IRRIGATION
Theory and Practice

Cover Design Courtesy:

M/s. Jain Irrigation Systems Pvt. Ltd., Jalgaon (Maharashtra)

IRRIGATION
Theory and Practice

SECOND EDITION
Revised and Enlarged

Dr. A.M. Michael, M.Tech., Ph.D.

Consultant, Water Management, Edappally North, Cochin

Formerly

Professor and Project Director, Water Technology Centre, IARI, New Delhi;
Director, Indian Agricultural Research Institute, New Delhi:
Vice-Chancellor, Kerala Agricultural University, Trichur

VIKAS® PUBLISHING HOUSE PVT LTD

VIKAS® PUBLISHING HOUSE PRIVATE LIMITED
E-28, Sector-8, Noida-201301 (UP), India
Phone: +91-120-4078900 • Fax: +91-120-4078999
VIKAS® Registered Office: A-27, 2nd Floor, Mohan Co-operative Industrial Estate, New Delhi-110044
E-mail: helpline@vikaspublishing.com • Website: www.vikaspublishing.com

SALES AND MARKETING OFFICES OF VIKAS® AND S CHAND AND COMPANY LIMITED

Chennai	: Ph: 044-2363 2120, chennai@schandpublishing.com
Guwahati	: Ph: 0361-2738 811, 2735 640, guwahati@schandpublishing.com
Hyderabad	: Ph: 040-4018 6018, hyderabad@schandpublishing.com
Jalandhar	: Ph: 0181-4645 630, jalandhar@schandpublishing.com
Kolkata	: Ph: 033-2335 7458, 2335 3914, kolkata@schandpublishing.com
Lucknow	: Ph: 0522-4003 633, lucknow@schandpublishing.com
Mumbai	: Ph: 022-2500 0297, mumbai@schandpublishing.com
Patna	: Ph: 0612-2260 011, patna@schandpublishing.com

Irrigation—Theory and Practice
ISBN: 978-81259-1867-7

Product Code: V3IRR46AGRI10ENAB07O

Second Edition 2007
First Published 1978
Reprinted several times between 1978 and 2022, 2023
Reprint 2024 (Twice)

PRINTED IN INDIA
By Vikas Publishing House Private Limited, Plot 20/4, Site-IV, Industrial Area Sahibabad, Ghaziabad-201010 and Published by Vikas Publishing House Private Limited, E-28, Sector-8, Noida-201301

Preface to the Second Edition

The author is greatly encouraged by the acceptance and popularity of the First Edition of *Irrigation Theory and Practice* by students, professional engineers, scientists and readers from various areas of water resources development and management in India and other developing countries of Asia and Africa as well as many institutions globally. The need for revising the book and enlarging it substantially was felt since long. The present volume incorporates the latest information on the subject and covers a number of new areas.

The new chapters included in this edition are 'Lift Irrigation Systems and their Design' and 'Economic Evaluation of Irrigation Projects and Water Pricing Policy'. The chapter on 'Water Conveyance and Control' of the first edition has been expanded and split into two separate chapters, namely, 'Open Channel Distribution in Command Areas', and 'Underground Pipeline System'. The earlier contents of the chapter on 'Soil-Plant-Water Relationships' have been enlarged and organized in two separate chapters, namely, 'Soil-Plant-Water Relationships' and 'Water Requirement of Crops and Irrigation Management'. Similarly, the earlier chapter 'Water Application Methods' has been substantially widened and presented in three new chapters, namely, 'Surface Methods of Water Application', 'Sprinkler Irrigation', and 'Drip Irrigation'. In addition to these, all the other chapters have been thoroughly updated with the inclusion of additional material covering the latest developments in the respective fields.

Short answer and Objective type questions, as also the matching exercises have been added at the end of each chapter to test the in-depth knowledge of students and help them prepare for various competitive examinations. To facilitate verification, answers to numerical and objective type questions are included in the Appendix at the end of the book. These are in addition to the solved examples and problems (with answers) provided along with the text.

The author is greatly indebted to many people, organizations and industries for the supply of material for inclusion in the new edition of the book, and for their support in the revision. The most valuable reference source has been the United Nations Food and Agriculture Organization, Rome. Prominent among the individuals who contributed substantially are Prof. T.P. George, former Professor and Dean, Faculty of Agricultural Engineering and Dr. Satish Babu, Agricultural Economist, and Mr. Sanjeev Pillai, Librarian, all from the Kerala Agricultural University, Trichur; Dr. Ashwani Kumar, Director and Dr. R.B. Singandhupe, Senior Scientists (Agronomy) of Technology Centre for Eastern Region, ICAR, Bhubaneshwar, Orissa; Dr. T.B.S. Rajput and Dr. A.K. Bhattacharya, Principal Scientists and Ms. Neelam Patel, Sr. Scientist of the Water Technology Centre, IARI, New Delhi; and Dr. R.K. Batta, Principal Scientist, ICAR Research Complex for Eastern Region, Patna, Bihar. Many leading industries provided valuable material for inclusion in the book. The support extended by the leading manufacturer of Drip and Sprinkler Irrigation Systems, 'Jain Irrigation Systems Ltd.', Jalgaon, Maharashtra has been extraordinary. Particular mention must be made here of Mr. Ajit B. Jain, Director and Mr. M.S. Sudhakar, Senior Manager, Projects, of this company. Mr. Vijay Jadhav, Manager, Irrigation Technology of EPC Industries Ltd., Nasik, Maharashtra provided valuable material. Every effort has been made to acknowledge the sources of information. Omissions, if any, are inadvertent and will be included, when pointed out. The author welcomes suggestions from the readers for improvements in the book.

The author wishes to make particular mention of the major contributions made by the publisher, Vikas Publishing House Pvt. Ltd., New Delhi in the processing and editing of the book as well as providing excellent marketing support.

A.M. Michael

Preface to the First Edition

This book is intended as a professional textbook for students of agricultural engineering degree programme. It will also serve as a text/reference for students at the undergraduate and postgraduate levels in civil engineering institutions and agricultural colleges. In addition to the student readership, the book will be a valuable reference to professional engineers and agricultural scientists working in the field of irrigation and on-farm water management.

In addition, the book will also be a valuable reference for the candidates appearing in competitive examinations for the Agricultural Research Services and for the examinations leading to the Associate Membership of the Institution of Engineers (India). Professional training institutions in soil conservation and water management, agricultural polytechnics, krishi vigyan kendras and rural institutes and similar other organisations having a program in water management would find this book useful.

Though written primarily to serve as a text/reference for students of agricultural and engineering institutions and practicing engineers and technologists in developmental organisations in India, it is hoped that the book will be valuable to similar groups in the developing countries of Asia and Africa.

Irrigation: Theory and Practice covers all the areas of irrigation related to agriculture. It is a comprehensive treatment on water resources in India, irrigation wells, irrigation pumps, measurement of water, land grading and field layout, soil-plant-water relationships, water conveyance and control, water application methods and salt problems in irrigation agriculture.

The details regarding the design and layout of irrigated practices in the field, including step-by-step procedures, and necessary equipment, are included wherever applicable. Class problems and examples have been included to emphasize the design principles and facilitate the understanding of the subject matter. Tables, figures and articles are numbered chapter-wise. A list of important references is given for each chapter. A set of problems for home work is provided with their answers at the end of each chapter. Suitable illustrations are provided wherever applicable so as to make easy the understanding of the subject matter. The entire book has been written using metric units of weights and measures. Conversion tables for change of units from one system to another are given in the Appendix.

The author is deeply indebted to many individuals and organisations for the supply of useful material for inclusion in the manuscript and the assistance provided at various levels for the preparation of the manuscript. Special mention is made of the valuable help received from L.V. Paliwal, Senior Scientist (Soil Salinity and Water Quality), Water Technology Centre, IARI, New Delhi, I Pai Wu, Professor of Agricultural Engineering, University of Hawaii, and R.K. Yadav, Agricultural Engineer, College of Agriculture, University of Udaipur, Jobner for the development of some of the specialised sections in the text. J.N. Luthin, Professor of Water Science and Engineering, University of California, Davis and Janardhan Prasad, Test Engineer, Tractor Training and Testing Centre, Budni, T.B.S. Rajput and Hanumantha Rao, Scientists, Water Technology Centre, IARI, New Delhi, were kind enough to review the manuscript and offer useful suggestions in improving the presentation of the subject matter. The assistance received from N.P. Singh, Irrigation Agronomist, P.B.S. Sarma, Senior Hydrologist, S.K. Sinha, Senior Plant Physiologist, R.K. Kaul, Irrigation Engineer, S.L. Pandey, Senior Scientist (Irrigation), A.K. Sinha, Soil Physicist and B.R. Yadav, Scientist of the Water Technology Centre and S.B. Hukkeri, Senior Agronomist of the Division of Agronomy of the IARI, New Delhi are gratefully acknowledged. While every effort has been made to acknowledge the sources of information, any omission therein is inadvertent.

The author would welcome suggestions from the readers to improve the text.

A.M. Michael

Contents

Preface to the Second Edition v
Preface to the First Edition vi

1. Utilization of Water Resources and Irrigation Development 1-123

Learning Objectives 1
Need for Sustainable Water Management 1
The Hydrologic Cycle 3
Global Water Scenario 4
Water Budget in India 8
Irrigation Development 11
Irrigation Development in the World: A Brief Historical Review 17
Irrigation Development from 1st to 20th Century A.D. 19
Irrigation Development in Former Soviet Union Countries 21
Water Resources Development in Monsoon Asia 21
South-East Asia 25
Major Island Nations of South-East Asia 27
Eastern Asia 30
Far East 31
Central Asia 32
Water Management Hazards and Environmental Degradation in Aral Sea Basin Countries of Central Asia 34
Irrigation Development in West Asia and the Middle East 35
Irrigation Development in Africa 39
Extent of Utilization of Irrigation Potential in Africa 55
Irrigation Development in India 57
Physiography, Climate, Soils, River Systems and Groundwater Regions of India Influencing Water Resources Utilization 69
Surface Water Sources of India 81
Groundwater Resources of India 83
Cropping Pattern and Irrigation Management 90
Canal Systems 91
Irrigation System Management 96
Participatory Irrigation Management 102
Floods and Droughts 104
Environmental Impact Assessment of Irrigation Projects 108
Inter-Basin Water Transfer 111
Summary 113
Questions 114
Further Reading References 119

2. Irrigation Wells 124-192

Learning Objectives 124
Groundwater and Aquifers 124
Hydraulics of Wells 132
Unsteady-State Flow 137
Pumping Tests of Wells 141
Design of Irrigation Wells 143
Well Construction Procedures 156
Drilled Tube Wells 161
Rotary Drilling 166
Summary 186
Problems 186
Questions 187
Further Reading References 190

3. Water Lifts and Pumps 193-248

Learning Objectives 193
Variable Displacement Pumps 208
Centrifugal Pumps 213
Vertical Turbine Pumps 224
Submersible Pumps 230
Propeller and Mixed Flow Pumps 232
Jet Pumps 234
Air-lift Pumps 235
Selection of Pumps 236
Power Requirements, Efficiency and Economics of Irrigation Pumping Plants 238

Summary 244
Problems 244
Questions 245
Further Reading References 247

4. Lift Irrigation Systems and their Design 249-289

Learning Objectives 249
Hydrology of Rivers in Plateau Regions Influencing Lift Irrigation Potential 255
Design of Lift Irrigation Systems 260
Summary 286
Questions 286
Further Reading References 289

5. Measurement of Irrigation Water 290-310

Learning Objectives 290
Methods of Water Measurement 291
Weirs and Flumes 296
Water Measurement Structures for Varying Flows in Earthen Channels 302
Orifices and Meter Gates 304
Tracer Methods of Water Measurement 306
Summary 307
Problems 307
Questions 308
Further Reading References 310

6. Open Channel Distribution in Command Areas 311-350

Learning Objectives 311
On-Farm Structures for Water Conveyance, Control and Distribution 331
On-Farm Water Conveyance and Distribution Systems in Irrigation Command Areas 343
Summary 347
Problems 347
Questions 348
Further Reading References 350

7. Underground Pipeline System 351-381

Learning Objectives 351
Underground Pipeline Structures 361
Underground Pipeline Water Distribution Systems in Community Irrigation Projects 373
Summary 378
Problems 378
Questions 379
Further Reading References 381

8. Land Grading and Field Layout 382-420

Learning Objectives 382
Land Grading Survey and Design 385
Construction Procedures 402
Contour Bench Levelling 403
Equipment for Land Grading and Field Layout 409
Summary 416
Problems 416
Questions 417
Further Reading References 420

9. Soil-Plant-Water Relationships 421-477

Learning Objectives 421
Soil Properties Influencing Irrigation Management 422
Soil Fertility Management 441
Water Relations of Soil 444
Measurement of Soil Moisture 459
Plant-Water Relationships 465
Summary 471
Problems 472
Questions 472
Further Reading References 475

10. Water Requirement of Crops and Irrigation Management 478-553

Learning Objectives 478
Evaporation 479
Transpiration 483
Evapotranspiration (ET) or Consumptive Use (CU) 483
Estimating Evapotranspiration Based on Climatological Approach 490
Rational and Quasi-Rational Approaches in Estimating Evaporation and Evapotranspiration 495
FAO Penman-Montieth Based Combination Formula for Estimating Reference Evapotranspiration, ET_o 501
Crop Evapotranspiration (ET_C) 510
Crop Water Requirements and Irrigation 515

Irrigation Scheduling 520
Irrigation Efficiency 523
Crops and Water 527
Planning Cropping Systems and Water Management in Irrigation Project Areas (Based on Crop Response to Water Availability) 530
Irrigation Requirements of Common Crops 532
Crop Management Under Rain-Fed Agriculture 545
Evaluating Lands for Irrigation Command Areas 546
Summary 548
Questions 548
Further Reading References 551

11. Surface Methods of Water Application 554-577

Learning Objectives 554
Surface Irrigation Methods 554
Border Irrigation 555
Check Basin Irrigation 564
Furrow Irrigation 568
Sub-Irrigation 574
Summary 574
Problems 575
Questions 575
Further Reading References 577

12. Sprinkler Irrigation 578-610

Learning Objectives 578
Development and Use of Sprinkler Irrigation 579
Very-Low-Pressure Sprinkler Irrigation Systems 599
Application of Fertilizers and Chemicals through Pressurised Irrigation Systems 601
Summary 606
Problems 607
Questions 607
Further Reading References 609

13. Drip Irrigation 611-660

Learning Objectives 611
Indigenous Micro-Irrigation Devices 616
Devlopment of Drip Irrigation: A Historical Perspective 619
Components of Drip Irrigation Systems 622
Drip Irrigation Water Distribution Systems 629
Application of Fertilizers and Chemicals through Drip Irrigation Systems 641
Installation, Operation and Maintenance of Drip Irrigation Systems 643
Summary 655
Questions 655
Further Reading References 658

14. Salt Management in Irrigated Agriculture 661-694

Learning Objectives 661
Application of Soil Amendments 683
Irrigation with Poor Quality Water 684
Summary 691
Problems 691
Questions 691
Further Reading References 693

15. Economic Evaluation of Irrigation Projects and Water Pricing Policy 695-727

Learning Objectives 695
Economic and Financial Feasibilities of Projects 695
Application of Discounting Methodology in Economic Analysis of Irrigation Projects 704
Water Pricing and Cost Recovery 716
Summary 724
Questions 725
Further Reading References 726

Appendices 728-756
Index 757-768

Chapter

1

Utilization of Water Resources and Irrigation Development

LEARNING OBJECTIVES

Global awareness on sustainable management of water resources. Global water scenario in terms of water availability and irrigation development. The hydrologic cycle and its various phases. Diverse uses of water and prioritization in water allocation to different sectors. India's water budget and scope for introducing greater efficiency in the use of surface and groundwater. Historical review of irrigation-based societies globally and the reasons for the collapse of many of the ancient civilizations due to faulty water management and unstable governments. Dependence on monsoons in water resources development in India and other countries of Asia. Highlights of physical resources, irrigation development and potential for increased availability and efficiency of water management in different countries of Asia and Africa. Special situation of water management in Central Asian countries, with particular reference to the problems of Aral Sea basin countries. Unique features of water resources development and irrigation management in the Middle East. Critical analyses of water resources availability and their efficient utilization in the different agro-climatic and ecological regions of Africa. Irrigation development in India and the emphasis on tank-based projects by ancient rulers. Causes of frequent occurrence of famines in India during the later part of the 19th century and pre-Independence period of the 20th century and emphasis on irrigation development to mitigate the malady. Rapid pace of irrigation development in India during the Five Year Plans. River systems and statewise potential and utilization of surface and groundwater in India. Physiography, climate, soils, river systems and groundwater regions of India in relation to irrigation development. Canal irrigation system management in India and scope for increased irrigation efficiency. Problems of night irrigation in canal command areas. Cropping systems and water management. Conjunctive use of surface and groundwater. Participatory irrigation management. Floods and droughts and measures to minimize their effect on crop production. Environmental impact assessment of water irrigation projects. Issues relating to large-scale inter-basin water transfer.

Water is a precious natural resource, a basic human need and a prime national asset. The extent to which water is plentiful or scarce, clean or polluted, beneficial or destructive, profoundly influence the extent and quality of human life. Rapid development of human civilization and advances of scientific and technological innovations are changing the condition of life on earth, giving rise to basic transformations of the environment. The relentless increase in population and the resulting spurt in the demand for water require careful planning and management of the limited water resources. In India, the growth of population is about 2% per year. It is essential that food production should increase about 2.5% per year to provide a better food intake. The available water resources are to be optimally harnessed and beneficially utilised with appropriate priorities of use. Long-term perspective planning of water resources is required to achieve the goal of economic prosperity and quality of life on a sustainable basis.

The competing demands for water are for irrigation, industrial production, domestic water supply, hydro-power generation and environmental management. Requirements of water for transport, recreation and fresh water fisheries also require adequate provision for water.

NEED FOR SUSTAINABLE WATER MANAGEMENT

Worldwide water use has increased more than tenfold during the 20th century. It has risen sharply in the second-half of the 20th century. The most frequently used criterion for assessing the availability of fresh water supply is the per capita supply. A supply of less than 500 m^3 per capita per annum is regarded as the critical lower limit, 1,000 m^3 per capita as very low and 2,000 m^3 as critical (Wolff, 1999). Countries which are in semi-arid and arid regions are often in a critical situation. Based on a supply of less than 500 m^3 per capita per annum, the most endangered regions of the world are North Africa and the Middle East. It is estimated that 22 countries in the world have renewable water supplies of less than 1,000 m^3 per capita per year.

Water is unevenly distributed all over the world and is usually scarce in many of the regions where the masses of people live. There are large seasonal variations. For example, large parts of India and Bangladesh suffer from major floods during the monsoon period, which is followed by a period of water shortage. Water scarcity is not a general phenomenon but is a regionally, locally and seasonally specific problem.

The UN Water Conference held at Mar del Plata (Argentina) in 1977 drew attention to the basic lacuna in water management in several countries. The Action Plan finalised at the Conference states as follows:

"In many countries water interests have been divided among numerous agencies without adequate coordination and without adequate links to other aspects of national planning. Institutional arrangements adopted in each country should ensure that the development and management of water resources take place in the context of national planning and that there is real coordination among all bodies responsible for the investigation, development and management of water resources. The problem of creating an adequate institutional infrastructure should be kept constantly under review and consideration should be given to the establishing of efficient water authorities to provide for proper coordination."

The 1992 International Conference on Water and the Environment in Dublin, Ireland, laid major emphasis on the need for intensive and serious consideration at different levels on the limited availability of water and the possibilities of exploiting and managing the earth's fresh water resources. In June 1992, the United Nations Conference on Environment and Development held in Rio de Janeiro, Brazil, building on the results of the Dublin Conference (1992), brought out the need for worldwide reforms in the water sector. Chapter 18 of Agenda 21 of the Conference states as follows:

"Sustainability of food production depends on sound and efficient water use and conservation practices consisting mainly of irrigation development and management, livestock water supply and inland fisheries. Achieving food security is a high priority item in many countries. Agriculture must provide food for the rising populations and save water for other uses. It is essential to develop and apply water-saving technology and management methods. Through capacity building, communities should be able to adopt new approaches, for both rain-fed and irrigated agriculture." (Wolff, P., 1999).

The Dublin Statement (1992) on Water and Sustainable Development, and Agenda 21 of the UN Conference (1992) in Rio de Janeiro stressed on the following emerging issues:

(*i*) Fresh water is a finite and vulnerable resource which is essential to sustain life, development and the environment on earth.

(*ii*) Water resource development and management should be based on a participatory approach, involving users, planners and policy-makers at all levels.

(*iii*) Women play a major role in the provision, management and safe-guarding of water and their effective involvement is vital in efficient water management.

(*iv*) Water has an economic value in all its competing uses and should be recognised as an economic good.

To give effect to the above principles, the Global Water Partnership (GWP) was established in 1996. The GWP stresses that to manage water sustainably for continued human development, the competing uses for water must be reconciled. This would be possible only if the parties competing for fresh water appreciate the needs and appropriately adjust their demands. There have been major international initiatives in the 1990s and early 21st century. The World Bank published a Policy Paper in 1993 defining new goals for the water sector. The FAO set up an International Action Programme on Water and Sustainable Agricultural Development. Other UN agencies also started programmes dealing with the problems of fresh water resources. The donor organisations, including non-government organisations (NGOs), have recently increased their activities in water resources development and management. The World Water Forum held at Kyoto (Japan) in March 2003, emphasised on the need for world water infrastructures, especially for developing countries to meet the present and future needs of the entire population. The forum stressed on the need for proper governance of water for sustained development and prevent environmental degradation. These are in view of the realization that water is an increasingly scarce and valuable resource and everything possible should be done to ensure sustainable use of fresh water.

The scope of water science and technology extend from the watershed to the farm and on to the drainage channel. The watershed yielding the rainfall run-off; the storage and conveyance of water; the management and distribution of water and the drainage problems arising from faulty water distribution are all equally important.

THE HYDROLOGIC CYCLE

Hydrology is the science that deals with the occurrence and movement of water on the earth in streams, rivers, on or below the land surface and in the atmosphere in all its states. It is concerned with the physical, chemical and hydrological reactions of water with the rest of the earth and its relation to the life on earth.

Hydrologic cycle is the circulation of water from the oceans or the sea, lakes, reservoirs, streams, rivers, land surface detention, underground water and the aqueous vapour in the atmosphere (Fig. 1.1). The processes involve cloud formation and movement, condensation, precipitation, interception by vegetation and the land surface, infiltration, run-off, soil moisture storage, percolation, subsurface flow, groundwater storage, evaporation and transpiration. *Hydrologic regime* is the behaviour of water in the hydrologic cycle, during a given period.

Hydrosphere is the aqueous envelope of the earth including the oceans, seas, lakes, streams, underground water and the aqueous vapour in atmosphere.

Climatic cycle is the cyclic recurrences of weather phenomena at more or less regular intervals, in response to long-range terrestrial and solar influences.

Agro-hydrology is the study of movement of water in the unsaturated zone of the soil and the shallow groundwater under the influence of rainfall, evaporation, irrigation and drainage. The study includes the effects of these processes on crop growth.

Hydrologic equation is the water balance equation (inflow = outflow + storage). It expresses the basic principle that during a given time interval the total inflow to an area must equal the total outflow plus the net change in storage.

Hydrometry is the measurement and analysis of the flow of water.

Cryology is the science of ice in all its forms.

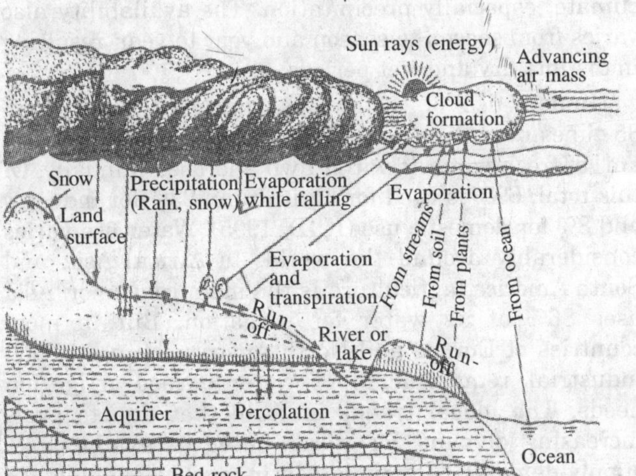

Fig. 1.1 Schematic diagram illustrating the earth's water cycle – the hydrologic cycle

The total amount of water contained in planet earth is constant and can neither be increased or decreased. However, the availability of water for the use of man is dependent on the earth's unending moisture cycle, the *hydrologic cycle*. It comprises of a gigantic distillation system operating in and on the land and the water bodies of the earth and in the atmosphere surrounding it, with the energy derived from the sun. Moisture is constantly circulating between the land, the ocean and the atmosphere. This complex series of phase changes and inter-connected flows are shown schematically in Fig. 1.1. The cycle has neither a beginning nor an end, but the concept of the hydrologic cycle commonly begins with the water of the oceans, since it holds the bulk of the water of the earth. Radiation from the sun evaporates water from the oceans to form water vapour. The water vapour rises and collects in the atmosphere to form clouds. The circulation of moisture from the ocean to the land is influenced by the winds, atmospheric pressure differences, the physiography of the continents, temperature differences and other factors. Under certain conditions, the cloud moisture condenses and falls back to the earth as precipitation (rain, hail, sleet or snow).

Precipitation from the atmosphere to the ground occurs under three specific conditions, viz., (*i*) the cooling of the incoming air until it is super-saturated with water vapour; (*ii*) condensation, comprising of the conversion of water vapour to liquid water in the form of cloud droplets which requires the presence of tiny nuclei, and (*iii*) the coalescence (fusion) of water droplets or of ice crystals to form particles which are large enough to fall to the ground as rain or snow.

Rainfall is classified according to the lifting mechanism of clouds into: (*i*) *orographical rain*, where the lifting is due to flow over a mountain barrier; (*ii*) *frontal rainfall*, when the lifting is due to the relative movement of two large air masses; and (*iii*) *convective rainfall* (thunderstorm) where the lifting is due to local heating of the ground, resulting in violent uplift of air over a localized area. The annual average precipitation that occurs on the earth is estimated as 1,030 mm. Taking oceans and the continents separately, the annual precipitation over the ocean surface is 1,140 mm, while that over the continents is 760 mm.

The average continental precipitation, evaporation and run-off are close to the world average in Europe, Asia and North America. In South America, the continental water balance estimates show that the precipitation is more than twice the world average. In Australia, the precipitation is only about 60% of the world average. In Africa, the evaporation is about the world average, but the precipitation is below world average and the run-off is only about half the world average.

Of the total annual precipitation in the world, 79% is over the sea and is not available to man. Of the remaining 21% falling on the land, about 13.4%

evaporates to the atmosphere, leaving 7.6% as run-off to rivers, soil moisture and groundwater recharge. The estimates available currently indicate that about half of the run-off is beyond the control of man and the remaining has the potential for development by man. However, the extent of utilization varies greatly amongst the continents and from country to country. Within the same country or a state, there are wide regional variations in the extent of water resources utilization. Precipitation reaching the earth's surface may be intercepted by vegetative material, may infiltrate into the ground, may flow over the land surface as run-off or may evaporate. Evaporation may be from the surface of the ground, from free water surface or from leaves of plants through transpiration.

A part of the precipitation, after wetting the foliage and ground surface, runs off over the land to streams and rivers and to the oceans. Another part infiltrates into the soil. Much of the water that enters the soil is retained in the plant root zone and is eventually drawn back to the atmosphere by evaporation or plant transpiration. Some of it, however, percolates below the plant root zone, and under the influence of gravity continue to move downward until it enters the groundwater reservoir.

On joining the body of groundwater, the percolating water may move through the pores of saturated subsurface materials and may reappear at the surface in areas at lower elevations as subsurface flow. Groundwater discharges naturally at such places in the form of springs and seeps which maintain the flow of streams in dry periods. The streams, carrying both surface run-off and subsurface flows, join the rivers and eventually flow back to the oceans. There are many opportunities to influence the hydrologic cycle and affect water conservation in agriculture.

The hydrologic cycle is not always punctual in delivering precipitation to the earth's surface, and drought so caused is one of the major deterrents to adequate food production for the world population. Although man has been able to modify, to some extent the pattern of availability of water supplies with respect to space and time, the total supply of water neither grows nor diminishes. Unlike other natural resources, water is not depleted through consumption. An evaluation of the water budget of any area of a country may lead to the conclusion that water can be used more efficiently than at present.

GLOBAL WATER SCENARIO

The oceans occupy about 70.8% of the earth's surface and only 29.2% is land. About 97.3% of the world water resource is in the oceans and is saline (Fig. 1.2). Of the remaining, 2.7% of the global fresh water resource, about 75.2% is in ice caps and glaciers and is not available for the requirements of mankind (Fig. 1.3). A major part of the balance amount of water (22.6%) occurs as groundwater, of which about 56% of the volume lies in water-bearing formations deeper than 800 m below ground surface and is not ordinarily available for economical development. The amount of water available for development by man is mainly the water in the streams, rivers, fresh water lakes and about 44% of the quantity of groundwater which occurs at depths less than 800 m from ground surface.

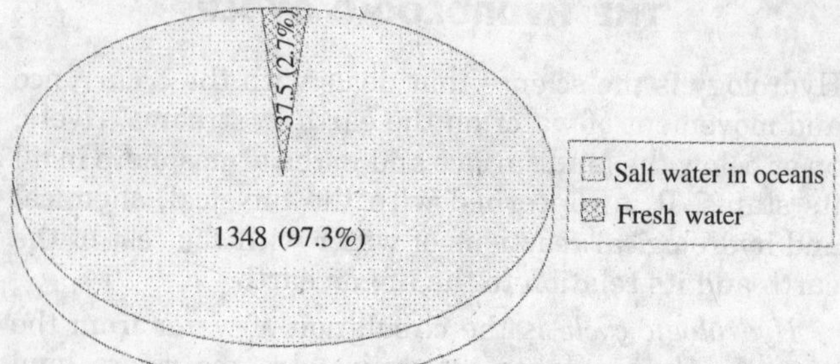

(*Note*: All units in million cubic km)

Fig. 1.2. Distribution of world's total water.
(***Source:*** Based on Central Water Commission, Govt. of India, 1996.)

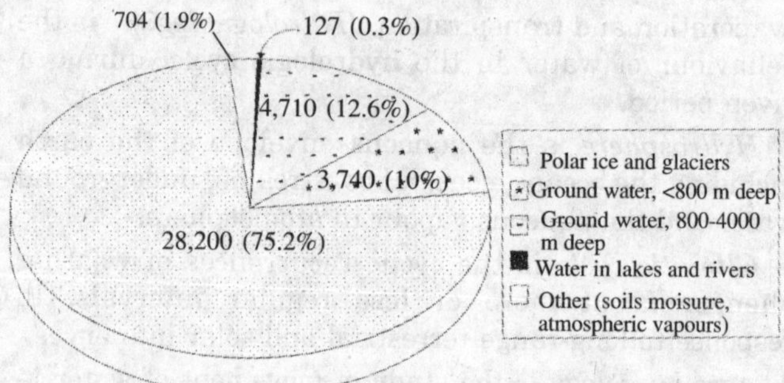

(*Note*: All units in thousand cubic km)

Fig. 1.3. Distribution of world's fresh water.
(***Source:*** Based on Central Water Commission, Govt. of India, 1996.)

Water availability is profoundly influenced by climate, especially precipitation. The availability also varies from season to season and year to year, resulting in distinct dry and wet periods.

Human use of water has increased more than 35 times over the past three centuries. Globally, 3,240 km^3 of fresh water is withdrawn and used annually. Of this total, 69% is used for agriculture, 23% for industry and 8% for domestic use (ICID, 1995). Water use varies considerably around the world. In Africa, Asia and South America, agriculture is the main consumer. Asia uses 86% of its water for irrigation. But in most countries of Europe and North America, domestic and industrial requirements of water exceed irrigation needs. The global trends indicate that due to ever-increasing demand for water, resource availability–supply-demand pattern is continuously changing. In almost all countries the demand-supply gap is widening.

The need for additional food supplies are causing the rapid expansion of irrigation throughout the world

(Fig. 1.4). The net irrigated area in the world in 1999 has been about 268 million ha.* India, China, the United States of America and Canada, Pakistan, Indonesia, Mexico, Iran, Thailand, Uzbekistan, Turkey and Bangladesh have the largest irrigated areas, amounting to about 55% of the world's total irrigated area (Fig. 1.4 and 1.5).

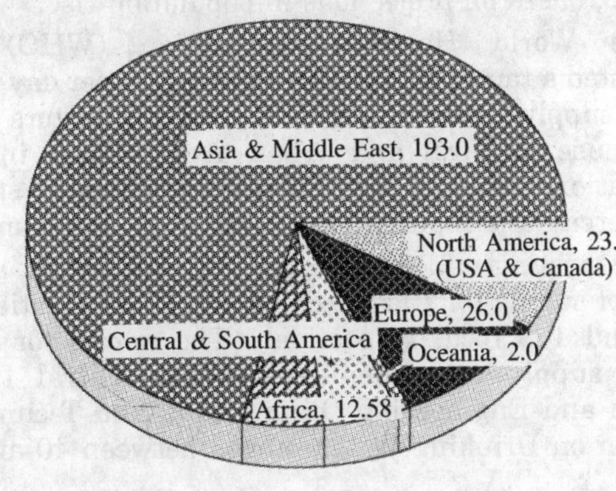

Oceania (Australia, New Zealand, Fiji, Papau, New Guinea)
All figures are in million ha of net irrigated area

Fig. 1.4. Irrigated area in the world, regionwise.
[***Source***: Based on Intl. Comm. on Irrig. and Drainage (1994), FAO (2000).]

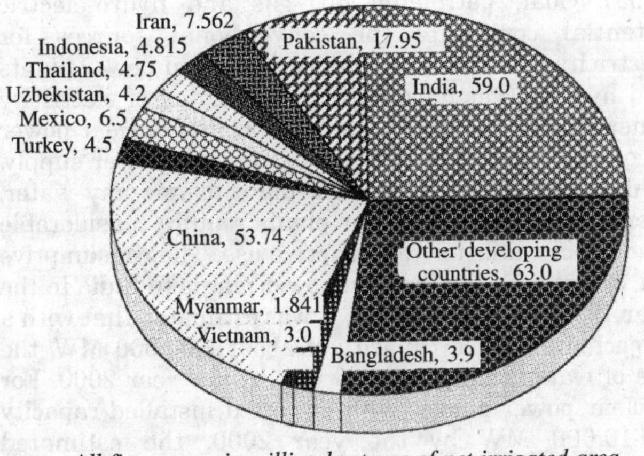

All figures are in million hectares of net irrigated area

Fig. 1.5. Irrigated area in developing countries, 1999.
[***Source:*** Based on Intl. Commn. on Irrig. and Drainage(1994), FAO (2000).]

Attempts to conserve water has challenges which are not there in other natural resources. In most parts of the world, water is not controlled by market mechanisms, nor is water a commodity that can be traded internationally like food or other merchandise. Whether people save or waste water in one region is of no concern to those who live in other regions. People need clean water close to their homes.

Mounting population and improved standards of living are increasing the demand for agricultural production. The available water resources will have to satisfy the increasing demand for irrigated agriculture. Urbanisation, which is likely to have over half of the world's population concentrated in urban settlements would demand greater water supplies. The rise in pollution and the consequent increases in water use are not the only factors leading to water shortage. Pollution is a major hazard. A unique quality of water is to cleanse itself in the hydrologic cycle, but there are limitations. Although fresh water is a renewable resource, increased use of fertilizers and pesticides and generation of new and complex wastes by industries and unscientific sewage disposal systems are adding to the extent and dimensions of water pollution. The demand for industrial uses of water is outpacing the demand for domestic uses. Pollution is becoming a problem in almost all the rivers, lakes and estuaries. Even deep sea waters are affected by pollution. Sufficient potable water for domestic consumption is not available in many regions of the world.

*[***Source***: UN Food and Agriculture Organization FAO, 2000).]

Diverse Uses of Water

The diverse uses of water can be broadly classified as (*i*) *off-stream use*, and (*ii*) *instream use*. Offstream use includes the use for irrigated agriculture, domestic and industrial water supply and thermal and nuclear power generation. Offstream use of water in India in 1999 has been 92% for irrigation, 5% for domestic use and 3% for industry. Instream use comprises of the use for hydro-electric power generation, fisheries, navigation and recreation. Water use for offstream use comprise the part returned to the surface water or to groundwater after use and the part consumed in use. An example of consumptive use is transpiration by vegetation and evaporation from wet soils.

Irrigation: Irrigation water is supplied to supplement the water available from rainfall and the soil moisture in the crop root zone and the contribution to soil moisture from groundwater. In many areas of the world the amount and timing of rainfall are not adequate to meet the moisture requirement of crops and irrigation is essential to raise crops necessary to meet the needs of food and fiber for the growing population. Irrigation is practised under the following major agro-climatic conditions:

1. In the arid and semi-arid regions of the world, with their characteristic low rainfall, where agriculture is mainly dependent on irrigation. The major source of irrigation water in these areas is rivers. These rivers have extensive catchments which contribute run-off from rainfall and/or snow-melt. A part of the water supply is from groundwater, the potential of which increases due to added recharge arising from irrigation.
2. In areas which have a marked seasonal shortage of rain, like monsoon Asia and regions with Mediterranean climate, irrigation is essential to grow crops during the dry season.

3. In humid areas where the rainfall is normally adequate to sustain crop production, irrigation is desirable as a safeguard against drought and to obtain high crop yields.

With the current global trend in population growth of about 90 million each year, the 21st century is set to face an uphill task of feeding the growing population with the limited resources. During the second-half of the 20th century, irrigation and drainage have contributed significantly to the increase in food production. About one-sixth of crop-land that is irrigated produces one-third of the world's harvest of food crops. Drainage has enhanced the productivity of another 10% of crop-lands.

Water requirement for irrigation: Irrigated agriculture has a predominent role in India's food production. The population growth in India is 2% per year. It is essential that the food production should increase by about 2.5% per year (to allow for higher food intake). Irrigated area should increase and the potential productivity of crop should be realised to the maximum possible extent. It is estimated that 70% of urban and 30% of rural water requirement would be met from surface water sources and the remaining from groundwater resources. Water use in irrigation at the beginning of the 21st century (2000 A.D.) was estimated at 501 billion cubic metres. To meet the increasing demand for additional food grain production, the demand for water in the irrigation sector will increase substantially (Report of the National Commission for Integrated Water Resources Development Plan (NCIWRDP), 1999).

Water requirement of animal population: The norms for water requirement of the cattle range from 18 to 30 litres per capita per day. Assuming a growth of 0.5%, the NCIWRDP (1999) has estimated the water requirement of animals at 4.8 km^3, 5.2 km^3 and 5.9 km^3, respectively for the years 2010, 2025 and 2050 A.D.

Water requirement for industries: The requirement of water in the industrial sector is high, especially in mining, chemicals, fertilizers, petrochemicals, paper and pulp, and textiles. Based on information obtained from the Central Pollution Control Board and from different associations of industries, the NCIWRDP (1999) has estimated the water requirement for industrial development in India as 37 km^3, 67 km^3 and 103 km^3 for the years 2010, 2025 and 2050 A.D. respectively, under the demand rates of water prevailing in 1998–99. However, the NCIWRDP (1999) presented an alternative figure of 81 km^3 for the year 2050, assuming that by then the industries would adopt major water-saving technology. It is possible to utilize 80% of industrial effluent after proper treatment. It is estimated that 70% of the industrial water requirement in India will be from surface water sources and the remaining 30% from groundwater.

Water requirement for domestic use: Water for drinking is a basic human need. In addition to meeting this basic need, community water supply has to meet the needs for cooking, washing, bathing and sanitation. Though the water requirement for community water supply is only a small fraction of man's total need for water resources, it is the most important requirement. The National Water Policy (1987 and 2002) of India has given the highest priority to domestic water supply.

The water use for community water supply in India in the year 1989 was 25 km^3. It was estimated to reach 30 km^3 by the year 2000. The estimate for the 2025 is 52 km^3, based on projections in population rise.

The World Health Organization (WHO) has suggested a target of 200 litres per capita per day (lpcd) water supply for urban areas. A variety of factors affect water use in rural and urban areas. These include population size of a habitat, economic status, commercial and manufacturing activities and climate.

The existing national average of water supply, at the point of supply for Class I cities and Class II cities are 182 and 103 lpcd, respectively. The norms for rural water supply recommended by the Central Public Health and Engineering Organization and Technology Mission on Drinking Water ranged between 40 and 70 lpcd.

Water requirement for power development: The major sources of power are thermal/nuclear power and hydro-electric power. In India the installed power generation capacity, which was 1,362 in 1947, has increased about 70 times to 92,281 MW by March 1999 (NCIWRD, 1999). Out of this only about 25% is the share of hydro-power (Ministry of Water Resources, 2003). Coal, petroleum oil, gas and hydro-electric potential constitute the conventional sources for electricity generation. Coal-based thermal power plants and hydro-power have been the mainstay of electricity generation in India. Oil, natural gas and nuclear power account for a smaller portion of the total power supply. While hydro-power plants hardly consume any water, thermal and nuclear power plants require considerable amount of water for consumptive use. The consumptive use of water by the thermal power plants in India in the year 1989 was about 4 km^3. It was estimated that with a projected installed capacity of about 1,00,000 MW, the use of water will go up to 6 km^3 by the year 2000. For nuclear power plants, with projected installed capacity of 10,000 MW by the year 2000, the estimated consumptive use has been about 0.3 km^3.

The total hydro-power potential of India, as assessed by the Central Electricity Authority is 84,000 MW at 60% load factor. The installed capacity as of 1989 was 17,798 MW. Apart from this, the Authority has also identified 56 potential pumped storage schemes with a possible installed capacity of 94,000 MW (Indian National Acad. of Engg., 1990). Hydro-electric source of power generation is low in cost, renewable, clean and pollution-free. The consumptive use of water for hydro-power generation is only by way of evaporation from the reservoirs. Only about 15% of India's hydro-power potential has been utilized till the year 2003 and another 7% is under various stages of development (Min. of Water Resources, 2003). Thus, there is tremendous scope for hydro-power development in India.

Inland water transport: In comparison to road and rail transport, inland water transport is the most inexpensive mode of transportation. The consumptive use of water for inland water transport is not significant.

Fisheries: There is considerable scope for substantially increasing fisheries production on scientific lines in inland water bodies, canal water conveyance systems and storage reservoirs, as instream use.

Water requirement for protection of environment and ecology: Increasing water shortage and water pollution are major ecological concerns. In India, the dense forest cover which was estimated at 40% of land area in the early part of the 20th century has reduced to less than 14% by the year 2000. Most of the rivers now carry polluted water during the non-monsoon periods. Protection, enhancement and restoration of water quality and prevention of water pollution are necessary for providing safe drinking water and maintaining human health. It is also important for sustaining land quality, agriculture and industrial production.

Water requirement for the purpose of environmental protection comprise (*i*) the requirement for afforestation and tree planting and, (*ii*) the requirement for abatement of water pollution and maintaining water quality safe for drinking and other uses in the streams/rivers and other water bodies. The water requirement for afforestation could mostly be met through effective soil and water conservation programmes in the catchment areas.

Rivers and other water bodies are being polluted by both point and non-point sources. Pollution at point sources are due to discharges of urban sewage and industrial effluents which contain organic pollutants, toxic chemicals and heavy metals. The non-point sources of pollution are mainly run-off from irrigated lands and urban solid waste dumps. The major contributor of water pollution is urban sewage. Industrial waste, though small in volume, contributes significantly in terms of the total pollutant loads. There are many instances of high pollution from irrigated lands using heavy doses of fertilizers and insecticides.

Estimation of the quantity of water required for managing ecological standards for all water bodies including lakes and rivers on sustainable basis is difficult at present. Information on quantity of water required for dilution and treatment of sewage in the rivers in the country is not available. The NCIWRDP (1999) has made a token provision of 5 km^3, 10 km^3 and 20 km^3 of water for ecological management for the years 2010, 2025 and 2050 A.D. respectively. It has suggested that periodic reassessment of this requirement should be made, taking into account all relevant data on sources and extent of pollution, technologies and institutional arrangements.

Participation of women: Globally, there is greater awareness of gender-related issues, However, it is still not being translated into concrete actions related to participation of women in decision-making in water management. The current developments in the Local Self Government (*Panchayat*) institutions is an encouraging step. Concerted efforts are needed to associate women with the village level committees for water supply and sanitation in the rural areas as well as policy decision at higher levels.

Sustainable development and management of water resources: Sustainable development and management of water resources would require a holistic approach and simultaneous attention to all aspects of water use and management. Some of the issues causing much concern and threaten sustainability relate to water quality management across the sectors, including ecology and environment, waterlogging, salt problems in agriculture and absence of on-farm drainage. Conjunctive use of surface and groundwater, utilization of canal flows in non-demand periods in agriculture, effective utilization of canal flows during the night, rapid silting up of reservoirs and increased flood damages due to uncontrolled development of catchment areas and watersheds, water quality management of lakes and tanks, recycling and proper treatment of industrial effluents. lack of adequate attention to preventive and regular maintenance of water infrastructure are the major issues to be addressed.

Wasteful use of water: A major challenge today is to introduce economy in agricultural, municipal and industrial water use. The efficiency level of water use in irrigation in the developing countries is often only 25 to 40% (Wolf, 1999). Major areas of waste of water in canal irrigation are under-utilization of night supplies, canal flow going waste in non-demand seasons due to lack of crop diversification and absence of storage tanks/reservoirs within the system (system storage), improper canal water releases, excessive seepage losses in water conveyance systems, field-to-field irrigation in rice crop and unscientific irrigation methods in upland crops. Conveyance losses are often high in urban water supply systems and is estimated at about 50% or more. Leaking pipelines, resulting from corrosion and incrustation in pipes and faulty pipe joints as well as intermittent water supply result in serious water pollution problems.

Soil degradation in irrigated areas: Soil degradation by waterlogging and salt problems is a serious threat to irrigated agriculture. The problem has worsened during the later part of the 20th century with the expansion of irrigation. Land areas which become unsuitable for cultivation each year due to waterlogging, salinization and alkalinization vary between 0.16 to 1.5 million hectares. Estimates on the total area worldwide under the influence of salinity varies between 20 to 46 million hectares, depending on the concepts used in defining salt problems in soils (Wolff, 1999).

Contamination of water with harmful substances. Contamination of water with harmful toxic and pathogenic substances is a major problem of the industrialized countries. In developing countries also, it is on the increase, Important factors influencing contamination are untreated or inadequately treated industrial effluents and sewage, and contamination by mining. Dumping municipal waste on lands at higher reaches of streams and rivers and its flow into waterways is a serious problem.

In India, it is reported that only few of the cities are having satisfactory waste water treatment plants. Poor quality drinking water, in combination with inadequate sanitary conditions at the domestic and municipal levels, are the major causes of the high incidence of diseases, especially among children in the developing countries. It is estimated that about 1 billion people in the world have no access to drinking water of satisfactory quality and that 1.7 billion people have inadequate sanitary facilities (Wolf, 1999). Unsafe water is the main cause of over 3 million deaths in the world every year. It also causes 900 million cases of illness each year (World Bank Policy Paper, 1993).

WATER BUDGET IN INDIA

India is a vast country with extremes of climate. Rainfall varies from 100 mm in western Rajasthan to over 11,000 mm at Cherrapunji in Meghalaya in the north east. The average annual rainfall isophyetes over the country are shown in Fig. 1.40. Variability of annual and seasonal rainfall at the same place is high, resulting in droughts and floods. One-third of India is drought-prone, while over one-eighth is flood-prone. India's topography comprises of towering mountain ranges, rolling hills, lofty plateaus and extensive plains traversed by major rivers. Bounded by the Himalayas in the north, India stretches southwards and tapers towards the Indian Ocean, with the Bay of Bengal in the east and the Arabian Sea in the west. The presence of the great mountain mass formed by the Himalayas and its spurs on the north and of the Indian ocean in the south, are the two major physical features influencing the climate of India. The Himalayas provide an impenetrable barrier to the influence of cold winds from Central Asia, and give the sub-continent, the elements of tropical type of climate. The Indian ocean is the source of cool moisture-laden winds reaching India and giving it the elements of the oceanic type of climate.

India has great diversity and a variety of climate, and extremes of weather conditions. The climate ranges from continental to oceanic, from extremes of heat and cold, and from aridity and neligible rainfall to excessive humidity and heavy rainfall. The climatic factors profoundly influence the availability and utilization of water resources.

Precipitation. The basic source of water is precipitation in the form of rainfall or snowfall. Run-off from precipitation drains through streams and rivers or collects in surface depressions forming tanks or ponds. Water of streams and rivers is stored in reservoirs or is diverted directly through canal systems for irrigation. Run-off water stored in tanks or ponds is also regulated for irrigation through suitable control and conveyance systems. A part of the rainfall is retained as soil moisture and stored as groundwater.

A proper understanding of water resources requires a careful analysis of the nation's water budget. Rainfall in India varies from place to place and from year to year. It is recorded by over 3,000 rain gauges set up by the Indian Meteorological Department and the State Governments. The country's average annual rainfall is about 1,200 mm, which when considered over the geographical area of over 328 million ha amounts to 390 million ha-metres. The contribution of snowfall is estimated to range from 80 to 100 mm. Thus, the total precipitation amounts to 400 million ha-metres. Rainfall in India is dependent in various degrees on the south-west and north-east monsoons, on shallow cyclonic depressions and disturbances and on violent local storms, which form in regions where cool humid winds from the sea meet hot dry winds from the land (which occasionally reach cyclonic dimensions). Most of the rainfall in India takes place under the influence of south-west monsoon between June to September. The south-west monsoon starting from the equatorial belt comes over the Indian sub-continent in two distinct wind currents known as the Bay of Bengal branch and the Arabian sea branch. Figure 1.38 shows the normal dates of onset and withdrawal of south-west monsoon. It has been estimated that during the four rainy months of June to September, the Arabian sea branch of the monsoon carries moisture amounting to about 770 million ha-m and the Bay of Bengal branch about 340 million ha-m of water. During the remaining eight months, there is a substantial amount of moisture over the country. It contributes to precipitation of the order of 100 million ha-m, mainly as north-east monsoon. The influence of the north-east monsoon is mainly in the southern states of Tamil Nadu, Andhra Pradesh and Kerala.

When rain falls, a portion of it soon evaporates from the ground or the vegetation that may intercept it, another portion infiltrates into the soil and the rest flows away over the land surface as run-off. When a shower is very light, say of the order of 2.5 mm or less, it moistens only a few millimetres of the top soil which evaporates rapidly. It neither contributes to surface water nor to groundwater. There are, on an average, 130 rainy days in a year in the country and on 75 days the rainfall is below 2.5 mm. Taking these factors into consideration, it is estimated that out of the average annual precipitation of 400 million ha-m, about 70 million ha-m is lost to atmosphere. Of the remaining 330 million ha-m, about 115 million ha-m flows as surface run-off and the rest, 215 million ha-m infiltrates into the ground, a part of which joins the streams as subsurface flow.

Assessment of surface water resources of India: The first attempt to assess the water resources of India was made by the First Irrigation Commission (1902–03). The data available to estimate surface flows were

Distribution of average water resources of India 1974 and (2025) AD

(All figures in million hectare-metre)

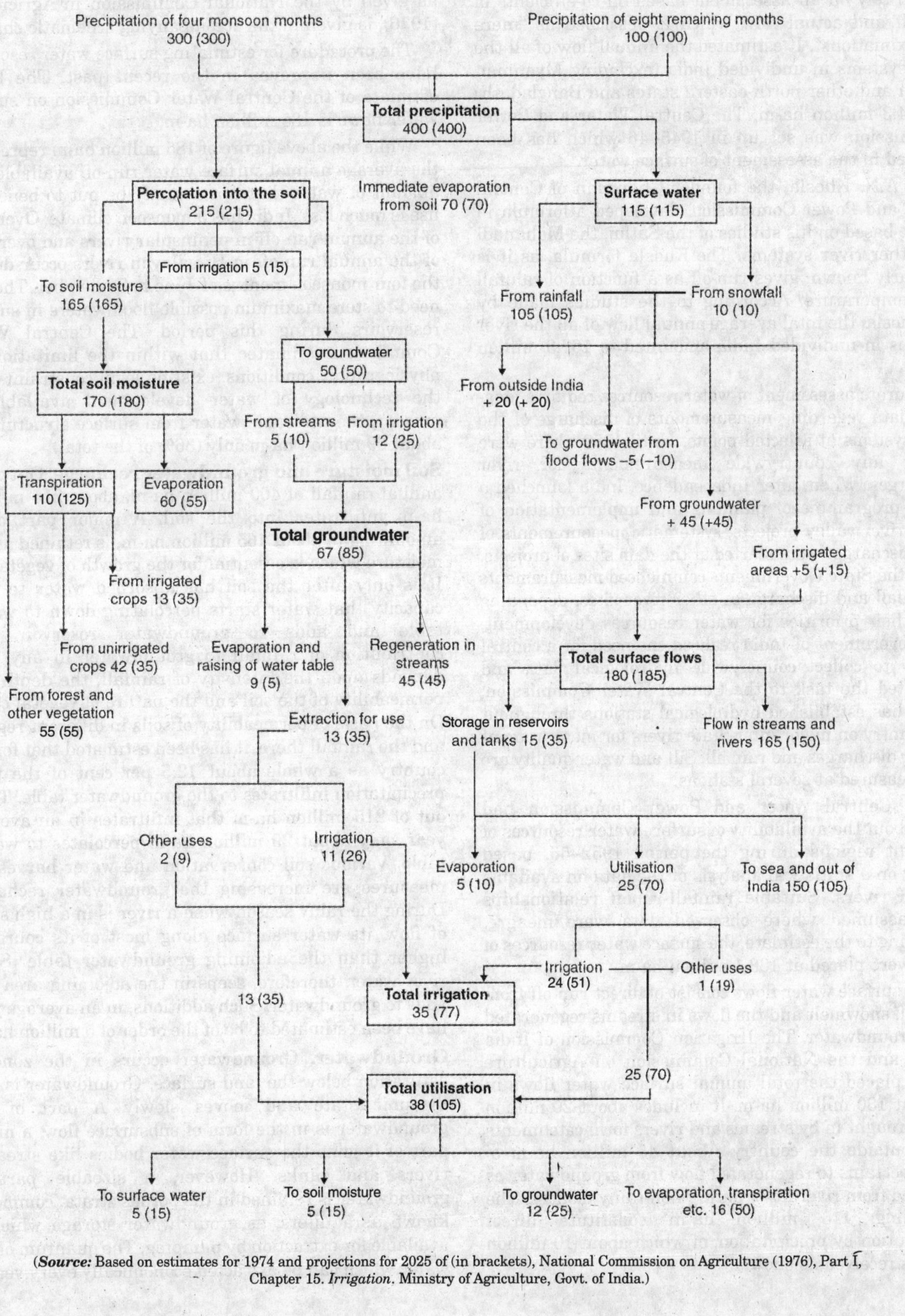

(***Source:*** Based on estimates for 1974 and projections for 2025 of (in brackets), National Commission on Agriculture (1976), Part I, Chapter 15. *Irrigation*. Ministry of Agriculture, Govt. of India.)

inadequate. While records on rainfall were available, data on river flows were not available. The Commission had to rely on an assessment based on co-efficients of run-off and actual flow which it described as "mere approximations". It estimated the annual flow of all the river systems in undivided India (excluding Myanmar, Assam and other north-eastern states and Bangladesh) as 144.3 million ha-m. The Central Water and Power Commission was set up in 1945–46 which has been engaged in the assessment of surface water.

Dr. A.N. Khosla, the founder Chairman of Central Water and Power Commission, developed a formula in 1940's, based on his studies of the Satluj, the Mahanadi and other river systems. The Khosla formula, as it is popularly known, gives run-off as a function of rainfall and temperature. According to the studies made by Dr. Khosla, the total average annual flow of all the river systems in undivided India amounted to 167.3 million ha-m.

Accurate assessment of water resources requires long-term data regarding measurements of discharge of the river systems at selected points. Until 1950, there were hardly any countrywide measurements for river discharges. When, after Independence, India launched a major programme of planning and implementation of major river valley projects, systematic measurements of river discharges were started at the dam sites of projects. Later, the State Governments commenced measurements of rainfall and discharge at other river sites, depending upon their priorities for water resources development. The Government of India realised the need for a central agency to collect countrywide hydrological data and entrusted the task to the Central Water Commission, which has established hydrological stations throughout the country on major inter-State rivers for measurement of river discharges and rainfall. Silt and water quality are also measured at several stations.

The Central Water and Power Commission had worked out the availability of surface water resources of different regions during the period 1952–56, based largely on a statistical analysis of the data on available flow of rivers. Suitable rainfall-runoff relationships were assumed where observed data were meagre. According to the estimate, the surface water resources of India were placed at 188.1 million ha-m.

The surface water flows consist of direct run-off from rainfall, snowmelt and the flows in streams regenerated from groundwater. The Irrigation Commission of India (1972) and the National Commission on Agriculture (1976) placed the total annual surface water flows in India at 180 million ha-m. It includes about 20 million ha-m brought in by streams and rivers form catchments lying outside the country. About 45 million ha-m of water pertains to regenerated flow from groundwater as assessed from river flows during non-rainy months. The remaining 115 million ha-m constitute direct contribution by precipitation, of which about 10 million ha-m is received as snowfall

Estimates on distribution of average annual water resources of India in 1974 and projection for 2025 A.D. as given by the National Commission in Agriculture (1976), is given in the accompanying schematic chart.

The procedure for estimating surface water resources have been improved in the recent past. The latest estimate of the Central Water Commission on surface water flows is 185 million ha-m.

While the above figure of 185 million ha-m represents the average annual surface water run-off available, the amount of water that can actually be put to beneficial use is much less. India has a monsoon climate. Over 90% of the annual run-off in peninsular rivers and over 80% of the annual run-off in Himalayan rivers occur during the four monsoon months of June to September. There is need to store maximum possible flood waters in surface reservoirs during this period. The Central Water Commission estimated that within the limitations of physiographic conditions, existing legal constraints and the technology of water development available at present, the utilisable water from surface structures is about 69 million ha-m only (36% of the total).

Soil moisture and groundwater recharge. Out of the annual rainfall of 400 million ha-m, about 215 million, ha-m infiltrates into the soil. A major part of it, amounting to about 165 million ha-m, is retained as soil moisture which is essential for the growth of vegetation. It is only after the soil has absorbed water to field capacity that water starts percolating down to water-table and adds to groundwater reservoir. The contribution of rainfall to groundwater in any area depends upon the intensity of rainfall, the depth and permeability of the soil and the nature of vegetal cover. On the basis of permeability of soils in different regions and the rainfall there, it has been estimated that for the country as a whole about 12.5 per cent of the total precipitation infiltrates to the groundwater table. Thus, out of 215 million ha-m that infiltrates in an average year, only about 29 million ha-m percolates to water-table. Various soil conservation and water harvesting measures are increasing the groundwater recharge. During the rainy season when a river is in a high stage of flow, its water surface along most of its course is higher than the adjoining groundwater table. Some river water, therefore, seeps in the adjoining area and adds to groundwater. Such additions, in an average year have been estimated to be of the order of 5 million ha-m.

Groundwater. Groundwater occurs in the zone of saturation below the land surface. Groundwater is in a dynamic state and moves slowly. A part of the groundwater is in the form of subsurface flow, a major part of it joins the surface water bodies like streams, rivers and tanks. However, a sizeable part of groundwater is retained in the porous strata, commonly known as aquifers, as groundwater storage which is available for extraction by pumping. The quantum of the water which can be extracted economically every year is considered as the groundwater potential. It implies that water extracted in any year should be returned to the aquifers in the form of recharge from the succeeding precipitations, so that over a cycle of 2, 3, or 5 years the groundwater table does not go down progressively. When there is an over-exploitation of groundwater, the water tables go down. Such over-exploitation has occurred at many places in India. This has induced saline waters of the sea into the coastal areas at several places or has resulted in lowering of water table below the economic level of pumping.

With shallow water aquifers, it is economical to utilize groundwater, as compared to surface water through storage reservoirs and canal systems. With the advent of electrical and oil engine operated pumps, groundwater extraction has greatly increased, specifically after the introduction of high yielding varieties of crops in the mid-1960s.

Scientific assessment of groundwater resources of India was not undertaken until a few decades ago. Over a period of more than three decades, the Central Groundwater Board has carried out countrywide explorations on the availability of groundwater and has arrived at a preliminary estimate of annual utilisable groundwater as 41.8 million ha-m.

IRRIGATION DEVELOPMENT

Irrigated agriculture has developed most extensively in the arid and semi-arid regions and areas having prolonged dry spells. The practice is essentially to supplement the available rainfall in an area. A review of the development of irrigation in the world is fascinating. It reveals the ingenuity and foresightedness of many of the ancient civilizations in harnessing the resources of nature for the benefit of man. It also shows the importance of careful planning of irrigation systems and efficient water management for maintaining the productivity of the land. The decline of irrigated agriculture in many countries has been due to unstable governments and lack of proper measures to ensure efficient water management.

The principal irrigation practice of the ancient times was diverting the flow of springs and streams with temporary barriers constructed across them and irrigating the adjoining fields. This practice is still common in small scale works owned by farmers or their associations (Fig. 1.6 to 1.8).

The practice of storing water in tanks was subsequently developed. Tanks of various sizes were in existence since the ancient times (Figs. 1.9, 1.10, 1.21, and 1.22). Major river diversion projects to harness waters of large rivers were constructed by ancient rulers using indigenous technology. However, modern well designed large scale irrigation systems were developed mainly after the 18th century. Irrigation technology made rapid strides in the 20th century.

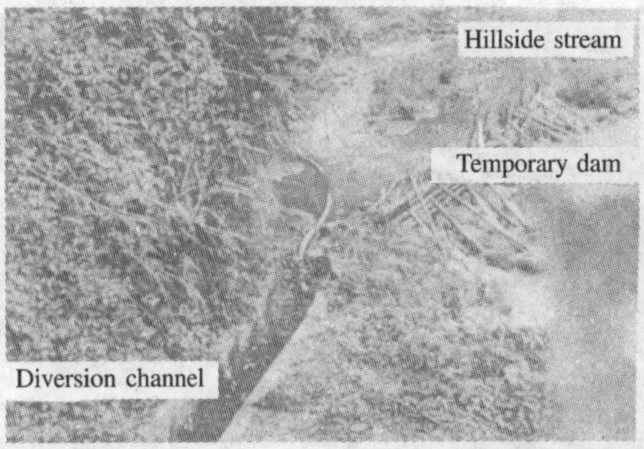

Fig. 1.6. An indigenous diversion near Jharnapani, Nagaland (1971).

Fig. 1.7. A temporary diversion dam near Chiang Mai in Thailand built across a small stream to irrigate the rice fields seen in the background (1973).

Fig. 1.8. An irrigation channel diverted from a hillside river (Kargil, Ladakh, 1975). Such diversion channels irrigate over one million hectares of land in India.

Fig. 1.9. A masonry tank for storing the flow from a hillside spring, near Solan in Himachal Pradesh. The tank is owned by a group farming society (1976). Note the lined channel which leads the spring flow to the tank.

Fig. 1.10. A minor irrigation tank at Sukhomanjri, near Chandigarh which has become a model for soil and water conservation (1976). Note the earth dam and the storage reservoir. In the foreground is a highly undulating terrain of the Shivaliks which is being brought under afforestation programme. In the background is the irrigation command area of the tank.

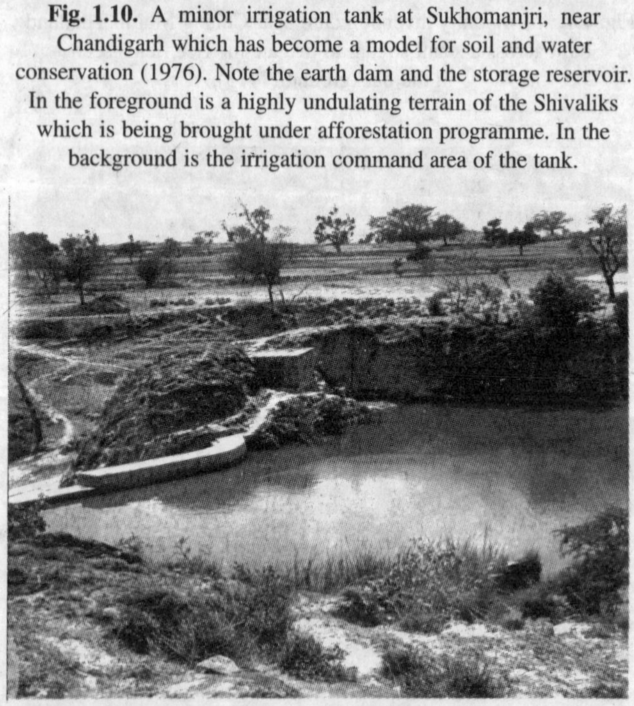

Fig. 1.11. A small masonry dam in a storage-cum-lift type minor irrigation scheme near Bilaspur, Himachal Pradesh. The area in the background is irrigated by installing a pumping set to lift water.

Irrigation Terminology

Storage and Diversion Schemes. Surface water irrigation schemes are of storage or diversion types. Under some situations, a storage-cum-lift scheme may be more suitable. Storage schemes include large and medium reservoirs, tanks and ponds which impound water of rivers and streams. Regions of hilly and undulating topography are suitable for the construction of reservoirs. They comprise of dams with appropriate spillways and sluicing structures (Fig. 1.12).

Fig. 1.12. Upstream view of7 a large masonry storage dam.

Gorge: A narrow passage between mountains which is a possible site for construction of a dam (Fig. 1.15 and 1.25).

Dam: A dam is a barrier, usually of large size, built across a river or stream for impounding water or creating a reservoir. It is used (i) to raise the water level, (ii) to divert water into a canal or conduit, (iii) to create a hydraulic head which can be used for hydro-power generation, (iv) to improve navigability of a river or canal through regulated releases of stored water, and (v) to retain debris.

According to the International Commission on Large Dams, a large dam is 15 m or more high (from the foundation). If dams are between 5-15 metres and have a reservoir volume of more than 3 million cubic metres they are classified as large dams. However, there are variations between countries in the size grouping of dams as large, medium and small.

Freeboard. The vertical distance between the maximum water level and the top of a dam.

Reservoir, tank. An artificially impounded body of water constructed for the purpose of storage, regulation and control of water (Fig. 1.12 to 1.20). Small reservoirs are called tanks in the Indian sub-continent.

The terminologies *tank* and *pond* are often used interchangeably. However, in irrigation practice in India, tanks are small reservoirs formed by constructing embankments across springs or streams. They may be of varying size and are used for irrigation, water supply and groundwater recharge. Tanks which are constructed exclusively for groundwater recharge are called *percolation tanks*. Tanks may be of the *impounding type* which collect and store the flow from springs or run-off of the contributing watershed. They could also be of the *dug out type*, or *built-in type* which are used to store water pumped from wells or diverted from canals.

Auxiliary reservoir. A reservoir constructed to supplement the storage of a main reservoir.

Balancing tank, balancing reservoir, compensating reservoir. A reservoir which is supplementary to the main reservoir. It is used for storing excess flow in the canals for use during high demand periods in irrigation.

Masonry dam. A dam made of bricks or stone or concrete (plain/reinforced).

Masonry-cum-earth dam. A dam comprising portions of earth and masonry along its length.

Arch dam, arched dam. A solid concrete or masonry dam, curved upstream in plan (Fig. 1.16). The arched design resists part of the pressure of the water in the reservoir by its own weight and obtaines a large measure of stability by transmitting the remainder of the water pressure or load by arch action into the canyon walls or other abutments.

Earth dam or embankment dam. A barrier of earth, including clay, sand or sand and gravel, or a combination of earth and rock across a river or stream.

Coffer dam. A temporary dam to exclude water from a work site.

Retention dams. Small earth dams constructed to retain water for short periods of time to control excessive run-off and soil erosion.

Diversion schemes. Diversion schemes provide gravity flow by diverting river or stream water without any appreciable storage. They are more economical than storage schemes, but their feasibility depend on the extent of flow in the river or stream during periods when irrigation is required. They comprise of a weir or barrage constructed across the river or stream for diverting water (Fig. 1.13 and 1.14).

Diversion weir. A low dam across a river or stream for the purpose of diverting all or part of the flow into canals or channels taking off from either or both its banks. The canals are named on the basis of the side from which it originates. Accordingly, they are right bank or left bank canals (Fig. 1.13).

Diversion weirs are called *anicuts* in South India, *bandhara* in Maharashtra and Gujarat and *thingal* in the Assam region.

The term *weir* ordinarily means a low dam or wall built across a stream to raise the upstream water level. When used for the purpose of measuring the flow, it is termed as *measuring weir* or *gauging weir*.

Pick-up-weir, regulating weir or regulating pondage weir. It is a weir built downstream of and as an adjunct to a dam.

Gravity weir. It is a watertight weir made of stone or brick masonry or concrete which depends solely on its

Fig. 1.13. A diversion weir in the Philippines showing component parts (1973). Energy dissipation structures are seen on the downstream. Note the head regulator at the intake of the main canal in the background.

weight for its stability. It may be of overfall or slope-weir type.

Slope weir. A weir with a sloping surface extending from the crest. The overflow water remains in contact with the weir surface till the downstream end and is guided by it, as desired (Fig. 1.13).

Solid slope masonry weir. An impermeable slope weir which may or may not have falling shutters for headwater control.

Permeable weir. A weir made of loose and permeable construction such as loose rock, stones, timber or brushwood.

Barrage. Barrage is a barrier across a river provided with a series of gates to regulate water surface level upstream and to divert flow into the canal on one or both its banks (Fig. 1.14). It is distinguished from a weir that it is gated over its entire length and may not have a raised sill. In Australia, barrages are primarily used for preventing salt water upstream in a tidal river.

Aqueduct. It is a cross-drainage work in which the canal is carried over a drainage canal or natural stream (Fig. 1.17). The bottom of the aqueduct is above the high flood water level in the river or drainage canal. An aqueduct is supported on piers. It is usually constructed of masonry or concrete. Small aqueducts for on-farm water conveyance may also be of timber or steel and are sometimes called *flumes*.

Canal syphon, inverted syphon, canal super-passage. A structure with a closed conduit designed to run full and usually under pressure to convey (1) canal water under roads, railway embankments, drainage canals and depressions, or (2) drainage of water under railway embankments, roads or other structures.

Canal overflow spillway. It is a by-pass structure usually designed to operate automatically, to discharge surplus and waste water into a waste water disposal channel. The spillway allows non-erosive passage of the excess water above a set elevation in a canal. In some designs it also helps to wash out weeds and other floating material.

Fig. 1.14. A view of the Sarda Barrage on river Sarda, Uttar Pradesh.

[*Courtesy:* Min. of Information and Broadcasting, Govt. of India. (PIB Photo File No. 31913).]

Canal supply regulation. The process of manipulating and distributing water supplies at the source of supply as per the demands of irrigation. This includes regulation at different off-taking canals or among branches and distributories on the same canal. The purpose is to provide equitable distribution of water and deliver the required quantity at various canal outlets. It is also designed to prevent or reduce the silt entering the different reaches in a canal system.

Wadi. A ravine or valley which is dry except in the rainy season. It may also refer to a stream running through such a valley. In-between the irregular floods, they are either dry or carry scanty base flow.

Irrigation canals. Canals meant for conveying and delivering water from headworks to the fields. They include main canals, branches, laterals, distributories and water courses.

Inundation canal. A canal in which the flow depends on the surface level of the water in the river above a certain limit.

Contour canal. A canal which is aligned to follow the contour. It can irrigate the land on its lower side (having lower elevations).

Ridge canal. A canal aligned along the ridge of a terrain. It can irrigate land on both its sides.

Command area. The area that can be irrigated from an irrigation canal or canal system by gravity flow due to its location higher than the surrounding land.

Ayacut. The term *'ayacut'* is often used in South India to mean the irrigation command area of an irrigation system.

Field command. It is the difference in elavation between the water level in the water course and the level of the highest point in the area to be irrigated.

Gross command area (GCA). It is the total area which can be irrigated by a certain channel or a project. It includes the area covered by roads, culverts, unculturable areas, villages, and other areas which are not available for cultivation within a command area.

Water right. A quantity of water per year to which each farm or user within an irrigated area is legally entitled. A permit to claim the use of water for beneficial purposes is limited by its economical use and subjected to other rights of older rate of use, called 'priority' or prior appropriation.

Catchment area, River basin, Drainage basin, Watershed. The area from which a river, stream or reservoir receives surface flow which originates as precipitation.

Conjunctive irrigation planning. Planning an irrigation project in a given area having undergroundwater resources in addition to surface water sources so that both the water resources are so planned as to obtain optimum and most economical use of water from the two sources.

Fig. 1.15. A narrow gorge on a stream near Bilaspur, Himachal Pradesh (1976). Note the lime marking indicating the location of a proposed dam.

Irrigation project. A project meant either solely or primarily for irrigation purposes, including development and improvement of land, although it may incidently serve other purposes.

Spate irrigation. A method of random irrigation using flood waters of a normally dry river system. It includes the construction of earthern diversion embankments across the river bed and canals leading to embanked (bunded) fields, where water is ponded till it is infiltrated.

Gross irrigated area. Area of land irrigated in a year (2 irrigation seasons counting as two).

Net irrigated area. The area of land that receives irrigation water in a year, irrespective of the number of crop seasons (e.g., 2 or 3 irrigation seasons in a year are counted as one).

Kharif. The summer south-west monsoon crop in the Indian subcontinent, with the onset of rains in May/June and withdrawal of rains mainly during September/October.

Rabi. Winter cropping season (usually October/November to March/April).

Irrigation scheduling. Planning the distribution of irrigation water, including specifying allocations and preparing timetables for the amounts of water and locations and making proper adjustments in them.

On-farm development. Land development below the canal outlet, including land leveling/grading, construction of field channels and drains, preventing soil degradation by erosion and providing for soil fertility development through appropriate crop rotations with leguminous crops intervening with cereals/oilseeds and other crops.

Performance improvement. Enhancing the well-being of an irrigation system, including productivity, financial costs and preventing adverse social, health, environmental and other effects.

Fig. 1.16. A view of the reinforced concrete arch dam at Idukki, Kerala—the largest multipurpose river valley project in the Western Ghats region.

(*Source:* Central Board of Irrigation and Power, Intl. Seminar on New Materials and Techniques in Dam Construction, March, 1987.)

Communal irrigation system. An irrigation system managed by a community of irrigators.

Farmer joint management. Irrigation system management on a segment of the canal system, involving both farmers and irrigation staff.

Water basin. A hydrologically delineated area that is drained by a river system.

Water resources. (*i*) Water available or capable of being made available for use in sufficient quantity and quality at a location and over a period of time appropriate for an identified demand. (*ii*) Supply of water in a given area or basin interpreted in terms of availability of surface and groundwater.

Integrated river basin planning. Formulating a course of action involving natural, agricultural, and human resources of a river basin, taking into account the social, economic and institutional factors in operating a river basin to achieve specific objectives. It signifies the interaction of various components with particular emphasis on certain components.

Watershed. An area drained by a river or stream system.

Watershed management. A process of planning and implementing a course of action that involves a region's natural and human resources and taking into account social, political, economic, environmental and institutional factors operating within a watershed and the surrounding river basin and other relevant regions to achieve the desired social goal.

Culturable command area (CCA). It is the portion of the GCA which is culturable or cultivable. The CCA is equal to the GCA minus the unculturable area in the GCA.

Intensity of irrigation. It is the ratio of irrigated to irrigable area. In North India and Pakistan this means the proportion of culturable command area proposed to be irrigated annually.

Irrigation potential. The irrigation potential of a project is the gross area that can be irrigated with the amount of water available under the envisaged cropping pattern. Utilisation is the gross area actually irrigated each year, which is generally assessed on the basis of construction of field channels in the command area of a project. Once a project is commissioned, i.e., the water

Fig. 1.17. An R.C.C. aqueduct beside a railway line in the Krishna Delta Irrigation System, Andhra Pradesh.

course up to the outlet level is ready, the irrigation potential as designed is said to have been created. Even if single watering is provided to a crop in one season and another single watering for another crop is provided in the same year, the irrigation potential created is twice the culturable command area of the project. The single watering may not be sufficient for the crop. In such cases, other sources of irrigation are to be tapped to meet the full requirement of the crop. The utilisation of irrigation potential is said to be full if the project is commissioned and field channels are constructed beyond the outlet level. Assumptions made in assessing irrigation potential vary from country to country. In most cases it is computed on the basis of available land and water resources, but economic and environmental considerations may also have been taken into account.

Permanent crops. Crops which occupy the land for long periods that do not need to be replanted after each harvest. This does not include woodland and forests which are classified separately.

Annual crops. Temporary (annual) crops which are grown during a crop season.

Harvested crops under irrigation. Total harvested irrigated area for the crop for a given year. Areas under double cropping is counted twice.

Cultivable area. Area of land potentially fit for cultivation. This term may or may not include part or all of the forests and permanent pastures.

Cultivated area. Area under annual and permanent crops. It refers to the area actually cultivated and does not include land which is fallow.

Agricultural water withdrawal. Annual quantity of water withdrawn for agricultural purposes including irrigation and livestock watering.

Groundwater depletion. Annual amount of water withdrawn from aquifers and which is not replenished (mining of rechargeable aquifers or extraction from fossil aquifers).

Safe yield of groundwater. The definition varies between countries. It may or may not include fossil water. For example, in Malta, the safe yield of groundwater is considered less than the renewable groundwater potential in order not to deplete the storage capacity of the main aquifer without causing salt water intrusion. In Qatar, it includes fossil water from the aquifers underlying a large part of the Arabian Peninsula (FAO, 2000b).

Environment. Environment is the combination of all external conditions affecting the life, development and survival of an organism. As applied to water resources projects, environment includes the earth resources of land, water, air and vegetation as well as man-made structures which surround or are directly related to a proposed structure.

Ecology. Ecology is the study of relationship of living things to one another and to their environment. It encompasses the pattern of relationship that exists between organism (plants, animal and human) and their environment. For instance, the location of a dam or canal within an environment should be done with due consideration of its effect on the fauna, especially fish and wildlife populations of the specific area.

Ecosystem. It is the interacting system of a biological community and its non-living surroundings which are linked by the cycling of materials and the flow of energy.

Ecological succession. It is the time succession (sequence) of distinctive plant and animal communities occurring within a given area.

Ecological balance. It is the natural equilibrium (balance) existing between the components of a natural community. Favourable equilibrium is maintained when the relative numbers of a natural community remain more or less constant, thus forming a stable ecosystem.

Ecological impact. It is the total effect of an environmental change, either natural or man-made, on the ecology of an area.

Environment parameters. They are the physical, chemical or biological components of an environmental system and their interactions which can be stated in quantitative terms.

Environmental standards. Limiting conditions of environmental quality, often expressed in numerical terms and usually with legal standing that has been set by a jurisdiction to protect man's health and well-being.

Environmental analysis. It is an analysis of determining whether a particular action under consideration may or may not have a significant effect on the environment. It is the evaluation of the short and long term, primary and secondary environmental effects of an action.

Environmental assessment. It is a written analysis describing the environmental impacts of the proposed actions.

Environmental impact. It is the effect of any kind on any component or the whole of the environment. Environmental impact assessment generally comprises of two major elements—a quantitative measure of magnitude and a qualitative measure of importance.

Effluent. A liquid flowing out of a containing space (e.g., factory).

Environmental elements. They are identified in three categories, namely physical, chemical and biological. They include the location, space, mass or resource complex which may directly to indirectly influence the environment by a particular action.

Environmental forecasting. It is the technique of predicting environmental effects of proposed developmental activities.

Pollution. Pollution is the presence of matter or energy whose nature, location or quantity, produces undesirable environmental effects. It could be any substance which is likely to create a nuisance problem or render the air and/or water supply injurious or detrimental to living organisms.

Pollution load. It is the quantity of material in a waste stream that requires treatment or exerts an adverse effect on the receiving system. *Primary pollutants* are emitted directly from a source. Toxic agents are substances which will kill or injure an organism through their chemical or physical action or by altering their environment. *Toxicity* is the degree of danger posed by a substance to animal or plant life.

IRRIGATION DEVELOPMENT IN THE WORLD: A BRIEF HISTORICAL REVIEW

Irrigation is as ancient as civilization. In the recorded history of ancient times, there are numerous references to the practice of irrigation from wells, tanks, canals and directly from rivers. In the *Vedas*, rituals were prescribed for the inaugural ceremonies of canals. Among the more notable irrigation works in the world were those of Egypt, Iraq, India and Pakistan, China, Peru and Mexico. The extensive irrigation works in these countries reflected the vision, planning and technical skill of the civilizations they served.

More than six major irrigation-based civilizations arose betwen 2,000 and 6,000 years ago. They include the Sumerians, Babylonians, and Assyrians, who thrived in the basins of the Tigris and Euphrates rivers of present-day Iraq. The longest-lasting irrigation society was developed in Egypt in the valley of Nile River. The Harappan civilization developed in the Indus River valley of present-day Pakistan. The Yellow River basin of north-central China saw major developments of irrigation works during the period. Sometime later, irrigation-based culture arose in Central Mexico, Coastal Peru, and the American Southwest, each of which saw the rise and fall of advanced societies dependent on irrigated agriculture.

The influence of irrigation on the development of the above societies and their vulnerability and environmental deterioration offer lessons for global irrigation societies of modern times. The overriding lesson is that most of the irrigation based societies of the ancient times failed due to faulty water management.

Rise and Fall of the Mesopotamian Civilization

The Sumerians migrated from the Mesopotamian highlands in around 4000 B.C. and settled in the lowland plains of southern Mesopotamia, called the 'Fertile Crescent'. They developed extensive irrigation facilities in the Mesopotamian plains, utilizing the water of the Tigris and Euphrates rivers. Sumerian farmers practiced check basin and furrow methods of irrigation, which are still used on about 90% of irrigated land in the world today. The Euphrates was easy to tap, because the river bed lay above the level of the surrounding plains due to heavy siltation of the river. Irrigators had only to breach the levee alongside the river to take water to the canals. The main canal would then supply water to branch canals, which in turn delivered water to the irrigation ditches from where water was turned to the farms (Sandra Postel, 1999).

Akkadian empire. In about 2300 B.C., one of the ministers of a Sumerian kingdom, Sargon of Akkad, took control of the independent cities and villages of Sumer and established the first known human empire. Akkadian empire encompassed about 200 km from the irrigated plains of southern Mesopotamia to the rainfed agricultural plains of north-eastern Syria. The empire prospered for only about 100 years. It became vulnerable to the strains of population pressures, water shortages, and land degradation.

Babylonian empire. With the rise of Babylon in the 18th century B.C., political power shifted towards northern Iraq. The former cities of Sumer which were the earliest urban societies were in ruins. King Hammurabi, who ruled during 1792-50 B.C. is known for his historic code of law, The Code of Hammurabi contained 285 laws, and several of them dealt with irrigation. The following provision was made in the Code:

> *"If anyone opens his irrigation canal to let in water, but is careless and the water floods the fields of his neighbour, he shall measure out grain to the latter in proportion to the yield of the neighbouring field."*

The irrigation works during Hammurabi's period revealed the greatness Babylon achieved. A large canal was built between the old Sumer city of Kish, south of Babylon, to the Persian Gulf. This greatly expanded the irrigated area of the empire and provided flood protection to the southern cities. Hammurabi's dynasty ended with an invasion by the Cassites. It took several hundred years before the next major empire of the Assirians could initiate the advancement of irrigated agriculture (Sandra Postel, 1999).

Assyrian empire. Like the Babylonians, the Assyrians took pride in building irrigation canals and boosting agricultural production. The Assyrians built an extensive network of canals, along with small reservoirs for storing run-off from the surrounding hills. Perennial irrigation systems were constructed for intensive croping and for orchards. The Mesopotamian society reached its peak of glory between the third and seventh centuries A.D. under the rule of the Sassanians. They planned and constructed many water resources schemes which expanded irrigation throughout the region. Heavy silt loads of Eupbrates and Tigris rivers resulted in the silting of many of the irrigation works. Inadequate drainage resulted in the build-up of salt in the once fertile areas.

In the centuries that followed, after the Arab conquest of Mesopotamia in A.D. 639, the population reduced and the tale of the great Mesopotamian societies virtually ended. By the sixteenth century (A.D.), the Fertile Crescent of Mesopotamia, from which the origin of human civilization could be traced was almost a salty wasteland. These societies had not developed a system of irrigated agriculture that could be sustained.

Ancient Irrigation Works of Yellow River Basin in China. The Chinese are known to have practiced irrigation since 2600 B.C. The earliest irrigation works in China relate to the development of the delta lands on the Yellow River. Extensive irrigation systems were planned by Li Ping, Governor of Chengtufu in 300 B.C. and were constructed after him by his son. The 1130 km long Imperial Canal (also called the Grand Canal), built in 700 A.D. for irrigation and navigation, is one of the greatest ancient irrigation works.

Diversion of water from the Yellow River to irrigate paddy lands dates back to the early Han Dynasty of 200 B.C. to A.D. 200. The Yellow River has been difficult to control. It carries huge quantities of silt each year which made the river prone to frequent changes in the direction of flow, channel meandering and flooding. It is reported that the Yellow River breached its dikes more than 1500 times over the 2000 years prior to A.D. 1950 (Sandra Postel, 1999). The Chinese had to make enormous efforts to continually raise and reinforce the dikes to hold back the flood waters. The ingenuity and hard work of the ancient Chinese could create an enterprising civilization in the valley in spite of the problems (Sandra Postel, 1999).

Egyptian Civilisation of Nile Valley. Early recorded history shows that irrigation was practiced in the Nile Valley of Egypt between 3000 B.C. and 2500 B.C. Technical developments recorded during the period include the construction of artesian wells, the Nilometer for guaging streams, and large canal systems for irrigation and navigation. Water lifting devices like the swing baskt, the *shaduf* and the *sakia*, all of which are still in use in the Nile valley originatd during the period.

The first Pharaoh of Egypt, King Menes in about 3000 B.C. introduced a system of basin irrigation along the Nile. In about 2000 B.C., diversion was taken from the Nile through a canal 19 km long into Lake Moeris. During the flood period, excess water was stored in the lake. The stored water was diverted to irrigated fields in the dry season. The Egyptians practiced water management through large basins. It comprised of a network of earthen banks that formed basins of various sizes. Regulated slucies were built to direct floodwater into a basin. The excess water, was drained to a lower level basin or to a drainage canal. The farmers of the drained plot, then undertook cropping operations. The Egyptian farmers practiced crop rotation to maintain soil fertility. They planted grain and legumes in alternate years, which helped to maintain soil productivity. Irrigated agriculture in the Nile Valley stood the test of time for the past over 5000 years without serious problems of sustainability.

Early Irrigation Works of Iran. Great irrigation works were developed in Iran (Persia) in about 500 B.C. under the rule of Darius. A system of *Kanatas* (tunnels), tapping groundwater from the recharge areas of surrounding mountains and leading it by gravity to the fields to be irrigated, is still used in Iran as it was practiced 2500 years ago.

Ancient Irrigation Systems of Japan and South-East Asia. Diversion of mountain streams by canals for irrigation was practiced in Japan in the Yayoi Era (before 600 B.C.). Irrigation by diversion works and direct flooding by rivers is known to have been extensively practiced in Thailand, Philippines and Indonasia.

Irrigation Works of Early Americas. Irrigated agriculture was prevalent in the Americas around 1000 B.C. The Incas of Peru had developed extensive irrigation works with considerable ingenuity. The Spaniards on their entry into Mexico and Peru, observed elaborate irrigation works which were in existence for many generations. Remains of an early civilization based on irrigated agriculture are evident in the United States of America. The remnants of ancient canals are visible along the Gila River in Arizona and in the valleys of Rio Grande. Development of irrigated agriculture based civilization in the western hemisphere was later in time and smaller in size than their counterparts in the orient. In coastal Peru, the ancient Chimu empire developed into an advanced economy from irrigation based agriculture. Beginning around 1000 B.C. and extending to A.D. 1500, it grew to a level in many ways comparable to that of ancient Mesopotamia (Sandra Postel, 1999).

Irrigation Development in Indian Sub-continent

The dependence of India's agriculture on the South-West monsoon and its consequent vulnerability have been recognised since ancient times. The importance given to irrigation and water management in ancient India is seen from the following excerpts from history:

> *"No grain is ever produced without water, but too much water tends to spoil the grain. An inundation is as injurious to growth as dearth of water."*
>
> *Narada Smriti XI, 19 (600-300 B.C.)*

Ancient Indian literature and history refer to wells, tanks and canals and the importance given to their construction by the rulers for the welfare of the people. The ruins of Indus valley civilization in Harappa, Mohenjodaro (of the Sindh province of Pakistan), Lothal (Gujarat), Inamgaon (Maharashtra) and other places in the northern and western India, provide evidence on irrigation practices in the form *of bundhs* (small dams) and canals.

By about 3000 B.C. the Mohenjodaro civilization of the Indus Valley, spanning part of present day Pakistan and India, had developed a variety of tanks and canals for irrigation. These works were destroyed by the invaders from Central Asia sometime around 2000 B.C. While the fall of the Mesopotamian civilization has been due to mismanagement of irrigation systems, the ancient Indus Valley civilization fell due to the wanton destruction by the invaders.

From the *Arthashastra* of Kautilya, the advisor of the legendary Gupta ruler Chandragupta, we get a vivid description of the administration, people and laws in ancient India. Irrigation systems were mostly state owned, though private irrigation facilities were also in existence. Farmers were required to pay tax for utilising the irrigation system. The amount of tax depended upon to extent of area irrigated and amount of crop produced. A quotation from Pali Works and Epics refers to the prosperous irrigated agriculture which prevailed in the Kingdom of Meghatha (Bihar) in 550 B.C.:

> *"The Khettas of Magadha (the land of Buddha) were intersected by network of canals and ridges...watering projects were undertaken by specialists who conducted the water as they pleased".*

The importance given to efficient water management in ancient India is seen from the following writings of Magasthenes, the Greek Ambassador in the Court of Emperor Chandragupta in about 300 B.C.:

> *"The whole country was under irrigation and very prosperous because of two crops grown in a year with irrigation facilities ... The District Officers measure the land and inspect the sluices by which water is distributed into the branch canals (water courses) so that every one may get his fair share of the benefit."*

Only the ruins of most of the ancient irrigation works are found now. The most famous one still functioning is the Grand Anicut on the Kaveri at the head of its delta in Tamil Nadu. A stone masonry weir (*anicut*) was originally built in A.D. 46 by the ancient Tamil King Karikala Chola. The diversion weir was subsequently enlarged in A.D. 1880. With the massive irrigation distribution system from the headworks, an area of 400,000 ha is being irrigated for centuries.

Lessons from the Past

There are several examples to be learnt from the experience of irrigation works of the ancient times. Irrigated agriculture which flourished in Mesopotamia in Iraq is almost extinct, but for the new works introduced by the Government in the 20th century. The loss of the once fertile lands may be attributed to faulty irrigation methods and inadequate drainage. There are also many other regions of the world which experienced a decline in productivity in irrigated agriculture. This has given rise to considerable debate on the long term effect of irrigation, unless adequate protective measures are provided.

The lessons of the past point to the need for (1) a strong government to ensure integrated planning of irrigation systems, and (2) control of irrigation water to prevent waterlogging and salt accumulation. The fact that good crop yields have been maintained under irrigation for a period of more than 4000 years in Egypt and China and about 2000 years in the Kaveri delta in South India, support the view that irrigation can be permanent and productive, provided the system is planned and operated efficiently.

IRRIGATION DEVELOPMENT FROM 1ST TO 20TH CENTURY A.D.

Brief Global Review

It took several centuries of the modern era before scientists could develop the principles of hydraulics and hydrology and for engineers to develop the modern technology to bring in a near revolution in irrigation system development. Much of the scientific and technical foundation of large irrigation system development was built during the second-half of 19th century and in the 20th century A.D. The progressive growth of irrigation in the world during the period 1800 to 2000 A.D. is presented in Fig. 1.18. Table 1.1 presents the net irrigated area in the top 20 countries having large scale irrigation works.

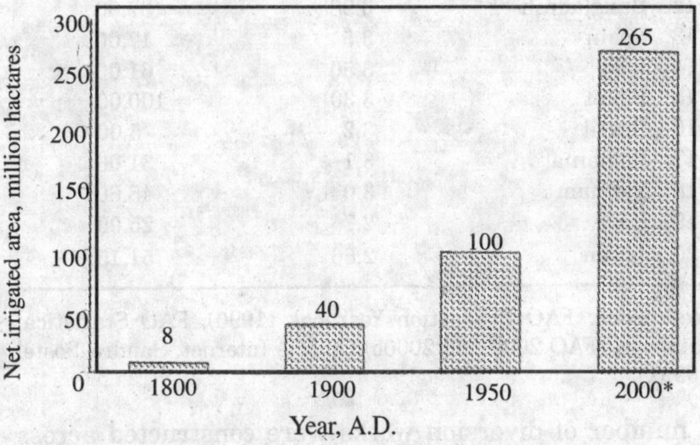

Fig. 1.18. Progressive increase in net irrigated area in the world during the period 1800–2000 A.D.
Estimated reported figure for 1995, 255.5 m ha.

Four countries namely, India, China, the United States of America and Pakistan account for over half of the world's irrigated land (Table 1.1). Major countries including India, China, Egypt, Indonesia and Pakistan depend on irrigated agriculture for more than half of their domestic food production. Crop yield from irrigated agriculture far exceeds that from unirrigated areas under the same climatic situation. With irrigation, farmers can grow two or three crops a year. Thus, irrigation has been a key driver in the size of food production in 20th century. About 40% of world's food is grown in the 17% of cropland that is irrigated.

The birth place of modern irrigation is considered to be Punjab, the *"Land of Five rivers"* the Satluj, Beas, Ravi, Chenab and Jhelam situated in northwest India and eastern Pakistan. The five rivers form the Indus River system and flow across gently slopping plains and joins the Arabian Sea. The river system derives its flow from the melting snow of the North Himalayan ranges and the monsoon rainfall. The annual flow of the Indus system is about 175 km^3 which is about twice that of the Nile River system. The major effort of the British during their rule of the Indian sub-continent was concentrated in the Indus basin.

Table 1.1. Net irrigated area in the 20 countries having large irrigated areas in the world, A.D. 1999

Country	Irrigated area (million ha)	% of net cropped land irrigated
1. India	59.00	34.70
2. China	53.74	10.09
3. United States	21.40	11.00
4. Pakistan	17.95	66.38
5. Iran	7.56	11.25
6. Mexico	6.10	22.00
7. Russia	5.40	4.00
8. Indonesia	4.81	11.41
9. Thailand	4.75	25.36
10. Turkey	4.50	11.54
11. Uzbekistan	4.20	15.21
12. Bangladesh	3.98	37.00
13. Spain	3.5	17.00
14. Iraq	3.50	61.00
15. Egypt	3.30	100.00
16. Brazil	3.2	5.00
17. Romania	3.1	31.00
18. Vietnam	3.0	45.60
19. Italy	2.7	25.00
20. Japan	2.66	51.15

[***Reference.*** FAO Production Yearbook (1996), FAO Statistical Database. (FAO 2000 and 2000b) On line Internet. Sandra Postel (1999).]

A number of diversion works were constructed across the Ravi and Satluj rivers during the first half of the 19th century. An important development during the period has been the introduction of barrages with mechanically controlled sliding gates that could raise the height of a river to the level needed for water to enter an irrigation canal at either or both of its banks. Gates at the head of the canal controlled how much water was admitted. By mid-1940s, the total irrigated area in British India was 20 million ha which was double than that of A.D. 1900. With the partition of India, about 30% of irrigated area, including a major part of the projects on the Indus system went to Pakistan.

Irrigation Development in the USA

The Hohokam, other native Indians, and Spanish settlers had been irrigating the drylands of the American West for centuries. By A.D. 1800, the Spanish communities had built more than 160 irrigation canals along the upper sections of the Rio Grande in New Mexico. Mormon settlers built many irrigation works to transform the valley of Great Salt Lake in the State of Utah. By 1890 more than 100,000 ha of land were under irrigation in Utah state. By about A.D. 1875, the center of irrigation development shifted to Colorado and California. By the end of the 19th century most of the land reclamation that could be accomplished with the then available technology and capital had been completed. By that time, a system of water rights called prior appropriation had spread throughout Western United States. It was a system of allocating water on a "first in time, first in right" basis.

There was the growing need for large size storage reservoirs and diversion works at feasible sites in many of the river systems in USA. Major reservior projects required financial support of the federal government. In 1902, the National Reclamation Act was passed, which opened a new era of irrigation development in Western United States. The Salt River Project in Arizona, which includes the Roosevelt Dam, was completed in 1912. The project created an irrigation potential of 1,00,000 ha. The Hoover Dam, on Colorado River, constructed in 1930s, was a great engineering feat. The design procedures developed in the project were adapted in the construction of large dams in other countries. During the first 50 years of existence, the Bureau of Reclamation of USA constructed 173 dams, including the four highest and largest dams in the world. These projects could irrigate about 2.7 million ha (Mha) of cropland. Irrigated area in the United States doubled from 10 Mha in 1950 to 21 Mha in 1995. During this period, there was substantial increase in the utilisation of groundwater, mainly through deep tubewells. East and Midwest United States have highly productive rainfed areas. Hence, USA is not so much dependent on irrigation as many other countries which have to depend mainly on irrigated agriculture.

Nile Valley Irrigation Development

The basin irrigation system of ancient Egypt was able to provide only seasonal irrigation to comparatively small areas in a predominently desert country. During mid-ninteenth century, series of diversion dams were constructed across the Nile at the head of the fan-shaped delta. A series of other dams were built after 1860, including the first Aswan Dam, completed in 1902. The Aswan Dam was enlarged twice by 1934. These works turned Egypt's irrigated agriculture from a seasonal system to a perennial one. The irrigation works which were completed by then could store about 10% of the average annual flow of the Nile river system. The Aswan High Dam which was completed in 1970 is a monument to Egypts' national pride. The reservior created by the dam has the capacity to store two full years of annual average flow of the Nile.

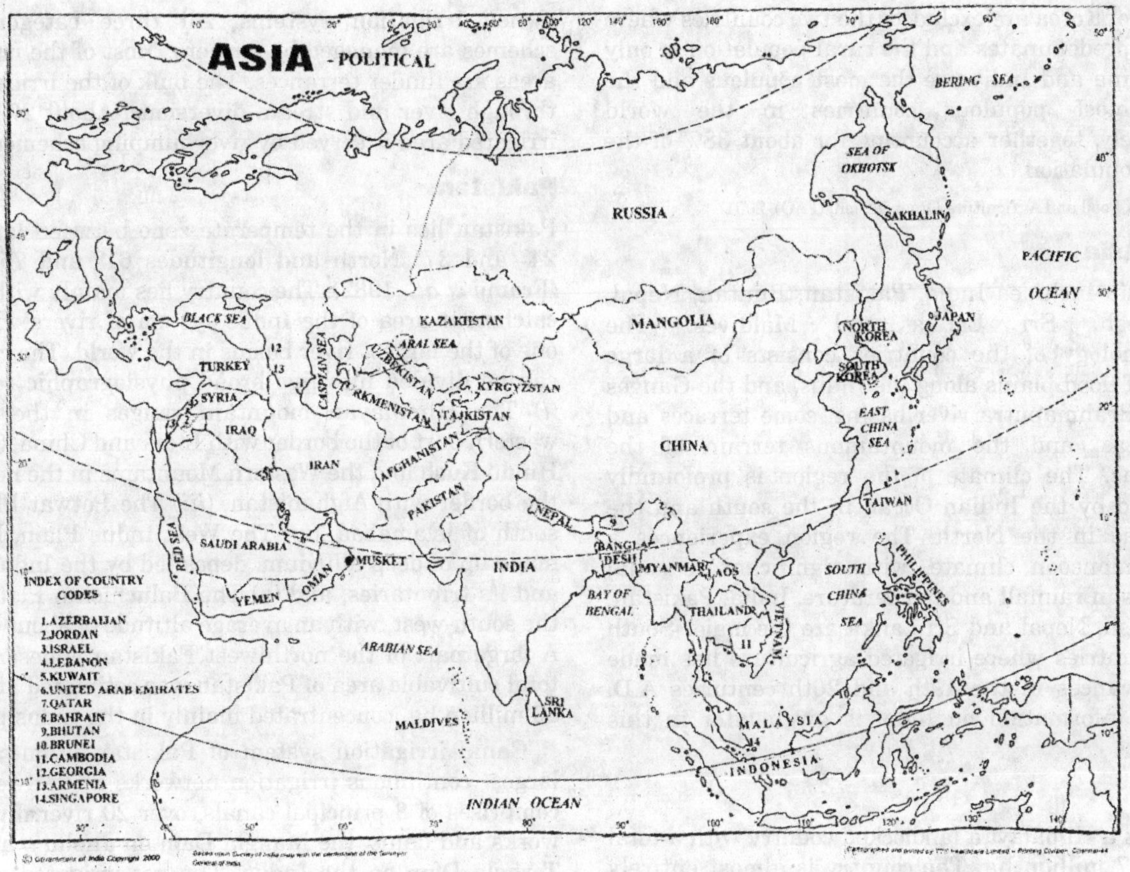

Fig. 1.19. Map of Asia, 2000.

(***Note:*** **International boundaries are not authenticated in some cases. The map is intended to show the locations of countries. Errors/omissions, if any, are unintentional and will be corrected, if brought to the notice of the author/publisher.)**

IRRIGATION DEVELOPMENT IN FORMER SOVIET UNION COUNTRIES

The total water resources of the countries of the former Soviet Union were abundant when all the countries were taken together. However, there is great variation in the distribution of water between countries and their regions. About 84% of run-off of rivers flowed into the Arctic and Pacific Oceans, away from the major population centers and arable lands. Hence, a top priority of Soviet leaders was to expand irrigation in regions with favourable climatic conditions for crop production, but with insufficient or unreliable rainfall. Two key areas which received major attention were the Central Asian republics, which accounted for about 40% of Soviet irrigated area prior to the nation's breakup in 1991, and the south-eastern European region, including parts of Russia and Ukraine. The deserts of Central Asia were converted into major cotton growing regions.

Massive water development took place in the central Asian region where irrigation expanded to 7 million ha by 1980s. Large diversions from the Amu Darya and Syr-Darya rivers that feed the Aral Sea, supplied most of the irrigation water. However, these diversions resulted in serious ecological deterioration. With the breakup of the Soviet Union, Russia and Uzbekistan now rank among the world's top 20 countries in irrigation development.

WATER RESOURCES DEVELOPMENT IN MONSOON ASIA

Monsoon Asia comprises the countries in South Asia, South-East Asian mainland and island countries, Eastern Asia and the Far East having climate, especially rainfall, influenced by the monsoons (Fig. 1.19). Heavy south-west monsoon rains concentrated during the period June to October and comparatively smaller north-east monsoon are the characteristics of major regions of Monsoon Asia. There are prolonged dry spells in the intervening periods, necessitating the conservation of the available rainfall run-off, mainly for use during the prolonged dry periods in most areas. The heavy rains frequently cause problems of waterlogging and accompanying salt build-up which require appropriate measures for prevention and control. The population of Monsoon Asia is predominantly rural: about 67% of the total, compared to 54% for the world as a whole. This percentage rises to 70%, if Japan and the

Republic of Korea are excluded (the two countries where industry predominates and the rural population is only 21%). China and India are the most populous and the second most populous countries in the world respectively, together accounting for about 38% of the world's population.

(***Source:*** UN Food and Agriculture Organization, (FAO), 2001).

South Asia

South Asia includes India, Pakistan, Bhutan, Nepal, Bangladesh, Sri Lanka and Maldives. The geomorphology of the countries consists of a large portion of flood plains along the Indus, and the Ganges and the Brahmaputra river basins, some terraces and hilly areas, and the mountainous terrain of the Himalayas. The climate of the region is profoundly influenced by the Indian Ocean in the south and the Himalayas in the North. The region experiences a tropical monsoon climate, with significant seasonal variations in rainfall and temperature. India, Pakistan, Bangladesh, Nepal and Sri Lanka are the major South Asian countries where irrigated agriculture has made rapid advances in the 19th and 20th centuries A.D. Detailed information on India is given later in this chapter.

Bhutan

Bhutan is a Himalayan landlocked country with a total area of 47 million ha. The country is almost entirely mountainous, with flat lands limited to the broader river valleys and along the foothills. Bhutan has three major landform features: the southern foothills, the inner Himalayas and the higher Himalayas. Owing to the extremely rugged mountainous terrain, only 3.6% of the total area is cultivated.

The south-west monsoon from mid-June to September, brings about 60 to 90% of the annual precipitation, depending upon the region. Precipitation ranges from 2,500 mm/year in the south to 500 to 1000 mm/year in the inner valleys, and less than 500 mm/year above 4000 m altitudes. The peak demand for supplementary irrigation is in April and May.

(***Source***: UN Food and Agriculture Organization (FAO), 2001).

Water Resources and Irrigation Development. Bhutan has swiftly flowing rivers and streams passing through its valleys. Except for a small river in the extreme north of the country which flows north, all rivers flow south to India. Bhutan has a large potential for hydro-power development, only a part of which is developed. The total net irrigated area of Bhutan in 1999 was about 40,000 ha (FAO, 2000). Irrigation schemes can be classified according to their origin into (*i*) Indigenous systems, with little or no government assistance; (*ii*) Farmer-initiated irrigation systems constructed by the beneficiaries with technical and other government assistance and (*iii*) Government owned irrigation systems. All three categories of schemes are managed by farmers. Most of the irrigated areas are under terrances. The bulk of the irrigation is through river and stream diversions. About 2% of the irrigated area is served by river jumping schemes.

Pakistan

Pakistan lies in the temperate zone between latitudes 24° and 37° North and longitudes 61° and 75° East (Framji *et al.*, 1982). The country lies mainly within the catchment area of the Indus system of rivers which is one of the largest river basins in the world. The country can be divided into five broad physiographic regions: (*i*) The Himalayan mountain ranges in the north-western part of the border with India and China, (*ii*) The Hindu Kush and the Western Mountains in the north on the border with Afghanistan, (*iii*) The Potwar Plateau, south of Islamabad, (*iv*) The West Indus Plain, largely made up of deep alluvium, deposited by the Indus river and its tributaries, and (*v*) The Baluchistan Plateau in the south-west, with an average altitude of about 600 m. A large part of the north-west Pakistan is desert. The total cultivable area of Pakistan was estimated at about 30 million ha, concentrated mainly in the Indus plain.

Canal irrigation system of Pakistan is one of the largest continuous irrigation networks in the world. It comprises of 8 principal canals, over 20 river diversion works and dams, the Mangla Dam on Jhelum, and the Terhela Dam on the Indus. The net irrigated area of Pakistan in 1999 was 17.95 million ha.*

After Independence, Pakistan made substantial development in expanding its irrigated area and increasing water management efficiency. The major irrigation projects completed after Independence include the Kalabagh Barrage (1947), the Kotri Barrage (1955), Guddu Barrage (1963) and Taunsa Barrage (1958). The Lower Sind Inundation Canals were converted into weir-controlled canals. The Indus Basin Irrigation System is the largest in Pakistan, covering areas in all provinces. In North-West Frontier Province (NWFP) about 8000 ha are irrigated through pump lifts. In the northern parts of NWFP, about 27,000 ha are irrigated by contour channels off-taking from locally available water sources. In Baluchistan, about 50,000 ha are irrigated from about 800 *karezes* (underground tunnels tapping an aquifer and perennial springs). They are generally small schemes of 50 and 400 ha each in area. About 130,000 ha are irrigated from infiltration galleries or small weirs. In 1989, over 30,000 tubewells were used for irrigation in Pakistan, most of them being privately owned.*

Spate irrigation can cover an area of over 1.4 million ha. It relies on the floods of the hill torrents. The run-off is harnessed for irrigation by weirs or temporary diversion structures. Farmers divert the spate flow onto their fields by constructing bunds. Flood recession

*(***Source***: UN Food and Agriculture Organization (FAO), 2000).

cropping areas (*Sailaba* cultivation) cover an area of about 1.23 million ha in Pakistan. It utilizes the moisture retained in the crop root zone after the flood subsides as well a capillary rise of groundwater.*

Irrigation is the mainstay of food security and sustainability in Pakistan, with its predominantly semi-arid and arid climates. The major challenges facing sustainability are the increasing trends in waterlogging and salinity, ensuring greater equity in water distribution to beneficiary farmers, and environmental management. Water losses from Pakistan's over 8,500 water courses have been estimated to be 35% of the water reaching the canal outlet (Framji *et al.*, 1982). Major emphasis has been laid on the improvement of water courses. With the introduction of large scale irrigation in the Indus Basin, the groundwater table started to rise menacingly. Waterlogging and salinity rendered extensive areas unproductive.

Intensive efforts to develop solutions to the drainage problem of Pakistan were initiated in 1954 when a comprehensive programme of investigation on hydrologic, groundwater and soil characteristics was started in the Upper Indus Plains. The investigation generated data for planning drainage and reclamation programmes under Pakistan's Salinity Control and Reclamation Project. It was the first of a series of similar large scale projects in which drainage was achieved by pumping groundwater for augmenting canal water supplies. Two major surface drainage projects were undertaken, namely, the Sukh-Beas Drainage Scheme and Left Bank Overfall Drain.

Bangladesh

Bangladesh is located between latitutdes 20° 30′ N and 26° 45′ N longitudes and 88° E and 92° 45° E (Framji *et al.*, 1981). The country is endowed with fertile soil, abundant water resources and a high ratio of cultivators to land. In spite of these attributes agricultural productivity is low, even though significant progress has been made in recent years. Bangladesh has 14.4 million ha of land area, of which the net cultivable area is estimated at 8.774 million ha. A greater part of Bangladesh lies within the combined Ganga-Brahmaputra-Meghana River system. The rivers overflow their banks with about 3.2 to 4 million ha getting flooded every year.

Irrigation development. A system of water control and irrigation existed in the Ganga delta for centuries. Large irrigation projects were almost unknown till the first-half of the 20th century. Since 1955 a number of irrigation, drainage and flood control projects have been initiated. The flood protection plan of Bangladesh envisages the construction of embankments along the major rivers, development of a network of drainage canals and the improvement of river sections to facilitate flow. The Coastal Embankment Project of Bangladesh is one of the largest of its kind in the world. It is to provide protection to about 1.38 million ha of land against tidal salt water intrusion and floods during the monsoon. The project envisages the construction of about 5000 km of earthen embankments along the banks of rivers, drainage canals and tidal estuaries. Automatic flap gates installed on the outlet structures prevent the intrusion of salt water into the adjoining land. Sluices are provided in the flood protection embankments to facilitate drainage.

The irrigated area in Bangladesh increased from 0.501 million ha in 1961–65 to 1.282 million ha in 1968 (Framji *et al.* (1981) and 3.985 million ha in 1999.*

A Master Plan Organization for water resources development of Bangladesh has been prepared in 1985 and is continuously being upgraded. It provides integrated plans for irrigation, flood control and drainage.

Groundwater potential. The groundwater storage in the alluvial plains of Bangladesh, comprising over 80% of its land area, is made of three components (a) upper layer of silty clay, (b) zone of fine to very fine sands averaging about 20 m thickness, and (c) thick column of medium to fine or medium to coarse sand with layers of clay and silt extending from a depth of about 150 m to over 800 m.

The principal shallow aquifer generally lies within 100 m of the surface under a cover of clay. The clay layer generally increases in thickness southwards from less than 2 m in Dinajpur and Rangpur districts to over 50 m in south and south-eastern parts of Bangladesh. Deep aquifers are present in most parts of Bangladesh at depths of 300 m and more. It provides the potential for high discharge deep tubewells. A single heterogeneous aquifer overlain by silty clays is the common groundwater reservoir of Bangladesh. Salinity is a serious problem in coastal Bangladesh. Low flows in rivers in the dry season allow salinity to penetrate far inland.

Planning policy of surface water irrigation systems. The single source planning concept generally adopted in Bangladesh is that there will be no groundwater development in areas served by surface water irrigation schemes. It is often argued that the same farmer should not take advantage from surface water and groundwater schemes.

In Bangladesh, the expansion of minor irrigation is vital. Irrigation through major canals covers only 6% of the total irrigated area, the remainder being classed as minor irrigation consisting of low lift centrifugal pumps drawing water from rivers, creeks and ponds, shallow tubewells, deep tubewells and manually operated water lifts. In 1991, the National Minor Irrigation Development Project was established to meet the requirement for increased private sector investment in irrigation.

*(***Source***: UN Food and Agriculture Organization (FAO), 2000).

The conjunctive use of surface and groundwater has the following specific applications to Bangladesh:

1. Inside large scale irrigation schemes, groundwater pumping can bring highland areas within the project under groundwater irrigation and lower the water table and prevent waterlogging.
2. Facilitate groundwater recharge from major rivers by lowering groundwater levels.
3. Use groundwater where it is plentiful to allow the scarce surface water to be used in areas with inadequate groundwater resources.

Nepal

Nepal lies between the latitudes 26° 20′ N and 30° 10′ N and longitudes 80° 15′ E and 88° 15′ E (Framji *et al.*, 1982). The country located entirely in the Ganges basin, has 14.8 million ha of land area. Three major physiographic units can be distinguished in Nepal: (*i*) The high Himalayas in the north, covering 23.4% of the land area. (*ii*) The mid-hills in the central region comprising 56.2% of the land area. (*iii*) The plains (Terrai and Inner Terrai) in the south covering 20.4% of the land area. Agro-climatic and socio-economic variations within the physiographic regions are wide, specially in respect of soils and topography. The land use for mountains and high-hills include pastures, for mid-hills fruits and forest development, and for plains food grain production. The mean annual rainfall is 1500 mm. The southwest mansoon (June to September) brings in more than 75% of the total rainfall.

There are more than 6,000 rivers in Nepal, all of which drain into the Ganga River. Nepal comprises of five river basins, namely (*i*) the Mohakali River basin, (*ii*) the Karnali River basin, (*iii*) the Gandaki River basin, (*iv*) the Kosi River basin, and (*v*) the southern river basin. The seasonal distribution of flow in the rivers varies widely between the monsoon and post-monsoon periods.

Groundwater resources. Much of the *terrai* physiographic region and some parts of Siwalik valleys are underlain by deep or shallow aquifers, many of which are suitable for exploitation. Installation of deep tubewells for irrigation is progressively on the increase.

Irrigation development. From ancient times farmers in Nepal, working individually or in groups used to harness water for irrigation by cutting canals from the banks of rivers. The first attempt at providing well designed irrigation system was in the district of Saptari in 1924. Many dams and diversion works were constructed since then. Important projects completed during the Five Year Plans period include Tikabharable Canal, the Mahadukhole Canal, the Bagmati Canal, the Tinau Project and Chandranadhar Project (Framji *et al.*, 1982). A number of major projects have been at various stages of completion in 1980s. The net area irrigated in 1999 was 1.135 million ha.*

Irrigated areas in Nepal are often classified as public systems and farmer managed systems. Public systems comprise about 25% of the total. They cover an area of over 283,000 ha, (91% dependent on surface water and the remaining on groundwater). In the Public scheme, river diversions comprise 76%, river pumping 13%, reservoirs 2% and groundwater 9%. Farmer managed irrigation systems amount to 75% of the total irrigated area, covering about 0.85 million ha.

*(***Source***: UN Food and Agriculture Organization (FAO), 1999)

Sri Lanka

Sri Lanka, a tropical island in the Indian ocean, lies between the latitudes 5° 55′ N and 9° 50′ N and longitudes 79° 42′ E and 81° 52′ E (Framji *et al.*, 1982). The country is located to the east of the southern end of India, and has a total area of 6,463 million ha. The island comprises of a central mountain range sloping on all sides and rivers flowing in different directions. The landforms comprises of plains, midland and the central highland regions. The coastal plains are narrow in the west, south and the east. It broadens to a vast terrain in the north. The average elevation of the coastal plains is 122 m and the midlands 488 m.

Rainfall in Sri Lanka is unevenly distributed and comprises both south-west and north-east monsoons. The land area can be broadly divided into two distinct regions—the *wet zone* and the *dry zone*. The wet zone receives most of the rainfall during the south-west monsoon. Agriculture in the dry zone depends heavily on irrigation. Tank irrigation has been common since the ancient times.

Surface water sources. Sri Lanka, with its central hill *massif*, has generally a radial pattern of the rivers. There are 103 rivers and streams draining the island. The largest and the longest river is the Mahaweli Ganga with a length of 327 km. All the rivers offer ample potential for utilisation for irrigation and power generation. Trans-basin water transfer is considered desirable from the wet zone to irrigate the dry zone.

Groundwater resources. Groundwater has been extensively used in Sri Lanka since ancient times for domestic purposes using shallow open wells. Sri Lanka's largest aquifer extends over 200 km in the northwestern and northern coastal areas. There were about 15,000 tubewells in 1990s.

Irrigation development. Irrigation works had flourished in Sri Lanka for over twenty centuries. Traditionally, rainwater was harvested by networks of tanks. Ancient irrigation works were concentrated in the dry zone. Many of the ancient tanks are functioning even now. The first dam in Sri Lanka is believed to have been constructed in 504 B.C. by King Panduwasa. From the time of Panduwasa up to the 12th century over 10,000 tanks were constructed. After the 12th century, many of the ancient tanks and irrigation works fell into ruin. In the nineteenth century, some of the tanks were restored (Framji et al., 1983).

In 1952, Gal Oya, a large multipurpose scheme was launched, followed in the 1960s by Mahaweli, the largest multi-purpose scheme. The Mahaweli project has been planned to irrigate over 300,000 ha land and generate 800 MW of hydropower.

In Sri Lanka, irrigation systems are almost exclusively designed for paddy irrigation. The net area irrigated in 1999 has been 0.662 million ha.* Flood control and drainage schemes have been incorporated into the irrigation system mainly in the lower reaches of rivers. In the coastal areas, saltwater exclusion schemes have been provided. Embankments and pumps are the main features in flood protection schemes, whereas gated regulators are adopted in saltwater exclusion schemes.

*[***Source***: UN Food and Agriculture Organization (FAO), 2000).]

SOUTH-EAST ASIA

South-East Asia, with its monsoon dominated rainfall characteristics and tropical climate, is one of the major agricultural regions of the world. Rice-based farming systems and predominance of plantation crops including coconut, rubber, spices and tropical horticultural crops are the major characteristics of the region. The monsoon rains are concentrated in a few months in a year, with dry weather during the rest of the year. The climatic conditions of the region necessitate the development of irrigation for sustainable farming. The region, in general, has made major strides in participatory irrigation management, especially in rice-based farming systems. The location of international institutions, namely the Asian Institute of Technology, Bangkok, Thailand and the International Rice Research Institute, Los Banos, Philippines and many national level institutions have contributed greatly in the development of water resources and rice-based farming systems in the region.

South-East Asian Mainland

The South-east Asian mainland region (Fig. 1.19) is composed of Myanmar and the riparian countries of the lower Mekong basin, namely, Thailand, Lao PDR, Cambodia, Vietnam and Malasia (Peninsular region). Mountains and hills are the main physiographical features, covering about two-thirds of the total area. The extensive plains along the Mekong, Red and Ayeyarwady rivers are frequently subject to flooding. The climate is characterized by the south-west monsoon with heavy rainfall and the north-east monsoon which is relatively short in most countries and the dry summer. About 75% of the total rainfall occurs during the south-west monsoon. The average annual rainfall of the region is 1877 mm, ranging from 500 mm in the central dry zone in Myanmar to over 4,000 mm in the mountains of Rakhine in Myanmar and Bac Quang in Vietnam.

(***Source***: UN Food and Agriculture Organization (FAO), 2001).

Myanmar

Myanmar, formerly known as Burma, lies between South Asia and the Mekong Basin countries of South-east Asian mainland. The total area of Myanmar is 67.658 million ha. Topographically, Myanmar can be divided into five regions: The northern and western mountains, the eastern plateaus, the central basin and the coastal strip. The climate is tropical, influenced by the monsoons and physiographic conditions. South-west monsoon extends from mid-May to mid-October. The rainless cool weather is from mid-November to mid-March and the summer extends up to mid-May. The average annual rainfall of the coastal regions of Arakan and Tennassero region is about 5000 mm while that of the central region ranges from 500 to 1,270 mm. The rainfall is generally adequate to meet the need of agriculture in lower Myanmar and parts of the central regions. However, in the vast plains of the country irrigation is required to supplement the rains for paddy cultivation.

Water resources. Myanmar has rich surface water sources. The major river basins are the Irrawady, Chindwin and Rangoon Rivers covering about 58% of the land area. Favourable sites exist on many of the rivers for the construction of multi-purpose water resources projects. Groundwater resources of Myanmar are substantial. The present use of groundwater is mainly for domestic water supply.

Irrigation development. The irrigation works constructed during the times of the Burmese kings, commencing the 11th century A.D., comprised river diversion works and tanks. Many storage reservoirs were constructed in the second-half of the 20th century. In addition, many minor works have been constructed and maintained by the farmers. A number of large projects are under various stages of development. The total area irrigated in Myanmar rose from 0.681 million ha in 1966 to 0.98 million ha in 1978 (Framji *et al.*, 1982). The net irrigated area in 1999 was 1.841 million ha.

(***Source***: UN Food and Agriculture Organization (FAO), 2000).

Myanmar has immense potential for additional development of its water resources for irrigation, hydropower generation and flood control.

Thailand

Physiographically, Thailand comprise the northern region which is mountainous with dense forests, the north-east plateau bordering the Mekong River; the extensive central plain which is often subject to flooding; and the southern peninsula. The Chao Phraya Basin is the largest and most important geographic unit of Thailand in terms of land and water resources development. It covers about 35% of the country. Next to it in size and development potential is the Mac Klong Basin of the Central Region. Another important geographic unit of the North and Central Region is the Bang Pa Kong Basin, a major part of which lies in the southeast of the Chao Phraya Plain. The Mekong River Basin occupies the entire North-east Region. The Southern Region extends from the west of the Mac Klong Basin to the Malaysian border.

Most of the irrigable area in Thailand are in the flood plains of the major rivers where heavy rainfall and frequent floods prevent the cultivation of crops other than paddy in the wet season. In the Chao Phraya Plain, heavy soils predominate. Paddy is the common crop in the dry season as well. It is estimated that about 3.4 million ha of land in Thailand is affected by salinity due to sea water intrusion, saline water table, and inadequate drainage and landscape characteristics. Salt affected soils are found along the coastal areas and in the North-eastern Region.

Climate. The climates is mainly governed by the alternation between the south-west monsoon, which brings heavy rainfalls (from May to October), and the northeast monsoon, which is comparatively dry and cool (from October to February). The average annual rainfall is estimated at 1,485 mm, ranging 1,100 mm in the central plain and the northeast to 4,000 mm in the southern peninsula near the Andaman Sea.

(***Source***: UN Food and Agriculture Organization (FAO), 2001).

Water resources. Thailand comprises of seven river basins with 25 sub-basins which are rich in water resources. The groundwater potential is good, especially in the plains. Thailand shares three rivers with its neighbours: (*i*) The Mekong River forms the border with Lao PDR in the north and east. (Thailand has been a member of the Mekong River Commission since its inception in 1957. The international agreement on the lower Mekong River was signed in 1995 by the four riparian countries: Thailand, Lao PDR, Cambodia and Vietnam. This agreement concerns cooperation for the sustainable development of the lower Mekong River basin), (*ii*) The Salawin River, on the northwestern border with Myanmar, and (*iii*) The Kolok River, on the southern border with Malaysia.

(***Source***: UN Food and Agriculture Organization (FAO), 2001).

Irrigation and drainage development. Irrigation in Thailand is an age old practice and comprised of diversion type works, small in size (Fig. 1.7). During 1930s, the Department of Irrigation initiated the construction of a network of navigation and drainage canals to cover more than 500,000 ha in the Southern Chao Phraya Plain. A number of irrigation projects were started in Central, North and North-east regions. They were designed for supplementary irrigation in the wet season. Limited dry season cropping was also possible. The Bhumiphol dam on Ping river was the first large multi-purpose storage project constructed in 1950s. Storage dams of various sizes were also built on several other rivers. Tank Irrigation projects for irrigation were constructed in the northeast region. Since 1976, small scale water resources projects were given emphasis. The Royal Irrigation Department (RID) completed more than 4,000 minor projects all over the Thailand. The rehabilitation works of irrigation projects were started in mid-1970s (Framji *et al.*, 1982). The net area irrigated in 1999 has been 4.75 million ha.* Large and medium-scale schemes nunbered about 650 in 1995 and covered 65% of the irrigated area. They are constructed by the RID and are concentrated mainly in the Chao Phraya river system in the Central Plain which is called the rice bowl of Thailand. Small-scale schemes, also called private schemes, have been designed and built by the RID, by other departments, by NGOs or by the private sector. They are farmer driven and managed. They cover 24% the met irrigated area.**

*(***Source***: UN Food and Agriculture Organization (FAO), 2000).

(*Source***: UN Food and Agriculture Organization (FAO), 2001).

Laos

Laos (Lao Peoples' Democratic Republic) is a landlocked mountainous country in the Indo-China peninsula with a total land area of 23.68 million ha. Physiographically, Laos comprises deeply dissected plateau country, generally 900 to 1,800 m in attitude. About 80% of the country consists of hills and mountains. The cultivable area is about 2 million ha, consisting of narrow valleys and the flood-prone plains of the Mekong River system.

Water resources. Laos is drained from north to south by the Mekong River, the 10th longest river in the world. In addition to the Mekong there are six small river basins which drain towards Vietnam. About 90% of the country is in the Mekong River basin. The underground water resources of Laos have not been fully assessed.

Irrigation and power development. The total area under irrigation in 1979 was 79,000 ha (Framji et al., 1982). The main irrigated crop is rice. Laos is rich in hydro-power potential.

Cambodia

Cambodia, formerly known as Kampuchea, is situated on the coast of Gulf of Thailand. The country can be divided into three main physiographical regions, namely (1) undulating plateau in the east, (2) the Mekong River and the Lake Tonle' Sap Lowland, in the central region and (3) the Cardamome Mountains in the south-west. Water bodies, including Lake Tonle' Sap, occupy about 2.2% the total area of the country. During high floods the Mekong river overflows in the direction of Tonle Sap Lake.

Cambodia has a unique hydrological system. The Mekong River and Lake Tonle Sap are connected by the Tonle Sap River which reverses its direction of flow twice a year. From July to the end of October, when the level of the Mekong is high, water flows into the Tonle' Sap River, which fills Lake Tonle' Sap, thereby increasing the size of the lake. In early November, when the level of the Mekong decreases, the Tonle' Sap River reverses its flow, and water flows from Lake Tonle' Sap to the Mekong River and then to the Mekong Delta.

About 86% of Cambodia's territory is in the Mekong River basin, the remaining 14% drain directly towards the Gulf of Thailand. Cambodia was a member of the Mekong River Committee between 1957 and 1975. In 1995, Cambodia, Lao PDR, Thailand and Vietnam signed an agreement for the development of the Mekong River.*

Currently, the use of groundwater resources is limited mostly for domestic purposes. There is considerable scope to develop groundwater for irrigation, especially in the alluvial plains.

*(**Source**: UN Food and Agriculture Organization (FAO), 2001).

Irrigation development. Different types of irrigation works have been in existence in Cambodia since the past several centuries. The total area irrigated in Cambodia in 1979 was 89,000 ha (Framji *et al.*, 1981). By 1999, it rose to 270,000 ha. No large-scale drainage works exist in Cambodia.

(***Source***: UN Food and Agriculture Organization (FAO), 2000).

Vietnam

Vietnam is located in the eastern part of the Indo-China peninsula. Mountains and hills cover more than 75% of the country, although over 70% of the land area is within 500 m above sea level. About 25% of the land area is covered by plains. The most important are the Bac Bo and Nam Bo plains along the courses of the Red and Mekong rivers, respectively. The climate varies from temperate and sub-tropical in the north to tropical in the south. Precipitation ranges from 2,000 to 2,500 mm in the mountainous areas and from 1,600 to 2,200 mm in the midlands and plains. The rainy season is from April/May to October/November. The driest periods are from December to March.

(***Source***: UN Food and Agriculture Organization (FAO), 2001).

River basins and water resources. There are 9 major river basins, each with a catchment area larger than 10,000 km^2 covering about 80% of the country's area and providing about 70% of its water resources. The largest basins are the Mekong and the Red River/Thai Binh. Vietnam has abundant surface water resources in terms of total run-off and potential for development. Red and Mekong rivers carry about 75% of the flow. The alluvial plains of the river basins are rich in groundwater. They have been exploited for irrigation to a limited extent only. In future, development of groundwater will be a major component of the irrigation water supply in Vietnam.

Irrigation by means of indigenous water lifting devices directly from rivers has been an ancient practice in Vietnam. In southern Vietnam, the topography is flat, almost at sea level, with many swampy areas. There are also numerous small rivers influenced by sea water intrusion. The tides and floods spread saline water over the surrounding land. Therefore, irrigation works in these lands have been combined with drainage and prevention of salt-water intrusion. Among the more recent water resources projects are the Caisan Project and the Phanrang Project.

Malaysia

Malaysia occupies the Malay Peninsula of the southern shores of Asian landmass and the States of Sabah and Sarawak in the east on the Borneo Island. The two regions are separated by the South China Sea. Peninsular Malaysia comprises mountain ranges, undulating land and coastal and riverine flood plains. About 61% of the peninsula is within 100 m above sea level and the land is generally suitable for cultivation. Permanent crops occupy about 90% of the cultivated area, while the remaining is under annual crops, mainly paddy. The agricultural sector is divided into large-scale plantations of rubber, oil palm and cocoa. Tropical rain forest occupies about 70% of Malaysia, concentrated in hills and mountains.

Malaysia lies in the equatorial zone. The climate is governed by the northeast and southwest monsoons.

River basins and water resources. Peninsular Malaysia is drained by a network of rivers and streams. There are about 150 major river basins. The longest is the Pahang River which drains into the South China Sea. Malaysia's longest river is the Rajang River (563 km) in Sarawak. On the west coast of peninsular Malaysia, the low gradient has resulted in large extensions of tidal plains and swamps.

(***Source***: UN Food and Agriculture Organization (FAO), 2001).

Development of irrigation and drainage facilities. Early water resources development works were for drainage and comprised of the construction of coastal embankments to prevent saline water intrusion and the provision of facilities for the removal of excess rainfall and run-off. Over 50,000 ha of land are provided with drainage facilities.

The first major irrigation project to be constructed in Malaysia is the Bukit Merah Scheme, commissioned in 1906 with a dam utilising the flow of the Sungai Merah and the Sungai Kurau rivers. A separate Department of Drainage and Irrigation was formed in 1932. The Department constructed a system of controlled drainage in paddy lands. Gradually, such schemes were improved to provide irrigation to paddy lands through diversion works and canal networks (Framji *et al.*, 1982).

During the 1960s and early 1970s, the introduction of double cropping in rice cultivation required the development of adequate water resources for the second cropping season. During 1980s, the priority was to strengthen irrigation facilities for rice cultivation. The Government formulated a policy to concentrate efforts in irrigation development in eight large irrigated areas, designated as *granary areas* of the country. By late 1990s, Malaysia had over 932 irrigation schemes covering a net area of 0.365 million ha.

(***Source***: UN Food and Agriculture Organization (FAO), 2001).

MAJOR ISLAND NATIONS OF SOUTH-EAST ASIA

The three important island countries of South-East Asia are Indonesia, Philippines and Taiwan located in the Indian Ocean and North Pacific Ocean (Fig. 1.19).

Indonesia

Indonesia is a tropical archipelago comprising more than 13,600 islands, of which 931 are habitated. The total area of Indonesia is about 190.45 million ha. It is situated between two oceans, the Pacific Ocean and the India Ocean, and between two continents, Asia and Australia. The major islands of Indonesia are Java and Madura, Sumatra, Kalimantan (Borneo), Sulawesi (Celebes), West Irian (West New Guinea), Moluccas and Nussa Tenggara or Lesser Sunda Islands (Bali, Lombok, Sumbawa and others). Most of the major islands have a mountain range running their entire length. The mountains are mostly of volcanic origin. The island of Java has a chain of volcanic mountains. From the foot of the mountains, extensive table-lands with elevations ranging from 500 to 1,500 m, slope down to the low lands of the coast. The table-lands and low lands comprise 70% of the land area of Java. The low lands are used for cultivation of paddy. Sumatra has a large mountain and vast stretches of low lands, consisting mainly of marshy alluvial plains. Kalimantan, the largest island of Indonesia, has a vast terrain of low lands, consisting of alluvial and swampy plains. Sulawesi is mountainous and hilly, except the central region. Bali and Limbok are highly fertile. In 1995, the total cultivated area of Indonesia was estimated to be about 35.6 million ha.*

Rainfall in Indonesia is typically tropical and most of it falls with high intensity. The dry season is from March to August and the wet reason from September. Sumatra and western Java have good rains almost throughout the year. The annual rainfall in general exceeds the potential evapotranspiration. Rainfall varies widely between 700 and 7,000 mm per year with an average of about 2,600 mm, while evapotranspiration is in the range of 1,200 to 1,400 mm per year which is nearly same all over Indonesia. *

Indonesia has vast stretches of fertile soils. In Kalimantan, Sulawesi and Sumatra the soils are comparatively poor due to excessive leaching caused by heavy rains. The extensive swampy soils of the alluvial plains of Sumatra, Kalimantan and West Irian require drainage for successful cultivation.

Water resources. Indonesia has over 5,590 rivers. The biggest rivers are in the islands of Sumatra, Kalimantan, Java and West Irian with perennial discharge. Most of the lakes in Indonesia are of volcanic origin. A part of the river flow has been developed for irrigation. Suitable dam sites exist in many parts of Indonesia.

Groundwater. Good water-bearing formations exist in several parts of Indonesia The best aquifers are lava streams located in the vicinity of volcanoes and limestones. The alluvial plains have substantial groundwater yield. Presently, groundwater is used mainly for domestic water supply.

Irrigation development. In the Java island of Indonesia, irrigation has been practiced from ancient times. Emphasis on irrigation development was given since the 18th century A.D. Irrigation is well developed in Java and Bali islands, but much less on the other islands. Almost all the irrigation works are designed to supply water to rice crops. The important irrigation works consist of diversion weirs, storage works and river pumping schemes. Till the end of the 19th century diversion projects were considered adequate. Reservoirs to store the vast monsoon flows for use during the dry period were constructed in the 20th century. Irrigated rice farming on the island of Bali differs from that in the other islands. In Bali irrigation schemes are constructed, operated and managed by co-operative societies, called *subaks*, rather than by the government.

The government laid major emphasis on irrigation development and reclamation of swampy areas with major emphasis on rice-based farming systems. The cultivated area in 1966 was 14 million ha and the area under irrigation was 0.379 million ha (Framji et al., 1982). The net irrigated area rose to 4.815 million ha in 1999.
(***Source***: UN Food and Agriculture Organization (FAO), 2000).

In Indonesia, irrigation systems are classified into four types: technical, semi-technical, simple and village managed (*i*) *Technical systems* serve an area of 3.328 million ha (1996). They have permanent canals and control structures. The government control of water distribution is up to the tertiary level. (*ii*) *Semi-technical systems* served an area of 1.1 million ha in 1996. They have only few permanent control or distribution structures and are managed by the farmers. (*iii*) *Simple systems* have only few permanent control or distribution structures and are managed by the farmers. (*iv*) *Village managed irrigation systems* are developed and managed by farmers.*

Coastal and inland swamps in Indonesia amount to 39.0 million ha. The extent of arable swamps is estimated at 7.5 million ha. In 1996, the swamp area used for irrigation (mainly for rice) was about 1.18 million ha.*

Field water management. A tertiary development programme was introduced in Indonesia for better water management at the farm level. The construction of tertiary networks has been the responsibility of the farmers who avail assistance from the government. For the efficient operation and management of tertiary networks, Water Users Associations have been organised.

Drainage. Extensive swampy lands exist in several islands of Indonesia, particularly in Kalimantan and Sumatra. The people of southern Kalimantan living in the tidal regions have developed their own method of opening up new rice fields. The land in these regions is flat and almost at sea level. New rice fields are opened by digging channels connecting the land with a tidal river. When the river rises at high tide, the water flows into the channels and inundates the surrounding rice fields for about two or three hours. When the river falls at low tide the water in the channels flows back to the river and carries with it salts which are harmful to

*(***Source***: UN Food and Agriculture Organization (FAO), 2001)

plants. In about two years what was formerly a marsh is transformed into cultivable land. A scheme for reclaiming 5.25 million ha of land of Kalimantan and Sumatra was taken up. By 1979 about 272,000 ha of tidal and swamp areas were reclaimed (Framji *et al.*, 1982).

Philippines

The Philippines is a tropical country comprising 7,017 islands with a total land area of about 30 million ha. About 92% of the area of Philippines lies in 11 large islands. Philippines has a varied topography with highlands and valleys. The four major lowland plains are the central plain and the Cagayan valley in Luzon, the Agusan valley and the Cotabato valley in Mindanao. The total cultivated area is estimated at 9.5 million ha, of which 56% is under annual crops. The average farm size is 2.2 ha.

(***Source***: UN Food and Agriculture Organization (FAO), 2001).

The climate of the Philippines is tropical and monsoonal. It is characterized by uniformity of temperature, high relative humidity (above 70%) low solar radiation, diversity of rainfall and high frequency of tropical cyclones. The rainfall distribution varies from one region to another, depending upon the direction of the moisture bearing winds and the location of the mountain systems. The mean annual rainfall varies from 965 to 4,064 mm in different regions. Typhoons have major influence on the climate and weather of the Philippines.

Water resources. There are 421 rivers in the Philippines, excluding small mountain streams. The rivers are an important means of transportation and a valuable source of water for irrigation. There are also 59 natural lakes and more than 100,000 ha of freshwater swamps.

Groundwater resources. There are major groundwater reservoirs offering considerable potential for development. Wells are widesprerd in rural areas for domestic puroses. Groundwater is used extensively for irrigation and municipal water supply.

Irrigation development. The history of irrigation in the Philippines dates back several centuries before the Spanish colonization (1521 A.D.). During the Spanish regime (1521-1898) irrigation systems were further improved and regulations regarding irrigation were introduced. Irrigation societies, came into existence. These societies built their own irrigation systems mostly with temporary brush and rock dams and earthen canals. During the American period of 1898-1941, laws governing water rights, water use, irrigation system development, formation of irrigation associations, and payment of irrigation charges were passed. After 1945, the government stepped up irrigation development. The total irrigated area increased from 48,900 ha in 1900 to 400,00 ha in 1949 and 960,000 ha in 1966, 1.27 million ha in 1979, and 1.55 million ha in 1999 (Framji *et al.*, 1982 and UN Food and Agriculture Organization (FAO), 2000).

Irrigation systems in the Philippines are of 3 types: (*i*) Pump irrigation, (*ii*) National irrigation, and (*iii*) Communal irrigation. The distinction between the types of systems is based on the ownership, operation and maintenance of the systems and the size of the command area. National irrigation systems are owned and operated by the government through the National Irrigation Administration (NIA). They irrigate areas of 1,000 ha and above. They are owned and operated by Irrigation Associations. Pump irrigation systems are owned by the government, Irrigation Associations and private persons.

Irrigation Associations are organized to ensure farmers' participation in the delivery, distribution and proper utilisation of irrigation water at the farm level. Farmers Irrigation Groups are the organizations of water users in the command area of every canal outlet to ensure equitable distribution of irrigation water. They serve as the basic unit of Irrigators' Associations. The service area of every outlet is a rotational area, which is further composed of rotational units to facilitate equitable water distribution.

Taiwan

Taiwan (Republic of China) is located in the Western Pacific Ocean between Japan and the Philippines. Taiwan comprises the island proper and 13 islands of the Taiwan Group and 64 islands of the Penghu Group. The Central Mountain Range, extending from north to south, divides the island into a rocky rugged region on the east and fertile plains on the west. The alluvial plains are mostly along the west coast.

Taiwan has an oceanic and sub-tropical monsoon climate which is influenced by its topography. Concentrated rainfall and high evaporation are the characteristics of Taiwan.

Taiwan has a total area of 3.598 million ha. Almost the entire arable area of 0.92 million ha is cultivated. About 2.224 million ha is covered by forests, and about 0.183 million ha is provided for buildings. The irrigated area in 1980 was 0.496 million ha (Framji *et al.*, 1981).

Irrigation development. Irrigation development in Taiwan can be traced back to the 14th century A.D. when emigrants from the Chinese mainland settled and started farming. The early works were small scale stream diversions and ponds (tanks). By 1895 irrigation made progress, and an area of over 2,00,000 ha was developed for rice crops. All irrigation systems, except the Tsao-Kung canal, were built by individuals. During 1896-1945 the old canals and ponds were renovated and new irrigation projects were constructed, (Framji *et al.*, 1981).

In 1901, the government initiated the development of irrigation systems which were supervised by local administration agencies. In 1922, they were reorganised into irrigation corporations. After World War II, considerable effort was made on rehabilitation works and on the revival of projects which had been started earlier. To achieve economy in water use, rotational irrigation system was introduced in 1954. Taiwan has achieved major strides in on-farm water management with special emphasis on (*i*) *Irrigation scheduling,* (*ii*) *Irrigation system improvement,* (*iii*) *Farmer participation in irrigation management.* Participation of beneficiary farmers in irrigation system management is through Irrigation Associations. The Association builds and operates irrigation and drainage systems, availing technical support from the Provincial Water Conservancy Bureau of the Government, (*iv*) *Water laws.* Legal provision on water resources utilization are inscribed in Taiwan's constitution. According to the constitution, all surface and groundwater resources belong to the State, irrespective of the proprietorship of the riparian land in which the resources exist.

EASTERN ASIA

The Eastern Asia region includes China, Mongolia and the Democratic People's Republic of Korea (Fig. 1.19). The region covers an area of 11,285 million ha, which is about 8.4% of the world area. Eastern Asia is mountainous with about 80% of the landmass lying at altitude of over 1000 m. Apart from east and south China, where the climate is determined by the monsoons, the region is generally characterized by long cold winters caused by the north and north-west winds from Siberia. Most of the rainfall is during the summer and early autumn. Large parts of south Mongolia and central China suffer from a very arid climate and are facing severe water scarcity problems. The average annual precipitation of the region is 597 mm.

(***Source:*** UN Food and Agriculture Organization (FAO), 2001).

China

The People's Republic of China is world's third largest country in territory. It lies between 20° N and 50° N latitudes and 75° E and 135° E longitudes. The average altitude ranges from over 4,000 m in the west to less than 100 m in the east.

Climate. Vast areas of east China and south China are influenced by the East Asian monsoon climate. Mountains and plateaus prevent the monsoon from penetrating deeply into the northwest of the continent, resulting in low precipitation. In winter, the mainland is generally under the influence of dry cold air masses from Siberia. The mean annual precipitation is 648 mm. In the coastal areas of the southeast and some parts the southwest, the mean annual precipitation exceeds 2,000 mm. It is less than 400 mm in parts of northeast China. April/May to July/August are rainy months in the south and June to September in the north.

(***Source***: UN Food and Agriculture Organization (FAO), 2001).

Water resources. The distribution of water resources in China, like in India, is highly uneven, both in time and space. About 80% of China's rivers flow in South China, while 60% of China's arable land is in the central and northern regions. There are more than 50,000 rivers, 1,500 of which have their basin areas exceeding 1,000 km^2. The rivers are of two categories: those discharging into seas, and the inland rivers which run to depressions into the interior. About 77% of the water resources developed are utilized for irrigation, 18% are for industrial purposes and 5% are for domestic use.

(***Source***: UN Food and Agriculture Organization (FAO), 2001).

Irrigation and drainage development. A historical review of irrigation development in China was presented earlier. By A.D. 1900, China had developed an irrigation potential of about 19 million ha (Framji *et al.*, 1981). After the establishment of the People's Republic of China in 1949; a massive programme was launched on the regulation of rivers, expansion of old irrigation works and the construction of major, medium and minor irrigation works with government investment and involvement of people. China has over 22,000 large dams, constituting about 45% of the large dams in the world. An estimated about 85,000 dams of all types are in operation in China, excluding small farm-scale irrigation dams and mini and micro-hydropower units. At least 90 dams of over 60 metres height were under construction in 2000 A.D. Major emphasis on groundwater development has been in North China. About 2.3 million water wells have been constructed.

Over 1,77,000 km of dikes have been constructed for flood control and prevention of sea water intrusion. These efforts raised the net irrigated area in China to 53.74 million ha in 1999.* In terms of scheme size the irrigated areas of China are divided into three categories. Large irrigation districts cover areas over 20,000 ha; medium districts cover areas between 667 and 20,000 ha, and small districts are under 667 ha. The large and medium irrigation districts are generally administered by special government Organizations. The small ones are usually farmer managed. Surface irrigation is the method practised (mainly for rice, wheat, millet, vegetables, maize and cotton) on about 99% of the total irrigated area. The remaining 1% is under sprinkler and micro-irrigation (mainly for fruit trees, vegetables and tea). About two-thirds of the food grain production in China is from irrigated areas.**

* (***Source***: UN Food and Agriculture Organization (FAO), 2000).

(*Source***: UN Food and Agriculture Organization (FAO), 2001).

Mongolia

Mongolia is located in the north of the Central Asian plateau. Mongolia comprises principally of inter-mountain plateau. A large part of south Mongolia is covered by the Gobi Desert steppe, while the eastern part consists of elevated plains. The total cultivated area is as low as 1.8 million ha. However, 80% of the total land area can be used for pastoral activities.*

Mongolia has long cold winters. The average annual precipitation is 251 mm. Many of Asia's major rivers, including the Yenisei, Irtysh and Selenge rise in Mongolia. Rainfall is the principal source of water for the rivers of the region, while water from melting snow makes up 15–20% of the annual run-off. From November to May many of the rivers in the north are frozen.*

Irrigation development. Irrigation in Mongolia can be traced to the period of the Huns in the first century A.D. Modern irrigation development started in the 1950s. In 1993, the total area equipped for irrigation was about 57,300 ha, of which 43,400 ha* were under sprinkler systems and the remaining under surface irrigation methods. In addition, an estimated 27,000 ha of pasture are benefited from traditional floodwater diversions. Of the total water resources developed 82% is from groundwater and 18% from surface water. About 53% of the water resources developed is used in irrigation.

There is considerable scope for increasing the irrigation potential of Mongolia and developing sound water management practices.

Democratic People's Republic of Korea

The Democratic People's Republic of Korea (DPR Korea) is located on the northern part of the Korean peninsula in the far east of Asia. Only 20% DPR Korea is cultivable. In 1994, the cultivated land was 2 million ha, of which about one-third was for rice and an equal area for maize. In the 1980s there were about 38,000 cooperative farms and 180 state farms. Cooperative farms included more than 90% of the total cultivated land. The average annual precipitation of DPR Korea is 1,054 mm, ranging from 810 to 1,520 mm.*

Water resources and irrigation. Most of the rivers run west to the Yellow Sea (Korea Bay). In 1976, the government launched a *nature re-making programme* with the following objectives: (*i*) to provide irrigation to non-paddy lands; (*ii*) to reclaim 100,000 ha of new land; (*iii*) to build 150,000-200,000 ha of terraced fields; (*iv*) to reclaim tidal land; and (*v*) to conduct afforestation and water conservation projects.

The net irrigated area in 1999 was 1.4 million ha. In 1999, the total land area under irrigation was estimated at 73% of the cultivated area, in which surface water sources covered 86% and groundwater sources 14%.*

FAR EAST

The Far East region includes Japan and the Republic of Korea (Fig. 1.19). Mountains cover almost 70% of the total area. Winds and the mountainous topography divide the landmass into two typical climatic zones: the Pacific coast zone, marked by the summer monsoon which blows from the Pacific Ocean bringing warmer temperatures and rain, and the continental zone, characterized by the winter monsoon from the Asian continent which brings freezing temperatures and heavy snowfalls. The average annual precipitation in the region is 1,634 mm, most of which falling during June to September. The region is often struck by typhoons which cause severe crop damage mainly during the summer and early autumn. The population density is the highest in the region, with an average of 358 inhabitants/km^2 due to the concentration of the population in the urban areas. The contribution of the agricultural sector to GDP is low, compared to the other countries of Asia, 2% in Japan and 6.5% in the Republic of Korea.*

Republic of Korea

Republic of Korea is a peninsula, comprising of many islands and with a total land area of 9.873 million ha. The major part of the cultivated area is under rice (Framji *et al.*, 1982). The mean annual rainfall in Korea is 1,274 mm, with about 900 mm in the east central part rising to about 1,400 mm in the southern and western coastal areas. Irrigation is necessary to supplement rainfall for successful paddy cultivation and for upland crops.

Water resources and irrigation development. There is a network of rivers in the country. Irrigation in Korea was developed mainly for rice cultivation. There were 2,781 reservoirs and many diversion weirs in 1909. (Framji *et al.*, 1982). The Irrigation Association Law was enacted in 1906. In 1940, the Korea Union of Irrigation Associations was founded. The Korea Land Development Corporation was set up for large scale projects. In 1965, the Agricultural Water Resources Development Plan was prepared which included large scale irrigation projects. The drought in 1967–68 led to extensive groundwater development, including nationwide construction of drilled wells, dug wells and infiltration galleries. (Framji *et al.*, 1982). During the second-half of the 20th century considerable effort has been made to develop water resources. Multi-purpose river basin schemes have been developed for flood control, irrigation, community water supply and hydropower production. The net area irrigated in 1999 was 1.159 million ha.

(***Source***: UN Food and Agriculture Organisaton (FAO), 2000).

Japan

Japan is an archipelago in East Asia stretching 2,600 km running northeast to southwest through the North Pacific. Five main islands, Honshu, Hokkaido, Kyushu, Shikoku and Okinawa, account for about 96% of the total land area, though there are more than 3,300 other islands. About 70 to 80% of the country is mountainous. Japan is a volcanic country. In 1996, the cultivated land area was 4.776 million ha, which is 96% of the total cultivable area. The cultivated land area is declining due to increasing demand for housing and industry. About 92% of the cultivated area consist of annual crops,

*(***Source***: Food and Agriculture Organization (FAO), 2001).

of which paddy occupy 2.56 million ha. The area under permanent crops is estimated at 0.392 million ha.*

Japan lies in the temperate zone, except some of the islands in the south. The average annual precipitation is 1728 mm, with wide variation between regions. The weather on the Pacific Ocean side of Japan is warm and humid in summer (June-August), while it is dry and windy in winter (December-February). The regions facing the Sea of Japan receive high precipitation from December to February.*

Water resources and irrigation. Japan has 2,721 major rever systems and abundant water resources. Due to steep topography and uneven rainfall distribution, proper development of water resources assumes importance. The history of dams and reservoirs in Japan dates back many centuries. A number of ancient earthen dams are still used for paddy irrigation. Since 1920s, dams and weirs with modern designs were constructed. These have contributed mainly to irrigation development and hydro-power generation. In A.D. 2000 there were 2675 large dams, which is about 6% of the world population of dams. In addition, there were about 18,000 small irrigation reservoirs. The net area irrigated in 1999 was 2.659 million ha.**

CENTRAL ASIA

Central Asia comprises Afghanistan and the five republics of the former Soviet Union, namely, Kazakhstan, Kyrgzytan, Tajikistan, Turkmenistan, and Uzbekistan (Fig. 1.19 and 1.19a).

Afghanistan

Afghanistan is a landlocked country characterized by its rugged terrain and an average elevation of 1,100 m above mean sea level. About 75% of the territory consist of mountains and hills. Lowlands include river valleys in the northern part, and desert regions in the southern and south-eastern parts. The cultivable area is estimated at 8 million ha.*

Climate and water resources development. Afghanistan has a continental climate. It is an arid country, with more than 50% of the territory receiving less than 300 mm of rain. The eastern border regions are at the far edges of monsoon influence and have about 50% of the precipitation in winter (January to March), much of which is in the form of snow in the central mountainous regions. About 30% of rainfall is in spring (April to June). The run-off from snow melt in the spring and summer months is the main source of water in Afghanistan. There is considerable potential of shallow groundwater. The groundwater is exploited through wells and *kareze* (*Kanats*) for irrigation and water supply. Construction of stone and brick masonry diversions across streams and storage dams have been practiced for generations. Framji *et al.* (1981) have reported that in many of the existing irrigation systems 25 to 50% of the command area often remains idle due to water shortage caused by insufficient capacity of canals and ancillary structures. In the lower reaches, rationing system of water release is being practiced in canal irrigated areas, under which each field reveives irrigation supply only every second or third year, depending upon the amount of water available.

Irrigation in Afghanistan can be grouped into the following four systems.**

(*i*) *Kereze systems.* A kareze (kanat) is an unlined tunnel in the hillside, bringing water by free flow from underground aquifers to be used for surface irrigation. There were, 6,470 *kareze systems,* irrigating 1,67,750 ha in 1967.

(*ii*) *Small-scale informal surface water systems.* These are the traditional irrigation systems.

(*iii*) *Large-scale informal surface water systems.* These are located mainly in the plains and along the main valleys. Their operation and maintenance are better planned. By late 1970s large-scale modern irrigation systems had been built and were in operation. The net irrigated area in 1999 was 2.386 million ha (UNFAO, 2000 and 2001).

There is great scope for expanding the irrigation facilities and irrigation efficiency in Afghanistan. On farm water management requires special emphasis.

CENTRAL ASIAN REPUBLICS

The five Central Asian countries which were in the former Soviet Union and became independent in 1991 are Kazakhstan, Kyrgyztan, Tajikistan, Turkmenistan and Uzbekistan together cover an area of 400 million ha. The relief in this region is extremely varied. In the east are the Tien Shan and Pamir mountain ranges. Much of the mountain region is permanently covered with ice and snow and there are many glaciers. On the border between the Kyrgyztan, Tajikistan and Uzbekistan is the Fergana valley, which is a major agricultural area in this region. In the southwest lies the Black Sand desert, one of the largest sand deserts in the world and covers 80% of Turkmenistan. The west of the region is dominated by the depressions of the Caspian Sea. The Aral Sea, in the central western part, is located on the border between Kazakhstan and Uzbekistan. The average annual precipitation in this region is 338 mm, varying from less than 70 mm in the plains and deserts to 2400 mm in the mountains of central Tajikistan. Many parts of the region face serious problems of salt and waterlogging. This makes drainage important in this region.*

*(***Source***: Food and Agriculture Organization (FAO), 2001).

(*Source***: UN Food and Agriculture Organization (FAO), 2000 and 2001).

Kazakhstan

Kazakhstan has about 80% of its land under deserts and steppes. The central part of the country consists of a sandy plateau with small hills. In the east and southeast, mountain chains alternate with depressions comprising sandy deserts. The cultivated area was estimated at 34.4 million ha in 1993, of which 99% consisted of annual crops. Fodder accounts for more than 10 million ha. There are more than 180 water reservoirs constructed in Kazakhstan, including. 19 large multi-purpose projects for hydro-power production, irrigation and flood control. The net irrigated area of Kazakhstan in 1999 was 2.35 million ha. This is about 10% of the cultivated area. About 20% of agricultural production is from irrigated areas. The irrigated area is mainly concentrated in the south of the country, particularly in the Syr Darya River basin (32% of the total).

(***Source***: UN Food and Agriculture Organization (FAO), 2000 and 2001).

Kyrgyztan

Kyrgyztan has a total area of 19.99 million ha. The country is largely mountainous. Much of the mountain region is permanently covered with snow. The cultivable land is estimated at 10.1 million ha. In 1994, the cultivated area was 1.34 million ha, of which about 97% consisted of annual crops and 3% permanent crops.

Irrigation is of vital importance to Kyrgyztan and covers 80% of the cultivated area. In 1990, there were a total of 1,346 irrigation schemes. Large schemes represented 60% of the irrigated area. Irrigation has been developed mainly in the Syr Darya River basin (42%), in the Talas and the Chu river basins (41%), and around Lake Issyk-Kul.

The net area irrigated in the country in 1999 was 1.072 million ha. Water is mainly supplied through river diversion.

(***Source***: UN Food and Agriculture Organization (FAO), 2000 and 2001).

Waterlogging, salinity and drainage development. In 1994, about 60,000 ha were saline by Central Asia standards (toxic ions exceed 0.5% of total soil weight). Another 34,200 ha are moderately saline and 25,800 ha highly saline. The waterlogged area was estimated at 89,200 ha in 1990. It is estimated that 7,50,000 ha of irrigated land would require drainage. In early 21st century, only 1,49,000 ha were having drainage facilities.

(***Source***: UN Food and Agriculture Organization (FAO), 2001).

Tajikistan

Tajikistan is a mountainous country in southeastern Central Asia with a total area of 14.3 million ha. The north of the country covers part of the Fergana valley, a major agricultural area in the region. The cultivable area of Tajikistan is estimated at 1.57 million ha. Cultivation is mainly in the flood-plains of the rivers. The average annual precipitation is 691 mm, ranging from less that 100 mm in the southeast up to 2,400 mm in the central part.

(***Source***: UN Food and Agriculture Organization (FAO), 2001).

Irrigation development. Irrigation in Tajikistan is vital for the development of agriculture. In 1960, the total ara equipped for irrigation was 4,08,000 ha. In 1994 it was 7,19,200 ha, which was 93.4% of the total cultivated area. About 33% of the total irrigated area is situated in the Syr Darya River basin and 67% in the Amu Darys River basin. In 1994, there were 19 dams in Tajikistan: 5 in the Syr Darya River basin and 14 in the Amu Darys River basin. The net irrigated area in 1999 was 0.716 million ha. In northern Tajikistan, irrigation is mainly done by pumping from the Syr Darya River. Surface irrigation, mainly furrow irrigation was practised in over 96% of the irrigated area. About 68,000 ha are irrigated from groundwater and about 25,000 ha from reused drainage water and waste water. River pumping schemes cover 34.8% of total area under irrigation, while elsewhere the water is gravity fed from river diversions or reservoirs. The major irrigated crops are cotton, fodder, cereals and vegetables. Out of the total irrigated area of 0.716 million ha, it is estimated that over 0.6 million ha require drainage. In 1994, drainage systems had been constructed in about 0.33 million ha; 58% was surface drainage, and 42% subsurface drainage. In 1994, about 16% of the irrigated land were classed as saline.

(***Source***: UN Food and Agriculture Organization (FAO), 2000 and 2001).

Turkmenistan

Turkmenistan has a total area of 48.81 million ha. The Kara Kum Desert covers 80% of its total area. In the southwest, along the border with Iran, lies the Kopetdag mountain chain. The cultivable area is estimated to be about 14% of the total area of the county. In 1994, the total cultivated area was about 1.75 million ha.

(***Source***: UN Food and Agriculture Organization (FAO), 2001).

Climate and water resources. The climate of Turkmenistan is sub-tropical desertic. The average annual precipitation is about 191 mm ranging from less than 80 mm in the northeast to 300 mm, in the Kopetdag mountain zone in the southwest. Precipitation occurs during the winter season, mainly between October and April.

The outflow of drainage water from canal irrigation systems has led to the creation of artificial lakes in many natural depressions. A major environment issue in Turkmenistan is the accumulation of pollutant salts in these lakes. In 1975, the total irrigated area was about 857,000 ha. In 1999, it rose to 1.8 million ha which was 99.4% of the total cultivated area. About 43,600 ha were irrigated with groundwater, while the remaining area with surface water.* The Kara Kum canal has caused massive waterlogging and salinization of the surrounding land. In 1994, about 37% of the irrigated area were classed as saline. It was estimated that over 1.222 million ha of irrigated land required drainage.

*(**Source**: UN Food and Agriculture Organization (FAO), 2000 and 2001).

Uzbekistan

Uzbekistan has a total area of 44.74 million ha. Physiographically, the country can be divided into three zones: (i) the desert (Kyzylkum), steppe and semi-arid region covering 60% of the country, mainly the central and western parts; (ii) the fertile valleys, including the Fergana valley, and (iii) the mountainous areas in the east. The cultivated land is estimated at 5.2 million ha, of which 87% is under annual crops and 13% under permanent crops. Uzbekistan is one of the world's largest cotton exporters. The climate of Uzbekistan is continental, desert areas cover 60% of the territory. The average annual rainfall is 264 mm, ranging from less than 97 mm in the northwest to 425 mm in the mountainous zone in the middle and southern parts.

(**Source**: UN Food and Agriculture Organization (FAO), 2001).

River basins and surface water resources. Its two river basins: Amu Darya basin in the south, covers 86.5% of Uzbekistan and Syr Darya basin in the north covers the remaining area. The drainage water outflow from irrigated areas has led to the creation of artificial lakes in many of the natural depressions. The groundwater potential of Uzbekistan is substantial, mainly due to recharge contribution from irrigation systems.

(**Source**: Food and Agriculture Organization (FAO), 2001).

Irrigation development. At the beginning of the 20th century, about 1.2 million ha were irrigated. Large-scale irrigation development started in the late 1950s, when the Soviet Union decided to specialize Uzbekistan in the production of cotton. In 1991, the net irrigation area was 4.281 million ha. Irrigated land accounts for more than 90% of crop production. About 44% of the total irrigated area is in the Syr Darya basin and 56% in the Amu Darya basin.

Waterlogging, salinity and drainage development. The total area of salt-affected soils in Uzbekistan is estimated at 10.9 million ha, of which 1.8 million ha is characterized with a high degree of soil salinization. In the Syr-Darya basin, salt-affected soils cover 3 million ha, of which 44% is moderate and highly saline. Salt-affected soils in the Amu Darya basin cover an area of 7.8 million ha. Since 1990, a reduction in the quantity of water allocated to each farm, lower water quality, and the decline of companies responsible for maintaining the drainage network have resulted in increased salinization. It is estimated that about 3.3 million ha of irrigated land would require artificial drainage. About 2.8 million ha are currently equipped with drainage infrastructure.

(**Source**: UN Food and Agriculture Organization (FAO), 2001).

WATER MANAGEMENT HAZARDS AND ENVIRONMENTAL DEGRADATION IN ARAL SEA BASIN COUNTRIES OF CENTRAL ASIA

The Aral Sea in Central Asia (Fig. 1.19a) has been the fourth largest inland seas in the world with a water spread area of 66,000 ha and volume of 1,060 km^3. It is a freshwater lake replenished by Amu Darya and Syr Darya rivers and their tributaries, which have a total catchment area of 69 million ha. The countries lying in the Aral Sea Basin are dependent on the water of Amu Darya and Syr Darya rivers and their tributaries which cover Kazakhstan, Tajikistan, Turkmenistan, Kyrghyztan, and Uzbekistan, all of which were in the former Soviet Union and became independent in 1991. The large scale expansion of irrigation in the Aral Sea basin area from 4.5 million ha in 1960 to about 7 million ha in 1990 by drawing water from the Amu Darya and Syr Darya river systems resulted in the shrinking of the Aral Sea. This was first observed in early 1970s. Salinity of the Aral Sea rose. Eco-system around the sea and in the deltas was adversely affected. The occurrence of diseases among local residents rose because of the increase of toxic salts and ecological deterioration. It has been estimated that if proper remedical measures are not taken in the river basins, the downsteam areas of the region would face further serious water shortage and continued environmental degradation in the coming decades (ICID, 1994).

Before 1960, the level of the Aral Sea was almost stable. The combined average discharge of the Amu Darya and Syr Darya rivers to the sea was about 47–50 km^3/year, to which could be added about 11 to 12 km^3/year of groundwater inflow and precipitation over the Aral Sea. The Aral Sea level was then fluctuating at

Fig. 1.19a. Location map of Aral Sea and its river basins (*Ref.* ICID, 1994).

around 50-53 m above sea level. The heavy withdrawal of water from the river basins of the Aral Sea has resulted in a sharp fall in the flow entering it. In mid-1990s, the annual flow entering the Aral Sea from the two river systems was estimated to average 3.5 km^3 with a continuing downward trend (ICID, 1994).

It is estimated that the Aral Sea's surface area would decrease from the 1987 level of 41 km^2 to 9.5 km^2 in 2015, unless concerted efforts are taken to drastically reduce the quantity of water being used in the catchment areas of the river basins. Effective steps are also to be adopted to manage the environmental issues in the basins of the two river systems. Various projects proposed to augment the flow into the Aral Sea are aimed at increasing the inflow to 16-18 km^3 in A.D. 2002 and 22-25 km^3 in A.D. 2010 (ICID, 1994).

The break up of the Soviet Union in 1991 left the newly independent countries to address the crisis on their own. Signing a water agreemnt in 1992 by the Heads of the five Republics was a significant achievement. The five countries agreed on joint measures for solutions of the problems of the Aral Sea region.

The Communities of the region face appalling health problems. At many locations drinking water supply is too low and polluted. The high contents of metals such as strontium, zinc and manganese cause diseases and prevent iron absorption, causing anaemia. The infant mortality rate is one of the highest in the world.

Aral Sea crisis has attracted world attention. It is considered as one of the major anthropogenic environmental degradations in the world. Many international organizations and experts have undertaken studies on the Aral Sea problems. The United Nations Environment Programme undertook a study of the region. The study report was issued in 1992. It provided a basis for analysis of the problem for developing effective solutions. In the same year, a joint mission of other UN agencies undertook a joint study of the problem area and made several proposals for action and a framework of activities to stabilise the Aral Sea disaster zone, and engage in strategic planning and comprehensive development and management of the water resources of the Aral Sea basin. The Aral Sea programme is large, complex, multi-country and long term. It involves redressing decades of mismanagement of water resources and destruction of the environment.

Measures adopted to mitigate environment problems. The control measures proposed and under implementation in the Aral Sea basin include the following: (*i*) Large scale adoption of pressurised water application methods like drip irrigation, coupled with underground pipeline water distibution and control. (*ii*) Increased use of greenhouse crop production. (*iii*) Large scale mulching to reduce evaporation losses. (*iv*) Development of irrigated pastures in the delta region. (*v*) Diversion of urban and industrial effluents to seperate evaporation basins. (*vi*) Recycling of effluent waters after proper treatment. (*vii*) Large scale adoption of appropriate cropping systems and cultural practices to economise on water use. (*viii*) Development of efficient drainage systems. (*ix*) Reuse of drainage water. (*x*) Development of shallow lakes in the delta regions for fisheries development. (*xi*) Automation of river regulation. (*xii*) Enhancement of water charges to reduce wastage and improve revenue. (ICID, 1994)

Massive development of irrigation on traditional lines to meet the needs of the growing population, but without predicting the environmental consequences, has been the main cause of the Aral Sea crisis in Central Asia. The five countries in the Aral Sea basin have considerable economic potential. They are endowed with substantial land and water resources, extensive irrigation networks, high reserves of oil and natural gas and hydro-power as well as a large labour force. Well planned and timely executed programmes are vital to redeem the region and ensure sustainable natural resource base for the present and future generations.

IRRIGATION DEVELOPMENT IN WEST ASIA AND THE MIDDLE EAST

Iran

Islamic Republic of Iran lies east of the Arabian Peninsula with a land area of 164.8 million ha. The physiography of Iran is characterized by plateau, mountains and coastal plains. About 52% of the country consists of mountains and deserts. The climate is hot and dry during the summer months, while sub-zero temperatures exist in the winter, except along the Persian Gulf. About 90% of Iran is predominantly semi-arid to arid. The average annual rainfall is 252 mm which decreases to 1200 mm in the south and south-east. About two-third of the country receives less than 250 mm of annual rainfall. The cultivated land in Iran is estimated at 63.265 million ha.

(***Source***: Framji *et al.*, 1982, UN Food and Agriculture Organization (FAO), 2000).

Water resources. Due to the low annual rainfall the surface water sources of Iran are limited. Iran can be divided into the following major river basins: the *Central Plateau* in the middle, the *Lake Orumich basin* in the north-west, the *Persian Gulf and the Gulf of Oman* basins in the west and south, the *Lake Hamoun basin* in the east, the *Kara-Kum basin* in the north-east and the *Caspian Sea basin* in the north. There are many large rivers. A few streams that empty into the Central Plateau dissipate into the saline marshes. All streams are seasonable. There exists substantial groundwater storage. Groundwater is tapped by *kanats,* shallow tubewells and deep tube wells.

(***Source***: UN Food and Agriculture Organization (FAO), 2001).

Irrigation development. Irrigation by *kanats* (underground tunnels) has been in practice in Iran since the ancient times. Presently, out of the 50,000 million villages in Iran 18,000 have *kanats* as their main source of irrigation. The construction of a system of *kanats* is undertaken after ascertaining the presence of springs in the area. A trial shaft of about 60 cm in diameter is dug until water has been reached. When water is located, a series of shafts are dug along the direction of flow of groundwater. Each shaft is then inter-connected by a tunnel or a gallery of about 130 cm in height and 60 cm in width. The distance between shafts depends on the elevation of the water surface in the gallery; the deeper the gallery the closer the shaft. Commonly used distance between shafts is from 25 to 50 m. The tunnel must be dug along the slope of the land to flat land to allow the water to reach the field by gravity flow.

Iran has built many dams and diversion weirs in the past several hundred years. The oldest dams are on the Karun River, more than 1,700 years ago. Two of the three dams are still in good condition. In 1943, an independent Irrigation Corporation (*Bongah*) was set up in Iran to develop irrigation.

Since 1953, 12 diversion dams and storage dams with an irrigation potential of 5,00,000 ha were built. By 1972, a number of projects, including the Nakichevan Dam Project, the Koda-Aferine Dam Project and the development of the Khuzestan basin were completed (Framji *et al.,* 1982). In 1994, 27 large dams were in operation mainly for irrigation and hydro-power generation.

Salinity hazards are high in many irrigation command areas in Iran. Drainage has been provided in some of them. For successful irrigated agriculture in Iran, drainage is complementary to irrigation. The new projects are being designed and constructed on this principle. The net irrigated area in 1999 was 7.562 million ha.

(***Source:*** UN Food and Agriculture Organization (FAO), 2000 and 2001).

Iraq

Iraq, situated in the middle East, has a total area of 43.382 million ha. Geographically, Iraq is divided into the northern mountainous region, the southern and western deserts and the great Mesopotamian plains traversed by the Euphrates and the Tigris rivers. About 25% of the land area is waterlogged. The total area which could be used for agriculture is estimated at 8 million ha. However, due to waterlogging, soil salinity, and other constraints only 3 to 5 million ha are actually cultivated annually.

The climate of Iraq is semi-tropical, arid and continental with dry hot summer and dry cold winter. The average annual rainfall in the north-east is 800 mm and in the middle and south about 150 mm. The average evaporation in summer is about 15 mm per day. The climatic condition of Iraq points to the need of well planned irrigation for successful farming (Framji *et al.*, 1982).

Surface water resources. The major water resources of Iraq are the twin rivers the Tigris and the Euphrates and their tributaries. Both the Euphrates and the Tigris originate from the high mountains in Turkey. The length of Euphrates in Iraq is 1,152 km, while that of Tigris is 1,418 km (Framji *et al.*, 1982).

Irrigation development. Iraq has been the cradle of irrigation in the ancient time. An account of the historical development of irrigation in the Mesopotamian region and the rise and the fall of irrigation based civilizations have been described in the earlier sections of this chapter. Most of the ancient works were destroyed by various invaders, the last being the Mongols in A.D. 1258. Irrigation works did not receive attention during the following seven centuries. Concerted efforts to improve some of the earlier projects and start new ones commenced in early 20th century. The Irrigation Department was established in 1918 and new schemes were undertaken and great expanses of lands were brought under irrigated agriculture. During the 1950s and 1960s, there was extensive development of lift irrigation. In 1946, the Government of Iraq constituted an Irrigation Development Commission to undertake a comprehensive study of irrigation and flood control in Iraq. During 1950s and 1960s, several major projects were constructed. The net area irrigated in Iraq in 1999 was 3.5 million ha.

(***Source***: UN Food and Agriculture Organization (FAO), 2000).

Drainage. While irrigation in Iraq is several thousand years old, hardly any drainage measures were provided in the irrigated areas. In the recent period, drainage works have been built on large projects in different parts of Iraq. Up to 1989, a total area of 1.7 million ha were provided with drainage facilities.* The drainage projects, however, did not include farm drains which were proposed initially, mainly due to inadequate financial provisions (Framji *et al.*, 1982).

*(***Source***: UN Food and Agriculture Organization (FAO), 2001).

Israel

Israel, situated on the eastern shore of the Mediterranean Sea has a total land area of 2.077 million ha. The physiography of Israel shows considerable variation in altitude within short distances. The land area can be divided into three longitudinal strips running from north to south; the Coastal Plain, the Hilly Zone and the Rift Valley. The climate of Israel is Mediterranean. It is influenced by the proximity to deserts and the variations in altitude. The average annual rainfall varies from 600 mm in the north to 150 mm in the south. The climatic conditions and the rainless long summer makes irrigation an essential requirement for intensive agriculture in Israel (Framji *et al.*, 1982).

Of the total land area of 2.077 million ha, the arable land is reported to be 0.652 million ha. The net irrigated area in 1999 has been 0.199 million ha.

(***Source***: UN Food and Agriculture Organization (FAO), 2000).

Water resources. The water resources of Israel comprise of the perennial surface flow of rivers, groundwater including springs, intermittent run-off through storm water drains and reclaimed waste water.

A part of the groundwater is salty. About 95% of of groundwater is being utilized. About two-third of irrigable water resources is groundwater. The use of reclaimed and treated municipal and industrial waste water is becoming increasingly important in Israel. The reclaimed water is used for irrigation and industrial uses.

Irrigation development. Irrigation in Israel is recent. In 1930s, local and regional water supply and irrigation schemes were launched. In 1937, the Mekoroth Water Co. Ltd. was established to execute and operate irrigation schemes utilizing latest available technology. A Master Plan was formulated in 1950s defining a water management policy and to adopt water legislation and increase the use of the available water resources. Simultaneously, work was started on the development of new irrigation methods to increase the efficiency of water utilization. The focus of the Master Plan was the construction of the National Water Carrier (NWC) and integration of all major regional projects together into a national water grid.

In 1959, a comprehensive water law was passed making water resources a public property. All major regional projects, including the National Water Carrier were completed by 1965. Subsequent projects were aimed at enlarging the distribution systems, incorporation of increasing amounts of intercepted run-off water and reclaimed waste water, and increasing on-farm water use efficiency. Due to scarce water resources, there has been concerted efforts in Israel to improve irrigation efficiency.

The National Water Carrier (NWC). It is the heart of the Israeli water system. Over 320 million m^3/year of water is pumped from Lake Kinneret to a distance of 130 km through a central conduit to Rosh Haayin near Tel-Aviv from where it branches out to carry water south-wards. The central conduit consists, mainly of large diameter pressure pipes, but includes small sections of open channels as well.

Drainage. Efforts on effective drainage of agricultural lands commenced in 1920s. The early efforts by the settlers were to draim swamps, control malaria and reclaim the land for agricultural use. The Layer Galilee Drainage Project, constructed during 1951–57, is the most important drainage project in Israel. It is to reclaim the Hula swamps. The project included the interception of springs and surface run-off from the hill slopes. Large scale drainage with surface canals and underground pipes to control high groundwater in the Zevulun Plain was constructed during the 1950s. Along the Northern Coastal Plain, a network of drainage canals which were dredged back from the sea enables the streams to drain freely. The swamps in the area have been drained. Control and regulation of local springs in the *Beth Shean Plain* which had caused serious drainage problems have been an integral part of the development of the local water supply system. In the Jordan Plain, soil salinization and high groundwater levels have been serious problems. A network of underground drainage pipes has been laid throughout the plain controlling the groundwater level (Framji *et al.*, 1982).

Syria

Syrian Arab Republic has a total land area of 18.518 million ha. Physiographically, Syria comprises of four distinct areas: the coastal area, the mountains and highland extending from north to south, the interior plains and the desert area of the southeast. The climate of Syria in characterised by a rainy winter and hot dry summer with two short intervening seasons. The annual precipitation ranges from 100 to 150 mm in the north-west, 150 to 200 mm from the south to the central region, 300 to 600 mm in the plains and 800 to 1000 mm along the coast.* Out of the total area, the arable land is 5.23 million ha (Framji *et al.*, 1983).

*(***Source***: UN Food and Agriculture Organization (FAO), 2001).

Water resources. The main rivers in Syria are the Tigris and the Euphrates and their tributaries. There are ten other small rivers. In addition, there are about seventy springs, having discharges exceeding 50 l/s each. Groundwater offers considerable potential for irrigation in many regions of Syria.

Irrigation development. Irrigation development in Syria dates back to over 4,000 years. Many of the old systems of canals are still used for irrigation. Some of the underground galleries built by the Romans are used for irrigation and water supply. Up to 1933, the irrigation development in Syria was carried out by private enterprise or from ancient works. The first irrigation project of Homs-Hama was completed in 1951. Since 1946, the number of private irrigation works have been increasing. The private works included *kanats*, wells, tubewells and river pumping schemes (Framji *et el.*, 1983).

There are over 140 dams in Syria. The largest is located at Al-Tabka on the Euphrates. In 1993, it was estimated that 60.2% of the area was irrigated from groundwater and the remaining from surface water. The net irrigated area in Syria in 1999 was 1.186 million ha.

(***Source***: UN Food and Agriculture Organization (FAO), 2000 and 2001).

Drainage. The drainage systems are generally mixed systems of surface and subsurface drainage. In 1993, 60,000 ha of irrigated land were estimated to be affected by salinization. About 5,000 ha in the Euphrates basin have been abandoned due to waterlogging and salinity. In the new irrigation schemes, open drainage systems have been installed on 90% of the irrigated land.

(***Source***: UN Food and Agriculture Organization (FAO), 2000).

Jordan

The Royal Kingdom of Jordan lies in southwest Asia, to the south of Syria and to the west and north of Iraq and

Saudi Arabia. The total cultivable area of Jordan is estimated to be 1.28 million ha. Total irrigated area in 1999 was 75,000 ha.* Physiographically, Jordan may be divided into three main regions (*i*) the *Ghors (lowlands)* in the western part of the country, (*ii*) the *highlands*, which run from north to south at an altitude of between 600 and 1,600 m above sea level; and (*iii*) the *desert region* (*Badiah*) in the east, which is an extension of the Arabian desert. The Jordan Valley is the most fertile area of the kingdom. (FAO, 2001).

The Jordan Valley enjoys a sub-tropical climate. The climate of other areas ranges from Mediterranean in the highlands to desert climate in the east. The average annual rainfall is less than 100 mm in the desert areas and 300 to 700 mm in the highlands. The rainfall is confined mainly to the months of December to March (Framji *et al.*, 1982).

*(**Source**: UN Food and Agriculture Organization (FAO), 2000).

Water resources. The main river in Jordan with its four tributaries possess great potential for irrigation development. Jordan has numerous springs throughout the country which are being systematically developed for irrigation. Jordan has subtantial groundwater potential, a part of which has been developed.

Irrigation development. Irrigation from springs, wells and small stream diversions and lifting water from rivers has been practiced in Jordan from the ancient times by individual farmers or groups of farmers. After 1954 the Government assumed the responsibility of developing major water resources in the country on a systematic basis.

The Central Water Authority was established in 1959 for "investigating, planning, establishing, promoting, implementing, constructing, and, to the extent necessary, managing and operating a comprehensive programme for development, conservation and control of the water resources of the Kingdom of Jordan for maximum feasible use for domestic, municipal and livestock, industrial, irrigation, hydro-electric power and other beneficial purposes." The Central Water Authority has completed a large number of small irrigation projects. (Framji *et al.*, 1982).

Saudi Arabia

The Kingdom of Saudi Arabia comprises 80% of the Arabian Peninsula in South-Western Asia. The four main physiographic units are: (*i*) *The Western Mountains,* called the Arabian Shield, intersected by deep valleys; (*ii*) *The Central Hills,* with elevation ranging between 900 and 1 800 m. (*iii*) *The Desert Regions,* east of the Central Hills, with elevations ranging between 200 and 900 m. (*iv*) *The Coastal Regions* along the Red Sea with a width of 16 to 65 km. The *Tahama Plain* in the south is generally wide and includes the region of oases. Though mostly desert, Saudi Arabia has stretches of agricultural lands in the Tahama and the mountains of Asir in the west, in Al-Hasa, in the eastern region and in the valuable oases in Hijaz and Neid (Framji *et al.*, 1983).

Saudi Arabia has no perennial rivers. Surface run-off occurs after storms. The water infiltrating into the *wadi* gravels and alluvium, sometimes reach the *Subkhas* where short period lakes are formed. In the South Tihama in the extreme south-west floods occur, from the rift escarpment mountains. The water flows into the Red Sea. In this area a few dams have been constructed for irrigation. In most parts of the country surface waters are too small, silting is a major risk and the evaporation losses are high. Groundwater constitutes an important source of water supply in Saudi Arabia. It has been exploited since the ancient times. In over half of Saudi Arabia conditions are favourable for groundwater development. The Ministry of Agriculture and Water has undertaken investigation of the water and agricultural resources of the Kingdom. Emphasis is laid on groundwater recharge (Framji *et al.*, 1983).

Desalination of water. Saudi Arabia is the largest producer of desalinated water from the sea. In 1965 the Saline Water Conversion Corporation was established. Various processes for desalination of sea-water were studied. However, desalination is expensive when used for irrigation.

Irrigation development. In the Qatif oasis of Saudi Arabia irrigation has been practised for the last about 1,800 years. Hand-dug wells have tapped relatively shallow aquifers. In some places there are flowing artesian wells. When artesian pressure was not adequate to bring water to ground surface, tunnels have been built on downhill grade which lead water to areas that could be irrigated by free flowing water from the wells. During the second half of the 20th century there have been large scale development of well planned irrigation systems. The Wadi Jizan Dam is the first irrigation work constructed by the Government. In 1969, the Government embarked upon a policy of holistic development of water resources to benefit all sections of the population (Framji *et al.*, 1983). By 1993, approximately 185 dams of various sizes had been constructed for flood control, irrigation and groundwater recharge. About 45 new dams were planned to be built. The net area under irrigation (1999) was 1.62 million ha.*

The entire cultivated area in Saudi Arabia is irrigated. Surface irrigation is practised on about 34% of the irrigated area. Sprinkler irrigation is practised on about 64% of the irrigated area. The central pivot sprinkler irrigation system covers practically all the lands under cereals. Vegetables and fruit trees are in general irrigated by micro-irrigation. Groundwater is used on almost 96% of the irrigated area. The cropped area increased more than three times between 1977 and 1992. The major irrigated crop is wheat. Other major crops are fodder, other cereals (mainly sorghum and barley), fruit trees and vegetables. Since 1988, self-sufficiency in wheat has been reached and part of the production is being exported. Vegetables, fruits, dates and fodder are also exported.

Extensive pumping of groundwater has resulted in a significant drop in the groundwater level. Further, only half the groundwater reserves are located near the cultivated areas. The coastal areas suffer increasingly from sea water intrusion. Waterlogging and drainage problems exist in the central and southern parts of the country, due to the shallow, impervious layers below the ground surface. The drainage systems consist mainly of open drainage canals. In several projects, such as the Al-Hassa Irrigation and Drainage Project, agricultural drainage water is reused for irrigation after mixing with fresh groundwater.

*(***Source***: UN Food and Agriculture Organization (FAO), 2000).

Yemen

The Republic of Yemen was formed in 1990s, as a result of the unification of the former Yemen Arab Republic and the People's Democratic Republic of Yemen. Yemen is located on the south-western edge of the Arabian Peninsula. Apart from the mainland, it includes many islands, the largest of which are Socotra in the Arabian Sea and Kamaran in the Red Sea. The total land use area in 1999 was estimated at 17.733 million ha. The net irrigated area in the same year was 0.49 million ha.* Yemen has a predominantly semi-arid to arid climate, with rainy seasons during spring and summer.

Yemen can be grouped into five main geographical-climatological regions: (*i*) *The Coastal Plains:* located in the west and south-west, with rainfall less than 50 mm/year. They contain important agricultural zones due to the numerous *wadis* that drain the adjoining mountainous and hilly areas. (*ii*) *The Yemen Mountain Massif:* a high zone of irregular and dissected topography. The western and southern slopes have rainfall ranging from 300-500 mm/year. (*iii*) *The Eastern Plateau Region:* covering the eastern half of the country. Elevations range from 1200-1800 m with average annual rainfall below 100 mm. However, flash floods following heavy rains are devastating. (*iv*) *The Desert:* Between the Yemen Mountain Massif and the Eastern Plateau lies the Ramlat al Sabatayn, a sand desert. (*v*) *The Island:* The most important of all the islands is Socotra, where exuberant flora and fauna can be found than in any other region in Yemen.**

*(***Source***: UN Food and Agriculture Organization (FAO), 2000 and 2001).
(*Source***: UN Food and Agriculture Organization (FAO), 2001).

Water resources. Yemen can be divided into four major drainage basins, regrouping numerous smaller *wadis*: (*i*) the Red Sea basin, (*ii*) the Gulf of Aden basin, (*iii*) the Arabian Sea basin and (*iv*) the Rub Al Khali interior basin. The floods of the *wadis* in Yemen are characterized by abruptly rising peaks that rapidly recede. Wadi Tiban, on which Aden and the surrounding areas depend, is an important source of water. Surface irrigation from floods and rains covers about half of the irrigated area in Yemen. In many regions of Yemen, undergroundwater is the main source of irrigation.

Irrigation development. In the past, irrigation used to be solely through individual efforts. Subsequently, the government initiated water resources development with a view to expand the area under irrigation. The farmers have reshaped their lands in the hilly and mountainous areas into nearly level fields surrounded by bunds which retain rainfall run-off. Intense rainstorms in the summer rainy season produce two or three short but intense floods. These floods with high peak discharge flow into the flat coastal plain. On each of the major streams, the lands adjacent to the streams are extensively developed for flood irrigation. Many dams are built for irrigation and domestic water supply and flood control. In 1999, the net irrigated area of Yemen was estimated at 0.49 million ha.

(***Souce***: UN Food and Agriculture Organization (FAO), 2000 and 2001).

Irrigation Development in the Other Countries of West-Asia, the Middle East and Near East

Critical reviews of some of the countries in the Middle East including Oman, United Arab Emirates, Kuwait, Bahrain, West Bank, Gaza Strip (Palestine) Cyprus and Lebanan could not be included. Water resources development has been substantial and is of vital importance in many of these countries. In the Middle East, Oman had 62,000 ha, Qatar 13,000 ha, Kuwait 7000 ha, Lebanon 120,000 ha, United Arab Emirates 76,000 ha, Bahrain 5000 ha and Gaza Strip (Palestine) 12,000 ha under irrigation in 1999.

(***Source***: UN Food and Agriculture Organization (FAO), 2000).

IRRIGATION DEVELOPMENT IN AFRICA

The 53 African countries could be grouped into seven regions on the basis of geographic and climatic homogeneity, which has a direct influence on irrigation. These regions are the Northern, the Sudano-Sahelian, the Gulf of Guinea, Central, Eastern, Indian Ocean islands, and Southern Africa. The population of Africa crossed the 500 million mark in 1985. The average density of population is 17 per sq km. However, the distribution of population varies from region to region even within a country. The current average rate of growth of population in Africa is about 3.2% annually. Rural population ranges from 60 to 80% in most countries. Food production in Africa has been increasing at an estimated 2.3% per year, while cereal crops have been increasing at only 1.8% per year. Hence, the food imports are rising alarmingly. A series of droughts in the region has been leading to acute food shortage in some of the countries. The situation calls for urgent and effective steps to increase agricultured production.

The diversity of water management situations in Africa suggest two distinct situations of irrigation. The land on which water is applied for the purpose of agricultural production is called the *water managed areas.* The term *irrigated areas* is limited to that part of

the water managed areas with hydraulic structures: full or partial control irrigation, equipped wetland or valley bottoms and areas equipped for spate irrigation. Water managed areas include cultivated wetland and valley bottoms which are not equipped for irrigation, and recession cropping areas. Full or partial control irrigation is the most widespread and constitutes about 81% of the area under irrigation. Among the other classes of water management, cultivated wetlands and valley bottoms and recession cropping area constitute 15% of the total water managed area. Recession cropping is used essentially on the flood plains of the river Niger and Senegal and their tributaries.
(***Source***: UN Food and Agriculture Organization (FAO), 2001).

Recession cropping. It is the practice of raising crops in flood plains adjoining rivers, after the recession of flood waters. The soil moisture stored in the crop rest zone during the floods sustains the crop. In Africa, water managed areas comprise 14.3 million ha (FAO, 2001). The net irrigated area in Africa in 1999 was 12.58 million ha, out of a total agricultural area of 1093.6 million ha.*

Irrigation development in Africa has been very uneven. The part of water managed areas in agriculture varies from less than 1% of the cultivated land in Zaire, Uganda, Ghana, Togo, and Comoros to 100% in Egypt and Djibouti, where agriculture is impossible without irrigation. In Equatorial Africa, where precipitation is high, rainfed agriculture is dominant. Irrigation is used only for winter cropping, rice cultivation and to produce value crops.**

About 42% of Africa's irrigated area is in Egypt and the Sudan. Morocco and Madagascar account for another 30% of the irrigated area (Table 1.4). The type of irrigation development has been uneven. Most of about two-third of the irrigated area is under 'formally organized' irrigation which are well designed medium or large-scale projects. They are concentrated mostly in Egypt and the Sudan. The remaining one-third of irrigated area is mostly traditional, small-scale systems. In these, there is often only a partial control of water.

About 85% of water withdrawals are for agriculture. However, it varies considerably from one region to another. The northern region, which is predominantly arid, represents more than half of the water withdrawal in agriculture in Africa. The humid regions have the lowest agricultural withdrawals.

The need to expand irrigation in Africa arises form the present food and agricultural production scenario. Because of the importance of irrigation to Africa's agricultural production, there is an immediate need for proper assessments of irrigation potential, particularly in the arid and semi-arid areas where irrigation development is of vital importance.

*(***Source***: UN Food and Agriculture Organization (FAO), 2000 and 2001).

(*Source***: UN Food and Agriculture Organization (FAO), 2001).

Historical perspective. Historically, the Nile Valley irrigation system of Egypt stands out as an example of the importance given to irrigation since the ancient times. Traditional irrigation practices were prevalent in other countries of North Africa as well. For instance, irrigation was practiced in the ages of Tanisia in the 10th century B.C., before its occupation by the Romans. The Romans laid emphasis on irrigation and introduced improved methods of water conveyance and distribution. After the fall of the Roman Empire, irrigation practice declined, in spite of efforts made to restore them from time to time by some of the enlightened dynasties like the Aglabites and later the emigrants from Spain.

Major Climatic Zones of Africa

The climatic factors that affect agricultural production most are temperature and rainfall. Their combined effect on crop growth can be expressed in the length of the growing period of crops. The following are the major climatic zones of Africa:*

(*i*) *Desert zone*, comprising 29% of land area with an average rainfall of about 100 mm per year.

(*ii*) *Arid zone*, forming 17% of land area with rainfall ranging from 100-400 mm/year.

(*iii*) *Semi-arid zone*, comprising 8% of land area, with rainfall ranging from 400 to 600 mm/year.

(*iv*) *Dry sub-humid zone*, comprising 10% of the land area, with rainfall ranging from 600-1,200 mm/year.

(*v*) *Moist sub-humid zone*, constituting 20% of land area, with rainfall ranging from 1,200-1,500 mm/year.

(*vi*) *Humid zone*, constituting 16% of land area with an average annual rainfall of 1500 mm/year.

There is deficient soil moisture for rained crop production in the desert, arid and semi-arid regions. Availability of soil moisture is usually adequate for rainfed agriculture in the dry sub-humid and moist sub-humid regions. Irrigation in these regions could increase the production of high value crops when water management programmes are scientifically planned. Usually there is adequate soil moisture for crop production in the humid regions. Dry sub-humid and moist sub-humid regions, constituting 30% of the land area, are ideally suited climatically for rainfed millets sorghum, maize and other staple food crops of Africa.

*(***Source***: Based on UN Food and Agriculture Organization (FAO), 1987).

Major Agro-Climatic and Ecological Regions of Africa

The Consultation Group of FAO on Irrigation in Africa * has suggested the classification of Africa into 6 major regions (Fig. 1.20), each region having broadly distinctive environmental features (Table 1.2). The major features of the six regions are as follows:

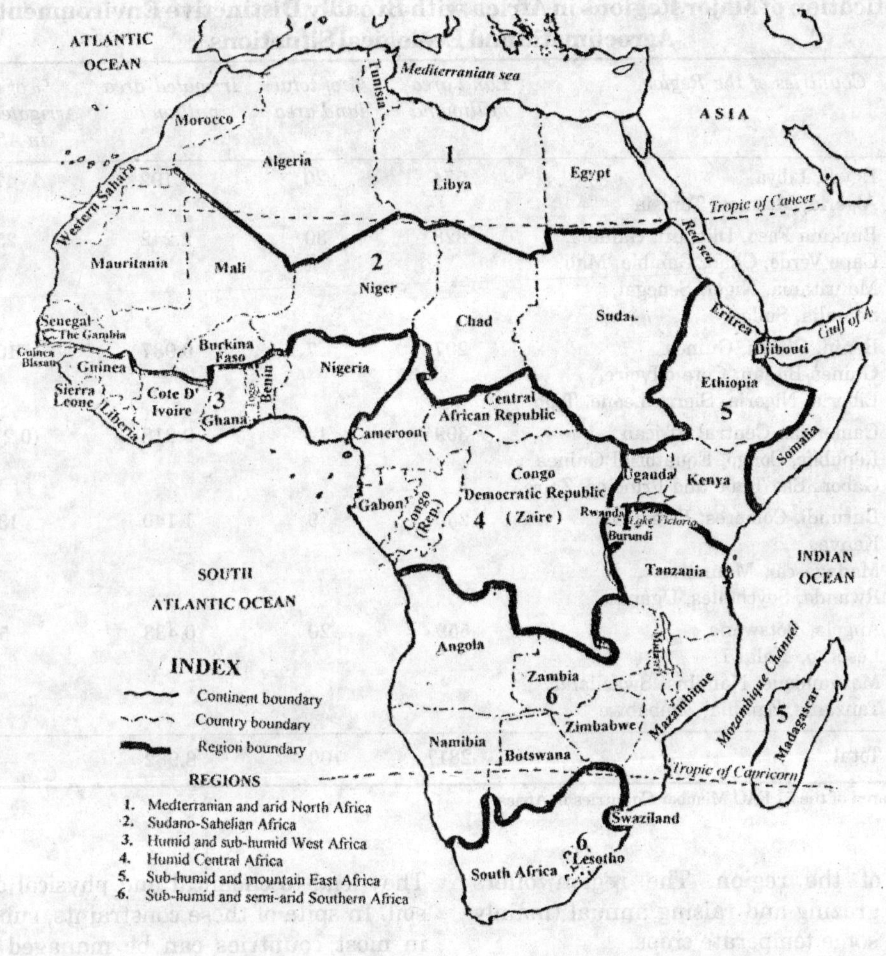

Fig. 1.20. Map of Africa showing broad agro-ecological regions with distinctive environmental features (not to scale). Boundary between Ethiopia and Eritrea is not shown.
[***Reference***: Framji *et al.*, (1981); FAO 1986–1987; *Oxford School Atlas* (2000).]

1. *Mediterranean and Arid North Africa.* The predominant part of the region forms the Sahara desert (86%). Adjoining arid region forms 7%. About 7% of the region, comprising semi-arid (3%), dry humid (2%) and moist sub-humid (2%) are suitable to raise rainfed temperate crops.
2. *Sudano-Sahelian Africa.* The region is predominantly desert (32%) and arid (30%). Semi-arid area (10%) forms extensive grazing land. Potential for tropical rainfed annual crops exists in the dry sub-humid (15%) and moist sub-humid (7%) areas.
3. *Humid and Sub-humid West Africa.* The predominant climatic zones of the region are moist sub-humid (47%) and humid (35%). They are suitable for a wide range of annual and perenial crops. Dry sub-humid zone comprise of 15%, while the remaining 3% is semi-arid.
4. *Humid Central Africa.* The region is dominated by humid (69%) and moist sub-humid (29%) climatic zones. They are suited for a limited number of annual and a wide range of perennial tropical crops. Extensive areas are under forests. Dry sub-humid conditions exists only in about 2% of the region.
5. *Sub-humid and Mountainous East Africa.* The region comprise the widest variety of environmental conditions, ranging from desert (10%) in eastern lowlands to humid areas (15%) in the cool highlands. Arid (25%) and semi-arid (13%) conditions prevail in the region. More favourable conditions occur in the dry sub-humid (11%), moist sub-humid (27%) and humid (14%) areas. The land in this region is under a wide range of uses. There is potential for annual and perennial, temperate and tropical crop production and grazing in the semi-arid parts of the region.
6. *Sub-humid and Semi-arid Southern Africa.* Desert (7%), arid (22%) and semi-arid (16%) conditions prevail in 38% of the region. More favourable environments exist in the dry sub-humid (19%), moist sub-humid (40%) and humid

Table 1.2. Classification of Major Regions in Africa with Broadly Distinctive Environmental Features and Agroclimatic and Ecological Situations

Region	*Countries of the Region*	*Land area million ha*	*% of total land area*	*Irrigated area million ha*	*% of total irrigated area in Africa*	*% of cropped area*
Mediterranean, arid North Africa,	Egypt, Libya, Algeria, Morocco, Tunisia	574	20	4.192	47	14.8
Sudano-Sahelian Africa	Burkina Faso, Djibouti, Gambia, Cape Verde, Chad, Gambia, Mali, Mauritania, Niger, Senegal, Somalia, Sudan	828	30	2.242	25	7.3
Humid and sub-humid West Africa,	Benin, Ghana, Guinea, Guinea-Bissau, Cote d'Ivoire, Liberia, Nigeria, Sierra Leone, Togo	207	7	0.937	10	2.1
Humid Central Africa	Cameroon, Central African Republic, Congo, Equatorial Guinea, Gabon, Sao Tome and Principe, Zaire	399	14	0.018	(0.2)	0.1
Sub-humid and mountainous East Africa	Burundi, Comoros, Ethiopia, Kenya, Madagascar, Mauritius, Rwanda, Seychelles, Uganda	250	9	1.140	13	4.1
Sub-humid and semi-arid Southern Africa	Angola, Botswana, Lesotho, Malawi, Mozambique, Namibia, Swaziland, Tanzania, Zambia, Zimbabwe	559	20	0.433	5	1.8
	Total	2817	100	8.962	—	5.2%

* Refers to the total land area of the 51 FAO Member Countries in Africa.
[***Reference:*** FAO (1987).]

(3%) parts of the region. The region offers potential for grazing and raising annual (mainly tropical) and some temperate crops.

*(***Source***: UN Food and Agriculture Organization (FAO), 1987)

Soils. The soils of Africa range from calcarious desert soils to deeply weathered and acid soils of the humid tropics. Dark clay soils and alluvial soils are usually very fertile. Saline soils, and shallow soils are suitable for agricultural production, only if appropriate soil management practices are adopted.

Out of the 2844 million ha of land area in Africa (including Western Sahara), deserts (620 m ha) occupy 21.8% sandy soils (577 m ha) 20.3%, saline soils (67 m ha) 2.3%, acid soils of tropical low lands (509 m ha) 17.9%, soils of tropical highlands (39 m ha) 1.4%, dark clay soil (99 m ha) 2.5%, Ferruginous tropical soils (194 m ha) 6.8%, Mediterranean soils (87 m ha) 3.1%, poorly drained alluvial soils (276 m ha) 9.7%, and shallow soils (376 m ha) 13.2% (Broad estimates, based on UN Food and Agriculture Organization (FAO), 1987). There are considerable areas in Africa, especially in the semi-arid and sub-humid regions with wide variations of soil degradation and accelerated soil erosion. Usually they are the result of overgrazing, deforestation, and inappropriate land use and cultural practices. Inappropriate land use is also leading to desertification. Poor water management practices are leading to excessive salinization in some of the irrigated areas. They lead to chemical and physical degradation of the soil. In spite of these constraints, substantial land area in most countries can be managed at high levels of productivity with proper inputs and appropriate soil and water management practices.

Irrigated crops. The most widespread crop is rice which represents more than one-third of the water managed crop area. Cultivated mostly in wetlands and valley bottoms, rice is the predominant crop in the humid zones of the Gulf of Guinea and Eastern Africa. It is also very much developed on the plateaus of Madagascar. In the Northern and Southern regions, it represents only a small portion of the total crops under water management. Among the other cereals, wheat and maize are cultivated and irrigated mostly in the countries of the North (Egypt and Morocco) and in South Africa, Sudan and Somalia. Sorghum and millets are cultivated in the Sudano-Sahelian region, mostly as recession cropping.

Vegetables (including root and tuber crops are present in all regions. They represent about 8% of the cultivated areas under irrigation in Africa. In Algeria, Mauritania, Kenya, Burundi and Rwanda, they are the most widespread crops under water management. Fruit crops, mostly citrus, is concentrated mostly in the northern region. Industrial crops are located mostly in Sudan and in the countries of the southern region (mostly cotton and oilseed crops). Sugarcane, coffee, cocoa, oilseeds, date palm, banana, tobacco and cut

Table 1.3. River basins of Africa with mean annual discharge of 10 km³ and above

River basin	*Major countries of the river basin*	*Basin area, km²*	*Mean discharge km³*	*Ocean/sea/inland to which flowing*
Congo (Zaire)	Congo, Congo Democratic Republic (Zaire)	4,000,000	132	South Atlantic
Ogooue	Gabon	203,500	149	South Atlantic
Sanaga	Cameroon	131,500	65	South Atlantic
Niger	Guinea, Mali, Niger, Nigeria	1,215,000	180	North Atlantic
Volta	Ghana, Burkina Faso	394,100	40	North Atlantic
Senegal	Guinea, Senegal, Mali, Mauritania	338,000	22	North Atlantic
Zambezi	Angola, Mozambique Zambia	1,250,000	103	Indian Ocean
Purfigi	Tanzania	1,58,000	30	Indian Ocean
Nile	Egypt, Ethiopia, Sudan, Uganda	2,800,000	84	Mediterranean Sea
Chari-Logone	Central African Republic, Niger, Chad, Nigeria		43	Inland
Omo	Kenya	77,200	16	Lake Turkana

Note: River Zaire has perennial flow. Almost all other rivers have high seasonal variations in their flow.
[***Reference:*** Framji *et al.*, (1981, 1982, 1983), UN Food and Agriculture Organization (FAO), 1986 and 1987).]

flowers). Apart from sugarcane, present in all regions except the Northern, the other crops are generally concentrated in a limited number of countries. Irrigation of fodder crops are important in the Northern and Southern region, and account for 15% of the water managed area in Africa. Irrigated berseem is cultivated almost exclusively in Egypt.

Root crops, particularly cassava, potatoes, sweet potatoes and yams are grown in humid tropical parts. Wheat and barley are grown in Morocco and Sudan and in other eastern and Southern African countries in areas where climate and altitude are favourable. Pulses, cotton, groundnut, sesame seed, sugarcane, cashewnuts, bananas and plantations of coffee, cocoa and tea are grown in various African countries. The percentage of irrigated areas under different crops are cereals 53%, fodder 15%, fibre crops 8%, vegetables (including potatoes) 7%, oilseeds crops 7%, fruit crops 6%, sugarcane 4%, and root crops 2%.

(***Source***: Based on UN Food and Agriculture Oganization (FAO), 1987 and 2001).

River Basins and Surface Water Sources

Africa has less surface water and a higher rate of evaporation, as compared to most other regions' continents. The flows of most of the main rivers, except the Zaire River, are seasonal. Most of the river systems would require substantial regulation for irrigation development.

In Africa, surface water is distributed highly unevenly. The Congo basin, which occupies almost 16% of the whole of Africa, south of the Sahara has 55% of the mean annual water discharge. Further, seven rivers namely the Niger, Ogoouc, Zambezi, Nile, Sanga, Chari-Lagone and Volta contribute another 25%. Only a few major rivers, namely Senegal, Niger and Nile flow through the drought-prone areas of the Sudano-Sahalion and Sahara regions where there are severe climatic restrictions to rainfed of Africa agriculture.

Nile is the most exploited river system in Africa. While the Nile discharges into the Mediterranean, ten major rivers flow into the Indian Ocean, three into the South Atlantic ocean and four into the North Atlantic Ocean. Two major rivers, the Niger and the Nile, pass through major inland swamps (the Niger Inland Delta and the Sudan Sudd). They lose a considerable amount of water through evaporation in the swamp areas. Several other basins are fully landlocked and eventually lose all their water by evaporation or percolation. Lake Chad system is the most important in this category. Particulars of the major river systems of Africa having mean annual discharge above 10 km³ is presented in Table 1.3.

Between the high rainfall tropical zone of Central Africa, and the arid zones of north and south lie large dry sub-humid or semi-arid areas. They contain many small rivers and streams, mostly with only seasonal flow, with considerable potential for small-scale irrigation development. In the Sahara region, there is very little run-off from rainfall. Hence, there are no significant surface water resources. In the Sudano-Sahelian region, extending from Senegal to Somalia,

Table 1.4. Irrigated Area in 12 Selected African Countries Having Major Programmes of Water Resources Development

Country in Africa	Net area under irrigation in 1999 million ha	% irrigation of cropped area of the country in 1982	% irrigation the country in Africa in 1999
1. Egypt	3.300	98.60	26.32
2. Sudan	1.950	14.10	15.55
3. South Africa	1.354	12.12	10.80
4. Morocco	1.305	9.50	10.41
5. Madagascar	1.090	32.00	8.69
6. Algeria	0.560	16.62	4.47
7. Libya	0.470	5.90	3.75
8. Tunisia	0.380	—	3.03
9. Nigeria	0.233	—	1.86
10. Somalia	0.200	—	1.60
11. Ethiopia	0.190	—	1.50
12. Tanzania	0.155	7.43	1.24
Total Africa	12.538	—	—

(***References:*** Compiled from Framji *et al.* (1981), UN Food and Agriculture Organization, 1987 and 2000).

the run-off is up to 10% of the rainfall. In the wet tropical highlands of Ethiopia, the run-off is more than 20% of the rainfall. In general, about 50% of total surface water resources are available for irrigation.*

A notable feature of the rivers in the African continent is the comparatively low sediment loads due to low density of population in the catchment areas. However, this picture is changing in areas which have lost their vegetation cover mainly due to overgrazing or deforestation.

*(***Source***: UN Food and Agriculture Organization (FAO), 1987).

Groundwater resources. About 20% of the total water resources of Africa comprise of groundwater. High yielding aquifers are present in about 10% of the land area. Until recently most of the studies of groundwater potential in Africa have been of general nature, limited to small areas. The water bearing formations in more than half of Africa consist of fractured, altered granitic, metamorphic or volcanic rocks containing small discontinuous aquifers. They have a low recharge rate which can meet relatively minor requirements of people, livestock and small scale irrigation.

Artesian aquifers occur in the desert of North Africa. However, their yields tend to fall due to inadequate recharge. Abundant shallow groundwater exists along alluvial river beds. Many coastal deltas and plains have sedimentary basins with shallow permeable horizons. At some places they are over-exploited. Some of the aquifers have been contaminated by intrusions of sea water.

*(***Source***: UN Food and Agriculture Organization (FAO), 1987).

Water quality. The quality of groundwater for irrigation is mostly good throughout Africa. However, there are localised problem areas due to salt water intrusion in the tidal areas and on mangrove swamps.

Irrigation development. In Africa the commonly used irrigation method is surface irrigation (check basin, furrow and border irrigation). Irrigation schemes obtain water from river diversions or reservoirs and use gravity-fed canal systems. Pumping from canals is used where gravity flow is not possible. Sprinkler and drip irrigation systems are used for large-scale sugarcane plantations and to a limited extent for orchards and vegetables. micro-irrigation is mostly limited to Egypt and South Africa. Sprinkler irrigation of sugarcane in common in Mauritius. The net irrigated area in 12 selected countries in Africa is presented in Table 1.4.

Classification of Irrigation Projects

Irrigation types and practices can be classified in various ways, namely based on the size of the scheme, degree of water control, level of technology or type of management. Both modern irrigation (full water control) and flood recession (little water control) can be found in some of the large and small schemes.

Based on the scheme size irrigation systems are classified into four main categories as follows :

(*i*) *Very large-scale schemes.* They are schemes having an irrigation potential of over 10,000 ha under government management. Examples are gravity schemes in the large river basins in Egypt, Sudan and Morocco. They are well designed systems.

(*ii*) *Large-scale schemes*. They are projects with irrigation potential ranging from 1,000 to 10,000 ha with adequate water control. They are generally under government or commercial management.

(*iii*) *Medium-scale schemes*. They are to irrigate 100 to 1,000 ha with full or partial water control. They are government managed or under government assisted cooperatives or private commercial estates.

(iv) *Small-scale schemes.* They irrigate areas up to 100 ha, and are usually controlled by farmers' groups, or single farmer. They are numerous in Kenya, Zimbabwe, Madagascar, and Tanzania. Pumping from lakes is common in small scale projects.

Based on the level of technology used in planning, construction and management, irrigation systems in Africa are classified as *formal* and *traditional*. Formal irrigation schemes are usually developed and managed by the government. They are typically medium, large or very large-scale projects. Traditional irrigation is usually small-scale. They are under local bodies or farmer groups and are controlled and operated by the community. The traditional irrigation schemes include the following:

(a) Irrigation based on the use of receding or advancing floods in river flood plains, inland valleys, lake shores and coastal swamps. The area planted by an individual or group of farmers. The aggregate area under such systems is large.

(b) Small-scale schemes using manual or animal operated water lifts, and power-driven pumps, using water from and wells, ponds, or streams.

(c) Small (temporary) rivers diversions or the development of swamps for wetland cultivation.

(d) Water harvesting tanks or diversions having earth bunds to collect run-off water or flash floods.

Traditional schemes often have only partial water control. They are constructed with locally available materials.

Out of the total irrigated area of 8,962 million ha in Africa in 1982, 56% was under formal irrigation systems and the remaining under traditional systems of irrigation. The concentration of traditional irrigation systems is in Madagascar, Nigeria, Morocco, Algeria and Tunisia.

(***Source***: UN Food and Agriculture Organization (FAO), 1986).

Egypt

Egypt, situated in the north-east of Africa has a total land area of 100.145 million ha. Physiographically, Egypt comprises three main regions, namely, the delta, the Nile valley and the deserts. The delta region, which was an estuary, is presently one of the most fertile and intensively cultivated regions of the world. The fertility of the area has been mainly due to the silt brought down by the Nile from the Ethiopian plateau. Out of the 6,700 km length of the Nile, the only river in Egypt, the terminal part of 1,550 km is in Egypt. In 1993, the total cultivated land was estimated to be 3.24 million ha, or 3.2% of the total area. The mean annual rainfall is estimated at 18 mm. It ranges from 0 mm in the desert to 200 mm in the northern coastal region.

(***Source***: UN Food and Agriculture Organization (FAO), 2001).

Water resources. The Nile river is the main source of water for Egypt. River Nile draws the greater part of its water from two regions, namely the Ethiopian Hills and the Equatorial Lake Plateau. The tributaries of the Nile which rise in Ethiopia are the Blue Nile, the Atbara and the Sobat, which together contributes about 85% of the annual discharge of the river. The water of the lake plateau region is Lake Victoria and partly of Lake Albert system. It flows into Southern Sudan and joins the Sobat to form the mainstream of the White Nile. The main source of recharge to groundwater is percolation from irrigated fields and seepage from the irrigation network. In the northern part of the Delta, groundwater becomes brackish to saline due to sea water intrusion.

Modern irrigation development. With the ever-increasing demand for agriculture to meet the needs of the rising population the need was felt to store all possible supplies in the Nile. This led to the construction of the Aswan Dam and Gebel Aulia Dam. The original Aswan Dam was constructed in 1902 A.D., to store 1000 million m^3 of water for the protection of perennial irrigated area in the Nile valley and for bringing an additional 1,80,000 ha area of basin irrigated area under perennial irrigation. In A.D. 1912, the height of the dam was increased by 6 m to raise the storage capacity of the reservoir to 2500 million m^3. The dam was again raised by another 8 m during 1929–33, increasing its capacity to 5,000 million m^3. The additional storage enabled the conversion of 80,000 ha of land under the old basin irrigation system to perennial irrigation and for irrigating additional 80,000 ha of new land. Under the 1959 Nile Waters Agreement between Egypt and Sudan, Egypt's share is 55.5 km^3/year. (Framji *et al.*, 1981 and UN Food and Agriculture Organization (FAO), 2001).

The construction of Aswan High Dam (Sadd-el-Aali Dam) Project was started in 1960 and completed in 1970. The High Dam, located 7 km upstream of the original Aswan Dam is a long term storage project for the full utilization of the waters of the Nile. It is a rock fill dam 110 m above the river bed with a storage capacity of 164,000 million m^3. The reservoir upstream of the river is one of the largest in the world with a length of 500 km and an average width 10 km. After the completion of the Aswan High Dam, the entire cultivated area in Egypt has been placed under perennial irrigation. Barrages in the Nile downstream divert water to main canals (district canals) and then to the branch canals and sub-branch canals.

Water is released to the main canals on 2 or 3 interval rotation system. The duration of the 'turn' depends on the estimated water requirement. The turns are usually of 4 and 7 days. From a branch canal water is supplied to farmers from outlets through a water course (*meska*). Pipe outlets are used. The design water elevation is located about 50 cm below the average elevation of the surrounding land to the irrigated. Each farmer (or group of farmers) is required to lift the water from the supply channel to the field with manually operated or animal operated water lifts or mechanically driven pumps. This practice is to discourage excessive use of water by the farmers (Framji *et al.*, 1981). In 1993, an area of 4,200 ha was irrigated from treated wastewater.

The irrigated area in Egypt in 1999 has been 3.3 million ha* comprising the entire cultivated area. The estimated irrigable area is 5 million ha. A Master Plan to increase the irrigation efficiency and irrigated area is being implemented. The Al Salam canal, under construction, is planned for the reuse of drainage water from two main drains in the Eastern Delta.**

*(***Source***: Un Food and Agriculture Organization (FAO), 2000).

(*Source***: UN Food and Agriculture Organization (FAO), 2001).

Drainage. An extensive National Drainage Programme has been carried out in Egypt to control waterlogging and salinity. The drainage system consists of open drains and sub-surface drains along with pumping stations. Of the total irrigated area, about 90% is drained, out of which about 57% is covered by surface drains and the remaining by sub-surface drains. Drainage water from agricultural land on both sides of the Nile Valley is returned to the Nile River in Upper Egypt and in the

Southern Delta. Soil salinity in Upper Egypt developed mainly after the construction of the Aswan High Dam and the conversion of flood irrigation into perennial irrigation. Most of the salt-affected agricultural land in Egypt has now been provided with drainage systems for leaching the excess salts.

(***Source***: UN Food and Agriculture Organization (FAO), 2001).

Sudan

Sudan, lying in north-east Africa has a total land area of 250.381 million ha. The area irrigated in 1999 was 1.95 million ha, which is 25% of the cultivated area.* Sudan consists of a flat internal plain, at about 325 m above sea level. It is crossed by the Nile river and its tributaries and by a number of mountains. In the south is the Sudd Region, the great wetland which comprises of channels, lakes and swamps. The Sudd area is nearly flat. The soils are predominantly clay, but poor in nutrients. The northern one-third of Sudan is covered by sandy desert.

The cultivable area of Sudan is estimated to be 105 million ha, or 42% of the total area. The cultivated land is 7.6 million ha, which is only 7% of the cultivable area.**

* (***Source***: UN and Agriculture Organization (FAO), 2000).

(*Source***: Food and Agriculture Organization (FAO), 2001).

Water resources. The most important surface water irrigation source in Sudan is the Nile river, including its tributaries. The main part of Nile is formed by the confluence of the Blue Nile (65%) and the White Nile (23%) in the capital Khartoum. Before flowing into Egypt, one more tributary, the Atbara river (12%) joins the Nile. The Atbara and the Blue Nile rivers originate in the Ethiopian catchment. Apart from the Nile system, there are also the seasonal rivers of Gash and Baraka in eastern Sudan. During the rainly period of July-September, the water flow is drawn off into canals and spreads to the land, forming a fertile delta area (spate irrigation).

(***Source***: UN Food and Agriculture Organization (FAO), 2001).

The Gezira triangle includes about 2.1 million ha of land lying between the White Nile and the Blue Nile. Of this about 0.735 million ha, are suitable for irrigated agriculture (Framji *et al.*, 1984).

There are four large dams in Sudan two on the Blue Nile, one on the Atbara river, and one on the White Nile.

(***Source***: UN Food and Agriculture Organization (FAO), 2001).

Irrigation development. Irrigation in Sudan may be classified as follows: (1) Major gravity irrigation with large scale storage reservoirs and river diversion projects. (2) River lift irrigation schemes involving large scale water lifting by pumps. (3) Flush irrigation in the Gash and Barak river deltas in eastern Sudan by inundation canals and in the Tokar Delta by natural flooding. (4) Basin irrigation in the Northern province which was started in the early 20th century. (5) Irrigation by small scale water lifting by manual and animal operated water lifts mainly on the Nile and its tributaries, but to a limited extent from shallow wells (Framji *et al.*, 1983).

Due to well planned development, the area under irrigation in Sudan increased from 1.250 million ha in 1969–71 to 1.750 million ha in 1980, and 1.95 million ha in 1999.

(***Source***: Framji et al, 1983 and UN Food and Agriculture Organization (FAO), 2000).

Use of groundwater. A large section of the population of Sudan depends heavily upon groundwater for domestic use. In the northern part of Sudan where rainfall is low, wells are the only sources of water. Several projects to use groundwater for irrigation are being implemented.

South Africa

The Republic of South Africa has a total area 122.1 million ha. The cultivable area is estimated to be 18 million ha, which is 18% of the total area. The net irrigated area in 1999 was 1.354 million ha.*

Topographically, South Africa may be divided into four zones: the Plateau, the Escarpment, the Folded Mountains, and the Coastal Plain. The Plateau is an extensive plain interrupted by small mountains. The Escarpment varies in appearance according to the elevation and extent of soil erosion. The average annual rainfall is 451 mm in the western deserts and rises to 1,200 mm/year in the eastern part.**

*(***Source***: UN Food and Agriculture Organization (FAO), 2000 and 2001).

(*Source***: UN Food and Agriculture Organization (FAO), 2001).

Water resources. There are two main river systems in South Africa: (i) The Orange River system, with its major tributary, the Vaal, drains almost the entire plateau, comprising about 48.5% of the total land area and contributing about 22.5% of the total run-off. (ii) The rivers draining the eastern slopes of the high escarpment. The dams command virtually all the run-off from the plateau. The untapped resources are concen-trated along the coast.*

The potential for full or partial control irrigation development, based on soil an water availability and suitability, is estimated at 1.5 million ha. The groundwater resources are estimated at 4.8 km^3/year. In 1980, about 1.79 km^3 were exploited (UNFAO, 2001).

*(***Source***: UN Food and Agriculture Organization (FAO), 2001).

Madagascar

Madagascar has a total land use area of 27.1108 million ha comprising the main Madagascar Island and minor islands in the Indian Ocean, off the south-east coast of Africa. It is separated from the mainland East Africa by the Mozambique Channel (Fig. 1.20). Madagascar consists of three parallel zones running longitudinally. They are the central Plateau of rugged highlands the Eastern Coastal alluvial belt of width about 48 km and a western zone of sedimentary formation of low plateau and vast plains. Irrigation is indispensible in the south-eastern region of the country while advantageous in the highlands and other parts.*

Madagascar has a number of important rivers. A number of dams have been built on these rivers for irrigation (Framji *et al.*, 1982).

****Source***: UN Food and Agriculture Organization (FAO), 2001).

Irrigation development: Irrigation started in Madagascar in about the 18th century on the plateau. Subsequently, it was extended to the coastal regions. Since the beginning of the 20th century large-scale irrigation works were built. Groundwater is used essentially for domestic water supply, except for an artesian aquifer in the south-west of Madagascar which irrigates 3,000 ha of land (Framji *et al.*, 1982).

The total area irrigated in 1982 was 0.96 million ha (Framji *et al.*, 1982). Almost 8,00,000 ha are covered under traditional irrigation system and the remaining under modern irrigation systems. The estimated irrigation potential is 1.2 million ha. The net irrigated area in 1999 has been 1.09 million ha.*

*(***Source***: UN Food and Agriculture Organization (FAO), 1986, 2000).

Nigeria

Nigeria, located in the tropical zone of West Aftrica, has a total area of 92.377 million ha. The net cultivated area is about 22.6 million ha. Wetland and bare-land represent 2.6% and 8.5% respectively, of the country's area. The net irrigated area in 1999 has been 0.233 million ha.*

Nigeria can be divided into six topographical units: (*i*) the low coastal belt, including the Niger Delta, (*ii*) adjoining valley region, (*iii*) the valleys of Niger and Benue rivers above their confluence at Lokoja, with altitude ranging from 76 to 183 m, (*iv*) the broad northern plateau comprising stepped plains, with altitudes ranging from 183 to 762 m, (*v*) the Jos Plateau of altitude ranging from 1,158-1,402 m, and (*vi*) the eastern highlands (Framji *et al.*, 1982).

The climate of Nigeria is semi-arid in the north, and humid in the south. Except for the coastal zone, where it rains all year round, rainfall is seasonal with distinct wet and dry seasons. Rainfall varies from under 500 mm/year in the northeast to over 3 000 mm/year on the coast.**

*(***Source***: Framji *et al.*, 1982, UN Food and Agriculture Organization (FAO), 1986, 2000 and 2001).

(*Source***: UN Food and Agriculture (FAO), 2001).

Water resources: There are four main river systems in in Nigeria: the Niger-Benue, the Chad basin area, the south-western littoral and the south-eastern littoral, with trans-border flows from Cameroon Nigeria is comparatively well endowed with surface water resources. The major rivers which flow through Nigeria are the Niger, the Benue and their tributaries. In the northeast region, the Benue, the Gongola, the Taraba, the Yobe and Ebeji are the major rivers. The other rivers have seasonal flows, confined to the summer months. The Niger river is the third longest river in Africa.

The other natural water resources are fresh water concentrated lakes, ponds and lagoons, mainly in the coastal areas. Lake Chad is a large land-locked, shallow, fresh water lake which is bounded by four countries, Niger, Nigeria, Cameroon and Chad. The Chad Lake is an important source of water for large-scale irrigation (Framji *et al.*, 1982).

In general, the availability of groundwater resources in Nigeria is limited. About 50% of the area of the country is have crystalline rocks underneath which are poor aquifers. The groundwater is exploited through springs and by constructing wells and bore holes. In parts of the Chad basin appreciable quantities of artesian and sub-artesian aquifers occur. Potential supplies of groundwater have also been located in the northern parts of the country. (Framji *et al.*, 1982).

Irrigation development: Well planned formal irrigation and drainage systems in, Nigeria are of recent origin, beginning in 1949. Since 1962, Nigeria has been a member of the Lake Chad Basin Commission consisting of the four countries bordering Lake Chad. (Framji *et al.*, 1982). Most of the irrigation projects in North-East Nigeria are river lift schemes using low head high discharge pumps.

According to the National Water Resources Master Plan Study, total irrigation potential of Nigeria would be about 3.14 million ha. The flood plain of the Niger river is considered the greatest untapped potential for future irrigation development in Nigeria. The most important irrigated crop is rice, which occupies about 75% of the total water managed area. A total of 60 large dams and 100 small dams have been built till 1999. The net irrigated area in 1999 was 0.233 million ha (UNFAO, 2000).

Irrigation systems can be divided into three categories: (*i*) Public irrigation projects which are schemes under government control (formal irrigation). (*ii*) Farmer-owned and operated irrigation projects (informal irrigation) which receive assistance from government (subsidies and training). (*iii*) Residual flood plains forming about 76% of the total water managed area where no government aid is given. In Nigeria, irrigation development is of great importance in view of the physical, economic and social conditions. Intensification of agricultural activities is mainly dependent on irrigation to grow crops like wheat, rice, cotton and sugarcane.

Morocco

Morocco, located in the north-west extremity of the African continent, has a total area of 44.65 million ha. Morocco can be divided into four physiographic units: (*i*) The Coastal Plains, extending along the entire Moroccan coastline, through which the majority of the rivers and valleys of the country crosses. (*ii*) The Northern Hills, running parallel to the Mediterranean Sea. (*iii*) The Central Hills, along the middle of the country. (*iv*) The Desert Hills, forming the southern slopes of the Upper and Lower Atlas Mountains.

The cultivable area has been estimated at 8 million hectares. In 1993, the total cultivated area was 7.23 million ha. The net irrigated area in 1999 was 1.305

million ha which is about 10.41% of the total irrigated area in Africa. The irrigated areas contribute about 45% to agricultural earnings. The average annual rainfall is 340 mm. It varies from more than 450 mm in the north, where rainfed agriculture is possible, to less than 150 mm towards the south-east, where irrigation is essential.

(***Source***: UN Food and Agriculture Organization (FAO), 2000 and 2001).

Irrigation development: Irrigation potential of Morocco is estimated at 1.65 million ha. Spate irrigation covers 1,65,000 ha. The remaining area is to be covered by full or partial control irrigation schemes. Since 1960, the emphasis has been on the construction of dams and on the development of large schemes. In addition to the areas covered by the large schemes, there are about 826,550 ha of medium and small schemes, including the spate irrigation area. The schemes in the mountains mostly use spring water, distributed by gravity. The net irrigated area in 1999 was 1.305 milion ha.

(***Source***: UN Food and Agriculture Organization (FAO), 2001).

Algeria

Algeria, with a total area of about 240 million ha, is located in the north-western part of the African continent, overlooking the Mediterranean Sea in the north. Algeria can be divided into three physiographic units: (*i*) The Atlas Mountains, consisting of two principal mountain chains separated by hills that contain a group of depressions known as El-Shatout. (*ii*) The South and South-east Hills, a group of isolated hills in the north of the Great Desert, the Sahara, lying south of the Atlas. (*iii*) The Hilly Coastal Depressions.

The cultivable area is estimated at about 10.74 million ha. In 1993, the cultivated area was 3.37 million ha. Agricultural development is mainly in the northern part of the country, where good soils are located and where the climatic conditions are more favourable. The climate varies from the tropical desert type in the Sahara region in the south to Mediterranean in the north. The average annual rainfall is about 68 mm, but varies from 0 mm in the southern desert up to 1, 500 mm in the north-eastern coastal area. Even in this area the dry season lasts about five months.

(***Source***: Framji *et al.*, 1981: UN Food and Agriculture Organization (FAO), 2001).

Water resources: The water resources, that are potentially available for use in the northern part of the country and the high plateaus have been estimated at 8.1 km^3/year, of which 6.5 km^3 is surface water to be regulated by dams and 1.6 km^3 is groundwater. The safe yield of fossil water in the Sahara varies between 2 and 5 km^3/year according to different estimates. The extraction of fossil water is expensive. By 1992, 79 dams had been constructed or were under construction with a total capacity of 4.3 km^3.*

Algerian rivers are usually seasonal and depend on the rains. Many of them diminish downstream or many get lost in inland basins or the Sahara. The Cheliff, the principal river, is more than 690 km long, with flow varying from 15 to 30 m^3/s during the rains to a few hundred litres/second in summer. The western rivers, whose basins are in the arid zone, contribute about 3,000 million m^3. However, these are important for development of irrigation in water scacre Algeria. (Framji *et al.*, 1982).

Good artesian aquifers occur in Oued R' Hir and Mitidja, many of which have been exploited. Groundwater is being used for irrigation around Oran and near Algiers. Forgaras structures groundwater galleries, similar to Kanats are also used to tap groundwater.

*(***Source***: UN Food and Agriculture Organization (FAO), 2001).

Irrigation development: Irrigation has been practiced in Algeria since ancient times. However, it is only during the later part of the 19th century that major irrigation systems have been developed. In 1920, a comprehensive programme to construct many dams was drawn up and many major dams were constructed since 1925. Since 1930, deep tubewells (100 m or more) have been dug in the Algerian desert. (Framji *et al.*, 1981). The net irrigated area of Algeria in 1999 was 0.57 million ha. Irrigation potential has been estimated at 1.24 million, based on soil resources potential. About two-third of the net irrigated area is under surface methods of water application, while about one-third area is under sprinkler irrigation.

(***Source***: UN Food and Agriculture Organization (FAO), 2000 and 2001).

Tunisia

Tunisia, located in north Mediterranean Africa, has a total area of 193.6 million ha, of which 7% (11.16 million ha) comprises of lakes and salty depression. The net irrigated area in 1999 was 0.38 million ha. Physiographyically, Tunisia can be divided into 6 different regions: (1) Mountainous regions, (2) Low coastal plains (Medjerdah, Matour, Mornag, Khanguet), (3) The High Tell (Maktar, Kof), (4) The Sahel, (5) The Steppes, and (6) The Sahara. The cultivable area is estimated at 8.7 million ha. In 1993, the cultivated area was estimated at 4.25 million ha. In the region close to the sea the climate is semi-arid, while in the south it is arid. Average annual rainfall is 207 mm, but almost half the country receives less than 100 mm/year. Only in the extreme north the rainfall is upto 1,500 mm/year. Evaporation losses are very high.

(***Source***: UN Food and Agriculture Organization (FAO), 2000, 2001).

Surface water sources: The greater part of runoff water, which could be stored in reservoirs is found to the north of the Dorsale. The availability of water resources are irregular due to the seasonal nature of most of the streams. Madjerda *wadi* in the north is the most important water course. This is the zone where the principal irrigation development and flood protection works have been carried out. By late 1990s there were 18 large dams and 22 hillside dams (FAO, 2001) (Framji *et al.*, 1982).

Groundwater resources are of great importance in Tunisia, especially central and southern regions. A part of the groundwater is fossil water. In many regions the groundwater is of poor quality. Thus, it is evident that the water resources of Tunisia are low, often of poor quality and have great irregularity.

(***Source***: Framji *et al.*, 1983; UN Food and Agriculture Organization (FAO), 2001).

Irrigation development: Irrigation has been practised in Tunisia since the ancient times, dating back to the 10th century B.C. At the time of the establishment of the French protectorate in the early 19th century, there existed about 100,000 ha of irrigation from natural springs in oasis and 3,000 ha of irrigation from streams. Up to 1930, the main effort was to provide water supplies to the towns. Major water resources projects were undertaken during the second-half of the 20th century (Framji *et al.*, 1983).

The net irrigated area in 1999 has been 0.38 million ha, including 0.35 million ha of full or partial control irrigation schemes and 30,000 ha of spate irrigation. Surface irrigation is mainly practiced on the schemes with supplementary irrigation. Sprinkler irrigation is used on a part of cereals in case of a severe rainfall deficit and on fruit trees during the early years of their development.

(***Source***: UN Food and Agriculture Organization (FAO), 2000 and 2001).

Mali

The Republic of Mali, is an inland State in north-west Africa, having a total area of 124.019 million ha. About 1.5 million ha were cultivated. In 1999, Mali had a net irrigated area of 0.138 million ha.*

The Niger River in the south is of vital importance to Mali. Between Segou and Timbuktu, there are many channels and lakes in Mali's inland delta. Much of the land area of Mali comprise of plains, 229 to 336 m high. There are mountains in the west and south. Rainfall exceeds 625 mm a year in the south. It drops rapidly towards the north. Rainfall and climatic conditions of the inland delta of Mali are ideal for the development of irrigation for cotton and rice. A number of irrigation schemes have been developed since 1929 (Framji *et al.*, 1982).

*(***Source***: UN Food and Agriculture Organization (FAO), 2000).

Water resources: The three important international rivers passing through Mali are the Niger, the Senegal and the Volta. The Niger has an inland delta in Mali which is being developed for irrigated agriculture. The government has initiated development of Mali's groundwater resources which are rather scarce. The Niger Scheme is the only large-scale irrigation project in Mali. It involves the development of the inland delta from Segou to Timbuktu. With the full development of the inland delta the irrigated area would be about one million ha. There is scope for substantial increase in irrigation using the flows of Senegal, Niger and Bani rivers (Framji *et al.*, 1982).

Tanzania

The United Republic of Tanzania is located on the south-eastern coast of Africa and comprises the mainland and two islands (Zanzibar and Pemba). Its total area is about 94.509 million ha. The cultivable area is about 40 million ha. The cultivated area in 1995 was 6.3 million ha, which is 7% of the total area and 16% of the cultivable area.*

The mainland has a coastline of 805 km and a coastal plain of 16 to 64 km wide. From the plain the area rises to the great central plateau having an average elevation of about 1220 m. Kilimanjaro, the highest mountain of Africa lies in the north-east of Tanzania, near the Kenyan border. Westwards lies the great lakes, namely, Lake Tanganyika, Lake Malawi and Lake Victoria (Framji *et al.*, 1983).

*(***Source***: UN Food and Agriculture Organization (FAO), 2001).

Climate: Tanzania has three climatic zones. The islands and the coastal belt of Tanzanyika are tropical. The central plateau is hot and dry. The mountainous area is semi-temperate. The average precipitation of Tanzania is 937 mm per year. The annual rainfall decreases from north to south.

(***Source***: Framji *et al.*, 1983; UN Food and Agriculture Organization (FAO), 2001).

Water resources: The important rivers of Tanzania flowing into the Indian Ocean are the Pangani, the Wami, Ruvu or Kingani, the Rufiji, the Matandu, the Mbemkuru, the Lukuledi and the Ruvuma. The Mori, the Maza and the Kagera flow into Lake Victoria. The River Malagarasi flows into Lake Tanganyika, and Songive and the Ruhuhu into Lake Malawi. Most of the river systems have considerable potential for irrigation development. Tanzania does not receive water from neighbouring countries, but shares three major lakes (Victoria, Tanganyika and Malawi) as well as the Rovuma River at the border with Mozambique.

(***Source***: Framji *et al.*, 1983; UN Food and Agriculture Organization (FAO), 2001).

Irrigation development: Irrigation in Tanzania comprises traditional and modern systems. Traditional irrigation is by diverting water from streams, rivers and springs. Traditional systems could be modernised by providing control structures, properly aligned and graded canals and appropriate drainage works. Storage could be provided wherever possible. (Framji *et al.*, 1983). The net area irrigated in 1999 has been 0.155 ha, forming 7.43% of the cultivated area of Tanzania The potential for irrigation development is estimated to be 0.828 million ha, based on soil and water availability

(***Source***: UN Food and Agriculture Organization (FAO), 2000 and 2001).

Libya

The Libya Arab Jamahiriya in North Africa has a total area of about 176 million ha, about 90% of which is desert. The fertile part is the coastal belt along the Mediterranean Sea. There are four distinct physiographic regions (*i*) The Coastal Plains, (*ii*) The Northern Mountains, which run close to the coastal plains, (*iii*) The Internal Depressions which cover the centre of Libya and include several oases, (*iv*) The Southern and Western Mountains. The landscape is characterized by broken country forming basins of inland drainage with intermittent streams and dry valleys (*wadis*). The deserts are both of sand and rock types. The cultivable area of Libya has been estimated at 3.80 million ha. In 1987, the total cultivated area was 2.28 million ha, of which 1.93 million ha consisted of annual crops and 0.35 million ha permanent crops.*

The climatic conditions of Libya are influenced by the Mediterranean Sea in the north and the Sahara desert in the south. Rainfall is essentially limited to the coastal and mountain areas for agriculture. The rainfall is less than 50 mm per year in about 90% of the country. The average annual rainfall is 26 mm. In general, the climatic and rainfall conditions in Libya are unfavourable to large-scale irrigation development. Agriculture is limited to the Mediterranean coast where shallow aquifers provide water for small-scale irrigation. Except small areas around regions of adequate rainfall, irrigation water is to be transported from distant areas.

*(*Source*: UN Food and Agriculture Organization (FAO), 2001).

***Water resources*:** There are no perenninal rivers in Libya. However, there are a large number of "*wadis*" (storm water drains) all over the country. Only during the rainy season these *wadis* flow, usually 3 to 5 times a year. The importance of *wadis* for agriculture in Libya was recognised since ancient times. Geographically, there are two groups of *wadis* in Libya: (1) The *wadis* which flow to the north to the Mediterranean sea, and (2) *The wadis* which disappear due to high evaporation and subsurface flow reaching sea. In addition to the *wadis,* there are about 3,000 springs in Libya, with yields ranging from 0.2 to 3,000 1/s. They are used for drinking water as well as agriculture (Framji et al, 1982).

Sixteen dams, with a total storage capacity of 387 million m^3, had been constructed by 1991. Additional dams are planned to increase the total storage capacity.*

Though the availability of surface water in Libya is limited the groundwater reserves are substantial. Libya has 17 groundwater provinces with aquifers having varying yields and water quality between regions. The largest groundwater basins are found in the desert zones of Kufra, Sarir, Murzuk and Calanshio. The 1980 Agricultural census had registered more than 80,000 wells, including many flowing artesian wells which have sustained the oasis since the past several centuries (Framji *et al.*, 1981).

In general, aquifers are recharged only in the northern region. Renewable groundwater resources of Libya are estimated at 800 to 1000 million m^3/year. It is estimated that about half of it is not available for use as they join the sea through subsurface flow or are lost into the evaporation areas (*subkhas*). Excess ground water withdrawal along the coast would lead to saline water intrusion from the sea. Most of the groundwater used in Libya comes from huge fossil reserves.

Concentration of population along the coast has led to excessive groundwater withdrawals. Limited recharge and high evaporation rates have caused drawdowns of the water table, salt water intrusion and deposition of salts on the soil surface. Appropriate water conservation and management are necessary to remedy the situation (Framji *et al.*, 1982).

*(*Source*: UN Food and Agriculture Organization (FAO), 2001).

***Desalinated water and treated wastewater*:** Since the late 1970s, there have been several attempts to introduce and expand sea water desalination plants and wastewater treatment facilities in Libya. However, the high cost of energy and spare parts have been major limitations.

(*Source*: UN Food and Agriculture Organization (FAO), 2001).

***Irrigation development*:** During the Greek and Roman times, a number of soil and water retaining dikes were constructed. At many sites they are still existing. During the 2nd century B.C., emphasis was laid on water conservation through a large number of underground water storage cisterns, dikes and dams. After independence in 1951, great importance has been attached to water resources development. Irrigation in Libya is mostly based on underground water resources. An important type of irrigation in Libya is water spreading from *wadis*. By the construction of dikes, the areas behind the dikes are benefited (Framji *et al.*, 1982).

The net area irrigated in 1999 was 0.47 million ha. The irrigation potential of Libya has been estimated at 0.75 million ha. However, the development of additional potential would have to rely mainly on the use of fossil water. Sprinkler irrigation is practised extensively because of the sandy soils in most areas of Libya.

(*Source*: UN Food and Agriculture Organization (FAO), 2000 and 2001).

Kenya

Kenya, lying on the east coast of Africa, has a total area of 58.265 million ha. The cultivable land is estimated at about 57 million ha, which is 98% of the total area. However, only about 10 million ha are classified as high-to-medium-potential land. About 80% of the land is semi-arid or arid. The cultivated area was 3.7 million ha in 1989, which was 38% of the high-to-medium-potential cultivable area.*

Kenya has four main physiographical regions: (1) Arid thorn-bush-covered plains in the north-east. (2) Low arid land, including lake Rudolf and a humid mountain zone in the north-west. (3) Dry, partly uninhabited bush land in the south-east. (4) A fertile plateau in the south-west where 85% of the population and major economic activities of Kenya are concentrated. The plateau is cut by the Rift Valley which is 48 to 65 km wide and has elevation ranging from 610 to 914 m above the land on either side (Framji *et al.*, 1982).

The mean annual rainfall is 572 mm, ranging from less than 200 mm in northern Kenya to 2,000 mm/year on the slopes of Mount Kenya. Kenya has a widely varying climate, caused mainly by variations in altitude. With the changes in altitude and rainfall, the climate changes from sub-tropical and temperate.*

Only about 3% of the total area of Kenya receives adequate rainfall to meet the full evapotranspiration demands of vegetation. In over 85% of the land, the rainfall is insufficient to meet the crop water needs and irrigation is highly beneficial (Framji *et al.*, 1982).

*(***Source***: UN Food and Agriculture Organization (FAO), 2001).

Water resources: The plateau land, receiving more than 760 mm rainfall, is the source of many springs and small streams which form perennial rivers. From the high rainfall zone of Mount Kenya and the Aberdare originate several rivers which combine to form the two largest river systems of Kenya, the Tana and the Athi which flow into the Indian Ocean. These river systems have substantial potential for irrigation development.

The rivers orginating from the Rift Valley feed the major lakes of Kenya. Some of the lakes have relatively fresh water while others are salty. The perennial rivers, many of them rising on the rim of the Rift Valley, flow into Lake Victoria, which is one of the sources of the Nile River. Lake Victoria is a fresh water lake and the rivers flowing into it constitute important sources for irrigation development. Two of the largest irrigation projects are situated on its shores, the Kano and the Yala (Framji et al, 1982).

Kenya possesses extensive groundwater resources, most of which are yet to be developed.

Irrigation development: Irrigation on a large scale is recent in Kenya. Some coffee and other plantations have been irrigated for many years. The net area under irrigation in 1999 was 67,000 ha (FAO, 2000). The National Water Master Plan estimates the total irrigation potential of Kenya to be 352,400 ha, About 44,610 ha were under surface irrigation (basin, furrow, hose), 21,000 ha under sprinkler, and 1,000 ha under drip and micro-sprinkler for commercial cut flower export market.

Three major types of irrigation are practiced: small-holder (28,000 ha), private commercial forms (26,600 ha) and government schemes (12,000 ha). Small-holder schemes grow mainly food crops and vegetables. The government-managed schemes grow mainly rice with tenant plot holders on 1.6 ha plots.

(***Source***: UN Food and Agriculture Organization (FAO), 2001).

Ethiopia

Ethiopia, dominated by a highland complex of mountains and plateau (above 1,500 m), is split from north-east to south-west by the East African Rift Valley, 40 to 60 km wide and occupied by a string of lakes. The total land area of Ethiopia is about 110 million ha. The cultivable land is about 13.2 million ha. The cultivated land is about 6 million ha. Small-scale farmers occupy 96% of the cropped area, while the remaining 4% are cropped by State farms and Producers' Cooperatives (UNFAO, 2001).

Physiographically, the land area of Ethiopea may be classified into four regions: (*i*) The Great Rift Valley in the centre of Ethiopia, known as the Ethiopian Plateau. (*ii*) The Ogaden Plateau. (*iii*) The Dauaki Lowlands, (*iv*) The Sudan lowlands. The Ethiopean Plateau which covers about half of Ethiopia have altitudes ranging from 2100 m to 2400 m (Framji *et al.*, 1981).

The average annual rainfall on the plateaus varies from 1,015 mm in the north to 2,030 mm in the south-west. The national average rainfall is 744 mm. The lowlands have less than 200 mm of rainfall (Framji *et al.*, 1981).

Rainfall variability is high, particularly in the lower rainfall areas of the north-east highlands. Recurrence of drought is a common phenomenon in Ethiopia. Climatic and rainfall conditions of Ethiopia show that irrigation is necessary in the eastern lowlands. Several areas in the highlands are also suitable for irrigation for diversification of crops and to overcome the effects of droughts (Framji *et al.*, 1981).

Water resources: Ethiopia has been popularly known as the water reservoir of Africa. The river systems are of vital importance to water resources development of the neighbouring countries as well. The major rivers include the Tekeze, the Blue Nile, the Sohat, the Awash, the Omo, the Shebelle and the Dawa. During the rainy seasons, the perennial streams flood vast areas of the plateaus and erode huge quantities of fertile top soils, to be deposited in the low-lying surrounding countries. Use of groundwater is mainly confined to domestic supplies (Framji *et al.*, 1981).

Irrigation development: The generally adequate rainfall to grow one crop in most of the populated parts of the country and subsistent nature of agriculture, have resulted in the slow pace of irrigation development. Small-scale irrigation by individual farmers by diversion from streams has been widely practiced. Modern irrigation schemes exist mainly along the Awash River Valley (Framji *et al.*, 1981).

The net irrigated area in 1999 was 0.19 million ha (FAO, 2000), and is concentrated in the Awash River Valley, which accounts for 48% of the total irrigated area and 92% of the large schemes built before 1990. There is vast potential for expansion of irrigation in the plains on the east and the west and in the highlands of Ethiopia (Framji *et al.*, 1981).

Eritrea

Eritrea, situated in North-eastern Africa, along the Red Sea, has a total area of about 12.2 million ha. The cultivable area is 1.6 million ha. The cultivated land is 0.439 million ha (26% of the cultivable area).

(***Source***: UN Food and Agriculture Organization (FAO), 2001).

Climate and water resources: The climate ranges from hot arid, adjacent to the Red Sea, to temperature sub-humid in isolated micro-catchments within the eastern escarpment of the highlands. Total annual rainfall increases from less than 200 mm at the northern border with Sudan to over 700 mm in the southern border with Ethiopia.

Eritrea has three main river systems: (*i*) The Mereb-Gash and Tekeze-Setit river system, draining into the Nile; (*ii*) The eastern escarpment and the Barka-Anseba river systems, draining into the Red Sea; (*iii*) The river systems of a narrow strip of land along the south-eastern border with Ethiopia, draining into the closed Denkel Basin. The most important river is the Tekezeo Setit, at the border with Ethiopia, with more than 90% of its catchment inside Ethiopia. The main sources of domestic water in rural areas are tubewells and shallow wells.

(***Source***: UN Food and Agriculture Organization (FAO), 2001).

***Irrigation development*:** The net area under irrigation in 1999 was 22,000 ha. The total water managed area in 1993 was about 28,124 ha, out of which about 12,500 ha were under partial control and 15,630 ha under spate irrigation.

(***Source***: UN Food and Agriculture Organization (FAO), 2000 and 2001).

Uganda

Uganda, East Central Africa, has a total area of 235.880 million ha, 15% of which is occupied by freshwater, swamps and lakes. The cultivable area was about 16.759 million ha in 1991. The cultivated area is estimated to be 4.6 million ha. More than two-thirds of the country is 1,000 to 2,500 m in elevation.*

Uganda has varied topography, having great plains, forests, swamps, lakes, desert and snow-capped peaks. The average elevation is 1,219 m. Some of the greatest lakes of Africa, including Lakes Victoria, Albert, Edward, George and Kyoga, lie wholly or partly within Uganda (Framji *et al.*, 1983).

Rainfall is fairly reliable and varies from 750 mm/year in the north-east to 1500 mm/year in the the shores of Lake Victoria and around the highlands. In spite of the lakes and the White Nile, which dominates the picture of surface water flow from about one-third of Uganda, it is in the water-scarcity zone. Besides the dry north-east, there are large areas of the southern, central and western parts which are deficient in water. Irrigation is necessary in these areas.

*(*Source*: UN Food and Agriculture Organization (FAO), 2001).

***Irrigation development*:** Major irrigation and drainage projects are yet to be developed in Uganda. The net irrigated area in 1999 has been 9000 ha.*

In general, the country is fertile and benefits from adequate rainfall, offering favourable natural conditions for agricultural production, with the exception of the far north and north-east, where the rainfall is confined to one rainy season and irrigation development is necessary.

*(***Source***: UN Food and Agriculture Organization (FAO), 2000).

***Surface water resources*:** The whole of Uganda lies within the Upper Nile catchment, with numerous rivers flowing into Lakes Edward, Victoria, Kyoga and Albert and also directly to the Nile.

Zimbabwe

Zimbabwe, situated in South Central Africa has a total area of 39.0759 million ha. Zimbabwe can be divided broadly into five main topographical zones: (*i*) The central watershed called the highveld which is a plateau, 650 km long and 80 km wide. (*ii*) On either side of the central watershed, between 700 m to 1,200 m above sea level, there is the middleveld, which is fairly deeply incised, (*iii*) The Zambesi Valley situated in the north, (*iv*) The lowveld, on the south-east of the middleveld escarpment, at an average altitude of less than 700 m and having streams with seasonal flow. (*v*) The Eastern Highlands situated in the eastern border (Framji *et al.*, 1983).

Out of the 39.076 million ha of the land area of Zimbabwe, 16.333 million ha are reserved as Tribal Trust Land for indigenous population. The allocation of Trust land to an individual family is by the tribal authorities. Another 16.827 million ha comprise of General Land with unrestricted ownership rights. Large-scale commercial farms have been developed in this category of land (Framji *et al.*, 1983).

Average annual rainfall is 656 mm. Rainfall ranges from 1,000 mm in the Eastern Highlands to around 350 to 450 mm/year in the lowveld in the South. About 90% of the precipitation occurs during mid-November to early April.

(***Source***: UN Food and Agriculture Organization (FAO), 2001).

***Water resources*:** Zimbabwe is bordered to the north by the Zambezi River and to the south by the Limpopo River, both flowing into Mozambique. Together with the Save in the east, they collect most of the streams originating from the Highveld. Most of Zimbabwe's rivers dry up a few months after the rainy season, with only few scattered pools. However, the rivers carry large volumes of flow. During drought years, infiltration galleries for extracting water from the sand layers in the river beds are a reliable form of water supply.

(***Source***: UN Food and Agriculture Organization (FAO), 2001).

The sedimentary formations have high potential for groundwater. Three extensive aquifers in these formations has been identified. They have been exploited only to a limited extent. About 60% area of Zimbabwe has hard rock formations which constitute the central watershed. In general, the ground water potential is low (Framji *et al.*, 1983).

***Irrigation development*:** The earliest development of modern irrigation in Zimbabwe was a diversion scheme from the Odzani River in the Eastern Districts in early 20th century. A steady growth in irrigation development took place during the succeeding years. By 1950 about 7,400 ha, were brought under irrigation. The early 1950s were marked by the construction of small water harvesting dams for numerous small-scale irrigation systems throughout the country. The Kyle Dam is the largest dam in Zimbabwe. An area of 26,500 ha, mainly planted with sugarcane, is irrigated in the estates and adjacent privately owned farms. The area irrigated in the main commercial farming areas increased by about 90,000 ha during 1960s, and 1970s. The Government has also developed irrigation schemes in the Tribal Trust Lands. By 1978, about 7,304 reservoirs of varying sizes were constructed in Zimbabwe. Of these 97 are large dams (Framji *et al.*, 1983). The net area irrigated in 1999 was 0.117 million ha. The gross irrigated area increased from 60,000 ha in 1968 to 1,91,000 ha in 1990, including double-cropped areas.

(***Source***: UN Food and Agriculture Organization (FAO), 2000 and 2001).

Angola

Angola, located on the west coast of Southern Africa, has a total area of 124.6 million ha. The arable land is 1.28 million ha, which is about 1% of the total area (Framji *et al.*, 1981). The net area irrigated in 1999 was 75,000 ha.

(***Source***: UN Food and Agriculture Organization (FAO), 2000).

Irrigation development: The total irrigation potential of Angola is estimated at 6.7 million ha. After independence in 1975, the Government emphasised on the development of irrigated agriculture. Hundreds of small dams and a few medium dams were constructed. Irrigation projects are more in the arid and semi-arid regions. There is an emphasis on irrigation development and exploitation of groundwater for irrigation and drinking water supply.

Zambia

Zambia, situated in Southern Central-Africa has a total area of 75.2715 million ha. The total cultivable land is estimated at about 16 million ha. Zambia comprises mainly of high plateau with elevations varying from 1,067 m to 1,372 m. The plateau is drained by the Zambezi River and its tributaries. Located in Zambia are the southern end of Lake Tanganyika, part of Lake Muweru and the shallow Lake Bangweulu. In the western province is the copperbelt, a highly industrialised region of Zambia. In general, the fertility of the soil decreases from east to west, as the soil changes from loam to coarse grained loose sands. A large part of this valley is a flood plain containing meandering rivers and seasonally waterlogged areas. Some areas have heavy dark clays. Zambia has large areas which are waterlogged (Framji *et al.*, 1983).

The mean annual rainfall of Zambia is 1011 mm. In the south it is 750 mm/year and increases uniformly to 1,400 mm/year in the north. Most of the cropped area in the central zone receives about 900 mm to 1,200 mm/year of rainfall. The climate and rainfall conditions of Zambia make irrigation desirable in regions with rainfall less than 760 mm.

(***Source***: UN Food and Agriculture Organization (FAO), 2001).

Water resources: Zambia has a number of major river systems with ample potential for irrigation development. Groundwater resources are estimated to be about 47 km^3/year. There are extensive limestone aquifers covering an area of 25,000 km^2, extending from Lusaka to the north-west.

(***Source***: UN food and Agriculture Organization (FAO), 2001).

Irrigation development: Irrigation in Zambia is recent. The total area under cultivation in Zambia in 1981 was 1.367 million ha, of which the irrigated area was 16,000 ha. The net area under irrigation in 1999 (full or partial water control) was 46,000 ha.*

The irrigation potential, based on water and soil resources, has been estimated to be 1.4 million ha. Almost the entire irrigation is from surface water surfaces, either from diversion works, pumping from rivers, or constructing dams. About 2,500 ha are irrigated from groundwater. Surface irrigation (furrow and basin) is practised on 28,400 ha. There are 17,200 ha under sprinkler and 800 ha with drip irrigation.**

*(***Source***: UN Food and Agriculture Organization (FAO), 2000 and 2001; Framji, 1983).

(*Source***: UN Food and Agriculture Organization (FAO), 2001).

Swaziland

Swaziland, located in Southern Africa, has a total area of 1.736 million ha. The cultivated land is 191,500 ha. Swaziland is divided into four physiographic regions extending longitudinally: (*i*) The mountainous highveld in the west, suitable mainly for afforestation. (*ii*) The middleveld, having good soil and plentiful rainfall. (*iii*) The lowveld which is gently undulating lowland, with fertile soils. (*iv*) The Lumbombo Plateau, mainly cattle grazing area but with scope for mixed farming (Framji *et al.*, 1983).

The highveld has a humid near-temperate climate; the Middleveld and Lubombo are sub-tropical; while the lowveld is tropical. The mean annual rainfall in Swaziland is 788 mm/year in the eastern lowlands to 1500 mm/year in the hilly, western half of the country. The climatic and rainfall conditions of Swaziland are such that irrigation is essential in the lowveld and the drier middleveld.

Water resources: Swaziland is rich in water resources. No quantitative assessment of groundwater resources has yet been made. The groundwater potential is considered to be limited.

Irrigation development: Development of irrigation in Swaziland is recent. The potential for irrigation development, based on available land resources, is estimated at 2,00,000 ha. The net irrigated area in 1999 was 67,000 ha (FAO, 2000). The irrigation sector in Swaziland consists maninly of large private farms. The main irrigated crop is sugarcane, followed by pineapple and grapefruit.

(***Source***: UN Food and Agriculture Organization (FAO), 2000 and 2001).

Botswana

Botswana is a semi-arid country in the centre of Southern Africa with a total area of 58.173 million ha. The cultivable area is about 6.2 million ha. Cultivated land was 0.3242 million ha in 1990. Agriculture (crop production and livestock) provide occupation to more than 70% of the population.* Botswana is divided into two topographical regions by a ridge of highland and hills extending in a north-northeasterly direction from near the south-eastern corner of the country to Zimbabwe's border. This ridge forms the western watershed of the Limpopo river system. On its eastern side rise a number of streams which flow generally in an easterly direction towards the Limpopo, or in the southern portion of the area into the Notwani and Marico rivers which are the tributaries of the Limpopo. To the north-west lies the Okavango Delta with a series of permanent and seasonal swamps, channels and lakes. The climate of Botswana is generally sub-tropical. The mean annual rainfall is 400 mm, ranging from 250 mm in the south-west to a maximum of 650 mm/year in the extreme north. (Framji *et al.*, 1981).

*(***Source***: UN Food and Agriculture Organization (FAO), 2001).

***Water resources and irrigation development*:** The Limpopo River System is the major source of surface water in Botswana. A few dams have been constructed on the tributaries of the river, mainly for urban water supplies. There is scope for the construction of more dams for irrigation and livestock and domestic supplies.

Groundwater is one of Botswana's important natural resources. It is estimated that nearly three-fourth of the human and cattle population of Botswana have been dependent on groundwater, mainly from boreholes (Framji *et al.*, 1981). The net irrigated area in 1999 was 1000 ha (FAO, 2000).

There are no drainage projects in Botswana. Considerable areas could be reclaimed by drainage of parts of the Okavango Delta Swamps. There is also potential for further development of the traditional recession-cropping.

Senegal

Senegal, situated in West Africa, covers an area of 19.672 million ha. Plains and plateau at altitudes not higher than 100 m area are the main topographic feature of Senegal. There are 4 major rivers, namely the Senegal, the Saloum, the Gambi, and the Casamance. The cultivated area of Senegal is 2.7 million ha. The estimated irrigation potential is 1,80,000 ha, of which 71,000 ha have been developed for irrigation by 1999. Rice and sugarcane are the principal crops of Senegal.

(***Source***: Framji *et al.*, 1983; UN Food and Agriculture Organization (FAO), 2000).

Somalia

Somalia, with a total area of 63.766 million ha has the longest coastline in Africa; in the north on the Gulf of Aden and in the east on the Indian Ocean. The cultivable area is estimated at about 8 million ha. In 1984 about 0.98 million ha were cultivated.*

The climate of Somalia is arid to semi-arid, with an average annual temperature of 27°C. The average annual rainfall is 253 mm with wide variations. Somalia is subject to periods of drought.

*(***Source***:UN Food and Agriculture Organization (FAO), 2001).

***Irrigation development*:** In spite of the importance of irrigation for the main cash crops in Somalia (bananas and sugarcane), the development of irrigation and drainage systems is low. The main irrigated areas are in the Juba and Shebelli valleys. In 1999, the net irrigated area was 0.2 million ha, of which only 50,000 ha had reasonably controlled irrigation and the rest is spate irrigation for maize production.

(***Source***: UN Food and Agriculture Organization (FAO), 2000 and 2001).

Ghana

Ghana lies in the Gulf of Guinea in Western Africa with a total area of 23.854 million ha. The cultivable area is about 10 million ha. The cultivated area is about 1.1 million ha.

The topography of Ghana is marked by two mountain ranges, the Akwapins-Kwahu and the Togo ranges. Bounded by the two mountain ranges is the Volta Plateau covering the whole of Northern Ghana and is generally flat. The rest of the country in the south-west of Akwapins-Kwahu range is undulating. The 563 km long coast consists mostly of scrub-land, much of which is extensively cultivated (Framji *et al.*, 1981).

***Climate and water resources*:** The average annual rainfall is 1,264 mm, but ranges from 890 mm/year in the coastal zone near Accra to 2,030 mm/year in the south-western rain forests. Its river network is divided into three main systems: the Volta river system, the south-western river basins, and the coastal rivers. The Volta river system is the largest in Ghana. The south-eastern part of the country is drained by a number of coastal rivers.

***Irrigation development*:** Development of irrigation in Ghana is recent. In 1953 a few small scale irrigation schemes were started in northern Savannah areas and in 1957 in the southern Savannah areas. The gross irrigated area in 1968 was 15,000 ha, which rose to 22,000 ha in 1978 (Framji *et al.*, 1981). However, the net irrigated area in 1999 was 11,000 ha.*

The potential for irrigated rice production in the Inkland Valley Swamps and river bed plains is estimated at 1.9 million ha in Ghana. Irrigation development in Ghana is dominated by the public sector.

*(***Source***: UN Food and Agriculture Organization (FAO), 2000).

Mauritius

Mauritius is an island country in the Indian Ocean with a total area of 0.204 million ha, consisting of the island of Mauritius representing most of the country's territory Rodriguez island and the distant Cargados Carajos shoals and Agalaga islands. In 1995 agriculture contributed 11% to GDP, of which 65% was from sugarcane.

(***Source***: UN Food and Agriculture Organization (FAO), 2001).

***Climate and water resources*:** The climate is tropical oceanic type in summer, and sub-tropical in winter. The average annual rainfall is 2,180 mm, but varies from 700 mm in the western coastal zone to 4,000 mm/year on the highlands of the central plateau. There are 25 main river basins. The largest are the Grand River South-west and the Grand River North-west.

(***Source***: UN Food and Agriculture Organization (FAO), 2001).

Table 1.5. Estimates of Irrigation Potential and Irrigated Areas (1982) in Different Countries Lying South of the Sahara in Africa.

(The countries for which descriptive information on natural resources and irrigation development have been described are not included)

Country	*Irrigation potential (million ha)*	*Modern irrigation (million ha)*	*Traditional irrigation (million ha)*	*Total million (ha)*	*% Utilization of irrigation potential*	*% of total irrigated area in Africa*	*Net irrigated area in 1999 (million ha)*
1. Zaire	4.0	0.004	0.02	0.024	0.60	0.27	0.011
2. Mozambique	2.4	0.066	0.004	0.070	2.90	0.78	0.107
3. Central African Republic	1.9	0.000	0.004	0.004	0.21	0.04	N.A.
4. Chad	1.2	0.01	0.040	0.050	4.20	0.56	0.020
5. Gabon	0.44	0.00	0.001	0.001	0.23	0.01	0.015
6. Burkina Faso	0.35	0.009	0.020	0.029	8.29	0.32	0.025
7. Congo, Republic of	0.34	0.003	0.005	0.008	2.35	0.09	0.010
8. Malawi	0.29	0.160	0.004	0.020	6.90	0.22	0.028
9. Cameroon	0.24	0.011	0.009	0.020	8.33	0.22	0.033
10. Guinea	0.15	0.015	0.015	0.045	30.00	0.50	0.095
11. Ivory Coast	0.13	0.042	0.010	0.052	40.00	0.58	N.A.
12. Niger	0.100	0.010	0.020	0.030	30.00	0.33	0.066
13. Sierra Leone	0.100	0.005	0.050	0.055	55.00	0.61	0.029
14. Togo	0.086	0.003	0.010	0.013	15.12	0.14	0.007
15. Benin	0.086	0.007	0.015	0.022	25.58	0.24	0.012
16. Gambia	0.072	0.006	0.020	0.026	36.1	0.29	0.002
17. Guinea Bissau	0.070	N.A.	N.A.	0.01	2.43	0.01	0.017
18. Burundi	0.052	0.002	0.050	0.052	100.00	0.58	0.074
19. Rwanda	0.044	0.000	0.015	0.015	34.10	0.17	0.005
20. Lesotho	0.008	0.000	0.001	0.001	12.50	0.01	0.001
21. Liberia	N.A.	0.003	0.016	0.019	N.A.	0.21	0.002

[*References:* Framji *et al.* (1981-83), FAO (1986, 1987, 2000).]

Irrigation development: Irrigation is mainly for sugarcane, which has one of the highest yields in the world. The net area under irrigation in 1999 was 20,000 ha. Sprinkler irrigation is the major irrigation method (83%).

EXTENT OF UTILIZATION OF IRRIGATION POTENTIAL IN AFRICA

The situation on land and water potential and irrigation development in Africa have been briefly described in the preceding sections. There are wide variations in resource potential and extent of development of the potential between regions and countries. Within the same country there are often widely varying situations in resource base, population pressure and other constraints which may confront efforts on development. Estimates on irrigation potential and the development of irrigation under modern and traditional systems of irrigation, relating to most of the countries in Africa, south of the Sahara in 1982 are presented in Table 1.5. The countries for which the resource base and irrigation situation have been described in the earlier sections are not included in Table 1.5.

During 1965 to 1975 the irrigated area in many of the African countries expanded at an average rate of about 4.3% per year and from 1975 to 1980 at 4.8% per year.* Due to climatic and demographic reasons, irrigation development in Africa has been concentrated in the north and north-east of the continent. All types of irrigation works can be found, from enormous storage to small-scale *wadi* control, from traditional surface irrigation system to modern and sophisticated sprinkler and drip systems. There are projects for the protection of crop land from flooding and the installation of extensive drainage systems for utilization of swamps and river valleys and for salinity control.

The average cropping intensity in North and North-East Africa is about 130%, as compared to about 110% in the other areas. Traditional irrigation has long been practised in most African countries. As population pressure on arable land is not so great in Africa as in other continents, the objective of modern large-scale irrigation in Africa has been mostly for commercial crop production, using modern technology and supported by efficient management input.

*(*Source*: UN Food and Agriculture Organization (FAO), 1986, 1987).

Major Issues in Land and Water Resources Management

Africa is a diverse continent. No single set of irrigation policies can be devised that is applicable throughout the continent. Problems and situations vary widely from area to area. Individual countries must develop their own policies to meet their specific requirements. It is essential to identify the roles of irrigated and rainfed cultivation in food production and in national development. In many arid regions irrigated agriculture may be the sole form of farming. In semi-arid areas, irrigation is to complement rainfed farming. It may be used to grow new crops in regions with moderate rainfall. Irrigation may be used to grow cash crops such as cotton or sugarcane to improve the national income.

It may also be used as an instrument of rural development, food self-sufficiency and employment.

The following are some of the specific issues in land and water resources development and management which are common to most of the situations in Africa, though they could be applied to many other developing countries as well:

1. **Need for holistic planning and development:** Irrigation in a country must be planned on a national or a watershed basis, and not as a series of disconnected individual projects. This is of special importance where watersheds/river basins are shared by two or more countries. It is necessary to ensure that traditional irrigation practices are given due priority, and are not neglected in favour of expensive and large projects. Soil erosion must be reduced in the catchments and excess of flows are to be stored in reservoirs or diverted into the river system for development downstream. The possibilities of large scale damage due to waterlogging, salinization and drying up of inland deltas are to be understood and preventive and remedial measures are to be adopted. Irrigation development programmes require appropriate institutions to plan and implement them.

The importance of irrigation in Africa is revealed from the study of the United Nations.* Although irrigation occupied only 6.5% of 124 million ha of cultivated land in 43 FAO participating African countries during 1970-80, it provided 20% of the value of all the agricultural crops grown in those countries. Irrigation increased the agricultural production, per hectare by more than three times.

The indispensability of irrigation development for the very existence of the civilizations in the arid North Africa has been recognized since ancient times. A number of countries in other regions, including Mauritania, Niger, Burkina Faso, Mali, Chad, Sudan, Ethiopia, Somalia, Kenya, Tanzania, Zimbabwe and Botswana have rainfed growing periods which are less than 120 days in over 25% of their land areas. Another ten countries have up to a quarter of their area in the semi-arid zone. In the high-risk zones of these countries irrigation is necessary for dependable farming. Irrigation will be beneficial in most of the other countries as well.**

*(***Source***: UN Food and Agriculture Organization (FAO), 1987).

(*Source***: UN Food and Agriculture Organization (FAO), 1987).

2. **Development of database:** Land and water resources development planning need adequate and reliable database. Information on the physical, social and manpower resources are essential for planning irrigation projects. An inventory of resources is critically important as a planning tool. Considerable attention is to be laid on collecting, compiling and analysing the physical resources database, such as soils and their characteristics, hydro-geological properties of underground formations, surface and groundwater resources and their characteristics and extent of development, geographical situation, climate, land use, population, availability of infrastructural facilities, agricultural and industrial situations, water table, soil salinity and socio-economic situation. Information on the nature and expectations of the societies to be affected by irrigation development is important. Planning and construction works of projects must be based on detailed information on the extent, type, location and state of all existing irrigation, and a reasonable assessment of the available potential.

3. **Decision on large versus small-scale irrigation systems:** Most African countries with experience of irrigation are against hasty adoption of large-scale irrigation in the absence of adequate databases on resources. There are many cases of poor performance of large systems. In general, it may be better for countries without much experience in irrigation to learn initially from small-scale schemes. African experience confirms the importance of small-scale irrigation and the rehabilitation of older schemes with poor performance. Such rehabilitation should not be limited to improvement of the structural components but should include sound land and water management practices and introduction of modern cropping systems and cultural practices. While support to small-scale irrigation schemes is to be considerably increased, the development of large-scale schemes should not be excluded. In each river basin the desirable mix of large and small-scale schemes could be established in relation to prevalent physiograhic features, capacity of government to handle schemes and social and economic conditions.

4. **Training and development of human resources:** Irrigation project planning and implementation require staff with a wide variety of knowledge and skills, including irrigation engineering, agriculture, economics, and social sciences. Shortage of trained and experienced staff at all levels is a major limitation in irrigation development in most African countries. Education and training programmes are essential to increase staff capability to deal with the various managerial and technical issues involved in irrigation development and management.

In many countries in Africa, many of the higher-level planning and technical work in irrigation project, is given to international consultants. As a result, only few countries are acquiring the benefits of high-level knowledge and skills they need to develop their own capability. There are instances of consultants and contractors resorting to expensive and sophisticated procedures which may not be appropriate to the social and economic situation in the country. Some of the consultants may not have the expertise on techno-socio-economic situations of the particular region/area. Hence, it is essential to develop manpower within a country to undertake and supervise projects.

Development of staff training programmes requires emphasis. The flow and exchange of information between nations and international institutions needs to be strengthened.

5. **The role of women in irrigated agriculture:** In many countries irrigation development has over-emphasised the role of men in irrigation system development and management. There is need to change the situation. Women contribute to about two-thirds of the total hours worked on the farm in most African countries. Practice of irrigation would change the role of women in the production of food and cash crops as well as animal production. For many reasons, including the seasonal migration of men to find work elsewhere, more and more women are heading households. About 22% of households in Africa south of the Sahara, are legally headed by women. Hence, it is important that women are given proper access to land and water management, equipment and production inputs, credit and marketing facilities, water users associations and training, research and transfer of technology.

6. **The role of non-governmental organizations:** Non-governmental Organizations (NGOs) are assuming increasingly important roles in irrigation development in Africa. They are able to work effectively with local population at the village level, promote self-reliance among producers and mobilize additional funds from donors. Reports of agricultural production in Burkina Faso, Senegal, Mali, Sierra Leone, Kenya, Mali, Tanzania, Niger and Chad testify the importance of NGOs in helping to develop small-scale irrigation. NGO irrigation projects usually base development on existing traditional technologies.

7. **Land tenure:** A constraint to the deep involvement of farmers of government controlled irrigation systems in many of the African countries is the short tenure of farming rights conferred by the Government. In many of the African countries the government allot irrigated land to farmers, giving them a one-year tenure. These schemes often fail to achieve their objective because tenants have no commitment to invest in land improvement, and cannot take up farming practices to suit their changing needs. The farmers must have assured rights to their irrigated land. Where irrigation is developed under customary law, the farmer cannot usually obtain credit as his title to the land is not recognized for security. For efficient irrigation farming tenants should be given land leases for long periods and should have a say in the choice of crops to be grown.

8. **Farmers' participation in irrigation system planning, development, operation and maintenance — lessons from South-East Asian countries:** There has been a major effort in South-East and Far East-Asian countries, particularly Indonesia, the Philippines, Thailand, Taiwan, Malaysia, and Korea, to involve farmers at various levels of planning, operation and management of irrigation systems. Farm level irrigation management has been given to a large extent to farmer organizations. These systems have produced substantial benefits in these countries. The main lessons learnt from South-East Asian countries are that farmers should begin to participate in irrigation schemes at the planning stage. The same agency should undertake construction of the project and the organization of farmers' participation. When the governments have to operate and maintain schemes, irrigation fees should be levied from users to lighten the government's financial burden. Farmers' participation is vital for the efficient operation of irrigation systems.

9. **Health control of irrigation command areas:** Practice of irrigation may have profound influence on health of men and animals. It is necessary to provide the population with good supplies of safe drinking water. Irrigation often introduces large scale water storage systems in an area. This can increase the incidence of water-related diseases, the most serious of which is malaria. Lymphatic filariasis, also transmitted by mosquitoes, is endemic over large parts of Africa and schistosomiasis transmitted by a water snail occurs in nearly all African countries. Irrigation development, therefore, requires that effective steps are taken to control diseases by prevention of water stagnation in irrigation command areas.

IRRIGATION DEVELOPMENT IN INDIA

First Century A.D. to Middle of 20th Century

Irrigation development in India in the first to third century A.D. has been mainly in southern India which was not influenced by the cultural ingression of tribes from Central Asia. The Sangam literature gives detailed description of the Tamil dynasties of Pandya, Chera, Chola, Pallava, Chalukya and Rashtrakuta and their contribution to development projects. Kings of these dynasties took great interest in promoting agriculture. Situated in the deltaic or valley regions, most of these kingdoms had fertile soil. With the expansion of irrigated agriculture, South India became the rice bowl of India. Sangam literature gives descriptions of the sites chosen for storing run-off water. Both tank and well irrigation were in practice in South India. There are distinct references to 'lift irrigation', where bullocks were used for lifting water from deep wells. An important irrigation work has been the construction of large embankments (both stone and earth) on Kaveri river in the delta region and the Vaigai river to store flood waters. Karikala, a Chola King (A.D. 180) constructed a 160 km long embankment along the Kaveri river primarily to protect his people from recurrent floods. The Grand *Anicut* in the Kaveri delta built in the second century A.D. is the first major irrigation system in the Indian sub-continent. He also patronised the construction of a large number of irrigation tanks.

***Gupta era* (A.D. 300 to 500):** The Gupta dynasty made substantial contribution by promoting irrigated agriculture in their empire. The Gupta empire extended over almost two-thirds of the Indian sub-continent. Dykes and embankments were constructed to divert river water for irrigation. Separate canals for irrigation and drainage have been reported.

Pallava Kingdom: The Pallava dynasty, founded in A.D. 550 promoted extensive tank irrigation throughout Tamil Nadu. The Parameswara tank and the Tiraiyaneri tank were contributions of this dynasty on irrigation. Vayiramega, Gudimallam and Ukkal tanks were constructed during the later part of the 8th century A.D. The large tanks at Kanakvallieri Solapuram and Kaveripak were constructed during the 9th century. The Kaveripak tank had a 6.4 km long earth dam and is still functional. The Chitramegha tank and the Tandalam tank with sluice gate are other major tanks of this era.

***Chola empire* (A.D. 985 to 1205):** Chola kings are known for their enormous contribution to irrigated agriculture. It led to unprecedented prosperity and wealth in the region. Among the famous ancient *anicuts* are those in the Thanjavur district. Huge stones were used to build this 329 km long, 12-18 m broad *anicut* built near the Srirangam islands. The Talkad *anicut* built downstream of the Kaveri river is another important structure. A number of channels were dug to cover large areas of land on both sides of the river bank to ensure production of crops throughout the year.

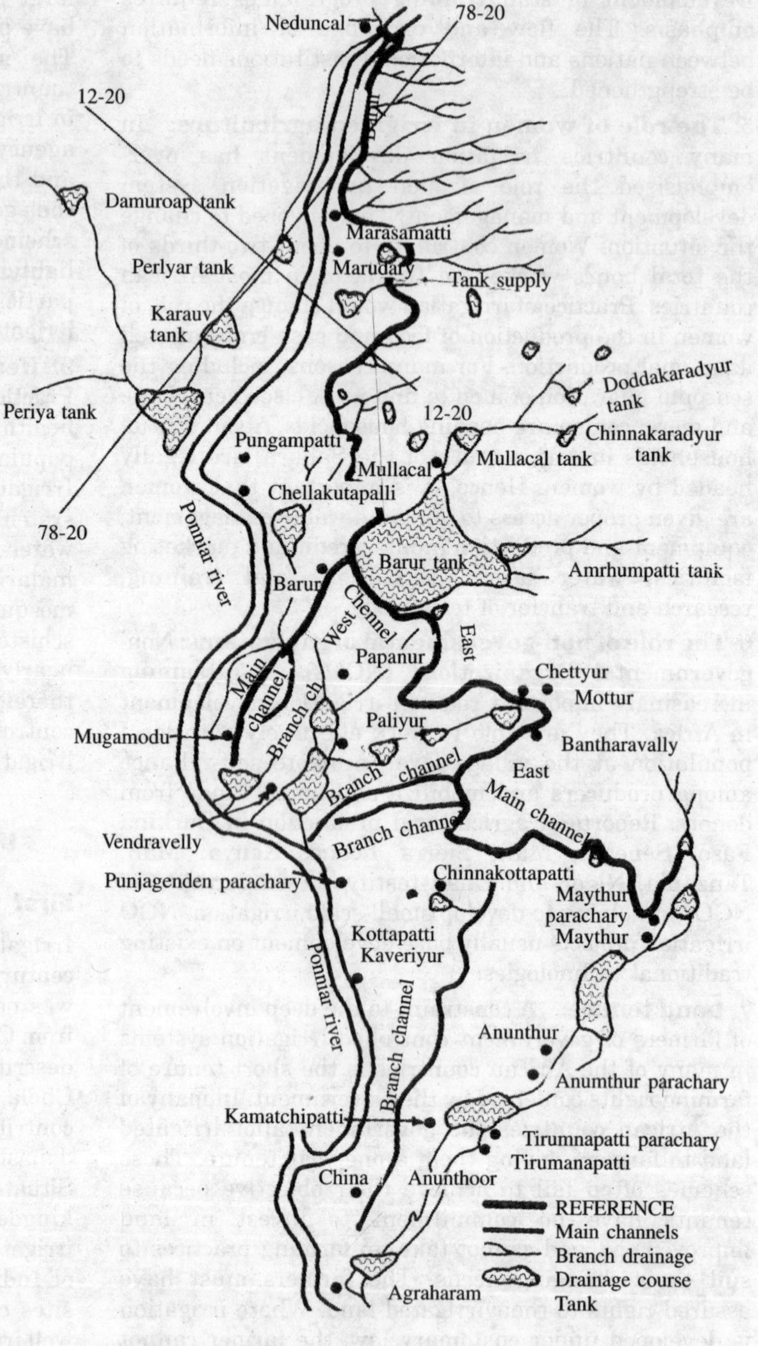

Fig. 1.21. Layout of system tanks (chain tanks). Barur Tank Project (1883), Salem district, Tamil Nadu.

Chain Tanks of Karnataka and Andhra Pradesh

Tanks for water storage and diversion were constructed in continuous series like links of an imaginary chain in many of the river valleys. The Telengana region of Andhra Pradesh was referred to as the land of thousand tanks. System tanks (chain tanks) are low dams which are constructed at the upper reaches of a stream to store water; its overflow is allowed to pass downstream to be stored in the next dam in the series. Such tanks have become common in Tamil Nadu, Karnataka and Andhra Pradesh (Fig. 1.21). Often, the chain tanks, are adequate to cover the whole watershed area. The disadvantage was that during heavy rains, if a breach occurred in an upstream tank, its water would come with tremendous force to the tanks on the lower reaches which consequently develop breaches unless the downstream tanks are provided with the required factor of safety.

Tanks of Tamil Nadu: Though initiated by the Pallava kings, the chain tanks of Tamil Nadu were extensively promoted by the Chola kings. Some of the important ones are the Cholavaridhi tank, Sodiyambakkam tank and a large number of tanks in Madurai, Salem and other districts (Figs. 1.21 and 1.22).

Fig. 1.22. An ancient minor irrigation tank in Madurai District, Tamil Nadu. On the right is a close view of the masonry spillway.

Tanks of Karnataka. A large number of tanks were built in Karnataka by the Chola kings during the 10th century A.D. They constructed tanks at every feasible point at regular intervals along a stream. Many of these are still in use. The ingenuity and precision with which the sites of tanks were selected are remarkable. The Pallar valley tank system had a chain of 1,000 tanks.

The neighbourhood of Delhi had at one time about 350 tanks (large and small) in use. Some of them like Surajkund are still in good condition. The greatness of a town or an area was based on the tanks it had for its use. The states of Andhra Pradesh, Karnataka, Madhya Pradesh, Maharashtra, Rajasthan, Tamil Nadu and West Bengal are still having a large number of old tanks. Separate management committees were formed at the village level to manage the tanks.

Irrigation development in Northern and Eastern India during the 9th to 12th century A.D.

Rajput rulers were reigning different parts of the Northern and North-Western India. King Krishnaraja made significant contribution in irrigation development. The Bhopal lake built during his period has 647 sq km of surface area and an extensive system of irrigation channels.

Vijayanagara Kingdom (1336 to 1500 A.D.). The Vijayanagara Kingdom was established by King Harihara I in A.D. 1336. Large-scale land reclamation and irrigation works were given boost by the kings. The most famous of these is the Anantaraja Sagar tank in Cuddapah district of Andhra Pradesh.

Mughal Period *(14th and 18th centuries A.D.).* Major efforts in irrigation development during the period was construction of numerous tanks and dug wells in Peninsular India and Rajasthan by local rulers. Amongst the Mughal emperors Ferozshah Tughlaq built the Western Yamuna Canal in A.D. 1355 to carry Yamuna water to his hunting ground at Safidem in Hisar district (Haryana). The canal was abandoned after his time. Later, Emperor Akbar renovated the canal in 1568 for irrigating lands in Hisar district. Emperor Shahjahan restored the silted and unusable Western Yamuna Canal and took a branch canal to irrigate the gardens in the Red Fort, in Delhi. Emperor Shahjahan built a 180 km canal from Ravi river through the Shalimar Gardens at Lahore. A branch of this canal carried water to the Golden Temple in Amritsar. A large part of Punjab and Kashmir were benefited by it. During the British period, this canal was replaced by the Upper Bari Doab Canal system. The Eastern Yamuna Canal system was constructed by Emperor Mohammad Shah in A.D. 1735. The canal was useful during the initial period, but got silted within a few decades. Development works did not make any major strides during the second-half of the eighteenth century, mainly due to the unstable governments in Delhi.

Irrigation Development in the 19th Century A.D.

There was substantial increase in irrigation in India in the nineteenth century. At the beginning of the 19th century, the net irrigated area was estimated at 0.8 million ha. This rose to 13.3 million ha at the close of the century when 16% of the gross sown area of 82.2 million ha was brought under irrigation. Of this, 45% was irrigated by canals, 35% by wells, 15% by tanks, and 5% by other sources.

Irrigation development in India in the nineteenth century, under the British rule was essentially a part of the famine relief works. The projects were designed to provide protective irrigation to as many areas as possible. The major irrigation projects constructed during the first half of the ninetenth century included the Kaveri delta system in Tamil Nadu in 1834 to irrigate 0.6 million ha, the Godavari *anicut* and canal system in Andhra Pradesh in 1846 to irrigate 0.4 million ha, and the Krishna delta system in Andhra Pradesh in 1852 to irrigate 0.5 million ha. The major credit for the massive delta projects goes to Sir Arthur Cotton, a great engineer. Cotton is held in high esteem in India for his spectacular contributions in irrigation development.

The other major irrigation projects of the 19th century are Upper Ganga Canal system in Uttar Pradesh commissioned in 1854 to irrigate 0.37 m ha, the Sone Canal in Bihar (1879) to irrigate 0.28 million ha and the Orissa Canal system on river Mahanadi (1895) to irrigate 0.1 million ha. The Upper Ganga Canal, with its headworks at Haridwar was the largest irrigation canal in the world at the time of its construction. It had a carrying capacity of 191 m^3/sec. It was later remodelled to carry a discharge of 288 m^3/sec. The Western Yamuna Canal in Haryana was remodelled in 1892 to provide an irrigation potential of 0.6 million ha. The lower Ganga Canal was constructed during 1872–78.

A definite policy for the construction of irrigation works was introduced in 1854 with the setting up of the Public Works Department by the then Government of India. The Government of India appointed the First Irrigation Commission in 1901 to report on irrigation as a 'means of protection against famine in India'. A few

Famine Commissions had been set up by the government and these Commissions had also recommended the development of irrigation works to contain the adverse impact of frequent famine in India. In its recommendations to the Government in 1903, the Irrigation Commission suggested definite policies regarding the selection, financing and maintenance of productive irrigation works. Apart from the remunerative works, the Commission recommended the extension of irrigation as a means of protection against famines in areas of insecure and precarious cultivation. The Commission observed that such irrigation works may not be directly remunerative to the Government, but are important to mitigate the miseries caused by famines.

In accordance with the recommendations of the First Irrigation Commission, several productive and protective irrigation projects were undertaken in the early part of the twentieth century. Prominent amongst them are the Sarda Canal in Uttar Pradesh (1926) to irrigate 0.6 million ha, Ganga Canal with dam and weir on Satluj river (1928) to irrigate 0.3 million ha in Rajasthan, Damodar Canal (1935) to irrigate 0.1 million ha in West Bengal, Ken Canal (1915) to irrigate 0.1 m ha in Uttar Pradesh; Nira Left Bank Canal (1906), and Right Bank Canal (1938) and Pawara Canal (1938) in Maharashtra with a total irrigation potential of 0.1 m ha; Mahanadi Canal (1931) and Kharung Canal Project (1931) in Madhya Pradesh with a total irrigation potential of 0.2 m ha.

The Triple Canal Project (1905–15), now in Pakistan was the first large-scale trans-basin diversion of water from one river basin to another. It comprised of two large weirs and one barrage across three different rivers and three canals to transfer water from the Jhelum across the Chenab and Ravi rivers to irrigate desert areas. The Satluj Valley Project constructed during 1921-31 was the largest irrigation project of the time. Major irrigation projects of South India constructed during 1921 and 1935 include the Krishnarajasagar Project with dam on Kaveri, the Nizamsagar Project with dam on Majira river, a tributary of Godavari in Andhra Pradesh and the Mettur Dam project with dam across Kaveri in Tamil Nadu constructed in A.D. 1927. A number of small works which could be taken up for famine relief in various provinces were recommended by the Irrigation Commission which gave considerable importance to the development of private irrigation works including wells in the tracts affected by famine.

In 1854, two categories of irrigation works were introduced, namely, minor works and major works. Major irrigation projects constructed during the second-half of the nineteenth century include the Sirhind Canal in Punjab (1873-82) consisting of 6000 km of canals to irrigate 1 million ha, the Lower Sohang and Para Canals (1872–77), now in Pakistan, the Lower Chenab Canal in Punjab, the Lower Ganga Canal, the Agra Canal in Uttar Pradesh (1873) with its headworks over Yamuna at Okhla in Delhi to irrigate 0.16 million ha, the enlargement of the Eastern Yamuna Canal in Uttar Pradesh to irrigate 0.19 m ha, the Periyar Dam (1887) in Kerala along with the Periyar Vaigai Canal system in Tamil Nadu to irrigate 60,000 ha and the Khadakvasla Dam (1869–79) and the Mutha Canals in Maharashtra. The Periyar Dam project is a remarkable feature involving the diversion of the west-flowing Periyar river with its catchment in the high rainfall Western Ghats region of Kerala to the rainshadow region of Madurai district in Tamil Nadu in the east.

Irrigation Works in India in the Pre-Independence Era of Twentieth Century

The indispensability of well planned and rapid development of irrigation was pronounced by the two great famines of 1897–98 and 1899–1900. The economic considerations were out-weighed for irrigation development by the need for increasing food production for sustaining the population.

The result of the large-scale expansion of irrigation works in the Indian sub-continent during the 19th and the first-half of the twentieth century has been largely due to the great efforts of irrigation engineers with the support of the governments, including the rulers of the princely states. The role of three great personalities are immemorable in the history of irrigation development during the period. They are Sir Arthur Cotton, Major P.T. Cautley, both in the 19th century and Sir M. Visveshwariah in the first-half of the 20th century. Major Cautley, a British military engineer assumed the major responsibility for the construction of the Upper Ganga Canal Project in Uttar Pradesh.

Table 1.6. Net Area Irrigated in the Indian Union from 1910 to 1950 (Quinquennial average)

(m ha = million hectares)

Period	Area irrigated from			Irrigated area as % of sown area
	Govt. canals m ha	Wells m ha	Total (all sources) m ha	
1910-11 to 1914-15	4.4	4.0	14.5	17.9
1915-16 to 1919-20	4.1	4.7	15.7	19.0
1920-21 to 1924-25	4.4	4.7	16.0	17.4
1925-26 to 1929-30	4.6	4.8	16.2	17.1
1930-31 to 1934-35	5.0	4.8	17.1	17.6
1935-36 to 1939-40	5.6	5.2	18.0	18.6
1940-41 to 1944-45	6.0	5.4	19.0	19.2
1945-46 to 1949-50	6.4	5.3	19.4	19.1

(***Source:*** Irrigation Commission, 1972).

Bharata Rathna M. Visveshwariah was largely responsible for the construction of Krishnarajasagar Scheme in 1911 A.D. on River Kaveri near the city of

Mysore. The dam was the largest reservoir built in India till then. The scheme created an irrigation potential of 40,000 ha and provided electricity to the people of Mysore and Bangalore. The project converted the Mandya district of Karnataka, which was once covered by shrub forests, into a prosperous region with high-yielding paddy and sugarcane (Bhavanishankar, 2000). Table 1.6 presents the irrigation development in India during A.D. 1910 to 1950.

Classification of Irrigation Works in India

Post-independence development of irrigation works in India were on the basis of classification of works into major, medium and minor projects. The classification has changed over the years from the basis of the extent of culturable command area of the project to that of expenditure involved in the project and back again to the basis of culturable command area. During the pre-plan period (prior to A.D. 1951), irrigation projects were classified on the basis of culturable command area of the project. During the period ranging from 1951 to 1978, the classification was based on the cost involved in the schemes. Since 1978, the basis of classification of irrigation projects has again been on the basis of the culturable command area of the project.

Accordingly, irrigation projects in India are presently classified into the following three categories:

(*i*) **Major irrigation projects.** All irrigation schemes with a culturable command area of 10,000 ha or more are classified as major irrigation projects. They are essentially surface water projects involving large-scale storage/diversion works.

(*ii*) **Medium irrigation projects.** Projects having a culturable command area of 2000 ha to 10,000 ha are classified as medium irrigation projects. Medium irrigation projects are also usually surface water projects, excepting a few large lift irrigation schemes.

(*iii*) **Minor irrigation projects.** All irrigation schemes having a culturable command area upto 2000 ha individually are classified as minor irrigation projects. Minor irrigation schemes are further grouped as (*a*) surface water minor irrigation schemes, and (*b*) groundwater minor irrigation schemes.

Jurisdiction of water resources. Under the Indian Constitution, water is the responsibility of the states. Thus, the federal states are primarily responsible for the planning, implementation, funding and management of water resources development projects. This responsibility in each state is borne by the Irrigation/Water Resources Departments. The Inter-State Water Disputes Act of 1956 provides a framework for the resolution of possible conflicts between states on the utilization of water resources.

At central level, there are three main institutions involved in water resources development and management:

(*i*) The Ministry of Water Resources, which is responsible for laying down policy guidelines and programmes for the development and regulation of the country's water resources. The ministry's technical arm, the Central Water Commission, provides general infrastructure, and technical and research support for water resources development at state level. The Central Water Commission is also responsible for the assessment of water resources.

(*ii*) The Planning Commission, which is responsible for the allocation of financial resources required for various programmes and schemes of water resources development to the states as well as to the Ministry of Water Resources. It is also actively involved in policy formulation related to water resources development at the national level; and the Ministry of Agriculture, which promotes irrigated agriculture through its Department of Agriculture and Cooperation. The research support at the national level is by the Indian Council of Agricultural Research. The Central Pollution Control Board is in charge of water quality monitoring, and the preparation and implementation of action plans to solve pollution problems.

International water treaties/agreements. Under the Indus Water Treaty (1960) between India and Pakistan, all the waters of the eastern rivers, i.e. the Satluj, Beas and Ravi rivers taken together, shall be available for the unrestricted use of India. All the waters, while flowing in Pakistan, or of any tributary which in its natural course joins the Satluj main or the Ravi main after these rivers have crossed into Pakistan, shall be available for the unrestricted use of Pakistan.

India controls the flow of the Ganges River through a barrage completed in 1974 at Farraka, 18 km from the border with Bangladesh. A treaty was signed in December 1996, under which Bangladesh is ensured a fair share of the flow reaching the Farraka barrage, in addition to the releases during the monsoon seasons. Appropriate agreements exist between Nepal and India for the exploitation of the Kosi River (1954, 1966) and the Gandak River (1959).

(***Source***: UN Food and Agriculture Organization (FAO), 2001).

Irrigation Development in India During the Five Year Plans

The partition of India in 1947 brought about a drastic reduction in the irrigated area in the post-independent India. A substantial portion of the irrigated area of undivided India went to Pakistan. In 1950–51, the net irrigated area was 20.85 million ha and the gross irrigated area 22.60 million ha. The Government of India gave priority to develop new irrigation projects, realising the urgent need for developing the available water resources of the country to meet the deficits in food production.

The increase in irrigation potential during the Five Year Plans, which commenced in 1951 and the Annual Plans was phenomenal and unprecedented. The gross irrigation potential rose from 22.6 ha in 1950–51 and was estimated to reach 106.61 million ha in 1997–2002. The period-wise break-up of development of irrigation-potential through major and medium irrigation projects and minor irrigation projects (surface water and groundwater) is shown Fig. 1.24. Table 1.8 presents the statewise number of major and medium irrigation projects in India till the year 1997. The net area under irrigation in 1999 was 59 million ha, which is the largest in the world.

(***Source***: UN Food and Agriculture Organization (FAO), 2001).

Fig. 1.23. A modern minor irrigation tank in Central Maharashtra.

In the First Five Year Plan (1951-56), India launched a major irrigation development programme. The construction of many multi-purpose major projects were taken up, such as the Bhakra Nangal (Figs. 1.25 and 1.26) the Damodar Valley (Fig. 1.29), Kosi, Chambal, Hirakud, Kakrapar and Tungabhadra. Simultaneously, numerous minor schemes, including groundwater, were developed. During the Second Five Year Plan (1956–61), Third Five Year Plan (1961–62) and the three Annual Plans (1966–69), more major, medium and minor irrigation and multipurpose schemes were taken up and those initiated earlier were completed and commissioned.

The most significant amongst the major irrigation projects undertaken and completed since the beginning of the Five Year Plans comprise the Bhakra Nangal, Rajasthan, Kosi, Gandhi Sagar, Ranapratap Sagar, Jawahar Sagar, and Kangrabati projects on Ganga river system; Kadna project on river Mahi, Ukai project on river Tapi, Hirakud dam project on river Mahanadi, Jayakwadi on river Godavari and Nagarjuna Sagar, (Fig. 1.27 and 1.28) and Tungabhadra, Malaprabha, Ghataprabha and Upper Krishna projects on Krishna river system.

At the start of the 20th century, India had 42 large dams. By 1950, a further 250 had been added. The rest

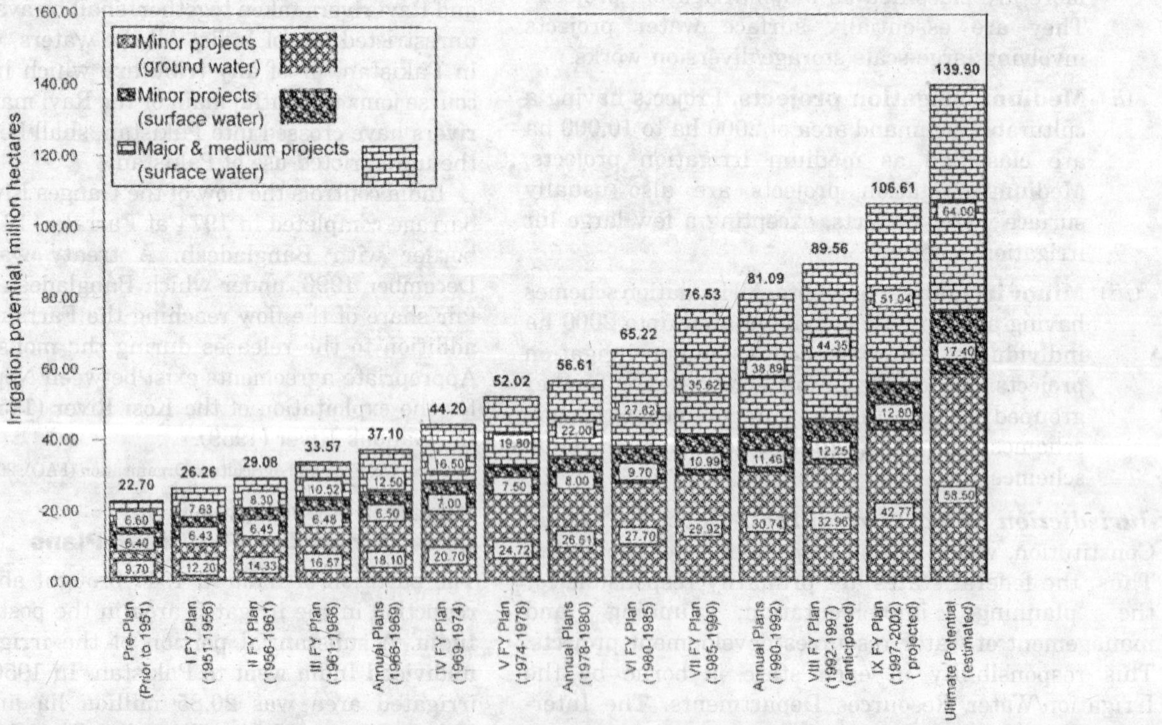

Fig. 1.24. Development of irrigation potential in India during the Five Year Plan periods.

(***Source:*** (i) Ministry of Water Resources, Central Water Commission, Govt. of India 1996. (ii) Reports of Working Gropus on IX Five Year Plans of various states, 1996-97. (iii) National Commission for Integrated Water Resources Development Plan, Report: Vol. 1. Ministry of Water Resources, Govt. of India, 1999).

Fig. 1.25. View of the Bhakra gorge—the site of the famous Bhakra dam before the construction work of the dam was started
(***Courtesy***: Press Inf. Bureau, Min. I and B, P.I.B., Govt. of India)

Fig. 1.26. Downstream view of the Bhakra dam, one of the world's highest straight gravity concrete dams. The dam has a height of 225.55 m and a width of 9.14 m at the top and 190.5 m at the base. The Gobind Sagar which is the Bhakra reservoir has a length of 96.56 km and a storage capacity of 9,87,000 ha-m.
(***Courtesy:*** Press Inf. Bureau, Min. of I and B, Govt. of India)

Fig. 1.27. A view of the Nagarjuna Sagar Dam, Andhra Pradesh.
(***Courtesy***: Mini of PIB, Govt. of India, Photo like No. NA–30666)

Fig. 1.28. In main canal taken through a tunnel across a rocky hillock in the Nagarjuna Sagar canal system in Andhra Pradesh.

Fig. 1.29. A view of Durgapur Barrage on River Damodar in West Bengal. The barrage is constructed to divert for irrigation the flow downstream of the systems of 5 major dams on Damodar river and its tributaries, as a part of the prestigious Damodar Valley Project, Flood control in the Damodar, earlier known as the river of sorrow, is a principal objective of the Project.
(***Courtesy***: PIB, Govt. of India.)

were undertaken in the second half of the 20th century. Half of the large dams were undertaken during 1970 to 1989. Statewise distribution of large dams shows that nearly half of the large dams are in the two States of Maharashtra and Gujarat, and that almost three-fourths lie within the three States of Gujarat, Maharashtra and Madhya Pradesh (including Chhattisgarh) (Rangachari, R. *et al.*, 2001).

The average annual runoff carried by the rivers in India, estimated by the Central Water Commission is around 1869 billion m^3. It has been revised by the National Commission for Integrated Water Resources Development Plan to 1953 billion m^3. The total live storage created under the various completed schemes (large and small) by the end of the 20th century was about 177 billion m^3. The scheme under construction is to provide an additional capacity of about 75 billion m^3 (Rangachari, R. *et al.*, 2001).

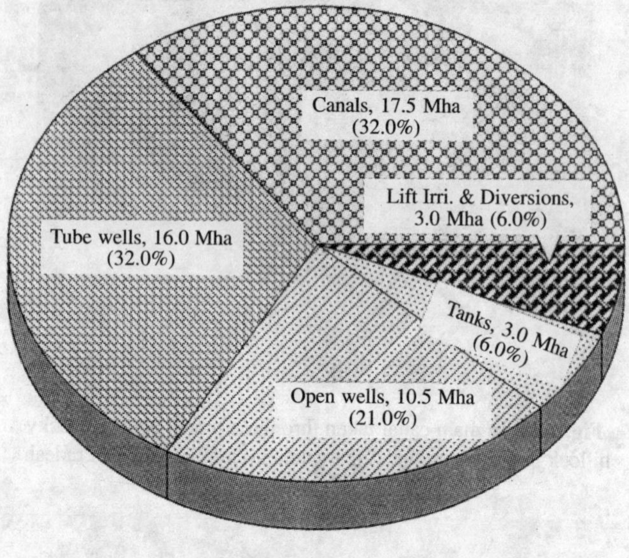

Fig. 1.30. Sourcewise distribution of net irrigated area in India, 1996.
(***Source:*** Based on Water and Related Statistics: Central Water Commission, 1996.)

During the Fourth Five Year Plan (1969–74), the emphasis was shifted to the completion of ongoing projects, integrated use of surface and groundwater, adoption of efficient water management techniques and modernization of existing schemes. Additional new projects were started. During the Fifth Plan (1974–78), Command Area Development Programme CADP was launched as a Centrally Sponsored Scheme with the objective of reducing the time-lag between the development of irrigation potential and optimum utilization of available land and water. The programme was conceived as a means of co-ordinating all related activities to meet with these objectives.

During the Annual Plans of 1978–80 and the Seventh Five Year Plan (1980–85), new projects were started. At the end of Seventh Plan, there were 182 major and 312 medium ongoing projects requiring completion. Increased emphasis was laid on completion of projects which were in the advanced stages of construction. This was continued during Annual Plans 1990–91 and 1991–92 and the Eighth Five Year Plan (1992–97). Rehabilitation and modernization of old irrigation schemes gained momentum during the period. User's participation in major and medium irrigation schemes received greater attention. Repairs and improvement to minor irrigation projects, and methods of irrigation also received support. Sprinkler and drip methods of water application and conjunctive use of surface and groundwater gained momentum. Figure 1.30 presents the sourcewise distribution of irrigated area in India in 1996. Table 1.6 presents the sourcewise and statewise gross irrigated area in India in 1996.

Irrigation development and management during the Ninth Plan has been with the following objectives:

(*i*) To improve water-use-efficiency by progressive reduction in conveyance and water application losses.

(*ii*) To bridge the gap between irrigation potential created and its utilisation, strengthening the Command Area Development Programme, institutional reforms and promoting farmers' involvement in irrigation management.

(*iii*) To complete the ongoing projects, particularly those which were started during the pre-Fifth Plan and Fifth Plan periods.

(*iv*) To restore and modernise the old irrigation systems.

(*v*) To introduce rational pricing of irrigation water, based initially on operation and maintenance costs.

(*vi*) To take definite steps towards integrated development of surface water resources and explore the possibility of a higher level of inter-basin transfer of water after ensuring environmental and ecological safety of the original river basin.

(*vii*) To promote adaptive research to ensure cost-effective and efficient development and management of irrigation systems.

(*viii*) To promote participatory irrigation system management with the involvement of farmers.

(*ix*) To encourage conjunctive use of ground and surface waters towards optimal utilisation of water resources.

(*x*) To promote the development of groundwater on sound technical, environmental and economic considerations.

Present-day Projection of Irrigation Potential

Based on the estimates of the Central Water Commission, the Central Groundwater Board and other agencies, the National Commission for Integrated Water Resources Development Plan (1999) has projected the ultimate irrigation potential under major and medium projects as 58.46 million ha. The estimated development upto 1996-97 has been 32.96 million ha of the potential created and 28.44 m-ha of potential utilized. A number of river valley projects are multi-purpose in character, envisaging irrigation and other benefits such as hydropower, flood control and water supply. Table 1.7 presents sourcewise and statewise gross irrigated area in India in 1992–93. Table 1.8 presents the statewise major and medium irrigation projects in India till 1997. Figure 1.31 presents the projections in major irrigation development during 1992 and 1997.

Surface water minor irrigation schemes. These schemes comprise mainly tanks, small diversion works and river lift irrigation schemes. They are more prominent in undulating and hilly regions and are spread all over the country. During the pre-plan period they were the principal source of irrigation. They are specially suitable in the plateau and semi-arid regions. They also facilitate recharging of groundwater. The total

Table 1.7. Sourcewise and Statewise Gross Irrigated Area in India (1996–97)

(Area in thousand hectares)

Sl. No.	*Name of the State / Union*	*Canals*				*Tanks*		*Wells*				*Other sources*		*Total gross irrigated area*
		Govt.	*Private*	*Total*	*%*	*No.*	*%*	*Tube-wells*	*Other wells*	*Total*	*%*	*No.*	*%*	
1.	Andhra Pradesh	2199	–	2199	38	969	17	1076	1316	2391	41	223	4	5782
2.	Bihar & Jharkhand	1368	–	1368	29	138	3	2238	142	2380	51	778	17	4664
3.	Goa	12	–	12	33	24	67	–	–	–	–	–	–	36
4.	Gujarat	720	–	720	20	35	1	906	1965	2871	79	17	–	3643
5.	Jammu & Kashmir	213	188	401	90	3	1	1	1	2	–	41	9	447
6.	Karnataka	1193	–	1193	41	272	9	471	567	1038	36	378	13	2881
7.	Madhya Pradesh and Chhattisgarh	1892	2	1894	29	197	3	1088	2514	3602	55	864	13	6557
8.	Meghalaya	–	–	–	–	–	–	–	–	–	–	53	100	53
9.	Mizoram	9	–	9	100	–	–	–	–	–	–	–	–	9
10.	Nagaland	–	–	–	–	–	–	–	–	–	–	73	100	73
11.	Punjab (1992-93)	2428	–	2428	34	–	–	4559	18	4577	46	137	2	7142
12.	Rajasthan	2200	–	2200	33	218	3	875	3391	4266	63	59	1	6743
13.	Tamil Nadu	930	1	931	28	720	22	277	1400	1677	50	19	1	3347
14.	Uttar Pradesh and Uttaranchal	4601	–	4601	26	107	1	11716	541	12257	70	503	3	17468
15.	W. Bengal (1992-93, Net Irri. Area)	717	–	717	38	263	14	689	23	712	37	219	11	1911
16.	D & N Haveli (UT)	1	–	1	3	–	–	34	–	34	–	1	25	4
17.	Delhi (UT)	2	–	2	50	–	–	25	–	25	–	–	–	37
15.	Pondicherry (UT)	9	–	9	24	–	–	–	1	1	–	2	6	34

Source: Central Water Commission, 1996, 2002 (data from Ministry of Agri., Directorate of Economics and Statistics).

Remarks: % denotes percentage to gross irrigated area.

Note: Similar information had not been received from other States.

ultimate irrigation potential from all sources of water is assessed as 139.89 million ha. Of this, a potential of 17.38 million ha is planned to be achieved through the surface water minor schemes.

The irrigation potential under surface water minor irrigation works has doubled during the second-half of the 20th century. These works serve an important means of recharge of groundwater through percolation tanks. They provide irrigation to the chronically drought affected areas. Where flow irrigation is not feasible, lift irrigation schemes are constructed.

Groundwater Minor Irrigation Schemes

Groundwater structures comprise of dug wells, dug-cum-bore wells, private tubewells and public tubewells using electrical pump sets or diesel pump sets. Groundwater development achieved a spectacular success with the potential increasing more than six times during 1950-2000. The development is primarily achieved through individual and co-operative efforts with the help of institutional finance or investment by farmers from their own savings and resources. The bulk of institutional finance is mobilised through land development banks, commercial banks and State co-operative banks with refinance from National Bank for Agriculture and Rural Development (NABARD).

Groundwater constitutes about 25% of the total irrigation potential of India. There was a spurt in groundwater development with the spread of rural electrification and the introduction of high yielding crop varieties in 1960s. Irrigation wells are predominantly privately owned and managed. The ratio of public tubewells to private tubewells is 1:73. The progressive development of open wells, shallow tubewells and deep tubewells since 1950–51 is shown in Fig. 1.32. Deep tubewells are generally owned by the government, barring a small number owned by farmers cooperatives and municipalities.

About 79% of the area irrigated under minor irrigation is from groundwater, which has been mainly through the construction of about 4.4 million dug wells and about 5 million shallow tubewells. The groundwater utilization in India has increased from 6.5 million ha at the end of the pre-Plan period to 53.9 million ha by 2000–01. However, the exploitation of groundwater has been uneven (Min. of Water Res., 2003). Gujarat, Tamil Nadu, Punjab and Haryana are amongst the states where large–scale groundwater development has taken

Table 1.8. Statewise Number of Major and Medium Irrigation Projects in India till 1997

Sl. No.	Name of the State	*Major irrigation projects*				*Medium irrigation projects*			
		Projects completed before the of Five Year Plans	Projects completed up to VIII Plan (1951-97)	Ongoing projects in IX Plan	New projects in IX Plan	Projects completed before the Five Year Plans	Projects completed up to VIII (1992–97)	Ongoing projects in IX Plan	New projects in IX Plan
1	2	3	4	5	6	7	8	9	10
1.	Andhra Pradesh	11	5	12	2	45	60	19	2
2.	Arunachal Pradesh	–	–	–	–	–	–	–	–
3.	Assam	–	2	4	–	–	8	5	–
4.	Bihar & Jharkhand	2	15	15	–	2	51	29	–
5.	Goa	–	–	1	–	–	1	1	–
6.	Gujarat	2	10	9	1	–	108	13	11
7.	Haryana	1	5	5	–	–	–	–	–
8.	Himachal Pradesh	–	–	1	–	–	2	2	2
9.	Jammu & Kashmir	1	–	1	–	6	10	8	–
10.	Karnataka	4	2	14	3	8	28	15	5
11.	Kerala	–	8	7	–	–	7	5	–
12.	Madhya Pradesh & Chhattisgarh	6	7	24	–	14	91	31	2
13.	Maharashtra	5	16	44	1	16	163	85	9
14.	Manipur	–	1	2	–	–	3	2	–
15.	Meghalaya	–	–	–	–	–	–	1	–
16.	Mizoram	–	–	–	–	–	–	–	–
17.	Nagaland	–	–	–	–	–	–	–	1
18.	Orissa	3	6	8	3	3	35	12	2
19.	Punjab	3	4	–	–	–	2	1	2
20.	Rajasthan	1	2	6	1	42	53	6	–
21.	Sikkim	–	–	–	–	–	–	–	–
22.	Tamil Nadu	17	5	–	1	7	39	2	–
23.	Tripura	–	–	–	–	–	–	3	–
24.	Uttar Pradesh & Uttaranchal	15	35	17	1	–	38	2	–
25.	West Bengal	3	1	3	–	–	9	17	–
	Total	74	124	171	13	143	708	259	37

(***Source:*** Central Water Commission, 2002).

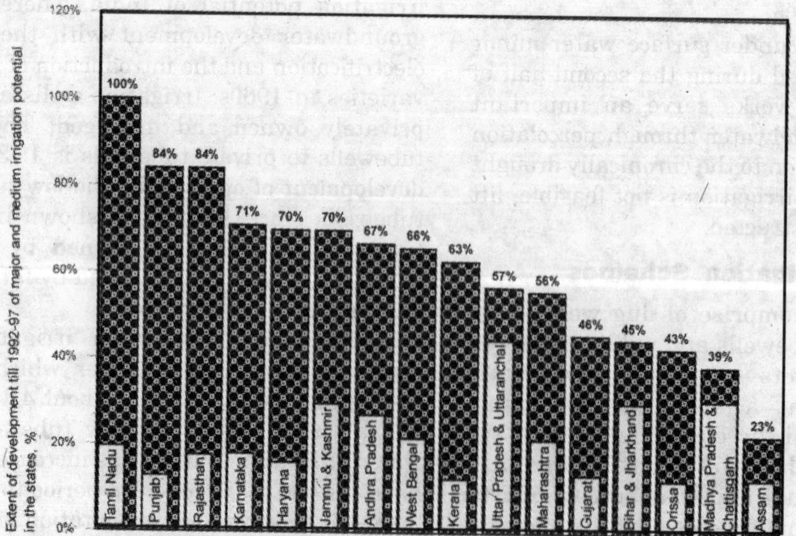

Fig. 1.31. Anticipated position of major and medium irrigation development at the end of VIII Five Year Plan (1992–97) in sixteen states of India.

(***Source:*** Report of the Working Group of Major and Medium Irrigation Projects for IX Five Year Plan, Govt. of India, Ministry of Water Resources, 1996.)

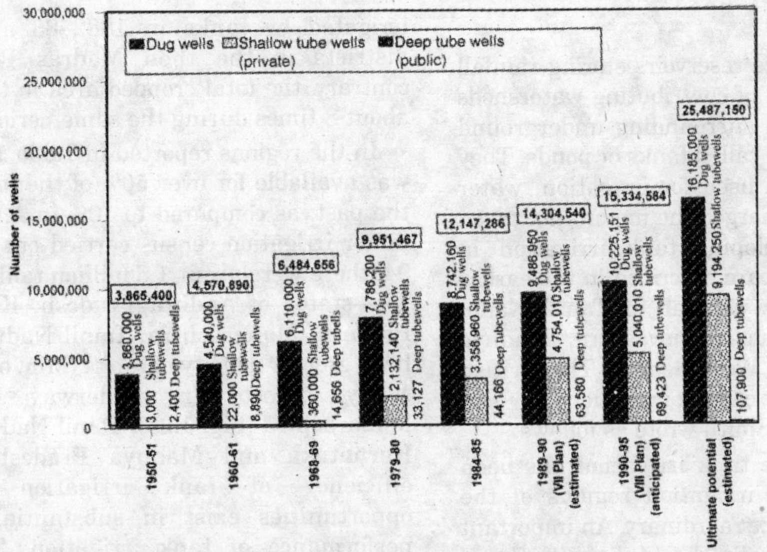

Fig. 1.32. Progressive development of dug-wells, shallow tubewells, and deep tubewells in India since 1950-51.

(***Sources:*** *(i)* Report of the Working Group on Minor Irrigation for the Formulation of VIII Five Year Plan (1990-95), Min. of Water Resources, Govt. of India, 1989. *(ii)* Central Ground Water Board, 1995).

place. In the eastern region, the development is low, namely, 8% in Orissa, 9% in Bihar, 20.5% in Assam and 34% in Eastern Uttar Pradesh. Groundwater in most parts of these states is available at shallow depths ranging from 5 to 10 m. Table 1.9 presents the estimates of the total number of groundwater structures in different states of India and projections of ultimate potential for development. (Fig. 1.32).

Table 1.9. Statewise Total Groundwater Structures in India till 1994 and Estimates of Ultimate Potential Development

States/Union	***Dug Wells***		***Shallow Tubewells***		***Deep Tubewells (Public)***	
	Constructed till 1994	*Ultimate potential (estimated)*	*Constructed till 1994*	*Ultimate potential (estimated)*	*Constructed till 1994*	*Ultimate potential (estimated)*
Andhra Pradesh	13,09,869	24,00,000	1,13,160	1,59,000	8,109	10,000
Arunachal Pradesh	–	500	45,597	2,00,000	2,307	15,000
Bihar	5,09,770	10,00,000	7,55,142	1,50,000	6,625	15,000
Goa	102	1000	–	100	105	125
Gujarat	7,09,070	8,00,000	83,300	50,000	5,588	10,000
Haryana	42,420	50,000	4,63,037	5,00,000	1,799	25,000
Himachal Pradesh	3,570	5,000	374	2000	289	500
Jammu and Kashmir	2,779	3,000	2,087	5000	172	300
Karnataka	5,45,156	10,00,000	37,837	2,00,000	–	100
Kerala	2,12,814	7,00,000	4,103	20,000	64	100
Madhya Pradesh and Chhattisgarh	13,07,070	30,00,000	23,886	50,000	1,940	2,000
Maharashtra	13,50,020	20,00,000	254	1000	–	100
Manipur	–	100	10	50	5	100
Meghalaya	–	1,000	780	1,000	3	50
Mizoram	–	–	–	50	–	108
Nagaland	450	5,000	–	50	40	100
Orissa	5,93,413	8,00,000	20,205	50,000	5,768	10,000
Punjab	93,470	1,00,000	6,22,600	7,00,000	2,002	2,500
Rajasthan	8,53,263	15,00,000	21,686	30,000	75	200
Sikkim	–	100	–	–	–	–
Tamil Nadu	14,70,807	16,00,000	1,71,305	2,00,000	100	–
Tripura	–	100	2,432	5,000	164	500
Uttar Pradesh and Uttaranchal	11,49,930	15,00,000	24,20,539	50,00,000	28,446	40,000
West Bengal	54,330	1,00,000	2,96,539	5,00,000	4,766	8,000
Total of States	1,02,08,294	1,61,50,000	50,13,927	91,64,250	68,626	1,07,400
Total Union Territories	16,857	–	26,083	30,000	797	1200
Grand Total	1,02,25,151	1,61,85,000	50,40,010	91,94,250	69,423	1,07,900

(***Source:*** Central Groundwater Board, 1995, Central Water Commision, 2002.)

Tank Irrigation in India

Small to medium size surface reservoirs storing rainfall runoff from catchment areas of contributing watersheds and collecting seepage from surrounding underground water bearing formations are called tanks or ponds. They are of varying size and are used for irrigation, water supply and groundwater recharge. The importance given by ancient rulers in developing tank irrigation in peninsular and central India are described in the earlier sections. Tank irrigation is popular in Tamil Nadu, Karnataka, Andhra Pradesh, Maharashtra, Madhya Pradesh, Chhattisgarh and West Bengal. The British were impressed by the extent of tank irrigation they saw in India. In A.D. 1853 Baird Smith wrote as follows:

"The extent to which the tank irrigation has been carried throughout all the irrigation regions of the Madras Presidency is truly extraordinary. An important record of the number of tanks in the 14 districts shows them to amount to not less than 43,000 in repair and 10,000 out of repair or 53,000 in all."

(Van Oppen and Subba Rao, 1980)

Table 1.10. Area Irrigated by Tanks and Total Cropped Area in Erstwhile Madras Presidency Region in 1882–83 and 1969–72

(hectares)

District	Average for 1882–83 Total cropped area ha	Average for 1882–83 Net area irrigated tanks ha	Average for 1969–72 Total cropped area	Average for 1969–72 Net area irrigated by tanks
Anantapur	45,600	35,600	9,00,000	31,200
Cuddapah	96,000	75,200	4,56,800	16,800
Kurnool	28,800	15,600	12,88,000	13,200
Bellary	28,800	21,200	6,02,800	8,000
Krishna	1,14,800	15,800	6,64,400	35,200
Nellore	90,000	60,400	6,58,400	90,400
Vishakapattanam	19,200	12,000	5,22,600	88,400
Salem	74,000	34,800	9,13,200	32,000
Coimbatore	56,800	17,800	4,27,200	4,800
Madurai	93,200	63,200	6,40,400	52,000
Chengelpet	1,92,000	1,26,000	42,800	16,200
North Arcot	1,35,200	80,400	6,35,800	1,04,000
South Arcot	1,58,800	85,200	7,14,400	1,07,200
Thanjavur	41,600	18,400	8,64,800	29,200
Tirunelveli	15,086	58,000	5,50,400	75,600
Tiruchirapalli	98,800	52,800	8,12,400	78,400
Total	12,88,686	7,72,400	1,06,94,400	7,82,600

(***Source:*** Data based on Tamil Nadu Season and Crop Reports, 1969-72, adapted from Van Oppen and Subba Rao, 1980.)

Unfortunately, tank irrigation has not received the attention it deserved during the large-scale irrigation development in the Five Year Plans which commenced in 1951. The extent of irrigation by tanks has remained nearly the same in most of the districts of the erstwhile Madras Presidency since the past one century. Table 1.9 presents the comparative figures on the extent of area irrigated by tanks in 1882–83 and 1969–72 in 16 districts in the then Madras Presidency. On the contrary, the total cropped area in the region increased about 8 times during the same period.

In the regions reported in Table 1.10, tank irrigation was available for over 50% of the total cropped area in the past, as compared to 10% in A.D. 2000. As per the minor irrigation census carried out for the year 1993–94, there were about 1.6 million tanks in various states. The states of Andhra Pradesh, Karnataka, Madhya Pradesh, Maharashtra, Tamil Nadu and West Bengal have greater density of tanks (Min. of Water Res., 2003). Many projects are underway in the states of Maharashtra (Fig. 1.23), Tamil Nadu, Andhra Pradesh, Karnataka and Madhya Pradesh to improve the efficiency of tank irrigation systems. Ample opportunities exist in substantially improving the performance of tank irrigation. There is scope to substantially increase the number of tanks to be constructed at suitable locations.

A nearly unutilized potential is the construction of *system storage reservoirs* in canal command areas to store the excess flows during the periods in the crop season when the demand for irrigation is low. Often there is considerable wastage of canal water supplies during night. These excess flows could be diverted to storage reservoirs (tanks) in depressions and other suitable sites constructed or existing in the command area. The water thus conserved can be recycled for use during periods of high water demand for irrigation or to irrigate additional land by gravity flow or by pumping to higher reaches.

The National Commission on Integrated Water Resources Utilization Plan (1999) recommended renovation and modernization of tanks and other local water resources projects as a priority task. The programme needs to be planned and implemented on a watershed basis, taking into account the comparative techno-economic feasibility of renovating existing tanks vis-a-vis construction of new tanks, upstream and downstream. Such a programme should focus on reforestation of the catchment areas of tanks; restoring unlined channels as well as removing encroachments, strengthening and improving tank bunds and structures, along with efficient water distribution network. All these activities are to be taken up on participatory mode. Under many situations it will be advantageous to construct *percolation tanks* which do not directly provide water for irrigation but serve to recharge the groundwater through seepage, benefitting surrounding areas.

Statewise Irrigation Development

The irrigated area is not evenly distributed over different states of India (Tables 1.7 and 1.8). Utilization of the estimated major and medium irrigation potential of 16 states of India is presented in Fig. 1.31. The highest utilization of available potential is in Tamil

Fig. 1.33. Views of a Kolhapur Type (KT) weir near Pune, Maharashtra. Note the piers with slots for inserting R.C.C. boards to control the river flow and store water in the upstream section of the rivers for lift irrigation and water suppy. The boards are removed during the periods of heavy rain, but are put in place well before the recession of the south-west monsoon.

Nadu which has attained nearly 100% utilization as per the present estimates. The lowest of 23% has been in Assam. Punjab and Rajasthan have utilized about 84% of the estimated ultimate potential in major and medium irrigation projects. The corresponding figures for Karnataka is 71% and those of Haryana and Jammu and Kashmir are 70%. Andhra Pradesh, West Bengal and Kerala have utilized about two-third of their available potential. In Uttar Pradesh (including Uttaranchal) the utilization has been about 57%, while that of Maharashtra 56%. The utilization in Gujarat, Bihar (including Jharkhand) and Orissa has been 46%, 45% and 43%, respectively. There is considerable scope for additional development in Madhya Pradesh and Chhattisgarh where the utilization is only 39% and in Assam with 23%. In the area of minor irrigation, both surface water and groundwater, all the states, offer immense scope for further development. Watershed based micro-level water resource development offer tremendous scope for local level irrigation and water supply. Careful investigation and appropriate technology in micro-level water resources development and on-farm water management offer immense scope for improvement and require much greater attention.

Kolhapur Type Pick-up Weirs

A comparatively new technology, originally developed in the Kolhapur region of Maharashtra state, with applicability, especially in the plateau region of India comprising over half the land area of the country and other similar geophysical regions, is the Kolhapur Type Weir, popularly known as KT weirs (Fig. 1.33). They are used to intercept and store the receding monsoon flows in streams and rivers. They comprise of masonry piers and removable R.C.C. boards. The R.C.C. boards have lengths ranging from 2 to 2.5 m, depending on the spacing of the piers, and depth of about 30 cm. The piers are designed and constructed to withstand the water pressure due to storage, and floods in the river. The weirs are kept open during the main rainy season. They are inserted, one by one into the slots built in the piers, during the receding phase of the monsoon to store the river flow. The water stored in the river course is pumped at suitable sites for lift irrigation and water supply upstream of the weir. Large areas of land on either side of the river course are provided with supplemental irrigation during the dry season. The boards are removed and stored before the onset of the monsoon and used again year after year.

PHYSIOGRAPHY, CLIMATE, SOILS, RIVER SYSTEMS AND GROUNDWATER REGIONS OF INDIA INFLUENCING WATER RESOURCES UTILIZATION

The origin, distribution and utilization of water resources of any country are profoundly influenced by its physiography, climate and soils. The Indian sub-continent lies between 8°4′ to 37°6′ North latitude and 68°7' to 97°25' East longitude, covering an area of 328.759 million hectares of land. India has a land frontier of 15,200 km and a coast line of 5,700 km. The diversity of land forms in India is matched by the diversity of its climate, its people and its cultures. Towering mountain ranges, rolling hills, lofty plateaus and extensive plains, have all played major roles in shaping the physical features of this great country and the cultural, economic, and political history of its people.

The total cultivable area is estimated at about 184 million ha, or about 56% of the total area. The total cultivated area was estimated at 142.5 million ha in 1995. The landholdings are predominantly of small size. Marginal size holdings of (<1 ha) with an average size of 0.4 ha constitute 59% and small holdings of 1 to 2 ha size constitute 19% of the total landholdings. Semi- medium holdings of 2 to 4 ha account for 13.2% and the rest are medium (7.2%) and large (1.6%). In the recent past, the evolution of cultivated land area has presented two distinct phases; from 1950 to 1970 the cultivated areas rose by 18%/year, while the cropping intensity increased from 111 to 118%. During 1970 to 1990 the cropping

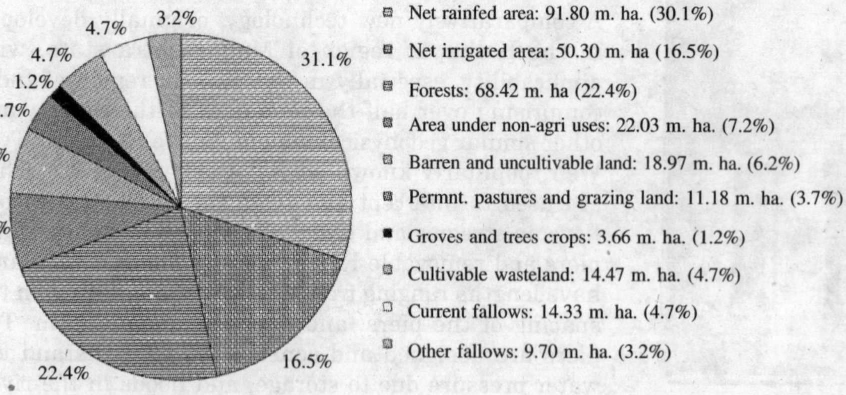

Total Geographical area : 328.53 m. ha.
Area reporting for land use classification : 304.86 m. ha. (92.80%)
Area not reporting : 23.67 m. ha. (7.2%)

Notes:
m. ha. Stands for million hectares and % indicates percentage of reporting area.

Fig. 1.34. Land use classification of India, 1993-94.

(***Source:*** Agricultural Statistics at a Glance, Directorate of Economics and Statistics, Ministry of Agriculture, Govt. of India, 1998).

intensity increased to 130%. The average cereal yield increased from 522 kg/ha in 1950 to 1457 kg/ha in 1992.

The net rainfed crop area of India is 91.8 million ha. Forests occupy 22.4% of the total geographical area. Figure 1.34 shows the land use classification of India, 1993-94.

India is the second most populous country in the world. The total population crossed 100.0 million in A.D. 2000. About 73% of the population is rural. The average population density is estimated at 287 inhabitants/km^2, but varies from fewer than 50 inhabitants/ km^2 to more than 500 inhabitants/km^2. The average annual population growth rate is estimated at 2.1%. Agriculture contributed about 29% of GDP in 1992–93. It employs more than 65% of the labour force and is the primary source of livelihood in rural areas.

(***Source***: UN Food and Agriculture Organization (FAO), 2001).

Principal Mountain Ranges, Plateaus and Plains

The mountain ranges of India greatly influence the rainfall over the sub-continent and the hydrology of the rivers which originate from them. The great Himalayan mountains of the north have a major role in influencing the climate of the country. The entire Himalayan range is about 2,500 km in length and 250 to 400 km in width. These mountains stretch continuously from Indus in the west to Brahmaputra in the east. The Himalayas consist of three parallel series of ranges, the northern most are the Great Himalayas with an average elevation of 6,000 m. South of these are the Middle Himalayas having elevations ranging from 3,200 to 4,500 m and the southern most are the Outer Himalayas or the Shivalik foothills, averaging 900 to 1,200 m in elevation. The entire Himalayan range is drained by a number of perenial streams and major rivers including the Indus, the Ganga and the Brahmaputra, along with their tirbutaries which flow through fertile valleys and plains. The North-Eastern ranges, constituting the Meghalaya plateau, are almost an extension of the Himalayas. They include the Garo, Khasi and Jaintia Hills of Meghalaya. They have an average elevation of about 830 m. The plateau rises abruptly from the plains of Bangladesh.

The hill ranges of peninsular India are the Eastern Ghats, Western Ghats, Aravallis, Vindhyas and Satpuras. The Eastern Ghats border the plateaus of peninsular India on the east and consist of a series of ranges and spurs cut into discontinuous blocks of hills by the large east flowing rivers. They run almost parallel to the east coast, leaving broad plains which stretch from the foot of the hills to the coast. The Eastern Ghats range in height from 500 to 1,000 m from the Godavari to the Mahanadi basins. The hills are low in the Krishna basin. In the Kaveri basin, they rise to heights of about 2,500 m at the Nilgiris where they meet the Western Ghats.

Along the Western coastline of India are the *Western Ghats* or the Sahyadris, extending 1,600 km from Kaniyakumari to Tapi Valley. They are in three distinct groups—the south, central, and northern ranges. The average elevation of the Western Ghats in the North is about 550 m, with a few peaks rising to 1,000 m. In the south of Mumbai, the average elevation is 1,400 m. Further south in Coorg, in Karnataka the peaks rise to more than 2,500 m. An important feature of the Western Ghats is that they give rise to all the important rivers of peninsular India, including Kaveri, Krishna and Godavari.

The Aravalli range is geologically the oldest of the mountain ranges in India, stretching a distance of about 800 km south-west from Delhi. They are highly eroded. The ravages of time have left only a few peaks, including Mount Abu (1,158 m), the rest having been reduced to low rugged hills and ridges, except the Mewar range.

The Vindhya range (1,050 km) traverse, almost the whole width of penisular India. Together with the Satpura range (900 km) which run nearly parallel to it, the Vindhyas form one of India's major watersheds, dividing North India from the Deccan. While the average elevation of the Vindhyas is only about 300 m, several of the peaks in the Satpura range rise above 1,000 m. The Satpura ranges have an average elevation of about 500 m for most of its stretch. The orientation of the mountain ranges with respect to the prevailing winds and their elevation have great influence on the distribution of rainfall in India.

The Plateaus. The plateaus are a prominent feature of the physiography of India. They range in elevation from 300 to 900 m. The surface of the plateaus usually consists of extensive plains, either flat or undulating. The major plateaus are the Malwa, the Vindhya, the Chhotanagpur (Ranchi, Hazaribagh and Kodarma), the Satpura (Betul and Maikala), the Deccan (Maharashtra, Karnataka and Telengana region of Andhra Pradesh), Ladakh and Meghalaya.

The Plains. A large part of the Indian sub-continent consists of extensive plains watered by great rivers and supporting a huge population. They comprise of the vast Indo-Gangetic plains, the coastal plains and the mountain plains such as the Kashmir Valley, the Imphal basin and Ladakh.

Physiographical Divisions of India

Physiographically, India can be divided into the following regions:

1. The northern mountains
2. The great plains
3. The central highlands
4. The peninsular plateau
5. The east coast belt
6. The west coast belt
7. The islands.

The Himalayan range in the north is the dominant geographical feature of India. It is a complex of several hill ranges intersected by valleys and plateaus.

The great *Indo-Gangetic Plains* have been built up in the west by the Indus and its tributaries and in the centre and east, by the Ganga and its tirbutaries. The thickness of the alluvium varies from area to area. It is the greatest in the Ganga plains and the least in the western plains. The great plains extend to an area of 6,52,000 sq km, of which about one-third is in the arid zone of western Rajasthan and another one-third in Uttar Pradesh. The remaining part is shared mainly by Punjab, Haryana, Bihar and West Bengal. The alluvium which forms the great plains was deposited by the rivers during successive geological eras. Of this, the older alluvium, locally known as *bangar*, forms the higher ground, while the newer alluvium, known as *khadar* in Uttar Pradesh and *bet* in Punjab forms the flood plains adjacent to the rivers. The elevation of the great plains varies widely, being at tide level at the mouth of the Ganga in West Bengal to over 200 m above sea level in Punjab

The *central highlands* lie between the great plains of North India and the Deccan plateau. It is a wide belt of hills and plateaus intersected by valleys and basins, largely covered by forests. It accounts for one-sixth of the total area of India. The northern part of the highland contains the Aravalli range, the east Rajasthan uplands, the Madhya Pradesh *pathar* and the Bundelkhand upland. The southern part contains the Malwa plateau, the Vindhya ranges and the Narmada valley.

The *peninsular plateaus*, lying between the Bay of Bengal in the east and the Arabian sea in the west and bordered by the Satpura ranges and the Eastern and the Western Ghats form the largest of the physiographical divisions of India. The area is drained by the major east-flowing rivers, namely, Godavari, Krishna and Kaveri and by the west-flowing Tapi.

The *east coast belt* bordering the Bay of Bengal. extends to a distance of about 1000 km from Kaniyakumari to the Krishna and Godavari deltas which form one of the most extensive paddy growing regions in the world. The belt has an average width of 100-130 km.

The *west coast belt,* bordering the Arabian sea extends a distance of 1500 km from Kaniyakumari to Broach. The coastal plains are confined to a narrow belt 10-25 km wide between Western Ghats and the sea. The two peninsulas of Kathiawar and Kutch and the wide Gujarat plains are at the northern end of the west coast.

The islands of India lie both in the Arabian sea and the Bay of Bengal. The Arabian sea islands include the Laccadive, the Amindivi and the Minicoy while the major ones in the Bay of Bengal are the Andaman and Nicobar islands.

Climate

India has a diverse climate and a variety of weather conditions. The climate changes from continental to oceanic, from extremes of heat to extremes of cold and from high aridity and scarce rainfall to excessive humidity and heavy rainfall. The great Himalayan ranges in the north and the Indian ocean in the south are the dominant features influencing India's climate. The Himalayan range, by providing an impenetrable barrier to the cold winds from Central Asia, provides the sub-continent with a tropical type of climate. The Indian ocean, which is the source of cool moisture-laden winds reaching India, provides the effects of an oceanic type of climate.

The varying climatic conditions of India result in diversity in agriculture. The country has regions experiencing cold and hot arid conditions to perhumid conditions across northern part. The peninsular region experiences tropical arid to humid conditions. Four distinct seasons are experienced in most parts of India, namely (*i*), winter from December to February, (*ii*) summer from March to May, (*iii*) monsoon from June to September, and (*iv*) post-monsoon from October to November. The seasons are identified based on the distribution of rainfall and temperature.

The distribution of crops and their productivity are profoundly influenced by climatic factors such as temperature and rainfall. Soil factors such as texture.

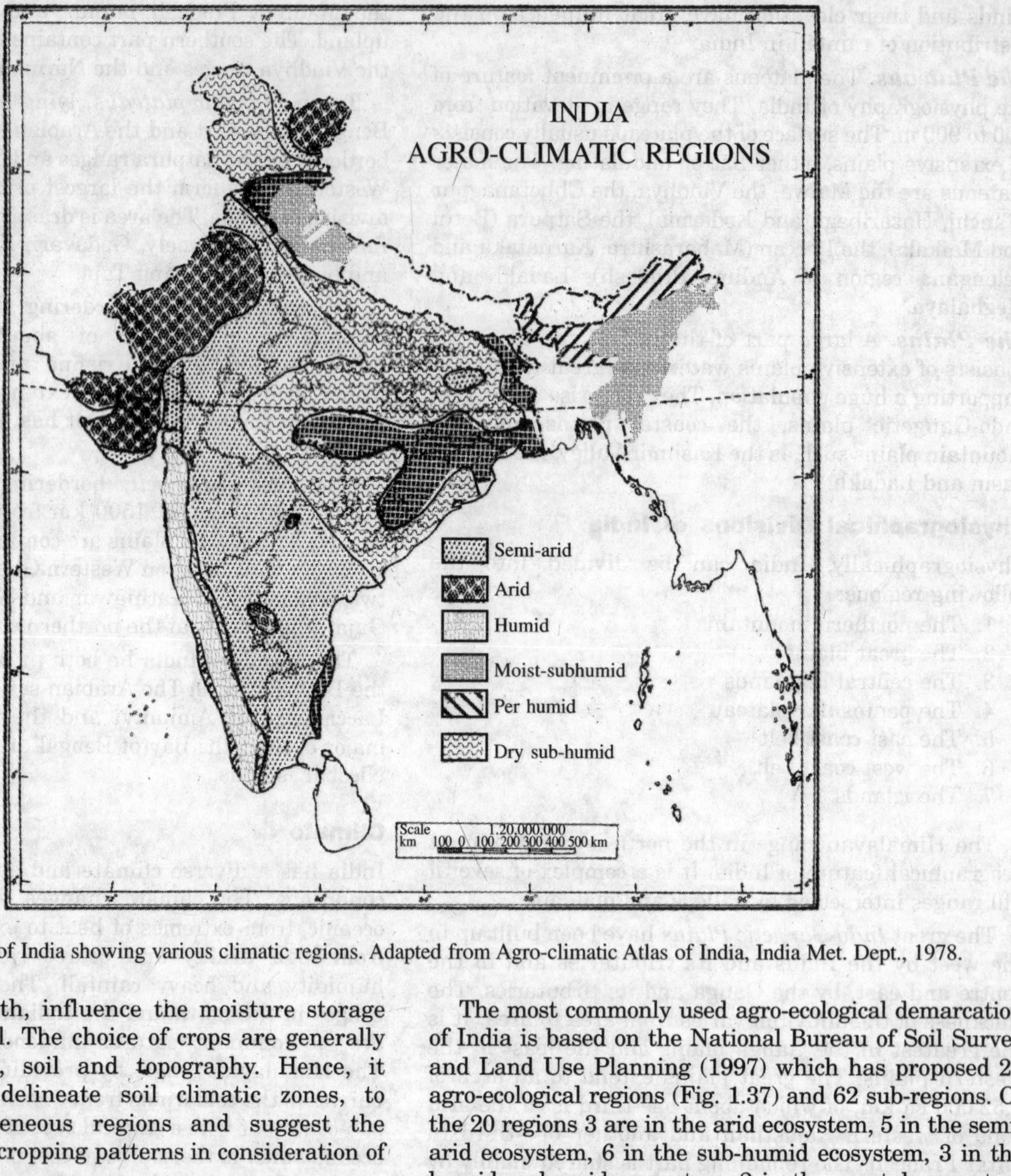

Fig. 1.35. Map of India showing various climatic regions. Adapted from Agro-climatic Atlas of India, India Met. Dept., 1978.

structure and depth, influence the moisture storage capacity of the soil. The choice of crops are generally based on climate, soil and topography. Hence, it is important to delineate soil climatic zones, to demarcate homogeneous regions and suggest the suitable crops and cropping patterns in consideration of ecological factors.

Agroclimatic zones and regions. Thornthwaite (1948) improved the climatic classification procedures adopted by earlier workers by introducing the water balance concept. Thornthwaite and Mather (1955) further revised the climatic classification. Using this modified method, Subramanyam *et al.* (1965) and Rao *et al.* (1972) classified India's climate into six major climatic types, viz., per-humid, humid, moist sub-humid, dry sub-humid, semi-arid and arid (Fig. 1.35). Krishnan (1988) has suggested the delineation of the country into 40 homogeneous soil climatic zones, Ghosh (1991) proposed the demarcation of the country into 12 agro-climatic sub-zones, based on soil, climate and physiography (Fig. 1.36). Sehgal *et al.* (1992) demarcated the country into 20 agro-ecological regions and further into 62 sub-regions.

The most commonly used agro-ecological demarcation of India is based on the National Bureau of Soil Survey and Land Use Planning (1997) which has proposed 20 agro-ecological regions (Fig. 1.37) and 62 sub-regions. Of the 20 regions 3 are in the arid ecosystem, 5 in the semi-arid ecosystem, 6 in the sub-humid ecosystem, 3 in the humid-perhumid ecosystem, 2 in the coastal ecosystem and one in the island ecosystem. The delineation map is based on physiography, soils, climate, growing period of crops, and taking into account available water holding capacity of the soil, soil depth and soil texture. Agro-ecological regions are carved out of agro-climatic regions, when it is super-imposed on land form and the kinds of soil and soil conditions that influence the length of growing period. Selection of crops, cropping system and cultural practices are profoundly influenced by agro-climatic and agro-ecological considerations.

Rainfall

Amongst all the climatic parameters, rainfall is the most important factor influencing the productivity of crops and cropping systems. With the exception of

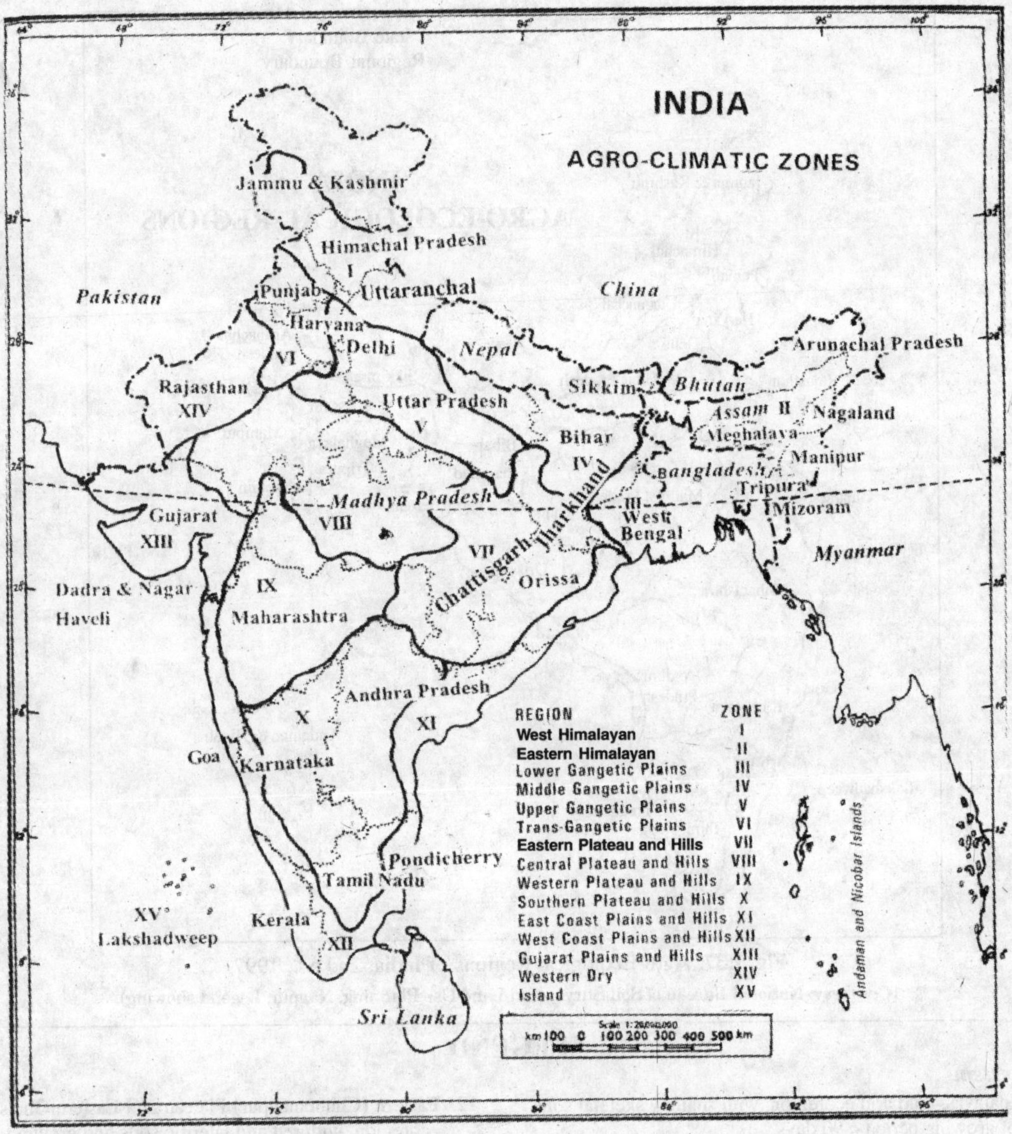

Fig. 1.36. Agro-climatic zones of India based on soil, climatic and physiography parameters.
(Source: National Land Use and Conservation Board, Mini. of Agriculture, Govt. of India (1991).

Jammu and Kashmir in the extreme north and Tamil Nadu in the peninsular south, about 74% of the rainfall in India occurs during the south-west monsoon season (June-September). In Jammu and Kashmir, a good part of the annual rainfall is contributed by winter precipitation due to western disturbances. In Tamil Nadu nearly half of the annual rainfall occurs during the north-east monsoon period of October-November.

India has a monsoon dominated rainfall pattern. South-west monsoon, north-east monsoon, cyclonic depressions and disturbances, violent local storms and occasional cyclones contribute to rainfall in different degrees in various rainfall regions of India. An outstanding feature of the wind system over the Indian ocean and the adjoining sea and land areas is the seasonal reversal of the *monsoons*. (*Monsoon* is termed as the periodic wind of the Indian ocean which is South-West from April to October and North-East during the rest of the year).

Due to the high temperature of the summer, moisture laden South-West monsoons originate from the equatorial region of the vast expanse of the Indian Ocean and enter the Indian sub-continent from the south-west. The physiographical features of peninsular India and the Western Ghats divert these monsoon winds into two branches, one the Arabian sea branch and the other the Bay of Bengal branch. The Bay of Bengal branch of monsoon winds moves into Assam by June when it strikes the Himalayas and precipitates heavily. A large portion of these monsoon winds are deflected towards the west by the Himalayas advancing the winds upto Delhi, Rajasthan and north-west India. These south-west monsoon winds increase from June to July and begin to weaken in September, specially in north India.

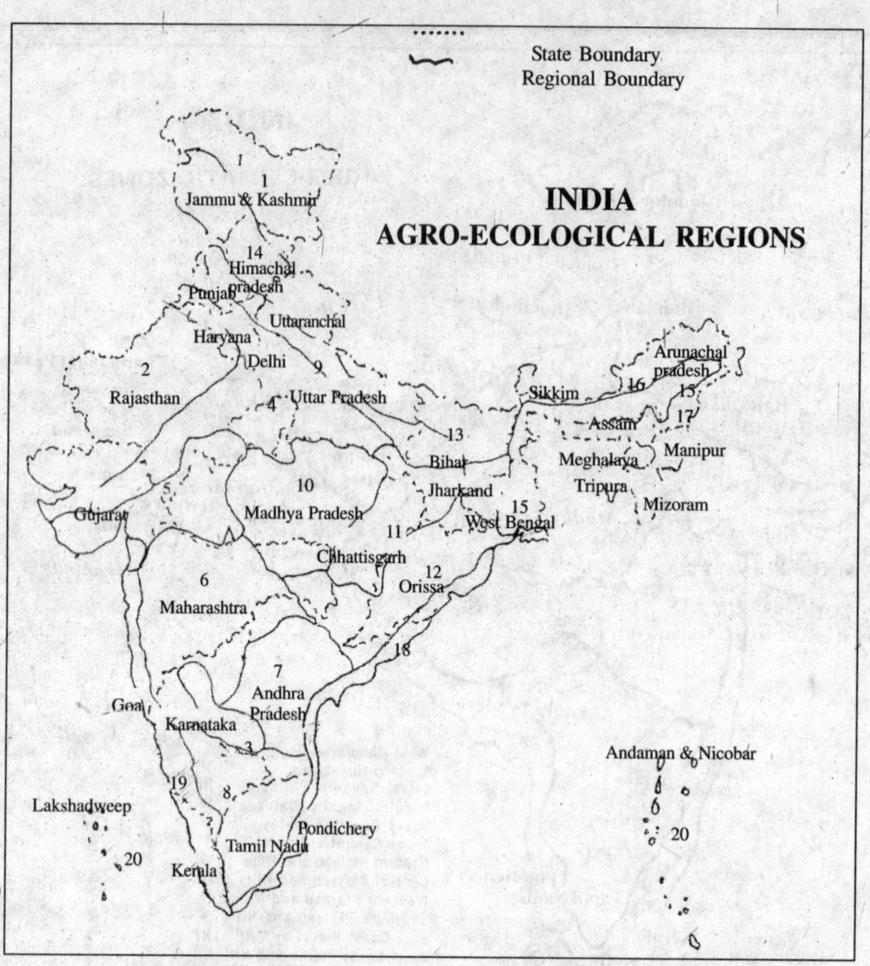

Fig. 1.37. Agro-Ecological Regions of India, 2nd ed., 1997.

(*Courtesy*: National Bureau of Soil Survey and Land Use Planning, Nagpur, Legend showing).

LEGEND

Arid ecosystem

1. Western Himalayas, cold arid ecoregion, with shallow skeletal soils and length of growing period < 90 days.
2. Western Plain, Kachchh and part of Kathiawar Peninsula, hot arid ecoregion, with desert and saline soils, and growing period < 90 days.
3. Deccan Plateau, hot arid ecoregion, with red and black soils and growing period < 90 days.

Semiarid ecosystem

4. Northern Plain and Central Highlands, including Aravallis, hot semi-arid ecoregion, with alluvium derived soils and growing period 90–150 days.
5. Central (Malwa) Highlands, Gujarat Plains and Kathiawar Peninsula, hot semi-arid ecoregion, with medium and deep black soils and growing period 90–150 days.
6. Deccan Plateau, hot semi-arid ecoregion, with shallow and medium (with inclusion of deep) black soils and growing period 90–150 days.
7. Deccan (Telangana) Plateau and Eastern Ghats, hot semi-arid ecoregion, with red and black soils and growing period 90–150 days.
8. Eastern Ghats, TN uplands and Deccan (Karnataka) Plateau, hot semi-arid ecoregion with red loamy soils and growing period 90–150 days.

Sub-humid ecosystem

9. Northern Plain, hot sub-humid (dry) ecoregion, with alluvium-derived soils and growing period 150–180 days.
10. Central Highlands (Malwa, Bundelkhand and Satpura), hot subhumid ecoregion, with black and red soils and growing period 150–180 (to 210) days.
11. Eastern Plateau (Chhattisgarh), hot subhumid ecoregion, with red and yellow soils and growing period 150–180 days.
12. Eastern (Chhotanagpur) Plateau and Eastern Ghats, hot subhumid ecoregion, with red and lateritic soils and growing period 120 180 (to 210) days.
13. Eastern Plain, hot subhumid (moist) ecoregion, with alluvium-derived soils and growing period 180–210 days.
14. Western Himalayas, warm subhumid to humid with inclusion of perhumid ecoregion with brown forest and podzolic soils and growing period 180–210 + days.

Humid-Perhumid ecosystem

15. Bengal and Assam Plains, hot subhumid (moist) to humid (inclusion of perhumid) ecoregion, with alluvium-derived soils and growing period 210 + days.
16. Eastern Himalayas, warm perhumid ecoregion with brown and red hill soils, and growing period 210 + days.
17. North-eastern Hills (Purvanchal), warm perhumid ecoregion with red and lateritic soils and growing period 210 + days.

Coastal ecosystem

18. Eastern Coastal Plain, hot subhumid to semi-arid ecoregion with coastal alluvium-derived soils and growing period 90–210 + days.
19. Western Ghats and Coastal Plain, hot humid to perhumid ecoregion with red, lateritic and alluvium-derived soils, and growing period 210 + days.

Island ecosystem

20. Islands of Andaman-Nicobar and Lakshadweep, hot humid to perhumid island ecoregion, with red loamy and sandy soils and growing period 210 + days.

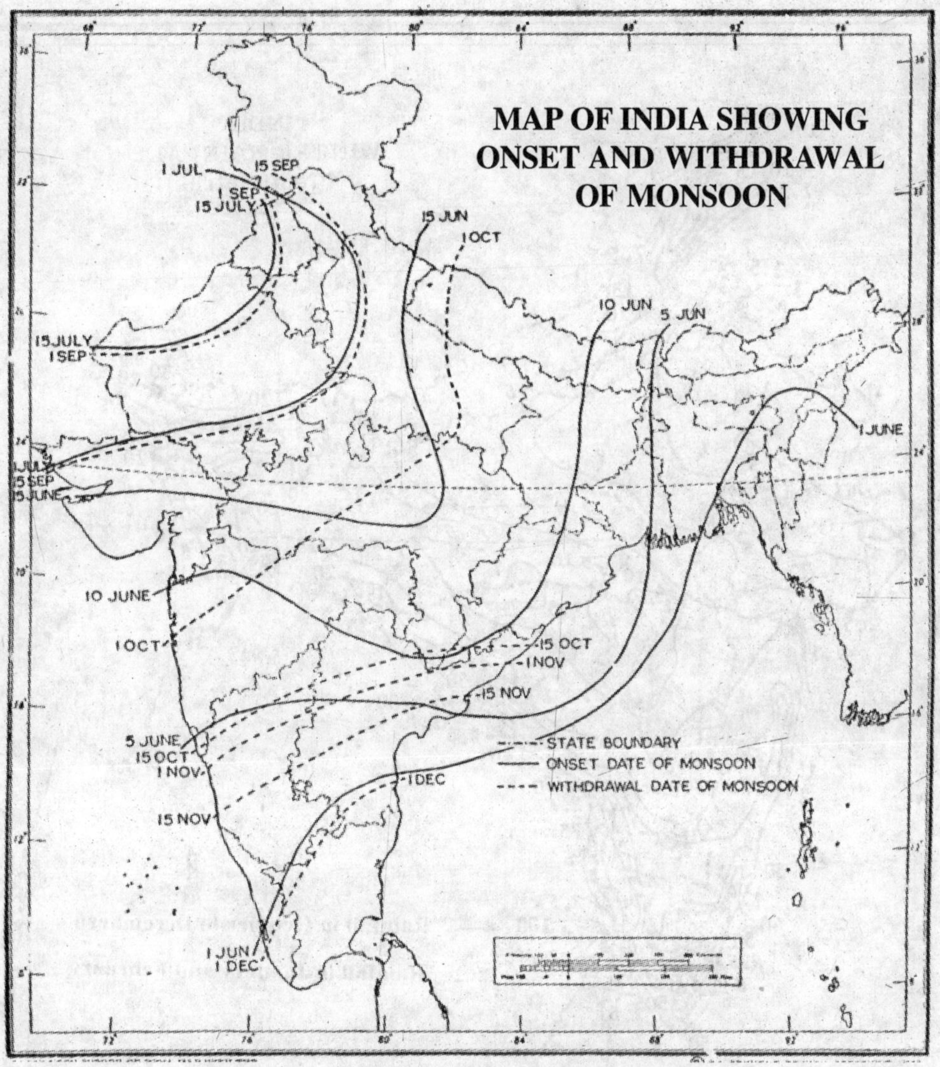

Fig. 1.38. Map of India showing the onset and withdrawal of monsoon.

(Source: *Climatological Atlas for Airmen,* India, Meteorological Department, Poona, 1943).

The Arabian sea branch of the monsoon strikes the Western Ghats and heavily precipitates along the west coast from Kerala to Gujarat. Normally, the onset of monsoon over Kerala coast takes place by 1st June. By the middle of June it covers Gujarat, Central India, North-east India, West Bengal, Orissa, Bihar and Eastern Uttar Pradesh. Almost the whole country comes under its sway by the end of June, except the Western Rajasthan region where it enters in the first fortnight of July. In most years, monsoon follows its normal time table for the regional onset over India. However, in some years there is delay and major variations. The normal dates of onset and withdrawal of south-west monsoon are shown in Fig. 1.38.

The rainy season crop, known as *kharif* in most parts of the country, is mainly dependent on the south-west monsoon rains. The monsoon season is not one of continuous rains in any part of India. Wet spells and periods of continuous dry spells may occur during the monsoon season. Provision of irrigation during the dry spells will result in substantial increase in crop production, especially rice. Wet spells may cause accelerated erosion of the soil, waterlogging and floods.

North-east monsoon. The north-east monsoon is associated with depressions in the Bay of Bengal. Under its influence, the retreating South-West monsoon currents are deflected towards the peninsula from the north-east, causing occasional showers on the east coast. The amount of rainfall decreases from the coast to the interior. During October and November cyclonic storms form in the Bay of Bengal which may bring heavy rains in the south and south-easterly regions. The coastal districts of Andhra Pradesh and Tamil Nadu receive rainfall in these months almost entirely from cyclonic storms or depressions. It contributes about 50% of the annual rainfall of this region. The productivity of crops in this region is very much dependent on the amount and distribution of north-east monsoon. The onset of north-

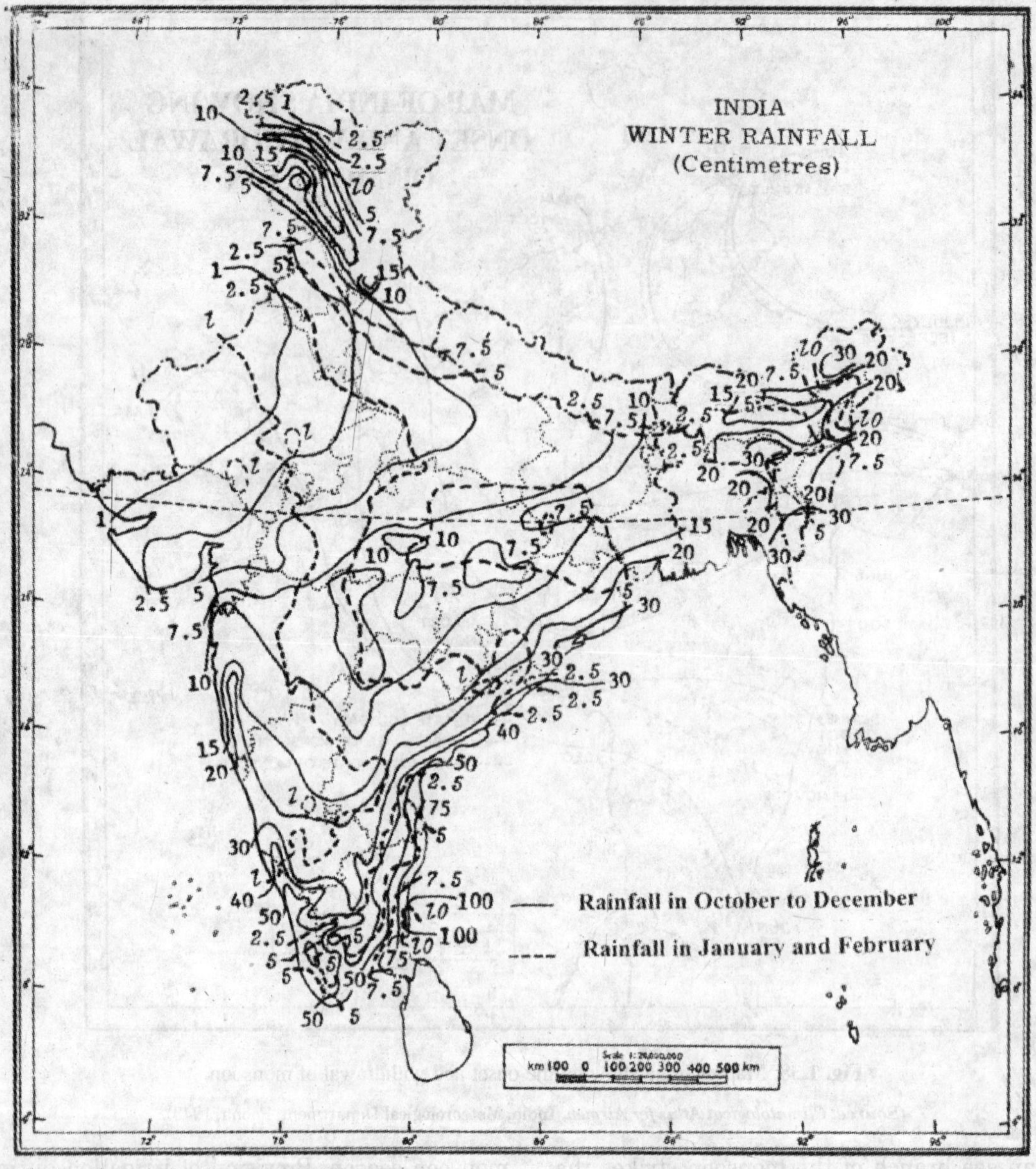

Based upon Survey of India outline map printed in 1979.

Fig 1.39. Map of India showing average rainfall during the winter crop (*rabi*) season of October-February.

*(**Source:** Based on data from India Meteorology Department).*

east monsoon normally takes place over the south peninsula (Tamil Nadu, south coastal Andhra the Pradesh, Rayalaseema and interior south Karnataka) by about 15 October. Withdrawal of north-east monsoon from south Peninsula is almost complete by mid-December. The highest seasonal rainfall of 600-800 mm occurs in north Tamil Nadu coasts; 400–600 mm in south coastal Andhra Pradesh, south Tamil Nadu and Kerala; 200–400 mm in the rest of coastal Andhra Pradesh, south interior Karnataka and coastal Karnataka; and 100-200 mm elsewhere in the south peninsula, Orissa and south-east Madhya Pradesh.

Western disturbances. During November to May, a series of western depressions, called western disturbances, enter India through the north-west frontier and Baluchistan and move eastwards across north India. These depressions cause cloudy weather and light rains in the plains of northern India and snowfall in the Himalayas. The western disturbances are important for the performance of winter *(rabi)* crops in northern and central India (Fig. 1.39).

Rainfall in the hot weather season. During March and April, western disturbances called Northwesters and their secondaries lead to limited rain over north India and central India. In May, the land surface gets heated appreciably, making the atmosphere unstable, resulting in dust-storms (in north-west India) and thunder storms (mostly in north-east India). In this season, the highest rainfall of 500–1000 mm occurs in south Assam, 300-500 mm in north Assam, Jammu & Kashmir and Kerala, and

Table 1.11. Seasonal Normal Rainfall in Different Meteorological Sub-divisions of India

S. No.	*Meteorological Sub-division*	*Area sq km*	*Normal rainfall, mm*						
			Jan. and Feb.	*Mar.-May*	*June-Sept.*	*Oct.-Dec.*	*Annual*	*% departure*	
								1997	*1998*
1	*2*	*3*	*4*	*5*	*6*	*7*	*8*	*9*	*10*
1	Andaman and Nicobar Islands	8,293	120	505	1595	744	2964	– 19	– 5
2	Arunachal Pradesh	82,578	119	597	2037	161	2914	– 10	30
3	Assam and Meghalaya	1,01,012	51	720	1901	201	2873	– 18	1
4	Nagaland, Mizoram, Manipur and Tripura	70.447	40	455	1247	191	1933	– 5	– 9
5	Sub-Himalayan West Bengal and Sikkim	28,924	40	437	2038	168	2683	– 11	22
6	Gangetic West Bengal	66228	37	176	1097	140	1450	– 5	– 10
7	Orissa	155782	37	118	1177	155	1487	3	– 10
8	Bihar Plateau	79638	44	90	1078	100	1312	15	23
9	Bihar Plains	94238	32	38	1005	76	1151	10	20
10	Uttar Pradesh East	146509	34	31	896	59	1020	3	13
11	Plains of Uttar Pradesh West	96732	40	31	774	47	892	0	23
12	Uttaranchal	51122	131	150	129	95	1671	– 12	15
13	Haryana, Chandigarh and Delhi	45821	41	31	513	30	615	38	49
14	Punjab	50362	55	51	499	42	647	39	31
15	Himachal Pradesh	55673	154	173	921	95	1343	1	3
16	Jammu & Kashmir	222236	181	294	474	117	1066	3	– 10
17	Rajasthan West	195086	9	15	283	9	316	1	24
18	Rajasthan East	147128	11	16	630	22	679	18	2
19	Madhya Pradesh West	229550	21	20	922	52	1015	15	– 4
20	Madhya Pradesh East	213291	42	46	1145	71	1304	4	14
21	Gujarat Region, Daman Dadra and Nagar Haveli	86597	3	10	1060	37	1110	25	26
22	Saurashtra, Kutch & Diu	109990	2	9	543	26	580	20	25
23	Konkan and Goa	34095	2	48	2804	147	3001	– 2	9
24	Madhya Maharashtra	115306	6	45	745	108	904	18	28
25	Marathwada	64525	6	36	717	85	844	– 3	45
26	Vidarbha	97537	24	40	978	73	1115	– 2	3
27	Coastal Andhra Pradesh	93045	241	95	609	311	1036	3	28
28	Telangana	114726	11	56	778	107	952	– 10	21
29	Rayalaseema	69043	12	79	372	211	674	16	41
30	Tamil Nadu and Pondicherry	1304549	49	141	323	478	991	21	5
31	Coastal Karnataka	18717	4	189	3178	283	3654	20	14
32	North Interior Karnataka	79895	5	87	476	314	702	– 1	33
33	South Interior Karnataka	93161	7	154	703	212	1076	13	14
34	Kerala	38864	32	416	2153	506	3107	3	0
35	Lakshadweep	32	35	189	943	334	1501	18	33
	All India	3286732	42	128	905	122	1197	5	9

(***Source:*** Central Water Commission, 2000, 2002).

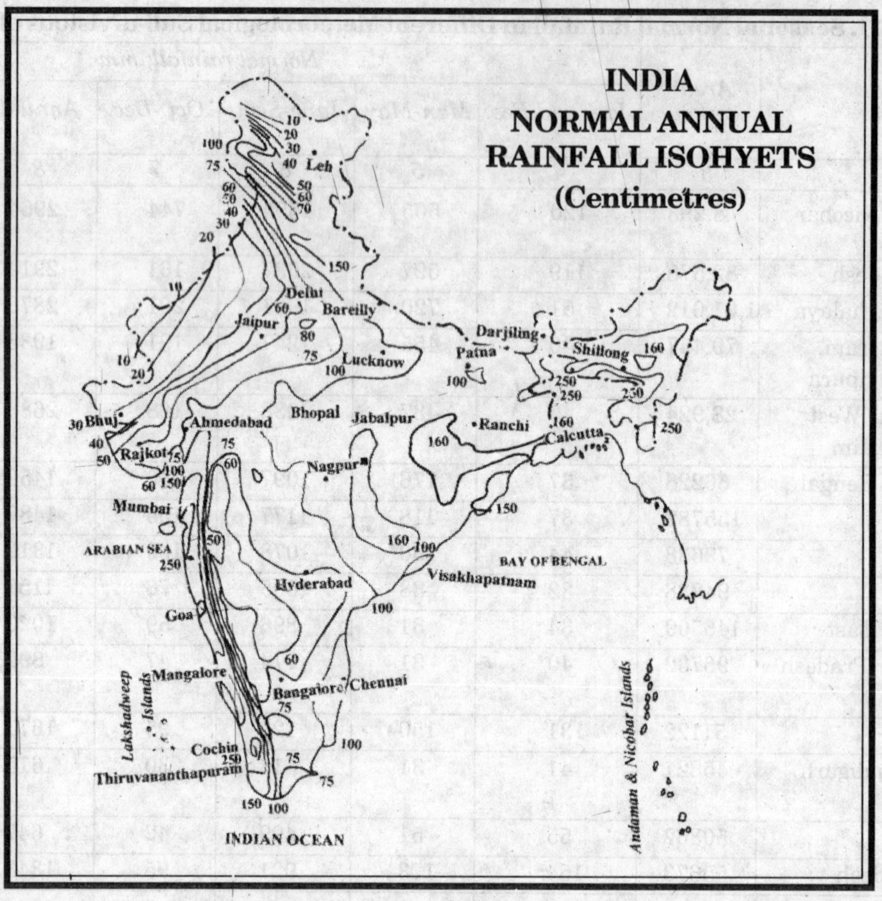

Fig. 1.40. India: Normal annual rainfall isophytes.

[***Sources:*** *(i)* Manual of Hydrometeorology, India Meteorological Department. (Based on rainfall data for the period 1901-50)
(ii) Water Management Perspective, Indian National Academy of Engineering (1990).]

100-200 mm in Bengal, Orissa and coastal Andhra Pradesh. Summer rains are important in West Bengal for the early spring crop of rice and for the tea crop in Assam and North Bengal.

All India average rainfall is 1,170 mm, but it varies respectively from 100 to 11,000 mm in Western deserts to North Eastern region. More than 50% of the precipitation takes place in about 15 days and less than 100 hours altogether in a year. The rainy days may be only about five in the deserts to 150 in the North East. The normal rainfall during the four seasons and also for the year as a whole is given in Table 1.11. Figure 1.40 shows the mean annual rainfall in India.

Temperature regime in different seasons. The growth and productivity of crops are profoundly influenced by temperature. Many of the physical and chemical reactions vital to crop growth are dependent on temperature regime. Generally, active growth of crops is confined to the temperature range of 5° to 40°C. Within this the optimum temperature range vary from species to species.

The temperature conditions of India are characterised by the oceanic type of climate of the coastal region, the moderate values of the sub-oceanic areas and large variations in the sub-continental and continental areas. The entire coastal belt, the Andaman and Nicobar islands and Lakshadweep islands of the Arabian sea are within the oceanic areas where the range in temperature is limited to 10–15°C. Beyond the coast are the sub-oceanic areas where the range is 15–20°C and the sub-continental areas with a range of 20–25°C. In the continental areas the range exceeds 25°C. During the winter season from November to February, due to the influence of continental winds over most parts of India, the temperature decreases from south to north. The minimum mean temperature varies from 24°C in the extreme south to below 5°C in the north. March to May is the period of continuous rise of temperature. The highest temperatures occur in the desert regions of north-west India where the maximum may exceed 46°C. With the advent of the south-west monsoon, there is a fall in the maximum temperature. It is during the monsoon season that nearly the entire country experiences almost uniform temperature. The temperature conditions of India are suitable for year round crop production in the whole of the country, except at high elevations in the Himalayas.

The number of days with mean maximum temperature exceeding 38°C is high in Deccan and central parts of India in March, April and May. The mean daily maximum temperatures are above 40°C in Vidarbha region in central India and Rayalaseema in Andhra Pradesh. In May many parts of India in central and north-east record mean daily maximum temperature above 40°C. High mean maximum temperatures in the form of heat waves are injurious to crops. In tea, coffee and cardamom gardens, farmers protect these sensitive crops by growing a permanent canopy of shade trees.

The northern states Punjab, Himachal Pradesh, Jammu and Kashmir, Rajasthan, Uttar Pradesh, Uttaranchal, Bihar, Jharkhand and Madhya Pradesh experience minimum temperatures less than 10°C. These regions are prone to occasional frost occurrence during the winter *(rabi)* crops season.

Sunshine. The number of days with bright sunshine hours in a year are least in north-east region (5.8 hrs) and highest in western Rajasthan (9.1 hrs). In most regions of India the sunshine hours during monsoon months are low. Konkan region experiences the least number of sunshine hours during July and August, compared to the other regions of India. The lower sunshine hours indicate lower availability of solar energy during the growing period of crops, which is a limiting factor for achieving high yields. Low sunshine hours and continued cloudiness provide conditions favourable for the multiplication of crop pests and diseases.

Wind speed and other weather parameters. Winds have varying effects on crops. High winds cause mechanical damage to crop plants and trees. Cold air can cause frost injury. If the wind is hot and dry, the rate of evaporation from reservoirs and transpiration from plants increases. Shelter belts and wind breaks reduce the adverse wind effects on crops.

Weather parameters like relative humidity, vapour pressure and evaporation influence plant growth and productivity. Regions suitable for the growth of a particular crop are determined by these factors. An understanding of the overall impact of various weather parameters on a long-term average basis is necessary to identify the climatic regions and their influence to specific crops and cropping patterns

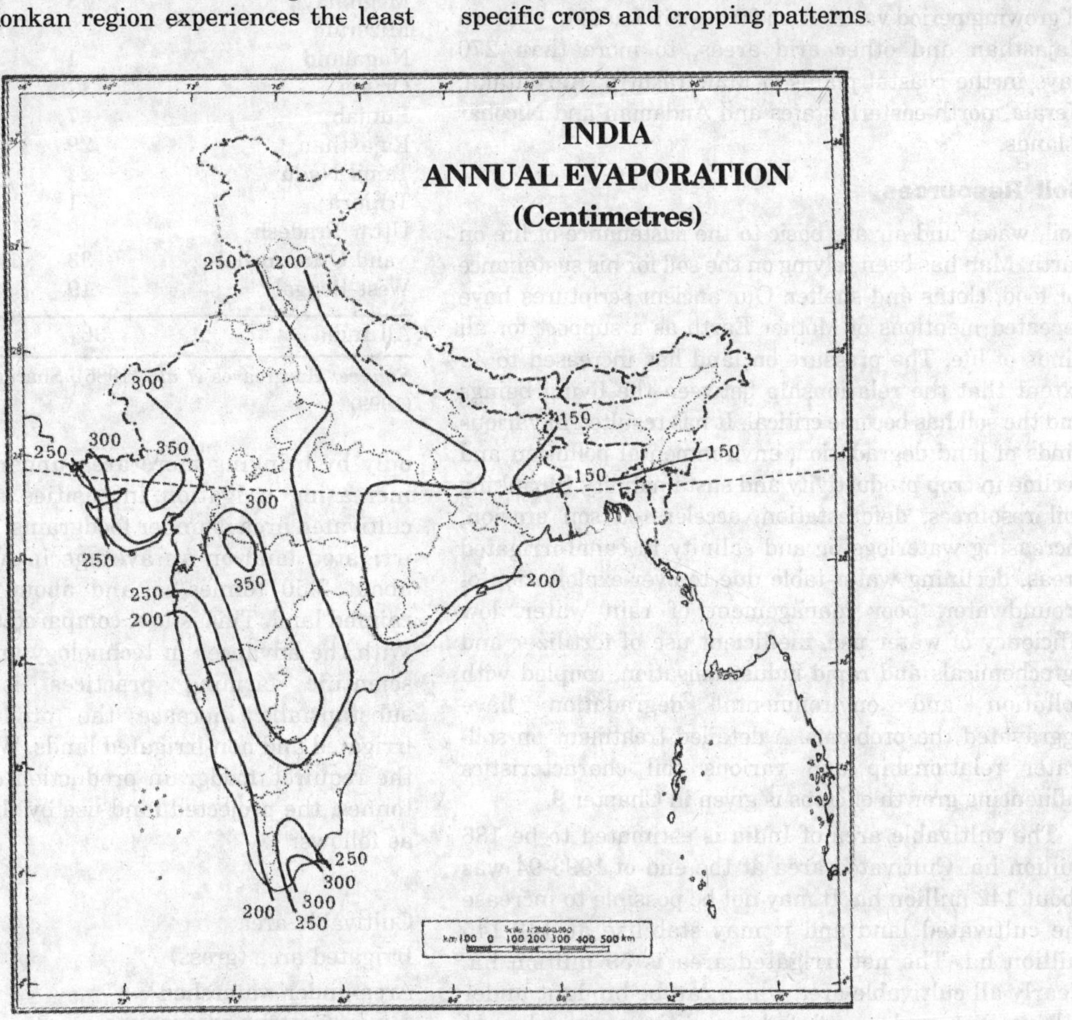

Fig. 1.41. Map of India showing annual evaporation.

(***Source:*** Based on data from India Meteorological Department).

Evaporation and Evapo-transpiration

The principal factors influencing evaporation are temperature, wind velocity, humidity and sunshine. Evaporation in India, varies greatly with seasons. In most parts of India, the rates of evaporation reach their peak in the summer months of April and May. Figure 1.41 shows the mean annual evaporation in India. Evapo-transpiration is a function of the availability of moisture at the earth's surface and solar energy to vapourise water and to transport the vapour. If there is no scarcity of water, evapo-transpiration from an extensive cover of vegetation will depend on the atmospheric conditions. The evapo-transpiration under such a situation is called the potential evapo-transpiration (PET). Potential evapo-transpiration (PET) is the realistic measure of water requirement of the vegetation. The annual potential evapo-transpiration in India is 1771 mm. It varies from a minimum of 1,239 mm in Jammu and Kashmir to a maximum of 2,052 mm in Andhra Pradesh (Table 1.12). The period of the year during which rainfall exceeds 50% of the PET is taken as the crop growth period. The length of growing period varies from 90 days in Ladakh, western Rajasthan and other arid areas, to more than 270 days in the coastal plains of Maharashtra, Karnataka, Kerala, north-eastern states and Andaman and Nicobar Islands.

Table 1.12. Average Annual Potential Evapo-transpiration of Different States of India

State	*Number of meteorological stations*	*Average annual potential evapo-transpiration (mm)*
Andhra Pradesh	39	2,051.82
Assam	11	1,741.82
Bihar and Jharkhand	22	1,851.36
Gujarat	25	1,992.08
Haryana	07	1,725.14
Himachal Pradesh	03	1,360.67
Jammu and Kashmir	13	1,239.38
Karnataka	27	1,929.19
Delhi	1	1,805.00
Goa	2	1,926.50
Kerala	8	1,915.50
Madhya Pradesh and Chhattisgarh	42	1,906.69
Maharashtra	30	1,966.57
Manipur	1	1,459.00
Meghalaya	3	1,520.33
Mizoram	2	1,533.50
Nagaland	1	1,459.00
Orissa	17	1.986.82
Punjab	7	1.795.14
Rajasthan	29	1,909..59
Tamil Nadu	24	1,944.04
Tripura	1	1,913.00
Uttar Pradesh and Uttaranchal	33	1,764.64
West Bengal	19	1,803.42
All India	367	1,770.84

Source: Hargreaves *et al.*, (1985), Sharma B.R, and Paul, D.K. (1999).

Soil Resources

Soil, water and air are basic to the sustenance of life on earth. Man has been relying on the soil for his sustenance for food, cloths and shelter. Our ancient scriptures have repeated mentions of Mother Earth as a support for all kinds of life. The pressure on land has increased to an extent that the relationship between the living beings and the soil has become critical. It has resulted in various kinds of land degradation, environmental pollution and decline in crop productivity and sustainability. Shrinking soil resources, deforestation, accelerated soil erosion, increasing waterlogging and salinity in canal-irrigated areas, declining water-table due to over-exploitation of groundwater, poor management of rain water, low efficiency of water use, inefficient use of fertilizer and agrochemicals and rapid industrialisation, coupled with pollution and environmental degradation have aggravated the problems. A detailed treatment on soil-water relationship and various soil characteristics influencing growth of crops is given in Chapter 9.

The cultivable area of India is estimated to be 186 million ha. Cultivated area at the end of 1993-94 was about 142 million ha. It may not be possible to increase the cultivated land and it may stabilize at 140-145 million ha. The net irrigated area is 59 million ha. Nearly all cultivable area which can be brought under cultivation under rainfed conditions is already cropped. Further increase in cropped area is possible only by bringing more area under irrigation and by increasing irrigation intensities. About 80% of the cultivated area is under foodgrains. The productivity of irrigated land on an average in A.D. 2000 has been about 2.50 tonnes/ha and about 0.7 tonnes/ha for rainfed land. This is low, compared to world standards. With the advances in technology and breakthrough in scientific farming practices it is possible to substantially increase the productivity levels of irrigated and non-irrigated lands. With a view to attain the required foodgrain production of over 450 million tonnes, the projected land use by the year is projected as follows:

		Million ha
Cultivable area	:	140–45
Irrigated area (gross)	:	150–160
Area under watershed management programme	:	10–15
Area to be protected from soil erosion	:	30–35

Increasing fragmentation of land-holdings is a serious problem which adversely affects crop production due to difficulty in adopting modern agricultural technology. The average operational holding size decreased from 2.1 ha in 1960–61 to 1.3 ha in 1991-92, with the number of marginal holdings (less than 1 ha) growing from 22 to 56 millions. The situation demands initiation of urgent steps for comprehensive land reforms.

Land Utilisation

The total geographical area of India is 328.53 million hectares. The net area sown is 142.1 million ha, which is about 43.2% of the total area. The area under forest in 68.42 million ha, which is about 22.8% of the total area. However, many forest lands are under varying degrees of degradation, in spite of stringent laws for their maintenance. The remaining area are grouped into area under non-agricultural uses (7.2%), barren and uncultivable land (6.2%), pasture and grazing land (3.7%), groves and tree crops (1.2%), cultivable wasteland (4.7%), current fallow (4.7%) and other fallows (3.2%). Figure 1.34 presents the land use classification of India in 1993–94. Table 1.13 presents the statewise distribution of land area, cultivated area and irrigated area in India during 1995–96.

SURFACE WATER SOURCES OF INDIA

Precipitation in the form of rainfall occurs only for a few days in a year. Hence, there is a need to conserve it in the soil profile, aquifers, ponds, lakes, reservoirs and rivers for use during the dry periods. The management of surface and groundwater resources requires a thorough understanding of the hydrology of the surface and groundwater systems, agro-ecosystem, soil-water-plant-environment interactions, existing and expected demands of water, availability of energy and social dynamics. For sustained agricultural production it is necessary to manage and utilize the water resources optimaly so that (i) both surface and groundwater supplies are maintained at the desired level, and (ii) the quality of land and water resources does not deteriorate with time.

Water for irrigation is obtained from natural streams or rivers, from surface reservoirs and from underground reservoirs. Flood water from rivers is collected in surface reservoirs by constructing dams at suitable sites (Figs. 1.11, 1.12, 1.15, 1.16, 1.23, 1.24) and by diverting water by constructing weirs or barrages at appropriate locations (Figs. 1.13, 1.14 and 1.26). Runoff water from small areas can also be collected by constructing ponds or tanks (Figs. 1.9, 1.10, 1.19 and 1.20). Water from underground reservoirs is utilized by constructing wells and installing pumps or other water lifts. Water from surface reservoirs or diversion points is conveyed through canals. Canals run from higher to lower elevations, and water flows in them by the force of gravity.

Based on whether the source of water is above or below the field surface, irrigation is classified into two types: (1) *Flow Irrigation* where the water reaches the field from the source by gravity flow; and (2) *Lift Irrigation* requiring the water to be raised or lifted from its source to the field surface.

River Systems of India

India is blessed with a large number of big and small rivers, the sweet waters of which can be put to many uses such as drinking, other domestic needs, agriculture and industrial uses, navigation, recreation, power generation and other uses. In the national strategy for water resource utilization, drinking water requirement is given priority over all other needs.

The river systems of India can be classified into four groups, namely (i) Himalayan rivers, (ii) Deccan rivers, (iii) Coastal rivers, and (iv) Rivers of the inland drainage basins. The Himalayan rivers derive their flows from rainfall runoff and by melting of snows and glaciers. Therefore, they have nearly continuous flow throughout the year. During the monsoon months, Himalayas receive heavy rainfall and rivers swell, causing frequent floods. The Deccan rivers are rainfed. Hence, they have major fluctuations in the volume of flow. Many of them are non-perennial. The coastal rivers, specially on the west coast, are short in length and have limited catchment areas. Most of them are non-perennial. The streams of inland drainage basins of Western Rajasthan are few and far apart. They drain into salt lakes like Sambhar or disappear in sands with no outlet to the sea.

The northern slopes of the Himalayas drain into Indus and Brahmaputra, while the southern slopes drain into the Ganga system. The large Indo-Gangetic plains lies between the Himalayas on the north and the plateau of the peninsula in the south and extends from the Bay of Bengal in the east and India's boundary with Pakistan on the west. These plains are fertile and support a major part of the population of the country. The heavy rainfall area of India, lying in the eastern and north-eastern regions, are drained by the Brahmaputra and its tributaries.

The Ganga has two main head reaches in Garhwal Himalayas, the Bhagirathi and the Alakhnanda which join at Devprayag to form Ganga. It traverses through Uttar Pradesh, Bihar, Jharkhand and West Bengal in India and thereafter enters Bangladesh. Its important tributaries are the Yamuna, the Ramganga, the Ghaghra, the Gandak, the Kosi, the Mahananda and the Sone. The rivers Chambal, and Betwa are the important sub-tributaries, which join the Yamuna, before it meets Ganga. The Brahmaputra rises in Tibet (China) and enters India in Arunachal Pradesh and after traversing through Assam and Bangladesh joins Ganga at Goalundo in Bangladesh. Its important tributaries are Dibang, Luhit, Subansiri, Manas and Tista. The Barak river, the head stream of Meghna, rises in the hills in Manipur. The Meghna river is the part of Ganga-

Table 1.13. Statewise Distribution of Land Area, Cultivated Area and Irrigated Area in India, 1995-96

S. No.	State Union Territory	Geographical area m ha	Gross area under cultivation m ha	Gross area irrigated m ha	Gross area irrigated to gross area cultivated %	Net area irrigated m ha
1	Andhra Pradesh	27.505	13.043	5.304	40.665	4.123
2	Arunachal Pradesh	8.374	0.244	0.036	14.754	0.036
3	Assam	7.844	3.938	0.572	14.525	0.572
4	Bihar and Jharkhand	17.388	10.019	4.581	45.723	3.68
5	Goa	0.37	0.165	0.036	21.818	0.023
6	Gujarat	19.602	10.082	3.655	36.253	3.002
7	Haryana	4.421	5.974	4.673	67.006	2.761
8	Himachal Pradesh	5.567	0.972	0.178	18.313	0.101
9	Jammu & Kashmir	22.224	1.073	0.44	41.007	0.386
10	Karnataka	19.177	11.958	2.845	23.792	2.302
11	Kerala	3.886	3.166	0.466	15.199	0.342
12	Madhya Pradesh and Chhattisgarh	44.345	25.04	6.178	24.673	5.928
13	Maharashtra	30.771	21.327	3.149	14.765	2.567
14	Manipur	2.233	0.192	0.075	39.062	0.065
15	Meghalaya	2.243	0.247	0.045	18.219	0.045
16	Mizoram	2.108	0.109	0.009	8.257	0.007
17	Nagaland	1.658	0.228	0.072	31.579	0.062
18	Orissa	15.571	9.668	2.629	27.193	2.09
19	Punjab	5.036	7.752	7.377	95.163	3.847
20	Rajasthan	34.224	19.672	6.361	2.335	5.232
21	Sikkim	0.71	0.142	0.016	11.268	0.016
22	Tamil Nadu	13.006	6.267	3.183	50.79	2.625
23	Tripura	1.049	0.426	0.06	14.085	0.035
24	Uttar Pradesh and Uttaranchal	29.441	25.793	16.972	65.801	11.675
25	West Bengal	8.875	8.972	2.491	27.764	1.911
26	Andaman and Nicobar	0.825	0.049	–	–	–
27	Chandigarh	0.011	0.004	0.002	50	0.002
28	Dadar and Nagar Heveli	0.049	0.027	0.005	18.519	0.005
29	Delhi	0.148	0.067	0.064	95.522	0.044
30	Daman and Diu	0.011	0.005	0.001	20	0.001
31	Lakshadweep	0.003	4	–	–	–
32	Pondicherry	0.049	0.045	0.035	77.778	0.02
	All INDIA	328.726	186.561	71.51		53.508

(*Source*: Central Water Commission, 2000; Directorate of Economic and Statistics, Min. of Agriculture, Govt. of India, 1998.)

Brahmaputra-Meghna system. The combined Ganga-Brahmaputra river meets Meghna in Bangladesh and this combined system flows as one river for some distance before joining the Bay of Bengal.

The river systems in the central and southern parts of India originate from mountains which are devoid of snow. These rivers are most stable as they flow through geologically stable areas. Rivers from the central plateau flow in all directions contributing to the Ganga in the north, Narmada and Tapi in the west, Damodar and Mahanadi in the east and the tributaries of Godavari in the south. Most of the southern rivers originate in the Western Ghats. The important river systems in Deccan are the Narmada and the Tapi which flow westwards into Arabian sea and the east flowing rivers—the Mahanadi, the Godavari, the Krishna and the Kaveri—which flow into the Bay of Bengal.

There are numerous coastal rivers which are comparatively small. There are as many as 600 rivers originating from the Western Ghats and joining the Arabian sea. The rivers draining the west coast are short in length, but have large velocities of flow due to the steep slopes. They carry large volumes of water due to the intensive monsoon rains. The rivers on the east coast are longer and have formed deltas which are extensive in area and are suitable for intensive cultivation with appropriate soil and water management practices. These deltas constitute the principal rice bowl of India.

Water resources regions can be planned on the basis of river basins, which are the natural units. Figure 1.42 shows the major river basins of India, excluding the island territories. A river basin has a defined watershed boundary. It has relationship with groundwater resources in most of the cases. The development of a balanced plan for water resources utilization requires detailed knowledge of the quantity, quality and distribution of water resources. Information on the changing pattern of land use in the entire watershed and its influence on the river flows are also required.

Table 1.14. Major River Basins of India

Sl. No.	Name or river	Catchment area in India, km^2	Catchment area falling in neighbouring countries, km^2
1.	Indus (Sindhu)	3,21,289	1,46,779
2.	(a) Ganga	8,61,404	1,88,569
	(b) Brahmaputra	1,87,113	3,92,887
	(c) Barak and other rivers flowing into Meghna	70,895	....
3.	Sabarmati	21,674	....
4.	Mahi	34,842	....
5.	Narmada	98,796	....
6.	Tapi	65,145	....
7.	Brahmani	39,033	....
8.	Mahanadi	1,41,589	....
9.	Godavari	3,12,812	....
10.	Krishna	2,58,948	....
11.	Pennar	55,213	....
12.	Kaveri	81,155	....
	Total	25,49,908	7,28,235

(***Source:*** Central Water Commission, 1996, 2000, 2003.)

The Ganga-Brahmaputra-Meghna basin is the largest in India, receiving water from an area which comprises about one-quarter of the total area of India. The second largest basin is that of Godavari which covers about 10% of the total area of India.

On the basis of catchments, the river basins of India could be divided into the following groups:

(*i*) *Large river basins* — River basins with catchment areas of 20,000 km^2 and above.

(*ii*) *Medium river basins* — River basins with catchment areas between 20,000 and 2,000 km^2.

(*iii*) *Minor river basins* — River basins with catchment areas below 2000 km^2.

Based on the above classification, there are 12 major river basins (Table 1.14). There are 45 medium river basins, 24 of which are of east flowing rivers, 17 of west flowing rivers, and 4 flowing into neighbouring countries in the north-eastern region. Together they contribute to over 90% of the rainfall-runoff occurring in India. The minor rivers account for about 10% of the runoff.

The average annual surface water flow in India has been estimated as 1869 billion m^3, of which only 690 billion m^3 can be utilised, if appropriate storages can be created. The reason for the vast difference between potential and availability for storage is the monsoonic climate as well as topographical and geological limitations. (Ministry of Water Resources, 2003).

Table 1.15 presents the availability of the water resources of India. The maximum potential for storage lies in the Ganga basin. Brahmaputra and Barak basin occupy the second place. The other basins where the live storage capacity of the projects is high are Krishna, Godavari, Narmada and Mahanadi.

In a majority of river basins, present utilization is significantly high and is in the range of 50% to 95% of utilizable surface water resources except in the rivers such as Narmada and Mahanadi basins where it is 23% and 34%, respectively. Major States like Andhra Pradesh, Gujarat, Karnataka, Madhya Pradesh, Maharashtra and Uttar Pradesh together account for more than 50% of total live storage capacity in India.

GROUNDWATER RESOURCES OF INDIA

Groundwater has a significant role in India's economy. The resource has gained importance as a source of drinking water and food security. Because the groundwater abstraction structures (wells) are usually under the direct control of the user, groundwater has become the preferred source for various uses. Groundwater contributes to about 50% irrigation water, 80% water for domestic use in rural areas and 50% water for urban and industrial areas.

The groundwater resource has two components, namely, *static* and *dynamic*. The static groundwater resources i.e. aquifer zones below the zone of water level fluctuation, of India have been estimated as 10,812 billion m^3. Generally, the static water resource is not regularly replenished on an annual basis. The dynamic resource is replenished annually, which has been assessed as 432 billion m^3. As per the National Water Policy, the development of groundwater resources is to be limited to the utilization of the dynamic component of groundwater. There is, however, scope for development of the static resources in periods of drought.

The annual availability of groundwater is predicted by the annual recharge of the aquifer. The water extracted from an aquifer in any year should be

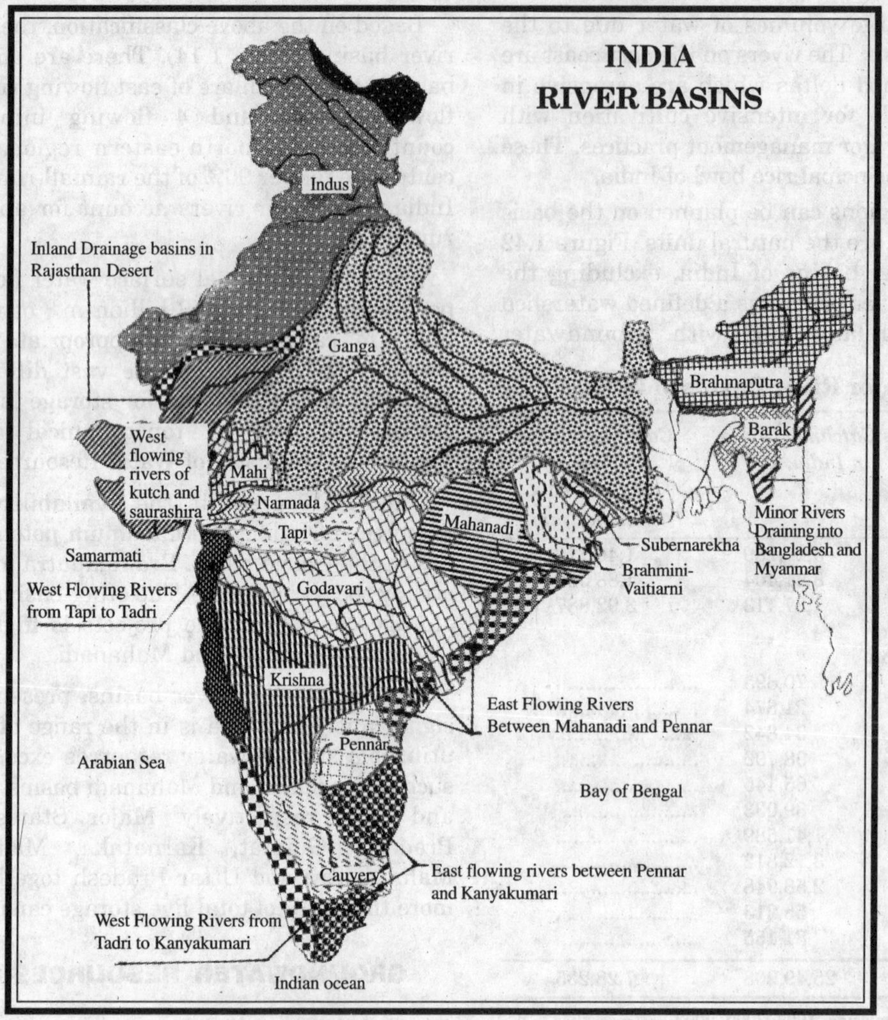

Fig. 1.42. River basins of India (excluding island territories).

(***Source:*** Based on Ministry of Water Resources, Govt. of India, 1999)

replenished through recharge from the succeeding precipitations (or other sources of recharge), so that over a cycle of 2, 3 or 5 years, the groundwater table does not go down. When there is over-exploitation of groundwater, known as *mining*, the water table goes down progressively. Such over-exploitations have occurred at several places in India, and has caused ingress of saline waters from the sea in some of the coastal areas.

Groundwater Regions of India

Based on the occurrence of groundwater in geological formations, nature and extent of aquifer bodies and flow characteristics, the Central Groundwater Board (1989) generalised the hydrological formations of India as follows:

(*i*) **Unconsolidated formations.** The quaternary rocks comprising recent alluvium, older alluvium and coastal alluvium. They are the important unconsolidated formations found in the Indo-Gangetic plains, coastal plains, Brahmaputra basin, inter-montane valleys of Kashmir, Himachal Pradesh, and intra-crotonic basins like Narmada, Tapi and Purna basins, aeolian sands of Rajasthan and Gujarat, arid tracts, including valley fill and buried channel areas.

(*ii*) **Semi-consolidated formations.** These are composed chiefly of shale, sandstone and limestone generally found in north-western region and the terrestrial deposits belonging to Gondwana group.

(*iii*) **Consolidated hard rock formations.** They occupy almost two-thirds of the area of India and can be classified in broad lithological units of sedimentary and meta-sedimentary, volcanic rocks and crystalline and metamorphic rocks.

Figure 1.43 presents the groundwater map of India showing the type of geological formation, scope for recharge and the yield potential of the regions. The unclassified crystalline formation of Archean age as

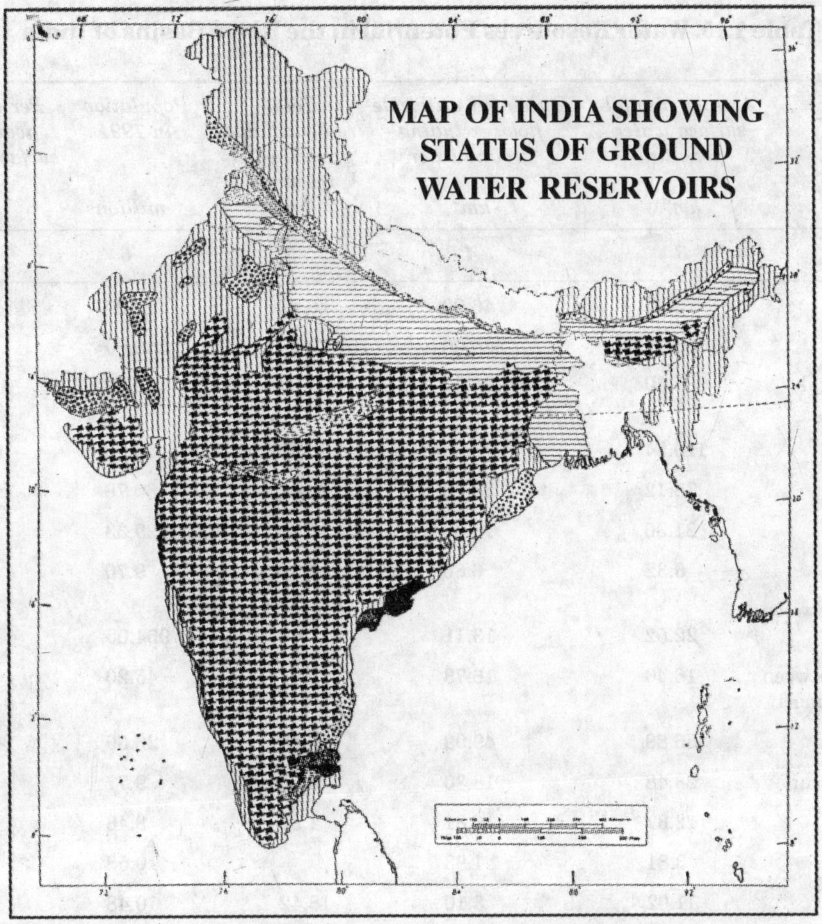

Fig. 1.43. Map of India showing groundwater reservoirs and their future management.

(***Source:*** Central Ground Water Board, 1989)

LEGEND

- Arid and semi-arid underground water reservoirs with limited recharge; calls for mangement on precise quantitative estimates of resources, both from quantitative and qualitative aspects and its budgeting for various needs (livestock, domestic and irrigation).
- Groundwater reservoirs with scope for recharge and occasionally coincide with surface drainage basins; calls for management to prevent waterlogging, and salinisation of soils.
- Groundwater reservoirs with ample scope for recharge, coincides partly or fully with surface drainage basins and fairly well developed; future management of reservoirs is hinged to the understanding of the inter-relationship of the surface and groundwater bodies and conjunctive use of the resources (areas with intensive tubewell irrigation (marked with dashed line).
- Groundwater reservoirs with ample scope for recharge along the sub-mountain tracts; future management of reservoirs depends on the economics of the lift involved.
- Status of information not enough on the reservoirs to define management.
- Artesian groundwater reservoirs with fairly large scope for recharge; future management of reservoirs depends on efficient well design, reclamation of land, and effective drainage measures.
- Coastal groundwater reservoirs; future management of basins depends on control of saline water intrusion into fresh groundwater bodies.
- Deltaic groundwater reservoirs under active exploitation; future management linked with shallow aquifer through wells or filter point wells coupled with return flows and drainage control.
- Groundwater reservoirs with fairly closed boundaries; future management calls for precise estimates on the quantitative determinations of usable storage/recharge, etc., components and methods of reservoirs operation.
- Groundwater basins in hard and consolidated rocks yet to be precisely delineated and quantitatively estimated; management of these limited supply basins depends largely on the geologic framework of groundwater reservoirs, permeability from weathering or fracturing and other structural features.

Table 1.15. Water Resources Potential in the River Basins of India

Sl. No.	*Name of the river and basin*	*Average annual surface water potential* km³	*Estimated utilisable flow excluding groundwater* km³	*Total replenishable groundwater resources* km³	*Population in 1991* millions	*Per capita available surface water* m³	*Per capita surface and groundwater* m³
1	*2*	*3*	*4*	*5*	*6*	*7*	*8*
1	Indus (up to Border)	73.31	46.00	26.49	41.90	1750	2382
2	(a) Ganga	525.02	250.00	170.99	356.80	1471	1951
	(b) Brahmaputra, Barak and other	585.60	24.00	53.91	35.24	16617	18147
3	Godavari	110.54	76.30	40.65	53.98	2048	2801
4	Krishna	78.12	58.00	26.41	60.78	1285	1720
5	Kaveri	21.36	19.00	12.30	29.33	728	1148
6	Pennar	6.32	6.86	4.93	9.70	652	1160
7	East flowing rivers between Mahandi and Pennar	22.52	13.11	23.60	954.00	1822	831
8	East flowing rivers between Pennar and Kanyakumari	16.46	16.73		45.20	364	
9	Mahanadi	66.88	49.99	16.46	26.60	514	3133
10	Brahamani and Baitarni	28.48	18.30	4.05	9.77	[illegible]15	3329
11	Subernarekha	12.37	6.81	1.82	9.46	[illegible]08	1500
12	Sabarmati	3.81	1.93		10.58	360	
13	Mahi	11.02	3.10	18.42	10.48	1052	1120
14	West flowing rivers of Kutch, Saurashtra including Luni	15.10	14.98		22.10	683	
15	Narmada	45.64	34.50	10.83	14.70	3105	3842
16	Tapi	14.88	14.50	8.27	14.80	1005	1564
17	West flowing rivers from Tapi to Tadri	87.41	11.94		25.80	3388	
18	West flowing rivers from Tadri to Kaniyakumari	113.53	24.27	17.69	32.60	3483	3744
19	Area of inland drainage, negligible in Rajasthan desert			7.10			
20	Minor river basins draining into Bangladesh and Myanmar	31.00			2.10	14762	14762
Total		1869.35	690.31	431.44	842.62	2218	2731

(***Source:*** Indian Water Resources Society (1999), Theme Paper on Water Vision 2050.)

well as Deccan trap formations are grouped together as hard rocks. The water yield of such areas is low. In the semi-consolidated formations which cover about 5 million ha, sedimentary rocks such as sandstones, limestones and conglomerates occur. The yield in these areas is moderate. These formations occur particularly in Rajasthan, Gujarat and Tamil Nadu. There are large areas of unconsolidated alluvial formations in India consisting mainly of sand, gravel, boulder, laterites, silt and clays. These areas include the Indus basin, Ganga-Brahmaputra-Meghna basin, coastal tracts and Kashmir valley. They yield substantial quantities of groundwater.

A scientific assessment of the groundwater potential of India was not undertaken until the initiation of the Five Year Plans in 1951. The groundwater mapping and groundwater explorations were sporadic and limited to specific areas. The Geological Survey of India was responsible for systematic investigation and mapping of groundwater in the country as a whole. With the increased emphasis of groundwater development during the Plan period, extensive exploration work was done in the desert areas of Rajasthan, the foothills of the Himalayas, coastal alluvial tracts of Orissa, Andhra Pradesh and Tamil Nadu and the basins of Narmada, Tapi and Purna. The Central Groundwater Board has estimated that the annual utilizable groundwater resource of India amounts to 423 km^3 per year and the present utilization at the end of the 20th century has been 100 km^3 per year.

There is a narrow band of artesian groundwater reservoir with high yields, just below the Himalayas adjoining the sub-mountain tracts. This band extends from Himachal Pradesh to Bihar through Uttar Pradesh. Such artesian aquifers exist at isolated locations in the coastal areas of Tamil Nadu.

The most potential groundwater reservoirs of India lie in the alluvial formations of the major river basins in the states of Punjab, Haryana, Uttar Pradesh, Bihar, West Bengal and Assam. These aquifers have ample scope for recharge. However, the future management of these reservoirs greatly depends upon the understanding of the inter-relationship of the surface and groundwater bodies and conjunctive use of the water resources. Some portions of this zone, especially in Punjab, Western Uttar Pradesh and Haryana have been developed intensively through tubewells.

Arid and semi-arid climatic areas exist in most parts of Rajasthan and major parts of Punjab, Haryana and Gujarat states. The groundwater reservoirs in this zone, are recharged to a limited extent. In general, the groundwater level is low and water quality tends to be saline at many places.

The hard rock formations lie south of the Indo-Gangetic plains and spread in the states of Rajasthan, Madhya Pradesh, Bihar, Gujarat, Maharashtra, Andhra Pradesh, Orissa, Karnataka, Tamil Nadu and Kerala. Groundwater is mostly pumped through open wells in these areas. The groundwater supply in these formations is limited and depends largely upon the geologic framework, permeability from weathering or fracturing, and other structural features of the formations.

There are important pockets of groundwater reservoirs along the eastern coast, in the deltas of Godavari, Krishna and Kaveri rivers, in which groundwater is under active exploitation, especially from shallow aquifers. Water is extracted through open wells and filter point wells.

For accurate assessment of groundwater resources, detailed and authentic information relating to hydrogeological formations and the water balance of the various river basins of the entire country are required. The current practice of estimating basinwise groundwater potential, unlike surface topographic divide, does not coincide with groundwater divide. There is considerable inter-basin or sub-basin transfer of groundwater in the aquifer system. The management of groundwater reservoirs of India requires detailed studies to regulate and maintain a hydrological balance so that groundwater is exploited on a sustainable basis.

Rainfall contribution to groundwater. Raghava Rao, *et al.* (1969) attempted to broadly assess the groundwater resources on the basis of rainfall contribution to the groundwater body and the contribution of seepage from canals, ponds, tanks, and other surface water bodies. The rocks and other geological formations of India have been classified in seven categories, as given in Table 1.16, for estimating the component of rainfall contribution to groundwater.

Contribution of seepage to groundwater. The underground aquifers are supplemented from sources other than rainfall, such as seepage from canals and field channels, ponds, tanks, influent drainage from rivers, and deep percolation from irrigated fields. Groundwater recharge could be increased through soil conservation measures such as large scale contour bounding and terracing and gully plugging and through intensive groundwater recharge programmes. In most of the canal irrigated areas of the country, a substantial component of the water applied to fields percolates below the root zone and contributes to the groundwater potential. Studies in Punjab have indicated that the magnitude of canal seepage is of the order of 115 hectare-metres per annum per kilometre length of the main canal and 38.24 hectare-metres per annum per kilometre length of distributory canal (Raghava Rao, *et al.*, 1969).

River basin based estimates of groundwater potential and utilization. Table 1.17 presents the river basin based availability of groundwater potential of India and the extent of their utilization. The Ganga basin has the maximum groundwater potential of about 171 km^3, which is over 38% of the total groundwater potential of India. Though the net groundwater draft is highest in the Ganga basin, the groundwater development is highest in the Indus basin at 77.7%.

The groundwater potential of the different states and Union Territories of India and the extent of utilization in each is presented in Table 1.18. The highest potential of groundwater is in Uttar Pradesh, which lies mostly in the Ganga basin. However, its level of groundwater utilization is only 37.67%. In Punjab the utilization is highest (94%), followed by Haryana (84%), Tamil Nadu (60.44%), and Rajasthan (50.63%). Groundwater utilization is very low in north-eastern states, except

Table 1.16. Percentage Rainfall Infiltration to Groundwater Body in Different Rock Types and Formations in India

Rock type formation	*Percentage rainfall infiltrating to groundwater body*
1. Hard rock formations and Deccan traps	10
2. Consolidated rocks (sandstone)	5–10
3. River alluvia	15–20
4. Indo-Gangetic alluvium	20
5. Coastal alluvia	10–15
6. Western Rajasthan dune sand	2
7. Intermontane valleys	15–20

[Raghava Rao *et al.* (1969).]

Manipur with 33.43%. Groundwater development in Orissa, Jammu and Kashmir and Goa are also low. In the remaining states, the utilization ranges from 15.28% to 41.45%, indicating substantial unexploited potential.

Legislative Measures for Utilization of Groundwater

The utilisable groundwater potential of India is limited and depends on the geo-hydrological conditions of different groundwater regions. The limited resources have to be utilised judiciously for sustained agricultural production and other uses. Indiscriminate exploitation will result in an imbalance in the underground hydrology and depletion of the valuable natural resource. Presently, the groundwater available at a place is considered as the property of the landowner. In many developed countries the ownership of the land does not permit the land owner to utilize the groundwater indiscriminately.

Table 1.17. River Basinwise Groundwater Potential, Availability for Different Uses and Percentage Development in India, 1995

River basin	*Total replenishable groundwater resources* km^3	*Provision for domestic, industrial and other uses* km^3	*Available groundwater resources for irrigation* km^3	*Net draft* km^3	*Balance ground water potential available for exploitation* km^3	*Level of ground water deve-lopement* %
Brahmain with Baitarni	4.05	0.61	3.45	0.29	3.15	8.45
Brahmaputra	26.55	3.982	2.56	0.762	1.80	3.37
Cambay composite	7.19	1.08	6.11	2.45	3.66	40.09
Kaveri	12.30	1.84	10.45	5.78	4.675	5.33
Ganga	170.99	26.03	144.96	48.59	96.37	33.52
Godavari	40.65	9.66	30.99	6.05	24.96	19.53
Indus	26.49	3.05	23.43	18.21	5.22	77.71
Krishna	26.41	5.58	20.83	6.33	14.50	30.39
Kutch and Sarurashtra composite	11.23	1.74	9.49	4.85	4.645	1.14
Madras and South Tamil Nadu	18.22	2.73	15.49	8.9	36.55	57.68
Mahanadi	16.46	2.47	13.99	0.97	13.02	6.95
Meghna	8.52	1.28	7.24	0.29	6.95	3.94
Narmada	10.83	1.65	9.17	1.99	7.182	1.74
Northeast composite	18.84	2.83	16.02	2.75	13.26	17.20
Pennar	4.93	0.74	4.19	1.53	2.66	36.60
Subernarekha	1.82	0.27	1.55	0.15	1.40	9.57
Tapi	8.27	2.34	5.93	1.96	3.97	33.05
Western Ghats	17.69	3.19	14.50	13.32	11.18	22.88
Total	431.43	71.08	360.35	124.36	255.15	31.97

(***Source:*** Central Groundwater Board, 1995).

Table 1.18. Groundwater Potential in the States and Union Territories of India

S. No	*Name of the State/ Union Territory*	*Total replenishable groundwater resource* km^3	*Provision for domestic, industrial and other uses* km^3	*Available ground-water resources for irrigation* km^3	*Net draft* km^3	*Balance ground-water potential available for exploitation* km^3	*Level of ground-water develop-ment* %
1	Andhra Pradesh	35.29	5.29	30.0	7.09	22.91	23.64
2	Arunachal Pradesh	1.44	0.22	1.22	0	1.22	–
3	Assam	24.72	3.71	21.01	0.94	20.07	4.48
4	Bihar	33.52	5.03	28.49	5.47	23.03	19.19
5	Goa	0.22	0.03	0.19	0.02	0.17	8.3
6	Gujarat	20.38	3.06	17.32	7.17	10.15	41.4
7	Haryana	8.53	1.28	7.25	6.08	1.17	83.88
8	Himachal Pradesh	0.37	0.07	0.29	0.05	0.24	18.04
9	Jammu & Kashmir	4.43	0.66	3.76	0.05	3.71	1.33
10	Karnataka	16.19	2.43	13.76	4.3	9.46	31.26
11	Kerala	7.9	1.31	6.59	1.01	5.58	15.28
12	Madhya Pradesh	50.89	7.63	43.26	7.13	36.12	16.49
13	Maharashtra	37.87	12.4	25.47	7.74	17.73	30.39
14	Manipur	3.15	0.47	2.68	Negl.	2.68	Negl.
15	Meghalaya	0.54	0.08	0.46	0.02	0.44	Negl.
16	Mizoram	Not Assessed					
17	Nagaland	0.72	0.11	0.62	Negl.	0.62	Negl.
18	Orissa	20	3	17	1.43	15.57	8.42
19	Punjab	18.66	1.87	16.79	15.76	1.03	93.85
20	Rajasthan	12.71	1.99	10.71	5.42	5.29	50.63
21	Sikkim	Not Assessed					
22	Tamil Nadu	26.39	3.96	22.43	13.56	8.87	60.44
23	Tripura	0.66	0.1	0.56	0.19	0.38	33.43
24	Uttar Pradesh	83.82	12.57	71.25	26.84	44.41	37.67
25	West Bengal	23.09	3.46	19.63	4.75	14.88	24.18
	ALL STATES	431.48	70.74	360.74	115.01	245.73	31.88
Union Territories							
1	Andaman and Nicobar	Not Assessed					
2	Chandigarh	0.03	–	0.03	0.02	0.01	–
3	Dadar and Nagar Heveli	0.04	0.01	0.03	0	0.03	12.81
4	Daman and Diu	0.01	Negl.	0.01	0.01	Negl.	–
5	NCT Delhi	0.29	0.18	0.11	0.12		
6	Lakshadweep	0	–		0	0	63.79
7	Pondicherry	0.03	0	0.02	0.01	0.02	24.34
ALL UTs.		0.41	0.19	0.21	0.17	0.06	76.19
ALL INDIA		431.89	70.93	360.96	115.17	245.79	31.92

(***Source:*** Based on data of Central Groundwater Board, 1995).

In view of the importance of groundwater, the government of India, Ministry of Agriculture, drafted a model bill called groundwater Control and Regulation Bill, 1970 and circulated it to all the states for possible adoption by the State Governments. However, only a few states have enacted any law providing for the rational use of groundwater. Even in these states, the enforcement of the law has not been adequate. The Irrigation Commission (1972) recommended that the State Government should enact legislation to have the legal power to regulate the utilisation of groundwater. The Commission, however, was of the view that water down to a particular depth, say 30 m, in alluvial plains could be exempted from control, as it felt that the upper aquifer in an alluvial formation was not likely to be over-exploited in the foreseeable future.

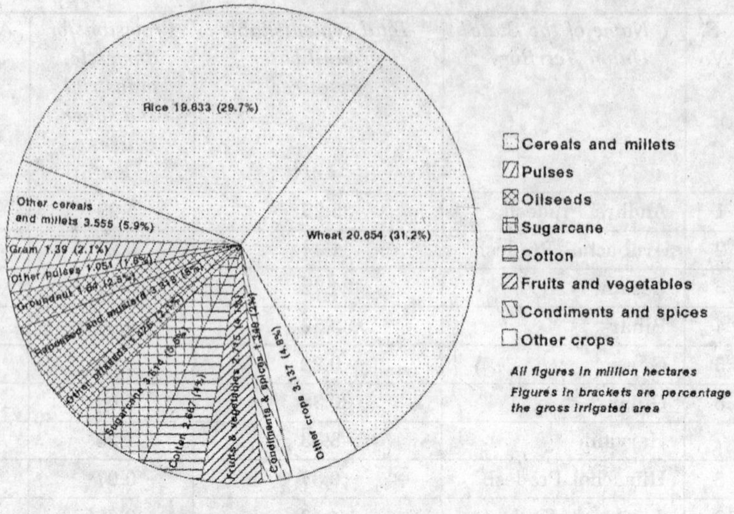

Fig. 1.44. Cropwise gross irrigated area in India, 1992–93.
(***Source:*** (i) Ministry of Agriculture, Directorate of Economics and Statistics, 1994, (ii) Central Water Commission, 1996.)

CROPPING PATTERN AND IRRIGATION MANAGEMENT

The cropping pattern in India, according to land use statistics shows that the total cropped area in 1992–93 was 185 million ha. Foodgrains cultivation formed 70% of the total cropped area, consisting of 57% under cereals and 13% under pulses. Among the cereals, rice and wheat formed 23% and 13%, respectively of the total cropped area. The percentage of cropped area under rice remained almost constant at 23% from the year 1950–51 to 1992–93. For wheat also, it remained constant at 13% for the period 1980–81 to 1992–93, while it rose from 8% to 11% during 1950–51 to 1970–71. There is a declining trend in the area under sorghum *(jowar),* pearl milet *(bajra),* barley and coarse cereals and millets. Declining trend has been observed in case of gram also which has come down to 3% in 1992–93 from 6% in 1950–51. The area under sugarcane increased from 1.3% in 1950-51 to 2.2% in 1992–93. It is observed that there is an appreciable increase in the proportion of area under fruits and vegetables from 1.7% in 1950–51 to 3.7% in 1992–93. The proportion of total area under oilseeds increased from 8% in 1950–51 to 14% in 1992–93. The area under cotton has remained constant at about 4%.

Area sown more than once was 10% of the total cropped area in 1950–51, which has gone up to 23% of the total cropped area in 1992–93. This could be attributed mainly to the spread of irrigation.

Irrigated area under principal crops. The gross irrigated area during 1992–93 was 66 million ha, of which foodgrain crops formed 70%. Among the cereals, wheat and rice were the main irrigated crops (Fig. 1.44). Among the cereals, the irrigated area under rice went up by 100%, while the increase has been six times in case of wheat during the period 1950–51 to 1992–93. Rice was having the largest area under irrigation at 44% of total irrigated area in 1950–51, against 15% for wheat during the same year. In 1992-93 their proportions were nearly equal (about 30%).

Punjab has the distinction of achieving the highest level of irrigation (95%) in the country, followed by Haryana (76%) and Uttar Pradesh (62%). The other states where this percentage is above the national level are Andhra Pradesh, Bihar, Jammu and Kashmir, Manipur and Tamil Nadu. The expansion of irrigation system along with the increased use of fertilizers, high yielding crop varieties, and scientific agronomic practices have increased the production of major crops between 1950–51 and 1996–97 (Table 1.19).

Table 1.19. Production of Major Crops in India in 1950–51 and 1996–97

Crop	*Production*	
	1950-51 (m tons)	*1996-97 (m tons)*
Rice	20.58	81.31
Wheat	6.46	69.27
Coarse cereals	15.38	34.28
Pulses	8.41	14.46
All foodgrains	50.82	199.32
Oilseeds	5.16	24.96
Sugarcane	57.05	277.25

(***Source:*** Indian Water resources Society (1999) (m = million)).

The population of India crossed the 1,000 million mark in A.D. 2000. In India, average foodgrain consumption as in A.D. 2000 was 550 gm per capita per day while the corresponding figures in China and USA are 980 gm and 2,850 gm, respectively. Present annual requirement on the basis of the consumption level of 550 gm for the country is about 200 metric tons which is almost equal to the annual production of foodgrains in India. A moderate rise in consumption level to 750 gm per capita per day is considered to be realistic for assessment of future needs. The annual foodgrain requirement of the country, thus works out to be 450 metric tons by the year 2050, which would subtantially increase the irrigation requirement. During this period there will be a major increase in water demand from other sectors, mainly industry, domestic water supply and environmental protection.

Fig. 1.46. A weir on river Mahi at Wanakbori near Vadodara, Gujarat. The weir is constructed to divert the river flow to the Mahi Right Bank Canal system. The regulator on the left is to release water to the original river course.

CANAL SYSTEMS

Surface water stored in reservoirs or diverted directly from rivers and streams is distributed for irrigation through the canal irrigation network. A canal takes off from the headworks of either a water storage dam (Fig. 1.45) or barrage or a diversion weir (Fig 1.46). Irrigation water flows through the canal system before it reaches the farmers' field to be irrigated (Fig. 1.47). The canal network consists of main canals, branch canals, distributories (Fig. 1.48), water courses, and field channels which are termed according to their capacity and orientation with respect to the headworks. In India, the canals and branches up to minor distributories are generally constructed and operated by the Government.

Headworks. It is the collective term used for all works (weirs or diversion dams) to regulate the flow at the head end to a canal system, including upstreams and downstream river training works.

Fig. 1.45. A view of the Kadana dam in Gujarat on river Mahi. The multi-purpose dam augments the irrigation water supply to the Mahi Right Bank Canal command area by releasing the stored water to the original river course and diverting the flow at the Wanakbori weir (Fig. 1.46) located 65 km downstream. The Mahi Left Bank Canal command area is fed directly from the Kadana reservoir.

Head regulator. Head regulator (Figs. 1.49 and 1.50) is a structure together with hydro-mechanical devices at the head-end of a canal that controls and regulates the water from the reservoir or diversion work to a canal in predetermined quantities. Regulators are also constructed at the off-take of branch and laternal canals.

Sluice off-take regulator. It is a regulator in which the discharge is regulated by means of a vertical sliding gate provided on the upstream side.

Silt selective distributory head. It is a distributory off-take regulator which is designed to adapt the approach section of a regulator without appreciable disturbance to exclude the entry of coarse silt into the distributory.

Closed conduit off-take regulator, box off-take regulator, or barrel off-take regulator. It is a regulator consisting of buried single or multiple barrel pipe or box section with gate control at its upstream end and a suitable transition into tailwater at the outlet (Fig. 1.49).

Cross regulator, check structure, or check. It is a structure built across a canal or ditch to control the water surface level and flow. It maintains a predetermined water depth on outlet or turnout structures, especially when the flow is small. It also enables to divert the flow into one or more directions (Figs. 1.54 and 1.55).

Sluicing outlet. It is an outlet in the body of the dam for sluicing out settled material from the reservior (Fig. 1.49).

Main canal. Main canal takes its supply directly from the reservoir or river diversion work (Figs. 1.51 and 1.52). The size of the main canal depends on the size of the irrigation system. The capacity of main canals in India usually varies from 280 to 425 cu m per second. Direct irrigation is usually not carried out from the main canals. Main canals are not usually

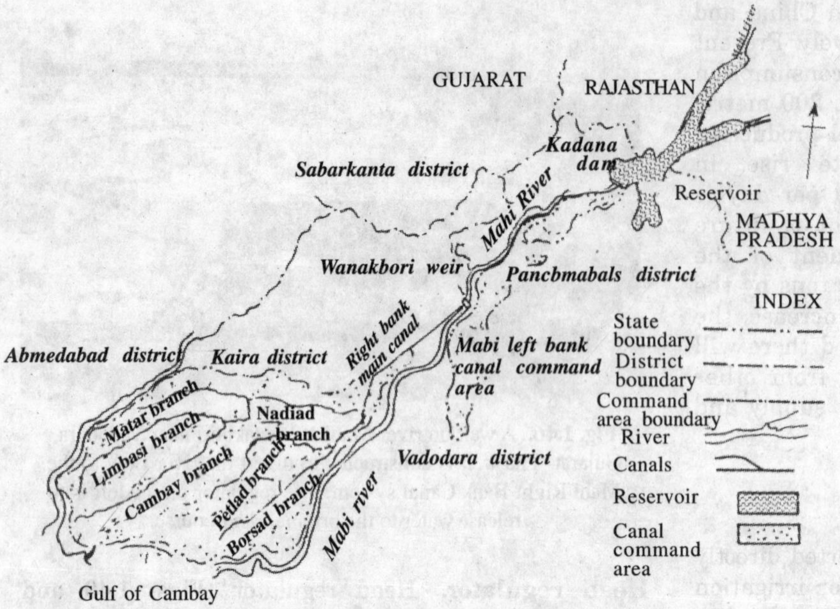

Fig. 1.47. Location map of the Mahi Irrigation Project in Gujarat. The map shows the Kadana reservoir, the Mahi Left Bank Canal command area, the downstream river course, the Wanakbori weir and head regulator, and the Mahi Right Bank canal command area.

lined. However, some of the main canals in the sandy and sandy loam areas are lined (Figs. 1.51 and 1.52).

Branch canals or laterals. Branch canals or 'branches' take off from the main canal and convey the water to different major parts of the irrigated areas (Figs. 1.48 to 1.55). Branch canals generally carry a discharge ranging from 4.0 to 8.5 cu m per second. Direct irrigation is generally not done from large branches. However, smaller branch canals may be provided with outlets for delivery of water to the fields.

Major distributory. Major distributories or 'distributories' take off from branch canals and sometimes from main canals and supply water to minor distributories or outlets. Major distributories generally carry discharges between 0.75 to 5.5 cu m per second. Distributories are usually unlined (Fig. 1.56), though they are sometimes lined to prevent excessive seepage losses (Fig. 1.57).

Minor distributory. Minor distributories or 'minors' are smaller channels taking their supply from major distributories and supplying water to outlets. The carrying capacity of minors is usually less than 750 litres per second. There may also be sub-minors branching off from minors (Figs. 1.56 and 1.57).

Canal outlet. Outlets are provided in irrigation canal system at suitable points. The water course originates from the canal outlet. The size of an outlet depends upon the irrigated area it commands. The common types of outlets used in canals in India are (*i*) adjustable proportionate module outlet (Fig. 1.58), (*ii*) open flume outlet (Fig. 1.59), (*iii*) pipe outlet (Figs. 1.60 and 1.61) and (*iv*) siphon outlet (Fig.1.62). The size of opening of the adjustable proportionate module (APM) outlet can be fixed to obtain the desired discharge from an outlet, depending on the culturable command areas. The discharge of the outlet varies with the level of water in the channel. The outlet capacity commonly fixed is at the full supply level (FSL) of the channel. In most of the older irrigation works in northern India, no device is provided to regulate and control the water flow. The outlet automatically starts flowing with the flow of the distributory and the rate of flow would depend upon the size of the opening and the head of water above the outlet. In some of the outlets in the

Fig. 1.48. Water distribution network of the Mahi Right Bank Canal command area in Gujarat. The map shows the layout of the main canal, branch canals and distributories.

Fig. 1.49. Head of the main canal originating from the closed conduit offtake regulator of the Neyyar dam near Trivandrum, Kerala. Note the shooting water jet emerging from the regulator.

Fig. 1.50. Head regulator of the Mahi right bank main canal at Wanakbori adjoining the pick-up weir.

Fig. 1.51. A view of the brick lined Upper Ganga Canal near Roorkee in Uttar Pradesh.

southern and eastern parts of India some regulating devices are provided, such as an iron gate with a few notches to set the gate to obtain a desired discharge rate.

Fig. 1.52. A section of the Mahi Right Bank main canal (low flow stage). The canal is lined with bricks laid in sandwitched cement mortar (1973).

Chak. The term is used in north-west India and Pakistan to define the irrigable area, normally under the control of the cultivators, and consisting of water courses, field channels (where they exist), fields and field drains. The *chak* is sometimes known as the unit command area. It is synonymous to the command area of a canal outlet.

Water course. A 'water course', as defined in the Northern India Canal and Drainage Act, means "any channel which is supplied with water from a canal but which is not maintained at the cost of the Government and all subsidiary works belonging to any such channels". The water courses pass through common land and are maintained by the farmers (Fig. 1.63).

Fig. 1.53. A branch canal taking off from a main canal in the Hisar Bhakra canal system in Haryana. Note the regulator to control the flow in the branch canal.

Fig. 1.54. A twin-gate cross regulator in a brick lined canal distributory in the Mahi Right Bank canal system in Gujarat.

Fig. 1.55. A multiple gate cross-regulator in an unlined branch canal.

Field channel. Field channels carry water to individual fields from the water course. In irrigated rice areas in many countries, field channels may be absent when the practice is to irrigate from field to field. The overflow from the upper field irrigates the adjoining lower field and so on. However, the practice results in loss of water and fertilizers.

Canal alignment. The alignment of a canal, including branches and minors is the location of its centre line in plan and fixing it on ground. There are two types of alignments, namely contour and ridge.

(i) *Contour alignment*. In this the canal follows on falling (or grade) contour. The main canal taking off from a reservoir often follows the contour alignment at its head reaches because the surroundings of the reservoir are in rough rolling tract. A contour canal usually has many cross drainage works. A contour canal can irrigate only one side, the other being higher.

(ii) *Ridge alignment*. The canal is laid along the highest or predominant ridge line in the command area. This is

Fig. 1.56. An unlined canal minor distributory during the close down period of canal supply in the West Yamuna Canal Circle near Dadra, Haryana (1973).

the most common and efficient alignment and provides the largest area under its command. At some locations, to avoid excessive cutting, the alignment is shifted to suit the prevailing land slope.

***Warabandi* (rotational water supply to fields by turn).** *Warabandi* is the delivery of water by turn to the farmers of the outlet command area. It is a common method of water delivery to field in northern India and Pakistan. In the *Warabandi* system the discharge of an outlet is usually fixed at 30 litres/sec (earlier 1 cusec), using APM type outlets. In some situations the outlets are set for higher discharges, ranging from 30 to 85 litres/sec. Discharge rate higher than 85 litres/sec is beyond the control of the farmers adopting *Warabandi* system.

***Shejpali* (Water allocation based on water requirement of crops).** In the *Shejpali* system also the discharge of an outlet is usually fixed at 30 ha/sec.

Intensity of Irrigation. *Intensity of irrigation* is the actual area irrigated in a year from an outlet. The actual irrigated area of an outlet usually falls short of the culturable command area in most of the canals of northern India. For instance, the sanctioned intensity of irrigation for canals under the Bhakra Canal Circle was 62 per cent for two crops a year, i.e., 27.6 per cent for *kharif* and 34.4 per cent for *rabi* crop in 1973.

In old irrigation projects, the general concept of providing irrigation water was for protective irrigation against failure of monsoon to prevent famines. Projects were designed only for about 50% irrigation intensity (including monsoon and winter crops). With the introduction of new crop varieties, it is now possible to raise economically two or more crops in a year from the same piece of land. Irrigation projects can have much higher irrigation intensities, i.e., about 200% or more, depending upon the availability of water.

Fig. 1.57. A canal distributory lined with single layer bricks laid in cement mortar near Bhiwani (Haryana).

Fig. 1.58. Adjustable proportionate module (APM) type canal outlet on a canal distributory. On the left is a close view of the outlet when the canal is dry. On the right is the view of the outlet at full flow stage.

Delta. It is the total depth of irrigation to a crop in centimetres. It can be calculated by dividing the volume of irrigation water by the area irrigated.

Base period. This is the period in days for which water is supplied to a crop. Therefore, it depends upon the duration of the crop season.

Duty of water. It is the ratio between the irrigated area and the quantity of water used. It is expressed in litres per second per hectare and indicates the continuous flow requirement per hectare of cropped area.

Water allowance. The quantity (rate of flow) of irrigation water allocated per unit area of the culturable command area of an outlet distributory. Its value is often a compromise between demand and supply of water. The carrying capacity of distributories and water courses is designed on the basis of this allowance. Whenever a distributory operates, it is expected to run full supply. The outlets to water courses are so constructed that all of them draw their authorised share of water concurrently. No distributory operates for all the days during a crop season. The ratio which the operating period of a distributary bears to the total period of the crop is called its capacity factor. It is separately designed for each of the sub-periods of a crop. This factor helps in the operation of the distributories while drawing water from the reservoir and the main canal. In case of Bakra Canal system in North India, the designed mean capacity factors for *kharif* and *rabi* are 0.8 and 0.72, respectively.

Fig. 1.59. Tail end of a 'minor' distributory in the West Yamuna Canal Circle near Dadra, Haryana. Note the inlets of three water courses originating from the end. The outlets are of 'open flume' type.

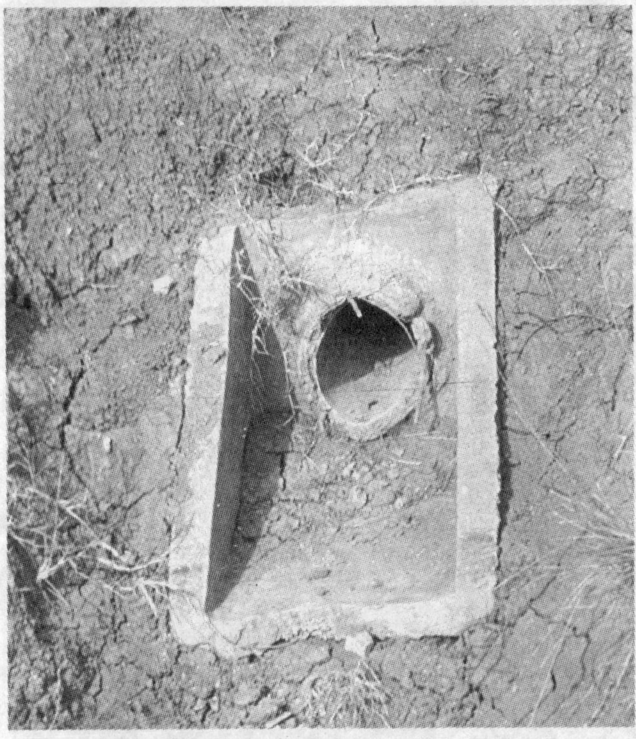

Fig. 1.60. The inlet end of an uncontrolled pipe outlet installed in a canal distributory to supply water to a water course.

IRRIGATION SYSTEM MANAGEMENT

Water resource projects cover a wide range of activities. Water is a scarce resource, especially in arid and semi-arid countries and is essential for all forms of life. It is essential that the efficiency in the use of water must be maximized. Maximization of water use efficiency requires careful planning and good project management. The paramount objective in the efficient management of an irrigation system is to ensure that water is distributed in adequate quantities and at proper time throughout the command area to meet the requirements of the crops grown. Proper regulation of the canals and the distribution of the required quantity of water are essential to obtain efficient use of irrigation water.

Fig. 1.61. A pipe outlet of a canal distributory near Cuttack, Orissa (1973). The discharge through the outlet is controlled by setting the gate at the inlet of the pipe opening (1973).

Fig. 1.62. A siphon outlet from the Bhakra canal near Nangal in Punjab (1972).

Fig. 1.63. Freshly removed soil which had accumulated in a canal water course.

The water from the reservoir/barrage/weir is released to the main canal through the head regulator. The main canal feeds two or more branches, which operate by rotation and may or may not run full, depending on the discharge available. Main canals run continuously throughout the irrigation season. Branch canals supply water to a large number of distributories which must run at full supply level by rotation. Distributories supply water to water courses from which it is let into the crop fields.

Regulation of canals. The process of distribution of irrigation water in a canal system is called *'regulation'* or *'rostering'*. 'Regulation' is specially required in canal systems where: (*i*) the demand is only on a part of the system and water is to be conveyed to that part only, or (*ii*) the demand is on the entire system but the available supply is not enough to meet the full demand. For instance, irrigation canals in the Bhakra Canal Circle run in rotation for a period of nine months in a year, while in the remaining three monsoonic months they run to their full discharge capacity. In the Bhakra Canal System, the channels in one group run for eight days and remain closed for 18 days when the water supply is directed to other groups of channels. The supply of the main canal is distributed to different 'branches' or 'distributories' in accordance with the demand on different channels. The main aim of rotation of water distribution is to regulate and evenly distribute the water over the command area of the canal system.

Canal system operation: Discharge vs. volume control. Operation of irrigation systems is aimed to control discharge, volume or a combination of discharge and volume. Discharge control is most common. Discharge control may be direct or, more commonly, indirect through control of water level. Some of the irrigation systems are designed and operated to control the volume of water in canal reaches. This technique requires the availability of storage, either on line storage capacity in the canal itself or in intermediate reservoirs or tanks to which the excess flow is released.

Dynamic regulation is an example of volume control methodology applicable to irrigation canals without intermediate storage reservoirs. For systems with intermediate storage within the canal, the control may be composite, i.e., discharge control for the canal reaches and volume control for intermediate reservoirs with closed loop feedback linked to the supply.

Types of controls of water releases. Most gravity irrigation systems are based on upstream control of water. All control structures are set according to the discharge to be released from the main intake structure. The objective is to maintain the water level upstream of each cross-regulator to control the backwater profile in the upstream reach. The backwater profile determines the head at off-takes in the reach.

Water control structures. Canal structures like gates/outlets combine the adjustability of the structures and the level of automation in operations. Some systems are fully adjustable, i.e., every structure has movable parts that can be set according to demands. Other systems have no provision for any manipulation, such as the fixed proportional distribution devices.

Manually operated gated structures. Manually operated systems have control regulators which are adjustable. Each structure has to be manipulated by irrigation staff when a change in the flow regime is scheduled or occurs. The difficulty in operating the systems is due to the numerous structures to be adjusted simultaneously when the flow regime is changing.

Fixed systems. These systems, known as 'structured systems' have been largely developed in India, Pakistan, and Nepal. Water delivery is organized around pulses of constant discharge with a varied frequency. Distribution is proportional and the structures are fixed permanently at the construction stage. The non-adjustable structures are limited to minor canals.

Automatic/semi-automatic systems. Hydraulically automated systems are equipped with control of the structures that control water levels in canals over a wide range of discharges. Usually, the control of water level is achieved by mechanical movements of regulator gates, driven by hydraulic forces without an external source of energy or human intervention. Variations of water level may still occur at locations remote from the control regulator.

Water distribution pattern at field level in canal irrigation systems. Different methods of water distribution are followed in canal irrigation. The following are the essential requirements of a proper water distribution system: (*i*) Ensuring the supply of designed flows in the minor distribution channel at each outlet; (*ii*) Developing proper distribution network of field channels up to each land holding in the command area of the outlet; (*iii*) Working out irrigation schedules, allocating a time and day for the flow of water to each individual farmer.

Different water distribution practices are being practised in India and other countries to ensure equity and to meet the crop demands.

The *warabandi* or *osrabandi* is a system of delivery of water in rotation amongst the cultivators sharing water from a canal outlet.

Wara/Osra means turn and *'bandi'* means fixation, i.e., *warabandi* means fixation of turns. In *warabandi* method, the available water is allocated to cultivators in proportion to their landholdings. *Warabandi* system has successfully held its sway over the canal systems of the Indo-Gangetic plains and in Pakistan during the past several decades. It has its main thrust on equitability of distribution of water. The basic principle underlying the *warabandi* system of managing irrigation water is that the available water, whatever be its quantum, is allocated to cultivators in equal proportion to their landholdings and not only to some of them to meet their total demand. This principle imposes water scarcity conditions, which necessarily lead to the adoption of more efficient water management practices by the farmer. Socially, it stands for equitable distribution of water to the largest number of farmers in the command area of an outlet/distubutory. In *warabandi* system the distribution of water is carried out by keeping the stream of water as constant and regulating the time of its flow. The entire time in a week is allocated to farmers. Time once lost cannot be regained. So, if for any reason, a farmer is unable to receive his share of water there is no way of compensating him. Similarly, if the flow of water is less than the authorised quantity due to some technical defects, the loss suffered on this account cannot be compensated. It, therefore, becomes necessary that distributories and water courses must flow as programmed. *Warabandi* is considered to be a powerful tool for promoting better water utilisation.

Roster of turns. In the rotation water supply (*warabandi*) system, a predetermined quantity of water is provided to each irrigator once a week. The duration of time the water supply is allowed per unit area of the irrigated land under the command of the outlet is determined by dividing the number of minutes in a week (10,080) by the area of land to be irrigated in hectares. Thus, in an outlet command area of 40 ha, the time allowed per ha to a farmer is 10,080 is 252 minutes or 4 hrs and 12 minutes in a week. A 2 ha farmer will get 8 hrs. and 24 minutes. The cycle of turns on a water course or its branch starts from its head, proceeds downwards and ends at its tail. The supply has to be cut-off from head when the last farmer is having his turn. The length of upper portion of water course which has been filled with common pool time (*bharai*) can be discharged only into the tail end fields and normally the total time spent on its filling is to be recovered from the tail end farmer. But the tail end farmer does not receive all this water at a constant rate. When the supply is cut-off from the head end, it starts depleting at tail and gradually approaches zero. Such a supply, beyond a limit is not efficient from the point of view of field application. The tail end former is to be compensated for it and this is done by allowing him a certain discount on the recovery of the filling time (*bharai*) time.

Thus, the lower reaches of a water course gains from the recession. However, in actual practice there is usually a discrimination in water supply in a water course. It is usually the tail end farmer who is usually deprived of water.

The seapage loss in a watercourse is not reflected in the *warabandi* system. The utilisation of irrigation supplies is left entirely to the cultivator, who is under no obligation to grow any particular crop in the area. In Andhra Pradesh, a 7-day rotation is worked out by which the farmer gets a weekly amount of water based on a maximum consumptive use calculation for an 'average' cropping pattern, a 65% overall irrigation efficiency, the irrigable area of his farm and the size of the rate of flow at the outlet (head end of the water course).

Shejpali and other systems of water distribution. The *Shejpali* and block system of water distribution in the outlet command area of canal systems is practised in Maharashtra, parts of Gujarat and Karnataka. Under this system, estimates of expected water availability are made. Applications are invited from the farmers seeking information on the crop to be grown and the area to be irrigated under each crop. Water is then sanctioned taking into account the total demand and the water availability. A schedule called *shejpali,* giving turns to different irrigators for the sanctioned crop area on outlets is prepared for each rotation of the canal system. In the block system, a long-term agreement for the supply of water for 6 to 12 years is done, especially in case of perennial crops. A system called *rigid shejpali*, in which definite dates and durations for the supply of water to a particular field area are recorded on the passes issued to the farmers for the sanctioned area has been introduced during the second half of the twentieth century. In the *shejpali* system, the farmers are supplied water, in turn, from tail to head, as per the schedule prepared and communicated in advance. The *shejpali* system gives consideration to the crops to be grown, contrary to the *warabandi* system. The canal outlets are equipped with gates to facilitate the operations.

In many of the irrigation projects of southern and north-eastern states, West Bengal, Orissa, Bihar, Jammu and Kashmir as well as many south and south-east Asian countries where paddy is the main crop, the irrigation flow below the canal outlet is allowed from one field to another through surface flooding. This system of water distribution is referred to as the *'localised system'* of irrigation for paddy areas. However, the practice usually results in insufficient water distribution and fertilizer use efficiency.

A system of *'zonal irrigation'* has been introduced in the Lower Bhavani Project in Tamil Nadu. In this system, the command area (*ayacut*) is divided into two-halves. Water is made available continuously in one-half of the area for one season for a period of 4 months in a year. The other half gets irrigation water sufficient for wetland crops the next year. This way, each half gets irrigation supplies for wet and upland irrigated crops (irrigated dry crops, as is sometimes known) on alternate years.

It may be seen from the preceding discussion that the water delivery schedules presently followed require substantial improvements. They are not adequate to meet the specific requirements of the crops. Except for the *warabandi* system, they do not take into account the filling time and the gain resulting from recession flow in water courses. None of the methods give consideration to conveyance losses in the water distribution system.

Cropping Systems and Water Management

The primary objective of irrigation is to increase agricultural production on a sustained basis. Plants vary greatly in their requirements for water and adaptability to environment. Irrigation systems are planned to meet the water requirement of crops under a specified cropping pattern. Alternatively, cropping patterns and cropping systems can be modified to suit, to a large extent, the availability of water in an area. Crop water requirements can be met from rainfall, from the soil moisture in the crop root zone, through capillary rise from shallow groundwater, by pumping groundwater from the area and through supplies from a canal system.

A major opportunity of increasing crop yields and water use efficiency in a particular environment is to select plant species adapted to the available amount and distribution of water. The plant characteristics influencing water requirements include the length of growing season, extensiveness of the root system, leaf area and the number, distribution and size of stomata. Species selection, crops selection and plant breeding have contributed greatly to increasing the efficiency of the water available in a given area. Plant breeders have contributed greatly by increasing the yield potential of crops and by introducing disease resistance characteristics. With the aid of suitable predictions on moisture availability from rainfall, groundwater and surface water sources, it is possible to arrive at suitable cropping patterns to make the maximum use of the available water.

Cropping pattern in a region is determined by a multiplicity of factors like types of soil, climate, rainfall and the farmers' choice. Assured water supply will motivate the farmer to change over to improved cropping patterns. However, there is a need to restrict crops with high water consumption, like rice and sugarcane, when the available water supply is low.

Night Irrigation in Canal Systems

Most of the canal irrigation systems in South and Southeast Asia and many other countries continue to flow throughout the day, and much of the night flows are inefficiently used or are wasted, unless effective procedures are adopted for their effective utilization. The importance of efficient utilization of night supplies of canal system has not received the right attention in many of the systems.

Darkness, cold, fear, departure from normal working hours, and desire for sleep deter irrigation staff, farmers and labourers from activities at night. Night irrigation usually requires extra labour and costs. It requires smaller stream flows and well graded fields. Paddy, tree crops and other widely spaced crops are more adaptable to night irrigation, as compared to other crops. Inefficient water application, breaches in channels, and wasted water flowing into drains are common during the night. Reuse of night drainage water lower down is sometimes possible. Substantial loss of water in the night has been reported by many workers. About the Chambal Canal project in India, Bottrail (1981) has observed that the farmers' unwillingness to practice night irrigation has contributed to high levels of water wastage. Similar observations have been made regarding Morna project in Maharashtra and Parambikulam Aliyar project in Tamil Nadu.

Loss of water during the night under the *warabandi* system of water distribution is negligible. *Warabandi* system does not distinguish between day and night. Farmers have accepted the system without complaints. Water is taken at night according to schedule and conflict is low. However, the major features favouring the adoption of the *warabandi* system are nearly level land, consolidated rectangular land holdings and water scarcity. But elsewhere, especially under undulating topography *warabandi* is rare and utilization of night water supply is usually inefficient.

Possible measures of combating the problems of managing canal flows during the night are (*i*) reducing irrigation at night—by regulating sluice releases or diversion of flows, providing intermediate storage in canal command areas, diversion of flows for travelling time, redistributing daytime water, and passing water to drains and escapes, and (*ii*) improving irrigation at night—by manageable streamflows, convenient field shaping and water application methods, choice of crops and distribution of flows in distinct zones.

Robert Chambers (1988) has made the flowing observations on the relative advantages of various resources to measure the efficiency of night irrigation: "It is difficult to estimate how much water is currently saved at night through intermediate storage in tanks or canals, diversion to 'travelling', or closure of headworks, but it is probably quite a small proportion. A reasonable estimate may be that 40% of the canal irrigation water on medium and major systems is either applied in night irrigation or sent into drains at night ... Night irrigation is often inefficient. Supervision is difficult and minimal ... Night flows are often diverted to crops which tolerate flooding, mainly paddy, or are allowed to flow into drains."

Topography and size of holding. The smaller and more sloping the field is, the more difficult night irrigation becomes. Some large flat fields can be left to flood all night. An extreme case is the Gezire Scheme in Sudan, where impervious soils and flat land permit water to be left to spread unattended. Likewise, some large-scale irrigation schemes in western countries with very large fields could be left unattended at night, and changes made during the day. At the other extreme are the sloping fields with small basins which are difficult to irrigate during the night.

Soil type. The ease, difficulty and efficiency of irrigation at night also depend on the type of soil. Night irrigation is easier on soils which are not sticky, and which make stable bunds. Night irrigation is often difficult in black cotton soils. Hence, the *warabandi,* as followed in northwest India cannot be adopted as such in black soils.

Size of irrigation stream. Handling stream flows is harder at night. The optimal night flow should normally be lower than the day supply, especially on difficult terrain. This requirement is often overlooked in irrigation planning and design.

Crop. The ease and efficiency of water application at night depend on the crop and its stage of growth, as well as the topography and the size of fields. The easiest crops are paddy and trees: paddy can be flooded. Over flows may not raise serious problems. Similarly, trees are tolerant to standing water for some hours. Crops which are widely spaced or in early stages of growth are easier to irrigate at night than those which are tall, dense, or in their later stages of growth.

Utilization of Waste Flow during the Night

When night irrigation is not possible the excess flow can be passed to drains and escapes, instead of canal outlets and then to crop fields. Uncontrolled large flows may cause soil erosion, especially at the tail end of canal minors. Drainage of excess flows at non-erosive velocities will be beneficial, especially when the aim is to avoid waterlogging and erosion.

Whether the water is wasted or not will depend on the possible utilization of drainage water which is needed lower down. In regions where there is a cascade or chain of tanks, wastage through night flows in the upper tanks may fill the tanks lower down. Such situations are possible in the plateau region of South India and the Dry Zone of Sri Lanka.

Soil erosion, especially in steep minors and field channels, can occur through large flows at night when extractions upstream cease. Detection and remedial action are difficult in the night. Flooding can also result from uncontrolled flows of water at night. Reducing irrigation at night has many potential advantages, including saving of water, convenience to all concerned, and limiting the damage (flooding, waterlogging, erosion) which night flows can cause.

Following are some of the ways in which irrigation at night can be reduced/improved:

(*i*) *Regulating sluice releases or river diversion flows*. Regulating the sluice gates/diversions on small schemes with command area less than 200 ha where the sluices are desirable. Where canals are long and large, sluice or off-take regulation to reduce night deliveries becomes infeasible.

(*ii*) *Intermediate night storage reservoirs/offtaking tanks*. Night storage in intermediate reservoirs, diverting part of night flows to tanks is a useful concept. It can reduce night irrigation, save water and introduce flexibility in canal system management. However, design of intermediate storages within the canal network needs careful consideration in relation to silting of canals due to the storage reservoir. Night storage of water can be at the field level as well, if the field conditions and crops permit. It is most feasible with paddy. It is a common practice in Sri Lanka to hold as much water as possible in the upper paddy basin and then release it to the lower fields in the following days.

(*iii*) *Zoning for night flows*. Where a farm or an outlet receives water throughout the day and night, the day flows can be used for the more difficult soils and crops, and the night flows for those which are easier. In South India, on the Sri Ramasagar Project, it is a common practice for an outlet's day flows to be used in a more controlled manner for upland irrigated crops in red soils near the outlet, and the night flows to go in a less controlled manner into low-lying paddy on black soils farther from the outlet.

(*iv*) *On-farm engineering measures*. Night irrigation can be substantially improved by providing appropriate structures in the water distribution network and adopting suitable water application methods.

(*v*) *Choice of crops*. Night irrigation can be improved by selecting crops which are tolerant to flooding and easy to irrigate. Paddy, trees and sugarcane at its early growth stages belong to this category.

Conjunctive Use of Surface Water and Groundwater

The basic source of water for both the surface and groundwater is the same, namely precipitation. Conjunctive use can be defined as management of all water resources in an area, including surface and groundwater in a harmonious manner such that the total water use in the area over a period of time is optimised. Integrated and coordinated development of surface and groundwater, commonly known as conjunctive use, is now widely recognised as the most suitable strategy for irrigation development, especially in alluvial plains. Normally under such use, the combined availability of water is more than the availability from separate components of surface and groundwater in a region. Conjunctive use mitigates the problems of high water table and salinity, resulting from canal irrigation and facilitates the use of high salinity groundwater by dilution with canal water. Several of the existing canal irrigation systems suffer from inadequate supplies. This calls for combined or conjunctive use of surface and groundwaters wherever possible. The conjuctive use of surface and groundwaters can take the form of full utilization of surface water supplies, supplemented by groundwater, or the direct use of groundwater during periods of low canal supplies or canal closures. It can also take the form of irrigating a part of the canal command area exclusively with groundwater. In the Western Yamuna Canal System in Haryana, tubewells have been constructed along the canal banks to augment canal supply and lower the groundwater table (Fig. 1.64).

In countries like Israel and USA, saline groundwater has been used for irrigation to a limited extent after diluting it with good quality surface water. There is scope for adopting, such practices in certain areas of Gujarat and Rajasthan where the groundwater is brackish. An outstanding example of obtaining substantial increase in rice production is the practice of raising rice nurseries, preferably through farmers' cooperation, utilizing groundwater and starting the transplantation of seedlings soon

Fig. 1.64. A tubewell discharging water to add to the flow of West Yamuna Canal near Panipat, Haryana. The conjunctive use of surface and groundwater has been planned to overcome the problems of water deficiency in the canal system and waterlogging of the area adjoining the canal.

after the onset of monsoon or when the canal water is released. This will gain 3 to 4 weeks of the growing period for the crops, resulting in higher crop yields and enabling timely planting of the succeeding (*rabi*) crops.

Conjunctive use of surface and groundwater will not only increase the irrigation potential but also mitigate the problems of waterlogging and land degradation. The technologies of irrigation from the two sources have to be integrated in a complementary manner to achieve sustainable optimum agricultural production. Such an integration can be brought about in one or more of the following ways:

(*i*) *Conjunctive use in space.* Some parts of the command area may be irrigated exclusively by surface water and the other parts by groundwater.

(*ii*) *Conjunctive use in time.* Parts of the command area may be irrigated by surface water at one time of the growing period or in one crop season, and by groundwater at another part of the growing period or in another crop season.

(*iii*) *Conjunctive use by augmentation.* Supplies from one source are augmented by those from the other source, e.g., augmentation tubewells (Fig. 1.64).

Efficient use of surface and groundwater requires careful investigations to quantify regionwise, surface and groundwater resources and the interaction of one source with the other. Though the conjunctive use of surface and groundwater is being advocated and is accepted as the most suitable means of irrigation development, not much headway has been achieved in its planning and implementation. All existing surface irrigation projects need to be operated in a manner to induce and encourage conjunctive use of available groundwater resources in the command area. Similarly, all future irrigation projects need to be planned for conjunctive use of the two resources.

Jurisdiction of Surface and Groundwater

Jurisdiction on the use of water varies from country to country. In many countries both surface water and groundwater are by law the properties of the state, to be used for the welfare and development of the nation. Under the Indian statutory law, as interpreted and applied by the courts, all surface water is public property, whether it is natural flow of a stream or river or storage behind a dam or natural lake. The responsibilities apportioned to the states and the centre (Union) by the Constitution fall in three categories, as listed in Schedule VII: The Union list (List I), the State List (List II) and Concurrent List (List III). Regulation and development of interstate rivers and river valleys fall under the Union Government to the extent that it is declared by the Parliament by Law in the public interest. Water supplies, irrigation and canals, drainage and embankments, water storage and water power are essentially within the power of state. The states have the pre-eminent position in relation to water resources development and utilization in India. Thus, not withstanding the power granted to the Union Government in Entry 56, water is perceived as a 'State Subject' as per Entry 17 of the Constitution of India. The legislation is substantially framed within the content of the state boundaries.

As river basins often extend over many states, disputes arising in inter-state river basin utilization are decided through interstate agreements or tribunal awards. Under the Constituion of India, Article 262, deals with the adjudication of disputes relating to waters of inter-state rivers or river valleys and Entry 17 of List II and Entry 56 of List I of Seventh Schedule deals with items that fall within the purview of the state legislation and the parliament, respectively. Article 262 states as follows:

"Parliament may by law provide for the adjudication of any dispute or complaint with respect to the use, distribution or control of the water of, or in, any inter-state river or river valley."

In case of groundwater, the legal and absolute rights rest with the owner of the overlying land. In many cases, this system has led to de facto rights to the land owner when he can construct a deep tubewell which will drain water from a much larger area of influence of groundwater than the land area belonging to the owner. This may deprive his neighbours from their fair share of water. Over-exploitation of groundwater leads to adverse effects.

The above situation on the rights of surface and groundwater brings to focus, in addition to development and environmental issues, the need for appropriate legislation to control water and provide for people's participation for efficient irrigation system management.

National Water Policy

India adopted a National Water Policy in 1987 for the planning and development of water resources to be governed by national perspectives. It emphasizes the need for river basin planning. Amongst the priorities in the allocation of water for various needs, drinking water is given the highest priority, followed by irrigation, hydro-power, navigation, and industrial and other uses. As water resources development is a State responsibility, all the states are required to develop their state water policies within the framework of the National Water Policy. Since the adoption of the policy in 1987, a number of new issues and challenges emerged in the development and management of water resources. Therefore, in 2002 the National Water Policy (1987) was reviewed and updated. The revised National Water Policy (2002) has the following priorities in water allocation: (1) Drinking water, (2) Irrigation, (3) Hydro-power, (4) Ecology, (5) Industries, (6) Navigation and other uses. However, it has been stipulated that the priorities could be modified or added if required by the area/region specific considerations (Ministry of Water Resources, 2002). The revised policy has stressed that in the planning, implementation and operation of a water resource project, the preservation of the quality of environment and ecological balance should be primary considerations. An integrated and multi-disciplinary approach to the planning, formulation, clearance and implementation of projects, including catchment area treatment, command area development and the rehabilitation of the people adversely affected by the project are envisaged. The drainage system should form the integral part of an irrigation project right from the planning stage. The involvement and participation of beneficiaries and other stakeholders should be encouraged from the project planning stage. Integrated coordinated development of surface and groundwater resources and their conjunctive use should be envisaged right from the project planning stage. There should be periodical assessment of the groundwater potential and its quality. Exploitation of groundwater should be regulated to prevent its use to exceed the recharging capabilities of aquifers and ensure social equity. There should be a close integration of water use and land use policies. Municipal and industrial effluents should be treated to acceptable levels and standards before discharging them into natural streams. Minimum flow should be ensured in the perennial streams for maintaining ecology and social considerations. Drought-prone areas should be made less vulnerable to drought-affected problems through soil moisture conservation measures and water harvesting procedures. The water sharing and distribution amongst states should be guided by a national perspective with due regard to water resources availability and needs within the river basin/sub-basin. Water resources development will have to be planned for a hydrological unit, multi-sectorally. Watershed management through extensive soil and water conservation in the catchment area, preservation of forests and increasing the area under the forests are to be given priority. Construction of check dams should be promoted wherever feasible.

Energy Requirement in Irrigation

Careful planning of energy sources and their efficient utilization are vital in meeting the water needs of agriculture. Agricultural production in developing countries has been seriously hampered by the non-availability of adequate power to lift water for irrigation from groundwater and surface water sources. In many countries there is lack of coordination between the agencies responsible for the development of water resources and those vested with energy development and allocation. Energy requirements in irrigation pumping may be broadly grouped into the following two categories: (1) Pumping from wells, and (2) River and canal pumping.

(1) ***Pumping from wells.*** Irrigation wells are classified into (*i*) Open wells, (*ii*) Shallow tubewells, and (*iii*) Deep tubewells. Water is lifted from open wells by manually operated and animal operated water lifts and pumps driven by diesel engines and electric motors. Water lifting from shallow tubewells and deep tubewells is almost exclusively done by pumps operated by diesel engines and electric motors.

(2) ***River and canal pumping.*** Pumping from canals and river-pumping for lift irrigation are almost exclusively done by diesel engine or electric motor driven pump-sets. The development of energy sources, especially electrical energy should match with the development of lift irrigation using groundwater and/or surface water sources. Whenever electric power is available, it is the most convenient and efficient source of energy in irrigation pumping. The reliability and efficiency of electrically operated pumping-sets are high, provided they are properly selected, installed and maintained.

PARTICIPATORY IRRIGATION MANAGEMENT

The traditional systems of irrigation the world over have had a built-in system of distribution of water among the beneficiaries, thereby linking utilisation with the creation of irrigation potential. In India, the management of tanks and small scale river diversion schemes have involved the irrigators in the maintenance, distribution as well as in modalilites of conflict resolution. There was an unwritten convention among the farmers under a tank to use water on a weekly rotation. One among the beneficiaries acted as the coordinator and leader for water distribution within the unit. Similarly, supply of water among fields on a particular day was decided by mutual agreement. The system on the whole, worked with minimal government intervention.

Against this background, major canal irrigation project planners of the 20th century did not concern themselves with developing a proper system of water sharing and infrastructure maintenance at the farm level. The responsibility of the irrigation department is deemed to end at the point where water leaves the outlet. The planners and executors of irrigation projects have tended to focus largely on engineering aspects with little attention to socio-economic dimensions and modern methods of water management. There has been a general lack of constructive and functional relationship between irrigation officials and the farmers. Thus, the farmer who had the most important stake in the proper management of irrigation water as the ultimate beneficiary, had little role to play in the management of irrigation water under the prevailing system. This realisation lead the government to emphasise the need for the participation of farmers in the management of irrigation water during the Sixth Plan and its reiteration during the Seventh Plan. In 1987, the Ministry of Water Resources, Govt. of India, issued guidelines to state governments to involve farmers' organizations to take over maintenance, management of water, allocation and collection of water charges and to develop participatory mode of management on distinct segments of the irrigation system (Joshi, 1997). Participatory irrigation management refers to the involvement of irrigation users in part or all the aspects of irrigation management and at part or all levels. This implies initial planning and design, construction, supervision, financing, decisions, rules, operation, maintenance, monitoring and evaluation of the system. 'All Levels' refers to the full physical limits of the irrigation project, comprising of harnessing water at the headworks, conveyance, distribution, and application systems (Singh, S.R., 1998).

Even though the intent to create farmers' organizations has been expressed during the Sixth and Seventh Plans, the visible efforts to involve farmers in the form of Water Users' Associations (WUAs) in different states have been made in late 1980s and early 1990s to manage irrigation water from outlets, minors, and distributories of the canal system. Farmers have also been involved in the management of irrigation water from tanks, and deep tubewells. These organizations in different states are registered under State Cooperative Acts or Irrigation Utilisation and Command Area Development Acts and their, variants. The water users' associations under the participatory irrigation management system are charged with the responsibilities (*i*) to improve the water delivery service through better operation and maintenance; (*ii*) to achieve optimum utilisation of available water form canals, tanks, and tubewells; (*iii*) to achieve equity in the distribution of water amongst water users; (*iv*) to promote economy in water use and to increase water and land productivity; (*v*) to optimise the conjunctive use of precipitation and groundwater with the water delivered from the canal system; (*vi*) to assist farmers in crop planning, and frequency and duration of water supply; (*vii*) to inculcate a sense of ownership and responsibility amongst the farmers and to ensure collection of water changes; (*viii*) to develop professional relationship between the water users and state irrigation agencies; (*ix*) to generate local resources in cash, kind or labour for operation and maintenance; and (*x*) to assist farmers' association and its members in other spheres of activities (e.g., agro-industries and marketing) for their upliftment. Thus, the water users associations are charged with the responsibilities of integrated management of water, including conveyance, distribution and application to optimally utilise the water resources for enhancing crop production (Singh, S.R., 1998).

International Status

Even though the irrigation systems are meant for the farmers, the countries around the world have promoted the development of irrigation projects primarily through the involvement of their governments. Only in 1960s and 1970s, several countries started providing incentives to their farmers to take over the management of operations and maintenance, leaving the government organizations to concentrate on the main system. During 1970s and 1980s, the work on irrigation systems' management in participatory mode involving farmers began in many countries, like United States of America, Japan, Italy, Mexico, Columbia, South Korea, Vietnam, Indonesia, China, Nepal, Pakistan, Sri Lanka, The Philippines, Turkey, Egypt, Sudan, Jordan, Morocco, Senegal, and Tanzania, with varying degrees of success. The United States of America, Philippines, Mexico and Turkey have developed their own models of involving farmers' organizations in the operation, fee collection, maintenance, and management of either complete irrigation projects or part of the systems commensurate with their political and socio-economic environment (Singh, S.R., 1998). The review on irrigation development in Asian and African countries presented earlier in this chapter gives an account of the efforts of different countries to accelerate people's participation in irrigation system development and management.

People's Participation in Managing Irrigation Systems in India

The first Water Users' Associations (WUAs) were established in the 1980s. About 4,400 WUAs had been formed by March 1997, managing an irrigated area of about 3,97,100 ha under major and medium irrigation schemes. For minor irrigation schemes (mostly tubewells), about 10,400 WUAs had been formed managing an irrigated area of about 50,000 ha.* The WUAs are formed and work on the basis of executive instructions/guidelines laid down by each state government. As of 1998, the only state which has passed legislation exclusively for farmer participation in the management of irrigation systems is Andhra Pradesh. In 1994, the Government of Andhra Pradesh promulgated Andhra Pradesh Irrigation and Command Area Development Act, 1984, which authorised the creation of Command Area Development Authorities and 'Pipe Committees'. The pipe committees, charged with the responsibilities of internal distribution of water below the outlet and maintenance of the water courses and field channels, were the first step in the direction of participatory irrigation management. However, there were problems due to the lack of coordination between the irrigation and command area development authorities, and poor reliability of water supply at the outlet. In 1995, the Govt. of Andhra Pradesh issued an order to promote Water Users' Associations (WUAs) for the management of irrigation systems at the minor level covering areas of 400 to 750 ha. The WUAs are to function as autonomous bodies on democratic lines and are responsible for the maintenance and operation of irrigation network within the area of their operation. To begin with, the WUAs were required to cover about 2,68,000 ha command area of Sriramsagar, and 65,000 ha command area of Srisailam Right Bank Canals.

In Gujarat, in 1997, 33 WUAs at the minor level of canal projects were formed. The Mohini Coperative Distributory Society which was formed in 1979 in the command area of Bhestan minor of the Ukai Kakrapar Irrigation Project made spectacular success due to government backing, effective local leadership, and an income generating cropping system. The Government of Gujarat has taken a policy decision to distribute water on volumetric basis to water users farmers' associations at the level of the minors of Sardar Sarovar Irrigation Project.

In Rajasthan, 3,483 'outlet committees' were formed in the command area of Indira Gandhi Nahar Pariyojana, each outlet commanding an area of 200-300 ha. The State of Madhya Pradesh has a provision for water *panchayats* in its Irrigation Act. Several WUAs have been formed in Karnataka by the Cooperative Department with the support of the Command Area Development Authority. The Government of Maharashtra, with the help of NGOs, organised farmers in the command area of two minors of major irrigation projects to evolve methodology and approach for involving in managing irrigation systems.

The state of Tamil Nadu has a tradition of contributing voluntary labourers for digging and maintenance of water courses and field channels below the minor. In 1982, the Centre for Water Resources, Anna University, Chennai initiated pilot projects on farmers' participation in rehabilitation and management of irrigation tanks. The farmers in the command area of each tank formed WUAs which were registered under the Tamil Nadu Society's Registration Act, 1975. The Agricultural Engineering Department, responsible for the Command Area Development Programme in Tamil Nadu, has initiated 2,770 Water Users' Farmers' Associations at the outlet level in Lower Bhavani, Kaveri, Periyar Vaigai, Parambikulam Aliyar and Sathanpur irrigation projects for distributing water, resolving conflicts and organising agricultural extension activities for the benefit of farmers.

In Uttar Pradesh, outlet committees have been constituted for the distribution of water in the Sharada Canal Command. Till 1992, the Government of Uttar Pradesh had constructed 28,626 deep tubewells for irrigation. In 1992, the government started pilot projects by transferring the operation and management of about 100 tubewells for a period of five years to Tubewell Cooperative Societies comprising farmers of the tubewell commands. The Government of West Bengal has implemented participatory irrigation management in the commands of low capacity and high capacity irrigation tubewells.

Involvement of farmers on different aspects and levels of water management varies in different states of India. All the states envisage the retention of ownership of the main systems from headworks to the branch canal of the major and medium projects. In different states of India, Water Users' Associations are envisaged at outlet, minor and distributory levels. These associations have different responsibilities in respect to maintenance, operation and collection of water charges. The rules for water allocation and distribution also differ form one state to the other. The performance of the programme of participatory irrigation management in India is yet to be evaluated.

*(***Source***: UN Food and Agriculture Organization (FAO), 2001).

FLOODS AND DROUGHTS

The highly seasonal nature of the rainfall over India gives rise to frequent heavy floods in may regions during the monsoons (Figs. 1.65 and 1.66). In sharp contrast, there are frequent failures of rains in one region or the other and prolonged dry period during the non-monsoon seasons. These give rise to serious drought conditions in many parts of India, especially the arid and semi-arid regions. Hence, the basic concept of irrigation development and water management in India centres essentially around the conservation and control of the excess rainfall runoff during the monsoon and its utilization for supplemental irrigation during periods of dry spells in

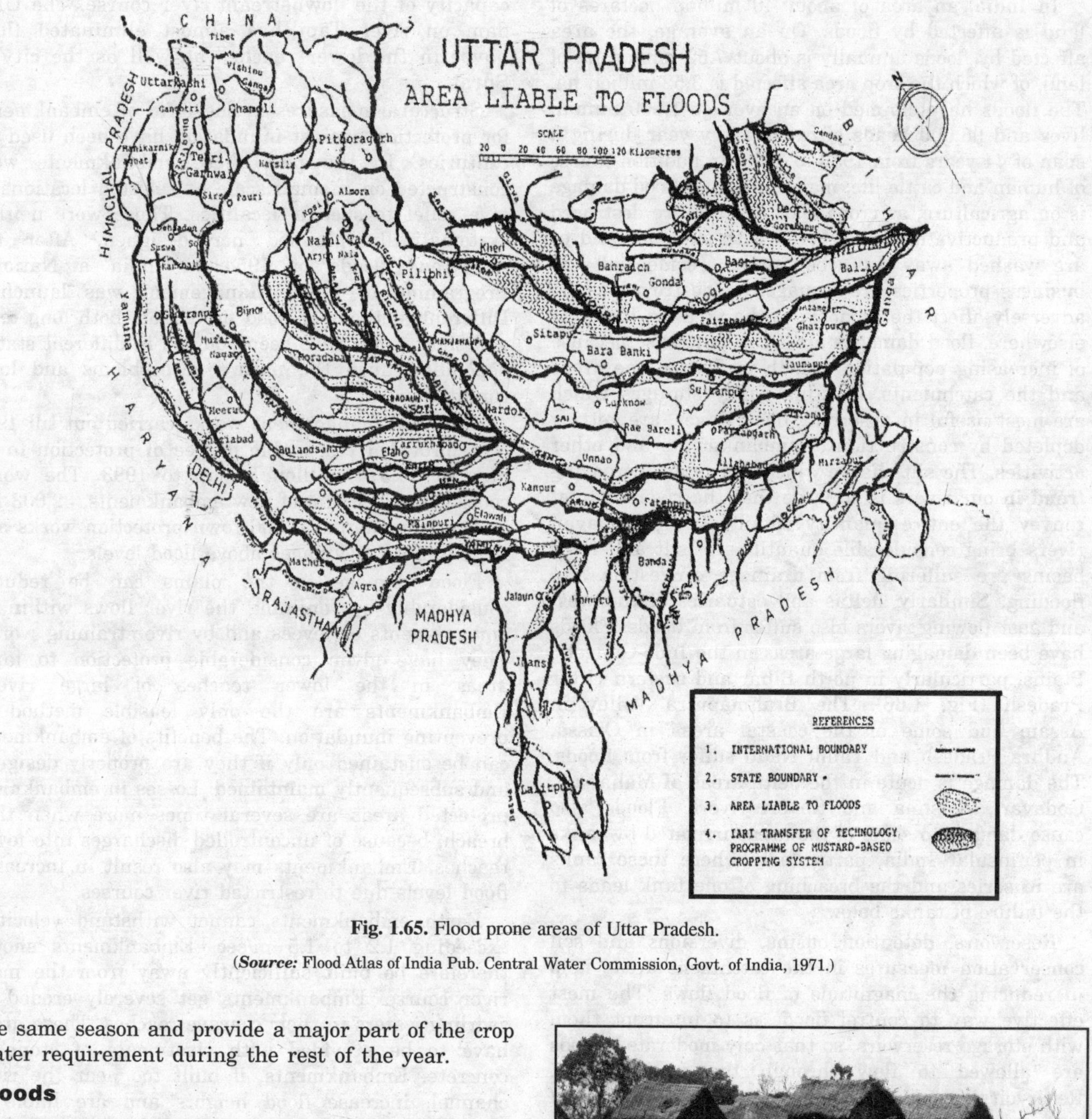

Fig. 1.65. Flood prone areas of Uttar Pradesh.

(*Source:* Flood Atlas of India Pub. Central Water Commission, Govt. of India, 1971.)

the same season and provide a major part of the crop water requirement during the rest of the year.

Floods

A river is stated to be in flood when its flow exceeds its capacity and spills over its bank. The magnitude of the floods depends mainly on the intensity and duration of the rainfall and the vegetational and physiographic characteristics of the watershed. Arid and semi-arid regions, where the mean annual rainfall is low, have poor drainage characteristics. Consequently, heavy rainfall, though limited to a few days in a year, often causes floods. The major factors influencing the severity of flood damage in the plains are the rate of erosion of the soil of the catchment area and the silting of the river beds. Silting progressively reduces the carrying capacity of rivers.

Fig. 1.66. A scene of vast desolation caused by floods in the Kosi River.

(*Photo courtesy:* Min. I. and B., P.I.B., Govt. of India)

In India, an area of about 40 million hectares of land is affected by floods. On an average, the area affected by floods annually is about 7.52 million ha of land, of which the crop area affected is 3.52 million ha. The floods have claimed on an average 1,500 human lives and 95,000 heads of cattle every year during a span of 44 years from 1953 to 1996. In addition to loss of human and cattle life, major impact of flood damage is on agriculture, as crops and livestock are destroyed and productive land is covered with silt and sand or are washed away. In urban areas, residential and business properties are damaged or destroyed which adversely affect the economy of the area. In India, as elsewhere, flood damages are on the increase because of increasing population growth in flood prone areas and the catchments. Existing valley storages which are most useful in absorbing flood waves, are getting depleted by construction of embankments and other activities. The silt charge which shows an increasing trend in our rivers is so large that the rivers cannot convey the entire quantity of silt. The Himalayan rivers bring considerable quantities of silt and their basins are suffering from drainage congestion and flooding. Similarly, deltas and estuaries of the west and east flowing rivers also suffer from floods. Floods have been damaging large areas in the Indo-Gangetic Plains, particularly in north Bihar and eastern Uttar Pradesh (Fig. 1.66). The Brahmaputra Valley in Assam and some of the coastal areas in Orissa, Andhra Pradesh and Tamil Nadu suffer from floods. The damage is acute in the delta areas of Mahanadi, Godavari, Krishna and Kaveri rivers. Floods also cause damage to some of the areas irrigated by tanks in peninsular India, particularly where these tanks are in series and the breaching of one tank leads to the failure of tanks below.

Reservoirs, detention basins, diversions and soil conservation measures in the catchment areas help in reducing the magnitude of flood flows. The most effective way to control floods is to intercept them with storage reservoirs, so that only moderated floods are allowed to flow through the valley below. Reservoirs constructed merely to detain and moderate floods may not be economically justifiable. However, multi-purpose projects could be planned to provide for flood control, irrigation and hydro-electric power generation. The works carried out by the Damodar Valley Corporation in the Damodar river basin in Jharkhand and the various benefits conferred by them are the result of such planning where the primary aim was flood control. Hirakud reservoir is another multi-purpose project which was primarily required for flood control.

In general, the reservoir is not filled to capacity till the middle of August so that it can absorb whatever floods may come in. Subsequently, water is released in such amounts which are within the capacity of the downstream river course. The Ukai dam on river Tapi has almost eliminated flood havoc in the lower reaches as well as the city of Surat.

Structural measures in the form of embankments for protection against inundation have been used for centuries. In the past these embankments were constructed on a small scale at selected locations to give relief to specific localities. These were neither systematically planned nor designed. After the disastrous floods of 1954 in India a National Programme of Flood Management was launched. Different methods of flood protection, both long term and short term, have been adopted in different states, depending upon the nature of problems and local conditions.

The flood management works carried out till 1993 have provided reasonable degree of protection to an area of 14.374 million ha up to 1993. The works consist of 16,199 km of new embankments, 32,003 km of drainage channels, 906 town protection works and raising of 4,721 villages above flood levels.

Flood damage in the plains can be reduced considerably by confining the river flows within its embankments or levees and by river training works. They have given considerable protection to large areas in the lower reaches of large rivers. Embankments are the only feasible method of preventing inundation. The benefits of embankments can be sustained only if they are properly designed and subsequently maintained. Losses in embankment protected areas are several times more when they breach, because of uncontrolled discharges into lower reaches. Embankments may also result in increased flood levels due to restricted river courses.

Earth embankments cannot withstand velocities exceeding 1.2 to 1.5 m/sec. Embankments should therefore be built sufficiently away from the main river course. Embankments get severely eroded by swirling waters in their concave reaches. These areas have to be provided with rivetments of stone or concrete. Embankments, if built too near the river channel, increase flood heights and are liable to breach. Except under special circumstances, flood embankments are to be constructed on both banks of the river, as otherwise it may cause severe damage to the unprotected side. Two of the major flood embankment works in India are the Kosi flood embankments and the Bagmati flood control embankments. The Kosi flood embankments benefit an area of 2,65,000 ha. The Bagmati flood control embankments protect about 57,000 ha. Extensive flood plains like the flood-prone areas of the Indo-Gangetic plains cannot be protected exclusively by constructing embankments. Suitable irrigation and agriculture technology to suit the flooding pattern of the area are essential.

The flood forecasting and warning of incoming floods have played a significant role in reducing the loss of life and property. The Central Water Commission issues forecasts from 157 forecasting stations covering all the major inter-state rivers. The Rashtriya Barh Ayog (National Flood Commission) was constituted in 1976 with the main terms of reference as under: (i) To review and evaluate the flood protection measures with special reference to the construction of embankments. (ii) To evolve a comprehensive approach to the problem of floods as a part of optimum and multipurpose utilization of water resources. (iii) Make recommendations on appropriate land use in flood plains. With the growing population and increased developmental activity, the flood plains are being increasingly occupied, resulting in more and more damages. It is desirable to supplement the construction of flood protection embankments by storage reservoirs or detention dams to detain the flows. Soil conservation and afforestation programmes of the catchment areas of projects would reduce the sediment load of rivers. Ensuring people's participation in flood management is vital to the success of the programme.

Droughts

The worst natural disaster that human race has to confront is drought. Its catastrophic effects are known from time immemorial. Generally, drought is a term applied when there is shortage of water either due to lack of precipitation or due to excessive rates of evapotranspiration. To a farmer, drought means shortage of moisture availability in the root zone of his crop (Fig. 1.67). Drought brings miseries for human life, creating shortage of soil moisture in agricultural lands and shortage of drinking water for human beings and cattle. It may be caused by either annual precipitation being less than the normal onset of monsoon, early withdrawal of monsoon or long dry spells coupled with very low surface and groundwater storage. History reveals that all over the world, famines have almost always been mainly a consequence of droughts. These are the periods of water shortage which affect biological activity (Veeriah, 2001).

The Irrigation Commission (1972) defined '*drought*' as a situation occurring in an area where the annual rainfall is less than 75% of the normal in 20% of the years examined. In case it has occurred in more than 40% of the years examined, these are considered as '*chronic drought years*'. A *taluk* (development block or an equivalent unit), where 30% or more of the cultivated area is irrigated, has been considered to sustain reasonably stable agriculture and to be treated as reasonably protected against drought. Classification of drought-prone areas on the basis of annual rainfall alone may often lead to erroneous conclusions. An immediate consequence of drought is loss in crop production and employment opportunities of people. The National Water Policy (NWP) of India, adopted in 1987 and revised in 2002 has recommended to combat drought-associated problems through soil-moisture conservation measures, water harvesting practices, minimization of evaporation losses, development of the groundwater potential and transfer of surplus water from surplus areas, where feasible and appropriate, to deficit ones. In planning water resources development projects, the needs of the drought prone areas should be given priority.

Fig. 1.67. Effect of drought on standing wheat crop in a black soil region.

About one-sixth of the area of India is drought prone. Recurrent droughts have been great deterrents to economic development of over one-third of the area. Nearly 30% of India's population is affected by drought. The states of Karnataka, Tamil Nadu, Rajasthan, Gujarat, Andhra Pradesh and Maharashtra have large areas prone to drought. The problems posed by drought vary from area to area, depending on the amount of rainfall and its variability and the extent to which irrigation has been developed. Assuming that the districts which receive less than 750 mm of rainfall per annum are liable to drought, there are 77 such districts in India, which account for over 34% of the net sown area. Some of the districts have developed adequate irrigation facilities. Still 50 districts, accounting for one-fourth of the cultivated area of the country, could be considered as vulnerable to drought. The chronic drought affected areas are mainly in West Rajasthan, particularly the districts of Jaisalmer, Barmer, Jodhpur, and Bikaner, and Kutch in Gujarat.

The problem of drought has socio-economic dimensions. There have been glaring disparities in income and living standards between the dry and drought-affected areas on the one hand, and the irrigated and high rainfall areas on the other. Providing irrigation to the maximum extent possible is the major step in relieving the drought affected areas from scarcity conditions. An example is the Ganga Canal in Rajasthan which has transformed parts of Ganganagar district, one of the intensely arid districts in the State, into a prosperous agricultural tract.

Unfortunately, most of the drought affected areas have only limited sources of water which could be harnessed for irrigation. In these areas, transportation of water from other basins and exploitation of groundwater to the extent possible and adopting measures for increased groundwater recharge will provide relief to the problem of drought. Soil moisture conservation would be of great value.

An essential requirement in combating drought is the adoption of small scale water harvesting structures in the form of tanks. Tanks are collection points of runoff, thereby moderating flash floods. They also facilitate recharge of groundwater, while serving as a source of drinking water. Several other water conservation structures/procedures like roof top water collection have been adopted traditionally in different parts of India. In more than one ways, the tanks constitute the primary water conservation work in the watersheds, making them the largest single component of an overall watershed development programme.

With the emphasis on large dams and canal systems and due to socio-economic changes in rural areas (the collapse of traditional power structure, and absence of cohesive village community), the emphasis on tanks declined over the years. Deforestation, mainly due to population pressure in the catchments of tank systems, their siltation, lack of maintenance and repairs and breaches of tank embankments are major deterrents in water harvesting through tanks.

ENVIRONMENTAL IMPACT ASSESSMENT OF IRRIGATION PROJECTS

Any direct human influence on environment shall have a positive objective, namely, to raise the living standards of a community. However, certain activities could cause undesirable side effects, leading to unexpected changes in the environment, unless they are carefully planned and executed. They may give rise to irreversible processes, causing deterioration of the desirable balance of nature.

All water resources development projects have social and environmental implications. The environmental impacts of irrigation projects vary with the physiography of the region, climate, soil, the characteristics of river flow and the size of the project.

Environmental Impacts Associated with Major Irrigation Projects

Irrigation projects invariably change the river flow characteristics and eco-system regimes of the downstream river course and the command areas. Different types of ecological impacts may be observed in the areas occupied by the reservoir at the headworks of the project, downstream from the headworks in the original river course, along the routes of the canal system and in the extensive irrigation command areas.

Global attention to environmental problems was drawn by the United Nations Conference on Environment at Stockholm in 1972. Following the establishment of the United Nations Environment Programme (UNEP), the international organizations like UNDP, UNICEF, FAO and WHO have engaged themselves in programmes of environmental improvement and the relationship between development and environment. The need to take measures to reduce the adverse environmental impacts of development activities have been stressed in many international conferences. International financing agencies like the World Bank, which were concerned earlier mostly with economic returns from projects, have been focusing on social and environmental issues.

Legislative measures. In India, interest in environmental issues began at the national level in early 1970s. Attention of the Government was drawn to the denudation of forests in the Himalayan range and other mountains and their adverse effects on soil erosion in different parts of the country. Article 48A of the Constitution of India, dealing with the Directive Principles of State Policy lays down that, "The State shall endeavour to protect and improve environment and to safeguard the forest and wildlife of the country." Article 51A(g) of the Constitution dealing with Fundamental Duties states that, "it shall be the duty of every citizen of India to protect and improve the natural environment, including forests, lakes, rivers and wildlife and to have compassion for living creatures." The Eleventh Schedule of the Constitution lists soil conservation, water management, watershed development, social and farm forestry, drinking water, fuel and fodder, non-conventional energy sources and maintenance of community assets in which specific functions should be devolved by state governments to *panchayats*. They are aimed at effective environmental protection. The 12th Schedule of the Constitution of India lists wildlife, protection of environment and protection of ecology in relation to water, sanitation and solid waste management, parks and gardens in which specific functions should be assigned by state governments to urban local bodies.

In order to ensure sustainable development of water resources, the Government of India has enacted various acts and legislations. In 1978, environmental clearance was made mandatory for all major irrigation, multi-purpose and flood control projects from the Central Ministry of Environment and Forests. The projects are considered by the Environmental Appraisal Committee which decides about the acceptance of the project from the environmental angle. Environmental safeguards are stipulated, which need to be implemented along with the construction activities of the project. The Environmental (Protection) Act, 1986, calls for the State Governments/Project Authorities to submit Environmental Impact Assessment Statements and Environment Management Plans for obtaining environmental clearance to a project. It is mandatory to get clearance from the forest angle whenever forest land is involved in the project. Other relevant Acts are Water (Prevention and Control) of Pollution Act, 1974 (amended in 1983), Water (Prevention and Control) of Pollution (Cess) Act, 1977 (amended in 1991), Forest Conservation Act, 1980 and Environmental Impact Assessment Notification of 1994 (amended in 1997).

The negative or detrimental effects of water projects are project specific and vary from case to case. A combination of some of the following impacts occur in every project:

(1) Project construction in remote and often pristine areas involves cutting and blasting, establishment of labour colonies and induction of large number of people from outside. These activities result in increased deforestation and creation of gullies on steep slopes leading to landslides. They disturb the lives of local inhabitants. During the process of environmental analysis of a proposed major irrigation projection in which the author was involved, it is suggested that a majority of the staff and the construction labours may be housed in suitable colonies at a place away from the work site near a town and the labourers and the staff transported to the work site daily, with a view to prevent cutting of trees for firewood and hunting of wild animals.

(2) Creation of large reservoirs and construction of canals lead to submergence of land and human settlements, displacement of people and their livestock, loss of occupation and other hardships.

(3) Standing water brings about changes in the ecosystem, including temperature stratification and variation in dissolved oxygen at different levels. These may have influence on aquatic and riparian life which originally existed in the project area.

(4) Some reservoirs pose danger for wildlife through possibilities of drowning or marooning of habitats. Groups and interdependent species could be split and food chains broken. Some species could disappear and this could affect other species. On the contrary, water bodies in general would provide drinking water for wildlife, especially during the dry season, facilitating the growth of wildlife. The Periyar wildlife sanctuary in Kerala is an example.

(5) Flora could be adversely affected through the construction processes, submergence and other factors.

(6) There could be reservoir-induced 'seismicity', leading to earthquakes for which adequate safeguards are required to ensure safety of dams.

(7) The construction of a dam across a river affects the river flow regime. Flows as well as the silt load and nutrient content downstream of the dam would be substantially reduced. This would have an impact on lives, occupations and livelihoods downstream. The reduction in flows could lead to deterioration of water quality and an increased concentration of pollutants in the river downstream of the dam.

(8) Dams are now being designed for the anticipated maximum flood. However, in older dams which are not adequately designed for peak flow, there may be overtopping and possible failure.

(9) The construction of canal networks could disrupt natural drainage, leading to drainage congestion, unless adequate drainage is provided. Canal irrigation can lead to waterlogging and salinisation of extensive areas in the command area.

Environmental impact assessment. To assess the likely changes in the environmental conditions due to water resources projects, it is necessary to carry out environmental impact study at the planning stage itself. Environmental impact assessment is to predict the environmental consequences of a development activity and to plan appropriate measures to eliminate the adverse effects and to augment positive effects. The study should be an important element in the process of project appraisal. The Central Water Commission and the Ministry of Environment and Forests, Govt. of India have laid down detailed guidelines on scope, coverage and methodology of the assessment.

Planning of a project has to include the formulation of detailed measures to remedy/mitigate the adverse effects of the project. A catchment area treatment programme for arresting the deterioration of the catchment areas is essential. Resettlement and rehabilitation of displaced people need advanced planning and timely implementation.

Forests. Forests are predominant components of the environment. They are important for soil and water conservation, protection of watersheds, conservation of ecological diversity (biodiversity) and moderation of regional environment. All aspects of environmental protection and social and economic developments, as they relate to forests, should be integrated. Degradation of forests gives rise to serious problems such as floods and drought.

The vital role of forests in protecting fragile ecosystems, and watersheds, freshwater resources and as store house of rich biodiversity is to be recognised. Trees, shrubs and other vegetation help to control temperature regimes by modifying solar radiation and due to increased relative humidity. Wetlands control flood, store and purify water and protect shores and hinterland, floral and fauna habitats, gene pools and recreation areas.

Reservoir site. The major ecological impacts of the reservoir site include the submergence of extensive areas of fertile land, dislocation of human habitats, changes in the scenic values, modification of the macro-climate of the region and alteration in the flora and fauna of the neighbourhood.

Resettlement and rehabilitation. One of the most sensitive and complex problem that arises in large water projects is that of rehabilitation of people displaced by submergence. Though, resettlement and rehabilitation of the population have been essential components of all river valley projects, their implementation have not always been satisfactory. Many States have formulated their own resettlement and rehabilitation policies. To help the State project authorities in formulating their policies, it will be desirable that the Government of India formulates a national policy with the objective that the people affected by the project should enjoy a better quality of life at the place of resettlement than that they enjoyed at the original habitat.

Reservoir sedimentation. Provision is kept in all reservoirs in the form of dead storage for trapping of sediments. Catchment area treatment, soil conservation and watershed management programmes are taken up in catchment areas of various water resources, projects. Effective implementation of these schemes will reduce the silt inflow into reservoirs and prolong their economic life.

Downstream river course. The ecological impacts downstream of the original river course influenced by the construction of irrigation works are many. The downstream hydrology of the river is disturbed. The beneficial impact is the reduction in the peak flows and the consequent reduction of floods in the plains adjoining the river course. The adverse effects of the reduced quantity of river flow downstream are the deterioration of water quality, reduction in the sediment carrying capacity of the river, changes in aquatic and riparian vegetation, modification of aquatic life, changes in the recreational potential of the river and alterations in the economic and social life of the people of the area. The capacity of the river to flush the pollutants which accumulate during the non-rainy season is greatly reduced. It may also result in abnormal and variable flows in the original river course due to different levels of regulation at the headworks to suit heavy rains and drought periods. Storage reservoirs often cause a deterioration in the quality of the water downstream of the original river course. Low flows may prevent survival of uniquely adapted species. Dams also interfere with the growth of migrating fish.

Minimum flows in the river. Dams create reservoirs to conserve flood flow for use during lean season. Often they result in the drying up of the river downstream of the dam, affecting riverine ecology, pisciculture and human habitations, unless an effective downstream river flow regulation scheme is evolved. In all new dams it is essential that low level outlets are provided to ensure minimum flows for sustaining ecology.

Canal routes. Ecological impacts are exhibited along the conveyance route of the canal system, including majors, branches, distributories and water courses. They include seepage along the unlined water conveyance systems and rise in water tables along their course, formation of stagnant water bodies in the borrow pits on either side of the canals, introduction of pests through the water conveyance system and changes in the landscape.

Command area. The ecology of the irrigation command areas gets modified substantially after the introduction of irrigation. Agriculture changes from that of rainfed farming to intensive irrigated farming. Fauna and flora get modified. The ill-effects attributed to poor water management in the command area include the rise of the groundwater table, build-up of salinity, leaching of nutrients and accelerated use of pesticides and their effect on the flora and fauna.

There is a strong view amongst a section of the scientists in India to attribute the increase in the incidence of malaria to the increase in irrigation. Even though irrigation by itself may not contribute to the spread of mosquitoes and thereby of malaria, extensive waterlogging, resulting from ill-planned irrigation projects facilitates breeding of mosquitoes.

Influence of insecticides and pesticides. Intensive farming resulting from the introduction of irrigation projects leads to increased use of pesticides, insecticides, fungicides and herbicides. Many of these agro-chemicals, if improperly used, adversely affect man, animals, birds and plants. There is need to adhere to concentration limits in pesticide application and to avoid their direct application to irrigation streams. Alternative to pesticides like inter-row cropping to improve the micro-climate in crop canopy are possible ways of crop management.

Influence of salinity of drinking water on cattle production. Saline drinking water influences the growth of cattle. High total soluble salts or specific ions, particularly magnesium and sulphates have been considered to influence the gain in body weight of sheep and cattle. Limiting magnesium content in drinking water for young animals is about 250 ppm. Normally, water containing more that 500 ppm is considered unsafe even for grown-up cattle.

Water weed problems in irrigation systems. Introduction of irrigation often results in a rapid increase in water weeds. Hence, study of the ecological factors which promote the growth of weeds is desirable to provide safeguards against their uncontrolled growth. Weeds may become serious hazards to agricultural lands and often block the drains.

Prevention and Control of Pollution and Wastewater Management

Pollution of streams, rivers and reservoirs due to effluents from domestic supplies, mining, toxic effluents from industries, pesticides, fertilizers and agricultural wastes is posing serious problems. Rivers, lakes and wetlands are the important natural sources of water. Maintaining water quality in them is vital for all life. Most of the major rivers in the developing countries suffer from pollution in one stretch or the other. A study conducted by the Central Pollution Control Board in 1984 revealed that nearly 75% pollution of the river was on account of discharge of untreated municipal sewage into the river from large and medium towns located along it and the remaining 25% pollution was caused by the discharge of partly treated/untreated effluents from the industry. The report also identified that some of the non-point sources of pollution include run-off from agricultural lands carrying chemicals and fertilizers, runoff from areas used for dumping of solid waste and open defecation, dumping of burnt/half burnt dead bodies and animal carcasses, washing ghats, cattle wallowing and unchecked mass bathing.

In 1998, a total of 12,145 million litres of wastewater per day was the estimated generation from 212 Class-I cities surveyed in India. Of this 40% is generated from 43 cities and about 60% from 12 metropolitan cities. The survey revealed that only 5 metropolitan cities have proper wastewater collection systems. In 142 Class-I cities the mode of disposal of effluents in 42 cities is on agricultural land, 44 cities into rivers, 32 cities both on agricultural land and into rivers and 24 cities discharging to other recipient systems such as sea, lakes and ponds. In 166 Class-II towns, the mode of disposal of wastewater in 66 towns is on agricultural land, 42 towns into rivers, 12 towns both on agricultural land and into rivers and 46 towns into sea, lakes and ponds. Due to indiscriminate discharges of untreated sewage and industrial effluent into water bodies, the quality of water is fast deteriorating. The Central and State Pollution Control Boards have been constituted by an Act of Parliament to prevent and control pollution of water.

The first major river cleaning programme in India was started with the launching of the Ganga Action Plan in June 1985 as a 100% Centrally Sponsored Scheme. The objective is to clean the river Ganga and its major tributary the Yamuna in a phased manner. The Phase-I of the Ganga Action Plan has been implemented and Phase-II has been taken up to cover the remaining works. The Government of India has initiated similar pollution control measures to clean up other important rivers under the National River Conservation Plan in a phased manner.

Sewage Irrigation

Using treated sewage water is a favourable proposition. The nutrients present in treated sewage water are an advantage for agricultural production. There are several examples of use of sewage water for irrigation. It can be successfully adopted wherever feasible. Sewage water is unsafe for irrigating leafy and root type vegetable crops where the edible part is consumed raw. Wastewater could be used for secondary domestic and municipal purposes such as flushing toilets and watering lawns. Only countries with a high level of technology for the operation and monitoring of advanced techniques of treatment of wastewater systems can consider reusing sewage water for domestic purposes. The quality of industrial and municipal wastewater requires to be constantly monitored, because such waters may have long term effects, both on the crops and on the health of the people.

INTER-BASIN WATER TRANSFER

An analysis of the water resources potential of any country would reveal that wide disparities exist in the availability of water resources in different regions. In India, during the monsoons, water pours down the giant rivers of north and east, flowing away to the sea, while major parts of southern, central and western India struggle for water due to low rainfall and/or prolonged dry periods. The central and western regions, being dominated by underground hard rock formations are poor in groundwater potential, as compared to the rich water bearing formations of the alluvial plains of north, east and the Brahmaputra valley of north east India. This calls for the need to explore the possibility of inter-basin transportation of water without environmental deterioration and other adverse effects within the donor basin and considering the overall benefit to the nation.

Historical perspective. The earliest works of linking of river basins in India are the Periyar Vaigai Project of late nineteenth century. Parambkulam Aliyar Project, Kurnool-Cudappa Canal and the recent Telugu Ganga Project are other examples. The inter sub-basin transfers in the Indus Basin from rivers Ravi to Beas, and Beas to Satluj have tackled difficult design and construction problems. Various proposals involving inter-basin transfer of water within India on a large scale have been studied since 1960s. In 1960s, Dr. K.L. Rao, the then Minister of Irrigation and Power, prepared proposals for the National Water Grid. His proposals, better known as the Ganga Kaveri Link, consisted of 2,640 km link from Ganga near Patna to transfer the Ganga waters during high flow to the south, based on some earlier work in the Central Water Commission. The objective was to provide for navigation and ameliorate spatial disparities between different river basins. Detailed examination showed that the proposal involved high energy costs in pumping and construction of extensive water conveyance systems. Captain Dastur, an air pilot, proposed a Garland Canal Scheme for inter-basin water transfer. The proposal, however, could not stand the test of technical scrutiny.

National perspective. The Ministry of Irrigation and the Central Water Commission in 1980 prepared a National Perspective Plan comprising of two components, namely (*i*) The Himalayan River Component, and (*ii*) The Peninsula River Component. It was decided to start studies on the Peninsula component, as that component could be implemented by India on its own. Consequent to these efforts, the National Water Development Agency (NWDA) was set up in 1982 by the Government of India. This Agency is charged with the detailed planning of the link proposals and preparing the feasibility reports of the various components. By 2002, the NWDA completed water balance studies and pre-feasibility studies of the schemes. It had completed six feasibility level studies and 18 other feasibility studies were in progress. The Himalayan Rivers Component envisages construction of storages on the main Ganga and Brahmaputra rivers and their principal tributaries in India and Nepal so as to conserve monsoon flows for flood control, hydro-power generation and irrigation. Interlinking canal systems are to be provided to transfer surplus flows of the Kosi, Gandak and Ghagra to the west. Surplus flows available on account of interlinking of Ganga and Yamuna are proposed to be transferred to the drought prone areas of Haryana, Rajasthan and Gujarat and also benefit large areas in South Uttar Pradesh and Jharkhand. The project will have provision to benefit Nepal, both in hydro-power generation and irrigation. In the peninsular component, the NWDA has formulated proposals for the Mahanadi-Godavari-Krishna-Pennar-Kaveri-Vaigai link running through the eastern region of peninsular India and inter-connecting all the major east-flowing rivers of southern India to benefit the drought prone areas of the region. The Government of India, in December, 2002 set up a Task Force to prepare action plans outlining the time schedules and options for funding and execution so as to facilitate implementation of the project by 2016 (Mohila, A.D., 2003).

The first generation of inter-basin water transfer projects in most countries were constructed without serious problems. Most of these projects were completed before environmental issues became important considerations. The new generation of water transfer projects are generally large and technical problems are not easy to resolve. They often span more than one state, and sometimes even more than one country. Hence, the institutional problems can become major issues. Environmental and social impacts assessment have become crucial issues now. Before inter-basin water transfers are taken up basin-wise, surplus and shortage of water resources have to be established. Legitimate projection of the demands of identified surplus basin states will have to be met. Proposals of inter-basin water transfer are likely to be opposed by the basin states, unless detailed studies have proved their feasibility from the environmental angle and the future needs of the basin states are fully met.

New focus on environmental impacts. In 1977, a Task Force comprising of international experts met at the International Institute of Applied Systems Analysis, Luxembourg, Austria, to consider issues relating to large scale water transfers (Goluber, G.N. and Biswas, A.K., editors, 1985). The Task Force identified the following major issues to be considered in appraising the feasibility of large scale inter-basin water transfers:

1. ***Political.*** (a) Power balance between the central and state governments. (b) Political boundaries versus hydrological boundaries. (c) Available channels for effective public participation.
2. ***Legal.*** (a) Existing water laws and water rights. (b) Inter-state water agreements and/or relations. (c) International water agreements and/or relations.

3. ***Institutional.*** (a) Central water authority. (b) Regional water authority. (c) Local authorities.
4. ***Environmental impacts.*** (a) Impacts of water transfer after the transfer. (b) Environmental impacts on exporting area/importing area. For each component impact should be considered on: (c) Physical system: (*i*) *Water quantity*: level; discharge; velocity; groundwater; storage losses. (*ii*) *Water quality*: sediments; nutrients; salinity and alkalinity; (*iii*) *Presence of toxic chemicals*; (*iv*) *Influence* on soil erosion; sedimentation; salinity; alkalinity; waterlogging; (*v*) Changes in land use patterns; (*vi*) Earthquake inducement; other hydro-geological factors, and (*vii*) Changes in micro- and macro-climate. (d) *Biological system*: (*i*) Aquatic life on flora and fauna, especially of the basin state. (*ii*) Socio-economic system: influence on agriculture; aquaculture; hydro-power; navigation; industries; recreation and mining. (*iii*) Socio-cultural: settlement of people; infrastructural developments.
5. ***Alternative strategies*:** (a) Efficient use of water. (b) Changes in water use, changes in allocation, changes in utilization (e.g., cropping pattern). (c) Demand control: economic, legal, institutional and technological (d) Reuse of water-sewage, drainage of waste water. (e) Desalination of brackish sea water.
6. ***Perception and attitude of people to inter-basin transfer.*** (a) Perception on need for water for social and economic development. (b) Political considerations. (c) Provision of reliable information on all aspects of water transfer and their effective dissemination. (d) Perception of politicians, bureaucracy, and various population groups (environmental, labour and business).

Many environmental impacts which were not considered serious earlier have come to focus now. There are a number of alternatives to large scale water transfers, both of an institutional and technical nature. Institutional, legal and political considerations play a major role in decisions on inter-basin water transfer. In a federal democratic set up, adequate legal safeguards are to be provided to ensure that water transfer issues are not influenced by political preferences. This is especially important when the political affiliation of the basin state and the recipient state are different and when the central government has political compulsions.

SUMMARY

Increased awareness on the efficient utilization of water resources globally, and the importance of rational allocation of available water to meet the competing demands of different sectors are stressed. The hydrologic cycle and its influence on the temporal and spatial availability of water resources are described. Global water scenario, with particular reference to different continents for irrigation and other uses are described. Analysis of the water budget of India focusing on the need for appropriate development and management of surface and groundwater. The historical review of irrigation development globally brings forth examples of the collapse of many irrigation-based societies due to faulty water management and unstable governments. The influence of monsoons on water resources development and irrigation management of different countries of Monsoon Asia, along with their physiographic and agro-climatic situations are described. The picture of Central Asian countries in relation to water management brings to focus the importance of integrated planning of water resources systems for sustained agricultural productivity. The physiographic and climatic situations of West Asia and the Middle East points out the need for situation-specific planning and management of the scarce water resources. Land drainage forms an essential component of integrated management of land and water resources in many countries: The status of water resources and their management in different countries of Africa in the six major agro-climatic and ecological regions are presented. Major issues of land and water management in Africa are examined. The regional variations in water availability and the requirements of irrigation as well as the extent of irrigation development in different countries of Africa are described. Specific details of Nile Valley irrigation system in Egypt are presented. The historical review of irrigation development in India, highlights the importance given by the ancient rulers on tank-based water resources development. Substantial progress in irrigation development could be achieved in India during the later part of the 19th century and pre-Independence period of the 20th century, with particular reference to famine relief and increased agricultural production. Rapid pace in irrigation development could be achieved during the second-half of the 20th century, mainly through the construction of multi-purpose river valley projects and emphasis on minor irrigation, especially groundwater development. The physiography, climate, soils, river systems and groundwater regions in relation to water resources availability and utilization in India are described. State-wise distribution of land area, cultivated area and irrigated area in India are presented. Specific details of canal system management in India are explained. Major losses of water in canal systems, especially in night irrigation and the need to improve efficiency in canal irrigation are pointed out. The importance of conjunctive use of surface and groundwater is highlighted. The vital role of people's participation in the management of irrigation systems and the status of its adoption in different countries are described. The occurrence of floods and droughts, the extent of areas affected and the means to minimize their adverse effects are described. The details of environmental impact assessment of irrigation projects are discussed. The critical issues to be examined in planning large scale inter-basin water transfer are pointed out.

QUESTIONS

1. What are the major issues in water resources development stressed in the international conferences and initiatives during 1990s and early 21st century?
2. What are the major objectives of Global Water Partnership?
3. What are the possibilities of human interventions in the hydrologic cycle to improve the availability of water resources on earth?
4. Which are the industries requiring high rate of water use? What are the different methods of economizing water use in the growing industrial sector?
5. What are the common point and non-point sources of pollution of water bodies? Discuss.
6. What are the factors affecting the requirement of water in the domestic sector? What are the commonly used criteria for water supply for domestic use in urban and rural sectors?
7. What is the estimated hydro-power potential of India? What is the present level of utilization of the potential?
8. What is the estimated surface water potential of India, based on the latest estimates? How much of it can be utilized?
9. What are the major lessons to be learnt from the rise and fall of the irrigation-based societies of ancient Mesopotamia in West Asia?
10. Define canal water regulation. What are the possible means of regulating canal supplies to achieve equity in water distribution and efficiency in canal irrigation?
11. Distinguish between contour canals and ridge canals. What are the topographic limitations in adopting them?
12. Define water right. What is 'prior appropriation' in water allocation?
13. Define spate irrigation. Which are the countries / geographical regions where spate irrigation is a common means of meeting the water requirement of crops?
14. Define 'safe yield' of aquifer. Does it include fossil water? Discuss.
15. Define environmental impact assessment. How is the term ecology different from environment?
16. What are the major lessons to be learnt from the review of the performance of ancient irrigation systems on the sustainability of water resources projects on agricultural production?
17. What are the special requirements in irrigation system management in the Indus basin of Pakistan?
18. How is conjunctive use of surface and groundwater vital in the integrated water resources development of Bangladesh?
19. Which is the major geographical region of Nepal requiring special emphasis on the development of surface and groundwater potential for irrigation? How much of the total irrigated area in Nepal is covered by 'public' and 'farmer managed' irrigation systems in Nepal?
20. What are the specific requirements of water resources development in the 'dry' and 'wet' zones of Sri Lanka?
21. How are the irrigation systems classified in India? Which class has the largest net area under irrigation? Discuss.
22. Which south-east Asian countries have made major strides in participatory irrigation management? What are the benefits achieved by them?
23. What are the important geographical regions of Thailand in relation to water resources planning and development? Discuss.
24. What are the unique features of the flow of Tole Sap River in Combodia?
25. Which are the basin countries of Mekong River in south-east Asia? How much of the geographical areas of each of the basin countries lie in the Mekong River basin?
26. What are the major physiographic and climatic features of south-east Asian countries in relation to water resources development?
27. What are the special features of 'granary areas' earmarked for concentrated development in Malaysia?
28. How are the irrigation systems of Bali Island in Indonesia different from those in other islands of the country?
29. How are the irrigation systems classified in Indonesia?
30. Describe the indigenous method of reclaiming salt-affected swampy lands in the southern Kalmantan island of Indonesia.
31. Discuss the major features of irrigation development in the Philippines.

32. What are the major activities of Irrigation Associations in Taiwan?

33. What are the major physiographic and climatic features of eastern Asia?

34. What are the two major characteristics of the river systems of China? How are they concentrated geographically?

35. Describe the procedure used for the location and construction of *kanats* in groundwater development? Which are the countries in which they are common?

36. What are the major river basins of Iran? Which are the important interventions introduced in the recent past to combat waterlogging and salt problems in major irrigation projects of Iran?

37. Discuss the major action points of the integrated development of water resources of Israel? What is the role of National Water Carrier?

38. What are the unique features of irrigation development in Saudi Arabia? Discuss.

39. Distinguish between the terms 'water managed area' and 'irrigated area', as commonly used in most of the African countries.

40. Define 'recession cropping'. Which are the countries in which the method is commonly adopted?

41. Which are the distinct climatic regions of Sudano-Sahelian Africa? How much of the land area in the region is covered by each of the distinct regions?

42. Discuss the commonly used classification of irrigation systems in Africa.

43. Which are the major rivers of Africa flowing through swampy areas that lose considerable amount of water through evaporation?

44. Which are the countries in the Humid Central Africa and what are the distinctive features of the water resources of the region in relation to irrigation development?

45. Discuss the main features of irrigation water distribution systems in the Nile Valley of Egypt?

46. What are the major features of drainage systems in the Nile Valley of Egypt?

47. Which are the major topographical units of Nigeria and what are their important features?

48. What are the major characteristics of *wadis* common in North Africa, Central Asia and the Middle East?

49. What are the major types of irrigation in Kenya and what is the extent of irrigation under each?

50. What are the major issues of land and water management in Africa?

51. Discuss the importance of Kaveri Delta project in India from the historical and sustenance of agricultural production points of view.

52. Describe the features of 'chain tanks' in Tamil Nadu, Karnataka and Andhra Pradesh.

53. What are the major lessons to be learnt from the development of tank irrigation in peninsular India?

54. When was the First Irrigation Commission appointed by the Government of India and what were its mandates? How long it took to submit its report?

55. Which are the major irrigation projects of India constructed in the 19th century and the pre-Independence period of the 20th century?

56. What are the major international treaties / agreements between India and its neighbouring countries?

57. What is the jurisdiction of the water resources of India? Discuss.

58. Briefly discuss the rapid development of irrigation potential in India during the Five Year Plans.

59. Describe the importance of groundwater development in the integrated development of the water resources of India.

60. Discuss the reasons for the slow pace of development of tank irrigation in India during the Five Year Plan periods.

61. What were the major points of emphasis on irrigation development during the 1st, 2nd, 3rd and the subsequent Five Year Plans in India?

62. Discuss the present day projections of irrigation potential in India.

63. Which are the most leading states of India in the utilization of irrigation potential? Discuss.

64. Discuss the adaptability, scope and operational procedures of Kolhapur Type Pick-Up welts.

65. Which are the major physiographic regions of India? What are their specific characteristics influencing water resources potential?

66. Which are the large west flowing rivers of India? Which are the major irrigation projects built on these rivers?

67. Which river in India was called the 'river of sorrow'? How could the serious floods of the river be controlled?

68. Describe the principal agro-climatic zones and regions of India with specific reference to crop production.

69. Which are the distinct agro-ecological regions of India? How do they influence agricultural production?

70. Define monsoon. What are the basic features of the monsoons?

71. What are 'western disturbances'? How do they influence precipitation?

72. Which are the states and regions of India profoundly influenced by North-East monsoon?

73. How do the wind characteristics influence crop growth and productivity?

74. What are the major constitutional provisions on the jurisdiction of water resources? Discuss.

75. What are the limitations in the present day legislation on groundwater rights?

76. Briefly describe the land utilization pattern of India.

77. What are the relative yields of groundwater in the major geo-hydrological regions of India?

78. Describe the river basin-wise potential of groundwater in India and the extent of utilization in each major river basin.

79. Define 'roster of turns' in the *warabandi* system of water delivery at the field level in canal systems. What are its limitations?

80. Under the *warabandi* system of water allotment, what is the weekly duration of irrigation allotted to a farmer having 3 ha of land in the command area of an irrigation outlet?

81. What are the major issues to be considered in planning large-scale inter-basin water transfer projects? Discuss.

82. Identify true and false statements from the following:

(*i*) Hydrologic regime is the behaviour of water in the hydrologic cycle in a specific period.

(*ii*) Water use in pollution control is an offstream use.

(*iii*) Common causes of soil degradation in irrigated agriculture are waterlogging and salt accumulation.

(*iv*) Water requirement in the power sector can be of offstream and instream types.

(*v*) Major reduction in water consumption in the industrial sector can be achieved by recycling the effluents after proper treatment, thus reducing the water requirement in the sector.

(*vi*) The Code of Hammurabi is the first recorded law to compensate the loss to a farmer by the improper management of a canal irrigation system.

(*vii*) Flood irrigation is a significant means of meeting the water requirement of crops in Yemen.

(*viii*) In the *Shejpali* system of water distribution in the command area of irrigation canal outlets, the 'turns' are adjusted to protect the interest of the tail-end fanner.

(*ix*) In the *warabandi* system of water delivery, due consideration is given to the type of crops grown by the fanner.

(*x*) The various patterns of water delivery to the farmers of an outlet command area include proper consideration to compensate the farmers for the losses in the water course.

(*xi*) The Nile Valley Irrigation System in Egypt and the Kaveri Delta Irrigation System stand out as examples of long term sustainability of irrigation works.

(*xii*) The collapse of the Harappan civilization of the Indus Valley could be traced to the mismanagement of its irrigation works.

83. Fill in the blanks:

(*i*) The 1992 International Conference on Water and Environment was held at ....

(*ii*) Agenda 21 of the United Nations Conference on Environment and Development held at .... in .... brought out the need for .... in the water sector.

(*iii*) The cyclic recurrence of the weather phenomena at more or less regular intervals in response to long range .... and influences is called ....

(*iv*) The measurement and analysis of the flow of water is called ....

(*v*) Rainfall is classified according to the lifting mechanism of clouds into (*i*) .... (*ii*) .... and (*iii*) ....

(*vi*) The average continental precipitation, evaporation and runoff are close to the world average in (*i*) .... (*ii*) .... and (*iii*) ....

(*vii*) Water use varies considerably around the world: In .... , ... and .... agriculture is the main user, in Europe and .... domestic and industrial requirements of water exceeds irrigation needs.

(*viii*) The net irrigated area in the world in 1999 was about .... ha.

(*ix*) Water use in irrigation in India at the beginning of the 21st century was estimated at .... billion cubic metres.

(*x*) The World Health Organization has suggested a target of .... lpcd of water supply for urban areas.

(*xi*) The total hydro-power potential of India, as assessed by the Central Water Authority is .... MW at .... % load factor.

(*xii*) The National Agriculture Commission (1976) estimated the total groundwater potential of India in 1974 at .... million ha and the projected potential for 2025 at .... million ha.

(*xiii*) An .... reservoir is constructed to supplement the storage of a main reservoir.

(*xiv*) The 'Imperial Canal' in China built in A.D. .... for irrigation and navigation is one of the greatest ancient irrigation works.

(*xv*) The Grand Anicut on River Kaveri was originally built in AD .... by the ancient Tamil King ....

(*xvi*) The Hoovar Dam on River .... in U.S.A., constructed in .... has been a great engineering feat, the design procedures of which formed the basis for large dams in many countries.

(*xvii*) The longest river in Malaysia is .... in Sarawak.

(*xviii*) The largest dam in Syria is at .... on River ....

(*xix*) The region of oasis in Saudi Arabia is in the .... Plain.

(*xx*) The average density of population of Africa is .... per sq. km.

(*xxi*) Rice is the principal crop in the .... and .... Africa.

(*xxii*) Dark clay soils form .... % of the soils of Africa.

(*xxiii*) Many of the rivers of Algeria diminish downstream or may be lost in .... or in the .... desert.

(*xxiv*) Fossil water is an important source of groundwater in three of the countries of North Africa, namely (*i*) ...., (*ii*) .... and (*iii*) ....

(*xxv*) The major part of Atlas Mountain lies in ....

(*xxvi*) The important international rivers passing through Mali are (*i*) ...., (*ii*) .... and (*iii*) ....

(*xxvii*) Amongst the African countries, ... is popularly known as the 'water reservoir' of Africa.

(*xxviii*) Irrigation in Mauritius is confined mainly to grow .... crop.

(*xxix*) The main irrigated areas of Somalia are in .... and .... valleys.

(*xxx*) The .... built on River .... in the second century A.D. is the first major irrigation project in the Indian sub-continent.

(*xxxi*) The Western Yamuna Canal was built in A.D. .... by Emperor ....

(*xxxii*) The major credit for the construction of the major delta irrigation projects in India goes to ....

(*xxxiii*) The most famous amongst the irrigation works of Vijayanagara Kingdom is the .... tank in .... district of Andhra Pradesh.

(*xxxiv*) The names of three great engineers of the 19th and early 20th centuries are immemorable in the history of irrigation in the pre-Independence period of India. They are (*i*) ...., (*ii*) .... and (*iii*) ....

(*xxxv*) At the beginning of the 21st century about half of the large dams in India were located in the two states of .... and ....

(*xxxvi*) The National Commission on Integrated Water Resources Development Plan (1999) has estimated the availability of average annual runoff of the rivers in India as .... billion cubic metres.

(*xxxvii*) The Himalayas comprise of three parallel series of ranges, namely, (*i*) The .... Himalayas, (*ii*) The .... Himalayas and (*iii*) The .... Foothills.

(*xxxviii*) In India, the average annual potential evaporation is highest in the state of ....

(*xxxix*) Amongst the different states of India, ... has the largest area under irrigation. The major geohydrological regions of India are (*i*) .... (*ii*) .... , and (*iii*) ....

(*xl*) The crops which are most adapted to night irrigation are .... and ....

(*xli*) The conjunctive use of surface and groundwater is achieved by any of the following ways: (*i*) Conjunctive use in ...., (*ii*) Conjunctive use in ...., and (*iii*) Conjunctive use in ....

(*xlii*) The Irrigation Commission (1972) defined 'drought' as a situation occurring in an area where the annual rainfall is less than .... % of the normal in .... % of the years examined.

84. Identify the most appropriate answer of the following questions from the four options:

(*i*) In south-east Asia, a major potential for increasing agricultural production lies in:
- (a) Reclamation of salt-affected areas
- (b) Provision for surface drainage
- (c) Reclamation of swampy and marshy lands
- (d) Provision for appropriate sub-surface drainage.

(*ii*) Drainage has increased agricultural production globally by about:
- (a) 10% (b) 15%
- (c) 20% (d) 25%

(*iii*) The utilization of hydro-power potential in India is estimated to be as low as:
- (a) 10% (b) 15%
- (c) 20% (d) 25%

(*iv*) The most significant historical evidence on the efforts of ancient rulers to ensure equitable distribution of irrigation water amongst the farmers of the command areas is:
- (a) The Code of Hammurabi in the ancient Mesopotamia.
- (b) The writings of Megasthenes, the Greek Ambassador in the Court of Emperor Chandra Gupta.
- (c) Pali Works and Epics referring to the prosperous irrigated agriculture in the Kingdom of Magadha (Bihar).

(*v*) In Indonesia irrigation is well developed in:
- (a) Kalimantan Island
- (b) Sumatra Island
- (c) Java and Bali islands
- (d) Sulaweski Island.

(*vi*) The major source of irrigation water in Israel is:
- (a) Surface water (b) Groundwater
- (c) Conjunctive use of surface and groundwater
- (d) Treated municipal and industrial effluents.

(*vii*) The climate of Israel is:
- (a) Tropical (b) Sub-tropical
- (c) Oceanic (d) Mediterranean.

(*viii*) The largest number of major rivers of Africa join the:
- (a) Mediterranean Sea
- (b) Indian Ocean (c) South Atlantic Ocean
- (d) North Atlantic Ocean.

(*ix*) Out of the total water resources of Africa the contribution of groundwater is estimated to be:

(a) 15% (b) 20%
(c) 25% (d) 30%

(*x*) The design elevation of water surface in the water distribution canals of Egypt is:

(a) At the level of the fields to be irrigated.
(b) 45 cm above the level of the fields.
(c) 30 cm below the average elevation of the irrigated fields.
(d) 50 cm below the average elevation of the irrigated fields.

(*xi*) The Aswan High Dam (Sadd-el-Ali Dam), considered to be the 'pride of Egypt' was constructed at:

(a) The site of the original Aswan Dam.
(b) Downstream of the original Aswan Dam.
(c) Upstream of the original Aswan Dam.
(d) In the Nile Delta region of Egypt.

(*xii*) The Hirakud Dam project in India is located on River:

(a) Godavari (b) Mahanadi
(c) Damoder (d) Krishna.

(*xiii*) In India, the irrigation potential increased during the second-half of the 20 century by:

(a) Double (b) Three times
(c) Four times (d) Five times.

(*xiv*) The largest area under tubewell irrigation in India is in:

(a) Punjab (b) West Bengal
(c) Uttar Pradesh
(d) Bihar and Jharkhand.

(*xv*) Amongst the different states of India, the largest gross irrigated area lies in:

(a) Maharashtra
(b) Uttar Pradesh, including Uttaranchal
(c) Madhya Pradesh, including Chhattisgarh
(d) Rajasthan.

(*xvi*) Amongst the 12 major river basins the largest catchment area in India is for:

(a) Indus (Sindhu) (b) Ganga
(c) Brahmaputra (d) Godavari.

(*xvii*) Provision to mitigate the disadvantages of the 'tail-end farmer' of a canal outlet is provided in the water delivery pattern under:

(a) *Warabandi* system (b) Block system
(c) *Shejpali* system (d) Localised system.

(*xviii*) In the National Water Policy (2002) the highest priority in the allocation of water resources of India is given to:

(a) Ecology (b) Irrigation
(c) Drinking water (d) Industries.

(*xix*) In the assessment of the feasibility of large-scale inter-basin transfer of water, the most vital factor to be considered is:

(a) Prevailing legal issues.
(b) Availability of appropriate institutions for the management of the water resource system.
(c) Environmental impact of the basin (donor) country/state.
(d) Perception and attitude of the people of the affected regions on inter-basin water transfer.

85. In the following, match each of the entries of the first column with the most closely linked entry in the second column:

(*i*)	(A) Rio de Janeiro	(I) Improper management of water resources
	(B) Climatic cycle	(II) Grand Anicut
	(C) Environmental degradation	(III) Delta irrigation systems in India
	(D) Sir Arthur Cotton	(IV) United Nations Conference on Environment and Development
	(E) Karikala Chola	(V) Long range terrestrial and solar influence
(*ii*)	(A) Sir M. Vishweshwariah	(I) South-East Asia
	(B) Granary Areas	(II) Estimation of surface water resources of India
	(C) A.N. Khosla	(III) Biological community
	(D) Ecosystem	(IV) Malaysia
	(E) Mekong River Basin	(V) Krishnarajasagar Project
(*iii*)	(A) Magasthenes	(I) Yellow River Delta Development
	(B) Code of Hammurabi	(II) Basin irrigation
	(C) Li Ping	(III) President Roosevelt
	(D) King Menes	(IV) Irrigation management in ancient India
	(E) National Reclamation Act of 1902	(V) Compensation for crop damage due to faulty irrigation management
(*iv*)	(A) Hoover Dam	(I) Nile River
	(B) Amu Darya	(II) Indus River
	(C) Tunsa Barrage	(III) Bangladesh
	(D) Master Plan Organization	(IV) Aral Sea
	(E) Aswan High Dam	(V) Colorado River

(*v*) (A) Mahaveli Ganga Basin Project — (I) Philippines
(B) Chao Phraya Basin — (II) Mekong River
(C) National Irrigation Administration — (III) Central Asia
(D) Lake Tonle'Sap — (IV) Sri Lanka
(E) Kanats — (V) Thailand

(*vi*) (A) National Water Carrier — (I) Tamil Nadu
(B) Swamp Rice Production — (II) Maharashtra
(C) Periyar Vaigai Project — (III) Upper Ganga Project
(D) Khadakvasla Dam — (IV) Israel
(E) Major Kautley — (V) Tanzania

(*vii*) (A) Mahi-Kadana Project — (I) Maharashtra
(B) Highest utilization of available water resources — (II) Telangana region of Andhra Pradesh
(C) KT Weirs — (III) Western Yamuna Canal
(D) Land of Thousand Tanks — (IV) Tamil Nadu
(E) Ferozshah Tughlaq — (V) Gujarat

FURTHER READING REFERENCES

Adhikari, K.D. *et al.*, eds. (2000). *Cooperation on Eastern Himalayan Rivers, Opportunities and Challenges.* Konark Publishers Pvt. Ltd., Delhi.

Asian Institute of Technology (1999). *Evaluation of Operational Performance Information and Communication in a Large Irrigation System.* Phitsanulok Irrigation System, Thailand. Monograph No. 1, Water Engineering and Management Program, School of Civil Engineering, A.I.T., Bangkok, p. 140.

Asian Productivity Organization, Tokyo (1997). *Farm Level Irrigation Management.* Report of the Study Meeting in Lahore, Pakistan, Feb. 7–18, 1989. Country Reports: Iran, Massoudi, H.; *Republic of Korea* (1): Kwun, S.K.; *Republic of Korea* (2): Ryu, K.S.; Malaysia: Ismail, A.B.; Nepal : Satyal, R.P.; Pakistan (1): Salim, A.M.; Pakistan (2): Choudhary, M.A.; Philippines: Jorge, O.F.; Thailand (1): Lekmanece, P.; Thailand (2): Sumitthemdhendra, W.

Bagchi, K.S. (1995). *Irrigation in India—History and Potentials of Social Management.* Uplabhi Trust for Developmental Initiatives, Delhi–110065, p. 149.

Bakkar, M. *et al.* (1999). *Multiple Uses of Water in Irrigated Areas*: A Case Study from Sri Lanka, International Irrigation Management Institute, Colombo, Sri Lanka, SWIM Paper, p. 48.

Bhatia, P.C. (1999). *Crops and Weather Resources.* In: Singh, G.B. and Sharma B.R. (eds.). 50 Years of Natural Resources Management Research, ICAR, New Delhi, pp. 49–62.

Bhavanishankar, B.S. (2001). Sustainable Water Resources Development and Management. Proc. National Workshop on Water Resources and Environmental Problems, 2001, Centre for Water and Environmental Studies, Kumaraguru College of Technology, Coimbatore, Tamil Nadu, p. 189.

Bhavanishakar B.S. (2001). *Dams, Drought and Distress.* Proc. Workshop on Dams and Droughts, October 2001. Organized by SAHAYOGA, Bangalore.

Biswas, A.K. and El-Habr, H.N. (1993). *Environment and Water Resources Management : The Need for a Holistic Approach.* Water Resources Development, Vol. 3, No. 2, pp. 117–125.

Biswas, A.K. and Qu Geping, eds. (1987). *Environmental Impact Assessment for Developing Countries.* Pub. for the United Nations University by Tycooly International, London, p. 232.

Biswas, A.K., Mohammad Jellali and Slout, G. eds. (1993). *Water for Sustainable Development in the Twentifirst Century.* Oxford University Press, Oxford, U.K.

Bottrail, A.F. (1981). *Comparative Study of Management and Organization of Irrigation Projects.* The World Staff Working Paper No. 458. The World Bank, Washington D.C. USA, p. 274.

Bruns, B.R. and Munzen R.S. eds. (2000). *Negotiating Water Rights.* Sage Publications (India) Pvt. Ltd., New Delhi.

Buckley, R.B. (1880). *The Irrigation Works of India and Their Financial Results.* Pub. W.H. Allen and Co., 13 Waterloo Place (Publishers to the India Office), p. 198. (*Rare Manuscript*, preserved and kept in the Central Library, IARI, New Delhi).

Buckley, R.R. (1905). *The Irrigation Works of India.* 2nd ed., E and F.N. Spon Ltd., London. (A.R. Choudhary).

Cantor, L.M. (1970). *A World Geography of Irrigation.* Oliver and Boyd, Edinburgh UK, p. 252.

Central Board of Irrigation and Power (1979). *Register of Dams in India.* Pub. C.B.I.P., New Delhi, p. 119.

Central Board of Irrigation and Power (1979). *Register of Water Resources Projects in India.* Publication No. 36, CBIP, New Delhi, p. 100.

Central Board of Irrigation and Power (1981). *Warabandi System and its Infrastructure.* CBIP, New Delhi, p. 78.

Central Ground Water Board (1982). *Ground Water Resources and Development Potential in India.* Ministry of Irrigation, Govt. of India, New Delhi, p. 112.

Central Ground Water Board (1995). *Ground Water Resources of India.* Ministry of Water Resources, Govt. of India, Faridabad, Haryana.

Central Water Commission, Government of India (1997). *Water Resources of India.* Min. of Water Resources, New Delhi, p. 36.

Central Water Commission (1996, 2000, 2002). *Water and Related Statistics.* Information System Directorate, Performance Overview and Management Information Organization, Min. of Water Resources. Govt. of India.

Centre for Water Resources, College of Engineering, Anna University of Technology, Chennai (1982). *Modernizing Tank Irrigation: Problems and Issues.* Pre-Workshop paper, February 10–12, 1982.

Centre for Water Resources Development and Management, Calicut (1984). *Inter-basin Transfer of Water*. Proc. National Workshop on Inter-basin Transfer of Water, 27–28 January, 1984, p. 114.

Chambers, R. and Carruthers, I. (1986). *Rapid Appraisal to Improve Canal Irrigation Performance: Experience and Options*. International Irrigation Management Institute, Colombo, Sri Lanka, p. 18.

Chambers, R. (1998). *Managing Canal Irrigation*. Oxford & IBH Pub. Co. Pvt. Ltd., New Delhi, p. 279.

Chitale, N.A. (2000). *A Blue Resolution*. Bharatiya Vidya Bhawan, Pune, p. 117.

Critchfied, H.J. (1974). *General Climatology*. 3rd ed., Prentice Hall Inc., New Jersey, U.S.A., p. 446.

Dahigaonkar, J.G. (1982). *A Text Book of Irrigation Engineering*. A.H. Wheeler & Co. (P) Ltd., Allahabad, p. 250.

Dakshinamurti, C., Michael, A.M. and Shri Mohan (1973). *Water Resources of India and Their Utilization in Agriculture*, Water Technology Centre, I.A.R.I., New Delhi, p. 400.

Dhawan, B.D. (1986). *Irrigation and Water Management in India : Perception of Problems and Their Solutions*. Ind. Jl. of Agri. Econ., Vol. 41, No. 3., pp. 271–282.

Dhawan, B.D. (1989). *Water Resource Management in India*. Ind. Jl. Agri. Econ., Vol. 44, No. 3, pp 233–241.

Dhawan, B.D. (1998). *India's Irrigation Sector : Myths and Realities*. National Bank, Mumbai, News Review, July-September 1998, pp. 34–46.

Directorate of Economics and Statistics, Ministry of Agriculture and Cooporation, Government of India (1998). *Land Use Statistics at a Glance. Part I.*

Dooge, J.C.I., Costin A.B. and Finkel, H.J. (1973). *Man's Influence on the Hydrological Cycle*. Food and Agriculture Organization of the United Nations, Rome Irrigation and Drainge Peper, *Special Issue*, p. 71.

Dougharty, T.C. and Hall, A.W. (1995). *Environmental Impact of Irrigation and Drainage Projects*. Overseas Div. Administration of the United Kingdom and Food and Agriculture Organization of the United Nations, Rome. Irrigation and Drainage Paper 53, p. 74.

Ellis, W.M. (1903). *College of Engineering Manual: Irrigation*. Controller. Govt. of Madras, Controller of Stationery and Printing, Madras, (Briefly revised in 1950), p. 455.

Engman, T. ed. (1992). *Irrigation and Drainage, Saving a Threatened Resource in Search of Solutions*. Am. Soc. of Civil Engs., New York, p. 124.

Food and Agricultural Organization of the United Nations (1978). *World Irrigation and Land Use Statistics*. FAO Production Yearbook, Vol. 32. FAO Statistics Series, No. 22. FAO, Rome.

Food and Agricultural Organization of the United Nations (1973). *Man's Influence on the Hydrological Cycle*. FAO, Rome, Special Issue. No. 17, p. 71.

Food and Agriculture Organization of the UN (1984). *Participatory Experiences in Irrigation Water Management*. Proc. Expert Consultation on Irrigation Water Management held in Yogyakarta and Bali, Indonesia, 16–22 July, 1984, Pub. FAO, Rome, p. 280.

Food and Agriculture Organization of the United Nations, Rome (1986). *Irrigation in Africa South of the Sahara*. FAO, Investment Centre Tech. Paper 5, p. 175.

Food and Agriculture Organization of the United Nations (1987). *Consultation on Irrigation in Africa*. FAO, Rome, Irrigation and Drainage Paper 42, p. 206.

Food and Agricultural Organization of the United Nations Statistical Database. FAO (2000). Online internet 18th July, 2000. From http://apps.fao.org/page/collections

Food and Agricultural Organizations of the United Nations (2001). AQUASTAT Database. FAO (2001). From http://www.fao.org/waicent/faoinfo/agricult/agl/aquastat/profile.htm

Food and Agricultural Organizations of the United Nations AGL-Global Network on Integrated Soil Management for Sustainable Use of Salt-affected Soils. On Line Internet. FAO, 2001. From http://www.fao.org/agl/agl/spush topics.htm

Framji, K.K. and Mahajan, I.K. (1969). *Irrigation and Drainage in the World, FAO, Production Year Book*, 1968, Vol. 22, International Commission on Irrigation and Drainage, Chanakyapuri, New Delhi.

Framji, K.K., Garg, B.C. and Luthra, S.D.L. (1981–83) Irrigation and Drainage in the World, A Global Review. Vol. I, 1981; Vol. II, 1982; Vol. III, 1983. International Commission on Irrigation and Drainage, Chanakyapuri, New Delhi–110021, p. 1667.

Goluber, G.N. and Biswas, A.K. eds. (1985). *Large Scale Water Transfers: Emerging Environmental and Social Experience*. Pub. for the United Nations Environment Programme, by Tycooly Publishing House Ltd., Oxford, England, p. 158.

Gopalaswamy, N. (1994). *Agro-climatic Classification. In: Agricultural Meteorology*. Rawat Publications, Jaipur and New Delhi, p. 92.

Government of Andhra Pradesh (1979). *Guidelines for Introduction of Warabandi System in Irrigation Projects of Andhra Pradesh*. Command Area Development Dept., Govt. of Andhra Pradesh, p. 16.

Government of Maharashtra (1979). *Traditional Water Supply on Girna Canal System*. Command Area Development Authority, Irrigation Dept., p. 20.

Government of Maharashtra (1983). *Manual of Minor Irrigation Works in Maharashtra State*. Director, Govt. Printing and Stationary, Mumbai, p. 250.

Gulati, N.D. (1972). *Development of Inter-State Rivers, Law and Practice in India*. Allied Publishers, Mumbai, p. 319.

Hafez M. and Shenouda, W.K. (1983). *Environmental Impacts of the Aswan High Dam*. Water World, Vol. V, No. 2, pp. 15–20.

Hiremath, C.R., (1987). *Cooperative Irrigation and Water Management Systems : Indian Experience*. Regionl Office for Food and Agricultural Organization of United Nations, Bangkok. RAPA Pub. 1987/6, p. 65.

Horst Lucas (1998). *The Dilemas of Water Diversion Considerations and Criteria for Irrigation System Design*. International Water Management Institute, Colombo, Sri Lanka and Wgeningen Agricultural University, The Netherlands, p. 123.

Ibrahim, A.M. (1983). *The Environmental Impact of Jonglei Canal Project in Sudan*. Water World, Vol. V, No. 2, pp. 9–14.

Indian National Academy of Engineering (1990). *Water Management Perspective, Problems and Policy Issues*. Pub. INAE, New Delhi, p. 88.

Indian Council of Agricultural Research (2000). *Natural Resources Management Research—Vision* 2020. I.C.A.R., New Delhi, p. 45.

Indian Water Resources Society (1997). *River Basin Management Issues and Options*. Theme Paper, Water Resources Day (1997). I.W.R.S., New Delhi.

Indian Water Resources Society (1998). *Five Decades of Water Resources Development in India*, Theme Paper, Water Resources Day, 1998. I.W.R.S., New Delhi, p. 53.

Indian Water Resources Society (1999). *Water : Vision 2050.* Theme paper, Water Resources Day, 1999. I.W.R.S. New Delhi, p. 74.

Institute of Scientific Cooperation (1975). *Natural Resources and Development.* Pub. Institute Fiir Wissenchafthche, Honover, Germany.

International Commission on Irrigation and Drainage (1975). *Multilingual Technical Dictionary an Irrigation and Drainage* English/French. Pub. ICID, Chanakyapuri, New Delhi–110021, p. 1262.

International Commission of Irrigation and Drainage (1994), *Report of the Special Session on Aral Sea Basin.* Varna, Bulgaria, 19 May 1994. ICID – CID, New Delhi, p. 74.

International Commission on Irrigation and Drainage (1995). *Watson Activities, Some Lessons and Experiences.* I.C.I.D, New Delhi, p. 32.

Inernational Hydrological Decade (1978). *World Water Balance and Water Resources of the Earth.* Pub. USSR Committee for the International Hydrological Decades, U.N. Educational, Scientific and Cultural Organization. UNESCO Press, Paris, p. 663.

Jancy Vijayan (1997). *Appraisal of a Proposal for Inter-Basin Water Transfer.* M.Tech. Thesis, Dept. of Civil Engg., Indian Instt. of Tech., New Delhi.

Joshi, L.K., 1997. Irrigation and Its Management in India, In: Joshi, L.K. (editor) "Management of Irrigation, A New Paradigm, Participatory Management", Ministry of Water Resrouces, Government of India, New Delhi.

Kanwar, J.S. ed. (1988). *Water Management—The Key to Developing Agriculture.* Proc. of National Seminar Organized by Indian National Science Academy, 28–30 April 1986. Pub. Agricola Publishing Academy, New Delhi, p. 789.

Khosla, A.N. (1949). *Appraisal of Water Resources— Analysis and Utilization of Data.* Proc. of the U.N. Scientific Conference on Preservation and Utilization of Resources UNESCO, New York.

Korten, F.F. (1982). *Building National Capacity to Develop Water Users' Associations: Experience from Philippines.* The World Bank Staff Working Paper No. 528 ARDD, The World Bank, Washington D.C., p. 69.

Krishnan, A. and Singh, M. (1968). *Soil-Climatic Zones in Relation to Cropping Patterns.* Proc. of Symposim on the Cropping Patterns of India. I.C.A.R., New Delhi, pp. 177–185.

Krishnan, A. (1988). *Delineation of Soil-Climatic Zones of India and Its Practical Application in Agriculture.* Fertilizer News 4, pp. 11–19.

Maria Saleth, R. (1996). *Water Institutions of India: Economics, Land and Policy.* Commonwealth Publishers, New Delhi (for Institute of Economic Growth, Delhi), p. 299.

Michael, A.M. (1986). *Conjunctive Use of Surface and Ground Water for Irrigation in Bangladesh.* Report (W.M. 6–86) submitted to Bangladesh Agricultural Research Council, Winrock Institute for Agriculture Development, Dacca, p. 42.

Michael, A.M. and Ojha, T.P. (1999). *Principles of Agricultural Engineering,* Vol. II, 3rd ed. Jain Brothers, East Park Road, New Delhi, p. 888.

Michael, A.M. and Walia, S. and Parmar B.S. (1993). *Emerging Water Quality Hazards and Their Management in Populated Alluvial Zones of India.* Proc. Stockholm Water Symposium, 10–14 August 1992, Stockholm, Sweden, pp. 253–265.

Michael, A.M. (2002). *Blue revolution—A Step towards Water Conservation and Management, small holder irrigation, drainage and water reuse to achieve sustainable public health engineering and food security and ushering in 2nd Green Revolution in India.* Theme Paper, Second International Conference on Sustainable Agriculture, Water Resources Development and Earth Care Policies. Bhoovigyan Vikas Foudnation, New Delhi, pp. 59–98.

Ministry of Environment and Forests, Govt. of India (1985). *Guidelines for Environmental Impact Assessment of River Valley Projects.* Dept. of Environment and Forests, Bikaner House, New Delhi.

Ministry of Information and Broadcasting (1955). *The Gazettes of India, Indian Union,* Vol. I, Country and People. Publications Div., Govt. of India, New Delhi.

Ministry of Irrigation, Govt. of India, New Delhi (1981). *Ground Water Development in India,* p. 66.

Ministry of Water Resources, Govt. of India (1987). *National Water Policy,* p. 15.

Ministry of Water Resources, Govt. of India (1995). *Annual Report* 1994–95 (p. 113), 1996–97 (p. 116).

Ministry of Water Resources, Govt. of India (1999). *Report of the National Commission for Integrated Water Resources Development Plan,* New Delhi, p. 541.

Ministry of Water Resources, Govt. of India (2002). *National Water Policy.* April 2002, p. 18.

Ministry of Water Resources, Govt. of India (2003). *Vision of Integrated Water Resources Development and Management.* Feb., 2003. p. 53.

Ministry of Water Resources, Govt. of India, New Delhi (1997). *Report of the Working Group on Participatory Irrigation Management for the IX Five Year Plan.* Command Area Development Wing, p. 77.

Mistry, J.F. (1991). *Large Dams vs. Small Dams. Theme Paper.* Symposium, Board of Irrigation and Power, New Delhi, December 2, 1991, p. 63.

Mohan Dharia (2002). *Management of Water, Development of Water, Land and other Natural Resources for Self Reliant India.* Keynote Address Second International Conference on Sustainable Agriculture, Water Resources Development and Earth Care Policies. Bhoovigyan Vikas Foundation, New Delhi, p. 6.

Mohile, A.D. (2003). *Inter-basin Transfer of Water in India — Prospects and Problems.* A Bridge Paper for discussion. Water Management Forum, Institution of Engineers (India). Feb. 2003, p. 23.

Naegamvala, J.P. (1973). *Development of Water Resources in India for Irrigation and Hydel Power Generation.* Proc. National Seminar on Water Resources of India and Their Optimum Utilization in Agriculture, Water Technology Centre, IARI, New Delhi, Feb. 2, 1973, New Delhi, pp. 1–24.

Nag, B.S. and Kathpalia, G.N. (1975). *Water Resources of India. In Water for Human Needs.* Proc. Second World Congress on Water Resources, Vol. II, Health and Planning, Organised by International Water Resources Association and Central Board of Irrigation and Power, New Delhi, pp. 373–382.

National Bureau of Soil Survey and Land Use Planning, Nagpur (1997). *Agro-Ecological Regions and Agro-Ecological Sub-Regions.* (Revised). Personal Correspondence.

National Land Use and Conservation Board, Ministry of Agriculture, Govt. of India, New Delhi (1991). *National Consulation on Perspective Plan for Conservation, Management and Development of Land Resources of the Country.* Discussion paper, Oct. 21–24, 1991, New Delhi.

Oxford University Press (2000). *Oxford School Atlas,* 30th ed. Oxford University Press, Jai Singh Road, New Delhi–110 001, p. 96.

Philip, J.R. (1977). *Water on the Earth.* Proc. Symposium on Water, Plants and People. Edited by McIntyre, A.K., Australian Academy of Science.

Planning Commission, Government of India (1972). *Report of the Task Force on Ground Water Resources,* New Delhi, p. 131.

Raghava Rao, K.V., Raju, T.S. and Ramesan, V. (1969). *An Estimate of Ground Water Potential of India-First Approximation.* Proc. ICAR Symposium on Soil and Water Management, Hisar, March 1969.

Rajput, T.B.S. and Michael, A.M. (1986) *Modification in Warabandi Rostering for Equitable Water Distribution.* Jl. of Irrigation and Power, pp. 59–67.

Ramdas, L.A. (1960). *Crops and Weather in India.* Pub. Indian Council of Agrl. Research, New Delhi, p. 127.

Ramdas, L.A. (1966). *Weather and Crops. In: Handbook of Agriculture.* Indian Council of Agrl. Research, New Delhi, pp. 1–31.

Randhawa M.S. (1980). *History of Agriculture in India,* Vols. I, II and III. Pub. Indian Council of Agricultural Research, New Delhi.

Rangachari R., Sengupta N., Iyer G.R. *et al.* (2001). Large Dams in India: An Overview Proc. Workshop on Dams and Droughts, 2000–2001. Organised by SAHAYOGA, Bangalore.

Rao, K.N., George, C.J. and Ramasastri, K.S. (1972). *Agro-climatic Classification of India.* Agricultural Materological Monograph, Met. 4 India Meteorology Department, Pune.

Rao, K.L. (1975). *India's Water Resources.* Orient Longmans Ltd., New Delhi, p. 255.

Renault, D. and Godaleyadda, G.G.A. (1999). *Generic Typology for Irrigation Systems Operation.* International Water Management Institute, Colombo, Sri Lanka, p. 22.

Saksena, K.K. (1995). *Design and Operation of Irrigation Systems for Conjunctive Water Management.* Doctoral Thesis, Dept. of Civil Engineering, Indian Institute of Technology, New Delhi, p. 333.

Saktivadivel, R. *et al.,* (1999). *Modernization, Using the Structured System Design of the Bhadra Reservoir Project, India : An Intervention Analysis.* Pub. International Water Management Institute, Colombo, Sri Lanka, Res. Report. 33, p. 24.

Sandra Postel (1999). *Pillar of Sand—Can the Irrigation Miracle Last?* W.W. Nortus & Company Ltd., Inc., New York, For the World Watch Institute, p. 313.

Schendal, U. (1975). *The World's Water Resource and Water Balance. In: National Resources Development,* Vol. I., Institute of Scientific Cooperation, Tubingen, Germany.

Seckler, D. *et al* (1998). *World Water Demand and Supply 1990–2025: Scenarios and Issues.* International Water Management Institute, Colombo, Sri Lanka, Res. Report 19.

Seghal J.L., Mandal, D.K., Mandal, C. and Vadivelu S. (1992). *Agro-Ecological Regions of India.* Pub. No. 24, National Bureau of Soil Survey and Land Use Planning, Nagpur, p. 130.

Shanmugham, C.R. (2000). *Tank Irrigation Management and Policy Issues.* Paper presented at the session on Water Resources Management in the International Conference on Managing Natural Resources for Sustainable Agricultural Production in the 21st Century. Feb. 14–18, 2000. I.C.A.R., New Delhi.

Shanmugham, C.R. and Karagavali, K. (2000). *Design of Supply Channels—Lifeline for Chain of Tanks.* Jl. Tank Cascade, Vol. 2, Nos. 1 and 2, pp. 65–76. Dhan Foundation, Madurai, Tamil Nadu.

Sharma, B.R. and Paul, D.K. (1999). *Water Resources of India. In: 50 Years of Natural Resource Management Research,* ICAR. New Delhi, pp. 31–48.

Singh, G.B. and Sharma, B.R. (1999). *Fifty Years of Natural Resource Management Research.* Indian Council of Agricultural Research. Division of Natural Resource Management. New Delhi, p. 644.

Singh, R.P. (1991). *Farm Ponds.* Pub. Central Research Institute for Dryland Agriculture, Hyderabad, p. 51.

Singh, S.R.; Chaudhary, T.N.; Bhalla G.K. *et al.* (1999). *Irrigation Water Management.* In Singh, G.B. and Sharma, B.R. 50 Years of Natural Resources Management Research. ICAR, New Delhi, pp. 265–370.

Singh, S.R. (1998). *People Managed Irrigation Systems in Partnership Mode-Status and Method for Performance Evaluation.* International Workshop on Peoples' Participation in Irrigation Systems Management, Div. of Natural Resources Management, ICAR, New Delhi, April 22–24, 1998, p. 14.

Subramanyam, V.P., Subba Rao, B.; and Subramanyam, A.R. (1965). *Koppen and Thornthwaite Climatic Classification as Applied to India.* Annals of Arid Zone, 49, pp. 46–55.

Swaminathan, M.S. (1995). *Population, Environment and Food Security Issues in Agriculture,* No. 7, Consultative Group on Intl. Agrl. Research.

Tawalchai Tingranchali, ed. (1993). *Environmentally Sound Water Resources Utilization.* Proc. Intl. Conference, 8–11 November, 1983, Vol. I and II. Water Resources Engineering Programme, School of Civil Engineering, Asian Institute of Technology. Bangkok, Thailand.

Thanh, N.C. and Biswas, A.K. (1990). *Environmentally Sound Water Management.* Oxford University Press, New Delhi, p. 276.

Thornthwaite, C.W. (1948). *An Approach Towards a National Classification of Climate* Geographical Review 38: pp. 55–94.

Thornthawiate, C.W. and Malter, J.R. (1955). *The Water Balance. In :* Climatology, Vol. 8, No. 1, p. 104. Drexel Institute of Technology, New Jersey, U.S.A.

Upkoff, N., Meinegen-Dick, R., Nancy St. Julien (1985). *Improving Policies and Programmes for Farmer Organization and Participation in Water Management.* Dept. of the Govt., Rural Sociology, in Regional Planning, Cornell University, Ithaca, New York, p. 41.

United Nations Educational, Scientific and Cultural Organization (1978). *World Water Balance and Water Resources of the Earth.* UNESCO Press, Paris, p. 663.

Van Oppen, M. and Subba Rao, K.V. (1980). *Tank Irrigation in India, Part I: Historical and Spatial Distribution.* International Crops Research Institute for Semi-Arid Tropics, Patancheru, Andhra Pradesh, p. 24.

Veiraih, L.H. (2001) *Dams, Droughts and Distress.* Proc. Workshop on Dam and Drought, Oct. 28, 2001, organised by SAHAYOGA, Bangalore.

Vincent, L. (1998). *Paticipation Paradigm and Their Prospects to Enhance Production Through Improved Irrigation Management.* Proc. Intl. Workshop on Peoples' Participation in Irrigation System Management, April 22–24, 1998. Div. of Natural Resources Megt., ICAR, New Delhi.

Walker, S.H.(1998). *International Experiences on Participatory Management of Canal Systems.* Proc. Intl. Workshop on Peoples, Participation in Irrigation Systems Management, April 22–24 1998. Div. of Natural Res. Management, ICAR, New Delhi, p. 15.

Water and Land Management Institute, Aurangabad (1989). *Water Distribution Practices in Maharashtra State.* 2nd ed. WALMI, Aurangabad, Maharashtra, Publication No. 22, p. 67.

Water Technology Centre and Indian Society of Water Management, I.A.R.I., New Delhi (1998). *Water Management for Sustainable Agriculture—Problems and Perspectives for the 21st Century.* Proc. of Natl. Seminar, W.T.C., I.A.R.I., April 15–17, 1998, p. 516.

Water Technology Centre, I.A.R.I., New Delhi (1983). *Resource Analysis and Plan for Efficient Water Management—A Case Study of Mahi Right Bank Canal Command Area, Gujarat,* p. 360.

Wittlesey, N.K. and Huffaker, R.G. (1995). *Water Policy Issues for the Twenty-first Century.* Am. J. Agric. Econ., Vol. 77, No. 5, pp. 1199–1203.

Wolff, P. (1999). *On the Sustainability of Water Use.* Natural Resources and Development, Vol. 49/50. The Institute of for Scientific Cooperation, Tiibingen, Federal Republic of Germany, pp. 9–27.

World Bank (1998). *India — Water Resources Management Sectoral Review: Report on The Irrigation Sector.* Report No. 18416. Rural Development Unit, South Asia Region, World Bank, New Delhi, p. 135.

World Bank (1998). *India — Water Resources Management Sectoral Review: Report on The Inter-Sectoral Water Allocation,Planning and Management, Vol. 1, Main Report No.* 18322. Rural Development Unit, South Asia Region, World Bank, New Delhi, p. 98.

World Bank (1998). *India—Water Resources Management Sectoral Review: Ground Water Regulation and Management Report.* Rural Development Unit, South Asia Region, World Bank, New Delhi, p. 98.

World Book Encyclopedia (1996). *Water and the Course of History.* Pub. The World Book International, World Book Inc, a Scott Fetzer Co., Chicago, U.S.A., pp. 68–83.

World Commisssion on Dams (2000). Dams and Development: A New Framework for Decision Making. Earthscan Publications Ltd., 120 Pentonville Road, London N1 9JN, U.K.

World Commission on Dams (2001). Global Dam Statistics "China". Region Consultations. Online. Internet. From http://www.dams.org/global/china.htm

Yadav, J.S.P. and Singh, G.B. (eds.) (2000). *Natural Resources Management for Agricultural Production in India.* Proc. of Intl. Conf. on Managing Water Resources for Sustainable Agricultural Production in the 21st Century, Feb. 14–18, 2000. Indian Soc. of Soil Science, New Delhi, p. 1022.

World Meteorological Organization (1975). *Drought and Agriculture.* Tech. Note No. 138. Secretariate of the World Meteorological Organization, Geneva, p. 127.

Chapter

2

Irrigation Wells

LEARNING OBJECTIVES

Characteristics of water bearing formations. Groundwater recharging. Groundwater exploration. Aquifer characteristics influencing yield of wells. Hydraulics of wells. Predicting well yield. Steady and unsteady state flows in confined, semi-confined and unconfined aquifers. Evaluation of hydraulic properties of aquifers. Pumping tests in wells—in hard rock and semi-consolidated formations. Type of tube well screens and their materials. Gravel packing in tube wells. Design procedures for gravel pack and slot size of well screens in tube wells. Spacing between wells. Hydraulics of multiple well systems, procedures. Direct rotary, reverse rotary and air rotary drills, including down-the-hole hammer drills and their adaptability. Drilling procedures for various types of drills. Well logs and their use. Installation of well screens. Procedures for gravel packing in tube wells. Sealing of brackish and saline water horizons. Procedures for development of tube wells. Testing wells for drawdown and yield. Common troubles in irrigation wells and their remedies. Protection of well screens affected by corrosion and incrustation. Rehabilitation of sick and failed wells.

A water well is a hydraulic hole in the earth down to a supply of water excavated for the purpose of bringing groundwater to the surface. Occasionally, wells serve other purposes, such as sub-surface exploration and observations and artificial groundwater recharge. Irrigation wells differ from those used to supply water for domestic purposes because of the large volumes of water that have to be pumped from them for irrigating even small farms. Efficient and economical utilization of groundwater through wells depend on the design of wells to best suit the characteristics of the water-bearing formations. Flow of groundwater into wells is influenced by the physical characteristics of the water bearing formations, the number and extent of these formations, design of wells and the methods used for constructing and developing the wells.

GROUNDWATER AND AQUIFERS

Types of Water-Bearing Formations

An understanding of groundwater occurrence requires a study of the vertical distribution of water in the sub-surface geologic formations. The earth's crust is called the lithosphere. The lithosphere is composed predominantly of rocks, consisting of disintegrated rock materials such as granite and sandstone. The lithology of a section through the earth's crust reveals the kind of rocks that occur in a succession of layers or strata below the surface that make up any part of the lithosphere.

The outer part of the earth's crust is usually porous to varying depths at different places. This is the zone of rock fracture. The pores or openings in this part of the lithosphere may be partially or completely filled with water. In the surface strata, the openings are only partially filled with water. This strata is called the *zone of aeration*. The layer below this, where the openings are completely filled with water, is called the *zone of saturation* (Fig. 2.1). The zone of aeration is divided into three zones—the soil water zone, the intermediate zone, and the capillary fringe. The zones vary in depth and their limits are not sharply differentiated by the differences in the physical properties of the earth materials. A gradual transition exists from one zone to another.

Formations or strata within the saturated zone below the ground surface, from which groundwater can be obtained for beneficial use, are called aquifers. Groundwater reservoirs or water-bearing formations are the other terms commonly used in place of the word aquifer. Aquifers are permeable geologic formations that permit appreciable amounts of water to move through them. Most aquifers extend to large areas and may be visualised as underground storage reservoirs. The thickness of the zone of saturation varies from a few metres to hundreds of metres.

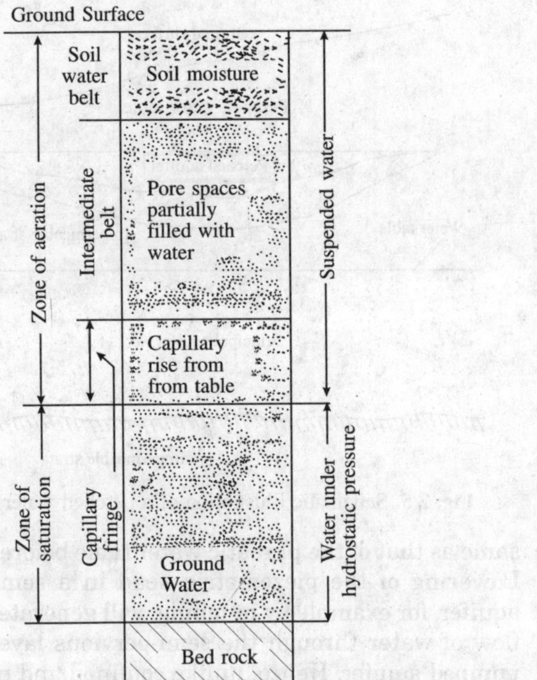

Fig. 2.1 Diagram illustrating soil-moisture relationship and groundwater in an underground strata.

The portion of the rock or soil not occupied by the solid material may be occupied by groundwater. These spaces are known as voids, interstices, pores or pore space. The porosity of a rock or soil is a measure of the contained interstices. It is expressed as the percentage of void space to the total volume of the mass as follows:

$$p = \frac{V_w}{V} \times 100 \qquad \text{... (2.1)}$$

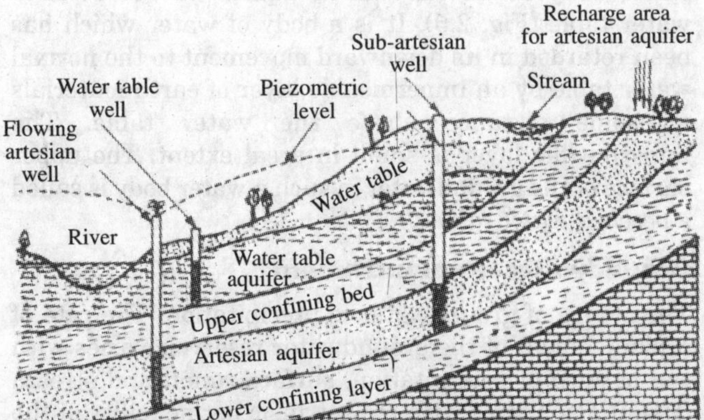

Fig. 2.2. Schematic sketch illustrating artesian and water table aquifers, water table well, flowing artesian well, sub-artesian well and potential recharge area for the artesian aquifer.

in which,

p = porosity (per cent)

V_w = volume of water required to fill or saturate the entire pore space (cm^3 or m^3)

V = total volume of rock or soil (cm^3 or m^3)

The size of the pores and the total pore space of an aquifer vary with the type of formation material. Individual pores in a fine grained material like clay are extremely small but the total pore space is usually large. While a clay formation has a large water-holding capacity, water cannot move readily through the tiny pores and is hence not an aquifer, even though it may be water saturated. On the other hand, a coarse material such as sand contains large pores through which water can move fairly easily. A saturated sand formation is an aquifer, as it can hold water and transmit it at a satisfactory rate when pressure differences occur.

Water enters a groundwater reservoir from natural or artificial recharge; it flows out under the action of gravity or is extracted by wells.

Unconfined aquifer: An unconfined aquifer (Fig. 2.2) is a permeable bed only partly filled with water and overlying a relatively impervious layer. Its boundary is formed by a free water table or phreatic level. It is also known as a free, phreatic water-table or non-artesian aquifer. The water in a well penetrating an unconfined aquifer does not, in general, rise above the phreatic level (Fig. 2.2 and 2.3).

The upper surface of the zone of saturation is called the *water table*. At the water table, the water in the pores of the aquifer is at atmospheric pressure. The hydraulic pressure at any level within a water table aquifer is equal to the depth from the water table to the point and is referred to as the *hydraulic head*. When a well is dug in a

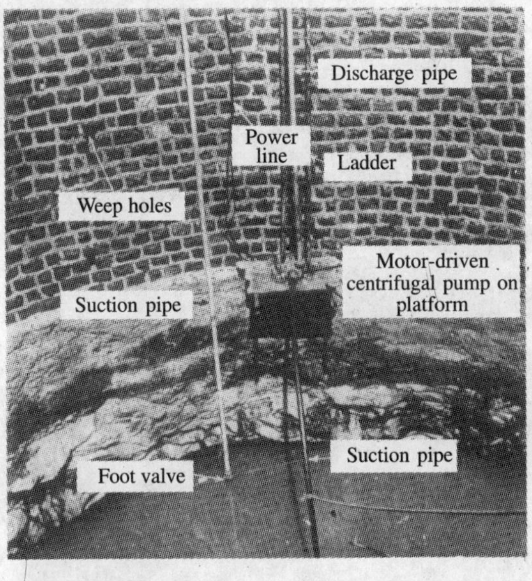

Fig 2.3. Inside view (in segment only) of a brick-lined open well. The flow of water from the aquifer into the well is through the crevices and weep holes.

water table aquifer, the static water level in the well stands at the same elevation as the water table (Figs. 2.2 and 2.3). In general, the shape of the water table tends to follow the topography of the ground surface. It is, however, not a stationary surface but rises or falls with the addition or withdrawal of water.

Confined aquifer: The zone of saturation may consist of permeable, impermeable and semi-permeable earth materials. An aquifer found between two impermeable layers is said to be *confined* (Fig. 2.2). It is also called an *artesian aquifer*. Because of the presence of an upper confining layer, the water in the pores of a confined aquifer is not open to atmospheric pressure. It is at pressures greater than the atmospheric pressure.

When a well is drilled into a confined aquifer, water rises in the well to some level above the top of the aquifer, depending on the pressure of water in it. The elevation to which the water level rises in a well which taps a confined aquifer is called the *piezometric level* (Fig. 2.2). An imaginary surface representing the hydrostatic pressure in a confined aquifer is called the *piezometric surface*. The piczometric surface is analogous to the water table in an unconfined aquifer. The hydrostatic pressure within a confined aquifer is sometimes great enough to cause the water to rise in a well and flow out, resulting in a flowing artesian well (Fig. 2.4).

Semi-confined aquifer: A semi-confined or leaky aquifer is a completely saturated aquifer that is bounded above by a semi-pervious layer and below by a layer that is either impervious or semi-pervious. A semi-pervious layer has low, though measurable, hydraulic conductivity. The elevation to which the water level rises in a well that taps a semi-confined aquifer is referred to as the peizometric level. The piezometric level is the

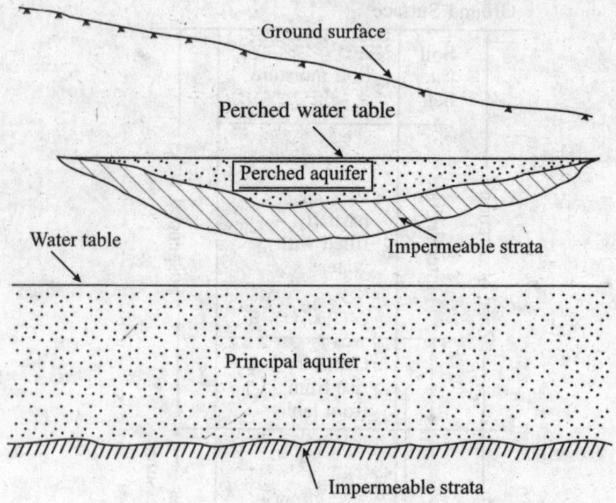

Fig. 2.5. Schematic illustration of a perched water table.

same as that of the phreatic water table before pumping. Lowering of the piezometric head in a semi-confined aquifer, for example by pumping, will generate a vertical flow of water through the semi-pervious layer into the pumped aquifer. Hence, unlike confined and unconfined aquifers, two types of hydraulic heads are developed. They are the phreatic head above the semi-pervious layer lying over the main aquifer and the piezometric head in the main aquifer. The piezometric head is always more than the phreatic head. The ratio of the heads depends on the hydraulic conductivity and the thickness of the semi-previous layer (Yadav, 1973).

In semi-confined aquifers, when the phreatic level is lowered almost simultaneously with the lowering of the piezometric level, the aquifer is called a 'semi-confined aquifer with prompt yield'. When there is a significant time lag between the lowering of the two water levels, the formation is called a 'semi-confined aquifer with delayed yield'.

Perched water table: A special case of a localised water body in an unconfined aquifer is the perched water table (Fig. 2.5). It is a body of water which has been retarded in its downward movement to the normal water table by an impermeable layer of earth materials at some distance above the water table. The impermeable layer is small in areal extent. The upper surface of the groundwater in such a water body is called a *perched water table*.

Groundwater Replenishment

The source of groundwater is precipitation. The rate of replenishment of the groundwater reservoir varies with the pattern of precipitation, surface runoff and stream flow. It also varies with the intrinsic permeability of the soil and other earth materials through which the water must percolate to reach the zone of saturation. In general, groundwater reservoirs are replenished by the rain falling directly on the land surface and by seepage from canals and tanks, provided the intervening layer is permeable.

Fig. 2.4. A free-flowing artesian well in the Thanjavar district of Tamil Nadu (1973).

Artificial groundwater recharge: Optimum development and sound management practices are vital to the sustained use of groundwater. For water bearing formations readily recharged from the surface, average pumpage may equal average recharge, but should not exceed it. For deep artesian aquifers, where a substantial amount of recharge occurs at a region away from the area where it is used, the pumpage may exceed recharge in the area. But groundwater utilization in such cases must be on the basis of quantitative estimates of how much of the total groundwater storage is being used. In coastal regions, however, the average pumping must be considerably less than the average recharge so as to allow the groundwater flow in the seaward direction, thus preventing the intrusion of sea water into well sites.

Methods of Artificial Recharge

Groundwater recharge may be increased by soil and water conservation measures and artificial recharge procedures. The amount of rainfall or ponded surface water infiltrated into the soil varies greatly with the condition of soil surface and the moisture content of the soil at the time of rainfall. The total amount of water infiltrated depends on the infiltration opportunity time, which depends mainly on the slope of the land and the field structures like percolation tanks, check dams, contour bunds and terraces, which tend to hold the runoff water over long periods on the land surface and as instream storage. In addition, a portion of the river flow which cannot be utilised for irrigation can be harnessed to recharge groundwater reservoirs. In recharge operation the water is allowed to stand on the field for a long period. Suitable diversion structures constructed across a stream channel cause the water level to rise on the upstream side, which can be diverted to the land surface. This type of application, which is generally called *water spreading,* can be broadly classified into (*i*) *basin method,* and (*ii*) *furrow or ditch method*. Of these, the basin method permits a maximum wetted surface area and is mostly recommended.

Basin method: In the *basin method,* water is diverted to the uppermost of a series of basins or percolation tanks separated from one another by low height levees. The levees generally run on contours. They are constructed with materials like stone or boulder, which will withstand the overtopping of water. In places where such materials are not available, suitable structures are constructed on the levees to allow water to drop from the higher basin to the lower. Basin method of recharge is particularly suitable where ground surface is irregular and infested with a number of shallow gullies and ridges.

Furrow method: In the *furrow method* of recharge, water is diverted from a main canal running along a contour into a series of parallel furrows. These furrows differ from irrigation furrows in that they are flat bottomed and shallow to allow maximum possible surface area to come in contact with water.

The selection of a particular method of water spreading depends upon factors like topography, slope, type of soil and cost of labour involved. Selection of sites for water spreading must be made with a knowledge of the geologic nature of the groundwater basin so as to obtain the most efficient utilization of the storage capacity and the transmissibility of the aquifers. This is necessary to ensure that the applied water reaches the zone intended to be replenished. The selection should be made with respect to the texture and structure of the surface soil to obtain the most rapid rates of water movement into and through the soil.

Wells and pits: In areas where the water bearing stratum is overlain by an impermeable layer, surface methods of recharge may not be suitable. Water may penetrate the top soil and as soon as it strikes the impermeable layer it may flow away as inter-flow and may not join the water table. Depending upon the depth of the impermeable layer, groundwater recharge can be done by large pits or wells, the former being used where the impermeable layer is encountered at large depths. Sometimes water is recharged into the soil through tube wells. Recharge by wells is also practised where the land is not available for surface methods of recharge because of cultivation requirements or is unsuitable for water spreading as in urban areas. The volume of water recharged by the well or pit method is low and the method is not practised where surface methods are feasible.

Optimum utilization of groundwater includes well planned pumping from groundwater reservoirs during dry periods in order to create a storage capacity to be filled during the monsoon season. Groundwater reservoirs, like surface reservoirs, are not utilised to the best advantage if they are always kept full or nearly full. Water spreading is done during the rainy season, when the demand for irrigation is low. Facilities for surface water distribution, like canals can often be utilised for spreading water without conflicting with irrigation.

Groundwater Exploration

Detailed study of groundwater and conditions under which it occurs may be made by sub-surface investigations.

Test drilling: The most reliable method to learn the character of the formations beneath the earth's surface is to drill through them, obtain samples while drilling, and record a log of the bore hole. *Well logging* consists of recording the characteristic properties of the samples of sub-surface materials obtained during the progress of drilling. Test drilling is done to satisfy two objectives: (*i*) to explore the groundwater resources of a region, and (*ii*) to obtain information on the geologic and groundwater conditions of a well site prior to installing a regular well. Almost any well drilling method may be used for test drilling; however, in unconsolidated formations, cable tool and hydraulic rotary methods are the most common. The cable tool method is slower but provides more accurate samples from the bailer. The rotary method is faster but it is sometimes difficult to determine the exact character of the formation. This is particularly true where fine-grained materials are encountered as they mix with the drilling fluid.

Geophysical Methods

Geophysical methods provide indirect evidence of the sub-surface formations that indicates whether the formations may possibly be aquifers. The methods do not directly measure the type of rock, porosity, intrinsic permeability or mass density of any formation. They measure other properties of the materials which determine whether a formation may be sufficiently porous and permeable to serve as an aquifer.

Geophysical methods are classified as surface-type and bore hole-type operations, depending on whether the measurements are made at or above the ground surface, or below the ground in drilled holes. The commonly used geophysical methods of groundwater prospecting are the electrical resistivity method and seismic refraction method.

Electrical resistivity method: Measurements of electrical resistivity show values that vary with certain aquifer characteristics. For instance, clean sand saturated with fresh water shows relatively high resistivity. A sand containing clay shows a lower resistivity. By measuring the electrical resistivities of the two materials, the higher electrical resistivity of the clean sand is interpreted to mean that it is apt to be a better aquifer than the sand having lower resistivity. Because various types of earth materials generally exhibit characteristic values of resistivity, strata of differing materials can be identified. Highly resistive sands, gravels, and sandstones may be differentiated from low resistive materials such as clays and shales.

Electric logging: The most common bore hole geophysical operation is electric logging (Fig. 2.6). An electric log consists of a record of the apparent resistivities of the sub-surface formations and the spontaneous potentials generated in the bore-hole, both plotted in terms of depth below the ground surface. These two properties are related indirectly to the character of the sub-surface formations and to the quality of water contained in them. They can be measured only in mud filled, uncased bore-holes.

Many types of electrode arrangements are employed for electric logging. Instrumentation for logging varies from simple hand operated portable equipment to truck-mounted, power-driven equipment. The type of equipment depends on the depth of the hole and its diameter. For wells 300 to 1000 metres deep, it is generally best to have power-driven equipment to do the job. Shallow wells can be logged readily by manually-operated portable units. Interpretation is affected by the diameter of the bore-hole, type of fluid in the hole, chemical quality of the formation water, porosity of the formation, degree of mud invasion into the formation, and type of electrode arrangement used in the operation.

One of the most important variables in electric logging is the chemical quality of the groundwater. The formation resistivity varies inversely with the dissolved solids content of the formation water. A clean sand saturated with water having 600 ppm dissolved solids will show a formation resistivity about half that of the same sand containing water with 300 ppm dissolved solids.

Gamma-ray logging: Gamma-ray logging is a bore-hole geophysical procedure based on measuring the natural radiation of gamma-rays from certain radioactive elements that occur in varying amounts in sub-surface formations. The log is a diagram showing the relative emission of gamma-rays, measured in counts per second, plotted against depths below the surface (Fig. 2.7). The curve obtained is similar in appearance to the resistivity curve of an electric log.

Equipment for gamma-ray logging is about the same as that used for electric logging, except for the well probe and detecting device. Logs can be made using a suitable counter for the down-the-hole sensing unit. The counting circuit at the surface converts the number of electric pulses per second received from the probe into voltages which are continuously recorded on a strip of tape or film as the probe is pulled up the hole.

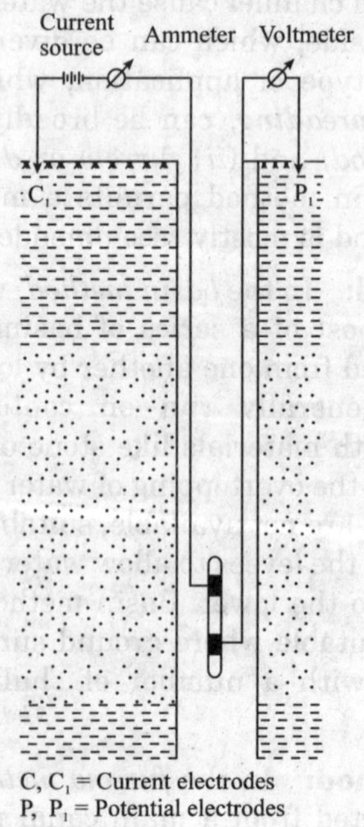

Fig. 2.6. Schematic diagram of electrode arrangement and electrical circuits for the two-electrode logging procedure.

Gamma-ray logs can be obtained in cased and uncased wells. The method can be used in situations where the presence of casing precludes electric logging. Changes in water quality have little effect on the gamma-ray log, so the log is valuable in identifying the position and thickness of clay formations where the alternating sand formations may contain brackish water.

Electrical resistivity surveying: Electrical resistivity surveying is a geophysical prospecting operation in which measurements of earth resistivity are made from the ground surface. The relative values of electrical resistivity are interpreted in terms of the general geology of the sub-surface at limited depths. Measurements of resistivity are made using four electrodes set in the ground. Electric current is applied to the ground through two of the electrodes and the potential drop across the other two is noted. The most common arrangement is with the electrodes equally spaced along a straight line (Fig. 2.8 and Fig. 2.9).

The apparent resistivity is computed from the potential drop, the applied current, and the electrode spacing. The apparatus is usually designed so that the ratio of the potential drop to the applied current is read directly as resistance in Ohms. The value thus obtained is referred to as the apparent resistivity of the lithosphere down to a depth which is proportional to the electrode spacing. Apparent resistivity is considered as being a weighted average of the real resistivities of the individual strata within the depth of penetration of the resistance measurement.

Seismic refraction surveying: The principle of seismic refraction surveying is based on the fact that shock waves travel through different earth materials at different velocities. The denser the material, the faster the waves travel. From measurement differences in velocity, the existence of differing layers of sub-surface materials is identified. The method is suitable where the velocity of the shock wave increases with depth.

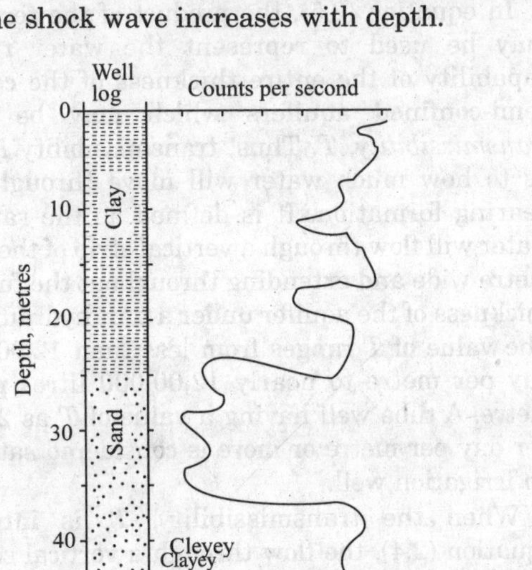

Fig. 2.7. Gamma-ray log and well log in a typical case.

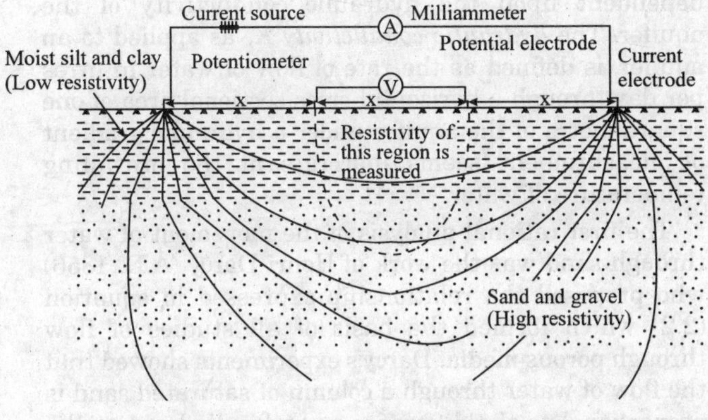

Fig. 2.8. Definition sketch illustrating prospecting of groundwater by electrical resistivity method.

Seismic or shock waves are produced by setting off an explosive charge in a shallow hole, or by striking the ground with a heavy hammer. A seismograph is used to detect and time the arrival of the shock wave over a measured distance from the shock point. The velocity of the first part of the shock wave is then calculated. Comparison of the velocities measured at various distances between the shock point and detector instruments provide a basis for estimating the sub-surface geologic conditions.

Aquifer Characteristics Influencing Yields of Wells

The properties of the aquifer that influence well performance are depth and areal extent of the aquifer, number of aquifers exposed to the well and the hydraulic properties of the aquifer. An aquifer performs two functions, viz., storage function and conduit function. The properties of the aquifer may be expressed in terms of its *hydraulic conductivity, transmissibility, storage coefficient and specific yield.* In case of semi-confined aquifers, two additional properties, viz., *leakage factor and hydraulic resistance* are also important.

Hydraulic conductivity: The rate of flow of groundwater in response to a given hydraulic gradient is

Fig. 2.9. Diagnosing groundwater potential by the electrical resistivity survey.

dependent upon the hydraulic conductivity of the aquifer. The *hydraulic conductivity K*, as applied to an aquifer, is defined as the rate of flow of water in litres per day through a horizontal cross-sectional area of one square metre of the aquifer under a hydraulic gradient of measurement seems inaccrate at the prevailing temperature of water.

The first rational analysis of the movement of water through sand was the work of Henri Darcy (A.D. 1856) who proposed the relationship expressed in equation (2.2) which formed the basis of all studies of flow through porous media. Darcy's experiments showed that the flow of water through a column of saturated sand is proportional to the difference in hydraulic head at the ends of the column, and inversely proportional to the length of the column. This is known as Darcy's law, expressed as

$$v = \frac{K(h_1 - h_2)}{l} \quad \text{... (2.2)}$$

in which,

v = velocity of flow (m/day)

K = hydraulic conductivity, depending upon the properties of the sand and the liquid (m/day)

$h_1 - h_2$ = difference in hydraulic head (m)

l = distance along the flow path between the points h_1 and h_2 (m).

Thus, Darcy's law states the rate of flow through a porous media is proportional to the loss head and inversely proportional to the length of the flow path. The difference in hydraulic head $(h_1 - h_2)$ divided by the distance, l, along the path of fluid flow is called the hydraulic gradient, i.

Thus,

$$v = Ki \quad \text{... (2.3)}$$

Often the quantity of flow may be of greater interest than the velocity. Hence, in terms of quantity of flow, Darcy's law may be expressed as follows:

$$Q = av = Kia \quad \text{... (2.4)}$$

in which,

Q = volume of water discharged in standard length of time, usually expressed as m^3/day.

a = cross-sectional area through which water moves (m^2).

Thus, the property of a water bearing formation that relates to its pipeline or conduit function is the hydraulic conductivity (K) and is termed as the capacity of a porous medium to transmit water. In the past, this property has been called as *coefficient of permeability*. Permeability, on the other hand, refers to the case with which any fluid moves through a formation. The term is more aptly applied to *intrinsic permeability*, k, which is a function of the medium alone and can be expressed as follows:

$$k = K\frac{\mu}{\rho g} \quad \text{... (2.4a)}$$

in which, K = hydraulic conductivity, μ = dynamic viscosity, ρ = fluid density, and g = acceleration due to gravity.

The concept of intrinsic permeability is used in petroleum industry because fluids of different phases (gas/liquid) are analysed for their trade of movement through rock formations.

Hydraulic conductivity is influenced by the size and shape of pores, the effectiveness of the inter-connection between pores and the physical properties of the fluid.

The value of K can be obtained by laboratory tests of the samples of the formation material. For field use however, it is more appropriate to determine the values of K from pumping test data using equation (2.8).

Transmissibility: The slope of the water table or the piezometric surface is the hydraulic gradient under which groundwater movement takes place. The magnitude of flow through any vertical section of an aquifer can be estimated if the thickness, average hydraulic conductivity of the aquifer and the hydraulic gradient at the section are known. The flow, q, through each metre width of the aquifer is,

$$q = Kbi \quad \text{... (2.5)}$$

in which,

K = the average hydraulic conductivity of the material from top to bottom of the aquifer (m/day)

b = thickness of the aquifer (m)

i = hydraulic gradient.

In equation (2.5), the product of the terms K and b may be used to represent the water transmitting capability of the entire thickness of the confined and semi-confined aquifers which may be termed as *transmissibility*, T. Thus, transmissibility Kb indicates as to how much water will move through the water bearing formation. It is defined as the rate at which water will flow through a vertical strip of the aquifer one metre wide and extending throughout the full saturated thickness of the aquifer under a unit hydraulic gradient. The value of T ranges from less than 12,000 litres per day per metre to nearly 12,00,000 litres per day per metre. A tube well having a value of T as 20,000 litres per day per metre or more is considered satisfactory as an irrigation well.

When the transmissibility, T, is introduced in equation (2.4), the flow through a vertical section of an aquifer may be expressed as follows:

$$Q = Tiw \quad \text{... (2.6)}$$

in which w is the width of the vertical section through which the flow occurs. The value of T can be determined from aquifer pumping tests by measuring in one or more wells the decline of head with time under the influence of a constant pumping rate.

Coefficient of storage: The storage properties of confined and semi-confined aquifers are expressed by the coefficient of storage S, which is defined as the volume of water the aquifer releases from or takes into storage per unit surface area of the aquifer per unit change in the component of head normal to the surface (Fig. 2.10). Water recharged to, or discharged from, an aquifer represents a change in the storage volume within the aquifer. In a confined aquifer, the coefficient of storage is a result of compression of the aquifer and expansion of the contained water as a result of reduced pressure due to pumping. The value of S, which is a dimensionless quantity, ranges from 0.00001 to 0.001 for confined aquifers. It may be determined from pumping tests of wells.

Specific yield: Specific yield S_y is the property of an unconfined aquifer. Specific yield is the ratio of the volume of water which a rock or soil, after being saturated will yield by gravity and other inherent soil moisture tensions, to the volume of rock or soil. It is expressed as:

$$S_y = \frac{\text{Volume of water drained}}{\text{Total of volume of the saturated rock or soil}} \times 100$$

When applied to soil the term has identical meaning as 'drainable porosity'. The concept of specific yield as applied to a confined aquifer is equivalent to its storage coefficient.

Under unconfined aquifer conditions, the groundwater is derived from the storage by the gravity drainage of the voids in the portion of the aquifer that has been unwatered by pumping (Fig. 2.11).

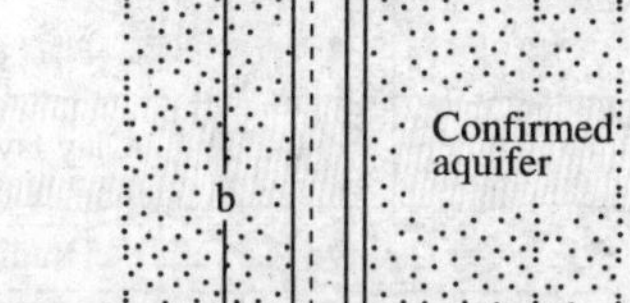

Fig. 2.10. Definition sketch of storage coefficient.

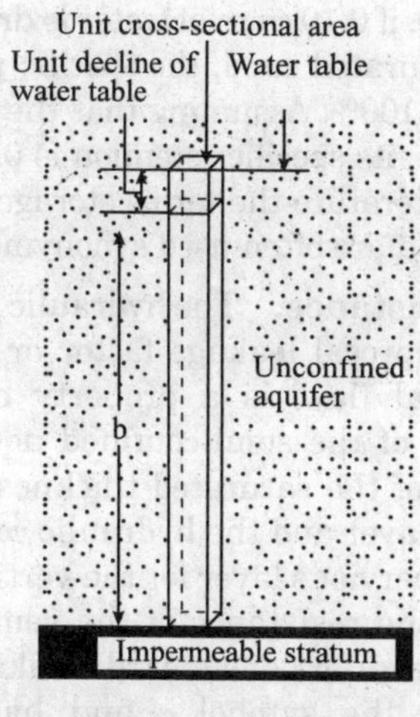

Fig. 2.11. Definition sketch of specific yield.

When water is drained from an aquifer by gravity force, only a part of the total volume stored in its pores is released. The quantity of water that a unit volume of the aquifer will yield when drained by gravity is called its specific yield. The part of the water that is retained in the aquifer mass is held against the force of gravity by molecular attraction and capillarity. The quantity of water that a unit volume of aquifer retains when subjected to gravity drainage is called its *specific retention*, S_r. Both specific yield and specific retention are expressed as fractions or percentages. The sum of the specific yield and specific retention equals the porosity of the aquifer.

Expresssed mathematically,

$$S_y = \frac{W_r}{V} \times 100 \qquad \text{... (2.6a)}$$

in which, W_r is the volume occupied by the retained water and V is the gross volume of the rock or soil. On the other hand, the specific yield S_y is the water that can be drained.

Therefore,

$$S_y = \frac{W_y}{V} \times 100 \qquad \text{... (2.6b)}$$

in which, W_y is the volume of water that is drained. As $w_r + w_y = w$, it is evident that:

$$\rho = S_r + S_y \qquad \text{... (2.6c)}$$

Thus, specific yield is a function of the porosity of an aquifer. The values of S_y depend upon the grain size, shape and distribution of pores, and compaction of the stratum. For uniform sand S_y may be as high as 30%, but in the case of most of the alluvial aquifers, the specific yield ranges from 10–20%. Values of specific yield for other formations encountered in water wells are: gravel 15–30%, sand and gravel 15–25%, sandstone 5–15%, shale 0.5–5%, and limestone 0.5–5% (Driscoll, F.G., 1986).

For example, if 0.10 cu m of water is drained from one cu m of saturated sand, the specific yield of the sand is 0.10 or 100%. Assuming that the porosity of the sand is 28%, its specific retention is 0.18 or 18%. (In American literature the terms storage coefficient and specific yield are often used synonymously).

Hydraulic resistance: The hydraulic resistance, also called reciprocal leakage factor or resistance against vertical flow, is a property of a semi-pervious layer of the semi-confined aquifer. It is the ratio, b/K of the saturated thickness b, of the semi-pervious layer and the hydraulic conductivity K, of the semi-pervious layer for the vertical flow. It characterises the resistance of the semi-previous layer to upward or downward leakage. It is designated by the symbol c and has reduced dimension of time. When the value of c equals infinity, the layer is considered to be impervious. When the value of c of any lithological section is equal to or near to zero, the layer is considered an aquifer. The value of hydraulic resistance of a semi-previous layer is evaluated on the basis of the pumping test data using the following equation:

$$c = \frac{B^2}{Kb} \quad \text{... (2.7)}$$

in which,

c = hydraulic resistance (days)
B = leakage factor (m)
K = hydraulic conductivity of the aquifer (m/day)
b = thickness of the horizontal pervious stratum confined between horizontal semi-pervious and impervious layers (m)

The value of c, of a semi-pervious layer generally ranges between 100 to 10,00,000 minutes.

Leakage factor: Leakage factor, $B = \sqrt{(Kbc)}$, is also a property of the semi-confined aquifer. It determines the distribution of leakage into or from the semi-pervious layer. In other words, it determines the origin of water withdrawn from a well tapping the aquifer. High values of leakage factor indicate a great resistance of the semi-pervious strata to flow, as compared to the resistance of the aquifer itself. This factor is designated by the symbol B and has the dimension of length, L. The values of B can be determined by the pumping test data of the semi-confined aquifer.

HYDRAULICS OF WELLS

Irrigation wells operate according to certain fundamental hydraulic principles. Water flows into the well from the surrounding aquifer because the pumping of the well creates a difference in pressure. Before pumping, the water in the well stands at a height equal to the static water level in the saturated sand around the well. When pumping starts, the water in the well is pulled down and the water starts to flow into the well from the water bearing formation because the water level or pressure inside the well during pumping is lower than it is in the aquifer outside the well. This pressure difference is the '*drive*' that causes the water to move through the pores of the sand towards the well, The closer the water gets to the well, the faster it has to move because the area through which it has to travel is continuously decreasing.

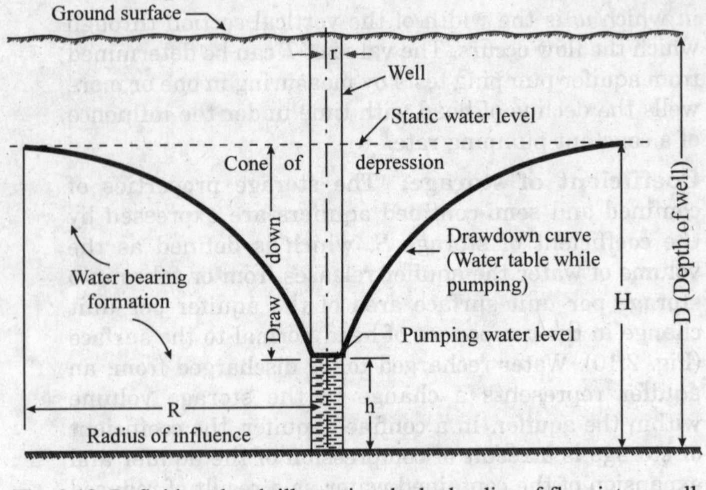

Fig. 2.12. Definition sketch illustrating the hydraulics of flow into a well penetrating an unconfined aquifer.

Definition of Terms

Static water level: The level at which the water stands in a well before pumping starts is called the static water level. It is generally the level of the water table, except in case of artesian wells where the static water level may be above the water table. The pressure of water at the static water level is atmospheric. Static water level is generally expressed as the distance from the ground surface to the water level in the well.

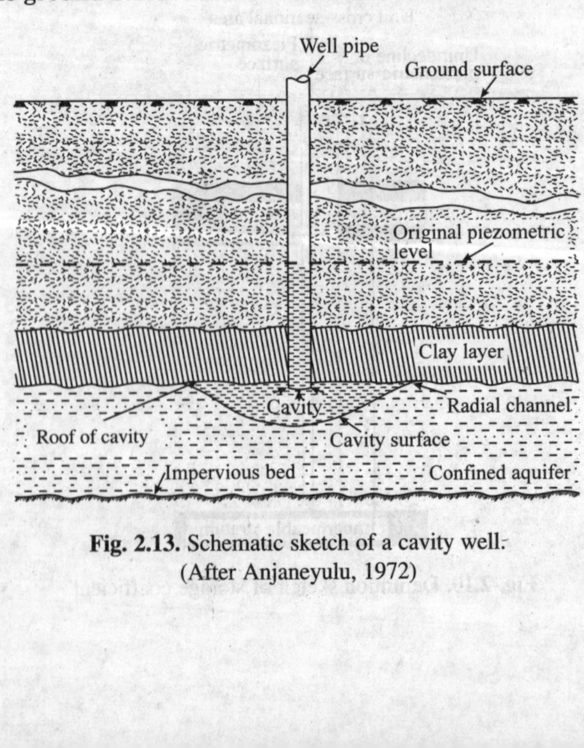

Fig. 2.13. Schematic sketch of a cavity well.
(After Anjaneyulu, 1972)

Piezometric surface: The piezometric surface is the height at which water will stand in a piezometer or pipe which is open at the end and which extends into the aquifer. The height h, to which the water will rise in the pipe above its bottom is equal to the pressure p at the bottom point of the pipe divided by the unit weight w of water, $h = p/w$.

Pumping water level: This is the level at which water stands in a well when pumped at any given rate. This level is variable and changes with the quantity of water being pumped.

Drawdown: Drawdown at any instant is the difference between the static water level and the pumping water level at that instant. Drawdown affects the yield of a well. The maximum practical drawdown in a well is limited to the pumping water level reaching the top of the well screen in the case of a tube well.

Area of influence: As water is pumped out of the well, it gets the supply from the surrounding formations. There is thus an imaginary inverted cone formed around the well having the static water level as base and pumping level as apex (Fig. 2.12). The area which gets affected by the pumping of the well is called the *area of influence*. The boundary of the area of influence is called the *circle of influence*. The radius of the circle of influence is called the *radius of influence*. As more and more water is pumped out of the well, it takes more water from storage and, as a result, the radius of influence gets extended till a position is reached when the rate of discharge from the well becomes equal to the rate of recuperation from the storage of the well. It is at this instant that the *cone of depression* gets stabilized. This equilibrium condition changes when the discharge rate is increased or decreased.

Well yield: The yield of a well is the volume of water discharged per unit of time from it. It is commonly expressed in litres per second or litres per minute.

Specific capacity: Specific capacity of a well is its yield per unit of drawdown. It is usually expressed as litres per minute per metre of drawdown. Dividing the yield by the drawdown, each yield measured at the same time, gives the specific capacity.

Open wells: Open wells are dug down to the water bearing strata. They derive water from the formations close to the ground surface. The large diameter of open wells permits the storage of large quantities of water in the well.

Tube wells: Tube wells are constructed by fixing a pipe below ground surface and passing through different geological formations consisting of water bearing and non-water bearing strata. Blind pipes are located at the non-water bearing strata and perforated pipes or well screens are placed against the aquifer. In *cavity wells* screens are not used. The well casing rests over a confined water bearing formation of sand and gravel. Water enters the well through the bottom only (Fig. 2.13).

Filter points: In deltaic regions, where the aquifer formations are of coarse sand and gravel, the tube wells are shallow and consist of a well screen and a short length of casing pipe. Such wells are called filter points.

Predicting Well Yield

Darcy's theory established the fundamentals of groundwater movement. Dupuit, a French hydraulic engineer established the analysis of seepage phenomenon and derived well equations based on the following assumptions:

1. The hydraulic gradient is the same at all points in a vertical section of an aquifer.
2. The hydraulic gradient at the water table (or at the piezometric surface in artesian flow) is equal to the slope of the water surface at that point.

Steady-state Flow to Wells in Unconfined Aquifer

Steady-state flow: The flow is said to be steady when the flow conditions at any instant are constant, *i.e.*,

$$\frac{dv}{dt} = 0$$

in which,

v = velocity of flow (m/sec)

t = time (sec).

Steady-state flow occurs when there is an equilibrium between the discharge of the pumped well and the recharge of the aquifer by an outside source. Such an outside source is, for example, a phreatic aquifer with constant water table overlying a semi-pervious layer. In practice, this condition is rarely fulfilled in groundwater flow through wells. However, for the ease of analysis, it is said that a steady state is reached if, in the observations wells, the changes in drawdown with time have become negligible or the hydraulic gradient has become constant.

Dupuit (1863) distinguished between unconfined and confined aquifers. To derive an equation of flow into an unconfined or gravity well, he considered the case of a cylindrical island of pervious sand surrounded by water at a constant level and underlain by a horizontal impervious base. Applying equation 2.4 to an arbitrary cylindrical surface surrounding a gravity or unconfined well located at the centre of the island and penetrating the horizontal base, and using plane polar coordinates with the well as the origin, the well discharge Q at any distance r equals—

$$Q = Kia = 2\pi rhK\frac{dh}{dr}$$

or,

$$hdh = \frac{Q}{2\pi K}\frac{dr}{r}$$

for steady radial flow to the well. Integrating for boundary conditions at the well, $h = h_w$ at $r = r_w$ and $h = H$ at $r = R$

$$\int_{h_w}^{H} h\,dh = \frac{Q}{2\pi K}\int_{r_w}^{R}\frac{dr}{r}$$

$$\text{or,} \quad H^2 - h_w^2 = \frac{Q}{\pi K}\ln(R/r_w)$$

$$\text{or,} \quad Q = \frac{\pi K (H^2 - h_w^2)}{\ln(R/r_w)}$$

$$= \frac{\pi K (H + h_w)(H - h_w)}{\ln(R/r_w)} \quad ...(2.8)$$

in which,

Q = constant discharge rate (m³/sec)

H = original elevation of the water surface measured from the impervious base (m)

h_w = depth of water in the well from the impervious base (m)

R = radius of island (m)

r_w = radius of well (m)

K = hydraulic conductivity of the aquifer (m/sec).

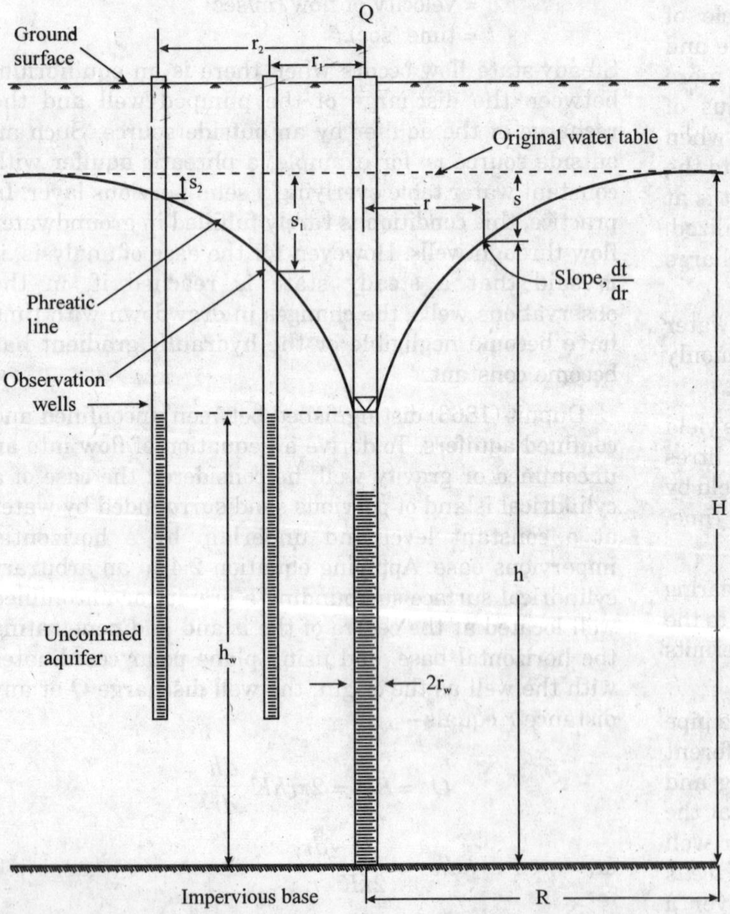

Fig. 2.14. Definition sketch illustrating steady-state flow into a well in an unconfined aquifer.

The practical applicability of equation (2.8) derived by Dupuit was established by Adolph Thiem (1870). Thiem showed that beyond a certain distance from the well, the drawdown of the phreatic surface from the original groundwater table became negligible. He observed that constant replenishment of the supply of groundwater by rainfall or irrigation tended to equalise the amount withdrawn by the well, thereby keeping R approximately constant. The Dupuit-Thiem theory stated above is important in well hydraulics. To evaluate the hydraulic properties of the aquifer, equation (2.8) can be used for steady-state conditions :

Let s_1 and s_2 be the steady-state drawdowns at two observation wells and r_1 and r_2 the distances of the observation wells from the centre of the test well, respectively, and s be the drawdown at a distance r from the test well (Fig. 2.14).

Since $h = H - s$, equation (2.8), viz.,

$$Q = \frac{\pi K (h_2^2 - h_1^2)}{\ln(r_2/r_1)}$$

can be transformed into

$$Q = \frac{\pi K [(H - s_2)^2 - (H - s_1)^2]2H/2H}{\ln(r_2/r_1)}$$

Replacing $s - s^2/2H$ by s', the corrected drawdown, yields—

$$Q = \frac{2\pi KH(s_1' - s_2')}{\ln(r_2/r_1)} \quad ...(2.9)$$

in which,

$$s_1' = s_1 - s_1^2/2H$$
$$s_2' = s_2 - s_2^2/2H$$

s_1 and s_2' are the corrected steady-state drawdowns at points 1 and 2, respectively.

$$\text{or,} \quad Q = \frac{2\pi T(s_1' - s_2')}{ln(r_2/r_1)}$$

in which,

$T = KH$ = assumed transmissibility of the aquifer (m²/sec).

$$\text{or,} \quad T = \frac{Q\,ln(r_2/r_1)}{2\pi(r_1' - s_2')} \quad ...(2.10)$$

Equation (2.10) can be used to estimate the values of transmissibility and hydraulic conductivity only when the drawdown in the aquifer is small in relation to the thickness of the saturated portion of the aquifer. Otherwise, the assumption that the thickness of the aquifer is constant is no longer satisfied.

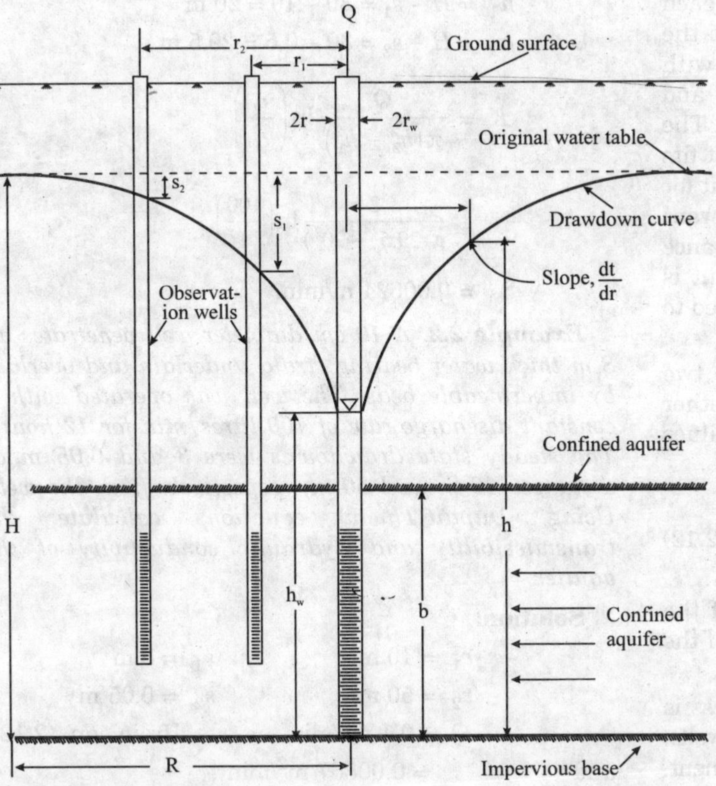

Fig. 2.15. Definition sketch illustrating the hydraulics of flow in a well penetrating a confined aquifer.

Steady-state Flow to Wells in Confined Aquifers

To derive the equation for radial flow in a confined aquifer (Fig. 2.15), Dupuit used equation (2.4) which he recognised in this case as the equation of piezometric surface. Using plane polar coordinates with the well as the origin, the well discharges Q at distance r, when the thickness of aquifers is b, is determined as follows:

$$Q = Kia = 2\pi rbK \frac{dh}{dr}$$

or,
$$dh = \frac{Q}{2\pi Kb} \frac{dr}{r}$$

Integrating for the boundary conditions at the well, $h = h_w$ at $r = r_w$ and $h = H$ and $r = R$ at the extremity of the area of influence,

$$\int_{hw}^{H} dh = \frac{Q}{2\pi Kb} \int_{rw}^{R} \frac{dr}{r}$$

$$H \,.\, hw = \frac{Q}{2\pi Kb} \ln (R/r_w)$$

After re-arranging,

$$Q = \frac{2\pi Kb(H - h_w)}{\ln (R/r_w)} \quad \text{... (2.11)}$$

in which b = thickness of the horizontal pervious stratum confined between two horizontal impervious strata, *m*.

The other variables are same as defined in equation (2.8). Equation (2.11) can be used to evaluate the hydraulic properties of an aquifer based on the measurements made during a pumping test. Thiem (1870), who worked independently of Dupuit, derived equations (2.8) and (2.11) based on the following assumptions which were more precisely defined than those of Dupuit:

1. The aquifer has a seemingly infinite areal extent.
2. The aquifer is homogeneous, isotropic and of uniform thickness over the area influenced by the pumping test.
3. The pumped well penetrates the entire thickness of the aquifer and receives water from the entire thickness of the aquifer by horizontal flow.
4. Flow to the well is in steady-state.

To evaluate the hydraulic properties of the water bearing formation in case of wells in a confined aquifer, any one of the following two procedures could be adopted:

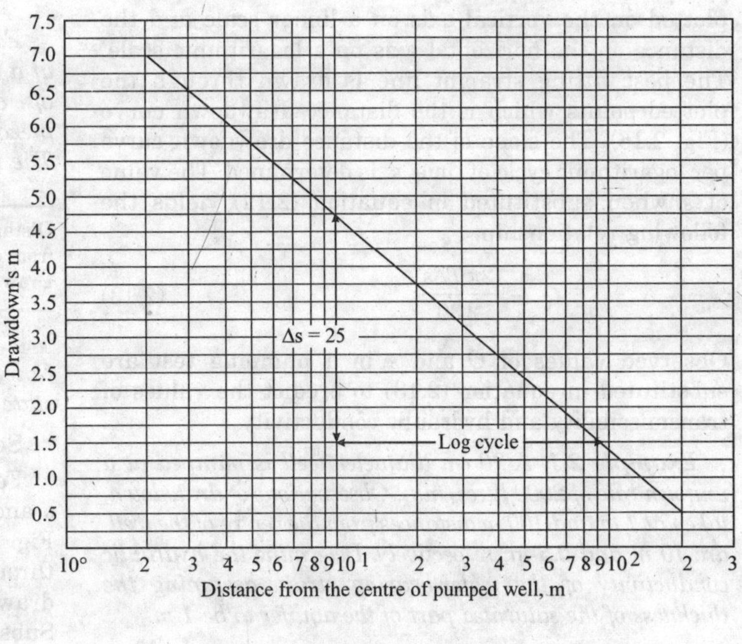

Scale: Horizontal 1 cm = 1 cm
Vertical 1 cm = 5 cm

Fig. 2.16. Distance-drawdown curve (on semi-log paper) for determining the transmissibility and hydraulic conductivity of aquifers.

Procedure 1. The observed drawdown in each piezometer or observation well is plotted against the corresponding time on a single logarithmic paper with time on the horizontal axis on a logarithmic scale and the drawdown on the vertical axis on a linear scale. The time—drawdown curve of each piezometer, that best fits through the points, is drawn. It will be observed that for the latter time data, the curves of the different piezometers run parallel and thus their mutual distance is constant. This means that the hydraulic gradient is constant and the flow in the aquifer can be considered to be in a steady-state.

The values of the steady-state drawdown of the two piezometers are substituted in equation (2.11) together with the corresponding values of r and the known values of Q and solved for the transmissibility, $T = Kb$.

$$Q = \frac{2\pi T\,(s_1 - s_2)}{\ln (r_2 / r_1)} \qquad \text{... (2.12)}$$

in which s_1 and s_2 are the values of drawdown of the piezometers at r_1 and r_2 distances from the centre of the well, respectively.

To get a more precise value of T, the same process is repeated for all possible combinations of observation wells. Theoretically, the results should show a close agreement. However, the results usually give more or less different values of T. The average is used as a final result.

Procedure 2. On a semi-logarithmic paper, the observed steady-state drawdown s of each observation well is plotted against the distance r between the pumped well and the piezometer. The drawdown is plotted on the vertical axis on a linear scale and the distance on the horizontal axis on a logarithmic scale. The best fitting straight line is drawn through the plotted points which is the distance drawdown curve (Fig. 2.16). The slope of the distance drawdown curve per logarithmic cycle of time, s, is determined. The value of s when substituted in equation (2.12) yields the following relationship:

$$Q = \frac{2\pi\, T\Delta s}{2.30} \qquad \text{... (2.13)}$$

Observed values of Q and s in a pumping test are substituted in equation (2.13) to predict the values of transmissibility and hydraulic conductivity.

Example 2.1. *A 20 cm diameter well is pumped at a uniform rate of 2000 litres/min. Observations of drawdown, taken at 1 m and 100 m distances from the centre of the well, are 10 m and 0.5 m, respectively. Determine the hydraulic conductivity of the water-bearing strata assuming the thickness of the saturated part of the aquifer to be 1 m.*

Solution:

$$Q = 2 \text{ m}^3\text{/min}$$
$$r = 1 \text{ m}$$
$$r_2 = 100 \text{ m}$$
$$h_1 = H - s_1 = 30 - 10 = 20 \text{ m}$$
$$h_2 = H - s_2 = 30 - 0.5 = 29.5 \text{ m}$$

$$K = \frac{Q}{\pi\left(h_2^2 - h_1^2\right)} \ln\left(\frac{r_2}{r_1}\right)$$

$$= \frac{2}{\pi(29.5^2 - 20^2)} \ln\left(\frac{100}{1}\right)$$

$$= 0.00624 \text{ m/min.}$$

Example 2.2. *A 10 cm diameter well penetrates an 8 m thick water bearing strata underlain and overlain by impermeable beds. The well was operated with a constant discharge rate of 100 litres/min for 12 hours. The steady state drawdowns were 3 and 0.05 m at distances 10 m and 50 m, respectively from the well. Using Dupuit-Thiem equation, calculate the transmissibility and hydraulic conductivity of the aquifer.*

Solution:

$r_1 = 10$ m $s_1 = 3$ m

$r_2 = 50$ m $s_2 = 0.05$ m

$Q = 0.1$ m^3/min [from eqn. (2.12)]

and $= 0.00867$ m^2/min

$$\text{Hydraulic conductivity} = \frac{\text{transmissibility (m}^2\text{/min)}}{\text{thickness of aquifer (m)}}$$

$$= \frac{0.00867}{8} = 0.00108 \text{ m/min.}$$

Example 2.3. *A well in a confined aquifer is pumped at a constant rate of 1200 litres/min. After 600 minutes of continuous pumping the drawdowns in the piezometers used for the pumping test were measured. The results are as follows:*

Distance of piezometer, from centre of well (m)	2	10	50	100	200
Drawdown (m)	7.25	4.75	2.5	1.50	0.50

Calculate the aquifer transmissibility, assuming steady-state drawdown.

Solution:

Following *Procedure 2* (Thiem method), the values of s and r are plotted on a single log paper as shown in Fig. 2.16. The slope of the best fit straight line drawn through the plotted points is measured to be equal to a drawdown difference of 3.25 m per log cycle of $r(s)$. Substituting the values of s and Q in equation 2.12,

$$T = \frac{2.30\,Q}{2\pi\Delta s} = \frac{2.30 \times 1.2}{2 \times \pi \times 3.25}$$

$$= 0.134 \text{ m}^2\text{/min.}$$

Example 2.4. *Using the table given in example 2.3, calculate the transmissibility values of different sections and average transmissibility, using the Thiem Procedure I.*

Solution:

$$T = \frac{Q}{2\pi(s_1 - s_2)} \ln(r_2 / r_1)$$

Section 1:

$r_1 = 2$ m, $r_2 = 10$ m

$s_1 = 7.25$ m, $s_2 - 4.75$ m

$$T = \frac{1.2}{2\pi(7.25 - 4.75)} \ln\left(\frac{10}{2}\right)$$

$= 0.132$ m^2/min

Section 2:

$r_1 = 10$ m, $s_1 = 4.75$ m

$r_2 = 50$ m, $s_2 = 2.5$ m

$$T = \frac{1.2}{2\pi(4.75 - 2.50)} \ln\left(\frac{50}{10}\right)$$

$= 0.1367$ m^2/min

Section 3:

$r_1 = 50$ m, $s_1 = 2.50$ m

$r_2 = 100$ m, $s_2 = 1.50$ m

$$T = \frac{1.2}{2\pi(2.50 - 1.50)} \ln\left(\frac{100}{50}\right)$$

$= 0.132$ m^2/min

Section 4:

$r_1 = 100$ m, $s_1 = 1.50$ m

$r_2 = 200$ m, $s_2 = 0.50$ m

$$T = \frac{1.2}{2\pi(1.50 - 0.50)} \ln\left(\frac{200}{100}\right)$$

$= 0.132$ m^2/min

Average transmissibility

$= 1/4\ (0.1320 + 0.1367 + 0.1320 + 0.1320)$

$= 0.133$ m^2/min.

UNSTEADY-STATE FLOW

The flow is said to be unsteady when the flow conditions at any moment are not constant, *i.e.*,

$$\frac{dv}{dt} \neq 0$$

The steady-state equations of Thiem are widely applied in determining the hydraulic conductivity and transmissibility of confined and unconfined aquifers. However, field conditions may be such that considerable time is required to reach the steady-state flow.

Unsteady-state Flow to Wells in Unconfined Aquifers

In unsteady-state flow in an unconfined aquifer with a declining water table, dewatering of the pore space is not instantaneous, but continues for some time after drawdown. The region above the water table, though unsaturated, goes on supplying water to the receding water-table. Thus, the specific yield increases at a diminishing rate with the duration of pumping. Hence, the saturated thickness of the unconfined aquifer is variable in magnitude. Assuming that the change in drawdown is negligible and almost constant in the dewatering area, the aquifer properties can be evaluated by the procedure adopted for unsteady-state flow in confined aquifer by assuming $s' = s - s^2/2H$, m, in which, s' is the drawdown component for the decrease in the saturated thickness of the unconfined aquifer.

Unsteady-state Flow to Wells in Confined Aquifers

Theis (1935) developed the non-steady-state analysis to determine the aquifer parameters by introducing the time factor and the storage coefficient. Theis noted that when a well penetrating an extensive artesian aquifer, is pumped at a constant rate, the influence of discharge extends outward with time. He showed that the rate of decline of head, multiplied by the storage coefficient and summed over the area of influence equals the discharge.

Theis' equation for non-steady state flow in aquifers, which was derived from the analogy between the flow of groundwater and conduction of heat, is based on the following assumptions which are in addition to the assumptions mentioned for the Thiem-Dupuit equations 2.8 and 2.11:

1. The aquifer is confined.
2. The flow to a well is in unsteady-state, *i.e.*, neither the drawdown differences with time are negligible nor is the hydraulic gradient constant with time.
3. The water removed from storage is discharged instantaneously with the decline of head.
4. The well diameter is very small, *i.e.*, the storage in the well can be neglected.

Jacob (1940) proposed the following differential equation governing the unsteady-state radial flow in a non-leaky confined aquifer in polar coordinate notation:

$$\frac{\partial^2 h}{\partial r^2} + \frac{1}{r}\frac{\partial h}{\partial r} = \frac{S}{T}\frac{\partial h}{\partial t} \quad \text{... (2.14)}$$

in which,

T = transmissibility of the aquifer (m^2/sec)

S = storage coefficient, dimensionless

r = radial distance of the piezometer from the centre of the pumped well (m)

t = elapsed time after pumping has started (sec).

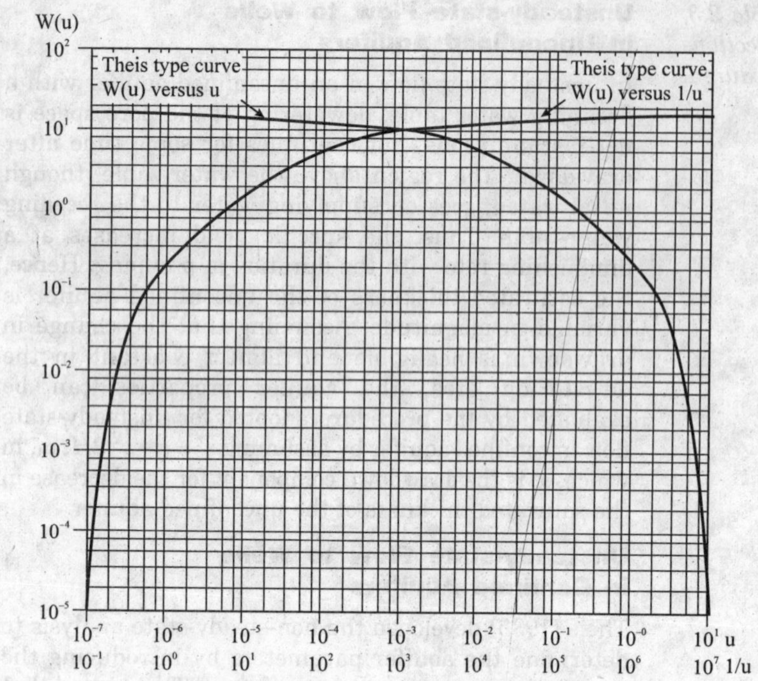

Fig. 2.17. Theis type curves for $W(u)$ versus u and $W(u)$ versus l/u (Based on values presented in Appendix A)

Theis (1935) obtained the solution of equation (2.14) based on the analogy between groundwater flow and heat conduction and for boundary conditions, $h = h_o$, before pumping and $h \to h_o$ as $r \to \infty$ as pumping begins ($t \geq 0$), which may be written as,

$$h_0 - h = \frac{Q}{4\pi T} \int_{r^2 S/4tT}^{\infty} \frac{e^{-u}}{u} du \quad \text{... (2.15)}$$

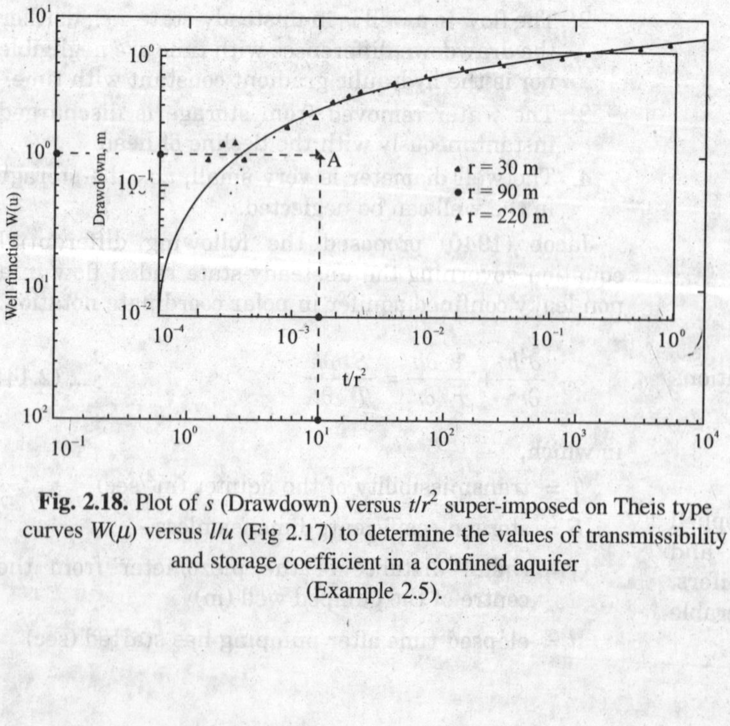

Fig. 2.18. Plot of s (Drawdown) versus t/r^2 super-imposed on Theis type curves $W(\mu)$ versus l/u (Fig 2.17) to determine the values of transmissibility and storage coefficient in a confined aquifer (Example 2.5).

in which,

$$u = \frac{r^2 S}{4Tt} \quad \text{... (2.16)}$$

and Q = constant discharge rate (m^3/sec)

The exponential integral is written symbolically as $W(u)$, which in this usage is generally read 'well function of u' or Theis' 'well function'.

Equation (2.16), in terms of Theis' well function, may be written as follows:

$$s = \frac{Q}{4\pi T} W(u) \quad \text{... (2.17)}$$

in which,

s = the unsteady-state drawdown (m)

Determining Hydraulic Properties of Confined Aquifer

1. Prepare a 'type curve' (Fig. 2.17) of the Theis well function on double logarithmic paper by plotting values of $W(u)$ against u, using table of functions (Appendix A) of Theis (1935).

2. Plot the values of s against t/r^2 on another double logarithmic paper on the same scale as that used for the type curve. Place the observed data plot over the type curve and keeping the coordinate axes of both data plot and type curve parallel, locate the position of best match between the data plot and type curve (Fig. 2.18).

3. Select an arbitrary match point A on the overlapping portion of the two sheets of graph paper and determine co-ordinates $W(u)$, l/u, s and t/r^2 for this match point. The calculations are greatly simplified if the point is selected where the coordinates of the type curve are $W(u) = 1$ and $1/u = 10$.

4. Substitute the values of $W(u)$, s and Q in equation (2.17) to obtain the value of transmissibility, T, i.e.,

$$T = \frac{Q}{4\pi s} W(u)$$

5. Calculate S by substituting the values of Kb, t/r^2 and u in equation (2.16). That is,

$$S = 4Tu \frac{t}{t^2}$$

Example 2.5. *Using this method calculate the hydraulic properties of an aquifer with the data presented in Table 2.1.*

Table 2.1. *Pumping Test Data, I.A.R.I. Farm, with Constant Rate of Discharge Q = 0.005 m^3/min*

Elapsed time, t, min	Drawdown, s, m — Distance of observation well from pumped well, r: 30 m	90 m	220 m
0	0	0	0
5	0.490	0.130	–
10	0.680	0.207	–
20	0.700	0.306	–
30	0.650	0.365	–
50	0.795	0.427	–
60	0.820	0.450	0.080
90	0.870	0.495	0.120
180	0.934	0.570	0.162
300	0.990	0.613	0.193
480	1.050	0.700	0.218
600	1.053	0.684	0.225

Solution: The plots of t/r^2 versus s of all observation wells are superimposed on the Theis type curve $W(u)$ versus l/u (Fig. 2.18). The match point A is so chosen that the value of $W(u) = 1$, and the value $u = 10$. The value of drawdown s on A = 0.15 m and t/r^2 on $A = 1.5 \times 10^{-3}$ min/m². Introduction of these values into equations (2.16) and (2.17) gives

$$T = \frac{0.005}{4 \times 3.14 \times 0.15} \times 1$$

$$= 0.00265 \text{ m}^2/\text{min}$$

and

$$S = 4 \times 0.00265 \times 10 \times 1.5 \times 10^{-3} = 1.59 \times 10^{-4}$$

In addition to the Theis (1935) method described above, Jacob (1940) and Chow (1952) have proposed procedures for determining the hydraulic properties of confined aquifers.

Steady-state Flow to Wells in Semi-confined Aquifers

The quantity of water moving through a semi-pervious layer is directly proportional to the difference between the phreatic and the piezometric heads and inversely proportional to the hydraulic resistance of the semi-pervious layer, *i.e.*,

$$q = \frac{(h_1 - h_2)}{c}$$

in which

q = discharge through the semi-pervious layer (m³/min)

h_1 = phreatic head of the semi-pervious layer (m)

h_2 = piezometric head in the main aquifer (m)

c = hydraulic resistance of the semi-pervious layer (min).

De-Glee (1930) and Jacob (1940) derived indepen-dently the following equation for steady-state drawdown in a semi-confined (leaky) aquifer:

$$s = \frac{Q}{2\pi Kb} K_0 \ (r/B) \qquad \text{... (2.18)}$$

in which,

s = steady-state drawdown in an observation well at a distance r from the pumped well, m

Q = steady-state constant discharge (m³/min)

$K_0(r/B)$ = modified Bessel function of the second kind and of order zero

B = $\sqrt{Kbc}$, leakage factor of the semi-pervious layer, m. The other terms are same as defined earlier.

The values of $K_0(r/B)$ for different values of (r/B) can be obtained from the table of functions provided by Hantush (1956) which is reproduced in Appendix *B*.

Evaluation of the Hydraulic Properties of Semi-confined Aquifer

The hydraulic properties of a semi-confined aquifer for steady-state condition can be evaluated by (*i*) De-Glee's method, and (*ii*) Hantush and Jacobs method. The De-Glee's method is described below:

De-Glee's method: The transmissibility T, leakage factor B and hydraulic resistance c of the semi-confined aquifer can be evaluated by adopting the following procedure:

(*i*) A type curve (Fig. 2.19) is prepared by plotting the values of K_0 (r/B) versus (r/B) on a double logarithmic tracing paper with (r/B on the x-axis), using the table of functions K_0 (r/B) and (r/B) (Appendix B) given by Hantush (1956).

(*ii*) The observed data curve is prepared on another double logarithmic paper on the same scale with the values of steady-state drawdown, s, of each observation well versus its corresponding distance r (r on the x-axis)

(*iii*) The observed data curve is superimposed over the type curve while keeping the coordinate axes parallel. By trial and error, adjustment is made until a position is obtained where most of the plotted points of the observed data curve overlap on a segment of the type curve.

(*iv*) An arbitrary point* A is selected on the overlapped portion of both the sheets and the values of the match point coordinates s, r, and K_0 (r/B) are determined for point A.

* To make the calculations easy the point A is selected on the graph sheet where K_0 (r/B) = 1.

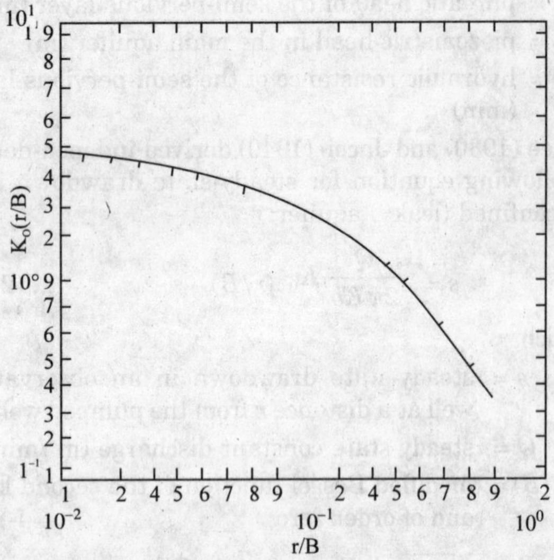

Fig. 2.19. De-Glee's type curve K_0 (r/B) versus r/B.

(*v*) The values of s, Q and K_0 (r/B) are substituted in equation (2.17) to obtain the value of transmissibility.

(*vi*) The leakage factor B and hydraulic resistance c are obtained by substituting the values of T, r and (r/B) in the following equation proposed by De-Glee:

$$c = \frac{B^2}{T} = \frac{1}{(r / B)^2} \times \frac{r^2}{T} \quad \text{... (2.19)}$$

Example 2.6. *A 10 cm diameter well penetrates a semi-confined aquifer which is underlain by an impervious layer and overlain by a semi-pervious layer. The depth of the well is 80 m and the thickness of aquifer is 20 m. The steady-state drawdowns measured at the radial distances of 1 m, 10 m, 20 m, 30 m, 40 m, and 50 m, were 3.40 m, 1.12 m 0.65 m, 0.40 m, 0.25 m, and 0.17 m, respectively. The constant discharge of the well during the test was 300 litres/min. Determine the hydraulic properties of the aquifer using the De-Glee's method*

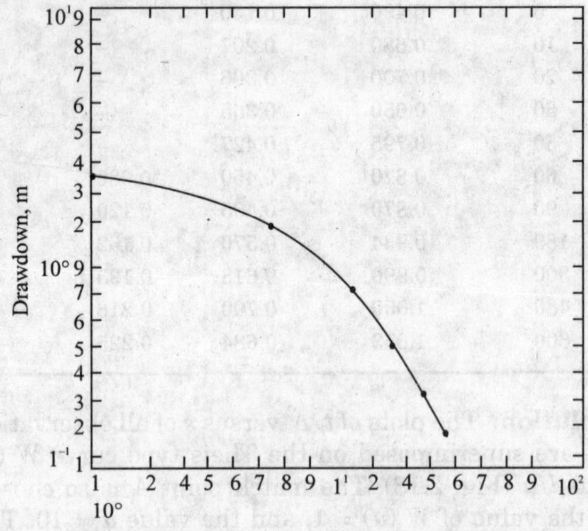

Fig. 2.20. Observed data curve of s versus r in Example 2.6.

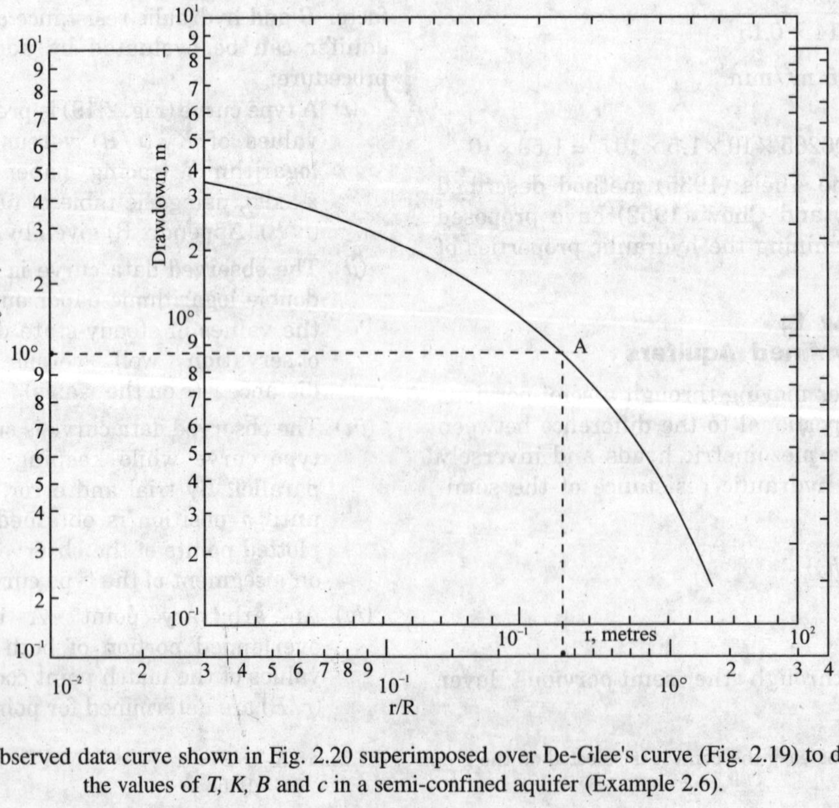

Fig. 2.21. Observed data curve shown in Fig. 2.20 superimposed over De-Glee's curve (Fig. 2.19) to determine the values of T, K, B and c in a semi-confined aquifer (Example 2.6).

Solution: The data of s versus r, plotted on double logarithmic paper is shown in Fig. 2.20. Figure 2.21 shows the superimposition of the observed data curve on De-Glee's type curve. From Fig 2.11.

$$K_0(r/B) = 1$$

$$r/b = 0.47$$

$$s = 0.84 \text{ m}$$

Substituting the above values in equation 2.18.

$$\text{Transmissibility, } T = \frac{0.3 \times 1}{2 \times 3.14 \times 0.84}$$

$$= 0.0568 \text{ m}^2/\text{min}$$

$$\text{Hydraulic conductivity, } K = \frac{0.0568}{20}$$

$$= 0.00284 \text{ m/min}$$

$$\text{Leakage factor, } B = \frac{15.58}{0.47} = 33.148 \text{ m}$$

$$\text{Hydraulic resistance, } c = \frac{(33.148)^2}{0.0568}$$

$$= 19344.9 \text{ min}$$

PUMPING TESTS OF WELLS

Pumping tests are conducted to obtain information on the characteristics of the water-bearing formations of the well. Such tests provide information on the hydro-geological properties of the well site such as the type of aquifer and its areal extent and thickness, water table gradients and recharge boundaries. The hydraulic properties of the aquifer such as hydraulic conductivity, transmissibility, storage coefficient, specific yield, leakage factor and hydraulic resistance could be obtained from pumping test data. A pumping test will also provide information about the drawdown and yield capacity of the well. This will help in selecting the pump, the characteristics of which will match best with those of the well.

Pumping Test Procedures

The step-by-step procedure commonly adopted in conducting pumping tests is enumerated below.

(1) Selection of the test site: In selecting the site of the pumping test the following points are kept in mind:

(*i*) The hydro-geological conditions of the test site should not change over short distances and should be representative of the area or a large part of the area under consideration.

(*ii*) The site should not be close to railway lines or highways with heavy traffic. Such sites may produce measurable fluctuations of the piezometric surface in case of confined aquifers.

(*iii*) The pumped water does not return back to the aquifer.

(*iv*) The gradient of water table or piezometric surface should be low.

(*v*) Man power, instruments and equipments to be used must be able to reach the site easily.

(2) Observation wells: Water level measurements during pumping test are made in observation wells installed close to the well or at some distance away from it. The installation and design of the observation well requires the following points to be considered:

(*i*) *Number of observation wells:* The number of observation wells depends on the amount of information desired and the degree of precision expected. Though a single observation well permits the determination of hydraulic properties, more precise values of these properties can be obtained by installing two or more observation wells at varying distances from the centre of the well. A large number of observation wells also provide information on distance-drawdown relationship which can be utilized for designing the spacing of wells.

(*ii*) *Spacing of observation wells:* In general, observation wells should be placed neither too far from the pumped well nor too close to it. Koul (1970) reported that there was no appreciable drawdown beyond 16 metres from the centre of a shallow tube well in the IARI farm in Delhi for a discharge capacity of about 450 litres/min. Krusseman and De-Ridder (1970) stated that placing the observation wells about 10 to 100 m from the centre of the pumped well will provide good results in most cases. Yadav (1974) reported that the nearest observation well, for a medium capacity shallow tube well in semi-confined aquifer, should be placed at about 1 m from the centre of the test well. The following are the main factors influencing the spacing of observation wells:

(*a*) *Type of aquifer.* In confined and semi-confined aquifers, a loss of head caused by pumping propagates faster than in an unconfined aquifer because the release of water from storage is due to the compressibility of the aquifer material and that of water. Hence, the nearest observation well should be placed a little farther in confined and semi-confined aquifers than in an unconfined aquifer for the same discharge rate of the test well. For the same discharge rate, the radius of influence is more in confined and semi-confined aquifers than in an unconfined aquifer. Therefore, the farthest observation well should be placed at a greater distance in these aquifers than in an unconfined aquifer to evaluate the boundary and the extent of the aquifer.

(*b*) *Hydraulic conductivity:* When the hydraulic conductivity of the aquifer material is high, the cone of depression produced by pumping will be wide and flat, which results in a larger radius of influence as well as a

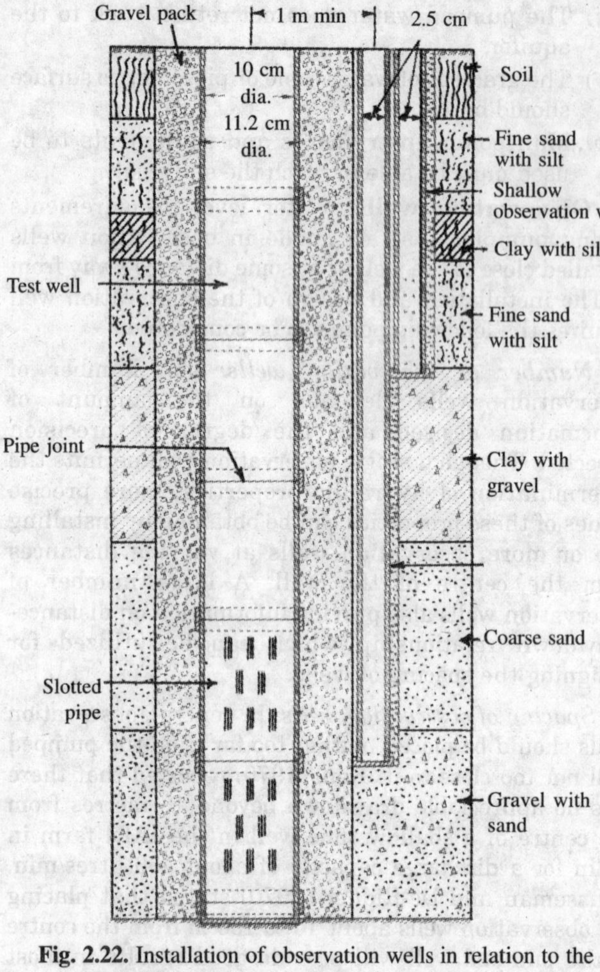

Fig. 2.22. Installation of observation wells in relation to the test well and the characteritics of the underground strata.

greater amount of turbulence. Therefore, observation wells should be placed farther in an aquifer with a high value of hydraulic conductivity as compared to an aquifer of lower conductivity. This applies to the observation well closest to the test well and those away from it.

(c) Length of well screen: The distance of the observation well adjacent to the test well is influenced by the length and depth of penetration of the well screen of the test well. The minimum distance of the observation well from the test well in partially penetrating confined and semi-confined aquifers should be greater than 0.5 to 2 times the thickness of the aquifer (Krussenman and De-Ridder, 1970). In case of an unconfined aquifer a lesser distance can be used.

(d) Stratification: Isotropic and homogeneous aquifers seldom occur in nature. As a result of stratification, the drawdown observed at a distance from the pumped well may be different at different depths within the aquifer because of the difference in the vertical and horizontal components of hydraulic conductivity. This effect neutralizes as the distance of the observation well from the test well increases. Although there is no fixed rule for determining the spacing of observation wells, a range of 1 m to 100 m for the minimum and maximum distances of the observation well, will usually be sufficient.

(iii) Depth of the observation well: In fully penetrating confined and semi-confined aquifers, the depth of the observation well should be up to the center of the well screen. For a partially penetrating well, the depth of the observation well should be the same as that of the test well.

In semi-confined aquifers with prompt yield and fully penetrating conditions, two observation wells should be installed at a particular location-one above the semi-pervious layer and the other reaching the middle of the well screen in the main aquifer (Fig. 2.22). When the layer below the main aquifer in a semi-confined formation is also semi-pervious, a third observation well should be installed reaching below the bottom of the semi-pervious layer. By doing so, the leakage from the main aquifer or to the aquifer is accounted for.

(iv) Diameter of the observation well: Precise measurements of the drawdown can be made in a small diameter observation well. In practice, however, it is difficult to fabricate and install an observation well less than about 2.5 cm diameter. The diameter of observation wells generally ranges from 2.5 to 5 cm (Fig. 2.23).

(v) Length of the perforated portion of the observation well: The portion of the observation well casing in the

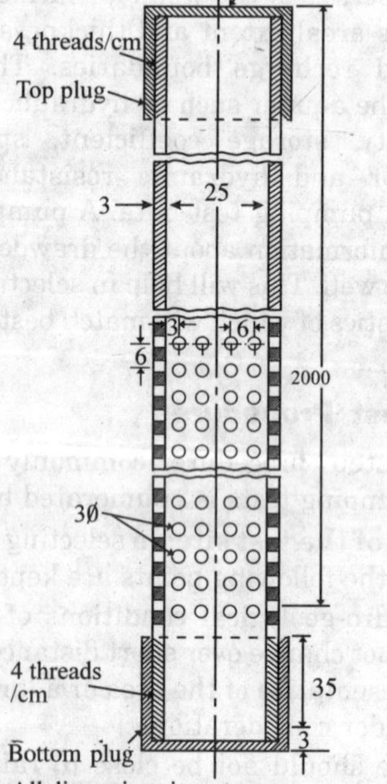

Fig. 2.23. Sectional view of a typical observation well.

aquifer should be generally perforated (Fig 2.23). However, a shallow observation well which is installed above or below a semi-confined aquifer may be perforated only about 1 to 2 m at the bottom.

(vi) Size of perforations of the observation well: The size of the perforations should be designed on the basis of the particle size distribution of the strata in which it is installed and the size of the drilling tool available. Since well development is not usually possible in an observation well the size of the perforations should be smaller than that of the well screen in order to protect it from clogging. However, perforations of size 2 to 5 mm covered with rope or PVC fibre may be used in coarse formations.

(vii) Installation of observation wells: The observation well is placed in the hole made by a soil augers or core drilling machine. After placing the observation wells, the annular space around the screen should be filled with either uniform coarse sand or gravel in order to facilitate the rapid entrance of water into it. When an observation well penetrates several aquifers or a semi-pervious or impervious layer, the annular space against the impervious or semi-pervious formation is filled with fine sand or clay to prevent leakage from one layer to another.

Duration of the pumping test: The duration of the pumping test depends on the type of aquifer and its hydraulic properties and the method to be used for analysing the pumping test data. Reliable information can be obtained if pumping continues till a nearly steady-state condition has reached. The condition is obtained in about 15 to 20 hours in a semi-confined aquifer. In a confined aquifer the duration should be about 24 to 40 hours. It is not always necessary to wait till steady-state condition is reached because even when the duration of pumping is small, methods are available for analysing the pumping test data.

Water level measurement: An important part of a pumping test is precise measurement of the depth of water in the observation well and, if possible, in the pumped well. These measurements must be taken many times during the pumping test. Since the water level falls very fast during the first few minutes to first few hours, the measurement should be made at close intervals during this period. The time interval may be increased with the progress of the pumping test. Different methods of measuring water levels in pumping wells and observation wells are described under 'Testing Well for Drawdown and Yield' in this chapter.

DESIGN OF IRRIGATION WELLS

Generally, wells are designed for the purpose of irrigation, drainage, and domestic and industrial uses. The design of each type of well for each purpose requires particular attention, taking into account its purpose. Designing a water well involves selecting the proper dimensional factors for the well structure and choosing the materials to be used in its construction. Good design aims to assure an optimum combination of performance, long service life and reasonable cost.

The hydraulic and hydro-geological characteristics of aquifers vary greatly. Irrigation wells should be designed and constructed to take advantage of the natural conditions at a given location. When this is done, and the materials of construction are properly selected, an economical and efficient well structure of long life can be achieved. Irrigation wells are usually designed to obtain the highest yield available from the aquifer, and the highest efficiency in terms of specific capacity. These factors bear directly upon operating costs. Two general principles influencing the design of both open wells and tube wells are the water requirement of the crops to be irrigated and the location of well.

Water requirement: The yield potential of a well is evaluated on the basis of hydrological conditions of the area — rainfall, runoff and recharge. When the yield potential of an area is not a limiting factor, a properly designed irrigation well should provide the required quantity of water to irrigate the entire area owned by the farmer, keeping in view the area to be irrigated, cropping pattern, and other needs such as domestic and livestock requirements.

When the yield potential of a well is more than the field water requirement of the farm under the control of the well owner, it is always economical to sell the additional yield to the neighbouring farmer, rather than limiting the capacity of the well to the requirement of a small or medium size holding. Often, it is not economical to have separate wells for individual small holdings. In an area with predominantly small holdings, it is essential that high discharge irrigation wells are owned jointly by a group of farmers or by a suitable public agency like the State Tube well Corporation.

Location of the well: The topography of the farm, the type of water conveyance system, the method of irrigation to be used, recharge possibilities of the area, safety of the pumping set and other installations are the main factors to be considered in deciding the location of the well. If the general slope of the farm does not exceed 1 per cent and surface methods of water conveyance and distribution are to be employed, the well should be constructed at a higher elevation. If underground pipeline water distribution is to be used or sprinkler or drip irrigation are used it is not necessary to locate the well at a high point. The location at a lower point of the farm usually facilitates better well yield.

Design of irrigation wells may be broadly divided into: *(i)* design of open wells in hard rock areas (consolidated formations) and unconsolidated formations and *(ii)* design of tube wells in unconsolidated formations.

Design of Open Wells

As seen in Chapter 1, a majority of irrigation wells in India are open wells, drawing their water mostly from shallow unconfined aquifers. The two common types of open wells are those located in hard rock areas and those in unconsolidated formations. Tube wells are the predominant groundwater structures in the alluvial formations of the Indo-Gangetic plains, the Brahamputra valley, the delta regions of the east flowing rivers of Kaveri, Krishna, Godavary, and Mahanadi and the alluvial and unconsolidated formations in river basins of the northwest India. Tube wells may be deep tube wells and shallow tube wells.

Open Wells in Hard Rock Areas

A major part of the geographical area of India, comprising of vast areas of Maharashtra, Madhya Pradesh, Gujarat, Andhra Pradesh, Karnataka and Tamil Nadu, is covered by consolidated underground formations, usually known as hard rock areas. The shallow groundwater reservoir is the only source of the limited supplies of groundwater in these areas. The aquifer is directly dependent on precipitation for recharge. Hence, the water table is prone to considerable fluctuations in relation to the incidence of rainfall. The shallow groundwater reservoir is constituted by the weathered mantle covering the unaltered rock and by the fracture porosity of the rock itself and is characterised by limited permeability. Hence, they are capable of yielding only limited quantities of groundwater. Due to their poor permeability, tube wells are usually unsuitable in such formations. It is therefore desirable in such formations to have open wells capable of storing fairly large supplies of water during a given recovery period. Thus, the available supplies of water can be obtained at small drawdowns in a relatively short period, thereby allowing for sufficient recuperation periods between successive periods of pumping. Open wells also expose a greater surface area of the aquifer for infiltration.

The hard rock areas are predominantly occupied by crystalline rocks like gneisses, granites, schists and compacted sediments. Rocks of precambrian sedimentary basins are largely covered in India, particularly in the States of Andhra Pradesh, Karnataka, Maharashtra, Madhya Pradesh, Rajasthan, Jharkhand, Uttaranchal and Assam. A major rock type in which design of open wells is important for optimum utilization of groundwater is the Deccan Traps. These are ancient volcanic flows of late Mesozoic era. There are a number of flows of different kinds, sometimes found with Inter-Trappean sedimentary formation. Vesicular types and weathered horizons and Inter-Trappean sediments are major sources of groundwater in these rocks. Important criteria for designing open wells are the pattern, density, and fabric of joints in these rocks. Vast areas in Maharashtra, Gujarat, Madhya Pradesh, Chattisgarh, Karnataka and Andhra Pradesh are covered with these rocks.

Dykes (thick and compact hard rock formations) which act as barriers to groundwater movement are often encountered intermittently in the granitic terrain. Dykes are actually dark coloured medium to fine-grained igneous rocks occurring as intrusions in wall like form. They have high specific gravity of about 3.0 and low porosity. Unless the rock is highly fractured and weathered, which is not common, the chances of striking water within a dyke are remote. Dykes act as sub-surface dams and effectively stop and/or change the lateral movement of groundwater. On the upstream side of groundwater flow by the side of a dyke one may get high-yielding wells while on the downstream side there may be no groundwater.

Sedimentary rocks occurring in many parts of the hard rock areas consist of sandstones, phyllites, and dolomites. The sandstones are compact and show good development of joints. Bedding and vertical joints and cleavages are well developed in shales and phyllites. There are number of fault zones in these rock formations. The rocks show good porosity and permeability. Groundwater occurs in these formations both under water table and semi-confined conditions. The occurrence and movement of groundwater is controlled by the grain size, degree of weathering and nature of bedding planes. The thickness of the weathered zones depends upon topography, climatic conditions and rock type.

In hard rock terrains where the groundwater occurrence is poor, the most important zone in which groundwater invariably occurs is the weathered zone. Therefore, study of the weathered zone profile, mode of weathering, structural features and correlation of all these features with the lithology of the area are useful. In the basement rock, one can expect good yield only if it is located in a shear zone or fracture zone.

Results of research at the Hyderabad Centre (Geology Dept., Osmania University, 1975) of the ICAR Coordinated Project on Wells and Pumps have revealed the following criteria for the design of open wells in hard rock areas:

(*i*) In the case of hard rock terrains, open wells rather than borewells are recommended. Open wells can tap the weathered zone better than the bore wells and they will be cheaper for shallow depths.

(*ii*) In the hard rock areas a large rectangular open well is preferable to the traditional circular shape well. Larger cross-sectional area will have chance to cut across fracture zones and increase the recharge into the well. This provides for storage of large amounts of water.

(*iii*) The longer side of the rectangular well should be normal to the trend of the fracture system to ensure better yield.

(*iv*) Dug wells in granitic areas should touch the basement rock. This is because the top of the basement rock is a more fruitful zone than the remaining part of the weathered zone.

(*v*) It is advisable to tap granites (for groundwater) rather than Deccan Traps when both are present in the same area.

(*vi*) In the case of sedimentary formations, open wells (or borewells) should penetrate the entire thickness of the aquifer.

(*vii*) Since dykes act as barriers for groundwater flow, in the vicinity of a dyke, wells should be located on the upstream side of the strata and not on the dyke or its downstream.

(*viii*) Beyond the weathered zone it is not advisable to deepen a dug well. Instead, one should go in for bores at its bottom or sides. These bores should be on the side where fissures or fractures are present. They can be radial, vertical or inclined, depending upon the local situation. As long as the pressure is sufficient, water will be seeping out of the bore wells.

Open Wells in Unconsolidated Formations

Open wells in unconsolidated formations are dug down about 7 to 10 m below the static water level in the dry season. The open excavation is usually circular in shape, the diameter varying from 1.5 to 4.5 m. The wells, in general, derive their water from unconfined aquifers.

Open wells in unconsolidated formations are usually provided with lining to prevent cave-in of the walls. The common materials used for lining are brick or stones laid in cement mortar or R.C.C. The lining consists of a wedge shaped well curb at the bottom which facilitates the sinking of the well and the steining which is the cylindrical wall forming the lining proper. The thickness of the steining may be designed by adopting the usual structural design procedures and assuming the well structure to be a thick cylinder with external pressure only. To make the design safe in case of drought, it is assumed that the well is empty and there is no internal pressure acting on it.

A graphical solution of the inter-relationship between depth of well and thickness of steining for wells of different diameters is presented in Fig. 2.24. The theoretical thickness of masonry for different depth intervals for a specific well diameter may be read from the nomograph and the same may be modified to suit the size of brick with a minimum of 23 cm for the first depth interval. The depth interval at which the thickness may change is be taken to be about 3 m.

Fig. 2.24. Graphical solution of the inter-relationship between the depth of well and thickness of steining in open wells of different diameters.

(*Source*: Khepar *et al.*, 1971.)

Design of Tube Wells

The design elements in tube wells include the well diameter, well depth, well screen, gravel pack and well spacing.

Well diameter: Choice of proper well diameter is important because it affects significantly the cost of the well. The diameter of a shallow tube well is usually the same from top to bottom. In deep wells, however, the well structure usually consists of two main elements. One element is the part of the well that serves as housing for the pumping equipment and the other as a vertical conduit through which water flows upwards to the pump. This is commonly the cased portion of the well, although some of its length may be uncased where the well is constructed in consolidated rock materials. The other main element is the intake portion of the well where water enters the well from the aquifer. A properly constructed well screen allows water to enter the well freely at low velocity, prevents sand from entering with the water, and serves as the structural retainer to support the loose formation material.

In a consolidated rock aquifer, the intake portion of the well is usually an open bore hole drilled into the aquifer to an adequate depth. The yield of such a well varies with the number, size, and continuity of the openings in the rock that are encountered in drilling the bore hole in the aquifer. The well diameter must be chosen to satisfy the following two requirements:

(1) The housing pipe must be large enough to accommodate the pump with proper clearance for installation and efficient operation.

(2) The diameter of the intake section of the well must be such as will ensure good hydraulic efficiency of the well.

In choosing the size of the housing pipe the controlling factor is usually the size of the pump that is required for the desired or potential yield of the well. The diameter of the well casing should be at least 5 cm larger than the nominal diameter of the pump.

Table 2.2 presents the recommended size of pump bowls (in vertical turbine pumps and submersible pumps) and the diameter of the housing pipe in tube wells.

Table 2.2. *Relationship between Well Yield and Recommended Size (diameter) of Pump Bowls and Diameter of Housing Pipe in Tube wells**

Anticipated yield of well (litres/sec)	*Nominal size of pump bowl (cm)*	*Optimum size of housing pipe (cm)*	*Minimum size of housing pipe (cm)*
6	10	15	12.5
5 to 11	12.5	20	15
10 to 25	15	25	20
22 to 40	20	30	25
37 to 56	25	35	30
53 to 82	30	40	35
75 to 112	35	50	40
100 to 190	40	60	50

*Shallow tube wells of size 10 cm diameter fitted with centrifugal pumps are frequently employed for discharges up to about 8 litres/second.

Well diameter (intake section) and yield: The yield of a well is a function of the diameter of its intake portion. Many cases are on record of costly large diameter wells which were put down because of the belief that "the bigger the well, the more the yield". It is true that the larger diameter tube well will yield some more, but the percentage of increase may be relatively small. Keeping the hydraulic properties of the aquifer as constant, doubling the diameter of the screen in a water-table well will increase the discharge only about 11% (Slichter, 1899 and Linsely *et al.,* 1964). Ahrens (1958) showed that doubling the diameter of the well screen, in a confined aquifer, will increase its yield by about 7%. The value of increase in yield in a confined aquifer is less than a water-table well because for the same discharge rate the radius of influence is larger for the former. Table 2.3 shows the approximate increase in yield in sand and gravel wells which may be expected by increasing the diameters, all other conditions remaining the same.

Table 2.3. *Effect of Increase in Well Diameter on Well Yield*

Well diameter (cm)	*10*	*15*	*20*	*30*	*45*	*60*	*90*
per cent increase	0	5%	10%	15%	23%	28%	38%
		0	5%	10%	18%	23%	33%
			0	5%	13%	18%	28%
				0	8%	13%	23%
					0	5%	15%
						0	10%

The yield of a well increases very rapidly with the coarseness of the formation.

Thickness of Casing Pipes and Well Screens

The thickness of pipes used in tube wells varies mainly with the depth of the well. Whereas thicker pipes are costlier, the thinner ones may be structurally weak to withstand stresses due to earth and water pressures and may also corrode earlier. A pipe may be considered adequately strong for use in tube wells if it has adequate tensile strength under the following conditions of use: (*i*) pipe assembly hanging freely in the bore hole, and (*ii*) pipe assembly being extracted out. The first criterion should be satisfied in all tube well assemblies, the weight of the pipe under water being taken as submerged. The second criterion should be satisfied only in such cases where the assemblies are intended to be recovered in the event of failure. In this case too, the weight of the pipe should be taken as submerged weight for the portion under water and as dry weight for the remaining portion. The tensile strength of the pipe should be able to overcome the weight of the assembly plus the frictional force caused by earth.

Metal casing should be of sufficient strength, toughness and thickness to withstand shear, tensile and compressive stresses imposed during and after installation. Heavy duty steel pipes have proved to be better for casing pipes to ensure durability. In areas where corrosion is a problem, corrosion resistant metals should be used. A rust inhibitor such as asphalt or coal tar may be used on black metals in corrosive water.

Studies at the Irrigation Research Institute, Roorkee (Sharma, 1972) have shown that in case of normal tube wells 100 metres deep and 10 to 12 cm diameter, using mild steel pipes of 1.0 mm thick are good enough for safe installation, operation and extraction. However, as the tube well pipes are usually joined together through sockets, an additional pipe thickness is required for cutting threads which is about 2.1 mm. Thus, the minimum thickness of the steel pipe at the ends may be fixed as (1.0 + 2.1) = 3.1 mm or near about. However, it would be desirable to cast the steel pipes of uniform thickness throughout so that the extra thickness of 2.1 mm in the rest of the body may be good enough to counteract the effect of corrosive agencies. The thickness will be more for larger diameter and deeper wells.

Well depth: The depth of a well is usually determined from the log of a test hole, from logs of other nearby wells in the same aquifer, or during the drilling of the production well. Generally, a well is drilled up to the bottom of the aquifer. This is desirable as more of the aquifer thickness can be utilized as the intake portion of the well, resulting in higher specific capacity. It also enables more drawdown, permitting greater well yield. Sometimes, however, the screen is centering between the top and bottom of the aquifer in order to make more efficient use of a given length of screen in a uniform artesian aquifer.

Length of well screens: The optimum length of a well screen is chosen in relation to the thickness of the aquifer, available drawdown and stratification of the aquifer. Edward E. Johnson Inc. (1966) has given the following recommendations for the length of a well screen under different aquifer conditions:

(1) In a homogeneous artesian aquifer, 70 to 80 per cent of the thickness of the water bearing sand should be screened, depending on aquifer thickness and assuming that the pumping water level is not expected to be below the top of the aquifer (Good design practice dictates that the maximum available drawdown in an artesian well should be the distance from the static water level to the top of the aquifer). Best results are obtained by centering the screen section in the aquifer or dividing the screen into sections of equal length and interspacing them with sections of blind pipe. It would thus be possible to obtain about 90 per cent or more of the maximum specific capacity that could be obtained by screening the entire aquifer.

(2) In a non-homogeneous artesian aquifer, it is best to screen the most permeable strata.

(3) In a homogeneous water-table aquifer, screening the bottom one-third of the aquifer provides the optimum design. In some wells, however, the bottom half of the aquifer may be screened in order to obtain higher specific capacity. Higher efficiency is more desirable than higher total yield in some cases.

For water-table wells, selection of screen length is something of a compromise between two factors. On the one hand, higher specific capacity is obtained by using as long a screen as possible. This reduces the convergence of flow and the entrance velocity, thereby obtaining higher specific capacity. On the other hand, more available drawdown results from using as short a screen as possible. *Available drawdown* is the distance between the static water level and the top of the well screen. A water-table well is usually pumped so that the pumping water level is maintained at or slightly above the top of the screen. The well screen is positioned in the lower or bottom portion of the aquifer.

(4) In non-homogeneous water-table aquifers the principle of well screen location is the same as in non-homogeneous artesian aquifers. In water-table aquifers, the screen or screen sections are located in the lower portions of the most permeable strata to obtain the maximum available drawdown.

Length of housing pipe, lower casing pipe and well screen: The length of the housing pipe of a tube well is governed by the lowest water level at which the intake of the pump is to be set. The top of the housing pipe is usually kept 60 cm above the ground level. The lowest static water level is that which exists at the end of the dry season. To this is added a pump clearance allowance which is the length of the suction pipe below the base of the pump. It usually varies from 1.5 to 2.7 m. Estimated drawdown during pumping is added to the sum of the above values. A depletion allowance, which takes into account the effect of groundwater abstraction of the area, as reflected in the dry season, is added to the above figure. A further factor of safety of 10% of the value, obtained by totalling the above values, is provided to arrive at the design depth of the housing pipe.

Lower well casing: All the possible aquifers in the area of influence of the well are not screened. Particles finer than medium sand containing more than 25% of fine sand are considered too fine. If they are screened, the well is likely to pump sand along with water. There are frequent clay and silt bands in sequence, against which screens should not be located in order to avoid sand pumping. Blank pipe lengths are placed against such materials. The length of blank pipe will depend on the thickness of the unscreenable layers. It is necessary to maintain 60 cm (minimum) offset between the ends of the blank pipe and the beginning and end of the unscreenable materials. This is to prevent mobilization of finer materials due to high entrance velocities close to the well screen.

Effect of Partial Penetration of Wells

The effect of partial penetration of wells on the flow lines is illustrated in Fig. 2.25. Within some distance from the centre of the test well (zone of effective drawdown) the flow lines in the aquifer are radial. Therefore, the drawdown, measured in the zone which is influenced by the vertical flow components, will be more in the case of a partially penetrating well than a fully penetrating well for the same discharge or for the same drawdown. The discharge obtained by the fully penetrating well will be more than a partially penetrating well.

The area in which the drawdown is affected by the vertical flow components is often reduced to an area with radius $r \approx b$ (Fig. 2.25). When b is unknown and the

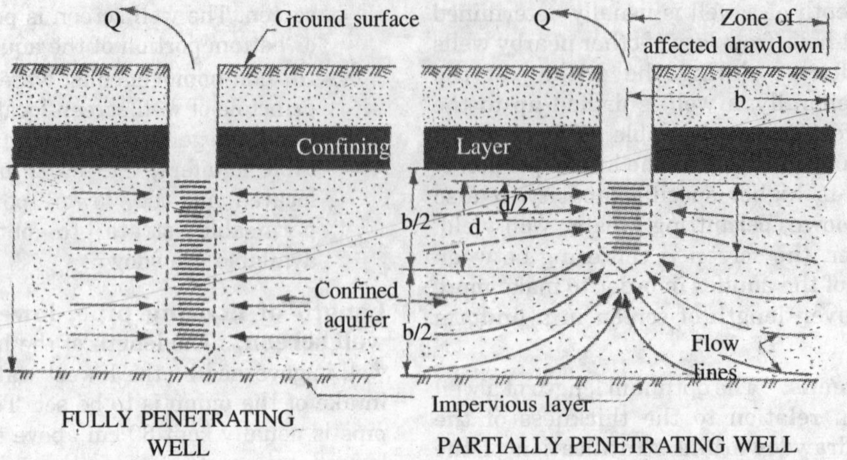

Fig. 2.25. Comparison of flow lines of a fully penetrating well and a partially penetrating well in a confined aquifer.

screened part of the well covers only a relatively small part of the aquifer thickness, it is better to accept the area with radius $r < 2\,b$ as being affected by the partial penetration effects.

Deep Well with Multiple Aquifers

Generally, deep tube wells encounter two or more aquifers. The aquifers may be of the same type or of different types. The possible combination of aquifers encountered by deep wells, depending on the hydrogeological characteristics of a particular area may be as follows:

(*i*) A series of confined aquifers
(*ii*) A series of semi-confined aquifers
(*iii*) A combination of confined and semi-confined aquifers
(*iv*) An unconfined aquifer underlain by a series of confined aquifers
(*v*) An unconfined aquifer underlain by a series of semi-confined aquifers
(*vi*) An unconfined aquifer underlain by a combination of confined and semi-confined aquifers.

The well should, under these circumstances, be designed on the basis of the characteristics of each water-bearing-formation in case they are not hydraulically interconnected. When aquifers are hydraulically interconnected, the resultant maximum permissible drawdown is the sum of the maximum permissible drawdown of all aquifers which are interconnected. In such a case, the well elements should be designed on the basis of the resultant maximum permissible drawdown.

Spacing of wells: Whenever a well is pumped, the greatest amount of lowering of the water table occurs within the well and the immediately adjacent areas are lowered almost as much. But as pumping continues, a hydraulic gradient, which may actually become an equilibrium gradient, is established, and the quantity of water entering the well from the surrounding zone will be approximately the same as that being pumped. The amount of lowering diminishes as the distance from the well increases, until a point is reached where pumping does not affect the level of the overall water-table (Fig. 2.14 and Fig. 2.15). If wells are spaced too closely together, the pumping of one may interfere with the yield of others. In other words, the yield of a well is decreased if its cone of depression is overlapped by the cone of an adjacent well. The hydraulic gradient may be short, or it may extend to over a kilometre, but experience and tests have shown that, in most cases, grouping of wells should not be made at intervals less than 70 metres. In shallow tube wells, however, the spacing can be closer. Koul (1970) observed that in shallow tube wells 20 metres deep, the drawdown curves were steep within 8 m from an individual well and flattened at farther distances from the centre of the test well.

Well Screens

A well screen is a strainer, which separates the groundwater from the granular material in whose pores it is contained. Generally all formations, except stable rock, require well screens. The quantity of water that can be tapped from a well depends primarily on matching the characteristics of the water-bearing formations to the elements of the well screen. The well screen elements refer to the length of the screen, its diameter, total open area and size, and arrangement of the slot openings.

The following are the basic requirements for any well screen:

(1) Resistance to corrosion, incrustation and deterioration,
(2) Enough structural strength to prevent collapse,
(3) Suitability to prevent excessive movement of sand into the well,
(4) Minimum resistance to flow of water into the well.

The selection of well screens is usually a matter of engineering judgement and experience. The recommendations on safe limit of entrance velocity of the flow into the well from the surrounding aquifer vary considerably. A criterion proposed by Bennison (1947) is that a velocity of 3 to 7.5 cm/sec through the individual openings of the screen will keep the sand movement and head losses to the minimum. Linsley and Frazani (1964) observed that the entrance velocity should be kept below 15 cm/sec in order to minimise sand movement and head loss. The variation in the two recommendations may be attributed to the differences in the particle size distribution of the aquifer materials investigated.

To prevent rapid clogging, the length of the well screen for a non-gravel pack well is designed on the basis of the following equation (Walton, 1962):

$$h = \frac{Q_0}{A_0\, V_e} \quad \text{... (2.20)}$$

in which,

h = length of the screen (m)

Q_0 = maximum expected discharge capacity of well (m^3/min)

A_0 = effective open area per metre length of the well screen (m^2)

V_e = entrance velocity at the screen (m/min).

Equation (2.20) is also used for determining the length of the screen in an artificial gravel-pack well. In this case the average value of the hydraulic conductivity of the aquifer and the gravel-pack is used to determine the entrance velocity at the screen.

The design procedure for the length of the well screen is as follows:

The optimum entrance velocity at the well screen is determined. Then, from the aquifer test the expected capacity of the well is calculated. From the information on the open area of the well screen per metre, the effective open area is determined as provided by the manufacturer. After providing a factor of safety of 2 to 5, the length of the screen is designed using equation (2.20).

Slot opening: Choosing the right slot width is one of the important steps in well design. Oversized slot will pump finer material (sand, silt and clay) indefinitely and clear water is difficult to obtain, while under-sized slots will provide more resistance to the flow of groundwater, resulting in more head loss. The fine slots are also blocked by small sand and silt particles in the long run which are carried up to the screen by suspension. The problem of clogging is reduced as the well screen openings are increased. Therefore, the well slot openings are used as wide as possible by matching the opening with the grain size distribution of the material surrounding the screen. The width of openings commonly used ranges from 1.5 to 5 mm, beyond which no substantial reduction in clogging is obtained.

Generally, horizontal slot openings give a better control of unconsolidated material than do vertical openings. The slots may be made in different ways, vertical or horizontal, continuous or intermittent. The width of the slot depends on the grain size distribution of the aquifer and varies in practice from values as low as 0.2 or 0.5 mm, depending on screen construction, to as large as 2 to 5 mm. The square openings in the wire mesh and circular drilled holes in the wall of the pipe are easily plugged by particles of nearly the same size as the openings.

Percentage open area: Water flows more freely through a screen with large open area than through one with limited open area. When the open area of the screen is large, the entrance velocity is slow and the head loss at the screen is minimum. Corey (1949) observed that little or no increase in well efficiency results when the open area is greater than 15 per cent of the total surface area of the screen. Open area larger than about 15 per cent affects the structural strength of the well screen.

When a screen is placed in an aquifer, sediment will settle around it and partially block the slot openings. Walton (1962) observed that on an average, about one-half of the open area of the screen is blocked by aquifer materials. Thus, it may be said that the effective open area averages about 50 per cent of the actual open area of the screen. Therefore, a new screen should be designed keeping this factor in view.

Well screen diameter: Screen diameter is selected to satisfy the essential basic principle that enough total area of the screen openings must be provided so that the entrance velocity of water will not exceed the design standard. Screen diameter is a factor that can be varied after the length of the screen and size of the screen openings have been selected. Screen length depends upon the thickness of the water-bearing sand; screen openings depend upon the gradation of the aquifer material. To a large extent, the naturai characteristics of the aquifer fix these dimensions, leaving the diameter as a factor that can be varied.

Laboratory tests and field experience show that if the screen entrance velocity is equal to or less than 3 cm per second, friction losses in the screen openings will be negligible and the rate of incrustation and corrosion will be minimum.

The entrance velocity is calculated by dividing the expected or desired yield of the well by the total area of the openings in the screen. If the value is greater than 3 cm per second, the screen diameter should be increased to provide enough open area so that the entrance velocity is less than 3 cm per second. If, on the other hand, the calculated entrance velocity is less than this figure the screen diameter may be reduced somewhat to achieve economy. Most well screen manufacturers provide tables that show the open area per metre of the screen for each size of the screen and for various widths of slot openings.

Well Screen Losses

The well screen losses are composed of the following:

(*i*) Loss of pressure head at the entrance due to size and shape,

(*ii*) Loss of head attributed to the turbulence of water in passing from the formation to inside the well screen,

(*iii*) Loss inside the well screen due to turbulence,

(*iv*) Loss of head due to partial perforations,

(*v*) Loss of head due to partial penetration,

(*vi*) Loss of head due to entrance (entrance losses).

Analytical formulae to predict the well screen losses have been developed by Peterson, *et al.* (1955), Saran (1961) and Huisman (1972).

Example 2.7. *A fully penetrating well in a confined aquifer has a maximum discharge of 1000 litres/min. The aquifer is overlain and underlain by impervious layers. The thickness of the aquifer is 20 m. Design the length of the well screen, assuming the percentage open area of the available strainer to be 30% and borehole diameter 20 cm.*

Solution: In equation 2.20

$$A_0 = \pi d\ 1$$
$$= 3.14 \times 0.06 \times 1$$
$$= 0.1884 \text{ m}^2$$

Assuming the safe entrance velocity to be 0.3 m/min, we have

$$h = \frac{1}{0.1884 \times 0.30}$$
$$= 17.692 \text{ m}$$
$$\approx 18 \text{ m}$$

This is the safe length of the well screen, because it provides 90% of the aquifer length to be perforated.

Material of the Well Screen

The well screen must be of sufficient structural strength to meet the requirements against corrosion and deterioration and to prevent collapse. This requirement can easily be met by choosing a strong metal such as steel of large wall thickness. If corrosion is anticipated the steel may be covered with a protective lining of chlorinated rubber paint, bitumen, hard rubber, plastic and similar materials.

When economy is not important, a screen may be used which is made entirely from corrosion resistant materials such as copper, bronze, nickel and stainless steel. When economy is important, locally available materials like bamboo and asbestos-cement pipe may be used as screen materials in shallow tube wells. When sanitation is important (medicinal wells), vitrified clay, porcelain, and porous glass (made of fibre glass) may be used as screen material.

Well Screen Types

Several types of well screens are manufactured to adopted standards and sold as commercial products. Edwarde E. Johson (1966) lists the following desirable features for a properly designed well screen:

(1) Openings in the form of slots which are continuous and uninterrupted around the circumference of the screen,

(2) Close spacing of slot openings to provide maximum precentage of open area,

(3) V-shape slot openings that widen inward,

(4) Single metal construction to avoid galvanic corrosion,

(5) Adaptability to different conditions by use of various materials,

(6) Maximum open area consistent with adequate strength,

(7) Ample strength to resist the forces to which the screen may be subjected during and after installation,

(8) Full series of accessories and fittings to facilitate screen installation and well completion operations.

The continuous-slot type of well screen (Fig. 2.26) is made by winding cold drawn wire, approximately triangular in cross-section, spirally around a circular array of longitudinal rods. At each point where the wire crosses the rods, the two members are securely jointed.

The louver or shutter-type well screen has openings that are really rows of louvers. The openings may be oriented either at right angles or parallel to the axis of the screen. They are produced in the wall of a welded tube by a stamping operation.

The continuous slot type and louver type screens, thought more efficient hydraulically, have not become popular in developing countries due to their high cost.

Common Types of Well Screens Used in India

Almost all the deep tube wells in India are of the gravel pack type using slotted pipes as screens. For shallow tube wells, however, a variety of strainers, some of them of low cost material, have been developed. These include coir strainers, PVC fibre strainers, bamboo strainers and agricultural strainers (Fig. 2.27).

Slotted pipes: Slotted pipes are usually made of medium and heavy duty mild steel pipes. Cement asbestos pipes are sometimes used for slotted pipes in shallow tube wells. Slotted pipes available in India are usually provided with vertical slots 1.5 to 4 mm wide and about 9 cm long. The minimum spacing between slots is 3 mm. Slots may be made in groups of 2 to 4 slots

Fig. 2.26. Continuous slot type well screen.

Fig. 2.27. Different types of well screens: *A.* Coir screen. *B.* Pipe-base screen with brass wire netting (agricultural strainer). *C.* Slotted pipe with vertical slots. *D.* Slotted pipe with horizontal slots.

(*Courtesy:* John Warwick Tockey, Wooloowar).

and are so arranged that the slots of one row are not in line with those of the adjacent row, so as to provide adequate strength for the screen section. The basic requirement regarding size of slot opening is that the slot should be wide enough to permit about 70 per cent of the formation material to pass through it. Non-availability in the market of a wide range of solt sizes to suit the varying characteristics of aquifers is often a limitation in designing efficient wells.

Steel pipes are not resistant to corrosion and may deteriorate soon under saline groundwater conditions. The Baroda Centre of the ICAR Coordinated Project on Wells and Pumps (Gujarat Engineering Research Institute, 1978 and 1988) has successfully demonstrated the use of rigid PVC slotted pipes under corrosive groundwater conditions. The design of the slots and the strength requirements of the pipe are the same as in metal pipes. The test well was 65 m deep and 20 cm in diameter.

Slotted pipes with pre-pack filters: Pre-pack filters (popularly known as Ashim filters, after its trade name) have their gravel pack on slotted pipes (Fig. 2.28). In the manufacture of pre-pack filters, graded hard gravel grains are coated and bonded together with water-proof chemicals over the outer surface of a slotted pipe. The gravel is fixed loose, providing a high value of permeability. The gravel size used varies, depending on the grain size distribution of the aquifer. Pre-pack filters are suitable for shallow and medium deep tube wells in areas having an aquifer composed of fine sand, sand and gravel. It is claimed that pre-packing provides economy in the quantity of gravel required and protects against dislodging of the gravel envelope.

Coir strainers: Coir strainer is a low cost strainer for shallow tube wells. Coir strainers are suitable when the aquifer is predominantly of coarse sand and gravel. This strainer is made of coir wound round a cylindrical steel frame. The frame is made of mild steel rods (sometimes, steel strips are also used) which are rivetted or welded with screwed pipe ends so that each length of the strainer can be joined to the next. There are no separate openings, and the whole surface of the coir acts as a screen, allowing the water to flow but preventing the sand from entering the tube well. This strainer is one of the cheapest strainers but its life is too short (five to ten years).

Bamboo strainers: The bamboo strainer is essentially a coir strainer having the longitudinal frame made of bamboo strips instead of steel rods. For making the strainer, bamboo strips are stretched lengthwise by fixing on 10 to 12 cm diameter steel rings placed at 30 cm interval. The coir rope is wound around this cylindrical frame. Originally developed in the Saharsa district of Bihar, bamboo strainers have become popular in many areas as low cost material for shallow tube wells. The adaptability of bamboo strainers is the same as that of coir strainer.

Polythene strainers: Polythene well strainers are, in general, similar to coir strainers in their construction, but are resistant to corrosion and deterioration (Fig. 2.28). They consist of high density polythene rings mounted close together on a steel framework. The frame has steel sockets fixed on either side. It is made of steel rods mounted on steel rings. The polythene rings vary in width from 6 to 10 mm. Water percolates through the passages of adjacent rings.

Fig. 2.28. Different types of well strainers.
(Photograph of pre-pack filter exposed view from Ashim Filters, Naini, Allahabad).

Agricultural strainers: Agricultural strainers are almost same as the coir strainer. It could, however, be used for deeper wells, but are more costly. The agricultural strainers are made of perforated mild steel pipes over which a copper wire mesh is soldered. The perforations on the mild steel pipe are round and are made with a drilling-machine.

The type of the copper mesh used depends upon the type of the sand strata encountered along the bore. Over the mild-steel pipe, iron strips, about 1.5 cm wide and 0.5 cm thick, are welded, taking care that they do not cover the perforations. The strainer sheet is then wrapped and soldered round these strips.

Gravel Packing

Gravel pack is a thin layer of coarse material (specially gravel) formed around the screened portion of a well (Fig. 2.29). Wells can be naturally gravel packed or artificially gravel packed. A naturally developed envelope can be produced by removing the fine sand and silt from the natural formations and transporting these fines through the well screen openings by surging and bailing. An artifical gravel envelope can be provided by keeping the bore of the well somewhat larger than the well screen, centering the screen in the hole and then filling the annular space around the screen with properly selected gravel (Fig. 2.30).

Analyses of Aquifer and Gravel Pack Samples

Sieve analysis of formation samples obtained during the drilling of test holes or production wells reveals the characteristics of the sub-surface formation. The results provide the basis for well screen specifications and gravel pack design.

The standard procedure for analysing sand/soil samples by the dry sieving method is adopted. A set of sieves conforming to Indian Standard (IS: 460-1962) or the particular standard adopted by the country is used (Fig. 2.31). The weight of the material retained on each sieve is recorded. These weights are then expressed as percentages of the total weight of the sample and a graph is plotted of the cumulative per cent of the sample retained on a given sieve and all the other sieves above it, versus the size of the given sieve, expressed in millimetres (Fig. 2.32). The "per cent retained" is plotted on the y-axis* and the size of the sieve opening or particle size is plotted on the x-axis. The size of the opening is considered to be the diameter of the smallest particle retained by each sieve.

Figure 2.32 presents the typical sand analysis curves for distinct classes of aquifer materials indicating the water bearing characteristics of the formation and the need for artificial gravel packing.

Fig. 2.29. Screen segment of a slotted pipe tube well with gravel packing developed with compressed air. Development removes the fine particles and leaves more permeable coarser particles around the well screen.

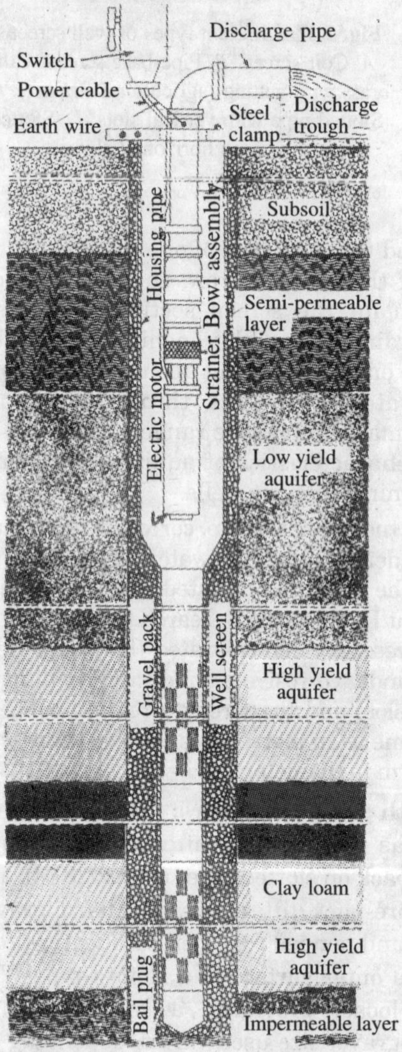

Fig. 2.30. Schematic sketch of a slotted pipe screen tube well assembly with submersible pump installed in the housing pipe. Note that the screen pipes are installed against the water bearing layers.

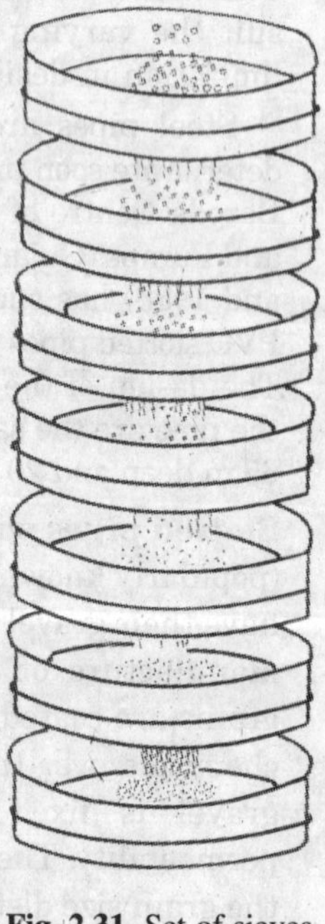

Fig. 2.31. Set of sieves for analysis of aquifer samples.

* It is a common practice to plot the graph on a semi-log paper, with the x-axis on the logarithmic scale (Fig. 2.33).

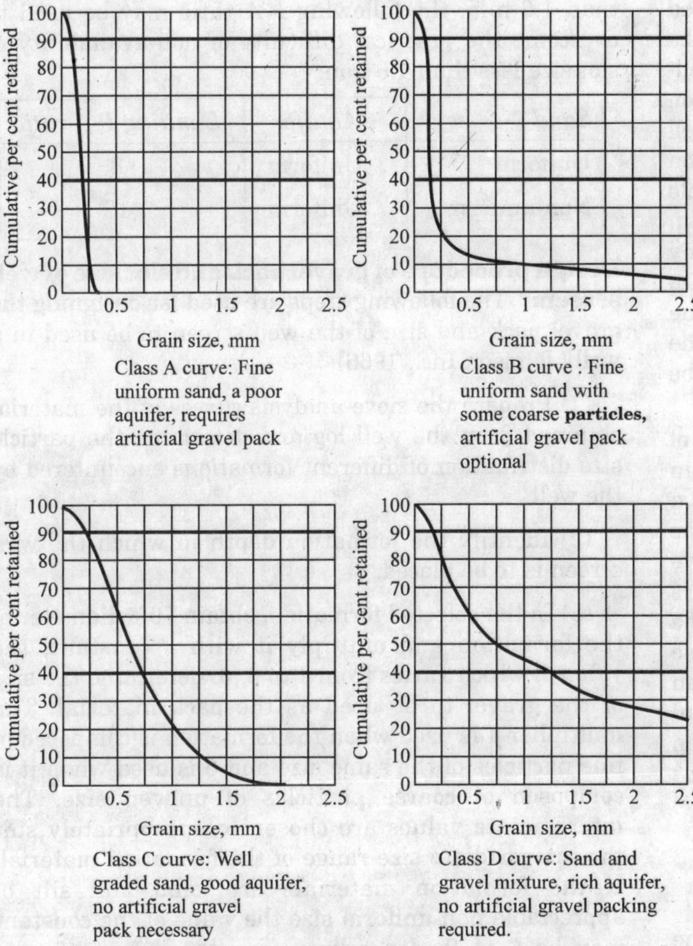

Class A curve: Fine uniform sand, a poor aquifer, requires artificial gravel pack

Class B curve : Fine uniform sand with some coarse **particles**, artificial gravel pack optional

Class C curve: Well graded sand, good aquifer, no artificial gravel pack necessary

Class D curve: Sand and gravel mixture, rich aquifer, no artificial gravel packing required.

Fig. 2.32. Typical sieve analysis curves of water bearing sands and gravels.

Effective size (D_{90}). The term effective size is defined as the particle size where 10% of the sand is finer and 90% is coarser. For example, the Class *C* curve of Fig. 2.32 shows that 90% of the sample consists of sand grains larger than 0.25 mm or that 10% is smaller than this size. Thus, the effective size of the formation material is 0.25 mm.

Uniformity coefficient **(C_u)**: This is a ratio expressing the variation in grain size of a granular material. It is the quotient obtained by dividing the sieve aperture that passes 60% of the material, by the sieve aperture that passes 10% of the material.

$$C_u = \frac{D_{60}}{D_{10}}$$

Thus, in Class *C* curve (Fig. 2.32) the uniformity coefficient is 0.75 mm divided by 0.25 mm or 3.0. This ratio was proposed by Hazen as a quantitative expression of the degree of assortment of water-bearing sand as an indicator of porosity. The value of the coefficient for complete assortment (one grain size) is unity; for fairly even-grained sand, it ranges between 2 and 3; for heterogeneous sand the value is high.

Formations Requiring Artificial Gravel Pack

Not all water bearing formations require artificial gravel packing. Generally, formations with an effective size of 0.25 mm and a uniformity coefficient of 2 or more can be safely developed without a gravel pack. As the formation becomes coarser, the desirability of the gravel pack decreases; however, exceptions to the above are common.

Artificial gravel pack construction is recommended where the natural formation consists of fine uniform sands and/or where the formation is extensively laminated (consists of alternating fine, medium, or coarse layers that are thin and difficult to locate precisely). These conditions are frequently et in most deep tube wells. An artificial gravel pack may also be used for an aquifer containing fine materials, when it is desirable to use larger screen openings than are indicated by the sieve analysis. This often occurs when the C_u is between 2.0 and 3.0 and the D_{40} size is less than 0.42 mm. In areas where incrustation is a problem it is desirable to use large screen openings.

The following are the advantages of the gravel pack wells:

(1) Gravel packing increases the effective diameter of the well, thus increasing its specific yield.
(2) It increases the yield of a well due to low resistance against flow at the well screen.
(3) It reduces incrustation. This is due to the large screen openings.
(4) When designed properly, gravel packs provide sand-free water, thus increasing the efficiency of the well and the pumping unit.
(5) It provides higher structural strength for the well screen.
(6) It prevents the caving-in of the formation material, thus reducing the danger of clogging of the well screen.
(7) It facilitates the extraction of the well casing and screens in shallow wells.

Gravel pack material: A good gravel pack material is clean gravel with well rounded, smooth, uniform grains. (These characteristics increase the permeability and porosity of the pack material). It should be hard, insoluble, siliceous material having less than 5% of calcareous particles (limestone). Particles of shale, anhydrite and gypsum are undesirable. The material should be uniform in size. A uniform material is one for which the uniformity coefficient (C_u) is 2.0 or less.

Two types of gravel packings are in general use—the uniform-grain size pack and graded-grain size pack. The former has been widely accepted in recent years, especially when manufactured screens are used because the size of the openings can be controlled. In the case of a graded pack, the formation material may invade a graded pack at the gravel-formation interface, partly filling the pores and resulting in reduced permeability. With a well-sorted (uniform) gravel pack, the fines of the formation can travel between the grains and be pulled into the well during development, thereby increasing the formation permeability while retaining the highly permeable nature of the pack. Lack of availability is the main drawback in adopting uniform pack material in many cases. An important physical property of uniform-grain-size material is the particle size as represented by the mean grain diameter which is 50% of the grain size. The maximum size of the particle in the gravel pack should be about 6.4 mm.

Design of gravel pack: To prevent the movement of formation material, it is necessary to provide a pack in which the mean grain size of the pack material bears a specific relationship to the mean grain size of the formation material. Figure 2.33 shows a typical sieve analysis curve of a water bearing formation which is classified as medium sand. The other curve is that of the gravel pack generally used, classified as very coarse sand to fine gravel. The mean grain size of the formation material is 0.38 mm and that of the gravel pack material is 1.8 mm, *e.g.*, the size ratio of gravel pack to formation is 4.8. This ratio of mean sizes is called the *gravel pack ratio* or pack-aquifer ratio (*P-A* ratio).

$$P\text{-}A \text{ ratio} = \frac{50\% \text{ size of gravel pack}}{50\% \text{ size of aquifer}}$$

Some authorities define the pack-aquifer ratio as the ratio of the 70% size of the pack and the aquifer. The 50% criteria is usually adopted.

In order to minimize the head loss through the gravel pack, the lower *P-A* ratio should usually be 4.0. *P-A* ratios exceeding 9.0 may allow the movement of sand. Smith (1954) reported that ratios of 4 to 5 were found satisfactory for the efficient design of wells with gravel packing. However, wells having gravel-pack ratios of 7 to 10 were found inefficient because of sand pumping.

The results of model studies conducted at Ludhiana (PAU, 1971) indicated the following upper limits of *P-A* ratios in order to maintain a stable filtering action:

Aquifer	*Gravel Limiting*	*P-A ratio*
Uniform	Uniform	6
Non-uniform	Uniform	9

It was further reported that in India the slotted pipes commonly available are in two-slot sizes, viz., 1.6 mm and 3.2 mm. Hence, it was recommended that if the size of the aquifer material is such that it requires a slot size less than 1.6 mm, the following *P-A* ratio may be used to overcome the practical difficulty of non-availability of slot size less than 1.6 mm:

Sand	*Aquifer*	*Limiting P-A ratio*
Uniform	Uniform	8
Non-uniform	Uniform	12

Design procedure of gravel pack and slot size of well screen: The following steps are used for designing the gravel pack and size of the well screen to be used in a well (Johnson Inc., 1966):

(1) Prepare the sieve-analysis curves of the material obtained from the well log and determine the particle size distribution of different formations encountered by the well.

(2) Identify the formation depth in which the well screen is to be placed.

(3) In the selected formation, obtain 70% diameter of the formation and multiply it with a constant, the value of which varies from 4 to 6, to determine the size of the gravel to be used as the pack material. The multiplier 4 is used when the formation is composed of fine particles of the same size and 6 is used when it is composed of coarse particles of uneven size. The intermediate values are chosen to appropriately suit the intermediate size range of the formation material. If the formation materials are sand and silt of appreciable non-uniform size the value of the constant may be 6 to 9, depending upon the extent of non-uniformity. Locate the gravel size, as computed above, on the particle-size distribution graph (Fig. 2.34).

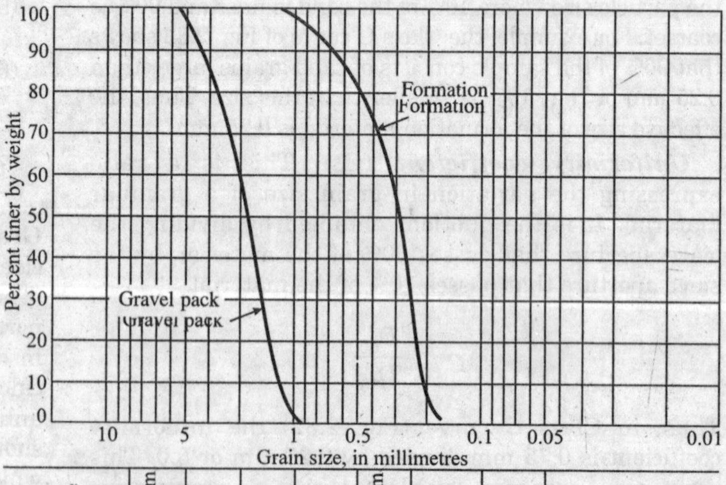

Fig. 2.33. Sample sieve analysis curve of aquifer material and gravel pack material (Adapted from Smith, 1954).

Through this point draw a curve parallel to the particle size distribution curve of the formation. The curve should be so drawn, by trial and error, that it should represent a uniformity coefficient equal to or less than 2.5. (Gravel material with a uniformity coefficient less than 2.5 is usually difficult to obtain. If the uniformity coefficient is higher than 2.5, it will result in the segregation of the gravel particles while placing the gravel. This will result in poor efficiency in the operation of the well.) The curve thus prepared will represent the size distribution of the gravels to be used for gravel packing. The particle size distribution curve of the gravel pack used in the well should more or less coincide with this graph.

(4) *Selection of the size of slot opening in the screen:* The slot size opening of the well screen should be determined on the basis of the particle size distribution curve of the gravel. On the particle size distribution curve of the gravel locate a point *C* indicating the 90% size of the gravel to be retained (Fig. 2.34). Through *C*, draw a line parallel to the *y*-axis, meeting the *x*-axis at point *D*, which indicates the slot size of the well screen. The size of the slot is fixed at ± 8% of the above size, depending on the size of the tool used in making the slots.

Example 2.8. *Table 2.4 shows the particle size distribution of the aquifer material from 10 to 12 m depth from the ground surface of a well penetrating a semi-confined aquifer in an alluvial terrain. Design the size of gravel pack and slot size opening of a slotted pipe well screen.*

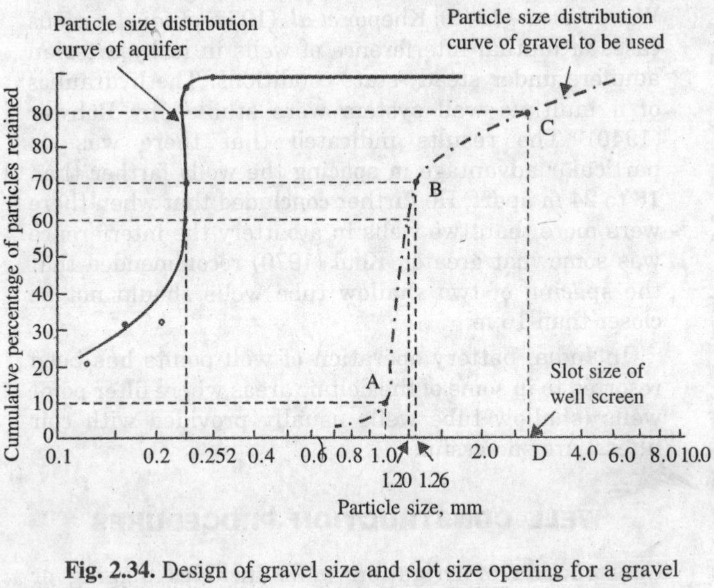

Fig. 2.34. Design of gravel size and slot size opening for a gravel pack tube well (Example 2.8)

Table 2.4. *Slieve analysis of lithological section between 10 to 12 m in Example 2.8*

Size of the sieve opening (mm)	*Cumulative retained (gm)*	*Cumulative percentage age retained*	*Cumulative percentage age finer*
> 2.54	0.00	0.00	100.00
1.80	5.00	1.00	99.00
0.30	16.00	3.20	95.00
0.25	318.00	63.60	32.00
0.21	3.00	0.60	31.60
0.16	50.00	10.00	21.00
0.12	38.00	7.60	14.00
< 0.12	70.00	14.00	0.00

Solution: In Fig. 2.34, curve *A* shows a plot of the particle size versus percentage retained of the formation material at depth 10 to 12 m from the ground surface. The 70% diameter of the formation material located on the curve is 0.252 mm. To obtain an uniform size of gravel, a multiplying factor 5 is used which gives the size of the gravel as 1.26 mm (0.252 × 5). The point is located at *B*. Through point *B*, a curve is drawn approximately parallel to the particle size distribution curve of the formation and representing a uniformity coefficient less than 2.5. The D_{60} size of the gravel is 1.2 mm and the D_{10} size 1 mm. This gives an uniformity coefficient of 1.2 which is well within the limit of 2.5 as recommended by Johnson Inc. (1966). The 90% diameter of the gravel is 3.01 mm. Thus, 3 mm size of slot is selected for the well screen.

Gravel pack thickness: The design theory of gravel pack gradation is based on the mechanical retention of the formation particles. Hence, a gravel pack thickness of only two or three grain diameters is adequate to retain and control the formation sand. Laboratory tests made by Johnson Inc. (1966) show that a gravel pack with a thickness of only a few mm successfully retains the formation particles, regardless of the velocity of water tending to carry the particles through the gravel pack. However, it is impractical to ensure complete coverage of the gravel pack around a well screen if it is only a few mm thick. To ensure that gravel will surround the entire screen, a minimum thickness of 7.5 cm is considered necessary under field conditions. Under most conditions, the gravel pack thickness should not exceed 20 cm. A thicker envelope does not appreciably increase the yield of a well. Too thick a gravel pack can make it difficult to develop a well properly.

Multiple Well Systems

A multiple well system is a group of closely located wells, usually connected to a common header pipe or a mainfold and pumped by suction lift (Fig. 2.35 and 2.36). Such a system is known as well point system/well battery. In areas where the water bearing materials are not appreciably thick and the deepening of well does not increase the well capacity, multiple well systems are used to derive water from aquifers at shallow depths. In a well battery system the drawdowns in the well do not usually exceed the suction limit of a centrifugal pump. In case where waterlogging is a problem, multiple well systems have shown their usefulness in controlling the water-level. A battery of wells is specially adapted under the following conditions:

(*i*) Where the water table is at shallow depth,

(*ii*) Where medium and deep tube wells are not economical,

(*iii*) Where the hydraulic properties of the aquifer are poor,

(*iv*) Where salts are present in the deeper layers of the aquifer,

(*v*) Wherever there is a problem of waterlogging,

(*vi*) Where the conditions are such that a constant level of water at some depth from ground surface is required to be maintained.

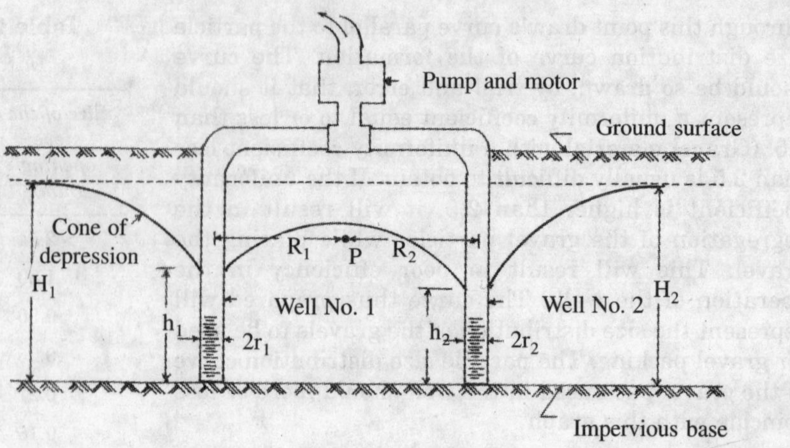

Fig. 2.35. Vertical section of two wells in an unconfined aquifer in a well battery.

As more wells are introduced into a given area the discharge of each of the existing wells may decrease with time, although the total discharge from the aquifer may increase. When two or more wells are installed in a manner such that one falls in the area of influence of the other, an overlapping of the cones of depression is experienced (Fig. 2.35) which results in an increased drawdown in each well. Consequently, there is a decrease in the discharge of an individual well. Thus, the relationship between discharge and time may not be unique for a well but may depend on the operation of the well itself as well as of neighbouring wells. The farther apart the wells are spaced, the lesser their mutual interference but greater the cost of connecting pipelines. The practice, therefore, is to install the wells such that each well is outside the predominant area of influence of the other. In a multiple well system, the wells may be arranged in various geometric patterns, either as isolated groups or as a continuous pattern extending over a large area. Generally, two wells in a straight line (Fig. 2.35), three wells in a triangular pattern (Fig. 2.36) and four wells in a square pattern are used for withdrawing the water from the shallow aquifers. The subject 'battery of wells' was first studied by Forchheimer (1898) with the introduction of theory of functions of complex variables for the analysis of gravity flow towards a group of wells. On the basis of theoretical investigations, Slichter (1899) stated that the amount of interference between wells in a battery depends on the diameter, depth, spacing and number of wells in the battery. Scott *et al.* (1967) studied the hydraulics of multiple shallow tube wells. They stated that the wells may be arranged in various geometric patterns, either as isolated groups or as a continuous pattern over large areas. Muskat (1937) studied analytically the hydraulics of multiple well systems arranged in different geometrical patterns. He deduced equations to compute the drawdown caused at certain specific points by the combined pumping action from three, four or more wells in unconfined and confined aquifers. Analysis of mutual interference of wells under steady-state flow for different geometrical patterns of wells was given by Muskat (1949), and for unsteady-state flow by Wenzel *et al.* (1943). Kheper *et al.* (1972) studied various cases of mutual interference of wells in leaky artesian aquifers under steady-state conditions. The hydraulics of a multiple well system were studied by Rohwer (1940). The results indicated that there was no particular advantage in spacing the wells farther than 16 to 24 m apart. He further concluded that when there were more than two wells in a battery the interference was somewhat greater. Koul (1970) recommended that the spacing of two shallow tube wells should not be closer than 16 m.

In India, battery operation of well points has been resorted to in some of the deltaic areas where filter point wells (shallow tube wells usually provided with coir filters) are successful.

WELL CONSTRUCTION PROCEDURES

Construction of irrigation wells includes digging or drilling operation, installation of casing and screens and well development. All these operations depend on the type of formations to be drilled or excavated, the size of wells and its depth and the discharge expected.

Fig. 2.36. Three-way manifold connecting three shallow tube wells to a common suction pipe of a centrifugal pump.

Equipment and Procedures for Open Well Construction

Open wells are usually dug by hand tools or mechanically operated equipment, while tube wells are drilled by a suitable drilling equipment.

Construction of open wells: Open wells are usually excavated using pick axe and shovels or power-driven shovels of various designs. Many open wells are still constructed manually. In the manual method, one or more men start to dig, throwing out the dirt till the hole becomes deep. Then a second man or a group of men standing at intermediate points relay the excavated material and dump it on the ground surface, and so on as the digging proceeds. Buckets and wheel barrows operated by suitable pulleys and lines (Fig. 2.37) are often used with advantage to lift the excavated material. After water is struck, hand excavated pits must be kept relatively dry by pumping. Dragline excavators using clamshell or orange-peel type buckets (Fig. 2.38) are sometimes employed to dig open wells in semi-consolidated formations. These buckets can excavate under water, unassisted as long as there is no cave-in.

When digging open wells in unconsolidated formations, safety requires that a carefully made continuous lining or cribbing is lowered as the hole is deepened. Unless this precaution is taken, the digger's life may be endangered by a cave-in occurring at any depth more than 1 to 1.5 metres from the surface. The common practice in concrete lined wells is to allow the concrete lining to sink as the well is deepened. The lining is made in sections of one metre or so which is cast each day over the previous section. It may often be necessary to add weight from the top or apply force to make the lining sink down as the digging proceeds. Temporary support may be provided with a crib made of wooden boards (Fig. 2.37). Dug wells used for irrigation must be deep enough to extend about 5 to 10 m below the water-table in dry weather.

Quicksand: Quicksand, also called running sand, is mobile fine grained loose sand super-saturated with water. It is sometimes encountered in excavations and occurs in localities where soils are underlain by lacustrine sediments or river alluvium. While digging open wells, the flow of quicksand can be largely prevented by removing the material with a bucket or pump and keeping the excavation partially full of water. This neutralises the hydraulic pressure in the quicksand and renders it comparatively solid. As the excavation proceeds, the edge of the well steining should be driven to 0.3 to 1.2 m below the bottom of the excavation. After the excavation is completed, further entry of quicksand at the bottom can be prevented by placing a thin layer of clean coarse sand and weighting it with several layers of sand or gravel in increasing coarseness; thereby creating a graded sand filter, each layer of which is held in position by the slightly coarser material above it.

Boring and Deepening of Open Wells

One of the major schemes launched during the Five Year Plans in India has been the boring and deepening of open wells to increase their yield. A large number of open wells in different parts of the country is amenable to improvement by boring and deepening operations. Boring of a well consists of drilling a small size bore through its bottom and extending the bore up to potential water bearing formations, with a view to tapping underlying aquifers. The bores may be of two types, according to the way they tap the water bearing strata. A cavity bore is to be preferred only where there is available a hard, thick and extensive layer of impervious stratum immediately overlying a comparatively thin bed of good water bearing layer. A thick and extensive layer of impervious stratum is necessary in order that it may support its own weight and that of the overlying strata by arch action. In case the water bearing stratum available below the impervious layer is thick, a more suitable arrangement for tapping the aquifer fully may be to have a bore fitted with a screen opposite almost the entire thickness of the water bearing layer. The suitable strata conditions under which one or the other types of bores may be expected to improve discharge of open wells are described below:

(1) *Good water bearing strata of small thickness underlying a layer of clay of requisite thickness* (Fig. 2.39): When a formation of this nature is available,

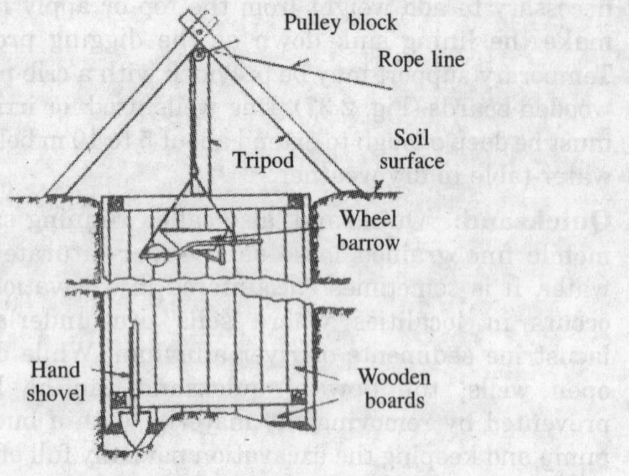

Fig. 2.37. Excavation of open well in an unconsolidated formation using hand-tools for digging and wheel barrow-operated, rope and pulley for lifting the excavated material. Note the wooden cribbing to retain the wall while the excavation proceeds.

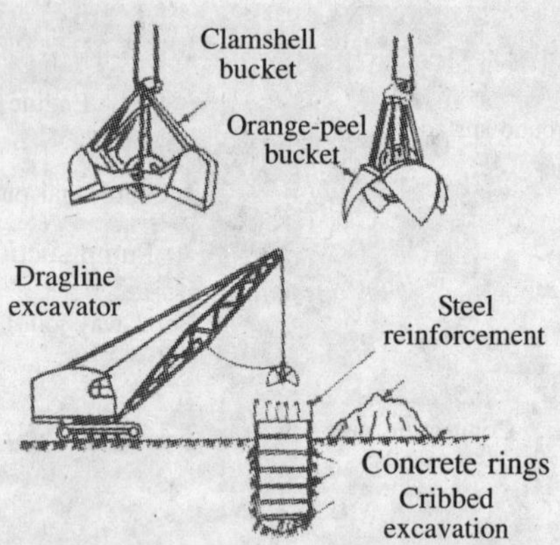

Fig. 2.38. Schematic sketch illustrating the construction of open wells using a dragline excavator equipped with orange-peel/clamshell bucket. Note the R.C.C. steining used for retaining the earth.

there are good prospects for increasing the well discharge by boring. A bore ending in a cavity is to be preferred because a hard impervious layer just overlying a good water bearing stratum is available and secondly the vertical thickness of the sandy water bearing stratum is small.

(2) *Good water bearing stratum available in substantial thickness below the bottom of the well* (Fig. 2.40): In such formations strainer-bore should provide good prospects for improving the well discharge. The strainer has to be provided against the sandy water bearing strata.

(3) *Two or more substantially thick layers of good water bearing layers available in the sub-strata below the bottom of the well:* In formations of this type, a high discharge can be obtained by tapping the available aquifer by installing strainers along the water bearing strata, with blind pipes provided along the non-water bearing strata.

(4) *A good water bearing stratum available below an intervening hard rock:* There are good prospects of getting additional water yields if the hard rock is pierced through and the bore is extended up to the water bearing strata, which may be tapped through a strainer. If, however, the water bearing stratum is immediately below the hard rock and is comparatively thin it may be economical to finish the bore just at the bottom of the hard rock and get the water supply through the cavity formed below the bore. Such a bore through the hard rock need not be protected by a pipe lining as the bore in the rock is able to stand on its own.

(5) *Artesian formation available below the bottom of the existing well:* In such a formation, there are not only chances that the boring would increase the water

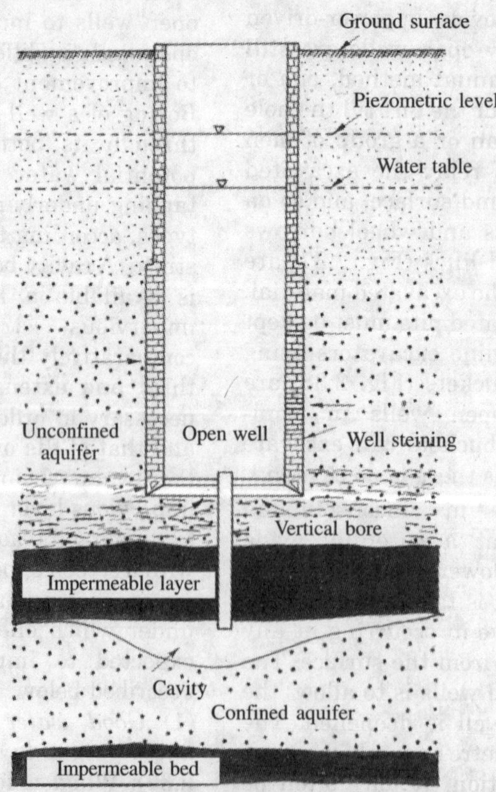

Fig. 2.39. Sectional view of an open well with a bore at the bottom ending in a cavity.

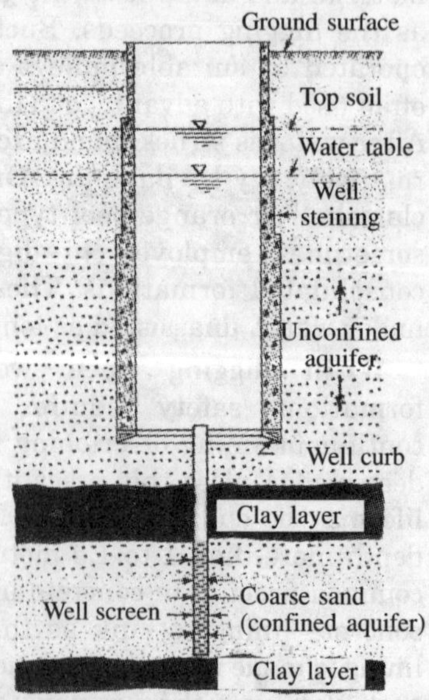

Fig. 2.40. Vertical boring provided with screen/strainer to augment the yield of an open well.

supply but also that the water level may rise considerably in the well. If the stratum of the artesian layer is thin and immediately below the intervening impervious stratum it may be tapped without providing a strainer; otherwise a strainer may be necessary.

Any of the common methods of well drilling, viz., percussion, rotary or down-the-hole drilling may be adopted for boring in open wells. The drilling method and type of equipment depend on the type of formation material encountered and the size and depth of bore.

Horizontal Boring in Open Wells

Studies have shown that in many groundwater regions, especially those having semi-consolidated formations and in hard rock areas the yield of existing open wells can be increased substantially by driving short lateral tunnels horizontally or at an angle below the watertable in the well. These borings are sometimes called revitalization holes. They open the paths of flow which may be otherwise restricted and also function as miniature horizontal wells or infiltration galleries leading to the main wells, thus increasing the well discharges. As a good part of the groundwater in hard rock areas is stored in cracks and fissures of the rocks, the lateral flow of water may be slow. At a certain distance from a well which has gone dry, the water-table may be considerably higher than the well itself. Because of this hydrostatic head, the water will often come out of the revitalization hole by its own pressure.

Fig. 2.41. Manually operated auger set for horizontal boring in open wells.

There is considerable local skill in Rajasthan and adjoining areas to make horizontal bores in semi-consolidated formations with indigenous boring tools consisting of sharp pointed or twisted bits and extension pieces consisting of 2 cm square rods in 1 to 1.5 m lengths with two bolt holes at each end for joining the adjacent piece (Fig. 2.41). Boring is done with a twisting and reciprocating motion of the auger set.

Radial borings may also be done by mechanically powered rock drilling equipment. The equipment consists of an air compressor and rock drilling kit for long hole drilling with extension equipment having pneumatic feed (Fig. 2.42). The rock drill unit is mounted on a feed beam built of steel channels. The cradle carrying the rock drill and the feed cylinder with retaining mechanisms move along the feed beam, thus giving the essential forward and backward movements of the rock drill. The unit commonly used in revitalising open wells is capable of drilling holes 4 to 5 cm in diameter and 25 to 30 m long.

In alluvial and other unconsolidated formations, unlined radial bores, if made, will collapse and it may sometimes be useful to install filters radially below the level of water-table in open wells. The equipment for making radial bores and installing filters consists of a jack screw unit (Figs. 2.43 and 2.44). The jack screw is cylinderical with external threads working in a pipe-based jack block with internal threads. Handles are fixed to the jack screw to rotate the screw and drive in the filter. The bore bit consists of an over-size conical pointed edge in front of the filter pipe. The filter is driven into the formation in successive lengths. The diameter of filter pipe usually ranges from 2.5 to 5 cm. While drilling horizontal bores the water level in the well is kept below the level of the bore. This is done by continuous pumping.

Fig. 2.42. Horizontal boring of an open well in a hard rock area in Vappani Village, Chengelpet District, Tamil Nadu, using mechanically powered rock drilling kit with pneumatic feed. (*Courtesy:* Dept. of Agrl. Engineering, Govt. of Tamil Nadu.)

Deepening of Open Wells

Deepening of wells consists of blasting the bottom of the wells to rocky sub-stratum by dynamite, in stages, up to the desired depths. This is resorted to in cases where boring is not likely to help. If the rocks dip more towards the vertical, borings are least likely to

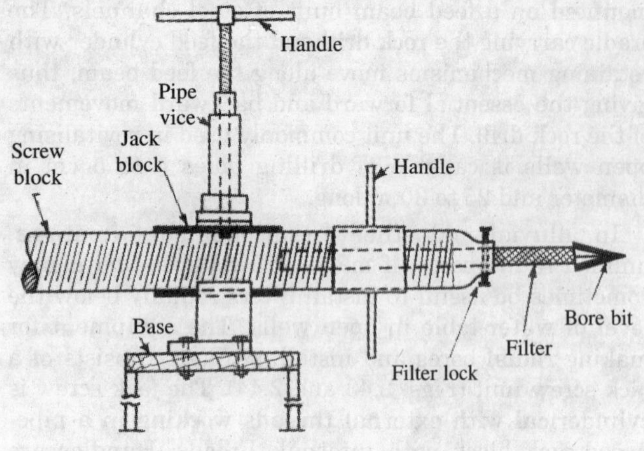

Fig. 2.43. Jack screw-based horizontal boring equipment (Khepar and Sondhi, 1973).

be useful. The objective of deepening is two-fold: (*i*) to expose additional water bearing joints, cracks and crevices, and (*ii*) to increase the storage capacity of the well. Some wells, however, require deepening in consequence of the lowering of water-table due to extreme drought conditions or over-pumping of groundwater.

Fig. 2.44. Horizontal boring with jack screw-based drill unit for installing filters in open wells (Khepar and Sondhi, 1973).

Rocks normally contain voids, interstices, cracks, joints and bedding planes which serve as receptacles holding the water in such formations. In most rocks, these are inter-connected so that a constant circulation of water is kept up under gravitation and hydraulic forces. In some rocks, however, these are generally isolated and there is little opportunity for water to permeate. In such type of rocks, where inter-connections are not sufficiently developed, percolation can be somewhat increased by deepening the existing wells.

The wells dug in alluvial soils can be deepened manually. There are tracts where soft rock appears after the first few metres of top soil have been removed. The farmer in such cases, normally tries to carry out further excavation with his own resources. He successfully carries out blasting up to a few metres of soft rock with the use of gun-powder and a fuse coil for igniting the gun-powder. The gun-powder is placed in holes which are drilled with crow bars, chisels and hammers. The efforts of the cultivator continue till some hard consolidated rock formation is reached or water is struck. At this stage, it is desirable that the work may be entrusted to organised blasting agencies, which may undertake the work systematically using explosives. The systematic operation for deepening wells consists in drilling holes in the bottom of the well be means of pneumatic tools operated by air compressors, filling these holes with suitable gelignite sticks and finally exploding the well bottom. After the well bottom has been cleared of debris, the process is repeated till the well has been deepened to the required depth.

The effect of the blast depends to a large extent on the position and inclination of the blast holes. The holes are usually drilled oblique with a view to minimise resistance to blast and thereby enabling the explosive force of the blast to be more effective in breaking the rock. Oblique holes also prevent the explosives from shooting out straight, as they sometimes do, if the holes are vertical and the charges not properly tapped. In soft rocks, the holes may be drilled straight as comparatively less explosive force is required to break the rock.

One of the guiding factors in the successful blasting of the stratum met with is to follow the track of moisture in the rock or water oozing out of the crevices. Sometimes, even horizontal blasting may be resorted to in search of a source of water suspected to exist. In no case the holes are made through crevices from which water oozes out, as otherwise the heavy shock of the blast may disturb the flow condition and sometimes choke the existing passages.

Tube Well Drilling Equipment and Procedures

With respect to methods of construction, tube wells may be classified as driven, jetted and drilled tube wells. Driven and jetted tube wells are relatively shallow and generally not over 15 m in depth. Drilled tube wells extend to deeper aquifers and are constructed with the help of well drilling machines.

Driven Tube Well

A driven well consists of a pipe and well point which are forced into the water bearing formation by driving with a wooden maul, or drop hammer. Driven tube wells are usually from 3 to 7.5 cm in diameter (Fig. 2.45). If the water level is 7 m or more below the land surface, the well pipe should be more than 5 cm in diameter so that a jet or cylinder may be inserted and submerged to permit the pump to function. A special device called a cap or drive head protects the top of the pipe during the driving operation. After each length of pipe is 'hammered' into the ground, the cap is removed and additional sections of pipes are attached and driven as required. The pipe is kept full of water all the time. In its descent through impervious material, the screen is sealed by the formations through which it travels. But as the well point contacts water-bearing sand or gravel some water flows out, and water remaining in the pipe drops to its static level. This is a signal to the driller that water-bearing material has been reached. To develop the well, the driller attaches a small pitcher pump to the pipe. A considerable amount of sandy water may be pumped for a short time, but if a good water bearing strata has been tapped, continued pumping should result in clear, sand-free water.

Driven tube wells develop small yields and their construction is limited to shallow depths in soft unconsolidated formations free from boulders and other obstructions. They are commonly used for domestic water supply.

Jetted Tube Well

A jetted tube well is constructed with hand operated equipment or power driven machines, depending upon the formation, and the size and depth of the well. A hole in the ground is made by the cutting action of a stream of water which is pumped into the well through a small diameter pipe and forced against the bottom of the hole through the nozzles of a jetting bit (Fig. 2.46). The hole is cased to prevent it from caving.

After the hole has been jetted down to the desired depth, the well assembly consisting of blank pipes and screen is lowered and the outer casing is pulled out. Jetted tube wells have small yields and their construction is possible only in unconsolidated formations.

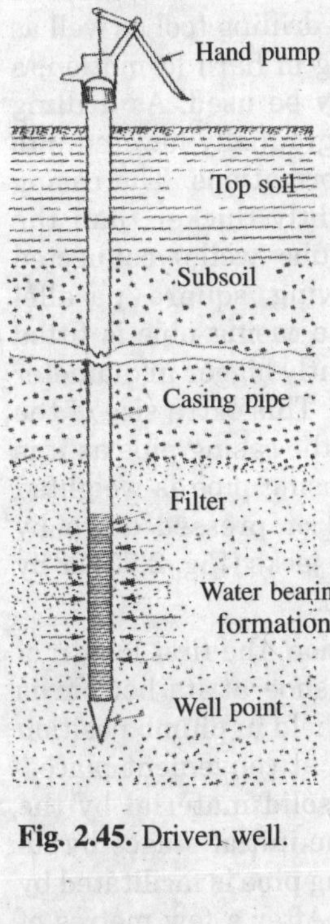

Fig. 2.45. Driven well.

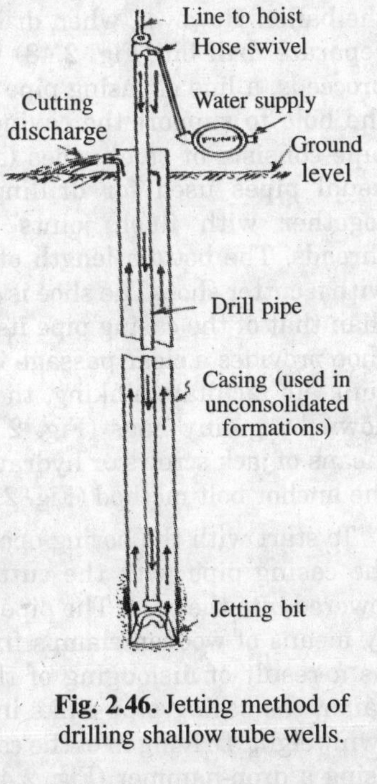

Fig. 2.46. Jetting method of drilling shallow tube wells.

DRILLED TUBE WELLS

Drilled tube wells are constructed by standard drilling procedures utilizing the type of equipment best suited to the formations encountered. Tube well construction involves four distinct operations. These include the drilling operation, installation of casing and well screen and developing the well for maximum yield and sand-free operation. Where artificial gravel packing is required, this forms a part of the well screen installation.

Classification of Drilling Equipment

Different types of drilling rigs are manufactured to suit different drilling conditions. They are usually classified under different categories according to the method of drilling employed. The different types of drilling equipment in common use are (*i*) percussion (cable tool) drills, (*ii*) rotary drills, including direct rotary, reverse rotary and air rotary, (*iii*) hammer drills, and (*iv*) core drills. Under each category, according to the method of drilling, rigs of various capacity ratings are available. From the capacity rating angle they may be sub-divided into three broad sub-categories, namely, light, medium and heavy duty drills. The choice of the capacity of the equipment is mainly governed by the size and the depth of the hole to be drilled. For catering to different mobility requirements the rigs may be mounted on a skid, a trailer, a truck or a tractor.

Cable-Tool Percussion Drilling

The cable tool drills operate by lifting and dropping a string of tools suspended on a cable. A bit at the bottom of the tool string strikes the bottom of the hole, crushing and breaking the formation material. The drill bit breaks hard rock into small fragments. When working in soft, unconsolidated formations, the drill bit loosens the material. In both instances, the reciprocating action of the tools mixes the crushed or loosened particles with water to form a slurry or sludge. The necessary water to form the slurry is put into the bore hole if no water is present in the formation being penetrated. The resulting slurry is removed at intervals from the bore hole by means of a sand pump or bailer. The accumulation of too much depth of sludge slows up the fall of the tools and retards penetration. This factor usually determines the frequency of bailing the sludge from the well.

Adaptability: Cable-tool method is best suited for drilling in boulders, slanted and fissured formations and lost circulation zones. The method is also suitable for drilling in alluvial formations but drilling rate in such formations, particularly those having clay, or sticky shale strata, is much lower as compared to direct or reverse rotary methods. Percussion drilling in hard rock is a slow process but better than rotary drilling.

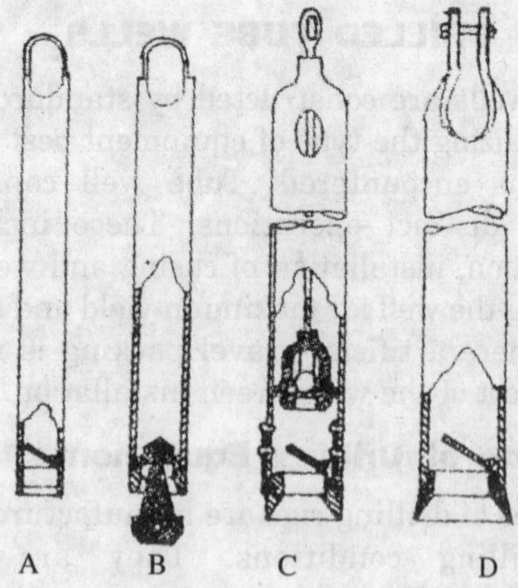

Fig. 2.47. Bailers with different types of valves and a mud scow: *A.* Flapper; *B.* Dart valve; *C.* Bayonet valve; *D.* Mud scow.

Relative advantages of percussion method are the collection of more precise data, detection of thin and weak water bearing zones, and facility of testing the quality and quantity of water in different formations, as the drilling proceeds. The method requires much less water for drilling as compared to direct rotary. It is widely applicable for boring in dug wells as the percussion drilling outfits can be conveniently installed on the side of the well (hand boring sets can be installed over the well) and carry out boring in the centre of the well.

Hand boring sets: The hand boring set is a simple manually operated outfit which works on the percussion principle. The equipment comprises of a tripod, a crab winch and bailers or sand pumps, cutter shoes, anchor bolt set, loading clamps, chain wrenches, pully blocks, flexible steel wire rope and tools such as spanners and wrenches (Fig. 2.53 and 2.54a and b). The bailer/sand pump (Fig. 2.47) which is equipped with cutting shoes at its lower end normally acts as the drilling tool as well as the bailer. However, when drilling in hard formation, a separate drill bit (Fig. 2.48) may be used. As drilling proceeds, a line of casing pipe is gradually lowered into the hole to support the caving formations. The casing pipe consists of steel tubes (slightly thicker than the usual pipes used for drilling tube wells) connected together with flush joints having square parallel threads. The bottom length of the casing pipe is fitted with a cutter shoe. The shoe is lightly bigger in diameter than that of the casing pipe itself. This extra size of the shoe provides a clear passage for the casing pipe as it is sunk. To facilitate sinking, the casing pipe is weighted down by gunny bags (Fig. 2.50) or pressed down by means of jack screws or hydraulic jacks (Fig. 2.73) or by the anchor bolt method (Fig. 2.51).

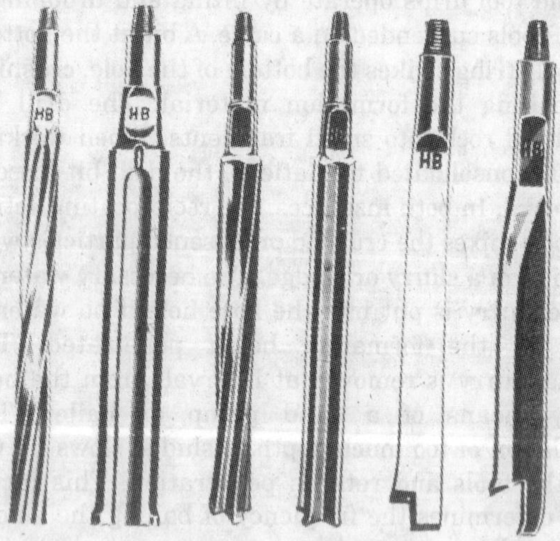

Fig. 2.48. Different types of drill bits used with percussion rigs: *A.* Deep-fluted twist drill bit (for hard formations). *B.* Flat-fluted drill bit (for soft rock strata). *C.* Twist star drill bit (specifically suited when the formation is at an angle). *D.* Straight star drill bit (for heavy duty jobs and straightening crooked holes.). *E* and *F.* Enlarging drill bits (suited to drill an enlarged hole below the casing shoe which in turn allows the bore case to follow).

(*Courtesy*: Hardwood Bagshaw Engg. Ltd.)

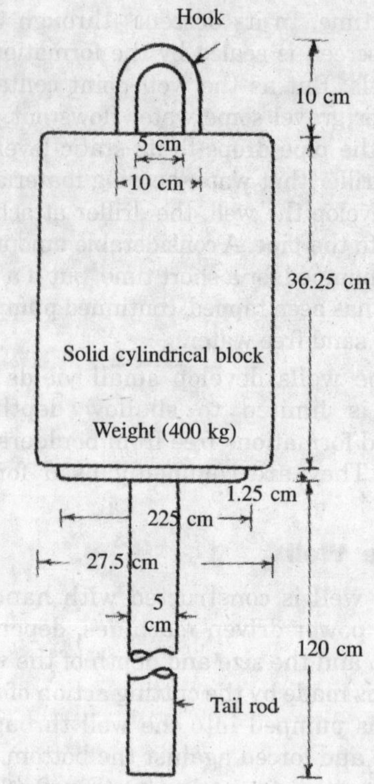

Fig. 2.49. Drop-hammer for driving in the casing pipe in cable tool drilling.

To start with the boring operation, the first length of the casing pipe with the cutter shoe at its bottom is lowered into the bore. The pipe is held in plumb position by means of wooden clamps fixed at convenient places. As a result of dislodging of the solid material by the bailer, the casing pipe sinks in the initial stages by its own weight. Driving in of the casing pipe is facilitated by using a drop-hammer (Fig. 2.49). After a few metres of

Fig. 2.50. Weighted drop hammer and sand bags used to drive in the outer casing pipe in cable-tool drilling.

sinking, the pipe is weighted down by gunny bags full of sand and the sand bags are placed over a platform made of wooden sleeper clamps fixed around the casing pipe at a convenient point (Fig. 2.50).

Anchor bolt method of sinking casing pipes: An improved method to sink casing pipes is by means of screws or hydraulic jacks (Fig. 2.51). The jacks are made to work against a wooden clamp *A* fastened rigidly around the pipe and to react from another wooden clamp *B* fitted loosely around the pipe and held rigidly in position by means of two bolts, which are securely anchored a few metres below the ground level. The anchor bolts consist of two mild steel rods about 4 cm in diameter and 5 m in length. They are positioned on opposite sides of the pipe, at a distance of about 10 cm from the pipe. The rods pass freely through the rigid clamp *A*, but are bolted firmly with the loose clamp *B* at their top. At the lower end, each rod is fastened by means of nuts on either side, with two pieces of steel joists about 1.2 m long to serve as anchors. The joists are placed in a cross form in a dug pit, which is subsequently filled up. As the jacks are operated, clamp *B*, held rigidly by the anchor bolts, remains stationary while clamp *A* moves down along with the pipe. After the pipe has sunk some distance, clamp *A* is moved up on the pipe and the jack re-adjusted. The process is repeated, sinking down the pipe a few metres each time. The method is suitable for hard strata for depths below about 30 m. During drilling, a periodical check on the verticality of the casing pipe is necessary. Hand boring sets are generally used to drill wells of size 10 cm to 20 cm in diameter and 50 m to 100 m deep.

Auger bits for handboring: Handboring sets are often equipped with augers (Fig. 2.52) to make shallow holes, especially in boring to install filter points. Augers of different shapes and sizes are made for the purpose. They are used to loosen and remove the soil material. It is turned into the hole until loaded and is then filled and cleaned. The hole is bored until caving formations are encountered or water table is reached. It is not suitable for drilling in water bearing sand and gravel, as these have a tendency to wash down the hole. The method is suitable for boring in shallow depths in unconsolidated formations which can be normally started without casing.

Percussion rigs: A standard percussion rig (Figs. 2.53 and 2.54a) consists essentially of (*a*) a mast (*derrick*), (*b*) a two-or three-line hoist which is the bull/drilling reel for operating drilling tools), the second (the sand reel) for operating the bailer/sand pump and the third (the casing reel) for operating the casing pipes, (*c*) a spudder for raising and dropping the tools, and (*d*) a diesel engine to provide power for these operations. The mast is made long enough to allow the longest string of tools/pipes to be hoisted out of the hole, the minimum height being about 10 m. The mast is telescoped to its shortest length and folded down at the top of the machine during transit. The reels are required to be large enough to hold sufficient cables for drilling the deepest holes which the rig is capable and strong enough to handle drilling and fishing tools and other equipment. The drilling process includes three types of operations: (*a*) making the hole, (*b*) bailing out drilled cuttings, and (*c*) casing the bored holes. These three operations generally follow one another till the bore is completed.

A full string of tools in ascending order consists of a bit of suitable type, a drill stem, jars and a swivel socket which is attached to the cable (Figs. 2.54a and 2.55). The drill stem gives additional weight to be bit and the effect of its added length helps to maintain a straight hole when drilling in hard rock. The jars consist of a pair of linked steel bars. The rope socket connects the string of tools to the cable. In addition, its weight supplies part of the energy of the upward blows by the jars when its use becomes necessary. It allows the tools to rotate slightly with respect to the cable.

The size of the drilling bit is determined by the diameter of the hole to be drilled. But the length of the bit may be varied according to the weight required to provide adequate capacity to the drilling machine. Figure 2.48 shows the common type of percussion drill bits and their adaptability. Different types of travel blocks and lighting devices in and clamps used in percussion drilling are shown in Figs. 2.55 and 2.56.

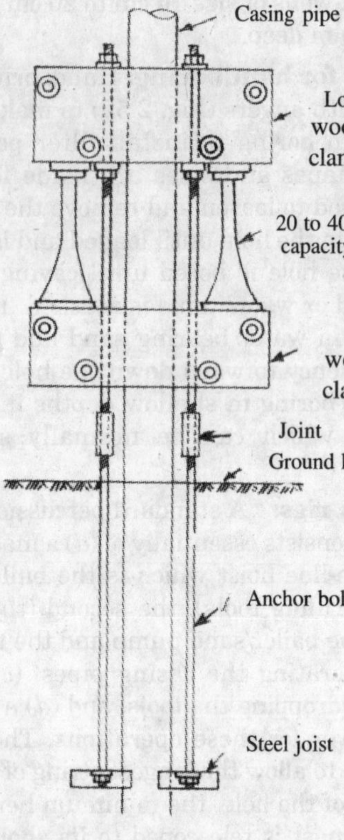

Fig. 2.51. Anchor bolt system of sinking casing pipes in percussion drilling.

Fig. 2.52. Auger used in a hand boring unit for shallow well drilling.

A bailer or sand pump is used for removing the drilled cuttings from time to time. A bailer consists of a section of pipe generally from 2 to 5 m long with a valve at the bottom and a bail at the top for attaching the cable. The diameter of the bailer is determined by the size of the hole being drilled, sufficient clearance being allowed so that the bailers will fall free. Bailers are generally equipped with flap, or dart valves (Fig. 2.47). The former has a flap valve which cannot be opened after the bailer is filled. This has to be emptied from the top by tipping the bailer. The design, however, enables this type of bailer to clean out closer to the hole bottom. The bailer with dart valve is emptied by striking the dart attached to the valve on something solid which forces the valve to open. It is used for the rapid bailing of the drilled cuttings which mix well with water during drilling and are kept in suspension. A sand pump is similar to a bailer except that it is equipped with a piston for the purpose of drawing the drilled cuttings from a well. It is used for the removal of heavy cuttings and is more effective than a flap valve bailer.

The up-and-down action is imparted to the drilling tools by a walking beam. The walking beam is pivoted at one end. Its outer end carries a sheave for the drill line. It is moved up-and-down by a pitman connected to a

Fig. 2.53. Basic components of percussion drilling rig.

crank gear. Both the length of stroke and its speed can be varied. The crank gear is driven by a pinion mounted on a friction clutch. The friction clutch and the drive pinion for the drill-line reel are all mounted on the same drive shaft assembly.

Fishing tools: Sometimes, during drilling, the tools get lost or jammed up in the bore hole due to a number of causes such as (*a*) snapping of wire rope, (*b*) loosening or unscrewing of tool joints, (*c*) breaking of pins or jars, (*d*) string of tools getting out of alignment in sloping formations, and (*e*) tools getting wedged by caving formations. A fishing job to recover the tools so jammed or lost, and to remove obstructions, becomes necessary under these circumstances. Special types of tools called fishing tools are used for the purpose. The fishing string is made like the drilling string except for two items, the drilling jars and bit which are removed (Fig. 2.57). At the bottom of the stem are attached the fishing jars which have extra long strokes and at the bottom of these are attached the fishing tools. The fishing tools include *rope, spears and crabs* for catching the loose wire rope; *rope knives* for swearing the loose rope from the 'fish', different types of sockets for gripping and holding; bumpers for loosening the stuck tools; and spuds for bringing the tools in alignment.

Cable-tool drilling rate: The speed of drilling and the rate of progress with a cable tool rig depends upon a number of basic factors: (*i*) the hardness of the rock, (*ii*) the diameter and depth of hole, (*iii*) the dressing of the bit, (*iv*) the weight of the tool string, and (*v*) the rate and length of the stroke. The capacity of a cable tool rig depends on the weight of tools which it can handle safely. The tools weight is determined largely by the diameter of the bit and the diameter and length of drill stem.

Capacity range of percussion rigs: Based on capacity range, percussion rigs are classified as light, medium and heavy. Light duty machines are capable of handling a total weight of tools ranging from 200 to 700 kg at the ground level. These drills are suitable for drilling 15 cm to 25 cm diameter holes up to a depth of 100 to 130 m or so in soft and medium hard formations. Their average performance may range from 1.5 to 5 m per day of 8 hours, depending on the formations. Medium rigs are capable of handling tool weights in the range of 800 to 1500 kg at the ground level. They are suitable for drilling 20 to 45 cm diameter holes up to depths of 30 to 200 m in soft, medium, hard and boulder formations. Their average performance may range from 1.5 to 5 m per day of 8 hours, depending on the size and depth of

Fig. 2.54a. Percussion drilling rig. The string of tools is seen to the right of the mast.

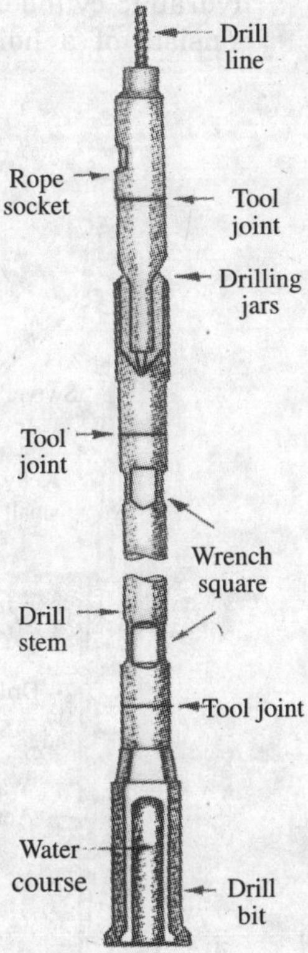

Fig. 2.54b. Drill tool assembly for cable tool percussion drilling.

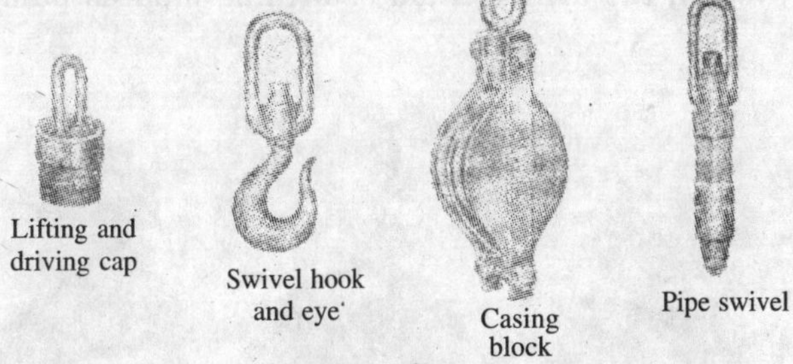

Fig. 2.55. Common types of travel blocks and lifting devices used in percussion drilling.
(***Courtesy***: Hardwood Bagshaw Engg. Ltd.)

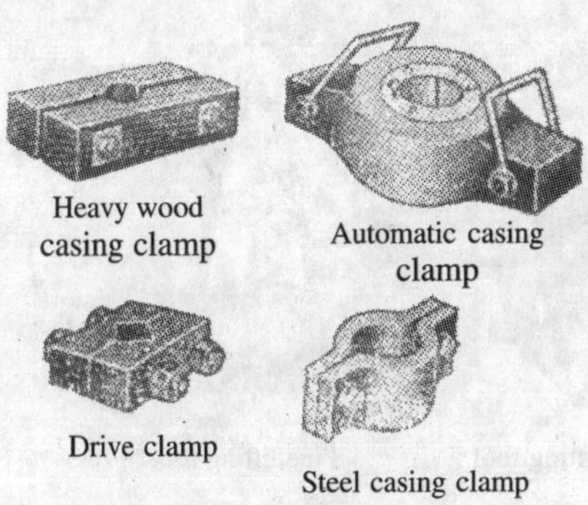

Fig. 2.56. Common type of clamps used in percussion drilling.
(***Courtesy***: Hardwood Bagshaw Engg. Ltd.)

bore and the type of formation encountered. In certain formations like sandy clay, soft sand strata, slate etc. the output may be much higher. Heavy duty percussion rigs are required especially for drilling deep holes in boulders or difficult formations.

ROTARY DRILLING

Rotary drilling is one of the fastest methods of drilling wells and is particularly adapted to drill large size holes. The drilling process involves boring a hole by using a rotating bit to which a downward force is applied. The bit is supported and rotated by a hollow stem, composed of high quality steel, through which a drilling fluid is circulated. The method is most suitable for drilling deep holes in unconsolidated (alluvial) formations. It is unsuitable for drilling in boulders and hard rocks due to slow progress and high bit cost. This method is also unsuitable for drilling in slanted and fissured formations. The hydraulic rotary drilling generally requires large quantities of water which may have to be brought from long distances, if not locally available. Because of adding large quantities of water and due to drill cuttings, the hydraulic rotary method is less suitable for obtaining accurate logs of the strata encountered.

Direct Rotary Drills

In the conventional direct circulation rotary drilling system, the drilling fluid or drilling mud, is pumped down through the drill pipe and out through the nozzles in the bit (Figs. 2.58 to 2.61). The mud fluid with the cuttings flows upward in the annular space around the drill pipe to the ground surface. At the surface, the fluid is lead into a settling pit through a channel and then into a storage pit where the cuttings settle. From the storage pit, the mud water is again pumped back into the drill pipe, thus repeating the cycle of operations.

The most important components in the rotary drilling method are the drill bit and the drilling fluid. They are essential in constructing and maintaining the bore hole. The component parts and the make up of the rotary drilling machine are designed to serve in the operation of the bit and continuous circulation of the drilling fluid.

Components of rotary rigs: A rotary rig (Figs. 2.58 and 2.59) consists essentially of a derrick, a hoisting mechanism for handling the tools and lowering the casing, a rotary table drill head for rotating the drill pipe and the bit; a pull down mechanism for applying a force or thrust load on the drill pipe; hydraulic system for hoisting the mast and adjusting the position of drill head; pump for handling mud; and a suitable source of power together with necessary shafts, gears and accessories required for operating the hoisting, rotating and pull down mechanisms and the pumps. Modern rigs carry the drill mast which forms an integral part of the drill frame and can be raised and lowered by means of hydraulic cylinders. The hoisting mechanism generally consists of a hoisting drum, a pick-up drum and a

Cable-cutting tool Impression block Pipe-locking tool

Cutting and Locking Tools and Impression Block

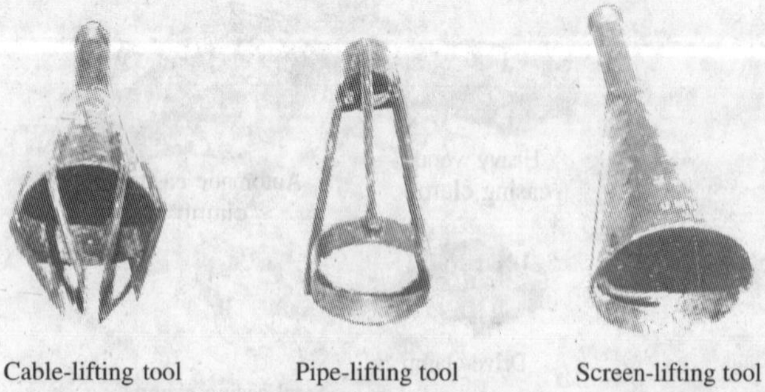

Cable-lifting tool Pipe-lifting tool Screen-lifting tool

Lifting Tools

Fig. 2.57. Commonly used of fishing tools in tube wells.

(***Source:*** ICAR Coord. Res. Project on Wells & Pumps, JNKVV, Jabalpur)

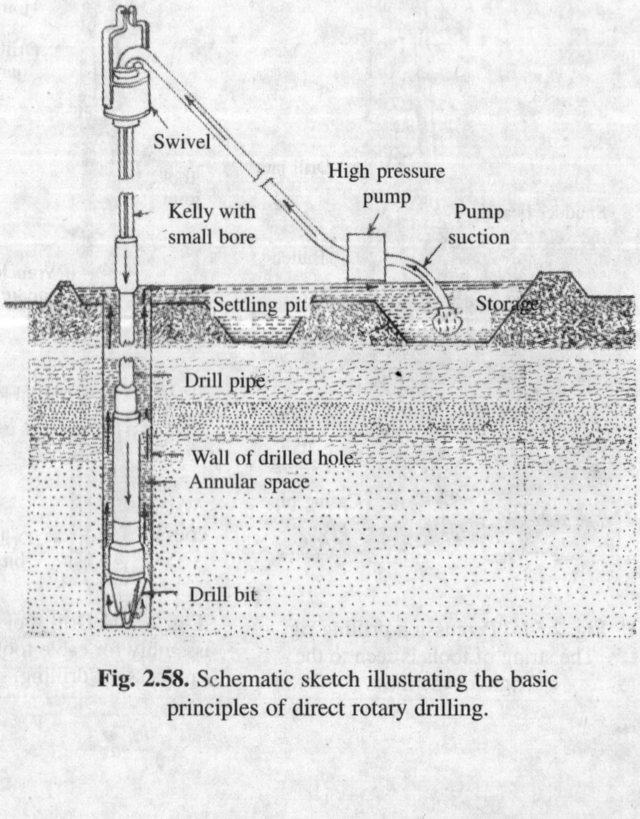

Fig. 2.58. Schematic sketch illustrating the basic principles of direct rotary drilling.

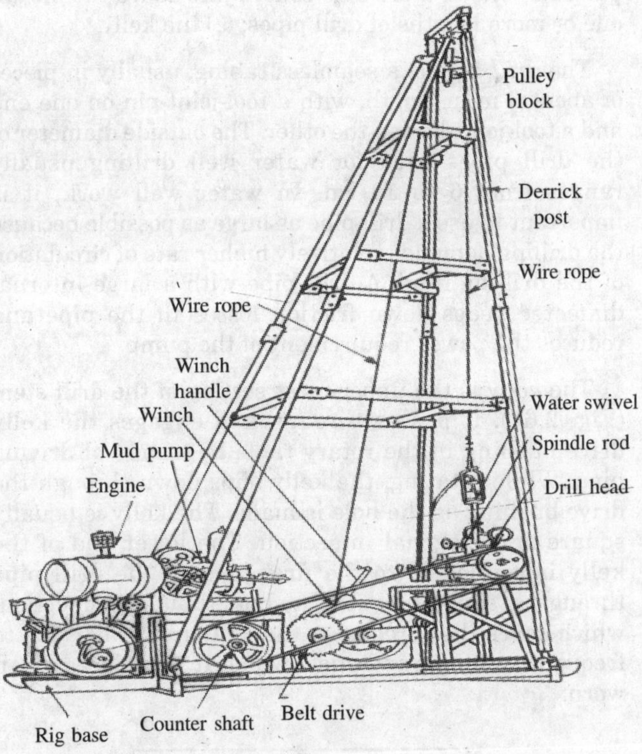

Fig. 2.59. Skid mounted light duty rotary drilling rig. (Adapted from Garlick & Co. Pvt. Ltd.)

cathead with their respective drives, lines and fittings. The rotary table is set on tool steel rollers and is operated by a spiral bevel gearing mechanism. The table is provided with bushing, which is grooved to fit similar grooves on the top joint of the drill pipe. This bushing, called the kelly drive bushing, is keyed to the kelly by means of driving pins, which fit in the grooves. The kelly can rotate and move vertically at the same time.

The pump used is a double acting piston type force pump capable of developing a high pressure (9 kg/sq cm or more) and delivering enough volume of the drill fluid as required for the operations. As the pumps handle mud-laden fluid, they are provided with a minimum number of valves. The pumps are connected to the drill pipe by means of a flexible hose.

The *power unit* in the rotary rig is an internal combustion engine (usually a diesel engine). The power of the engine varies according to the size and capacity of the rig.

Drilling string: The drill string (Fig. 2.62) for rotary drilling consists of a drill bit, drill stem and a swivel. The proficiency in drilling rests largely on the bit. The bits may be divided into two broad classes: (*a*) *the drag bit* and (*b*) the *rolling cutter type bits*, also called rock bits (Fig. 2.63). The drag bits have a shearing and scraping action and are adapted to drilling in soft and unconsolidated formations. They have short blades or wings, which are forged and dressed to form thin cutting edges. Wearing surfaces are hard-faced and dressed with hardened inserts set in the corners. The body of the bit is hollow with holes properly positioned for directing the drill mud to keep the blades clean and cool and at the same time assist in penetration by means of a jetting action against the bottom of the hole. The drag bit may be a two-way bit—called the fish tail bit because of its resemblance with fish tail, a three-way bit, or a six-way bit—called the pilot bit. The two-way bit having two blades is adapted specially to loose sands and clays. The three-way bit is much like the fish tail in design, but has three cutting blades instead of two. The three blades make its action smoother in shattered or irregular formations. It also makes a better hole in semi-consolidated formations and has a lesser tendency for deflection. In soft formations, it cuts a little slower than the fish tail. The pilot bit is of the same general type as the fish tail and three-way bits. It has six cutting edges, each with a hole for circulating fluid. Three of these cutting blades cut a hole of about half the diameter of that is cut by the other three. This bit is more suited to shattered formations and cemented sands than either the fish tail or the three-way bits. It will cut somewhat harder formations, and with less irregular strain on the drill rods.

Roller type bits are specially adapted to drilling in hard formations. They exert a crushing and chipping

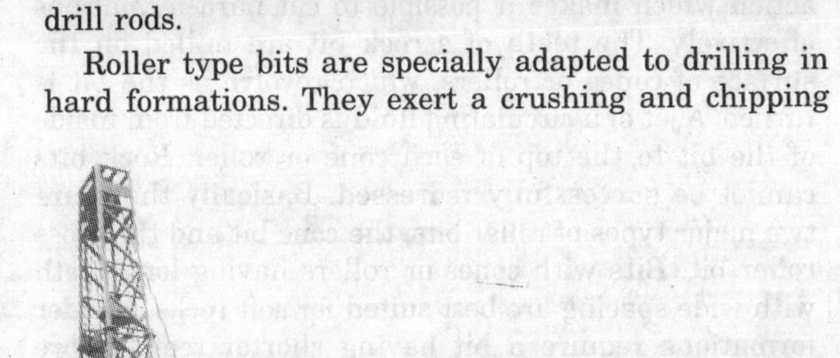

Fig. 2.60. View of a truck mounted heavy duty rotary drilling rig (***Courtesy***: Garlick and Co. Pvt. Ltd.)

Fig. 2.61. A truck mounted direct rotary drill rig. Note the casing pipes in the foreground. At the centre of the row of pipes is the lower most housing pipe to which is fixed the taper joint.

action which makes it possible to cut hard formations effectively. The teeth of a rock bit are milled on the surface of cones or rollers, which revolve as the bit is turned. A jet of a circulating fluid is directed from inside of the bit to the top of each cone or roller. Rock bits cannot be successfully redressed. Basically there are two major types of roller bits, the cone bit and the cross roller bit. Bits with cones or rollers having long teeth with wide spacing are best suited for soft rocks. Harder formations require a bit having shorter teeth, more closely spaced. A cone type bit has three-cones, a forged-steer body, and a cone axle or pin, which form a part of the body. The cones have roller bearings fitted at the time of assembly. The three-cone construction provides smooth operation. Bits with different shapes of cutting surfaces and designs of teeth are available to suit different formations. All cutting surfaces are flushed by the circulating fluid. The tricone bit is more commonly used due to its structural strength and balanced cutting action.

Cross roller rock bits constitute a broad class of bits that have their main cutting elements disposed on a right angle cross. The cross roller bit usually consists of four cutters. The cutters are arranged so that all four cut the bottom of the hole, but only two of them cut the hole to gauge. As in the cone bits, the teeth of this type of bit vary with the formation to be cut. The cross-roller bit is generally used in formations where there is a tendency for it to become crooked. It is also used to straighten crooked holes.

The *drill stem* is a hollow rotating shaft, which operates the bit at the lower end. It forms the link between the bit and the surface. It usually consists of 3 parts: (*a*) one or more drill collars just above the bit; (*b*) one or more lengths of drill pipes; (*c*) the kelly.

The *drill pipe* is a seamless tubing, usually in pieces of about 6 m in length, with a tool-joint-pin on one end and a tool-joint-box on the other. The outside diameter of the drill pipes used for water well drilling usually ranges from 6 to 12 cm. In water well work, it is important to use a drill pipe as large as possible because the drilling demands relatively higher rate of circulation of the drilling fluid. A drill pipe with a large internal diameter keeps down friction losses in the pipe and reduces the power requirement of the pump.

The *kelly* is the upper-most section of the drill stem (Fig. 2.62). It passes through and engages the kelly drive bushing in the rotary table by means of driving pins. While rotating, the kelly slips down through the drive bushing as the hole is made. The kelly is usually square or hexagonal in section. The lower end of the kelly is connected to the first joint of the drill pipe through a '*sub*'. The sub is a short tubular connection, which saves the threads on the kelly from wear due to frequent screwing and unscrewing. It is replaced when worn.

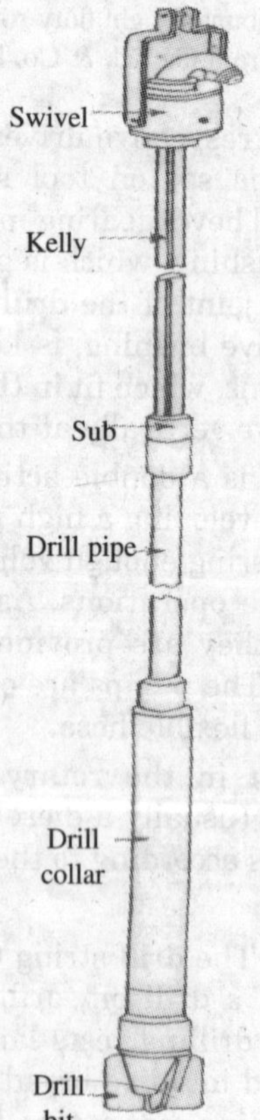

Fig. 2.62. Drill tool assembly for rotary drill. (Adapted from Garlick & Co. Pvt. Ltd.)

The upper end of the kelly connects the *water swivel* (Fig. 2.62). The swivel is so built that the weight of the entire drill string is supported on freely-moving bearings. This permits the drill stem and the bit to rotate with the driving mechanism during the drilling operation, while the upper part of the swivel remains stationary. The swivel also provides the mud hose connection. This permits free passage of the drilling fluid as drilling proceeds. The drilling fluid passes through the swivel to the drill pipe and to the bit. The swivel is hooked to a pulley block assembly called the travelling block.

Operation of the drill: When the drilling machinery is completely rigged up and ready to operate, the drilling bit is fastened to the drill pipe, which in turn is screwed on to the end of the kelly. The gripping bushing is set around the kelly, the pump is started and the drill string positioned in the hole by means of the drilling hoist. The drilling fluid is circulated from the mud pit for a few minutes, the rotary table and drill pipe are set in motion and the bit is slowly lowered. When the bit touches the ground, drilling begins. As the drill pipe is rotated, adequate bit pressure against the material cut is maintained. The speed of rotation and bit pressure are regulated according to the type, form and size of the bit, the formation to be drilled and the strength of the drill pipe. These factors are adjusted for penetrating the formation at the highest rate, consistent with the protection of the cutting edge of the tool from undue wear, drilling a straight hole and the torsional and fatigue limits of the drill pipe. The drilling speeds usually range from 30 to 150 rpm and the bit pressures from 100 to 200 kg per sq cm per cm diameter of the bit.

Drilling mud circulation: As the bit rotates, it cuts chips and abrades the formation into small particles. The removal of the drill-cutting is effected by the mud-laden fluid kept under constant circulation by the mud pump. The drilling mud moves under considerable pressure, from the mud pump to the drill pipe and to the bit. It passes out through the jetting nozzles of the bit at a high velocity and picks up the drill cuttings. Deflected upwards from the bottom of the hole, it travels through the annular space between the drill pipe and the walls of the drilled hole. At the ground surface, the drilling mud with the cuttings is led away from the drilled hole through a channel to a settling pit. It over-flows from the settling pit to a suction pit where it is picked up in the suction hose of the mud pump to repeat its original course. Most of the cuttings settle down in the drain and the settling pit.

The drill fluid performs a number of functions in the drilling operation. The properties of the fluid, such as viscosity, gel strength and density have to be controlled and varied according to the conditions met with in drilling

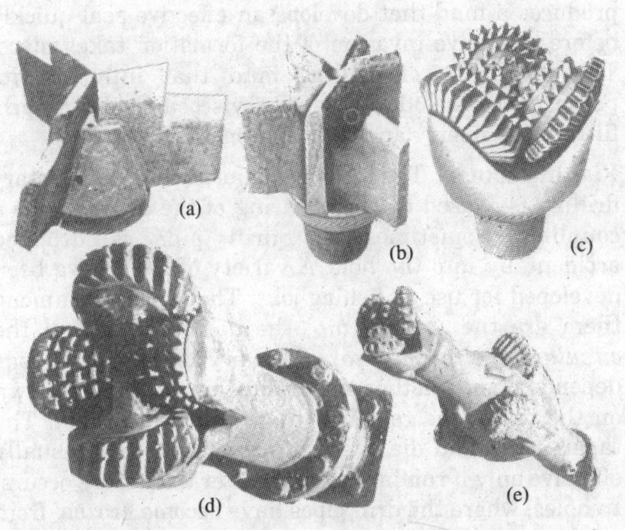

Fig. 2.63. Drill bits for direct rotary drills: (a) Three-way drag bit, (b) Six-way drag bit, (c) Three-cone roller bit, (d) Cross-roller bit, (e) Reamer bit provided with rollers for pilot hole and reaming.

different formations and in working at different depths, in order that the fluid may perform its functions effectively.

The *return* or *annular* velocity at which the drilling fluid flows upwards in the space between the wall of the hole and the drill stem is important in keeping the bit clean and the hole washed clear of cuttings. The ideal return velocity ranges between 40 and 60 m per minute. Should the pump be incapable of delivering sufficient volume of fluid with desirable return velocities, suitable adjustment in the quality and characteristics of the mud fluid may be necessary.

Forming the filter cake: As drilling progresses, a mud cake or filter cake builds up on the wall of the bore hole. It is a lining of silt, clay and colloids formed by a filtration effect when the pressure of the drilling fluid squeezes some of the water out of the fluid. The mud cake works as a plaster and holds the loose particles in place. It helps to protect the well wall from being eroded by the upward stream of drilling fluid. It tends to seal the well wall and minimises the loss of drill fluid into permeable formations. The filter cake can hold loose particles in place because it is sticky and plastic; but it cannot prevent outside pressure from collapsing the hole. Resistance to collapse results from the outward hydrostatic pressure of the drilling fluid.

Drilling fluid: Drilling fluid may vary from simple muddy water to a skilfully prepared, viscous mixture of special-purpose materials like bentonite. It is necessary for well drillers to prepare a drilling fluid which produces adequate sealing effect with little invasion into the formation and yet is not too viscous for pumping with standard equipment. A high-grade bentonite produces a mud that develops an effective seal quickly before extensive invasion of the formation takes place. The properties of drilling mud that influence its performance include its density, viscosity, gel strength, filtration property and sand content.

Fishing tools: The most frequent fishing job in rotary drilling is caused by the shearing of the drill pipe or a coupling. Sometimes the drill pipe is dropped accidentally into the hole. A variety of tools have been developed for use in fishing jobs. The important among them are the *tapered tap*, the *die overshot* and the *circulating slip overshot*. Recovery of lost drill rods depends upon whether the driller can set the tool down on the top of the rods and then connect to them. The tapered tap and die overshot are simple tools usually effective only if run immediately after the failure occurs. In holes, where the drill pipes have become 'frozen' from cuttings settling around them, circulation of drill fluid to the bottom of the hole is helpful in loosening them. The circulating slip overshot which, after gripping the top of the pipe, enables circulation of drilling fluid through the pipe is the best tool for the purpose. In some cases use of a rotary jar immediately above the overshot becomes essential to exert jarring blows on the jammed object.

Capacity range of rotary rigs: Based on the capacity, direct rotary rigs may be classified into: (i) light rotary rigs, (ii) medium rotary rigs and (iii) heavy rotary rigs. Rotary rigs having the rated maximum hook capacity up to about 9,000 kg and capable of drilling slim (pilot) holes with 6 cm drill pipe up to about 380 m is considered as light rigs. The rigs are suitable for drilling and completion of light capacity tube wells (up to 30 litres/second discharge) to depths up to 60 to 140 m or so in alluvial and semi-consolidated formations. Their average performance may range between 5 to 30 m per day, depending on the size and depth of the bore and the type of the formation, the rate of progress being slow in hard formations.

Medium capacity direct rotary rigs have a rated maximum hook-load capacity up to 13,000 to 18,000 kg and are capable of drilling pilot holes up to depths of 450 to 700 metres using 6 to 7 cm drill pipes. These rigs are suitable for drilling and completion of medium capacity tube wells (30 to 60 liters/second discharge) to depths up to 30 to 250 m or so in alluvial or semi-consolidated formations.

Heavy duty rotary rigs have a maximum hook load capacity of 22,000 to 45,000 kg and are capable of drilling pilot holes up to 750 to 1,000 m using 9 cm drill pipes. These rigs are capable of drilling heavy capacity tube wells (60 to 150 litres per second discharge) up to depths of 225 m and above in alluvial and semi-consolidated formations.

Reverse Circulation Rotary Drills

A reverse circulation rotary drill works on the same principle as the direct rotary method except that the drilling fluid is circulated in reverse to the direction of that of the direct rotary drilling method (Figs. 2.64 and 2.65). The drill fluid, which is generally clear water in this case, is fed from the tank down the annular space between the hole and the drill pipe and is recovered along with the drill cuttings through the drill pipe, generally by means of a centrifugal pump and discharged into the tank where the cuttings are allowed to settle. The process is continuous. The hole is not cased during drilling and is prevented from cave-in by the hydrostatic pressure of the hole and also due to the film of fine grained material deposited on the wall of the hole by the circulating water. Considerable supply of water in the range of 3 to 9 litres per second (depending on the size and depth of the hole) is required from the time the drilling is started till the gravel packing of the well has been completed. Mud is added in the water if the formation starts caving. The drilling operation has to be carried out round the clock. For drilling deeper holes up to 300 m or so, compressed air is used for circulating the drilling fluid.

The reverse circulation rigs differ from the direct rotary rigs in the following respects: (*a*) the rotary table has lower speed range and fewer number of speeds, (*b*) the drill pipes used are usually larger in size and are

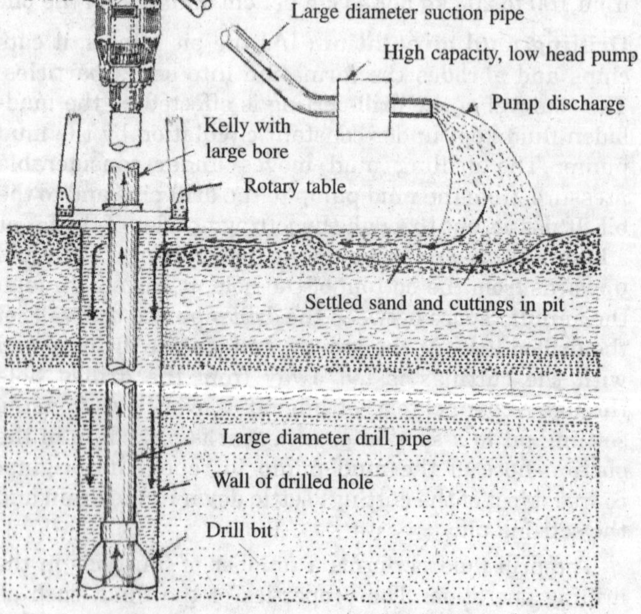

Fig. 2.64. Schematic sketch illustrating basic principles of reverse rotary drilling.

Fig. 2.65. Four-way fish tail bit with large diameter opening for reverse rotary drilling.

flange jointed, (*c*) the pump used is centrifugal provided with non-clog impellers, and (*d*) air lift method is used in heavier rigs for drilling deeper depths. The drilling fluid is essentially muddy water rather than drilling mud. The suspended clay and silt which recirculate with the drilling fluid are the fine materials, picked up from the subsurface formations as drilling proceeds. The following are the basic points to be adhered to while drilling with reverse rotary drills in alluvial formations.

(*i*) Static water tables are to be between 2 and 12 m below ground level. Static water levels less than 2 m could cause problems as insufficient hydrostatic pressure may lead to collapse of the well.

(*ii*) Velocity of water inside the drill pipe should be as follows:

Minimum	:	42 m/min for lifting fine cuttings.
Desirable	:	60 m/min for lifting fine cuttings.
Desirable	:	145 m/min for lifting coarse cuttings.

(*iii*) Velocity of water outside the drill pipe (annular down-hole velocity) should not exceed 18 m/min to avoid wall erosion and possible eventual collapse of the well.

(*iv*) When drilling too slowly, the possibility of borehole collapse will increase.

(*v*) When drilling too fast, the mud film or fines skin would be poorly formed, again resulting in possible borehole collapse.

(*vi*) Drilling rates will depend upon the type of formation and the speed of rotation, combined with the weight on the drilling bit.

During drilling, the drill string should always be suspended. If the full weight of the string is placed on the bit, the bit is likely to shift and produce a non-vertical hole. By suspending the drill string and using rotation, gravity will assist in producing a vertical hole. Periodic checks on the verticality of the rig and mast are required, as the rig may settle on its foundation as the drilling progresses. Bentonite or other additives are usually not added to water to make a viscous fluid in case of reverse rotary drilling. To prevent cave-in of the hole as the drilling proceeds, the fluid level must be kept at ground level at all times. The hydrostatic pressure of the water column together with the inertia of the body of water moving downward outside the drill stem support the wall of the bore hole. Usually the well wall does not erode because the velocity of fluid in the annular space is low.

Adaptability: Reverse circulation rotary drilling is the cheapest way to drill large diameter holes in soft unconsolidated formations. It is generally used to drill holes of size 35 cm and above. Its use is restricted to locations where ample water supply is available, as it requires about five times the amount of water that is normally needed for direct rotary drilling. The method is also not suitable in formations where caving is prominent. It is suitable only for the construction of gravel pack tube wells. During the reverse circulation process of drilling, the flow of water into and out of the aquifer actually results in preliminary development of the well. As clear water is used, there is no mud invasion. After completion of drilling, the resultant well is cleaned and requires only a minimum of development to ensure maximum yield.

Air Rotary and Down-the-Hole Hammer Drilling

Percussion and rotary methods of well drilling are usually uneconomical in water well drilling in hard rock areas due to the slow operation rate, high cost of drill bit and high maintenance cost of the machinery. Air rotary drilling equipment uses compressed air as the drilling fluid, rather than drilling mud. Air is circulated through the drill pipe, out through ports in the drill bit, and upward in the annular space around the drill pipe (Figs. 2.66 and 2.67). Air, moving at high velocity in the annular space, carries the cuttings to the surface or blows them out into rock crevices.

Air-operated down-the-hole (DTH) drills and air rotary drills have proved to be suitable for the construction of water wells in hard rock areas. Drilling with DTH machines combines the percussive action of percussion drilling and rotary action of rotary drilling. The tool is essentially a pneumatic hammer attached at the end of a drill string (Figs. 2.66, 2.67, 2.68 and 2.70). The hammer tool comprises of an air-operated piston, striking directly on the bit at a high frequency, while the entire assembly is hydraulically pressed against the rock. By rotating the drill string at the same time, a hammering action is imparted to the full surface of the

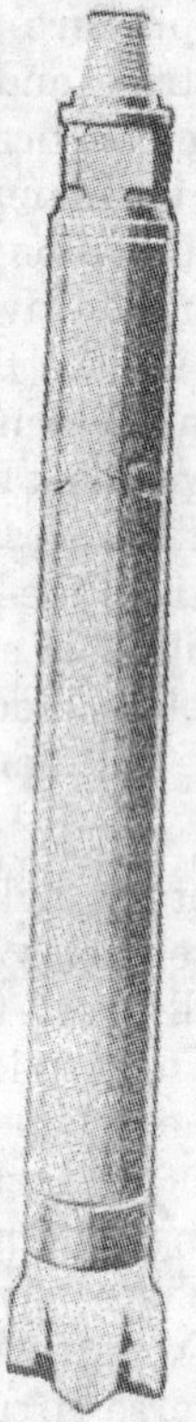

Fig. 2.66. Down-the-hole hammer drill. (*Courtesy:* Atlas Copco)

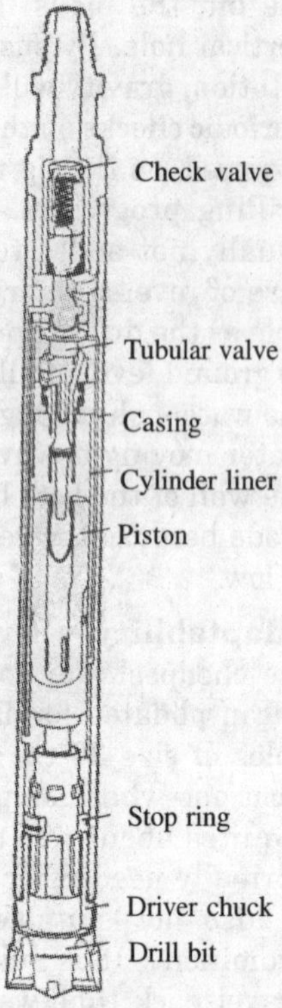

Fig. 2.67. Longitudinal section of down-the-hole drill (Adapted from Atlas Copco).

bottom of the hole, which pulverises the formation. The air for the hammering action is allowed to escape at the bottom of the hole to lift the cuttings up to the surface through the annular space between the hole and the drill string (Fig. 2.69).

With DTH drilling the penetration is faster because the hammer at the bottom of the hole allows the piston to transmit the blows directly on the bit, without losing its energy through the drill string. The hammer may drill granite 2-3 times faster than the air-rotary method and several times faster than the percussion method.

The operation of the drill bit depends upon the capacity of the air compressor, the design of the hammer, the type of bit and the type of coolant used in the drilling operation. An air compressor continuously provides the required quantity of air to operate the hammer as well as clean the cuttings. A straight hole is ensured by the short rapid blows which minimise the effect of chipping and breaking of the formation. DTH drilling is limited to constructing small diameter wells. The largest drill bit commonly used is 15 cm. However, the well size can be larger (about 30 cm) when rock-roller bits are used.

The drill bit is an alloy steel hammer with heavy tungsten carbide inserts which provide the cutting surfaces. Tungsten carbide is extremely resistant to abrasion, but drill bits become dull with continued use. The inserts are sharpened by grinding when the operating conditions indicate that the bit is not cutting properly.

The drillability of rock depends on the hardness of the minerals of which it is composed and the size of the grains. Quartz, which is the most common amongst the rock forming minerals is the hardest to drill, causing heavy wear on the drill bit. However, rocks with large calcite centent are easily drilled and cause little drill bit wear.

Rock drilling methods: The basic method of rock drilling comprises of—(*i*) Parcusscim drilling, (*ii*) Rotary drilling, (*iii*) Rotary cutting, and (*iv*) Rotary abrasive drilling. The percussion method incorporates a single cylinder with, reciprocating air piston driven by the circulating air. The extremely rapid hammer blows delivered with relatively light weight on the bit, increase bit life and help to control deviation while maintaining relatively high penetration rates. The pressure required to operate the hammer ranges between 7 to 17 kg per sq. cm and the blows per minute range from 600 to 1200. The hammer hits the bit while it is also being rotated along with the tool string. Since the hammer moves down the hole, the piston blows are transmitted directly to the bit without losing percussive energy through the drill string. The drill cuttings are lifted by compressed air which passes through the bit and keeps the hole clean. To remove the cuttings effectively, the annular air velocity is required to be maintained between 900 and 1500 m/min. Much faster rate of penetration result from abrasive percussion action.

To start drilling, air is turned on and the bit rotates and lowers slowly until it starts hammering on the formation. Initially just enough pull down pressure is applied to start breaking the rock and the pressure is increased gradually till smooth operation is obtained. The recommended pull down pressure is in the range of 30 to 150 kg per cm diameter of the bit. Application of higher pressure may decrease the effectiveness in hard rock drilling. As the hole gets deeper, pull down pressure is required to be adjusted to compensate for the additional weight of drill pipe added to the drill string. With deeper holes, instead of additional pull pressure, counter balancing of the weight of the drill string may be necessary. The optimum rotational speed vary between

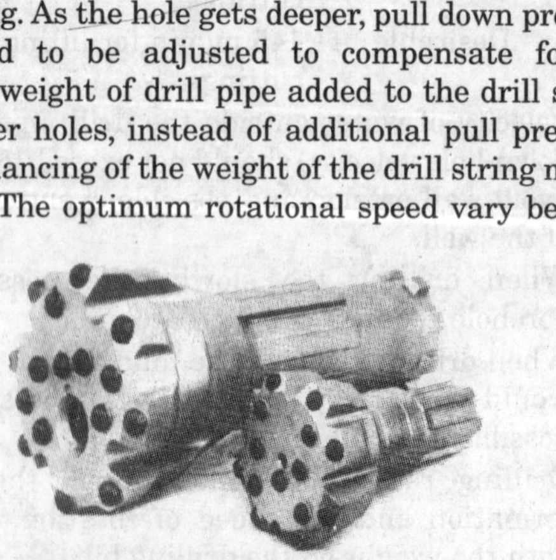

Fig. 2.68. Button bits used in down-the-hole air-percussion drilling.
[*Courtesy:* Atlas Copco (India) Ltd.]

15 to 50 rpm. The harder or more abrasive the rock, the lower the rate of rotation.

One of the latest developments in down-the-hole percussion air drilling is a button bit (Figs. 2.76 and 2.77) with replaceable, tungsten carbide buttons. The bit has buttons set on its face to a predetermined height above it. The buttons when crushing, are removed with a special punch through access holes at the back of the bit. *Rotary crushing* procedure, originally developed for oil well drilling, is being increasingly used in water well drilling as well. The procedure is used for drilling in soft rock formations. It is mainly used in groundwater prospecting when a core of the formation is required for testing. The drilling is done with a diamond core bit.

In *percussive crushing* and *cutting drilling,* the rock is broken by being subjected to high pressure from buttons or inserts of cemented carbide. Since the button is sufficiently pointed, the stress in the rock will gradually increase and cause the rock to be broken into chips. A uniform and high feed rate will cause the button to continue penetrating. If the button is blunt, more force will be required to cause penetration and breaking. Eventually, the breaking phase will stop and penetration will cease, despite the button being subjected to high loads. The bit buttons or inserts must be reground before the penetration rate has decreased excessively.

In percussive drilling, energy is transmitted from a rock drill through a drill steel and cemented carbide inserts to the rock, where it is used for crushing. In down-the-hole drilling energy is transmitted in the same way, but the piston of the hammer works directly on the drill bit. When drilling is done by rotary crushing, energy is transmitted to the bit through a pipe which is rotated and which presses the bit against the rock. The cemented carbide buttons are pressed into the rock and break off chips in the same way as in percussive drilling. In the rotary cutting method, energy is transmitted to the insert through a drill tube which rotates and which presses the inserts against the rock. The edge of the insert then generates a pressure on the rock and cracks off chips.

To enable drilling to proceed effectively, it is essential that the bottom of the hole is kept clean and the cuttings are continuously removed from the drilled hole. This is done with a flushing medium which may be air, water or foam, which is forced down to the bottom of the hole through a central flushing hole in the drill rod and flushing passages in the drill bit. The cuttings, mixed with the flushing medium, are forced out of the drill hole through the space between the rod and the wall of the drill hole. To obtain effective cleaning of the hole, the flushing air must flow at a high velocity.

Down-the-hole drills are available for various types of uses. These include the light-duty in-well drills, wagon drills, and truck/tractor-mounted drills of various capacities. The drill unit can be powered by a separate

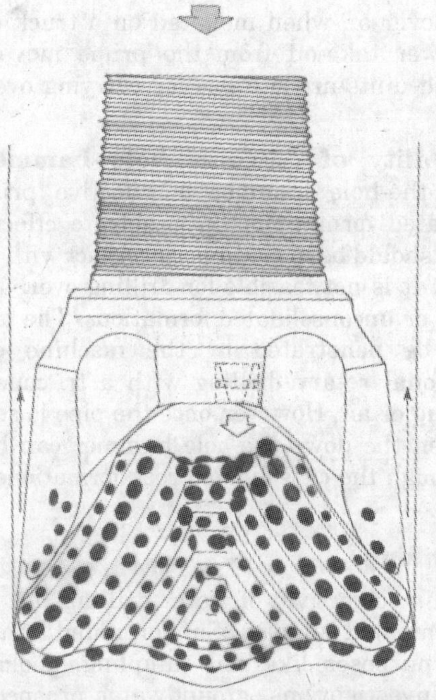

Fig. 2.69. Sketch illustrating the details of operation of a pneumatic rock roller botton bit. The downward arrows at the centre show the direction of the main air flow for the operation of the bit. The upward arrows on the side show the direction of flow of the material cut by the bit.

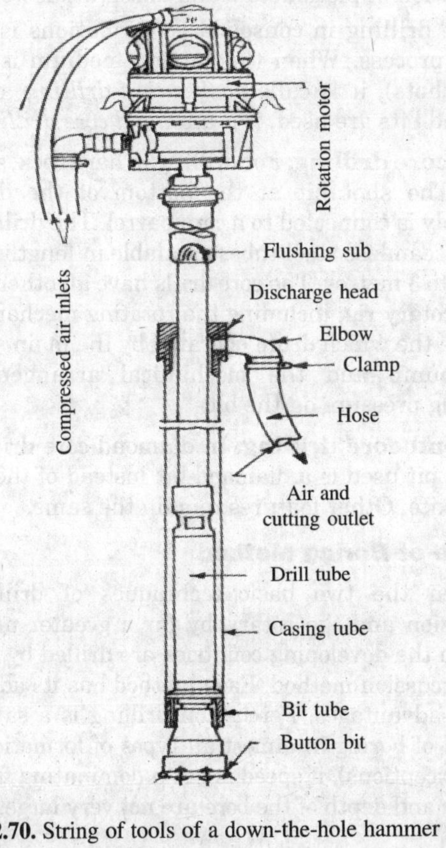

Fig. 2.70. String of tools of a down-the-hole hammer drill.

prime mover or, when mounted on a truck or tractor, with power take-off from the prime mover. Tractor amounted units are suitable for carrying over difficult terrains.

Adaptability of down-the-hole hammer drills: A down-the-hole hammer is effective primarily in consolidated formations. To penetrate effectively, the hammer should be in continuous contact with the face of the rock. It is not feasible for drilling overburden with boulders or unconsolidated formations. The overburden has to be penetrated by the machine employing conventional rotary drilling with a tri-cone drag bit using mud or air. However, once the pipe is set into bed rock, then the down-the-hole hammer can be used to drill through the consolidated rock formations at a very fast rate.

Core Drilling

The main objective of core drilling is to obtain uncontaminated samples of underground formations for various purposes like soil sampling, geological and mining investigations, groundwater prospecting, and foundation engineering. Rigs employing this method are occasionally used for water well drilling (mostly for drinking water and sometimes irrigation) in rocky areas, if no alternative and more efficient equipment are available. Engine operated hydraulic core drills are useful in drilling wells in soft formations. The equipment is commonly used in soil survey work and for installation of piezometers and observation wells.

Core drilling in consolidated formations is a direct rotary process. When the cutting medium is calyxite (steel shots), it is called *calyx core drilling* and when diamond bits are used, it is *diamond core drilling*.

Calyx core drilling: For drilling in hard rock, shot bit is used. The shot bit at the bottom of the drill pipe assembly is connected to a core barrel. The drill rods are hollow seamless steel tubes available in lengths varying from 1 to 3 metres. The core drills have all other features of the rotary rig, including the rotating mechanism, the derrick, the winch drum operated by the main shaft, the mud pump and the mechanical arrangement for applying pressure on the bit.

Diamond core drilling: In diamond core drilling, the cutting bit used is a diamond bit instead of the shot bit or calyxite. Other features remain the same.

Choice of Boring Method

Between the two basic techniques of drilling, the percussion and the rotary, by far a greater number of wells in the developing countries are drilled by the cable tool percussion method. Each method has its advantages and disadvantages. Percussion drilling is a satisfactory method of boring in almost all types of formations (with some exceptions), if speed is not a dominating factor and the size and depth of the bore are not very large. Cost per metre of drilling is lower for relatively shallow holes in unconsolidated materials. A percussion drill is lighter and can be easily transported in rough countryside.

In certain types of formations, such as dune sand, quick sand and unconsolidated river gravels, cable tools are unsatisfactory, particularly for drilling deep, small diameter wells. The pounding of the drill, the low fluid level, and the lack of mud sealing, contribute to slumping and caving in such formations.

For drilling large size holes, particularly where speed is an important factor, rotary drilling is most suitable. The method is also advantageous for drilling test holes so that if the prospects are found to be poor, the hole can be abandoned at a minimum expense without the trouble of pulling the casing.

For drilling in low lands with clay and silt or sticky shale formations and saline or alkaline conditions, the rotary rig is a much faster method. Dunes, wind blown sand and quick sand are handled easily by rotary machines. However, rotary drills are unsatisfactory in extremely permeable lost circulation zones. The circulating fluid is lost to a large extent in such formations and drilling becomes at times almost impossible. Rotary method also may not drill successfully in alluvial formations with boulder beds and cobbles. The removal of large size cobbles with circulating mud fluid becomes difficult.

Under certain conditions, the cost of drilling large diameter holes by the rotary method does not go up in the same proportion as it does in cable tool work. Thus, where favourable conditions exist, large diameter holes can be drilled more economically by the rotary method.

The percussion method requires much less drilling water than the rotary method. It is, therefore, more suitable for boring in areas where sufficient water is not available within a reasonable distance. Under the drilling technique adopted in the percussion method, a more accurate sample of the formation is obtained than in the rotary method. In percussion drilling, it is possible to test the quality of water in each stratum as the drilling proceeds. Zones of poor quality water can often be shut off by locating blank casing pipes against such strata. Such an information becomes difficult in case of the rotary method due to the mixing of the dislodged soil material and water with large quantities of surface water.

Among the other methods of drilling, the auger method has a specific limited use for boring shallow wells in unconsolidated formations which can normally stand without a casing. Core drilling is a suitable method only where it is necessary that the soil sample should be extracted in the form of a core for investigation purposes. The method is expensive for water well drilling. Down-the-hole hammer drill is advantageous for drilling small diameter wells in hard rock formations.

Common Drilling Difficulties

Some of the common difficulties encountered in rotary drilling are lost circulation, crooked holes, bailing up, and stuck-up drill pipes.

Lost circulation: Formations with joints and fissures and those with sands and gravels absorb enough of the drilling mud to hinder the drilling process seriously. Jointed and fissured shale, on the other hand, seldom drains off an excessive amount of the circulating fluid. However, the mud absorbed in this case causes the shale to swell and heave, filling up the drill hole. These difficulties can usually be overcome and satisfactory drilling progress continued by suitably altering the character and properties of the circulating fluid.

Crooked holes: Crooked holes that would seriously damage the pump may frequently be drilled without noticeable effect on the drilling machinery or equipment. To avoid crooked holes, the bits should be of a form and size that prevent undue eccentricity during rotation. They must be sharp and dressed to proper gauge. The drill collar which holds the bit to the lower end of the drill pipe must be large enough in diameter to hold the pipe centrally in the hole and to prevent the bit from working off to one side. Excessive bit pressures may be avoided. Some of the considerations which help in straight hole drilling are that the diameter of the hole may not exceed twice the diameter of the drill pipe and that the speed of bit rotation may be at least 150 rpm.

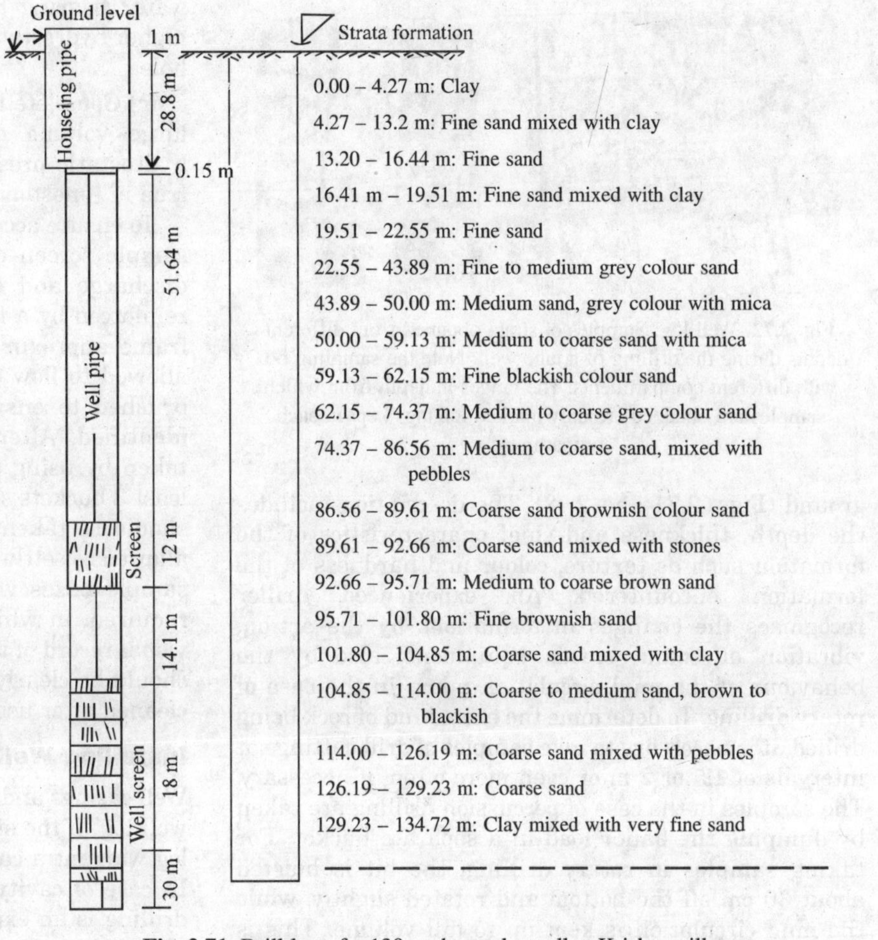

Fig. 2.71. Drill log of a 130 m deep tube well at Krishna village, Ranghat Block, West Bengal (1988).

Bailing up: Bailing up is the accumulation of soft, sticky shale or clay around the bit and drill collar. Occasionally mud collars are formed which are forced up the hole by the pump action. This bailing up, if allowed to continue, forms a coating around the drill collars which sticks to the drill pipe securely when it is raised off the bottom. The usual causes for bailing up is a high rate of penetration combined with a speed of rotation insufficient to mix the drill cuttings thoroughly. When such conditions are encountered, the drill pipe is 'spudded' frequently by raising it off the bottom and dropping it 1 to 1.5 m at a time, while the pipe is being rotated rapidly. If this is done and if the rate of penetration is held at a speed which gives the circulating fluid time to mix the drill cuttings thoroughly, this source of trouble can be held to a minimum.

Stuck-up drill pipe: Lost-circulation causes the drill pipe to stick when a crevice or formation which drains away the drilling mud is penetrated. The sudden reversal of the circulation in the hole deposits the suspended drill cuttings around the drill pipe, sticking to it tightly. This often happens so suddenly that there is little or no time to remove the pipe. When a drill pipe is so stuck, recovery may be effected by pulling the pipe with jacks.

Well Log

Sampling of the formations encountered in drilling is a major source of information essential for lithological evaluation and design of a tube well. The distribution of sand, silt and clay particles will profoundly influence the location of the well screen. Thus, accurate sampling and description of the strata are essential if the well is to be designed properly. Lack of care in either sampling or description may cause screens to be placed in wrong positions and can lead to well failure. The well log is the recorded description of the materials encountered in sequence throughout the drilling, with formations accurately noted as distances from the surface of the

Fig. 2.72. Well log: Samples of strata obtained from different depths during the drilling of a tube well. Note the sampling box with different compartments. The ranges in depth from which samples are obtained are duly marked ion tags kept in each compartment.

ground (Figs. 2.71 and 2.72). The description includes the depth, thickness and chief characteristics of the formation such as texture, colour and hardness of the formation encountered. An experienced driller recognises the changes in formations by the action, vibration or sound of his equipment; also by the behaviour of the mud circulation pump in the case of rotary drilling. To determine the exact kind of rock being drilled, it is necessary to take samples of drill cuttings at intervals of 1.5 or 2 m or even more often, if necessary. The samples in the case of percussion drilling are taken by dumping the bailer load in a separate bucket. For taking samples in rotary drilling, the bit is hoisted about 30 cm off the bottom and rotated slightly, while the mud circulation is kept up to full volume. This is continued until the mud is comparatively free of drill cuttings. After this operation is completed drilling is resumed for another 1.5 m, after which the drill pipe is raised slightly and the mud circulated until then is again cleaned of drill cuttings. The mud is screened carefully as it flows over the top of the well casing and the samples of the strata are obtained.

It is possible to detect water-bearing beds in percussion holes being drilled under ordinary conditions, because the comparatively clear water in the hole does not clog the pores in the water-bearing formation. Hence, a sudden rise or decline in the water level usually indicate that a permeable bed has been struck. Crevices or soft streaks met with while drilling hard formations, are often water-bearing. Sand gravel, sand stones and lime stone usually produce large quantities of water.

In rotary drilling, the following criteria, based on changes in the character and circulation of the mud indicate some of the most important means of recognising water producing zones:

(a) A loss of circulation fluid, much in excess of the normal or usual loss, indicates the presence of water-bearing stratum.

(b) A noticeable increase in the volume of the returning circulation fluid invariably means that a water bearing formation has been met which has a higher hydrostatic pressure than the mud column in the hole.

(c) Complete loss of circulation usually means that a large volume of water with a comparatively low hydrostatic pressure has been met. This is especially true in limestone.

To ensure accurate logging in case of rotary drilling a sample screen could be used to take samples of the discharge and cuttings. It consists of a fine mesh, reinforced by a larger mesh wire screen contained in a frame approximately 1.5 m square. The discharge is allowed to flow over the screen, which is continuously brushed to ensure that any change in formation is identified. Alternatively, formation samples may be taken by using buckets of about 14 litres capacity. At least 3 buckets should be ready at the site. The samples should be taken directly from the discharge point and allowed to settle prior to decanting and description. Two sample boxes with about 50 compartments each are required, in which the samples are kept to provide a visual record of the strata penetrated. The sample boxes should be clearly marked with each depth interval and cleaned after use.

Installing Well Screens

Well casings and screens are installed on the basis of the well log. If the strata conditions as revealed by the well log warrant a cavity bore, no screens will be necessary. In case of cavity wells, if the casing pipe used during drilling is an expensive one, a string of ordinary blank pipes of smaller diameter may be inserted inside the casing pipe, which may be pulled out later. The annular space between the tube and the sides of the bore hole in the clay layer should be plugged with care so that no soil particle can enter the cavity from the upper layers. On the other hand, if the casing used consists of ordinary pipes, it may be left as such in the bore to tap the cavity. It is desirable that the cutting shoe at the lower end of the tubes in cavity bores should be left about 60 cm above the bottom of the clay layer. This enables the formation of a bell-mouth at the entrance of the pipe, thereby reducing friction and increasing discharge for the same drawdown.

Procedures for installing well screens vary with the design of the well and the method employed in drilling. The screens have to be located opposite the water bearing strata in order to draw water from the strata (Fig. 2.71). This is done by assembling together the whole length of the screens and the blank pipes in exactly the same order and lengths in which they are to be lowered in the bore. Every pipe, socket and strainer are numbered serially. The bottom end of the assembly is fitted with about one to two metre length of blind pipe with a cap to close the bottom. The cap is known as *bail plug* and has an 'eye' fixed inside, which can be connected with a hook whenever it is necessary to pull out the string of pipes. The blank pipe at the bottom serves to collect the sand particles that enter the bore-hole during the development of the well, thus preventing choking the lower portion of the screen.

The strainers and plain pipes are lowered down the bore, one by one, starting from the bottom end. To start with, the bottom pipe is fixed with a pair of clamps and is lowered in the bore hole till the pipe is supported by the clamp over the bore. The next piece is then screwed on to the end of the bottom one and a second set of clamps is fixed at the top end of the upper piece. The pipe assembly is then lifted slightly by means of the drilling cable and the lower clamp is removed. The drilling cable is gradually slackened, and the tubes lowered in the bore and the whole operation is repeated as before till the entire line of pipes is lowered into the bore. The complete string is kept suspended from the top and the outer casing pipe is jacked up and extracted one by one. As the casing pipe is withdrawn, the surrounding material gets loosened and grips the strainers and plain pipes and holds them in position.

Verticality: Upon successful installation of the pipes and screens, the housing pipe must be checked for verticality. This is to ensure that the pump will hang vertically in the well. The verticality of the hole and of the assembled casing string in it shall be such that both before and after placing the gravel filter, no point on the actual centre line of the upper well casing shall deviate appreciably from an imaginary plumb line passing through the centre of the top of the upper well casing. The test shall be carried out in the well, both before and after the insertion of the gravel pack.

The method of measuring verticality consists of a cylindrical plumb bob, suspended from a tripod aligned over the center of the well. Prior to any measurements, the north direction is marked, using a compass, so that the correct direction of any deviation can be noted. The point of suspension for the verticality test should be at least 3 m above the top of the casing. The tripod is moved so that the bob hangs freely in the centre of the well top. The bob is then lowered 3 m and the position of the wire, relative to the casing, is measured. The amount and direction of any deviation from the centre should be noted. The bob is then lowered an additional 3 m and another measurement taken. This is continued until the full length of the housing pipe has been tested. The deviation, if any should be within permissible limits.

Gravel Packing

The procedure for installing well screens in gravel pack wells involves centering the screen in an over-size bore hole, followed by placement of the gravel. The double-casing method commonly used in wells drilled by the cable tool equipment uses one string of casing of a size corresponding to the outside diameter of the gravel pack and a second string of casing that is nominally the same size as the screen.

The inner casing and the well screen are centered in the outer casing. The annular space between the two casings is filled with gravel poured from the ground level up to about 3 metres above the top of the screens. The casing pipe is then raised to about 5 cm above the bottom of the screens and the development of the well is started. As development takes place, some of the displaced sand may be filled with the gravel from the surface. The gravel pack is maintained at the same top level throughout the course of development, by adding more gravel, if necessary. While the processes of development and gravel pouring continue, the casing pipe is gradually withdrawn by means of jacks (Fig. 2.73). The extraction of the casing pipe should be done slowly and without jerks in order to ensure that the screen tube is not dragged up with the casing pipe.

When a gravel packed well is constructed with a rotary rig, the hole should be thoroughly washed clean of cuttings and finally reconditioned with fresh mud of good viscosity before lowering the casing assembly. The hole should be kept full of mud during lowering so that it does not cave in. After lowering, the assembly should be suspended in tension and suitably guided in order that it will be centered at all points in the hole.

Before gravel feeding, water is introduced into the system and mud weight in the annular space is reduced substantially. Gravel is then introduced against the circulating stream at the surface or pumped through tubing into the annular space. The circulating stream serves to remove the fine materials from the gravel and prevents possible bridging. During gravel pack, the top

Fig. 2.73. Ejecting the outer casing pipe with screw jacks after installing the strainer and well casing in a well drilled by percussion method.

of the well assembly is sealed so that the circulation returns through the annular space. In the construction of a gravel packed well with reverse rotary method, the well assembly is held in tension and guided for centering in the hole. The hole is kept full of water during the lowering of assembly and gravel is fed into the annular space from the surface or pumped through a tubing.

Sealing of brackish and saline aquifer horizons: Brackish and saline water occurring in some of the aquifer horizons may sometimes be a serious problem in well construction. Techniques are available to 'seal' these horizons, once they are properly identified and located. The *drill stem test* is carried out to demarcate saline horizons. In this method the bore hole is reamed with a drill bit of the packer size in order to consolidate the well wall up to firm strata such as clay above the aquifer. Slotted pipes are placed opposite the aquifer and water is pumped by the air lift method. Water samples are collected and analyzed to determine its quality. The process is repeated for every aquifer, one after the other.

Sealing saline top zones: In some cases the top horizons may have saline water while good quality water exists in the lower horizons. To seal such zones the well pipe assembly is lowered in such a way that the blank pipe is placed opposite the saline aquifer. The well is gravel packed up to a few metres below the saline aquifer where there is an impervious strata at least 6 m below the saline aquifer. Cement slurry is pumped into the annular space between the casing and the well bore and around the casing pipe in order to fill up the space. The cement slurry is allowed to settle there for at least 72 hours. Thus, the upper aquifer is sealed off from the remaining bottom aquifers and cannot contaminate the water below. The annular space over the cement plug can be filled with clay.

Sealing of intermediate saline aquifers: Sometimes intermediate aquifers are found to be saline and these have to be sealed off from top zones as well it from the bottom aquifers. This is a somewhat complicated job and it requires considerable precaution and care. To seal the horizon the assembly of pipes is lowered as usual, taking care that the saline aquifers are faced against blank pipes. The well is gravel packed up to a few metres below the saline aquifer. It is essential to ensure that proper gravel packing is done up to such a point. This may be possible only by surging the well against the lower aquifer in order to get the gravel settled properly. After a proper gravel pack is ensured, cement slurry is pumped at the top of the gravel pack and below the saline zone. If the saline layer is not very thick the slurry may be pumped up to the top of the saline aquifer so that the entire aquifer is sealed off. In case the aquifer is very thick then the sealing has to be done at the bottom of the aquifer and at its top and clay and other material may be filled in between the two cement slurry sealings. The cement slurry is allowed to remain about 72 hours and again gravel packing is done above the top cement seal for the upper aquifers. The same process can be repeated if there is more than one saline aquifer.

Sealing the bottom zones: There are cases where only the bottom zones are found saline. In such cases the sealing is easy and may be done with or without cement. In most cases the saline aquifers can be ignored and only the top aquifers which are suitable can be tapped.

The consistency of the cement slurry requires special attention. It is necessary to estimate the drilling fluid pressure at the various depths where cement slurry has to be pumped. The density of the slurry has to be the same as that of the drilling mud. In order to have the desired density of the slurry, various additives are used. These include gel, hydrocarbons, barite, and sodium chloride.

Well Development

Tube wells are developed to increase their specific capacity, prevent sand pumping and obtain maximum economic well life. Development means the stabilization of the walls of a well adjacent to the screen by a process which removes fine particles from the formation immediately surrounding the well screen, leaving coarser particles to contact and surround the screen (Figs. 2.29 and 2.30). The main objectives of development are: (1) to correct any damage to or clogging of the water bearing formation, (2) to increase the porosity and permeability of the water bearing formation in the vicinity of the well, and (3) to stabilize the sand formation around a screened well so that the well will yield sand-free water. Development is necessary in all gravel packed wells and other screened wells, except when the screen is formed of fine wire mesh or coir or other closely knit filters located in a highly permeable formation.

The basic principle in the development operation is to cause reversals of flow through the screen openings that will rearrange the formation particles. This is essential to break down interlocked groups of particles. Figure 2.74 shows how small particles can bridge between large particles and across screen openings when the flow of water is in one direction only. Reversing the direction of flow by surging overcomes the tendency

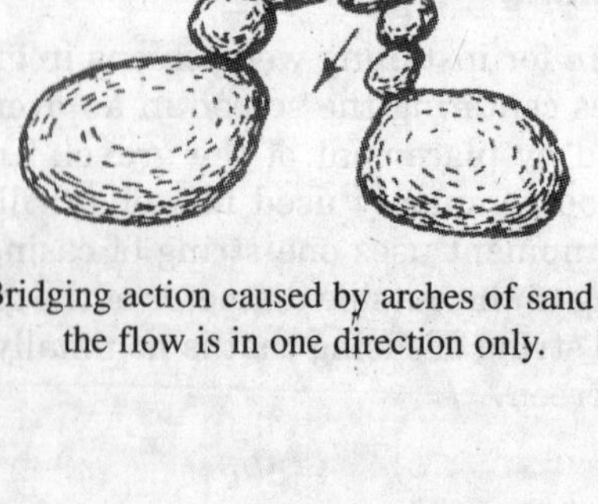

Fig. 2.74. Bridging action caused by arches of sand grains when the flow is in one direction only.

of bridging of particles. The outflow portion of the surge cycle breaks down bridging while inflow portion moves the fine material toward the well screen and into the well.

The methods commonly employed for well development are surging, pumping, use of compressed air, jetting and use of dispersing agents (chemicals).

Development by surging: It is the working of a block or plunger up and down in the well so that water is alternately forced out into the surrounding formation and then allowed to flow back into the well. This action loosens fine sand or gravel particles near the screen and carries finer particles into the well where they can be removed.

The tool normally used in called a surge plunger or surge block (Fig. 2.75). A heavy bailer may be used to produce the surging action, but is not as effective as a close-fitting surge block. The surge block is especially favoured in cable tool drilling. Surge blocks may be of two types: (*i*) a solid plunger, and (*ii*) a plunger with valve openings. The valve-type plungers give a lighter surging action than the solid block. This is an advantage in developing tight formations since it is always good to start surging lightly and to increase the force of the surge as development proceeds. A solid-type plunger is preferred when greater surging is necessary.

Fig. 2.75. Surge plunger used in developing tube wells (solid type).

Simple surge plungers are made by wrapping sacking, old rope, or some such material around the drill stem or bailer. A mud-scow type bailer is a satisfactory surge plunger. For developing formations requiring more thorough treatment, specially built surge plungers may be used.

Surge plungers may be operated by the spudding mechanism of boring outfits using long strokes. The tools are lowered into the casing until the plunger is 3 to 5 m in water. Care is taken not to lower the tools so far that they batter the top of the screen. Surging is started slowly and the speed increased until it reaches the fastest limit at which the tools will drop and rise without too much slap of the cable. After the surging has been continued for some time and sufficient sand accumulates in the screen, tools are hoisted and the bailer is run into the well to clean it thoroughly. Surging is again resumed after cleaning, and the procedure is repeated until the well is fully developed as evidenced by the fact that little or no more sand can be pulled into the screen. Two or three days are usually required to develop a deep well by this method.

Development with pumps. Tube wells may be developed by pumping, either by ove-pumping or by a process called *rawhiding the well* (Fig. 2.76). Over-pumping means pumping the well with excessive drawdown. This is not a very effective method of developing a well, particularly because the flow of water remains in one direction and, thus, encourages the phenomenon of bridging. The process of rawhiding the well consists of starting and stopping the pump intermittently to produce relatively rapid changes in the head of the well. While this may be done with any kind of pump, it is most effective with a turbine pump installed without a foot valve (Fig. 2.76). There are two main types of developing reactions which may be obtained by operating the pump in different ways:

(1) The pump is operated at its full capacity until it has produced the greatest possible drawdown of water. It is then stopped, the water recedes rapidly out of the column of the pump, and the well is permitted to regain its full static water level. This process is repeated until the well shows no further improvement.

(2) The well is pumped to obtain maximum drawdown and the pump is then stopped and started alternately at short intervals. This has the effect of holding down the water level in the well and frequently agitating the formation by the backwash of water in the pump column.

The time required to develop a well by pumping varies. Sometimes a few hours only are required but usually at least a day is necessary. The method is inexpensive, speedy and fairly effective, but not vigorous enough for wells requiring heavy development.

Fig. 2.76. Developing a tube well using a vertical turbine pump.

One of the objections to development by pumping is the damage which it can cause to a permanent pump. If considerable sand is being pumped with the water, the pump will be subjected to excessive wear, which reduces its efficiency. Under severe conditions, it is possible that the pump may get sand-locked. Should sand-locking occur, the pump must be pulled, disassembled, and cleaned carefully before being placed back into service.

Backwashing by pumping: One of the methods of developing a well or cleaning a clogged strainer by means of pumps is that of 'backwashing'. Water under pressure is forced in the casing. It eventually passes out into the formation, cleaning the strainer and carrying with it clogging sand particles. The operation of backwashing is followed by bailing or pumping and the process is repeated to develop the well fully.

Development by compressed air: The most common method of well development is by compressed air (Figs. 2.77 and 2.78). There are two general methods: the backwashing method; and the surging method. A combination of these methods may also be used for more effective development.

(a) Backwashing method: The principle of the backwashing method is to force the water back out of the well through the screen and into the water bearing formations by means of compressed air introduced into the well through the top of the casing after it has been closed. The method cannot be used unless water in the well stands at a considerable height above the screen section of the casing. It is best used after the well has been cleared as completely as possible with a bailer. When this method is used, the top of the casing is sealed with an air tight cap through which the air line extends (Fig. 2.78). The air line is equipped with a three-way valve so that the pressure in the well can be released at any time. When compressed air is turned into the air line, the water in the well is forced out through the screen. As soon as the air begins to escape through the openings in the screen, the air stream is shut off and the pressure in the well is released by opening the three-way valve. This is done to prevent the danger of air-logging in the water bearing formation. When the pressure is released, the water will flow back into the well through the screen and bring fine sand from the area surrounding the well. This process is repeated until no more sand is brought in. The cap is then removed and the accumulated sand bailed out.

(b) Surging method: The use of the surging method, sometimes called open well method, is limited to wells in which the depth of water exceeds two-thirds of the total depth of the well. Well development is accomplished under this method by a combination of surging and pumping. The inrush of air creates a powerful surge within the well. The process loosens the fine material surrounding the perforations which may then be brought into the well by continuous air injection. The operation is repeated at intervals along the screened section of the well until sand accretion becomes negligible. The principle of pumping is the same as described under the section on air lift pumps (Chapter 3).

Combination of backwashing and surging methods: A combination of backwashing and pumping methods of developing wells by means of air is more effective than either of the two methods used separately. When the

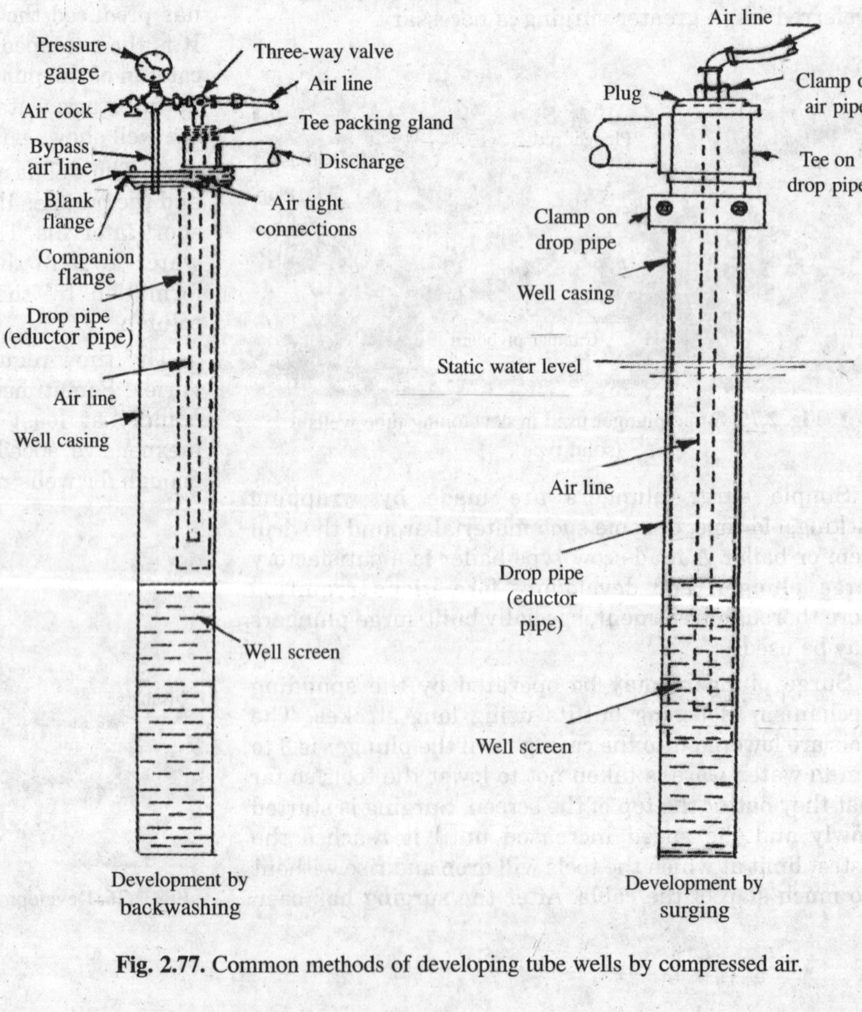

Fig. 2.77. Common methods of developing tube wells by compressed air.

combination method is used, the top of the casing is closed air tight and a three-way valve is used as in the backwashing method. The drop pipe is run through the air tight cap on the casing. The air line is inserted into the drop pipe as in the surging method and a separate line with an air cock is run through the cap. The drop pipe is carried down into the well to a point about 30 to 60 cm above the top of the screen. To develop the well, the air cock is opened and air is turned into the air line in the drop pipe. Water is pumped out of the well through the discharge pipe and washes some of the fine sand material. When the water becomes clear, the supply of air is cut off and the water in the well is allowed to regain its static level. The air cock is then closed and a three-way valve turned to direct the air supply down the other air line. This forces the water out of the casing and back through the screen, agitating the sand and breaking down the bridges of sand grains. When the water has lowered to the bottom of the drop pipe, the hydraulic connection is broken which makes it impossible to do further development.

When the air is heard escaping out of the discharge pipe, and when the pressure stops increasing, the supply of air is cut off and the air cock is reopened to allow water to reach the static level. The three-way valve is turned on and the air supply is again directed down the air line to pump water in the well. The process of alternately pumping and backwashing is continued until the sand accretion becomes negligible.

Development by jetting: Jetting with water at high velocity is an effective method of well development. A simple jetting tool, a high pressure pump, and the required length of hose and piping are the main items of equipment needed. The high velocity water jet applied through the well screen agitates and rearranges the sand and gravel particles of the aquifer around the screen. The mud cake deposited on the bore hole in rotary drilling is broken and dispersed. This enables the drilling mud to be pumped out easily. By slowly rotating the jetting tool and gradually raising or lowering it, the entire surface of the screen gets the action of the jet. Fine sand, silt and clay are flushed out of the aquifer. The turbulence created by the jet brings the fine materials into the well through the screen openings.

Development by dispersing: Adding a small quantity of a suitable poly-phosphate to the drilling mud to the water used in backwashing and jetting and to the water standing in the well helps in mud removal. The phosphates disperse the clay particles in the drilling mud. Dispersal induces the movement of particles. A suitable mixture is formed by adding 2 kg of the chemical to about 400 litres of water.

Development of a Cavity Well

Development of a cavity well is the drawing out of sand from below the clay roof. As sand is drawn out, a cavity is formed in the water bearing stratum at the bottom of the clay roof. The depth of the cavity is small, only a few cm but the circular area of the cavity is large; hence the strain on the roof. The area of the formation developed is much larger than the area of the cavity, and it is on the development of this formation that the yield of the cavity well depends.

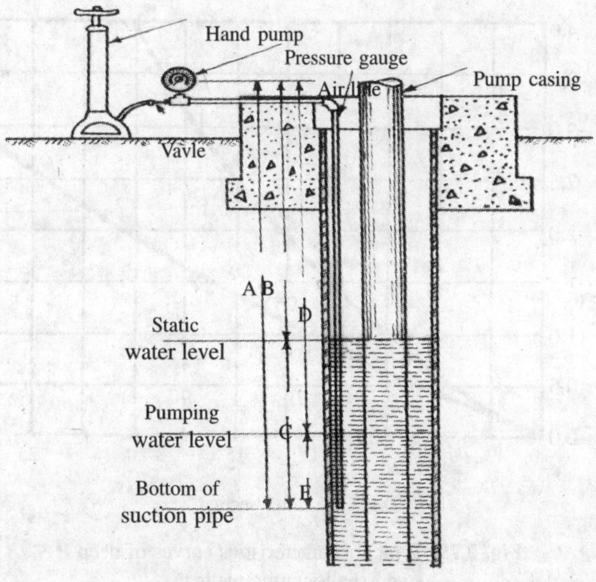

Fig. 2.78. Air line installed in a tube well for measuring water level.

Air compressors or turbine pumps may not develop a cavity well satisfactorily, unless a small cavity has previously been created by means of a sludger. A centrifugal pump is usually better suited for developing a cavity well than an air compressor. The cavity should be developed slowly and with great care. Otherwise, sand is likely to rush into the casing and choke it. This may cause such a disturbance in the formation that the satisfactory development of a cavity may become impossible even after removing the initial mass of sand.

To begin with, the discharge is kept low. When the discharging water has become clear, the depression may be increased slightly, which may result in more sand being drawn out. This process is repeated till the normal depression is reached and the discharge is clear. The pumping operation may then be stopped for an hour or so and then resumed. The discharge after restarting may again contain sand. The operation may be continued till the water is clear again. The procedure may be repeated till the well is fully developed, which is evidenced by the discharge being clear even on resumption of pumping after an interval.

Testing Wells for Drawdown and Yield

Following the development of a new well, it should be tested to determine it's yield and drawdown. This information provides the basis for determining the yield of the well in order to select the type of pump to be installed. The information to be obtained from a well test includes location of the static water level, the yield

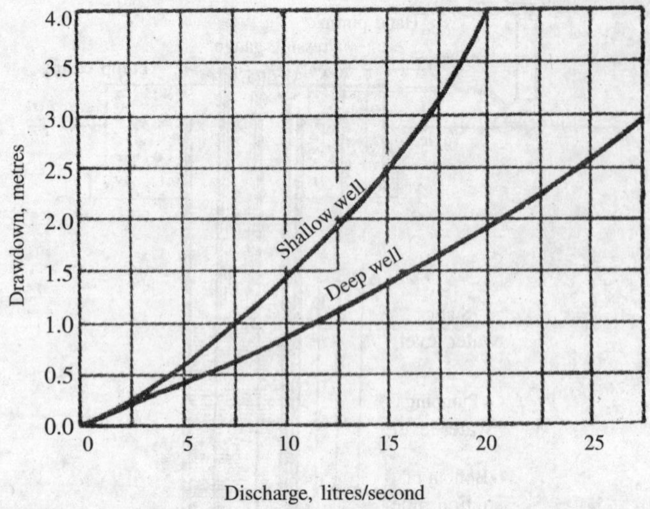

Fig. 2.79. Typical characteristics curves of deep and shallow tube wells.

and the depth of water when the well is being pumped at different rates. From these data, the drawdown, the yield and the specific capacity (the yield per metre of drawdown) can be computed. In deep wells tapping substantial thickness of the water bearing medium, the yield is almost directly proportional to the drawdown. This, however, is not true for shallow wells. The well characteristics curve is obtained by plotting the data on discharge versus drawdown. Typical characteristic curves of deep and shallow wells are shown in Fig. 2.79.

Measurement of discharge and drawdown: Well discharge can be measured by any of the conventional water measuring devices like weirs, Parshall flumes and orifices. The depth of water in most wells may be measured with reasonable accuracy by using a steel tape with a weight to make it hang straight. The tape is chalked at the lower end and lowered into the well until about 30 to 50 cm of the chalked lower end is submerged. The wetted length of the tape, which shows up clearly on the chalked portion of the tape, is subtracted from the total length lowered below the reference point; this gives the depth to water. The difference in the depth of water at any stage of pumping and the depth of static water level will give the drawdown. Depth to water table measurement can also be made by the electrical depth gauge (Fig. 2.80).

In the case of horizontal centrifugal pumps directly connected with the well casing, the drawdown can be determined by measuring the suction head with a vacuum gauge screwed on the suction side of the pump. The reading on the gauge will indicate the suction head from the centre of gauge. The suction head will include the head due to friction in the suction pipe, which may be estimated. Making adjustments for the friction head and the distance between the static water level and the gauge, the drawdown can be estimated. In the case of bore-hole pumps, where the casing is not directly connected to the suction side of the pump, drawdown may be measured by means of an 'air line' lowered into the annular space between the pump column and the well casing and fitted at the upper end with a pressure gauge (Fig. 2.78).

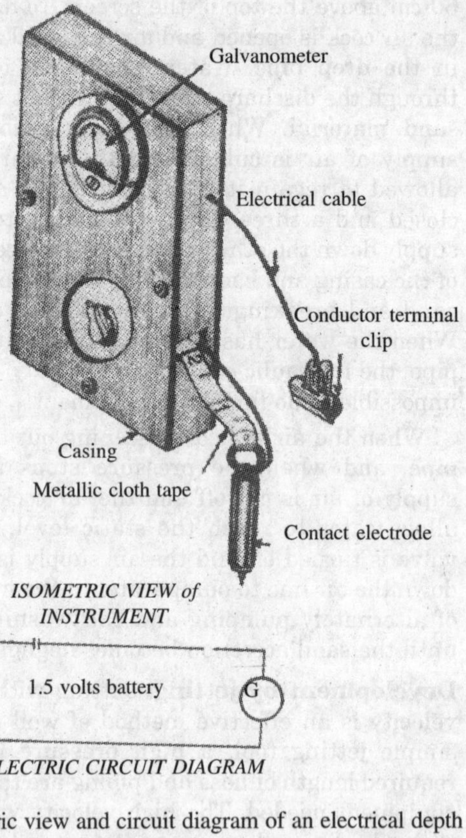

Fig. 2.80. Isometric view and circuit diagram of an electrical depth gauge. The electrical circuit is complete when the electrode dips in water which is indicated by the galvanometer. The corresponding depth is read on the tape.

The air line should extend about 3 m below the limit of drawdown. Air is pumped into the line until the maximum possible pressure is reached. Normally the air line is full of water up to the level of water in the well (static or pumping level). When air is forced into the line, it creates a pressure which forces the water out of the lower end until it is completely expelled and the line is full of air. If more air is pumped in, air instead of water is expelled and it is not possible to increase the pressure further. The head of water C or E (Fig. 2.78) above the end of the line maintains this pressure and the gauge shows the pressure or head above the end of the line. If the gauge is graduated in metres of water, it registers directly the amount of submergence of the end of the line. This reading subtracted from the length A of the line gives the water level B or D (static or pumping level). If the gauge is graduated in kg/sq cm the reading must be multiplied by 10 to get the submergence in metres.

Common Well Troubles

After a tube well has been constructed and pumped for some time, the user may find that it is loosing its yield capacity. This condition may become progressively worse and result in complete well failure. For a detailed treatment on rehabilitation if rock and failed wells, reference is invited to the book entitled. *Well and Pump Engineering*, listed at the end of this chapter.

The following are the common troubles in tube wells:

(1) Inflow of fine sand, silt and muddy water.
(2) Low discharge due to improper development and improper size of gravel.
(3) Low discharge due to improper design of screens.
(4) High rate of pumping resulting in depletion of water in the aquifers.
(5) Corrosion or incrustation problems.
(6) Mechanical faults caused during lowering of tube well assembly.
(7) Non-vertical wells.

1. *Inflow of fine sand, silt and muddy water:* The trouble is mainly due to the improper selection of screen to suit the aquifer characteristics and wrong placement of screens, improper development of the tube well and improper gravel pack.

2. *Low discharge of a tube well due to improper development and improper size of gravel:* A tube well has to be developed properly with due consideration to the method of drilling adopted, the characteristics of the aquifer encountered and the type of strainer used. If a gravel packed well is developed immediately after drilling, there are less chances of having any sort of choking of aquifer.

3. *Low discharge of a tube well due to improper design of tube well assembly and use of unsuitable screens:* Before a tube well assembly is lowered, the particle size distribution of the aquifer should be carefully studied and a suitable size of screen selected.

4. *High rate of pumping which results in the depletion of water in the aquifers:* After pumping for some time the static water level of a tube well may gradually go down due to over-pumping which indicates the depletion of aquifers. Over-pumping should be avoided. Before a pumping plant is purchased the quantity of water which can safely be withdrawn is determined so that depletion does not take place.

5. *Well failure due to incrustation and corrosion:* Well failures are often caused due to incrustation and corrosion.

Corrosion and Incrustation of Well Screens

Reduction in the yield of wells constructed according to proper standards is often caused by corrosion or incrustation (Fig. 2.81 and 2.83). *Corrosion* is a chemical attack on any material, which changes its form or breaks it down. It is a common cause of failure of buried metallic pipes and mild-steel tube wells. *Incrustation*, on the other hand, is the accumulation or deposition of external materials. Incrustation results in clogging or stoppage of the well screen and the surrounding water-bearing formation.

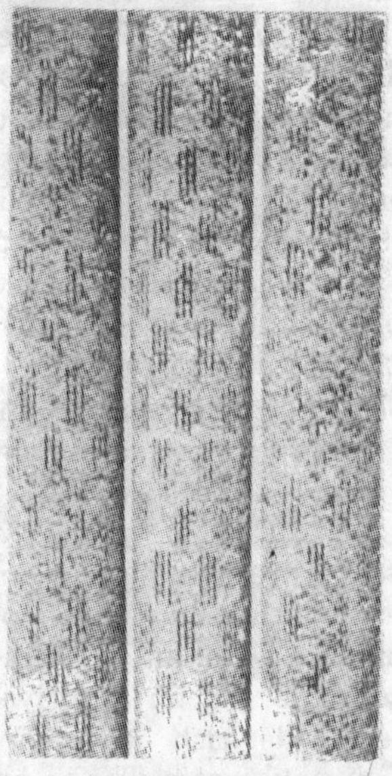

Fig. 2.81. Corrosion and incrustation in mild steel well screen.

Corrosion is a chemical action on the well casing or screen, exerted by external factors which result in the erosion of the metal. Corrosion can severely limit the life of tube wells as follows:

1. Screen-slot openings get enlarged, resulting in the loss of metal and sand pumping.
2. The strength of the well screen and casing get reduced, resulting in their collapse.
3. The well yield gets progressively reduced by deposition of corrosion products and blockage of screen.
4. The pump components get corroded, leading to their failure.

Metals are corroded due to various factors, including acidity of water, high velocity of entrance of water into the well and high temperature. Corrosion rate doubles due to every increase of 2°C in temperature of water. Other causes of corrosion are high value of dissolved oxygen content, total dissolved salts in water, dissolved carbon dioxide, electrical conductivity and presence of hydrogen sulphide and sulphur dioxide as well as presence of slime forming organisms.

Corrosion may be inhibited in various forms:

(*i*) *Direct chemical corrosion:* It is an electro-chemical phenomenon involving oxidation, resulting in the loss of electrons and ionization of the metal. Direct chemical corrosion result in the uniform loss of metal with occasional perforations in some areas. Increasing the rate of flow of corrosive water across a metal surface intensifies corrosion. The characteristics of the soil influence corrosion substantially. Poor aeration, high acidity, low EC and high salt and moisture contents are characteristics of corrosive soils. Presence of chlorides and sulphates in soils increases the corrosion rate of ferrous and some non-ferrous metals.

(*ii*) *Galvanic corrosion:* Whenever two different metals are electrically connected in a conducting solution, galvanic corrosion results due to their forming an electrolytic cell. The section of the cell being corroded is the anode, and the section on which the protective coat is being formed is the cathode. A well screen made of two different materials is easily damaged as a result of galvanic corrosion.

(*iii*) *Microbiological corrosion:* Sulphur and carbon, when present in soils, are attracted by bacteria and form acidic products which increase corrosion. The microbes causing severe corrosion are sulphate-reducing anaerobic bacteria which reduce the sulphates present in the soil or in water and convert metallic ion into the sulphide. Bacteria thrive well in poorly drained soils rich in organic matter and containing sulphates. The optimum range in pH for bacterial corrosion is 5 to 9.

(*iv*) *Dezincification:* Dezincification is sometimes called selective corrosion. It is the removal of zinc from a bimetallic alloy containing zinc (e.g., brass). The residue left behind is weak and porous.

(*v*) *Graphitization:* Graphitization is a selective corrosion process. Graphitization of low carbon steel and cast iron is an electro-biological process in the anaerobic condition. It is usually associated with sulphate-reducing bacteria in the well.

(*vi*) *Pitting or localised corrosion:* Pitting is defined as the formation of small holes in the blind pipe, positioned between the static water level and pumping water level, by the oxidising action of air and the corrosive action of groundwater. This results in a considerable decrease in well discharge.

(*vii*) *Fatigue or corrosion cracking:* Fatigue is caused by alternating stresses set up by vibrations or other periodic mechanical changes occurring in the pipes or pump. Fatigue, combined with corrosive water, causes cracks in the casing pipe.

Protection of well screens against corrosion: While installing a new tube well, one should determine the conditions under which the screens are to be operated and the type of corrosion anticipated. Adequate protective measures against corrosion should be provided in the design and management of a tube well. A mind steel well screen is liable to be corroded quickly under saline groundwater conditions. Hence, wherever possible, corrosion resistant materials are used for well screens in corrosive groundwater. Rigid PVC pipes, fibre-glass reinforced plastic pipes and stainless steel pipes are used to overcome corrosion. Under corrosive conditions, well screens should be designed with large open area so as to reduce the entrance velocity of water. The thickness of the casing pipe and screen should be comparatively large. Management practices like decreasing the rate of pumping or decreasing the drawdown could be undertaken.

Corrosion-resistant paints and coatings: Painting of mild steel pipes with anti-corrosive paints greatly reduce corrosion. Some commonly used paints/coatings to reduce corrosion are aluminium, asphalt, red lead and coal tar. However, paints have the disadvantage of forming scales on the metal surface. The scales are subsequently removed by the flowing water. Hence, the effects of the paints do not last long.

Cathodic protection of mild steel tube wells in corrosive groundwater: The principle of cathodic protection of tube wells is as follows:

Whenever two different metals are connected together in a conducting solution, the electric current set up due to the metals forming an electrolytic cell causes galvanic corrosion. The section being corroded is the anode and the section on which deposition takes place is the cathode.

The commonly used method of applying cathode protection to tube wells is through sacrificial anode. In case of the sacrificial anode system of cathodic protection, a metal of higher negative potential than that of the materials of the pipe to be protected is used as the anode. The metal pipe acts as the cathode and the intervening water as electrolyte, thus establishing the flow of electrons from the anode to the cathode. During electrolysis, the anode gets dissolved slowly and the metal ions in the solution are deposited at the cathode. Thus, the main pipe is protected from corrosion by sacrificing the metal of the anode. The anodes may be made of magnesium, zinc, aluminium, tin or their alloys. They are available in diameter ranging from 1.5 and 8 cm, 1 to 3 m long.

Incrustation: The following are the different forms of incrustations.

(*i*) Precipitation of carbonates, sulphates and silicates of calcium and magnesium.

(*ii*) Precipitation of compounds of iron and manganese.

(*iii*) Slime clogging produced as a result of iron bacteria and other slime forming organisms.

(*iv*) Mechanical incrustation formed by deposits of silt and clay in well screen openings.

Rehabilitation of incrusted tube wells: The following are the major management procedures adopted to reduce incrustation in tube wells:

(*i*) The drawdown while pumping is kept low. This will control the release of CO_2 so that precipitation of carbonates in the well screen is reduced.

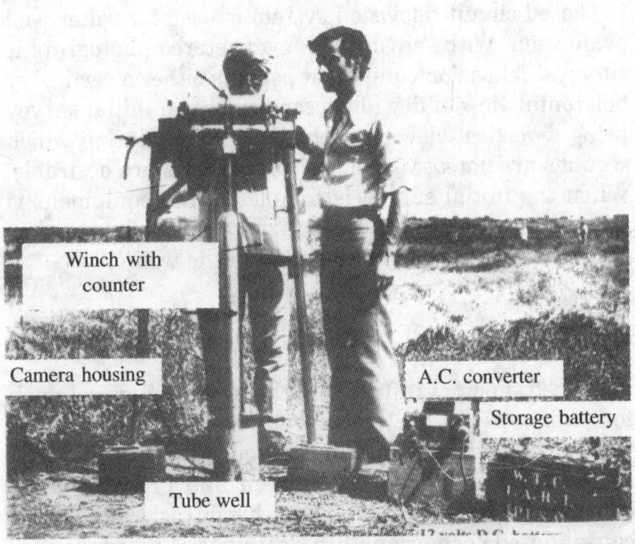

Fig. 2.82. The three-dimensional stereo camera being lowered into a tube well.

(*ii*) Entrance velocity of water into tube wells is kept low by providing greater area of opening in the screen section.

(*iii*) Well screens are cleaned periodically by subjecting the tube well to development, once in two or three years.

When the yield of wells get decreased in spite of the preventive measures stipulated above, it is necessary to undertake suitable rehabilitation procedures. These include acid treatment, using hydrochloric acid, sulphamic acid, glassy phosphates, chlorine or other material, depending on the cause of incrustation. Acid treatment is effective in removing carbonates. When calcium and magnesium silicates or sulphates form the basic binder, they are not treated by the acid treatment method as acid has no action on them. Glossy phosphates are effective in removing iron and manganese as well as silt and clay. Bacterial growth and slime can be removed by chlorine treatment. For detailed procedures for various treatments, reference is invited to the book *Water Well and Pump Engineering* listed at the end of this chapter.

Photographic Evaluation of Well Failures

With the expansion of photographic technology, photographs can be produced to clearly show specific problems that might exist in a well. Study and evaluation of these photographs assist in the selection of appropriate corrective methods. Well repair expenses can be significantly reduced if downhole photographs are used. The equipment consists of a bore hole camera provided with a flash arrangement which can be lowered into a tube well to take photographs (usually, stereo-photographs) at any desired depth (Fig. 2.82). Photographs may be made from black and white or colour films. The colour photographs have been used in uncased holes to identify formation materials. The best colour photographs of formation materials are obtained in dry holes. However, colour photographs of bore holes containing clear water can also be used for formation identification. Stereo-photographic surveys made of new wells provide a visual inspection of the entire well. These surveys suggest to the controlling agency and well owner, whether the well has been completed in accordance with guaranteed plans and specifications or not. However, the surveys cannot show detailed features of grouting, gravel pack and screen placement versus formation changes. Such photographic records of new wells can be used for comparison with photographs taken on a later date;

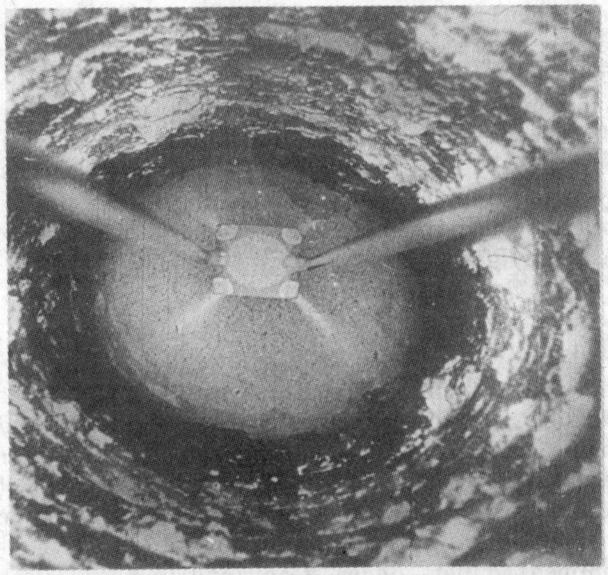

Fig. 2.83. Underwater view showing the inside of a screen tube well which is clogged by incrustation. Note the fluffy deposits blocking the slot openings of the well screen.

Fig. 2.84. Inside view of a tube well showing partial blocade by foreign material. Note the broken screen and the rod blocking the well.

conclusions may be drawn for well rehabilitation. These comparisons also provide a basis for determining the endurance of the casing and screen materials. Subsequent surveys can show the apparent effectiveness of the chemical rehabilitation of the well screen. Well failures by blocade can be clearly identified by photographic surveys, including their precise location which would suggest the remedial measures to be taken (Figs. 2.83 and 2.84).

Closed circuit television system is used for water well evaluation with advantage over stereo-photographic surveys. Television equipment permits either a vertical or horizontal view of downhole conditions. An initial survey using a vertical-viewing probe is common, during which sections are noted, where horizontal views are desirable. When the initial survey is completed, the equipment is hoisted to the surface; the horizontal viewing mirror is attached to the cable and inserted in the well.

SUMMARY

Importance of efficient management of groundwater is highlighted. Types of water bearing formations and their characteristics are explained. Groundwater recharge procedures are described. Methods of groundwater exploration are introduced. Hydraulic characteristics of unconfined, confined and semi-confined aquifers under steady and unsteady state conditions are described. Procedures for determining hydraulic characteristics of aquifers are presented along with appropriate mathematical relationships and examples using data from field tests. Methods of conducting pumping tests of obtain information on the characteristics of water bearing formations are described with requisite procedures. Design procedures for open wells in hard rock and semi-consolidated formations are explained. Criteria for the design of tube wells including component parts are described. Various types of well screens and their adaptability are explained. Well screen location in relation to aquifers are presented. Effects of partial penetration of wells influencing well yield in confined aquifers are explained. Specifications of well screens to suit varying aquifer characteristics and materials for their construction are described. Factors influencing the design of gravel pack for tube wells are described along with related procedures. Adaptability and principles of multiple well systems are described. Equipment and procedures for construction of open wells in hard rock and semi-consolidated formations are described. Procedures for boring and deepening of open wells are described. Tube well drilling equipment and procedures for the construction of tube wells in varying types of groundwater formations are explained. Components of cable tool percussion drills, direct and reverse rotary drills, and air rotary drills including down-the-hole hammer drills are described. Well construction procedures using different types of drilling and boring equipment are explained. Well log based installation of well screens are described. Procedures for dealing with brackish and saline aquifer formations are described. Equipment and procedures for the development of tube wells are described. Procedures for testing of wells for yield and drawdown are explained. Common troubles in the functioning of wells are presented along with the requisite management and rehabilitation procedures, including photographic evaluation of wells.

PROBLEMS

2.1. A well penetrating aquifer which is underlain and overlain by impermeable layers was tested with a uniform discharge of 1000 litres/min. The steady-state drawdowns measured in two observation wells which were at 1 m and 10 m radial distances from the centre of the pumped well were 13.40 m and 4.2 m, respectively. Determine the hydraulic properties of the aquifer, if its saturated thickness is 10 m.

[**Ans.** Transmissibility = 0.039 m^2/min.
Hydraulic conductivity = 0.00397 m/min.]

2.2. A 10 cm diameter well was pumped at a uniform rate of 500 litres/min, while observations of drawdown were made in an observation well located at a distance of 50 m from the well. The original head of water, measured from the top of the impervious layer, was 25 m. The hydraulic conductivity of the aquifer was 1.83 ञ 10^{-3} m/min. Determine the drawdown at the face of the well, using Dupuit-Thiem equation and assuming that the flow to the unconfined aquifer is under steady-state.

[**Ans.** 20 m]

2.3. A semi-confined aquifer was tested with a constant discharge of 550 litres/min. The thickness of the semi-confining layer was 8.0 m. Evaluate the hydraulic properties of the aquifer from the following information and assuming that the phreatic drawdowns were negligible compared to piezometric drawdowns:

(*i*) Size of well = 10 cm
(*ii*) Piezometric level measured from the ground surface = 7 m
(*iii*) Hydraulic conductivity of the phreatic aquifer above the semi-impervious layer = 1.5 cm/min
(*iv*) Match point co-ordinates:

$K_0\ (r/B) = 1$
$(r/B) = 1$
$r = 500$ m
$s = 0.10$ m

[**Ans.** T = 08763 m^2/min; K= 0.1095 m/min
B = 500 m; c = 2852.9 min]

2.4. A well penetrating an aquifer (b = 20 m) which is overlain by a semi-confining layer and underlain by an impermeable layer (aquiclude) was pumped at a constant discharge by super-imposing the observed data sheet over De-Glee's type curve:

K_0 (r/B) = 0.01

(r/B) = 1

r = 1,000 m

s = 0.5 m

Evaluate the hydraulic properties of the aquifer and the semi-confining layer.

[Ans. T = 0.2 m^2/min

K = 0.1 m/min

B = 1000 m

c = 5,00,000 min]

2.5. A 15 cm diameter well penetrating an artesian aquifer was tested with a constant discharge of 800 litres/min. The drawdowns measured in three observation wells spaced 10 m, 50 m, and 100 m away from the well are given below:

Time since pumping started (min)	*r = 10 m Drawdown (m)*	*r = 50 m Drawdown (m)*	*r = 100 m Drawdown (m)*
1	0.595	—	—
2	0.605	—	—
3	0.650	—	—
5	0.800	—	—
10	0.905	0.450	0.100
20	0.955	0.500	0.300
40	1.312	0.550	0.400
60	1.565	0.605	0.450

Determine hydraulic properties of the aquifer using Theis' method

[Ans. T = 0.3302 m^2/min,

$S = 198 \times 10^{-4}$]

2.6. An aquifer overlain by a semi-pervious layer and underlain by an impervious layer was tested with a uniform discharge of 0.3 m^3/min. The diameter of the well was 10 cm and the depth of the water bearing formation was 0.7 m. The measurements of the drawdowns made after steady-state conditions were obtained in various observation wells were as follows :

Distance of the observation well from the centre of the test well (m)	10	50	100	200	400
Drawdown in deep observation well (m)	0.254	0.202	0.1510	.100	0.070
Drawdown in shallow observation well (m)	0.080	0.050	0.005	0.0	0.0

Evaluate hydraulic properties of the aquifer and of the semi-confining layer using De-Glee's method of analysis.

[Ans. T = 0.873 m^2/min

B = 1150 m, c= 1514897.3 min]

2.7. A pumping test was conducted in an aquifer in which the upper confining layer was semi-pervious and the lower layer was impervious. The constant discharge of the well was 250 litres/min. The thickness of the semi-pervious layer was 10 m and that of the aquifer was 15 m. The drawdowns at 0.5, 1, 2, 3 and 5 minutes duration were 0.15, 0.20, 0.30 and 0.50 metres respectively. The drawdowns were obtained in an observation well which was at 10 m distance from the centre of the test well. Assuming that the drawdowns in the layer above the semi-imperious layer were negligible, evaluate the hydraulic properties of the aquifer, using Hantush Jacob's methods of analysis.

[Ans. T = 0.639 m^2/min

$S = 5.6 \times 10^{-5}$

B = 200 metres

c = 62597.8 min]

QUESTIONS

1. Distinguish between:
   (*i*) Lithosphere and Lithology
   (*ii*) Zone of saturation and zone of aeration
   (*iii*) Soil water zone and capillary fringe
   (*iv*) Confined and unconfined aquifers
   (*v*) Water table and piezometric surface
   (*vi*) Artesian and water table aquifers
   (*vii*) Phreatic head and piezometric head
   (*viii*) Electric logging and gamma ray logging
   (*ix*) Electrical resistivity surveying and seismic refraction surveying
   (*x*) Hydraulic conductivity and transmissibility
   (*xi*) Specific retention and specific yield
2. Define porosity of an aquifer. If a sample of 1 m^3 of sandstone contains 0.12 m^3 of water, what is the porosity of the formation?
3. An aquifer comprising of sand and gravel has a porosity of 20%. A 1 m^3 volume of the aquifer yields 0.11 m^3 of water. What are the values of specific yield and specific retention?
4. What is a perched water table?
5. Define well logging. What is its purpose?
6. Define apparent resistivity of a geological formation.
7. What are the specific requirements in locating open wells in hard rock areas where dykes are encountered?
8. What is the major consideration in choosing the diameter of housing pipes in tube wells?
9. How is the width of slot determined in the design of a tube well?
10. What is the recommended relationship between the total surface area and the area of opening of a well screen?

11. What are agricultural strainers? What are their advantages and limitations?

12. Distinguish between naturally gravel packed and artificially gravel packed tube wells.

13. What are the basic requirements of the underground formations which limit the adaptability of cavity wells?

14. In the analysis of aquifer particles, what is the effective size?

15. Which groundwater formations require gravel pack? What is P-A ratio?

16. How is mudcake formed in rotary drilling? How does it help the drilling process?

17. How are lost circulation zones formed during well drilling? How are they overcome in well drilling?

18. Distinguish between static water level and pumping water level.

19. What is leakage factor in an aquifer? Is it applicable to all types of aquifers?

20. What are the specific properties of underground formations favouring deepening and boring of open wells to increase well yield?

21. How is quicksand condition, which may occur in certain formations, stabilized during the construction and operation of open wells?

22. How does the diameter of tube well influence well yield?

23. What are the various types of losses in well screens?

24. What is uniformity coefficient, as applied to design of gravel packs in wells? How is the concept used in the design of gravel packs?

25. In open wells, what formations favour horizontal boring to increase well yield?

26. Describe the anchor bolt method of sinking casing pipes in percussion drilling.

27. What is 'finishing job' in well drilling? What are the common types of fishing tools?

28. What are the basic principles of direct rotary well drilling?

29. What are the formation and site conditions favouring the use of reverse rotary drills for the construction of tube wells?

30. Distinguish between air rotary drilling and down-the-hole hammer drilling. What are the underground formations where down-the-hole drilling is to be preferred?

31. What are the common problems encountered in well drilling? How are they overcome?

32. What are the commonly used procedures for the development of screen type tube wells?

33. What are specific procedures for the development of cavity wells?

34. How do the performance characteristics of a shallow well differ from that of a deep well?

35. What are the common troubles experienced in the operation of tube wells? How are they overcome?

36. What is cathodic protection of mild steel tube wells in corrosive groundwater?

37. What is incrustation in tube wells? What are the rehabilitation procedures to protect tube wells against incrustation?

38. How do photographic evaluation help in rehabilitation of sick and failed wells?

39. Identify the *True* and *False* statements from the following:

(*i*) The increase in the yield of a tube well by increasing the diameter of the well is more in a confined aquifer than in a water table aquifer.

(*ii*) The drawdown while pumping a tube well is identical in partially penetrating and fully penetrating wells.

(*iii*) In gravel-packed tube wells, the entrance velocity of water is determined by taking the average values of the hydraulic conductivity of the aquifer and the gravel pack.

(*iv*) In the analysis of aquifer materials the 'per cent retained' is plotted on the x-axis and the 'particle size' on the y-axis.

(*v*) Roller type drill bits used in direct rotary drilling are specially adapted to drilling in hard formations.

(*vi*) A limiting factor in the use of reverse rotary drilling is the need for having a fairly large supply of water for drilling.

(*vii*) Quartz, which is most common amongst rock forming minerals, is the hardest to drill, causing heavy wear on the drill bit.

(*viii*) To solve steady state flow to wells in confined aquifers, the assumptions made by Dupuit have been more refined than those made by Thiem.

(*ix*) The hydraulic properties of a semi-confined aquifer for steady state conditions can be evaluated either by the De-Glee's method or the Hanthush and Jacob's method.

(*x*) In the operation of a direct rotary drill, the 'used drilling fluid', which is collected in the settling pit, is pumped back to the drilling system for reuse.

(*xi*) The collapse of a drilled, hole by external pressure during well drilling with a rotary drill is prevented by the upward stream of drilling fluid.

(*xii*) In reverse rotary drilling, the desirable velocity of water inside the drill pipe to lift coarse cuttings is about 60 m/min.

40. Identify the most appropriate answer of the following questions from the four options:

(*i*) The geophysical method of groundwater exploration which is suitable for both cased and uncased wells is:

(a) Electrical resistivity method
(b) Electric logging
(c) Gamma ray logging
(d) Seismic refraction surveying

(*ii*) Keeping the hydraulic properties of the aquifer as constant, doubling the diameter of the tube well will increase the well yield by about:
(a) 7% (b) 9%
(c) 11% (d) 15%

(*iii*) In a homogeneous artesian aquifer, the total length of the well screen as a percentage of thickness of the aquifer being tapped should range between:
(a) 50 – 60% (b) 60 – 70%
(c) 70 – 80% (d) 80 – 90%

(*iv*) In the design of a tube well, the velocity of water entering a well screen should be kept about:
(a) 1 cm/sec (b) 3 cm/sec
(c) 5 cm/sec (d) 7 cm/sec

(*v*) In drilling tube wells with direct rotary drills, the drilling speeds range from:
(a) 30 to 150 rpm (b) 150 to 300 rpm
(c) 300 to 500 rpm (d) 500 to 750 rpm.

(*vi*) In reverse rotary drilling, the drilling, fluid is recovered using:
(a) Reciprocating pump
(b) Centrifugal pump
(c) Rotary pump
(d) Propeller pump

(*vii*) In reverse rotary drilling, the desirable velocity of water inside the drill pipe to lift fine cuttings is about:
(a) 40 m/sec (b) 60 m/sec
(c) 80 m/sec (d) 100 m/sec

(*viii*) The drilling fluid used in air rotary drilling is:
(a) Clear water
(b) Water mixed with bentonite
(c) Water mixed with other additives
(d) Compressed air.

(*ix*) The most suitable equipment for well drilling in hard rock areas is:
(a) Cable tool drilling
(b) Direct rotary drilling
(c) Down-the-hole air rotary drilling
(d) Reverse rotary drilling.

(*x*) The optimum length of a well screen for a tube well in a confined aquifer should extend to:
(a) Full depth of the aquifer
(b) 50–60% of the depth of the aquifer
(c) 60–70% of the depth of the aquifer
(d) 70–80% of the depth of the aquifer.

(*xi*) The slot width of a well screen should be limited to:
(a) 1.0 – 1.5 mm (b) 1.5 – 5.0 mm
(c) 5.0 – 7.5 mm (d) 7.5 – 10.0 mm

(*xii*) The open area of a well screen as a percentage of the surface area of the pipe is:
(a) 10% (b) 15%
(c) 20% (d) 25%

(*xiii*) The drilling speed in a rotary drilling machine ranges from:
(a) 30 – 150 rpm (b) 150 – 250 rpm
(c) 250 – 500 rpm (d) 500 – 750 rpm

(*xiv*) The ideal return velocity of the fluid in the annular space of a reverse rotary drilling machine is:
(a) 20 – 40 m/min (b) 40 – 60 m/min
(c) 60 – 80 m/min (d) 80 – 100 m/min

(*xv*) The type of drill bit most suitable to be used with a direct rotary drilling machine to cut through rock formations, with a tendency to become crooked is:
(a) Three-way drag bit
(b) Six-way drag bit
(c) Three-cone roller bit
(d) Gross roller bit

(*xvi*) For fishing, a drill pipe dropped accidentally during rotary drilling, a suitable fishing tool is:
(a) Pipe locking tool
(b) Tapered tap
(c) Die overshot
(d) Circulating slip overshot

(*xvii*) The maximum hook load capacity of a medium duty rotary drill ranges from:
(a) 8000 to 13,000 kg
(b) 13,000 to 18,000 kg
(c) 18,000 to 23,000 kg
(d) 23,000 to 28,000 kg

41. In the following, match each of the entries of the first column with the most closely related entry in the second column:

(*i*)	(A) Lithosphere	(I) Confining formations
	(B) Zone of aeration	(II) Localised water body
	(C) Perched water table	(III) Geophysical method of groundwater exploration
	(D) Electric logging	(IV) Earth's crust which is only partially filled with water
	(E) Artesian aquifer	(V) Disintegrated rock material
(*ii*)	(A) Specific capacity	(I) Water storage properties of confined and semi-confined aquifers
	(B) Transmissibility	(II) Property of unconfined aquifer
	(C) Specific yield	(III) Semi-pervious layer of a semi-confined aquifer

	(D) Hydraulic resistance	(IV) Yield per unit depth of drawdown
	(E) Coefficient of storage	(V) Rate of movement of water through a vertical strip of the aquifer
(*iii*)	(A) Henri Darcy (1856)	(I) Hydraulic properties of aquifers
	(B) Dupuit (1863)	(II) Unsteady state flow to wells in confined aquifers
	(C) Adolf Theim (1870)	(III) Unsteady state radial flow in non-leaky confined— aquifers in polar coordinates
	(D) Thies (1935)	(IV) Hydraulic conductivity
	(E) Jacob (1940)	(V) Unconfined and confined aquifers
(*iv*)	(A) Stratification	(I) Safe limit of entrance velocity of flow into wells
	(B) Dykes in hard rock formations	(II) Slotted pipes with pre-pack filters
	(C) Bennison (1947)	(III) Negative indicator of occurrence of groundwater
	(D) Ashim filters	(IV) Size ratio of gravel pack to aquifer
	(E) Pack-aquifer ratio	(V) Difference in vertical and horizontal components of hydraulic conductivity of aquifers
(v)	(A) Dragline excavator with orange-peel type bucket	(I) Sinking of casing pipes in cable tool method of well drilling
	(B) Revitalization hole	(II) Straightening crooked well holes in rotary drilling
	(C) Anchor bolt method	(III) Holding of loose particles in place while drilling with rotary drills
	(D) Cross-roller rock bit	(IV) Augmentation of yield of open wells
	(E) Filter cake	(V) Construction of open wells

FURTHER READING REFERENCES

Ahrens, T.P. (1958). *Water Well Engineering.* Water Well Jl., 12, December 1958.

Ahrens, T.P. (1970). *Basic Considerations in Well Design,* Parts I, II and III. Water Well Jl. April, May and June 1970.

Anjaneyulu, B. (1972). *Hydraulics of Cavity Wells.* Ph.D. Thesis, Dept. of Agrl. Engg., Indian Instt. of Technology, Kharagpur.

Anonymous (1964). *Steady Flow of Groundwater Towards Wells.* Proc. Commonwealth Hydraulic Research T.N.O. No. 10.

Anonymous (1966). *Procedure Used in Gravel Packing Wells.* Johnson Drillers Jl. 38 : 2.

Appa Rao, V. (1971). *Groundwater Aquifers in Cuddapah Basin.* Bulletin of the National Geophysical Research Institute, Hyderabad-7, Vol. 9, Nos. 3 and 4, NCRI Centrobution No. 293. pp. 113–122.

Atlas Copco A.B., Stockholm, Sweden, Anderson SB. et al (Eds.) Atlas Copco Manual, p. 585.

Bangladesh Agricultural Development Corporation (1984). *Well Design and Construction Manual Guide to Supervision Staff.* Sir M. McDonald & Partners Ltd. and Hanting Technical Service Ltd. for IDA Tube Well II, Project II, Irrigation Mangt. Programme, p. 82.

Bennison, E.W. (1947). *Groundwater.* E.E. Johnson, St. Paul Minn., U.S.A., p. 500.

Borewell Equipment Co. Hyderabad (1983). *BEC Model 160 Inwell Rig (2p.), BEC Torpedi (2p.) BEC Model 500 Challenger, Buttom bits (4p.) Leaflets.*

Boulton, N.S. (1963). *Analysis of Data from Non-equilibrium Tests Allowing for Delayed Yield from Storage.* Proc. Instt. Civ. Eng., Vol. 26: 469–482.

Campbell, M.D. and J.H. Lehr (1973). *Water Well Technology.* McGraw-Hill Book Co.

Central Groundwater Board (1986). *Manual on Analysis of Pumping Test Data of Large Diameter Wells.* Ministry of Water Resources, Govt. of India, Faridabad, Tech. Series M, No. 2, p. 33.

Chakraborty, M.N. (1976). *Information on Polythene Filters.* Chatterjee Filter Industries, Calcutta.

Chow, V.T. (1952). *On the Determination of Transmissivity and Storage Coefficients from Pumping Test Data.* Am. Geophys. Union Trans., Vol. 33: 397–404.

Clark, F.E. (1980). *Corrosion and Incrustation in Water Wells—A Field Guide for Assessment, Prediction and Control.* Pub. Food and Agriculture Organization of the United Nation FAO Irrigation and Drainage Paper 34. FAO, Rome, p. 95.

Cooper, H.H. and Jacob, C.E. (1946). *A Generalised Graphical Method for Evaluating Formation Constants and Summarising Well Field History.* Am. Geophys. Union Trans., Vol. 27: 526–534.

Corey, C.L. (1949). *Hydraulic Properties of Well Screens.* M.S. Thesis (Unpublished) Colorado Agri. and Mech. College, Fort Collins, U.S.A.

Dakshinamurthi, C., Michael, A.M. and Shri Mohan (1973). *Water Resources of India and Their Utilisation in Agriculture.* Water Technology Centre, I.A.R.I., New Delhi, p. 398.

Darcy, H.C. (1956). *Less fontaines publiques de la ville de D. Yon.* V. Dalmont, Paris, p. 647.

Das, S.K. (1983). *Equipment—Indian Scene.* Paper presented in 2nd All India Conf. on Drilling and Drilling Equipment. Institution of Public Health Engineers (India), New Delhi, p. 13.

De-Glee, G.J. (1930). *Over groundwaterstromingen bij water onttrekking door middle van putten.* J. Waltman, Delf (The Netherland), p. 175.

Department of Agriculture, Ministry of Food and Agriculture, Govt. of India (1968). Brochure on Drilling Equipment, p. 46.

Department of Army and Air Force, USA (1957). *Manual on Wells. Pub.* U.S. Govt. Printing Press, Washington D.C., U.S.A., p. 264.

Drillmark India Ltd. (1971). Instructions for Reverse Rotary Drilling, p. 10.

Driscoll, F.G. (1986). *Groundwater and Wells.* 2nd ed. Pub. Johnson Division, St. Paul, Minnesota U.S.A., p. 1989.

Dupuit, J. (1963). *Etuds theoriques et pratiques sur le movement dex eaux dans les canaux de couverts et a travers les terrains permeables.* Zeme edition; dunot, Paris, p. 304.

Dylla, A.S. and Muckel, D.C. (1967). *Experimental Development of Shallow Groundwater Wells.* Pub. Univ. Nevada and USDA, p. 23.

Edward, E. Johnson Inc. (1966). *Groundwater and Wells.* Pub. Edward E. Johnson Inc., Saint Paul, Minnesota, U.S.A., p. 440.

Forchheimer, P. (1898). *Grundwasserspiegel bei brunnesan lagen.* Zeitschrift, Der Osterreichesche Ingenieur unel 629–35, 645–49.

Garden, R.W. (1958). *Water Well Drilling with Cable Tools.* Pub. Bucyrus-Erie Co., South Melwankie, Wisconsin, p. 230.

Garg, S.P. (1978). *Grand Water and Tube wells.* Oxford and IBH Pub. Co., New Delhi, p. 333.

Garlick and Co. Pvt. Ltd. (1973). *Information on Well Drilling Equipment.* Personal correspondence.

Geology Dept., Osmania University, Hyderabad (1975). *Optimisation of Groundwater Utilization through Wells and Pumps.* Annual Technical Progress Report, p. 62.

Gujarat Engg. Res. Instt., Vadodara (1978 and 1988). Annual Progress Reports of ICAR Cood Res. Project on Wells and Pumps.

Gupta, S.N. (1964). *Design of Gravel Envelopes for Tube wells,* T.M. No. 34–RR (G–69) U.P. Irri. Res. Instt., Roorkee.

Hall, H.P. (1954). *A Historical Review of Investigations on Seepage Towards Well.* Jl Boston Soc. Civ. Engrs. Vol. 41, No. 3.

Hantush, M.S. (1956). *Analysis of Data from Pumping Tests in Leaky Aquifers.* Am. Geophys. Union Trans., Vol. 37: 702–714.

Hantush, M.S. and Jacob, C.E. (1955). *Non-steady Radial Flow in an Infinite Leaky Aquifer.* Am. Geophys. Union Trans., Vol. 36: 95–100.

Hazen, A. (1893). *Some Physical Properties of Sands and Gravels with Special Reference to Their Use in Infiltration.* 24th Annual Report of the State Board of Health of Massachusetts, 1892, Public Document No. 34, Write and Potter Printing Co., U.S.A., p. 553.

Huisman, L. (1972). *Groundwater Recovery.* Macmillan Press, London, p. 336.

Indian Standards Institution (1985). *Requirements for Well Screens and Slotted Pipes* IS:8110–1985. ISI, Manak Bhawan, New Delhi.

Indian Standards Institution (1983). *Specification for Steel Tubes Used for Water Wells.* IS: 4270–1983. ISI, Manak Bhawan, New Delhi.

Indian Standards Institution (1979). *Code of Practice for Construction and Testing of Tube wells:* Part I *Construction*; Part II *Testing* IS: 2800 (Part I and II)– 1979. ISI, Manak Bhawan, 9 Bahadur Shah Zafar Marg, New Delhi–110002.

Indian Standards Institution (1985). *Method of Tube well Development* IS:11:89–1985. ISI Manak Bhawan, New Delhi.

Jacob, E. C. (1940). *On the Flow of Water in a Plastic Artesian Aquifer.* Am. Geophys Union Trans., Vol. 72, Part II: 574–586.

Jahnke, E. and F. Embde (1945). *Tables of Functions with Formulas and Curves.* Dover Publications, New York, p. 306.

Johnson, C.N. (1954). *Irrigation Well and Well drilling Methods in California.* Uni. of California, Agrl. Expt. Stn., Berkeley, Circular No. 361, p. 37.

Johnson Division (1968). *Principles and Practical Methods of Developing Water Wells.* Bull. No. 1033, St. Paul, Minnesota, p. 24.

Johnson Division (1967). *Corrosium and Incrustation of Well Screens.* UOP Johnson Division, St. Paul, Monnesota, U.S.A.

Kennedy, M.R. (1970). *Bore Wells in the Hard Rock and the Equipment Used.* Proc. Summer Instt. on Tube well Technology, Div. of Agrl. Engg., I.A.R.I., New Delhi.

Khepar, S.D. and Sondhi, S.K. (1973). *Equipment for Radial Boring in Open Wells.* Jl. Agl. Engg., Vol. 10, No. 4, pp. 42–43.

Khepar, S.D., Arora, S.K. and Sondhi, S.K. (1971). *Studies on Engineering Aspects of Design and Constructions of Open Wells and Tube wells.* Annual Progress Report: Dept. of Agrl. Engg., Punjab Agrl. Uni., Ludhiana, p. 48.

Khepar, S.D., Prakash, A. and Boundurant, B. (1972). *Mutual Interference of Wells In Leaky Aquifers.* J. Agril. Engrs. 9: 40–49.

Koegel, R.G. (1971). *Self Help Wells.* Pub. Food and Agriculture Organisation of the United Nations FAO, Irrigation and Dramage, 30, p. 78.

Koul, R.K. (1974). *Hydraulics of Open Wells and Shalow Tube wells.* M.Sc. Thesis, Div. of Agrl. Engg., I.A.R.I., New Delhi, p. 145.

Krishna Rock Drills (Pvt.) Ltd. Secunderabad, Andhra Pradesh (1983). *Information on down-the-hole hammers drills.* Leaflets.

Linsley, R.K. and Franzini, J.B. 1964. *Water Resources Engineering.* Pub. Mc-Graw Hill Book Co. Inc. New York, p. 65.

Micheal A.M. and Khepar, S.D. (1989). *Water Well and Pump Engineering.* Pub. Tata McGraw Hill Publishing Co. Ltd., West Patel Nagar, New Delhi–110008

Ministry of Food and Agriculture, Govt. of India (1963). *Handbook on Boring and Deepening of Wells,* p. 115.

Ministry of Water Resources, Govt. of India (1986). *Groundwater Development in India.* p. 73.

Mohandas B. (1982). *Corrosim Studies in the Coastal Artesian Belt of Tamil Nadu.* M.E. Thesis, College of Engineering, Guindy, p. 120.

Murty, V.V.N., Jain, K.C. and Pandey, R.N. (1973). *Tube-Wells.* Pub. Punjab Agricultural University, Ludhiana, p. 57.

Muskat, M. (1937). *Flow of Homogenous Fluids Through Porous Media.* McGraw Hill Book Co Inc., New York.

Muskat, M. (1949). *Physical Properties of Oil Production.* McGraw Hill Book Co. Inc., New York.

National Bank for Agriculture and Rural Development (1993). *Borewells in Jodhpur District, Rajasthan—An Ex-Post Evaluation Study.* Evaluation Study Series, Jaipur–4, p. 52.

Pandey, D. (1974). *Construction of Water Wells—Various Methods and Machinery.* Central Groundwater Board. Mimeo., p. 23.

Peterson, J.S., Rohwer, C. and Albertson, M. L. (1955). *Effect of Well Screen on Flow Into Well.* Trans. Am. Soc. Civil Engrs., 120: 563–606.

Ram Krishnan, S. (1998). *Groundwater.* Pub. S. Bawakreyhnan H 34/F TNHB Colony, Chennai, Tamil Nadu, p. 761.

Raman, R.J, Rangaswamy, M., Chandrasekaram V. Ramalingam, M. (1984). *Analysis of Design Requirement of Open Wells in Tamil Nadu.* ICAR Cord. Project on Wells and Pumps, Institute of Hydraulics and Hydrology, Pondicherry, Tamil Nadu.

Rao, J. Vijayam, B.E., Raju, S. and Papi Reddy, M. (1975). *Optimum Design of Dug Wells in Hard Rock Terrains.* Indian Farming.

Rewathi Machinery Traders, Coimbatore (1970). *Instruction Manual for MAH - 1200 Rig,* p. 40.

Richards, L.S. (1952). *Report of the Sub-Committee on Permeability and Infiltration.* Committee on Terminology, Soil Science Society of America. Soil Science, Vol. 16: 85–88.

Rohwer, C. (1940). *Putting Down and Developing Wells for Irrigation.* U.S. Dept. of Agri., Washington, D.C., Cir. No. 546, p. 85.

Sahai, M.M. (1975). *Bamboo Boring—A Cheap Irrigation Device.* Paper presented in Annual Conf. of Indian Soc. of Agrl. Engrs. Hyderabad.

Saran, R.C. (1961). *Selection of Well Screens and Gravel Packs, and the Development of Model Technique for Design of Tube-Wells.* T. M. 31, R. R. (G-41). Irrigation Res. Instt., Roorkee.

Sankarnaryana Reddy, N. (1972). *Drilling Equipment and Well Drilling Methods.* The Madras Agrl. Jl. Vol. 59, No. 2, pp. 94–102.

Sharma, H.D. (1972). *Research Studies on the Engineering Aspects of Design and Construction of Open Wells and Tube* wells. Annual Prog. Report 1971–72, U.P., Irrigation Report Research Institute, Roorkee, p. 21.

Sharma, H.D. and Chawla, A.S. (1977). *Manual of Groundwater and Wells.* Tech. Rep. No. 18, Cent. Bd. of Irrigation and Power, p. 200.

Shah, S.M., Patel, S.D. and Joshi, A.R. (1974). *Development of New Screens for Tube wells.* Report No. TW4/1974 Tube well Division, Jl. Gujarat Engineering Research Institute, Baroda, p. 25.

Slichter, C.S. (1899). *Theoretical Investigation of the Motion of Groundwaters.* 19th Annual Report, U.S.G.S., 295–384.

Smith, H.F. (1954). *Gravel Packing is Water Wells.* Water Well Jl., 8 (1): 31–34.

Srivastava, R. (1969). *Evaluation of Hydraulic Characteristics Influencing Shallow Tube wells.* M.Sc. Thesis, Div. of Agrl. Engg., P.G. School, I.A.R.I., New Delhi, p. 104.

Tandon, R.N. (1970). *Design of Tube wells in Alluvial Strata.* Proc. Summer Instt. on Tube well Technology, Dev. of Agrl. Engg., I.A.R.I., New Delhi.

Theis, C.V. (1935). *The Relation Between Lowering of Piezometric Surface and the Rate and Duration of Discharge of a Well Using Groundwater Storage.* Am. Geoph. Union Trans., Vol. 16: 513–524.

Thiem, A. (1870). *Die Ergiebigkeit artesischer Bohrlocher, Schachtbrunnen, and Filter Gallerign.* Journal fur Casbelevchtung and Wasserversorgung, Munich, Vol. 14, pp. 450–467.

Todd, D.K. (1959). *Groundwater Hydrology.* Pub. John Wiley & Sons, Inc., New York, p. 336.

Tribeni Sales Corporation, New Delhi (1974). *Ashim Filters.* Leaflet.

United Nations Educational, Scientific and Cultural Organisation Regional Office, New Delhi (1981). *Groundwater in Hard Rocks.* Proc. Intl. Regional Seminar. Coimbatore, Dec. 1999. Pub. UNESCO Regional Office for South and Central Asia, New Delhi, p. 203.

United States Department of the Interior, Bureau of Reclamation (1985). *Groundwater Manual.* U.S. Govt. Printing Office, Denver, U.S.A.

Walton, W. C. (1962). *Selected Analytical Methods for Well and Aquifer Evaluation.* Illinois State Water Survey, Bull. No. 49, p. 81.

Walton, W.C. (1970). *Groundwater Resource Evaluation.* McGraw-Hill Book Co., New York, p. 664.

Water Development Society, Hyderabad (1983). *Water Well DTH Rig. The Faster Driller, WDM 100, Supertrack WDC 150.*

Wenzel, L.K. (1942). *Methods for Determining Permeability of Water Bearing Materials with Special Reference to Discharging Well Methods.* U.S. Geological Survey, Water Supply Paper, 887.

Wenzel, L.K. and Greenlee, A.L. (1943). *A Method for Determining Transmissibility and Storage Coefficients by Test of Multiple Well System.* Trans. Am. Geophy. Union 24: 547–64.

Widia India Ltd. Bangalore (1983). *Widia Button Bits Regrinding Manual* (41 p.), Widia Rock Roller Bits for Blast Hole Drilling (8 p.) Widia Countersunk Cutter Picks Leaflets.

Yadav, R.K. (1973). *Hydraulics of Shallow Tubewells in Semi-Confined Aquifer under Multiple Operation.* Ph.D. Thesis (unpublished), P.G. School, Indian Agricultural Research Institute, New Delhi, p. 344.

Chapter

3

Water Lifts and Pumps

LEARNING OBJECTIVES

Principles of operation and adaptability of indigenous water lifts and pumps. Commonly used indigenous water lifts. Application of non-conventional energy sources and appliances in pumping—windmill, hydraulic ram, solar photovoltaic pump, biogas engine powered pumps. Positive displacement pumps—shallow well and deep well reciprocating pumps and their performance. Variable displacement pumps. Specific speed of pumps and their significance in the design of variable displacement pumps. Pump characteristic curves and their use in the selection of pumps for particular applications. Criteria for the design of centrifugal pumps. Principles of operation of centrifugal pumps. Types of centrifugal pumps and their adaptability. Medium lift submersible centrifugal pumps with hydraulic drive—their principles of operation and adaptability. Electrical connections and devices used in the installation of pumping sets. Vertical turbine pumps—their principles of operation and component parts. Oil lubricated and water lubricated vertical turbine pumps and their adaptability. Submersible pumps—their principles of operation, component parts and adaptability. Installation, operation, maintenance and trouble shooting of centrifugal pumps, vertical turbine pumps and submersible pumps. Propeller and mixed flow pumps—their principles of operation, component parts and adaptability. Jet pumps—their operating principles and adaptability. Airlift pumps—their operation, adaptability and performance. Criteria for selection of irrigation pumps. Fitting pumps and wells based on their performance characteristics. Power requirement, efficiency and economics of irrigation pumping plants. Pumping plant design. Economical pipe size selection.

Lift irrigation requires that water be raised from its source to the field surface. Whatever the source of water in a lift irrigation project, the efficiency of the system depends on the application of sound principles in the design and construction of the utilization structure, usually the well and the characteristics of the water lifting device in relation to the source of water. Devices for irrigation water lifting range from age-old indigenous water lifts to highly efficient pumps. Pumps operated by electric motors or engines have come into prominence in all large scale lift irrigation schemes. This is because high output and efficiency levels can be attained and controlled, using mechanically-powered water lifts.

Basically, there are four principles involved in pumping water. These are atmospheric pressure, positive displacement, centrifugal force, and movement of columns of water caused by the difference in specific gravity. Numerous kinds of pumps are available, all of which embody one or more of these principles. The classification of commonly used irrigation water lifts is shown in the schematic chart on the following page.

Selection of a suitable water lifting device for a particular situation depends on the characteristics of the source of water and the lifting device, the amount of water to be lifted, the depth to the pumping water level, type of power available and the economic status of the farmer.

Man and animal power are the major energy sources in small-scale irrigation pumping and domestic water supply in many regions of developing countries. From a purely economic standpoint, the cost of raising water with man or animal power is substantially higher than those which are driven by electric motors and engines. Studies have revealed that the cost of raising a unit quantity of water is the lowest in case of pumps operated by electric motors. The relative cost for engine-operated pumps is about 15 to 25 per cent more than that of electric motor-driven pumps. The cost of lifting water by animal power is about 10 to 20 times more than with electric motors. Manually-operated water lifts are the costliest, being 3 to 4 times more than animal-operated devices. In spite of the economic disadvantage, man and animal power continue as important energy sources in domestic water supply and small-scale irrigation in developing countries. Low economic status of the rural masses often necessitate the use of man and animal power in water lifting in developing countries.

Indigenous Water Lifts

Several types of indigenous water lifts are in use in India. They may be manually-operated or animal-operated. Table 3.1 presents the performance characteristics and adaptability of the different types of indigenous water lifts commonly used. Based on the optimum range in the height of lift, they may be grouped under devices for low lift, medium lift and high lift.

Hierarchy of Pumps

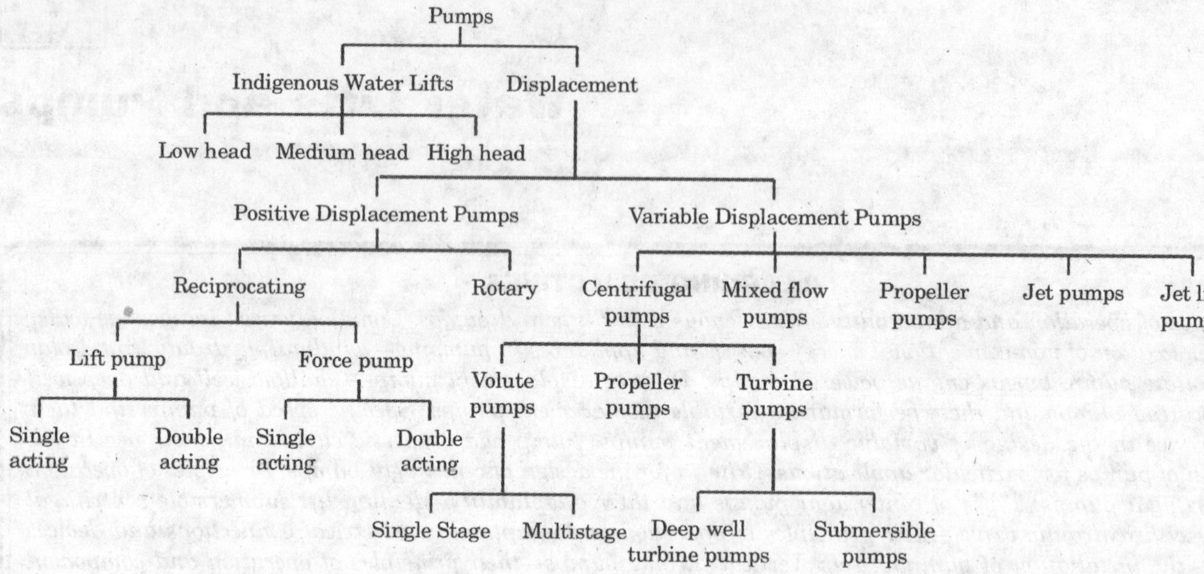

Table 3.1. Performance and adaptability of common types of indigenous water lifts.

Name of device	*Kind of power*	*Optimum lift (m)*	*Average discharge (litres/hr)*	*Remarks*
Swing basket	Two men	0.9 to 1.2	14,000–19,000	Occasionally used in the rice growing belts of eastern and southern India
Don	Single man	0.5 to 1	9,000–13,000	Sometimes used in the eastern region of India
Archemedian screw	Single man	0.5 to 1.2	14,000–19,000	Used in some parts of the deltaic regions of Andhra Pradesh, and in Egypt
Water wheel	One pair of bullocks or buffaloes and one man	1 to 1.2	40,000–60,000	Used in some parts of the canal irrigated areas of north India to lift water from water courses which run below field level
Persian wheel (*Rahat*)	One pair of bullocks or buffaloes or single camel and one man	5 to 10	14,000–18,000	The most common water lift in northern India during the first-half of twentieth century
Chain pump (washer *rahat*)	One pair of bullocks or buffaloes and one man	3 to 6	15,000–20,000	Used in some parts of U.P. Manually operated chain pumps are also available, though not popular
Self-emptying type rope-and-bucket lift (*Sundia charas*)	One pair of bullocks or buffaloes and one man	4 to 6	10,000–15,000	Used in some parts southern India, Deccan region and parts of Rajasthan
Circular two-bucket lift	Single bullock/buffalo and one man	4 to 5	12,000–14,000	Used in some parts of Tamil Nadu.
Counterpoise-bucket lift (*dhenkli, piccottah*)	Single man	1.2 to 4	8,000–11,000	Used in some parts southern India, Bihar and the Deccan region
Rope-and-bucket lift	Two pairs of bullocks/ buffaloes	10 to 30	6,000–10,000	Used in some parts of Rajasthan, Maharashtra and other areas with deep water-table

Fig. 3.1. Two swing baskets in operation.
Note the farmer couple on the stream bank. The operators on the other side are not visible in the picture.
(Photo: Ministry of Information and Broadcasting, Govt. of India).

Fig. 3.2. *Don* used to lift water from a shallow stream. *Dons* are commonly used in West Bengal and parts of Bihar and Orissa as a manually operated water lifting device.

Low Head Water Lifts

The swing basket, *Don*, Archemedian screw, and water wheel are suitable when the depth to water surface from the ground does not exceed 1.2 m.

Swing basket: One of the most ancient water lifts is the swing basket consisting of a basket or shovel-like scoop to which four ropes are attached (Fig. 3.1). Two persons stand facing each other and swing the basket to fill in water. The basket is raised and discharged into the field channel.

Fig. 3.3. An Archemedian screw in operation on a farm in the Godavari delta of Andhra Pradesh.

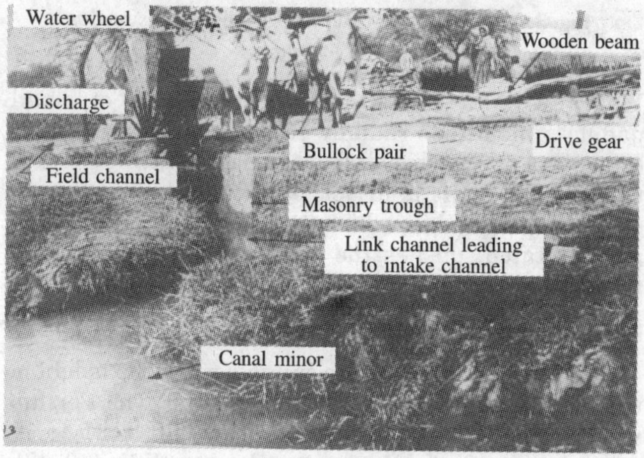

Fig. 3.4 (a). A bullock operated water wheel (*Jalaar*) lifting water from a canal minor near Rohtak, Haryana.

***Don*:** The *don* is a manually operated boat-shaped trough, closed at one end and open at the other (Fig. 3.2). The closed end of the trough is tied with a rope to a long wooden pole which is pivoted as a lever on a post. A weight—a large stone or a ball of dried mud—is fixed to the shorter end of the lever. The open end is hinged to the discharge point. The trough is dipped into the water by applying the body weight and the force of the operator. Water is lifted by the counter-weight on the beam and is emptied automatically.

Archemedian screw: The device is manually operated and consists of a wooden or metal drum with interior partition in the form of a screw, rotated by means of a handle fixed to a central spindle [Fig. 3.3(a)]. The spindle projects from both ends and is supported near its extremities by posts. The drum is placed at an angle of less than 30°, with its lower end in water. When the handle is turned, the water moves up through the drum and discharges through the upper end. To obtain high efficiency in operation, the lower end of the drum should be only half submerged in water. The Archemedian screw is more efficient compared to the swing basket and the *don*.

Fig. 3.4 (b). Close view showing details of construction of a water wheel.

Water wheel: The water wheel consists of small paddles mounted radially on a horizontal shaft [Fig. 3.4 (a)]. The wheel is fixed on a close-fitting concave trough. The shaft is driven by a bullock-wheel drive similar to the drive used in a Persian wheel. The wheel, when rotated, pushes the water to the field surface through the trough.

Medium Head Water Lifts

Medium head water lifts are suitable when the height of the lift is within the range of 1.2 to 10 m. The Persian wheel, chain pump, leather bucket lift with self-emptying bucket, circular two-bucket lift and the counterpoise-bucket lift belong to this category.

Persian wheel: The Persian wheel consists of a chain and a row of buckets mounted on an open-spoked drum and provided with a suitable driving mechanism (Fig. 3.5). The capacity of the bucket ranges from 7 to 14 litres. Two parallel loops of chain, joined by spacing bars and having metal or earthern-ware pots attached to them at intervals, pass over the drum and loop into the water in the well. A horizontal shaft extends from the central axis of the drum to a small vertical cogwheel or gear. The teeth of this vertical gear mesh with the teeth of a large horizontal wheel gear. Fitted to the shaft of the horizontal gear is a long horizontal beam to which the animals are hitched. In some models, the long horizontal shaft is eliminated and the bullocks go round the well, instead of moving on a separate platform.

When the Persian wheel is operated, the bucket-drum revolves, causing the buckets at the lower end of the chain to get filled with water and carried to the top with their open mouth upward. The buckets get filled by dipping into the water in the well. Therefore, a certain minimum length of the chain must remain immersed in water. When the buckets pass over the top of the drum, they spill their contents into a trough, from where it is

Fig. 3.5. A Persian wheel operated by a camel. (Photo: Directorate of Pub. Rel., Govt. of Rajasthan)

conveyed through the irrigation channel. Each bucket has a small hole at the bottom for drainage when the lift stops working. A ratchet arrangement prevents the wheel and buckets from being turned backward by the weight of the filled buckets.

The Persian wheel lifts water over a height of upto about 10 m. But its efficiency is considerably reduced when the lift exceeds 7.5 m. With the increase in height, the number of loaded buckets is increased and this puts a heavy strain on the animals.

At places where the depth to water level does not exceed about 5 m, two rows of buckets, mounted on separate drums and fixed on a common shaft could be used with advantage. This arrangement almost doubles the discharge.

In a Persian wheel, water is lifted considerably above the ground level before it is discharged from the buckets into the receiving trough. The extra energy required in this is especially significant when the water table is close to the ground.

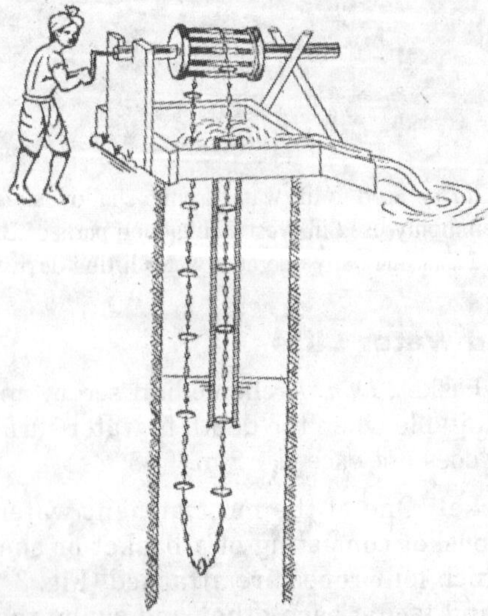

Fig. 3.6 (a). Sketch showing the construction details of a manually operated chain pump.

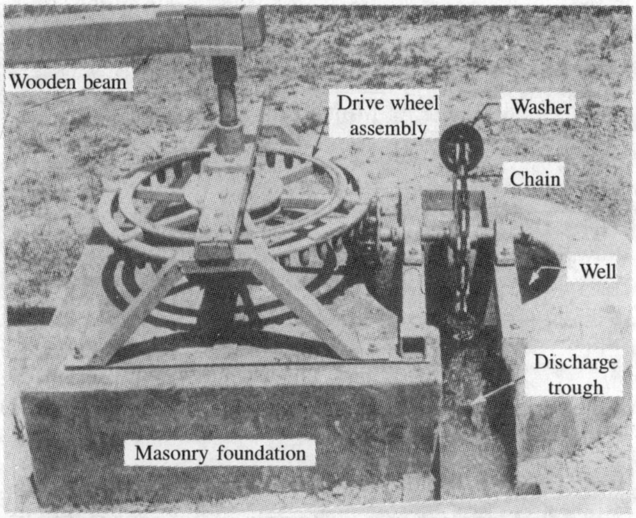

Fig. 3.6 (b). Close view of an animal-operated chain pump showing the details of the drive wheel and chain with washer. Note the discharge of water into the delivery channel.

Chain pump: The chain pump consists of an endless chain which is provided with leather discs or washers spaced at intervals of about 25 cm [Fig. 3.6 (a) and Fig. 3.6(b)]. In some of the recent designs, leather washers, strengthened by iron washers, are being used to make the discs. The chain passes over a notched wheel mounted on a suitable platform fixed on top of the well. On one side of the chain is a pipe of about 10 cm diameter, having a flared opening at its bottom and connected to a trough at the top. The bottom of the pipe is submerged about 60 to 90 cm below the surface of water. The discs have the same diameter as the inside of the pipe. When the wheel is turned, each disc brings up a volume of water.

Chain pumps may be either animal-driven or manually operated. In case of animal-driven chain pumps, the arrangement of the driving mechanism is the same as in a Persian wheel. In manually operated chain pumps, a handle is fixed directly to the drive shaft over which the chain works. An advantage of the chain pump is that it eliminates the extra lift which is encountered in case of a Persian wheel.

Rope and bucket lift with self-emptying bucket: The self-emptying type rope-and-bucket lift is suitable for operation with a pair of average size bullocks. The device consists of a sheet metal or leather bucket, having a capacity of about 100 to 150 litres. At the bottom of the bucket is fixed a leather tube or spout (Fig. 3.7). The bail of the bucket is attached to a heavy rope which passes over a pulley. A second lighter rope is fastened to the lower end of the spout. This second rope passes over a roller fixed to the lip of the receiving through on the ground surface. Both the ropes are tied together and then attached to the bullock yoke. Their lengths are so adjusted that the spout folds up along the side of the bucket while it is being raised from the well.

Fig. 3.7. Water discharging through the spout of a self-emptying type rope-and-bucket lift.

When the bucket reaches the top of the well the spout follows its own rope over the roller, while the bucket continues to rise upwards towards the pulley. When the spout is extended downwards, water is discharged through it into the trough. After the water is discharged, the bullocks are made to walk up a ramp to the place of starting.

The self-emptying bucket has the advantage that it saves the labour of one man. However, some time is lost in operation when the bullocks are made to walk back on the ramp. A toggle is often used to disengage the rope from the yoke when the bucket is empty. The bullocks could then be made to turn around and walk up the ramp, thus speeding up the operation to some extent.

Circular two-bucket lift: The device uses two buckets which are alternately raised, emptied, lowered and filled. While one bucket is filled and lifted, the other is lowered empty into the well. A rope and pulley arrangement along with a central rotating lever permit

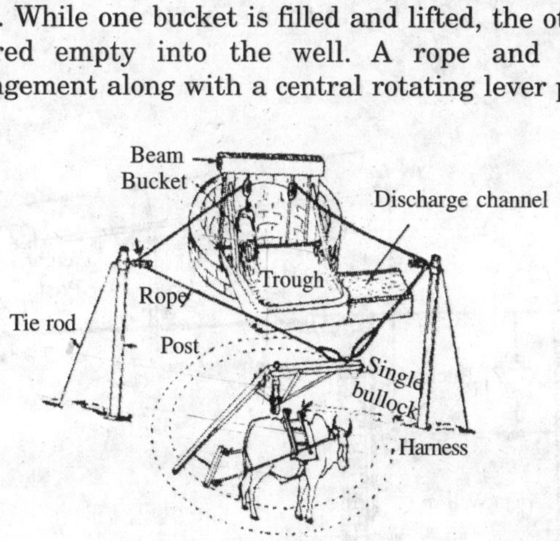

Fig. 3.8. Schematic view of the two-buckets circular lift showing the arrangement of rope and pulleys and the self-emptying bucket.

reciprocating action while the bullocks move in a circular path (Fig. 3.8). The rope arrangement is such that at no time are both the buckets filled and drawn together. Each bucket is of about 70 litres capacity and is provided with a flap valve at the bottom for filling. As the bucket is lowered into the water in the well, the valve opens and the bucket is filled automatically. When it moves up, the weight of water keeps the valve closed. Two vertical rods fixed to the well walls guide the buckets in their up and down movements. When the bucket comes to the top, it is tilted automatically due to the loop on the rod and water is discharged into the receiving trough.

Counterpoise-bucket lift: This device consists of a long wooden pole which is pivoted as a lever on a post. A weight, usually a large stone or a ball of dried mud or a basket full of stones, is fixed to the shorter end of the pole. This weight serves as a counterpoise to a bucket suspended by a rope or a rod attached to the long arm of the lever. To operate the lift, a man pulls down the rope or rod, using his body weight and strength until the bucket is immersed in the water and filled. The bucket is lifted up by the counter weight. As the bucket reaches the ground level, it is tipped into a trough. Alternatively, a rod can be fixed to the well wall on which the bucket is made to slide. This enables the bucket to be emptied automatically into a trough, as it reaches the top of the well.

The counterpoise is adjusted to balance the weight of the full bucket. The proper weight of counterpoise can be computed as follows:

Referring to Fig. 3.9, let

W_1 = weight of bucket (kg)
W_2 = weight of water in the bucket (kg)
W_3 = weight of counterpoise (kg)
Z_1 = lift or motion of load (m)
Z_2 = motion of effort (m)

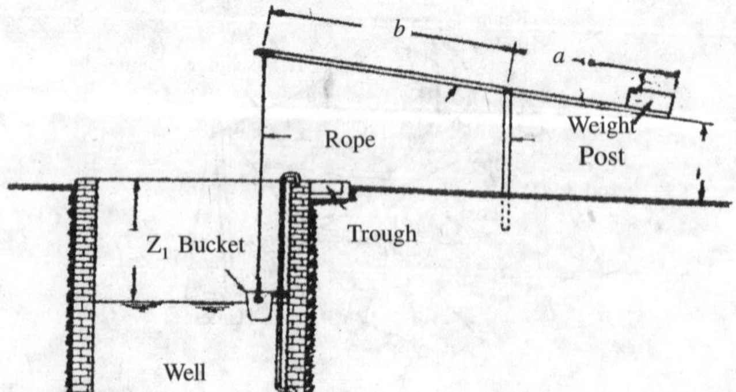

Fig. 3.9. Sketch illustrating the principle of working of a counterpoise-bucket lift.

Velocity ratio

$$= \frac{\text{Motion of effort}}{\text{Motion of load}}$$

$$= \frac{Z_2}{Z_1} = \frac{a}{b}$$

Now, $(W_1 + W_2)b = W_3 a$

or $$\frac{W_1 + W_2}{W_3} = \frac{a}{b} = \frac{Z_2}{Z_1}$$

$$W_3 = (W_1 + W_2)\frac{b}{a} \qquad \text{... (3.1)}$$

High Head Water Lifts

Rope-and-bucket lift: The only indigenous water lift suitable for deep wells is the rope-and-bucket lift (*charas*) operated by bullocks. The device may be operated singly or in multiples of two or more working simultaneously, depending on the yield of the well and the requirement of irrigation water. It consists of a bucket or bag having a capacity of about 150 to 200 litres and made of leather or galvanised iron sheet. The bucket is fixed on a stout iron ring with an iron frame-work at the top and is tied to a long rope. The rope passes over a pulley set on a wooden frame-work fixed on top of the well. A pair of bullocks, hitched to the other end of the rope, provide the power to lift the bucket. The bullocks, while pulling up the full bucket, walk down on an earthen ramp sloped at an angle of 5 to 10 degrees, thereby obtaining some advantage of their body weight in exerting the necessary force to lift the bucket.

The rope-and-bucket lift can be worked by one pair of bullocks and two men. More commonly, however, two pairs of bullocks and three men are used, one man for emptying the bucket and two others for driving the bullock teams. When two pairs of bullocks are used, one pair walks up the slope while the other is drawing up the bucket.

Application of Non-conventional Energy Sources in Pumping

Fossil fuels (petrol/diesel) and electricity generated at centralised grid power systems have been the conventional energy sources in pumping for irrigation and water supply. In spite of their outstanding merits, fossil fuels have some major limitations. They are non-renewable and their supply is bound to reduce and approach exhaustion with continued exploitation. The cost of diesel/petrol and their distribution to remote areas may sometimes be prohibitive in many regions. Their indiscriminate use result in environmental hazards like atmospheric pollution. Likewise, the cost of generating electricity at hydro-electric and thermal stations, transmitting it over long distances and distribution to individual small landholdings for operating pumps is becoming increasingly high. Hence,

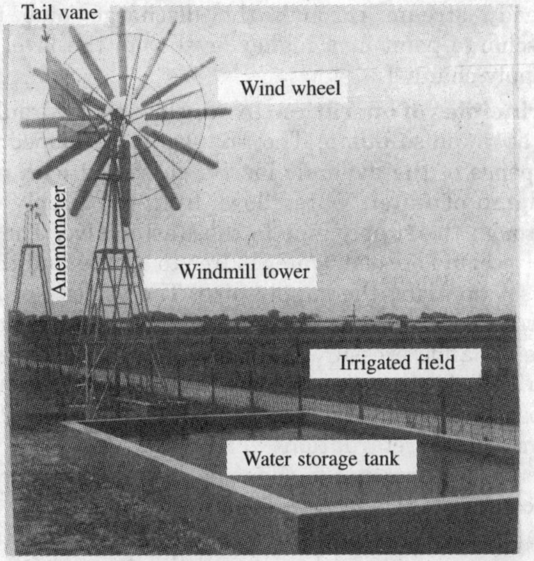

Fig. 3.10. A typical windmill installation for irrigation.

there is an ever increasing effort to adopt non-conventional energy sources in pumping, wherever feasible.

The four major non-conventional energy sources which have established their feasibility in water pumping are wind, micro hydro-power, sunlight and biomass. All these energy sources are renewable, are available to the user at the pumping point and are environmentally safe. The pumping units suitable for non-conventional energy sources are the windmill-operated reciprocating pumps, micro-water power operated hydraulic rams, photo-voltaic power operated centrifugal pumps and biogas-engine operated centrifugal pumps. All of them are currently limited to small scale use.

Wind-powered Water Lifts

The windmill (Fig. 3.10) utilises the natural power from wind to lift water. It consists of a large diameter vane wheel or rotor fixed on the top of a high steel or wooden tower and a reciprocating pump at its bottom. The upper end of the piston rod is fixed eccentrically to the vane wheel. The pump cylinder, or at least the lower part of the suction pipe, is always immersed in water in the well. When a strong wind blows, it rotates the vane wheel, which in turn works the pump. The delivery pipe rises to a reservoir where the water is stored to be used whenever required (Fig. 3.10). The discharge from a windmill is usually intermittent and seasonal. Its capacity depends on wind velocity and frequency. The device is suitable to lift water from shallow and moderately deep wells, depending on the wind velocity.

Wind is a free source of power that can be harnessed in areas where winds of sufficient velocity can be relied upon. The operational cost of lifting water is almost eliminated when wind power is utilised. The seasonal nature of winds, the low energy output of windmills and the high cost of commercially produced mills are considered major limitations in the large-scale adoption of windmills in pumping. In general, the most suitable applications of windmills for irrigation are at places which are windy and away from the source of supply of electrical power, and remote regions where petroleum fuels are not available at a reasonable cost.

The minimum wind speed required for the operation of a windmill usually ranges from 6 to 8 km/hr depending on the design. The anticipated windiness of a location is to be studied before deciding on the installation of windmills. The relative windiness of different regions could be broadly compared from the average wind speeds. The energy from the wind varies as the cube of the wind velocity and the duration of wind at a particular velocity. Therefore, two places having the same monthly average wind speed may vary greatly in their wind-energy potential. The characteristics of wind in India show that it is a potential source of power only in limited areas. The regions of high wind velocity are almost limited to the coastal areas in the western and eastern regions of India. Wind farms with clusters of windmills with power generation facility have made substantial growth in the south-eastern region of Tamil Nadu, including Tinneveli and adjoining districts where wind power has great potential. Windmills are used exclusively for electricity generation or to supplement the electric power supplied by the State Electricity Board to industries. Many textile mills are investing on windmills to supplement their power requirement. Surplus power generated by these windmills are purchased by the State Electricity Board and are fed to its power grid. This facility is known to be available only in Tamil Nadu at present. States like Gujarat could also make feasibility studies on such systems, especially for the Saurashtra region.

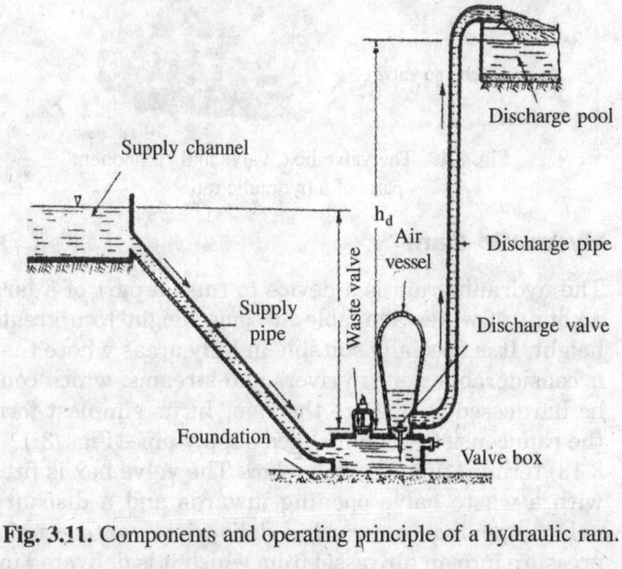

Fig. 3.11. Components and operating principle of a hydraulic ram.

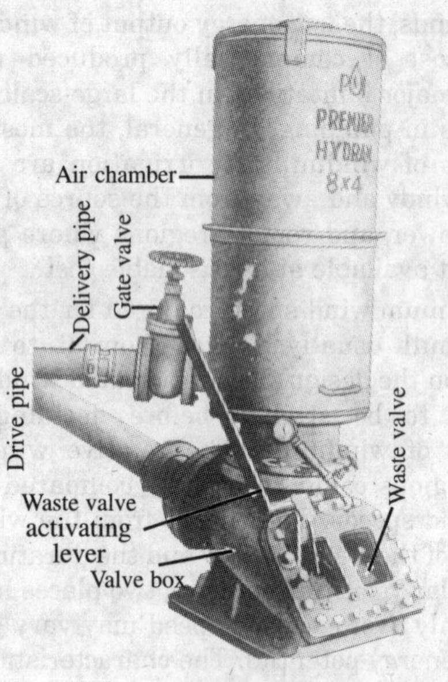

Fig. 3.12. A hydraulic ram of size 20 cm × 2 cm

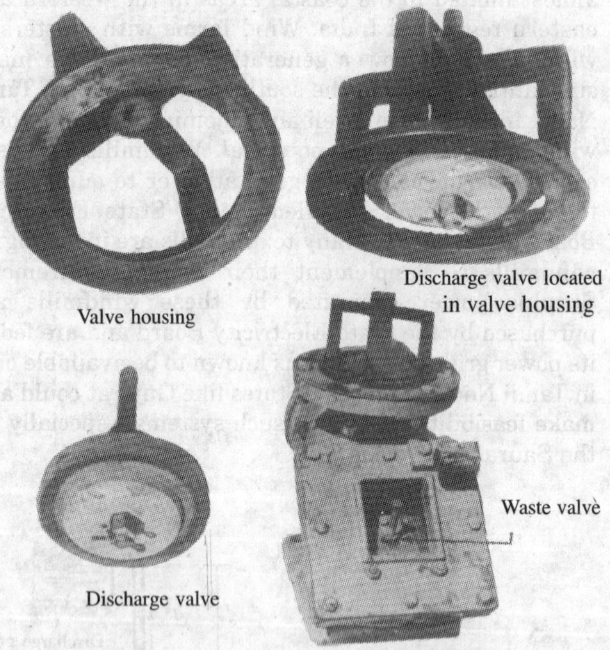

Fig. 3.13. The valve box, valve and component parts of a hydraulic ram.

Hydraulic Ram

The hydraulic ram is a device to raise a part of a large amount of water available at some height to a greater height. It is specially suitable in hilly areas where there is considerable slope in rivers and streams, which could be harnessed to operate the ram. In its simplest form, the ram consists of an inclined supply pipe (Figs. 3.11 to 3.13) terminating in a valve box. The valve box is fitted with a waste valve opening inwards and a discharge valve opening outwards, delivering water under pressure into an air vessel from which it is delivered in a steady stream through the discharge pipe to the discharge point at a higher level than the level of the supply channel.

Principles of operation: In principle, the hydraulic ram is an impulse pump. The impulse is developed at the expense of the dynamic inertia possessed by a moving column of water. Water flows from the supply source through the supply pipe to the waste valve. The waste valve being opened, water is free to escape and the flow is set up along the supply pipe. The velocity of flow is increased under the influence of the supply head until the dynamic pressure on the side of the valve becomes sufficiently great to overcome its weight. The valve now closes rapidly and the supply column suffers a consequent retardation which gives rise to a rapid increase of pressure in the valve box until this pressure becomes sufficiently great to open the delivery valve. Water then escapes through this valve into the air vessel, compresses the air and flows out through the discharge pipe. As soon as the momentum of the supply column is destroyed the delivery valve closes and a backward motion of water is instituted. This motion, once set up, can only be checked by a reduction of pressure in the valve box below that corresponding to the static head and consequently the pressure in the valve box is reduced rapidly until at some instant the waste valve reopens, and the whole cycle of operations is repeated.

In order to develop maximum impulse, the supply pipe should be as long as possible. Installation of a ram close to the source of supply will reduce the impulse and consequently the delivery head.

Advantage is taken of the reduction of pressure to keep the air vessel charged. By the introduction of a shifting valve, air is drawn through this valve into the valve box, when the pressure falls below that of atmosphere. The whole cycle, which may only take a fraction of a second to complete, may be divided into four periods, during which the waste valve is respectively opening, wide open, closing and closed.

Efficiency. The efficiency of the ram may be considered from two points of view. If q is the volume of water delivered by the ram, and Q that is escaping through the waste valve, H being the effective supply head and h_d the effective delivery head measured from the level of the waste valve and including the friction in the delivery pipe line, the total input of energy to the ram is $(Q + q)H$ and the total output is qha, known as D'Aubuission's efficiency ratio, then gives the efficiency of the ram as a machine.

$$\frac{qh_d}{(Q+q)H} \qquad \text{... (3.2)}$$

The main advantage of the hydraulic ram is that once installed, it needs hardly any running cost. The machine can work continuously for all the 24 hours and provide regular water supply.

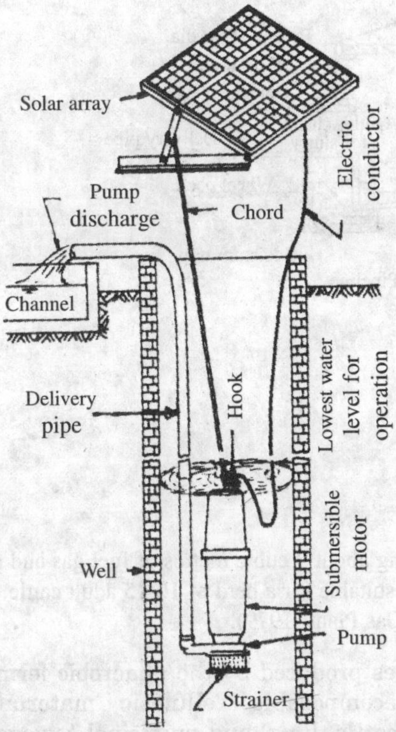

Fig. 3.14. Schematic sketch illustrating the installation of a photo-voltaic pumping set using submersible motor and pump installed in a shallow open well.

Solar Photo-voltaic Pumping System

Solar energy, with its virtually infinite potential and free availability, represents a non-polluting and inexhaustible energy source which can be developed to meet the energy needs of mankind in a major way. The solar-pump unit consists essentially of a solar array, a direct-current electric motor and a pumping unit (Fig. 3.14). The other components are the electrical control and some mechanism for tracking the array against the sun. Two types of pumping sets are used with photo-voltaic systems, a vertical centrifugal pump coupled to a submersible DC electric motor (Figs. 3.14 and 3.15), or an ordinary volute centrifugal pump close-coupled to a horizontal DC electric motor. However, the submersible pump unit is more suitable for the photo-voltaic system. This arrangement eliminates the suction pipe and foot valve and results in higher efficiency of the pumping unit. The submersible pump is made leak-proof by a mechanical seal. In case a volute centrifugal pump is used, care is taken to limit the pump suction within about 4.5 m to maintain a high level of pump efficiency.

Fig. 3.15. A solar photo-voltaic pump in operation.

Photo-voltaic power generation: Photo-voltaic cells, frequently referred to as solar cells, convert the light part of the solar spectrum (sunlight) into electricity. They are the most rapidly expanding energy sources in the world. The solar cell operates on the principle of the photo-voltaic effect—the creation of charge carrier within a material by the absorption of energy from the incident solar radiation. The solar cell is made of two thin layers of a semi-conductor material appropriately doped with impurity atoms so as to give one layer a negative electrical bias (*n*-bias) and the other a positive bias (*p*-bias). Sunlight falling on the cell reaches the junction between the two layers in the form of photons of energy and knocks electrons across the junction. This results in the development of a potential difference across the two layers. Direct current (DC) electricity can be drawn across the two layers through an external circuit.

The efficiency of solar cells in converting incident solar energy into electrical energy depends on the illumination spectrum intensity, material of construction and design of the cell, atmospheric temperature and dustiness of the sky around the solar panel. Silicon is the most commonly used material in manufacturing solar cells. A solar cell behaves like a low voltage (0.5 volt) battery whose charge is continuously replenished at a rate proportional to the incident solar radiation. Connecting multiples of cells into series-parallel configuration results in photo-voltaic modules or solar arrays with high currents and voltages. The power developed by a solar array ranges from 80 to 120 watts/sq m area of the panel. The output of the solar array varies with the incoming radiation, orientation of the panel and other factors. The photo-voltaic power can be utilized to operate conventional electrical appliances, including direct current (DC) motors. The solar panel is mounted on a suitable frame which can adjust the array to the direction of the sun rays. The angle of the array has to be changed 3 or 4 times a day.

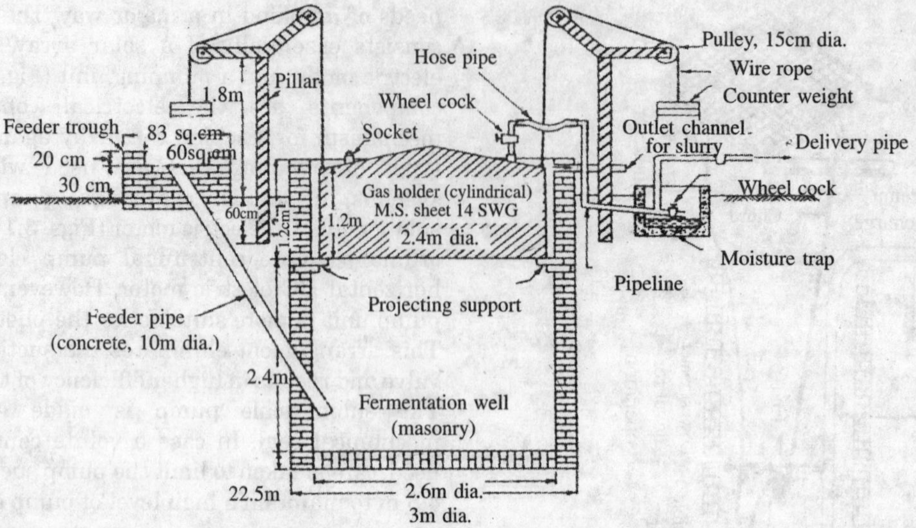

Fig. 3.16. Sectional view of a medium size drum type biogas plant producing about 9 cubic metres of fuel gas and requiring about 150 kg of cattle dung per day. The design specifications are suitable for a herd of 10-15 adult cattle. *Adapted from* IARI Tech. Bul. Cow Dung Gas Plant (1979).

Biogas Engine Powered Pumping Plants

The potential of biomass as an energy source is being increasingly realised. Biomass constitutes a significant, clean and renewable energy source. The biomass in the biological system may be classified in two broad categories: terrestrial biomass (organic residues and higher plants) and aquatic biomass (fresh-water aquatic plants, seaweeds, micro-algae and floating marine plants). Biogas is a mixture of gases containing methane, carbon dioxide, hydrogen and traces of a few other gases produced by the anaerobic fermentation of easily decomposable cellulosic materials. Animal manure (cattle dung) and municipal sewage have been the main materials used for producing biogas. The process has the advantage that animal and human waste can be used to generate energy while, at the same time, retaining their nutrient value for use as organic fertilizer. The production of methane gas from crop residues and aquatic plants like water hyacinth have also been attempted with considerable success. Water

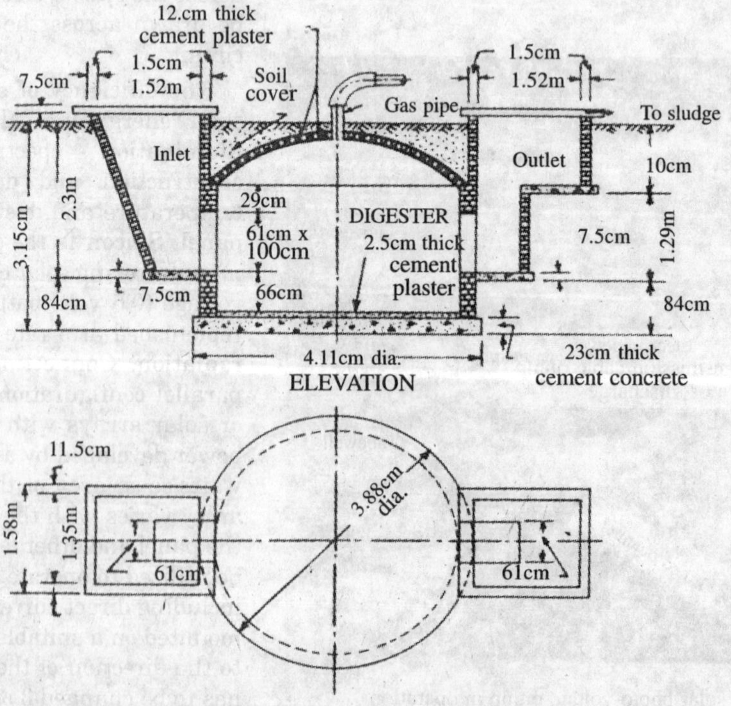

Fig. 3.17. Views of a drumless type biogas plant.

hyacinth is one of the major weeds leading to the blocking of many drainage canals and even irrigation canals and tanks. Their removal and use as a source of energy provide a substantial energy source and aid in preventing blocking of water bodies.

Biogas plants: Biogas plants may be classified into two, namely drum type and drumless type. The conventional biogas plant, originally developed at the Indian Agricultural Research Institute, New Delhi in 1935, is the drum type. It consists of a masonry digester (fermentation tank), with an inlet pipe on one side for feeding cattle dung mixed with water into the plant and an outlet pipe on the other side for discharging the spent slurry (Fig. 3.16). The gas collects in a gas holder or drum made of mild steel. The gas holder is inverted over the slurry and moves up and down with the accumulation and discharge of gas.

Operation of biogas plants: To operate the biogas plant, a mixture of cattle dung or other animal excreta and water, in the ratio 1 : 1 is added as slurry to fill the digester. In a new plant, the production of gas may start in 5 to 10 days in summer, and 15 to 20 days in winter. When fresh dung is added into the digester, the digested slurry overflows into a collection pit.

Biogas engine pumping set: Ordinary (petrol/gasoline) engines can be adapted to run on biogas. In case of compression-ignition engines (diesel) adapted to run on biogas, a small part of the fuel continues to be diesel. A popular commercial make of biogas-run internal combustion engine working on the diesel cycle is designed to use about 80 per cent biogas and 20 per cent diesel oil. Minor modifications are made in the combustion chamber of the conventional diesel engine to adapt it for use with biogas. The engine is started on diesel oil and after warming, is made to run on the biogas-diesel mixture for continued operation. The gas consumption of biogas engines is about 450 litres (0.45 m^3) per brake horse power per hour of operation.

Pumping unit: Any of the common types of pumps—centrifugal or propeller — can be used with biogas engines. The pump can be coupled directly to the engine or operated by a belt drive. Low-head, high-discharge pumps are usually preferred for use with biogas engines.

Positive Displacement Pumps

Positive displacement pumps discharge the same volume of water regardless of the head against which they operate. This type of pump must be powered to meet the maximum load resulting from its discharge capacity and the greatest head under which it will operate. As the capacity is small, these pumps are not very popular in irrigation and drainage. However, they are commonly used in home water supply, well drilling and under special situations in irrigation pumping.

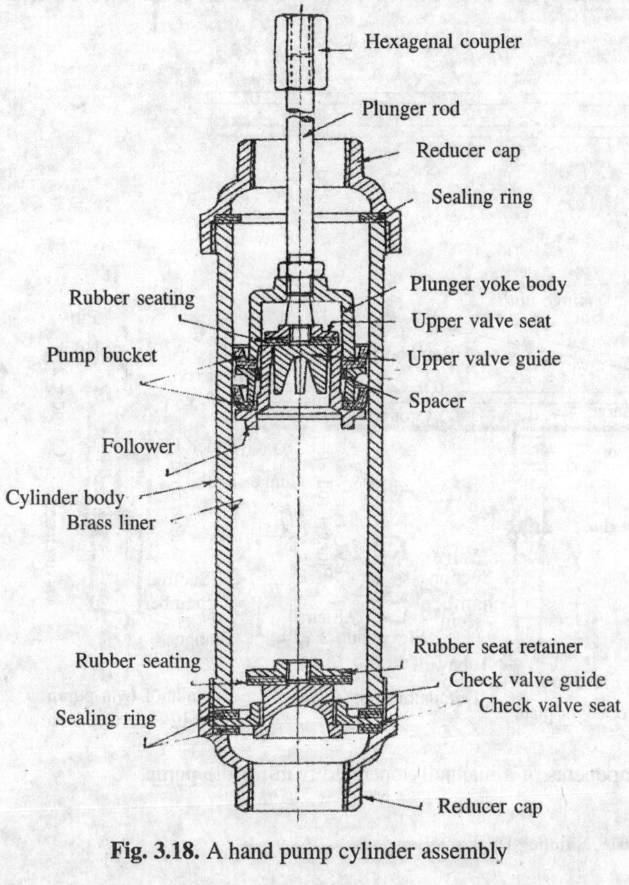

Fig. 3.18. A hand pump cylinder assembly

Fig. 3.19. A twin treadle pump being operated by a small boy for irrigating tobacco near Dinapur, Bangladesh. It is a common sight in the region to see women operating the pump.

Reciprocating Pumps

Reciprocating pumps, sometimes called piston or displacement pumps, function by means of a piston movement which displaces water in a cylinder (Fig. 3.18). The flow is controlled by valves. A *piston or plunger* is a cylindrical piece which moves backward and forward inside a hollow cylinder. The capacity of the reciprocating pumps depends on the size of the cylinder chamber and the length and speed of the stroke. Numerous forms of packings are used to prevent leakage past a piston. Cup-leathers are usually used for packing in case of pump cylinders. The pressure of the water acting inside the cup presses the leather outwards against the cylinder, thus preventing leakage.

Shallow Well Reciprocating Pumps

Reciprocating pumps for shallow wells are usually of the lift type, using atmospheric pressure to raise water in the pump column. The piston is in the pump body, near the handle. The piston when moved up and down by the movement of the handle displaces air from the pump column. This creates a vacuum which permits the force of atmospheric pressure to push the water from the well up into the pump column and from it to the outside. Theoretically, it can lift water from a height of upto one atmosphere or 10.33 m. But in practice, the height reached is only about 6.5 to 7 m due to friction and other losses.

Animal powered reciprocating type pump: An animal powered duplex reciprocating pump developed by Khepar *et al.* (1975) is specially suitable for pumping water from shallow tubewells. The conventional bullock gear of the type commonly used with Persian wheels is used to transmit the power of the pair of bullocks or other draft animals to operate the pump. The pumping unit consists of a pair of ordinary piston pumps. Each pump has a cylinder diameter of 30 cm with pistons having a stroke of 11 cm. The suction ends of the two pumps are connected by bends to a T-joint to which a common suction pipe is connected. A foot valve is fixed at the bottom of the suction pipe, as usual. The suction pipe which is smaller in diameter than the tubewell, is lowered into the well. The power transmitted through the bullock gear is applied to operate the piston through a flywheel to which are connected the pump handles. The linkage mechanism is designed to obtain the desired piston stroke. The drive provides one discharge stroke in each pump for each revolution of the flywheel. The suction and discharge strokes of the pair of pumps alternate with each other, i.e., when there is suction in one pump there is discharge in the other. The flywheel provides the inertia required for the smooth running of the pump. The pump has a discharge of about 7 litres per second against a head of 4 metres.

Manually-Operated Twin Treadle Pump

A foot-operated twin pump developed by RDRS, Bangladesh*, called twin treadle pump, has been a major contribution to small scale lift irrigation through low-cost shallow wells (Fig. 3.19). Over 20,000 units of the pump were manufactured and installed in northern Bangladesh during 1981-84. The main features of the pump are the twin plunger pump and the use of the body weight of the operator in pumping.

The components of the pump are shown in Fig. 3.20. The suction pipe at the bottom of the pump is connected to the water source, such as an open well, tubewell, stream, or pond. The pump sucks water through the suction pipe into a manifold through which are connected the two pump cylinders. Each cylinder is equipped with a foot valve which prevents the back-flow of water from the cylinder to the suction pipe during the return stroke of the pump. The foot valve could be a simple rubber flap made from a used inner tube of a truck tyre. The pump plunger consists of two round

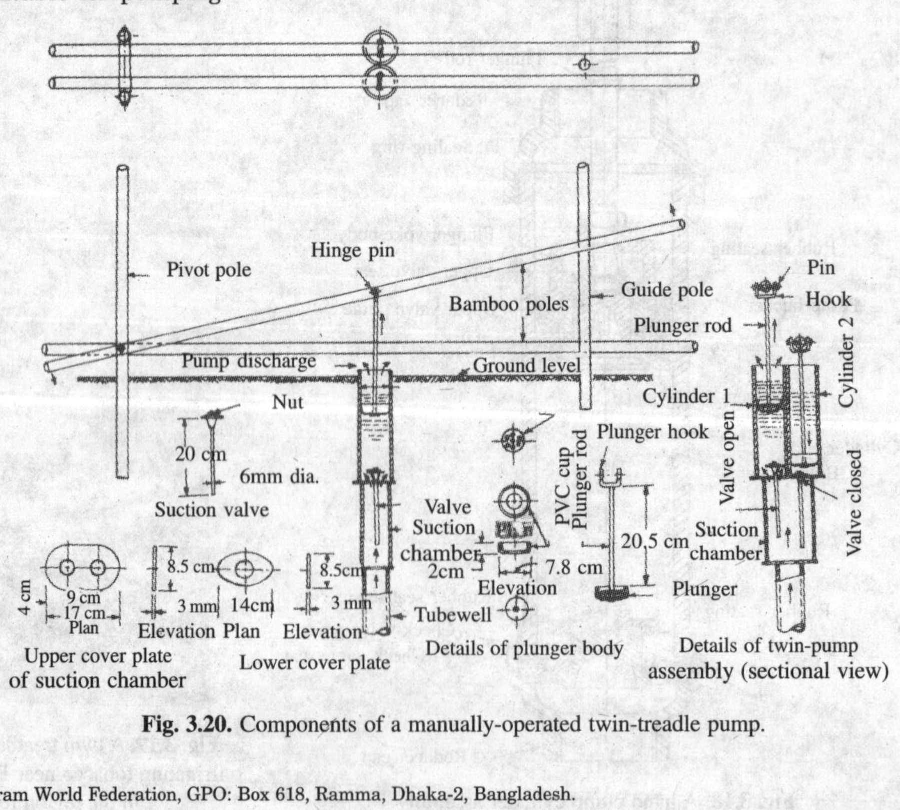

Fig. 3.20. Components of a manually-operated twin-treadle pump.

*Rangpur Dinapur Rehabilitation Service, Lutheram World Federation, GPO: Box 618, Ramma, Dhaka-2, Bangladesh.

disks fastened to a rod and a moulded rubber or PVC cup or bucket. The upper disk has holes which allow water to pass through the plunger during the downward return stroke. On the upward stroke, the bucket is pressed against both the lower disk and the cylinder wall. This provides a seal that prevents water from passing by the piston. It also creates a partial vacuum that sucks water into the cylinder from the manifold and the suction pipe. The plunger assembly is easy to fabricate with simple tools. It utilizes the buckets of standard domestic water hand pumps, which are widely used in developing countries. The plunger rods are connected to the treadles (generally bamboo poles) by means of a hinged joint. The pump cylinders are generally made from standard size of steel pipes or of cast iron. It could also be made of mild steel sheet (16 gauge) using sheet metal. During operation, it is possible to adjust the position of the feet of the operator to obtain maximum efficiency in pumping. The twin-treadle pump has a capacity of about 3 lit/sec for a lift of 3 metres and 2 lit/sec for a lift of 4 metres.

Deep Well Reciprocating Pumps

By introducing the pump cylinder with its plunger and valve into the water in a well, water can be lifted to almost any height required in practical use. The plunger is connected to the pump handle, or other operating devices like a mechanically powered crank shaft. The details of construction of a deep well reciprocating pump are illustrated in Fig. 3.21. As the plunger in which the upper valve is located moves upward, the water on the top of the valve is forced upward through the delivery pipe and another charge of water fills the space between the valves. The cycle is repeated in each upward stroke.

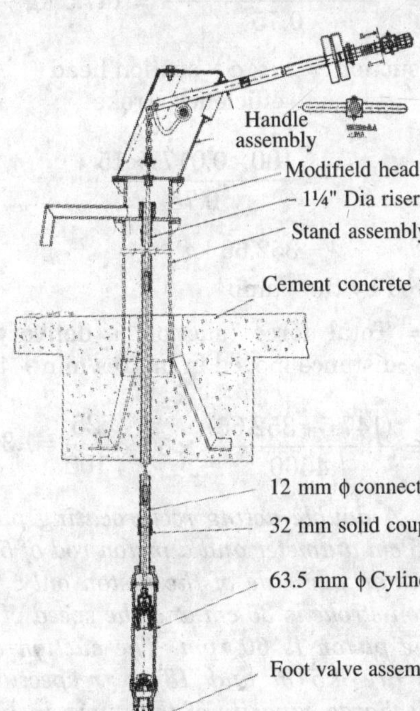

Fig. 3.21. Deep well hand pump assembly and installation.

Deep well reciprocating pumps may be manually operated or mechanically driven. Manually operated deep well pumps are suitable only for lifts upto about 45 m, beyond which manual pumping becomes progressively difficult due to increasing weight of the water column and the connecting rod that must be lifted with each stroke. Engine or motor operated reciprocating pumps usually employ a crank shaft arrangement called the working head, placed at the top of the well to change the rotating motion of the power to that of reciprocating motion at the piston.

Installation of mechanically powered deep well reciprocating pumps: Engines or electric motors may be used to operate mechanically driven deep well reciprocating pumps. As the pump has to work at a low speed, a speed reduction device fixed to a working head is used in the power transmission system. The working head is usually provided with fast and loose pulleys. The head is fixed on a frame placed over the top of the well. Power is transmitted to the working head by belt drive.

Types of Reciprocating Pumps Based on Construction and Operating Features

While the basic principles of operation apply to all piston pumps, there are many modifications in design, which adapt these pumps to specific uses. Piston pumps may be either single acting or double acting.

Single acting pumps (Fig. 3.18) have one discharge stroke for every two strokes of the piston. Thus, the water is delivered during alternate strokes of the piston. The flow through the delivery pipe is therefore intermittent. An air vessel (Fig. 3.22) may be fixed over or near the delivery valve to remedy this. During the delivery stroke of the piston, the air in the air vessel is

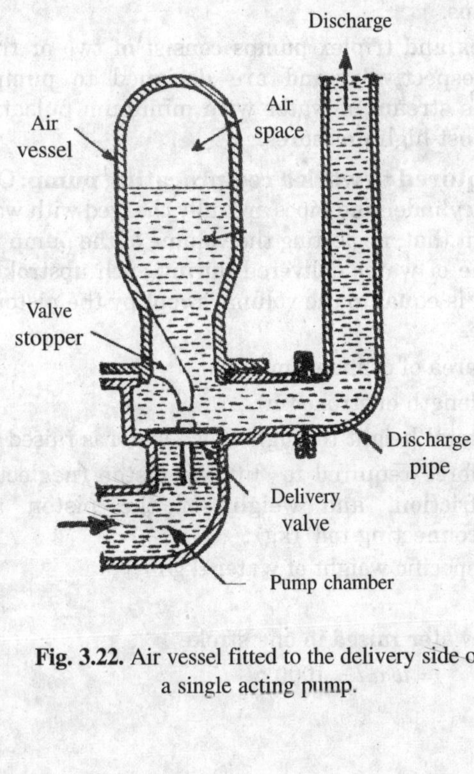

Fig. 3.22. Air vessel fitted to the delivery side of a single acting pump.

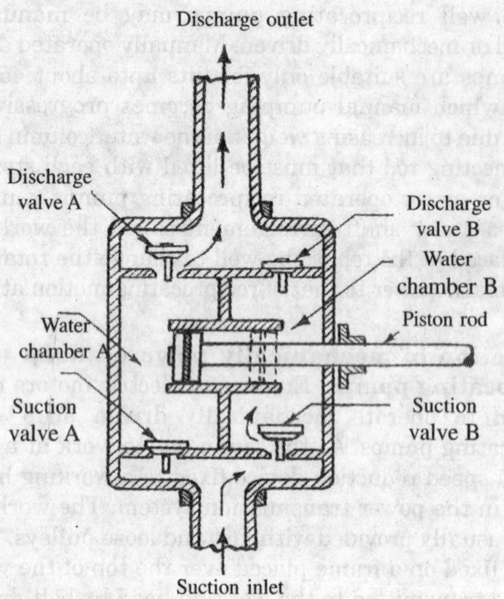

Fig. 3.23. Sketch illustrating the principle of working of a double acting reciprocating pump.

compressed to a greater pressure than that corresponding to the head of water at the bottom of the delivery pipe. During the suction stroke, the pressure of the air in the vessel maintains the flow of water through the delivery pipe, thereby ensuring nearly continuous discharge.

Double acting pump s are constructed with piston and valves so arranged that water is pumped on both the inward and the outward movements of the piston (Fig. 3.23). Though the arrangement is more commonly used in lift pumps, it is also sometimes incorporated in force pumps.

Duplex and triplex pumps consist of two or three pistons respectively, and are designed to pump a continuous stream of water with minimum pulsation, often against high pressures.

Force required to work a reciprocating pump: Once the pump cylinder and pipes are fully charged with water it is evident that, neglecting the volume of the pump rod, the volume of water delivered during each upstroke of the piston is equal to the volume swept by the piston in one stroke.

Let, a = area of cylinder (m^2)

l = length of stroke (m)

h = total height through which water is raised (m)

p = force required to lift the piston (neglecting friction, and weight of the piston and connecting rod) (kg)

w = specific weight of water (kg/m^3)

Then,

Weight of water raised in one stroke

$= w\,a\,l = 1000\,al$

Work done in one upstroke $= 1000\,a\,l\,h = P\,h$

Therefore, $P = 1000\,al$ ... (3.3)

Thus, the pull on the pump rod is equal to the weight of a column of water whose base is equal to that of the piston and whose height is the total head. Hence, when friction is neglected, P is independent of the diameters of the suction and delivery pipes.

***Example 3.1**. A single acting reciprocating pump has its piston diameter 15 cm and stroke 25 cm. The piston makes 50 double strokes per minute. The suction and delivery heads are 5 m and 15 cm, repectively. Find, (i) the discharge capacity of the pump in litres per minute, (ii) the force required to work the piston during the suction and delivery strokes if the efficiency of suction and delivery strokes are 60% and 75%, respectively, and (iii) the H.P. required by the pump for its operation*

Solution:

(*i*) Area of piston,

$$a = \frac{\pi}{4}d^2 = \frac{3.15}{4}\times\frac{15}{100}\times\frac{15}{100} = 0.0177\ \text{m}^2$$

Volume swept by piston per stroke

$$= al = 0.0177 \times \frac{25}{100} = 0.0044$$

Discharge of pump $= 0.0044 \times 50 = 0.22\ \text{m}^3/\text{min}$

$= 0.22 \times 1000 = 220$ litres/min

(*ii*) Average force of suction = $w \times a \times$ suction head ÷ efficiency stroke

$$= \frac{1000 \times 0.0177 \times 15}{0.75} = 147.5\ \text{kg}$$

Average force of suction = $w \times a \times$ suction head ÷ efficiency stroke

$$= \frac{100 \times 0.0177 \times 15}{0.75}$$

$= 352.66$ kg

(*iii*) H.P. required by the pump

= Total force (suction + delivery) × distance moved in metres/min ÷ 4560

$$= \frac{(147.5 + 352.66)}{4560} \times 50 \times \frac{25}{100} = 1.371$$

***Example 3.2.** A double acting reciprocating pump has a piston of 25 cm diameter and a piston rod of 5 cm diameter which is on one side of the piston only. The length of the piston stroke is 35 cm and the speed of the crank moving the piston is 60 rpm. The suction and discharge heads are 4.5 m and 18 m, respectively. Determine the discharge capacity of the pump in litres per minute and the H.P. required to operate the pump.*

Solution:

Cross-sectional area of piston,

$$a = \frac{\pi d^2}{4} = \frac{3.14}{4} \times \frac{25}{100} \times \frac{25}{100}$$

$$= 0.049 \text{ m}^2$$

Cross-sectional-area of piston rod,

$$al = \frac{3.14}{4} \times \frac{5}{100} \times \frac{5}{100} = 0.00196 \text{ m}^2$$

(a) Force required to work the piston during 'in' stroke

(i) For suction $= w \times a \times$ suction head

$= 1000 \times 0.049 \times 4.5 = 220.50$ kg

(ii) For delivery $= w(a - a_1) \times$ delivery head

$= 10000(0.49 - 0.00196) \times 18$

$= 846.72$ kg

Total force during 'in' stroke

$= 220.50 + 846.72$

$= 1067.22$ kg

(b) Force required to work the piston during 'out' stroke

(i) For suction $= w(a - a_1) \times$ suction head

$= 1000(0.049 - 0.00196)4.5$

$= 211.680$ kg

(ii) For delivery $= w \times a \times$ delivery head

$= 1000 \times 0.049 \times 18$

$= 882$ kg

Total force during 'out' stroke

$= 211.68 + 882.00$

$= 1093.68$ kg

Discharge during 'in' stroke $= a \times l \times$ rpm

$$= 0.0049 \times \frac{35}{100} \times 60$$

$= 1.029 \text{ m}^3\text{/min}$

$= 1029$ litres/min

Discharge during 'out' stoke

$= (a - a_1) \times 1 \times$ rpm

$$= 0.04704 \times \frac{35}{100} \times 60$$

$= 0.98784 \text{ m}^3\text{/min}$

$= 987.84$ litres/min

Total quantity of water raised by the pump

$= 1.29 + 987.84$

$= 2016.84$ litres/min

H.P. required by the pump

$$= \frac{\text{Total force (kg)} \times \text{distance moved (min)}}{4560}$$

$$= \frac{1093.68 \times 35}{4560 \times 100} \times 2 \times 60 = 10.73.$$

Hand Pumps for Domestic Water Supply

Properly installed hand pumps serve as a means of providing clean drinking water in rural areas. Handpumps are important in villages which are not provided with piped water supplies. The practice of drawing water from exposed open wells with a rope and bucket, coupled with bathing and washing near wells and in ponds used for drinking water supplies, are health hazards in rural areas of developing countries. Most governments have launched large programmes to provide potable water through handpumps and piped water supplies.

The type of handpump to be selected will depend on the depth of ground-water, the yield of the well, constructional features and efficiency of the pump, and availability of facilities for maintenance and repair. The per capita provision of drinking water under government-sponsored domestic water supply projects in most states in India, vary from 25 to 70 l/day/person. The discharge of handpumps varies from 500 to 1500 litres/hour, depending mainly on the depth to groundwater. India Mark II hand pumps* [Fig. 3.21] to suit various pumping situations are widely used in developing countries for domestic water supply. They are comparatively trouble, free.

Handpump installation: Handpumps may be installed in bore wells (Fig. 3.21), filter points and open wells. If the pumping water level is within 6 m during the dry season, it is advantageous to install a shallow-well lift pump, with the pump chamber located at the ground surface on top of the well. A deep-well lift pump, with its cylinder submerged in water, is selected when the ground-water level is below 6 m. While installing a handpump in an open well, it is essential to close the top of the well with a removable slab made of reinforced concrete. Well covers made of wooden boards or stone slabs are also used where an R.C.C. slab is considered expensive. The cover should slope towards the periphery of the well in order to drain away the waste water. The well cover will prevent the entry of pollutants from the ground surface into the well.

Whenever a hand pump is installed in a bore well or filter point, it is essential to provide a permanent platform, about 1 m square, around the well. The platform, may be of reinforced concrete or masonry.

* Manufactured by Richardson and Crudas Ltd. (A Government of India Undertaking), Rajaji Salai Chennai-600 001, Tamil Nadu.

Brick or stone masonry may be used, depending on their relative costs. Provision of suitable drainage through a pipe or lined channel of semi-circular or rectangular cross-section is essential in a handpump installation. The drain should lead the waste water to a natural channel or ditch.

Maintenance and repair of handpumps: A large number of hand-pumps remain out of order due to inadequate arrangements for repair and maintenance. There is a need for training village youth, including artisans, in the repair of handpumps. An adequate stock of parts exposed to wear and tear should be maintained within easy reach of a group of hand pumps. The parts of handpumps most exposed to wear and tear are the cup-leather washers of the plunger, valve seats, gaskets, bolts and nuts, cotter pins and sockets. A stock of a few sections of the pump rod and arrangements for threading the rods are also essential.

Rotary Pumps

Though commonly used in pumping lubricating oils, rotary pumps are sometimes used as boosters in irrigation pumping. Figure 3.24 illustrates the construction details of a rotary pump. The designs commonly encountered are those using cams or gears.

The pump body is a plain housing with inlet and outlet pipes and openings for shafts which carry the gears or cams. One of the gears is the driving gear and is rotated by an outside source of power. The other is the 'idler' gear driven by the driving gear. The gears are fitted closely in the housing and mesh with minimum clearance. They rotate in the direction shown and force the water out through the discharge opening. This creates a partial vacuum and brings in a replacement supply of water along the inlet side. Such an operation creates an even, continuous flow. The capacity delivered is constant, regardless of pressure. Due to the necessity for close clearance and metal to metal contact, rotary pumps work best and last long when pumping liquids having lubricating qualities.

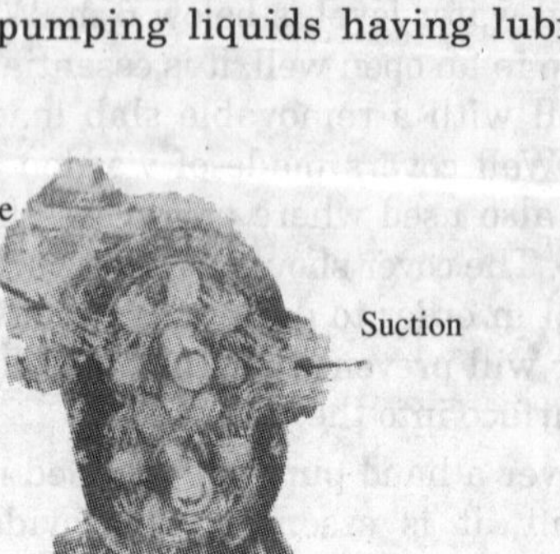

Fig. 3.24. Exposed view of a spur gear type rotary pump showing the principle of working

(***Courtesy:*** Goulds Pumps, Inc).

VARIABLE DISPLACEMENT PUMPS

The distinguishing feature of variable displacement pumps is the inverse relationship between the discharge rate and the pressure head. As the pumping head increases, the rate of pumping decreases. Unlike positive displacement pumps, variable displacement pumps require the greatest input of power at a low head because of the increase in discharge as the pumping head is reduced.

Variable displacement pumps of the impeller type, including centrifugal, mixed flow and propeller pumps are predominantly used in irrigation pumping. They use a rotating impeller to pump water. In general, they range from pumps with small discharges and high heads to large discharges with low heads.

Specific Speed of Pumps

Specific speed is often used as an index to the operating characteristics of pumps. It expresses the relationship between speed, discharge and head. The index, originally developed for FPS units, is the speed in revolutions per minute at which a theoretically and geometrically similar pump would run if proportioned to deliver one gallon per minute against one foot total head at its best efficiency. The specific speed of a pump, as originally developed, is calculated from the following formula:

$$n_s = \frac{n\,Q^{1/2}}{H^{3/4}} \qquad \text{... (3.4)}$$

in which, n_s = specific speed (rpm)

n = pump speed (rpm)

Q = pump discharge, US gallons (per min)

H = Total head (ft)

In metric units, spec ific speed may be defined as the speed of a geometrically similar pump when delivering one cubic metre/second of water against a total head of one metre (Church and Jagdish Lal, 1973). Expressed mathematically,

$$n_s = \frac{n\,Q^{1/2}}{H^{3/4}}$$

$$n_s = \frac{n\,Q^{1/2}}{H^{3/4}} \qquad \text{... (3.4a)}$$

in which, n_s = specific speed (rpm)

n = pump speed (rpm)

Q = pump discharge (m^3/sec)

H = total head (m)

The Bureau of Indian Standards (BSI) has defined the specific speed as the speed expressed in revolutions per minute of an imaginary pump geometrically similar in every respect to an actual pump under consideration and capable of raising 75 litres of water per second to a height of one metre. Expressed mathematically,

$$n_s = \frac{3.65 \ nQ^{1/2}}{H^{3/4}} \qquad \text{... (3.4b)}$$

in which all terms are the same as in Eq. 3.4a. If the discharge is expressed in litres per minute, the expression for specific speed, as per IBSI definition is

$$n_s = \frac{0.01\,49nQ^{1/2}}{H^{3/4}} \qquad \text{... (3.5)}$$

***Example 3.3.** A centrifugal pump at its best point of efficiency discharges 0.03 cubic metres of water per second against a total head of 40 m when the speed is 1450 rpm. Compute the specific speed of the pump.*

Solution:

$$\text{Specific speed} \quad n_s = \frac{nQ^{1/2}}{H^{3/4}}$$

$$= \frac{1450 \times (0.03)^{3/4}}{40^{1/2}} = 15.9 \text{ rpm}$$

The normal range in specific speeds encountered in single suction impeller designs is from 500 to 15,000 when measured in FPS units and 10 to 300 when measured in metric units. Figure 3.25 illustrates the characteristics of impeller pumps as a function of the specific speed in FPS units. For a multi-stage pump, the specific speed is calculated on the basis of the head per stage. When the specific speed of a double suction impeller is compared with that of a single suction impeller, the capacity of the first should be divided by 2 or its specific speed should be divided by $\sqrt{2}$.

Generally, the lower the specific speed, the higher the head that can be developed per stage by the pump. For given total head and suction lift conditions, it is often specified that the specific speed of a pump should be below a certain limit for successful operation. From equations 3.4 and 3.4a, it follows that for the same head-capacity requirements, higher specific speed pumps will run at a higher speed and will be of smaller physical dimensions. Further, for the same speed and capacity, a higher specific speed pump will operate at a lower head, or, for the same speed and head, a higher specific speed pump will operate at a lower head, or, for the same speed and head, a higher specific speed pump will deliver a higher capacity (Fig. 3.25).

Pump Characteristics

A pump operates most satisfactorily under a head and at a speed for which it is designed. The operating conditions should therefore be determined, as accurately as possible, to select pumps well adapted to the particular conditions of operation.

Terminology

Capacity is the volume of water pumped per unit time. It is generally measured in litres per second. Small capacities, however, may be stated in litres per minute or litres per hour and large capacities in cubic metres per second.

Suction lift exists when the source of water supply is below the centre line of the pump.

Static suction lift is the vertical distance from the free suction water level to the centre line of the pump.

Total suction lift is the sum of static suction lift, friction and entrance losses in the suction piping.

Suction head exists when the source of water supply is above the centre line of the pump, as is the usual case in a turbine pump. There will, however, be no suction head in a volute centrifugal pump, unless it is operated as a vertical pump.

Static suction head is the vertical distance from the centre line of the pump to the free level of water to be pumped.

Total suction head is the vertical distance from the centre line of the pump to the free level of the liquid to be pumped minus all friction losses in suction pipe and fittings, plus any pressure head existing on the suction supply. (Total suction head, as determined in a pump test, is the reading of a gauge connected to the pump suction, expressed in metres of water and corrected to

Specific speed (Approximate)	*Impeller cross section*	*Type of pump*	*Head-discharge characteristics*
500	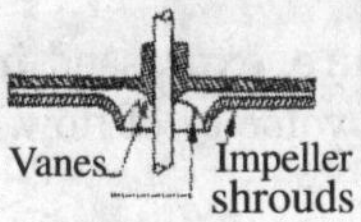	Centrifugal (Radial flow)	High head, low discharge
1000		-do-	Medium to high head and low to medium discharge
2000		-do-	Intermediate head and discharge
5000		Mixed flow	Intermediate to low head and medium to high discharge
10,000		Propeller (Axial flow)	Low head, high discharge

Fig. 3.25. Relationship between impeller design, specific speed and operating characteristics. (Specific speed relates to FPS system).

the pump centre line, plus the velocity head at the point of gauge attachment).

Static discharge head is the vertical distance from the centre line of the pump to the discharge water level.

Total discharge head is the sum of the static discharge head, friction and exit losses in the discharge piping plus the velocity head and pressure head at the point of discharge.

Total static head is the vertical distance from suction water level to discharge water level or the sum of static suction lift and static discharge head. (*Note:* If there is static suction head, total static head is the static discharge head minus the static suction head).

Friction head is the equivalent head, expressed in metres of water required to overcome the friction, caused by the flow through the pipe and pipe fittings.

Pressure head is the pressure, expressed in metres of water, in a closed vessel; from which the pump takes its suction or against which the pump discharges.

Expressed mathematically,

$$H_p = \frac{p}{w} \quad \text{... (3.6)}$$

in which, H_p = pressure head (m)

p = pressure inside the vessel (kg/m^2)

w = specific weight of water (kg/m^3)

Total head is the energy imparted to the water by the pump. It is the sum of total discharge head and total suction lift when *suction lift* exists. It is the total discharge head minus the suction head where *suction head* exists.

Velocity head is the pressure, expressed in metres of water, required to create the velocity of flow.

Expressed mathematically,

$$H_v = \frac{v^2}{2g} \quad \text{... (3.7)}$$

in which H_v = velocity head (m)

v = velocity of water through the pipe (m/sec)

g = acceleration due to gravity (m/sec^2) (usual value 9.81 m/sec^2)

Net positive suction head (NPSH) is the total suction head, determined at the suction nozzle (corrected to pump centre line) minus the vapour pressure of water at the pumping temperature, both expressed in metres. In the pumping of liquids, the pressure at any point in the suction line must not be reduced to the vapour pressure of a liquid. The vapour pressure of a liquid at any given temperature is that pressure at which it will vapourise if heat is added to the liquid or, conversely, that pressure at which vapour at the given temperature will condense into liquid, if heat is subtracted. (Vapour pressure of water at different temperatures is given in standard physical tables). The NPSH is a characteristic of the pump and is usually furnished with the pump characteristic curves.

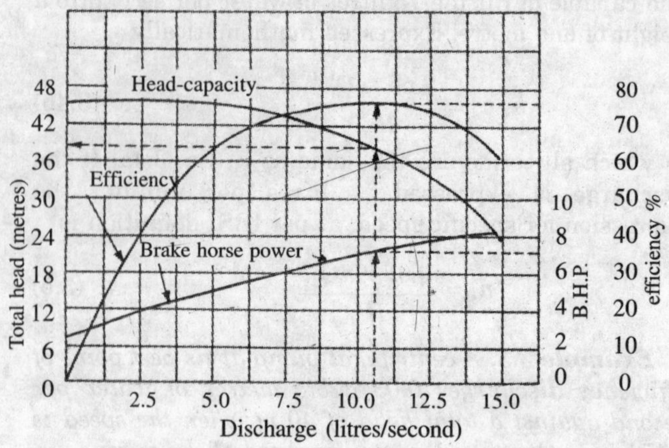

Fig. 3.26. Typical characteristic curves of a centrifugal pump.

Maximum practical suction lift of pumps: For the operation of a centrifugal pump without cavitation, the suction lift plus all other losses must be less than the theoretical atmospheric pressure. The maximum practical suction lift can be computed by the equation,

$$H_s = H_a - H_f - e_s - \text{NPSH} - F_s \quad \text{...(3.8)}$$

in which,

H_s = maximum practical suction lift, or elevation of the pump centre line minus the elevation of the water surface (m)

H_a = atmospheric pressure at water surface (m)

H_f = friction losses in the strainer, pipe, fittings, and valves on the suction line (m)

e_s = saturated vapour pressure of water (m)

NPSH = net positive suction head of the pump including losses at the impeller and velocity head (m)

F_s = factor of safety, which is usually taken as about 0.6 m.

The approximate correction of H_a for altitude is a reduction of 0.36 m for each 300 m of altitude. Friction losses and suction lift should be kept as low as possible. For this reason the suction pipe is usually larger than the discharge pipe, and the pump is placed as close as possible to the water supply. The head loss equivalent due to the vapour pressure of water must be considered to prevent cavitation but it does not add to the total suction head when the pump is operating.

Example 3.4. *Determine the maximum practical suction lift for a pump having a discharge of 38 litres per second. The water temperature is 20°C. The total friction loss in the 10 cm diameter suction line and fittings is 1.5 m. The pump is operated at an altitude of 300 m above sea level. The NPSH of the pump, as obtained from the characteristic curve supplied by the manufacturer is 4.7 m.*

Solution:

e_s at 20°C = 0.24 m (From standard physical tables)

F_s is assumed to be 0.6

Atmospheric pressure = 10.33 – 0.36 = 9.97 m

$H_s = H_a = H_f - e_f - \text{NPSH} - F_s$

$= 9.97 - 1.5 - 0.24 - 4.7 - 0.6 = 2.93$ m

Water horse power (WHP) is the theoretical horse power required for pumping. It is the head and capacity of the pump expressed in terms of horse power.

$$\text{WHP} = \frac{\text{Discharge in litres per sec} \times \text{total head in metres}}{76}$$

$$= \frac{\text{Discharge in cubic metres per hour} \times \text{total head in metres}}{273}$$

Shaft horse power is the power required at the pump shaft.

$$\text{Shaft horse power} = \frac{\text{Water horse power}}{\text{Pump efficiency}}$$

Shaft horse power is used up in the pump in water horse power, disc friction, circulation losses, stuffing box and bearing friction and hydraulic losses (friction, shock and turbulence). It is always greater than water horse power.

Efficiency is the ratio of the power output to power input.

$$\text{Pump efficiency} = \frac{\text{Water horse power}}{\text{Shaft horse power}}$$

Brake horse power is the actual horse power required to be supplied by the engine or electric motor for driving the pump.

(*i*) With direct driven pump (drive efficiency 100%):
Brake horse power = Shaft horse power

(*ii*) With belt or other indirect drives:
Brake horse power

$$= \frac{\text{Water horse power}}{\text{Pump efficiency} \times \text{drive efficiency}}$$

Horse power input to electric motor

$$= \frac{\text{Brake horse power} \times 0.746}{\text{Motor efficiency}}$$

Kilowatt input to electric motor

$$= \frac{\text{Brake horse power} \times 0.746}{\text{Motor efficiency}}$$

Characteristic Curves

The characteristic curves, also called performance curves, show the interrelations between capacity, head, power and efficiency of a pump. The knowledge of pump characteristics enables one to select a pump which is best adapted to particular conditions of operation and thus obtain a relatively high value of efficiency with low operating cost.

It is usual to plot the head, the power input and efficiency as ordinates against capacity as abscissa at a constant pump speed, as shown in Fig. 3.26. The net positive suction head, when shown, is also plotted as ordinate. About 6 to 12 values are taken during the pump test to plot the points. Smooth curves are drawn joining the points. The head-capacity curve shows how much water a given pump will deliver at a given head. As the discharge increases, the head decreases. The resulting efficiency is observed to increase from 0 when the discharge is 0 to a maximum and then decreases. The brake horse power curve for a centrifugal pump usually increases over most of the range as the discharge increases, reaching a peak at a somewhat higher rate of dischage than that which produces maximum efficiency. The curves vary with the speed of the pump. Hence, speed must be considered while selecting a pump to obtain maximum efficiency. Each of the curves also varies with the type of pump.

Several curves, representing different pump speeds or impeller diameters, may be drawn on the same graph. This type of graph shows a number of head-capacity curves for one impeller diameter and different speeds or head-capacity curves for different impeller diameters and one speed (Fig. 3.27). A curve of this type is known as a composite characteristic curve. When this is done, iso-efficiency lines are plotted by joining points of equal efficiency on the head-capacity curves.

The method of reading a performance curve is the same as for any other graph. For example in Fig. 3.26, it is desired to find the head, horse power and efficiency at a capacity of 10.7 litres per second. By reading the graph as illustrated, it is observed that at 10.6 litres per second the pump will develop 38 m head, require 7.1 BHP and has an efficiency of 75.5%. The curve could also by entered from any other scale if one of the other values had been given.

The composite characteristic curve shown in Fig. 3.27 are read in much the same way as the simple curves shown in Fig. 3.26. For example, suppose it is required to select a pump and power unit capable of delivering 16 litres per second against a head of 30 metres. It is important to choose a pump that will operate near its

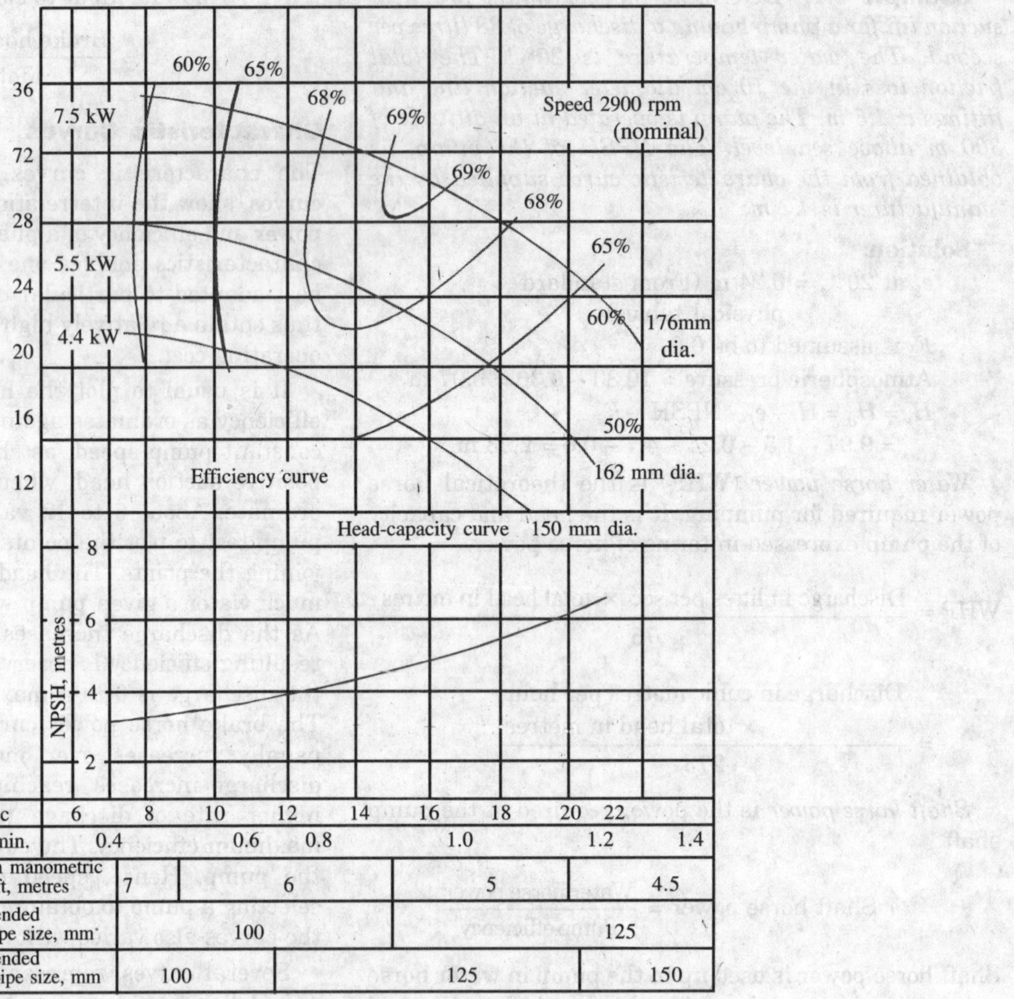

Fig. 3.27. Composite characteristic curves of a centrifugal pump with different impeller diameters (Adapted from Kirloskar Bros. Ltd., Kirloskarwadi, Dist. Sangli, Maharashtra)

highest efficiency most of the time. From the graph, it is observed that the above conditions are met by selecting a pump with an impeller diameter of 176 mm. It will then operate at its highest efficiency of 69%. (By interpolation, however, it may be seen that actual efficiency level at this point is 69.2%). For determining the power requirement, it is observed that the point is above 5.5 kW and below 7.5 kW. Hence a 7.5 kW motor is selected. A more exact value of the power required may be computed from the values of head, capacity and efficiency, using the relationships expressed earlier. The corresponding suction lift is read as 5 m, and size of delivery and suction pipes as 125 mm. If the composite graph presents the pump characteristics at different speeds, the best operating speed along with the other factors are selected in the same way as explained above. Many operating points will fall between various curves. Therefore, it is necessary to understand a composite curve well enough to interpolate and find the approximate values required.

Effect of Speed and Impeller Diameter on Pump Performance

The performance of centrifugal pumps can be changed by changing the impeller diameter or speed.

Effect of change of pump speed: When the speed of a centrifugal pump is changed, the operation of the pump is changed as follows:

(*i*) The capacity varies directly as the speed.

(*ii*) The head varies as the square of the speed.

(*iii*) The brake horse power varies as the cube of the speed.

Expressed mathematically,

$$Q = Q_1\left(\frac{n}{n_1}\right)$$

$$H = H_1\left(\frac{n}{n_1}\right)^2$$

$$P = P_1\left(\frac{n}{n_1}\right)^3$$

or
$$\frac{n}{n_1} = \frac{Q}{Q_1} = \sqrt{\frac{H}{H_1}} = 3\sqrt{\frac{P}{P_1}} \qquad \text{... (3.9)}$$

in which,

n = new speed desired (rpm)
Q = capacity at the desired speed n (litres/sec)
H = head at the desired speed n for capacity Q (m)
P = BHP at the desired speed n at H and Q
N_1 = speed at which the characteristics are known (rpm)
Q_1 = capacity at speed n_1 (litres/sec)
H_1 = head at capacity Q_1 and speed n_1 (m)
P_1 = BHP at speed n_1 at H_1 and Q_1

Effect of change of impeller diameter: Changing the impeller diameter has the same effect on the pump performance as changing the speed. Therefore, the following relationships apply :

(*i*) The capacity varies directly as the diameter.
(*ii*) The head varies as the square of the diameter.
(*iii*) The brake horse power varies as the cube of the diameter.

Expressed mathematically,

$$Q = Q_1 \times \left(\frac{D}{D_1}\right)$$

$$H = H_1 \times \left(\frac{D}{D_1}\right)^2$$

$$P = P_1 \times \left(\frac{D}{D_1}\right)^3$$

or
$$\frac{D}{D_1} = \frac{Q}{Q_1} = \sqrt{\frac{H}{H_1}} = 3\sqrt{\frac{P}{P_1}} \qquad \text{... (3.10)}$$

in which,

D = changed diameter of impeller (mm)
D_1 = original diameter of impeller (mm)

The other terms are analogous to those used in equation (3.9).

When pumps are driven by belts, or variable speed drivers, it is possible to change the operating speed. Many pumps are directly coupled to electric motors and must run at constant speed. In this case, it is necessary to change the impeller diameter to alter pump performance.

CENTRIFUGAL PUMPS

Amongst modern pumps, centrifugal pumps are most widely used in irrigation practice. They are simple in construction, easy to operate, low in initial cost and produce a constant steady discharge. The wearing parts are few. They are adapted to direct motor or engine drives without the use of expensive gears. This type of pump is well adapted to usual pumping services such as irrigation, water supply and sewage service. Having no valves, the pump can handle liquids having solids in suspension, provided it is constructed to suit such conditions.

Principles of Operation of Centrifugal Pumps

A centrifugal pump is a rotary machine consisting of two basic parts—the rotary element or impeller and the stationary element or casing (Fig. 3.28 to 3.30). The impeller is a wheel or disc mounted on a shaft and provided with a number of vanes or blades usually curved in form. The vanes are arranged in a circular array around an inlet opening at the centre. In some pumps, a diffuser consisting of a series of guide vanes or blades, surrounds the impeller (Fig. 3.29). The impeller is secured on a shaft mounted on suitable bearings. The shaft usually has stuffing box or seal where it passes

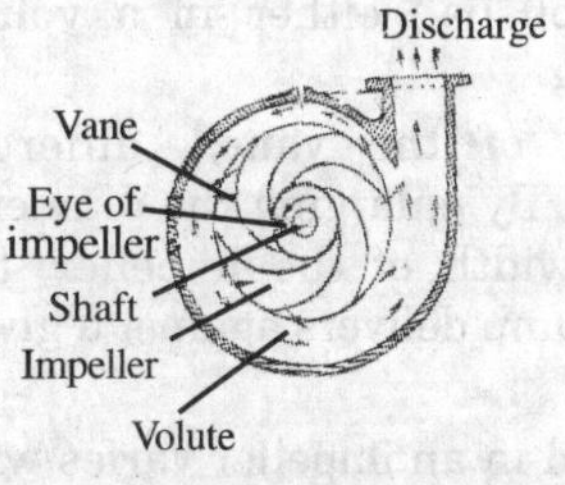

Fig. 3.28. Volute type centrifugal pump.

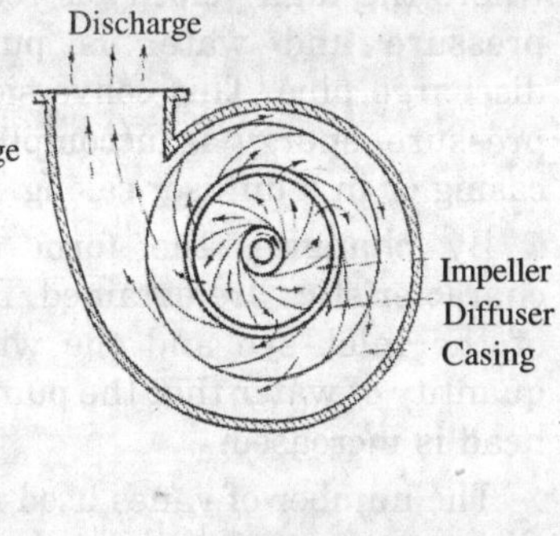

Fig. 3.29. Diffuser type centrifugal pump.

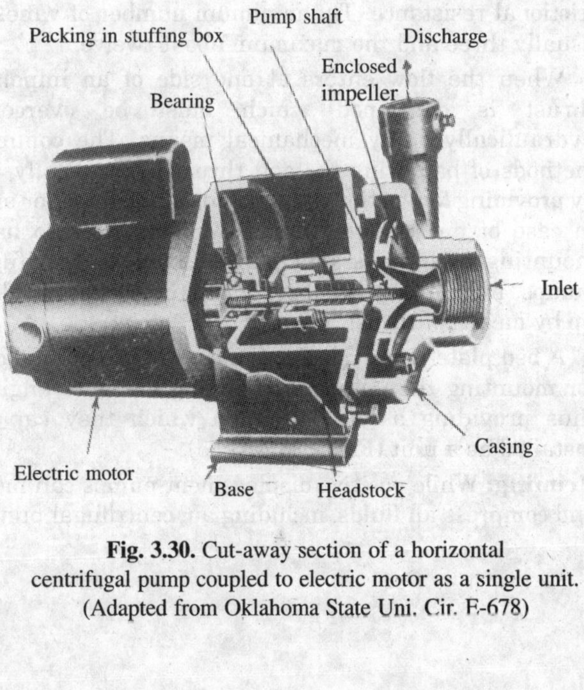

Fig. 3.30. Cut-away section of a horizontal centrifugal pump coupled to electric motor as a single unit. (Adapted from Oklahoma State Uni. Cir. E-678)

through the casing wall (Fig. 3.30). Stuffing box packings are generally made of materials such as asbestos or organic fibre. The casing surrounds the impeller and is usually in the form of a spiral or volute curve with a cross-sectional area increasing towards the discharge opening.

A centrifugal pump may be defined as one in which an impeller rotating inside a close-fitting case draws in the liquid at the centre and by virtue of centrifugal force throws out the liquid through an opening at the side of the casing. The underlying hydraulic principle in the design of an impeller is the production of high velocity and the partial transformation of this velocity into pressure head. In operation, the pump is filled with water and the impeller is rotated. The blades cause the liquid to rotate with the impeller and in turn impart a high velocity to the water. Centrifugal force causes it to be thrown outward from the impeller into the casing. The outward flow through the impeller reduces the pressure at the inlet, allowing more water to be drawn in through the suction pipe by atmospheric pressure or an external pressure. The liquid passes into the casing where the high velocity is reduced and converted into pressure and water is pumped out through the discharge pipe. This conversion of velocity energy into pressure energy is accomplished either in a volute casing or in a diffuser casing.

By changing the form of the vanes, different characteristics are obtained. By enlarging the diameter of the inlet eye and the width of the impeller, the quantity of water that the pump delivers against a given head is increased.

The number of vanes used in an impeller varies with the operating conditions. In general, the higher the head, the more the vanes used; and greater the rate of pumping, the fewer the vanes. Too few vanes provide poor guidance for the water; too many cause excessive frictional resistance. The minimum number of vanes is usually three and the maximum about twelve.

When the flow enters at one side of an impeller, thrust is developed which must be overcome hydraulically or by mechanical means. The common methods of balancing the end thrust hydraulically are by providing for suction at both ends of the impeller and, in case of multi-stage pumps by using back to back mounting of impellers in pairs. In most centrifugal pumps, including vertical turbines, the thrust is taken up by means of thrust bearings.

A bed plate is provided at the base of the pump body for mounting the pumps and driving motor or engine, thus providing a foundation on which they can be installed as a unit (Fig. 3.33 & 3.35).

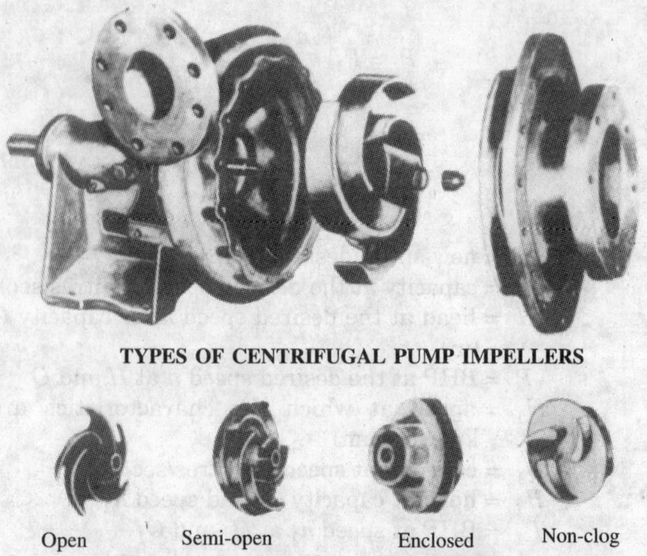

Fig. 3.31. Components of a volute centrifugal pump. At the top is an exploded view. An open type impeller is seen in the middle. Different types of impellers used in volute centrifugal pumps to match specific pumping conditions are shown at the bottom.

Priming: While positive displacement pumps can move and compress all fluids, including air, centrifugal pumps are very limited in their capacity to do so. Hence, they are to be primed, or filled with water upto the top of the pump casing to initiate pumping.

Many devices and techniques are used for priming centrifugal pumps and maintaining the primed condition. In general, they involve one or a combination of the following: (*i*) a foot valve to hold the water in the pump (Fig. 3.32), (*ii*) an auxiliary piston pump to fill the pump casing and suction line with water, (*iii*) connection to an outside source of water under pressure for filling the pump, and (*iv*) use of a self-priming construction. The self-priming construction retains water for priming in an auxiliary chamber which forms a part of the pump body. The foot valve retains water in the pump and suction line. The need for priming each time the pump is started is therefore eliminated. The valve is installed at the bottom of the suction pipe. A strainer is usually fixed to the foot valve to prevent foreign material from entering the pump.

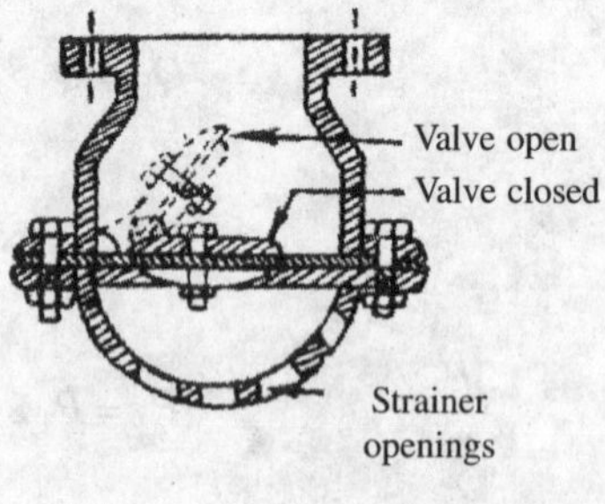

Fig. 3.32. Foot valve with strainer.

Centrifugal Pump Classification

Centrifugal pumps are built in a wide variety of types. They are primarily classified on the following basis:

1. Type of energy conversion
   (*a*) Volute (*b*) Diffuser or turbine.
2. Number of stages
   (*a*) Single stage (*b*) Multi-stage.
3. Impeller types
   (*a*) Open (*b*) Semi-open
   (*c*) Closed (*d*) Non-clog.
4. Type of suction inlet
   (*a*) Single suction (*b*) Double suction
5. Construction of casing
   (*a*) Vertically split (*b*) Horizontally split
6. Axis of rotation
   (*a*) Horizontal (*b*) Vertical
7. Method of drive
   (*a*) Direct connected (*i*) Coupled (*ii*) Close-coupled or Uni-built
   (*b*) Belt-driven.

Volute centrifugal pumps. The volute type pump (Figs. 3.31 and 3.33) has a casing made in the form of a spiral or volute curve. The volute casing starts with a small cross-sectional area near the impeller periphery and increases gradually to the pump discharge. The casing is proportioned to reduce the velocity of water gradually, as it flows from the impeller to the discharge, thus changing velocity head into pressure head. Most of the irrigation pumps are of the volute type.

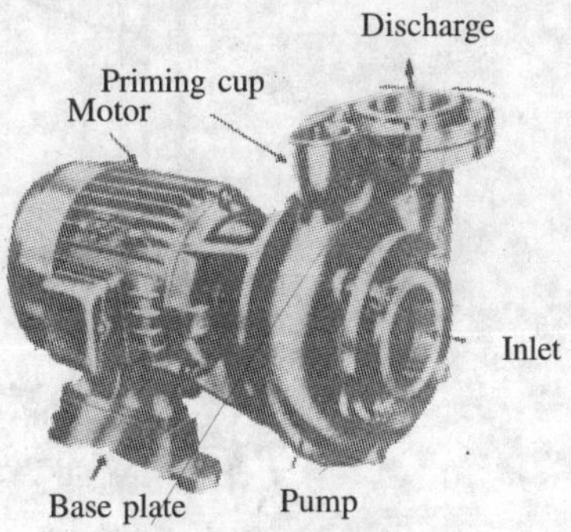

Fig. 3.33. A close-coupled horizontal volute centrifugal pump and electric motor (***Courtesy:*** Jyoti Ltd.)

Diffuser or turbine pump: In the turbine type pump, the impeller is surrounded by diffuser vanes. The diffuser vanes have small openings near the impeller and enlarge gradually to their outer diameter where the liquid flows into the chamber and around to the pump discharge. A major part of the conversion of velocity into pressure takes place between the diffuser vanes. The diffuser vane casing was introduced from water turbine practice where diffusion vanes are indispensable. Hence, these pumps are often called turbine pumps.

The choice between volute type and turbine type pumps varies with the conditions of use. Ordinarily, the volute type pump is preferred for large capacity, low head applications. Turbine pumps are usually used for high head conditions. Turbine pumps are most popular in deep tubewells because of its design advantage where the diameter of the pump is small.

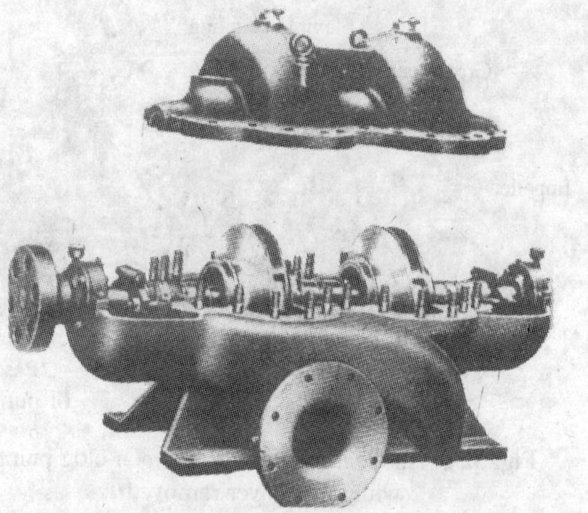

Fig. 3.34. A two-stage horizontal centrifugal pump with top cover lifted up (*Courtesy*: Lee Howl and Co. Ltd.)

Single stage and multi-stage pumps: A single stage pump is one in which the total head is developed by a single impeller. A multi-stage pump has two or more impellers on a common shaft, acting in series in a single casing (Fig. 3.34). The liquid is conducted from the discharge of the preceding impeller or stage to the suction of the following impeller, causing a pressure or head increase as it passes through each stage. The use of multi-stage pumps is a standard practice in volute as well as turbine type pumps for operation under high heads. For a given type of impeller, the characteristics exhibited by a multi-stage pump are as follows:

1. The head and power requirement increase in direct proportion to the number of stages (impellers).
2. The discharge capacity and efficiency are almost the same for a single stage of the pump operating alone.

Types of impellers: The design of the impeller greatly influences the efficiency and operating characteristics of centrifugal pumps. Centrifugal type impeller used in irrigation practice may be open, semi-open, or closed (Fig. 3.31). An open impeller consists essentially of a series of vanes attached to a central hub. It is used to pump water with a limited amount of small solids. A semi-open or semi-enclosed impeller has a shroud or side wall on one side only, usually on the back. It can be used to pump water having some amount of suspended sediments. In an enclosed impeller, the vanes are enclosed between shrouds or side walls on either side. It is designed to pump clear water. Enclosed impeller develops higher efficiencies, especially in high pressure pumps.

For pumping ordinary water, centrifugal pump impellers may be made of bronze or cast iron. To handle brackish or salt water, gun metal impellers are commonly used.

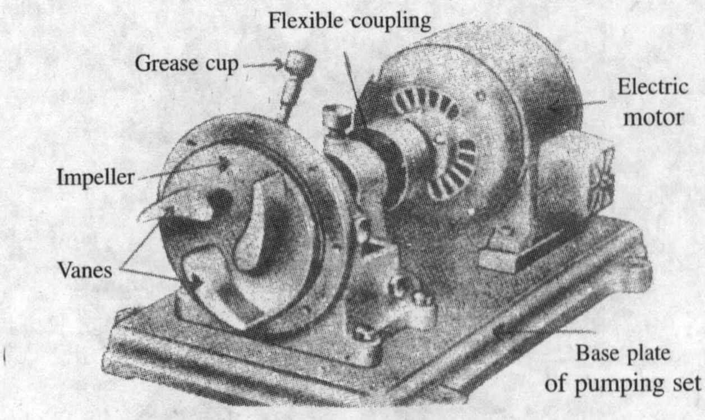

Fig. 3.35. An electric motor driven non-clog pump with outer cover removed
(***Courtesy***: Abang Engg. Co. Ltd.)

Non-clog impeller (Fig. 3.35): Non-clog impellers are specially designed for sewage service. They have vanes which are well rounded at their entrance ends and have large passage-ways between the vanes. They can handle sewage water containing solid particles, rags and other impurities.

Single suction and double suction pumps: In a single suction pump, the liquid enters the impeller from one side (Fig. 3.33). In a double suction pump, it enters from both sides. The double suction impeller is similar to two single suction impellers cast back to back. They are theoretically in axial hydraulic balance making a thrust bearing unnecessary. However, due to manufacturing difficulties, double suction pumps are not as common as single suction pumps.

Horizontal centrifugal pumps: A horizontal centrifugal pump has a vertical impeller mounted on a horizontal shaft (Fig. 3.30). This type of pump is most commonly used in irrigation. It costs less, is easier to install and is more accessible for inspection and maintenance. The pump should be installed so that it is always above the water surface, but as close to it as possible. For satisfactory operation, the suction lift of the pump should not exceed 4.5 to 6 m.

Vertical centrifugal pumps: The vertical centrifugal pump has a horizontal impeller mounted on a vertical shaft (Fig. 3.36). This type of pump has the advantage that it can be lowered to the depth required to pump water and the vertical shaft is extended to the surface where the power is applied. The volute type vertical centrifugal pump may be either submerged or exposed. The exposed pump is set in a sump at an elevation that will accommodate the suction lift. In the submerged pump, the impeller and suction entrance remain submerged below the water level. Thus, the pump does not require priming. However, the arrangement is not popular in irrigation practice due to the difficulty in lubricating the bearings. Volute type vertical centrifugal

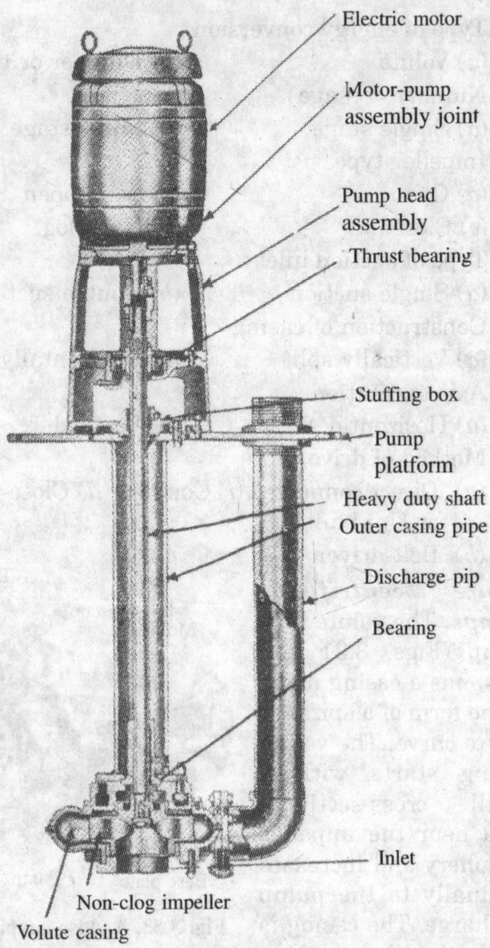

Fig. 3.36. Cut-away sectional view of a vertical centrifugal pump which is commonly used as a sump pump in sewage service.
(***Courtesy:*** Aurora Pumps)

pumps are usually restricted to pumping heads upto about 15 metres and are commonly used to pump from sumps or pits.

Close-coupled or unibuilt pumps: Close-coupled pumps are built with a common shaft and bearings for the pump and driver so as to form a single compact unit (Fig. 3.33). They are commonly used with electric motor driven pumping sets of small to medium capacity.

Direct connected pump: In this type, the pump is mounted on a base plate and is connected directly to its driver through a flexible coupling (Fig. 3.37). Flexible couplings are most commonly used to connect the pump shaft to motor or engine shaft. They permit minor misalignment of the shafts.

Belt-driven pump: The pump is provided with a pulley head for belt drive (Fig. 3.37). It is suitable when the power source is located away from the pump. Usually a set of two pulleys–one fast pulley and a loose pulley—are provided to facilitate engaging and disengaging the pump.

Medium Lift Submersible Centrifugal Pump with Hydraulic Drive

A comparatively recent development in irrigation pumps is the medium lift (submersible) pump developed by Volvo Flygmotor AB, Sweden. The unit, which is designed to lift water from pumping depths between 5 and 25 m comprises of one internal combustion engine (or less commonly an electric motor) and a submersible centrifugal pump. These two are connected together by an oil hydrostatic transmission. The engine drives a hydraulic pump which, through hydraulic hoses, feeds a hydraulic motor connected to the centrifugal pump (Fig. 3.38). The two main advantages of the pump are the easy and efficient transmission of power from engine to pump, using the hydraulic drive, and the ability to operate without any suction lift limits, as the pump works submerged below the water level. It is also easily portable. It can lift about 4 to 33 litres of water per second with a total head ranging from 5 to 15 m. The medium lift submersible pump is especially adapted to deep open wells in areas where electricity is not available and pumping from rivers and streams having steep banks. The pump can be used in either stationary or mobile applications.

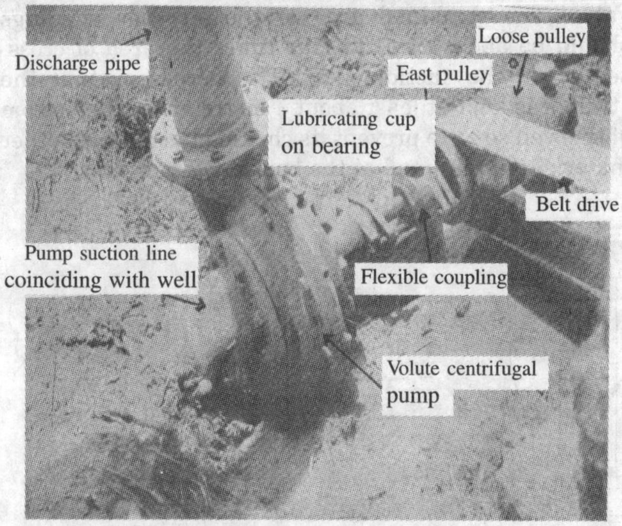

Fig. 3.37. A belt-driven centrifugal pump coupled directly to a shallow tubewell.

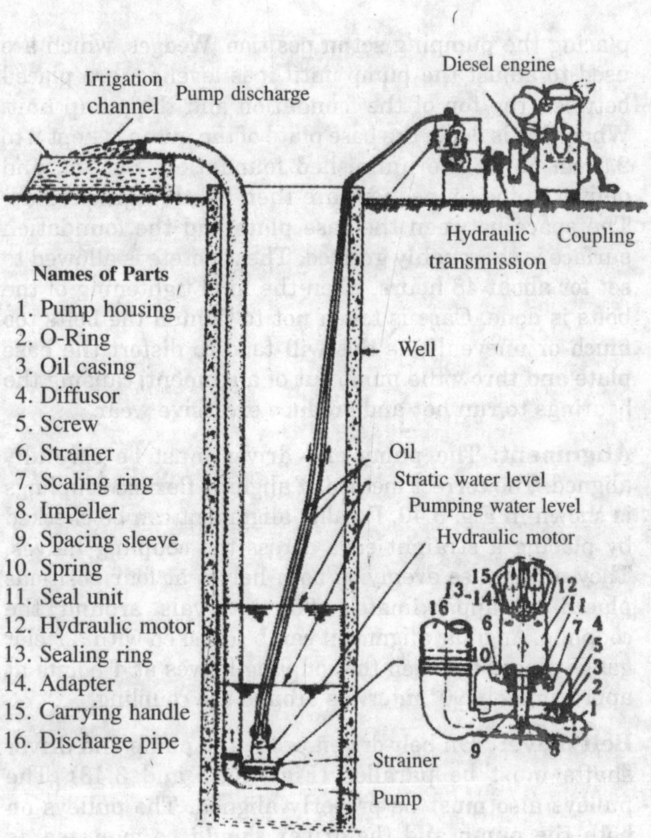

Fig. 3.38. Medium lift submersible centrifugal pump with hydraulic drive.

Casing: Centrifugal pump casings are usually of the horizontally split type (Fig 3.34) or vertically split type (Figs. 3.31 and 3.33). Horizontally split easings are also called axially split. In this, both suction and discharge openings are usually on the lower half of the casing. The upper half can be lifted up for inspection. Vertically split casings are commonly used in close coupled designs. Casings are usually made of cast iron.

Installation of Horizontal Centrifugal Pumps

Proper installation of a centrifugal pump ensures its prolonged trouble-free service. The pump is installed as close to the water surface as possible. It is located at an easily accessible place in clean, dry, well ventilated surroundings. To ensure maximum capacity, the site selected should permit the use of the shortest and most direct suction and discharge pipes.

Foundation: The pump is installed on a foundation rigid enough to absorb all vibrations (Fig. 3.39 and Fig. 3.41). A cement concrete base of adequate dimensions, steel girders, and wooden beams are used as pump foundations. Pumps driven by electric motors do not require any special foundation but are bolted securely to support and anchor the pump independent of the piping. The support to which the motor-driven pumping set is bolted should be stiff and rigid enough to prevent bending. While installing a motor-driven pump, shims are inserted where necessary, between the base plate and foundation, so as to give an even bearing on all the four corners.

Direct connected pumping sets are always mounted in a horizontal position on a level foundation. Concrete foundations, with foundation bolts imbedded in the concrete, are in general, a satisfactory arrangement. The recommended method of anchoring bolts in concrete foundations is to place them in pipe sleeves (Fig. 3.39). The inside diameter of the sleeve is kept approximately two and half times the diameter of the bolts. Large washers placed on bolts to hold them from slipping through the pipe sleeves. The top of the foundation should bc roughened, cleaned and dampened before placing the pumping set in position. Wedges, which are used to adjust the pump until it is levelled, are placed between the top of the foundation and the pump unit. When this is done, the base plate of the pump is kept 2 to 2.5 cm above the unfinished foundation. Suction and delivery pipe connections are then lightly tightened up. The space between the base plate and the foundation surface is thoroughly grouted. The concrete is allowed to set for about 48 hours, when the final tightening of the bolts is done. Care is taken not to tighten the bolts too much or unevenly, as this will tend to distort the base plate and throw the pump out of alignment, causing the bearings to run hot and produce excessive wear.

Alignment: The pump and driver must be carefully aligned. The correct method of aligning flexible couplings is shown in Fig. 3.40. Parallel alignment can be checked by placing a straight edge across the coupling halves. They must raise evenly on both halves at four positions placed at approximately 90° intervals around the coupling. Angular alignment can be checked with a feeler gauge placed between the coupling halves at 4 points at approximately 90° intervals around the coupling.

Belt drive: On belt-driven units, the pump and driver shafts must be parallel. (Figs. 3.37 and 3.43). The pulleys also must be properly aligned. The pulleys on both the pump and the driver should be mounted as close to the bearing housing as possible to minimise overhang and also allow sufficient clearance for the end play of the shafts. On flat belt and V-belt drives, the tight side of the belt should be at the bottom. A vertical drive is avoided on flat belt drives. An angle of 45° or less between the line of shaft centres and the horizontal is desirable. Normally, the belt speed should not exceed about 1,500 metres per minute. The ratio of the diameters of the pulleys should not exceed 5 to 1. The belt tension is adjusted just sufficient to prevent slippage. Excessive tension overloads the bearings.

Supports for piping: Suction and delivery pipes should be supported independently of the pump without being strained into position (Fig. 3.41).

Suction piping: The suction piping should be as direct and short as possible. It should have a minimum of fittings so as to avoid excessive friction losses. Sharp angle bends should be avoided. Particular care is taken to see that the suction pipe and its joints are absolutely air-tight. The size of the suction pipe should be equal to or larger than the suction opening of the pump. The foot valve is located at the bottom of the suction pipe with a minimum submergence of 60 cm below the pumping water level. The size of the suction pipe is so selected that the velocity of water on the suction side does not exceed 3 metres per second. Suction velocities in excess of this may cause cavitation. There should be ample openings on the strainer which is fixed below the foot valve. The combined area of the strainer openings should be about three to four times the area of cross-section of the suction pipe. It is necessary that the strainer is kept at least about 1 metre above the bottom of the well so as to prevent its choking by mud and other material accumulated at the bottom of the well.

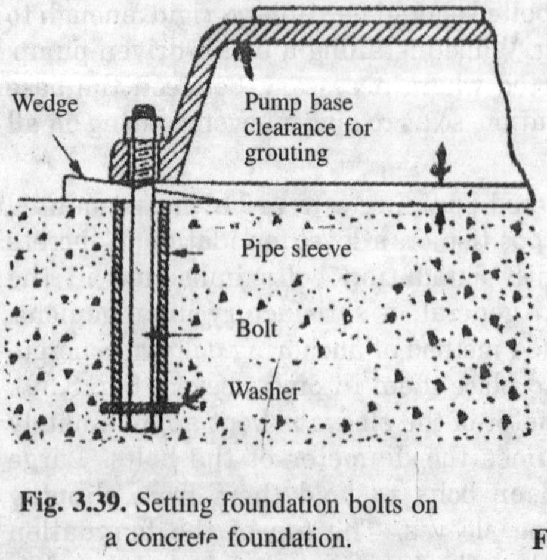

Fig. 3.39. Setting foundation bolts on a concrete foundation.

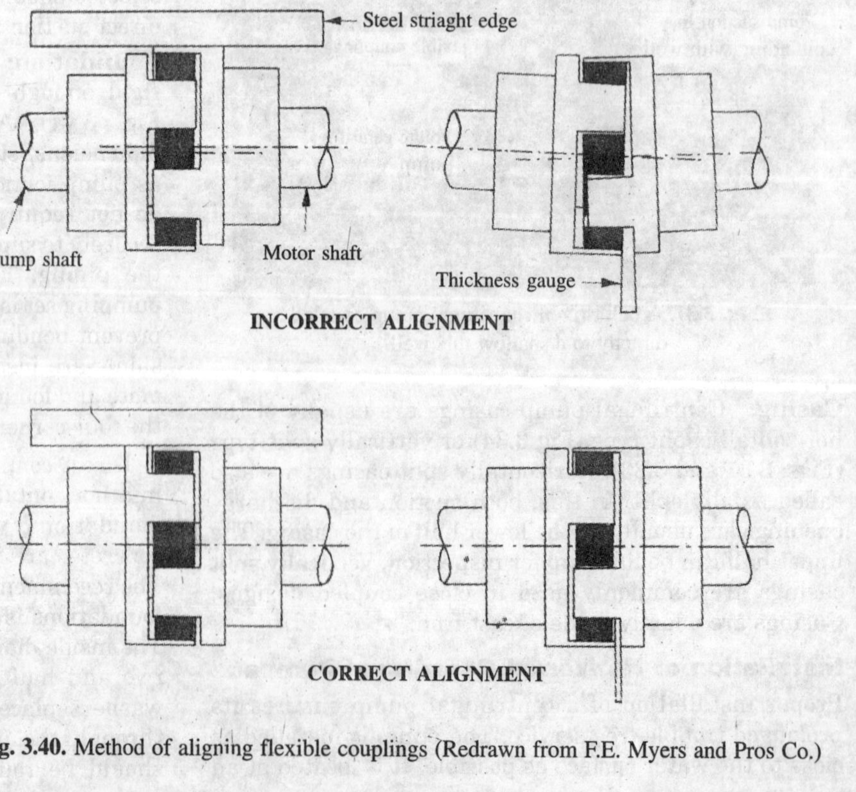

Fig. 3.40. Method of aligning flexible couplings (Redrawn from F.E. Myers and Pros Co.)

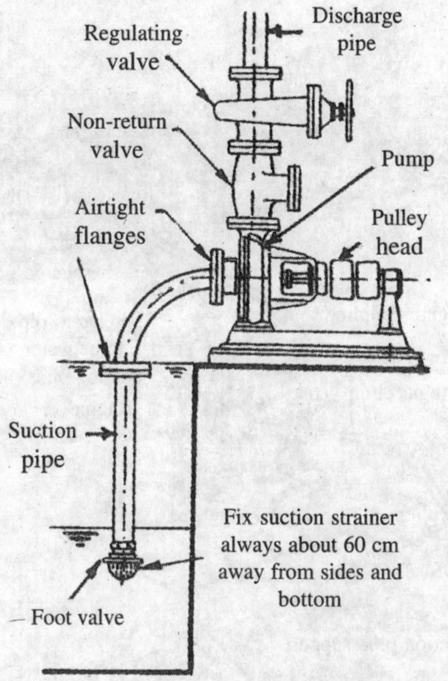

Fig. 3.41. Correct arrangement of suction piping and valves on the discharge side of a centrifugal pump.

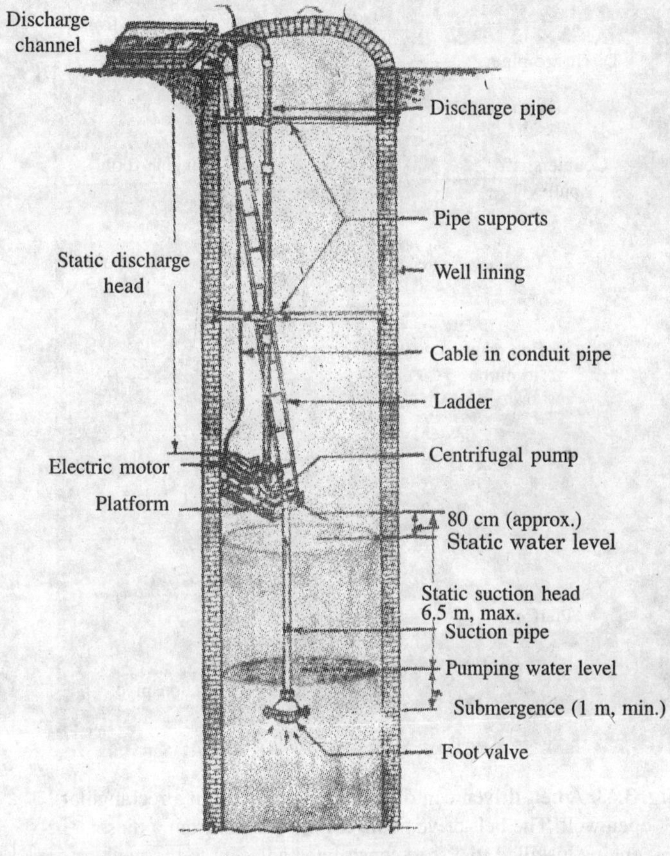

Fig. 3.42. Schematic sketch illustrating the details of installation of a motor-driven centrifugal pump in a deep open well.

Delivery piping: A pipe of suitable size to carry the normal discharge of the pump without excessive frictional resistance should be selected. Use of bends, elbows, tees, and other fittings is kept to the minimum to reduce head loss in the discharge line. However, when pumping to distant places or under high heads, it is desirable to have a reflux or non-return valve fixed on the delivery side close to the pump. This will overcome water hammer which may occur at the time of stopping the pump. Sometimes, a sluice valve is fitted immediately after the reflux valve. The sluice valve helps in regulating the discharge rate and thus creates controlled working conditions. It is also useful for disconnecting the pump for inspection and repairs without disturbing the delivery pipeline.

Installation of Centrifugal Pumps in Deep Open Wells

The horizontal centrifugal pump should be installed so that it is always above the water surface, but is also as near to it as possible. When driven by electric motor, the pump and motor coupled to it, can be conveniently installed close to the static water level of deep open wells (Fig. 3.42). It is especially important to avoid submergence of electric motors. At places where the water table varies greatly between seasons during the year, it may be necessary to change the location of pumps during seasons. In such cases, two or more platforms may be necessary in the well, the upper platform to be used during the period of a comparatively high water table. The pumping set is lowered to the lower platform when the water table falls. When the depth of water column in the well exceeds the pump suction limit substantially, it may be necessary to install the centrifugal pump on a float and provide a minimum length of flexible pipe on the delivery line.

The centrifugal pump is not efficient in deep-wells when it is driven by an engine, as it involves excessive power loss with long vertical belt drives used in transmitting power from the engine placed at the ground surface to the pump installed in the well (Fig. 3.43). To reduce the distance of belt drive, the engine is usually installed in a pit. A better arrangement is the use of a vertical shaft drive to transmit power from the engine placed at the ground surface to the pump placed close to the static water level in the well. Short length quarter-twist belt drives are used to transmit power from the engine to the vertical shaft and from the shaft to the pump.

When used to pump water from a tubewell with a deep water table, it is necessary that the centrifugal pump is installed in a pit such that the suction lift does not exceed about 6 m. Since the construction of deep water-tight pits is often difficult and costly, and since the draw-down in tubewells usually exceeds the limit of 6 m, volute centrifugal pumps have very limited application in deep tubewells.

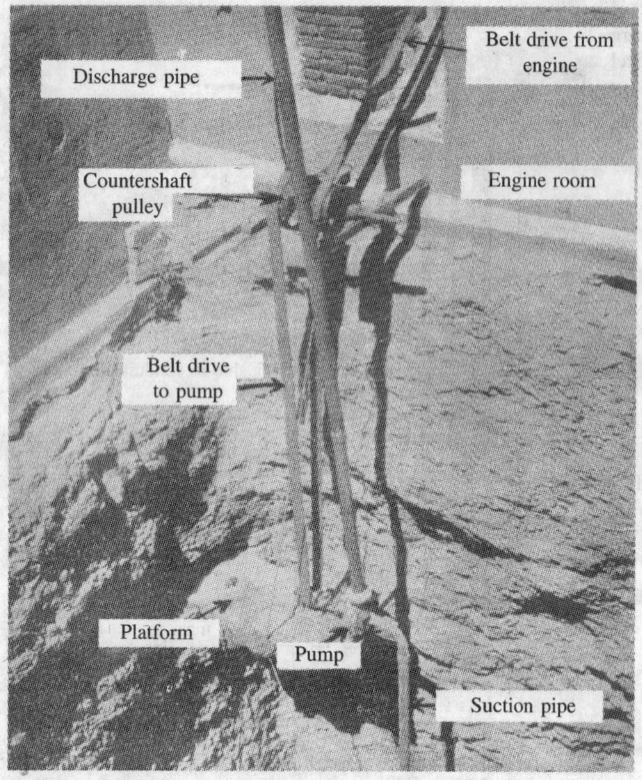

Fig. 3.43. A belt-driven centrifugal pump installed in a rectangular open well. The belt-drive to the countershaft is from a diesel engine installed in a shed constructed adjacent to the well.

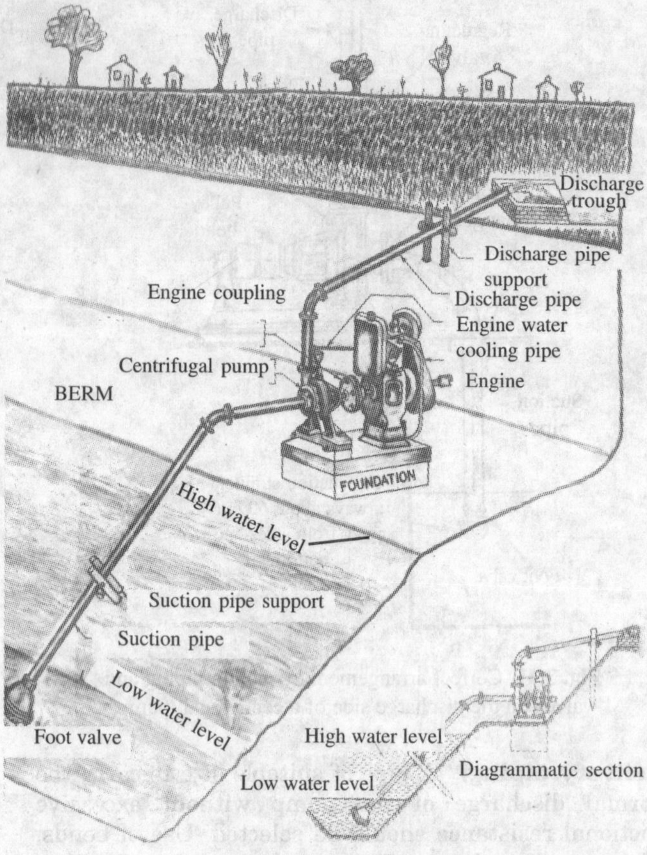

Fig. 3.44. Schematic sketch illustrating the installation of a centrifugal pump coupled to a vertical diesel engine to lift water from a canal.

Pumping from canals and rivers: Centrifugal pumps are commonly used for river and canal pumping (Fig. 3.44). The pump, coupled to an engine or electric motor, may be fixed on a permanent foundation. Engine-operated pumping sets may also be installed on a float and used as a portable unit. Flexible pipes are preferred for suction and discharge lines on portable units. When installed on a permanent foundation, the suction and delivery lines are made of rigid pipes which may be kept at an angle, conforming to the slope of the canal bank. The pipes are supported on posts at suitable points. The foot valve is so located that the minimum water level during the pumping period will be at least 60 cm above it.

Portable Pumping Sets

Centrifugal pumps may be used as portable units for pumping from several wells, one after the other, or from streams, canals or rivers. The essential requirement for portable units is that the pumping water level should not be deeper than about 6 metres from the ground surface. Portable pumping sets are usually engine-driven. The pump and engine are mounted on a common base plate which is fixed to a trolley made of steel girders or wooden beams. The trolley is mounted on steel wheels or rubber tyred wheels.

Fig. 3.45. A portable centrifugal pump fixed to the 3-point power transmission linkage system of a wheel tractor.

Fig. 3.46. A case of improper operation of power unit—a tractor operating a small size centrifugal pump installed to lift water from a shallow tubewell.

Tractor-driven centrifugal pumps: Tractor power can be used with advantage to operate centrifugal pumps. Direct shaft drive or belt drive may be used for power transmission from tractor to pump (Fig. 3.45 and 3.46). With a belt drive, power is transmitted from the tractor belt pulley to the pump pulley by a flat belt. The size of the pump pulley should be selected in relation to the speed and size of the tractor pulley so that the pump may operate at the rated speed. With shaft drive it is necessary to install a speed increasing mechanism between the tractor power take-off shaft and the pump.

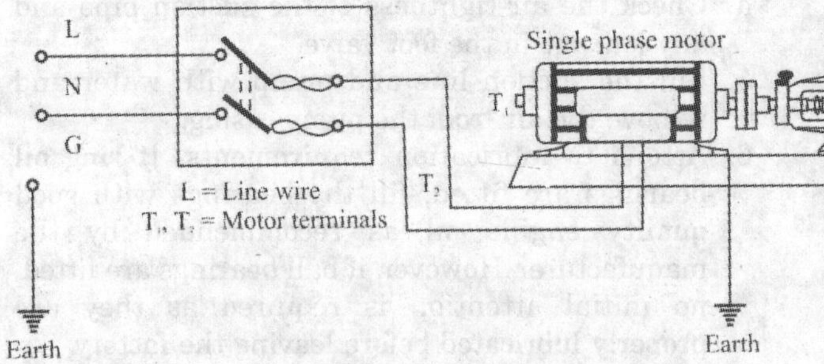

Fig. 3.47. Wiring diagram for a single phase motor with hand operated switch.

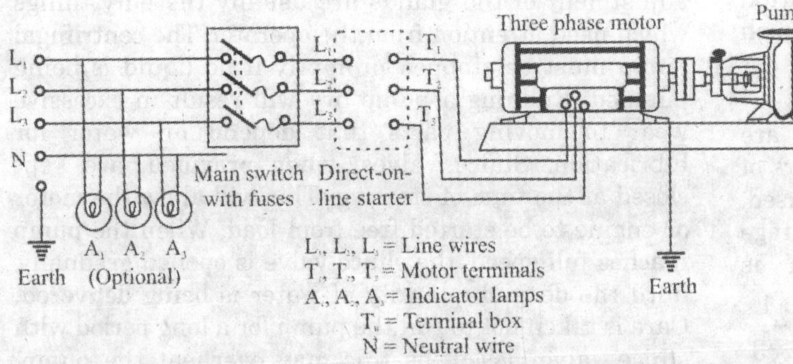

Fig. 3.48. Wiring diagram for 3-phase motor with direct-on-line starter. The indicator lamps shown in the diagram are optional.

This is because the speed of the tractor shaft is standardised at 540 ± 10 rpm and the pump speed usually exceeds 1,450 rpm. With portable units, the pump and the speed increasers are mounted as a unit on a two-wheel trailer hitched to the tractor drawbar. The desired pump speed is obtained through the use of a gearbox. Care is taken to properly align the shaft. A telescopic shaft, using universal joints, is usually used. The power-take-off shaft is usually shielded as a safety measure.

Electrical Connections

Efficient use of electric motors in irrigation pumping requires the selection of the right type of motor for the job, its proper installation, and use of suitable controls for the operation and protection of the motor. Motor and control wiring and grounding should be carefully done in accordance with the Electric Code in force in the area. All wiring jobs should be done, in accordance with the existing wiring rules, by a competent, authorised electrician. Wiring should be of ample capacity, based on a maximum drop of 2 per cent of line voltage at full load current.

Alternating current, 50 cycles, 400-440 volts, 3-phase electric power supply is usually available on the farm. Four wire, 3-phase supply is taken from the local sub-station. Three of the wires are phase wires and the fourth is the neutral wire. The 3-phase system is usually used for operating 3-phase electric motors. However, single phase 230 volts line could also be tapped from the 3-phase system for lighting, running single phase motors and other uses. The wiring is so done that single phase loads are equally divided between all the phases of the 3-phase system. In the 2-phase system the neutral wire of the service entrance to all buildings must be grounded for safety reasons. Grounding of a circuit involves the permanent contact of one of the circuit wires with damp earth (Figs. 3.47 to 3.49). Usually an uninsulated single wire is driven into the ground.

AC motors: The most common type of electric motors used in irrigation pumping are the 3-phase squirrel cage induction motors. They are low in initial and running costs, smooth in running and have a long life, if maintained properly. Sometimes, however, single phase induction motors are also used with pumps under low head and low discharge conditions.

The following accessories are used in the electrical connections using 3-phase motors: (*i*) Energy meter, (*ii*) Volt meter and ammeter, (*iii*) Indicator lamps, (*iv*) Main switch, and (*v*) Starter.

The energy meter, voltmeter and ammeter are useful indicators of the line

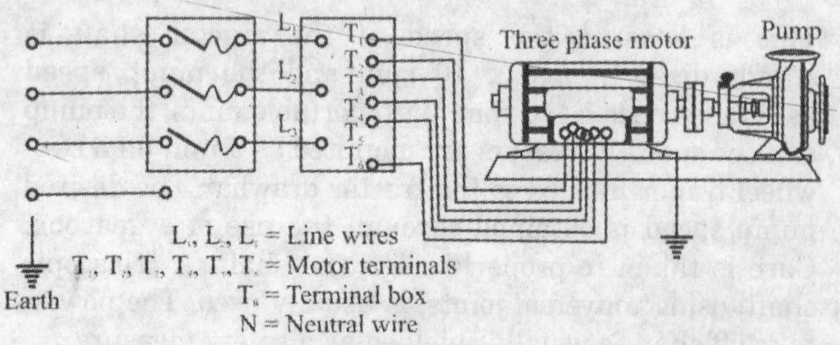

Fig. 3.49. Wiring diagram for a 3-phase motor with star-delta starter.

supply but are not essential in every installation. The energy meter gives the amount of energy in kilowatt hours consumed by the motor. The voltmeter and ammeter indicate the line voltage and current, respectively. Indicator lamps show whether all the three phases are live.

Main switch: The main switch connects the motor to the supply lines. It also includes fuses of appropriate capacity to protect the supply lines and instruments from faults in the motor. The switch contains a moving system connected to an insulated lever. In a single phase switch, two fuses are provided, one on the phase wire and the other on the neutral. In a 3-phase switch, fuses are provided for the three phase wires. The fuses are intended to give protection against any persistent fault condition in the event of the starter becoming defective.

AC motor starters: If an AC motor is started on full line voltage, it will draw about four times its normal running current. This may damage the motor and cause line disturbances affecting the operation of other motors on the same line. For single phase small motors, a hand-operated switch can be used to control the motor (Fig. 3.47). On three phase motors it is necessary to insert a starter in the line which will reduce the starting current. *Direct-on-line* or *push button starters* are commonly used for motors up to 5 H.P. Star-delta starters are used for motors above 5 H.P. (Figs. 3.48 and 3.49).

The direct-on-line starter connects the motor directly to the line (Fig. 3.48). It is simple in construction and is comparatively lower in cost. The starter contains an overload relay to enable it to shut off in case of overloads. It may have handle control or push button control. The push button type starter has two push buttons, one for starting (usually green) and the other for stopping (usually red) the motor.

Star-delta starters of the air break type are commonly used with motors of 5-15 H.P. For motors of higher capacity, oil-immersion type starters are used. Star-delta starter (Fig. 3.49) uses the reduced voltage starting principle. While starting, the motor is connected star and the line voltage is reduced to $\frac{1}{\sqrt{3}}$ times the phase voltage. The reduced phase voltage reduces the phase current in the same proportion. Six terminals are taken from the motor so that they can be changed to 'star' at starting time and 'delta' in the running condition. In starting, as the starting lever is pressed down, the motor starts rotating as a star connected machine. After the motor accelerates, the starting lever is raised up, the connection is changed to 'delta' and the motor runs on full voltage.

Starters are usually equipped with overload releases to protect the motor from 'single phasing' and overload. Overload releases may be of the magnetic type or thermal type and could be adjusted to 'trip off' and disconnect the motors in case of a power failure in any of the phases. The overload releases are set to trip off at about 10 per cent above the maximum motor current.

Operation, Maintenance and Trouble-Shooting of Centrifugal Pumps

Proper care and maintenance of centrifugal pumps ensure their efficient and prolonged service.

Starting: Prior to starting the pump for the first time special attention is paid to the following points:

1. Check the alignment of the pump. Any misalignment is corrected by placing shims under the pump or driver.
2. Make sure that the engine or motor will drive the pump in the direction indicated on the pump body.
3. Make sure that the gland is tightly and evenly adjusted and the pump shaft revolves freely when turned by hand.
4. Check the air-tightness of the suction pipe and any leakage in the foot valve.
5. Fill the suction line and pump with water and remove the air from the pump casing.
6. Attend to lubrication requirements. If ring oil bearings are fitted, fill the bearings with good quality engine oil as recommended by the manufacturer. However, if ball bearings are fitted, no initial attention is required as they are properly lubricated before leaving the factory.

Operation: On account of its simple construction, the centrifugal pump requires practically no attention while running. Proper lubrication of the bearing and adjustment of the glands are usually the only things which need attention from the operator. The centrifugal pump must be stopped promptly if no liquid is being pumped. Running a pump dry will result in excessive wear to moving parts that depend on water for lubrication. Sluice valves, when provided, are kept closed at the time of starting. This will allow the motor or engine to be started free from load. When the pump reaches full speed, the sluice valve is opened gradually, until the desired quantity of water is being delivered. Care is taken not to run the pump for a long period with sluice valve closed, as this may overheat the pump.

Table 3.2. Check-list for centrifugal pump troubles and their causes

1. *No water delivered*
   - (*a*) Pump not primed
   - (*b*) Speed too low
   - (*c*) Discharge head too high
   - (*d*) Suction lift too high
   - (*e*) Impeller or suction pipe completely plugged
   - (*f*) Wrong direction of rotation
   - (*g*) Air pocket in suction line
   - (*h*) Air leak in suction line or stuffing box
   - (*i*) Insufficient NPSH available.
2. *Not enough water delivered*
   - (*a*) Air leak in suction line or stuffing box
   - (*b*) Speed too low
   - (*c*) Discharge head higher than anticipated
   - (*d*) Suction lift too high
   - (*e*) Impeller or suction pipe partially plugged
   - (*f*) Wrong direction of rotation
   - (*g*) Insufficient NPSH available
   - (*h*) Foot valve too small
   - (*i*) Insufficient submergence of suction inlet
   - (*j*) Wearing rings worn.
3. *Not enough pressure developed*
   - (*a*) Speed too low
   - (*b*) Excessive amount of air or gas in liquid
   - (*c*) Wrong direction of rotation
   - (*d*) Viscosity of liquid higher than anticipated
   - (*e*) Wearing rings worn
   - (*f*) Impeller diameter too small.
4. *Pump works for a while and then loses prime*
   - (*a*) Air leak in suction line or stuffing box
   - (*b*) Excessive amount of air or gas in liquid
   - (*c*) Air pocket in suction line
   - (*d*) Water seal tube clogged
   - (*e*) Water seal ring improperly located
   - (*f*) Suction lift too high
   - (*g*) Insufficient submergence of suction inlet.
5. *Pump requires excessive power*
   - (*a*) Speed too high
   - (*b*) Head lower than anticipated; pumps too much water
   - (*c*) Specific gravity or viscosity of water too high
   - (*d*) Wrong direction of rotation
   - (*e*) Misalignment
   - (*f*) Stuffing box too tight
   - (*g*) Rotating element rubbing or binding
   - (*h*) Bent shaft
   - (*i*) Wearing rings worn.
6. *Stuffing box leaks excessively*
   - (*a*) Packing is worn or improperly lubricated
   - (*b*) Packing improperly installed
   - (*c*) Incorrect type of packing
   - (*d*) Shaft sleeve scoured
   - (*e*) Bent shaft.
7. *Pump noisy or vibrates*
   - (*a*) Suction lift too high
   - (*b*) Insufficient NPSH available
   - (*c*) Impeller or suction pipe partially plugged
   - (*d*) Misalignment
   - (*e*) Foundation not rigid
   - (*f*) Lack of lubricant
   - (*g*) Bearing worn
   - (*h*) Rotating element out of balance
   - (*i*) Bent shaft.

Before stopping the pump, the discharge valve is closed slowly in order to prevent water hammer. This is especially important in high head pumps.

Pump maintenance: Operating conditions of pumps vary widely and so do the maintenance requirements. The following maintenance schedules apply to most pumps under average operating conditions:

1. *Every month:* Check bearing temperature. Bearing may run hot due to lack of lubrication or excess lubricants.

2. *Every three months:* Drain lubricants in oil ring bearings and wash out oil wells and bearings with kerosene. In case of sleeve bearing, check to see that oil rings are free to turn with the shaft. Refill with the lubricant recommended by the manufacturer. Check the wear in the bearings and replace, if excessive.

3. *Every six months:* Replace gland packing. Check alignment of pump and driver and add shims, if required. If misalignment occurs frequently, the entire piping system may have to be checked and corrected.

4. *Every year:* Thoroughly inspect the unit once a year. Remove bearings, clean, and examine for flaws. Clean bearing housings. Remove packing and examine the wear in the shaft sleeve or shaft. Disconnect coupling valves and check alignment. Inspect foot valve and check valves.

Pump Troubles

When the centrifugal pump fails to operate or the discharge or pressure drops, the cause of trouble should be investigated and steps taken to eliminate it. A list of the most common troubles and their causes are given in Table 3.2. The remedies are evident from the cause of the trouble.

Cavitation: The term cavitation refers to the formation of cavities filled with liquid vapour due to a local pressure drop and their collapse as soon as the vapour bubbles reach regions of high pressure. In a centrifugal pump, as the water flows through the suction line and enters the eye of the impeller, an increase in velocity takes place. This increase in velocity is accompanied by a reduction in pressure. If the pressure falls below the vapour pressure, corresponding to the temperature of water, the water will vapourise and the flowing stream will consist of water plus pockets of vapour. Flowing further through the impeller, water reaches a region of high pressure and the cavities of vapour collapse. The collapse of vapour pockets cause the noise incident of cavitation.

Cavitation is the result of improper design, installation and operation of pumps. It may be very mild to severe. A pump may be operating noiselessly and yet be cavitating mildly. The only effect may be a slight drop in efficiency. On the other hand, severe cavitation will result in noise and vibration, and may destroy the impeller and other parts of the pump. To prevent cavitation, the following precautions are taken in operating a centrifugal pump:

1. Avoid operating the pump at heads and capacities much lower than the design value.
2. Excessive suction lifts and high values of water velocity on the suction side are avoided.
3. Pump speed is not allowed to exceed the manufacturer's recommendation.

Water hammer: Water hammer or hydraulic shock occurs when water flowing through a pipe undergoes a sudden change in velocity. The kinetic energy of the flowing liquid, under a sudden change in the velocity of flow, is converted to a dynamic pressure wave which may produce a substantial impact in rebounding back and forth in the pipeline. Water hammer sometimes reaches destructive magnitudes, especially in long pipelines. It has been found to occur in starting or stopping a pump, at an abrupt change in pump speed, and in case of a power failure or rapid closure of a valve in the piping system. Destructive water hammer must be eliminated to reduce stress on the pipes and fixtures and to prevent reversal of the pump, which may be dangerous to the motor or engine. Reflux valve and surge pipes are installed in long pipelines to reduce the effect of water hammer. Rapid closure of valve is avoided.

VERTICAL TURBINE PUMPS

A vertical turbine pump, also called a deep well turbine pump, is a vertical axis centrifugal or mixed flow type pump comprising of stages which accommodate rotating impellers and stationary bowls possessing guide vanes. The bowl assemblies are nearly always located beneath the water surface and hence the deep-well turbine pumps are adapted to seasonal fluctuations in water level in the well. They are specially adapted to tubewells where the pumping water level is below the practical limits of a volute centrifugal pump. The comparatively small diameter of turbine pumps suit their installation in tubewells.

Vertical turbine pumps are adapted to high lifts and have high efficiencies under optimum operating conditions. They have, however, higher initial cost and are more difficult to install and repair, as compared to volute pumps.

As with all centrifugal pumps, the pressure head developed depends on the diameter of the impeller and the speed at which it is rotated. In the deep-well turbine pump, the diameter of the bowl and that of the impeller inside it are restricted by the relatively small diameter of the tubewell. Hence, the pressure head developed by a single impeller is not large. Additional head is obtained by adding more bowl assemblies or stages.

Pump Construction and Characteristics

The turbine pump has three main parts: the pump element, the discharge column and the discharge head (Figs. 3.50 to 3.52).

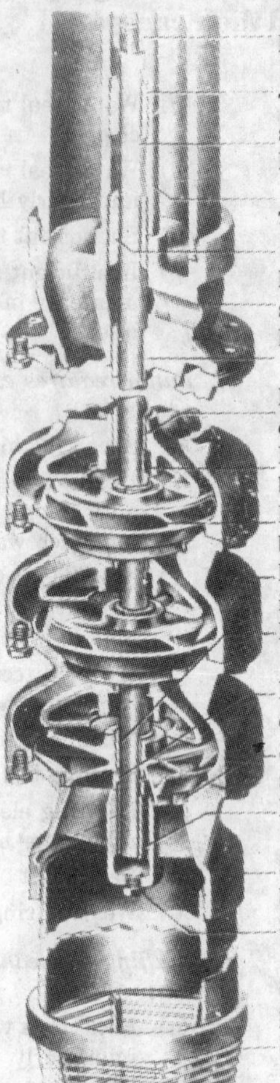

Fig. 3.50. Cut-away section of the bowl assembly of a vertical turbine pump (water-lubricated type). Note that the cut-away section of the impeller is shown in the bottom while the upper two impellers are shown in full (*Courtesy*: Johnston Pumps India Ltd.)

Pump element: The pump element is made up of one or more bowls or stages (Fig. 3.50). Each bowl assembly consists of an impeller and a diffuser and a bearing. It has a screen or strainer fixed to its bottom to keep coarse sand and gravel from entering the pump. Impellers used are either enclosed or semi-open type.

Discharge column: The discharge column (Fig. 3.51) connects the bowl assembly and the pump head assembly and conducts water from the former to the latter. It consists of the discharge or column pipe and the line shaft with its couplings and bearings.

Discharge head: The head assembly (Fig. 3.52) consists of the base from which the discharge column, bowl assemblies and shaft assemblies are suspended, the discharge head elbow which directs the water into the delivery piping system and the driving head. Electric motors, right angle gear drives, flat or V-belt pulleys and suitable combinations of these drives may be used on the

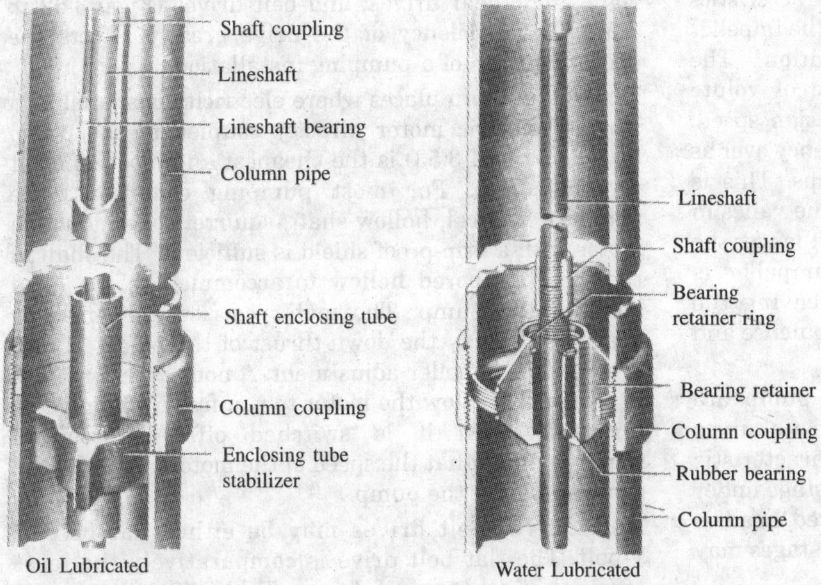

Fig. 3.51. Cut-away views showing the details of the column assembly of oil lubricated and water lubricated vertical turbine pumps.
(*Courtesy:* Johnston Pumps India Ltd.)

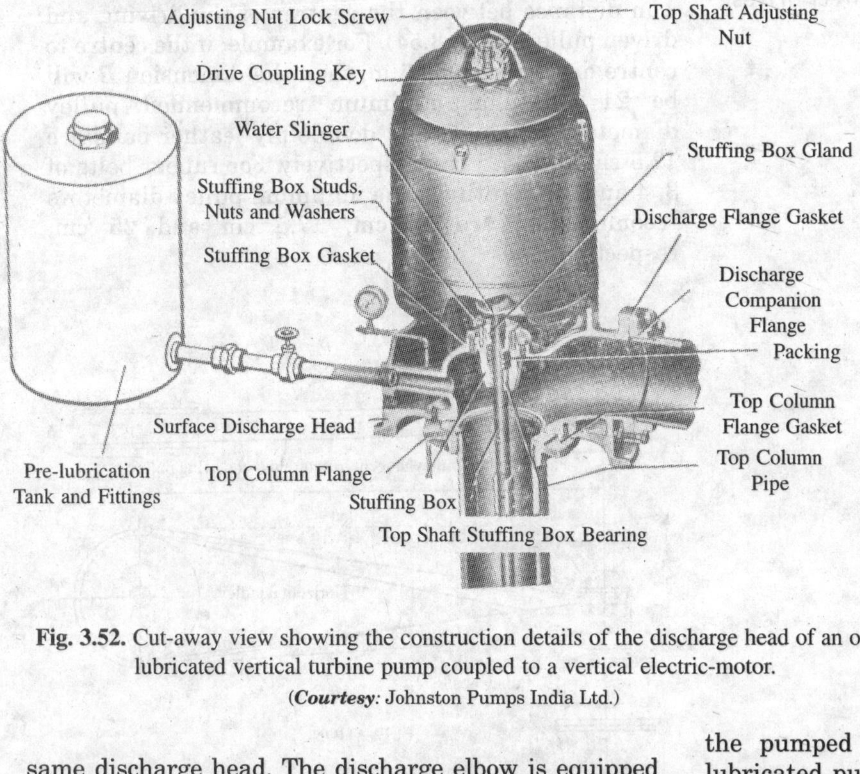

Fig. 3.52. Cut-away view showing the construction details of the discharge head of an oil lubricated vertical turbine pump coupled to a vertical electric-motor.
(***Courtesy:*** Johnston Pumps India Ltd.)

same discharge head. The discharge elbow is equipped with flanges to which the discharge pipe is fixed.

Oil lubricated and water lubricated pumps: Based on the method of lubrication of the pump, turbine pumps are classified into two types, oil lubricated and water lubricated. Figure 3.51 illustrates the constructional features of the column assemblies of the two types. Column pipes are made of steel and are joined by threaded sleeve type couplings. The drive shaft is located in the centre of the discharge pipe. It is made of cold finished, carbon steel of proper size to operate the pump without distortion or vibration.

When oil lubricated, the drive shaft is enclosed in a cover pipe which guides the oil to proper parts. Bronze bearings are provided at every 1.5 m. The bearings also serve as couplings for the cover pipe. The shaft and cover pipe assemblies are supported and aligned by reinforced rubber spiders spaced in the column pipe at about every 15 m. The oil lubricated column construction has the advantage that the shafting is protected from any abrasive or corrosive elements which may enter the pump. This type of construction provides a more positive system of shaft and tube alignment.

In water lubricated pumps, the bearings, lubricated by the water being pumped, are of natural or synthetic rubber, set in bronze retainers. A renewable stainless steel sleeve, which eliminates shaft wear, shrouds the line shaft at the points of bearing contact. The cover pipe is not necessary. The shaft is held in line by the rubber bearings placed at about 3 m intervals. Rubber bearings are safely lubricated only when wet. Water lubricated columns are provided with a pre-lubrication system consisting of a water storage tank placed near the discharge head. Water is made to run through an inlet provided in the discharge head down to the discharge column before starting the pump each time. This wets the rubber bearings, so that they will not become overheated and get damaged.

The choice between oil lubricated and water lubricated pumps depends on the conditions of operation. When the pumped water contains fine soil particles, oil lubricated pumps waste some oil into the water. It is desirable that water lubricated pumps are used. In cases where the pump has to stand idle frequently for considerable time, it is desirable to have an oil lubricated pump, as the rubber bearings of the water lubricated pumps may get out of shape and may be damaged. In general, the cost of a water lubricated pump is slightly lower than an oil lubricated pump.

Pump characteristics: The operating characteristics of turbine pumps depend upon the design of the impeller and the bowl and the speed of rotation. The characteristic curves are similar to those of volute centrifugal pumps when operated at the design speed. Turbine pumps, cannot operate at high efficiency over as wide a range of speed as can volute pumps. This is because a high efficiency is possible only if the vanes in the bowl are in line with the flow of water as it leaves the tip of the impeller. If the speed of the impeller is changed, the direction of flow of water leaving the impeller is also changed. This will cause turbulence and will result in reduced efficiency.

Operating characteristics of a turbine pump are determined by an inspection of the characteristic curves furnished by the manufacturer. The characteristic curves represent the performance of a single stage under various operating conditions. If the required head is more than that for a single stage, additional stages may be added.

Pump Drives

One of the main advantages of the turbine pump is its adaptability to different types of power sources and drives. In a vertical turbine pump, the driver is located at the ground surface. The drives may be direct drives, right angle gear drives, and belt drives (Figs. 3.52 to 3.54). The efficiency of the drive greatly affects the operating costs of a pumping installation.

Fig. 3.53. Sketch illustrating the installations of a vertical turbine pump in a tubewell.

Direct drive: In places where electricity is available, a vertical electric motor directly coupled to the pump (Figs. 3.52 and 3.53) is the cheapest and most efficient type of drive. For most pumping conditions, the standard vertical, hollow-shaft, squirrel cage induction motor with a drip-proof shield is sufficient. The shaft of the motor is bored hollow to accommodate the head shaft of the pump. The motor carries a bearing or bearings to carry the down-thrust of the pump. It also provides for impeller adjustment. A non-reverse ratchet is provided to allow the motor to run free in the reverse direction when it is switched off. An essential requirement is that the speed of the motor should be the same as that of the pump.

Belt drive: Belt drives may be either flat belt or V-belt. The flat belt drive is comparatively the least efficient of all types of drives (Table 3.3). Slippage is the main cause of loss of efficiency. V-belt drive is more dependable and efficient. Belt drives for vertical turbine pumps require a quarter-twist arrangement for the belt (Fig. 3.54). To allow for belt sag, the centre of the face of the driving pulley is raised 6 cm for each 1 m distance between the centres of the driving and driven pulleys (Fig. 3.54). For example, if the centre to centre distance A is 3.5 m, then the dimension B will be 21 cm. The minimum recommended pulley diameters for single and double ply leather belts are 12.5 cm and 17.5 cm, respectively. For rubber belts of 3, 4 and 5 ply ratings, the minimum pulley diameters recommended are 15 cm, 17.5 cm and 25 cm, respectively.

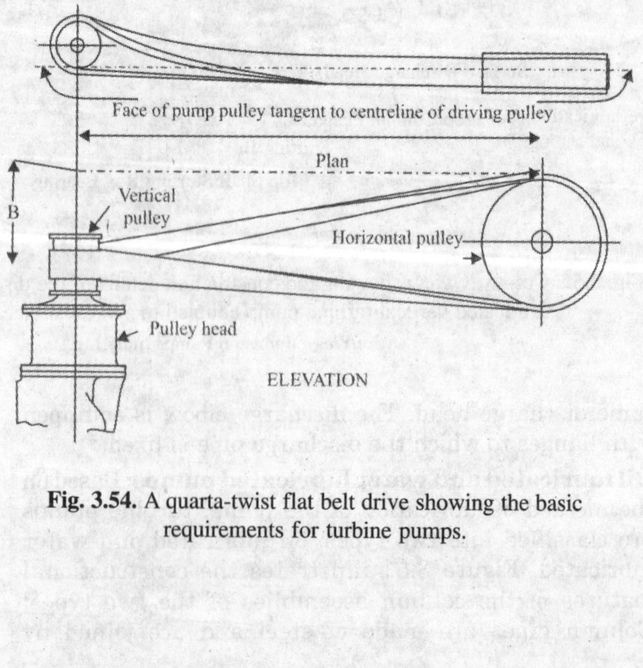

Fig. 3.54. A quarta-twist flat belt drive showing the basic requirements for turbine pumps.

Table 3.3. Specifications of Flat Belts for Quarter-Twist Drives

H.P. transmitted	Leather belts Width (cm)	Leather belts Ply	Rubber belts Width (cm)	Rubber belts Ply	Minimum centre distance 'A' Motor drive (m)	Minimum centre distance 'A' Engine drive (m)
10	10.0	1	12.5	3	3	4.5
15	12.5	2	12.5	4	3.6	5.5
20	15.0	2	15.0	4	4.2	6.5
30	20.0	2	20.0	5	4.8	7.5
50	25.0	2	25.0	6	5.4	9.0

The pulley head has a shaft which may be bored hollow to receive the head of the pump shaft. In case of a hollow shaft construction, the impeller adjustment is made at the top of the pulley head. The pulley head is provided with a bearing to carry the total down-thrust as well as a bearing to carry the belt pull.

Right angle gear drive: The right angle gear drive provides the most dependable and efficient method of transmitting power from horizontal prime movers to vertical pump shafts (Fig 3.55). The assembly consists essentially of a horizontal and a vertical gear mounted on shafts. The vertical shaft, if bored hollow, accommodates the head shaft of the pump. The horizontal shaft receives power from the prime mover. Impeller adjustment is possible at the top of the gear head. A bearing is provided on the hollow shaft for taking up the down-thrust. The gear drive is usually provided with a non-reverse clutch to prevent the pump turning in the opposite direction when the power is shut off.

Fig. 3.55. An engine driven vertical turbine pump equipped with right angle gear drive.

Power from the engine or motor is transmitted to the gear drive by a flexible shaft or a belt. The flexible shaft connects the horizontal shaft of the gear drive to the engine shaft. The belt transmits power from engine to gear drive with the help of pulleys mounted on both. Sometimes a combination of direct drive, using a vertical electric motor mounted on a right angle gear drive is used, so that the pump may be connected to the engine when there is a power failure, thus ensuring continuous water supply.

The selection of gear drive is based on the horse power required by the pump and the thrust conditions of the pump. The total pump thrust is the combined weight of the rotating parts of the entire pump plus the hydraulic thrust. The thrust load varies with the size and number of stages, shaft size, and total head. In very deep wells or where high head conditions exist, the thrust loads may be very high and special bearings are required.

Installation of Vertical Turbine Pumps

Most of the installation features discussed under centrifugal pumps also apply to turbine pumps. Turbine pumps are relatively much more bulky than centrifugal pumps with their long column pipes, drive shafts and bowl assembly. The installation is, therefore, a special job requiring heavy equipment like a derrick or a tripod, chain pulley blocks, supporting clamps, heavy duty pipe wrenches and chain tongs. Reputed manufacturers provide detailed instructions regarding the installation and maintenance of their products. Only the general principles of installation are discussed below.

Vertical turbine pumps must be in correct alignment with the well and the power unit (Fig. 3.53). The pump must be aligned in the well casing so that no part of the pump assembly touches the well casing. It must be mounted on a good foundation so that its alignment between pump and drive and pump and well casing will be maintained at all times. Concrete foundations are usually preferred. The foundation must be large enough so that the pump and drive assembly can be securely fastened. It should have at least 30 cm of bearing surface on all sides of the well. In the case of a gravel packed well, this 30 cm clearance should be measured from the outside edge of the gravel packing. When the pump is installed in a gravel-packed well, at least two openings should be provided in the foundation on opposite sides of the well to permit refilling with gravel as the gravel-pack settles.

Several precautions are taken while installing turbine pumps. Care must be exercised in handling and installing all parts, particularly the shaft and tubes. These are machined to close tolerances for accurate alignment and must not be sprung or bent by mistreatment. All joints are threaded, and great care is taken so that all threads are wiped clean from dirt of any sort before they are lubricated with lead compound and turned down in place. Care is also taken that the threads start accurately matched so that no cross threading is permitted to injure the starting turns. Every joint is set up tight.

The bowl assembly is placed below the static water level in the well. It must be low enough so that the drawdown while pumping does not exceed the suction limit below the impellers. Preferably, the bowls must be placed below the lowest drawdown.

When the pump discharges into a long line of discharge pipes, a large amount of water could run back down through the pump at the time it is shut off, unless a reflux valve is provided. Water running down the pump column tends to turn the pump backwards. This rotation can unscrew the drive shaft. Results of unscrewing the drive shaft are mechanically disastrous. Therefore, in such installations, a reflux valve is to be fitted on the delivery pipe near the pump outlet.

The pump, after installation, must be adjusted so that the impellers turn freely without touching either the top or the bottom of the bowls. Adjustments in position are made by raising or lowering the impeller shaft. Provision for adjustment is made in the threaded section of the top end of the drive shaft and consists of two hexagonal-nuts supported by a heavy duty ball bearing with a thrust plate. Impeller adjustments are of special importance in semi-open impellers which have their lower sides open, or enshrouded. With this type, wear on the exposed vanes can be compensated for by lowering impellers by a fraction of a millimetre, and efficiencies often respond by an increase by 10 to 20%.

Operation, Maintenance and Trouble-Shooting of Vertical Turbine Pumps

Before starting, the sluice valve is kept closed. It must be ensured that there are no tools or rags near the pump. The mains are then switched on to start the pump. The starting handle of the starter should be kept in the 'START' position till the pump picks up full speed and then in one motion thrown over to the 'RUN' position. The following points are checked with the sluice valve fully closed:

1. Oil is flowing into the pump. If not, there may be something wrong with either the solenoid coil, or the sight feed valve may have some dirt. For manual lubrication, adjust the set screw of sight feed valve for the rate of oil feed desired.
2. Line voltage is as specified on the motor. The variations should not be more than ± 5%.
3. Check the amperage. This should not exceed the ampere rating stamped on the motor.
4. Check for water leakages at the tube tension nut and plate (stuffing box in case of water lubricated pumps), pressure gauge line connection and connection of discharge head and sluice valve. If leakages are noted they, will have to be sealed off by packing, with lead compound and jute.

The sluice valve is opened slowly, letting the water discharge in open. If the water being pumped is dirty and has substantial sand content, the sluice valve is opened only partially. The pump is allowed to operate with throttled flow for 10 to 15 minutes. The water should clear out in this period if the tubewell has been properly developed. After the valve is fully open and clear water is being pumped, check the line voltage, amperage, any leakage in tube tension plate or nut or stuffing box in case of water lubricated pumps. Flow of oil to the pump and any vibration or excessive noise in operation are checked. The sluice valve is closed completely before stopping the pump.

Maintenance: For the safe and efficient operation of the pump, the lubrication of the different parts should be carried out strictly according to the manufacturer's instructions. The following maintenance points for the various parts of the pump apply to most makes of turbine pumps:

Vertical belt drive head

Each year, or 2000 hours' running:

(*a*) Remove drain plug and filler plug from the oil filler cup, and allow oil to drain.

(*b*) Replace drain plug and fill drive head with the recommended grade of lubricating oil, up to the top of the oil filler cup.

(*c*) After a few minutes, check oil level, and add oil, if necessary, to bring oil level up to the top of the oil filler cup.

(*d*) Replace filler plug.

Right angle gear head

Each year, or after 2000 hours' running:

(*a*) Remove drain plug from the oil filler cup, and allow oil to drain.

(*b*) Replace drain plug and fill drive head with the recommended gear oil to the top of the oil filler cup.

(*c*) After a few minutes, check oil level and add oil, if necessary, to bring oil level up to the top of the oil filler cup.

Electric motor drive head

Each 500 hours' running:

(*a*) Unscrew cap from the thrust bearing lubricator and fill the cap with grease.

(*b*) Replace cap and screw it down one turn. (One cap full of grease is sufficient for 500 hours' running. The lubricator cap should be usually screwed down each day, so that the amount of grease in the cap will last 500 hours approximately.)

Shaft packing gland

As required:

(*a*) Unscrew the cap from the shaft packing gland lubricator and fill the cap with grease.

(*b*) Replace the lubricator cap, and screw it down two turns. Each day of operation, screw the cap down one turn.

Pump troubles and remedies: When properly installed and maintained, turbine pumps give trouble-free service. Common troubles of vertical turbine pumps and suggested remedies are listed below:

(1) *Pump does not start:* (*i*) Faulty electrical connections of motor; (*ii*) Excessive bearing friction due to misalignment of pump, insufficient lubrication or improper grade of lubricating oil, or dry rubber bearing in case of water lubricated pumps; (*iii*) Impellers are locked, improper impeller adjustment; and (*iv*) Discharge packing is excessively tight. The remedies, in general, are evident from the cause of the trouble. Raising or lowering the impeller may release them. Backwashing with clear water may clean the bowls. When other methods fail it may be necessary to pull out the pump and set right the defect.

(2) *Pump does not deliver water*: (*i*) Pump suction broken, (*ii*) Speed too slow (Check pump speed and see whether motor is getting full voltage), (*iii*) Water level is lowered such that the pump bowls are not submerged, (*iv*) Impellers plugged (this will happen when the tubewell is not fully developed and the pump discharges clay and sand at the start). The bowl assembly will have to be pulled out and flushed with a hose. It may also be necessary to open the bowl assembly and clean thoroughly, (*v*) Impellers jammed on account of foreign matter in the bowls. In the case of open wells, there is a likelihood of small pieces of wood or rags getting through the strainer and getting caught in the impeller, (*vi*) Strainer clogged, (*vii*) Wrong direction of rotation, (*viii*) Pumping head too high, (*ix*) Pump speed too slow, and (*x*) Mechanical failures like broken shaft or bowl assembly, loose impellers or leakage in column pipe joints.

(3) *Insufficient water*: (*i*) Speed too slow, (*ii*) Water lowered so that all bowls are not submerged, (*iii*) Worn out pump bowl and impellers, (*iv*) Impellers loose or damaged, (*v*) Column joint defective and leaking, (*vi*) Impellers or strainer partially clogged, and (*vii*) Improper impeller adjustment–an important factor in case of semi-open impellers where ¼ turn of the adjustment nut makes considerable difference in discharge.

(4) *Insufficient pressure:* (*i*) Speed too slow, (*ii*) Air in water, (*iii*) Pump bowls not submerged, (*iv*) Worn out bearings, (*v*) Impellers damaged, (*vi*) Column joints leaking, and (*vii*) Wrong impeller adjustment.

(5) *Pump takes excessive power:* (*i*) Operating conditions differ from those for which the pump is designed, (*ii*) Poor lubrication, (*iii*) Impellers not correctly adjusted (They may be too high, thus binding against the top of the bowls, or too low, thus rubbing against the bowl seats. (*ii*) Pumping sand-impeller partially clogged and rubbing against sand deposited on bowl seats, (*iii*) Too much oil in the oil tube or oil used is too heavy, (*iv*) Pump out of alignment, (*v*) Shafts bent, (*vi*) Worn out or tight bearings, (*vii*) Dirty oil used in lubrication — impurities may bind the shaft against the bearings, and (*viii*) Improper greasing or oiling of bearings in the pump drive.

(6) *Pump vibrates:* (*i*) Speed too high, (*ii*) Pump improperly aligned and leveled, (*iii*) Crooked well, (*iv*) Bent shaft, (*v*) Impellers rubbing against bowl seats on account of improper adjustment, (*vi*) The pump may not be properly bolted to foundation, (*vii*) Pump not resting on solid foundation, (*viii*) Bearings in drive not properly lubricated, (*ix*) Impellers loose, (*x*) Imbalance due to foreign material lodged between impeller or bowl passage, vibration of drives or excessive wear in rotating parts, and (*xi*) Poor suction conditions.

(7) *Capacity too low:* (*i*) Head too high, (*ii*) Speed too slow, (*iii*) Poor suction conditions caused by turbulence, eddying and vortexing at the pump suction, air leaks in suction pipe, (*iv*) Leakage due to loose threads of flanges, worn gaskets or packing, (*v*) Incorrect impeller adjustment and (*vi*) Mechanical failures like worn impellers, wearing rings, or bearings; or impellers loose on shaft.

(8) *Water enters shaft tubes or oil mixes with water:* (*i*) Tube joints not tightened flush, (*ii*) Line shaft bearings broken, (*iii*) Discharge head gland not properly tightened, (*iv*) Dents or cuts on the tube faces, and (*v*) Shaft enclosing tube leaking or broken.

(9) *Components wear out quickly:* (*i*) Bearings run dry or insufficient lubrication, (*ii*) Pump bearings misaligned, (*iii*) The tubewell crooked, (*iv*) Shafts bent, (*v*) Sand in water, and (*vi*) Vibrations due to misalignment or improper levelling of discharge head.

SUBMERSIBLE PUMPS

A vertical turbine pump close-coupled to small diameter submersible electric motor is termed a submersible pump. The motor is fixed directly below the intake of the pump. The pump element and the motor operate entirely submerged. Such an installation eliminates the long vertical shaft in the column pipe. The performance characteristics of the submersible pump are similar to the vertical turbine pump. Efficiency is increased by the direct coupling of the motor and its effective cooling by submergence in water.

The principal advantage of the submersible pump is that it can be used in a very deep tubewell where a long shaft would not be practical. These pumps are also less affected by deviations in vertical alignment of wells. As the submersible pump has no above-ground working parts, it can be used where flooding may be a hazard. It is also adaptable to locations like public grounds where an above-ground pump house would be inconvenient.

Submersible Pump Construction

The submersible pump consists of a pump and motor assembly, a discharge column, a head assembly and a waterproof cable to conduct the electric current to the submerged motor (Figs. 3.56 and 3.57).

Pump element: The construction of the pump element is similar to the vertical turbine pump. The propelling shaft, usually made of stainless steel, is very short and on it are mounted the bronze impellers. Impellers are usually made of bronze and may be closed or semi-open. The closed-impeller is generally used where it is necessary for the pump to develop high pressure. Water enters the pump through a screen located between the motor and the pump.

Electric motor: The submersible electric motor has the same diameter as the pump bowl but it is much longer than an ordinary motor. It is a squirrel cage induction motor which may be of the dry or wet type. The dry motor is enclosed in a steel case filled with a light oil of high dielectric strength. A mercury seal placed directly above the armature prevents oil

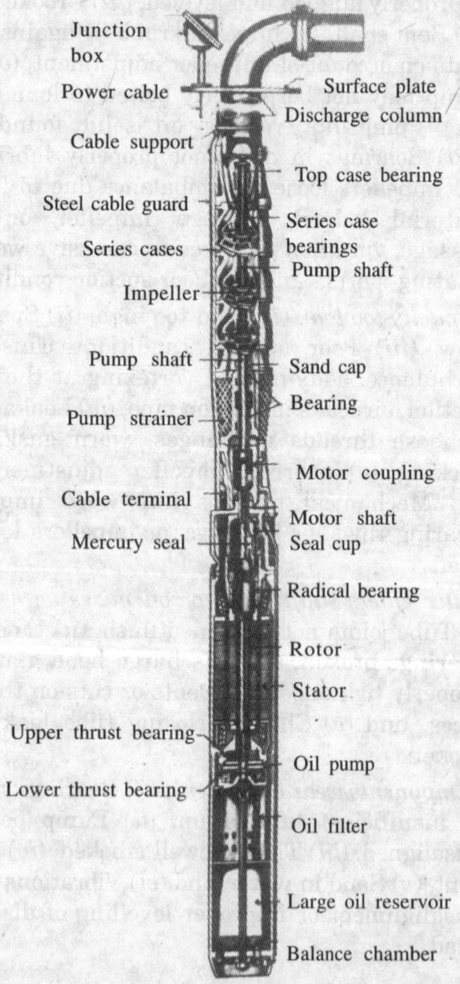

Fig. 3.56. Sectional view (schematic) of a submersible pump showing parts.

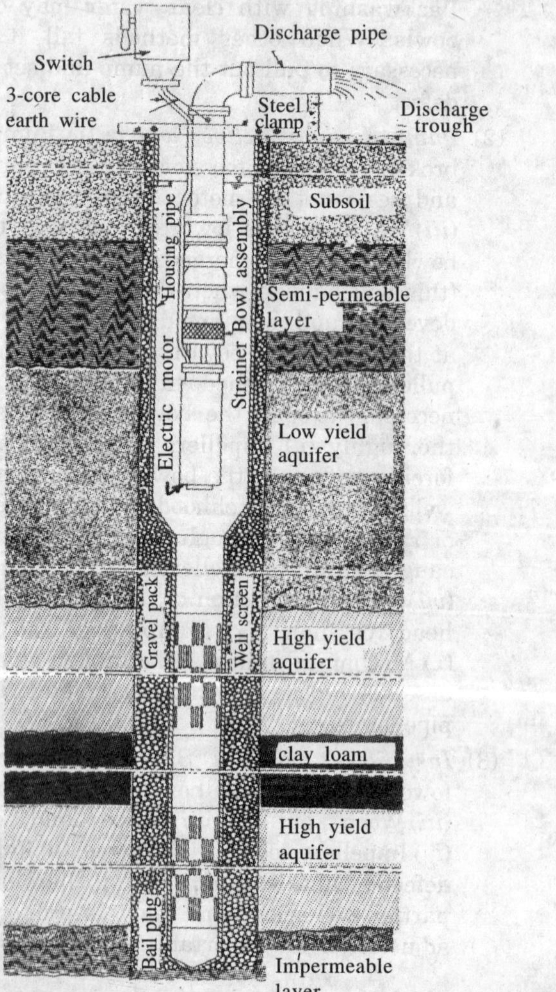

Fig. 3.57. Sketch illustrating the installation of a submersible pump in a tube well.

leakage or water entrance at the point where the drive shaft passes through the case to the impellers.

Wet motors are those in which the well water has access to the inside of the motor, with the rotor and bearings actually operating in water. In this type of motor, windings of the stators are completely sealed off from the rotor by means of a thin stainless steel inner liner. A filter around the shaft is required to prevent the entrance of abrasive material into the motor. The wet type motor should be filled with water during installation so that the bearings will have sufficient lubrication when the motor is first started.

The stator windings are continuous for the whole length of the motor. The rotors are made in sections on a continuous shaft, with bearings between them to guide the shaft and maintain correct alignment. The electric cables, leading from the motor to the starting box on the ground surface, are water proofed and are placed outside the discharge pipe.

Fig. 3.59. A submersible pump installed in a tubewell.

Installation, Operation and Maintenance of Submersible Pumps

The ease of installation is an outstanding feature of the submersible pump. The pump and motor assembly is lowered first into the well, adding pipe lengths as required. When lowering the pump and piping, care is taken not to damage the outside sheathing of the water proof cable with pipe tools or against the top edge of the well casing. The cable is taped to the column pipe at every 2 metres. A pipe clamp is used at the top of the well to support the pump and the delivery pipe (Fig. 3.58). A sluice valve is usually fitted at the discharge end of the pump (Fig. 3.59). Since the complete pump and motor assembly is in the well, no pump house is required. The control board consisting of the switch, starter and meter should be enclosed in a water proof box (Fig. 3.59).

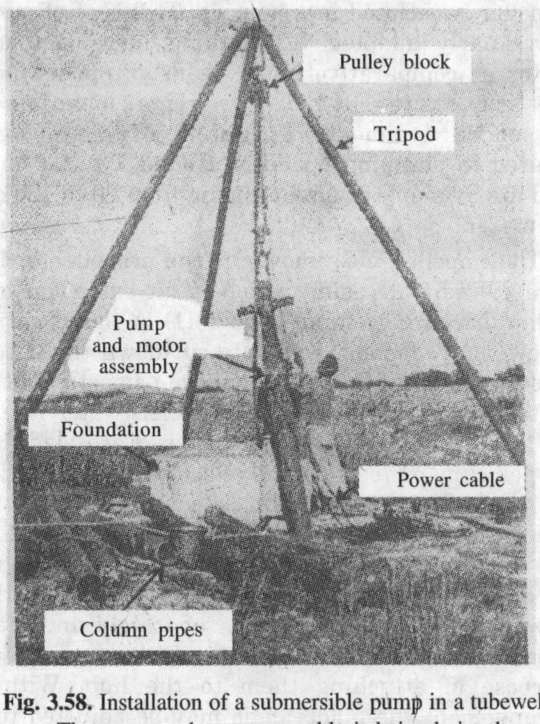

Fig. 3.58. Installation of a submersible pump in a tubewell. The motor and pump assembly is being hoisted for lowering into the tubewell.

Starting and initial run: The pump is started with the sluice valve closed or slightly open. During the initial run, it is observed whether the water pumped is clear or muddy, and whether any impurities are being pumped. If the water contains sandy or gritty particles or other impurities, care is taken not to stop pumping, otherwise the particles will settle down inside the pump and on top of the non-return valve (if fitted) and may choke or cease the pump. The sluice valve is kept partly closed and the pumping continued until the water clears. When the water is reasonably clear, the valve is opened a little more, checking whether the increased flow of water brings up more impurities. If it does, the valve is adjusted so that impurities are kept to the minimum. The pumping is continued in this manner until it is possible to pump at full open valve, without having a large proportion of sand and grit in the water at any time. When the water is running clear, the pump may be stopped and restarted, as required.

When pumping first, most new tubewells produce water laden with a certain amount of sand, silt, grit and other debris from the drilling operation, but the impurities gradually lessen until the water runs clear. In a few cases, mostly bore holes in sandstone and similar formations, a certain amount of running sand is present continuously. This can be troublesome, as the sand content may cause undue wear on the pump parts, but all that can be done after the well is completed is to pump at such a rate that the sand content is kept to a minimum. The object of initially pumping with the valve partly closed is to ensure that the pump parts do not suffer undue wear, as might be the case if all the loose sandy material present in the tubewell is passed through the pump in the first few minutes after switching on.

Maintenance: When correctly installed and operated under suitable working conditions, the submersible pump requires little maintenance. After about 6000 hours of operation (or two years of service life), it may be necessary to withdraw the pump from the bore hole and overhaul it, adopting the instructions of the manufacturer.

Common Troubles and their Remedies

The common troubles experienced in the operation of submersible pumps and their possible causes are as follows:

(1) *Pump fails to start:* (*i*) Fuses blow, (*ii*) Overload causes short-circuits, and (*iii*) Heavy load on pump.

(2) *Pump starts and the discharge is steady but reduced amount of water, or discharges no water at all:* (*i*) Motor is running in reverse direction, (*ii*) Pump is operating against a head greater than intended, (*iii*) Pump suction is blocked by foreign matter, mineral salts deposit, collapse of bore, etc., (*iv*) Pump is air-locked (no discharge at all), (*v*) Reflux valve above pump is jammed/closed, or riser pipe is closed by obstruction, and (*vi*) Voltage is lower than required.

(3) *Pump runs and discharges steadily, but discharge is below normal*: (*i*) Fault in power supply, (*ii*) Mechanical friction in pump or motor, (*iii*) Riser pipe has developed a hole, or leak has developed in the system below ground level, and (*vi*) Pump is worn by sand, as well as increase in mechanical friction.

(4) *Pump runs and discharges intermittently, air bubbles are present in discharge*: The pumping rate is greater than the rate of flow into the well bore. If allowed to persist this will cause damage to pump, shaft keys, couplings, and possibly to the motor.

PROPELLER AND MIXED FLOW PUMPS

Propeller and mixed flow pumps are used for low head and high discharge operations. At very low heads, a much higher efficiency can be obtained from a propeller pump than from a centrifugal pump. For slightly higher heads mixed-flow pumps are suitable than centrifugal pumps. Both of these pumps are extensively used for drainage pumping. They are also suitable to pump water for irrigation from canals, rivers or streams.

The characteristic difference in the principles of operation of the pumping elements of propeller and mixed flow pumps, as compared to turbine pumps, is illustrated in Fig. 3.60. In turbine pumps, as well as in other centrifugal pumps, the pressure head is developed principally by the action of centrifugal force. But a propeller pump develops most of the head by the propelling or lifting action of the impeller

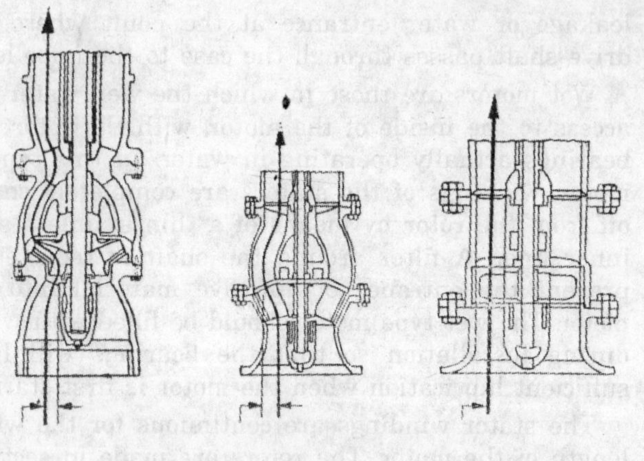

Fig. 3.60. Characteristic differences in the principle of operation of turbine, mixed flow and propeller pumps.

vanes on the liquid. The mixed flow pump combines some of the features of both the turbine pump and the propeller pump.

The principal parts of propeller and mixed flow pumps are similar to the deep well turbine pump in that they have a head, an impeller and a discharge column. A shaft extends from the head, down the centre of the column to drive the impeller.

Construction and Operation of Propeller Pumps

The flow through the impeller of a propeller pump is parallel to the axis of the drive shaft rather than radial. Therefore, propeller pumps are often referred to as axial flow pumps. The principle of operation is similar to that of a boat propeller, except that the impeller is enclosed in a housing. This type of pump is particularly adapted to handling large volumes of water at comparatively low heads. Propeller pumps are not easily clogged by foreign materials and suspended sediments in water. They are usually limited to pumping heads of around 1 to 2.5 metres and are available in sizes ranging from 20 to 120 cm in diameter.

The impeller, also known as the propeller, operates in a cylindrical casing which is an extension of the pump discharge column (Fig. 3.61). A flared entrance below the propeller is used to cut down the entrance losses, and guide vanes above smooth out the disturbances caused by the propeller. Water is moved up by the lift of the propeller blades. Each blade of the propeller helps to impart a velocity in the direction of the shaft. The number of blades is usually 3 to 5. They are set on the shaft at angles determined according to the head and speed. The propellers are usually made of bronze and are cast in one piece. Sometimes, however, the blades are cast separately and may be threaded on their hub ends to receive a nut for the purpose of attaching them to the hub. With this arrangement the blade angle may be adjusted to suit the operating conditions. The propeller is balanced

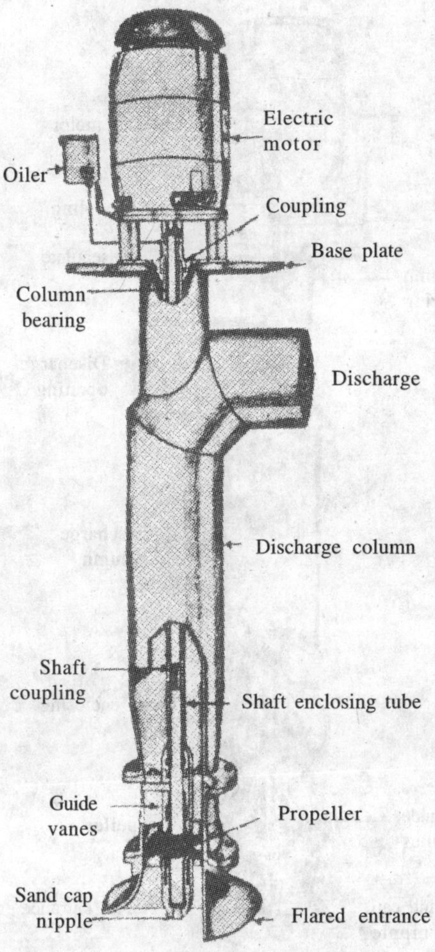

Fig. 3.61. View of a propeller pump with the propeller and lower shaft assembly exposed.

hydraulically and statically. The blades are carefully cast and scraped to reduce skin friction. They are keyed to the drive shaft and are accurately positioned by a locking collar and nut. A cone-shaped cover is usually installed over the locking nut to eliminate eddies and to prevent the entrance of sand or grit into the lower pump bearings. Diffusion casings are usually provided on propeller and mixed flow pumps to convert into pressure the tangential component of the velocity of discharge from the impeller. The discharge column, discharge head and pump drive of propeller pumps are, in general, similar to the vertical turbine pump. Propeller pumps are commonly of the oil-lubricated type.

Operating Characteristics

A slight increase in the pumping head causes a large decrease in the quantity of water delivered by a propeller pump. From the standpoint of efficiency, it is necessary that the propeller pump is operated at the rated head as far as possible. The power required is increased in a propeller pump as the head is increased, and the capacity is reduced. Overload is likely to occur when the discharge valve is nearly closed instead of open, in contrast to the centrifugal pump. Because of the steep horse power curve at shutoff, propeller and mixed flow pumps are started against an open discharge.

Propeller Pump Installation

The propeller pump should be installed on a firm, strong foundation. The foundation should be able to support the weight of the pump evenly on all sides of the base plate and allow the unit to hang vertically.

For drainage installations, the pump is usually installed in a sump with an automatic float control. The automatic operation of the electric motor is by means of start and stop collars on a float rod. Automatic switches are provided for the start and stop of the pump at the required flood stages. For large drainage installations, two pumps with different capacities may be necessary, one for handling surface runoff during the rainy seasons and the other for seepage or flow from tile drains.

When the source of water contains appreciable amounts of floating material, some type of strainer or screen should be installed to exclude the floating material that may damage the propeller if drawn into the pump. Provision is made to attach a small strainer that can be attached to the suction bowl. These strainers are satisfactory when the water pumped is comparatively free of floating vegetation and small debris. When the source of water supply contains this type of foreign material, the small strainer is likely to get clogged. Under such situations, it is desirable to construct some type of screen around the inlet, so as to increase the area for straining out the small debris. Sometimes, it may be necessary to provide trash racks at some distance away in the drainage channel.

Propeller pumps are not suitable for suction lift. The impeller bowl must be submerged with the pump operating at the proper submergence depth, the submergence being the distance from the pumping water level to the lowest part of the suction pipe. The minimum submergence required varies from 65 to 90 cm for pumps of size 20 to 35 cm. Failure to observe the required submergence depth may result in cavitation. It is also important that proper clearances are maintained between the end of the suction pipe and the side walls and the bottom of the pit or pump intake bay. The minimum clearance between the pump and the side wall of the intake structure varies from 30 to 50 cm, and from the bottom of the sump floor to pump inlet end between 20 to 30 cm, for pump size of 20 to 30 cm. Inadequate clearances can result in the lowering of pump efficiency.

Mixed Flow Pumps

A mixed-flow pump combines some of the features of both the vertical turbine pump and the propeller pump (Fig. 3.64). The head is developed partly by the

Fig. 3.62. A portable engine-operated propeller pump lifting water from a stream to a field channel.

centrifugal force as in a conventional centrifugal pump and partly by the lift of the impeller vanes on the liquid as in a propeller pump. It is applicable for high discharge medium head conditions. The head usually varies from 5 to 10 metres.

The main difference between mixed-flow and propeller pumps is in the construction of the impeller. The impeller blades of the mixed-flow pumps are designed to give an outward thrust to the water, in addition to imparting to it a velocity upward. Above the impeller, curved vanes guide the water to the column and straighten the flow. The mixed flow pump can work satisfactorily with two or three stages at the bottom of the shaft. The construction of the column assembly and pump drive is similar to the propeller pump. When clear water is being pumped, a water lubricated column assembly is suitable. Oil-lubricated pumps are needed when the water contains suspended sediments and impurities.

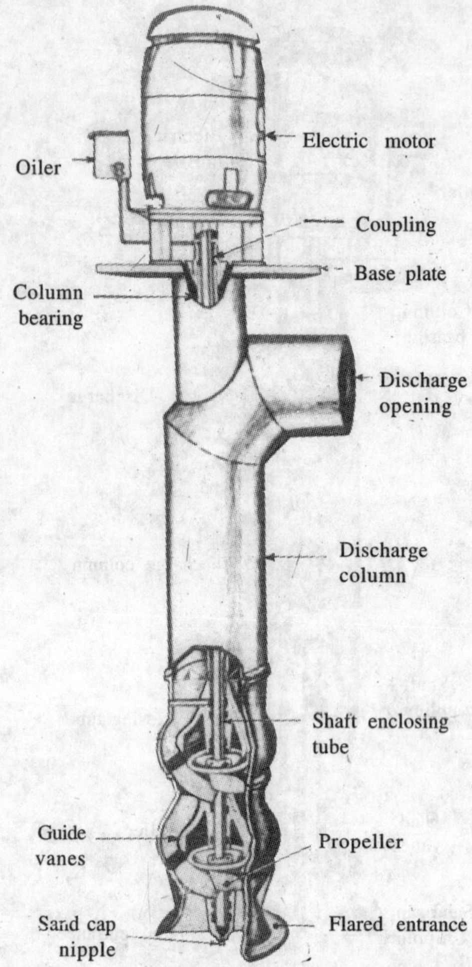

Fig. 3.64. View of a two-stage mixed flow pump with the pump element exposed.

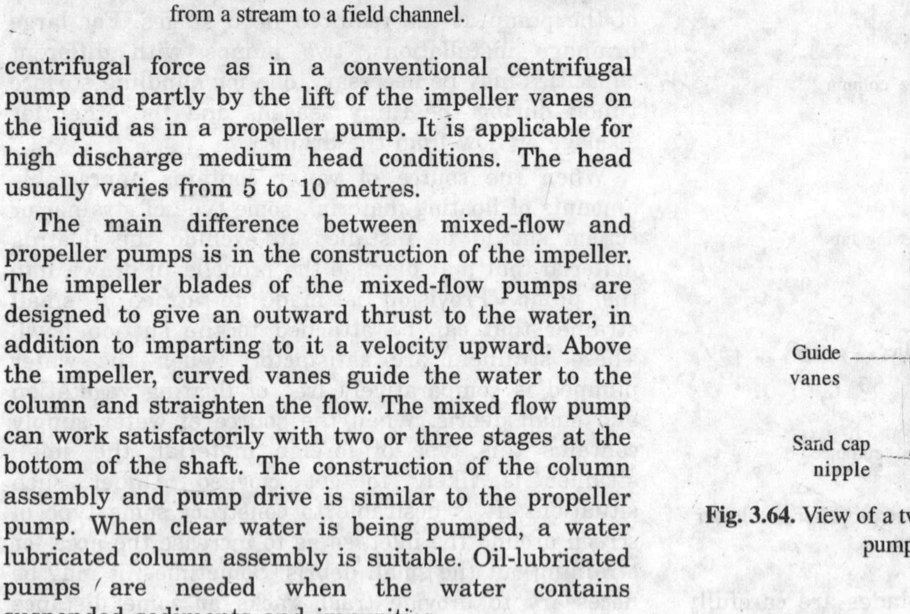

Fig. 3.63. Installation of a propeller pump to lift water from a canal over an embankment, taking advantage of the siphon action to reduce lift.

JET PUMPS

Jet pumps which are used in irrigation practice consist of a combination of a centrifugal pump and a jet mechanism, or an ejector (Fig. 3.65). The centrifugal pump, coupled with the electric motor or engine, is placed at the ground surface and furnishes the driving head and capacity for the jet placed in the well below the water surface. The main parts of a jet pump are a nozzle and a venturi. At the pump delivery, a portion of the high pressure water returns through the pressure pipe to activate the nozzle in the ejector.

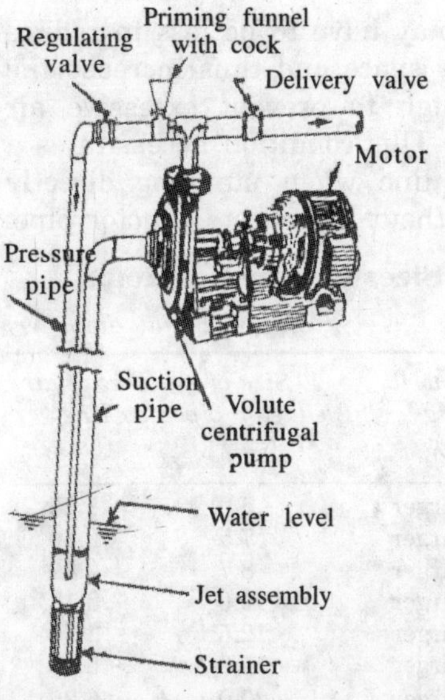

Fig. 3.65. Jet pump assembly with electric motor driven centrifugal pump placed at ground level. The Jet assembly is located below the water level.

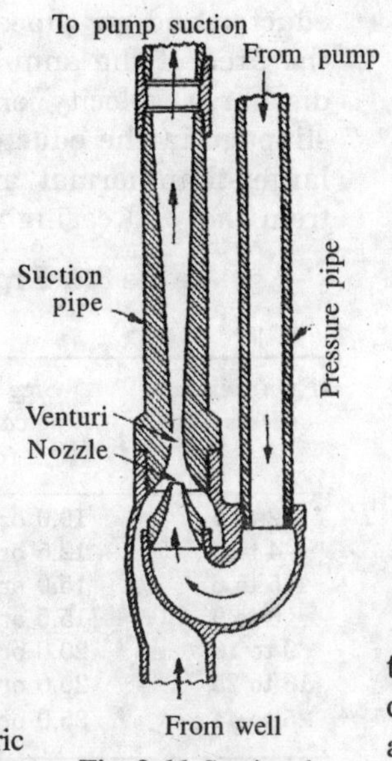

Fig. 3.66. Sectional view of jet assembly.

The nozzle is shaped so that it smoothly but abruptly reduces the area through which the flow must pass, thus increasing the velocity of flow. This creates a low pressure area around the venturi which draws more water from the well. The gradual enlargement in the venturi tube to the full diameter of suction pipe reduces the velocity of water with a minimum of turbulence. The vacuum created by the impeller of the pump placed at the ground surface draws the flow through the delivery pipe at the desired pressure. The additional supply of water which is obtained from the well is discharged past the control valve while the volume required for producing the flow is recirculated through the pressure pipe. The control valve is set to maintain the necessary pressure to produce flow at the existing pumping head.

The most important factor governing the capacity and efficiency of a jet pump is the selection of the jet to correspond to both the type of the pump and depth of low-water level at the source. Jet pumps are by nature inefficient. The highest efficiency usually obtained is about 35%. The efficiency of the pump is influenced mainly by the nozzle throat ratio. For small jet pumps used in domestic water supply, the spacing between the nozzle and the throat is equal to one nozzle diameter. The length of the throat is about six throat diameters.

Adaptability: Jet pumps are generally used for low capacity deep well pumping where it is necessary to locate the moving parts of the pump and the prime mover at the ground surface. The following are the main advantages of a jet pump:

(1) Its adaptability to be installed in wells as small as 5 cm inside diameter. (When the well diameter is too small to admit two pipes, a single pipe ejector may be used to connect the piping to the pump head. When this is done, the jet should always be submerged and the well casing extended to the low water level).

(2) Its high lift which cannot be reached by ordinary centrifugal pumps. The suction lifts usually obtained with low to medium size jet pumps vary from 12 to 18 metres.

(3) Accessibility of all moving parts which are located at the ground surface.

(4) Simplicity combined with relatively low cost and easy maintenance.

(5) Adaptability to being installed with the moving parts offset from the well.

In some locations, where water levels are subject to large seasonal variations, or where severe corrosion or plugging at the nozzle occurs, jet pumps are not very satisfactory.

AIR-LIFT PUMPS

Principle of Working of Air-lift Pumps

An air-lift pump operates by the injection of compressed air directly into the water inside a discharge or eductor pipe at a point below the water level in the well. The operating principle of the pump is shown in Fig. 3.67. The injection of the air results in a mixture of air bubbles and water. This composite fluid is lighter in weight than water so that the heavier column of water around the pipe displaces the lighter mixture forcing it upward and out of the discharge pipe. The piping assembly used for air-lift pumping from a well consists of a vertical discharge pipe called eductor pipe and a smaller air pipe.

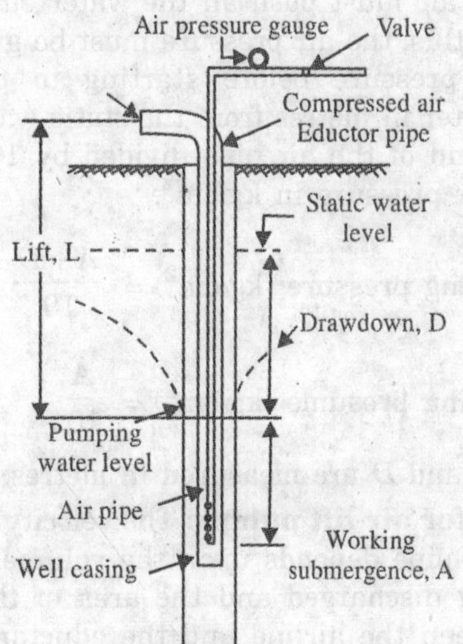

Fig. 3.67. Principle of operation of an air-lift pump.

An air-lift pump for pumping or developing a well is arranged commonly with the air pipe inside the eductor pipe as shown in Fig. 3.67. It is also possible to locate the air pipe outside the eductor pipe if there is space in the bore hole, as otherwise there is considerable friction loss when the air pipe is located inside a small diameter eductor pipe. Both the eductor pipe and the air pipe must be submerged in water in the well with 40% or more of their lengths extending below the pumping level. The energy that is available to operate the air-lift pump is that which is contained in the compressed air.

The well casing itself can be used for the eductor pipe if the diameter of the casing is not much larger than the air line. This provides a practical way to pump sand and mud from the bottom of a well during development and cleaning operations. When test-pumping a well, however, it is better to use a separate eductor pipe since the pumping level can then be measured with steel or electronic tapes.

The two most important factors in air-lift pumping are the per cent submergence of the airline and the relative sizes of the air and eductor pipes. Generally, the air-lift pump works best with a submergence of 60% or more. It is the length of airline that is submerged below the pumping level divided by the distance from the level of water discharge to the lower end of the airline and multiplied by 100, to give the result as percentage. Referring to Fig. 3.67,

$$\text{Per cent submergence} = \frac{A}{L - A} \times 100$$

The length of air pipe below the static water level is significant only for determining the air pressure required to start the air-lift pump. Before air can discharge from the lower end of the air pipe, the compressed air must push all the water out of the air pipe. To do this, the air pressure must be greater than the water pressure before starting to pump. The depth of water in metres from the static water level to the lower end of the air pipe divided by 10 gives the required air pressure in kg/cm^2.

$$\text{Thus, starting pressure (kg/cm}^2) = \frac{A + D}{10}$$

$$\text{Working presume (kg/cm}^2) = \frac{A}{10}$$

in which A and D are measured in metres.

Pipe sizes for air-lift pumps: The velocity of water in the eductor pipe depends upon the volume of air and water being discharged and the area of the annular space between the airline and the eductor. Table 3.4 gives the size of airlines for various sizes of eductor used under most conditions. When the water yield is unusually low, the difference between the sizes of eductor and air pipes may have to be less to reduce the area of the annular space and thus increase the discharge velocity enough to prevent excessive air slippage in the eductor. This condition necessitates a larger-than-normal air line when pumping directly from the well casing without a separate eductor pipe.

Table 3.4. Pipe Sizes for Air-lift Pumps

(internal diameter)

Pumping rate (litres/sec)	*Size of well casing (cm)*	*Size of eductor pipe (cm)*	*Size of air line (cm)*
2 to 4	10.0 or larger	5.0	1.25
4 to 5	12.5 or larger	7.5	2.50
5 to 6	15.0 or larger	8.7	2.50
6 to 9	15.0 or larger	10.0	3.13
9 to 16	20.0 or larger	12.5	3.75
16 to 25	20.0 or larger	15.0	5.00
25 to 44	25.0 or larger	20.0	6.25

Best operation of an air-lift pump requires good regulation of the amount of air injected. Too much of air causes excessive friction in the pipe lines and waste of air. Too little air results in reduced yield and in surging, intermittent discharge. When developing a well with an air-lift pump, the discharge should be started at a very low rate and brought up gradually. The air flow should be slowly increased in proportion to about what appears to be the increase in water flow into the well from the water-bearing formation.

Air-lift pumping is extensively used in the development and preliminary testing and cleaning of tubewells. It is also sometimes used in crooked tubewells where vertical turbine pumps or submersible pumps cannot be installed. The advantages of air-lift pumps are: simplicity, tubewell need not be perfectly straight or vertical, and impure water will not damage the pump. The main disadvantage is its low efficiency. The maximum efficiency usually obtained is about 30 per cent. The initial cost of the pump, including the air compressor, is high and it requires an extra depth of water for proper submergence.

SELECTION OF PUMPS

Irrigation wells and pumps are costly installations which require efficient utilization. A major part of the energy used in agriculture is in pumping. The total energy utilized in irrigation pumping in India in the year 1973-74 was about 8,400 million kilowatt-hours. It was estimated that the requirement would double in the next two decades. Efficient utilization of limited energy resourse calls for the selection of the most suitable pump, keeping in view the requirement of irrigation, characteristics of the well or other sources of water, kind of power available, economic conditions of the farmer and other factors.

Criteria and Procedures for Selection of Irrigation Pumps

The main factors influencing the selection of pumping sets are: (*i*) the requirement of irrigation water by the crops to be irrigated, (*ii*) yield of the sources of water (open wells, tubewells, streams, rivers, ponds), and (*iii*) availability and cost of the pump and kind of energy.

Determining the Discharge Capacity of Pumps

The data on the safe discharge rate of the well (or other source of water) and the discharge rate required for the crops to be irrigated with a particular cropping pattern are estimated. If necessary, the cropping pattern can be adjusted according to the safe yield of the well in different seasons. In many cases, such adjustments may not be possible on one's own farm since the holdings may be smaller than the command area required by a pump set of moderate size. In some cases, the holdings may be fragmented and only a few fragments belonging to the farmer are within the service orbit of the pump set installed in a well. In such cases, any one of the following alternatives could be adopted:

(1) Design the pumping plant according to the safe yield of the well and use it to irrigate the farm of an individual farmer, keeping the spare capacity idle. But it leads to high investment and under-utilization of installed capacity and invested funds.

(2) Design the pumping plant for the safe yield of the well and use the installed capacity to irrigate the land of the well owner and sell the remaining water to irrigate other farmer's land, based on such demands. This alternative is better than the first and leads to the full utilisation of the water resource potential developed at a high cost.

Discharge capacity of pump based on crop requirements: The pump discharge should meet the peak demand of water for the selected croping pattern. The rate of pumping depends on the area under different crops, the water requirement of the crops, rotation period (interval between two successive irrigations of a crop) and the duration the pump is operated each day. It may be computed by the following relationship:

$$q = \frac{Ay}{RT} \times \frac{1000}{36}$$

$$= 27.78\ \frac{Ay}{RT} \quad \text{... (3.11)}$$

in which,

q = rate of discharge of pump (litres/sec)
A = area of land under the crop (hectares)
y = depth of irrigation (cm)
R = rotation period (days)
T = duration of pumping (hours/day)

Example 3.5. *A farmer wishes to have his own pump set for the following cropping pattern to be followed in his holding of three hectares and his brother's holding of 2 hectares. Calculate the right size of the centrifugal pump he should have.*

Season	*crop*	*Area to be irrigated (ha)*	*Intensity of irrigation (cm)*	*Rotation period (days)*	*Period work (hrs/day)*
Writer	*1. Wheat*	*2.0*	*7.5*	*12*	*10*
(rabi)	*2. Cotton*	*0.4*	*7.5*	*20*	*10*
	3. Vegetables	*0.4*	*7.5*	*10*	*10*
	4. Mustard	*2.2*	*5.0*	*40*	*10*

Solution: Discharge rate of pump

$$= 27.78\ \frac{Ay}{RT}$$

$$= 27.78\left(\frac{2\times7.5}{12\times10}+\frac{0.4\times7.5}{20\times10}+\frac{0.4\times7.5}{10\times10}+\frac{2.2\times5}{40\times10}\right)$$

$$= 5.49 \text{ litres/second}$$

Fitting Pumps and Wells

The drawdown-discharge characteristics of the well influence the selection of a pump for a particular well. The pump that best suits a well can be selected by matching the characteristics of the pump with that of the well.

Well characteristics: The yield of a well depends on how much the water table is lowered by pumping. The well yield is the volume of water discharged from a well per unit of time. The discharge-drawdown relationship of a well is known as the well characteristics. Data on this relationship may be plotted on a graph paper with discharge as the abscissa and total depth to water level, as measured from the ground surface, as the ordinate (Fig. 3.68).

Matching pump and well characteristics: The characteristic curves of the well and the pump enable the selection of the pump which best suits the well. It is usual to draw the well characteristic curve on a tracing paper in the same scale as that of the pump curve. The tracing paper is placed on the graph showing the pump characteristics. The characteristics of pumps manufactured by leading manufacturers are matched with

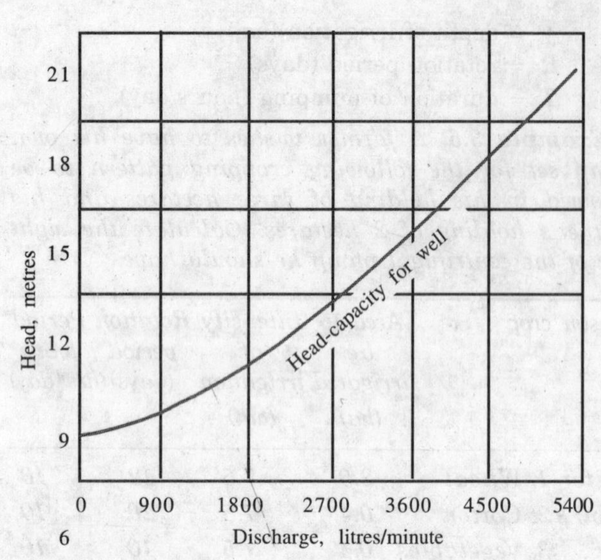

Fig. 3.68. A typical well characteristic curve.

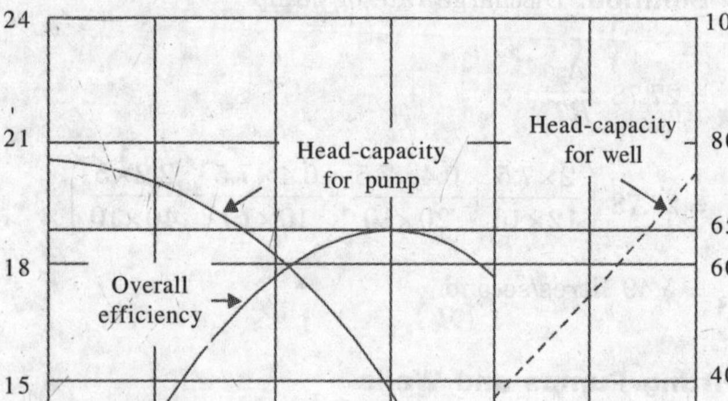

Fig. 3.69. Matching the pump characteristic curves with the well characteristic curve shown in Fig. 3.68.

the well characteristics. This will lead to the selection of the pump which gives the maximum efficiency and the desired discharge at the estimated total head. Figure 3.69 shows the matching of the characteristic curves of a pump with the well characteristics shown in Fig. 3.68. The point of intersection of the head-capacity curves of the pump and well provides a point of selection. At this point, the pump will give a discharge of 31,500 litres per minute against a total head of 13 metres and an efficiency of 65%. The power input is 12 horse-power. By careful selection from a wide range of products, it is possible to select a pump which will give the desired discharge rate at the estimated total head and provide an efficiency of 75 to 80% or more.

Some pump manufacturers provide information of their products as tabular data. Though precise matching of pump and well characteristics is not possible with tabular data, by checking the data of a large number of pumps, it is possible to select an efficient pump.

Pump selection based on types: The relative merits and applicability of different types of pumps were discussed earlier. Volute centrifugal pumps are adapted to a wide range of head-discharge conditions, as compared to turbine pumps and propeller pumps. The efficiency of vertical turbine pumps, including submersible pumps and propeller pumps, drop rapidly when applied under conditions different from those for which they are designed.

POWER REQUIREMENTS, EFFICIENCY AND ECONOMICS OF IRRIGATION PUMPING PLANTS

To determine the horse power of the electric motor or engine used in driving a pump, it is necessary to know the efficiency of the pump, the type of drive, the type of power unit, the head under which the pump operates and the losses in the piping system.

Friction Head in Pipe Systems

The head losses pipes and accessories include the energy or head required to overcome the friction in pipes and fittings (including strainer, foot valve, elbows, bends, reducing sockets, tees, valves, etc.) in the pumping system. Friction head exists on both the suction and discharge sides of a pump and varies with the rate at which the water is being pumped, pipe size, condition of the interior of the pipe and the material of which the pipe is made.

Loss of head due to friction in a pipe: Loss of head due to friction in a pipe, between two points at a distance l apart, is given by the formula:

$$h_f = \frac{4flv^2}{2gd} \qquad ...(3.12)$$

in which, f = coefficient of friction for pipe, m

l = length of pipe, m

d = diameter of pipe, m

The friction coefficient f is not simply a coefficient of friction between the water and the surface of the pipe, but includes a coefficient of resistance due to eddying motions in the water itself. It further depends upon the smoothness or roughness of the pipe surface. The value of f is less if the pipe is new and smooth. The approximate values of f are given by the following empirical relations:

(1) Old pipes:

$$f = 0.01\left(1+\frac{1}{4.725\,d}\right) \qquad ...(3.13)$$

(2) New pipes:

$$f = 0.005\left(1+\frac{1}{4.725\,d}\right) \qquad ...(3.14)$$

Since the factor $\frac{1}{4.725d}$ is usually very small, it is often neglected and the value of f is assumed to be 0.01 and 0.005 for old pipes and new pipes, respectively.

Example 3.6. *Compute the head loss due to friction in a pipe 7.5 cm in diameter and 120 m long when the water is flowing at a velocity of 1.8 m/sec. The value of f may be assumed to be 0.005.*

Solution:

d = 0.075 m $\qquad l$ = 120 m

v = 1.8 m/sec $\qquad g$ = 9.81 m/sec^2

Head loss due to friction, $h_f = \frac{4flv^2}{2gd}$

$$= \frac{4 \times 0.005 \times 120 \times 1.8 \times 1.8}{2 \times 9.81 \times 0.075} = 5.28 \text{ m}$$

Friction loss in pipe fittings and pump accessories. Head loss in the strainer and foot valve of a pump are obtained from the following formulae:

Strainer,

$$h_f = K_2\,\frac{v^2}{2g} \qquad ...(3.15)$$

Foot valve,

$$h_f = K_f\,\frac{v^2}{2g} \qquad ...(3.16)$$

in which,

h_f = head loss due to friction (m)

K_2 and K_f are constants.

The value of K_s usually assumed is 0.95 and that of K_f = 0.80.

Head loss in valves and pipe fittings could be determined from the nomographic solution presented in Fig. 3.70. The dotted line in the figure illustrates the use of the nomograph for determining the head loss due to sudden contraction in a pipeline when the ratio between the small diameter d and large diameter D is 1:2 and the value of d is 80 mm. The point 80 mm on the pipe size scale is joined to the point on the nomograph representing sudden contraction of d/D = ½. The point of intersection of the line on the equivalent pipe length scale gives the desired resistance value as equivalent to the friction loss in a 0.9 m long straight pipe of 80 mm size. The

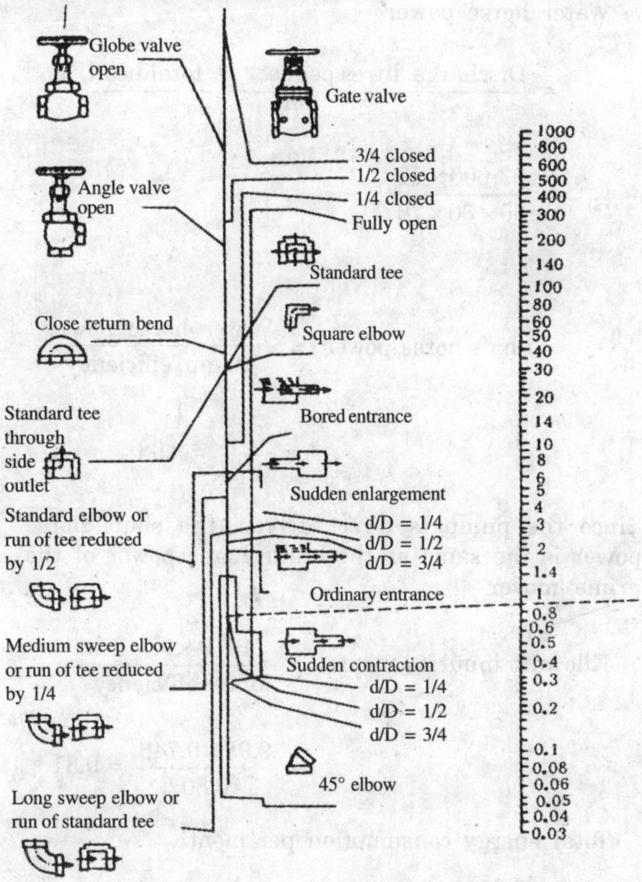

Fig. 3.70. Nomograph for determining friction losses in pipe fittings.
(Adapted from Kirloskar Bros. Ltd. '*Engineering Data*'.)

actual head loss in an 80 mm diameter pipe at the given discharge rate may be computed using equation (3.13.) It may be noted that for sudden enlargements or sudden contractions, the smaller diameter d is used on the pipe size scale.

Pumping Plant Design

The power requirements and efficiency of pumping plants can be computed using the relationships expressed earlier in this chapter.

Example 3.7. *A pump lifts 93,600 litres of water per hour against a total head of 21 metres. Compute the water horse power. If the pump has an efficiency of 72%, what size prime mover is required to operate the pump. If a direct drive electric motor having an efficiency of 80% is used to operate the pump, compute the cost of electrical energy in a month of 30 days. The pump is operated for 12 hours daily for 30 days. The cost of electrical energy is 50 paise per unit.*

Solution:

Water horse power

$$= \frac{\text{Discharge, litres per sec} \times \text{total head, m}}{76}$$

$$= \frac{93{,}600 \times 21}{60 \times 60 \times 76}$$

$$= 7.18$$

$$\text{Shaft horse power} = \frac{\text{Water horse power}}{\text{Pump efficiency}}$$

$$= \frac{7.18}{0.72} = 9.98$$

Since the pump is direct driven, the shaft horse power is the same as the brake horse power of the prime mover.

$$\text{Kilowatt input to motor} = \frac{\text{BHP} \times 0.746}{\text{Motor efficiency}}$$

$$= \frac{9.98 \times 0.746}{0.80} = 9.31$$

Total energy consumption per month

$= 9.31 \times 12 \times 30$

$= 3351.6$ kilowatt-hours (electrical units)

Cost of electrical energy per month

$$= 3351.6 \times \frac{50}{100}$$

= Rs. 1675.80

Example 3.8. *A direct driven centrifugal pump coupled to a 3-phase electric motor is installed in a deep open well. The discharge rate of the pump is 18 lit/sec. The pump efficiency is 67%. The centre line of the pump is 60 cm vertically above the static water level and 6.2 m above the pumping water level. The suction pipe is 7.5 m long and 3 cm in diameter. A foot valve with strainer is fixed at the bottom of the suction pipe. The suction line is connected to the pump inlet by a long sweep bend of the same size as the suction pipe. The pump discharges water to the top of a pump stand of an underground pipe line water distribution system. The vertical distance between the top of the stand and the centre line of the pump is 16 m. The total length of the 7 cm diameter discharge pipeline is 24 m. The pipe fittings on the discharge side are three long sweep bends, one gate valve and one reflux valve, all of the which are of the same size as the discharge pipe.*

From the above data, compute–

(1) Total head,

(2) Water horse power, and

(3) Brake horse power of the motor driving the pump.

Solution:

$$\text{Area of suction pipe} = \frac{\pi}{4} \times \frac{8}{100} \times \frac{8}{100} = 0.005, \ \text{m}^2$$

Velocity of water in suction pipe

$$= \frac{Q}{a_1} = \frac{18}{1000 \times 0.005} = 3.6 \text{ m/sec}$$

Cross-sectional area of discharge pipe

$$= \frac{\pi}{4} \times \frac{7}{100} \times \frac{7}{100} = 0.0038 \text{ m}^2$$

Velocity of water in discharge pipe

$$= \frac{Q}{a_2} = \frac{18}{1000 \times 0.0038} = 4.74 \text{ m/sec}$$

(*i*) Total head = Total suction lift + total discharge head

Static suction lift = 6.2 m

Head loss in suction pipe 8 cm dia, 7.5 m long
= 2.25 m

Head loss in long sweep bend, 8 cm dia.
= 0.41 m
(Fig. 3.70 and Eq. 3.12)

Head loss in strainer

$$= K_3 \frac{v_1^2}{2g} = 0.95 \times \frac{3.6 \times 3.6}{2 \times 9.81}$$

$= 0.63$ m

Head loss in foot valve

$$= K_f \frac{v_1^2}{2g} = 0.80 \times \frac{3.6 \times 3.6}{2 \times 9.81}$$

$= 0.53$ m

Velocity head (suction line)

$$= \frac{v_1^2}{2g} = \frac{3.6 \times 3.6}{2 \times 9.81} = 0.66 \text{ m}$$

Total suction lift

$= 6.20 + 2.25 + 0.63 + 0.53 + 0.66 + 0.41$

$= 10.68$ m

Static discharge head = 16.00 m

Head loss in discharge pipe 7 cm dia., 24 m long

$= 14.4$ m (Eq. 3.12)

Head loss in 3 long sweep bends, 7 cm dia.

$= 2.07$ m

Head loss in gate valve, 7 cm dia.

(Fig. 3.70 and Eq. 3.12)

$= 0.33$ m

Head loss in reflux valve, (using Eq. 3.17 for foot valve)

$$= K_f \frac{v_1^2}{2g} = 0.8 \times \frac{4.74 \times 4.74}{2 \times 9.81} = 0.92 \text{ m}$$

Velocity head (discharge line)

$$= \frac{v_1^2}{2g} = \frac{4.74 \times 4.74}{2 \times 9.81} = 1.14 \text{ m}$$

Total discharge head = 16.00 + 14.40 + 2.07 + 0.33 + 0.92 + 1.14

$= 34.86$ m

Total head $= 10.68 + 34.86 = 45.54$ m

(*ii*) Water horse power $= \frac{18 \times 45.54}{76} = 10.78$

(*iii*) Brake horse power of motor driving the pump

$$= \frac{10.78}{67} \times 100 = 16.10$$

Example 3.9. *A deep well turbine pump installed in a tubewell has a discharge of 22 litres per second. The discharge pipe is 12.5 cm in diameter. The pumping water level is 24 m below the centre line of the discharge head. The delivery pipe is 15 m long, measured from the end of the discharge head. A sluice valve and two long sweep bends, each of 12.5 cm inside diameter, are fitted on the delivery pipe. The delivery pipe outlet is 4.6 m above the discharge end of the pump. Determine the brake horse power of the directly connected vertical electric motor required to operate the pump if the pump efficiency is 68%.*

Solution:

Area of the delivery pipe

$$= \frac{\pi}{4} \times \frac{125}{100} \times \frac{12.5}{100} = 0.0122 \text{ m}^2$$

Velocity of water in the delivery pipe

$$= \frac{Q}{a} = \frac{22}{1000 \times 0.0122} = 180 \text{ m/sec}$$

Static head

= Pumping-head + static discharge head

$= 24 + 4.6 = 28.6$ m

Friction head in the delivery pipeline of 12.5 cm dia. and 15 m long

$= 0.615$ m (Eq. 3.12)

Head loss in sluice valve (gate valve) 12.5 cm dia.

$= 0.045$ m (Fig. 3.70)

Head loss in two long sweep bends

$= 0.197$ m

Velocity head at the end of delivery pipe (Fig. 3.70)

$$= \frac{v^2}{2g} = \frac{1.8 \times 1.8}{2 \times 9.81} = 0.165 \text{ m}$$

Total head $= 28.6 + 0.615 + 0.045 + 0.197 + 0.165$
$= 29.622$ m

Water horse power $= \frac{22 \times 29.622}{76} = 8.57$

Brake horse power of electric motor

$$= \frac{8.57}{0.68} = 12.61$$

Economics of Pumping

Estimates of cost of pumping are necessary to compare the relative costs of different types of pumping installations and evaluate the economics of irrigation. Cost of pumping includes fixed costs and operating costs.

Fixed Costs

1. *Interest:* Interest is calculated on the average value of the installation at the prevailing interest rate.

Annual interest cost

$$= \frac{(\text{Value of installation} - \text{salvage value}) \times \text{interest rate}}{2}$$

2. *Depreciation:* Depreciation is the loss in the value of the pumping plant due to operation or age

Annual depreciation

$$= \frac{\text{Orignal cost} - \text{salvage value}}{\text{Useful life of pumping unit in years}}$$

The estimated useful life of pumping plants and allied installations are presented in Table 3.6.

Operating Costs

1. *Fuel/energy consumption:*

(*a*) *Electric motor*: Efficiencies of electric motors usually vary from 80 to 90 per cent.

Energy consumption, kilowatt-hours

$$= \frac{\text{Brake horse power}}{\text{Motor efficiency}} \times 0.746$$

(*b*) *Engine:* An estimate of the rate of fuel consumption for a given engine can most accurately be made if the manufacturer's fuel consumption curve for that engine is available.

Approximate value of the rate of fuel consumption of diesel engines which are commonly used in irrigation pumping is 0.23 litres per brake horse power hour.

Cost per hour of operation

$$= \text{BHP} \times \begin{Bmatrix} \text{Fuel consumed} \\ \text{in litres} \\ \text{per hour} \end{Bmatrix} \times \begin{Bmatrix} \text{Cost of fuel} \\ \text{per litre} \end{Bmatrix}$$

2. *Lubricating oil and grease:*
   (*a*) Negligible for electric driven plants.
   (*b*) Diesel and petrol engines:
   4.5 litres per 1000 hp-hours.
3. *Pump maintenance and repairs:*
   Centrifugal pump:

$$\text{Yearly cost} = \frac{\text{Total cost}}{\text{Estimated life in years}}$$

Turbine pump:

$$\text{Yearly cost} = \frac{\text{One-half of the total cost}}{\text{Estimated life in years}}$$

4. *Engine maintenance and repairs:*

Difficult to predict, a nominal amount may be added towards this item.

5. *Operator's wages:*

At prevailing rates.

Table 3.5. Estimated Useful Life of Pumping Plants and Allied Installations

Item	*Estimated useful life*
Mild steel pipe	25 years
Well and casing	20 years
Turbine pump:	
Bowl (about 50 per cent of cost of the pump unit) 16,000 hours or 8 years	
Column, pipe, etc.	32,000 hours or 16 years
Centrifugal pump	32,000 hours or 16 years
Power transmission:	
Gear head	30,000 hours or 15 years
V-belt	6,000 hours or 3 years
Flat belt, rubber and fabric	10,000 hours or 5 years
Flat belt, leather	20,000 hours or 10 years
Electric motor	50,000 hours or 25 years
Diesel engine	28,000 hours or 14 years

(From SCS National Engineering Handbook, Section 15, Ch. 8, USDA)

Example 3.10. *The water horse power of a centrifugal pump, installed in an open well and operated by a direct coupled 3-phase electric motor, is 2.3. The efficiencies of the pump and electric motor are 68% and 76%, respectively. The pump is operated for a total of 2,600 hours in 210 days a year. Estimate the annual cost of operating the pump. The cost of the pump and the motor are Rs. 800 and Rs. 2,200, respectively. The total cost of the suction and discharge pipes, pipe fittings, foot valve and strainer amount to Rs. 950.00. The cost of electrical accessories including starter, switch and wiring is Rs. 820.00. The cost of electricity is 18 paise per unit. Prevailing interest rate is 8 per cent. The salvage value of pump and motor may be assumed to be Rs. 20 and Rs. 30, respectively. The salvage values of other items are negligible. The operator's time spent on the pumping set is 1 hour daily. The wage of the operator is Rs. 8 per day including his other duties. (Note: Costs and wage relate to early 1970s).*

Solution:

Fixed costs

1. Annual interest cost

$$= \frac{(4770 - 50) \times 8}{2 \times 100} = \text{Rs. } 188.80$$

2. Depreciation

$$(a)\ \text{Pump} = \frac{800 - 20}{16} = \text{Rs. } 48.75$$

$$(b)\ \text{Motor} = \frac{2200 - 30}{25} = \text{Rs. } 86.80$$

(*c*) Pipe & fittings $= \dfrac{950-0}{25}$

= Rs. 38.00

(*d*) Electrical accessories

$= \dfrac{820-0}{25}$

= Rs. 32.80

Total = Rs. 206.35

Operating costs

1. Annual energy consumption

$= \dfrac{2.3}{0.68 \times 0.76} \times 0.746 \times 2600$

= 86332 kilowatt-hours

2. Cost of electrical energy

= 8632 × 0.18 = Rs. 1533.76

3. Pump maintenance and repairs

$= \text{Rs. } \dfrac{800}{16} = 50.00$

4. Operator's wages $= \dfrac{1}{8} \times 210 \times 8$ = Rs. 210.00

Total annual cost of operation

= Fixed cost + operating cost

= 395.15 + 1813.76 = Rs. 2208.91

Economical Pipe Size Selection

The selection of pipe size of pumping plants and pipelines should be based on careful economic analysis. A small pipe may require a lower initial investment but the head loss due to friction is greater and this increases the power cost. A larger pipe in many cases will save more in power cost than the additional investment. Further, the larger pipe may so reduce the total pump head that a lighter and lower priced pump and power unit may be used.

Example 3.11: *A pump is to supply water at the rate of 7.5 litres per second through a pipe 300 m long. The overall efficiency of the pump and the motor is 70%. The cost of electricity 18 is paise per kilowatt-hour. Select the most economical size of pipe from out of the pipe sizes commonly available in the market. Prevailing rate of interest on investment may be assumed to be 7%. The pump is to be operated for a total period of 2,600 hours in a year.*

Solution:

1. The possible sizes of pipes from which the selection could be made are assumed to be of 5 cm, 6 cm, 7 cm, 8 cm, 10 cm and 12.5 cm diameter.
2. The cost of 300 m of pipe of each size is estimated.
3. Interest on the cost of investment on 300 m length of pipe of each size is assumed.
4. Cost of depreciation in the value of pipe is computed in each category.

   Head losses due to friction in the different sizes of pipes are computed.
6. Horse power consumed in overcoming the friction head and the consequent cost of the electric power consumed is calculated from the following relationship:

   (*i*) Energy consumed annually, kilowatt hours

   $$= \frac{\text{discharge (litres/sec)} \times \text{friction head (m)} \times 746 \times \text{hrs of pumping}}{76 \times \text{overall efficiency}}$$

   (*ii*) Cost of energy

   = Energy consumption × cost of electric energy

It is seen that the annual operating cost is the lowest in case of a 10 cm pipe. However, the initial cost of this pipe is Rs. 19,200, as compared to Rs. 15,126 for a pipe of 8 cm diameter. Evidently, the choice will fall between the sizes of 8 cm diameter and 10 cm. The actual cost of the installation based on the cost of pump and electric motor, taking into account the friction head under the two cases, should be estimated and the choice of the pipe size finalized.

The results are tabulated below:

Size of pipe (*cm*)	*Cost of pipe 300 m long* (*Rs*)	*Interest on investment* (*Rs*)	*Depreciation charges in 300 on long pipe* (*Rs*)	*Head loss due to friction in pipeline* (*m*)	*Annual power consumption in overcoming friction* (*kW-hrs*)	*Cost of energy consumed* (*Rs*)	*Total cost per year* (*Rs*)
5	8,760	306.00	350.40	186.0	50,859.8	9154.77	9811.77
6	10,800	378.00	432.00	72.0	19,687.7	3543.79	4353.79
7	13,104	458.40	524.16	31.8	8,695.4	1565.17	2547.97
8	15,126	529.41	605.04	15.6	4,265.7	767.83	1902.28
10	19,200	672.00	768.00	4.5	1,230.5	221.49	1661.49
12.5	24,450	855.75	978.00	1.41	385.6	69.40	1903.15

SUMMARY

Basic principles of indigenous water lifts and pumps are introduced with their adaptability and relative economics. Performance and adaptability of common types of indigenous water lifts are presented under low head, medium head and high head operating conditions. Feasibility of utilizing non-conventional energy sources in pumping are explained along with the types of appliances to suit wind, water, solar photovoltaic, and biogas energy. Principles of operation, component parts and application of different types of positive displacement pumps are explained with their functional relationships. Importance of hand pumps in rural water supply in developing countries and their installation procedures are described. Principles of operation of variable displacement pumps are presented with their basic design criteria and adaptability. Importance of pump characteristics curves are explained, including family of curves. The functional relationship between pump speed and impeller diameter on pump performance are explained. Procedures for installation, operation, maintenance and troubleshooting of centrifugal pumps under widely varying operating conditions are described. Causes of cavitation and water hammer in pumping systems are discussed and their remedial measures explained. Types of vertical turbine pumps and their adaptability are explained. Construction and characteristics of vertical turbine pumps, including their installation, operation, maintenance and trouble-shooting are described. Adaptability, merits and limitations of submersible pumps are explained, including the requirements for their installation, operation and maintenance. Characteristic differences in the principles of operation and pumping elements of turbine, propeller and mixed flow pumps are explained. Construction, illustration, operation and maintenance of mixed flow and propeller pumps are described. Principles and adaptability of jet pumps and airlift pumps are described, including their installation operation and maintenance requirements.

PROBLEMS

3.1. A deep well single acting piston pump has a cylinder diameter of 10 cm and stroke 22.5 cm. The pump makes 300 stroke per minute. Calculate the discharge of the pump in litres per hour. If the total head is 36 m and the pump efficiency is 60%, determine the horse power required to drive the pump.

[**Ans.** 15900 litres/hr, 3.5 HP]

3.2. A centrifugal pump is running at a speed of 1800 rpm at its best efficiency point and delivers 50 litres of water per second at a total head of 30 metres. Compute the specific speed of the pump.

[**Ans.** 31.4 rpm]

3.3. Calculate the desired discharged rate of a centrifugal pump to be used by a farmer for the following irrigation schedule.

Season	*Crop*	*Area to be irrigated ha*	*Intensity of irrigation cm*	*Rotation Period days*	*Period work hrs/day*
Kharif	Paddy	1.6	10.0	15	10
	Cotton	2.0	10.0	20	12
Rabi	Wheat	3.2	7.5	12	12
	Cotton	2.0	7.5	20	12
	Vegetables	1.6	7.5	10	12
Hot weather	Bajra	0.8	7.5	8	10

[**Ans.** 9.14 litres/second]

3.4. Determine the annual consumption of electrical energy by a motor driven centrifugal pump installed in a shallow tubewell. The pump discharge is 16 lit/sec against a total head of 7 m. The pump efficiency is 70% and the motor efficiency 84%. The drive efficiency may be assumed to be 100%. The pump is operated for 3500 hours per year.

[**Ans.** 6546 kilowatt-hours]

3.5. A volute centrifugal pump installed in a deep open well is driven by a diesel engine using a two-stage belt drive. The pump discharge is 20 litres per second. The suction pipe is 10 cm in diameter and is 7 m long. The pumping water level is 5.8 m below the centre line of the pump. The suction pipe is connected to the pump by a long sweep bend of 10 cm inside diameter. A 10 cm diameter foot valve is fixed at the intake of the suction pipe. The static discharge head is 12 m. The discharge pipe is 8 cm in diameter and 16 m long. Three long sweep bends and a reflux valve, each of 8 cm diameter, are fixed on the discharge line. The pump discharges into a pressure tank, the pressure inside which is 0.6 kg/cm^2. Determine (*i*) the total head, (*ii*) brake horse power of the engine if the pump efficiency is 75% and drive efficiency is 80%, (*iii*) if the specific fuel consumption of the diesel engine is 0.23 litres per kilowatt-hour, what is the monthly requirement of diesel oil to run the engine if the pump works for 10 hours each day.

[**Ans.** (*i*) Total head, 34.52 m
(*ii*) Brake horse power, 15.33
(*iii*) Diesel, 799.40 litres]

QUESTIONS

1. Discuss the adaptability and relative advantages of manually operated and animal operated indigenous water lifts.
2. Compute the weight of the stone required to be attached to the pole of a counterpoise water lift working on a post with its length extending 1.5 m from pivot point at the load end and 3 m on the bucket end. The weight of water bucket is 10 kg .
3. Briefly describe the components and operating principles of a hydraulic ram.
4. Define D' Aubuission's efficiency ratio.
5. What is the principle of photo voltaic power generation?
6. Describe the operating principle of a twin-treadle pump developed in Bangladesh in 1980s.
7. Distinguish between single acting and double acting type positive displacement pumps.
8. Briefly describe the installation procedures of India Mark II handpumps.
9. Define specific speed of pumps. What is its significance in the design of centrifugal pumps?
10. Define net positive suction head in centrilugal pumps.
11. What is the importance of characteristic curves in the selection of centrifugal pumps?
12. Express the relationships between pump speed and impeller diameter in the performance of centrifugal pumps?
13. Discuss the functioning of diffuser when used in centrifugal pumps.
14. How will the shape and number of vanes influence the performance of centrifugal pumps?
15. What are the common methods of balancing the end thrust developed in a centrifugal pump?
16. What is priming, as applied to centrifugal pumps? What are the different methods used in priming?
17. What are the special pumping situations requiring the use of non-clog pumps?
18. How are centrifugal pumps classified?
19. What are the special applications of medium lift submersible pump with hydraulic drive? Describe its operating principles.
20. List the basic requirements in installing volute centrifugal pumps.
21. What are the common types of starters used with electric motors for operation with pumps? What are the operating principles of a star-delta starter.
22. What is the function of an overload release in the electrical system of a pumping set? How does it function?
23. What is cavitation? How does it affect the performance of a pumping system?
24. How is cavitation prevented in a centrifugal pump?
25. What are the operating situations requiring the use of (i) Oil-lubricated vertical turbine pumps (ii) Water-lubricated vertical turbine pumps?
26. What is the function of non-reverse ratchet in a turbine pump assembly?
27. What are the basic requirements in the installation of a vertical turbine pump?
28. What are the major causes of vibration during the working of a vertical turbine?
29. What are the major advantages of a submersible pump in comparison to a vertical turbine pump?
30. What are the components of the pump element in a submersible pump?
31. What are the major features of a submersible electric motor used in a submersible pump?
32. Distinguish between propeller and mixed flow pumps. What are the components which are common in the two types?
33. What is net positive suction head? How is it computed?
34. What are the basic hydraulic principles of a centrifugal pumps?
35. Distinguish between cavitation and water hammer in pumping. How are they overcome?
36. What are the specific situations requiring the use of air-lift pumps?
37. Define the concept of 'fitting of wells and pumps'.
38. Describe the graphical procedure for matching the characteristics of wells with those of the pump.
39. Identify the true and false statements from the following:

    (*i*) The four basic principles involved in pumping water are:

    (a) Atmospheric pressure,

    (b) Positive displacement,

    (c) Centrifugal force,

    (d) Movements of columns of water caused by their difference in specific gravity.

    (*ii*) To obtain high efficiency in the operation of an Archemedian screw, the lower end of the screw should be just half submerged in the source of water.

    (*iii*) Water wheel is the most efficient amongst the manually operated indigenous water lifts.

    (*iv*) In a Persian wheel, the water is lifted considerably above the ground level. The percentage increase in the energy required in water lifting is significant, when the water table is deep.

    (*v*) Silicon is the most common material used in manufacturing solar cells.

(*vi*) The capacity of a twin-treadle pump is about 3 litres/sec for a lift of 4 m.

(*vii*) The discharge rate of a volute centrifugal pump is directly proportional to the pressure head.

(*viii*) In centrifugal pumps, the lower the specific speed, the higher is the pressure head developed.

(*ix*) A volute centrifugal pump is preferred for large capacity, high speed operation.

(*x*) The discharge capacity and efficiency are almost similar in single stage and multi-stage centrifugal pumps.

(*xi*) The impellers used in centrifugal pumps designed to pump brackish water are commonly made of gun metal.

(*xii*) The bowl assembly of a vertical turbine pump is located to remain submerged in water in the well during pumping.

(*xiii*) The minimum pulley diameter for using a 4-ply rubber belt in a turbine pump installation is 15 cm.

(*xiv*) In a centrifugal pump, the conversion of velocity energy into pressure energy is accomplished either in the volute casing or in the diffuser casing.

(*xv*) In the design of a centrifugal pump, the greater the rate of pumping, the more is the number of vanes provided in its impeller.

(*xvi*) The used of multi-stage pumps is a standard practice in volute and turbine pumps operating under high heads.

(*xvii*) Compared to volute centrifugal pumps, the efficiency of a turbine pump remains high over a wide range of speeds.

40. Identify the most appropriate answer of the following questions from the four options:

(*i*) Archemedian screw is suitable to lift water from open water bodies to heights ranging from:

(a) 0.5 to 1.2 m (b) 1.2 to 3 m
(c) 2 to 3 m (d) 3 to 4 m

(*ii*) The gas consumption of a biogas engine per brake horse power per hour is:

(a) 0.25 m^3 (b) 0.45 m^3
(c) 0.65 m^3 (d) 0.85 m^3

(*iii*) The minimum wind speed required for the operation of a windmill usually ranges from:

(a) 4–6 km/hr (b) 6–8 km/hr
(c) 8–10 km/hr (d) 10–12 km/hr.

(*iv*) The energy output of a windmill varies:

(a) directly with the wind speed
(b) as the square of the wind speed
(c) as the cube of the wind speed
(d) inversely proportional to the wind speed.

(*v*) Overload releases in electric motor starters are set to trip off when the motor current exceeds:

(a) the current rating of the motor
(b) 5% above the motor current
(c) 10% of the motor current
(d) 15% of the motor current

(*vi*) For optimum performance, the per cent submergence of an air lift pump is:

(a) 50% (b) 60%
(c) 70% (d) 80%

40. In the following, match each of the entries of the 1st column with the most closely related entry in the 2nd column:

(*i*)	(A) Rotary pump	(I) Godavari Delta region of Andhra Pradesh, Egypt
	(B) Twin-treadle pump	(II) Hydraulic ram
	(C) Archemedian screw	(III) Silicon
	(D) D' Abuisson's efficiency ratio	(IV) Bangladesh
	(E) Solar cells	(V) Positive displacement
(*ii*)	(A) Methane gas	(I) Operating characteristics of variable displacement pumps
	(B) Richardson Crudas Ltd.	(II) Centrifugal pump
	(C) Specific speed	(III) Sewage water pumping
	(D) Stuffing box	(IV) Hand pumps for drinking water supply
	(E) Non-clog impeller	(V) Biogas plant
(*iii*)	(A) Cavitation	(I) Turbine pumps
	(B) Water hammer	(II) Centrifugal pumps
	(C) Bowl assembly	(III) Medium lift submersible centrifugal pump
	(D) Diffuser	(IV) Collapse of water vapour pockets
	(E) Hydraulic drive	(V) Kinetic energy of flowing fluids
(iv)	(A) Overload release	(I) Vertical turbine pump
	(B) Rubber bearing	(II) Mixed flow pump
	(C) Right angle gear drive	(III) Jet pump

(D) Mild centrifugal action	(IV) Water lubricated pump		(B) Pump selection	(II) Reverse flow in pipelines
(E) Ejector	(V) Single phasing of electric motors		(C) Gate valve	(III) Centrifugal pump
			(D) Reflux valve	(IV) Air lift pump
(v) (A) Working submergence	(I) Piping system accessory		(E) Volute casing	(V) Matching pump characteristics

FURTHER READING REFERENCES

Addison, H. (1959). *A Treatise on Applied Hydraulics*, 4th edn, Asia Publishing House, Bombay, p. 724.

Brauman, E. and Fuller, R.J. (1984), *Manual Irrigation in Bangladesh*. Swiss Development Cooperation, Mirpur Agricultural Workshop and Training School, Dhaka, Bangladesh, Mimeo, p. 38.

Brooks, J.B. (1953). *Electric water systems for farms*. Univ. of Kentucky, College of Agr. and Home Eco., Cir, 452, p. 39.

Bureau of Indian Standards, *Specifications for Deep Well Hand Pumps*, (1982). IS: 9301–1982. New Delhi, p. 8.

Chanhan, H.S., Sewa Ram, Trivedi, S.K. and Sharma, H.C. (1979). *Selection, Installation and Operation of Horizontal Centrifugal Pumps for Irrigation*. Dept. of Agrl. Engineering, College of Technology, G.B. Pant University of Agriculture and Technology Pantnaga, Uttaranchal, p. 129.

Church, A.H. and Jagdish Lal (1973). *Centrifugal pumps and blowers*. Pub. Metropolitan Book Co. Pvt. Ltd., New Delhi.

Dakshinamurti, C., Michael, A.M. and Mohan, S. (1973). *Water resources of India and their utilization in agriculture*. Water Technology Centre, IARI, New Delhi, pp. 269–285.

Das, G. and Michael, A.M.(1967). *Evaluation of performance characteristics of circular two-bucket lift*. Jl. of Indian Soc. of Agrl. Engrs., Vol. 4, No. 2, pp. 1–8.

Ebara Solar Powered Pumping System. Ebara Corporation, Tokyo, Japan, (1982), p. 5.

Fogel, M.M. and Myles, G.A. (1962). *Pumping from irrigation wells*. Univ. of Nevada, Bul. No. 110, p. 23.

Fraenkel, P.K. (1986). *Water lifting devices,* FAO Irrigation and Drainage Paper 43, Pub. Food and Agriculture Organization of the United Nations, Rome, p. 295.

Fritz, M. (1982) *Relevant criteria for harnessing new and renewable sources of energy in the developing world,* Natural Resources and Development, 15. Institute of Scientific Cooperation, Tibingen, Fed. Rep. of Germany.

Ghate P. (1977) *Scope for irrigation through hydraulic rams in the hills of India*, Agri. Engg. Today, I (XII), Indian Soc. of Agril. Engrs.

Gibson, A.H. (1924). *Hydraulics and its applications*. Pub. Constable and Co. Ltd., London.

Griffin, R.E. *et al.* (1982). *Irrigation pumping planning guide*, Water Magt. Synthesis Project, Colorado State Uni., Fort Collins, USA, p. 23.

Horsey, H.R. (1983). *A technical and economic analysis of low-lift irrigation pumping in Egypt*. M.S., Thesis (unpublished), Colorado State University, Fort Collins, Colorado, USA, p. 213.

Indian Standards Institution (1970). *Indian standard specification for sewage and drainage pumps*. IS: 5600–1970. ISI, New Delhi, p. 8.

Indian Standards Institution (1970). *Indian standard specification for vertical turbine pumps for clear, cold, fresh water*. IS: 1971–72, ISI, New Delhi, p. 14.

Indian Standards Institution (1972). *Indian standard specification for horizontal centrifugal pumps for clear, cold, fresh water for agril. purposes.* IS:6595–1972, ISI New Delhi, p. 8.

Indian Standards Institution (1980). *Code of Practice for the Selection, Installation, Operation and Maintenance of Horizontal Centrifugal Pumps for Agricultural Applications*, Part I *Selection* (15.9694–1080), Part II *Installation* (15–9694–1980), Part III *Operation* (15–9694–1980). Part IV *Maintenance*. Bureau of Indian Standards, Manak Bhavan, Bahadur Shah Zafar Marg, New Delhi.

Industrial and Allied Sales Pvt. Ltd. *NSALA India Mark II Deep Well Hand Pump Installation and Maintenance Manual*. Industrial and Allied Sales Pvt. Ltd. Surya Kiran, Kasturba Gandhi Marg, New Delhi–110001.

Israelson, O.W. and Hansen, V.E. (1950). *Irrigation principles and practices*., 3rd ed. John Wiley and Sons, Inc., New York, pp. 52–74.

Johnson Inc. (1966). *Ground water and wells*. Edward E. Johnson, Inc., Minnesota. pp. 375–394.

Johnson, C.N. (1950). *Irrigation pumps, their selection and use*. Uni. of California, Agr. Exp. Stn., Cir. 415, p. 54.

Jyoti Ltd., Vadodara (1982), "Jyoti low head tubular turbines" (1982). *Commercial literature*, p. 8. (*ii*) "Jyoti microhydel units", p. 10.

Khandewal, K.C. and Mahdi, S.S. *Biogas Technology–A Practical Handbook*, Tata McGraw Hill Book. Co., New Delhi.

Khepar, S.D. Kaushal, M.P. and Katyal, A.K. (1975). *Development of an animal—drawn lift irrigation pump*. Jl. of Agr. Eng. Vol. XII, No. 3–4.

Kul, D. (1973). *Pumping water by the air lift method has practical applications*. The Johnson Drillers Jl. Nov.–Dec.

Kindel, E.W., (1977). *A Hydraulic ram for village use*, Volunteers in Technical Assistance, 3706, Rhode Island Ave., Mt. Rainier, Maryland, USA.

Kirloskar Borthers (1962). *Pump bulletins*. Nos. 1 to 15. Pub. Kirloskar Brothers Ltd., Kirloskarvadi.

Kirloskar Borthers (1962). *Pump bulletins*. Nos. 16 to 24, Kirloskar Brothers, Kirloskarvadi.

Kirloskar Brothers Ltd. (1967). *Engineering data*. Kirloskar Borthers Ltd., Kirloskarvadi.

Kumar, A., Rajput, T.B.S. and Michael, A.M., (1981), *Optimal water use in windmill system.* Paper Presented in the XVIII Annual Convention of Indian Society of Agri. Engineers, p. 6.

Lal, J. (1969). *Hydraulic machines*, Metropolitan Book Co. (P) Ltd., New Delhi, p. 631.

Low, D.A. (1949). *Applied mechanics*. Longmans, Green and Co., London, pp. 499–519.

Lysen. E.H., (1982). *Introduction to Wind Energy*, SWD Steering Committee, Wind Energy for Developing Countries, Amersfort. The Netherlands, p. 310.

Manual of Information of Rife Rams, Rife Hydraulic Engine Manufacturing Co., Millburn, New Jersey, USA (1969), p. 16.

Merriam, M.G. (1972). *Is there a place for the windmill in the less developed countries?* Working Paper Series, No. 20, East-West Technology and Development, Honolulu, Hawai, USA.

Micheal, A.M. and Ojha, T.P. (1999). *Principles of Agricultural Engineering*, Vol. II. 3rd ed. Jain Brothers, 16/873. East Park Road, Karol Bagh, New Delhi-110005, p. 888.

Micheal, A.M. and Khepar, S.D. (1989). *Water Well and Pump Engineering*. Tata McGraw Hill Publishing Co. Ltd., West Patel Nagar, New Delhi, p. 929.

Molenear, A. (1956). *Water lifting devices for irrigation*, Agril. Development Paper, No. 60. Food and Agriculture Organization of the United Nations, Rome, p. 71.

Myles, R.M. (1981). *Construction of a 2 cubic metre demonstration-cum-training model*. Janta biogas plant, Mimeo, Action for Food Production, New Delhi, p. 14.

National Acadamy of Sciences, Washington DC (1976). *Energy for Rural Development*—Renewable Resources and Alternative Technologies for Developing Countries, Commission on International Relations, National Academy of Sciences, 2101 Constitution Avenue, NW Washington DC, 20418, USA, p. 306.

Patel S.M. (1970). *Conditions for designing ideal lift irrigation plants for farmers*. Agriculture and Agro-industries Jl., December 1970.

Peter Stern (1979), *Small-scale irrigation*, Intermediate Technology Publication Ltd., London, and International Irri. Inf. Centre, Volcani Centre, Bet Dagan, Israel, p. 152.

Rastogi R.A. (1966). *Performance characteristics of a washer rahat*. Jl. of Indian Soc. of Agrl. Engrs., Vol. 3, No. 2, pp. 23–27.

Richardson and Crudas Ltd. (1972). *India Mark II Deep Well Hand Pump Leaflet*. Richardson & Cruddas Ltd. 23, Rajaji Salai, Chennai–600001.

Schwab G.O., Frevert, R.K., Edminister, T.W. and Barnes, K.K. (1966). *Soil and water conservation engineering*. 2nd ed. John Wiley and Sons, Inc., New York, pp. 483–505.

Sharma S.C., Agarwal, R.L. and Lal, R. (1968). *Analysis of power consumption in Persian wheel*. Journal of Irrigation and Power, Vol. 24, No. 4, pp. 375–382.

Stepanoff, A.J. (1957). *Centrifugal and axial flow pumps*. 2nd ed. John Wiley & Sons, Inc., New York, p. 462.

Sherman, M.M. (1977). *Report on the practical application of wind-power water pumps*. Economic and Social Commision for Asia and the Pacific, Workshop on Biogas and Other Rural Energy Sources, Sna. Fiji, p. 27.

Singh, D. (1978). *Comparative study of hydram sprinkler irrigation system with other water lifting electric and diesel operated devices in the hills*, Tech. Report, Planning, Research and Action Division, State Planning Institute, UP, Lucknow, p. 31.

Singh D. (1977). *Hydraulic rams, their design, installation and operation* (1977) Agril. Engg. Today, 1, (XII), Indian Soc. of Agril. Engrs.

Solar photovoltaic project (1980). *Annual Rep.*, No. 4, 1979–80, Central Electronics Limited, Industrial Area 4, Shaibabad, UP.

Srivastava, K.P. and Sarkar, T.K. (1981). *Field evaluation and testing of solar pump*, Report, Water Technology Centre, IARI, New Delhi, p. 40.

Srivastav, N.C. and Michael, A.M. (1965). *Evaluation of performance characteristics of Archimedian screw*, Unpublished Report, College of Agriculture, Jobner, Rajasthan.

"SVOAN wind generator 0.5" (1982). *Commercial literature*, Swedish Windpower Industries Ltd., p. 4.

Talbert, S.G., Fischer, R.D., Alexander, D.H., Frieling, D.H. and Eibling, J.A. (1978). *The development of a 37 kW solar powered irrigation system*, Mimeo, BATELIE Columbus Laboratories, Columbus, Ohio, USA, p. 5.

Taneja, D.S. and Sandhi, S.K. (1988). *Handbook of Indian pumps for irrigation*. India Soc. of Agrl. Engineering, Punjab Agrl. University, Ludhiana, p. 287.

Tewari, S.K. (1977). The utilization of wind energy and wind regimes of arid areas in India, *Proc. Intl. Symposium on Desertification and Its Control, Nairobi, Kenya*. Aug. pp. 291–300.

Thind, B.S. and Sardana, V.K. (1982). *Hydraulic ram pumps*. Tech. Inf. Service No. 1/82, Action for Food Production, Community Centre, New Delhi, p. 36.

UNICEF- India. (1984). *Procedures for maintenance of India Mark II Deep Well Hand Pump*. UNICEF House II, Jor Bagh, New Delhi.

United States Dept. of Agri. (1959). *Irrigation pumping plants*. SCS National Engg. Handbood, Section 15, Irrigation. Soil Cons. Service, U.S. Dept. of Agri., Washington, D.C., pp. 8–1 to 8–70.

Van de Ven, N. (1977). *Construction manual for Cretan windmill*, SWD Steering Committee, Wind Energy for Developing Countries, Amersfort, The Netherlands, p. 59.

Volunteers in Technical Assistances (1977). *Using water resources* VITA, 3706, Rhode Island Ave. Mt. Rainier, Maryland, USA, p. 143.

Water Development Project. (1976). *A study of tubewell installations in Betul district of Madhya Pradesh, India*, E.L.C. Water Development Project, Betul, M.P.

White, R.G. and Kidder, E.H. (1962). *Tractor power for power-take-off driven pumps*. Extn. Ser., Michigan State Univ., East Lansing, Cir. E–23, p. 23.

Wolf, J.W. (1950). *Pumping manual for irrigation and drainage*. Oregon State College, Agri. Expt.. Stn., Corvallis, Bull. 481, p. 39.

Wood, I.D. (1950). *Pumping for irrigation*, Soil Conservation Service, U.S. Dept of Agr., Washington DC, p. 40.

Wood, A.D., Ruff, J.F. and Richardson, E.V. *Pumps and water lifts for rural development*, Colorado State University, Fort Collins, Colorado, USA, p. 189.

World Health Organization. (1976). *Typical designs for engineering components in rural water supply, South-East Asia Series No. 2*, WHO Regional Office, New Delhi (1976), p. 49.

Chapter

4

Lift Irrigation Systems and their Design

LEARNING OBJECTIVES

Importance and scope of river and canal lift irrigation systems. Assessment of feasibility of lift irrigation projects. Riparian rights and environmental considerations. River hydrology and characteristics influencing the location of intake points of river lift irrigation schemes. Hydrology and characteristics of rivers in alluvial plains and plateau regions influencing the location and design of lift irrigation schemes. Water lifting from drainage canals. Water lifting permission. Types of river and canal pumping schemes. Intake wells in sandy river beds. Design and installation of intake pipes of river lift irrigation schemes. Design of pumping wells and pump houses. Installation details of horizontal centrifugal pumps and vertically driven turbine type pumps in lift irrigation schemes. Design, layout and installation of rising mains. Optimization analysis. Water hammer control devices. Pumping stages and booster pumps. Intake and delivery chambers. Irrigation water distribution systems in lift irrigation command areas. Economic water rates in lift irrigation.

Historically, lift irrigation is as old as civilization. The Hanging Gardens of Babylon in Iraq, one of the wonders of the ancient world, was irrigated by lifting water from the river Euphrates through a series of manually operated counterpoise water lifts. The potential for water lifting from rivers, streams, canals, tanks and reservoirs for irrigation and water supply continues to be under-exploited. Surface water pumping schemes play vital roles in areas where the groundwater potential is low and the topography of the area does not permit direct-flow irrigation through diversion or storage works.

In lift irrigation projects, water is lifted directly from rivers or streams or from suitable points at different segments of a canal system. River lift irrigation schemes are feasible in regions where the stream flow is adequate during the peak period of irrigation demand by the crops and where electricity or diesel oil is available at reasonable cost. Canal lift irrigation schemes are beneficial in many areas where sizable land area adjoining a canal, but lying at higher elevations cannot be irrigated through gravity flow from the canal. Often such areas can be brought under irrigation by installing

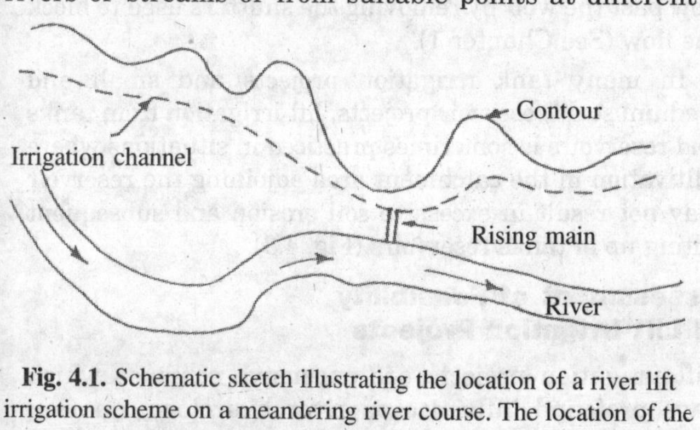

Fig. 4.1. Schematic sketch illustrating the location of a river lift irrigation scheme on a meandering river course. The location of the pumping station is on the straight section of the river. The rising main has the minimum possible length to serve the command area.

Fig. 4.2. A high capacity multi-stage horizontal centrifugal pump in a river pumping scheme near Vadodara, Gujarat. Note the suction pipe coupled to the pumping unit. The panel board in which the electrical accessories are housed is seen in the background.

low and medium head, high discharge pumping sets. The water can be pumped to one or more distribution chambers located at suitable points in the command area from where it is distributed through appropriate water distribution systems.

There has been a spurt in the growth of river lift irrigation schemes in India since 1970s. The total number of lift irrigation schemes in India was estimated at almost 5,00,000 in 1987.* Lift irrigation schemes are being implemented mostly by cooperative societies, state owned lift irrigation corporations and minor irrigations departments of state governments. The schemes in the cooperative sector usually avail credit facilities of the National Bank for Agriculture and Rural Development (NABARD), land development banks and commercial banks, availing refinance facilities of the NABARD.

The state of Maharashtra has made significant progress in the development of financially viable lift irrigation projects in the cooperative sector, availing institutional financing, mainly through the NABARD. Other states where lift irrigation systems have been developed as a significant source of irrigation are Andhra Pradesh, Karnataka, Madhya Pradesh, Bihar, Orissa, West Bengal and Tamil Nadu. Most state governments have set up separate statutory corporations to undertake minor irrigation works, including river lift irrigation systems.

Direct lifting of water from rivers, streams and canals is extensively practiced in rural water supply schemes where it may be the only source of potable water. River pumping for irrigation and water supplies is practiced extensively in south Asia, south-east Asia and in many other developing countries. In most countries, however, it is still an under-exploited water resource potential for irrigation.

A characteristic feature of traditional canal irrigation systems in some of the tropical regions with predominance of plantation crops is that flow irrigation is available only to paddy lands in the lower reaches of the watersheds, while orchards and upland crops, and plantation crops, which usually have high field economic values, are not provided with irrigation. Lack of irrigation not only results in low yields of crops, but in drought years they result in permanent wilting of orchard and plantation crops. A system of flow-cum-lift irrigation in such areas would greatly increase agricultural productivity and save valuable crops, including orchards and plantations. In developing country situations, a lift irrigation system is usually not economical to be owned by a single small farmer. The system is usually designed to serve a group of farmers sharing water from a common source. The project is owned and operated by government agencies or by farmers' cooperatives. The cooperatives usually enjoy the benefit of technical advice from the government or consultancy agencies and credit from financing institutions.

Fig. 4.3. The intake of a lift irrigation system pumping from a lake near Ranchi, Jharkhand (1979). Note the two independent suction pipes connected to the respective horizontal centrifugal pumps installed in the pump house.

It is often possible to develop integrated lift irrigation projects comprising of storage reservoirs which store the rainfall runoff in the upper reaches of a river course, with a series of river pumping schemes downstream. The stored water can be released into the original river course during dry periods when there is acute water scarcity, to irrigate areas adjoining the river, either on one or both banks, and provide drinking water supplies by installing suitable pumping systems (Fig. 4.2 and 4.3).

The state of Maharashtra has made rapid progress in lift irrigation through farmer's cooperatives. Many of them attained high levels of economic viability. A recent development (mainly after 1970) has been the large scale construction of *pick-up weirs*, called *Kolhapur type weirs or KT weirs* to impound river flows during the receding monsoon periods. When the weir is in operation, the water impounded upstream of it is used almost exclusively for lift irrigation and water supply. The river flow during high rainfall periods is allowed to flow past the weir by removing the shutters used to block the flow (See Chapter 1).

In many tank irrigation projects and small and medium scale reservoir projects, lift irrigation from tanks and reservoirs is sometimes practiced in situations where cultivation in the catchment area adjoining the reservoir may not result in excessive soil erosion and subsequent silting up of tanks/reservoirs (Fig. 4.3).

Assessment of Feasibility of Lift Irrigation Projects

Lift irrigation projects are expensive schemes which have profound influence on river hydrology, riparian rights, the characteristics of the command area,

*Report of the Working Group on Minor Irrigation for the Formulation of the Ninth Five Year Plan (1996). Ministry of Water Resources, Govt. of India, New Delhi.

cropping systems and other site specific requirements. They are usually single purpose projects which have to be sustained by the returns from the increased agricultural production of the command areas. It is therefore essential that the techno-economic feasibility of the project should be established before the project is sanctioned or undertaken.

The techno-economic feasibility report of a lift irrigation/water supply project should include the following:

1. Location of the scheme.
2. Availability of water in the river or the stream, with respect to irrigation seasons.
3. Number of existing irrigation and water supply projects on the same river or stream in both upstream and downstream sections of the proposed scheme.
4. Potential for additional pumping from the river without adversely affecting the water requirement of existing schemes/riparian users.
5. Suitability of water for irrigation/water supply.
6. Proposed cropping pattern/cropping intensity in the command area/water supply demands of the population.
7. The extent of land development work needed in the command area
8. On-farm water conveyance system proposed (open channels/underground pipelines).
9. The method of water application proposed (surface/sprinkler/drip).

The economic feasibility of the scheme must be determined by estimating the expected farm income from the command area and/or the returns and social benefits of the scheme. These should be compared with the investment cost of the scheme, including the equipment and accessories and the cost of operation and maintenance. The total cost of the scheme consists of the following:

1. Fixed cost or investment cost.
2. Operating cost.
3. Maintenance cost.

The investment cost or the capital cost of a water lifting scheme includes the cost of the following activities:

1. Preliminary works (investigation, surveying, design of the system, site clearance and construction of approach roads).
2. Land acquisition, if applicable.
3. Civil construction works, including water conveyance and distribution system.
4. Mechanical works, including machinery and equipment, and special tools and plants during the time of construction of projects.
5. Electrification charges.
6. Miscellaneous expenses (contingencies, work charge, and unforeseen expenses) during implementation.
7. Land development work in the command area, if applicable.

The economic viability of a lift irrigation scheme should be determined on the basis of the net incremental income expected to be generated as a result of the implementation of the scheme. The water rate to be charged from the beneficiaries should be estimated on the basis of the annual fixed and recurring costs of the scheme and the investments and returns from the farming enterprise. For detailed discussions on economic evaluation and water pricing, reference is invited to Chapter 15 of this book.

Water Availability

The assessment of the amount of river or stream flow which could be safely harnessed for the proposed lift irrigation/water supply scheme is the most important factor in establishing the technical feasibility of a river pumping project. The availability of water in the river should be determined from the observed minimum discharges during the periods when the demand for water is maximum. The river discharge data should be for a sufficiently long period, preferably a ten-year period, if possible. River flow records are usually available from the concerned Executive Engineer of the Government or from local bodies.

If relevant data on river discharges are not available, reasonable estimates are to be made on the quantum of flow during the dry seasons of the year. The number of irrigation and water supply projects existing on the river, upstream and downstream of the project site, should be ascertained. Provision to meet the continued requirements of water for domestic use and other riparian requirements and existing contracts for water supply, like those of industries are to be ensured.

Lift irrigation schemes which are designed and constructed by government agencies usually provide for a certain risk, especially when the water availability is scarce. A confidence level of 75 per cent is usually adequate for schemes in the peninsular region. Lift irrigation schemes depend on the minimum flow (base flow) during the crop period. There is a likelihood of the base flow depleting in certain years. Hence, it is necessary to provide for this contingency.

Riparian Rights

The right for use of the river/stream flows downstream of a proposed water resource project is called the riparian right. It includes the quantum of water already granted to industries, water supply and sanitation schemes, irrigation schemes and public recreation areas. These rights may be formally granted by the government or local administration/social organization or acquired due to continued use over long periods. It is necessary that the essential downstream flow to meet the riparian requirements is ensured, while deciding the quantity of water which can be pumped from a river or stream. Riparian rights are of major significance when the river flow is low and there is competition for the use of the available water. The requirements of the users within close reaches downstream of a project are of greater significance than those farther down. This is because of the gradual enrichment of downstream river flows through subsurface ground-water flow and accretion of surface flow from downstream tributaries of a stream or a river. There would, of course, be added requirements of riparian users. In short, a holistic view of riparian requirements and possible accretions to downstream flows is required while estimating the quantity of water that can be pumped from a point, especially during periods of low flow.

Water Lifting from Canals

In canal pumping schemes (Fig. 4.4), the capacity of the pumping sets and the periods of pumping have to be fixed after careful consideration of the irrigation requirements of downstream users, including those receiving their supplies from gravity flow and/or are allowed to pump canal flows at specified time periods. The basic requirements of both these categories of users should not be adversely affected by a new pumping scheme. However, in many situations there are ample opportunities to provide adequate supplies for pumping at several points along a canal. They arise due to a change in the cropping system of the command area served by the gravity flow system, in order to combat the problem of excessive rise in ground water table after the introduction of canal irrigation. Canal pumping schemes will help in the effective utilization of canal flows during periods of low demand in the gravity command area. With the present day advances in plant breeding, there is ample scope for selecting crops and their varieties which could be grown successfully during this period. In many regions, the practice of night irrigation is not well established, when surface methods of water application are practiced. Sizable quantities of canal supplies during the night are not effectively utilized. It is desirable to permit pumping during the night to utilize the under-utilized canal flows during the period. There are also situations where the initial reaches of a canal system cannot be brought under gravity flow irrigation due to the unfavourable topography of the surrounding areas. In such cases canal pumping assumes importance as they add to the culturable command area of the canal system. Utilizing canal flows in its upstream reaches results in a major reduction in the total seepage loss of a canal system.

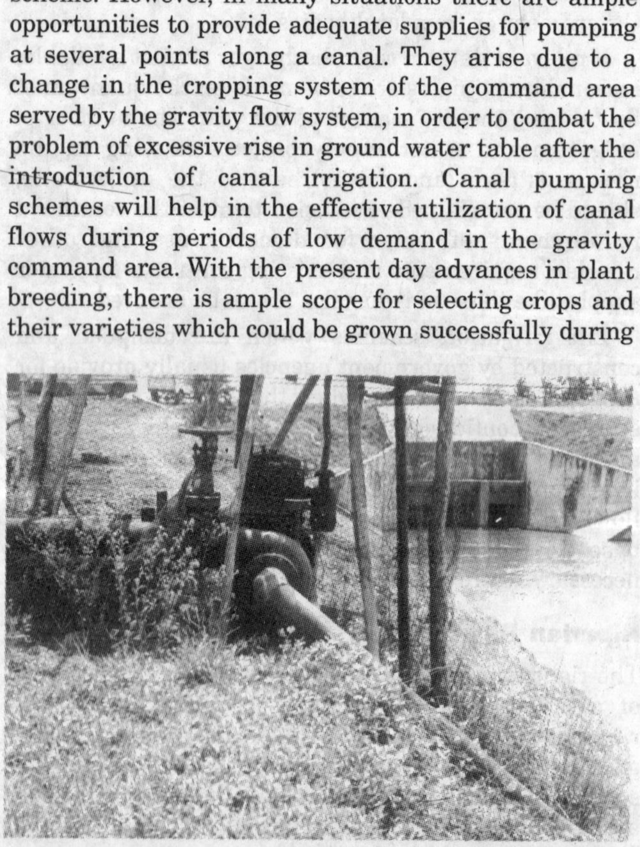

Fig. 4.4. A medium capacity canal pumping system comprising of an engine-driven volute centrifugal pump. Note the suction and delivery pipes and the sluice valve on the delivery side.

Fig. 4.5. Pumping from a canal in the Godavari delta, Andhra Pradesh (1973). A volute centrifugal pump is operated by a diesel engine using a belt drive. Note the high discharge of the pump due to the proper selection of the pump and the engine.

Canal pumping schemes are most economical when the lift command area is close to the canal and the pumping heads are low (Fig. 4.6). In general, lift irrigation results in more efficient use of the canal supplies, as compared to gravity flow systems, as the farmers are more conscious of the cost of pumped water supplies than those obtained from a gravity flow system.

Environmental Considerations

Many river lift water resources schemes require proper assessment at the planning stage to identify changes in the environment that can result from the project. Often it is an essential requirement for sanctioning the scheme. The environmental inventory for lift irrigation development provides the characteristics of the environment which exist prior to development. It will include base line data for assessing changes that will result from the development. The environmental enquiry should include mainly the following:

(*i*) Physical and chemical characteristics of downstream reaches—surface and groundwater qualities, sanitation and proliferation of pathogens causing water-borne diseases in humans and animals, and leaching of pesticides and other chemicals from irrigated areas.

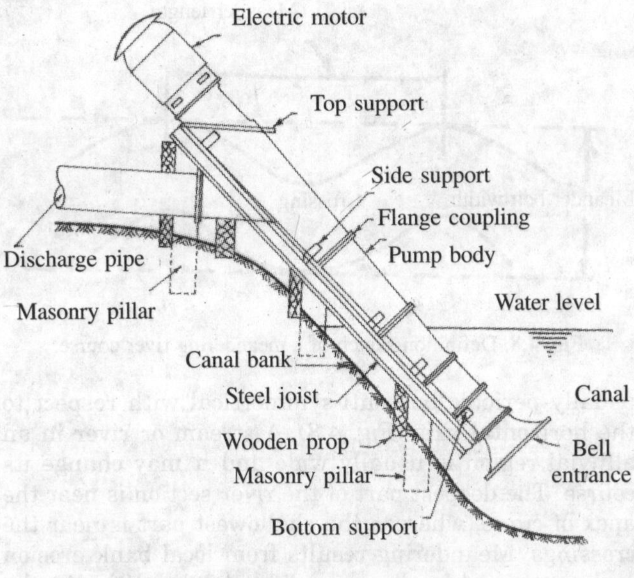

Fig. 4.6. View of an angle-mounted propeller pump installed to lift water from an unlined canal/stream.

(*ii*) Influence on flora and fauna, including riparian fish life, due to the changes in water quality and salt problems in the soil profile.

In canal lift schemes, environmental consequences are usually not significant. However, there may be special situations like the possibility of pesticide residues from irrigated areas leaching to lower reaches and entering the canal system. Appropriate pest management, involving minimum use of chemicals in agriculture and resorting to biological plant protection measures, and resorting to increased use of organic fertilizers will alleviate the problem.

River Characteristics Influencing Location of Intake Points for Water Supply

The terrain of the river course and the resulting river flow characteristics have profound influence on the location and design of lift irrigation systems. The boundaries of natural rivers vary, depending on

(*i*) the existence of rock and boulders,

(*ii*) whether they pass though gravel flood plain deposits or sand and silt deposits, and

(*iii*) what material is in transport, if any.

The above boundary conditions are distinctly different during the progression of the river from the mountains, through the hills, through the upper alluvial plains and the nearly flat delta regions. The river bed may comprise of rock, boulder, cobbles, gravel, sand, silt or clay, depending on the location of the river reach. These materials may be stationary, or in transport, depending on the flow velocity. Most rivers have the following four stages of flow:

1. Rocky stage. In this stage, the river is flowing through steep and rolling hill slopes. The bed slope and flow velocities are high. The bed and sides of the river are mostly made up of rock or very large boulders which seldom move. There is also little evidence of regular movement of cobbles or gravel.

Fig. 4.7. A river pumping system in a river in the boulder stage. Note the intake channel, jack well and the pump house. The main river course is seen in the background.

2. Boulder stage. As the river emerges from the hills, its bed slope and flow velocity are reduced. The bed and sides are composed of boulder or gravel (Fig. 4.7). The stream cross-section gets gradually well defined. The flow is confined between non-submersible banks on either side. The subsurface flow in this region is high because of the permeable bed.

3. Trough stage. After the boulder stage, the river enters the alluvial plain created by itself. It flows through the formation made up of alluvial sand and silt. The bed slope is small and the velocity of flow low during high floods. The river spreads out over a wide area as banks higher than the high flood level are usually far away from the main current of the river.

4. Delta stage. As the river approaches the sea or the ocean, the bed slope and the flow velocity get reduced. These result in the deposition of large quantities of sediment which constitute the delta.

Hydrology of Rivers in Alluvial Plains

The sediment carrying capacity of a river is greatly influenced by its flow velocity. A reduction in the flow velocity of a river, through the construction of a dam, weir or barrage, will result in the deposition of large quantities of silt. Likewise, wherever river water is charged with silt spills and spreads over a wide surface, it deposits silt, raising the level of the water spread area. The river silt thus deposited is called alluvium. The flat lands formed by the deposition of alluvium are the alluvial formations.

Formation of alluvial land plains. When a river or a stream enters a depression, forming a lake, its velocity is reduced and it deposits silt in a fan shape over the lake bottom, thus raising the lake bed sufficiently high to provide for the growth of vegetation. This process leads to accelerated deposition of sediment from the river. The raised areas gradually become dry land through which the river flows in one or more branches, still pushing out into the lake further raised beds. Thus, in course of time, the whole lake gradually gets transformed into an alluvial plain through which the river flows, keeping open waterway of areas for the main river course and its branches such that the combined cross-section of the main river and its branches are sufficient to carry the normal discharge.

Development of deltas. The hydrologic process involving increased silt deposition continues as the bed slopes of rivers get progressively reduced and nearly flattened at sea mouths. The land formed by the deposit of the silt carried by rivers at sea mouths results in the formation of deltas. The name '*delta*' is derived from the Greek letter Δ, since such deposits generally take a triangular shape. A delta is formed by successive layers of river-borne silt deposited in the sea which gradually rises in its level and finally reaches the surface. They continue to grow by repeated deposition of silt and partly by the growth and decay of vegetation, until an extensive plain is formed. The seaward end of the delta gradually pushes out into the sea and continues to grow, unless checked by tidal or other currents carrying away the silt.

After the alluvial river enters a delta, it generally ceases to receive tributaries. Further, it shows a tendency to subdivide into branches which themselves tend to further sub-divide. The deposit of silt from the flood overflow on either side of a river is the heaviest close to the river banks, and progressively decreases as the distance from the river bank increases. Thus, on either side of each river or branch, an alluvial plain is formed.

Characteristics of Rivers in the Alluvial Plains

Based on the river flow patterns and silt deposition characteristics, rivers in alluvial plains may be classified into the following three types:

(*i*) Meandering type
(*ii*) Aggrading type
(*iii*) Degrading type

It is likely that all the three types may be found at different reaches of the same river course. It is also possible that the same reach may change type, depending on a change of factors influencing river flow characteristics.

Meandering type rivers. A meandering type river flows in consecutive curves connected by short straight reaches called crossing (Fig. 4.8). Meandering is a self-induced plan deformation of a stream or river that is

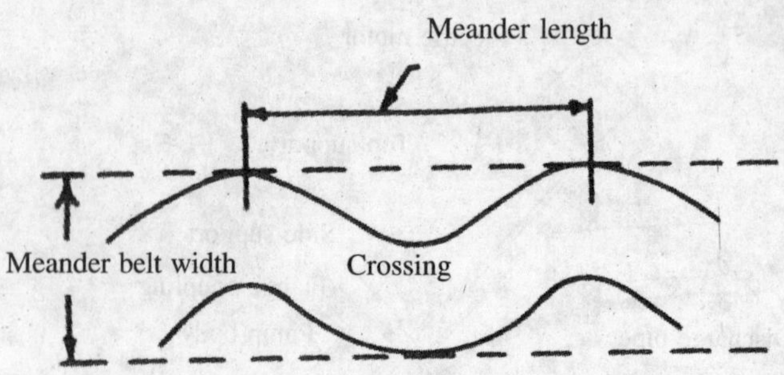

Fig. 4.8. Definition sketch of a meandering river course.

usually periodic but anti-symmetrical with respect to the horizontal axis (Fig. 4.8). A stream or river in an alluvial region is usually wide and it may change its course. The deepest part of the river section is near the apex of curve, whereas the shallowest part is near the crossings. Meandering results from local bank erosion and consequent local over-loading and deposition by the river of the heavier sediments which move along the river bed. The river cannot maintain symmetrical axial flow when there is a heavy bed load. This causes unevenness of the river bed. It would develop cross flow in the stream, leading to scouring on one side and deposition of silt on the other until an equilibrium is attained. The parameters influencing meandering increase with the increase in sediment load. The major factors influencing meandering are (1) type of river bed material, (2) bed slope of the river, (3) silt content, (4) hydraulic properties of the river channel, and (5) hydrologic conditions of the river flow.

Any section of a bank which is more friable than the rest is likely to be eroded, causing the river to deflect into a small curve. This induces a more rapid current on the concave side, and slower flow on the convex. It further accentuates the curvature by increased erosion on the concave side and silt deposition on the convex (Fig. 4.8 and 4.9).

By reflux action, there is a tendency for the current, after passing such a curve, to form a similar concave reach formed by erosion downstream on the opposite side of the river, with a similar convex silt deposit on its opposite side. The action is repeated along the course of the stream.

A meander, when fully developed, has a definite pattern of curvature, length, depth, and width of channel. The pattern of the meander changes whenever there is a change in river discharge or the type of the material transported by the stream. In general, the size of bends, their length, width as well as the degree of sinuosity (the ratio between the curved length and the straight distance) increases with an increase in the sediment load of the river. However, a river flowing in a tortuous rocky terrain or in a rigid sinuous flume, whose curvilinear pattern has not been created by the stream itself, is not meandering.

Aggrading type rivers. An aggrading type of river is in the process of building up its bed to a nearly definite slope. The instability in channel bed may have caused due to excessive sediment entering a river with a sudden reduction of slope on the plains, with the construction of a barrage or dam across the downstream reach of a river, or a sudden influx of sediment from a river tributary. The aggrading reach of a river is usually straight and wide with shoals in the middle, resulting in divided flow.

Degrading type rivers. When the sediment load of a river is blocked by a dam upstream on the river, a degrading reach may develop downstream. The river starts losing its bed gradually in the form of loss of sediment. Degrading reaches are also developed due to river training works, when a loop of a meandering river is cut off by a straight river channel.

Channel Flow Patterns

The following flow patterns are usually observed in a river channel (Fig. 4.8):

(*i*) *Straight.* Straight sections of a river channel may have *braided* channels, branch rivers in deltas and structurally controlled channels. (*Braiding* is the splitting of an alluvial channel into a multitude of channels. Like meandering, it is a self-induced form of an alluvial stream flow. *Braiding* is also influenced by large-scale horizontal turbulence.)

(*ii*) *Sinuous.* Channel reaches with light curvature where channel bed width is less than two channel widths.

(*iii*) *Irregular.* No repeatable pattern, many braided channels fall in this group.

(*iv*) *Irregular meanders.* A repeated pattern can be vaguely distinguished in the channel plan. Irregular meanders are usually observed in sand bed channels with high bed loads.

(*v*) *Regular meanders.* Regular meanders are characterised by distinct repeated patterns. The angle between the channel and the general valley trend is less than 90°. Confined meanders are usually regular. Freely meandering gravel bed channels on comparatively steep slopes are also usually regular.

(*vi*) *Tortuous meanders.* Tortuous meanders exhibit a nearly repeated meandering pattern characterised by angles larger than 90° between the channel axis and the valley trend. This pattern is usually associated with free meanders of sand-bed streams in flood plains influenced by aggradation.

HYDROLOGY OF RIVERS IN PLATEAU REGIONS INFLUENCING LIFT IRRIGATION POTENTIAL

Major plateau regions of India are the Deccan plateau, the Chottanagpur plateau and the Malwa plateau. The entire peninsular India, lying east of the Western Ghats and south of the Vindlya and Satpura hill ranges, is traversed by four major river systems, namely, Mahanadi, Godavari, Krishna and Kaveri. They offer considerable potential for lift irrigation. In addition, there are several fast flowing medium and minor rivers and their tributaries with substantial lift irrigation potential. The west flowing rivers of the Western Ghats region also offer scope for installing lift irrigation systems, mainly in the lower and middle reaches. Installation of lift systems with augmentation of flow in these rivers through storage reservoirs and pick-up weirs remains an under-exploited potential for irrigation and water supply, especially in the west flowing rivers of the Western Ghats region. The peninsular river systems can be classified in the following groups, in relation to their lift irrigation potential:

(*i*) Major perennial rivers.

(*ii*) Medium and minor rivers.

(*ii*) Streams with regenerated flow from irrigation command areas in the upstream reaches.

(*iii*) Drainage channels of irrigated areas.

In case of major perennial rivers, elaborate gauging is usually not required for planning lift irrigation systems, as the quantity of water lifted by these units in usually insignificant, as compared to their discharge. However, it should be ensured that adequate flow is available at the lift sites during peak periods of irrigation. Keeping in view the cropping systems of seasonal crops adopted in most parts of the region, water availability up to the middle of March is usually considered while designing lift irrigation systems. While selecting the site for installing the pumping sets, it should be ensured that the river should flow close to the bank where the intake of the lift system is constructed.

In case of lift irrigation systems proposed in the foreshore of major reservoirs such as Pochampad, Nagarjunasagar and Srisilam in Andhra Pradesh, the question of availability of water is usually not examined, as the quantity of water required for lift irrigation in the catchment areas is negligible, when compared to the total storage in the reservoirs.

In case of many of the medium and minor rivers, the flow is scanty in the post-monsoon dry periods. Estimates of the river flow which can be tapped during different periods of the crop growing season is desirable in terms of standard weeks, especially during the dry season. An additional potential in many of the rivers is the deep sandy bed with substantial groundwater. Hence, intake wells located in sandy river beds should be provided with window filters or percolation holes so that groundwater of the river bed is tapped (Fig. 4.14 and Fig. 4.15). River gauging in such rivers should quantify the flow during the dry post-monsoon period at periodic intervals. When summer flows are available or the river bed groundwater could be tapped, there exists the possibility of short duration upland irrigated crops. These crops require only a limited number of irrigations, usually a pre-sowing irrigation and one or two irrigations during the crop growing season.

In proposals for lift irrigation from streams/rivers having dry season regenerated flow, originating mainly from deep percolation and lateral flow of irrigation command areas in the upper reaches, it is necessary to collect field data for predicting the available flow. Such data should be for a minimum period of one dry season, and preferably 2 or 3 consecutive dry seasons.

With increasing systematic utilization of groundwater in irrigation command areas, the quantity of water available from regenerated groundwater flow into streams may decrease over a period of time. However, this aspect is not significant in most irrigation command areas currently, as accelerated groundwater utilization in most of these areas is yet to be developed.

Water lifting from drainage channels. While planning lift irrigation schemes from drainage canals, important factors to be considered are realistic estimates of the flow potential and the quality of the drainage water. The extent of the command area from which the drainage water is likely to enter into the drain is to be assessed. Water quality is to be monitored at different periods during the crop growing season.

Water Lifting Permission

Water availability in the river/stream during the periods of low flow, previous commitments for the utilization of water, for the increases in demand for drinking water, increasing need of the industry and provision for environmental protection are the major factors influencing the quantity of water which could be allotted to a particular project. In almost every country, it is obligatory under law to obtain the water availability certified and approved by the concerned executive engineer of the state irrigation/public health engineering department or other designated authority, before undertaking the project on water lifting for irrigation/water supply. In most states in India, it is the irrigation department which is given the authority to issue water lifting permission. In some states including Maharashtra, there are two categories of rivers for the purpose of water lifting permission: *notified rivers* and *non-notified* rivers. Irrigation department and command area development authorities (CADA) are vested with the authority to grant permission in case of notified rivers, while the revenue authorities are empowered to issue permits in case of non-notified rivers. Local bodies are sometimes empowered to issue water lifting permission in case of minor rivers and streams. The following guidelines are usually followed for issue of water lifting permission:

1. Water lifting permission should ordinarily be issued by only one agency, usually the irrigation department of the state. In cases where the permission is issued by the revenue department or any other agency authorised by the state, the concurrence of the irrigation department is to be obtained about the availability and adequacy of water.
2. Water lifting permission could be issued to individuals, registered farmers' associations and cooperative societies.
3. On receiving an application for a scheme, and after initial scrutiny, it is a common practice that a letter of intent is issued by the designated authority to the applicant. The letter of intent is usually valid for a period of one year. During this period the applicant is required to complete the detailed project proposal, register the cooperative society and indicate the anticipated cash flow to establish the bankability of the project.
4. Water lifting permission should indicate the volume of water permitted to be lifted season-wise and the cropping pattern and the validity period which should normally be a minimum of 15 years.
5. The period of completion of the lift irrigation project is normally two years. It could be extended in deserving cases.
6. In case of projects financed by banks, the permission may not be withdrawn without the consent of the financing institution.
7. It is desirable that the beneficiary/state government installs a volumetric measuring device near the headworks so that season-wise records of the quantity of water lifted are maintained.

The competent authority usually prescribes the cropping pattern to be followed by the beneficiary in the overall public interest. The state of Maharashtra, for instance, has been adopting the work of restricting the area under sugarcane, which is a high water consuming crop to 15 per cent of the command area of the scheme and 85 per cent under other crops in the rainy season (*kharif*) and 42.5 per cent area under the winter (*rabi*) crop for granting water lifting permission.

Dug wells near river banks. In the plateau region of peninsular India, underlain by hard rock underground strata, there is often the need to restrict pumping from wells in the areas close to the river. These wells receive a major part of their recharge from the water which flows through fissures and cracks in the strata adjoining the river. It is desirable that the state government regulate installation of high capacity pumping sets in such wells. The working committee of the NABARD (1991) has recommended that water lifting schemes from dug wells within 1 km from the river banks may not be financed under the institutional financing programme.

Site Selection for Locating River Lift Pumping Stations

Location of sites for river pumping stations depends on the site specific requirements of the pumping system. Sites for lake or reservoir intake structures are governed by such factors as shoreline, topography, unstable soils or rock, depth of water, aquatic growth, fish protection and surface trash. For rivers, the factors include their flow characteristics with special reference to scour or deposition of bed or bank material (Fig. 4.9), distance from unstable channel reaches, adverse effects or nearby bridge piers or other hydraulic structures, the stability of river banks, the location of river bends and the hazards from floods. With such a wide range of river types, stages of river flow, and sediment carrying characteristics and scale and nature of the pumping schemes, there cannot be any standard design of intake to suit the widely varying river characteristics.

Influence of River Characteristics on Intake Site Selection

(a) Rivers in rocky and boulder stages. Mountain rivers in rocky and boulder stages provide favourable conditions for intake to pumping stations in terms of durability and lack of interference from sediments under normal conditions. However, these sites are vulnerable to high flow velocities, turbulence and movement of boulders during floods. In many situations the choice of the intake site is dominated by the presence of rock outcrops. Rock outcrops prevent migration of the river channel and hold the deepest part of the channel against it for long periods, before some major event provides sufficient geomorphic energy to cause a significant change in the pattern of the channel. In steep mountain rivers, it is often advantageous to avoid an exposed intake structure by having the intake behind the shelter of a rock outcrop. This, however, has the disadvantage that the water levels upstream of such rock outcrops is lower than its downstream. In other cases there may be an opening between a large boulder and rock wall which can be used as a natural intake.

Steep hill rivers in narrow valleys with bed material of rock and boulders, or gravel or cobbles are more amenable to the choice of site. Such situations facilitate the location of permanent intake structures taking off from the side of the river channel or as an angled or frontal intake built into the channel.

Fig. 4.9. An intake well located on the concave side of a meandering river section, resulting in stream bank erosion downstream of the well and silt deposition on the opposite side. The intake structure of a river pumping system is to be located on the straight length of the river course.

(b) Gravel bed river. Gravel bed reaches follow the rocky and boulder stages of mountain river sections. They are less steep with beds formed mainly of cobbles and gravel, together with some boulders. General principles of siting intakes under such situations are the same as discussed in (a) above, but they usually offer suitable sites for the construction of head walls. For control of river bank erosion in meandering type of gravel bed rivers, bank erosion and shoal development progress downstream in a rather irregular manner. In such cases, intakes are sited downstream of the concave bank on a firm strata, preferably utilizing a rocky outcrop on the bank. Such intakes will require protection against bank erosion, both upstream and downstream. If meandering is very active, the protection may have to be extended and modified from time to time.

(c) Meandering streams in alluvial formations. Alluvial streams usually are on sand beds and are usually of the meandering type. The more active the meanders, the less suitable is the river reach for locating intakes. Often, river training works may be required for locating a permanent intake point. Sometimes, there may be a firm bank or a rock valley wall which provides good locations for intakes. The general guidelines for locating intakes in sand bed rivers are:

(*i*) avoid an intake inside of a bend on the river course (Fig. 4.9),

(*ii*) take advantage of the deeper clearer water (and permanent flow) downstream of a bend where the effect of secondary currents has fully developed, and

(*iii*) locate the intake structures on the straight length of the river course and never on curves.

Fig. 4.10. A tractor power take-off driven volute centrifugal pump mounted on a two-wheel trolley, suitable for river and canal pumping. Adapted from Cir., 23, Coop. Extension Service, Michigan State University, Each Lansing, USA.

Fig. 4.12. A floating barge type pumping station in a river lift irrigation project near Varanasi in Uttar Pradesh.

Fig. 4.11. The intake section of a river pumping system in the Brahmaputra valley near Jorhat in Assam (1974).

(***Note***: the pump house supported on pillars raised in the river bed and the extensive growth of water weeds on the river bank.)

(d) Rivers in delta stage. Rivers in delta stage have highly fluctuating water levels, with river flow shifting far away from the bank during low flows. Portable pumping units (Fig. 4.10) which may be installed on temporary foundations (Fig. 4.11) or on floating barges are ideally suited to such situations (Fig. 4.12). The intake pipe should be of portable nature, provided with quick coupling joints. In case of floating barge, the initial length of the delivery pipe should be flexible. The main limitation of river pumping projects in delta areas is the damage to the pumping sets during floods. Great care is required to shift the pumping sets, including floating barges to flood-free zones well before the onset of the rainy season. They should be reinstalled for operation during the dry season.

(e) Rivers in non-alluvial plateau regions. Rivers in plateau regions have firm banks. The river flow during the dry seasons is usually along one bank. Therefore, the site for the pumping plant has to be located on the bank where the water flows. If the water is on the opposite bank, it will necessitate laying pipelines in the river bed, which would not only be expensive but may also give rise to problems of scouring or silting. Structures for river pumping in rivers in plateau regions usually require an intake well in the river bed or close to the river bank (Fig. 4.20) and a jack well away from the high flood levels. The intake well and jack well are connected together by an intake pipe (Fig. 4.20). The site for the jack well should be such that it necessitates a short inlet pipe and is above the high flood level, wherever possible.

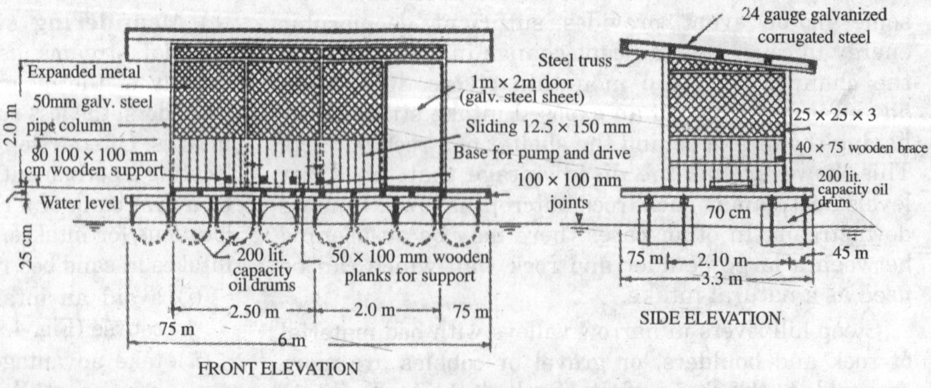

Fig. 4.13. Views of a floating barge made of empty drums with the platform and pump house fixed to them.

Topography and Location of Command Areas Influencing Site of River Pumping Stations

Lift irrigation efficiencies and economics are influenced in a major way by the location of the pumping sites with reference to the topography of the command area. A topographical map of the area comprising the river course, proposed irrigation command area and adjoining watershed is an essential requirement in planning a major lift irrigation project. The contour lines in the topo-sheet are to be at close intervals. As already discussed, a point on a straight length of a river is to be selected for locating the pump intake structure. Examination of the contour lines on the side of the river course where the scheme is proposed provides the most economical location from the point of view of the minimum length of the rising main. Figure 4.1 illustrates a suitable site for locating the intake works at a point where the top of the contour loop comes close to the river. At such a location the length of the rising main is short.

The site should be either in the centre of the irrigation command or at the upstream end. From a central position, field channels can be taken on either side so that the length of each side is small and irrigation can be better managed. If the site is situated towards the downstream end of the command area, it would be unnatural to run the channel opposite the direction of flow of the river, as the command area would normally be rising from the downstream end, towards upstream.

Types of River and Canal Pumping Systems

River and canal pumping systems may range from small-size portable or fixed pumps driven by diesel engines or electric motors to large-scale pumping systems irrigating hundreds of hectares of land and/or meeting the domestic water requirements of large communities.

Fig. 4.14. An intake well in an alluvial river bed. Note the percolation holes (window filters) provided for direct entry of river water into the well from above the river bed, when the water level in the river rises.

Small-scale Pumping Units

The pumping units used for irrigating small areas by individual farmers may be either portable or fixed uses.

Portable pumping sets. To irrigate small areas, portable pumping units coupled with engines are feasible, as they provide considerable flexibility and mobility for location at different points along a stream or canal (Fig. 4.5). The pumping set may be mounted on a small trolley provided with rubber tyres or steel wheels, Tractor-operated portable pumps have a gear box to provide the required increase in pump speed, a telescopic shaft, a flexible coupling, and a hitch at the front of the frame (Fig. 4.10). For operation on level ground, an economical engine-operated pump of small size consists of the unit mounted on a pair of wooden beams.

Fig. 4.15. Close view of the *window filter* (percolation hole) of an intake well in a river bed.

Fixed installation. Permanent installations of small-scale pumping sets consist of the units installed on stable foundations above the high flood level in the river or the full supply level of a canal. The type of pump depends on the head and discharge requirement. In case of centrifugal pumps, the suction line may be a flexible pipe or a rigid pipe laid along the bank. Propeller and mixed flow pumps could be fixed to rest on supports (Fig. 4.6). The pump intake or the foot valve is so located that it remains submerged at least about 60 cm below the low water level in the stream/canal. The delivery pipe is laid along the canal/river bank and may be supported at one or more places. The delivery pipe discharges into a masonry trough at the head end of the field channel.

Installation of small scale pumping units are relatively simple (Fig. 4.5). A firm foundation to support the pumping set, comprising the engine or electric motor and the pump, pump house, and supporting structures for the pumping system and accessories are adequate.

Large-Scale River and Canal Pumping Schemes with Permanent Locations

In river and canal pumping schemes, a permanent location for the pumping plant is advantageous at most sites. However, permanent locations are not possible when the rivers have highly fluctuating water levels and the river course shifts away from the banks during low flow, or when a suitable reach is not available along the river course close to the area to be irrigated. In general, permanently located pumping stations are most suitable for rivers in the rocky and boulder stages (Figs. 4.7, 4.8, 4.17 and 4.20). Permanent locations are usually necessary for rivers in the plateaus.

River pumping schemes in the plateau regions usually consist of a suitable intake structure, an intake pipe, a jack well (pump well) in which is installed the pump, the rising main, delivery chamber, water distributions system and allied accessory components and structures (Fig. 4.20). While pumping from streams and rivulets, it is often desirable to construct a small dam, either temporary or permanent, at a suitable site to head up and store water which can be pumped. It will also help to provide the required submergence depth for the pump intake point. In lift irrigation schemes, pumps are installed at or near the water intake points located along the river course or canal reaches. Water is conveyed to the irrigation command area through a common delivery pipe, called the *rising main*. Pumping is done in a single or in two or more stages, depending on the site conditions. Water distribution in the command area is by open channels or through underground pipeline systems.

DESIGN OF LIFT IRRIGATION SYSTEMS

The design of the various components of lift irrigation systems are highly situation specific. Detailed investigation of the site, including river flow characteritics, topography, and climatic conditions, mainly rainfall characteristics will enable the development of the system which best suits the requirements and provides maximum possible economy in installation and operating costs. The main structural components of lift irrigation systems are (*i*) intake structure, (*ii*) intake pipe, (*iii*) jack well/pump well, (*iv*) pump house, (*v*) rising main and accessories, (*vi*) delivery chamber, (*vii*) distribution chamber, and (*viii*) on-farm water distribution system. The pumping system equipment comprise of pumps, electric motors/engines, electrical accessories and fittings, transformer and power connection and protective devices.

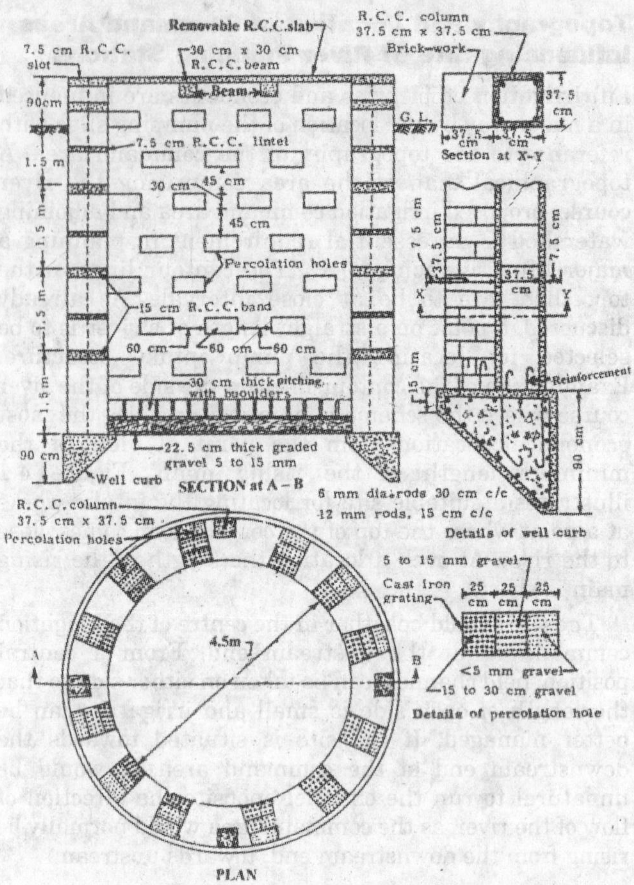

Fig. 4.16. Views of an intake well in an alluvial river bed to tap the groundwater from the sandy river bed during the period when the river is dry.

Intake Structure

The intake structure of a large-scale lift irrigation project comprises of a head wall or intake well on the river bank from where the intake pipe leading to the jack well originates (Fig. 4.17 and 4.18). The head walls provides a direct intake from the river. For lift irrigation schemes on canals an intake well is not necessary. The intake pipe is fixed in a head wall constructed on the canal bank. Intake well is a small diameter well constructed in the river near its bank to direct river water and deliver it to the jack well through an intake pipe. The intake pipe opening is located below the low water level in the river, keeping in view the possible scouring in the river bed. Suitable grating is provided at the mouth of the intake pipe to keep away floating material from entering the pipe (Fig. 4.19).

The intake well, when provided, serves to keep away silt and debris from the intake pipe and the jack well. It also facilitates the flow of groundwater into it from the stratum below the river bed. It is desirable to locate the intake structure at a place where there is flow even in low water conditions and preferably where a stable foundation is possible. Where the stream flows in a

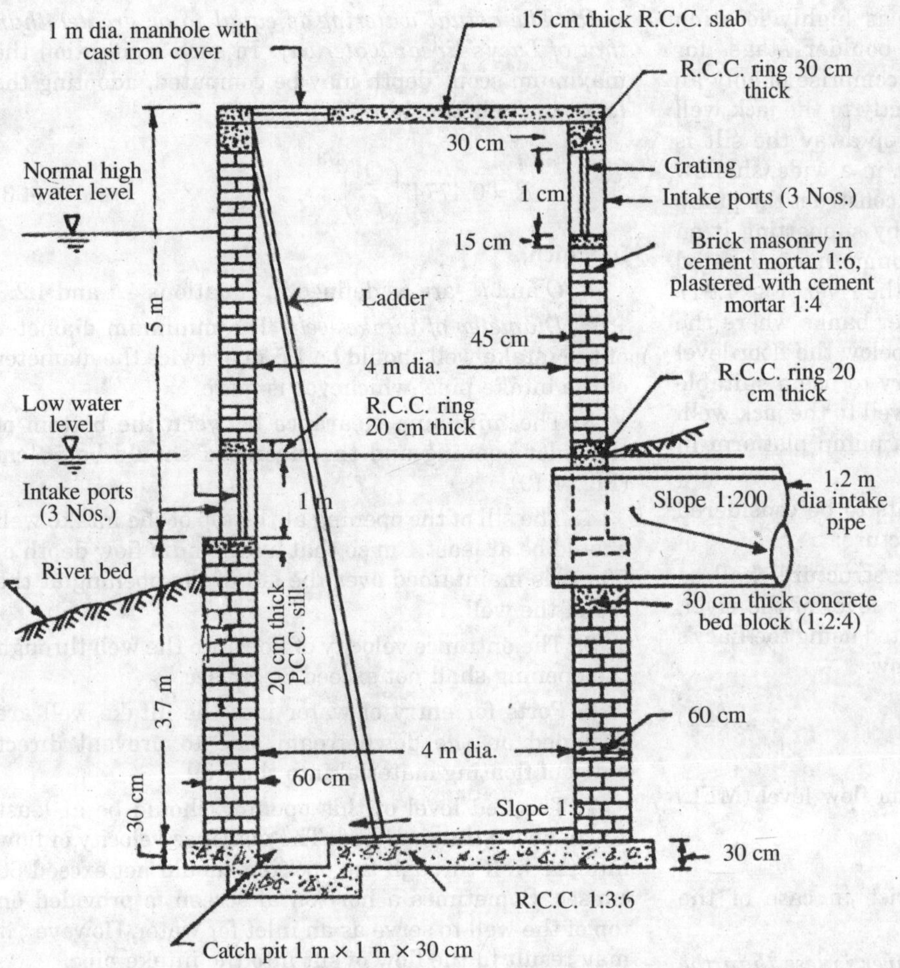

Fig. 4.17. Vertical section of an intake well located on the bank of a river.

curve, the concave side of the reach is more suitable for locating the pump intake, as the stream course adheres to this side of the curve during low flows.

Direct inlet type intake wells. In non-alluvial river beds, the intake wells need not be taken into the river beds. They may be provided with direct-entry ports (Figs. 4.17 and 4.19), for the purpose of minimizing silt accumulation. In order to prevent silt and floating material from entering, inlets are provided at two levels, low flow stage and high flow stage (Fig. 4.17). At each level, one or more ports permit the direct entry of water into the well. The entry of water is usually limited to the downstream section only. The lower layer of ports permits the direct entry of water at the low-flow stage of the river, while the upper layer meets the requirement of high-flow stages. All intake ports are provided with screens made of mild-steel bars fixed to rigid angle iron frames (Fig. 4.19). The section between the screens is made of RCC and is usually shaped as a hydrofoil to allow smooth entry of water into the well. The lower section of the intake well below the level of the intake pipe, serves as the space for the sediment to accumulate. Desilting of the sediment is carried out periodically to keep the intake pipe free of silt.

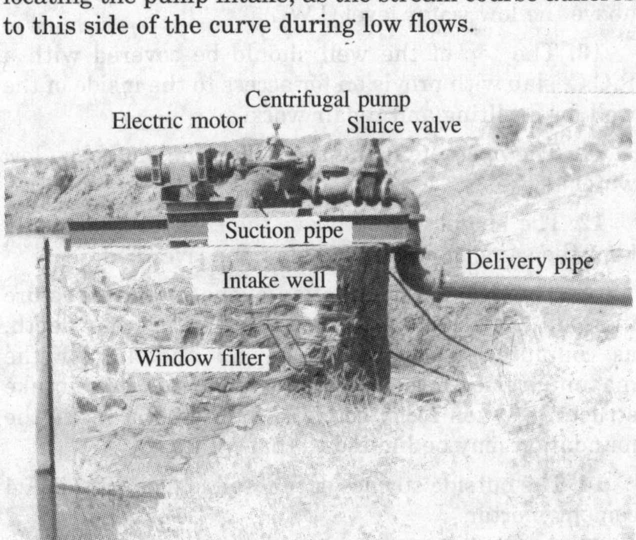

Fig. 4.18. An electric motor-driven centrifugal pump installed on top of an intake well in a lift irrigation project in the plateau region near Akola, Maharashtra (1977).

Fig. 4.19. Mouth of an intake pipe under construction. Note the steel grills which prevent the entry of floating objects into the intake pipe.

The design of intake structure is highly location specific. When a river is in the boulder stage, for instance, the intake structure may comprise of only an intake channel to divert water directly to the jack well (Fig. 4.19). A suitable device to keep away the silt is often necessary. On the other hand, in a wide, shallow river course, it may be necessary to construct the pump house directly on the river course by supporting it on pillars, with the floor level of the pump house located above the high flood level (HFL) of the river (Fig. 4.11). In certain situations, like steep river banks where the water level in the intake well falls below the floor level of the jack well, it may be necessary to use a suitable pump to lift water from the intake well to the jack well. This will require the provision of a pump platform in the intake well structure.

The following are the main points to be considered while designing an intake well/structure:

1. The foundations of the intake structure shall be provided below the maximum scour level of the river. The maximum scour depth is computed using the Lacy's linear waterway formula stated below:

$$L = 4.75\sqrt{Q} \quad \text{... (4.1)}$$

in which,

L = linear waterway at maximum flow level (MFL) (m)

Q = Discharge at MFL (m^3/sec)

Two specific situations may arise in case of the actual waterway in the river:

(a) The actual waterway in the river is less than the Lacy's waterway, as defined in Eq. 4.1: In such a situation, the maximum scour depth may be computed adopting the following formula:

$$R = 1.35\left(\frac{q^2}{f'}\right)^{1/3} \quad \text{... (4.2)}$$

in which,

R = Normal scour depth below MFL (m)

q = Discharge per unit width of actual waterway (m^3/sec/m)

f' = Lacy's silt factor values (Table 4.1)

Table 4.1. Values of Lacy's Silt Factor

S. No.	*Type of silt*	*Value of f'*
1	*Silt :*	
	(*i*) Medium	0.85
	(*ii*) Standard	1.00
2.	*Sand :*	
	(*i*) Medium	1.25
	(*ii*) Coarse	1.50

(b) The actual watering is equal to or greater than that of Lacy's linear waterway: In such a situation the maximum scour depth may be computed, adopting the following relationship:

$$R = 0.4774\left(\frac{Q}{f'}\right)^{1/3} \quad \text{... (4.3)}$$

in which,

R, Q and f' are as defined in equations 4.1 and 4.2.

2. *Diameter of intake well:* The minimum diameter of the intake well should be 1.5 m or twice the diameter of the intake pipe, whichever is more.

3. The minimum clearance between the bottom of the inlet opening and the river bed should be 50 cm (Fig. 4.19).

4. The sill of the opening at the top of the intake well should be at least 1 m so that a minimum flow depth of 50 cm is maintained over the sill of the opening at the top of the well.

5. The entrance velocity of flow into the well through the opening shall not exceed 60 cm/sec.

6. Ports for entry of water into the intake well are provided on the downstream side to prevent direct entry of floating material into the well.

7. The bed level of this opening should be at least 15 cm above the river bed. The entrance velocity of flow into the well through the opening should not exceed 60 cm/sec. Sometimes a horizontal screen is provided on top of the well to serve as an inlet for water. However, it may result in the flow of silt into the intake pipe.

8. Grating up to 20 to 50% should be provided at the entrance to the intake pipe.

9. The top of the intake well is kept at least 50 cm above the low water level (LWL).

10. The top of the well should be covered with a R.C.C. slab with provision for access to the inside of the well for desilting and repair work.

11. The opening at the top is kept 50 cm below low water level.

12. The height of well above the river bed is usually kept the same as its depth below the river bed.

13. The depth of foundation of the intake structure in erodible soils should be 1.25 times the scour depth, as computed using Equations 4.2 and 4.3 for the maximum flood level situation at the site of the intake structure. When rocky bed is encountered at a site the foundation may be limited to that strata.

14. The outside surface of the well is plastered in 1:4 cement mortar.

When an intake well is located away from the river bank with a lead channel directing the river flow into it, there is no possibility of scour of river bed affecting the intake well. Hence, the well foundation need not be taken to the full scour depth of the river.

The intake structure may be constructed of stone or brick masonry laid in cement mortar. Usually it is necessary to provide protection to the river bed through rivetments around the well or along the side slopes in case of intake wall. Rivetments are usually of rough stone laid in dry packing, 30 cm thick.

Intake wells in alluvial river beds: Intake wells constructed on river beds in the alluvial plains are provided with adequate means for entry of surface water when there is river flow and percolation of ground water from the river bed aquifer during dry periods (Fig. 4.14). The conventional infiltration gallery is suitable as an intake structure in lift-irrigation schemes as well. A simpler structure comprises of a series of *weep holes*, often called *window filters* throughout the depth of the intake well below the river bed (Figs. 4.14 to 4.16). Such holes are also provided at points above the bed level of the river to permit surface flow into the well. The number of weep holes are decided on the basis of the yield of the aquifer.

The details of an infiltration type intake well suitable for sandy river beds carrying a considerable amount of sediment load is shown in Fig. 4.16. The well is 4-5 metres in diameter. The window filters are staggered to tap the flow above the river bed and the groundwater in the saturated zone below it. No top cover is required if the wall extends above the high flood level (HFL) of the river. The window filters consist of perforated cost-iron plates at the outer and inner faces (Fig. 4.17). The space between the plates is filled with graded gravel in three or more distinct layers, the layer of fine gravel being laid at the outer periphery and the coarsest at the inner. The construction procedure includes the provision of RCC well curbs and RCC rings of about 60 cm length added from the top. Wherever the bed material is not very deep, the infiltration well is taken to the bed rock.

Pumping from reservoirs: Pumping from reservoirs require the determination of the shore line at different reservoir stages when pumping is done. The pump suction point may be located directly below the lower level in the reservoir, wherever feasible. Major structures are usually avoided in such situations. The suction point of the pump should be provided with a suitable device to keep away the debris and trash. A cage type filter, usually cylindrical, provided with wire mesh in two or more layers is adequate. The wire meshes used in the filter are of decreasing order in terms of the size, the smallest near the inner end. The size of the filter should be sufficiently large to permit unrestricted flow of the design discharge into the pump intake.

Intake Pipe

The intake pipe connects the intake well to the jack well or sump well (Fig. 4.20). RCC pipes, usually of the non-pressure type, are used as intake pipes, The intake pipe is laid with a gentle slope (usually 1 to 400) towards the jack well. The slope should not exceed 1:100 and not less than 1:1000. The diameter of the intake pipe depends on the design discharge of the pumping plant. The pipe size should be such that the flow velocity does not exceed about 1.2 m/s. The diameter of an intake pipe should not ordinarily be less than 7.5 cm. When the pipe length is larger than 15 m, the minimum diameter of the intake pipe is kept 90 cm in case of major installations. Smaller pipes may get chocked and, in the worst cases, may have to be removed and the entire pipeline relaid. Watertight joints are to be ensured while laying concrete pipes especially with collar joints. Leakage of intake pipe due to jointing have resulted in the failure of many of the river pumping schemes. Procedure for making watertight joints in concrete pipe is described in Chapter 7.

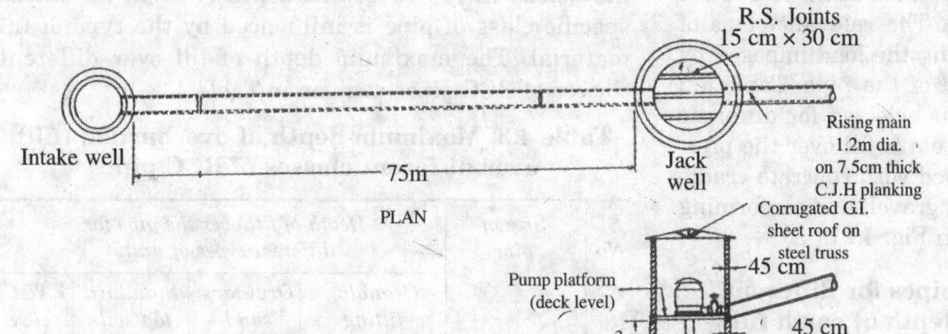

Fig. 4.20. Sectional views of the intake well, jack well and intake pipe in a river lift irrigation project in a plateau region.

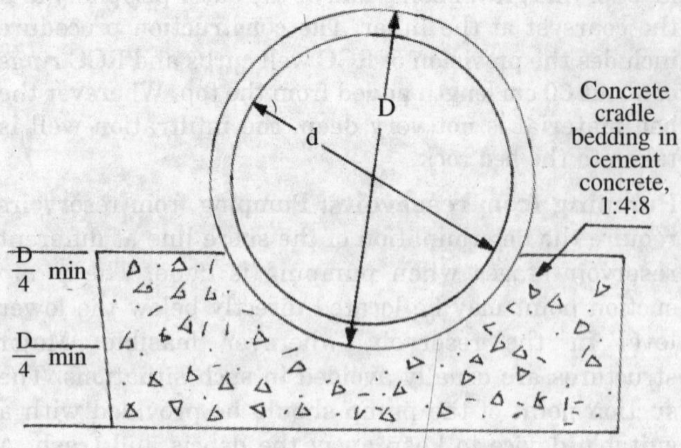

Fig. 4.21. Concrete cradle for laying RCC pipe.

The intake pipes shall be RCC pipes of suitable class. The class of pipe shall be based on site conditions. Standard specifications of different classes of pipes are given in Bureau of Indian Standards IS:458-1971. They include the sizes and corresponding test strengths of pipes. The pipes may be pressure and non-pressure (NRCC) pipes of different classes. All cement concrete pipes, except NP 1 class are provided with reinforcement. Earth pressure acting on buried pipeline is the predominent force acting on the pipe. Pipes of classes above NP 1 are preferred in lift irrigation schemes because the pipeline is also subjected to a certain amount of hydrostatic pressure. The choice is usually limited to NP 2 and NP 3 classes of RCC pipes.

Pipe laying procedures are specified in the Bureau of Indian Standards IS:783-1959. The relevant class of pipe is selected after determining the load imposed by the earth fill and the strength of the pipe. Table 4.2 presents the class of pipe to the selected for different hydrostatic heads and depth of earth fill over the pipe. Intake pipes are usually provided with concrete cradle bedding of 1:4:8 (cement : sand : gravel) and conforming to the specifications indicated in Fig. 4.21.

Table 4.2. Class of RCC pipes for different hydrostatic heads and depth of earth fill

Hydrostatic head (m)	*Depth of earth fill over the pipe*				
	2 m	*4 m*	*6 m*	*8 m*	*10 m*
2	NP 2	NP 3	NP 3	NP 3	NP 3
4	NP 2	NP 3	NP 3	NP 3	NP 3
6	P 1	NP 3	NPC	PSC	PSC
8	P 1	PSC	PSC	PSC	PSC
10	P 1	PSC	PSC	PSC	PSC

(***Note:*** NP indicates non-pressure pipe, P 1 pressure pipe and PSC means pre-stressed concrete pipes.)

Assumptions: Diameter of pipe 90 cm.

Unit width of saturated earth, 2080 kg/m^3

Unit weight of dry earth overfill on pipe, 1850 kg/m^3 Ref. IS:458-1971 and IS:783-1959).

The diameter of the intake pipe is computed using the Mannings formula to suit the low flow situations in the stream. At full flow stage, Darcy-Weisbach formula or William Hauzen formula may be used to estimate the velocity of flow in the pipe (see Chapter 7). The area of cross-section of flow multiplied by the velocity provides the discharge capacity of the pipe. It is necessary that the intake should draw the discharge required for the scheme for all flow conditions of the stream on which lift irrigation schemes are located. In cases where the water levels in the stream and the sump well (jack well) are such that the required discharge can be carried even with a horizontal pipeline, it is desirable to provide a nominal slope to facilitate cleaning of the pipeline.

Whenever the length of the intake pipe is more than 100 m, a man-hole is necessary. The position of the man-hole is usually the middle point of the intake pipeline. In long pipelines, a man-hole each is provided at nearly 100 m intervals for inspection and cleaning.

Determination of the Supporting Strength and Depth of Overburden of Intake Pipes

The procedure for determining the supporting strength of pipes and the intensity of load imposed on it by the earth fill over it are provided in Bureau of Indian Standards IS:783-1959. The depth of earth fill on the specific class of pipe is influenced by the type of fill material. The maximum depth of fill over different classes of RCC pipes is given in Table 4.3.

Table 4.3. Maximum depth of overburden (fill) over different classes of RCC pipes

Sl. No.	*Size of pipe*	*Depth of fill over the pipe for different types of units*			
		Granular filling (soil without cohesion)	*Ordinary sand*	*Standard top soils*	*Wet clay*
	NP 2 Class				
1	90 cm	2.9 m	2.6 m	2.5 m	2.4 m
2	120 cm	2.2 m	2.1 m	2.1 m	2.0 m
	NP 3 Class				
3	90 cm	19.5 m	19.5 m	19.5 m	13.5 m
4	120 cm	24.5	24.5 m	13.2 m	9.8 m

(***Source:*** Based on A.P. State Irrigation Dev. Corpn. (1980).

The following approximations are made for estimating the load factors on RCC intake pipes:

(*i*) Unit weight of earth fill: 2080 kg/m^3

(*ii*) Average *strength* factor for concrete cradle bed: 2.8 (Strength factor for concrete cradle ranges from 2.2 to 3.4. An average value of 2.8 is usually assumed).

(*iii*) Factor of safety: 2.5.

(*iv*) Live load on the pipe (over the earth fill): Negligible.

(*v*) Bearing test strength of pipes, adopting the 3-edge bearing test method (i.e., load to produce 0.25 mm crack) is as follows:

Class of pipe	*3-edge bearing test load kg/m length*
1. *90 cm diameter pipe:*	
(a) NP_2	2500
(b) NP_3	6760
2. *120 cm diameter pipe:*	
(a) NP_2	2880
(b) NP_3	9000

Vertical load on pipe: The vertical load imposed on intake pipes may be under trench conditions or culvert conditions.

Trench formula: Under the situation of concrete cradle bedding, the vertical load acting on the pipe by the soil fill is estimated by the following relationship:

$$W = C_t\, w\, B_t^2 \quad \text{... (4.4)}$$

in which.

W = vertical external load under trench conditions due to filling material (kg/m length)

C_t = trench coefficient

w = weight of filling material (kg/m^3), and

B_t = breadth of trench just below the top of the pipe (m)

The value of B_t is obtained as follows:

B_t = Diameter of pipe + (2 × thickness of pipe) + (2 × space between outer face of pipe and trench edge)

The value of the coefficient C_t depends on the characteristics of the fill material and the ratio H/B_t. The values of C_t for various fill materials and H/B_t ratios are given in Bureau of Indian Standards IS:583-1959.

Culvert formula: The vertical load on the pipe under culvert condition by the materials in the embankment is estimated from the following relationship:

$$W = C_c\, w\, B_c^2 \quad \text{... (4.5)}$$

in which,

W = vertical external load on pipe under culvert condition due to embankment material (kg/m length)

C_c = culvert coefficient as a function of H/B_c value (to the obtained from IS:783-1959)

m = weight of embankment material (kg/m^3), and

B_c = horizontal width on the outside of the pipe, i.e., external diameter in case of circular pipes, m

Hydrostatic head on intake pipe: The intake pipe is to withstand the resultant stress due to dry earth backfill from the top and upward hydrostatic thrust within the pipe. The RCC non-pressure pipes are capable of withstanding a test pressure of 0.7 kg/cm^2, or 7 m head (IS:458–1971). The design pressure is not to exceed two-thirds of the test pressure. The limiting conditon is expressed as follows:

Total upward maximum thrust—downward load due to earth back fill = Net safe upward hydrostatic head allowable.

Example 4.1. *A lift irrigation system is to carry a discharge of 400 litres/sec. The intake pipe is to be located at an average depth of 5 m from the ground surface. The earth fill over the pipe comprises of red and black soils and gravel. Design the intake pipe and check for limiting conditions.*

Solution:

Considering the site conditions as specified, NP 2 Class pipe of 90 cm diameter laid on a gradient of 1:1000 is proposed.

Assumptions:

Weight of saturated earth fill : 2080 kg/m^3

Strength factor of concrete cradle bedding: 2.8

Factor of safety: 1.5

Three-edge bearing test load for 90 cm NP 2 Class pipe: 2500 kg/m length

Thickness of pipe: 5 cm

Clearance space between pipe and trench face: 15 cm

(i) *Vertical load acting on the pipe:*

(a) Trench formula approach,

$$W = C_t\, w\, B_t^2$$

B_t = Diameter of pipe + (2 × thickness of pipe) + (2 × clearance space)

$$= \frac{90}{100} + 2 \times \frac{5}{100} + 2 \times \frac{15}{100} = 1.3 \text{ m}$$

$$C_t = \frac{H}{B_t} = \frac{5}{1.3} = 3.85$$

For the soil conditions specified and the ratio of H/B_l of 3.85, the value of C_t, as given in IS:783-1959 is 2.006

$$w = 2080 \text{ kg/m}^3$$

$$W = 2.006 \times 2080 \times 1.3^2$$

$$= 7051 \text{ kg/m length}$$

(b) Culvert formula approach,

$$W = C_t\, w\, B_c^2$$

$$B_c = \frac{90}{100} + 2 \times \frac{5}{100} = 1.0 \text{ m}$$

$$H/B_c = \frac{5}{1} = 5.0$$

Value of C_c for H/B_c of 5.0 and the site conditions as specified = 3.0 (Ref. IS:783-1959)

$$W = 3 \times 2080 \times 1^2$$

$$= 6240 \text{ kg/m length}$$

The value of the load estimated by the Culvert formula is less than that by the trench formula. Hence, the lower value of 6240 kg/m length is considered as the effective load due to soil fill on the pipe (Ref. IS:783-1959)

(*ii*) *Check for the supporting strength of the pipe:*

$$\frac{\text{Three-edge bearing strength}}{\text{Factor of safety}} = \frac{W \text{ due to filling material/m length}}{\text{Corresponding strength factor}} + \frac{W \text{ due to surface loads (kg/m length)}}{\text{Strength factor for surface loads}}$$

In the present case, the surface loads are considered negligible. Strength factor of concrete cradle is assumed to be 2.8.

Applying the above relationship,

$$\frac{2500}{1.5} = \frac{W}{2.8}$$

$$\therefore \quad W = \frac{2500 \times 2.8}{1.5} = 4667 \text{ kg/m length}$$

The value of 4667 kg/m length for W is against the requirement of 6240 kg/m length. Hence, NP 2 Class pipe is insufficient in the present case. The next higher value of NP 3 Class pipe of 90 cm diameter is considered.

Three-edge bearing test load for = 6760 kg/m length, 90 cm diameter NP 3 Class pipe (IS:458-1971)

Thickness of pipe = 10 cm

(*i*) *Vertical load acting on the pipe:*

(a) Trench formula approach,

$$W = C_t\, w\, B_t^2$$

$$B_t = \frac{90}{100} + 2 \times \frac{10}{100} + 2 \times \frac{15}{100}$$

$$= 1.4 \text{ cm}$$

$$H/B_t = 5/1.4 = 3.57$$

Value of C_t for the value of H/B_t of 1.57 = 1.94 (Ref. IS:783-1959)

$$W = 1.94 \times 2080 \times 1.4^2$$

$$= 7909 \text{ kg/m length}$$

(b) Culvert formula approach,

$$W = C_c\, w\, B_c^2$$

$$B_c = \frac{90}{100} + 2 \times \frac{10}{100}$$

$$= 1.1 \text{ m}$$

$$H/B_c = \frac{5}{1.1} = 4.55$$

The value of C_c for H/B_c of 4.55 = 2.8 (IS:783-1959)

$$W = 2.8 \times 2080 \times 1.1^2$$

$$= 7047 \text{ kg/m length}$$

This is less than the value of 7909 kg/m length obtained by the trench formula approach.

Hence, the value of W = 7047 kg/m length is accepted as the effective load on the pipe.

(*ii*) *Check for the supporting strength of the pipe:* Adopting the precedure as described in case of NP 2 Class pipe,

$$= \frac{6760}{1.5} = \frac{W}{2.8}$$

$$\therefore \quad W = 12{,}619 \text{ kg/m length}$$

The value of W = 12,619 > 7047 kg/m length strength of the pipe proposed.

Hence, NP 3 Class of 90 cm pipe is safe.

(c) *Check for hydrostatic load for the pipe:* RCC NP class pipes are capable of withstanding a test pressure of 0.7 kg/cm^2 (7 m head), vide IS:458-1971. The design pressure is not to exceed two-thirds of the test pressure, i.e., $7 \div \frac{2}{3}$ = 4.67 m head. With a factor of safety of 10%, the working pressure head will be 4.67/1.1 = 4.2 m head.

NP class pipes can withstand 4.2 m head of water when there is no overburden on the pipe. If H' is the hydrostatic head when the pipe is simultaneously subjected to hydrostatic head and overburden due to back-fill on the pipe, the hydrostatic pressure on the pipe is H' kg/cm^2.

Total upward thrust in the pipe per m length of the 90 cm diameter pipe

$$= 100 \times D \times H'$$

$$= 100 \times 90 \times H'$$

$$= 900\, H' \text{ kg/m}$$

Net allowable upward thrust at 0.42 kg/cm^2

$$= 42\, D = 0.42 \times D \times 100$$

$$= 0.42 \times 90 \times 100$$

$$= 3780 \text{ kg/m}$$

The following limiting condition is assumed:

Total upward maximum thrust-Downward load due to earth back-fill

= Net safe available upward hydrostatic thrust,

i.e., $100 \times D \times H' - W = 42\,D$

or, $100 \times D \times H' - 0.42 \times D \times 100 = W$

or, $$H' = 0.42 + \frac{W}{100\,D} \quad \text{... (4.6)}$$

in which, D = diameter of pipe, cm

H = head, m or kg/cm^2

In the present case,

$9000\,H' = 6268 = 3780$

or, $9000\,H' = 3780 + 6268 = 10{,}048$

or, $H = 1.12$ kg/cm^2 (11.2 m head)

Hence, it should be ensured that the depth of water at the entrance of the intake pipe does not exceed 11.2 m. If it exceeds this limit, it may be necessary to select pressure pipes, since non-pressure pipes can withstand pressures only up to 0.7 kg/cm^2 hydrostatic pressure (7 m head).

Open channel intake: Even though the intake to jack well is usually an RCC pipe, there may be situations where sheet rock or hard terrain is available right from the top level of the ground surface and the length of intake pipe is small, less than about 20 m. In such situations it will be feasible and economical to have an intake channel, with bell mouth and trash screen. The longitudinal slope of the open channel shall be about 1:1500. The side slopes of the channel under different channel conditions are as follows:

(*i*) Soils: 1.5 horizontal to 1 vertical

(*ii*) Fraclured rock formation: 1 horizontal to 1 vertical

(*iii*) Compact rock formation: 0.5 horizontal to 1 vertical

The excavated material should be disposed off away from the intake channel.

Trash screen: Trash screens are provided at the entrance and top of the intake well/cistern/mouth of intake pipe or intake channel to prevent entry of floating materials of the river into the intake and damaging pump impellers (Fig. 4.19). The screen is so designed that the minimum velocity of flow of 60 cm/sec is maintained in the intake pipe. The horizontal screen over the top of an intake well/cistern serves the same purpose when the water level in the river rises above the top of the structure. The screens on the face of the entry point is usually made of iron bars (square or round) and that at the top of the structure may be RCC bars or steel bars. In all cases, the screens are to be anchored deep into the walls of the intake structure through anchor bolts.

Operation of stop logs: Immediately behind the trash screen is provided stop logs of suitable size inserted in grooves formed on the sides and bottom of the intake opening. The sill in the intake well/cistern is cleared during the low water level situations by operating the stop logs in the grooves. When the water level in the river starts rising, the stop logs are inserted in the grooves, one after another, starting from the sill level of the inlet opening. When the depth of flow in the river is over the top of the intake well, all stop logs are inserted in the grooves, thus blocking the entry of water through the vertical opening of the intake. The flow into the well is then only from the top of the well which is provided with a strong metal or RCC screen. This arrangement will minimise the entry of bed load silt in the intake cistern.

When the water level in the river falls below the top of the intake cistern, the logs are removed from the notch, one after another, starting from the top most log, thus maintaining the required flow of water to the jack well/sump well. This will ensure the flow of water into the cistern from higher levels and prevent bed load silt from entering the cistern. During the idle period, the stop logs are kept in safe custody.

Pumping Well and Pump House

The water diverted from the river through the intake pipe in collected in a *jack well* or *sump well* from where it is pumped to the irrigation command area. The term jack well is used when the pump house is located directly over the well. In such cases vertical turbine pumps are mostly used. When the pump house is

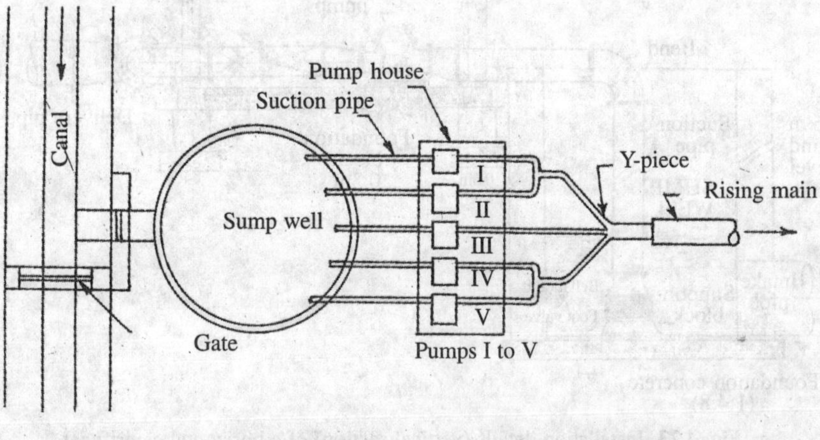

Fig. 4.22. Schematic view illustrating the installation of multiple pump sets connected through Y-piece joints in a canal pumping system.

located on the ground close to the well the term sump well is used. For permanent lift irrigation schemes, properly designed pumping wells are provided for installing the pumping sets (Fig. 4.22 to 4.27). The arrangement of shifting the pumping sets to platforms constructed at different levels, as the water level rises or falls, is not recommended for large projects. Such an arrangement is cumbersome and, if the management is not alert and active, much damage could be caused by accidental submergence of the electric motors or engines. In the case of electric motor drive, the electrical control system should be located above the high water level in the pump house/jack well. A common practice is to locate the pump house on the upper section of the jack well (Fig. 4.26).

Sump Wells

Sump wells, also called dry wells, are constructed on the banks of rivers to protect pumping units from flood water. They are usually used to house centrifugal pumps on rivers or streams in which the flood levels are not high and the water level in the river is within the suction lift of centrifugal pumps. A sump well should be plastered both on the inside and outside to make it watertight. The well has the disadvantage that if it is not made water proof, water may leak and damage the motors. A small dewatering pump is usually fitted to keep the well dry during floods.

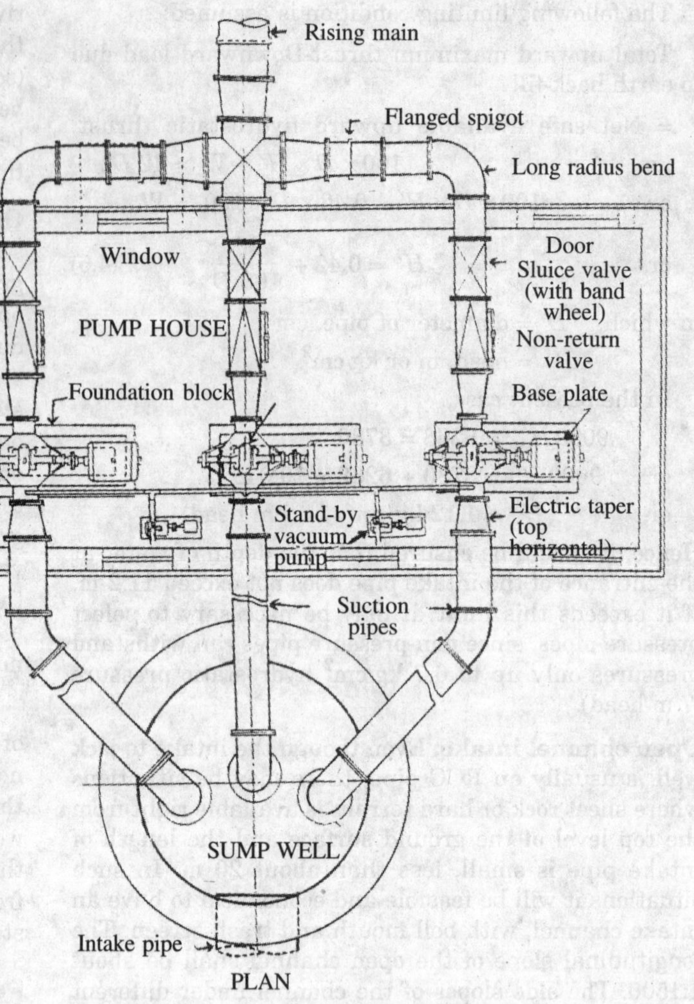

Fig. 4.24. Schematic sketch (Plan) illustrating the layout of a multi-stage (2 + 1 system) horizontal centrifugal pumping plant in a lift irrigation project.

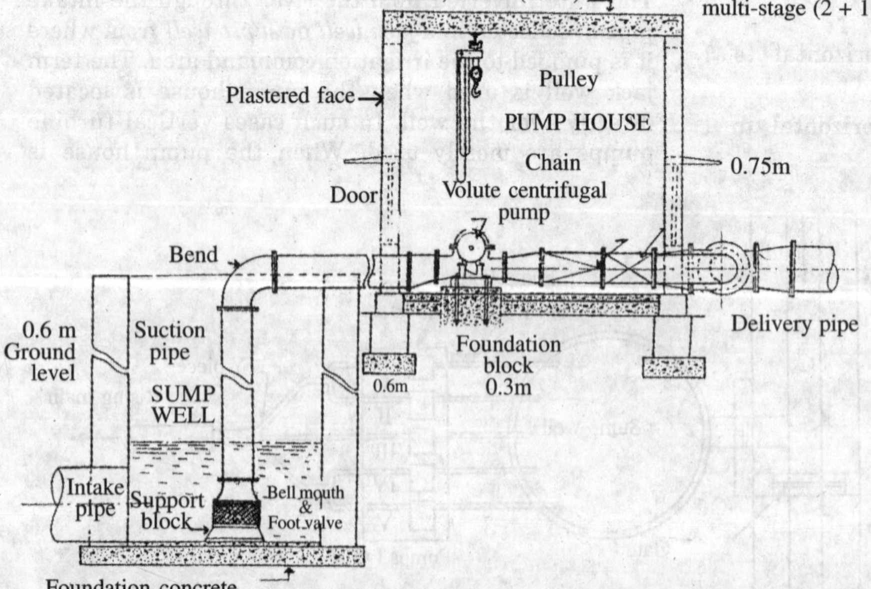

Fig. 4.23. Installation details (vertical section) of a horizontal centrifugal pump in a lift irrigation project provided with a rectangular sump well and pump house with ancillary equipment.

Horizontal centrifugal pumps are used in lift irrigation schemes, where the suction lift is within the permissible limits (Fig. 4.23). The suction lift of a horizontal centrifugal pump should not generally exceed 4.5 m. The actual value depends on the net positive suction head required (Chapter 3). When the suction lift is more than the permissible limit, vertical turbine pumps or submersible

Fig. 4.25. A clogged intake pit of a lift irrigation system. Note that the foot valve of one of the two pumps is fully submerged in the material settled in the pit, while the inflow to the other foot valve is restricted by the material settled around it. Further, the foot valves are located too close to the sides and bottom of the tank which would adversely influence the flow to the valves.

pumps are to be used. The number of pumps in a scheme depends on the discharge capacity required for providing irrigation to the command area of the system. A standby pump set is usually required to ensure uninterrupted flow in the system. Generally, two pump sets with one standby pump (2 + 1) system is found suitable for power ranges up to 400 h.p. (Fig. 4.24).

All the pumps, together with the electrical installation, panel board and service crane are to be located in a pump house to protect them from interference by unauthorised persons and the effect of sun and rain.

In case of vertical turbine pumps and submersible pumps, it is feasible and economical to provide the pump house on top of the jack well or sump well itself (Fig. 4.27). In such cases, the size or the pump house is usually kept the same as that of the well. In case of schemes having more than one stage, a sump well is required at each pumping stage.

Size of sump wells: The standard diamensions of a sump well to house horizontal centrifugal pumps will depend on the spacing between the suction pipes and the size of bell mouth or foot valve, the clearance of the foot valve from the sump bottom and side walls, the minimum depth of submergence of the foot valve and the maximum water level in the sump well.

Ordinarily, the shape of the sump well is kept circular if the total horse power is limited to about 45 h.p. For higher values of horse power, rectangular shape is generally preferred. The minimum diameter of a circular sump well is estimated by the following relationship:

$$D = 4 \times \frac{\text{No. of pump sets} \times d}{\pi} + 3d \qquad \text{... (4.7)}$$

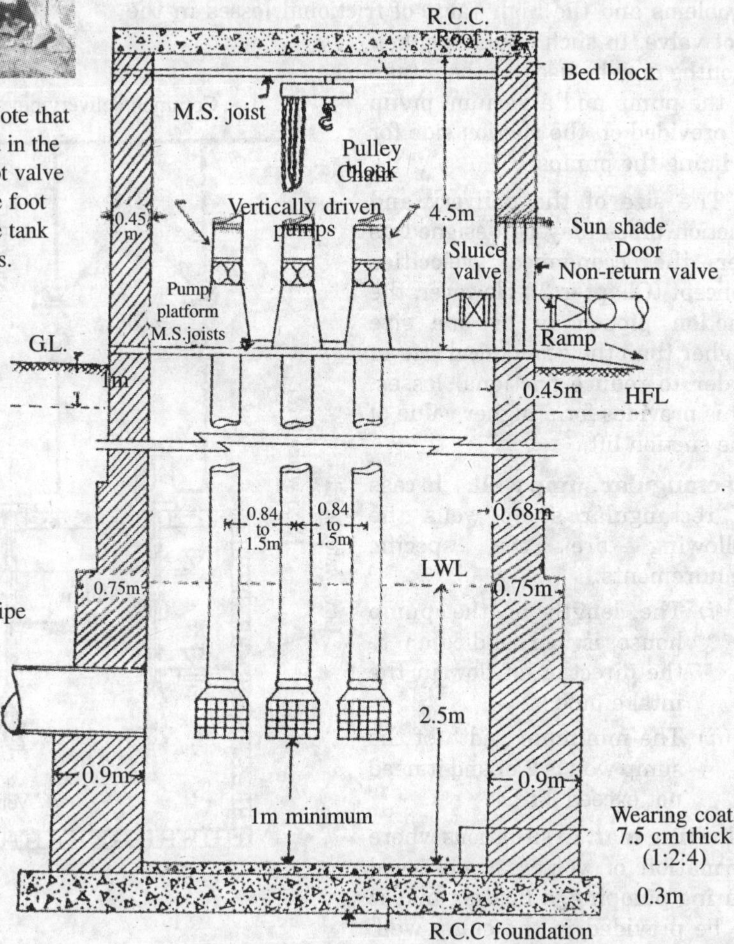

Fig. 4.26. Schematic view (vertical section) of a jack well (circular) with pump house built over it. The pump spacing of vertically driven pumps and valve arrangement in the delivery section are also shown.

(Adapted from National Bank for Agriculture and Rural Development, 1989)

in which,

D = diameter of sump well (m)

d = diameter of foot valve (m)

The following are the other specifications usually adopted:

(i) The minimum clearance between the sump well wall and the outer face of the foot valve is d.
(ii) The minimum depth of sump bottom from the bottom of the foot valve is 60 cm.
(iii) The minimum submergence level between the foot valve and the low water level in the sump is 1 m.
(iv) Freeboard above the maximum water level in the sump well and the ground surface is 1 (m).
(v) The minimum height of sump well above ground level is 60 cm. (This height is required to prevent animals and human beings from falling into the well).

For discharges less than 150 litres/sec, low frictional loss globular foot valves are used (Chapter 3). The use of foot valves, may be dispensed with when the discharge values exceed about 150 litres/sec, in view of priming problems and the high value of frictional losses in the foot valve. In such situations, bell mouths may be used at the intake of the pump and a vacuum pump is provided on the suction side for priming the pumps.

The size of the delivery and suction pipes may be designed as per the economical velocities concept (Chapter 3). However, the suction pipe may be one size higher than the economical size in order to reduce frictional losses. This provides for a higher value of the suction lift.

Rectangular sump well: In case of rectangular sump wells the following are the specific requirements:

(i) The length of the pump house is perpendicular to the direction of flow in the intake pipe.
(ii) The minimum width of the sump well is 2 m and it need not exceed 3 m.

Floating mat: In situations where formation of vortex is observed during pumping, a floating mat is to be provided in the sump well. The mat may be made of wooden reapers of size 1.25 × 1.25 cm with 15 × 15 cm openings.

Retention period: In case of multi-stage pumping system lift irrigation projects, the retention period of inflow into the sump well is kept about 5 minutes. In some cases, when economy is a prime consideration, lower retention periods are also adopted in design. It is essential to provide erosion-proof disposal of overflow from sump wells in case of a sudden breakdown of pumping units in a multi-stage river pumping scheme.

Jack Wells

Jack wells are usually associated with high-head vertical turbine pumps and submersible pumps or mixed flow pumps. They are usually preferred in major lift-irrigation schemes (Fig. 4.26). A jack well is constructed on high ground. It is to be located as close to the river as possible. Although it would be an advantage to locate it on high ground above the highest flood level (HFL), sometimes, it may not be possible without substantially increasing the length of the intake pipe. Consequently, considering the cost of the system it may be economical to locate the jack well near the water body during the low-flow stage, even if a good portion of its height remains above

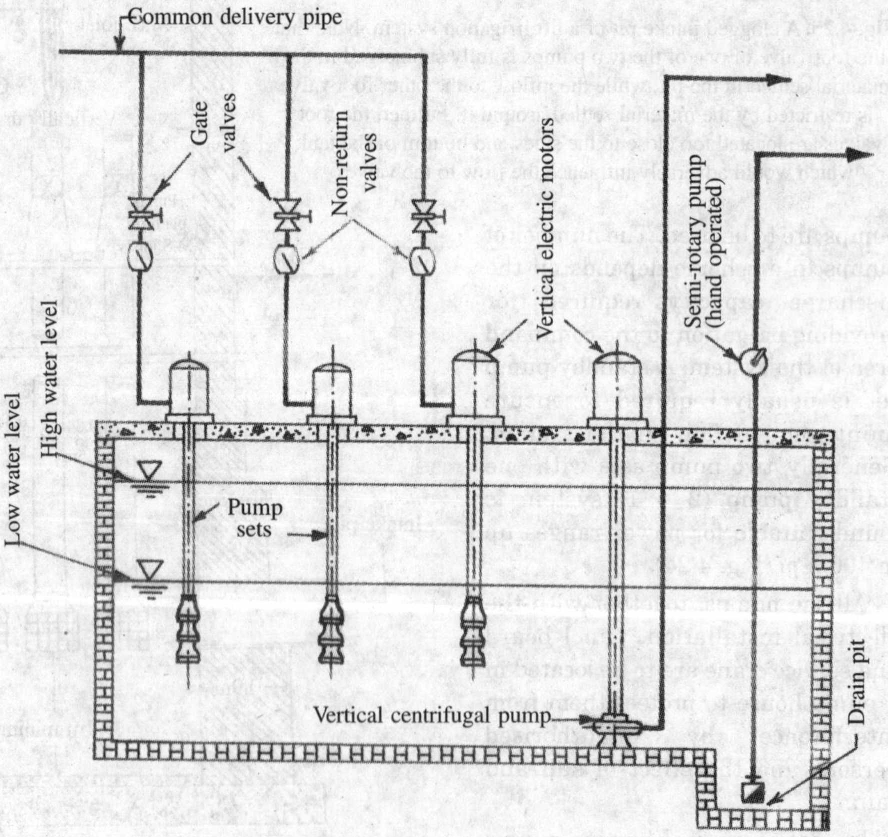

Fig. 4.27. Schematic sketch illustrating the installation of a series of vertical turbine pumps in a masonry sump functioning as the jack well in a lift irrigation project. Note the ancillary pumping sets (vertical centrifugal pump and semi-rotary pump) for dewatering of the sump.

ground level. Access should be provided by a rigid ladder.

The thickness of the well steining below the ground level is governed by the earth pressure and the superimposed load. This load consists of the pumps, motors, valves, other accessories and the live load on the pump-house. To distribute this load evenly over the entire steining, an RCC ring of 30 cm thickness is provided on top of the steining masonry. The thickness of the steining above ground level is controlled by the requirements of stability. Generally, the thickness of steining below ground level should be 75 cm to 90 cm and between 38 cm to 45 cm above ground level.

The jack well should be located on hard strata with a bearing capacity of 4.5 kg/cm^2 or more. If such a strata is not available, an RCC raft may be provided at the bottom. Even when the jack well is dug through rocky strata, it may be desirable to smoothen the sides or provide a masonry lining. This will facilitate access to the bottom of the well for maintenance and desilting operations. For installing turbine pumps, the bottom of the well should be about 3 metres below the lowest water level in the well in order to accommodate the pump bowl assembly, suction pipe and strainer and provide for a silt pocket.

Mild-steel bars of 20 mm diameter, bent in the form of *U* and embedded in masonry, is adequate to provide access into the well. Two mild-steel girders should be provided at mid-height between the low water level (LWL) and floor level. These will enable the erection of a platform for the installation and maintenance of the pumping set and ancillary equipment. The girders are to be fixed in RCC blocks. The following specifications are maintained in the design of the jack well:

(*i*) The minimum clearance between the base plate on which the pumps are installed and the wall of the jack well is 50 cm.

(*ii*) The top of the jack well is kept 1 m above the high flood level (HFL) in the river or 0.3 m above the ground level, whichever is more.

(*iii*) The minimum submergence of the pump bowl assembly below the low water level (LWL) in the jack well is 2 m in case of vertical turbine and submissible jumps and 1.5 m in case of centrifugal jumps.

(*iv*) The minimum clearance between the bottom of the pump and the floor of the jack well is 1.5 m.

(*v*) The bottom of the jack well should be at least 50 cm below the bottom of the intake pipe.

Silt removal: Wherever necessary, a slurry pump of suitable capacity may be provided to remove the silt and discharge it downstream of the jack well.

Pump House

A circular pump house could be conveniently located on top of the jack well. A rectangular or square shape of the pump house is more convenient and economical than a circular shape. A rectangular pump house gives more space for locating the pumping sets and ancillary equipment. A rectangular or square pump house can be constructed over the circular steining by providing girders over the top of the steining and cantilevering them suitably. In case of a sump well, however, much space is not required in the pump house, as the motors are not located at the same level. In such cases, it is more convenient to continue the steining masonry with the circular shape. A lightening conductor should be provided on top of the pump house.

Pump House for Centrifugal Pumps

The pump house for centrifugal pumps shall be rectangular in shape. The pumps, motors, panel box for electrical components, starters and other accessories are located in the pump house separately. The wall of the pump house is about 45 cm thick and the exposed surface is plastered with cement mortor. The dimensions of the pump house, namely length, breadth and height are based on the specific requirements of the installation. The length of the pump house is the total length along the shaft of horizontal centrifugal pumps, the spacing between pumps, the spacing between the pump house walls, the outer ends of the first and the last units. The shaft length of the pumping set (pump with motor coupled to it) is obtained from the specifications of the pumping unit proposed. The spacing between each pumping unit depends on the h.p. of the unit and ranges from 60 cm to 1 m. For pumps up to 30 h.p.: 60 cm, 31 to 60 h.p.: 70 cm, 61-90 h.p.: 80 cm, 91-120 h.p.: 90 cm and above 120 h.p.: 1 m. The same values may be adopted for the clearance between the wall and the first and the last pump in the row.

The width of the pump house is computed on the basis of the requirements of the following:

(*i*) The width of the pump, including the length between the suction side flange and the delivery side flange of the pump.

(*ii*) The clearance between the wall and the suction flange of the pump is kept 60 cm in all cases. This includes 15 cm as the clearance from the wall to the eccentric taper flange and 45 cm as the length of the eccentric taper (Fig. 4.23). When the length of the eccentric taper exceeds 45 cm, the width of the pump house has to be correspondingly increased.

(*iii*) The length of the non-return valve on the delivery side.

(*iv*) The length of the sluice valve on the delivery side.

(*v*) The clearance between the sluice valve flange and the pump house wall is 30 cm for pumps up to

50 h.p., 45 cm for those from 50 to 100 h.p., and 60 cm for units above 100 h.p.

The height of the pump house is fixed, taking into account the following factors:

(*i*) The height of the foundation block of the pumping set above the floor of the pump house is usually kept at 10 cm.

(*ii*) The thickness of the base plate of the pumping unit is usually kept at about 15 cm.

(*iii*) The height of the pumping set above the base plate is obtained from the specifications of the pump proposed.

(*iv*) Clearance required between the top of the pumping unit and the bottom of the gantry hook (Fig. 4.23).

(*v*) The depth of the I-beam supporting the gantry.

(*vi*) *The depth of the gantry unit:* This is the distance between the bottom of the gantry girder and the centre of the hook.

The sum of the above items provides the height of the pump house.

Pump House Structure

(*i*) *Foundation:* The depth of foundation of the pump house depends on the properties of the soil in which the foundation is constructed. It is 60 cm below ground level in hard soils and 90 cm in black soils or below the summer crack level of the soil of the foundation. The foundation is constructed with concrete (1:6:10) for a minimum thickness of 30 cm. The width of the foundation is to be designed on the basis of the safe bearing capacities of the soil. The maximum bearing capacity is limited to 5 tonnes/m^2.

(*ii*) *Basement:* The floor level of the pump house shall be at least 1 m higher than the maximum flood level (MFL) of the river or stream from which the water is pumped or 30 cm above the ground level, whichever is higher.

The material of construction of the basement is RR (random rubble) masonry in cement mortar (1:8) or brick masonry in cement mortar (1:6), whichever is more economical at the site.

(*iii*) *Flooring:* The common material for floor is cement concrete (1:3:6). The floor thickness is usually 10 cm. Where good quality flat stones are available the same could be used, depending on cost considerations. The floor concrete is isolated from the foundation blocks of the pumping units by providing a mastic pad of bitumenous joint to prevent transmission of vibration effect of the pumps when they are in operation.

(*iv*) *Superstructure:* The walls of the pump house which support its roof, are made of 35 cm thick brick masonry in cement mortar (1:6) or RR stone masonry 45 cm thick in cement mortar (1:6), whichever is more economical. The walls are plastered in cement mortar (1:4). The thickness of the plaster is 1.2 cm in case of brick masonry and 2 cm in case of stone masonry.

(*v*) *Roof:* The roof of the pump house is usually 15 cm thick RCC (1:2:4). It could also be corrugated *GI* or asbestos cement sheets supported on steel trusses. The details of design of RCC slabs are given in IS:456-1978 of Bureau of Indian Standards.

(*vi*) *Doors, windows and ventilators:* The number of windows of the pump house walls along the length is usually three. One door, one window and a ventilator are provided in the wall on the delivery side. The common size of window is 90 cm wide and 1.25 m high and that of the door 2.1 m high and 1.25 m wide. Slit type ventilators are usually provided close to the roof level of the doors and windows. The common size of ventilators is 60 cm × 15 cm.

(*vii*) *Foundation blocks:* The foundation blocks of the pump sets are made of cement concrete (1:3:6) and 10 cm thick above the floor of the pump house. As already stated, the foundation blocks are separated from the pump house floor with mastic pad or bitumen in order to localise the effect of vibration to the foundation blocks only. The foundation blocks are made larger than the base plates of the pump to provide a margin of 15 cm on all its sides. The length of the foundation bolts are about 70 cm, out of which 10 cm are in the bed block and 60 cm in the floor.

The minimum depth of the pump foundation is 60 cm, with an all-round offset of 15 cm. This is considered safe as the weight of the pumping set is usually less than the permissible bearing capacity of 5 tonnes/m^2 in almost all types of soils.

(*viii*) *Gantry arrangement:* The gantry is the overhead structure comprising of a girder and crane of adequate capacity, depending on the weight of the pumps, motors and other accessories. Provision of suitable gantry arrangement is essential for the maintenance and repair of pump sets.

(*ix*) *Pump delivery system:* The delivery pipe of the pump may discharge into a cistern located close to the pump house or into a rising main leading to the distribution chamber. The delivery pipe of the system, which discharges into the cistern, is located close to the pump house. It is provided with suitable long radius bends. In such a system, provision of a reflux valve is unnecessary and the width of the pump house could be proportionately reduced. When the pump delivery pipe is continued by the rising main, a non-return valve with by-pass arrangement will have to be provided on the delivery side (Fig. 4.23 to 4.29).

(x) *Location of starters:* Air break contactor type star-delta starters are usually used in motor-driven pumping systems with capacity up to 50 h.p. For higher horse power units, oil immersed starters are prefered. Starters of smaller units may be mounted on a common panel board. Oil immersed starters are heavy and are installed separately on the floor close to the motor so as to reduce the cost of cables. Starters for vacuum pumps are mounted on the common panel board.

Example 4.2. *A lift irrigation pumping system is to be designed for a total discharge capacity of 570 litres/sec. The total dynamic head is 18.5 m and the length of the rising main is 1000 m. Static suction lift is 5 m. The frictional losses on the suction side is 0.5 m. Determine (i) the capacity of the motors required, (ii) the dimensions of the pump house, and (iii) the net positive suction head (NPSH).*

Solution: *Selection of pumps and motors:*

Select a 2 + 1 pumping system, with one pump to serve as stand-by:

(*i*) The capacity of each pump = $\frac{570}{2}$ = 285 lps

(*ii*) Assuming a pump efficiency of 80%, h.p. of motor

$$= \frac{Q \times H}{75} \times \frac{100}{80}$$

$$= \frac{285 \times 185}{75} \times \frac{100}{80} = 87.875 \text{ h.p.}$$

A cushioning effect of 10%, i.e., 8.78 h.p. is added to compensate for low voltages.

Hence, the corrected value of motor capacity

= 87.875 + 8.787 = 96.66 h.p.

The nearest available size of electric motor is 100 h.p. (75 kW).

Pump House Design

(*i*) *Length of pump house*

Assumptions:

Length of each pumping set (from the pump specifications) : 2.30 m

Clearance between pump sets (see text) : 0.9 m (90 cm) (for 100 h.p. pumping set)

∴ Length of pump house = (2.3 × 3) + (0.9 × 4) = 10.50 m

(*ii*) *Width of pump house*

Assumptions:

Flange to flange width: 1.14 m

Length of reflux valve: 1 m

Length of sluice valve: 0.75 m

Length of eccentric taper: 0.65 m

Clearance between pump house wall and sluice valve: 0.5 m

Length of taper on suction side + 15 cm clearance: 0.6 m

Total width = 1.14 + 1 + 0.75 + 0.65 + 0.5 + 0.6 = 4.64 m

(*iii*) *Height of pump house*

Assumptions:

Height of top of pump house from floor level: 1.6 m

Clearance between crane hook to pump top: 1.85 m

Depth of crane hook below roof: 1 m

Total height = 1.6 + 1.85 + 1 = 4.45 m, say 4.5 m

Net positive suction head

Assumptions:

Saturated vapour pressure at the locality: 0.3 m

Total suction lift = 5 + 0.5 = 5.5 m

Entrance losses = 0.1 m (given)

Total suction list = 5.5 + 0.3 + 0.1 = 5.9 m

Atmospheric pressure at the location (assumed) = 9.9 m

NPSH available = 9.9 – 5.9 = 4 m

NPSH required for the pump may not exceed the NPSH available minus the vapour pressure, i.e., 4 – 0.3 = 3.7 m.

Hence, a pump whose NPSH is under the limit of 3.7 m is selected.

Note: In case a suitable pump whose NPSH value is within the above limit is not available, the alternative is to suitably reduce the suction lift to permissible limits. There is little scope for reducing the entrance loss, if the foot valve/bell mouth entrance is well designed.

Pump House for Vertical Driven Pumps

Vertical driven pumps for lift irrigation schemes include vertical turbine pumps, submersible pumps, mixed flow pumps and propeller pumps. Proper arrangement of pumps, sluice and reflux valves, delivery manifold, flow training devices in the jack well for ensuring hydrostatic efficiency and gantry arrangement should be designed properly to ensure smooth functioning of the pumping system. Gantry arrangement with chain pulley block is provided at a height of 4 m above the floor of the pump house. The capacity of the gantry should be adequate to lift the maximum weight of a single unit which usually varies from 0.5 to 3 tonnes, depending upon the size and horse power of the pump set. Cost considerations are equally important as the performance requirements.

Specific arrangement of the various ancillary components depends on the type of pumps and their capacities.

In case of horizontal centrifugal pumps the sump well and the pump house are generally located separately in order to be economical. However, in the case of vertically operated pumps, the pump house is superimposed on the jack well. The specific requirements in case of vertical turbine pumps, submersible pumps and other vertically operated pumps are as follows:

(*i*) Provision of suitable hydraulic inflow training devices to ensure hydraulic efficiency.

(*ii*) Layout of the pumps in the pump house.

(*iii*) Location of reflux and sluice valves.

(*iv*) Arrangement of delivery manifold.

(*v*) Specifications of the pump house and the supporting girders for the pumps.

(*vi*) Gantry arrangement.

Number of pumps: The number of pumps to be installed in a particular system and the stand-by capacity required depend on the discharge capacity of the rising main and relative cost of various alternatives. The larger the number of pumping sets the more will be the size of the jack well. A common arrangement is 2 + 1 system, with two pumping sets working at a time and the third serving as a stand-by.

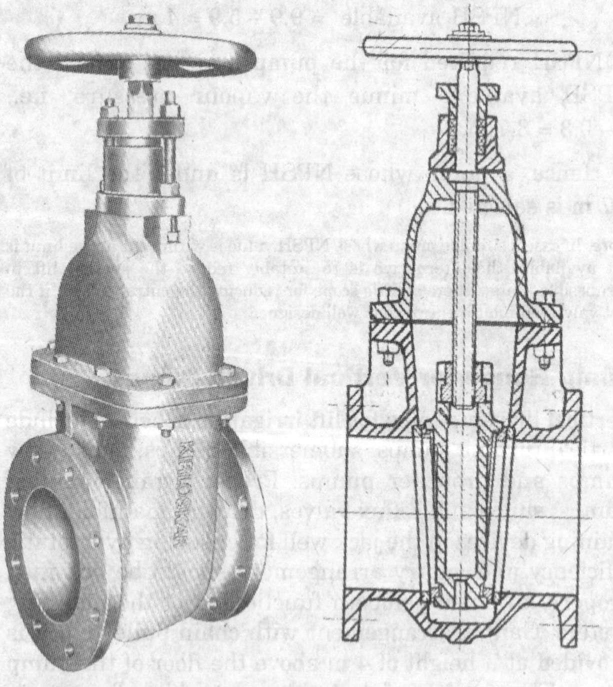

(a) General View (b) Sectional View

Fig. 4.28. Cast iron sluice valve suitable for high pressure pipelines in lift irrigation schemes.
(***Courtesy:*** Kirloskar Brothers Ltd., Pune.)

Specifications of jack wells: The internal diameter of the jack well depends on the number of pumps to be installed and clearances between the pumps and from the well walls as well as the bottom of the pump and that of the jack well. The space between the adjacent pump sets will depend on the size of the base plates for pumps of different capacities, hydraulic efficiency considerations for inflow of water from intake pipe to jack well, and the working space required for the installation, operation and maintenance of the pumps. The other major specifications are as follows:

(*i*) The steining of sump well shall be of R.R. masonry or brick masonry laid in cement mortar (1:5), adopting usual design procedures. The maximum earth pressure on steining is when the well is dry. The minimum thickness of steining is 60 cm.

(*ii*) The floor level of the pump house is kept greater than 30 cm above the ground level or 1 m above the high flood level in the river or streams, whichever is higher.

(*iii*) The floor slab of jack well is designed to withstand the possible loads due to the weight of the pumps and various accessories as well as the live load during various stages of installation and operation. An impact factor of 10% is usually added to the above values. Although the floor can be of RCC slab with suitable opening for lowering of the pump assembly, it is desirable to construct it with wooden planks fixed with nuts and bolts on mild steel girders. This minimises the chance of electric leakages.

(*iv*) The roof slab of pump house is kept at 5 m. It is designed to withstand a live load of 200 kg/m^2 and the dead load which it supports, including its own weight. They are usually 75 cm thick R.C.C. slabs in cement concrete (1:2:4). The height of the jack well is fixed to facilitate easy handling of one column pipe of 3 m length at a time.

(*v*) Bed blocks are provided for the bearings of girders carrying the pumps and the gantry system to ensure safe working loads, on the wall of the jack well. An impact factor of 20% is usually added to the total computed loads. Bed blocks are constructed of cement concrete (1:3:6) with nominal reinforcement.

(*vi*) Lintel over the doors and windows on the circular portions of the pump house are designed to withstand the torsional effect due to their location in the circular areas. They are designed for load imparted by an equilateral triangle of the masonry. They are constructed of RCC (1:2:4). In case of lintels located below the gantry girders the maximum possible load from the gantry system is considered. Lintels over doors and windows on straight walls are designed, adopting the usual procedure for design of lintels.

(*vii*) One door and two windows are usually provided in case of a jack well. Doors are usually of size 1.2 m × 2.1 m. One window is provided for the jack well and another for the valve room. The windows are usually 1.2 m wide and 1.2 m high. Ventilators of size 1.5 cm wide and 60 cm high are provided above the door and each of the windows.

(*viii*) The minimum depth of foundation of the wall of the valve room is 90 cm in case of black soils and the summer crack depth in case of all types of soils, whichever is more. The thickness of foundation concrete shall be a minimum of 22.5 cm (1:4:8). If rock is available at shallow depth, the depth of foundation may be limited to the rock level. The maximum load on foundation in made-up soils (fill areas) may not exceed 5 tonnes/m^2.

(*ix*) Suitable bracings are provided for the girders supporting the pumping sets to minimise vibration. In case of centrifugal pumps, each pump set should have a foundation block constructed in 1:3:6 cement concrete. To protect against vibration effects the foundation block is isolated from the floor by a layer of bitumen.

(*x*) When the column pipe length of the pump exceeds 10 m, suitable brackets are provided on RS joists at the mid-height of the column pipe.

Pump chamber: If all the sluice valves and reflux valves are located within the jack well, the diameter of the jack well will be more and this will therefore be expensive. To minimise the cost of jack well, a separate pump room, rectangular in shape, could be constructed above the ground level but adjoining the jack well. Foundations for this may be separated from the jack well.

Suitable sizes of man-holes, frames and covers are provided on the roof adjacent to the gantry girder, vertically above the pump for easy handling of component parts.

The height of the roof of the pump house, when it is constructed separately adjoining the jack well, is kept at about 3 m above the floor level. Construction-joints are provided at the junction of the rectangular and circular chambers. The floor level of the pump house is fixed with a minimum freeboard of 1 m above maximum water level.

Hydraulic efficiency of pump chamber: To provide for high pump efficiency, the water that flows into the jack well is to be streamlined so as to provide equal distribution of flow towards all the pumps located in the well and avoiding turbulence and vortices. Baffles and diaphragms are provided for the above purpose. The designs of these are to be in conformity with the standards as given by the American Hydraulic Research Institute, Denver or the Irrigation Research Institutes of state irrigation departments. Baffle walls and diaphragms are constructed in RCC (1:2:4) with a minimum thickness of 10 cm and provided with nominal reinforcement. Brick masonry, if used, will occupy more space in the jack well. RCC can withstand vibration effect due to turbulent flow at the initial stage, when water impounds in the jack well.

Silt removal. A slurry pump of suitable capacity is usually provided to facilitate the removal of silt and discharge it to the river course downstream of the jack well.

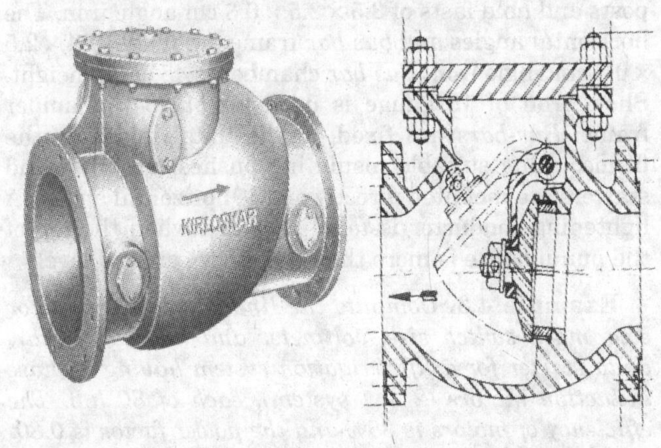

(a) General View (b) Sectional View

Fig. 4.29. A cast iron swing-type non-return valve (a) General view, (b) Sectional view showing the working of the non-return valve.

(***Courtesy:*** *Kirloskar Brothers Ltd., Pune.*)

Electrical Equipment and Accessories

Pumps for lift irrigation schemes are driven by a.c. induction motors. Various component parts are required for the proper control of electric motors and to ensure their trouble-free service. The requirements as prescribed by the State Electricity Boards are to be adhered to. Controls for various ranges in the capacity of motors are to be provided with matching accessories.

A common arrangement in lift irrigation systems is the 2 + 1 type pumping sets with two pumps operating at a time. However, the incoming cables and switch fuse unit/air circuit breaker are designed to provide for all 3 pumping sets operating simultaneously. This also provides the opportunity for operating all the pumps to work together to meet emergencies. Similarly, the transformer capacity is also designed to suit all 3 pumps working together.

Cable size: The cross-sectional areas of conductors are determined to meet the current carrying capacity. The current carrying capacity of a power supply system is computed adopting the following relationship:

$$\text{Line current} = \frac{\text{h.p. of motor} \times 735}{\text{line voltage} \times 1.732 \times \eta_m \times P_f} \quad \text{... (4.8)}$$

in which,

The line voltage, usually assumed to be 415 V

η_m = efficiency of motor (usually assumed as 0.85)

P_f = power factor (usually assumed as 0.80)

Suitable conductor size is selected from handbooks and reference tables to match the line current. For the incoming cable to the panel board, the total h.p. of all the pumps is considered. However, for cable connections of individual motors the h.p. of each motor, only is considered. Copper conductors are superior but are expensive. Hence, good quality aluminium cables are often substituted.

Switch fuse units, air circuit breakers and fuses: The selection of switch fuse unit/air circuit breakers is done on the basis of the current rating required for each motor. The rewirable fuses which were common in earlier days are being replaced with modern high rupturing capacity fuses (HRC) which are capable of successfully interrupting fault currents. The main function of an HRC fuse is to limit the short-circuit current both in magnitude and duration which can pass through the circuit and the devices. Switch fuse unit/air circuit breakers are selected on the basis of the current ratings of each motor. Air circuit breakers, in addition to functioning as switch fuse units, have relays for the protection of electrical equipment and circuits. While the switch fuse unit needs frequent replacement, the circuit breaker lasts long. The choice between the two is based mainly on relative costs. It is generally seen that air circuit breakers are economical for systems with ampere range of 100 and above and will last long with little repair. For current ratings below 100 amperes, it is economical to have HRC type switch fuse units.

Contactor type air circuit breakers are more advantageous than oil immersed air circuit breakers. They last long, almost the full service period of a motor.

Voltmeters and ammeters: Selection of voltmeters and ammeters depends on the anticipated range in voltages and current. Voltmeters of range 0-600 V are usually selected. The starting current is usually 2 to 3 times the line current. Hence, the ammeter is selected based on possible current in starting.

Capacitors: Capacitors are used in the circuit to provide high value of the power factor (P_f). Suitable capacitors are selected, based on standard tables to ensure economical operation of power supply in the circuit.

Starter: Direct-on-line starters are used for motors up to 5 h.p. Contactor type fully automatic star-delta starters are used for motors above 5 h.p. Contactor type star-delta starters are less expensive and lighter in weight, as compared to oil immersed type. They are used for motors in the range of 5 to 50 h.p. For motors of 50 h.p. and more, oil immersed type star-delta starters are preferred, even though they are bulky and more expensive.

Single phase preventer: Single phase preventer provides protection to 3-phase motors and other 3-phase equipment when there is a failure in any one of the phases. When any phase fails in the 3-phase circuit, a relay provided in the preventer trips the starter. Single phase preventers can be provided for each of the starters of the motors. In some starters this device is built-in and a separate unit is not required.

Panel board connections: In case of high tension supply, metering of the current is done on the high tension side. In such cases, the low tension side of the transformer is connected to the air circuit breaker or switch fuse unit, as the case may be. In case of low tension supply the pole wire is connected to the LT kWh metre board kept inside the pump house. The kWh meter is connected to the incoming side of the main switch-fuse unit. The outgoing wires of the main switch fuse unit/air circuit breaker is connected to the *bus-bars*. From the *bus-bars*, connections are made to the 3 sub-switch fuse units/air circuit breakers. From them connections are made to the starters of the motors. Connections to the capacitors are made through the sub-switch fuse units only, thus eliminating a separate switch for the capacitor. Connections are made from the *bus-bar* to the single phasing preventer, from which connections are made to the starters. All connections are made by using lugs.

The panel board frame is fabricated with vertical posts and hold fasts of 3.5 × 3.5 × 0.3 cm angle iron. The horizontal angles and *bus-bar* frame are made of 4 × 2.5 × 0.3 cm angle iron. *Bus-bar* chamber is 45 cm in height. Sheet iron of 18 gauge is used for *bus-bar* chamber frame. *Bus-bars* are fixed on the flats welded to the frame, using suitable insulating bushes. Switches and starters are mounted directly on the horizontal angles. A lightening conductor is to be provided when the top of the pump house is more than 5 m above ground level.

Example 4.3. *Compute the line current, conductor size, main switch size, voltmeter and automatic star-delta starter for a lift irrigation system having 3-phase induction motors (2 + 1 system), each of 80 h.p. The efficiency of motors is 85% and the power factor is 0.80.*

Solution: Total horse power of the system working at a time

= 3 × 80 = 240 h.p.

Line voltage = 415 V (assumed)

Percentage variation of voltage fluctuation is assumed as 10%

Design voltage $= 415 - 0.1 \times 415$

$= 373.5$ V

Line current (amperes)

$$= \frac{240 \times 735}{373.5 \times 1732 \times 0.85 \times 0.8}$$

$= 401$ A

The sizes of main switch and conductor are selected from standard tables to suit the current load of 401 amperes.

A voltmeter of 0 – 600 V capacity and 150 mm dial and individual ammeters of 0 – 410 A, 150 mm dial along with voltage selector switch could be selected.

Line current of each motor:

Line current rating per motor

$$= \frac{401}{3} = 133.67$$

= say 134 A

Air circuit breaker of 150 A is selected as sub-switch.

Contactor type fully automatic star-delta starters of 80 h.p., capacity may be selected for mounting on the panel box. Alternatively, oil immersed type starter may be selected for each motor and mounted separately near each of them, depending on cost considerations.

Transformer: Whenever the connected load is more than 75 h.p. it is mandatory on the part of the consumer to use high tension (HT) power supply. In most lift irrigation systems the power requirement exceeds this value and a transformer of appropriate capacity is required.

State Electricity Boards maintain distribution transformers, mostly of less than 250 KVA in large numbers. They have trained staff and well equipped worshops to assist in prompt repairs/replacements when the transformers go out of order. Similarly, Public Health Engineering Departments and Rural Electricity Corporations also extend such services.

Procedure for Starting and Stopping of Pumps

Specific procedures in the operation of the pumps will ensure trouble free functioning of the system. For the purpose of starting and stopping of the pumping sets the pumps may be classified into two groups:

1. Radial flow pumps and mixed flow pumps of low specific speed (specific speed < 4000)
2. High specific speed mixed flow pumps and axial flow propeller type pumps (specific speed > 4000)

1. Low specific speed pumps: The procedure for starting and stopping of low specific speed pumps (both horizontal and vertical shaft) are described below:

(*i*) *Starting procedure:*

(*a*) Close the sluice valve on the delivery side (in case of water-lubricated turbine pumps, the sluice valve is kept slightly opened).

(*b*) Prime the pumps.

(*c*) Open the valve in the cooling water supply to the bearings (in the case of water-lubricated turbine pumps).

(*d*) The valve in the cooling water supply is opened, in case of pumps with stuffing boxes which are water cooled.

(*e*) Check the voltage in each phase.

(*f*) Start the motors (when the pumping system consists of a number of pumps in parallel, the pumps are started one after another, adopting the above procedure, keeping the sluice valves closed).

(*g*) Open slowly the sluice valve on the delivery side, after starting all the motors.

(*h*) Observe leakage in the stuffing boxes and carefully adjust the sealings, if required.

(*i*) Observe the general mechanical functioning of the pumping system.

(*j*) Avoid starting an individual pump while others are in operation.

(*ii*) *Stopping procedure:*

(*a*) Close the sluice valve slowly.

(*b*) Stop the motor.

(*c*) Close the valve on the pipe to supply cooling water for the bearings and to water-cooled stuffing boxes, if provided.

2. High specific speed pumps: In case of axial flow pumps (propeller type) operated at low heads, there is excessively high power consumption when the pump is started with the delivery valve closed. Hence, an axial flow pump shall not be started with the sluice valve closed. Sometimes, it may be necessary to provide a by-pass to direct the flow back into well or stream to prevent "shut off" in case the gate valve is to be kept closed. However, if the delivery pipeline is relatively small and the pump discharges close to the delivery end, there is no need for a by-pass arrangement. However, care is taken to see that the delivery sluice valve is fully open when the pump is started. Whenever by-pass is provided, the same is located after the reflux valve and the delivery sluice valve.

1. Starting procedure

(*i*) *By-pass is not provided:*

(a) Open the delivery sluice valve fully.

(b) When reflux valve is provided, open the by-pass of the reflux valve and sluice valve and empty the delivery pipe. After emptying. close the by-pass of the reflux valve.

(c) Start the motor.

(ii) *By-pass is provided:*

(a) Open the delivery sluice valve and by-pass.

(b) Start the motor.

(c) Close the by-pass valve.

2. Stopping procedure

(i) *By-pass is not provided:*

(a) Stop the motor.

(b) Close the sluice valve in the delivery side.

(ii) *By-pass is provided:*

(a) Open the by-pass valve.

(b) Stop the motor.

(c) Close the sluice valve.

When the pumping system comprises of pumps without by-pass arrangement and those with them, it is necessary to start the pumps one after the other. The sluice valves are kept fully open. In case of pumps with by-pass arrangements, full load is given to the pump sets only after all the pump sets are started. In such situations the sluice valves on the delivery side are opened fully, only after starting all the pump sets.

Piping System in Pump Chamber

The piping system in the pump chamber comprises of suction and delivery pipes and accessories, such as bends and valves. A well designed piping system results in the satisfactory performance of the system and reduction in friction losses, with consequent savings in energy.

Suction and delivery pipes: Correct sizes of suction and delivery pipes are selected to keep the flow velocities within permissible limits. The limiting velocity in the suction pipe is 1.5 m/sec while that in the delivery pipe is 2 m/sec. When velocities higher than the recommended permissible limits are encountered in design, it is necessary to increase the pipe size to the next higher value. However, increase in pipe sizes beyond economical limit will increase the capital cost of the project disproportionate to the savings.

It may often be advantageous to have larger diameter pipes, especially on the suction side, as compared to the suction and discharge openings in the pump, and connect the pipes to the pump suction and delivery ends by taper pieces (Fig. 4.23). The taper pieces are called reducers, when applied to the suction side and enlargers, when applied to the delivery side. The optimum arrangement of the piping system is to have the pipes run straight with minimum of long radius bends and avoiding elbow joints, using only the minimum number of fittings.

Operation of multiple pumping sets in parallel: Lift irrigation pumping systems usually comprise 3 or more pumping sets working in parallel and discharging into one or two main delivery pipes, referred to as *rising mains.* Parallel operation of pump sets requires a well-designed manifold to deliver water from each pump to the main discharge pipe (Fig. 4.22). The delivery pipes of the pumps should be connected to the rising main in such a manner that the flow is smooth and the dismantling of connections for maintenance and repairs easy. Multiple pump sets installed in parallel should be connected to a single main. With centrifugal pumps, they are connected through a Y-piece (Fig. 4.22). When there are more than two units, a multiple Y-piece is required. The Y-piece can be made of cast iron or mild steel. Cast iron manifolds specially cast for high pressure pipeline system, have the advantage of reduced cost. They are free from rusting and incrustation.

With turbine pumps, the pipe connections can be made in two ways. In the first method, the delivery pipes are taken out of the jack well and connected to a Y-piece fitted at the bottom end of the rising main. In the second case, the rising main itself is taken into the jack well and the delivery pipes of the pumps are directly connected to it. The latter is a better layout and the head losses are minimum as bends are avoided. However, it may be more expensive. For deep well turbine and submersible pumps, the delivery pipe can be taken out at any level. It is convenient to take it at the level of the rising main so that bends are avoided.

The piping is connected to the pumps only after the grout on the foundation is set hard and the foundation bolts tightened up. Flexible type of connection using flanged spigot and flanged socket is provided for each pump set. After the installation of the pump set the delivery manifold is connected to the rising main, keeping the socket joint open. After all the joints and bolts are tightened the socket is completed by pouring lead and caulking the same. The discharge passing through the parts of the manifold and the accessories, excluding control valves, are to be taken into account and the frictional losses computed along the path of flow.

Control valves (Fig. 4.28 and 4.29): One set of control valves, comprising a reflux valve and a sluice valve, is usually provided for every pump. The reflux valve prevents emptying of the rising main and the sluice valve permits regulation of discharge. In case the rising mains are taken in different directions, depending on the topography of the area to be irrigated, separate sluice valves are provided near the head-end of each branch line, a few metres downstream of the junction point (usually one pipe length of about 6 m). Thus, every pump should be able to supply any part of the system singly or in combination with others. Further, any pump or main branch can be isolated from the system for repairs, without affecting the remaining part of the system. The reflux valve (non-return valve) is provided close to the pump and the gate valve immediately after it. Where the discharge pipe is larger than the delivery end of the pump, the reducer socket should be attached directly to the pump, followed by the control valves which are of the size of the discharge pipe.

Suction and delivery pipes of the pumps should preferably be placed below the pump house floor level in accessible concrete trenches covered with removable gratings. The valve stems with hand wheels should project above the floor for convenient operation.

Installation of pumping sets: The general procedure for the installation of different types of pumps and their accessories apply to river pumping schemes as well. The pump should never be throttled from the suction side. A control valve may, however, be required for pumps with a positive suction head, as may sometimes be the case in booster pumps. Excessive branching and too many accessories in the delivery pipeline results in heavy head losses due to friction.

To prevent vortex movement in the sump (jack well), the intake pipe should approach the side walls of the sump radially at right angles. To prevent free cascading flow into the water surface, the intake pipe should preferably be extended by a long sweep bend to discharge below the water level. In case of round sump wells, vertical baffle boards are installed opposite the pump entrance to prevent vortex formation around the pump suction. The following are a few guidelines to be observed in the installation of pumping sets:

1. Two pumps should never be connected rigidly together.
2. The pump should not be connected rigidly to the suction or delivery pipes. These pipes should be supported independent of the pumps.
3. The prime movers (electric motors/engines) used in river pumping schemes should be connected to the pumps through flexible couplings.
4. All pump components and pipe fittings should be easily accessible. Each one could be removed without affecting the continuous operation of others.
5. All moving parts should be provided with guards to protect the operating crew.

Rising Main

The rising main is the pipeline conveying the pump discharge to the distribution chamber under hydrostatic pressure (Fig. 4.30).

Choice of pipes: RCC pipes of class P_1, P_2, P_3 and P_s (prestressed) cement concrete and cast iron pipes and PVC pipes are commonly used for the pipeline system in lift irrigation schemes. Steel pipes are usually not preferred due to corrosion and incrustation problems and high cost and weight. Under special situations like steep slopes, steel pipe could be used for limited lengths. RCC pipes are being increasingly used for rising mains in lift irrigation schemes. The class of pipe selected depends on the hydraulic pressure encountered in the pipe, the site conditions, and the possibility of water hammer effects. The adoption of these pipes shall conform to IS:458-1971 and IS:784-1956 of Bureau of Indian Standards or other relevent codes of practice. Cast iron pipes are commonly used in water supply schemes. They are durable and are not subjected to corrosion. They have good strength, low maintenance cost and can be easily tapped for making service connections due to the availability of a wide variety of pipe specials like T-s, Y-s and other tapping specials. They have the disadvantage of heavy weight, high transport cost, high cost in pipe laying and jointing, low tensile strength, liability to defects, such as blow holes, and rough internal surface, resulting in higher friction losses. They are suitable for working pressures up to 130 m and up to 1 m diameter. The durability of cast iron pipes are limited to about 30 years. They are attacked by soft peaty waters. Hard waters tend to cause incrustation of calcareous matter which offer considerable resistance to the flow of water, resulting in reduced discharge and high energy requirement. However, a lining of cement mortar protects the internal surface of the pipes from corrosion. Specifications of cast iron pipe specials are given in IS:1537-1960 and IS:1538-1976 of Bureau of Indian Standards.

Mild steel pipes are used for water distribution network in water supply systems. Steel pipes could be used for exposed sections of rising mains, inverted syphons, and on bridges and other structures where tensile strength and lighter weight are required at high pressure ratings (> 7 kg/cm^2). Due to greater strength and ductility, the wall thickness of steel pipes is much lower than that of cast iron pipes. Thus, there is saving in transportation, pipe laying and jointing costs. However, they are liable to action by acid and alkalies in water and even traces of these will produce rust and incrustation. Steel pipes are subject to corrosion than cast iron pipes. They are to be frequently painted to protect both the inner and outer surfaces. Buried steel pipes are subject to damage by corrosion and incrustation, unless they are properly protected by suitable coatings. The coatings should be resistant to scratching during transport and laying, and chemical attack. A popular lining material is bitumen (3 to 5 mm thick) or coal tar enamel (2 to 3 mm thick). For applying the lining material, the pipe is cleaned and the lining is applied by brush or spray or dipping or spinning. The recommended thickness of lining varies with the type of paint. The usual thickness is 0.33 mm, in 3 to 4 layers.

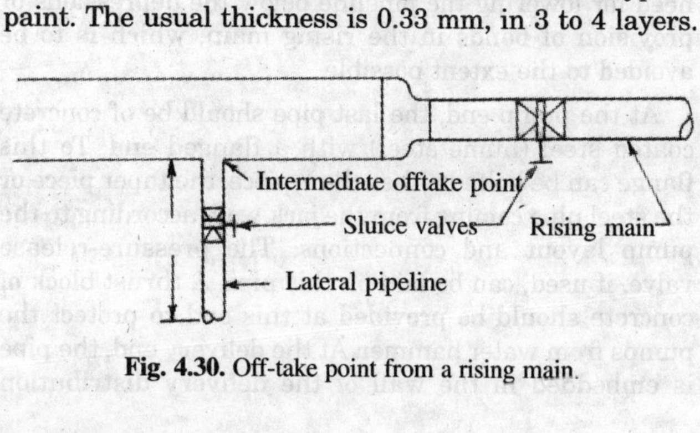

Fig. 4.30. Off-take point from a rising main.

In case of large diameter pipes, cement mortar lining, 5 to 10 mm thick, could be applied centrifugally by spinning the pipe. The maintenance cost of mild steel pipes is high and durability low. Therefore, steel pipes are seldom used for rising mains in lift irrigation systems.

Piping system manifolds at the discharge end of pumps in lift irrigation systems are made of cast iron or mild steel. Cast iron fittings are more flexible and lower in cost by 20 to 28 per cent, as compared to mild steel. Hence, cast iron pipe fittings/specials are preferred in suction and delivery pipes and the delivery manifold in lift irrigation schemes. However, they are to be provided with adequate supports.

PVC pipes of appropriate strength are increasingly being used in lift irrigation projects. Both PVC and HDPE (high density polyethylene pipes) are popular. These pipes are free from corrosion and incrustation, have smooth internal surface, and resulting low friction loss, and are light in weight. Laying and jointing of these pipes are much easier, as compared to concrete pipes. Their adoption will depend on the relative costs. The cost is usually high for use in large capacity lines. PVC pipes with Bureau of Indian Standard (BIS) specifications are available up to 60 cm diameter. A problem encountered in lift irrigation systems is when PVC/HDPE pipes are fitted to metal valves and fittings. The joints often give way due to water hammer in pipelines, unless proper protective measures are adopted.

Layout of rising main: As far as possible, the rising main should be laid on a continuously rising straight line. However, where the ground profiles do not permit laying the pipe on a straight line, bends will have to be provided in the vertical plane at the change of gradients. These bends should preferably be of mild steel and coated with concrete, if possible, or provided with a thick coating of paint. The bends should have plain ends connected to concrete blocks. The blocks are placed over the bend to withstand the thrust due to change of direction of the flowing water. At summit points, air release valves should be provided (Fig. 4.31).

The alignment of the rising main should be such that it does not traverse any depressions or high mounds. These will require either excessive cutting, due to the need for lowering the pipeline below the depressions or provision of bends in the rising main, which is to be avoided to the extent possible.

At the pump end, the last pipe should be of concrete coated steel (hume steel) with a flanged end. To this flange can be bolted either the Y-piece, the taper piece or the steel pipe coming from the jack well, according to the pump layout and connections. The pressure-release valve, if used, can be laid on this pipe. A thrust block of concrete should be provided at this end to protect the pumps from water hammer. At the delivery end, the pipe is embedded in the wall of the delivery distribution chamber and left open. The rising main should preferably be buried in the ground with a soil cover of at least 60 cm on top.

Intermediate off-take points (Fig. 4.30): The topography of the irrigated area may sometimes require one or more off-take points from the rising main. Such points should be kept to the minimum, as there is considerable pressure drop in the main line when an off-take valve is opened and water does not reach the upper delivery point. Hence, off-takes at two levels cannot be operated simultaneously. Full discharge has to be run at both the points for a short period and larger irrigation channels are required at both ends. Moreover, the pumps are required to function at two radically different heads. Therefore, the overall efficiency would be low. The range of heads over which the pumps can normally function satisfactorily are given by the manufacturers. If the heads at the end and intermediate off-take fall within this range, the loss in efficiency may not be much, otherwise the pump would be working at every low efficiency when a lower off-take is operating. In such a situation, it may be more economical to provide a separate rising main of short length to serve the lower area. Decisions on multistage pumping are to the taken on the basis of detailed economic analysis of the entire system, mainly the type of water distribution system which could be adopted.

Bifurcation of pumping mains: For diverting a part of the flow from the rising main, pipes are sometimes designed telescopically according to the lower discharge beyond the bifurcation point. In such situations, if the bifurcated pipe is closed or partly operative, the increased discharge with higher flow velocity may endanger them in case of water hammer. Hence, diversion of part of the total flow at intermediate stages in the rising main should normally be avoided.

Permissible velocity of flow in the rising main: The maximum permissible velocity of flow in the rising main ranges from 1.35 to 1.5 m/sec. The normal limit followed in the design is 1.37 m/sec. However, design velocities up to 1.50 m/sec are permissible. Design velocity and permissible friction loss in pipes are inter-related. The design criterion usually followed is to limit friction loss in concrete pipes to 3 m/100 m length of pipe. Slightly higher values could be used in case of PVC pipes. Details of computation of friction losses in pipelines and fittings, using the Darcy-Weisbach formula, is explained in Chapters 3 and 7.

Optimisation analysis: The cost of the rising main is a major component in the capital cost of lift irrigation projects. Hence, it is desirable that proper optimisation analysis is carried out on a scheme specific basis while designing lift schemes with long length of pipeline and having large command area. The optimum diameter of rising main will be that diameter at which the present value of the sum of capital cost and operation and maintenance charges is minimum. Capital cost includes cost of pipes, cost of laying the pipe and the cost of pump sets. The operation and maintenance (O&M) charges comprise of energy charges, and establishment costs pertaining to rising mains. Since the horse power of the pump sets depends upon frictional loss, energy charges are also a function of the pipe diameter. The formula for determining the present value for the purpose of optimisation analysis is given below:

Present value

$$= \text{Capital cost} + (\text{O\&M charges}) \times \frac{(1+r)^n 1}{r(1+r)^n} \quad \text{... (4.9)}$$

in which,

r = Rate of interest as applicable, and

n = life of the project (usually assumed as 25 years)

By minimising the present value, optimum diameter is estimated and results for different discharges and types of pipes are computed and compared for their relative costs. The diameter of the pipe selected should be nearest to the optimum size thus arrived at.

Optimum static head per stage of lift irrigation schemes: The total static head per stage of a lift irrigation scheme should be about 50 m. This is to limit the pressure head and to ensure that the scheme is safe against water hammer pressures that may arise due to long length of rising mains. It also enables irrigation at intermediate stages and online inspection at each stage. In case of long rising mains, a stage can be provided in-between at a convenient location. Any variation from this norm should be justified on technical, financial and operational considerations. However, too many stages would increase the operation and maintenance costs considerably.

Water Hammer Control Devices

Pressure release valves: There may be situations in lift irrigation systems to guard against water hammer which may occur due to sudden change in the velocity of flow by sudden closure of sluice valves, power failure or rupture of the pipe. Depending upon the rate and magnitude of the velocity change, water hammer can lead to high pressure fluctuations in pipelines. Knocking sound and vibrations occur in the pipeline and when their magnitudes are high the pipe might burst. However, in well designed and carefully operated lift irrigation systems, such situations are rarely encountered in the rising main. But, under unfavourable situations, it may be necessary to analyse the system behaviour and provide for safety devices such as surge pipes/tanks and relief valves. Water hammer control devices are available as compressed air vessels and zero energy air cushion valves. In irrigation systems having discharges larger than 150 litres per second and/or having a rising main of over 100 m it is highly desirable to provide a water hammer control device. Only one water hammer control device per rising main of a lift scheme is adequate. Zero energy air cushion valves are less expensive and are preferred.

Pressure relief valves can be used in the system to minimise water hammer pressures. For small pipelines a valve of the type shown in Fig. 4.31, which permits the escape of a relatively small amount of water, will alleviate the problem. Either a spring or a counter weight may be used to set the opening of the release valve. When water flowing in a closed pipe suddenly stops due to tripping of motor, the pressure in the pipeline suddenly rises much above the normal pressure head due to sudden change of kinetic energy to pressure head. This pressure which is of a dynamic nature in transmitted along the pipe and travels the entire length of the pipe in the form of pressure wave. Such instant pressure changes are called water hammer.

Automatic pressure reducers of the type shown in Fig. 4.32 are used in case of large scale systems. In this device a pilot control valve is adjusted manually for any desired pressure in the delivery side of the main valve. Whenever the delivery pressure lowers below the adjusted value, the spring above the pilot diaphragm causes the pilot valve to open and permits the water above the piston to escape through valves to the delivery side. The pilot valve opens due to the high pressure under it. Detailed information on the location of air release valves/vent pipes are given in Chapter 7.

Booster Pumps

While irrigating hilly areas, it is often desirable to provide one or more booster pumps, taking off from a delivery chamber located at an intermediate point. It is highly uneconomical and wasteful of energy to lift water to a hill top and provide field channels and drop structures to irrigate the lower areas with delivery chamber located at the top of the hill. A contour map of a hill area will reveal that the total area between the higher contours is only a small fraction of the area lying between those representing the lower elevations. Hence, for hillside irrigation, intermediate delivery chambers are to be provided, which also serve as reservoirs for locating the inlet of the booster pump. In such a system, high-discharge medium or high head pumps are required to lift water from the intermediate delivery chamber, while one or more low-discharge high-head pumps are required to lift water from the intermediate delivery chamber to the top chamber. Such an arrangement will require a much smaller size prime mover for the pumping system, as well as a smaller diameter rising main between the intermediate and top delivery chambers.

In-line boosters having the downstream section of the rising main connected directly to the suction side of a volute centrifugal pump could also be used in river pumping schemes as per requirements.

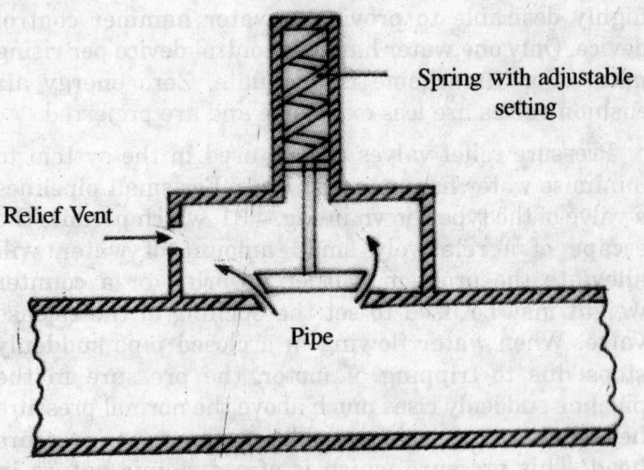

Fig. 4.31. Sketch of a pressure release valve suitable for low-head water distribution systems.

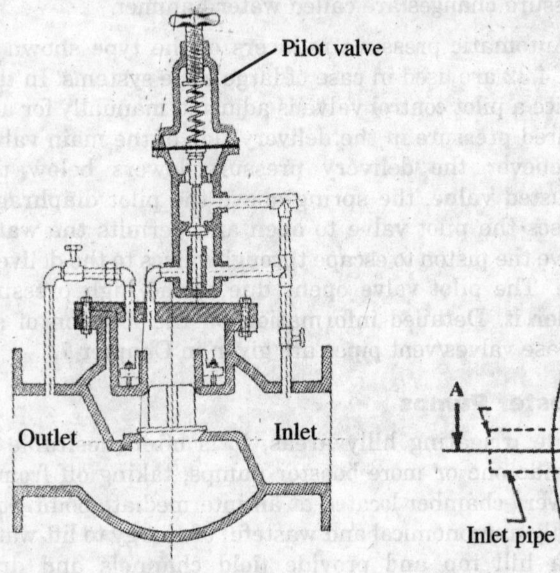

Fig. 4.32. Sectional view of an automatic pressure reducing valve in closed position suitable for high head water distribution systems.

Delivery Chamber

The rising main should discharge into a masonry or concrete delivery chamber (Fig. 4.33). When the water distribution is through underground pipelines, the delivery chamber comprises a junction box provided with control valves at the head end of the pipelines.

The size of the delivery chamber will depend upon the rate of supply of water pumped into it. A common practice is to provide the chamber adequate capacity to store 2 to 5 minutes supply. The pumping chamber may be square or rectangular.

The delivery chamber receives water from the rising main and distributes it to various chambers or distribution pipes taking off from the chamber. It is desirable to provide water distribution systems at all intermediate points when valves are provided to take off water from the rising main.

A maximum of three off-takes could be located on the 3 sides of the chamber, leaving the fourth side to locate the pumping mains. The off-takes are usually provided with rectangular notches to deliver the required quantity of water to the respective distributory channel taking off from the chamber. The length of the notch is usually kept 3 times larger than the maximum depth of water over the notch (Chapter 5). The minimum space between two pumping mains is one metre. The pumping main shall be laid with a minimum earth cover of 60 cm, up to the distribution chamber. The side walls of the chamber beyond the rectangular notch shall not be less than 2.5 times the depth of flow over the notch. The rectangular notch shall be made of 6 mm thick mild steel plate painted with two coats of water proof anti-

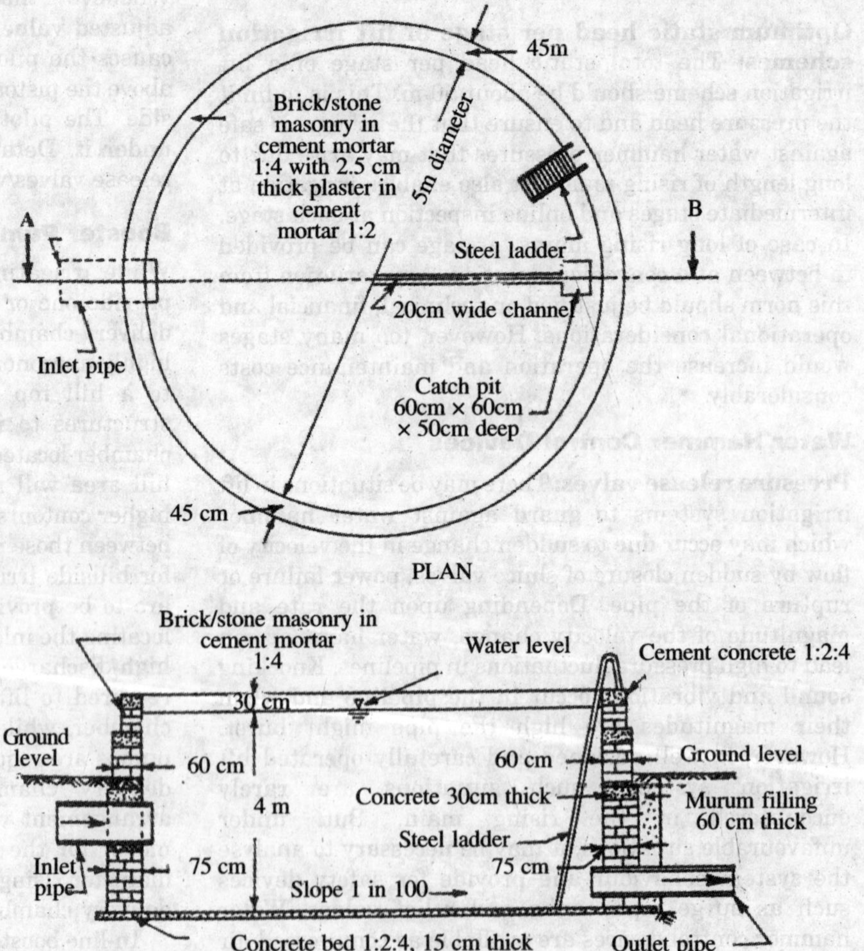

Fig. 4.33. Views of a typical distribution chamber in a lift irrigation system.

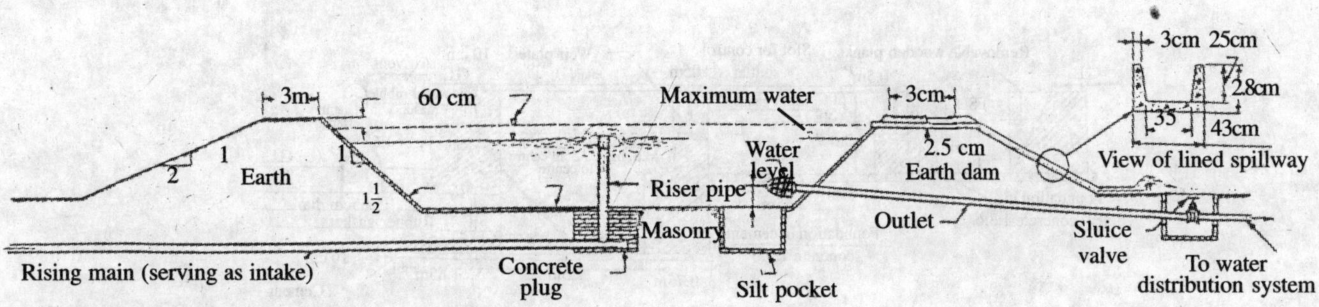

Fig. 4.34. Sectional view of a typical storage-cum-distribution tank with riser pipe outlet for a major lift irrigation project near Akola, Maharashtra (1978). Any masonry material or buried plastic film may be used to prevent seepage in the tank. The large capacity of the tank is to provide storage capacity of the water pumped during the night.

corrosive paint and firmly anchored to the side walls. The edges are made sharp by chambering from the downstream side only (Chapter 5).

Specifications of the distribution chamber: The depth of foundation shall be deeper than the zone of summer cracks in the soil of the locality. The common value of depth in black soils is 90 cm and in other soils 60 cm. The floor level of the chamber is kept below the bed level of the channel taking off from it. All channels taking off from the chamber are lined for a minimum distance of 10 m from the point of off-take.

The thickness of the walls is designed to provide safeguard against the earth pressure acting on them when the chamber is empty. When there is no earth pressure acting on the walls, the same is designed to withstand the water pressure of the chamber.

The following are the other specifications required:

(*i*) The foundation shall be of cement concrete (1:4:8).

(*ii*) The wall shall be either brick masonry or RR masonry in cement mortar (1:5), depending on the cost.

(*iii*) The base coat on the floor of the chamber shall be of cement concrete (1:2:4), 7.5 cm thick.

(*iv*) The exposed faces of the chamber shall be plastered with cement mortar (1:4), 2 cm thick for RR masonry and 1.2 cm thick for brick masonry walls.

Use of water storage tanks in lift irrigation schemes: In large scale lift irrigation schemes it may sometimes be necessary to provide for the storage of water during the days when there is no demand for irrigation and when the pumps are operated during the night, if electric power is not available during the day time or during periods of power failure (Fig. 4.34). In such situations, the rising main is taken to the tank, with its outlet above the maximum water level in the storage tank. Off-take pipes are taken from one or more points above the bottom of the tank. Suitable screens are provided to prevent silt and debris entering the off-take pipes. It is essential that storage tanks are lined with suitable lining material like bricks, stone pitching, both in cement mortar or polyethylene sheets (provided with an earth cover of about 25 cm thick) to prevent seepage. Concrete lining is usually expensive.

On-farm water distribution system for irrigation originates from the distribution chamber of a storage tank (Figs. 4.33 and 4.34). It compromises of an open channel water distribution system (Chapter 6) or underground pipeline water distribution system (Chapter 7) or a combination of both. Some of the structures used in large scale lift irrigation system are illustrated in Fig. 4.35 to Fig. 4.39.

Norms for Fixing Economic Water Rates in Lift Irrigation Schemes

Lift irrigation projects are generally undertaken in areas where water is scarce and no other alternative source of irrigation, such as groundwater, is available. As surface water availability is limited, competition for exploiting the lift irrigation potential is on the increase. In such a situation, apart from technical feasibility, the financial viability of a lift irrigation project becomes crucial. The economic water rates to be charged for recovering the capital cost, operation and management costs and other charges will have to be within the repayment capacity of beneficiaries, many of whom will be small farmers. Economic water rates are to be determined from the incremental income generated out of the cropping pattern sustained by the lift irrigation project. Fixing and recovering of capital cost and other charges by lift irrigation societies through water rates or otherwise is important (also see Chapter 15).

The cost of delivery of a unit quantity of water will have to be determined for each scheme to know the cost of irrigation. This will vary from scheme to scheme depending mostly upon the capital cost. The capital cost depends on the designed discharge, distance of command area from the lift point, and other factors. This is independent of the optimum cropping pattern which will be feasible in terms of the quantity of water allowed to be lifted under the water lifting permission.

The basic consideration is that the cost of delivery of unit quantity of water should be recovered from the beneficiaries. Therefore, their capacity to repay this cost is an important consideration in project appraisal.

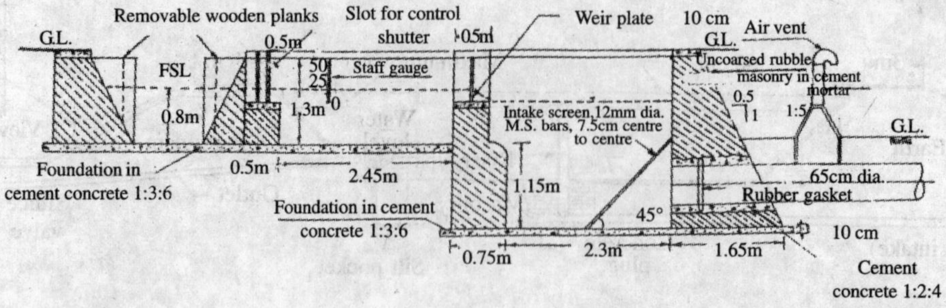

Fig. 4.35. Sectional view of an intake chamber with weir box intake structure for underground pipeline on-farm water distribution system in the Shahapur Lift Irrigation Scheme, Konkan Region, Maharashtra. (Adapted from Command Area Development Authority, Irrigation Dept. Maharashtra).

The following norms are usually followed for fixing the economic water rates of lift irrigation schemes:

(*i*) For lift irrigation schemes, the economic water rate should be estimated based on the capital cost component and the operation and maintenance and other charges required to deliver a unit volume of water.

(*ii*) The principle underlying the collection of water charges is that each beneficiary pays in full the cost of water availed by him.

(*iii*) The operation and maintenance and other charges comprising establishment charges, maintenance of works, energy and other charges should be determined realistically.

(*iv*) The capital cost should be apportioned either on the basis of the lands of beneficiaries under the command of a scheme or recovered as part of the water rates. While in the case of state owned corporations, the capital cost will have to be necessarily recovered by way of water charges, the private/cooperative sector could follow either the cropwise water rates or on the basis of the volume of water used. The operation and maintenance and other charges, will have to be recovered by way of water charges in both the cases.

(*v*) The annual liability, including operation and management and other charges, should be within

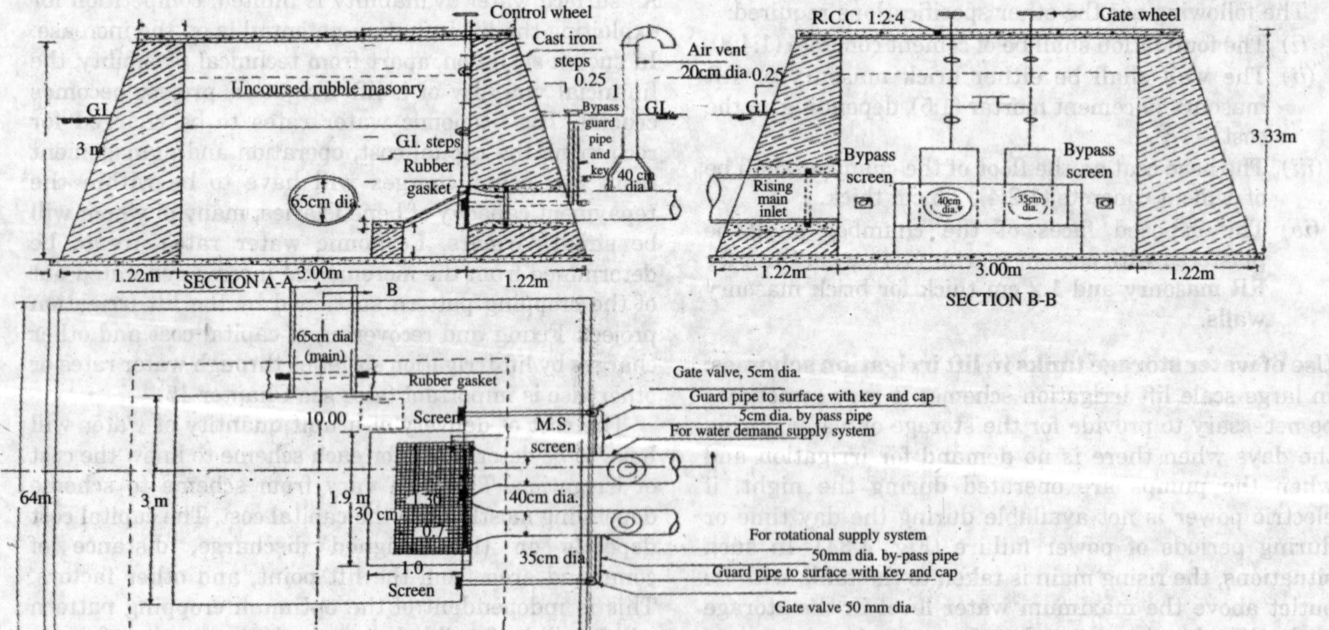

Fig. 4.36. View of an intake chamber for a twin pipe underground pipeline water distribution system in the Shahapur Lift Irrigation Project, Konkan Region, Maharashtra. (Adapted from Command Area Development Authority, Irrigation Dept., Maharashtra)

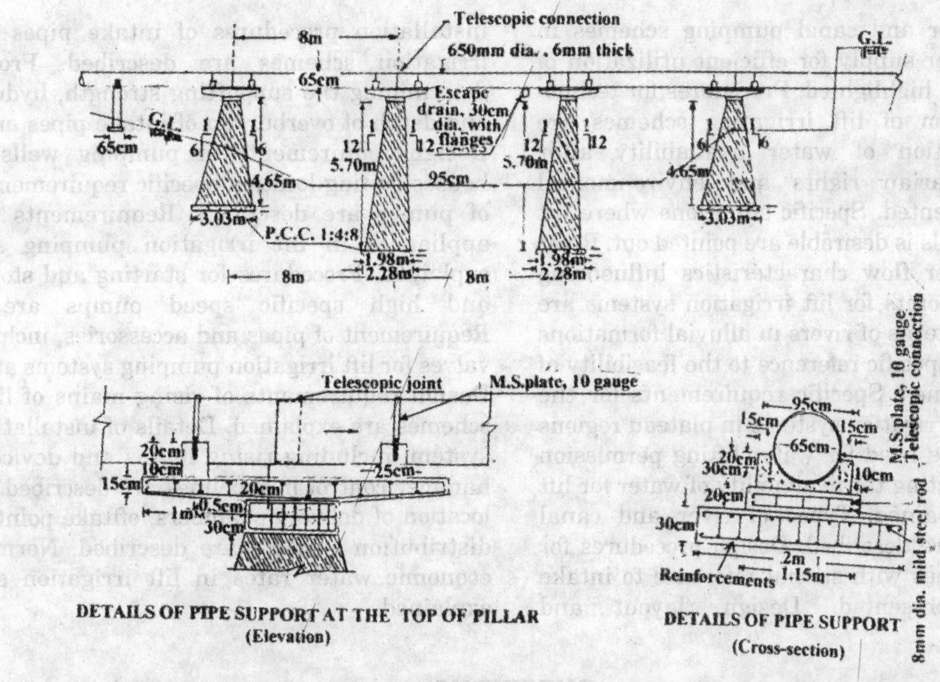

Fig. 4.37. An aqeduct for crossing a natural drain/depression in a large capacity rising main of a river lift irrigation system at Shahapur, Konkan Region, Maharashtra (Adapted from Command Area Development Authority, Irrigation Dept., Maharashtra).

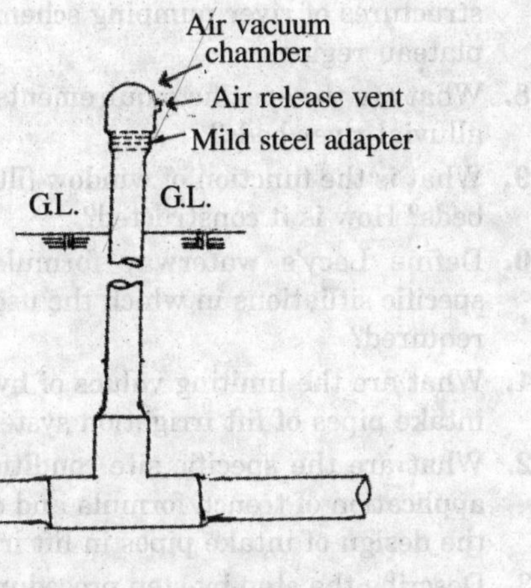

Fig. 4.38. A typical air vent type pressure release valve fixed to a pipeline.

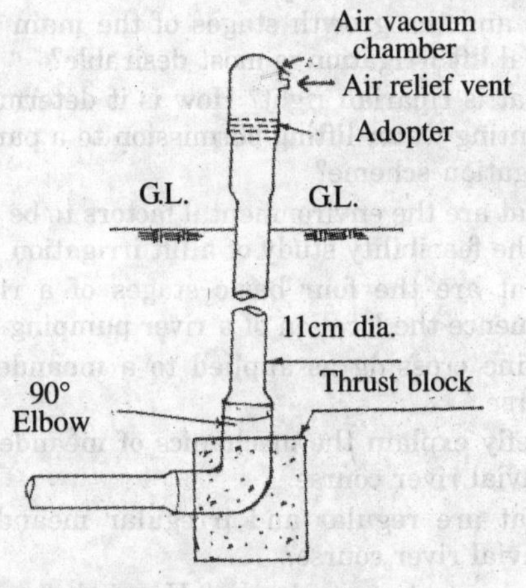

Fig. 4.39. An air vacuum type pressure release valve fixed at the end of a pipeline.

the repayment capacity of individual borrowers, usually at 50% to 75% of their net incremental income.

(*vi*) The financial liability of the project should be determined for the scheme as a whole on the basis of realistic cropping pattern for the entire command area and for the individual beneficiaries.

(*vii*) Charging of water at a flat rate power tariff results in pumping surface water far in excess of that permitted as per the cropping pattern given in the water lifting permission. This will result in substantial loss to the state.

SUMMARY

Importance of river and canal pumping schemes in irrigation and water supply for efficient utilization of water resources are highlighted. Procedures for techno-economic evaluation of lift irrigation schemes are described. Estimation of water availability after providing for riparian rights and environmental protection are presented. Specific situations where lift irrigation from canals is desirable are pointed out. River hydrology and river flow characteristics influencing location of intake points for lift irrigation systems are described. Flow patterns of rivers in alluvial formations are described with specific reference to the feasibility of lift irrigation schemes. Specific requirements for the design of river lift irrigation systems in plateau regions are pointed out. The need for water lifting permission and the factors affecting the availability of water for lift irrigation are explained. Types of river and canal pumping systems are described. Design procedures for lift irrigation schemes with special reference to intake structures are presented. Design, layout and installation procedures of intake pipes of river lift irrigation schemes are described. Procedures for determining the supporting strength, hydrostatic head and depth of overburden of intake pipes are presented. Design requirements of pumping wells and pump houses suiting location specific requirements and types of pumps are described. Requirements of electrical appliances in lift irrigation pumping stations are explained. Procedures for starting and stopping of low and high specific speed pumps are presented. Requirement of pipes and accessories, including control valves for lift irrigation pumping systems are explained. Design requirements of rising mains of lift irrigation schemes are explained. Details of installation of piping system, including rising mains and devices for water hammer control in pipelines are described. Design and location of delivery chambers, offtake points and water distribution systems are described. Norms for fixing economic water rates in lift irrigation schemes are explained.

QUESTIONS

1. What is the scope of lift irrigation from canals in relation to the topography of the adjoining irrigation command areas?
2. What are the time periods in relation to day/night use and the growth stages of the main crop when canal lift irrigation is most desirable?
3. What is riparian right? How is it determined while granting water lifting permission to a particular lift irrigation scheme?
4. What are the environmental factors to be considered in the feasibility study of a lift irrigation project?
5. What are the four basic stages of a river which influence the location of a river pumping scheme?
6. Define crossing as applied to a meandering river course.
7. Briefly explain the mechanics of meandering in an alluvial river course.
8. What are regular and irregular meanders in an alluvial river course?
9. What is tortuous meander? How is it formed?
10. Distinguish between aggrading and degrading type river flow patterns and silt deposition characteristics of rivers in alluvial plains.
11. How does an aggrading section develop in an alluvial river course?
12. What are the hydrologic processes involved in the development of deltas?
13. Define braided channels in an alluvial river course.
14. What are the major factors to be considered in planning lift irrigation from drainage canals?
15. What are notified and non-notified rivers, as they relate to water lifting permission?
16. What are the geohydrological characteristics requiring the imposition of restrictions in pumping from wells near river banks where lift irrigation projects are located?
17. What are the specific requirements of intake structures of river pumping schemes in non-alluvial plateau regions?
18. What are the specific requirements of intake wells in alluvial river beds?
19. What is the function of window filters in sandy river beds? How is it constructed?
20. Define Lacy's waterway formula. What are the specific situations in which the use of the formula is required?
21. What are the limiting values of hydrostatic head on intake pipes of lift irrigation systems?
22. What are the specific site conditions requiring the application of trench formula and culvert formula in the design of intake pipes in lift irrigation systems?
23. Describe the step-by-step procedure in the design of intake pipes in lift irrigation schemes?
24. Under what situations is an open channel intake preferred over a pipe inlet in a lift irrigation scheme?
25. What are the design requirements of the sump well in a lift scheme?
26. How is the diameter of a sump well determined?
27. What are the functions of a jack well in a lift irrigation scheme?
28. How are the design dimensions of pump houses for installing horizontal centrifugal pumps and vertically driven turbine type pumps determined?
29. What are the specific requirements of foundation blocks of pump sets used in lift irrigation projects?
30. What is the function of bed blocks in jack wells?

31. What are the functions of baffles and diaphragms in a pump chamber?

32. What is the function of single phase preventer in the installation of motor-driven pumping sets?

33. What is water hammer? How is it controlled in the pipeline of lift irrigation schemes?

34. How are the specifications of the distribution chamber of lift irrigation schemes determined?

35. What are the usual norms for determining economic water rates in a lift irrigation project?

36. Identify the true and false statements from the following:

(*i*) River lift irrigation systems form an integral part of a Kolhapur type pick-up weir.

(*ii*) In assessing the technical feasibility of a lift irrigation project, the average flow of the river during the period of irrigation demand in the command area is considered in estimating the amount of water, for which water lifting permission is granted.

(*iii*) Canal lift irrigation is an effective means of beneficial utilization of excess canal flow during the night, in regions where rotation irrigation system (*warabandi*) is not feasible.

(*iv*) Canal pumping schemes are economical in areas where the lift command area is close to the canal and the pumping head is low.

(*v*) Environmental considerations are significant in assessing the technical feasibility of both river and canal lift irrigation schemes.

(vi) In the boulder stage of a river, the flow is confined between non-submersible banks on either side.

(*vii*) When the sediment load of a river is blocked by a dam upstream on the river course, a degrading reach may develop downstream.

(*viii*) In rivers, in rocky and boulder stages, rock outcrops are not safe to locate intake points of lift irrigation schemes.

(*ix*) Oil-immersed type star-delta starters are used for electric motors of capacity of 50 h.p. or more.

(*x*) In alluvial river courses, the pattern of meander changes whenever there is a change in river discharge or the type of material transported by the stream.

(*xi*) In an intake well the entry of water into the well is usually limited to the upstream section only.

(*xii*) A catch pit for silt deposition is an integral component of an intake well.

(*xiii*) In the construction of window filters for intake wells in sandy river beds, using sandwiched graded gravel fill between the inner and outer cover plates, the coarsest grade of gravel is laid at the outer periphery and finer gravel at the inner.

(*xiv*) The commonly adopted capacity of a delivery chamber in a lift irrigation system is to hold 2 to 5 minutes supply of the line on which the chamber is located.

37. Identify the most appropriate answer to the following questions from the four options:

(*i*) In the meandering course of a river in alluvial formation, the intake of a lift scheme is to be located on:

(a) concave side of the river
(b) convex side of the river
(c) at a point where the water flow is deep
(d) on a straight length of the river

(*ii*) Intake structures for lift irrigation from rivers in delta regions generally comprise of:

(a) portable pumping sets
(b) floating barges and their adaptations
(c) specially designed intake structures suiting the river flow stage
(d) a combination of (b) and (c).

(*iii*) The most important requirement in the installation of RCC intake pipe is:

(a) providing the required bed slope in the pipeline
(b) ensuring watertight pipe joints
(c) provision of man-holes where the length of the pipeline exceeds about 100 m
(d) provision of suitable concrete cradles as bottom support for the pipes.

(*iv*) The commonly used limit of suction lift of horizontal centrifugal pumps used in lift irrigation schemes is:

(a) 4 m (b) 4.5 m
(c) 5 m (d) 5.5 m

(*v*) The usually adopted value of retention period of inflow into a sump well is:

(a) 3 min. (b) 5 min.
(c) 7 min. (d) 9 min.

(*vi*) The minimum submergence of the pump bowl assembly of a turbine / submersible / mixed flow pump in a jack well is:

(a) 1 m (b) 2 m
(c) 3 m (d) 4 m

(*vii*) The minimum clearance between the bottom of a vertically driven turbine type pump and the floor of the jack well is:

(a) 1 m (b) 1.5 m
(c) 2 m (d) 2.5 m

(*viii*) The minimum spacing between each pumping unit in the power range of 30 to 60 h.p. is:

(a) 60 cm (b) 70 cm
(c) 80 cm (d) 90 cm

(*ix*) The common type of pipe used for rising main in lift irrigation scheme is:

(a) PVC (b) Cast iron
(c) RCC (d) Mild steel

(*x*) Pipe specials commonly used in lift irrigation schemes are generally made of:

(a) Cast iron (b) Mild steel
(c) PVC (d) RCC

(*xi*) The total static head per stage of a lift irrigation system is limited to about:

(a) 30 m (b) 40 m
(c) 50 m (d) 60 m

(*xii*) The diameter of the intake pipe of a lift scheme flowing partially full is computed using:

(a) Darcy Weisbach formula
(b) William Razen formula
(c) Manning's formula
(d) Chezy's formula

38. Arrange the following in proper sequences:

(*i*) The various stages in the flow of a river course:

(a) Boulder stage (b) Rocky stage
(c) Delta stage (d) Trough stage

(*ii*) Mechanics of the initial stage of meandering in an alluvial river course:

(a) local bank erosion,
(b) scouring of river bed on one side and deposition of silt on the other until an equilibrium is attained,
(c) local overloading and deposition of heavier sediments in the river bed,
(d) development of uneven river bed.

(*iii*) The various events in the meandering process of an alluvial river course are:

(a) deployment of rapid current on the concave side and slower flow on the convex,
(b) erosion of the river bank that is more friable than the other,
(c) accentuation of curvature by increased erosion on the concave side and silt deposition on the convex,
(d) repetition of formation of concave and convex reaches.

(*iv*) The various types of valves installed in a lift irrigation system, commencing from the pumping point are:

(a) sluice valve
(b) sluice valve
(c) foot valve
(d) intermediate take-off valve.

(*v*) The components of the piping system in a lift irrigation scheme, commencing from the discharge point of the pumping system are:

(a) sluice valve (b) reflux valve
(c) delivery chamber (d) rising main.

39. In the following, match each of the entries of the 1st column with the most closely related entry in the 2nd column:

(*i*)

(A) Maximum utilization of lift irrigation potential amongst the states in India	(I) Rivers flowing on steep slopes on rocky beds and sides
(B) Riparian rights	(II) Rivers in mid reaches passing through boulders and gravel
(C) Rocky stage	(III) Rivers flowing through alluvial formations
(D) Boulder stage	(IV) Right for the continued use of the flow downstream
(E) Trough stage	(V) Maharashtra

(*ii*)

(A) Delta formation	(I) River training works
(B) Crossing	(II) Sand bed channels with high bed loads
(C) Degrading reach	(III) Repeated meandering patterns in a river course
(D) Irregular meander	(IV) Deposition of large quantities of sediment as the river approaches the sea or ocean
(E) Tortuous meander	(V) Straight reaches in a meandering river course

(*iii*)

(A) Floating pumping stations	(I) Intake wells in sandy river beds
(B) Flexible type delivery pipe	(II) High head vertical turbine pumps or submersible pumps
(C) Window filters	(III) Silt collection at the bottom of the intake pipe
(D) Jack well	(IV) Floating barge
(E) Catch pit	(V) Rivers with highly fluctuating water levels

(*iv*)

(A) Rising main	(I) Pipeline system in lift irrigation
(B) Intermediate off-take points	(II) Take-off point of on-farm water distribution in lift irrigation
(C) Pressure release valve	(III) Large-scale water storage in lift irrigation schemes
(D) Delivery chamber	(IV) Water distribution in lift irrigation system
(E) Storage-cum-distribution tank	(V) Water conveyance in lift irrigation water distribution systems

FURTHER READING REFERENCES

Agricultural Finance Corporation Limited, Mumbai (1981). *Appraisal of Faridkhanwadi Lift Irrigation Project,* Ugarkhurd, Belgaum District, Karnataka, p. 146.

Agricultural Finance Corporation, Mumbai. (1981). *Rehabilitation of Mahatma Gandhi Lift Irrigation Project,* Gaddipalli, Nalgonda District, Andhra Pradesh, Main Report & Appendix 1.

Andhra Pradesh State Irrigation Development Corporation Ltd. (1980). *Type Designs and Standards for Lift Irrigation Schemes,* Hyderabad, p. 184.

Bureau of Indian Standards (1959). *Code of Practice for Laying Concrete Pipes*: IS:783–1959. BS9, Manak Bhawan, Bahadur Shah Zafar Marg, New Delhi.

Bureau of Indian Standards (1971). IS:458–1971, *Specification of Concrete Pipes (with and without reinforcement):* BS9, New Delhi, Revised Version, IS: 458–1978.

Central Ground Water Board (1992). *Ground Water Statistics.* Ministry of Water Resources, Govt. of India, New Delhi.

Central Water Commission (2000). *Water Related Statistics.* Ministry of Water Resources, Govt. of India, New Delhi.

Design Bureau for Hydramtic Planning of the National Water Authority (1966). *Project Report for River Lift Irrigation Systems in India.* Budapest, Hungary, p. 8.

Ellis, W.H. (1963). *Irrigation.* College of Engineering Manual, Govt. of Madras, p. 455.

Government of Maharashtra, Law and Judiciary Department (1977). *The Maharashtra Irrigation Act, 1976.* Printed at the Government Central Press, Mumbai, pp. 9621-9673.

Government of Maharashtra (1980). *A Note on Preparation of Lift Irrigation Schemes.* Dept. of Irrigation, Mumbai, Mimeo, p. 48.

Kundu, P.S. (1981). *Requirement of Lift Irrigation Schemes,* Mimeo, p. 10.

Madhya Pradesh Lift Irrigation Corporation Limited Bhopal (1981). *Soyat-Khurd Lift Irrigation Scheme,* District Shajapur, Madhya Pradesh.

Maria Saleth, R. (1996). *Water Institutions in India, Economics, Law, and Policy.* Commonwealth Publishers, New Delhi for Institute of Economic Growth, Delhi, p. 299.

Michael, A.M. and Khepar, S.D. (1989). *Water Well and Pump Engineering*, Pub. Tata McGraw Hill Publishing Company Ltd., West Patel Nagar, New Delhi, p. 929.

Minor Irrigation Dept., Govt. of Bihar (1978). *Outlines of Lift Irrigation Schemes in Chhotanagpur and Santhal Pargana,* Mineo, p. 7.

Ministry of Irrigation, Govt. of India (1981). *Development of Irrigation in India Through Surface Water Minor Irrigation Schemes,* Minor Irrigation Division, New Delhi, p. 23.

National Bank for Agriculture and Rural Development (1989). *Technical Aspects of Agricultural Projects.* Vol. I. NABARD, Shrineketan, Dr. Annie Basant Road, Worli, Mumbai, pp. 158–214.

National Bank for Agriculture and Rural Development (1995). *Lift Irrigation Schemes in Maharashtra–Tecbino-Economic Evaluation.* Study No. 15, Department of Economic Analysis and Research, Mumbai, p. 129.

National Bank for Agriculture and Rural Development, (1991). *Report of the Working Group on Lift Irrigation Schemes.* Mumbai, p. 90.

National Sample Survey Organization (1985). *Sarvekshana,* Vol. 9, No. 1.

Varshney, R.S., Gupta, S.C. and Gupta R.C. (1979). *Theory and Design of Irrigation Structures I.* Pub. Nem Chand Brothers, Roorkee, p. 452.

Wagner, E.G. and Lanoia, J.N. (1969). Water *Supply for Rural Areas and Small Communities.* World Health Organization, Geneva, p. 340.

Waterhouse, J. (1982). *Water Engineering for Agriculture.* Pub. Batsford Academic and Educational Ltd., London.

World Health Organization Region for South-East Asi, New Delhi (1976). *Typical Designs for Engineering Components in Rural Water Supply,* p. 49.

Chapter

5

Measurement of Irrigation Water

LEARNING OBJECTIVES

Importance of measurement of irrigation water. Units of measurement. Flow measurement in streams and channels. Methods of measuring flow through pipes—water meters and venturi meters. On-farm water measuring devices for field channels—weirs, Parshall flumes, Dethridge meter, orifices and meter gates. Specific requirements in the use of weirs in farm water measurement—rectangular weirs, Cipoletti weir and triangular weir. Parshall flumes, and cut-throat flumes, and their application. Flow measurement under free flow and submerged flow conditions in flumes and orifices. Water measurement structures for varying flows in earth channels in canal water distribution systems—broad-crested rectangular weirs, truncated flumes and triangular throated flumes. Tracer methods of flow measurement. Principles of application of radio-isotopes in the measurement of flow.

Water is the most valuable asset of irrigated agriculture. Accurate measurement of irrigation water permits intelligent use of this valuable natural resource. Such a measurement reduces excessive waste and allows the water to be distributed among users according to their needs and rights. Systematic water measurements properly recorded, interpreted and used constitute the foundation upon which increasing efficiencies of water conveyance, application and use must be based. Accuracy in water measurement is, therefore, of prime importance in the operation of any water distribution system.

Units of Measurement of Water

Water is measured under two conditions—at rest and in motion. Water at rest—that is, in reservoirs, ponds, tanks, soil—is measured in units of volume such as litre, cubic metre, hectare-centimetre, hectare-metre. Measurement of water in motion, that is, flowing in rivers, canals, pipelines, field channels, and channel structures—is expressed in rate of flow units such as litres per second, litres per hour, cubic metres per second, hectare-centimetres per hour, and hectare-metres per day.

Litre. A volume equal to one cubic decimetre or 1/1000 cubic metre.

Cubic metre. A volume equal to that of a cube 1 metre long, 1 metre wide and 1 metre deep (1 cubic metre = 1000 litres).

Hectare-centimetre. A volume necessary to cover an area of 1 hectare (10,000 sq m) up to a depth of 1 centimetre (1 hectare-centimetre = 100 cubic metres = 100,000 litres).

Hectare-metre. A volume necessary to cover an area of 1 hectare up to a depth of 1 metre (1 hectare-metre = 10,000 cubic metres = 10 million litres).

Litre per second. A continuous flow amounting to 1 litre passing through a point each second (generally used to denote the discharge of a pump, small stream or pipeline).

Cubic metre per second. A flow of water equivalent to a stream 1 metre wide and 1 metre deep, flowing at a velocity of 1 metre per second.

Example 5.1. *A Persian wheel with an average discharge of 230 litres per minute irrigates 1 hectare wheat crop in 50 hours. What is the average depth of irrigation?*

Solution: Discharge in 50 hours

$$= 230 \times 60 \times 50 = 6{,}90{,}000 \text{ litres} = 690 \text{ m}^3$$

Depth of irrigation, cm

$$= \frac{\text{Volume of water, m}^3}{\text{Area of land, m}^2} \times 100$$

$$= \frac{690}{10{,}000} \times 100 = 6.9 \text{ cm}$$

Example 5.2. *Wheat crop requires 45 cm of irrigation water during 120 days irrigating period. How much land can be irrigated with a flow of 20 litres per second for 22 hours a day?*

Solution: Total discharge during the irrigation period

$$= \frac{20 \times 60 \times 60 \times 22 \times 120}{1000} = 1{,}90{,}080 \text{ m}^3$$

Irrigation requirement per hectare

$$= \frac{45}{100} \times 10{,}000 = 4{,}500 \text{ m}^3$$

$$\text{Area Irrigated} = \frac{\text{Volume of water available, m}^3}{\text{Volume of water requ. per hectare, m}^3}$$

$$= \frac{190080}{4500} = 42.24 \text{ hectares.}$$

METHODS OF WATER MEASUREMENT

Several devices are used for measuring irrigation water on the farm. They can be grouped into four categories: (*i*) volumetric measurements, (*ii*) velocity — area methods, (*iii*) measuring structures (orifices, weirs and flumes), and (*iv*) tracer methods (dilution).

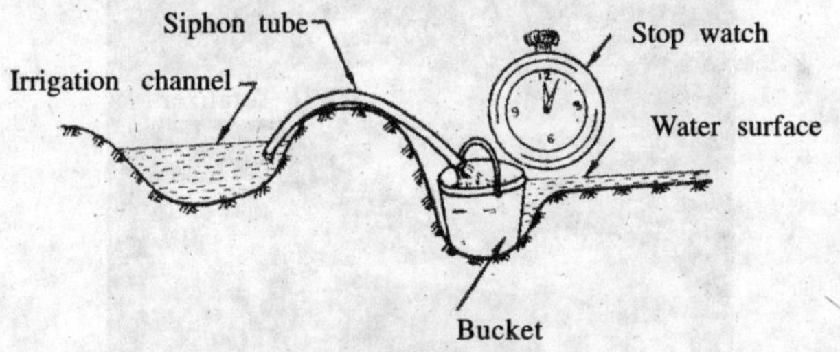

Fig. 5.1. Field installation for volumetric measurement of small flows.

Volume method of water measurement. A simple method of measuring small irrigation streams is to collect the flow in a container of known volume for a measured period of time (Fig. 5.1). An ordinary bucket or barrel is used as the container. The time required to fill the container is reckoned with a stop watch or the second's hand of a wrist watch. The rate of flow is measured by the formula

Discharge rate, litres/sec

$$= \frac{\text{Volume of container, litres}}{\text{Time required to fill, seconds}}$$

For example, if a 24 litre capacity bucket is filled in 12 seconds by the discharge from a Persian wheel, then the rate of flow is 2 litres per second or 120 litres per minute.

The method can be used to determine the discharge rate of pumps and other water lifts like Persian wheel and leather bucket lift if a barrel of 150 to 200 litres capacity is used to collect the flow. The method is also suitable when the stream is discharged into a reservoir of sufficient capacity. The volume can be calculated by direct measurements with a tape. The flow is determined by noting the time required for the reservoir to fill to a certain depth, or for the water surface to rise from one level to another.

Velocity-Area Methods

The rate of flow passing a point in a pipe or open channel is determined by multiplying the cross-sectional area of the flow section at right angles to the direction of flow by the average velocity of water.

$$\text{Discharge} = \text{Area} \times \text{Velocity}$$

$$Q = a \times v \qquad \text{... (5.1)}$$

in which, Q = discharge rate (m^3/sec)

a = area of cross-section of channel or pipe (m^2)

v = velocity of flow (m/sec)

The cross-sectional area is determined by direct measurements. The velocity is generally measured with a current metre. Approximate values of rate of flow may also be obtained by the float method.

(1) Float method. The float method of making a rough estimate of the flow in a channel consists of noting the rate of movement of a floating body. A long-necked bottle partly filled with water or a block of wood may be used as the float. A straight section of the channel about 30 metres long with fairly uniform cross-section is selected. Several measurements of depth and width are made within the trial section to arrive at the average cross-sectional area. A string is stretched across each end of the section at right angles to the direction of flow. The float is placed in the channel, a short distance upstream from the trial section. The time the float needs to pass from the upper to the lower section is recorded. Several trials are made to get the average time of travel.

To determine the velocity of water at the surface of the channel, the length of the trial section is divided by the average time taken by the float to cross it. Since the velocity of the float on the surface of the water will be greater than the average velocity of the stream, it is necessary to correct the measurement by multiplying it with a constant factor which is usually assumed to be 0.85. To obtain the rate of flow, this average velocity (measured velocity × coefficient) is multiplied by the average cross-sectional area of the stream.

(2) Current meter method. The velocity of water in a stream or river may be measured directly with a current meter and the discharge estimated by multiplying the mean velocity of water by the area of cross-section of the stream. The current meter is a small instrument containing a revolving wheel or vane that is turned by the movement of water

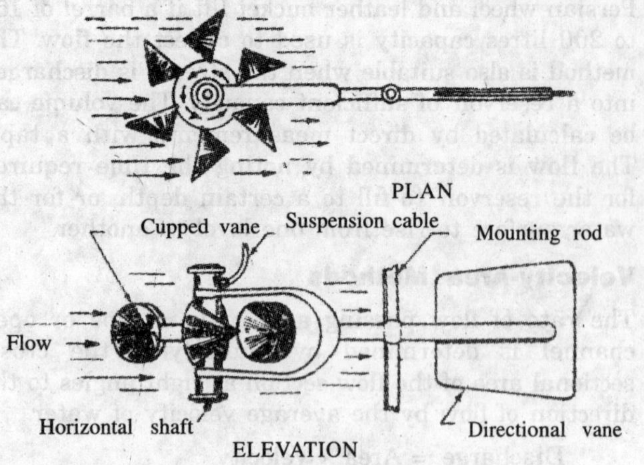

Fig. 5.2. A cup type current meter for flow measurement in streams/rivers/canals. Note the rod mounting and cable suspension.

(Fig. 5.2). It may be suspended by a cable for measurements in deep streams or attached to a rod in shallow streams. The propeller is suspended facing the centre of the flow in the stream (or the pipe). The propeller is rotated by the flow of water. The speed of the propeller (r.p.m.) is proportional to the average velocity of flow in the conduit. The number of revolutions of the wheel in a given time interval is obtained and the corresponding velocity is reckoned from a calibration table or graph of the instrument.

Current meter measurements in canals and streams are generally made at metering bridges, at cableways, or at other structures giving convenient access to the stream. The channel at the measuring section should be straight, with a fairly regular cross-section. Structures with piers in the channel are avoided as far as possible.

When the mean velocity of a stream is determined with a current meter, the cross-section of the flow is divided into a number of sub-areas and separate measurements are made for each sub-area. The width of the sub-areas may vary from 1 m to 6 m, depending on the size of the stream and the precision desired. It has been found that the average of readings taken at 0.2 and 0.8 of the depth below the surface is a fairly good estimate of the mean velocity in a vertical plane.

Both the float and the current meter methods, however, have very limited application in farm irrigation practice.

Flow through Pipelines

Water meters. Water meters utilise a multi-blade conical propeller made of metal, plastic or rubber, rotating in a vertical plane and geared to a totalizer in such a manner that a numerical counter can totalise the flow in any desired volumetric units (Figs. 5.3 to 5.7). Water meters are available for a range of sizes suiting the pipe sizes commonly used on the farm.

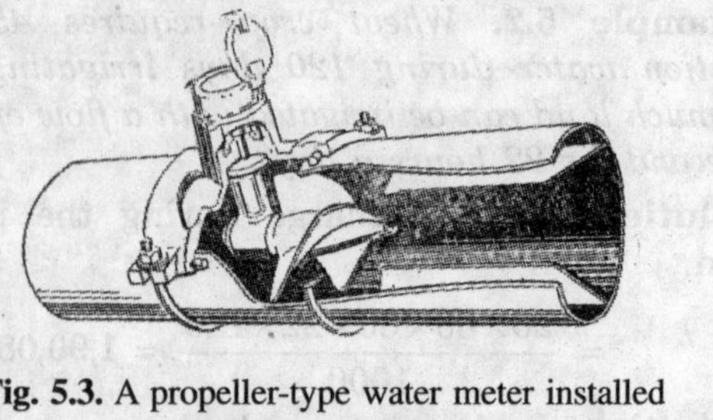

Fig. 5.3. A propeller-type water meter installed in a pipeline. Note the flow straightening vanes installed in the pipeline to eliminate turbulance.

(***Source:*** Scott, V.H. and Houston, C.E. (1959), Uni. of California Exp. Stn. Cir. 473:, USDA, SCS, *National Engg. Handbook,* Section 15.)

There are two basic requirements for accurate operation of the water meter: (1) The pipe must flow full at all times, and (2) the rate of flow must exceed the minimum for the rated range. Meters are calibrated in the factory and field adjustments are usually not required. When water meters are installed in open channels, the flow must be brought through a pipe of known cross-sectional area. Care must be taken that no debris or other foreign material obstruct the propeller.

Fig. 5.4. A water meter with internal parts exposed.

Fig. 5.5. Water meter installed to measure the flow through a portable aluminium pipeline.

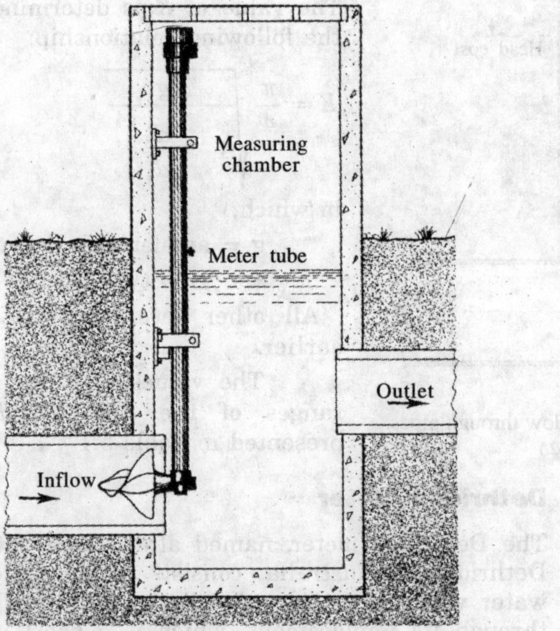

Fig. 5.6. An open-flow type water meter installed to measure the flow in an underground pipeline water distribution system.
(Source: USDA, SCS, National Engg. Handbook, Ch. 15, 1962.)

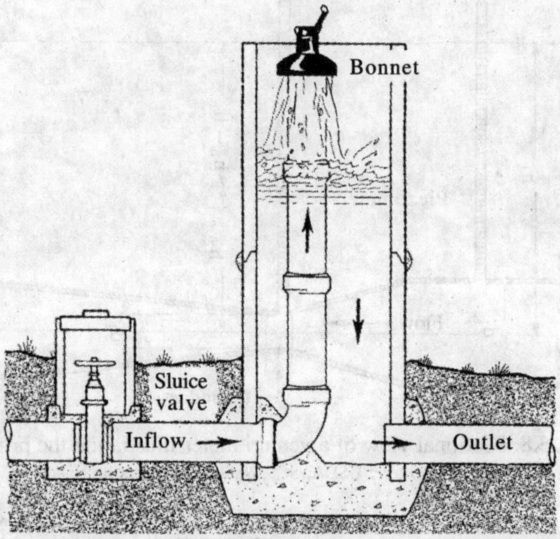

Fig. 5.7. Typical installation of a vertical flow water meter in a pipeline.
(Source: USDA, SCS, National Engg. Handbook, Sec. 15, 1962.)

Three basic types of water metres are: (*i*) Low-pressure line meters, (*ii*) Open-flow meters and (*iii*) Vertical-flow or hydrant-type meters.

Low-pressure line meters. They are used in underground pipeline water distribution systems (Fig. 5.6). The principal use of the low pressure line meter is to measure the flow delivered to individual farms.

Open-flow meters. They are similar to low pressure-line meters and are used to measure the flow in open channels or gravity-flow in closed-conduit systems (Fig. 5.6). The meter is suspended from a wall or other support structure into the centre of a full flowing submerged discharge end of the pipe, culvert, or siphon which serves as the meter tube. Open-tube meters may be installed permanently or may be portable. They do not interrupt the normal flow of water. Smaller size meters are used to measure the flow at farm turnouts (Fig. 5.6). Larger sizes are used to measure large volumes of flow like those in canals.

Vertical-flow meters. They are used to measure the flow in vertical pipes (Fig. 5.7). They are identical to the other two types, except that the propeller is vertical. Water in the riser pipe, flowing unrestricted in an upward direction through the meter tube, activates the propeller. The flow is then deflected downwards by a bonnet or cover to the outside of the pipe. The bonnet is designed to provide adequate area to permit the maximum flow. It also prevents tampering with the propeller. Vertical-flow meters are used to measure the flow to individual large farms or groups of small farms through pipe turn-outs or underground pipeline systems. They are usually of portable type and can be used on any number of hydrants or pipe turn-outs. They can be placed in the open end of any vertical riser-type discharge pipe.

Water meters are usually costly devices to measure irrigation water in surface water distribution systems. Their application is usually limited to measure the flow in pipelines.

Venturi meters. Venturi meters measure the flow of water in pipes under pressure. They work on the venturi principle in that the flow passing through a constricted section of a pipe is accelerated and its pressure head is lowered. By determining the cross-sectional areas of the pipe and the constricted section as well as the pressure drop between the two sections, the quantity of water passing the pipe can be computed.

Figure 5.8 is the sectional view of a venturi meter showing the principle of flow measurement. The meter is inserted between two flanges of a pipeline. The inside diameter of the meter at its upstream end is the same as that of the pipeline. A hole is drilled near the upstream end of the meter to connect a piezometer tube for measuring the static pressure (head). The upstream end is short and straight. There is a conical entrance section after the short upstream end. The entrance section leads to a short constricted section or cylindrical throat. The entrance cone has an angle of 21°. The cylindrical throat section is provided with a side hole for fixing a piezometer tube. The diameter of the throat is about one-fourth to one-half the diameter of the pipeline. A conical exit section with an angle of 5° to 7° follows the throat section ends at the flange in the pipeline. The downstream section is longer than the upstream section.

The rate of flow through a venturi meter is computed in metric/FPS system of units from the following relationship, using consistent units:

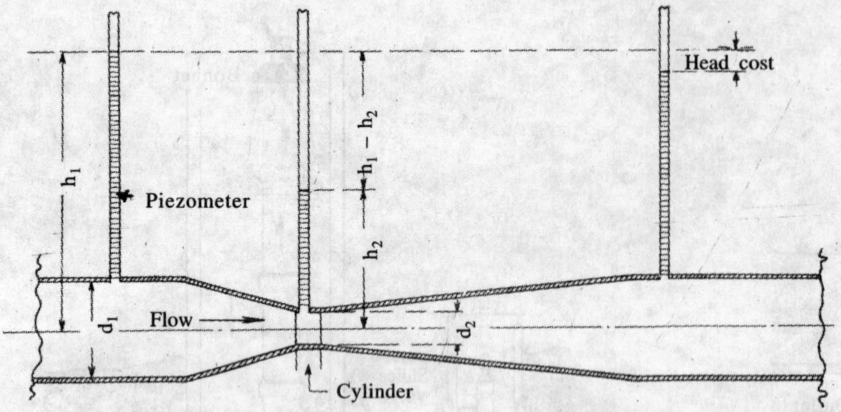

Fig. 5.8. Sectional view of a venturi meter illustrating the principle of flow through pipes.
(***Source:*** USDA, SCS, National Engineering Handbook, Sec. 15, 1962.)

$$Q = CKd\sqrt{h_1 - h_2} \quad \text{... (5.2)}$$

in which,

Q = rate of flow, L^3/T.

C = discharge coefficient (ranging in values from 0.935 to 0.988).

(The lowest values are for pipes of 2.5 cm dia. and the highest for that of 45 cm dia. The values also increase with the increase in the velocity of flow in the throat section. However, the value of C is usually almost equal to unity and this value is often used in computations).

K = a factor corresponding to the ratio of the throat diameter to the diameter of the entrance section (d_2/d_1).

d_1 = diameter of the entrance section, L

d_2 = diameter of the throat section, L

h_1 = pressure head measured above the axis of the meter, L

h_2 = pressure head measured above the axis of the meter at the throat, L

The value of K is determined from the following relationship:

$$K = \frac{\pi}{4}\sqrt{\frac{2g}{1-\left(\frac{d_2}{d_1}\right)^4}} \quad \text{... (5.3)}$$

in which,

g = acceleration due to gravity, L/T^2

All other terms are as defined earlier.

The values of K for varying ranges of the ratio of d_2/d_1 is presented in Table 5.1.

Dethridge Meter

The Dethridge meter, named after its inventor, J.S. Dethridge of Australia, consists of an under shot water wheel turned by the discharge water passing through its implacement, which is a short concrete outlet specially formed to provide only the minimum practicable clearance of the lower half of the wheel at its sides and around its circumference. (Figs. 5.9 and 5.10). The meter gives a positive measurement of discharge, recorded directly in volumetric units by a revolution counter driven by the wheel axle. Under normal working conditions, the accuracy of measurement is within five per cent.

Two standard sizes of meters are in common use: 1.5 m diameter 'large' meter for discharges from 45 to 150 litres per second and the 'small' 1.2 m diameter for discharges from 30 to 75 litres per second.

The main advantages of the Dethridge meter are that it operates with small head loss; it is simple and robust in construction, and it will pass ordinary floating debris without damage to or stoppage of the wheel.

The meter, though widely used in Australia to measure the quantity of water diverted from supply

Table 5.1. Values of K in Equation 5.3 for venturi meters

$\frac{d_2}{d_1}$	K	$\frac{d_2}{d_1}$	K	$\frac{d_2}{d_1}$	K	$\frac{d_2}{d_1}$	K	$\frac{d_2}{d_1}$	K
.20	6.31	.33	6.34	.46	6.45	.59	6.72	.72	7.37
.21	6.31	.34	6.34	.47	6.46	.60	6.75	.73	7.45
.22	6.31	.35	6.35	.48	6.47	.61	6.79	.74	7.53
.23	6.31	.36	6.35	.49	6.49	.62	6.82	.75	7.62
.24	6.31	.37	6.36	.50	6.51	.63	6.86	.76	7.72
.25	6.31	.38	6.37	.51	6.52	.64	6.91	.77	7.82
.26	6.31	.39	6.37	.52	6.54	.65	6.95	.78	7.94
.27	6.32	.40	6.38	.53	6.56	.66	7.00	.79	8.06
.28	6.32	.41	6.39	.54	6.59	.67	7.05	.80	8.20
.29	6.32	.42	6.40	.55	6.61	.68	7.11	.81	8.35
.30	6.32	.43	6.41	.56	6.64	.69	7.17	.82	8.51
.31	6.32	.44	6.42	.57	6.66	.70	7.23	.83	8.69
.32	6.32	.45	6.43	.58	6.69	.71	7.30	.84	8.89

(***Source:*** USDA, SCS, National Engg. Handbook, Sec. 15, 1962.)

Fig. 5.9. A Dethridge wheel installed at the inlet of a branch water course.

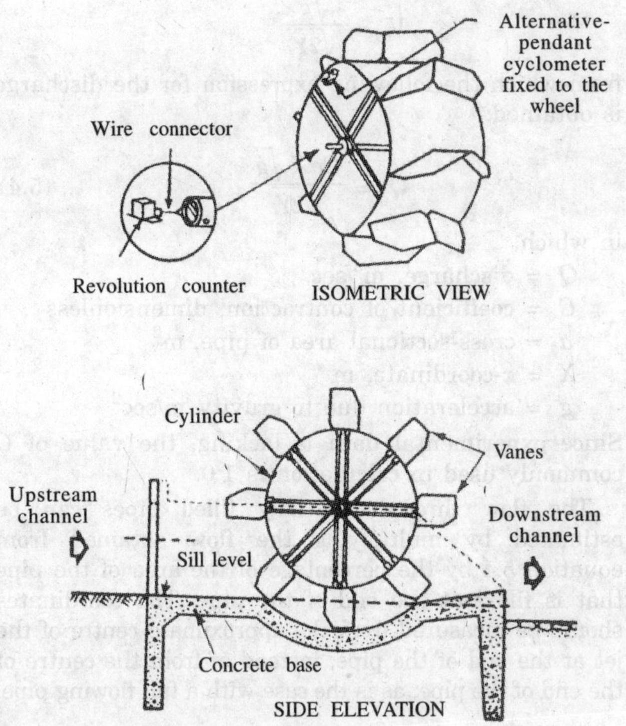

Fig. 5.10. Views of Dethridge meter showing parts details and installation.

channels (canal minors) to water courses and field channels, has not yet been introduced in other countries to any great extent.

Coordinate Method of Measuring Discharge from Pipes

Coordinate methods are used to estimate discharge from flowing wells discharging vertically or from small pumping plants discharging horizontally. These methods have limited accuracy owing to the difficulty in making accurate measurements of the coordinates of the jet. They are used only where facilities for making more accurate measurements by other methods are not available. Since estimates of discharge from vertical pipes have very limited practical application on the farm, the discussion in this text is confined to horizontal pipes.

For pipes discharging horizontally, it is necessary to measure both a horizontal and a vertical distance from some point at the end of the pipe to a similar point in the jet. These horizontal and vertical distances are called the X and Y coordinates, respectively. The coordinates are measured from the centre of the end of the pipe to the centre of the jet, as shown in Fig. 5.11.

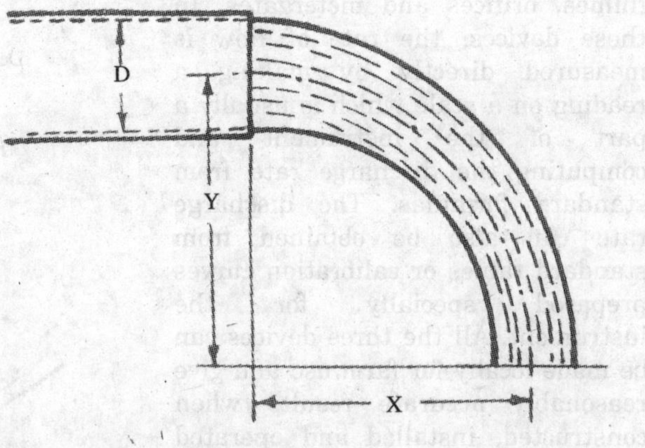

Fig. 5.11. Definition sketch illustrating the coordinate method of measuring flow through a full flowing pipe.

The expression for the flow from completely filled pipes is derived as follows: The distance Y that a particle falls in time, t, after issuing from the end of a horizontal pipe is

$$Y = \frac{gt^2}{2}$$

and the horizontal distance, X, it travels is $X = Vt$, in which V is the initial horizontal velocity. Combining these two equations by eliminating t and solving for V,

$$V = \frac{\sqrt{g}}{\sqrt{2Y}}$$

from which the following expression for the discharge is obtained:

$$Q = \frac{CaX\sqrt{g}}{\sqrt{2Y}} \quad \text{... (5.4)}$$

in which,

Q = discharge, m^3/sec
C = coefficient of contraction, dimensionless
a = cross-sectional area of pipe, m^2
X = x-coordinate, m
g = acceleration due to gravity, m/sec^2

Since experimental data is lacking, the value of C commonly used in calculation is 1.0.

The flow through partially filled pipes can be estimated by multiplying the flow obtained from equation 5.4 by the percentage of the area of the pipe that is filled at the end of the pipe. The coordinates should be measured from the approximate centre of the jet at the end of the pipe, instead of from the centre of the end of the pipe, as is the case with a full flowing pipe.

WEIRS AND FLUMES

In farm irrigation practice, the most commonly used devices for measuring water are weirs, Parshall flumes, orifices and metergates. In these devices, the rate of flow is measured directly by making a reading on a scale which is usually a part of the instrument and computing the discharge rate from standard formulas. The discharge rate can also be obtained from standard tables or calibration curves prepared specially for the instrument. All the three devices can be made locally for farm use and give reasonably accurate results when constructed, installed and operated properly. The choice between one or the other depends on the expected flow rates and site conditions.

Weirs

Weirs are used to measure the flow in an irrigation channel, or the discharge of a well or canal outlet at the source. A weir is a notch of regular form through which the irrigation stream is made to flow. In its simplest form, a weir consists of a weir wall of concrete, timber or sheet metal. In most case, a sheet metal weir plate fixed to the weir notch on the upstream side. Weirs may be built as stationary structures or they may be made portable. The notch may be rectangular, trapezoidal or triangular. Rectangular weirs and 90° V notch weirs are commonly used on the farm. It is desirable to install the weir at a point where there is a drop in the elevation of the channel bed. This will make the flow through the weir to fall freely to the downstream channel.

Terms Used

Weir pond. Portion of the channel immediately upstream from the weir.

Weir crest. The bottom of the weir notch.

Head. The depth of water flowing over the weir crest measured at some point in the weir pond.

Sharp crested weir. A weir having thin-edged crest such that the over-flowing sheet of water has the minimum surface contact with the crest.

End contraction. The horizontal distance from the ends of the weir crest to the sides of the weir pond.

Weir scale or gauge. The scale fastened on the side of weir or on a stake in the weir pond to measure the head on the weir crest.

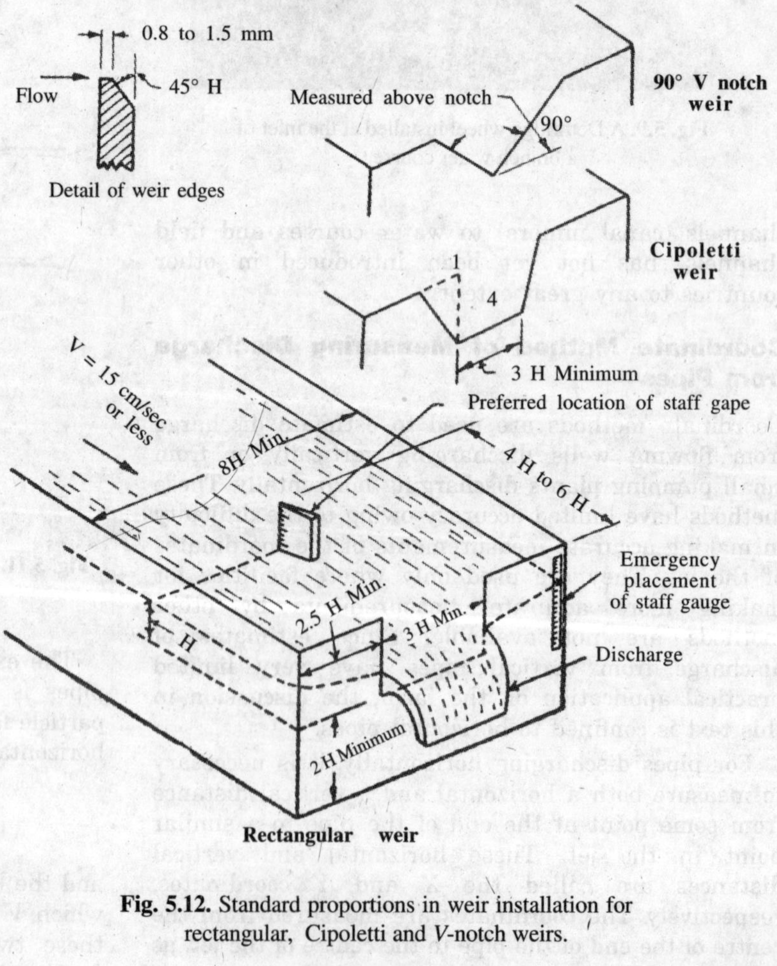

Fig. 5.12. Standard proportions in weir installation for rectangular, Cipoletti and V-notch weirs.

Nappe. The sheet of water which overflows a weir.

Weirs may be divided into two general classes: (*i*) sharp crested, and (*ii*) broad crested. Sharp crested weirs are commonly used for water measurement on the farm. They may be divided into weirs with end contractions and those without end contractions. End contractions are produced when the crest length, or width of the notch is less than the channel width. In suppressed weirs, the crest extends across the full width of the channel, and no contractions are produced. When the water surface downstream from the weir wall is far enough below the crest so that air has access around the nappe, the flow is said to be free, otherwise it is submerged.

The discharge through a weir notch is proportional to the head on the crest and is affected by the condition of the crest, the contraction, the velocity of the approaching stream and the elevation of the water surface downstream from the weir.

The basic formula for calculating discharge through a weir is:

$$Q = CLH^m \quad \text{...(5.5)}$$

in which,

Q = discharge

C = a coefficient dependent on the nature of the crest and approach conditions

L = length of crest

H = head of crest

m = an exponent, depending upon the weir opening.

The values of C and m are determined by carefully conducted calibration tests. Figure 5.12 shows the standard proportions of rectangular, trapezoidal and 90° V notch weirs.

Rectangular Weirs. The rectangular weir takes its name from the shape of the notch. They are used to measure comparatively large discharges. It has a horizontal crest and vertical sides (Figs. 5.13 and 5.14). They may be either contracted rectangular weirs or suppressed rectangular weirs. They have a sharp crest and are bevelled on the downstream side only. The sides are not bevelled.

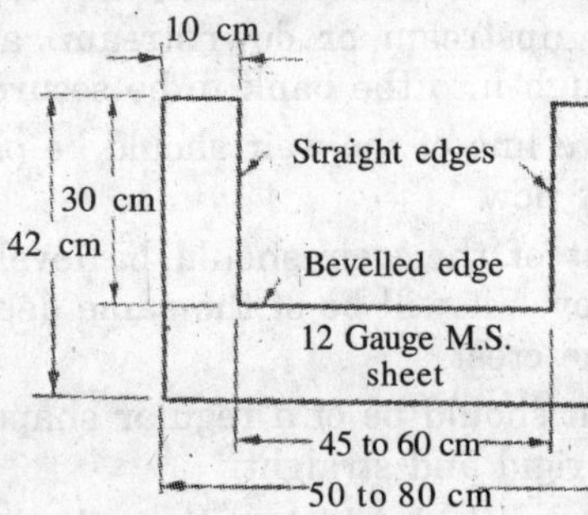

Fig. 5.13. Dimensions of commonly used rectangular weirs in average farm practice.

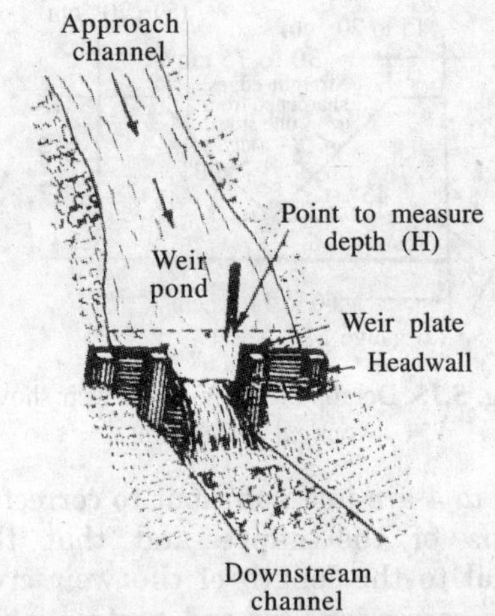

Fig. 5.14. Schematic sketch showing the installation of a rectangular weir with end contractions in an unlined field channel.

The discharge through rectangular weirs may be computed by the Francis' formula stated below:

(*i*) *Suppressed rectangular weir*

$$Q = 0.0184\ LH^{3/2} \quad \text{... (5.6)}$$

in which,

Q = discharge, litres/second

L = length of crest, cm

H = head over the weir, cm

(*ii*) *Contracted rectangular weir* (with end contractions at both ends)

$$Q = 0.0184\ (L - 0.2H)\ H^{3/2} \quad \text{... (5.7)}$$

Example 5.3. *Using Francis' formula, compute the discharge of a rectangular weir 45 cm long with a head of 12 cm, under the following conditions:*

(*i*) *with no end contraction.*

(*ii*) *with one end contraction.*

(*iii*) *with two end contractions.*

Solution:

(*i*) With no end contraction:

$$Q = 0.0184\ LH^{3/2}$$
$$= 0.0184 \times 45 \times 12^{3/2}$$
$$= 34.4 \text{ litres/second}$$

(*ii*) With one end contraction:

$$Q = 0.0184\ (L - 0.1H)\ H^{3/2}$$
$$= 0.0184\ (45 - 0.1 \times 12)\ 12^{3/2}$$
$$= 33.5 \text{ litres/sec}$$

(*iii*) With two end contractions:

$$Q = 0.0184\ (L - 0.2H)\ H^{3/2}$$
$$= 0.0184\ (45 - 0.2 \times 12)\ 12^{3/2}$$
$$= 32.6 \text{ litres/sec.}$$

Cipoletti Weir. The Cipoletti weir is a contracted trapezoidal weir, in which each side of the notch has a slope of 1 horizontal to 4 vertical. It is named after its inventor Cesare Cipoletti, an Italian engineer. Its popularity rests largely upon the belief that side

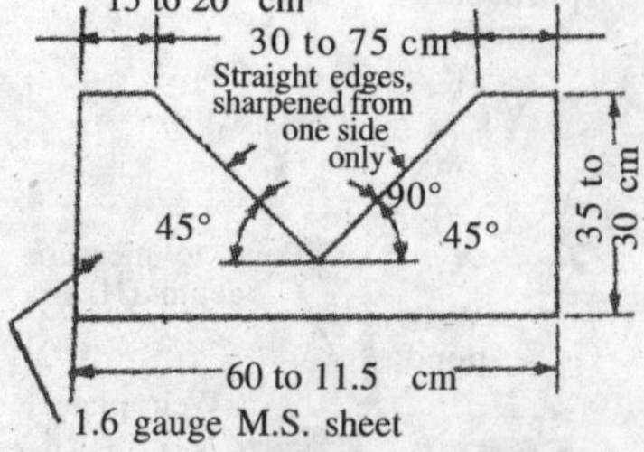

Fig. 5.15. Details of 90°-V-notch weir showing typical dimensions.

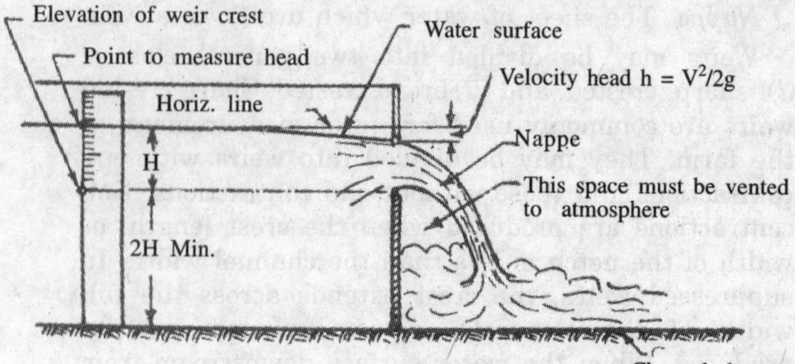

Fig. 5.17. Sectional view (vertical) of a weir installation showing essential requirements.

slopes of 1 to 4 are just sufficient to correct for the end contractions of the nappe and that the flow is proportional to the length of the weir crest. It does not require corrections for end contractions. The weir has sharp crest and sharp sides, bevelled from the downstream side only. It is commonly used to measure medium discharges.

The discharge through a Cipoletti weir is computed by the following formula:

$$Q = 0.0186\ LH^{3/2} \quad \text{... (5.8)}$$

in which,

Q = discharge, litres/second

L = length of crest, cm

H = head over the crest, cm.

V notch weirs. The 90° V-notch weir is commonly used to measure small and medium size streams. The advantage of the V-notch weir is its ability to measure small flows accurately. Typical dimensions are given in Fig. 5.15. It has both its sides sharp, bevelled from the downstream side only.

The discharge through a 90° V-notch weir may be computed by the following formula:

$$Q = 0.0138\ H^{5/2} \quad \text{... (5.9)}$$

in which,

Q = discharge, litres/second

H = head, cm

For heads lower than 5 cm, the weir should preferably be calibrated to obtain the discharge.

Fig. 5.16. A 90° V-notch weir installed in a field channel. Note the position of the scale used to measure the head.

General Requirements for the Setting and Operation of Weirs

The weir, when properly constructed and installed, is one of the simplest and most accurate devices for measuring irrigation water. Under ideal conditions, it will be accurate within 2 to 3%. However, improper setting and operation may result in large errors in discharge measurement with weirs. In general, the approaching flow should be the same as tranquil flow in long straight canals (without obstructions) of the same size. Any deviation from a normal horizontal or vertical flow distribution may cause error in measurement. Sand, gravel or sediment bars submerged in the approach channel, weeds or other obstructions along the banks or in the flow area can cause unsymmetrical approach flow. Standard weir installations are illustrated in Figs. 5.16 and 5.17. To ensure reliable measurements, the following precautions are necessary in the use of weirs:

(1) The weir should be set at the lower end of a long pool sufficiently wide and deep to give an even, smooth flow of the stream. The velocity of the approaching flow should not exceed 15 cm per second. Baffles may be put in the weir pond, if necessary, to reduce velocity and equalise the flow. To measure the discharge from pumps, a portable weir pond made of sheet metal or a built-in weir pond made of brick or stone masonry is constructed and the weir plate is fixed at its outlet.

(2) The weir wall or bulkhead must be vertical (not leaning to the upstream or downstream) and must extend far enough into the bank to be secured.

(3) The centre line of the weir should be parallel to the direction of flow.

(4) The crest of the weir should be level so that water passing over it will be of the same depth at all points along the crest.

(5) The notch should be of a regular shape and its edges must be rigid and straight.

(6) The weir crest should be above the bottom of the approach channel (weir pond) by at least about twice the depth of water flowing over the weir. A

minimum distance of about twice the depth of flow should be maintained between the end of the weir crest and the bank of the upstream channel.

(7) The crest of the weir is placed high enough so that water will fall freely below the weir, leaving an air space under the overfalling sheet of water.

(8) The depth of water flowing over the rectangular weir should not be less than about 5 cm, and not more than about two-thirds the crest width.

(9) The scale or gauge used for measuring the head should be located at a distance of about four times the approximate head. It should be far enough to one side so that it will be in comparatively still water. The zero of the scale should be exactly on the same level as the crest of the rectangular or Cipoletti weir or the apex of a V-notch weir.

Limitations in the use of weirs. Although weirs are easy to construct and convenient to use, they are not always suitable. They are not accurate unless proper conditions for their measurements are maintained. They require a considerable loss of head, often not available in channels on flat grades. They are not easily combined with turnout structures. Weirs are unsuitable for water carrying silt, which deposits silt load in the approach channel and destroys the proper conditions for weir measurement.

Parshall Measuring Flume

The Parshall flume is an open channel type measuring device that operates with a small drop in head, adopting the venturi principle. The loss of head for free flow limit is only about 25 per cent of that for weir. It is a self-cleaning device. Sand or silt in the flowing water does not affect its operation or accuracy. Parshall flumes allow reasonably accurate measurement even when partially submerged. The velocity of the approaching stream has very little influence on its operation (Figs. 5.18, 5.19 and 5.20).

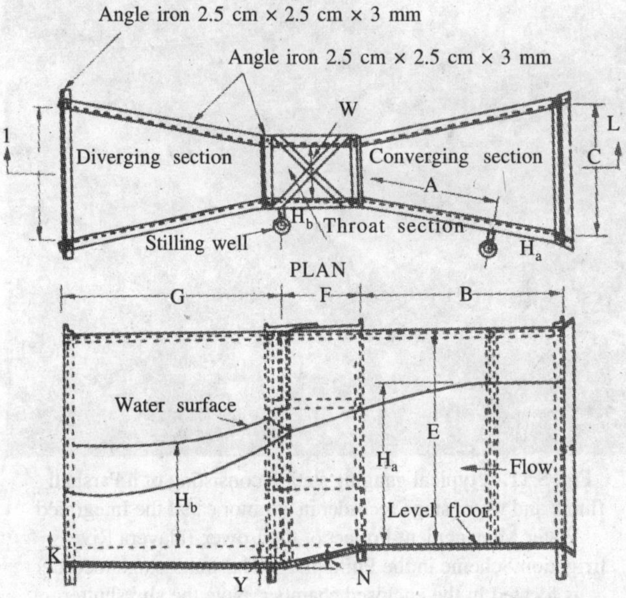

Fig. 5.19. Plan and elevation of a sheet metal Parshall flume showing component parts.

A Parshall flume consists of three principal sections: (*i*) A converging or contracting section at its upstream end leading to, (*ii*) a constricted section or throat, and (*iii*) a diverging or expanding section downsteam. The floor of the upstream section of the flume is level and the walls converge toward the throat section (Figs. 5.19 to 5.22). The walls of the throat section are parallel and the floor is inclined downward. The walls of the downstream section diverge toward the outlet and the floor is inclined upward. Large size flumes have an approach section and wing walls at the upstream end. Small flumes are made of sheet metal (Fig. 5.18 and 5.19), while large ones are made of concrete (Figs. 5.20 to 5.22). The size of the flume is determined by the width of its throat. Sizes ranging from 7.5 cm to several metres in

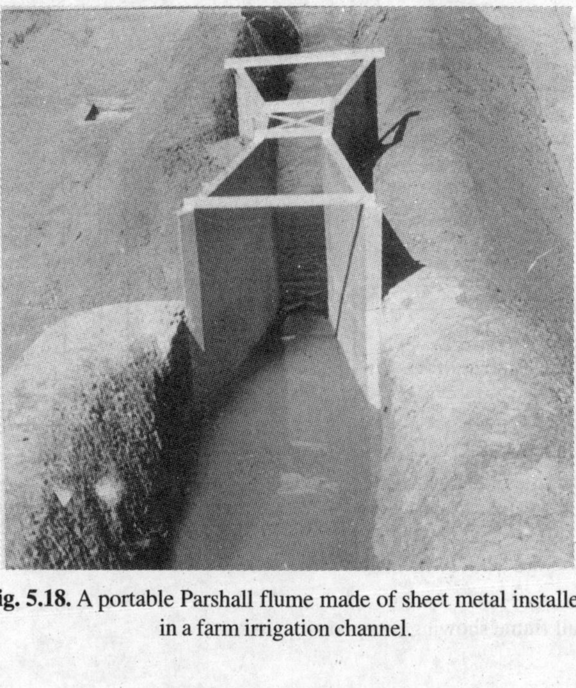

Fig. 5.18. A portable Parshall flume made of sheet metal installed in a farm irrigation channel.

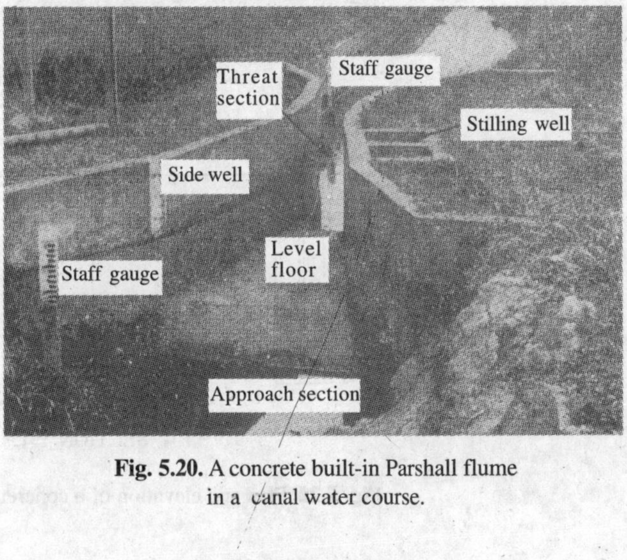

Fig. 5.20. A concrete built-in Parshall flume in a canal water course.

Fig. 5.21. A typical gauging station consisting of a Parshall flume and water stage recorder in a minor canal the Integrated Water Management Project of the Lower Talavera River Irrigation Scheme in the Philippines. The water stage recorder is located in the enclosed chamber. Note the slip shutter placed on the flume wall top.

throat width have been calibrated and formulas and tables developed. Stilling wells can be used to measure the water surface elevations in the flume, if turbulence is present. When the flow is smooth a plastic scale fixed to the inside of the flume will be sufficient to measure the head. The zero of the scales should be set at the same level as the floor level of the converging section.

Standard dimensions of Parshall flumes and discharge values are presented in Tables 5.2. and 5.3. There is no need to have a pond above the flume, as the velocity of approach has little effect on the accuracy of measurement.

Discharge through the flume can occur under either free flow or submerged flow conditions. When the elevation of the water surface downstream from the flume is high enough to retard the rate of discharge, the flow is considered submerged. To determine the discharge, two scales, H_a and H_b are provided at the upstream and downstream sections of the flume. Only H_a needs to be measured under free flow conditions. Free flow conditions are satisfied if the degree or percentage of submergence, as represented by the ratio H_b/H_a, is within the following limits:

Width of throat	*Free flow limit* H_b/H_a
2.5 to 7.5 cm	0.5
1.5 to 22.5 cm	0.6
3.0 to 24.0 cm	0.7

It is desirable to develop calibration curves of the flume to avoid the risk of inaccuracies in the dimensions of the flume. One of the important characteristics of the Parshall flume is its ability to withstand a relatively high degree of submergence over a wide range of backwater conditions downstream from the structure, and still maintain flow conditions.

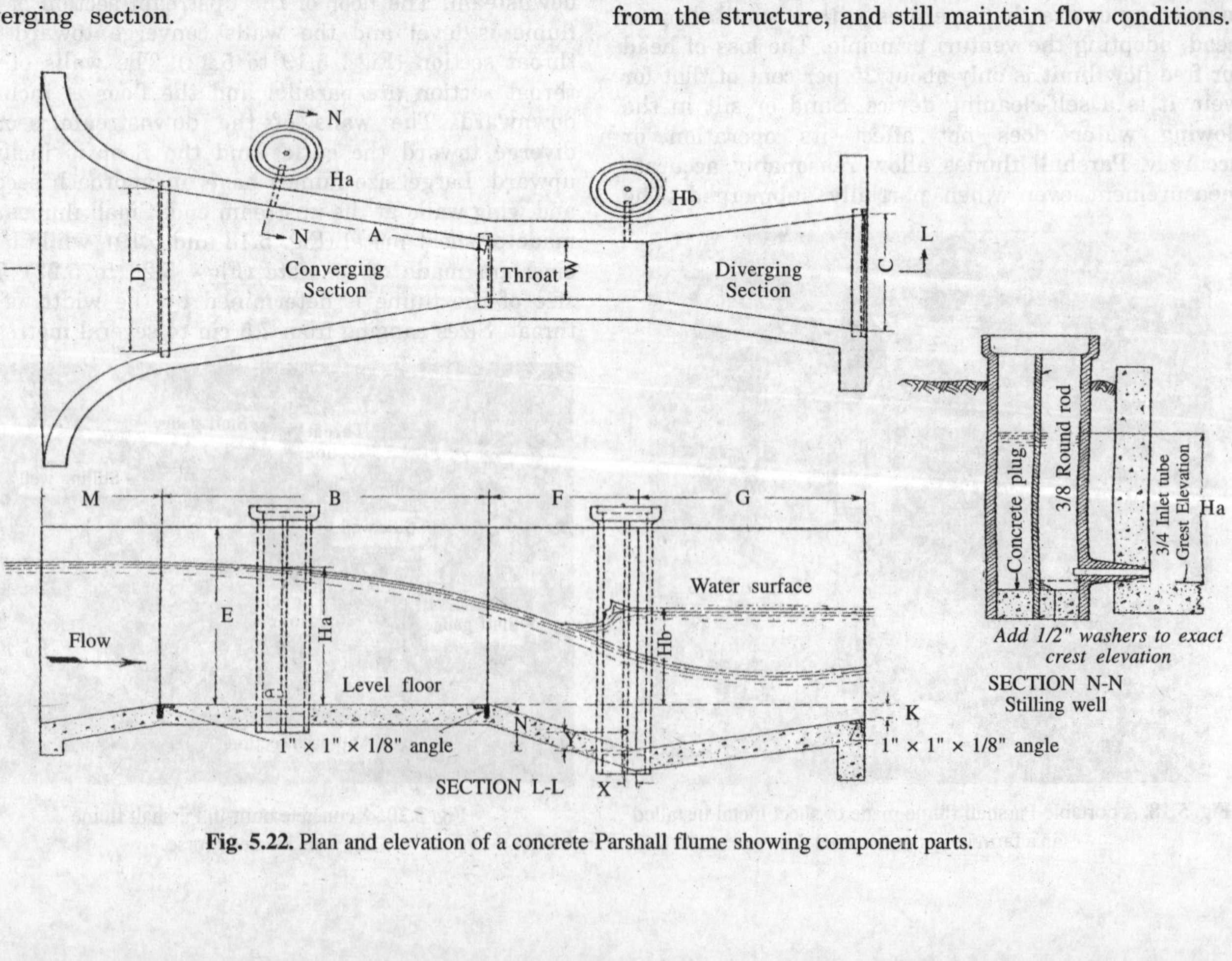

Fig. 5.22. Plan and elevation of a concrete Parshall flume showing component parts.

Table 5.2. Dimensions and capacities of Parshall flumes of various sizes

(Letters, refer to Fig. 5.22)

Throat width	*A*	*B*	*C*	*D*	*E*	*F*	*G*	*K*	*N*	*X*	*Y*	*Free-flow capacity*	
												Minimum, litres/second	*Maximum, litres/second*
cm	*cm*	*cm*	*cm*	*cm*	*cm*	*cm*	*cm*	*cm*	*cm*	*cm*	*cm*		
7.5	31	45.5	17.8	26	45.5	15	30.5	2.5	5.7	2.5	3.8	0.85	28.4
15	41.4	61	39.4	39.7	61	30.5	61	7.6	11.5	5.1	7.6	1.4	110.8
23	58.8	86.4	38.1	57.5	76	30.5	45.5	7.6	11.5	5.1	7.6	2.5	253
30	91.5	134	61	84.5	91.5	61	91.5	7.6	23	5.1	7.6	3.13	456.6

Cut-throat Flume

The cut-throat flume (Fig. 5.23) is an attempt to improve on the Parshall flume, mainly by simplifying the construction details. The details of the flume are shown in Fig. 5.25. The flume has a flat bottom, vertical walls, and a zero length throat section. Since it has no throat section, it was given the name 'cut-throat' by the developers (Skogerboe, Hyatt, Anderson and Eggleston, 1967). The major advantage of a cut-throat flume is economy, due to simplicity in construction. Any flume length between 45 cm to 3 m can be used, while throat widths between 2.5 cm to 1.8 m have been investigated.

The cut-throat flume can operate either as a free flow or a submerged flow structure. Under free flow conditions, critical depth occurs in the vicinity of minimum width, w, which is called the flume throat or the flume neck. The attainment of critical depth makes it possible to determine the flow rate, knowing only an upstream depth, h_a. This is possible because whenever critical depth occurs in the flume the upstream depth, h_a, is not affected by changes in the downstream depth, h_a. For free flow, the ratio of inlet flow depth h_a to flume length should preferably be less than 0.4.

The relationship between flow rate Q and upstream depth of flow h_a, in a cut-throat flume under free flow conditions is given by the following experimental relationship:

$$Q = C_1 h_a^{\,n} L \quad \text{... (5.10)}$$

in which,

Q = flow rate.

C_1 = free flow coefficient, which is the value of Q when h_a is 1.0 foot, which is the slope of the free flow rating curve when plotted on logarithmic paper.

n = exponent, whose value depends only on the flume length L.

The value of n is a constant for all cut-throat flumes of the same length, regardless of the throat width W.

Submerged flow analysis. When the flow conditions are such that the downstream flow depth, h_b, is raised to the extent that the flow depths at every point through the structure become greater than the critical depth, resulting in a change in upstream depth (Fig. 5.24), the flume is operating under submerged flow conditions and requires that the two flow depths be measured, one upstream (h_a) and the other downstream (h_b) from the flume neck. The submergence, S, is defined as the ratio, often

Fig. 5.23. A cut-throat flume installed in a field channel.

Fig. 5.24. Water surface profiles in a cut-throat flume showing free flow and submerged flow conditions.

Table 5.3. Free-flow discharge values for Parshall flumes

Head	Discharge, litres-per second			
	Throat width			
cm	*7.5 cm*	*15 cm*	*23 cm*	*30 cm*
3	0.8	1.4	2.6	3.1
4	1.2	2.3	4.0	4.5
5	1.7	3.3	5.5	7.0
6	2.3	4.4	7.2	9.6
7	2.7	5.4	8.5	11.4
8	3.4	7.2	11.1	14.4
9	4.3	8.5	13.5	17.7
10	5.0	10.2	15.9	21.1
11	5.8	11.6	18.1	23.8
12	6.7	13.5	21.1	27.5
13	7.5	15.0	23.3	31.0
14	8.5	17.3	26.7	35.0
15	9.4	19.2	29.5	38.7
16	10.4	21.2	32.5	42.7
17	11.4	23.2	35.6	46.6
18	12.4	25.3	39.0	51.2
19	13.6	27.8	42.5	55.0
20	14.3	30.0	45.8	59.7
21	15.8	32.7	49.3	64.7
22	17.1	35.2	53.3	69.8
23	18.2	37.7	56.8	74.0
24	19.4	40.1	60.5	79.0
25	20.7	42.7	64.5	84.1
26	22.0	45.7	69.3	89.0
27	23.3	48.1	72.4	94.3
28	24.8	51.5	76.7	100.0
29	26.0	54.0	80.7	105.1
30	27.5	57.3	85.2	111.0

expressed as a percentage, of the downstream depth to the upstream depth.

$$S = h_b/h_a \quad \text{... (5.11)}$$

For detailed procedure for computing the discharge rate under submerged flow conditions, reference is invited to Skogerboe, *et al.* (1973).

The cut-throat flume should be installed in a straight section of the channel. It should not be located too near a farm gate because of unstable and surging effects which might result from the gate

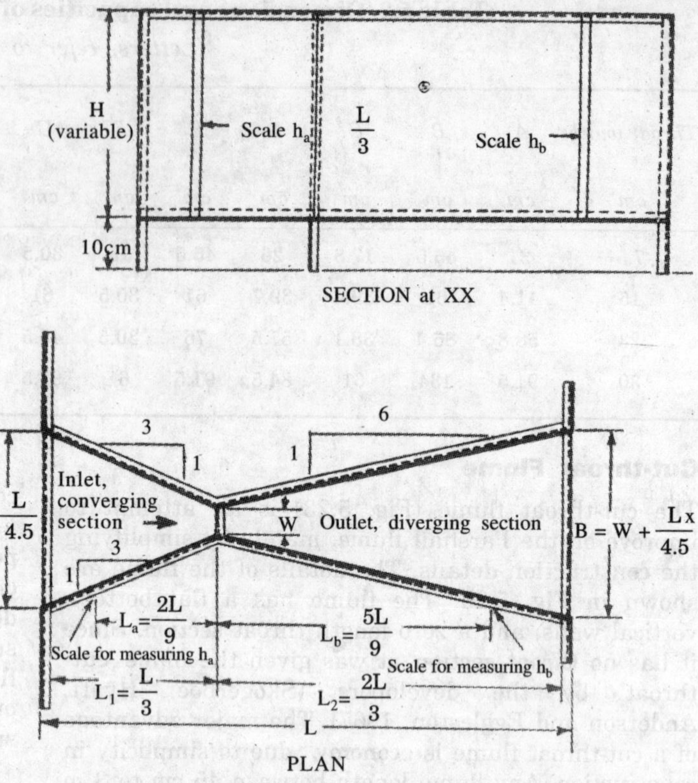

Fig. 5.25. Views of cut-throat flume showing dimensions.

operation. Whenever feasible, it is better to have a flow measuring device operate under free flow conditions.

WATER MEASUREMENT STRUCTURES FOR VARYING FLOWS IN EARTHEN CHANNELS

Flow conditions vary widely in earthen irrigation and drainage channels. Earth channels are usually wider and shallower than concrete or masonry lined canals. They have comparatively low velocities of flow. Bos, Replogle and Clemens (1991) have described various flow measuring flumes for open channel systems. They are usually concrete or masonry structures built at the desired measuring points. A number of alternative shapes of broad-crested weirs and long throated flumes have been described (Figs. 5.26 to 5.28). They are commonly used to measure the discharge of large canals, as compared to field channels.

In a triangular flume, the control section is triangular (Fig. 5.28). They are particularly suited to measure smaller flows like those in drainage channels and streams. Trapezoidal shapes are also available, depending on site conditions.

Broad-crested rectangular weirs. The following are the basic components of weirs and flumes used in earth channels: entrance to approach channel, a converging transition section, throat, a diverging transition section and stilling basin (Fig. 5.26). A suitable riprap section (usually of stone paving) is provided in the downstream channel section to prevent the erosion of channel bed and sides. The approach section has vertical side walls and are of the same width as the throat of the flume.

Truncated flumes. Shortening of the full length structure is possible by deleting the diverging section and the tailwater section if the head loss over the section exceeds 0.4 times the head causing the flow, as measured from the crest of the throat section. Such a structure is called a truncated flume (Fig. 5.27). In such a situation, it is not necessary to account for the velocity head. The rectangular tailwater section could be avoided if at maximum flow the Froude Number at the entrance of the channel is less than 1.7. Whenever site conditions permit, the choice of a truncated flume will greatly reduce the cost of the measuring structure.

Discharge through broad-crested weirs. In a rectangular broad-crested weir with a

$$Q = bq \qquad \ldots (5.12)$$

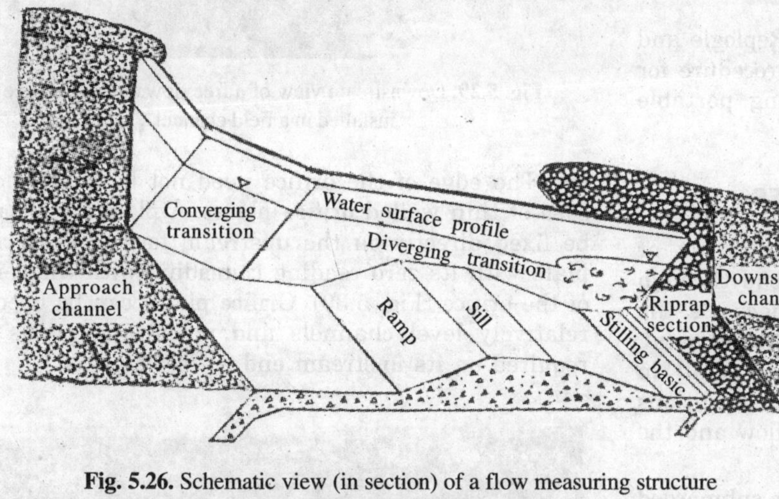

Fig. 5.26. Schematic view (in section) of a flow measuring structure using rectangular control section. (*Ref.* Bos, Replogel and Clemens, 1991).

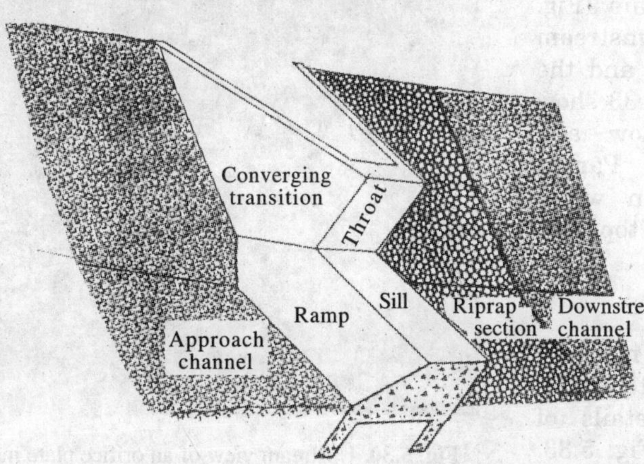

Fig. 5.27. Schematic view of a truncated flume for earthen-channel. (***Ref.*** Bos, Replogle and Clemens, 1991).

in which,

Q = discharge over the weir, m^3/sec

b = width of the weir, m

q = discharge per metre width of the weir, m^3/sec

Bos, Replogle and Clemens (1991) have given the required specifications for the design of earth channel weirs and flumes. They have provided theoretical analysis for estimating the flow over the weirs. Ratings for varying flow conditions are also provided. The basic requirement for obtaining reliable estimates of the flow over the weirs is that there must be a drop in the energy head across the flume so that critical flow will occur in the throat. This is necessary to maintain the required head-discharge relationship for the flume which is not influenced by the downstream tailwater level. Design procedures and theoretical considerations for truncated weirs are also presented in the reference cited above.

Triangular throated flumes. Triangular throated flumes (Fig. 5.28) are adapted to measure a wide range of discharges, including low flows like return flows to drainage systems and operational spillage from irrigation systems. It also permits, wide variations in the rate of flow.

The ratio between the maximum and minimum flow (Q_{max} /Q_{min}) is high in triangular throated flumes. The drop in water surface, required to obtain a constant discharge of water passing over the flume structure (modular structure), irrespective of the water levels in the upstream and downstream of the channel, is as low as 10% of the design head of flow over the structure.

Portable flumes for earthen channels: Portable long-throated flumes are designed to measure comparatively low flows such as those of irrigation furrows and on-farm surface water distribution systems when a permanent point for measurement is not required. They are made of sheet metal with suitable bracing. The common material used is 1 mm thick galvanised

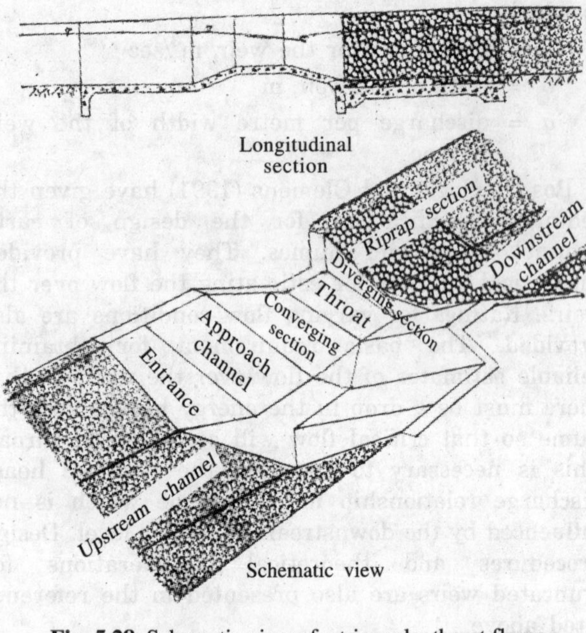

Fig. 5.28. Schematic view of a triangular throat flume.
(*Ref.* Bos, Replogle and Clemens, 1991)

steel sheet with angle iron bracing. Bos, Replogle and Clemens (1991) have presented detailed procedure for installation and flow measurement, using portable flumes with selected base width.

ORIFICES AND METER GATES

Orifices in open channels are usually circular or rectangular openings in a vertical bulkhead through which water flows (Fig. 5.29). The edges of the openings are sharp and often constructed of metal. The cross-sectional area of the orifice is small in relation to stream cross-section. These conditions allow complete contraction of the stream flow and the velocity of approach becomes negligible.

Orifices may operate under free flow or submerged flow conditions. Under free flow conditions, the flow from the orifice discharges entirely into air (Fig. 5.29). In submerged flow orifices, the downstream water level is above the top of the opening and the flow discharges into water. Figures 5.30 to 5.33 show how measurements of head on free flow and submerged flow orifices are made. Partial submergence occurs when the downstream water surface is between the elevations of the top and bottom of the orifice.

Free flow orifices

Free flow orifice plates can be used to measure comparatively small streams like the flow into border strips, furrows or check basins. The details of construction of orifice plates are given in Fig. 5.33. They consist of sheet iron, steel or aluminium plates that contain accurately machined circular openings or orifices usually ranging in diameter from 2.5 cm to 7.5 cm. The edge of the orifice need not be sharpend in case of thin walled orifice plates. A plastic scale may be fixed directly on the upstream face of the orifice plate with its zero reading coinciding with the centre of the orifice (Fig. 5.30). Orifice plates can be used on relatively level channels and no stilling ponds are required on its upstream end as in weirs.

Fig. 5.29. Downstream view of a free flow circular orifice installed in a field channel.

Fig. 5.30. Upstream view of an orifice plate installed in a field channel. Note the plastic scale fixed to the orifice plate. The zero reading of the scale coincides with the centre of the orifice.

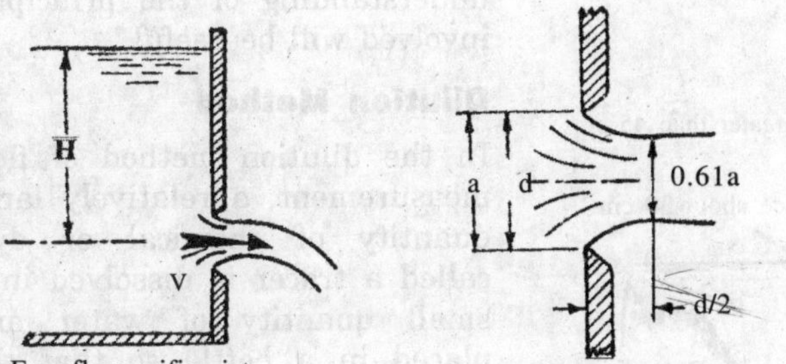

Free flow orifice jet

Contraction of jet after passing through orifice

Fig. 5.31. Schematic sketches illustrating flow through free flow orifice.

Figure 5.31 illustrates the flow through orifice plates. The discharge through an orifice is calculated by the formula

$$Q = 0.16 \times 10^{-3}\ a\sqrt{2gH} \qquad \text{... (5.13)}$$

in which,

Q = discharge through orifice, litres/second.

a = area of cross-section of the orifice, cm^2.

g = acceleration due to gravity, cm/sec^2 (981 cm/sec^2).

H = Depth of water over the centre of the orifice (on the upstream side) in case of free flow orifice, or the difference in elevation between the water surface at the upstream and downstream faces of the orifice plate in case of submerged orifice, cm.

Submerged Orifices

Submerged orifices may be of two types: (*i*) those having orifices of fixed dimensions, and (*ii*) those in which the height of opening may be varied. A standard submerged orifice has fixed dimensions. The opening is sharp edged and usually rectangular, with the width being 2 to 6 times the height. The discharge through a standard submerged orifice may be obtained using equation (5.11). The adjustable submerged orifice is one in which the height of opening and head may be varied to suit the requirements. The ordinary form is a simple head gate used to control flow from one channel to another. Adjustable submerged orifices should be calibrated to obtain the discharge under varying head conditions.

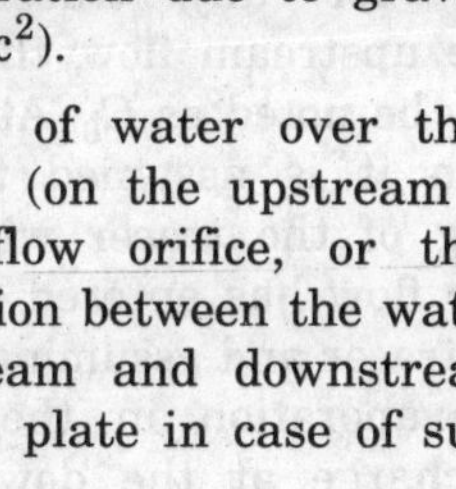

Fig. 5.32. Measuring head through sub-merged orifice.

Meter Gate

A meter gate is basically a modified submerged orifice so arranged that the orifice is adjustable in area (Fig. 5.34). Meter gates are manufactured commercially and are used to control the water flowing from one channel to another. They may serve to measure the rate of flow, if the head and area of opening can be determined and the gate has been calibrated. Meter gates are commonly used by canal irrigation agencies in USA and many other countries at canal outlets where irrigation water is charged on volumetric basis. Their use, however, has not become popular in India.

The head, as applied to meter gates, is the same as that described under submerged orifice. Although it is possible to compute the discharge using formulas, it is rarely done because there are usually too many departures from standard definable conditions for which correction coefficients are not known.

Normally the gate and the opening into the outlet are circular and they can be determined easily if the gate is fully open. However, under most conditions the gate is only partially open, leaving a crescent-shaped area which is difficult to measure. Consequently, most gates are calibrated and tables supplied, giving the rate of flow as a function of the head and the degree of gate opening as measured by the displacement of the gate stem. The pipe outlet with control gates of canal distributories (Chapter 1) used in some parts of India, is functionally similar to a meter gate, but not so precise. The adjustable proportionate module type canal outlet, commonly used in northern India, works on the orifice principle and can also be used to measure water by calibrating it under different operating heads.

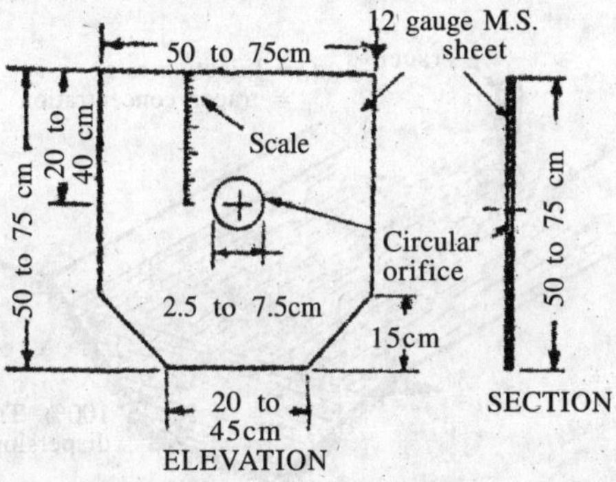

Fig. 5.33. Details of a circular orifice. The dimensions shown are the values commonly used in farm irrigation practice.

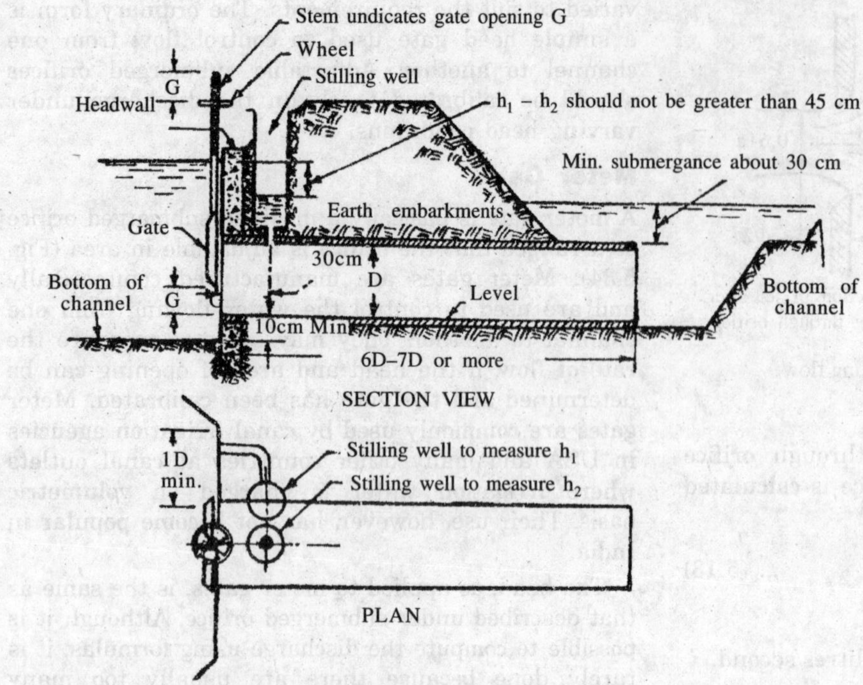

Fig. 5.34. Installation of a typical meter gate.

TRACER METHODS OF WATER MEASUREMENT

The tracer methods of water measurement are independent of stream-cross-section and are suitable for field measurements without installing fixed structures. In these methods, a substance (tracer) in concentrated form is introduced into the flowing water and allowed to thoroughly mix with it. The concentration of the tracer is measured at a downstream section. Since only the quantity of water necessary to accomplish the dilution is involved, there is no need to measure velocity, depth, head, cross-sectional area, or any other hydraulic factors usually considered in discharge measurement. Though the methods have yet to become popular, an understanding of the principles involved will be useful.

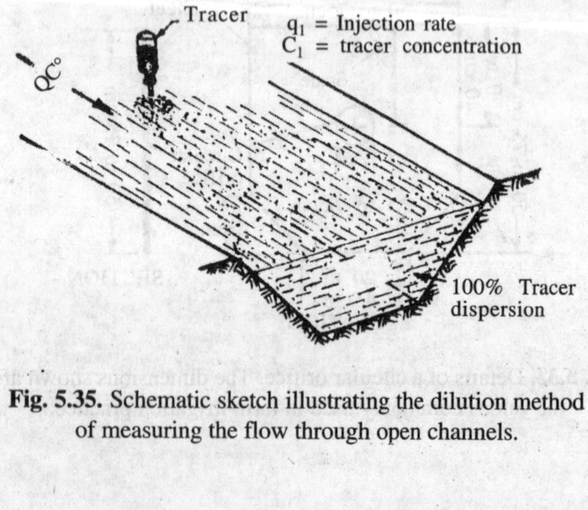

Fig. 5.35. Schematic sketch illustrating the dilution method of measuring the flow through open channels.

Dilution Method

In the dilution method of flow measurement, a relatively large quantity of chemical or dye, called a tracer is dissolved in a small quantity of water and placed in a bottle so that the tracer solution can be discharged at a known rate into the water flowing in a channel or pipe. Figure 5.35 illustrates the method of flow measurement using the dilution method of measurement in an open channel.

The concentration given here $\left(\dfrac{\text{weight of tracer}}{\text{weight of water}}\right)$ of tracer in the bottle is a ratio C_1. The rate of injection is q_1. To account for any tracer which might already be in the upstream flow, the original tracer concentration may be noted as C_0. At the downstream measuring station it is assumed that a thorough transverse mixing of the tracer with the flow has taken place. If no flow has entered the canal during the mixing of the tracer and assuming no loss of water by seepage or evaporation in the section of the channel, the discharge at the downstream, is the discharge Q at the upstream plus q_1 which is the quantity of tracer added. The concentration of tracer at the downstream station may be designated as C_2.

The above relationship may be expressed mathematically as follows:

$$QC_0 + q_1C_1 = (Q + q_1)C_2$$

or

$$QC_0 + q_1C_1 = QC_2 + q_1C_2$$

Rearranging the terms,

$$QC_0 - QC = q_1C_2 - q_1C_1$$

Simplifying,

$$Q(C_0 - C_2) = q_1(C_2 - C_1)$$

or

$$Q = q_1\left(\frac{C_2 - C_1}{C_0 - C_2}\right) \quad \text{... (5.14)}$$

Salt concentration, as such, is difficult to measure directly. However, over a useful range, the electrical conductivity bears a linear relationship with the salt concentration and it is easily measured.

Many different substances have been used in dilution tests. Chemicals, including common salt (NaCl) have been used and the dilution has been

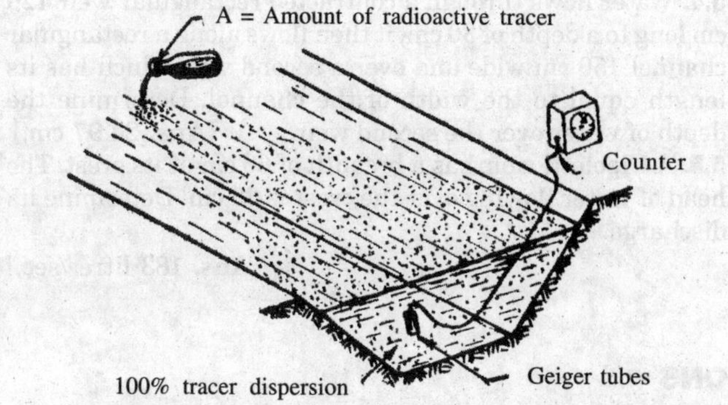

Fig. 5.36. Schematic sketch illustrating the radioisotope method of making discharge measurement.

evaluated by the change in the ability of the flowing water to conduct electric current. Sodium dichromate is another chemical used in dilution tests. Its concentration can be determined by analysing water samples by chemical analysis or flame photometer determination. Dyes have been used to colour the water and calorimeters and comparators are used to determine the dilution by comparing the samples with standard concentrated or diluted solutions.

Radioisotope method. Radioisotopes may be used in place of chemical or dye tracer and the degree of dilution determined by counting the gamma ray emissions from the diluted isotope solution (the downstream flow) using Geiger counters or scintillation counters. The method is illustrated in Fig. 5.36. In the '*pulse*' or 'total' count method, a known amount of radioisotope is introduced into the flow in a relatively short time. At the measuring station downstream, where the isotope is thoroughly mixed, the concentration of the radioisotope tracer is determined from the gamma ray emissions detected and counted by the counter. The following mathematical relationship may be derived:

$$Q = \frac{FA}{N} \qquad \text{... (5.15)}$$

in which,

Q = volume of water flowing in the channel per unit time.

F = counts per unit of radioactivity per unit volume of water per unit of time.

A = total units of radioactivity to be introduced for each discharge measurement.

N = total counts.

In preparing A, a sufficient quantity of isotope for the desired measurements (Gold-198-salts dissolved in aqua regia) is divided into parts known as aliquots, using a portable field laboratory containing a standardised counting system. The individual parts, A, are usually contained in plastic bottles and are transported to the test site in numbers sufficient for the discharge measurements. Thorough mixing of isotope with the flowing water is of primary importance in accurate measurements. Turbulent water aids in dispersion and thorough mixing. Studies are underway in many parts of the world to obtain satisfactory mixing. Geiger or scintillation counters and portable decade scalers, for totalising the gamma ray emissions, are used in the measurement.

Radioactive material is to be handled with great care. Only licensed personnel are allowed to handle it. Permission of the concerned public health agencies is always obtained before any tests are run with radioactive tracer.

SUMMARY

Need for proper measurement of water in irrigation water conveyance and application is highlighted. Various methods of water measurement under pipe flow and open channel flow are introduced. Commonly used devices for measuring flow of water in farm water distribution systems, including weirs, flumes and orifices are described. Methods of estimating the discharge rates under free flow and submerged flow conditions in weirs, flumes and orifices are described. Specific requirements in the installation and operation of various flow measuring devices are described. Water measurement structures for measuring varying flows in earth channels of canal water distribution systems are presented. Scope and applicability of tracers and radio-isotope methods of flow measurement are introduced.

PROBLEMS

5.1. A centrifugal pump has a discharge of 38 litres per second. One hectare of potato crop is irrigated in 6 hours. What is the average depth of irrigation ?
[**Ans.** 8.208 cm.]

5.2. Berseem requires 92 cm of water in 122 days of its irrigating period. How much land will be irrigated with a flow of 30 litres per second for 12 hours daily, assuming that there is no rainfall concentration during the period.

[**Ans.** 17.185 ha.]

5.3. A farmer having 10 hectares of land grows wheat in 5 hectares, potatoes in 2 hectares, peas in 1 hectare and mustard in 2 hectares as winter crops. The irrigating season is 4 months. The water require-ments of wheat, potatoes, peas and mustard are 45 cm, 50 cm, 15 cm and 10 cm, respectively. What size of irrigation stream will be required from a source which can supply for 8 hours each day during the season? No appreciable rainfall occurs during the period. [**Ans.** 10.416 lit/sec.]

5.4. Water flows through a contracted rectangular weir 120 cm long to a depth of 30 cm, it then flows along a rectangular channel 150 cm wide and over a second weir which has its length equal to the width of the channel. Determine the depth of water over the second weir. [**Ans.** 24.97 cm.]

5.5. A Cipoletti weir has a breadth of 60 cm at its crest. The head of water flowing over the crest is 30 cm. Determine its discharge.

[**Ans.** 183 litres/sec.]

QUESTIONS

1. Distinguish between:

(*i*) Volume method and velocity-area method of measurement of water.

(*ii*) Open-flow meters and vertical flow meters in flow measurements.

(*iii*) Water meters and venturi-meters.

(*iv*) Free flow and submerged flow in flumes.

(*v*) Free flow and submerged flow type orifices.

(*vi*) Sharp-crested and broad-crested weirs.

(*vii*) Orifices and water meters.

(*viii*) Broad-crested rectangular weirs and truncated flumes used for flow measurement in earth channels.

(*ix*) Broad-crested rectangular long-throated weirs and triangular-throated weirs for flow measurement in large earth channels with varying flows.

(*x*) Dilution method and radio-isotope method of measurement of water in open channels.

2. What are the specific requirements for the proper operation of a water meter?

3. What are the basic principles of operation of a venturi meter?

4. What are the advantages and limitations in the use of the coordinate method for estimating flow through horizontal pipes?

5. What are the specific requirements to be adhered to in the installation of weirs?

6. What are the limitations in the use of weirs in flow measurement?

7. What are the conditions under which the Cipoletti weir is preferred over rectangular weirs?

8. What are the specific field conditions under which a Parshall flume is preferred over weirs in measuring open channel flows?

9. What are the advantages claimed for a cut-throat flume over a Parshall flume?

10. Where will the critical flow depth occur in a cut-throat flume working under free flow conditions?

11. What are the common types of water measuring structures for measuring varying flows in earthen channels?

12. What are the basic components of weirs and flumes used in flow measurement under varying flows in earth channels?

13. What are the situations where broad-crested rectangular weirs are preferred over other types of weirs and flumes?

14. What is the limiting value of Froude Number in the use of a truncated flume?

15. Briefly describe the application of tracer method in open channel flow measurement?

16. How is the discharge through broad-crested weirs computed?

17. What is the concentration of tracer in flow measurement in open channels?

18. Express the mathematical relationship to estimate the rate of flow in open channels using the tracer method.

19. What is the basic principle of flow measurement in open channels using the dilution method?

20. What is a radio-isotope? Express the functional relationship to estimate the rate of flow in open channels using the radio-isotope method.

21. Identify the True and False statements from the following:

(*i*) The velocity of water at the surface of a stream is greater than the average velocity of the stream.

(*ii*) The current meter provides a fairly accurate means of the rate of flow in a field channel.

(*iii*) The average of the readings taken at 0.2 and 0.8 of the depth below the surface of a river gives a fairly good estimate of the velocity of flow.

(*iv*) The large type model of a Dethridge meter is suitable to measure the rate of flow in open channels with capacities up to 500 litres/sec.

(*v*) In a rectangular weir only the crest of the weir plate is bevelled and not the sides.

(*vi*) The Cipoletti weir could either be of the suppressed type or contracted type.

(*vii*) The V-notch weir has both the sides sharp, bevelled from the downstream side only.

(*viii*) A sharp crested rectangular weir is bevelled from the downstream side only.

(*ix*) The weir crest should be above the bottom of the approach channel by at least about twice the depth of water flowing over the weir.

(*x*) A Parshall flume should not be located near a head gate in a canal water distribution system, as the turbulent water jet downstream of the outlet may cause surging and unbalanced flow.

(*xi*) In a truncated flume used for measuring varying flows in earth channels, the rectangular tail water section could be avoided if at maximum flow, the Froude Number at the entrance of the channel is less than 1.7.

(*xii*) Critical depth in a cut-throat flume under free-flow conditions occurs in the vicinity of the minimum width.

(*xiii*) Stilling basins upstream of the orifice are desirable to obtain accurate measurements while using orifice plates for flow measurements in open channels.

(*xiv*) The ratio between maximum and minimum flows is low in a triangular-throated flume as compared to a long-throated rectangular weir.

(*xv*) Meter gates are usually used by canal irrigation agencies in many countries at canal outlets where irrigation water is charged on volumetric basis.

(*xvi*) Flow measurements, using meter gates, are usually done using calibration curves/tables developed for the gate.

(*xvii*) In flow measurements using radio-isotops, the mixing of the isotope with the flowing water is of primary importance.

22. Identify the most appropriate answer of the following questions from the four options:

(*i*) In measuring the rate of flow in irrigation canals using current meters, the measurements are taken at two locations below the water surface and the average of the two values provides an estimate of the average rate of flow. To obtain fairly accurate results the measurement points are located at:

(a) 0.1 and 0.9 of the depth of flow
(b) 0.2 and 0.8 of the depth of flow
(c) 0.3 and 0.7 of the depth of flow
(d) 0.4 and 0.6 of the depth of flow

(*ii*) The depth of flow over a sharp crested rectangular weir should not be more than about:

(a) half the crest width
(b) two-thirds of the crest width
(c) three-fourth of the crest width
(d) the width of the weir

(*iii*) The gauge used to measure the depth of flow over a weir is located upstream of the weir at a distance of about:

(a) twice the approximate head
(b) thrice the approximate head
(c) four times the approximate head
(d) five times the approximate head.

(*iv*) The appropriate device for the measurement of flow in irrigation furrows is:

(a) V-notch weir
(b) Dethridge meter
(c) Portable Parshall flume
(d) Submerged orifice

(*v*) The most suitable device for measuring large flows in earth channels in a canal water distribution system is:

(a) In-built Parshall flume
(b) Suppressed rectangular weir
(c) Broad-crested rectangular weir
(d) Truncated flume.

(*vi*) The principle of working of a meter gate is:

(a) free flow rectangular orifice
(b) free flow circular orifice
(c) submerged flow circular orifice
(d) submerged flow rectangular orifice.

23. In the following, match each of the entries of the 1st column with the most closely linked entry in the 2nd column:

(*i*)	(A) Current meter	(I) Trapezoidal weir
	(B) J.S. Dethridge	(II) Measurement of flow through pipes
	(C) Cisare Cipoletti	(III) Flow measurement at the discharge end of horizontal pipes
	(D) Coordinate method	(IV) Measurement of flow in rivers and canals
	(E) Water meter	(V) Measurement of flow in water courses
(*ii*)	(A) Francis' formula	(I) Flow measurement in open channels
	(B) Nappe	(II) Cut-throat flumes
	(C) Parshall	(III) Measurement of varying flows in earth channels
	(D) Skogerboe	(IV) Rectangular weir
	(E) Bos, Replogle & Clemens	(V) Sheet flow through a weir
(*iii*)	(A) Critical depth	(I) Modified submerged orifice
	(B) Truncated flume	(II) Flow measurement in Parshall flumes
	(C) Meter gate	(III) Flow measurement in open channels
	(D) Stilling wells	(IV) Flow measurement structures for varying flows
	(E) Tracer	(V) Cut-throat flume

FURTHER READING REFERENCES

Anonymous (1962). *Measurement of Irrigation Water.* Chapter 9 in SCS National Engineering Handbook, U.S. Dept. of Agriculture, Washington, D.C., p. 72.

Bos, M.G., editor (1976). *Discharge Measurement Structures. Working Groups in Small Hydraulic Structure.* International Institute for Land Raclamation and Instrument, Wageningen TZ, Netherlands, p. 464.

Bos, M.G. 1977a. *The use of long-throated flumes to measure flows in irrigation and drainage canals. Agricultural Water Management,* Vol. 1. No. 2, pp. 111–126.

Bos, M.G., and Reinink, Y, 1981. "Head loss over long-throated flumes". *Journal of the Irrigation and Drainage Division,* American Society of Civil Engineers, Vol. 107, No. IR1, pp. 87–102.

Bos, M.G. Replogle, J.A. and Clemens, A.J. (1991). *Flow Measuring Flumes for Cyclen Channel Systems.* American Society of Agricultural Engineering, St. Joseph Michigan, U.S.A., 11, p. 321.

Christiansen, J.E. (1935). *Measuring Irrigation Water.* University of California, Agri. Expt. Stn. Berkeley, p. 91.

Gupta, S.N. (1964). *Measurement of discharge at state tubewells by steel V-notch.* U.P. Irrigation Research Institute, Roorkee, p. 12.

Michael, A.M. and Ojha, T.P. (2003). *Principles of Agricultural Engineering,* Vol. II. Jain Brothers, (New Delhi), Karol Bagh, New Delhi, p. 888.

Michael, A.M., Arora, D.R., Bhattacharya, A.K., Mandal, A. and Gupta, A.K. (1970). *Handbook of farm irrigation structures.* Indian Agricultural Research Institute, New Delhi, p. 82.

Parshall, R.L. (1926). *The improved venturi flume.* Transactions, American Society of Civil Engineers, Vol. 89, pp. 841–880.

Peterka, A.J. (1967). *Water Measurement Procedures.* U.S. Bureau of Reclamation, Denver, Colorado, Report No. Hyd–557, p. 153.

Replogle, J.A. (1970). "Flow Meters for Water Resource Management". *Water Resources Bulletin.* Vol. 6, *No. 3,* pp. 345–374.

Replogle, J.A. (1971). *Critical-depth flumes for determining flow in canals and natural channels.* Transactions of the American Society of Agricultural Engineers. Vol. 14, No. 3, pp. 428–433.

Replogle, J.A. (1977a). "Discussion of 'Venturi flumes for circular channels' *by M.H. Disking". Journal of the Irrigation and Drainage Division.* American Society of Civil Engineers, Vol. 103, No. IR3, pp. 385–387.

Replogle, J.A. (1977b). "Compensating for constraction errors in critical flow flumes and broad-crested weirs". In: *Flow Measurement in Open Channels and Closed Conduits.* L.K. Erwin, ed., National Bureau of Standards Special Publication No. 434, Vol. 1, U.S. Government Printing Office, Washington DC, pp. 201–218.

Replogle, J.A. 1978. *"Flumes and broad-crested weirs: mathematical modelling and laboratory ratings".* In: *Flow Measurement of Fluids.* H.H. Dijstelberger and E.A. Spencer, Eds., North-Holland Publishing Co., Amsterdam, The Netherlands, pp. 321–328.

Replogle, J.A. and Bos, M.G. (1982). "Flow measurement flumes: Application to irrigation water management, In: Advances in Irrigation. Vol. 1, D. Hillel, Ed., Academic Press, New York, pp. 147–217.

Replogle, J.A. and Clemens, A.J. (1979). *Broad-crested weirs for portable flow metering.* Transactions of the American Society of Agricultural Engineers, Vol. 22, No. 6, pp. 1324–1328.

Replogle, J.A., and Clemmens, A.J. (1980). *Modified broad-crested weirs for lined canals.* In: *Irrigation and Drainage–Today's Challenges,* Specially Conference Proceeding, American Society of Civil Engineners. Boise, Idaho, July 23–25, pp. 463–479.

Replogle, J.A. Clemmens. A.J., Tanis, S.W., and McDade. J.H. (1983). *Performance of large measuring flumes in main canals.* In: Advances in Irrigation and Drainage: Surviving External Pressures. Speciality Conference Proceedings, American Society of Civil Engineers, Jackson, Wyoming, July 20–22, pp. 530–537.

Stock, E.M. (1955). *Measurement of irrigation water.* Bul. No. 5, Utah State Engg. Expt. Stn., p. 80.

Scott, V.H. and Houston, C.E. (1959). *Measuring Irrigation Water.* California Agrl. Expt. Stn. Circular 473, p. 52.

Skogerboe, G.V., Bennet, R.S. and Walker, W.R. (1973). *Selection and installation of cut-throat flumes for measuring irrigation and drainage water.* Colorado State University, Expt. Stn., p. 74.

Skogerboe, G.V. and Hyatt, M.L. (1967). *Rectangular cut-throat flow measuring flumes.* Jl. of the Irrigation and Drainage Dn., ASCE, Vol. 98, No. IR 4, pp. 569–583.

Skogerboe, G.V. and Hyatt, M.L., Anderson, R.K. and Elgleston, K.O. (1967). *Design and calibration of submerged open channel flow measurement structures: Part 3, Cut-throat flumes.* Report W G31–4, Utah Water Research Laboratory, College of Engineering, Utah State University, Logan, Utah.

Skogerboe, G.V. and Hyatt, M.L. (1967). *Analysis of submergence in flow measuring flumes.* Jl. of the Irrigation and Drainage Division, ASCE, Vol. 93, No. IR4, Dec., pp. 1–13.

Skogergoe, G.V., Bennett, R.S. and Walker, W.R. (1972). *Generalized discharge relations for Cut-throat flumes.* Jl. of the Irrigation and Drainage Division, ASCE, Vol. 98, No. IR4, Dec., pp. 569–583.

United States Department of Agriculture, Soil Conservation Services (1962). *Measurement of Irregation Water,* Section 15, Chapter 9, p. 92.

Woffinden, D.S. and Clyde, C.G. (1971). *A Portable direct reading open channel flow measurement device.* Paper No. 2–6–190 presented at the Symposium on Flow: Its Measurement and Control in Science and Industry. Pittsburg, Pennsylvania, May 10–14, 1971, p. 15.

Chapter

6

Open Channel Distribution in Command Areas

LEARNING OBJECTIVES

Importance of appropriate design and layout of water conveyance and distribution systems in irrigated areas. Need for lining of water conveyance systems in permeable soils to minimize seepage losses. Factors influencing canal seepage. Estimation of conveyance losses in different sections of a canal system. Economic advantages of lining of different segments of a canal water distribution system in permeable soils. Measurement of seepage in canal sections. Hydraulics of open channel flow. Design of open channels. Estimating velocity of flow and discharge capacity of open channels. Materials for lining canals and on-farm water conveyance systems. Procedures for canal lining with different materials. Anti-weed treatment in canals. Lining of water courses and field channels—materials and methods. Use of pre-fabricated non-reinforced concrete channel sections in lining field channels. Use of semi-elliptical R.C.C. canalets in canal minors, water courses and field channels. Optimum length of water course to be lined. Structures to control erosion in on-farm water conveyance systems—open drop structures, chute spillways and pipe drop spillways. Water control structures. Use of synthetic rubber seal strips in check gates. Structures for diversions and turnouts in water courses and field channels. Structures at channel crossing—flumes, culverts, inverted siphons. Survey and design of water distribution systems in outlet command areas of canal irrigation systems—command areas with consolidated land holdings and those with irregular boundaries and scattered holdings.

A properly designed water distribution system will make irrigation easy and efficient. Several types of structures are used to convey, divert and control irrigation water. Good irrigation structures are an essential part of an irrigation layout. Conveyance of irrigation water from the headworks at a reservoir, barrage or weir to the farms is essentially through a system of open channels and allied structures. Open channels also form the predominant means of water conveyance in lift irrigation projects and irrigation by wells. In many regions of the world, a substantial amount of water conveyed through the canal system is lost through seepage, evaporation, transpiration from the weeds in the canals and banks, breaches in the canal banks, leakage through gates and turnouts and operational losses. Operational losses include return flow to drains due to faulty canal regulation, inadequate arrangement for irrigation in the night and unused flows during periods when there is no demand for irrigation. While some loss in the canal system is unavoidable, a major part of it could be saved by proper design and management of the system. Operational waste in most canal systems could be reduced considerably by diverting excess canal flow to neighbouring irrigation tanks and temporary storage in equalising reservoirs (system storage) along the course of canals from which water could be withdrawn to supply local demands.

Fig. 6.1. A small diversion canal in a highly permeable soil near Leh, District Ladakh, Jammu and Kashmir (1975). It is estimated that about 70% of the water diverted to the canal is lost by seepage in a length of about 5 km of the canal.

Fig. 6.2. An unlined water course infested with weeds in a sandy tract. High seepage losses are inherent in such soils. Breaches and weed growth obstruct the flow and contribute to additional loss of water.

The major component of water loss in the irrigation conveyance system is seepage (Fig. 6.1 and 6.2). Seepage control measures, consisting essentially of resorting to lining of the permeable reaches of a canal system, provide the greatest opportunity in saving the water supplies developed at a high cost. The water thus saved can be used for irrigating additional area or for more intensive irrigation to the existing area. Little can be done to save the evaporation loss in channels, which in any case is very small. Losses due to transpiration through bank vegetation is reduced by keeping the banks clear of weeds (Fig 6.1). Lining with masonry material eliminates weeds in a canal almost completely. In addition to the major objective of saving irrigation water, seepage control measures contribute to solving the problems of waterlogging and salt accumulation in land areas adjoining the canal system. Water which seeps through canals and laterals collects in lower-lying lands. The problem of waterlogging in an irrigated area could often be attributed to the faulty design of the water distribution system and poor water management. There are also other benefits resulting from minimizing the water loss in the conveyance system. Unlined channels are to be designed with wide beds and nearly flat slopes, in order to make them stable. In a lined channel much higher velocities (nearly three times that of unlined channels) are allowed. A deeper and narrower channel section is possible, resulting in saving in land cost and earth-work. Lining results in a general reduction in the size and hence in the cost of canal structures including cross-drainage works, bridges, falls and diversions. Lined channels are easier and less costly to maintain, and have safer banks. In case of an on-farm system, the bed levels of minor canals and channels are often raised which result in an increase in the command area of the system. Canal systems, originating from reservoirs are often infested with weeds, which greatly reduce their capacity and make canal operations difficult. Lining such canals have the added advantage of eliminating the problem.

Lining assumes special importance in areas, which on introduction of irrigation, would be liable to waterlogging and salinity.

Farm irrigation distribution systems for surface irrigation methods can be classified into two groups—surface channels and underground pipelines. Facilities for the conveyance and control of irrigation water are different for each type. Field channels may be either unlined earth channels or lined channels.

Seepage in Canals

Estimates of conveyance losses in canal systems in different countries of the world reveal that about one-third of all the water diverted for irrigation is lost in conveyance from the headworks to the farm fields. The following are the major factors influencing seepage from a canal:

1. Characteristics of the soil traversed by the canal system (The dominant soil characteristics influencing seepage is its hydraulic conductivity).
2. Area wetted by the canal.
3. Extent of deposition of silt in the canal systems, especially for canals in the alluvial plains.
4. Location of water table relative to the canal.
5. Frequencies of canal usage.
6. Age of the canal system, particularly for canals in the alluvial plains in which the seepage loss gets progressively reduced due to the natural sealing of the wetted area by the fine silt carried by the canal flow.

Estimates on Canal Losses

The problem of substantial loss of water in the canal system has been attracting the attention of irrigation engineers throughout the world. In India, measurement of seepage losses in a canal system was made as early as in the later part of the nineteenth century (1875 A.D.). In 1881, Kennedy reported that in the Upper Bari Doab Canal, nearly 47% of the supply entering the canal head was lost in conveyance (20% in main canals and branches, 6% in distributories and 21% in water courses) (ICID, 1967). The corresponding loss in transmission in the Ganga canal system was 44% (15% in the main canals and branches, 7% in disributories and 22% in water courses). Values close to these were obtained in 1907 in case of canals in Khandesh in Maharashtra (ICID, 1967). Observations by Malhotra (1970) on seepage losses showed that about 25% of the water entering a canal system was lost in the water courses. The following are the commonly accepted figures of conveyance losses in the different segments of a canal system in the alluvial plains of North India (ICID, 1967; Irrg. Commission, 1972):

1. Main canals and branches	17% of head discharge
2. Distributories and minors	8% of head discharge
3. Water courses	20% of head discharge
Total	45%

The above figures do not include the losses in the field itself which is estimated to be about 30% of the water supply reaching the field or about 16.5% of the discharge at the head end of the canal system. The corresponding values of conveyance loss are substantially higher in the sandy and sandy loam soils of the arid and semi-arid regions.

Estimates of seepage loss in the canal systems in lower Egypt and upper Egypt are 8% and 10%, respectively. In Greece, where the canals are ordinarily unlined, seepage losses are reported to be of the order of 40 to 50% of the supply at the headworks. It may thus be seen that the transmission loss in canal systems vary widely, depending on the major factors influencing seepage. Hence, decisions on seepage control works in a canal system should be based on a careful evaluation of the physical and economic factors of the project under consideration.

A case study on the economics of lined and unlined canals in Northern India revealed that on the basis of the cost of construction alone a lined canal costs 3.25 times as much as an unlined canal to carry the same flow. However, when the value of the water lost in the unlined canal is taken into account, the unlined canal costs about 1.08 times as much as the lined canal. When the additional advantages of reduced maintenance costs, prevention of waterlogging and increased agricultural production resulting therefrom are concerned, lining would prove to be economical in many cases. However, in canal reaches where natural sealing of earthen canal sections is expected with fine particles of silt, lining may not be economical.

There will be considerable economic advantage if the decision on lining or otherwise of a canal section is made at the planning stage, rather than at some later date when the need becomes apparent during operation. Thus, the estimation of the possible seepage in a canal reach is an important part of the investigations required in planning an irrigation project. Broad estimates on the extent of seepage in any particular reach of a canal may be made after identifying the soil textural group along the route of the canal system.

Various approaches are in use for estimating the seepage loss in canals. The practice most commonly used in canal irrigation works in India is to assume a seepage loss varying from 1.828 cubic metres per second per million square metres (6 cusec per million square feet) of wetted area for a canal with a discharge of 3.4 cubic metres per second (120 cusec) to 2.45 cubic metres per second per million square metre (8 cusec per million square feet) for a canal with a discharge of upto 56.6 cubic metres per second (2000 cusec).

In the United Arab Republic (UAR), the following empirical formula is used to estimate the losses in a canal section (ICID, 1967):

$$S = CLP\sqrt{R} \qquad \text{... (6.1)}$$

in which,

S = Seepage and evaporation losses in a canal of length L, m^3/sec
L = Length of canal, m
P = Average wetted perimeter, m
R = Hydraulic radius of the canal, m
C = A constant, the value of which depends on the nature of the soil. The value of C is 0.0015 in stiff clay soils and rises to 0.003 in very sandy soils.

Evidently, estimates based on empirical formulas are highly situation specific. Often they may not provide reasonable values of the loss in a canal section. For reliable estimates of seepage in a proposed canal length, it is necessary to estimate the loss based on field measurements at selected segments of the canal.

Measurement of Seepage in a Canal Section

The following are the commonly used methods adopted for measuring the quantity of water lost by seepage in a section of an existing canal:

1. Ponding method
2. Inflow-outflow method
3. Seepage meter method.

1. Ponding Method: The ponding method is the most dependable means of determining seepage in a canal length. In this method, temporary water-tight dikes or walls are used to isolate the specific reach of the canal under investigation. Water is ponded between the two dikes and the time rate of drop of the water surface and the dimensions of the ponded reach of the canal are used to compute the seepage loss.

Seepage rate is usually expressed in cubic metres per square metre of the wetted area in 24 hours. To obtain satisfactory results, the ponded reach to be tested must be selected so as to avoid any inflow or outflow within the reach which cannot be accurately measured.

For precise measurements of seepage loss, water is added to the ponded reach to maintain a constant water surface elevation. The volume of water added is accurately measured. This represents the seepage loss and the elapsed time provides the rate of loss. However, this is often difficult in the field. In areas where the seepage loss is low and under arid and semi-arid climates, it is necessary to measure the rate of evaporation at the site, using a standard pan evaporimeter. In case there is any rain during the test, necessary correction must be made. The ponding method is best adapted to locations where irrigation is seasonal in nature. Ponding tests are done during the off season. The method can be used to investigate the seepage characteristics of short reaches of canals in contrast to the inflow-outflow method which is adapted to long reaches of a canal.

2. Inflow-Outflow Method: The inflow-outflow method of seepage measurement utilises measurements at the

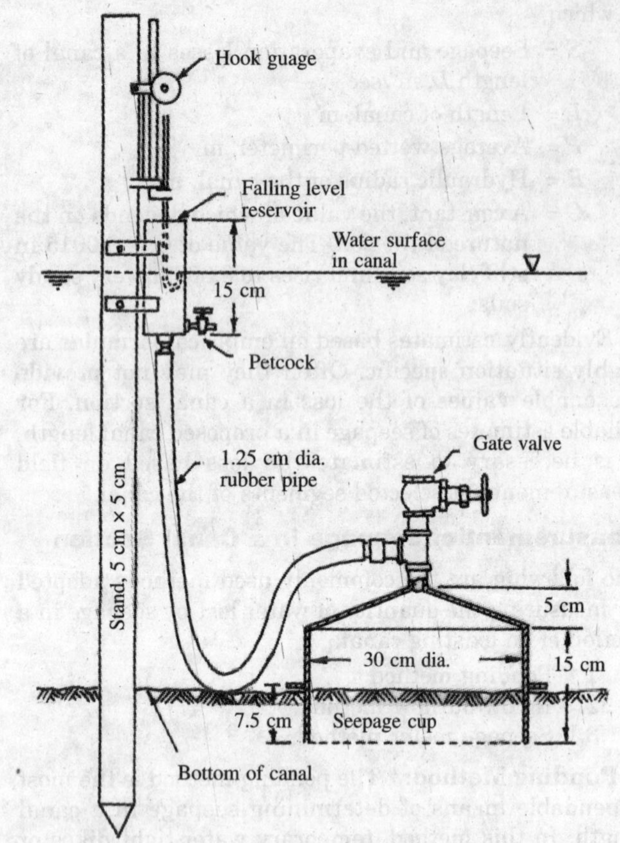

Fig. 6.3. Details of a falling head seepage meter. Adapted from Skogerboe, G. (1977).

upstream and downstream ends of the canal reach being investigated. The accuracy of the method depends on the precision obtained in the measurements. The quantities of water flowing into and out of the test length of the canal are measured as accurately as possible for a predetermined length of time and the difference is attributed to seepage. The rate of seepage is established by the elapsed time. Reaches in which there are the least number of outlets and diversions and no appreciable inflow from higher lands should be selected.

Existing calibrated weirs and Parshall flumes in a canal system may be used with advantage for measuring flows. Where these are not available, temporary gauging stations may be constructed. Current meters are sometimes employed in case of large canals. Water flowing out of the test section through outlets or diversions should be carefully monitored and accounted for. If the tests are of long duration, it is advantageous to install a water stage recorder to record the height of flow in the flume as a function of the elapsed time. The inflow-outflow method is not sufficiently accurate for the precise determination of seepage in short reaches of a canal.

3. Seepage Meter Method: The seepage meter (Fig. 6.3) is a modified form of the constant head permeameter. It is mainly used to determine locations of relatively high seepage losses. An indication of the average rate of seepage in a given stretch of a canal can be made if the tests are made at close intervals along the section. Seepage meters can be used during normal canal operation. Their use is limited to unlined canals, whereas the ponding and inflow-outflow methods can be used both in lined and unlined canals.

Open Channels

An open channel is defined as any conduit in which water flows with a free-water surface. Rivers, canals and uncovered flumes are open channels. Pipes, tile drains and sewers act as open channels when flowing partially full. The following are some of the common terms describing flow in open channels.

Steady flow. This refers to the condition in which the flow characteristics at any point do not change with time (i.e., $\frac{\partial v}{\partial t} = 0$, $\frac{\partial y}{\partial t} = 0$ etc.) in which v denotes the velocity of flow, y denotes the flow depth and t denotes time. Flows in lined irrigation channels tend to be steady under normal flow conditions.

Unsteady flow. When the flow characteristics change with time at a channel cross-section, the flow is unsteady (i.e., $\frac{\partial v}{\partial t} \neq 0$, $\frac{\partial y}{\partial t} \neq 0$, etc.) Intermittent runoff into terrace outlets, diversion channels etc., is generally unsteady.

Uniform flow. When the channel flow is steady and the mean velocity is the same at each succeeding cross-section, the flow is uniform (i.e., $\frac{\partial v}{\partial l} = 0$, in which l denotes the length of the channel section under consideration). Irrigation channels must have a constant cross-section and slope. In designing field channels, the flow is considered to be uniform at the maximum discharge.

Non-uniform flow. When the mean velocity changes from cross-section to cross-section, the flow is non-uniform $\left(\frac{\partial v}{\partial a} \neq 0\right)$. Flow in the upper portions of steep channels near the entrance is non-uniform. In a channel of changing cross-section the flow is non-uniform.

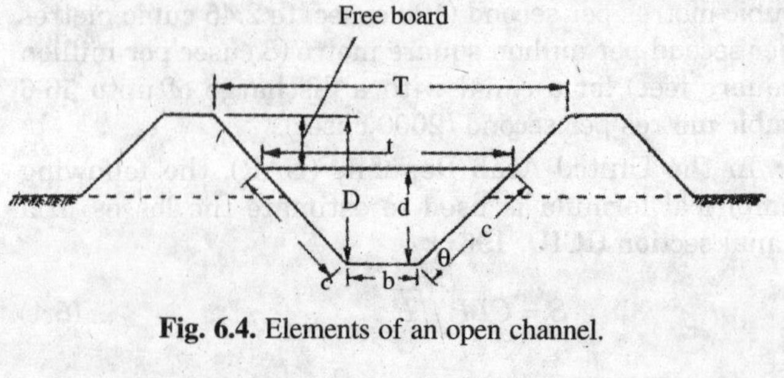

Fig. 6.4. Elements of an open channel.

Design of Open Channels

The elements of an open channel are shown in Fig. 6.4. Referring to Fig. 6.4,

T = top width of the channel
t = width of surface when the water is at depth d
D = depth of channel after free board is added
d = depth of flow in channel
b = bottom width of channel
c, c = wetted sides of channel
θ = angle between the sloping side and the horizontal

Definitions

Wetted perimeter. Wetted perimeter is the sum of the lengths of that part of the channel sides and bottom which are in contact with water.

$$p = b + c + c \quad \text{... (6.2)}$$

Frictional resistance offered to the flow of water in channels depends upon the area of the surface passed over, which in turn is proportional to the wetted perimeter.

Area of cross-section. Cross-sectional area a refers to the area of the wetted section of the channel:

$$a = \frac{(b+t)d}{2} \quad \text{... (6.3)}$$

Hydraulic radius. Hydraulic radius R is the ratio between the cross-sectional area of the stream and its wetted perimeter:

$$R = \frac{a}{p} \quad \text{... (6.4)}$$

The hydraulic radius is an important parameter in determining the velocity of water flowing in channels. Velocity of flow in channels is directly proportional to the square root of the hydraulic radius (i.e., $v \propto \sqrt{R}$). The hydraulic radius decreases with the increase in wetted perimeter for a given cross-sectional area.

Hydraulic slope. Hydraulic slope S of a channel is the ratio of its vertical drop h for a length l of the channel.

$$S = \frac{h}{l} \quad \text{... (6.5)}$$

The velocity of flow in channels varies directly with the square root of the hydraulic slope (i.e., $v \propto \sqrt{S}$).

Free-board. Free board is the vertical distance between the highest water level anticipated in the design and the top of the retaining banks. It is provided to prevent overtopping of structures because of wave action or damage due to trampling or other unforeseen conditions.

Angle of repose. If a mass of loose dry soil is allowed to fall freely to the ground, the soil heap formed by it will attain a conical shape. The maximum angle of the sloping sides of the cone, measured from the horizontal, is called the angle of repose. Channel sides will be stable if its slope does not exceed the angle of repose of the soil in which it is made. Clay soils have steeper angles of repose, while in sandy soils they are flatter.

Estimating Velocity of Flow in Open Channels

The design of channels is based on the mean velocity of flow which is obtained by standard formulae. These formulae express the mean velocity of flow as a function of the hydraulic roughness of the channel, the hydraulic radius and the hydraulic slope.

From the foregoing discussions on the influence of hydraulic radius and hydraulic slope on the mean velocity of flow, the following relationship is obtained:

$$v \propto \sqrt{R} \,.\, \sqrt{S}$$

or

$$v = C\sqrt{RS} \quad \text{... (6.6)}$$

in which C is a constant, depending on the size, character of slope, depth of flow of stream and nature of surface of the channel. Equation (6.6) is known as Chezy's formula. Determination of the value of C in equation (6.6) is comparatively difficult. A more workable and satisfactory formula, carefully examined experimentally is the Manning's formula stated below:

$$v = \frac{R^{2/3}S^{1/2}}{n} \quad \text{... (6.7)}$$

in which,

v = mean velocity of flow in the channel, metres per second
R = hydraulic radius, metres
S = hydraulic slope, dimensionless
n = roughness coefficient of the channel

The Darcy-Weisbach formula stated below is also sometimes used to determine the mean velocity of flow in open channels:

$$v = \sqrt{\frac{8gRS}{f}} \quad \text{... (6.8)}$$

in which,

g = acceleration due to gravity, metres/sec^2
f = Darcy-Weisbach roughness coefficient or friction factor, dimensionless

Other factors are as in equation (6.7).

The resistance coefficients C, n and f in equations (6.6), (6.7) and (6.8), respectively, can be related to each other as follows:

$$\frac{C}{\sqrt{g}} = \sqrt{\frac{8}{f}} = \frac{R^{1/6}}{n\sqrt{g}} \quad \text{... (6.9)}$$

The Manning's formula is frequently used in the design of water courses and field channels. The values of the roughness coefficient n for Manning's formula have been determined experimentally for various conditions of the channel sides and bottoms. Table 6.1 presents the values of n under the conditions commonly met with in irrigation channels.

Table 6.1. Values of Manning's 'n'

Type and description of channel	*Manning's n*
Earth channels	
Straight and uniform	0.023
Winding, sluggish	0.025
Stony bed, weeds on bank	0.035
Small drainage ditches	0.04
Lined channels	
Concrete	0.015
Masonry, rubble	0.017 to 0.030
Metal, smooth	0.011 to 0.015
Wooden	0.011 to 0.014
Vegetated waterways	0.020 to 0.040
Pipes	
Cast iron	0.012 to 0.013
Clay or concrete drain tile	0.011
Steel	0.015 to 0.017
Vitrified sewer pipe	0.013 to 0.015

Discharge Capacity of Channels

The discharge capacity of channels is obtained from the formula:

$$Q = a\,v \qquad \text{...(6.10)}$$

in which

Q = discharge capacity, cubic metres per second

a = cross-sectional area of the wetted section of the channel, square metres

v = mean velocity of flow in channel, metres per second.

Unlined Field Channels

Unlined earth channels are frequently used in water conveyance on the farm (Fig. 6.5). The advantages of earth channels are that they are understood and accepted by farmers. They can be built and maintained by unskilled persons and require no special equipment or materials. The low initial cost constitutes the major advantage of earth channels.

Earth channels should be built with stable side slope and with banks strong enough to carry the required flow of water safely. They should have ample capacity to carry the design discharge at non-erosive velocities (Table 6.2). Side slopes should be flat enough so that the banks will neither cave in nor slide when they are saturated with water. Earth channels should not ordinarily have side slopes steeper than 1½ horizontal to 1 vertical.

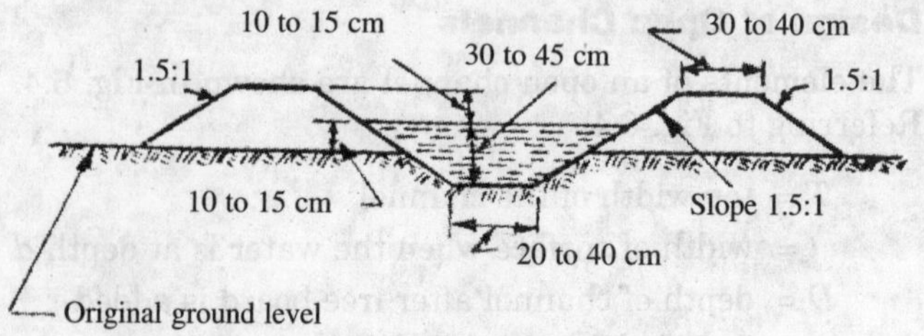

Fig. 6.5. Typical cross-section of an unlined field channel.

In stiff clay, however, steep slopes upto 1/2 to 1 are possible. Loose sand should have flat slopes of about 2 to 1. Main water supply channels that are built up higher than the surrounding field level should have banks large enough so that they will not be damaged by slippage or by trampling.

Example 6.1. *Assume an earth channel on a grade of 0.10%, depth of water 40 cm, bottom width 60 cm and side slopes 1½ to 1. Calculate the velocity of flow and carrying capacity of the channel.*

Solution:

Side slopes $1\frac{1}{2}$ to 1 of flow depth of 40 cm

Length of wetted side, c $= \sqrt{60^2 + 40^2}$

$= 72$ cm $= 0.72$ m

Wetted perimeter, p $= 72 + 40 + 72$

$= 184$ cm $= 1.84$ m

Top width, T $= 60 + 40 + 60$

$= 160$ cm $= 1.6$ m

Area of cross-section, a $= \dfrac{(40+160)40}{2}$

$= 4{,}000\ \text{cm}^2 = 0.4\ \text{m}^2$

Hydraulic radius, R $= \dfrac{t}{p} = \dfrac{0.4}{1.84} = 0.217$ m

Hydraulic slope, S $= \dfrac{0.1}{100} = 0.001$

Manning's n (from Table 6.1)

$= 0.025$

Mean velocity of flow $= \dfrac{R^{2/3}.S^{1/2}}{n}$

$= \dfrac{(0.217)^{2/3} \times (0.001)^{1/2}}{0.025}$

$= 0.456$ m/sec

Carrying capacity $= a\,v = 0.4 \times 0.456$

$= 0.182\ \text{m}^3/\text{sec}$

$= 182$ litres/second.

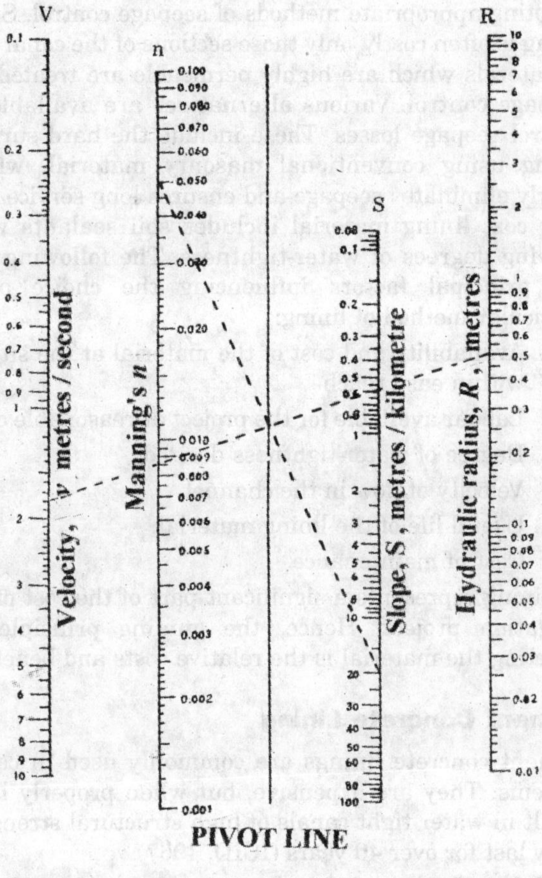

Fig. 6.6. Nomograph showing the solution of Manning's formula. The dotted lines show the determination of v at 1.8 m/sec when $R = 0.5$ metres, $S = 20$ m/km and $n = 0.05$.

(Manning's formula $v = \dfrac{1}{n} R^{2/3} S^{1/2}$)

Figure 6.6 presents the nomographic solution of the Manning's formula for different values of R, S and n. The nomograph permits rapid determination of the mean velocity of flow when the values of R, S and n are known. To use the nomograph, a line is drawn to join the S and n values on the respective scales, and passing through the pivot line. The point of intersection of this line with the pivot line is the pivot point. A line originating from the known value on the R scale and passing through the pivot point when extended to meet the velocity scale, provides the required value of velocity. The nomograph may also be used to determine the value of any one of the factors R, S, n and y when any three of the factors are known.

Permissible slope of earth channels. The natural slope of the land is usually the deciding factor in determining the channel bed slope. The steeper the channel, the more will be the velocity and more the discharge for the same cross-section. But excessive gradients produce very high velocities which cause erosion of the sides and bottom of the channel. Normally, a channel should slope about 0.1%. Silting may take place in the channel if the slope is less than 0.05%. For determining the bed slopes of channels the velocities should be checked and the maximum grade designed so that the velocities do not exceed the limits given in Table 6.2. Where earth channels are to be used on steep slopes, it is necessary to control the gradients and thus the velocity by constructing drop structures or by building up the channel bed.

Table 6.2. Permissible velocities for various soil textures

Type of soil	*Maximum velocity permissible cm / second*
Bare channels	
Sand and silt	45
Loam, sandy loam, silt loam	60
Clay loam	65
Clay	70
Vegetated channels	
Poor vegetation	90
Fair vegetation	120
Good vegetation	150

A change in the direction of the channel should be brought about by smooth curves and not abruptly. Similarly, when the cross-section is changed, the change should be gradual.

Channel cross-sections. The cross-section of a channel which carries large flows should be such that it will give the largest possible hydraulic radius under the natural surface conditions. Theoretically, the most efficient cross-section for an open channel is a semi-circle, as the wetted perimeter would then be minimum and its hydraulic radius maximum. In the case of rectangular cross-section, half a square is the best discharging rectangular channel, i.e., when depth = 1/2 × breadth (i.e., $d = b/2$). But semi-circular and rectangular cross-sections are possible only in small size *lined channels*.

Most lined channels which are designed to carry large flows, have a trapezoidal section and earth channels some approximation thereof. The best hydraulic cross-section of a trapezoidal channel under favourable structural condition is:

$$b = 2d \tan \theta/2 \quad \text{...(6.11)}$$

in which,

b = bed width

d = depth of flow of water

θ = the angle between the side and the horizontal.

But local topographic and soil conditions and also practical problems of construction and maintenance necessitate considerable departure from the above relationship. For instance, in lining canal water courses of comparatively smaller sections, a rectangular cross-section having a depth of about half its width has been found economical. This is because the land area occupied by a trapezoidal section of the same capacity is more than that of the rectangular section as the former requires wider embankments and its cost of construction is higher.

Lining Canals

Earthen channels are lined with impervious materials to prevent excessive seepage and growth of weeds. A large portion of the water harnessed at a high cost, through the canal irrigation network or through wells and pumps, is lost by seepage from unlined conveyance systems comprising of the main canal, branches, distributories, water courses and field channels. In permeable soils like sand and sandy loam, the losses in earth channels may be as high as 20 to 40 per cent of the water delivered to the channel. The loss in the field distribution system is even more significant in areas irrigated by wells.

Water losses in water courses and field channels may occur by seepage, by breaches along the channel through rat holes, by ponding of water in depressions and irregularities in the channel sections (Fig. 6.2) and by evaporation. The loss by seepage depends on the length of the water course, its wetted perimeter and the intrinsic permeability of the strata through which the channel passes. Seepage is usually designated in litres/hr per sq. metre of the wetted area of the channel.

One of the main problems in the use of earth channels is the control of weeds. Weeds in a channel obstruct the flow of water. If the weeds are allowed to grow upto maturity, their seeds may spread over the farm through the irrigation water. Earth channels require continuous maintenance to control moss and weed growth, and to repair damage by livestock, rodents and erosion.

There are several other problems which confront a farmer sharing water from unlined water courses. Usually, the farmer at the head of the water course gets nearly the full supply of water due to less seepage in the segment close to the canal outlet. The water available to the farmer at the tail end is usually much less than that is due to him. Lining of water courses will provide nearly equal quantity of water per unit area of the landholding to all the farmers in the culturable command area. Lining will reduce the labour cost in maintaining the water course. It will also almost eliminate waterlogging caused by seepage in water courses. More land will be available for cultivation since the cross-sectional area of the channel required to carry a given discharge through a lined channel is considerably less than the unlined channel.

Materials for Lining Large Canals and Field Channels

The substantial loss of water in the conveyance system of an irrigation project can be successfully overcome by adopting appropriate methods of seepage control. Since lining is often costly, only those sections of the canal and its laterals which are highly permeable are treated for seepage control. Various alternatives are available to control seepage losses. These include the hard surface lining using conventional masonry material, which nearly eliminates seepage and ensures long service life. Low cost lining material includes soil sealants with varying degrees of water-tightness. The following are the principal factors influencing the choice of a particular method of lining:

1. Availability and cost of the material at the site or within easy reach.
2. Labour available for the project at reasonable cost.
3. Degree of water-tightness desired.
4. Velocity of flow in the channel.
5. Useful life of the lining material.
6. Cost of maintenance.

Lining represents a significant part of the cost of an irrigation project. Hence, the guiding principle in selecting the material is the relative costs and benefits.

Cement Concrete Lining

Cement concrete linings are commonly used in canal systems. They are expensive, but when properly laid, result in water tight canals of high structural strength. They last for over 40 years (ICID, 1967).

The concrete used in lining a canal should be plastic enough for thorough consolidation and stiff enough to stay in place on the side slopes. Close control of the workability and consistency of the concrete is essential. Investigations at the irrigation research institutes in India have revealed that the degree of impermeability of a concrete mixture is improved and their costs reduced by adding *kankar lime* and *surkhi* in various proportions. (*Kankar lime*, similar to *hydraulic lime* reported in US literature, is obtained by burning *kankar*, a locally available calcarious material. *Surkhi* is an ingredient in concrete or mortar in the form of powder obtained by grinding burnt bricks or specially burnt lumps of earth.) However, field and/or laboratory tests are essential to determine the percentage of lime or *surkhi* which could be added, based on the depth of water in the canal and the extent of cracking that is expected under the prevailing climatic conditions.

Subgrade preparation. Like any other concrete structure, the subgrade (base) on which the concrete lining is placed should be well consolidated. In areas where backfilling is necessary, the material used should be well graded to prevent its piping. Where concrete lining is placed over expansive clays, a clay layer of adequate thickness is removed and replaced by sand or gravel material before placing concrete linings.

Concrete placement. Cement concrete for canal lining can be placed by a variety of methods. Jobs involving considerable lengths of large canals usually use slip forms drawn lengthwise. The slip form is supported on rails placed along both berms of the canal or on track-laying machines. Concrete placement by hand laying is usually used for lining short lengths of canals, distributories, minors and water courses. Fixed forms or templates are used in guiding the finish and ensuring correct dimensions (Fig. 6.7).

Practices vary with regard to the thickness of concrete used and the mixes adopted. In Tamil Nadu, mix of concrete range from 1:4:7 (cement : sand : coarse aggregate) to 1:6:10 and thickness of lining vary from 6 to 10 cm. A mortar layer of 1:4 (cement : sand) is applied as plaster over the concrete to obtain water tightness. In Andhra Pradesh, 10 to 23 cm thick plain cement concrete of mix ranging from 1:4:8 to 1:6:10 has been generally adopted. In Punjab, in the Nangal Hydel Canal and Bhakra canals a mix of 1:3:6 and thickness ranging from 13 to 18 cm has been used. Lining with *kankar lime* concrete 15 cm thick has been used on the Gang Canal in Rajasthan. In the Thal Canal in Pakistan, the concrete lining is 10 cm thick. The linings are laid in compartments of size not exceeding 6 m × 6 m. Alternate blocks are laid and cured before laying concrete in blocks in between. In Greece, the thickness of concrete linings vary from 7 cm to 12 cm. In Australia, the standard adopted for concrete lining varies from 5 to 7.5 cm in thickness. Joints are provided at 2.5 m interval to prevent uncontrolled cracking (ICID, 1967).

Joints and grooves. The main cause of failure of concrete lining is the development of cracks. Cracks develop by the initial contraction and subsequent tension produced by seasonal temperature changes and by the temperature gradient between the exposed surface and the bottom of the concrete layer. It may also result from inadequate curing of the concrete after laying, which results in quick drying of the surface of the concrete layer.

Both transverse and longitudinal grooves are provided in concrete lining to absorb cracking (Fig. 6.8). The groove is placed at the top one-third thickness of the concrete layer. The width of the groove ranges from 0.6 to 1.2 cm in concrete layers of 5 to 10 cm thick. The groove is usually wedge shaped with the top end about 3 mm wider than its bottom.

Satisfactory curing of the freshly laid concrete layer by sprinkling of water or placing damp covers is essential to prevent quick drying and consequent cracking. The subgrade should be moistened well prior to placing the concrete. In exceptional cases, the lining has to be properly reinforced. Methods of repairing cracks in concrete lining include the use of pneumatically applied (shotcrete) mortar, spraying asphalt cement and application of asphaltic matting. Careful laying and proper curing will minimise maintenance problems in concrete lining.

Fig. 6.7. Concrete lining of a trapezoidal water course. Note the metal frame used to obtain the correct dimensions.

Cement Mortar Lining

Cement mortar has many of the characteristics of cement concrete. The mortar is usually pneumatically applied, that is, shot into place by pneumatic pressure. Hence, it is often referred to as shotcrete. Mortar linings

Fig. 6.8. A trapezoidal concrete lined canal water course. Note the grooves in the lining.

have given satisfactory service in many irrigation projects where the temperature variations are not high and the canals are kept full all the year round and where the subgrade material is stable. It has the advantage of being more easily placed over rougher subgrades than concrete lining. Hence, they are particularly adapted for use in rock cuts where trimming to exact line and grade is expensive.

Mortar lining may or may not be reinforced with steel, even though it has been observed that simple reinforcement with wire mesh greatly increases the life of the lining. Shotcrete lining of 2.5 cm thickness, reinforced with light wire mesh has been used extensively in the Salt River Valley in Arizona (USA) with varying degrees of success. The thickness of the mortar layer, usually adopted in practice, ranges from 4 to 5.5 cm. The usual mortar mix is 1:4.5 (cement: good quality sand). It is desirable to provide expansion joints or grooves as in the case of concrete. Curing is important in mortar linings. The maintenance requirements of mortar lining include the filling of cracks with elastic material soon after they occur.

Bricks (or tiles) with sandwiched mortar linings. Bricks (or tiles), sandwiched in cement mortar, is the most popular lining material used in the plains in India (Fig. 1.52, 1.53 and 1.54, Chapter 1). It was first used in the Haveli canal and a part of the Thal canal (both now in Pakistan), and later on the Sarda canal, the Bhakra canal, the Rajashan canal and other canal systems. In the plains, stones and sand have often to be transported from long distances for concrete lining; while for sandwiched mortar lining, bricks can be made near the construction site. The lining consumes much less cement and sand as compared to concrete. It can be laid by skilled manual labour without the use of machinery. The lining is economical where labour is cheap. When properly constructed, it has almost the same degree of impermeability and service life as with cement concrete lining. The joints in bricks in the sandwiched mortar linings provide numerous hair cracks and hence do not develop major cracks due to temperature variations as in cement concrete linings.

Sequence of construction. The canal is constructed to design specifications and the subgrade prepared suitably. Bricks (or tiles) with their top ends at correct formation levels are placed at, say, 7.5 m centres in the bed of the canal along the centre line. The bottom line of the side slopes (tangent points) are similarly marked by setting out from the centre line tile. The top of the bottom layer of brick lines of side are also fixed to correct formation levels. Templates of appropriate dimensions are used to obtain the correct profile of the side slopes. A layer of mortar about 1 cm thick is laid on the compacted and wetted subgrade. The first layer of bricks is laid with their face having the groove upwards to obtain grip on the sandwitched plaster. The bricks on the bed are laid with their lengths at right angles to the centre line of the canal, while those on the sloping sides are laid parallel to it.

To lay the bricks, mortar is applied to the side of one tile and the next tile struck after putting it adjacent to the first so that the mortar adheres to it. The excess mortar is squeezed out and cleaned by brushes before it hardens. In canal linings as in any good masonry construction, strength increases by making the mortar joints uniform and as small in width as possible. While laying the bricks, it may be remembered that canal lining is a water retaining structure where a defect in one part is likely to ruin the whole structure.

The first layer of bricks is allowed to cure for 3 to 4 days by keeping it wet by sprinkling water. The sandwiched mortar layer of thickness about 1.5 cm is then placed over the bricks layer. It is important to maintain uniform thickness in the sandwich layer. A wooden form is usually used to guide the finishing of the sandwich layer to the correct thickness. The finished surface is roughened with a brush to obtain proper bond with the upper layer of bricks.

The sandwiched mortar layer is kept wet and allowed to cure for a day. The top layer of tiles is then laid. It is bedded in a 0.5 cm layer of cement mortar. The procedure in laying the top layer of bricks is the same as the lower layer, but with greater care. However, the grooved faces of the bricks are laid against the sandwich mortar layer to obtain the grip and maintain a smooth surface at the top. The top layer is kept wet by sprinkling water for a day and is cured for 3 to 4 weeks by retaining water between earthen bunds constructed in the bed and sprinkling water on sides. Curing of sides is best done by constructing a perforated drain on the top banks and keeping it filled from a tank at a higher level. It can also be done by laying portable gated aluminium or PVC pipes and maintaining a low pressure head. Wet gunny bags are sometimes employed to prevent rapid drying by direct exposure to the sun.

As with the concrete lining, the success of cement mortar sandwich linings depends on the degree of compaction of the subgrade. To prevent the subgrade absorbing moisture from a newly laid lining, it should be kept thoroughly moist.

Single layer brick (or tile) lining. Single layer brick (or tile) lining can be used in minor canals, distributories (Chapter 1) and water courses. Even in case of large canals, the bottom could often be made with a single layer of bricks while the sides are lined with the sandwitched layer. Single layer bricks are laid on a compacted 1 cm thick 1:5 cement mortar bed. The bricks are also laid in 1:5 cement mortar. The procedure in brick laying is the same as in the sandwich lining.

Hot Applied Asphalt Membrane Lining

Hot applied asphalt membrane serves as a low cost lining in irrigation canals and tanks. It is reported to have been first used on a large scale in the Klamath Project in California, USA in 1947. When protected from weed growth the material provides satisfactory protection against seepage for decades. The lining consists of a 6 mm layer of high-softening point asphalt sprayed in-place at a high temperature (204°C) on a prepared subgrade. The asphalt layer is usually protected against damage and weathering by a layer of earth or gravel. The subgrade is given a smooth finish by roller or plank (dragger) to provide for the application of a uniformly thick layer of asphalt. The quantity of asphalt required is about 6 litres per sq m of the surface to be covered. Hot asphalt is applied at a pressure of about 3.5 kg/sq cm through slot type spray nozzles, using hand sprays or multiple spray bars mounted on a distributor.

Asphaltic concrete is sometimes used in preference to asphalt membranes. A 4 to 5 cm thick layer of asphaltic concrete provides a nearly water-tight surface. The mix consists of well sieved aggregates with about 7 to 10% asphalt content. The mix is laid by spreading with shovels and rakes or mechanically by spreader box or slip form. Pre-fabricated asphaltic linings have been developed for lining canals. They are easy to lay and do not require any cover for protection.

Lining with LDPE Film

Low density polyethylene (LDPE) film as a lining material in medium size perennial canals was introduced in India in 1957. Its use continued to develop, and it is being used as one of the approved types of lining in Gujarat, Madhya Pradesh, West Bengal, Rajasthan, Punjab, Haryana and some other States. However, certain properties of LDPE film such as ageing, resistance to sunlight, and microbiological action, stability in side slopes, bond with cement concrete/cement mortar continue to be under investigation. The polyethylene film used is of 400 g (100 microns), black in colour available in 35 to 40 m lengths and 3 m width. The following procedures are adopted in lining with LDPE film:

(*i*) The canal section is over-excavated by 15 cm and the excavated soil is carried on to the embankments.

(*ii*) For affording proper cushion to the polyethylene film, a layer of 1.2 cm thick sand is laid.

(*iii*) The polyethylene film 400 g, is laid along the canal over the bed and slopes to the shape.

(*iv*) This is then covered by soil to a thickness of 25 cm. Light tamping is done with the help of wooden mallets.

The surface of the subgrade need not be very smooth, but it should be free from any sharp and protruding object which might puncture the films. The black polyethylene film 400 g manufactured by M/s. Union Carbide India Ltd., Mumbai-1 is frequently utilised. Any two widths of the polyethylene film can be joined by heat sealing which can be done electrically. However, where polyethylene films are covered with about 30 cm earth or with hard rigid material like bricks or concrete slabs, joining may not be necessary.

LDPE film lining with soil cover has been used with advantage in the Kangsabati Irrigation Project in West Bengal. Apart from the non-availability of construction materials, the main problem was the scarcity of water for construction of hard surface lining. The project area lies in highly porous lateritic zones, having high rate of seepage both in the field and in the canals. The inherent high permeability of the soil does not permit the storage of water in tanks.

The earthwork related to the excavation for film laying and placement over the film comprises the major cost of the burried membrane lining. Obviously, for economy, the material should be locally available and placed as early as possible for adequate protection of the membrane. IS:9698-1980 'Code of Practice for Lining of Canals with Low Density Polyethylene Film' lays down that the soil used for cover material should be such that it is not likely to damage the film and it would not be easily eroded by water. For effecting economy, as far as possible the soil of the canal excavation could be used as covering material.

In actual practice, locally available soil is used as the cover material, as the transportation of material from long distances is not economical. Should the soil of the canal excavation be found unsuitable, it would be advisable to go in for a rigid cover.

Cover thickness and side slopes. No uniform criteria is being followed to decide the cover thickness. IS:9698-1980 lays down that minimum thickness of cover over the film shall not be less than 30 cm in bed and 60 cm on the sides, for canals with velocities less than or equal to 60 cm/sec. Studies have indicated that the minimum side slope of a channel for LDPE film lining is 2:1. Flat slide slopes are required to prevent the slipping of the cover material in the membrane.

Sand cushion. Specifications laid down by the Bureau of Indian Standards, IS:9698-1980 prescribe that a layer of fine sand shall be laid over the subgrade (bed) to facilitate working conditions over the film. About 25 mm thick sand cushion is provided in bed and on sides for LDPE film lining with rigid cover in Gujarat. It is justified on the ground that black cotton soil subgrade has imperfections, irregularities and projections which, in the absence of the sand cushion, may puncture the film. Site specific investigations are to be carried out on the various design details and economic advantages before large scale lining works using LDPE film are undertaken.

Anti-weed Treatment

A major factor influencing the durability of hot applied asphalt membranes as well as asphaltic concrete linings is excessive weed growth. Weeds pierce through the lining and slowly disintegrate it. Weed damage is high in the layer above the water surface. Weeds get established in the flow section during the periods when the canal is not in use. Treatment of the subgrade with a soil sterilent would minimise the problem. Sodium chlorate, as a 5% solution in water at 2 litres per sq m is a commonly used sterilent. Pentachloro-phenol in oil distillates and chlorates and boron compounds in water solution are amongst the other sterilents used. There are also other chemicals with varying degrees of effectiveness. The soil is to be left undisturbed for 4 to 5 days after spraying the sterilent.

The feasibility of asphaltic lining in irrigation canals in a given region would depend on the relative cost of the materials, the severity of the weed problem and the extent of use of the canal system.

Earth Linings

Various types of earth materials have been tried to reduce seepage in canals. The simplest form of earth lining consists of a silt layer deposited in the flowing stream. It can be cast loosely compacted in the canal subgrade. Puddling the subgrade with silt and sediment increases the effectiveness of the earth lining. Buried bentonite or clay membrane type linings have been adopted for reducing seepage in canals. Bentonite and bentonitic type clays are characterised by high water absorption, followed by imperviousness. Their effectiveness depends on the quality of the material used and the method of placement. Field trials conducted in USA with 2 to 5 cm thick bentonite layers covered with 15 to 30 cm thick earth layers have shown satisfactory results.

Resins, chemicals and petro-chemicals have been used in the stabilization of earth type linings. The use of chemically stabilised soil for canal linings is still in an experimental stage.

Lining of Minor Distributories, Water Courses and Field Channels

General principles and procedures for lining large canals apply to smaller channels including minor distributory, water course and field channels (Fig. 6.8 to 6.13). Concrete and brick or stone masonry are the commonly used materials for lining small channels. Well mixed and well made cement concrete lining, and bricks or stones laid in cement or lime mortar, provide virtually water proof channel lining. Limited use has also been made of bituminous mixtures, soil cement, chemical sealants, and impervious earth materials. Most of these are susceptible to damage by trampling by livestock, attack by insects, weed growth and erosion by high velocity flows. Their short life often makes them uneconomical unless special protection is provided.

Concrete linings. Concrete may be placed in forms or may be applied as a plaster to the sides and bottom of the channel.

An embankment on which the channel lining is to be placed should be thoroughly compacted by tamping, rolling or water soaking. When new embankments are made, they should be compacted in approximately 15 cm layers. The top width of the embankment should be about 40 to 45 cm or more. A free board of at least 5 cm should be allowed. The side slopes can be steeper than in earth channels. They should not normally exceed slopes of 1 to 1. However, when concrete is hand placed and the channel is over 60 cm deep, side slopes can sometimes be as steep as 3/4 horizontal to 1 vertical. When steeper slopes or vertical side walls are desired, forms are necessary to hold the concrete in place until it sets. Non-reinforced concrete linings with vertical sides are not suitable for depths more than about 40 cm.

A mixture of one part cement, three of sand and four or five of graded gravel of 1 to 3 cm size, makes a strong and durable lining. In alkali soil, an alkali-resistant cement should be used. A water cement ratio of 25 litres per bag of cement has been found to give good results. The linings should not be less than 3.5 cm thick. A wooden template made to fit the inside dimensions of the channel will be useful in obtaining a uniform cross-section of the lined channel (Fig 6.7).

Transverse contraction joints are used in concrete channel linings to control transverse cracking (Fig. 6.8). Cracking results from shrinkage during volume changes caused by changes in temperature or moisture content. The spacing between transverse contraction joints to reasonably control cracking should be 2 to 3 metres. The lining is laid in lengths of 2 to 3 metres. Contraction joints are formed by constructing a vertical groove in the top third of the concrete. For example, if the thickness of

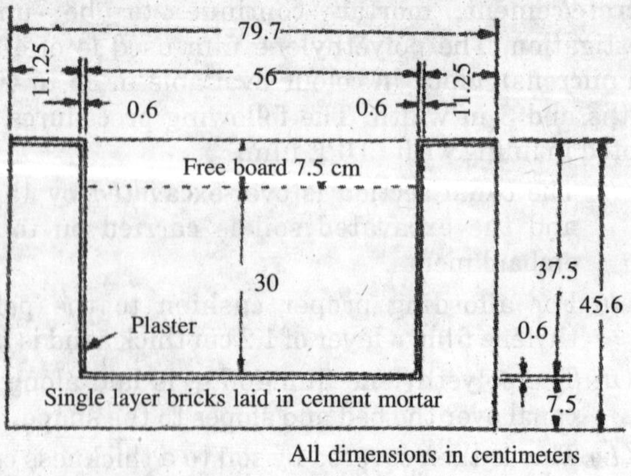

Fig. 6.9. Cross-section of a brick lined channel in Example 6.2.

Fig. 6.10. A rectangular water course in Haryana lined with bricks on sides and stone slabs at the bottom.

Fig. 6.11. A rectangular brick-lined water course constructed on a freshly formed embankment. The lining was badly damaged due to settlement of embankment in the following monsoon season.

the lining is 4.5 cm the groove should be 1.5 cm deep from the upper surface. The groove can be made of a flat steel piece about 4 to 6 mm thick. This groove should be filled with a suitable sealing compound like bitumen before the channel is placed in operation.

Proper curing is necessary to obtain the maximum strength and durability of the concrete. Curing also prevents rapid drying of the concrete, thereby reducing the possibility that shrinkage cracks will occur while the concrete is still plastic. Curing may be done by keeping the concrete damp by sprinkling water frequently or by covering with wet gunny bags.

Brick or stone masonry lining. Single layer brick or tile or rubble stone may be used to obtain satisfactory lining of field irrigation channels (Fig. 6.9 to 6.13). The bricks, tile or stones are laid flat on the compacted sides and bottom of the channel and the joints are filled with cement mortar. The mortar should have a cement-sand ratio of 1:4. The choice between brick, flat clay tile or stones depends mainly on their availability and relative costs.

Example 6.2. *Estimate the mean velocity of flow and carrying capacity of a lined canal water course, rectangular in section with a bottom width of 50 cm and depth of flow of 25 cm (inside). Single layer bricks are laid in cement mortar with a cement plaster 8 mm thick. The slope of the channel bed is 2 metres per kilometre.*

Fig. 6.12. Zig-zag alignment of a water course in a canal command area resulting from inadequate mobilisation of the beneficiary farmers who did not agree to a straight alignment.

Solution:

Area of cross-section of the channel,

$$a = \frac{50}{100} \times \frac{25}{100}$$

$$= 0.125 \text{ m}^2$$

Wetted perimeter, p $= 0.5 + 0.25 + 0.25$

$= 1$ m

Hydraulic radius, R $= \frac{a}{p} \quad \frac{0.125}{1}$

$= 0.125$ metres

Hydraulic slope, S $= \frac{2}{1000} = 0.002$

Manning's n $= 0.015$ (Table 6.1)

Mean velocity of flow, $v = \frac{R^{2/3}\, S^{1/2}}{n}$

$= \frac{(0.125)^{2/3}\,(0.002)^{1/2}}{0.015}$

$= 0.746$ m/sec

Discharge capacity, $Q = a \times v = 0.125 \times 0.746.$

$= 0.093$ cu m/sec

$= 93$ litres/sec

Precast Concrete Channel Sections

Precast concrete is a term applied to concrete units that are manufactured at a central place and hauled to the job site. Their low cost and operating advantages make them particularly suitable for handling comparatively low discharges. Precast concrete channel sections may be made in semi-circular or U-sections. The semi-circular channels or 'half pipes' are made especially for irrigational use by commercial manufacturers of concrete pipes. They are made in lengths of one metre or more and are generally 30 to 45 cm in diameter. The area of land occupied by the channel is greatly reduced when precast concrete sections are used since side embankments are unnecessary.

Low cost concrete channel sections can be made in a U-shaped wooden form (Fig. 6.14) by tamping in a concrete mixture made with a bare minimum of water. The wooden form can be made by a local carpenter. Hard wood boards 2.5 cm thick are used to make the form. Removable pins at corners permit removal of forms without disturbing the newly cast section. The size of the form will depend on the size of the irrigation stream to be conveyed. The quantity of water that may be conveyed by a channel of a given section will depend on the slope of the channel. They are usually suitable for flows upto 30 litres/per second.

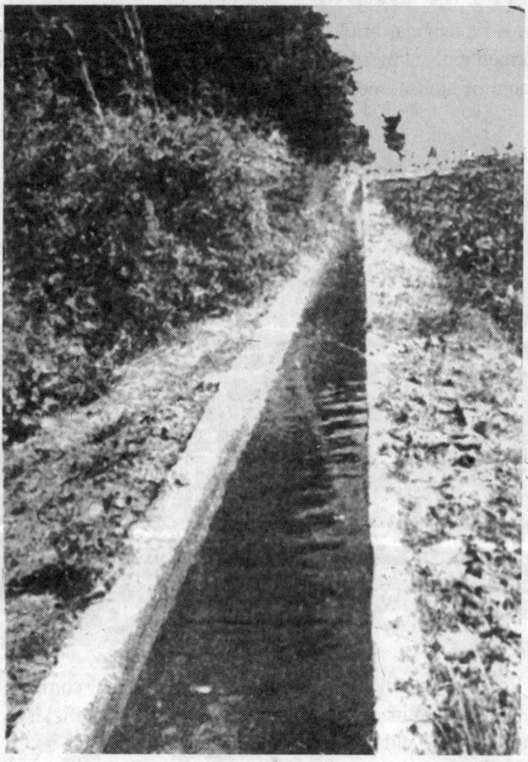

Fig. 6.13. A brick lined canal water course near Faridkot, Punjab. Lining of this section of the channel has saved about 20 per cent of water that was otherwise being lost by seepage. The channel bed which has been given a uniform slope has increased the command area of the channel, thus reducing land levelling cost.

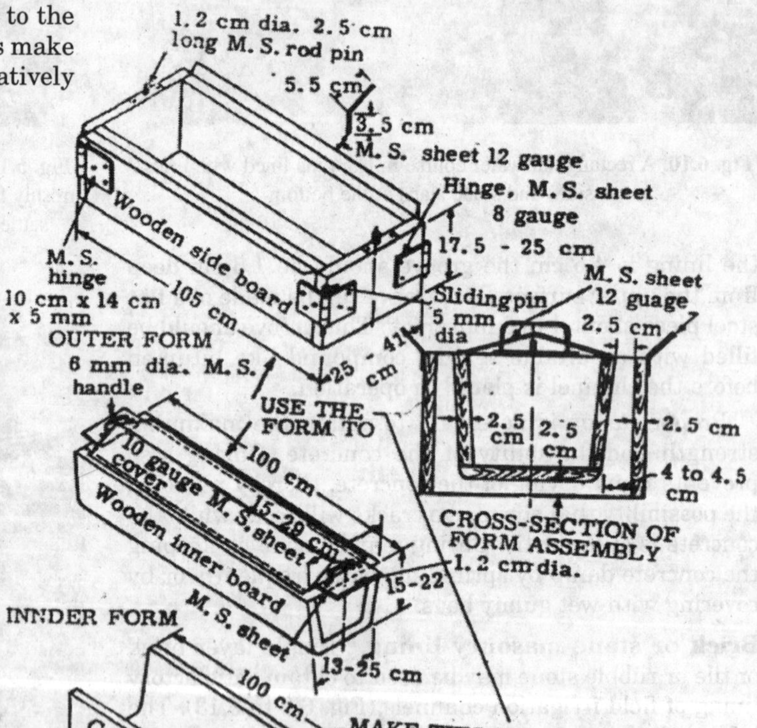

Fig. 6.14. Wooden form to make precast concrete channel sections. Any suitable size within the range indicated may be used, depending on the size of the irrigation stream.

A mixture of cement, sand and gravel in the ratio of 1 cement: 2½ sand: 3½ gravel is suitable to make precast concrete channels. The gravel size should not exceed about one centimetre. Only the minimum quantity of water necessary to make the mixture workable should be added.

To make a concrete section, the outer form is first placed over a thin layer of sand spread on a level ground. A 2.5 cm thick concrete layer is first put to form the bottom and tamped. The inner form is then placed over the concrete bed and the side gaps are filled with concrete. The concrete is tamped into the form with a steel rod (Fig. 6.15). To remove the form, after the gap of a few hours (if possible), the inner form is lifted up slowly (Fig. 6.16). The pins of the outer form are then removed and the side boards are slid out slowly without damaging the freshly cast section (Fig. 6.17). The new concrete section is protected from sun and rain for about 24 hours and then cured for over a week by sprinkling water frequently. After curing, the inside of the channel section is coated with a thin cement slurry to prevent leakage through the micro-pores in the concrete section (Fig. 6.18).

Fig. 6.15. Casting concrete channel section (non-reinforced), using the wooden form illustrated in Fig. 6.14. Note the partially filled sides being tamped thoroughly before adding subsequent layers of concrete.

Fig. 6.16. The inner form is lifted and removed after the concrete is tamped in the form.

Fig. 6.17. The side boards are slid out slowly after the inner form is removed.

Fig. 6.18. Coating the inside of a precast channel section with cement slurry.

For laying the concrete sections, a suitable small earth embankment is made (Fig. 6.19 and 6.20). The embankment is compacted while moist, and graded to obtain the desired bed slope. The precast concrete sections are then laid in straight lines. After laying the sections, their joints are filled with cement mortar

Fig. 6.19. Laying precast concrete channel sections. Note the compacted channel bed. The joints are made with cement mortar.

Fig. 6.20. A well-laid out water course lined with prefabricated concrete channel-sections and provided with masonry turnouts with check gates at the head-end of each pair of fields in a cooperative farming system organised by a group of small farmers in the Alipur Block in Delhi State, (1976).

(1 cement : 2 sand). It is necessary to cure the joints by sprinkling water frequently for three or four days or by ponding water in the newly laid channel.

Water Conveyance through Pre-fabricated Canalets

Pre-fabricated canalets of reinforced concrete construction (RCC) are becoming increasingly popular for water conveyance in many countries. Canalets are usually semi-elliptical in shape and are often made in 5 m long sections (Fig. 6.21 and 6.22). They can be made

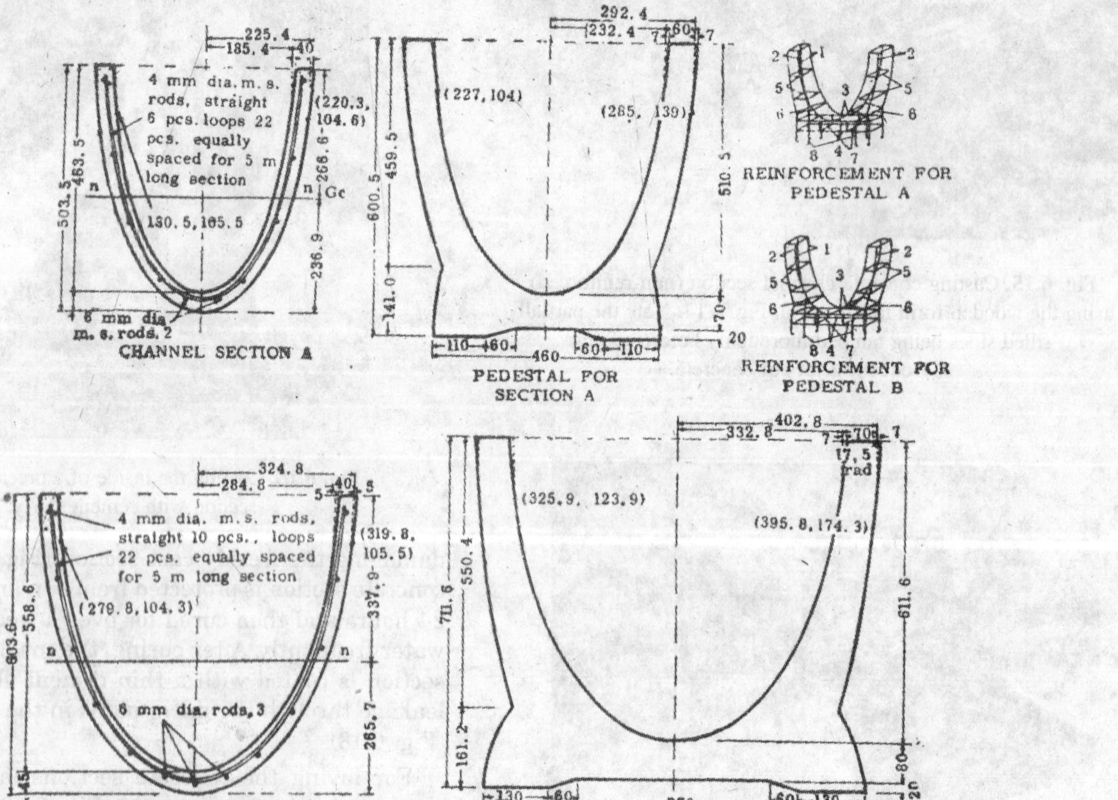

Fig. 6.21. Two typical sections of canalets and their pedestals suitable for canal minors and water courses. The discharge capacity of section A varies from 25 lit/sec at 0.1% channel slope with a water depth of 42 cm and a free-board of 4.35 cm. Channel section B has a capacity ranging from 58.6 lit/sec at 0.1% slope to 185.2 lit/sec at 1% slope with 50 cm water depth and 5.86 cm free-board.

Table 6.3. Specifications of reinforcements for the pedestals shown in Figures 6.21 and 6.22

Channel section A						*Channel section B*						*Channel section C*					
Piece No. (see fig.)	*Quantity*	*Dia. (mm)*	*Length of piece (cm)*	*Total length (cm)*	*Weight (kg)*	*Piece No. (see fig.)*	*Quantity*	*Dia. (mm)*	*Length of piece (cm)*	*Total length (cm)*	*Weight (kg)*	*Piece No. (see fig.)*	*Quantity*	*Dia. (mm)*	*Length of piece (cm)*	*Total length (cm)*	*Weight (kg)*
1	2	8	122	244	0.956	1	2	8	155	310	1.215	1	2	8	210	420	1.646
2	4	5	68	272	0.416	2	4	5	82	328	0.501	2	4	5	100	400	0.612
3	2	5	43	86	0.131	3	2	5	74	148	0.226	3	2	8	107	214	0.838
4	2	6	43	86	0.190	4	2	8	74	148	0.580	4	2	12	107	214	1.887
5	6	5	28	168	0.257	5	6	5	32	192	0.293	5	6	5	36	216	0.330
6	2	5	46	92	0.140	6	4	5	58	232	0.354	6	4	5	64	256	0.391
7	3	5	26	78	0.119	7	5	5	30	150	0.229	7	8	5	40	320	0.489
8	2	5	47	94	0.143	8	2	5	53	106	0.162	8	2	5	67	134	0.205
			Total		2.352						3.560						6.398

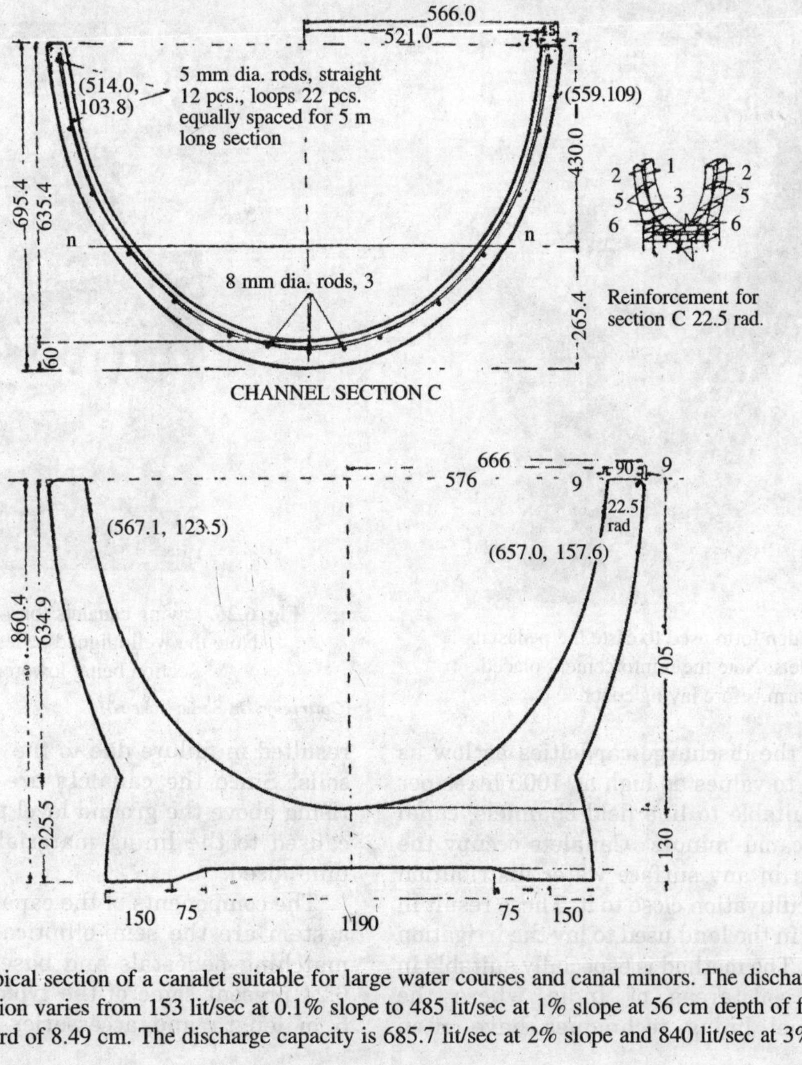

Fig. 6.22. Typical section of a canalet suitable for large water courses and canal minors. The discharge capacity of the section varies from 153 lit/sec at 0.1% slope to 485 lit/sec at 1% slope at 56 cm depth of flow and a free-board of 8.49 cm. The discharge capacity is 685.7 lit/sec at 2% slope and 840 lit/sec at 3% slope.

Fig. 6.23. Fabrication of a semi-elliptical canalet using a wooden form on which is cast the ferrocement canalet at IARI, New Delhi. (1976).

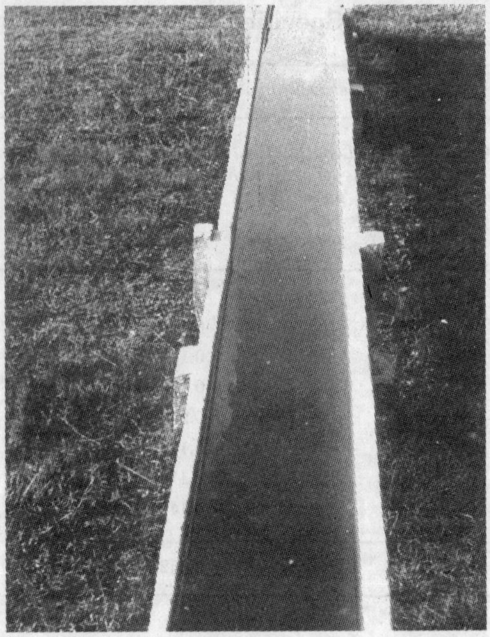

Fig. 6.25. A canal minor made of prefabricated reinforced concrete canalets mounted on R.C.C. pedestals.

Fig. 6.24. A wooden form used to caste the pedestals for raised canalets. Note the reinforcement placed in the form before laying concrete.

Fig. 6.26. Laying canalets for a large canal water course. Note the well-aligned pedestals and the channel section being lowered with a crane

(***Courtesy***: Dr. Sadiq Toksos).

of any size suiting the discharge capacities as low as 20 litres per second to values as high as 1000 litres per second. They are suitable to line field channels, canal water courses and canal 'minors'. Canalets occupy the minimum land area in any surface water distribution system and permit cultivation close to it. These result in considerable saving in the land used to lay the irrigation distribution system. The method is especially suitable in the black cotton soil areas of India where the conventional channel lining techniques have often resulted in failure due to the extensive cracking of the soils. Since the canalets are supported on pedestals rising above the ground level the damage that could be caused to the lining material due to soil cracking is minimised.

The components of the canalet irrigation distribution system are the semi-elliptical RCC channel sections, matching pedestals and base blocks. Figures 6.21 to 6.24 present some of the typical designs of canalets of 5 m length and accessories. Table 6.3 presents the specifications of reinforcements for the pedestals shown in Fig. 6.21 to Fig. 6.22. Figure 6.25 shows canal water courses formed of canalets supported on RCC pedestals and base blocks. The discharge capacity of the canalets can be estimated using the Manning's formula. The channel sections and pedestals are formed of good concrete mix of 1:2:4 (cement : sand : coarse aggregate) ratio. Wooden or metal forms of matching sizes are used to make the channel sections and pedestals (Fig. 6.23 and 6.24). Factory-made sections are more competitive in price and have better strength.

While installing the canalets on the farm, care is taken to align the pedestals properly and lay them and with a uniform gradient (Fig. 6.25). Pulley blocks mounted on tripods or small cranes are used in laying large sections (Fig. 6.26). An advantage of the canalet water distribution system is its suitability for laying on uneven land with RCC props erected on base blocks to support the pedestals (Fig. 6.27 and 6.28).

Cost Analysis of Surface Irrigation Distribution System

The costs of lining irrigation channels and constructing the allied structures vary with the prevailing cost of materials, labour, and other charges. The cost components include the expenditure on earth work, masonry materials, laying the lining, plastering, and on engineering and supervision. The cost is usually expressed on a unit length basis such as cost per metre length. In estimating the total cost, the proportionate expenditure on ancillary structures is added to the cost of lining the channel. Example 6.3 shows the cost estimate of lining a water course, on the basis of the average rates prevailing in Haryana in 1973.

Fig. 6.27. Pre-fabricated canalets installed on matching pedestals. Note the base blocks supporting the pedestals. (***Courtesy***: Dr. Sadiq Toksos).

Fig. 6.28. A canal water course formed of prefabricated canalets crossing a natural driangeway. RCC props of suitable, heights resting on base blocks support the pedestals holding the channel sections. (***Courtesy***: Dr. Sadiq Toksos).

Example 6.3. *Design a brick-lined rectangular water course to carry a discharge of 42.5 litres per second. Estimate the cost incurred in lining the channel. The water course has a uniform slope of 0.02 per cent.*

Solution:

$$q = 42.5 \text{ litres/sec} = 0.0425 \text{ cu m/sec.}$$

As a first approximation, on the basis of the available size of bricks and their possible arrangements in laying, the depth and width of the channel are assumed to be 30 cm and 56 cm, respectively.

$$a = b \times d = 0.30 \times 0.56 = 0.168 \text{ sq m}$$

$$p = b + 2d = 0.56 + (2 \times 0.30) = 1.16 \text{ m}$$

$$R = \frac{a}{p} = \frac{0.168}{116}$$

$$= 0.1.448 \text{ m}$$

$$S = 0.0002$$

$$n = 0.015 \qquad \text{...(from Table 6.1)}$$

$$v = \frac{R^{2/3}S^{1/2}}{n}$$

$$= \frac{(0.1447)^{2/3} \times (0.0002)^{1/2}}{0.015}$$

$$= 0.26 \text{ m/sec} = 26 \text{ cm/sec}$$

$$q = a \times v = 0.168 \times 0.26$$

$$= 0.0435 \text{ cu m/sec} = 43.5 \text{ litres/sec}$$

The section is suitable as the velocity is within permissible limits and the discharge capacity is sufficient.

Cost estimate per metre length of channel

(*i*) Earth work quantity, average value of 0.77 cu m
Cost of earthwork @ Rs 22/cu m = Rs 1.71

(*ii*) Brick work
Sides: 2 × 0.456 × 0.1125 × 1 = 0.1026 cu m
Bottom: 1 × 0.56 × 0.075 × 1 = 0.0420 cu m
0.1446 cu m
Cost of brickwork @ Rs 250 per cu m = Rs 12.00

(*iii*) Plastering
Sides: 2 × 1 × 0.375 = 0.750 sq m
Bottom: 1 × 1 × 0.56 = 0.560 sq m
Top: 2 × 1 × 0.1125 = 0.225 sq m
1.535 sq m
Cost of plastering @ Rs 31.75 per 10 sq m = Rs 4.88

(*iv*) Cost of gates, inverted siphons, survey, etc. (proportionate

= Rs 2.30 per metre length cost)

Sub-total of items (*i*) to (*iv*)

= 1.71 + 12.00 + 4.88 + 2.30

= Rs 20.89

(*v*) Cost of design, supervision and other miscellaneous costs @ 27°% of the sub-total of items (*i*) to (*iv*)

= Rs 5.75

Total cost = Rs 20.89 + 5.75

= Rs 26.64

Inside surface area of channel (including free-board) per metre length of channel

$= (2 \times 0.375 + 0.56) \times 1$

= 1.31 sq m

Cost per sq m of inside surface area

$$= \frac{26.64}{1.31} = \text{Rs } 20.33.$$

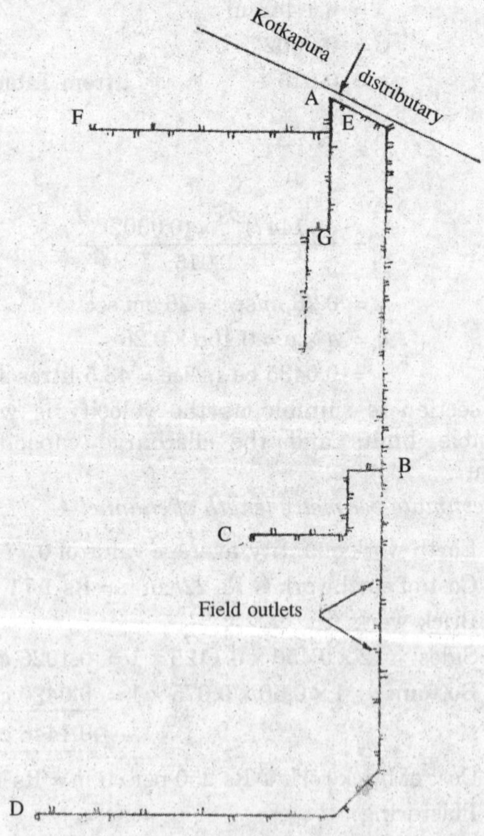

Fig. 6.29. Layout of water course R.D. 34790 in Kotkapura distributory in Punjab.
(***Source:*** Khepar, S.D. and Rai, A. (1979)).

Optimum Length of Water Course to be Lined

The seepage loss in a water course is directly proportional to the length and time for which the water flows in each reach. Different reaches of a water course run for different periods. The upper reaches run for a much longer time than the lower ones. Therefore, the seepage loss per unit length of a water course is larger in the upper reaches than the lower ones. However, the cost of lining of a unit length of the water-course remains constant. Studies have revealed that by lining the upper two-third length of a water course, over 80% of the seepage loss in a water course is saved. A decision on the length of the water course to be lined could be based on economic consideration, i.e., benefit-cost ratio or social compulsions. Khepar and Rai (1979) suggested that normally the length of the water course to be lined could be decided on the basis of benefit-cost ratio of 1.

The cost of lining per unit length (Rs/m) may be estimated adopting the procedure generally used for determining annual cost (i.e., considering fixed cost (its life, interest on the investment, repair and maintenance). Then the benefit-cost ratio may be determined as

$$B.C. \text{ ratio} = \frac{BLi}{Ci \times Li} \quad ...(6.12)$$

where $B.C.$ ratio = Benefit-cost ratio

BLi = Benefit from lining of length L of ith arm of the watercourse (Rs.)

C = Cost of lining per unit length (Rs/m)

Li = Length of the water course of the ith arm (m)

The benefit-cost ratio may be determined for lining of different lengths of different sections of a water course. Decisions on length of lining of a watercourse could be taken based on the desired benefit-cost ratio to be applied. There may, however, be social compulsions to line the full length of a main water course when the beneficiary farmers of a water course at its lower reaches may not agree to limit the lining to upper reaches only.

Kheper and Rai (1979), in a case study of water course No. 34790 of the Kotakapura distributory in Punjab (Fig. 6.29), observed that only the lining of the main water course could be economically justified. They studied the weekly operation schedule of water supply in distinct reaches of the water course on a weekly rotational irrigation supply system (*warabandi*). Thus, in Fig. 6.29, lining of section AD of the water course, namely, the trunk line could be justified, while the branch water courses FE and BC were uneconomical.

ON-FARM STRUCTURES FOR WATER CONVEYANCE, CONTROL AND DISTRIBUTION

Appropriate structures are integral components of an effective and efficient water conveyance and distribution system at the farm level. In many of the Command Area Development Projects, component structures which are essential for a lined water conveyance system are not given the due attention. On-farm strucutres are equally important as lining the water conveyance system. These structures include structures to check channel erosion, to maintain the desired water level in the operating section of the channel, for crossing obstructions and for the delivery of the desired quantity of water to the fields.

Structures to Control Erosion in Irrigation Channels

Often it is necessary to build irrigation channels on land slopes so steep that the water will attain erosive velocities. Severe erosion will occur in earth channels if structures to control the slope are not provided. Drop structures and chute drops are used to prevent erosion in field channels.

Drop Structures

Drop structures are used to discharge water in a channel from a higher level to a lower one. They may be open type drops or pipe drops.

Open drop structures. Open drop structures can be made of timber, concrete, or brick or stone masonry. Timber is usually not preferred due to its short life. Low cost drop structures can be built by using discarded drums or barrels of the type used to transport coal tar (Fig. 6.30).

The components of a drop structure are the inlet, the vertical overfall section, and the outlet. Figure 6.32 shows the basic components and typical dimensions of an open drop structure. The check gate provided at the inlet of the drop structure is used to control the water

Fig. 6.30. A low cost drop structure made from a discarded drum. *Photo*: G.C. Kneirim.

Fig. 6.31. A farmer operating the wooden shutter fixed at the inlet of a masonry drop structure. Note the stilling basin downstream.

surface elevation on the upstream stretch of the channel. The minimum width of the inlet of a drop structure is equal to the bottom width of the irrigation channel. Water enters the structure through the inlet which is in the form of a weir or notch in a wall. Vertical walls, known as cut-off walls, extend down into the soil under the inlet in order to prevent water seepage under the structure. Similar walls extending from the inlet to prevent seepage around the ends of the structure are called headwall extensions. Water falls into a stilling basin, which is an essential part of an erosion control structure. The stilling basin controls the erosive force of water. The length of the stilling basin is nearly twice the height of the drop. A small cross wall, the end still, about 10 to 12 cm high, placed at the end of the basin increases the efficiency of energy dissipation. Instances of serious channel erosion and improper design of drop structures are frequently encountered in many of the irrigation command areas (Fig. 6.33 and 6.34).

Drop structures often set up eddy currents in the irrigation stream and these currents tend to cause erosion of the channel section immediately downstream from the structure. Stones or brick-bats placed over a length of one to two metres from the structure help to prevent channel erosion near erosion control structures. There are many instances where farm irrigation structures do not receive the required attention from the design point of view (Fig. 6.35). In Figs. 6.33 and 6.34 the absence of an end sill at the downstream end of the drop structure will induce erosion in the immediate lower reach of the structure.

When earth channels are to be built on steep slopes or on bench terraced lands, it is necessary to construct a series of drop structures (Fig. 6.36) to flatten the channel grade. The section between succeeding drop structures is nearly flat.

Pipe drop structures. Sometimes construction of an open drop structure is not possible without disturbing an existing bund or dam. In such cases water can be safely discharged from a higher level to a lower one by providing a pipe drop as shown in Figs. 6.37 and 6.38. This type of structure allows the discharge of water through a pipeline, leaving the bund or dam undisturbed. Vitrified sewer pipes or concrete pipes made with bell joints or corrugated metal pipes are used as the conduit. A water-tight lid at the inlet will function as a check gate. The corner in the pipeline is made of a large radius bend. A

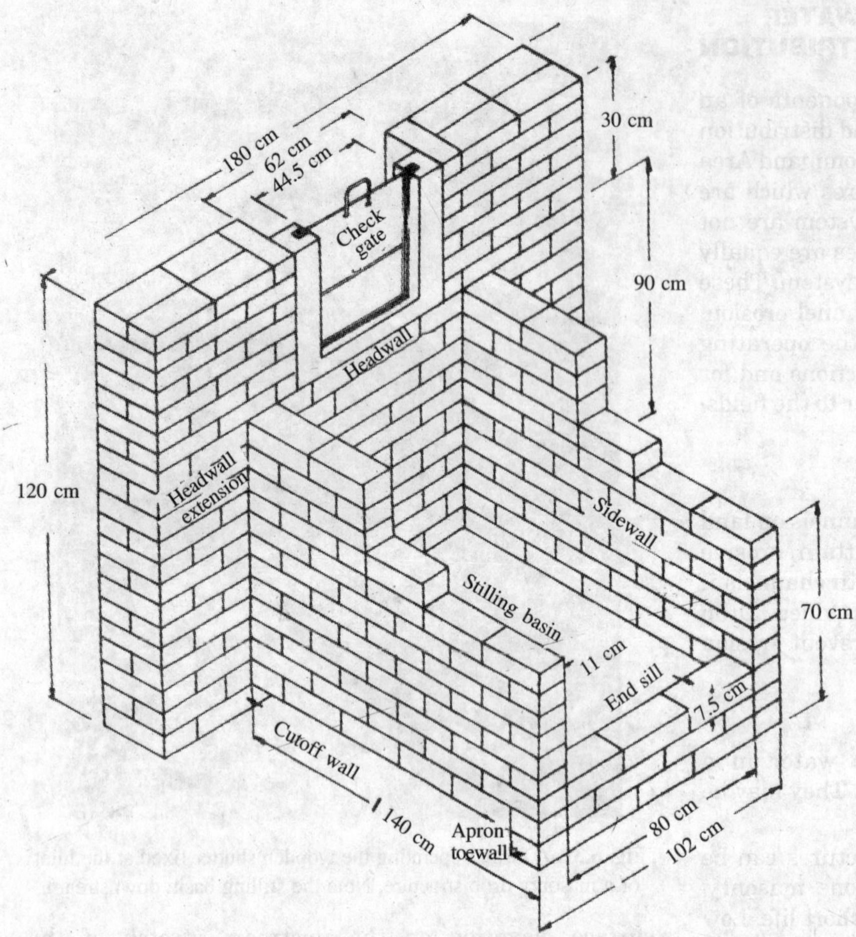

Fig. 6.32. Isometric view of a typical brick masonry drop structure for a drop of 90 cm.

Fig. 6.33. A drop structure in an unlined water course in a major irrigation project area. The short length of the stilling basin and the absence of an end sill at the downstream end are basic defects in the structure.

Fig. 6.34. A simple drop strucutre formed by placing pieces of stone slabs in the black cotton soil region of a major irrigation command area. The absence of a properly constructed stilling basin to prevent soil erosion is a basic shortcoming of the structure.

Fig. 6.35. An inadequately designed irrigation channel and drop structures washed away by monsoon runoff in a black cotton soil area of a canal command near Dharwar Karnataka, (1973).

stilling basin, made of brick or stone masonry, or concrete is provided at the outlet of the pipe conduit to dissipate the energy of the incoming stream. A masonry or concrete apron is provided around the inlet end of the pipe to prevent seepage around it.

The velocity of flow of water in pipe drop spillways using different size pipes may be calculated from the following relationship obtained by applying the Bernoullis' theorem:

Available head = Frictional loss in the pipeline + Velocity head + head loss at the entrance of the pipe + Head loss at the bend

i.e., $$H = 4\frac{flv^2}{2gd} + \frac{v^2}{2g} + K_1\frac{v^2}{2g} + K_2\frac{v^2}{2g} \quad \text{... (6.13)}$$

Fig. 6.36. Series of masonry drop structures constructed on an unlined field channel on a steep slope at Jobner, Rajasthan (1965). Note the use of rubber siphons for diverting water from the channel to the field. Improved structures and devices make irrigation easy, efficient and a pleasure to the farm family to operate.

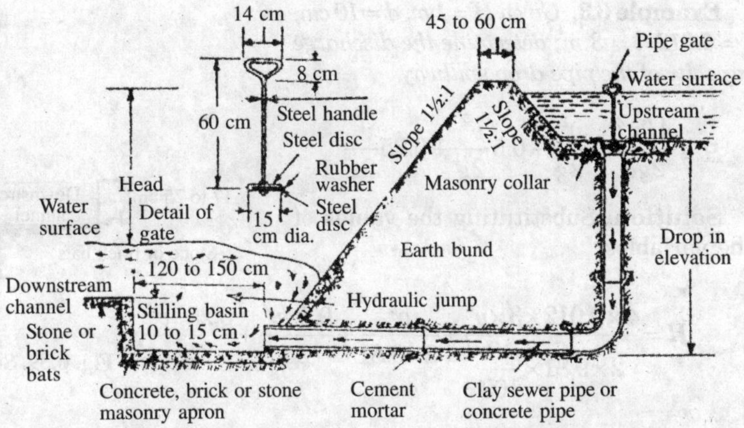

Fig. 6.37. Sectional view of a pipe drop spillway.

in which,

H = difference in elevation between the water level at the upstream and downstream ends of the structure (m)

v = velocity of flow in the pipe (m/sec)

f = coefficient of friction for the pipe (usually assumed to be about 0.01)

l = length of pipe (m)

g = acceleration due to gravity (m/sec^2) (9.81 m/sec^2)

d = diameter of pipe (m)

K_1, K_2 = constants (for pipe drops K_1 = 0.5 and K_2 = 0.25

The discharge capacity of the pipe drop structure may be determined by the relationship

$$q = av$$

in which,

q = discharge (m^3/sec)

a = area of cross-section of the pipe (m^2)

v = velocity of flow (m/sec)

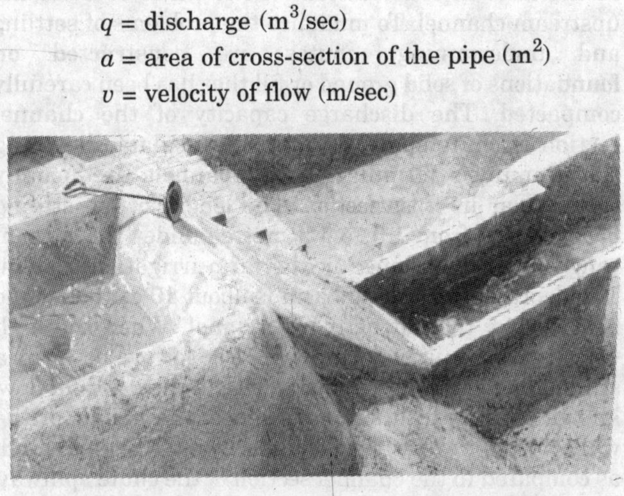

Fig. 6.38. View of a pipe drop spillway in a field demonstration system at the Water Technology Centre, IARI, New Delhi. Note the whirlpool developed at the inlet upstream and the turbulent flow in the stilling basin downstream. A check gate made of two circular metal plates with a rubber washer in between and the mild steel rod handle is seen kept on the upstream bank of the structure.

Example 6.3. *Given H = 1 m, d = 10 cm, f = 0.012, l = 3 m; determine the discharge capacity of the pipe drop spillway.*

$$H = \frac{4flv^2}{2gd} + \frac{v^2}{2g} + 0.5\frac{v^2}{2g} + 0.25\frac{v^2}{2g}$$

Solution: Substituting the values of the variables

$$H = \frac{4 \times 0.012 \times 3 \times v^2}{2 \times 9.81 \times \frac{10}{100}} + \frac{v^2}{2 \times 9.81} + \frac{0.5\ v^2}{2g} + \frac{0.25\ v^2}{2g},$$

$$v = \sqrt{6.26} = 2.5 \text{ m/sec}$$

$$a = \frac{\pi}{4}d^2 = \frac{\pi}{4} \times \frac{10}{100} \times \frac{10}{100} = 0.0078 \text{ sq m}$$

$$Q = av$$

$$= 0.0078 \times 2.5 = 0.0195 \text{ cu m/sec}$$

$$= 19.5 \text{ litres/sec.}$$

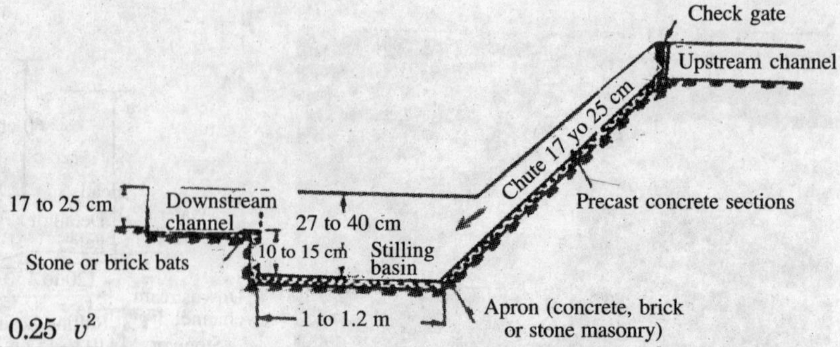

Fig. 6.39. Sectional view of a low cost chute spillway made of precast concrete channel sections and a check gate.

Chute Spillways

Chute spillways carry the flow down steep slopes through a lined channel rather than by dropping the water in a free overfall (Fig. 6.39). On steep slopes, chutes are more economical than a series of drop structures required to take the flow down the slope. The chute spillway consists of an inlet, channel section and outlet. The structure may be made of concrete, or stone or bricks laid in cement mortar. A low cost chute spillway can be made of precast concrete channel sections with a stilling basin at the outlet (Figs. 6.39 and 6.40). It is frequently desired to install a check gate at the inlet in order to control the water level in the upstream channel. To minimise the problems of settling and undermining, chutes are constructed on foundations of solid ground or fill that has been carefully compacted. The discharge capacity of the channel section of chute spillways may be calculated by using the Manning's formula. Chute channels are usually rectangular in cross-section. The length of the stilling basin varies from 1.0 to 1.5 metres under the normal ranges of flow obtained in the farm irrigation system. The depth of the stilling basin is about 10 cm below the bed level of the downstream channel. When the high velocity water is slowed down to a low velocity in a stilling basin, there is a sudden rise in the depth of flow, known as a hydraulic jump. Hence, the height of the walls in the downstream channel should be increased, as compared to the channel section of the chute spillway, in order to prevent overflow. Usually it is sufficient to keep the same depths for the upstream and downstream sections of the channel and to deepen the stilling basin floor level.

Water Control and Diversion Structures

Water control and diversion structures are necessary to give easy and effective control of irrigation water on the farm. Good control will reduce the labour required to irrigate and check erosion and water loss. The structure include check gates, portable check dams, diversion boxes, turnout boxes, siphons and pipe turnouts.

Check gates. Check are placed in an irrigation channel to form an adjustable dam to control the elevation of the water surface upstream (Fig 6.53 to 6.58). To apply water from channels into a field it is often necessary to raise the water level. The water level in the channel should be at least about 8 to 12 cm above the ground surface in order that siphon tubes or pipe turnouts may be used efficiently. Check gates placed at intervals along the channel keep a satisfactory water level for applying water to the field.

A check structure consists of canvas, metal or masonry walls built across the channel and provided with a suitable check gate or outlet device. The masonry wall is built in place while the check gate is usually precast and fixed to the wall. The crest of the check gate is at the same level as the bottom of the upstream channel.

Precast check gates. Precast reinforced concrete check gates can be made in a form made of sheet metal and provided with slot formers. Figure 6.41 shows the details of the form. Accurately, machined slot formers will provide suitable slots for inserting water tight rubber seal strips on the sides and at the bottom (Fig. 6.42). Removable plates facilitate the filling of concrete and the removal of the form after the concrete is set. Care is taken that the two sides are in proper alignment. By varying the length and height of the form, check gates of any size may be made. Suitable reinforcement may be fabricated from 6 mm diameter mild steel bars (Fig. 6.43). Two or four rods may be used, depending on the size of the gate. A suitable concrete mix for check gates is 1 : 2 : 3 (cement : sand : aggregate) made with minimum water, consistent with the need to compact the concrete thoroughly.

Fig. 6.40. A chute spillway made with precast concrete channel sections. Note the precast check gate at the inlet and the stilling basin at the downstream end.

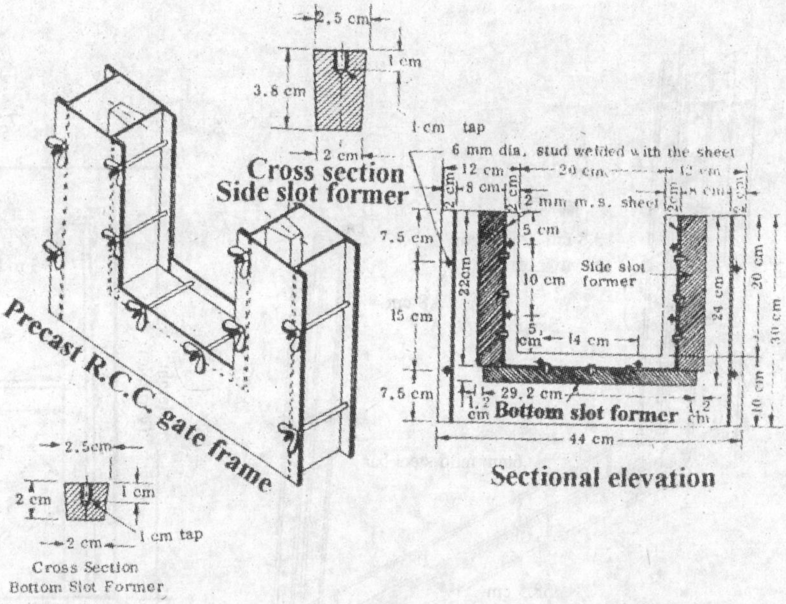

Fig. 6.41. Details of the form for casting reinforced concrete check gate frame. By changing the length and breadh suitably, frames of any size to suit the field channel may be constructed.

To cast the gate frame, the form is placed in inverted position and the reinforcement is placed in position. The form is filled with concrete and thoroughly tamped. When the concrete has set, about 8 to 12 hours after pouring, the form is removed. The concrete must be cured for about a week to prevent cracking. After curing, the check gate can be transported to the field and installed. Hooked bars which may be fixed on either side of the frame at the time of casting will ensure bondage of the gate with the masonry check wall to which it is fixed.

The gate. Plate steel gates or wooden boards may be used as stops for check gates. Wooden boards inserted in the slots are usually leaky and the farmer usually shovels earth around them to reduce leakage. Metal gates, together with specially designed rubber seal strips, provide watertight gates. Figure 6.44 shows the design of the metal gate. It is provided with one or two handles, depending on the width of the gate. For effective sealing, it is important that the bottom edge of the gate is perfectly plane and smooth (Fig. 6.45).

Watertight seal for check gates. To obtain a watertight seal, specially designed rubber strips are used. Two types of sealing strips are used—the guide strip, in which the vertical edge of the gate engages and the still strip in which the bottom edge of the gate makes plane surface contact (Fig. 6.46). The designs for the strips, developed by the State Rivers and Water Supply Commission, Victoria, Australia, can be adopted with advantage to suit almost any field situation. Figure 6.46 presents the details of the seal strips. The length of the strip may vary with the size of the gate. Figure 6.47 shows the relative position of rubber seal strips in the slots of check gate.

Fig. 6.42. Close view of the slot formers fixed to the inside of the cover plates used in the form shown in Fig. 6.41. The bottom slot former for the sill strip (bottom) is at the centre and the two side slot formers are on either side.

Seal strips are manufactured commercially by an extrusion process. Suitable metal dies could be designed for manufacturing the strips locally by a technician (Fig. 6.49 and 6.50). Seal strips can be made with a synthetic rubber called neoprene or with commercial raw rubber used for rubber casting works. The synthetic rubber is preferred as it contains additives to give anti-oxidation properties and abrasion resistance.

For bonding the seal strips into the concrete check gate the surfaces of the slots are cleaned to remove loose

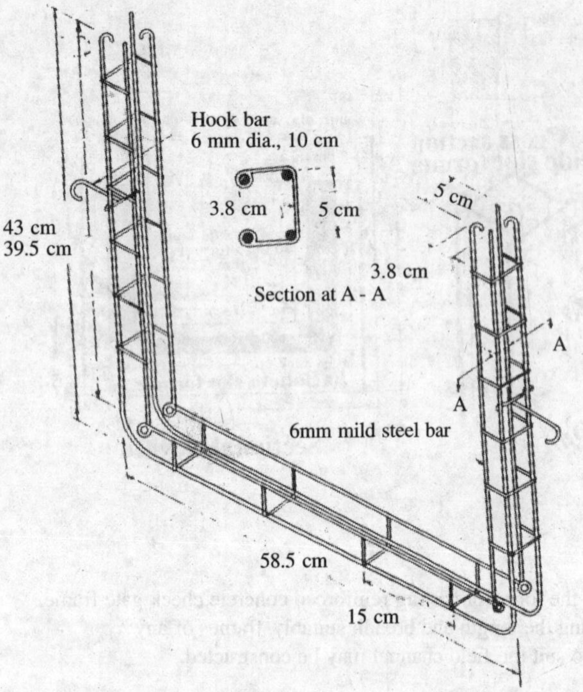

Fig. 6.43. Steel reinforcement for a medium size RCC check gate. The height and width of the reinforcement and the number and size of steel rods used will vary on the basis of the size of the frame.

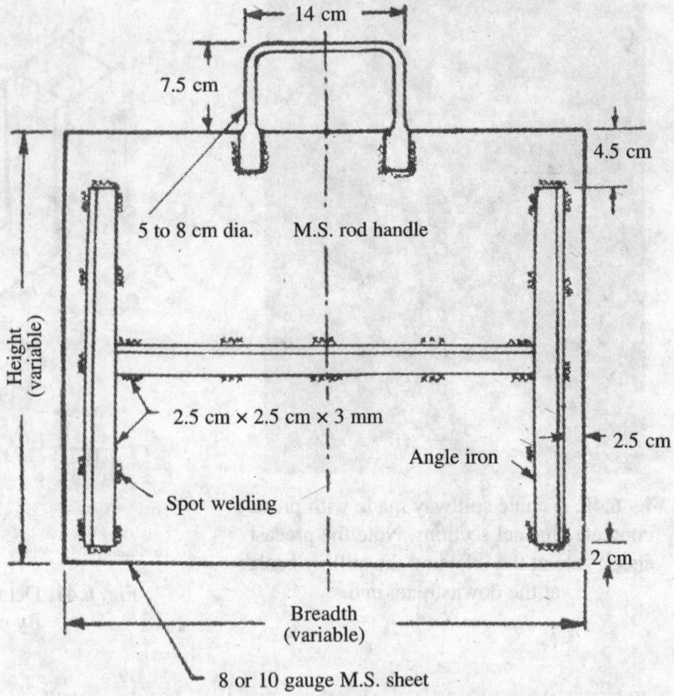

Fig. 6.44. Metal slide gate to fit into watertight rubber seals. The length and breadth of the gate may be varied to suit the size of the check gate frame. For wide gates the size of the members may be increased appropriately and two handles may be provided.

Fig. 6.45. Casting concrete frame for check gate. Note the precast check gate with watertight rubber seal and metal slide gate beside the form.

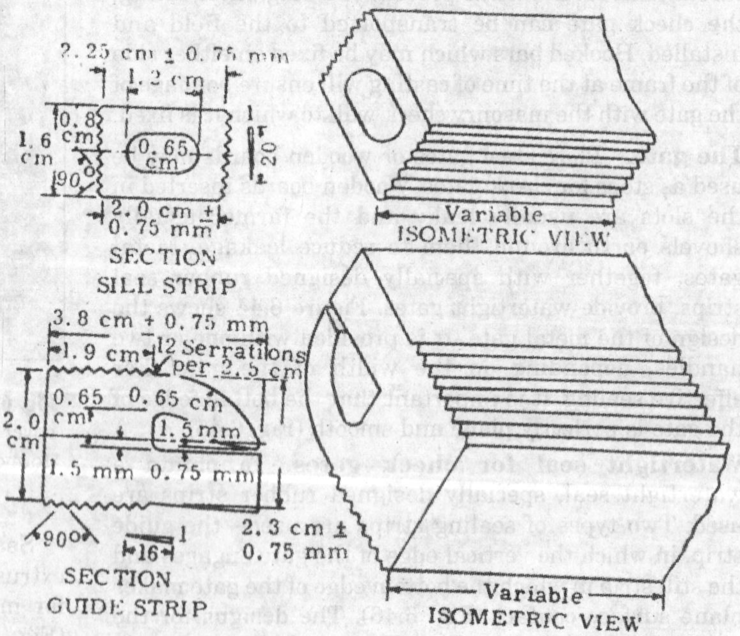

Fig. 6.46. Details of water-tight rubber seal strips for check gates.

particles and coated with an adhesive glue (Fig. 6.52). This coating is allowed to dry. The serrated surface of the sill strip and guide strips are then coated with the adhesive and the strips inserted to the full depth of the slots (Fig. 6.52). With the guide strip, the cut lengths should be the same as the vertical slots. With the sill strip there is some elongation on insertion and it is necessary to allow for this by cutting the strip shorter than the length of the horizontal slot by about 3 to 6 millimetres, depending on the size of the check gate. Suitable metal dies may be fabricated to cast waterlight rubber (or other synthetic) seal strip for check gates

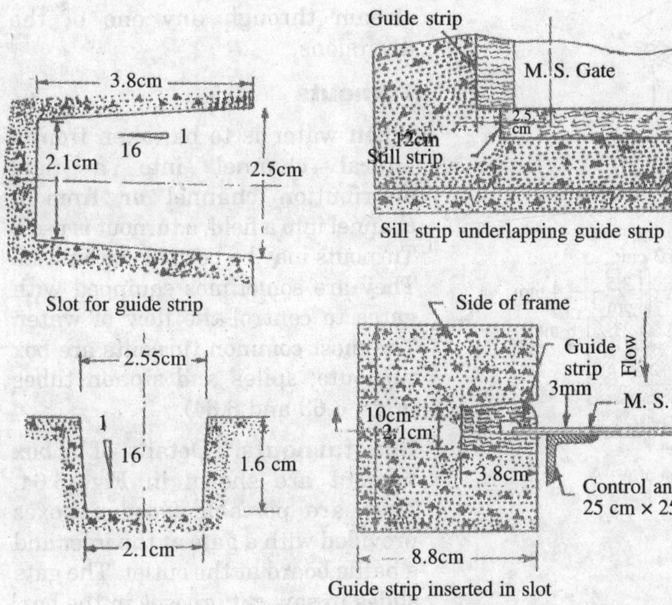

Fig. 6.47. Details of slots and relative positions of rubber seal strips inserted into the slots.

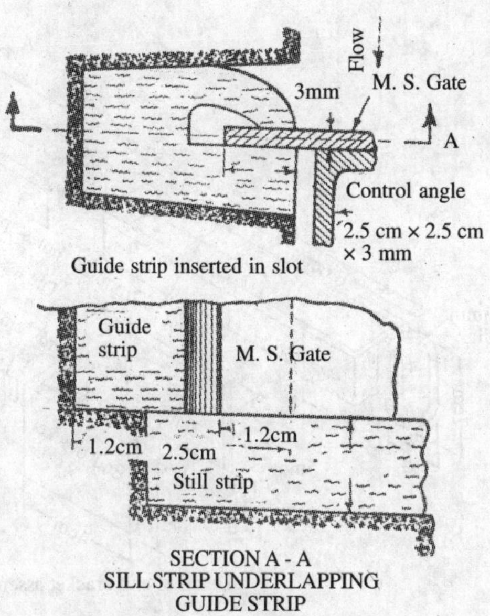

Fig. 6.48. Relative position of rubber seal strips in the slots of check gate.

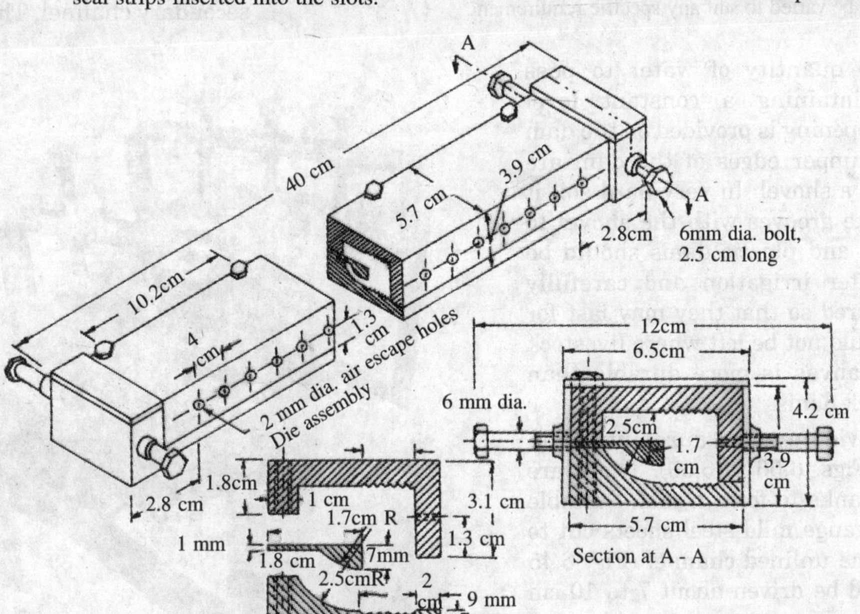

Fig. 6.49. Isometric and sectional views with typical dimensions of a mild steel die set for casting the guide strips for the sides of water-tight check gates. The length of the die may be varied suitably to suit the dimensions of the check gate.

(Figs. 6.49 and 6.50). (See also Fig. 6.51). To prevent leakage at the corners, care must be taken to obtain a good bond between the lower end of the guide strip and the surface of the sill strip. It is essential to cut the guide strip accurately at right angles, apply adhesive to the cut surface and ensure firm contact with the sill strip on insertion (Fig. 6.52).

Portable Check Dams

Canvas, plastic and sheet metal dams are suitable to check water and raise or control water surface elevations (Figs. 6.53 to 6.60). Canvas and plastic dams are supported on a pipe or wooden cross-piece. A loop is usually provided at the bottom to anchor the dam to a small post. A sleeve made near the bottom of a canvas

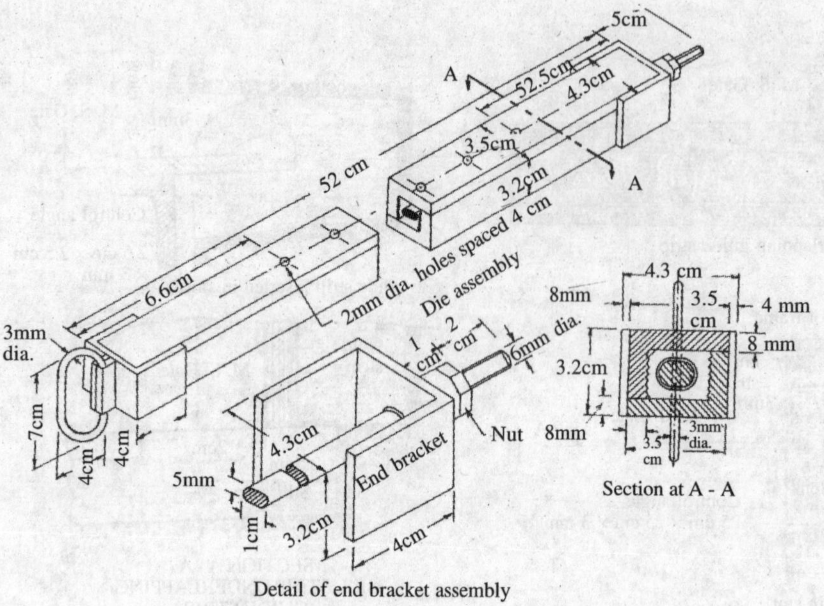

Fig. 6.50. Views of a die set (mild steel or other metal) for casting the seal strips for inserting at the bottom of water tight check gates. The length of the die may be varied to suit any specific requirement.

dam allows any desired quantity of water to pass downstream, while maintaining a constant level upstream. Sometimes an opening is provided on the dam to divide a stream. The upper edges of the dam are pressed into the soil with a shovel. In very hard soil it may be necessary to make grooves with the shovel to insert the edges. Canvas and plastic dams should be removed immediately after irrigation and carefully washed out, dried and stored so that they may last for several seasons. They should not be left where livestock will trample on them. Canvas is more durable than plastic when used for check dams.

Sheet metal dams provide greater durability than canvas or plastic dams (Figs. 6.56 to 6.60). These are driven into the channel banks to form a seal. Suitable dams can be made of 18 gauge mild steel sheets cut to suit the cross-section of the unlined channel (Fig. 6.45 and 6.46). The dam should be driven about 7 to 10 cm deep into bank to prevent seepage around the sides and bottom. One or two supports made of angle iron with pointed bottom provide greater strength and a better anchor. Sheet metal dams may be equipped with watertight check gates in the same way as masonry check dams. To prevent rusting, metal dams are always painted.

Diversions. In carrying water to different parts of the command area it is necessary for the water course or the main channel to divert the stream into the proper channel (Fig. 6.61 and 6.62). Two-way or three-way and four-way diversions made of check gates at the channel junction will permit the operation of diverting all or any predetermined portion of the stream through any one of the diversions.

Turnouts

When water is to be taken from a lateral channel into a field distribution channel or from a channel into a field, a turnout is used. Turnouts may be portable or built-in. They are sometimes equipped with gates to control the flow of water. The most common turnouts are box turnouts, spiles and siphon tubes (Figs. 6.63 and 6.64).

Box turnouts. Details of a box turnout are shown in Fig. 6.64. They are portable wooden boxes provided with a gate at the inlet and a baffle board at the outlet. The gate slides in saw-cut grooves in the box. It can be adjusted to divert any desired flow into a field or a secondary channel. The baffle board divides the stream and reduces its velocity (Fig. 6.65). The box should be long enough to extend completely through the channel bank. A collar made of wooden pieces fixed around the box will prevent seepage around the box turnout. The size of the box turnout will depend on the size of the stream to be diverted.

Fig. 6.51. A simple die-cast equipment for fabricating water tight rubber seal strips for check gates. Larger dies may be designed for factory production of seal strips.

Fig. 6.52. Inserting rubber seal strip in the side slot of an RCC check gate frame. Note the glue used to coat the surface of the slots and the seal strips before insertion.

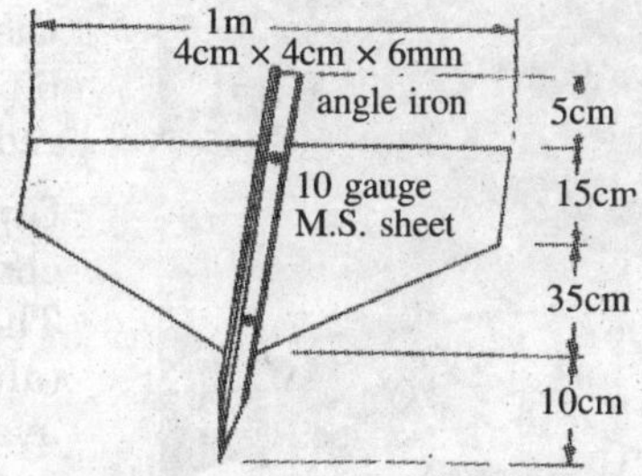

Fig. 6.56. Portable sheet metal dam suitable for small unlined channels.

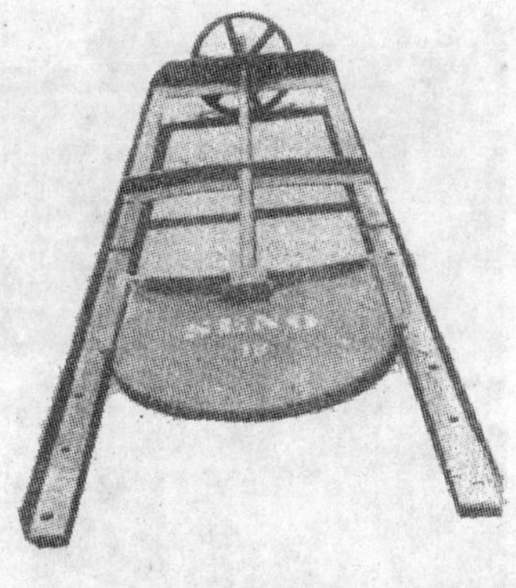

Fig. 6.57. A metal sluice gate suitable for water control and diversion.

Fig. 6.53. Watertight sheet metal check gate fixed to the inlet of a drop structure.

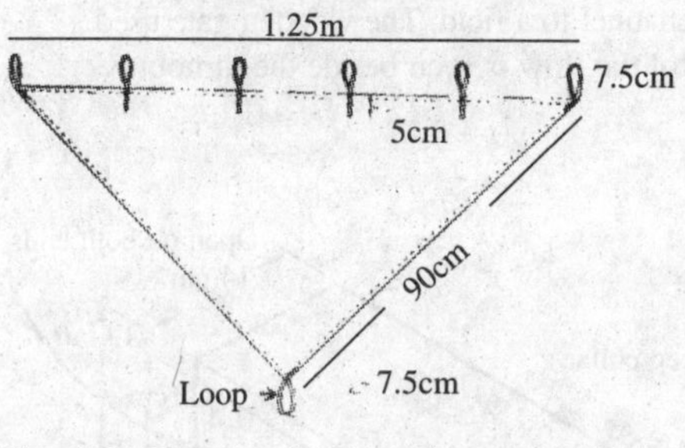

Fig. 6.54. Details of a triangular canvas dam to control water level in small earth channels.

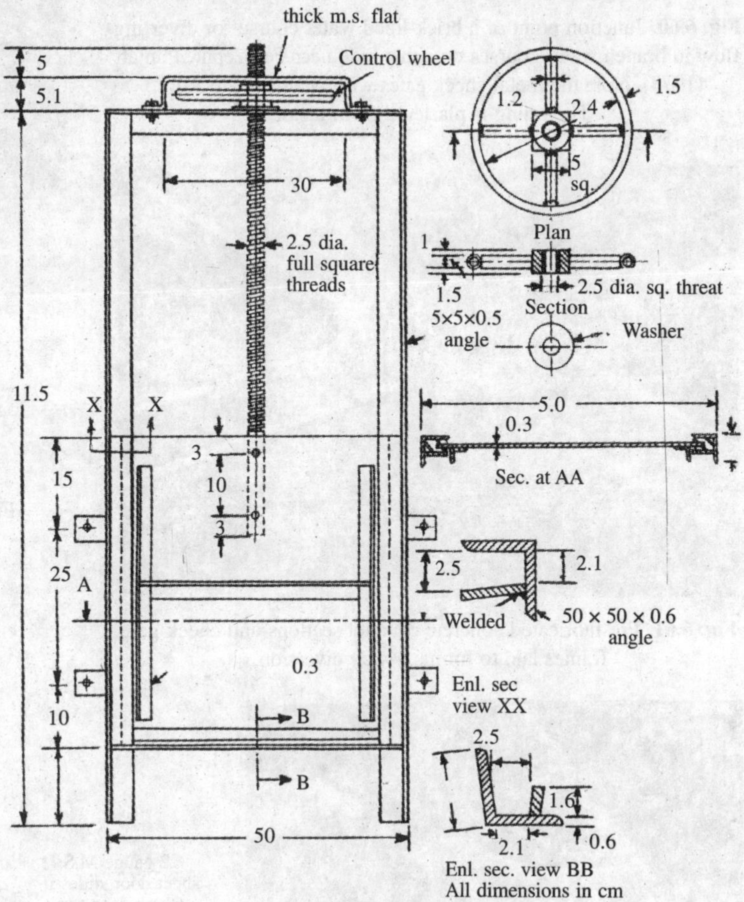

Fig. 6.58. Details of a portable steel check gate (meter gate) with rubber seal strips.

Fig. 6.55. Portable canvas dam with coat-sleeve by-pass. The quantitiy of by-pass water can be controlled by the string on the sleeve.

Fig. 6.59. A two-way masonry diversion fitted with metal sluice gates.

Fig. 6.60. Junction point of a brick-lined water course for diverting flow to branch water courses on either side near Ferozepur, Punjab (1973). Note the metal check gates with chains for locking. The lining is plastered from inside.

Fig. 6.61. Pre-fabricated concrete channel sections and check gate frames laid to form a 4-way diversion.

Fig. 6.62. A 4-way diversion for a riser of an underground pipeline irrigation distribution system to distribute water to rice fields.

Spiles. Spiles are made of bamboo, concrete, or baked clay pipes. The size of the spiles ovary from 2.5 cm to 10 cm or more, depending on the flow of water desired and the depth of water at the inlet and outlet of the pipe (Fig. 6.68). The spile should be placed low in the bank to reduce erosion at the outlet.

Siphon tubes. Siphon tubes convey water over a channel bank into a field or furrow (Figs. 6.67 and 6.69). They are made of plastic, rubber or sheet metal (aluminium or mild steel) and are commercially available in different sizes, usually ranging from 2.5 cm to 7.5 cm. Each siphon tube is about one metre or more in length, depending on the height of the channel bank

Fig. 6.63. Wooden turnout installed to divert water from a supply channel to a field. The wooden gate used to control the flow is seen beside the turnout.

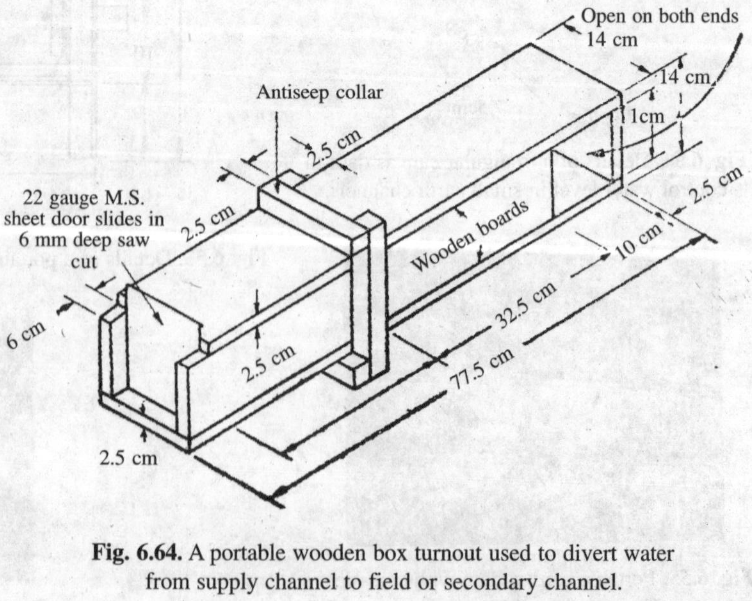

Fig. 6.64. A portable wooden box turnout used to divert water from supply channel to field or secondary channel.

Fig. 6.65. Wooden box turnout installed to divert flow from an earth channel. Note the spiles placed in the secondary field channel to uniformly distribute water to furrows.

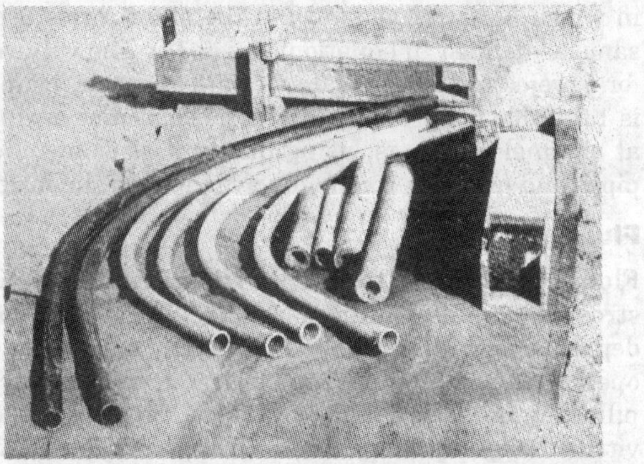

Fig. 6.66. Assorted irrigation syphon tubes, spiles, and turnouts.

Fig. 6.67. Plastic siphon tubes used to divert water from a field channel lined with non-reinforced precast concrete sections.

above the field. In using a siphon tube, it is dipped in water and filled completely. One end of the tube is then closed tightly with the palm of hand, keeping the other end dipped in water, the closed end is taken out and released at the ground surface.

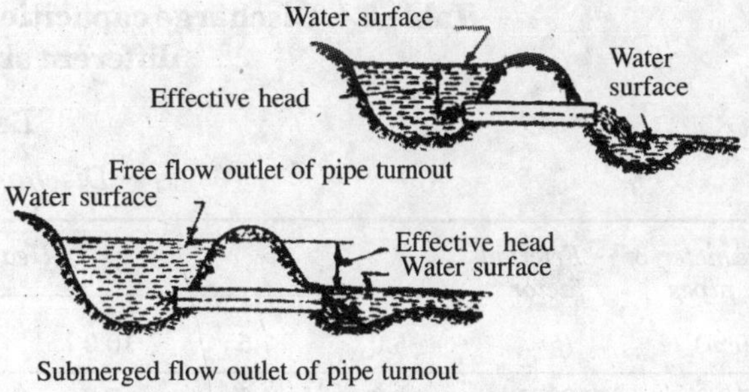

Fig. 6.68. Schematic sketch illustrating flow through spile tubes under free flow and submerged flow conditions.

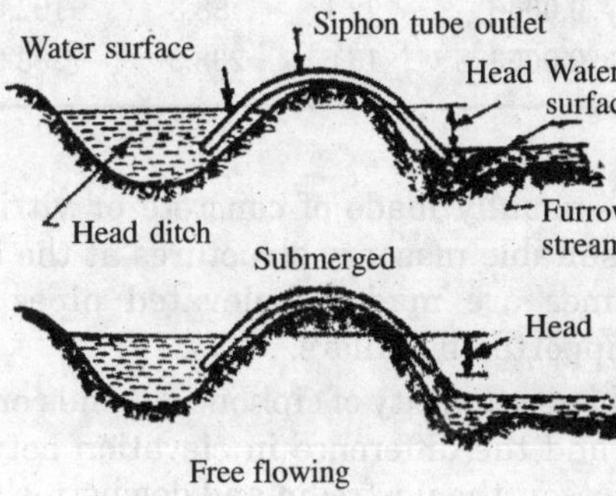

Fig. 6.69. Schematic sketch illustrating flow through siphon tubes.

The discharge from siphon tubes depends on the size of the tube and the difference in elevation between the water surfaces at the upstream and downstream ends of the tube. It may be estimated by the following formula:

$$Q = 0.65 \times 10^{-3}\, a \sqrt{2gH} \qquad \text{... (6.14)}$$

in which,

Q = discharge from siphon tube (litres/sec)

a = area of cross-section (inside of tube) (sq cm)

g = acceleration due to gravity (cm/sec^2) (981 cm/sec^2)

H = effective head causing flow (cm)

If the outlet is not submerged, the effective head is the vertical distance from the water level over the inlet end to the centre of the discharge end.

Structures at Channel Crossings

It is often necessary to carry irrigation channels across roads, hillsides and natural depressions or drainageways. Siphons are used to carry irrigation channels across natural farm obstructions. For the average farm job they are usually in the form of a straight pipe or an inverted siphon. When the siphon is laid over a drainageway or depression it is often called a flume (Fig. 6.70). When it crosses a ridge, highway, or bund it is called a culvert. Where the road surface lies too close to the field surface to permit using a culvert, or if the channel surface is higher than the road, an inverted siphon is constructed for crossing the road.

Table 6.4. Discharge capacities of siphons at channel crossings using different sizes of concrete pipes

Length = 5 m

Discharge, litres / second

Diameter of pipes	*Friction factor*	*Head (Centrimetres)*								
(cm)	*(f)*	5.0	7.5	10.0	12.5	15.0	17.5	20.0	25.0	30.0
10.0	0.0095	3.9	4.8	5.6	6.2	6.1	7.4	7.9	8.8	9.6
15.0	0.0091	9.8	12.0	13.8	15.4	16.9	18.3	19.5	21.8	23.9
22.5	0.0071	24.3	29.7	34.3	38.4	42.0	45.0	48.5	54.3	59.4
30.0	0.0060	45.2	55.3	63.9	71.4	78.2	84.5	90.3	101.0	110.6
37.5	0.0056	72.1	88.3	102.0	114.0	124.9	134.9	144.2	161.3	176.6
45.0	0.0053	16.5	233.3	233.9	260.9	285.7	308.6	330.0	368.9	404.1

Siphons are usually made of concrete or vitrified clay pipes with suitable masonry structures at the inlet and outlet. Flumes are made of elevated pipes or open channels supported by pillars.

The discharge capacity of siphons depends on the size of the pipe and the difference in elevation between the water surfaces at the upstream and downstream ends of the structure. The capacities of siphons may be estimated in the same way as those of pipe drop spillways. For siphons using straight pipes the Bernoulli's theorem may be stated as follows:

$$H = \frac{4flv^2}{2gd} + \frac{v^2}{2g} + K_3 \frac{v^2}{2g} \qquad ...(6.14)$$

Fig. 6.70. Semi-circular metal flume on wooden supports to carry irrigation water across a drainageway. *(Colourado State Univ. Expl. Stn. Bil. 465-S-S)*

in which K_3 is a constant and the other terms are the same as defined in equation (Fig. 6.13). The value of K_3 for a projecting pipe entrance, commonly used in siphons is 1.0. Table 6.4 presents discharge capacities of siphons at channel crossings, using different sizes of concrete pipes and at heads commonly available on the farm.

Flumes

Flumes are used to carry irrigation water across streams, canals, gullies, ravines or other natural depressions (Figs. 6.70 and 6.71). They may be made of open channels or pipes which are often supported by pillars or may be fixed to bridges. Steel, concrete or vitrified clay pipes are used. The open channels, when used, are semi-circular metal pipes or rectangular or trapezoidal wooden channels. The supporting structure may be of timber, steel or concrete. If timber is used, it should be treated with a preservative like creosote oil to extend its life. Steel structures should be painted. When pipes are used in flumes, they should be placed below the water surface on the upstream end to ensure that they flow full. The capacities of open channels used in flumes may be estimated using the Manning's formula. The discharge capacity of pipe flumes (flowing full) may be determined using equation (Fig. 6.14).

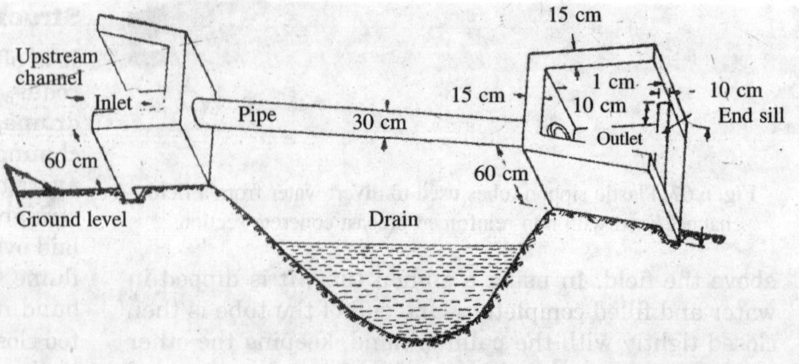

Fig. 6.71. Details of a pipe flume for carrying an irrigation stream across gullies or drainageways.

Culverts

Culverts are commonly used at channel crossings when the road fill is sufficiently high and the channel bed lies on the field surface on either side. The structure consists of masonry headwalls at the inlet and outlet ends connected together by a buried pipeline. The earth covering over the pipe should not be less than 30 cm, but preferably about 45 cm. At the outlet of the culvert a stilling basin of the type described under erosion control structures is usually required to prevent erosion downstream. Low-cost culverts made of pipes of adequate capacity may be constructed for ditch crossing (Fig. 6.72).

Inverted Siphons

The inverted siphon is economical when a channel has to cross a wide depression or where the road surface lies close to the filed surface. The structure consists of an inlet and an outlet tank connected together at their bottom by a pipe (Figs. 6.73 and 6.74). A check gate is usually provided at the inlet end to control the water surface level in the upstream channel. Since the tanks of an inverted siphon act as stilling basins and slow down the velocity of the flowing water, silt carried by the water from upstream erosion is likely to be deposited in the tank. To collect silt the bottom of the tanks is kept about 15 cm below the bottom of the pipe. Occasional cleaning of the tanks is required. Hence, the tanks must be of sufficient size to permit access into it. The size commonly used is 60 cm × 60 cm. As with culverts, the pipes of an inverted siphon must be protected against traffic damage by an earth cover of 30 cm or more. In lined channels with adequate depth, the inlet and outlet tanks may be avoided.

ON-FARM WATER CONVEYANCE AND DISTRIBUTION SYSTEMS IN IRRIGATION COMMAND AREAS

On farm water conveyance and distribution systems in canal irrigation project areas comprise of open channels and ancillary structures designed to ensure efficient and effective supply of water to the farmers' fields in the areas under the command of irrigation outlets (see Chapter 1). Till the early part of the twentieth century irrigation project development was essentially to provide water at the outlet. The farmers were given the responsibility to manage water below the outlet. This resulted in numerous problems and the water distribution efficiency in the command area was poor. Problems of unequal distribution of water to the beneficiary farmers of an outlet such as waterlogging and soil salinity cause serious problems. Pioneering efforts in alleviating the problem were done during 1960s at some of the regional centres of the Central Soil Conservation Research Institute, Dehradun and at the College of Agriculture, Jobner of Rajasthan Agricultural University. The first effort of a major public sector

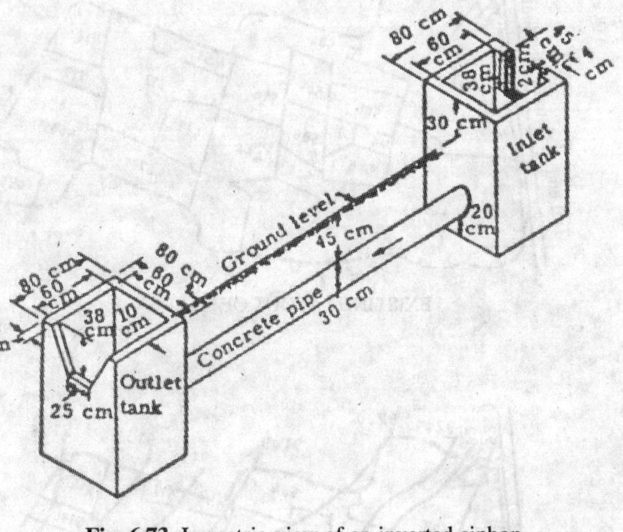

Fig. 6.73. Isometric view of an inverted siphon.

Fig. 6.72. A low cost pipe culvert.

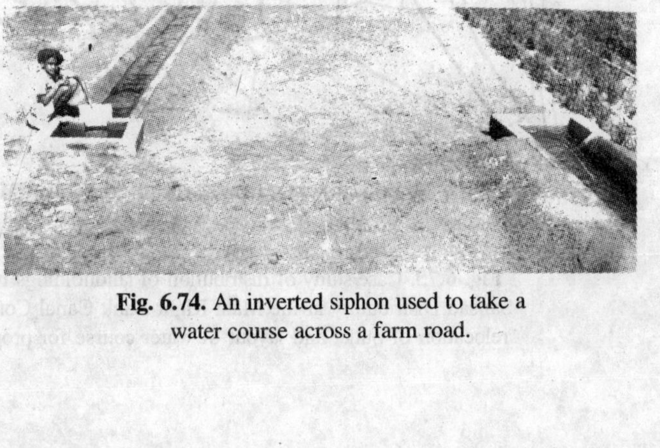

Fig. 6.74. An inverted siphon used to take a water course across a farm road.

involvement in improving watercourses was the lining of watercourses in Haryana and Punjab with the support of the Agricultural Finance Corporation (presently NABARD). The first national level effort to focus attention on the problems of canal irrigation command areas has been the establishment of the Command Area Development Programme in the mid-1970s by the Government of India and its adoption by most of the state governments. Simultaneously, many of the research institutes of State Irrigation Departments, Command Area Development Authorities, Agricultural Universities and ICAR Research Institutes, and Water and Land Management Institutes at the State level concentrated their efforts in developing technology and manpower in command area development. The Water Technology Centre of Indian Agricultural Research Institute, New Delhi has contributed substability to the development of technology for on-farm water conveyance and distribution and other problem areas in irrigation command areas. A policy decision was taken by the Government to reach canal water supplies to sub-outlets which commands an area exceeding 8 ha. In open channels the flow is under gravity. Hence, the supply channels are to be laid along the ridge lines in a command area so that water can flow by gravity on either side of the channel. In case of sloping land the channels are laid around the slope on the higher reach of the field and water flows to the downfield area which is levelled to non-erosive grade (Fig. 6.75). In underground pipeline systems water is under pressure and the pipeline can be laid along a convenient route on as short a distance as possible to reduce the cost of the system. The pipe outlets, however, are to be located at commanding elevations so that water can flow from it and reach every part of the field it serves.

Survey and Planning of Water Distribution Systems in Command Areas

A watercourse originates from the canal outlet to supply water to a group of farmers having their landholdings within the command area of the outlet. Its ownership belongs to the beneficiary farmers whose land fall within the reach of the watercourse so that water can reach each field plot by gravity flow. Watercourses are usually maintained by the beneficiary farmers. When some of the field plots cannot be served directly from the main watercourse, branch watercourses are taken out to supply water to these plots (Fig. 6.75).

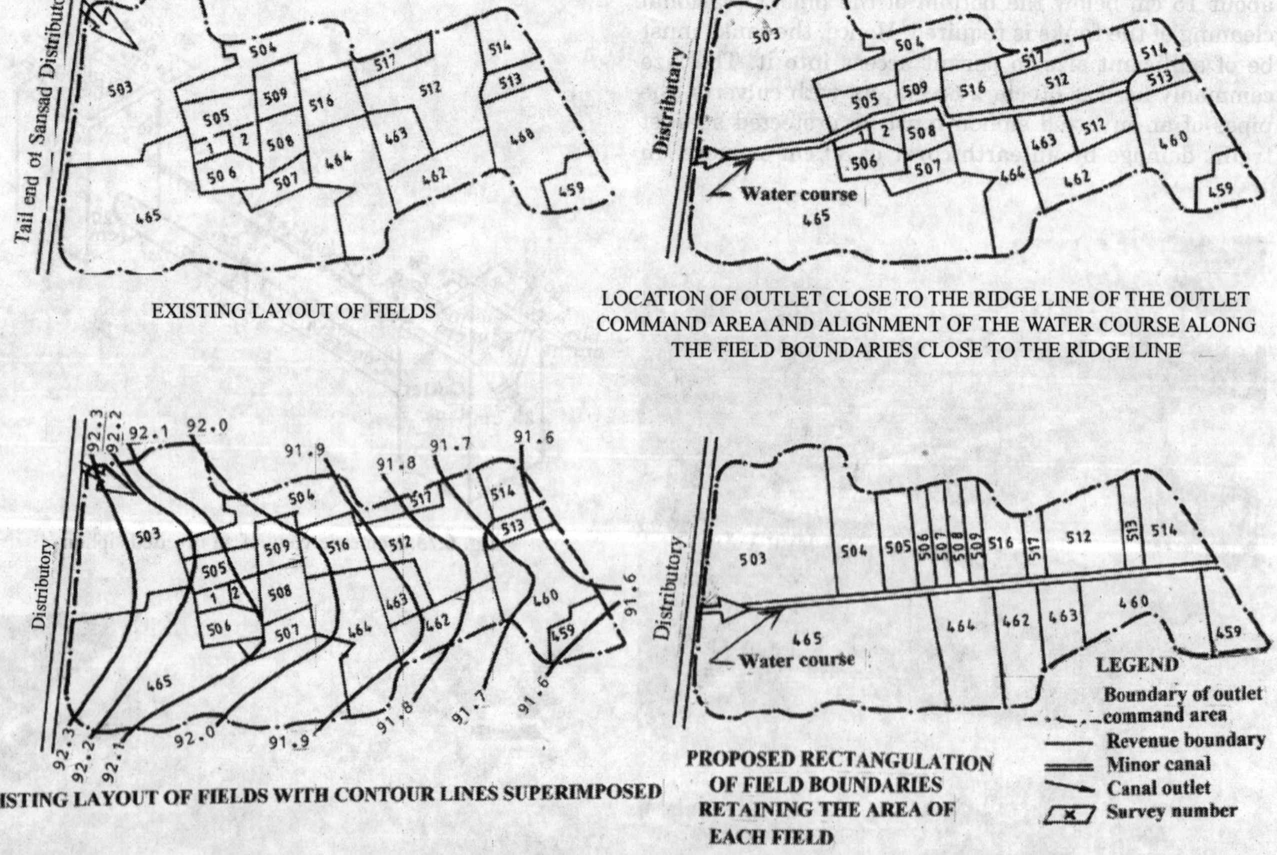

Fig. 6.75. Case study of distribution of landholdings in the command area of Outlet No. 04/L of the tail minor of Sansad Distributory in the Mahi Right Bank Canal Command Area in Gujarat in 1982 and proposed programme of relocation of outlet and layout of water course for proper distribution of water amongst the different landholdings.

Arrangement of field plots vary from region to region. In states like Punjab, Haryana and Western Uttar Pradesh, where land consolidation and rotational water delivery system (*warabandi*) are common, the fields are divided into rectangular blocks and the water distribution is comparatively easy (Fig. 6.76). The quantity of water which a farmer gets is proportional to the size of his landholding. In most other states in India land consolidation has not been effective and the fields are usually of widely varying sizes and the boundaries are not properly aligned. In such situations also it is necessary to supply water to all the farmers who have their landholdings within the command of the outlet (Fig. 6.77 and 6.78).

Topographic survey and preparation of field maps showing contour lines, field boundaries and all structures and other features of the command area are basic requirements to plan efficient irrigation water distribution and drainage systems and soil conservation works at the farm level in an outlet command area. Most of these works involve group efforts and mutual cooperation of the beneficiary farmers and frequently those of other farmers in the upper and lower reaches of a watershed. Detailed procedures in land survey and field layout are presented in the book Principles of Agriculture Engineering, Vol. II listed at the end of this chapter.

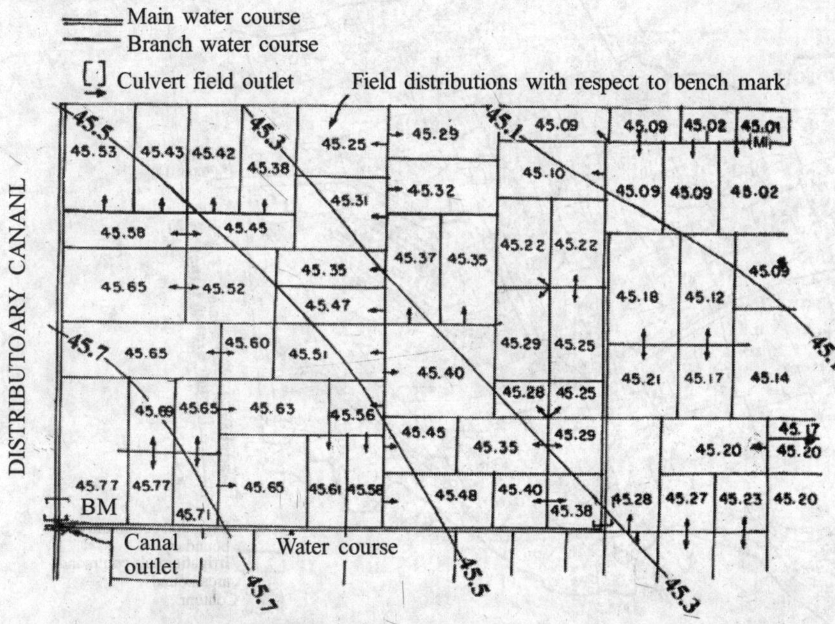

Fig. 6.76. Topographic map of the command area of a canal outlet in a region where land consolidation and rectangulation have been done. The figure shows the canal distributory on which the outlet is located, layout of field plots, water course, field outlets, contour lines and other field structures used in water conveyance and control. The average elevations of the field plots are shown in each plot.

Survey and Design of Water Distribution System in Outlet Command Areas with Rectangular Landholdings

In this survey, the arrangement of all bunded units with their elevations and the location of canal outlets (demarcated with earth ridges), culverts, roads, and other important features of the watercourse command area are determined and recorded. A map obtained from the local revenue officer serves as a guide for locating the field plots and recording their land revenue numbers. The topographic survey is started from the nearest benchmark and the field notes are recorded. The elevation of each bunded unit should be measured and recorded. The survey is closed and error of closure is computed. Contour lines are drawn on the map to indicate general slopes and as a rough check of the measurements and calculations. Coutour lines represent locations of equal elevations on the ground surface. Each line represents an equal increment of elevation change. All existing and proposed channels and structures are indicated in the map. Branch water courses are clearly identified to ensure that each shareholder of the canal outlet gets his fair share of water.

Survey and Design of the Water Distribution System in Unconsolidated Holdings with Irregular Boundaries

The survey procedure is the same under such situations as well. Field elevations are established and contour lines are drawn as described earlier. The location of the outlet need careful consideration. It should be at a commanding elevation from where the water course, together with branch water courses, would distribute water to all the field plots within the command area. The water course is to be laid along the ridge line to the extent possible, but preferably along field boundaries so that an individual landholding is not disturbed (Fig. 6.75). At many points branch water courses are to be located so that water could flow into those fields which are not within the direct reach of the main watercourse.

Drainage channels are laid along the valley line. Identifying the main and branch water courses at the planning stage itself will avoid future disputes. Figure 6.75 shows a typical case of layout of an irrigation distribution network in the command area of a canal outlet comprising irregular boundaries and varying sizes of landholdings.

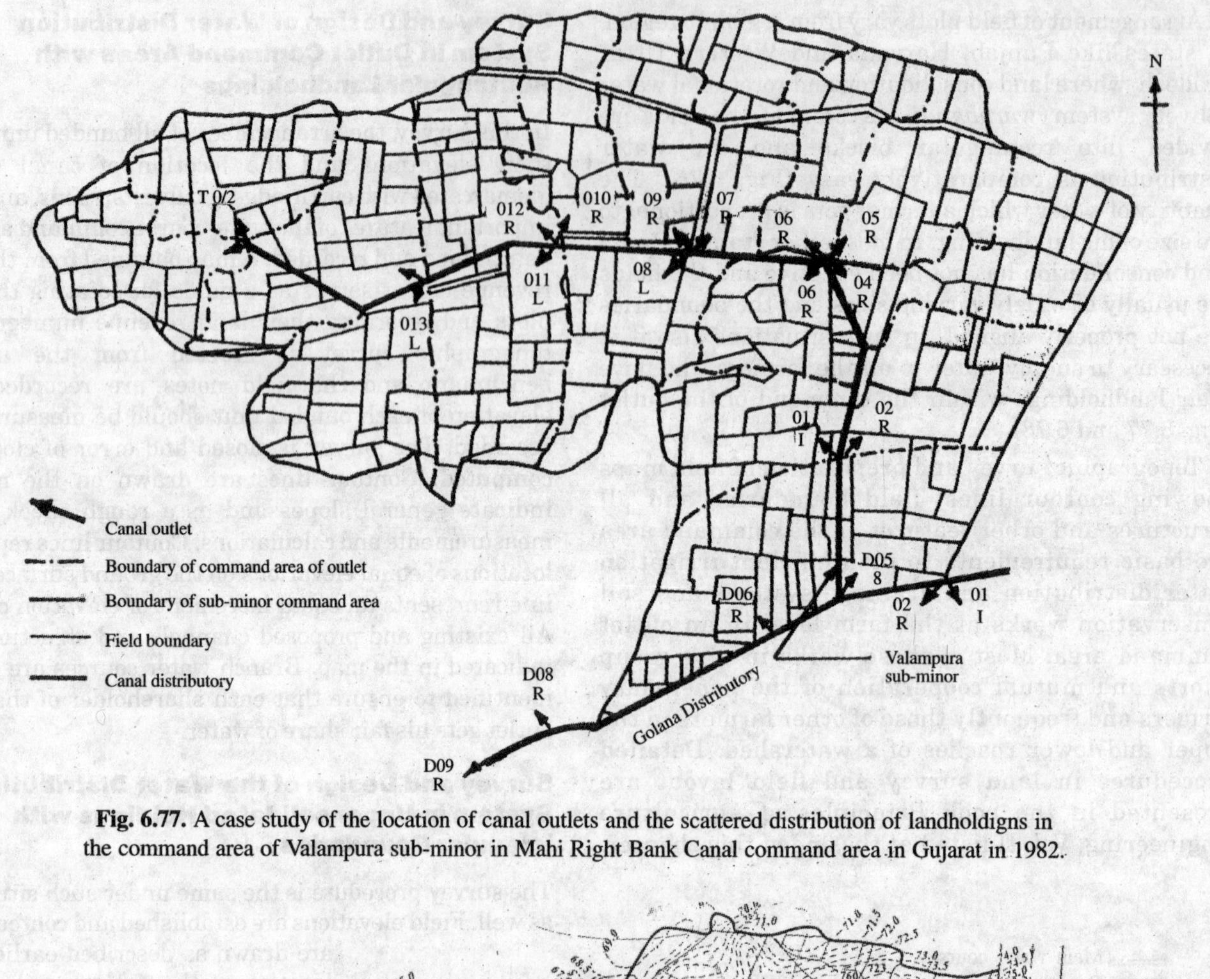

Fig. 6.77. A case study of the location of canal outlets and the scattered distribution of landholdigns in the command area of Valampura sub-minor in Mahi Right Bank Canal command area in Gujarat in 1982.

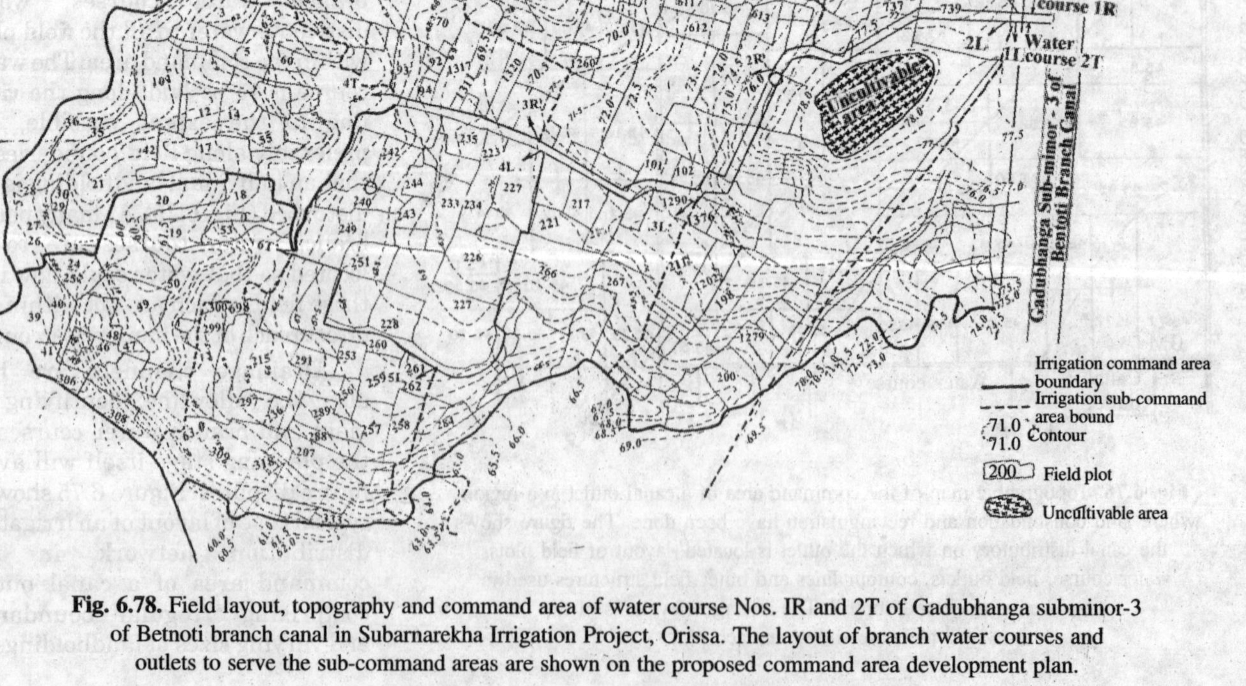

Fig. 6.78. Field layout, topography and command area of water course Nos. IR and 2T of Gadubhanga subminor-3 of Betnoti branch canal in Subarnarekha Irrigation Project, Orissa. The layout of branch water courses and outlets to serve the sub-command areas are shown on the proposed command area development plan.

SUMMARY

Common problems encountered in open channel water conveyance and distribution systems in irrigation command areas are presented. Importance of lining the vulnerable sections of canal reaches and on-farm water conveyance systems are explained. Factors influencing seepage in water conveyance systems, including estimates of seepage losses in different segments of a canal system are described. Methods of measuring seepage losses in canal systems are described. Flow regimes in open channels are explained. Methods of estimating the velocity of flow and discharge capacities of open channels are described. Procedures for design of earth channels and lined channels are explained. Materials and methods used in lining of canals and on-farm water conveyance systems are described. Use of pre-fabricated non-reinforced concrete channel sections for field channels are described with design of forms and step-by-step procedure for fabrication of channel sections. Role of pre-fabricated reinforced concrete canalets in canal minors and farm water distribution systems are described with appropriate designs and installation techniques. Cost analysis of open channel water conveyance systems in irrigation command areas are presented. Optimum length of water courses to be lined are explained with results of case studies. Structures to control erosion in farm irrigation channels, including open drop structures, pipe drop spillways and chute spillways are described, with design requirements of each. Water control and diversion structures, including portable and built-in check gates are described. Application of synthetic rubber seal strips in check gates are described, including design requirements and fabrication techniques. Use of siphon tubes and spiles are presented. Structures at channel crossings, including flumes, culverts and inverted siphons are described, with design procedures for each. Procedures for design of on-farm water conveyance and distribution systems in irrigation command areas are described, including importance of appropriate location canal outlets as illustrated by the results of case studies.

PROBLEMS

6.1. Determine the velocity of flow and discharge capacity of an unlined canal distributory on a grade of 1 metre on 100 metres. The depth of flow is 1.2 m, bottom width 60 cm and side slopes 1 : 1.

(**Ans.** v = 0.912 m/sec, q = 1.97 cu m/sec)

6.2. Estimate the carrying capacity of a pre-fabricated concrete channel section of specifications: Bottom width 17.5 cm, top width 20 cm, height 17.5 cm and bed slope 0.25%. (**Ans.** 17.16 litres/sec)

6.3. Design an earth channel to carry a flow of 55 litres per second in loamy sand soil. The land slope favours a channel's bed slope of 1 metre per 1000 metres. Manning's n = 0.0225.

(**Ans.** d = 25 cm, free board = 5 cm, b = 25 cm, side slope 1.5:1)

6.4. Design a rectangular water course lined with single layer brick masonry to carry a discharge of 70 litres/sec. The section is to be adjusted to suit the available brick size of 7.5 cm × 11.25 cm × 22.50. The available bed slope of the water course is 0.2 m/km.

(**Ans.** Outside dimensions: Width 90 cm, depth 53.1 cm,
Inside dimensions: Width 66.3 cm, depth 45 cm,
including 7.5 cm freeboard)

6.5. Determine the optimum bottom width, depth of flow and mean velocity of flow in a channel with 1.5:1 side slopes for a flow of 1.416 cu m/sec. Assume S = 0.2 m/km, n = 0.04.

(**Ans.** v = 0.289 m/sec., b = 2.3 m, d = 1.25 m)

6.6. Estimate the discharge from 10 siphon tubes, each of 10 cm diameter. The elevation of the water surface in the channel is 98.87 m and that in the field 98.32 m.

(**Ans.** 13.4 lit./sec.)

6.7. A field channel having a discharge of 25 lit/sec has to cross a natural drainageway. A semi-circular metallic flume is to be constructed at the crossing. Determine the size of the flume. The width of the drainageway is 20 m. There is a fall of 10 cm between the upstream and downstream ends of the structure.

(**Ans.** 25.1 cm diameter)

6.8. A field channel of capacity of 34 lit/sec has to cross a ridge. The length of the culvert is 12 m. The difference in elevation between the upstream and downstream water surface is 25 cm. Determine the size of the concrete pipe to be used. (**Ans.** 22.5 cm)

QUESTIONS

1. What are the approximate values of seepage loss in different segments of major canal command areas in the sandy loam soils of India/Pakistan as per available estimates?

2. What are the factors influencing seepage in canals?

3. What are the different methods commonly used in measuring seepage loss in canal segments?

4. What is the principle of working of a seepage meter?

5. Distinguish between:

(*i*) Steady flow and unsteady flow in open channels.

(*ii*) Uniform flow and non-uniform flow in open channels.

(*iii*) Perimeter and wetted perimeter of a channel.

(*iv*) Hydraulic radius and hydraulic slope.

(*v*) Earth channels and lined channels.

6. State the mathematical relationships linking the resistance coefficients of commonly used mathematical formulae for estimating the velocity of flow in open channels.

7. What is angle of repose? How does it influence the design of an earth channel?

8. What are the usual causes for the formation of cracks in concrete lining?

9. How are the cracks in concrete-lined open channels repaired?

10. What are the advantages of using pre-fabricated RCC canalets for canal minors, courses and field channels?

11. In the lining of water courses originating from a canal outlet what is the optimum length to be lined, considering the benefit and cost?

12. What are the common types of drop structures used in on-farm water distribution? What are the specific situations under which each type is particularly adapted?

13. Why are check gates used in on-term water distribution systems? What are the different kinds of check gates and their distinct advantages and limitations?

14. What is the function of precision machined slot formers in the form for casting RCC frame for check gates?

15. What are the advantages of synthetic rubber seal strips in check gates?

16. What is the purpose of coat sleeve by-pass in portable canvas dams in farm water distribution systems? How is the by-pass flow controlled using a string?

17. What is the function of a baffle board in a box turnout?

18. What is an anti-seep collar? What is its function?

19. Distinguish between spiles and siphon tubes used as turnouts in farm channel. Are both of them suitable for use in lined channels?

20. Distinguish between culverts and inverted siphons used in channel crossings.

21. What are the factors influencing the location of a canal outlet to best serve its command area?

22. How is the alignment of a water course determined?

23. How are isolated land holdings of a command area served from a water course?

24. Identify the true and false statements from the following:

(*i*) The velocity of flow of an open channel is directly proportional to its hydraulic radius.

(*ii*) Lining results in an increase in the command area of an outlet.

(*iii*) The lighter the type of soil, the larger will be the value of its angle of repose.

(*iv*) The most efficient hydraulic cross-section of a lined water course is rectangular with its depth of water equal to half its breadth.

(*v*) Both longitudinal and transverse grooves are provided in concrete linings to minimize cracking.

(*vi*) In lining canals with brick (or tiles) sandwiched material the usual procedure is to have the sandwiched layer both on the sides and at the bottom of the canal.

(*vii*) Single layer bricks or rubble stones with cement mortar joint are considered as the most appropriate lining material for water courses and field channels.

(*viii*) Placing lining material like bricks or stones on an unconsolidated soil bed is a hazardous practice.

(*ix*) In case of lined field channels, a trapezoidal section is preferred over a rectangular section.

(*x*) A major advantage of precast RCC canalets is its suitability for laying in undulating terrains with RCC pedestals and props of suitable size.

(*xi*) Short length of stilling basins and absence of end sills are commonly observed shortcomings of drop structures on the farm.

(*xii*) The velocity of flow of pipe drop spillways is computed using Manning's formula.

(*xiii*) In fixing synthetic rubber seal strips in the slots of a check gate frame, both the sill strip and the guide strips are cut exactly to the dimensions of the corresponding slots.

(*xiv*) Bernoulli's theorem is applied to design the pipe sizes of pipe drop spillways, inverted siphons and culverts.

(*xv*) In a canal command area when it is observed that an outlet is located far away from the ridge line of the outlet command area, the appropriate step to be taken is to relocate it suitably.

25. Identify the most appropriate answer to the following questions from the four options:

(*i*) The permissible value of mean velocity of flow in an earth channel in loamy soil is:

(a) 40 cm/sec (b) 60 cm/sec
(c) 80 cm/sec (d) 1 m/sec

(*ii*) The most commonly used material for lining canals in India and other South Asian countries is:
(a) cement concrete
(b) cement mortar
(c) low density polyethylene film
(d) bricks or tiles in sandwiched cement mortar

(*iii*) The minimum side slope of an earth channel for polyethylene lining is:
(a) 15:1 (b) 2:1
(c) 2.5:1 (d) 3:1

(*iv*) The policy for lining of water courses for on-farm water distribution systems in irrigation command areas is to provide sub-outlets to serve every:
(a) 6 ha (b) 8 ha
(c) 10 ha (d) 12 ha

(*v*) In a programme on the improvement of an outlet command area the important activities include:
(a) location of canal outlet at the most appropriate point
(b) alignment of the water course along the ridge line or as close to it as possible
(c) taking branch water courses to reach isolated field plots
(d) rectangulation of land holdings, subject to topographic limitations and consent of beneficiary farmers.

26. Re-arrange the events in the following procedures in their appropriate sequences:

(*i*) The events in the lining of canals with bricks (or tiles) in cement mortar:
(a) marking of bottom line of the sloping sides
(b) laying of bricks (or tile) with their top ends at the correct phonation levels
(c) laying of bricks (or tiles) on beds with their length at right angles to the centre line of the canal
(d) the placement of the first layer of bricks with their face having the groove upwards.

(*ii*) The events in the lining of canals with low density polyethylene film:
(a) laying of polyethylene film along the sides and bottom to the canal.
(b) placing a layer of 1.2 cm thick sand to provide proper cushion to the polyethylene film
(c) excavation of the canal and transport of excavated soil to the canal embankment
(d) covering the film with soil to a thickness of 25 cm.

(*iii*) Procedures for fabricating non-reinforced concrete sections for lining field channels:
(a) removing the side boards from the channel section
(b) lifting the inner form after concrete is tamped in the form
(c) coating cement slurry inside precast channel section
(d) laying the base layer and tamping concrete in the side space of the form.

(*iv*) Construction of field channels with precast concrete channel sections:
(a) laying precast channel sections along its alignment
(b) phonation of earth embankments and its compaction
(c) curing the joints of concrete channel sections by sprinkling water
(d) filling the joints with cement mortar.

(*v*) Construction of portable RCC frames for check gates of water courses and field channels:
(a) filling the form with concrete
(b) placing the form in inverted position and placing the reinforcement
(c) curing freshly cast frames
(d) removal of components of the form after the concrete is set.

27. In the following, match each of the entries in the 1st column with the most closely linked entry in the 2nd column:

(*i*)

(A) Kennedy (1881)	(I) Open channel flow
(B) Seepage meter	(II) Friction factor
(C) Chezy	(III) Design of open channels
(D) Darcy-Weisbach	(IV) Constant head permeameter
(E) Manning's formula	(V) Estimates on seepage loss in canal systems

(*ii*)

(A) Contraction joint	(I) Cross wall at the downstream end of a stilling basin
(B) Cut-off wall	(II) Erosion control structure down earth embankments
(C) End sill	(III) Seepage control around masonry structures
(D) Pipe drop structure	(IV) Rubber seal strips
(E) State River and Water Supply Commission, Victoria, Australia	(V) Vertical groove in the top third of concrete lining

(*iii*)

(A) Coat-sleeve by-pass	(I) Seepage control around turnout box
(B) Baffle board	(II) Inverted syphon
(C) Anti-seep collar	(III) Semi-elliptical prefabricated concrete channels
(D) Inlet tank	(IV) Canvas dam
(E) Canalet	(V) Turnout box

FURTHER READING REFERENCES

American Society of Testing Materials (1965). *Standard Specification for Concrete Sewer, Storm Drain and Culvert Pipes*. ASTM: C 14–65.

American Society of Testing Materials (1966). *Tentative Specification for Concrete Pipes for Irrigation or Drainage*. ASTM: C 118–66 T.

Asopra, V.N. and Shingi P.M. (1987). *Command Area Development Programme Strategies for Improving Performance in India*. Centre for Management in Agriculture, Indian Institute of Management, Ahmedabad, p. 108.

Booher, L.J. (1967). *Surface Irrigation*. Food and Agriculture Organization of the United Nations, Rome, p. 161.

Campbell, D.E. (1980). *Canal Systems and Land Development for Irrigation in India*. Collected Papers (Mimeo). D.E. Campbell, Investment Centre, FAO of UN, Rome.

Central Board of Irrigation and Power (1987). *Use of Plastics for Lining of Canals*. Proc. of Workshop, Gujarat Engg. Res. Instt., Govt. of Gujarat, Vadodara, May 1987, p. 24.

Code, W.E. (1961). *Farm Irrigation Structures*. Colorado State University Bul. 496–S, p. 59.

Food and Agriculture Organization of the United Nations (1988). *Design and Optimization of Irrigation Distribution Networks*. FAO Irrigation and Drainage Paper 44, FAO, Rome, p. 247.

Haryana State Minor Irrigation (Tubewell) Corporation, Limited (1980). *Design Manual for Modernization of Water Courses in Haryana*. Pub. H.S.M.I.T.C., Sector 17–B, Chandigarh, p. 34.

Haryana State Minor Irrigation (Tubewell) Corporation, Chandigarh (1983). *Focus on Lining of Water Courses in Canal Command Areas*. Proc. National Seminar on Lining of Water Course, Chandigarh, December 1983, p. 181.

International Commission on Irrigation and Drainage (1967). *Controlling Seepage in Irrigation Canals*. Pub. I.C.I.D., 14, Nyaya Marg, New Delhi–110 021, p. 100.

Irrigation Department, Govt. of Maharashtra (1987). *Manual of Design for On-Farm Development Works*. Govt. Central Press, Mumbai.

Khepar, S.D. and Rai, A. (1979). *Optimum Length for Lining of Canal Water Courses — A Case Study*. J. Agrl. Engg., Indian Soc. of Agrl. Engrs., Vol. XVI, No. IV, pp. 173–181.

Kraatz, D.B. (1971). *Irrigation Canal Lining*. Irrigation and Drainage Paper 2, Food and Agriculture Organization of the United Nations, Rome, p. 170.

Kraatz, D.B. and Mahajan, I.K. (1975). *Small Hydraulic Structures*. Food and Agriculture Organization of the United Nations, Rome, Irrigation and Drainage Paper 26/1 to 26/2.

Kwun Soonkuk (1975). *Design of Irrigation Structures*. Consortium for International Development, Colorado State University, Fort Collins, Colorado, USA Tech. Rep. No. 33, p. 123.

Malhotra S.P. (1970). *The Untold Story of Water Courses in Hisar Bhakra Canal Circle*. Booklet, p. 11.

Michael, A.M., Arora, D.R., Bhattacharya, A.K., Mandal, A. and Gupta, A.K. (1970). *Handbook of Farm Irrigation Structures*. Indian Agricultural Research Institute, New Delhi, p. 82.

Michael, A.M. and Ojha, T.P. (2003). *Principles of Agricultural Engineering*, Vol. II. 4th ed. Pub. Jain Brothers, 873 East Park Road, New Delhi–110005, p. 888.

Michael, A.M. (1973). *Report of the Technical Evaluation of the Scheme for Lining of Water Courses of Bhakra and West Yamuna Canal Systems in Haryana*. Mimeo. Haryana State Minor Irri. and Tubewell Corp., Chandigarh, p. 38.

Ministry of Food, Agriculture and Cooperatives, Govt. of Pakistan, Islamabad (1980). *Water Course Improvement*, Vol. III, On-Farm Water Management Field Manual. Dec. 1980, Mimeo, p. 97.

Normal Belonarme Ve Ilkel Gerilmeli Retonarme Kanallar-Istinat Engerlen-Ayaklar-Temeller (1964). *Prefabricke Sulaona Kanali Pipleri*. CILT II. Ankara, Turkey DSI Mathaasi.

Portland Cement Association (1969). *Concrete Pipe for Irrigation*. American Concrete Pipe Association, USA, p. 24.

Portland Cement Association (1957). *Lining Irrigation Canals*. Portland Cement Association, USA, p. 33.

Raginato, R.J. and Myers, L.E. (1966). *Repair of Cracks in Concrete Channel Linings*. ARS 4–126, U.S. Dept. of Agriculture, U.S. Water Conservation Laboratory, Phoenix, Arizona, USA, p. 8.

Rajput, T.B.S. and Michael, A.M. (1987). *An Equitable Water Distributions System for Tubewell Command Areas*. Jl. Indian Water Res. Society. Vol. 7, No. 4, pp. 11–20.

Rao, B.K. and Gandhi, K.B. (1977). *Evolving Low Cost Canal Linings*. Proc. 46th Session of Central Board of Irrigation and Power, New Delhi, Vol. III, pp. 79–98.

Robinson, E.P. (1963). *Watertight Seals for Flow Control Structures*. Australian Civil Engineering and Construction.

Robinson, A.R., Lauritzen, C.W. and Muckel, D.C. (1963). *Distribution, Control and Measurement of Irrigation Water on the Farm*. U.S.D.A., Misc. Pub. No. 926, p. 27.

Robinson, A.R. (1982). *Farm Irrigation Structures Planning Guide No. 4*. Water Management Synthesis Project, Agriculture and Irrigation Engineering, Utah State University Logan, USA, p. 29.

Robinson, A.R. (1983). *Farm Irrigation Structures*. Water Management Synthesis Project, Handbook No. 2. University Services Centre, Colorado State University. Fort Collins, Colorado, USA, p. 89.

Skogerboe, G. (1977). *Assessment and Improvement of On-Farm Irrigation Systems*. Participant Manual International. Training Office, US Dept. of Agriculture, Dept. of Agricultural Engineering, Colorado State University, Fort Collins, Colorado, USA.

Trout, T.J. and Kemper W.D. (1980). *Water Course Improvement Manual*. Water Management Research Project, Engineering Research Centre, Colorado State University, Fort Collins, Colorado, USA Mimeo, p. 244.

United States Dept. of the Interior. (1963). *Linings for Irrigation Canals*. Pub. Bureau of Reclamation, Denver, Colorado, USA, p. 149.

United States Dept. of Agriculture, Soil Cons. Service, Engineering Division (1964). *Planning and Desing of Open Channels*. Tech. Release No. 25, U.S.D.A., Washington, D.C.

U.S. Dept. of Agriculture (1966). *Handbook of Channel Design for Soil and Water Conservation*. Stillwater Hydraulics Laboratory, Stillwater, Oklahoma, USA.

Water and Power Consultancy Services (India). Ltd., New Delhi (1985). *Economics of Plastics in Canal Lining*.

Westfall D.G. ed. (1980). *Training Manual for Agricultural Water Management*. Water Magt. Tech. Rep. No. 60, Consortium for Intl. Development, Colorado State University, Fort Collins, Colorado, USA.

Chapter

7

Underground Pipeline System

LEARNING OBJECTIVES

Adaptability and advantages of underground pipeline water conveyance and distribution systems. Types of pipes for underground pipeline systems—reinforced and non-reinforced concrete pipes, high density polyethylene pipes and asbestos cement pipes. Hydrostatic pressure requirement of irrigation pipes. Testing of pipes. Estimating discharge capacity of pipelines. Procedures for installing concrete pipelines—pipes with faucet and spigot ends, pipes with tongue and groove joints, pipes with collar joints. Repair of leakage in concrete pipeline joints. Installation of PVC pipes. Jointing of rigid PVC and asbestos cement pipes. Structures for underground pipeline water conveyance and distribution systems—pump stands, gravity inlets, gate stands, air vents, end plug and end wall. Riser valves, hydrants and gated pipes. Design of underground pipeline irrigation systems. Common troubles encountered in the operation of underground pipelines and measures to overcome them. Cost analysis. Underground pipeline systems in community irrigation projects—types of pipeline networks, elevated water tank, outlet valves and surge risers. Rotational water supply in community irrigation projects equipped with underground pipelines.

An underground pipeline water distribution system consists of buried pipes for conveying water to different points on the farm and allied structures required for the efficient functioning of the system. The system offers many advantages over open channels in water conveyance and distribution. Since the pipes are laid underground, cultivation can be done above the pipeline; no culverts or other structures are required. Open channels often take 2 to 4% of the land area out of cultivation, which is saved by adopting the underground pipeline water distribution system. The pipelines do not interfere with farming operations. Efficient use can be made of modern farm machinery because fields are not cut into short runs by channels. When properly installed, the pipelines have long life and low maintenance costs. They are essentially leakproof. Their placement below the ground level prevents damage to pipes and eliminates water loss by seepage and evaporation. Water supply to the field plots are controlled precisely with leakproof riser valves used in the distribution system.

Since underground pipelines operate under pressure, they can be laid uphill or downhill, thus permitting delivery of irrigation water to areas not accessible to open channels (Fig. 7.1). With an underground pipeline system, wells need not be located at the high point of the farm, but may be at a location that provides the best water supply. The water pumped from wells may be delivered directly into the pipeline system. The pipelines are not clogged by vegetation and wind blown material. They are ideally suited to undulating topography to reach water to fragmented and isolated locations on the farm and in areas where land is costly.

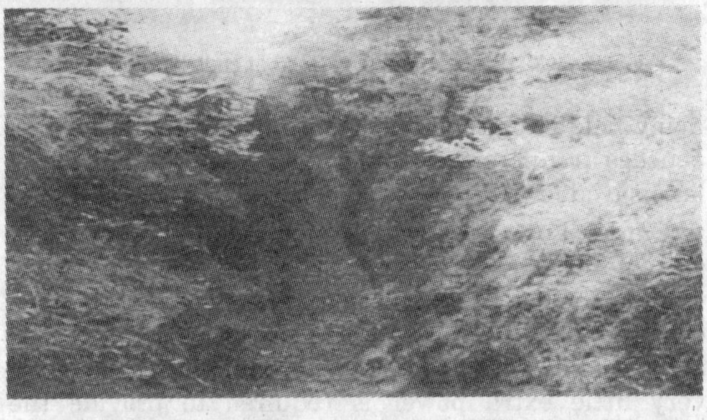

Fig. 7.1. A canal water course cut across a hill on a sandy tract. The heavy loss of water through seepage in the channel section and frequent blocking of the channel due to soil falling from the sides could have been avoided by resorting to underground pipeline water conveyance.

Underground pipeline irrigation distribution systems have a higher initial cost, as compared to lined open channels. They may, however, be more economical under

Fig. 7.2. Home-made concrete pipe with bell end. Note the inner form and one half of the outer form placed on either side of a freshly cast pipe. The outer form and the wooden base are kept with the pipe.

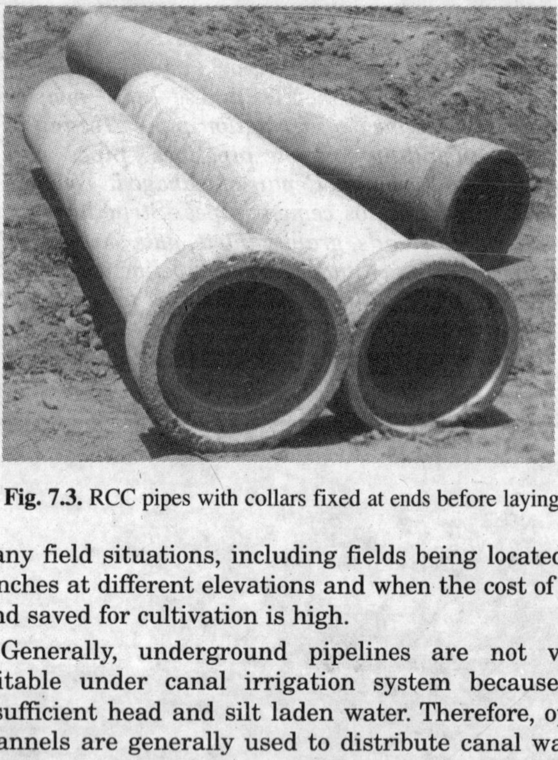

Fig. 7.3. RCC pipes with collars fixed at ends before laying.

many field situations, including fields being located in benches at different elevations and when the cost of the land saved for cultivation is high.

Generally, underground pipelines are not very suitable under canal irrigation system because of insufficient head and silt laden water. Therefore, open channels are generally used to distribute canal water supplies. Pipelines are suitable for pumping schemes, as very little extra power is required to provide the necessary head required to distribute water through them.

Pipes for Underground Water Distribution Systems

Underground pipeline water distribution systems are usually constructed of concrete (Fig. 7.2) or PVC pipes. Occasionally, however, vitrified clay pipes and cement asbestos pipes are used in small size landholdings. Concrete pipelines are usually constructed with factory-made reinforced concrete pipe. Plastic or PVC pipes are being used in tubewell command areas (Fig. 7.4). In case of small farms, where high pressures are not involved, concrete pipe may be constructed of non-reinforced home-made (Fig. 7.2) or factory-made concrete pipes (Fig. 7.4).

The choice of the material for pipes depends on the conditions under which the pipe is to operate and the relative costs. Under most field conditions, concrete pipes are the most economical. Where soils contain salts in large quantity, a chemical analysis should be made to determine the salts which are present. If sulphates are high, concrete should be avoided or the pipe should be made with special sulphate-resistant cements. Vitrified clay tiles with bell joints and PVC pipes are suitable almost under any kind of field environment.

Asbestos cement pipes can be easily installed but are often subject to leakage, unless carefully handled; they have a long service life and are adapted to a wide range of water pressures. Plastic pipes and pipes made of thermoplastic resins are especially suitable in the smaller diameters for high pressure underground pipelines used for sprinkler irrigation, and water distribution systems in public or cooperative irrigation systems.

Non-reinforced Concrete Pipes

Non-reinforced concrete pipes are commonly used when the pipe is not subject to high pressures. Normally, they are suitable for operating heads (pressures) not exceeding 6 metres (0.6 kg/sq cm). Non-reinforced concrete pipes are cheaper than reinforced pipes and could be used with advantage in small to medium size farms. They may be either hand made or machine made. Hand-made pipes are cast in lengths of about one metre and may have bell joints (Fig. 7.2) or tongue and groove joints. Steel or wooden forms are used to cast concrete pipes. Typical specifications of non-reinforced concrete pipes are given in Table 7.1.

Table 7.1. Dimensions and strength requirements for standard and faucet and spigot non-reinforced concrete pipes

Internal diameter (cm)	*Thickness of wall (minimum) (cm)*	*Inside diameter at mouth of socket (cm)*	*Depth of socket (cm)*	*Minimum strength, three-edge bearing method (kg/metre length)*
10	1.5	15.0	3.8	1500
15	1.7	20.6	5.1	1640
20	2.0	27.4	5.7	1940
25	2.2	33.0	6.4	2085
30	2.5	38.4	6.4	2235
37.5	3.5	47.6	6.4	2600
45	4.0	56.6	7.0	2980
52.5	4.5	65.4	7.0	3280
60	5.5	75.0	7.6	3580

(Based on American Society of Testing Materials standard specifications, ASTM C14-65)

The concrete used for making pipes should be made of good quality cement, sand and coarse aggregate so proportioned and mixed as will produce a homogeneous concrete mixture of the required strength. A mixture made of 1 part cement, 2 parts sand and 3 parts coarse aggregate, with a minimum amount of water to make it workable, is suitable to make pipes of adequate strength.

Reinforced Concrete Pipes

The majority of the RCC pipes used in India are manufactured according to Indian Standard, IS:458. They are available in sizes of 15 cm, 22.5 cm, 30 cm and 60 cm. Their lengths vary from 2 to 2.5 metres.

The concrete mixes used are 1 : 1½ : 3, 1 : 2 : 2, 1 : 2½ : 2, 1 : 2 : 3, and 1 : 2 : 4 (cement : sand : coarse aggregate). The sand used varies from fine sand to coarse sand and is to be free from organic material and other impurities. The size of the coarse aggregate usually ranges from 10 to 15 mm. The reinforcement consists of woven steel wire conforming to standard specifications. It is so embedded that the steel and the concrete act together. The pipes are usually made by the spun process. In the case of RCC spun pipes, there is a recession at the end of the pipe for jute braiding. Often a reinforced concrete collar is used with this pipe. Tables 7.1 and 7.2 present the physical strength requirements and specifications of standard factory-made concrete irrigation pipes. Concrete pipes are available with faucet and spigot ends, tongue and grooved ends with collars for jointing.

Table 7.2. Specifications for standard concrete irrigation pipes

Internal diameter (cm)	*Wall thickness (cm)*	*Minimum internal hydrostatic pressure (kg/sq cm)*	*Minimum three-edge bearing load (kg/linear metre)*
15	2.2	3.5	1935
20	2.5	3.5	2010
25	2.9	3.5	2090
30	3.2	3.2	2235
35	3.5	3.2	2380
37.5	3.8	3.2	2456
40	3.8	3.2	2530
45	4.4	3.2	2680
50	5.0	2.8	2750
52.5	5.4	2.8	2825
60	5.7	2.8	2980

(Based on American Society of Testing Materials Standard specifications, ASTM C118-66T)

High density polyethylene pipes. The PVC (Poly Vinyl Chloride) is a thermoplastic material produced by the polymerisation of vinyl chloride as a sole or principal monomer. The main raw materials required for the manufacture of PVC pipes are acetylene and hydrogen-chloride or ethylene-chloride. The PVC pipes are produced by extrusion of thermoplastic in which the PVC contents are between 85-90% and the rest are stabilisers in the form of lubricants, fillers, impact strength modifiers and pigments. PVC pipe fittings (Fig. 7.4) are produced by injection moulding methods

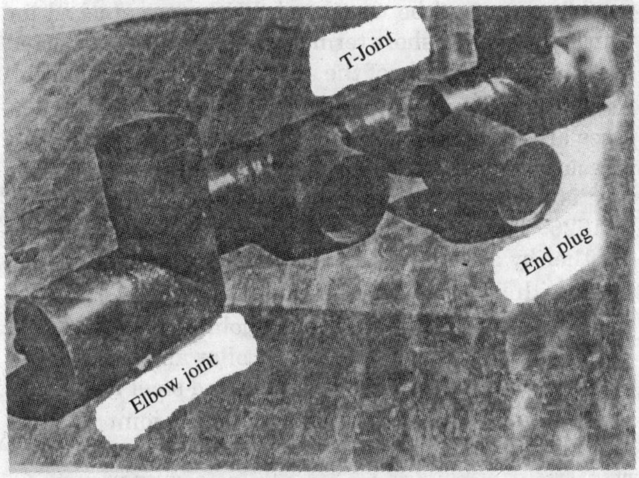

Fig. 7.4. Accessories for PVC buried pipeline system.

up to 110 mm diameter. For large diameters, the fittings are fabricated.

Extrusion of PVC pipes. Extrusion moulding is employed to form thermoplastic material into continuous piping. In extrusion, dry plastic compound, thoroughly mixed, is loaded into a hopper, then fed into a long heating chamber through which it is moved by the action of a continuously revolving screw. At the end of the heating chamber the molten plastic is forced out through a small opening or the die of the desired size. The plastic extrusion from the die is fed into a conveyor belt where it is cooled by immersing in water. The pipe is pulled through a traction unit or hauled off and finally cut to length by a circular saw.

Injection moulding of PVC pipes. Injection moulding is the principal method of forming thermoplastic material. In this, plastic material is put into a hopper which feeds into a heating chamber. A plunger pushes the plastic through this long chamber where the material is softened to a fluid state. At the end of the chamber there is a nozzle which abuts firmly against an opening into a cool, closed mould. The fluid plastic is forced at high pressure through this nozzle into the cold mould. As soon as the plastic cools to a solid state, the mould opens and the finished PVC fitting is ejected from the press.

The Bureau of Indian Standards has given detailed specifications for high density polyethylene (HDPE) pipes, vide IS:4984:1995 and for unplastic pipes, vide IS:4985:1988. Specifications for choice of materials and general recommendations are provided in IS:7634 (Part 1):1975.

PVC should not be subjected to rough handling, particularly during loading and unloading operations. Precautions shall be taken to prevent deformation and degradation due to heat, pressure and ultraviolet radiation. PVC pipes stored for more than three months should not be exposed to open sky and to temperatures above 49°C. The pipe should be covered with a heavy flexible cover and be ventilated. PVC pipes stacked for transportation or short term storage shall be supported below and at the side of the stack at one metre interval.

Asbestos cement pipes. Two types of asbestos cement pipes are used for irrigation pipelines. One type is a pressure pipe made according to IS:1592. The other type is a non-pressure pipe manufactured according to IS:1626. The asbestos cement pressure pipe is joined together by any one of the following three ways:

(*i*) with asbestos cement couplers,

(*ii*) with a cast iron detachable joint (The detachable joint consists of one centre collar and two flanges of cast iron, two rubber rings and steel bolts),

(*iii*) with a concrete collar. The collar joint is used under low heads only.

The asbestos cement non-pressure pipes (building pipes) are made with a faucet and spigot. The joint is made with a rope dipped in cement slurry caulked into the space between the socket and spigot. The space is then caulked with cement mortar and the joint is pointed.

Testing Irrigation Pipes

The standard tests for irrigation pipes are tests for water absorption and for strength. The water absorption in standard irrigation pipes should not exceed eight per cent of the original dry weight of the pipe. To test the water absorption, samples of the pipe are boiled for about five hours and the gain in weight by moisture absorption is noted.

The hydrostatic pressure requirement prescribes the minimum hydrostatic pressure the pipe may sustain without excessive leakage through the shell of the pipe. Figure 7.5 shows the layout for the three-edge bearing test to determine the strength of concrete pipes. The tests are made on a standard testing machine. The load is applied parallel to and along a vertical plane passing through the axis of the pipe. The lower bearings consists of two wooden strips 2.5 to 3.8 cm in height and not more than 5 cm wide at the base, having their interior top corners rounded to a radius of approximately 1.2 cm. The strips are straight and are securely fastened to a rigid uniformly supported wooden block at least 15 cm wide. The upper bearing is a rigid wooden block, free of knots and straight and uniform from end to end. The upper and lower bearings extend to the full length of the pipe. The load is applied continuously and the load at which the pipe is crushed is also noted. The load per linear metre of the pipe is calculated by dividing the total recorded load by the length of the pipe.

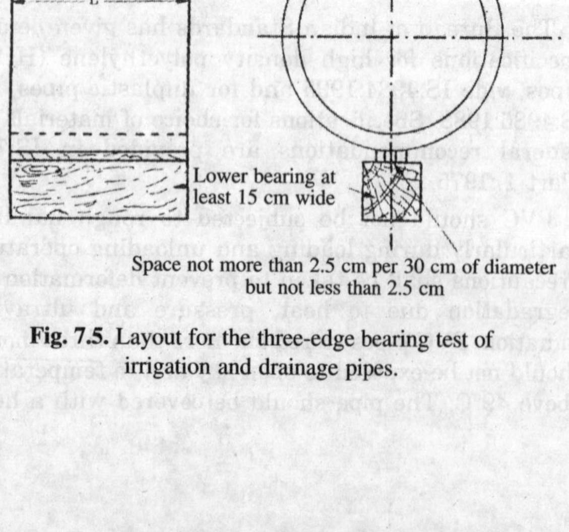

Fig. 7.5. Layout for the three-edge bearing test of irrigation and drainage pipes.

Special Fittings for Concrete Pipes

Special fittings for concrete pipes such as branches, saddles, matching pieces, bends and tees are usually supplied in cast iron or mild steel (concrete lined). Sluice valves, reflux valves and air valves are supplied in cast iron, fabricated to suit the pipes. The fittings are usually made with faucets to suit the concrete pipe. Joints of tightly caulked lead and hemp are generally used. When pouring lead joints, care is taken to ensure that the molten lead is not too hot. Overheating of lead may result in cracking of the concrete pipe.

Discharge Capacity of Pipelines

The discharge through a pipeline can be determined by applying Darcy's equation

$$V = \sqrt{\frac{Hdg}{2fl}} \quad \text{... (7.1)}$$

in which,

V = velocity of flow of water through the pipe (cm/sec)

H = available head causing flow (difference in elevation between water level in the pump stand and the outlet point) (cm)

d = diameter of pipe (cm)

g = acceleration due to gravity (cm/sec^2)

l = length of pipe (cm)

f = Darcy's roughness coefficient

Table 7.3 presents the values of the Darcy's roughness coefficient f for different kinds of pipes at various values of velocity of flow.

Example 7.1. *Determine the discharge capacity of an underground concrete pipeline from the following data: Diameter of pipe 15 cm, length of pipeline 150 metres, difference in elevation between water levels at pump stand and discharge point 2 metres.*

Solution: The value of f is assumed to be 0.009 (Table 7.3, with an assumed velocity of flow of 90 cm/sec).

Table 7.3. Values of coefficient 'f' in Darcy's equation for pipe flow

Diameter of pipe (cm)	Velocity, centimetre per second														
	30	6	90	120	150	30	60	90	120	150	30	60	90	120	150
	Galvanised iron pipe					Cast iron pipe					Concrete pipe				
15	0.0056	0.0047	0.0043	0.0042	0.0038	0.0061	0.0059	0.0058	0.0057	0.0055	0.0102	0.0094	0.0090	0.0086	0.0084
22.5	0.0051	0.0044	0.0041	0.0038	0.0036	0.0057	0.0055	0.0054	0.0053	0.0052	0.0080	0.0074	0.0070	0.0067	0.0064
30	0.0047	0.0042	0.0038	0.0035	0.0034	0.0054	0.0052	0.0051	0.0050	0.0048	0.0067	0.0062	0.0059	0.0056	0.0054
45	0.0044	0.0037	0.0035	0.0033	0.0032	0.0052	0.0051	0.0050	0.0048	0.0047	0.0056	0.0051	0.0048	0.0046	0.0044
60	0.0042	0.0035	0.0033	0.0032	0.0030	0.0050	0.0048	0.0045	0.0043	0.0041	0.0050	0.0046	0.0044	0.0042	0.0040
90	0.0037	0.0033	0.0031	0.0029	0.0029	0.0042	0.0040	0.0037	0.0036	0.0034	0.0044	0.0040	0.0038	0.0036	0.0034
120	0.0035	0.0032	0.0029	0.0028	0.0027	0.0036	0.0034	0.0032	0.0032	0.0031	0.0039	0.0037	0.0035	0.0033	0.0030

Applying Darcy's equation,

$$V = \sqrt{\frac{Hdg}{2ft}} = \sqrt{\frac{200 \times 15 \times 981}{2 \times 0.009 \times 15000}}$$

= 100 cm/sec = 1.0 m/sec

Area of cross-section of pipe,

$$a = \frac{\pi d^2}{4}$$

$$= \frac{\pi}{4} \cdot 0.15^2$$

$$= 0.0176 \text{ sq m}$$

Discharge capacity, $Q = a\ v$

$= 1.0 \cdot 0.0176$

0.0176 cu m/sec

= 17.6 litres/sec.

Spacing of Concrete Pipelines

Pipelines are spaced to suit the field layout and soil type. One main line is usually laid along the upper boundary of the field. When the land slopes in nearly opposite directions, the pipeline may be laid in the middle, so that one line can command fields on either side. When the prevailing land slope is in one direction only, the spacing between pipelines varies from 100 to 200 metres. The closer spacings are suitable for sandy soils and the farther spacings for heavy clay soils. It is not necessary to lay the pipelines along the field boundaries. Often it may be possible to lay it diagonally across the fields, thus reducing the total length and the cost.

Installation of Concrete Pipelines

Proper installation of the pipes is the key to trouble-free service of an underground pipeline irrigation system. The process of laying the pipeline involves the following steps that follow in succession: *(i)* Selection of depth and grade of laying, *(ii)* Digging to proper depth and grade, *(iii)* Lowering the pipe and squeezing, *(iv)* Sealing the joints, and *(v)* Backfilling the trench.

Fig. 7.6. Checking the trench gradient with a dumpy level.

To prevent damage of pipelines from loads acting on it, concrete pipes should be laid with their upper surface at a depth of at least 45 cm to 60 cm from the ground surface. The width of the trench should be adequate to provide sufficient working space. The grade of the pipeline is kept uniform and is usually the same as the average land slope along the line where the pipes are laid. To obtain uniform depth and grade, the elevations of the trench bottom are fixed with a dumpy level (Fig. 7.6). The layout of the pipeline is decided to suit the irrigation requirement of different fields.

The route of the proposed pipeline is cleared and smoothened with suitable tools and equipment. A tractor-operated blade terracer, if available, will make the operation easy and efficient. On the dressed route, a base line is established. The base line is aligned with the help of ranging rods and rope. Elevations of the different points on the base line along the route of the pipeline is determined with a dumpy level. Stakes are fixed along the base line, with marks indicating the extent of cut required at the point. It is ensured that the pipeline is laid at a uniform gradient. The minimum cover over the top of the pipe should be 60 cm, so that the pipelines will not interfere with farming operations.

Lines are marked on either side of the base line, marking the border of the trench in which the pipeline is laid. This facilitates digging of the trench. For a 30 cm pipeline, the minimum width of the trench should be about 75 cm. The trench lines are marked with lime.

The trench is dug manually with spades or with trenching machines. For speedy operation a tractor mounted backhoe digger could be used. The trench is dressed to ensure a uniform section. The trench bottom is prepared to ensure laying of pipes on a firm and smooth bed. At all pipe joints the trench is further cut to a depth of about 15 cm and a width of 30 cm to provide working space for proper jointing of the pipes (Fig. 7.7).

Pipes and collars, if used, are laid on one side of the trench. They are inspected for any damage, including cracks. Defective pipes or collars, if any, are replaced.

The pipes are lowered gently with the help of a strong rope passing through the pipe and held on both ends. Similarly, the collar is lowered and slipped inside the respective pipe end. Pipes are set on firm trench bed in sets of 4 to 5 pipes at a time with the respective collars laid over it. The horizontality of each pipe is checked with a spirit level to maintain a uniform slope for the pipeline. After a series of pipes are laid and set, the gradient of the line is rechecked with a dumpy level.

Each pipe should be laid on firm, undisturbed soil. The bottom of the trench should be smooth, with large stones, clods, or other irregularities removed. The fill should be properly compacted to produce a firm uniform foundation. When the pipelines are to be placed in soft, unstable soil, the trench should be over-excavated and backfilled to grade with suitable bedding for the pipe. The bed below each joint is scrapped to a small pit so that jointing work is done conveniently. Before lowering the pipes, holes for fixing riser pipes and air vents are made on the appropriate pipes. The pipes are first strung along the trench and then tilted into the trench. Heavy RCC pipes may be wrapped at the ends by ropes and then lowered by gradually releasing the ropes till the pipe has rested on the bottom of the trench. The pipes are placed in a straight line at the centre of the trench.

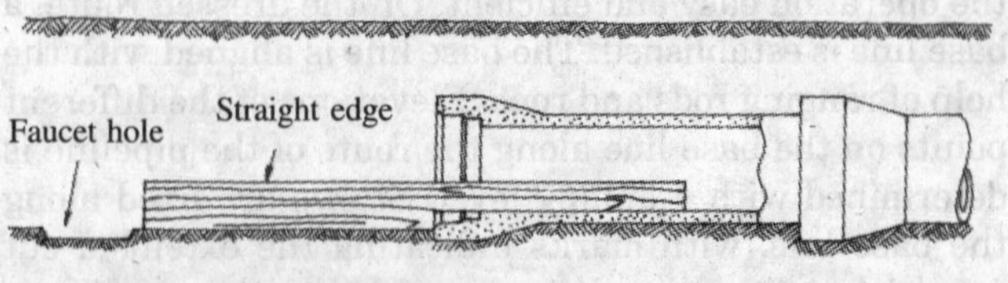

Fig. 7.7. Pipe laying procedure: Preparation of bed for the next pipe to be laid.

(*Courtesy:* The Indian Hume Pipe Co. Ltd., Mumbai.)

Faucet and Spigot End Pipes

Pipes with faucet and spigot ends (Fig. 7.8) are laid with their faucet (enlarged) ends facing upstream. The joints may be made using jute or hemp rope or rubber rings. In case of joints made with hemp ropes, the ends of the pipe are cleaned and wetted with a brush. Jute or hemp rope dipped in a cement paste is wrapped round the plane end of each pipe. The rope is just thick enough to be inserted into the socket of the pipe already in position. Before the pipe is inserted in the socket, mortar made of one part cement and two parts sand is applied to a thickness of about one cm over the spigot end of the pipe and also inside the socket. The pipe is placed into the socket and care is taken that the packing is not pressed through the pipe. Steel chisels are used to ram the hemp rope tight when the pipes are in position. After ramming the hemp rope tight with a steel tool, the remaining space in the socket is filled with cement mortar (1:2 mix) and finished with a bead on the outside.

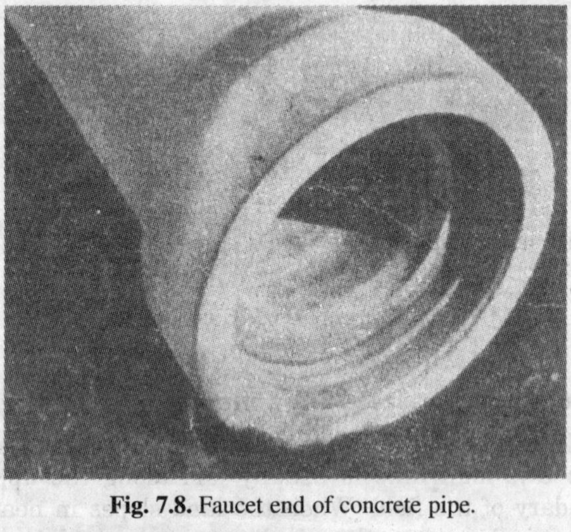

Fig. 7.8. Faucet end of concrete pipe.

Courtesy: The Indian Hume Pipe Co. Ltd., Mumbai.)

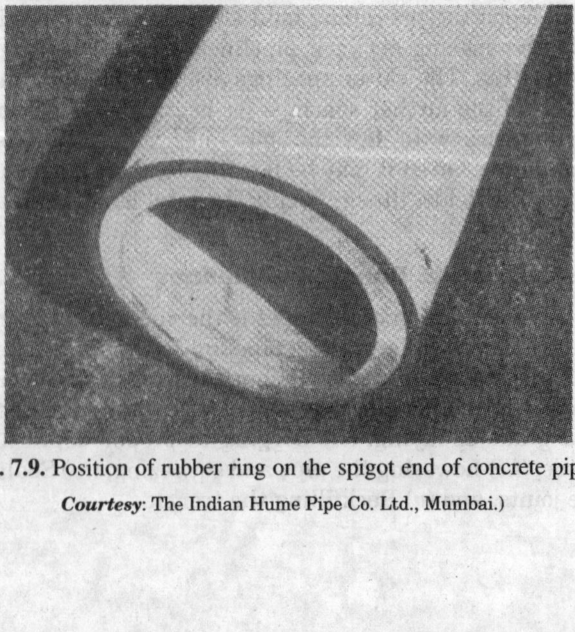

Fig. 7.9. Position of rubber ring on the spigot end of concrete pipe.

Courtesy: The Indian Hume Pipe Co. Ltd., Mumbai.)

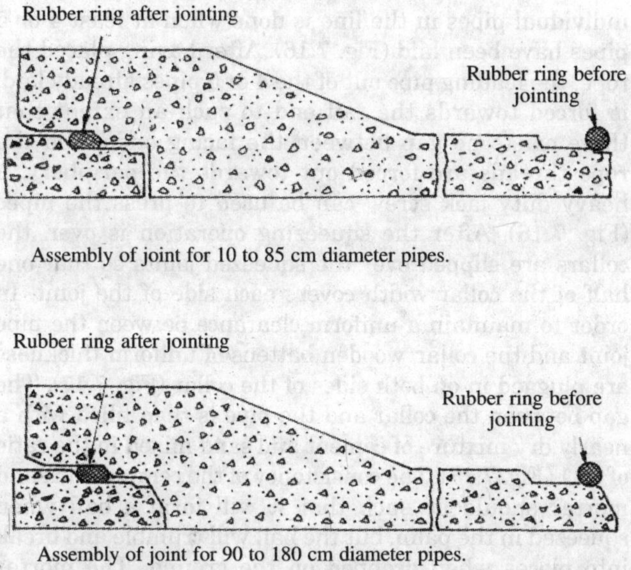

Fig. 7.10. Sectional view showing the location of rubber ring in jointing faucet and spigot end concrete pipes.
(***Courtesy:*** The Indian Hume Pipes Co. Ltd., Mumbai).

Pipe joints using rubber rings. Concrete pipes joined with rubber rings provide a certain amount of flexibility. They are to be anchored at places where there are unbalanced forces acting on the pipe, such as bends, tees and valves. The most suitable anchorage is obtained by casting a block of concrete around or against the filling. Care is taken that the connector block is bearing on solid ground.

Laying and jointing of pipes. Before being laid, each pipe is carefully examined for cracks or other damage which may have been due to cartage, handling or other causes. The size of rubber rings are checked. Larger pipes are generally lowered into the trench by use of a three-legged derrick or a mobile crane, and the smaller ones by hand, using ropes. Handling methods will depend on the size of the pipe, the number of men available and the conditions generally prevailing at the site. After lowering the pipe into the trench, it is ensured that both the spigot and faucet are clean and dry. Any dirt in the faucet of the preceding pipe is brushed out.

As the pipe is pushed into position, the rubber ring rolls along the spigot and at the same time is squeezed between the faces of the spigot and faucet ends to form a watertight joint (Fig. 7.9 and 7.10). For pipes of 22.5 cm diameter and smaller, this operation can be done by hand, but for larger diameter pipes, crowbars are required for forcing the pipe forward (for larger pipes 60 cm to 75 cm two bars are necessary) and another one to raise the spigot end of the jointing point. A chain is placed around the pipe approximately 7.5 cm back from the spigot end and the point of the crowbar is passed through the loop of the chain to rest on a block on the faucet end of the pipe already laid. The pipe layer, who is

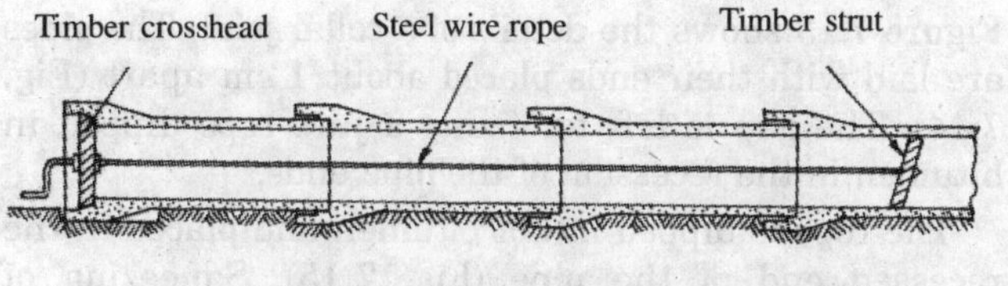

Fig. 7.11. Pipe laying procedure: Take-up device for pulling in large pipes.
(***Courtesy***: The Indian Hume Pipes Co. Ltd., Mumbai).

standing in the trench, has an effective leverage and can easily manage to lift and maneuver pipes up to 75 cm in diameter. When the pipes are pushed by means of one or more crowbars, the bars should bear against stout pieces of wood placed across the faucet ends, so as to prevent any damage to the faucet (Fig. 7.11). It should be ensured that the pipe ends are brought together closely, and internal inspection should be made after each pipe is laid to see that this has been carried out. If a pipe is inclined to creep out of its joint for no apparent reason, it is usually an indication that the rubber ring has not rolled evenly. In such a case, the pipe should be pulled out and the joint made again.

Pipes with Tongue and Groove Joints

Pipes with tongue and groove joints are laid with the groove ends facing upstream. The tongue or spigot end of the first section of pipe is cleaned and wetted with a brush. Sufficient mortar to form the lower section of the outside band or collar is placed in a depression at the lower side of the pipe joint. The groove end of the next section is wetted and filled with mortar. This section is then tipped over carefully and the mortar end is pushed in to make a snug, tight joint (Fig. 7.12). Excess mortar will be squeezed out of the joint on both the sides.

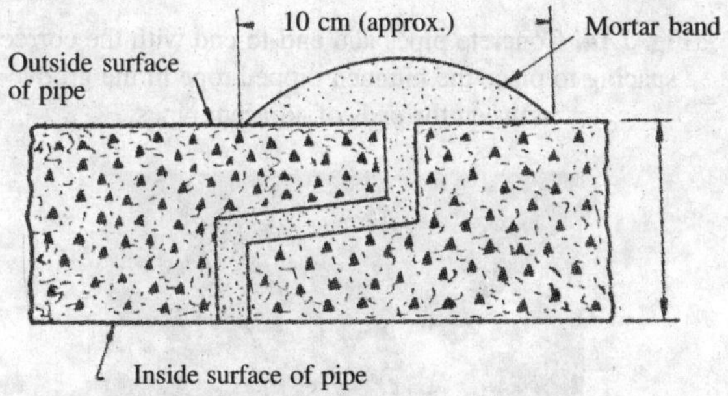

Fig. 7.12. Details of mortar joint for concrete pipes with tongue and groove joint. The outside mortar band is approximately 10 cm wide. The inside is brushed smooth.

Reinforced Concrete Pipes with Collar Joints

Reinforced concrete pipes are usually available in lengths of 2.5 to 3 metres. The pipes are lowered into the trench with the help of ropes by two men standing on the trench bank on either side of the pipe. Before an adjacent pipe is laid, a collar is slipped to its end.

Figure 7.13 shows the details of a collar joint. The pipes are laid with their ends placed about 1 cm apart (Fig. 7.14). The gap is left for fixing a jute rope dipped in bitumen in the recession of the pipe ends.

The rope is dipped in hot bitumen and placed in the recessed end of the pipe (Fig. 7.15). Squeezing of individual pipes in the line is done when at least 4 or 5 pipes have been laid (Fig. 7.16). After having placed the rope, the leading pipe out of the 4 or 5 pipes already laid, is forced towards the tail end to such an extent that there exists no gap between the facing ends, and the rope threads are forced out towards the periphery. A heavy duty jack screw can be used to press the pipes (Fig. 7.16). After the squeezing operation is over, the collars are slipped over the squeezed joints so that one half of the collar width covers each side of the joint. In order to maintain a uniform clearance between the pipe joint and the collar, wooden battens of uniform thickness are plugged in on both sides of the collar (Fig. 7.17). The gap between the collar and the pipe is then filled with a nearly dry mixture of cement and sand mixed in the ratio of 1 : 1 (Fig. 7.18). The consistency of the cement and sand mortar should be such that it will form a ball when squeezed in the palm, but the ball will crumble and break into pieces when dropped on the ground. The mortar which is filled in the space between the collar and the pipe is thoroughly tamped into the space with a flat steel chisel and hammer. Once the gap is sealed satisfactorily the collar is lined with 1.5 cm thick plaster (1 : 2 ratio) and bevelled off at an angle of 45° with the outside edge of the collar. The finishing of the joints should be at least 5 sections behind the laying operation.

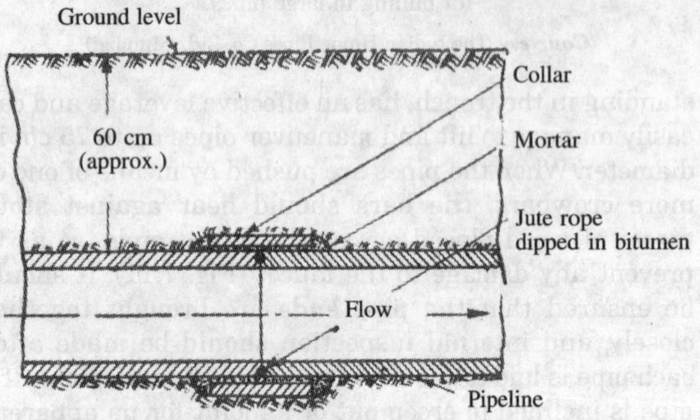

Fig. 7.13. Collar joint for reinforced concrete pipes.

Fig. 7.14. Concrete pipes laid end-to-end with the correct spacing to place the bitumen dipped rope in the groove between the ends of adjacent pipes.

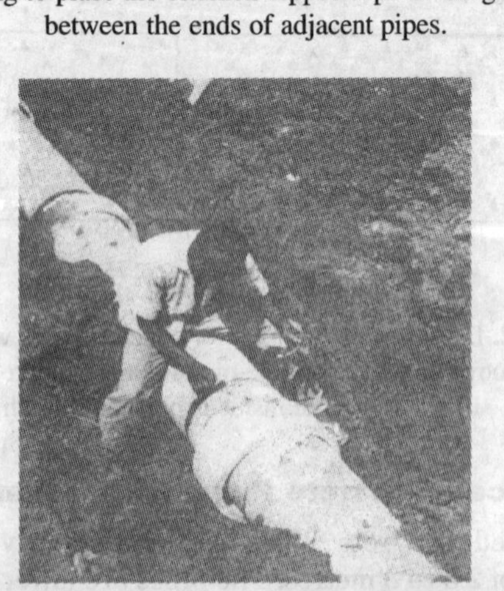

Fig. 7.15. Placing the bitumen-dipped rope in the groove between the ends of adjacent pipes before squeezing (Michael, A.M., *et. al.*, 1970).

Fig. 7.16. Squeezing the pipe with a screw jack. Note the wooden supports and the concrete end plug.

Fig. 7.17. Correct placement of collar at the joint. The concentric space is of uniform thickness to tamp in the concrete. Note the wooden spacers to maintain uniform concentric space.

Openings in all concrete pipelines should be covered to prevent air circulation, except when the work is in progress. Such openings should be kept closed until the pipeline is filled with water. There should be an initial backfill of soil around the pipe (Fig. 7.20). The soil backfill should cover the pipe to a depth of at least 15 cm for the full width of the trench and not more than seven sections behind the laying. In backfilling, the space on either side of the pipe is filled first to avoid injury to the freshly applied mortar bands and not to dislocate the pipes.

Mortar joints should be protected from drying out (Fig. 7.20). After initial setting, curing is continued for seven to ten days. This can be done by covering the joints with earth and adding water to the earth cover to keep it wet. Alternatively, gunny bags may be put over the joints and water sprinkled on it to keep them wet. Negligence in curing may lead to poor setting of cement

Fig. 7.18. The space between the collar and pipe ends at the joint is filled with *a nearly dry mixture* of cement and sand (1:1 mix.). The mixture is thoroughly tamped into the space with a flat steel chisel and hammer. The joint is finished with a band of cement mortar bevelled to an angle of 45°, placed on either side of the collar. (Michael, A.M. *et al.*, 1970)

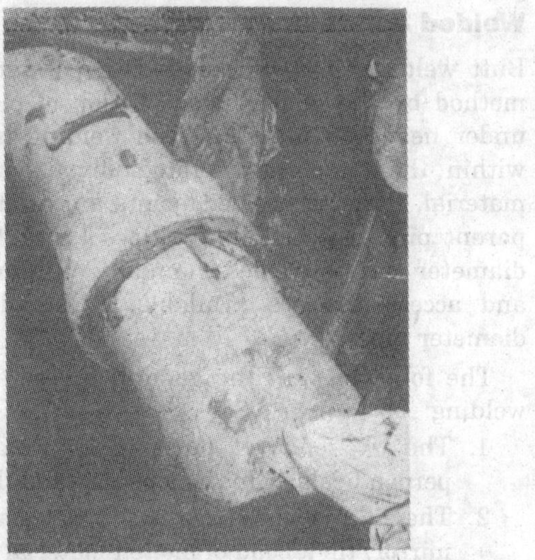

Fig. 7.19. Hemp or jute rope dipped in wet cement-sand mixture (1:1 ratio) placed on either end of the collar after filling the concentric space with cement-sand mixture.

Fig. 7.20. A long line of concrete pipes with collar joints being cured before backfilling. Note the mounds of soil at each joint for holding the water added for curing.

mortar and resultant leakage. Sometimes to facilitate the jointing process, collars are fixed to the faucet end of pipes and cured before laying in trench.

After the joints are cured, the pipeline is tested at low head (about 1 metre). Any leakage has to be carefully observed and exact points needing repair are marked. Leaky joints are repaired after the water in the pipeline is drained. The surface is exposed, cleaned and then a thick band of mortar is applied. After the pipeline is tested, the trench is backfilled with earth.

Repair of leakage is concrete pipeline joints. Pipelines may leak due to faulty jointing. *It is better not to install an underground pipeline system if watertight joints cannot be provided.* However, there may be cases of a few leakes due to cracks in the line or other unforeseen reasons. The normal symptoms of leakage is continued wetness of the soil cover over the pipeline.

To repair the pipeline at the point of fault, the wet area is marked with lime and soil is dug to expose the leaking joint or crack in the pipeline. A good working space is formed around the joint. Water is drained from the pipeline by releasing it from a suitable outlet at the downstream end. A hole of size equal to that of an arm is made in the pipe to enable the placement of mortar to plug the leaky part. Reinforcing wires in the pipe are cut and the accumulated debris removed. A layer of rich cement mortar in the ratio of 1:1 is put all around the internal periphery of the joint. It is properly finished and cured. After allowing the first layer of cement mortar to set, another layer is put in a similar manner. Once the cement mortar has set, the pierced hole should be capped by covering it by galvanised iron sheet plates. For this two plates of size equal to that of the pierced portion of the pipe section are made with holes at each end. One plate is inserted inside. It is held in position by another plate over its top. The two are joined by bolts and nuts.

Installation of PVC Pipes and Fittings

Rigid PVC pipes are easy to install. There are few simple rules to follow to obtain optimum results:

(*i*) *Installation above or below ground*. Changes in temperature can cause the 'snaking' of pipes laid above ground due to expansion and contraction. For example, a 100 metres long line will expand 60 mm for every 10°C rise in temperature.

(*ii*) *Burying of PVC pipes*. Although it is preferable to bury rigid PVC pipes to a depth of about 45 cm, there are many applications where it is safe to bury them close to the surface. Where the external loading is expected, the pipes are to the laid at 45 cm or deeper.

Jointing of rigid PVC pipes. The following systems of jointing are used for rigid PVC pipes:

(*i*) *Solvent cement jointing*. Solvent cement jointing is used to fit spigot and faucet joint bounded by a solvent cement comprising of a solution of plastic material in a strong solvent.

(*ii*) *Rubber ring jointing*. This is based on the concept that the rubber ring in compression are used to remain in compression such that it excerts a sealing pressure.

(*iii*) *Flanged jointing*. Flange joints are used for jointing PVC pipes, provided with flanged ends. Their principal application is for the connection of pipes to valves and vessles where strength in tension is required.

(*iv*) *Threaded jointing*. The cutting of a screw thread into PVC pipes is not recommended. However, moulded threaded fittings may be used for jointing to metal connections.

Solvent cement jointing. To make solvent cement joint in rigid PVC pipes the following procedures are adopted:

(*i*) Ensure that the pipe ends are undisturbed after storage or transportation.

(*ii*) Ascertain that the pipes enter into the socket and fit properly.

(*iii*) For spigot and faucet joints, scour with an emery paper the surfaces which are to be cemented.

(*iv*) Wipe the pipe with a clean cloth if it is heavily contaminated with grease or oil; use thiner to remove grease or oil.

(*v*) Use brush and apply the solvent cement recommended by the manufacturer evenly on the outside of the surface of the pipe and the inside of the socket, elbow, or other fitting.

(*vi*) Enter the pipe socket within a minute after applying the solvent cement. Thereafter, give a quarder turn (90°) to ensure even distribution of solvent cement with the mating surface. The excess cement which is pushed out of the socket is removed immediately with a clean cloth. A minimum of 5 to 8 minutes setting time is allowed without disturbing the pipe. To obtain nearly full strength, the setting time is 30 minutes. Full strength is attained in about 24 hours.

(*vii*) In case of any leakage observed during testing after 24 hours of jointing the last pipe piece, the leaking portion of the pipe is cut with a hacksaw, the defective portion is removed and the pipes rejointed, adopting the same procedure as described above.

Solvent cement is highly inflamable and should not be kept near an unguarded flame. It should be stored in a cool dry place. The lid of the container must be tightly screwed when not in use so as to prevent evaporation of the solvent.

Detailed procedure for jointing of PVC pipe and fittings are given in IS:7634 (Part-2) and IS:7634 (Parts-3 to 6) of the Bureau of Indian Standards listed at the end of this chapter.

Welded Joints in PVC Pipes

Butt welding of HDPE pipes is a permanent joining method by the homogeneous fusion of pipe material under heat and pressure. The welding is performed within the temperature range for fusion of HDPE material. Perfectly welded joints are as strong as the parent pipe material itself. Pipes up to 11 cm outer diameter can be welded manually. Welding machines and accessories are available for small and large diameter pipes.

The following are the general procedures in butt-welding:

1. The two ends of pipes to be welded are cut perpendicularly to the pipe-axis and then cleaned.
2. They are passed against a hot-plate (heating mirror) till a bead of molten material is formed.
3. They are pressed together after removing the heating mirror and allowed to cool down.

Specific Procedures in Butt Welding

Preparation of welding surfaces:

1. *Cutting the pipe.* The pipes to be butt-welded should be cut perpendicular to the pipe axis. For cutting pipes up to 110 mm outer diameter, the use of wooden jigs is useful. Straightness of the cut for pipes larger than 110 mm diameter can be achieved only by holding and clamping the pipe on a welding jack.
2. *Planing pipe ends after cutting.* After cutting the pipe, the ends to be welded must be planed. The pipes are brought in axial alignment. The prepared ends of the pipes should be brought in contact with each other (face to face). The gap, if any, should not exceed the limits given below:

Pipe outer diameter	*Gap*
60 – 200 mm	0.30 mm
225 – 400 mm	0.50 mm

3. *Inspection before welding.* Before the ends of the pipes are welded, they must be inspected for the following:
   (*i*) Internal or external damage: The damaged portions should be removed.
   (*ii*) The difference in wall thicknesses of pipes to be welded should not exceed 5% of nominal wall thickness with a maximum of 0.5 mm for pipes up to 200 mm outer diameter and 3% with a maximum of 1 mm for pipes up to 400 mm outer diameter.
   (*iii*) The pipe ends to be joined must be checked for cleanliness. Any impurity on the welding surface will act as separating agents for the joint.

   The following precautions are to be observed:

   (*a*) Before welding, the surfaces should be mechanically cleaned again with a blade or scraper.
   (*b*) Shavings should be removed.
   (*c*) Traces of grease or oil, which may fall from tools on the welding surface should be removed.
   (*d*) In difficult situations, an additional method of cleaning the pipes immediately before the start of welding is to use a cloth dampened with methylated spirit.
   (*e*) After cleaning, the surfaces may not be touched and the welding is done immediately.
   (*f*) The alignment of the prepared ends of the pipes should not be disturbed till the welding is over.

Heating plates or mirror: Heat supply to the ends of pipes to be butt welded is usually effected by using a metallic plate which, in general is electrically heated. It is called *heating mirror*. For sites without electricity, mirror heated by blow torches can be used for small diameter pipes. For butt-welding, the heating mirrors can be circular or square. The contact surfaces of the heating mirror have to be parallel to each other. The flatness of the mirror can be checked with the aid of a steel "dummy" pipe. A maximum permissible gap between dummy pipe and heating surface should be as follows:

Mirror suitable for welding of pipes of outer diameter	*Gap*
60 – 200 mm	0.15 mm
225 – 400 mm	0.25 mm

The contact surface of the heating mirror should be covered with PTFE cloth which is commercially known as 'Teflon'. PTFE covered mirrors have the advantage that the molten material does not stick to its surface. The Teflon covered heating surface should be wiped with clean cloth before and after the heating operation to get rid of any adhered impurities.

The heating should be such that the temperature of the front and back of the heating mirror and at various spots of the contact surface does not differ from the nominal plate temperature to a value of more than ± 5°C.

UNDERGROUND PIPELINE STRUCTURES

Specialised structures are used with underground pipeline systems to control the water and protect the pipeline from damage. These include the structures at the inlet, water control and diversion structures, air release vents and end plugs.

Inlet Structure

Water may enter a pipeline by gravity from an open channel or it may be pumped from a well or any other source. An inlet structure is required to develop the full flow capacity of the pipeline, to maintain water surface elevations sufficient to distribute water to the different points on the farm, to ensure that excessive pressures do not cause damage, and keep trash from entering the pipe.

Pump stands. A vertical pipe extending above the ground and connected to the underground pipeline system is known as a pump stand (Figs. 7.21 to 7.23). Pump stands are located at the inlet of an underground pipeline system. It is larger than the size of the pipeline. The larger capacity of the stand permits dissipation of the high velocity stream and release of entrapped air before the water enters the pipeline. To enable the mason to make mortar joints from inside the stand, its

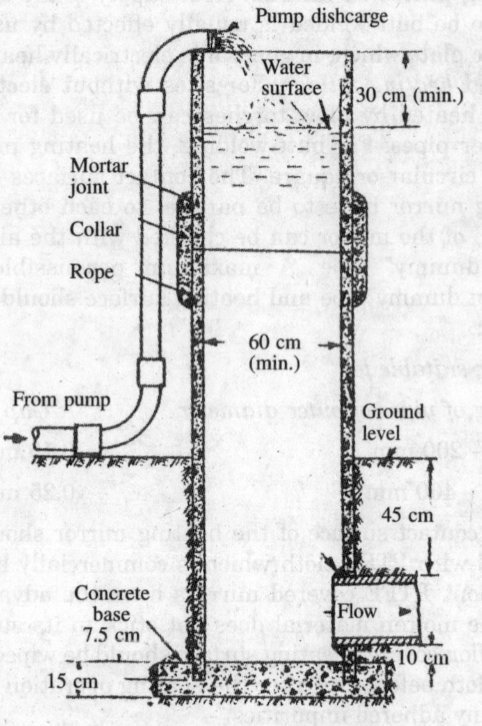

Fig. 7.21. Pump stand for underground pipeline.

minimum inside diameter should be about 60 cm. Any air entrained by the high velocity stream coming from the pump will have an opportunity to escape at the pump stand. Entry of air into the pipeline can cause air pockets which restrict the flow of water, surging flow conditions and development of excessive pressures.

Pump stands must extend upward to a point where water will not overflow, except when unusual pressures occur. This sometimes happens when a pump is started, and the pipeline could be broken if the stand does not overflow. The elevation of the water surface in the stand must be sufficient to permit the discharge of water through the outlets at any point on the farm. This should also include the pressure head required to overcome friction in the pipe and valves.

To install the pump stand a concrete bed is made. The thickness of the bed should be about 10 cm for pump stands up to 3 metres high and 15 cm when the height is more than 3 metres. The diameter of the concrete bed should be at least 30 cm more than the outer diameter of the pipe used as pump stand. After laying the concrete and before the bed hardens, the pipe is placed vertically in the centre of the bed and a concrete of 1:2:4 mix is added to a thickness of about 5 cm.

In small scale installations, pump stands consist of simple masonry tanks of the required height, about one metre square. When the pump discharges water containing appreciable amount of sand or silt, it is necessary to construct an auxiliary desilting tank of the same height as the pump stand. In such cases, the pump discharge is first allowed into the desilting tank. The desilting tank is connected to the pump stand by a pipe fixed near the top. The desilting tank is occasionally cleaned.

Fig. 7.22. Jointing of the pump stand pipes.

Gravity inlets. When water enters the pipeline from an open channel, a gravity inlet is used (Fig. 7.24). A screen is fixed to the inlet to keep trash out of the pipeline. The top of the structure is usually provided with a cover to prevent accidents and to keep trash from blowing in.

If the irrigation water contains appreciable quantities of sand, a trap can be built into the inlet structure to remove most of the suspended material. When the stand functions as a sand trap also, it has an extra-large diameter to ensure low velocity of water and its bottom is set at some distance (about 60 cm) below the bottom of the pipeline.

When the canal water contains considerable quantities of debris and weed seed, a debris screen is

Fig. 7.23. Inlet of an underground pipeline water distribution system laid in a hilly tract in Ropar district, Punjab (1973). Note the tall pump stand and the breather pipe beside it. The delivery pipe of the pump installed in the nearby open well discharges water into the pump stand.

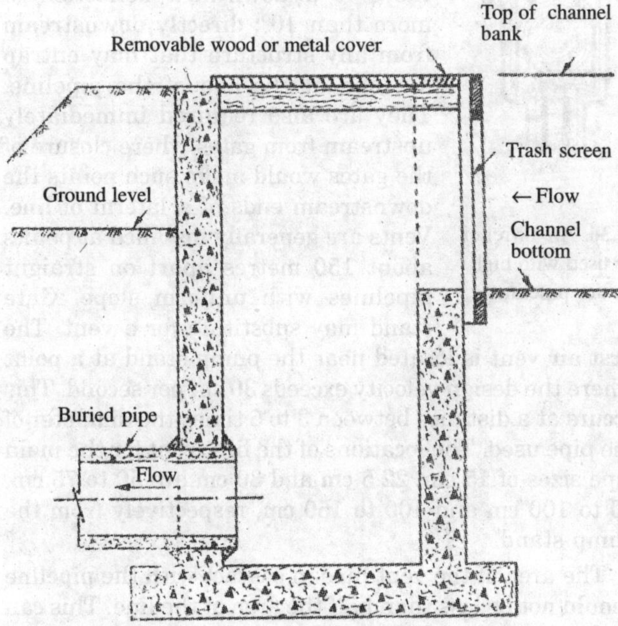

Fig. 7.24. Sectional view of an inlet for taking water from a minor canal into an underground pipeline. A separate desilting tank is necessary at the upstream end of the inlet structure when the incoming stream carries appreciable amounts of silt.

provided at the inlet to clean the water before it enters the pipeline. Commonly, the stream is allowed to fall through a fine screen into a gravity inlet. The screen will need frequent cleaning.

Fig. 7.25. Joining the pipeline to a pump stand from which two lines of underground pipes are taken out. Bitumen-dipped rope is first inserted and the joint finished with cement mortar. The joint is made both from inside and outside the stand. Note the air vent installed lose to the stand in one of the lines and the opening left in the pipe for fitting the same on the other line.

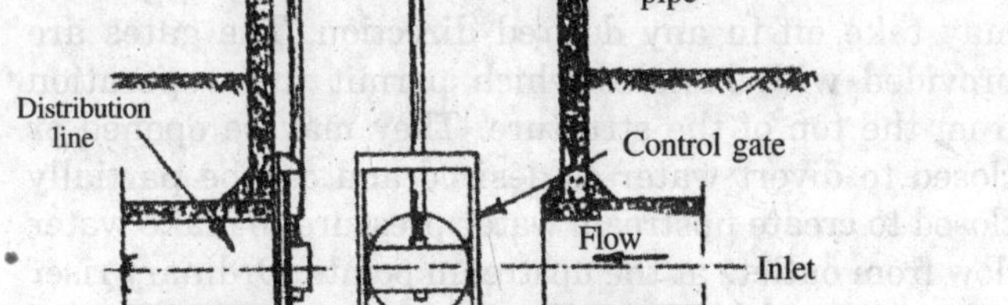

Fig. 7.26. Details of a concrete gate stand to control the flow into different laterals.

Gate Stands

Gate stands are provided to control the flow into laterals when pipelines branch off in different directions. They also prevent high pressures and act as surge chambers.

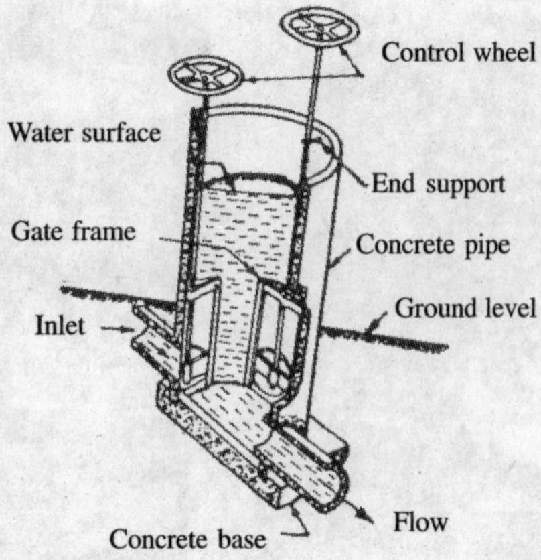

Fig. 7.27. Isometric view (in section) of a gate stand used to control the flow into several laterals.

Fig. 7.31. Constructing the end of a pipeline. Note the end plug and the space made for constructing the end wall. An air vent and riser valve are always installed near the end of a pipeline.

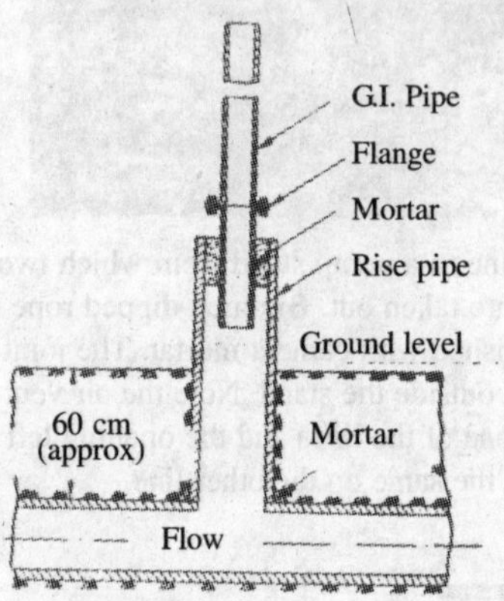

Fig. 7.28. Sectional view of an air vent for an underground pipeline.

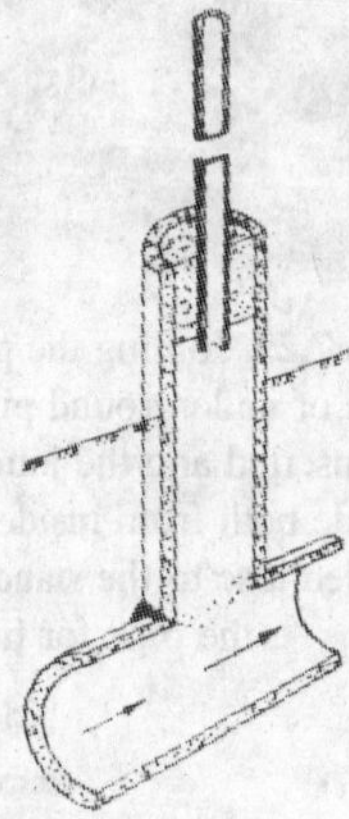

Fig. 7.29. Isometric view (in section) showing the details of an air vent.

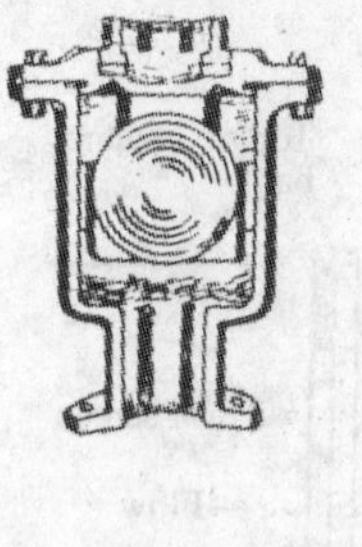

Fig. 7.30. An air relief valve used with high pressure pipelines.

They are made in the same way as pump stands, but are provided with slide gates or valves to control the flow into the laterals (Figs. 7.25 to 7.27). Lateral pipelines may take off in any desired direction. The gates are provided with handles which permit their operation from the top of the structure. They may be opened or closed to divert water as desired and can be partially closed to create upstream water pressure to make water flow from outlets at the upstream points. Ordinary riser valves, placed horizontally are often used in place of gates to construct low cost gate stands. They are operated manually. The gate stand must be large enough to accommodate the gates to be used and to permit access for maintenance and repair.

Air Vents

Air vents are vertical pipe structures to release air entrapped in the pipeline and to prevent vacuum. Entrapped air must be removed to permit an even flow and avoid the danger of water hammer. They are installed near the pump stand, at all high points in the line, at sharp turns, at points where there is a downward deflection of more than 10°; directly downstream from any structure that may entrap air, and at the end of the pipeline. They are also required immediately upstream from gates where closure of the gates would make such points the downstream ends of a lateral of line. Vents are generally installed at points about 150 metres apart on straight pipelines with uniform slope. Gate stand may substitute for a vent. The first air vent is located near the pump stand at a point where the design velocity exceeds 30 cm per second. This occurs at a distance between 3 to 5 times the diameter of the pipe used. The locations of the first vent for the main pipe sizes of 15 cm, 22.5 cm and 30 cm are 50 to 75 cm, 70 to 100 cm and 100 to 150 cm, respectively from the pump stand.

The area of the vent pipe in contact with the pipeline should not be less than half the area of the line. This can be further reduced (Figs. 7.28 to 7.31) until the area of the small pipe is not less than one-sixtieth of the area of the main line. In no case should the diameter of the small pipe be less than 5 cm. The vents should be high enough to provide a free board of at least 60 cm above the maximum height to which the water will rise at that point under normal operation.

Pipelines designed to operate under a higher pressure than that can be conveniently served by open air vents, should have air relief valves (Fig. 7.30) placed at the critical points along the line. Air relief valves are fixed to the vertical vent pipe at some convenient height above the ground surface. These are usually spring operated valves which are set to open whenever excess pressure develop in the pipeline. Such a valve will permit air to escape or enter but will not allow water to pass. Air relief valves of the type shown in Fig. 7.30 should not be used in any location where it may be necessary to relieve momentary high pressure surges.

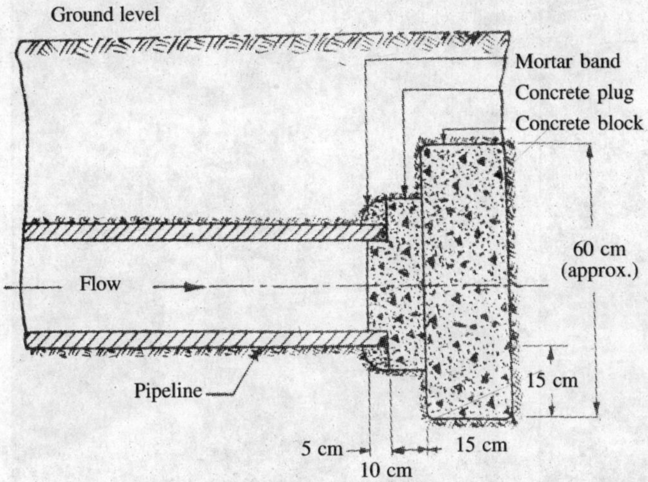

Fig. 7.32. Concrete end plug and end wall at the end of a concrete pipeline.

End Plug

End plug is provided where the line terminates. The function of an end plug is to close a line and to absorb the pressure developed at the end of the line, on account of water hammer. The plug is backed by a masonry block which provides sufficient strength to meet unexpected high pressures developed due to sudden opening or closing of valves. Figure 7.32 gives the constructional details of an end plug and the supporting concrete end wall.

Outlets

Outlet structures are necessary to provide controlled delivery of water to the fields at any desired location. These should be easy to open and close, be of proper size to provide the flow required, and be so constructed that the water released will not cause soil erosion. The most common outlet consists of a concrete riser pipe to bring the water from the buried pipeline to the ground surface and an attached valve to control the flow of water. The diameter of the riser pipe is usually the same as the main pipe. This will permit the entire flow of the pipeline to be released through the valve. In large-sized pipelines, when a number of outlets are to discharge water simultaneously, the size of the riser pipe is smaller than the main pipe.

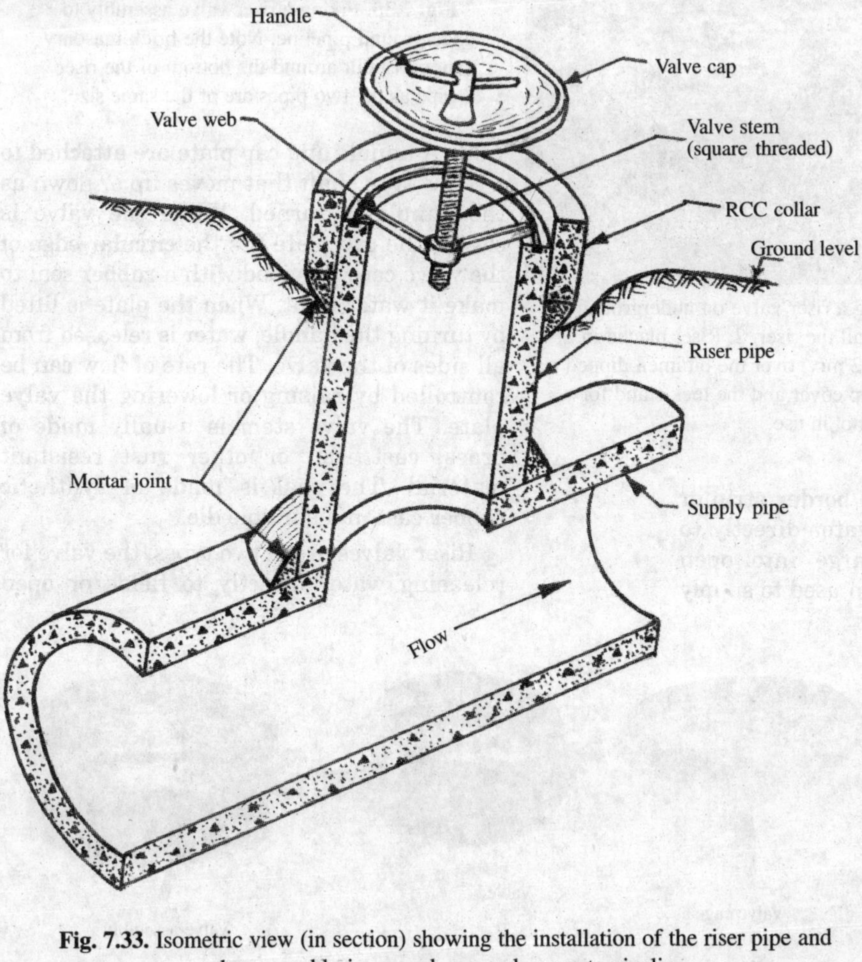

Fig. 7.33. Isometric view (in section) showing the installation of the riser pipe and valve assembly in an underground concrete pipeline.

The riser pipes are installed on the pipeline by cutting a hole on the upper surface of the pipe (Fig. 7.33). They are saddled over the openings in the pipeline and permanently attached to it with cement mortar joints after inserting a bitumen-dipped rope around the joint. The joint is made after the pipe ends are cleaned by grouting in a mixture of cement and sand (1:1 ratio) and placing a mortar band all around the outside to obtain a water-proof joint (Figs. 7.33 and 7.34). It is a good practice to tie together one or two of the reinforcing rods or wires of the main pipe and the riser pipe.

Fig. 7.34. Step-by-step procedure in installing a riser valve on underground pipeline. 1. Hole made in the pipeline to install the riser. 2. Riser placed in position. 3. Joint made with cement mortar (1:2 mix) over the bitumen dipped rope ring. 4. Trench, back-filled. Note the flap cover and the lock stand for locking the valve, when not in use.

Fig. 7.35. Fixing a riser valve assembly to underground pipeline. Note the brick masonry support built around the bottom of the riser pipe as the two pipes are of the same size.

The outlet may open directly into a border strip or check basin or at a point to deliver water directly to several furrows. It may also discharge into open channels or portable surface pipes. When used to supply water directly to the field, the spacing between adjacent risers is 15 to 35 metres. When the risers deliver water to open channels or to gated pipes, the spacing varies from 60 to 100 metres. The outlet valves should be constructed so that the top of the riser or valve is at or slightly below the ground surface.

Riser Valves

Riser valve is a screw valve installed on top of the riser pipe (Figs. 7.35 to 7.38). A handle and cap plate are attached to a screw type shaft that moves up or down as the handle is turned. When the valve is closed, the cap plate fits the circular edge of the valve case provided with a rubber seal to make it water tight. When the plate is lifted by turning the handle, water is released from all sides of the valve. The rate of flow can be controlled by raising or lowering the valve plate. The valve stem is usually made of brass, cast iron or other rust resistant material. The seal is made of synthetic rubber cast in a suitable die.

Riser valves are of two types; the valve for releasing water directly to fields or open

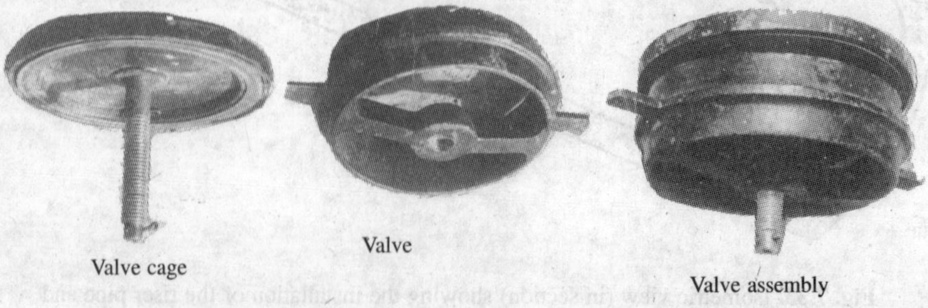

Valve cage

Valve

Valve assembly

Fig. 7.36. Views of a riser valve used in the underground pipeline system.

Fig. 7.37. A low cost riser valve fitted to a concrete riser pipe.

Fig. 7.38. A partially opened riser valve.

channels and hydrant valves which are adapted for fixing a protable hydrant on it. The hydrant valves are provided with a cage to seat the hydrant. The entire riser body may be built in a single unit and may be fixed to the top of the riser pipe with cement mortar (Fig. 7.37). Ribs or projections are provided on the body of the valve to securely fix it with the pipe. When RCC pipes are used as risers, it is desirable to embed the projections within the reinforcement network. In some types of valves, the cage is not fixed to the valve body. In such cases, an over-sized RCC collar is necessary to fix the valve to the riser pipe (Fig. 7.34 and 7.40).

Hydrants

Hydrants are devices placed over riser valve outlets as a means of connecting portable gated pipes to the pipeline (Fig. 7.39 to 7.41). They are portable so that they can be moved from one valve outlet to another to serve the portion of the field which is being irrigated at a particular time. Portable hydrants are constructed of cast aluminium or cast iron. They have a double screw arrangement which is connected to the handle on the riser valve. One of the screws permits forming a tight seal between the hydrant and the top of the riser pipe, and the other screw is used for opening the riser valve after the hydrant is in place. Hydrants may have a single outlet which permits the flow of water in only one direction, or may have outlets on two sides. The size of the outlet corresponds to the size of the gated pipe used. When the size of the outlet is larger than the size of the pipe, an adapter may be used for connecting the pipe to the hydrant. Hydrants can also be used for connecting the suction hose of a pump to the water supply carried in the pipeline under low pressure, so that the pump can develop the high pressure needed for operating sprinkler or drip irrigation systems or for lifting water to higher elevations.

Gated Pipes

Gated pipes are portable lines with uniformly spaced outlets used for releasing irrigation water to furrows, border strips or check basins (Figs. 7.42 to 7.44). When connected to buried pipelines through hydrants, they permit conveyance of the water in an enclosed system

Fig. 7.39. View of a hydrant.

Fig. 7.40. A portable hydrant installed on a riser valve as a means of connecting portable gated pipes for water application. The farmer is seen inserting the gated pipe into the hydrant opening.

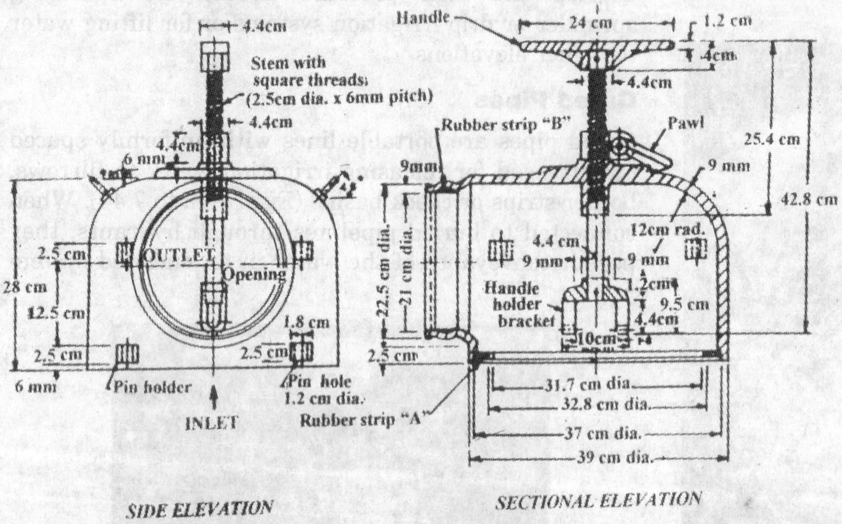

Fig. 7.41. Views of a hydrant suitable for fitting water distribution pipes.

from the source to the head end of the irrigation strip. Seepage losses are reduced to a minimum. They also provide a convenient means of regulating the flow of water into the furrows or border strips.

Gated pipes are usually constructed of aluminium or light weight steel tubing. Small finger tip operated rubber or metal slide gates are placed along the side of the pipe at a uniform distance apart (Fig. 7.42). The spacing of the gates is usually the same as the spacing of the furrows so that the water is discharged directly into them. The gates are placed in a rubber frame which provides a water-lubricated bearing surface for the slide gates. The flow of water can be regulated by the degree of opening of the gates. Gated pipes are usually manufactured in length of six to ten metres. Their diameter ranges from 10 to 30 cm. The individual pipes can be connected to the hydrant or to each other by means of quick-coupling connections (Fig. 7.42) to obtain any desired length of the line. A portable end plug with a locking device (Fig. 7.43) is fitted at the end of the line.

Flexible rubber or PVC pipes may also be used as gated pipes. The pipes may have slide gates similar to those used with a metal pipe, or they may have small diameter flexible pipe sleeves about one metre in length connected to the tubing through which the water is released into the furrows or field strips. The sleeves can be folded back to stop the flow of water, or some type of clamp can be used to regulate the flow.

The outlets on the gated pipe operate as sharp edge or long-tube orifices. The flow through them is a function of the pressure within the pipe and the size of the opening. A pressure of 30 to 150 cm of water is usually required at the hydrant to operate the gated pipe.

The velocity of flow in the pipe depends on the pipe size, the initial flow in the pipe and the number of gates that open upstream from a given point. The quantity of water flowing past a given gate depends upon the quantity introduced into the pipe, the discharge from each gate upstream, and the number of gates open upstream (Fig. 7.44).

Pressure must be provided at the hydrant to overcome friction losses in the gated pipe. It is desirable to limit the length of the line and use a pipe of sufficiently large diameter so that the friction losses can be kept within about 60 cm of water. The risers and outlet valves are generally spaced along the buried line so that not over 50 to 70 metres of gated pipe will be needed to irrigate the entire field.

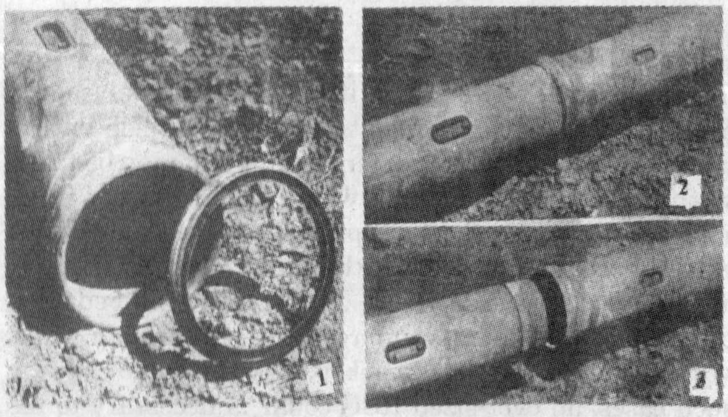

Fig. 7.42. Gated pipe joints. 1. End view of a flared joint. 2. Joints closed. 3. Joints open.

Fig. 7.43. End plug fitted to the tail-end of a line of gated pipes. Note the position of the lock.

Fig. 7.44. Gated aluminum pipes used to irrigate from an underground pipeline water distribution system. The gates are adjustable to obtain the desired flow rates.

Design of Underground Pipeline Irrigation System

An underground pipeline irrigation system must be properly designed to handle efficiently the required flow throughout the system (Fig. 7.45). If pipelines are too small in diameter, pumping costs are increased and the capacity of the system may be seriously limited. On the other hand, pipelines larger than necessary add to the cost of the system and may cause uneven flow. Pump stands must be high enough to allow sufficient operating head for the pipeline. Stands higher than necessary may permit high heads of water to build up, leading to excessive line pressures.

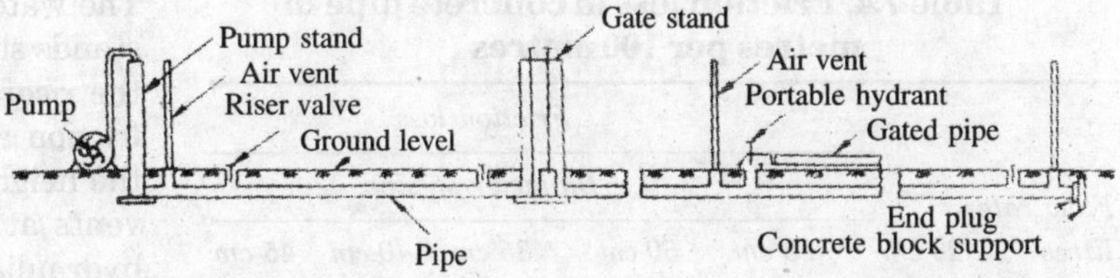

Fig. 7.45. Design elements of an underground pipeline irrigation system.

The pipeline sizes and heights of pump stand and control stands must be determined so as to obtain a balanced water distribution and provide trouble-free, economical operation. The working pressures in the pipeline are kept within one-fourth the internal bursting pressures of the pipe. When it is necessary to design pipelines with higher heads, reinforced concrete pressure pipes are used. The size of the outlets are selected to suit the flow required at diversion points.

In designing an underground pipeline system, the head loss due to friction in the pipeline and other factors must be taken into account. Loss of head due to friction in the pipe between two points in a pipe at a distance l apart in given by the formula:

$$h_f = \frac{flv^2}{2gR} \qquad ...(7.2)$$

in which

f = coefficient of friction for pipe (dimensionless)

l = length of pipe (metres)

v = velocity of water in pipe (metres per second)

g = acceleration due to gravity (metres sec^2) (9.81 m/sec^2)

R = hydraulic radius (metres)

$$R = \frac{a}{p}$$

in which,

a = area of cross-section of pipe (sq m)

and, p = wetted perimeter (m)

$$a = \frac{\pi d^2}{4}$$

and, $p = \pi d$, in which d = diameter of pipe (metres).

Also,

$$R = \frac{d}{4}$$

Therefore,

$$h_f = \left(\frac{2f\,lv^2}{gd}\right)$$

Table 7.3. presents the values of f for the common types of irrigation pipes. Table 7.4 gives the values of friction less in concrete pipes at various discharge rates.

Table 7.4. Friction loss in concrete pipe in metres per 100 metres

Flow rate, litres per second	*Friction loss*					
	Diameter of pipe					
	20 cm	*25 cm*	*50 cm*	*35 cm*	*40 cm*	*45 cm*
2.8	.01					
5.7	.03					
8.5	.07					
11.3	.12	.04				
14.2	.19	.06				
17.2	.26	.09	.04			
22.7	.45	.15	.06			
28.3	.68	.23	.09	.04		
34.0	.95	.32	.13	.06		
39.6	1.23	.43	.18	.08	.04	
45.3	1.62	.55	.22	.11	.06	
51	2.02	.68	.28	.13	.07	
57	2.45	.83	.34	.16	.08	.05
68	3.43	1.16	.48	.22	.12	.07
79	4.57	1.54	.63	.30	.16	.09
91	5.85	1.97	.81	.38	.20	.11
102	7.27	2.46	1.01	.48	.25	.14
113	8.84	2.98	1.03	.58	.30	.17
127		3.71	1.53	.72	.38	.21
142		4.51	1.86	.88	.46	.26
156		3.38	2.21	1.05	.55	.31

Entrance losses in pipelines may be calculated from the following relationships:

(*i*) Pipe end flush with the wall of pump stand

$$h_e = 0.5 \frac{v^2}{2g} \qquad \text{...(7.3)}$$

(*ii*) Pipe end protruding from wall of pump stand

$$h_e = 1.0 \frac{v^2}{2g} \qquad \text{...(7.4)}$$

(*iii*) Pipe with bell entrance

$$h_e = 0.1 \frac{v^2}{2g} \qquad \text{...(7.5)}$$

in which,

h_e = head loss at pipe entrance (metres), and v and g are the same as defined earlier.

The other losses which may be significant are the velocity head in the pipe and the head loss through riser valves. The velocity head due to flow in pipes is obtained from the formula:

$$h_v = \frac{v^2}{2g} \qquad \text{...(7.6)}$$

in which,

h_v = velocity head (metres), and v and g are the same as defined earlier. Head loss through riser valves is given in Table 7.5.

The water surface elevations in pump stand and control stands should be sufficient to distribute water through all the risers in the field and to overcome the losses due to friction and other causes. The hydraulic gradient shows the height to which water will rise in open standpipes or vents at any place along the line for a given flow. The hydraulic gradient line at any point in the pipeline may be obtained by deducting the total losses in the pipeline up to that point from the elevation of the water surface in the stand pipe. Changes in slope, sharp bends, and other factors will cause a shift in the hydraulic gradient and should be considered. The designer should plan the system so that the pipeline flows full and the hydraulic gradient remains above the pipeline to prevent surges and trouble due to entrapped air. In case gated pipes are used, the pressure required to overcome the friction in them is to be added to the system requirement.

Table 7.5. Head loss through riser valves

Flow rate, (litres per second)	*Pressure loss in centimetres of water*					
	Diameter of opening					
	40 cm	*15 cm*	*20 cm*	*25 cm*	*30 cm*	*35 cm*
10	2	1				
20	10	3	1			
30	26	7	3	1		
40	50	13	5	2	1	
50	83	22	8	4	2	1
60	126	34	12	6	3	2
80	250	64	24	11	5	3
100		105	40	18	9	5
125		165	65	29	15	7
150		250	97	44	22	12
175			136	65	30	17
200			185	83	41	22
225			245	100	55	30
250				137	69	37
275				170	83	45
300				210	100	56
350				300	145	77
400					195	103
450					250	135
500						170
550						210

Example 7.2. *A concrete pipeline is to carry water across a distance of 700 metres from a pump which discharges 100 litres per second. The land has a downfield slope of 0.1 per cent along the line the pipes are laid. It is desired that the maximum pressure in the pipeline should not exceed 5 metres of water. What size pipe is be used? How high should be the pump stand?*

Solution: Pressure available in overcoming friction = permissible pressure in the pipeline + gain in pressure head due to downfield gradient = 5 + 0.100 × 7 = 5.7 metres.

5.700 ÷ 7 = 0.814 metres per 100 metres

Table 7.4 shows that a 35 cm diameter pipe will deliver a discharge of 102 litres per second and the friction loss will be 0.48 metres per 100 metres. Hence, a 35 cm diameter pipe is selected.

The water will rise to a height of 7 × 0.48 – 0.7 = 2.66 metres. Allowing a freeboard of 30 cm and a head of 50 cm at the entrance end, the height of stand pipe will be 2.66 + 0.30 + 0.50 = 3.46 metres, or say 3.50 metres.

Example 7.3. (*i*) *The installation of an underground pipeline on a hilly area is shown in Fig. 7.46. A pump discharge of 110 litres per second is allowed into the pump stand A and is carried in the pipeline between stand A and gate stand B. The top of stand A is four metres above the ground surface. A minimum water surface elevation of 30 cm above ground surface is needed in gate stand B. What size of pipe is required between stands A and B?*

(*ii*) *A quantity of 43 litres per second is diverted at B, leaving a flow of 67 litres per second in the 30 cm pipeline between stands B and D. How high will the water rise in stand D?*

(*iii*) *Air vents are required at C because of the high point in the pipeline and at the end of the pipeline at E. What should be the heights of the vent pipes to prevent overflowing? The pipeline is level between stand D and vent E. The size of pipeline between D and E is 30 cm.*

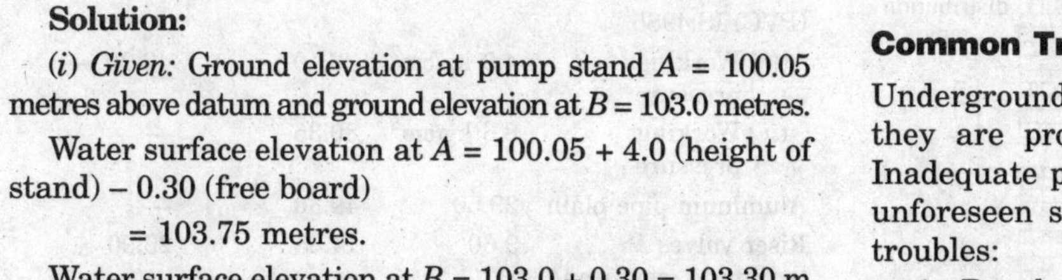

Fig. 7.46. Illustrative sketch of design Example 7.3.

Solution:

(*i*) *Given:* Ground elevation at pump stand *A* = 100.05 metres above datum and ground elevation at *B* = 103.0 metres.

Water surface elevation at *A* = 100.05 + 4.0 (height of stand) – 0.30 (free board)

= 103.75 metres.

Water surface elevation at *B* = 103.0 + 0.30 = 103.30 m

Head difference between *A* and *B* = 103.75 – 103.30

= 0.45 metres per 280 metres of pipeline

∴Allowable friction loss per 100 metres of pipeline

$$= \frac{2.45}{2.80} \; 0.161 \text{ metres}$$

From Table 7.4, it is seen that a 45 cm pipe will have a friction loss of 0.16 metres per 100 metres of pipeline. Hence, the pipe to be used between section *A* and *B* should be of 45 cm diameter.

(*ii*) For a flow of 68 litres per second in a 30 cm pipeline, Table 7.4 shows the friction loss to be 0.48 metres per 100 metres of line. Since the line between *B* and *D* is 412 metres long, the total friction loss will be 0.48 · 4.12 = 1.98 metres. Hence, the hydraulic gradient will drop by this amount and the elevation of the water surface in stand *D* will be 103.30 – 1.98 = 101.32 metres. Since the ground elevation at stand *D* is 99.5 m, the height of stand should be 101.32 – 99.5 + 0.30 (freeboard) = 2.12 metres above ground surface.

(*iii*) At *C*, the hydraulic gradient between *B* and *C* will drop by an amount equal to the friction loss or 0.48 metres per 100 metres of line. Since vent *C* is 200 metres apart from stand B, the total friction loss will be 0.96 metres. Therefore, the hydraulic gradient will be at an elevation of 103.30 – 0.96 = 102.36 metres. Since the ground elevation is 101.30 metres at *C*, the water surface in the vent will be 102.34 – 101.30 = 1.04 metres above ground level. Allowing a free board of 30 cm, the top of vent *C* should be 1.34 metres above ground level.

Elevation of water surface at stand *D* is 101.32 metres. The loss of head due to friction in the pipeline between stand *D* and vent *E* is $\frac{0.48 \times 320}{100}$ = 1.54 metres. Therefore, the elevation of water surface in vent *E* is 101.42 – 1.54 = 99.88 metres. Since, the ground elevation at *E* is 99.50 metres, the water level in vent *E* will be 28 cm above ground surface. Allowing a freeboard of 30 cm, the height of vent should be 58 cm above ground level. It may be noted that the pressure head available at *E* for distribution of water through the outlet is 28 cm which is about the optimum head necessary for the satisfactory operation of an irrigation system, if gated pipes are used for water distribution. There is no appreciable residual energy at the end of the pipeline.

Common Troubles of Underground Pipelines

Underground pipelines operate without trouble when they are properly designed and correctly installed. Inadequate procedures in design and installation and unforeseen situations may give rise to the following troubles:

1. Development of longitudinal cracks in the pipe, usually at the top or both at top and bottom
2. Telescoping of sections
3. Pushing of the pipe into the stands
4. Development of circumferential cracks
5. Surging or intermittent flow of water.

The first four troubles are caused mainly by the expansion and contraction of concrete pipes by wetting and drying. Concrete is also affected by temperature; it expands when heated and contracts when cooled. Circumferential cracks are caused by a drop in water level or soil temperature or by drying of the pipe.

The axial stress set up in pipe by a natural restraint to longitudinal expansion may cause longitudinal cracks. This may also cause telescoping and pushing of pipes into stands. The longitudinal cracks normally occur within a few days after the pipe is laid. Wetting stresses may also occur as a result of change in moisture gradient through a pipe wall, or from a wetter condition around the bottom than around the top of a pipe. Air circulation through a pipe may also develop stress. The prevention of occasional failure lies not in eliminating wetting expansion, but in keeping it from becoming excessive, and in minimising moisture gradients around the pipe. Failure of underground pipelines can be avoided by adopting proper procedures in pipe laying. Using moist soil for the initial back-fill after laying the pipe minimises the circumferential stress and also reduces excessive longitudinal stress. Another precaution is to minimise air circulation through a line when laying the pipe and when the line is not in use. Pushing of pipes into stands is not a serious trouble as the joint can be remade with cement mortar.

Surges in pipelines. Surging is a common trouble in pipelines when adequate vents are not provided. Air gets entrained in water as it over-pours into the pump stand.

Fig. 7.47. Typical pump house of a public tubewell showing the installation of submersible electric motor driven pump and fittings.

The entrained air is carried in the pipeline. Because of turbulence of the flow, the tendency for the air to separate out is minimised. After a short interval, the upstream portion of this reach of water becomes lighter than the downstream portion. This may cause a reversal of the hydraulic gradient until the water with entrained air flows back to the stand and the air is dissipated. Provision for adequate vents will eliminate the problem.

Cost Analysis of Underground Pipelines

The cost of laying pipelines varies considerably. It may vary with the type of pipe, labour charges, depth of laying and type of joint used; all of which may differ from place to place and over the years. Therefore, exact cost figures which will hold good for every locality are difficult to arrive at. Table 7.6 presents the comparison of approximate costs of different types of pipes and pipe accessories.

Table 7.6. Cost comparison of various types of irrigation pipes*
[Price per meter length, in Rs. (Approx.)]

Type of pipe	*Diameter of pipe*		
	15 cm	*22.5 cm*	*30 cm*
Reinforced concrete pipe	7.0 to 8.50	10.00 to 11.00	15.00 to 18.00
Collars for above pipe	1.00 to 1.65	2.00 to 2.50	2.80 to 3.75
Non-reinforced concrete pipe, hand made	3.00 to 4.00	5.00 to 7.00	8.00 to 11.00
Asbestos-cement pipe (5 kg/cm^2, IS:1952)			
(*i*) Pipe without collar	11.35	21.65	38.90
(*ii*) Pipe with cast iron detachable joint	18.90	27.75	49.90
(*iii*) Pipe with asbestos cement collar	17.85	33.45	—
Plastic pipe: Polyvinyl chloride** (PVC) IS:4985			
(*i*) Working pressure	4.0 kg/cm^2	33.10	43.20 —
(*ii*) Working pressure	8.6 kg/cm^2	80.35	— —
Aluminum pipe plain	29.50	49.30	—
Riser valves (ordinary), cast iron	9.60	14.85	23.90
Riser valves (hydrant), cast iron	24.00	30.00	54.00

* The rates are based on the rates prevailing in 1972-73.

** Rates refer to the year 1970.

The components to be included in the estimate are cost of earthwork including trenching, blinding and backfilling, cost of pipes and accessories, cost of installation of pipes, cost of structures used with underground pipelines and cost of engineering and supervision.

UNDERGROUND PIPELINE WATER DISTRIBUTION SYSTEMS IN COMMUNITY IRRIGATION PROJECTS

The water distribution system of deep tubewells, which are usually owned by the government or cooperatives or group farming societies (Fig. 7.47 to 7.49), usually comprise of open channels or buried pipes with outlets at suitable points in the tubewell command area. Buried pipeline water distribution systems, though comparatively more expensive, have major advantages over surface water distribution systems comprising of a network of field channels. The following are the major advantages:

1. The farmers get water at or near their fields.
2. Water conveyance losses through seepage, evaporation and breaches in the channels are avoided.
3. The quantity of water delivered from each outlet remains the same, irrespective of the elevation of the outlet.
4. The pipeline can be laid with complete freedom to best suit the requirements of water supply and cost of lines, irrespective of the topographic features of the tubewell command area.
5. Outlet valves can be provided wherever desired in the pipeline, as determined in the interest of minimising the distance from the outlet to the field, the number of cultivators served by an outlet and ensuring gravity flow from outlet to the fields.
6. Water is supplied to each field plot either directly or through a field channel of short length originating from the outlet.
7. Maintenance cost of the water distribution system is very low.
8. There is full control of the water supply to the fields within a tubewell command area.

Both reinforced concrete pipes and PVC pipes are used for constructing water distribution systems in the command areas of wells. PVC pipes are often preferred because of the ease of installation and to ensure leakproof joints. Other factors favouring its use are speed of laying and greater resistance to internal pressure, including surges. However, PVC pipes are more expensive, as compared to concrete pipes of a given diameter, to convey large quantities of water. However, when skill and adherence to proper procedure in laying the pipes and accessories can be ensured in installation, concrete pipes could be used more economically and with equal efficiency, as compared to PVC pipes on plain land. Under undulating topography, PVC pipes have distinct advantages over concrete pipes.

Fig. 7.48. Water storage tank and electrical power supply system in a Public Tubewell in the Gangetic plains of Uttar Pradesh.

Survey Procedures of the Command Area

An essential requirement for the design, layout and construction of a buried pipeline water distribution system for a deep tubewell is the preparation of a contour map of the area at about 30 cm contour interval on the village cadastral map of the Revenue Department. The map should show field boundaries, streams, rivers, tanks, earth embankments, roads, wells and village boundary and other major features of the area which will come under the command of the tubewell and the area immediately adjoining it. Field numbers, according to revenue records, are indicated in the map. The map also specifies the location of the area with respect to the north and its boundaries and the name of the revenue division and sub-division in which the area is located. On this map, the location of the tubewell and elevated water distribution tank are indicated. Detailed procedures for the survey of irrigation command areas are presented in the book *Principles of Agricultural Engineering*, Vol. II (2003), authored by A.M. Michael and T.P. Ojha (Pub. Jain Brothers, New Delhi) listed at the end of this chapter.

Based on the inspection of the field contours and the various features on the ground, the alignment of the buried pipeline water distribution system, the location

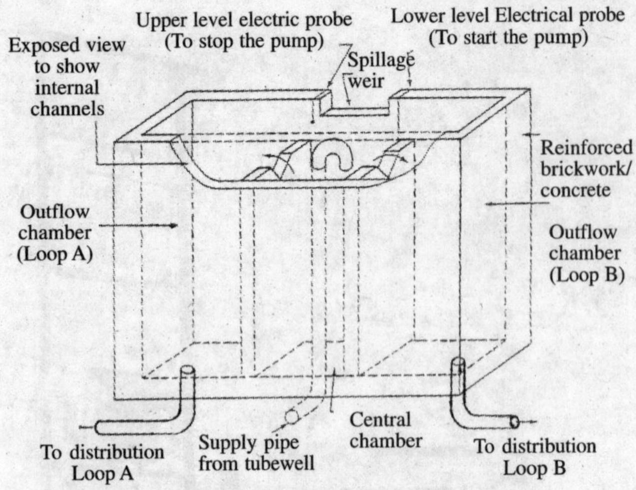

Fig. 7.49. Elevated water distribution chamber in a public tubewell.

of the valves, the alignment of the common earthen field channels and essential field structures like inverted siphons are decided. Profiles along each pipe and channel route are surveyed and plans prepared showing depth, gradient and earth work along their length and the locations of the structures. The locations of the outlets are reviewed with the farmers and necessary adjustments are made. The routes of the pipeline are marked and the adjustments are made. The routes of the pipeline are marked and the outlet points are flagged. The alignments of field channels to convey water from an outlet to each field plot are marked in consultation with the farmers.

Systems of pipeline networks. Pressurised water supply through the pipeline could be of the following types, according to the layout and interconnection of the pipes:

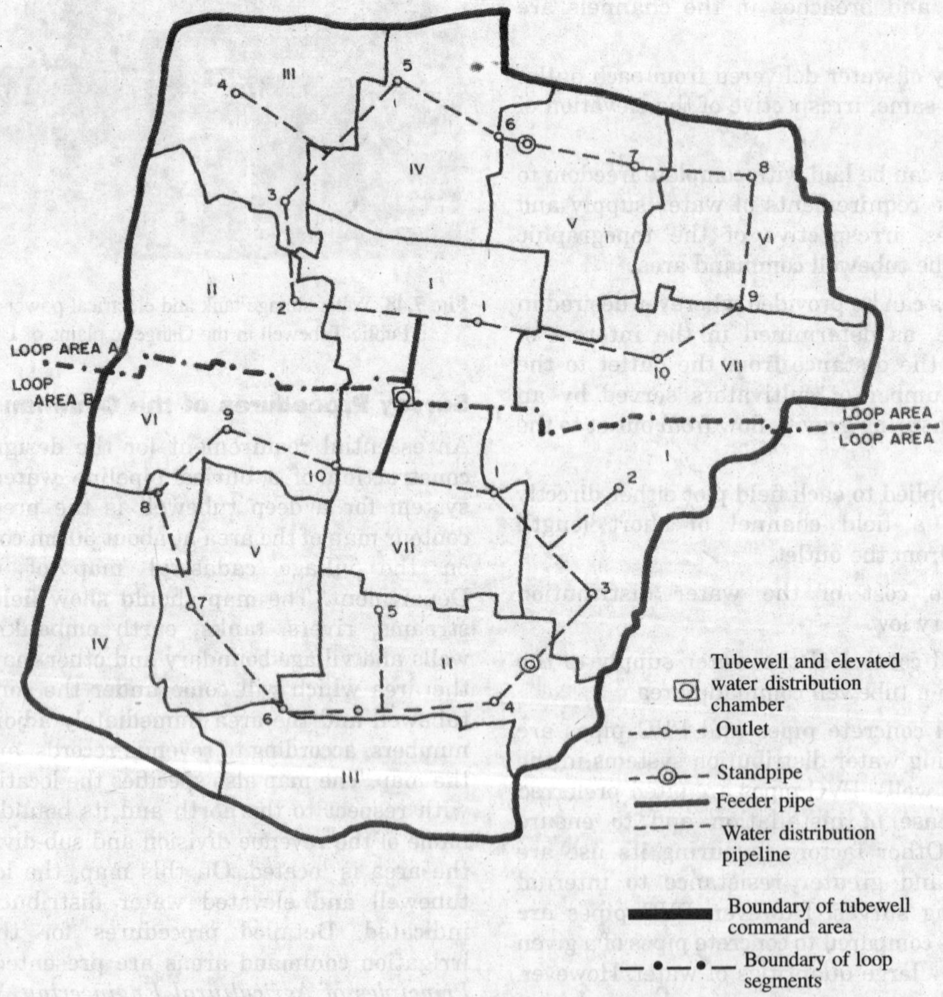

Fig. 7.50. Layout of a buried pipeline water distribution system for a deep tubewell. The pipes are laid in two loop segments. Outlet Nos. 2 and 5 of the bottom segment and No. 4 of the top segment are branches from the loop. The *Arabic numbers* (1, 2, 3, ...) denote serial number of the pipe outlets in each loop. The elevations of the field plots in the blocks served by them are shown below the corresponding outlets. The *Roman numbers* (I, II, III, ...) denote the field block to be served by an outlet on a particular day of the week in a weekly rotation irrigation regulation system. For example, in the bottom segment, Field Block III is to be supplied water on the third day of the week, from the outlet No. 6.

(1) Branching network. This system is used in areas with undulating topography or on steep slopes. In a branched network only the main line and the particular branch on which the operating valve is located contributes to conveyance of water.

(2) Looped network. These are mostly used in plain areas where the difference in elevation between different fields in the command area is not high. In a closed loop system, both arms of the loop contribute to conveyance of water, regardless of the location of the valve in operation (Figs. 7.50 and 7.51). The system is designed to have only one valve in a loop to operate at a time. Since the water conveyance to an outlet is from both arms of the outlet, only a smaller size pipeline is required, about 30% smaller in size as compared to a branching line system. The cost of the pipeline is about 50% more for the branching system than for the loop system.

Elevated Water Tank

An elevated water tank constructed near the tubewell serves as the pump stand of the PVC pipeline system (Figs. 7.48 and 7.49). The elevation of the water surface in the tank will determine the pressure in the pipeline. A stand pipe located near the elevated tank serves as a pressure release structure and prevents water hammer in the pipeline.

The distribution chamber maintains the operating head in the piped water distribution system. It also enables automatic operation of the pumping unit. The latter is achieved by means of level operated probes which activate the power supply according to the water level in the chamber and in response to irrigation demand. The water level operated probes also enable water distribution at less than the discharge rate of the pump without spillage. The design of the distribution chamber ensures equal division of the design discharge from the well into the two separate segments of the water distribution system (Fig. 7.48).

The water distribution chamber is the link between the tubewell and delivery pipeline system (Fig. 7.49). The major function of the chamber are to regulate tubewell discharge to match the capacity of the pipe distribution system and to ensure equal streams to each of the loops in the system. This is regardless of differences in elevation or length of the lines and as to which outlet (alfa-alfa type valve) is operating on the respective pipelines at any particular time. Thus, the discharge remains constant regardless of how far the outlet in operation may be from the chamber.

Equal division of the flow between the loops is achieved by delivering the tubewell discharge into a central inlet chamber (Fig. 7.49). From the inlet chamber, water flows over two equal weirs into outlet chambers serving the respective pipelines. The water level in the outlet chamber may be at any distance below the weir without influencing its discharge. In operation, the water level in a chamber varies with the location of the particular outlet valve in use at any time. Water levels in the two chambers may also differ considerably from one another but the division of flow remains unaffected.

The elevation of the weir in the distribution chamber is a major factor influencing the capacity of the delivery pipelines. Other critical factors which influence the design are the diameter and length of the pipe distribution system and the elevation of the outlet valve farthest from the tubewell. These factors together govern the line capacity. However, the desired line capacity (75 m^3/h) had been pre-determined in the public tubewells of the World Bank supported projects in Uttar Pradesh, Bihar and West Bengal in 1980s and early 1990s. The size of the command area to be served by each loop line system was kept at about 50 ha. The shape of the service area, the length of pipeline, and the difference in ground elevations between the distribution chamber and the farthest outlet valve are fixed, based on site conditions. The remaining variables are pipe diameter and elevation of the weir and the height of the distribution chamber. The smaller the pipe diameter, the higher the pumping head and the higher the distribution chamber required, and vice versa. In practice, the elevation of the bottom of the weir is usually about 3 m above ground level. The tubewell output may be reduced to match the capacity of the pipeline system by adjustment of the main inlet valve. Alternatively, the difference in capacity between the tubewell and the distribution system may be accommodated by periodic automatic on/off operation of the electrical probes which respond to maximum and minimum permitted water levels in the distribution chamber.

An electric probe shuts off the pump when the water level in the distribution chamber rises to an upper limit. This may be necessary if all cultivators cease irrigating and all the outlet valves are closed. The water level in the distribution chamber would then rise towards the overflow level. The level probes also cycle the pump on-and-off periodically, if cultivators cease irrigating on one of the two pipelines/loops, which reduces the water demand from the well by one-half. The design discharge of a public tubewell of say 150 m^3/h is divided into two equal streams, each feeding a separate loop line. Each loop line has a command area of about 50 ha. The pipe distribution system for the 150 m^3/h capacity well usually consists of two loops, each of 2 km length comprising of 160 mm PVC pipes, each connected to the outlet point of the elevated water distribution chamber by a short run (150 to 250 m) of 200 m PVC the pipe which is the feeder pipeline. The total length of the pipe is generally about 40 m per ha command area served. The PVC pipe is of 2.5 kg/cm^2 working pressure or 25 m of hydraulic head.

The water distribution system is designed to have only one valve in a loop to operate at a time. Each loop line could have command areas extending up to about 50 ha. There are no air vents in the pipeline system, or vacuum valve, since each delivery valve acts as an air release valve, if the release of air is required at the time of initial filling of the pipeline. In normal service, however, the pipeline remains full even when the pump is not in operation. The outlet compartments of the distribution chamber serve as air vents and there is little chance of air bubbles being carried downwards to the pipelines.

Laying of PVC pipelines. The pipelines are designed to remain full all the time without the possibility of gravity drainage of the line. This permits immediate delivery of water to fields after a period of shut down of the well.

The pipeline is laid at a depth of about 1 m below the ground surface and hence, gravity drainage of the well is usually not possible. If emptying of the pipeline is required for any reason, this can be done by removing the head cover of the lower most outlet valve and inserting a suction line from a portable centrifugal pump or a positive displacement pump. Pipe fittings such as bends, T-joints, and couplers are selected to suit the field conditions.

The following are the three main features of a buried pipe layout of the type described above:

(*i*) *There are no air or air/vacuum valves* since each delivery valve can act as an air release valve if release of air is required, as during the initial filling of the line. Vertical bends in the line (the usual location of airlock problems) are so small that any air entering the line would be swept downslope to the nearest valve in operation. However, when a vertical bend is essential, it is made at a delivery valve of the distribution chamber.

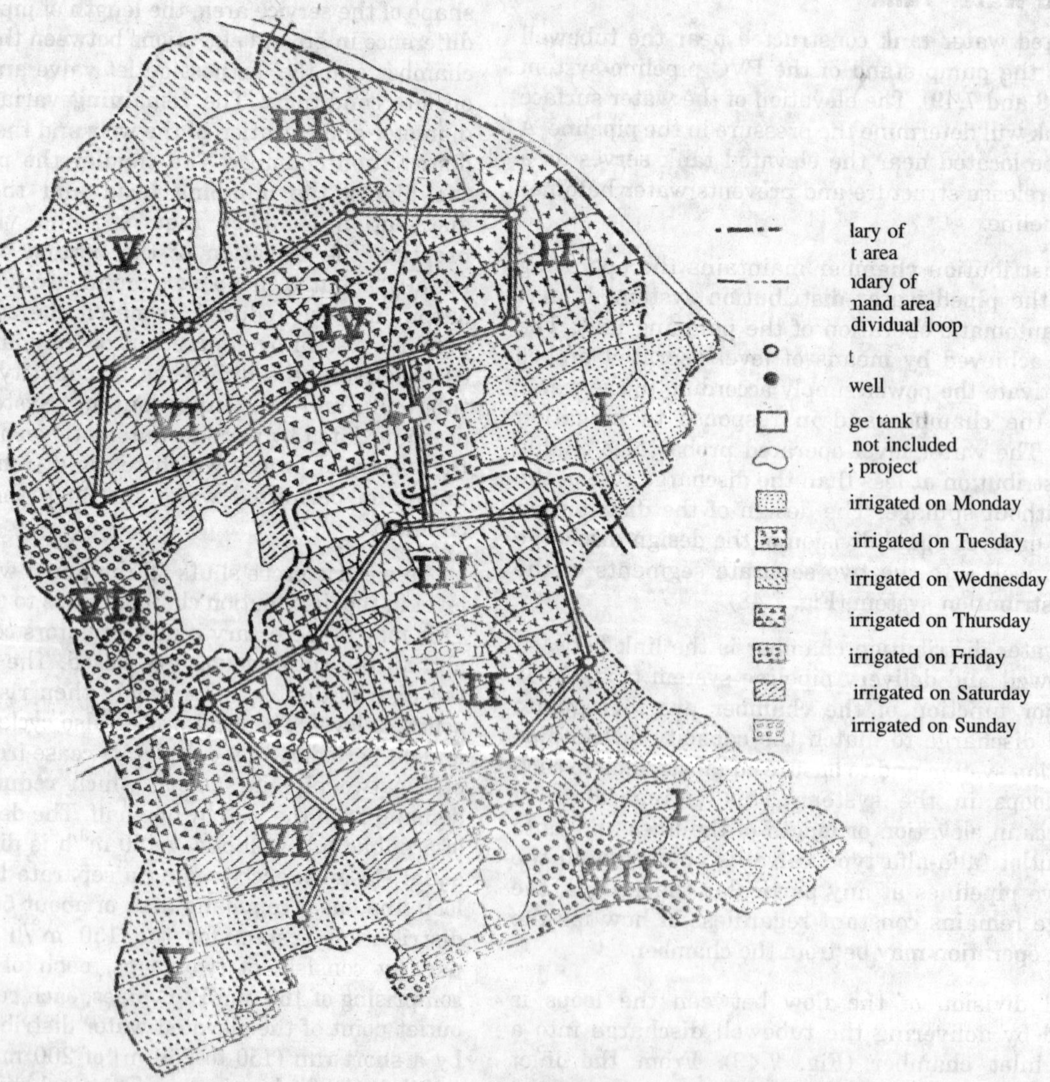

Fig. 7.51. A typical integrated water distribution system in a Public Tubewell command area in Uttar Pradesh (1990).

(ii) There is only one surge riser/vent located near the outer end of the loop. This is with regard to surge or water-hammer during valve closure and the time taken for the pressure waves to travel and the minimum pressure rise that can be expected.

(iii) The line is designed to remain full at all times, without the possibility of gravity drainage. This permits immediate resumption of delivery to valves after a period of shutdown of the well since there is no 'filling time'.

The maintenance requirements of PVC pipeline is low and comprise mainly of the removal of sand which may accumulate in the pipeline. With the distribution chamber operating as a sand-trap, removal of sand or silt from the pipeline will not be high. Pressure rubber hose of 2.5 to 3.5 cm diameter with specially shaped radial nozzles or rotary brush could be progressively fed down the length of the line between the two valves.

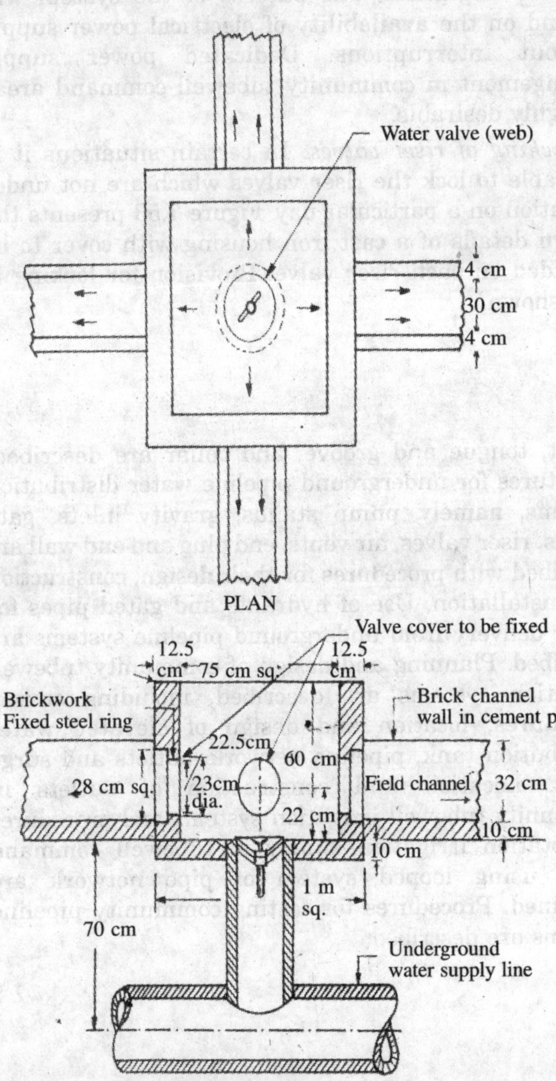

Fig. 7.52. A water diversion box constructed at the outlet of an underground pipeline water distribution system. Valves (alfa-alfa type) fixed horizontally or slide gates could be used to control the flow.

Use of lateral lines from loops. Lateral lines are usually avoided in loop network. However, in some cases the shape of the tubewell command area may necessitate the provision of one lateral line or a maximum of two in exceptional cases. It may be necessary to take a distribution line aside a pond or a village area or when the command area is of very irregular shape. Lateral lines are fitted directly to the loop line at a convenient point. They are of the same diameter as the loop network. The length of a lateral line is not to exceed about 100 m. An outlet valve is fixed at the outer end of the lateral line.

Depth of pipeline below ground. About 80 cm deep earth cover is usually maintained over the PVC pipe. In case of uneven land the pipe is laid about 1 m deep. At places where an earth cover of at least 60 cm cannot be maintained, as in case of crossings of drains, the pipe is to be encased in a RCC pipe.

The appropriate time for making the trenches and laying the pipeline is the dry season. In the rainy season the trenches get filled with water, making the construction difficult. It is also necessary to avoid digging in fields having standing crops.

Outlet Valves

Outlets are provided along the underground pipeline to supply water to each field plot (Fig. 7.52). The command area of each outlet could extend up to about 5 ha on plain land. It is a good practice to locate the outlet on the boundary between two or more landholdings. Wherever possible, they are located near the centre of their respective command areas. The valves are the conventional riser valves of the *alfa-alfa* type and are of the size which match the size of the riser pipe (usually 15 to 20 cm diameter). The valves and a section of the riser pipe are embedded in masonry or concrete. Water distribution boxes are provided to distribute water in desired directions (Fig. 7.52).

Surge Riser

Each loop line is provided with one surge riser or air vent located near the centre of the loop. This is to protect the pipeline against spillover surge or water hammer during valve closure and the time taken for the pressure waves to travel and the minimum pressure rise that can be expected. A 20 cm diameter RCC pipe is usually used as the surge pipe. The pipe is provided with a perforated PVC cap. The pipe is embedded in masonry at the bottom. The surge pipe is located at the highest point near the middle of each loop. The top of the pressure release pipe (surge riser) should be about 50 cm above the hydraulic grade line to prevent spillover of water.

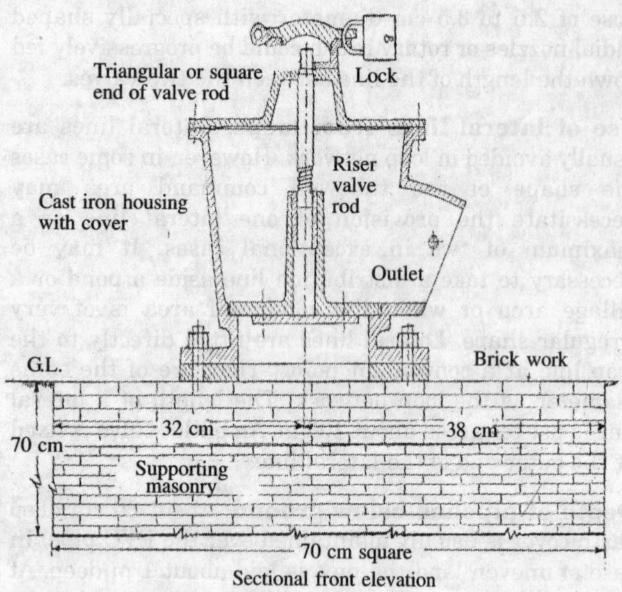

Fig. 7.53. Sectional view of a riser pipe outlet valve with locking arrangement suitable for fixing to community owned underground pipcline systems wherever required.

Testing of Pipeline After Laying

After laying of PVC pipes and backfilling of the trenches and construction of the outlet structure, the tubewell is run to fill the pipeline. All air in the pipe should escape through the pressure release pipe. For 3 to 4 days, all the outlets are kept open for 3 to 4 hours, with the pump in operation, in order to check if there is any leakage in the pipe or in any of the outlets. If leakage is noticed, repairs are done after draining the entire pipeline system.

Rotational Water Delivery System

Buried pipeline water distribution systems are ideally suited for rotational water supply to the beneficiary farmers of the command area. Figure 7.51 illustrates a water distribution programme in a tubewell command area having loop segments in the distribution network. The system ensures reliable supply to the beneficiary farmers once a week in each of the command areas of the loop segments. The success of the system will depend on the availability of electrical power supply without interruptions. Dedicated power supply arrangement in community tubewell command areas is highly desirable.

Locking of riser valves. In certain situations it is desirable to lock the riser valves which are not under operation on a particular day. Figure 7.53 presents the design details of a cast iron housing with cover to be provided on each riser valve. Provision for locking is also shown.

SUMMARY

Adaptability, advantages and limitations of underground pipeline water conveyance and distribution for on-farm irrigation systems are described, with related economic considerations. Types of pipes for underground pipeline systems, including non-reinforced and reinforced concrete pipes with different types of joints, high density polyethylene pipes and asbestos cement pipes are described with their relative merits and limitations. Standard specifications of concrete irrigation pipes and procedures for testing them are presented. Fabrication of HDPE pipes by extrusion and injection moulding processes are explained. Installation procedures, care and maintenance of HDPE pipes are described. Methods of estimating the discharge capacity and spacing of pipes for underground pipelines are explained. Detailed procedures for the design and installation of concrete pipes with different types of joints, including faucet and spigot, tongue and groove, and collar are described. Structures for underground pipeline water distribution systems, namely, pump stands, gravity inlets, gate stands, riser valves, air vents, end plug and end wall are described with procedures for their design, construction and installation. Use of hydrants and gated pipes for water delivery from underground pipeline systems are described. Planning and design of community tubewell irrigation systems are described, including survey procedures, location and design of elevated water distribution tank, pipeline network, outlets and surge risers. Location and construction of outlets in community tubewell irrigation systems and procedures for rotation irrigation supply in tubewell command areas using looped system of pipe network are explained. Procedures for testing community pipeline systems are described.

PROBLEMS

7.1. A concrete pipeline is to be installed to carry a discharge of 100 litres per second. The friction loss in the pipeline should be equal to the gain in head due to the downfield slope of the land. The land slope along the line the pipes are to be laid is 1%. What pipe size should be selected?

(**Ans.** 30 cm)

7.2. An underground pipeline is to be used to deliver 90 litres per second of water to a distance of 500 metres. The land has an upward slope of 0.3 per cent. It is desired to limit the pressure in the pipeline to 6 metres of water pressure Select a suitable size of pipe. What is the maximum pressure developed in the pipeline?

(**Ans.** 30 cm, 5.55 m)

QUESTIONS

1. What are the specific field situations favouring the use of underground pipeline water conveyance and distributions system in preference to open channel systems?

2. What are the limitations in the large scale adoption of underground pipelines in on-farm irrigation management systems?

3. What are the different kinds of pipes used in underground pipelines? State the particular adaptability of each.

4. What are the kinds of pipes which can tolerate salt-affected soils with high sulphate content?

5. Distinguish between extrusion moulding and irrigation moulding of PVC pipes.

6. What are the general precautions to be taken while using PVC pipes for water distribution on the farm?

7. What are the two distinct types of asbestos cement pipes suitable for use in underground pipeline water distribution systems? What are the limitations of asbestos pipes for use in underground pipeline systems?

8. What is the procedure for conducting three-edge bearing test of irrigation pipes?

9. What are the fittings commonly used in concrete pipes? How are they fabricated?

10. What are the specific procedures to be adopted in making cement mortar joints in concrete pipes with tongue and groove joints?

11. What are the requirements for laying concrete pipes with faucet and spigot ends, using rubber rings?

12. How are concrete pipes with faucet and spigot ends and using jute or hemp ropes laid?

13. What is a 'take-up' device for pulling in large concrete pipes?

14. What is the function of heating mirror in butt welding of HDPE pipes?

15. What is the purpose of an auxiliary pump stand sometimes used along with the regular pump stand in an underground irrigation system?

16. What is the function of gate stands in underground pipeline systems?

17. What are the different types of air vents used in underground pipelines? What are the specific situations under which each one is particularly adapted?

18. Which are the locations where air vents are installed in an underground pipeline system?

19. What are the functions of end plug and end wall in an underground pipeline system using concrete pipes?

20. What are the two common types of riser valves (alfa-alfa valves) used in underground pipelines? What is the adaptability of each?

21. What is the function of a hydrant valve in an underground pipeline system?

22. What are the factors affecting the rate of flow of each gate in gated pipes used in underground pipelines?

23. What are the basic factors influencing the design of an underground pipeline irrigation system?

24. What are the common troubles observed in the functioning of underground pipeline systems? How are they overcome?

25. What is the usual cause for developing longitudinal cracks in underground pipelines? How are they controlled?

26. What are the common types of pipeline networks used in underground pipeline systems? What is the special adaptability of each type?

27. What are the advantages of PVC pipelines with closed loop network in community tubewell irrigation projects?

28. How is the pressure head of both the loops maintained uniformly in a double loop PVC pipe based underground pipeline system of a community tubewell project.

29. Describe the design requirements of an elevated water tank in a community tubewell irrigation project with PVC double loop pipe network?

30. What is the location of a surge riser in an underground pipeline loop network?

31. In a community tubewell irrigation system served by a loop system of pipe layout, how is the rotation irrigation distribution planned?

32. What is the procedure for testing an underground pipeline after laying a loop network?

33. Identify the True and False statements from the following:

(*i*) After a concrete pipeline is laid, holes for fixing riser pipes and air vents are made at appropriate locations.

(*ii*) Under most field conditions concrete pipes are the most economical for use in underground pipeline water distribution systems.

(*iii*) Cement concrete pipes are unsuitable for laying in soils with high sulphate content.

(*iv*) PVC pipes are specially adapted to situations where large diameter pipes are to be used under high pressure.

(*v*) The spacing between the pipes in underground pipeline systems is mainly based on the type of soil of the field. Closer spacings are suitable for heavy soils and farther spacings for light soils.

(*vi*) While laying concrete pipes with faucet and spigot ends, the faucet ends are laid facing downstream.

(*vii*) Concrete pipe joints using rubber rings are to be anchored, preferably by casting a block of concrete at places where there are unbalanced forces acting on the pipe such as bends, tees and valves.

(*viii*) While laying concrete pipes with faucet and spigot ends and using rubber rings, the common cause for the tendency of the pipes to creep out of their joints is when the rubber rings have not rolled evenly while laying the pipes.

(*ix*) While laying concrete pipes with tongue and groove joints, the tongue end is laid facing downstream.

(*x*) While laying RCC pipes with collar joints, the adjacent pipes are laid with their ends placed about 1 cm apart to provide space for inserting a rope dipped in hot bitumen in the recession at the pipe ends.

(*xi*) The most vital operation in laying RCC pipes with collar joints is filling the space between the collar and the pipe with cement mortar made in the ratio of 1:1, adding a bare minimum of water to obtain a consistency at which the mortar will form a ball when squeezed in the hand, but which will crumble when dropped, and tamping the mortar thoroughly with a flat steel chisel and hammer.

(*xii*) Joining HDPE pipes adopting butt welding procedure is done by the homogeneous fusion of the pipe material under heat and pressure.

(*xiii*) In an underground pipeline system, the cross-sectional area of the vent pipe in contact with the pipeline is kept the same as that of the pipeline.

(*xiv*) In the design of an underground pipeline system to prevent surges in the pipeline, the hydraulic gradient is maintained above the pipeline.

(*xv*) Equal distribution of flow in the two loop segments of a community tubewell project is achieved at the control inlet chamber of the elevated tank.

(*xvi*) Air release valves / vacuum valves at points of change in elevation of the pipeline and near turning points in the pipeline form integral parts of a loop system of pipeline network.

34. Identify the most appropriate answer of the following questions from the four options:

(*i*) To prevent damage of pipeline from the load acting on it, concrete pipes should be laid with their upper surface at a depth of about:

(a) 30 cm (b) 45 cm
(c) 60 cm (d) 75 cm

(*ii*) The limiting operating pressure which can be sustained by a non-reinforced concrete pipe in irrigation water conveyance is:

(a) 4 m (b) 6 m
(c) 8 m (d) 10 m

(*iii*) The most commonly used concrete pipe for underground pipeline water distribution system is:

(a) pipes with bell ends
(b) pipes with tongue and groove joints
(c) pipes with faucet and spigot joints
(d) pipes with collar joints.

(*iv*) To enable the workman to enter into the stand pipe to make joints inside it, the diameter of the stand pipe should have a minimum value of:

(a) 50 cm (b) 60 cm
(c) 70 cm (d) 80 cm

(*v*) The most important factor influencing the capacity of the delivery pipeline using loop system of pipe layout in a community tubewell irrigation system is:

(a) the diameter of the pipe
(b) the length of the pipeline
(c) elevation of the weir in the water distribution chamber of the elevated water tank
(d) elevation of the outlet farthest from the tubewell.

35. Rearrange the events in the following procedures in their proper sequences:

(*i*) Processes in the laying of different types of concrete pipes:

(a) Digging the trench to proper depth and grade
(b) Selection of depth and grade
(c) Lowering the pipes and squeezing them
(d) Sealing of joints.

(*ii*) Procedure for laying concrete pipes with collar joints:

(a) Lowering the pipes into the trench
(b) Squeezing the pipes after 4 or 5 pipes are laid
(c) Inserting hot bitumen dipped jute rope in the grooved pipe ends
(d) Slipping of RCC collars over the squeezed joints
(e) Filling the gap between the collars and the pipe with cement and sand mortar of 1:1 ratio
(f) Placing wooden battens between the collar and pipe to ensure uniform space between them
(g) Lining the collar ends with 1.5 cm thick cement and sand mortar and bevelling off the plaster at an angle of 45°

(*iii*) Procedure for the repair of a leaky joint in a concrete pipeline:

(a) Making a hole of size equal to that of an arm in the pipe to enable the placement of cement mortar to plug the hole.
(b) Draining the water in the pipeline by releasing it through a suitable outlet at the downstream end
(c) Capping the pierced hole by covering it at the bottom and top with galvanized steel plates
(d) Filling the pierced hole with cement mortar of 1:1 mix in distinct layers.

(*iv*) The process of butt welding of HDPE pipe end:

(a) Cutting the pipe ends perpendicular to the pipe axis and cleaning.
(b) Pressing together the pipe ends after removing the heating mirror.
(c) Passing pipe ends against a hot plate (heating mirror) till a bead of molten material is formed.

(*v*) Survey procedure for the design and layout of underground pipeline water conveyance and distribution in a community tubewell project with pipe loop network:

(a) Alignment of the pipelines and location of outlets, lateral lines and surge riser
(b) Procurement of village cadastral map and identification of tubewell command area and field plots along with their boundaries and ownership.
(c) Profile survey of the routes of the pipeline system and random determination of elevations of fields.
(d) Review of the design of water distribution system and proposed water release patterns with the beneficiary farmers of the tubewell command area.

36. In the following, match each of the entries in the 1st column with the most closely inked entry in the 2nd column:

(*i*)	(A) Spun process	(I) Formation of thermoplastic material into continuous pipeline
	(B) Impact strength modifiers	(II) PVC pipes
	(C) Extrusion moulding	(III) Load-bearing strength of pipes
	(D) Injection moulding	(IV) Additives in PVC pipes
	(E) Three-edge bearing test	(V) RCC pipes
(*ii*)	(A) Darcy equation	(I) Filler material in the space between RCC collar and pipes
	(B) Faucet and pipe joint	(II) PVC pipes
	(C) Nearly dry cement-sand mixture	(III) PVC pipe jointing process
	(D) Solvent cement jointing	(IV) RCC pipes
	(E) Heating mirror	(V) Velocity of flow in pipes
(*iii*)	(A) Gravity inlet	(I) Connecting portable gated pipes to underground pipelines
	(B) Concrete end plug	(II) Community tube-well irrigation project
	(C) Hydrant	(III) Undulating topograpby
	(D) Branching network	(IV) Laying underground pipelines
	(E) Elevated water tank	(V) Entry of open channel flow into underground pipelines

FURTHER READING REFERENCES

American Society of Testing Materials. (1965). *Standard specifications for concrete sewer, storm drain and culvert pipes.* ASTM: C 14-65.

American Society of Testing Materials. (1966). *Tentative specifications for concrete pipe for irrigation or drainage.* ASTM : C 118-66 T.

Bhatnagar, R.N. (1982). *Use of PVC Pipes in Agriculture.* Proc. Seminar on use of plastics for lining of water conveyance systems. National Commission in Plastics in Agriculture. Paper No. 9, pp. 219-213.

Bureau of Indian Standards. (1999). *Specifications for High Density Plyethylene Pipes for Potable Water Supplies* (4th revision): IS:4984–1995 Specifications for unplasticised PVC pipes: IS:4985–1988. Bureau of Indian Standards, Manak Bhawan, New Delhi.

Bureau of Indian Standards. (1975). *Code of Practice for Plastic Pipes Work for Potable Water Supplies* IS:7634: 1975.

Part-1 : 1975. *Choice of Materials and General Recommendations* (1975);

Part-2 : *Laying and Jointing Polyethylene (PVC) Pipes* (1987);

Part-3 : *Specifications for Injection Moulded PVC Fittings with Solvent Cement Joints for Water Supplies.* (1987).

Part-4 : *Specific Requirements for* 90° *Ts*;

Part-5 : *Specific Requirements for* 45° *Ts*;

Part-6 : *Specific Requirements for Sockets;*

Part-7 : Specific Requirements *for Caps.*

Bureau of Indian Standards. (1976): Specifications for Injection Moulded HDPE Fittings for Potable Water Supplies: IS: 157634.

Part-1 : *General Requirements;*

Part-2 : *Specific Requirements for 90° Bends;*

Part-3 : *Specific Requirements for 90° Ts;*

Part-4 : *Specific Requirements for Reducers;*

Part-5 : *Specific Requirements for Ferrul;*

Reducers, Part-6 : *Specific Requirements for Pipe Ends*;

Part-7: *Specific Requirements for Sandwitch Flanges.*

Indian Hume Pipe Company Limited: *General Laying Instructions for Reinforced Concrete Pipes with Rubber Ring Joints.*

Climax Pipes Private Limited: *Jointing of HDPE Pipes–Butt Welding.*

Koluvek, P.K. (1970). *Draft report on irrigation pipelines in India,* U.S.A.I.D., New Delhi.

Michael, A.M., Arora, D.R., Bhattacharya, A.K., Mandal, A. and Gupta, A.K. (1970). *Handbook of farm irrigation structures.* Indian Agricultural Research Institute, New Delhi, p. 82.

Michael, A.M. and Ojha, T.P. (2003). *Principles of Agricultural Engineering,* Vol. II. Jain Brothers, 873 Joshi Lane, Karol Bagh, New Delhi-110005, p. 888.

Portland Cement Association (1969). *Concrete pipe for irrigation.* American Concrete Pipe Association, USA, p. 24.

The Concrete Association of India (1969). *Concrete pipes for irrigation.* New Delhi, p. 8.

Chapter

8

Land Grading and Field Layout

LEARNING OBJECTIVES

Basic requirements for land levelling. Land levelling procedures – rough grading, land levelling and land smoothening. Layout of fields and field sub-divisions based on contour patterns. Layout of farm roads and irrigation and drainage channels. Methods of land levelling design – plane method, profile method, plan inspection method and contour adjustment method. Determination of centroid of fields, formation levels and cuts and fills. Computing cut-fill ratio, slope of plane of best fit and earthwork quantities. Construction procedures in land levelling jobs. Contour bench levelling — design, layout, and construction of contour benches. Equipment for land grading and field layout — animal operated and mechanically powered machinery. Procedures for operation of equipment for land grading, land smoothening and construction of ridges and ditches.

Land grading is reshaping of the field surface to a planned grade. It is necessary in making a suitable field surface to control the flow of water, to check soil erosion and provide for surface drainage. Usually the field is not graded to a truly level surface, but a gentle uniform slope is maintained to meet the requirements of irrigation and drainage. The uniform grades allow fields to be laid out into irrigation runs of the proper length.

Irrigated areas benefit greatly from land grading as the ground in its natural state is seldom suited for the efficient application of water. When an uneven land is irrigated, the high spots are watered too little and the low spots too much (Figs. 8.1 and 8.2). This results in

Fig. 8.1. A poorly laid rice field with inadequate land levelling and poor drainage.

Fig. 8.2. Ponding of runoff water in a depression at the lower end of a field near the head reaches of the Mahi Right Bank Canal command area, Gujarat (1982). Improper land development has been one of the principal causes of inefficient irrigation resulting from non-uniform water distribution over fields and excessive loss of irrigation water.

Fig. 8.3. Good stand of rice crop in a well laid out field with adequate drainage, good land preparation and proper irrigation in the Kaveri Delta Canal Project, Tamil Nadu (1973).

uneven crop, reduced yields, and loss of water and fertilisers. A properly graded land surface ensures unobstructed smooth flow of water into the land, without eroding the soil and ensures uniform distribution of water throughout the field.

Land grading is beneficial in unirrigated areas to conserve moisture, reduce soil erosion and provide for surface drainage. In low rainfall areas, land grading will result in a smooth uniform land surface which reduces runoff and induces a greater amount of the rainfall to infiltrate the soil and assures even moisture distribution. On sloping ground, levelling eliminates small depressions, cuts and furrows where water would otherwise concentrate and start to run off, causing damage by erosion year after year. Proper land grading, coupled with surface drainage measures, can reclaim unproductive poorly drained areas.

Land levelling operations may be grouped into three phases: rough grading, land levelling and land smoothing. *Rough grading* is the removal of abrupt irregularities such as mounds, dunes, and ridges, and filling of pits, depressions and gullies. *Rough grading* is followed by *land levelling* which reshapes the land surface to a planned grade. (The terms land levelling, land grading, land forming and land shaping are synonymous). *Land levelling* requires moving large quantities of earth over considerable distances. Levelling operation leaves an irregular surface due to dumping of the loads. These irregularities are removed and a plane surface obtained by *land smoothing* which is the final operation in land levelling. (The terms land smoothing, land floating and land planning are synonymous). In addition, land smoothing is often done prior to seeding as a regular land preparation practice.

Criteria for Land Levelling

Criteria for land grading are influenced by the characteristics of the soil profile, prevailing land slope, rainfall characteristics, cropping pattern, methods of irrigation, and other special features of the site, including the preferences of the farmer.

Soil profile conditions. A soil survey of the area to be levelled is necessary before undertaking the levelling work. The soil survey map will show the nature of the top soil, the subsoil, and the materials under the subsoil. It will show the depth to sand, gravel, hard pans, rock or other material that might limit the depth of cut, as well as the extent of such areas. Alkali spots will be outlined and depth to water table shown. The information will aid in planning the best possible layout for land levelling. Soils that are shallow have limitations in the depth of the cut that is allowable. Deep cuts in such cases may expose inert materials which may not grow crops. In deep soils, however, cuts into the subsoil may not be harmful because organic matter and fertilisers can be added to build up its fertility quickly.

Where deep cuts are unavoidable and the soil is shallow, the harmful effects of top soil removal may be mitigated by scraping and storing the top soil, which is then replaced by the new grade after the movement of the subsoil material. In such cases, the fill areas also must be stripped of the top soil, the fills partly made with the subsoil material brought from the high spots, and the top soil replaced. The process involves moving the same soil twice. Hence, it is an expensive procedure which should be justified. Bench levelling, in which the field is levelled in small strips, minimises the harmful effects of severe top soil removal.

Soil surveys furnish information relative to infiltration characteristics and hydraulic conductivity of the soil. If the subsoil and substratum are sand or sandy loam, and the infiltration rates are high, the irrigation runs must be shorter than if there is clay loam or clay underlying the surface.

Land slope. If the land is very steep and undulating, and the soils are shallow, it may not be possible to shape the surface to uniform slopes on good irrigation grades. Such areas should be left without any major land grading operations being done and should be kept in pasture for as much of time as possible, in order to prevent severe soil erosion.

The development of a uniform non-erosive grade in the direction of irrigation and the removal of excess slopes should be the aim of a high quality levelling job. This is particularly true for surface irrigation methods like border strip. Safe limits of longitudinal slope of fields, based on soil types are given in Table 8.1.

Table 8.1. Recommended safe limits of land slope for efficient irrigation

Type of soil	*Longitudinal slope, %*
Heavy (clay) soils	0.05 to 0.20
Medium (loamy) soils	0.20 to 0.40
Light (sandy) soils	0.25 to 0.65

A good land grade, designed in accordance with the infiltration characteristics of the soil, the size of the irrigation stream available, the crops to be grown and the erosion hazard from rainfall, permits uniform water distribution and high irrigation efficiencies.

Usually, excessive cuts are necessary to eliminate cross slopes. To reduce the extent of cuts, the field is divided into parts and the levelling is done in strips at different elevations, separated by low ridges (Fig. 8.6). The practice is commonly called bench levelling, especially if there is considerable difference in elevation between adjacent strips. The width of benches will be influenced by the farming equipment to be used and the earthwork involved. The amount of earthwork is governed by the magnitude of the diagonal slope at right angles to the direction of irrigation. The extent of cross slope permissible depends on the method of irrigation. While border strips and check basins do not permit any cross slope within the strip, furrow and corrugation methods of irrigation permit a certain degree of cross slope within a field.

Rainfall characteristics. The rainfall characteristics of an area are important in determining the maximum and minimum grades allowable for a field. Minimum grades must meet the drainage requirements which are substantially influenced by the amount of rainfall and its intensity, frequency and season of occurrence. Likewise, the minimum permissible grade for a given soil is determined by the maximum grade which is non-erosive under the prevailing rainfall conditions.

Cropping pattern. The kind of crops to be grown should be considered in selecting the irrigation or drainage method and the resulting land grading criteria. A high value crop, or a crop with high labour requirement, may justify a high degree of levelling to reduce labour and production costs. Intensively cultivated crops such as vegetables may justify a high levelling cost whereas a fodder crop may warrant a much smaller investment.

Irrigation methods. Each method of irrigation has its own limitations as to the permissible cross slope and downfield slope. The criteria for land grading must specify slopes within these limits. When several methods of irrigation are to be used on the same field, the requirements of the most restrictive method must be adapted. Graded and level borders and check basins require precise grading.

Other considerations. Field subdivision, based on natural topographical boundaries, should be considered for the entire farm in the initial plan itself, even though only a part of the farm may be levelled in the first year. Often it may be possible to subdivide the farm area into relatively narrow strips on the approximate contour to reduce the slope to safe limits and avoid deep cuts of the surface soil. Each strip is considered as a separate field for land levelling design. Plans for levelling should include the use of waste soil from drains and the requirements for earth fill in constructing irrigation channels. As far as possible, the irrigation and drainage systems for a particular field should be constructed at the same time as the adjacent areas are levelled.

Land Clearing

Prior to making the land grading survey, it is advantageous to remove heavy vegetative growth from the land. Land clearing consists of removing some or all of the trees, brush, vegetation, trash and boulders from the area specified for land grading. The operation also includes *grubbing* which is the process of removing stumps, sunken stones, and boulders and all roots over a minimum size and upto a specified depth as may be required. Where the land to be levelled is covered with large brush and timber, all merchantile timber should be cut and removed before starting the clearing operations. Some areas may be strewn with loose or partially buried boulders. These should be removed by digging and pushing. Land clearing may be done with manual labour or with heavy earth moving equipments like bulldozers, root rakes, stumpers, root cutters, rotary choppers or other appropriate machinery.

Layout of Fields, and Irrigation and Drainage Systems

Prior to levelling design, the land development programme must be planned so that the location of the field boundaries, irrigation water supply system, drains and farm roads are known. The levelling plan for an individual field must provide for furnishing field material or absorbing the excavated earth from adjacent areas. It must also provide for the proper ratio between excavation and fill.

A topographical map of the farm area is necessary in planning the field layout, water conveyance system and field drains. The topography is a major factor in selecting the method of irrigation, estimating the number and kind of water control structures needed, and determining the need for land levelling. The relative elevations of the source of water supply, the land surface between the water source and the area to be irrigated, the different parts of the farm area to be irrigated and the drainage outlets must be known to properly plan a farm irrigation system.

Field channels and underground pipelines to supply irrigation water to the field are located along the upper reach of the irrigation runs. They are usually perpendicular to the direction of irrigation for surface irrigation methods. The water surface in field channels should be 20 to 30 cm higher than the ground to be irrigated. If possible these channels should be nearly level (less than 0.1 per cent slope) so that water can be checked up for a maximum distance. Where the topography permits, the water supply lines may be so located as to serve the fields on either side of it. The location of the supply lines is determined by standard route-type survey. A profile of each proposed centre line is needed to compute the amount of excavation or fill required.

Field arrangement. Laying out fields of workable size and shape is important to successful irrigation farming. The fields are laid out as nearly rectangular as possible. Sharp turns in field boundaries should be avoided as far as possible in order to facilitate the use of modern farm equipment. The field length is based on the maximum allowable length of run for the irrigation method selected. Table 8.2 gives the range of lengths for border strip and furrow methods of irrigation. The field length may be equal to a single run length or a multiple of the run lengths. Alternatively, the field lengths may be limited by ownership boundaries.

Table 8.2. Recommended length of run for border strip and furrow methods of irrigation

Soil type	*Length of run*
Sandy and sandy loam soils	60 to 120 m
Medium loam soils	100 to 180 m
Clay loam and clay soils	150 to 300 m

The width of the field depends on cropping system, operations schedule and type of farm equipment used. With border and check basin methods of irrigation, the width of a field is a multiple of the width of the irrigation strip. The width of border or basin is based on the size of the irrigation stream and the type of equipment used in land preparation.

Subdivision of fields is based on ownership boundaries, obstructions, soil boundaries, land slopes and cropping systems. It is important to retain the existing facilities of irrigation, drainage and farm roads as much as possible. In most cases, the contour map of the area indicates the most advantageous way to subdivide the land for grading. The following contour patterns may warrant subdivision of land for successful irrigation farming:

(*i*) An abrupt major difference in contour interval implies a sharp change in slope. Separation of fields may be desirable along the line of slope change.

(*ii*) Sharp bends in otherwise nearly straight contours indicate an important change in the direction of maximum slope. Separation of fields at the bend is desirable for surface irrigation methods.

(*iii*) Contour lines either close together or far apart may imply that the average natural slope is either too steep or too flat to be satisfactory for the intended land use and that it must be graded to a lesser or greater slope. In such cases, the lengths of fields are kept to the minimum required for efficient irrigation so as to reduce the amount of cuts and fills required.

(*iv*) Great irregularity in spacing and direction of contours may show that the topography is non-uniform to the extent that it fails to show where a separation into fields can be made advantageously. Such areas may be set apart to be graded individually as units.

(*v*) Excessively irregular and/or closely spaced contours indicate that the cost of levelling may be prohibitive. Such areas may be planned for irrigation with sprinkler or drip methods.

Field road system. It is essential to provide a field road system for ready access to all areas of the farm for farm equipment, transportation of farm produce, and easy operation of the irrigation system. Field roads are provided above irrigation channels and below field drains.

Drainage. Good drainage, both surface and subsurface, is essential for successful irrigation. Provision should be made to drain the excess rainfall promptly and safely. Surface drainage may be needed to prevent or modify saline-alkali conditions in a soil by leaching. If the land is not naturally well drained, artificial drainage must be established at the same time the irrigation system is installed. Seepage from over-irrigated areas at higher elevations and irrigation canals can damage lands in the low-lying area. Interceptor drains may be necessary at the upper boundaries of the low-lying area to divert the seepage and prevent waterlogging. Integrated irrigation and drainage planning is often necessary for laying out a farm area for efficient water use.

LAND GRADING SURVEY AND DESIGN

Survey and Staking

With the field boundaries considered and established, the next step is to survey the area for land levelling design. The general practice is to establish a grid system over the field and set stakes at the grid points (Fig. 8.4). The usual grid spacing is 25 m in each direction. Other spacings such as 30 × 30 m, 20 × 20 m, and 15 × 15 m are also sometimes used, depending on the nature of surface, relief of the area and the precision required in levelling. Each grid point is at the centre of the grid square and represents nearly equal area. For convenience in identification, the row lines are lettered and the column lines numbered (Fig. 8.4). Often ridges or low spots fall between grid corners. Then, additional stakes should be set at high and low spots.

In locating the grid points, the usual practice is to establish two or more base lines in each direction and then to sight in the rest of the stakes. Thus, in Fig. 8.5, line B might first be established by measurement at a distance of one and a half times the grid interval (37.5 m), parallel to the south boundary of the field. Lines 4 and 5 are likewise located at right angles to line *B*. Line *C* should also be located by measurement. Other stakes can then be located by sighting. For instance, stake *D*-3 can be located at the intersection of the line of sight *D*-4 and *D*-5 on *B*-3 and *C*-3.

If two adjacent sides of a field are straight lines which form a 90° angle as in the case illustrated in Fig. 8.5, the staking problem is relatively simple. For crooked-sided fields, there is the added problem of establishing the two straight lines and making them conform as nearly as possible to the boundaries of the area.

The stakes are usually of size 1 cm by 4 cm by 1 m. They should be sharpened and driven into the ground far enough to ensure that they will withstand strong wind.

After all the points have been staked, levels are run to determine the ground elevation at each stake. The

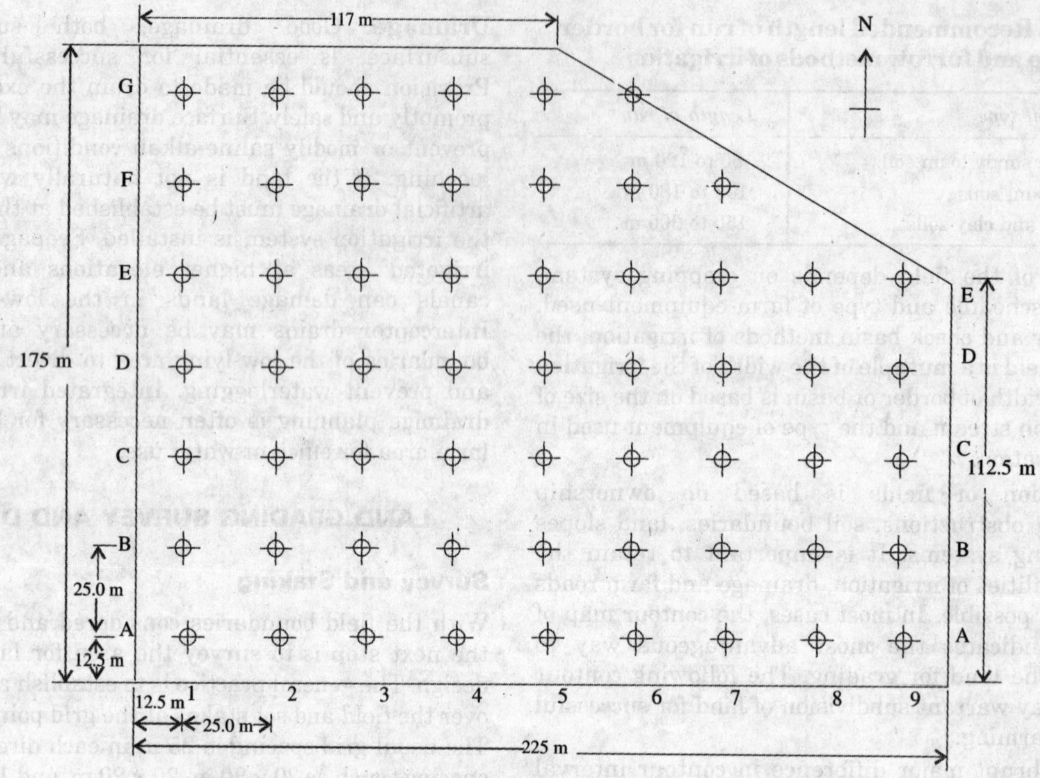

Fig. 8.4. Grid pattern used for staking a field which is to be graded.

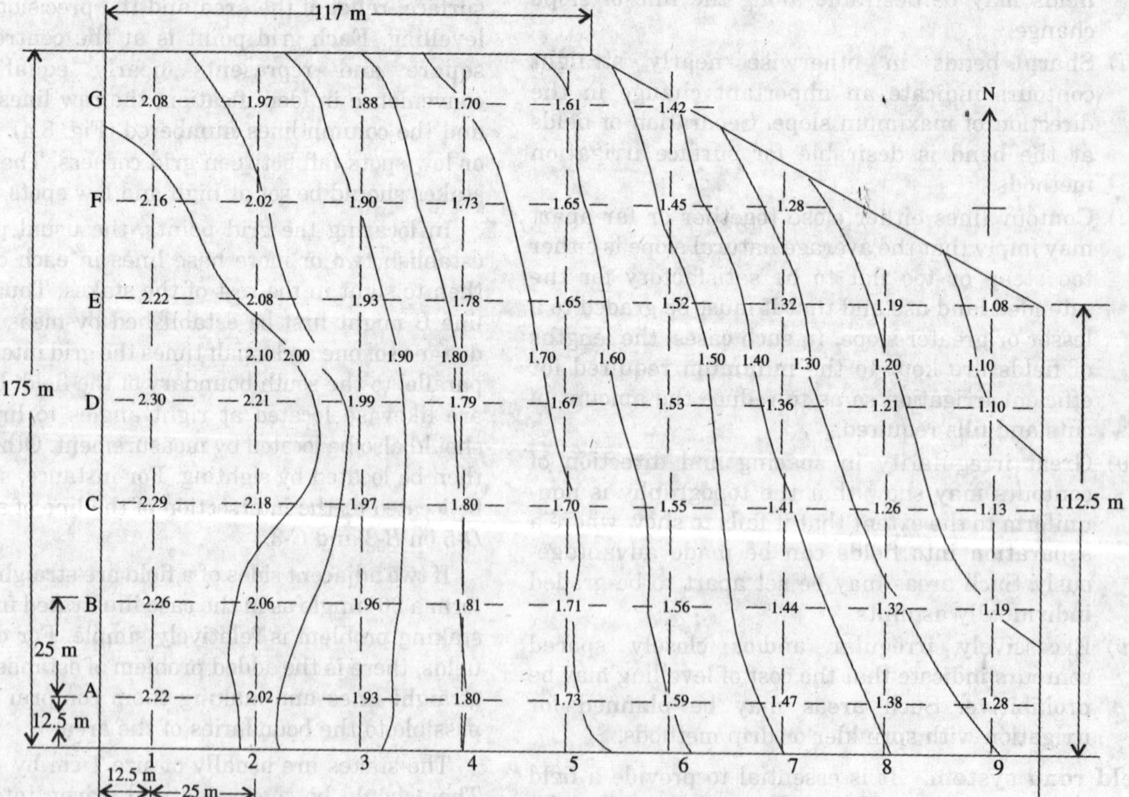

Fig. 8.5. Grid point elevations of the field marked in Fig. 8.4. Note the contour lines drawn at 10 cm intervals on the basis of the grid point elevations.

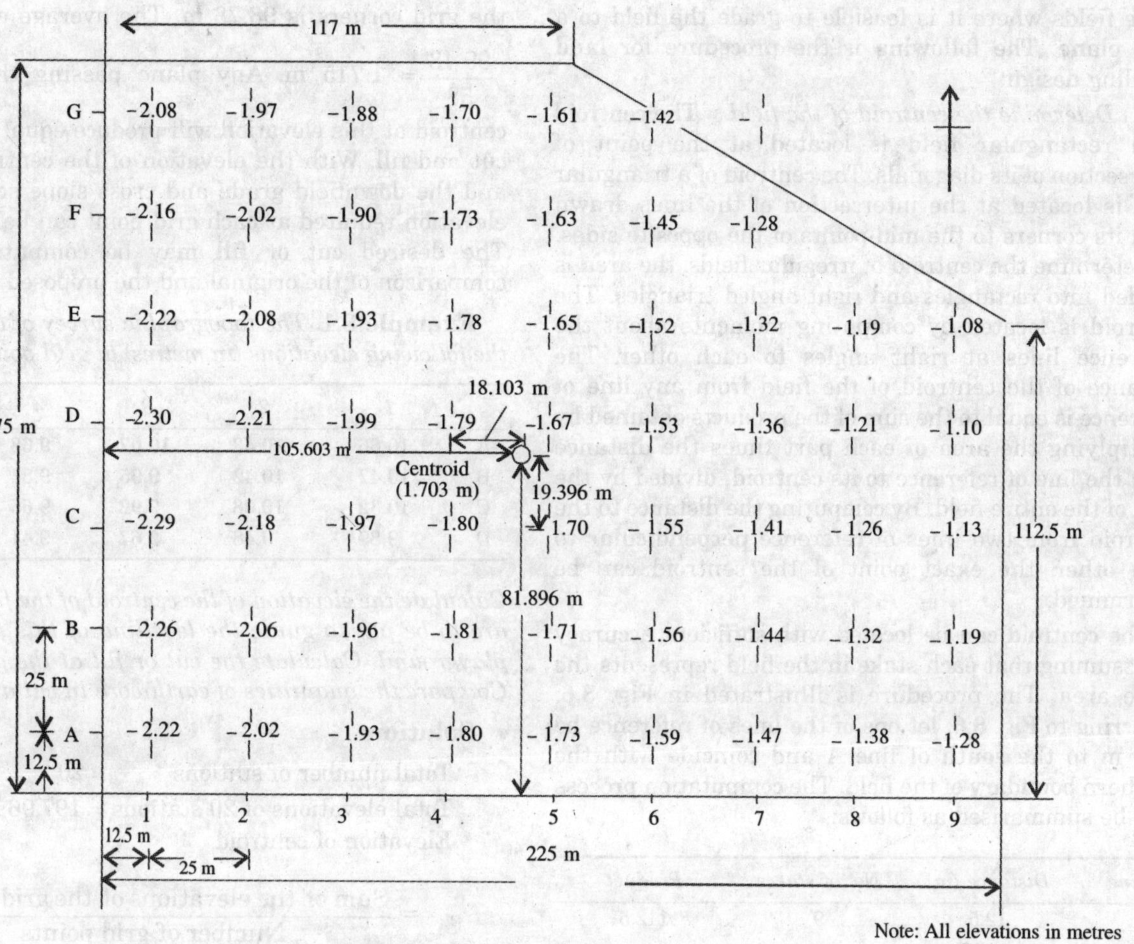

Fig. 8.6. Location of the centroid of the field shown in Fig. 8.5.

water surface elevation of the water supply and the drainage outlet are also determined to facilitate the layout of irrigation and drainage systems.

The elevations are determined with a dumpy level or farm level and a level rod, adopting the usual survey procedures. The first rod reading is made on a bench mark (BM) of known or assumed elevation. Any permanent point on or near the farm may be taken as the bench mark. In case of surveys involving large areas, it is desirable to establish the elevation of the bench mark with reference to mean sea level (MSL) by running a line of differential levels from a point of known elevation, like a nearby railway platform or other points, or the G.T.S. (Grand Trignometrical Survey) bench marks which have been established by the Survey of India or other agencies. For small scale works the elevation of the bench mark is assumed to be 0.00, 10.00 or 100.00 metres from an imaginary datum. When only a single field is to be surveyed, the first grid point may be taken as the bench mark.

Based on the rod readings, the reduced levels (RL) of the grid points are computed. The values of RL may be entered in a tabular form or may be indicated directly on the base map at the respective grid points (Fig. 8.5).

To make the survey information more readily understood and studied, contour lines are drawn at suitable intervals. The contour intervals are usually based on the average land slope. The recommended contour intervals for land slopes ranging between zero to one per cent, one to two per cent, two to five per cent and five to ten per cent are 6 to 15 cm, 15 to 30 cm, 30 to 60 cm and 60 to 150 cm, respectively. The contour lines clearly indicate the direction and degree of slope, ridges, depressions and other topographical features. With the help of the contour map the land is divided into fields that can be graded and irrigated individually to the best advantage.

Land Levelling Design Methods

The basic methods of land levelling design are: (1) Plane method, (2) Profile method, (3) Plan inspection method, and (4) Contour-adjustment method. Each method has its advantages and disadvantages and is best adapted to specific site conditions.

Plane Method

The plane method is the most commonly used method of land levelling design. Its use, however, is restricted to those fields where it is feasible to grade the field to a true plane. The following is the procedure for land levelling design:

(1) *Determine the centroid of the field.* The centroid of a rectangular field is located at the point of intersection of its diagonals. The centroid of a triangular field is located at the intersection of the lines drawn from its corners to the mid-points of the opposite sides. To determine the centroid of irregular fields, the area is divided into rectangles and right angled triangles. The centroid is located by computing moments about the reference lines at right angles to each other. The distance of the centroid of the field from any line of reference is equal to the sum of the products obtained by multiplying the area of each part times the distance from the line of reference to its centroid, divided by the area of the entire field. By computing the distance to the centroid from two lines of reference perpendicular to each other, the exact point of the centroid can be determined.

The centroid can be located with sufficient accuracy by assuming that each stake in the field represents the same area. The procedure is illustrated in Fig. 8.6. Referring to Fig. 8.6, let one of the lines of reference be 12.5 m to the south of line *A* and coincide with the southern boundary of the field. The computation process may be summarised as follows:

Line	*Distance (m)*	*No. of stakes*	*Product*
A	12.5	9	112.5
B	37.5	9	337.5
C	62.5	9	562.5
D	87.5	9	787.5
E	112.5	9	1012.5
F	137.5	7	962.5
G	162.5	6	975.0
	Total	58	4750.0

The distance of the centroid from the reference line is obtained by dividing the sum of the products by the total number of stakes, or $\frac{4750}{58}$ = 81.896 m. Another line of reference may be assumed at 12.5 m to the left of line 1 and the location of the centroid in the east-west direction is computed as 105.603 m from the reference line in the same way. With the two dimensions, the centroid point is located at 105.603, 81.896. It may be noted that the centroid point on the *x*-axis is 105.603 – 87.500 = 18.103 m from point *D*4. On the *y*-axis it is 81.896 – 62.500 = 19.396 m from point *C*5 (Fig. 8.6).

(2) *Determine the average elevation of the field.* This is obtained by adding the elevations of all the grid points in the field and dividing the sum by the number of points. Thus, in Fig. 8.6, the total of the 58 elevations on the grid corners is 98.78 m. The average elevation is $\frac{99\ 48}{58}$ = 1.715 m. Any plane passing through the centroid at this elevation will produce equal volumes of cut and fill. With the elevation of the centroid known and the downfield grade and cross slope selected, the elevation required at each grid point can be calculated. The desired cut or fill may be computed from a comparison of the original and the proposed elevations.

Example 8.1. *The topographic survey of a field gave the following elevations (in metres) at grid points:*

	1	*2*	*3*	*4*	*5*
A	10.65	10.43	10.07	9.68	9.67
B	10.47	10.42	9.95	9.84	9.75
C	10.32	10.08	9.92	9.65	9.48
D	9.89	9.48	9.67	9.41	9.13

Calculate the elevation of the centroid of the field. Stakes are to be put to guide the levelling of this field into a playground. Calculate the cut or fill at the grid points. Compare the quantities of earthwork in cut and fill.

Solution:

Total number of stations = 20

Total elevations of 20 stations = 197.96 m

Elevation of centroid

$$= \frac{\text{Sum of the elevations of the grid points}}{\text{Number of grid points}}$$

$$= \frac{197.96}{20} = 9.898 \text{ m}$$

Since the field is to be used as a playground, it is to be levelled without any slope in any direction. The cuts and fills at various grid points are obtained by subtracting the elevation of the corresponding grid point from the elevation of the centroid.

Thus, the cut/fill at grid point A1 is 9.898 – 10.65 = – 0.752 m. The – sign indicates the cut, while + sign indicates fill. The cuts and fills (in metres) at the different grid points are computed which are tabulated below:

	1	*2*	*3*	*4*	*5*
A	– 0.752	– 0.532	– 0.172	+ 0.218	+ 0.228
B	– 0.572	– 0.522	– 0.052	+ 0.058	+ 0.148
C	– 0.442	– 0.182	– 0.022	+ 0.248	+ 0.418
D	+ 0.008	+ 0.418	+ 0.228	+ 0.488	+ 0.768

Check : Σ cut = Σ fill = 3.28 m.

Example 8.2. *A topographic survey of a rectangular field was done for planning a land levelling programme. Grid points were set at intervals of 25 m. The elevations of the points are given as follows:*

Stations	Elevation (m)				
	Line 1	Line 2	Line 3	Line 4	Line 5
A	8.26	8.75	9.30	9.52	10.44
B	7.94	8.12	8.90	8.80	9.62
C	7.12	7.86	8.35	8.60	8.42
D	6.74	7.28	7.94	8.16	7.80

Determine the elevation of the centroid of the field. The field is to have a downfield slope of 0.2%. Determine the formation levels at the grid points and also the amount of cut or fill at each grid point.

Solution: Total number of stations = 20

Sum of the elevations of the 20 stations = 167.92 m

$$\text{Elevation of the centroid} = \frac{167.92}{20} = 8.396 \text{ m}$$

The formation levels at the grid points are computed at each grid point with reference to the elevation of the centroid. For example, station A1 is 50 m from the north-south line passing through the centroid. At 0.2 per cent slope, the elevation of this point should be 0.1 m below the centroid, or 8.296 m. Similarly, the formation level at each grid point is computed. The results are tabulated below:

	Formation levels at grid points (m) 0.2% slope				
	1	2	3	4	5
A	8.296	8.346	8.396	8.446	8.496
B	8.296	8.346	8.396	8.446	8.496
C	8.296	8.346	8.396	8.446	8.496
D	8.296	8.346	8.396	8.446	8.496

The cut/fill is computed by subtracting the original elevation of the point from the formation level at the grid point. Thus, at station *A*1, the cut/fill is 8.296 – 8.260 = + 0.036 m. The results are tabulated below:

	Cuts and fills (m)				
	1	2	3	4	5
A	+ 0.036	– 0.404	+ 0.096	– 1.074	– 1.944
B	– 0.356	+ 0.226	– 0.504	– 0.354	– 1.124
C	+ 1.176	+ 0.486	+ 0.046	– 0.154	+ 0.076
D	+ 1.556	+ 0.066	+ 0.456	+ 0.286	+ 0.696

Check: Σ cut = Σ fill = 5.558 m.

(**Note:** Fill heights when loose and before settlement should, depending on the soil type, be about 10 to 15 per cent more than the values given above, in order to allow for settlement.)

(3) *Compute the slope of the plane of best fit.* The slope of any line in the *x* or *y* direction of the plane which fits the natural ground surface, can be determined by the least squares method. In a rectangular area this can be represented by the following equation:

$$S = \frac{\Sigma(DH) - (\Sigma D)(\Sigma H)/n}{\Sigma(D^2) - (\Sigma D)^2/n} \quad ...(8.1)$$

in which,

S = slope of line in a plane (dimensionless*)

D = distance from the reference line (L)

H = elevation of the grid point (L)

n = number of grid points

The slope of the plane can be determined on both *x*- and *y*-axes, and the elevation of any point calculated from the elevation of the centroid. Although equation (8.1) is valid only for rectangular fields, it is often satisfactory to enclose most of the fields within an arbitrary rectangular boundary and extend the slopes of the plane to the remaining areas.

The grades determined by equation (8.1) must be checked to see if they fall within the safe limits selected for cross slope and downfield grade (Table 8.1). If they exceed these values, it will be necessary to use the nearest values that satisfy the criteria. Example 8.3 illustrates the computation procedure for determining the plane of best fit.

Example 8.3. *Determine the plane of best fit for the field shown in Fig. 8.7.*

Solution: Assuming the southern boundary of the field as the *x*-axis, the slope of the best fit plane in the *y*-direction is determined by applying equation (8.1). The distances of the stations from the *x*-axis on the different lines and the number of stations on each line are given below:

Line	Distance, D (m)	No. of stations
A	12.50	9
B	37.50	9
C	62.50	9
D	87.50	9
E	112.50	9
F	137.50	7
G	162.50	6
		Total 58

Values of *H*, the elevations of each station are indicated on Fig. 8.7.

Each factor in equation (8.1) is computed as follows:

$$\begin{aligned}\Sigma(D_yH_x) = {} & 12.5\,(2.22 + 2.02 + 1.93 + ... + 1.28) \\ & + 37.5\,(2.26 + 2.06 + 1.96 + ... + 1.19) \\ & + 62.5\,(2.29 + 2.18 + 1.97 + ... + 1.13) \\ & + 87.5\,(2.30 + 2.21 + 1.99 + ... + 1.10) \\ & + 112.5\,(2.22 + 2.08 + 1.93 + ... \\ & + 1.08) + 137.5\,(2.16 + 2.02 + 1.90 + ... \\ & + 1.28) + 162.5\,(2.08 + 1.97 + 1.88 + ... \\ & + 1.42) = 8116.27\end{aligned}$$

* To obtain the slope in percentage, the quantity *S* should be multiplied by 100.

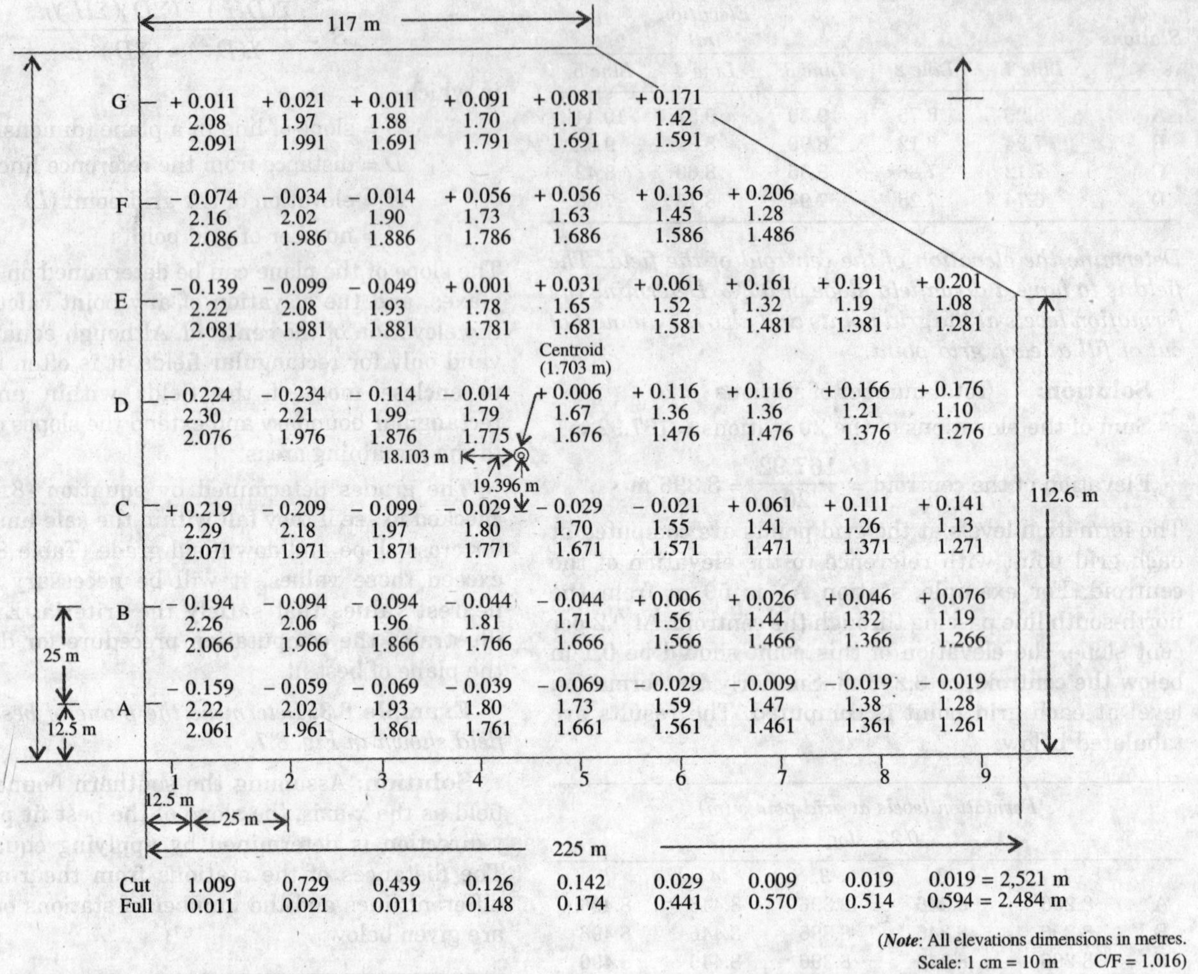

Fig. 8.7. Determination of formation levels, cuts and fills for land levelling design of the field shown in Fig. 8.5, adopting the plane method. Note that at each grid point the middle figure shows the original ground elevation, the top figure the cut/fill and the bottom figure the formation level.

$$(\Sigma D_y)(\Sigma H_x)/n = [9 \times 12.5 + 9 \times 37.5 + 9 \times 62.5 + ... + 9 \times 162.5]\,[2.22 + 2.02 + 1.93 + ... + 1.61 + 1.42] \div 58$$

$$= (4750.0 \times 98.78)/58 = 8089.741$$

$$\Sigma(D_y)^2 = 9 \times 12.52 + 9 \times 37.52 + 9 \times 62.52 + ... + 7 \times 137.52 + 6 \times 162.52$$

$$= 522812.50$$

$$(\Sigma D_y)^2/n = [9 \times 12.5 + 9 \times 37.5 + 9 \times 62.5 + ... + 7 \times 137.5 + 6 \times 162] \div 58$$

$$= \frac{(4750.0)^2}{58} = 389008.62$$

$$S_x = \frac{\Sigma(D_y H_x) - (\Sigma D_y)(\Sigma H_x)/n}{\Sigma(D_y)^2 - (\Sigma D_y)^2/n}$$

$$= \frac{8116.27 - 8089.741}{522812.50 - 389008.62}$$

$$= \frac{25.502}{133803.88}$$

$$= 0.00019$$

$$= 0.019 \text{ per cent}$$

From similar calculations and taking the western boundary of the field as the reference line on the y-axis,

$$S_z = -0.007$$

$$= -0.7 \text{ per cent}$$

It is seen that S_x is negative. This means that the plane reaches a lower elevation as the distance from the y-axis increases. Since S_y is positive, the plane of best fit rises towards north.

The values of S_x and S_y reveal that the slope of the best fit plane in the y-direction is negligible and can be tolerated as the cross slope, while that in the x-direction is substantial. The recommended safe limits of land slope presented in Table 8.1 show that the slope of the best fit plane along the x-direction exceeds the safe values in all the three soil types indicated in the table. Assuming that the soil is loamy, the maximum permissible land slope is 0.4 per cent. Hence, the value of S_x is to be limited to – 0.4 per cent or – 0.004.

(4) *Compute the formation levels, cuts and fills.* With the elevation of the centroid determined, the formation level of any point (the elevation which the point should attain after land grading operation) may be determined, using the computed or assumed values of S_x and S_y. For the convenience in computation, a grid point close to the centroid is selected and its elevation computed with reference to the centroid elevation, using the values of S_x and S_y. For example, in Fig. 8.7, the grid point $C4$ may be selected as the reference station. The distance of $C4$ from the centroid is 18.103 m in the x-direction and 19.396 m in the y-direction. Since the plane of best fit must pass through the centroid, the elevation of any point on the x-direction should increase at the rate of S_x to the left of the centroid and reduce by an equal amount to the right of the centroid. But since S_y is positive, the elevation in the y-direction increases at the rate of S_y from the centroid towards the north while it decreases at the same rate towards the south.

In Fig. 8.7, the elevation of a point on the x-axis, 18.1 m to the left of the centroid is

$$1.703 + \frac{18.103 \times 0.4}{100} = 1.775, \text{ m}$$

Similarly, the decrease in elevation of a point on the y-axis 19.396 m from the centroid, towards the south, is $\frac{19.396 \times 0.02}{100} = 0.0039$ m. Hence, the elevation of the grid point $C4$ = 1.775 – 0.0039 = 1.7711 m. The elevations of all other points on line C can be determined by adding $\frac{0.4 \times 25}{100} = 0.1$ m to each grid point to the left of $C4$ and subtracting an equal amount from each station to the right of $C4$. For example, the elevation of the point $C3$ is 1.771 + 0.1 = 1.871 and that of $C5$ = 1.771 – 0.1 = 1.671 and so on. The elevations of the points on line 4 can be obtained by progressively adding $0.02 \times \frac{25}{100} = 0.005$ m to the elevations of grid points in the north direction and subtracting the same amount from the elevations of the preceding grid points in the south direction from $C4$. Thus, the elevation of $B4$ is 1.771 – 0.005 = 1.766 m, of $D4$ = 1.771 + 0.005 = 1.776, of $E4$ = 1.776 + 0.005 = 1.781 and so on. Similarly all the design elevations are determined (Fig. 8.7).

The desired cut or fill at each point may be obtained from a comparison of the original and the proposed elevations. When the proposed elevation is less than the original elevation of the ground surface, the difference will indicate the depth of cut at the point. When the formation level is higher than the original ground surface, the difference will show the depth of fill required at the point. The cuts/fills at each of the grid points are shown in Fig. 8.7.

(5) *Determine the cut-fill ratio.* Experience in land grading with modern earthmoving equipment has shown that the cut-fill ratio should be greater than one. This means that a greater volume of cut than fill must be allowed. Compaction from equipment in the cut area which reduces the volume and also the compaction in the fill area which increases the fill volume needed, are the principal reasons for this effect. Another reason for the need to have additional fill material results from imperfections in the levelling operation. An argument usually put forth is that on level ground between stakes, the operator has an optical illusion of a dip in the middle, and therefore in filling, crowning often occurs, which requires additional fill material than what is estimated.

It is rarely possible to estimate exactly the cut-fill ratio. It usually varies from 1.2 to 1.6. In extreme cases of heavy or light textured soils and deep or shallow excavation, the ratio may be as low as 1.1 and as much as 2.0. In case the soil is composed principally of organic material such as peat or muck the ratio must be about 2.0. The type of soil, quantity of earthwork involved and the type of equipment to be used should be considered in deciding the cut-fill ratio.

With the plane method of computing cuts and fills, a settlement correction for the whole field is more convenient to apply. The settlement allowance or amount of lowering of elevation may range from 0.3 to 1 cm for compact soils and 1.5 to 4.5 cm for loose soils. A small change in elevation will cause a considerable change in the cut-fill ratio.

Since precise estimates on the additional cut volume cannot be made in advance, it may often be necessary to first set the grades tentatively. After the work has made progress, the levels are checked again and the stakes reset to obtain the desired grade as closely as possible.

In instances where borrow is to be taken from the field for construction works such as roads, irrigation channels and filling of pits, the plane surface can be lowered by the amount of earthwork required. Similarly, if the earth excavated from drainage channels or other sources is to be utilised in making land levelling fills, the elevation of the plane surface should be raised by the quantity of earth available. These can be done by dividing the volume of the borrow or excavation by the area of the field. The adjustment of the elevation of the plane can be made at the centroid by subtracting the value of the borrow taken from the field and adding the value of the excavated earth brought to the field. The elevations of other points are then computed, and the amount of cut or fill at each point determined. The cuts and fills are marked to the nearest cm since construction tolerances do not warrant a higher degree of precision.

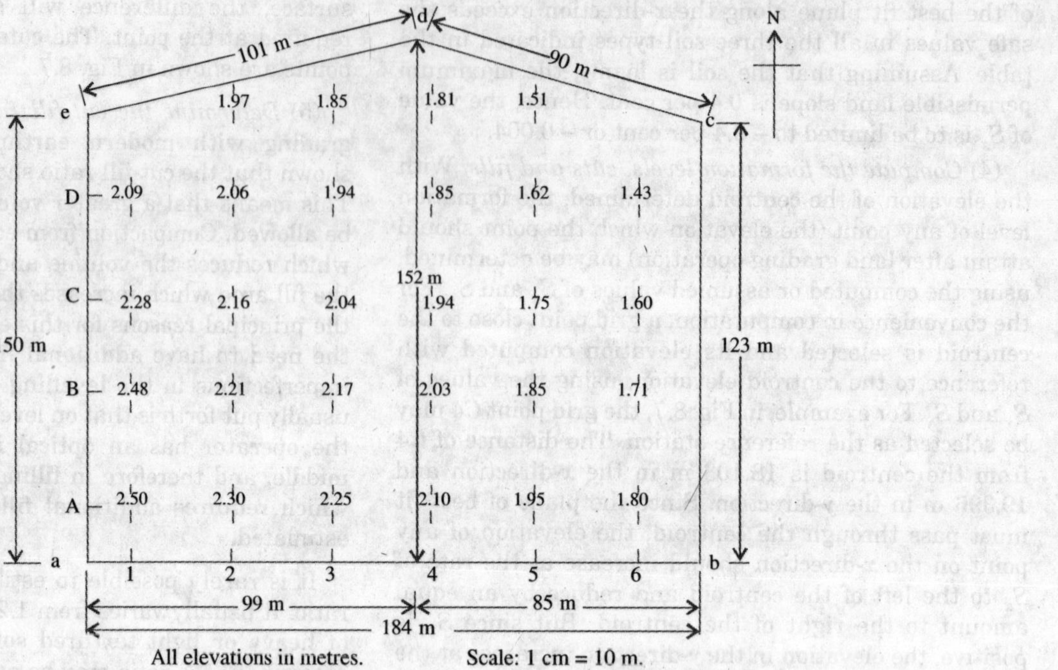

Fig. 8.8. Plan of field in Example 8.4 with grid point elevations marked.

Example 8.4. *The field shown in Fig. 8.8 is to be levelled to a downfield gradient of 0.4 per cent and a cross slope of 0.02 per cent. The following additional provisions may be made while designing the land levelling work:*

1. *A farm road is to be constructed along the boundary ab. The dimensions of the road are: bottom width 4.2 m, top width 3.5 m, and average height of embankment 30 cm. The entire earth fill for the farm road is to be taken from the field.*
2. *A drainage canal is to be dug outside the field along the side cd. The canal is 1.5 m deep, 1 m wide at the bottom and has side slopes 1.5:1. The excavation from the canal is to be spread in the field.*
3. *An allowance of 1 cm may be provided to account for the shrinkage of the cut and fill areas.*

Determine the design elevation of each grid point. Compute the cut and fill at each grid point and the cut-fill ratio.

Solution:

(1) Weighted average elevation of the grid points

$$= \frac{\Sigma H_i}{n} = (2.50 + 2.48 + ... + 1.43)/28)$$

$$= 1.973 \text{ m}$$

(2) The location of the centroid is determined by taking moments about the reference line *ox* and *oy*. (Note that the reference line *ox* is drawn parallel to the field boundary *ab* and is spaced 15 m away from the field boundary line *ab*. The line *ox* is laid at right angles to *oy* and is 15 m from the field boundary line *ae*. It is usually advantageous to locate the reference lines outside the field boundary, and space at least one of it at half grid spacing from one of the field boundary lines. This is especially important in the land grading design of fields with irregular boundaries).

Referring to Fig. 8.9, the location of the centroid on the plane on the *x*-axis is (6 × 30 + 6 × 60 + 6 × 90 + 6 × 120 + 4 × 150)/28 = 85.71 m from line *ox* or 85.71 − 15 = 70.71 m from line *ab*. Similarly, the location of the centroid on the plane on *y*-axis is (4 × 30 + 5 × 60 + 5 × 90 + 5 × 120 + 5 × 150 + 4 × 180) ÷ 28 = 105.0 m from line *oy* or 105.0 − 15 = 90 m from line *ae*.

(3) The slopes of the best fit planes on the *x*- and *y*-axis are determined by using equation (8.1)

$$(a) \quad S_x = \frac{\Sigma(D_x H_y) - (\Sigma D_x)(\Sigma H_y)/n}{\Sigma(D_x)^2 - (\Sigma D_x)^2/n}$$

$$\Sigma(D_x H_y) = 15(2.50 + 2.48 + ... + 2.09) + 45(2.30 + 2.21 + ... + 1.97) + ... 165(1.80 + 1.71 + ... + 1.43)$$

$$= 4663.050$$

$$\Sigma D_x = 4 \times 15 + 5 \times 45 + 5 \times 75 + ... + 4 \times 165 = 2520.00$$

$$(\Sigma D_x)(\Sigma H_y)/n = \frac{2520 \times 55.25}{28} = 4972.500$$

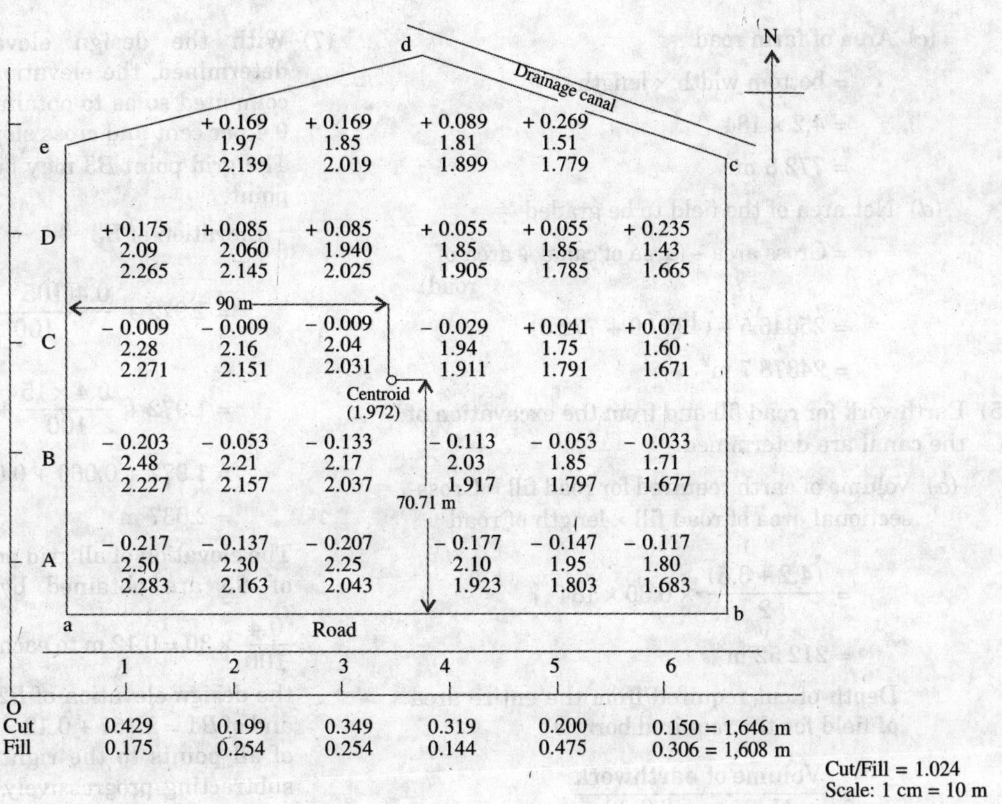

Fig. 8.9. Determination of design elevations, cuts and fills of the field shown in Fig. 8.8 adopting the plane method.

$$\Sigma(D_x)^2 = 4 \times 15^2 + 5 \times 45^2 + \ldots + 4 \times 165^2$$

$$= 294300.000$$

$$= (\Sigma D_r)^2/n = \frac{(2520)^2}{28}$$

$$= 226800.00$$

$$\therefore \quad S_x = \frac{4663.050 - 4972.500}{294300 - 226800}$$

$$= \frac{-309.450}{67500} = -0.00458$$

$$= -0.458 \text{ per cent}$$

(b) $$S_y = \frac{\Sigma(D_y H_x) - (\Sigma D_y)(\Sigma H_x)/n}{\Sigma(D_y)^2 - (\Sigma D_y)^2/n}$$

$$\Sigma(D_y H_z) = 15(2.50 + 2.30 + \ldots + 1.80) + 45(2.48 + 2.21 + \ldots + 1.71) + \ldots + 135(1.97 + 1.85 + \ldots + 1.51) = 3754.350$$

$$\Sigma D_y = 6 \times 15 + 6 \times 45 + \ldots + 4 \times 135 = 1980$$

$$(\Sigma D_y)(\Sigma H_z)/n = \frac{1980 \times 55.25}{28}$$

$$= 3906.964$$

$$\Sigma(D_y)^2 = 6 \times 15^2 + 6 \times 45^2 \ldots + 4 \times 135^2$$

$$= 186300$$

$$(\Sigma D_y)^2/n = \frac{(1980)^2}{28}$$

$$= 140014.286$$

$$S_y = \frac{3754.350 - 3906.964}{186300 - 140014.286}$$

$$= \frac{-152.614}{46285.714} = -0.003297$$

$$= 0.3297 \text{ per cent}$$

Since $S_x > S_y$, the downfield slope is taken in the x-direction and the cross slope in the y-direction.

(4) Net area of the field which is to be graded is determined.

(*a*) Gross area of field

$$= \frac{(130 \times 152)}{2} \times 99 + \frac{(152 + 123)}{2} \times 85$$

$$= 13959 + 1168.5$$

$$= 25646.5 \text{ m}^2$$

(*b*) Area of the drainage canal

$$= \text{top width} \times \text{length}$$

$$= 5.5 \times 90$$

$$= 495 \text{ m}^2$$

(*c*) Area of farm road

= bottom width × length

= 4.2 × 184

= 772.8 m^2

(*d*) Net area of the field to be graded

= Gross area – (area of canal + area of road)

= 25646.5 – (495.00 + 772.8)

= 24378.7 m^2

(5) Earthwork for road fill and from the excavation of the canal are determined.

(*a*) Volume of earth required for road fill = Cross-sectional area of road fill × length of road

$$= \frac{(4.2 + 3.5)}{2} \times 0.30 \times 184$$

= 212.52 m^2

Depth of cut required from the entire area of field for the required borrow

$$= \frac{\text{Volume of earthwork}}{\text{Net area of field}}$$

$$= \frac{212.52}{24378.7}\ \text{m}$$

= 0.0087 m

(b) Volume of excavated earth from drainage canal

= Cross-sectional area × length

$$= \frac{(5.5 + 1)}{2} \times 1.5 \times 90$$

= 438.75 m^3

Height of earth fill resulting from the spreading of the excavated earth in the net area of field.

$$= \frac{\text{Volume of excavation}}{\text{Net area of field}}$$

$$= \frac{438.75}{24378.7}$$

= 0.018 m

(6) Design elevation of centroid

= Average elevation of field – depth of shrinkage – depth of borrow + height of excavated earth

= 1.973 – 0.010 – 0.0087 + 0.018

= 1.972 m

(7) With the design elevation of the centroid determined, the elevation of the grid points are computed so as to obtain a downfield gradient of 0.4 per cent and cross slope of 0.02 per cent. The grid point *B*3 may be taken as the reference point.

Elevation of *B*3

$$= 1.972 + \frac{0.4(105 - 90)}{100} + \frac{0.02(85.71 - 60)}{100}$$

$$= 1.972 + \frac{0.4 \times 15}{100} + \frac{0.02 \times 25.71}{100}$$

= 1.972 + 0.060 + 0.005

= 2.037 m

The elevation of all grid points on line *B* to the left of *B*3 are obtained by progressively adding $\frac{0.4}{100} \times 30 = 0.12$ m to each succeeding point. Thus, the design elevation of *B*2 = 2.037 + 0.12 = 2.157 and of *B*1 = 2.157 + 0.12 = 2.277 m. The elevations of all points to the right of *B*3 are obtained by subtracting progressively. 0.12 m from each grid point from the elevation of the previous point on the left. Thus, the design elevation of *B*4 = 2.037 – 0.12 = 1.917 m, of *B*5 = 1.917 – 0.12 = 1.797 m and so on.

The design elevations of points on line *A* can be determined by adding $\frac{0.02}{100} \times 30 = 0.006$ m to the elevation of the corresponding point on line *B*. Similarly, the design elevation of all points on line *C* can be determined by subtracting 0.006 m from the elevations of the corresponding points on line *B*. The elevations of points on line *C* are obtained by subtracting 0.006 m from the corresponding elevations on line *D* and so on. The design elevations, thus obtained, are shown in Fig. 8.9.

(8) The cut or fill at each grid point is determined by a comparison of the original and design elevations. The values are entered at the corresponding grid point (Fig. 8.9). The total cuts and fills for each column are shown below the corresponding columns. Referring to Fig. 8.9,

$$\Sigma\ \text{Cut} = 1.646$$

$$\Sigma\ \text{Fill} = 1.608$$

$$\text{Cut-fill ratio} = \frac{1646}{1.608}$$

$$= 1.024$$

(**Note:** If a higher value of the cut-fill ratio is desired, a larger value of shrinkage depth may be assumed and the problem solved again.)

Profile Method

The profile method of land levelling design consists of plotting the profiles of the grid lines and then laying the desired grade on the profiles. Some designers prefer to plot the profiles at approximate right angles to the direction of irrigation or drainage; others plot them downfield. The method is especially adapted to levelling design for very flat lands or land with undulating topography on which it is desired to develop a fairly uniform surface relief.

There are many variations of the profile method, but essentially it consists of a trial and error method of adjusting grades on plotted profiles until the irrigation criteria are met with and the earthwork balance is attained. The method is frequently employed by many experienced workers, as it is relatively easy to select grades on a profile that will provide balanced cut and fill with a relatively short haul distance.

The field in Fig. 8.6 is again shown in Fig. 8.10 to illustrate the profile method. The base map is drawn to a (horizontal) scale of 1 cm = 10 m. The original grid point elevations are the same as shown in Fig. 8.6. The original grid profiles are drawn in the x-direction with a datum elevation of 1.5 m for each line A, B, etc. (It is possible to assume any suitable elevation for the datum line so that the corresponding profile line will fall below the adjacent grid line. Likewise, it is possible to assume different datum elevations for the different lines in case of undulating topography.) For example, the original ground elevation at station $A1$ is 2.22 m. The point on line for original ground elevation is 2.02 – 1.50 = 0.72 m = 72 cm. With the scale of 1 cm = 40 cm, this point is marked at a distance of 1.8 cm from point $A1$. Similarly, the original ground elevation at grid point $A2$ is marked on the vertical scale at a distance of 2.22 – 1.50 = 0.52 = $52 \times \frac{1}{40}$ = 1.3 cm from the point on line 2. Similarly, the

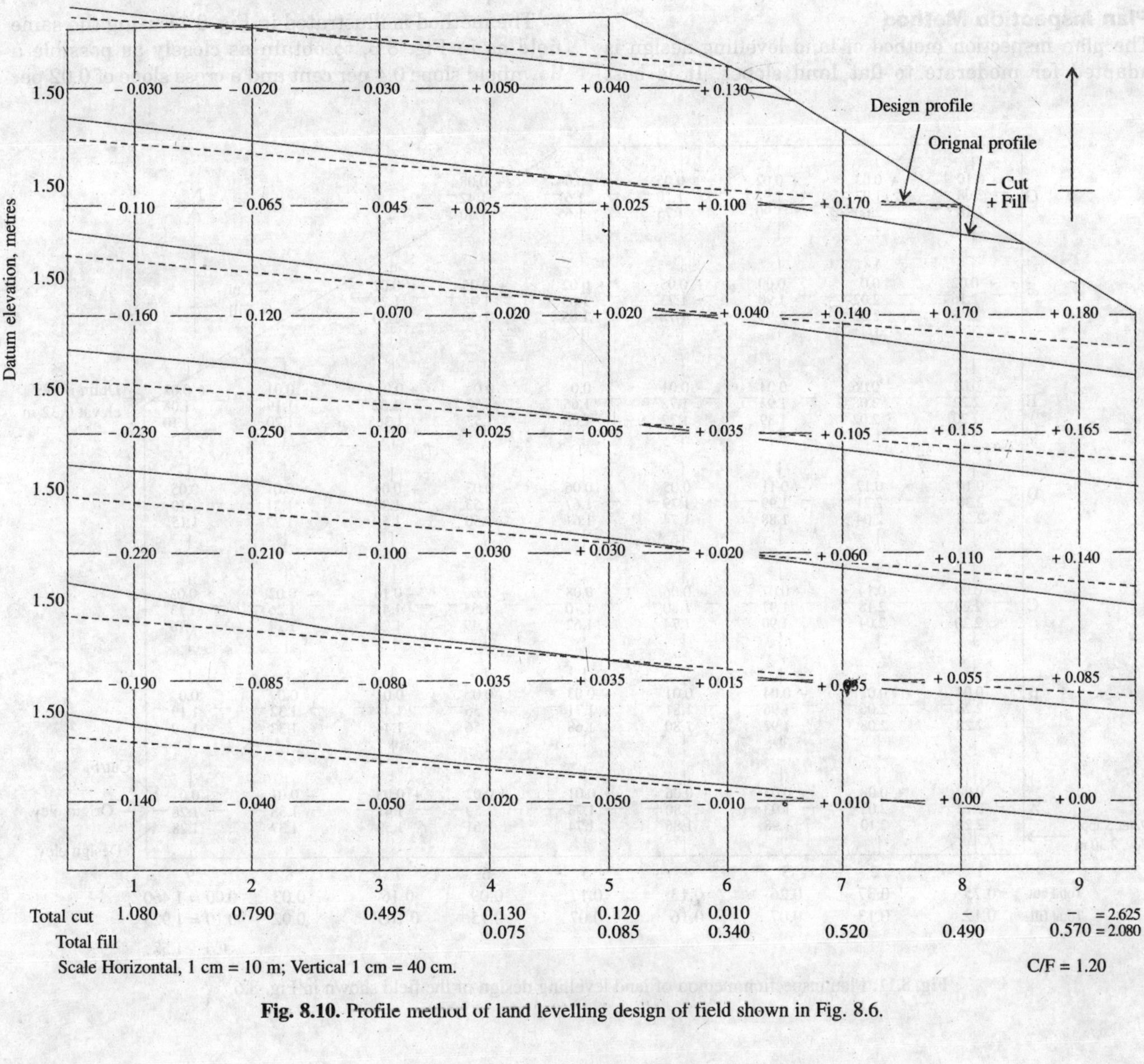

Fig. 8.10. Profile method of land levelling design of field shown in Fig. 8.6.

profile points on the vertical scale are located for lines *A*, *B*, *C*, *D*, *E*, *F*. Profile lines are drawn, connecting the points plotted above. Design profiles are then drawn by trial and error, with a view to obtain a downfield gradient of 0.4 per cent, a cross slope not exceeding 0.02 per cent and involving the minimum earthwork. The design profiles thus drawn, are shown in dotted lines in Fig. 8.10. Having drawn the design profiles, the cuts and fills are estimated graphically by comparing the original and design profiles. For example, on line 1, corresponding to grid point *A*1, the distance between the point of intersections of original and design profiles is 0.35 cm which represents an elevation difference of 0.35 × 40 = 14 cm = 0.14 m. Since the design elevation is less than the original ground elevation, this quantity shows the cut at *A*1. The amount of cut is posted as – 0.14 at *A*1. Similarly, the cut or fill at each grid point is determined. Referring to Fig. 8.11, Σ cut = 2.625 and Σ fill 2.080 which give a cut-fill ratio of 1.26.

Plan Inspection Method

The plan inspection method of land levelling design is adapted for moderate to flat land slopes. It is best adapted to the use of experienced personnel in those areas where land levelling is widely practiced. In this method, the grid point elevations are noted on the plan, and the design grade elevations are determined by inspection after a careful study of the topography. It is largely a trial and error procedure. Usually it is necessary to assume trial elevations for one or two lines of stakes. In selecting the elevations (formation levels) the designer must simultaneously consider the downfield slope, cross slope, earthwork balance and haul distance. The desired cut-fill ratio and volumes of earthwork are estimated from the summation of all cuts and fills. The grades are frequently adjusted to obtain a favourable earthwork balance and to maintain the downfield and cross slopes within safe limits. Although the method does not assume minimum cuts and fills or the shortest haul distance, it is a rapid procedure. The quality of the design depends on the success of the designer in weighing all factors as he selects the grades.

The method is illustrated in Fig. 8.11 using the same field as in Fig. 8.6, to obtain as closely as possible a downfield slope 0.4 per cent and a cross slope of 0.02 per

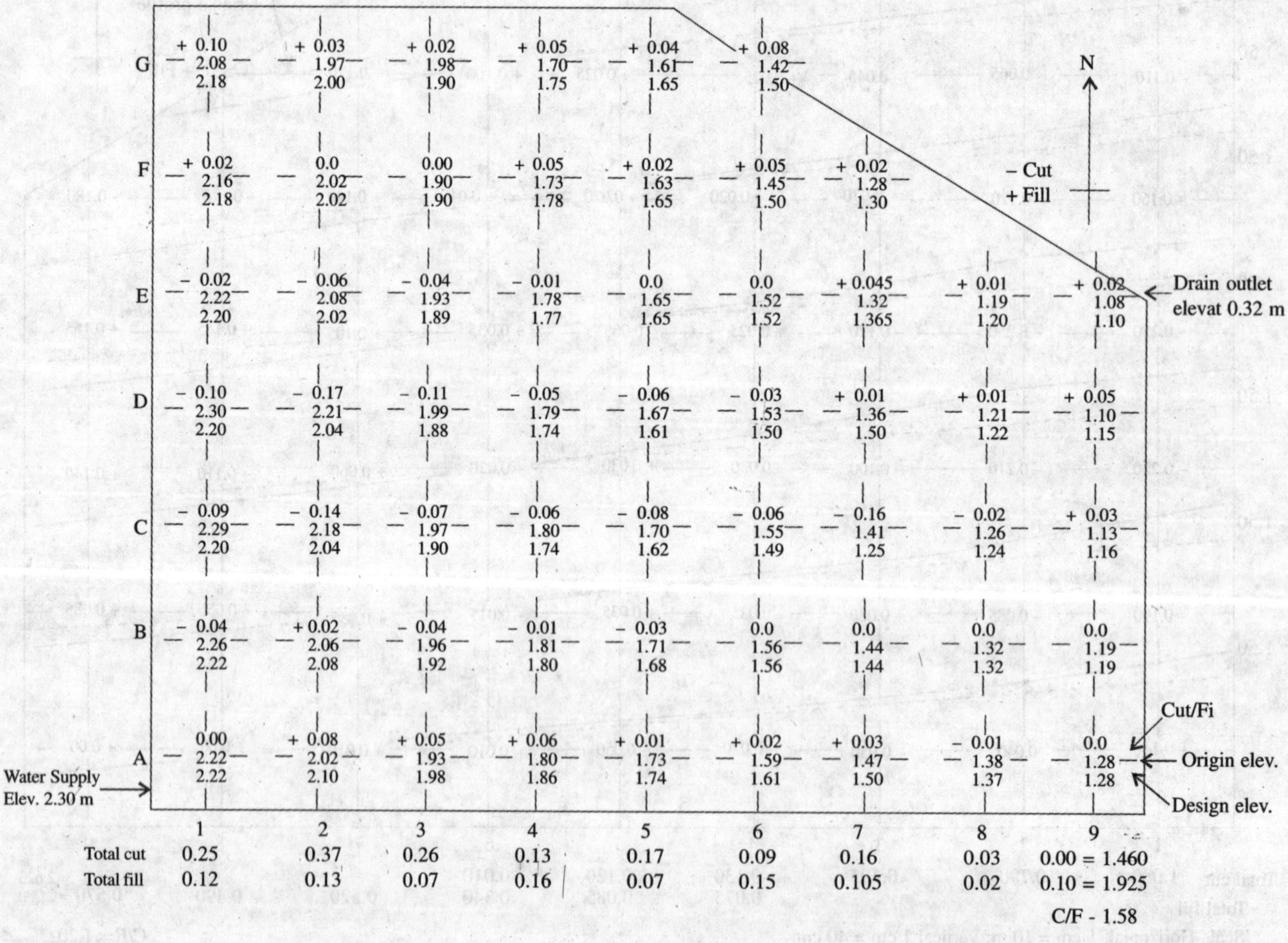

Fig. 8.11. Plan inspection method of land levelling design of the field shown in Fig. 8.6.

cent. With the original elevations of grid points known, the design elevations are so decided as to obtain the desired requirements. The difference between design elevation and original ground level is computed at each grid point to obtain the cut or fill. The summation of cuts and fills for each column of grid points is entered below the corresponding column. Referring to Fig. 8.11, Σ cut = 1.460 and Σ fill = 0.925. Therefore, the cut-fill ratio is

$$\frac{\Sigma \text{ Cut}}{\Sigma \text{ Fill}} = \frac{1.460}{0.925} = 1.58.$$

Contour Adjustment Method

The contour adjustment method of land levelling design consists of a trial and error adjustment of the contour lines on a plan map. The method is especially adapted to the smoothening of steep lands that are to be irrigated. Some workers also find it useful on moderate slopes. A contour map is drawn and the proposed ground surface is shown on the same map by drawing new contour lines. The uniformity of downfield slope is controlled by the uniformity of the horizontal spacing between contours, and the cross slopes can be examined by scaling the distance between contours at right angles to the direction of irrigation. A balance between cut and fill can be approximated by maintaining the proposed contour in an average position with reference to the original contour at the same elevation. The design elevation of grid points can be determined by interpolation between the design contours. A close approximation of the

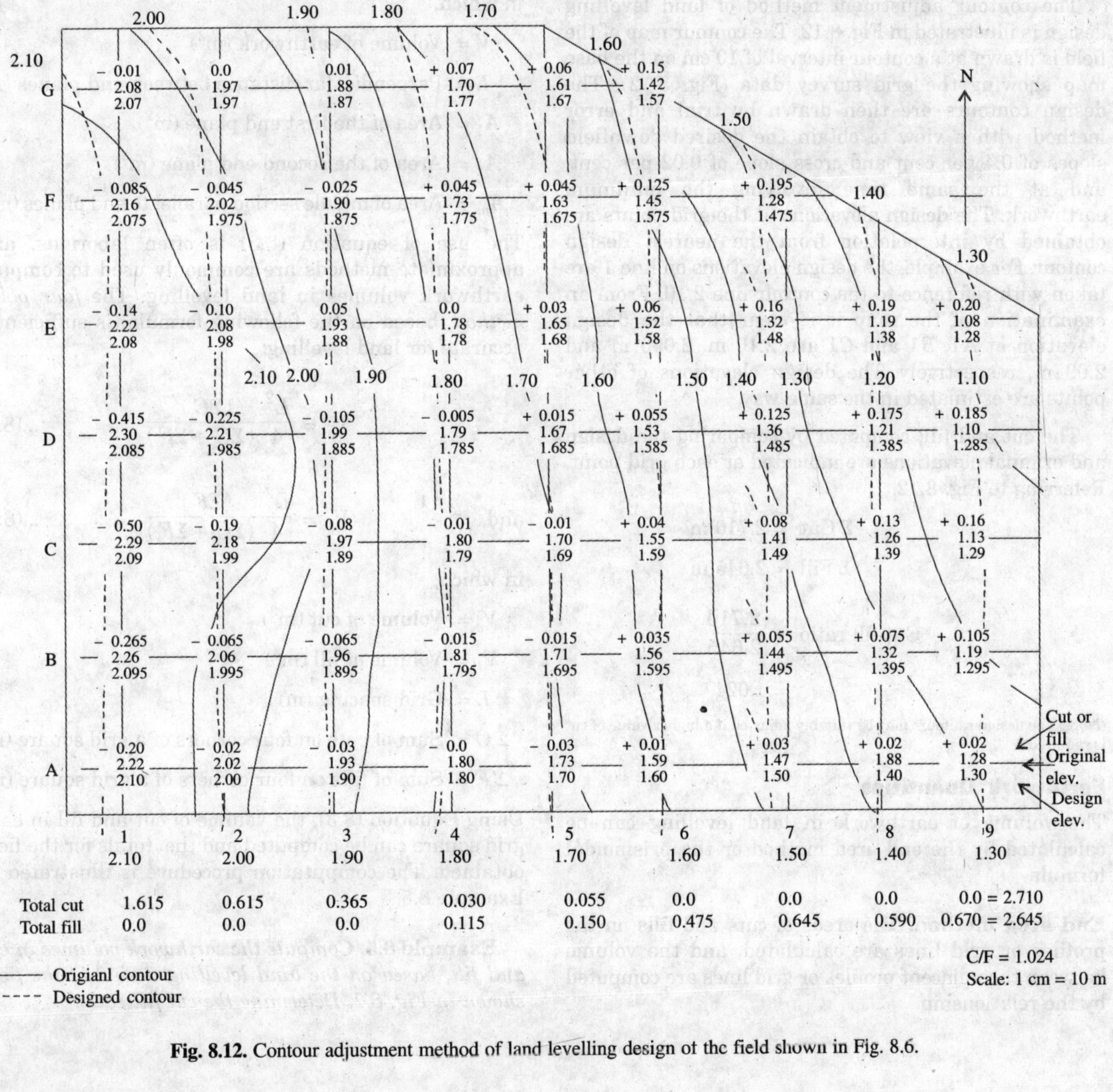

Fig. 8.12. Contour adjustment method of land levelling design of the field shown in Fig. 8.6.

earthwork balance can be obtained by summing the design cuts and fills from the staked points and comparing the totals.

The contour adjustment method demands considerable judgement on the part of the designer to keep the earthwork and haul to a minimum. Like the profile method and the plan inspection method, success in its use depends on the ability of the designer to recognise the relative importance of all the factors involved and to select a solution which satisfies the levelling criteria and requires the least earthwork. On steep slopes, particular attention must be given to the maximum permissible depth of cut that the soil will permit. On moderate slopes, the length of haul between the cut and fill area is often the point which requires the closest attention.

The contour adjustment method of land levelling design is illustrated in Fig. 8.12. The contour map of the field is drawn at a contour interval of 10 cm on the base map showing the grid survey data (Fig. 8.12). The design contours are then drawn by trial and error method with a view to obtain the desired downfield slopes of 0.4 per cent and cross slope of 0.02 per cent, and at the same time involving the minimum earthwork. The design elevations at the grid points are obtained by interpolation from the nearest design contour. For example, the design elevations on line 1 are taken with reference to the contour line 2.10. From an examination of the map it is clear that the design elevation at $A1$, $B1$ and $C1$ are 2.10 m, 2.095 m and 2.09 m, respectively. The design elevations of other points are estimated in the same way.

The cut and fill, computed by comparing the design and original elevations are indicated at each grid point. Referring to Fig. 8.12.

$$\Sigma \text{ Cut } = 2.710 \text{ m}$$

$$\Sigma \text{ Fill } = 2.645 \text{ m}$$

$$\text{Cut-fill ratio } = \frac{2.710}{2.645} = 1.024$$

(*Note:* The design contours may be suitably adjusted if a higher value of cut is required.)

Earthwork Quantities

The volume of earthwork in land levelling can be calculated by the end area method or the prismoidal formula.

End area method. The areas of cuts and fills on the profiles or grid lines are calculated, and the volume between the adjacent profiles or grid lines are computed by the relationship

$$V = L\,\frac{(A_1 + A_2)}{2} \quad \text{... (8.2)}$$

in which,

V = Volume of cut or fill (m^3)

L = Distance between profiles or lines (m)

A_1 = Area of cut or fill in the first profile or line (m^2)

A_2 = Area of cut or fill in the second profile (m^2)

Prismoidal formula. A precise method of computing the volume of earthwork in land levelling makes use of the prismoidal formula

$$V = \frac{L}{6}\,(A_1 + 4A_m + A_2) \quad \text{...(8.3)}$$

in which,

V = Volume of earthwork (m^3)

L = Perpendicular distance between end planes (m)

A_1 = Area of the first end plane (m^2)

A_2 = Area of the second end plane (m^2)

A_m = Area of middle section parallel to end planes (m^2)

The use of equation (8.3) is often laborious, and approximate methods are commonly used to compute earthwork volumes in land levelling. The *four point method*, based on the following formula, is sufficiently accurate for land levelling:

$$V_e = \frac{L^2}{4}\,\frac{(\Sigma C)^2}{(\Sigma C + \Sigma F)} \quad \text{...(8.4)}$$

and,

$$V_f = \frac{L^2}{4}\,\frac{(\Sigma F)^2}{(\Sigma C + \Sigma F)} \quad \text{...(8.5)}$$

in which,

V_e = Volume of cut (m^3)

V_t = Volume of fill (m^3)

L = Grid spacing (m)

ΣC = Sum of cuts on four corners of a grid square (m)

ΣF = Sum of fills on four corners of a grid square (m)

Using Equation (8.3), the volume of cut and fill in each grid square can be computed and the totals for the field obtained. The computation procedure is illustrated in Example 8.5.

Example 8.5. *Compute the earthwork volumes in cut and fill, based on the land levelling design of the field shown in Fig. 8.7. Determine the cut-fill ratio.*

Solution: Referring to Fig. 8.7, the volumes of earthwork in cut and fill are calculated separately for each square grid, using equations (8.4) and (8.5). For the boundary areas outside the square grids, the method is applied by taking the area of the individual figures and assuming the amount of cut or fill to be the same as that of the nearest grid point. The earthwork volumes in the individual areas are computed as follows:

The square grid formed by rows A and B and columns 1 and 2 may be taken and named $AB12$. For this,

$$\Sigma C = 0.132 + 0.187 + 0.077 + 0.032$$

$$= 0.428 \text{ m}$$

$$\Sigma F = 0$$

$$V_c = \frac{L^2 (\Sigma C)^2}{4(\Sigma C + \Sigma F)}$$

$$= \frac{25^2 \times 0.428^2}{4 \times 0.428} = 66.875 \text{ m}^3$$

Similarly, the cut and fill volumes of all square grid areas are calculated. The results are presented in Table 8.3.

The earthwork in the area lying between the left side and bottom boundaries of the field and row A and column 1 may be designated as A-1. The area of this field is

$$12.5 \times 12.5 = 156.25 \text{ m}^2$$

Assuming that all corners of this field have cuts equal to that of grid point A-1,

$$\Sigma C = 0.132 \times 4$$

$$= 0.528 \text{ m}$$

$$\Sigma F = 0$$

Since the term L^2 in equations (8.4) and (8.5) represents the area of the square grid, this term may be substituted by the actual area of the section of the field.

$$\therefore \quad V_e = \frac{156.25\,(0.528)^2}{4 \times 0.528}$$

$$= 20.625 \text{ m}^2$$

Similarly, the volumes of earthwork in all the boundary areas are calculated. The results are included in Table 8.3.

Table 8.3. Computation of earthwork volumes of field shown in Fig. 8.7

Sub-area	ΣC *(m)*	ΣF *(m)*	$\frac{(\Sigma C)^2}{(\Sigma C + \Sigma F)}$ *(m)*	$\frac{F^2}{(\Sigma C + \Sigma F)}$ *(m)*	V_e *(m³)*	V_f *(m³)*
1	2	3	4	5	6	7
Square areas (Area: $25 \times 25 = 625$ m²)						
1. AB12	0.506	—	0.506	—	79.062	—
2. BC12	0.716	—	0.716	—	111.875	—
3. CD1 2	0.886	—	0.886	—	138.437	—
4. DE12	0.696	—	0.696	—	108.750	—
5. EF12	0.346	—	0.346	—	54.062	—
6. FG12	0.108	0.032	0.083	0.007	12.969	1.094
7. AB23	0.316	—	0.316	—	49.375	—
8. BC23	0.496	—	0.496	—	77.500	—
9. CD23	0.656	—	0.656	—	102.500	—
10. DE23	0.496	—	0.496	—	77.500	—
11. EF23	0.196	—	0.196	—	30.625	—
12. FG23	0.048	0.032	0.029	0.013	4.531	2.031
13. AB34	0.246	—	0.246	—	38.437	—
14. BC34	0.266	—	0.266	—	41.563	—
15. CD34	0.256	—	0.256	—	40.000	—
16. DE34	0.177	0.001	0.176	—	23.50	—
17. EF34	0.063	0.057	0.033	0.027	5.203	4.219
18. FG34	0.014	0.158	0.001	0.145	0.1562	2.656
19. AB45	0.196	—	0.196	—	30.625	—
20. BC45	0.146	—	0.146	—	22.812	—

(Contd ...)

1	2	3	4	5	6	7
21. CD45	0.072	0.006	0.067	0.00046	10.469	0.072
22. DE45	0.014	0.038	0.004	0.027	0.625	4.219
23. EF45	—	0.144	—	0.144	22.500	—
24. FG45	—	0.284	—	0.284	44.375	—
25. AB56	0.142	0.006	0.136	0.00024	21.250	0.037
26. BC56	0.073	0.027	0.053	0.007	8.281	1.094
27. CD56	0.029	0.073	0.008	0.052	1.250	8.125
28. DE56	—	0.144	—	0.144	—	22.500
29. EF56	—	0.284	—	0.284	—	44.375
30. FG56	—	0.444	—	0.444	—	69.375
31. AB67	0.038	0.032	0.021	0.015	3.281	2.344
32. BC67	—	0.114	—	0.114	—	17.812
33. CD67	—	0.244	—	0.244	—	38.125
34. DE67	—	0.384	—	0.384	—	60.000
35. EF67	—	0.564	—	0.564	—	88.125
36. AB78	0.028	0.072	0.008	0.052	1.250	8.125
37. BC78	—	0.244	—	0.244	—	38.125
38. CD78	—	0.454	—	0.454	—	70.937
39. DE78	—	0.634	—	0.634	—	99.062
40. AB89	0.038	0.122	0.009	0.093	1.4061	4.531
41. BC89	—	0.374	—	0.374	—	58.437
42. CD89	—	0.594	—	0.594	—	92.812
43. DE89	—	0.734	—	0.734	—	114.687
End area (non-square)						
1. A-1 12.5 × .5 = 156.25	0.636	—	0.636	—	24.844	—
2. A-B-1 25 × 12.5 =312.5	0.706	—	0.706	—	55.156	—
3. B-C-1 25 × 12.5	0.826	—	0.826	—	64.531	—
4. C-D-1 25 × 12.5	0.886	—	0.886	—	69.219	—
5. D-E-1 25 × 12.5	0.726	—	0.726	—	56.719	—
6. E-F-1 25 × 12.5	0.426	—	0.426	—	33.281	—
7. F-G-1 25 × 12.5	0.148	0.022	0.129	0.003	10.078	0.234
8. G-1 25 × 12.5 = 156.25	—	0.044	—	0.044	—	1.719
9. G-1-2 25 × 12.5	—	0.064	—	0.064	—	5.000
10. G-2-3 25 × 12.5	—	0.064	—	0.064	—	5.000
11. G-3-4 25 × 12.5	—	0.204	—	0.204	—	15.937
12. G-4-5 25 × 12.5	—	0.344	—	0.344	—	26.875

(Contd ...)

1	*2*	*3*	*4*	*5*	*6*	*7*
13. G-5-6 $\frac{1}{2}(4.5 + 25) \times 12.5$ = 184.375		0.414		0.414		19.083
14. F-6-7 $\frac{1}{2}(9 + 25) \times 25$ = 425	—	0.719	—	0.719	—	76.394
15. $\frac{1}{2} \times 16 \times 9 = 72$	—	0.618	—	0.618	—	11.124
16. E-F-7-8 $\frac{1}{2}(19 + 9) \times 8 = 152$	—	0.794	—	0.794	—	30.174
17. E-F-7-8 $\frac{1}{2}(11 + 25) \times 25$ = 450	—	0.749	—	0.794	—	84.262
18. E-8-9 $\frac{1}{2}(6 + 20) \times 25$ = 325	—	0.784	—	0.784	—	63.700
19. E-9 $\frac{1}{2} \times 6 \times 12.5$ = 37.5	—	0.603	—	0.603	—	5.653
20. D-E-9 25 × 12.5 = 312.5	—	0.754	—	0.754	—	58.906
21. C-D-9 25 × 12.5	—	0.634	—	0.634	—	49.531
22. B-C-9 25 × 12.5	—	0.434	—	0.434	—	33.906
23. A-B-9 25 × 12.5	0.038	0.152	0.076	0.122	0.594	9.531
24. A-9 12.5 × 12.5	0.076	—	0.076	—	2.969	—
25. A-8-9 25 × 12.5	0.076	—	0.076	—	5.937	—
26. A-7-8 25 × 12.5	0.056	—	0.056	—	4.376	—
27. A-6-7 25 × 12.5	0.076	—	0.076	—	5.937	—
28. A-5-6 25 × 12.5	0.196	—	0.196	—	15.312	—
29. A-4-5 25 × 12.5	0.216	—	0.216	—	16.875	—
30. A-3-4 25 × 12.5	0.216	—	0.216	—	16.875	—
31. A-2-3 25 × 12.5	0.256	—	0.256	—	20.000	—
32. A-1-2 25 × 12.5	0.436	—	0.436	—	34.062	—
				Total	1604.933	1379.946

$$\frac{\Sigma \text{Cut}}{\Sigma \text{Fill}} = \frac{1604.933}{1379.946} = 1.163$$

CONSTRUCTION PROCEDURES

After the land levelling design is complete, the desired cuts and fills are marked on the grid stakes. The levelling job is done with suitable equipment and adopting the techniques for obtaining high efficiency in levelling operation.

Construction and Maintenance

Cut and fill stakes for construction purposes must be established to guide the equipment operator. When a grid type survey exists, the stakes indicating cuts and fills are located at the grid points. Sometimes, the cut or fill may be marked directly on the grade stakes.

Setting Grade Stakes

The amount of cut or fill required to bring the ground surface to a planned grade is indicated on grade stakes.

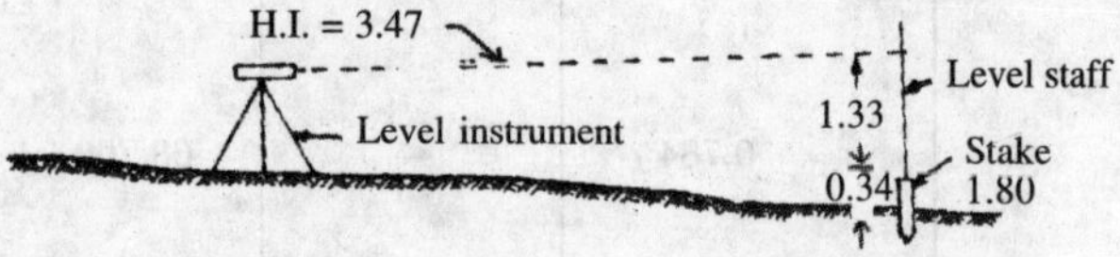

Fig. 8.13. Setting grade stakes for land grading.

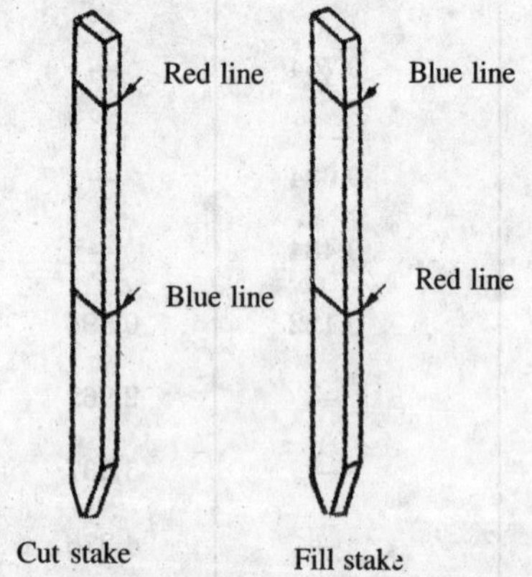

Fig. 8.14. Marking 'Cut' and 'Fill' stakes for 'blue top levelling'.

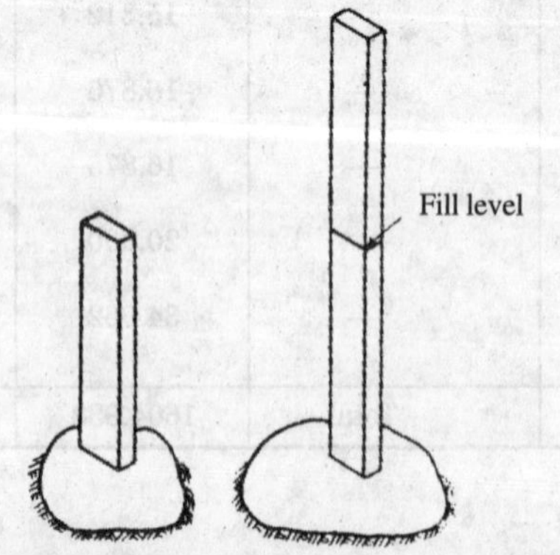

Fig. 8.15. Indicating 'Cut' and 'Fill' by short and long stakes.

The following are the common methods of setting grade stakes:

(1) Each stake may be driven so that its top is at grade. If a point *C*1 is to be established on a given grade, then let the formation level of *C*1 be 2.14 m and the original ground elevation 1.80 (Fig. 8.13). The height of the instrument H.I. is established with reference to a known bench mark elevation. Let the value of H.I. be 3.47 m.

Now the grade rod (reading on rod)

= H.I. – formation level

= 3.47 – 2.14 = 1.33 m

First set the stake at *C*1 with its top above grade level (only partly driven). Take a rod reading on the top of the stake. Adjust the stake level by driving it deeper until the desired reading of 1.33 m is obtained. The top of the stake is 2.14 – 1.80 = 0.34 m above the ground surface. indicating the depth of fill required at the point.

(2) Each stake has two lines marked on it—one blue and the other red. The blue line indicates the formation level and the red line the ground level at the stake point (Fig. 8.14). The blue line will be above the red line at 'fill' points and below it at 'cut' points. The distance between the red and blue lines indicates the amount of cut or fill, as the case may be. The desired grade is obtained when the level of the work (formation level) coincides with the blue lines. This method is frequently employed in farm work and is referred to as *'blue top levelling'*.

(3) Each stake may be driven to locate a point with little regard to its height. A mark may be made across, on one side of it, to indicate the depth of cut or height of fill required at the point.

(4) Cut and fill points are marked by short and long stakes, respectively. This helps in differentiating easily the cut and fill areas (Fig. 8.15). The amount of cut or fill at each point is marked on the corresponding stake. As in 'blue top levelling' the formation levels may be marked with the blue line above or below, as the case may be, of the original ground level indicated by the red line.

Job Planning

The following points are to be considered in deciding the job plan for land levelling:

(1) *Time available for completing the job.* The land levelling job must fit the cropping sequence. Often the number of days the field is open for land levelling is limited. When the levelling job can be spread over a long period, it is possible to use comparatively smaller machines and manual labour. But, when the time is pressing, it is advantageous to use heavy earth moving equipment that can do the job in a short time.

(2) *Skill of the operator.* Heavy earthmoving equipments are to be used by experienced operators.

With inexperienced operators, it is better to do the job with small equipments. The slow rate of progress allows more time to size up the work and take decisions.

(3) *Haul distance.* The haul distance should be reduced to the minimum by selecting the areas of cut as close as possible to the fill area. When the earth is to be moved a considerable distance, rubber tyred equipments are to be selected as far as possible.

Construction Procedures

A construction map is a useful aid in deciding the best approach to a levelling operation. To keep the haul distance to a minimum, it is important to construct the fills from the nearest excavation available. It is often advantageous to make a detailed study of the construction and prepare balance areas in which there is just sufficient excavation to make the required fill. Each balance area is levelled as a unit, ensuring short haul distance.

In performing the levelling operations, care is taken not to disturb the grid stakes. In opening a lane for levelling, it is usually preferred to first cut and fill a strip of one scraper width adjacent to the stake line. After this is brought to the design elevation, work begins on the intermediate area. By working at a slight diagonal, the areas between stakes can be almost completely worked out until there remains only small islands around each grid stake. These are allowed to remain until the field has been checked.

When the depth of fill is more than 15 cm, it is necessary to build them in layers of about 15 cm deep, to avoid excessive settling. Sometimes, it may be necessary to provide extra fill heights upto about ten per cent to compensate for settlement.

Preservation of top soil. In some soil conditions it may be necessary to stockpile the topsoil from an area, over-excavate the cuts and replace the top soil. The fill areas also must be stripped, the fill made partly with the materials available, and the top soil replaced. When these operations are essential, the top soil from one lane is stockpiled first on an area requiring little cut or fill. The cuts and fills in this lane are completed and the top soil from the adjacent lane is stripped and used for dressing the surface of the first lane. Thus, progressively across the field, the top soil is moved to the adjacent lane as the levelling is completed, until the last lane is dressed with the top soil stockpiled from the first lane.

Construction tolerance. Normally, earthwork operations have a permissible tolerance of ± 3 cm. On flat slopes it is necessary to make an additional restriction prohibiting reverse grades in the direction of irrigation. It is impractical to accomplish a high finish with scraper equipment. The finishing job is always done with land floats or land planes. The scraper work should be accomplished to the degree that two trips with the float or plane will produce the desired finished surface.

Checking. The final check must be made before the grid stakes are removed. The grade between stakes must be uniform. Necessary corrections are marked. After the field has been checked and found correct the mounds and depressions left at the grid stakes are levelled to match the adjacent surface.

Finishing. After the earthmoving work has been completed, the land surface is smoothened with a float or land plane to remove minor irregularities. This is usually done by working first at a 45° angle to the grid lines and again at right angles to the first operation.

Although a land has been levelled correctly, sometimes, after irrigation water is applied, the area will settle unequally. These irregularities are to be removed by repeating the floating operation prior to the sowing of the succeeding crop.

Maintenance. A levelled field requires maintenance to preserve its surface. Soil erosion and improper use of tillage implements may render a smooth field into an uneven area. Although annual scraping is not usually required, irrigated fields should be smoothened with a float or plane prior to seeding as a regular land preparation practice.

CONTOUR BENCH LEVELLING

Contour bench levelling is suitable for grading steep and undulating fields where irrigation is often difficult and hazardous. It also controls rainfall erosion. In this

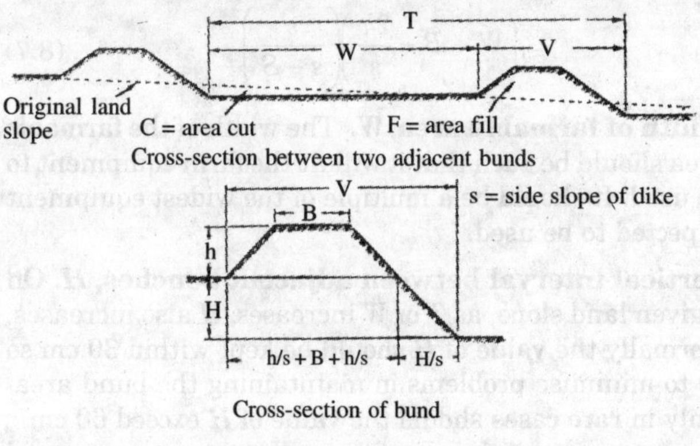

Fig. 8.16. Design relationships for contour bench levelling.

method the field is divided into a series of strips on the approximate contour, and each strip is levelled as an independent area. Thus, a series of steps is formed in successive elevations around the slope. Within the benches border strips, furrows and check basins can be used to apply irrigation water efficiently. The nearly flat benches provide increased opportunity for considerable quantity of water to infiltrate into the field, thereby reducing the irrigation requirement.

Design, Layout and Construction of Contour Benches

Selecting the proper cross-section of contour benches is one of the important steps in planning contour bench levelling. The elements of the bench cross-section are shown in Fig. 8.16.

Referring to Fig. 8.16,

T = Overall width of bench (m)

W = Width of cultivable strip (m)

B = Top width of bench dike (m)

h = Height of dike (m)

H = Vertical interval between two adjacent benches (m)

S = Original land slope, dimensionless

s = Side slope of dike, dimensionless

The following relationships between W, T and H are evident:

$$H = TS \quad \text{...(8.6)}$$

$$T = W + \frac{h}{s} + B + \frac{h}{s} + \frac{TS}{s}$$

$$T - \frac{TS}{s} = W + B + \frac{2h}{s}$$

$$T\left(1 - \frac{S}{s}\right) = W + B + \frac{2h}{s}$$

$$T = \left(W + B + \frac{2h}{s}\right)\left(\frac{1}{1 - S/s}\right)$$

$$= \left(W + B + \frac{2h}{s}\right)\left(\frac{s}{s - S}\right) \quad \text{...(8.7)}$$

Width of farmable area, *W*. The width of the farmable area should be such that it will fit the farm equipment to be used. It should be a multiple of the widest equipment expected to be used.

Vertical interval between adjacent benches, *H*. On a given land slope, as T or W increases, H also increases. Normally the value of H should be kept within 30 cm so as to minimise problems in maintaining the bund area. Only in rare cases should the value of H exceed 60 cm.

Height of dike, *h*. The height of dike should be sufficient to safely contain both the normal irrigation stream and the rainfall runoff. After computing the depth of flow, a free board of about 10 cm is given above the maximum water level expected. In sandy soils and pasture lands the free board should be about 15 cm.

Side slopes of dikes. The dike should have stable side slopes. Normally it should not be steeper than 1½ : 1. The usual slope provided is 2 : 1. In areas where stones are present in the field, they could be used to support the bund and a steeper side slope could be used.

Top width of the dike, *B*. The top width of the dike should be sufficient to prevent lowering of its height by trampling or by other sources. It is a usual practice to keep the top width of dike equal to the vertical interval between benches, i.e., $B = H$.

Longitudinal slope of bench. The benches are level in the direction of the original land slope but are given a gentle downfield gradient along their length to ensure efficient irrigation and drainage of rainfall runoff at non-erosive velocities. The safe limits of longitudinal slope for different soils, given in Table 8.1, are applicable to contour benches also.

Example 8.6. *Determine a suitable top width and vertical interval for a contour bench for the following conditions: the maximum width of the implement to be used is 3 m; the land slope is 4 per cent; the side slope of the dike is 1½:1 and the top width of dike is 36 cm. The maximum depth of irrigation stream is 12 cm. The depth of runoff produced by the most intense storm is 20 cm.*

Solution: The width of farmable area of the bench, W, should be a multiple of the width of the widest implement used in farming. Since $H = ST$, and the value of H should be about 30 cm and should not exceed 60 cm, the value of W may be taken as three times the width of the implement or 9 m. This gives a value of H (for the first approximation) at $\frac{4}{100} \times 9 = 0.36$ m = 36 cm. (The value of W could also be taken as twice the implement width or 6 m, but in that case the width of land will be too narrow which may effect farming efficiency. In general, the widest strip within the limits of the other factors involved should be selected).

Assuming a free board of 15 cm, the height of dike,

$$h = 20 + 15 = 35 \text{ cm}$$

$$S = \frac{4}{100} = 0.040$$

$$s = \frac{1}{1.5} = 0.666$$

$$B = 36 \text{ cm}$$

Overall width of the bench,

$$T = \left(W + B + \frac{2h}{s}\right)\left(\frac{s}{s - S}\right)$$

$$= \left(0 + 0.36 + \frac{2 \times 0.35}{0.66}\right)\left(\frac{0.666}{0.666 - 0.04}\right)$$

$$= 11.04 \text{ m}$$

Vertical interval between adjacent benches, H

$= ST$

$= 0.04 \times 11.04$

$= 0.44$ m

$= 44$ cm

Bench Location

The location of contour benches should be considered as a part of the basic farm planning in which the field boundaries, irrigation and drainage systems and farm roads are located. For levelling, each bench is regarded as a separate field. Since the earthwork involved in contour benches is often great, the benches should conform as closely as possible to the original topography of the area, as farming operations and other considerations will permit. Contour benches may be located either on a contour map or a grid map of the area.

Bench location based on contour map. The method is suitable when the contour map of the field to be levelled in contour benches is available. The procedure is illustrated in Fig. 8.17. By inspection of the contour map, the whole farm area is divided into key areas having nearly uniform topographic features. The key area is bounded by two contour lines, one on its upper boundary and the other at the lower. In the field shown in Fig. 8.17, one key area may be considered between the 0.75 and 1.75 contour lines and another between 0.25 and 0.50 contour lines.

Guidelines are located which represent an average condition of each key area. This is done by locating a line that represent the average shape of the contour lines in the key area and then computing the location of the guideline to the grade desired.

In Fig. 8.17, random lines 0, 1, 2, ... and 9 are drawn. On these lines the distances from a reference line to the contour lines within a key area are measured. Line bc is taken as the reference line. The distances of the contour lines 1.75, 1.50, 1.25, 1.00 and 0.75 on random line 0, measured from the reference line, are 10.4, 16.8, 21.6, 28.0 and 32.8 metres, respectively. The average distance is:

$$\frac{10.4 + 16.8 + 21.6 + 28.0 + 32.8}{5} = 21.92 \text{ m}$$

Similarly, the locations of the average shape of the contour lines in the two key areas are computed. The results are presented in Table 8.4. The line showing the average shape of the contours in the respective key areas is shown in Fig. 8.17.

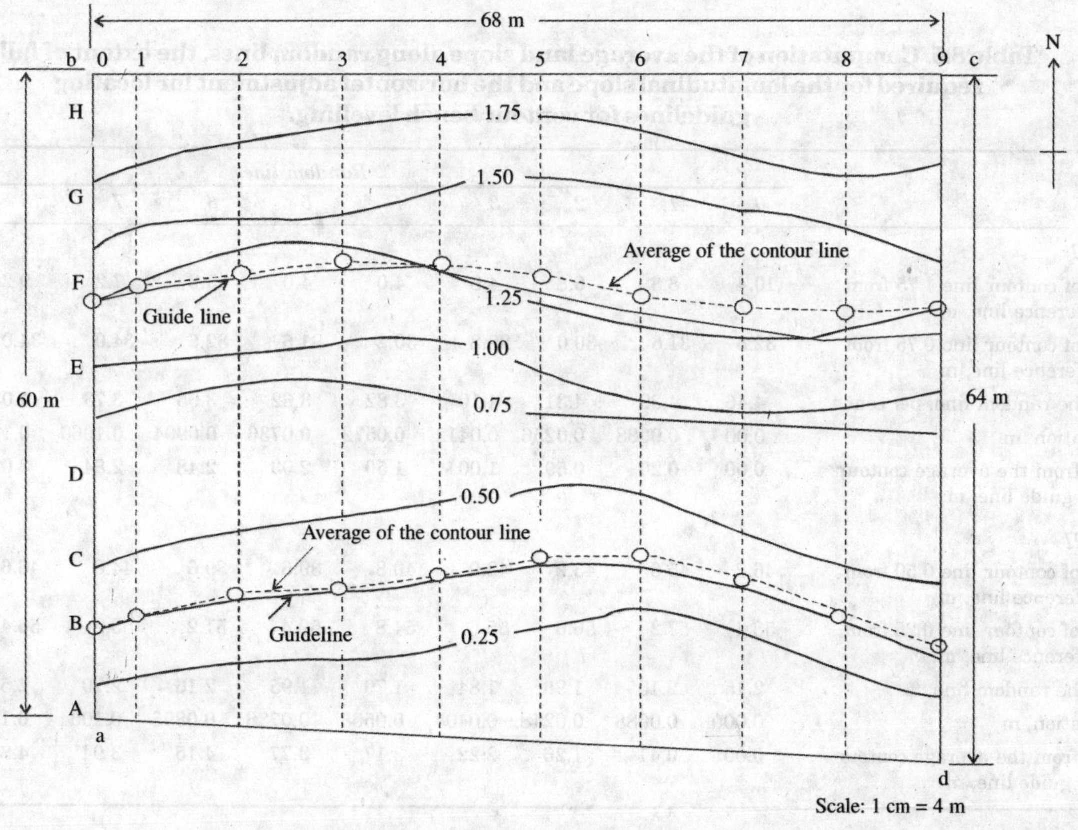

Fig. 8.17. Establishing guidelines for contour bench levelling from grid map

Table 8.4. Computation of average shape of contour lines in key areas for locating the guideline in contour bench levelling, based on the contour map of the field shown in Fig. 8.17.

Contour				*Random line*						
	0	*1*	*2*	*3*	*4*	*5*	*6*	*7*	*8*	*9*
	Distance from line bc (metres)									
Key area I										
1.74	10.4	8.8	6.8	4.8	4.0	4.0	5.2	7.2	9.2	8.0
1.50	16.8	14.8	13.2	12.8	10.8	9.6	11.2	13.2	14.8	17.6
1.25	21.6	20.9	16.8	16.4	19.0	21.2	24.0	24.8	24.0	22.0
1.00	28.0	28.8	26.4	26.0	25.0	27.6	31.0	32.0	31.6	30.0
0.75	32.8	31.6	30.0	29.2	30.2	31.6	32.6	34.0	34.0	33.6
Average	21.92	20.40	18.64	17.84	17.8	18.8	20.8	22.24	22.72	22.24
Key area II										
0.50	46.8	45.6	43.2	42.0	40.8	39.6	39.6	44.0	46.6	48.0
0.25	58.4	57.2	56.0	55.6	54.8	52.4	51.2	53.2	56.4	58.8
Average	52.6	51.4	49.6	48.8	47.8	44.6	45.4	48.6	51.5	53.6

The average land slope along the random lines in each key area is computed. Usually it is sufficiently accurate to determine this slope from the bottom edge to the top edge of the key area. The average slope is calculated by dividing the difference in elevation by the difference in distances from the reference line 0. For example, in Fig. 8.17, the average slope along random line 0 in key area *i* is

$$\frac{1.75-0.75}{32.8-10.4} \times 100 = 4.46\%$$

Similarly, the per cent slopes on the other random lines for key area *i* and for all lines for key area II are computed. The results are presented in Table 8.5.

After plotting the line representing the average of the contours and computing the land slope along the

Table 8.5. Computation of the average land slope along random lines, the extent of fall required for the longitudinal slope and the horizontal adjustment for locating guidelines for contour bench levelling.

	Random lines									
	0	*1*	*2*	*3*	*4*	*5*	*6*	*7*	*8*	*9*
Key area I										
Distance of contour line 1.75 from the reference line, m	10.4	8.8	6.8	4.8	4.0	4.0	5.2	7.2	9.2	8.0
Distance of contour line 0.75 from the reference line, m	32.8	31.6	30.0	29.2	30.2	31.6	32.6	34.0	34.0	33.6
Slope of the random line, per cent	4.46	4.39	4.31	4.10	3.82	3.62	3.65	3.73	4.03	3.91
Fall at station, m	0.00	0.0088	0.0256	0.0412	0.0572	0.0736	0.0904	0.1060	0.1216	0.1376
Distance from the average contour line to guide line, m	0.00	0.20	0.59	1.00	1.50	2.03	2.48	2.84	3.02	3.52
Key area II										
Distance of contour line 0.50 from the reference line, m	46.8	45.6	43.2	42.0	40.8	39.6	39.6	44.0	46.6	48.4
Distance of contour line 0.25 from the reference line, m	58.4	57.2	56.6	55.6	54.8	52.4	51.2	53.2	56.4	58.8
Slope of the random line, %	2.15	2.15	1.96	1.84	1.79	1.95	2.16	2.70	2.5	52.40
Fall at station, m	0.000	0.0088	0.0248	0.0408	0.0568	0.0736	0.0896	0.1064	0.1240	0.1408
Distance from the average contour line to guide line, m	0.00	0.41	1.26	2.22	3.17	3.77	4.15	3.94	4.86	5.87

random lines, the guidelines for laying out the contour benches are located. For the field shown in Fig. 8.17, let the contour benches have a longitudinal slope of 0.2 per cent. The direction of the longitudinal slope is given along the direction of the maximum cross slope of the field. The guide line is located to obtain this gradient in the desired direction.

Referring to Fig. 8.17, the point of intersection of the line representing the average of the contours with random line 0 may be selected as the origin of the guideline. The distance between random lines 0 and 1, measured along the line of the average contour, is 4.4 m. The guideline will have a fall of $\dfrac{4.4 \times 0.2}{100} = 0.0088$ m from the point of intersection of the average contour line on random line 1. But random line 1 has an average slope of 4.39% (Table 6.5). Hence, to provide this fall, the horizontal measurement on random line 1 from the line of average contour is

$$\frac{0.0088 \times 100}{4.39} = 0.2004 \text{ m}$$

or say, $\quad$ 20 cm

The position of the guideline on random line 1 is located at a distance of 20 cm to the south of the line representing the average values for the contours. Similarly, the location of the guideline on random line 2, 3, 4,... and 9 are determined. The results, computed for the two key areas, are presented in Table 8.5. The guideline is drawn joining the points so determined.

Contour bench location based on grid point elevations. The method is faster than the one described above, but is suitable only when the ground surface elevations at grid points are available. The procedure, illustrated in Fig. 8.18 is as follows:

(*i*) The average elevations of the grid lines are determined: This is done by taking the sum of the grid point elevations on a line and dividing by the number of points. Thus, the average elevation of grid line 1 is

$$\frac{2.0 + 1.4 + 0.7 + 0.5}{4} = 1.15 \text{ m}$$

The average elevations of the other grid lines are computed in the same way. The results are presented in Table 8.6.

(*ii*) The average grade on each grid line is determined. This is usually done by subtracting the elevation on the lower edge of the field at a grid line from the elevation on the upper edge and dividing by the distance between the two points. Thus, the average grade of line 1 is

$$\frac{2.0 - 0.5}{24} = 0.625$$

$$= 6.25\%$$

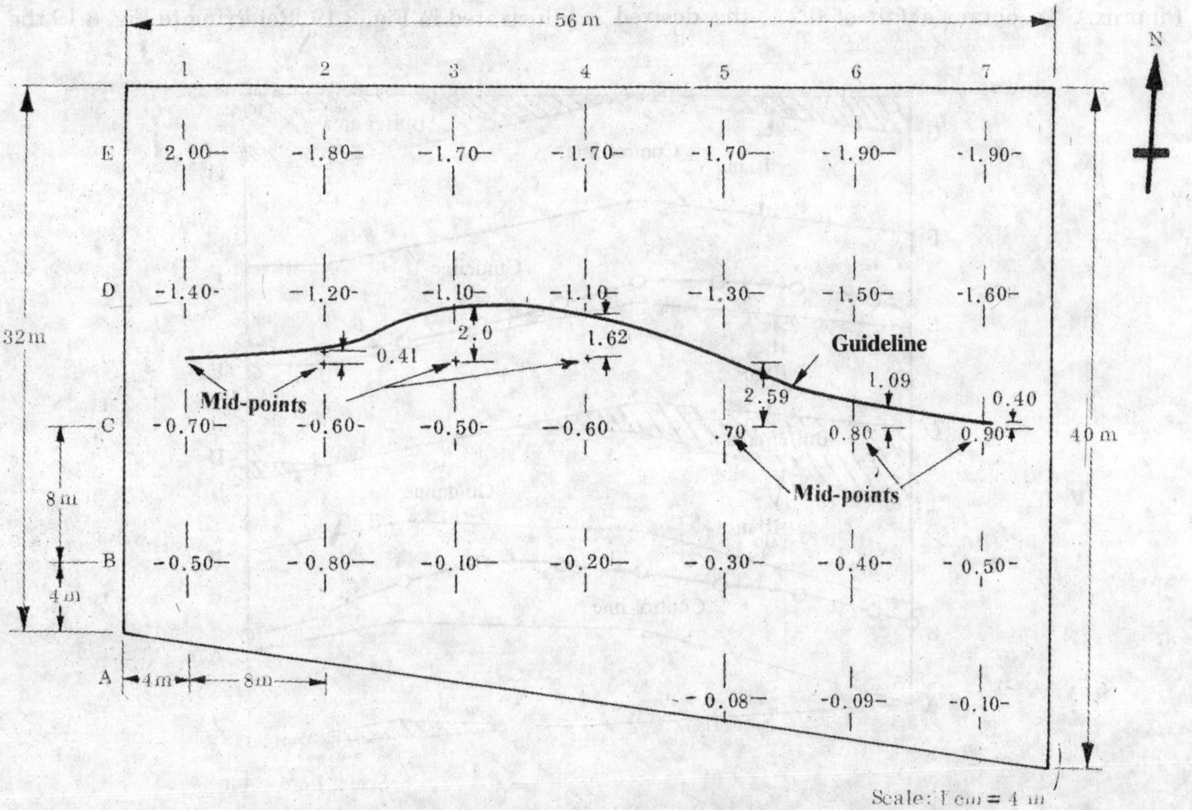

Fig. 8.18. Establishing control line for contour bench levelling from grid map.

Similarly, the average grade of lines 2, 3,... and 7 are computed. The result are given in Table 8.6.

Table 8.6. Computations for locating guideline for contour bench levelling from grid map.

	Grid line						
	1	*2*	*3*	*4*	*5*	*6*	*7*
Sum of elevations (m)	4.6	4.4	3.4	3.6	4.08	4.69	5.0
Average elevation of grid line (m)	1.15	0.10	0.85	0.90	0.816	0.938	1.0
Average grade of grid line (%)	6.25	4.17	6.67	6.25	5.06	5.66	5.63
Elevation of guideline (m)	1.15	1.134	1.118	1.102	1.084	1.068	1.052
Horizontal adjustment (m)	0.00	0.82	4.02	3.23	5.30	2.30	0.92

(*iii*) The midpoints of grid lines are determined. The average elevations are considered to lie on the midpoint of each line. For lines 1, 2, 3 and 4 the midpoint lie midway between row lines *C* and *D*. For lines 5, 6 and 7 the midpoints lie on line *C*.

(*iv*) The location of the guideline with reference to the midpoints is determined. The longitudinal gradient of the benches, and hence the guideline, is assumed to be 0.2 per cent. Let the guideline start at the midpoint of line 1 which has an average elevation of 1.15 m (Table 8.6). Between lines 1 and 2 the guideline is 8 m long (approx.). To obtain a fall of 0.2%, the desired elevation of the point on line 2 for the guide line is

$$1.15 - 8 \times \frac{2.0}{100} = 1.134 \text{ m}$$

Since the average elevation of line 2 is 1.10 m, the desired point will be above the midpoint at a vertical distance of 1.134 – 1.10 = 0.034 m.

As the slope of line 2 is 4.17%, the horizontal adjustment required is

$$0.034 \times \frac{100}{4.17} = 0.815 \text{ or say, } 0.82 \text{ m}$$

The position of the guideline is located at a distance of 0.82 mm from the midpoint of line 2 towards the north. Similarly, the positions of the guidelines on other grid lines are computed and the line joining the points is drawn. The results are presented in Table 8.6.

Planning Bench Locations

With the guidelines drawn, the contour benches are located on the map. The closer the guidelines are followed, the lesser will be the earthwork required. The following points are considered in planning the benches:

(*i*) Sharp bends are avoided in order to facilitate tillage operations.

(*ii*) Normally the benches should continue completely across the field.

(*iii*) Avoid pointed areas that are difficult to construct and maintain.

The procedure for laying out contour benches with the help of the guidelines shown in Fig. 8.17 is illustrated in Fig. 8.19. Referring to Fig. 8.19 the width

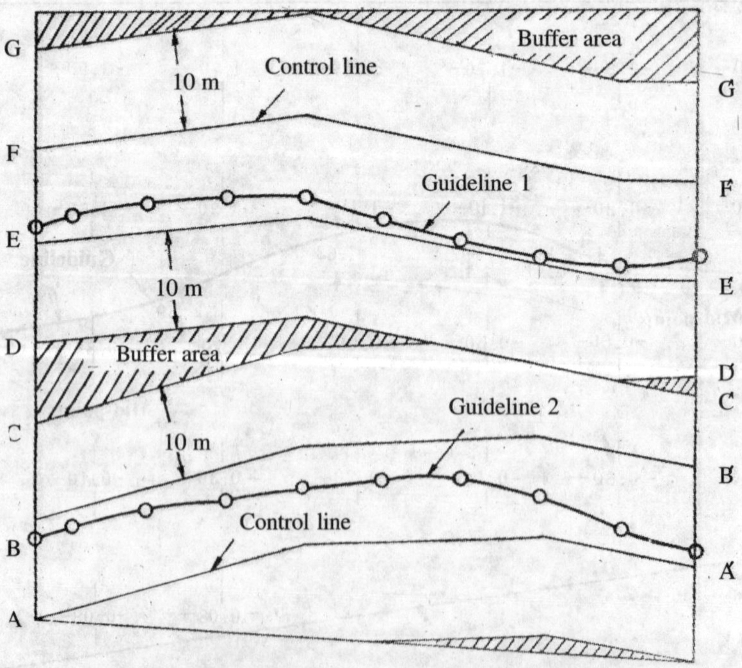

Fig. 8.19. Contour bench levelling layout.

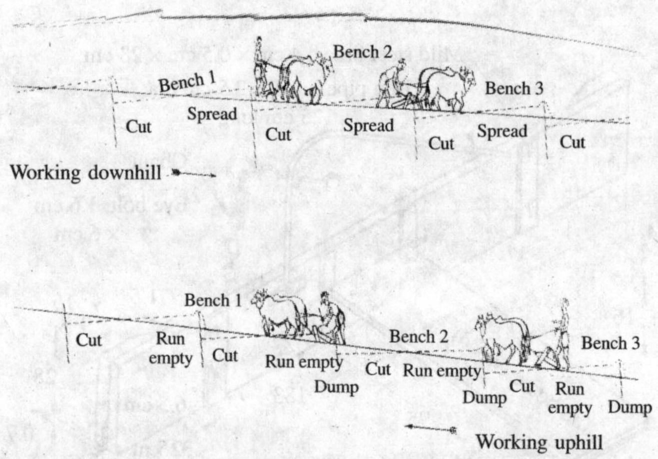

Fig. 8.20. Sequence of jobs in uphill and downhill operations in contour bench levelling.

of bench is assumed to be 10 m. The control line AA' is drawn nearly parallel to guideline 2. The control line is so shaped as to provide the minimum buffer area. The other line BB' is drawn nearly parallel to AA' and at a spacing of 10 m. Similarly, the other bench lines are drawn to cover the entire field and matching the guide lines in the respective key area. The unavoidable irregular portions, shown as buffer areas, are included in the corresponding bench while levelling the field.

Staking

The bench layout, as planned on the contour or grid map, must be transferred to the field. Usually one or two bench boundary lines are selected as control lines and these are located on the ground by horizontal measurement from the points of reference. If a grid survey is used for planning, the bench boundaries may be established by measurement from existing grid stakes.

Computation of Cuts and Fills

For the purpose of land levelling, each bench is regarded as a separate field. Any of the methods of land levelling design can be employed for computing cuts and fills. For levelling individual benches, allowance for shrinkage, dikes, irrigation and drainage channels and farm roads should be made. After computing the cuts and fills, the stakes are marked with a 'fill' on one side for the area above and a 'cut' on the other for the lower bench.

Contour Bench Construction

Contour benches can be constructed with animal-drawn or tractor-drawn equipment. The sequences of operation involved are illustrated in Fig. 8.20. In levelling land in contour benches, it is advantageous to scrape directly across at right angles to the benches. The steep slope is changed into benches by scraping the upper portion of each strip and filling the lower portion as the scraper is worked down the hill. On the return trip uphill (Fig. 8.20) the ridges are formed from the earth cut below and dumped exactly on the dike. Land smoothing is done by a wooden float or land leveller.

Provision for the disposal of rainfall runoff is essential at the downstream end of benches. A grass waterway, reinforced with drop or chute structures, is necessary to carry the runoff at non-erosive velocities.

EQUIPMENT FOR LAND GRADING AND FIELD LAYOUT

Land grading is done by tractor-drawn and animal-drawn equipment. Rough grading is done by bulldozers, tractor-drawn carrier-type scrapers, small scrapers drawn by tractor and by animal-drawn buck scrapers (Figs. 8.21 to 8.28). Land smoothing is done by tractor-drawn land planes or animal-drawn wooden floats (Figs. 8.29 to 8.33).

Both crawler and rubber-tired wheel type tractors are used for land grading. They vary in size from medium size farm tractors to heavy crawler tractors. The crawler tractor is frequently employed when heavy earth movement is involved. Its main distinguishing feature, the crawler track, gives superior traction on the varying soil and ground conditions encountered in land grading. It is more capable than wheel tractor in loading an attached scraper, and for cutting and pushing of earth when equipped with dozer blade. Since its speed is limited to about 8 km per hour it is not very suitable when long haul distances are involved. Rubber-tyred tractors are used for earth-moving, when the emphasis is on speed rather than drawbar pull. They can be used with advantage on land grading jobs requiring long distance hauling.

Fig. 8.21. A bulldozer in use in a major land levelling operation involving huge earth movement.

Fig. 8.22. Land levelling with 'terracer blade' mounted on a medium size wheel tractor.

Fig. 8.23. Carrier-type scraper drawn by a crawler tractor hauling earth in a land development project.

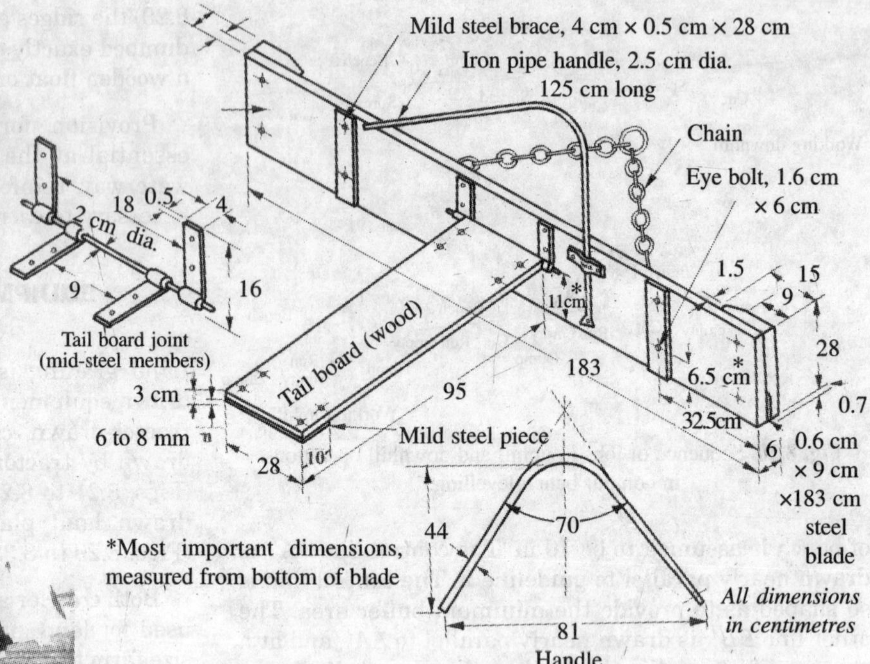

Fig. 8.24. Isometric view of an animal-drawn wooden buck scraper showing design details.

Bulldozers. Bulldozers, consisting of crawler tractors equipped with dozer blades, are frequently employed in cutting and pushing earth to short distances (Fig. 8.21). They are suitable for rough grading when the haul distance does not exceed about 25 metres.

Tractor-drawn scrapers. There are a variety of scrapers in land grading. They range in size from the terracer blade (Fig. 8.22) to heavy carrier-type scrapers (Fig. 8.23). The carrier-type scraper is widely used for large scale land grading operations. It consists of a bowl or bucket mounted on rubber-tyred wheels with a blade and apron across its front end for cutting, scooping and retaining soil. To load, the bowl is lowered and the apron is partly lifted. In hauling position the apron is closed and the bowl is lifted clear of the ground. To dump or spread, the apron is lifted and the load is pushed forward through the open end of the bowl by an ejector member which normally rests against the rear of the bowl. The machine cuts to grade, hauls the load for fairly long distances and spreads the soil evenly at the desired location. The capacity of carrier-type scrapers range from 1 to 2.5 cubic metres.

Wheeled scrapers and leveller blades are frequently used for medium and small scale levelling jobs (Fig. 8.23). They are also called 'bottomless scrapers'. As the name implies, the bowl or bucket has little or no bottom and earth movement is accomplished with them by scraping up a load, dragging it a short distance and dumping it. To operate these scrapers, it is necessary to loosen the ground with a plough or harrow.

Animal-drawn buck scraper. Animal-drawn buck scraper (Figs. 8.24 to 8.28) is the most efficient implement for land grading when animal power is to be used in the job and the fields are of small to medium size. It may be used to move the soil loosened by ploughing or other tillage practices. It can be operated satisfactorily when the haul distance does not exceed about 50 metres. The essential components of the buck scraper are the front board, the tail board, joints, handle and hitch. The implement incorporates a number of automatic features when constructed and operated properly. As indicated in Fig. 8.26, the most important dimensions of the implement are the location of the hinge point of the tail board at the back of the front board. The centre of hinge pin which should be exactly 11 cm from the bottom of the front board and the position of 'eye bolt' or hitch of the chain on the front board. The hitch points on either side are to be fixed 6.5 cm from the bottom of the front board.

Although the buck scraper is a simple implement, certain exact procedures must be followed to make it

Fig. 8.25. Buck scraper operation, Step 1: Placing the right foot of the operator on the tail end of the buck board. (As the animals move forward the buck board swings back and the implement comes to operating position.)

Fig. 8.26. Buck scraper operation (Step 2). The correct position of feet of the operator to load the implement. The operator places his right foot on the buck board and the left foot on the tail board with his body wieght acting on the buck board to load the implement with soil.

work effectively. To bring the scraper to loading position the operator approaches the tail board from the left and places his right foot at the end of the tail board as the scraper is pulled forward (Fig. 8.25). To load the scraper, the operator stands on the scraper by placing his one foot on the front board and the other foot on the tail board in such a way that his body weight acts directly over the cutting edge (Fig. 8.26). After loading, the operator can move back and walk beside the implement, since the scraper will usually carry a full load without the operator riding or holding the handle (Fig. 8.27). To unload, the handle is lifted and the load dumped (Fig. 8.28). Uniform spreading of the earth may also be done by slowly raising the handle which allows the soil load to be distributed in a thin layer.

Land plane. Land levelling jobs are finished by land planes, levellers and floats. The land plane (Fig. 8.29) is

Fig. 8.27. Buck scraper operation (Step 3). The operater steps out and walks beside the implement as the load of soil is transported to the point of "fill." (*There is no need to hold on to the handle during the transport of the soil.*)

Fig. 8.28. Buck scraper operation, Step 4: Unloading of the soil load at the place of "fill". The operator lifts the handle and throws the blade (buck board) to the foreground. The load of soil is dumped at the "fill site". (It is also possible to spread the soil uniformly over a distance by slowly lifting the handle upwards progressively till the load of soil is fully emptied.)

essentially a bottomless scraper provided with a long frame and mounted on four wheels. The blade is adjustable and is located at about the centre of the frame. The frame is at least 15 metres long and this great length makes it possible to finish a field to a uniform smooth surface. When in use, the blade is permanently set at a level which will maintain about

Fig. 8.29. A land plane in use for the final shaping (smoothening) of land to a uniform grade after the land levelling operation.

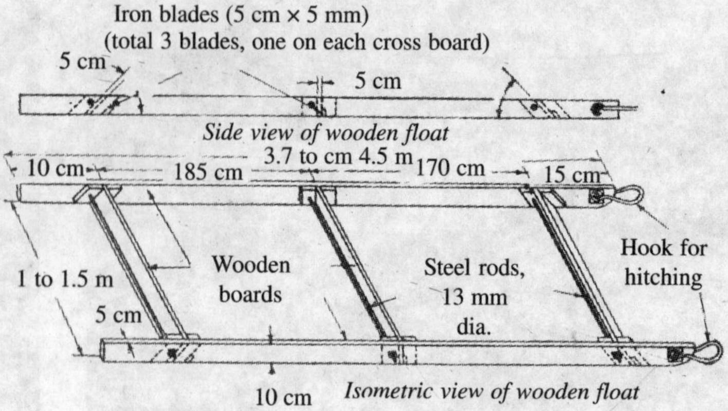

Fig. 8.31. Views of a sledge-type wooden float used for land smoothing with animal power.

Fig. 8.30. Rear view of a two-wheel automatic leveller.

Fig. 8.32. Land smoothing with an animal-drawn wooden float.

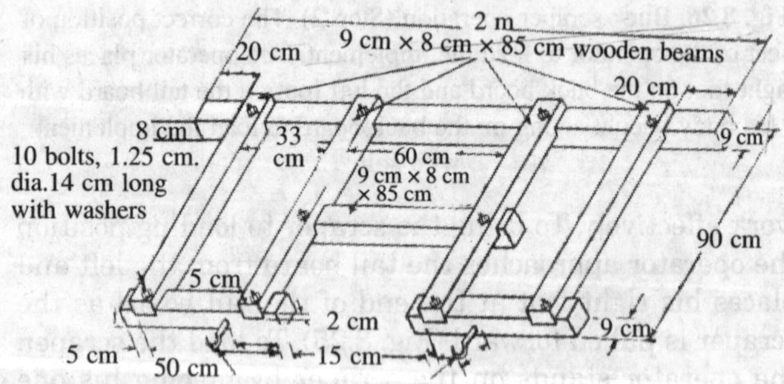

Fig. 8.33. Wooden U-leveller for smoothening of small fields.

one-third of the load in the bowl. With this adjustment properly made the machine will, as it is drawn across the field, automatically remove high spots and fill depressions. The machine is operated in diagonal directions and downfield. The machine, however, is suitable only to grade large fields.

Levellers. The two-wheeled automatic type leveller (Fig. 8.30) is used for the fine grading of small and medium size fields. It can be operated by a medium size tractor. The machine has an adjustable blade which is so constructed that it will drag a considerable volume of earth.

Wooden float. The float is a long sled-like drag (Fig. 8.31 and 8.32) which is used for land smoothing with animal power. Large size floats are sometimes used with tractors also. The machine operates on the principle of a carpenter's plane. It has three blades—the cutting blade in the front, the spreading blade in the centre and the covering blade at the back. As the float is pulled forward, the cutting blade removes the high spots and pushes this soil into the low areas ahead. The other blades assist in obtaining a uniform field surface. For best smoothing on flat fields, it is desirable to float the surface three or more times by going across the field lengthwise, crosswise and diagonally.

As the float is being drawn forward, there is a tendency of the front to be lifted slightly. This defect can be overcome by hitching the animals farther away from the float. It can also be partially overcome by having the operator stand on the front cross-bar.

A wooden U-leveller (Fig. 8.33) is a useful implement for smoothening small fields for check basin method of irrigation.

Fig. 8.34. An animal-drawn bund former. Note the adjustable collecting mould boards.

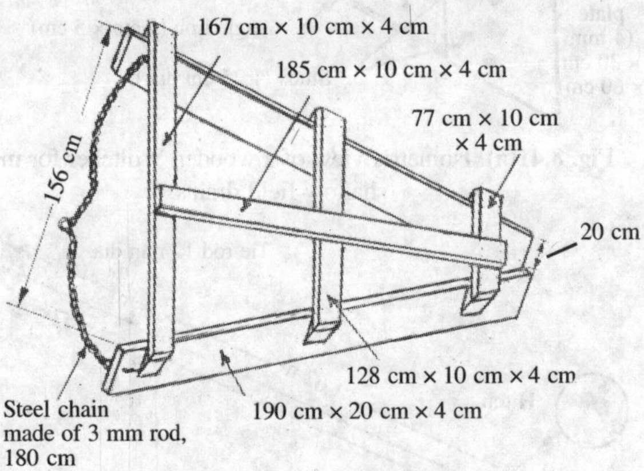

Note. All boards to be of good strong timber.

Fig. 8.35. View of a wooden A-frame ridger showing design dimensions.

Fig. 8.36. Making bund with an A-frame ridger.

Fig. 8.37. Border ridge formed with a tractor-drawn disc ridger.

Equipment for Making Ridges and Ditches

Layout of fields for efficient irrigation requires the construction of ridges for borders and check basins and ditches for field channels and field drains.

Equipment for ridge making. Ridges at the outer boundaries of irrigation border strips and check basins are prepared with animal-drawn bund former, A-frame ridger, and tractor-drawn disc ridger. All these implements work on well prepared soils.

The bund former (Fig. 8.34) is suitable for making small ridges suitable for check basins. The implement has two collecting mould boards which form the ridge when the implement is pulled forward.

The wooden A-frame ridger (Fig. 8.35) is a suitable implement for making ridges for border strips and check basins. The ridger is similar in principle to the bund former. However, it gathers soil from a larger area and can make larger bunds (Fig. 8.36). The operator usually rides on the ridger, thus applying his weight to get a more effective job. The operator has a choice of standing positions on the cross planks on top of the ridger. The largest ridge is formed when the operator stands on the cross plank nearest to the animals.

The tractor-drawn disc ridger (Fig. 8.37) is an effective implement for making border ridges. It consists of four discs in two gangs so assembled as to gather the soil into a ridge as the implement is pulled forward. The disc gangs can be moved in or out on the tool bar to make larger or smaller ridges.

Making furrows. Animal-drawn lister or middle breaker (Fig. 8.35) and the tractor-drawn multi-row furrowers are commonly used for making irrigation furrows. While using the single-row animal-drawn lister, the rows are spaced uniformly by making the

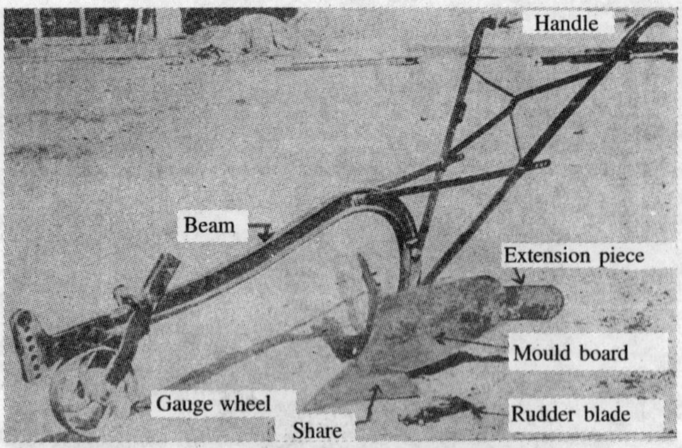

Fig. 8.38. An animal-drawn lister (double mould board plough). Extension pieces fixed to the mould boards are used for making wide furrows. Note the rudder blade placed beside the lister which could be fixed to the lister bottom to ensure steady movement of the machine.

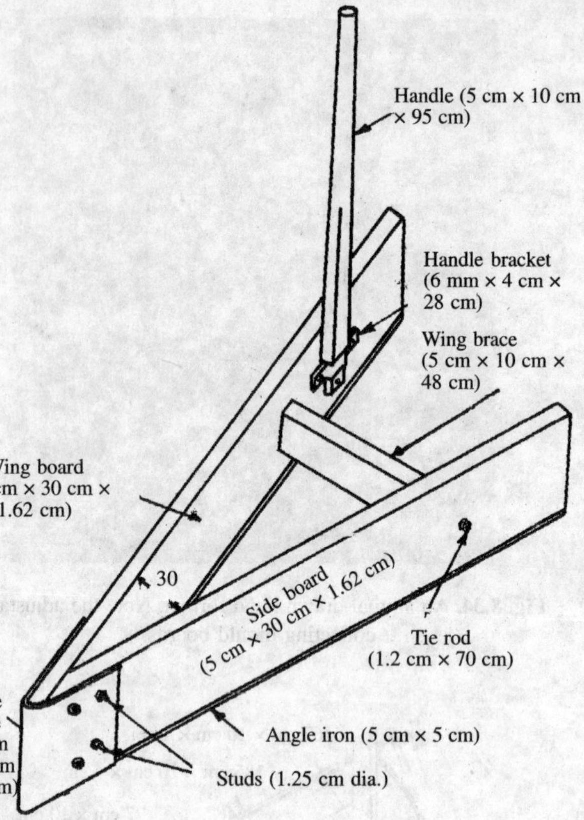

Fig. 8.41(a). Isometric view of a wooden V-ditcher for making shallow field drains.

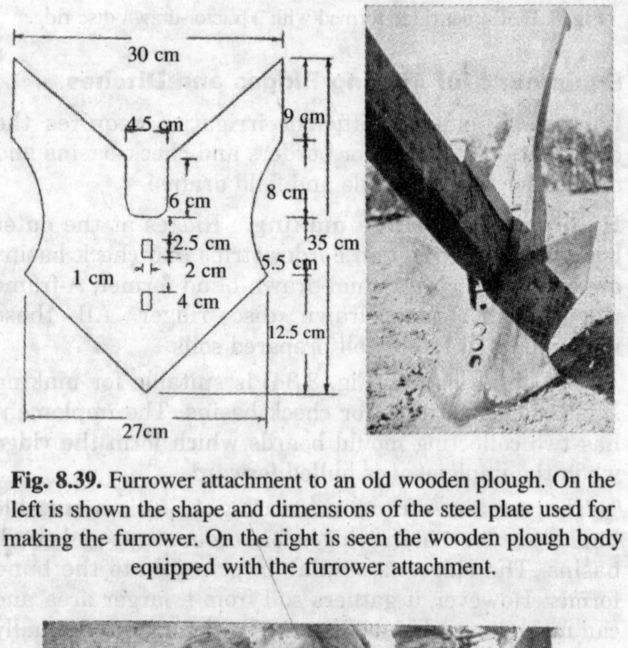

Fig. 8.39. Furrower attachment to an old wooden plough. On the left is shown the shape and dimensions of the steel plate used for making the furrower. On the right is seen the wooden plough body equipped with the furrower attachment.

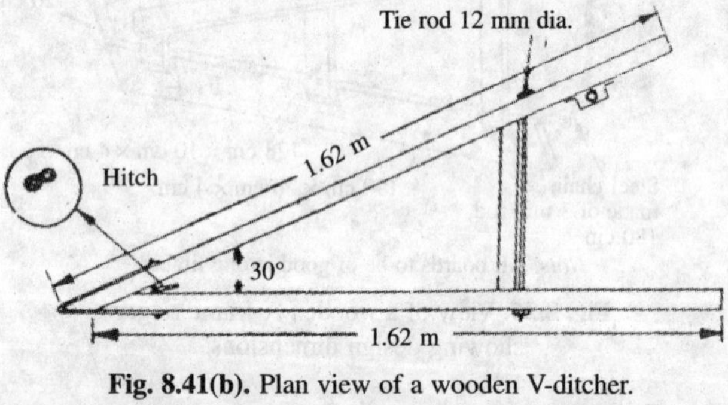

Fig. 8.41(b). Plan view of a wooden V-ditcher.

Fig. 8.40. A two-row lister fixed to the frame of a tractor-drawn tiller making deep furrows.

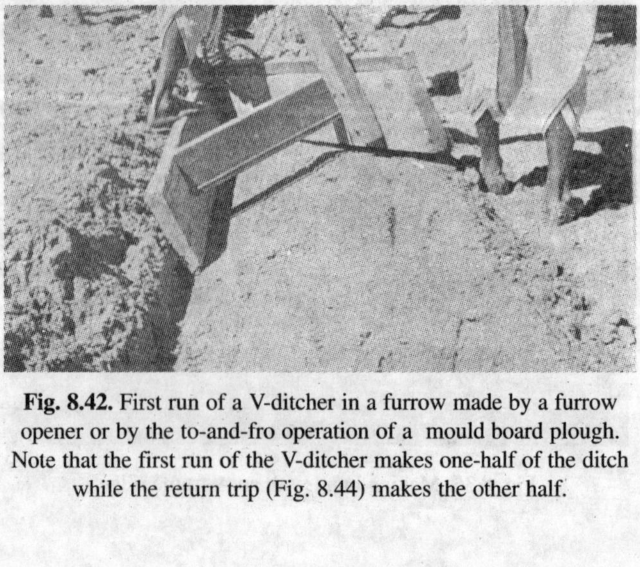

Fig. 8.42. First run of a V-ditcher in a furrow made by a furrow opener or by the to-and-fro operation of a mould board plough. Note that the first run of the V-ditcher makes one-half of the ditch while the return trip (Fig. 8.44) makes the other half.

Fig. 8.43. Construction of field drain with an animal-drawn V-ditcher.

Fig. 8.44. An Eversman D7 ditcher used with 40–50 h.p. wheel tractor. The ditcher is capable of construction of field drains of about 2 m width and 90 cm depth.

(***Source:*** Bligh, F.L., The Eversman Manufacturing Co., Denver, Colorado, USA.)

Fig. 8.45. A dragline excavator equipped with V-shaped ditcher bottom used to construct a drainage canal. The same equipment with appropriate accessories can be used for desilting and cleaning of existing drains.

Fig. 8.46. A dragling excavator with bucket type cutter head cleaning a drainage canal. Note the spoil dumped on the canal bank.

animals walk in the previously laid furrow. A yoke of the exact length for proper row spacing is essential. An alternative to the conventional lister is a steel furrower attached to a wooden plough bottom (Fig. 8.39).

Tractor cultivators may be equipped with furrower bottoms to make irrigation furrows (Fig. 8.40). Two-row and four-row furrowers are commonly used with medium size tractors.

Making field channels. The V-ditcher (Figs. 8.41(a) and (b) to 8.43) is a useful implement for the construction or cleaning of field channels for irrigation and drainage. The operation of the V-ditcher in constructing a channel is much like that of an over-size mould board plough. The channel line is first laid out. Then a mould board plough is used to make a furrow where the line has been staked. The plough is run again on the same furrow, but in the opposite direction, throwing the soil out on the other side. The V-ditcher is then used in the furrow to enlarge it into a channel (Figs. 8.42 and 8.43). Tractor operated wheel type ditchers can be used to make larger ditchers (Fig. 8.44).

Dragline excavators equipped with V-shaped cutter blades (Fig. 8.45) and tractor-mounted digger loaders are frequently employed for constructing large size irrigation and drainage canals. Dragline excavators with bucket type cutter heads are used both for digging canals and for cleaning them (Fig. 8.46). Self-propelled trenching machines are used for making trenches for laying tile lines and underground pipeline irrigation systems.

The choice of equipment for land grading and field layout depends on the quantity of earthwork involved, type of power available, economic feasibility of the machine and the size of the farm. While animal-drawn equipments are suitable for small jobs, heavy earth moving equipment is needed for major works involving large areas.

SUMMARY

The need for appropriate land grading for surface methods of irrigation water application and soil and water conservation in unirrigated and irrigated areas are highlighted. The three distinct operations in land levelling, viz., rough grading, land levelling and land smoothening are identified. Criteria for land levelling, including soil profile considerations, land slope, rainfall characteristics, cropping patterns and irrigation methods to be used are explained. Layout of fields and on-farm irrigation and drainage systems and farm roads are described. Land grading procedures, including survey and staking, location of grid points and preparation of contour maps are explained. Land levelling design methods, including plane method, profile method, plan inspection method and contour adjustment method are described in detail. Procedures for determination of the centroid of fields, plane of best fit, design elevations of grid points, cuts, fills, cut-fill ratios and computation of earthwork quantities are explained. Land levelling procedures, including job planning, setting grade stakes, equipment operations, preservation of top soil, maintenance of construction tolerances, land smoothening and checking of the field job after completing the levelling operation are described. The need for maintenance of field grades, especially through land smoothening operations as a regular seed bed preparation practice is highlighted. Adaptability of contour bench levelling in undulating lands and fields with moderate to steeper slopes, including design and layout of contour benches and specific steps in equipment operation for constructing contour benches are presented. Adaptability and application of animal-drawn and mechanically operated equipment for land grading and field layout are explained. Design, construction and operation of animal-drawn equipment in land levelling, land smoothening, and preparation of ridges and ditches are described. Adaptability and general principles of construction and operation of mechanically operated machinery for earth moving, land grading and field channel construction and maintenance are presented.

PROBLEMS

8.1. A topographic survey of a field was done for planning land levelling. Grid points were selected at intervals of 25 m. The elevations of the points are given below:

Stations	Elevation *m*					
	Line No. 1	*Line No. 2*	*Line No. 3*	*Line No. 4*	*Line No. 5*	*Line No. 6*	*Line No. 7*
A	8.41	9.75	8.27	9.38	—	—	—
B	9.06	9.35	8.68	8.79	9.01	—	—
	No. 1	*No. 2*	*No. 3*	*No. 4*	*No. 5*	*No. 6*	*No. 7*
C	8.75	9.97	9.13	8.89	8.80	9.11	—
D	8.89	9.12	8.78	8.78	7.97	8.37	8.56
E	8.95	9.10	8.98	9.07	9.17	8.77	9.14
F	9.21	8.95	7.98	9.13	8.97	8.87	8.12

Calculate the position of the centroid of the field and the average elevation of the field. The field is to have a downfield slope of 0.3%. Determine the formation levels at grid points and the amount of cut or fill at each grid point.

(**Ans.** Position of centroid 77.75 m from left side boundary and 67.25 ma frombottom boundary of field. Average elevation of the field is 8.867 m)

Formation levels at grid points, 0.3% slope

	1	2	3	4	5	6	7
A	9.063	8.988	8.913	8.838	—	—	—
B	9.063	8.988	8.913	8.838	8.763	—	—
C	9.063	8.988	8.913	8.838	8.763	8.688	—
D	9.063	8.988	8.913	8.838	8.763	8.688	8.613
E	9.063	8.988	8.913	8.838	8.763	8.688	8.613
F	9.063	8.988	8.913	8.838	8.763	8.688	8.613

Cuts and fills

	1	2	3	4	5	6	7
A	+ 0.653	− 0.762	+ 0.643	− 0.542	—	—	—
B	+ 0.003	− 0.362	+ 0.233	+ 0.048	− 0.247	—	—
C	+ 0.313	+ 0.018	− 0.217	+ 0.052	− 0.037	− 0.422	—
D	+ 0.173	− 0.132	+ 0.133	+ 0.058	+ 0.793	+ 0.318	− 0.053
E	+ 0.113	− 0.112	− 0.067	− 0.232	− 0.407	− 0.082	− 0.527
F	− 0.147	− 0.038	+ 0.933	− 0.292	− 0.207	− 0.182	+ 0.493

(+ sign indicates fill and – sign indicates cut)

Check: Σ cut = 5.028,

Σ fill = 5.016

(***Note:*** The minor difference in the amounts of cut and fill is due to the rounding of digits at the third place of decimal.)

8.2. Draw the base map of a field *a b c d e*, based on the following data:

ab	= 213 m,	∠*eah*	= 90°
bc	= 91.5 m,	∠*abc*	= 110°
cd	= 130 m,	∠*bcd*	= 122°
de	= 170 m,	∠*cde*	= 132°

The side *ea* lies in the east-west direction. The field is laid out in grid pattern on a grid spacing of 30 m. Row lines *A*, *B*, *C*, *D*, *E*, *F*, and *G*, are drawn parallel to the field boundary *abcde*. Column lines 1, 2, 3, 4, 5, 6 and 7 are drawn perpendicular to the row lines. Line *A* is 15 m from the field boundary *ae*. Column 1 is 15 m to the right of the field boundary *ab*. The elevations of the grid point are as follows:

	1	*2*	*3*	*4*	*5*	*6*	*7*
A	3.22	3.24	3.27	3.31	3.35	3.34	3.40
B	3.10	3.12	3.15	3.20	3.25	3.27	3.30
C	3.03	3.03	3.08	3.10	3.13	3.15	3.15
D	2.85	2.90	2.92	2.95	2.98	3.10	3.05
E	2.75	2.80	2.85	2.85	2.90	2.92	2.95
F	2.60	2.65	2.68	2.72	2.75	2.78	2.80
G	2.50	2.52	2.55	2.58	2.60	2.64	—
H	—	2.40	2.45	2.40	—	—	—

A farm road is to be constructed inside the field along the boundary line *ae*. The top width of the road is 4 m and bottom width 4.8 m. The average height of the road embankment is 40 cm.

An irrigation channel is to be laid beside the road, all along its length, using the road embankment as one of the side embankments of the channel. The channel is trapezoidal in cross-section and is 40 cm deep. The side slopes of the inside section of the channel and of the channel embankment are each 1 : 1. The bottom width of the channel is 25 cm.

The land is to be graded to a downfield slope of 0.2% and cross-slope of 0.01%. Earth fills for the road and irrigation channel are to be obtained from the field itself. A shrinkage allowance of 1.5 cm may be allowed.

Using the plane method of land levelling design determine the following:

(*i*) Area of field in hectares.
(*ii*) Average elevation of the grid points.
(*iii*) Location of the centroid.
(*iv*) Slopes of the planes of best fit in *x* and *y* directions.
(*v*) Design elevation of centroid.
(*vi*) Design elevations of the grid points.
(*vii*) Cuts and fills at grid points.
(*viii*) Cut-fill ratio.

(**Ans.** (*i*) 4.48335, (*ii*) 2.933 m, (*iii*) 101.47 m, 119.70 m (*iv*) S_x = 0.0007 per cent and S_y = – 0.414 per cent, (*v*) 2.909 m)

(*vi*) *Design elevations of grid points*

	1	*2*	*3*	*4*	*5*	*6*	*7*
A	3.11	3.12	3.12	3.12	3.12	3.13	3.13
B	3.05	3.06	3.06	3.06	3.06	3.07	3.07
C	2.99	3.00	3.00	3.00	3.00	3.01	3.01
D	2.93	2.93	2.94	2.94	2.94	2.95	2.95
E	2.87	2.87	2.88	2.88	2.88	2.89	2.89
F	2.81	2.81	2.82	2.82	2.82	2.83	2.83
G	2.75	2.75	2.76	2.76	2.76	2.77	—
H	—	2.69	2.70	2.70	—	—	—

(*vii*) *Cut and fill at grid points*

	1	*2*	*3*	*4*	*5*	*6*	*7*
A	– 0.11	– 0.12	– 0.15	– 0.19	– 0.23	– 0.21	– 0.27
B	– 0.05	– 0.06	– 009	– 0.14	– 0.19	– 0.20	– 0.23
C	– 0.04	– 0.03	– 0.08	– 0.10	– 0.13	– 0.14	– 0.14
D	+ 0.08	+ 0.03	+ 0.02	– 0.01	– 0.04	– 0.15	– 0.15
E	+ 0.12	+ 0.07	+ 0.03	+ 0.03	– 0.02	– 0.02	– 0.06
F	+ 0.21	+ 0.16	+ 0.14	+ 0.10	+ 0.07	+ 0.05	+ 0.03
G	+ 0.25	+ 0.23	+ 0.23	+ 0.18	+ 0.16	+ 0.13	—
H	—	+ 0.29	+ 0.25	+ 0.30	—	—	—

Σ *cut* = 3.33

Σ *fill* = 3.12

(*viii*) 1.067

8.3. Compute the earthwork volumes in cut and fill, based on the land levelling design shown in Figure 8.9. Determine the cut-fill ration.

(**Ans.** Σ volume cut = 1452.434 m^3,
Σ volume fill = 1451.345 m^2,
cut-fill ratio = 1.001)

8.4. Determine a suitable top width and vertical interval for a contour bench for the following conditions: the maximum width of the implement to be used is 2.5 m; the land slope is 3%; the side slope of the dike is 1½:1 and top width of the dike is 30 cm. The maximum depth of irrigation stream is 15 cm and that of intense storm is 25 cm.

(**Ans.** Overall width of bench, *T* = 12.24 m
Vertical interval, *H* = 0.37 m)

QUESTIONS

1. What are the factors influencing land levelling requirements?
2. Distinguish between land levelling and land smoothening with reference to equipments and procedures.
3. How do the soil profile influence the extent of cuts and fills in a land levelling job?
4. How are the harmful effects of deep cut in shallow soils mitigated in a land levelling job?

5. What are the different operations in land clearing?
6. Define bench levelling. Under what topographic situations the practice is necessary?
7. What is 'grubbing' in a land levelling job?
8. What are the specific contour patterns which warrant sub-division of land for successful irrigation farming?
9. How is the elevation of the centroid of a field determined?
10. Define 'slope of the plane of best fit'. How is it determined?
11. How are the formation levels computed in a land levelling job?
12. What is 'cut-fill ratio'? What are the common values of cut-fill ratios in the land levelling job of an area with average loamy soil?
13. Why is 'plane method' of land levelling design preferred in small scale levelling jobs?
14. Define formation level. How is it determined?
15. How are the earthworks for road fills and excavations for channels accounted in land levelling design?
16. When is 'profile method' adopted in land levelling design?
17. What is the scope of 'plan inspection method' of land levelling design?
18. What is the specific situation where 'contour adjustment method' of land levelling design is suitable?
19. How are cuts and fills distinguished while setting stakes for land levelling?
20. What is 'blue top levelling'?
21. How is the width of farmable area determined in bench levelling?
22. How are contour benches located in a contour bench levelling job?
23. Describe the step-by-step procedures in contour bench levelling using an animal-drawn buck scraper.
24. What are the step-by-step procedures in the operation of an animal-drawn wooden buck scraper?
25. What is a land plane? How does it function in a land levelling job?
26. What is the principle of operation of an animal-drawn wooden float?
27. How is the centroid of a field with irregular boundaries determined?
28. Distinguish between end area method and prismoidal formula for computing earthwork quantities in land grading.
29. What is construction tolerance in land grading? What are the permissible levels of tolerances usually adopted?
30. What are 'key areas' in contour bench levelling design?
31. What are the usually adopted locations for setting grade stakes marking cuts and fills in a land levelling job?
32. What are the basic principles adopted in the design of contour benches?
33. What is the basis for determining the longitudinal slope of a contour bench?
34. What are the measures required to prevent soil erosion at the tail-end of a contour bench where there is a concentration of flow?
35. What are the major points to be considered in locating contour benches?
36. What are the limitations in the use of bull dozers and tractor-drawn scrapers in small scale land levelling jobs?
37. What are the common equipments used in completing a land levelling job? What are the relative advantages of each?
38. What is the principle of working of a wooden float? What are the specific operating principles of each of the three blades in a wooden float?
39. What are the common machines used for making ridges in a well prepared seed bed? What are the relative advantages of each?
40. What is the principle of working of a V-ditcher?
41. Identify the True and False statements from the following:
    (*i*) In a land levelling job the volume of cut is more than the volume of fill.
    (*ii*) The vertical interval between adjacent benches in contour bench levelling is kept at 1 m interval to facilitate proper maintenance of the bund.
    (*iii*) The usual practice in a contour bench levelling job is to keep the top width of the dyke equal to the vertical interval between benches.
    (*iv*) The width of the farmable area of a bench in contour bench levelling is a multiple of the widest implement used in farming operations.
    (*v*) A common cause of breach of contour benches at its tail-end is the non-provision of erosion control measures / structures.
    (*vi*) Rubber-tyred wheel-type tractors equipped with scrapers are inefficient when applied to land levelling jobs.
    (*vii*) Carrier-type scrapers drawn by crawler-type tractors have limited application in small scale land levelling jobs.

(*viii*) A wooden float used in land smoothening operation works on the principle of a carpenter's plane.

(*ix*) The most important dimensions of an animal-drawn wooden buck scraper which are to be exactly adhered to are: (*i*) the centre of the hinge pin on the back of the front board should be 11 cm above its bottom, and (*ii*) the location of the 'eye bolt' or the hitch of the chain on either side of the front board are fixed at 6.5 cm from the bottom of the board at its front face.

(*x*) An animal-drawn V-ditcher is capable of making a field channel in a single run.

(*xi*) By computing the distance to the centroid from two lines of reference perpendicular to each other, the exact point of the centroid of a field can be determined.

(*xii*) Plan inspection method of land grading design is largely a trial and error method and is adapted to moderate to flat land slopes.

(*xiii*) The operation of a land plane or float is carried out along the downfield slope of the field.

(*xiv*) A mechanically operated carrier-type scraper mounted on rubber-tyred wheels is able to cut, load, and carry the soil to comparatively long distances.

(*xv*) For the proper operation of wheeled scrapers and leveller blades, it is necessary to loosen the ground with a plough or harrow.

(*xvi*) The tendency of the wooden float to lift up when it is being pulled forward can be overcome by hitching the animals farther away from the float or loading its front by the weight of the operator who stands on the front blade.

42. Identify the most appropriate answer to the following questions from the four options:

(*i*) The recommended safe limits of land slope in loamy soils range from:

(a) 0.01 to 0.05% (b) 0.05 to 0.10%
(c) 0.10 to 0.2% (d) 0.10 to 0.5%

(*ii*) The side slope of the dyke in a contour bench levelling job is kept about:

(a) 1:1 (b) 1.5:1 (c) 2:1 (d) 2.5:1

(*iii*) The most efficient equipment for making ridges in the layout of fields for irrigation borders and large size check basins is:

(a) bund former (b) A-frame ridger
(c) disc ridger
(d) double mould board plough

(*iv*) The most commonly adopted land levelling design in small scale jobs is:

(a) contour adjustment method
(b) profile method (c) plane method
(d) plan inspection method

(*v*) The most appropriate wooden equipment for land smoothening operation is:

(a) leveller board (b) buck scraper
(c) float (d) U-leveller

(*vi*) The most appropriate mechanically operated equipment for medium scale land grading and smoothening is:

(a) terracer blade (b) land plane
(c) two-wheeled automatic leveller
(d) carrier-type scraper

(*vii*) In a land plane the scraper blade is provided at:

(a) the front of the frame
(b) about the centre of the frame
(c) the back of the frame
(d) one-third distance from the front of the frame

43. Rearrange the events in the following procedures in their proper sequences:

(*i*) The proper procedures in the levelling operation of a field:

(a) Providing for construction tolerance
(b) Preservation of top soil
(c) Development of a construction map
(d) Checking the completed job
(e) Finishing the job

(*ii*) The procedure for the location of contour benches in the contour map of a field comprise of:

(a) Location of guidelines to represent specific average conditions of a key area
(b) Dividing the whole field area into key areas having uniform topographic features
(c) Identifying two contour lines forming the boundary of a key area
(d) Location of reference lines to the contour lines within a key area
(e) Computation of average land slope along the random lines, the extent of fall required and the horizontal adjustment for locating guide-lines
(f) Establishing control lines for contour bench levelling.

(*iii*) The procedures for the operation of a wheeled scraper in land grading comprise of:

(a) Scooping of soil (b) Cutting of soil
(c) Dumping and spreading of soil
(d) Hauling of soil load

(*iv*) The exact procedures which are to be observed in the operation of an animal-drawn buck scraper are:

(a) To load the scraper the operator stands on it by placing his one foot on the front board and the other on the tail board in such a way that his body weight acts directly on the cutting edge.
(b) To bring the buck scraper to loading position the operator approaches the tail board from the left and places his right foot at the end of the tail board, as the scraper is pulled forward.
(c) To unload, the handle is lifted and the load dumped or spread.
(d) After loading, the operator moves out and walks beside the implement.

(*v*) The procedures for the construction of field channels using an animal-drawn V-ditcher comprise of:

(a) Using a mould board plough to make a furrow along the line of the channel

(b) Running the plough again on the furrow in the opposite direction, throwing the soil in that direction

(c) Laying out the channel line in the field

(d) Using the V-ditcher in the furrow in both the directions to enlarge the furrow into a field channel.

44. In the following; match each of the entries in the 1st column with the most closely linked entry in the 2nd column:

(*i*)	(A) Centroid of the field	(I) Elevation after land grading operation	
	(B) Plane of best fit	(II) Settlement allowance	
	(C) Formation level	(III) Smoothening steep lands	
	(D) Cut-off ratio	(IV) Least square method	
	(E) Contour adjustment method	(V) Plane method of land levelling design	
(*ii*)	(A) End-area method	(I) Estimation of earthwork volume	
	(B) Four-point method	(II) Use of differential colour codes in setting grade stakes	
	(C) Prismoidal formula	(III) Contour bench levelling	
	(D) Blue top levelling	(IV) Computation of volume of earthwork in land levelling	
	(E) Guidelines	(V) Earthwork volume between adjacent profiles	
(*iii*)	(A) Carrier-type scraper	(I) Finishing land levelling job	
	(B) Critical dimensions	(II) Construction of bunds	
	(C) Land plane	(III) Sled-like drag	
	(D) Disc ridger	(IV) Long distance hauling of earth	
	(E) Wooden float	(V) Animal-operated buck scraper.	

FURTHER READING REFERENCES

Gandhi, R.T., *et al.* (1971). *Handbook of irrigation water management.* Water Management Div., Dept. of Agr., Mini. of Agri., Govt. of India, New Delhi, pp. VII–1–27.

Kellog. F.H. (1956). *Construction methods and machinery.* Prentice Hall, Inc., New York, p. 415.

Marr, J.C. (1957). *Grading land for surface irrigation.* Uni. of California, Agrl. Expt. Stn. Cir.–438, p. 53.

Michael, A.M. and Knierim, G.C. (1965). *Cultivation practices in irrigation farming.* University of Udaipur, College of Agriculture, Jobner, Extn. Bull. No. 3, p. 32.

Michael, A.M., Knierim, G.C. and Reeser, R.M. (1964). *Simple bullock-drawn implements for efficient irrigation farming.* University of Udaipur, College of Agriculture, Jobner, Extn. Bul. No. 1, p. 16.

Michael, A.M. and Ojha, T.P. (2003). *Principles of Agricultural Engineering,* Vol. II, Jain Bros, Karol Bagh, New Delhi, p. 888.

Ministry of Irrigation and Power (1951). *Handbook of earth moving machinery.* Central Water and Power Commission, Govt. of India, p. 258.

Phelan, J.T. (1961). *Land levelling and grading—in Agricultural Engineer's Handbook,* edited by Richey, C.B. *et al.* McGraw Hill Book Co., N.Y., pp. 347–355.

United States Department of Agriculture (1961). *Irrigation,* Chapter 12, *Land levelling.* Section 15, SCS National Engineering Handbook. U.S. Dept. of Agri., SCS, Washington, D.C., p. 58.

Chapter

9

Soil-Plant-Water Relationships

LEARNING OBJECTIVES

Importance of soil-plant-water relationships in sustainable agricultural production. Basic information on soils, their origin, classification, and components. Soils properties influencing irrigation management. Mechanical composition of soils. Soil—physical and chemical properties. Soil structure and its management. Volume and mass relationships of soil constituents. Soil profile characteristics. Soil resources survey and classification. Soils resources of India. Soil taxonomy. Major soil orders, their characteristics and distribution. Crop production technology in Alfisols and associated soils of Semi-Arid Tropics. Vertisols and associated soils, their distribution, characteristics and management. Processes of soil degradation. Extent of soil degradation in India. Management of degraded lands. Soil fertility and its management. Macro-elements and their availability in the soils of India. Methods of increasing fertilizer use efficiency. Micro-nutrients and other beneficial elements in soils. Soil biological fertility. Water relations of soils. Kinds of soil water. Movement of water into and within soils. Infiltration characteristics of soils and their measurement. Soil moisture retention and movement. Viscosity, surface tension and capillarity. Soil moisture tension. Soil water potential. Soil moisture characteristics and constants. Soils moisture movement under saturated and unsaturated conditions. Water vapour movement. Measurement of soil moisture. Soil moisture potential and its measurement. Plant water relationships. Plant structure. Water movement along soil-plant-atmospheric system. Moisture stress and plant response. Drought tolerance of plants. Movement of water and nutrients in plants. Rooting characteristics and moisture use of crops.

Soil-plant-water relationships relate to the properties of soil and plants that affect the movement, retention and use of water. Soil provides the room for water to be used by plants through the roots present in the same medium. Water, as such and also as a carrier of large amount of nutrients, is required in a large measure for the successful growth of crops. Due to inadequate and/or uneven distribution of rainfall during the growth span of a crop, it becomes essential to apply additional water to the soil for plant use in the form of irrigation. The rate of entry of water into the soil and its retention, movement and availability to plant roots are all physical phenomena. Hence, it is important to know the physical properties of soils in relation to water for efficient management of irrigated agriculture.

Soil

The soil is a granular porous medium. It is a natural body of mineral and organic constituents differentiated into horizons of variable depths. It differs from the material below in morphology, physical make-up, chemical properties and composition as well as biological characteristics. In agricultural science the term soil is applied only to the upper part of the weathered rock penetrated by the roots of plants. Soil is a store-house of plant nutrients, a habitat for soil micro-organisms, an anchorage for plants, and a reservoir that holds the water needed for plant growth. Soil is a natural body of mineral and organic constituents. In the traditional concepts, soil is considered as a natural medium for plant growth. The concept of soil as an independent natural body was introduced towards the end of the 19th century. Accordingly, soil is defined as an independent natural body differentiated into horizons of layers of various depths, each with a unique morphology resulting from the combination of climate, living matter, parent rock material, relief and time. The modern scientific definition of soil is that it is a collection of natural bodies on the earth's surface containing living matter and supporting or capable of supporting plants. The word *horizon* is not included in the modern concept of soil because all soils do not have distinct horizons. However, the presence or absence of the horizons is important in soil classification. The nature, kind and arrangements of soil horizons in a profile have significant influence on soil-water-plant relationships.

Definitions

Soil profile is a vertical section of the soil from the surface through all its horizons.

Pedology is the science which deals with the origin, fundamental nature and classification of soils.

Pedosphere is the outer soil layer of the earth in and on which organic life exists.

Soil survey is the systematic classification of soils in an area. Standard procedure for soil surveys and their classification are available in USDA (1975).

Soil class is a group of soils which has a definite range of properties, or a group in relation to texture, structure, acidity, drainage properties and such other characteristics.

Soil classification is the systematic grouping of soils based on profile characteristics such as thickness, colour and structure of different horizons, drainage and parent materials.

Soil map depicts the distribution of soil types or soil mapping units in relation to the prominent physical and cultural features of the earth's surface.

Soil microbiology deals with the study of micro-organisms in the soil such as bacteria, algae, fungi and protozoa.

Soil morphology deals with soil and arrangements of horizons or separate natural layers in a soil profile. It is the index for identifying the genetic processes in soil development attributing to unique properties of the soil.

Soil genesis is the mode of origin of the soil, particularly the processes which are responsible for the development of the upper part of the soil profile (solum) from the parent material.

Soil type denotes a group of soils which have genetic horizons of similar characteristics (including texture and arrangement in the soil profile) and developed from a particular type of parent material. In a taxonomic classification it is a sub-division based primarily on soil texture.

Soil tilth denotes the soil structure or soil physical conditions favourable for the germination and growth of crop plants. It is developed by proper tillage and soil building and management practices.

Soil series is a group of soils closely similar in properties, except the texture of the surface soil. It expresses soil profile characteristics such as colour, structure and thickness of horizons, type of parent material and drainage.

Soil depth is the potential depth of the root zone of crops in a soil profile.

Soil deterioration is the loss of potential productive capacity of a soil resulting from destructive processes accelerated by activities of man, e.g., soil erosion, waterlogging and accumulation of salts.

Soil degradation is the result of excessive leaching, bringing about replacement of sodium by hydrogen.

Soil building is the formation of soil from parent material by natural processes. It also refers to increasing the productivity of soil by manuring, fertilizing, growing of soil building crops, and soil conservation measures.

Soil building crops are those which have a high capacity for adding nutrients, especially legumes which have nodules of nitrogen fixing bacteria in their roots.

Soil depleting crops tend to deplote the nutrients and organic matter in the soil and induce deterioration of the soil structure.

Soil climate denotes the moisture, air and temperature existing within the soil.

Soil colloids comprise of inorganic and organic matter of very small size having relatively large surface area per unit of mass.

Soil conditioners are synthetic chemicals which can stabilise soil aggregates and improve soil structure or tilth.

Soil aeration is the movement of air into the soil as water is drained out. It is the process by which air and other gases in the soil are removed. The rate of aeration depends largely on the size and number of soil pores and the soil moisture content. Large pores facilitate aeration. Waterlogging and continued high levels of soil moisture content restrict aeration, resulting in unfavourable soil physical condition for plant growth. Restricted soil aeration is a major limiting factor in the development of extensive root systems in crop plants. Restricted aeration impairs the essential process of respiration of an established root system which retards water and nutrient absorption by plant roots. It also adversely influences the essential biological processes in the soil. Hence, it is important to know how much air the soil contains, the composition of the air and the method of renewal of soil air.

Base exchange capacity, also called exchange capacity, is a measure of the adsorptive capacity of soil for bases. A soil with a high exchange capacity will retain more plant nutrients and will be less subject to leaching, as compared to the one with a low exchange capacity.

SOIL PROPERTIES INFLUENCING IRRIGATION MANAGEMENT

The amount of water a soil can hold in available form for plant use is determined by its physical properties. This amount determines the length of time a plant can survive without water being added. It determines both the frequency of irrigation and the capacity of irrigation system needed to ensure continuous crop growth.

Soil is a three-phase system comprising of the solid phase made of mineral and organic matter and various chemical compounds, the liquid phase called the *soil moisture* and the gaseous phase called the *soil air.* The main component of the solid phase is the soil particles, the size and shape of which give rise to pore spaces of different geometry. These pore spaces are filled with water and air in varying proportions, depending on the amount of moisture present. The presence of solid particles, liquid (soil solution) and gas (soil air) constitute a complex polyphasic system. The volume composition of the three main constituents in the soil system varies widely. In addition to the three basic components, soil usually contains numerous living organisms such as bacteria (the most dominant), micro-organisms, fungi, algae, protozoa, insects and small animals which directly or indirectly affect soil structure and plant growth. The most important soil properties influencing irrigation are its infiltration characteristics and water holding capacity and other soil properties such as texture, or structure, capillary conductivity, soil profile conditions, and depth of water table.

The inorganic and organic materials in the solid phase of the soil varies markedly in different soils. The inorganic material consists of mineral particles ranging in size from coarse sand grains to fine clay. The coarser particles are derived primarily from the physical weathering of rocks. They represent primary minerals of the rocks or material derived therefrom, such as quartz, mica, and felspars, which have not undergone any chemical change in their composition. The finer particles are mostly secondary minerals which are the products of chemical alterations of primary minerals. Largely they are represented by a variety of clay-minerals together with colloidal iron oxides and silica. The finer clay particles form the most important reactive constituent of a soil. Availability of plant nutrients, water, air and resistance of a soil to degradation are profoundly influenced by the minerals together with decomposed organic matter. The organic matter in the soil is derived largely from the remains of plants and animals. These include grasses, shrubs, trees, bacteria, fungi, rodents, earthworms, city wastes, and sewage.

The liquid phase of soil constitutes water that occurs in the pores and on the surface of the particles and is held by the forces of adhesion, cohesion and surface tension. In an ideal condition, half of the soil pores are filled with water and the remaining with air. Plants need water for their existence. Water helps maintaining turgidity, formation of food material in presence of sunlight (photosynthesis) and acts as solvent in which plants take their nutrients through roots.

The gaseous phase of soil consists of soil air. The amount of soil air varies inversely with the amount of water present. Restricted soil aeration is the major limiting factor in the development of extensive root system in crops. Both water and nutrient absorption of plants are governed by soil aeration. The composition of soil air as percentage by volume are (*i*) oxygen 20%, (*ii*) nitrogen 78.6%, carbon dioxide 0.5%, and argon 0.03%. The percentage of oxygen and nitrogen in atmospheric air and soil air are nearly same (21% and 78.03%, respectively for atmospheric air). The carbon dioxide content of soil air is 15 to 20 times more than atmospheric air. The argon content in atmospheric air is 0.94%, as compared to 0.03%, in soil air. The larger content of CO_2 in soil air is mainly because of the respiration of plant roots and soil organisms. When the pore space is filled with water the concentration of CO_2 in soil air rises, resulting in unfavourable conditions for biological activities. Toxic substances to plants such as methane, hydrogen sulphide, and other reduced compounds are produced under waterlogging conditions due to restricted diffusion of fresh air.

Soils vary in their capability for successful crop production. Some are suited to agriculture, others are not. Even amongst those suited to agriculture, marked variations exist in the yield of different crops from different soils. The variations in different soils arise due to differences in their genetic history, and the nature, composition and arrangement of their mineral and organic components. Proper use of soil necessitates an intimate knowledge of the soils encompassing its various properties

Mechanical Composition of Soil

Mechanical composition of a soil refers to its solid phase composed of the mineral fraction. Knowledge of the mechanical composition of a soil gives an insight into its physical, chemical and biological potential. It is a vital aid in managing the soil for plant growth as well as engineering (construction) purposes.

The mineral particles are the chief components of most soils on a volumetric basis, except in an organic soil such as peat, and are the most stable. They consist of rock particles developed *in situ* by weathering or deposited in bulk by wind or water. Mineral soil particles are usually classified according to their size. The minerals of size between 2 mm and 20 mm diameter is called gravel, while materials smaller than 2 mm in diameter is fine earth. Only the latter is considered in the chemical and mechanical analyses of soils. The components of fine earth are sand, silt and clay. The size limits of the soil-particle fractions have been established by various national and international organisations. The most commonly used classifications of soils are those proposed by the United States Department of Agriculture (USDA) and by the International Soil Science Society (ISSS) which are given in Table 9.1a.

Clay is sometimes sub-divided into coarse clay of 0.002 to 0.0002 mm in diameter and colloidal clay of less than 0.0002 mm in diameter. The term 'colloidal' refers to the size of matter. The colloidal size is intermediate between matter that is visible under an optical microscope and invisible molecules.

Table 9.1 Soil textural classification

Soil particle fraction	*Particle diameter*	
	USDA	*ISSS*
Gravel	> 2 mm	> 2 mm
Very coarse sand	1 to 2 mm	—
Coarse sand	0.5 to 1 mm	0. to 2 mm
Medium sand	0.25 to 0.5 mm	—
Fine sand	0.1 to 0.25 mm	0.02 to 0.2 mm
Very fine sand	0.05 to 0.1 mm	—
Silt	0.002 to 0.05 mm	0.002 to 0.02 mm
Clay	< 0.002 mm	< 0.002 mm

Sand and silt particles are approximately spherical or cubical in shape. Soil that is formed in place and has not been transported contains sand and silt particles with comparatively sharp edges. The edges of the soils deposited by wind or water are smoothed off in transportation. Chemical decomposition and physical influences such as freezing and thawing tend to make the soil particles smooth. Unlike sand and silt, clay particles are plate or lath shaped, depending on the crystal structure of the specific type of clay.

Soil Texture

Soil texture is determined by the size of the particles that make up the soil. The relative proportion of sand, silt and clay determines the soil texture. Texture is designated by using the names of the predominant size fractions and the word 'loam' whenever all three major size fractions occur in sizable proportions. Thus, the term 'silty clay' describes a soil in which the clay characteristics are outstanding and which also contains a substantial quantity of silt. A 'silty clay loam' is similar to silty clay except that it contains sand in a sizable proportion. Sandy soils are classified as coarse-textured, loam soils as medium-textured and clay soils as fine textured. Figure 9.1 presents the textural classification chart for 12 main textural classes of soils.

The least complex textural group is sand which contains less than 15 per cent silt and clay. Such soils form relatively simple capillary systems with a large volume of non-capillary pores spaces, which ensure good drainage and aeration. Sandy soils are relatively inert chemically, are loose and non-cohesive, and have a low water-holding capacity. They have usually a low cation exchange capacity.

Clay soils are at the other extreme with reference to the size of particles and complexity. They contain more than 40% of clay particles and 45% sand or silt. The clay particles are usually aggregated together into complex granules. Because of their plate-like shape, clay particles have a much greater surface area than cubes or sphere of similar volume. Their extensive surface enables clay particles to hold more water and minerals than sandy soils. There are three major types of clay minerals: kaolinite, montmorillonite and illite. Unit crystals or micelles of kaolinite consists of silica and alumina platelets in a 1:1 ratio. Soils composed chiefly of kaolinite show little swelling or shrinking with change in hydration, Montmorillonite and illite micelles are composed of silica and alumina platelets in a 2:1 ratio. In illite, potassium ions occurring between silica platelets of adjacent micelles form chemical bonds strong enough to prevent separation and swelling. Such bonds do not exist between micelles in montmorillonite. Hence, soils containing a large proportion of montmorillonite swell and shrink with changes in hydration. Such soils develop broad, deep cracks during prolonged droughts.

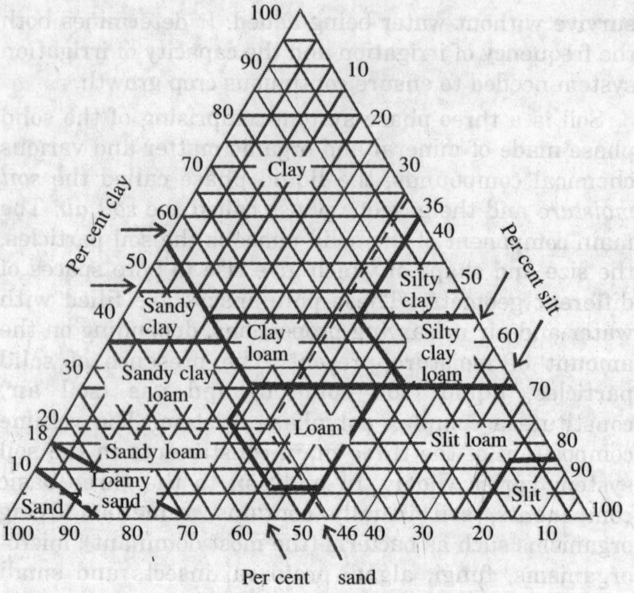

Fig 9.1. USDA soil textural classification chart. The dotted lines indicate the method of determining the textural class of a soil containing 18% clay, 36% silt and 46% sand. The three lines join at point A which lies within the class 'loam.'

Clay micelles are negatively charged, chiefly by replacement of silicon and aluminum ions with other cations within the crystals lattice. The intensity of the negative charge determines the cation exchange capacity or the ability of soils to hold cations. The intensity of hydrogen ions the clay particles hold, relative to basic ions such as calcium, potassium and sodium, determines whether a soil is acidic or alkaline (Fig. 9.2).

Loam soils contain more or less equal amounts of sand, silt and clay. They have properties which are intermediate between those of sand and clay. Such soils are considered most favourable for plant growth because

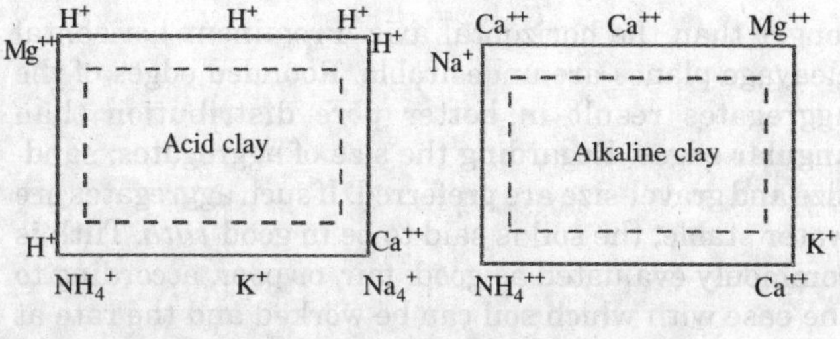

Fig. 9.2. Attraction of cations to the surface of negatively charged clay micelles. Acid soils contain a high proportion of hydrogen ions. Alkaline soils have most of the exchange positions held by basic ions such as Ca^{++}, K^{+} and Na^{+}.

they hold more available water and cations than sand and are better aerated and easier to work than clay.

Textural classification has only an approximate relationship to the behaviour of a soil as a medium for plant growth. Textural properties may be modified appreciably by organic matter content and the kinds of clay minerals present and kinds of ions associated with them. For example, aggregation effects of organic matter tend to give a fine-textured soil high in clay some of the pore space properties of a coarser-textured soil. Similarly, colloidal effects of organic additions to a coarse-textured sandy soil give it some of the moisture and cation retention characteristics of fine textured soil.

The soil texture forms the basic matrix. The geometry of voids created in this soil matrix is dependent on the class of soil texture. The soil texture, therefore, influences considerably the other phases (water and air) contained in the spaces in the soil matrix. The texture of a soil is almost constant and does not change with tillage or other practices.

Chemical nature of soils. The mineral components of the soil are made largely of silica and silicates. Although the chemical composition of soil particles varies from profile to profile, generally the larger particles are highest in silica and the finer particles contain more potassium, calcium felspars in fine sand and silt; and mica; vermiculite, montmorillonite, kaolinite and amorphous colloids in clay.

Specific surface area of soil particles. The specific surface area is an important characteristic of the soil system. It is defined as the ratio of the surface area to the volume of the soil system. It is expressed as square centimetres per cubic centimetre or as square metres of surface per gram of soil. It is a measure of the total interfaces in the system. It varies inversely with the size of the soil particles. The specific surface of clay is several times larger than silt and sand. Specific surface is important since most chemical and physical reactions of soil occur at the surfaces and hence the amount of these reactions is approximately proportional to the specific surface.

Shrinking and swelling of soils. Dry soils expand when they are wetted. Conversely, a wet soil shrinks upon drying. The amount of swelling and shrinking depends on the amount and type of clay in the soil. The more montmorillonite the soil contains, the greater is the swelling. To a certain extent, cycles of shrinking and swelling induce a good soil structure. However, if the amount of shrinking and swelling is large, they adversely influence the soil system. A swollen soil is relatively impermeable to the movement of water. The cracks in a shrunken dry soil promote the rate of further drying. When a dry soil is wetted under constrained conditions it exerts excessive pressure which may result in the cracking of roads, building foundations and earth dams.

Mechanical analysis. The determination of the particle size distribution of soils is called mechanical analysis. The process of mechanical analysis involves the separation of all particles from each other—complete dispersion into ultimate particles—and the measurement of the amounts of each size group in the sample. For detailed procedure for mechanical analysis, reference is invited to standard books on soil science or soil physics.

Soil Structure

The arrangement of individual soil particles with respect to each other into a pattern is called *soil structure*. Soil structure is an important characteristic used to classify soils and it heavily influences agricultural productivity. Aggregated soil types are desirable for plant growth. Soil structure is unstable and can change with climate, biological activity and soil management. Structureless soils may be either fine grained individual unattached particles like in sand dunes or massive individual particles adhered together without regular cleavage like clay pans or hard pans. Grouping of particles into structural units occurs in all soils. However, the strength of the bond, the size and shape of the structural units and the proportion of the soil particles involved in the units differ considerably among soils. Two distinct phases in the formation of soil structure are recognised: (*i*) development of interparticle bonds, related to aggregate stability, and (*ii*) the separation of structural units from each other, related to size and shape characteristics of the aggregates. Aggregates vary greatly according to their resistance to the impact of rain drops or under submergence. The stability of aggregation depends upon the clay content, the extent of floculation, the organic—inorganic linkages, the microbial glue of the aggregates, and the presence of mineral cementing materials, such as iron and aluminium oxides.

Based on the shape of aggregates, soil structure may be classified into (*i*) simple structure in which natural cleavage planes are absent or indistinct, and (*ii*) compound structure in which natural cleavage planes are distinct. Simple structure may be a *single-grain structure* or *massive structure.* Single-grain structure occurs normally only in sands and silts of low organic matter content. In sandy soils, single grain structure facilitates aeration and capillary movement of soil moisture. In other soils, single grain structure is undesirable because it results in the absence of large pores needed for aeration. Massive structure is similar to single grain structure, except that it is coherent. Examples of massive structure are dense soil crusts and plough pans. A *compound structure* may be described according to the relative lengths of their vertical and horizontal axes and by the contours of their edges. Accordingly, they may be grouped into (*i*) *cube-like structure* in which the vertical and horizontal axes are almost of equal length, (*ii*) *columnar structure* with vertical axes longer than the horizontal, (*iii*) *platy structure* with horizontal axes longer than vertical, (*iv*) *angular structure* in which the corners and edges are sharp, (*v*) *subangular structure* with sharp edges but rounded corners, and (*vi*) *granular structure* with both the corners and the edges rounded.

The size of aggregates is a valuable criterion of soil structure. Sand-sized aggregates are more favourable for plant growth than very small and very large ones. For instance, a soil made up exclusively of silt-size aggregates cannot be drained by gravity, since the pores are too small.

Since the pore spaces are as important as the solid particles, soil structure is sometimes defined as the arrangement of small, medium and large soil pores into a structural pattern. Such a definition highlights one of the primary effects of soil structure, namely, its effect on porosity.

Soil structure has a pronounced effect on such soil properties as erodibility, porosity, hydraulic conductivity, infiltration characteristics and water-holding capacity. The quality of the soil structure may be expressed in terms of porosity, aggregation, and cohesiveness or permeability for water or air. Porosity is of great importance, as the chemical and biological processes occur in the pores. Large pores induce aeration and infiltration, medium-sized-pores facilitate capillary conductivity, and small pores induce greater water-holding capacity. The effect of pore size on crop response depends on the crop species, climatic conditions and depth to water table. Under conditions of plentiful water supply through irrigation or rainfall, water storage pores (small pores) are not so important, but aeration pores are essential for optimum plant growth. Under dry farming conditions with restricted rainfall, both aeration pores and storage pores are needed to facilitate infiltration and moisture retention. In humid climate, large, medium and small pores should occupy nearly equal volumes.

The soil aggregates should be stable to withstand rainfall impact and temporary submergence. The vertical axis of the aggregate should be as long as or longer than the horizontal axis. Prominent horizontal cleavage planes are undesirable. Rounded edges of the aggregates result in better pore distribution than angular edges. Regarding the size of aggregates, sand-size and gravel-size are preferred. If such aggregates are water stable, the soil is said to be in good *tilth.* Tilth is commonly evaluated as good, fair, or poor, according to the ease with which soil can be worked and the rate at which it takes in water. Soils in good tilth are mellow, crumbly, and easily worked. Soils in poor tilth are generally hard, cloddy and difficult to work. Dispersed clay represent the most unfavourable soil structure.

Cohesiveness of soils is related to its moisture content. Soil should be in friable condition, but not too loose. A massive compact soil restricts aeration and root spread. For optium crop growth, soil structure should be such that the infiltration capacity is large, the percolation capacity is medium and aeration is sufficient, without being excessive.

Measurement of soil structure. The four properties of soil structure, namely, porosity, aggregation, cohesiveness and permeability can be quantitatively measured. The pore space distribution of a soil can be determined by saturating the soil sample with water and removing the water by successively increased tension. The volume weight of a soil whose true density is known enables the calculation of the total pore space. The distribution of aggregates is determined by dry saving. The stability and the size distribution of water stable aggregates is determined by wet sieving. The shape of aggregates is determined approximately by the measurement of the vertical and horizontal axes.

Managing soil structure. Soil structure management is aimed at obtaining the maximum yields under the given conditions of climate, moisture availability and nutrient supply. Another objective is to conserve the soil by decreasing its detachability and transportability by water or wind and to increase its infiltration and percolation capacities so that runoff and erosion are minimised. Common methods of soil structure management are proper land use, adoption of suitable tillage practices at optimum soil moisture content, subsoiling, addition of organic matter, crop rotation, application of optimum levels of fertilizers, mulching, drainage, controlled irrigation, soil conservation practices, protection from compaction of wet soils, and, occasionally, the use of soil conditioners.

Volume and Mass Relationships of Soil Constitutents

A schematic diagram of soil shown in Fig. 9.3 may be useful to define volume and mass relationships of the three soil phases. The diagram shows the presence of the three phases in relative proportions both in masses and volumes.

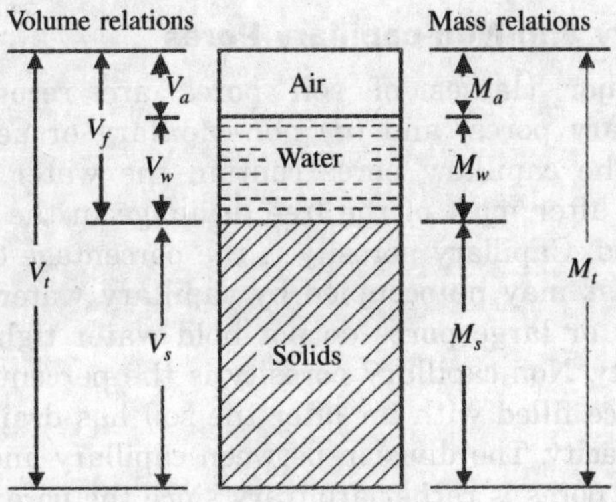

Fig. 9.3. Schematic diagram of the soil as a three-phase system.

Referring to Fig. 9.3, let

M_a = Mass of air (negligible)
M_w = Mass of water
M_s = Mass of solid
M_t = Total mass ($M_a + M_w + M_s$)
V_a = Volume of air
V_w = Volume of water
V_f = Volume of pores ($V_a + V_u$)
V_s = Volume of solids
V_t = Total volume ($V_f + V_s$)

Density of soils solids, ρ_s. The density of soil solids is defined as the ratio of mass of solid to its volume:

$$\rho_s = \frac{M_s}{V_s \rho_w} \quad ...(9.1)$$

in which ρ_w = density of water at 4°C.

Soil density is the mass per unit volume of the soil particle. The density of dominant soil minerals such as quartz and several of felspars is nearly 2.65 grams/cc. Consequently, the particle matter is 2.65 grams/cc. Alternatively, it may be said that the true specific gravity (also called real specific gravity or particle density) of soil is about 2.65. However, the density of soil organic matter is usually between 1.3 and 1.5 grams/cc.

Dry bulk density, ρ_b. The dry bulk density of soil is defined as the ratio of the mass of dried particles to the total volume of soil (including particles and pores).

$$\rho_b = \frac{M_s}{V_t} = \frac{M_s}{V_s + V_a + V_w} \quad ...(9.2)$$

The term dry bulk density and *apparent specific gravity* are often used synonymously. Whereas the term apparent specific gravity denotes a dimensionless quantity, bulk density is expressed in grams per cc or mass per unit volume. However, since 1 gram of water fills a volume of 1 cc at normal temperatures, the two terms have equal numerical values. The apparent specific gravity is influenced by the structure, texture and compactness of the soil. It is an important soil physical property, considering its influence on the water-holding capacity of soils and its hydraulic conductivity. For instance, when the bulk density of medium to fine-textured subsoils exceeds about 1.7 grams/cc, the hydraulic conductivity values will be so low that drainage may become difficult.

Total (wet) bulk density. Wet bulk density is the mass of moist soil per unit volume:

$$\rho_t = \frac{M_t}{V_t} = \frac{M_s + M_w}{V_s + V_a + V_w} \quad ...(9.2a)$$

Porosity. Porosity can be defined as the ratio of the volume of pores (voids) to the total soil volume:

$$n = \frac{V_a + V_w}{V_s + V_a + V_w} \quad ...(9.3)$$

Porosity is an index of the relative volume of pores. It is influenced by the textural and structural characteristics of the soil. The porosity of sandy soils usually ranges from 35 to 50 per cent, while that of clay soils from 40 to 60 per cent. The more finely divided are the individual soil particles, the greater is the porosity.

Void ratio. The quantity expressing the ratio of the volume of pores to the volume of solids is termed as void ratio or relative porosity:

$$e = \frac{V_a + V_w}{V_s} \quad ...(9.4)$$

This index has certain advantages over porosity (n). In the case of void ratio, the total soil volume changes with volume change of voids, whereas in the case of porosity the volume of pores may change without change in the volume of solids. Therefore, knowledge of the dynamics of pores due to external load is possible with the information on void ratio. The term is commonly used in engineering works relating to the compaction of foundations and embankments. The following relationships exist between porosity and void ratio to apparent and true specific gravity:

$$\rho_b = \rho_s \left(1 - \frac{n}{100}\right) \quad ...(9.5)$$

$$\rho_s = \rho_b (1 + e) \quad ...(9.6)$$

Soil Wetness

The term soil wetness refers to the relative water content in the soil. It can be expressed in the following ways:

Mass wetness. It is the ratio of mass of water to the mass of soil and is often referred to as the 'gravimetric water content.'

$$\text{Mass wetness} = \frac{M_w}{M_s} \quad ...(9.7)$$

This is a commonly used term and is often called soil moisture content. It may be expressed in decimals or, more commonly, as a percentage by multiplying by 100. In mineral saturated soil, it can range between 0.25 to 0.60 (or 25% to 60%) mass wetness, depending upon the bulk density.

Volume wetness. The relative water content of soil is expressed on volume basis of water and soil. Volume wetness is the ratio of volume of water to the total soil volume:

$$\text{Volume wetness} = \frac{V_w}{V_t} = \frac{V_w}{V_s + V_f} \quad ...(9.8)$$

It is also termed as volumetric water content. The use of the term volume wetness is preferred over mass wetness as the former is more directly adaptable to the computation of fluxes and water quantities. The volume wetness at saturation varies from 40-50 per cent in sandy soils to 60 per cent in clayey soils.

Degree of saturation. This relationship, also termed as degree of sturation, refers to the volume of water present in the total pore volume.

$$\text{Degree of saturation} = \frac{V_w}{V_f} = \frac{V_w}{V_a + V_w} \quad ...(9.9)$$

Determination of Soil Bulk Density

The usual method of determining the bulk density or apparent specific gravity of a soil is to obtain an uncompacted soil sample of known volume. Core samplers are commonly used to take undisturbed soil samples (Fig. 9.4). The cylinder of the core sampler, which has its cutting edge, is driven into the soil and an uncompacted core obtained within the tube. The samples are carefully trimmed at both ends of the core cylinders. They are dried in an oven at 105°C for about 24 hours, until all the moisture is driven off and the sample is weighed again. The volume of the soil core is the same as the inside volume of the core cylinder. The weight of the soil in grams divided by the volume of the soil core in cc is the bulk density or the numerical value of its apparent specific gravity. The bulk density of most soils ranges from 1.0 gram/cc for clays to 1.8 gram/cc for sands (Example 9.1).

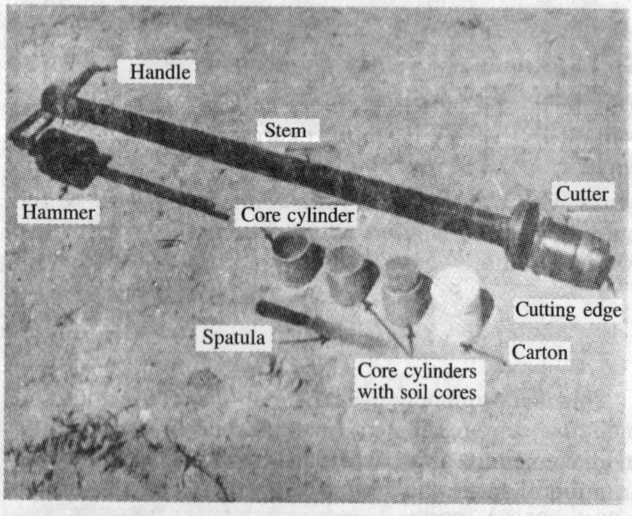

Fig. 9.4. Soil core sampling unit showing component parts.

Capillary and Non-capillary Pores

Two major classes of soil pores are recognised: (*i*) capillary pores, and (*ii*) non-capillary or aeration pores. The capillary pores contain the water which remains after most of the free drainage in the soil is completed. Capillary porosity is the percentage of pore space that may be occupied by capillary water. Non-capillary or large pores do not hold water tightly by capillarity. Non-capillary porosity is the percentage of pore space filled with air after the soil has drained to field capacity. The division between capillary and non-capillary pores is rather arbitrary since the pore space, water retention, and drainage characteristics are linked dynamically. However, this concept is usually used as an initial subdivision of total porosity and has considerable significance with respect to plant growth. The large non-capillary porosity of sandy soils result in better drainage and aeration, but it also results in a lower water holding capacity than clay soils, which have a larger proportion of small capillary pores. An ideal soil has its pore space about equally divided between capillary and non-capillary pores. Such a soil has enough large pores to permit adequate drainage and aeration and enough small pores to give adequate water-holding capacity.

Soil Consistency

Soil consistency describes the characteristics of the soil at various moisture contents when influenced by the physical forces of *cohesion* and *adhesion*. Cohesion (moleular attraction) is brought about by the surface charges of clay paticles, by the broken bonds at the edge of clay particles and by the attraction of particle to particle. The effect of molecular attraction is greatest in a dry soil and it sharply decreases as water enters between particles, thus keeping them apart. Adhesion is due to surface tension. Its effect on holding soil particles together depends on the presence of both water and air. Surface tension decreases as the soil moisture increases. In a soil fully saturated with water, surface tension ceases to exist.

Soil consistency varies with texture, structure, organic matter, percentage of colloidal material, and the shape and type of clay mineral. Several soil consistency terms express the various conditions of the soil mass, such as plasticity, hardness, friability, stickiness, and resistance to compression and shear. Since the consistency concepts are applied mostly in construction works, and are not of much consequence in farm irrigation practice, a detailed discussion on the subject is not attempted in this book.

Soil Profile

Soil profile is a vertical section through the soil mass. Significant changes in soil texture and structure with depth are observed in most soils. The soil column in most cases is composed of a series of layers of varying degrees of distinctness. These layers are nearly parallel with the soil surface and are called *soil horizons.*

Collectively, the horizons make up the soil profile. Three distinct horizons usually recognised are *A, B* and *C* horizons with sub-groups for *A* and *B* horizons. '*A horizon*' is the zone of maximum biological activity. The '*B horizon*' includes the lower layers of a soil column that have been influenced strongly by the factors of soil formation. In soils developing on fresh geological formations the *B* horizon may not be distinct. The *B* horizon is differentiated from the *A* horizon above and the *C* horizon below by colour and change in structure, and the content of clay and organic matter. The *B* horizon has a blocky or prismatic structure that is not shared by *A* and *C* horizons. The organic matter is low in the *B* than in the *A* horizon. In cool, humid regions the *B* horizon is a zone of accumulated clay, sesquioxides and organic matter moved downward in varying proportions from the *A* horizon. The *C* horizon is a layer of unconsolidated material that is affected relatively little by the action of soil organisms. The *C* horizon usually represents the type of material from which at least part of the overlying horizons have developed. Individual soils may have one or more of these horizons but need not contain all of them. In addition, letter '*D*' or '*R*' is used to designate the stratum below *C* horizon, such as hard rock, sand or clay which is not the parent material but may have some significance to the overlying soil.

In recently deposited alluvium, however, soils may be of uniform texture and structure upto several metres. Changes in soil characteristics with depth are often found in older soils where much downward leaching has occurred. As a result, undisturbed soils usually possess a characteristic profile consisting of definite horizons or layers of soil which differ in their properties. It is, therefore, necessary to sample at least as deeply as plant roots penetrate and determine the properties of the different horizons penetrated by roots in order to understand the soil-plant-water relationships.

Soil Resources Survey and Classification

Soils support the plant life on earth. Land refers to a delineated part of the earth's surface, including mountains, plains, river, swamps, deserts and coastal areas. Soil comprising mineral and biological materials is the basic feature of the earth's surface. Land characteristics are influenced by the climate, the soil and underlying geology, the topography, the hydrology, the plant and animal populations, and the human activities. Soil-water-plant relationships are profoundly influenced by the type of soil and its properties.

The soils are derived from a wide variety of parent rocks and materials. Their composition varies widely, depending on the nature and kind of mineral and organic matter present in them and in their proportions. The origin of the parent rocks is varied and the soils have been formed over different geological periods.

In India, systematic approach to the collection of information of extensive areas began in early 20th century. The early surveys were on the fertility status of soils, involving studies on nutrient status. The map prepared from such surveys showed the distribution of soils having different levels of total and available nitrogen, phosphoric acid and potash (Govinda Rajan and Gopala Rao, 1978). Soil surveys of irrigated areas were started in 1928 in Punjab, followed by Uttar Pradesh. These surveys extended to depths of 2 to 3 m for examining the physico-chemical properties influencing movement of salt and water, leading to salinity and waterlogging problems. The surveys enabled the grouping of soil into those suited and those unsuited to irrigation. During 1930s and later pre-irrigation soil surveys were conducted in irrigation command areas in central and south India. Subsequently, post-irrigation soil surveys were conducted in many of the irrigation project areas. In the subsequent stages of soil classification, zones of accumulation of soluble salts and various morphological characteristics of soil were included. During 1938-47 soil research was given strong orientation towards the genetic aspects of soil development. Valuable data were collected on the main group of soils of the Indian subcontinent.

The second-half of the 20th century (post Independence period) witnessed substantial strengthening of soil survey works at project levels in the states and at the all India level. An All India Scheme on Soil Surveys and Fertilizer Experiments on Farmer's Fields was initiated with the coordinated efforts of the Central Goverment and linking various states. Subsequently, two national level organisations were established by the Government of India to conduct detailed studies on various soil profile characteristics. These are the All India Soil Survey Unit of the Department of Agriculture and the National Bureau of Soil and Land Use Survey of ICAR. These two organisations, with distinct responsibilities have been engaged in extensive survey and mapping of soil groups of India.

Over the Indian subcontinent, three main geological regions are recognised, namely, the peninsular region, the Indo-Gangetic plains and the Himalayan region. The peninsular region comprises of the area located to the south of the Vindhyas and bordered by the range of hills along the western and eastern limits of the Peninsula which end at Kanyakumari at the southern-most tip of the Peninsula. The Himalayan region at the northern limits of the subcontinent represents the highly folded sedimentary rocks running from the west in the Punjab and Himachal Pradesh towards north in Kashmir Valley and Ladakh, and then turning east in a south-eastern direction through Uttar Pradesh and ending in the North-Eastern region. The third region, viz., the great Indo-Gangetic plain, representing the largest extent of most intensively populated area in the world lies between the two regions mentioned above.

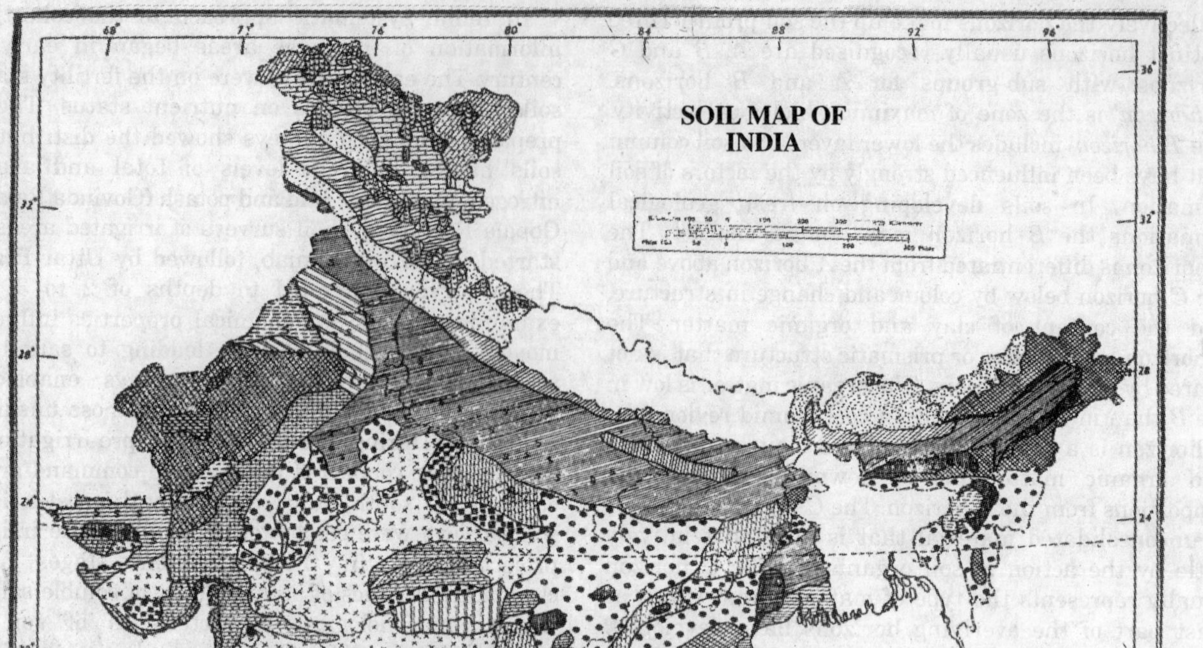

BASED UPON SURVEY OF INDIA MAP WITH THE PERMISSION OF THE SURVEYOR GENERAL OF INDIA.

THE TERRITORIAL WATERS OF INDIA EXTEND INTO THE SEA TO A DISTANCE OF TWELVE NAUTICAL MILES MEASURED FROM THE APPROPRIATE BASE LINE.

THE BOUNDARY OF MEGHALAYA SHOWN ON THE MAP IS INTERPRINTED FROM THE NORTH-EASTERN AREAS (REORGANISATION) ACT, 1971, BUT HAS YET TO BE VERIFIED.

Fig. 9.5. Soil map of India.

(***Source:*** Adapted from All India Soil and Land Use Survey, IARI, New Delhi.)

Soil Resources of India

The following are the predominant soil groups of India: *red* (105.5 million ha), *black* (73.5 million ha), *alluvial* (58.4 million ha), *laterite* (11.7 million ha), *desert* (30 million ha) and *hill* and *terai* soils (26.8 million ha) (Singh, 1999). It is estimated that an area of 174.2 million ha is potentially exposed to various degradation forces like water erosion (153.2 million ha) and wind erosion (15.0 million ha). About 40.0 million ha area is subjected to floods and 22.0 million ha is not reclaimable for cultivation. An area of about 25.0 million ha is affected by acidity having pH below 5.5, and 33.7 million ha is degraded because of specific problems like waterlogging (8.5 million ha), sodicity (3.6 million ha), salinity (5.5 million ha), shifting cultivation (5.0 million ha), ravine and gullies (4.0 million ha) and riverine and torrent lands (2.73 million ha).

With the wide range of geological materials forming the surface of the Indian subcontinent and highly varying climatic factors and associated variations in vegetative cover have resulted in the development of a wide variety of soils on the land surface. The earliest soil map of India was prepared in 1932 which was revised in 1944 and further improved in 1952 by Raychaudhary (Govindarajan, 1988). Based on further information collected by various states and the All India Soil Survey, Govindarajan and Dutta Biswas prepared a revised soil map of India in 1968 which was subsequently revised by the National Bureau of Soil Survey and Land Use Planning, Nagpur in 1996 (Fig. 9.5). Twenty three different soil groups are identified in the soil map. Brief description of the major broad groups of soil represented in Fig. 9.5 is given in the following paragraphs.

Black Soils

They are characteristically black in colour, ranging from dark brown to deep black and occur extensively in the central and south central parts of India. They are spread over the states of Andhra Pradesh, Gujarat, Madhya Pradesh, Chhattisgarh, Tamil Nadu, Maharashtra and Karnataka. The parent materials on which these soils are formed in Maharashtra, parts of Karnataka and Andhra Pradesh, are traps, granites, and granite-gneiss, particularly those rich in lime feldspars. These are also found along the river courses and valleys in Gujarat, Madhya Pradesh and Maharashtra. In Maharashtra, the black soils occur in the river valleys of Tapi, Godavari, Bhima and Krishna. As a group, these soils are heavy textured and their clay content generally ranges from 40 to 60%. They contain mostly montmorillonitic clay mineral, on account of which they show severe swelling and shrinkage with changing moisture conditions, causing heavy fissuring and deep cracking on drying. They vary in pH from 8.0 to 9.0 or higher in different regions.

Black soils are grouped into: (*i*) *shallow black soils* which have a depth of 30 cm or less, (*ii*) *medium black soils* which are 30 to 100 cm in depth, and (*iii*) *deep black soils* which are over 1 m deep. Deep black soils are also referred to as *black cotton soils* since cotton is the most important commercial crop of the area.

Shallow black soils. The black soils are derived from basalts, mostly of the Deccan traps. The soil is usually silty loam to clay in texture, and the colour at the surface range from dark-brown to dark yellowish-brown. The structure is granular or blocky. Lime in the form of fine grains or nodules is usually present and the soil is well drained.

Medium black soils. These are developed from a variety of rocks. The texture ranges from silty clay to clay. The soils are moderately rich in organic matter and are fairly well drained. They contain lime in varying proportions. Some of them have thick layers of calcareous nodules (20 to 30 cm thick) in the C-horizon.

Deep black soils. These soils are derived from basaltic traps and represent a large part of the black soils of the Deccan plateau, commonly identified as the *Regur* soils. The texture of the surface and sub-surface layers is nearly uniform and ranges from silty clay to clay. The percentage of clay vary from 40 to 60% or more. Distribution of lime, in the form of irrigular-shaped nodules of varying sizes, may be uniform throughout the profile with heavier accumulations at depths of over 1.5 m. The soil reaction is moderately alkaline (pH 8.0 to 8.5).

In India, black soils occur mostly in the peninsular region on gently sloping lands at low elevations. They have high cation exchange capacity with the advantage for nutrient and moisture retention capacity. Moisture shortage under rain-fed conditions and waterlogging, salinity and alkalinity under irrigated conditions adversely affect the productivity of the soils. Though regarded as fertile, the average productivity of black soils is generally low. Black soils possessing more water holding capacity are relatively fertile and less drought prone. On saturation they become impervious and are more erodible. When wet, they are difficult to cultivate and cropping is concentrated more during the post-rainy season, unless the timing of tillage operations are properly planned.

Mixed red and black soils. In some states, black and red soils occur side by side, the former on the uplands and the latter at the bottom of the slopes. They are of varying sizes and patterns. Even though the mechanical composition of some samples taken from red and black soil tracts may show identical proportions of sand, silt and clay, the type of clay mineral assumes greater importance with respect to soil moisture storage characteristics.

In black soils the cropping patterns are recommended on the basis of the availability of water. As the soils are fairly rich in plant nutrients and have a high cation exchange capacity, the productivity can be high if they are properly managed. Under efficient water management and appropriate agronomic practices, it is possible to work with these soils both under wet and dry conditions. The disadvantage of these soils is their poor drainage and consequent waterlogging.

Red Soils

These soils have red, yellowish red, reddish brown and yellow colours with sandy loam, loam and silty loam textures. Soil reaction is usually acidic with pH ranging from 4.5 to 6.5, while a few have a pH on the alkaline side. The soils contain, gravels, quartz and concretions of iron and aluminium in their profiles. Usually, the soils are non-calcareous. The clay mineralogy is predominantly kaolinitic. The cation exchange capacity and water-holding capacity are low. The soils are generally poor in nitrogen, organic matter and phosphorus.

Red soils are found in Andhra Pradesh, Karnataka, Tamil Nadu, Orissa, Kerala, Manipur, Tripura, Nagaland and western portion of West Bengal, Jharkhand, parts of Madhya Pradesh and small areas in the States of Assam, Maharashtra, Uttar Pradesh, Gujarat and Rajasthan. The rice soils of West Bengal are red soils with a pH between 6.2 and 6.6.

Red soils are classified into (*i*) *red loam soils,* and (*ii*) *red sandy soils*. This differentiation, in addition to the textural group, is also related to their genesis from geological materials of distinct compositional differences. In general, the red soils as they occur in Peninsular India, are derived from granites and gneisses.

The red sandy soils have a moderately deep A-horizon of pale reddish brown and red colour with textures varying from fine sand through loamy sand to coarse sand. The B-horizon is dark reddish-brown loam or sandy clay loam of varying depths. The depth varies from 15 cm to 30 cm or more. The C-horizon is a weathered parent material of disintegrating coarse grained granite or granatoid gneiss. The texture may very from gritty coarse sandy clay to gravelly loam.

Laterite Soils

Laterite soils in India extend to an area of about 24.8 million ha, occurring mainly in the hills and plateaus of Karnataka, Kerala, Tamil Nadu, Madhya Pradesh, Orissa, Maharashtra, West Bengal, Bihar and Assam. The formations may be massive and firm or it may be in the form of loose aggregates. If the formation is massive and under favourable conditions of moisture content and consistency, it may be quaried and cut up into blocks like bricks which on exposure dehydrate and become hard. The laterite soils which are generally acidic have an average pH between 5.0 and 6.0.

Lateritic Soils

These soils are associated with and derived from laterites. In lateritic soils there is a larger proportion of the primary kaolinitic minerals than in laterites. Though usually occurring at high elevations, laterites may be carried to lower levels by streams where they are redeposited and form into a compact mass. Lateritic soils are generally poor in lime and magnesium and deficient in nitrogen. They are usually poor in available phosphorus and calcium and have relatively low organic matter content. However, the high level and low level laterites of the west coast and Tamil Nadu are rich in plant nutrients. They support extensive paddy fields on the lower reaches and rich plantation crops like tea, rubber, coconut and arecanut at higher elevations.

Red and lateritic soils are characterised by dry to humid tropical and subtropical climate. These soils occur on gently sloping to undulating lands. They are good in sesquioxides, low in organic matter and have a large proportion of kaolinite and non-expanding illitic clay fractions. Due to the nature and quality of clay minerals, red and associated soils are less water retentive and are more drought-prone. These soils are structurally unstable. Except under irrigation or in regions with mean annual rainfall exceeding 1000 mm, red soils are generally cropped only during the rainy season. The constraints to crop production in red and lateritic soils are soil erosion, hardening of soil, low water-holding capacity, high rates of drought related stress, low cation exchange capacity, low organic matter, high acidity, iron and aluminium toxicity, and poor fertility status.

Red and Yellow Soils

In the east-central and in the north-east central parts of India, occurs soils which are characteristically yellow in colour. The colour may range from reddish-yellow to yellowish-brown, and the soils are usually fine textured. In some cases, these soils are associated with soils developed from laterites or overlying laterites. Otherwise, they are derived from a variety of rocks underlying in the profile at depths varying from 1 to 2 m. The soils have a pH around nutrality or else slightly on the acidic side. The *A*-horizon of these soils is loam to silty loam in texture and moderately rich in humus.

Alluvial Soils

Alluvial soils are formed by transportation in streams and rivers and are deposited in flood plains or along coastal belts. The character of these soils varies a great deal and reflects the nature of soils that occur in the region from where they are transported. They are generally deep soils. Most of the fresh alluvia show little or no horizonation, while older alluvium shows distinct profile development. The alluival soils occur in the Indo-Gangetic basin of North India. The soils of the Indo-Gangetic plains are deposits of the great rivers Yamuna and Ganga and their tributaries Gandak, Gomti, Ghagra and others which flow out of the Himalayas. The soils found in the Godavari and Krishna river basins, although of alluvial origin, are black in colour. In Gujarat, alluvial soils of the coast and the Rann of Kutch, exhibit grey and black colours with salt patches here and there and pH ranging between 7.0 and 9.0. Coastal alluvium runs along the west and east coasts. These soils have fairly good water-holding capacity and are easily manageable from tillage and water management points of view. Unless badly affected by salts and poor drainage, these soils are fertile and respond well to manuring.

Coastal alluvium. The alluvial soils of the coastal belt all along the peninsular region, and extending over varying widths between the sea and the range of hills along the east and the west coasts, are characterised by being relatively recent deposits of alluvial origin. The texture of these soils is extremely variable and range from sandy to silty clay. Except in the older deposits, these soils do not show a prominent horizonal differentiation. The soils are usually deep and the colours range from bright reddish-brown and yellowish-brown to grey and dark-grey.

The composition and mineralogy of these soils is greatly influenced by the parent materials of the catchment areas. Alluvial soils derived from calcareous materials are usually composed of dark-coloured heavy clay, while in areas where the red soils from the granitic and granite gniessic rocks are dominant, the alluvial soils are poorer in fertility. They are often medium or light-textured and relatively rich in kaolinitic clays. Frequently, these soils in areas which are not much above sea-level get inundated by sea-water and suffer from problems of salinity.

Coastal sand. In certain areas along the coast in the peninsular region occur soils which are sandy, deep, and lacking in profile development. The lack of profile development is due to the coarse nature of the parent material. The topography varies from flat to gently undulating with occasional dunes. Some areas in the low-lying flat lands can be marshy and saline, in which case, the swampy condition makes the areas unfit for cultivation. These sandy stretches are low in organic matter content. They can be used for raising fuel trees, particularly, casuarina.

Deltaic alluvium. The soils of the deltaic alluvium represent the heterogeneous types of sediments brought by rivers and deposited at their mouths. The east coast of peninsular India is characterised by the formation of deltas at the mouths of the major rivers of the subcontinent that flow into the Bay of Bengal. Of these rivers, the Ganga and the Brahmaputra flow from the Himalayas and drain the extensive plains of northern India; the Mahanadi, the Godavari and Krishna, in the Deccan plateau and the Central India plains, and the Kaveri in the southern part of the peninsula. These rivers carying the alluvium of the extensive areas they traverse, deposit it at the regions they join the sea. These deposits form the alluvial soils of these deltas. The drainage conditions in most of the deltaic soils, other than the Gangetic plains, are satisfactory. The deltas of the eastern region are highly suited to paddy cultivation.

Calcareous alluvial soils. These are alluvial soils which occure along the north-eastern districts of Uttar Pradesh and extending to the north-western parts of Bihar. These calcareous soils are developed on the alluvium brought by the river Gandak, flowing from the Himalayas in a north-west to south-east direction and joining Ganga. The main characteristic of the soil is the high content of $CaCO_3$ (10 to 40%) which is distributed throughout the depth of the profile. The soils are light-coloured, and lack in horizonal differentiation. Their texture varies from sandy loam to loam. The pH of the soils is on the alkaline side and the contents of available phosphoric acidic and potash contents are low.

Calcareous sierozemic soils. The old Indo-Gangetic alluvium of the north-west region of India has been subject to varying climatic conditions, ranging from humid to arid as the distance from the Himalayas increases. These soils, like the other soils of the region, are deep and clayey. There is a clear indication of the movement of $CaCO_3$ throughout the profile, and a layer of $CaCO_3$ accumulation is reached below 150 cm. The pH of the soil is neutral to slightly alkaline. The soils are deficient in phosphate, with the medium status in organic matter and nitrogen.

Terai soils. The soils of the Terai region, lying at the foot of the Himalayan range, possess certain characteristics of their own. These soils occur all along the foothills in the northern parts of Uttar Pradesh, Bihar and West Bengal, and are fairly deep and moderately fertile. These soils are deposited as a result of their movement by water erosion of the Himalayan range. The excessive soil moisture and high fertility of the soil have resulted in excessive growth of vegetation and weeds, but once drainage is improved and cultivation adopted, the soils become highly productive.

Desert Soils

Desert soils in India cover an area 14.2 million ha in the arid and semi-arid regions of Rajasthan and parts of Punjab and Haryana. Soil development in the region is inhibited by a mantle of sand blown from the coastal region and the Indus valley, and the sand formed locally by weathering. These soils are composed of sand to a depth extending beyond 50 cm. The *A*-horizon is weakly developed or absent, as in shifting sand dunes. The *B*-horizon is also weak or absent. Some of these soils are high in soluble salts in the lower horizon and are alkaline, with varying amounts of $CaCO_3$ and low organic matter content.

Grey Brown Soils

These soils have developed under semi-arid conditions. The texture of the soils is usually light, being sandy loam with a preponderance of coarse sand. The *B*-horizons a distinctly heavier textured. $CaCO_3$ is present throughout the depth of the profile.

Brown Hill Soils

These soils occur in hilly regions, under moderately heavy vegetation. The soils are formed over a variety of parent rocks. They form a system of low foot hills corresponding to the Shiwalik ranges. The surface soils are dark-brown in colour and loam to silty-clay loam in texture and moderately rich in organic matter.

Submontane soils. These soils are found in the sub-Himalayan region under forest vegetation of the coniferous type. The rock types are hard and soft sandstones and shales, occurring in the upper Himalayan regions. The top surface layer which may extend from 10 to 15 cm is of dark-brown and black sandy loam with loose, undecomposed organic matter. The next layer extending to a depth of about 50 cm is a dark-brown sandy clay loam rich in humus. The pH of both these layers is around 5.0.

Mountain Meadow Soils

In the top Himalayan region at elevations where conifers or other trees do not grow, and also on the slopes with a southern aspect at lower elevations exist soils of moderately shallow depth, developed from sandstones and shales. The soils are relatively thin formations with poor profile development. The vegetation is mainly grass. The growth of grass helps to bind the soil material to prevent its loss through wind and snow action. The build-up of organic matter from the grass roots is assisted by the relatively low temperatures in the region.

Saline and Alkali Soils

Saline soils are characterised by the occurrence of a high proportion of soluble salts, usually chlorides and sulphates. They are found among the groups of red, black and alluvial soils. In the case of saline soils occurring close to the sea, the salinity is due to the action of sea water and the salts therein. In other cases, the enhanced salinity is due to the removal of salts from soils in higher topographic situations and their accumulation is low-lying areas in an extensive manner. The soils affected by alkali have high pH which may range between 9.0 and 10.5, and the exchangeable sodium may be over 15% of the total exchange capacity. The physical condition of the soils is affected considerably by the dispersion of the clay colloids.

Peaty and Peaty Saline Soils

The peaty saline soils are developed from brackish water sediments and contain a good amount of sulphides, principally composed of pyrites (FeS_2). They may be classed as very acid alluvial soils and are found to occur in the backwater areas of Kerala. When not drained, their pH may be only slightly acid. By oxidation and hydrolysis of the sulphur compounds contained in the soil, sulphuric acid is formed in varying amounts and the pH goes down to 3.0 to 4.0, and in extreme cases, a low pH of 2.0 is not unusual. They have high content of free aluminium and iron, which are formed by the breakdown of clay through the action of sulphuric acid. Accumulation of organic matter in layers of peaty matter results in the lack of oxidation and poor drainage conditions. Small areas of peat accumulation at the surface occurs in the northern parts of Bihar, and they are characterised by the presence of a high amount of organic matter, and low pH.

Skeletal Soils

The skeletal soils are so designated because of the poor thickness of the soil and the lack of profile development in them. Such soils usually occur over sandstones of the Vindhyan formation. The soils are coloured pale-brown to dark-brown and may have a thickness varying from 7 to 15 cm. The soils are susceptable to erosion. The texture of the soils is usually light, being sandy loam to loam, and reflects the origin of the soils from highly siliceous rocks. The poor nutrient supply from such soils sustains only a hardy type of natural vegetation, and the soils have low value for cultivated crops. The skeletal soils in the Himalayan and Ladakh regions are thin soils overlying fragmented sandstone and shale material. The low rainfall, poor vegetation, and unfavourable climate that induce intensive weathering have prevented the development of deep soils.

Soil Type Distribution of India for Land Use Planning

Within any ecosystem the formation of soils can be attributed to the interaction of climate and vegetation on parent rock as conditioned by topography over a period of time. Depending on the variations in one or more soil forming factors or processes, a great variety of soils with widely different properties are formed. Soil properties, such as soil depth, texture and mineralogy control the available water-holding capacity for plants which is of vital significance in land use planning.

The distribution of major soils in India is available in published soil maps (Govindrajan and Dutta Biswas, 1968 and NBSS & LUP, 1996). The National Bureau of Soil Survey and Land Use Planning superimposed the informations on soils on the physiographic map to yield a soil-scape map which provides the soil types for land use planning (Fig. 9.6).

Figure 9.6 correlates the soil characteristics and the physiography. Twenty one distinct soil types and their geographic distributions are presented in the map. The identification of the soil types and their specific distribution statewise/regionwise is given in the index to the map.

Classification of Soils by Properties (Soil Taxonomy)

A comprehensive system for soil classification, called *soil taxonomy* was introduced by the United States Department of Agriculture (USDA in 1975), based essentially on variations in soil properties distinguished by soil profile characteristics. According to the presence or absence of well defined diagnostic horizons and other profile characteristics, soils are alloted to 10 orders, with internationally recognised names derived from Latin and Greek. The orders are divided into 47 suborders, 185 great groups and a number of subgroups, families and series. The following are the 10 orders:

1. **Alfisols.** Soils with a clay-enriched horizon, through translocation, and medium to high base saturation and susceptable to erosion hazard. They are predominant in the semi-arid tropics (SAT).
2. **Vertisols.** Shrinking and swelling dark clay soils which in dry season develop deep, wide cracks. Material from upper part usually falls into the cracks and cause partial inversion of soil.
3. **Entisols.** Soils with no differentiated horizons. They are the common soils of alluvial areas and steep slopes. The common feature is the lack of profile development.
4. **Inceptisols.** Soils in which the horizons are weekly developed. The profile development is more advanced than that of entisols but less than that of other orders.

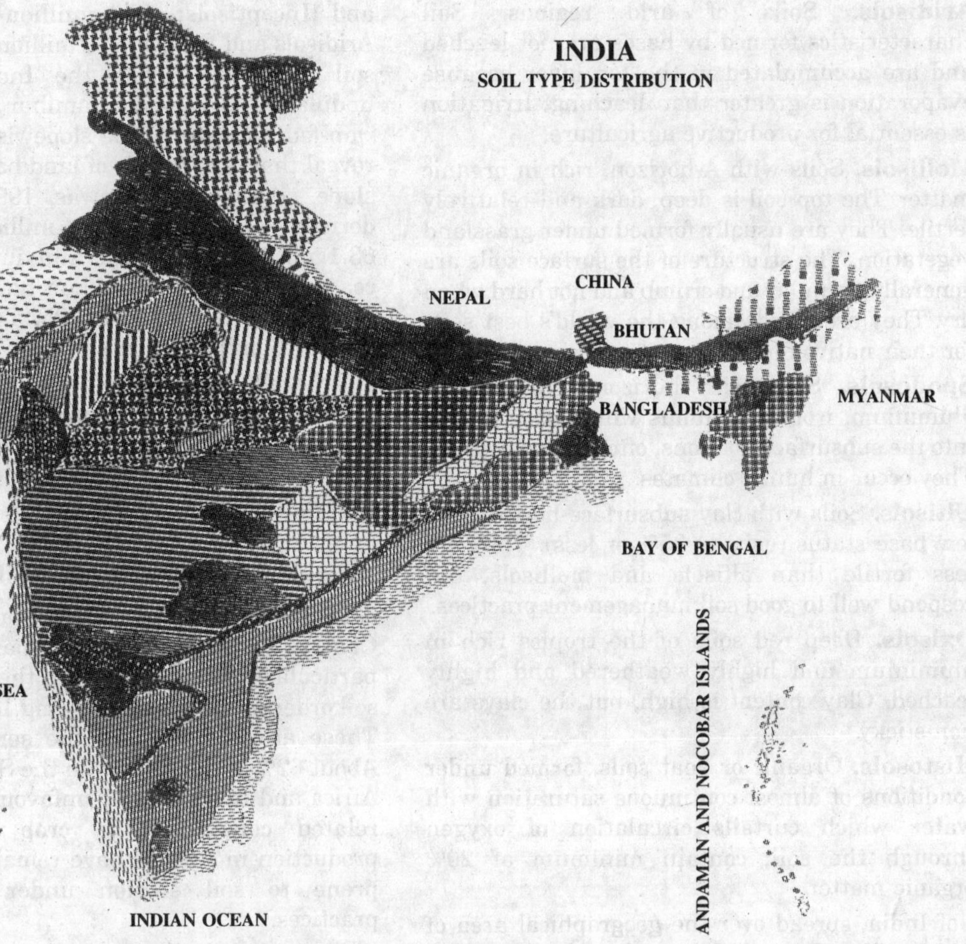

Fig. 9.6. Soil type distribution map of India (not to scale)

(***Source:*** Adapted from N.B.S.S. Pub. No. 35, National Bureau of Soil Survey and Land Use Planning, Nagpur, 1999)

Index to Fig. 9.6

- Brown Forest and Podzolic Soils of North-West Himalayas.
- Alluvium Soils of Western, Northern and Eastern Indo-Gangetic Plains and Bengal and Assam Plains.
- Terai Soils of the foot hills of Central and Eastern Himalayas, including Darjeeling and Jalpaiguri and northern hilly strips of Assam.
- Desert and Saline Soils of Western Rajasthan, Kachchh and pockets of Haryana and Punjab.
- Mixed Red and Black Soils of Deccan Plateau and including Telengana Plateau, western part of Eastern Ghats, Bellary and Anantapur Region of North Karnataka Plateau.
- Medium and Deep Black Soils of Central Highlands and Narmada Valley, including Malwa Plateau, Bundelkhand Upland and Kathiawar Peninsula (Gujarat) and small strips in the Eastern Ghats region.
- Red and Lateritic Soils of Deccan and Chhotanagpur Plateau, and North-Eastern hills.
- Red and Yellow Soils of Eastern Plateau adjoining the Central Highlands and covering the states of Madhya Pradesh, Chhattisgarh and Orissa.
- Red Loamy Lateritic and Brown Hill Soils of Eastern Himalayas and North-Eastern Hills.
- Deep Black Soils of Maharashtra, Andhra Pradesh, Karnataka, Madhya Pradesh and Gujarat.
- Red and Yellow Soils of Kachchh and Kathiawar Peninsular Region.
- Mixed Red and Black Soils of Decean Plateau.
- Red Loamy Soils of Eastern Ghats and Deccan Plateau.
- Alluvial Soils of East Coast and Kachchh and Kathiawar Peninsular Region
- Medium and Deep Black Soils of Eastern Ghats and Kachchh and Kathiawar Peninsular Region.
- Shallow Skeletal Soils of Ladakh plateau and ranges of Kashmir Himalayas.
- Brown and Red Hill Soils of Eastern Himalayas (Sikkim).
- Sandy and Littoral Soils of Andaman and Nicobar Islands, Lakshadweep and Minicoy Islands, and Daman and Diu.
- Red and Lateritic Soils of Western Ghats.
- Red Lateritic Soils of West Coast Region.
- Saline and Alkali Soils of Kachchh and many pockets of Haryana, Uttar Pradesh, Punjab, Maharashtra, Karnataka, Tamil Nadu, Goa and Gujarat.

5. **Aridisols.** Soils of arid regions. Soil characteristics formed by bases are not leached and are accumulated in the top layer because evaporation is greater than leaching. Irrigation is essential for productive agriculture.
6. **Mollisols.** Soils with A-horizon, rich in organic matter. The top soil is deep, dark and relatively fertile. They are usually formed under grassland vegetation. The structure of the surface soils are generally granular and crumb and not hard when dry. They are rated among the world's best soils for their native fertility.
7. **Spodosols.** Soils with horizons enriched in aluminium, iron and humus which have moved into the subsurface horizons, often called *podzols*. They occur in humid climates.
8. **Ultisols.** Soils with clay subsurface horizon and low base status (usually 35% or less). They are less fertile than alfisols and mollisols, but respond well to good soil management practices.
9. **Oxisols.** Deep red soils of the tropics rich in aluminium and highly weathered and highly leached. Clay content is high, but the clays are non-sticky.
10. **Histosols.** Organic or peat soils, formed under conditions of almost continuous saturation with water which curtails circulation of oxygen through the soil; contain minimum of 20% organic matter.

Soils of India, spread over the geographical area of nearly 328.7 million ha could be broadly classifed under 8 soil taxonomic orders as follows:

(*i*) **Entisols**, 80.1 million ha (24.37% of the total land area), (*ii*) **Inceptisols,** 95.8 million ha (29.13%), (*iii*) **Vertisols**, 26.3 million ha (8.02%), (*iv*) **Aridisols,** 14.6 million ha (4.47%), (*v*) **Mollisols**, 8 million ha (2.43%), (*vi*) **Ultisols,** 0.8 million ha (0.24%), (*vii*) **Alfisols,** 79.7 million ha (24.25%), and (*viii*) **Oxisols**, 0.3 million ha (0.08%). There are also Histosols in isolated areas of Kuttanad in Kerala and the Dal Lake area in Kashmir. About 23.1 million ha (7.01%) of land are not classified (Prasad and Biswas, 1999). The National Bureau of Soil Survey and Land Use Planning, Nagpur, has developed a detailed soil map of India at 1:7 million scale at soil suborder association level (totalling about 103 soil suborders) based on analysis of 80,000 profile samples collected at 10 km grid. The soils were further classified following soil taxonomy up to family level. Nearly 500 soil series have been identified, out of which 180 series have been given the status of established series and 78 series a tentative status.

From the viewpoint of soil groups based on Soil Taxonomy, the soils of India comprises of red and laterite soils (*Alfisols, Oxisols, Ultisols, Inceptisols*: 117.2 million ha), black soil (*Vertisols* and their associations: 73.5 million ha), alluvial soils (Entisols and Inceptisols: 58.4 million ha), desert (mostly Aridisols and Entisols: 30 million ha) and the rest other soil types. Excepting the Indo-Gangetic alluvium, undulating terrain is a familiar feature particularly in rain-fed ecosystem. The slopewise distribution of India reveal that nearly 60% of land belongs to more than 3% slope. (Prasad and Biswas, 1999). Broadly, alluvium derived soils extend to 73.4 million ha, inland alluvium 63.1 million ha, deltaic alluvium 6.8 million ha and coastal alluvium 4.4 million ha. The black soils cover 73.2 million ha, red soils 105. 5 million ha and laterite soils 87.6 million ha, desert soils 30 million ha and soils of the Himalayan and Shivalik region 28.7 million ha.

Amongst the taxonomic soil groups. Alfisols and Vertisols are the most important from the point of view of efficient soil management to protect them from soil degradation.

Alfisols and Associated Soils

Alfisols occur extensively in South Asia, Western and Central Africa and in many parts of South America, particularly Brazil. They are the third most important soil order in the world, covering 13.1% of the world area. These are abundant in the semi-arid tropics (SAT). About 62% of the Alfisols in the SAT are located in West Africa and India. Due to unfavourable weather and soil related constraints to crop growth, agricultural production in Alfisols have remained low, unstable and prone to soil erosion under traditional farming practices.

Most of the areas of Alfisols (red and related soils) in the semi-arid tropics of India fall in three states, namely, Andhra Pradesh (18.8 million ha), Karnataka 11.5 million ha and Tamil Nadu (10.8 million ha). They are also spread over significant areas in Maharashtra, Kerala, Goa, Chhattisgarh, Madhya Pradesh, Orissa, Jharkhand, West Bengal and Assam.

The red soils of semi-arid tropics (SAT) of India fall mainly in 5 orders of Soil Taxomony, viz., Alfisols, Inceptisols, Oxisols, Ultisols and Entisols. But Alfisols are predominant in SAT. SAT environments are identified at the sub-order level in Soil Taxonomy by the denotation of the *Ustic* moisture regime. This term implies dryness during part of the year but moisture is present at a time when conditions are suitable for plant growth. The specific definition of *Ustic* is based on mean annual soil temperature and duration of the period in which the control section of the soil profile remains moist or dries out. This occurs in tropical regions with rainy season lasting 3 months or more.

Alfisols possess an argilic horizon within the profile. The clay content increases with depth, although shallow and gravelly Alfisols are common as a result of soil erosion. The predominance of coarse particles in surface layers is assumed to be the result of clay migration with percolating water, termite activity or selective removal of fine particles by erosion. The soils may contain distinct layers of gravel and weathered rock fragments at lower depths (called *murrum*). Effective rooting depths of crops may be restricted by the presence of limited soil depth or by the compact argillic horizon. Restricted root development on these soils prevent many crops from withstanding even moderate droughts.

Cultivated Alfisols of semi-arid tropics are characterised by lack of structural development, a consequence of low content of fine particles in the surface horizon, clay minerals of low activity (Kaolinate with small proportion of 2 : 1 layer lattice clays and sesquioxides) and small amount of decomposed organic matter.

A major consequence of lack of aggregation or unstable aggregation is the tendency of these soils to display rapid surface sealing following rainfall, and crusting with subsequent drying cycle. Crusting adversely affects seedling emergence and plant establishment. Alfisols also depict the phenomenon of slumping, slaking, packing and compaction.

Alfisols of SAT have low nutrient and moisture retention capacity (0.03-0.1 cm/cm). Since the rooting depth is generally restricted to 20-30 cm the crops suffer from water stress. Because of crusting and surface sealing the runoff in these soils, particularly in the early part of growing season, is high.

Alfisols are generally well drained, particularly after the establishment of a crop, leading to leaching of water soluble nutrients. The soil erodibility is high. They have low moisture and nutrient retention capacity. Hence, they are cropped only in the rainy season, unless irrigation facilities are available. The surface soil layers are often dominated by coarse particles. The soils may contain distinct layers of weathered rock fragments at lower depths (often called *murrum*). Effective rooting depth of crops could often be restricted either by the limited soil depth or by the compact argellic horizon (with heavy clay concentration) which may restrict root penetration. Restricted root development on these soils prevents many crops from withstanding even moderate droughts.

Alfisols of SAT are characterised by the lack of structured development, resulting from a low content of fine particles in the surface horizon, presence of clay minerals of low activity and the tendency to stabilise only minute amounts of decomposed organic matter within the soil mass. The increased clay content with depth distinguishes these soils from other 'sandy' soil groups (e.g., Entisols). Crusting can adversely affect the establishment of the plant. Crusting often extends deeper than the immediate soil surface, resulting in consolidation of the soil profile. Alfisols permit easy tillage when wet, but become hard and difficult to till when dry. However, tillage operations when the soil is too wet may result in excessive compaction. Thus, only a limited soil moisture range is available to produce optimum tilth.

Alfisols usually possess low water retention characteristics because of their mineralogical composition. Shallow depth of the potential crop root zone further reduces the water-holding capacity of the soil. Lack of moisture storage, combined with mechanical impedance problems, result in limited crop root proliferation. Soil structural instability and surface sealing and crusting induce excessive runoff, and adversely affect the emergence of seedlings. The adverse effects are more pronounced in crops with small seeds, such as pearl millet, finger millet and sorghum. Localised droughts are also likely in the seed environment in the ridges. Entry of water by infiltration is restricted by surface sealing and poor moisture retention characteristics. Rainfed crops of Alfisols face constant threat of soil moisture stress even during short dry spells.

In many SAT Alfisols, soil workability in the dry state and the presence of stones are deterrants to effective land preparation. Since the soils are usually well drained, leaching of water soluble fertilizers is common with heavy rains. The losses by leaching are enhanced by low nutrient-retention capacity. The cation exchange capacity range between 1 and 10 milli-equivalents per 100 grams.

The lack of active clays in Alfisols indicates the importance of organic matter in influencing the soil's structural characteristics. Hence, in humid tropics the management of Alfisols for sustained agricultural productivity is easier than in the semi-arid tropics. In humid tropics, abundant and uniformly distributed raifall allows continuous biological activity and build-up of organic matter. Alfisols display wide deversity in spatial variability.

Rainfed farming in SAT Alfisols is practiced only during the rainy season. The growth period for sequential cropping in Alfisols is extended into the post rainy season in areas where facility for supplemental irrigation is available. The main traditional crops of SAT Alfisols include sorghum, pearl millet, finger millet, groundnut, pigeonpea, cotton, castor, green gram (mung bean) and black gram. They are grown as monocrops or as intercrops, depending on farmers' choice and other factors. Inter-cropping of cereals with pulses or oilseeds is common. In the SAT Alfisols of India, supplemental irrigation facilities from tanks, wells, and canal irrigation projects are common.

Under the traditional farming system, the fields of SAT are relatively bare when the monsoon begins. The unstable structure of these soils enhances their tendency to develop surface seals and reduce infiltration and soil profile recharge. These surface seals harden into crusts during the dry periods. They prevent establishment of protective crop cover during the peak monsoon period, resulting in heavy runoff and soil loss.

Most SAT Alfisols have much higher production potential than that is realized under traditional farming system. Land smoothening and drainage facilities are essential for improved management of Alfisols. Tillage operations, especially primary tillage should be appropriate to provide sustained productivity. A favourable zone for root proliferation will enhance infiltration and soil moisture storage. Secondary tillage operations are necessary for seadbed proparation and weed centrol. Breaking of soil crusts through well designed rolling crust breakers with spokes mounted at predetermined positions can substantially improve seadling emergence under suitable soil moisture conditions. Deep ploughing, once in 3 years, is a useful practice for soils having a compact subsoil. Addition of organic mulch increases crop yields in Alfisols. Building up of organic matter levels is important to develop favourable soil physical and chemical conditions. Harvesting runoff or development of other water resources for supplemental irrigation will increase crop productivity in Alfisols substantially.

Water-nutrient interactions play an imporant role in drought response of crops. Crop water use efficiency could be improved through appropriate soil and water management practices, selection of appropriate crop varieties and cropping systems. Graded bunding for shallow to medium Alfisols and graded contour border strips for deep soils are suitable soil and water conservation measures. Sowing flat on grades not excessing 0.5% and ridging later after the seedlings establish are appropriate practices, especially for crops like sorghum. Moderate and balanced fertilizer use increase and stabilize productivity. Integrated nutrient management for the nutrient deficient Alfisols is needed. Other improved practices are proper crop establishment through timely operations, use of high quality seeds, precision fertilizer placement, timely weed control and effective plant protection measures (Singh and Das, 1983). The International Crop Research Institute for Semi-Arid Tropics (ICRISAT), the Central Research Institute for Dryland Agriculture (CRIDA), Hyderabad. All India Coordinates Project for Dryland Agriculture and the State Agricultural Universities located in the semi-arid tropics have developed valuable information suitable for large scale field adoption in the Alfisols of SAT.

All India Coordinated Research Project for Dryland Agriculture (AICRPDA) was started in 1970 at 23 centres and historical rainfall data were obtained for all the stations and analysed, particularly on (*i*) the date of start and termination of rainy season and its variability, and (*ii*) the ditribution of rainfall within the rainy season, and (*iii*) assessing inevitable runoff. Suitable crops and their varieties for normal and late commencement of rains, and suitable inter-cropping systems have been identified.

Vertisols and Associated Soils

Vertisols and associated soils (Vertic sub-groups) constitute one of the most important and widely occurring soils of the world in general and the Indian subcontinent in particular. Their unique morphology and high coefficient of shrinking and swelling properties result in their specific genetic implications. In India, Vertisols and associated soils occupy 72.9 million ha, or 22.2% of the total geographical area. They are commonly called as black cotton soils. The associated soil groups are contiginuous to Vertisols. Globally, Vertisols and associated soils are located between 39°58' N and 36°57′ S latitudes and 74°30′ W and 154°41′ E longitudes. The major concentration of these soils is in the eastern half of Australia and the central-north-western parts of peninsular India.

Of the total area of 72.9 million ha in India under black cotton soils, Vertisols constitute 26.3 million ha, amounting to 8.02% of the geographical area. The Vertisols are 70 cm or more deep. The rest, the associated soils, also called Vertic soils are Vertic Inceptisols and Inceptisols. Medium deep (30–70 cm) Vertic soils are called Vertic Inceptisols. Shallow black soils having less than 30 cm depth are Inceptisols.

In India, Vertisols and other associated soils are mostly concentrated in peninsular India extending from 8°45′ to 26°0′ N latitudes and 68°0′ to 83°45′ E longitude. These soils are distributed in the states of Maharashtra, Madhya Pradesh, Gujarat, Andhra Pradesh, Karnataka, Tamil Nadu, Rajasthan, Orissa, Bihar and Uttar Pradesh. Maharashtra and Madhya Pradesh account for a major portion of the Vertisols and their associates. Maharashtra has 7.9%, Madhya Pradesh 5.1%, Gujarat 2.6%, Andhra Pradesh 2.2%, Karnataka 2.1%, Tamil Nadu 1.0%, Rajasthan 0.7%, Orissa 0.4%, and Bihar 0.2% of the total geographical area of the country under Vertisols and associated soils. The climatic conditions of the Vertisol regions, though predominantly semi-arid, also include arid, dry sub-humid and moist sub-humid regions.

Vertisols and associated soils occur on plateaus and are confined mainly to elevations of 300 to 900 m above mean sea level with high summer and mild winter temperatures. The factors contributing to the development of Vertisols and associated soils are: (1) landforms in lower topographical situations, (2) medium to low rainfall confined to specific period, (3) high temperature, (4) alternate wet and dry periods, (5) basic character of rocks and parent materials rich in alkaline earths, and (6) restricted drainage. (Murthy *et al.*, 1982).

The soilscape of Vertisols and vertic subgroups are closely associated with three major landforms: (*i*) gently sloping extensive erosional surface, (*ii*) level to gently sloping piedmont plain and floodplain, and (*iii*) delta and coastal plains. The climate is influenced by variations in microtopographic units causing locally arid and humid soil climate in accordance with slope variations. In the Vertisol dominated peninsular India rainfall varies from 60 mm to over 1200 mm and is mainly confined to the south-west monsoon period. In Gujarat it is about 400 mm in the Vertisol areas.

Soil texture of Vertisols is generally clay loam, silty clay loam, silty clay, and clay. The clay content ranges from 40 to 60%, but it may sometimes go as high as 80%. The organic carbon is low, ranging from 0.3 to 0.7%. The distribution of clay and organic matter is nearly uniform through depths up to 1 to 1.6 m. The distribution of clay in associated vertic subgroups increases up to a particular depth and thereafter decreases. Organic carbon decreases regularly with depth, indicating the dominance of the process of horizontal layering of *A*, *B* and *C* horizons.

The bulk density of soils is normally high in Vertisols. In India, the normal range, is 1.5 to 1.8. The cation exchange capacity (CEC) of Vertisols in normally high ranging from 35 to 50 me/100 g, mainly due to relatively large fine clay < 0.2 μ fraction. (Murthy *et al.*, 1982).

Montmorillonite is the dominant clay mineral in black soils. Significant relationships exist in the composition of the clay fraction, silt and fine sand of black soils and red soils. Varying rainfall and topography influence the development of Vertisols in basaltic terrain.

Soil and Crop Management in Vertisols and Associated Soils

The beneficial factors of Vertisols and their associated vertic subgroups are: favourable surface structure, deeper *A* and *B* horizons in the profile, high water-retentive capacity and high cation exchange capacity. An essential requirement of improved crop production in Vertisols and associated soils is effective soil and water conservation practices and appropriate agronomic practices. The major difference between Vertisols and associated Vertic soils lies in their effective soil depth. This affects their water-holding capacity, which in turn influences their management for dryland crop production.

The rainfall environment and crop growing periods of Vertisols and associated soils may be grouped into two distinct classes: (*i*) Regions with dependable rainfall, where the annual rainfall exceeds 750 mm, and (*ii*) Regions with undependable rainfall where the annual rainfall in less than 750 mm. The first group has great potential for crop production, but is prone to high rates of soil erosion, unless appropriate soil conservation measures are practiced. The crop growing season for rainfed farming ranges from 180–210 days in Vertisols, 130–150 in Vertic Inceptisols and 90–120 in Inceptisols. Inceptisols have only medium potential for crop production, but are less susceptible to soil erosion, as compared to Vertisols and Vertic Inceptisols. In the second group of undependable rainfall where the annual rainfall is less than 750 mm, the Vertisols have growing period ranging from 90-110 days. The crop productivity is medium, while soil erosion in high. In Vertic Inceptisols in this group, the growing period is around 90 days, the productivity is low and erosion hazard is medium. Inceptisols in this group have low productivity and are less susceptible to erosion.

1. Management of vertisols and associated soils. Vertisols of dependable rainfall areas located in the flood plains, valley bottoms and plateaus often have severe drainage problems. Successful farming in these areas require adequate drainage provision. The following are the major recommended practices in Vertisols and associated soils, based on the studies by ICRISAT and CRIDA, as reported by Virmani *et al.*, (1986):

(*i*) Adoption of land management practices that reduce runoff and erosion, and provide improved surface drainage leading to better aeration and workability of the soils.

(*ii*) Cropping systems and crop management practices are to be aimed at establishing a crop at the very beginning of the rainy season which will make efficient use of moisture throughout the rainy and post-rainy seasons.

(*iii*) Use of appropriate implements for cultivation, sowing and fertilizer application.

(*iv*) Adoption of adequate plant protection practices, as required.

2. Improved management of vertisols and vertic inceptisols in undependable rainfall areas. These areas are located mainly in rain-shadow regions of Maharashtra and Karnataka. The average annual rainfall ranges between 500 and 750 mm and is characterised by undependable onset and frequent dry spells during the monsoon. Hence, cropping in these regions is delayed till sufficient moisture is available in the soil profile which results in delayed or extended monsoon season (*Kharif*) cropping, utilising only the stored soil moisture.

3. Management of vertic inceptisols. In the Vertisols region of India there is considerable area under associated shallow Vertic soils which presents a serious challenge for the production of economically viable crops. Such areas are prone to high degree of soil erosion. Because of shallow soil depth and other physical constraints, the Vertic Inceptisols do not provide a good plant nutrient base. In such soils agroforestry, silvi-pastoral land use and forestry may often be more advantageous.

Soil Degradation

Land degradation is a human induced or natural process which results in drastic or gradual reduction in the productivity of the soil. It is one of the greatest challenges facing mankind. The term soil degradation has been used rather vaguely. In 1982, UNEP defined it as "decline in soil quality caused through its misuse by humans". They further clarified that it refers to "decline in soil productivity through adverse changes in nutrient status and soil organic matter, structural attributes, and concentration of electrolytes and toxic chemicals. In other words, it refers to domination of the soils' current and/or potential capability to produce quantitative or qualitative goods or services as a result of one or more degradative processes" (Lal and Stewart, 1990). Bhumbla and Khare (1983) gave the following definition of wastelands. "Wastelands are the lands in which the productive capacity of the soil has been reduced due to loss of organic matter and nutrient rich surface soil, or loss of soil structure or accumulation of toxic substances in the root zone". Soil degradation could be considered as a process(s) resulting in loss of production potential of a soil.

Processes of soil degradation. Soil degradation processes include the following:

A. Physical: (*i*) Erosion by water (*ii*) Erosion by wind, (*iii*) Compaction, (*iv*) Loss of soil structure, (*v*) Water-logging, (*vi*) Hardsetting.

B. Chemical: (*i*) Salinization, (*ii*) Sodification, (*iii*) Acidification, (*iv*) Fertility depletion, (*v*) Accumulation of toxic compounds, (*vi*) Laterization.

C. Biological: (*i*) Decline in soil organic matter and (*ii*) Decline in microbial population.

The following are the factors affecting soil degradation: (*i*) Land (Topography, vegetation, geology, hydrology); (*ii*) Climate (Rainfall, temperature, humidity, wind velocity); (*iii*) Land use (Cultivated crops, grasslands, shrubs and forests); (*iv*) Socio-economic situation (land tenure, human and livestock population, government policies); and (*v*) Urbanisation.

Extent of Soil Degradation in India

The extent of soil degradation in India has been assessed by various agencies. The estimates show wide reliability, probably due to the different approaches in defining degraed soils or variation in criteria of delineation. Precise information about the extent of soil degradation in India is not available. The estimates vary from about 60 million hectares to 189 million hectares (Parsad and Biswas, 1999). The estimated area under different soil degradation problems as reported by the Ministry of Agriculture, Govt. of India are as follows: Water erosin: 104.4 million ha; Wind erosion, 40.0 million ha; Water-logged area, 8.5 million ha; Alkali lands, 3.6 million ha; Saline and coastal sand, 5.5 million ha; Ravine and gully, 4.0 million ha; Riverine, 2.7 million ha; and Shifting cultivation, 4.9 million ha (Total 173.6 million ha).

It has been estimated that a total of more than 5,000 million tonnes of top soil is being eroded every year. Of this, about 30% of the total eroded mass is permanently getting lost to the sea (Singh, G.B., 1999).

The chief cause of land degradation is human intervention, resulting from population pressure. They lead to loss of vegetation through deforestation, in man's effort to bring more land under cultivation. Shifting cultivation leads to serious soil erosion. Lack of appropriate soil conservation measures and conservation farming practices, non-judicious use of fertilisers and pesticides, improper irrigation system planning and water management, discharge of industrial effluents and sewage/sludge are the other factors leading to soil degradation.

Among the soil groups which are prone to serious degradation are the red soils (mostly Alfisols, Inceptisols and Ultisols), the black soils (Vertisols and Vertic sub-groups) and the mountain soils (Alfisols and Inceptisols). Vertisols of Peninsular India are highly susceptible to erosion because of heavy runoff (about 40 to 50% of total rainfall), poor vegetative cover and erosive land slope. The red soils which have low infiltration due to crusting are prone to erosion. The lateritic soils suffer from severe soil erosion in the absence of proper soil conservation measures.

Wind erosion is a serious problem in the arid and semi-arid regions of the Western Rajasthan and parts of Gujarat, Haryana and Punjab. Wind erosion is also prevalent in the coastal areas having predominantly sandy soils. Soil erosion in directly linked to soil health, which in turn affects productivity and sustainability.

Chemical degradation of soils occur mainly due to the loss of nutrients and organic matter and through accumulation of salts and pollutants. Depletion of nutrients and organic matter adversely affect soil fertility and microbial biomass. Excessive leaching of basic cations leads to soil acidification. Among the soil types, the red and laterite soils, representing Alfisols, Ultisols and Oxisols, are more prone to chemical degradation, especially in the sub-tropical regions and areas under shifting cultivation. Soils suffering from salinity and alkalinity problems are prevalent in arid, semi-arid, sub-humid, and coastal regions.

Physical degradation of soils are mainly due to waterlogging. Low lying areas, canal irrigated areas without provision for drainage and flood prone areas usually suffer from physical degradation. The adverse effects of temporary/long term waterlogging are reflected in ecology degradation, reduced agricultural productivity, limited choice of crops and a series of adverse socio-economic conditions.

Soil degradation is a major concern to our planned effort to increased production and sustainability. To achieve sustainability in agricultural production it is of vital importance to halt further deterioration of soils and maintain improved soil quality/health through alternative agricultural practices such as efficient soil and water management, adoption of farming systems with increased emphasis on soil building and soil maintaining crops, adoption of appropriate crop rotations, conservation tillage practices integrated nutrient and water management and effective pest management.

Development of degraded loads is one of the major opportunities in most of the developing countries to enhance food production and to restore the fragile ecosystem. The information on the extent and spatial distribution of degraded lands is a prerequisite for formulation of development action plans and monitoring. The wide variation in the estimates on the extent of degraded lands in India by different agencies has been due to lack of accepted methodology and criteria for characterising degraded lands. The All India Soil and Land Use Survey (AISLUS) of the Ministry of Agriculture, Government of India initiated land degradation mapping on district basis, using remote sensing technique during the Eighth Five-Year Plan (1990-95). The objective has been to generate realistic information on degraded lands and their spatial distribution for use in developmental planning. Suitable methodology for land degradation mapping, using remotely sensed data has been developed. The mapping has been based on the kind and degree of degradation and degradation under major land use and landforms. The various causative factors include both natural and biotic degradation processes, including water and wind erosion, waterlogging, salt affliction, shifting cultivation and mining. Remote sensing from satellite in conjunction with ground truth verification provides a means of speedy survey of degraded lands. For the purpose of mapping, the whole country has been divided into four major landscape classes, namely, plain land, undulating land, rolling land, and hilly/mountainous land, based on slope ranges. Land degradation mapping on 1:50,000 scale, using remotely sensed data provides the first hand information for macro-level planning. Detailed planning, however, will require micro-level data on soils and associated features of the land. Cadastral maps would be needed for depiction of processed field data on soils, vegetation and climate. Development of database on degraded land, using remotely sensed and Geographic Information System (GIS) data are of vital importance, in development planning.

The International Conference on Land Resources Management for Food, Employment and Environmental Security held in New Delhi in November 2000, under the auspices of the Soil Conservation Society of India, adopted the **New Delhi Declaration 2000 A.D.** on Land Resources Management containing far reaching policy guidelines on "the conservation, enhancement and sustainable and equitable use of the land resources by the people of all nations, both individually and collectively, for the benefit of the present and future generations. It states that common property resources on which the rural poor survive are highly degraded and encroached upon. The land base per person is shrinking, posing the challenge of maintaining the soil health, while obtaining more and more from less and less area and quality of land."

SOIL FERTILITY MANAGEMENT

Soil fertility is the inherent capacity of a soil to supply essential nutrients to plants in adequate quantity and right proportion for their optimum growth. Proper management of soil fertility demands identification of current nutrient deficiencies and monitoring changes in soil fertility under varying management levels. The deficiencies are to be removed through sound and proven practices of nutrient, water, crops, and soil management practices.

The requirements of soil fertility management for sustainable productivity vary from sol to soil. The problems include acidity in hill and laterite soils, nutrient leaching in sandy soils, nutrient fixation in red, lateritic and clayey-soils, restricted drainage in black soils, and volatilisation losses of nitrogen (N) from calcareous soils. Zinc deficiency is common in sandy soils with low organic matter content, high pH and high water table. Boron deficiency is often observed in calcareous and leached acid soils. Soils with wheat and rice cropping systems frequently encounter problems of Fe and Mn deficiency, mainly in sandy soils. Salt-affected soils frequently have Ca, N and Zn deficiencies. The N/K relationships in sodic soils require careful management. In acid soils Ca, P and B deficiencies and Al taxicities are common fertility constraints.

Continuous cultivation of the same crop result in poor fertility, unless soil building cropping systems and crop management are practiced in rotation. Soil fertility and crop response studies of ICAR have revealed that most soils in India have low to medium available P and 60% soils have medium K status and N is deficient in most soils. About 47% of soils are deficient in Zn, 12% in Cu and 4% in Mn. Phenomenal increase in S deficiency has been observed in recent years (Takkar and Biswas, 1999).

The nutrient requirement of crops can be met through various sources, but their effectiveness varies with the method of application. Fertiliser (inorganic salts), synthetic compounds, organic manures, composts, bio-fertilizers and organic wastes are the main nutrient sources for soils. The materials vary in their nutrient content, effectiveness and cost for different soil and crop conditions. The consumption of macro-nutrients (NPK) in India increased to 14.3 million to tonnes per annum in 1999. On all India basis, the fertilizers use efficiency is estimated to be low, at 30 to 40%. Continuous cereal to cereal rotation has resulted in increased nutrient deficiency (Singh, G.B., 1999).

Nitrogen. Nitrogen is the most important plant nutrient influencing the fertility level of soils in India. Its status is low in 62%, and medium in 36.6% of the soil samples tested. In general, higher values of N are found in the surface soil layer. The N content decreases sharply at depths below the surface. N level in soils is influenced by the rainfall, temperature and elevation of the region. Continuous farming decreases the N status, unless, leguminous crops are included in crop rotation. Addition of organic matter increases N status. The response of crop, especially rice and wheat, to N application are usually very good in the soils of north India (Takkar and Biswas, 1999).

Various procedures have been developed to improve the nitrogen use efficiency of soils. They include root zone placement of N, use of urea coated neem-cake/rock phosphate and gypsum. Synchronizing N application with peak period requirement of crops has been beneficial. The number of N applications usually vary with the soil texture. In fine textured soils a single basal application is adequate. In coarse textured soils split dozes of N are more efficient. Placement of N, particularly in rice at a depth of 5 to 8 cm below the surface has been found to be beneficial (Takkar and Biswas, 1999).

Studies of the ICAR institutes and projects have revealed that the application of nitrate containing fertilizers to lowland rice or waterlogged soils usually results in loss of N due to denitrification. Similar is the situation in coarse textured soils in high rainfall regions. In light textured acid soils, losses in the form of ammonium and nitrate through surface runoff and leaching are common.

Phosphorus. Deficiency of P is widespread. Both low and medium P soils require P application to maintain soil fertility. Phosphorus use efficiency is influenced by P-fixing capacity of soils, crops, as well as the method, rate and time of P application. Phosphorus fertilizers or materials of high water solubility are better in neutral or alkaline soils. In general, phosphatic fertilizers are low in efficiency due to chemical fixation in soil. Further, native P is usually not available to crops due to low solubility. Thus, there is build-up of insoluble phosphates in soil. Introduction of P solubilizers in the rhizosphere of crops and soils increases the availability of P from insoluble sources of phosphorus and its use efficiency (Chonkar, 1994). Fertilizer P applied to a crop is not fully utilized by the crop. A part of the applied P gets fixed in the soil. Management of the fixed P or built up fertility is important.

Potassium. Potassium is absorbed by plants in larger amounts than any other elements, receipt nitrogen. Apart from that added in fertilizers, potassium contained in soils originates from the disintegration of rocks containing potassium-bearing minerals. Unlike phosphorus, potassium is present in large quantities in most soils. In tropical soils the total content of potassium is low because of their origin, high rainfall and continued high temperatures. High rainfall and temperature facilitate the release and subsequent leaching of soil potassium over the years. The long term fertilizer experiments conducted by the ICAR reveal increasing deficiency of K in most Indian soils. Even under optimum rates of NPK application, the K balance under most of the soil-cropping systems have been low. The low K fertility status of the soils often limit the responses of crops to N and P as well. For proper management of soil K fertility it is necessary to monitor the exchangeable and non-exchangeable status of K in the soil and add adequate quantity of K from fertilizers, manures and crop residues to compensate for the gradual loss of soil K fertility.

Sulphur. Sulphur is considered as the 13th most abundant element on the earths' crust. The average sulphur content range from 0.06 and 0.1%. Sulphur occurs in the element form and as sulphides, sulphates, and in organic combinations with carbon and nitrogen. Sulphur in most arable land is in the form of organic matter, sulphates in the soil solution or adsorbed on the soil complex. Sulphur is a component of protiens. When protein materials are returned to the soil and converted to humus, a large fraction of the sulphur remains in organic combination. Much of the total sulphur in the surface horizon of humid region soils is in the organic form. In arid soils, the sulphates of Ca, Mn, Na and K are often precipitated in large quantities in the soil profile. Appreciable quantities of exchangeable SO_4^{2-} may be present in subsoils that contain 1:1 clays and hydrous oxides of iron and aluminium, (Tisdale *et al.*, 1985).

S deficiency is widespread in many soils in India. Studies by ICAR revealed that intensively cultivated coarse-textured soils of the Indo-Gangetic alluvial plain are deficient in S. Crops like groundnut, rapeseed-mustard, chickpea, soyabean, wheat, rice sugarcane and potato respond markedly to S application. S deficiency has been widespread in fine textured black soils as well. In neutral to calcareous soils, three kinds of applications have proved effective in correcting S deficiency, namely, (*i*) broadcast, (*ii*) broadcast and mixing, and (*iii*) drill and split applications. On Alfisols, application of S, 75% at sowing as broadcast and mixing and 25% S as broadcast at flower-initiation stage, enhance crop yields (Takkar and Biswas, 1999).

Calcium. Calcium is absorbed by plants as the ion, Ca^{2+}, from the soil solution. The calcium concentration of the earth's crust is about 3.64% which is more than most other plant nutrients. Soils vary widely in Ca content. Sandy soils of the humid regions are low in Ca. Calcium deficiencies, leading to yield reductions are not common in soils, since most soils contain sufficient amounts of Ca required by crops.

Magnesium. Magnesium a absorbed by plants from the soil solution as the ion Mg^{2+}. Like Ca, the availability of Mg in soils is usually adequate for plant growth. Magnesium, like Ca can easily be leached from soils. Magnesium deficiency occurs in the acid soils. The MgO content in the acid soils of India is low, varying between 0.1 and 0.6% (Takkar and Biswas, 1999).

Plant species and varieties differ in their magnesium requirement. Pastures, maize, potatoes, oil palm, cotton, citrus, tobacco and sugarbeets are among the crops which are highly responsive to magnesium.

Micronutrients and Other Beneficial Elements in Soils and Fertilizers

Boron, cobalt, copper, iron, manganese, molybdenum and zinc are commonly classified as micronutrients. Sodium, chlorine and silicon are less known, but are beneficial for the growth of some plants.

Boron. Boron is the only non-metal among the micronutrient elements. Boron occurs in low concentration in the earth's crust and in most igneous rocks and soils. Boron concentration in soils usually ranges from 7 to 80 ppm. Large amounts of boron can retard the growth of plants and can also be toxic. Normally, boron does not occur in toxic quantities in most arable soils, unless added in excessive quantities in commercial boron fertilizers. In arid region, it can occur naturally or may develop due to irrigation with water containing high amounts of boron. Boron deficiency is usually associated with dry weather and low soil-moisture. Plants differ in their born requirements. The crops which are most sensitive to born deficiency include sugarbeets, apple and many vegetable, crops, including cabbage and cauliflower, and sunflower. Genetic variability of boron sensitive plants greatly influences boron uptake and concentration. Semi-tolerant crops include, barley, grapes soyabeans, peas, maize, lucern, tobacco, tomato, cotton, turnips and sugarbeet (Tisdale *et al.,* 1985).

Cobalt. Cobalt is required in symbiotic N_2 fixation. Its role in higher plants is yet to be established. Cobalt deficiencies are often associated with forages produced in soils containing less than 5 ppm cobalt (Tisdale *et al.,* 1985).

Copper. Copper concentration on the earth's crust usually ranges from 55 to 70 ppm. Copper deficiency is often noticed in newly reclaimed acid Histosols. Vegetable crops such as onion and carrots, when grown on Histosols often show copper deficiency. Crops which are highly responsive to copper include wheat, rice, lucern, citrus, onion, carrots, and lettuce. There are major genotypic differences in the copper nutrition of plants. Like most micro-nutrients, copper in large quantities is toxic to plants.

Iron. Iron comprises about 5% of the earth's crust. Most of the soil iron is in primary mineral, clays, oxides, and hydroxides. Plant genotypes differ in their capability to extract iron from a growth medium. Traces of iron are found in most rocks. The average iron content of soils is about 600 ppm.

Manganese. Manganese occur in soils in the form of oxides and hydroxides. Plants and their varieties differ in their sensitivity to manganese deficiency. Soyabean is highly susceptable to Mn deficiency. The other crops include small grains, some vegetable crops and fruit trees.

Molybdenum. Molybdenum occur in soils in extremely small quantities. Its concentration is less than 1 ppm in plants. In soils it ranges from 0.2 to 5 ppm. Plant availability of molybdenum decreases with increasing soil acidity. Application of lime to correct soil acidity will improve molybdenum nutrition of plants. (Tisdale *et al.,* 1985).

Zinc. Zinc content of the earth's crust is estimated to be about 80 ppm. Zinc deficiencies are widespread in many regions of India and other countries. Soil conditions conducive to zinc deficiencies are: acid sandy soils low in total zinc; neutral soils, especially calcareous soils; soils high in clay content and silt, soils high in available phosphorus; some organic soils and subsoils exposed by land levelling operations, or by soil erosion. Zinc availability to plants is highly dependent on its pH value. Most pH induced Zn deficiencies occur within the range of 6.0 to 8.0 and calcareous soils are especially prone to this problem (Tisdale *et al.,* 1985). Species and varieties of plants differ in their susceptibility to Zn deficiency Common crops which are highly sensitive to Zn deciency are: beans, citrus, maize, deciduous fruit trees, grapes, rice, and soyabeans. Several zinc-containing sources are used to correct Zn deficiency. Zinc sulphate containing 35% Zn is a popular fertilizer.

Sodium. Sodium content in soils ranges from 0.1 to 1%, the average concentration being 0.63% (Tisdale *et al.,* 1985). The major adverse effect of sodium on soil properties is its dispersing action on clay and organic matter, resulting in a breakdown of soil structure and impairment to air and water due to the blocking of large soil pores. Penetration of crop roots is impeded, soil clods become hard and seedbed preparation becomes difficult. Surface crust restricts emergence of seedlings. The adverse effect of high levels of exchangeable Na^+ depends on the soil texture and the type of soil clays. High concentration of sodium are toxic to some plants. Growth of most plants is severely reduced on alkali soils. High pH can create deficiencies in micronutrient cations. Reclamation of sodic soils is by replacement of the excess exchangeable Na^+ with Ca^+, followed by leaching of the displaced Na^+ as sodium sulphate. Calcium-containing substances such as gypsum are used to supply Ca^{2+}. On soils with calcium carbonate, acid-forming sulphur compounds can be used to stabilize calcium.

Chlorine. Chlorine is a minor constituent in the earths' crust. Chloride uptake inplants is restricted by concentrations of NO_3^- and SO_4^{2-}. Chloride is beneficial for plants such as spinach, beet root, cabbage and to some extent in potatoes.

Silicon. Silicon is the second most abundant element in the earth's crust. Silicon plays an important role in the weathering of rocks and soil development. Plant uptake of silicon is influenced by the content of iron and aluminium oxide, liming and nutrient supply. High soil moisture content encourages silicon uptake in crops like rice.

Solenium. Solenium is not needed by common crops, but must be present in fodder crops as it is essential for animals. The solenium concentration in most soils is in range of 0.1 to 2 ppm with an average of 0.3 ppm. Plant availability of solenium is influenced by the pH of soils, inter-relationships of solenium with other elements and climate. Solenium deficiencies of animals can be overcome by solenium fertilization of forages.

Soil Biological Fertility

Soil micro-organisms and their biological processes increase the availability of nutrients under favourable physical and chemical soil conditions. Consequently, they lead to increase in soil fertility and productivity. The soil micro-organisms act as agents of nutrient-element transformations as well as store house of carbon and mineral nutrients, specially N, P and S, in their own living biomass. Soil microbial biomass carbon and soil enzyme activity are favourably influenced by crop management practices and environmental conditions. Organic matter plays numerous roles in soil. The most important is as a source of nutrients and energy to the soil biota. In agricultural soils, physical and chemical properties are influenced by the nature and amount of soil organic matter.

The maintenance of soil organic matter benefits the sustainability of agricultural production systems by improving soil physical properties and protecting the soil surface from erosion, providing a reservoir of plant nutrients, enhancing cation-exchange capacity, and stimulating biological activity. Soil-organic matter decomposition depends on the quality of organic matter inputs and the environmental conditions, particularly the soil water and temperature. In the tropics, organic matter decomposes fast because of high temperature regime.

Soil Physical Fertility

Crop yields are limited more by the physical conditions of the soil than its nutrient status. Poor physical conditions may lead to retardation of the crop through its effect on soil organisms. The basic requirements for crop growth in terms of soil physical environment are adequate supplies of moisture and oxygen, optimum soil temperature and freedom from mechanical stress. Soil-water potential controls seed germination and seedling emergence. Water relations of soil depend upon the size distribution and stability of soil pores. Management of soil for root growth is essential for creating and stabilising the right sizes of pores in the soil. Soil temperature influences seed germination, seedling emergence and early plant growth. Soil temperature shows considerable interaction with other factors such as moisture content and planting depth.

Tillage, mulching, irrigation, rotation of crops and incorporation of organic matter and amendments like gypsum on sodic soils, and liming of acid soils are major management practices for creating soil-physical environment for crop growth. Tillage loosens the soil, reduces clod size distribution, increases surface roughness and soil porosity and destroys weeds. They affect soil-water regime, resistance to erosion, mechanical impedance, aeration and soil temperature. Covering the soil surface with crop residues and mulches changes its radiant energy balance and determines the seed-zone temperature. Soil physical properties like bulk density, aggregate stability, water retention capacity and hydraulic conductivity are affected by long-term manuring and intensive cropping. Physical properties of soil modify the irrigation water-fertiliser interactions. Nitrate-nitrogen moves freely with water. Heavy irrigation at greater intervals leaches large amount of soluble fertilizers, compared to frequent light irrigations. Farm yard manures, green manures, legumes, organic wastes, and inoculation of plants with efficient strains of bacteria *Rhyzobium*, *Azotobactor*, *Azospirillum*, and blue-green algae enhance the nutrient status and soil physical conditions substantially.

WATER RELATIONS OF SOIL

The mineral and organic compounds of soil form a solid (though not rigid) *matrix*, the interstices of which consists of irregularly shaped pores with a geometry defined by the boundaries of the matrix (Fig. 9.7). The pore space, in general, is filled partly with soil air and liquid vapour and partly with the liquid phase of soil water. Soil moisture is one of the most important ingredients of the soil. It is also one of its most dynamic properties. Water influences many physical and chemical reactions of the soil as well as plant growth.

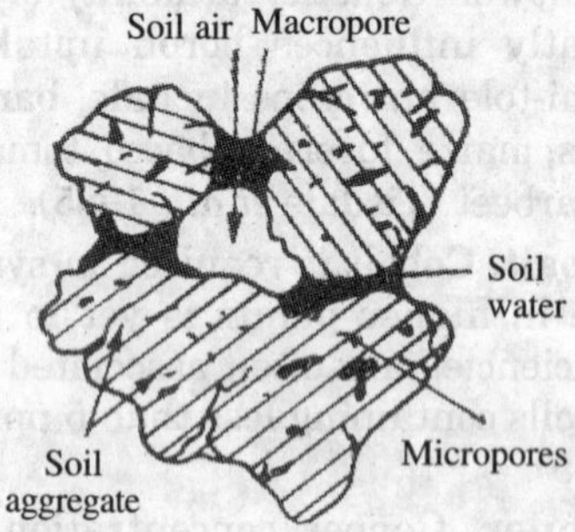

Fig. 9.7. Soil moisture within soil aggregates and suspended between adjacent aggregates.

The properties of water can be explained by the structure of its molecule. Two atoms of hydrogen and one atom of oxygen combine to form a molecule of water. The space occupied by an individual water molecule is largely determined by that of the oxygen ion. Water molecules do not exist individually. The hydrogen in the water serves as a connecting link from one molecule to the other.

Soil serves as the storage reservoir for water. Only the water stored in the root zone of a crop can be utilized by it for its transpiration and build-up of plant tissues. When ample water is in the root zone, plants can obtain their daily water requirements for proper growth and development. As the plants continue to use water, the available supply diminishes, and unless more water is added, the plants stop growing and finally die. Before the stage is reached when crop growth is adversely affected, it is necessary to irrigate again. The amount of water to be applied in each irrigation and the frequency of irrigation are dependent on the properties of the soil and the crop to be irrigated.

Kinds of Soil Water

When water is added to a dry soil either by rain or irrigation, it is distributed around the soil particles where it is held by adhesive and cohesive forces; it displaces air in the pore spaces and eventually fills the pores. When all the pores, large and small, are filled, the soil is said to be saturated and is at its maximum retentive capacity. As soil dries out, the water films become thinner and thinner and more tightly bound.

The following are the three main classes of soil water:

(*i*) *Hygroscopic water.* Water held tightly to the surface of soil particles by adsorption forces.

(*ii*) *Capilary water.* Water held by forces of surface tension and continuous films around soil particles and in the capillary spaces.

(*iii*) *Gravitational water.* Water that moves freely in response to gravity and drains out of the soil.

Adhesion is the attraction of solid surface for water molecules. Adhesion is operative only at the solid-liquid interface, Hence, the film of water established by it is very thin. *Cohesion* is the attraction of water molecules for each other. This force makes possible a marked thickness of the films of water established by hydration until they attain microscopic size. As the film gets thicker and thicker the forces of gravity act and water flows downward through the large pores. Such water is loosely held. Thus, when a soil is near saturation it is

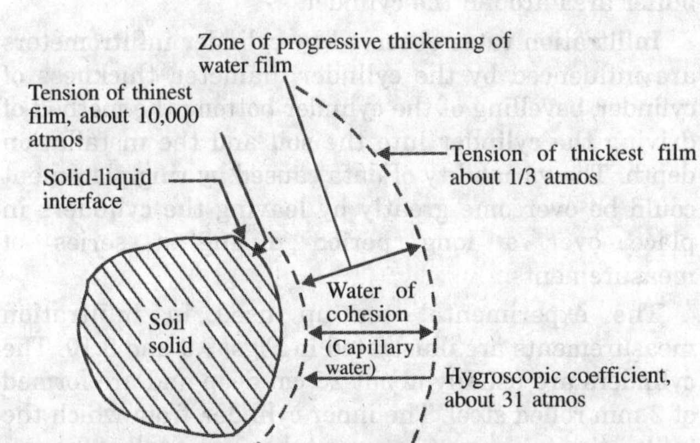

Fig. 9.8. Diagramatic representation of the progressive thickening of water film in a macropore and the corresponding decline in tension at which the surface molecules are held.

easy to remove an increment of water, but as moisture becomes less and less in the soil, the greater will be the force required to remove a unit amount of moisture. Figure 9.8 is the diagramatic illustration of the above concept.

When a dry soil sample is exposed to water vapour, it will take up moisture. The amount adsorbed depends on the nature and magnitude of the surface exposed, the temperature, and the degree of humidity. The moisture thus adsorbed is the water of hydration, water of adhesion, or commonly called the hygroscopic water. When the air saturation is 100% the maximum amount of such moisture will be acquired.

The capillary water is held between tensions of about 31 atmospheres and one-third atmosphere (Fig. 9.8). Between 31 and 15 atmospheres, capillary adjustment is very sluggish. Comparatively easy movement does not occur until the water film thickness and pressures measuring one-third atmosphere are reached. As a result of its energy relations, the capillary water is the only fluid water bearing solutes, that remain, in the soil for any length of time, if drainage is satisfactory. Thus, it functions physically and chemically as the soil solution. The principal factors influencing the amount of capillary water in soils are the structure, texture and organic matter. The finer the texture of the mineral soil particles, the greater is likely its capillary capacity. Granular soil structure produces higher capillary capacity. Presence of organic matter increases the capillary capacity.

Water held in the soil at tensions of one-third atmosphere or less will respond to gravity and move downward, hence the name gravitational water. Capillary water is present in the non-capillary (large) pores of the soil. Of the three forms of water, only capillary and gravitational water are of interest to the irrigationist, since hygroscopic water is not available to plants.

Movement of Water into Soils

The movement of water from the surface into the soil is called *infiltration.* The infiltration characteristic of the soil is one of the dominant variables influencing irrigation. *Infiltration rate* is the soil characteristic determining the maximum rate at which water can enter the soil under specific conditions, including the presence of excess water. It has the dimensions of velocity. The actual rate at which water is entering the soil at any given time is termed the *infiltration velocity.*

The infiltration rate decreases during irrigation. The rate of decrease is rapid initially and the infiltration rate tends to approach a constant value. The nearly constant rate that develops after some time has elapsed from the start of irrigation is called the basic *infiltration rate.*

Accumulated infiltration, also called *cumulative infiltration,* is the total quantity of water that enters the soil in a given time. Infiltration rate and accumulated infiltration are the two parameters commonly used in evaluating the infiltration characteristics of soil. For design purposes, the relationship between accumulated infiltration and elapsed time are usually expressed by the following empirical equations:

$$y = at^{\alpha} \quad \ldots(9.10)$$

$$y = at^{\alpha} + b \qquad t \neq 0 \quad \ldots(9.11)$$

in which, y = accumulated infiltration in time t, cm

t = elapsed time or infiltration opportunity time, min.

and a, α, b are characteristic constants.

Field experimental data on accumulated infiltration versus time, when plotted on an ordinary co-ordinate paper, give a parabolic curve. When the data are plotted on a log-log paper a linear relationship is indicated. But some data on the initial stage of the experiment usually fall along a slight curved line. Since the constant b rectifies this slight deviation from the linear log-log relationship, equation (9.11) is considered superior to equation (9.10) to express the accumulated infiltration-time relationship. The values of a, α and b usually range between 0 and 1.

The infiltration rate at any time t is obtained by differentiating equation (9.11) as follows:

$$y = at^{\alpha} + b$$

$$\frac{dy}{dt} = a^{\alpha}t^{\alpha-1} \quad \ldots(9.12)$$

Factors affecting infiltration rate. The major factors affecting the infiltration of water into the soil are the initial moisture content, condition of the soil surface, hydraulic conductivity of the soil profile, texture, porosity, degree of swelling of soil colloids, organic matter, vegetative cover, duration of irrigation or rainfall and viscosity of water. The antecedent soil moisture content has considerable influence on the initial rate and total amount of infiltration, but decreasing as the soil moisture content rises. The infiltration rate of any soil is limited by any restraint to the flow of water into and through the soil profile. The soil layer with the lowest permeability, either at the surface or below it, usually determines the infiltration rate. Infiltration rates are also affected by the porosity of the soil which is changed by cultivation or compaction. Cultivation influences the infiltration rate by increasing the porosity of the surface soil and breaking up the surface seals. The effect of tillage on infiltration usually lasts only until the soil settles back to its former condition of bulk density because of subsequent irrigations.

Infiltration rates are generally lower in soils of heavy texture than in soils of light texture. The influence of water depth over soil on infiltration rate was investigated by many workers. It has been established that in surface irrigation, increased depth increases initial infiltration slightly but the depth of application has negligible effect after prolonged irrigation. Infiltration rates are also influenced by the vegetal cover. Infiltration rates on grassland is substantially higher than bare uncultivated land. Additions of organic matter increase infiltration rate substantially. The hydraulic conductivity of the soil profile often change during infiltration, not only because of increasing moisture content, but also because of the puddling of the surface caused by reorientation of surface particles and washing of finer materials into the soil. Viscosity of water influences infiltration. The high rates of infiltration in the tropics under otherwise comparable soil conditions is due to the low viscosity of warm water.

Measurement of infiltration. Three methods of estimating infiltration characteristics of soil for the design of irrigation systems have been recognised. They are: (*a*) the use of cylinder infiltrometers, (*b*) measurement of subsidence of free water in a large basin, and (*c*) estimation of accumulated infiltration from the water front advance data. The use of cylinder infiltrometers, described below, is the most common method.

Infiltration characteristics of soils may be determined by ponding water in a metal cylinder installed on the field surface and observing the rate at which water level is lowered in the cylinder. In the earlier studies only a single cylinder was used and many of the data indicated a high degree of variability. The variability was mainly due to the uncontrolled lateral movement of water from the cylinder after the wetting front reached the bottom of the cylinder. After the initiation of infiltration, while the wetting front is in the cylinder, the water subsidence rate corresponds to the infiltration rate. When the wetting front passes below the cylinder, a more or less divergence of flow will occur. The lateral movement of water from cylinders is minimised by pending water in a guard cylinder or buffer area around the cylinder.

Infiltration rates observed by cylinder infiltrometers are influenced by the cylinder diameter, thickness of cylinder, bevelling of the cylinder bottom, the method of driving the cylinder into the soil and the installation depth. The variability of data caused by ring placement could be overcome greatly by leaving the cylinders in place over a long period during a series of measurements.

The experimental set up used in infiltration measurements are illustrated in Figs. 9.9 and 9.10. The cylinders are usually about 25 cm deep and are formed of 2 mm rolled steel. The inner cylinder, from which the infiltration measurements are taken, is usually 30 cm in diameter. The outer cylinder, which is used to form the buffer pond is about 60 cm in diameter. The cylinders are installed about 10 cm deep in the soil. Care is taken to keep the installation depth of the cylinders the same in all experiments. This is accomplished by marking the outside of the cylinders at the 10 cm level and driving the cylinders upto the mark. The cylinders are driven

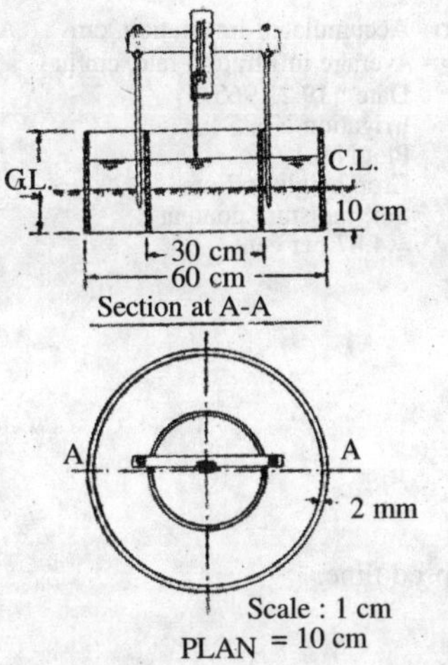

Fig. 9.9 Plan and elevation (in section) of the experimental set up for measuring infiltration with a concentric cylinder infiltrometer.

into the ground by a falling weight type hammer striking on a wooden plank placed on top of the cylinder or by light blows with an ordinary hammer and using a short wooden plank to prevent damage to the edges of the metal cylinder.

The water level in the inner cylinder is read with a field type point gauge (Fig. 9.9) or hook gauge. (Sometimes an ordinary plastic scale placed in the inner cylinder or a manometer fixed to the outside of the cylinder are also used, but for accurate work a point gauge or hook gauge should be used). The point rod is set at the desired level to which water is to be added. Water is added to the inner cylinder from a container of known volume and a graduated jar (Fig. 9.10). It is added by pouring on to a piece of folded jute matting. The matting is used to prevent pudding and sealing of the surface soil. After filling the cylinder to about three-fourths of the desired level the matting is removed.

A stop watch or the second's hand of a wrist watch is used to note the instant the addition of water begins and the time the water reaches the desired level. The total quantity of water added to the inner cylinder is determined by counting the number of full containers of water and the fractional volume in the jar, which is added last. Care is taken to fill the container completely each time before adding water to the cylinder. The difference between the quantity of water added and the volume of water in the cylinder at the instant it reached the desired point is taken as the quantity of water that infiltrates during the time interval between the start of filling and the first measurement.

Fig. 9.10. Determining infiltration with a concentric cylinder infiltrometer.

After the initial reading, point gauge measurements are made at frequent intervals to determine the amount of water that has infiltrated during the time interval. Water is added quickly after each measurement so that a constant average infiltration head could be maintained. Point gauge readings are taken before the water level recedes more than 1 cm. The average depth of water maintained in the cylinder is 7 to 12 cm, which is approximately equal to the water level expected in the irrigation border or basin during irrigation. The buffer pond is filled with water immediately after filling the inner cylinder. Water levels in the inner cylinder and the buffer pond are kept approximately the same. The experiments are continued beyond the estimated time the water would stand in borders or basins during an irrigation. The data are tabulated on standard forms, as shown in Table 9.2.

It is necessary to conduct replicated tests at suitable locations in the field. The average values of accumulated infiltration y and average infiltration rates are plotted as a function of elapsed time (Fig. 9.11).

Curve fitting. The functional relationship between y and t is best represented by the equation $y = at^{\alpha} + b$. The values of the constant a, α, and b may be determined by the method of averages using the procedure suggested by Davis (1943). The first step is to plot y against x and choose two points (x_1, y_1) and (x_2, y_2) on and near the extremities of the smooth curve representing the data. Now a point $x_3 = \sqrt{x_1 x_2}$ is chosen, y_3 is read against x_3. The value of b is determined by using the following equation:

$$b = \frac{y_1 y_2 - y_3^2}{y_1 + y_2 - 2y_3} \quad \ldots(9.13)$$

Example 9.1. *Using the field data presented in Table 9.1 the procedure in determining the values of a, α and b, and evaluating the goodness of fit, are described below.*

Solution: From the plot of y against t (Fig. 9.11)

For $t_1 = 5$ minutes, $y_1 = 1.73$ cm, and

$t_2 = 130$ minutes, $y_2 = 18.43$ cm

Adopting the procedure suggested by Davis (1943), the rectifying value of t is found from the following relationships:

$$t_3 = \sqrt{t_1 \times t_2} \quad \ldots(9.13a)$$

$$= \sqrt{.5 \times 130} = 25.55 \text{ minutes}$$

The corresponding value of y_3, as determined from Fig. 9.11 is 5.6 cm. The value of the constant b is obtained as follows:

$$b = \frac{y_1 y_2 - y_3^2}{y_1 + y_2 - 2y_3} = \frac{1.73 \times 18.43 - (5.6)^2}{1.73 + 18.43 - 2 \times 5.6} = 0.05$$

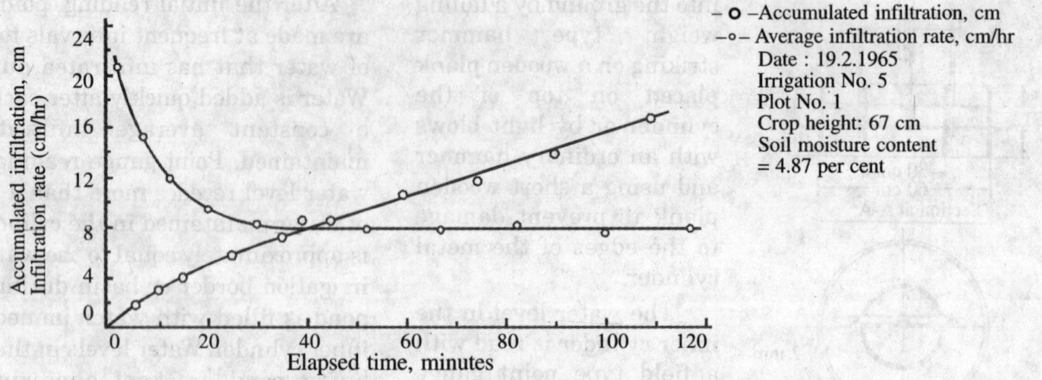

Fig. 9.11. Plots of accumulated infiltration and average infiltration rate against elapsed time, based on the experimental data presented in Table 9.1.

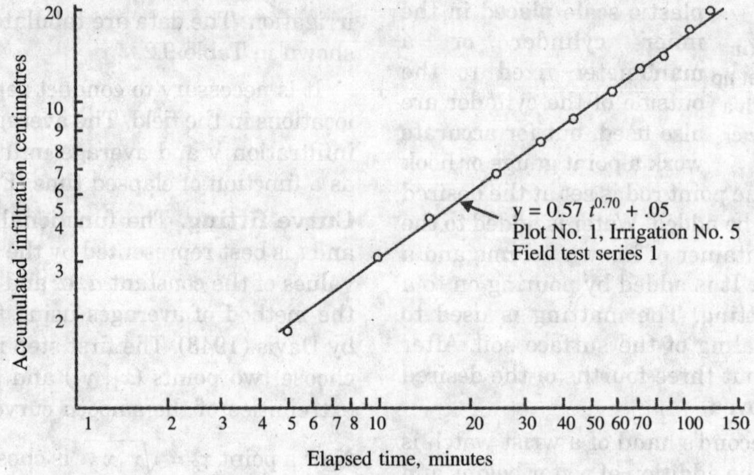

Fig. 9.12. Plot on log-log paper of accumulated infiltration—time relationship, based on the experimental data presented in Table 9.2.

Table 9.2. Sample data of cylinder infiltrometer tests

Date: 19.2.1965
Irrigation No. 5
Plot No. 1

Crop height = 65 cm
Soil moisture content = 4.87 per cent

Elapsed time	*Cylinder No. 1*					*Cylinder No. 2*					*Cylinder No. 3*					*Average infiltration rate*	*Average accu. infiltration*
	Distance of water surface from ref. point		*Infiltration during period*			*Distance of water surface from ref. point*		*Infiltration during period*			*Distance of water surface from ref. point*		*Infiltration during period*				
	Before filling	*After filling*	*Depth*	*Ave. rate*	*Accu. infilt.*	*Before filling*	*After filling*	*Depth*	*Ave. rate*	*Accu. infilt.*	*Before filling*	*After filling*	*Depth*	*Ave. rate*	*Accu. infil.*		
(min)	*(cm)*	*(cm)*	*(cm)*	*(cm/hr)*	*(cm)*	*(cm)*	*(cm)*	*(cm)*	*(cm/hr)*	*(cm)*	*(cm)*	*(cm)*	*(cm)*	*(cm/hr)*	*(cm)*	*(cm/hr)*	*(cm)*
–	–	11.0	–	–	–	–	11.0	–	–	–	–	11.0	–	–	–	–	–
5	9.3	11.0	1.7	20.4	1.7	9.5	11.0	1.5	18.0	1.5	9.0	11.0	2.0	24.0	2.0	20.80	1.730
10	10.0	11.0	1.0	12.0	2.7	9.7	11.0	1.3	15.6	2.8	9.5	11.0	1.5	18.0	3.5	15.20	3.000
15	10.1	11.0	0.9	10.8	3.6	10.0	11.0	1.0	12.0	3.8	10.0	11.0	1.0	12.0	4.5	11.60	3.970
25	9.5	11.0	1.5	9.0	5.1	9.4	11.0	1.6	9.6	5.4	9.5	11.0	1.5	9.0	6.0	9.20	5.500
45	9.9	11.0	2.2	6.6	7.3	9.5	11.0	3.0	9.0	8.4	9.5	11.0	3.0	8.0	9.0	8.20	8.230
60	9.5	11.0	1.5	6.0	8.8	9.0	11.0	2.0	8.0	10.4	9.0	11.0	2.0	8.0	11.0	6.66	10.067
75	9.5	11.0	1.5	6.0	10.3	9.0	11.0	2.0	8.0	12.4	9.0	11.0	2.0	8.0	13.0	7.33	11.900
90	9.5	11.0	1.5	6.0	11.8	9.0	11.0	2.0	8.0	14.4	9.0	11.0	2.0	8.0	15.0	7.33	13.067
110	9.0	11.0	2.0	6.0	13.8	8.5	11.0	2.5	7.5	16.9	8.5	11.0	2.5	7.5	17.5	7.00	16.067
130	9.0	11.0	2.0	6.0	15.8	8.5	11.0	2.5	7.5	19.4	8.5	11.0	2.5	7.5	20.0	7.00	18.430

The value 0.05 of b is subtracted from each value of y in Table 9.2. The logarithms of $(y - 0.05)$ and t are taken. The variables are related by the expression

$$y - 0.05 = at^{\alpha}$$

the logarithmic form of which is

$$\log (y - 0.05) = \log a + \alpha \log t \qquad \text{...(9.14)}$$

(If the functional relationship between y and t can be expressed by the equation $y = at^{\alpha}$, it is not necessary to determine the value of the rectifying factor b. The logarithmic form of the expression will then be $\log y = \log a + \alpha \log t$. The other steps in the computation of the values of a and α, and the verification of the goodness of fit are same in both the types of equations).

Substituting the data on average infiltration y and elapsed time t presented in Table 9.2 into equation 9.14 yields the following eleven equations:

$$0.2253 = \log a + 0.6990\ \alpha$$
$$0.4698 = \log a + 1.0000\ \alpha$$
$$0.5933 = \log a + 1.1761\ \alpha$$
$$0.7364 = \log a + 1.3979\ \alpha$$
$$0.8338 = \log a + 1.5441\ \alpha$$
$$0.9128 = \log a + 1.6532\ \alpha$$
$$1.0004 = \log a + 1.7782\ \alpha$$
$$1.0738 = \log a + 1.8751\ \alpha$$
$$1.1142 = \log a + 1.9542\ \alpha$$
$$1.2044 = \log a + 2.0414\ \alpha$$
$$1.2642 = \log a + 2.1139\ \alpha$$

Adding the first six and last five equations

$$3.7714 = 6 \log a + 7.4703\ \alpha$$
$$5.6570 = 5 \log a + 9.7628\ \alpha$$

Multiplying (*ii*) by 6 and (*i*) by 5

$$33.9420 = 30 \log a + 58.5768\ \alpha$$
$$18.8570 = 30 \log a + 37.3515\ \alpha$$

Solving simultaneously

$$\alpha = 0.70$$
$$\log a = -0.2429$$
$$a = 0.57$$

Figure 9.12 is the plot of y against t on log-log paper. To determine the goodness of fit the values of y are calculated by substituting the values of a, α and b in the equation $y = at^{\alpha} + b$ for each observed value of t. The values of t are substituted in the equation as follows in the logarithmic form:

$$\log (y - b) = \log a + \alpha \log t$$

We have, at

$t = 5$ minutes

$$\log (y_1 - 0.05) = (-0.2429 + 0.70 \times 0.699)$$
$$y_1 = 1.766 + 0.05 = 1.816 \text{ cm}$$

$t = 10$ minutes

$$\log (y_2 - 0.05) = (-0.2429 + 0.70 \times 1.00)$$
$$y_2 = 2.865 + 0.05 = 2.915 \text{ cm}$$

$t = 15$ minutes

$$\log (y_3 - 0.05) = (-0.2429 + 0.70 \times 1.1761)$$
$$y_3 = (3.806 + 0.05 = 3.856 \text{ cm})$$

$t = 25$ minutes

$$\log (y_4 - 0.05) = (-0.2429 + 0.70 \times 1.397)$$
$$y_4 = 5.4339 + 0.05 = 5.489 \text{ cm}$$

$t = 35$ minutes

$$\log (y_5 - 0.05) = (-0.2429 + 0.70 \times 1.544)$$
$$y_5 = 6.885 + 0.05 = 6.935 \text{ cm}$$

$t = 45$ minutes

$$\log (y_6 - 0.05) = (-0.2429 + 0.70 \times 1.6532)$$
$$y_6 = 8.247 + 0.05 = 10.09 \text{ cm}$$

$t = 60$ minutes

$$\log (y_7 - 0.05) = (-0.2429 + 0.70 \times 1.7782)$$
$$y_7 = 10.04 + 0.05 = 10.09 \text{ cm}$$

$t = 75$ minutes

$$\log (y_8 - 0.05) = (-0.2429 + 0.70 \times 1.8751)$$
$$y_8 = 11.74 + 0.05 = 11.79 \text{ cm}$$

$t = 90$ minutes

$$\log (y_9 - 0.05) = (-0.2429 + 0.70 \times 1.9542)$$
$$y_9 = 13.34 + 0.05 = 13.39 \text{ cm}$$

$t = 110$ minutes

$$\log (y_{10} - 0.05) = (-0.2429 + 0.70 \times 2.0414),$$

or
$$y_{10} = 15.36 + 0.05 = 15.41 \text{ cm}$$

$t = 130$ minutes

$$\log (y_{11} - 0.05) = (-0.2429 + 0.70 \times 2.1139)$$
$$y_{11} = 17.25 + 0.05 = 17.3 \text{ cm}$$

The results are given in Table 9.3 along with the percentage deviation in each case. The average percentage deviation is 2.37.

Table 9.3. Determination of goodness of fit of equation $y = at^a + b$. [*Data from Table 9.2*]

y observed cm	*t min*	*y – 0.05*	*log (y – 0.05)*	*log t*	*y calculated cm*	*Deviation %*
1.730	5	1.680	0.2253	0.6990	1.816	+ 4.96
3.000	10	2.950	0.4698	1.0000	2.915	– 2.83
3.970	15	3.920	0.5933	1.1761	3.856	– 2.87
5.600	25	5.450	0.7364	1.3979	5.489	– 0.20
6.870	35	6.820	0.8338	1.5441	6.935	+ 0.94
8.230	45	8.180	0.9128	1.6532	8.297	+ 0.64
10.067	60	10.017	1.0004	1.7782	10.090	+ 0.23
11.900	75	11.850	1.0728	1.8751	11.790	– 0.92
13.067	90	13.017	1.1142	1.9542	13.390	+ 2.47
16.067	110	16.017	1.2044	2.0414	15.410	– 4.09
18.430	130	18.380	1.2642	2.1139	17.300	– 6.13
						2.37 av.

The instantaneous infiltration at any time t after the beginning of the test may be obtained from the formula

$\frac{dy}{dt} = a\,\alpha\,t^{\alpha-1}$ (Equation (9.12)].

Measurement of infiltration in irrigation furrows. Infiltration characteristics vary with the method of irrigation. In border strip and check basin methods of irrigation the land surface is completely flooded. Water moves into the soil vertically and the boundary effects are negligible. The infiltration rate is usually expressed in cm per hour. The concentric cylinder method of infiltration measurement is suitable under these conditions. The method is also considered suitable for sprinkler irrigation. In furrow irrigation, however, only a part of the soil surface is flooded. Water moves both downward and outward from a furrow, and the infiltration rate depends on the capacity of the soil to transmit water, both vertically and laterally. It also depends on the wetted perimeter of the furrow. The infiltration rate is expressed in litres per minute per 100 metres length of furrow (other units of furrow length, say, 25 m, 30 m or 50 m may also be used). The inflow-outflow method (Chapter 11) is the most commonly used method of infiltration measurement in furrow.

Soil Moisture Retention and Movement

The moisture content of a sample of soil is usually defined as the amount of water lost when dried at 105°C, expressed either as the weight of water per unit weight of dry soil or as the volume of water per unit volume of bulk soil. Although useful, such information is not a clear indication of the availability of water for plant growth. The difference exists because the water retention characteristics may be different for different soils.

The forces that keep soil and water together are based on the attraction between the individual molecules, both between water and soil molecules (adhesion) and among water molecules themselves (cohesion). In the wet range, surface tension is the most important force, while in the dry range adsorption is the main factor. Thus, the higher the moisture content, the smaller is the attraction of the soil for water. The energy of water tension in a soil depends on the specific surface as well as the structure of the soil and on its solute content. When water is present in the capillaries, the energy with which it is attached is a function of surface tension and capillary size but when it is present in bigger pores, it is bound loosely to the soil and can be acted upon by gravity. When salts dissolve in water, they decrease the free energy of water. Soil water, by virtue of the salts dissolved in it, has a lower free energy than pure water. Further, soil water that is bound to solid particles as hygroscopic water is tightly held by the surface of contact and has a low free energy by virtue of binding forces. Thus, there are two types of interactions which decrease the free energy of water, namely, (*i*) due to the solubility of salts, and (*ii*) due to the interaction of water and solid surface. Both these add together in decreasing the energy of soil water. Thus, the retention of water in the soil and the tendency of water to move in the soil are consequences of energy effects.

Viscosity, Surface Tension and Capillarity Viscosity

Viscosity of a liquid or a gas is its internal friction that brings about resistance to flow. Viscosity of water, as of most liquids, changes greatly with temperature. The value of the dynamic viscosity for the measured temperature of water may be obtained from the following relationship*:

$$\mu = \frac{0.0179}{1 + 0.03368T + 0.000221T^2} \quad ...(9.15)$$

in which,

μ = dynamic viscosity of water, dyne-sec/cm^2 (poises), and

T = temperature of water, °C

Viscosity is measured in poises. (A poise is a dyne-second per square cm. One-hundredth of a poise is called a centipoise. The viscosity of water at 20°C is almost exactly 1 centipoise). The relationship between dynamic viscosity and kinematic viscosity may be expressed as follows:

$$\nu = \frac{\mu}{\rho} \quad ...(9.16)$$

in which,

ν = kinematic viscosity of water, cm^2/sec, and

ρ = mass density of water, gm/cm^3

Due to hydrogen bonding, the viscosity of water is high for a liquid of so low a molecular weight. Viscosity is an important factor in infiltration and percolation of water. Viscosity also influences the characteristics of fluid flow in open channels and pipes.

Surface tension. The molecules of all liquids are held together by intermolecular attraction (cohesion) which enables the liquid to withstand a small stress. When dissimilar substances make contact at an interface, the inequalities of molecular attraction tend to change the shape of the interface until the potential energy of the whole molecular system attains a minimum value. Surface tension is a force pulling inward at the surface of a liquid, tending to make the surface area as small as possible.

The nature and dimensions of surface tension may be analysed by considering the force that holds a soap bubble together. At an imaginary central plane of a soap bubble, the upper half is held to the lower half by surface tension. Surface tension acts all along the length of the circle that forms contact between the upper half and the lower half of the bubble. Since the surface tension acts on the outside and the inside of the bubble, the total length over which it acts is twice as long as the circumference of the bubble, i.e., $4\pi r\sigma$, in which σ is the

*From Gibson, A.H. (1930). *Hydraulics and Its Applications*. 4th ed. Constable and Co. Ltd., London. Computed values of viscosity are presented in standard physical tables.

surface tension and r is the radius of the bubble. (Surface tension is the energy per unit area and is measured in erg/cm^2). Let P_b be the pressure on the inside of the bubble and P_a the atmospheric pressure on the outside of the bubble. The difference in pressure $(P_b - P_a)$ acting over the entire area of cross-section of the bubble (πr^2), is the force counteracting the inward pull of surface tension. Equating the two,

$$4\pi r\sigma = (P_b - P_a)\,\pi r^2 \quad \ldots(9.17)$$

or

$$P_b - P_a = \frac{4\sigma}{r} \quad \ldots(9.18)$$

Equation (9.18) shows that the bigger the radius of the bubble, the smaller is the difference in pressure. In other words, the greater is the curvature of the bubble, the greater is the inside pressure. Within the temperature limits normally encountered in the soil, the surface tension of water does not change appreciably with temperature.

Capillarity. One of the important phenomenon of surface tension in soil is capillarity, the attraction of water into 'hair-like' openings or capillary pores. Capillarity depends both on cohesion of the liquid and on the adhesion of the liquid to solid walls. The height to which the liquid rises is determined by the surface tension and the weight of the liquid column. Figure 9.13 is the schematic illustration of the rise of water in a tube of capillary dimensions. A pressure difference exists across a curved air-water interface, with the pressure under the concave meniscus smaller than the pressure under the plane meniscus of the same liquid at the same height. This causes the water to rise in the capillary tube. The sides of the capillary tube attract the water so that a concave meniscus is formed as a result of surface tension. Such capillary attraction or capillary lift occurs only in a medium in which the walls of the capillary have more attraction for the liquid than the molecules of the liquid have for each other. This is the case for water in most soils and in clean glass.

Referring to Fig. 9.13, the weight of the column of water elevated to a height h is supported by the vertical

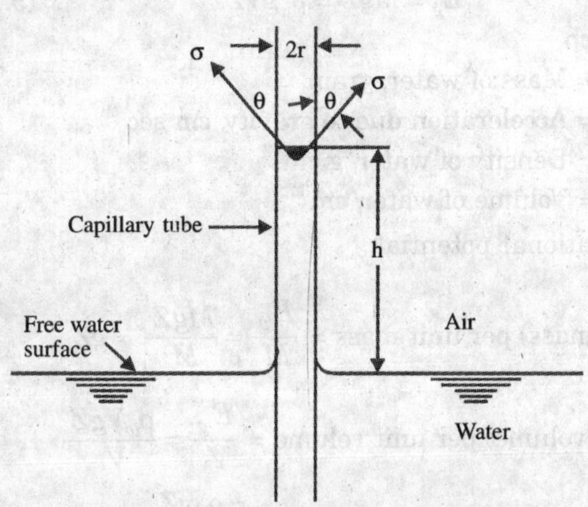

Fig. 9.13. Rise of water in a capillary tube.

component of surface tension acting around the perimeter of the tube. The vertical component of the surface tension force is $2\pi r\sigma \cos\theta$, in which $2\pi r$ represents the line of contact between the liquid and the tube, $\sigma$is the surface tension of water, and θ is the angle of contact. The weight of the column of water of height h and diameter $2r$ is $g\,\pi r^2 h$, in which ρ is density of water and g is the acceleration due to gravity. Equating the two,

$$2\pi r\sigma \cos\theta = g\rho\pi r^2 h$$

$$h = \frac{2\sigma\cos\theta}{g\rho r} \quad \ldots(9.19)$$

If the contact angle is zero, $\cos\theta = 1$ and equation (9.14) becomes

$$h = \frac{2\sigma}{g\rho r} \quad \ldots(9.20)$$

For water the angle of contact is so small that $\cos\theta$ is nearly equal to 1. Equation (9.19) is used to calculate the rise of water in a soil for which the size of the largest effective pore space is known. The relationship bet-ween the diameter of soil pores and the height to which water rises above a free water surface is shown in Fig. 9.14.

Soil pores are rarely uniform. The height of capillary rise will therefore depend on the largest opening that the water encounters. Once the soil is saturated with water, and then the water is allowed to drain away, some water is held by the smaller capillaries, even though the larger openings are freed of water. The larger the diameter of the capillaries, the faster is the rate of rise. In a fine-textured soil, the rise is so slow that the plants seldom gain from the presence of groundwater if its level is about 80 cm or more below the roots. The capillary rise on the 'dry range' of soil moisture content is much slower than in the 'wet-range'.

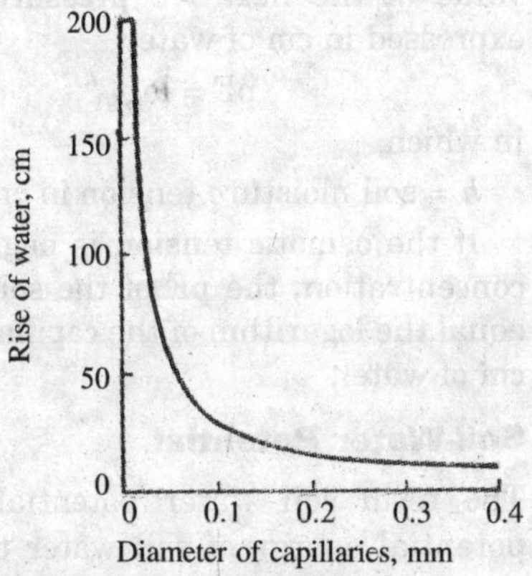

Fig. 9.14. Relationship between the diameter of capillaries and the rise of water in soil.

Soil Moisture Tension

Soil moisture tension is a measure of the tenacity with which water is retained in the soil and shows the force per unit area that must be exerted to remove water from a soil. The tenacity is measured in terms of the potential energy of the water in the soil, usually with respect to free water. It is usually expressed in atmospheres, the average air pressure at sea level. Other pressure units like cm of water or cm or mm of mercury are also often used (1 atmosphere = 1036 cm of water or 76.39 cm of mercury). It is also sometimes expressed in bars (1 bar = 10^6 dynes/cm^2 = 1023 cm of water column. 1 millibar = 1/1000 bar).

Soil moisture tension is brought about at the smaller dimensions by surface tension (capillarity), and at the higher dimensions by adhesion. Buckingham (1907) introduced the concept of 'capillary potential' to define the energy with which water is held by soil. This term, however, does not apply over the entire moisture range. In a wet soil, as long as there is a continuous column of water, it might be called 'hydrostatic potential'. In the intermediate range the term 'capillary potential' is appropriate. In the dry range, the term 'hygroscopic potential' would be suitable. However, the term 'soil moisture potential', 'soil moisture suction' and 'soil moisture tension' are often used synonymously to cover the entire range of moisture (Khonke, (1968).

pF of soils. Schofield (1935) suggested the use of the logarithm of soil moisture tension and gave the symbol pF to this logarithm which is an exponential expression of a free-energy difference (based on the height of a water column above free-water level in cm). The pF function, analogous to the acidity-alkalinity scale pH, is defined as the logarithm to the base 10 of the numerical value of the negative pressure of the soil moisture expressed in cm of water.

$$\text{pF} = \log_{10}{}^{h} \qquad \text{...(9.21)}$$

in which,

h = soil moisture tension in cm of water

If the osmotic tension is negligible, *i.e.,* at low salt concentration, the pF of the soil moisture may nearly equal the logarithm of the capillary tension expressed in cm of water.

Soil-Water Potential

The term soil water potential denotes the specific potential energy of soil water to that of water under standard reference state. The major forces that act on water in the soil are (*i*) gravity, (inertial or body forces), (*ii*) hydrostatic forces (forces associated with pressure gradients, (*iii*) osmotic forces (related with solutes) and (*iv*) adsorption forces. If temperature of the soil water system is not constant, an additional field due to temperature will also be acting. However, isothermal conditions are usually assumed in soil water systems. The total soil-water potential can be defined as the amount of work that must be done per unit quantity of pure water in order to transport reversibly and isothermally an infinitesimal quantity of water from a pool of pure water at a specified elevation at atmospheric pressure to soil water at the point under considerations (Aslyng, *et al.,* 1963). Total potential includes both the chemical potential and the potential of external force fields. When expressed per unit amount of water, the potential energy thus defined is referred to as the potential of water (ψ_k) in the particular force field, or

$$\psi_k = -\int F_k dl \qquad \text{...(9.22)}$$

in which,

F_k = Force per unit mass due to field 'K'.

As the potential is a scalar quantity, the total potential can be determined by taking the algebraic sum of component potentials corresponding to the different force fields, or

$$\psi = \Sigma_\lambda \psi_\kappa = -\Sigma_k \int F_k \, .dl \qquad \text{...(9.22}a\text{)}$$

in which,

ψ = Total potential, erg g^{-1}

The soil water is under the influence of different force fields and accordingly the total potential of soil water can be considered as the sum of the separate contributions of the various forces as follows:

$$\psi = \psi_g + \psi_{p(m)} + \psi_0 \qquad \text{...(9.22}b\text{)}$$

in which,

ψ = Total potential, erg g^{-1}

ψ_g = Gravitational potential, erg g^{-1}

$\psi_{p(m)}$ = Pressure (or metric) potential, erg g^{-1}

ψ_o = Osmotic potential, erg g^{-1}

Gravitational potential (ψ_g). Gravitational potential is the potential attributable to the gravitational force field and is dependent on the elevation. It is the amount of work that a unit quantity of water in an equilibrium soil-water (or plant-water) system at an arbitrary level is capable of doing when it moves to another equilibrium identical in all respects except that it is at a reference level. As the gravitational potential of soil water at each point is determined by the elevation of the point relative to some arbitrary reference level, it is customary to set the reference level at the elevation of a pertinent point within the soil or below the soil profile being considered, so that this potential can always be taken as zero or positive.

Considering a point at height Z above the reference point. the gravitational potential energy,

$$E_g = MgZ = \rho_w gVZ \qquad \text{...(9.23)}$$

in which,

M = Mass of water, gram

g = Acceleration due to gravity, cm sec^{-2}

ρ_w = Density of water, g cm^{-3}

V = Volume of water, cm^3

Gravitational potential,

$$\psi_g \text{ (mass) per unit mass} = \frac{E_g}{M} = \frac{MgZ}{M} = gZ$$

$$\psi_g \text{ (volume) per unit volume} = \frac{E_g}{V} = \frac{\rho_w VgZ}{V} = \rho_w gZ$$

$$\psi_g \text{ (weight) per unit weight } = \frac{\rho_w VgZ}{\rho_w Vg} = Z$$

Pressure potential (ψ_p). Pressure potential is the amount of work that a unit quantity of water in an equilibrium soil-water (or plant-water) system is capable of doing when it moves to another equilibrium system identical in all respects except that it is at a reference pressure. In the field, the pressure potential at a point is a direct result of the overlying water, *i.e.*, where the air pressure on a soil water system is determined by the depth *h* of the point below the water table. This is also referred to as submergence potential. The hydrostatic pressure *P* of water with reference to atmospheric pressure is

$$P = \rho_w gh \qquad \ldots(9.23a)$$

The potential energy of the water is *Pdv*, in which *dv* is the infinitesimal volume of water.

Pressure potential,

$$\psi_D \text{ (volume) per unit volume} = \frac{Pdv}{dv} = p$$

$$\psi_p \text{ (mass) per unit mass} = \frac{\rho_w gh}{\rho_w} = gh$$

$$\psi_D \text{ (weight) per unit weight} = \frac{\rho_w gh}{\rho_w g} = h$$

Metric potential (ψ_m). The metric potential of soil water is defined as the amount of work that a unit quantity of water in an equilibrium soil-water (or plant-water) system is capable of doing when it moves to another equilibrium system identical in all respects except that there is no matrix present. It is the portion of the total water potential that is attributable to the solid colloidal matrix of the soil system. It is the negative pressure potential which results from the capillary and adsorptive forces emanating from the soil matrix. The terms matrix suction, soil water suction and capillary potential have been in use for metric potential.

Considering an infinitesimal volume *dv* of water, with pressure deficit, *p*

Metric potential,

$$\psi_m = pdv \qquad \ldots(9.23b)$$

Metric potential per unit volume, ψ_m(volume)

$$= \frac{pdv}{dv} = p = \rho_w gh$$

Metric potential per unit mass, ψ_m (mass)

$$= \frac{\rho_w gh}{\rho_w} = gh$$

Metric potential per unit weight, ψ_m (weight)

$$= \frac{\rho_w gh}{\rho_w g} = h$$

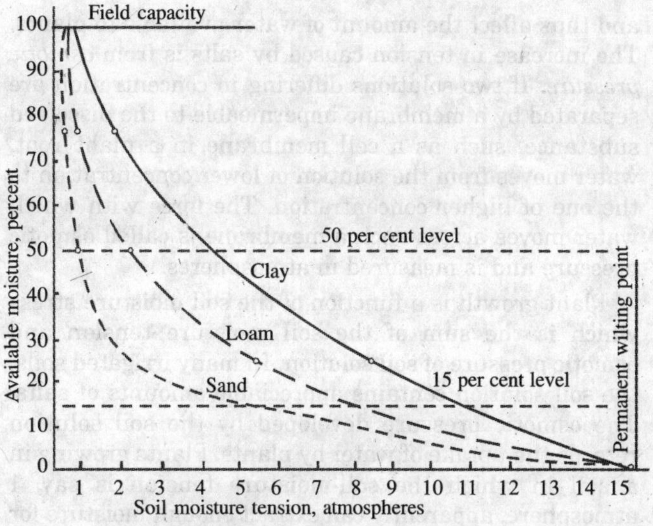

Fig. 9.15. Typical moisture characteristic curves of clay, loam and sandy soils.

(Adapted from USDA, SCS, National Engg. Handbook, Sec. 15, Chapter 1).

Osmotic potential (ψ_o). Osmotic potential can be defined as the amount of work that a unit quantity of water in an equilibrium soil water system is capable of doing when it moves to another equilibrium system identical in all respects except that there are no solutions. Osmotic potential which is also termed as solute potential is the portion of water potential that results from the solute present in the soil system;

$$\psi_o = -\pi \qquad \ldots(9.23c)$$

in which, π = osmotic pressure due to dissolved salts.

Soil Moisture Characteristics

Soil moisture tension is not necessarily an indication of the moisture content of the soil nor the amount of water available for plant use at any particular tension. These are dependent on the texture, structure and other characteristics of the soil and must be determined separately for each soil. Generally, sandy soils drain almost completely at low tension, but fine-textured clays still hold a considerable amount of moisture even at such high tensions that the plant growing in the soil may wilt. Moisture extraction curves, also called moisture characteristic curves (Fig. 9.15), which are plots of moisture content versus moisture tension, show the amount of moisture a given soil holds at various tensions. A knowledge of the amount of water held by the soil at various tensions is required to understand the amount of water that is available to plants, the water that can be taken up by the soil before percolation starts, and the amount of water that must be used for irrigation.

Soil Moisture Stress

Soil-moisture tension as discussed in the preceding paragraphs is based on pure water. Salts in soil water increase the force that must be exerted to extract water and thus affect the amount of water available to plants. The increase in tension caused by salts is from *osmotic pressure.* If two solutions differing in concentration are separated by a membrane impermeable to the dissolved substance, such as a cell membrane in a plant root, water moves from the solution of lower concentration to the one of higher concentration. The force with which water moves across such a membrane is called osmotic pressure and is measured in atmospheres.

Plant growth is a function of the soil moisture stress which is the sum of the soil moisture tension and osmotic pressure of soil solution. In many irrigated soils, the soil solution contains appreciable amounts of salts. The osmotic pressure developed by the soil solution retards the uptake of water by plants. Plants growing in a soil in which the soil-moisture tension is say, 1 atmosphere, apparently can extract enough moisture for good growth. But if the osmotic pressure of the soil solution is, say, 10 atmospheres, the total stress is 11 atmospheres, then the plants cannot extract enough water for good growth. Thus, for successful crop production in soils having appreciable salts, the osmotic pressure of the soil solution must be maintained as low as possible by controlled leaching.

Soil Moisture Constants

Soil moisture is always being subjected to pressure gradients and vapour pressure differences that cause it to move. Thus, soil moisture cannot be said to be constant at any pressure. However, it has been found experimentally that certain moisture contents described below are of particular significance in agriculture and these are often called soil moisture 'constants'.

Saturation capacity. When all the pores of the soil are filled with water, the soil is said to be under saturation capacity or *maximum water-holding capacity.* The tension of water at saturation capacity is almost zero and it is equal to free water surface.

Field capacity. The field capacity of soil is the moisture content after the drainge of gravitational water has become very slow and the moisture content has become relatively stable. This situation usually exists for one to three days after the soil has been thoroughly wetted by rain or irrigation. The terms field capacity, field-carrying capacity, normal moisture capacity and capillary capacity are often used synonymously. At field capacity, the large soil prose are filled with air, the micro pores are filled with water and any further drainge is slow. The field capacity is the upper limit of available moisture range in soil moisture and plant relations. The soil moisture tension at field capacity varies from soil to soil, but it generally ranges from $\frac{1}{10}$ to $\frac{1}{3}$ atmospheres.

Field capacity is determined by ponding water on the soil surface in an area of about 2 to 5 sq m and permitting it to drain for one to three days, with surface evaporation prevented. Evaporation may be prevented by spreading a polyethylene sheet or a thick straw mulch on the ground surface. One to three days after the soil is thoroughly wetted, soil samples are collected with an auger from different soil depths at uniform intervals throughout the wetted zone. The moisture content is determined by the gravimetric method.

Moisture equivalent. Moisture equivalent is defined as the amount of water retained by a sample of initially saturated soil material after being subjected to a centrifugal force of 1000 times that of gravity for a definite period of time, usually half an hour. To determine the moisture equivalent, a small sample of soil is whirled in a centrifuge with a centrifugal force of 1000 times that of gravity. The moisture remaining in the sample is determined. This moisture content, when expressed as moisture percentage on oven dry basis, gives the value of the moisture equivalent. The concept is sometimes used to estimate field capacity. In medium textured soils, the values of field capacity and moisture equivalent are nearly equal. In sandy soils, the field capacity exceeds the moisture equivalent. In very clayey soils, the field capacity is generally lower than the moisture equivalent.

Permanent wilting percentage. The permanent wilting percentage, also known as permanent wilting point or wilting coefficient, is the soil moisture content at which plants can no longer obtain enough moisture to meet transpiration requirements; and remain wilted unless water is added to the soil. At the permanent wilting point the films of water around the soil particles are held so tightly that roots in contact with the soil cannot remove the water at a sufficiently rapid rate to prevent wilting of the plant leaves. It is a soil characteristic, as all plants whose root systems thoroughly permeates the soil will wilt at nearly the same soil moisture content when grown in a particular soil in a humid atmosphere.

The moisture tension of a soil at the permanent wilting point ranges from 7 to 32 atmospheres, depending on soil texture, on the kind and condition of the plants, on the amount of soluble salts in the soil solution, and to some extent, on the climatic environment. Since this point is reached when a change in tension produces little change in moisture content, there is little difference in moisture percentage, regardless of the tension taken as the permanent wilting point. Therefore, 15 atmospheres is the pressure commonly used for this point.

The *wilting range* is the range in soil-moisture content through which plants undergo progressive degrees of permanent or irreversible wilting, from wilting of the oldest leaves to complete wilting of all leaves. At the permanent wilting point, which is the top of this range, plant growth ceases. Small amounts of water can be removed from the soil by plants after their growth ceases, but apparently the water is absorbed only slowly and is enough only to maintain life until more water is available. The moisture content at which the wilting is complete and the plants die is called the

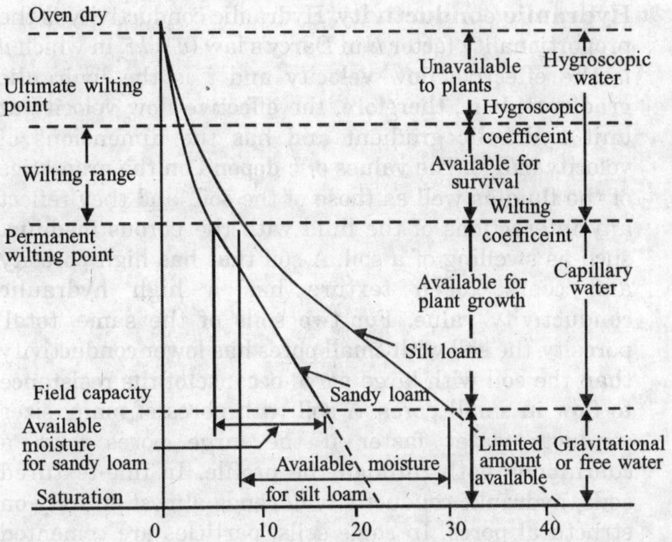

Fig. 9.16. Kinds of water in the soil and difference in available moisture content between a sandy loam and a silt loam soil. Note that the silt loam contains more than twice as much readily available water.

(Adapted from USDA, SCS Handbook Sec. 45, 1964)

ultimate wilting point. Although the difference in the amount of water in the soil between the two points may be small, there may be a big difference in tension. At the ultimate wilting point soil-moisture tension may be as high as 60 atmospheres.

The most common method of determining the permanent wilting percentage is to grow indicator plants in containers, usually in small cans, holding about 600 grams of soil. Sunflower plant is commonly used as an indicator plant. The plants are allowed to wilt and are then placed in a chamber with an approximately saturated atmosphere to test them for permanent wilting. The residual soil moisture content in the container is then calculated, which is the permanent wilting percentage. The determination of moisture content at 15 atmosphere tension, which is the usually assumed value of permanent wilting point, can be done by the pressure membrane apparatus (Richard, 1947).

Available water. Soil moisture between field capacity and permanent wilting point is referred to as readily available moisture. It is the moisture available for plant use. In general, fine-textured soils have a wide range of water content between field capacity and permanent wilting point than coarse texured soils. In contrast, sandy soils with their larger proportion of non-capillary pore space release most of their water within a narrow range of potential because of the predominance of large pores.

The available water of a soil may be computed from the following relationship: Available water = Depth of root zone, cm – (Field capacity, % by volume – Permanent wilting point, % by volume).

Theoretically, plants are able to obtain water from the soil whenever soil water contents exceed the permanent wilting point. However, the rate at which they transpire decreases in response to the decline in soil moisture content, due to increased stomatal resistance. The decline in plant's actual transpiration is small till the soil moisture content reaches a critical value. After the critical value, which varies with plant types, the transpiration declines sharply and crop growth suffers adversely. The volume per unit surface area of soil water between field capacity and critical soil moisture content is called *readily available water*, the value of which is computed from the following relationship:

Readily available water = Depth of root zone, cm (Field capacity % by volume – Critical soil moisture content, % by volume)/100

The concept of maximum available *moisture deficiency* is sometimes used to estimate the amount of water to be used without adversely affecting the plant:

Maximum available moisture deficiency = Readily available water, cm/Available water, cm

In most crops, the maximum available deficiency is about 0.65. In general, vegetable crops are less tolerant to soil moisture stress. Their normal depth of rooting is limited to about 60 cm. Their maximum available moisture deficiency is about 0.5. Potatoes have the least tolerance to moisture stress, with a tolerance value of 0.3.

Figure 9.16 illustrates the three kinds of soil water and the difference in available water between typical sandy loam and silt loam soils. Table 9.4 presents the range of available water-holding capacities of different soil textural groups. For irrigation system design, the total available water is calculated for a soil depth based on the root system of a mature plant of the crop to be grown.

Table 9.4. Range of available water-holding capacity of soils

	Percent moisture, based on dry weight of soil		*Depth of available water per unit depth of soil*
	Field capacity	*Permanent wilting percentage*	*cm per metre depth of soil*
Fine sand	3–5	1–3	2–4
Sandy loam	5–15	3–8	4–11
Silt loam	12–18	6–10	6–13
Clay loam	15–30	7–16	10–18
Clay	25–40	12–20	16–30

(Adapted from Booher, 1967)

Movement of Water within Soils

The movement of water within the soil controls not only the rate of infiltration but also the rate of supply of moisture to plant roots and the rate of underground flow to springs and streams and recharge of groundwater. Water in the liquid phase flows through the water filled pore spaces under the influence of gravity. In the films surrounding soil particles (under unsaturated conditions), it moves under the influence of surface tension forces. Water also diffuses as vapour through the air-filled pore spaces along gradients of decreasing vapour pressure. In all cases, the movement is along gradients of decreasing water potential.

Terminology

Water intake. The movement of irrigation water from the soil surface into and through the soil is called water intake. It is the expression of several factors, including infiltration and percolation.

Percolation. Percolation is the downward movement of water through saturated or nearly saturated soil in response to the force of gravity. Percolation occurs when water is under pressure or when the tension is smaller than about ½ atmosphere. *Percolation rate* is synonymous with infiltration rate with the qualitative provision of saturated or near saturated conditions.

Interflow. Interflow is the lateral seepage of water in a relatively pervious soil above a less pervious layer. Such water usually reappears on the surface of the soil at a lower elevation.

Seepage. Seepage is the infiltration (vertically) downward and lateral movements of water into soil or substrata from a source of supply such as a reservoir or irrigation canal. Such water may reappear at the surface as wet spots or seeps or may percolate to join the groundwater or may join the subsurface flow to form springs or streams. Seepage rate depends on the wetted perimeter of the reservoir or the canal and the capacity of the soil to conduct water both vertically and laterally.

Permeability. (1) *Qualitative.* It is the characteristic of a pervious medium relating to the readiness with which it transmits fluids.

(2) *Quantitative.* The specific property governing the rate or readiness with which a porous medium transmits fluids under standard conditions. According to this definition, equations used for expressing flow, which take into account the properties of the fluid, should give the same soil permeability value for all fluids which do not alter the medium.

The term *intrinsic permeability* is used as permeability factor independent of the fluid. It must, however, be remembered that the factors which tend to change the permeability of the soil matrix to water will influence this value and prevent its use unless they can be measured or evaluated separately.

Hydraulic conductivity. Hydraulic conductivity is the proportionality factor k in Darcy's law ($v = ki$, in which v is the effective flow velocity and i is the hydraulic gradient). It is, therefore, the effective flow velocity at unit hydraulic gradient and has the dimensions of velocity (LT^{-1}). The values of k depend on the properties of the fluid as well as those of the soil, and they reflect any interactions of the fluid with the porous medium, such as swelling of a soil. A soil that has high porosity and coarse open texture has a high hydraulic conductivity value. For two soils of the same 'total' porosity, the soil with small pores has lower conductivity than the soil with large pores because of the resistance to flow in small pores. A soil with pores of many sizes conducts water faster if the large pores form a continuous path through the profile. In fine-textured soils, hydraulic conductivity depends almost entirely on structural pores. In some soils, particles are cemented together to form nearly impermeable layers commonly called *hardpans.* In other soils, very finely divided or colloidal material expands on absorbing water to form an impervious gelatinous mass that restricts the movement of water.

Hydraulic head. Hydraulic head is the elevation with respect to a standard datum at which water stands in a riser pipe or manometer connected to the point in question in the soil. This will include elevation head, pressure head, and velocity head, if the terminal opening of the sensing element is pointed upstream. For non-turbulent flow of water in soil the velocity head is negligible. In unsaturated soil, a porous cup must be used for establishing hydraulic contact between the soil water and water in a manometer. Hydraulic head has the dimensions of length (L).

Hydraulic gradient. Hydraulic gradient is the rate of change of piezometric or hydraulic head with distance. Hydraulic gradient of groundwater records the head consumed by friction in the flow in unit distance because in groundwater flow the velocity heads are generally negligible.

Hydraulic equilibrium of water in soil. It is the condition for zero flow rate of liquid or film water in the soil. This condition is satisfied when the pressure gradient force is just equal and opposite to the gravity force.

Moisture Movement under Saturated Conditions

Saturated flow in soils occurs when the water is not under tension. All or most of the pores are completely filled with water.

Poiseuille's law. The law of Poiseuille express the flow of water in a narrow tube:

$$q = \frac{P\pi r^4}{8l_\mu} \qquad \ldots(9.24)$$

in which,

q = volume of flow per unit time, cm^2/sec

P = pressure difference between two ends of a tube of length 1, $dynes/cm^2$

r = radius of tube, cm

l = length of tube, cm

μ = viscosity of liquid, dyne sec/cm^2 (poises)

Poiseuille's law forms the basis for a number of different equations which have been developed for determining the hydraulic conductivity of the soil from a knowledge of its pore-size distribution. Equation (9.24) shows that the pore size is of outstanding significance, as its fourth power is pro-portional to the rate of saturated flow. This indicates that saturated flow under otherwise identical conditions decreases as the pore size decreases. Generally, the rate of flow in soils of various textures is in the following sequence:

Sand > loam > clay

Darcy's law. The theory of water movement in soils is based on Darcy's law or a generalized form of it. Darcy's law states that the quantity of water passing a unit cross-section of soil is proportional to the gradient of hydraulic head. In mathematical symbols, Darcy's law is

$$q = kia \qquad ...(9.25)$$

in which,

q = volume of flow per unit time, cm^3/sec

i = hydraulic gradient, dimensionless

a = cross-section of flow area, cm^2

k = hydraulic conductivity, cm/sec

The hydraulic gradient is the difference in hydraulic head, divided by the distance between the points measured along the path of flow.

$$i = \frac{h_1 - h_2}{l} \qquad ..(9.26)$$

in which h_1 and h_2 are the hydraulic head at points of measurement 1 and 2, respectively and l is the distance between the points. (see also Chapter 2). Darcy's law can be of the form

$$\frac{q}{a} = ki = v \qquad ..(9.27)$$

in which v is the velocity of flow, cm/sec. The term $\frac{q}{a}$ is also called the velocity flux v, which is defined as the flow per unit area. The validity of Darcy's law is limited by the condition that the flow through the porous medium must be laminar. The usual index used to determine the tendency of flow to be laminar is the Reynold's number R_n, defined as $\rho v d/\mu$, in which ρ is the density of fluid, v is the velocity of flow, d is the mean diameter of soil particles and μ is the dynamic viscosity of the fluid. Darcy's law applies as long as the velocity of flow and the size of soil particles are such that the value of R_n is less than 1. This condition is valid for most cases of water flowing through soils and for groundwater flow problems involving the flow of water towards wells.

In equations, 9.26 and 9.27 if the hydraulic head is replaced by the difference in total potential ψ, the resulting equation is

$$v = -\frac{k\Delta\psi}{l} \qquad ..(9.28)$$

in which l is the distance along the path of greatest change in potential. Equation (9.28) states that the rate of movement is proportional to the potential gradient $\frac{\Delta\psi}{\Delta l}$. The negative sign is introduced because the movement is in the direction of decreasing potential. Equation (9.28) is the generalized expression for soil moisture movement. The nature of the potential ψ and of the conductivity k depends on the soil moisture range in which movement is occurring. It may be seen that equation (9.28) is analogous to Ohm's law for flow of electricity and to the Fourier's law for flow of heat.

Laplace's equation. Application of Darcy's law and the equation of continuity to three-dimensional flow of an incompressible fluid (water) through a porous medium (soil) results in the derivation of the Laplace's equation:

$$\frac{\partial^2\psi}{\partial x^2} + \frac{\partial^2\psi}{\partial y^2} + \frac{\partial^2\psi}{\partial z^2} = 0 \qquad ..(9.29)$$

Equation (9.29) states that the second partial derivatives of the potential with respect to x, y and z directions sums to zero. Solutions of this differential equation have been developed for many problems related to soil moisture movement (Muskat, 1937; Scheidegger, 1960).

Considering both the Poiseuille's equation and Darcy's law, it is recognised that the conductivity of water does not depend upon the total amount of pore space. For example, the conductivity of clay, through a porous material, is so low that it makes one of the best confining materials. The conductivity of sand and gravel depends much upon the sorting and grading of the particles composing the materials; those having grains of uniform size being the most permeable. In most soils, the value of hydraulic conductivity is not constant but changes with time and management history. Since the pore space distribution determines the permeability characteristics, any factor which affects pore space, and particularly the degree of swelling of the soil colloids, influences hydraulic conductivity. Entrapped gases also affect the hydraulic conductivity, primarily by reducing the effective cross-sectional area available for flow.

Moisture Movement under Unsaturated Conditions

As drainage proceeds in a soil and the larger pores are emptied of water the contribution of hydraulic head or the gravitational component to total potential becomes progressively less important and the contribution of the metric potential ψ_m becomes more important. The effect of pressure is generally negligible because of the continuous nature of the air space. The solute potential (osmotic potential) ψ_0 does not affect the potential gradient, unless there is unusual concentration of salt at some point in the soil. The negligible effect of solute potential is due to the fact that both solutes and water are moving. Thus, in case of moisture movement under unsaturated conditions, the potential ψ (Equation 9.28) is the sum of the metric potential ψ_m and, to some extent the gravitational potential ψ_g. In horizontal movement, only ψ_m applies. Under conditions of downward movement, capillary and gravitational potentials act together. In upward capillary movement, ψ_m and ψ_g oppose one another. For unsaturated flow, Equation 9.28 may be rewritten as:

$$v = -k \frac{\Delta(\psi_m + \psi_g)}{\Delta l} \quad \text{...(9.30)}$$

The direction of l is the path of greatest change in $(\psi_m + \psi_g)$.

Under unsaturated conditions, Darcy's law (Equation 9.28) is still applied but with some modifications and qualifications. It is applicable to unsaturated flow if k is regarded as a function of water content, *i.e.*, $k(\theta)$ in which θ is the soil moisture content. As the soil moisture content and soil moisture potential decreases, the hydraulic conductivity decreases very rapidly (Fig. 9.17), so that ψ_{soil} is – 15 bars, k is only 10^{-3} of the value at saturation. According to Philip (1957*a*), the rapid decrease in conductivity occurs because the larger pores are emptied first, which greatly decreases the cross-section available for liquid flow. When the continuity of the films are broken, liquid flow no longer occurs.

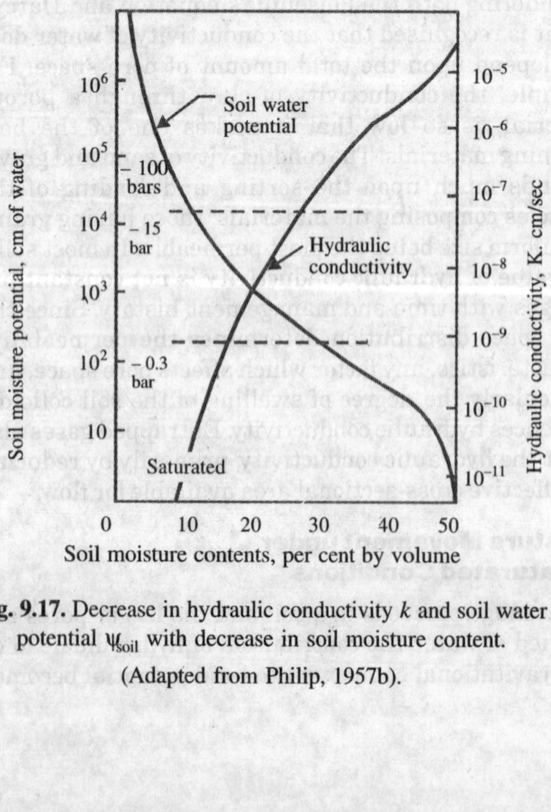

Fig. 9.17. Decrease in hydraulic conductivity k and soil water potential ψ_{soil} with decrease in soil moisture content.
(Adapted from Philip, 1957b).

Unsaturated sil moisture movement, also called capillary movement, k (Equation 9.28) is often termed as *capillary conductivity,* though the term hydraulic conductivity is also frequently used. The unsaturated conductivity is a function of soil moisture content as well as number, size and continuity of soil pores. At moisture contents below field capacity, the capillary conductivity is so low that capillary movement is of little significance in relation to plant growth. Many investigations have shown that capillary rise from a free water table can be an important source of moisture for plants only when free water is within 60 to 90 cm of the root zone.

Movement of unsaturated flow ceases in sand at a lower tension than in finer textured soils, as the water films lose continuity sooner between the larger particles. The wetter the soil, the greater is the conductivity for water. In the 'moist range', the range of unsaturated flow in soils of various textures is in the following order:

Sand < loam < clay

It may be noticed that this is the reverse of the order encountered in saturated flow. However, in the 'wet range' the unsaturated conductivity occurs in the same or similar order as saturated conductivity.

Water Vapour Movement

Movement of soil water in unsaturated soils involves both liquid and vapour phases. Although vapour transfer is insignificant in high soil water contents, it increases as void space increases. At a soil moisture potential of about – 15 bars, the continuity of the liquid films is broken and water moves only in the form of vapour. Diffusion of water vapour is caused by a vapour pressure gradient as the driving force. The vapour pressure of soil moisture increases with the increase in soil moisture content and temperature, it decreases with the increase in soluble salt content.

Water vapour movement is significant only in the 'moist range'. In the 'wet range' vapour movement is negligible because there are few continuous open pores. In the 'dry range' water movement exists, but there is so little water in the soil that the rate of movement is very small.

Water vapour movement goes on within the soil and also between soil and atmosphere, for example, evaporation, condensation and adsorption. The rate of diffusion of water vapour through the soil is proportional to the square of the effective porosity, regardless of pore sizes. The finer the soil pores, the higher is the moisture tension under which maximum water vapour movement occurs. In a coarse textured soil, pores become free of liquid water at relatively low tensions and when the soil dries out there is little moisture left for vapour transfer. But a fine textured soil retains substantial amounts of moisture even at high tensions, thus permitting vapour transfer. It is interesting to note that maximum water vapour movement in soils occurs just before the permanent wilting point is reached. At such a stage vapour movement is of greatest importance for the growth and survival of plants.

MEASUREMENT OF SOIL MOISTURE

The importance of the moisture content in the soil in relation to plant growth has resulted in the development of many methods for measuring soil moisture. Soil moisture measurements are important in scheduling of irrigations and estimating the amount of water to apply in each irrigation. Measurement of changes in soil moisture storage with time is important in estimating evapotranspiration. There are also many experimental situations where careful measurement and control of soil moisture is necessary, if the results of investigation on soil-plant-water—relationships are to be interpreted properly.

The principal methods of expressing soil moisture are: (*i*) by the amount of water in a given amount of soil, and (*ii*) the stress or tension under which the water is held by the soil. The relationship between these two properties throughout the entire moisture range gives a good deal of insight into the physical properties of a soil. This relationship can be shown in the soil moisture characteristic curves (Fig. 9.15).

Expressing the amount of soil moisture. The amount of moisture that is held by a certain mass or volume of soil can be expressed as weight per cent or volume per cent. Soil moisture on weight basis is based on the dry weight of the sample:

Soil moisture content, per cent by weight

$$= \frac{(\text{Weight of moist sample} - \text{weight of oven dry sample})\,100}{\text{Weight of oven dry sample}}$$

Expression of moisture content as a percentage of dry weight may not indicate the amount of water available to plants, unless the soil moisture characteristic curve or the field capacity and permanent wilting point are known. For some purposes it is more useful to convert moisture content per unit weight into moisture content per unit of volume. Also, additions and losses of water from the soil are often measured in centimetres which, on an area basis becomes the volume. The conversion from weight to volume units can be made by multiplying the percentage by weight times the bulk density of the soil under study.

Moisture content, per cent by volume = Moisture content, per cent by weight × Bulk density

The per cent by volume of moisture content obtained by the above relationship is numerically equal to cm of water per metre depth of soil. For example, if the bulk density (apparent specific gravity) is 1.6 gm/cc and the moisture content is 14 per cent, on dry weight basis, the soil moisture is $1.6 \times 14 = 22.4$ cm per metre depth of soil.

Location of soil moisture measurement. The location of soil moisture measurements is important. It is generally recommended that one location should be near the side of the field where irrigation is to be started, as a reference point for starting the irrigation cycle. For row crops the measurements should be made in the row or near the plants. In sprinkler irrigation, the measuring stations should be between the sprinkler heads and three to four metres away from the lateral. For tress, measurements are made generally about one to two metres away from the *trunk*. Soil moisture measurements should be made in that part of the soil from which plant roots extract their moisture and according to the moisture extraction pattern of the particular crop. One measurement is made in the upper quarter of the root zone and one or two measurements at lower levels.

Gravimetric Method of Soil Moisture Measurement

In the gravimetric method, basic measurements of soil moisture are made on soil samples of known weight or volume. Soil samples are collected with a soil auger or sampling tube (Fig. 9.18). The samples are taken from the desired depth at several locations for each soil type. They are collected in air tight aluminium containers. The soil samples are weighed and they are dried in an oven at 105°C for about 24 hours, until all the moisture is driven off. After removing from oven they are cooled slowly to room temperature and weighed again. The difference in weight is the amount of moisture in the soil. The method, though accurate, is used mainly for experimental work.

Several methods have been developed to reduce the time required for oven drying and to simplify the determination of soil moisture. One technique is to mix the soil sample with a known volume of methyl alcohol and then measure the change in specific gravity of the

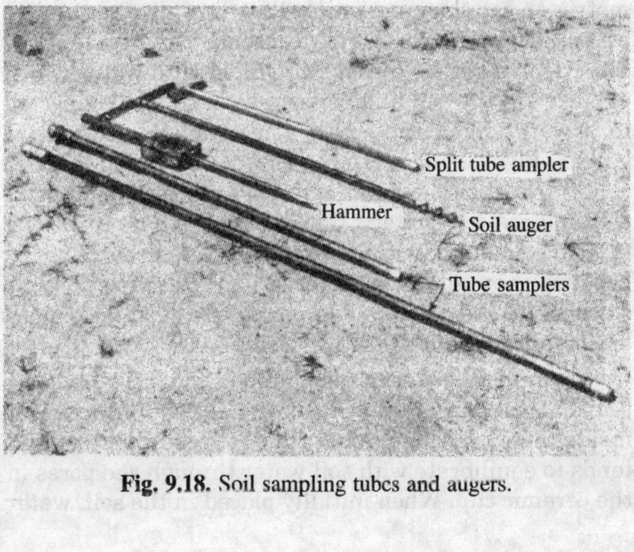

Fig. 9.18. Soil sampling tubcs and augers.

alcohol with a hydrometer. Another method is to mix the soil with a known amount of calcium carbide. Calcium carbide reacts with water and removes it in the form of acetylene gas. The moisture is determined by the difference in weight. More recently, infrared radiation is utilized for drying soil samples quickly. However, neither this nor any other suggested short-cut procedures have come into common use.

Example 9.2. *Undisturbed soil sample was collected from a field two days after irrigation, when the soil moisture was near field capacity. The inside dimensions of the core sampler were 7.5 cm diameter and 15 cm deep. Weight of the core sampling cylinder with moist soil was 2.76 kg and the weight with oven dry soil was 2.61 kg. The weight of the core sampling cylinder was 1.56 kg. Determine the water depth in centimetres per metre depth of soil.*

Solution: Weight of moist soil = 2.76 – 1.56 – 1.20 kg

Weight of oven dry soil = 2.61 – 1.56 – 1.05 kg

$$\therefore \text{ Moisture content} = \frac{1.20 - 1.05}{1.05} \times 100 = 14.28\%$$

$$\text{Volume of core sampler} = \frac{\pi}{4}d^2h = \frac{\pi}{4} \times 7.5 \times 7.5 \times 15$$

$$= 662 \text{ cu cm}$$

Apparent specific gravity

$$= \frac{\text{Wt. of dry soil in grams}}{\text{Volume of soil (sampler) in cu cm}}$$

$$= \frac{1.05 \times 1000}{662} = 1.58$$

$\therefore$ Moisture content in linear units

= Ap. sp. gr. × moisture content

= 1.58 × 14.28 = 22.56 cm per metre depth of soil

Tensiometers

Tensiometers provide a direct measure of the tenacity with which water is held by soils. They measure the metric or capillary potential, they can also be used to estimate the soil moisture content. The tensiometer consists of a porous ceramic cup filled with water which is buried in the soil at any desired depth and connected to a water filled tube with a manometer or vacuum gauge (Figs. 9.19 to 9.21). The scales are generally calibrated in either hundredths of an atmosphere or in centimetres of water. Tensiometers that utilize a mercury manometer are preferred as research tools because they afford great precision. Because of their simplicity, tensiometers equipped with vacuum gauges are better suited for practical field use.

When the tensiometer is placed in the soil where the tension measurement is to be made, the bulk water inside the porous cup comes into hydraulic contact and tends to equilibrate with soil water through the pores in the ceramic cup. When initially placed in the soil, water

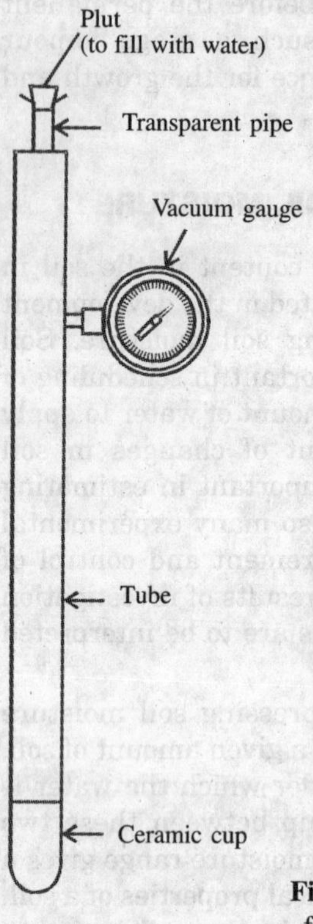

Fig. 9.19. Tensiometer fitted with vacuum gauge.

Fig. 9.20. Tensiometers installed in a wheat field to determine the moisture tension at different depths within the root zone during crop growth.

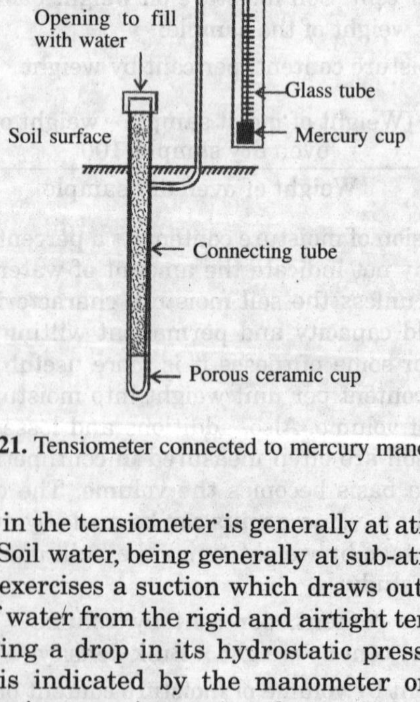

Fig. 9.21. Tensiometer connected to mercury manometer.

contained in the tensiometer is generally at atmospheric pressure. Soil water, being generally at sub-atmospheric pressure, exercises a suction which draws out a certain amount of water from the rigid and airtight tensiometer, thus causing a drop in its hydrostatic pressure. This pressure is indicated by the manometer or vacuum gauge. Any increase in tension that occurs as the soil dries causes the gauge or manometer reading to increase, which can be read above ground. Conversely, an increase in soil-water content reduces tension and lowers the reading.

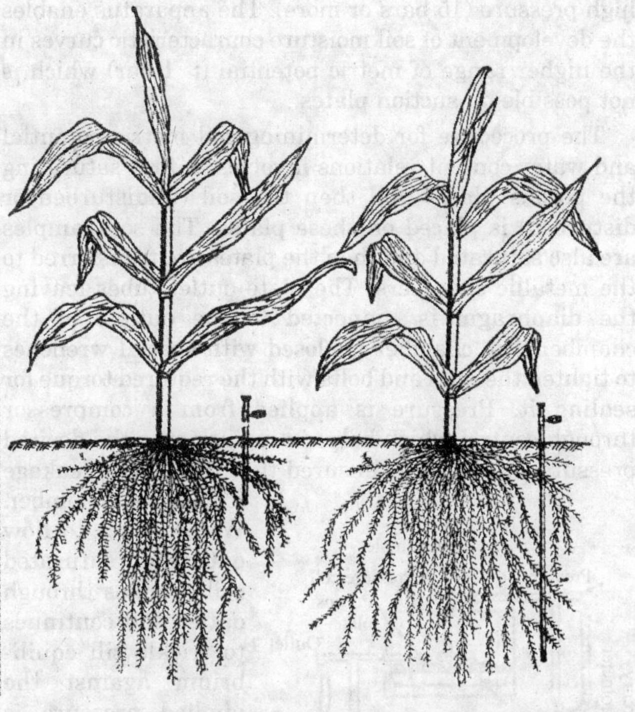

Fig. 9.22. Installation of tensiometers to measure soil moisture tension at different depths in the crop root zone.

Since the porous cup walls of the tensiometer are permeable to both water and solutes, the water inside the tensiometer assumes the same solute composition and concentration as soil water, and the instrument does not indicate the osmotic suction of soil water (unless equipped with some type of an auxiliary salt sensor).

Tension measurements by tensiometers are generally limited to metric suction values of below 1 atmosphere. This is due to the fact that the vacuum gauge or manometer measures a partial vacuum relative to the external atmospheric pressure, as well as to the general failure of water columns in macroscopic systems to withstand tensions exceeding one atmosphere. Further, as the ceramic material is generally made of permeable and porous material, too high a suction may cause air entry into the cup, which would equalize the internal pressure to the atmospheric. Under such conditions, soil moisture tension might continue to increase even while the tensiometer fails to show it. In practice, however, the useful limit of most tensiometers is at about 0.85 atmospheres. The limited range of tension measurable by the tensiometer is not serious since it generally encompasses the greater part of the soil wetness range.

Tensiometer readings reflect soil-moisture tension only; that is, they indicate the relative wetness of the soil surrounding the porous cup. They do not provide direct information on the amount of water held in the soil. Tension measurements are useful in deciding when to irrigate, but they do not indicate how much water

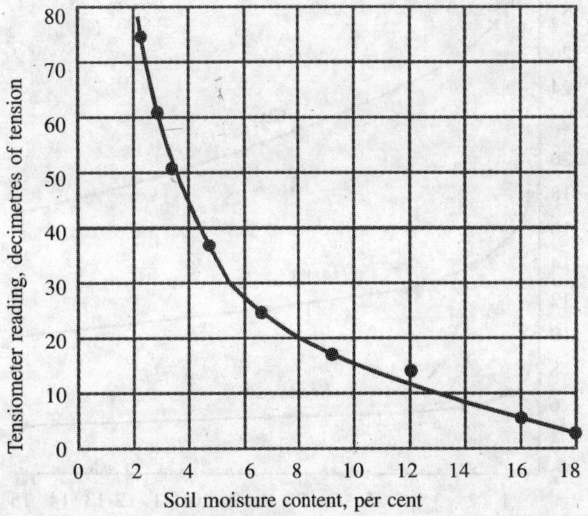

Fig. 9.23. Calibration curve of a tensiometer in a sandy loam soil.

should be applied. A special moisture characteristic curve for the particular soil is needed to convert moisture-tension measurements into available-moisture percentages (Fig. 9.23).

Tensiometers do not satisfactorily measure the entire range of available moisture in all soil types. But they probably are the best field instruments to use to determine moisture conditions in the wet range. They are best suited for use in sandy soils since in these soils a large part of the moisture available to plants is held at a tension of less than 1 atmosphere. Tensiometers are not well suited for use in fine-textured soils in which only a small part of the available moisture is held at tensions of less than 1 atmosphere. A general practice is to place a tensiometer at few soil depths representing the root zone, and to irrigate when the tensiometer indicates that the metric suction has reached some prescribed value. The use of several tensiometers at different depths can indicate the amount of water needed in irrigation (with the help of calibration curves). They can also allow calculation of the hydraulic gradients in the soil profile.

Some experience is required to use a tensiometer. If air enters the tensiometer through any leaks at the rubber connections, measurements will not be reliable. Air leaks can result from faulty cups. They may occur also at the contact points of the setscrews used to secure the porous cup to the metal support. Some manufacturers provide a test pump that can be used to test the gauge and to remove air from the instrument. After an irrigation or rain, it is desirable to fill the instrument with water.

Care is needed in installing the tensiometer properly. For good installation, part of the tensiometer containing the ceramic cup and tube is inserted into a prepared hole with a near-close fit so that the walls of the ceramic cup are in close contact with undisturbed soil and roots. The

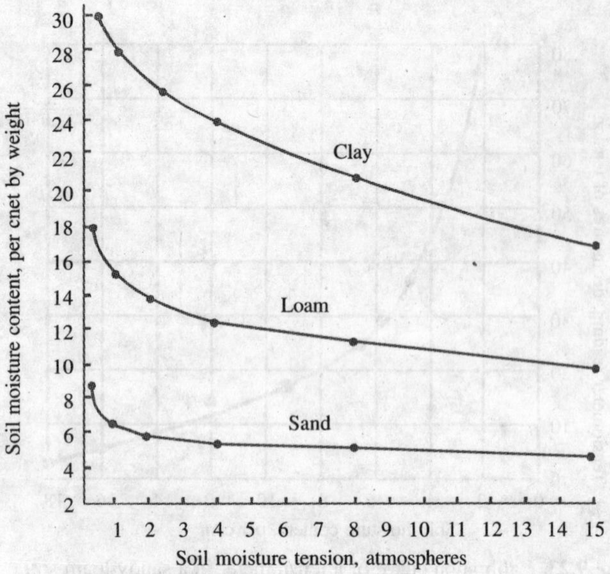

Fig. 9.24. Moisture characteristics of common soil types. (Adapted from USDA, SCS Handbook Sec. 15, 1964).

Fig. 9.25. Pressure membrane apparatus with manifold.

hole can be made by driving a steel rod of diameter equal to the tensiometer and provided with a tapered driving point. At the ground surface the soil is pressed around the tensiometer and it is piled slightly so that water does will not collect and seep down along the tube of the tensiometer.

Pressure Membrane and Pressure Plate Technique

Laboratory measurements of soil moisture potential are mostly made with the pressure membrane and pressure (or suction) plate equipment (Figs. 9.25 and 9.26) developed primarily by Richards (1949, 1954). The apparatus consists of ceramic pressure plates or membranes of high air entry values contained in air-tight metallic chambers strong enough to withstand high pressure (15 bars or more). The apparatus enables the development of soil moisture characteristic curves in the higher range of metric potential (> 1 bar) which is not possible on suction plates.

The procedure for determining soil metric potential and water content relations involves in first saturating the porous plates and then the soil (undisturbed or disturbed) is placed on these plates. The soil samples are also saturated and then the plates are transferred to the metallic chambers. The plate outlet tubes leaving the diaphragm is connected to the outlets of the chamber. The chamber is closed with special wrenches to tighten the nuts and bolts with the required torque for sealing it. Pressure is applied from a compressor through control which helps in maintaining the desired pressure. It should be ensured that there is no leakage from the chamber. Water starts to flow out from saturated soil samples through outlet and continues to trickle till equilibrium against the applied pressure is achieved. After that the soil samples are taken out and oven-dried for determining the moisture content, either on weight basis (disturbed soil), or volume basis (undisturbed soil). Similarly, the moisture content of the soil can be determined against other pressure values. For a large number of soils the pressure values chosen are 1/10, 1/5, 1/3, 1, 3, 5, 7, 10 and 15 bars. The pairs of pressure and moisture content data so obtained are used to construct the soil moisture characteristics curve.

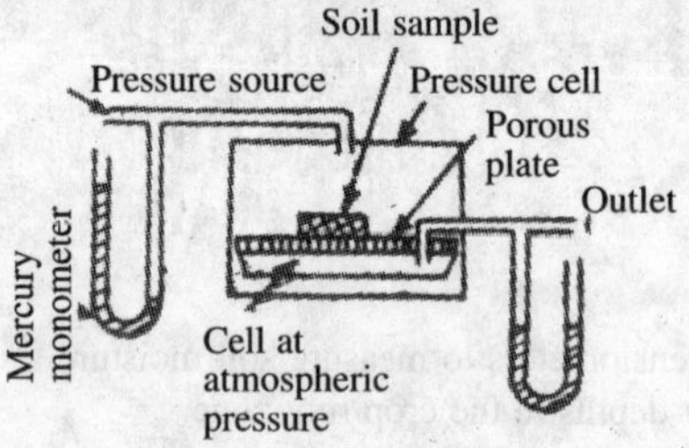

Fig. 9.26. Schematic diagam showing main parts of a pressure plate apparatus.

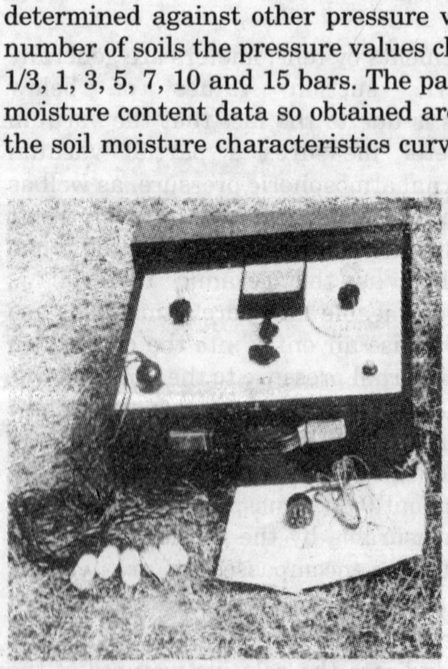

Fig 9.27. A soil moisture-cum-temperature meter. Note the gypsum blocks and thermisters kept on the foreground. (Instrument developed by Dr. M.C. Oswal)

Fig. 9.28. Electrical resistance block.

Measurement of Solute (Osmotic) Potential

Field methods of measuring soil moisture with porous blocks and tensiometers can be used to estimate the metric component, but not the osmotic component, of the soil water potential. The pressure and gravitational potentials can usually be neglected, but the osmotic potential may be important, especially in heavily fertilized and salt affected soils. Unfortunately, there are at present no entirely suitable methods for routine measurement of the total soil water potential in the field. Measurements of solute potential or osmotic pressure of the soil solution can be made after the solution is extracted from soil samples. Extraction involves making up the water content to a specified supersaturated level and then preparing a "saturation extract" (Richards, 1954). Osmotic pressure is determined by using vapour pressure psychrometers (Slatyer, 1967). The electrical conductivity of the soil solution is often measured because there is usually a close agreement between electrical conductivity and osmotic potential of the saturation extract.

Electrical Resistance Method

The dependence of electrical conductivity of a porous solid on the amount of water forms the basis of this method. Porous blocks containing electrodes inside (Fig. 9.28), when embedded in soil, the moisture content of the porous body comes in equilibrium with soil moisture. The electrical resistance between the two electrodes varies with the moisture of the porous block which is in equilibrium with soil moisture. The resistance is measured by a resistance meter. The blocks are calibrated against a range of moisture levels and are installed in the field at the required depths. The electrode lead wires are taken out of the soil surface and well secured and protected against rain. When required, the resistance meter is carried to the field and through the electrode leads the resistance is measured to know the moisture content from the calibration curve.

The resistance blocks are made of various materials like gypsum, nylon, fibre glass, plaster of paris or a combination of these materials. The blocks are generally rectangular. However, cylindrical blocks having concentric electrodes are also common. The pair of electrodes is usually made of 20 mesh stainless steel wire screen soldered to the copper lead wire. The common dimensions of screen electrodes are 3.75 cm long and 1.25 cm wide. The usual spacing between the electrodes is 2 cm. A smaller block is always sensitive. A convenient size of a gypsum block is 5.5 cm long, 3.75 cm wide and 2 cm thick.

Nylon and fibre-glass units are more sensitive in the higher ranges of soil moisture than plaster of paris blocks, but often their contact with soil that is alternately wet and dry is not very good. Nylon units are most sensitive at a tension of less than 2 atmospheres. Plaster of paris blocks function most effectively at a tension between 1 and 15 atmosphere, and fibre-glass units operate satisfactorily over the entire range of available moisture. A combination of fibre glass and plaster of paris provides sensitivity in both the wet and dry range and provides good contact between the soil and the unit. The conductivity blocks must be properly installed in order to obtain readings which are representative of the area. It is necessary that the blocks must have close contact with undisturbed soil. A satisfactory procedure is to force the blocks into undisturbed soil along the sides of a hole dug for placement of blocks (Fig. 9.29). After placement, the hole is back-filled quickly.

The electrodes in the blocks are attached by wires to a resistance or conductance meter that measures the changes in electrical resistance in the blocks. The blocks are buried at the desired depths (Fig. 9.29). When the units are buried in the soil, they become almost a part of the soil mass and respond to changes in soil moisture content. The most commonly used electrical resistance instrument was developed by Bouyoucos (1949). Hence, they are sometimes called Bouyoucos moisture meters. The instrument uses a specially designed wheatstone bridge for resistance measurement. Resistance measurements vary with soil type. Since the same reading may indicate different amounts of available

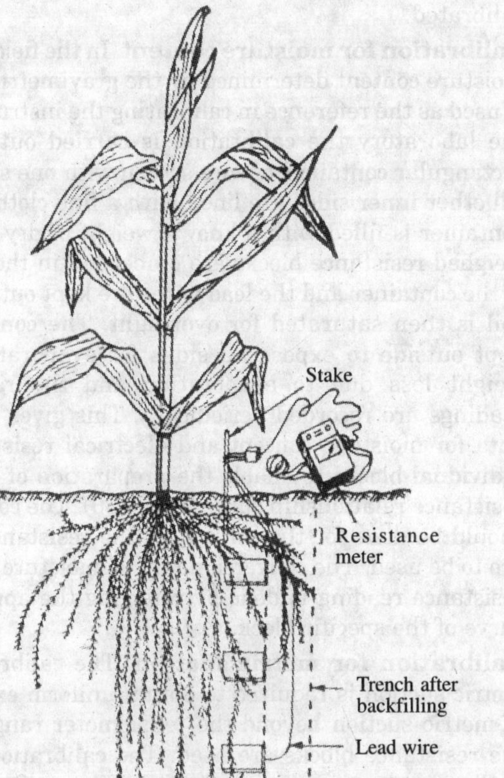

Fig. 9.29. Schematic sketch showing the use of electrical resistance unit for soil moisture measurement.

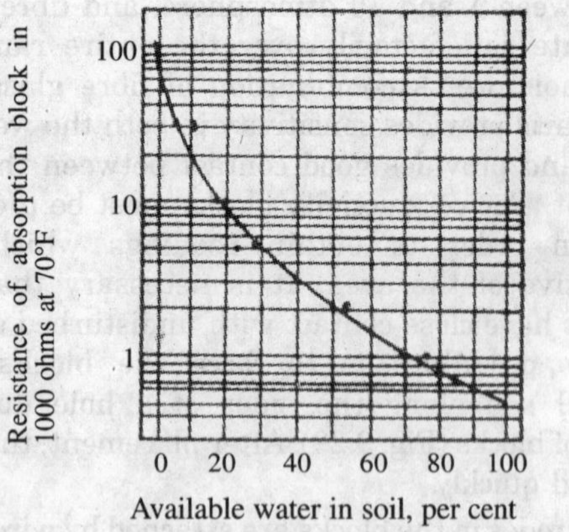

Fig. 9.30. Calibration curve of a plaster of paris resistance block for a silt loam soil (Bouyoucos, 1954).

moisture for different soil textures, the instrument must be calibrated for the soil in which it is to be used.

Calibration of Electric Resistance Blocks

It is important to calibrate individual blocks for moisture content and moisture tension. In other words, a calibration curve of resistance versus moisture per cent/moisture tension has to be done. It is difficult to maintain exactly equal spacing of electrodes in all the blocks. Therefore, it is desirable that each block is calibrated.

Calibration for moisture content. In the field, the soil moisture content determined by the gravimetric method is used as the reference in calibrating the instrument. In the laboratory the calibration is carried out using a rectangular container of wire screen with one side open. All other inner sides are lined with a fine cloth and the container is filled with a known weight of dry soil. The weighed resistance blocks are embedded in the dry soil of the container and the lead wires are kept outside. The soil is then saturated for overnight. The container is kept outside to expose all sides for evaporation. The weight loss due to evaporation and the resistance readings are recorded periodically. This gives a pair of data for moisture content and electrical resistance for individual blocks to enable the preparation of moisture resistance relationship curve (Fig. 9.30). The calibration should be done for the soil where the resistance blocks are to be used. The conversion for soil moisture from the resistance reading is done by referring the appropriate curve of the specific block.

Calibration for metric suction. The calibration for metric suction is required to obtain uniform expression of metric suction beyond the tensiometer range where the resistance blocks are used. The calibration can be easily done in a pressure plate apparatus. Soil passed through a 2 mm sieve is kept on a saturated pressure plate. The required number of resistance blocks are embedded in the soil and the whole thing is saturated overnight. The plate containing blocks is placed inside the pressure plate chamber bringing out the lead wires

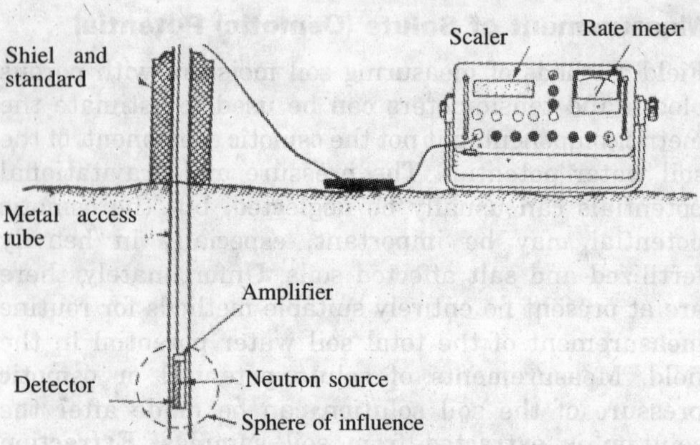

Fig. 9.31. Schematic sketch illustrating the use of neutron moisture meter.

Fig. 9.32. Neutron moisture meter for soil moisture measurement. Note the storage battery, counter and neutron probe.

of the resistance blocks through "electrical lead through", the passage provided for taking the lead wires out. The required pressure is applied and at equilibrium resistance, readings are recorded. These observations for resistance are obtained for a range of pressure values to enable the preparation of the calibration curve between metric potential and resistance readings.

Limitations of electrical resistance blocks. Resistance blocks are not useful for saline soils, since the resistance reading is affected by salt concentration. However, gypsum blocks work better even in saline conditions because of buffering action. But nylon and fibre glass blocks cannot be used. Fibre-glass units are more sensitive than plaster of paris. Their readings are also affected by concentrations of fertilizer. Where the fertilizer is spread in bands, the unit should be placed to one side of the band. Generally, the resistance blocks should not be used for wet soil moisture range. Resistance measurements change somewhat with temperature and thermistors are sometimes installed with the blocks to measure the temperature (Fig. 9.27).

Neutron Moisture Meter

The neutron scattering method is a rapid means of making *in situ* measurement of soil moisture. The principle of the neutron method is based on the measurement of the number of hydrogen nuclei that are present in a unit volume of soil, their number being a direct function of the number of water molecules contained in that same volume. This measurement is made by inserting a source of fast neutrons and by counting the slow neutrons produced.

The neutron moisture meter consists of two principal parts: a probe, which is lowered in the access tube inserted vertically in the soil and which contains a source of fast neutrons and a detector of slow neutrons; (*ii*) a scaler or ratemeter (usually battery-powered and portable) to monitor the flux of slow neutrons, which is proportional to the soil water content. The fast-neutron source may be a 2–5 millicurie mixture of radium and beryllium or a mixture of americium and beryllium. In use, the probe is lowered into the access tube, and counting rates are determined at the desired depths. The count rates, adjusted for background and standardized to counts made in a tank of pure water, are then calibrated against volumetric determinations of soil moisture content (Fig. 9.33). (For detailed treatment on the method see Van Bavel *et al.,* 1963; Holmes, 1956).

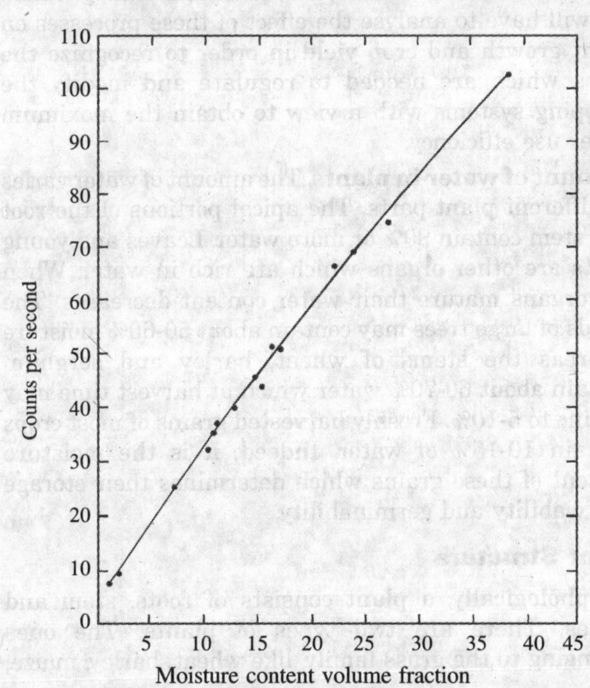

Fig. 9.33. A typical calibration curve of a neutron moisture probe.

The effective volume of soil in which the water content is measured depends upon the concentration of the hydrogen nuclei, *i.e.,* upon the pre-volume wetness of the soil as well as upon the energy of the emitted fast neutrons. With the commonly used radium-beryllium sources, the soil volume measured is in effect a sphere, which in a wet soil is about 15 cm in diameter, but in a relatively dry soil it may be as great as 50 cm or more. The relatively large volume monitored can be an advantage in water-balance studies, as such a volume is generally more representative of the field soil than a small sample. Another advantage is the absence of a lag period while the soil water equilibrates with a sensing instrument. However, it involves the disturbing influence of the access tube, and the large volume of the sample prevents sampling at a point, or near the surface unless special precautions are taken. Further, the results are influenced by other sources of hydrogen such as organic atoms, matter, and by other elements, notably chlorine, iron and boron.

Improper use of the neutron moisture meter can be hazardous. The danger from exposure to radiation depends upon the strength of the source, the distance from the source to the operator, and the duration of exposure. A protective shield (generally consisting of lead and paraffin or polyethylene in a cylindrically shaped unit with a central hole to accommodate the probe) is an essential component of the equipment. It also serves as a standard absorber for checking the readings. With reasonable care and attention to safety rules, the equipment can be used safely.

PLANT-WATER RELATIONSHIPS

Almost every plant process is affected directly or indirectly by the water supply. Water constitutes about 80 to 90 per cent of most plant cells and tissues in which there is active metabolism. It forms a continuous liquid phase through the plant from the root hairs to the leaf epidermis. Plant roots take up water and transfer it through the root tissues to the conducting vessels of the plant. These conducting vessels pass it on to the mesophyll tissues of the leaves. After moving through these tissues, the water reaches the evaporation sites which are primarily the walls of the sub-stomatal cavities. All these flows are in the liquid phase along a potential gradiant. The final transfer of water from the sub-stomatal cavities through the stomata of the leaves is a process of molecular diffusion of water vapour. The factors influencing the water relations of plants, and thus their growth and yield responses, may be grouped into the following:

(*i*) Soil factors—soil moisture content, texture, structure, density, salinity, fertility, aeration, temperature and drainage.

(*ii*) Plant factors—type of crop, density and depth of rooting, rate of root growth, aerodynamic roughness of the crop, drought tolerance and varietal effects.

(*iii*) Weather factors—sunshine, temperature, humidity, wind and rainfall.

(*iv*) Miscellaneous factors—soil volume and plant spacing, soil fertility, and crop and soil management. Figure 9.34 is a schematic sketch of soil-water-float relationship and the basic morphology of a typical plant.

Water as a Plant Component

The metabolic activity of cells and plants is closely related to their water content. Growth of plants is controlled by the rates of cell division and enlargement, and by the supply of organic and inorganic compounds required for the synthesis of new protoplasm and cell walls. A decreasing water content is accompanied by a loss of cell turgor* and wilting, cessation of cell enlargement, closure of stomata reduction of photosynthesis and interference with many basic metabolic processes. Eventually, continued dehydration

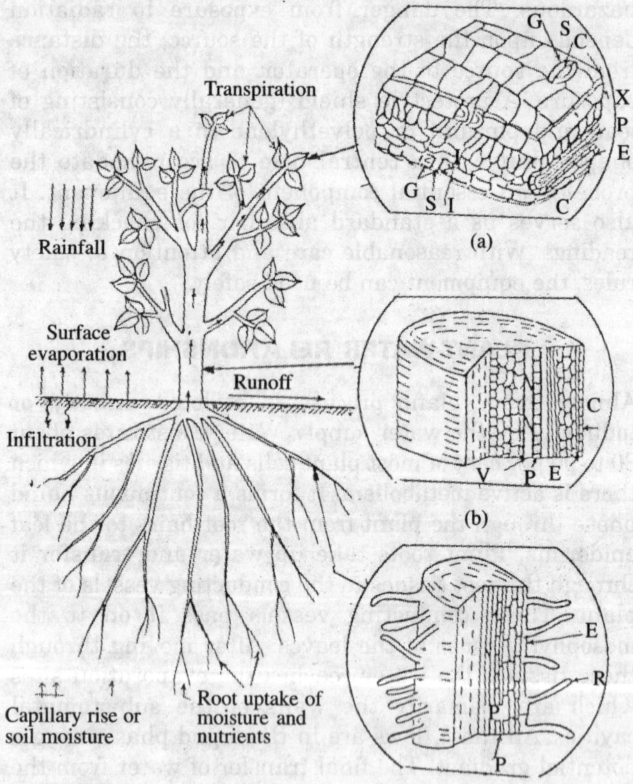

Fig. 9.34. Schematic sketch of water-soil-plant relationship. Enlarged cut-away views of the parts of a plant which are concerned with the absorption, conduction, conservation, and loss of water are shown: (*a*) is an enlarged portion of leaf cut-away to show internal structure: (*b*) an enlarged segment of stem; (*c*) enlarged segment of root: *G*, is a guard cell; *S*, Stomata; *C*, cuticle; *E*, epidermal cell; *V*, vein; *P*, phloem; and *R*, root hair.

causes disorganization of the protoplasm and death of most organisms. The relation of water content to physiological processes is shown strikingly in seeds, where respiration and other physiological activities increase manifold as the water content increase. In photosynthesis, water is as important a reagent as carbon dioxide. An essential function of water in plants is of the solvent through which gases, minerals, and other solutes enter plant cells and move from cell to cell and tissue to tissue within the plant. The permeability of most cell walls and membranes to water results in a continuous liquid phase extending throughout the plant in which translocation of solutes occurs. Water is a reactant or reagent in many physiological processes including photosynthesis and hydrolytic processes such as the hydrolysis of starch to sugar. Water is essential to maintain sufficient turgidity for growth of cells and maintance of the form and position of leaves and new shoots.

The total quantity of water required for the essential physiological functions of the plant is usually less than five per cent of all the water absorbed. Most of the water entering the plant is lost in transpiration, directly contributing little to its growth. However, failure to replace the water lost by transpiration results in the loss of turgidity, cessation of growth, and eventual death of the plant from dehydration.

The following are the main areas of water-plant relationship: (1) water absorption, (2) water conduction and translocation, and (3) water loss or transpiration.

In determining the importance of water in crop productivity we have to understand clearly all the three processes—absorption, translocation and transpiration. We will have to analyse the effect of these processes on plant growth and crop yield in order to recognize the steps which are needed to regulate and modify the cropping systems with a view to obtain the maximum water use efficiency.

Amount of water in plants. The amount of water varies in different plant parts. The apical portions of the root and stem contain 90% or more water. Leaves and young fruits are other organs which are rich in water. When the organs mature their water content decreases. The woods of large trees may contain about 50-60% moisture whereas the stems of wheat, barley and sorghum contain about 60-70% water which at harvest time may decline to 5-10%. Freshly harvested grains of most crops contain 10-15% of water. Indeed, it is the moisture content of these grains which determines their storage life, viability and germinability.

Plant Structure

Morphologically, a plant consists of roots, stem and leaves. There are two types of plants. The ones belonging to the grass family, like wheat, barley, maize,

*Turgor or turgidity is the distension (steel out or dilation by pressure from within) of the cell wall and protoplasmic layer of plants by fluids. It is essential for plant growth.

sorghum, millets, etc. which are called monocotyledonous. The second group of plants named dicotyledonous includes peas, red gram, green gram, chickpea, mustard, seasamum, tomatoes, potatoes, tapioca, etc. Once these plants grow in the field, it is easy to distinguish them on the basis of their stem. The monocotyledonous wheat, sorghum, etc. give an appearance of leaves directly emerging from the soil because these plants have a very condensed stem during their early stages of growth. A typical stem can be noticed during the later stages of growth. In mustard, tomatoes, beans, castor etc., which are dicotyledonous plants, a distinct stem is visible from the very beginning.

The leaves are borne throughout the stem in all the plants. These organs are mainly responsible for the loss of water. The internal structure of a typical leaf is shown in Fig. 9.35. The path of water in a leaf is shown in Fig. 9.36. The stomata regulate the loss of water as vapours and exchange of carbon dioxide in the leaf and

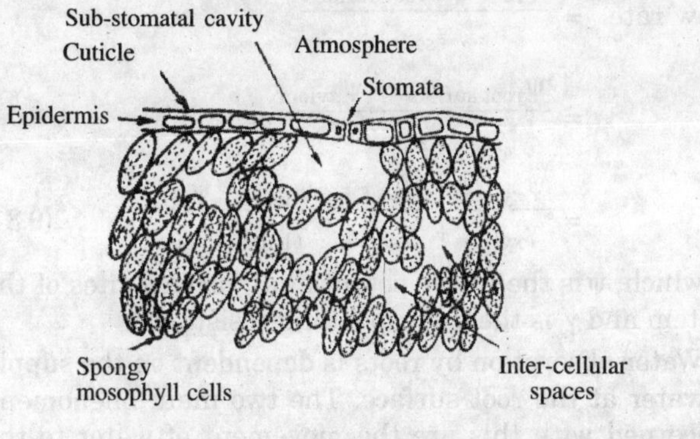

Fig. 9.35. Portion of a leaf cross-section adjacent to stomate.

other organs. It is the efficiency of these structures which determines the water loss from the plant. The efficiency of the stomata depends upon their size and number per unit area.

The leaves maintain their continuity of structure with the stem which has conducting tissues called xylem and phloem. The former are the main channels of water transport. The stem maintains its continuity with the root which eventually is in contact with the water source, the soil.

The structure of a typical root is shown in Fig. 9.37. The root apex has three distinct zones namely, (*i*) apical meristem with root cap, (*ii*) root hairs, and (*iii*) root maturation zone. It is the second zone which is largely involved in water uptake. A transverse section of root brings out the relationship of root hairs with the remaining structure of the root (Fig. 9.38). It can be seen from this diagram that an outermost cell or an epidermal cell usually gets elongated into a long hair which has its continuity with the remaining cells. It must be emphasized that other epidermal cells are also capable of absorbing water but the root hair provides the

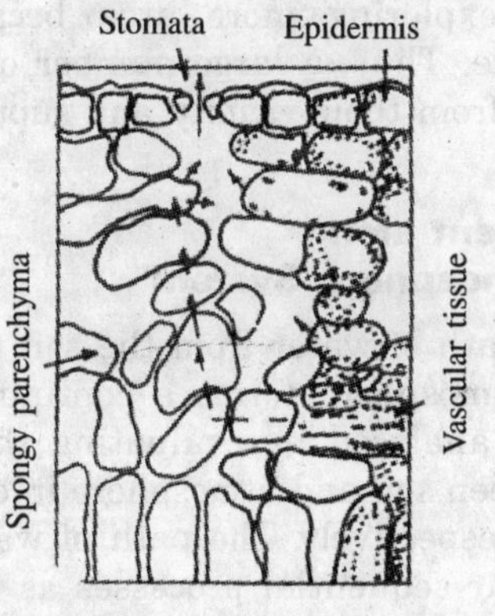

Fig. 9.36. Path of water vapours in a leaf.

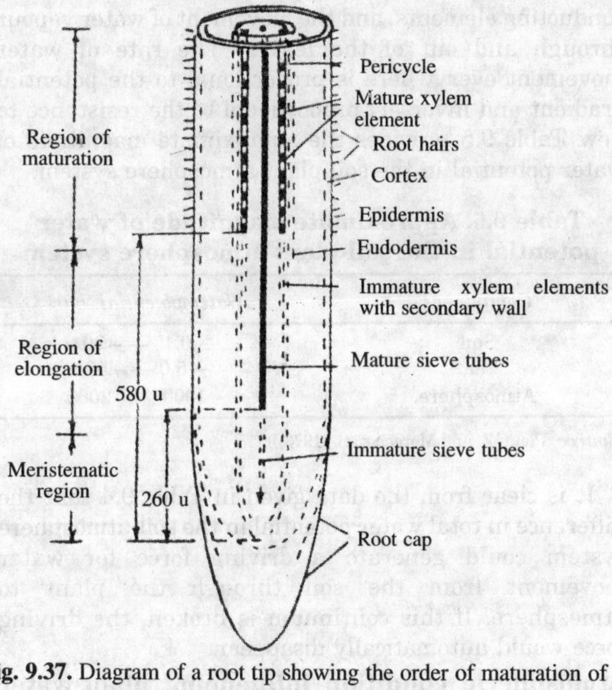

Fig. 9.37. Diagram of a root tip showing the order of maturation of various tissues (After Kramer, 1969).

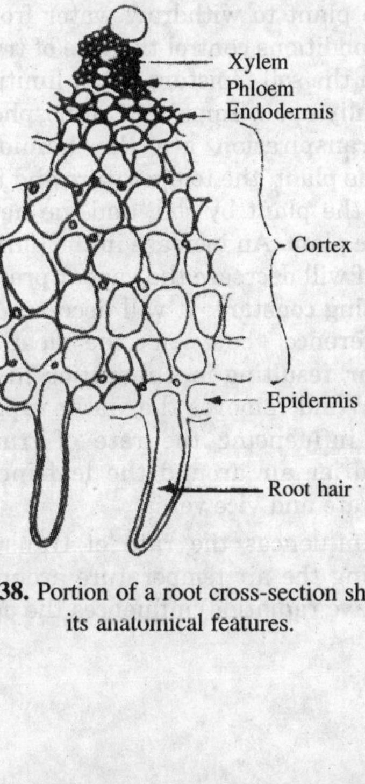

Fig. 9.38. Portion of a root cross-section showing its anatomical features.

advantage of exploring more area because of its enlarged surface. Thus, a large number of root hairs draw moisture from their vicinity and supply water to the cortex.

Water Movement along Soil-Plant-Atmosphere System

The complete path of water from the soil through the plant to the atmosphere forms a continuous system which may be analysed by evaluating the potential difference between soil and atmosphere in contact with root and leaf, respectively. The path of water may be divided into four sequential processes as follows: the supply of liquid water to the root surface, the entry of water into the root, the passage of water in the conducting elements, and the movement of water vapour through and out of the leaves. The rate of water movement everywhere is proportional to the potential gradient and inversely proportional to the resistance to flow. Table 9.5 provides the approximate magnitude of water potential in the soil-plant-atmosphere system.

Table 9.5. Approximate magnitude of water potential in the soil-plant-atmosphere system

Component	*Water potential, bars*
Soil	– 0.1′ — – 20
Leaf	– 5.0 — – 50
Atmosphere	– 1000 — – 2000

(***Source:*** Plout, Z. and Moreshet, S. (1973).)

It is clear from the data given in Table 9.4 that the difference in total water potential in the soil-atmosphere system could generate a driving force for water movement from the soil through the plant to atmosphere. If this continuum is broken, the driving force would automatically disappear.

Atmospheric condition influencing plant-water relationships. The atmosphere provides the energy needed by the plant to withdraw water from the soil. Atmospheric conditions control the rate of transpiration of plants when the soil moisture is not limiting and the stomata are fully open. Important atmospheric factors influencing transpiration are the humidity of air surrounding the plant, the temperature and humidity of air carried to the plant by soil, and the net radiation available to the plant. An increase in the humidity of air around the leaf will decrease the vapour pressure, other factors remaining constant. It will decrease the vapour pressure difference between the leaf and the surrounding air, resulting in a reduction in the rate of transpiration. Wind removes the water vapour around the air, thus influencing the rate of transpiration. Warmer and drier air around the leaf increases the transpiration rate and vice versa.

Radiation influences the rate of transpiration of plants by raising the air temperature around the leaf. Light (short wave radiation) influences the opening and closing of stomata. In most plants the stomata are open during the day and closed at night.

Absorption of water in soil-plant-atmosphere system. The root system is extremely variable in different crop plants. The variability exists in rooting depth, root length and horizontal distribution of roots. These are further influenced by environmental factors and the genetic constitution. Nevertheless, both the properties of soil and the roots determine the water uptake by roots. The roots of cereals, apparently occupy more surface area of the soil than other crops. For example, it has been shown that cereal roots extend to 200-4000 cm/cm^2 of soil surface area as against 15-200 cm/cm^2 for non-graminaceous plants.

It is desirable to consider water absorption in the total soil-plant-atmosphere system instead of the roots alone. In this system, one can partition the system in such a manner that the involvement of different plant parts is taken into account. The flow rate of water in this system is given by the following equation:

$$\text{Flow rate} = \frac{\psi_{soil} - \psi_{root\ surface}}{\gamma_{soil}} = \frac{\psi_{root\ surface} - \psi_{xylem}}{r_{root}} = \frac{\psi_{xylem} - \psi_{leaf}}{\gamma_{xylem} + \gamma_{leaf}} = \frac{\psi_{leaf} - \psi_{air}}{\gamma_{leaf} + \gamma_{air}} \quad \ldots (9.31)$$

in which, ψ is the water potential at various sites of the system and γ is the corresponding resistance.

Water absorption by roots is dependent on the supply of water at the root surface. The two main phenomena concerned with this are the movement of water to the root surface and the growth of roots into the soil mass. As the soil dries out from a saturated state, the rate of water movement in the soil decreases rapidly. The water

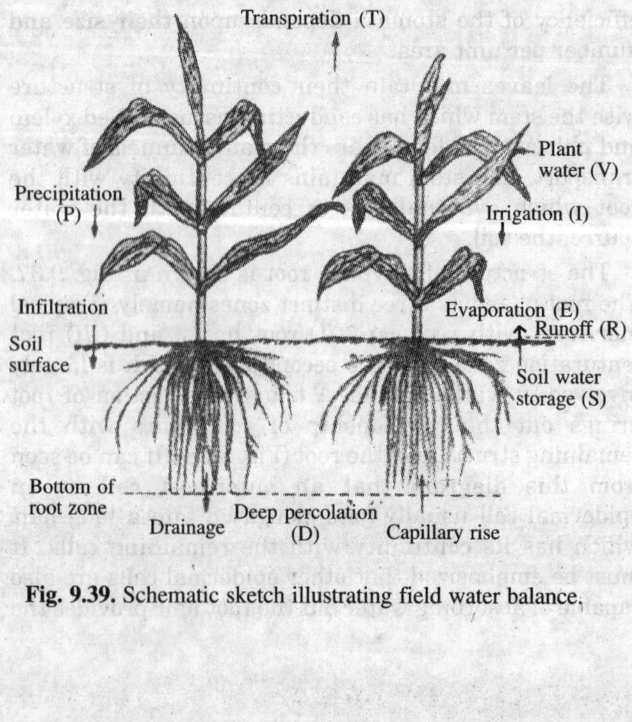

Fig. 9.39. Schematic sketch illustrating field water balance.

movement in the soil drier than field capacity controls the distance in the soil from which roots can extract water. Thus, under conditions where the water extracted by roots is not frequently replaced by rain or irrigation, it is important that the root system must expand continuously or else have already occupied a large enough volume of soil to provide the plant with sufficient water to replace the transpiration losses. Hence, all the factors which affect root growth or the occupation by roots of a large enough soil volume, will also affect the absorption of water by plants.

The actual entry of water into the roots is affected by the extent of the absorbing zone of the roots, the permeability of the root cortex to water movement and the water potential at the root surface. The movement of water through the root and conducting elements of the leaves' xylem to the leaves is initiated and largely controlled by the transpiration from the leaves in response to the water potential gradient extending from the soil water, through the plant to the atmosphere. The water moves from the xylem strands of the leaf across the mesophyll tissue and through the cell walls bordering the sub-stomatal cavities where the liquid vapourises and diffuses out of the leaves through the stomatal openings. Transpiration, though an energy-controlled process, is modified by the soil, plant and atmospheric factors which govern the potential gradients in the various parts of the water path to the leaf surface.

Moisture Stress and Plant Response

As mentioned previously, plant-water relations consist of a group of interrelated and interdependent processes. Thus, the internal water balance or degree of turgidity of a plant depends on the relative rates of water absorption and water loss, and is affected by the complex of atmospheric, soil and plant factors that modify the rates of absorption and transpiration.

Water moves in response to a potential gradient. When the plant roots are in equilibrium with the soil water potential, and the soil water potential gradients are near zero, a base level of leaf turgor or plant water potential is reached. Under the conditions of low evaporative demand during the night and early morning (prior to sunrise) the values of water potential are often at or near this level. An increase in the rate of transpiration, coincident with the increase in evaporation during the day, causes a decrease in the turgor pressure of the upper leaves and the development of water potential gradients through the plant from the evaporating surface of the leaves to the absorbing surface of the roots. Conditions are often such that the rate of water loss exceeds the rate of water absorption, causing an internal water deficit to develop in the plant. It is this internal water deficit, through its influence on many of the physiological processes in the plant, that is directly responsible for the growth and yield of a crop under the prevailing conditions.

The yield of a crop is the integrated result of a number of physiological processes. Water stress can affect photosynthesis and respiration. It can also affect growth and reproduction. Reduction in leaf area, cell size and inter-cellular volume are common under water stress. Dehydration of protoplasm may be responsible for decreasing several physiological processes. Water stress produces important changes in carbohydrate and nitrogen metabolism of plants. Water stress at certain critical stages of plant growth causes more injury than at other stages.

Drought Tolerance of Plants

Plants survive the periods of water stress by various means. Short duration crop varieties that avoid extensive drought period may be better drought resistant than other varieties. Researches on producing ideotype plants with leaf and stomatal characteristics suitable for drought resistance may result in developing suitable drought resistant varieties. Plants exhibit drought tolerance either because of factors affecting their intake or loss of water or because the plants are able to survive tissue desiccation. Drought-hardy plants usually have smaller cells than those living in moist habitats. When desiccated, small cells undergo a much smaller proportionate reduction in volume than do large cells and therefore do not suffer as large disturbances as the latter. In general, increased osmotic values are characteristic of plants having superior drought hardiness. The higher osmotic values not only increase the ability of cells to retain water, but also may have an additional effect by increasing the resistance of the protoplasm to dehydration. One of the most effective safeguards against drought injury is a deep, wide spreading root system such as that of sorghum. Plants with shallow, sparsely branched root system like potatoes, onions etc. suffer sooner than deep rooted species like lucern and maize.

Terminology

Chemical potential of water. The chemical potential of a substance in a system is a measure of the capacity of that substance to work. In a simple solution of a non-electrolyte in water, the chemical potential of water depends on the mean free energy per molecule and the concentration of water molecules.

Osmotic pressure. When a solution is separated from pure solvent by a membrane permeable only to the solvent, there tends to be a net flux of solvent into the solution since the chemical potential of the solvent is higher in the pure phase than in the solution. This process is called osmosis and the pressure difference which must be applied to the solution to prevent a net flux of solvent is called osmotic pressure. In general, dehydration is accompanied by an increase in osmotic pressure. Osmotic pressure, however, is not sufficiently sensitive to be used as an indicator of small changes in water balance.

Diffusion pressure deficit. Diffusion pressure deficit (usually abbreviated as *DPD*) of water in a cell or solution is defined as the amount by which its diffusion pressure is lower than that of pure water at the same temperature and under same atmospheric pressure. The *DPD* of a cell may be regarded as a measure of the pressure with which water would diffuse into the cell when immersed in pure water. The equation for the *DPD* in a cell is:

$$DPD = OP - TP \quad \ldots(9.32)$$

in which *OP* is the osmotic pressure of the cell contents and *TP* is the turgor pressure within the cell. The *DPD* is a more sensitive indicator of the degree of turgidity. It is comparable to the total soil moisture stress (moisture tension + osmotic pressure) of a soil. A fully turged tissue has a zero *DPD*. The *DPD* increases as the water deficit increases until it equals the osmotic pressure, except when a negative wall pressure or tension develops which may even exceed the osmotic pressure. This value is especially useful because it is expressed in atmospheres and can therefore be compared with soil moisture tension and osmotic pressure of the soil solution. The main difficulty in the extensive use of *DPD* is the difficulty in measuring it. Another objection to its usage is that it is no longer used in physical chemistry and/or thermodynamics, and that diffusion is no longer considered as the only mechanism by which water will move under the influence of free energy of chemical potential gradients. The term water potential is often used instead of *DPD*.

Cell water potential. Water potential is a measure of the free energy status of water. As applied to plant cells under isothermal equilibrium conditions, the various factors involved in cell water relations can be summarized by the following equation:

$$\psi_{cell} = \psi_s + \psi_p + \psi_m \quad \ldots(9.33)$$

in which ψ_{cell} is the potential of water in the cell and the other terms express the contribution to ψ_{cell} by solutes (ψ_s), pressure (ψ_p) and metric forces (ψ_m). ψ_s and ψ_m are both negative while ψ_p is positive. ψ_s expresses the effect of solutes in the cell solution, and ψ_m expresses the effect of water-binding colloids and surfaces in the cell. The sum of $\psi_s + \psi_p + \psi_m$ is a negative number, except in fully turgid cells when it becomes zero. In this case the positive pressure potential ψ_p balances the sum of the negative osmotic and matric potentials. It may be shown that *DPD* is numerically equal to ψ_{cell}, but opposite in sign (Kramer, 1969), i.e.,

$$\psi_{cell} = -DPD \quad \ldots(9.34)$$

The potential of water in a cell is less than that of pure water, *i.e.*, it is negative whereas *DPD* is positive because it is defined as a deficit.

Movement of Water and Nutrients in Plants

Water and solutes in cells are in continuous motion, moving within cells, from cell to cell and from tissue to tissue. Several forces are involved in the movement of water and solutes. These may be classified into two groups: (1) passive movement by diffusion and mass flow, and (2) active transport, depending on expenditure of metabolic energy. Movement of materials by mass flow occurs when force is exerted on the moving substance by an external agent so that the molecules tend to move in the same direction in mass, whereas diffusion results from random movement of individual molecules or ions by their own kinetic energy. The movement of water from cell to cell is usually described as occurring by osmosis. However, in recent years it has been claimed that osmosis really involves mass flow of water through the pores of differentially permeable membranes.

The absorption of salts by plants involves movement of ions from soil to root surfaces, ion accumulation in root cells, movement of ions from root surfaces into the xylem and finally their translocation from roots to shoots. The kinds and amounts of ions accumulated by root cells vary widely among different species of plants, with the metabolic activity of roots, the kinds and concentrations of ions in the soil, soil temperature, and aeration. Mobile ions, such as nitrate and potassium, move toward roots in the soil solution but immobile ions, such as phosphates, are available only from the soil in the immediate vicinity of roots. Root extension is therefore an important factor in ion absorption. Ions which reach the xylem sap of the roots are usually carried to the shoots in the transpiration stream. However, much salt is removed from the xylem sap by adjacent living cells, and the salt concentration of xylem decreases from roots to shoots.

Rooting Characteristics and Moisture Use of Crops

The amount of soil moisture that is available to a plant is determined by the moisture characteristics of the soil, the depth to which the plant roots extend and the proliferation or density of the roots. Soil moisture characteristics, such as field capacity and wilting percentage are peculiar to a soil and are a function of the texture and organic matter. Little can be done to alter these limits to any great extent. Greater possibilities lie in changing the characteristics of the plant and the root zone environment, thus enabling the crop to extend its rooting system deeper into the soil, thereby enlarging its reservoir of water. The density of root proliferation is important. Water in an unsaturated soil moves very slowly. To utilize effectively the moisture stored in the soil profile, roots must continue to proliferate into unexploited zones throughout the plant's growth cycle. During favourable growing periods, roots often elongate so rapidly that satisfactory moisture contacts can be maintained even when the soil moisture content declines. Where a good root system has developed during favourable growing periods, a plant can draw its moisture supply from deeper soil layers.

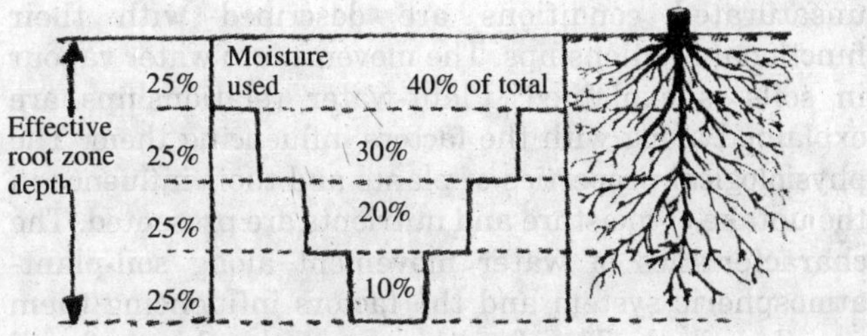

Fig. 9.40. Average moisture-extraction pattern of plants growing in a soil without restrictive layers and with an adequate supply of soil moisture.

Plants vary genetically in their rooting characteristics. Vegetable crops, such as onions and potatos, have a sparse rooting system and are unable to use all the soil water within the root zone. Forage grasses, sorghum, maize and such other crops have very fibrous, dense roots. Lucern has a deep root system. Whether a plant is an annual or perennial is another factor affecting its moisture relations. An annual plant must extend its roots down into the soil to make available all the water it can potentially use. A perennial plant has an already established root depth, and needs only to extend its small roots and root hairs to be able to utilize the entire amount of available soil water.

Plants may be limited in their rooting by factors other than genetics. High water table, shallow soils, and an impermeable formation near the ground surface restrict the depth of rooting. Fertility and salt status of the soil influence the rooting of plants. Crop management practices, such as cutting the top growth at different physiological stages, alter rooting habits. The rooting pattern of common crop plants vary widely from soil to soil. For example, roots of maize crop have been found to extend as deep as 1.5 metres in medium to coarse textured soils while in a fine textured soil the crop has a shallower root system.

Effective root zone. Effective root zone is the depth from which the roots of an average mature plant are capable of reducing soil moisture to the extent that it should be replaced by irrigation. It is not necessarily the maximum root depth for any given plant, especially for plants that have a long tap root. Table 9.6 presents the approximate values of effective root zone depth of some of the common crops grown in deep well drained soils.

Table 9.6 Effective root zone depth of some common crops*
(grown on very deep, well drained soils)

Rooting characteristics			
Shallow rooted	*Moderately deep rooted*	*Deep rooted*	*Very deep rooted*
Depth of root zone			
40 cm	90 cm	120 cm	180 cm
Rice	Wheat	Maize	Sugarcane
Potato	Tobacco	Cotton	Citrus
Cauliflower	Castor	Sorghum	Coffee
Cabbage	Groundnut	Pearl millet	Apple
Lettuce	Muskmelon	Soyabean	Grapevine
Onion	Carrots	Sugar beet	Safflower
	Pea	Tomato	Lucerne
	Bean		
	Chilli		

Moisture extraction pattern within root zone. The moisture extraction pattern shows the relative amount of moisture extracted from different depths within the crop root zone. Figure 9.40 shows the moisture extraction pattern of average crop plants growing in deep uniform soils. It may be seen that about 40 per cent of the total moisture used is extracted from the first quarter of the root zone, 30 per cent from the second, 20 per cent from the third and only 10 per cent from the last quarter. This indicates the importance for making soil moisture measurements at different depths (at least two) within the root zone in order to have a fair estimate of the soil moisture status.

SUMMARY

Basic principles of soil-water-plant relationships are explained along with their importance in sustainable and efficient crop production. Soil properties influencing irrigation management are presented. Soil texture, its classification and procedures for determination are described. Role of montmorillonite type clay minerals influencing swelling and drying characteristics of soils is explained. Soil structure, its measurement, significance in crop productivity and methods of management are described. Volume and mass relationships of soil constituents and procedures for their determination are described. Soil profile characteristics and their importance in irrigation system planning are explained. Procedures for soil resources survey and their classification are presented. Soil resources of India, their characteristics and distribution are described. Soil taxonomical orders, their properties and distribution are presented. Management procedures of Alfisols and Vertisols and their associated soil groups are explained. The processes of soil degradation, the extent of soil degradation in India due to various factors and the methods of managing degraded soils are described. Soil fertility and its management practices are described. The availability and requirements of macro- and micro-nutrients in the soils of India are presented. Water

*Adapted from Gandhi *et al.* (1970). The values are approximate and may be used for making estimates in the absence of better information from local research stations.

relations of soils are explained. Movement of water in soils, including infiltration, percolation, seepage and interflow are described with methods of their measurement. The phenomena of viscosity, surface tension and capillarity are explained. Soil moisture tension and its relationships to crop growth and productivity are presented. Methods of measuring soil moisture tension are described along with equipment and procedures. Soil-water potential and their physical relationships are presented. Soil moisture characteristics and their relationships to plant growth are described. Soil moisture movement in saturated and unsaturated conditions are described with their functional relationships. The movement of water vapour in soils is explained. Plant-water relationships are explained along with the factors influencing them. The physiological properties of plants and their influence in the uptake of moisture and nutrients are presented. The characteristics of water movement along soil-plant-atmospheric system and the factors influencing them are described. The functional relationships of soil moisture stress and plant response are presented. Rooting characteristics of plants and soil moisture use, and drought tolerance of crops are described.

PROBLEMS

9.1 In the determination of the soil moisture content, adopting the gravimetric method, the following observations were made:

Weight of moist soil, including box = 180.04 gm
Weight of oven dry soil, including box = 157.84 gm
Weight of moisture box alone = 42.2 gm
Calculate the soil moisture content. [**Ans.** 10.3%]

9.2 A tube well having a capacity of 4000 litres/hr operates for 20 hours each day during the irrigation season. How much area can it command if the irrigation interval is 20 days and depth of irrigation is 7 cm? [**Ans.** 22.86 ha.]

QUESTIONS

1. Distinguish between:
   (*i*) Pedology and pedosphere.
   (*ii*) Soil classes and soil classification.
   (*iii*) Soil morphology and soil genesis.
   (*iv*) Soil type and soil tilth.
   (*v*) Soil degradation and soil deterioration.
   (*vi*) Soil colloids and soil conditioners.
   (*vii*) Soil building crops and soil depleting crops.
   (*viii*) Soil climate and soil aeration.
   (*ix*) Primary minerals and secondary minerals.
   (*x*) Inorganic and organic materials in the solid phase of a soil system.
   (*xi*) Course clay and colloidal clay.
   (*xii*) Columnar structure and platy structure in soils.
   (*xiii*) Water-holding pores and aeration pores in soils.
   (*xiv*) Porosity and void ratio of soils.
   (*xv*) Subangular structure and granular structure in soils.
   (*xvi*) Mass wetness and volume wetness of soils.
   (*xvii*) Massive structure and compound structure in soils.
   (*xviii*) Dry bulk density and wet bulk density of soils.
   (*xix*) Capillary and non-capillary pores in soils.
   (*xx*) Cohesion and adhesion in soils.
   (*xxi*) Soil horizons and soil profiles.
   (*xxii*) Laterite soils and lateritic soils.
   (*xxiii*) Alfisols and Vertisols.
   (*xxiv*) Coastal alluvium and deltaic alluvium.
2. Define base exchange capacity of soils.
3. What are the most important soil properties influencing irrigation?
4. What are the forces that hold soil moisture in the soil?
5. Define mechanical composition of soils.
6. What are the three most important types of clay minerals?
7. What it the specific property of illite type clay minerals which restricts the tendency to swell?
8. Why are loam soils considered more favourable to crop production?
9. Define specific surface area of a soil system. How is it expressed?
10. What is mechanical analysis of soils? Outline briefly the procedure involved.
11. How does soil structure influence the productivity of soils?
12. Define soil tilth. How does good soil tilth influence the productivity of the soil?
13. What are the practices for managing proper soil structure?
14. Define the degree of saturation of soils.
15. Distinguish the specific characteristics of A-, B-, and C- horizons in a soil profile.
16. What are skeletal soils? What are their special characteristics?
17. What are the major characteristics of Alphisols? What are the important practices required for the management of Alphisols?

18. What are the causes for low level of structural development in Alphisols?

19. What are the consequences of lack of aggregation and/or unstable aggregation in Alphisols?

20. Define soil degradation.

21. What are the physical, chemical and biological processes involved in soil degradation?

22. What are the estimated areas in India under different processes of soil degradation?

23. What are the major causes of chemical and physical degradation of soils?

24. What are the biological factors influencing soil degradation?

25. Explain the application of remote sensing and Geological Information System (GIS) data in mapping degraded lands.

26. What are the major points stressed in the New Delhi 2000 A.D. Declaration on Land Resources Management?

27. Define soil fertility.

28. What are the common causes for the decline of soil fertility?

29. What are the measures to maintain and enhance the fertility of soils?

30. What are the major macro-nutients? Enumerate their availability in the soils of India.

31. What are the procedures to increase the efficiency in the use of N, P and K in soils?

32. What are the common micro-nutrients and what is the status of their availability/requirements in the soils of India?

33. What is the role of cobalt in the soil-plant-atmospheric system?

34. How does the presence of large amounts of boron influence the plant growth?

35. What are the soil conditions conducive to zinc deficiency in soils?

36. What are the soil moisture tension relationships of hygroscopic water, capillary water and gravitational water?

37. What are the principal factors influencing the amount of capillary water in soils?

38. Define basic infiltration rate of soils.

39. What are the major forces which act on water in the soil?

40. How do the forces of adhesion and cohesion operate in the soil system?

41. Define soil water potential.

42. Distinguish between:

(*i*) Infiltration rate and cumulative infiltration.

(*ii*) Capillary potential and hydrostatic potential of soils.

(*iii*) Soil moisture tension and soil moisture suction.

(*iv*) pF of soils and capillary tension.

(*v*) Gravitational potential and pressure potential.

(*vi*) Osmotic forces and adsorption forces.

(*vii*) Metric potential and osmotic potential.

(*viii*) Soil moisture constants and soil moisture equivalent.

(*ix*) Saturation capacity and field capacity.

(*x*) Moisture equivalent and available water.

(*xi*) Permanent wilting point, wilting range and ultimate wilting point.

(*xii*) Percolation, interflow and seepage.

(*xiii*) Permeability and hydraulic conductivity.

(*xiv*) Hydraulic head and hydraulic gradient.

(*xv*) Available water and maximum available moisture deficiency.

43. What is the limiting factor for the application of Darcy's law in flow through soils and aquifers?

44. Define Reynold's Number.

45. What is the generalized expression for soil moisture movement?

46. What are the basic factors influencing soil moisture movement under unsaturated conditions?

47. What are the principles of operation of a tensiometer? What are its limitations?

48. What are the specific applications of pressure plate apparatus in soil moisture measurement?

49. How is the osmotic potential of soils measured?

50. What are the limitations in the use of electrical resistance meter in soil moisture measurements?

51. What is the principle of working of a neutron moisture meter?

52. What are the basic soil, plant and atmospheric factors influencing the water relations of plants?

53. What are the approximate ranges in the magnitudes of water potential in the soil-plant-atmosphere system?

54. Define diffusion pressure deficit.

55. Identify the *True* and *False* statements from the following:

(*i*) The amount of soil air varies inversely with the amount of water present.

(*ii*) The percentage of oxygen and nitrogen in atmospheric air and soil air are distinctly different.

(*iii*) The argon content in the soil air is insignificant when compared to the argon content of atmospheric air.

(*iv*) When the pore space of soils is filled with water the concentration of CO_2 in the soil air decreases, resulting in unfavourable conditions for bio-logical activities.

(*v*) Toxic substances to plants such as methane, hydrogen sulphide and other reduced compounds are produced under waterlogging conditions due to restricted diffusion of fresh air.

(*vi*) The variations in different soils in their capability for successful crop production arise due to their genetic history, and the nature, composition and arrangement of their mineral and organic components.

(*vii*) Because of their plate-like shape, clay particles have much greater surface area as compared to sand or silt of identical volume which are spherical or cubical.

(*viii*) Amongst the three types of clay minerals the montmorillonite is the most resistant to swelling

(*ix*) Montmorillonite and illite types of clay minerals are composed of silica and alumina platelets in 1 : 1 ratio.

(*x*) The intensity of the negative charge in clay soils determines its cation exchange capacity.

(*xi*) Alkaline soils have most of the exchange positions held by the basic ions such as Ca^{++}, K^+ and Na^+.

(*xii*) The specific surface area of a soil system varies directly with the size of the soil particles.

(*xiii*) The extent of drying and swelling in soils depend on the amount and type of clay in it.

(*xiv*) The more montmorillonite the soil contains, the greater is its swelling when wet.

(*xv*) Sand-sized soil structure are more favourable for plant growth than very small and very large ones.

(*xvi*) Soil structure has profound influence on soil properties like erodibility, porosity, hydraulic conductivity, infiltration characteristics and water-holding capacity.

(*xvii*) Prominent cleavage planes are desirable in soil aggregates.

(*xviii*) Surface tension in soils decreases with the decrease in soil moisture content.

(*xix*) The major cause of high level water stress experienced by the crops grown in the semi-arid tropics is the low nutrient, moisture retention capacity and shallow soil depth.

(*xx*) Vertisols and associated soils occur in plateaus and are confined mainly to elevations of 300 to 900 m above mean sea level with high summer and mild winter temperatures.

(*xxi*) Chemical degradation of soils occur mainly due to the loss of nutrients and organic matter and through the accumulation of salts and pollutants.

(*xxii*) The availability of the micro-nutrient molybdenum to plants increases with increasing soil acidity.

(*xxiii*) The finer is the texture of the mineral soil particles, the greater is likely its capillary capacity.

(*xxiv*) Field capacity is the upper limit of available moisture range in soil and plant relationship.

(*xxv*) For successful crop production in soils having appreciable salts, the osmotic pressure of the soil solution must be maintained as low as possible by controlled leaching.

(*xxvi*) In sandy soils the moisture equivalent exceeds field capacity while in clayey soils the moisture equivalent is lower than field capacity.

(*xxvii*) The rate at which the plants respire decreases in response to the decline in soil moisture content.

(*xxviii*) Potato crop has the least tolerance to soil moisture stress with a maximum available moisture deficiency of 0.3.

(*xxix*) The rate of diffusion of water vapour through the soil is proportional to the square of the effective porosity, regardless of the pore space.

(*xxx*) The permeability of most cell walls and membranes in plants to water results in a continuous liquid phase extending throughout the plants in which translocation of solute occurs.

56. Identify the most appropriate answer to the following questions from the four options:

(*i*) The most desirable soil structure is:
(a) granular structure
(b) single grain structure
(c) massive structure
(d) compound structure

(*ii*) Among the predominant soil groups of India the largest area is under:
(a) laterite soils (b) alluvial soils
(c) red soils (d) black soils

(*iii*) In India the maximum land area under Alfisols is in:
(a) Karnataka (b) Tamil Nadu
(c) Andhra Pradesh (d) Maharashtra

(*iv*) Though not an essential micro-nutrient of plants, the availability of selenium in the harvested produce of the crop is highly desirable in:
(a) cereal crops (b) pulses
(c) oilseeds (d) fodder crops

(*v*) Tensiometers installed in soils measure:
(a) osmotic suction of soil moisture
(b) capillary potential of the soil
(c) soil moisture content
(d) soil moisture constants

(*vi*) Electrical resistance blocks made of nylon are most sensitive at soil moisture tensions less than:
(a) 15 atmospheres (b) 10 atmospheres
(c) 5 atmospheres (d) 2 atmospheres

57. In the following, match each of the entries in the 1st column with the most closely related entry in the 2nd column:

(*i*)	(A) Pedosphere	(I) Mode of origin of the soil
	(B) Soil morphology	(II) Moisture, air and temperature in soils
	(C) Soil genesis	(III) Measure of adsorption of a soil for bases

	(D) Soil climate	(IV)	Kind of arrangement of separate natural layers
	(E) Base exchange capacity	(V)	Outer soil layer of the earth in which organic life exists
(*ii*)	(A) Shrinking and swelling of soils	(I)	Zone of maximum biological activity
	(B) Structureless soils	(II)	Molecular attraction
	(C) Cohesion	(III)	Predominant soil group of India
	(D) A – horizon	(IV)	Cracking of roads, building foundations
	(E) Red soils	(V)	Fine grained unattached soil particles
(*iii*)	(A) Vertisols	(I)	Deep red soils of the tropics, weathered and leached
	(B) Entisols	(II)	Soils with a clay enriched horizon, susceptible to erosion, predominant in semi-arid tropics
	(C) Oxisols	(III)	Organic or peaty soils
	(D) Alfisols	(IV)	Shrinking and swelling dark grey soils
	(E) Histosols	(V)	Lack of profile development, common in alluvial soils
(*iv*)	(A) New Delhi Declaration, 2000 AD.	(I)	Non-metal among micro-nutrients
	(B) Nitrogen	(II)	Symbiotic Nitrogen fixation
	(C) Boron	(III)	Needed for fodder crops
	(D) Cobalt	(IV)	Most important plant nutrient
	(E) Solenium	(V)	Soil Conservation Society of India
(*v*)	(A) Viscosity	(I)	Attraction of water into minute pores
	(B) Surface tension	(II)	Negative pressure resulting from capillary and adsorption forces
	(C) Capillarity	(III)	Lateral seepage of water
	(D) Matric potential	(IV)	Force pulling inwards at the surface of a liquid
	(E) Interflow	(V)	Internal friction that brings about resistance to flow

FURTHER READING REFERENCES

Agriculture Refinance and Development Corporation (1979). *Irrigated Agriculture.* ARDC. (now NABARD), Mumbai–18, p. 197.

Aslyng, H.C. (1963) *Soil physics terminology*. Int. Soc. Soil Sci. Bull, 23:1–4.

Bavel, C.H.M., Nixon, P.R. and Hanser, V.L. (1963). *Soil Moisture Measurement with Neutorn Method.* US Dept. Agr. ARS–48–70, p. 39.

Bhumbla, D.R. and Khare, A. (1983). *Estimates of Wastelands in India.* Society for Promotion of Wasteland Development, New Delhi.

Bhumbla, D.R. and Michael, A.M. (1992). *Degradation in Asplisols of the Indian SAT:* A Proposal for ICAR/ICRISAT Jt. Research. ICRISAT, Potencheri, Andhra Pradesh, p. 43.

Black, C.A. (1957). *Soil-plant-relationship.* John Wiley and Sons, New York, p. 332.

Bouyoucos, G.J. (1949). *Nylon electrical resistance units for continuous measurement of electrical resistance in the field.* Soil Science 67, 319–331.

Bouyoucos, G.J. (1954). *New type electrode for plaster of paris moisture blocks.* Soil Science 78, 339–342.

Buckingham, E. (1907). *Studies on the movement of soil moisture.* US Dept. Agr. Bur. Soils Bull. 38.

Chonkar, P.K. (1994). *Mobilization of Soil Phosphorous through Microbes: Indian Experience. In: Phosphorous Researches in India,* pp. 120–195.

Dakshinamurti, C., Michael, A.M. and Mohan, S. (1973), *Water resources of India and their utilization in agriculture.* I.A.R.I. Monograph No. 3, Water Technology Centre, I.A.R.I., New Delhi, p. 392.

Dastane, N.G. and Saraf, C.S. (1966). *Evaluating lands for irrigation commands.* Proceedings of Sixth New East-South Asia Regional Irrigation Practices Seminar, Amman, Jordan, pp. 393–397.

Davis, D.S. (1943). *Empirical equations and monography.* McGraw Hill Book Co., New York, p. 200.

Directorate of Economics and Statistics, Dept. of Agri. and Cooperation, Ministry of Agri., Govt. of India, New Delhi (1988). *Indian Agriculture in Brief,* 22nd ed., p. 406.

Doneen, R.D. (1972). *Irrigation Practice and Water Management.* FAO Irrigation and Drainge Paper 1, Food and Agriculture Organization of United Nations, Rome, p. 84.

Food and Agriculture Organization of United Nations (1984). *Guidelines: Level Evaluation for Rainfed Agriculture.* Soils Bulletin No. 52, FAO, Rome, p. 84.

Food and Agriculture Organization of United Nations (1985). Guidelines: *Land Evaluation for Irrigated Agriculture.* Soils Bulletin No. 55, FAO, Rome.

Gandhi. R.T. *et al.* (1971). *Handbook of Irrigation Water Management.* Dept. of Agri., Ministry of Agriculture, Govt. of India.

Ghildayal, B.P. and Tripathi, R.P. (1987). *Soil Physics.* Wiley Eastern Ltd., New Delhi, p. 656.

Govinda Rajan, S.V. (1968). *Soils of India.* All India Soil and Land use Survey. I.A.R.I., New Delhi, p. 22.

Govinda Rajan, S.V. and Gopala Rao., H.G. (1978). *Studies on Soils of India.* Vikas Publishing House Pvt. Ltd., New Delhi, p. 425.

Hagen, R.M., Haise, H.R. and Edminister, T.W. (1967). *Irrigation of Agricultural Lands.* American Soc. of Agronomy, Madison, Washington, USA, p. 1180.

Hillel, D. (1971). *Soil and Water: Physical principles and processes.* Academic Press, New York, p. 288.

Hinkel, H.J. (1983). CRC *Handbook of Irrigation Technology, Vol. 1.* CRC Press, Inc. Boca Raton, Florida, USA, p. 169.

Holmes, J.W. (1956). *Calibration and field use of the neutron scattering method of measuring soil water content.* Australian Jl. of Appl. Sci. 7, 45–58.

Indian Farmers Fertiliser Cooperative Ltd., (1990). *Soil Fertility and Fertilizer Use,* Vol. IV: *Nutrient- Management and Supply System for Sustaining Agriculture in the 1990s.* PR Dept. Marketing Dn., IFFCO, Nehru Place, New Delhi, p. 357.

International Crops Research Institute for the Semi-Arid Tropics (1987). *Alfisols in the Semi-Arid Tropics.* Proc. of the Consultants Workshop on the State of Art and Management Alternative for Optimizing the Productivity of SAT Alfisols and Related Soils, 1–3 December 1983, ICRISAT, Patancheru, Andhra Pradesh, p. 188.

Jensen, M.V. (1967). *Improving Irrigation Efficiency, in Irrigation of Agricultural Lands.* American Society of Agronomy, Madison, Washington, USA, pp. 112–113.

Kanwar, J.S. ed (1988). *Water Management the Key to Developing Agriculture.* Proc. National Seminar on Water Management the Key to Developing Agriculture, Indian National Science Academy. Agricol Publishing Academy, New Delhi, p. 789.

Khonke, H. (1968). *Soil Physics.* McGraw Hill Book Co., New York, p. 224.

Klingebiel, A.A. and Montgomery, P.H. (1961). *Land Capability Classifications.* Agriculture Handbook No. 210. Soil Conservation Service, US Department of Agriculture, Washington, D.C.

Kramer, P.J. (1969). *Plant and soil water relationships: A modern synthesis.* McGraw Hill Book Co., New York, p. 482.

Lal, R. and Stewart, B.A. (1990). *Advances in Soil Science.* Vol. II, Soil Degradation. Springer-Verlag, New York, p. 345.

Means, R.E. and Parchar, J.V. (1963). *Physical Properties of Soils.* Charles E. Merril Books Inc., Columbus, Ohio, USA, p. 469.

Michael, A.M. and Ojha, T.P. (1999). *Principles of Agricultural Engineering.* Vol. II. 3rd ed. Jain Brothers, East Park Road, Karol Bagh, New Delhi–110005, p. 888.

Murhty, R.S., Hirekerur, L.R., Deshpande, S.B., Venkata Rao, B.V. and Shankaranarayana, H.S. eds. (1982). *Benchmark Soils of India, Morphology, Characteristics and Classification for Resources Management.* National Bureau of Soil Survey and Land Use Planning, ICAR, Nagpur, p. 374.

Murhty, R.S. and Pandey, S. (eds.) (1983). *Soil Map of India* (1:7 million), NBSSLUP, I.C.A.R., Nagpur.

Murhty, R.S. *et al.* (1982). *Distribution, Characteristics and Classification of Vertisols.* Vertisols and Rice Soils of the Tropics, pp. 3–22.

Muskat, M. (1937). *The flow of homogeneous fluids through porous media.* McGraw Hill Book Co., New York.

National Bureau of Soil Survey and Level Use Planning (1999). *Agro-Ecological Subregious of India for Planning and Development.* NBSSULP, Nagpur No. 30.

National Land Use and Conservation Board (1988). *National Land Use Policy:* Outline and Action Points. Dept. of Agriculture and Cooperation, Govt. of India, New Delhi.

Philip. J.R. (1957). *The physical principles of soil water movement during irrigation cycle.* Proc. 3rd International Congress on Irrigation and Drainage 8, 125–154.

Philip. J.R. (1966). *Plant water relations: some physical aspects.* Annual Review of Plant Physiology, Vol. 7, 245–268.

Plaut, Z. and Moreshet, S. (1973). Transport of water in plant-atmosphere system (In Arid Zone Irrigation). Chapman and Hall, Ltd.

Prasad, R.N. and Biswas, P.P. (1999). *Soil Resources of India.* In 'Fifty Years of National Resources Management Research'. Indian Council of Agricultural Research, New Delhi, pp. 13–30.

Rege, M.D., Karale, R.L. and Boli, Y.P. (1979). *Know Your Soils.* Agricultural Refinance Corporation (now NABARD), Mumbai, p. 29.

Richards, L.A. (1947). *Pressure membrane apparatus—construction and use.* Agri. Engg. 28 : 5451–54.

Richards, L.A. (1949). *Methods of measuring soil moisture tension.* Soil Science 68, 95–112.

Richards, L.A. (ed.) (1954). *Diagnosis and improvement of saline and alkali soils.* Agriculture Handbook No. 60, US Dept. of Agr., Washington, D.C.

Rose, C.W. (1966). *Agricultural physics.* Pergamon Press, New York, p. 226.

Russel, M.B.—Coordinator (1959). *Water and its relation to soils and crops.* Academic Press, New York, p. 131.

Saini, K.M. and Das, S.N. (2000). *Rapid Inventorying of Degraded Lands Using Space-Borne Remotely Sensed Data.* Advances in Land Resources Management for 21st Century. Read Paper of the International Conference on Land Resource Management for Food, Employment and Environmental Security, November 9–13, 2000. New Delhi. Soil Conservation Society of India, New Delhi, pp. 41–53.

Scheidegger, A.E. (1960). *The Physics of flow through porous media.* University of Tornoto Press, London.

Schofield R.K. (1935). *The p^F of the water in soil.* Third Intl. Congress on Soil Science, Vol. II, pp. 37–48.

Schwab, G.O., Fangmeier, D.D., Elliot W.J. and Frevert, R.K. (1996). *Soil and Water Conservation Engineering.* John Wiley Sons, Inc., New York, p. 507.

Sehgal, J.L. *et. al.* (1990). *Red and Laterite Soils of India—Resource Approsal and Management.* National Bureau of Soil Survey and Land Use Planning, Nagpur, p. 356.

Sehgal, J.L. Venugopal, K.R. and Kalbande, A.R. (1992). *Micromorphology of Soils of India.* National Bureau of Soil Survey and Land Use Planning, Nagpur, NBSS 34, Res. Bull., p. 76.

Sehgal, J.L. and Mandal, D.K. (1973). *Soil Moisture Regimes of India.* National Bureau of Soil Survey and Land Use Planning, Nagpur. NBSS 43, Tech. Bul.

Sehgal J.L., and Mandal, D.K. (1994). *Soil Temperature Regimes of India.* National Bureau of Soil Survey and Land Use Planning, Nagpur NBSS. 41. Tech. Bul.

Singh, G.B. (1999). *Natural Resources Management for Sustainable Agriculture. In Fifty Years of Natural Resources Management Research*, Indian Council of Agricultural Research, New Delhi.

Singh, N.P. and Sinha, S.K. (1976). *Water use efficiency in crop production.* I.A.R.I. Mono. No. 4, Water requirement and irrigation management of crops in India. Water Tech. Centre, I.A.R.I., New Delhi.

Singh, R.P. and Das S.K. (1983). *Limitations and Prospects of Crop Production in SAT Alfisols.* Proc. Consultants Workshop on the State of Art and Management Alternatives for

Optimizing the Productivity of SAT Alfisols and Related Soils, 1–3 December, 1983, ICRISAT, Patancheru, Andhra Pradesh, pp. 123–125.

Sinha, A.K. (1976). *Soil-water relationships.* I.A.R.I. Mono. No. 4, Water requirement and irrigation management of crops in India. Water Tech. Centre, I.A.R.I., New Delhi.

Sinha, S.K. and Singh, N.P. (1976). *Plant-water relations* I.A.R.I. Mono, No. 4, Water requirement and irrigation management of crops in India. Water Technology Centre, I.A.R.I., New Delhi.

Sivakumar, M.V.K., Piara Singh and Williams, J.H. (1983). *Agro-climatic Aspects in Planning for Improved Productivity of Alfisols*. Proc. Consultants Workshop on the State of the Art and Management Alternatives for Optimizing the Productivity of SAT Alfisols and Related Soils, ICRISAT, Patancheru, Andhra Pradesh.

Slatyer, R.O. (1967). *Plant-water relationships.* Academic Press. New York, p. 366.

Soil Conservation Society of India, New Delhi (2000). *Advances in Land Resources Management for 21st Century*. Lead Papers of the International Conference on Land Resources Management for Food, Employment and Environmental Security, November 9–13, 2000, New Delhi, p. 655.

Sree Ramulu, US (1998). *Management of Water Resources in Agriculture*. New Age International (P) Ltd., New Delhi, p. 159.

Takkar, P.N. and Biswas, A.K. (1999). *Fertility Management of India Soils. Fifty Years of Natural Resources Management Research*, I.C.A.R. New Delhi, pp. 115–144.

Tisdale, S.L., Nelson, W.L. and Beaton, J.D. (1985). *Soil Fertility and Fertilizers*. Pub Macmillan Publishing Company, New York, p. 754.

Thomas Varghese and Byju, G. (1993). *Laterite Soils*. State Committee of on Science, Technology and Environment, Thiravananthapuram. Tech. Monograph No. 1, p. 116.

United States Department of Agriculture (1960). National Engineering Handbook, Section 15. Ch. 1, *Soil-plant-water relationships*, USDA, SCS, Washington D.C., p. 72.

United States Department of Agriculture (1975). *Soil Taxomony—A basic system of soil classification for making and interpreting soil surveys*. Agriculture Handbook No. 436. USDA, Washington, D.C., p. 754.

Velayutham, M. (1999). *Mapping Soil Resources for Optimal Land Use Planning. In Fifty Years of Natural Resource Management Research*, I.C.R.A. New Delhi, pp. 84–114.

Virmani, S.M., Sharma, D. and Venkateswarulu, J. (1986). *Management of Vertisols and Associated Soils,* Joint Paper by ICRISAT and CRIDA, presented in the INSA National Seminar on Water Management, New Delhi, April 28–30. 1986 (Mimeo), ICRISAT, Patancheru, Andhra Pradesh.

Water and Land Management Institute, Aurangabad. (1989). *Application of Soil Survey in Irrigation Water* Management, 2nd ed. WALMI, Aurangabad, p. 187.

Chapter

10

Water Requirement of Crops and Irrigation Management

LEARNING OBJECTIVES

Concepts of water requirement for crops, evaporation, evapotranspiration and consumptive use and the factors influencing them. Procedures for estimating evaporation from water bodies and land surfaces. Meteorological factors influencing evapotranspiration. Estimating crop evapotranspiration, adopting crop coefficient approach. Rational quasi-rational and empirical approaches in estimating evapotranspiration. Penman formula, Modified Penman formula and FAO Penman-Montieth formula and their applications. Estimating missing climatic data for application in FAO Penman-Montieth equation. Basic concepts on field water balance and irrigation requirement. Effective rainfall and factors influencing it. Principles governing irrigation scheduling. Irrigation efficiency concepts and their application. Harvest index and yield response of crops. Crops response to soil moisture availability at different stages of its growth. Irrigation requirement of common crops. Planning cropping systems and irrigation management in irrigation project areas. Crop management under rain-fed agriculture.

The basic resources for crop production are climate, water and soil. An efficient utilization of these is essential for optimum crop production. Climate determines the suitability of a crop as a habitat for different flora and fauna as well as the availability of water for crop production and other uses. Soil serves as the reservoir for water and nutrients. It also provides anchorage for plants. All plants require water in order to live, grow and provide satisfactory yields to meet the human needs. The water required by crops may be provided by precipitation, soil moisture storage in the crop root zone or obtained as seepage from higher reaches, capillary contribution of moisture from shallow water table or by irrigation. The relationship between crops, climate, soil and water involves many biological, physical and chemical processes.

The maximum potential yield of a crop is determined primarily by the climate and the genetic potential of the crop. The extent to which the upper limit of crop's productivity is reached is profoundly influenced by the availability of water in quantity and time, including periods of water shortage during the critical periods in its growing phase. The irrigation system is to be designed and managed to meet the crop's water requirement in quantity and time to attain optimum yields. A critical appraisal of the water requirements of crops, including different crop combinations is necessary for scheduling of irrigation and in planning farm irrigation systems.

Water requirement of crops defines the quantity of water needed to meet the water losses through evapotranspiration of a disease-free crop under non-restricting soil conditions, including soil, water and fertility and achieving the full within potential under a given soil environment in a given time. It is usually expressed in water-depth units or depth-area units per unit area. Water requirements include evapotranspiration and unavoidable losses of water like deep percolation. It also includes the requirement of pre-sowing irrigation and the leaching of salts, whenever required. Irrigation water requirement is the quantity of water, exclusive of precipitation and contribution of soil moisture stored in the crop root zone and the upward flow of water to the root zone from a saturated zone below, that is required for normal crop production.

The basic requirement of water is a function of the soil moisture deficits, as influenced by evapotranspiration and soil and plant characteristics. In many crops there are criticial periods in their physiological growth stages when inadequate soil moisture would be detrimental to their growth. Crop productivity is also affected by other factors like crop variety, nutrient availability, presence of salts in the crop root zone, pests and diseases, and cultural practices. Cultural practices include time and method of sowing/planting, soil tilth, weed control and crop growth period and duration.

Evaporation, Transpiration, Evapotranspiration or Consumptive Use

Evaporation, transpiration, evapotranspiration and consumptive use are important factors in estimating irrigation requirement and planning irrigation systems.

Terminology

Evaporation. Evaporation is the process by which liquid water is converted into water vapour (vaporization) and removed from the evaporating surface (vapour removal). The process of evaporation of water in nature is one of the fundamental components of the hydrologic cycle by which water changes to vapour through the absorption of heat energy. This is the only form of moisture transfer from land and oceans into the atmosphere.

Transpiration. Transpiration is the process by which water vapour leaves the living plant body and enters the atmosphere.

Evapotranspiration (ET). Evapotranspiration denotes the quantity of water transpired by plants during their growth, or retained in the plant tissue, plus the moisture evaporated from the surface of the soil and the vegetation.

Potential evapotranspiration (PET). The concept of potential evapotranspiration PET was suggested by Thornthwaite (1948) who defined it as the evapotranspiration from a large vegetation covered land surface with adequate moisture at all times. He felt that since the moisture supply was not restricted the PET depended solely on available energy. Penman (1947) defined PET as the ET from an actively growing short green vegetation completely shading the ground and never short of moisture availability. Though Penman's definition specifies the important characteristic of reference vegetation, it does not specify the name of the vegetation. Jensen (1968) assumed PET as the upper limit of ET that would occur with a well watered agricultural crop having an aerodynamically rough surface such as lucerne with 30 to 50 cm of top growth.

Apparently, the concept of PET can be visualised as the integral effect of all the climatic factors governing the evapotranspiration process. It may be defined as the ET that occurs when the ground is completely covered by actively growing vegetation and where there is no limitation in the soil moisture. It may be considered to be the upper limit of evapotranspiration for a crop in a given climate.

Consumptive use. Consumptive use (CU), often called evapotranspiration (ET), is the amount of water used by the vegetative growth of a given area in transpiration and building up of plant tissues and evaporation from soil and plant surfaces. Thus, CU exceeds ET by the amount of water used in photosynthesis, transport of minerals and photosynthates, digestion of plant food, plant growth and structural support to plants. Since the difference between ET and CU is usually less than 1%, ET and CU are usually assumed to be equal.

EVAPORATION

Evaporation from natural surfaces, such as open water, bare soil or vegetative cover is a diffusive process, by which water in the form of vapour is transferred from the underlying surface to the atmosphere. The essential requirements in the evaporation process are: (1) the source of heat to vaporise the liquid water, and (2) the presence of a gradient of concentration of water vapour between the evaporating surface and the surrounding air. The source of energy for evaporation may be solar energy, the air blowing over the surface or the underlying surface itself. The energy required for evaporation, regardless of the surface where evaporation is taking place is 590 calories per gram of water evaporated at 20°C. Evaporation can occur only when the vapour concentration at the evaporating surface exceeds that in the overlying air.

Evaporation is influenced by the soil surface, the degree of shading of the crop canopy and the amount of water available at the evaporating surface. Frequent rains, irrigation and upward movement of water in a soil from a shallow water table wet the soil surface. Where the soil is able to supply water fast enough to satisfy the evaporation demand, the evaporation from the soil is determined solely by the meteorological conditions. However, where the interval between rains and irrigation becomes large, the water content in the top soil drops and the soil surface dries out. Under these circumstances the limited availability of water exerts a controlling influence on soil evaporation. In the absence of any supply of water to the soil surface, evaporation decreases rapidly.

The fundamental principle of evaporation from a free surface was enunciated by Dalton in the year 1882. Dalton stated that evaporation is a function of the difference in the vapour pressure of the water and the vapour pressure of the air:

$$E = C(e_s - e_d) \quad \text{...(10.1)}$$

in which,

E = rate of evaporation

e_s = saturated vapour pressure at the temperatures of the water surface (mm Hg.)

e_d = actual vapour pressure of the air (e_s times the relative humidity) (mm Hg).

C = a constant, which is a function of the wind velocity and atmospheric pressure.

Many researchers have developed empirical relationships to estimate the value of C in Equation 10.1 in order to determine the rate of evaporation from open water bodies like tanks, ponds, and reservoirs.

Rower (1931), based on his studies proposed the following empirical relationship to determine the value of C in Equation 10.1:

$$C = (0.44 + 0.073\ W)\ (1.465 - 0.00073\ p) \quad ...(10.2)$$

in which,

W = average wind velocity in km/hr at a height of 15 cm.

p = atmospheric pressure in mm Hg at 0°C.

The value of E, estimated on the basis of the above relationship, is in mm/day. To estimate the evaporation from reservoirs, the value of E as calculated on the basis of Equations 10.1 and 10.2 should be multiplied by 0.77.

Meyer (1942) proposed the following empirical relationship to estimate the value of C from evaporation pans and shallow ponds in which E is computed in mm/month:

$$C = 15 + 0.93\ W \quad ...(10.3)$$

and for small reservoirs and lakes,

$$C = 11 + 0.68\ W$$

in which,

W = average wind velocity in km/hr for the period at a height of 7.6 m. The vapour pressure e_d should also be measured at 7.6 m height. The air temperature is the average of the daily minimum and maximum values.

Example 10.1. *Estimate the evaporation in a month from a shallow pond having an average surface water temperature of 15.6°C and average wind velocity of 5.3 km/hr. The average temperature and relative humidity at 7.6 m height over the area are 21.1°C and 42%, respectively.*

(***Note***: The saturated vapour pressure at 15.6°C and 21.1°C are 13.0 mm Hg and 18.5 mm Hg, respectively, based on the values given in Meteorological Tables.)

Solution: Substituting the data in Equations 10.1 and 10.3 (Meyer equation),

$$E = (15 + 0.93 \times 5.3)\ (13.0 - 18.5 \times 42/100)$$
$$= (15 + 4.929)\ (13.0 - 7.77)$$
$$= 104.23 \text{ mm/month.}$$

Energy balance approach in evaporation studies. The requirement of the source of heat to vapourise the liquid in the evaporation process provides the basis of the energy balance approach to the study of evaporation. Solar energy arrives at the upper limit of the earth's atmosphere at the rate of about 2 calories per square cm per minute. A considerable fraction of this is reflected or scattered by the atmosphere back into space, and does not affect the energy balance at the earth's surface. The thermal balance can be expressed as follows:

$$R_s - R_r - R_{lw} = H_a + H_s + H_e \quad ..(10.4)$$

in which,

R_s = incoming solar radiation

R_r = reflected solar radiation

R_{lw} = net outgoing longwave radiation

H_a = sensible heat flow into the air

H_s = heat flow into the soil, and

H_e = evaporation heat or latent heat flow into the air.

Evaporation Measurement

Evaporation measurements may be made by the Piche evaporometer or, more commonly, by the pan evaporometer. The Piche evaporometer consists of a graduated tube, 30 cm long, with one open end covered by a dry paper held in place by a metallic device. The tube is filled with water and turned upside down. The water slowly wets the paper and evaporates. Water loss is read on the tube, as a measurement of evaporation. The Piche unit is installed in a conventional shelter with other weather instruments. Because of the small size of the Piche unit the values obtained from the instrument are higher than the water use by crops. Multiplying Piche readings by 0.7 gives a reasonable estimate of comparable pan values.

Standard USWB Class A Pan Evaporimeter

The Standard US Weather Bureau Class A pan is the most widely used evaporation pan (Fig. 10.1). It is circular and is made of 22 gauge galvanized iron (or other materials), 10.7 cm in diameter and 25 cm deep. The pan is painted white. It is mounted on a white painted wooden open frame which is 15 cm above the ground level. The open frame permits free movement of air beneath the pan. The pan is kept level. It is filled with clean water up to 5 cm below the rim of the pan. The water level is not allowed to drop more than 7.5 cm below the rim. The water in the pan is changed regularly, at least once in a week to eliminate turbidity. Water is added each day to bring the water level to a fixed point. Often the pan is provided with a standard wire mesh (Fig. 10.1) to prevent loss of water from the pan by birds and animals wherever such losses cannot be prevented by other means. However, it is not an ideal practice as it has some shade effect. The pan is installed in an area which is protected by providing woven wire fence about 20 m × 20 m around it. The water temperature is measured by a thermometer which is immersed just below the water level in the pan.

The level of water in the pan is measured daily at a fixed time in the morning, using a hook gauge fixed over a cylindrical metallic stilling well, about 10 cm in diameter and 20 cm deep provided with a small hole at its bottom (Fig. 10.1). The site where the pan is located should preferably be under grass and open on all sides, except for the fencing, to permit free circulation of air. It is better to locate the pan in the middle of a weather station located on the leeward side of a large crop field.

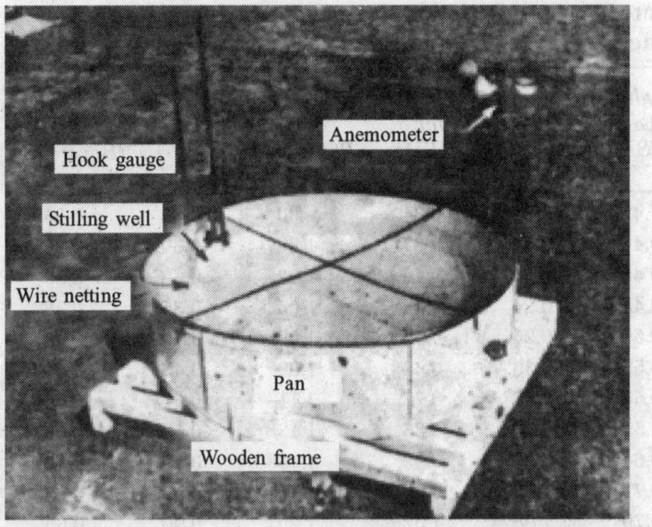

Fig. 10.1. Open pan evaporimeter (US Weather Bureau type)

The pan site has a marked influence on evaporation from the pan. Hagan *et al.* (1967) quoted the observation of Pruitt and Angus that evaporation from a Class A pan was 130 cm during June to December when located in a large crop field, 150 cm when the crop was surrounded by 14.6 m fetch of grass and 175 cm in an open dry land.

Evaporation pans have higher rates of evaporation than a large free water surface, and a factor of about 0.7 is usually recommended for converting the observed evaporation rate to those of large water surface areas. This factor is called *pan coefficient.*

Evaporation pans are usually used to predict ET for periods of 10 days and longer. Pan evaporation is related to reference evapotranspiration by an empirically derived coefficient:

$$ET_o = K_p\,E_{pan} \quad ..(10.5)$$

in which,

ET_o = reference evapotranspiration (mm/day)

K_p = pan coefficient (fraction)

E_{pan} = pan evaporation (mm/day)

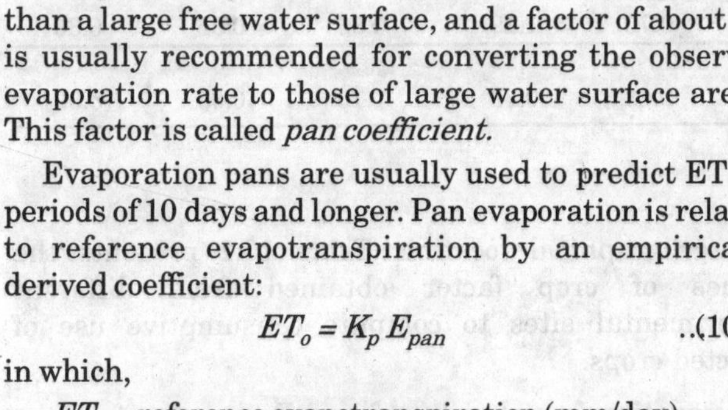

Fig. 10.2. Cross-sectional view of a sunken screen pan evaporimeter. (Sharma and Dastane, 1968)

Sunken screen pan evaporimeter. The sunken screen evaporimeter (Fig. 10.2) developed by Sharma and Dastane (1968), provides a simple device to make an estimate of consumptive use. As compared to the USWB pan in which the ratio between evaporation and evapotranspiration (E_o/ET) was 1.2 to 1.5, the same ratio in a sunken screen pan evaporimeter was observed to be 0.95 to 1.05. The device, however, has not been extensively tested under varying climatic conditions and crops.

Pan coefficient. The pan coefficient of a Class A pan depends on the location of the pan, the surrounding ground cover, the distance of the place of measurement from the pan and the climate, particularly the relative humidity. The UN FAO Irrigation and Drainage Papers 24 (1977) and 56 (1998) provide information on estimated values of pan coefficient of Class A pans placed under two distinct surroundings, viz., pans placed in a short green cropped area and those placed in a dry fallow area. The values are for distinct ranges in the mean relative humidity (RH mean), wind speed and windward side distance of green cropped/dry fallow area. The values coefficient for pans located in green cropped area are given in Table 10.1.

When the pan is located in a dry fallow area, the corresponding values of the pan coefficient are reduced by values ranging from 0.05 to 0.10. For example, under moderate wind conditions (wind speeds ranging from 2

Table 10.1. Estimates of the values of pan coefficient of USWB Class A pan at varying locations from the pan and at different values of relative humidity

	Place of measurement	*Relative humiity (RH mean)*		
		40%	40 to 70%	70%
(*i*)	*Low wind speed (< 2m/sec):*			
	10 m from the pan	0.65	0.75	0.85
	100 m from the pan	0.70	0.80	0.85
	1000 m from the pan	0.75	0.85	0.85
(*ii*)	*Moderate wind speed (2 to 5 m/sec):*			
	10 m from the pan	0.60	0.70	0.75
	100 m from the pan	0.65	0.75	0.80
	1000 m from the pan	0.70	0.80	0.80
(*iii*)	*High wind speed (5 to 8 m/sec):*			
	10 m from the pan	0.55	0.60	0.65
	100 m from the pan	0.60	0.65	0.70
	1000 m from the pan	0.65	0.70	0.75
(*iv*)	*Very high wind (> 8 m/sec):*			
	10 m from the pan	0.45	0.55	0.60
	100 m from the pan	0.50	0.60	0.65

Source: Based on UNFAO Irri. and Drainage Paper 24 (1977) and 56 (1998).

Table 10.2. Some suggested values of crop factor to compute consumptive use from pan evaporation data at selected places (measured or computed).

Stage of crop growth in % of growth season	*Wheat (Ludhiana)*	*Wheat (Pune)*	*Maize (Ludhiana)*	*Maize (Albama, USA)*	*Sorghum (Albama, USA)*	*Cotton Pune)*	*Sugarcane (Hawaii USA)*	*Rice (Los Banos, Philippines)*	*Groundnut (Israel)*	*Small vegetables, (Hargreaves data)*
0	0.14	0.30	0.40	0.40	0.42	0.22	0.34	1.00	0.30	0.25
5	0.17	0.40	0.42	0.40	0.44	0.22	0.37	1.02	0.30	0.28
10	0.23	0.51	0.47	0.43	0.46	0.23	0.40	1.03	0.32	0.30
15	0.33	0.62	0.54	0.46	0.48	0.24	0.44	1.05	0.35	0.38
20	0.45	0.73	0.63	0.52	0.50	0.26	0.50	1.07	0.40	0.45
25	0.60	0.84	0.75	0.59	0.52	0.35	0.60	1.09	0.49	0.50
30	0.72	0.92	0.85	0.67	0.56	0.58	0.72	1.11	0.60	0.55
35	0.81	0.96	0.96	0.76	0.59	0.80	0.86	1.13	0.69	0.58
40	0.88	1.10	1.04	0.85	0.64	0.95	0.93	1.16	0.78	0.60
45	0.90	1.10	1.07	0.93	0.71	1.03	0.98	1.18	0.85	0.63
50	0.91	1.00	1.09	1.02	0.79	1.08	1.02	1.20	0.90	0.65
55	0.90	0.91	1.10	1.09	0.92	1.08	1.05	1.21	0.94	0.65
60	0.89	0.80	1.11	1.14	1.01	1.07	1.07	1.22	0.96	0.65
65	0.86	0.65	1.10	1.19	1.07	1.05	1.10	1.22	0.95	0.63
70	0.83	0.51	1.07	1.21	1.09	1.00	1.13	1.21	0.94	0.60
75	0.80	0.40	1.04	1.23	1.09	1.93	1.16	1.19	0.91	0.58
80	0.76	0.30	1.00	1.22	1.05	0.85	1.19	1.16	0.86	0.55
85	0.71	0.20	0.97	1.16	0.99	0.73	1.20	1.10	0.79	0.50
90	0.65	0.12	0.89	0.06	0.91	0.62	1.20	0.03	0.72	0.45
95	0.58	0.10	0.81	0.95	0.82	0.50	1.19	0.96	0.64	0.38
100	0.51	0.10	0.70	0.75	0.40	0.70	1.19	0.86	0.55	0.30
Seasonal average value										
k	0.61	0.61	0.86	0.86	0.75	0.68	0.89	1.10	0.68	0.50

Source: B based on Water Management Division, Govt. of India, 1971.

to 5 m/sec), the value of pan coefficient at a distance 10 m from the pan on a bare dry soil is 0.55, as against 0.60 in a green cropped field.

Estimating Evapotranspiration from Pan Evaporation Data

It has been observed that a close relationship exists between the rate of consumptive use by crops and the rate of evaporation from a properly located evaporation pan. The standard US Weather Bureau Class A open pan evaporometer described earlier, or the sunken screen open pan evaporimeter (Sharma and Dastane, 1968) or other types of evaporometer may be used for the measurement.

Evaporation pans have proved their utility and have been used to estimate reference evapotranspiration by observing the evaporation loss from the pan water reference and applying empirical coefficients to relate pan evaporation to ET. The relationship between evapotranspiration and pan evaporation is given by the *crop factor*:

Evapotranspiration = pan evaporation × crop factor.

The value of crop factor for any crop depends on foliage characteristics, stages of growth, environmental and geographical location. Table 10.2 presents the values of crop factor obtained from different experimental sites to compute consumptive use of selected crops.

Evaporation from land surfaces. Evaporation from the land surface is affected mainly by the degree of saturation of the soil surface, temperature of air and soil, humidity and wind velocity. Several of these factors are greatly influenced by the vegetative cover. The density of vegetation is an important factor influencing evaporation from land surface. Because of differences in soil moisture and characteristics of soils, it is difficult to generalize on the amounts of evaporation from land surfaces. At high moisture contents, evaporation from land surfaces is nearly equal to that from free water surface. At lower moisture levels, evaporation may decrease in proportion to the water content in the soil. Evaporation from land surfaces is usually confined to shallow depths and its rate reduces rapidly with the increase in depth below the surface. Mulches are effective for a few days after a rain. The mulch restricts air movement, maintains a high air vapour pressure near the soil surface, and shields the soil from solar energy, all of which reduce evaporation.

TRANSPIRATION

Transpiration is the process by which water vapour leaves the living plant body, principally the leaves, and enters the atmosphere. During the growing period of a crop there is a continuous movement of water from the soil into the roots, through the stem and out through the leaves to the atmosphere. The process includes *cuticular transpiration,* or direct evaporation into the atmosphere from moist membranes through the cuticle, and *stomatal transpiration*, or outward diffusion into the atmosphere through the stomate and lenticels, of water vapour previously evaporated from imbibed membranes into intercellular spaces within the plant.

Transpiration is basically an evaporation process. However, unlike evaporation from a water or soil surface, transpiration is modified by the plant structure and stomatal behaviour operating in conjunction with the physical principles governing evaporation.

The atmosphere provides the energy required by the plant to withdraw water from the soil. Under conditions of unlimited soil moisture and when the stomata are fully open, the atmospheric conditions control the rate of transpiration. The major atmospheric factors influencing transpiration are the humidity of the air surrounding the plant, the temperature and humidity of air carried by wind to the plant, and the net radiation available to the plant. When the humidity of the air surrounding the leaf is increased, there will be a decrease in the vapour pressure difference between the leaf and the surrounding air. This will reduce the rate of transpiration. The water vapour surrounding the leaves is swept away by wind. If the wind brings in cooler and/or humid air the rate of transpiration will decrease. Radiation influences transpiration substantially. It raises the leaf's temperature above the surrounding air, resulting in an increase in the rate of transpiration. This is because the saturation vapour within the leaf increases as the leaf's temperature rises. Leaf temperatures in the sunlight are usually 5 to 10°C above the air temperature. The presence of sunlight stimulates the opening or closing of stomata. In most plants the stomata are open during the day and closed at night.

Transpiration is the dominant factor is plant-water relations because evaporation of water produces the energy gradient which causes the movement of water into and through the plants. The rate of transpiration depends on the supply of energy to vaporise water. The water vapour pressure or concentration ɛ adient in the atmosphere constitutes the driving force and the resistance to diffusion in the vapour pathway.

The climate, the soil and the plant factors influence transpiration. The important climatic factors affecting transpiration are light intensity, atmospheric vapour pressure, temperature, and wind. The soil factors are those governing the water supply to the roots. The plant factors include the extent and efficiency of the root systems in moisture absorption, the leaf area, leaf arrangement and structure, and stomatal behaviour.

In comparing the rate of evaporation from a free water surface to the transpiration of a crop, the interval over which evaporation is considered is important. Vegetation virtually ceases to transpire at night due to stomatal closure but a water surface continues to evaporate, though at a slower rate. Thus, over a period of, say, a month, evaporation from a free water surface may exceed the transpiration from a crop. Over a few hours during the day the crop may evaporate more than a water surface.

Transpiration ratio. The effectiveness of the plants in the use of water was often given (in the comparatively earlier literature) in terms of its transpiration ratio. This is the ratio of the weight of water transpired by a crop in its growth to the weight of dry matter in the plant. It varies with the same factors as does transpiration. Approximate values of transpiration ratio for some of the common crops are 500 for wheat, 640 for potatoes, 250 for sorghum, 350 for maize and 1000 for lucerne.

In the earlier works on evaluating transpiration ratio, plants were mostly grown in pots in green houses and the evaporation was minimised by sealing the soil surface. The size of the pots and the criteria for irrigation were not standardised. These factors considerably influenced the results. By the term '*dry matter*' some workers took shoot weight, while others meant only the weight of economic produce of the crop. Now, it is established that the transpiration and dry matter production by a plant are not related in a simple manner as in the transpiration ratio. Several soil, plant and atmospheric factors influence the growth of crops. Thus, the knowledge of transpiration ratio is of little value in planning irrigation practices.

EVAPOTRANSPIRATION (ET) or CONSUMPTIVE USE (CU)

In designating water use by crops, evaporation and transpiration are combined into one term called evapotranspiration (ET), as it is difficult to separate these two losses in cropped fields. The relative amounts of direct evaporation from land and water surfaces and transpiration depend on the amount of ground covered by crops. For most crops covering the whole soil surface only a small amount of water is lost from the ground surface.

Under field conditions the incoming solar radiation supplies the energy for the evapotranspiration process. Usually a close relationship exists between net incoming solar radiation and evapotranspiration. In general, the factors influencing evaporation and transpiration also govern the evapotranspiration process.

Evaporation and transpiration occur simultaneously and it is not easy to distinguish between the two processes. Apart from the water availability in the top soil, the evaporation from a cropped soil is mainly determined by the fraction of the solar radiation reaching the soil surface. It decreases over the growing period as the crop develops and the crop canopy shades more and more of the ground area. During the initial stages of the crop, water is lost mainly by soil evaporation but once the crop is well developed and completely covers the soil, transpiration is the predominant process.

Other factors being equal, the stage of growth of the crop has a considerable influence on its consumptive use rate. This is particularly true for annual crops which generally have three rather distinct stages of growth. These are: (*i*) emergence and development of complete vegetative cover, during which time the consumptive use rate increases rapidly from a low value and approaches its maximum; (*ii*) the period of maximum vegetative cover, during which time the consumptive use rate may be maximum, if abundant soil moisture is available, and (*iii*) crop maturation stage, when for most crops, the consumptive use rate decreases.

The interception of solar energy by row crops is a function of their orientation with respect to the sun. Depending on the surrounding crop areas, particularly those upwind of the crop, air masses passing across the field will vary in humidity and temperature. Thus, if the surrounding fields are bare, more water will be lost than if they were cropped and irrigated. This, therefore, stresses on the need for developing compact field blocks for irrigation.

The rate of evapotranspiration is normally expressed in mm per unit of time. It is the amount of water lost from a cropped surface in units of depth. The unit of time may be an hour, day, week, month, crop season or year. For example, one hectare area is 10,000 sq m and 1 mm is equal to 0.001 m. An evapotranspiration value of 1 mm of water corresponds to a loss of 10 cubic metres of water per ha. This, 1 mm per day is equivalent to 10 cu m/ha per day (10 m^3/ha/day).

Evapotranspiration in terms of depth of water can also be expressed in terms of energy received per unit area. The energy is the heat required to vaporize free water. It is known as latent heat of vaporization (λ) which is a function of the water temperature. For example, at 20°C, λ is about 2.45 megajoule per kg (MJ/kg). In other words, an energy input of 2.45 MJ are needed to vaporize 1 kg or 0.001 cu m of water. Thus, an energy input of 2.45 MJ per m^2 will vaporize 0.001 m or 1 mm of water. The rate of evapotranspiration is represented in units of MJ per square metre per day, the latent heat flux.

Seasonal consumptive use. The total amount of water used in evapotranspiration by a cropped area during the entire growing season is called seasonal consumptive use. It is expressed as the depth of water in cm or volume in hectare-cm per hectare. Seasonal consumptive use values are required to determine the seasonal irrigation water supply.

Peak period consumptive use. The average daily water use rate during the few days of the highest consumptive use of the season is called the peak period consumptive use rate. This is the design rate to be used in planning an irrigation system. The peak-use period for various crops in a given area may occur at different times in the crop season. In irrigation project design, the peak period of consumptive use is the period during which the weighted average daily rate of consumptive use of the various crops grown in the project area is at a maximum. In planning the cropping pattern for an area it may be possible to have such crop combinations which have their peak consumptive use rates spread out in such a way as to use the available supply of water in the most efficient manner. Table 10.3 presents the typical values of the peak rate of soil moisture use by crops under different climatic conditions. For precise estimates, however, the peak-use rates of different crops should be established on the basis of experimental data from the concerned or similar agro-climatic region.

Table 10.3. Maximum rates of soil moisture use by crops under different climatic conditions

Climatic conditions	*Peak rate of soil moisture removal (mm/day)*
Cool, humid	3
Cool, dry	4
Moderate, humid	4
Moderate, dry	5
Hot, humid	5
Hot, dry	8

(*Source*: Adapted from Molenar, 1960)

Basic Concepts

Reference crop evapotranspiration, ET_o. It is the evapotranspiration rate from a reference surface. The FAO Expert Consultation on Revision of FAO Methodologies for Crop Water Requirements (1990) adopted the following definition for the reference surface:

"A hypothetic reference crop with an assumed crop height of 0.12 m, a fixed surface resistance of 70 s per m^{-1} and an albedo of 0.23."

The reference surface closely resembles an extensive area of actively growing green grass of uniform height completely shading the ground and with unlimited availability of soil moisture. The concept of reference evapotranspiration, ET_o provides information on the evaporation demand of the atmosphere, independent of the type of crop, its stage of development and the management practices. As soil moisture is abundantly available at the reference surface, the soil factors do not affect ET_o. Relating ET to a specific surface enables the reference to which ET from other surfaces can be related. ET_o is influenced only by climatic factors. Hence, it can be computed from weather data. ET_o expresses the evaporating power of the atmosphere at a specific location and time of the year.

Crop Evapotranspiration under Standard Conditions (ET_c)

Crop evapotranspiration, ET_c is the evapotranspiration of a crop under standard conditions. The standard conditions assume a disease-free and well fertilized crop grown in large plots under optimum soil moisture conditions and obtaining optimum production under the prevailing climatic conditions. Crop evapotranspiration is the amount of water that is lost through evapotranspiration. Crop water requirement is the amount of water that is to be applied. Irrigation water requirement is the difference between crop water requirement and effective rainfall/precipitation. Irrigation water requirement includes the leaching requirement and other special needs like pre-sowing irrigation.

Crop Evapotranspiration under Non-standard Conditions (ET_c adj)

ET_c under non-standard conditions refer to ET_c of crops grown under management and environmental conditions which are different from the standard conditions mentioned earlier. These include conditions such as pests and diseases, low soil fertility, salt-affected soils, waterlogging and high soil moisture stress due to limited availability of water. ET_o under non-standard conditions is calculated by using water stress coefficients K_s or by adjusting K_c for distinct types of stresses and environmental constraints. ET_c under such situations is called ET_c adj.

Meteorological Factors Influencing Evapotranspiration

The meteorological factors influencing evapotranspiration are those which provide energy for the vaporization of water and removal of water vapour from the evaporating surface. The principal factors are solar radiation, air temperature, humidity and wind speed. The meteorological factors are profoundly influenced by the location of the place. The essential factors to be known to specify the location are the *altitude*, which is the elevation of the place above mean sea level, and the *latitude* which is the angular distance of the place with reference to the equator (degrees north or south or radians, i.e., $\pi/180$).

Solar Radiation

Evapotranspiration process depends on the energy available to vaporize water. The energy source is solar energy. The amount of radiation which reaches the evaporating surface at a particular place depends on its location in terms of the latitude and the time of the year. It is also influenced by the turbidity of the atmosphere and cloudiness. Solar radiation is also called shortwave radiation, because the sun emits energy through electromagnetic waves of shortwave length. Solar radiation received at the top of the earth's surface on a horizontal surface is referred to as extra-terrestrial (solar) radiation denoted by the symbol R_a. When the sun is directly overhead, the angle of incidence of sun's rays is zero and the value of R_a is 0.0820 megajoules per square metre per minute (MJ m^{-2}/min). With a change in the season, the angle of incidence of sun's rays also changes. The amount of solar radiation, denoted by the symbol R_s (and/or R_a) is a function of the latitude and the date and time of the day. On a cloudless day, R_s is about 75% of R_a. Even in a very cloudy day about 25% of R_a may still reach the earth's surface mainly due to diffused sky radiation.

Relative shortwave radiation. It is the ratio of the solar radiation R_s to the clear-sky solar radiation, usually denoted as R_{so}, i.e., R_s/R_{so}. R_s is the radiation that actually reaches the earth's surface at a particular period, while R_{so} refers to the solar radiation that would occur at the same location and time under cloudless conditions. Thus, R_s/R_{so} expresses the cloudiness of the atmosphere; the cloudier the sky is the smaller is the ratio R_s/R_{so}. In general, the ratio R_s/R_{so} varies from 0.33 under dense cloudy conditions to 1.0 on a clear day.

Relative sunshine duration. It is the ratio of the actual sunshine duration, n to the maximum possible duration of sunshine or daylight hours, N. When there are no clouds, $n = N$ and the ratio is one. As with R_a, the value of N which depends on the day length, is a function of the latitude and date.

Albedo. Albedo is a fraction of the solar radiation, usually denoted as $\propto$, which is reflected by the earth's surface. The value of albedo depends on the type of surface, the angle of incidence of the sun's rays and the slope of the ground surface. Green vegetative cover has an albedo usually ranging from 0.2 to 0.5. On a freshly fallen snow it may be as high as 0.95, while on a bare wet soil it may be as low as 0.05.

Net solar radiation, R_{ns}. It is a fraction of the solar radiation R_s that is not reflected from the earth's surface:

$$R_{ns} = (1 - \propto)\, R_s \qquad \text{...(10.6)}$$

Net long wave radiation. The earth which absorbs solar radiation emits radiation energy through long wave radiation, as the earth is at a much lower temperature than the sun. This terrestrial radiation is called long wave radiation which is absorbed by the atmosphere or is reflected back to the earth. The difference between the outgoing and incoming long wave radiation is the *net long wave radiation, R_{nl}*.

Net radiation. The net radiation, denoted as R_n is the difference between the incoming and outgoing radiation, including both short and long wave lengths. The total daily value of R_n is positive during the day and negative during the night.

Soil heat flux. Soil heat flux, usually denoted as *G*, is the energy utilized in heating the soil. The value of *G*, when compared to R_n is very small.

Units of radiation. The standard international (SI) units of radiation is megajoule per square metre per day (MJ m^{-2} day^{-1})

Conversion factors:

equivalent evaporation (mm/day):

$1 \text{ mm day}^{-1} = 2.45 \text{ MJ m}^{-2} \text{ day}^{-1}$

joule per cm^2 per day (J cm^{-2} day^{-1}):

$1 \text{ J cm}^{-2} \text{ day}^{-1} = 0.01 \text{ MJ m}^{-2} \text{ day}^{-1}$

calorie per cm^2 per day (cal cm^{-2} day^{-1}):

$1 \text{ cal} = 4.1868 \text{ J} = 4.1868\ 10^{-6} \text{ MJ}$

$1 \text{ cal cm}^{-2} \text{ day}^{-1} = 4.1868\ 10^{-2} \text{ MJ m}^{-2} \text{ day}^{-1}$

watt per m^2 (Wm^{-2}):

$1 \text{ W} = 1 \text{ J s}^{-1}$

$1 \text{ W m}^{-2} = 0.0864 \text{ MJ m}^{-2} \text{ day}^{-1}$

Pressure

1 mbar	=	0.1 kPa (kiloPascal)

Radiation

1 cal cm^{-2} d^{-1}	=	0.041868 MJ m^{-2} d^{-1}
1 MJ m^{-2} d^{-1}	=	23.884 cal cm^{-2} d^{-1}
	=	0.408 mm d^{-1}
1 mm d^{-1}	=	2.45 MJ m^{-2} d^{-1}
	=	58.6 cal cm^{-2} d^{-2}
1 Wm^{-2}	=	0.0864 MJ m^{-2} d^{-1}
	=	2.064 cal cm^{-2} d^{-1}

Measurement of radiation. Solar radiation is measured by pyranemometers, solarimeters or radiometers. These instruments contain sensors placed on horizontal surfaces which measure the intensity of total radiation. Total solar radiation includes both direct and diffuse radiation from cloudy conditions. Glass domes are placed to protect the sensors and keep them in dry atmosphere. They are wiped and cleaned periodically. In net radiometers the glass domes are replaced by polyethylene domes which have a transmission range for both shortwave and longwave radiation.

At places where pyranemometers and other precision instruments are not available, solar radiation is estimated by the duration of sunshine. The sunshine recorder is usually of the Campbell Stokes type comprising a spherical glass ball of 7.5 cm diameter mounted on a stand with a paper card underneath. The sun's rays are concentrated at a focal point that burns a hole in a specially treated card mounted concentrically with the sphere. The length of the burn indicates the duration of sunlight. The width of the burn gives a qualitative measure of the intensity of radiation. The position of the paper chart is to be changed according to the season of the year. The amount of heat conducted into the soil is measured by a system of soil heat flux plates and thermocouples or thermisters.

Wind Speed

The removal of water vapour depends mainly on wind and air turbulence. During vaporization, the air above the evaporating surface gets saturated with water vapour. If the saturated air is not continuously replaced with drier air, the rate of evaporation decreases. High humidity of air and clouds reduce the rate of evapotranspiration.

Velocity and direction are the basic characteristics of wind. Wind vane indicates the wind direction. Wind velocity is the factor which influences ET. Wind velocity is expressed in metres per second (ms^{-1}) or kilometres per hour or day.

Conversion factors:

Standard (SI) unit: metre per second (m s^{-1})

kilometre per day (km day^{-1})

$1 \text{ km day}^{-1} = 0.01157 \text{ m s}^{-1}$

nautical mile/hour (knot)

$1 \text{ knot} = 0.5144 \text{ m s}^{-1}$

foot per second (ft s^{-1})

$1 \text{ ft/s} = 0.3048 \text{ m s}^{-1}$

Wind speed measurement. Anemometers are used to measure the wind speed. Cup type or propeller type anemometers are commonly used in weather stations. Cups are of the hemispherical shape, usually 3 in number. Cups/propellers are turned by the force of the wind. The rotations are recorded by a counter. By counting the number of rotations over a given time period, the average wind speed during that period is computed.

Wind profile. Wind speeds vary with the height above the ground surface. It is the lowest at the ground surface and increases with height. Hence, anemometers are placed at standard heights, which is 10 m in meteorology and 2 or 3 m in air-meteorology. The standard height in computing ET is 2 m above the ground surface.

Air Temperature

In the computation of ET and other agro-meteorological studies, the air temperature near the level of the crop canopy is important. Solar radiation absorbed by the atmosphere and the heat emitted into the atmosphere by the earth increases the air temperature. ET is greater in sunny warm weather, as compared to cloudy cool weather. The average daily maximum and minimum temperatures in degree Celsius (°C) are required in computing ET. The mean daily air temperature is defined as the mean of the daily maximum and minimum temperatures:

$$T_{\text{mean}} = \frac{T_{\text{max.}} + T_{\text{min.}}}{2} \qquad \text{...(10.7)}$$

Units:

Standard SI Unit: degree Celsius (°C)

degree Centigrade (°C) = (°F – 32) $\frac{5}{9}$

degree Fahrenheit (°F) = 1.8°C + 32

Kelvin (K) = (°C) + 273.16

Measurement of temperature. At the weather stations, air temperature is measured with instruments located inside the conventional wooden box shelters having openings formed by a series of inclined battens (louvered construction), called Stevenson's screen or radiation shelter. The instruments are placed as per the specifications prescribed by the World Meteorological Organization (WMO) at 2 m above the ground surface. The location at 2 m height is to protect the instruments from direct exposure to solar heating. Air temperature is measured with thermometers, thermisters or thermocouples. The maximum and minimum temperatures measured are for a 24-hr period. Thermographs, when provided, plot the instantaneous temperature over a day or a week. (Soil thermometers are used to measure soil temperature at specified depths below the ground surface. Usually they are installed at depths of 5 cm, 15 cm and 30 cm below the ground surface). The time of observation of the instruments should adhere to the national standards. In India, the Indian Meteorological Department has specified the timings for observing the readings of various weather instruments. Agro-meteorological stations are sited in cropped areas where the instruments are exposed to atmospheric conditions similar to those of the surrounding crops.

Air Humidity

Air humidity expresses the amount of water vapour in the air. It is the mass of water vapour per unit volume of space. It also refers to the gaseous pressure exerted by water vapour present in space. Air humidity is expressed in terms of vapour pressure, dew point temperature and relative humidity.

Vapour pressure. The pressure of water vapour in the air contributes to the total atmospheric pressure. The amount of water vapour in the air is directly proportional to the partial pressure exerted by it in the air. Hence, water vapour pressure, is a measure of the water vapour content. In the air column above a water surface, an equilibrium is reached between the water molecules escaping and returning to the water surface. When air is saturated it cannot store any more water molecules at the corresponding temperature. The vapour pressure at this point is called saturation vapour pressure usually denoted e^o (T). The higher the air temperature, the higher is the capacity to store water molecules, and the higher the saturation vapour pressure. When the air is saturated, the actual vapour pressure e_a is lower than saturation vapour pressure. The difference between the saturation vapour pressure and the actual vapour pressure is the *vapour pressure deficit* or *saturation deficit* $(e_s - e_a)$. Vapour pressure deficit indicates the evaporative capacity of the air.

Dew point temperature. It is the temperature to which the air is to be cooled to make it saturated.

Relative humidity. It is the ratio of the mass of moisture actually present in any volume of air of a given temperature to the maximum possible at that temperature and pressure. It is expressed in percentage by the following relationship:

$$RH = \frac{e_a}{e^o(T)} 100 \qquad \text{...(10.8)}$$

in which,

RH = relative humidity, %

e_a = actual vapour pressure (kPa) (kiloPascal)

e^o (T) = saturation vapour pressure (kPa)

The actual vapour pressure is nearly constant throughout the day. However, the relative humidity fluctuates between a maximum at about sunrise to a minimum around early afternoon. This is because the saturation vapour pressure depends on the air temperature which changes during the day.

Units of pressure (air pressure, vapour pressure):

Standard (SI) unit: kiloPascal (kPa)

millibar (mbar)	1 mbar	= 0.1 kPa
bar	1 bar	= 100 kPa
		= 1.013 atm
		= 76 cm Hg
mbar	1 mbar	= 0.76 mmHg
centimetre of water (cm)	1 cm of water = 0.09807 kPa	
millimetre of mercury (mmHg)	1 mmHg = 0.1333 kPa	
atmospheres (atm)	1 atm = 101.325 kPa	
pound per square inch (psi)	1 psi = 6.896 kPa	
	1 inch H_2O = 2.49 mbar	

Measurement of vapour pressure. Vapour pressure is derived from relative humidity or dew point temperature. The direct measurement of relative humidity is with hygrometers. The common type of hygrometer is the hair hygrometer operated as mechanical hygrographs. The accuracy of measurements with hair hygrometers decline with dust and ageing of hairs. Modern hygrometers use a dielectric polymer film that changes its dielectric constant with changes in surface moisture. The most common method of measuring relative humidity is with *dry and wet bulb thermometers.* The bulb of a wet bulb thermometer is covered with a wick which is kept constantly saturated with distilled water. Evaporation of water from the wick lowers the temperature of the wet bulb thermometer.

The drier the air layer is, the larger is the evaporative cooling and larger the temperature drop. The difference between the dry and wet bulb thermometer readings is a measure of the air humidity.

The dew point temperature, T_{dew} is measured with dew point meters. Most dew point meters comprise of a mirror-like metallic surface that is artificially cooled. When dew forms on the surface, the corresponding temperature is the dew point temperature.

All weather instruments require proper installation and maintenance. Instruments for measuring vapour and dew point temperature, which require comparatively greater care in maintenance are installed in the conventional Stevenson's screen. Details regarding agro-meteorological instruments, their installation and maintenance are contained in FAO Irrigation and Drainage Paper 27 (Doorenbos, 1976), Agro-meteorological Field Stations, WHO Guide to Agro-meteorological Instruments and Practices (WHO, 1983) and meteorology handbooks. Data from weather stations, other than meteorological stations, need to be examined as to their applicability in crop-weather studies. For example, meteorological data collected at airports usually give data relating to urban environment and may not be directly applicable to agro-meteorological studies.

Measuring Evapotranspiration

The principal methods for direct measurement of evapotranspiration are: (*i*) lysimeter experiment, (*ii*) field experimental plots, (*iii*) soil moisture depletion studies, and (*iv*) water balance method. The methods are, however, laborious, costly and time consuming. Even though direct measurements of ET are usually not possible for routine measurements, they are important for the evaluation of ET estimates obtained by indirect methods.

Lysimeter experiments. Lysimeter studies involve the growing of crops in buried soil filled large containers (lysimeters) and measuring their water loss and gains. A lysimeter can be defind as a device in which a volume of soil planted with vegetation is located in a container to isolate it hydrologically from the surrounding soil. Mainly there are two types of lysimeters: (*i*) non-weighing type, and (*ii*) weighing type. Non-weighing type lysimeters give less accurate results, compared to weighing type. To get more accurate results as well as short period estimates, weighing type lysimeters have been developed. Many designs of weighing type lysimeters are reported.

The soil and crop conditions in the lysimeters should be close to the natural conditions. From the irrigation point of view, weighing lysimeters are set up to enable the operator to measure the water balances: water added, water retained by the soil, and water lost through all sources—evaporation, transpiration and deep percolation. These measurements involve weighing which may be made with scales or by floating the lysimeters in water or a suitable heavy liquid, in which case the change in liquid displacement is computed against water loss from the tank. The technique yields a measurement of total water loss and is useful as an indicator of field water loss, provided suitable precautions are taken. The tanks must be permanently buried in the ground and surrounded by a large area of crop of the same height if the readings made are to bear relation to losses from the crop in the field. The water table is maintained at a specific depth in the tank by connecting it to a supply tank provided with a float mechanism which has an arrangement for receiving excess water that tends to build up in the tank. Water is applied in measured amounts to the lysimeter, as irrigation is applied to the surrounding cropped area. The overflow and deep percolation, if any, are measured. The water received either from the storage tank or precipitation, excluding the outflow, constitutes the water used by the crop.

Though lysimeters provide the means of precise and direct measurement of the amount of water supplied to and lost by the crop, often they encounter a number of problems. The major limitations are the reproduction of physical conditions such as temperature, water table, soil texture, structure and density within the lysimeter, comparable to those outside in the field. To ensure proper drainage, the bottom of an isolated soil column will often require artificial application of a moisture suction equivalent to that present at the same depth in the natural soil.

Field experimental plots. Measurements of water supplies to the field and changes in soil moisture contents of field plots are sometimes more dependable for computing seasonal water requirement of crops than measurements with lysimeters which do not simulate field conditions. The seasonal water requirements are computed by adding measured quantities of irrigation water, the effective rainfall received during the season, and the contribution of moisture from the soil. Field water balance may be expressed by the following relationship:

$$WR = IR + ER + \sum_{i=1}^{n} \frac{M_{bi} - M_{ei}}{100} . A_i . D_i \qquad ...(10.9)$$

in which,

WR = seasonal water requirement (mm)

IR = total irrigation water applied (mm)

ER = seasonal effective rainfall (mm)

M_{bi} = moisture percentage at the beginning of the season in the ith layer of the soil

M_{ei} = moisture percentage at the end of the season in the ith layer of the soil

A_i = apparent specific gravity of the ith layer of the soil

D_i = depth of the ith layer of the soil within the root zone (mm)

n = number of soil layers in the root zone D.

The method requires that the amount of water applied to a field is measured accurately. This method, though satisfactory for computing seasonal water requirements, does not provide information on intermediate soil moisture conditions, short term use, profile use, deep percolation losses and peak use rate of water by the crop.

Soil moisture depletion studies. The soil moisture depletion method is usually employed to determine the consumptive use of irrigated field crops grown on fairly uniform soils when the depth to the groundwater is such that it will not influence the soil moisture fluctuation within the root zone. These studies involve measurement of soil moisture from various depths at a number of times throughout the growth period. Greater the number of measurements, more is the information obtained from such studies. Consumptive use (Cu/ET) is calculated from the change in soil water content in successive samples from the following relationship:

$$u = \sum_{i=1}^{n} \frac{M_{1i} - M_{2i}}{100} . A_i . D_i \qquad ...(10.10)$$

in which,

u = water use from the root zone for successive sampling periods or within one irrigation cycle (mm)

n = number of soil layers sampled in the root zone depth D

M_{1i} = soil moisture percentage at the time of the first sampling in the ith layer

M_{2i} = soil moisture percentage at the time of the second sampling in the ith layer.

A_i = apparent specific gravity of the ith layer of the soil

D_i = depth of the ith layer of the soil (mm)

Seasonal consumptive use ($Cu = \Sigma u$) is calculated by assuming consumptive use values of each sampling interval. A correction is made by adding PET value(s) for accelerated water loss for the interval(s) just after irrigation(s) and before soil moisture sampling.

Water balance method. The water balance method, also called the inflow-outflow method, is suitable for large areas (watersheds) over long periods. It may be represented by the following hydrological equation:

Precipitation = Evapotranspiration + surface runoff
+ sub-surface drainage
+ change in soil water contents

This method necessitates adequate measurement of all factors, except evapotranspiration. The value of evapotranspiration is computed from the measured data.

Evaporation units:

Standard (SI) unit: millimetre per day (mm day^{-1})

(*i*) m^3 per hectare per day ($m^3\ ha^{-1}\ day^{-1}$)

$1\ m^3\ ha^{-1}\ day^{-1} = 0.1$ mm day^{-1}

(*ii*) litre per second per hectare ($1\ s^{-1}\ ha^{-1}$)

$1\ s^{-1}\ ha^{-1} = 8.640$ mm day^{-1}

(*iii*) equivalent radiation in megajoules per square metre per day (MJ m^{-2} day^{-1})

1 MJ m^{-2} day^{-1} = 0.408 mm day^{-1}

Calculating Crop Evapotranspiration: Crop Coefficient Approach

All the methods of calculating crop evapotranspiration involve the following relationship:

$$ET_c = K_c\, ET_o \qquad ...(10.11)$$

in which,

ET_c = Evapotranspiration of a specific crop (mm/day)

K_c = Crop coefficient (dimensionless)

ET_o = Potential evapotranspiration (or reference crop evapotranspiration) (mm/day)

The influence of various weather factors are incorporated in ET_c estimates. Thus, ET_c is an index of climatic demand. Crop coefficient K_c is influenced mainly by the specific characteristics of the crop, including its stage of growth. To a limited extent K_c is also influenced by the climate. The value of K_c for a crop is determined experimentally and reflects the physiology of the crop, the degree of crop cover, the location from where data are obtained and the method used to compute ET_c. K_c represents the integrated value of four major factors, viz., (*i*) *crop height*, which influences aerodynamic resistance, (*ii*) *albedo* (reflectance) of crop-soil surface, (*iii*) *canopy resistance*, and (*iv*) *evaporation from the soil*, especially exposed soil. Albedo is affected by the extent of coverage of the ground surface by the crop as well as the extent of wetness of the soil surface. Canopy resistance of the crop to water vapour transfer is influenced by the leaf area, which indicate the number of stomata, age, condition of the leaf and stomatal properties. Canopy resistance influences the surface resistance, r_s. The rate of water vapour transfer from the soil is high in a wet soil, especially when the crop canopy provides incomplete cover on the ground surface. Closely spaced plants, tall crops and the surface of full grown crop field may raise the values of K_c larger than 1, sometimes up to 1.2. ET of full grown crops covering the entire ground surface comprises essentially the transpiration component, as evaporation from the ground surface is negligible. After rainfall or irrigation the influence of evaporation is high, especially, when the crop is at seedling stage without much ground cover. Changes in wind velocity alter the aerodynamic resistance of crops and hence the K_c values for crops taller than the reference crop of grass. Influence of climate on K_c is high in arid and windy conditions. Under humid and calm wind conditions, K_c is less dependent on the aerodynamic components of ET_o and K_c values do not usually exceed the reference value of 1.

Influence of Crop Growth Stages on Crop Coefficient

The values of crop coefficient K_c varies greatly with the stage of growth of the crop, as the crop develops the ground cover, crop height and leaf area change. The growing period of crops can be divided into four distinct stages: (*i*) *initial stage,* (*ii*) *crop development stage,* (*iii*) *mid-season stage,* and (*iv*) *late season stage*. The length of growth periods vary with the crop and its variety, the planting date, and climate. The trends in the values of K_c at different growth stages of crops are shown in Fig. 10.3. For annual crops the initial stage starts from planting to the canopy coverage of about 10% of the ground surface. Crop development stage is from the end of the initial stage to the attainment of about 70 to 80% effective ground cover. Mid-season stage, which is usually the longest, continues from the end of the growth stage to the time of starting of maturity. Late season stage is from the end of mid-season stage to full maturity or harvest. The values of K_c for seasonal crops increase from initial low to plateau to a peak plateau and then decline during the late season stage. (Fig. 10.3). In case of perennial crops the planting date is replaced by the period of initiation of new leaves. During the initial stage, the leaf area and the transpiration are low. K_c values during the initial period is profoundly influenced by the wetness of the soil (due to irrigation or rainfall). K_c values are high when the soil is wet and reduce sharply as the soil becomes dry.

As the crops grow and cover more and more of the ground, the component of evaporation in ET reduces and that of transpiration increases. At this stage K_c values correspond to the extent of ground cover reached and the development of the plant. The mid-season stage is the longest for perennials and many of the annual crops, except short duration vegetable crops. The values of K_c reach the maximum during the mid-season and is relatively constant (Fig. 10.3). The computation of K_c and ET_c end when the crop is harvested or dries out naturally or the leaves drop (in case of perennial crops). K_c value at the end of the late season depends on crop management and water management practices. K_c value is comparatively high when the crop is irrigated frequently until harvested fresh (e.g., fodder crops). The values of K_c will be low when the crop is allowed to dry in the field before harvest (most of the grain crops).

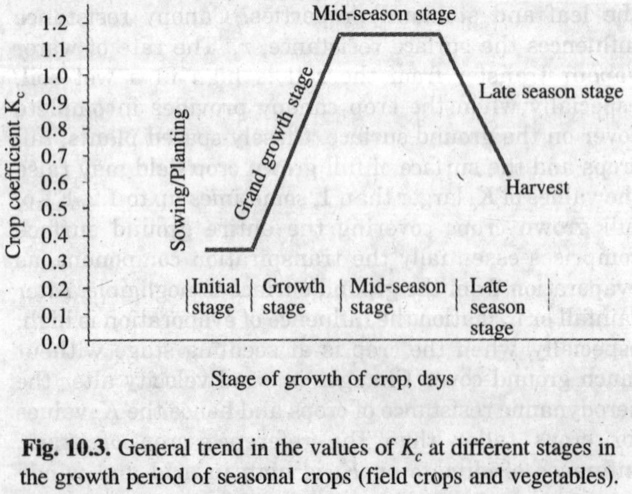

Fig. 10.3. General trend in the values of K_c at different stages in the growth period of seasonal crops (field crops and vegetables).

Climatic Data Sources

Every country has its own national agency for systematic collection, storage and retrieval of weather data. The Indian Meteorological Department, Pune is the repository of weather data in India. Climatic data specific to crop-weather studies are collected and compiled in the agro-meteorological stations at national and state levels in research institutions, various centres of agricultural universities and other scientific and technical institutions. Commencing 1984, the FAO of the United Nations has published mean monthly agro-climatic data from 2300 stations under FAO Plant Production and Protection Series. There are separate volumes for different countries. FAO Irrigation and Drainage Paper No. 46 contains monthly data from 3262 climatic stations, grouped by country and by continent. Monthly averages of maximum and minimum temperatures, mean relative humidity, wind speed, sunshine hours, radiation data, rainfall, and ET_o (calculated using the Penman-Montieth method) are listed in diskets (FAO, 1998: Allen, R.G. *et al.*). FAOCLIM provides user friendly information on compact discs as an extension of the above referred publications. The number of weather stations have later been increased from 2300 to about 19,000. However, the coverage in relation to all the parameters is not complete. Other electronic databases for some parts of the globe have been published by the International Water Management Institute, Colombo, Sri Lanka. They include daily and monthly values of temperature, precipitation and ET_o, based on Hargreave's ET_o equation (UNFAO, 1998).

ESTIMATING EVAPOTRANSPIRATION BASED ON CLIMATOLOGICAL APPROACH

Owing to the difficulty in obtaining accurate direct measurement of pan evaporation under field conditions, evapotranspiration is often predicted on the basis of climatological data. The approaches followed are to relate the magnitude and variation of evapotranspiration to one or more climatic factors (temperature, day length, humidity, wind, sunshine, etc.). Broadly, these approaches fall in two classes: purely empirical attempts to correlate evapotranspiration with one or more climatic factors, or the application of a more theoretical approach. Most prediction formulae use a differentiation between climate and crop. Often such formulae have to be used under climatic and agronomic conditions different from those for which they were originally developed. It is therefore necessary to test the adaptability of the formulae before using them under a new set of conditions.

The more commonly used empirical formulae in estimating evapotranspiration are: (*i*) Blaney-Criddle method, (*ii*) Thornthwaite method and (*iii*) Hargreave's method. The methods based on climatic parameters have been calibrated for 10-day and monthly calculations and not for daily or hourly calculations. The Blaney-Criddle method is recommended for periods of one month or longer.

Blaney-Criddle Method

Blaney and Criddle (1950) observed that the amount of water consumptively used by crops during their growing seasons was closely correlated with mean monthly temperatures and daylight hours and the length of the growing season. The correlation coefficients are then applied to determine the ET for other areas where only climatic data are available. Blaney–Criddle formula is one of the best known procedures for estimating potential evapotranspiration (*PET*) and is widely used. The popularity of the procedure is due to its simplicity and its use of readily available data. It requires the use of only two factors, namely, temperature which is readily available from weather stations and information on day light hours which is a factor based purely on the latitude of the place. Using Blaney-Criddle approach, *PET* can be expressed as follows, in metric units:

$$PET = 0.46\,p\,(T + 17.8) = p\,(0.46\,T + 8.188) = p\,(0.46\,T + 8) \quad ...(10.12)$$

in which,

PET = potential evapotranspiration, mm of water per day (mean value over the month)

p = monthly per cent of total day time hours of the year

$$= \frac{\text{monthly day time hours} \times 100}{\text{Total annual day time hours}}$$

T = mean monthly temperature in °C (average of daily maximum and minimum values)

(***Note***: *p* is entirely a factor of latitude and can be determined from Climatological Tables).

(Appendix D, Table D1)

In English units,

$$PET = \frac{TP}{100} \quad ...(10.13)$$

in which,

PET = monthly potential evapotranspiration (inches/day) (mean value over the month)

P = mean daily percentage of maximum possible annual day light hours

T = monthly mean temperature (°F)

In order to convert the PET values determined by using Eq. 10.12, to actual evapotranspiration or consumptive use *CU* (or the older expression *duty of water*), it is necessary to multiply the *PET* values obtained based on the above relationship by a *crop factor* (*K* when used for seasonal values and *k* when used for monthly values). The values of *K* or *k*, as the case may be, are determined experimentally.

The seasonal consumptive use of a crop can be determined from the following relationship.

$$U = KF = \sum kf = \sum u \quad ...(10.14)$$

in which,

U = seasonal consumptive use of water by the crop for a given period (mm/inches).

u = monthly consumptive use (mm) (Eq. 10.12), inches (Eq. 10.13) applied to obtain monthly values.

K = empirical seasonal consumptive use cofficient for the growing season.

F = sum of the monthly consumptive use factors (f) for the growing season.

k = empirical consumptive use crop coefficient for the month (u/f).

f = value of monthly *PET* in mm (Eq. 10.12) or inches (Eq. 10.13) applied to obtain monthly values).

The monthly percentage of daylight hours (p) of the year for different latitudes are given in Appendix D. Table D1. (India lies between 8°4′, and 37°6′ North latitude).

Equation 10.12 uses the monthly values, while Eq. 10.13 uses the daily values. The value of p for use in Eq. 10.13 can be taken directly from Table D1, Appendix D. The values of mean daily percentage of annual day time hours for different latitudes can be obtained from Table D1 by dividing the monthly values by the number of days in the month. For example, in Table D1 (Appendix D), the monthly percentage of day time hours in the year for the month of January for 60°N latitude is 4.46. The corresponding daily value for the month is 4.67/31 = 0.15.

The mean monthly temperature of a locality is obtained from the long term weather data. In the absence of a more reliable information based on local experimental data, the values of consumptive use crop coefficients (k) for common irrigated crops given in Table 10.4 may be used in computations.

Table 10.4. Monthly consumptive use crop coefficients (k) for use in Blaney-Criddle formula

Crop	*Months*											
	Jan.	*Feb.*	*Mar.*	*Apr.*	*May*	*June*	*July*	*Aug.*	*Sep.*	*Oct.*	*Nov.*	*Dec.*
Rice				0.85	1.00	1.15	1.30	1.25	1.10	0.90		
Maize				0.50	0.10	0.70	0.80	0.80	0.60	0.50		
Wheat	0.50	0.70	0.75	0.70								
Sugarcane	0.75	0.80	0.85	0.85	0.90	0.95	1.00	1.00	0.95	0.90	0.85	0.75
Cotton				0.50	0.60	0.75	0.90	0.85	0.75	0.55	0.50	0.50
Vegetables	0.50	0.55	0.60	0.65	0.70	0.75	0.80	0.80	0.70	0.60	0.55	0.50
Berseem	0.50	0.70	0.80	0.90	1.00				0.60	0.65	070	0.60
Citrus	0.50	0.55	0.55	0.60	0.60	0.65	0.70	0.70	0.65	0.60	0.55	0.55

(***Source***: Dastane, 1972).

The method of computing *PET*, applying Equations 10.12 and 10.13 is illustrated in Example 10.2. The method of estimating consumptive use for a particular crop (e.g., sugarcane in Delhi) is shown in Example 10.4).

Example 10.2. *Compute the ET value in the month of July at a place 28°N latitude if the long term data of temperature for the month of July is 30.5°C. The value of consumptive use coefficient for the month may be taken as 0.62. Compute the values separately, using Equations 10.12 and 10.13.*

Solution:

(*i*) *Metric units* (Equation 10.12). From Appendix D, Table D1 the monthly percentage of day time hours in the year for latitude 28°N is 9.58. Substituting the values in Eq. 10.12.

$$PET = p\,(0.46\,T + 8.188)$$
$$= 9.58\,(0.46 \times 30.5 + 8.188)$$
$$= 9.58 \times 22.218 = 212.71 \text{ mm/month}$$
$$CU = 0.6 \times 212.71 \text{ mm/month}$$

(*ii*) *English units* (Eq. 10.13):

$$PET = \frac{TP}{100}$$

Computing temperature in °F,

$$°F = \left(°C \times \frac{2}{5}\right) + 32$$
$$= 54.9 + 32 = 86.9°F \approx 87°F$$

From Table D1, Appendix D, the monthly percentage of day time hours in the year at 28°N latitude is 9.58. Hence, the daily percentage for the month is 9.58/31 = 0.31.

Substituting the value of *T* and *p* in Eq. 10.13,

$$PET = \frac{8.7 \times 0.31}{100}$$
$$= 0.27 \text{ inches per day}$$
$$= .27 \times 31 = 8.37 \text{ inches for July}$$

Expressed in mm, it is 25.4 × 8.37 = 212.6 mm.

(**Note:** The small difference in decimals in the values obtained by using Equations 10.12 and 10.13 is due to the rounding of the decimals.)

The Blaney-Criddle formula has generally given sufficiently accurate estimates of seasonal consumptive use owing to the inclusion of locally developed crop coefficient factor (K_c). Basically, it assumes that the consumptive use of water is dependent only on temperature and day length which is not true. Crop water requirements vary widely between climates having similar air temperature but different humidity and wind conditions. The consumptive use crop coefficient K_c will, therefore, vary not only with crop but also with climatic conditions. Doorenbos and Pruitt (1977) have not agreed to with the use of crop coefficients (K_c) normally applied in Blaney-Criddle approach as originally developed, because (*i*) the original crop coefficients (K_c) are heavily dependent on local conditions, and wide variations of K_c values reported in literature make the selection of this value rather difficult; (*ii*) the relationship between Blaney-Criddle *f*-values and *PET* can be adequately described for a wide range of temperatures for areas having minor variations in relative humidity, sunshine and wind velocity, and (*iii*) once *PET* has been determined by any standard method one set of crop factors (K_c) can be used to determine crop ET.

Doorenbos and Pruitt (1977) pointed out that the Blaney-Criddle method to calculate the mean daily *PET* should normally be applied for periods not shorter than one month and for each calender month for each year of record, instead of using mean temperature from several years record. According to them the formula is not suitable for use in:

(*i*) equatorial regions where temperatures remain fairly constant, but other weather parameters change,

(*ii*) small islands where air temperature is affected by the surrounding sea temperature, showing little response to seasonal change in radiation,

(*iii*) high altitudes where day time radiation is practically independent of night temperature, and

(*iv*) climates with a high variability in sunshine hours during transition months.

Example 10.3. *Calculation of the consumptive use of water (CU) for sugarcane crop at Delhi (28°N) by using Blaney-Criddle formula (Equation 10.13)*

Month	*Mean monthly temperature (°F T)*	*Monthly crop co-efficient (k) from Table 10.4*	*Per cent day light hrs (p) computed from Appendix D, Table D1*	*Monthly consumptive use (inches) (3 × 2 × 4)*
1	2	3	4	5
Jan.	55.7	0.75	7.38	3.08
Feb.	60.4	0.80	7.02	3.39
March	69.3	0.85	8.39	4.94
April	80.6	0.85	8.69	5.95
May	88.0	0.90	9.48	7.51
June	92.3	0.95	9.41	8.25
July	87.1	1.00	9.60	8.36
August	84.2	1.00	8.18	7.73
Sept.	82.8	0.95	8.33	6.55
Oct.	76.6	0.90	8.01	5.52
Nov.	66.0	0.85	7.25	4.07
Dec.	56.7	0.75	7.24	3.08

$$\text{Annual value of } PET \sum_{i=1}^{12} \text{ of monthly values}$$

$$= 68.43 \text{ inches} = 173.8 \text{ cm}$$

(***Source:*** Michael, Hukkeri and Singh, 1977).

Thornthwaite Formula

Thornthwaite (1948) assumed that an exponential relationship existed between mean monthly temperature and mean monthly consumptive use. The relationship was based largely on experience in the central and eastern United States of America. No allowance was made for different crops or varying land uses. The formula was originally developed for the purpose of a rational classification of the broad climatic patterns of the world. Suitable coefficients should, therefore, be developed locally for reliable estimation of crop ET values. Thornthwaite proposed the following formula:

$$e = 1.6\,(10t/I)^a \quad \ldots(10.15)$$

in which,

e = unadjusted potential evapotranspiration (cm per month) (month of 30 days each and 12 hours day time).

t = mean air temperature (°C)

I = annual or seasonal heat index, the summation of 12 values of monthly heat indices (i) when,

$$i = (t/5)^{1.514}$$

a = an empirical exponent computed by the equation $a = 0.0000006751I^3 - 0.0000771I^2 + 0.01792I + 0.49239$

The unadjusted values of e are corrected for actual day light hours and days in a month. The correction

Example 10.4. *Calculation of the potential evapotranspiration (PET) for Delhi (28°4′ N) by using Thornthwaite formula (Equation 10.15).*

Month	*Mean monthly temperature (°C) (t)*	*Monthly heat index from Appendix Table D-3 (i)*	*Unadjusted PET (cm) e = 1.6(10t/I)^a*	*Correction factor for 'e' from Appendix Table D-2*	*Corrected values of PET (cm) (4 × 5)*
1	2	3	4	5	6
January	12.6	4.05	1.36	0.91	1.24
February	15.8	5.70	2.72	0.87	2.36
March	20.7	8.59	6.39	1.03	6.58
April	27.0	12.85	14.27	1.07	15.26
May	31.1	16.04	22.04	1.17	25.78
June	33.5	17.73	27.64	1.16	32.06
July	30.6	15.53	20.91	1.19	24.88
August	29.0	14.32	17.74	1.13	20.04
September	28.2	13.65	16.30	1.03	16.78
October	24.8	11.30	10.19	0.98	9.98
November	18.9	7.49	4.72	0.90	4.24
December	13.7	4.60	1.75	0.89	1.55
		131.85, or 132 = *I*			160.75 cm

(***Note***: $a = 0.000000675(132)^3 - 0.0000771(132)^2 + 0.01792(132) + 0.49239 = 3.07$).

factors of e and monthly values of i are given in Appendix D, Tables D-2 and D-3, respectively. The method to use the formula is shown in Example 10.4.

The Thornthwaite formula gives a reasonable estimate of *PET* in the temperate, continental climate of North America where the formula was originally derived because there the temperature and radiation are strongly correlated. In other parts of the world this approach has been less successful. The drawbacks of the formula as enumerated by Chang (1968) are as follows:

(*i*) temperature alone is not a good indication for the energy available for evapotranspiration,

(*ii*) air temperature of a place lags behind radiation,

(*iii*) according to this formula ET will cease when mean temperature is below 0°C which by no means is true, although the amount of evaporation will be very small,

(*iv*) the formula does not take into account the wind effect which might be an important factor in some areas, and

(*v*) it does not consider the effect of warm and cool air on the temperature of a place.

Hargreaves' Equation

Hargreaves, based on his work on data from grass lysimeters, proposed the following relationship to estimate ET (Hansen *et al.* 1979)

$$PET = 0.0135\,(t + 17.78)\,R_s \qquad \ldots(10.16)$$

PET = reference crop potential consumptive use, well watered grass in langleys/day

t = average daily temperature in °C

R_s = incident solar radiation in langleys/day

Hargreaves suggested that a reference crop like lucerne would have a consumptive use value 20% more than grass. Thus, ET lucerne will be 1.2 times *PET*. Equation 10.16 does not have any adjustment for site specific conditions. Hargreaves suggested that at places where measured values of R_s are not available, a fair estimate of R_s is

$$R_s = 0.10\,R_{\text{reference grass}}\,(S)^{1/2} \qquad \ldots(10.17)$$

in which,

S = per cent possible sunshine.

Conversion factor for reference crop from Langlays per day to millimeters per day: The units of α, langley (calories per square centimetre) when multiplied by 10 and divided by the latent heat of water (λ) are converted to mm of water per day, which is the common unit for consumptive use computations.

Hargreaves' Radiation Formula

Hargreaves' radiation formula, adjusted and tested at a large number of weather stations under varying climatic conditions, is expressed as follows (FAO, 1998):

$$R_s = k_{R_s}\sqrt{(T_{\max} - T_{\min})}\,R_a \qquad \ldots(10.18)$$

in which,

R_a = extraterrestrial radiation [MJ m^{-2} d^{-1}],

$T_{\max}$ = maximum air temperature [°C],

$T_{\min}$ = minimum air temperature [°C],

k_{R_s} = adjustment coefficient (t)

The factor $\sqrt{(T_{\max} - T_{\min})}$ is closely related to the daily solar radiation at a given location. The adjustment coefficient k_{R_s} is empirical. It differs between inland (interior) regions and coastal regions. For interior regions where the landmass is predominant and air masses are not significantly influenced by a large water body (sea/ocean/large lake) k_{R_s} is approximately equal to 0.16. For coastal locations, situated or adjacent to the coast of a large landmass, when the air masses are influenced by the water body, k_{R_s} is approximately equal to 0.19 (FAO, 1998).

Under island conditions where radiation data from another comparable location in the island are not available, an estimate of the monthly solar radiation can be obtained from the following relationship (FAO, 1998).

$$R_s = 0.7\,R_a - b \qquad \ldots(10.19)$$

R_s = solar radiation [MJ m^{-2}/day^{-1}]

R_a = extraterrestrial radiation [MJ m^{-2}/day^{-1}],

b = empirical constant, equal to 4 [MJ m^{-2}/day^{-1}]

Equation 10.19 is applicable only for low altitudes (from 0 to 100 m).

An alternative Hargreaves' equation for *PET* (ET$_o$) when some of the weather data are missing

When the weather data are inadequate, *PET* (ET_o) can be estimated with reasonably satisfactory results, using Hargreave's alternative equation, expressed as follows (FAO, 1998):

$$ET_o = 0.0023\,(T_{\text{mean}} + 17.8)\,(T_{\max} - T_{\min})^{0.5}\,R_a \qquad \ldots(10.20)$$

All the parameters in Eq. 10.20 have been previously defined. The units of both ET_o and R_a are in mm/day. Equation 10.20 should be verified in each new region by comparing with estimates of ET_o obtained by the FAO Penman-Montieth equation (discussed later) at places where data on radiation, air temperature, humidity and wind speed are available. It is desirable to compute ET_o on a monthly or seasonal basis (see Examples 10.9 and 10.10 for computations using Eq. 10.20).

RATIONAL AND QUASI-RATIONAL APPROACHES IN ESTIMATING EVAPORATION AND EVAPOTRANSPIRATION

Radiation Method

Radiation method is an adaptation of the relationship developed by Makkink (1957). It is recommended for areas where the weather data are not sufficient to use the Penman formula. The essential climatic data include air temperature and sunshine or cloudiness or radiation. Only general levels of wind and humidity, which could be estimated from local weather stations are adequate.

ET_o is estimated from the following relationship (FAO, 1998):

$$ET_o = c\,(W - R_s)\ \text{(mm/day)} \qquad \ldots(10.21)$$

in which,

ET_o = reference crop evapotranspiration in mm/day for the period considered

R_s = solar radiation at the ground level, expressed in equivalent evaporation in mm/day

W = weighting factor, which is a function of the temperature and altitude

c = adjustment factor which depends on mean humidity and day time wind conditions.

Detailed procedures to calculate the values of R_s and W as well as to select proper values of the adjustment factor are given in UNFAO 24 (1977). The radiation method is considered superior to Blaney-Criddle method, particularly in equatorial regions, on small islands and at high altitudes, even if recorded data on sunshine or cloudiness are not available. In such situations solar radiation maps which are available is most parts of the world are considered adequate to provide the required radiation data. The radiation method has proved valuable, particularly in humid regions.

Penman Formula

Penman (1948) proposed an equation for evaporation from open water surface, based on a combination of energy balance and sink strength which is given below with changes in certain symbols in view of the recent trends:

$$E_o = \frac{\Delta Q_n + \gamma E_a}{\Delta + \gamma} \qquad \ldots(10.22)$$

in which,

E_o = evaporation from open water surface (mm/day)

Δ = slope of saturation vapour pressure *vs.* temperature curve (de_a/dT) at the mean air temperature T_a mm Hg per °C (Fig. 10.4).

Q_n = net radiation (mm of water) = $Q_A\,(1 - r)(0.18 + 0.55\,\frac{n}{N})$ $- \sigma T_a^{\,4}\,(0.55 - 0.092\,\sqrt{e_d}\,)(0.10 + 0.90\,\frac{n}{N})$

e_a = saturation vapour pressure of the evaporating surface (e_s) in mm Hg at mean air temperature T_a (Fig. 10.5). [Here e_s is considered equal to e_a by assuming zero temperature gradient between surface(s) and air temperatures.]

T_a = mean air temperature in °K = 273 + °C

r = reflection coefficient of evaporating surface, 0.06 for open water surface

Q_A = Angot's value of mean monthly extra-terrestrial radiation (mm of water/day) (Table D-4, Appendix D)

$\frac{n}{N}$ = ratio between actual and possible hours of bright sunshine (Appendix D, Table D-5 for N)

σ = Stefan-Boltzman constant (Table 10.4 for $\sigma T_a^{\,4}$)

e_d = saturation vapour pressure of the atmosphere, in mm Hg, at dew point temperature = $\frac{RH_{\text{mean}}}{100} \times e_a$, in which RH is the mean relative humidity

γ = psychrometric constant or the ratio of specific heat of air to the latent heat of evaporation of water (0.49 for 0°C and mm Hg)

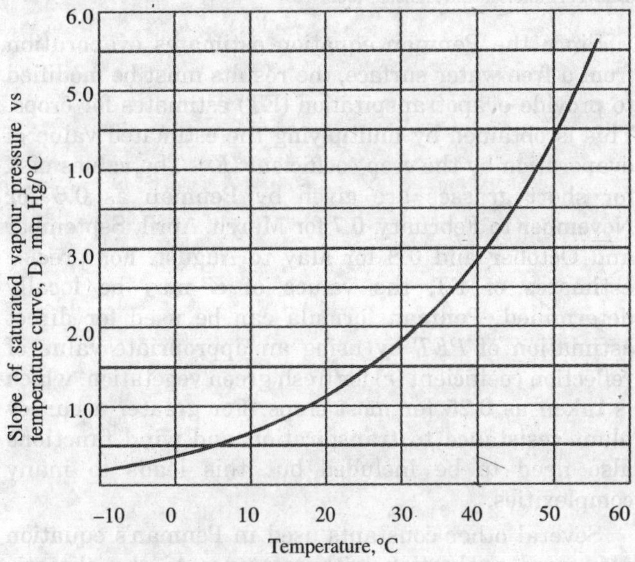

Fig. 10.4. Slope of saturated vapour pressure vs. temperature curve.

[***Source***: Michael A.M., Hukkeri S.B. and Singh, N.P. (1977)]

E_a = an aerodynamic component in which e_s is considered equal to $e_a = 0.35\,(e_a - e_d)\,(1 + 0.0098\,u_2)$

u_2 = wind speed in miles/day at 2 metres height

$$= u_1\left(\frac{\log 6.6}{\log h}\right)$$

in which, u_1 is wind speed in miles per day at any other height h in feet.

Table 10.5. Values of σT_a^4 for various temperatures when computing evapotranspiration by the Penman method (after Criddle)

Temperature (°K)	σT_a^4 *(mm water/day)*
270	10.73
275	11.51
280	12.40
285	13.20
290	14.26
295	15.30
300	16.34
305	17.46
310	18.60
315	19.85
320	21.15
325	22.50

(***Note:*** Heat of vaporisation was assumed to be constant at 590 cal/gm of water)

(***Source***: Israelsen and Hansen, 1962)

Since the Penman equation estimates evaporation from a free water surface, the results must be modified to provide evapotranspiration (ET) estimates for crops. This is obtained by multiplying the estimated value of evaporation by the crop coefficient (K). The values of K for short grasses are given by Penman as 0.6 for November to February, 0.7 for March, April, September and October and 0.8 for May to August. For precise estimates of ET, the values of K may be locally determined. Penman formula can be used for direct estimation of *PET* by using an appropriate value of reflection coefficient (r) for fresh green vegetation, which is taken as 0.25 for most crops. For greater accuracy plant resistance to transpiration and wind functions also need to be included but this leads to many complexities.

Several other constants used in Penman's equation also need evaluation with reference to the climatic conditions of a place. The values of a and b in the equation, $Q = Q_A\,(a + b.n/N)$, for instance, vary from place to place (Doorenbos and Pruitt, 1977). Gangopadhyay, *et al.*, (1970), has worked out these

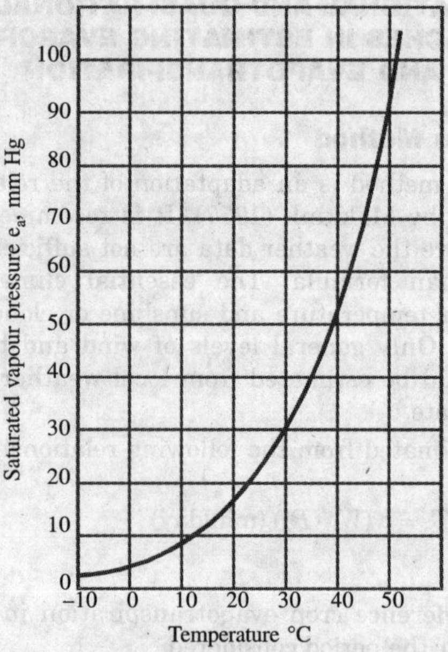

Fig. 10.5. Saturated vapour pressure versus temperature.

[***Source:*** Michael, A.M., Hukkeri, S.B. and Singh, N.P. (1977)]

values for selected locations in India which are presented in Table 10.6.

Chakravarti and Sastry (1977) observed large variations in the values of E_o obtained by Penman method at Delhi due to advective effect. The method to use the Penman formula is given in Example 10.6. The Penman's method has been found to be satisfactory in both humid and arid regions under calm weather conditions. Its main drawbacks are that many meteorological parameters are needed and it involves tedious computations.

Table 10.6. Values of constants, 'a' and 'b' in Penman formula

Place	*Latitude °N*	*Value of 'a'*	*Value of 'b'*
New Delhi	28.4	0.31	0.46
Jodhpur (Rajasthan)	26.3	0.31	0.49
Ahmedabad (Gujarat)	23.1	0.42	0.30
Nagpur (Maharashtra)	21.1	0.16	0.68
Pune (Maharashtra)	18.5	0.35	0.40
Hyderabad (Andhra Pradesh)	17.4	0.14	0.55
Chennai (Tamil Nadu)	13.1	0.30	0.44
Bangalore (Karnataka)	13.0	0.18	0.62
Trivandrum (Kerala)	8.5	0.37	0.38
Shillong (Meghalaya)	24.6	0.18	0.66

[***Source:*** Gangopadhyaya *et al.* (1970)]

Example 10.5. *Calculation of evaporation from open water surface (E_o) in mm/day for Delhi (28°4′ N), using Penman's formula (Equation 10.22) for the period of August 1-10, 1970.*

A. Data (Period 1–10 Aug., 1970)

1.	Mean air temperature	28.7°C
2.	Mean relative humidity, %	88.5
3.	Mean sunshine hours, n	3.43
4.	Possible sunshine hours (N from Table D-5, Appendix D)	13.1
5.	Value of the ratio n/N	0.26
6.	Wind speed at 3 m height in miles/day (u_1) = 90.02 miles/day and at 2 m height = $u_1 \times 0.93 = u_2$	83.7
7.	Extra-terrestrial radiation in mm/day (Q_A from Appendix D, Table D-4)	15.7
8.	Reflection coefficient (r)	0.06

B. Solving expression Q_A $(1-r)(a + b\,.\,n/N)$

9. $(1 - r) = (1 - 0.06) = 0.94$
10. $(a + b.\ n/N) = (0.18 + 0.55 \times 0.26) = 0.32$
11. Item Nos. $(7 \times 9 \times 10) = 4.72$

C. Solving expression $\sigma T_a^{\ 4}$ $(0.56 - 0.092\ \sqrt{e_d}\,)(0.10 + 0.90\ n/N)$

12. Vapour pressure (*i*) e_a (from Fig. 10.5) = 29.2
 (*ii*) e_d (recorded) = 25.5
 (*iii*) $\sqrt{e_d} = 5.04$
13. $\sigma T_a^{\ 4}$ (from Table 10.5) = 16.72
14. $(0.56 - 0.092\ \sqrt{e_d}\,)$
 $= (0.56 - 0.092 \times 5.04) = 0.10$
15. $(0.10 + 0.90 \times n/N)$
 $= (0.10 + 0.90 \times 0.26) = 0.23$
16. Item (Nos. $13 \times 14 \times 15) = 0.38$

D. Solving for Q_n = Item No. 11 – Item No. 16 = $(4.72 - 0.38) = 4.34$

E. Solving for $E_a = 0.35\ (e_a - e_d)(1 + 0.0098\ u_2)$
$= 0.35\ (29.2 - 25.5)(1 + 0.0098 \times 83.7) = 2.37$

F. Solving for $E_o = \dfrac{\Delta Q_n + \gamma E_a}{\Delta + \gamma}$

$$= \frac{1.65 \times 4.34 + 0.49 \times 2.37}{1.65 + 0.49} = 3.89 \text{ mm/day}$$

(For Δ value, see Fig. 10.4)

G. (*i*) $PET = E_o \times K$ for August
 $= 3.89 \times 0.8 = 3.11$ mm/day

(*ii*) *PET* by using 0.25 as the value of r for fresh green vegetation in equation (10.21) = 3.15 mm/day

(*iii*) *PET* by using in equation (10.21) the value of $r = 0.25$ and the values of a and b in item 10 as 0.31 and 0.46, respectively for Delhi given by Gangopadhyay, et al. (1970) = 4.24 mm/day

(*From* Michael, A.M., Hukkeri, S.B., and Singh, N.P., 1977).

Modified Penman Formula

The original Penman equation (1948) predicts evaporation losses from an open water surface (E_o). Experimentally determined crop coefficients, ranging from 0.6 in winter months to 0.8 in summer months related E_o to grass evapotranspiration for the climate in England. The Penman equation consists of two terms: the radiation term and the aerodynamic (wind and humidity) term. The relative importance of the terms varies with climatic conditions. Under calm weather conditions the aerodynamic term is less important than the energy term and the original Penman E_o equation using a crop coefficient of 0.8 predicts ET_o closely in humid and semi-arid regions worldwide. Under windy conditions and in arid regions the aerodynamic term is more important and errors can result in predicting ET_o when using 0.8 E_o.

Based on intensive studies of the climatic and measured grass evapotranspiration data from various research stations in the world and the available literature on prediction of ET or PET, Doorenbos and Pruitt (1975, 1977) proposed a modified Penman method, for estimating fairly accurately the reference crop ET and gave tables and graphs to facilitate the necessary computations. The modified Penman equation involves a revised wind function term. The method uses mean daily climatic data. Since day and night time weather conditions affect the level of evapotranspiration, an adjustment for this is included. The formula contains components which are to be derived from measured climatic data when no direct measurements of needed variables are available. For places where no direct measurements of net radiation are available, these can be obtained from measured solar radiation, sunshine duration or cloudiness along with measured humidity and temperature.

The Modified Penman Equation is stated as follows (FAO 24, 1977):

$$ET_o = c\ [\underbrace{W\,.\,R_n}_{\text{Radiation term}} + \underbrace{(I - W)\,.f(u).(e_a - e_d)}_{\text{Aerodynamic term}}] \quad \ldots(10.22\text{ a})$$

in which,

ET_o = the reference crop evapotranspiration in (mm/day)

e_a = saturation vapour pressure in mbar at the mean air temperature in °C (Appendix D, Table D-6)

e_d = mean actual vapour pressure of the air in mbar

$= e_a \times \dfrac{RH_{\text{mean}}}{100}$, in which, RH = relative humidity. This can also be determined from dry and wet bulb temperatures or dew point temperature (as stated earlier)

$f(u)$ = a wind related function (Appendix D, Table D-7)

$(1 - W)$ = a temperature and elevation related weighting factor for the effect of wind and humidity on ET_o (Appendix D, Table D-8)

W = a temperature and elevation related weighting factor for the effect of radiation on ET_o (Appendix D, Table D-9), and

R_n = net radiation (same as $Q_n = R_{ns} - R_{nl}$)

in which,

R_{ns} = the net incoming shortwave solar radiation = $R_A (1 - \alpha) (0.25 + 0.50\, n/N)$ in which R_A is same as Q_A or extra-terrestrial radiation, expressed in equivalent evaporation in mm/day, given in Appendix D, Table D4; n/N is the same as explained in equation (10.22), and α is same as r or reflection coefficient; the value of which is taken as 0.25 for most crops (Appendix D, Table D-10) gives conversion factors for R_A to R_{ns} for a given reflection of 25 per cent and ratios of n/N), and

R_{nl} = the net long wave radiation = $f(t) . f(ed)$. $f(n/N)$, the values of which are given in Appendix D, Tables D-11, D-12 and D-13, respectively.

c = adjustment factor to compensate for the effect of day and night weather conditions (Approximate values) for different areas given by Doorinbos and Pruitt (FAO, 1975) are given in Appendix D, Table D-7.

Vapour pressure is to be expressed in mbar. If e_s is given in mm Hg, it is to be multiplied by the factor 1.32 (i.e., 1013/766) to convert the value in mbar.

The wind factor $f(u)$ is given by the following relationship (FAO, 1977).

$$f(w) = 0.27 \left(1 + \frac{u_2}{100}\right) \quad \text{...(10.22 b)}$$

in which u_2 is the total wind run in km/day at 2 m height. Where wind data are not collected at 2 m height, suitable corrections for the wind measurements taken at different heights are to be made. Appendix D, Table D-7 presents the values of the crop factors given by Doorenbos and Pruitt (FAO, 1975, 1977). The computation technique to use Eq. 10.22b is shown in Example 10.6.

Example 10.6. *Computation of reference, crop ET_o (PET) in mm/day for Delhi (28°4′ N and 228 m altitude) using Modified Penman Formula (equation 10.22) for the period of August 1-10, 1970.*

A. Data (Period 1–10 Aug., 1970)

1. Mean air temperature 28.7°C
2. Mean relative humidity, % 88.5
3. Mean sunshine hours, (n) 3.43
4. Possible sunshine hours (N from Appendix D, Table D-5) 13.1
5. Value of the ratio n/N 0.26
6. Extra-terrestrial radiation, mm/day (Table D4) = R_a 15.7
7. Wind speed at 3 m height in km/day (u_1) = 145.2.
   Therefore, u_2, i.e., wind speed at 2 m height = $u_1 \times 0.93$ = 135.0
8. Reflection coefficient (α) 0.25

B. Solving for aerodynamic term $(1 - W) . f(u) . (e_a - e_d)$

9. e_a mbar from Appendix D, Table D-6 39.4
10. e_d mbar = $\frac{RH_{mean}}{100} \times e_a = 0.885 \times 39.4 = 34.9$
11. $(e_d - e_d) = (39.4 - 34.9) = 4.5$
12. $f(u)$ from Appendix D, Table D-7 0.635
13. $(1 - W)$ from Appendix D, Table D-8 = 0.22
14. Item Nos. 11 × 12 × 13 = 4.5 × 0.635 × 0.22 = 0.63

C. Solving for radiation term, $W . R_n = W(R_{ns} - R_{nl})$

15. $R_{ns} = R_a \times$ factor from Table D-10 = 15.7 × 0.284 = 4.46
16. $R_{nl} = f(t) . f(e_d) . f(n/N)$ from Appendix D, Table D-10 = 16.5 × 0.08 × 0.34 = 0.45
17. $R_n = (R_{ns} - R_{nl})$ = Item Nos. 15 – 16 = 4.46 – 0.45 = 4.01
18. W from Appendix D, Table D9 = 0.78
19. $W. R_n$ = Items 17 × 18 = 4.01 × 078 = 3.13

D. ET_o (unadjusted) = $W. R_n + (1 - W) . f(u) (e_a - e_d)$ = Items 14 + 19 = 3.76

E. Adjustment factor c based on the approximate relationship explained above ~ 1.09

$c \times ET_o = 1.09 \times 3.76 \sim 4.10$ mm/day

Note: The value of ET_o (4.1 mm/day) by the above method is close to the value of *PET* (4.24 mm/day) obtained in Example 10.6.

(Adapted from Michael, A.M., Hukkeri, S.B., and Singh, N.P., 1977)

Modified Penman Method (FAO No. 24, 1977) described above was being widely used in irrigation development and management projects. Detailed investigations by FAO Expert groups revealed that the method frequently overestimated ET_o, sometimes even up to 20% for low evaporative conditions (FAO No. 56, 1998). It was observed that the Penman methods may require local calibration of the wind function to achieve satisfactory results.

Formulation of Penman-Montieth Equation

Original Penman-Montieth equation. The 1948, Penman formula has been the combined energy balance method with the mass transfer method to compute the evaporation from an open water surface from standard climatological data of sunshine, temperature, humidity and wind speed. It was further developed by Montieth and other researchers with a view to extend to cropped areas by introducing resistance factors, including surface resistance and aerodynamic resistance (Montieth 1965, 1981).

The original form of the Penman-Montieth equation can be written as follows (UN FAO 1990, 1998).

$$\lambda ET_o = \frac{\Delta(R_n - G) + \rho C_p\,(e_a - e_d)/r_a}{\Delta + \gamma(1 + r_c/r_a)} \quad ...(10.23)$$

in which λET_o : latent heat flux of evaporation [kJ m^{-2} s^{-1}]

R_n : net radiation flux at surface [kJ m^{-2} s^{-1}]

G : soil heat flux [kJ m^{-2} s^{-1}]

ρ : atmospheric density [kg m^{-3}]

C_P : specific heat of moist air [kJ kg^{-1} °C^{-1}]

$(e_a - e_d)$: vapour pressure deficit [kPa]

r_c : crop canopy resistance [s m^{-1}]

r_a : aerodynamic resistance [s m^{-1}]

Δ : slope of vapour pressure curve [kPa °C^{-1}]

γ : psychrometric constant [kPa °C^{-1}]

λ : latent heat of vaporization [MJ kg^{-1}]

To facilitate computations the Panel of Experts of FAO Consultations held in 1990 defined the aerodynamic and radiation terms in Eq. 10.23 as follows:

$$ET_o = ET_{rad} + ET_{aero} \quad ...(10.23a)$$

in which,

ET_o : reference evapotranspiration of standard crop canopy [mm d^{-1}]

ET_{rad} : radiation term [mm d^{-1}]

ET_{aero} : aerodynamic term [mm d^{-1}]

Observations of the Revision of FAO Methodologies for Crop Water Requirements

Studies conducted by various agencies worldwide after the publication of the Modified Penman Method (Eq. 10.22a) in Irrigation and Drainage Paper 24 (United Nations, FAO, 1977) showed that in many cases the prevailing methods for estimating evapotranspiration (ET_o), though are suitable for many regions, could not be applied worldwide without local level calibration to develop correction factors. For instance, the Modified Penman Method (Eq. 10.22a) recommended in [(FAO, 24 (1977)] was found to frequently overestimate ET_o while the other methods presented in the same paper, namely, the Radiation, Blaney. Criddle and Pan Evaporation methods showed variable adaptability to the grass evapotranspiration. The Modified Penman Method (Eq. 10.22a) was considered to offer comparatively satisfactory results in relation to living grass reference crop. The reliability of the Evaporation Pan Method would greatly depend on the location of the pan. Local level calibration required to obtain correction factors for various equations were often difficult and expensive.

To evaluate the performance of the various ET_o estimation procedures under different climatological conditions, a detailed study was undertaken under the auspices of the Committee on Irrigation Water Requirements of the American Society of Civil Engineers (ASCE). The study analysed the performance of 20 different methods, and compared the results to carefully screened lysimeter data from 11 locations with variable climatic conditions. The study showed the widely varying performance of the methods under different climatic conditions. The results of the study were reproduced in FAO Consultation (UNFAO, 1990). An identical study commissioned by the European Community through various institutions in Europe showed similar variations in the prediction of ET_o. The following are the major conclusions resulting from the two studies and the observations of other research stations elsewhere:

1. The Penman methods may usually require local calibration of the wind function to obtain satisfactory results.

2. The radiation methods show good results in humid climates where the aerodynamic term is relatively small, but their performance in arid conditions is erratic and tends to underestimate evapotranspiration.

3. Temperature methods are essentially empirical and require local calibration to achieve satisfactory results, except possibly the Hargreaves' method (Eq. 10.20) which has shown reasonable ET_o values with a global validity.

4. Pan evapotranspiration methods reflect the shortcomings of predicting crop evapotranspiration from open water evaporation. The methods are susceptible to the microclimatic conditions under which the pans are operating and the level of weather station maintenance. Their performance proved erratic.

5. The relatively accurate and consistent performance of the Penman-Monteith approach in both arid and humid climates has been indicated in both the ASCE and European studies.

To improve on the methodologies the FAO solicited the collaboration of a wide range of resource persons and institutions worldwide with a view to review the methodologies and update procedures for estimation of crop water requirements. They included the International Commission for Irrigation and Drainage, the American Societies for Civil and Agricultural Engineers and the World Meteorological Organization. With this background the FAO organized a consultation of eminent experts and researchers which was held is May 1990 in Rome, Italy, in collaboration with the International Commission on Irrigation and Drainage and the World Meterorological Organization. The Consultation was to review the FAO methodologies in crop water requirements and to advice on the revision and update of the procedures. Considering the limitations in the availability of reliable climatic data to many users, it was concluded that several ET_o methods should be recommended, based on type of climatic data available. It is necessary that a clear relationship should be established among the various methods, allowing meaningful comparisons. The following methods and concepts were therefore recommended by the group of experts:

1. **The FAO Modified Penman Method (UNFAO, 1977).** It provides an estimation of the evapotranspiration of a reference crop from measured climatic data on temperature, humidity and wind. However, there was evidence that the method was somewhat overpredicting under non-advective conditions. Comparative studies carried out under the auspice of the American Society of Civil Engineers and the European Commission have shown convincing performance under varying climatic conditions of the method based on the Penman-Monteith approach. The consultation agreed to recommend climatic formula based on the Penman-Montieth approach as the best performing equation. Details on the various parameters to be used for estimation of the Reference Evapotranspiration were developed in the Consultation.

2. **Radiation method.** Several authors introduced equations for estimating reference evapotranspiration based on measured solar radiation data and temperature only. Doorenbos and Pruitt (1977) proposed the FAO Radiation method based on the Makkink equation (1957), introducing a correction factor based on estimates of wind and humidity conditions to compensate for advective conditions. The radiation method has proved useful in particular under humid conditions. It was suggested that the original FAO radiation method could possibly be improved by introducing long wave radiation as part of the radiation term and by including a simplified aerodynamic term of the combination formula (Eq. 10.24) with standardized wind and humidity conditions. However, additional studies are required to arrive at firm conclusions on the suggested procedure.

3. **Temperature method.** Many users have access only to data on temperature and rainfall. The temperature method was included amongst the various methods, although its results will be affected by local conditions and calibrations. The FAO Temperature Method recommended by Doorenbos and Pruitt (1977) was based on the Blaney-Criddle method, introducing a correction factor based on estimates of humidity, sunshine and wind. The Consultation, recognised the historical value of the Blaney-Criddle method and its validity where regionally calibrated correction factors were used. The method was found useful for areas having only mean daily temperature data and where appropriate correction factors could be obtained. It was felt that possibly a more satisfactory temperature method could be developed by using maximum and minimum temperature measurements. Differences in maximum and minimum temperatures may be correlated with humidity conditions and net solar radiation. In this respect reference was cited of the Hargreaves method (Eq. 10.20). Further studies were recommended to develop a satisfactory temperature method under different climatic conditions.

4. **Pan evaporation method.** Extensive use and testing of the evaporation from standardized evaporation pans such as the Class A pan have shown, the great sensitivity of the daily evaporation of the water in the pan, influenced by a range of environmental conditions such as wind, soil-heat flux, vegetative cover around the pan, painting and maintenance conditions and the use of screens over the pan. The introduction of the pan-factor related to various advective and environmental conditions by Doorenbos and Pruitt (1977) in FAO No. 24 was recommended to be maintained. The use of the pan evaporation method in particular in comparative studies and for practical irrigation scheduling was recognised. However, use of the pan evaporation method for estimating reference evapotranspiration could be recommended only if instrumentation and the site are properly calibrated and managed.

5. **Parameters and climatic variables.** To obtain uniformity in calculation procedures, the various parameters in the different equations used were reviewed. It was agreed that the various parameters and the units could better be expressed in the SI system. The algorithms to determine the various parameters were reviewed and a uniform procedure was proposed to determine each parameter. Values for several parameters were recommended based on their general and global validity.

Details on the various parameters, their algorithm, recommended values and units which they are expressed were included in the report the Consultation (UNFAO, 1990). The various climatic variables were reviewed. The importance of proper observations, and maintenance of the recording devices and practical procedures for data processing and satisfiable analysis were stressed.

6. **Crop water requirements.** The sound and practical methods of determining crop water requirements as introduced by Doorenbos and Pruitt (1977) who provided a list of crop coefficients for a range of crops at various growth stages were to a large extent valid.

The FAO Expert Consultation on Revision of FAO Methodologies for Crop Water Requirements accepted the following definition for the reference surface:

"A hypothetical reference crop with an assumed height of 0.12 m, a fixed surface resistance of 70 s/m and an albedo of 0.23."

The Consultation stressed the need to update the crop coefficients, in particular trees and fruit crops and many of the perennial crops. The values of crop coefficients at the initial stage of the crop growth are reviewed by evaluating soil evaporation and basal crop transpiration separately. Similarly, the effect of climate and advective conditions on the crop factor need to be reviewed. There is also need to review and update the length of various growth stages of crops by possibly introducing growth function of crops relating crop growth parameters.

Like leaf area index, dry matter production and crop height with climatic variables like temperature, relative humidity and day length need further study. The importance of organizing reviews on factors like effective and dependable rainfall, salinity control and leaching as well as irrigation scheduling were identified in the Consultation.

FAO PENMAN-MONTIETH BASED COMBINATION FORMULA FOR ESTIMATING REFERENCE EVAPOTRANSPIRATION, ET_o

The combination formula recommended by the Expert Consultation on FAO Methodologies for Crop Water Requirements (UNFAO, 1990) derived by combining the equation of aerodynamic resistance and surface resistance is stated below:

$$ET_o = \frac{0.408\,\Delta(R_n - G) + \gamma\frac{900}{T+273}u_2\,(e_s - e_a)}{\Delta + \gamma(1 + 0.34u_2)} \qquad \text{...(10.24)}$$

in which,

ET_o = reference evapotranspiration [mm day^{-1}]

R_n = net radiation at the crop surface [MJ m^{-2} day^{-1}]

G = soil heat flux density [MJ m^{-2} day^{-1}]

T = mean daily air temperature at 2 m height [°C]

u_2 = wind speed at 2 m height [m s^{-1}]

e_s = saturation vapour pressure [kPa]

e_a = actual vapour pressure [kPa]

$e_s - e_a$ = saturation vapour pressure deficit [kPa]

Δ = slope of vapour pressure curve [kPa $°C^{-1}$]

γ = psychrometric constant [kPa $°C^{-1}$]

The figure 900 is a conversion factor (UNFAO, 1990). When no measured radiation data are available, the net radition can be estimated as follows (UNFAO, 1990):

$$R_n = R_{ns} - R_{nl} \qquad \text{... (10.25)}$$

$$R_{ns} = 0.77\left(0.25 + 0.50\frac{n}{N}\right)R_a \qquad \text{...(10.26)}$$

$$R_{nl} = 2.45\,.\,10^{-9}\left(0.9\frac{n}{N} + 0.1\right)\left(0.34 - 0.14\sqrt{e_d}\right)\left(T_{kx}^4 + T_{kn}^4\right) \qquad \text{...(10.27)}$$

$$G = 0.14\,(\text{T}_{\text{month}\,n} - \text{T}_{\text{month}\,n-1}) \approx 0 \qquad \text{...(10.28)}$$

in which,

R_n : net radiation [MJ m^{-2} d^{-1}]

R_{ns} : net shortwave radiation [MJ m^{-2} d^{-1}]

R_{nl} : net long wave radiation [MJ m^{-2} d^{-1}]

R_a : extraterrestrial radiation [MJ m^{-2} d^{-1}]

n/N : relative sunshine fraction [ ]

Tk_x : maximum temperature [K]

T_{kn} : minimum temperature [K]

e_d : actual vapour pressure [kPa]

G : soil heat flux [MJ m^{-2} d^{-1}]

$T_{\text{mean}\,i}$: mean temperature of the month, °C

$T_{\text{mean}\,(i-1)}$: mean temperature of the previous months, °C

Equation 10.24 determines the ET_o from an assumed grass reference surface and serves as a standard to which evapotranspiration in different periods of the year and in other regions can be compared and to which ET_o of different crops can be related.

Based on the recommendations of the Expert Consultation of FAO Methodologies for Crop Water Requirements (UNFAO, 1990) and the results of other studies, the United Nations Food and Agriculture Organization (Allen, R.G. *et al.*, UNFAO, 1998) brought out a comprehensive treatise in FAO Irrigation and Drainage Paper 56, entitled Crop Evapotranspiration, Guidelines for Computing Crop Water Requirements. Clear solutions of the various factors in Eq. (10.24) and related climatic parameters and their functional relationships have been brought out and detailed procedures for computing ET_o based on Eq. (10.24) are presented with appropriate examples and explanatory notes. They are summarized below to facilitate the solution Eq. 10.24. Procedures for arriving at close approximations of ET_o when some of the data in Eq. 10.24 are missing are also given in the above cited reference.

1. **Wind pressure at different heights from ground surface.** Wind pressures measured at different heights above ground surface vary. Friction at the ground surface slows down the wind speed. Wind speed increases with the height above the ground surface. Anemometers are usually placed at height of 10 m in meteorological observatories and 2 or 3 m in agro-meteorological stations. To compute ET_o the wind speed measured at 2 m above the ground surface is necessary. To adjust the wind speed measured at heights other than 2 m, the following relationship is used:

$$u_2 = u_z \frac{4.87}{\ln(67.8z - 5.42)} \quad \text{... (10.29)}$$

u_2 = wind speed at 2 m above ground surface [m s^{-1}]

u_z = measured wind speed at z m above ground surface [m s^{-1}]

z = height of measurement above ground surface [m]

2. **Atmospheric pressure, *P*.** The atmospheric pressure P, as exerted by the weight of the atmosphere. P is computed using the following relationship:

$$P = 101.3\left(\frac{293 - 0.0065z}{293}\right)^{5.26} \quad \text{...(10.30)}$$

in which,

P = atmospheric pressure [kPa]

z = elevation above mean sea level [m]

3. **Slope of saturation vapour pressure.** To compute ET_o, the slope of the line showing the relationship between saturation vapour pressure and temperature is required. It is obtained from the following relationship:

$$\Delta = \frac{4098\left[0.6108\exp\left(\frac{17.27T}{T+237.3}\right)\right]}{(T+237.3)^2} \quad \text{...(10.31)}$$

in which,

Δ = slope of saturation vapour pressure curve at air temperature T [kPa °C^{-1}]

T = air temperature [°C]

exp [...] = 2.7183 (base of natural logarithm) raised to the power [...]

4. **Psychrometric constant (γ).**

The psychrometric constant, γ, is given by

$$\gamma = \frac{c_p P}{\varepsilon\lambda} = 0.665 \times 10^{-3} P \quad \text{...(10.32)}$$

in which,

γ = psychrometric constant [kPa °C^{-1}]

P = atmospheric pressure [kPa]

λ = latent heat of vaporization, 2.45 [MJ kg^{-1}]

c_p = specific heat at constant pressure, 1.013 10^{-3} [MJ kg^{-1} °C^{-1}],

ε = ratio of molecular weight of water vapour/dry air = 0.622.

The specific heat at constant pressure is the energy required to increase the temperature of a unit mass of air by one degree at constant pressure. Its value depends on the composition of the air, i.e., its humidity. Under normal atmospheric conditions a value c_p = 1.013 10^{-3} MJ kg^{-1} °C^{-1} is used. As an average atmospheric pressure is used for each location the psychrometric constant is kept constant for each location.

5. **Mean saturation vapour pressure.** Saturation vapour pressure is related to air temperature and can be expressed by the following relationship:

$$e^0(T) = 0.6108\exp\left[\frac{17.27T}{T+237.3}\right] \quad \text{...(10.33)}$$

in which,

$e^0(T)$ = saturation vapour pressure at the air temperature T [kPa],

T = air temperature [°C]

exp [...] = 2.7183 (based of natural logarithm) raised to the power [...]

The mean saturation vapour pressure for a day, week or longer period is computed as the mean between the saturation vapour pressure at the mean daily maximum and minimum air temperatures for that period:

$$e_s = \frac{e^0(T_{max}) + e^0(T_{min})}{2} \quad \text{...(10.34)}$$

The use of mean air temperature instead of daily minimum and maximum temperatures results in lower estimates for the mean saturation vapour pressure. The corresponding vapour pressure deficit (for expressing the evaporating power of the atmosphere) will also be smaller and the result will be underestimation of the reference crops evapro-transpiration. The mean saturation vapour pressure should be calculated as the mean between the saturation vapour pressure at both the daily maximum and minimum air temperature.

6. **Vapour pressure deficit.** *Derivation of actual vapour pressure (e_a) from relative humidity data*: The actual vapour pressure can be computed from the relative humidity. Depending on the availability of the humidity data, different equations are used:

(*i*) For RH_{max} and RH_{min}:

$$e_a = \frac{e^0(T_{min})\frac{RH_{max}}{100} + e^0(T_{max})\frac{RH_{min}}{100}}{2} \quad \text{...(10.35)}$$

in which,

e_a = actual vapour pressure [kPa]

$e^0(T_{min})$ = saturation vapour pressure at daily minimum temperature [kPa]

$e^0(T_{max})$ = saturation vapour pressure at daily maximum temperature [kPa]

RH_{max} = maximum relative humidity [%]

RH_{min} = minimum relative humidity [%]

For periods of a week or a month, RH_{max} and RH_{min} are obtained by dividing the sum of the daily values by the number of days in that period.

***(ii)* For RH_{max}:**

While using weather instruments where errors in estimating RH_{min} can be large, or when RH data are in doubt, only RH_{max} should be used:

$$e_a = e^0(T_{min})\frac{RH_{max}}{100} \quad ...(10.36)$$

Vapour pressure deficit ($e_s - e_a$) is the difference between the saturation (e_s) and actual vapour pressure (e_a) for a given time period. When the time period is a week or a month e_s is computed from Equation 10.34 using the T_{max} and T_{min} averaged over the time period and similarly the e_a is computed with equations 10.35 using average measurements over the period.

7. **Extraterrestrial radiation for daily periods (R_a).** The Extraterrestrial radiation, R_a, for each day of the year and for different latitudes can be estimated from the solar constant, the solar declination and the time of the year by the following relationship:

$$R_a = \frac{24(60)}{\pi} G_{sc} d_r \,[\omega_s \sin(\varphi)\sin(\delta) + \cos(\varphi)\cos(\delta)\sin(\omega_s)] \quad ...(10.37)$$

in which,

R_a = extraterrestrial radiation [MJ m^{-2} day^{-1}]

G_{SC} = solar constant = 0.0820 MJ m^{-2} min^{-1}

d_r = inverse relative distance Earth-Sun (Equation 10.39)

ω_s = sunset hour angle (Equation 10.44) [rad]

φ = latitude [rad] (Equation 10.38)

φ = solar declination (Equation 10.40) [rad]

R_a can be expressed in mm day^{-1} by multiplying by a conversion factor 0.408 (equal to the inverse of the latent heat of vaporization). The latitude, φ, expressed in radians is positive for the northern hemisphere and negative for the southern hemisphere. The conversion from decimal degrees to radians is given by the following:

$$[\text{Radians}] = \frac{\pi}{180}\,[\text{decimal degrees}] \quad ...(10.38)$$

The inverse relative distance Earth-Sun, d_r, and the solar declination, δ, are given by the following:

$$d_r = 1 + 0.33\cos\left(\frac{2\pi}{365}J\right) \quad ...(10.39)$$

$$\delta = 0.409\sin\left(\frac{2\pi}{365}J - 1.39\right) \quad ...(10.40)$$

in which,

J is the number of the day in the year between 1 (1 January) and 365 or 366 (31 December).

The sunset hour angle, ω_s, is expressed as follows:

$$\omega_s = \arccos\,[-\tan(\varphi)\tan(\delta)] \quad ...(10.41)$$

8. **Daylight hours, *N*.** The daylight hours are given by the following:

$$N = \frac{24}{\pi}\omega_s \quad ...(10.42)$$

in which,

ω_s is the sunset hour angle in radians given by Equation 10.41.

Solar radiation (R_s). In situations when the solar radiation, R_s, is not measured, it can be calculated with Angstrom formula, which relates solar radiation to extraterrestrial radiation and relative sunshine duration as follows:

$$R_s = \left[(a_s + b_s)\frac{n}{N}\right]R_a \quad ...(10.43)$$

in which,

R_s = solar or shortwave radiation [MJ m^{-2} day^{-1}]

n = actual duration of sunshine [hours]

N = maximum possible duration of sunshine or daylight hours [hours]

n/N = relative sunshine duration [–]

R_a = extraterrestrial radiation [MJ m^{-2} day^{-1}]

a_s = regression constant, expressing the fraction of extraterrestrial radiation reaching the earth on overcast days (n = 0)

$a_s + b_s$ = fraction of extraterrestrial radiation reaching the earth on clear days ($n = N$)

The corresponding value of R_s (equivalent evaporation) in mm day^{-1} is obtained by multiplying R_s by 0.408. Depending on atmospheric conditions (humidity, dust) and solar declination (latitude and month), the Angstrom values a_s and b_s will vary. At places where actual solar radiation data are not available the values a_s = 0.25 and b_s = 0.50 are usually assumed.

Clear-sky solar radiation R_{so}. Computation of the clear-sky radiation, R_{so}, when $n = N$, is required for obtaining net long wave radiation:

(i) *At near sea level or when calibrated values for a_s and b_s are available:*

$$R_{so} = (a_s + b_s)\, R_a \qquad \text{...(10.44)}$$

in which,

R_{so} = clear-sky solar radiation [MJ m^{-2} day^{-1}]

$a_s + b_s$ = fraction of extraterrestrial radiation reaching the earth on clear-sky days ($n = N$)

(ii) *When calibrated values for a_s and b_s are not available:*

$$R_{so} = (0.75 + 2 \times 10^{-5} \times z)\, R_a \qquad \text{...(10.45)}$$

in which,

z = station elevation above sea level [m]

9. **Net solar or net shortwave radiation R_{ns}**. The net shortwave net solar radiation resulting from the difference between incoming and reflected solar radiation is given by the following relationship.

$$R_{ns} = (1 - \alpha)\, R_s \qquad \text{...(10.46)}$$

in which,

R_{ns} = net solar or shortwave radiation [MJ m^{-2} day^{-1}]

α = albedo or canopy reflection coefficient, which is 0.23 for the hypothetical grass reference crop (dimensionless)

R_s = the incoming solar radiation [MJ m^{-2} day^{-1}]

10. **Net longwave radiation, R_{nl}**. The rate of longwave energy emission is proportional to the absolute temperature of the earth's surface raised to the power 4. It is expressed quantitatively by the Stefan-Boltzmann law. However, the net energy flux leaving the earth's surface is less than that emitted as given by the Stefan-Boltzmann law due to the absorption and downward radiation from the sky. Water vapour, clouds, carbon dioxide and dust are absorbers and emitters of longwave radiation. Their concentrations should be known to assess the net outgoing flux. Humidity and cloudiness play important roles in the longwave energy expression. Hence, the Stefan-Boltzmann law is corrected by these two factors when estimating the net outgoing flux of longwave radiation. It is assumed that the concentrations of the other absorbers are constant. R_{nl} is expressed by the following formula:

$$R_{nl} = \sigma\left[\frac{T_{\max,K^4} + T_{\min,K^4}}{2}\right]\left(0.34 - 0.14\sqrt{e_a}\right)\left(1.35\frac{R_s}{R_{so}} - 0.35\right) \qquad \text{...(10.47)}$$

in which,

R_{nl} = net outgoing longwave radiation [MJ m^{-2} day^{-1}]

σ = Stefan-Boltzmann constant [4.903 10^{-9} MJ K^{-4} m^{-2} day^{-1}]

$T_{\max, K}$ = maximum absolute temperature during the 24-hour period [K = °C + 273.16]

$T_{\min, K}$ = minimum absolute temperature during the 24-hour period [K = °C + 273.16]

e_a = actual vapour pressure [kPa]

R_s/R_{so} = relative shortwave radiation (limited to ≤ 1.0)

R_s = measured or calculated (Equation 10.35) solar radiation [MJ m^{-2} day^{-1}]

R_{so} = calculated (Equations 10.35 or 10.36) clear-sky radiation [MJ m^{-2} day^{-1}]

The above relationship of T is commonly used for 24-hour time steps. The term $(0.34 - 0.14\sqrt{e_a})$ expresses the correlation for air humidity, and will be smaller if the humidity increases. The effect of cloudiness is expressed by $(1.35\, R_s/R_{so} - 0.35)$. The smaller the correction terms, the smaller the net outgoing flux of longwave radiation. The R_s/R_{so} term in Equation 10.47 must be limited so that $R_s/R_{so} \le 1.0$.

11. **Net radiation R_n**. The net radiation (R_n) is the difference between the incoming net shortwave radiation (R_{ns}) and the outgoing net longwave radiation (R_{nl}):

$$R_n = R_{ns} - R_{nl} \qquad \text{...(10.48)}$$

12. **Soil temperature**. Soil temperature remains lower than air temperature. Hence, the average temperature for a period should be considered when assessing the daily soil heat flux, Δt. The depth of penetration of the temperature wave is determined by the length of the time interval. The effective soil depth, Δz, is only 0.10–0.20 m for a time interval of one or a few days but might be 2 metres or more for monthly periods. The magnitude of the soil heat flux when computed for a day is small and hence is often ignored.

Estimation of Missing Climatic Data

Estimation of reference evapotranspiration, ET_o with Penman-Montieth Method (Eq. 10.24) requires mean daily, ten-day or monthly maximum and minimum air temperature ($T_{\max}$ and $T_{\min}$), actual vapour pressure (e_a), net radiation (R_n) and wind speed measured at 2 m (u_2). If some of the required weather data are missing or cannot be computed, UN FAO (1998) has recommended the following procedure to make a reasonable estimate of the missing climatic data:

Estimating missing humidity data. Where the humidity data are lacking, an estimate of actual vapour pressure, e_a, can be obtained by assuming that dew point temperature (T_{dew}) is near the daily minimum temperature (T_{min}). This statement implicitly assumes that at sunrise, when the air temperature is close to T_{min}, the air is nearly saturated with water vapour and the relative humidity is nearly 100%. Substituting T_{min} to represent T_{dew},

$$e_a = e^0(T_{min}) = 0.6108 \exp\left[\frac{17.27\ T_{min}}{T_{min} + 237.3}\right] \quad ...(10.48a)$$

The assumption that $T_{dew} \approx T_{min}$ is satisfactory for locations where the cover crop of the station is well watered. However, mainly in the arid regions, the air might not be saturated when its temperature is at its minimum. Hence, T_{min} might be greater than T_{dew}. In such situations, "T_{min}" in Equation 10.47 may be better approximated by subtracting 2-3°C from T_{min}. Where Equation 10.47 is validated for a region, it can be used for daily estimates of e_a.

Estimating missing radiation data. Net radiation measuring devices are not commonly installed in many of the agrometeorological stations. In the absence of a direct measurement, longwave and net radiation can be derived from the commonly observed weather parameters, namely solar radiation or sunshine hours, air temperature and vapour pressure. When solar radiation is not available, it may have to be estimated from measured hours of bright sunshine. In situations when daily sunshine hours (n) are not available, solar radiation may be estimated with the data from a nearby weather station.

Solar radiation data from a nearby weather situation. The method is based on the assumption that for the same month and often for the same day, the variables influencing incoming solar radiation, R_s, and sunshine duration, n, are similar throughout a given region. The following requirements must be satisfied to use data from a neighbouring weather station to arrive at a reasonable estimate: (*i*) the size of the region is small; (*ii*) the air masses governing rainfall and cloudiness are almost identical within the region; and (*iii*) the physiography of the region is almost homogeneous.

Data from nearby stations may not be suitable for adoption in mountain and coastal areas where rainfall variations are high. For instance, there is substantial difference in rainfall characteristics between the windward and leeward sides of a mountain/hill. Use of Equation 10.47 is recommended for monthly calculations of ET_o. For the estimation of daily values of ET_o careful analysis of the importing and exporting meteorological stations is required to verify whether the two stations are in the same homogeneous climatic region.

Solar Radiation Data Derived from Air Temperature Differences

The difference between T_{max} and T_{min} is related to the cloudiness of the sky. Clear sky result in high values of air temperature during the day and low temperatures in the night. In cloudy weather a significant part of the incoming solar radiation never reaches the earth's surface and is absorbed and reflected by the clouds. During the night, the cloud cover acts as a blanket and decreases the net outgoing long wave radiation.

Missing Wind Speed Data: Wind Speed Data from Weather Stations

Adopting wind speed readings of nearby stations for estimating ET_o is governed by the same considerations as for radiation data. Wind speed data of a nearby station can be used when making monthly estimates of ET_o. Computation of daily ET_o can be justified when it is utilized as a sum or average over a several day period such as a week or longer.

Minimum Data Required for Estimating ET_o

There are procedures for estimating the values of solar radiation, vapour pressure and wind data, when any one of them is not available. However, there is no satisfactory method of estimating ET_o without information on the maximum and minimum daily temperature data. Hence, measurements of maximum and minimum temperatures are basic requirements of a weather station. To apply for ET_o evaluation, Equation 10.20 (Hargreaves' ET_o equation) can be used to make reasonable estimates of ET_o when information on daily values of T_{max} and T_{min} only are available.

Example 10.7. *Estimate the value of ET_o for the month of March based on the following monthly average climatic data for the 25 year period (1957 to 1981) at the Regional Research Station of Kerala Agricultural University at Pattambi, District Palakkad, Kerala, India. The Station is located at 10°48 N′ latitude and 76°12′ E longitude. The altitude of the Station is 25 m above mean sea level. The values of daily maximum and minimum temperatures for each month during the period are available at the weather station attached to the Regional Research Station, Pattambi:*

Values & Units

Daily (average) maximum temperature = 36.470 °C

Daily (average) minimum temperature = 23.220 °C

Maximum relative humidity (RH_{max}) = 87.480 %

Minimum related humidity (RH_{min}) = 39.420 %

Wind speed measured at 3 metre height = 4.520 km/h

Actual hours of sunshine (n) = 9.830 hours

Wind speed at 3 metre height in metre/second (U_2) = 1.256 m/s

Solution:

Wind speed at 2 metre height

$$U_2 = U_Z\left(\frac{4.87}{\ln(67.8\times z - 5.42)}\right) \quad \text{(Eq. 10.29)}$$

z is the height at which wind speed is measured

$$= 1.256\left(\frac{4.87}{\ln(67.8\times 3 - 5.42)}\right) = 1.157 \text{ m/s}$$

$$P = 101.3\left[\frac{293 - 0.0065\,Z}{293}\right]^{5.26} \quad \text{(Eq. 10.30)}$$

Z is the elevation above mean sea level

$$= 101.3\left[\frac{293 - 0.0065\times 25}{293}\right]^{5.26} = 101.005 \text{ kPa}$$

$$T_{mean} = (T_{max} + T_{min})/2$$

$$= (36.470 + 23.220)/2 = 29.845\ °C$$

$$\Delta = 4098\left[\frac{0.6108\exp\left(\frac{17.27T}{T+237.3}\right)}{(T+237.3)^2}\right] \quad \text{(Eq. 10.31)}$$

$$= 4098\left[\frac{0.6108\exp\left(\frac{17.27\times 29.845}{29.845+237.3}\right)}{(29.845+237.3)^2}\right]$$

$$= 0.241 \text{ kPa/°C}$$

$$\gamma = 0.665\times 10^{-3}\times P \quad \text{(Eq. 10.32)}$$

$$= 0.665\times 10^{-3}\times 101.005 = 0.067 \text{ kPa/°C}$$

$$1 + 0.34U_2 = 1 + 0.34\times 1.157 = 1.393$$

$$\frac{\Delta}{[\Delta + \gamma(1 + 0.34\,U_2)]} = \frac{0.241}{(0.241 + 0.067\times 1.393)} = 0.721$$

or

$$\frac{\gamma}{[\Delta + \gamma(1 + 0.34\,U_2)]} = \frac{0.067}{(0.241 + 0.067\times 1.393)} = 0.200$$

Thus

$$\frac{900}{(T_{mean} + 273)}\times U_2 = \frac{900}{(29.845 + 273)}\times 1.157 = 3.438$$

$$e°T_{(max)} = 0.6108\exp\left[\frac{17.27\times T_{max}}{T_{max} + 237.3}\right] \quad \text{(Eq. 10.33)}$$

$$= 0.6108\exp\left[\frac{17.27\times 36.47}{36.47 + 237.3}\right] = 6.096 \text{ kPa}$$

$$e°T_{((min)} = 0.6108\exp\left[\frac{17.27\times T_{min}}{T_{min} + 237.3}\right]$$

$$= 0.6108\exp\left[\frac{17.27\times 23.22}{23.22 + 237.3}\right] = 2.847 \text{ kPa}$$

$$e_s = \frac{e°T_{max} + e°\,T_{min}}{2}$$

$$= \frac{6.096 + 2.847}{2} \quad \text{(Eq. 10.34)}$$

$$= 4.472 \text{ kPa}$$

$$e_a = \frac{e°T_{min}\times\frac{RH_{max}}{100} + e°\,T_{max}\times\frac{RH_{min}}{100}}{2} \quad \text{(Eq. 10.35)}$$

$$= \frac{\frac{2.87\times 87.48}{100} + \frac{6.096\times 39.42}{100}}{2} = 2.447 \text{ kPa}$$

Vapour pressure deficit $= e_s - e_a$

$$= 4.472 - 2.447 = 2.025 \text{ kPa}$$

J for day 15, month March (Eq. 10.39)

= 74 days

$$R_a = \frac{24\,(60)}{\pi}G_{Sc}\,dr\,(\omega_s\sin\phi\sin\,\delta + \cos\phi\cos\,\delta\sin\omega_s) \quad \text{(Eq. 10.37)}$$

$$dr = 1 + 0.033\cos\left[\frac{2\pi}{365}J\right] \quad \text{(Eq. 10.39)}$$

$$= 1 + 0.033\cos\left[\frac{2\pi}{365}\times 74\right] = 1.010 \text{ radians}$$

$$\delta = 0.409\sin\left[\frac{2\pi}{365}\times J - 1.39\right] \quad \text{(Eq. 10.40)}$$

$$= 0.409\sin\left[\frac{2\pi}{365}\times 74 - 1.39\right] = -\,0.047 \text{ radians}$$

ϕ = latitude (10°48′) in radians = 0.188 radians

$$\omega_s = \text{arc}\cos(-\tan\phi\times\tan\delta) \quad \text{(Eq. 10.41)}$$

$$= \cos^{-1}(-\tan 0.188\times\tan(-\,0.047)) = 1.562 \text{ radians}$$

Day light hours

$$N = \frac{24}{\pi}\,\omega_s = \frac{24}{\pi}\times 1.562 \quad \text{(Eq. 10.42)}$$

$$= 11.933 \text{ hours}$$

$$\sin\phi\times\sin\delta = \sin 0.188\times\sin - 0.047$$

$$= -\,0.009$$

$$\cos\phi\times\cos\delta = \cos 0.188\times\cos - 0.047$$

$$= 0.981$$

G_{Sc} (Solar constant) = 0.082 $MJm^{-2}d^{-1}$

$$R_a = \frac{24(60)}{\pi}\times 0.082\times 1.01\times[1.562\times - 0.009 + 0.981\times\sin 1.562] \quad \text{(Eq. 10.37)}$$

$$= 36.706 \text{ MJm}^{-2}\text{d}^{-1}$$

$$R_s = (a_s + b_s)\,\frac{n}{N} \times R_a \quad \text{(Eq. 10.43)}$$

$$= \left(0.25 + 0.5 \times \frac{9.83}{11.933}\right) \times 36.706$$

$= 24.295\ MJm^{-2}d^{-1}$

Net solar radiation, $R_{ns} = (1 - \alpha)\,R_s$ (Eq. 10.46)

$= (1 - 0.23)\ 24.295$

$= 18.707\ MJm^{-2}d^{-1}$

$R_{so} = (0.75 + 2 \times 10^{-5} \times Z)\ R_a$ (Eq. 10.45)

$= (0.75 + 2 \times 10^{-5} \times 25)\ 36.706$

$= 27.548\ MJm^{-2}d^{-1}$

$$R_{nl} = \sigma\left[\frac{T_{\max},K^4 + T_{\min},K^4}{2}\right](0.34 - 0.14\sqrt{e_a})$$

$$\times \left(1.35\,\frac{R_s}{R_{so}} - 0.35\right) \quad \text{(Eq. 10.47)}$$

σ = Stefan-Boltzmann constant

$= 4.903 \times 10^{-9}\ MJ\ K^{-4}\ m^{-2}\ day^{-1}$

$(K = °C + 273.16)$

$\sigma T_{\max}K^4 = 4.903 \times 10^{-9}$

$= 45.064\ MJm^{-2}d^{-1}\ (36.47 + 273.16)^4$

$\sigma T_{\min}K^4 = 4.903 \times 10^{-9}\ (23.22 + 273.16)^4$

$= 37.382\ MJ\ m^{-2}d^{-1}$

$$0.34 - 0.14\sqrt{e_a} = 0.34 - 0.14\sqrt{2.447} = 0.121$$

$$\left(1.35.\frac{R_s}{R_{so}} - 0.35\right) = 1.35 \times \frac{24.295}{27.548} - 0.35 = 0.841$$

$$R_{nl} = \left(\frac{45.064 + 37.832}{2}\right) \times 0.121 \times 0.841$$

$= 4.218\ MJm^{-2}d^{-1}$

$R_n = R_{ns} - R_{nl} = 18.707 - 4.218$

$= 14.489\ MJm^{-2}d^{-1}$ (Eq. 10.48)

$T_{\text{month}}\ i - 1$ = mean temperature of the previous month (i.e., February) = 28.205 °C

Soil heat flux, G:

$G = 0.14\ (T_{\text{month}}\ i - T_{\text{month}}\ i - 1)$ (Eq. 10.28)

$= 0.14\ (29.845 - 28.205)$

$= 0.230\ MJm^{-2}d^{-1}$

$R_n - G = 14.489 - 0.23 = 14.259\ MJm^{-2}d^{-1}$

Grass reference evapotranspiration, ET_o

$$ET_o = \frac{0.408\,\Delta\,(R_n - G) + \gamma\,\dfrac{900}{(T+273)}\,U_2(e_s - e_a)}{\Delta + \gamma\,(1 + 0.34\,U_2)} \quad \text{(Eq. 10.24)}$$

$$0.408\,(R_n - G) \times \frac{\Delta}{(\Delta \times \gamma)(1 + 0.34\,U_2)}$$

$= 0.408 \times 14.259 \times 0.721 = 4.195$ mm/day

$$\frac{900}{(T+273)} \times U_2 \times (e_s - e_a) \times \frac{\gamma}{(\Delta + \gamma)(1 + 0.34\,U_2)}$$

$$= \frac{900}{(29.845 + 273)} \times 1.157 \times 2.025 \times 0.200$$

$= 3.438 \times 2.025 \times 0.200 = 1.392$ mm/day

$ET_o = 4.195 + 1.392 = 5.587$ mm/day

Determination of ET_o with missing data:

Example 10.8. *Given the monthly average daily maximum and minimum temperature of March at Delhi 28°35′ N latitude, altitude 216 m. No other climatic data are available.*

Values & Units

Monthly average daily maximum temperature $T_{\max}$ = 28.610 °C

Monthly average daily minimum temperature $T_{\min}$ = 13.970 °C

Solution:

Estimation of wind speed:

2 m/s is used as an estimate = 2.000 m/s

Due to relatively small crop height of 0.12 m of the reference crop and the appearance of U_2 on both numerator and denominator of FAO Pennman Montieth equation ET_o is not highly sensitive to normal range of wind speed

$$T_{\text{mean}} = \frac{T_{\max} + T_{\min}}{2}$$

$$= \frac{28.610 + 13.970}{2} = 21.290°C$$

$$\Delta = 4098\left[\frac{0.6108\exp\left(\dfrac{17.27\,T}{T + 237.3}\right)}{(T + 237.3)^2}\right]$$

$$= 4098\left[\frac{0.6108\exp\left(\dfrac{17.27 \times 21.29}{21.29 + 237.3}\right)}{(21.29 + 237.3)^2}\right]$$

$= 0.155$ kPa/°C

Altitude = 216.00 m

$$P = 101.3\left[\frac{293 - 0.0065 \times Z}{293}\right]^{5.26}$$

$$= 101.3\left(\frac{293 - 0.0065 \times 216}{293}\right)^{5.26} = 98.773 \text{ kPa}$$

$\gamma = 0.665 \times 10^{-3} \times P = 0.665 \times 10^{-3} \times 98.773$

$= 0.066$ kPa/°C

$\Delta \div [\Delta + \gamma\ (1 + 0.34\ U_2)]$

$= 0.155 \div [0.155 + 0.066 \ (1 + 0.34 \times 2)]$

$= 0.583$

$\gamma \div [\Delta + \gamma \ (1 + 0.34 \ U_2]$

$= 0.066 \div [0.155 + 0.066 \ (1 + 0.34 \times 2)]$

$= 0.248$

$$\frac{900}{T_{\text{mean}} + 273} \times U_2 = \left(\frac{900}{(21.29 + 273)}\right) \times 2$$

$= 6.116$

Estimation of humidity data

Assume $T_{dew} \approx T_{min} = 13.970$ °C

$e_a = e°T_{(dew)}$

$$= 0.6108 \exp\left[\frac{17.27 \times 13.97}{13.97 + 237.3}\right] = 1.595 \text{ kPa}$$

$$e°T_{(max)} = 0.6108 \exp\left(\frac{17.27 \times 28.61}{(28.61 + 237.3)}\right)$$

$= 3.916$ kPa

$$e°T_{(min)} = 0.6108 \exp\left(\frac{17.27 \times 13.97}{(13.97 + 237.3)}\right) = 1.595 \text{ kPa}$$

$e_s = (e°T_{max} + e° \ T_{min}) \div 2$

$= (3.916 + 1.595) \div 2 = 2.756$ kPa

$e_s - e_a = 2.756 - 1.595 = 1.161$ kPa

$$R_{H\ min} = 100 \times \frac{e_a}{e° T_{max}} = \frac{100 \times 1.595}{3.916} = 40.730 \%$$

$$R_{H\ max} = \frac{100 \times e_a}{e°T_{min}} = 100 \times \frac{1.595}{1.595} = 100.000 \%$$

$R_{Hmean} = (R_{H\ max} + R_{H\ min}) \div 2 = 70.365 \%$

(R_a can be taken from climatological table or computed)

$= 1.967$ MJm^{-2}d^{-1}

$$R_s = k \ R_s \ \sqrt{T_{max} - T_{min}} \ \times R_a$$

$$= 0.16 \ \sqrt{28.61 - 13.97} \ \times 31.967$$

$= 19.570$ MJm^{-2}d^{-1}

$R_{so} = (0.75 + 2 \times 10^{-5} \times Z) \ R_a$

$= (0.75 + 2 \times 10^{-5} \times 216)31.967$

$= 24.113$ MJm^{-2}d^{-1}

$$\frac{R_s}{R_{so}} = \frac{19.570}{24.113} = 0.812$$

$R_{ns} = (1 - 0.23) \ R_s = 0.77 \times 19.570$

$= 15.069$ MJm^{-2}d^{-1}

$\sigma T_{max} \ K^4 = 4.903 \times 10^{-9} \ (28.61 + 237.16)^4$

$= 40.660$ MJm^{-2}d^{-1}

$\sigma T_{min} K^4 = 4.903 \times 10^{-9} \ (13.97 + 273.16)^4$

$= 33.325$ MJm^{-2} d^{-1}

$$R_{nl} = \left[\frac{\sigma T_{max} \ K^4 + \sigma \ T_{min} \ K^4}{2}\right](0.34 - 0.14 \ \sqrt{e_a})$$

$$\left[1.35 \times \frac{R_s}{R_{so}} - 0.35\right]$$

$$= \left[\frac{40.660 + 33.325}{2}\right][0.34 - 0.14\sqrt{1.595}]$$

(Eq. 10.47)

$[1.35 \times 0.812 - 0.35] = 36.993 \times 0.163 \times 0.746$

$= 4.498$ MJm^{-2} d^{-1}

$R_n = R_{ns} - R_{nl} = 15.069 - 4.498 = 10.571$ MJm^{-2} d^{-1}

Assume $G = 0$

$R_n - G = 10.571 - 0 = 10.571$ MJm^{-2} d^{-1}

$0.408 \ (R_n - G) = 0.408 \times 10.571 = 4.313$ mm/day

Grass Reference Evapotranspiration:

$$0.408 \ (R_n - G) \times \frac{\Delta}{\Delta + \gamma \ (1 + 0.34 \ U_2)} = 4.313 \times 0.583$$

$= 2.514$ mm/day

$$\frac{900}{(T + 273)} \times U_2 \times (e_s - e_a) \times \frac{\gamma}{\Delta + \gamma \ (1 + 0.34 \ u_2)}$$

$6.116 \times 1.161 \times 0.248 = 1.761$ mm/day

$ET_o = 2.514 + 1.761 = 4.275$ mm/day

Alternate method of computation based on Hargreaves' equation:

$$ET_o = 0.0023 \ (T_{\text{mean}} + 17.8) \ (T_{\text{max}} - T_{\text{min}})^{0.5} \times R_a \quad \text{(Eq. 10.20)}$$

In the Hargreaves equation ET_o and R_a are expressed in mm/day. The unit of R_a computed using the formula suggested in FAOs Penman-Monteith method is in *MJm^{-2}d^{-1}*. To convert this into mm/day the value of R_a obtained by this method is divided by the conversion factor 2.45.

$$ET_o = 0.0023 \ (T_{\text{mean}} + 17.8) \ (T_{\text{max}} - T_{\text{min}})^{0.5} \times \frac{R_a}{2.45}$$

$$ET_o = 0.0023 \ (21.29 + 17.8) \ (28.61 - 13.97)^{0.5} \times \frac{31.967}{2.45}$$

$= 4.488$ mm/day

Determination of ET$_o$ with missing data:

Example 10.9. *Given the monthly average daily maximum and minimum temperature for Pattambi 10° 48′ N altitude 25 m (Example 10.7) and no other climatic data are given.*

Solution:

Values & Units

Monthly average daily
maximum temperature (T_{max}) for March = 36.470 °C

Monthly average daily
minimum temperature (T_{min}) for March = 23.220 °C

$T_{\text{mean}} = (36.47 + 23.22) \div 2 = 29.845$ °C

Wind speed (assumed) = 2.000 m/s

$$\Delta = 4098 \left[0.6108 \exp\left(\frac{17.27 \ T}{T + 237.3}\right)\right] \div (T + 237.3)^2 \quad \text{(Eq. 10.31)}$$

$$= 4098 \left[0.6108 \exp\left(\frac{17.27 \times 29.845}{29.845 + 237.3}\right)\right]$$

$$\div\ (29.845 + 237.3)^2 = 0.241 \text{ kPa/°C}$$

$$P = 101.3 \left(\frac{293 - 0.0065\,Z}{293}\right)^{5.26} \quad \text{(Eq. 10.30)}$$

$$= 101.3 \left(\frac{293 - 0.0065 \times 25}{293}\right)^{5.26} = 101.005 \text{ kPa}$$

$$\gamma = 0.665 \times 10^{-3} \times P = 0.665 \times 10^{-3} \times 101.005$$

$$= 0.067 \text{ kPa/°C}$$

$$\Delta \div [\Delta + \gamma\,(1 + 0.34\,U_2)] = 0.241 \div [(0.241 + 0.067\,(1 + 0.34 \times 2)]$$

$$= 0.682 - \gamma \div [\Delta + \gamma\,(1 + 0.34\,U_2)]$$

$$= 0.067 \div [0.241 + 0.067\ \ (1 + 0.34 \times 2)]$$

$$= 0.190 \quad \frac{900}{(T + 273)} \times U_2 = \frac{900}{(29.845 + 273)} \times 2 = 5.944$$

Estimation of humidity data:

Assume $T_{dew} \approx T_{min} = 23.220$ °C

$$e_a = e°T_{\text{dew}} = 0.6108 \exp\left[\frac{17.27 \times 23.22}{23.22 + 237.3}\right] \quad \text{(Eq. 10.48a)}$$

$$= 2.847 \text{ kPa}$$

$$e°T_{\max} = 0.6108 \exp\left[\frac{17.27 \times 36.47}{36.47 + 237.3}\right] = 6.096 \text{ kPa}$$

$$e°T_{\min} = 0.6108 \exp\left[\frac{17.27 \times 23.22}{23.22 + 237.3}\right] = 2.847 \text{ kPa}$$

$$e_s = (e°T_{\max} + e°T_{\min}) \div 2$$

$$= (6.096 + 2.847) \div 2 = 4.472 \text{ kPa}$$

$$e_s - e_a = 4.472 - 2.847 = 1.625 \text{ kPa}$$

$$R_{H\,\max} = \frac{100\,e_a}{e°T_{\min}} = \frac{100 \times 2.847}{2.847} = 100.000\ \%$$

$$R_{H\,\min} = \frac{100\,e_a}{e°T_{\max}} = \frac{100 \times 2.847}{6.096} = 46.703\ \%$$

R_a can be taken from the table or calculated = 36.706 $MJm^{-2}d^{-1}$

Hargreaves Radiation formula: (Eq. 10.18)

$$R_s = kR_s \sqrt{(T_{\max} - T_{\min})} \times R$$

$$0.16 \sqrt{(36.47 - 23.22)} \times 36.706 = 21.378 \text{ MJm}^{-2}\text{d}^{-1}$$

$$R_{so} = [0.75 + 2 \times 10^{-5}\,Z]\,R_a$$

$$= (0.75 + 2 \times 10^{-5} \times 25) \times 36.706$$

$$= 27.548 \text{ MJm}^{-2}\text{d}^{-1}$$

$$\frac{R_s}{R_{so}} = \frac{21.378}{27.548} = 0.776$$

$$R_{ns} = (1 - 0.23)\,R_s = (1 - 0.23)\ 21.378$$

$$= 16.461 \text{ MJm}^{-2}\text{d}^{-1}$$

$$R_{nl} = \sigma \left[\frac{T_{\max}\ K^4 + T_{\min}\ K^4}{2}\right] (0.34 - 14\sqrt{e_a})$$

$$\left(1.35\ \frac{R_s}{R_{so}}\ 0.35\right)$$

$$\sigma T_{\max}\,K^4 = 4.903 \times 10^{-9} \times (36.47 + 273.16)^4$$

$$= 45.064 \text{ MJm}^{-2}\text{d}^{-1}$$

$$\sigma T_{\min} K^4 = 4.903 \times 10^{-9} \times\ (23.22 + 273.16)^4$$

$$= 37.832 \text{ MJm}^{-2}\text{d}^{-1}$$

$$0.34 - 0.14\sqrt{l_a} = 0.34 - 0.14\ \sqrt{2.847} = 0.104$$

$$1.35 \times \frac{R_s}{R_{so}} - 0.35 = 1.35 \times 0.776 - 0.35 = 0.698$$

$$\left(\frac{45.064 + 37.832}{2}\right) \times 0.104 \times 0.698 = 3.009 \text{ MJm}^{-2}\text{d}^{-1}$$

$$R_n = R_{ns} - R_{nl} = 16.461 - 3.009$$

$$= 13.452 \text{ MJm}^{-2}\text{d}^{-1}$$

$$G = 0 \text{ (Assume)}$$

$$R_n - G = 13.452 - 0 = 13.452\ \text{MJm}^{-2}\text{d}^{-1}$$

$$0.408\,(R_n - G) = 0.408 \times 13.452 = 5.488 \text{ mm/day}$$

Grass reference evapotranspiration:

$$0.408\,(R_n - G)\,\Delta \div [\Delta + \gamma\,(1 + 0.34\,U_2)\ 5.488 \times 0.682$$

$$= 3.743 \text{ mm/day}$$

$$\frac{900}{(T + 273)} \times U_2\,(e_s - e_a) \times \frac{\gamma}{(\Delta + \gamma)(1 + 0.34\,U_2)}$$

$$5.944 \times 1.625 \times 0.190 = 1.835 \text{ mm/day}$$

$$ET_o = 3.743 + 1.835 = 5.578 \text{ mm/day}$$

Alternative method of computation based on Hargreaves equation:

$$ET_o = 0.023\,(T_{\text{mean}} + 17.8)\,(T_{\max} - T_{\min})^{0.5} \times R_a \quad \text{(Eq. 10.20)}$$

In the Hargreaves equation, ET_o and R_a are expressed in mm/day. The unit of R_a in FAO Penman-Montieth method is in MJ $m^{-2}d^{-1}$. To convert this into mm/day the value of the R_a obtained by this method is divided by the conversion factor, 2.45.

$$ET_o = 0.0023\,(T_{\text{mean}} + 17.8)(T_{\max} - T_{\min})^{0.5} \times \frac{R_a}{2.45} \quad \text{(Eq. 10.20)}$$

$$ET_o\ = 0.0023\,(29.845 + 17.8)$$

$$(36.47 - 23.22)^{0.5} \times \frac{36.706}{2.45} = 5.976 \text{ mm/day}$$

Example 10.10. *Determination of ET_o for March from pan evaporation data of Pattambi, Kerala (Examples 10.8 and 10.9)*

Daily mean pan evaporation for the month of March =7.01 mm

Average relative humidity = 63.45 %

Average wind speed at 2 m height (U_2) = 1.157 m/s

At Pattambi, evaporation pan is placed in an area surrounded by green crops. The windward side distance of green crops is about 100 m. In tables for such pan sitings, relative humidity and wind speed, the recommended K_p values is 0.8.

Solution:

$$ET_o = K_p \times E_{pan} \quad \text{(Eq. 10.5)}$$
$$= 0.8 \times 7.01$$
$$= 5.61 \text{ mm}$$

Estimating Crop Water Requirement:

Example 10.11. *Determination of monthly average K_c values and Monthly water requirements of a groundnut crop grown in the rice fallows at Pattambi, Kerala, India (Examples 10.7, 10.9)*

Solution: Date of sowing = 1 January

Date of harvest = 20 April

Duration = 110 days

Stage	Length of growth stage	K_c value	Remarks
Initial stage	25 days	0.5	Assumption
Crop development	30 days	0.5 to 1.15*	*Values suggested in FAO Irrigation and Drainage paper 56 (1998)
Mid-season stage	30 days	1.15*	
Late season stage	25 days	1.15 to 0.6*	

K_c values for January

1–25 January for 25 days, K_c value (Initial stage) = 0.5

From 26 January K_c value increase linearly from 0.50 to 1.15 in 30 days during the crop development stage

$$\text{Daily rate of increase} = \frac{1.15 - 0.5}{30} = 0.0217$$

Cumulative increase for 6 days from 26th to 31st January

$$= 6 \times 0.0217 = 0.13$$

K_c value on 31st January = 0.5 + 0.13 = 0.63

Average K_c value for 6 days

$$= \frac{0.5 + 0.63}{2} = 0.565$$

Average K_c value for January

$$= \frac{25 \times 0.5 + 6 \times 0.565}{31} = 0.51$$

K_c value for February

K_c value at the beginning of February = 0.63

K_c value on 24th February = 1.15

(End of crop development stage)

K_c value from 25th to 28th February (Mid-Season) = 1.15

Average K_c value

$$\text{for February} = \frac{\left(\frac{0.63 + 1.15}{2}\right) 24 + 1.15 \times 4}{28} = 0.93$$

K_c value for March

K_c value from 1st to 26th March (March 26 days) - mid-season stage = 1.15

From 27th March to 20th April (25 days) K_c value decreases from 1.15 to 0.6 in the late season stage.

The rate of decrease/day

$$= \frac{1.15 - 0.6}{25} = 0.022$$

K_c value on 31st March

$$= 1.15 - 0.022 \times 5 = 1.04$$

Average value from 26th March to 31st March (5 days)

$$= \frac{1.15 + 1.04}{2} = 1.095$$

Average value for March

$$= \frac{1.15 \times 26 + 1.095 \times 5}{31} = 1.14$$

K_c value for April

K_c value at the beginning of April = 1.04

K_c value on 20th April = 0.6

Average value for 1 – 20 April

$$= \frac{1.04 + 0.6}{2} = 0.82$$

Month	*January 31 days*	*February 28 days*	*March 31 days*	*April 20 days*
K_c value	0.51	0.93	1.14	0.82
*ET_o mm/day	4.60	5.00	5.59	5.20
ET crop (mm/day)	2.35	4.65	6.37	4.26
ET crop Total for the month mm	72.85	130.20	197.47	85.20

*ET_o values are computed using the values for Pattambi, Kerala, India (Example 10.7)

Total water requirements of the crop = 485.72 mm

Irrigation requirements for the above months are obtained by subtracting monthly effective rainfall from ET_o crop for the respective months.

CROP EVAPOTRANSPIRATION (ET_C)

The procedures for estimating reference crop evapotranspiration. ET_o were explained in the earlier sections.

Equation 10.11, $ET_c = K_c\, ET_o$. Alternatively, the ratio ET_c to ET_o is K_c.

i.e., $$K_c = ET_c / ET_o \qquad \text{...(10.48)}$$

The concept of crop evapotranspiration ET_c and its application under standard and non-standard conditions were introduced in the initial part of this chapter. The influence of various weather conditions on evapotranspiration are included in the section on estimating ET_o. The value of K_c is influenced by the specific characteristics of the crop at various stages of its growth. The following are the basic characteristics that distinguish a crop from reference grass:

(*i*) **Crop height.** The crop height influences the aerodynamic resistance term R_a, of the FAO Penman-Monteith equation and the turbulent transfer of vapour from the crop into the atmosphere.

(*ii*) **Albedo (reflectance) of the crop-soil surface.** The albedo is influenced by the fraction of ground covered by vegetation and by the soil surface wetness. The albedo of the crop-soil surface influences the net radiation of the surface, R_n, which is the main source of energy exchange for the evaporation process.

(*iii*) **Canopy resistance.** The resistance of the crop to vapour transfer is affected by leaf area (number of stomata), leaf age and condition, and the degree of stomatal control. The canopy resistance influences the surface resistance, R_s.

(*iv*) **Evaporation from soil, essentially exposed soil.** The wetness of the soil and the extent of ground cover by vegetation influence the surface resistance *r*. Vapour transfer from a wet soil is high, especially from a bare ground.

The K_c value in Eq. 10.48b, predicts ET_c under standard conditions. This represents the conditions where no limitations are placed on crop growth or evapotranspiration due to water shortage, crop density, disease, weed, insect or salinity. The ET_c predicted by K_c is adjusted, if necessary, to non-standard conditions, $ET_{c\ adj}$ when limiting conditions exist.

Factors Influencing Crop Coefficient, K_c

The value of crop coefficient K_c integrates the effect of the characteristics that distinguish a field crop from reference grass. Different crops will have different K_c coefficients. The changing crop characteristics under various stages of growth also influence K_c values substantially. Conditions affecting soil evaporation will also influence K_c values.

Crop type. As crops vary in albedo, plant height aerodynamic properties and the properties of leaf and stomata, the evapotranspiration from well watered crop differs from ET_o. The close spacing of plants and taller canopy heights of many crops cause them to have K_c values more than the reference crop. Typical K_c values of selected crops are given in Table 10.7. They are the common values under standard climatic conditions which is defined as sub-humid climate with average daytime hours, relative humidity less than 45% and having calm to low wind.

Climate. Variations in wind change the aerodynamic resistance of crops and their K_c values, especially for those crops that are much taller than the hypothetical grass reference. The effect of the change in aerodynamic properties between the grass reference surface and crops is not only crop specific but also varies with the climatic conditions and crop height. Because aerodynamic properties are greater for many crops, as compared to the grass reference, the ratio of ET_c to ET_o (*i.e.*, K_c) for many crops increases as wind speed increases and as relative humidity decreases. More arid climates and higher wind speed will have higher values for K_c. More humid climates and conditions of lower wind speed will lower K_c values.

3. **Soil evaporation.** Differences in soil evaporation and crop transpiration and the reference grass are integrated within the crop coefficient. The value of K_c for full-cover crops primarily reflects differences in transpiration, as the contribution of soil evaporation is small. After rainfall or irrigation, the effect of evaporation is significant when the crop is small. For low-cover conditions, K_c is influenced mainly by the frequency with which the soil surface is wetted. Where the soil is wet the evaporation from its surface is considerable and K_c may exceed 1. When the soil surface is dry, evaporation is restricted and K_c will be small.

4. **Crop growth stages.** The stage of growth of a crop profoundly influence K_c values. The crop growing period can be divided into four distinct growth stages: (*i*) *the initial stage* (from sowing to about 10% ground cover), (*ii*) *crop development stage* (from 10% to about 70% ground cover), (*iii*) *mid-season stage* (including flowering and grain setting and yield formation stage), (*iv*) *late season stage* (including ripening and harvest). In general, out of the four growth stages (Fig. 10.3) the mid-season stage is the most sensitive to water shortages, as it is the period of the highest crop water need. Water shortage during the mid-season will reduce crop yields substantially.

Table 10.7 presents the commonly observed values of the duration of different growth stages in crops at various experiment stations in different climatic/ geographic regions along with the months of planting. The data are indicative of the general conditions but variations in agro-climatic conditions will result in similar variations in K_c values. Hence, the data presented in Table 10.7 could be used only under situations when more reliable data are not available.

Table 10.7. Length of crop development stages of selected crops during different stages of growth in some of the climatic regions, days

Crops	*Region*	*Initial*	*Develop. stage*	*Mid-stage*	*Late stage*	*Total*	*Planting Month*
Vegetable crops							
1. Carrots	(*i*) Arid	20	30	30-50	20	100	Oct./Jan.
	(*ii*) Mediterranean	30	40	60	20	150	Feb./Mar.
2. Cauliflower	Cali. Desert, USA	35	50	40	15	140	Sep.
3. Cabbage	Cali. Desert, USA	40	60	50	15	165	Sep.
4. Lettuce	(*i*) Mediterranean	20	30	15	10	75	April
	–do–	30	40	25	10	105	Nov./Jan.
	–do–	35	50	45	10	140	Feb.
	(*ii*) Arid Region	25	35	30	10	100	Oct./Nov.
5. Onion, dry	(*i*) Mediterranean	15	25	70	40	150	April
	(*ii*) Arid (Cali., USA)	20	35	110	45	210	Oct./Jan.
Onion, green	(*i*) Mediterranean	25	30	10	5	70	April/May
	(*ii*) Arid	20	45	20	10	95	October
6. Spinach	(*i*) Mediterranean	20	20	15/25	5	60/70	Apr./Sep./Oct.
	(*ii*) Arid	20	30	40	10	100	November
7. Radish	(*i*) Med., Europe	5	10	15	5	35	Mar./Apr.
	(*ii*) Arid	10	10	15	5	40	Winter seas.
8. Egg plant	(*i*) Mediterranean	30	45	40	25	140	May/June
	(*ii*) Arid	30	40	40	20	130	October
9. Tomato	(*i*) Arid	30	40	40	25	135	January
	–do–	35	45	70	30	180	Oct./Nov.
	(*ii*) Mediterranean	30	40	45	30	145	Apr./May
10. Cucumber	(*i*) Arid	20	30	40	15	105	June/Aug.
	–uo–	25	35	50	20	130	Nov./Feb.
Roots and tubers							
11. Casava 1st yr.	Tropic regions	20	40	90	60	210	Rainy seas.
2nd yr.		150	40	110	60	360	
12. Potato	(*i*) Semi-arid	25	30	30/45	30	115/130	Jan./Nov.
	(*ii*) Continental climate	25	30	45	30	130	May
	(*iii*) Europe	30	35	50	30	145	April
13. Sweet potato	(*i*) Mediterranean	20	30	60	40	150	April
	(*ii*) Tropical reg.	15	30	50	30	125	Rainy seas.
14. Sugarbeet	(*i*) Mediterranean	25	35	50	50	160	May
	–do–	45	75	80	30	230	Nov.
	(*ii*) Arid	35	60	70	40	205	Nov.
	(*iii*) Cali., USA	30	45	90	15	180	March
	–do–	25	30	90	10	155	June
Legumes							
15. Beans, green	(*i*) Mediterranean, Califor.	20	30	30	10	90	Feb./Mar.
	(*ii*) Cali., Egypt, Lebanon	15	25	25	10	75	Aug./Sep.
16. Beans, dry	(*i*) Continental climate	20	30	40	20	110	May/June
	(*ii*) Pakistan, Cali., USA	15	25	35	20	95	June
17. Green gram, cowpea	(*i*) Mediterranean	20	30	30	20	110	March

(*Contd ...*)

Crops		Region	Initial	Develop. stage	Mid-stage	Late stage	Total	Planting Month
18. Groundnut	(i)	West Africa	25	35	45	25	130	Dry season
	(ii)	High Latitudes	35	35	35	35	140	May
	(iii)	Mediterranean	35	45	35	25	140	May/June
	(iv)	South-west India (Kerala)	25	30	30	25	110	January
19. Lentil	(i)	Europe	20	30	60	40	150	April
	(ii)	Arid	25	35	70	40	170	Oct./Nov.
20. Peas	(i)	Europe	15	25	35	15	90	May
	(ii)	Mediterranean	20	30	35	15	100	March/April
21. Soyabeans	(i)	Tropics	15	15	40	15	85	December
	(ii)	Central USA	20	30/35	60	25	140	May
	(iii)	Japan	20	25	75	30	150	June
Fibre crop								
22. Cotton	(i)	Egypt, Pak., Cali, USA	30	50	60	55	195	March/May
	(ii)	Texas, USA	30	50	55	45	180	April
Oil crops								
23. Castor beans	(i)	Semi-arid	25	40	65	50	180	March
	(ii)	Indonesia	20	40	50	25	135	Nov.
24. Safflower	(i)	Cali., USA	20	35	45	25	125	April
	(ii)	High Latitudes	25	35	55	30	145	March
	(iii)	Arid region	35	55	60	40	190	Oct./Nov.
25. Sesame		China	20	30	40	20	100	June
26. Sunflower		Medit., Cali., USA	25	35	45	25	130	April/May
Cereals								
27. Wheat/	(i)	Cent. India	15	25	50	30	120	November
Barley/Oats	(ii)	East Africa	15	30	65	40	150	July
Winter wheat	(iii)	Cali., USA	20	60	70	30	180	December
–do–	(iv)	Mediterranean	30	140	40	30	240	November
28. Maize, grain	(i)	East Africa	30	50	60	40	180	April
	(ii)	Arid regions	25	40	45	30	140	Dec./Jan.
	(iii)	Nigeria (humid)	20	35	40	30	125	June
	(iv)	India (dry, humid)	20	35	40	30	125	October
29. Millet	(i)	Pakistan	15	25	40	25	105	June
	(ii)	Central USA	20	30	55	35	140	April
Sugarcane								
30. Sugarcane,	(i)	Low Latitudes	35	60	190	120	405	
virgin	(ii)	Tropics	50	70	220	140	480	
	(iii)	Hawaii, USA	75	105	330	210	720	
31. Sugarcane,	(i)	Low Latitudes	25	70	135	50	280	
ratoon	(ii)	Tropics	30	50	180	60	320	
	(iii)	Hawaii, USA	35	105	210	70	420	
Tropical fruits and trees								
32. Banana, 1st yr		Mediterranean	120	90	120	60	390	March
2nd yr			120	60	180	5	365	February
33. Pineapple		Hawaii, USA	60	120	600	10	790	

(Contd ...)

Crops	Region	Initial	Develop. stage	Mid-stage	Late stage	Total	Planting Month
34. Grapes	(*i*) Low Latitudes	20	40	120	60	240	April
	(*ii*) California, USA	20	50	75	60	205	March
	(*iii*) High Latitudes	20	50	90	20	180	May
	(*iv*) Mid-Latitudes (wine)	30	60	40	80	210	April
Fruit trees							
35. Citrus	Mediterranean	60	90	120	95	365	January
36. Olives	Mediterranean	30	90	60	90	270	March
37. Walnuts	Utah, USA	20	10	130	30	190	April

The 'region' shown in the table indicates the location of the source of data. Based mainly on UNFAO (1977), Doorenbos and Pruitt; UNFAO (1998), Allen *et al.*

Table 10.8. Values of crop coefficient K_c of selected crops under well managed conditions in sub-humid climates (RH_{min} 45% and wind speed 2 metres per second) for use in FAO Penman-Monteith equation to determine ET_o.

Crops	Initial stage	Mid-stage	Late stage
1	2	3	4
Vegetable crops			
1. Cabbage, carrot, cauliflower		1.05	0.95
2. Cucumber	0.60	1.00	0.90
3. Egg plant		1.05	0.90
4. Garlic		1.00	0.70
5. Lettuce		1.00	0.95
6. Onion – dry		1.05	0.75
– green		1.00	1.00
7. Spinach		1.00	0.96
8. Radish		0.90	0.85
9. Watermelon	0.4	1.00	0.75
Root and tubers			
10. Casava 1st year	0.3	0.80	0.30
2nd year	0.3	1.10	0.50
11. Sweet potato		1.15	0.65
12. Sugarbeet	0.35	1.20	0.70
13. Turnip		1.10	0.95
Legumes			
14. Beans (green)	0.5	1.05	0.90
15. Beans (dry) and pulses	0.4	1.15	0.35
16. Chick pea		1.00	0.35
17. Green gram and cowpea		1.05	0.35-0.60
18. Groundnut		1.15	0.60
19. Lentil		1.10	0.30
20. Peas – fresh	0.50	1.15	1.10
– dry/seed	0.50	1.15	0.30
21. Soyabeans		1.15	0.50
Fibre crops			
22. Cotton		1.15-120	0.50-0.70
Oil crops			
23. Castorbean		1.15	0.55
24. Rapeseed		1.0-1.15	0.35
25. Safflower		1.0-1.15	0.25
26. Sesame		1.10	0.25
27. Sunflower		1.0-1.15	0.35
Cereals			
28. Barley		1.15	0.25
29. Oats		1.15	0.25
30. Spring wheat		1.15	0.25-0.4
31. Winter wheat – non-frozen soil	0.7	1.15	0.25-0.4
32. Maize (grain)		1.20	0.35-0.60
33. Millet		1.00	0.30
34. Sorghum (grain)		1.00	0.55
35. Rice	1.05	1.20	0.60-0.90
Sugarcane			
36. Sugarcane	0.40	1.25	0.75
Tropical fruits and trees			
37. Banana 1st year	0.50	1.10	1.00
2nd year	1.00	1.20	1.10
38. Cocao	1.00	1.05	1.05
39. Coffee-bare soil	0.90	0.95	0.95
40. Date palms	0.90	0.95	0.95
41. Palm trees	0.95	1.00	1.00
42. Pineapple-bare soil	0.50	0.30	0.30
43. Rubber trees	0.95	1.00	1.00
44. Tea – shaded	0.95	1.00	1.00
– non-shaded	1.10	1.15	1.15
Grapes			
45. Grapes – table	0.30	0.85	0.45
– wine	0.30	0.70	0.45
Fruit trees			
46. Almond – no ground cover	0.40	0.90	0.65
47. Apple, Cherry, Pear	0.45-0.8	0.90-1.15	0.65-0.85
48. Apricot, Peach			
(*i*) no ground cover – no frosts	0.55	0.90	0.65
(*ii*) active ground cover – no frosts	0.80	1.15	0.85
49. Avocado – no ground cover	0.60	0.85	0.75

(*Contd ...*)

Crops	*Initial stage*	*Mid-stage*	*Late stage*
1	2	3	4
50. Citrus—no ground cover, 70% canopy	0.70	0.65	0.70
51. Conifer trees	1.00	1.00	1.00
52. Pistachios – no ground cover	0.40	1.10	0.45
53. Walnut	0.50	1.10	0.65

Based on Doorenbos and Kassam (1979; Doorenbos and Pruitt UNFAO (1977); Allen *et al.*, UNFAO (1998), Pruitt (1986); Wright (1981, 1982).

ET_c is computed by multiplying ET_o by K_c which is a coefficient expressing the difference in evapotranspiration between the cropped and reference grass surface. The difference can be combined into a single coefficient, or can be split into two factors describing separately the differences in evaporation and transpiration between both surfaces. The selection of the approach depends on the purpose of the calculation, the accuracy required and the climatic data available.

Table 10.8 presents the approximate values of the crop coefficient K_c of some of the common crops at sub-humid climate (relative humidity RH_{min}, 45% and wind speed 2m/sec) at 3 different growth stages, based on selected studies.

Single crop coefficient approach (K_c). In this approach the effect of crop transpiration and soil evaporation are combined into a single K_c coefficient. The coefficient integrates differences in the soil evaporation and crop transpiration rate between the crop and the grass reference surface. As soil evaporation fluctuates almost daily as a result of rainfall or irrigation, the single K_c coefficient expresses only the time-averaged (multi-day) effects of crop evapotranspiration. The approach is used to compute ET_c for weekly or longer periods. The time averaged single K_c coefficient is valid for irrigation management, especially for surface and sprinkler irrigation methods of water application.

Dual crop coefficient approach ($K_{cb} + K_e$). In this approach the effects of crop transpiration and soil evaporation are computed separately. Two coefficients are used: the basal crop coefficient (K_{cb}) to describe plant transpiration, and the soil water evaporation coefficient (K_e) to describe evaporation from the soil surface. The single K_c coefficient is replaced by the following relationship:

$$K_c = K_{cb} + K_e \qquad \text{...(10.49)}$$

in which,

K_{cb} = basal crop coefficient

K_e = soil water evaporation coefficient.

The basal crop coefficient K_{cb} is the ratio of ET_c to ET_o when the soil surface layer is dry but average soil moisture content of the root zone is adequate to sustain full plant transpiration. The soil evaporation coefficient, K_e, describes the evaporation component from the soil surface. When the soil is wet, following rain or irrigation, K_e may be large. However, the sum of K_{cb} and K_e cannot exceed a maximum value, $K_{c\ max}$, determined by the energy available for evapotranspiration at the soil surface. As the soil surface becomes drier, K_e reduces and reaches zero when no water is left for evaporation. The estimation of K_e requires a daily water balance computation of the water content remaining in the upper top soil.

The dual coefficient approach requires more numerical calculations than the single time-averaged K_c coefficient. The dual procedure is best for real time irrigation scheduling, for soil water balance computations, and for research studies on day-to-day variations in soil surface wetness. This is useful for high frequency irrigation with micro-irrigation systems.

The largest difference between K_c and K_{cb} is in the initial growth stage where evapotranspiration is predominantly through soil evaporation and crop transpiration is small. Because crop canopies are near full ground cover during the mid-season stage, soil evaporation beneath the canopy has less effect on ET and the value of K_{cb} will be nearly the same as K_c. Depending on the ground cover, the basal crop coefficient during the mid-season may be only 0.05-0.10 lower than the K_c value (UNFAO, 1998).

The calculation procedure for crop evapotranspiration, ET_c, consists of (*i*) identifying the crop growth stages, determining their lengths, and selecting the corresponding K_c coefficient; (*ii*) adjusting the selected K_c coefficients for the frequency of wetting or climatic conditions during the growth stages, (*iii*) constructing the crop coefficient curve to determine K_c values for any period during the growing period, and (*iv*) calculating ET_c as the product of ET_o and K_c. For detailed procedures on computation of K_c values under different stages of growth in multiple cropping systems and overlapping situations of crops, reference is invited to United Nations Food and Agriculture Organization Report 56 (1998). Procedures for estimating K_c under single and dual crop coefficient approach ($K_{cb} + K_e$) under standard (ET_c) and non-standard (ET_{cadj}) conditions are also available in the above cited reference.

CROP WATER REQUIREMENTS AND IRRIGATION

The estimation of the water requirements (*WR*) of crops is one of the basic needs for crop planning on the farm and for the planning of any irrigation project. Water requirement may be defined as the quantity of water, regardless of its source, required by a crop or diversified pattern of crops in a given period of time for its normal growth under field conditions at a place. Water requirement includes the losses due to evapotranspiration (ET) or consumptive use (C_u) plus the losses during the application of water (unavoidable losses) and the quantity of water required for special operations like land preparation, pre-sowing irrigation and transplanting.

It may thus we expressed as follows:

$WR = ET$ or C_u + application losses + special needs.

Water requirement is, therefore, a 'demand' and the 'supply' would consist of contributions from any of the sources of water, the major source being the irrigation water (IR), effective rainfall (ER) and soil profile contributions (S) including that from shallow water tables. Numerically, therefore, water requirement is given as:

$$WR = IR + ER + S \quad \text{...(10.50)}$$

Irrigation Requirement

The irrigation requirement of a crop is the amount of water to be supplied by irrigation to a disease-free crop growing in a large field with adequate soil water and fertility and achieving full production potential under the prevailing soil and climatic conditions. Irrigation requirement includes the evapotranspiration of the crop (ET_c) (or consumptive use C_u), maintaining a favourable salt balance within the root zone of the crop. Irrigation requirement does not include the water which is available to a crop through precipitation, soil moisture storage and soil moisture accretion through capillary movement of groundwater or seepage from such surrounding areas.

With the estimated value of ET_c the irrigation requirement of a crop can be computed, adopting the following relationship (using identical units of depth, i.e., mm/cm/inches):

$$I = ET_c - P + RO + DP + L + D_r\,(\theta_f - \theta_i) \quad \text{...(10.51)}$$

in which,

I = Irrigation water requirement

ET_c = Evapotranspiration of the crop

RO = Runoff from the crop field from irrigation and/or rainfall

DP = Deep percolation in the field from irrigation and/or rainfall

L = Leaching requirement

D_r = Depth of the crop root system

θ_f = Root zone soil moisture content at field capacity level

θ_i = Root zone soil moisture content prior to irrigation.

The field irrigation requirement of a crop, therefore, refers to the water requirement of crops, exclusive of effective rainfall and contribution from soil profile, and it may be given as:

$$IR = WR - (ER + S) \quad \text{...(10.52)}$$

The farm irrigation requirement depends on the irrigation needs of individual crops, their area and the losses in the farm water distribution systems, mainly by way of seepage. The irrigation requirement of an outlet command area includes the irrigation requirement of individual farm holdings and the losses in the conveyance and distribution system.

Field Water Balance

The water balance of a field is an itemized statement of all gains, losses, and changes of storage of water occurring in a given field within specified boundaries during a specified period of time. The task of monitoring and controlling the field water balance is vital to the efficient management of water and soil. A knowledge of the water balance is necessary to evaluate the possible methods to minimize loss and to maximize utilization of water which is so often the limiting factor in crop production.

Gains of water in the field are generally due to precipitation and irrigation. Occasionally, there may be gains due to accumulation of runoff from higher tracts of land, or to capillary rise from below (especially where a water table is present at some shallow depth). Losses of water include surface run-off from the field, deep percolation out of the root zone (drainage), evaporation from the soil surface, and transpiration from the crop canopy. The change in storage of water in the field can occur in the soil as well as in the plants. The total change in storage must equal the difference between the sum of all grains and the sum of all losses.

Accordingly, the water balance equation may be stated as follows:

(Gains) – (Losses) = (Change in storage)

$$(P + I) - (R + D + E + T) = (\Delta S + \Delta V) \quad \text{...(10.53)}$$

in which, P is precipitation, I irrigation, R runoff from the field, D downward drainage out of the root zone, E evaporation from the soil, T transpiration by the crop canopy, ΔS the change in soil water content of the root zone, and ΔV the change in plant water content. All of these quantities are usually expressed in terms of water depth per unit of land area (ha-cm) or units of depth (cm).

An important consideration in water balance studies is the period, or time interval, for which the balance is made. Too short a period might be impractical, while too long a period might mask the occurrence of short-term critical stages. At such critical stages as flowering and fruit-set even temporary imbalance in crop-water status can be have a lasting effect. It is necessary to ensure that the crop water balance is maintained positive continuously during the growing season.

Measurements of field water balance is usually conducted by the use of weighing or floating lysimeters. However, they cannot be used when the mixed nature of the species composition, the spatial distribution of the vegetation, the characteristics of the root systems, the size of plants, or other factors make it impossible to simulate the natural environment in the lysimeter itself. In such cases, determination of soil water storage at different points in the community provide the only means for evaluating the water balance.

Effective Rainfall

In its simplest sense, effective rainfall means useful or utilizable rainfall. Rainfall is not necessarily useful or desirable at the time, rate, or amount in which it is received. Some of it may be unavoidably wasted while some may even be destructive. Just as total rainfall varies, so does the amount of effective rainfall.

The term effective rainfall has been interpreted differently by specialists in different fields. To a canal irrigation engineer, the rain which reaches the storage reservoir directly and by surface runoff from the surrounding area is the effective portion. Geo-hydrologists would define as effective rainfall that portion of rainfall which contributes to groundwater storage. Agriculturists consider as effective that portion of the total rainfall which directly satisfies crop water needs. In the field of dry land agriculture, when the land is left fallow, effective rainfall is that which can be conserved for the following crop.

Thus, it may be seen that even though the concept is the same in all the above cases, the values of effective rainfall are different for different agencies for the same total rainfall. Since there are such varied interpretations of what may be regarded as effective rainfall, it is difficult to develop a definition to suit all the interested disciplines. Rainfall which is ineffective according to one may be effective according to another. From the point of view of the water requirement of crops, the Food and Agricultural Organisation of the United Nations (Dastane, 1974) has defined the annual or seasonal effective rainfall as that part of the total annual or seasonal rainfall which is useful directly and/ or indirectly for crop production at the site where it falls, but without pumping. It therefore includes water intercepted by living or dry vegtetation, that is lost by evaporation from the soil surface, the precipitation lost by evapotranspiration during growth, that fraction which contributes to leaching, percolation or facilitates other cultural operations either before or after sowing without any harm to yield and quality of the principal crops. Consequently, ineffective rainfall is that portion which is lost by surface runoff, unnecessary deep percolation losses, the moisture remaining in the soil after the harvest of the crop and which is not useful for next season's crop. The growing season is counted from the start of the first tillage operation until the harvest.

The above concept of effective rainfall is suggested for use is planning and operation of irrigation projects. The irrigation water supply in a given year should be planned to complement rainfall. Since annual rainfall varies from year to year, an irrigation project is to be planned based on data, over a long period to calculate the effective rainfall on the basis of probability of occurrence.

Factors Influencing Effective Rainfall

(i) Rainfall characteristics. A soil has a definite and limited infiltration and moisture holding capacity. Hence, greater amount and intensities of rainfall normally reduce the effective fraction of rainfall and increase runoff. A well distributed rainfall in frequent light showers increases the effective rainfall and is conducieve to crop growth than heavy downpours. In India the intensity, frequency and amount of rainfall are high during July and August and hence the effective fraction is low. From November to April most of the rainfall is effective due to its low intensity, frequency and amount.

(ii) Land slope. The slope of the land has a profound influence on the time available for the rain water to infiltrate into the soil (infiltration opportunity time). Water stays longer on flat and levelled land and thus has a longer opportunity time than on slopping land where there is a rapid runoff.

(iii) Characteristics of the soil. The soil properties influencing infiltration and moisture retention influence the effective fraction of rainfall. High values of infiltration rates and hydraulic conductivity of the soil enhance the infiltration and reduce runoff. The moisture content of the soil at the time of occurrence of the rain (initial moisture content), affects the effective rainfall considerably. The higher the moisture content, the lower the infiltration rate and higher the surface runoff which reduces the effective rainfall.

(iv) Management practices. Any management practice which influences runoff, infitration, hydraulic conductivity or evapotranspiration also influences the degree of effective rainfall. Bunding, terracing, contour tillage, ridging and mulching reduce runoff and increase effective rainfall.

(v) Crop characteristics. Crops with high water consumption create greater deficits of moisture in the soil. The effective rainfall is directly proportional to the rate of water uptake by the plant. Crop characteristics influencing the rate of water uptake are the degree of ground cover, rooting depth and stage of growth. Soil moisture stored in deeper layers can be tapped only when roots penetrate to these depths. Deep-rooted crops, therefore, increase the proportion of effective rainfall in a given area.

(vi) Carry-over soil moisture. It is the moisture stored in the crop root zone between cropping seasons or before the crop is planted. This moisture is available to meet the consumptive water needs of the crop. The contribution of a fairly good rain occurring just prior to sowing may be equivalent to one full pre-sowing irrigation. An additional benefit of excess pre-sowing season rain is the leaching of salts accumulated in the root zone in the summer season. Carry-over of soil water is of great importance in rainfed farming in water scarcity areas.

(vii) Groundwater contribution. Soil moisture contribution from the groundwater table is determined by the depth of groundwater below the root zone and the capillary and conductive properties of the soil.

Measurement of effective rainfall. Evaluation of effective rainfall involves the measurement of rainfall and/or irrigation losses by surface runoff, percolation beyond root zone and soil moisture use by crops. Precise measurements are often made by weighing type lysimeters.

Groundwater contribution. The contribution to the root zone soil moisture from the groundwater table is determined by the depth of groundwater below the root zone, the capillary and conductive properties of the soil and the soil moisture content in the root zone. Both the rate and the distance of water movement are important criteria. For heavy soil the distance of movement is high and the rate low, while for light textured soil the distance of movement is low and the rate high.

Surface and sub-surface in- and out-flows. Computation of surface inflow normally does not apply, except for areas subject to occasional flooding. Under efficient irrigation practices surface outflow is small. Management losses and waste of water due to technical faults are normally accounted for in irrigation efficiency. Sub-surface inflow is only of local significance in areas where there is upward movement of water from deeper subsoil caused by seepage from reservoirs and canals. Sub-surface inflow may also occur locally on or near the toe of sloping lands.

Deep percolation. Deep percolation is that amount of water which passes below the root zone of the crop and can continue for a long time after field capacity has been reached in the crop root zone, following irrigation or heavy rain. Total water loss by deep percolation in irrigated conditions can account for 20% or more of the total amount of water applied. However, soil water movement in and below the root zone, after an initial downward outflow, can later be reversed to an upward inflow from the wet sub-soil to the drying root zone above.

Net Irrigation Requirement

The net irrigation requirement is the depth of irrigation water, exclusive of precipitation, carry-over soil moisture or groundwater contribution or other gains in soil moisture, that is required consumptively for crop production. It is the amount of irrigation water required to bring the soil moisture level in the effective root zone to field capacity (Fig. 10.6). Thus, it is the difference between the field capacity and the soil moisture content in the crop root zone before starting irrigation.

This may be obtained by the relationship given below:

$$d = \sum_{i=1}^{n} \frac{(M_{fci} - M_{bi})}{100} . A_i . D_i \qquad ...(10.54)$$

in which,

d = net amount of water to be applied during an irrigation, cm

M_{fci} = field capacity moisture content in the ith layer of the soil, per cent

M_{bi} = moisture content before irrigation in the ith layer of the soil, per cent

A_i = bulk density of the soil in the ith layer

D_i = depth of the ith soil layer within the root zone cm, and

n = number of soil layers in the root zone D.

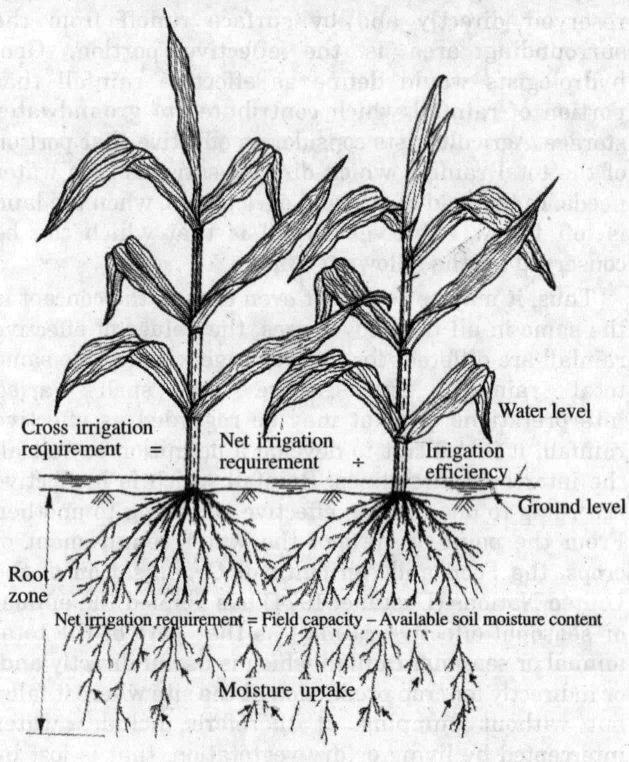

Fig. 10.6. Schematic sketch illustraing the concept of gross and net irrigation requirements.

In drawing up the seasonal or monthly net irrigation requirements for a given crop or cropping pattern the main variables composing the field water balance include: (*i*) crop water requirements as determined by climate and crop characteristics, ET_c (*ii*) contribution from precipitation, and groundwater, and (*iii*) carry-over of soil moisture. The deficit in the soil water balance is met by the net irrigation requirement.

Gross Irrigation Requirement

The total amount of water applied through irrigation is termed as 'gross irrigation requirement'. In other words, it is the net irrigation requirement plus losses in water application and other losses. The gross irrigation requirement can be determined for a field, for a farm, for an outlet command area or for an irrigation project, depending on the need, by considering the appropriate losses at various stages of growth of the crop.

Gross irrigation requirement (in field)

$$= \frac{\text{Net irrigation requirement}}{\text{Field efficiency of system}}$$

For example, if the net amount of irrigation is 8 cm and the field efficiency is 75 per cent, the gross amount of water to be aplied to the field is

$$\frac{8}{0.75} = 10.66 \text{ cm}$$

The gross irrigation requirement (*IR*) at the field head, for instance, can be determined as follows:

$$IR = \sum_{1}^{n} \frac{d}{E_{(\text{application})}} \quad ...(10.55)$$

IR = seasonal gross irrigation requirement at the field head (cm)

d = net amount of water to be applied in each irrigation (cm)

$E_{(\text{application})}$ = water application efficiency, and

n = number of irrigations in a season.

Irrespective of the method of irrigation, no system is 100 per cent efficient and not all the water applied during an irrigation enters the soil and is held in the root zone. Unavoidable losses are caused through seepage and leaks in the water conveyance system, non-uniform distribution of water over a field, percolation below crop root zone and wastage due to surface runoff at the end of borders and furrows. In case of sprinkler irrigation, additional losses may be caused by evaporation from the spray and by the retention of water on the plant foliage.

Irrigation frequency. Irrigation frequency refers to the number of days between irrigations during periods without rainfall. It depends on the consumptive use rate of a crop and on the amount of available moisture in the crop root zone. It is a function of crop, soil and climate. Sandy soils must be irrigated more often than fine-textured deep soils. Moisture-use rate varies with the kind of crop and climatic conditions and increases as the crop grows larger and the days become longer and hotter. In general, irrigation should start when about 50 per cent and not over 60 per cent of the available moisture has been used from the zone in which most of the roots are concentrated. The stage of growth of the crop with reference to the critical periods of growth is also kept in view while designing irrigation frequency.

In designing irrigation systems, the irrigation frequency to be used is the time (in days) between two irrigations in the period of highest consumptive use of the crops grown. Irrigation frequency depends on how fast soil moisture is extracted when a crop is transpiring at its maximum rate. The average moisture-use rate during this period is used to plan irrigation systems. For an irrigation system to be adequate, it must have sufficient capacity to supply the water required during this period. The design irrigation frequency may be computed as folows:

Design frequency (days)

$$= \frac{\begin{array}{c}\text{Field capacity of the soil in the effective crop root zone}\\ -\text{ Moisture content of the same zone at the time of starting of irrigation}\end{array}}{\text{Peak period moisture use rate of crop}}$$

Irrigation period. Irrigation period is the number of days that can be allowed for applying one irrigation to a given design area during the peak consumptive use period of the crop being irrigated. It is the basis for irrigation system capacity and equipment design. The irrigation system must be so designed that the irrigation period is not greater than the irrigation frequency.

Irrigation period

$$= \frac{\text{Net amount of moisture in soil between start of irrigation and lower limits of moisture depletion}}{\text{Peak period moisture use rate of crop}}$$

Example 10.11. *A Persian wheel discharges at the rate of 11,000 litres per hour and works for eight hours each day. Estimate the area commanded by the water lift if the average depth of irrigation is 8 cm and irrigation interval is 15 days.*

Solution:

The discharge of Persian wheel in 15 days

$$= \frac{11,000 \times 8 \times 15}{1000} = 1320 \text{ m}^3$$

Water required to cover one hectare to a depth of 8 cm

$$= 10,000 \times \frac{8}{100} = 800 \text{ m}^3$$

Area commanded by Persian wheel

$$= \frac{\text{Water available}}{\text{Water required per unit area}}$$

$$= \frac{1320}{800} = 1.65 \text{ hectares.}$$

Example 10.12. *The following data were obtained in determining the soil moisture content at successive depths in the root zone prior to applying irrigation water:*

Depth of sampling cm	*Wt. of moist soil sample gm*	*Oven dry wt. of soil sample gm*
0–25	*134.60*	*126.82*
25–50	*136.28*	*127.95*
50–75	*122.95*	*115.32*
75–100	*110.92*	*102.64*

The bulk density of the soil in the root zone was 1.50 gm/cc. The available moisture holding capacity of the soil was 17.8 cm/m depth. Determine (i) the moisture content at the different depths in the root zone, (ii) moisture content in the root zone at the time of irrigation, (iii) net depth of water to be applied to bring the mositure content to field capacity, and (iv) gross irrigation requirement at an estimated field irrigation efficiency of 70 per cent.

Solution:

(*i*) Soil moisture content at different depths

0–25 cm: $\frac{(134.60 - 126.82)100}{126.82} = 61.4\% = 6.14 \times 1.50$

$= 9.20$ cm/m depth $= 9.20 \times \frac{25}{100} = 2.30$ cm

25–50 cm: $\frac{(136.28 - 127.95)\ 100}{127.95} = 6.52\% = 6.52 \times 150$

$= 9.76$ cm/m depth $= 9.76 \times \frac{25}{100} = 2.44$ cm

50–75 cm: $\frac{(122.95 - 115.32)100}{115.32} = 6.6\% = 6.60 \times 1.50$

$= 9.90$ cm/m depth $= 9.90 \times \frac{25}{100} = 2.47$ cm

75–100 cm: $\frac{(110.92 - 102.64)100}{102.64}$

$= 8.06\% = 8.06 \times 1.50$

$= 12.10$ cm/m depth $= 12.10 \times \frac{25}{100} = 3.02$ cm

(*ii*) Moisture content in the root zone at the time of irrigation

$= 2.30 + 2.44 + 2.47 + 3.02 = 10.23$ cm

(*iii*) Net irrigation requirement

$= 17.80 - 10.25 = 7.55$ cm

(*iv*) Gross irrigation requirement

$= \frac{7.55}{70} \times 100 = 10.78$ cm

IRRIGATION SCHEDULING

Irrigation scheduling is the process of determining the time to irrigate and how much water is to be applied (irrigation depth) in each irrigation. Proper scheduling is essential for the efficient use of water and other inputs in crop production. Irrigation schedules are planned to either fully or partially provide the estimated water requirement of the crop.

Amount of water to apply per irrigation. With 'full irrigation' the amount of water required per irrigation is computed as follows:

$$1 = D_r(f_c - f_m)/\text{Irrigation efficiency} \quad \text{...(10.56)}$$

in which,

D_r = Depth of root zone

f_c = Soil moisture content at field capacity

f_m = Soil moisture content prior to irrigation.

Full irrigation provides adequate water to meet the entire irrigation requirement and is aimed at achieving the maximum production potential of the crop. Excess irrigation, however, reduces crop yields by adversely influencing soil physical properties like soil aeration, soil temperature and the microbial activities in the soil. Full irrigation is justified when there is no scarcity of water and the cost of irrigation is low.

Deficit irrigation meets the water requirement of crops only partially. It is economically justified when reducing the amount of irrigation supply below the full level causes production costs to decrease faster than the decline in the value of crop harvest. Deficit irrigation is practiced when there is water scarcity or when the irrigation system capacity is limited. With 'deficit' irrigation the crop root zone is not always filled to the field capacity moisture content level. In locations with appreciable amounts of precipitation during the irrigation season, it is possible to fill the root zone only partially so that some precipitation can be stored in the crop root zone. The recent concepts on the management of 'deficit irrigation'. i.e., replacing only a part of the water that is used by the plant and achieving maximum possible production per unit quantity of water and/or land require careful analysis and planning. For example, it has been observed that the yield and fruit quality of matured apple trees are not affected by partial replenishment of the water used during the period prior to fruiting. Full replacement is required during the period following fruiting. Many research studies indicate that with increasing water shortage globally it may be more economical to plan irrigation systems to obtain economical crop production per unit of land. However, care should be taken to ensure that there is no water shortage during the 'critical stages' in the growth of the crop.

Ideally, the irrigation depth and/or the irrigation interval vary with the stage of growth of the crop. At the beginning of the growing season the depth of irrigation is kept low, but the applications are frequent. This is due to the low value of ET_c of the growing seedling and their shallow root system. During the mid-growth stage of the crop the depth of irrigation is increased and irrigations are given less frequently. Controlling irrigation depths is not difficult under sprinkler and drip methods of irrigation. However, with surface irrigation methods the scope for changing irrigation depths is limited. The amount of effective rainfall which may occur during the crop season is deducted from the amount of irrigation water estimated to be applied in the irrigation which follows. Alternatively, provision may be made to allow for the precipitation which is expected during the crop growing period.

Irrigation Indices

Scientific irrigation must be based on an understanding of the soil-water-plant-atmospheric relationship. Irrigation needs of crops depend on the evaporation demand of the ambient atmosphere, soil-water regime in the crop root zone and the plant foliage.

Visual plant symptoms. Visual symptoms in plants to decide on the time to irrigate are changes in the colour of plants, curling of the leaves and tendencies of wilting. These are observed by looking at the crop as a whole and not individual plants. When the crop comes under water stress the appearance changes from vigorous growth to retarded growth. Under conditions of water stress the number of emerging leaves get reduced, and often there is a marked change in the colour of the leaf which gets darker and sometimes grey. Successful interpretation of crop stress requires keen observation and experience in crop fields. However, symptoms may sometimes be misleading due to changes in crop varieties and nutritional disorders as well as insect and pest incidence. Another limitation in resorting to the appearance of crop foliage is that by the time the symptoms are evident irrigation has already been withheld for too long a period for most crops and the yield losses have become substantial.

Plant water content and water potential. The status of water in the plant is generally indexed from the measurement of leaf-water content and leaf-water potential. Their values at any time of the day will depend on the lag between the evaporative demand of the atmosphere and the moisture uptake rate by the crop. When their values fall below certain critical limits specific to the plant species and their growth stages, important physiological and growth factors are adversely affected. Hence, these values can serve as reliable indicators for irrigating crops.

Plant temperature. Almost the entire radiation energy received on the leaf surface during the daylight hours is utilized for evapotranspiration (*ET*). The remaining energy, if any, is used to heat the leaf tissues and the ambient air. When *ET* reduces due to water deficit in the plant, the energy saved in the process is partly used to raise the leaf temperature. Many investigations have shown that the leaf canopy temperature is a sensitive index in crops like soyabean, oats, barley, lucerne (alfalfa), wheat, sorghum and maize (Prihar, S.S. and Sandhu, B.S., 1987).

Soil-water regime. A commonly recommended method to decide on irrigation schedules is soil moisture measurements in the field. When the soil moisture content has dropped to a certain critical level, say about 50% of field capacity level in the crop root zone, irrigation is applied. Normally irrigation is not delayed beyond 60% field capacity moisture content.

Irrigation schedules based on prediction of *ET*. Irrigations can be suitably scheduled on a farm if the allowable maximum water depletion in the crop root zone and the evapotranspiration (*ET*) for short periods during the growing period are known.

Irrigation scheduling based on pan evaporation. Prihar *et al*. (1974) suggested a 'practical approach' relating irrigation requirement to the cumulative evaporation during a time period as measured in a standard open pan. The relationship is expressed as IW/E_{pan}, in which IW is the depth of irrigation water and E_{pan} is the cumulative open pan evaporation. The rainfall during the period, if any, is subtracted from the value of E_{pan}. The validity of the procedure will depend on the proper installation of the evaporation pan and the rain gauge and the precision in measurements of pan evaporation and rainfall. Further, the suitability of the method is situation-specific and limited to the particular variety of a crops.

General Guidelines on Planning Irrigation Schedules

The characteristics of different soil groups and soil depth influencing irrigation scheduling are as follows:

(*i*) *Shallow and/or sandy soil.* In sandy soils in general and soils with limited depth (due to the presence of hard pan close to the ground surface) the soil moisture holding capacity is low. These soils require frequent irrigations and the amount of water applied in each irrigation is kept low.

(*ii*) *Loamy soils.* Loamy soils can store more water than sandy soils or shallow soils. Hence, irrigation is applied less frequently with larger depth of water in an irrigation.

(*iii*) *Clayey soils.* The moisture holding capacity of clayey soils is more than loamy soils. Hence, irrigations are less frequent and the quantity of water in an irrigation is higher.

The estimated irrigation schedule of major field crops during peak water use periods is given in Table 10.9. The data in Table 10.9 is based on the estimated value of ET_c and root depth for each crop. The information is given for three representative climates, viz., *Climate 1* when ET_o ranges from 4 to 5 mm/day. *Climate 2* when ET_o ranges from 6 to 7 mm/day and *Climate 3* when ET_o ranges from 8 to 9 mm/day.

Adjusting irrigation depth for non-peak periods. The irrigation schedule shown in Table 10.9 is for peak period moisture use situations. During the early growth stages of crops it is desirable to keep the irrigation interval at shorter intervals and reduce the quantity of water used in irrigation. If the irrigation interval is kept far apart due to the low value of ET_o, the young plants may suffer from water shortage as their roots are not able to take up water from the lower layers of the root

Table 10.9. Estimated irrigation schedules of common field crops during peak periods in water requirement

	Shallow and/or sandy soil				*Loamy soil*				*Clayey soil*			
	Irri. int. (days)			*Net irri. depth (mm)*	*Irri. int. (days)*			*Net irri. depth (mm)*	*Irri. int. (days)*			*Net irri. depth (mm)*
Climate*	1	2	3		1	2	3		1	2	3	
Alfalfa (Lacerne)	9	6	5	40	13	9	7	60	16	11	8	70
Banana	5	3	2	25	7	5	4	40	10	7	5	55
Barley/Oats	8	6	4	40	11	8	6	55	14	10	7	70
Beans	6	4	3	30	8	6	4	40	10	7	5	50
Cacao	9	6	5	40	13	9	7	60	16	11	8	70
Carrot	6	4	3	25	7	5	4	35	11	8	6	50
Citrus	8	6	4	30	11	8	6	40	15	10	8	55
Coffee	9	6	5	40	13	9	7	60	16	11	8	70
Cotton	8	6	4	40	11	8	6	55	14	10	7	70
Cucumber	10	7	5	40	15	10	8	60	17	12	9	70
Cabbage/Cauliflower	3	2	2	15	4	3	2	20	7	5	4	30
Egg plant	6	4	3	30	8	6	4	40	10	7	5	50
Fruit trees	9	6	5	40	13	9	7	60	16	11	8	70
Grains (small)	8	6	4	40	11	8	6	55	14	10	7	70
Grapes	11	8	6	40	15	11	8	55	19	13	10	70
Grass	9	6	5	40	13	9	7	60	16	11	8	70
Groundnut	6	4	3	25	7	5	4	35	11	8	6	50
Lentil	6	4	3	30	8	6	4	40	10	7	5	50
Lettuce	3	2	2	15	4	3	2	20	7	5	4	30
Maize	8	6	4	40	11	8	6	55	14	10	7	70
Melons	9	6	5	40	13	9	7	60	16	11	8	70
Millet	8	6	4	40	11	8	6	55	14	10	7	70
Olives	11	8	6	40	15	11	8	55	19	13	10	70
Onions	3	2	2	15	4	3	2	20	7	5	4	30
Peas	6	4	3	30	8	6	4	40	10	7	5	50
Potatoes	6	4	3	30	8	6	4	40	10	7	5	50
Radish	4	3	2	15	5	4	3	20	7	5	4	30
Safflower	8	6	4	40	11	8	6	55	14	10	7	70
Sorghum	8	6	4	40	11	8	6	55	14	10	7	70
Soyabeans	8	6	4	40	11	8	6	55	14	10	7	70
Spinach	3	2	2	15	4	3	2	20	7	5	4	30
Sugarbeet	8	6	4	40	11	8	6	55	14	10	7	70
Sugarcane	7	5	4	40	10	7	5	55	13	9	7	70
Sunflower	8	6	4	40	11	8	6	55	14	10	7	70
Tea	9	6	5	40	13	9	7	60	16	11	8	70
Tobacco	6	4	3	30	8	6	4	40	10	7	5	50
Tomatoes	6	4	3	30	8	6	4	40	10	6	5	50
Wheat	8	6	4	40	11	8	6	55	14	10	7	70

(***Note:*** Climate 1: ET_o ranges from 4 to 5 mm/day; Climate 2 : ET_o ranges from 6 to 7 mm/day; Climate 3: ET_o ranges from 8 to 9 mm/day)

[***Reference:*** Irrigation Water Management Training Manual No. 4. *Irrigation Scheduling.* Brown, C. and Prins, K. (1980).]

zone. On the other hand, crops which are harvested dry like maize (grain) need much less water at the late season stage. Hence, the frequency of irrigation during the late season is reduced.

Adjusting irrigation schedule to rainfall occurring during the growing season. The data presented in Table 10.9 is on the assumption that no rainfall occurs during the crop growing season If the contribution from rainfall is significant during the period the schedule is to be suitably adjusted. This is usually done by making the irrigation interval longer or reducing the net irrigation depth.

Simple procedure to arrive at approximate solutions to irrigation schedule. A simple procedure to make estimates leading to approximate values for scheduling irrigation comprise of the following steps:

1. Estimate the net and gross irrigation depths (mm).
2. Compute the irrigation water need over the total growing season of the crop (mm).
3. Compute the total number of irrigations over the growing season.
4. Calculate irrigation interval (days).

The number of irrigation applications over the growing season of a crop is obtained by dividing the irrigation water need during the growing season by the net irrigation depth per application. The interval between two irrigations is obtained by dividing the number of days of the growing season by the number of irrigations. To be on the safe side the irrigation interval is rounded of to the lower whole figure.

Example 10.13. *Compute the gross depth of water application, the total number of irrigations and the irrigation interval of tomatoes of 5 months growing period in loamy soil. The irrigation water need of tomatoes for the whole growing period, estimated monthwise is 720 mm.*

Solution: From Table 10.9 the net depth of irrigation for tomatoes in loamy soil is 40 mm.

Therefore, the gross irrigation depth $= \frac{720}{40} = 18\,\text{mm}$

The growing season = 5 × 30 = 150 days

Irrigation interval $= \frac{150}{18} = 8.33$ days, rounded of to 8 days.

In using the above simple procedure the effective rainfall received during particular intervals are to be deducted from the depth of water to be applied in the next irrigation. To overcome the risk of water shortage in the peak growing period it may sometimes be necessary to base irrigation interval during the peak season, particularly to the peak period irrigation requirement rather than generalizing it to the requirement of the full growing period.

Soil-Water Constants, Root Depth of Crops and Irrigation Scheduling

Scientific, planning of irrigation schedules requires a sound knowledge of the related soil-water constants and rooting characteristics of the crops in relation to their stages of growth (see Chapter 9).

Available water (AW). The amount of soil moisture held between field capacity and permanent wilting point is the *available water*. Available water in a crop root zone is computed by the following relationship:

AW = Root zone depth, cm (Field capacity, % by vol. – Perm. wilting point, % by volume (see Chapter 9).

Readily available water (RAW). The volume per unit surface area of soil water between field capacity and critical soil moisture content is called *readily available water.* (*RAW*). *RAW,* expressed in identical units, is given as follows:

RAW = Depth of root zone of the crop (Field capacity–Critical moisture content).

Maximum available deficiency (MAD). Another concept of importance in irrigation scheduling is the maximum available deficiency (MAD) of soil moisture. It is the ratio of RAW and AW and is expressed as follows:

$$MAD = RAW/AW \quad \text{... (10.56)}$$

MAD for most vegetable crops is estimated to be about 0.5 and for other crops (peas, grain crops, fruit plants, sugarcane and oilseeds) it is 0.65. The exception is potatoes which has an *MAD* value of about 0.30.

The root depth of a crop depends on the type of soil horizon in which the crop is grown, its physiological characteristics and its stage of growth. When the root zone depth is not restricted by the presence of impermeable layers the roots can extend freely. The maximum root depths attained by crops like lucern (alfalfa), apples, pears, plums, safflower and sunflower is about 180 cm. Crops like beans, carrot, sorghum and wheat have roots extending up to about 90 cm. Shallow rooted crops with maximum root depth limited to about 60 cm are onion, potatoes, radish and spinach. Maize (corn) and cucumber extend their roots up to about 120 cm. The above values are for a fully grown crop with unrestricted soils. Hard pans close to the land surface restrict root growth. Further, the plant root system grows with the stage of growth of the plants and attain the full depth and spread at the full growth stage of the plant.

When the *MAD* concept is used in irrigation scheduling, the *RAW* (expressed in identical units) is computed as follows:

RAW = (*MAD*) × (root depth of the crop) × (Field capacity–Permanent wilting point).

Average root depth of a crop can be determined by digging test pits and observing the roots. However, it is not very reliable, as the part of the root system which is most active are the tiny root hairs which are difficult to be identified and would often be destroyed by digging. In agricultural experiment stations root distribution is determined by observing the soil moisture deficits in the successive layers of the root zone soil.

IRRIGATION EFFICIENCY

Irrigation efficiency E_i is the ratio, usually expressed as per cent, of the amount of water used to meet the consumptive use requirement of the crop plus that necessary to maintain a favourable salt balance in the crop root zone to the total volume of water diverted, stored or pumped for irrigation. Irrigation systems are designed and operated to supply the irrigation requirement to each field of the farm. The performance of an irrigation system is determined by the efficiency with which water is stored in the surface reservoir at the headworks, diverted and conveyed to the irrigated area through the conveyance system and applied to the fields and by the adequacy and uniformity of water application in each field. The overall efficiency of an irrigation system is defined as the per cent of water supplied to the farm that is beneficially used for irrigation on the farm. Overall system efficiency, often known as irrigation efficiency, is expressed as follows:

$$E_i = 100\,(S - P - R - O)/S \quad \text{...(10.56a)}$$

in which,

E_i = Irrigation efficiency %

I = Irrigation requirement (i.e., ET_c–Effective rainfall)

L = Leaching requirement

S = Amount of water supplied to the farm

P = Total deep percolation on the farm

R = Total runoff from the farm

O = Operational losses in channels and pipelines

Example 10.14. *Compute the efficiency of an irrigation system based on the following data:*

1800 litres/min of water is diverted to the farm each day of 24 hrs. Each day 0.8 ha of maize and 1 ha of wheat are irrigated. The irrigation requirement of maize is 8 cm and that of wheat 12 cm.

Solution: Applying Equation 10.56 and assuming that the leaching requirement is nil

$$S = 1800 \text{ litres/min} = \frac{(1800 \times 60 \times 24)}{1000}$$

$$= 2592 \text{ cubic metres}$$

Volume of water required to irrigate maize

$$= \left(\frac{8}{100}\right) \times 0.8 \times 1000 = 640 \text{ cubic metres}$$

Volume of water required to irrigate wheat

$$= \left(\frac{12}{100}\right) \times 1 \times 1000 = 1200 \text{ cubic metres}$$

Total water use by maize and wheat

$$= 640 + 1200 = 1840 \text{ cubic metres}$$

$$E_i = \left(\frac{1840}{2592}\right), 100 = 70.99\%, \text{ say } 71\%.$$

In Example 10.14, 71% of the water applied is beneficially utilized while 29% is lost as deep percolation, runoff, evaporation and spillage.

Water losses in an irrigation system can be considered in two groups, namely, (*i*) the losses in the storage reservoir (in water storage projects) and the conveyance system, and (*ii*) the losses in the field. Losses in the storage reservoir are due to evaporation from the water surface in the reservoir, seepage losses from the sides and bottom of the reservoir and spillage from the reservoir. Conveyance losses include seepage losses in the canal system, spillage due to overtopping of the canals and breaches and rat holes in the canal bunds, runoff to drains from the canals and evaporation from the water surface in the canal network. Field losses are mainly due to deep percolation below the crop root zone and runoff from irrigated fields in case of land served by surface irrigation methods and evaporation during water application in case of land served by sprinkler irrigation.

In evaluating the performance of irrigation systems it is useful to examine the efficiency of each component of the system. This allows identification of the components which are not performing well and taking suitable measures to improve them. The overall irrigation system efficiency may be expressed as follows:

$$E_i = \left(\frac{E_r}{100}\right)\left(\frac{E_c}{100}\right)\left(\frac{E_a}{100}\right)(100) \quad \text{...(10.57)}$$

in which,

E_i = Irrigation efficiency (%)

E_r = Reservoir storage efficiency (%)

E_c = Water conveyance efficiency (%)

E_a = Irrigation application efficiency (%)

1. Reservoir storage efficiency E_r. The term defines the efficiency with which water is stored in a reservoir. Reservoir storage efficiency is expressed as follows:

$$E_r = 100\left(1 - \frac{V_e + V_s}{V_i}\right) = 100\left(\frac{\Delta_o + \Delta S}{V_i}\right) \quad \text{...(10.58)}$$

in which,

V_e = Evaporation volume from the reservoir

V_s = Seepage volume from the reservoir

V_i = Inflow to the reservoir during a time interval

V_o = Volume of outflow from the reservoir during a time interval

ΔS = Change in reservoir storage during the time interval.

The ΔS term is often neglected when long time periods are considered.

Example 10.15. *Compute the reservoir storage efficiency for a 24-hr period when 3100 liters/minute of water are being diverted from the reservoir based on the following data:*

The rate of inflow into the reservoir of 4420 litres per minute.

ΔS = 410 cubic metres (the quantity of water to be removed to restore the initial water level in the reservoir.

Solution:

V_i = 4420 litres/minute = $\dfrac{(3790 \times 60 \times 24)}{1000}$

= 6364.8 cubic metres per day

V_o = 3790 litres/minute = $\dfrac{(3790 \times 60 \times 24)}{1000}$

= 5457.6 cubic metres per day

ΔS = 410 cubic metres (given)

$$E_r = \frac{100\,(5457.6 + 410)}{6364.8} = 92.2\%$$

2. Water conveyance efficiency. This term is used to measure the efficiency of water conveyance systems associated with the canal network, water courses and field channels. It is also applicable where the water is conveyed in channels or pipelines from the well to the individual fields. It is expressed as follows:

$$E_c = \frac{W_f}{W_d} \times 100 \qquad \text{...(10.59)}$$

in which,

E_c = Water conveyance efficiency (per cent)

W_f = Water delivered to the irrigated plot (at the field supply channel)

W_d = Water diverted from the source.

3. Water application efficiency. After the water reaches the field supply channel, it is important to apply the water on the land as efficiently as possible. A measure of how efficiently this is done is the water application efficiency, defined as follows:

$$E_a = \frac{W_s}{W_f} \times 100 \qquad \text{...(10.60)}$$

in which,

E_a = Water application efficiency (per cent)

W_s = Water stored in the root zone of the plants

W_f = Water delivered to the field (at the field supply channel)

Water application efficiencies below 100 per cent are due to seepage losses from the field distribution channels and deep percolation below the crop root zone. Sometimes, in case of very long fields, there may be runoff losses at the tail end of borders and furrows. In general, water application efficiency decreases as the amount of water applied during each irrigation increases. However, very small irrigations may not fill the root zone adequately and may reduce crop yields, and in the long run give rise to salt problems due to inadequate leaching.

4. Water storage efficiency. It has been stated that small irrigations may lead to high water application efficiencies, yet the irrigation practice may be poor. The concept of water storage efficiency is useful to evaluate this problem. This concept relates how completely the water needed prior to irrigation has been stored in the root zone during irrigation. It is defined as,

$$E_s = \frac{W_s}{W_n} \times 100 \qquad \text{...(10.61)}$$

in which,

E_d = water storage efficiency (per cent)

W_s = water stored in the root zone during irrigation

W_n = water needed in the root zone prior to irrigation.

Water storage efficiency becomes important when water supplies are limited or when excessive time is required to secure adequate penetration of water into the soil. Also, when salt problems exist, the water storage efficiency should be kept high to maintain a favourable salt balance.

5. Water distribution efficiency. Not only the application of the right amount of water to the held but also its uniform distribution over the field is important. Permissible lengths of irrigation runs are controlled to a large extent by the uniformity of water distribution which is possible for a given soil and irrigation management practice. Water distribution efficiency indicates the extent to which water is unformly distributed in the field.

It is defined as

$$E_d = \left(1 - \frac{\bar{y}}{\bar{d}}\right) \times 100 \qquad \text{...(10.62)}$$

in which,

E_d = water distribution efficiency (per cent)

$\bar{d}$ = average depth of water stored along the run during the irrigation

$\bar{y}$ = average numerical deviation from $\bar{d}$.

(**Note: It may be seen that Equation 10.62 is identical to the expression used for determining *uniformity coefficient* described in chapter 12 under sprinkler irrigation.**)

Water Use Efficiency

The term water use efficiency denotes the production (of crops) per unit of water applied. It is expressed as the weight of crop produce per unit depth of water over a unit area, *i.e.*, kg/cm per hectare. Two distinct terms are used in expressing water use efficiency.

(*i*) *Crop water use efficiency.* It is the ratio of crop yield (Y) to the amount of water depleted by the crop in the process of evapotranspiration (ET).

$$\text{Crop water use efficiency} = \frac{Y}{ET} \qquad \text{..(10.63)}$$

(*ii*) *Field water use efficiency.* It is the ratio of crop yield (Y) to the total amount of water used in the field (WR)

$$\text{Field water use efficiency} = \frac{Y}{WR} \quad \text{...(10.64)}$$

Crop yield itself can be defined in terms of total growth (*i.e.,* dry matter production) or in terms of marketable product. Of the two indices for water use efficiency, the concept of crop water use efficiency is of fundamental interest while field water use efficiency is of greater practical importance. Water use efficiency obtained in a mixed cropping system is often more than a monocrop system, especially in supplemental irrigation farming system.

Referring to Equation (10.63) water use efficiency can be increased either by increasing the numerator (crop yield) or by decreasing the denominator. The numerator, being plant production, depends on such plant factors as gains due to photosynthesis verses losses due to diseases and pests. Hence, water use efficiency can be influenced by such means as pest and disease control, the choice of crop and the genetic improvement (by selection and breeding) of its productivity and adaptation to the particular environment; as well as by improvement of the water, air and nutrient supply to the roots, and of light and carbon dioxide supply to foliage.

Economic (irrigation) efficiency. Economic efficiency in irrigation is the ratio of the total production (net or gross profit) attained with the operating irrigation system, compared to the total production expected under ideal conditions. This parameter is a measure of the overall efficiency, because it relates the final output to input.

Example 10.16. *An area of 20 hectares is to be irrigated by a pump working for 12 hours a day. The available moisture holding capacity of the soil is 16 cm/m and the depth of root zone is 1 m. Irrigation is to be done when 50 per cent of the available moisture in the root zone is depleted. Water application efficiency is 70 per cent. Peak rate of moisture use by the crops is 4 mm (weighted average). Losses in water conveyance are negligible. Determine the irrigation period, net depth of water application, depth of water pumped per application, and the required capacity of the irrigation system in hectare-cm/day and litres per second.*

Solution:

$$\text{Net depth of water application} = 16 \times \frac{50}{100} = 8 \text{ cm}$$

$$\text{Irrigation period} = \frac{\text{Net irrigation requirement}}{\text{Peak use rate}}$$

$$= \frac{8}{0.4} = 20 \text{ days}$$

Depth of water pumped per application

$$= \frac{8}{0.7} = 11.4 \text{ cm}$$

Required capacity of irrigation system

$$= \frac{11.4 \text{ cm} \times 20 \text{ ha}}{20 \text{ days}}$$

$$= 11.4 \text{ ha-cm/day}$$

$$= \frac{11.4 \times 10{,}000 \times 1000}{100 \times 12 \times 60 \times 60}$$

$$= 26.4 \text{ litres/sec.}$$

Example 10.17. *A stream of 135 litres per second was diverted from a canal and 100 litres per second were delivered to the field. An area of 1.6 hectares was irrigated in eight hours. The effective depth of root zone was 1.8 m. The runoff loss in the field was 432 cu m. The depth of water penetration varied linearly from 1.8 m at the head end of the field to 1.2 m at the tail end. Available moisture holding capacity of the soil is 20 cm per metre depth of soil. Determine the water conveyance efficiency, water application efficiency, water storage efficiency and water distribution efficiency. Irrigation was started at a moisture extraction level of 50 per cent of the available moisture.*

Solution:

(*i*) Water conveyance efficiency,

$$E_c = \frac{W_f}{W_d} \times 100 = \frac{100}{135} \times 100 = 74\%$$

(*ii*) Water application efficiency,

$$E_a = \frac{W_s}{W_f} \times 100$$

Water delivered to the plot

$$= \frac{100 \times 60 \times 60 \times 8}{1000} = 2880 \text{ cu m}$$

Water stored in the root zone

$$= 2880 - 432 = 2448 \text{ cu m}$$

Water application efficiency

$$= \frac{2448}{2880} \times 100 = 85\%$$

(*iii*) Water storage efficiency,

$$E_s = \frac{W_s}{W_n} \times 100$$

Water holding capacity of the zone

$$= 20 \times 1.8 = 36 \text{ cm}$$

Moisture required in the root zone

$$= 36 - \frac{36 \times 50}{100} = 18 \text{ cm}$$

$$= \frac{18}{100} \times 1.6 \times 10{,}000 = 2880 \text{ cu m}$$

$$\text{Water storage efficiency} = \frac{2448}{2880} \times 100 = 85\%$$

(*iv*) Water distribution efficiency,

$$E_d = 100\left(1 - \frac{y}{\bar{d}}\right)$$

$$\bar{d} = \frac{1.8 + 1.2}{2} = 1.5\,\text{m}$$

Numerical deviation from depth of penetration:

At upper end = 1.8 – 1.5 = 0.3

At lower end = 1.5 – 1.2 = 0.3

$$\text{Average numerical deviation} = \frac{0.3 + 0.3}{2} = 0.3$$

$$E_d = 100\left(1 - \frac{0.3}{1.5}\right)$$

$$= 80\%$$

Increasing Water Use Efficiency by Crop and Soil Management Practices

Water use efficiency (crop yield/evapotranspiration of crop area) is influenced by crop and soil management practices. The numerator of this formula, namely, crop yield can be changed appreciably by management practices. The evapotranspiration or denominator of the formula is more difficult for man to control because it is dependent on climate and the availability of water for the crop. Water use efficiency is not closely dependent on the water available if the supply is within the evapotranspiration limit. The crop yields depend on the adequacy of the water supply. Most crops are sensitive to water stress at the critical phases of their growth.

Plant species adaptation. One of the primary ways of increasing crop yield and water use efficiency in a particular environment is to select plant species adapted to the total amount and distribution of water in an area. Plant species vary greatly in water requirement. Species selection, plant selection and plant breeding have contributed greatly to increasing the efficiency of the water available in an area.

Planting pattern. Planting patterns have a direct effect on yield, solar energy capture, and evaporation and thus an indirect effect on water use efficiency. Two important planting pattern practices are plant density and row spacing.

Planting date. In arid and semi-arid regions planting date is an extremely important cultural practice in efficient water use. The main reason for choosing the optimum dates for sowing is to ensure good germination by placing the seed in the optimum moisture zone, neither too deep nor very shallow. After the seeds germinate, the moisture should be optimum for root growth and root penetration and to envelope maximum soil volume for nutrient uptake. Planting dates also indicate the right type of climate for the shoot growth and optimum utilisation of moisture by the roots under normal rainfall conditions. Short season and early maturing crop species are often grown to escape the draught part of the season.

Weed control. One of the main management means of obtaining more efficient water use is the elimination of weeds in crops. Weeds compete with crops for soil nutrients, water and light. Technology for efficient weed control, both mechanical and chemical, have been developed which should be availed in increasing water use efficiency of crop plants.

Disease and insect pests. An often overlooked means of increasing water use efficiency lies in the area of disease and pest management of crops. It is possible to select crop varieties which are resistant to pests and diseases common to an area. Efficient crop and water management also includes the judicious use of chemical pesticides and selecting appropriate cropping patterns and cropping sequences. Growing the same crop year after year on a land tend to increase disease incidences in crops. Planting dates could influence pest incidence. For example, in North India the recommended sowing date for mustard crop is in the early part of the last week of September. Delayed sowing will result in delayed flowering which usually coincides with the incidence of aphids (small insects of the family Ahidae which may completely destroy the crop by sucking plant juice or by producing secretions which attract ants and cause heavy damage).

CROPS AND WATER

Water is essential for crop production and every effort must be made to get the maximum possible efficiency in water utilization by plants. A proper understanding of the effect of water—rainfall and/or irrigation—on crop growth and yield under different agro-climatic conditions is essential in planning, design and operation of irrigation schemes. The relationships involved are those between crop, climate, water and soil. Many biological, physiological, physical and chemical processes are involved. It is possible to analyse the influence of water supply on crop yields. The upper limit of crop production depends on the climatic conditions and the genetic potential of the crop. The extent to which this limit is reached will depend on the quality of engineering input in irrigation system design and operation to tune the irrigation water supply to meet the biological needs for water in crop production.

The relationship between crop yield and water supply is determined when crop water requirements and crop water deficits as well as maximum potential crop yield and actual crop yield are known. Water deficits in crops and the resulting water stress on the plant influence the evapotranspiration, ET and the yield, *Y* of the crop.

Water stress in the plant can be determined by the rate of actual evapotranspiration (ET_a) in relation to the rate of maximum evapotranspiration (ET_m). When water supply meets the full water requirement of the crop $ET_a = ET_m$. However, when water supply is insufficient, $ET_a < ET_m$. Water deficits beyond a certain limit can adversely affect crop yields. Actual crop yield is usually denoted as Y_a and the maximum yield potential is denoted as Y_m. The manner in which water deficits influence the crop growth and the yield varies with the crop species and stage of growth of the crop.

Water utilization efficiency and harvest index. Water requirement and yield of crops vary between crops and their varieties. The total dry matter and yield produced per unit quantity of water (kg per cubic metre) vary with the crop. This is expressed as the water utilization efficiency in kg per cubic metre. Usually only a part of the dry matter like grain, oil content or sugar content is useful. The ratio between the total dry matter and the harvested yield is known as the harvest index. The harvest index varies with the crop and its variety.

Yield response factor of a crop. The response of crop yield to water supply is quantified using the yield response factor k_y, which relates relative yield decrease $(1 - Y_a/Y_m)$ to relative evapotranspiration deficit $(1 - ET_a/ET_m)$. Results of research on actual crop yield Y_a in relation to maximum potential yield Y_m under different water supply regimes have resulted in the following empirical relationship (Doorenbos and Kassem, UNFAO, 1979):

$$(1 - Y_a / Y_m) = k_y (1 - ET_a/ET_m) \quad ...(10.65)$$

in which,

k_y = Yield response factor of a crop
Y_a = Actual harvested yield of the crop
Y_m = Maximum (potential) harvested crop yield
ET_a = Actual evapotranspiration
ET_m = Maximum potential evaporation

In addition to the availability of water in terms of the amount, time, and crop growth stage, crop yields are influenced by its variety, density of plants, nutrient availability, soil salinity, and control of crop pests and diseases.

In deriving the values of k_y of crops it is usually assumed that the relationship between relative yield (Y_a/Y_m) and relative evapotranspiration (ET_a/ET_m) is linear and holds good up to 50%, i.e., $1 - (ET_a/ET_m) = 0.5$. The results of selected field experiments (Doorenbos and Kassem, FAO, 1979) show the values k_y as 0.2 during the early part of the vegetative period for crops such as cabbage, peas and tobacco, 0.25 for sunflower while it is 0.45 for potato and watermelon. During the flowering period, k_y values are 1.1 for beans and tomato, 1.0 for sunflower; 0.8 for groundnut, soyabean and watermelon, 0.6 to 0.65 for wheat; 0.55 for safflower and sorghum. The coresponding values are much lower in the ripening period, namely, 0.2 for bean, groundnut, potato, pea and sorghum, 0.25 for cotton; 0.3 for watermelon and 0.4 for tomato. The values of k_y averaged for the entire growing period are 1 to 1.35 for banana; 1.25 for maize, 1.2 for sugarcane; 1.15 for bean, pea and spring wheat; 1.1 for potato, pea and watermelon; 1.0 for winter wheat; 0.95 for sunflower; 0.9 for sorghum and tobacco; 0.85 for cotton, grape and soyabean; 0.8 for potato; and 0.7 for groundnut.

The yield response of a crop to deficient water availability at different stages of its growth depends on the sensitivity of the crop to water stress at that period. In general, crops are more sensitivie to water deficits during the emergence of the seedling, flowering and early yield formation. Thus, after the emergence of the seedling, the mid-season stage which includes flowering and grain setting or yield formation is most sensitive to water shortage in most crops. Water shortage during the mid-season will greatly reduce crop yields as compared to other periods. The growth stage which is least sensitive to water shortage is the late season stage which includes ripening and harvest, especially in crops which are harvested dry (e.g., grain crops). Thus, in an irrigation command area with a single crop, which is planted in different periods (staggered planting), and there is water shortage, it is advisable to give priority to irrigating those fields on which the crop has reached the mid-season stage. It is interesting to note that some crops react favourably to water shortage during the crop development stages. They develop deeper root systems which are beneficial to the crops during the later stages. In general, for the total growing period of a crop the decrease in yield is proportionally lower for crops like lucerne (alfalfa), groundnut and safflower, while it is proportionally higher in crops like banana, maize and sugarcane.

The yield response to water deficit can vary among different varieties of the same crop. In general, high yielding varieties of a crop are more sensitive to water, fertilizer and other agronomic inputs than the traditional low yielding varieties. Crops grown for their fresh leaves and fruits are usually more sensitive to water stress than those grown for their seeds or dry fruits.

Under situations of deficient water supply, crops vary in their response to water shortage. In some crops there is an increase in water utilization efficiency, E_y under limited water supply, whereas in other crops E_y decreases with increase in water shortage. For example, in maize with water deficit spread equally over the growing season E_y will decrease whereas in sorghum it will increase marginally. The yield per unit area (kg/ha) will decrease in both the cases, but the yield reduction will be greater in maize than in sorghum. When water shortage occurs during a particular period in the growing season of a crop, the yield response to water shortage will depend mainly on how sensitive that stage is to deficiency in water availability and the degree of deficiency. In planning cropping systems in irrigation project areas the application of the yield response factor will be of great value in estimating water supply requirements and irrigation scheduling. Under situations of deficient water supply spread equally over the entire crop growing period with crops with different k_y values, the crop with a higher k_y value will suffer a greater yield loss than a crop with a lower k_y value.

Crop Response to Water at Different Stages of Growth

It has been observed that water requirement of crops varies with the stage of its growth. When water supply is limited, it is necessary to take into account the critical stages of crop growth with respect to moisture. The term *critical stage* is commonly used to define the stage of growth when plants are most sensitive to shortage of water. Each crop has certain critical stages at which if there is shortage of moisture, yield is reduced drastically. When there is shortage of water, it is better to take care of the critical stages first to obtain increased water use efficiency. The criticial stages of crops growth are required to be studied for different crops.

Growth stages of cereals in relation to irrigation. The following terms are used to describe the growth and developmental stages of grain crop in relation to irrigation (Salter and Goode, 1967).

Stage	*Details*
Germination	the appearance of the radicle.
Tillering	the formation of tillers, i.e., branches produced from the base of the stem.
Jointing	the stage when 2 nods can be seen, i.e., the beginning of shooting. (During this stage the internodes of the stem are elongating).
Shooting	the stage of elongation of internodes.
Booting	end of shooting stage and just prior to the emergence of ears.
Heading (earing)	the emergence of the ear from the tube formed by the leaf sheaths.
Flowering	the opening of the flowers.
Grain formation	the period of grain development from fertilization until maturity, which can be further subdivided into: (*i*) 'milk stage'–grain contents have a milky consistency. (*ii*) 'dough stage'–grain contents have a doughy consistency. (*iii*) 'waxy-ripe'–grain contents have a waxy appearance. (*iv*) 'full-ripe'–grain contents are hard. (*v*) 'dead-ripe'–grain ripe for harvesting.

Table 10.10 presents the sensitivity of common crops to water shortage.

Table 10.10. Sensitivity of common field crops to deficient water availability

Crops	*Sensitivity*		
High	*Medium to high*	*Low to medium*	*Low*
Green vegetables	Cabbage	Wheat	Cotton
Rice	Beans	Lucerne	Millet
Potato	Onion	Groundnut	Sorghum
Sugarcane	Tomato	Sunflower	Pigeon pea
Banana	Maize	Soyabean	Casava
	Peas	Citrus, Grape	

Table 10.11 presents the values of evapotranspiration (ET_m) of some of the important crops. The values presented in Table 10.11 are broad ranges. The actual values will depend on the climate, soils, type of crops and their varieties as well as the total growth periods. Field studies have revealed that during the growth period in some of the crops the availability of optimum soil moisture during particular periods produces the greatest benefits. They are jointing to soft dough stage in wheat, first bloom to full formation of boll in cotton, tuber formation to maturity in potato, bud to pod fill period in soyabean and boot to soft dough stage in sorghum.

Table 10.11. Growth stages of crops in which adequate availability of soil moisture is highly desirable for growth and yield

Crop	ET_m	*Growth periods sensitive to water deficit*
Vegetable crops		
1. Bean	300 to 500 mm	Flowering and pod filling
2. Cabbage	380 to 500 mm	Head enlargement and ripening
3. Onion	350 to 550 mm	Bulb enlargement
4. Pea	350 to 500 mm	Flowering and yield formation
5. Potato	500 to 700 mm	Stolonization and tuber formation
6. Tomato	400 to 600 mm	Vegetative period during and just after transplanting, flowering, yield formation
Foodgrains		
7. Maize	500 to 800 mm	Just prior to flowering, grain filling
8. Rice	500 to 700 mm	Tillering, flowering

(*Contd ...*)

Crop	ET_m	Growth periods sensitive to water deficit
9. Sorghum	450 to 650 mm	Flowering, grain formation (booting to heading)
10. Wheat	450 to 650 mm	Crown root initiation (20 to 25 days after sowing), flowering (80 to 85 days after sowing). Next in importance are jointing, tillering and yield formation
Oilseed crops		
11. Groundnut	500 to 700	Flowering, pod setting
12. Mustard and rapeseed	—	Pre-flowering (about 40 days after sowing), pod filling (about 80 days after sowing)
13. Safflower	600 to 1200 mm	Flowering, seed filling
14. Sunflower	600 to 1000 mm	Late vegetative stage, floweing, yield formation
Fibre crop		
15. Cotton	700 to 1300 mm	Flowering, bud formation, boll formation
Sugar crops		
16. Sugarcane	1500 to 2500 mm	Vegetative period (tillering and stem elongation)
17. Sugarbeet	550 to 750 mm	First month after emergence
Fruit crops, trees and other crops		
18. Banana	1200 to 2200 mm	Throughout the growing season, particularly first part of vegetative period, flowering, yield formation
19. Citrus	900 to 1200 mm/yr	Flowering and fruit set, followed by fruit enlargement
20. Grape	500 to 1200 mm	Vegetative period
21. Pineapple	700 to 1000 mm	Period of vegetative growth
22. Olive	400 to 600 mm	Just prior to flowering, yield formation, stone hardening
23. Watermelon	400 to 600 mm	Vegetative period, particularly flowering and fruit filling
24. Tobacco	400 to 600 mm	Period of rapid growth, yield formation and ripening

Based mainly on Brouwer and Prins (1980), Doorenbos and Kassem, FAO (1979), Michael and Pandey (1970) and Hukkeri and Pandey (1976).

PLANNING CROPPING SYSTEMS AND WATER MANAGEMENT IN IRRIGATION PROJECT AREAS

(Based on Crop Response to Water Availability)

In the design and operation of irrigation projects, the cropping systems are to be planned and water supplies regulated keeping in view the physical resources which are available, mainly the climate, soil and water resources (rainfall, soil moisture storage, groundwater and possible quantity of irrigation supplies at different periods of the crop season in different segments of the project area). Further, there are many technical, social and managerial issues to be considered to meet the production objectives and make the project economically and financially viable. When the available water resources are adequate and fully meet the crop water requirements of the project area, the production potential is maximum, keeping other crop production inputs at optimum levels. The irrigation supplies are determined on the basis of the crops selected, the length of the growing period of each crop and the area to be irrigated under each. Under conditions of water scarcity, crop production will depend on the management of irrigation supplies to best suit the crops in terms of their yield response to water availability. An equally important opportunity is to select cropping patterns and crop varieties aimed at maximizing the production within the limits of available inputs in crop production.

The cropping pattern in the irrigation command area, namely the sequence of crops grown in regular order on any particular field or fields is important in maintaining sustainability in agricultural production. It is always advisable to include useful leguminous crops in the sequence to improve soil productivity. Alternating shallow and deep rooted crops is beneficial for efficient utilization of the nutrients in the root zone. The application of the yield response factor k_y in the design and operation of irrigation projects permits quantification of water supply and water use in terms of the yield and production of crops in a particular area of a project. For example, with limited water availability spread over the total growing period, the yield decrease for maize (k_y = 1.25) will be greater than for sorghum (k_y = 0.9). When these crops are grown in a project area, maize should have priority in water supply, with emphasis to meet the full requirement during the flowering and yield formation periods. When the aim is to obtain the maximum possible production from an area and land is not a limiting factor, the available water should be supplied to fully meet the requirement of maize in a limited area and increase the area under sorghum which has lower water requirement and can tolerate water deficiency within certain limits, depending on its growth stage. In general, the yield response to deficient water availability during specific growth periods of crops is of great value in the scheduling of limited water supply with the aim of maximizing the yields. Allocation of limited water supply should be to meet the full water requirements of the crops during the most sensitive growth periods for water rather than spreading the available limited supply equally over the total growing period.

For the efficient operation of an irrigation project, the most suitable crop calendar must be outlined well before the start of the growing season. Depending on the available water resources of an area, a preliminary irrigation schedule is made. To obtain a high level of water utilization efficiency and maximum possible production, the schedule is adjusted according to the changes in crop water requirements over the growing season. Step-by-step procedures for crop selection, areas under different crops and irrigation scheduling of the available supplies of water are presented under the two different situations, viz., (*i*) under adequate water supply, conditions, and (*ii*) under deficit water availability. Under adequate water supply situations $ET_a = ET_m$ and $Y_a = Y_m$, other inputs maintained at optimum levels. Land and water management in such a situation requires optimum utilization of water and other inputs to achieve maximum production per unit of land for the selected crops over the available area. Under limited water supply conditions the available supply of water may be used either to meet the full requirement over a limited area under a crop or partially meet the water requirements over an extended area which is available for growing the same crop.

1. Adequate water supply. Under such a situation, the selection of crops and the area under each crop are based mainly on the suitability of the climate of the area for the crop and the availability of other inputs in crop production. The cost of the irrigation system and its management are also factors to be considered. The design and operation of the water supply and distribution system of the project would take into account limitations in the capacity of the system, the quantity of water to be supplied on a monthly, 10-day or weekly basis, the area covered under a crop, cropping pattern adopted and the irrigation schedule to be followed. To reduce excessive demand on the capacity of the irrigation system, adjustments in the crop calendar may be necessary.

The other factors include the optimum use of the rainfall during the crop season and available soil moisture storage from pre-season rains and reducing the peak requirements in water by shifting the sowing dates so that peak demand periods of different crops/crop varieties are spread over a long period. The design and operation of an irrigation project must meet the need to deliver the required amount of water to the different segments of the project area at pre-determined intervals during the crop season. To arrive at the water supply schedule in units of flow (cubic metres/second or litres/second) and flow duration in hours, the water supply requirements of individual fields are weighted and determined for a block of fields and for areas served by the different segments of an irrigation distribution system. The calculation procedure includes the determination of soil water balance, total water available in the root zone, interval and depth of irrigation application, and operation of the water supply and distribution system.

2. Limited water availability. With limited water supply, the selection of crops and the area under each crop are based on the yield of the crops as influenced by the extent to which crop water requirements (ET_m) are met by the available supply during the growing season. Under this situation it is necessary to consider the effect of the limited water supply during different growth stages of the crops. In case of crops which are very sensitive to water deficits, irrigation applications are scheduled to meet the full water requirement of the crops. Crops which are not very sensitive to water deficits and can be grown with economically viable yields without meeting the full water requirements, water supplies are scheduled minimizing water deficits during the crop growth stages which are most sensitive to water shortage. The considerations in the selection of crops and cropping patterns for a project area, the area under different crops and their sequences as well as the water requirement of crops and irrigation scheduling are generally same as under the situations of adequate water supply conditions. In addition, the selection of crops and the area under each crop must consider the crops that are most efficient in water utilization (a high E_y) under adequate water supply as well as those which can maintain an acceptable yield level under limited water supply (a low k_y). Under conditions of limited water supply, a preliminary estimate of actual crop yields and total production for the project area can be made by considering the yield response factor k_y for the total growing period of the crops with high E_y values. As k_y values are used along with the relative evapotranspiration deficit $(1 - ET_a/ET_m)$, the available and the required water supplies are expressed in terms of ET_a and ET_m.

Example 10.18 *Compute the relative evapotranspiration deficit and relative yield decrease in a maize crop with 4 months growing period. Water requirements for the maize crop are 860 mm. Effective rainfall during the growing period is 30 mm. The water supply available at the head works of the project is 135 cm/ha. The overall irrigation efficiency of the project is 45%.*

Solution:

Water available at the project headworks

$$= 135 \text{ cm/ha} = 1350 \text{ mm}$$

Water available to the crop at 45% project efficiency

$$1350 \times \frac{45}{100} = 607.5 \text{ mm}$$

Effective rainfall = 3 mm

Total water available for the crop

$$= 607.5 + 30 = 637.5 \text{ mm}$$

Relative evapotranspiration deficit $(1 - Et_a/ET_m)$

$$= \frac{1 - 657.5}{860} = 1 - 0.74 = 0.26$$

k_y for maize (from the text) = 1.25

$$k_y\,(1 - ET_a/ET_m) = 1.25 \times 0.26 = 0.325$$

$$= 1 - Y_a/Y_m$$

$$Y_a/Y_m = 67.5\%.$$

IRRIGATION REQUIREMENTS OF COMMON CROPS

Water requirement of crops vary with their type and variety and stage of growth, properties of the soil and the climate. The tolerance of the crop to soil moisture stress is influenced by the genetic make up of the crop, the rooting characteristics and the properties of the stomata. Water management for efficient water use is profoundly influenced by the selection of suitable crops and their varieties based on the amount of water available in the area during different periods of the crop growing season.

Adequate soil moisture, temperature and air are the essential requirements for the germination of seeds. The seeds are to be planted at shallow depth (usually 5 to 7 cm) below the ground surface on firm moist soil to ensure proper germination. When the soil moisture at the time of sowing is inadequate a pre-sowing irrigation is necessary. Additional water over and above the crop water requirements may be necessary for leaching of excessive salts in order to maintain proper salt balance in the crop root zone. In wetland rice cultivation, puddling of the seed bed is necessary to minimize water loss due to deep percolation below the root zone.

Rice. Rice is the most important cereal crop grown. It is the staple food for over half of the population of the world. It occupies about 40% of the irrigated area in India and is grown extensively in other south Asian countries, the entire south-east Asia and the Far East as well as many countries in other continents. It is a semi-aquatic plant and hence its water requirements are many times more than most other food crops. It is, therefore, a major consumer of the water and thus needs careful management in order to increase its efficiency of water use.

Rice is grown under varied soil and climatic conditions and one to three crops per annum are taken in India and other countries in monsoon Asia. Though rice could be grown on a variety of soils, it grows best on clay loams to clays since these are retentive of moisture and have low percolation rates (1-5 mm/day). It can also grow on soils with relatively high lime or alkali content. The crop thrives best under conditions of high temperature and humidity in North India as well as temperate regions. At high altitudes, the crop is grown only in the warm part of the year.

The cultural practices of rice vary widely, depending upon the variety and the local soil and climatic conditions. The conditions under which rice is grown could broadly be grouped into two, namely, wetland or lowland rice and upland rice. *Floating rice* is a particular variety of wetland rice which is grown in areas subject to deep flooding, with water depths ranging from 60 cm to 6 m. It is a long-stemmed variety which is sown before the onset of rains and starts growing quickly with the early rains. When flooding commences it grows rapidly and the plants float on the water surface.

Upland rice or dryland rice is directly sown by drilling or by broadcasting the seeds and it mostly depends on rain as the source of water. It constitutes only a small percentage (5%) of the total area under rice. Lowland rice culture is a widely accepted practice for high production. Under lowland conditions rice is generally transplanted on puddled soils and land is kept under submerged conditions, covering most of its growth period by rain and/or irrigation. Puddling reduces percolation losses, checks weed growth and regulates soil and water temperatures.

Wetland rice cultivation has many advantages. Crop growth in the nursery, which occupies only 5 to 10% of the normal crop area, saves considerable amount of water and provides the time required to prepare the seedbed for transplanting. Weed problem is low in a transplanted crop field. The transplanted crop under predominantly submerged condition during their growth provides the required environment for growth.

Wetland rice production comprises of the following operations: (*i*) Preparation of the rice nursery and sowing good quality seeds, (*ii*) Preparation of rice fields (usually level basins in rectangular or contour benches), (*iii*) Transplantation of the seedlings when it is about one month old in the nursery, (*iv*) Water control, (*v*) Weeding, (*vi*) Fertilization, (*vii*) Pest control, (*viii*) Harvesting, threshing and storage.

On germination of the rice seed, the radicle (the rudimentary root of the embryo) develops from the base of the grain, quickly followed by two additional roots. Subsequently, short lateral roots develop from the nodes of the stem below the ground level. Tillers are produced at the nodes and further adventitious roots are produced from the lower nodes and the plant develops a mass of adventitious roots. The tiller produced directly from the stem is the 'primary tiller', which is followed by further tillers. The roots tend to develop horizontally more than in the vertical direction. Hence, the nutrient uptake of the rice crop is mainly from the top layer of the root zone. A remarkable feature of the root system of rice plant is that it does not suffer from lack of oxygen under submerged condition. Using large basins is most suitable in wetland rice cultivation. For the distribution of water between basins, field channels are constructed between basins to deliver water to individual fields on either side, in case of gentle slopes (keeping field plots level), and to the down-slope fields in case of contour channels. The traditional practice of 'field to field' irrigation, in which water is turned to the adjoining lower basins after those at higher elevations are filled reduces the water distribution efficiency substantially and results in loss of fertilizer and nutrients.

In case of individual field plots, the traditional practice has been to keep the field submerged with water to a depth of about 8 to 10 cm or more. Field experiments conducted at various locations in India and other countries on wetland rice reveal that the practice of keeping the soil under shallow depth of submergence (about 5 $\pm$ 2 cm water) throughout the crop growth period is conducive to higher yields. The practice of shallow submergence directly saves considerable amount of water, as compared to deep submergence. These results, in general, hold good under different climatic conditions. The practice of continuous shallow submergence, however, is possible only when the water supplies are adequate and assured. Land needs to be scrupulously levelled to facilitate uniform spreading of water. Weeds, especially the grassy types, need to be controlled.

Experimental results are available to show that it is not always necessary to follow the usual practice of continuous submergence. The practice of intermittent submergence during the critical stages of initial tillering and/or flowering and maintenance of saturation to field capacity during the rest of the stages give yields comparable to those obtained under continuous shallow submergence. The water supplies, if limited, could be curtailed during the non-critical stages of crop growth. The shortage of water during initial tillering and flowering reduces the yield considerably, while the stages of tillering, grain formation and maturity tolerate water stress to a great extent. The water is drained completely in about 7 days before harvesting to facilitate the use of mechanical equipment in harvesting.

Drainage needs of rice. When continuous land submergence by ponding is practised, it is necessary to drain the soil either through horizontal (surface) or through vertical (sub-surface) drainage, once or twice during the growth period, especially on poorly drained clayey soils having the rate of percolation less than 2.5 mm/day. This practice is beneficial in removing the toxic substances such as sulphides and in supplying oxygen to the root system. The drainage period could last from four to eight days, depending upon the type of the soil. In case of sandy soils, the drainage period is about three days while in clay soils it is about five to seven days.

The following growth periods can be identified in a wetland rice crop:

(*i*) *Growth period in the nursery* (usually one month).

(*ii*) *Vegetative stage.* From transplanting to panicle initiation (initiation of the inflorescence with its main stem and sub-divided branches). This period may range from 45 days to 90 days, depending on the variety of the crop.

(*iii*) *Mid-season stage*. From panicle initiation to flowering, and it includes stem elongation, panicle extension and flowering.

(*iv*) *Late season or ripening stage*. Period from flowering to full maturity (duration about one month).

Transplanted rice may be of short duration, medium duration or long duration, depending on their growing period in the field after transplanting. The growth period of short duration rice usually ranges from 90 to 100 days, medium duration crops 100 to 130 days and long duration crops 130 to 160 days after transplanting.

Components of water loss in rice culture. Considerable amount of water delivered to the rice field in wetland cultivation is lost through deep percolation, evaporation from the water surface in the field and field channels, seepage from the bunds and field channel embankments, overtopping of the bunds, breaches and rat holes in the bunds, and runoff to drains. Vamadevan and Dastane (1968) observed that out of 1960 mm of water needed by rice, 1200 mm were lost through deep percolation (about 70%) on the sandy loam soils of Delhi and only 480 mm were actually used consumptively.

Great economy in water use by rice crop could be achieved if suitable measures are adopted to reduce the deep percolation losses. Some of the practices to improve water use in lowland rice are as folows (Dakishinamurthi, Michael and Shri Mohan, 1973).

1. Selecting heavy soils or those with hard pans in subsurface depths to reduce permeability.

2. Growing crop in large blocks instead of in isolated small holdings.

3. Scrupulous land levelling.

4. Puddling to reduce permeability.

5. Compacting soil, embedding polythene, sheet, applying chemicals such as bitumen, asphalt, etc. (these practices, however, are expensive and are not usually followed).

6. Treating fields by adding one to two per cent bentonite with small quantities of sodium chloride and sodium carbonate.

Wheat. Wheat (*Triticcum spp.*) is the most widely grown of all the cereal crops in the world. Wheat is grown in almost all temperate and most of the sub-tropical zones worldwide. It is also grown in some tropical countries at higher elevations. Wheat varieties can be broadly grouped into *winter wheat* and *spring wheat*. Winter wheat requires a cold period during early growth. Winter wheat has a growing period of 120 to 160 days. For winter and spring wheat, the minimum daily temperature for growth is about 5°C. The mean daily temperature for optimum growth and tillering is in the range of 15°C and 20°C. In regions of severe winter and frost winter, wheat requires a period of dormancy which may extend up to about 90 days, depending on the temperature conditions.

In India, wheat is grown on an area of about 19-20 million hectares and nearly 54 per cent of this area is irrigated. The best wheat producing region in India is the Indo-Genetic plain where cool and dry climate prevails during the growing season (*rabi*). Areas that are always moist and warm are not suited to wheat. Well drained clay loams, loams and sandy loams are the best soils for wheat. Wheat is commonly sown around the middle of October under rainfed conditions and around the middle of November under irrigated conditions. The crop takes 4 months to mature in the northern region, 3 months in the central, and 3 in the southern regions.

Wheat farming in India has been almost revolutionised with the introduction of the high yielding dwarf wheat varieties in mid-1960s. Thereafter, a number of dwarf varieties have been evolved in the country which are highly responsive to irrigation and fertilizer applications. A short, stiff, strong stem is the main advantage a dwarf wheat has over the taller varieties. The shorter stem allows the plant to utilize irrigation water without lodging, and results in higher yields.

Wheat is grown as an irrigated crop or rainfed crop, depending on the occurrence of rains during the growing period. In many temperate regions it is grown as a rainfed crop. The need for pre-sowing irrigation depends on the adequacy of soil moisture for germination of the wheat seed.

Considerable research information on the water management of wheat, using various criteria for irrigation scheduling based on soil, plant and climatic factors, has been obtained in wheat growing areas worldwide. The most practical criterion commonly adopted for scheduling of irrigations to wheat is the one based on the physiological growth stages critical in demand for water. In wheat, different growth stages such as crown root initiation, tillering, jointing, booting, flowering, milk, and dough could be well delineated. Experiments conducted to study the important stages critical in their demand for water have clearly indicated that some stages can tolerate moisture stress to a certain extent. Most of the workers have observed that in case of dwarf varieties of wheat, irrigation at the crown root initiation stage (20-25 days after sowing) resulted in the maximum production per unit of water applied and, therefore, this stages was considered as the most critical stage for irrigation. It has been observed that regardless of the depth of planting, the crown root develops about 2 cm below the surface of the soil (Fig. 10.7). If the soil zone around the crown is dry at the time of the crown root initiation, crown roots do not develop properly and only a few tillers are produced. An irrigation at the crown root initiation stage stimulates root development and tillering. The second important stage is that of flowering which takes place 80-85 days after sowing. Among the other stages, jointing and milk rank third, followed by tillering stage, and then dough stage. Irrigation is applied at the initiation of the stage so that the moisture in the soil is available during the stage in question. In general, wheat requires 3 to 5 irrigations in North India.

Fig. 10.7. Seedling of dwarf wheat showing crown root. Irrigation is critical at the crown root initiation stage.

There has been substantial contribution by the Plant Breeders of the Indian Agricultural Research Institute (IARI) and the Agricultural Universities in Indian since 1980s in developing stress tolerant wheat varieties which could be grown successfully under limited irrigation and rainfed conditions. The 'Kundan' variety of wheat developed at IARI in 1980s is recommended for cultivation under limited irrigation conditions in the northern plains zone of India. It is resistant to soil moisture stress and is maintaining its resistance since the past 25 years (Pandey, H.N., 2005). It can be grown successfully with 2 to 3 post-emergence irrigations.

Predominant drought situation prevails in many parts of Central India where wheat is cultivated under rainfed or limited irrigation availability. For such situations the IARI Regional Station, Indore has developed a number of *Aestivum* and *Durum* varieties of wheat which are listed below (Pandey, H.N., 2005):

Aestivum varieties	*Durum varieties*
HW 224 (Amar)	HD 4672 (Malav Ratan)
HI 1500 (Amrita)	HI 8627 (Malav Kirti)
HI 1531 (Proposed for release in 2005)	HI 8638 (to be released)

The above *Aestivum* varieties can be cultivated under rainfed as well as under one or two supplementary irrigations after germination and yield 30 to 40 quintals/ha. The *Durum* varieties mentioned above are semi-dwarf and under 2 support irrigations can yield 45 to 50 quintals/ha under October sowing.

Maize. Maize, also called as corn, is an important cereal crop grown in climates ranging from temperate to tropical during the period when mean daily temperatures are above 15°C and frost-free. About half of maize production in the world is from USA.

Maize is grown almost all over India under varied soil and climatic conditions. However, it is essentially a warm and humid season crop and in areas of mild climate it can be grown throughout the year. It grows best on fertile, well drained loamy soils. In North India maize is grown mostly during the monsoon period (*kharif*) of June-October as rainfed or irrigated and on a smaller scale in the summer season (Feb./Mar.-May/June) under irrigated conditions. Three crops of maize could be grown in a year (June-September, October-January and February-May) in southern India because of favourable weather conditions. Most of the improved varieties of maize require 100-120 days to mature, except in the northern hills where they take 10-20 days more.

When maize seed germinates, four seminal roots appear. They are soon superceded by adventitious roots which develop from the lowest nod of the stem, below ground level. They begin to spread out laterally in the upper layers of the root zone soil. Subsequently, they grow downward as the plant grows. In about 10 to 12 weeks after sowing the entire root zone of 75 to 120 cm is filled with a mass of fibrous roots. However, a major part of the maize roots lie within 75 cm from the ground surface.

The irrigation requirements of maize vary with the type of soil and the season in which they are grown, depending upon the rainfall received. In regions where rainfall is inadequate maize can be grown with supplemental irrigation. In northern India, the crop is normally sown after a pre-sowing irrigations as the monsoon arrives late. During the monsoon, supplemental irrigation may be required whenever soil moisture falls below the desired level. The early vegetative stage (20-40 days after sowing) and tasselling and silking stage (45-60 days after sowing) have been observed as critical in the demand for water by maize. Maize is very sensitive to excess water and hence, it is advisable to plant it on ridges or make ridges in the field after its establishment.

Sorghum. Sorghum (*jowar / cholam*) is next in area and importance to rice and wheat is India. It is grown for grain and fodder in the relatively dry tracts of central and southern India. Sorghum grows best in semi-arid areas with well distributed rainfall of about 300 to 1000 mm. The temperature in the growing season should be within the range of 16 to 40°C. Optimum temperature for growth is around 25°C. Sorghum has many features which make it a drought resistant crop. This is mainly due to the extensive root system, effective control of ET and stomata with an ability to recover rapidly after a period of water stress. In dry areas with low and/or erratic rainfall the crop can respond well to supplementary irrigation. Under conditions of deficient water supply it is advantageous to spread the water application to larger areas of sorghum and provide for irrigation at the critical periods of growth.

In India sorghum crop is grown mostly as a monsoon season crop (*kharif,* June-November). It is also grown in the October-March season on the moisture retentive deep black soils of South Gujarat and in those parts of Andhra Pradesh, Karnataka and Tamil Nadu which benefit from the North-East monsoon rains. The crop takes 110-130 days for maturity, dependng upon the variety.

Sorghum is grown on a variety of soils ranging from clay to sandy loam. When grown on clay to clay loams the irrigation requirement is low and on lighter soils, irrigation requirement is higher. Initial seeding, pre-flowering, flowering, and grain formation stages coinciding with 2-4, 12-14, 14-16 and 17 weeks after sowing, respectively, have been found critical for water demand of sorghum.

Pearl millet. Pearl millet (*bajra*) is well suited to warm areas of low rainfall (200-500 mm). In India it is commonly grown in Rajasthan, Haryana, Punjab, Uttar Pradesh, Gujarat, Karnataka and Tamil Nadu. It is mostly grown as a rainy season crop (*kharif*) under rainfed conditions and it takes 85-90 days for maturity. Pearl millet is preferred when the rainfall is inadequate for maize and sorghum, as it is more tolerant to drought. Observations, however, have revealed that pearl millet (*bajra*) responds well to the applications of supplemental irrigations.

The irrigation requirement of the crop vary from 150 to 200 mm in most parts of India. Experiments have revealed that the crop tolerates about 75% soil depletion from 0-30 cm depth of soil on heavy clay soils. On sandy loam soils, however, best yields are obtained with irrigations at 50% depletion level.

Finger millet. Finger millet (*Eleusine coracane* Gaertn.), commonly known as *ragi* in India is one of the important grain crops of Karnataka. It is also grown on a fairly large scale in Tamil Nadu, Andhra Pradesh and Maharashtra and on a smaller scale in Bihar, Uttar Pradesh and Gujarat. It is one of the drought tolerant crops suited for dry farming areas. The crop is grown as rainfed in the rainy season (June-October) and as an irrigated crop all the year round, particularly in Karnataka and Tamil Nadu. It can be grown on a variety of soils, including to some extent saline and alkali soils. The crop takes about 60 days for vegetative growth and matures in about 100-120 days.

Studies on the heavy black clay soils of Siruguppa (Karnataka) have shown that irrigations given at 50 per cent available moisture depletion in the top 30 cm soil layer were necessary to obtain high yields of *ragi* sown in the summer season while that sown in the (*kharif*) season did not require any irrigation (Patel, *et al.*, 1969). Similar results have been reported from the sandy loam soils of Tirupati (Andhra Pradesh).

Barley. Barley is commonly grown as a *rabi* crop (October-March) mainly in the States of Uttar Pradesh, Rajasthan and Haryana. It needs less water than wheat for its growth. It is grown as a rainfed crop in areas having rainfall of 400-500 mm in the growing season. In drier areas the crop needs irrigation for good growth. It is generally grown on sandy loam soils of medium fertility level. The crop has a moderate level of tolerance on saline and alkaline soils. The moisture sensitive period occurs at the end of the shooting stage and during earing. Studies have shown that during this period, drought conditions have their maximum effect in reducing yield, and irrigations have the greatest effect in increasing yield.

Pulses. Pulses, or grain legumes, by virtue of their deep tap root system, utilize soil moisture efficiently and require less number of irrigations as compared to cereals. These crops, therefore, have been introduced in the cropping pattern in areas with limited water resources. Legumes in monsoon season (*kharif*) are normally grown without any irrigation in most parts of India, unless the rainfall distribution is very unfavourable. However, legumes grown in winter (*rabi*) and summer seasons, depend on irrigation. Most of the legumes do not require irrigation at the early vegetative stage, as it may do more harm than good by interfering with the nodulation and restrict oxygen requirement of roots at this stage. These crops profit from irrigations during flowering and podding.

Bean. Common bean (*Phaseolus vulgaris*), also known as French bean or kidney bean, is grown as a vegetable crop for fresh pods or as a pulse crop for dry pods. Common bean grows well in areas with medium rainfall, but the crop is not suited to the humid wet tropics. Excessive rains and continued hot weather cause flower and pod drops and diseases incidence. Bean is sensitive to salinity. Optimum temperature range for the growth of beans is 15°C to 20°C. The minimum temperature tolerated is 10°C. The length of the growing period is 60 to 90 days for green beans and 90 to 120 days for dry beans.

The tap root of the bean plant may reach depths of 1 to 1.5 m. The lateral root system is extensive and is mainly concentrated in the first 30 cm from the ground surface. At emergence the root depth is about 7 cm, at the start of flowering 30 to 40 cm and at maturity 1 to 1.5 m. Water uptake takes place mainly from the first 30 to 70 cm depth.

Water management requirements for green and dry beans comprise the maintenance of soil moisture around 50% of the field capacity level. When the crop is grown with supplemental irrigation, water application is directed to meeting the water requirement during the crop establishment period, flowering and yield formation periods. When bean is grown for dry seeds the water deficit in the root zone should not exceed 60% to 70% of the field capacity level. For green gram, water supply is continued till just prior to the last picking. When harvested for dry pods, irrigation is discontinued 20 to 25 days before crop harvest.

Gram. Gram (*Cicer arietinum* L.), also known as chick pea (*chana*), is extensively grown in northern India in the winter season (*rabi*) on alluvial sandy loam to loam soils. In central and southern India, the crop is grown on water retentive clay loam soils and hence is not normally irrigated. The crop is best suited for low rainfall areas. However, it has been found to respond to irrigation in the absence of winter rains, especially on the lighter soils of northern India. The crop generally requires only one or two irrigations when grown in *rabi* in northern India. Pre-flowering (45 days after sowing) and flowering (70 days after sowing) stages have been found to be most responsive to irrigation.

Pigeon pea. Pigeon pea [*Cajanus cojan* (L). Millsp.], commonly known as red gram (*arhar*), is grown in practically every state of India. The crop can be grown on almost all types of soils, ranging from alluvial sandy loam to heavy black clay soils. It is mostly grown in the rainy season (June-November). It is grown as a *rabi* crop in the south where mild winter prevails, in which case the sowing is done in September-October, and harvesting in March-April. The crop benefits from irrigation if dry spells prevail for long periods. Flower initiation (about 75 days after sowing) and pod filling (about 100 days after sowing) are found to be most responsive to irrigation for short duration red gram.

Green gram. Green gram (*mung*) or mungbean (*Vigna radiata*), is cultivated almost all over India. The major producing states being Uttar Pradesh, Madhya Pradesh, Punjab, Andhra Pradesh, Rajasthan, Karnataka and Tamil Nadu. It is largely grown as a rainfed *kharif* crop. In some parts of Maharastra, Andhra Pradesh, Karnataka and Tamil Nadu it is also grown in the *rabi* season as a second crop. The crop takes two and half to three months for maturity. In *rabi* and summer, the crop needs irrigation for higher production. The crop when grown in *rabi* or summer generally requires two to four irrigations for high yields. A pre-sowing irrigation is generally found to be beneficial.

Oilseeds. Oilseeds occupy approximately 25 per cent of the total cropped area of India. Like grain legumes pulses, oilseeds are also efficient in water use and hence require less water for their growth. These crops are mostly taken as rainfed but their yields could be boosted substantially if grown under limited irrigation conditions, especially in the arid and semi-arid areas and on lighter type of soils. Irrigation requirement is mainly based on soil moisture depletion levels, depending on the rainfall during the growing season. Flowering and fruiting stages are observed to be most responsive to irrigation.

Mustard and rapeseed. These are edible oilseed crops mainly grown in North India. These are crops of tropical as well as temperature zones and require cool weather for satisfactory growth. In India, these are grown in the winter season (*rabi*), mostly on sandy loam to clay loam soils. The optimum sowing time for the common varieties is the last week of September. Sowing after September end will delay flowering and result in attack by insects. Being a crop with well developed tap root system, mustard and rapeseed can utilize soil moisture from deep layers in the root zone and can produce high yields without any irrigation after germination. The crop has been found to be highly successful in the flood-prone areas of the Indo-Gangetic plains (covering an area of about 2 million ha) when sown in moist soils in the last week of September, with yields ranging from 20 to 30 quintals/ha. The technology sponsored by the IARI, New Delhi, revolutionized mustard cultivation in India in late 1980s (Fig. 10.8). These crops benefit from irrigation if the winter rains are inadequate. Pre-flowering (about 40 days after sowing) and pod filling (about 80 days after sowing) stages have been observed to be the most responsive to irrigation.

Fig. 10.8. Rich growth of high yielding mustard in a compact community farming block of the unirrigated flood plains of Khora Kanak, Fatehpur district, Uttar Pradesh (1989), under the IARI Transfer of Technology Programme.

Groundnut. Groundnut (*Arachis hypogaea*), also called peanut, has high oil content and nutritive value. The oil content varies from 48 to 51% and the protein content is about 25%. The crop is highly adapted to well-drained, loose, friable medium textured soils. The growing period of groundnut ranges from 90 to 115 days for sequential branched varieties and 120 to 140 days for alternately branched varieties. For optimum growth of groundnut the mean daily temperature ranges from 22 to 28°C. The crop yield gets reduced at temperatures above 33°C and below 16°C. The germination gets delayed at temperatures below 20°C. The crop is planted with a row spacing of 75 cm to 1 m. Groundnut is moderately sensitive to soil salinity. Being a legume, groundnut can fix nitrogen from the atmosphere.

Groundnut is commonly grown in semi-arid regions both under rainfed and irrigated conditions. It has a well developed tap root system extending to about 1.8 m depth. The major part of the root system is within the top 50 to 60 cm soil layer. The flowering period is most sensitive to water stress. Water deficiency during flowering causes flower drop and improper pollination. Next is the yield formation period. Water deficit during this stage results in reduced pod weight. In general, water deficiency during the vegetative period causes delayed flowering and harvest, accompanied by reduced growth and yield. To obtain higher values of total production per unit of water applied it is better to cover larger areas under the crop and reduce the quantity of water applied, taking care of the growth periods sensitive to water stress.

Sunflower. Sunflower is an important oilseed crop widely grown as a rainfed crop and under supplemental irrigation. The crop is photo-insensitive and can be grown in almost any season. It takes about 3 to 4 months for its maturity. Sunflower is sensitive to frost. Mean daily temperature for good growth is about 18 to 25°C. It is grown on a wide range of soils. Optimum soil pH is in the range of 6.0 to 7.5. At lower values liming may be required. Sunflower is particularly sensitive to boron deficiency. Under erratic and low rainfall conditions, a deep soil with good water-holding capacity is required. Sunflower has deep root system which may extend up to 2 to 3 m in deep soils. Hence, the plant can extract moisture from deep layers.

The water requirement of sunflower is in the range of 600 to 1000 mm, which is high when compared to most crops. The ET of sunflower may be as high as 12 to 15 mm/day at the flowering stage. Despite the high water requirement the plant can withstand short periods of severe water stress. However, severe water stress during the early vegetative stage results in reduced plant height, but deeper root depth. Adequate soil moisture during the later part of the vegetative period is necessary for proper bud development. Flowering period is most sensitive to soil moisture stress which may result in considerable yield reduction, as only fewer flowers come to full development. Yield formation stage is the next most sensitive to water deficit, causing severe reduction in yield and oil content. Supplemental irrigation is highly desirable during periods of failure of rainfall at these stages.

Safflower. Safflower (*Carthamus tinctorius*) is an important oilseed crop in India, the Mediterranean basin, China, Japan, Mexico and USA. In India, safflower is grown largely in Maharashtra, Madhya Pradesh, Chhattisgarh, Andhra Pradesh and Karnataka. The crop is grown in the winter season (*rabi*) as a rainfed crop, especially on water retentive clay soils.

Safflower requires a fertile, fairly deep and well drained soil. The crop is not suited to lowland humid tropics. It is a draught resistant deep rooted crop. In deep soils, the roots may grow as deep as 3 m or more. The crop can withdraw water from depths as deep as their root system. Safflower is moderately resistant to soil salinity. However, high yields are produced on soils with neutral reaction. When the pH is lower than 6, liming may be required. The row-to-row spacing of the crop varies from 50 to 80 cm, with 15 to 35 plants per metre length on each row.

Safflower is susceptible to diseases under excess soil moisture conditions. Excessive humidity may cause head rot. Heavy rains during flowering may adversely influence pollination and seed filling. Soil moisture deficits during the early and late vegetative periods cause reduction in growth and prolong the growing period. Safflower is fairly tolerant to soil moisture stress. However, flowering and yield formation stages are sensitive to water shortage. The water uptake gets reduced when about 60% of the available soil moisture has been abstracted. Under conditions of limited water availability, the overall production is increased by increasing the area under the crop and partially meeting the crop water requirements rather than meeting the full water requirements over a limited area.

Cotton. Cotton is the most important fibre plant in agriculture. Most of the cotton grown in the world is rainfed, except in regions with low annual rainfall where the crop is grown under irrigated conditions. Cotton crop falls into two groups: (*i*) *Gossypium arboreumum,* and *Gossypium herbaceum* which have short fibres which have low commercial value and are being progressively replaced, (*ii*) *Gossypium hirsutum* which has medium length and strong fibres. Varieties of this type are most widespread in the world. *Gossypium barbadense* is a cotton variety of this group with long fibres which are smooth and fine and of high commercial value. The Egyptian cotton belongs to this variety.

In northern India, the crop is sown in the month of May after pre-sowing irrigation and at least two to three irrigations are required prior to the start of monsoon. During the rainy season, normally no irrigation is required, unless rainfall distribution is very erratic. In central and south India, sowing is coincided with the onset of monsoon in the month of June. In parts of Tamil Nadu and Karnataka, the crop is also sown in August-October as these regions receive rainfall during this period and the temperature remains high and favourable, unlike the cold temperatures in northern India.

Cotton is very susceptible to frost. The total growing period ranges from 150 to 180 days. Depending on the variety and temperature conditions, bud formation takes about 50 to 60 days from planting, 25 to 30 days for flower formation and 50 to 60 days from the opening of flowers to maturity. The optimum temperature for germination ranges from 18 to 30°C. Temperatures between 27 and 30°C are optimum for boll formation. Above 38°C the yield gets reduced. Cold and strong winds harm the young seedlings. Such winds at maturity will blow away fibres from open bolls.

The cotton plant has a deep and extensive root system when grown on deep, well drained, and fertile soil. The root system develops rapidly, and within 70 to 80 days may extend to a depth of about 1.8 m and laterally to the same extent. Rooting depth is affected by soil and drainage conditions, as well as by irrigation frequency. In general, rooting depths are shallower on clay soils, or with frequent irrigations. Cotton can extract most of the soil moisture to the depth of rooting and this characteristic has an important influence on irrigation of cotton.

Pre-planting irrigation is important unless rainfall has brought the soil to field capacity to a depth of about 1 m. This will ensure good germination and rapid plant development. Under extremely unfavourable conditions, a light irrigation after planting may be needed to ensure germination.

Sufficient water should be added to the soil, either from rainfall or irrigation, to ensure that the entire root zone is brought to field capacity early in the growing season; say within 40 days after planting. Cotton is relatively tolerant to short periods of waterlogging. However, waterlogging and heavy rains may lead to lodging of the plants. The crop is relatively tolerant to soil salinity.

Adequate soil moisture is needed for rapid growth and good budding and formation of healthy bolls. However, excess water early in the growing period will restrict root and crop development. Cotton requires adequate water supply, particularly just prior and during bud formation. Severe water deficits during flowering may restrict the growth, but with subsequent water application the crop growth recovers. In the vegetative period, irrigation is scheduled when about 60% of the available moisture in the top 75 cm of the root zone has been taken up by the crop. During flowering the soil moisture depletion of about 70% could be tolerated without adversely affecting the yield.

Irrigation after flowering and fruit setting period is needed only to supply sufficient soil moisture for fibre and seed development. The soil moisture levels maintained during the flowering and fruit setting period are usually adequate for this period. Particular attention should be given to avoid over-irrigation during this period as this may encourage delay in opening of bolls. The timing of the last irrigation will depend on the water-holding capacity of the soil. In general, it is desirable to allow the cotton crop to mature on stored soil moisture, and extract almost all available soil moisture at the end of the season. Thus, in north India, irrigation of cotton on deep loamy soils is not generally needed after late September, while on drier soils, it may be necessary to irrigate till mid-October.

Jute. Jute is an important fibre crop of India and Bangladesh. It is grown on a large scale is West Bengal and to some extent in Bihar. Uttar Pradesh, Orissa and Andhra Pradesh. The crop requires warm and humid climate, with temperature ranging from 24°C to 35°C and relative humidity of the order of 90 per cent. The crop requires irrigation if rains are untimely and not well distributed. Sowing of jute in lowland is done is February, on midlands in March-April and on uplands in May-June. In some parts of Bihar and Uttar Pradesh, the sowing may continue upto mid-July. Irrigation schedules based on the maintenance of a soil moisture deficiency level of 50 to 60 per cent have been reported to be most favourable for jute.

Sugarcane. Sugarcane (*Saccharum officinarum*) is the most important source of sugar in the world. The crop requires a long, warm growing season with a high incidence of radiation and adequate soil moisture, followed by dry, sunny and fairly cool but frost-free ripening and harvesting periods. Nearly 70% of the area under sugarcane in India is irrigated but this constitutes only about five per cent of the irrigated area of the country. Sugarcane usually occupies the land for about 10 to 18 months and thus necessitates adequate irrigation for realising its potential yields.

In North India, planting is mostly done with the commencement of spring season (February-March). In the States of Tamil Nadu and Andhra Pradesh in the south, planting is done during December-February. In Maharashtra and parts of Karnataka, the crop is planted in December-February, October-November and July-August for 12, 15 and 18 months crops, respectively.

Sugarcane is propagated by planting stem cuttings from mature plants. Optimum temperature for sprouting (germination) is 32 to 38°C. Roots emerge quicky from the original cutting. After a time, secondary shoots emerge quickly from the original cutting, and they send down shoot roots which eventually develop into a mature root sysem.

Optimum growth in sugarcane is achieved between 22 and 30°C. Minimum temperature for active growth is about 20°C. However, for ripening a relatively lower temperature in the range of 20 to 10°C, which has a significant influence on the reduction of vegetative growth and enrichment of sucrose in the cane, is preferable. The growth period ranges from 9 months with harvest before winter frost to 24 months in Hawaii, USA. In general, the growing period is between 15 to 16 months. The first crop is usually followed by 2 to 4 ratoon crops, each taking about a year to mature. Adequate soil moisture throughout the growing period is important to obtain high yields.

Sugarcane has an extensive fibrous root system, with roots most extensive in the upper 60 to 90 cm of soil. Some roots may extend to depths as much as 240 cm in well drained, deep loamly soils. Moisture extraction is greater in the upper 120 cm, decreasing rapidly below that depth. Rooting patterns will vary with soil type and drainage conditions.

Basically, three methods of water application are adopted: furrow, sprinkler and drip. The frequency and depth of irrigation vary with the growth periods of the crop. During the emergence and establishment of young seedlings, light and frequent irrigations are required. During the early vegetative period, the tillering is in direct proportion to the adequacy of soil moisture in the root zone, which requires frequent irrigations. During stem elongation the irrigation interval can be extended, but the depth of irrgation is increased. The response of sugarcane to irrigation is greater during the vegetative and yield formation stages.

Average soil moisture in the first 60 cm should generally be kept at a level of above 66 per cent of total available moisture. However, during winter time in North India, temperatures are too low for growth, and available soil moisture levels need only to be maintained at about 50 per cent level in the root zone. Just before the harvest of the crop, irrigations may be somewhat delayed as the practice results in the accumulation of sucrose in the cane.

Sugarbeet. Sugarbeet *(Beta vulgaris),* is a sugar enriched root crop, which provides about 40% of sugar production in the world. Sugarbeet needs a relatively long growth period, ranging from 140 to 200 days. The crop is grown under a variety of climatic conditions. The minimum temperature for seed germination is 7 to 10°C. During the vegetative growth period a higher temperature is preferred. High sugar yields are obtained when the night temperatures are in the range of 15 and 20°C and daytime temperatures between 20 and 25°C. During the later part of the growing period, day temperatures above 30°C decrease the sugar yield substantially. Sugarbeet is grown in a variety of well-drained soils with medium to slightly heavy textures. Seeds are planted 1 to 2 cm deep in single or double rows. The row-to-row spacing of the crop range from 50 to 70 cm. Crust formation on the soil surface during seed germination can lead to poor crop stand. When the plant attains 4 to 8 leaves, thinning is done either manually or by machines.

Sugarbeet has a relatively high tolerance to soil salinity. However, soil pH below 5.5 is unfavourable for growth. The process of photosynthesis results in the formation of sugar in the leaves, a part of which is utilized in the metabolic processes in the plant and the rest is stored in the fleshy part of the roots, particularly during the growing period when the vegetative growth slows down. The sugar yield of the crop is determined by the size of the roots and their sugar concentration. With the rapid growth of the storage roots, sugar concentration reaches a high value, depending on the climatic factors, soil moisture content and the nitrogen level in the soil. Nitrogen is given in split applications, a small part at the time of planting and the rest after the thinning of the crop. However, excess application of nitrogen as well as application during the late growing season will reduce the sugar yield. Sugar content is also influenced by the crop variety and plant spacing. Often the sugar content in the root is greater than 15% of the fresh root weight. When grown for sugar, flowering and seed formation stages are avoided. When the crop is grown for seeds several weeks at low temperatures near 4°C are required to induce flowering.

Three distinct growth stages are recognized in sugarbeet. The first, ranging from 25 to 35 days is from the time of sowing to the emergence of seedlings. The second, lasting 75 to 115 days is from the emergence of the seedling to the point of constant growth. During this period the leaf surface develops faster than the root. In the third period, which extends from 40 to 50 days, the rate of growth is more constant and strong root develops along with a large canopy of leaves. A well-developed plant has a deep tap root and a set of well-branched lateral roots which fill the upper 35 cm of the soil, extending to a radius of about 50 cm.

Sugarbeet is grown both under rainfed and irrigated conditions, depending on the availability of rains. Sugarbeet is particularly sensitive to soil moisture deficits during the emergence of the seedlings and a period of about one month after emergence. Frequent light irrigations are desirable during these periods. On the other hand, over-irrigation will retard leaf development and can encourage flowering in the first year (bolting) which is harmful when the crop is grown for sugar. Water deficits during the mid-growing period (vegetative and yield formation periods) will adversely influence the sugar yield more than during the ripening stage. Soil moisture deficits, together with deficiency in nitrogen towards the end of the growing period, may reduce the growth of roots but increase the sugar concentration. Thus, except during the emergence and early growth periods, the crop is less sensitive to soil moisture deficits.

Tobacco. Tobacco *(Nicotiana tabacum)* is grown under a wide range of climates. The crop requires a frost-free period of 90 to 120 days from transplanting to last harvest of leaves. The optimum temperature range for growth is 20 to 30°C. A dry period is required for ripening and harvest of leaves. Tobacco is raised in nurseries (soil beds) and is transplanted 40 to 60 days after sowing when the plant is about 15 cm tall.

Tobacco is grown under a variety of soils and climatic conditions. Cigarette tobacco is normally grown on light soils and mainly in *rabi* (winter season) in the southern parts of India. *Bidi* tobacco is grown as a rainfed crop during *kharif* and *rabi* seasons, mostly in Gujarat, Karnataka and Maharashtra. *Hookah* tobacco is mostly grown in the alluvial soils of Uttar Pradesh, Bihar, West Bengal, Haryana and Punjab.

Tobacco when grown on heavy black clay soils normally does not require irrigation especially in the moonsoon season (*kharif*), but the crop grown on light soils and during winter season (*rabi*) requires irrigation. Furrow irrigation is more efficient than other surface methods of irrigation.

Tobacco has a well developed tap root with extensive and well developed secondary roots. However, about 75% of the moisture uptake is from the top 30 cm of the root zone. When grown under irrigated conditions tobacco requires careful scheduling of irrigation applications. In the nursery the seedlings are given light frequent irrigations. After 30 to 40 days of sowing the seedlings require less water in order to obtain robust plants. The period of maximum water requirement is 50 to 70 days after transplanting. This is followed by a decrease in water requirements. Moderate water deficits during the early vegetative period may enhance root development. However, water deficits during the mid-vegetative period (period of rapid growth) may result in reduced growth and smaller leaves. Severe water deficits during the yield formation and ripening periods adversely affect leaf weight and chemical composition which will influence the fire-holding capacity. However, a mild water deficit during ripening may be favourable as it will restrict the growth of new young leaves. On the other hand, excess water due to rains or irrigation will result in leaves of low quality. Waterlogging for two or more days will severely damage the plant.

Fodder Crops

Important fodder crops of the monsoon season (*kharif*) are maize, cowpea, pearl millet and sorhum. The important *rabi* fodders are oats and berseem or Egyptian clover. Some fodder crops like napier and lucerne are grown as perennial crops.

The objective in the irrigation of fodder crops generally is optimum vegetative growth of a certain quality. Since seed production is not generally the objective, the timing of irrigation with respect to grain yields does not apply. For example, in maize and sorghum (*jowar*) grain yields are not appreciably affected by early season stress (prior to bloom) if there is no severe wilting. Such a stress will have a marked affect on vegetative growth and fodder yield. Lucerne, for example, generally produces good seed yields only under moisture stress. As a general rule, optimum forage production is possible only with a continuous high moisture level. A safe rule to follow is to irrigate so as to keep the available soil moisture level above 50 per cent at all times.

Lucerne (Alfalfa). Lucerne (*Medicago sativa*), also known as alfalfa, is an important fodder crop of high nutritive value. It is a perennial crop planted from seed and reaches its maturity in the first year. The highest yields are obtained in the second year of growth. It is a legume which grows well on a variety of well-drained soils with pH values above 7. Lucerne is moderately sensitive to soil salinity. When properly innoculated it develops nitrogen fixing bacteria in the root nodes. It is a dense crop grown in closely spaced rows. It is cut for fodder several times during each growing season and may remain productive for 3 to 7 years or more. Lucerne is also grown as a short season annual crop.

An important feature of the crop from the point of view of irrigation is the well-develped tap root system. The tap root penetrates up to 4 m in deep soils. To stimulate growth, lucerne is to be irrigated frequently. Root development is adversely affected by dryness. To promote growth, irrigation is normally applied just after each cutting. Adequate soil moisture is maintained in the root zone throughout the growing period by irrigation when the rainfall is inadequate. However, excess irrigation may cause reduced soil aeration which is harmful to the crop.

Vegetable Crops

Vegetables form an important component of human diet, supplying major amounts of vitamins A and C and numerous nutrients, protein and fat. Vegetable crops are grown over a wide variety of soils and climatic conditions. One or the other vegetable crop is grown at any time of the year, depending upon the availability of irrigation water. The important monsoon season (*kharif*) and summer vegetable crops are lady finger, tomato, brinjal, clusterbean, sweet potato, and cucurbitaceous crops. The important winter (*rabi*) vegetables are cabbage, cauliflower, knolkhol, beetroot, radish, turnip, carrot, potato, onion and leafy vegetables like lettuce and spinach. Most of the vegetable crops are of short duration and quick growing in habit and need frequent irrigations for their maximum production. The irrigation requirements will vary, depending upon the duration of the crop and the season when grown.

Vegetable crops include perennials such as asparagus (*Asparagus officialis*) which have extensive and deep root system and hence are draught resistant. Another perennial vegetable crop is globe artichoke *(Cynara scolymus)*, which also has extensive root system but requires frequent irrigations throughout the year. Soil moisture deficiency, high temperature particularly during the time buds are forming, results in loose buts which are not marketable. The crop does not tolerate waterlogging. Almost all other vegetable crops are seasonal. They include potherbs or greens, including spinach, and kole which are shallow rooted. The salad crops include lettuce, celery, endive, and parsley which are relatively long season crops with shallow root systems. Soil crust on the soil surface adversely affects the emergence of seedlings of lettuce. The other group of important vegetable crops are the crucifers, namely cabbage, cauliflower, broccoli and brussels sprouts. Important root crops amongst the vegetable crops are beets, carrot, radish and turnip which are rapid growing crops which readily respond to irrigation. Bulb crops include onions and garlic. Beans and green peas are the prominent legumes amongst the vegetable crops. Solanaceous vegetable crops include tomatoes which are deep rooted with high water requirements and high yields. Cucurbits include cucumbers, muskmelon, watermelon, pumpkin and squash which are crops with medium or deep root systems.

Adequate soil moisture availability in the root zone is of prime importance in the cultivation of vegetable crops. In arid and semi-arid climates irrigation is a basic requirement in the cultivation of vegetable crops. Supplemental irrigation is common in most climatic regions, including humid and sub-humid regions. In surface irrigated crops, furrow irrigation is usually preferred. Under conditions of limited water availability both the drip and sprinkler methods of water applications are suitable, though during the recent times drip irrigation is preferred in many situations.

The water requirement of vegetable crops and the frequency of irrigation vary with the type of crop, soil and climatic conditions. However, as a general rule, applicable to most of the vegetable crops, the available soil moisture in the crop root zone should be replenished when 50% of it has been depleted. Specific characteristics of some of the common vegetable crops influencing irrigation are presented in the following pages.

Cabbage. Cabbage *(Brassica oleragea)* requires a cool climate for growth. The total growing period depends on the growing season. Spring season crop usually has a growing period of 90 days, while autumn crop has an extended growth period up to 200 days, depending on climatic conditions, the variety of the crop and the date of planting. Normally, the growing period ranges from 120 to 140 days. Most varieties can tolerate short periods of frost. Optimum growth occurs at mean day time temperatures around 27°C. Mean relative humidity may be in the range of 60 to 90%. In general, heavy loam soils are preferred. Under heavy rainfall conditions, sandy and sandy loam soils are preferable as they have better drainage properties.

Cabbage has an extensive shallow root system. The majority of the roots are in the top 40 to 50 cm of the root zone. The transpiration of the crop increases with the increase in the growth, with the peak towards the end of the growing season. During the vegetative period, the development of the crop is slow and the water requirements low. During the yield formation stage, which is the period of rapid growth, the availability of adequate moisture in the root zone is important. Depending on the climate, stage of development of the crop and soil type, the frequency of irrigation varies from 3 to 12 days.

Cauliflower, Brocolli and Brussels Sprouts. Plant characteristics of cauliflower, broccoli and brussels sprouts are generally similar to cabbage. However, cauliflower plants will not produce marketable heads when grown under deficit soil moisture. Brocoli is more tolerant to soil moisture stress, compared to cauliflower. Brussels sprouts require adequate soil moisture throughout the growing season. Maintenance of 50% of available moisture in the top 60 cm of the root zone is the recommended practice in all these crops.

Carrots. Carrots grow best in deep sandy and sandy loam soils. Heavy, compacted and cloddy soils hinder smooth development of carrots. The crop requires adequate soil moisture throughout the growing season.

Pea. Pea (*Pesum sativum)* is grown both for fresh and dried seed. The crop varieties range from tall, climbing plants to small bush types. Bush type plants have shorter growing period. Pea prefers cool climate with optimum mean daily temperatures around 17°C, with a minimum of 10°C and a maximum of 23°C. Young plants can tolerate light frost, but flowers and green pods are injured by frost. Peas are grown on well-drained soils with pH in the range of about 5.5 to 6.5. Peas have tap root systems with many lateral roots. Root depths range from 1 to 1.5 m. However, moisture uptake is usually limited to the 60 cm to 1 m depth of the root zone. The sensitive periods for soil moisture deficits are flowering and yield formation. For high yields soil moisture depletion should not exceed 60% of the total available moisture and 40% during flowering and yield formation periods.

Radish. Radishes are very short season crops, maturing in 4 to 6 weeks. They are grown on a variety of soils. Since the crop is very shallow rooted, frequent irrigations are required. The top 15 cm of root zone should be maintained at about 0.3 atmosphere tension.

Onion. Onion is the most important bulb crop grown under a variety of climates from temperate to tropical. The crop flourishes in mild climates without extremes of temperature and without excessive rainfall. During the initial growth period, cool weather and adequate soil moisture are advantageous for the proper establishment of the crop. During the crop ripening period, warm, dry weather is desirable for high yields of good quality bulbs. The mean daily temperature for optimum growth of the crop ranges from 15 to 20°C. The length of the growing period varies with the climate, but usually ranges from 150 to 175 days, from sowing to harvest. The crop is usually grown in the nursery and transplanted after 30 to 35 days. Direct seeding is also practiced. The optimum soil temperature for germination is 15 to 25°C. For bulb production the plant should not flower, as it adversely affects the yield. For the initiation of flowering, temperatures lower than 14 to 16°C and low humidity are desirable. Onion is sensitive to soil salinity. Optimum pH is in the range of 6 to 7. Onion, like most other vegetable crops, is sensitive to soil moisture deficits. The crop is shallow rooted with roots concentrated in the top 30 cm of the soil, requiring frequent and light irrigations. When the soil is kept relatively wet, root growth is reduced and this favours the enlargement of the bulb. Irrigation is discontinued when the crop approaches maturity.

The growth periods of onion crop are the *establishment period*, reckoned from sowing to transplanting which takes 30 to 35 days, the *vegetative* period comprising 25 to 30 days, the *yield formation* (bulb enlargement) period of 50 to 80 days and the *ripening* period of 25 to 30 days. Onion is most sensitive to soil moisture deficit during the yield formation period, particularly rapid bulb growth which occurs about 60 days after transplanting. It is also equally sensitive during the planting. For a seed crop, the flowering period is very sensitive to soil moisture deficit. However, over-irrigation will reduce growth and yield. Frequent irrigations are required to prevent cracking of the bulb and to ensure good quality bulbs.

Potato. Potato *(Solanum tuberosum)* is grown in most regions throughout the world, but is particularly important in the temperate climates. The optimum mean temperatures are around 18 to 20°C. Night temperatures below 15°C is desirable for tuber initiation. Temperatures below 10°C and above 30°C adversely affect growth. Potato varieties can be grouped into *early* (90 to 120 days), *medium* (120 to 150 days and *late* (150 to 180 days). Early varieties are bred for temperate climates with long day lengths of 15 to 17 hrs, while late varieties result in good yields under long and short day conditions. For tropical climates varieties tolerant to short days are required.

Potato is grown in 3 or more year rotations with other crops such as maize, bean and lucerne to maintain soil productivity, to check weed growth and prevent crop damage due to insects, pests and diseases, particularly soil-borne diseases. Potato requires well drained soils with adequate aeration, with pH ranging from 5 to 6. Under irrigations, the crop is grown on ridges. The crop is moderately sensitive to soil salinity.

Potato requires adequate soil moisture for optimum growth. The available soil moisture should not be depleted beyond 30 to 50%. Soil moisture depletion more than 50% reduces the yield of the crop. Soil moisture deficits during the period of stolonization, tuber initiation and yield formation are particularly sensitive to potato. Early vegetative and ripening stages are relatively tolerant to potato. However, there is considerable variation between varieties in crop response to soil moisture availability. Potato has a shallow root system and about 70% of the total moisture uptake is from the top 30 cm of the root zone and nearly 100% from the top 40 to 60 cm of the root zone. Under irrigated conditions, irrigation scheduling should be based on avoiding soil moisture deficits during the periods of stolonization and yield formation. To obtain water savings greater level of soil moisture depletion can be allowed during the ripening period. This will also help in advancing the maturity of the crop.

Tomato. Tomato *(Lycopersicon esculentum)* is the second most important vegetable, next only to potato. It is a rapidly growing crop with a growing period of 90 to 150 days. The crop is neutral to day length. Optimum mean daily temperature is in the range of 18 to 25°C with night temperatures between 10 and 20°C. The crop is sensitive to frost. The crop can be grown on a wide range of soils but prefers well-drained light loam soils with pH from 5 to 7. Waterlogging induces diseases, particularly bacterial wilt.

The seed is usually grown in nursery plots and seedlings are transplanted after 25 to 35 days. Tomato is moderately sensitive to soil salinity. The most sensitive period to salinity is during germination and early vegetative periods. The plant produces flowers from bottom to top.

The growth period of tomatoes are recognized as follows: *nursery stage* (25 to 35 days), *flowering stage* (20 to 30 *days), yield formation stage* (20 to 30 days) and *ripening stage* (15 to 20 days). Highest yields are obtained by frequent, light irrigations. Tomato is most sensitive to soil moisture stress during and immediately after transplanting, during flowering and yield formation. Soil moisture deficit during the flowering period causes flower drop. Moderate moisture deficit during the vegetative period enhances root growth. Tomato has fairly deep root system. In deep soils roots may penetrate up to 1.5 m. About 80% of the soil moisture uptake is from the top 50 to 60 cm of the root zone soil. For high yields of good quality, tomato needs controlled supply of water throughout the growing period. However, excessive irrigation during the flowering period induces flower drop and reduces fruit set. It may also lead to excessive vegetative growth and delay in ripening. Highest demand for water is during flowering. Irrigation during and after fruit set should be limited to the extent that it will not induce stimulation of new vegetative growth at the cost of fruit development. Furrow irrigation is the most commonly adopted method of surface water application. The crop is highly suited to drip irrigation. With drip irrigation even moderately saline water could be used to irrigate the crop. However, tomato is not suited to sprinkler irrigation. Under sprinkler irrigation fruit set may be reduced with an increase in fruit rotting.

Cucurbits. Cucurbits, mainly cucumber *(Cucumis sativus)*, muskmelon (*Cucumis melo*), water melon (*Citrullus vulgaris*), pumpkin *(Cucumbita pepo)* and squash (*Cucurbita maxima*) are long season crops with medium or deep root systems. As in other vegetable crops irrigation increases the yield and fruit quality. Excess irrigation reduces the quality and shelf life of cucurbits. Watermelons grow well on sandy loam soils with pH of 5.6 to 7.2. It is moderately sensitive to salinity.

In watermelons the growth period of 80 to 110 days can be distinguished into the *establishment period* (10 to 15 days), the *vegetative period* (20 to 25 *days*), *flowering period* (15 to 20 days), *yield formation period* (20 to 30 days) and *ripening* (15 to 20 days). Irrigations must be scheduled at the start of the growing period, the flowering period and the yield formation period. During these stages soil moisture deficit must not exceed 50% of the available moisture level. During the ripening period, relatively dry soils are preferred to increase sugar content and to avoid flesh becoming fibrous and less juicy. During the growing season, one heavy irrigation may be sufficient to mature the crop. Furrow irrigation and drip irrigation are most common in cucurbits. Watermelons can also be grown under spate or flood irrigation in basins.

Fruit Crops

Fruit crops may be of evergreen type or deciduous type. The deciduous trees shed their leaves during winter and remain dormant for 3-4 months. A typical example is the grapevine. Amongst the evergreen fruit trees are citrus, mango, banana, olive, pineapple, papaya and date palm. All fruit crops require adequate soil moisture during their establishment period of three to four years. Soil moisture should be maintained above 50% availability in the surface 1 m depth of soil. Later, on full development, the crops may be irrigated when two-thirds of available soil moisture is depleted during blossoming, fruit setting and fruit enlargement periods. At other periods and during dormancy in the case of deciduous trees, irrigations may be applied when the soil moisture depletion reaches below 30% of the available soil moisture in the root zone. Papaya and banana are shallow rooted and comparatively shortlived species. They need to be irrigated when the available moisture in the top 30 cm soil layer is depleted to about 75 to 80 per cent of availability. This will amount to irrigations at 8 to 12 day intervals during rainless periods in a tropical climate. Date palm can withstand high temperature and low humidity but needs irrigations during flowering and fruiting stages to obtain good yields.

Banana. Banana (*Musa* spp.) is an important tropical fruit crop. A mean daily temperature of about 27°C is optimum for banana. The growth gets restricted when temperatures fall below 16°C. Banana plant is multiplied by planting suckers which originate from the underground stem (corm or rhizome) which bears several sprouts or suckers which are removed for planting, except for one or two which provide the ratoon crop. Nutrients must necessarily be applied at the time of planting and at the start of the ratoon crop. The plant has a sparse, shallow root system. The feeding roots are mostly spread laterally near the surface. The root depth is usually limited to about 75 cm. In general, the soil moisture uptake is from the top 80 cm of the root zone.

The growth of the banana plant can be distinguished into *vegetative stage* (from planting to shooting) which takes about 7 to 9 months and *flowering and yield formation stages* (from shooting to harvest) which take about 2 to 3 months. The number of ratoons varies with the variety. The average life of a ratoon plantation is over 3 years. Some varieties are replanted after each harvest. In irrigated plantations ample and frequent supply of water is necessary. Water deficits in excess of 35% adversely affect growth and yield. Establishment period, early phase of the vegetative period and flowering period are vital in terms of water requirement. Water deficit during the yield formation period adversely affect the fruit size and quality. Under limited water availability, the total production will be higher when full water requirements are met over a limited area than when it is met partially over a large area. This is unlike many other crops described earlier. Furrow and basin methods are the commonly adopted methods of surface irrigation in bananas. Both drip and sprinkler methods of irrigation are suitable in bananas. Drip irrigation results in considerable savings in water.

Citrus. Citrus is a widely cultivated perennial fruit tree of various species, including orange (about 70% of the total citrus crop), lemon, grape fruit and mandarin. Citrus is cultivated up to 1800 m altitude in the tropics and 750 m in the sub-tropics. For large scale cultivation the crop is usually not suited to the humid tropics. Optimum daily temperatures for growth are in the range of 23 to 30°C. There is a great reduction in growth at temperatures above 38°C and below 13°C. Strong wind is harmful to citrus, as they cause drop of flowers and young fruits.

Citrus is grown in well-drained deep soils which allow the tap root to penetrate to depths of 1 to 2 m. Most citrus species develop a single tap root. The lateral roots are spread laterally and form the feeding roots. Light, and medium soils without impervious layers and not subject to waterlogging are preferable. The soil pH in the range of 5 to 8 is preferred. Citrus is sensitive to high concentration of salts in the soil. The plant is propagated by bud grafting.

Adequate soil moisture is necessary for unrestricted growth of citrus. Water deficit retards growth and induces leaf curl and drop; young fruits fall and mature fruits are of inferior quality. Water deficits are avoided during periods of vigorous vegetative growth. However, prior to flowering, and fruit set too vigorous and luxuriant growth may impair production of high quality fruits. The flowering period is highly sensitive to soil moisture stress.

Citrus plant requires well planned irrigation scheduling. The plant requires good aeration and over-irrigation is harmful, particularly in young trees. Too frequent and heavy irrigations will affect root development and yield as well as leaching of nutrients. Surface irrigation of citrus are by furrow and basin methods. Both sprinkler and drip methods of irrigation are suitable and will result in uniform distribution of water and savings in water. Sprinkler irrigation is helpful in protecting the crop from frost.

Grape. Grape is an important fruit crop requiring a long, warm, dry summer and cool winter. Grapes are adapted to a wide range of well-drained soils. Largest vines and high yields are obtained in deep, fertile soil. Grapes are moderately sensitive to soil salinity.

Most grape varieties are propagated by cuttings grown in nurseries to produce roots. Grape vines are deep rooted with roots penetrating to depths of 2 to 3 m. Grapes can adjust moderately to limited water supply. Prior to and during flowering adequate soil moisture is essential for flower development. During the vegetative period, flowering and early yield formation stages soil moisture depletion should not exceed 0.35 to 0.45 when the ET_m is 5 to 6 mm/day. Later in the growth period, moisture depletion can be at a higher level, while towards and after harvest a high level of soil moisture depletion is required.

When the rainfall during the growth period is insufficient, irrigation should be scheduled to fill the root zone to field capacity at regular intervals, particularly during the periods when the crop is sensitive to moisture stress. In such situations irrigation should be applied before the vegetative growth starts. In shallow and light soils properly scheduled irrigations are required till harvest. Furrow method of surface irrigation is common in grape cultivation. Drip irrigation is the most suitable irrigation method under limited water availability. Sprinkler irrigation may also help in protecting the crop against frost. However, sprinkling during the ripening period increases the chance of bunch rot.

Pineapple. Pineapple (*Ananas comosus*) is grown mainly in regions with high relative humidity. For good growth pineapple requires a mean daily temperature of of 22 to 26°C. Mean daily maximum and minimum temperatures are 30 and 20°C, respectively for the whole growing period. Pineapple can grow on a wide range of soils, but well-drained sandy loam soils are most suitable. The lime content of the soil should be low. The crop is sensitive to waterlogging.

Pineapple is usually grown in double rows on raised beds. The plant population is about 50,000/ha. Shading is sometimes used to protect the crop from scorching when temperatures and radiation are high. The growing period, from planting to harvest, ranges from 1 to 2 years and in ratoon crops 9 to 18 months, depending on the planting material and climatic conditions. The root system of pineapple is shallow and sparse. In deep soils the maximum root depth may be about 1 m, but roots are generally concentrated in the top 30 to 60 cm from where the full water requirements are extracted by the plant. Flower initiation is induced by low temperature, water deficit or hormone sprays. Hormone spray results in uniform fruiting and harvest periods.

Pineapple can tolerate long dry periods due to its ability to retain water in the leaves, which is used by the plant during periods of water scarcity. However, pineapple is sensitive to soil moisture stress, particularly during the vegetative period. Soil moisture deficits retard growth, flowering and fruiting. However, soil moisture deficits during flowering is less harmful. On the other hand, it may hasten fruiting. Frequent irrigations or rains at the time of harvest may cause poor quality of fruits, and make the crop susceptible to fungal attack, causing heart rot. Under deficient water availability, irrigation scheduling is based on preventing water deficits during the period of vegetative growth. Water supply can be restricted during the period of ripening. Soil moisture depletion up to 75% of the available level can normally be allowed, except during the vegetative growth period. About a month prior to harvest, irrigation is discontinued. Furrow, sprinkler and drip methods of water application are suitable in pineapple cultivation. Pineapple fruit contains about 80 to 85% water and 10 to 14% sugar. Frequent irrigations prior to harvest decrease the sugar content.

Spices and Condiments

Important crops in this group are turmeric, ginger, cardamom, cumin, coriander, betal vine, black pepper and chillies. Only limited information is available on the water management practices of these crops. Turmeric, ginger and chillies should be grown on broad ridges and irrigated to maintain over 50 per cent of available moisture in the zone of maximum root spread which is the top 60 cm layer of soil. Coriander and cumin are winter crops and need irrigation at intervals of 10 to 12 days on light soils and 15 to 20 days on heavy soils. The water requirements of these crops are around 500 to 600 mm.

Lawns and Gardens

In parks and gardens, seasonal flowering plants need to be irrigated frequently to maintain the surface 30 cm soil layer moist. These plants should receive irrigations at an interval of five to eight days depending upon the severity of the climate. A lawn may be irrigated at an interval of eight to ten days in summer and 15 to 20 days during winter to maintain its good growth. In lawns, one has often to resort to surface irrigation methods and hence proper consideration should be given to levelling at the time of planting to ensure that water will spread uniformly. The young shrubs and live-fences should be irrigated along with the lawns. Fully developed ornamental trees may be able to tap deep percolated water and hence no special care may be necessary for their growth.

Plantation Crops

Apple, coconut, date palm, arecanut, black pepper, cocoa, coffee, tea and rubber are amongst the major plantation crops grown in many regions of the world. There are many other fruit crops and spices which are important in specific regions. Plantation crops are usually high price crops. Many of the plantation crops are developed in favourable environments so that irrigation may not be a major limiting factor in their production at the global level. However, most plantation crops respond well to irrigation during periods of water scarcity at specific growth periods.

Coffee. Coffee crops are normally of two species, namely, *Coffee Arabica* and *Coffee robusta.* Only Coffee Arabica is irrigated on a significant scale. Coffee Robusta commands a much lower price in the world market and is usually grown in hotter and wetter areas. Coffee Arabica provides the highly aromatic variety, forming the bulk of the commercial blends of coffee, which is usually grown in moderately high altitudes. Mean monthly minimum temperature for coffee plantations is 10°C.

Coffee varieties have a vigorously developed root system and can withstand soil moisture stress to a great extent. Properly controlled irrigations, however, increase the yield and bean quality of Coffee Arabica. Limited soil moisture availability controls flowering and hence better period of picking. Continuous replacement of soil moisture through irrigation is not advisable in the monsoon regions. Amongst the surface methods of water application, furrow irrigation (usually contour furrows) is preferred. Both sprinkler and drip irrigation are suitable. Drip irrigation is preferred in regions with limited water availability.

Tea. Tea *(Thea sinensis)* is one of the important crops of India and Sri Lanka which together produces over half of the world production. It is also grown in many regions of the tropics and sub-tropics under cool and high altitude conditions.

The major root system of tea plant is in the top 70 cm of the soil. Irrigation is highly desirable in regions with prolonged drought periods. Sprinkler and drip methods of water application are suitable in tea plantations. Furrow irrigation, particularly contour furrows is also suitable.

Cocoa. Cocoa *(Theobroma cocoa)* is a tree crop grown in wet, tropical lands around the world. The climatic requirements are high humidity and comparatively high temperature. Cocoa is a shallow rooted plant with most of its active roots spread in the top 60 cm of the soil. The crop is sensitive to soil moisture stress. Soil moisture stress below 60% of the field capacity moisture level affects the plant growth. Furrow irrigation is the most commonly used surface irrigation method. Drip irrigation is suitable and will result in substantial savings in water.

CROP MANAGEMENT UNDER RAIN-FED AGRICULTURE

Selection of crops and their varieties adapted to the total amount and distribution of water available and adoption of suitable agronomic practices are important factors in the success of agriculture in the rain-fed areas (Fig. 10.9). The crops commonly grown in unirrigated areas are grain legumes including groundnut, millets, cotton and oilseeds. There is also considerable area under cereals, grasses and several perennial trees. Wide variations in the adaptability exhibited by crops to drought prone conditions, *viz.,* uncertain and erratic pattern of rainfall could be utilised in the selection of crops and their varieties. In North India the more efficient dryland crops are gram, mustard, barely, safflower and linseed. Wheat is the least efficient under unirrigated conditions. Both linseed and safflower appear promising in the east. In the Deccan plateau and in south India, sorghum, safflower and cotton are efficient. The principal pulse crops grown in unirrigated areas are gram (*Cicer arietinum*), *arhar* (*Caganas cagan*), *urad* (*Phaseolus mungo*), *mung* (*P. auneus*), beans (*Dolichos sp.*), moth bean (*Phaseolus aconitifylius*) and groundnut (*Arathis hypogaea*).

Fig. 10.9. Result of choice of crops in dryland agriculture. In the foreground is a crop of wheat withered due to lack of moisture, while in the background stand lush green and profusely flowering mustard crop grown at the same field moisture level (Photograph: IARI).

With the development of short-duration varieties of castor, mustard, cotton, pegion pea (*arhar*) and other crops, several double-croppnig and mixed-cropping patterns can be standardised for different regions. The rotation chosen should be such that plants having the same pests or diseases in common should not be grown in succession. A deep rooted crop may be succeeded by a shallow rooted one, so that nutrients and moisture could be tapped from different zones of the soil.

Whether in irrigated or unirrigated areas, there is a relationship between the supply of inputs and the output of the crops. Hence, arrangements will have to be made to feed plants grown in dry farming areas with nutrients, if the yield is to be increased. Most of the soil in such areas are deficient in nitrogen and hence a good response ought to result from the supply of nutrients particularly nitrogen.

EVALUATING LANDS FOR IRRIGATION COMMAND AREAS

The primary objective of evaluating lands for introducing irrigation projects is to assess its capacity to withstand intensive irrigation without any deterioration and to assess its perpetual productivity per unit of water applied. The main technological factors influencing the choice of command areas for irrigation projects may be grouped into physical, chemical and climatological groups as follows (Dastane and Saraf, 1966):

Physical	*Chemical*	*Climatological*
1. Texture	1. Soluble salt content	1. Aridity
2. Permeability	2. Exchangeable sodium	2. Humidity
3. Non-capillary porosity	3. pH	
4. Water-table depth		
5. Soil depth		
6. Topography		
7. Water-holding capacity		

Loamy soil texture is ideal for irrigation, and deviation on either side lowers the value of the land. Clayey texture hinders leaching and results in poor aeration, while sandy texture increases water application losses. High values of available water-holding capacity, depth of soil and non-capillary porosity are desirable characteristics from both irrigation and drainage points of view. As regards the permeability of water, moderate values are the best and the extremes on either side are harmful. The deeper the water table the beter the land, as it can be drained satisfactorily and hazards such as waterlogging, salinity and alkali problems are not involved. A good levelled land is the most suitable for even distribution of water on the surface and in the soil profile. The steeper the slope, the more the land is wasted in bunds and channels in surface irrigation methods and the land development cost is higher.

Higher contents of soluble salts result in salinity problems and those of exchangeable sodium in alkali problems. Hence, the values of both these should be as low as possible for high rating of the land for irrigation. Values of pH beyond 9 is undesirable on account of alkali hazards. Neither is the pH below 4.5 good, as the availability of nutrients to plants becomes a limiting factor. The limits for the various physical and chemical characteristics are shown in Table 10.12. The land could be classified into four classes, namely, excellent, good, fair and poor, according to the values of the characteristics. The land in the last class is generally unsuitable or uneconomical to bring under irrigation.

Table 10.12. *Limits for different characteristics of land for surface irrigation.*

Description	*Texture of surface soil and sub-soil*	*Available water-holding capacity (mm of water per metre soil depth)*	*Topography (slope per cent) Depth from the surface level*	*Possibility of water rise after introduction of irrigation*	*Minimum permeability in root zone (mm/hr)*	*Minimum non-capillary porosity (per cent)*	*Maximum soluble salt content in root zone (per cent)*	*Maximum exchange-able sodium percentage in root zone*	*pH*	*Depth of soil (cm)*
Excellent	Loam	175	0—1	<300 cm	15—25	>20	<0.20	<5	6.5—7.5	>150
Good or	Clay loam sandy-loam	100—175	1—3	<200 cm	10—25 or 25—50	15—20	0.20—0.30	5—10 150	7.5—8.5 or	50—
Fair	Clayey or fine sandy	25—100	3—10	<100 cm	5—10 or 50—125	5—15	0.30—0.50	10—15	8.5—9.5 or 4.5—5.5	10—50
Poor	Coarse sandy or gravelly	<25	>10	<100 cm	<5 or >125	0—5	>0.50	>15	>9.5 or <4.5	<10

Dastane and Saraf (1966)

For assessment of the permeability, non-capillary porosity, soluble salt content and exchangeable sodium percentage, the values of the limiting layer in a root zone are to be taken into account. Thus, if the surface soil is permeable and sub-soil impermeable, the later will be taken into account.

It is necessary to have a thorough knowledge about the various physical and chemical characteristics of the soil and meteorological factors and economic conditions of the inhabitants before putting a land under irrigation, in order to avoid hazards of development of waterlogging, salinity and alkali problems. Pre-irrigation surveys are, therefore, conducted to give complete information of the proposed command areas. After a pre-irrigation survey, only the most suitable areas are brought under command. Post-irriation surveys are then conducted after the operation of the project for some time to examine the changes that have taken place in various physical and chemical properties of the soil profile. If any problems have developed then proportionate increased care is necessary.

SUMMARY

Concepts on evaporation, transpiration, evapotranspiration and consumptive use are discussed. Methods of measurement of evaporation are described. Factors influencing evapotranspiration are explained. Procedures for estimating evapotranspiration using open pan evaporimeter are presented. Influence of crop growth stages on crop coefficients are explained. Procedures for estimating crop evapotranspiration, adopting crop coefficient approach, are presented. Methods of estimating evapotranspiration based on climatological approach and using empirical formulae are explained. Use of Penman formula and Modified Penman formula for estimating evaporation are described. Observations of the 1990 FAO Consultation on FAO methodologies on estimating crop water requirement are presented. F AO Penman-Montieth based Combination formula for estimating reference evapotranspiration is described along with procedures for computing the various factors included in the formula. Methods of making broad estimation on the values of the missing data when measured field data are not available are presented. Case studies utilizing the FAO Penman-Montieth equation using field data are presented. The applicability of the Hargreaves' Alternate formula when temperature and radiation data only are available is illustrated. Factors influencing crop coefficients in estimating crop evapotranspiration are described. The concepts of crop water requirement, irrigation requirement and field water balance are explained. The component of effective rainfall and factors influencing it are presented. The principles of scheduling irrigation applications and the guidelines for planning irrigation schedules are described. The various terminologies used in expressing irrigation efficiencies along with the methods of their application are presented. Yield response factor and harvest index of crops are presented. Crop response to soil moisture availability at different stages of their growth are described. Irrigation requirement of common crops are presented. Principles of planning cropping systems and water management practices in irrigation project areas are described. Criteria for evaluation of lands for irrigation command areas are explained. Principles of crop management in rainfed areas are explained.

QUESTIONS

1. Distinguish between:

(*i*) Evapotranspiration and potential evapotranspiration

(*ii*) Cuticular transpiration and stomatal transpiration

(*iii*) Transpiration and transpiration ratio

(*iv*) Potential evapotranspiration and consumptive use

(*v*) Reference crop evapotranspiration, ET_o and crop evapotranspiration, ET_c

(*vi*) Reference evapotranspiratrion and pan evaporation

(*vii*) Crop evapotranspiration under standard and non-standard conditions

(*viii*) Relative short wave radiation and relative sunshine radiation

(*ix*) Single crop coefficient and dual crop coefficient

(*x*) Water requirement and irrigation requirement of crops

(*xi*) Net irrigation requirement and gross irrigation requirement

(*xii*) Infiltration and deep percolation

(*xiii*) Available water and readily available water

(*xiv*) Deficit irrigation and full irrigation

(*xv*) Reservoir efficiency and water storage efficiency

(*xvi*) Water application efficiency and water distribution efficiency

(*xvii*) Crop water use efficiency and field water use efficiency

(*xviii*) Overall efficiency and economic efficiency of an irrigation system

(*xix*) Harvest index and yield response function of a crop

(*xx*) Tillering and jointing stages in the growth of cereal crops

(*xxi*) Shooting and heading stages in the growth of cereal crops

2. Define energy balance in evaporation.
3. What are the components which increase the value of CU over ET?
4. What are the commonly used empirical equations to estimate the value of C in Dalton's theory of evaporation from a free water surface?
5. What are the standard procedures to be adopted in the installation and use of USBR Class A Pan Evaporometer?
6. How does soil moisture content influence evaporation from land surface?
7. What is the energy input, expressed in SI units, required to vaporize 1 mm of water?
8. Define latent heat flux.
9. Define albedo. What is the common range in the values of albedo of a green vegetative cover?
10. What is the functional relationship between solar radiation and clear-sky solar radiation?
11. What are the limitations in the use of Blaney-Criddle formula in estimating evapotranspiration?
12. What are the assumptions on which the Thornthwaite formula is based? What are the essential requirements for its use at different places?
13. What are the basic characteristics that distinguish a specific crop from reference grass in evapotranspiration studies? .
14. Define effective rainfall. What are the principal factors influencing effective rainfall from crop production point of view?
15. Define irrigation scheduling.
16. What is the maximum available deficiency allowed in planning irrigation schedules?
17. How is the amount of water required to be applied in an irrigation computed?
18. What are irrigation indices? What are the commonly used indices in scheduling irrigation?
19. Define the yield response factor of a crop.
20. How can the water requirement of a crop be increased by crop and soil management practices?
21. Name the draught tolerant wheat varieties recommended for cultivation in the plains and plateau regions of north and central regions of India.
22. Which are the common soil building crops suitable for conditions of limited irrigation supplies?
23. What are the specific soil profile characteristics to be investigated in the pre-irrigation soil survey of irrigation command areas to assess their suitability for sustained agricultural production?
24. What are the limiting values of soil pH to introduce irrigated agriculture in the command area of an irrigation project?
25. Identify the True and False statements from the following:

(*i*) The basic requirement of water in plants is a function of the soil moisture deficits as influenced by evapotranspiration and soil and plant characteristics.

(*ii*) Dalton's theory, Rowher's equation and Meyer's equation are the basic equations to express the evaporation from a free surface.

(*iii*) Precipitation, contribution of soil moisture in the crop root zone and upward flow of moisture to the crop root zone are included in the term irrigation requirement of a crop.

(*iv*) The essential requirements in the evaporation process are (*i*) source of heat to vaporize the liquid water, and (*ii*) the presence of a gradient of water vapour pressure between the evaporating surface and the surrounding air.

(*v*) The values of pan coefficient increases with wind speed.

(*vi*) The pan coefficient increases with the increase in relative humidity.

(*vii*) The two major factors influencing evapotranspiration are the moisture content of the soil surface and the fraction of solar energy reaching the land surface.

(*viii*) Blaney-Criddle formula is applicable for use in high altitudes where day time radiation is practically independent of night temperature.

(*ix*) What is the main assumption on which the Thornthwaite formula is based?

(*x*) The Radiation method, proposed in FAO (1998) is considered superior to Blaney-Criddle method, particularly for equatorial regions, on small islands and at high altitudes.

(*xi*) FAO Consultation on Revision of FAO methodologies for estimating evapotranspiration observed that the Penman methods require local calibration of the radiation data to achieve satisfactory results.

(*xii*) Studies have shown that the Modified Penman method proposed in FAO 24 (1997) frequently underestimates ET_o under non-advective conditions.

(*xiii*) In the different growth stages of common field crops, the late season stage is usually longer than other growth stages.

(*xiv*) The water use efficiency of a mixed crop system is usually lower than a mono-crop system.

(*xv*) The upper limit in the productivity of a crop is influenced mainly by the availability of soil moisture at critical stages in the growth of the crop.

(*xvi*) In the cultivation of wetland rice, puddling reduces percolation losses, checks weed growth and regulates soil and water temperatures.

(*xvii*) While planning to provide irrigation at the critical stages in the growth of crops, it is desirable to apply water at the initiation of the critical stage so that moisture is available in the soil during the stage.

(*xviii*) The irrigation requirement of sorghum is low when grown in sandy soils, as compared to clay loam soils.

(*xix*) Pulse crops sown in moist soils in winter do not generally require irrigation at the early vegetative stage, as it may do more harm than good by adversely affecting nodulation and reduce oxygen requirement at this stage.

(*xx*) To obtain high values of total production of groundnut under conditions of limited water availability, it is desirable to cover large areas under the crop and reduce the quantity of irrigation water applied, taking care of the growth periods sensitive to soil moisture stress.

26. Identify the most appropriate answer of the following questions from the four options:

(*i*) In general, the major components of water requirement of crops is:
- (a) pre-sowing irrigation
- (b) evapotranspiration
- (c) leaching requirement
- (d) deep percolation from crop root zone

(*ii*) Generally, the most significant factor in determining the productivity of a crop is:
- (a) non-availability of adequate soil moisture in the root zone at the critical stages of their growth
- (b) crop variety
- (c) nutrient availability
- (d) pests and diseases

(*iii*) The most important factors influencing evaporation from land surfaces is:
- (a) degree of saturation of soil surface
- (b) temperature of air and soil
- (c) humidity and wind velocity
- (d) density of vegetative cover

(*iv*) Amongst the four distinct growth stages of seasonal crops the maximum value of crop coefficient K_c is during:
- (a) initial stage
- (b) crop development stage
- (c) mid-season stage
- (d) late season stage

(*v*) The most important factor influencing the evapotranspiration of a crop is:
- (a) physiology of the crop
- (b) stage of growth of the crop
- (c) degree of crop cover
- (d) the location from where the data are obtained

(*vi*) The principal meteorological factor influencing evapotranspiration is:
- (a) air temperature (b) solar radiation
- (c) humidity (d) wind speed

(*vii*) The more commonly used empirical formulae for estimating evapotranspiration, namely, Blaney-Criddle method, Thornthwaite method and the Hargreaves method have been calibrated for:
- (a) daily calculations
- (b) hourly calculations
- (c) 10-day and monthly calculations
- (d) seasonal calculations

(*viii*) Amongst the most commonly used empirical formulae to estimate evapotranspiration, the one which has been found to provide reasonable approximation at most locations is:
- (a) Blaney-Criddle formula
- (b) Thornthwaite formula
- (c) FAO Radiation method
- (d) Hargreaves' alternative method

27. In the following, match each of the entries in the 1st column with the most closely linked entry in the 2nd column:

	1st column	2nd column
(*i*)	(A) Hargreaves equation	(I) Weather data
	(B) Stefan-Boltsman constant	(II) Solar radiation
	(C) Albedo	(III) Value of constant C in Dalton's equation
	(D) Rower	(IV) FAO Penman-Montieth formula
	(E) FAOCLIM	(V) Evapotranspiration
(*ii*)	(A) Thornthwaite formula	(1) Extra-terrestrial radiation
	(B) Water vapour	(II) FAO Penman-Montieth formula
	(C) Agnots value	(III) Relative humidity
	(D) Soil heat flux density	(IV) Evapotranspiration
	(E) Vapour pressure deficit	(V) Air humidity
(*iii*)	(A) Field water	(I) Peak period consumptive use balance
	(B) Dastane (1974)	(II) Plant water content and water potential
	(C) Irrigation indices	(III) Irrigation scheduling
	(D) Pan evaporation	(IV) Effective rainfall
	(E) Irrigation frequency	(V) Change in storage
(*iv*)	(A) Soil water constants	(I) Irrigation scheduling
	(B) Leaching requirement	(II) Plant species selection
	(C) Water needed in the root zone	(III) Water storage efficiency
	(D) Increasing water use efficiency	(IV) Available water

	(E) Maximum available deficiency of soil moisture	(V) Irrigation efficiency
(*v*)	(A) Tillering	(I) Yield response factor
	(B) Crown root initiation stage	(II) Potato
	(C) Limiting values of soil pH	(III) Pre-irrigation soil survey
	(D) Stolonization	(IV) Wheat
	(E) Relative yield of a crop	(V) Critical stage of growth in cereal crops

FURTHER READING REFERENCES

Allen, R.G., Pereira, L.S., Raes, D. and Smith, M., (1998), *Crop Evaporation: Guidelines for Computing Crop Water Requirements,* FAO Irrigation and Drainage Paper 56. Food and Agriculture Organization of the United Nations, Rome, p. 300.

Allen, R.G., Jensen, M.E., Wright, J.L. and Burman, R.D., (1989), *Operational Estimates of Reference Evapotranspiration,* Agronomy Jl. 81: 650–662.

Allen, R.G. and Pruit, W.O., (1986), *Rational Use of the FAO Blaney - Criddle Formula,* J. Irri. and Drainage Engg., ASCE, 112 (IRR2): 139–155.

Allen, R.G. and Pruit, W.O., (1991), *FAO–24 Reference Evapotranspiration Factors.* J. Irri. and Drainage Engg., ASCF 117(5): 758–773.

Aslyng, H.C. (1963), *Soil physics terminology,* Int. Soc. Soil Sci. Bull, 23: 1–4.

Bavel, C.H.M., Nixon, P.R. and Hanser, V.L., (1963), *Soil moisture measurement with neutron method.* US Dept. Agr. ARS–48–70, p. 39.

Black, C.A. (1957). *Soil-plant-relationship.* John Wiley and Sons, New York, p. 332.

Blaney, H.F. and Criddle, W.D., (1950), *Determining water requirements in irrigated areas from climatological and irrigation data,* US Dept., Agr. SCS-TP 96, p. 44.

Bouyoucos, G.J., (1949), *Nylon electrical resistance units for continuous measurement of electrical resistance in the field,* Soil Science 67, 319–331.

Bouyoucos, G.J., (1954), *New type electrode for plaster of paris moisture blocks,* Soil Science 78, 339–342.

Bredero, T.J. (1988), *Concepts and Guidelines for Crop Water Management Research,* World Bank, New Delhi, p. 106.

Brouwer, C. and Prins, K., (1980), *Irrigation Scheduling,* Irrigation Water Management Training Manual No. 4, FAO Land and Water Management Div., Food and Agriculture Organization of the United Nations, Rome, p. 36.

Burman, R.D., Jensen, M.E. and Allen, R.G., (1987), *Thermodynamic Factors in Evapotranspiration,* In: Irrig. and Drainage Spl. Conf. James, L.G. Euglish, M.J. eds., ASCE, Portland, OR., USA, July, p. 28–30.

Buckingham, E. (1907), *Studies on the movement of soil moisture,* US Dept. Agr., Bur. Soils Bull. 38.

Burman, R. and Pochop, L.O., (1994), *Evaporation, Evapotranspiration and Climatic Data,* Elsavier Science, B.V., Amsterdam, The Netherlands.

Chakravarti, N.V.K. and Sastry, P.S.N., (1977), *Seasonal variation of pan to potential evapotranspiration in semi-arid region,* Indian J., Power and Rivery Valley Development (in press).

Chang Jen-Hu, (1968), *Climate and Agriculture – An Ecological Survey,* Aldine Publishing Co., Chicago, p. 304.

Chaudhary, T.N., (1977), *Water Management in Rice for Efficient Production,* Directorate of Water Management Research (ICAR), Patna, Res. Bull. No. 1, p. 63.

Christiansen, J.E., (1968), *Pan evaporation and evapotranspiration from climatic data,* J. Irrigation and Drainage, 94: 243–265.

Dakshinamurti, C., Michael, A.M. and Mohan, S., (1973), *Water resources of India and their utilization in Agriculture,* I.A.R.I., Monograph No. 3, Water Technology Centre, I.A.R.I., New Delhi, p. 392.

Dastane, N.G. and Saraf, C.S., (1966), *Evaluating land for irrigation commands,* Proceedings of Sixth New East-South Asia Regional Irrigation Practices Seminar, Amman, Jordan, pp. 393–397.

Dastane, N.G., Singh, M., Hukkeri, S.B. and Vamadevan, V.K. (1970), *Review of work done on water requirement of crops in India,* Navabharat Prakashan, Poona, p. 106.

Dastane, N.G., (1972), *A practical manual for water use research in agriculture,* Navbharat Prakashan, Poona, p. 120.

Dastane, N.G., (1974), *Effective Rainfall in Irrigated Agriculture,* FAO Irrigation and Drainage Paper 25, Food and Agricultural Organization of the United Nations, Rome, p. 62.

Davis, D.S., (1943), *Empirical equations and nomography,* McGraw Hill Book Co. New York, p. 200.

Department of Agricultural Research and Education, Govt. of India (1975), *Annual Report 1974–75,* Govt. of India, Ministry of Agriculture and Irrigation, New Delhi.

Doneen, I.D., (1972), *Irrigation Practice and Water Management.* FAO Irrigation and Drainage Paper I. United Nations Food and Agriculture Organization, Rome, p. 84.

Doorenbos, J. and Kassam, A.H. with Bentvelsen, C.I.M. *et al.* (1979), *Yield response to water,* FAO Irrigation and Drainage Paper 33, Food and Agriculture Organization of the United Nations, Rome, p. 193.

Doorenbos, J. and Pruitt, W.C., (1975), *Guidelines for Predicting Crop Water Requirements,* Irri. and Drainage Paper 24, FAO, Rome, p. 197.

Doorenbos, J and Pruitt, W.O., (1977), *Guidelines for Predicting Crop Water Requirements,* FAO Irrigation and Drainage Paper 24 (Revised), Food and Agriculture Organization of the United Nations, Rome, p. 144.

Duru J.O., (1984), *Blaney–Morin–Nigeria Evapotrans-piration Model,* Jl. Hydrol., 70: 71–83.

Echeverria, H.G., (1967), *Evaporation and evapotranspiration in Venezuela,* M.S., Thesis, Utah State University, Logan, Utah.

Finkel, H.J., (1983), *CRC Handbook of Irrigation Technology, Vol. I & II.,* CRC Press Inc., Boca Raton, Florida, USA.

Food and Agriculture Organization of the United Nations (1986), *Early Agrometeorological Crop Yield Assessment.* FAO Plant Production and Protection Paper 73, pp. 114–149.

Food and Agriculture Organization of the United Nations (1990), *Expert Consultation on Revision of FAO Methodologies for Crop Water Requirements.* Report, FAO, Rome, May 28–31, 1990, p. 90.

Frera, M. and Popov, G.F. (1979), *Agrometeorological Crop Monitoring and Forecasting.* FAO Plant Production and Protection Paper 17, FAO, Rome, p. 38–43.

Frevert, D.K., Hill, R.W. and Braten, B.C. (1983), *Estimation of FAO Evapotranspiration Coefficients,* Irrig. and Drainage Engg., ASCE 109 (IR2): 265–270.

Gandhi, R.T., Gupta, P.C., Joseph, A.P., Rege, N.D., Coover, J.R., Jones, D.F., Phelan, J.T., and Pope, E.T. (1970), *Handbook on irrigation water requirement,* Water Management Div., Dept. of Agr., Ministry of Agriculture, New Delhi, p. 129.

Gangopadhayay, M., Datar, S.V. and George, C.J. (1970), *On the global solar radiation, climate and evapotranspiration estimates in India,* Indian J. Met. Geophys., 20: 23–30.

Government of India, Ministry of Irrigation, Water Management Division, New Delhi (1984), *A Guide for Estimating Irrigation Water Requirements,* Tech. Series No. 2 (Revised), p. 114.

Government of India, Ministry of Water Resources, Water Management Dn., New Delhi (1985), *A Water Management Manual,* Tech. Series No. 2 (Revised), p. 281.

Govinda Rajan, S.V. (1968), *Soils of India.* All India Soil and Land Use Survey. I.A.R.I., New Delhi, p. 22.

Hagan, R.M., Haise, H.R., Edminster, T.W. editors and Dinauer, R.C. Mng. Editor (1967), *Irrigation of Agricultural Lands,* American Society of Agronomy, Madison, Wisconsin, USA, p. 1180.

Hansen, V.E. Israelson, O.W. and Stringham, G.E. (1979), *Irrigation Principles and Practices,* 4th ed. John Wiley and Sons Inc., New York, p. 417.

Hargreaves, G.H., Patwardhan, M.M., Prasad, U.K., Pawar, D.H., Swani, Z.A. and Bhola, A.M. (1985), *A Crop Water Evaluation Model for India,* The International Irrigation Centre, Dept. of Agricultural and Irrigation Engineering, Utah State University, Logan, Utah, USA, p. 155.

Hargreaves, G.H. (1956), *Irrigation requirements based on climatic data,* Paper 1105, Proc. ASCE, Jl. Irrig. and Drainage Div., 82 (IR–3) 1–10.

Hendricks, D.W. and Hansen, V.E. (1962), *Mechanics of evapotranspiration,* Jl. of the Irri. and Drainage Division, Am. Soc. of Civil Engineers. IR 2, pp. 67–82.

Hillel, D. (1971), *Research on soil and water use efficiency,* Report to the Govt. of Cyprus, October 1971, p. 74.

Holmes, J.W. (1956), *Calibraton and field use of the neutron scattering method of measuring soil water content,* Australian Jl. of Appl. Sci. 7, pp. 45–58.

Hukkeri, S.B. and Pandey, S.L., (1977), *Water Requirement and Irrigation of Crops,* In: Water Requirement and Irrigation Management of Crops in India. Water Technology Centre, I.A.R.I., New Delhi, pp. 163–285.

Indian Society for Spices, (1997), *Water and Nutrient Management for Sustainable Production and Quality of Spices,* Proc. National Seminar on Water and Nutrient Management for Spices, 3–6 October, 1997, Madikeri, Karnataka. Pub. Indian Institute of Spices Research, Calicut, Kerala, p. 182.

Jagtab, S.S. and Alabi, (1997), *Evaluation of Evapotranspiration by Adjusted Penman Formula by Class A Pan at Humid and Sub-humid Regions of Nigeria and Republic of Benin Using Long Term Data,* Sustain Africa, 3, 5–12.

James, L.G. (1993), *Principles of Farm Irrigation System Design,* Krieger Publishing Co., Malabar, Florida, USA, p. 543.

Jensen, M.E. (1968), *Water consumption of agriculutral plants: Water deficits and plant growth* 2: 1–22, Academic Press, New York.

Jensen, M.E., Burman, R.D. and Allen, R.G. (1990), *Evaporation and Irrigation Water Requirement,* American Soc. of Civil Engineers Manuals and Reports on Engineering Practices, No. 70, ASCE, New York, USA, p. 360.

Jensen, M.E. (1973), *Consumptive Use of Water and Irrigation Water Requirements,* A Report of the Technical Committee on Irrigation Water Requirements of the Irri. and Drainage Div. of Amer. Soc. of Civil Engrs., New York, p. 256.

Kerala Agricultural University (1993), *Package of Pracices Recommendations 'Crops' 93,* Directorate of Extension, Kerala Agrl. Uni., Mannuthy - 680 651, Thrissur, Kerala, p. 237.

Khonke, H. (1968), *Soil Physics.* McGraw Hill Book Co., New York, p. 224.

Kramer, P.J. (1969), *Plant and soil water relationships: A modern synthesis,* McGraw Hill Book Co., New York, p. 482.

Makkink, G.F., (1957), *Testing the Penman Formula by Means of Lysimeters,* J. Inst. Water Engg., 11(3): 277–288.

Martin, D.L. and Gillary, J.K. (1993), *Irrigation Water Requirements,* Chapter 2, SCS National Engineering Handbook, US Dept. of Agriculture, Soil conservation Service, Washington, D.C., p. 284.

McCulloch, J.S.G., (1965), *Tables for rapid computation of the Penman estimate of evaporation,* East African Agri. and Forestry J., 30: 286–95.

Meyer, A.F. (1942), *Evaporation from Lakes and Reservoirs,* Minnesota Resources Commission, St. Paul, USA.

Michael, A.M., Hukkeri, S.B. and Singh, N.P. (1977), *Estimating Water Requirement of Crops,* In: Water Requirements and Irrigation Management of Crops in India, Water Technology Centre, I.A.R.I., New Delhi, pp. 91–162.

Michael, A.M., Mohan, S. and Swaminathan, K.R. (1972), *Design and evaluation of irrigation methods,* I.A.R.I. Monograph No. 1, Water Technology Centre, I.A.R.I., New Delhi, p. 208.

Michael, A.M. and Pandey, S.L. (1970), *Time your irrigation for top yields,* Intensive Agriculture, May 1970.

Molenaar, A. (1960), *Irrigation by sprinkler,* Food and Agriculture Organisation of the United Nations, Rome, p. 91.

Michael, A.M. and Ojha, T.P. (2003), *Principles of Agricultural Engineering,* Vol. II, 4th ed., Jain Brothers 16/873, East Park Road, Karol Bagh, New Delhi, p. 888.

Montieth, J.L. (1981), *Evaporation and Surface Temperature,* Quarterly Jl. Royal Meteo. Soc. 107: 1–27.

Montieth, J.L. (1965), *Evaporation and Environment,* In: The State of Movement of Water in Living Organisms, 19th Symposium of the Soc. of Environmental Biology, Cambridge University Press, Cambridge, U.K., 19: 205–234.

Montieth, J.L. and Unsworth, M.H. (1990), *Principles of Environmental Physics,* 2nd ed. Edward Arnold, London.

Muskat, M. (1937). *The flow of homogeneous fluids through porous media.* McGraw Hill Book Co., New York.

Oswal, M.C. (1972). *Effect of different tillage practices and moisture conservation and water budgeting under dryland agriculture.* Ph.D. Thesis, P.G. School, I.A.R.I., New Delhi.

Olufayo, A.A. and Ognutunde, P.G. (2000). *Estimating Potential Evapotranspiration under Humid Tropical Conditions of Africa.* Jl. Agriculutral Engineering, Vol. 36. Indian Society of Agricultural Engineers, CIAE, Bhopal.

Pai, A.A. and Hukkeri, S.B. (1969). *Manual on Irrigation Water Management.* Govt. of India, Ministry of Agriculture and Irrigation, Dept. of Agriculture, New Delhi, p. 176.

Pandey, H.N. (2005). *Development of Rainfed and Limited Irrigation Wheat Cultivation Technology.* I.A.R.I. Regional Research Station, Indore, Personal correspondence.

Penman, H.L. (1963). *Vegetation and Hydrology.* Tech. Comm. No. 53, Commonwealth Bureau of Soils, Harpenden, England, p. 125.

Penman, H.L. (1948). *Natural evapotranspiration from open water, bare soil and grass.* Proc. Royal Soc. of London 193, pp. 120–145.

Penman, H.L. (1956). *Estimating evapotranspiration.* Trans. Am. Geophy. Union 37, 43–46.

Perriar, A. (1985). *Updated Evapotranspiration and Water Requirement Definitions.* In: Perriar, A. and Riou, C. eds. Crop Water Requirement (ICID Intl. Conf., Paris. Sept. 1984). INRA, Paris: 885–887.

Philip, J.R. (1957). *The physical principles of soil water movement during irrigation cycle.* Proc. 3rd Intl. Congress on Irri. and Drainage 8, pp. 125–154.

Philip, J.R. (1957*b*). *Evaporation and moisture and heat fields in the soil.* Jl. Meteorology 14: 354–366.

Philip, J.R. (1966). *Plant water relations: some physical aspects.* Annual Review of Plant Physiology, Vol. 7, pp. 245–268.

Pierre, W.H., Kirkham, D., Pesek, J. and Shaw, R.J. Editors (1965). *Plant environment and efficient water use.* Amer. Soc. Agronomy and Soil Sc. Soc. of Amer, Wisconsin, USA, p. 295.

Prihar, S.S., Gajri, P.K. and Narang, R.S. (1974). *Scheduling Irrigation to Wheat Using Open Pan Evaporation.* Indian J. Agrc. Sci. 44: 567–71.

Prihar, S.S. and Sandhu, B.S. (1987). *Irrigation of Field Crops: Principles and Practices.* I.C.A.R., Krishi Anusandhan Bhavan, Pusa, New Delhi, p. 142.

Pruitt, W.O. (1960). *Relation of consumptive use of water to climate.* Transactions, Am. Soc. Agrl. Engrs. 3(1): 9–13, 17.

Richards, L.A. (1947). *Pressure membrane apparatus–construction and use.* Agri. Engg. 28: 451–454.

Richards, L.A. (1949). *Methods of measuring soil moisture tension.* Soil Science 68, 95–112.

Richards, I.A. – Editor (1954). *Diagnosis and improvement of saline and alkali soils.* Agriculture Handbook No. 60, US Dept. of Agr., Washington, D.C.

Rijtema, P.E. (1965). *Analysis of Actual Evapotranspiration.* Agri. Res. Report No. 69, Centre for Agriculture Publications and Documentation, Wageningen, The Netherlands.

Rose, C.W. (1966). *Agricultural Physics.* Pergamon Press, New York, p. 226.

Rowher, C. (1931). *Evaporation from Free Water Surfaces.* USDA Tech. Bul. 271, GPO, Washington, D.C.

Russel, M.B. – Coordinator (1959). *Water and its relation to soils and crops.* Academic Press, New York, p. 131.

Saksena, K.K. (1987). *Design and Operation of Irrigation Projects.* Dept. of Civil Engineering, Indian Institute of Technology, New Delhi, p. 372.

Salter, P.J. and Goode, J.E. *Crop responses to water at different stages of growth.* Commonwealth Agricultural Bureaux. Farnham Royal, Bucks, England.

Scheidegger, A.E. (1960). *The Physics of flow through prorous media.* University of Toronto Press, London.

Schofield, R.K. (1935). *The* pH *of the water in soil.* Third International Congress on Soil Science, Vol. II, pp. 37–48.

Schwab, G.O., Fangmeir, D.D., Elliot, W.J. and Frevert, R.K. (1996). *Soil and Water Conservation Engineering,* 4th ed. John Wiley and Sons Inc., New York, p. 507.

Sharma, R.G. and Dastane, N.G. (1968). *Use of screened evaporimeters in evapotranspirametry.* Proc. Water Management Symposium, Udaipur (1966). Indian Soc. of Agronomy, New Delhi, pp. 66–67.

Singh, N.P. and Sinha, S.K. (1976). *Water use efficiency in crop production.* I.A.R.I., Mono No. 4, Water requirement and irrigation management of crops in India. Water Tech. Centre, I.A.R.I., New Delhi, pp. 57–90.

Sinha, S.K. and Singh, N.P. (1976). *Plant-water relations* I.A.R.I., Mono. No. 4, Water requirement and irrigation management of crops in India. Water Technology Centre, I.A.R.I., New Delhi.

Slatyer, R.O. and Mcllroy, L.C. (1961). *Evaporation and the Principles of its Measurement.* In: Practical Meteorology, CSSRI (Australia) and UNESCO, Paris.

Slatyer, R.O. (1967). *Plant-water relationships.* Academic Press, New York. 366, p. 104.

Sinha, A. K. (1976). *Soil - water relationships.* I.A.R.I. Mono, No. 4, Water requirement and irrigation management of crops in India. Water Tech. Centre, I.A.R.I., New Delhi, pp. 9–56.

Smith, M., Allen, R.G., Motieth, J.L., Perrier, A., Pereira, L. and Segeran, A. (1992). *Report of the Expert Consultation for Revision of FAO Methodologies for Crop Water Requirements.* Food and Agricultural Organization of the United Nations, Rome, p. 60.

Sri Ramulu, US (1998). *Management of Water Resources in Agriculture.* Pub. New Age International (P) Ltd., New Delhi, p. 159.

Thornthwaite, C.W. (1944). *Report of the Committee on Transpiration and Evaporation.* Trans. Am. Geophysical Union, 25: p. 687.

Thornthwaite, C.W. (1948). *An approach toward a rational classification of climate.* The Geographical Review, Vol. 38, No. 1, pp. 55–94.

United States Department of Agri. (1960). National Engineering Handbook, Section 15. Ch. 1, Soil-plant-water relationships. USDA, SCS, Washington, D.C., p. 72.

United States Dept. of Agri., Soil Conservation Service, Engineering Div., Washington, D.C. (1967). *Irrigation Water Requirements.* Tech. Series No. 21, p. 83.

USDA (ARS). (1962). *Determining consumptive use and irrigation water requirements.* United States Dept., Agri. Res. Serv. Techn. Bull. 1275: p. 59.

Vamadevan, V.K. and Dastane, N.G. (1968). *Drainage needs of rice.* IL RISO, Anno. XVI, 3: 243–50.

Water Management Division, Ministry of Agriculture, Department of Agri., Govt. of India. New Delhi (1971). Hanbook of Irrigation Water Mangement, Part III. *Scheduling Irrigation to Meet Crop Needs.* III-1–6.

Wright, J.L. (1982). *New Evapotranspiration Crop Coefficients.* J. Irri. and Drain. Div., ASCE 108 (IR2): 57–74.

World Meteorological Organization, Geneva (1983). *Guidelines to Meteorological Instruments and Observing Practices,* 5th ed.

Chapter 11

Surface Methods of Water Application

LEARNING OBJECTIVES

Different methods of surface irrigation and their adaptability to the specific characteristics of soil, topography and crops. Basic variables influencing the hydraulics of flow in surface methods of water application and their influence on their design. Rational and quasi-rational approaches in the design of borders, check basins and furrows laid along the prevailing land slope and on contours. Infiltration opportunity time and its signficance in surface irrigation design. Concepts of quarter-time rule and cut-back stream in check basin and furrow methods of irrigation, respectively. Sub-irrigation and its applicability.

Irrigation water may be applied to crops by flooding it on the field surface, by applying it beneath the soil surface, by spraying it under pressure or by applying it in drops. The common methods of irrigation are indicated schematically below:

The water supply, the type of soil, the topography of the land and the crop to be irrigated determine the correct method of irrigation to be used. Whatever the method of irrigation, it is necessary to design the system for the most efficient use of water by the crop. Effective agronomic practices are essential components of irrigated agriculture. Management of soil fertility, cropping systems selection, appropriate crop rotations and plant protection measures are equally important as irrigation itself. Irrigation in the arid and semi-arid areas provides two essential requirements: (*i*) moisture supply for plant growth which also transports essential plant nutrients, and (*ii*) the water required to leach or dilute salts in the soil.

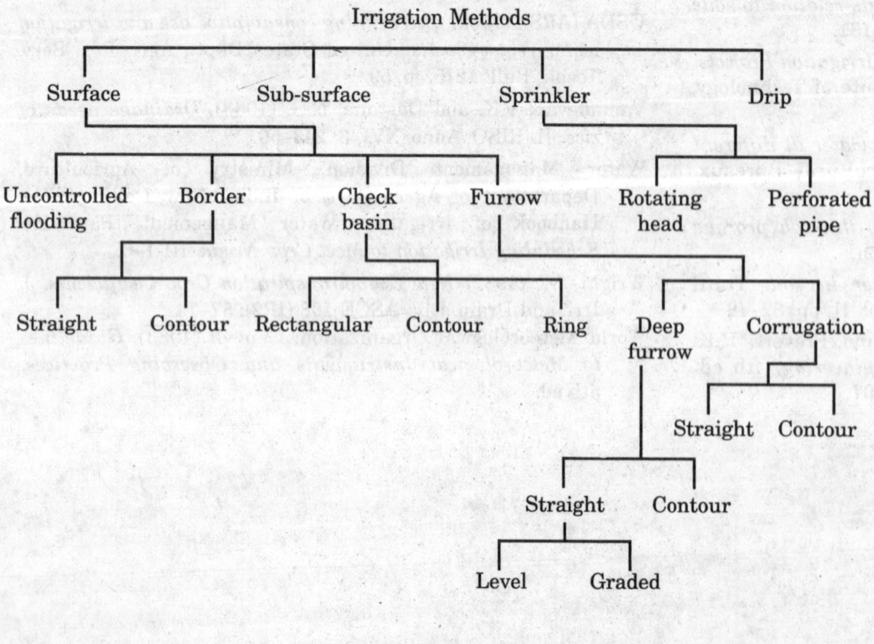

SURFACE IRRIGATION METHODS

In the surface methods of irrigation, water is applied directly to the soil surface from a channel located at the upper reach of the field. Water may be distributed to the crops in border strips, check basins or furrows. Two general requirements of prime importance to obtain high efficiency in surface methods of irrigation are: properly constructed water distribution systems to provide adequate control of water to the fields and proper land preparation to permit uniform distribution of water over the field. They are also designed to minimize labour and capital requirements. Effective management practices are dependent on the type of irrigation, and the climate and topography of the region.

Uncontrolled Flooding

There are some situations in which crop lands are irrigated without regard to efficiency or uniformity. These are situations where spate irrigation is prevalent, and the purpose of application of flood flows is to recharge the crop root zone's soil moisture to grow a crop after the flood season, or when the value of the crop is small or the field is used for grazing. The practice is often called *free flooding*. Irrigation of individual trees with ring basins would also sometimes fall in this category. Water may be brought to the field through field ditches, often contour field ditches or large furrows, and water application is done by cutting openings in the ditch at regular intervals to provide for full coverage of the field. No specific design criteria is followed in this method of water application. The method is wasteful of water and many times results in soil erosion and non-uniform application of water in a field.

BORDER IRRIGATION

Basic Details of Border Method of Irrigation

The border method of irrigation makes use of parallel ridges to guide a sheet of flowing water as it moves down the slope. The land is divided into a number of long parallel strips called borders that are separated by low ridges. The border strip has little or no cross slope but has a uniform gentle slope in the direction of irrigation (Fig. 11.1). The essential feature of border irrigation is to provide an even surface over which the water can flow down the slope with a nearly uniform depth. Each strip is irrigated independently by turning in a stream of water at the upper end. The water spreads and flows down the strip in a sheet confined by the border ridges. The irrigation stream must be large enough to spread over the entire width between the border ridges without overtopping them. When the advancing waterfront reaches the lower end, a few minutes before or after that the stream is turned off. The water temporarily stored in the border moves down the strip and infiltrates, thus completing the irrigation. Borders are of two types: (*i*) *Open-end borders* and (*ii*) *Blocked-end borders*. Open-end borders are limited to large size fields Blocked-end borders are adapted to smaller size of field which are most common in developing countries.

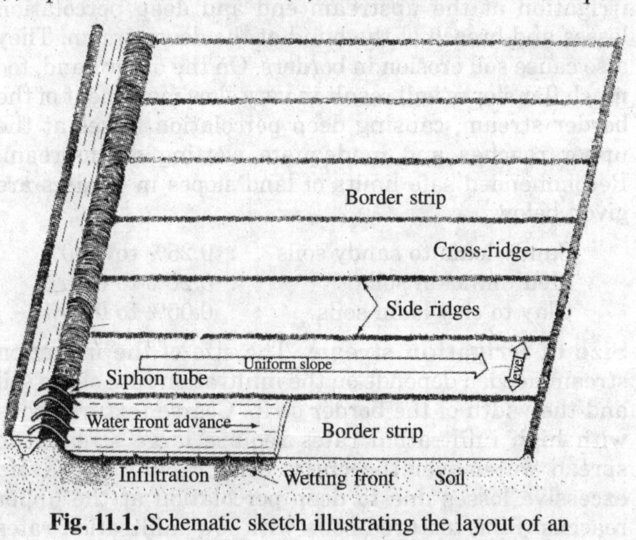

Fig. 11.1. Schematic sketch illustrating the layout of an impounding type border irrigation system.

The border method of irrigation is adapted to most soils where depth and topography permit the required land levelling at a reasonable cost and without permanent reduction in soil productivity. It is, however, more suitable to soils having moderately low to moderately high infiltration rates. Usually it is not used in coarse sandy soils that have very high infiltration rates because of the stringent limitations in design. It is also not well suited to soils having a very low infiltration rate, since to provide adequate infiltration opportunity time, without surface run-off at the lower end, the irrigation stream may be too small to completely cover the border strips. The size of irrigation stream per unit width of border strip must be larger as compared to other methods of surface irrigation.

The border method is suitable to irrigate all close-growing crops like wheat, barley, fodder crops and legumes. It is, however, not suitable for crops like rice which requires standing water during most parts of its growing season.

Fig. 11.2. Uniform spread of water in blocked-end border strips.

The border method has a number of advantages. They include: (*i*) border ridges can be constructed economically with simple farm implements like a bullock-drawn A-frame ridger or *bund former* or tractor-drawn disc ridger, (*ii*) labour requirement in irrigation is greatly reduced as compared to the conventional check basin method of irrigation, (*iii*) uniform distribution and high water application efficiencies are possible if the system is properly designed, (*iv*) large irrigation streams can be efficiently used, (*v*) operation of the system is simple and easy, and (*vi*) adequate surface drainage is provided if outlets are available.

Straight and Contour Borders

Borders may be laid along the general slope of the field (straight or down the slope borders), or may be laid across the general slope of the field (contour borders).

Fig. 11.3. Wheat growing in border strips 135 metres long and 7.5 metres wide. Note the unoformity of growth and large earheads resulting from uniform distribution of water.

When fields can be levelled to desirable land slopes economically and without affecting its productivity, graded borders (Figs. 11.1 to 11.3) are easier to construct and operate. When land slope exceeds safe limits, fields are undulating and levelling is not feasible, borders may be laid across the slope and are called contour borders (Fig. 11.4). The design criteria of a contour border is the same as a straight border. Each contour border is level crosswise and has a uniform longitudinal gradient as a straight border. The width and length of a contour border are identical to that of a straight border for a particular set of conditions.

In laying contour borders, the field is divided into a series of strips on the approximate contour, and each strip is levelled as an independent area. Thus, a series of steps are formed in successive elevations around the slope. The vertical interval between adjacent benches should, as far as possible, be limited to 30 cm, but should not exceed 60 cm. The height of bund (ridge) should be sufficient to contain both the normal irrigation stream and storm runoff. Since the earthwork involved is often large, contour borders should be laid out as closely to the original topography of the land, as farming operations and other considerations will permit. In laying out contour borders the irrigation channel is laid down the land slope, providing sufficient number of drop structures. At the downstream end of borders, a drainage channel is provided to drain away storm runoff. The drainage channels, which run down the land slope, should be provided with erosion control structures.

Fig. 11.4. Field laid out for contour border irrigation.

Border Specifications and Stream Size

Successful operation of the border method requires that the system is designed and operated properly. Proper design requires consideration of the hydraulics of flow in borders.

Width of border strip. The width of a border usually varies from 3 to 15 metres, depending on the size of the irrigation stream available and the degree of land levelling practicable. When the size of the irrigation stream available is small, the width is reduced. It is, however, not economical to keep the width less than about three metres; as otherwise, too many ridges will have to be formed per unit area of the field surface.

Border length. The length of the border strip depends upon how quickly it can be wetted uniformly over its entire length. This in turn depends on the infiltration rate of the soil, the slope of the land, and the size of the irrigation stream available. For moderate slopes and small to moderate size irrigation streams, the following border lengths are suggested:

Sandy and sandy loam soils	:	60 to 120 metres
Medium loam soils	:	100 to 180 metres
Clay loam and clay soils	:	150 to 300 metres

Border slope. The borders should have a uniform longitudinal gradient. Excessive slopes will make the water run to the lower end quickly, causing insufficient irrigation at the upstream end and deep percolation losses and breach of the bund at the downstream. They also cause soil erosion in borders. On the other hand, too much flat slopes will result in very slow movement of the border stream, causing deep percolation losses at the upper reaches and inadequate wetting downstream. Recommended safe limits of land slopes in borders are given below:

Sandy loam to sandy soils	:	0.25% to 0.60%
Medium loam soils	:	0.20% to 0.40%
Clay to clay loam soils	:	0.05% to 0.20%

Size of irrigation stream. The size of the irrigation stream needed depends on the infiltration rate of the soil and the width of the border strip. Coarse textured soils with high infiltration rates require large streams to spread water over the entire strip rapidly and avoid excessive losses due to deep percolation at the upper reaches. Fine textured soils with low infiltration rates require smaller streams to avoid excessive loses due to runoff at the downstream end and deep percolation at the lower reaches.

The size of the stream delivered into each border is the controlling factor in border irrigation which can be varied after the irrigation system has been installed. It can, therefore, be used to partly compensate for inadequacies in width or length of borders which have been selected and for changes which might take place in soil infiltration rates or depths of rooting of crops. The size of the stream required, however, should be determined as accurately as possible as a part of the design of the system.

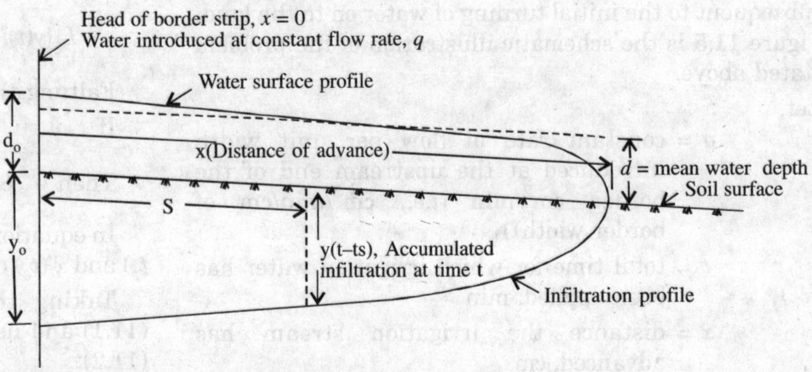

Fig. 11.5. Diagram illustrating the infiltration–advance problem in border irrigation.

It is often convenient to express the requirement of the irrigation stream in terms of the rate of water flow per unit width of the border, such as in litres per second per metre of border width. This value multiplied by the width of the border is the size of the irrigation stream that should be delivered into each border.

The depth of water applied to the soil can be regulated by the size of the irrigation stream. A larger stream is used to apply at a shallower depth, and a smaller stream to apply at a greater depth water. This is because the amount of water entering the soil is related to the infiltration opportunity time which, in turn, is related to how fast the entire area of the border can be covered with the flow of water. Thus, by varying the size of the irrigation stream, it is possible to vary the depth of water applied.

Table 11.1 presents some typical values of stream sizes for different soil types and field slopes.

Table 11.1. Some typical values of stream sizes to suit varying soil characteristics and border slopes.

Soil type	*Border slope %*	*Flow per metre width of border strip, litres per second*
Sandy soil, infiltration rate 2.5 cm per hr	0.20–0.40	10–15
	0.40–0.65	7–10
Loamy sand, infiltration rate 1.8 to 2.5 cm per hr	0.2–0.40	7–10
	0.4–0.60	5–8
Sandy loam, infiltration rate 1.2 to 1.8 cm per hr	0.2–0.40	5–7
	0.4–0.60	4–6
Clay loam, infiltration rate 0.6 to 0.8 cm per hr	0.15–0.3	3–4
	0.30–0.4	2–3
Clay, infiltration rate 0.2 to 0.6 cm per hr	0.1–0.2	2–4

Hydraulics of Border Irrigation

Efficient water application by border irrigation depends on the knowledge of the hydraulic characteristics of the irrigation stream. The flow in a border strip is a case of spatially varied unsteady open channel flow with decreasing discharge. The discharge rate decreases downstream due to infiltration. The flow phenomenon is affected by several variables that must be determined before proper criteria for the design of border irrigation system are developed. The dominant variables influencing border irrigation are—size of irrigation stream, slope of the land surface, infiltration characteristics of the soil and the resistance to flow offered by the soil surface and vegetative cover. The advance of the water front as a function of the elapsed time, infiltration characteristics of the soil, hydraulic resistance and land slope, are the most important factors to be determined in designing an efficient border irrigation system.

A complete analysis of border irrigation would require information on the effect of land slope, infiltration, roughness of soil surface, vegetation, and depth of water on the velocity of flow down the slope. Careful consideration, however, shows that the effect of slope, hydraulic resistance offered by the soil surface roughness and vegetative retardance, and the size of the irrigation stream are reflected in their influence on the depth at which the water flows. If the vegetation or the size of the irrigation stream are increased or the slope is decreased, the water will slow down until a great enough quantity of water has accumulated to overcome these factors. The reverse conditions will result in flow taking place at a shallower depth. Thus, the prediction of advance rate of the border stream consists of finding the advance of the water front as a function of the elapsed time, when the infiltration-time function $y\ (t)$, the size of entrance stream q, and the depth of flow d, are known.

As irrigation water flows into a field and progresses down its length, an ever-decreasing portion of the total volume of water applied flows above the ground, while the remainder infiltrates into the soil and composes the subsurface storage. Any rational approach to predict surface irrigation flows must equate the total volume of water discharged at the supply channel to the sum of surface storage and subsurface storage; moreover, this volume balance is obtained at every instant of time, subsequent to the initial turning of water on to the land. Figure 11.5 is the schematic illustration of the problem stated above.

Let,

q = constant rate of flow per unit width introduced at the upstream end of the border, cm^2/min (*i.e.*, cm^3/min/cm of border width)

t = total time for which irrigation water has been applied, min

x = distance the irrigation stream has advanced, cm

d = average depth of water over the ground surface, cm

t_s = value of t at which $x(t) = s$, min.

$y(t - t_s)$ = accumulated infiltration at the point $x = s$ at time t_s, cm

s = value of x at $t = t_s$, cm, and

$x'(t_s)$ = the value of $\frac{dx}{dt}$ at $t = t_s$.

Referring to Fig. 11.5.

Total water admitted in time t per unit width = qt

Volume of water stored on the ground surface per unit width = dx

Volume of water infiltrated into the soil =

Applying the volume balance relationship,

$$qt = d.x +$$

y is a function of $(t - t_s)$, and

x is a function of t_s

$$\int_0^x y ds = \int_0^t y(t - t_s) x'(t_s) dt_s$$

$$\therefore \quad qt = dx + \int_0^t y(t - t_s) x'(t_s) dt_s \qquad \text{...(11.1)}$$

Equation 11.1 was originally proposed by Lewis and Milne (1938).

Philip and Farrel (1964) analysed the physical restrictions on x and y in equation (11.1). The equation is valid if x is a monotonic increasing function of t. This condition on $x(t)$ places a restriction on the form of $y(t)$ for which the analysis is valid. Sufficient conditions are

$y \geq 0, \frac{dy}{dt} \geq 0,$ and $\frac{\partial^2 y}{\partial t^2} \geq 0.$ Generally the physical restrictions on y are such that these conditions are realized and equation (11.1) is valid.

Philip and Farrel (1964), using the Faltung or convolution theorem of Laplace transformation, obtained the general solution of equation (11.1) as follows:

The following definition of Laplace transformation is used:

$$L\{y(t)\} = \int_0^\infty e^{-st} y(t) dt = f(s) \text{ and } L^{-1}\{f(s)\} = y(t)$$

Faltung theorem:

If $\quad L\{F(t)\} = f(s)$ and $L\{G(t)\} = g(s)$

$$\text{Then} \quad L^{-1}\{f(s) g(s)\} = \int_0^t F(\lambda) G(t - \lambda) d\lambda \qquad \text{...(11.2)}$$

In equation (11.2), $G(t - \lambda)$ and $F(\lambda)$ correspond to $y(t - t_s)$ and $x'(t_s)$ of equation (11.1).

Taking the Laplace transformation of equation (11.1) and using the relationship obtained in equation (11.2).

$$= dL\{x\} + L\{x'\} L\{y\}$$

Now, since $x(0) = 0$, equation (11.1) becomes

$$\frac{q}{s^2} = dL\{x\} + sL\{x\} L\{y\} = L\{x\} [d + sL\{y\}]$$

$$\therefore \quad L\{x\} = \frac{q}{s^2[d + sL\{y\}]}$$

$$L\{x\} = \frac{q}{[ds^2 + s^3 L\{y\}]}$$

$$\text{or} \quad \frac{x}{q} = L^{-1}\left[\frac{1}{s^3 L\{y\} + ds^2}\right] \qquad \text{...(11.3)}$$

Equation (11.3) represents the general solution of equation (11.1) in terms of the Laplace transformation. It can be applied to those particular forms of $y(t)$ of interest.

Field tests on infiltration at pre-sowing and post-emergence irrigations indicate that the functional relationship between accumulated infiltration and elapsed time can be expressed best by the following empirical formula:

$$y = at^\alpha + b \qquad 0 \leq a \leq 1, t \neq 0 \qquad \text{...(11.4)}$$

in which,

y = accumulated infiltration at time t, cm

t = elapsed time, minutes, and

a, α and b are characteristic constants.

Michael (1968) obtained the following particular solutions of equation (11.3) to predict the water front advance when the accumulated-infiltration-time relationship is expressed by equation (11.4):

Taking the Laplace transformation of equation (11.4)

$$L\{y\} = \frac{a\Gamma(\alpha + 1)}{s^{\alpha+1}} + \frac{b}{s}$$

(Γ denotes gamma function)

Equation (11.3) then becomes

$$\frac{x}{p} = L^{-1}\left[\frac{1}{s^3\left\{\frac{a\Gamma(a+1)}{s^{\alpha+1}} + \frac{b}{s}\right\} + ds^2}\right]$$

Let $a\Gamma(\alpha+1) = k$, then

$$\frac{x}{p} = L^{-1}\left[\frac{1}{ks^{2-\alpha}+(b+d)s^2}\right]$$

Case 1: Small values of t

Let $\frac{k}{b+d} = \beta$, then

$$\frac{1}{ks^{2-\alpha}+(b+d)s^2} = \left[\frac{1}{1+\frac{k}{b+d}s^{-\alpha}}\right]\frac{1}{(b+d)s^2}$$

$$= \frac{1}{(b+d)s^2}(1+\beta s^{-\alpha})^{-1}$$

$$= \frac{1}{(b+d)s^2}\sum_{n=0}^{\infty}(-\beta s^{-\alpha})^n \quad \text{for } s>\beta^{1/\alpha}$$

$$= \frac{1}{(b+d)}\sum_{n=0}^{\infty}(-\beta)^n s^{-(n\alpha+2)}$$

$$\therefore \quad \frac{x}{q} = L^{-1}\left[\frac{1}{ks^{2-\alpha}+(b+d)s^2}\right]$$

$$= \frac{1}{b+d}L^{-1}\sum_{n=0}^{\infty}(-\beta)^n s^{-(n\alpha+2)}$$

$$= \frac{1}{b+d}\sum_{n=0}^{\infty}(-\beta)^n\frac{t^{n\alpha+1}}{\Gamma(n\alpha+2)}$$

$$\frac{x}{q} = \frac{1}{b+d}\sum_{n=0}^{\infty}\frac{(-\beta t^{\alpha})^n}{\Gamma(n\alpha+2)}$$

$$\therefore \quad x = \frac{qt}{b+d}\left[\frac{1}{\Gamma(2)}-\frac{\beta t^{\alpha}}{\Gamma(2+\alpha)}+\frac{\beta^2t^{2\alpha}}{\Gamma(2+2\alpha)}-\frac{\beta^3t^{3\alpha}}{\Gamma(2+3\alpha)}+\frac{\beta^4t^{4\alpha}}{\Gamma(2+4\alpha)}-\ldots\right] \quad \ldots(11.5)$$

Case 2: Large values of t

$$\frac{1}{ks^{2-\alpha}+(b+d)s^2} = \frac{1}{ks^{2-\alpha}}\left[\frac{1}{1+\left(\frac{(b+d)}{k}-s^{\alpha}\right)}\right]$$

$$= \frac{1}{ks^{2-\alpha}}(1+\beta^{-1}s^{\alpha})^{-1} = \frac{1}{ks^{2-\alpha}}\sum_{n=0}^{\infty}(-\beta^{-1}s^{\alpha})^n \quad \text{for } s<\beta^{1/\alpha}$$

$$= \frac{1}{k}\sum_{n=0}^{\infty}(-\beta^{-1})^n\frac{1}{s^{[2-(n+1)\alpha]}}$$

$$\frac{x}{q} = L^{-1}\left[\frac{1}{ks^{2-\alpha}+(b+d)s^2}\right]$$

$$= \frac{1}{k}L^{-1}\sum_{n=0}^{\infty}(-\beta^{-1})^n\frac{1}{s^{[2-(n+1)\alpha]}}$$

$$= \frac{1}{k}\sum_{n=0}^{\infty}\frac{t^{[1-(n+1)\alpha]}(-\beta^{-1})^n}{\Gamma[2-(n+1)\alpha]}$$

$$= \frac{1}{\beta(b+d)}\sum_{n=0}^{a}\frac{t^{[1-(n+1)\alpha]}(-\beta^{-1})^n}{\Gamma[2-(n+1)\alpha]}$$

$$= \frac{-t}{b+d}\sum_{n=0}^{a}\frac{t^{[1-(n+1)\alpha]}(-1)^{n+1}(\beta)^{-(n+1)}}{\Gamma[2-(n+1)\alpha]}$$

$$\frac{x}{q} = \frac{-t}{b+d}\sum_{n=1}^{\infty}\frac{(-\beta t^{\alpha})^{-n}}{\Gamma(2-n\alpha)}$$

$$\text{or} \quad x = \frac{qt}{b+d}\left[\frac{1}{\beta t^{\alpha}\Gamma(2-\alpha)}-\frac{1}{(\beta t^{\alpha})^2\Gamma(2-2\alpha)}+\frac{1}{(\beta t^{\alpha})^3\Gamma(2-3\alpha)}-\frac{1}{(\beta t^{d})^4\Gamma(2-4\alpha)}+\ldots\right] \quad \ldots(11.6)$$

Fig. 11.6. Water front advance and depth of flow measurement in evaluating a non-vegetated border strip.

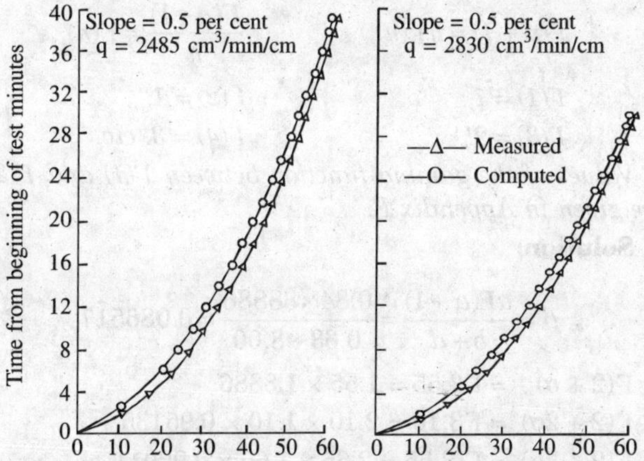

Fig. 11.7. Comparisons of measured advance curves with those predicted by using equations (11.5) and (11.6) in field tests in pre-sowing irrigation.

Michael (1968) obtained the following limiting conditions for the use of equations (11.5) and (11.6):

Equation (11.5) is suitable when

$$\frac{\beta t^{\alpha}}{\Gamma(2+\alpha)} < 1, \text{ and}$$

Equation (11.6) is suitable when ...(11.7)

$$\frac{\beta t^{\alpha}}{\Gamma(2+\alpha)} > 1$$

The validity of the limiting conditions expressed in equation (11.7) is evident from an examination of the limiting values of the terms in equations (11.5) and (11.6), all of which should necessarily be less than unity. It is also evident that the critical limit for the use of the two equations is influenced by the elapsed time t, the infiltration characteristics of the soil which determine the value of a, α and b, and the average depth of flow d.

Equations (11.5) and (11.6) may be used to predict watcr front advance in broder irrigation on the basis of the infiltration characteristics of the soil, the size of the irrigation stream and average depth of flow. Figure 11.7 shows two sample plots of measured and computed values of the advance distance and elapsed time in field tests on irrigation borders. Examples 11.1 and 11.2 illustrate the computation process involved in applying the prediction equations (11.5) and (11.6) to evaluate the field test data.

Example 11.1. *In a field test of border irrigation in a wheat field, the following observations were made:*

$q = 1600\ cm^3/min/cm$, $d = 8$ cm,

$a = 0.84$, $\alpha = 0.55$

b - 0.63.

Using equation (11.5), determine the distance advanced by the water front 54 minutes after the start of irrigation.

Note: *The following relationships of the gamma function, G(n), in which n is real and positive, are used in the solution of the problem:*

$$\Gamma(n+1) = n\Gamma(n), \qquad \frac{\Gamma(n+1)}{n} = \Gamma(n),$$

$$\Gamma(1) = 1, \qquad \Gamma(2) = 1,$$

$$\Gamma(3) = 2!, \qquad \Gamma(4) = 3! \text{ etc.}$$

Values of the gamma function between Γ(1) and Γ(2) are given in Appendix E.

Solution:

$$\beta = \frac{a\Gamma(\alpha+1)}{b+d} = \frac{0.84\times 88886}{0.63+8.00} = 0.086517$$

$\Gamma(2+\alpha) = \Gamma 2.55 = 1.55 \times 1.8886$

$\Gamma(2+2\alpha) = \Gamma 3.10 = 2.10 \times 1.10 \times 0.95135$

$\Gamma(2+3\alpha) = \Gamma(3.65 = 2.65 \times 1.65 \times 0.90011$

$\Gamma(2+4\alpha) = \Gamma 4.20 = 3.20 \times 2.20 \times 1.20 \times 0.91816$

$\Gamma(2+5\alpha) = \Gamma 4.75 = 3.75 \times 2.75 \times 1.75 \times 0.91906$

$\Gamma(2+6\alpha) = \Gamma 5.30 = 4.30 \times 3.30 \times 2.30 \times 1.30 \times 0.89747$

$\Gamma(2+7\alpha) = \Gamma 5.85 = 4.85 \times 3.85 \times 2.85 \times 1.85 \times 0.94561$

Substituting the above values in equation (11.5)

$$x = \frac{1600\times 54}{0.63+8.00}\left[1 - \frac{0.086517\times 54^{0.55}}{1.55\times 0.8886} + \frac{0.086517^2\times 54^{1.10}}{2.10\times 1.10\times 0.95135} - \frac{0.086517^3\times 54^{1.65}}{2.65\times 1.65\times 0.90011} + \frac{0.086517^4\times 54^{2.20}}{3.20\times 2.20\times 1.20\times 0.91816} - \frac{0.086517^5\times 54^{2.75}}{3.75\times 2.75\times 1.75\times 0.91906} + \frac{0.086517^6\times 54^{3.30}}{4.30\times 3.30\times 2.30\times 1.30\times 0.89747} - \frac{0.086417^7\times 54^{3.85}}{4.85\times 3.85\times 2.85\times 1.85\times 0.94561}\right]$$

$$= \frac{1600\times 54}{0.63+8.00}\left[\begin{matrix}1-0.58031+0.29087-0.12985+0.05268\\ -0.01969+0.00685-0.00224\end{matrix}\right]$$

$= 6534.06$ cm $= 65.34$ cm.

Example 11.2. *The following data were obtained in a field test in a border strip at pre-sowing irrigation in a sandy loam soil:*

$q = 2880\ cm^3/min/cm$, $d = 2.345$ cm

$a = 2.7$ $\alpha = 0.49$

$b = 0.142$

Compute the distance advanced by the water front 26 min. after the start of irrigation, using equation 11.6.

Solution:

$$\beta = \frac{2.27\Gamma(1+0.49)}{0.142+2.345} = \frac{2.27\times 0.9279}{0.142+2.345} = 0.8469$$

$$\Gamma(2-\alpha) = \Gamma 1.51 = 0.88659$$

$$\Gamma(2-2\alpha) = \Gamma 1.02 = 0.98884$$

$$\Gamma(2-3\alpha) = \Gamma 0.53$$

$$= \frac{1.53}{0.53} = \frac{0.88756}{0.53}$$

$$\Gamma(2-4\alpha) = \Gamma 0.04$$

$$= \frac{\Gamma 1.04}{0.04} = \frac{0.97843}{0.04}$$

$$\Gamma(2-5\alpha) = \Gamma -0.45$$

$$= \frac{\Gamma 1.55}{-0.45\times 0.55} = \frac{0.88886}{-0.2475}$$

$$\Gamma(2-6\alpha) = \Gamma -0.94$$

$$= \frac{\Gamma 1.06}{-0.94\times 0.06} = \frac{0.96874}{-0.0564}$$

Substituting the above values in equation (11.6)

$$x = \frac{2880\times 26}{0.142+2.345}\left[\frac{1}{0.8469\times 26^{0.49}\times 0.88659}\right.$$

$$-\frac{1}{(0.8469\times 26^{0.49})^2\times 0.98884}$$

$$+\frac{0.53}{(0.8469\times 26^{0.49})^3\times 0.88756}-\frac{0.04}{(0.8469\times 26^{0.49})^4\times 0.97843}$$

$$\left.+\frac{0.2475}{(0.8469\times 26^{0.49})^5\times 0.8886}-\frac{0.0564}{(0.8469\times 26^{0.49})^6\times 0.96874}\right]$$

$$= \frac{2880\times 26}{0.142+2.345}\left[\begin{array}{l}0.26984-0.05788+0.00817\\ -0.00013+0.00021-0.00001\end{array}\right]$$

$= 6617.71$ cm $= 66.17$ metres.

Recession Flow

After the irrigation stream is cut off, the tail water recedes downstream. The rate of recession of the tail water is determined by noting the times at which water just disappears from the upstream end and recedes downstream past the border strip. An accurate description of flow involves the use of unsteady state flow equations. Rational formulae to predict the recession flow under field situations are not known to have been developed.

In plotting advance and recession curves, the distance down the border (or furrow) is plotted on the x-axis and the elapsed time on the y-axis. Both the advance and the recession curves are plotted on the same graph (Fig. 11.8). Parallelism of advance and recession curves ensures uniform distribution of water throughout the border.

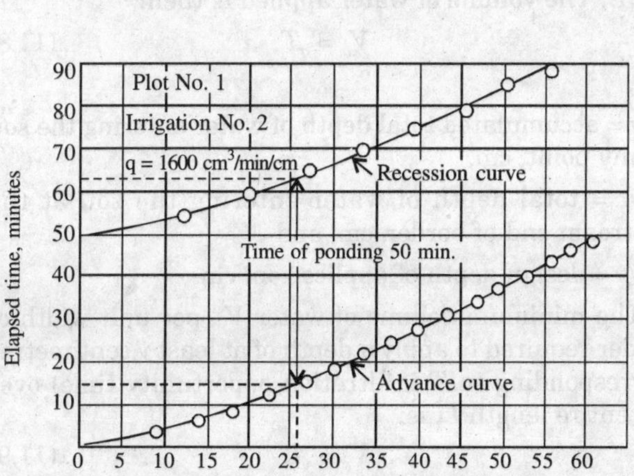

Fig. 11.8. Advance and recession curves obtained in the field test of a border strip. The parallelism of the two curves indicates uniform distribution of water throughout the length of the border.

(***Source:*** Michael, A.M., 1968.)

Infiltration opportunity time (time of ponding). The difference between the time the water front reaches a particular point along the border (or furrow) and the time at which the tail water recedes from the same point is the infiltration opportunity time, or the time of ponding. The infiltration opportunity time at any point along the border (or furrow) is the vertical distance (in time scale) between the advance and recession curves at the point. For example, in Fig. 11.8 the infiltration opportunity time at a distance 26 metres from the upstream end of the border is 50 minutes.

Design of Border Irrigation Systems

A well designed border irrigation system provides for the application of the desired amount of water nearly uniformly throughout the field. To obtain the maximum water application efficiency the water should remain on the surface sufficiently long to allow just the desired amount of water to infiltrate into the soil. The required infiltration opportunity time is obtained using the accumulated infiltration-time relationship for the soil in question. Water distribution efficiency is governed mainly by the uniformity of the infiltration opportunity time obtained throughout the border length. When the behaviour of the irrigation system can be predicted from the measurable characteristics of the site, it is possible to obtain a nearly uniform infiltration opportunity time throughout the border by suitably adjusting the entrance stream size, and the length of run or slope of the border. In an existing field layout, where the length or slope of the borders cannot be altered, the main controllable variable is only the entrance stream size.

The following procedure is suggested for the design of an efficient border irrigation system:

(1) The length, width and slope of the border strip are determined.

(2) The depth of water required to replenish the soil moisture in the root zone of the crop to field capacity is estimated. This may be done by actual measurement of the soil moisture content in the root zone before irrigation, or predicted from the consumptive use requirements of the crop under the prevailing agro-climatic conditions.

(3) The accumulated infiltration-time relationship of the soil under the existing soil conditions and vegetation is determined. The relationship can be established by actual measurements with cylinder infiltrometers before each irrigation; but it can also be obtained from previously established relationships.

(4) The desired infiltration opportunity time is determined. It is the time necessary for the soil to absorb the estimated depth of irrigation water. It is obtained from the accumulated infiltration-time relationship of the soil as determined in item (3).

(5) The hydraulic resistance is estimated on the basis of the soil surface roughness and the hydraulic characteristics of the crop. The hydraulic resistance, expressed as the Manning's n, may be estimated by the relationship expressed in Chapter 6. An approximate value of the stream size is selected first and the value of n is calculated. More accurate values of n are obtained by successive approximations on the basis of the stream size ultimately selected.

(6) The average depth of flow is estimated. For a given entrance stream size q, hydraulic resistance n, and water surface slope s, a depth d_o, will be reached at the upstream end of the border. This depth is obtained directly from Manning's equation, and represents the minimum steady state flow depth. The flow depths d_1, d_2, ... d_n at successive points along the border may be calculated by considering the distance advanced at equal time increments. Initially, since the depth d_o is small compared to the distance the water front will move during the first time increment, the value of d_o, obtained using the true water surface slope s, will be nearly that obtained using the ground slope. The water depth necessary for flow past successive points down the border will decrease since the total flow past these points decreases. The water depth is thus a function of the distance down the border. The stream size q_o, at any point down the border is equal to the difference between the entrance stream size q, and the average rate of the volume of water q_i infiltrating over the area upstream from the point under consideration.

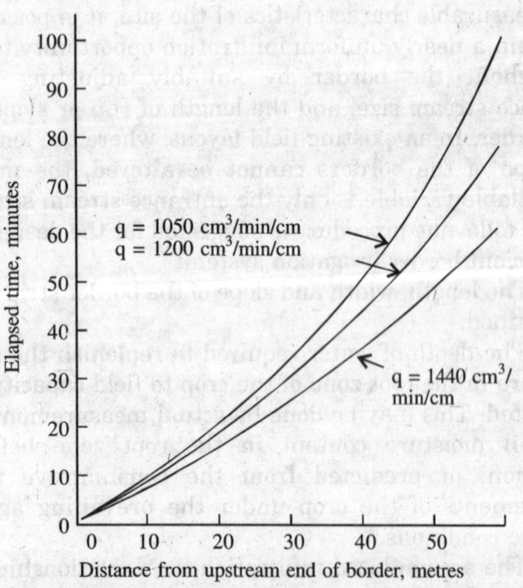

Fig. 11.9. Water front advance curves for different rates of discharge and accumulated infiltration elapsed time function $y = 0.77\ t^{0.61} + 0.35$.
[***Source***: Michael, A.M. (1968).]

With short borders and low gradients the slope of the water surface may differ from that of soil surface due to differences in the depth of water at various distances from the upstream end, but with borders of considerable length and moderate slope, the water surface slope is essentially the same as that of the ground surface.

(7) The water front advance is predicted. The water front advance-time relationship as a function of the entrance stream size, average depth of flow, and infiltration characteristics, may be predicted with reasonable accuracy using equation (11.5) and/or equation (11.6).

To arrive at an efficient irrigation system design it is necessary to calculate a family of advance curves using different values of q and possible border lengths.

(8) The irrigation system is designed to obtain the optimum water application efficiency and border length. The procedure proposed by Hall (1960) may be adopted to arrive at a suitable design. This procedure is discussed below:

It is presumed that the limits on non-erosive stream and range of optimum land slope are known. The first step in the procedure is to calculate a family of advance curves as discussed under step 7 (Fig. 11.9). In Fig. 11.9, time is plotted as the ordinate and distance as abscissa and rates of discharge as the parameter.

The volume of water V_a admitted to the border strip per unit width is equal to the entrance stream size q, multiplied by the length of time the water flows into the border. For the preliminary design, it is supposed that this period of time is also equal to T, the time required for the advance front to reach the end of the border, $X = L$. The volume of water applied is then:

$$V_a = {}_qT \qquad \text{...(11.8)}$$

Let,

y = accumulated total depth of water entering the soil at any point, cm,

y_o = total depth of water entering the soil at the upstream end of border, cm, and

y_d = design depth of application, cm.

The minimum volume of water V_r, per unit width of border required to apply a depth of at least y centimetres (corresponding to T infiltration opportunity time) over the entire length L, is:

$$V_r = {}_yL \qquad \text{...(11.9)}$$

The volume of water admitted to the border must at least be equal to the volume required by the soil. Therefore, the first design criterion is:

$$_qT = {}_yL \qquad \text{...(11.10)}$$

As the second step of the design procedure, a volume balance function $B(q, t, y, L)$ is developed, which expreses the difference between V_a and V_r, presuming that a uniform depth y_o, corresponding to an infiltration opportunity time, T is added to the entire field.

By definition, $$B = {}_qT - y_oL \quad ...(11.11)$$

The function B, is then constructed for each possible and probable field lengths $L_1, L_2, L_3, ...L_n$. This procedure involves the determination of time T, required for each entrance stream size to reach a distance L_1 using the advance curves (Fig. 11.9). This determines the values of T and y_o for each value of q. The corresponding values of B for $L = L_1$ are calculated from equation (11.11) and plotted as the ordinate against y_o as the abscissa (Fig. 11.10). The process is repeated for the different lengths $L_2, L_3 ... L_n$, giving a family of curves B versus y_o with L as the parameter.

Only those values of B which are greater than zero represent solutions which may possibly apply to a depth equal to y_o over the entire border length. To allow for non-uniformities and non-parallel advance and recession curves, only values of B greater than some minimum value B_d can be expected to accomplish adequate irrigation. Then, by definition,

$$B_d = f_{1y}L \quad ...(11.12)$$

in which f_1 represents a factor of safety or a measure of the excess water. This function is then plotted as a series of straight lines on the same graph as B. The lines are plotted with B_d as the ordinate and y_o as the abscissa and they pass through the origin (0, 0). The intersection of B and B_d for a given length represents the minimum value of B which can be accepted for that length, as lower values presumably would not provide adequate irrigation to parts of the field. It also represents the point of optimum application efficiency for a particular border length, as larger values of B represent a greater than necessary amount of water admitted to the border. Thus, the intersection of B and B_d nearest to, yet larger than the design y_d will represent the optimum length.

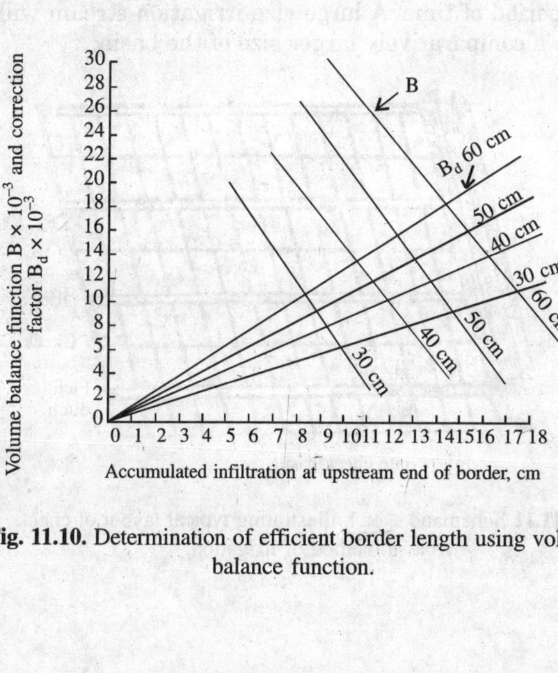

Fig. 11.10. Determination of efficient border length using volume balance function.

Example 11.3. *In the following sample calculations, using the field data obtained from a field test illustrate the procedure described in the preceding pages:*

Given

$$y = 0.77\, t^{0.61} + 0.35, \text{ and}$$

(i) $q = 1050\ cm^3/min/cm,\ d = 7.27\ cm$

(ii) $q = 1200\ cm^3/min/cm,\ d = 7.52\ cm$, and

(iii) $q = 1440\ cm^3/min/cm\ d = 7.93\ cm$.

Solution: The predicted advance curves for each value of q, using equation (11.5), are presented in Fig. 11.9. Let the border lengths selected for the design trial be 30 m, 40 m, 50 m and 60 m. Table 11.2 shows the values of T, y_o, B and B_d for the different border lengths and entrance stream sizes. Fig. 11.9 is the plot of B and B_d against y_o with L as parameter. The values of B_d are obtained for the corresponding trial border lengths and a constant factor of safety f_1 of 0.2.

The intersection of $B(y_o)$ and $B_d(y_o)$ for a given length represents the minimum value of B which is acceptable for that length, as lower values presumably would not provide adequate irrigation to all parts of the field. It also represents the point of optimum application efficiency as larger values of B represents a greater than necessary amount of water admitted to the border. The intersection nearest to, yet larger than the design y_d represents the optimum length.

In Fig. 11.10, if the design depth of application of each irrigation is 13 cm, the optimum border length is 50 m. If a depth of 12 cm is required, the optimum length is 40 m, and so on.

With the optimum border length determined, the optimum entrance stream size q and the corresponding water application efficiency may be computed. The optimum value of q is calculated from equation (11.12) using the design value of y_d, the optimum length, and the advance time corresponding to y_d, as determined from the accumulated infiltration function.

The water application efficiency E_a, is equal to the ratio of the desired uniform volume of application y_dL to the actual volume of water admitted to the border.

$$E_a = \frac{y_dL}{qT}$$

or, from equation (11.11)

$$E_a = \frac{y_dL}{(B + y_dL)} \quad ...(11.13)$$

In the example presented above, if y_d is 13 cm and L is 50 m, the corresponding value of B is 15,500 and the application efficiency is

$$E_a = \frac{13 \times 5000}{(15{,}500 + 13 \times 5000)} \times 100 = 80.7 \text{ per cent}$$

Table 11.2. Values of T, y_o, B and B_d for different trials of border lengths and entrance stream sizes in Example 11.3

Border lenth	Entrance stream size 1050 $cm^3/min/cm$				1200 $cm^3/min/cm$				1440 $cm^3/min/cm$			
	T	y_o	B	B_d	T	y_o	B	B_d	T	y_o	B	B_d
m	min	cm	cm^3	cm^3	min	cm	cm^3	cm^3	min	cm	cm^3	cm^3
30	34.50	7.03	14925	4218	30.00	6.48	16560	3888	25.00	5.84	18480	3504
40	51.00	8.33	18230	7064	43.50	8.04	20040	6432	35.50	7.14	22560	5712
50	71.00	10.72	20950	10720	59.50	9.66	23100	9660	47.75	8.43	26610	8430
60	101.50	13.24	27135	15828	78.50	11.37	25980	13644	61.75	9.87	29700	11844

If the solution obtained by this method results in unsatisfactory efficiency for every economic length, it may be necessary to modify the land slope to get more satisfactory results.

Additional experimental research is needed to determine the exact nature of the factor of safety f_1. In practice, f_1 could be determined by experience and from the factors affecting the lack of parallel advance and recession curves. Thus, the method is not precise but depends on experience and empirical observations. Secondly, the length L and time T may be regarded only as approximations, for, in many cases, more efficient irrigation may be accomplished by turning off the entrance stream before the water front reaches the downstream end.

CHECK BASIN IRRIGATION

Check basin irrigation, often referred to as basin irrigation is the most common method of irrigation in India and in many other countries. This is the simplest in principle of all methods of irrigation. There are many variations in its use, but all involve dividing the field into smaller unit areas so that each has a nearly level surface (Figs. 11.11 and 11.12). Bunds or ridges are constructed around the areas forming basins within which the irrigation water can be controlled. A basin is typically rectangular or square in shape but exists in all sorts of irregular configurations in undulating and sloping fields. It may be furrowed or corrugated, may be raised beds in case of certain crops, but as long as the inflow into the basin is not specifically directed and not controlled into these field modifications, it remains a basin. The basins are filled to the desired depth and the water retained in them until it infiltrates into the soil. When irrigating rice or ponding water for leaching salts from the soil, the depth of water may be maintained for considerable periods of time by allowing water to continue to flow into the basins.

Check Basin Types, Adaptability and Design Considerations

The distinguishing features of the various uses of the check basin method of irrigation involve the size and shape of the basins and whether irrigation is accomplished by intermittent or continuous ponding of water in the basins. The ridges or bunds may be temporary for a single irrigation as in the pre-sowing irrigation of seasonal crops, or may be for a cropping season as in post-emergence irrigations. They may be semi-permanently constructed for repeated use, as in the case of paddy fields. The size of the ridge will depend on the depth of water to be impounded as well as on the stability of the soil when wet.

Water is conveyed to the field by a system of supply channels and lateral field channels. The supply channel is aligned on the upper side of the area and there is usually one lateral for every two rows of check basins (Fig. 11.11). Water from the laterals is turned into the beds and is cut off when sufficient water has been admitted to the basin. Water is retained in the basin until it seeps into the soil. The size of the irrigation stream is not critical as long as it is sufficient to provide a coverage of the entire strip in a relatively small portion of the time required to apply the desired amount of water into the soil. As the infiltration rate of the soil increases, the stream size must be increased or the size of the basins reduced in order to cover the area within a short period of time. A large size irrigation stream will permit a comparatively larger size of the basin.

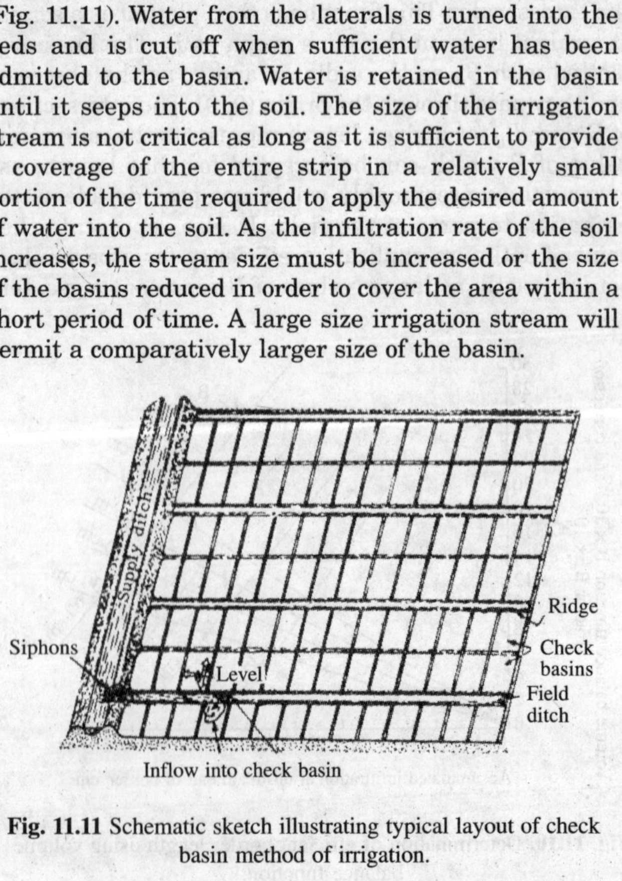

Fig. 11.11 Schematic sketch illustrating typical layout of check basin method of irrigation.

Fig. 11.12. Field laid out for check basin method of irrigation.

Fig. 11.14. Ring method of basin irrigation in an orchard. The field channel is laid in between the rows of trees and the link channels connecting the basins are formed for each tree.

Types of Check Basins, Based on Size and Shape

The size of check basins may vary from 1 sq m, used for growing vegetables to as large as one or two hectares or more, used for growing rice under wetland conditions. When the land can be graded economically into nearly level fields, the basins are rectangular in shape. In rolling topography the ridges follow the contours of the land surface (Fig. 11.13). The contour ridges are connected by cross ridges at intervals. The vertical interval between contour ridges usually varies from 6 to 12 cm in case of upland irrigated crops like wheat and 15 to 30 cm in case of lowland irrigated crops like rice.

The size of basins in rice fields varies with the size of the irrigation stream available and the size of the landholding. Large irrigation streams and large holdings permit large basins. The size also depends on the infiltration characteristics of the soil. Sandy and sandy loam soils with high infiltration rates permit only small size basins while clay soils having low infiltration rates allow large basins.

In rice fields, water is held at the desired level in the basins by check gates. The height of the shutter of the check gate is equal to the depth of submergence desired. Drainage may be provided by placing pipe outlets in the levees, allowing excess water from one area to spill over into the area immediately below.

In irrigating orchards, square or contour basins may be used as in other crops. When the plants are widely spaced, the *ring method* of basin irrigation (Fig. 11.14) may be adopted. The rings are circular

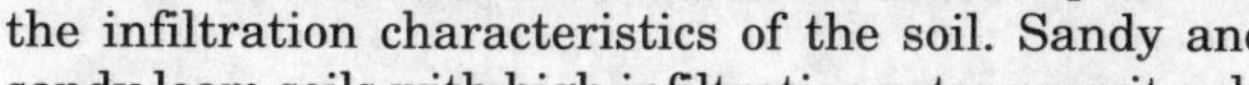

Fig. 11.13. Schematic sketch illustrating the layout of contour basin irrigation of large rice fields.

basins formed around each tree. The ring basins are small when the plant is young. The size is increased as the plant grows. An advantage of the ring method is that the entire area is not flooded, thus obtaining high water use efficiency. Usually there is one basin to a tree. In rectangular and contour basins, however, there may be one basin to a single tree or two or more trees.

Adaptability. Check basin irrigation is suited for smooth gentle and uniform land slopes and for soils having moderate to low infiltration rates. Steep slopes require complex layouts and heavy land levelling. Both row crops and close-growing crops are adapted to be used with basins as long as the crop is not affected by temporary inundation or is planted in raised beds so that it will remain above the water level. The method is especially adapted for irrigation of grain and fodder crops in heavy soils where water is absorbed slowly and is required to stand for a relatively long time to ensure adequate irrigation. It is also suitable in very permeable soils which must be covered with water rapidly to prevent excessive deep percolation losses at the upstream end.

Many different kinds of crops can be grown in sequence in the same field without making major changes in the design, layout or operating procedures. Practically, all irrigated paddy is grown in basins. Check basins are useful when leaching is required to remove salts from the soil profile. The method enables the conservation of rainfall and reduction in soil erosion by retaining a large part of the rain in the basin to be infiltrated gradually, without loss due to surface runoff. The method usually results in high water application and distribution efficiencies if the desired net depth of irrigation can be estimated adequately and if the size of the irrigation stream is measured properly. The low efficiencies obtained in check basin irrigation are due to inadequate land levelling and uncontrolled water application.

Limitations. The principal disadvantage of the check basin method of irrigation is that the ridges interfere with the movement of animal-drawn or tractor-drawn implements for interculture or harvesting of crops. Considerable land is occupied by ridges and lateral field channels and crop yields are substantially low on the ridge and in the lateral channels. The method impedes surface drainage. Precise land grading and shaping are required. Labour requirements in land preparation and irrigation are much higher in check basin irrigation as compared to other methods, except when the basins are very large. The method is not suitable for irrigated crops which are sensitive to wet soil conditions around the stems of plants. Plants which are damaged by these conditions are usually irrigated by furrows.

Hydraulics of Check Basin Irrigation

Hydraulics of flow in check basins may be considered to comprise of four stages: (*i*) initial spreading of the entrance stream to cover the full width of the basin and simultaneous advance of the irrigation stream, (*ii*) advance of the water front after the initial spreading, (*iii*) rise of water level after the advancing stream reaches the downstream end, and (*iv*) subsidence of water after the irrigation stream is stopped. The variables influencing the hydraulics of check basin irrigation are, in general, the same as in border irrigation. The essential differe-nces in the phenomenon are in the initial spreading of the entrance stream to cover the full width of the basin and in the characteristics of the recession flow.

If the check basin irrigation system is properly designed, it is possible to apply the right amount of water nearly uniformly throughout the basin. The problem of efficient irrigation by check basin consists essentially of having the right size of basin to suit the available stream size for a particular set of soil and crop conditions.

(*i*) *Spreading the entrance stream in a check basin.* The water front advance in a check basin differs from borders in the initial stages at the upstream end of the basin. The entrance stream spreads on either side as it advances forward till the entire width of the basin is covered. When water is introduced into a check basin from an orifice or other inlets, the flow is non-linear. In non-linear flow, the paths of flow may diverge or converge along the flow line.

Since the hydraulic characteristics of non-linear water flow patterns have not been developed yet, the differential equation which relates the parameters of water flow at the initial spreading stage in a check basin is not well established. In the absence of suitable mathematical models to predict the spreading pattern in basins, on the basis of measurable variables, check basins are evaluated by observing the spreading pattern in the field. The desired water distribution efficiency is obtained by adjusting the entrance stream size or the size of the check basin or both. In long basins, the initial spreading time is very small as compared to the total time required for the water front to reach the downstream end of the basin. In such cases, check basins may be designed using the procedure suggested for borders.

(*ii*) *Water front advance in check basins.* The advance of the water front after the initial spreading in check basins may be predicted using equations 11.5 and 11.6, as in borders. The dominant variables are the same as in borders, namely, entrance stream size, infiltration characteristics of the soil, hydraulic resistance offered by the soil surface and vegetation, water surface slope and elapsed time. Figure 11.15 shows the pattern of water spread in a check basin. The water spread pattern is shown by drawing iso-time lines at uniform time intervals based on field data of the measured values of time when the water front reaches each station.

(*iii*) *Water storage and rate of rise in check basins.* Ponding occurs after the water front reaches the downstream end of the check basin. The volume of storage above the soil surface in a given time period is equal to the difference between the volume of water admitted into the basin during the period and the volume infiltrated into the soil. This may be expressed as:

$$V_s = qt_s - I_s t_s = A_c d_s \qquad \text{...(11.14)}$$

in which,

V_s = volume of water stored in a given time t_s, cm^3

q = average size of the entrance stream, cm^3/min

t_s = storage time, min

I_s = average infiltration rate during the storage time, cm/min

A_c = area of check basin, cm^2, and

d_s = depth of storage during t_s, cm.

$$\text{Rate of rise of water level} = \frac{d}{dt_s}(d_s)$$

(*iv*) *Recession of water in check basins.* Recession in a check basin may be taken as the subsidence of water due to infiltration. The recession rate can be estimated with the help of the infiltration equation $y = at^{\alpha} + b$ (Equation 9.12, Chapter 9). Figure 11.16 shows the definition sketch of recession in a check basin, and Figure 11.17 shows typical advance and recession curves plotted from the data obtained from a check basin irrigation test.

Design of check basin irrigation system. Efficient irrigation by the check basin method depends on the knowledge of the hydraulics of flow in the basin. Careful evaluation of the irrigation system under representative soil and crop management practices is necessary to obtain the criteria for the design of efficient check basin irrigation systems.

In check basins it is possible to apply precise amounts of water required to fill the root zone moisture to field capacity if provisions are made to measure the rate of flow into the basin and control the duration of flow of the irrigation stream. This will result in high values of water application and water storage efficiencies. Water

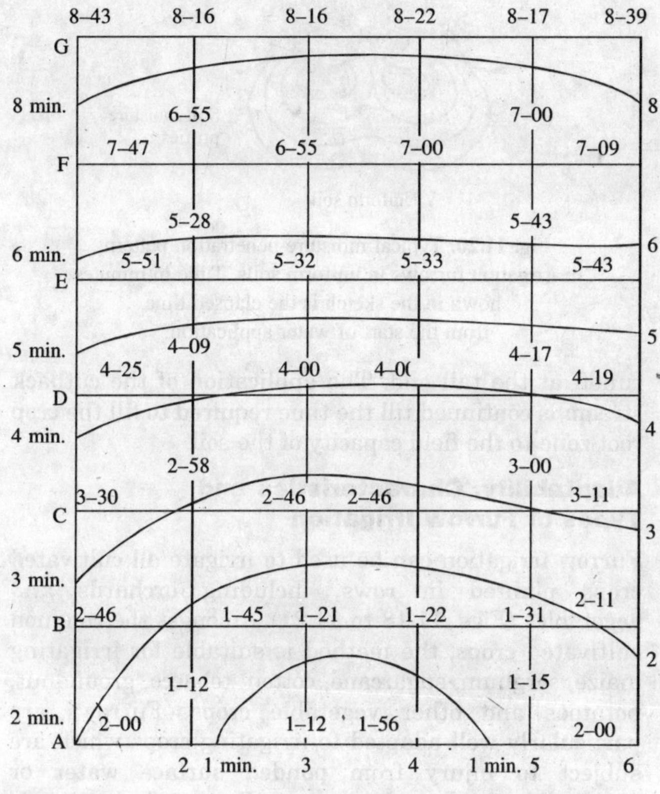

Fig. 11.15. Iso-time profiles of water spreading and advance of water front in a check basin, based on field data from a test basin of size 10 m × 12 m in a sandy loam soil. The total time of water application has been 10 minutes, average moisture content 14.5% and depth of irrigation 3.79 cm.

distribution efficiency is usually the limiting factor. A useful thumb rule in check basin design is that the water spread in the entire basin should be covered in one-fourth of the time required to infiltrate the net depth of irrigation water. This is often called the *quarter-time rule.* By careful selection of the size of the check basin and the stream size to suit the infiltration characteristics of the soil, it is possible to arrive at designs of check basins which give high values of water distribution efficiency.

Quasi-rational designs of check basins may be developed using the procedure suggested for border irrigation. The procedure is simplified in case of level check basins as the recession rate can be predicted with reasonable accuracy using the infiltration equation. In

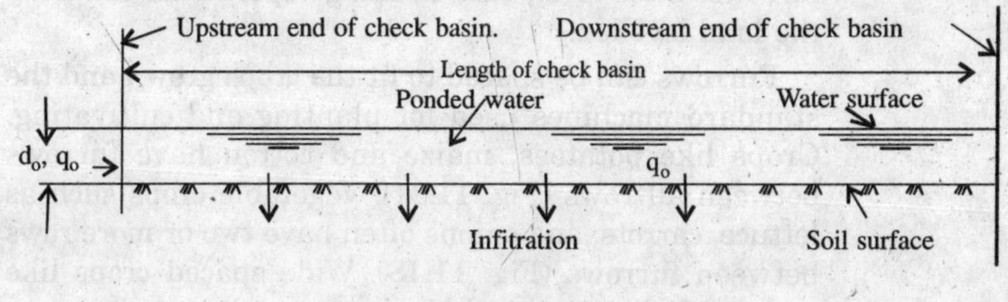

Fig. 11.16. Definition sketch illustrating recession in check basins.

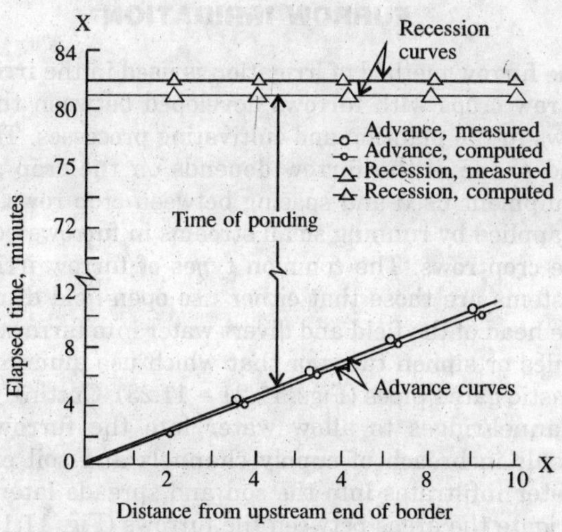

Fig. 11.17. Typical advance and recession curves in a check basin. Note that the recession curve is nearly a straight line, almost parallel to the *x*-axis, as the recession occurs nearly at constant time.

comparatively short basins, the time required for the lateral spread of water at the upstream end is considerable. Results of data from representative soil and crop conditions will be useful in designing efficient check basins.

Example 11.4. *An irrigation stream of 27 litres per second is diverted to a check basin of size 12 m × 10 m. The water holding capacity of the soil is 14%. The average soil moisture content in the crop root zone prior to applying water is 6.5%. How long should the irrigation stream be applied to the basin to replenish the root zone moisture to its field capacity, assuming no loss due to deep percolation. The average depth of crop root zone is 1.2 m. The apparent specific gravity of the root zone soil is 1.50.*

Solution: Net irrigation requirement = Field capacity – Available soil moisture content

$$= 14 - 6.5 = 7.5\%$$

$$= 1.5 \times 7.5 = 11.25 \text{ cm/m of soil depth}$$

$$= 11.25 \times 1.2 = 13.5 \text{ cm}$$

Total volume of water required in the check basin

= Area of basin × net depth of irrigation

$$= 12 \times 10 \times \frac{13.5}{100} = 16.2 \text{ cu m}$$

$$= 16{,}200 \text{ litres}$$

∴ Required duration of irrigation

$$= \frac{16200}{27} = 600 \text{ seconds}$$

$$= 10 \text{ minutes.}$$

FURROW IRRIGATION

The furrow method of irrigation is used in the irrigation of row crops with furrows developed between the crop rows in the planting and cultivating processes. The size and shape of the furrow depends on the crop grown, equipment used and spacing between crop rows. Water is applied by running small streams in furrows between the crop rows. The common types of furrow irrigation systems are those that either use open-field ditches at the head of the field and divert water into furrows using spiles or siphon tubes or that which use aluminium or plastic gated pipes (Figs. 11.21 – 11.23). Cutting of field channel/ridges to allow water into the furrows will result in breach of supply channels and soil erosion. Water infiltrates into the soil and spreads laterally to irrigate the areas between the furrows (Fig. 11.18). The length of time the water is to flow in the furrows depends on the amount of the water required to replenish the root zone and the infiltration rate of the soil and the rate of lateral spread of water in the soil (Figs. 11.19 and 11.20). Both large and small irrigation streams can be used by adjusting the number of furrows irrigated at any one time to suit the available flow. In areas where surface drainage is necessary, the furrows can be used to dispose off the runoff from rainfall.

Cutback stream in furrow irrigation. A common method to obtain the optimum opportunity time to apply a given depth of irrigation in a furrow system is to allow the initial design stream to flow the full length of the furrow and then reduce the stream size by adjusting outlet gates or by reducing the number or size of spiles or systems to permit reduced stream size. The furrow stream in such a situation is called the *cutback stream*. The size of the cutback stream is adjusted to match the cumulative infiltration in a furrow so that there is no

Fig. 11.18. Schematic sketch illustrating furrow irrigation of a vegetable crop with one furrow for each two rows of the crop.

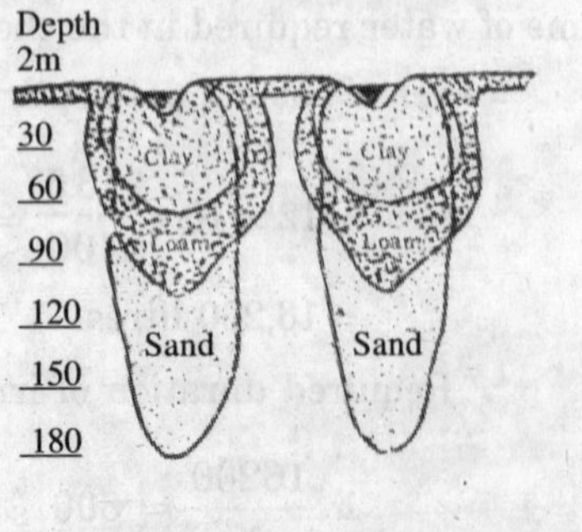

Fig. 11.19. Profiles showing the downward movement of water in different soil types.

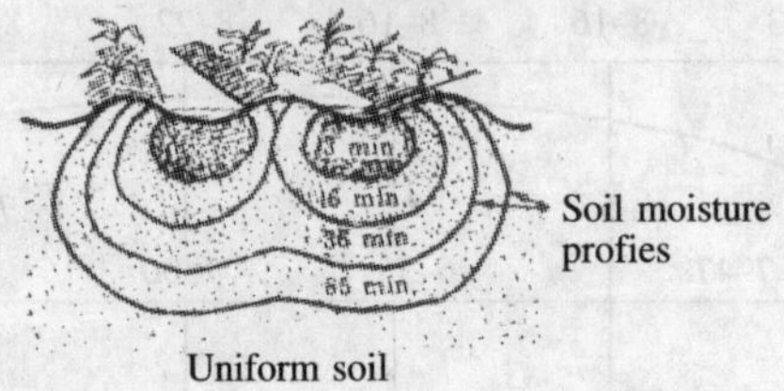

Fig. 11.20. Typical moisture penetration pattern in irrigation furrows in uniform soils. Time in minutes shown in the sketch is the elapsed time from the start of water application.

runoff at the tail end. The application of the cutback stream is continued till the time required to fill the crop root zone to the field capacity of the soil.

Adaptability, Characteristics and Types of Furrow Irrigation

Furrow irrigation can be used to irrigate all cultivated crops planted in rows, including orchards and vegetables (Figs. 11.18 to 11.21). Amongst the common cultivated crops, the method is suitable for irrigating maize, sorghum, sugarcane, cotton, tobacco, groundnut, potatoes and other vegetable crops. Furrows are particularly well adapted to irrigating crops which are subject to injury from ponded surface water or susceptible to fungal root rot. Furrow irrigation is suitable to most soils except sands that have a very high infiltration rate and provide poor lateral distribution of water between furrows.

As compared to other methods of surface irrigation, the furrow method has several distinct advantages: (*i*) Water in the furrows contacts only one-half to one-fifth of the land surface, thereby reducing puddling, and crusting of the soil, and evaporation losses, and (*ii*) earlier cultivation is possible which is a distinct advantage in heavy soils. The furrow method may be adapted to use, without erosion, on a wide range of natural slopes by carrying the furrows across a sloping field rather than down the slope. The method reduces labour requirements in land preparation and irrigation. Compared to check basin method, there is no wastage of land in field ditches.

Furrow irrigation requires proper land grading. The land must be graded so that water can travel the entire length of the row without ponding. This means that the high and low spots must be removed and the land must be given enough slope to let the water flow down the furrows. Once level, care in tillage operations can keep the land levelled.

Furrows can be spaced to fit the crops grown and the standard machines used for planting and cultivating. Crops like potatoes, maize and cotton have furrows between all rows (Fig. 11.21). Vegetable crops such as lettuce, carrots, and onions often have two or more rows between furrows (Fig. 11.18). Wide spaced crops like melons, fruits trees and berries generally require more than one furrow between crop rows. Furrows should be spaced close enough to ensure that water spreads to the sides into the ridge and root zone of the crop before it moves down below the root zone.

In general, small plants require small furrows; larger plants require large furrows. Furrows of 7.5 to 12.5 cm depth are appropriate for vegetable crops, while some row crops and orchards require much deeper furrows.

In soils that absorb water slowly, a wide, relatively shallow furrow is preferable since it gives more area for the water to infiltrate. Where furrows are long and the soil is permeable, narrow deep furrows may be used to discourage excessive percolation at the upper end. In general, in furrow irrigation, a wide and shallow furrow in preferable.

Irrigation furrows may be classified into two general types based on their alignment. They are: (*a*) *straight furrows* (Fig. 11.21) and (*b*) *contour furrows* (Fig. 11.22). Based on their size and spacing, furrows may be classified as *deep furrows* and *corrugations*.

Fig. 11.21. Furrow irrigation of a vegetable crop. The crop is grown on top of the ridge with furrows between each crop row. Note the siphon tubes used to divert water from the field channel to each furrow.

Straight Furrows

Straight furrows, like borders, are laid down the prevailing land slope (Fig. 11.21). They are best suited to sites where the land slope does not exceed 0.75 per cent. In areas of intense rainfall, however, the furrow grade should not exceed 0.5 per cent so as to minimise the erosion hazard. The ranges in furrow slopes for efficient irrigation in different soil types are the same as those recommended for borders.

In fine textured soils having very low infiltration rates, the furrows are usually level lengthwise. With *level furrows*, the same stream size is maintained until the required amount of water is applied. The water is ponded in the furrows until it is absorbed by the soil.

Contour Furrows

Contour furrows carry water across a sloping field rather than down the slope. Contour furrows are curved to fit the topography of the land (Fig. 11.23 and 11.24). The furrows are given a gentle slope along its length as in the case of graded borders. Field supply channels run down the land slope to feed the individual furrows and must be provided with erosion control structures.

Contour furrow method can be successfully used in nearly all irrigable soils. The limitations of straight furrow irrigation are overcome by contouring to include sloping lands. Light soils can be irrigated successfully across slopes upto 5 per cent.

The topography must be uniform enough to permit a head ditch that can feed the entire area of contour-furrows. It must also be regular enough to permit installing a waste ditch to carry off surplus irrigation water and rain water. Fields should be smoothed as much as possible by land grading and by using a float or land plane, before contour plantings are laid out.

Contour furrows for row crops are installed annually. But in orchards and other permanent contour plantings, the same furrows and ditches are used every year, and they may be considered as a permanent part of the water handling system. Furrows and waste ditches should be cleaned and prepared for the rainy seasons immediately after the crops are harvested.

Contour furrows may be used on most soil types, except on light sandy soils and soils that crack. The ridges between furrows in sandy soils may break and wash out, overloading the furrow below, which also breaks. This may continue all the way down the slope causing heavy erosion damage. Soils that crack provide channels for water, causing similar downslope furrow breaks. Hence, contour furrows have only limited use on steep slopes in sandy soils and heavy black soils.

Fig. 11.22. Large bamboo spiles taking out water from field channel to furrows for pre-sowing irrigation in a black soil region.

Fig. 11.23. Pre-sowing irrigation with contour furrows in black cotton soil.

Fig. 11.24. Young cotton growing in rows 250 metres long with 0.1% slope, in heavy clay soil. Note the contour furrows laid within contour borders. Any desired gradient may be obtained by changing the direction of the rows.

In heavy rainfall areas the length of furrows should be short enough to dispose off the runoff safely without breaking the furrows. Erosion control structures are needed to carry the surplus water down the slope. Contour furrow irrigation used in conjunction with contour bunding and terracing provides an insurance against furrow breaks. Land which is too steep for contour furrows may sometimes be graded to levelled strips (bench terraces) across the slope on which the furrows can be constructed. Land with slopes upto 25% or more can be benched to permit the production of irrigated crops.

Furrow Irrigation in Black Cotton Soils

In heavy black soils, a combination of border strip irrigation with furrows is usually preferred (Fig. 11.24). These soils crack badly, so it is desirable to have furrows within the border strips, so cracking will not allow the water to escape from furrow to furrow. The border ridges are sufficiently large to confine the flow within the strip and the furrows distribute it uniformly. Another advantage is that the furrows provide uniform drainage during heavy rains and erosion is kept at a minimum.

Corrugation Irrigation

Corrugation irrigation consists of running water in small furrows, called corrugations which direct the flow down the slope. It is commonly used for irrigating non-cultivated, close-growing crops such as small grains and for pasture growing on steep slopes. Corrugations may be used in conjunction with border irrigation on lands with relatively flat slopes to help in obtaining uniform coverage with water. The main point of difference from regular furrow irrigation is that more, but smaller furrows are utilised and the crop rows are not necessarily related to the irrigation furrows.

The soil may be prepared and the crop planted without regard for irrigation layout, using corrugation method. After the seed has been sown in the soil, but before germination has taken place, a corrugator is used, making small furrows or corrugations to aid in controlling irrigation water. The corrugations can be made with a simple bamboo corrugator (Fig. 11.25) or cultivators equipped with small furrowers or other similar implements.

Corrugations are V-shaped or U-shaped channels about 6 to 10 cm deep, spaced 40 to 75 cm apart. The entire soil surface is wetted slowly by the capillary movement of the water which flows in the corrugations. This method of wetting the soil minimizes the crusting effect on the surface soil which may be a problem when the entire surface is flooded. For this reason, corrugation irrigation is sometimes used for germinating seeds which have been drilled or broadcast. Border or check basin methods are used later, after the plants have established.

Corrugation irrigation is most suitable in loamy soils in which the lateral movement of water takes place readily. Clay soils with low infiltration rates are difficult to irrigate by this method, unless the slopes are quite flat, or water can be held in the corrugations for a considerable length of time. The method is also not suitable in deep sandy soils because of the excessive loss of water by deep percolation before the entire soil surface is wetted.

The corrugation method is not recommended on saline soils or when the irrigation water has a high salt content. The capillary movement and subsequent evaporation of the water will tend to concentrate the salts in the surface soils.

Fig. 11.25. Corrugations made by a simple manually operated bamboo corrugator in a field immediately after the sowing of wheat seeds.

The corrugations are usually made in the direction of the steepest slope. They may also be laid across the slope as in contour furrows. They often become blocked by soil or plant debris, causing the water to overflow the corrugations. If the corrugations are placed across the land slope, the overflow of water may move down into the lower corrugations and may cause soil erosion. The maximum slope of the corrugations depends on the type of soil and the crop being grown. The permissible length of corrugations varies from about 50 m in light textured soils with slopes of 2% to 4% to about 150 m in heavy textured soils upto 2% slope.

Furrow Irrigation Hydraulics

Efficient water application in furrow irrigation depends on the knowledge of the hydraulics of flow in furrows. Like in borders, the flow phenomenon in furrow irrigation is a case of unsteady open channel flow with decreasing discharge. Furrow irrigation, however, differs from borders and basins in the pattern of wetting the soil, because the water which soaks into the soil spreads laterally to the adjacent areas. The dominant variables influencing the rate of flow in furrows are the entrance stream size, infiltration rate, size and shape of wetted section of furrow, furrow slope and hydraulic resistance. The hydraulic resistance to flow may be due to the combined effect of the roughness offered by the wetted surface of the furrow and the resistance offered by the crop.

As in borders, the time of ponding or infiltration opportunity time in a furrow should be the time period required for the net depth of water to infiltrate so that the crop root zone is filled to its field capacity. The criterion for providing uniformity in moisture penetration is the value of the time of ponding and should, as far as possible, be uniform throughout the length of the furrow. The time of ponding is demarcated by the difference between the advance and recession times in the furrow.

Rational procedures for predicting the water front advance and tail water recession in furrows which are applicable to field designs have not been adequately developed. A number of quasi-rational procedures have been proposed by various workers with varying degrees of adaptability.

Evaluation of Furrow Irrigation

Infiltration of water into the furrow is the most important variable affecting the characteristics of flow in furrows. The infiltration rates in furrows may be determined by the gravimetric method, by specially designed furrow infiltrometers or, more commonly, by the inflow-outflow method. In the gravimetric method, the difference in moisture content in the soil before and after irrigation is determined by soil sampling and moisture determination. The method, though accurate, is time consuming as it involves sampling at many locations. The furrow infiltrometer techniques consists of blocking the furrows at their two ends so as to assess the volume storage difference in the furrow in relation to time. The instrument includes a float mechanism and a water stage recorder. The inflow-outflow method, also known as volume balance method, is considered to be the most satisfactory one because it gives the average infiltration value by compensating various errors, inherent in the furrows, arising out of soil heterogeneity, furrow cross-sectional difference, cracks and puddling effects.

Inflow outflow method. In the inflow-outflow method, the furrow is divided into a number of sections and Parshall flumes or other suitable water measuring devices are installed at each station to measure the flow rate. The furrow cross-section profile is determined at representative locations in the test section with a point gauge before admitting water. The furrow spacing is measured from the centre of the furrow to the centre of the adjacent furrow. The rate of advance of water in the test section and the depth of flow at different points at definite lime intervals are measured. From the above measurements it is possible to obtain the furrow cross-sectional area and wetted perimeter. The average value of the wetted perimeter multiplied by the length of the test section area gives the wetted area. Furrow infiltration is determined from the following relationships:

$$\text{Accumulated infiltration} = \text{Accumulated inflow} - \text{Accumulated storage (volume)}$$

$$\text{Accumulated infiltration (depth)} = \frac{\text{Accumulated infiltration (volume)}}{\text{Wetted area of test section}}$$

Example 11.5. *The data obtained from a test furrow in a sandy loam soil are given as follow.*

Stream size litres/min	Distance metres	Advance time minutes	Wetted perimeter cm	Furrow cross-sectional area corresponding to depth of flow cm^2
92.00	20	1.75	25.39	60.00
	40	5.75	25.82	93.00
	60	10.91	26.39	103.00
	80	17.83	26.70	108.40
	100	23.67	27.11	111.65
	110	27.75	27.42	112.28

Compute the furrow infiltration.

Solution: As a first step, the accumulated inflow is computed. This is accomplished by multiplying the stream size and advance time. As a second step, the accumulated storage volume is computed. This is accomplished by multiplying the furrow cross-sectional area and the distance traversed by the furrow stream. The third step is the computation of the accumulated wetted area. This is the wetted perimeter multiplied by the distance traversed.

Sample calculations: For distance 40 m,

Accumulated inflow = 92 × 5.75
= 529.00 litres

Accumulated storage = 93 × 4000
= 372000 cm^3 or 372 litres

Accumulated wetted area = 25.82 × 4000
= 103280 cm^2

The results are tabulated below:

Accumulated inflow litres	Accumulated storage litres	Wetted area cm^2
161.00	120.00	40780
529.00	372.00	103280
1003.72	618.00	157980
1640.36	867.20	213600
2176.72	1116.50	271100
2553.00	1235.08	301620

The second phase of the computation involves the determination of accumulated infiltration volume and depth. Continuing the sample calculation for distance 40 m,

Accumulated storage = 372 litres

Accumulated inflow = 529 litres

Accumulated infiltration (volume) = 529 – 372
= 157 litres = 1,57,000 cm^3

Area of wetted furrow = 1,03,280 cm^2

Accumulated infiltration depth

$$= \frac{157000}{103280} = 1.52\,\text{cm}$$

Similarly, for infiltration depths, another length along the furrow are computed and are summarised below:

Advance time minutes	Accumulated infiltration volume litres	Accumulated infiltration depth cm
1.75	32.00	0.63
5.75	157.00	1.52
10.91	385.72	2.44
17.83	773.16	3.62
23.67	1060.22	3.91
27.75	1317.92	4.37

Furrow Irrigation Design Considerations

Efficient irrigation by the furrow method is obtained by selecting proper combinations of spacing, length, and slope of furrows, suitable size of the irrigation stream and duration of water application.

Furrow spacing. Furrows can be spaced to fit the crops grown and the type of machines used for planting and cultivation. Furrows should be spaced close enough to ensure that water spreads to the sides into the ridge and the root zone of the crop, to replenish the soil moisture uniformly. The lateral movement of water from the furrow in soils with uniform profiles depends primarily upon the texture of the soil, with a broader wetting pattern occurring in clays than in sandy soils. To obtain complete wetting of sandy soils to depths of 1 to 1.5 metres, the furrows should not be spaced more than 50 to 60 cm apart. In uniform clay soils, complete wetting to the same depth may be obtained with a furrow spacing of one metre or more.

Furrow length. The optimum length of a furrow is usually the longest furrow that can be safely and efficiently irrigated. Long furrows are an advantage in intercultivation. If the length is too long, water soaks in too deep at the head of the furrow by the time the stream reaches the lower end. This results in over-irrigation at the upper end or under-irrigation at the lower end. Short furrows require field supply channels to be spaced too close, with consequent loss of land and increase in labour requirement.

Proper furrow length depends largely on the hydraulic conductivity of the soil. Furrows must be shorter on a porous sandy soil than on a tight clay soil. The length of furrow which can be efficiently irrigated may be as short as 45 m on soils which take up water rapidly, or as much as 300 m or longer on soils with low infiltration rates. The length of the furrow may often be limited by the size and shape of the field. If only a small area is to be irrigated the length of furrow may be determined by the length of the field. If the area is large, it may be desirable to have the furrow lengths equal to an even fraction of the total length of the field.

With gentle slopes (less than 0.3%), the length of the run can be increased with the increase in the furrow slope. Furrow lengths can be increased as the average depth of water to be applied increases. Since the depth of irrigation needed is related to the water holding capacity of the soil and the depth of rooting of the crop, much longer furrows can be used with deep rooted crops growing on clay soils than for shallow rooted crops growing on sandy soils. Table 11.3 presents the recommended maximum lengths of furrows for different soils, slopes and depths of irrigation.

Table 11.3. Recommended lengths of furrows for different soil types, furrow lengths and depths of irrigation.

Furrows slope %	*Furrow length, metres*						
	Slope type						
	Clays		*Loams*		*Sands*		
	Net depth of water application						
	7.5 cm	15 cm	5 cm	10 cm	5 cm	7.5 cm	10 cm
0.05	300	400	120	270	60	90	150
0.10	350	440	180	330	90	120	190
0.20	370	470	220	370	120	190	250
0.30	390	500	280	400	150	225	280
0.50	380	500	280	370	120	190	250
1.00	270	400	250	300	90	150	220

Furrow slope. The slope or grade of the furrow is important because it controls the speed at which water flows down the furrow. A minimum slope of 0.05% is needed to ensure surface drainage. In general, the ranges in slope recommended for borders apply to furrows also.

As the furrow grade increases, the rate of infiltration slows down and the side spread of water into the crop ridge decreases, so that wastage may occur at the end of the furrow. With highly permeable soils, these factors may not be limiting. However, steeper grades lead to higher water velocities and more soil erosion. If the slope of the land is too steep for effective irrigation, the furrows should lead around the hill rather than straight down the slope. As with border irrigation, a uniform slope is desirable.

Furrow stream. The size of the furrow stream is one factor which can be varied after the furrow irrigation system has been installed. The size of the furrow stream usually varies from 0.5 to 2.5 litres per second. To obtain the most uniform irrigation, the largest stream of water that will not cause soil erosion is used in each furrow at the beginning of irrigation. Its purpose is to wet the entire length of each furrow as quickly as possible, thus enabling the soil to absorb water evenly through the entire furrow length. After the water reaches the lower end of a furrow, the stream is reduced or cut back so that it will just keep the furrow wet throughout its length with a minimum waste at the end. This *cut back stream* flows until the required amount of water has been applied. With level furrows, however, the initial stream is continued from the beginning to the end of irrigation. The water is ponded in the furrow until it is absorbed by the soil.

The maximum size of the irrigation stream that can be used at the start of the irrigation is limited by considerations of erosion in furrows, overtopping of furrows and prevention of runoff at the downstream end. The maximum non-erosive flow rate in furrows is estimated by the following empirical equation:

$$q_m = \frac{0.60}{s} \quad ...(11.15)$$

in which,

q_m = maximum non-erosive stream, litres per second

s = slope of furrow expressed as a per cent

The average depth of water applied during an irrigation can be calculated from the following relationship:

$$d = \frac{q \times 360 \times t}{w \times L} \quad ...(11.16)$$

in which,

d = average depth of water applied, cm

q = stream size, litres per second

t = duration of irrigation (elapsed time), hours

w = furrow spacing, metres

L = furrow length, metres

Example 11.6. *Furrows 90 m long and spaced 75 cm apart are irrigated by an initial furrow stream of two litres per second. The initial furrow stream reached the lower end of the field in 50 minutes. The size of the stream was then reduced to 05 litres per second. The cut back stream continued for 1 hour. Estimate the average depth of irrigation.*

Solution: Average depth of irrigation by the initial furrow stream

$$= \frac{q \times 360 \times t}{w \times L} = \frac{2 \times 360 \times 50}{0.75 \times 90 \times 60} = 8.88 \text{ cm}$$

Average depth of irrigation by the cut back stream

$$= \frac{0.5 \times 360 \times 1}{0.75 \times 90} = 2\,66 \text{ cm}$$

Net average depth of irrigation

$$= 8.88 + 2.66 = 11.54 \text{ cm}$$

Flow into furrows can be carefully regulated for uniform water distribution and efficient irrigation. In the absence of satisfactory rational equations to predict advance and recession in furrows, furrow specifications and stream sizes are selected on the basis of field evaluation data obtained under representative field conditions.

Fig. 11.26. Trench dug for sub-irrigation of a maize field.

Fig. 11.27. Sub-irrigation of coconut palms and tapioca through open ditch in the organic soils reclaimed from lake bed in Kuttanad, Kerala.

SUB-IRRIGATION

In sub-irrigation, water is applied below the ground surface by maintaining an artificial water table at some depth, depending upon the soil texture and the depth of the plant roots. Water reaches the plant roots through capillary action. Water may be introduced through open ditches (Figs. 11.26 and 11.27) or underground pipelines such as drains or mole drains. The depth of open ditches or trenches vary from 30 to 100 centimetres and they are spaced about 15 to 30 metres apart. The water application system consists of field supply channels, ditches or trenches suitably spaced to cover the field adequately and drainage ditches for the disposal of excess water.

The sub-irrigation method requires rather special site conditions as it is necessary to have complete control of the water table through controlled water application and drainage. Each site requires special investigations to determine its adaptability to sub-irrigation. The method is suited to soils having reasonably uniform texture and permeable enough for water to move rapidly both horizontally and vertically within and for some distance below the crop root zone. The soil profile must also contain a barrier against excessive losses through deep percolation, either a nearly impermeable layer in the substratum or a naturally high water table on which a perched or artificial water table can be maintained throughout the growing season. Topography must be smooth and nearly level.

Sub-irrigation can be used for soils having a low water holding capacity and a high infiltration rate where surface methods cannot be used and sprinkler irrigation is expensive. In sub-irrigation, it is possible to maintain the water level at optimum depths for crop needs at different growth stages. Evaporation losses from ground surface are held to a minimum. Since the method requires an unusual combination of natural conditions, it can be used in only a few areas. Water having a high salt content cannot be used. In India sub-irrigation is practised to a limited extent for growing vegetable crops around Dal lake in Kashmir and for irrigating coconut palms in the organic soils of Kuttanad in Kerala (Fig. 11.27).

SUMMARY

Different methods of surface irrigation along with their sub-groups and adaptability are introduced. Basic features of border strip method of irrigation are explained. Hydraulics of border irrigation with special reference to the advance of the water front under pre-sowing and post-emergence irrigation situations are described. Procedures for the design of border irrigation systems using rational and quasi-rational approaches are explained. Basic features of check basin and furrow methods of irrigation are described along with procedures for their design. The concept of quarter-time rule in check basin irrigation is explained. Applicability of cut-back stream approach to obtain high efficiencies in furrow irrigation are described. Procedures for field evaluation of border strip, check basin and furrow methods of irrigation are explained. The applicability and advantages of corrugation irrigation are described. Sub-irrigation, its applicability and limitations are described.

PROBLEMS

11.1. The following data were obtained in a field test in a border strip in the third irrigation of a wheat crop in a sandy soil:

$$q = 1150 \text{ cm}^3/\text{min/cm}$$
$$d = 7.47 \text{ cm}$$

The functional relationship between accumulated infiltration and elapsed time is represented by the equation.

$$y = 0.52t^{0.72} + 0.66$$

Determine the distance advanced by the water front 15 min, 30 min, 45 min, 60 min and 75 min after the start of irrigation.

[**Ans.** 21.75 m, 35.00 m, 45.50 m, 53.50 m, 59.75 m.]

11.2. A wheat crop is to be irrigated using the check basin method. The size of each basin is 10 m × 8 m. The size of the available irrigation stream is 18 liters per second. The water holding capacity of the root zone soil is 16%. The apparent specific gravity of the soil is 1.58. Soil moisture content before irrigation is 8.5%. Determine the irrigation duration if the water application efficiency is 96%. The depth of root zone is 80 cm. [**Ans.** 6 min 57 sec.]

11.3. The field data obtained from a test furrow are as follows:

Stream size	*Distance from upstream end of furrow*	*Time of advance*	*Wetted perimeter*	*Furrow cross-sectional area corresponding to depth of flow*
liters/min	*metres*	*minutes*	*cm*	*cm*
74.00	60	11.75	22.10	91.50
	80	17.42	22.50	92.50
	100	23.75	22.90	92.40
	110	26.83	23.20	91.91

Compute the depth of infiltration at the different values of elapsed time. [**Ans.** 2.42 cm., 3.05, 3.64 cm, 3.82 cm.]

11.4. Furrows 110 m long and spaced 90 cm apart and having a slope of 0.2 per cent are irrigated for 45 minutes by an initial stream of size equal to the maximum non-erosive stream. The stream size is then reduced to half and continued for 1 hour 20 minutes. Determine the average depth of irrigation. [**Ans.** 15.45 cm.]

QUESTIONS

1. Classify the different methods of surface irrigation and their sub-groups.

2. What are the specific conditions under which free flooding is practiced? What are its limitations?

3. What are the basic features of a border strip?

4. What is the limiting value of the vertical interval between adjacent contour borders?

5. What are the specific situations where open-end and blocked-end borders are suitable?

6. What are the limiting values of the slope of border strips for moderate size irrigation streams in different soil textural groups?

7. What is the limiting minimum width of a border strip?

8. How is the size of irrigation stream expressed in border irrigation?

9. Define infiltration opportunity time. How is it determined in the graph of water front advance and tail water recession in a border strip?

10. What are the limitations in the adaptability of check basin method of irrigation?

11. What are the four basic stages in the hydraulics of check basin irrigation?

12. What is quarter-time rule, as applied to check basin irrigation?

13. What is the principle of cut-back stream in furrow irrigation?

14. What are the specific soil and land slope conditions to prefer the use of corrugation irrigation over regular furrow irrigation?

15. What are the variables influencing the hydraulics of flow in furrow irrigation? How do they differ from those influencing border irrigation?

16. What are the methods used to measure infiltration of water in furrows?

17. Why is the inflow-outflow method of evaluation of furrow irrigation considered superior to other methods?

18. What are the parameters in the design of furrow irrigation system?

19. In irrigating furrows laid on a longitudinal slope of 0.5%, what is the maximum size of the non-erosive irrigation stream? Express the value in litres/sec.

20. What are the special field situations in which sub-irrigation is suitable?

21. What is the technology adopted in irrigating crop lands through sub-irrigation?

22. Identify the True and False statements from the following:

(*i*) In the layout of border strip irrigation systems uniform gentle gradient is maintained both downfield and across the border strip.

(*ii*) In the layout of a field for contour border irrigation, irrigation and drainage channels are provided along the contour to prevent erosion of the channel bed.

(*iii*) Border strip irrigation of fine textured soils with low infiltration rates require smaller streams to avoid excessive losses due to runoff at the downstream end and deep percolation losses at the lower reaches.

(*iv*) By varying the size of the irrigation stream it is possible to vary the depth of water applied.

(*v*) Parallelism of advance and recession curves in a graph showing distance from the upstream end of the border versus elapsed time indicates uniform distribution of water along the border strip.

(*vi*) The flow of water in a border strip is a case of spatially varied steady open channel flow with decreasing discharge.

(*vii*) The vertical interval between contour ridges in check basin irrigation is low in case of lowland rice, as compared to upland irrigated crops like wheat.

(*viii*) In check basin irrigation sandy and sandy loam soils permit the use of large size basins.

(*ix*) In check basin irrigation, as the infiltration rate of the soil increases, the stream size must be increased or the size of the basins reduced, in order to cover the area within a short period of time.

(*x*) The water front advance in a check basin is a case of linear flow with decreasing discharge, as in case of border strips.

(*xi*) When the concept of cut-back stream approach is applied to furrow irrigation the size of the cut-back stream is so adjusted to match the cumulative infiltration in a furrow so that there is no runoff at the tail end.

(*xii*) In furrow irrigation operating on the cut-back stream approach, the cut-back stream continues to flow in the furrow until the quantity of water required to fill the crop root zone to field capacity has been applied.

(*xiii*) Furrow irrigation reduces the labour requirements in land preparation and irrigation.

(*xiv*) Corrugation irrigation is recommended on saline soils or when the irrigation water has a high salt content.

(*xv*) Sub-irrigation is suitable in soils having low water-holding capacity and high infiltration rate where surface irrigation methods cannot be used and sprinkler irrigation is expensive.

23. Identify the most appropriate answer of the following questions from the four options:

(*i*) The controllable variable influencing the advance of the water front in an irrigation border is:
- (a) slope of the border strip
- (b) infiltration characteristics of the soil
- (c) size of the irrigation stream
- (d) hydraulic resistance offered by field surface roughness and vegetative retardance

(*ii*) The overland flow in an irrigation border strip is a case of:
- (a) steady flow
- (b) unsteady flow
- (c) steady flow with decreasing discharge
- (d) unsteady flow with decreasing discharge

(*iii*) The cut-back stream concept cannot be used in:
- (a) graded furrows laid along the slope
- (b) contour furrows laid on gently sloping grade
- (c) level furrows on heavy soils
- (d) corrugations

(*iv*) The minimum land slope required in land irrigated with surface methods of water application is:
- (a) 0.01% (b) 0.05%
- (c) 0.10% (d) 0.15%

(*v*) The size of the furrow stream usually varies from:
- (a) 0.1 to 0.5 litres/sec
- (b) 0.5 to 1.0 litres/sec
- (c) 0.5 to 2 litres/sec
- (d) 0.5 to 2.5 litres/sec

(*vi*) In sub-irrigation water reaches the plant roots through:
- (a) deep percolation
- (b) lateral spread of water from open ditches
- (c) capillary action
- (d) subsurface flow

24. In the following, match each of the entries of the 1st column with the most closely linked entry in the 2nd column:

(*i*)

(A) Uncontrolled flooding	(I) Subsurface storage of soil moisture
(B) Limiting land slope	(II) Solution of water front advance function in border irrigation
(C) Infiltration opportunity time	(III) Design of border irrigation system
(D) Gamma function	(IV) Spate irrigation
(E) Recession flow	(V) Irrigation border strips

(*ii*)

(A) Philip and Farrel	(I) Irrigation of orchards
(B) Hall	(II) Check basin irrigation
(C) Ring basins	(III) Furrow irrigation
(D) Quarter-time rule	(IV) General solution of volume balance function in border strips
(E) Cut-back stream	(V) Border irrigation system

(*iii*)

(A) Fungal root rot	(I) Evaluation of furrow irrigation
(B) Minimum land slope	(II) Sub-irrigation
(C) Organic soils	(III) Evaluation of check basin irrigation hydraulics
(D) Iso-time profiles	(IV) Furrow irrigation
(E) Inflow-outflow method	(V) Surface drainage

FURTHER READING REFERENCES

Booher, L.J. (1967). *Surface irrigation.* Food and Agriculture Organization of the United Nations, Rome, p. 161.

Chauhan H.S. and Singh, P. (1974). *Surface Irrigation Design.* Part 1, Section 1, Surface Irrigation, in Handbook of Agricultural Engineering. Dept. of Agricultural Engineering, College of Technology, G.B. Pant University of Agriculture and Technology. Pantnagar, Uttaranchal, p.16.

Criddle, W.E., Davis, S., Pair, C.H. and Shockley, D.G. (1956). *Methods of evaluating irrigation systems.* Agriculture Handbook No. 82, Soil Cons. Service, US, Dept. of Agriculture, Washington, D.C., p. 24.

Davis, J.R. (1961). *Estimating rate of advance for irrigation furrows.* Trans. Am. Soc. Agr. Engrs. Vol. 4, No. 1.

Deshmukh, M.T. (1970). *Hydraulics of flow in irrigation furrows.* M.Sc. thesis, Division of Agri. Engg., Post Grad. School, I.A.R.I., New Delhi, p. 157.

Hall, W.A. (1960). *Design of irrigation border checks.* Agricultural Engineering, Vol. 41, pp. 439–442.

Kataria, D.P. (1970). *Hydraulics of flow in check basin irrigation.* M.Sc. thesis, Division of Agri. Engg., Post Grad. School, I.A.R.I., New Delhi, p. 135.

Lewis, M.R. and Milne, W.F. (1938). *Analysis of border irrigation.* Agri. Engg., Vol. 19, pp. 267–272.

Michael, A.M., Reeser, R.M. and Kñierim, G.C. (1965). *How to improve irrigation farming in India.* University of Udaipur, College of Agriculture, Jobner, Extn. Bull. No. 2, p. 60.

Michael, A.M. and Ojha, T.P. (2003). *Principles of agricultural engineering,* Vol. II, 3rd ed., Jain Brothers, Karol Bagh, New Delhi, p. 888.

Michael, A.M. (1968). *Evaluation of hydraulic characteristics influencing border irrigation design.* Ph.D. thesis, Dept. of Agri. Engg., Indian Institute of Technology, Kharagpur, p. 420.

Michael, A.M. and Pandya, A.C. (1971). *Hydraulic resistance relationship in irrigation borders.* Journal of Agri. Engineering Research, 16(1), pp. 72–80.

Michael, A.M. and Pandya, A.C. (1971). *Water front advance in irrigation borders.* Journal of Agricultural Engineering Research, 16(1), pp. 62–71.

Michael, A.M., Shri Mohan and Swaminathan, K.R. (1972). *Design and evaluation of irrigation methods.* I.A.R.I., Monograph No. 1, Water Technology Centre, I.A.R.I., New Delhi, p. 208.

Philip. J.R. and Farrel, D.A. (1964). *General solution of the infiltration—advance problem in irrigation hydraulics.* Journal of Geophysical Research, Vol. 69, No. 4, pp. 621–631.

Slabers, P.J. (1971). *Design criteria for basin irrigation systems.* FAO Irrigation and Drainage Paper 3, Food and Agriculture Organization of the United Nations, Rome, p. 15.

United States Dept. of Agriculture, Soil Conservation Service (1969). *Contour levee irrigation.* SCS Natl. Engg. Handbook, Section 15, Ch. 6, p. 30.

Walker, W.R. (1989). *Guidelines for Designing and Evaluating Surface Irrigation Systems.* FAO Irrigation and Drainge paper 45, Food and Agriculture Organization of the United Nations, Rome, p. 135.

Williams, G.S. and Hazen. A. (1960). *Hydraulic tables,* 3rd ed., John Wiley and Sons, Inc., New York.

Chapter

12

Sprinkler Irrigation

LEARNING OBJECTIVES

Critical appraisal of the adaptability, advantages and limitations of sprinkler irrigation systems. Sprinkler types and components. Moisture distribution patterns and uniformity co-efficient of sprinkler systems influencing sprinkler irrigation design. Sprinkler irrigation system design under varying topographic situations and wind characteristics. Sprinkler system operation and maintenance. Types and adaptability of very low pressure sprinkler irrigation systems. Principles of fertigation. Fertigation system components and their operation. Selection of fertilizers for use in pressurized irrigation systems.

Modern irrigation equipment and technology like sprinkler, micro-sprinkler and drip systems have the capability of applying the desired quantity of water with high precision and uniformity. Consequently, there is considerable saving in the use of water as compared to surface irrigation, sometimes to the extent of 25% to 50%, depending on the type and design of the system, the spacing of the nozzles/drip emitters, plant-to-plant and row-to-row spacing of the crop and the physiological characteristics of the plant. Research studies on sprinkler irrigation conducted in India and other countries have shown that sprinkler method leads to appreciable saving of water and results in achieving high crop yields, as compared to surface irrigation. Sprinkler irrigation is well suited to water scarcity areas and undulating topography where surface irrigation is difficult and expensive.

Sprinkler irrigation system is composed of a network of pipes and sprinklers. The pipes convey water and supply it to all the operating sprinklers at the appropriate pressure (Fig. 12.1). At the nozzles of the sprinklers the pressure head is converted to velocity head. Water flows out of the nozzle in the form of a jet which breaks down into drops of water which fall on the land surface and the foliage of growing plants like rain drops. The area of land wetted by a sprinkler depends on the velocity head given to the water jet, the angle of flow of the jet, the type of sprinkler and its design and the wind conditions during irrigation. The efficiency of a sprinkler system depends on the pressure head available at each of the sprinklers in the irrigated area and the proper spacing of the sprinklers, according to their type and design, to suit the soil type and crop characteristics as well as the wind conditions during irrigation. The quantity of water applied can be adjusted to precisely meet the water requirement of the crop. When properly designed, sprinkler irrigation can achieve high values of water application and water distribution efficiencies. It is possible to apply light and frequent irrigations to maintain favourable soil moisture regime in the crop root zone, resulting in higher crop yields.

Adaptability of sprinkler irrigation. Sprinkler irrigation is suitable to almost all soil types, topographic situations and all types of crops, especially close growing crops, except rice and jute. It is, however, not usually suitable for heavy clay soils with very low infiltration rates. Extensive field surface channels, which occupy considerable land area (often 2 to 3%) for water distribution are not necessary. There is substantial saving of labour in irrigation application. The method is particularly suited to sandy soils that have a high infiltration rate. Soils too shallow to be levelled properly for surface irrigation methods, can be irrigated safely by sprinklers. The flexibility of the sprinkler equipment and its efficient control of water application make this method adaptable to most topographic conditions without extensive land preparation. It is especially suitable for steep slopes or irregular topography (Fig. 12.2). If soil erosion is a hazard, sprinkler irrigation can be used in conjunction with contour bunding, terracing, mulching and strip cropping.

Land levelling is not essential for irrigation with sprinklers. Some smoothing or grading is, however, advisable if surface drainage is a problem or to provide a more uniform surface for sowing, tillage and harvesting. Small streams of irrigation water can be used efficiently, and well-designed sprinklers distribute water better than in other methods. Surface runoff of irrigation water can be eliminated. The amount of water can be controlled to meet crop needs, and light applications can be made efficiently on seedlings and young plants.

Soluble fertilizers, herbicides and fungicides can be applied in the irrigation water economically and with little extra equipment. Penetration of fertilizers into the soil can be controlled by applying the fertilizer at selected times during the application of water. Sprinkler irrigation can be used to protect crops against frost and against high temperatures that reduce the quantity and quality of harvest. Labour costs are usually less than for surface methods on soils having a high infiltration rate and on steep and rolling land. The irrigation method does not interfere with the movement of farm machinery.

Limitations. Wind distorts sprinkler patterns and causes uneven distribution of water. Ripening soft fruits must be protected from the spray. A stable water supply is needed for the most economical use of the equipment. The water must be clean and free of sand, debris and large amounts of dissolved salts.

The sprinkler method usually requires high initial investment as compared to surface irrigation methods, except where extensive land levelling is necessary for surface irrigation. Power requirements are usually high since sprinklers operate with a water pressure of 0.5 (micro-sprinklers) to more than 10 kg/cm^2 (giant nozzle sprinklers). Fine-textured soils that have a low infiltration rate cannot be irrigated efficiently in hot windy areas. If water is applied at the low rate required for these soils, the percentage lost by evaporation and wind drift increases.

DEVELOPMENT AND USE OF SPRINKLER IRRIGATION

Sprinkler irrigation was developed in the early years of the twentieth century. However, the widespread introduction of the system was after the Second World War with the introduction of light weight, portable aluminium pipes. Development of improved sprinklers and quick couplers along with other fittings facilitated convenient coupling and uncoupling of pipes and their easy installation. Presently, sprinkler and drip methods of irrigation are fast replacing surface irrigation in many developed countries and other countries in the water scarcity regions. The system became popular in the Central Asian countries, the USA, UK, Zimbabwe, Italy, Turkey and other European countries. By 1990 the total area under sprinkler irrigation in the world was about 21.58 million ha (INCID, 1998). Table 12.1 presents the estimated area under sprinkler irrigation in countries with more than 100,000 ha under sprinklers. In addition, the following countries have sizable areas under sprinkler irrigation : Algeria (40,000 ha), Austria (46,000 ha), Canada (65,000 ha), Iran (47,200 ha), Kenya (21,000 ha), Lebanon (21,000 ha), Norway (69,500 ha), Tunisia (55,000 ha) and Yugoslavia (49,192 ha, 35% of irigated area).

Table 12.1. Estimated area under sprinkler irrigation in top 20 countries having more than 100,000 ha under the system.

Country	*Reporting year*	*Area under sprinkler irrigation, ha*	*% of total irrigated area*
1. Afghanistan	1967	114,000	
2. Bulgaria	1980	496,000	50%
3. Denmark	1980	295,000	
4. Egypt	1993	313,000	
5. Germany (GDR)	1980	437,000	
6. Greece	1980	380,000	
7. Hungary	1980	333,802	
8. France	1980	600,000	
9. India	1995	658,000	
10. Italy	1980	517,000	17%
11. Kyrgystan	1990	141,000	13%
12. Libya	1990	470,000	100%
13. Moroco	1989	103,200	9.4%
14. Saudi Arabia	1992	1,029,000	64%
15. Turkey	1994	263,849	6.5%
16. USSR	1980	6,000,000	35%
17. United Kingdom	1980	117,000	90%
18. USA	1991	8,572,621	35%
19. Zimbabwe	1980	110,000	70%

Based on INCID (1998).

In India, the area irrigated by the sprinkler system is about 0.66 million ha, which is less than 1% of the total area under irrigation. The sprinkler system was first introduced in the mid-1950s by the plantation owners in the hills to irrigate tea, coffee and cardamom plantations. In mid-1970s progressive farmers in the Narmada valley in Madhya Pradesh, Southern region of Haryana and North Eastern part of Rajasthan and parts of Punjab started using sprinkler irrigation to overcome the problem of water scarcity. The adoption of sprinkler system later spread to more areas in these states and in the states of Maharashtra and Karnataka. It is estimated that about 135,000 sprinkler sets were in use in India in 1997 (INCID, 1998). About 65% of the area under sprinkler irrigation is under field crops like cereals, pulses, oilseeds, cotton, sugarcane and vegetables and the rest 40% under tea, coffee and cardamom plantations in the Western Ghats region and in the North Eastern states. The popularization of sprinkler irrigation in India received significant support from the various schemes involving subsidies by the Central and State Governments.

Fig. 12.1. A rotating head sprinkler line irrigating a row crop.

Types of Sprinkler Irrigation Systems

Sprinkler systems are of the following two major types, on the basis of the arrangement for spraying irrigation water:

1. **Rotating head system.** Small size nozzles are placed on riser pipes fixed at uniform intervals along the length of the lateral pipe (Figs. 12.1 and 12.2). The lateral pipes are laid on the ground surface. They may also be mounted on posts above the crop height and rotated through 90° to irrigate a rectangular strip. In rotating type sprinklers, the most common device to rotate the sprinkler heads is a small hammer activated by the thrust of the water striking against a vane connected to it.

Fig.12.2 A panoramic view of sprinkler irrigation on a hill slope.
(***Courtesy***: Perrot-Regnerbau GMBH & Co.)

2. **Perforated type system.** This method consists of holes perforated in the lateral irrigation pipe in a specially designed pattern to distribute water fairly uniformly. The system is usually designed for low operating pressures of about 0.5 to 2.5 kg/cm^2. The pressure is so low that the system can be connected to an overhead tank to obtain the necessary pressure head. The sprays are directed on both sides of the pipe and can cover a strip of land from 6 to 15 metres wide. The water is applied at a relatively high rate, and therefore, it is suitable for soils having moderately high infiltration rates. This system is suited for irrigation of lawns, gardens and small vegetable crop fields and other plants when the height does not exceed 40 to 60 cm. The water should be cleaned through a filter to prevent clogging of the small perforations.

Rotating Head Type Sprinkler Systems

Based on portability, sprinkler systems are classified into the following types:

1. **Portable system.** A portable system has portable main lines, and laterals and a portable pumping plant. It is designed to be moved from field to field or to different pump sites in the same field (Figs. 12.1 and 12.2).

The portable system may be designed to be moved manually or by mechanical power. The *hand move system* in which the laterals and mains are moved manually is the type that has been introduced in India. Comparatively, the initial investment on this type of sprinkler unit is not high but the labour cost is more. The systems using mechanical power for moving the laterals are usually of the *wheel move system*. In this system, the lateral is mounted on wheels and is moved as a unit instead of one pipe at a time (Fig. 12.4). The wheel move system involves higher capital investment but the labour requirement is less.

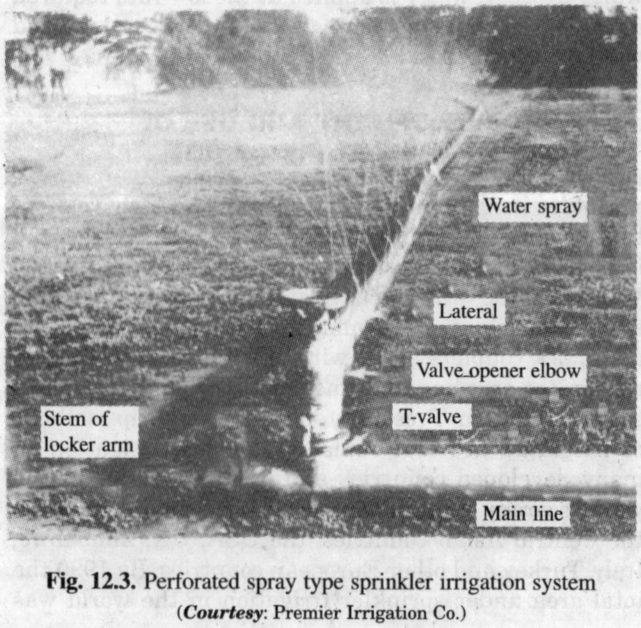

Fig. 12.3. Perforated spray type sprinkler irrigation system
(***Courtesy***: Premier Irrigation Co.)

Fig. 12.4. A portable sprinkler irrigation unit using the wheel move system for moving the laterals in the field.

2. **Semi-portable system.** A semi-portable system is similar to a fully portable system except that the location of the water source and pumping plant are fixed. Such a system may be used on more than one field where there is an extended main line, but may not be used on more than one farm unless there are additional pumping plants.

3. **Semi-permanent system.** A semi-permanent system has portable lateral lines, permanent main lines and submains, and a water source and pumping plant. The main lines and submains are usually buried, with risers for nozzles located at suitable intervals.

4. **Solid-set system.** A solid-set system has enough laterals to eliminate their movement. The laterals are positioned in the field early in the crop season and remain there for the season. The system is used for crops requiring short and frequent irrigations.

5. **Permanent system.** A fully permanent system consists of permanently laid mains, submains and laterals and a water source and pumping plant. Mains, submains and laterals are usually buried below plough depth. Sprinklers are permanently located on each riser. Such systems are costly and are suited to automation of the system with moisture sensing devices. Sprinkler installations in orchards are usually of the permanent type (Fig. 12.5).

Fig. 12.5. A permanent overhead sprinkler system irrigating an orchard.

Centre pivot system. The center pivot system of sprinkler irrigation consists of a pipeline rotating around a pivot point and supported by a large number of self-propelled towers (Fig. 12.6). The water is supplied at the pivot point and flows outward through the pipeline and is distributed through rotating sprinklers located on the pipeline at regular intervals. A part-circle sprinkler (end gun) is usually placed at the end of the pipeline to extend the diameter of coverage to the maximum extent possible. The towers are powered by a hydraulic or electric motor. The center pivot system of sprinkler irrigation is fully automated. Each pivot point covers a large circular area of land. The system is suitable in large scale farming on nearly plain land which does not have obstructions to block the circular movement of the pivot system. The portion of the land left uncovered by the circular area by the adjoining pivot points is usually ignored, considering the advantage of large area coverage and fully automatic operation and substantial savings in labour. To obtain uniform application of water the sprinkler spacing may be constant or with a combination of spacing and discharges. However, the system is not adaptable to small scale intensive farming systems and on irregular land holdings.

Components of the Sprinkler System

The parts of all sprinkler systems are similar in most respects. They consist of the pump to provide the needed pressure, the main pipelines and laterals, risers and sprinkler heads (Fig. 12.7).

The materials of construction of sprinkler irrigation systems vary. Amongst the main parts, aluminium, polyethylene and steel pipes are used for the mains and laterals. The advantage of aluminium are light weight and portability. Polyethylene pipes have better durability and flexibility. Steel pipes are preferred for permanently laid mains. The riser pipes on which the sprinkler heads are mounted are usually of mild steel. Sprinkler heads are made of brass or other alloys.

Pumping set. The pump usually lifts water from the source and pushes it through the distribution system and the sprinklers. It may also be used to boost the water in an existing water distribution line (Fig. 12.8) to force it through the sprinkler system at the desired pressure. In all cases it is important that the pump should be designed to lift the required amount of water from the source of supply to the highest point in the field and maintain an adequate operating pressure.

Both volute centrifugal pumps, deep well turbine pumps and submersible pumps are used to operate sprinkler systems. The centrifugal pump, which is more popular is usually used when the source of water is an open well, river, canal or pond. Booster pumps are almost exclusively of the centrifugal type. The turbine pump is used to pump from deep tubewells. It should generally be set at a fixed location. But, centrifugal pumps may be portable or in permanent installations.

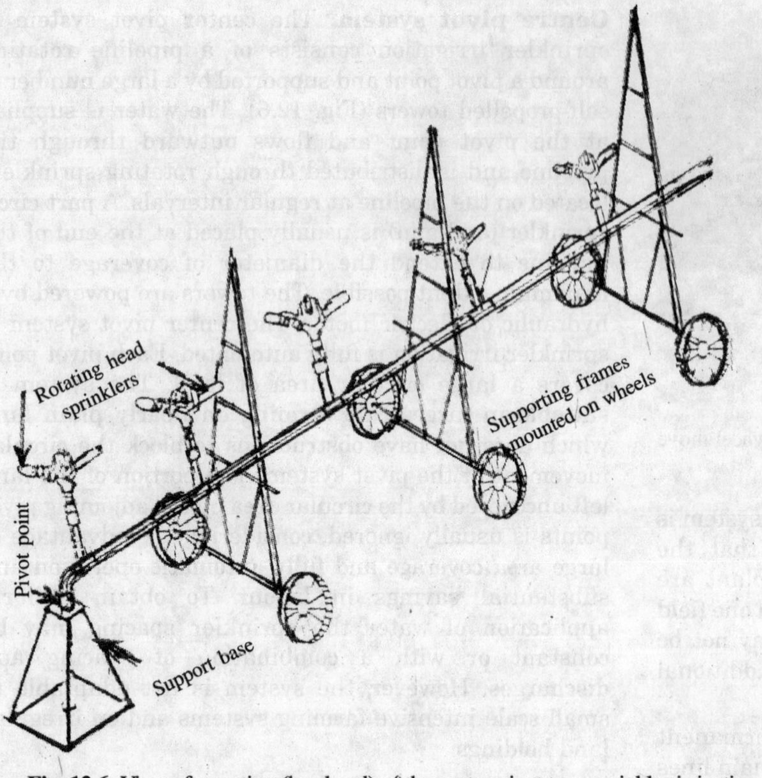

Fig. 12.6. View of a section (head end) of the center-pivot type sprinkler irrigation system. Adapted from Khepar, S.D. *et al* (1981); Fry, A.W. and Gray, A.S. (1970)

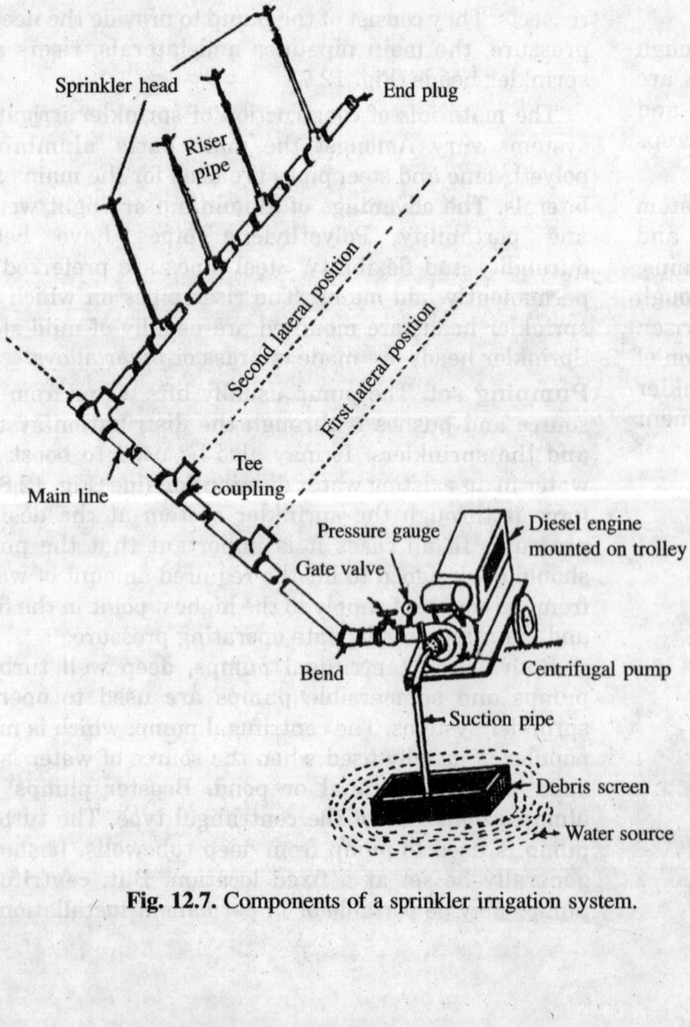

Fig. 12.7. Components of a sprinkler irrigation system.

Electric motors and internal combustion engines are used to drive the pumps operating the sprinklers. Electric motors are better suited for fixed installations. They have low initial and running costs and are easier to maintain. Engines are used when the pumping unit is to be portable and at places where electricity is not available.

Main lines. Main lines may be permanent or portable. Permanent mains are used to best advantage on farms where field boundaries are fixed and where crops require full-season irrigation. Portable mains are more economical when a sprinkler system is to be used on any of a number of fields. Steel pipe is used for most permanent main lines. Asbestos, cement and PVC pipes are also used, but concrete-pipelines are not adapted to use in high pressure sprinkler systems. The permanent lines should be buried so as to be out of the way of farming operations. Light-weight aluminium pipes with quick couplers are used for most portable main lines.

Movable mains generally have a lower initial cost and do not provide obstruction to field operations. Water is taken from the mains either through a valve placed at each point of junction with a lateral or in some cases through an L- or T-section that has been supplied in place of one of the couplings on the mains (Fig. 12.9).

Lateral lines. The lateral lines usually are portable. Buried permanent laterats are, however, used for some orchards, tree nurseries, and for other special sites. Quick-coupled aluminium pipe is best for most portable laterals. The lateral pipes are usually available in lengths of 5, 6 or 12 m. Each length has quick couplings (Figs. 12.9 and 12.10). The rubber gasket in the female portion of these couplings has a U-shape. The water pressure forces the outside of the 'U' to form a watertight seal. When the water is turned off, the seal is broken and water drains from the pipe, making it easier to uncouple and move.

Figure 12.11 illustrates a commonly used pipe coupling using a hooking device. Figure 12.12 presents different types of couplings used in portable sprinkler irrigation systems. A cut-away

Fig. 12.8. A portable engine-pumping set drawing water from an underground pipeline water distribution system to operate a portable sprinkler system.

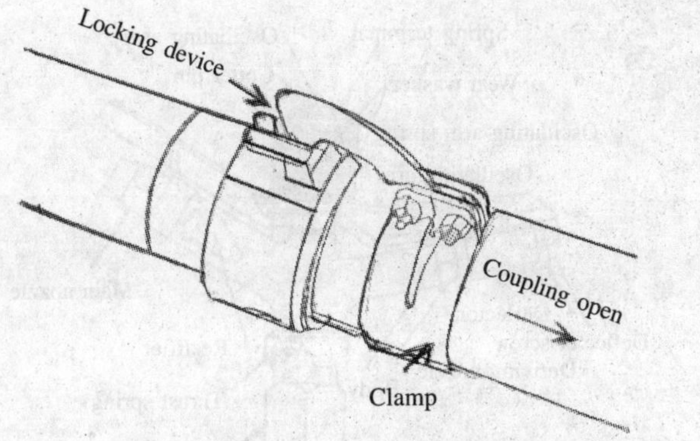

Fig. 12.11. Coupling of portable irrigation pipes. Note the hook arrangement. Adapted from Khepar, S.D. (1981).

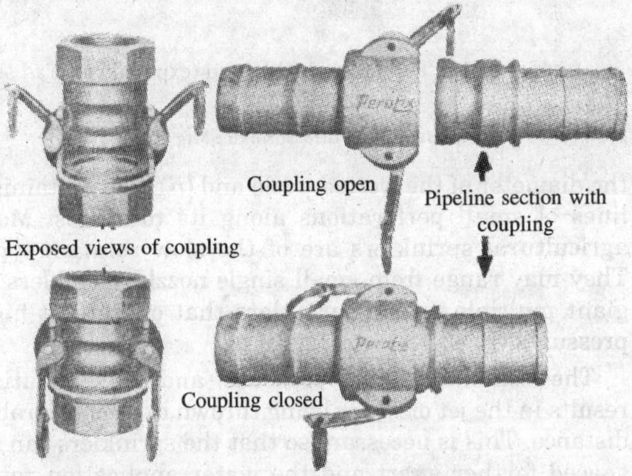

Fig. 12.9. Cardan-joint coupling of sprinkler line.

(***Courtesy***: Perrot-Regnerbau, GMBH & Co.)

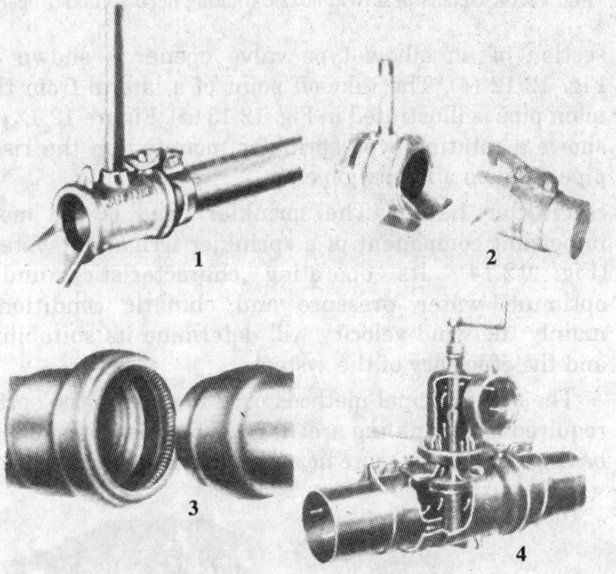

Fig. 12.12. Different types of couplings and elbow-type valve opener used in portable sprinkler irrigation systems: 1, 2 & 3, couplings; 4. Cut-away section of an elbow-type valve opener.

[***Source***: Scott, V.H. (1956)]

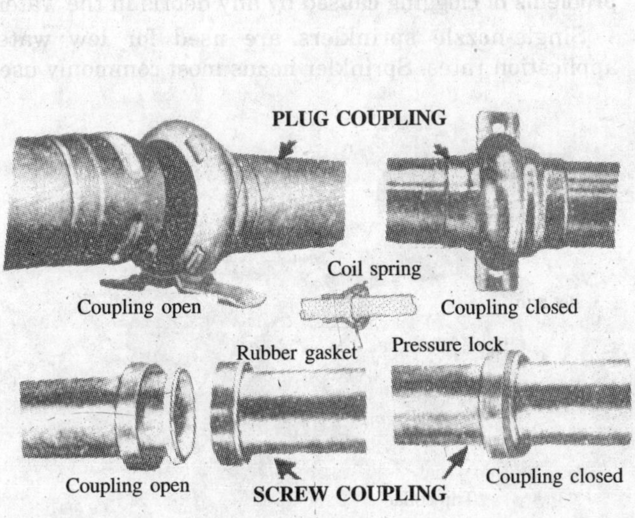

Fig. 12.10. Views of plug type and screw type couplings of sprinkler line.

(***Courtesy***: Perrot-Regnerbau GMBH & Co.)

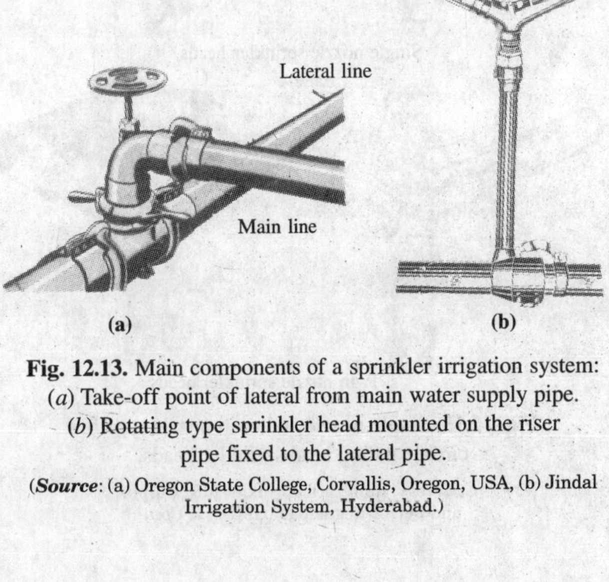

Fig. 12.13. Main components of a sprinkler irrigation system: (*a*) Take-off point of lateral from main water supply pipe. (*b*) Rotating type sprinkler head mounted on the riser pipe fixed to the lateral pipe.

(***Source***: (a) Oregon State College, Corvallis, Oregon, USA, (b) Jindal Irrigation System, Hyderabad.)

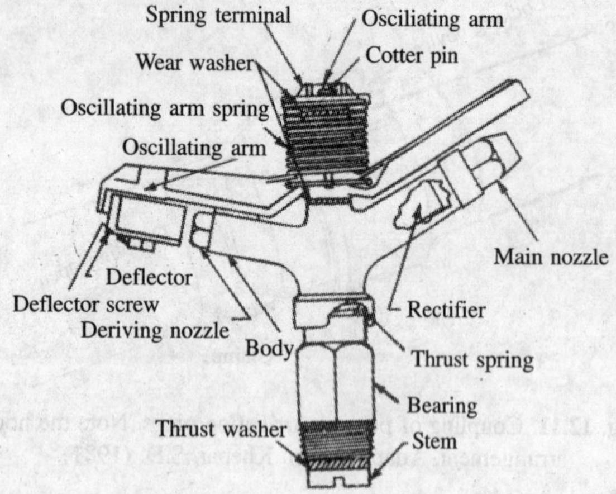

Fig. 12.14. Details of a twin-nozzle rotating head sprinkler head.

section of an elbow type valve opener is shown in Fig. 12.12 (4). The take-off point of a lateral from the main pipe is illustrated in Fig. 12.13 (*a*). Figure 12.13 (*b*) shows a rotating type sprinkler mounted on the riser pipe fixed on a lateral pipe.

Sprinkler heads. The sprinkler head is the most important component of a sprinkler irrigation system (Fig. 12.14). Its operating characteristics under optimum water pressure and climatic conditions, mainly the wind velocity, will determine its suitability and the efficiency of the system.

The two principal methods used to develop the spray required for sprinkling are: (*i*) revolving head sprinklers having one, two or more nozzles, depending primarily on the diameter of the wetted circle and (*ii*) pipe containing lines of small perforations along its top sides. Most agricultural sprinklers are of the slow rotation type. They may range from small single-nozzle sprinklers to giant multiple nozzle sprinklers that operate at high pressures (Figs. 12.15 to 12.18).

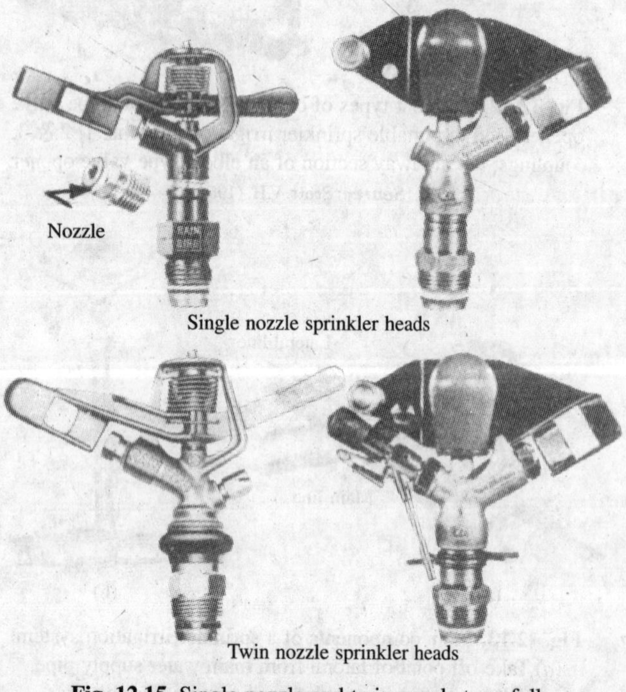

Fig. 12.15. Single nozzle and twin nozzle type full circle rotating head sprinkler heads.

[***Courtesy***: Rain Bird Sprinkler Mfg. Corpn. and Perrot-Regnerbau (GMBH & Co.)]

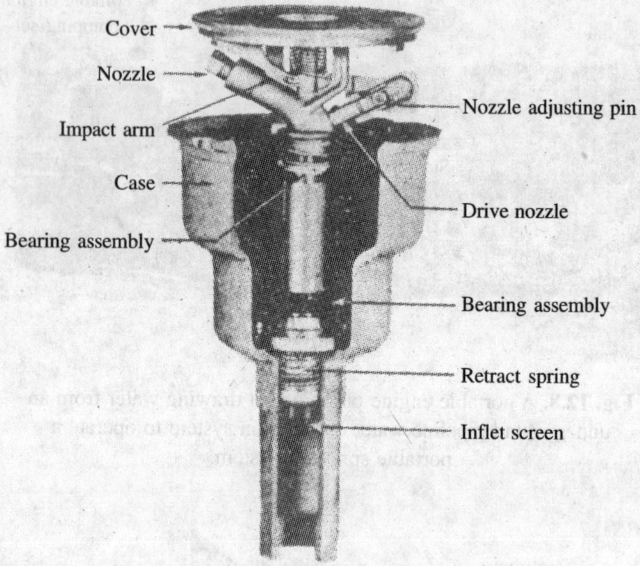

Fig. 12.16. 'Pop-up' sprinkler head (exposed view) suitable for lawns.

(***Courtesy***: Rain Bird Sprinkler Mfg. Corp.)

The combination of pressure and slow rotation results in the jet of water being thrown to a considerable distance. This is necessary so that the sprinklers can be spaced farther apart and the water application rates with an overlapping pattern can be low enough to be commensurate with the infiltration capacity of the soil. Further, it permits the use of larger orifices to minimize problems of clogging caused by any debris in the water.

Single-nozzle sprinklers are used for low water application rates. Sprinkler heads most commonly used

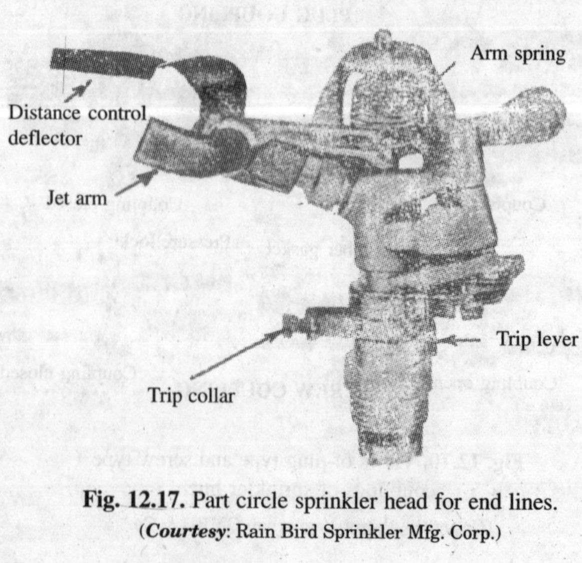

Fig. 12.17. Part circle sprinkler head for end lines.

(***Courtesy***: Rain Bird Sprinkler Mfg. Corp.)

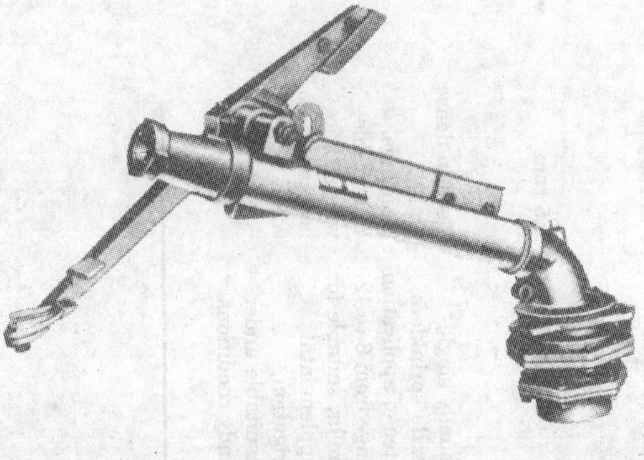

Fig. 12.18. Giant nozzle high pressure sprinkler head (rain gun)
(*Courtesy*: Rain Bird Sprinkler Mfg. Corp.)

have two nozzles, one to apply water at a considerable distance from the sprinkler and the other to cover the area near the sprinkler centre. Two-nozzle sprinkler systems cannot apply water at as low a rate as single-nozzle sprinkler systems can, but they provide good uniformity of application at low pressures. 'Pop-up' sprinkler heads (Fig. 12.16) are especially adapted to irrigation of lawns. The sprinkler head is housed in a casing which is provided with a cover at the top. The top of the cover is almost at the land surface. The cover opens by means of a retract spring when the sprinkler is in operation. The cover remains closed when the sprinkler is shut off. Giant sprinklers (Fig. 12.18) usually have one large nozzle with smaller supplemental nozzles to fill in pattern gaps. They are adaptable to close growing high plantings, such as sugarcane where closer spacing of sprinklers is difficult. They are limited to soils with high infiltration rates and are not suitable in regions of appreciable wind. Giant sprinklers usually have a low water distribution efficiency.

Revolving head sprinklers are classified according to pressure range and their position in relation to irrigated crops. Table 12.2 present the ranges in sprinkler operating pressures, sprinkler characteristics and adaptability of common types of sprinklers. It is important that the proper pressure be provided for the sprinkler heads to be used. Pressures too low or too high for a particular spinkler head will cause poor water distribution patterns.

Of the devices used to rotate the sprinklers, the most usual type, taps the sprinkler head with a small hammer activated by the force of water striking against a small vane connected to it. Some sprinklers, particularly the high pressure giant sprinklers, are rotated by water-activated gear drives.

Debris screens. Debris screens are usually needed when surface water is used as the source of irrigation (Fig. 12.7). Details vary, but the function of screens is to keep the system free of trash that might plug the sprinkler nozzles. Screens should be fine enough to catch weed seeds and other small particles. They may also be required to prevent sticks, plant stalks, and other large items of debris from entering the system. Two or more screens, of progressively finer mesh, can be used when heavy loads of debris are expected. The accumulated trash must be removed from the screens before water flow to the pump becomes restricted.

Desilting basins. Desilting basins may be required to trap sand or suspended silt when the water comes from streams, open ditches or well water having silt. Sometimes desilting basins and debris screens are built as a combination structure. Desilting basins should be large enough to provide protection for at least one full day. Larger capacities are desirable.

Booster pumps. Booster pumps should be used when a sprinkler irrigation system is used with an existing pumping system installed in a well and the pump capacity is insufficient to force the water through sprinklers. They should also be used to provide adequate pressure for small areas that lie at elevations above the principal area to be irrigated. The use of the booster pump under such situations removes the need to carry high pressures from the main pumping plant for the relatively small fraction of the total discharge that is needed on the high area.

Take off-valves. Take-off valves [Fig. 12.19 (g)] generally are needed to control pressures in the lateral lines. The valves should always be used in systems where there are significant differences in main line pressures at the various lateral take-off points. In multiple-lateral systems, they also permit moving individual laterals without shutting down the entire system.

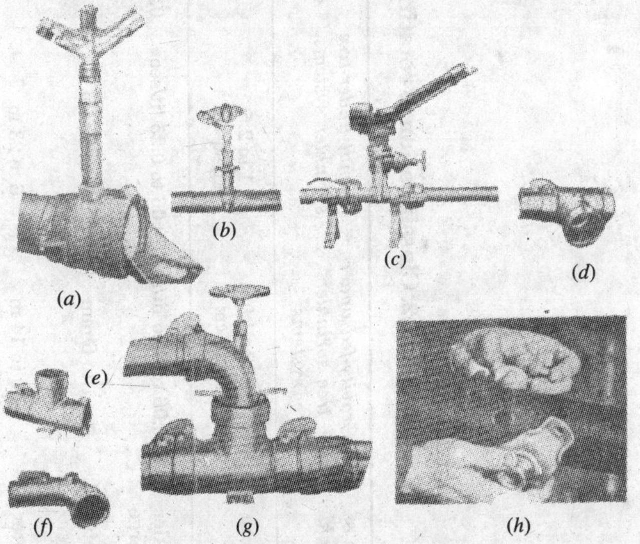

Fig. 12.19. Sprinkler line mountings and accessories: (*a*) Sprinkler head mounted on lateral coupler, (*b*) Sprinkler head mounted directly on pipe, (*c*) 'Rain gun' mounted on Tee-joint and provided with control valve, (*d*) Tee and reducer, (*e*) Tee with valve, (*f*) Elbow, (*g*) Tee valve and valve opener elbow, (*h*) Pipe clamp joint for riser pipe.
(*Courtesy*: Spramotor Ltd. and Perrot-Regenerbau GMBH & Co.)

Table 12.2. Classification of rotating head sprinklers, their characteristics and adaptability

Type of sprinkler	*Gravity-fed under-tree sprinkler systems*	*Normal under-tree sprinkler systems*	*Permanent overhead systems*	*Small overhead systems*	*Low pressure systems*	*Intermediate pressure systems*	*High pressure systems*
Pressure range	0.7 to 1.0 kg/sq cm	1 to 2.5 kg/sq cm	3.5 to 4.5 kg/sq. cm	2.5 to 4 kg/sq cm	1.5 to 2.5 kg/sq m	2.5 to 5 kg/sq cm	5 to 10 kg/sq cm
Sprinkler discharge	0.06 to 0.25 lit/sec	0.06 to 0.25 lit/sec	0.2 to 0.6 lit/sec	0.6 to 2.0 lit/sec	0.3 to 1 lit/sec	2 to 10 lit/sec	10 to 50 lit/sec
Diameter of nozzles	1 to 6 mm	1.5 to 5 mm	3 to 6 mm	6 to 10 mm	3 to 6 mm	10 to 20 mm	20 to 40 mm
Diameter of coverage	10 to 14 m	6 to 23 m	30 to 45 m	25 to 35 m	20 to 35 m	40 to 80 m	80 to 140 m
Range of sprinkler spacings (square)	–	–	18 to 30 m	9 to 24 m	9 to 18 m	24 to 54 m	54 to 100 m
Recommended speed of sprinkler rotations	–	0.5 to 1 rpm	1 rpm	0.67 to 1 rpm	0.5 to 1 rpm	0.7 rpm	0.5 rpm
Adaptability	Usually uses single nozzle sprinkler heads, used as under-tree systems in uplands, has low uniformity of coverage.	Usually used in closely spaced orchards with full low hanging branches, single nozzle slow rotation sprinklers are often used.	Used for orchards. Triangular spacing necessary for low application rates (1.5 to 3 mm/hr).	Commonly used for low rate of application (3.5 to 6 mm/hr) and to help reduce the effects of wind. High risers are used for orchards and lower risers for field crops.	Two-nozzel sprinklers can be used with lower pressures than single-nozzle sprinklers. More overlap is required. Rate of application tends to be high.	Usually single-nozzle sprinklers, rates of application range from 6 to 12 mm/hr, suitable for supplemental irrigation, unsuitable under windy conditions.	Used in dense high plantings. Unsuitable where there is appreciable wind.

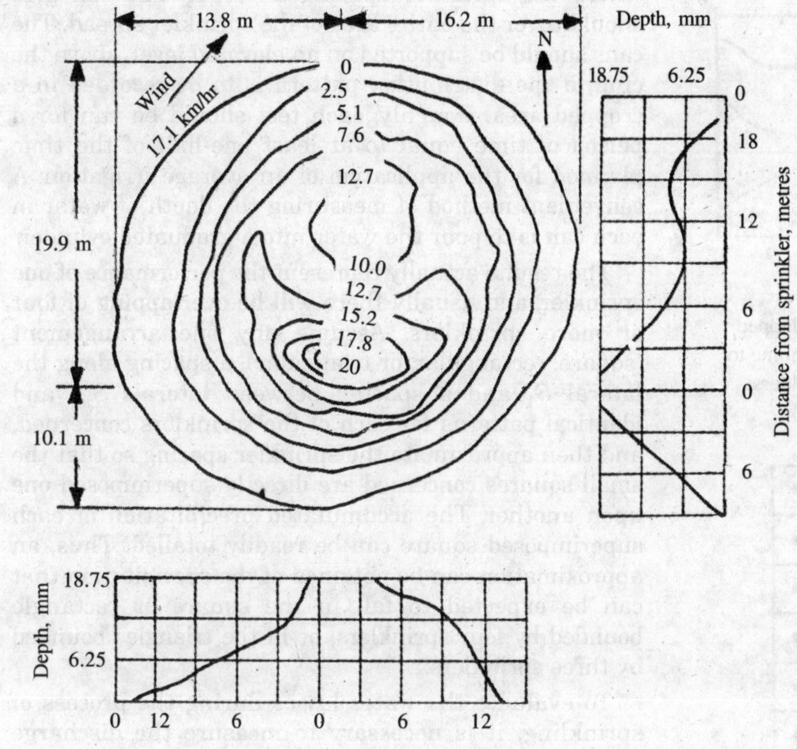

Fig. 12.20. Moisture distribution pattern of a rotating head sprinkler under windy conditions. (Adapted from USDA, SCS Handbook, 1960)

Flow control valves. Flow control valves to regulate the pressure and discharge of individual sprinklers may be helpful when unevenness of the ground causes an unequal distribution of pressure along the lateral (Fig. 12.19). They are seldom needed on level fields or smooth slopes.

Moisture Distribution Patterns and Uniformity of Coverage

The basic objective of sprinkler irrigation is to simulate rainfall and to apply uniformly a calculated depth of water at a predetermined application rate. The irrigation efficiency of sprinklers will depend upon the degree of uniformity of water application. The water spray distribution characteristics of sprinklers and their spacing will regulate the uniformity of water application. The spray distribution characteristics of sprinkler heads are typical and change with nozzle size and operating pressure. With lower pressures, the drops are larger and the water from the nozzle falls in a ring away from the sprinkler. With higher pressures, the water from the nozzle breaks up into very fine drops and falls close to the sprinkler. Wind also distorts the application pattern. The higher the wind velocity, greater the distortion, and this factor should be considered when selecting the sprinkler spacing under windy conditions (Fig. 12.20). The distribution pattern from sprinklers operating under favourable conditions of pressure and wind is given in Fig. 12.21. In this distribution pattern, the depth of water applied surrounding the sprinkler, decreases as the distance from the sprinkler increases. The absorption of water in the soil is also in the same pattern because the rate of water application at any point under the sprinkler coverage is lower than the infiltration rate of the soil. To obtain uniformity of water application, the wetted circles of the adjacent sprinklers should be overlapping so as to add water to areas of the adjoining sprinklers. The aggregate depth of distribution obtained by overlapping thus becomes nearly uniform as shown in Fig. 12.22.

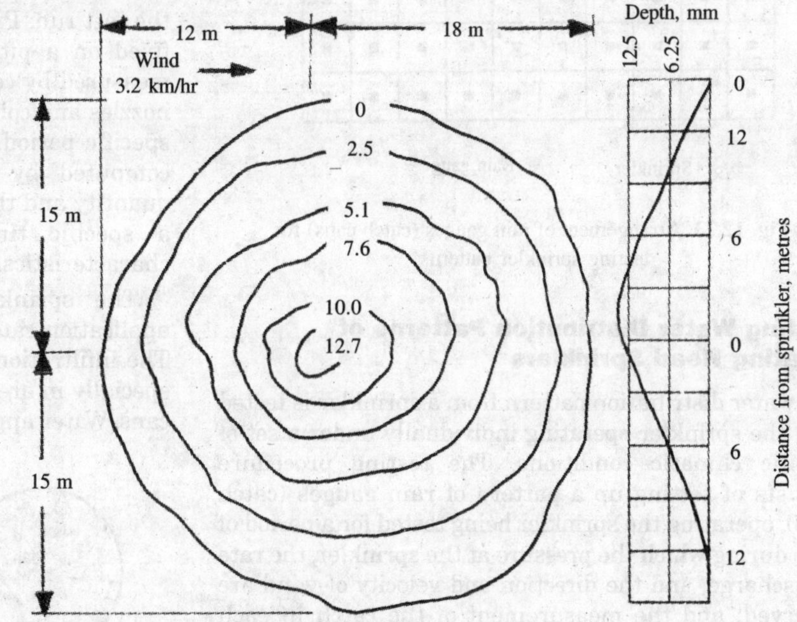

Fig. 12.21. Moisture distribution pattern from a rotating head sprinkler under favourable conditions of pressure and wind. (Adapted from USDA, SCS Handbook, 1960)

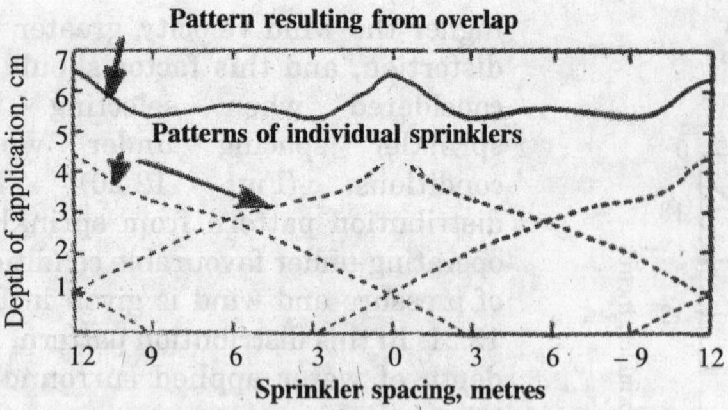

Fig. 12.22. Moisture distribution pattern from overlapped sprinklers showing how the overlapping patterns combine to give a relatively uniform distribution between sprinklers. (Adapted from USDA, SCS Handbook, 1960)

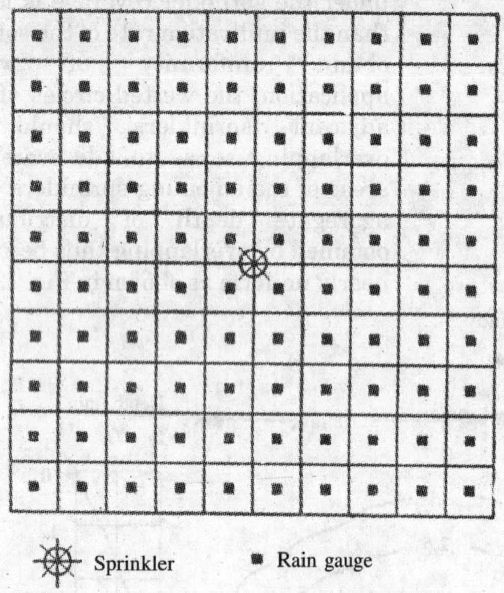

Fig. 12.23. Arrangement of rain gauges (catch cans) for testing sprinkler patterns.

Testing Water Distribution Patterns of Rotating Head Sprinklers

The water distribution pattern from a sprinkler is tested with the sprinkler operating individually under a set of specific climatic conditions. The testing procedure consists of setting up a pattern of rain gauges (catch cans), operating the sprinkler being tested for a period of time during which the pressure at the sprinkler, the rate of discharge, and the direction and velocity of wind are observed; and the measurement of the catch in each catch-can at the conclusion of the period. Figure 12.23 illustrates an arrangement of rain gauges for such a test.

The rain gauges used are normally ordinary tin cans of about 1 litre capacity. The area around the sprinkler is divided into squares of equal area. A can is placed at the centre of each square which represents the precipitation falling on that area. These cans should be about two metres apart where the sprinklers are spaced about 10 metres apart, and about three metres apart where the sprinkler spacing is larger. The can grid should cover the entire area of the sprinkler spread. The cans should be supported on an elevated level, above the crop, if the distribution pattern is to be recorded in a cropped area. Roughly, each test should be run for a period of time equal to at least one-half of the time planned for the application of an average irrigation. A convenient method of measuring the depth of water in each can is to pour the water into a graduated cylinder.

The results actually represent the performance of one sprinkler, and usually there will be overlapping of four or more sprinklers. Assume any one arrangement (square, rectangular, or triangular), a spacing along the lateral S_1 and a spacing between laterals S_w, and identical patterns for each of the sprinklers concerned, and then approximate the sprinkler spacing so that the small squares concerned are directly superimposed one upon another. The accumulated precipitation of each superimposed square can be readily totalled. Thus, an approximation can be obtained of the precipitation that can be expected to fall in the square or rectangle bounded by four sprinklers, or in the triangles bounded by three sprinklers.

To evaluate the water losses during the process of sprinkling, it is necessary to measure the discharge from the sprinkler head under the operating pressure. The pressure and discharge of the sprinkler head, on either side of the nozzles, is measured before and after the test run. Pressure is measured by a pressure gauge fixed on a pitot tube (Fig. 12.24). The discharge is measured by connecting flexible tubes on the sprinkler nozzles and collecting the discharge in a container for a specific period. The losses in evaporation can also be computed by finding the difference between this quantity and the quantity received by the catch-cans in a specific time period, while testing the spray characteristics.

The sprinkler nozzles are evaluated for their application rate and suitability for the soil of the area. The infiltration characteristics of the soil are observed, specially in areas around the fastest filling spray gauge cans. Water applied by the revolving sprinkler at a point

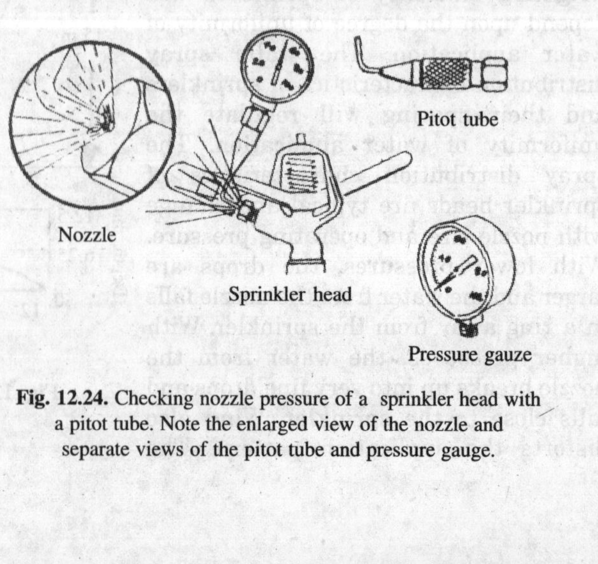

Fig. 12.24. Checking nozzle pressure of a sprinkler head with a pitot tube. Note the enlarged view of the nozzle and separate views of the pitot tube and pressure gauge.

should disappear before water is applied again at the same point by the sprinkler in its next round. There should be no runoff for movement of water over the soil surface. Any movement of water is unsatisfactory and is indicative of too high an application rate. These infiltration observations should be made when the soils are saturated after a few hours of operation of the sprinklers.

While testing the sprinklers, other observations such as wind direction, wind velocity, temperature and humidity should be recorded. Crop damage is also to be checked, as it could be due to drop impact, unfavourable jet trajectory or inadequate riser height.

Uniformity Coefficient

A measurable index of the degree of uniformity obtainable for any size sprinkler operating under given conditions is known as the uniformity coefficient (C_u). This uniformity coefficient is affected by the pressure-nozzle size relations, by sprinkler spacing, and by wind conditions. The coefficient is computed from field observations of the depths of water caught in open cans placed at regular intervals within a sprinkled area. It is expressed by equation 12.1, developed by Christiansen (1942):

$$C_u = 100\left(1.0 - \frac{\Sigma X}{mn}\right) \quad \text{...(12.1)}$$

in which,

m = average value of all observations (average application rate), mm

n = total number of observation points

X = numerical deviation of individual observations from the average application rate, mm

A uniformity coefficient of 100 per cent (obtained with overlapping sprinklers) is indicative of absolutely uniform application, whereas the water application is less uniform with a lower percentage. A uniformity coefficient of 85 per cent or more is considered to be satisfactory.

Example 12.1. *Determine the uniformity coefficient from the following data obtained from a field test on a square plot bounded by four sprinklers:*

Sprinkler – 4.365 × 2.381 mm nozzles at 2.8 kg/cm²

Spacing — 24 m × 24 m

Wind — 3.5 km/hr from south-west

Humidity — 42 per cent

Time of test—1.0 hour

S	*8.9*	*7.6*	*6.6*	*S*
8.1	*7.6*	*9.9*	*10.2*	*8.3*
8.9	*9.1*	*9.1*	*9.4*	*8.9*
9.4	*7.9*	*9.1*	*8.6*	*9.1*
S	*7.9*	*6.6*	*6.8*	*S*

(***Note:*** S indicates location of sprinklers.)

Solution: The computations are shown in the following tabular form:

Observation	*Frequency*	*Application rate × frequency*	*Numerical deviations*	*Frequency × deviations*
10.2	1	10.2	1.6	1.6
9.9	1	9.9	1.3	1.3
9.4	2	18.8	0.8	1.6
9.1	4	36.4	0.5	2.0
8.9	3	25.7	0.3	0.9
8.6	1	8.6	0.0	0.0
8.3	1	8.3	0.3	0.3
8.1	1	8.1	0.5	0.5
7.9	2	15.8	0.7	1.4
7.6	2	15.2	1.0	2.0
6.8	1	6.8	1.8	1.8
6.6	2	13.2	2.0	4.0
	21	mn = 178.0		ΣX = 17.4

$$\text{Mean} = \frac{178.0}{21} = 8.48$$

$$C_u = 100\left(1.0 - \frac{\Sigma X}{mn}\right)$$

$$= 100\left(1 - \frac{17.4}{178.0}\right) = 100\,(1 - 0.0977)$$

$$= 100 \times 0.9023 = 90.23\%.$$

The data on uniformity coefficient are useful as a basis for selecting the combination of spacings, discharge, nozzle size and operating pressure to obtain high values of irrigation efficiency at specific operating conditions.

The leading manufacturers of sprinkler irrigation equipment usually provide information on the diameter of coverage of each nozzle size and the pressure required at the sprinklers. Some manufacturers give C_u values for each sprinkler, for each set of nozzles, each common spacing and for each pressure.

Design of Sprinkler Irrigation Systems

A sprinkler irrigation system, to suit the conditions of a particular site, is specially designed in order to achieve high efficiencies in its performance and economy. The step-by-step procedure in the planning and design of a sprinkler irrigation system is enumerated below.

1. **Inventory of Resources and Conditions.** *Map of the area.* It is essential that a map of the area concerned is prepared and drawn to scale with sufficient accuracy to show all dimensions so that lengths of main and laterals can be scaled therefrom. It should be a contour map or, at least, should show all relevant elevations with respect to water source, pump location, and critical elevations in the fields to be irrigated. The elevation

differences, together with friction losses in the mains and laterals and the pressure requirements of the sprinklers, determine the pressure that must be developed by the pump.

Water supply—source, availability and dependability. It is important that an irrigation system of sufficient size is available to meet the maximum demand of crops. The quantity available should also meet the seasonal and annual requirements of the crops and the area to be irrigated. The water should be chemically suitable for irrigating the crops and soils of the area. It should not have any corrosive effect on the equipment. The water should be relatively clean and free of suspended impurities so that the sprinkler lines and nozzles are not clogged.

Climatic conditions. The consumptive use of a crop depends upon the climatic parameters such as temperature, radiation intensity, humidity and wind velocity. Details to obtain consumptive use information have been provided in Chapter 10. Sprinkler system is designed for the daily peak rate of consumptive use of the crops irrigated by it. Guidelines for determining the approximate peak rates of consumptive use on the basis of different climatic conditions are described in Chapter 10. A peak demand in the range of 2 to 10 mm depth per day is equivalent to a continuous flow of 0.23 to 1.16 litres/second/hectare.

Depth of irrigation. The depth of irrigation is calculated on the basis of available moisture holding capacity of the soil in its different layers and the soil moisture extraction pattern of the crop in its root zone depth. Detailed explanation and the procedure to work out the irrigation requirement are given in Chapter 10.

Irrigation interval. From the point of view of sprinkler design, the irrigation interval is the length of time allowable between successive irrigations during the peak consumptive use of the crop. It is interesting to note that the irrigation interval can also be varied by altering the hours of set. For instance, a system designed to supply 100 mm with 20-hour sets at 20-day intervals can also apply 50 mm with 10-hour sets at 10-day intervals. Such practices are common during the early stages of crop growth when the root system has not been fully developed.

Water application rate. The rate of water application by sprinklers is limited by the infiltration capacity of the soil. Application at rates in excess of the infiltration capacity of the soil results in runoff, with accompanying poor distribution of water, loss of water and soil erosion. Table 12.3 presents some suggested values of maximum rates of water application for various soil conditions. These values may be used where reliable local information is not available.

Table 12.3. Suggested maximum application rates for sprinklers for average soil, slope and tilth.

Soil texture and profile	*0–5% slope cm/hr*	*5–8% slope cm/hr*	*8–12% slope cm/hr*	*12–16% slope cm/hr*
1. Coarse sandy soil upto 2 m	5.0	3.7	2.5	1.3
2. Coarse sandy soils over more compact soils	3.7	2.5	2.0	1.0
3. Light sandy loams upto 2 m	2.5	2.0	1.5	1.0
4. Light sandy loams over more compact soils	2.0	1.3	1.0	0.8
5. Silt loams upto 2 m	1.3	1.0	0.8	0.5
6. Silt loams over more compact soils	0.8	0.6	0.4	0.3
7. Heavy textured clays or clay loams	0.4	0.3	0.2	0.1

Adapted from Hurd (1969).

Sprinkler spacing. To achieve uniform sprinkling of water, it is necessary to overlap the area of influence of the sprinklers. The overlap increases with the increase in wind velocity. Table 12.4 may be used as a guide in the design of sprinkler overlap under different wind conditions.

Power source. The source of power to operate the pump is to be known in advance. Electric power is most convenient when the pump is stationary. Electric pumping sets are cheaper in initial cost and maintenance cost. Portable diesel pumping sets are the most suitable and practical for fully portable sprinkler systems.

Table 12.4. Maximum spacing of sprinklers under windy conditions.

Sl. No.	*Average wind speed*	*Spacing*
1.	No wind	65% of the diameter of the water spread area of a sprinkler
2.	0–6.5 km/hr	60% of the diameter of the water spread area of a sprinkler
3.	6.5 to 13 km/hr	50% of the diameter of the water spread area of a sprinkler
4.	Above 13 km/hr	30% of the diameter of the water spread area of a sprinkler

Adapted from Hurd (1969).

2. **Types of Systems and Layout.** The common types of sprinkler systems were described earlier, Table 12.2 presents the characteristics and adaptability of different types of rotating head sprinklers. With the type of sprinkler determined, based on pressure limitations, application rates, cover conditions, crop requirements, and availability of labour, the next step is to determine the location of the pumping unit, the

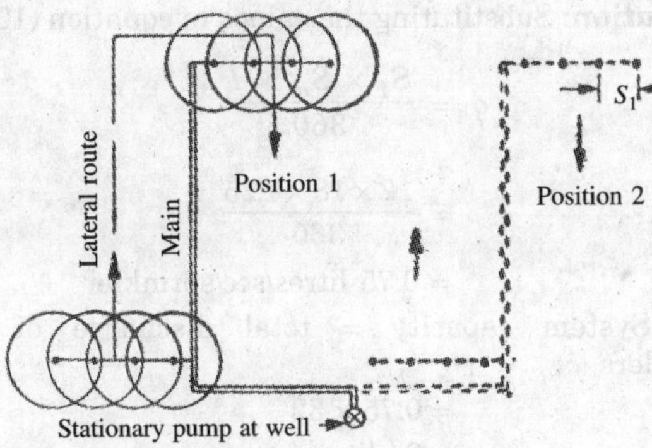

Fig. 12.25. Layout plan for sprinkler irrigation system. The laterals are moved to successive positions up on one side of the main and then down on the other.

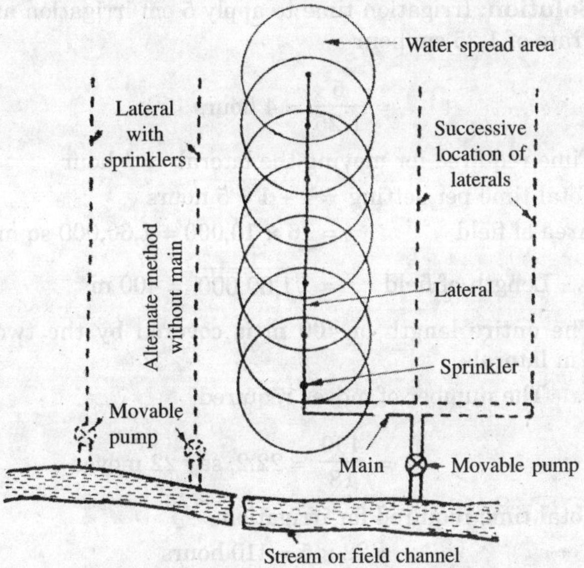

Fig. 12.26. Typical field layouts for fully portable sprinkler units drawing water from streams or field channels.

orientation of mains and laterals, sprinkler spacing, operating pressure, and nozzle sizes that will most nearly provide the optimum water-application rate with the greatest degree of uniformity of distribution.

Location and nature of water supply. The source of water supply for sprinkler irrigation is usually a tubewell or open well located on the farm. Sometimes, however, surface water sources may also be used with sprinkler irrigation systems. The well may be located at a high corner or, more likely, at the centre of the farm to minimize the distance the water must be pumped (Fig. 12.25). The layout of the mains will depend on the location of the well. It is advantageous if the pump for the well is designed to lift the water and provide necessary pressure to overcome the friction loss in the pipelines and to operate the sprinklers.

Sometimes it may be necessary to adopt a sprinkler irrigation system with an already laid underground pipeline water distribution system or field channels. In case of underground pipelines, a portable pumping set can be used with the suction attached to the hydrants mounted on the pipe outlet (Fig. 12.8). In case of field channels running along one edge of the farm, a portable pumping set and sprinkler unit with the lateral extending to the field may be used to draw water directly from the channel and distribute it through the sprinklers (Fig. 12.26). Alternatively, the channel can be run down the centre of the field. The laterals will then be only half as long so that smaller length laterals could be used, but the channel may interfere with tillage operations and may result in some reduction in the net area cropped. Another alternative is to have a permanent pumping plant at the source and distribute the water in buried pressure pipelines. These pipelines will usually run down the centre of the field so that the outlets offer little hindrance to tillage and other farm operations.

Orientation of laterals. To obtain a reasonable degree of uniformity in the discharge of each sprinkler, the mains should be located in the general direction of the steepest slope, with the laterals at right angles thereto and as close as is practical to the contour. The usual design is based on the lateral being level. If the lateral slopes upgrade appreciably, it is difficult to design for a reasonable length. If it slopes downgrade, the length can be longer than usual, but rarely does the slope remain uniform for each setting.

Any system in which the laterals are moved should be planned for successive irrigations in strict rotation, so that the interval between irrigations is the same for each portion of a field. This means that, with a main located along the middle of a field, a given lateral will normally be moved to successive positions up one side and then down the other. Otherwise, when the irrigation of one side has been completed, the lateral would have to be carried all the way back to the lower end in order to start over again.

The number of possible arrangements for mains, laterals and sprinklers are many. The arrangement selected should provide for a minimal investment in irrigation pipe, have low labour requirement and provide for the application of water over the total area in the required period of time. The most suitable layout can be determined only after a careful study of the conditions to be encountered. The choice will depend to a large extent upon the types and capacities of the sprinklers and their operating pressures.

3. **Sprinkler Selection and Spacing.** The actual selection of the sprinkler is based largely upon design information furnished by the manufacturers of the equipment. The choice depends mainly on the diameter of coverage required, pressure available and sprinkler discharge. The data presented in Tables 12.3 and 12.4

may serve as guidelines in selecting the pressure and spacing desired. The best combination of sprinkler spacing and lateral moves, suiting the application rate for the soil and wind conditions should be selected.

The required discharge of an individual sprinkler is a function of the water application rate and the two-way spacing of the sprinklers. It may be determined by the following formula:

$$q = \frac{S_1 \times S_m \times I}{360} \quad \text{...(12.2)}$$

in which,

q = required discharge of individual sprinkler, litres/second

S_1 = spacing of sprinklers along the laterals, metres

S_m = spacing of laterals along the main, metres

I = optimum application rate, cm/hr.

Height of sprinkler riser pipes. Sprinklers are located just above the crops to be irrigated and, therefore, the height of the risers depends upon the maximum height of the crop. To avoid excessive turbulence in the riser pipes the minimum height of riser is 30 cm when the riser pipe is of 2.5 cm diameter and 15 cm when it is of 1.8 to 2 cm diameter.

4. **Capacity of Sprinkler Systems.** The required capacity of a sprinkler system depends on the size of the area to be irrigated (design area), the gross depth of water applied in each irrigation, and the net operating time allowed to apply water to this depth. The capacity of the system may be calculated by the formula:

$$Q = 2780 \frac{A \times d}{F \times H \times E} \quad \text{...(12.3)}$$

in which,

Q = discharge capacity of the pump, litres/sec

A = area to be irrigated, hectares

d = net depth of water application, cm

F = number of days allowed for the completion of one irrigation

H = number of actual operating hours per day

E = water application efficiency, per cent.

In equation (12.3) it may be noted that F and H are of major importance in that they have a direct bearing on the capital investment per hectare required for equipment. From the formula it is clear that the greater the product of these two factors (operating time) the smaller is the system capacity (and thereby the cost) for a given area.

Example 12.2. *Determine the required capacity of a sprinkler system to apply water at the rate of 1.25 cm/hr. Two 186 metres long sprinkler lines are required. Sixteen sprinklers are spaced at 12 metre intervals on each line. The spacing between lines is 18 metres.*

Solution: Substituting the values in equation (12.2)

$$q = \frac{S_1 \times S_m \times I}{360}$$

$$= \frac{12 \times 18 \times 1.25}{360}$$

$$= 0.75 \text{ litres/sec/sprinkler}$$

$\therefore$ System capacity = total discharge of all sprinklers

$$= 0.75 \times 32$$

$$= 24 \text{ litres/sec.}$$

Example 12.3. *Allowing 1 hour for moving each 186 metre sprinkler line described in Example 12.2, how many hours would be required to apply a 5 cm irrigation to a square 16-hectare field? How many days are required assuming 10-hour days?*

Solution: Irrigation time to apply 5 cm irrigation at the rate of 1.25 cm/hour

$$= \frac{5}{1.25} = 4 \text{ hours}$$

Time required for moving the lateral = 1 hour

Total time per setting = 4 + 1 = 5 hours

Area of field = 16 × 10,000 = 1,60,000 sq m

$\therefore$ Length of field $= \sqrt{1,60,000} = 400$ m

The entire length of 400 m is covered by the two 186 m laterals.

$\therefore$ The number of moves required

$$= \frac{400}{18} = 22.2, \text{ say 22 moves}$$

Total time required for irrigation

$$= 22 \times 5 = 110 \text{ hours}$$

$$= \frac{110}{10} = 11 \text{ days.}$$

Example 12.4. *Determine the system capacity for a sprinkler irrigation system to irrigate 16 hectares of maize crop. Design moisture use rate is 5 mm per day. Moisture replaced in soil at each irrigation is 6 cm. Irrigation efficiency is 70 per cent. Irrigation period is 10 days in a 12-day interval. The system is to be operated for 20 hours per day.*

Solution:

Given: $A = 16, F = 10, H = 20, d = 6$

System capacity (Eq. 12.3),

$$Q = 2780 \frac{A \times d}{F \times H \times E}$$

$$= 2780 \times \frac{16 \times 6}{10 \times 20 \times 70}$$

$$= 19 \text{ litres/sec.}$$

Example 12.5. *A sprinkler irrigation system is to be designed to irrigate 8 hectares of vegetable crops in deep silt loam soil in moderate dry climate. The field is flat. Determine the limiting rate of application, the irrigation period, the net depth of water per application, the depth of water pumped per application and the required system capacity in hectare-centimetres per day. Assuming that the system is operated for 15 hours each day, determine the pump capacity in litres/sec.*

Solution: From Table 12.3 the limiting application rate is 1.3 cm/hr. The moisture holding capacity of the soil may be assumed to be 9.5 cm/metre depth, and the depth of root zone is 60 cm. The total available moisture capacity is thus $9.5 \times \frac{60}{100} = 5.7$ cm. Assuming that irrigation is started at 50 per cent moisture depletion level, the net depth of irrigation is $\frac{5.7}{2} = 2.85$ cm. Assuming a water application efficiency of 75 per cent, the depth of water pumped per application is

$$\frac{2.85}{0.75} = 3.8 \text{ cm.}$$

The peak rate of moisture use by crops is assumed to be 5 mm/day. Irrigation period is thus $\frac{2.85}{0.5} = 5.7$, say 6 days. To cover the field in 6 days the system must be able to pump and discharge $\frac{3.8 \times 8}{6} = 5.05$ hectare-centimetres per day.

The pump capacity, $$Q = 2780 \times \frac{A \times d}{F \times H \times E}$$

$$= 2780 \times \frac{8 \times 2.85}{6 \times 15 \times 75}$$

$$= 9.4 \text{ litres/second.}$$

5. **Hydraulic Design of Sprinkler Systems.** The hydraulic design of sprinklers is aimed at obtaining a uniform irrigation coverage, the desired rate of application, the break-up of sprinkler drops necessary to minimize structural deterioration of the soil surface, and the efficiency desired to reduce the energy requirement in operating the system and to maximise the area of coverage. The main hydraulic principles involved in a sprinkler system design are given below.

Discharge of sprinkler nozzle. The discharge of a sprinkler nozzle may be computed from the following orifice flow formula derived by Toricelli.

$$q = Ca\sqrt{2gh} \quad ..(12.4)$$

in which,

q = nozzle discharge, m^3/sec

a = cross-sectional area of nozzle or orifice, m^2

h = pressure head at the nozzle, metres

g = acceleration due to gravity, m/sec^2

C = coefficient of discharge which is a function of friction and contraction losses (the coefficient C of good nozzles should be 0.95 to 0.96)

Water spread of sprinkler. The area covered by a rotating head sprinkler may be estimated by the following formula suggested by Cavazza (Pillsbury, 1968):

$$R = 1.35\sqrt{dh} \quad ..(12.5)$$

in which,

R = radius of wetted area covered by the sprinkler, metres

d = diameter of nozzle, metres

h = pressure head at the nozzle, metres.

Maximum coverage is attained when the jet emerges from the sprinkler at an angle of 30° to 32° above the horizontal. Most rotating sprinklers are standardised at 30°.

Break-up of jet. Some break-up of the jet of water is necessary to attain uniformity of coverage and to minimize the presence of excessively large drops. The larger drops lose less velocity and travel farther. They also strike the soil surface at a higher velocity, with more energy to cause structural deterioration of the soil surface. There is need for some compromise among distance of throw, uniformity of coverage, and effect on the soil surface. There is a natural tendency of jets to break-up because of air resistance, and the break-up increases with pressure. The break-up is usually increased by having slots in the nozzle or by a pin impinging on the jet. Slow rotation sprinklers, which make about 0.67 to 1 rpm for small sprinklers and 0.25 to 0.5 rpm for large sprinklers, provide good coverage. The following empirical formula suggested by Tanda (Pillsbury, 1968) provides an index of jet break-up:

$$P_d = \frac{h}{(10q)^{0.4}} \quad ..(12.6)$$

in which,

P_d = index for jet break-up

h = pressure head at nozzle, metres

q = sprinkler discharge, litres/sec

It has been noticed if P_d is greater than 2, the condition of drop size is good; if 4, the condition of drop size is best; and if greater than 4, pressure is being wasted (Pillsbury, 1968).

Rate of application. The average rate of application, often called "precipitation intensity", for a single sprinkler may be estimated by the following formula:

$$R_a = \frac{q}{360 \times A} \quad ..(12.7)$$

in which,

R_a = water application rate, cm/hour

q = rate of discharge of sprinkler, litres/sec

A = wetted area of sprinkler, m^2

Design of Sprinkler Laterals

A sprinkler lateral has equally spaced sprinklers along its length. In the beginning the flow will equal the combined discharge of all the sprinklers. This will decrease along the line until, at the farthest end, there will be only the flow of the last sprinkler. Therefore, the friction loss in a lateral will be much less than if the total flow were carried through the entire length of the line. However, many authorities advise against 'tapering' of pipe diameters in laterals, as it then becomes necessary to keep the various pipe sizes in the same relative position.

It is recommended that the total pressure variation in the laterals, when practicable, should not be more than 20 per cent of the higher pressure. If the lateral runs uphill or downhill, allowance for this difference in elevation should be made in determining the variation in head. If the water runs uphill, less pressure will be available, at the nozzle; if it runs downhill, there will be a tendency to balance the loss of head due to friction.

It is necessary to compute the friction loss in the lateral line where flow decreases along its length. A simple approximate procedure, devised by Christiansen (1942), has become standard. First, ascertain the friction loss with full flow to the end of the line, using appropriate formulae or the values given in Appendix F, Table F1 to F5 (based on Scobey's formula). Next, refer to Table 12.5 for *"FI"* factors by which the friction loss with full flow to the end of the line are multiplied to obtain the approximate friction loss in the lateral. By this method and through trial and error, it will be possible to select a lateral of the proper size, so that the friction loss will not exceed that which is allowable.

Example 12.6. *Determine the friction loss H_f in 396 metres of a 15 cm diameter aluminium pipe with couplers, made of 12-metre sections having 22 sprinklers spaced 18 metres apart, each discharging 1.58 litres per second. The first sprinkler is 18 metres from the main.*

Solution: Total discharge of the system

$$= 22 \times 1.58 = 34.7 \text{ litres/sec}$$

From Appendix F, Table F1 the friction loss in a 15 cm aluminium pipe with a discharge of 34.7 litres/sec = 2.52 metres/100 metres.

From this, 3 per cent is deducted for 12-metre lengths of sections (see footnote on Appendix F, Table F1).

From Table 12.5, the value of F factor for 22 sprinklers, when the first sprinkler is one sprinkler interval from main = 0.357.

$$\therefore \quad H_f = \frac{396}{100} \times 2.52 \times \frac{97}{100} \times 0.357 = 3.45 \text{ metres.}$$

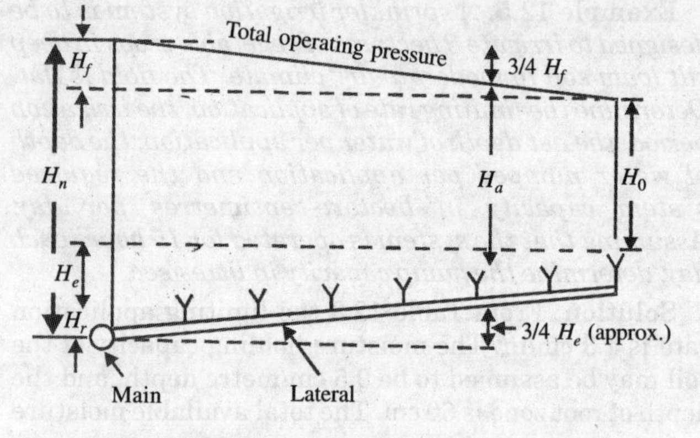

Fig. 12.27. Pressure profile in sprinkler irrigation lateral laid uphill.

The design capacity for sprinklers on a lateral is based on the average operating pressure. Where the friction loss, H_f, in the laterals is within 20 per cent of the average pressure, the average head, H_a, for design in a sprinkler line can be expressed approximately by

$$H_a = H_0 + \tfrac{1}{4} H_f \qquad \text{...(12.8)}$$

in which H_0 is the pressure at the sprinkler on the farthest end (Figs. 12.27 and 12.8). Thus, the average pressure is equal to the pressure at the farthest end plus one-fourth the friction loss. Where the lateral is on nearly level land or on the contour, the head (pressure) H_n at the main is

$$H_n = H_0 + H_f \qquad \text{...(12.9)}$$

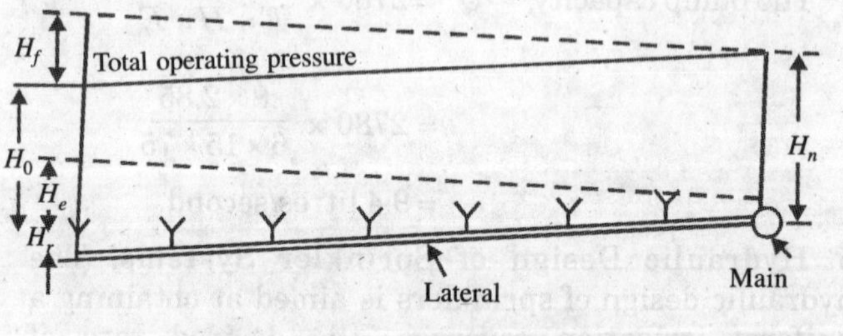

Fig. 12.28. Pressure profile in sprinkler irrigation line laid downhill.

By solving for H_0, in equation 12.8, by substituting in equation 12.10 and by making an allowance for differences in elevation along the lateral

$$H_n = H_a + \tfrac{3}{4} H_f + \tfrac{3}{4} H_e + H_r \qquad \text{...(12.10)}$$

in which,

H_e = maximum difference in elevation between the first and last sprinklers on the lateral, metres, and

H_r = the riser height, metres.

The term $\frac{3}{4} H_e$ is positive if the lateral runs up the slope (Fig. 12.27) and negative if it runs down the slope

(Fig.12.28). This term is an approximation since the allowance for the difference in elevation varies with the number of sprinklers on the lateral.

Table 12.5. Correction factor 'F' for friction losses in aluminium pipes with multiple outlets.

No. of sprinklers on lateral	*1st sprinkler is one sprinkler interval from main*	*1st sprinkler is 1/2 sprinkler interval from main*	*No. of sprinklers on lateral*	*1st sprinkler is one sprinkler interval from main*	*1st sprinkler is 1/2 sprinkler interval from main*
1	1.000	1.000	16	0.365	0.345
2	0.625	0.500	17	0363	0.344
3	0.518	0.422	18	0.361	0.343
4	0.469	0393	19	0.360	0.343
5	0.440	0.378	20	0.359	0.342
6	0.421	0.369	22	0.357	0.341
7	0.408	0.363	24	0.355	0.341
8	0398	0.358	26	0.353	0.340
9	0.391	0.355	28	0.351	0.340
10	0.385	0.353	30	0.350	0.339
11	0.380	0.351	35	0.347	0.338
12	0.376	0.349	40	0.345	0.338
13	0.373	0.348	50	0.343	0.337
14	0.370	0.347	100	0.338	0.337
15	0.367	0.346	> 100	0.335	0.335

Main Line Pipe Size

The function of the main lines and submains is to convey the required quantity of water at the desired pressure to all lateral lines under maximum pressure conditions. The selection should be based on economic considerations. Main line friction loss of about 3 metres for small systems and 12 metres for large systems may be allowed. The friction losses in main lines of the portable aluminium pipe can be determined from Appendix F, Tables F1 and F3. Semi-rigid plastic irrigation pipelines manufactured of PVC or asbestos compound are also used in sprinkler irrigation. In designing sprinkler irrigation, water may be pumped through low pressure reinforced cement concrete (RCC) pipes to a booster pump employed to pump water directly in the laterals. Appendix F Tables F4 and F5 provide, respectively, the friction loss in PVC or asbestos compounds and asbestos cement pressure pipes. Figure 13.77 (Chapter 13) presents the nomograph to determine the fraction loss in plastic pipes.

Pumps and Power Units

In selecting a suitable pump, it is necessary to determine the maximum total head against which the pump is working (see Chapter 3). This may be determined by

$$H_t = H_n + H_m + H_j + H_s \qquad ...(12.11)$$

in which,

H_t = total design head against which the pump is working, m

H_n = maximum head required at the main to operate the sprinklers on the lateral at the required average pressure, including the riser height, m

H_m = maximum friction loss in the main and in the suction line, m

H_t = elevation difference between the pump and the junction of the lateral and the main, m, and

H_s = elevation difference between the pump and the source of water after drawdown, m.

The amount of water that will be required is determined by multiplying the number of sprinklers by the capacity of each. When the total head and rate of pumping are known, the pump may be selected from rating curves or tables furnished by the manufacturer (Chapter 3).

Cost Estimation

After the complete design of the system, an estimate of cost is required. Suggested items are (*i*) well, (*ii*) pump house, (*iii*) power—(*a*) electrical installation, transmission line, transformer: motor, switches etc., (*b*) engine, (*iv*) pump unit, including foot valve, non-return valve, suction and discharge pipes, (*v*) main line and submains, (*vi*) lateral lines, (*vii*) sprinklers and risers, (*viii*) special equipment, if any, (*ix*) installation charges, (*x*) value of farm labour and equipment, and (*xi*) other costs. The information presented in Table 12.6 may be used as a guide in estimating the depreciation of sprinkler equipment.

Table 12.6. Guide for estimating depreciation of sprinkler equipment.

1. Water supply	
Well and casing	25 years
2. Pump, turbine	
(*a*) Bowl (estimated 50% of pump cost)	15,000 hrs or 7 years
(*b*) Column	25,000 hrs or 12 years
3. Pump, centrifugal	25,000 hrs or 12 years
4. Power, transmission	
(*a*) Gearhead	25,000 hrs or 12 years
(*b*) V-belts	2,000 hrs or 1 year
(*c*) flat belt, rubber, fabric	4,000 hrs or 2 years
5. Electric motors	30,000 hrs or 15 years
6. Diesel engines	20,000 hrs or 10 years
7. Pipe	
(*a*) Aluminium (main and laterals)	10 years
(*b*) P.V.C.	6 years
8. Sprinkler heads	7 years
9. Misc. fittings aluminium (valves, couplers, elbows, etc.)	10 years
10. Misc. fittings P.V.C.	6 years

In considering the economics of sprinkler irrigation, all costs and benefits should be included. Initial costs are important because the purchaser must finance this usually heavy capital investment. However, annual costs per hectare compared to annual returns per hectare are the best measure of the economics of an irrigation system. The annual cost per hectare may be estimated from the cost of the various components including the cost of equipment, labour, energy, depreciation and interest detailed above. When considering the returns to be expected from a new irrigation system, credit should be shown for any savings resulting from—(*i*) Increased yield, (*ii*) Less land out of production, (*iii*) Reduction in land preparation, tillage, and harvesting costs, (*iv*) Saving in labour, repair and maintenance, and operating costs, and, (*v*) Saving in water and power costs.

Example 12.7. *Determine the size of sprinklers, laterals, pump and power unit for the sprinkler system layout shown in Fig. 12.27 with the following conditions:*

H_a *= 30 m,* H_j *= 0.5 m,* H_e *= 2.2 m,* H_s *= 3.5 m*

I = 1.25 cm/hr, maximum length of main = 60 m

S_1 *= 15 m,* S_m *= 20 m and allowable variation in pressure in the lateral is 20 per cent.*

Solution:

1. Determine the sprinkler and the lateral capacities:

From equation 12.2, the required discharge capacity of each sprinkler,

$$q = \frac{S_1 \times S_m \times I}{360}$$

$$= \frac{15 \times 20 \times 1.25}{360}$$

$$= 1.04 \text{ litres/sec}$$

From equation 12.4,

$$q = Ca\sqrt{2\,gh}$$

Area of sprinkler nozzle,

$$a = \frac{1.04 \times 10^{-3}}{0.95\sqrt{2 \times 9.8 \times 30}}$$

$$= 0.45 \times 10^{-4} \text{ m}^2$$

Diameter of sprinkler nozzle,

$$d = \sqrt{\frac{4 \times 0.45 \times 10^{-4}}{\pi}}$$

$$= 0.75 \text{ cm}$$

Based on the catalogue of the manufacturer, a suitable nozzle can be selected to meet the requirement.

Lateral capacity for 10 sprinklers

$$= 1.02 \times 10 = 10.2 \text{ lit/sec.}$$

2. Determine the diameter of the lateral and the main:

Total allowable variation of pressure in laterals

$$= 0.2 \times \frac{30}{10.18}$$

$$= 0.59 \text{ kg/cm}^2$$

Allowable variation of pressure due to friction

$$= 0.59 - H_e$$

$$= 0.59 - 0.216$$

$$= 0.374 \text{ kg/cm}^2$$

From Appendix F, Table F1, H_f for 7.5 cm diameter aluminium lateral at 10.2 lit/sec discharge is approximately 8 m/100 m.

From Table 12.5, the correction factor *F* for 10 sprinklers on one lateral is 0.353. The friction losses in laterals and main for various possible sizes of pipes are shown below:

Outside diameter of pipe (cm)	*Lateral, 150 cm* $H_f \times F$ *(m)*	*Main 60 m* H_m *(m)*
7.5	4.236	9.75
10	1.034	2.33
12.5	0.337	0.80

3. Select 7.5 cm diameter lateral.

4. Select 10 cm diameter main.

From equation 12.10, assuming H_r = 0.5 m

$$H_n = 30 + \quad \times 4.236 + \tfrac{3}{4} \times 2.2 + 0.5$$

$$= 35.327 \text{ m}$$

5. Determine pump size:

From equation 12.28,

$$H_f = 35.327 + 4.8 + 0.5 + 3.5 = 44.13 \text{ m}$$

Select, from manufacturer's guide, a pump capable to deliver 10.4 litres/sec at a head of 45 m. Assuming an efficiency of 70 per cent

$$\text{h.p.} = \frac{1 \times 1 \times 10^3 \times 10^{-3} \times 45}{75 \times 0.7} = 9.43$$

Select an electric motor of 10 h.p.

Example 12.8. *Design a sprinkler irrigation system for a square, 10-hectare field to irrigate the entire area within 10 days period. Not more than 16 hours per day are available for moving the pipe and sprinkling. The required depth of irrigation is 5 cm and the water application rate is not to exceed 0.75 cm/hr. A 30 m deep well located in the centre of the field will provide the following discharge-drawdown relationship:12.5 lit./sec at 15 m and 15.8 lit./sec at 20 m. Design the system for an average pressure of 3 kg/cm² at the sprinkler nozzle. The highest point in the field is 1.25 m above the well* *site and 1 m risers are needed for the sprinklers. Assuming a pump efficiency of 60 per cent and supposing that the engine will furnish 70 per cent of its rated output for continuous operation, determine the rated output for a water cooled internal combustion engine.*

Solution: Time $= \dfrac{5}{0.75} + 1.33$ hrs (assumed for moving)

$= 8$ hrs

Thus, 2 moves/day are possible.

Since 10 ha area is to be irrigated within 10 days, the area to be covered in one move,

$$= \frac{10 \times 100 \times 100}{10 \times 2}$$

$= 5000$ m^2 (or 100 m × 50 m)

Select 12.5 m × 20 m spacing.

Length of main = 100 m

Length of laterals = 50 m

No. of laterals per set = = 5

No. of sprinklers per lateral $= \dfrac{50}{12.5} = 4$

Total No. of sprinklers/set = 5 × 4 = 20

From equation (12.2)

$$q = \frac{S_1 \times S_m \times I}{360}$$

$$= \frac{12.5 \times 20 \times 0.75}{360}$$

= 0.52 lit./sec.

Total discharge = 0.52 × 20

Q = 10.4 lit./sec.

Size of nozzle

From equation (12.4) $q = C_a \sqrt{2\,gh}$

$$a = \frac{0.52 \times 10^{-3}}{0.95\sqrt{2 \times 9.8 \times 29.6}}$$

$= 0.22$ cm^2

Diameter of nozzle $= \sqrt{\dfrac{4 \times 0.22}{\pi}} = 0.51$ cm.

Select sprinkler nozzles of 0.5 cm diameter single nozzle.

Select 10 cm diameter main.

From Appendix F, Table F3, for q = 10.4 lit./sec. and 10 cm diameter aluminium pipe when 6 m pipe sections are used, friction loss

$$H_f = 2.32 \times \frac{7}{100} + 2.32 \times \frac{100}{100}$$

= 2.48 m

Select 5 cm diameter lateral.

Discharge of one lateral $= \dfrac{10.4}{5} = 2.08$ lit/sec.

From Table F1, for a discharge of 2.08 lit/sec and 5 cm diameter aluminium pipe when 9 m pipe sections are used.

Friction loss = 3 m per 100 m

However, since 6 m pipe sections are used, increase the value by 7%.

Hence, friction loss $= \left[3 + 3 \times \dfrac{7}{100}\right] \dfrac{50}{100}$

From Table 12.5 for 4 sprinklers on a lateral when the first sprinkler is placed one sprinkler spacing from main

$F = 0.469$

Friction loss in one lateral = 1.61 × 0.469

= 0.755 m

Total friction loss in 5 laterals

= 0.755 × 5

= 3.775 m

From equation (12.11)

H_t = 29.6 + 3.775 + 1.61 + (1.25 + 1) + 12

= 49.24 m

Select a pump that will deliver 10.4 lit/sec at a head of 49.24 m, say 50 m,

Engine horse power $= \dfrac{1 \times 49.24 \times 10.4}{0.6 \times 75}$

= 11.4 h.p.

Engine output = = 16.3 h.p.

Operation and Maintenance of Sprinkler Systems

Proper design of a sprinkler system does not in itself ensure success. It should be ensured that the prime mover and the pump are in alignment, particularly in the case of tractor-driven pumps. For these the drive shaft and the pump shaft should lie at nearly the same height to prevent too great an angle on the universal shaft. Service and installation procedures of the pump and power unit should be strictly observed.

While laying the main and lateral pipes, always begin laying at the pump. This necessarily gives the correct connection of all quick coupling pipes. While joining couplings, it is ensured that both the couplings and the rubber seal rings are clean.

In starting the sprinkler system, the motor or engine is started with the valves closed. The pump must attain the pressure stated on type-plate, or otherwise there is a fault in the suction line. After the pump reaches the regulation pressure, the delivery valve is opened slowly. Similarly, the delivery valve is closed after stopping the power unit.

The pipes and sprinkler-lines are shifted as required after stopping. Dismantling of the installation takes place in the reverse order to the assembly described above.

Maintenance. A sprinkler system, like any other farm equipment, needs maintenance to keep it operating at peak efficiency. Parts of the system subject to wear are the rotating sprinkler heads, the pumping set, the couplers and the pipeline. Details regarding the maintenance of the pumping unit are given in Chapter 3. General principles regarding the maintenance of the pipes and fittings and sprinkler heads are given below:

1. *Pipes and Fittings.* The pipes and fittings require virtually no maintenance but attention must be given to the following procedures:

(*a*) Occasionally clean any dirt or sand out of the groove in the coupler in which the rubber sealing ring fits. Any accumulation of dirt or sand will affect the performance of the rubber sealing ring.

(*b*) Keep all nuts and bolts tight.

(*c*) Do not lay pipes on new damp concrete or on piles of fertilizer. Do not lay fertilizer sacks on the pipe.

The pipes are automatically emptied and ready to be moved. When the pump is first started and before the pressure has built up in the system the seals may give a little leakage. With full pressure in the system the couplers and fittings will be effectively leak-free. If, however, there is a leakage, check the following:

(*a*) There is no accumulation of dirt or sand in the groove in the coupler in which the sealing ring fits. Clean out any dirt or sand and refit the sealing ring.

(*b*) The end of the pipe going inside the coupler is smooth, clean and not distorted.

(*c*) In the case of fittings such as bends, tees and reducers ensure that the fitting has been properly connected into the coupler.

2. *Sprinkler heads.* The sprinkler heads should be given the following attention:

(*a*) When moving the sprinkler lines, make sure that the sprinklers are not damaged or pushed into the soil.

(*b*) Do not apply oil, grease or any lubricant to the sprinklers. They are water lubricated and using oil, grease or any other lubricant may stop them from working.

(*c*) Sprinklers usually have a sealed bearing and at the bottom of the bearing there are washers. Usually it is the washers that wear and not the more expensive metal parts. Check the washers for wear once a season or every six months—this is especially important where water is sandy. Replace the washers if worn.

(*d*) After several season's operation the swing arm spring may need tightening. This is done by pulling out the spring end at the top and rebending it. This will increase the spring tension.

In general, check all equipment at the end of the season and make any repairs and adjustments and order the spare parts immediately so that the equipment is in perfect condition to start in the next season.

Storage. The following points are to be observed while storing the sprinkler equipment during the off season:

(*a*) Remove the sprinklers and store in a cool, dry place.

(*b*) Remove the rubber sealing rings from the couplers and fittings and store them in a cool, dark place.

(*c*) The pipes can be stored outdoors in which case they should be placed in racks with one end higher than the other. Do not store pipes along with fertilizer.

(*d*) Disconnect the suction and delivery pipe-work from the pump and pour in a small quantity of medium grade oil. Rotate the pump for a few minutes. Blank off the suction and delivery branches. This will prevent the pump from rusting. Grease the shaft.

(*e*) Protect the electric motor from the ingress of dust, dampness and rodents.

Troubleshooting

The following are the general guidelines to identify and remove the common troubles in the sprinkler systems:

1. *Pump does not prime or develop pressure.*

(*a*) Check that the suction lift is within the limits. If not get the pump closer to the water body.

(*b*) Check the suction pipeline and all connections for air leaks. All connections and flanges should be air tight.

(*c*) Check that the strainer on the foot valve is not blocked.

(*d*) Check that the flap in the foot valve is free to open fully.

(*e*) Check the pump gland(s) for air leaks. If air leaks are suspected tighten the gland(s) gently. If necessary, repack the gland(s) using a thick grease to seal the gland satisfactorily.

(*f*) Check that the gate valve on the delivery pipe is fully closed during priming and opens fully when the pump is running.

(*g*) Check that the direction of rotation of the pump is correct.

2. *Sprinklers do not turn.*

(*a*) Check pressure.

(*b*) Check that the nozzle is not blocked. Preferably unscrew the nozzle or use a small soft piece of wood to clear the blockage. Do not use a piece of wire or metal as this may damage the nozzle.

(*c*) Check that the sprinkler bearing is quite free and smooth. During operation, the sprinkler can usually be pushed down towards the riser pipes so that the water pressure flushes out the bearing. If the bearing is still stiff dismantle and then clean it. Do not use oil, grease or any lubricant.

(*d*) Check the condition of washers at the bottom of the bearing and replace them if worn or damaged.

(*e*) Check that the swing arm moves freely and that the spoon which moves into the water stream is not bent by comparing it with a sprinkler which is operating correctly. If it is bent then very carefully bend it back into position.

(*f*) Adjust the swing arm spring tension. Usually it should not be necessary to pull up the spring by more than about 6 mm.

3. *Leakage from coupler or fittings.* The sealing rings in the couplers and fittings are usually designed to drain the water from the pipes when the pressure is turned off. This ensures that the pipes are automatically emptied and ready to be moved. When the pump is first started and before the pressure has built up in the system, the seals may give a little leakage. With full pressure in the system the couplers and fittings will be effectively leak-free. If, however, there is a leakage, check the following:

(*a*) There is no accumulation of dirt or sand in the groove in the coupler in which the sealing ring fits. Clean out any dirt or sand and refit the sealing ring.

(*b*) The end of the pipe going inside the coupler is smooth, clean and not distorted.

(*c*) In the case of fittings such as bends, tees and reducers ensure that the fitting has been properly connected into the coupler.

Standardization, Quality Control and Testing of Sprinkler Irrigation Systems

All major nations have well established institutions for standardization and quality control of equipment and products. In India the Bureau of Indian Standards (Manak Bhawan, 9, Bahadur Shaw Zafar Marg, New Delhi - 110002) is responsible for developing standards to ensure the quality requirements and related procedures for industry, agriculture and other areas. Indian standards for various facets of pressurized irrigation systems, including sprinkler irrigation are formulated by the Irrigation and Farm Drainage Equipment System Sectional Committee of FAD-54 (Food and Agriculture Division) of the Bureau of Indian Standards. The standards available for various components of sprinkler irrigation systems (including very low pressure sprinklers, usually grouped under micro-irrigation) are listed in the section under references at the end of this chapter. The standards are designated as follows: IS 12786 : 1989 etc. in which IS stands for Indian Standard, 12786 is the reference number and 1989 shows the year in which the standard was issued. Manufacturers, traders, consumers and other users are advised to ensure that the various products manufactured and used conform to the Standard. Revisions of the standards which may be made subsequently are notified by the Bureau.

VERY-LOW-PRESSURE SPRINKLER IRRIGATION SYSTEMS

Very-low-pressure (VLP) sprinkler irrigation is becoming increasingly popular for many specialized applications. The most important application is in greenhouse agriculture to raise high value vegetable crops and flower plants under controlled environment. VLP sprinkling is valuable to raise seedlings of plantation crops, vegetable crops and fruit trees and in floriculture. There are many field conditions which offer major situation specific advantages to adopt VLP sprinkler irrigation as an energy saving irrigation method. These include hose-pull and permanent under-tree orchard sprinklers, perforated pipes for orchards and row-crops, sprinkler systems operated with low gravity pressure (which would otherwise require pumping sets), and systems which utilize available pressures in underground pipeline irrigation systems. VLP sprinkler system is especially advantageous to energy deficit areas and in small scale intensive agriculture for under-tree applications. However, unlike the common sprinkler irrigation systems, VLP sprinkler systems are seldom used for irrigating common field crops.

Sprinkler irrigation systems are usually classified according to the nozzle operating pressure. The classification system used by the Irrigation Association in USA [Pair, C.H. *et al.* (1975)] is as follows: (1) *low pressure systems* (1/3 to 2 atmospheres), (2) *intermediate pressure systems* (2 to 4 atmospheres), and (3) *high pressure systems* (above 4 atmospheres). In the early period of sprinkler irrigation development and application, most systems were operated at nozzle pressures higher than 2 atmospheres. It was considered that with lower pressures the distribution of the sprinkler discharge is less uniform, the intensity of water application is greater and the sprinklers must be spaced closer together which increases the investment cost on the system. However, in the recent years, even in countries having low cost energy, there are situations where VLP sprinkler systems are desirable (Burt, C.M. and Keller, J., 1977). Burt and Keller (1977), during the performance evaluation of VLP sprinkler systems

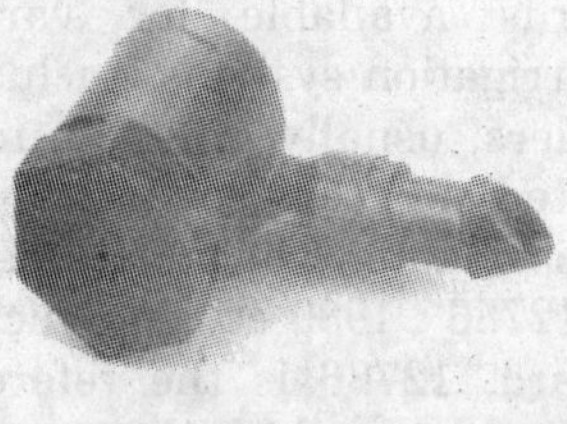

(*a*) Fogger

(*b*) Bubbler

(*c*) Mist sprayer

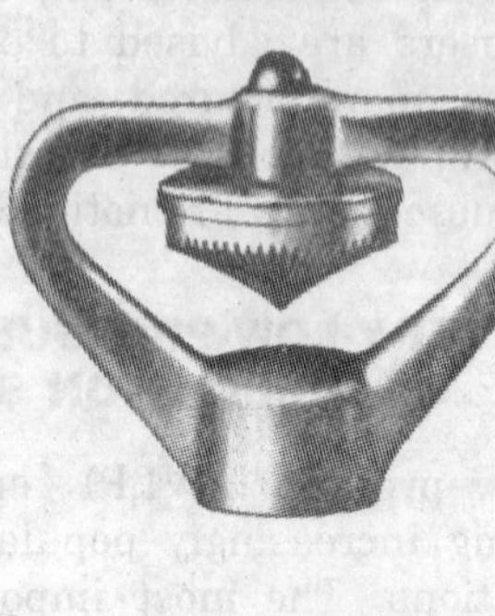

(*d*) Spray head sprinkler (Minnesmann Model "MR")

(*e*) Spray

(*f*) Micro-sprinkler

Fig. 12.29. Various types of spray devices and micro-sprinkler used in micro-irrigation.

(**Source:** (a), (b), (c), (e) *and* (f) – *Dr. Ashwani Kumar;* (d) *Burt, C.M. and Keller, J., 1977)*

conducted in Pakistan in 1970s, concluded that an ideal irrigation system for many of the crops grown in the undulating sandy soils of the Indus basin would be properly designed VLP sprinkler system.

Spray mode jet nozzle

Atomizer type spray nozzle

Mini sprinkler system

Fig. 12.30. Different types of spray nozzles and micro-sprinkler irrigation systems in operation.

(***Source***: Jain Irrigation Systems Ltd., Jalgaon, Maharashtra)

Fig. 12.31. Schematic sketch illustrating perforated pipe spray irrigation of an orchard.

[***Source***: University of Florida, Agricultural Extension Service (1962)]

Types of VLP Sprinklers

The basic types of VLP sprinklers include non-rotating spray heads, rotating type mini sprinklers and perforated pipelines. According to the means of rotation, rotating sprinklers are classed as reaction drive and impact drive.

Spray irrigation. The simplest VLP sprinkler is the stationary spray head which has no rotating mechanism. The application of water is by a small spray, mist or fog to the soil surface. The spray of water is in the form of fine jets or in a fan shape (Fig. 12.30). They are frequently used for residential sprinkling systems in many of the developed countries. As the spray heads do not have any moving parts, these sprinklers are durable and simple to operate and maintain. However, they cover only small wetted diameter and have narrow passages which could get clogged easily. They require relatively high water/application rates. In spite of these limitations the 'Minnesman Model MR' spray head [Fig. 12. 29 (*d*)], tested by Burt and Keller (1977) gave good results in water distribution tests. The Minnesman Model spray head has a slotted cone held above a large vertical nozzle with no narrow passages to clog. In addition to residential sprinkling jobs this type of spray head is also used in small farms in Germany and other countries.

Bubbler irrigation. In bubbler irrigation the application of water to the soil surface is as a small stream or fountain. The discharge rates for point-source bubbler emitters are greater than for drip emitters but is generally less than 225 litres/hour. Because the bubbler discharge rate may sometimes exceed the infiltration rate of the soil, a small basin is often required to contain the water.

Micro-sprinkler system. Where overhead or over the crop water application is desirable, micro- or mini-sprinklers having a rotor to distribute the water cover a large diameter with fine, evenly spread droplets. Micro-sprinklers are well suited to nurseries raising seedlings. The components of micro- or mini-sprinkler systems (Figs. 12.29 and 12.30) are the same as in the conventional rotating head sprinkler irrigation system. However, they are designed to work under low pressure. Rotating type mini-sprinklers are of two types:

(1) *Reaction drive sprinklers* and (2) *Impact drive sprinklers.*

Reaction drive sprinklers rely on the jet thrust action of the water as it leaves the nozzle to provide the force required for rotation. However, many reaction type sprinklers turn very rapidly and hence, are called spinners. At very low pressures reaction type sprinklers encounter operating problems. As the pressure is reduced, the turning force is also reduced because it is a function of both the discharge rate and the jet velocity. A few grains of sand in the rotating collars can cause a sprinkler to stick, regardless of the pressure. Reaction type sprinklers usually have adjustable nozzles or arm angles to help the irrigator to adjust the speed of rotation or the angle of trajectory. However, a minor adjustment of the nozzle angle might disturb the water distribution pattern and rate of application. Therefore, optimum angle adjustment should be determined, preset and locked in position. The rotating collars must be relatively loose to prevent jamming.

Impact drive sprinklers. The impact sprinklers are driven by the sharp impact of the driving arm which is activated by the water jet. The rotating mechanism is powered by the water thrust on the impact driving arm. Consequently the rotation becomes erratic when the pressure is too low. However, the VLP sprinkler under the trade name The Rain Bird Model 20L-TNT. Low Pressure Impact Sprinkler (Fig. 12.29) tested by Burt and Keller (1977) operated fairly well at a pressure head of 0.32 atmosphere (3.3 m).

Perforated Pipe VLP Sprinkler System

Perforated pipe sprinkler systems spray water from 1.6 mm or smaller holes drilled along the top and sides of the lateral pipeline. The holes are sized and spaced so as to uniformly apply water between adjacent lines of perforated pipeline. The water issues from the holes and produces a rain-like application over a rectangular strip (Fig. 12.31). Each hole emits a jet of water, which in rising and falling break into small drops that are spread over the irrigated area by air turbulence. The width covered ranges from about 7.5 to 15 m which increases as the pressure increases. Such systems can operate effectively over the entire VLP range. However, they can only be used on soils having high infiltration capacities such as loamy sands and coarser textured soils, as the water application rate is about 19 mm or greater. Another limitation of perforated pipe sprinkling is that the small holes are very sensitive to clogging. Most perforated pipe laterals have been made by drilling holes in portable aluminium pipes. Rigid PVC pipes may also be a suitable material if the holes are made accurately and properly de-burred.

APPLICATION OF FERTILIZERS AND CHEMICALS THROUGH PRESSURISED IRRIGATION SYSTEMS

Pressurised irrigation systems (both sprinkler and drip) provide a convenient means of supplying nutrient materials to the crop. The term *fertigation* is often used for the method of fertilizer application through pressurized irrigation systems. Besides fertilizers, other chemical products can also be applied through pressurized irrigation systems for the following purposes: (1) for improving the chemical properties of water (*e.g.,* lowering the pH) or of the soil (use of chelates to control certain chloroses of fruit trees); (2) for controlling certain crop diseases (insecticides, nematicide, systemic fungicides); (3) cleaning the irrigation equipment network and removing from it calcareous deposits or disinfecting it in order to prevent clogging resulting from the proliferation of certain micro-organisms. In almost all pressurized irrigation systems the fertigation component is an integral part. This is because fertigation allows more efficient use of fertilizers (water application limited to the crop root zone), resulting in higher crop yields, and to prevent the problem of clogging. Most water soluble fertilizers and liquid fertilizers can be applied efficiently to almost any crop through irrigation water, at any time of the growing season. The reason why fertigation has become the 'state of art' in vegetable cultivation is that nutrients can be applied in the correct dosage at the time appropriate for the specific stage of plant growth.

The fertilizers most commonly used are nitrogen, phosphorus and potassium (N, P and K). Micro-nutrients are to be applied when they are deficient in the crop root zone.

Nitrogen application. Nitrogen, the plant nutrient most commonly deficient for crop production, is often applied in pressurized irrigation systems. Nitrate nitrogen moves readily in the soil with irrigation water and can be applied separately or in mixture with such compounds as ammonium sulphate, urea, calcium ammonium nitrate, and, ammonium nitrate. Calcium nitrate can also be used when bicarbonates are low. Anhydrous ammonia, aqua ammonia and ammonium phosphate in most instances cause clogging problems. Nitrogen source selection should be based on its possible reactions with the irrigation water and the soil. In general, first five nitrogen fertilizers mentioned will cause few, if any, precipitation problems in irrigation systems.

Phosphorus application. Phosphorus has not generally been recommended for application with irrigation water because of its tendency to cause clogging and its limited movement in soil. If irrigation water is high in calcium and magnesium, precipitate of insoluble calcium and magnesium phosphate may result from the application of inorganic phosphate. The application of phosphoric acid along with short pulses of sulphuric acid keeps the water pH low in drip irrigation system and solves the precipitation or clogging problems. Organic phosphates will not precipitate unless the compound is being hydrolyzed to inorganic phosphate in water or the water pH is high. The studies also show that phosphorus was delivered to greater soil volume when applied as ortho-phosphoric acid through a drip system than triple super phosphate applied as a soil amendment beneath each emitter. The ortho-phosphoric acid lowered the pH of irrigation water enough to minimize clogging problems from phosphate precipitation over 3 and 24 days irrigation period.

The most common phosphate fertilizers are triple super-phosphate, mono-ammonium phosphate, di-ammonium phosphate and ammonium polyphosphate. The ammonium phosphate is also a good source of nitrogen. Mono-ammonium phosphate and ammonium polyphosphate, either alone or with some added potassium, makes excellent starter fertilizers because of their high P to N ratios, high water-solubility, and low in free ammonia. Di-ammonium phosphate (DAP) is not recommended as a starter material. However, many starter fertilizers contain DAP. Thus, it is critical that the starter is accurately placed and high rates are avoided. Hence, the common practice is to apply the phosphorus separately and not through the irrigation system.

Potassium application. There are no problems associated with potassium application through micro-irrigation systems. Most potassic fertilizers are water-soluble and quick acting such as potassium chloride or muriate of potash, potassium sulphate, potassium magnesium sulphate, also known as sulphate of potash magnesia.

The K ions are adsorbed in the soil and thus remain available, and largely protected against leaching. However, split application is advisable where higher leaching losses may be expected. Some immobilization into clay lattice layers reduces availability but strong fixation into completely unavailable forms is limited to a few special soil types.

All types of fully-water soluble granular and liquid fertilizers are suitable for fertigation. However, for higher yield and quality, chloride-free fertilizers such as Multi-K (potassium nitrate), Mono Ammonium Phosphate and Mono Potassium Phosphate are preferable. Soluble dry fertilizers containing N, P and K in different combinations are also available in the market. Liquid fertilizers with varying N, P and K contents are also available but these are more expensive. Mostly, Nitrogen (N). Potassium (K), or both are injected. But phosphorus does not move freely in the soil. Commercially prepared liquid fertilizers for fertigation are also good. These are usually combinations of N and K.

Water temperature is an important factor when determining solubility. The solubility of most fertilizers decreases with decreasing temperatures. Most granular nitrogenous fertilizers will absorb heat from the water when they are mixed, since cold water will not hold the same amount of soluble material as warm water. Dry fertilizer products that significantly lower the temperature of the water in which they are dissolved include urea, ammonium nitrate, calcium nitrate and potassium nitrate.

Fertilizer compounds differ greatly in their solubility in water. These differences are usually unimportant for application in the solid form. High solubility, however, is one of the major considerations for the farmers who purchase dry granular fertilizers for dissolving in irrigation water or for application in foliar sprays.

Micronutrients application. Manganese, zinc, iron, copper, etc., may be applied as soluble salts through the irrigation system. These should each be injected separately and apart from other fertilizers and chemicals to avoid chemical interaction and precipitation in emitters. Micronutrients such as iron, zinc, copper and manganese can be applied as chelates or sulphate salts in drip irrigation systems. Normal plant requirements for these nutrients are very low and their application through drip irrigation requires careful and precise metering.

Frequency of fertilizer application. Fertilizers can be injected into the irrigation system at various frequencies once a day or once every two days or even once a week. The frequency depends on system design, irrigation scheduling, soil type, nutrients requirement of the crop and the farmer's preference. It is also important to monitor the application of water, as fertilizer application is linked to water application (Locascio and Smajstrla, 1989). In any case, it is extremely important that the nutrients applied in any irrigation are not subject to leaching either during the same irrigation or during subsequent irrigations.

When applying the fertilizer through the sprinkler system, it is desirable to operate the system long enough to wet the soil and the plant foliage. The fertilizer is then injected through the system in a solution of 1 kg of fertilizer to about 30 kg of water and timed in such a way as to complete the injection in about 30 minutes. After this, the system is worked for 20 to 30 minutes to flush it from any toxic effects of the fertilizer solution on the sprinkler tubings.

Computation of the quantity of fertilizer to be applied. To determine the quantity of fertilizer to be injected into the system for each setting, the area irrigated in each setting of the lateral line is obtained by multiplying the length of the lateral coverage and the move of the lateral. The quantity of fertilizer to be injected is calculated for this area according to the recommended rate of fertilizer application in kilograms per hectare, using the following formula:

$$W_F = \frac{D_s \times D_e \times N_s \times W_f}{10,000} \quad \text{...(12.12)}$$

in which,

W_F = amount of fertilizer per setting, kg
D_s = distance between sprinklers, metres
D_e = distance between laterals, metres
N_s = number of sprinklers, and
W_f = recommended fertilizer dose, kg/ha

Example 12.9. *A lateral has 12 sprinklers spaced 14 metres apart. The laterals are spaced 20 metres on the main line. Determine the amount of fertilizer to be applied at each setting when the recommended fertilizer dose is 80 kg/ha.*

Solution: Using Eq. 12.12, D_s = 14 m, D_e = 20 m = N_s = 12 and W_f = 80 kg/ha

$$W_f = \frac{12 \times 20 \times 14 \times 80}{10,000} = 26.88 \text{ kg.}$$

Equipment and Methods for Fertilizer Injection

Fertilizers can be injected into pressurised irrigation systems by selecting a wide variety of available pumps, valves, fertilizer tanks, venturies and aspirators. Fertilizers may be injected into sprinkler/drip irrigation system by a differential pressure system, venturi injector, or by pumping under pressure into the irrigation pipeline. Other chemicals such as acids require a constant concentration rate of the fertilizer solution entering the irrigation water. They shall be injected by constant rate injection devices only. An injector that will operate within a range of injection rates for a number of chemicals is desirable. The concentration of chemicals to be injected in the irrigation water is low and usually ranges from 200 to 500 ppm for fertilizers and 0.5 to 10 ppm for bactericides. It shall be a regular practice to measure the chemical concentration at reasonable intervals. Measurements are to be made after the filters and before the main pipelines and occasionally at the end of the last lateral line to ensure that the entire system is working.

The commonly used components of the equipment to apply fertilizers in pressurized irrigation systems, in addition to the irrigation system, include a fertilizer tank, fertilizer dissolver, fertilizer injection device, a filter, a pressure gauge, check valves, and a pressure regulator.

Fertilizer tank. The fertilizer tank is used for mixing the fertilizer (Fig. 12.33). The fertilizer tank should have a sufficiently large capacity to contain the entire fertilizer solution required for the cropped area for any application. It should be made from materials that withstand corrosivity of the fertilizers. Large, low cost tanks constructed from epoxy-coated metal, plastic or fiberglass are useful when injection pumps are used. For a pressure differential injection system (Fig. 12.33), the high pressure rated chemical tank should have enough capacity for a complete irrigation application.

Fertilizer dissolver. Fertilizer dissolver is a device that can be used for preparation of fertilizer solution. After filling the tank with water, the pump injects the water into an inner strainer that contains the dry fertilizers. The water gradually dissolves the fertilizer into a solution by circulating the solution within the dissolver.

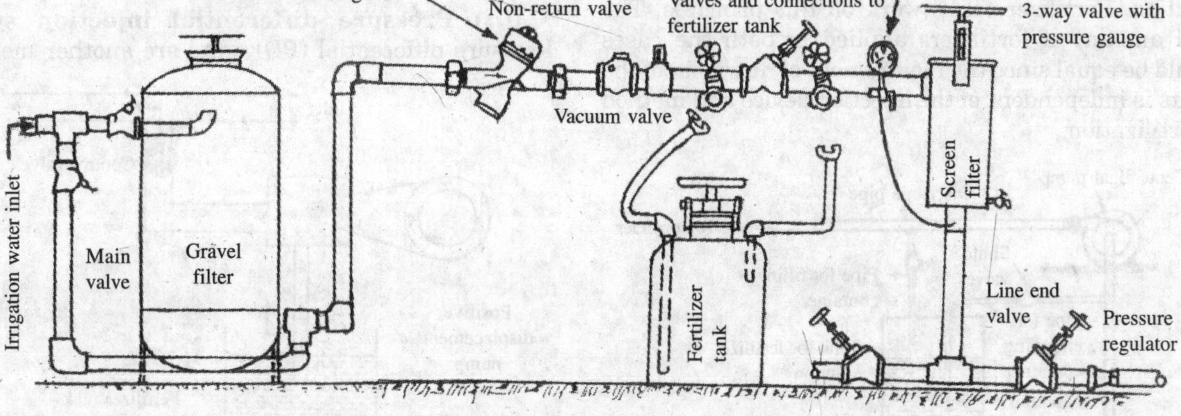

Fig. 12.32. Diagram illustrating the control head assembly of a drip irrigation system provided with fertigation equipment, filters, control valves and other accessories.
[***Source***: Adopted from Sapir, E. and Yadav, E. (1995)]

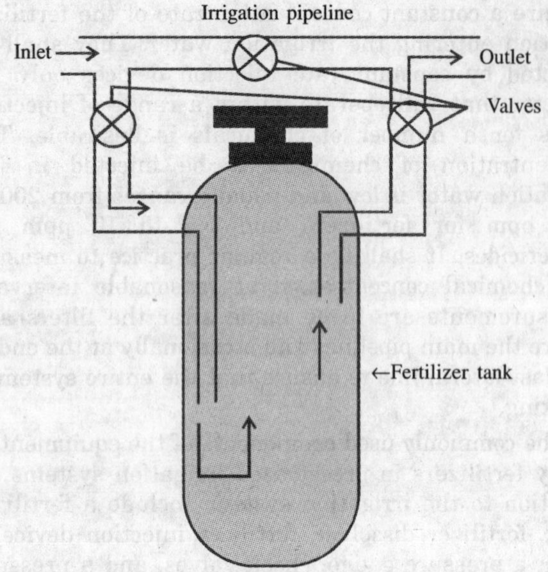

Fig. 12.33. Schematic diagram of a pressure differential fertilizer injection system.

(***Courtesy***: Dr. Ashwani Kumar)

Fertilizer Injection Devices

Pressurized irrigation methods require fertilizer injector for injection of fertilizers, into the irrigation water. The following are the two basic concepts on the basis of which fertilizer injectors are designed:

(*a*) *Proportional concept* is characterized by constant concentration of the fertilizer solution in irrigation water throughout the irrigation duration. Fertilizer injection devices such as venturi pump and fertilizer injection pump operate on this principle. It enables the delivery of constant concentration during the entire irrigation duration.

(*b*) *Quantitative or non-proportional concept* is characterized by change in the concentration of fertilizers in irrigation water. The concentration of fertilizer decreases gradually with the irrigation duration. Fertilizer tanks work on this principle. The total amount of fertilizers applied in both the cases should be equal since the requirement of nutrients of the plants is independent of the injection device and method of fertilization.

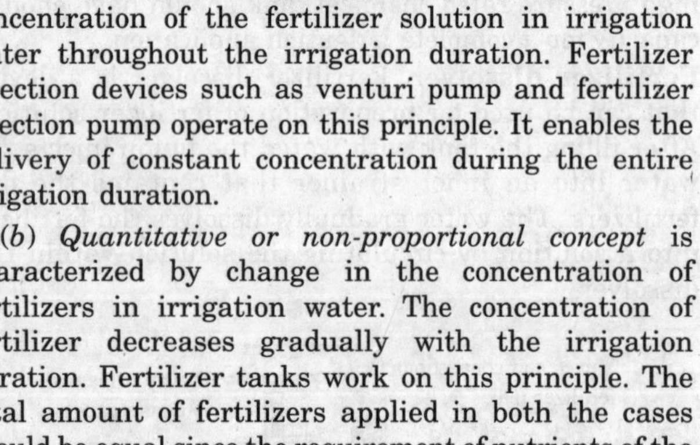

Fig. 12.34. Method of adding fertilizer in solution through the suction side of a centrifugal pump.

***(i)* Fertilizer injection through a volute centrifugal pump.** Fertilizer is introduced into the system from the suction side of the pump through a pipe and regulated by a valve (Fig. 12.34). Another pipe is connected from the discharge side of the pump to the fertilizer container for the required water supply in the tank. This system is relatively simpler, but pump impellers are likely to get corroded due to the fertilizer solution, unless the impellers are made of corrosion resistant materials.

***(ii)* Fertilizer injection pump system.** In this method a pump is used to draw the fertilizer stock solution from the storage tank and inject it under pressure into the irrigation stream. The injection rate can be set to create a desirable mixing ratio. The fertilizer solution is normally pumped from an unpressurised storage tank.

Pumping is a common method of injecting fertilizer into a drip irrigation system. Injector energy is provided by electric motors, internal combustion engines and water driven hydraulic motors. Positive injection pumps include single or multiple piston pumps, diaphragm pumps, gear pumps, and roller pumps (Fig. 12.35). Where two or more different types of fertilizers are required, multiple pump units can be used to avoid or reduce precipitation problems. All of the injection pumps can be regulated to achieve the desired or required rate of application, usually by adjusting the length of stroke of the piston pump or by selecting the appropriate pulley diameter. Another means of adjusting fertilizer application is with variable-speed motors. The system should flush itself with clean water at least once after injecting the fertilizer solution. The major advantages of the system are flexibility and high discharge rate, the system does not add to the head loss in a pressurized irrigation system and that it maintains a constant concentration of nutrients throughout the period of fertilizer application. The disadvantages are the high cost of the equipment and higher operation and maintenance costs.

***(iii)* Pressure differential injection system.** Pressure differential (PD) units are another method of

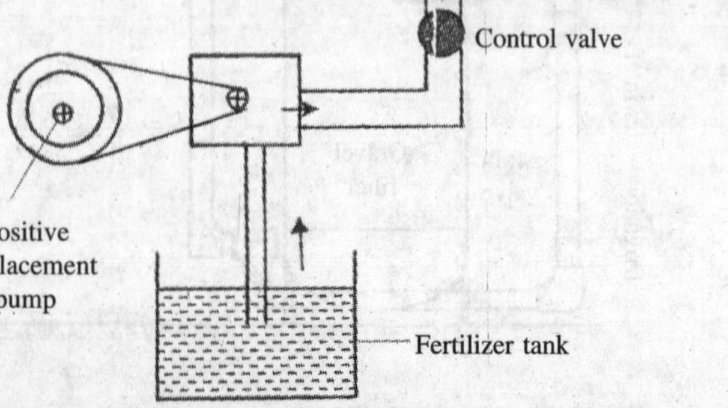

Fig. 12.35. Schematic sketch illustrating the layout of a fertilizer injection system using a positive displacement pump.

(***Courtesy***: Dr. Ashwani Kumar)

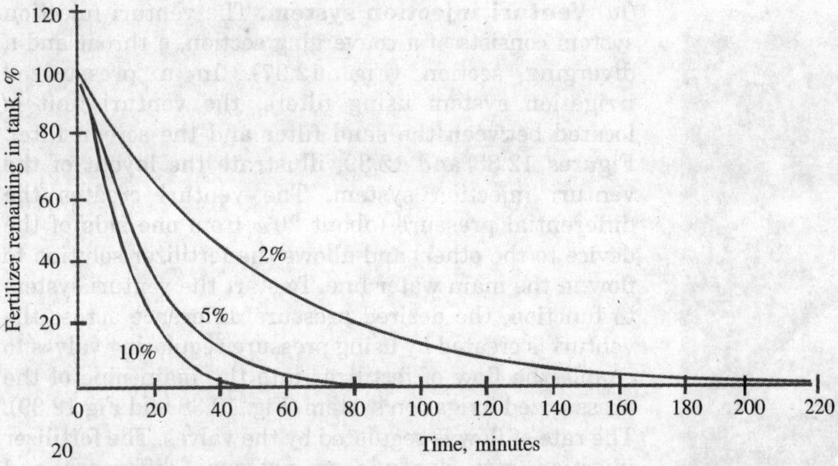

Fig. 12.36. The % fertilizer remaining in the fertilizer tank as a function of time in the pressure differential system of fertilizer injection.

(***Courtesy***: Dr. Ashwani Kumar)

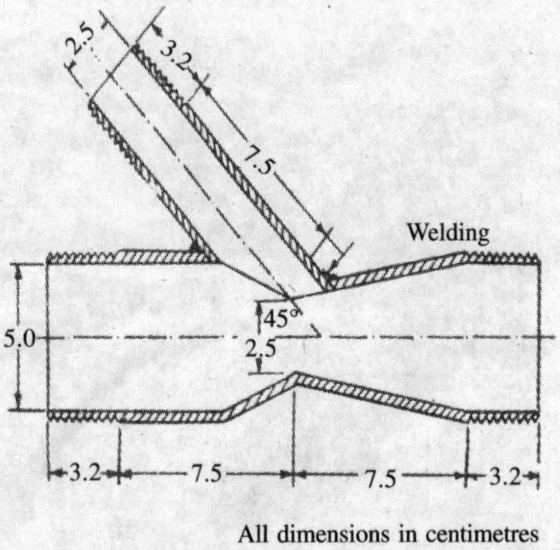

Fig. 12.37. Venturi used with fertilizer applicator (sectional view).

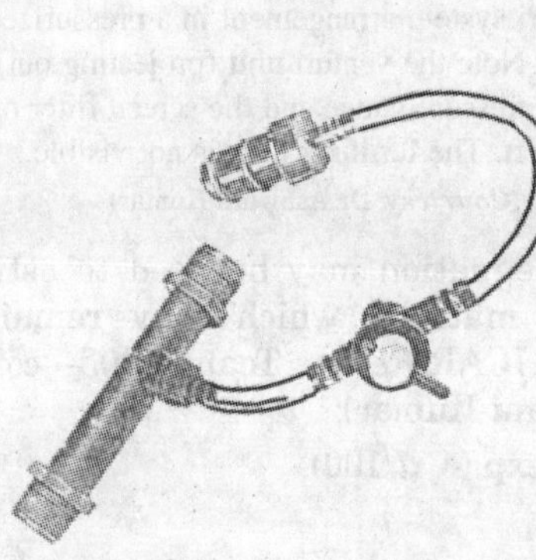

Fig. 12.38. Ventury assembly for fertigation.

(***Source***: Jain Irrigation systems Ltd., Jalgaon)

injecting fertilizer into drip irrigation systems. The PD unit takes advantage of the system's pressure-head differences. Valves, venturi, elbows, or pipe friction can develop pressure differences. The main advantage of the PD applicators is the absence of moving parts. They are simple in operation and require no electric, gasoline, or water-powered pumps. The main disadvantage of the PD units is that the rate of application is not constant and changes continuously with time; thus, a uniform concentration of a nutrient cannot be maintained. Figure 12.36 is the schematic diagram of a pressure differential fertilizer injection system.

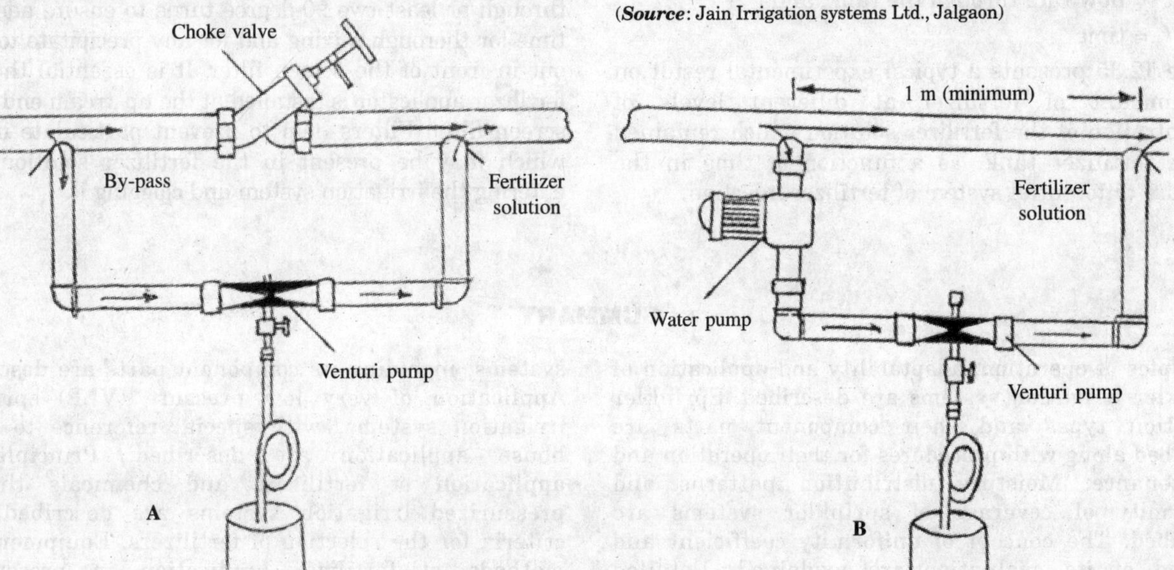

Fig. 12.39. Methods of assembling a venturi pump: **A.** Using choke valve. **B.** Using pump.

[***Source***: Sapir, E. and Yagev, E. (1995)]

Fig. 12.40. Venturi system arrangement in a pressurized irrigating system. Note the venturi unit (projecting out), valves, pressure gauge and the screen filter on the left. The fertilizer tank is not visible.
(***Courtesy***: Dr. Ashwani Kumar)

The following equation may be used to estimate the percentage of material which may remain in the fertilizer tank (ICAR Review Team, 2003—contribution from Dr. Ashwani Kumar):

$$n = 100 \exp(-xt/100) \quad ...(12.13)$$

in which,

n = % mixture remaining in the fertilizer tank,
exp = exponential function,
x = flow rate through the tank, and
t = time

Figure 12.35 presents a typical experimental result on the amount of fertilizer at different levels of concentration of the fertilizer solution which remained in the fertilizer tank, as a function of time in the pressure differential system of fertilizer injection.

(*iv*) **Venturi injection system.** The venturi injection system consists of a converging section, a throat and a diverging section (Fig. 12.37). In a pressurized irrigation system using filters, the venturi unit is located between the sand filter and the screen filter. Figures 12.38 and 12.39, illustrate the layout of the venturi injection system. The venturi creates the differential pressure (about 20% from one side of the device to the other) and allows the fertilizer solution to flow in the main water line. To start the venturi system to function, the desired pressure difference across the venturi is created by using pressure regulating valves to enable the flow of fertilizer into the main pipe of the pressurized irrigation system (Fig. 12.38 and Fig 12.39). The rate of flow is regulated by the valves. The fertilizer injection rate depends on pressure difference and pressure fluctuations in the system change the injection rate. The main advantages of the venturi system of fertilizer injection are the simplicity in operation, low cost and good control over the fertilizer concentration. The disadvantages are heavy power loss (i.e., about 1/3 of the total operating pressure), the relatively low discharge rate, difficulty in precise regulation of flow and the need to operate the system at full capacity prior to injecting the fertilizer solution (Rajput, T.B.S. and Patel, N., 2003).

Location of Fertilizer Injection Systems

In a pressurized irrigation system the fertilizer injection unit is located, between the sand filter (when used) and the screen filter (Figs. 12.38 and 12.40). The general recommendation is that the fertilizer solution should pass through at least two 90-degree turns to ensure adequate time for thorough mixing and for any precipitate to come out in front of the screen filter. It is essential that the fertilizer applicator is installed at the upstream end of the screen filter. Filters help to prevent particulate matter which may be present in the fertilizer solution from entering the irrigation system and clogging it.

SUMMARY

Principles of operation, adaptability and application of sprinkler irrigation systems are described. Sprinkler irrigation types and their component parts are described along with procedures for their operation and maintenance. Moisture distribution patterns and uniformity of coverage of sprinkler systems are described. The concept of uniformity coefficient and method of its evaluation are explained. Detailed procedures for the design of sprinkler irrigation systems, including its component parts are described. Application of very low pressure (VLP) sprinkler irrigation systems, with special reference to green house application are described. Principles of application of fertilizers and chemicals through pressurized irrigation systems are described with criteria for the selection of fertilizers. Equipment and methods of fertilizer application in pressurized irrigation systems are described.

PROBLEMS

12.1. The data obtained from a field test on determining the water distribution pattern of a sprinkler is given below:

Sprinkler:	7.10 × 4.76 mm—95.76 litres per min, 4.22 kg/cm
Spacing:	18.2 × 18.2 metres
Wind:	3.86 km/hr from the south-east
Humidity:	36 per cent
Time of test:	1 hour

S	15.70	13.20	13.20 North	16.70	17.0	S
15.0	15.0	15.70	15.2	17.5	19.5	17.8
14.5	12.7	16.7	16.0	17.8	11.9	14.2
W 15.0	15.5	18.3	19.0	16.2	11.9	12.4 E
17.3	18.8	16.0	15.2	10.7	14.2	16.0
12.2	20.3	14.5	12.4	14.7	10.1	12.9
S	15.5	13.20	13.2 South	16.5	16.0	S

S indicates the location of sprinklers.

Determine the uniformity coefficient.

[Ans. 87.4 per cent]

12.2. At what rate in cm/hr, would a sprinkler discharging at the rate of 0.95 litres per second apply water if the sprinkler spacing is 18 m × 24 m ? **[Ans.** 0.791 cm/hr]

12.3. Determine the discharge in litres per second for a sprinkler operating at 2.8 kg/sq cm and having 4 mm × 2.8 mm diameter nozzles with a discharge coefficient of 0.96.

[Ans. 0.418 lit/sec]

12.4. A 12-hectare field is to be irrigated at a maximum rate of 1 cm/hr with a sprinkler system. The root zone is 90 cm deep. The available moisture holding capacity of the soil is 16.5 cm/metre depth. The water application efficiency is 70 per cent. The field is to be irrigated when 45 per cent of the available moisture capacity is depleted. The peak rate of moisture use is 5 mm per day. Determine the net depth of application per irrigation, depth of water to be pumped in centimetres, days to cover the field and the area irrigated per day.

[Ans. 8.167 cm, 11.66 cm, 1.45 days, 8.275 ha]

12.5. Compute the total friction loss for a sprinkler irrigation system having a 10 cm diameter main, 244 m long and one 7.5 cm diameter lateral, 116 m long. The pump delivers 8.5 litres of water per second and there are 10 sprinklers on the lateral. Portable aluminium pipes 16 guage, 1.3 mm and couplers are used throughout.

[Ans. 5.57 m]

12.6. Determine the total pumping head for a sprinkler irrigation system on a level land. The average operating pressure at the nozzle is 3 kg/cm^2. The friction loss in the main is 6.5 m and in the lateral is 4 m; the drawdown of the well is 5 m at the required discharge of 31.6 lit/sec; the riser height is 1.5 m and friction loss in all valves is 3.25 m. Determine the horsepower requirements of the pump if it operates at 65 per cent efficiency.

[Ans. 48.85 m, 31.66 hp]

12.7. Assuming a triangular depth-distribution pattern for a rectangular placement of sprinklers and a diameter of coverage of 42 m, compute the uniformity coefficient for a 14 m × 21 m sprinkler spacing using 7 m grids for depth measurements. Assume that the depth of application at the sprinkler is 2.5 cm and zero at 18 m.

[Ans. 94.8%]

12.8. If the pressures at the opposite ends of a sprinkler lateral are 3 kg/cm^2 and 2.6 kg/cm^3, what would be the discharge of the farthest sprinkler, provided the sprinkler at 3 kg/cm^2 end discharges 0.76 lit/sec.

[Ans. 0.696 lit/sec]

12.9. If 100 sample cans are uniformly spaced in the area covered by four sprinklers, and the average depth of water caught in a given time is 1.25 cm with the average variation from the mean 0.2 cm, what is the uniformity coefficient? Assuming that the infiltration rate did not exceed and the water did not penetrate below the root zone, what is the application efficiency?

[Ans. 84%, 100%]

QUESTIONS

1. Distinguish between:

(i) Basic differences in the methods of water application in sprinkler and drip methods of irrigation.

(ii) Solid set system and permanent system of sprinkler irrigation.

(iii) Semi-portable and semi-permanent systems of sprinkler irrigation.

(iv) Plug type and screw type couplings of a sprinkler pipeline.

(v) Single nozzle and double nozzle sprinkling heads.

(vi) Over-head sprinklers and under-tree sprinklers.

(vii) Low pressure sprinklers and high pressure sprinklers.

(viii) Spray and bubbler types of micro-irrigation.

(ix) Reaction drive sprinklers and impact type sprinklers.

(x) Proportional concept and quantitative concept in fertilizer injection.

2. Specify the distinct advantages of sprinkler irrigation over other methods of irrigation water application.

3. Amongst the two methods of pressurized irrigation systems, namely, sprinkler and drip, which are the specific situations where sprinkler irrigation is more advantageous than drip irrigation?

4. What are the limitations in the large-scale adoption of sprinkler irrigation in many of the developing countries?

5. Under which cropping system conditions is the adoption of solid set system of sprinkler irrigation preferred?

6. What are the field situations favouring the adoption of centre-pivot system of sprinkler irrigation?

7. What are the principal methods used to develop the spray required for sprinkling?

8. What are the factors influencing the spray distribution pattern of a rotating head sprinkler?

9. What are the different types of valves used in a sprinkler irrigation system?

10. Define uniformity coefficient. How is it determined?

11. What is the purpose of overlapping of sprays in a sprinkler system?

12. How are the successive locations of sprinkler laterals determined?

13. What is the basis for the proper orientation of main line and lateral lines in a sprinkler system to obtain a reasonable degree of uniformity in the discharge of each sprinkler?

14. How is the required discharge of an individual sprinkler in a sprinkler system computed?

15. Express the mathematical relationship to compute the capacity of a sprinkler irrigation system.

16. How is the total head of a pump used in a sprinkler system determined?

17. What are the step-by-step procedures for the maintenance of sprinkler heads?

18. What are the points to check if the pump of a sprinkler system does not prime or develop pressure?

19. How is leakage from couplers of sprinkler lines detected and rectified?

20. How are the standardization and quality control of sprinkler systems controlled and enforced?

21. Define fertigation. What are its advantages over conventional methods of fertilizer application?

22. How is the quantity of fertilizer to be applied in fertigation units computed?

23. What are the common methods of fertilizer injection in pressurized irrigation systems?

24. What are the commonly used fertilizer injection devices in pressurized irrigation systems?

25. What are the limitations in fertilizer application through a volute centrifugal pump?

26. Identify the True and False statements from the following:

(*i*) Fine textured soils with low infiltration rates cannot be irrigated effectively in hot windy areas.

(*ii*) In the center-pivot system of sprinkler irrigation, the area of land left over between adjoining pivot points is usually ignored, considering the area coverage and the fully automatic operation of the system.

(*iii*) Giant sprinklers usually have two or more large nozzles with smaller supplemental nozzles to fill in pattern gaps.

(*iv*) The spray distribution characteristics of sprinkler heads change with the nozzle size and operating pressure.

(*v*) The moisture distribution pattern of a rotating head sprinkler is profoundly influenced by the available pressure head and the wind conditions.

(*vi*) Phosphorous has not generally been recommended for fertigation.

(*vii*) Most potassium fertilizers are difficult to dissolve in water.

(*viii*) In pressurized irrigation systems the fertilizer injection unit is located between the sand filter and the screen filter.

27. Identify the most appropriate answer of the following questions from the four options:

(*i*) Amongst the top 20 countries with large-scale adoption of sprinkler irrigation the maximum percentage coverage under sprinklers is in:

(a) USA (b) Saudi Arabia
(c) Libya (d) United Kingdom

(*ii*) The pressure head in a sprinkler system is converted into velocity head at:

(a) the sprinkler head (b) the sprinkler jet
(c) the nozzle (d) the riser pipe

(*iii*) The major factor influencing the efficiency of a well designed sprinkler irrigation system is:

(a) operating pressure
(b) wind velocity
(c) sprinkler spacing
(d) diameter of sprinkler lateral

(*iv*) In a sprinkler irrigation system the most commonly used pipe for a permanently laid main line is:

(a) asbestos cement pipe
(b) aluminium pipe
(c) steel pipe
(d) PVC pipe

(*v*) The width of land covered by a perforated spray type sprinkler system ranges from:

(a) 4 to 8 m (b) 5 to 10
(c) 6 to 15 m (d) 8 to 20 m

(*vi*) The diameter of the area covered by a rotating type permanent overhead type of sprinkler system ranges from:

(a) 15 to 20 m (b) 20 to 30 m
(c) 30 to 45 m (d) 40 to 50 m

(*vii*) The minimum satisfactory value of uniformity coefficient under normal conditions of pressure and wind velocity is:

(a) 80% (b) 85%
(c) 90% (d) 95%

(*viii*) The maximum water application rate of a sprinkler in light sandy loam soils on land with slope ranging from 1 to 5% is:

(a) 1.5 cm/hr (b) 2 cm/hr
(c) 2.5 cm/hr (d) 3 cm/hr

28. In the following, match each of the entries in the first column with the most closely linked entry in the second column:

(*i*)	(A) Pivot point	(I) Measurement of nozzle pressure of sprinkler heads
	(B) Carbon joint coupling	(II) Uniformity coefficient
	(C) Pitot tube	(III) Discharge of sprinkler nozzle
	(D) Christiansen	(IV) Sprinkler lateral
	(E) Toricelli	(V) Center-pivot sprinkler system
(*ii*)	(A) Covassa	(I) Rate of water application of sprinkler
	(B) Tanda	(II) VLP sprinkler system
	(C) Precipitation intensity	(III) Minnesan Model sprinkler head
	(D) Bubbler	(IV) Index for jet break-up in sprinklers
	(E) Burt and Keller	(V) Water spread area covered by a rotating head sprinkler
(*iii*)	(A) Mono-Ammonium phosphate	(I) Drip irrigation system
	(B) Choke valve	(II) Micro-irrigation in plastic house nurseries
	(C) Pressure differential injection system	(III) Monitoring fertilizer application
	(D) Fogger	(IV) Fertilizer injection system
	(E) Locasio and Smajstrla	(V) Fertigation

FURTHER READING REFERENCES

Bureau of Indian Standards (1989), Indian Standard IS 12786: 1989 (with Amendments 1992 & 1999), *Irrigation Equipment-Polyethylene Pipes for Irrigation Laterals.* Bureau of Indian Standards, Manak Bhawan, New Delhi, p. 9.

Bureau of Indian Standards (1991), Indian Standard IS 13062: 1991. *Irrigation Equipment and Systems—Evaluation of Field Irrigation Efficiencies Guidelines*. Bureau of Indian Standards, Manak Bhawan, New Delhi, p. 13.

Bureau of Indian Standards (1992), Indian Standard IS 13479: 1992, ISO 3501: 1976. *Assembled Joints Between Fittings and Polyethelene (PE) Pressure Pipes—Test of Resistance to Pull Out,* Bureau of Indian Standards, Manak Bhawan, New Delhi, p. 2.

Bureau of Indian Standards (1994), Indian Standard IS 12785: 1994, *Irrigation Equipment-Strainer Type Filters—Specification*, Bureau of Indian Standards, Manak Bhawan, New Delhi, p. 5.

Bureau of Indian Standards (1994), Indian Standard IS 14178: 1994, *Pressurised Irrigation Equipment—Terminology*, Bureau of Indian Standards, Manak Bhawan, New Delhi, p. 12.

Bureau of Indian Standards (1995), Indian Standard IS 12232 (Part 2): 1995 (with amendment 1998), *Irrigation Equipment—Rotating Sprinkler: Part II, Test Method for Uniformity of Distribution*, Bureau of Indian Standards, Manak Bhawan, New Delhi, p. 6.

Bureau of Indian Standards (1996), Indian Standard IS 12232 (Part 1): 1996/ISO7749–1: 1995, *Irrigation Equipment—Rotating Sprinkler Part I, Design and Operational Requirements,* Bureau of Indian Standards, Manak Bhawan, New Delhi, p. 12.

Bureau of Indian Standards (1997), Indian Standard IS 14483 (Part 1): 1997, *Fertilizer and Chemical Injection System,* Bureau of Indian Standards, Manak Bhawan, New Delhi, p. 4.

Bureau of Indian Standards (1998), Indian Standard IS 14606: 1998, *Irrigation Equipment-Media Filters—Specifications,* Bureau of Indian Standards, Manak Bhawan, New Delhi, p. 5.

Bureau of Indian Standards (1998), Indian Standard IS 14605: 1998, *Irrigation Equipment—Micro-Sprayers Specifications,* Bureau of Indian Standards Manak Bhawan, New Delhi, p. 9.

Bureau of Indian Standards (1999), Indian Standard IS 10799: 1999, *Irrigation Equipment—Design, Installation and Field Evaluation of Micro-Irrigation, Systems—Code of Practice*, Bureau of Indian Standards, Manak Bhawan. New Delhi, p. 16.

Bureau of Indian Standards (1999), Indian Standard IS: 1999, *Irrigation Equipment—Sprinkler Pipes—Specifications, Part I, Polyethelene Pipes*, Bureau of Indian Standards, Manak Bhawan, New Delhi, p. 9.

Bureau of Indian Standards (1999), Indian Standard IS 14151: 1999, *Irrigation Equipment—Sprinkler Pipes, Part 2, Quick-Coupled Polyethylene Pipes*, Bureau of Indian Standards, Manak Bhawan, New Delhi, p. 3.

Burt, C.M., and Keller, J., (1977), *Very Low Pressure Sprinkler Irrigation*, Dept. of Agricultural and Civil Engineering, Utah State University, Logan, Utah, USA, p. 24.

Cook, R.L., Davis J.R. and Kiddler, E.H. (1954), *Fertilizing Through Irrigation Water,* Michigan State University, Cooperative Extension Service, East Lansing. Michigan, USA, p. 23.

Finkel, H.K., ed. (1983), *CRC Handbook of Irrigation Technology,* CC Press Inc. Boca Raton, Florida, USA, pp. 193–245.

Fry, A.W. Gray, (1970), *Sprinkler Irrigation Handbook,* Rain Bird Sprinkler Mfg. Corp., Glendora, California, p. 42.

Hurd, C.J., (1969), *Sprinkler Irrigation Guide Book,* US Agency for International Development, Washington DC.

Imperial Chemical Industries of Australia and New Zealand Ltd. (1970), *Polyptot hydraulic design for multi-nozzle irrigation system,* ICIANZ Ltd. Nicholson Street, Melbourne, Australia.

Indian Council of Agricultural Research (2003), *Report of the Quinquinnial Review Team* on ICAR. All India Coordinated Research Project on Application of Plastics in Agriculture, Vol. II: *Micro-Irrigation System—Past, Present and Future*, Central Institute of Post Harvest Technology, Ludhiana, pp. 30-40.

International Committee on Irrigation and Drainage (1998), *Sprinkler Irrigation in India,* Ministry of Water Resources, Government of India, New Delhi, p. 134.

Jain Irrigation Systems Limited, Leaflet, *Liquid Fertilizers for Drip Irrigation,* Jain Irrigation System Ltd., Jalgaon, Maharashtra, p. 2.

Jindal Irrigation Systems Limited, Leaflet, *Sprinkler Irrigation Systems,* Jindal Irrigation System Ltd., Kushaiguda, Hyderabad, p. 7.

Khepar, S.D., *et al.,* (1981), *Sprinkler Irrigation.* Communication Centre, Punjab Agricultural University, Ludhiana, p. 78.

Molenaar, A., (1960), *Irrigation by Sprinker,* Food and Agriculture Organization of the United Nations, Rome, p. 91.

Moncada, A., (1970), *Evaluation of Sprinkler Performance from Can Catches Set in Different Patterns.* M.Sc., thesis, Utah State University, Logan, p. 215.

Oregon State College Bulletin, *Sprinkler Irrigation.* Oregon State College, Corvallis, Oregon, The State College of Washington, Pullman, Washington and University of Idaho, Mosco, Idaho, USA, p. 15.

Pair, C.M. *et al.* (1975), *Sprinkler Irrigation,* 4th ed. Sprinkler Irrigation Association, Silver Springs, Maryland, USA, p. 615.

Patel, N., and Rajput, T.B.S., (2003), *Effect of Fertigation on Growth and Yield of Onion (Allium cepa),* In Singh, H.P., *et al.,* (2003), Micro-irrigation, Central Board of Irrigation and Power, New Delhi, pp. 451–454.

Pillsbury, A.F., (1968), *Sprinkler Irrigation,* Food and Agriculture Organization of the United Nations, Rome, p. 179.

Polyolefins Industires Ltd., (1987), *Sprinkler Irrigation Systems—Use of Plastic Pipes,* A Case Study Based on HASTI High Density Polyethylene Pipes, Pl:, Mafatal Centre, Nariman Point, Mumbai, p. 13.

Premier Irrigation Equipment Pvt. Ltd., (1971), *Operating Instruction Manual for Premier Irrigation System,* No. T. 26, Issue No. 1, p. 35.

Premier Irrigation Equipment Limited, Leaflet, *Towards Greener World,* Premier Equipment Ltd., Alypore Road, Kolkata.

Rajput T.B.S., and Patel, N., (2003), *Fertigation Theory and Practice,* Water Technology Centre, I.A.R.I., New Delhi, p. 54.

Schwab G.O., Fangmeier, D.D., Eliot, W.J., and Frevert, R.K. (1996), *Soil and Water Conservation Engineering,* 4th ed., John Wiley & Sons, Inc., New York, USA, pp. 425–451.

Scott, V.H., (1956), *Sprinkler Irrigation,* California Agricultural Experiment Station Extension Service, University of California, Davis, California, USA, p. 42

Singh, H.P., *et al.* (2001), *Micro irrigation,* Pub. Central Board of Irrigation and Power, New Delhi, p. 774.

United States Department of Agriculture (1960), SCS National Enginering Handbook, Chapter II, *Sprinkler Irrigation, USDA, SCS, Washington* D.C., p. 83.

University of California (1969), *Solid Set Sprinklers for Starting Vegetable Crops,* AXT 294, University of California Agricultural Extension Service, p. 10.

University of Florida (1962), *How to Buy an Irrigation System,* Circular No. 173A, Agricultural Extension Service, University of Florida, Gainesville, Florida, USA, p. 23.

Voltas Limited, Leaflet, *Sprinkler Irrigation Systems,* The Voltas Ltd., Dr. Ambedkar Road, Mumbai.

Chapter 13

Drip Irrigation

LEARNING OBJECTIVES

Appraisal of basic principles, adaptability, advantages and limitations of drip irrigation systems. Review of indigenous drip irrigation devices. Drip irrigation system components. Types of drippers and their use under location-specific situations. Filtration systems, their adaptability, functioning and maintenance. Fertigation procedures. Design, operation, performance evaluation and maintenance of drip irrigation systems.

Drip or trickle irrigation is an efficient irrigation method of recent origin, which is becoming increasingly popular in areas of water scarcity and poor quality irrigation water. In drip irrigation, water is applied frequently at low rates from a low pressure delivery system comprising of small diameter plastic pipes fitted with outlets, called emitters or drippers, directly to the land surface close to the plant where the roots grow (Fig. 13.1). Drip irrigation is best suited to water scarcity areas and regions where irrigation water quality is marginal or low (saline water). Due to the frequent application of irrigation water, salts are pushed to the periphery of the moist zone, away from the root spread area (Fig. 13.2). It is suitable to almost all types of soils. In clay soils with low infiltration rates, water is to be applied slowly to prevent surface water ponding and runoff. In sandy soils with high infiltration rates, higher dripper discharge rates will be required to ensure adequate lateral wetting of the soil. Drip irrigation is advantageous on lands with undulating topography without undertaking major land levelling

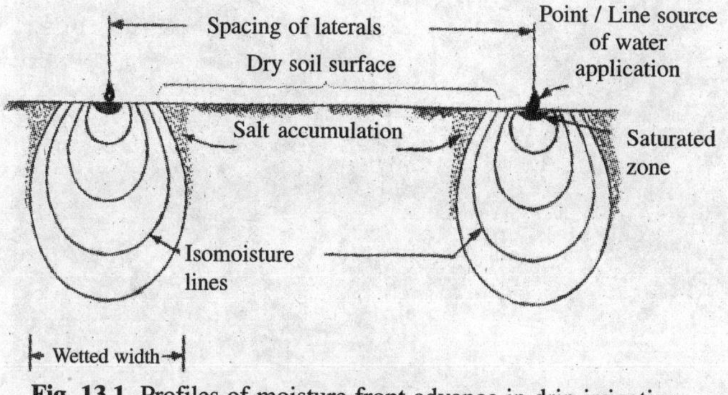

Fig. 13.1. Profiles of moisture front advance in drip irrigation in medium textured soils.

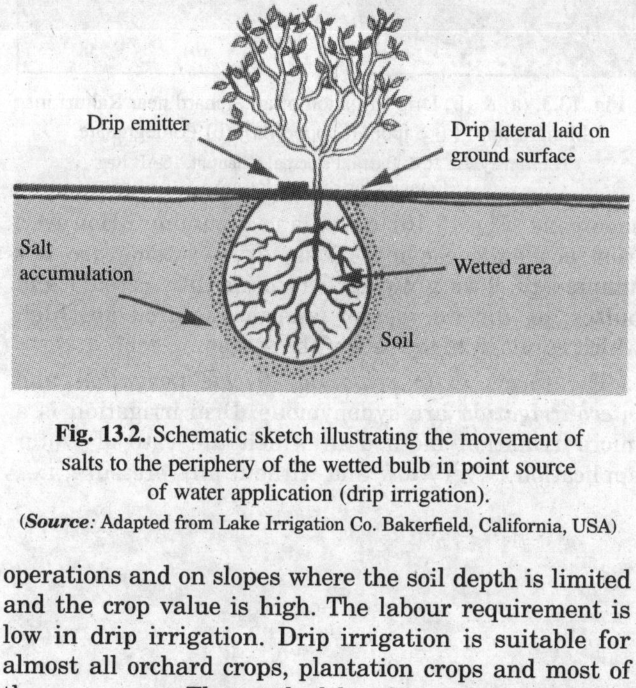

Fig. 13.2. Schematic sketch illustrating the movement of salts to the periphery of the wetted bulb in point source of water application (drip irrigation).
(***Source:*** Adapted from Lake Irrigation Co. Bakerfield, California, USA)

operations and on slopes where the soil depth is limited and the crop value is high. The labour requirement is low in drip irrigation. Drip irrigation is suitable for almost all orchard crops, plantation crops and most of the row crops. The method has been observed to be highly economical and beneficial in water scarcity areas to grow orchard and plantation crops (Fig. 13.3 to 13.6) like coconut, tea, coffee, cardamom, citrus, grapes, banana, papaya, mango, guava, pineapple and pomegranate; row crops like sugarcane, cotton, groundnut, sapota, strawberry and vegetable crops including tomato, potato and other widely spaced vegetable crops, and flower plants (Figs. 13.7 to 13.9). It is highly suitable for nursery raising and establishment of forestry plantations, especially under the wasteland development programme. It is also suitable to irrigate

Fig. 13.3. (a) & (b) Drip irrigation of an orchard near Rahuri in Maharashtra: (a) Sapota (Sappodilla), (b) Pomegranate.

(*Courtesy:* Dr. R.K. Batta, Principal Scientist, ICAR Res. Complex for Eastern Patna).

sugarcane (Fig. 13.10), cotton and groundnut. However, from economic considerations, drip systems are not adapted to close growing field crops like cereals and pulses, as the number of laterals required are high which result in high costs of the system.

The terms *drip irrigation, trickle irrigation* and *micro-irrigation* are synonymous. Drip irrigation is a micro-irrigation method in which the rate of water application is very low and without any pressure, i.e., drop by drop. While drip irrigation and trickle irrigation have identical meaning, the term micro-irrigation encompasses other low pressure water application methods like microsprinkler, bubbler and spray systems in which water is emitted in the form of small sprinkling, tiny streams or miniature sprays through fixed diffusers or applicators, localizing the water on the soil in separate spots (described in Chapter 12), as well as indigenous methods of micro-level application of water to plants. The term *localized irrigation* is sometimes used in place of drip/trickle/micro-irrigation, as the systems cause wetting which is localized to the plant root zone. The devices for water application are in the form of emitters, strip tubings, jets or mini-sprinklers. The service pressure of the system is of the order of 1 bar or less. Fertilizers and other chemicals can be applied efficiently to individual plants or a group of plants, using the drip system. Although the terms drip irrigation and trickle irrigation are used interchangeably, the term drip irrigation is mostly used in this book.

Drip irrigation is based on the basic concept of irrigating only the root zone of the crop, rather than the entire land surface on which the crop is grown. Water flows from the emission points through the soil by capillarity and gravity. The soil moisture content of the crop root zone is maintained at near optimum levels to facilitate optimum crop growth and production. This will involve frequent applications of small quantities of water. Irrigating only a part of the land surface minimizes evaporation losses, reduces weed growth—especially in widely spaced crops and minimizes interruption of cultural operations like inter-cultivation and harvesting.

Major advantages of drip irrigation systems. The major advantages of drip irrigation, as compared to other methods of irrigation include the following: (*i*) *Higher crop yields.* Studies on crop performance with

Fig. 13.4. Young grapevine plants irrigated by drip system at the Research Farm of Water Technology Centre. IARI, New Delhi (1975).

Fig. 13.5. A coconut garden under drip irrigation, using microtubes.

Fig. 13.6. A papaya orchard under drip irrigation.
[***Source:*** Jain Irrigation Systems Ltd. (2005 b)]

drip irrigation, in comparison to other irrigation methods have consistently shown yield increases ranging from 20 to 50% or more in drip irrigated crops, namely, fruit crops (banana, grapes, citrus, pomegranate, papaya, pineapple, watermelon and most other fruits), vegetable crops (cabbage, cauliflower, okra, tomato, potato, onion, chilli, radish, brinjal, bottle gourd, French bean and most other vegetable crops grown in the tropical, sub-tropical and temperate regions) and commercial crops like cotton and sugarcane. The increase in crop yield is mainly due to the maintenance of optimum soil moisture in the crop root zone throughout the growing period of the crop with the better control of irrigation water application in the drip system. (*ii*) *Improved quality of the harvested produce of the crop.* Drip irrigation results in improved quality of crop produce due to the maintenance of optimum soil moisture conditions in the root zone throughout the growing season, resulting in uniform crop growth. In addition, damage and loss due to water contact with foliage are eliminated, resulting in better quality of produce. (*iii*) *Savings in irrigation water.* With the elimination of water application losses such as deep percolation, runoff and evaporation and reduction in the volume of the soil which is wetted (limited to the crop root zone), there is substantial saving in the amount of water required for irrigation, as compared to other methods of irrigation. Loss of water from the land surface which is not occupied by the crop is eliminated. As compared to sprinkler irrigation, the loss of water due to evaporation from the plant foliage is eliminated. Further, the losses in water conveyance such as seepage is eliminated with the adoption of drip method of water application, as compared to the unlined open channel water conveyance system commonly used in surface irrigation methods. The savings in the quantity of irrigation water resulting from the adoption of drip

Fig. 13.7 View of an onion field under drip irrigation.
(*Source:* Jain Irrigation Systems Ltd.)

irrigation usually range from 20 to 50% or more, when compared to surface methods of water application and to a lesser extent when compared to sprinkler irrigation. Further, the irrigation water requirement of crops under drip irrigation is substantially low during the initial stages of crop growth. The saving in water by introducing large scale application of drip irrigation in a project area increases the irrigation command area of the project and provides opportunity for diversification of crops and introduction of high value crops in the cropping system. (*iv*) *Increased efficiency in fertilizer use.* A drip irrigation system is an effective means of applying fertilizers and other plant nutrients into the crop root zone where they are needed. The efficiency in fertilizer application increases greatly when fertilizers and nutrients are applied with the drip system, as compared to other methods of water application. (*v*) *Reduced energy consumption.* Drip irrigation results in reduced energy requirement in pumping, compared to sprinkler irrigation, due to the low operating pressure required by the drip system. Further, the quantity of water required for irrigation is low under the drip system, which further reduces the energy requirement.

Fig. 13.8 Drip irrigation of cabbage.
(***Source:*** Jain Irrigation Systems Ltd.)

(*vi*) *Tolerance to windy atmospheric conditions.* Wind has no adverse influence on drip irrigation, as compared to the sprinkler system. In the drip system, the laterals with the emitters are laid at the land surface and only the point of water application is wetted which is not influenced by the wind. The evaporation loss from the land surface is also negligible, as the wetted area is under the plant foliage. (*vii*) *Reduced labour costs.* In irrigated agriculture using the drip system, the labour requirements are low due to reduced cost of field preparation, elimination of fertilizer application as a separate operation, low requirement of weeding and fewer harvesting rounds due to more uniform ripening of the crop. (*viii*) *Improved disease and pest control.* Plant diseases and pests are lower in drip irrigated crops, as compared to other methods of water application. Bacteria, fungi and other organisms are reduced as the 'above ground' plant parts remain dry. As the application of water is limited only to the crop root zone, the areas surrounding the plant remain dry and hence the weed growth in the crop field is greatly reduced. With little or no weed growth, the intensity of pest and disease incidence is reduced. With the elimination of surface runoff resulting from drip irrigation, the spread of disease causing organisms by water movement is reduced. (*ix*) *Feasibility of irrigating undulating terrain and sloppy lands.* Hilly and undulating terrain cannot be irrigated with surface irrigation methods without extensive land levelling operations. Drip and sprinkler methods can be designed to operate efficiently on almost any topography without incurring the expenditure in land levelling. Drip irrigation can be designed to suit changing field gradients. (*x*) *Suitability for problem soils.* In heavy clay soils, with low infiltration rates, low rates of water application can be selected with the drip system to match the soil characteristics. On the other hand, in sandy soils with high infiltration rates, water can be applied in small quantities frequently, avoiding deep percolation losses. (*xi*) *Improved tolerance to salinity.* With the adoption of drip irrigation systems, it is possible to use more saline water for irrigation, as the root zone soil column is kept at higher level of moisture content, resulting in reduced soil moisture stress. The contact of water with the plant is minimized, compared to other methods of irrigation. In general, drip irrigation will reduce the sensitivity of most crops to saline soils and waters, due to its ability to maintain low water tension in the root zone. The frequent application of water with drip irrigation reduces the concentration of salts in the root zone by moving salts away from the root zone to the edges (perimeter) of the wetted bulb formed by the point source of water application in the drip system (Fig. 13.2). This process, called *micro-leaching,* prevents the harmful combination of high soil salinity and low moisture availability. (*xii*) *Promotes congenial soil physical conditions in the root zone.* The maintenance of soil moisture at nearly constant and optimum levels by renewing the water supply to the root zone nearly at the same rate as it is used by the plant results in low soil suction and facilitates water and nutrient uptake by the plant and high soil hydraulic conductivity. The soil, on the other hand, is never saturated in a properly managed drip irrigation system, and adequate aeration is maintained throughout the growing period of the crop. The low rate of water application makes drip irrigation suitable even in soils with low infiltration characteristics. In comparison to sprinkler irrigation, where the losses due to over-irrigation and irrigating areas not occupied by crops, including field edge losses, (may be as high as 20% or more), these types of losses are almost eliminated in drip irrigation. Further, no water is lost due to the transpiration of weeds, as irrigation is limited to the crop. The drip system facilitates better control of the irrigation water supply, as compared to the sprinkler irrigation system. There is no formation of soil crust in drip irrigation. (*xiii*) *Promotes better uniformity in irrigation water application.* The large number of irrigation points (emitters) in a drip system provide better uniformity of application of water. Low discharges and low pressure heads in the water distribution network enable the use of smaller pipes of lower pressure ratings at reduced costs. Fertilizers, pesticides and other chemicals may be injected into the system and applied in small quantities, as needed, with irrigation water. (*xiv*) *Easy in operation.* Drip irrigation system is comparatively easy to operate. Irrigation can be continued throughout the day and the night, regardless of wind, daytime temperature or cultural practices. (*xv*) *Facilitates automation.* Automation could be easily incorporated into the drip irrigation system. (*xvi*) *Adapted to irrigate crops in green houses.* Drip irrigation system is well adapted to irrigate crops grown in covered green houses with no wetting of the walls or the cover.

Problems associated with drip irrigation systems. In spite of the major advantages described earlier, there

Fig. 13.9. Capsicum under drip irrigation.
[***Source:*** EPC Industries Ltd. (2005).]

are several problems associated with drip irrigation systems:

(*i*) *Emitter clogging.* A major problem associated with drip irrigation systems is the clogging of emitters, unless preventive measures are taken. Emitter outlets are very small and can become clogged easily by suspended materials (sand and silt), precipitated dissolved salts (mainly carbonates), rust and other iron oxides and organic matter (including plant roots, algae and other minute animals). Clogging can reduce emission rates and cause poor uniformity of water application. Prolonged clogging can cause severe damage to the crop. Clogging also increases the maintenance costs, as it becomes necessary to check, repair or replace clogged emitters and other components. Chemical treatment and proper filtration of water can prevent or correct emitter clogging. The first requirement is to select emitter devices which may require less maintenance. Well-designed filtering systems, when properly maintained, can solve the problem caused by suspended material (sand and silt) and in some cases the organic matter. An effective means of dissolving carbonates is by means of dilute acids (mostly HNO_3 or HCl). However, it will require an interruption in irrigation water application. Flushing by water or air under pressure is feasible if the pipes and joints can withstand the pressure. *Pulse irrigation,* namely, irrigating at a higher rate for a very short time, is effective.

(*ii*) *Restricted root development of crops.* Drip irrigation normally wets only a part of the root zone Hence, the root distribution is almost limited to the moist zone. Many factors, involving soil and plant characteristics, management practices (amount and frequency of water application) and the design of the irrigation system (number of emitters per plant, placement and discharge rate of emitters) influence the root development of crops. The concentrated distribution of roots may reduce the plants ability to withstand strong winds. Further, the ability of the plant to withstand drought, resulting from any breakdown in the irrigation system, is greatly reduced as water of the wetted zone gets depleted soon and the surrounding region is dry. Careful planning and operation of the irrigation system and prevention of breakdowns will reduce the magnitude of the problem.

(*iii*) *Salt accumulation at the root zone periphery.* Where high salinity water is used for irrigation in arid regions, salts tend to accumulate towards the periphery of the wetted zone and the interface between the irrigated and non-irrigated zones in the soil. As the root zone is kept constantly at high moisture level, there may not be any major impact on the crop, but in the next growing season these salts, if not leached away, may damage the succeeding crop if planted on the interface having the salt patch. In regions where drip irrigation is practiced in the dry season with rainfall exceeding about 300 mm in the following rainy season, the salts will generally be leached. Like in other methods of irrigation, good natural or artificial drainage is needed in drip irrigated areas as well. Otherwise, artificial leaching is to be provided once every one or two years.

Fig. 13.10. Layout of drip laterals in a large field to irrigate row crops grown on ridges. Note the sugarcane crop in the background. [*Courtesy:* EPC Industries Ltd. (2005).]

(*iv*) *Damage from rodents and other animals.* There are several kinds of burrowing animals which can cause damage to surface or buried polyethylene laterals. They include rats, mice, squirrels, dogs and rabbits. When present in large numbers, these animals can cause heavy damage to drip irrigation systems by chewing holes in the lateral lines. Use of repellents to keep the animals away and baiting or trapping them are possible control measures. Rats and squirrels can be controlled by adopting subsurface drip system. Rats generally never cross a wet surface. Keeping the system moist is a possible remedy to control it. A chemical repellent which tastes or smells bad to the animal can be injected through the system or laid down with the laterals during installation. There are a number of chemicals available which are obnoxious to animals, including anhydrous or aqua ammonia and a number of insecticides which can be injected into the drip system. Rodent damage can also be prevented by the use of polyvinyle chloride (PVC) laterals. Ants and beetles can damage thin-walled polyethylene tubing. Ant damage typically takes the form of holes chewed through the sides of strip tubing and enlargement of orifices in strip tubing. Some types of insects will build a cocoon inside emitter outlets in water distribution tubing. Cocoons are built when the system is shut down for a period of time. These insects are to be controlled only during prolonged shut down periods. Ant damage can be successfully controlled with chlorinated hydrocarbon insecticides. However, these chemicals are usually toxic and persist in the environment and should be used with great care. A common recommendation is to use lateral tubes with sufficient wall thickness (not less than 15 mils (0.38 mm) to reduce damage by ants and insects.

(*v*) *High cost of drip irrigation systems.* The cost of drip irrigation system is high when compared to surface or portable sprinkler irrigation systems. (Many studies have shown that the cost *of solid state* sprinkler systems and the drip system are nearly same). Drip irrigation systems are expensive because of their requirements of large quantities of piping and filtration equipment to clean the water. However, the cost of drip irrigation systems varies considerably depending on the crop and terrain. Steep terrain may require several pressure regulators in the system. In many situations, the benefits of the drip system will usually overweigh the cost of the system, when compared to other methods of irrigation. Selecting widely spaced crops of high market value is necessary to increase the benefit-cost ratio in drip irrigation. Under average field conditions, the major components of the total cost of the system lie in the number of laterals/and emitters of the system. Evidently, both these items will be low in orchards and other widely spaced row crops.

Soil-Plant-Water Considerations in Drip Irrigation

Drip irrigation systems are designed and managed to deliver light, frequent applications of water that wet only a section of the soil. The irrigation requirements described in Chapter 10 of this book are to be adjusted for drip irrigation. Under the surface and sprinkler methods of water application, the irrigation needs for depth, frequency and salinity controls are based on maximum moisture storage in the root zone. However, to meet the objective of drip irrigation, water application is based on moisture replacement in a small area of the soil. This requires determining the wetted area, wetting pattern, and horizontal and vertical soil moisture movements in the soil. The values of water requirement, consumptive use, and frequency of irrigation are adjusted accordingly. Most soils have layers of various densities. On sloping fields, the wetting pattern distorts in favour of the downslope direction. Hence, the most reliable method of determining the wetted area in a drip irrigation system is to conduct field tests in which test emitters are operated at a few representative sites in the field and the wetting patterns are observed.

Wetting pattern in a drip irrigation system is influenced mainly by the soil characteristics, discharge rate of the emitter, and the duration of water application. Figure 13.11 illustrates the typical wetting patterns in heavy, medium and light soils. Koul (1979) studied the wetting pattern resulting from point source irrigation in sandy loam soils. He observed that the soil moisture in the wetted zone manifested itself by a rapid increase in the moisture content in the soil layer close to the land surface. This zone was identified to extend to about 15 cm deep and 20 cm diameter. It was observed that the moisture content stabilized to about the saturation moisture content of the soil and continued at this level till the remaining part of the application time. It was further observed that the soil moisture profile at any instant during irrigation was a function of the water application rate. The ultimate profile, however, was eventually a function of the total volume of water applied. Field studies on vertically downward movement of water in sandy loam soils have shown that on-and-off trickling of 1 to 2 hours, if applied after an initial continuous trickling, would reduce the downward movement of water and the moisture availability to the crop would increase (based on review by Rema Devi, 1980).

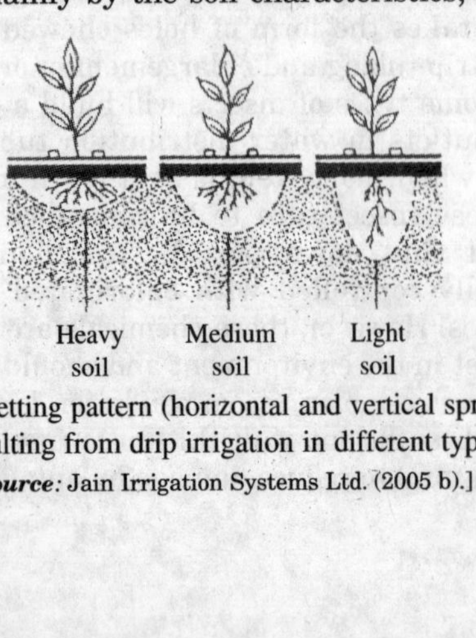

Fig. 13.11. Wetting pattern (horizontal and vertical spread of soil moisture) resulting from drip irrigation in different types of soils. [***Source:*** Jain Irrigation Systems Ltd. (2005 b).]

Plant response is about the same to drip irrigation as to other methods of water application. Even mature orchards that have been irrigated by surface or sprinkler methods of irrigation can be converted to drip irrigation. The root system of most trees can adapt to the smaller wetted area in a few months. Optimum soil moisture levels are easily maintained with a well designed drip irrigation system. Therefore, drip irrigation systems are often run daily, on alternate days, or twice weekly, depending on crop needs and agronomic practices. It is important to wet a relatively large part of the potential root system to ensure some degree of moisture reserve to cope with a temporary failure of the drip irrigation system. It is also important to have a large enough volume of moist soil to promote root extension and water uptake.

INDIGENOUS MICRO-IRRIGATION DEVICES

Pitcher Method

An indigenous method of micro-irrigation is the use of earthen pots equipped with cotton wicks or plastic tubes. Earthen pots of capacity ranging from 15 to 20 litres are commonly used. The pots are placed in the ring basin formed around the plant. When cotton wicks are used, each pot usually serves one plant. For use with wicks, a 1 mm diameter hole is made about 2 cm above the bottom of the pot into which is inserted the wick which permits the slow release of water from the pot to the crop root zone. The pots are filled periodically when they get dry. The pots are covered with a lid to prevent evaporation of water in the pot. The basin around the plant is covered with straw, dry leaves, coconut husks or plastic sheets to prevent evaporation from the land surface. Fertilizers can be added to the water in the pot.

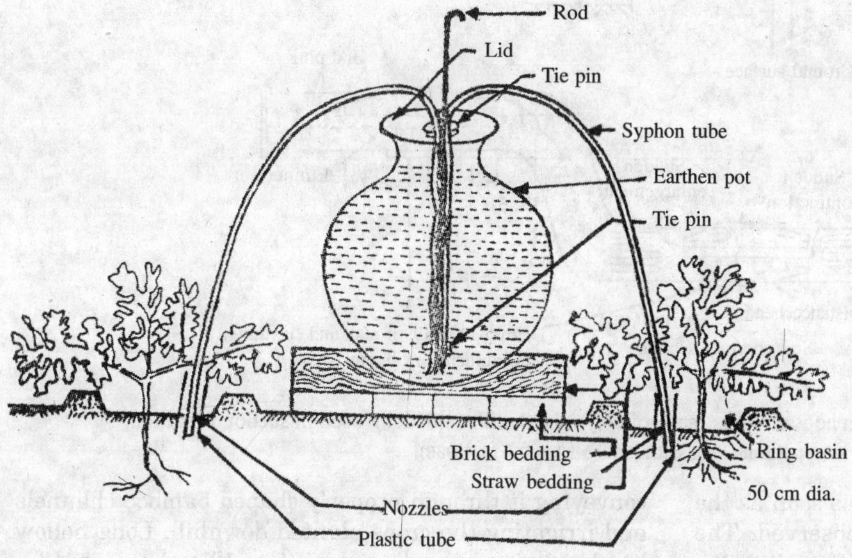

Fig. 13.12. Schematic sketch illustrating the use of earthen pot for irrigating young watermelon crops.

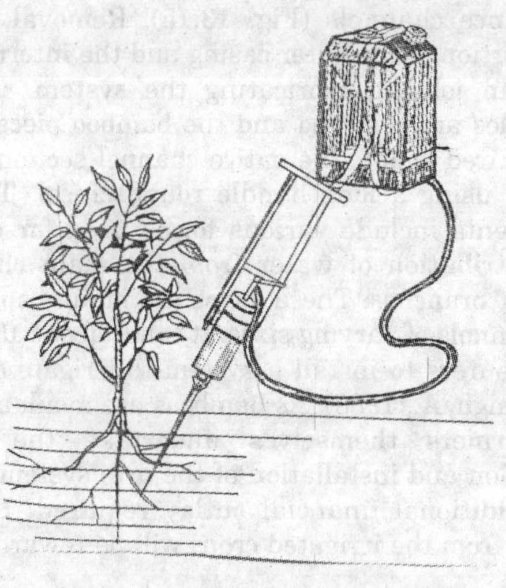

Fig. 13.13. A sub-soil injector for crop life-saving irrigation in dry periods in regions having acute water scarcity. (Adapted from Vamadevan, V.K., 1985).

A modified form of pitcher irrigation, adapted to widely spaced vegetable crops and vines, consists of pitchers provided with plastic siphon tubes of 4 mm diameter. The tubes are provided with nozzles, the ends which dip in the basin around each plant (Fig. 13.12). The nozzle is made of hard plastic tube about 2 cm long and having the same diameter (4 mm) as that of the siphon tube. At the end of the tube, a rectangular hole is made, the opening of which could be adjusted as per the discharge requirement of the nozzle. The adjustment is done with the help of a 1 cm long sliding tube which also connects the nozzle tube to deliver about 2 to 2.5 litres of water in 24 hours, as per the irrigation requirement. The pitcher is filled daily while irrigating vegetable crops. The nozzle could be sucked by mouth to draw water for siphoning.

Each set could irrigate 4 plants at a time. The plastic tubes are inserted about 5 cm below the ground level beside the plant it irrigates. One end of the siphon tube is dipped into the earthen pot and held in position with pins on a rod which is held in a vertical position. The other end with nozzle is placed in a hard tube in the ring basin as soon as the nozzle starts functioning. The pitcher is placed on a platform which is about 25 cm above the ground level in order to develop hydrostatic pressure. The platform is made of bricks over which is laid a straw bedding.

Sub-soil Injector

Sub-soil injector, also called root injector, is an equipment used for sub-soil irrigation by injecting water directly to the root zone of the plants at a depth of 15 to 20 cm from the ground surface (Fig. 13.13). The equipment consists of a hollow tube with a piston arrangement connected to a container of about 15 to 20 litres capacity. A bare pipe connects the tube with the water container. A needle is provided to inject water at the lower end of the tube. The equipment is used by inserting the needle into the soil in the root spread region. The irrigator moves from plant to plant and takes about 30 seconds to irrigate one plant at the desired depth with a predetermined quantity of water. The method is specially suited in porous soils on sloppy land for establishing plantation crops and trees. It is, however, not suitable in clay soils. It is also labour intensive.

Suction Irrigation

Suction irrigation works on the principle of utilizing the energy present in the soil in the form of soil moisture tension, or soil moisture suction, for irrigating the crop root zone. Figure 13.14 illustrates the layout of a suction irrigation system. The principal components of the system are the emitters made of clay soil mixed with 30 to 120 % fresh cattle dung on weight basis. They can be fabricated locally. The water emission rates of emitters depend on the relative proportion of clay soil and cattle dung. The emitters are about 2 cm outside diameter and 3 mm diameter inside.

The soil water suction at field capacity ranges from 1/10 to 1/3 atmosphere. A negative pressure head of more than 1 m in magnitude exists in the soil or plant root zone, depending upon the type of soil, plant and the moisture content at which the plant is to be irrigated.

The emitters are embedded below the root zone in order to utilise capillary water. They are connected to a water tank by means of PVC pipes. The air in the pipe and emitters is removed by filling water in the system.

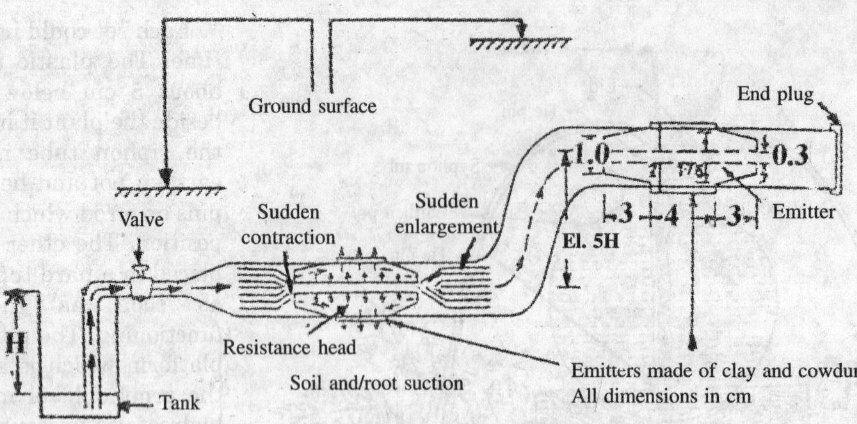

Fig. 13.14. Schematic diagram of indigenous emitters made of clay mixed with cattledung used in suction irrigation.
[*Source:* Adapted from Yadav, R.K. and Michael, A.M. (1980)]

The water tank is lowered in the ground as soon as the hydraulic connection of the emitter is observed. The water is allowed to pass through the emitters into the crop root zone. The static head 'H', is the height of the water column which is required to cause the flow in the mains, submains, laterals and emitters of the system. This head is required only at the initial stage of installation for a few minutes. By regulating the static head 'H', which is negative in magnitude, the rate of emission of water from the emitter can be regulated to nearly match the evapotranspiration rate of the plant. The number of emitters in the system will depend on the size and the spacing of the crops. The joints of the emitters and pipes are made air-light. As the designed hydraulic resistance or 'leakage' of the emitters is less than the pressure head of water at field capacity, the system does not require any external energy for its operation. Water of higher salt content can also be used, because of the high level of soil moisture which is maintained constantly in the root zone throughout the growing period of the crop. The salt is constantly being moved away from plant roots to the upper parts of the wetted soil volume. At places where water is scarce and the entire field is not to be wetted, suction irrigation provides good scope for utilization.

Bamboo Drip Irrigation System

Bamboo drip irrigation is an indigenous technique of irrigation evolved and used by tribal farmers of the Jaintia Hills District of Meghalaya state in the North-Eastern Region of India, since the past over 200 years (Singh, A., 1988). The system is used on steep slopes on rugged terrain for irrigating betel leaf and black pepper crops planted with areca nut. (It could be used to irrigate other crops as well). Natural perennial streams are the main source of water for domestic use and irrigation in the hills of the Region. Irrigation water conveyance through earth channels down steep slopes is not feasible. Using bamboo, which is locally available, the tribal farmers have gained skill and expertise in diverting part of the flow of the hillside stream, conveying it through properly shaped bamboo channels and irrigating the crops planted downhill. Long hollow bamboos of varying diameters (5 to 10 cm) are used for making channels. Indigenous tools have been developed to cut and shape the bamboo to serve as efficient water conveyance channels (Fig. 13.15). Removal of about 1/3rd portion of the outer casing and the internodes are the main jobs in fabricating the system. Once the internodes are removed and the bamboo piece is given the required shape, the entire channel section is made smooth, using a long handle round chisel. The other components include various forms used for diversion and distribution of water from the main channel to different branches. These forms are mainly small pipes and channels of varying sizes. It takes about 15 days for two labourers to install a system to irrigate one ha of land (Singh, A., 1988). As bamboos are available locally and farmers themselves undertake the jobs of fabrication and installation of the drip system, there is little additional financial outlay required. Enhanced returns from the irrigated crops will be rewarding.

Fig. 13.15. View of a network of bamboo channels in a bamboo drip irrigation system.
[*Source:* Singh, A., ICAR Res. Complex for NEH Region (1988).]

The water conveyance system is made of a multi-stage network of bamboo channel sections, properly shaped and jointed (Fig. 13.16). The channel sections convey water efficiently at the rate of 18 to 20 litres/minute or more, over a distance of several hundred metres. Elevation difference between the water source and the irrigated area may be as high as 100 metres. Water is carried to the irrigated plot site and distributed into branches with the use of different forms of bamboo pipes/channels. The unique feature of the system is diversion of water without leakage and the control of water flow in any branch as per requirement. The farmers have gained the necessary skill to suitably adjust the intake pipe positions. Depending on the points where water is to be applied (plant locations), water is diverted through the network of bamboo channels. Reduced channel section units are used at the last stage of water application. The last channel section enables the application of water drop by drop (Fig. 13.16). The system works round the clock, once commissioned. Once installed, most components of the system will last 2 to 3 years or more, except for minor items which may require more frequent replacement (Singh, A., 1988). Bamboo drip irrigation, though location specific, is a living example of the ingenuity, self-reliance and cooperation prevailing in the tribal community.

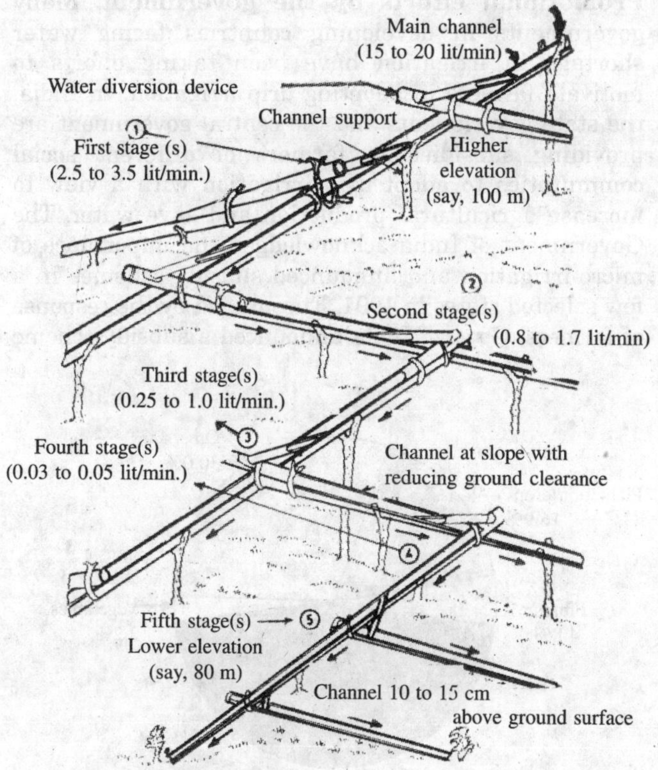

Fig. 13.16. Schematic sketch illustrating the distribution of water diverted from a hillside stream in bamboo drip irrigation system.
[***Source:*** Singh, A., ICAR Res. Com. for NEH Region (1988).]

DEVLOPMENT OF DRIP IRRIGATION: A HISTORICAL PERSPECTIVE

The use of drip irrigation can be traced back to the ancient custom in some parts of India of keeping a *tulsi* plant in the courtyard and irrigating it with water stored in an earthen pot which is hung above the plant. A minute hole is made at the bottom of the pot through which water drips (quoted from Rajput, T.B.S. and Patel, N., 2001). Pitcher irrigation, using earthen pots, is known to have been in practice to irrigate fruit trees and spices in some parts of India and in Egypt since ancient times. In India, micro-irrigation was practiced through indigenous methods such as bamboo pipes, perforated clay pipes and pitcher/porous cup irrigation. In Maharashtra, perforated earthenware pipes were used and their efficiency and benefit-cost ratio have been elaborated for popularizing them. Earthen pitchers and porous cups have also been used for growing vegetable crops in Rajasthan. The technique envisages embedding of earthen cups of 500 ml capacity at the site of seedlings. The cups are filled with water at an interval of 4 to 5 days (ICAR, 2003).

Internationally, micro-irrigation was developed originally as a sub-irrigation system and the basic idea underlying micro-irrigation can be traced back to experiments in Germany. Early attempts to develop a subsurface irrigation system were made in 1860 by German researchers, who pumped irrigation water into existing underground drainage systems. In the nineteen twenties and thirties, various types of porous pipes were used by a number of researchers. The first work in micro-irrigation in the United States of America was a study carried out by House in Colorado in 1913. An important development reported from Germany way back in 1920 was the introduction of perforated pipe micro-irrigation.

During the early 1940s, Symcha Blass, an Israeli Engineer, observed that a big tree near a leaking pipe exhibited a more vigorous growth than other trees in the area, which were not reached by water from tap. This led him to the concept of an irrigation system that would apply water in small quantity, literally drop-by-drop. In Israel, the first extensive research on drip irrigation was conducted in the Arava and Nageve deserts, where adverse conditions of climate, very sandy alkaline soils and saline water had produced good results on crops grown with drip irrigation.

The breakthrough which facilitated easy adoption and large scale manufacture of the drip irrigation system was when polyethylene was produced in a British laboratory in 1935. Polyethylene, or PE, is produced when the monomer ethylene, a component of natural gas and other fossil fuels, is subjected to heat and pressure. In the early years of its development, only the low density form of polyethylene was known. Low Density Polyethylene (LDPE), produced at relatively high pressures (up to 3500 kg/cm^2), was found to possess resistance to most chemicals and extraordinary electrical insulating properties. However, the LDPE pipes were soon found to be cracking due to degradation by UV light from the sun in open fields and there was need for a material with good characteristics to withstand the degradation from UV light. In 1948, the discovery of high density polyethylene (HDPE), produced at low pressures (20 kg/cm^2), broadened the scope of applications for polyethylene. In 1977, the polyethylene industry achieved a significant breakthrough with the development of a low-pressure process for polyethylene production, which lowered the cost and improved the quality of polyethylene resins. The state-of-the-art in plastics development has continued to advance, with the recent introduction of linear low density polyethylene (LLDPE), which exhibits greatly improved strength and stress-crack resistance. Polyethylene is a tough and flexible material, with a wide range of physical properties depending on the size and arrangement of the polyethylene molecules.

Polyethylene manufactures currently produce polyethylene resins which are tailored specifically for use by makers of film (for bags, plastic sheeting), by injection moulders (for emitters) and by makers of extruded products (drip laterals, emitting pipes, porous pipes). When suitable additives, including antioxidants, stabilizers and carbon black, are added to the clear plastic, a durable and economical product for micro-irrigation lateral pipes is produced. The large scale commercial use of micro-irrigation began in the 1960s and early 1970s, Interest in micro-irrigation was most keen in Australia, Southern California, USA, Israel and the Middle East areas which have traditionally suffered from scarcity of irrigation water.

Since 1970, there has been, rapid growth in the adoption of drip irrigation in different countries. During 1997-98, the area covered under drip irrigation in countries having large areas under the system, other than India which is described later, are as follows: United States of America (10,50,000 ha); Spain (230,000 ha); Israel (181,000 ha); South Africa (144,000 ha); France (140,000 ha); Egypt (104,000 ha); Italy (80,000 ha); Jordan (38,300 ha); China (34,000 ha); Cyprus (25,000 ha).

(***Note***: *The figures are based on the report of the National Committee on Plasticulture Application in Horticulture, 2001 listed at the end of this chapter*). There has since been substantial increase in the area under drip irrigation in all the above countries and in other countries, especially in West Asia.

Application of Drip Irrigation in India

In India, drip irrigation was introduced in the early seventies at the agricultural universities and other research institutions. Significant development has taken place in 1980s. The growth of micro-irrigation has gained momentum in recent years. From a mere 1500 ha in 1985, the area under drip irrigation has grown to 462,300 ha in 2003 (ICAR, 2003). These developments have taken place mainly in areas of acute water scarcity. The area-wise distribution of drip irrigation in different states of India are Maharashtra (1,94,000 ha); Andhra Pradesh (59,500 ha); Karnataka (58,500 ha); Tamil Nadu (46,500 ha); Rajasthan (41,500 ha); Gujarat (20,500. ha); Madhya Pradesh (8800 ha); Kerala (8500 ha); Uttar Pradesh (4500 ha); Orissa (3900, ha); Haryana (3400 ha); Punjab (2200 ha); West Bengal (800 ha); Assam (800 ha) and others (8900 ha), in 2003 (contribution by Dr. Ashwani Kumar).

During the year 2004, the area under drip irrigation in India was 540,000 ha (Rajput, T.B.S. and Patel, N., 2005). Figure 13.17 presents the percentage of the area irrigated by the drip system in India. Amongst the horticultural crops, coconut covers 19%, banana 11%, grapes 10%, mango 9.4%, citrus 7.9% and pomegranate 6.2% (Rajput, T.B.S. and Neelam Patel, 2001).

To promote the concept, efforts have been made at research level by Indian Council of Agricultural Research, Agricultural Universities, National Committee on Use of Plastics in Agriculture, Ministry of Water Resources and Drip Manufacturers Association. Ministry of Agriculture and the Ministry of Water Resources of the Government of India and the State governments have sponsored promotional activities for drip irrigation.

Promotional efforts by the government. Many governments in developing countries facing water shortage in irrigation have been taking efforts to motivate farmers in adopting drip irrigation. In India, the state governments and the central government are providing subsidies to farmers of different social communities to adopt drip irrigation with a view to increase agricultural production and save water. The Government of India acknowledged the importance of micro-irrigation and announced subsidy schemes in a few selected states in 1991. Encouraged by the response of farmers, Govt. of India announced a subsidy scheme

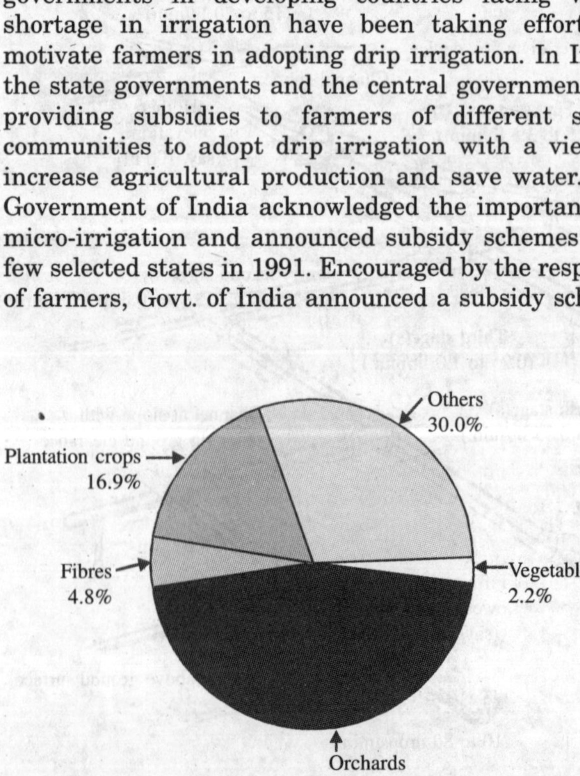

Fig. 13.17. Crop coverage under drip irrigation in India [***Source:*** Rajput, T.B.S. and Neelam Patel (2001).]

of Rs 250 crores during VIII plan (1992-1997). During IXth plan, plasticulture application got a further thrust with an outlay of Rs. 375 crores. Considering the importance of plasticulture in horticulture in particular, the Government of India reconstituted the National Committee on Plastics in Agriculture (NCPA) to National Committee on Plasticulture Applications in Horticulture (NCPH) in May 2001. Till early 2005, the assistance of the Government of India under the "Centrally sponsored scheme" on development of horticulture through "plasticulture interventions" is available for all types of micro-irrigation systems. The assistance covers all farmers growing horticultural crops like fruits, vegetables including potato, onion and other root and tuber crops, spices and medical and aromatic plants. (Rajput, T.B.S and Neelam Patel, 2005).

Surface and Subsurface Drip Irrigation Systems

Based on the installation and location of drip laterals with respect to the depth of installation, drip irrigation systems can be classified as follows:

(*i*) *Surface drip irrigation systems.* Surface systems are the conventional drip irrigation systems. In this the drippers and the lateral lines are laid on the land surface. (*ii*) *Sub-surface drip irrigation systems.* In the sub-surface systems the drippers and the lateral lines are laid below the ground level within the root zone of the crop. Sub-surface installations are sometimes preferred in semi-permanent and permanent installations, particularly for crops in which ratooning is practiced in lawns and landscapes. The depth of installation of sub-surface laterals usually ranges from 20 to 60 cm. The mains and sub-mains are laid underground at depths below 60 cm to avoid any damage due to lateral pressure. Spacing of laterals range from 0.25 to 5 m, depending on the spacing of the crops. Sub-surface drip system was introduced on commercial basis only in mid-1980s in some countries, though they are still not widely used. An advantage of the sub-surface system is that the tubes remain underground for years together and does not interfere with tillage or other cultural operations. Perhaps the greatest advantage is that the system can be automated easily. Sub-surface drip systems can effectively provide water and nutrients for many crops, even in areas with high seasonal rainfall. Other advantages claimed are labour savings, avoiding evaporation losses from the land surface and better water distribution efficiency. The method prevents the formation of soil surface crusts. Multiple cropping and ratoon cropping will be easier because laterals remain in place.

However, in India and many developing countries, surface drip irrigation systems are more prevalent than subsurface systems owing to the ease of installation and also because de-clogging of drip systems with chemicals is not practiced widely. Farmers prefer cleaning of emitters and laterals manually rather than injecting chemicals into the systems. Manual cleaning is not feasible in subsurface drip systems.

Low-Cost Small Scale Drip Irrigation Systems

Apart from the conventional pressurized drip irrigation systems, small scale gravity-operated drip irrigation systems are also available. These systems are suitable for kitchen gardens and very small fields. They are operated from the gravitational head available from an elevated water source like a bucket or a drum. Thin-walled drip tapes and cloth filtration systems reduce the cost of these systems. The *bucket kit* (Fig. 13.18) is popular amongst housewives to raise small kitchen gardens. It has two lateral lines of about 5 m each, fitted with about 40 microtubes and a 20-litre bucket. Depending on the type of crop, about 40–200 vegetable plants which are usually sufficient to provide vegetables for a small family could be grown with a bucket kit. The bucket needs to be filled twice daily. The *drum kit* (Fig. 13.19) is suitable for small commercial vegetable gardens. It is similar to the bucket kit in principle, but is

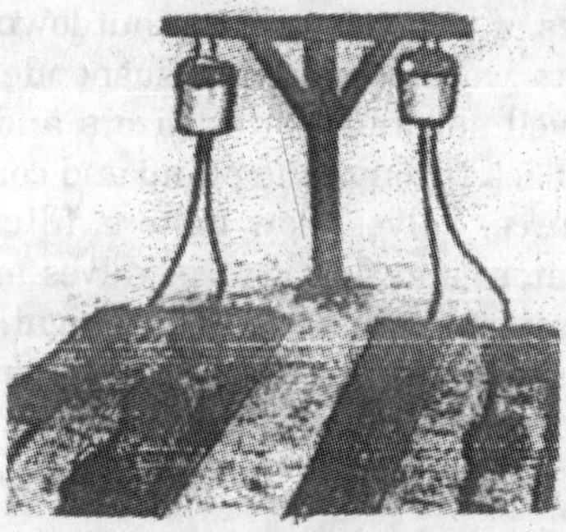

Fig. 13.18. Two-bucket kit balanced on a small beam mounted on a wooden post. Note the two lateral lines fitted to the bottom of each bucket. About 20 microtubes are fitted to each of the laterals.
[***Source:*** Jain Irrigation Systems Ltd. (2005 c).]

Fig. 13.19. A 200-litre capacity drum mounted on a platform for use in a drum kit type drip irrigation system.
[***Source:*** Adapted from Rema Devi, A.N. (1980).]

about 5 times larger in size. The system is designed to be fitted to a drum of water placed at about 1 m height from the ground surface. It can also be fitted with running water from a stream or pipe. The drum is to be filled once or twice daily, depending on the crop and weather conditions. Usually it has 5 lateral lines of about 10 m each fitted with a total of 200 microtubes and can irrigate about 150 sq. m of area. Depending on the type of crop, it can irrigate about 200 – 800 vegetable plants. However, both the bucket kit and the drum kit are labour intensive and are not suitable for common farm fields.

COMPONENTS OF DRIP IRRIGATION SYSTEMS

Water is pumped into drip irrigation systems and flows through valves, filters, mainlines, sub-mains or manifold lines and laterals before it is discharged into the field through point-source or line-source emitters. A check valve is usually installed just downstream of the pump which is open during the normal operation but gets closed to prevent backflow of water towards the pump at times of pump failure or shutdown. The check valve prevents water containing suspended materials and/or dissolved substances (fertilizers and chemicals) from flowing back through the pump and contaminating the water source. Valves, flow meters, filters, pressure gauges, regulators and flow control valves are the usual additional components used in drip irrigation systems. Figure 13.20 shows the layout of a typical drip irrigation system, including major components of the system.

Drippers

The dripper is an emitter fitted to a drip irrigation lateral and intended to emit water in the form of drops or continuous flow at emitter rates not exceeding 15 litres/hr., except during flushing. The emitter controls the flow from the lateral. The emitter inlet is the point at which water enters the emitter. The emitter inlet end ("barb") is usually a sharp point extending backward on an arrow inside the lateral pipe (Fig. 13.29). The emitter outlet is the opening or group of openings in an emitter from which water is emitted to a particular location. There are various types and sizes of drippers, based on different operating principles.

Drippers function as energy dissipaters, reducing the inlet pressure head in the lateral pipeline, which generally range from 0.3 to 1.5 atmospheres to zero atmosphere at the outlet. The pressure loss by an emitter is achieved by small openings, long passageways, vortex chambers, manual adjustment or other mechanical devices. The basis of the design of a dripper is to produce an emitting device which will provide a specified constant discharge, which docs not vary much with pressure changes and does not get clogged easily. Drippers are generally manufactured from poly-propylene or polyethylene material. There are many types and designs of drippers available commercially (Figs. 13.21 to 13.44). Drippers may be classified on the basis of various criteria, the most important of which are the operating pressure, discharge rate, design, flow regime, and cross-section of the flow area.

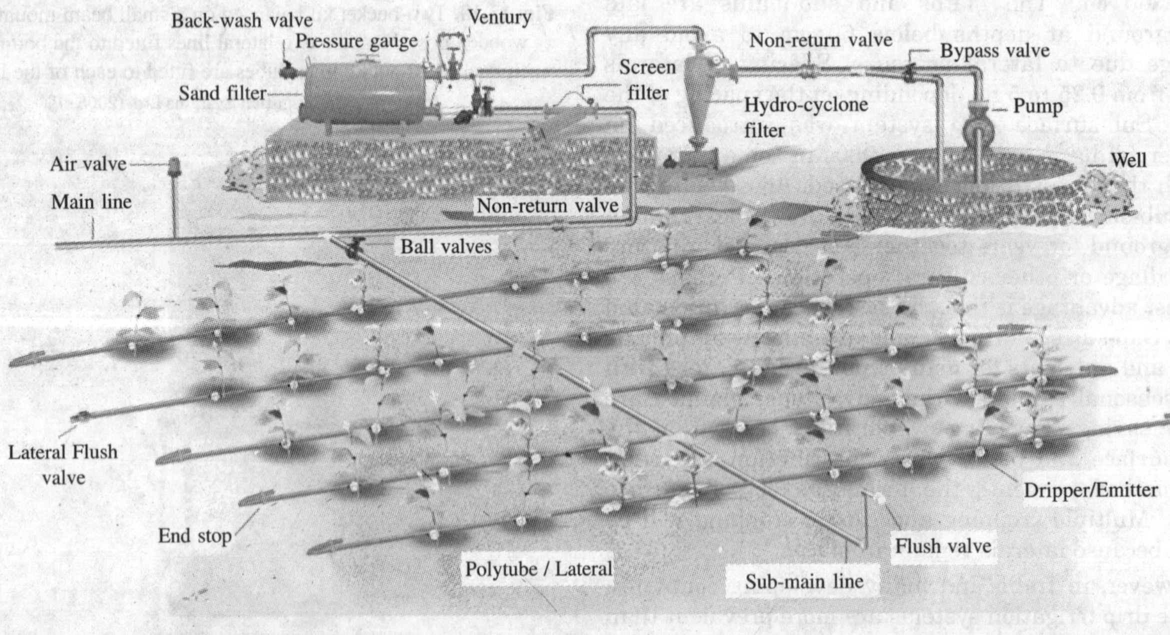

Fig. 13.20. Schematic layout of a typical drip irrigation system. Note the major components of the system including well, pump, fertigation system, filters, main line, sub-main, and laterals with emitters and accessories.
(*Source:* Jain Irrigation Systems Ltd., Jalgaon).

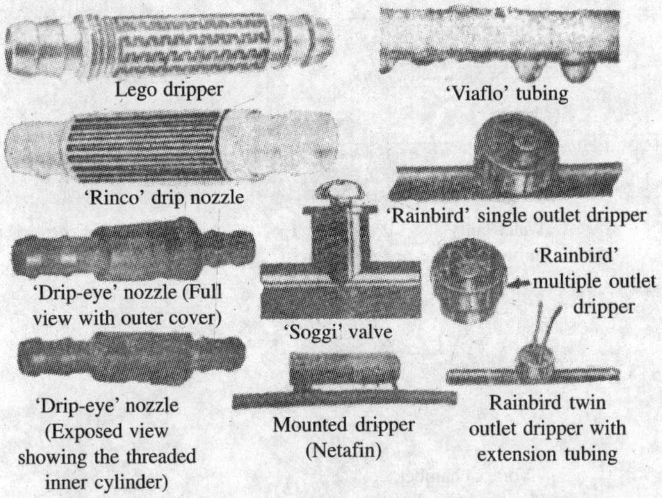

Fig. 13.21. Some of the common types of emitters and tubings used in drip irrigation systems.

(***Courtesy:*** (1) M-Lemelshtrich Ltd., (2) Rinko Irrigation Systems Inc., (3) Drip-eye systems, (4) Aqua-Data Co., and (5) Netafim Irrigation Eqt.)

Operating pressure: Based on the operating pressure in the lateral, drip irrigation systems may be grouped into (*i*) *Low pressure* (below 0.8 m), (*ii*) *Medium pressure* (2 to 5 m), and (*iii*) *High pressure* (8 to 15 m).

Path cross-section: Based on the relative sensitivity to clogging, drip irrigation systems may be grouped into (*i*) *Low* (below 0.8 mm), (*ii*) *Medium* (0.8 to 1.5 mm), and (*iii*) *Wide* (above 1.5 mm).

Discharge rates: Based on the discharge rate of the emitters, drip irrigation systems are grouped as (*i*) *Low* (4 litres/hr), (*ii*) *Medium* (4 to 10 litres/hr) and (*iii*) *High* (15 litres/hr or more).

The major types of emitters and their definitions, based on American Society of Agricultural Engineers (ASAE) are as follows: (*i*) *Emitter:* Small micro-irrigation devices designed to dissipate pressure and discharge a small uniform flow in drips of water at a constant discharge, which does not vary significantly because of minor differences in pressure head. (*ii*) *Compensating emitter* (Fig. 13.22). Designed to discharge water at a constant rate over a wide range of operating pressures. (*iii*) *Continuous flushing emitter* (Fig. 13.23): Designed to continuously permit passage of large solid particles while operating at a drip flow, thus reducing filter fineness requirements. (*iv*) *Flushing emitter* (Fig. 13.24). Water is discharged from closely spaced perforations, emitters, or porous wall along the tubing. (*v*) *Long path emitters* (Fig. 13.25). Employs a long capillary-sized tube or channel to dissipate pressure. (*vi*) *Multi-outlet emitter* (Fig. 13.26): Supplies

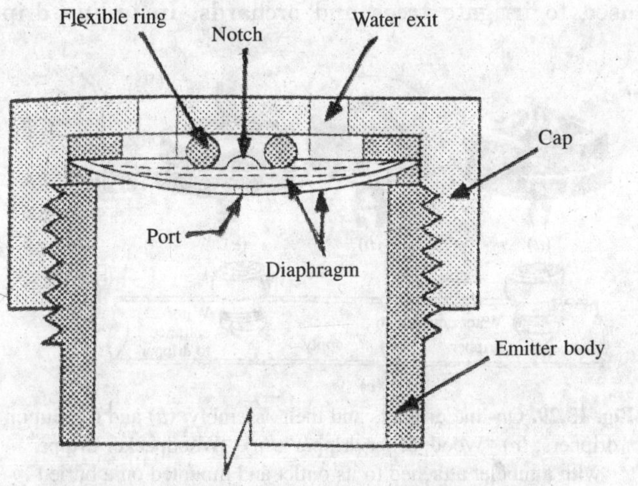

Fig. 13.22 Sectional view of a flushing type emitter.

(***Note:*** The diaphragm is shown in relaxed position. The dotted line shows the diaphragm in operating position.)

[***Source:*** United States Soil Conservation Service. (1984)]

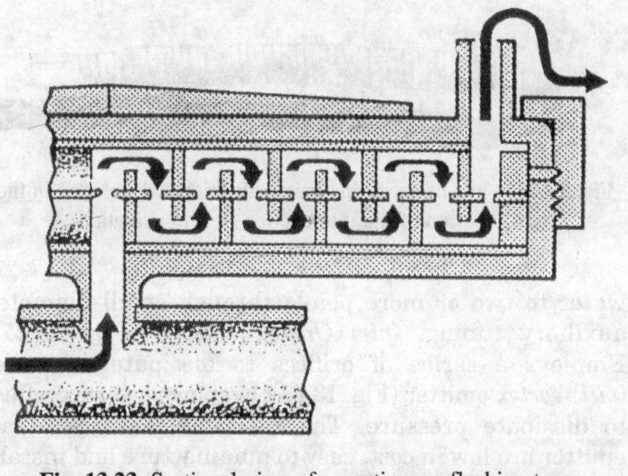

Fig. 13.23. Sectional view of a continuous flushing type emitter showing the flow path of water from the lateral pipe through the emitter and the emitter exit.

[***Source:*** United States Soil Convservation Service (1984).]

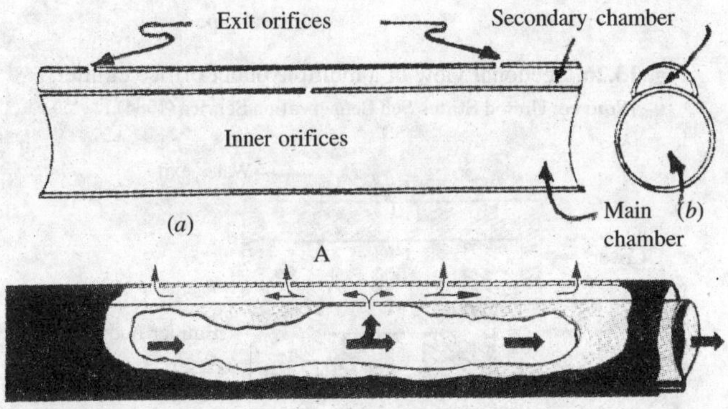

Fig. 13.24. Sectional views of double/twin wall emitter laterals: (A) Views of a twin wall lateral pipe showng emission points: (*a*) Longitudinal view, (*b*) Cross-section.

(***Source:*** United States Soil Conservation Service (1984).

(B) Partially exposed view of a double wall pipe used in drip irrigation

(***Note:*** The path of flow of water in the inner and outer tubes).

(***Source:*** Chapin Watermetics Inc., 368, N. Colorado Ave., Watertown, New York-13601.)

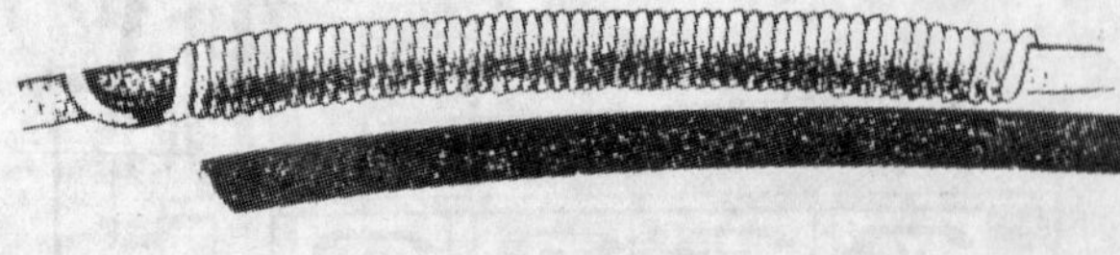

Fig. 13.25. View of an in-line long path dripper developed in the early period of the introduction of drip irrigation.
[***Source:*** Sapir, E. and Yagev, E. (1995).]

water to two or more points through small diameter auxiliary tubing. (*vii*) *Orifice emitter* (Fig. 13.27): Employs a series of orifices to dissipate pressure. (*viii*) *Vortex* emitter (Fig. 13.28): Employs a vortex effect to dissipate pressure. The desirable features of an emitter are low in cost, easy to manufacture and install, resistant to clogging, uniformity in discharge and reliable performance. Emitters are located at predetermined spacings on the lateral and are connected by various means. (Fig. 13.29). On the basis of their design, drippers are classified into three types, namely, point source, line source and disc source.

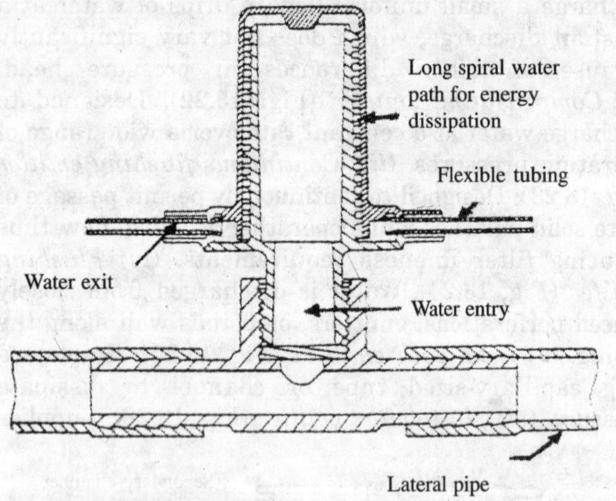

Fig. 13.26. Sectional view of a multiple outlet orifice emitter.
[***Source:*** United States Soil Conservation Service (1984).]

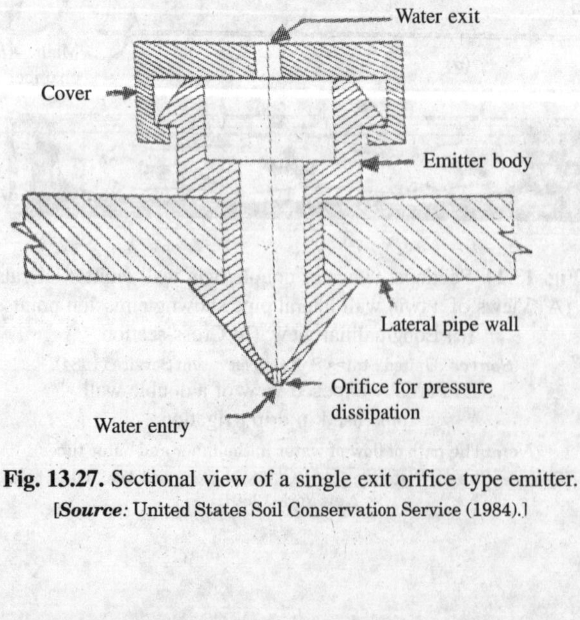

Fig. 13.27. Sectional view of a single exit orifice type emitter.
[***Source:*** United States Soil Conservation Service (1984).]

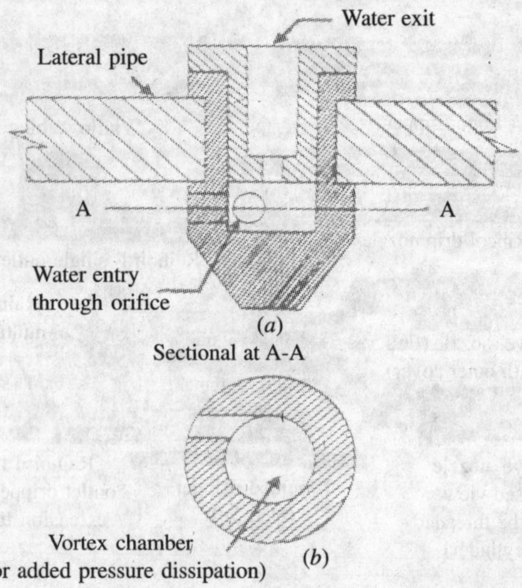

Fig. 13.28. Views of an orifice type vortex emitter: (*a*) Sectional view of emitter mounted on the wall of the lateral, (*b*) Section at A-A showing vortex chamber.
[***Source:*** United States Soil Conservation Service (1984).]

Point Source Drippers

The point on or beneath the ground at which water is discharged from an emitter is called the emission point. Drip irrigation with water discharged from emission points that are individually and widely spaced–usually about 1 m or more is called point source application. Point source drippers (emitters) discharge water from individual or multiple outlets. Most point source drippers are either *on-line* (Figs. 13.26 to 13.29) or *in-line* (Figs. 13.30 to 13.33) emitters. On-line emitters are installed in the wall of the irrigation lateral by punching suitable holes at locations specific to the crop being irrigated (Fig. 13.29). Wetting is confined to the root zone beneath the emitter. On-line emitters are widely used to irrigate trees and orchards. In in-line drip

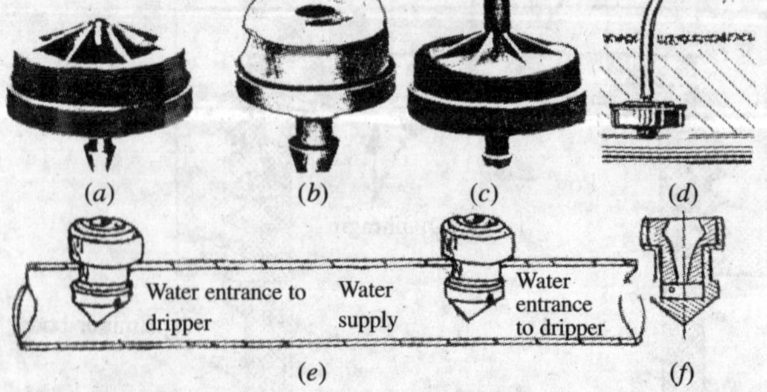

Fig. 13.29. On-line drippers and their assembly: (*a*) and (*b*) Button drippers; (*c*) "Woodpecker dripper". (*d*). "Woodpecker dripper" with a tubelet attached to its outlet and mounted on a buried lateral line in a drip system; (*e*). Longitudinal section of a drip lateral line with vortex type drippers mounted on it; (*f*). Sectional view of a vortex dripper.
[***Source:*** Adapted from Sapir, E. and Yagev, E. (1995).]

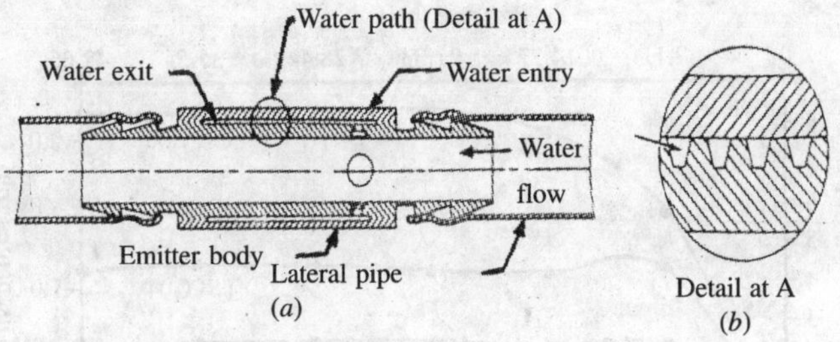

Fig. 13.30. Views of an in-line single exit long path emitter. (*a*) Longitudinal section of emitter fitted to drip lateral (*b*) Enlarged view of water path.
[***Source:*** Adapted from United States Soil Conservation Service (1984).]

Fig. 13.31. An in-line emitter fitted between two lengths of polyethylene tubing (drip lateral). Emitter and fittings feature self-seating compression attachment
(*Source: LOC-CZE Drip Irrigation System*: Reed Irrigation System, 585 Vermon Way, P.O. Box X, El Cajon, CA91022, USA)

irrigation systems the drippers are usually factory installed within the drip lateral. The emitters are installed between two lengths of an irrigation lateral (Figs. 13.30 to 13.33). In-line emitters are used primarily to irrigate small fruits, vegetable crops and other closely spaced crops. The primary difference between an on-line and an in-line emitter is that the entire flow required downstream of the emitter passes through an in-line emitter. The pressure drop in the lateral is usually more with on-line emitters than with in-line emitters, since the 'barbs' of on-line emitters obstruct flow and create additional loss of pressure in the lateral. On the other hand, on-line emitters can be replaced more easily when they are defective or become permanently clogged. Often it is necessary to shut off flow to the lateral and cut the pipe to replace a malfunctioning in-line emitter. The choice amongst the various types of drippers depends on the pressure – discharge relationship of the dripper (Fig. 13.34 and 13.35), type of crop, topography and soil type of the area.

On-line pressure compensating type emitters. Pressure compensating drippers are designed and constructed to obtain nearly uniform discharge from each dripper, in spite of varying operating pressures. A flexible internal diaphragm or disc inside the emitter changes shape at higher pressures to create greater restriction to flow as the pressure rises (Fig. 13.22). As a result, the discharge remains constant over the operating pressure range. At low pressure, such as at system start-up, the internal diaphragm does not change shape, leaving the flow path and/or orifice inside the emitter at maximum size. At this pressure, the emitter is operating in 'flushing' mode. Hence, these drippers are also called as 'self-flushing' type. Pressure compensating type emitters are especially useful in undulating and hilly terrains where larger fields can be irrigated using lesser number of sub-mains and fewer pressure regulating devices. The use of pressure compensating emitters simplifies the design of drip irrigation systems and improves the uniformity of water application, especially in undulating and hilly areas. The disadvantages include higher investment costs and higher operating pressures and hence higher operating costs, compared to non-pressure compensating drippers.

Non-pressure compensating type emitters. Non-pressure compensating emitters do not have the compensation feature explained earlier. Hence, their discharge tends to vary widely with the operating pressure, as compared to pressure compensating emitters. These drippers are less expensive and are widely used in fields with little or no undulations and almost flat terrains.

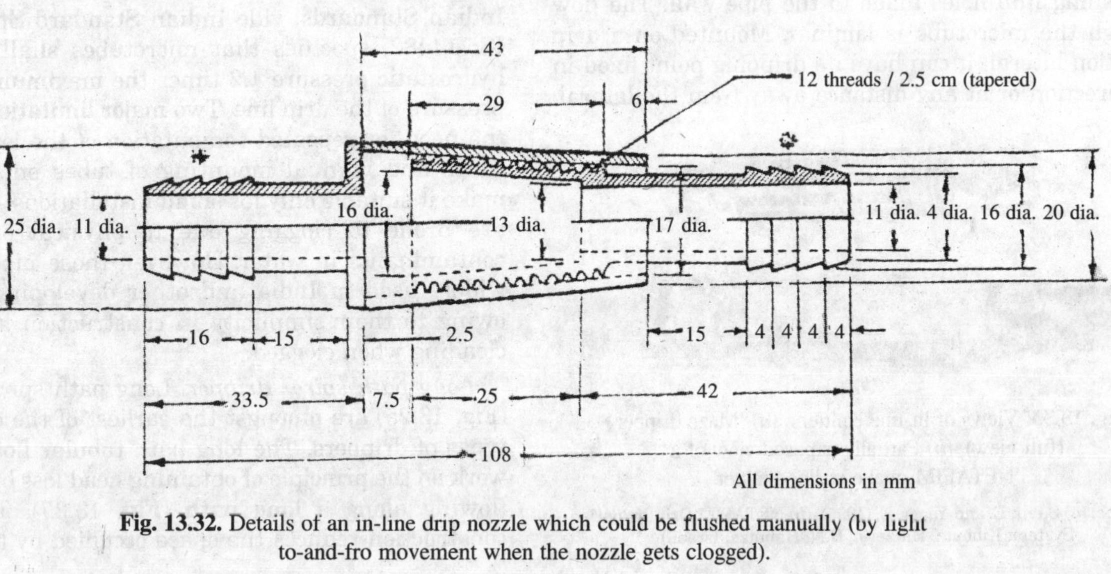

Fig. 13.32. Details of an in-line drip nozzle which could be flushed manually (by light to-and-fro movement when the nozzle gets clogged).

Point source emitters are also classified as *long path (laminar flow), turbulent flow or partially turbulent flow* type emitters:

Long path emitters. In the earlier days of the commercial introduction of drip irrigation, the drippers worked on the principle of high head loss by water flowing along a long path. The friction along the long path reduced the pressure so that at the end of the path there was a low laminar flow of about 2 litres/hr. Long path drippers are of the following types:

(*i*) Micro-tubes (tube type laminar dripper).

(*ii*) Spiral long path dripper.

(*iii*) Pressure compensating spiral laminar dripper.

(*iv*) In a smooth long path dripper, the head loss in the flow path in which the flow remains laminar.

The length of path required to bring down the pressure can be computed by the following relationship:

$$Ie = 4hgd/128q \quad \ldots(13.1)$$

in which,

Ie = length of path, m

h = working pressure, m

d = diameter, m

q = discharge, cu. m./sec

Microtube (Fig. 13.35 and 13.36). Microtubes (or "spaghetti" tubes) are the earliest forms of emitters during drip irrigation development. A microtube consists of a simple tube made of low density polyethylene (LDPE) or linear low density polyethylene (LLDPE) tube of internal diameters ranging from 0.5 to 1 mm and outer diameter of about 3.5 mm diameter. The longer the microtube, the lower will be the discharge from it. The usual range of discharges from microtubes varies from 1 litre/hr for a length of about 2 m to 10 litres /hr for a length of 15 mm. The microtube is pushed into the lateral pipe of the drip system, usually by pushing into holes made in the pipe wall. The flow through the microtube is laminar. Mounted on a drip irrigation lateral, it can have its dripping point fixed in any direction or at any distance away from the lateral.

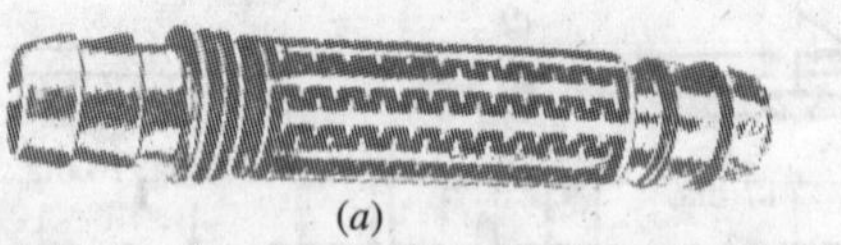

(*a*)

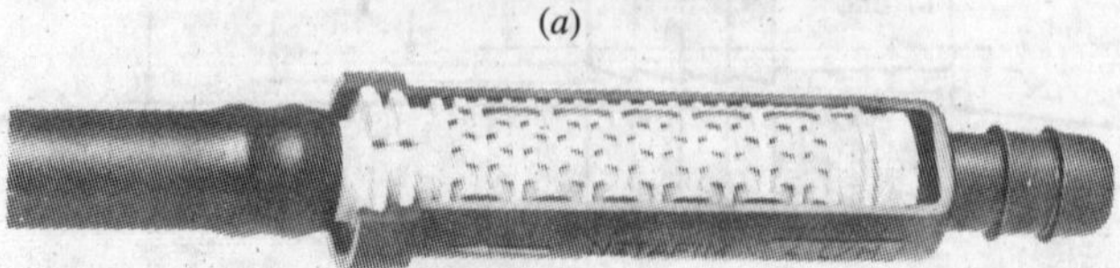

(*b*)

Fig. 13.33. Views of in-line emitters: (*a*) 'Maze dripper' (full view). (*b*) Partially exposed view of a 'NETAFIM, make in-line emitter.

(***Source:*** (*a*) Sapir, E. and Yagev, E. (1995). (*b*) NETAFIM Drip Irrigation System, Kibbutz, Hatzerin, D.N. Habutza, Israel.)

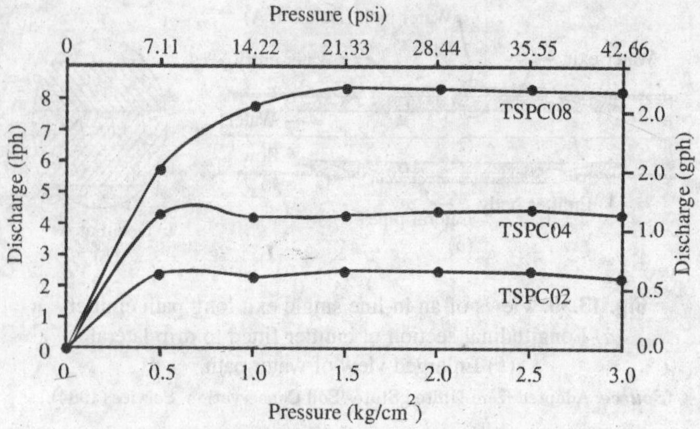

Fig. 13.34. Performance graph of a *'Turbo Seal Emitter'* under standard testing conditions.

[***Source:*** Jain Irrigation Systems Ltd. (2005 *a*).]

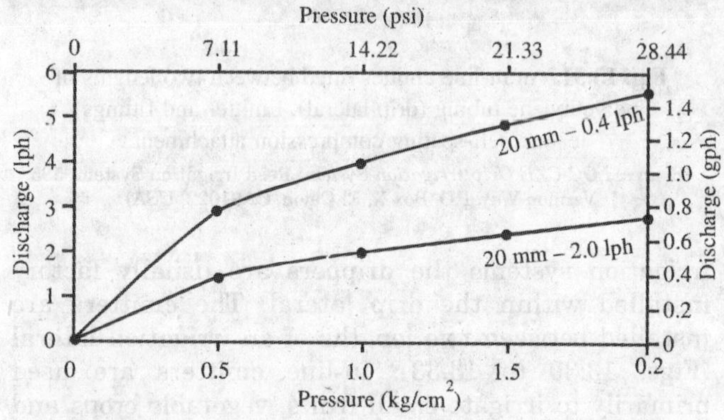

Fig. 13.35. Performance graph of *'J-Turbo Line'* of 20 mm outer diameter, as per performance Class I of IS:13488.

[***Source:*** Jain Irrigation Systems Ltd. (2005 *b*).]

Microtubes can be used for orchard trees and for row crops. For row crops, the outlets are spaced at 60 to 90 cm intervals along each lateral line, while in orchards, one or more outlets are inserted into the lateral line at each tree. The length of the microtube can be calculated to obtain a nearly constant discharge. The Bureau of Indian Standards, vide Indian Standard Specification, IS: 14482, specifies that microtubes shall withstand hydrostatic pressure 1.2 times the maximum operating pressure of the drip line. Two major limitations, namely, the need for repeated computation of the length of the tubes and manual mounting of tubes on the lateral make it suitable only for small installations. Microtubes are prone to clogging due to physical or chemical contaminants in water. However, these pipes are still widely used in India and other developing countries owing to their simplicity in construction and ease of cleaning when clogged.

Long path spiral dripper. Long path spiral drippers (Fig. 13.25) are amongst the earliest of the commercial types of drippers. The long path tubular flow passages work on the principle of obtaining head loss by the water flowing along a long path (Fig. 13.37). The 'spiral' construction reduces the space occupied by the dripper.

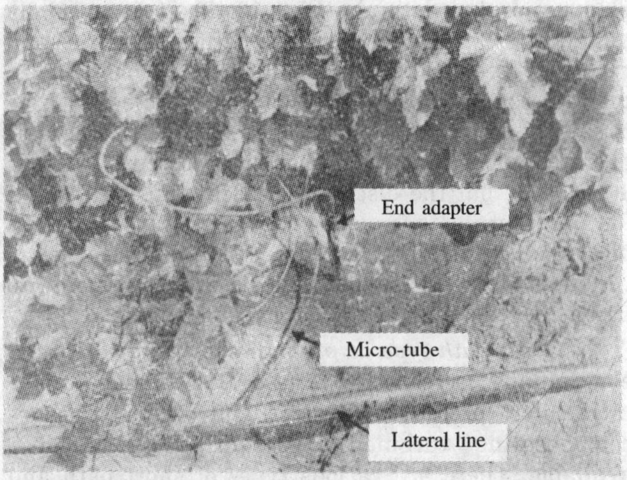

Fig. 13.36. Micro-tube fixed to a drip lateral irrigating a grape vine plant at the Research Farm of the Water Technology, IARI, New Delhi (1975). Note the end adapter of micro-tube and the water spray from the tube. The adapter protects the end of the tube from getting clogged with soil.

However, the spiral effect of flow may create considerable turbulence. *Pressure compensating spiral long path drippers* are constructed to yield a nearly constant discharge over a wide range of pressures.

Turbulent flow drippers. Turbulent flow drippers may be of three types: *tortuous dripper, short path type dripper* and *orifice type dripper.*

(*a*) *Tortuous dripper.* In the beginning of 1970s, many manufacturers concentrated on the production of tortuous drippers. The 'maze' emitter [Fig. 13.33 (a)], the 'Netafim' make emitter [Fig. 13.33 (a) and (*b*)] and the manually flushing type in-line emitter (Fig. 13.30 and 13.32) are some examples. It is a modification of long path dripper and is often known as a 'labyrinth dripper'. In tortuous emitters, the path of flow has changing directions, angles and diameters. The flow is turbulent and rapid, with small whirlpools along the path. The advantage of the principle is high head loss in a shorter path with a larger flow path diameter, usually more than 1 mm, with less chances of clogging. There are also on-line (inserted) drippers like the "button dripper' [Fig. 13.29 (*a*) and (*b*)] the 'woodpecker dripper' [Fig 13.29 (*c*) and (*d*)] working on this principle. The woodpecker drippers' have small flexible tubes attached to the outlets as part of the dripper unit. The drip line lateral is buried and only the outlet is above ground. This dripper was developed to prevent wood peckers from making holes in the laterals. It is also convenient when using mechanical harvesters on the farm (Sapir, E. and Yagev, E., 1995).

A number of emitters with distinct advantages in design for high quality performance have been introduced by Jain Irrigation Systems Ltd., Jalgaon, India in late 1990s and early 21st century. They include *Turbo Seal Emitter* (Fig. 13.38) *J-Sc-PC-Plus Emitter* (Fig. 13.39) J-*Turbo Key Plus Dripper* (Fig. 13.40). *Turbo*

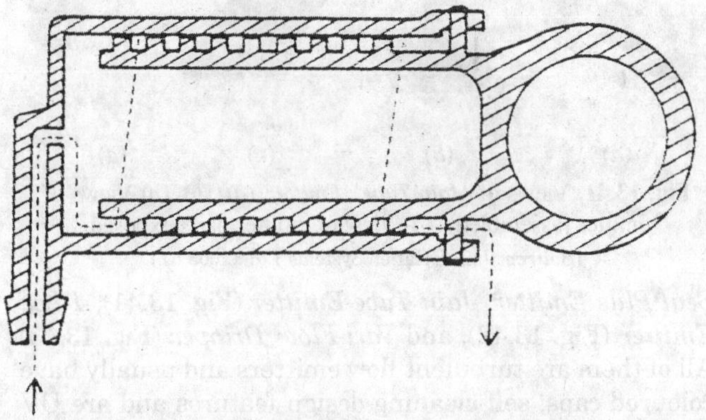

Fig. 13.37. Sectional view of a long path emitter that can be opened for easy cleaning.

[*Source:* United States Soil Conservation Service (1984).]

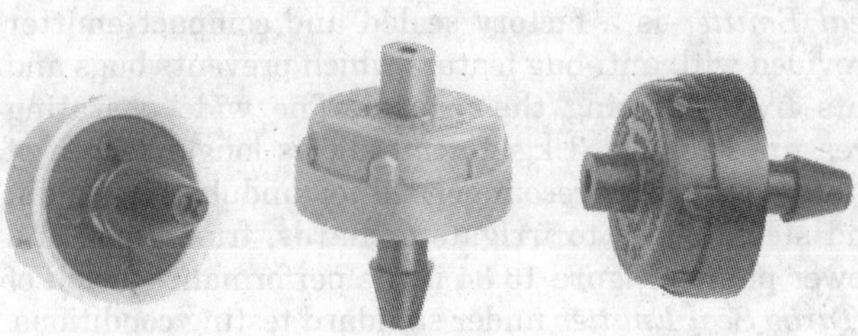

Fig. 13.38. Views of *'Turbo Seal' Emitter.*

[*Source:* Jain Irrigation Systems Ltd. (2005 b).]

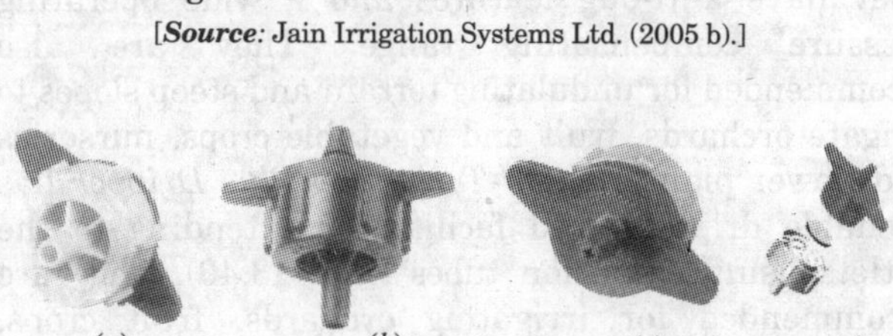

Fig. 13.39. Views of *'J–SC–PC–Plus' Emitter.* (*a*), (*b*) and (*c*) Views of emitter assembly, (*d*) Exposed view of emitter parts

[*Source:* Jain Irrigation Systems Ltd. (2005 *a*).]

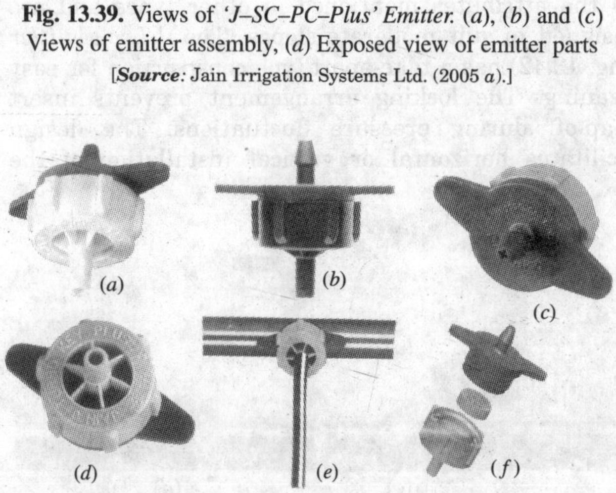

Fig. 13.40. Views of *J–Turbo Key Plus Emitter.* (*a*), (*b*), (*c*) and (*d*) Different views of emitters (assembled), (*e*) Emitter fitted with extension tube mounted on the lateral, (*f*) View of emitter parts exposed.

[*Source:* Jain Irrigation Systems Ltd. (2005 *a*).]

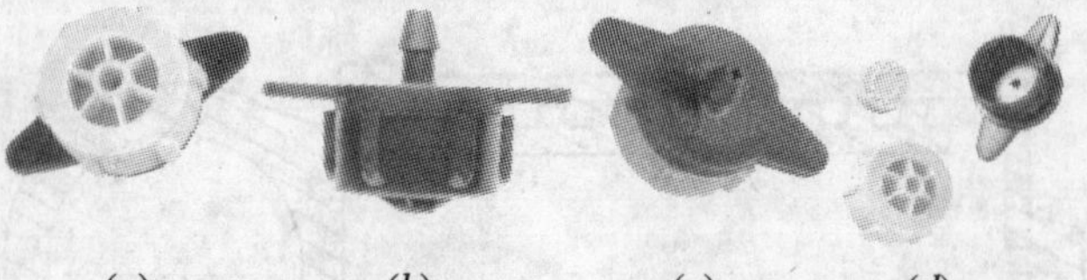

Fig. 13.41. Views of *'Jain Tube' Emitter.* (*a*), (*b*), (*c*) Views of emitter (assembled), (*d*) View of emitter parts exposed.
[*Source:* Jain Irriation Systems Ltd. (2005 *a*).]

Seal Plus Emitter, Jain Tube Emitter (Fig. 13.41), *J-Loc Emitter* (Fig. 13.42), and *Vari-Flow Dripper* (Fig. 13.43) All of them are turbulent flow emitters and usually have coloured caps, self-cleaning design features and are UV stabilized. Silicon rubber diaphragms provide longer service periods. Narrow cross shaped inlets act as filters. Coloured caps facilitate easy identification of flow rates. Self-cleaning design ensures cleaning at starting, shut down and during operation. The *Turbo Seal Emitter* is a factory sealed and compact emitter provided with anti-bug feature which prevents bugs and ants from entering the dripper. The wide operating pressure range (1-3 kg/sq. cm) allows longer length of laterals. They are recommended for undulating terrain and steep slopes to irrigate orchards, fruit crops and flower plants. Figure 13.34 is the performance graph of a *Turbo Seal Emitter* under standard testing conditions. The *SC-PC-Plus Emitter* can be opened for cleaning. They have anti-bug features and a wide operating pressure compensating range. They are also recommended for undulating terrain and steep slopes to irrigate orchards, fruit and vegetable crops, nurseries and flower plants. The *J-Turbo Key Plus Dripper* has openable drippers and facility for extending to the outlets using extension tubes (Fig. 13.40). They are recommended for irrigating orchards, fruit crops, vegetables, sugarcane and cotton. The *Jain Emitter* has all the attributes mentioned in other types and are designed to suit moderate slopes. The *J-Loc Emitter* (Fig. 13.42) has a take-apart type construction for easy cleaning. The locking arrangement prevents insert snap-off during pressure fluctuations. The design facilitates horizontal or vertical installation of the dripper on the lateral. (Fig. 13.42) It is specially designed to suit closely spaced row crops, nurseries and green houses. The *Vary Plow Dripper* has rugged construction and is provided with adjustable discharge to suit the water requirement and crop growth stages. It operates under a nominal operating pressure of about 1 kg/sq.cm, but the discharge can be adjusted even at low pressure. Its unique design feature prevents blockage of dripper in desert conditions. It is recommended for sandy soils/desert conditions to irrigate trees having high water requirements.

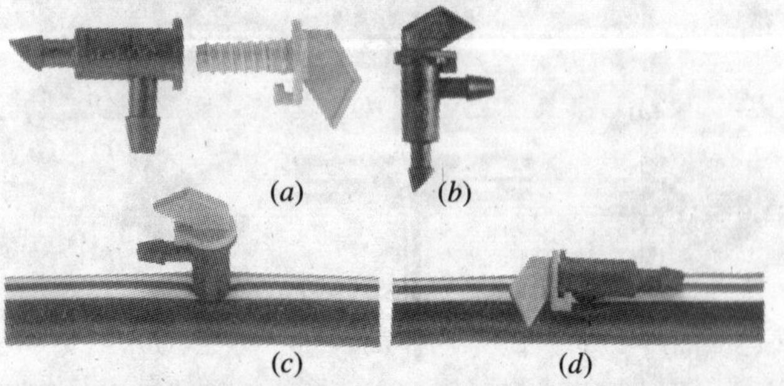

Fig. 13.42. Views of *'J–Loc' Emitter* and their mountings on lateral tubes : (*a*) Emitter parts separated, (*b*) Emitters (assembled), (*c*) Emitter fitted vertically on the lateral, (*d*) Emitter fitted horizontally on the lateral.
[*Source:* Jain Irrigation Systems Ltd. (2005 *a*)]

(*b*) *Short path type dripper.* In short path drippers, elastic flaps decrease the area of flow passage under increasing pressure. 'Grove and flap' pressure compensating turbulent type drippers are also available. There are various types of short path and orifice type drippers with distinct features produced commercially. The Bureau of Indian Standards, vide IS : 13487 : 1992 has specified that emitters shall be tested for resistance against hydrostatic pressure and leakage at pressure twice the maximum working pressure.

(*c*) *Orifice drippers.* In orifice drippers, the pressure in the pipeline is converted into velocity. The jet emitted at the outlet is broken by a jet breaker to be converted into drop-by-drop flow. The flow at the outlet is fully turbulent. The most common orifice drippers are the simple orifice (Fig. 13.27), the *vortex dripper* (Fig. 13.28), and the multiple orifices in series. In a simple orifice dipper, the jet of water strikes a cap which functions as a pressure regulator. In a vortex dripper, water enters the dripper at a tangent (Fig. 13.28), causing a whirlpool, thereby losing energy. Water is discharged at the top. Micro-sprinklers, micro-sprayers and bubblers (Chapter 12) work on the orifice principle.

Discharge of Drippers. The discharge of any dripper may be expressed by the following relationship:

$$q = kh^x \qquad \ldots(13.2)$$

in which,

q = discharge of the dripper, volume/time (L^3/T)
P = operating pressure, force/area (F/L^2)
k, x = constants for a specified emitter

In Equation 13.2, the exponents can be determined by measuring the slope of the log-log plot of pressure head versus discharge. With x known, k can be determined from Equation 13.2. The above relationship provides considerable insight on the performance of a particular emitter: (*i*) When x approaches 1, the dripper is a *long path* or laminar flow type one. Because discharge and operating pressure are linearly related, the discharge of these emitters is sensitive to variations in operating pressure. Further, the discharge can be very sensitive to the temperature of water, as the theoretical value of k is a function of the viscosity of water. (*ii*) An orifice type point source emitter has an x value of about 0.5. The emitter discharge is less sensitive to operating pressure and viscosity (*iii*) Emitters with exponents less than 0.5 is a pressure compensating type emitter. The insensitivity of discharge to pressure increases as the value of x approaches zero. Pressure compensating type emitters are especially useful in reducing emitter discharge variations resulting from major variations in operating pressure due to undulating terrain.

Line source emitters. Three types of line source tubings are used in line source drip irrigation: *Porous-wall tubing, Single tubing, and Double chamber tubing.* Porous-wall tubing is a small diameter hose or tape with a uniformly porous wall (Fig. 13.24). The pores are of capillary nature when under pressure and discharge water along their entire length. Laterals equipped with line source emitters are generally used to irrigate row crops, though they are also sometimes used to irrigate other crops as well. Porous tubes are usually made of polymer compounds with minute pores through which water seeps out of the pipe drop-by-drop. Porous tubes are sensitive to clogging and hence complete filtration of the irrigation water is essential. The tubes are buried in the ground and plants can draw water through soil suction.

Mono-walled and bi-walled pipes. Mono-walled and bi-walled perforated polyethylene pipes are commonly used as line source emitters. Water seeps from mono-walled pipe through perforations (orifices) in the pipe wall. Du Pont "Viaflow" porous plastic tubing is a unique product made from a sheet structure based on linear polyethylene (Fig. 13.21). With the application of water pressure from within, the tubing inflates and allows the passage of water through micron-size pores in the wall. The flow is uniform along the tubing. Double wall pipes used in drip irrigation are of two pipes, one within the other (Fig. 13.24). Water flows inside the inner tube at high discharge rates and pressures varying from 3 to 10 m. The flow passes through small perforations in the inner pipe to the space between the two, from where it flows out at low pressure (Fig. 13.24) through larger perforations in the outer wall. The discharge per outlet is a function of the pipe diameter, inlet and the ratio of the number of holes between the inner and outer pipes. The discharge rates usually range from 1 to 5 litres/hr. Double wall pipes are comparatively inexpensive and last for 2 or 3 crop seasons.

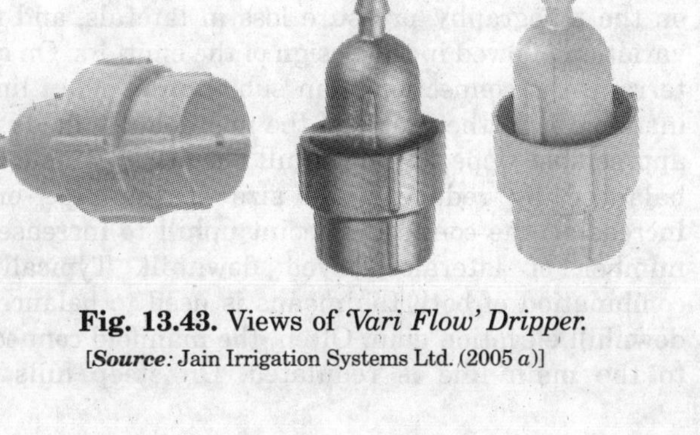

Fig. 13.43. Views of *'Vari Flow' Dripper.*
[***Source:*** Jain Irrigation Systems Ltd. (2005 *a*)]

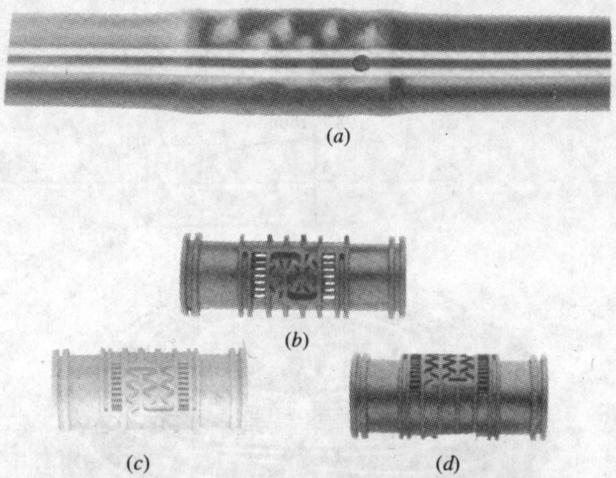

Fig. 13.44. *'Turbo Line' Emitter:* (*a*) Emitter fitted to the drip irrigation lateral. (*b*), (*c*) and (*d*) Views of the dripper.
[***Source:*** Jain Irrigation Systems Ltd. (2005 *a*)]

Typhoon Dripper. Typhoon thin-walled dripper lines are single chamber polyethylene tubes with injection moulded drippers welded to the inner wall of the tube. A wide water passage effectively prevents clogging.

J-Turbo Line. The Turbo Line developed by Jain Irrigation Systems Ltd., Jalgaon, India is a thin-walled polyethylene tube to which are fitted injection molded cylindrical drippers at short intervals (Fig. 13.44). The line is UV stabilized with clog-resistant turbulent flow path. The cylindrical shape permits wide flow path cross-section to avoid clogging due to salt precipitation. Built-in filters provide large filtration area to avoid clogging due to physical contaminations. The *Turbo Line* is available in standard dripper spacings of 30, 40, 50, 60, 75 and 90 cm. *Turbo Line* is suitable for irrigating closely spaced row crops like sugarcane, closely-spaced vegetable crops and flower plants. They are suitable for surface as well as subsurface drip irrigation systems.

DRIP IRRIGATION WATER DISTRIBUTION SYSTEMS

The water distribution network for a drip irrigation system will typically consist of the following components: (*i*) *Main lines,* which transport water within the fields, (*ii*) *Sub-mains,* which distribute water evenly to a number of lateral lines, and (*iii*) *Lateral lines,* which distribute water uniformly along their length by means of drippers or emitters (Fig. 13.20). The drip irrigation distribution system must be rugged because of the severe conditions under which they usually operate – fire, vandalism, movement of heavy machinery, rats, squirrels, birds, ants and chemicals.

Main and sub-main pipes. Rigid PVC and high density polyethylene (HDPE) pipes are used as main pipes to minimize corrosion and clogging. It is a good practice to lay the main pipes underground. Pipes of 65 mm diameter and above with a pressure rating of

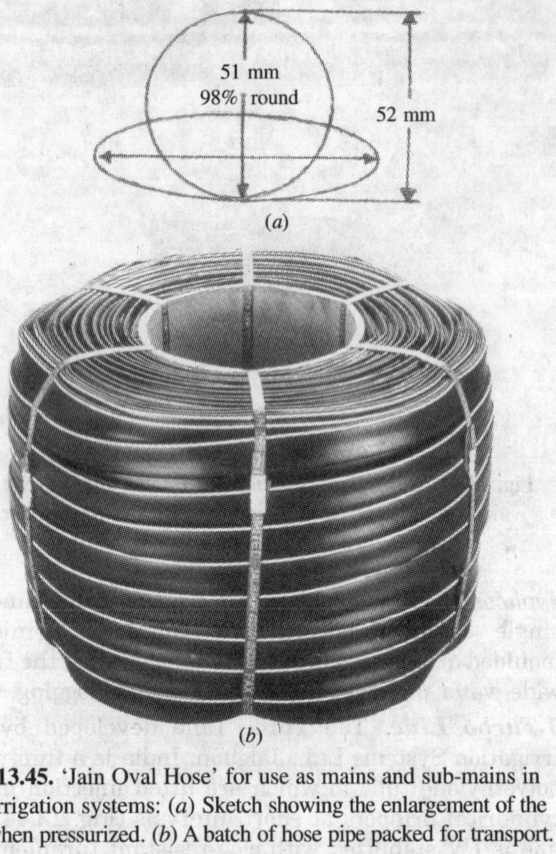

Fig. 13.45. 'Jain Oval Hose' for use as mains and sub-mains in drip irrigation systems: (*a*) Sketch showing the enlargement of the hose when pressurized. (*b*) A batch of hose pipe packed for transport. [***Source:*** Jain Irrigation Systems Ltd. (2005 *a*).]

4 to 10 kg/sq. cm are recommended for main pipes. Rigid PVC, HDPE and low density polyethylene (LDPE) are used as sub-main pipes. Pipes having an outer diameter of 32 mm to 75 mm with a pressure rating of 6 kg/sq. cm are usually recommended for use as sub-main pipes. The diameter of main and sub-main pipes will be based on the quantity of water to be conveyed. Hose pipes (Fig. 13.45) manufactured from LLDPE using extrusion technology and UV stabilization and with carbon black content provide durable tubing with close dimensional tolerances. In Fig. 13.45 (*a*), the oval shape, which attains 98% round shape when pressurized, reduces freight and storage cost. Larger diameter hoses are used in main lines and sub-mains and smaller diameter hoses are used as laterals in drip irrigation systems. Main lines are sized on the basis of irrigation sectioning or sequencing.

A sub-main is usually connected to the main line through a control valve assembly (Fig. 13.20). The function of the assembly is to regulate the water pressure in the submain, which is best achieved by a throttling valve. The function of a sub-main is to distribute water uniformly to a number of lateral lines. Uniform distribution requires that the pressure along the sub-main should be relatively constant, with a minimum of pressure variation. On a sub-main, additional components such as air and vacuum relief

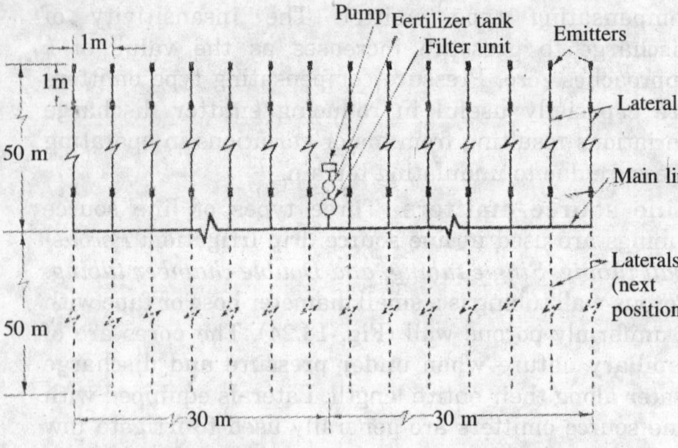

Fig. 13.46. Field layout of drip irrigation system for crop row spacing of one metre. In case of fields sloping in opposite directions laterals could be placed alternately on either side of the main, thus commanding double the area normally covered.

and pressure relief valves, pressure gauges, filtration and automatic on-off control are usually provided. At the end of each sub-main is fitted a valve for flushing the sub-main to remove contaminants and physical and chemical sediments (Fig. 13.20).

Laterals. Laterals (Figs. 13.20 and 13.46) are pipes on which drippers are mounted or within which they are inserted. Laterals are made of polyethylene (LDPE and LLDPE), butylenes or PVC materials. They usually range in diameter from 10 to 20 mm with wall thickness varying from 1 to 3 mm with pressure ratings of 2.5 kg/sq.cm. Lateral pipes should be flexible, non-corrosive and resistant to ultra-violet (UV) rays. LLDPE gives better protection against UV rays and longer service life than LDPE. Each lateral is connected to the sub-main pipe, or to the main pipe, usually through a manifold. Twin-line hoses are also used as drip irrigation laterals, using line source emission devices. Drip line laterals are designed to maintain an acceptable variation of emission device discharge along their length. Pressure differences due to pipe friction, minor losses and changes in ground surface elevation are the major reasons for along-the-lateral variation in emitter discharge.

Manifold. The manifold, or header, connects the main line to the laterals. Pressure loss in manifolds depends on the topography, pressure loss in laterals, and total variation allowed in the design of the emitters. On a flat terrain, the connection from sub-main or main line to manifold is in the centre of the manifold. In fields with appreciable slope, the downhill elevation gains can be balanced by reducing the size of the pipe or by increasing the connection point uphill to increase the number of laterals served downhill. Typically, a combination of both the means is used to balance the downhill elevation gain. Often, the manifold connection to the main line is regulated. On steep hills, one

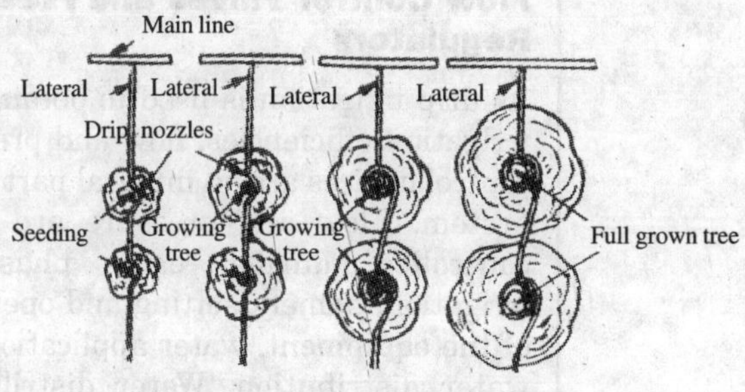

Fig. 13.47. Schematic sketch illustrating progressive increase in the number of nozzles required to cover adequately the root spread area of a growing tree ('Snake' layout).

pressure regulating valve can serve only one lateral. In such cases several pressure and flow regulating valves may be required. In general, one regulating point may serve 2 to 5 laterals, depending on the degree of slope.

The layout of lateral lines should be such that it provides the required emission points for the crop to be irrigated. Sometimes, two laterals per row of trees are required. Other means of obtaining more emission points per tree are zigzag and "snake" layouts (Figs. 13.47 and 13.48). Microtubes may also be used to obtain multi-outlet emission points.

Survey and mapping of the area to be irrigated to decide on the type and layout of the pipe system. For the proper layout of pipe lines in a drip irrigation system, properly conducted topographic survey of the area to be brought under irrigation is essential. A contour interval of 1 m will be usually necessary. The source of water, pumping point, availability of electricity and its timing, operational details of the pump, details of construction and non-planted area, details of the crops and their spacing and details of the climate are included in the survey and mapping. Soil samples have to be collected and analyzed for the physical and chemical properties, including pH, EC, soil texture, nutrient status and other properties. Water samples have also to be collected and analyzed with respect to the clogging potential of the water and its suitability for drip irrigation as well as for designing the filtration system.

Pipeline accessories and fittings. PVC pipes can be joined by using PVC resin and HDPE pipes by using heat-butt welding. Polyethylene tubing, which are normally not cementable, are connected with barbed or compression fittings. The common fittings required in drip irrigation systems are elbows, tees, crosses, unions, and plugs or end caps [Fig. 13.49 (*a*) and 13.49 (*b*)]. Similarly, for joining PVC/HDPE pipes, fittings of various sizes such as couplers, tees, elbows, bends, threaded adapters, tail pieces, collars and reducers are required. Accessories include take-outs, rubber grommet, end plug, joints, and manifolds. They are usually available in 4, 10, 12, 16, and 20 mm sizes. Take-outs are used for taking out laterals from sub-main

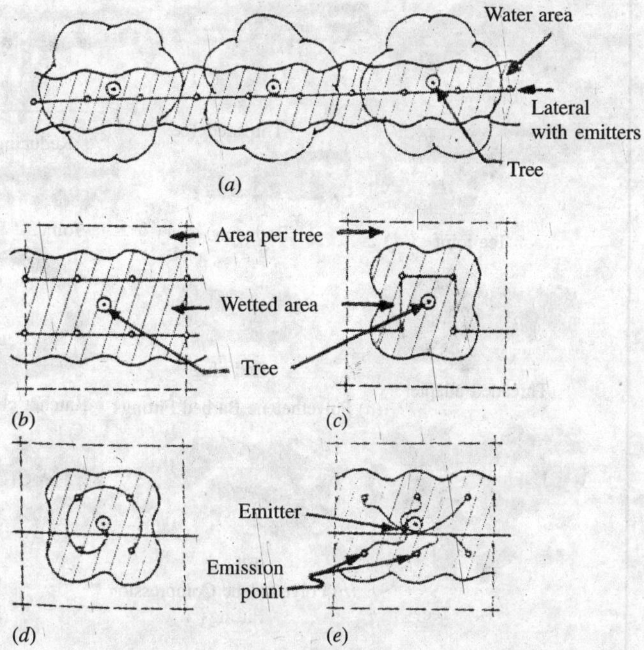

Fig. 13.48. Views of layout of laterals for widely spaced tree crops: (*a*) Single lateral for each row of trees, (*b*) Two laterals for each row of trees, (*c*) Zigzag layout of a single lateral for each tree in the row, (*d*) 'Pigtail' layout of emitters at each tree. (*e*) Multi-outlet emitter with distribution tubing.

[***Source:*** United States Soil Conservation Service (1984).]

connector. Tees, bends and end caps are used in drip laterals. They can be selected based on the sizes of pipes used in the distribution network.

Field failure of polyethylene tubings and hoses. Polyethylene is subject to failure in the field due to two main causes, namely, degradation due to exposure to ultra-violet light (UV degradation), and environmental stress-cracking.

UV degradation. UV degradation of polyethylene tubings and hoses occurs when ultraviolet light penetrates the plastic and causes molecular changes to occur, resulting in progressive deterioration of the plastic. The symptoms of UV degradation are cracking of the tube or the hose, usually along the top half which is exposed to sunlight. The standard method of preventing UV degradation is by the incorporation of a small amount of "carbon black" into the resin prior to extrusion (about 2%). Carbon black colours the clear polyethylene black and prevents degradation. UV degradation is the result of insufficient levels of carbon black in the tubing/hose, and can be prevented by careful manufacturing controls.

Environmental stress-cracking. Environmental stress-cracking is a broad term which refers to splitting or cracking of tube or hose which has been exposed to the environment. Usually, the failure occurs at points where

Fig. 13.49 (a) Different types of polyethelene and PVC fittings used in drip irrigation systems.

(***Source:*** Jain Irrigation Systems Ltd., Jalgaon.)

the material has been stressed. An example of environmental stress-cracking is the cracking of polyethylene hose at the location where it is attached to an insert fitting or in-line emitter. Extruded polyethylene hose and tubing is tested for its environmental stress-cracking resistance through the test procedure prescribed by the Bureau of Indian Standards (IS: 12786: 1989, ANNEX D, Clause 7.4) Other Standards, as applicable in the country where used may also be followed. It is observed that failures are likely to occur within six months to one year of installation of the material which has not passed these tests.

Flow Control Valves and Pressure Regulators

As drip irrigation is used to obtain high irrigation efficiencies, flow and pressure control devices are an integral part of the system. Flow and pressure are to be controlled during each phase of irrigation, namely, setting and operation of the equipment, water application and water distribution. Water distribution can be by hand-operated pressure controls and on-off valves, sequential operation, or partial or full automation.

Hand-operated pressure controls and on-off valves. The discharge rate is controlled by adjusting the pressure with manual valves set to balance flow rates among the sub-units of the system. It is important to check and adjust the valves to keep emitter discharges uniform. Another method of flow control is the use of pressure or flow regulators at the inlet to each lateral or header feeding a small group of laterals. They are usually preset for a given pressure or flow rate and often cannot be adjusted or reset. These valves must be incorporated into the system design because only a limited selection of pressures or flow rates is available with the small, low-cost valves. *Jumper tubes* of various diameters and lengths can be used to connect each lateral to the manifold. The tubes can be cut to the length that provides the pressure loss required to produce uniform lateral inlet pressures along a manifold with non-uniform pressures.

Sequential operation. Parts of the system can be operated sequentially with volumetric control valves that are interconnected by hydraulic control lines. As each valve closes, the next valve opens. When the sequencing operation is completed, the valves must be re-adjusted. The first valve must be activated manually to start the cycle again. It is also desirable/essential in steep areas to plan the irrigation so that valve activation proceeds from lower to higher plots.

Partial automation. Volume control is well suited to drip irrigation. Volume can be controlled simply with some automation by the use of volumetric or mechanical time-clock valves. Semi-automatic volumetric control valves can be placed at the head of each sub-unit or a single such valve can be used at the control head along with ordinary valves controlling each sub-unit. The volumetric valve requires manual opening and

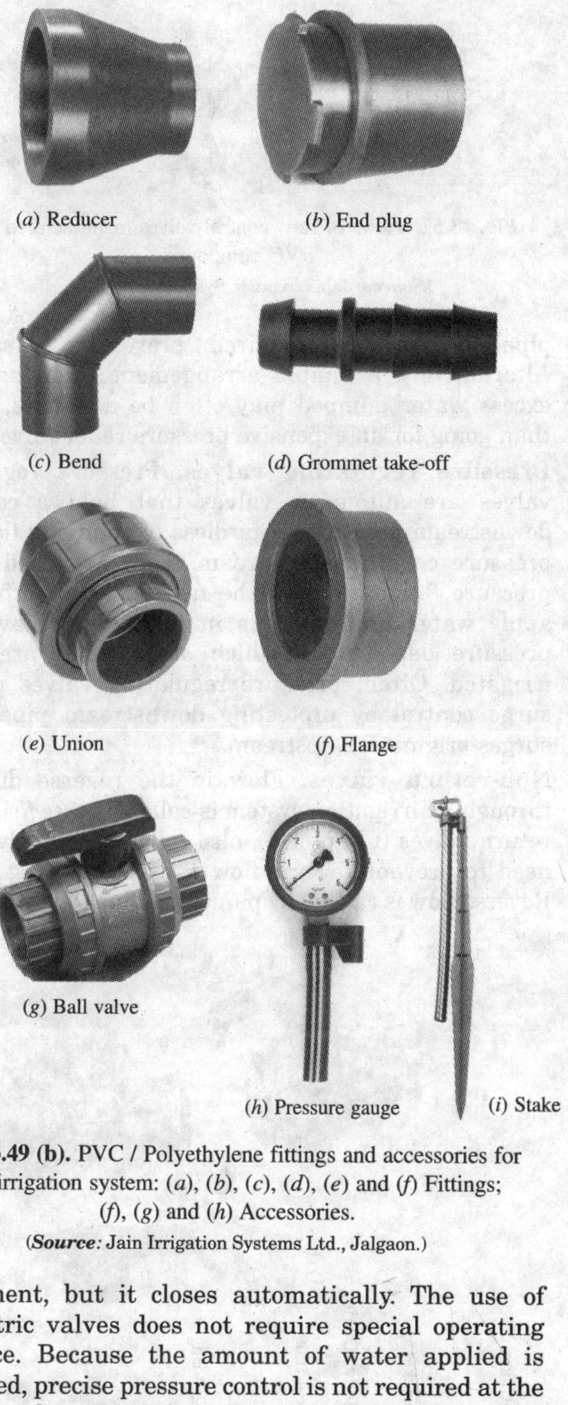

Fig. 13.49 (b). PVC / Polyethylene fittings and accessories for drip irrigation system: (*a*), (*b*), (*c*), (*d*), (*e*) and (*f*) Fittings; (*f*), (*g*) and (*h*) Accessories.
(***Source***: Jain Irrigation Systems Ltd., Jalgaon.)

adjustment, but it closes automatically. The use of volumetric valves does not require special operating sequence. Because the amount of water applied is measured, precise pressure control is not required at the inlets to volumetric valves. Pressure control is required if mechanical time clock valves are used.

Full automation. Drip irrigation can be fully automated either by using a central controller operated on a time or volume basis or by soil moisture sensing. Automation on a time or volume basis requires a control system operating either hydraulic or electric valves. The controller automates the irrigation for an unlimited number of cycles. The order in which the valves operate can be altered from one cycle to the next. Both the operating time of each valve and the quantity of water distributed can be changed easily at the control panel. In time-based automation systems, the operation of the valves is based on prefixed irrigation start times and end times and the sequence of valve operations required. The controller in these systems can be programmed with different valve sequences and irrigation times. These controllers activate/deactivate the low voltage electric solenoid valves in the field, as per the programme set into them. In sensor-based automation systems, the operation of valves could be based on the various parameters sensed from the field, such as soil moisture, leaf temperature, weather equipment and electronic water meters. Any of these sensors either individually or in combination can be attached to the controller for activating and deactivating the low voltage electric solenoid valves and pumps, based on the sensor reading. Usually a tensiometer is used as the moisture sensor. The tensiometer measures the soil moisture tension and signals the valve controlling each sub-unit, and the valve opens or closes. Because each valve operates automatically and is not connected to any other valve, the order of operation is not planned in advance. Therefore, the circuitry must pass through some type of control panel to eliminate the simultaneous opening of more than the desired number of valves. The use of automation equipment is growing worldwide and there exists great potential for these systems.

Valves

Valves are integral parts of pressurized irrigation systems, including drip irrigation. The valves used in drip irrigation systems include air release and vacuum relief valves, pressure regulating valves, flow regulation valves, non-return valves and on-and-off valves.

Air release and vacuum relief valves. Presence of air in the pipeline restricts the flow of water and increase pumping costs. Air may enter the pipeline during filling, where water enters the pipeline at the pump or gravity inlet point and through leaky joints and fittings. Air can be prevented from entering the pipeline by careful design of pump or gravity inlet, maintaining the average water velocity between 0.3 to 0.6 m/sec during pipeline filling, and laying pipe to a grade that results in a minimum number of high points. In order to remove any air that may enter the pipeline, properly sized air and vacuum relief valves are to be installed at all high points of main lines, sub-mains, control risers and other points in the pipeline system (Fig. 13.50). *Air and vacuum release valves* (Fig. 13.51) are designed to exhaust air at various pressure conditions during normal pipeline operation while restricting the outflow of water. Exhaust ports for such valves are usually of diameter 1.6 to 6.4 mm. Air release valves allow air to escape when filling pipelines with water and remove air pockets at high points in the system. *Vacuum relief valves,* which have orifices of size 25 to 200 mm in diameter, are designed to exhaust large volumes of air during pipe filling and to close when the filling stops. Further, the vacuum relief valves allow air to enter

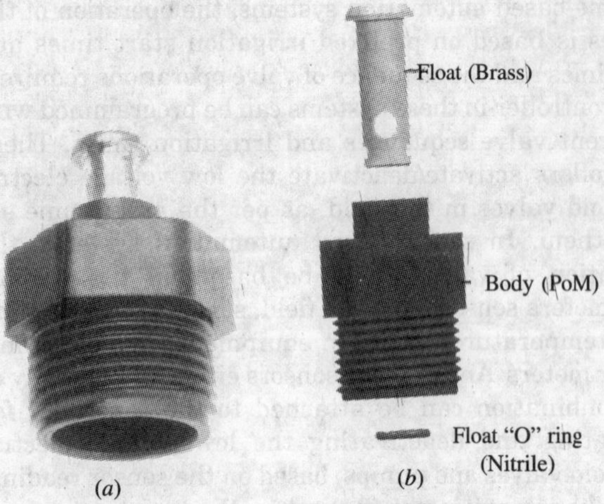

Fig. 13.50. Air release valve: (*a*) View of valve assembly (*b*) Exposed view of parts.
[***Source:*** Jain Irrigation Systems Ltd. (2005 *a*)]

Fig. 13.51. Air and vacuum release valve.
(***Source:*** Jain Irrigation Systems Ltd., Jalgaon.)

when draining pipelines and to prevent negative (suction) pressure in laterals, after the system shuts down. Both air release and vacuum relief valves are closed either by the impact of water against the valve closure element or by a float.

Pressure relief valves. Pressure relief valves may be installed at any point where the possibility exists for excessively high pressures, either static or surge pressures, to occur. Abnormally high pressure conditions may result from sudden opening or closing of a valve, starting or stopping of a pump, failure of a pressure regulator valve and sudden closure of an air vent under high pressure. Many of these potential problems can be prevented by proper design and careful operation of the system. Closing of manually operated valves slowly, selection of proper size and number of air vents and control of static pressure conditions in the

Fig. 13.52. Views of flow control valves manufactured from PVC compound.
[***Source:*** Jain Irrigation Systems Ltd. (2005 *a*)]

pipeline are the required preventive measures. Alternatively, a simple arrangement to by-pass the excess water pumped may often be adequate, rather than going for an expensive pressure relief valve.

Pressure regulating valves. Pressure regulating valves are automatic valves that hold a constant downstream pressure, regardless of changing flow and pressure conditions upstream. They are used where pressure fluctuations in the field make it difficult to apply water uniformly, as in fields where available pressure depends on which other fields are being irrigated. Often, pressure regulating valves provide surge control by protecting downstream pipes from surges originating upstream.

Non-return valves. Flow in the reverse direction through an irrigation system is called *reverse flow.* Non-return valves (Chapter 3), also called check valves, are used to prevent reverse flow or backflow (Fig 13.20). Reverse flow is caused by pump, pipeline or valve failure

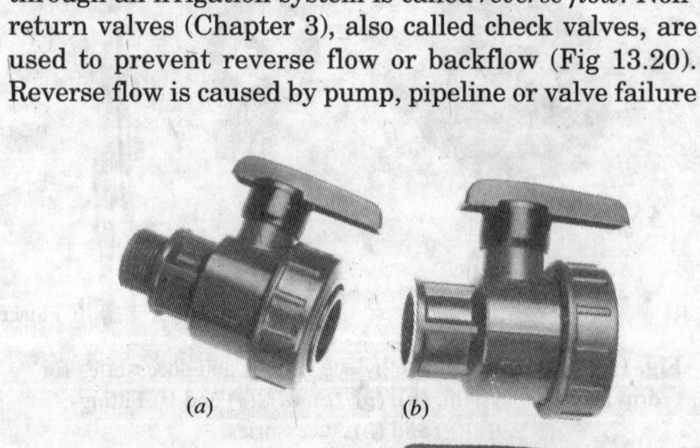

Fig. 13.53. Views of ball valve: (*a*) and (*b*) Views of valve assembly, (*c*) Exploded view showing parts.
(***Note:*** All parts are made of PVC compound, except "O" ring of Nitrile, seat ring of HDPE/PTFE, and handle of ABS.)
[***Source:*** Jain Irrigation Systems Ltd. (2005 *a*)

and can result in irrigation system damage and contamination of the water source. Backflow protection is often required by law when fertilizers and other agricultural chemicals are injected into the irrigation system, so as to prevent contamination of the water source, due to reverse flow. Non-return valves are installed immediately downstream of the water source. Pressure relief valves, manual flow control vales and vacuum relief valves provide protection from backflow when the system is off.

Flow control valves. Flow control valves, also called *on-off valves*, may be manually operated (Fig. 13.52) or automatic. They are usually made of rigid PVC compound. Chemical resistant nitrile rubber "O" rings (Fig. 13.53) provide leak proof operation. Control valves are usually available in 20 to 90 mm sizes. They are used for a variety of purposes. When they are located at the upstream end of a pipe, they provide on-off service to pipes downstream. This allows water to be cycled to different parts of the system to meet irrigation requirement of each portion of the farm. It also allows portions of the system to be isolated for maintenance and repair works. Control valves, when located at the downstream end of permanently laid pipelines allow flushing of pipes of sediment and debris.

Filration Systems

Filtration is the key to the success or failure of a drip irrigation system. Drip irrigation can cause clogging of the drippers and other components mainly due to the small diameter water passage in the drippers, dirt in the irrigation water, and the relatively low operating pressure of the system. When planning a drip irrigation system, it should be kept in mind that the water quality does not remain constant and may change over a period or different seasons in a year.

Causes of clogging. Generally, the reasons for clogging are: (*i*) *Mineral particles.* Mineral particles of sand, silt and clay, and debris that are too large to pass through the small openings of filters and emission devices. (Sometimes, silt and clay particles that are usually much smaller than the smallest passages may be deposited in the low velocity areas of the laterals where they coagulate to form masses to clog emitters. Coatings of clay particles in filters and emitters may also adversely influence the flow rate in the system). (*ii*) *Organic matter.* Organic contaminants including organic matter, algae and microorganisms like bacteria. Favourable environmental conditions within the drip irrigation system can cause rapid growth of several species of algae and bacteria. These microorganisms often become large enough to cause complete clogging of the system. Biological oxidation processes involving certain species of bacteria and water with even very low concentrations of ferrous and manganese ions can produce precipitates of iron and manganese oxides which can cause clogging. Small animals such as ants, spiders and crustaceans can also cause clogging in the drip system. *(iii) Chemicals.* Precipitates which form within the drip system and the outside surface of emitters are major chemical causes of clogging. When irrigation water contains soluble salts, crusts of salt often form on emitters as water evaporates between irrigations. Unless salts are dissolved during subsequent irrigations or rain, crust accumulation will continue and clog the system. Carbonates may accumulate within the drip system and cause clogging when the concentration levels of calcium, magnesium, and bicarbonates present in irrigation water are high. When well waters with dissolved iron and manganese are exposed to the atmosphere, precipitates of manganese oxide are formed and cause clogging of drip irrigation systems. Irrigation water with high contents of sulphides may produce insoluble compounds which cause clogging. In addition to naturally occurring compounds, some types of liquid fertilizers and other chemicals injected into the drip irrigation system may form precipitates and lead to clogging of drip systems.

Removal of the particles/objects causing clogging is essential for the satisfactory performance of a drip irrigation system. Settling basins, sand or media filters, screen filters, centrifugal sand separators and disc filters are the basic devices used to remove suspended and other materials clogging drip irrigation systems.

Settling Basins

Settling Basins or ponds can remove large volumes of sand and silt carried by streams which are used as the source of water for drip irrigation, usually through diversion works. The minimum size of particle that can be removed depends on the time that the sediment laden water is detained in the basin. The pond should be designed to provide the required settling time. Long and narrow ponds are preferred. If the construction area is limited, baffles may be constructed. A minimum of 15 minutes will be required for the settling of mineral particles larger than 80 microns. Longer detention time is required to remove smaller particles. Clay-sized particles will require several days to settle, unless flocculating agents like alum or polyelectrolites are used. For settling tanks to be effective, the intake pipe to the drip irrigation system should be so located that water from the upper level of the pond enters the irrigation system. However, sedimentation alone may not provide the desired water quality. Algal growth and other wind-blown contaminants in the settling pond may cause serious problems. Chemical treatments with chlorine or copper sulphate may be required to control algae. Because of these problems, settling basins are not used in drip irrigation using ground water. In canal irrigation also, settling basins are used only under conditions of heavy sediment loads in the source of water. Settling basins are to be cleaned several times a year, when large quantities of water with heavy sediment loads are passed through them.

Gravel/Sand Media Filters

Sand Media Filters, frequently called media filters, consist of layered beds of graded sand and gravel placed in one or more cylindrical tanks. Media filters are essential for the primary filtration of irrigation water. The diameter of filter tanks ranges from 20 to 200 cm, depending on the capacity of the system (Fig. 13.55). They may be of vertical or horizontal types and are usually made of steel and coated with 70 to 80 micron thick blue polyester powder from both inside and outside surface for protection against corrosion and weather effects. Media filters are relatively of low cost and easy to operate. Each media filter is contained within a pressure vessel or tank having service openings at the top and side, inlet, outlet and back-flush connections usually with a gravel bed to facilitate back-flushing (Fig. 13.56 and 13.57). The filter tank houses all filter components, except the control valves and the connecting pipes. Removable covers provided in the filter tank (Fig. 13.54 and 13.57) permit connection and removal of inlet water distributor, filtered water collection system, and filling and removal of media material (sand/gravel). The cover also acts as a deflector for uniform spread of water over the media when it is located at the center of the tank. A drain plug is provided at the bottom of the filter which should be large enough (about 32 to 50 mm diameter) to flush out relatively large particles. An inlet deflector is provided in the tank to reduce the velocity of the incoming raw water, to distribute water evenly over the filter bed and to facilitate collection of back-wash water.

Fig. 13.54. View of a sand filter: Capacity 18m^3/hr, steel tank of 1000 kPa, requires 150 kg of silica crushed sand media (0.7 to 1.2 mm), manually flushing type. (***Source:*** Jain Irrigation Systems Ltd., Jalgaon).

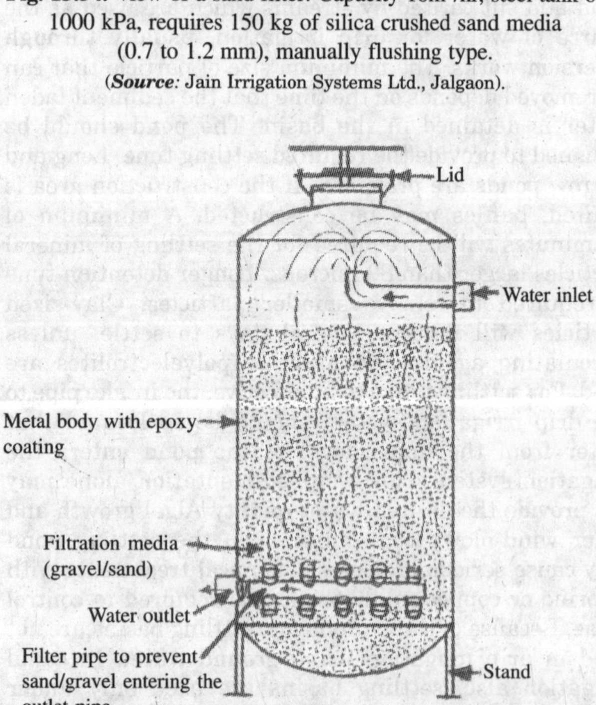

Fig. 13.55. Schematic sketch illustrating the principle of working of gravel filter.
Adapted from Sapir, E. and Yagev, E. (1995).

Fig. 13.56. Sand filter of a drip irrigation system installed to irrigate sugarcane crop.
[***Courtesy:*** EPC Industries Ltd. (2005).]

Generally, the filter media consists of crushed granite, silica or quartz sieved to specific particle sizes (generally of 0.7 to 1.2 mm size). The sand media used in most drip irrigation system filters are designated by numbers. Numbers 8 and 11 are crushed granite, and numbers 16, 20 and 30 are silica sands. The mean granule size is about 1900, 1000, 825, 550 and 340 microns, respectively for numbers 8, 11, 16, 20 and 30 (US Soil Conservation Service, 1984). It is a common practice to select the smallest medium possible. However, a larger medium may sometimes be desirable as it causes less pressure drop and has a slower buildup of particles. The quality of filtering required in a given drip irrigation system will depend on the configuration of the system and the type and size of the emission

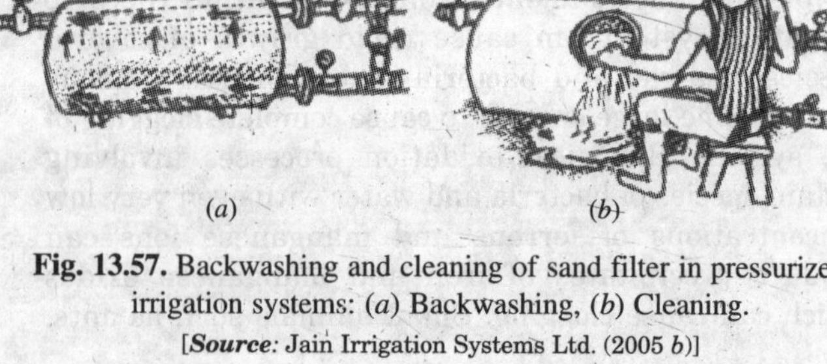

Fig. 13.57. Backwashing and cleaning of sand filter in pressurized irrigation systems: (*a*) Backwashing, (*b*) Cleaning.
[***Source:*** Jain Irrigation Systems Ltd. (2005 *b*)]

devices used. A usually successful rule of thumb is to filter the water to the equivalent of a 74 micron particle size. This corresponds to a media filter with a No. 11 sand media bed.

The sand media should be free of calcium carbonate. Selection of proper quality of sand is important in media filter performance. The ideal sand is sharp-edged and of a uniform size. River bed sand, usually found in nature is often rounded, poorly graded and full of fine particles which seriously impair filter performance. Water enters from the top and leaves through a bed of coarser gravel laid at the outlet. *Media grade* is the classification of the filtering media with respect to particles and their size distribution (% of total mass). *Minimum operating flow* is the rate of flow in the filter, as specified by the manufacturer, that ensures proper filtration and back-washing functions.

In a media filter, the water flows from the top of the sand bed, down through the bed and out at the bottom of the filter, leaving soil particles and other impurities trapped inside the sand bed. When too much foreign material has accumulated within the sand bed, it becomes difficult to force the water through the filter, and the head loss across the filter is increased. This increased head loss indicates that the filter needs to be cleaned by backwashing. *Safe maximum pressure drop* is the maximum allowable pressure difference between the inlet and outlet pressures across the filter when the filter media has clogged to a level which requires cleaning or backwashing. Pressure gauges are located at the inlet and outlet of the media filter to monitor the pressure drop and determine the need for backwashing. The pressure drop usually allowed is about 2 m; when it exceeds the limit the filter needs cleaning. *Backwashing* is a procedure that flushes clean water through a filter tank in reverse direction to the normal operating direction, in order to remove the intercepted organic material and mineral particles from the filter media. Backwashing must be frequent enough to hold the pressure drop within the prescribed design limits. It is important that the backwash flow rate be high enough to vigorously fluidize and agitate the sand bed, yet low enough, so that only the filtered material is washed from the filter bed and not the filter media itself. Media filters allow relatively low flow rate ranging from 5 to 50 cu.m/hr. Hence, in many cases a battery of filters connected in parallel is installed at the head of a drip irrigation system. The Bureau of Indian Standards, vide IS 14606: 1998 has prescribed the specifications for media filters.

Media filters effectively remove suspended minerals, organic materials, and most other suspended substances originating from surface or ground water sources. However, they do not remove very fine substances like silt and clay as well as bacteria. Filter cleaning involves the removal of the filtered substances which have accumulated in the pores of the filter material. The need for cleaning can be determined by monitoring the pressure drop across the filter.

Screen Filters

Screen Filter (Fig. 13.58 and 13.59) is fitted in series with the media filter in order to remove the solid impurities like fine sand and dust from the irrigation water after it has passed through the media filter. The parts of the screen filter consist of body, one or two filtration elements, gaskets, cover, inlet, outlet and drainage valves. Screen filters are characterized by the following parameters (Fig. 13.60):

(*i*) ***Size.*** The size of a screen filter is defined by the inlet and outlet diameter, which generally range from 2 to 30 cm.

(*ii*) ***Filtration area.*** It is the area of the filtration screen.

(*iii*) ***The size of openings***. The size of the screen hole is usually expressed in "mesh". Originally, this term was used in weaving and means the number of threads per inch (2.54 cm). Usually 120 mesh is used for screen filters used in drip irrigation, although 140 – 200 mesh are also used. A more exact measure is the size of the space between the threads, designated in microns: 1 micron = 0.001 mm.

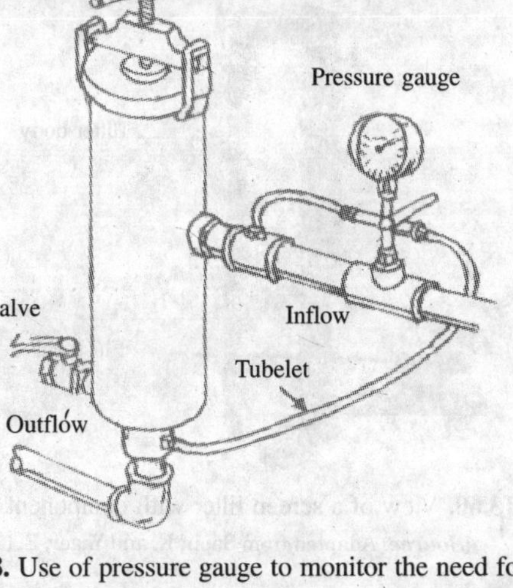

Fig. 13.58. Use of pressure gauge to monitor the need for flushing a screen filter.
(***Source:*** Sapir, E. and Yagev, E. (1995).

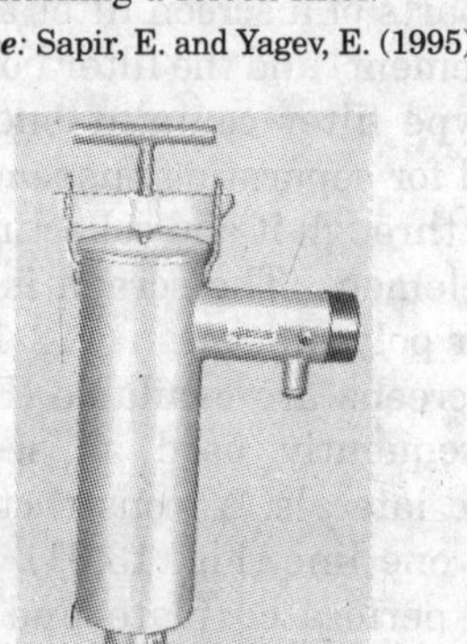

Fig. 13.59. View of a screen filter: Flow rate 40 m^3/hr. Stainless steel/ PVC 150 mesh screen, 1000 kPa max. Operating pressure, manually flushing type.
(***Source:*** Jain Irrigation Systems Ltd., Jalgaon.)

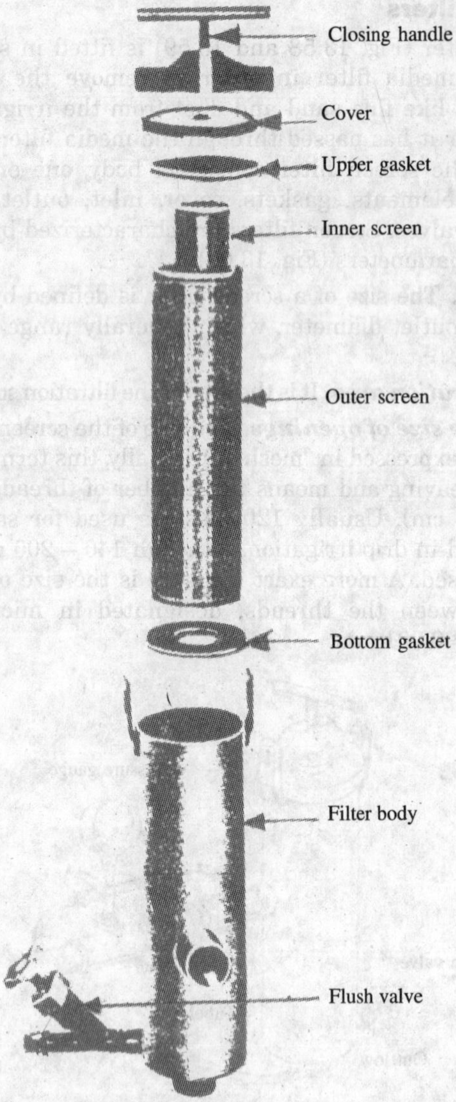

Fig. 13.60. View of a screen filter with component parts exposed.
[***Source:*** Adapted from Sapir, E. and Yagev, E. (1995).]

The basic parts of a screen filter are the filter screen or strainer (element) and the filter housing. The strainer in a screen-type filter contains one or more filtering elements used for separating suspended solids from the water flowing through it and collecting them on the face of the filter element. The screen is made of stainless steel, nylon, or polyester mesh (Fig. 13.59). Wye (Y) and basket type screens are available (Fig. 13.61). 'Y' type filters are frequently used as secondary filters in sub-mains and laterals. 'Y' construction keeps the inlet and outlet in one line (Fig. 13.61). The filter element consists of a perforated plate and fine mesh screen intended to retain solid contaminants larger than a given size from the water flowing through the component. The standard mesh size of screen filters is 100 microns. Other sizes are also available. The *'turbo-clean'* mechanism shown in Fig. 13.61 creates continuous vortex inside the elements preventing contaminants from settling on the filtration surface. The flow direction is from the inside of the element to outside. The strainer housing is the part of the strainer that contains all the strainer components, except the control valves. The material of construction of the filter housing and the filter element usually may be mild steel or plastic. Mild steel, when used as filter housing, is coated with 70 to 80 micron thick deep blue coloured polyester, both inside and outside, as protection against corrosion and weather effects.

Screen filters require periodic cleaning of the screen element for removing the filtered materials from the surface of the screen and from inside the housing. Arrangements for drain valves and ports for measuring pressure on both inlet and outlet of screen filter are essential (Fig. 13.62 and 13.63). Careful attention to maintenance and cleaning is necessary with screen filters. Manually cleaned screens must be regularly disassembled to clean the elements (Fig. 13.64). Failure to do so will lead to build-up of material on the screen and an excessive pressure drop across the filter. An excessive pressure loss across the screen may force out particles, particularly organic material, through the screen, where they can pass into the irrigation system and cause clogging problems. Because most screens must be disassembled for cleaning, they are not practical for use with waters containing high levels of contaminants, as it will require frequent disassembly

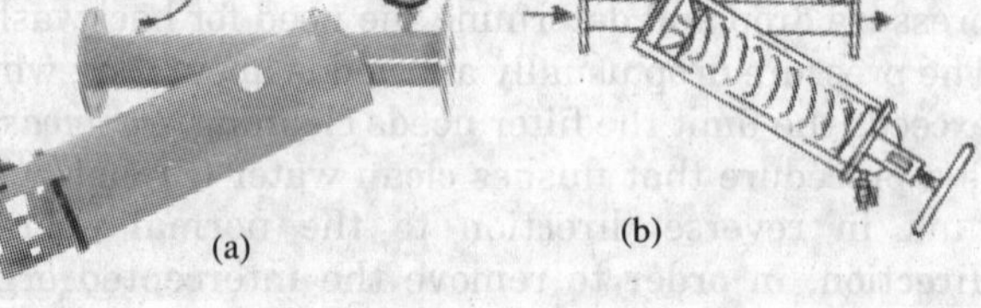

Fig. 13.61. 'Turbo-Clean' 'make Y-type screen filter used for secondary filtration in drip irrigation systems:
(*a*) View of filter assembly. (*b*) Sketch showing the vortex flow of water inside the screen element.
[***Source:*** Jain Irrigation Systems Ltd. (2005)]

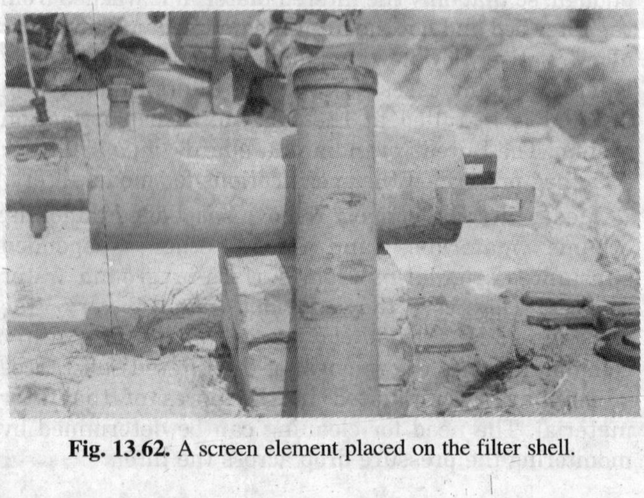

Fig. 13.62. A screen element placed on the filter shell.

(*a*) (*b*)

Fig. 13.63. Flushing and cleaning of a screen filter in a drip irrigation system: (*a*) Flushing, (*b*) Cleaning.
[***Source:*** Jain Irrigation Systems Ltd. (2005 *b*)]

and cleaning. However, like media filters, screen filters can be cleaned automatically as well. One type of automatic system has dual cylinder filters, one of which operates while the other is being cleaned. The flow of water is switched from cylinder to cylinder automatically on a predetermined pressure drop level, as with media filters. Cleaning is accomplished by opening a valve on the outlet end of the screen. It is important that every drip irrigation system should incorporate a screen filter as a minimum requirement, irrespective of the presence or absence of other types of filters.

Screen filters can be selected on the following basis: (*i*) The diameter of the housing cylinder, (*ii*) The flow rate in the main pipe of the system, which may vary from 3 to 50 cu.m/hr, (*iii*) By the total surface area of the filter and the active or 'net filter area'. The net filter area is usually about one-third of the total filter area. The ratio between the net filter area and the cross-sectional area of the inlet pipe should be at least 1 : 8. In screen filters, the mesh size and the total open area determine their efficiency and operational limits.

Hydrocyclones or Centrifugal Sand Separators

Centrifugal filters, also called hydrocyclones or cyclonic separators, use vortex motion and centrifugal force to remove suspended substances with specific gravities greater than 1.2 (Fig. 13.65). The filtration capacity of hydrocyclones is greater for high density particles than for lower density ones. Because these filters effectively remove large quantities of sand particles, they are placed upstream of other types of filters. The water is let in tangentially at the top of the cone and creates a circular motion resulting in a centrifugal force, which throws the heavy suspended particles towards the wall of the filter body. The separated particles move towards the bottom to the collecting tank (Fig. 13.65 and 13.66). The collected filtrate (sand) is emptied manually or through a flushing valve. The clean water rises up in a spiral motion to the outlet. The diameters of the top and bottom of the cone are designed in proportion to the rate of flow of water which varies from 3 to 300 cu.m/hr. The pressure loss in the filter ranges from 4 to 8 m. If the sediment concentration is high in the water source, then more filters can be installed in series. The hydrocyclone has proved its superiority in separating large quantities of sand from water, as compared to other types of filters. They are particularly effective for primary filtration of irrigation water.

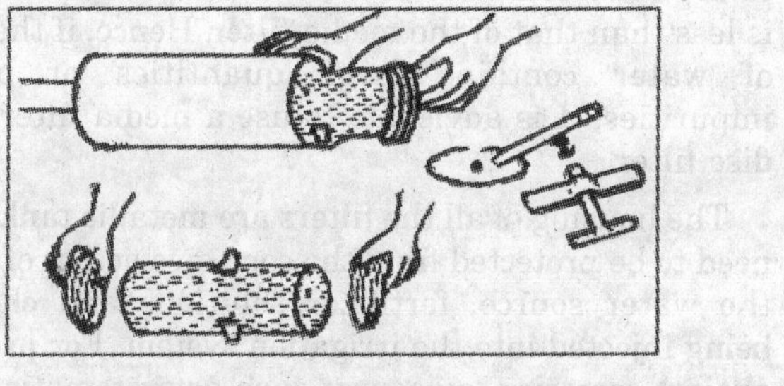

Fig. 13.64. Cleaning of screen filter.
[***Source:*** Rajput, T.B.S. and Neelam Patel (2001).]

Disc Filters

Disc filters can remove organic material and algae and other extremely fine particles which cannot be removed by other types of filters. Disc filters are often the final filtering component before water enters the drip line. Disc filters (Fig. 13.67) contain stacks of grooved, ring-shaped discs that capture the contaminating particles. During filtration the discs are pressed together. In the back-flushing mode, the direction of flow is reversed. The discs are loosened and start a spinning motion, ensuring complete removal of retained particles (Fig. 13.68). The back-flushing can be carried out

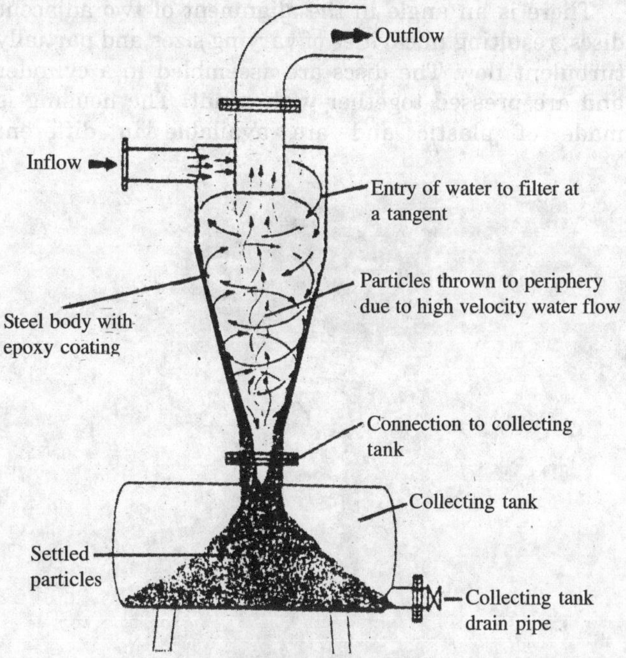

Fig. 13.65. Diagram illustrating the operation of a hydro-cyclone filter.
[***Source:*** Adapted from Sapir, E. and Yagev,.E. (1995)]

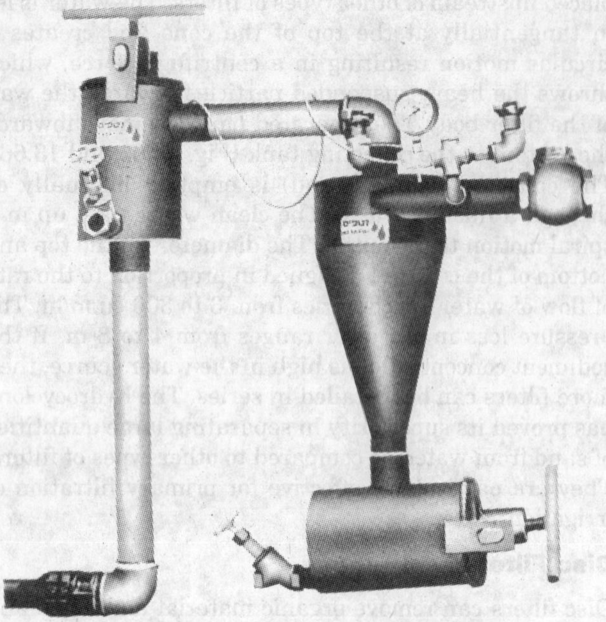

Fig. 13.66. Assembly of screen filter and hydrocyclone in a drip irrigation control head.
(***Source:*** NETAFIM Drip Irrigataion Systems, Kibbutz, Hatzerin, D.N. Habutza, Israel.)

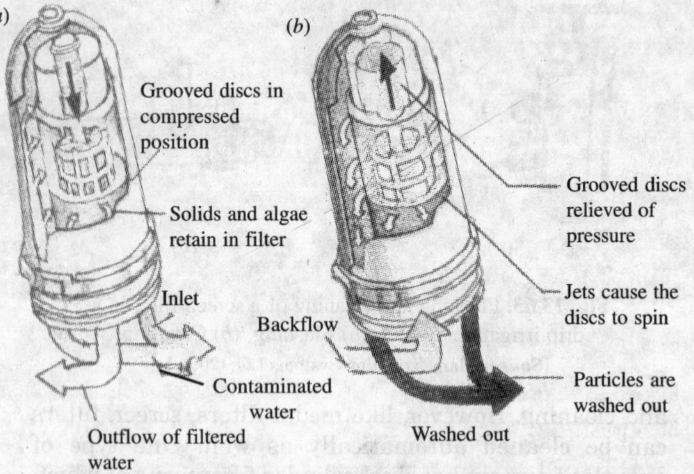

Fig. 13.68. Exposed views (schematic) showing the principle of operation of a disc filter: (*a*) Filtration, (*b*) Backwashing.
[***Source:*** Adapted from Rajput, T.B.S. and Neelam Patel (2001).]

manually by stopping the water flow, opening the disc housing and directing a water jet to start the spinning motion. It can also be done automatically, as the water flow is reversed and a drainage valve is opened when the pressure difference reaches a predetermined value.

There is an angle in the alignment of two adjacent discs, resulting in cavities of varying sizes and partially turbulent flow. The discs are assembled in a cylinder and are pressed together with a nut. The housing is made of plastic and are available in different sizes, mainly 2 cm to 7.5 cm. The flow rates vary from 4 to 30 cu.m/hr. They can be installed in batteries in parallel for very large flow rates.

Fig. 13.67. Disc filter assembly.
[***Courtesy:*** EPC Industries Ltd. (2005).]

Selection of filters and their protection against corrosion. The types, specifications and location of filters used in drip irrigation systems will depend on the quality of irrigation water, the flow rate and the configuration of the system layout. In general, hydrocyclone filters, being primary filters, are installed at or near the head-end of the system, followed by the sand filters. Screen filters and/or disc filters are usually installed at the head-end of the laterals, whenever required. Any one or more of the filters may be adequate, depending on the quality of irrigation water. If sand is of particular concern for a given system, choice of disc filters may not be correct. During back-flushing, when the discs are separated, sand may get lodged between the discs and filtration becomes less effective. In such cases where sand is a major contaminant, use of hydrocyclone combined with a sand and screen filter combination may be the most effective. The quantum of physical contaminant that can be held by a disc filter may be somewhat higher than the screen filter of the same capacity. However, it is less than that of the media filter. Hence, if the source of water contains large quantities of physical impurities, it is advisable to use a media filter than a disc filter.

The housing of all the filters are metallic tanks which need to be protected from the corrosive action of salts in the water source, fertilizer solutions and chemicals being injected into the irrigation system. For protection against corrosion, measures such as systematic surface cleaning and powder coating with pure polyester colours should be followed by the manufacturers.

APPLICATION OF FERTILIZERS AND CHEMICALS THROUGH DRIP IRRIGATION SYSTEMS

Water soluble fertilizers can be effectively and efficiently applied through drip irrigation systems (Fertigation). Reduced labour, equipment and energy costs and higher fertilizer use efficiency are the major benefits of fertigation, compared to the conventional methods of water application. The success of drip irrigation, to a good degree, is due to the improved supply of nutrients to plants, which is a unique capability of fertigation. Proper control of the time of application, the concentration of fertilizers/chemicals, proportion of the nutrients and the location of the point of application are possible with the fertilization systems used in pressurized irrigation systems. Plant protection chemicals can also be applied effectively using the same facility. However, materials such as inorganic forms of phosphorous that form chemical precipitates and cause clogging should not be injected into drip systems.

A detailed treatment on the subject of fertigation is presented in Chapter 12. Figure 13.69 illustrates the layout of a drip irrigation system with fertilization unit. Figure 13.70 illustrates the control head assembly of a drip irrigation system provided with fertilization equipment, filters, control valves and other accessories in a drip system. [Figure 13.71 (*a*) and (*b*) are the views of the fertilizer tank].

The requirement of fertilizer application, including the type of fertilizers to be injected should be considered in designing a drip irrigation system. Some types of fertilizers are not suitable for application through drip systems, because of volatalization of gaseous ammonia, low water solubility and problems with the chemical quality of irrigation water. Therefore, fertilizer injection equipment should be designed with an understanding of the chemical composition of the fertilizer to be used. Nitrogen is relatively problem free.

Nitrogen. Anhydrous ammonia and aqua-ammonia can be injected into the irrigation water, but volatalization is a major problem. Another problem with ammonia injection is the rise of hydroxide ion concentration in water. Ammonia increases the pH, which causes soluble calcium and magnesium to precipitate in irrigation water and coat the inside of the pipes and plug emitters. This problem, however, can be solved by injecting a water softener ahead of the ammonia gas. However, the process may increase the cost of fertilization.

Ammonium sulphate and ammonium nitrate are very common fertilizers. In the former, all the nitrogen is in the ammonium form, and in the latter about 28% by weight of the fertilizer is ammonium nitrogen and 8% is nitrate nitrogen. Urea is a soluble nitrogen fertilizer. It does not react with water to form ions. Urea and ammonium nitrate are mixed in water to give a fairly concentrated liquid fertilizer mixture. All of these nitrogen materials can be injected with no side effects in the water or irrigation system.

Both urea and nitrate nitrogen stay in solution and move with the soil water. However, ammonium nitrogen behaves differently. Because it is a positively charged ion, it enters into cation exchange reactions in the soil. A small change in either soluble constituents alters the relative amount of the ions in exchangeable form. In the exchangeable form, ammonium is immobile. Because cation exchange reactions are very rapid, ammonia applied in irrigation water is immobilized almost instantly on contact with soil and remains on or near the soil surface (US Soil Cons. Service, 1984).

Ammonium applied in water readily converts into exchangeable ammonium and generates an equivalent amount of cations in solution. In semi-arid and arid regions, soils are usually neutral to alkali (pH 7 to 8.20),

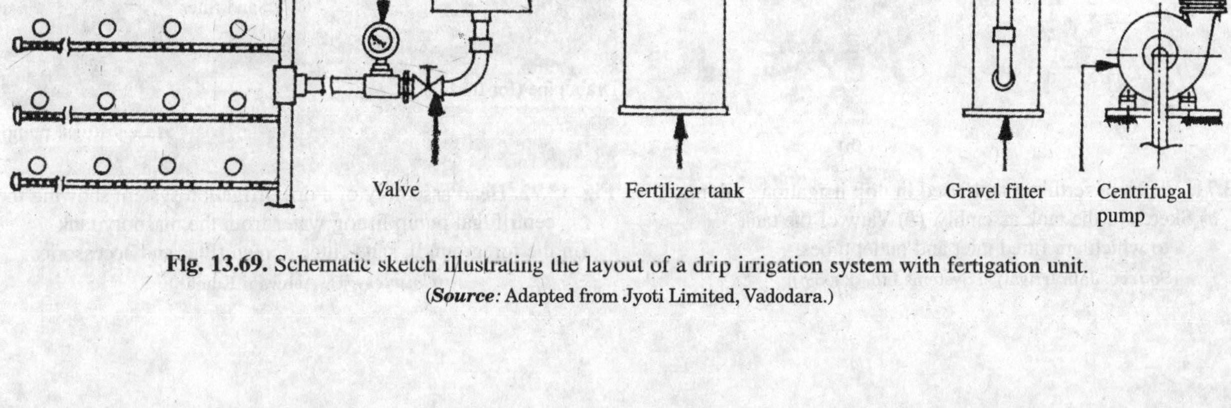

Fig. 13.69. Schematic sketch illustrating the layout of a drip irrigation system with fertigation unit.
(***Source:*** Adapted from Jyoti Limited, Vadodara.)

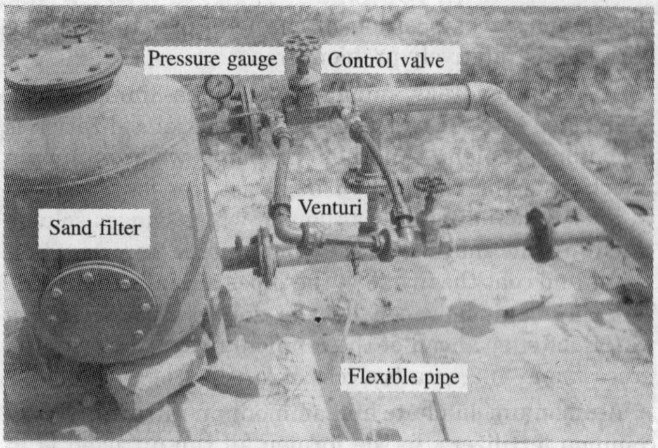

Fig. 13.70. View of venturi system assembly in a pressurized irrigation system. The venturi is projecting from the pipe at the top. Note the transparent plastic pipe connecting the venturi system to the fertilizer tank (not seen).

(***Courtesy:*** Dr. Ashwani Kumar.)

depending on how much free lime or calcium carbonate is present. In such soils, any exchangeable ammonium that exists at the soil surface is likely to volatize. Ammonium is sensitive to temperature and moisture. Water evaporates rapidly from soil after irrigation, and ammonium is susceptible to gaseous loss at this time (US Soil Cons. Service, 1984).

Phosphorous. Phosphorous is difficult to apply by injection into the drip system. Treble-superphosphate, which is commonly used is usually classified as water soluble, but is only moderately soluble. Hence, it is usually not recommended for injection in irrigation water, unless special precautions are taken. Several kinds of ammonium phosphates are soluble in water. Ammonium phosphate sulphate, ammonium phosphate and di-ammonium phosphate are suitable for injection in drip systems when the soil requires nitrogen and phosphorous. Phosphoric acid is another form of soluble phosphorous. The quality of irrigation water is to be determined before injecting phosphorous into the drip system. If the irrigation water has a pH above 7.5 and a high calcium content, the injected phosphorous will precipitate as calcium phosphate and may clog the emitters and restrict the flow in the pipeline. Under such situations, phosphoric acid could be used to meet phosphorous needs. Flushing the drip irrigation system with a solution of either sulphuric acid or hydrochloric acid immediately after applying phosphoric acid prevents clogging. Organic phosphate compounds like glycerol-phosphoric acid can also be injected into the drip irrigation system without any precipitation problem.

Potassium. Potassium is easy for injection through drip irrigation systems. The commonly used potassium oxide is very soluble. The fertilizer moves readily into the soil and is not easily leached away.

Trace Elements. The common trace elements like magnesium, zinc, boron, iron, and copper can be applied through drip irrigation systems. The application rates of trace elements should be based on proper analysis of soil and water, as trace elements applied in excessive quantities can react with salts in the water and can be toxic to plants.

Pumping Unit

Drip irrigation systems have pumping units to lift water from wells, streams, canals, ponds and reservoirs and provide pressure for the operation of the system. The pumps can be powered either by electric motors or internal combustion engines. The types of pumps used in drip irrigation include volute centrifugal pumps, vertical turbine pumps and submersible pumps. Figure 13.72 is a view of the head assembly of a drip irrigation system showing a centrifugal pump lifting water from a water storage tank, sand filter, screen filter and other accessories. Details regarding various types of pumps

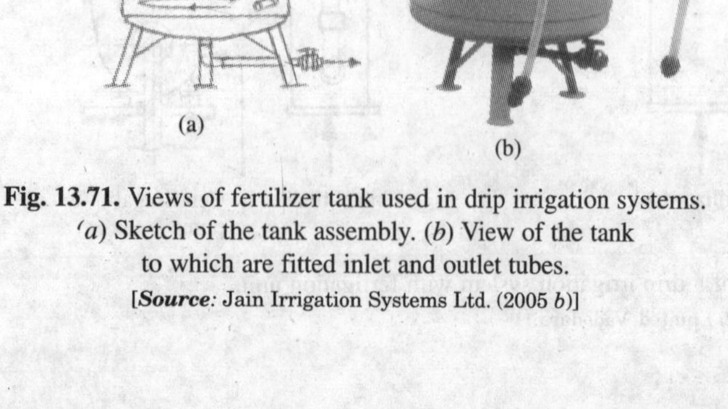

Fig. 13.71. Views of fertilizer tank used in drip irrigation systems. (*a*) Sketch of the tank assembly. (*b*) View of the tank to which are fitted inlet and outlet tubes.

[***Source:*** Jain Irrigation Systems Ltd. (2005 *b*)]

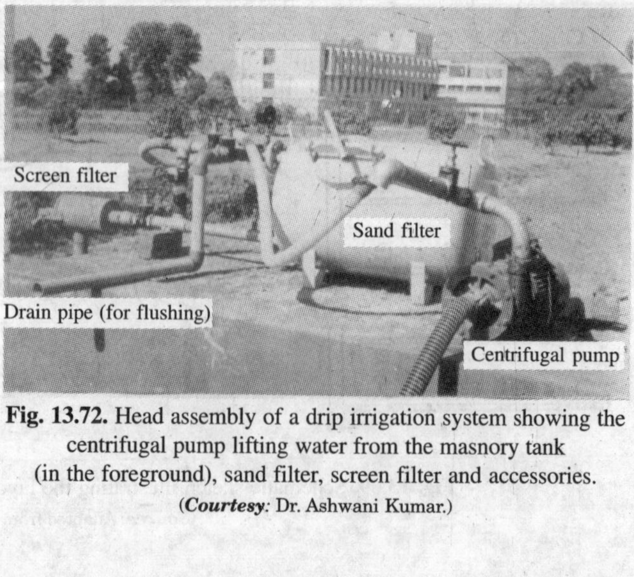

Fig. 13.72. Head assembly of a drip irrigation system showing the centrifugal pump lifting water from the masonry tank (in the foreground), sand filter, screen filter and accessories.

(***Courtesy:*** Dr. Ashwani Kumar.)

and prime movers and accessories, and their selection and application are presented in Chapter 3 of this book.

INSTALLATION, OPERATION AND MAINTENANCE OF DRIP IRRIGATION SYSTEMS

Many problems with drip irrigation systems arise out of poor installation. The main items in the installation of drip irrigation systems are the installation of the head assembly (Control head), comprising the pumping set, non-return valve, water meter, filters, fertilization equipment, and flow control, air release and pressure release valves. The network of water distribution pipes including the main line, sub-mains, and laterals require careful attention in joining and installation. General requirements in installation, operation and maintenance were included in the description of the various components of drip irrigation systems presented earlier.

Installation of filters and fertilization unit. The major points to be considered in the installation of the filtering system are the use of minimum number of accessories like elbows and reducers. Schematic sketch showing the assembly of various components of the system is shown in Fig. 13.20. Normally, the pump delivery pipe should be connected directly to the hydrocyclone or the media filter, followed by the fertilization equipment, and the screen filter, all of which are installed in the main pipe. Arrangements for backwashing of the filter and the disposal of the contaminated water are essential requirements. Suitable facility to collect or to dispose off the bypass water is necessary. Sufficient space is to be provided for the operation of filter valves and other valves. Hard surface or masonry/concrete foundation is to be provided for the sand filter and the hydrocyclone, so that they will not collapse due to vibration and load. For a screen filter, a strong support by using G.I. fittings is necessary to avoid its vibration due to load. Use of 'hold-tight' over the threads of G.I. fittings and application of proper mixture of 'M-seal' over the joints uniformly are necessary to prevent leakages.

Installation of mains and sub-mains. It is a desirable practice to lay the mains and sub-main pipes underground, at a depth not less than 45 cm to avoid damage due to vehicles and equipment. The United States Soil Conservation Service has recommended the following minimum cover of earth over for various pipe sizes (Fred Hamisch, 1977):

Pipe size	*Depth of earth cover*
1.2 to 6 cm diameter	45 cm
6 to 10 cm diameter	60 cm
Over 60 cm diameteı	75 cm

The PVC pipes are usually joined by solvent welding. Pipe sizes of 4 cm diameter and above are usually connected by gasketed joints. Thermal properties of PVC pipes require that an allowance for expansion and contraction be made when the pipeline is assembled. This is normally accomplished by 'snaking' of the PVC pipe in the trench. PVC pipe is affected by sunlight and heat. The pipe must be buried, or otherwise protected from these elements. Otherwise, it will become brittle and lose many of its valuable properties. The pipe should be laid in the trench and allowed to come to within a few degrees of the lowest possible temperature before backfilling. When the trench contains rock, hard pan, or boulders, which might damage the pipe, it should be surrounded with good clean fill material to a depth of 15 cm over the pipe with no stones larger than 2.5 cm (Fred Hamisch, 1977). The remaining portion of the trench may be filled with the available excavated material, avoiding large stones.

The assembly procedures for PVC solvent weld pipes are available from the manufacturers of PVC pipes. An important point to be kept in mind in making a good joint is to avoid leaving any excess solvent on the joint. No excess solvent should be allowed on the outside or the inside of the joint. The solvents used in making the joints are flammable and toxic. Hence, care must be taken when handling them. Solvent jointing requires a certain amount of skill which can be acquired.

The installation procedures for gasket jointed pipes are the same as for solvent-weld PVC pipes. However, the assembly procedures are simplified, as gasket joints do not require solvent cement application. In both the cases, thrust blocks should be used at all changes of direction, tees and bends, changes of pipe size, dead-ends and valves. The size and shape of the thrust block will depend on the operating pressure, pipe size and soil type. Normally, a thrust block is a stiff concrete poured between the pipe and the undisturbed bearing wall of the trench.

The main points to be kept in mind while connecting mains and sub-mains are the removal of mud and other impurities in the pipe before fitting, fitting a gunmetal gate valve at the start of the sub-main, and providing flush valves at the end of sub-mains (Fig. 13.20). The main and sub-mains are made firm by putting soil near the flush valve and at different places so that the pipe

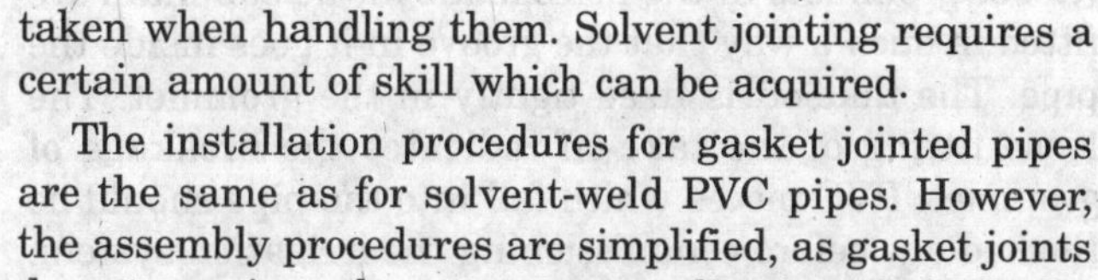

Fig. 13.73. Installation of drip irrigation laterals/thin walled drip tape in drip irrigation systems.
[***Source:*** Jain Irrigation Systems Ltd. (2005 *b*)]

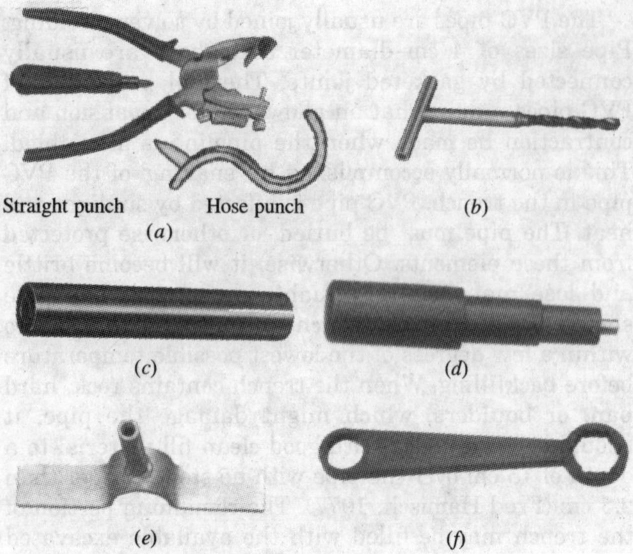

Fig. 13.74. Common tools used with drip irrigation systems: (*a*) Dripper punch, Straight punch, Hose punch, (*b*) Drill, (*c*) Punch for grommet take off, (*d*) Steel pin punch, (*e*) Drill guide, (*f*) 'J-Turbo Key' spanner [***Source:*** Jain Irrigation Systems Ltd. (2005 *b*)]

will not move while making a drill hole into it. The rubber grommets in the holes made in the sub-main are fitted in such a way that the groove in it goes inside the pipe. The take-off is fixed tightly in the grommet. The loose fitting of the take-off indicates the breakage of grommet. PVC pieces which fall into the pipe should be flushed out before commissioning the irrigation system.

Punching of lateral and fixing of drippers. *Punching of laterals* (Fig. 13.73). The following procedures may be followed in the punching of laterals: (*i*) Pass water through the polytube so that it gets bulged and easy for punching. (*ii*) Punch the lateral sideways. (*iii*) The dripper position is fixed according to design. (*iv*) All the drippers should come on the same straight line. (*v*) Do not fix the drippers until a complete lateral line is punched. (*vi*) Punching should start from the sub-main.

Fixing of drippers: (*i*) In case of drippers having arrow marks, the arrow head should be towards the sub-main. (*ii*) While fixing the dripper, push it into the lateral and pull it slightly. (*iii*) Close the end of the lateral by fitting end cap. (ICAR, 2003 – contribution by Dr. Ashwani Kumar). Figure 13.74 shows the common hand tools used in punching holes in laterals and fixing drippers. However, in medium and large scale installations the drippers should preferably be installed in the factory.

Testing of the drip irrigation system. After the irrigation system is installed, it is tested by adopting the following procedure: (*i*) Backwash the filter till clean water comes out through its flush valve. (*ii*) Ensure that all the valves are open before the testing. (*iii*) Close the flush valve after the sub-main is completely flushed. (*iv*) When the laterals are completely flushed, close them with end caps. (*v*) Check the pressure on the gauge installed on the filter. (*vi*) If excess pressure is observed, open the by-pass valve slowly till the desired pressure is obtained. (*vii*) Measure the discharge at a minimum of 3 different places. For this the volumetric method may be used. (*viii*) Check the working of the air release valve on the sub-main. (*ix*) After fitting the entire system, install the venturi on the filter (ICAR. 2003 – contribution from Dr. Ashwani Kumar).

Operation of the Drip Irrigation System

The proper operation of a drip irrigation system involves the following steps for the owner/operator : (*i*) Acquiring complete information and instructions from the designer and dealer. (*ii*) Determining when and how long to irrigate. (*iii*) Checking the water meter readings and recording the figures. (*iv*) Accurately setting the hydraulic metering valve. (*v*) Operating the head valve to begin irrigation. (*vi*) Checking the system along all components for proper operation, beginning with pressure readings at the header. (*vii*) Checking the emitters, at least on a random basis. (*viii*) Setting the chemical and fertilizer injection equipment (US Soil Cons. Service, 1984).

Maintenance of Drip Irrigation Systems

Reliable performance of a drip irrigation system depends on preventive maintenance that includes proper filtration, pipe flushing, field check of leakage in any component of the system, and monitoring the effective functioning of mechanical devices. General maintenance includes periodic checking of drippers, including their discharge, wetting zone, and leakage of pipes and filter gaskets in the lids, flushing valves and fittings. The placement of drippers should be periodically inspected. In case the placement is disturbed, they should be put at the proper location. Filter is the heart of the drip irrigation system. Filter failure will lead to clogging of the entire system. Various methods for cleaning of filters were explained earlier.

Flushing of sub-mains and laterals. After installation/repairs, the system should be flushed systematically, beginning with the main line and proceeding to the sub-mains, manifolds, and laterals (Figs. 13.76 and 13.77). The main lines and then the sub-mains should be flushed one at a time with the manifold or riser valves turned off. Closing the valves on all lines, except the one being flushed, allows a large flow of water. The manifolds should be flushed with all the lateral riser valves turned off. Finally, the lateral hoses should be connected and flushed for about an hour at each operating station.

Sometimes silt escapes through the filters and settles in sub-mains and laterals. Further, some algae and bacteria lead to the formation of slimes/pastes in the pipe and laterals. To remove these silts and slimes, opening the flush valve should flush the sub-mains.

(a) (b)

Fig. 13.75. Flushing of main/sub-main pipe in a drip irrigation system. (*a*) Drain pipe fitted with cap; (*b*) Opening of the cap for flushing.
[*Source:* T.B.S. and Neelam Patel (2001).]

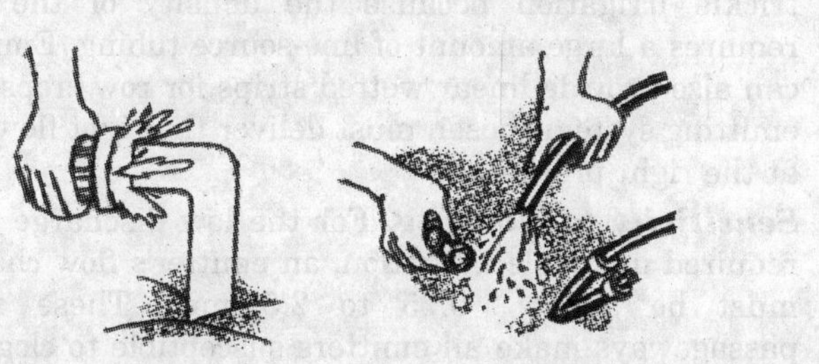

(a) (b)

Fig. 13.76. Flushing of sub-main and laterals in a drip irrigation system: (*a*) Flushing of sub-main, (*b*) Flushing of drip lateral tube/drip tape.
[*Source:* Jain Irrigation Systems Ltd. (2005 b)]

Removing the end stops flushes the lateral lines. By flushing, even the traces of accumulated salts will also be removed. Stop the flushing once the water going out is clean.

Fine sand, silt and clay tend to settle in the low-velocity section of the system, at the ends of manifolds and laterals. Emitters receiving high concentrations of fine contaminants are susceptible to clogging; therefore, periodic flushing is a recommended part of a good maintenance program. Annual flushing is enough for many systems, but some water and emitter combinations require more frequent flushing to control clogging.

Chemical treatment. Clogging or plugging of drippers will be due to precipitation and accumulation of certain dissolved salts like carbonate, bicarbonate, iron, calcium and manganese salts. Clogging is also due to the presence of microorganisms and the related iron and sulphur slimes due to algae and bacteria. Clogging or plugging is usually avoided/cleared by chemical treatment of water. Chemical treatments commonly used in drip irrigation systems include addition of chloride and/or acid to the water supply.

Acid treatment. Hydrochloric acid (HCl) is injected into drip systems, at the rate suggested. The acid treatment is performed till a pH of 4 is observed and the system is shut for 24 hours. Next day the system is flushed by opening the flush valve and lateral ends.

Chlorine treatment. Chlorine treatment in the form of bleaching powder is performed to inhibit the growth of organisms like algae and bacteria. The bleaching powder is dissolved in water and this solution is injected into the system for about 30 minutes. Then the system is shut off for 24 hours. After 24 hours, the lateral ends and flush valves are opened to flush out the water with impurities. Bleaching powder is directly added into the source at a rate of 2 mg/litre.

Standardization and Quality Control of Drip Irrigation Systems

Indian standards for different facets of drip irrigation systems are formulated by Irrigation and Farm Drainage Equipment and System Sectional Committee of the Bureau of Indian Standards. They are listed at the end of this chapter. Most of the Indian Standards on irrigation equipment are based on corresponding international standards, which have been adopted after suitable modifications to meet indigenous requirements.

Design of Drip Irrigation System

Design of a drip irrigation system broadly involves the estimation of water requirement of the crops to be irrigated, number of drippers and laterals, diameter of the main, sub-main and lateral pipes, and the size of the pumping unit. The basic principles of soil-water-plant relationships, and water requirement of crops are the same as explained in Chapters 9 and Chapter 10, respectively. The major difference between drip irrigation and other methods of irrigation is that not all the area is irrigated. Under normal conditions in widely spaced crops, a minimum of 30% of the area is irrigated when the drip system is adopted. For mature trees, about 75% of the area may have to be irrigated, while nearly the entire area is irrigated in case of closely-spaced crops. The necessary data to be collected includes soil type, infiltration characteristics of the soil, type of crops, consumptive use of water by crops, water quality, and climate, Class A pan evaporation data, and the availability of funds. A topographical map of the area to be irrigated, with contour lines at a spacing of about one metre is necessary in planning and layout of the system.

A check list of procedures in designing a drip irrigation system is given below:

(*i*) Prepare on inventory of available resources and operating conditions. (*ii*) Determine the water requirement to be met by the drip system. (*iii*) Determine the appropriate type of the system. (*iv*) Determine the type and design of emitters (on-line/in-line and the specific type required). (*v*) Determine the capacity of the pumping system. (*vi*) Decide on the economic sizes of the piping system. (*vii*) Determine the maximum and minimum operating pressures and the minimum efficiency requirement. (*viii*) Determine the appropriate filtering system to suit the site conditions. (*ix*) Determine the requirement of the fertilization system. (*x*) Plan field evaluation. (*xi*) Prepare drawings, specifications, cost estimates, schedules, and instructions for proper layout, installation, operation, and maintenance.

Emitter Selection Criteria

Emitter design and selection procedures require an estimation of discharge, spacing, and the type of emitter to be used. The efficiency of the drip irrigation system depends mainly on the selection of the type of emitter and its design. The following are the major characteristics of the emitter that influence the efficiency of the irrigation system:

(*i*) Discharge rate variations caused by emitter variations within manufacturing tolerance. (*ii*) Closeness of discharge-pressure relationship to design specifications. (*iii*) Emitter discharge exponent. (*iv*) Possible range of suitable operating pressures. (*v*) Pressure loss on lateral lines caused by the connection of emitters to the lateral. (*vi*) Susceptibility to clogging, siltation, or buildup of chemical deposits. (*vii*) Stability of discharge-pressure relationship over a long period.

The choice of emitters depends not only on emitter physical characteristics, but also on emitter placement, type of operation, diameter of laterals and user's preference. Selection requires four steps: (1) Evaluate and choose the general type of emitter that best meets the need in the area to be wetted. (2) Choose the specific emitter needed to meet the required discharge, spacing and other planning considerations. (3) Determine the average emitter discharge (q_a) and pressure-head (h_a) requirements. (4) Determine the allowable sub-unit pressure-head variation (ΔH_3) for the desired emission uniformity (EU).

Critical items in emitter selection are the % area wetted (P_w) and the emitter reliability (resistance to clogging and malfunctioning). The greater the P_w, the longer the system can be down or an emitter can be plugged before the plants become excessively stressed (US Soil Cons. Service, 1984).

Initially, emitter selection depends on the soil, plant water requirement, emitter discharge, water quality, emitter cost and terrain of a particular location. It is necessary to have emission points to wet at least one-third and up to one-half of the potential horizontal cross-section of the root system. There are some interactions between the emitter discharge rate and area wetted per emission point; but the density of emission points required to obtain $P_w \geq 33$ % can usually be based on a 1-gph emitter discharge rate by using the procedures described under 'Area Wetted'. The water required for plant growth increases until the plant reaches its peak-use growth stage (Chapter 10).

General suitability. General emitter suitability means how well the emitter fits into the particular design and matches the size and water requirements of the crop. Emission devices are available that will emit water at individual point locations or along the length of a line. Point source emitters are available with single or multiple outlets. With more than one outlet, distribution tubing is generally used to deliver the water from the emitter to the desired discharge location. Single-outlet emitters can be used to irrigate small individual areas or can be arranged around larger plants to provide multiple-outlet emission points. Multiple-outlet emitters are generally used in orchards, where each tree may require several emission points. For row crops such as vegetables, line-source tubing fits well with the cropping pattern because it provides the linear wetted strip desired. Cost is especially important in row-crop trickle irrigation because the density of the crop requires a large amount of line-source tubing. Emitters can also provide linear wetted strips for row crops. The emitting system chosen must deliver the right flow rate at the right pressure.

Sensitivity to clogging. For the low discharge rates required in trickle irrigation, an emitter's flow channel must be about 0.25 to 2.5 mm. These small passageways make all emitters susceptible to clogging. Therefore, careful filtration of irrigation water is required. Filtering to remove particles 10 or more times smaller than the emitter passageway is a typical recommendation. Self-flushing type emitters require less filtration. Long-path emitters, which have the largest passageways for a given flow rate, may still require filtering of even the smaller particles to prevent clogging.

Manufacturing Variation in Emitters. The variations in passage size, shape, surface and finish that occur are small in absolute magnitude but represent a relatively large percent variation. Also, some emitters use an elastomeric material to achieve a pressure-compensating or flushing ability, and such materials are inherently difficult to prepare with consistent dimensions and characteristics. The amount of difference to be expected varies with the design of the emitter materials used in its construction, and the care with which it is manufactured. The emitter coefficient of manufacturing variation (v) is used as a measure of the anticipated variations in discharge in a sample of new emitters. The value of C_v should be available from the manufacturer, or it can be estimated from the measured discharges of a sample set of at least 50 emitters operated at a reference pressure head. The value of C_v can be computed in FPS units as follows (US Soil Cons. Services, 1984):

$$C_v = \frac{S}{q}$$

$$= \frac{\sqrt{q_1^2 + q_2^2 \ldots + q_n^2 - n(q)^2}}{q\left(\sqrt{n-1}\right)} \quad \ldots(13.3)$$

in which,

C_v = emitter coefficient of manufacturing variation.

$q_1, q_2 \ldots q_n$ = individual emitter discharge-rate values, gallons per hour.

n = number of emitters in sample.

q = average discharge rate of the emitters sampled, gallons per hour.

S = standard deviation of the discharge rates of the sample.

The value of C_v is a useful characteristic with consistent physical significance, because the discharge rates of emitters at a given pressure are essentially normally distributed.

The recommended ranges in the values of C_v in point source and line source emitters are as follows:

(i) Point source emitters	Less than 0.05 is rated as *excellent*, 0.05 to 0.07 is rated as *average*, 0.07 to 0.11 is rated as *marginal*, 0.11 to 0.15 as *poor*, and above 0.15 is rated as *unacceptable*.
(ii) Line source emitters	Less than 0.10 is rated as *good*, 0.10 to 0.20 is rated as *average*, and less than 0.2 as *marginal* to *unacceptable*.

System coefficient of manufacturing variation

The system coefficient of manufacturing variation (v_s) is a useful concept because more than one emitter or emission points may be used per plant. In such a situation, the variations in flow rate for each emitter around the plant partly compensate for one another.

The value of v_s can be computed by following equation (US Soil Cons. Service, 1984):

$$v_s = \frac{v}{\sqrt{e'}} \qquad \ldots(13.4)$$

in which,

v = emitter coefficient of manufacturing variation.

e' = minimum number of emitters per plant, or 1 if one emitter is shared by more than one plant.

Emission Uniformity in Drip Irrigation

The coefficient of uniformity proposed by Christiansen (1942) is widely accepted for estimating the uniformity in water application in sprinkler irrigation systems (Chapter 12). The Christiansen formula, when applied to drip irrigation systems will give only a broad indication of the uniformity of water application, as the emission characteristics of droppers used in drip systems are significantly different from those of sprinkler nozzles. The water application uniformity of emitters (except pressure compensating type emitters) is influenced by the operating pressure, emitter spacing, land slope, size of the pipe line, emitter discharge rate, and emitter discharge variability. The emitter discharge variability is due to the variations in operating pressure and temperature, manufacturing variability (coefficient of variation, C_v), clogging, and aging of the emitters.

Design emission uniformity (Uniformity coefficient). Karmeli and Keller (1975) were the first to define an empirical design emission uniformity percentage EU for the evaluation of the performance of drip irrigation systems (Hillel, D., 1982). Based on it the following equation is commonly used to estimate the design emission uniformity in point source and line source drip irrigation systems (Karmeli and Keller, 1975; Hillel, D. 1982; Schwab *et al.*, 1996 and Bureau of Indian Standards, 1999):

$$EU = 100\left[1.0 - \frac{1.27\,C_v}{\sqrt{n}}\right]\frac{q_m}{q_a} \qquad \ldots(13.5)$$

in which,

EU = design emission uniformity, %

n = (i) for point source emitter on a perennial tree crop, number of emitters per plant,

(ii) for a line source emitters on a perennial tree crop, either the rooting diameter of the plants divided by the same unit length of lateral used to compute C_v or 1, whichever is greater.

C_v = the manufacturer's coefficient of variation for point source or line source emitters,

q_m = the minimum emitter discharge rate for a minimum pressure in the section (sub-unit), litres/hr.

q_a = the average or design emitter discharge for the section (sub-unit), litres/hr.

The commonly recommended ranges in design emission uniformity are as follows:

(i) Point source emitters for tree crops with spacing larger than 4 m:

(*a*) Land slope less than 2% – 90 to 95%

(*b*) Land slope greater than 2% – 85 to 90%

(ii) Point source emitters on tree crops and bushes spaced less than 4 m:

(*a*) Land slope less than 2% – 85 to 90%

(*b*) Land slope greater than 2% – 80 to 90%

(iii) Line source emitters on annual or perennial crops:

(*a*) Land slope less than 2% – 80% to 90%

(*b*) Land slope greater than 2% – 75 to 85%

Example 13.1. *The following data were obtained in a field test to determine the emission uniformity (uniformity coefficient) of a drip irrigation lateral:*

$C_v = 0.07$, $q_{min} = 45$ litres/hr, $q_{ave} = 50$ litres/hr, land slope 1.5%.

Solution:

$$EU = 100\left[1.0 - \frac{1.27 \times 0.07}{\sqrt{1}}\right]\left(\frac{45}{50}\right)$$

$$= 82.008\%.$$

The value of EU in Example 13.1 is lower than the recommended value stated above. The EU could be improved by reducing the difference between q_{min} and q_{ave} (by using larger diameter and / or shorter laterals or by using pressure compensating emitters) or by using an emitter with a lower C_v.

Relationship between Pressure Head and Discharge

The relationship between changes in pressure head and discharge is an important characteristic of emitters. In Equation 13.2, pressure compensating emitters have a low value of the exponent x. However, since they have some physical part that responds to pressure, their long-range performance requires careful consideration. The compensating emitters usually have a high coefficient of manufacturing variation (v), and their performance may be affected by temperature, material fatigue, or both. On undulating terrain, the design of a highly uniform system is usually constrained by the pressure sensitivity of the average emitter. Compensating emitters provide the solution. Emitters of various sizes may be placed along the lateral to meet pressure variations resulting from changes in elevation. In laminar-flow emitters, which include the long-path, low-discharge devices, the relation between the discharge and the operating pressure is linear, i.e., doubling the pressure doubles the discharge. Therefore, the variations in operating pressure head within the system are often kept to within ± 5 percent of the desired average. In a turbulent-flow emitter, the change in discharge varies with the square root of the pressure head, *i.e.*, $x = 0.5$, and the pressure is to be increased four times to double the flow. Therefore, the pressure head in drip irrigation system with turbulent flow emitter is often allowed to vary by ± 10% of the desired average (US Soil Cons. Service, 1984).

Layout of Drip Irrigation System

The layout of a drip irrigation system comprise of the alignment and lengths of main, sub-mains and lateral lines connected to the pumping system installed at the water source. The usual practice is to lay the mains and sub-mains across the slope, while the laterals are laid along the slope.

Irrigation Water Requirement

The irrigation water requirement of an area depends on the type of crop, weather data (mainly the mean monthly evaporation data obtained from the Class A pan evaporimeter, excluding the rainfall in the month), type of soil, and area under cultivation. The monthly irrigation water requirement can be estimated on the basis of the monthly pan evaporation data and the crop coefficient. Detailed procedures for estimating the water requirement of crops, which differ from place to place, are given in Chapter 10. The procedures described in Chapter 10 are for universal application, based on the geographic location of a place. Table 13.2 presents the normal monthly pan evaporation based on the data from the meteorological stations maintained by the India Meteorology Department as well as the site specific data at Delhi (IARI). The crop coefficients of most of the major crops are given in Chapter 10.

Table 13.1 Normal monthly pan evaporation data.

Month	*All India (average) mm*	*Delhi*
January	99.2	82.9
February	119.6	114.2
March	176.3	202.8
April	210.2	272.1
May	245.4	365.3
June	198.8	312.9
July	145.6	211.3
August	134.6	237.1
September	134.6	222.7
October	144.6	161.2
November	112.2	109.7
December	94.4	87.8

[*Source:* Rajput, T.B.S. and Neelam Patel, 2001 (based on IMD data)]

The monthly irrigation water requirement can be estimated on the basis of monthly pan evaporation data and crop coefficient by using the following equation (Rajput, T.B.S. and Neelam Patel, 2001):

$$V_m = K_c \times K_p \times C_c \times E_p \times A \qquad ...(13.6)$$

in which,

V_m = Monthly irrigation water requirement, L

K_c = Crop coefficient

C_c = Canopy factor (C_c = 1.0, for closely spaced field crop, C_c = wetted area/plant area for orchards and vegetable crops)

K_p = Pan evaporation factor (generally 0.8)

E_p = Normal monthly pan evaporation, mm

A = Area to be irrigated, m^2

Capacity of Drip Irrigation System

The capacity of a drip irrigation system is based on irrigation water requirement of the irrigated area, the duration of operation of the system per day, irrigation interval and irrigation efficiency. The duration of operation of each section (block) of a drip irrigation system is usually limited to 1.5 to 2 hours, so as to avoid deep percolation losses. In most cases, the irrigation interval may be on the basis of irrigating daily, every alternate day, or once in 3 days. Irrigating intervals larger than 3 days may cause soil moisture stress in some crops irrigated by the drip system. The capacity of a drip irrigation system may be computed using the following relationship (Rajput, T.B.S. and Neelam Patel, 2001:

$$Q = V_d \times T\,(\eta \times t) \qquad \text{...(13.7)}$$

in which,

Q = Capacity of drip system, Lph

V_d = Daily water requirement, L

T = Irrigation interval days

n_a = Water application efficiency (in fraction)

t = Duration of each irrigation in hrs

The discharge required per plant (Q_p) can be estimated by dividing the capacity of the drip system (Q) by the number of plants (n) in the area irrigated (i.e., $Q_p = Q/n$).

Number of Laterals Required

In case of vegetable crops, usually there is one lateral for each row of crops. For orchards the number of laterals usually range from 1 to 2 per row of trees, depending on the plant-to-plant and row-to-row spacing of the plants to be irrigated and the type of soil. The number of drippers required per plant can be estimated by dividing Q_p by the discharge of the selected dripper. An *emission point* is a location where a concentration of water is applied by one or more emission devices or drippers. The total number of drippers or emission points required in the area can be estimated by multiplying the number of drippers per plant with the number of plants to be irrigated. Detailed procedure for emitter spacing in an orchard is given in the following section.

In close-growing field crops, the drippers are to wet the whole area and act as a line source of water rather than a point source. Hence, in close growing crops a large number of drippers are required, if point source emission points are selected. This may sometimes create problems. Therefore, line source emitters comprising emitting pipes or laterals with built-in drippers at 30 to 40 cm spacing are often used.

Emitter Spacing

Emitter spacing and the number of emitters required depends on the wetting pattern of emission points and the spacing of the plants. Closely spaced drippers result in a line source drip irrigation system, while islands of wetted areas are created by locating the drippers apart. Due to the heterogeneity of the soil in most fields, it is advisable to conduct field tests at several representative sites in the field to be brought under drip irrigation and collect data on the horizontal and vertical movement of water from point source and line source emitters for deciding the emitter spacing. Under situations where data based on field measurements are not available, estimates of the maximum horizontal wetted diameter from a single outlet can be obtained from Table 13.2.

Table 13.2. Estimated maximum diameter of the wetted circle formed by an emission point in a drip irrigation system comprising point source drip outlets at a discharge rate of 4 lit./hr.,

(***Note***: For line source application, the values should be multiplied by a factor of 0.8.)

Depth of root zone and soil texture	*Homogeneous soil layers*	*Soil profiles of varying textural groups*	
		Mostly of low density	*Mostly of moderate density*
	m	*m*	*m*
Root zone depth, 0.75 m:			
Coarse soil	0.45	0.75	1.05
Medium soil	0.90	1.20	1.50
Fine soil	1.05	1.50	1.80
Root zone depth, 1.5 m:			
Coarse soil	0.75	1.40	1.80
Medium soil	1.20	2.10	2.70
Fine soil	1.50	2.00	2.40

[***Source:*** Adapted from United States Soil Cons. Service, (1984).]

When the system is used for irrigation with saline water, the spacing between the laterals and between the drippers will have to be suitably reduced. The estimated area mentioned in Equation 13.9 is given as a rectangle with the wetted width equal to the maximum expected diameter of the wetted circle and the optimum emitter spacing equal to 80% of the wetted circle and the optimum emitter spacing equal to 80% of that diameter. (Multiplying the lateral spacing and the emitter spacing will give about the same area as that of a circular wetted area). If the quality of irrigation water is saline, the spacing is further reduced another 80%, which provides overlap of the wetting patterns both along and between laterals and reduces salt accumulation within the wetted area. The closer spacing thus obtained will reduce the dry areas between emitters, where salts are likely to accumulate.

$$\text{Number of drippers per plant} = \frac{\%\ \text{total area shaded by the tree} \times \text{Area per tree}}{\text{Effective area wetted by a single emitter}} \qquad \text{...(13.8)}$$

The total area shaded by the tree will depend on the spacing between the trees. The effective area covered by an emitter will depend on the wetted diameter, and water quality.

Example 13.2. *In an orchard with fully grown trees, determine the number of emitters required per tree if 50% of the area is to be irrigated under the following conditions:*

(i) Good quality irrigation water,

(ii) Saline irrigation water,

(iii) The estimated area as given in Equation 13.9.

The soil is of medium texture and the soil layers are of medium density. The effective depth of root zone is 1.5 m. The tree-to-tree spacing in the orchard is 3.5 m and the spacing between laterals is 6 m.

Solution: Using Equation 13.8 and referring to the values presented in Table 13.2.

The diameter of the wetted circle = 2.4 m

(*i*) The number of emitters required for irrigation with good quality water.

$$= \frac{0.5 \times 6 \times 3.5}{0.8(2.4) \times 1.5} = \frac{10.5}{2.88} = 3.646 \text{, or 4 emitters/tree}$$

(*ii*) The number of emitters required for irrigation with saline water

$$= \frac{0.5 \times 6 \times 3.5}{0.8(2.4) \times 0.8(1.5)} = \frac{10.50}{2.304} = 4.56 \text{ or 5 emitter / tree}$$

Area irrigated by a dripper. The area irrigated by an emission device or dripper is computed using the following relationship:

$$A_i = \frac{L \times S \times P}{100 \times N_e} \quad ...(13.9)$$

in which,

A_i = Area irrigated, m^2

L = Spacing between adjacent plant rows, m

S = Spacing between emission points, m

P = % of cropped area to be irrigated

N_e = Number of drippers at each emission point

The term P is 100% when the entire area is irrigated. The value of P generally varies from 30% to 100%, depending on the crop and its stage of growth. It is large in mature crops and crops with close row spacing. For widely spaced crops like vine, bush and tree crops, 30 to 60% of the horizontal cross-section of the root zone is irrigated and a part of the land surface remains dry. In areas which have significant rainfall during the irrigation season, a design that irrigates less than one-third of the horizontal cross-section of the root system is preferred. Determination of P and N_e for the desired wetting pattern requires data on the horizontal and vertical movement of water in the soil, preferably based on field tests.

Loss of Pressure Head in Pipelines

The pipe line in a drip irrigation system is of plastics (PVQ polyethylene) and comprise of the main line, sub-mains and laterals. Drip irrigation system design must ensure nearly uniform discharge of the drippers in each section that is controlled by a valve and irrigated as a unit of the system. The maximum pressure difference allowable in a system is 20% and the maximum difference in pressure between the head end and the tail end of a lateral should not exceed 10%. The relation between pressure and discharge can be obtained from the manufacturers catalogues. The pressure loss can be estimated from nomographs, and the physical relationships expressed in the form of equations.

Graphical procedure using nomograph. Figure 13.77 is the flow nomograph for polyethylene pipes of size ranging from 15 mm to 500 mm. The nomograph has 4 scales, namely, internal diameter of the pipe (Scale A), flow rate in litres/second and litres/minute (Scale B), flow velocity in metres/second (Scale C), and hydraulic gradient, representing the loss of pressure head due to friction in the pipe in metres/100 m length of the pipe (Scale D). To use the nomograph, at least 2 values on the lines should be known. Joining the two values on the lines and extending the line formed by the two points will provide the desired values.

Example 13.3. *The internal diameter of a 50 m long drip irrigation main pipe is 50 mm and the discharge rate, as obtained from the flow meter installed in the main line is 0.4 litres/sec. Determine (i) The flow velocity, (ii) The head loss due to friction in the main pipe.*

Solution: Referring to Fig. 13.77,

Internal diameter of the pipe = 50 mm,

Discharge rate = 0.4 lit/sec.

Joining the two points on the respective scales A and B and extending the straight line so formed to meet scales C and D.

Flow velocity = 0.21 lit/sec.

Friction head loss in the pipe = 0.14 m/ 100 m = 0.07 m (for the 50 m pipe).

Use of hydraulic formulae. The loss of pressure head due to friction in the mains and sub-mains of drip irrigation systems can be estimated using the Darcy-Weishbach equation, the Blassius equation, or William and Hazen equation, or combination of Darcy-Weishbach and Blassius equations. The application of Darcy. Weisbach equation for estimating friction loss in pipe lines has been explained in Chapter 7.

Diameter of Drip Irrigation Lateral Pipe

Laterals convey water from main lines and sub-mains to the drippers or emission devices. Drip laterals are designed to maintain an acceptable variation in the discharge rates of the drippers along their length. The

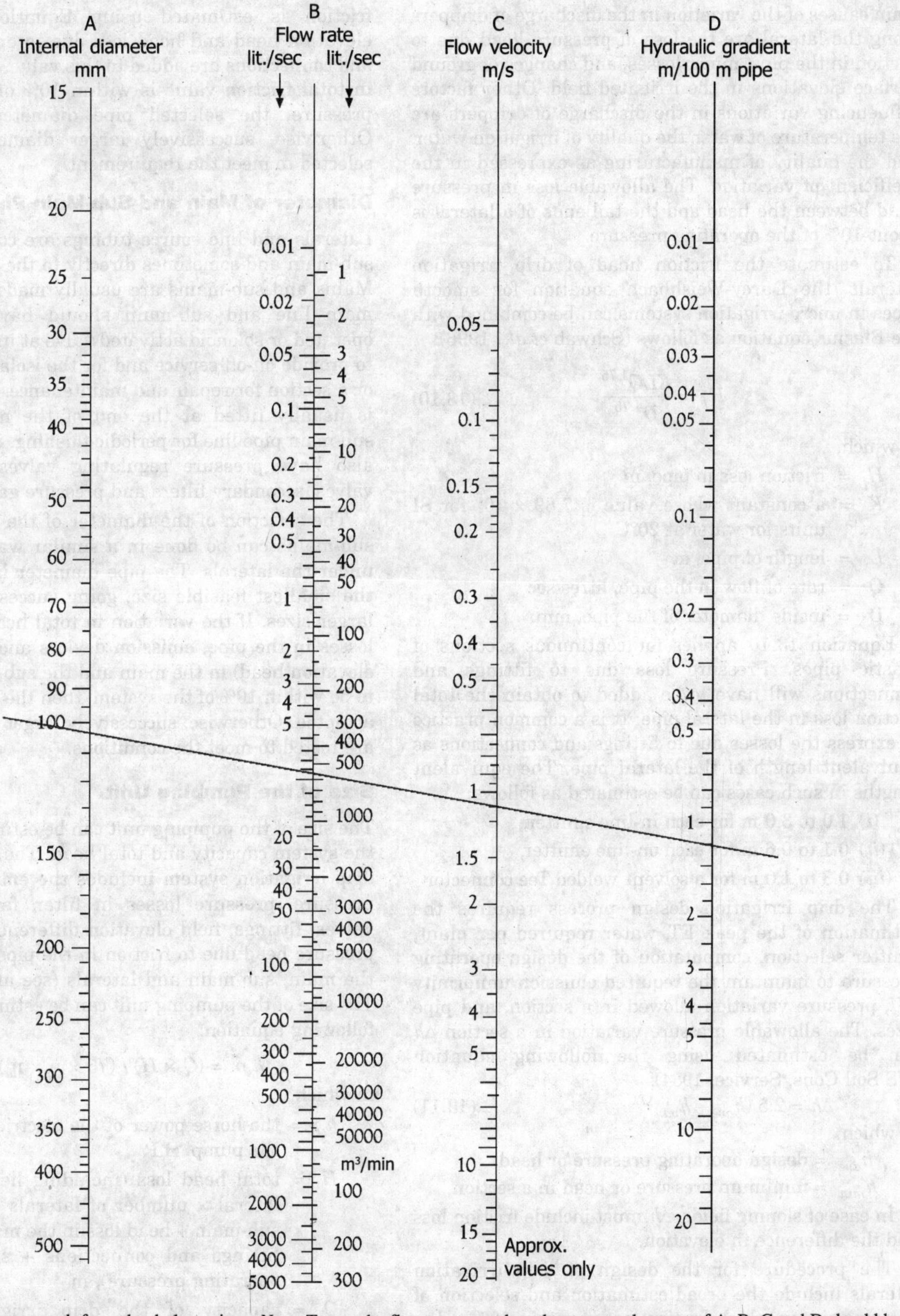

Fig. 13.77. Flow nomograph for polyethylene pipes. *Note:* To use the flow nomograph, at least two values out of A, B, C and D should be known. Joining the two values on the lines and extending the line formed by the two points will provide the desired values. (*Example :* Flow 10 lit/sec, pipe size 110 mm outer dia. (104 mm inner dia.), velocity 1.2 m/sec., hydraulic gradient 1.2 m/100 m. *See the darkened line in the nomograph*)

[***Source:*** Jain Irrigation Systems Ltd. (2005 a)]

main causes of the variation in the discharge of drippers along the lateral are the loss of pressure head due to friction in the pipe, minor losses, and changes of ground surface elevations in the irrigated field. Other factors influencing variations in the discharge of drippers are the temperature of water, the quality of irrigation water, and the quality of manufacturing as expressed in the coefficient of variation. The allowable loss in pressure head between the head and the tail ends of a lateral is about 10% of the operating pressure.

To estimate the friction head of drip irrigation laterals, the Darcy-Weishbach equation for smooth pipes in micro-irrigation systems can be combined with the Blasius equation as follows (Schwab *et al.*, 1996):

$$H_f = \frac{KLQ^{1.75}}{D^{4.75}} \qquad ...(13.10)$$

in which,

H_f = friction loss in pipe, m

K = a constant whose value is 7.89×10^5 for SI units for water at 20°C

L = length of pipe, m

Q = rate of flow in the pipe, litres/sec

D = inside diameter of the pipe, mm

Equation 13.10 applies for continuous sections of plastic pipes. Pressure loss due to fittings and connections will have to be added to obtain the total friction loss in the lateral pipe. It is a common practice to express the losses due to fittings and connections as equivalent length of the lateral pipe. The equivalent lengths in such cases can be estimated as follows:

(*i*) 1.0 to 3.0 m for each in-line emitter,

(*ii*) 0.1 to 0.6 m for each on-line emitter,

(*iii*) 0.3 to 1.0 m for a solvent welded Tee connector

The drip irrigation design process requires the estimation of the peak ET, water required per plant, emitter selection, computation of the design operating pressure to maintain the required emission uniformity *EU*, pressure variation allowed in a section, and pipe sizes. The allowable pressure variation in a section Δh can be estimated using the following equation (US Soil Cons. Service, 1984):

$$\Delta h = 2.5\,(h_{ave} - h_{min}) \qquad ...(13.11)$$

in which,

h_{ave} = design operating pressure or head,

h_{min} = minimum pressure or head in a section

In ease of sloping fields, Δh must include friction loss and the difference in elevation.

The procedure for the design of drip irrigation laterals include the broad estimation and selection of one pipe size in which friction loss is within 10% of the operating pressure of the system. The smaller diameter pipe is first selected with a view to reduce the total cost of the system. Using this value, the head loss due to friction is estimated using Equation 13.10. The elevation head and head loss due to emission devices and connections are added to this value. If the variation in total friction value is within 10% of the operating pressure, the selected pipe diameter is accepted. Otherwise, successively larger diameter pipes are selected to meet the requirement.

Diameter of Main and Sub-Main Pipes

Laterals and line source tubings are connected to the sub-main and sometimes directly to the main pipe line. Mains and sub-mains are usually made of PVC. Each main line and sub-main should have a manually operated or solenoid activated valve at its upstream end to provide on-off service and for the isolation of the unit or a section for repair and maintenance. An on-off valve is usually fitted at the end of the main and each sub-main pipe line for periodic flushing. Sub-mains may also have pressure regulating valves, flow control valves, secondary filters and pressure gauges.

The selection of the diameter of the main pipe and sub-mains can be done in a similar way as described under the laterals. The pipe diameter is selected from the smallest feasible size, going successively towards larger sizes. If the variation in total head loss (friction losses in the pipe, emission devices and connections + elevation head) in the main and the sub-mains is found to be within 10% of the system, then the pipe selected is accepted. Otherwise, successively larger diameter pipes are tested to meet the conditions.

Size of the Pumping Unit

The size of the pumping unit can be estimated based on the system capacity and total head. The total head in a drip irrigation system includes the emitter operating pressure, pressure losses in filter, fertigation unit, valves, fittings, field elevation differences, and loss of pressure head due to friction in the pipe line including the main, sub-main and laterals (see also Chapter 3). The size of the pumping unit can be estimated using the following equation:

$$h.p. = (Q \times H) / (75 \times \eta_p \times \eta_m) \qquad ...(13.12)$$

in which,

h.p. = the horse power of the electric motor driving the pump, H.P.

H = total head loss (including head loss in one lateral × number of laterals + head loss in sub-main + head loss in the main + loss in all fittings and connections + suction head + operating pressure), m

Q = capacity of the drip irrigation system, litres/sec.

η_p = pump efficiency, fraction

η_m = motor efficiency, fraction

Estimation of Drip Irrigation System Flow

For a pump of a given horse power, the following relationship is possible, assuming that the efficiency remains same in both the cases:

$$Q_1 \times H_1 = Q_2 \times H_2 \qquad ...(13.13)$$

in which,

Q_1 = discharge of the existing pump, litres/sec

H_1 = total head of the existing pump, m

Q_2 = discharge planned for the drip irrigation system, litres/sec.

H_2 = total head required while using the drip irrigation system, m

Equation 13.13 can be rewritten as follows:

$$Q_1 \times H_1 = Q_2 \times (H_1 + H_d) \qquad ...(13.14)$$

in which,

H_d = the additional head required for operating the drip irrigation system, m

A broad estimate of the additional head required for operating the drip irrigation system can be obtained using Equation 13.14. The additional head required is based on field conditions. A common thumb rule for arriving at the first approximation in design is to assume a value of 25 m for H_d.

Based on Equation 13.14, the design discharge that can be considered for the drip irrigation system to be operated by the existing pump can be rewritten as follows:

$$Q_2 = \frac{(Q_1 \times H_1)}{(H_1 + H_d)} \qquad ...(13.15)$$

For example, consider a pump with a discharge capacity of 3 litres/sec. at a total head of 30 m which is being used for surface irrigation. If the same pump is to be used for operating a drip irrigation system requiring an additional head of 25 m, the design discharge of the drip irrigation system can be estimated as follows:

$$Q_2 = \frac{3 \text{ lit./sec.} \times 30 \text{ m}}{(30 \text{ m} + 25 \text{ m})} = 1.636 \text{ litres/sec.}$$

Examples on Design of Drip Irrigation Systems

Example 13.4. *A dripper is to deliver a discharge of 5 litres / hr. The distance between the drippers is 1 m and the distance between the laterals is 4 m. Determine the discharge rate of the dripper in mm / hr.*

Solution: Discharge rate required of the dripper

$$= \frac{5 \text{ lit./hr}}{1 \text{ m} \times 4 \text{ m}} = 1.25 \text{ mm/hr}$$

Example 13.5. *Plan a suitable drip irrigation system for an orchard on a nearly flat land with medium heavy soil. The dimensions of the field are 375 m × 145 m. The source of water is a tube well located at the top corner of the farm. The tree spacing is 5 m × 4 m. Emitters are spaced 1 m apart in each lateral. The monthly evaporation rate observed with a Class A pan is 240 mm. Irrigation is to be applied daily. Determine the discharge capacity of the drip irrigation system.*

Solution: Drip irrigation system planning:

Area of the field = 375 m × 145 m – 54375 sq. m

= 5.4375 ha

The orchard is divided into 3 segments of 125 m width with laterals laid down the gentle slope and each segment is to be irrigated one after the other.

Laterals are spaced at 5 m interval.

The total number of laterals required = 375 / 5 = 75.

The number of laterals in each segment (irrigation block) 125 / 5 = 25.

Effective length of each lateral = 145 m.

Actual length of lateral, allowing for 30% additional length to provide for multiple emitters to cover each tree on a nearly circular path around the tree

= 145 + (145 × 30/100) = 188.5 m, say 188 m.

Total length of laterals in each section

= 188 × 25 = 4700 m.

Total length of laterals required for the entire system

= 4700 × 3 = 14100 m.

Number of trees per lateral = 145/4 = 36.25, say 36

Total number of trees in each irrigation segment

= 36 × 25 = 900.

Total land area irrigated by each segment = 125 × 145

= 18125 sq.m = 1.8125 ha.

Area covered by a dripper, using the values presented in Table 13.2.

Maximum diameter of the wetted circle formed by a dripper with a discharge rate of 4 litres/hr = 2.7 m.

Therefore, area covered by each dripper $\frac{\pi}{4} d^2$ = 5.72 sq.m.

Effective area shaded by a mature tree = 80% of the area under the tree = 5 × 4 (80/100) = 16 sq.m.

Number of drippers per tree, using Eq. 13. 8

= 5 × 4 (80/100)

= 16/5.72 = 2.8, say 3 trees

Total number of drippers in each segment

= Number of trees × 3 = 900 × 3 =2700.

Water distribution efficiency (assumed) = 90%.

The monthly irrigation water requirement is estimated using Eq. 13.6, assuming crop coefficient for orchard to be 0.80, and pan factor 0.8,

V_m = 0.8 × 0.8 × 0.8 × 240 × 54375 = 6681600 litres

Daily water requirement, V_d = 6681600/31

= 215535 litres

Therefore, irrigation requirement per 2 hour shift

= 215535/12 = 1796 litres.

Capacity of the drip irrigation system, using Equation 13.7.

Assuming the water application efficiency to be 90%,

$Q = V_d \times T\ (\eta_a \times t)$ = 1766 × 1 / (0.9 × 2) = 1796 / 1.8 = 997.8 litre/hr = 997.8 / (60 × 60) = 997.8/3600 = 0.277 litres/sec.

Thus, a drip irrigation system with a capacity of 0.277 litres/sec is designed for the layout suggested above.

Example 13.6. *Design a drip irrigation system for a 3 ha citrus orchard in the Delhi region. The field is rectangular with a length of 300 m along the head end and a width of 100 m. The field is nearly flat and the soil is sandy loam. The irrigation water source is a river flowing close to the top corner of the field. The crop spacing is 3 m × 3 m.*

Solution:

(i) Layout of drip irrigation system:

The main line is laid along the length of the field. The laterals are connected directly to the main. The total length of the main pipe is 300 m. The laterals are spaced 3 m apart on each row of the plants, covering the full length of the field.

(ii) Irrigation water requirement, V_m*:*

May is the hottest month and the maximum monthly pan evaporation is 245.4 mm (Table 13.1). The crop coefficient is assumed to be 0.6. The average canopy coverage of the plant ranges from 5 to 6 sq.m. Hence, the canopy coefficient is taken as 0.6.

Thus, C_e = 0.6, K_p = 0.8 (the usual value),

E_p = 245.4 mm, and the total area of the field,

A = 300 × 100 = 30,000 sq.m.

Using Eq. 13.6, the monthly irrigation water requirement, V_m

$= K_c \times K_p \times C_c \times E_p \times A$

= 0.6 × 0.8 × 0.6 × 245.4 × 30000

= 2120256 litres

Therefore, daily water requirement V_d

= 2120256 /31 = 68395.4 litres

(iii) Capacity of the drip system, Q:

Assuming water application efficiency of 90%, and irrigation daily with operation duration of 2 hr/day, the capacity of the drip system can be computed using Eq. 13.7, T = 1 day, t = 2.0 hr, η_a = 0.9,

$Q = V_d \times T / (\eta_a \times 2)$

= 68395.4 × I / (0.9 × 2) = 37997 litres /hr

= 37997/60 = 633.3 litres/min

= 633.3/60 = 10.55 litres/sec.

(iv) Number of laterals and drippers:

Number of laterals = 300 / 3 = 100

Length of each lateral is 100 m

Number of plants per lateral = 100 /3 = 33

Total number of plants to be irrigated at a time

= 33 × 100 = 3300

Discharge required per plant per day, Q_p

= 37997 / 3300 = 11.52 litres / hr.

Drippers with a discharge rate of 6 litres/hr is selected.

Number of drippers required/plant 11.52 / 6

= 1.92, say 2.

Total number of drippers required = 2 × 3300

= 6600.

(v) Diameter of lateral pipe:

Assume that the diameter of the lateral pipe is 20 mm. sign discharge rate of a lateral with 33 × 2 = 66 drippers with a discharge of 6 litres/hr = 396 litres/hr = 6.6 litres/min. Referring to Fig. 13.77 (nomograph), joining the corresponding points on scale A and B, and extending the straight line to scale D, the friction head loss in a 20 mm diameter pipe with a flow rate of 6.6 litres/min = 0.8 m/100 m length of pipe. There is no appreciable elevation head in the lateral, as the land is nearly flat. Thus, the head loss in one lateral of 0.8 m is less than 10% of the usually assumed operating head of 1 atmosphere or 10 m. Hence, the selected size of 20 mm internal diameter for the lateral pipe is accepted.

(vi) Diameter of the main pipe, 300 m long:

Design discharge rate of the main pipe with 100 laterals and 6600 drippers with a discharge rate of 6 litres/hr = 6600 × 6 = 39600 litres/hr = 660 litres / min, or 11 litres/sec. The diameter of the main is assumed to be 125 mm. Referring to Fig. 13.77 (nomograph), a pipe diameter of 125 mm is taken on scale A and the flow rate of 660 litres / min on scale B. Joining the two points by a straight line and extending the line to intersect scale D, the head loss due to friction is observed as 1 m or 3 m for the 300 m length of the main pipe which is very high. Further, the cost of a 125 mm pipe be very high for a 300 m long main pipe.

As the head loss due to friction is very high and the cost of the system prohibitive with the set up described above, it is proposed to divide the drip system into 3 segments of 100 m length each and providing 3 sub-main pipes of 100 m length each to which the laterals are fixed. The length of the main pipe could be reduced to 200 m, as the flow in the third segment will be one-third of the total flow. Irrigation is applied in each segment separately, one after the other. This will reduce the flow rate by one-third, namely, 220 litres / min. With this assumption, the diameter of the main pipe is assumed to be to 80 mm. The resulting head loss due to friction is 0.7 m /100 m or 1.4 m for 200 m length of pipe. As the head loss is still on the high side, the next higher size of 85 mm is assumed. From Fig. 13.77, the resulting friction head loss is 0.5 m/100 m or 1 m /200 m length which is accepted.

(vii) Diameter of sub-main pipes, 100 m long each:

Based on the results of the earlier trials, the diameter of the sub-main pipe is assumed to be 75 mm. Using Fig. 13.77, the head loss in the sub-main pipe is 0.9 m for the 100 m long pipe segment. Hence, the diameter of 75 mm is accepted for the 3 sub-main pipes, 100 m long each.

(viii) Size of the pumping unit:

The suction head is assumed to be 5 m and head loss in fittings and accessories is assumed to be 7 m. The pump efficiency is assumed to be 65% and the efficiency of the electric motor is assumed to be 80%. The operating pressure is taken as 10 m.

The total length of laterals in each segment

= 33 × 100 = 3300 m.

The total head loss in the laterals of each segment

= 33 × 0.8 = 26.4 m

The total head = Head loss in main pipe + Head loss in sub-mains + Head loss in laterals + Head loss in accessories and fittings + Operating pressure

= 1.0 + 0.9 + 26.4 +5 + 7 + 10 = 50.3 m

Applying Equation 13.12

Horse power of the electric motor, *hp*

= (10.55 × 50.3) / (75 × 0.65 × 0.80)

= 530.66 / 39 = 13.61, say 15 horse power.

A drip irrigation system of the above configuration and specifications is acceptable. However, greater efficiency and a more cost effective system would be possible by selecting a pump and a motor which provide still better levels of efficiency. Similarly, better performance and lower costs of emitters and pipes may probably be possible by more trials based on the above procedure.

SUMMARY

Application, adaptability, advantages and limitations of drip irrigation systems are described. Various types of indigenous micro-irrigation devices are described. Different components of drip irrigation systems are explained along with their materials of construction. Types of drippers, their performance and adaptability are described. Different types of filters used in drip irrigation systems are described along with their adaptability, performance and maintenance requirements. Application of fertilizers and chemicals through drip irrigation are explained with the suitable types of fertilizers to be used. Procedures for the design of various components of drip irrigation systems are described. Installation, operation, performance and evaluation of drip irrigation systems are described.

QUESTIONS

1. Distinguish between:
   (*i*) Drip irrigation and micro-irrigation
   (*ii*) Drip irrigation and trickle irrigation
   (*iii*) Drip irrigation and localized irrigation
   (*iv*) LDPE and HDPE pipes
   (*v*) Surface and sub-surface drip irrigation systems
   (*vi*) Low pressure, medium pressure and high pressure drip irrigation systems
   (*vii*) Compensating emitter and continuous emitter
   (*viii*) Long path emitter and short path emitter
   (*ix*) On-line emitter and in-line emitter
   (*x*) Orifice emitter and tortuous flow emitter
   (*xi*) Mono-walled and bi-walled perforated polyethylene pipes
   (*xii*) Pressure relief valves and vacuum relief valves
   (*xiii*) Pressure regulation valves and flow regulation valves
   (*xiv*) Non-return valves and on-and-off valves
   (*xv*) Media filter and screen filter
2. What are the specific advantages of drip irrigation over sprinkler irrigation?
3. What are the usually recommended control measures to prevent damage to drip irrigation pipelines from rodents and other animals?
4. What is the principle of working of a sub-soil injector?
5. What are the special characteristics of Linear Low Density Polyethylene (LLDPE) pipes?
6. What are the different promotional efforts of the Government of India in promoting adoption of drip irrigation in India?
7. How is the line pressure controlled in a bamboo irrigation device?
8. Why is the pressure loss in an on-line emitter more than in an in-line emitter?
9. What is the principle of working of a pressure-compensating dripper?
10. Under which site situations are pressure compensating emitters particularly suitable? What are their limitations?
11. What are the different types of long path emitters?
12. In a long path emitter, how is the length of path required to bring down the pressure in a drip line computed?
13. Express the mathematical relationship for computing the discharge of a dripper.
14. How is automation achieved in drip irrigation?
15. Define time-based automation in a drip irrigation system.

16. Which are the points in a drip irrigation pipeline system where air relief valves are to be located?

17. What are the causes for abnormally high pressure conditions in a pipeline?

18. What are the causes for reverse flow occurring in a pipeline?

19. Which is the point in a pipeline where the non-return valve is to be installed?

20. What is the purpose of locating a flow control valve at the end of a pipeline?

21. What are the common causes for clogging of drippers?

22. What are the common organic contaminants which cause clogging in drippers?

23. What are the common chemical causes resulting in clogging of drippers?

24. What are the common types of filtration devices used in drip irrigation systems using surface water? What is the specific adaptability of each of them?

25. What are the common types of coatings used in media filters for protection against corrosion and weather effects?

26. Define minimum operating flow in a media filter.

27. What is the advantage of providing a battery of media filters in a drip irrigation system?

28. What is the role of screen filters in a drip irrigation system?

29. What is the material of construction of filter elements in screen filters?

30. What is the coating used in mild steel screen filter housing to protect it from corrosion and weather effects?

31. What are the points which form the basis for selecting a screen filter?

32. What is the principle of working of a hydro-cyclone filter?

33. What is the specific function of a disc filter in a drip irrigation system?

34. What is the principle of operation of a disc filter?

35. Which are the commonly used fertilizers which are suitable for injection in drip systems when the soil requires nitrogen and phosphorous?

36. What are the steps necessary to prevent leakage in drip irrigation pipelines?

37. What is the purpose of snaking of PVC pipes in the trench while laying?

38. What are the step-by-step procedures required in punching of laterals and fixing of drippers?

39. List the step-by-step procedures required to test drip irrigation systems after their installation.

40. What is the purpose of chlorine treatment as a part of the maintenance requirement of a drip irrigation system?

41. How is acid treatment done in a drip irrigation system?

42. What is the important precaution to be taken in the solvent jointing of PVC pipes?

43. What are the critical items in the selection of emitters for a drip irrigation system?

44. Define emitter coefficient of manufacturing variation. How is it computed?

45. What is system coefficient of manufacturing variation in emitters?

46. How is design emission uniformity in a drip irrigation system different from uniformity coefficient in sprinkler irrigation?

47. How is design emission coefficient of a drip irrigation system determined?

48. What are the commonly recommended ranges in the values of design emission coefficients in point source and line source emitters?

49. What is the relationship between pressure head and discharge in a turbulent flow emitter?

50. What are the major factors influencing the design capacity of a drip irrigation system?

51. Identify the True and False statements from the following:

(*i*) In drip irrigation, the soil moisture content is maintained at near optimum levels to facilitate optimum growth and production of crops.

(*ii*) Selection of widely spaced crops of high market value is necessary to increase the benefit-cost ratio in drip irrigation.

(*iii*) In drip irrigation, continuous trickling is more advantageous to obtain uniform soil moisture distribution and reduce deep percolation losses than on-and-off trickling.

(*iv*) Plant response to irrigation is better in drip irrigation as compared to other methods of irrigation.

(*v*) Wetting pattern in a drip irrigation system is influenced mainly by soil characteristics, discharge rate of emitters and the duration of water application.

(*vi*) Cracking due to UV degradation is the main limitation in the use of HDPE pipes laid on the ground surface in a drip irrigation system.

(*vii*) The wide range of physical properties in polyethylene depends on the arrangement of polyethylene molecules.

(*viii*) The longer the micro-tube, the lower will be its discharge.

(*ix*) A 'woodpecker dripper' is a common type of in-line emitter.

(*x*) In an orifice type emitter, the discharge is less sensitive to operating pressure and viscosity.

(*xi*) Laterals in a drip irrigation system are usually made of HDPE pipes.

(*xii*) Pressure regulating valves provide surge control by protecting downstream pipes from surges originating upstream.

(*xiii*) Reverse flow protection is often required by law when fertilizers and other agricultural chemicals are injected into the irrigation system so as to prevent the contamination of the source of water in wells and streams used for domestic water supply.

(*xiv*) When well water with dissolved iron and manganese are exposed to the atmosphere, precipitates of manganese oxide are formed which may cause clogging of emitters.

(*xv*) For settling tanks to be effective in a drip irrigation system, the intake pipe should be located close to the bottom of the tank essential for the primary filtration of water in a drip.

(*xvi*) Media filters are essential for the primary filtration of water in a drip irrigation system.

(*xvii*) The usual direction of flow of water in a screen filter is from outside of the filter element to inside.

(*xviii*) The net filter area in a screen filter is usually about half that of the total filter area.

(*xix*) The filtration capacity of a hydro-cyclone filter is greater for high density particles than for lower density ones.

(*xx*) Because hydro-cyclone filters effectively remove large quantities of sand particles, they are placed upstream of all other types of filters.

(*xxi*) Disc filters, when used, form the final filtering component, before water enters the drip line.

(*xxii*) Anhydrous ammonia and aqua-ammonia are ideally suitable for use in fertigation.

52. Identify the most appropriate answer of the following questions from the four options:

(*i*) In drip irrigation, water flows from the emission points through the soil by:

(a) infiltration (b) percolation
(c) capillarity
(d) capillarity and gravity

(*ii*) Ant damage in drip laterals can be controlled by using pipes not less than:

(a) 5 mils. (b) 10 mils.
(c) 15 mils. (d) 20 mils.

(*iii*) High density polyethylene pipe is produced at pressures of about:

(a) 20 kg/sq. cm (b) 30 kg/sq. cm
(c) 40 kg/sq. cm (d) 50 kg/sq. cm

(*iv*) Globally, the maximum area under drip irrigation is in:

(a) Israel
(b) United States of America
(c) Saudi Arabia (d) France

(*v*) The largest crop area covered by drip irrigation in India is under:

(a) Plantation crops (b) Orchards
(c) Vegetable crops (d) Fiber crops

(*vi*) Under normal working conditions, the maximum rate of emission by drippers is:

(a) 5 litres/hr (b) 10 litres/hr
(c) 15 litres/hr (d) 20 litres/hr

(*vii*) The internal diameter of a microtube ranges from:

(a) 0.3 to 0.7 mm (b) 0.7 to 1.2 mm
(c) 1.7 to 2.2 mm (d) 2.2 to 2.7 mm

(*viii*) The minimum time required for the settling of mineral particles, larger than 80 microns, in a settling basin is:

(a) 5 min. (b) 10 min.
(c) 15 min. (d) 20 min.

(*ix*) In a media filter, the size of the filter media (crushed granite, silica or quartz) is sieved to particle sizes ranging from:

(a) 0.3 to 0.7 mm (b) 0.7 to 1.2 mm
(c) 1.2 to 1.7 mm (d) 1.7 to 2.2 mm

(*x*) The pressure drop usually allowed in a media filter is about:

(a) 1 m (b) 2 m
(c) 3 m (d) 4 m

(xi) The usual mesh size of the screen hole in a screen filter is:

(a) 80 mesh (b) 100 mesh
(c) 120 mesh (d) 140 mesh

(*xii*) The minimum specific gravity of the suspended particle which can be removed by a hydro-cyclone filter is:

(a) 1.0 (b) 1.2
(c) 1.4 (d) 1.6

(*xiii*) In the chlorine treatment of irrigation water, the quantity of bleaching powder added in the water source is:

(a) 1 mg/litre (b) 2 mg/litre
(c) 3 mg/litre (d) 4 mg/litre

(*xiv*) The minimum recommended earth cover for the mains and sub-mains in a permanently laid drip irrigation system for pipe sizes ranging from 6 to 10 cm diameter is:

(a) 40 cm (b) 50 cm
(c) 60 cm (d) 75 cm

53. In the following, match each of the entries of the 1st column with the most closely-linked entry in the 2nd column:

(*i*)

(A) Suction irrigation	(I) Indigenous irrigation device
(B) Symcha Blass	(II) Low density polyethylene pipes
(C) Bamboo irrigation	(III) Plastic sheeting
(D) UV radiation	(IV) Drip irrigation
(E) Polyethylene resins	(V) Soil moisture tension

(*ii*)

(A) Long path emitter	(I) On-line emitter
(B) Button emitter	(II) Viscosity of water
(C) Turbo-flow emitter	(III) Mono-walled perforated pipe
(D) Constant k in estimating emitter discharge	(IV) In-line emission device
(E) 'Viaflow' porous tubing	(V) Microtube

(iii)

(A) Ultraviolet light	(I) Splitting of polyethylene pipes
(B) Pig-tail layout	(II) Connection of drip lateral to manifold
(C) Volumetric control valves	(III) Sequencing operation in irrigation
(D) Jumper tube	(IV) Drip irrigation in orchards
(E) Environmental stress cracking	(V) Degradation of PVC pipes

(iv)

(A) Grommet take-off	(I) Hydro-cyclone
(B) Solenoid valve	(II) Disc filter
(C) Filtration of organic matter	(III) Media filters and screen filters
(D) Flush valve	(IV) Accessory of drip irrigation
(E) Primary filtration of large quantities of sand particles	(V) Automation in pressurized irrigation systems

(v)

(A) Turbo-clean mechanism	(I) Control valve
(B) Nitrile rubber ring	(II) Fertigation
(C) Solvent welding	(III) Anhydrous ammonia
(D) Volatalization	(IV) Joining of PVC pipes
(E) Water soluble fertilizers	(V) Screen filter

(vi)

(A) M-seal	(I) Drip irrigation system performance
(B) Snaking	(II) Bureau of Indian Standards
(C) Quality control of pressurized irrigation systems	(III) Friction loss in pipes
(D) Darcy-Weishbach	(IV) Joining of PVC pipes
(E) Emission uniformity	(V) Installation of PVC pipes in a trench

FURTHER READING REFERENCES

Brahmanand, P.S. and Singandhupe, R.B., (2001), *Drip Irrigation in India – Current Status and Future Needs,* Proc. Intl. Conf. on Micro- and Sprinkler Irrigation Systems, 8–10, Feb., 2001, Pub. Central Board of Irrigation and Power, New Delhi, pp. 44–51.

Bucks, D.A. and Myers, L.E., (1973), *Trickle Irrigation – Application Uniformity from Simple Emitters, Trans. ASAE,* Vol. 16, No.6. pp. 1108–1111, Pub. Am Soc. of Agrl. Engrs, St. Joseph, Michigan, USA.

Bucks, D.A., Nakyama, P.S. and Warrick, A.W., (1982), *Principles, Practices and Potentalines of Trickle (Drip) Irrigation,* In Advances in Irrigation, Vol.1, ed. Hillel, D. Academic Press, New York, pp. 219–297.

Bureau of Indian Standards (1984), **Indian Standard IS 10799: 1984,** *Code of Practices of Design and Installation of Trickle Irrigation,* Pub. Bureau of Indian Standards, Manak Bhawan, New Delhi.

Bureau of Indian Standards (1985), **Indian Standard IS 4985: 1985,** *Specifications for unplasticised PVC Pipes,* Pub. Bureau of Indian Standards, Manak Bhawan, New Delhi.

Bureau of Indian Standards (1989), **Indian Standard IS 12786: 1989** (with Amendments 1992 and 1999), *Irrigation Equipment–Polyethylene Pipes for Irrigation Laterals,* Pub. Bureau of Indian Stan-dards, Manak Bhawan, New Delhi, p. 9.

Bureau of Indian Standards (1991), **Indian Standard IS 13062: 1991** *Irrigation Equipment and Systems Evaluation of Field Irrigation Efficiencies–Guidelines*. Pub. Bureau of Indian Standards, Manak Bhawan, New Delhi, p. 13.

Bureau of Indian Standards (1992), **Indian Standard IS 13479: 1992, ISO 3501: 1976,** *Assembled Joints Between Fittings and Polyethelene (PE) Pressure Pipes – Test of Resistance to Pull Out.* Pub. Bureau of Indian Standards, Manak Bhawan, New Delhi, p. 2.

Bureau of Indian Standards (1992), **Indian Standard IS 13487: 1992.** *Irrigation Equipment – Emitters – Specification,* Pub. Bureau of Indian Standards, Manak Bhawan, New Delhi, p. 5.

Bureau of Indian Standards (1992), **Indian Standard IS 13488: 1992.** *Irrigation – Equipment – Emitting Pipe Systems – Specification,* Pub. Bureau of Indian Standards, Manak Bhawan, New Delhi, p. 7.

Bureau of Indian Standards (1994), **Indian Standard IS 12785: 1994**. *Irrigation Equipment – Strainer Type Filters–Specification,* Pub. Bureau of Indian Standards, Manak Bhawan, New Delhi, p. 5.

Bureau of Indian Standards (1994), **Indian Standard IS 14178: 1994.** *Pressurized Irrigation Equipment – Terminology,* Pub. Bureau of Indian Standards, Manak Bhawan, New Delhi, p. 12.

Bureau of Indian Standards (1997), **Indian Standard IS 14482: 1997,** *Polyethylene Micro Tubes for Drip Irrigation—Specification,* Pub. Bureau of Indian Standards, Manak Bhawan, New Delhi, p. 2.

Bureau of Indian Standards (1997), **Indian Standard IS 14483 (Part I): 1997,** *Fertilizer and Chemical Injection System,* Bureau of Indian Standards, Manak Bhawan, New Delhi, p. 4.

Bureau of Indian Standards (1997), **Indian Stanoard IS 14483 (Part I): 1997,** *Fertilizer and Chemicals Injection System, Part I: Ventury Injector,* Pub. Bureau of Indian Standards, Manak Bhawan, New Delhi.

Bureau of Indian Standards (1998). **Indian Standard IS 14606: 1998.** *Irrigation Equipment – Media Filters —Specification.* Bureau of Indian Standards, Manak Bhawan, New Delhi, p. 5.

Bureau of Indian Standards (1998), **Indian Standard IS 14605: 1998,** *Irrigation Equipment – Micro-Sprayers – Specification.* Pub. Bureau of Indian Standards, Manak Bhawan, New Delhi, p. 9.

Bureau of Indian Standards (1999), **Indian Standard IS 10799: 1999,** *Irrigation Equipment – Design, Installation and Field Evaluation of Micro-Irrigation Systems, Code of Practice,* Pub. Bureau of Indian Standards, Manak Bhawan New Delhi, p. 16.

Bureau of Indian Standards (1999). **Indian Standard IS 14743: 1999,** *Irrigation Equipment – Hydro-cyclone Separators–Specification,* Pub. Bureau of Indian Standards, Manak Bhawan, New Delhi.

Burt, C., O'Connor, K. and Ruehr, T. (1995), *Fertigation,* Pub. Irrigation Training and Research Center, California Polytechnic, State University, San Luis Obispo, CA 93407, USA, p. 293.

Camp, C.R. (1998), *Subsurface Drip Irrigation: A Review,* Trans. Amer Soc. of Asgrl Engrs., Vol. 41(5): 1353–1367.

Chapin Watermatics Inc. **Leaflet,** *Chapin Irrigation Products – Twin Wall Hose,* Chapin, Watermatics Inc., 368 N. Colorado Ave., New York.

Chauhan, H.S.(2007), *Elements of Microirrigation and Its Crop Application.* Pub. International Book Distribution Co. Ltd., Meerabhai Marg, Lucknow, p. 811.

Dov Nir (1983), *Drip Irrigation,* In Finkel, H.J. editor, Handbook of Irrigation Technology, Pub. CRS Press, Florida, USA, pp. 247–298.

E.I. du Pont de Nemorse and Co. Inc. (1973), *DUPONT 'VIAFLO' Plastic Tubing,* Tech. Bull. E.I. du Font and Co Inc., Development Department, Wilmington, Delaware, p. 4.

E.I. du Pont de Nemours and Co. Inc. (1973), *DUPONT "VIAFLOW" Porous Tubing for Row Crop Irrigation,* Tech. Bull., E.I.du Pont de Nemours and Co., Inc. Development Department, Wilmington, Delaware.

EPC Irrigation Ltd., **Leaflet,** *EPC Drip Irrigation,* EPC Irrigation Ltd., B–20 MIDC, Ambad, Nasik, Maharashtra, p. 18.

EPC Industries Ltd., (2005), *Material on Drip and Sprinkler irrigation: Views of irrigation systems in the field and system components,* Photographs supplied on request for inclusion in Irrigation: Theory and Practice, EPC Industries Ltd., H–109, MIDC, Ambad, Nasik, Maharashtra.

Gilbert, R.G., Nakyama, F.S, Bucks, D.A.; French. O.F., and Adamson, K.C., (1981), Trickle *Irrigation: Emitter Clogging and Other Flow Problems,* Agrl. Water Management 3, (1980/1981) 159–178–Elsevier Scientific Pub. Co., Amsterdam, The Netherlands, pp. l59–177.

Goldberg, D. and Shmueli, M., (1970), *Drip Irrigation–a method used under arid and desert conditions of water and soil salinity,* Trans. Am. Soc. of Agrl. Engrs. Vol. 13, No. 1, pp. 38–41.

Hamish, F., (1977), *Main line installation,* in Drip/Trickle Irrigation No. 5, Vol. 2, No. 2, 1977, Pub. International Drip Irrigation Asso., P.O. Box 288, Bloomington, California–92316 (714) 877–4405, p. 12.

Indian Council of Agricultural Research (2003), *Status Report on Major Areas of Plasticulture, Vol. II.* Central Institute of Post Harvest Technology, Ludhiana, Punjab, pp. 5–29.

Indian National Committee on Irrigation and Drainage (1994), *Drip Irrigation in India.* Pub. Water and Power Consultancy Services (India) Ltd., New Delhi, p. 176.

International Drip Irrigation Association, (1977), *Drip/Trickle Irrigation,* IDIA, Vol. 2, No. 2 – March/April, 1977, p. 30.

I-Pai Wu and Gittin, H.M., (1974), *Design of Drip Irrigation Lines,* Tech. Bul. No. 96, Hawaii Agricultural Experiment Station, College of Tropical Agriculture, University of Hawaii, Honolulu, Hawaii; USA, p. 27.

Jain Irrigation Systems Limited *Profile, Leaflet Micro-Irrigation A Review,* Pub. Jain Irrigation Systems Ltd., Jalgaon, Maharashtra.

Jain Irrigation Systems Limited, *Product Catalogue,* Pub. Jain Irrigation Systems Ltd. Jalgaon, Maharashtra.

Jain-Kemira Fertilizers Ltd., (1996), *Fertigation Manual,* Pub. Jain Irrigation Systems Ltd., Plastic Park, P.O. Box 72, NH6, Jalgaon, Maharashtra, India, p. 82.

Jain Irrigation Systems Ltd., (2005a), *Jain Products Catalogue,* Jain Irrigation Systems Ltd., Jain Plastic Park, P.O. Box 72, N.H. No. 6, Jalgaon, Maharashtra, India.

Jain Irrigation Systems Ltd. (2005b), *Maintenance Manual for Micro-Irrigation Systems,* Jain Plastic Park, P.O. Box 72, N-H. No. 6, Jalgaon, Maharashtra, India.

Jain Irrigation Systems Ltd., (2005c), *Principles, Products, Types: Advantages and Benefits of Drip Irrigation,* Drip–1, amm. doc. and Drip–2– amm.doc Personal Communications, May 4, 2005, Jain Irrigation Systems, Jain Plastic Park, P.O. Box 72, NH 6, Jalgaon, Maharashtra, India.

Jain Irrigation Systems Ltd. (2005 d), *Design, Operation and Maintenance of Drip Irrigation Systems, Part I initial pages.* Personal correspondence. April/May, 2005, Jain Irrigation Systems Ltd., Plastic Park, P.O. Box 72, NH 6, Jalgaon, Maharashtra, India, p. 10.

James, L.G. (1988), *Principles of Farm Irrigation System Design,* Pub. Krieger Publishing Company, Malabar, Florida, USA.

Jyoti Limited, Vadodara. *Leaflet, 'Jyoti' Drip Irrigation System,* Jyoti Ltd., Industrial Area, P.O. Chemical Industries, Vadodara, Gujarat, p. 4.

Kaul, R.K. (1979), *Hydraulics of Moisture Front Advance in Drip Irrigation,* Ph.D. Thesis Division of Agrl. Engg., IARI, New Delhi.

Karmeli, D. and Keller, J., (1974), *Evaluation of Drip Irrigation Systems,* Proc. Second Drip Irrigation Congress, San Diego, California, USA.

Karmeli, D. and Keller, J., (1975), *Trickle Irrigation Design,* Rain Bird Sprinkler Manufacturing Corporation, Glendora, California, USA.

Lake Company (1982), *How to Evaluate Drip Irrigation Systems,* Lake Company, P.O. Box 2248, Bakersfield, California, USA, p. 6.

Madhava Chandran, K.L. and Sushanth, C.M. (1992), *Handbook on Drip Irrigation,* Pub. Centre for Water Resources Development and Management, Kozhikode, Kerala, p. 28.

Michael, A.M. and Cjha, T.P. (2003), *Principles of Agricultural Engineering, Vol. II.,* Pub. Jain Brothers, 16/873, East Park Road, New Delhi.

Naidu, N. Chandrababu and Singh, H.P., (2004), *Report of the Task Force on Micro-irrigation,* Ministry of Agriculture, Department of Agriculture and Cooperation, Government of India, New Delhi.

Nakyama, F.S. and Bucks, D.A. ed., (1986), *Trickle Irrigation for Crop Production: Design, Operation and Management.* Pub. ELSEVIER, Amsterdam, The Netherlands, p. 383.

National Bank for Agriculture and Rural Development (1992), *Model Scheme for Financing Drip Irrigation for Availing Bank Finance,* Pub. NABARD, Mumbai, p. 30.

National Committee on Plasticulture Application in Horticulture (2001), *Progress Report – 2001,* Department of Agriculture and Coop., Ministry of Agriculture, Govt. of India, New Delhi, p. 98.

National Committee on Plasticulture Application in Horticulture (2004), *Review Meeting of Plasticulture Development Centres in North India,* Department of Agriculture and Coop., Ministry of Agriculture, Govt. of India, New Delhi.

Netafim Irrigation Equipment and Drip Systems, *Leaflet, Netafim Drip Irrigation Systems.* Netafim Irrigation Equipment and Drip Systems, Kibbutz Hatzerim Israel, p. 4.

Patel, V.C. and Angadi, S.S. (1997), *Drip Irrigation,* Pub. Agronomy Club, University of Agricultural Sciences, Dharwad, Karnataka, p. 140.

Rajput, T.B.S. and Michael, A.M. (1984), *Drip Irrigation – Its Methodology and Uses,* Souvenir of National Seminar on PVC Pipes, pp. 115–124.

Rajput, T.B.S. and Neelam Patel, (2001), *Drip Irrigation Manual* Pub. Water Technology Centre, IARI, New Delhi, p. 53.

Rajput, T.B.S. and Neelam Patel (2002), *Fertigation Theory and Practice,* Water Technology Centre, IARI, New Delhi, p. 53.

Rajput, T.B.S. and Neelam Patel (2005), *Development of Micro-Irrigation Technology.* Proc. Fifth International R and D Conference of Central Board of Irrigation and Power, Bangalore. Feb. 15–19, 2005, Vol. 4, pp. 15–19.

Rema Devi A. N., (1980), *Influence of Soil Moisture and Salt Distribution Characteristics on Crop Growth and Yield under Point Source of Water Application,* Ph.D. Thesis. Div. of Agrl. Engg., IARI, New Delhi, p. 159.

Ravinder Singh, Kundu, D.K., James, B.K. and Verma, H.N. (2000), *Progress of Micro-irrigation Research in India.* Proc. National Seminar on Micro-irrigation Research in India: Status and Prospects for the 21st Century. Pub. Water Technology Centre for the Eastern Region (ICAR), Bhubaneswar, Orissa, p. 327.

Schwab, G.O. Fangmeier, D.D., Elliot, W.J. and Frevert, R.K. (1996), *Soil and Water Conservation Engineering,* Pub. John Wiley and Sons, Inc., New York, pp. 452–466.

Singh, H.P., Kaushik, S.P., Ashwani Kumar, Murthy, T.S. and Samuel, J.C. (2001), *Micro-irrigation,* Proc. International Conference on Micro- and Sprinkler Irrigation System, 8–10, 2001, Central Board of Irrigation and Power, Chanakyapuri, New Delhi, p. 774.

Sivshankar, K., ed. (1995), *Micro-Irrigation and Fertigation.* Proc. of the National Workshop on Micro-Irrigation and Fertigation, Sep. 21 and 22, 1995, Bangalore, Pub. University of Agricultural Sciences, Bangalore, Kemira, OY Finland and Jain-Kemira, Jalgaon, p. 96.

Singh, A., (1988), *Bamboo Drip Irrigation System,* Pub. ICAR Research Complex for N.E.H. Region, Shillong-793004, Meghalaya, p. 17.

Sivanappan, R.K., Padma Kumari, O, and Kumar, V. (1998), *Drip Irrigation,* Pub. Keerthi Publishing House Pvt. Ltd., Coimbatore, Tamil Nadu, p. 412.

Sivanappan, R.K., Rao, A.S. and Dikshit, N.K. (1994), *Drip Irrigation in India.* Indian National Committee on Irrigation and Drainage, Jolly Reprographics, New Delhi, p. 176.

United States Department of Agriculture, Soil Conservation Service (1984), Trickle Irrigation. US Dept. of Agriculture, Soil Conservation Service, *National Engineering Handbook.* Chapter 15, Section 15. U.S.D.A., S.C.S., Washington, D.C., p. 129.

Vamadevan, V.K. (1985), *The Potential of Micro-Irrigation Techniques for High Value Tree Crops,* Proc. of Short Term Course on Sprinkler and Drip Irrigation, 25th Feb. to 16th March, 1985, Training Unit. WAPCOS (India) Ltd., New Delhi, pp. 169 – 202. Mimeo.

Vermieren, I. and Jobling, G.A. (1980), *Localized Irrigation,* Pub. United Nations Food and Agriculture Organization, Rome. Irrigation and Drainage Paper 36, p. 203.

Water and Land Management Institute (1988), *Drip Irrigation.* Pub. No. 34, WALMI, Aurangabad, Maharashtra, p. 114.

Williams, G. S. and Hazen, A., (1960), *Hydraulic tables,* 3rd ed., John Wiley and Sons, Inc., New York.

Wu, I, Pai (1975), *Design of drip irrigation mainlines* (1975), Jl. of the Irrigation and Drainage Division, ASCE, Vol. 101, No. IR 4, Proc. Paper 11803, pp. 265–278.

Wu, I. Pai and Gitlin, H. M. (1973), *Hydraulics and uniformity for drip irrigation,* Jl. of the Irrigation and Drainage Division, ASCE, Vol. 99, No. IR 3, Proc. Paper 9786, pp. 157–168.

Wu, I. Pai and Gitlin H. M. (1974), *Drip irrigation design, based on uniformity,* Trans. Am. Soc. Agrl. Engrs., Vol. 17, No. 3, pp. 157–168.

Wu, I. Pai and Gitlin, H. M. (1975), *Energy gradient line for drip irrigation laterals.* Jl. of the Irrigation and Drainage Division, ASCE, Vol. 101, No. IR 4, Technical Notes, Proc. Paper 11750, pp. 323–326.

Yadav, R.K. and Michael, A.M. (1980), *Indigenous Drip Emitter.* Proc. Third Afro-Asian Regional Conf. ICID, New Delhi, pp. 297–310.

Yagev, E. and Sapir, E. (1995), *Drip Irrigation.* Pub. Ministry of Foreign Affairs, Center for International Cooperation, State of Israel.

Yella Reddy, K., Appa Rao, D., Satya Prasad, A. and Madhava, M. (2000), *Evaluation of On-line Trickle Irrigation Emitter Constants,* Jl. of the Institution of Engineers (India) Vol.81, June 2000, pp. 24–26.

Chapter

14

Salt Management in Irrigated Agriculture

LEARNING OBJECTIVES

Nature and extent of salt problems in irrigated and unirrigated areas and their influence in crop growth and production. Causes of waterlogging and salt accumulation. Influence of management practices on soil salinity and alkali. Soil physical and chemical properties as influenced by salt concentration and their determination. Irrigation water quality—classification and characteristics. Irrigation management with poor quality water. Reclamation of saline, saline-alkali and alkali soils. Plant response to saline and alkali soils and appropriate soil and crop management practices.

In large parts of the world, including India, the production from irrigated agriculture is below the potential. Crop production in irrigated areas can be enhanced by adopting a proper mix of activities complementary to irrigation such as drainage, soil management, crop selection and appropriate agronomic practices. In India, the near stagnation of the cropped area under food grains and a decline in the productivity and yield of crops in the recent past is to be viewed with concern. Sustainability of agricultural production is the ability of the production system to enhance productivity towards the potential and maintain it, without degrading the resource bases of land and water.

Extensive areas of land in the arid and semi-arid regions have gone out of cultivation due to the rise of water table and accumulation of salt (Fig. 14.1 and 14.2). Excessive irrigation and poor water management are the chief causes of waterlogging and salt build-up. Development of soil salinity is a challenge to the sustenance of irrigated agriculture. Accumulation of salts in soils leads to unfavourable soil-water-air relationships and decreases crop production. Gradually, the land goes out of cultivation, unless remedial measures are undertaken.

In India, salt-affected soils are designated by different names according to their morphological, chemical and physical characteristics and the language of the region. Most commonly, the term *usar*, which is derived from the Sanskrit word *ushtra* meaning sterile or barren, is collectively used for all kinds of saline and

Fig. 14.1. Stunted wheat crop in a highly saline area in the Najafgarh block of Delhi State (1973).

Fig. 14.2. Extensive salt incrustation on agricultural land in the Najafgarh Block of Delhi State (1973). Note the stunted and patchy growth of the wheat crop.

alkali soils. The other words used in different regions of the country are *reh, thur kallar rakkar* and *kulrati* in Rajasthan, UP, and Punjab; *karu, chopan* and *lona* in Maharashtra; *karu, uippu, choudu* and *palachoudu* in Tamil Nadu; *phodus* in Andhra Pradesh and *kari, pokhali* or *kaipad* in Kerala. However, these terms are not associated with any particular limit of the indices measuring salinity or alkalinity of the soils.

Plant roots require a favourable environment to extract the nutrient solutions and water to meet the plant's requirement. For most crops, soil moisture ranging from field capacity to 50 per cent of the field capacity in the root zone is considered ideal. Only a few crops such as rice and jute need standing water in the field at certain stages of their growth. Chemically, a neutral and non-saline soil is ideal for proper growth and yield of most of the food crops. Excess water or high salt concentration in the root zone or at the land surface does not allow the plant roots to function normally. As a result, the plant growth and yield are adversely affected. In the extreme cases of waterlogging and salinity, the seeds may not germinate and the plants may wilt permanently (Fig. 14.2). Agricultural land drainage is both a preventive and a curative measure for the prevention of physical and chemical degradation of soil and for the reclamation of the already degraded lands.

Definitions. Salt affected soils have commonly been grouped into saline, saline-alkali and alkali soils. Considering their specific management aspects, these soils are usually identified in two main classes: (*i*) saline soils and (*ii*) alkali soils.

Saline soils. These soils contain soluble salts, mostly chlorides and sulphates of sodium, calcium and magnesium, in quantities enough to interfere with the growth of most crop plants (Fig. 14.3). The soil pH of saturated paste of these soils is less than 8.5, electrical conductance (EC_e) more than 4 deci-Siemens/m (dS/m) and Exchangeable Sodium Percentage (ESP) less than 15.

Alkali soils. Alkali soils contain salts dominated by bicarbonates, carbonates and silicate of sodium, capable of alkaline hydrolysis or have sufficient exchangeable sodium to interfere with the growth of most crops (Fig. 14.4). The soil pH is greater than 8.5, electrical conductance (EC_e) of saturation extract less than 4 dS/m and Exchangeable Sodium Percentage (ESP) of 15 or more. Excess sodium on the exchange complex imparts these soils poor physical properties such as low infiltration rates. As a result, most of the rainfall received in areas dominated by these soils goes as run-off.

Fig. 14.4. An extensive stretch of alkali land in the Pilot Project area of CSSRI in Haryana, prior to reclamation. (*Courtesy:* CSSRI, Karnal)

Fig. 14.3. Extensive saline barren land in the Pilot Project area of CSSRI in Haryana, prior to reclamation. (***Courtesy***: CSSRI, Karnal)

In scientific literature, alkali soils are often called *sodic soils*. In India, alkali soils are known as *usar* in Uttar Pradesh and *kallar* in the states of Punjab and Haryana.

Saline-alkali soil. A soil which shows EC of the saturation extract greater than 4 millimhos/cm at 25° C and its ESP value is greater than 15. The pH of the saturated soil paste may somewhat exceed 8.5.

Non-saline alkali soils. A soil which has exchangeable sodium percentage (ESP) above 15 and EC of the saturation extract less than 4 millimhos/cm. The pH value of such soils is more than 8.5 and ranges between 8.5 to 10.

Cation exchange. Interchange of a cation in solution with another cation on a surface-active material.

Cation exchange capacity (CEC). The total quantity of cations which a soil can adsorb by cation exchange, usually expressed as milli-equivalents per 100 grams. Measured values of the cation exchange capacity depend somewhat on the method used for its determination.

Dispersed soil. Soil in which the clay readily forms a colloidal suspension. Dispersed soils usually have low permeability. They tend to shrink, crack and become hard on drying and plastic on wetting.

Electrical conductivity (EC). It is the reciprocal of the electrical resistivity. Quantitatively the electrical resistivity is the resistance, in ohms, of a conductor, metallic or electrolytic, which is one cm long and has a cross-sectional area of 1 sq cm. Hence electrical conductivity is expressed as the reciprocal of ohm-cm or mhos per cm. The terms "electrical conductivity" and specific electrical conductance have identical meanings. For convenience in units, millimhos/cm (10^{-3} mhos/cm) or micromhos/cm (10^{-6} mhos/cm) are used.

Symbol	*Term*	*Unit*
EC	Electrical conductivity	mhos/cm
m mhos/cm	Millimhos/cm	10^{-3} mhos/cm
μ mhos/cm	Micromhos/cm	10^{-6} mhos/cm

Equivalent weight. It is the combining capacity of an element or radical with hydrogen. It is the weight in grams of an ion or compound that combines with or replaces one gram of hydrogen

$$\text{Equivalent weight} = \frac{\text{Atomic weight}}{\text{Valency}}$$

For example,

$$\text{Equivalent wt. of } Na^+ = \frac{23}{1} = 23; \quad Cu^{++} = \frac{40}{2} = 20$$

$$Cl^- = \frac{35.5}{1} = 35.5 \text{ and } SO_4^{\bullet\bullet} = \frac{96}{2} = 48$$

$$\text{Milliequivalent wt.} = \frac{\text{Equivalent weight}}{1000}$$

Equivalent weights of some common salts and chemical amendments are given in Table 14.1.

Exchangeable cation. A cation that is adsorbed on the exchange complex and which is capable of being exchanged with other cations.

Exchange complex. The surface-active constituents of soil (both inorganic and organic) that are capable of cation exchange.

Exchangeable sodium percentage (ESP). It is the degree of saturation of the soil exchange complex with sodium and may be calculated by the formula

$$ESP = \frac{\begin{array}{c}\text{Exchangeable sodium}\\ \text{(milliequivalent per 100 gm soil)}\end{array}}{\begin{array}{c}\text{Cation-exchange capacity}\\ \text{(milliequivalent per 100 gm soil)}\end{array}} \times 100$$

Leaching. The process of removal of soluble material by the passage of water through soil.

Leaching requirement. The fraction of the water entering the soil that must pass through the root zone in order to prevent soil salinity from exceeding a specified value. Leaching requirement is used primarily under steady-state or long-time average conditions.

Milliequivalent (meq). One thousandth of an equivalent.

Milliequivalent per litre (meq/l.). A milliequivalent of an ion or a compound in 1 litre of solution.

Molar solution. It is the solution having a salt concentration equal to one gram molecular weight dissolved per litre (molar sol. = gram mol. wt. per litre). Millimole is one-thousandth salt concentration of the molar solution (Millimole = Molar solution/1000).

Table 14.1. Chemical symbols equivalent weights and common names of some common salts and chemical amendments

Chemical symbol or formula	*Equivalent weight (grams)*	*Common name*
Ions		
Ca^{++}	20.04	Calcium
Mg^{++}	12.16	Magnesium
Na^+	23.00	Sodium
K^+	39.10	Potassium
Cl^-	35.46	Chloride
SO_4^{--}	48.03	Sulphate
CO_3^-	30.00	Carbonate
HCO_3^{--}	61.01	Bicarbonate
Salts		
$CaCl_2$	55.50	Calcium chloride
$CaSO_4$	68.07	Calcium sulphate
$CaSO_4.2H_2O$	86.09	Gypsum
$CaCO_3$	50.04	Calcium carbonate
$MgCl_2$	47.62	Magnesium chloride
$MgSO_4$	60.19	Magnesium sulphate
$MgCO_3$	42.16	Magnesium carbonate
NaCl	58.45	Sodium chloride
Na_2SO_4	71.03	Sodium sulphate
Na_2CO_3	53.00	Sodium carbonate
$NaHCO_3$	84.01	Sodium bicarbonate
KCl	74.56	Potassium chloride
K_2SO_4	87.13	Potassium sulphate
K_2CO_3	69.10	Potassium carbonate
$KHCO_3$	100.11	Potassium bicarbonate
Chemical amendments		
S	16.03	Sulphur
H_2SO_4	49.04	Sulphuric acid
$Al_2(SO_4)_3.18H_2O$	111.07	Aluminium sulphate
$FeSO_4.7H_2O$	139.01	Iron sulphate (ferrous)

Parts per million (ppm). The results of a chemical analysis of water are usually reported in parts per million of the various substances present in the sample. One part per million (ppm) means one part in a million parts. As commonly measured and used, parts per million is numerically equivalent to milligrams per litre.

Saturation extract. The solution extracted from a soil at its saturation percentage.

Saturation percentage. The moisture percentage of a saturated soil paste expressed on dry weight basis.

Sodium adsorption ratio (SAR). A ratio for soil extracts and irrigation waters used to express the relative activity of sodium ions in exchange reactions with soil

$$SAR = \frac{Na^+}{\sqrt{(Ca^{++} + Mg^{++})/2}}$$

in which the ionic concentrations are expressed in milliequivalents per litre.

Soil extract. The solution separated from a soil suspension or soil at a particular moisture content.

pH. The pH value of a soil or natural water is a measure of its alkalinity or acidity. More accurately stated, the pH value is a measure of the hydrogen-ion concentration in water. Water molecules (H_2O) have a slight tendency to breakdown into ions. Ions are atoms or groups of atoms carrying positive or negative electrical charges. The chemical formula H_2O may also be written HOH. When a water molecule breaks down, it divides into a positive hydrogen ion (H^+) and a negative hydroxyl ion (OH^-). In distilled water, the number of H^- ions formed is such that their concentration is expressed by a pH value of 7 (Mathematically this is the logarithm, to the base 10, of the reciprocal of the hydrogen-ion concentration of pure water). Thus, a pH value of 7 indicates a neutral solution, neither alkaline nor acid. A pH value of 7.5 to 8.0 usually indicates the presence of carbonates of calcium and magnesium, and a pH of 8.5 or above usually indicates appreciable exchangeable sodium.

Specific ion effect. Any effect of a salt constituent in the substrate on plant growth that is not caused by the osmotic properties of the substrate.

Specific surface. The surface area per unit weight of soil, commonly expressed as square metres per gram of soil (m^2/gram).

Osmotic pressure (OP). It is the equivalent negative pressure that influences the rate of diffusion of water through a semi-permeable membrane. Its direct experimental value for a solution is the pressure difference required to equalize the diffusion rates between the solution and pure water across a semi-permeable membrane. Osmotic pressure in atmospheres may be calculated from the freezing point depression (ΔT) in °C, by the formula

$$OP = 12.06\ \Delta T - 0.021\ \Delta T^2.$$

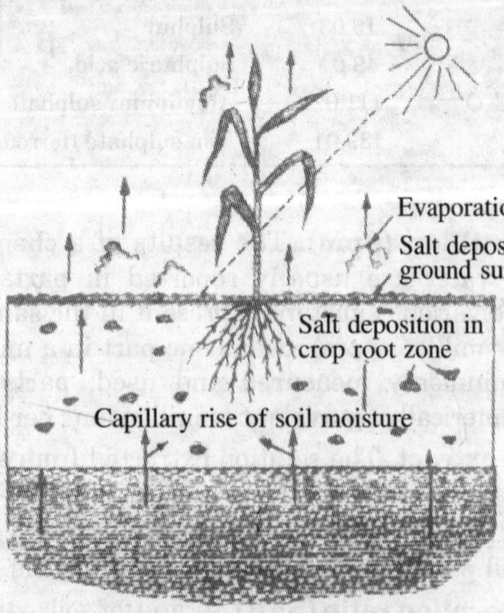

Fig. 14.5. Schematic illustration of salt build-up in soil due to high water table.

Fig. 14.6. A drainage canal blocked with water hyacinth in the Godavari delta of Andhra Pradesh (1973).

Causes of Waterlogging and Salt Problems

Waterlogging and salt problems on agricultural lands occur due to natural causes or human interventions. The important natural causes are high rainfall during the rainy seasons or due to over-irrigation, unfavourable topography, backwater entry from rivers, sea water intrusion, high evaporation during long dry periods and the salts present in the soil (Fig. 14.5). The important human factors are unscientific management of land and irrigation water, use of poor quality water for irrigation, adoption of unscientific and non-sustainable cropping pattern, sand blockage of natural outlets due to construction of roads, railway lines and urbanization (Fig. 14.6 and 14.7). Waterlogging is a situation of a cropland when the crop root zone has been saturated. Such a condition restricts normal air circulation, reduces the oxygen level and increases the carbon dioxide level in the root zone. Salt-affected soils are those in which the concentration of salts in the root zone adversely affects the normal root activity.

UN Food and Agricultural Organization (UNFAO, 1973) defined waterlogged areas as those where soils are temporarily saturated or where the groundwater table is too shallow such that capillary rise of water encroaches upon the crop root zone and may even reach the soil surface. The National Commission of Agriculture, Govt. of India (NCA, 1976) has defined an area as waterlogged when the water table causes saturation of crop root zone soil, resulting in restriction to air circulation, decline in oxygen and increase in carbon dioxide levels.

Following are the common approaches adopted in India to express the water table depth below the soil surface:

Fig. 14.7. Waterlogging and salt accumulation on agricultural land bordering a road embankment. Non-provision of adequate drainage outlets has given rise to the problem. Borrow pits on either side of the road serve as breeding grounds for mosquitoes.

(*i*) Pre-monsoon (April/May) depth to water table.

(*ii*) Post-monsoon (October/November) depth to water table.

(*iii*) Seasonal (monsoon/winter/summer) or annual average depth to water table.

(*iv*) Sum of the number of days in a year when the water table is shallower than a specified depth.

After reviewing the suggestions made by the different agencies, Bhattacharya and Michael (2003) suggested the following norms for the classification of different categories of waterlogged areas in India and other South Asian countries:

(*i*) Water table within 2 m from soil surface during pre-monsoon (April/May).

(*ii*) Water table within 1 m from soil surface during post-monsoon (October/November).

The problem of waterlogging on account of surface water ponding, primarily during the monsoon season, is due to topographical conditions, rainfall and related (back water-flow) causes. Its seriousness in relation to crop depends on the type of crops grown, its stage of growth and its physiological characteristics, duration of rainfall and soil characteristics. The productivity of sensitive crops like pigeon pea and maize reduces by 30 to 50 per cent of the potential, when they are either inundated or remains under a saturated soil condition for 24 to 48 hours.

Salt build-up in soils. Soluble salts in the parent rocks, which have weathered to form soil, seawater intrusion and high evaporation are the major natural causes of salinizing agricultural lands. Under a monsoon climate, much of the accumulated salts are washed or leached out during the rainy season. However, high evaporation during the remaining dry and hot months in the year draws up the saline groundwater at shallow depths towards the surface. Salts are left behind after the water evaporates (Fig. 14.5). The important man-made causes for salinity development are the use of poor quality and excess application of irrigation water.

Irrigation projects involving inter-basin transfer of water without adequate drainage has disrupted the equilibrium between the groundwater recharge and discharge, resulting in accretions to the groundwater table. This is the major cause for waterlogging and secondary salinization in various irrigation command areas. Millions of hectares of agricultural lands in India and other countries have become too saline to economic agricultural production. Although salt infestation results in gradual yields reduction, the severity of the problem is not fully realized till the crop suffers heavily due to excessive salts. In most such cases, the land is abandoned as the land has become non-productive due to salinization.

Irrigation with water of low salinity but with dominant HCO_3^- anion and migration of sodic salts in arid climate promote salinity. The main causes for soil salinity and sodicity are: (*i*) irrigation mismanagement; (*ii*) poor land levelling; (*iii*) leaving land fallow during dry periods, especially in regions of shallow water table; (*iv*) improper use of heavy machinery, resulting in soil compaction; (*v*) leaching without adequate drainage, and (vi) adoption of improper cropping patterns and crop rotations. In irrigated agriculture, scientific water and land management is the key to avoid waterlogging and salinity problems.

Saline or alkali soils are seldom formed *in situ* by residual weathering of primary rocks. On the other hand, hydrological conditions contribute substantially to the development of soil salinity and alkalinity. The hydrological processes release the salt constituents from the primary and secondary minerals to the surrounding waters. In arid regions, due to high evaporative conditions, salt concentration is gradually increased in the water sources.

In humid regions, salts thus released are transported to lower layers and ultimately by streams to the ocean. Moreover, the salts deposited by irrigation, are leached down to lower layers of the soil by rainfall. On the contrary, in arid regions, the released salts may not be transported far away because of insufficient rainfall and whatever leaching of salts may occur, it would mostly be local. However, high evaporating conditions of the arid regions tend to decrease further the limited available groundwater for leaching the salts. Moreover, the salts which move down due to rain, come back again to the surface by capillary action during the hot season.

Effect of Irrigation, Soil Management and Crop Management on Soil Salinity and Alkalinity

The nature and extent of salt accumulation and the degree of soil salinity and alkalinity depend on the quality of irrigation water, frequency of irrigation, soil type and its permeability, salt tolerance characteristics of the plant and climatic conditions. Generally, soils of light texture are less salinised than those of medium and heavy texture. The soils irrigated with waters having the same concentration of salts develop more alkalinity with waters having higher proportion of sodium and bicarbonate ions. The presence of a hard pan of lime or clay further enhances the degree of alkalinity. The situation becomes more complex to predict when the depth of water table is shallow and the quality of irrigation water is poor. Under such situations, the final salt balance near the root zone is, governed by the combined effect of the quality of irrigation water, irrigation management, climate and water transmission properties of the soil.

Salt accumulation is associated with the relief of the land. In flood plains, deltas and coastal belts and in areas of shallow water table, the salt concentration is usually high. In such situations, surface runoff is low and the drainage water evaporates, leaving the salts on the surface. In areas of shallow water table, salts move up under thermal gradient and are deposited on the surface. Saline groundwater and shallow water table are factors favouring soil salinity. However, the rate of salinisation by capillary action decreases when the depth of water table is large, say beyond 1.5 metres and the soil moisture content is also low.

During irrigation, large volumes of water percolate down to lower layers, as a result of which the depth to groundwater table regularly decreases. Under such a situation, salts present in the water move up by capillarity and are deposited on the surface as a result of evaporation.

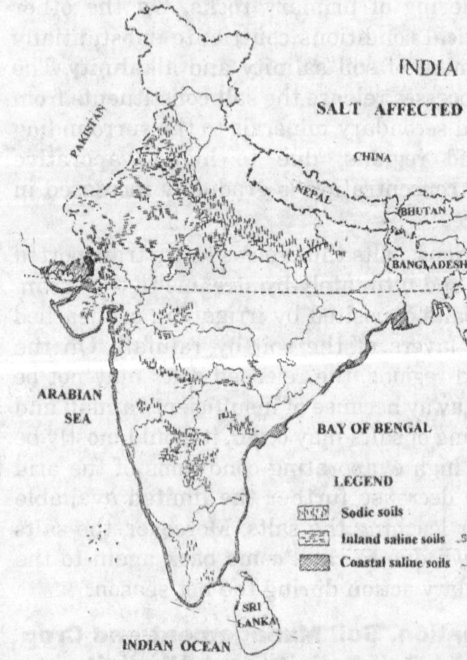

Fig. 14.8. Distribution of salt-affected soils of mainland India (not to scale)

[*Source:* Abrol, I.P. and Gupta, R.K. (1991), National Training Course, Lecture Note 1, CSSRI, Karnal.]

In addition to the above factors, salinity and alkalinity problems develop in areas where drainage facilities are adequate but irrigation is insufficient to meet both the crop needs and leaching requirement of the soil. Inadequate drainage is another factor intimately associated with the development of soil salinity. Sometimes, in spite of good quality irrigation water and good irrigation practices, soil salinity develops because of the poor soil physical conditions which impede drainage.

Nature and Extent of Salt Problems

A broad indication of the waterlogged and salt affected areas in some of the states of India are presented in Table 14.2, which gives the total waterlogged areas both within and outside the irrigated regions, including the coastal saline lands. The distribution of salt-affected soils in mainland India is shown in Fig. 14.8. The United Nations Food and Agricultural Organization made the following broad estimates of the salt affected areas in different continents/sub-continents (UNFAO, 2000):

South Asia, 87.6 million ha; Central and North Asia, 211.7 million ha; Australia, South-East Asia and East Asia, 357.3 million ha; Africa, 80.5 million ha; South America, 129.2 million ha; Mexico and Central America 2 million ha; North America 15.8 million ha and Europe 50.8 million ha.

Table 14.2 Geographical, waterlogged and salt affected areas of some states in India

State	*Geographical area* (*million ha*)	*Waterlogged area* (*million ha*)	*Salt-affected area* (*million ha*)
Andhra Pradesh	27.44	0.339	0.813
Bihar	17.40	0.363	0.400
Gujarat	19.60	0.484	0.455
Haryana	4.22	0.275	0.455
Karnataka	19.20	0.036	0.404
Kerala	3.89	0.012	0.026
Madhya Pradesh	44.20	0.057	0.242
Maharashtra	30.75	0.111	0.534
Punjab	15.54	0.196	0.400
Orissa	5.04	0.199	0.520
Rajasthan	28.79	0.348	1.122
Tamil Nadu	12.96	0.128	0.340
Uttar Pradesh and Uttaranchal	29.40	1.980	1.295

Compiled from: Ghosh, 1991 and Tyagi, 1999.
(***Source:*** Bhattacharya and Michael, 2003.)

Heavy monsoon rain causes accumulation of runoff. Its ill-effects are more severe in flat terrains. In the absence of drainage, natural or man-made (Fig. 14.9), the accumulated water gets more opportunity time to percolate down and causes the water table to rise. Often valuable field operation time is lost due to excess wetness of the soil, particularly in heavy textured soils. Delay in the sowing of winter crop after the withdrawal of monsoon is a common feature which reduces the crop yield.

The topography of an area profoundly influences the drainage conditions. Areas with adverse topography from the drainage point of view are known by different local names such as *bhal* (as flat as the forehead) lands in Gujarat, *diara* (concave) and *tal* and *chaur* (flood plains of low elevation) lands in Bihar and Uttar Pradesh. There are low-lying lands with severe restrictions of natural drainage in several parts of India. Examples are the large areas in the Indo-Gangetic plain, the delta regions of the eastern and south-eastern India and the below-sea level rice farms in the Kuttanad region of Kerala. In many of such lowlands or land with adverse topography, the monsoon cultivation season is lost due to high level of land inundation following incessant heavy rains, flood water accumulation and backwater flow.

Seepage from unlined and ill-maintained canal systems in the irrigated areas causes long duration wet condition over the adjoining agricultural lands and rise in water table and development of salinity and alkali conditions. Other problems leading to waterlogging and salt build-up are ill-planned construction of roads and embankments (Fig. 14.8), rail lines and railway embankments. Absence of culverts or under-capacity culverts leads to restrictions in the natural flow of water, resulting in waterlogging. Adoption of unscientific cropping practices such as growing more water requiring crops against the planned cropping pattern in a canal irrigated area results in applying more than the anticipated quantity of water over relatively smaller areas. This causes excess water problems in certain parts and water deficits over the other parts.

Brackish ground and surface water when used in irrigation, for want of adequate quantity of fresh water, adds undesirable salts to the agricultural lands. Disposal of untreated municipal and industrial effluents on the land and into the natural streams cause serious surface and groundwater quality problems. High salt content in the soil due to properties of the parent rock material, poor water transmission characteristics in heavy soils and high evapo-transpiration demands in a major part of the year lead to salinity and alkali problems. This is more conspicuous in the semi-arid regions.

The salt affected areas of India can be classified into three categories: (*a*) arid and semi-arid regions with low rainfall and irrigated with poor quality groundwater, (*b*) regions of high water table and areas influenced by temporary waterlogging and seepage from canals and other sources, and (*c*) coastal regions influenced by sea water intrusion.

Fig. 14.9. An outlet drain near Faridkot, Punjab maintained by the Drainage Section of Irrigation Department, Punjab (1968). The main drain also serves as outlet for field drains in the catchment area.

Salt problems in the arid and semi-arid regions having rainfall below 450 mm per annum are spread over the states of Gujarat and Rajasthan and in the districts of Agra and Mathura in Uttar Pradesh, Ferozepur and Sangrur in Punjab and Rohtak, Gurgaon and Mohindergarh in Haryana. In these areas, soil salinity is primarily due to the use of poor quality irrigation water. The severity of the salinity problem in these areas is further increased by the arid climate.

In Rajasthan, out of a total area of about 1.1 million hectares under irrigation with groundwater in 1970s, about 57 per cent of the area is affected by problems of salinity and alkalinity (Paliwal, 1972). The salt affected area is about 70 per cent of the irrigated area in the districts of Bikaner, Jaisalmer, Pali, Jodhpur, Bharatpur, Barmer, Nagaur, Jaipur and Bhilwara.

The salt problems in some of the areas of Punjab, Haryana, Uttar Pradesh and Delhi are due to the combined effect of poor quality irrigation water, shallow water table and inadequate drainage. In Punjab, the quality of irrigation water does not seem to pose a serious problem in the districts of Gurdaspur, Kapurthala, Hoshiarpur and Ludhiana, while in Amritsar, Patiala and Sangrur districts, it is of moderate level and the serious problem is in the district of Ferozepur. In Haryana, the problem is serious in the districts of Rohtak and Gurgaon. However, the salt problems in some of these areas are due to low permeability of the soils, rather than high water table or poor quality water. In the Union Territory of Delhi, all the 5 blocks, *viz*., Alipur, Shahdara, Nangloi, Najafgarh and Mehrauli are influenced by the problem of soil salinity and poor quality of well waters. The irrigated areas have shallow water table and poor drainage.

The salt-affected lands in the central and southern states are mostly confined to the black cotton soil areas in the semi-arid regions comprising parts of Maharashtra, Karnataka and Andhra Pradesh. Saline, alkali and saline-alkali soils in varying gradations are observed in these areas. Nearly all the deep black soils and soils in the low-lying areas are potentially saline or alkali, depending upon the depth and nature of alkalisation. Improper use of irrigation water has resulted in the rise of the subsoil water table and ultimately by secondary salinisation, salt-affected soils have developed. In some of these areas, gypsum is found in the B horizon which helps in keeping the soil in good condition. Almost all the salt-affected black cotton soils of the south have a good reserve of lime throughout the profile. This may act as a source of calcium and help in their better utilization.

The saline-alkali soils of the Kaveri delta have been classified into *Old delta* and *New delta.* The *Old delta* soils are dark gray, clayey, low-lying and poorly drained, with a water table at about 1.5 metres depth. The *New delta* soils are light textured, more dispersed and have low permeability.

Salt-affected soils of coastal lands. Along the coastal line of India, thousands of hectares of lands are salt affected by sea water intrusion. Large areas of land along the mouths of rivers and creeks in West Bengal, Orissa, Andhra Pradesh, Tamil Nadu, Maharashtra and Gujarat are subjected to periodic tidal waves and are inundated with sea water. Such areas are devoid of all types of vegetation.

Studies at the Land Reclamation, Irrigation and Power Research Institute, Amritsar, showed a rise of water table by 2.6 to 6.6 metres during the period 1940-60 (Uppal, 1962). It was observed that the salt problem in high water table areas is mostly governed by the amount and intensity of rainfall. In the monsoon season, when water table rises and comes near the surface, the land is temporarily waterlogged due to which the salinity is reduced. But, after the rains and with the onset of winter, water is lost by evaporation and salts are deposited on the surface (Fig. 14.5). During the winter crop (*rabi*), such lands are irrigated and the salts present in the soil profile move up and down. Consequently, the total soil salinity is controlled by the period of waterlogging, water transmission characteristics of the soil, the salt content of the groundwater and the evaporative conditions of the region. Topography of the land also modifies soil salinity.

Chemistry of Salt-Affected Soils

The main processes occurring in soils while irrigating with poor quality waters are: (*i*) ion exchange between cations in irrigation water and those present on the soil exchange complex, (*ii*) dissolution and precipitation of $CaCO_3$, (*iii*) weathering of minerals, (*iv*) hydration and dehydration of the soil as a result of fluctuation in soil moisture, (*v*) leaching down of ions, (*vi*) upward movement of ions through capillary activity, and (*vii*) mineral nutritional characteristics of the crop grown. Amongst these processes, cation exchange is the most important process, which governs the accumulation of excessive sodium during irrigation with saline water and reclamation of alkali soils during treatment with gypsum.

The finest fractions of soil particles, called clay, have a certain charge deficit. This charge is balanced by the charge of the adsorbed cation taken from the soil solution. These adsorbed ions on the clay phase are replaceable by any other cation. However, the rate of reaction depends upon its chemical affinity to the soil, its valency and size. These replaceable ions are called exchangeable cations. The maximum adsorption capacity of the soil for cations is called cation exchange capacity, which is generally expressed as meq./100 gram of soil. An order of replaceability of any ion present in the soil can be expressed as

$$NH_4^+ > Mg^{++} > Ca^{++} > K^+ > Na^+$$

This means that Ca^{++} will replace Na^+ from the exchange complex more easily, while the replacement of Ca^{++} by Na^+ is more difficult.

In normal soil, generally, Ca, Mg, K and NH_4 constitute 70, 25, 3 and 2 per cent, respectively, of the exchangeable cations. Such soils are dominated by Ca, followed by Mg, K and NH_4. But in case of alkali soils, the proportion of exchangeable Na is more and increases with the pH of soil. The pH of alkali soils generally increases with the degree of Na saturation. Exchangeable sodium per cent (ESP) value of the soil is further influenced by the nature of the anion associated with the sodium salt in the solution phase.

A clay particle with sodium and calcium ions attached tends to hydrolyze. When a sodium ion is exchanged for a hydrogen ion and the sodium ion combines with a molecule of water, sodium hydroxide (NaOH) is formed.

$$\mathrm{Ca}\overset{\mathrm{Na}}{\underset{\mathrm{Na}}{\boxed{\text{Clay Particle}}}}\mathrm{Na} + H_2O \rightarrow$$

$$\mathrm{NaOH} + \mathrm{H}\overset{\mathrm{Na}}{\underset{\mathrm{Na}}{\boxed{\text{Clay Particle}}}}\mathrm{Ca}$$

When carbon dioxide (CO_2) is present in the soil air, it readily reacts with the water to form hydrogen carbonate (H_2CO_3). However, the sodium hydroxide (NaOH) reacts readily with the hydrogen carbonate (H_2CO_3) to form sodium carbonate (Na_2CO_3).

$$H_2CO_3 + 2NaOH \rightleftharpoons Na_2CO_3 + 2H_2O$$

The Na_2CO_3 is gradually removed with extensive leaching and the soil is left with hydrogen ions having replaced sodium ions. This increase of hydrogen ions is reflected in a lower pH.

Where soils contain calcium carbonate ($CaCO_3$) or gypsum, calcium is dissolved into the soil solution. This available calcium is exchanged for sodium during the leaching process to obtain a normal soil.

$$Ca\,\overset{Na}{\underset{Na}{\boxed{\text{Clay Particle}}}}\,{}^{Na}_{Na} + CaCO_3 \rightleftharpoons Na_2CO_3 + Ca\,\overset{Na}{\underset{Na}{\boxed{\text{Clay Particle}}}}\,Ca$$

Effect of salts on soils. The physical and chemical properties of irrigated soils depend largely on the chemical composition of the irrigation water, soil type and drainage characteristics and climatic conditions. The main factors are:

(*i*) Concentration of salts in the soil solution and

(*ii*) Nature and degree of exchangeable cations.

Under field conditions, it is difficult to separate the two effects completely because the process of ion exchange begins simultaneously, as the concentration of soil solution increases. However, the physical properties of the soil are more deteriorated in the presence of excessive exchangeable sodium.

Salinity hazard. When the salt distribution in the soil profile becomes excessive, crop growth is deteriorated. Such soils are often recognized by the presence of a white surface crust (Fig. 14.1). In saline water irrigated areas, particularly in the arid regions of Rajasthan, the EC of the saturation extract of the soils is usually about three times and even more than that of irrigation water. In general, salts concentration is maximum on the surface and decreases with the increase in depth. In irrigated soils, the extent of salinity is more related to the process of salt water movement within the soil column than whether the salts are moving downwards with the leaching effect of irrigation water or are moving up by capillarity under the thermal gradient. In a well-drained light textured soil immediately after irrigation, the soil salinity, as measured on the basis of saturation extract of the soil sample, would be less at surface layers and it would be more at deeper layers. On the contrary, in heavy soils accumulation of salts would always be more on the surface due to impeded drainage. Quantitatively, under similar soil water conditions, more salinity would be observed in arid than in humid regions and in heavy textured soils than in light soils.

Sodium hazard. In irrigated soils the cations present in irrigation water as soluble salt take part in an exchange reaction with the soil. The main cation exchange which takes place is between sodium ion in the irrigation water and calcium ion in the exchange complex of the soil:

$$Ca\,\boxed{\text{Clay Particle}}\,Ca + 2Na^+ \rightleftharpoons Ca\,\overset{Na}{\boxed{\text{Clay Particle}}}\,Na + Ca^{++}$$

According to this process, a soil may or may not be changed into an alkali soil, depending upon the relative proportion of sodium and calcium ions in the equilibrium solution. However, the process of alkalization could be very quick and can come to an equilibrium condition, if the relative proportion of Na: Ca + Mg, as expressed by the sodium adsorption ratio is high. The exchangeable sodium percentage increases with the total salt concentration and/or SAR of irrigation water or soil solution.

In arid and semi-arid regions, as soil solution becomes concentrated through evaporation or transpiration, the solubility limits of calcium carbonate and magnesium carbonate are exceeded. As a result of this they are precipitated, with a corresponding increase in the relative proportion of sodium in the soil solution. Hence, a part of exchangeable calcium and magnesium is replaced by sodium, increasing the sodium content. This leads to the conclusion that any chemical process in the soil, causing a decrease in calcium, either in solution or exchange phase, will lead to enhancement of alkalinity.

In addition to the above process, the extent of degree of sodium saturation will be governed by the chemical affinity of the ion to the soil, its replacing power in comparison to other ions and its own replaceability by other ions. Fortunately, amongst sodium, calcium and magnesium, the replacing power, both of calcium and magnesium, and their chemical affinity to soil are more than sodium. Hence, a higher proportion of sodium is necessary in the soil solution to replace calcium. The proportion of sodium to calcium and magnesium should be above 50 per cent in the soil solution before it is adsorbed in significant amounts. Moreover, more Na is adsorbed at the same salt concentration and the same Na content when Mg is present rather than Ca, as the replaceability of Mg is more than that of Ca. It may be summarised that the process of alkalization, and its extent is influenced by the total salt concentration, ionic composition, relative proportion of sodium to other cations, nature and total amount of clay and their ion exchange characteristics.

Magnesium hazard. Sometimes more magnesium is present in irrigation water than calcium. This increases the degree of magnesium saturation which also deteriorates the soil structure. Irrigation waters having relatively more magnesium than calcium are likely to decrease soil productivity. About 73 per cent of the 4162 samples of poor quality well waters of Rajasthan were observed to have more Mg than Ca (Paliwal, 1972)

Influence of Salts on the Physical Properties of Soils

The physical properties of soils may be improved or deteriorated in the presence of salts, depending upon the nature and amount of salts, the reaction product and the initial physical and chemical conditions of the soil. The main physical properties influencing the air-water relationships in irrigated agriculture are dispersibility of the clay particle and permeability of the soil.

These properties are markedly influenced by the nature and amount of exchangeable cations and swelling characteristics of the soil. The extent of dissociation of exchangeable cations depends upon the nature of the cations, *i.e.*, their charge and size including the water of hydration as well as the concentration of salts and pH.

The degree of dispersion increases with the increase of the easily dispersible clay and the degree of sodium saturation, and decreases with the presence of cementing agents and salt. The degree of dispersion may be calculated as follows:

$$\text{Degree of dispersion} = \frac{\text{Amount of clay dispersed}}{\text{Total clay}} \times 100$$

However, various types of soils need different levels of sodium saturation for their complete dispersion. Generally, light textured soils need a higher degree of sodium saturation for the same degree of adverse effect than heavy soils.

The degree of dispersion of soil is controlled by the actual salt concentration in contact with the soil phase, as sometimes, a saline-alkali soil irrigated with good quality water or less saline water generally shows a higher degree of dispersion. These factors are of great practical importance in the management of alkali soils. Irrigation with highly saline water, particularly having divalent cations, is recommended to decrease the dispersibility of the soil and improve its physical condition, though the salinity of such soils may slightly increase.

Permeability. Permeability of soil is defined as the readiness with which a soil transmits fluids. A high permeability is essential for leaching down the excessive salts or removal of Na released by the application of gypsum in alkali soil. Again, the permeability of soil decreases with the increase of dispersibility and exchangeable Na percentage of the soil. Generally, water moves slowly downwards through a highly dispersed soil because the micropores are clogged by the swelling of dispersed clay. However, when the applied water contains sufficient divalent cations, it suppresses the diffused layer of clay particles and increases soil permeability. Soil permeability also depends on soil texture, structure and depth of the water table. The efficiency of soil reclamation processes by way of gypsum and other chemical amendments depends largely on its permeability. Application of some chemical amendments carrying soluble Ca not only increases the proportion of Ca in soil but also the infiltration rate of water, by reducing the exchangeable Na status and dispersibility of the soil. Good tillage practices improve soil permeability.

In order to predict the success of irrigation in a particular situation, it is essential to have adequate information on the quality of irrigation water, its effect on the physical and chemical properties of the soil, water transmission characteristics and salt build-up.

Determination of the Properties of Saline and Alkali Soils

Determinations of the chemical and physical properties of saline and alkali soils serve as a basis for their diagnosis, treatment and management. The important properties to be determined for the diagnosis of salt-affected soils are the pH, presence of soluble salts, exchangeable sodium, cation exchange capacity and gypsum requirement.

pH. The pH of soil in a 1:2.5 soil water suspension and irrigation water can be determined using a glass electrode with a standard pH meter (Fig. 14.10).

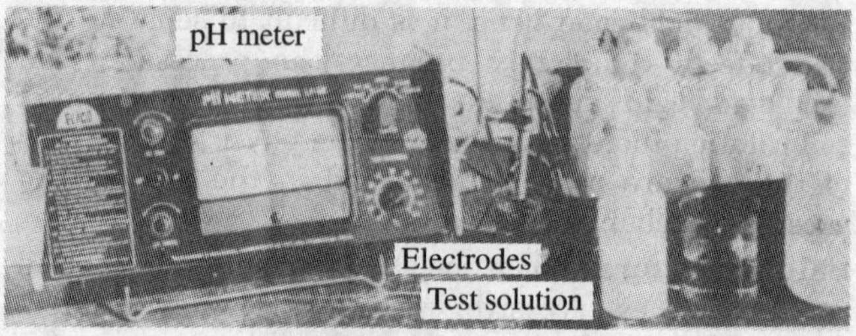

Fig. 14.10. pH meter for determining the concentration of H^+ ions in irrigation water.

Fig. 14.11. Electrical conductivity bridge for measuring the total salt concentration in irrigation water.

Soluble salts. The presence of soluble salts in soils may be estimated quite rapidly, though approximately, by determining the electrical conductivity of the saturation extract of the soil sample or irrigation water on an electrical conductivity meter (Fig. 14.11). Since a high degree of correlation exists between EC, the total cations and osmotic pressure of soil-water extract, the following relationships may be employed for the evaluation of salt concentration:

1. Salt concentration, mg/1 or ppm
   = 640 × EC, millimhos/cm
2. Total cation concentration, meq/l
   = 10 × EC millimhos/cm

   when EC is measured up to the range of 5 millimhos/cm at 25°C

3. Osmotic pressure, atmospheres

   = 0.36 × EC millimhos/cm

   For soils, EC value of the saturation extract could be used for this purpose.
4. ppm/Eq wt = meq/1
5. Milliequivalents per litre to ppm

   = Sum of the product of the milliequivalent of each ion times its equivalent weight.

Example 14.1. *Express 6400 ppm salt concentration in micromhos, millimhos and mhos/cm.*

Solution: Salt concentration, mg/*l* or ppm

$$= 640 \times \text{EC, millimhos/cm}$$

$$\text{EC} = \frac{\text{ppm}}{640} = \frac{6400}{640}$$

$$= 10 \text{ millimhos/cm}$$

$$= 10 \times 1000 = 10{,}000 \text{ micromhos/cm}$$

$$= \frac{10}{100} = 0.01 \text{ mhos/cm}$$

Example. 14.2. *Convert 1.5 millimhos/cm EC into osmotic pressure in atmospheres.*

Solution: Osmotic pressure, atmospheres

$$= 0.36 \times \text{EC, millimhos/cm}$$

$$= 0.36 \times 1.5 = 0.54 \text{ atmospheres}$$

Example 14.3. *Express 1170 ppm sodium chloride salt concentration in meq/1.*

Solution:

$$\frac{\text{ppm}}{\text{Eq wt}} = \text{meq/1}$$

Equivalent weight of NaCl = 58.45

$$\therefore \quad \text{Salt concentration} = \frac{1170}{58.5}$$

$$= 20 \text{ meq/l.}$$

Example 14.4. *A 100 g sample of oven dry soil requires 40 g of water to saturate and 25 g of water to bring it to field capacity. A soil-water suspension is prepared by adding the 100 g oven dry sample in 250 g of water. The suspension is thoroughly stirred and then filtered. The electrical conductivity of the filtrate was measured as 0.5 dS/m. Find the electrical conductivity of the soil-water at saturation and at field capacity.*

Solution: EC at any moisture content = EC of suspension filtrate × Ratio of water used to prepare the suspension to the water needed to obtain the given moisture content, all for the same soil mass.

Hence, $EC_{Saturation} = 0.5 \times (250/40) = 3.125$ dS/m

and $EC_{Field\ capacity} = 0.5 \times (250/25) = 5$ dS/m.

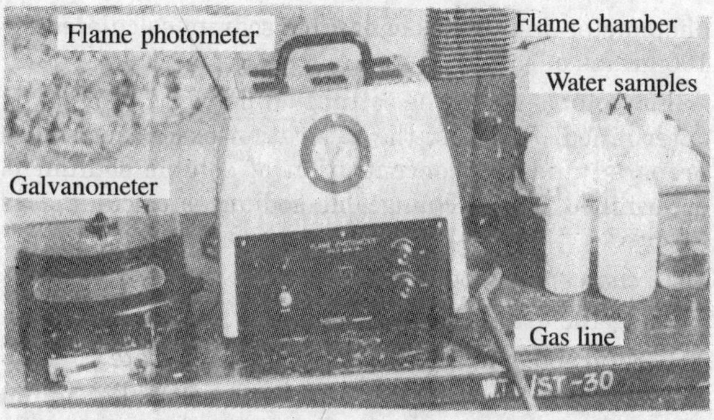

Fig. 14.12. Flame photometer for determining the total Na^+, K^+ and other elements in irrigation water.

Cation Exchange Capacity (CEC). Reagents: (1) Sodium acetate (1 N), (2) Ammonium acetate (1 N), and (3) Isopropyl alcohol (90%). The CEC may be estimated by saturating the soil, first with sodium ions and then replacing the same with ammonium ions as described below:

A 1 *normal* sodium acetate solution is prepared by dissolving 136 grams of the salt in 1 litre of water. The pH of this solution is adjusted at 8.2 by adding a few drops of acetic acid or sodium hydroxide. A 1 *normal* solution of ammonium acetate is also prepared and its pH is adjusted at 7. About 4 to 6 grams of soil is weighed and the sample is transferred to a 50 ml centrifuge tube. Thirty-three ml of sodium acetate is added to the solution and the contents are shaken for 5 minutes on a mechanical shaker after putting the stopper on the tube. The liquid is decanted off and the extraction process is repeated 3 times. The sample is then washed in an identical manner as described above with 33 ml of isopropyl alcohol. This adsorbed sodium should be replaced in the same way as described above with the 33 ml portion of ammonium acetate solution three times. The decanted fraction is kept in a 100 ml volumetric flask and the volume is made up to 100 ml with ammonium acetate. The sodium concentration is estimated by flame photometer (Fig. 14.12). The result, expressed in meq/100 gm of oven dry soil, is the CEC of the soil.

Exchangeable sodium. Reagents: 1 *normal* ammonium acetate solution adjusted to pH 7, concentrated nitric acid, hydrochloric acid 0.1 N and acetic acid 0.1 N. About four to six grams of soil is taken and 33 ml of ammonium acetate solution is added to a centrifuge tube and the suspension is shaken mechanically in the centrifuge for five minutes at 1000 rpm until the supernatant liquid is clear. The supernatant liquid is transferred into a 100 ml volumetric flask. The above process is repeated three times and the volume is made up to 100 ml. The solution is evaporated and the residue treated with a few drops of nitric acid to oxidise any organic matter. The solution is extracted in water to a known volume. The sodium concentration is determined on a flame photometer and the exchangeable sodium calculated for 100 grams of soil.

For soluble slats, the saturation percentage of soil is determined. For this, the saturation extract of soil is prepared and the concentration of soluble sodium is determined. The exchangeable sodium is calculated as follows:

Exchangeable sodium,
(meq/100 grams soil)
= Total sod. extracted by ammonium acetate
—Soluble Sodium
(meq/100 grams soil) (meq/100 grams soil)

Example 14.5. *Analysis of 5 gm soil sample showed that the total concentration of exchangeable cations was 1.25 meq and that of exchangeable sodium was 0.8 meq. Express the cation exchange capacity in meq / 100 grams soil and exchangeable sodium in per cent.*

Solution: In 5 grams soil, the total exchangeable cations

= 1.25 meq

∴ In 100 grams soil, the total exchangeable cations

$$= \frac{1.25 \times 100}{5}$$

Hence, CEC = 25 meq/100 grams soil

Exchangeable Na per cent, ESP

$$= \frac{\text{Exchangeable Na}}{\text{Total exchangeable cations}}$$

$$= \frac{0.8 \times 100}{1.25} = 64$$

or, Exchangeable Na/100 gm soil

$$= \frac{0.8 \times 100}{5} = 16 \text{ meq.}$$

Fig. 14.13. Soil salinity bridge and sensor unit.

The total salt concentration (EC) of the soil solution can be measured *in situ* with the help of a salinity sensor, which consists of a salt-sensitive ceramic electrical conductivity cell (Fig. 14.13).

Gypsum requirement (GR). *Schoonover's method:* A saturated solution of gypsum is prepared by adding 5 grams gypsum in 1 litre of water and shaking mechanically for 10 minutes. The suspension is then filtered. Five grams soil is weighed and 100 ml solution of saturated gypsum is added. The mixture is shaken for 5 minutes on a mechanical shaker or 30 minutes by hand. It is filtered and Ca + Mg is estimated by a standard versenate solution.

[Ca concentration of added gypsum solution, GR, meq/100 gm soil = (meq/1) — (Ca + Mg) concentration of filtrate, meq/l] × 2

It is essential that the soil should be free from soluble salts for GR estimation.

Example 14.6. *A quantity of 100 ml of gypsum solution, having 32 meq / 1 concentration as calcium, on reacting with 5 grams of an alkali soil showed 29.5 meq / l of Ca + Mg concentration in the filtrate. Estimate the gypsum requirement in meq / 100 grams soil.*

Solution: Wt. of soil = 5 grams

Volume of gypsum solution added = 100 ml

Concentration of Ca in gypsum solution = 32 meq/l

Concentration of Ca + Mg after shaking with soil
= 29.5 meq/l

Gypsum requirement (meq/100 grams soil)
= 2 (32 – 29.5)
= 5 meq/100 grams soil

Gypsum requirement = 5 meq Ca/100 grams soil

Note: 1 meq Ca/100 gm soil
= 1.72 metric tonnes per hectare for 15 cm depth of soil

Salt Balance

The relationship between the quantity of soluble salts brought into an area by the irrigation water and the quantity removed from the area by the drainage water has been called the salt balance of the area. Therefore, in an irrigated area, a favourable salt balance—a condition wherein the output of salts equals or exceeds the input—must be maintained if irrigated agriculture is to be permanent. This concept can be applied to a given piece of land or to an entire watershed.

Maintenance of a favourable salt balance in the soil requires proper water management practices. Soluble salts increase or decrease in the root zone, depending on whether the net downward movement of salt is lower or greater than the net salt input from irrigation water and other sources. The salt balance in soil is influenced by the quantity and quality of irrigation water and the effectiveness of leaching and drainage. If the amount of salt added is more than the salt removed, there would be excessive salt in the soil, which would adversely affect crop growth.

The salt balance of soils is expressed by the following relationship:

$$V_i C_i + S_m - V_d C_d - S_v - S_c = 0 \qquad ...(14.1)$$

in which,

V_i = volume of irrigation water

C_i = salt content of irrigation water

S_m = amount of salt dissolved from soil minerals

V_d = volume of drainage water

C_d = salt content of drainage water

S_p = the amount of salt precipitated over the soil surface

S_c = the amount of salt removed by the crop

S_m, S_p and S_c are generally very small in relation to other major quantities, except when C_i is low. Hence, these may be neglected for the evaluation of salt balance. If D_d and D_i are volumes per unit of area of equivalent depths of drainage and irrigation waters, respectively, equation (14.1) be written as follows:

$$\frac{D_d}{D_i} = \frac{C_i}{C_d} \qquad ...(14.2)$$

Example 14.7. *Water of electrical conductivity 1 dS/m is used for irrigation in a command area at the rate of 500 mm per year. If 50 per cent of the salts accumulated in a year is washed out and leached down during the monsoon season, find how many tons of salt would get added to 1 hectare of land after 10 years of irrigation.*

Solution:

Annual irrigation = 500 ha-mm = 5000 m^3

= 5 × 10^6 litre.

Salt concentration = 1 × 640 = 640 mg/l.

Salt added annually = 640 × (5 × 10^6) mg = 3200 kg

= 3.2 tons.

Salt washed out and leached down = 1.6 tons.

Net salt added in one year = 1.6 tons.

Salt added in 10 years = 16 tons.

Quality of Irrigation Water

All irrigation waters contain some dissolved salts from as low as 100 ppm to as high as 7,500 ppm (Paliwal, 1972; Rhoades, 1974). The canal water qualities are generally the same as the waters of their parent rivers, unless it is contaminated while passing through a salt-affected area. The water quality of the Indian rivers is generally good, except in the dry season in some of the rivers and when the river water gets contaminated by polluting water from different sources.

The salt concentration is expressed in ppm (parts per million) or by EC (electrical conductivity) of the salt solution, expressed in dS/m or µmhos/cm or µmhos/cm (micromhos per centimetre). Numerically, 1 dS = 1 µmhos/cm. The salt load due to irrigation will be over and above the salts already present in the soil. The salinity of the waters in canals fed from the rivers such as Ganga, Yamuna and Kosi are in the range from 0.2 to 0.44 dS/m (128 –282 ppm).

Groundwater and runoff water stored in tanks and used for irrigation have great variations in their salinity. In many cases, they are more saline than river or canal waters. Whatever may be the source of irrigation water, *viz.*, river, canal, tank, open well or tubewell, some soluble salts are always dissolved in it. The nature and quantity of dissolved salts depend upon the source of water and its course before use. The main soluble constituents in water are calcium, magnesium, sodium and sometimes potassium as cations and chloride, sulphate, bicarbonate and sometimes carbonate as anions. However, ions of some other elements such as lithium, silicon, bromine, iodine, copper, nickel, cobalt, fluorine, boron, zirconium, titanium, vanadium, barium, rubidium, ceasium, arsenic, antimony, bismuth, berylium, chromium, manganese, lead, molybdenum, selenium and phosphate, and organic matter, are present in minor quantities. These elements usually do not affect the quality of irrigation water as far as the total salt concentration is concerned, but some ions as selenium, molybdenum and fluorine, if absorbed by plants in excessive amounts may prove harmful to the animal life when taken by them through drinking water, feed or forage. Among the soluble constituents, calcium, magnesium, sodium, chloride, sulphate and bicarbonate, and boron are of prime importance in determining the quality of irrigation water and its suitability for irrigation. However, other factors as texture and structure of the soil, its drainage characteristics, nature of the crop grown and climatic conditions are equally important in determine the suitability of irrigation water in agriculture.

The suitability of water for irrigation will be determined by the amount and kind of salts present. With poor water quality, various soil and cropping problems can be expected to develop. Special management practices may then be required to maintain crop productivity.

The following are the most common problems that result from using poor quality irrigation water:

(*i*) *Salinity.* A salinity problem related to water quality occurs if the total quantity of salts in the irrigation water is high enough for the salts to accumulate in the crop root zone to the extent that yields are affected. If excessive quantities of soluble salts accumulate in the root zone, the crop has difficulty in extracting enough water from the salty soil solution.

This reduced water uptake by the plant usually results in slow or reduced growth.

(*ii*) *Reduced hydraulic conductivity*. A low permeability or hydraulic conductivity problem related to water quality occurs when the rate of water infiltration into and through the soil is reduced by the effect of specific salts in the water to such an extent that the crop is not adequately supplied with water and yield is reduced. The poor soil permeability makes it more difficult to supply the crop with water and may greatly add to cropping difficulties through crusting of seedbeds, waterlogging of surface soil and accompanying disease, salinity, weed, oxygen and nutritional problems. It is evaluated first from total salts in the water, since low salt water can result in poor soil permeability. Further, carbonates and bicarbonates can also influence soil permeability. The adverse influence of sodium on soil permeability has been recognized since long. But in many cases the evaluation of sodium influence alone has been found to be incorrect. This is because the long-term influence of water on soil permeability depends on the interaction of (1) sodium content relative to calcium and magnesium; (2) bicarbonate and carbonate contents, and (3) the total salt concentration of water.

(*iii*) *Toxicity*. A toxicity problem occurs when certain constituents in the water are taken up by the crop and accumulate in amounts that result in reduced yield. This is usually related to one or more specific ions in the water, namely, boron, chloride and sodium.

(*iv*) *Miscellaneous*. Various other problems related to irrigation water quality occur with sufficient frequency and should be specifically noted. These include excessive vegetative growth, lodging and delayed crop maturity resulting from excessive nitrogen in the water supply, white deposits on fruit or leaves due to sprinkler irrigation with high bicarbonate water and abnormalities indicated by an unusual pH of the irrigation water.

Chemical Analysis of Irrigation Water

Three general methods of quantitative determinations are used in water analysis—gravimetric, volumetric and colorimetric. The gravimetric method consists of separating single substances from the water sample by chemical and mechanical means, weighing each substance separately, and calculating the proportion of each with respect to the total sample. This method is generally used to determine the amount of suspended (undissolved) solids in a sample. It is also a convenient means for determining the total amount of dissolved solids but is more laborious than the other methods for determining the amount of each mineral constituent.

The volumetric analysis is performed by adding to measured portions of the water sample, the exact volumes of standard strengths of chemical solutions required to bring about a specific chemical change. The procedure is called *titration*. The point at which the chemical change takes place as the standard solution is added to the water being analyzed is the *end point* of the titration. A few drops of certain coloured chemical indicators such as phenolphthalein and methyl-orange are usually added before beginning a titration and a change in colour of the indicator used shows the *end point*. The amount of each mineral constituent being analyzed in the water sample is estimated directly from the volume of each standard chemical solution required.

The calorimetric analyses are those whose results are based on comparison of colours produced in the water sample with those of prepared standards. A most common calorimetric determination in water analyses is for iron. A given volume of the water sample is treated with certain chemical reagents and if iron is present, the sample becomes coloured. The colour is compared with that of prepared sample containing known quantities of iron, thus determining within close limits the amount of iron present in the sample. The photometric water analyzer (Fig. 14.14) is a precise photoelectric colour measuring instrument. It comprises of a constant voltage transformer, a prefocussed bulb, a holder for precise, optically polished glass cells and a sensitive photo cell. In line with the light beam is a rotatable disc containing accurately designed colour filters. These filters serve to accentuate the instrument's ability to measure accurately minute colour differences. In addition to the major and minor elements, the water analyzer is also used for determining the pH, turbidity and oxygen demand index of water samples.

In order to use saline irrigation water most efficiently, it is of primary importance to have its detailed chemical analysis done before use. The analysis should include total salt concentration or electrical conductivity, cationic and anionic composition, viz., calcium, magnesium, sodium, potassium, chloride, sulphate, bicarbonate, carbonate, nitrate and boron. Some of the methods for the analyses of soil salinity and

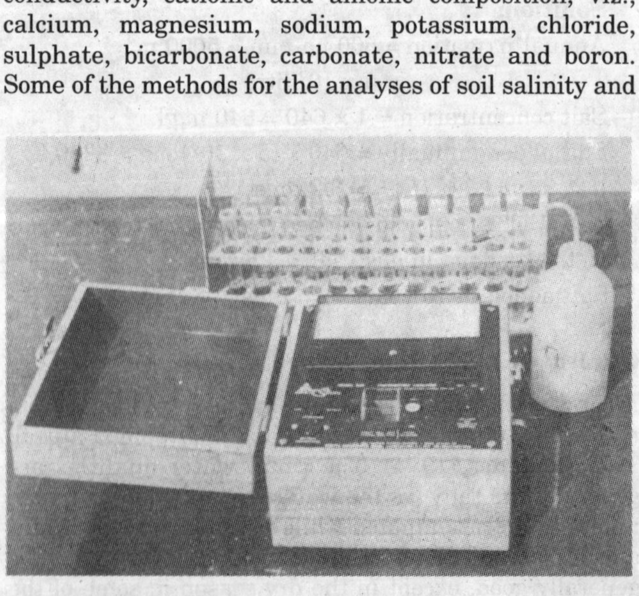

Fig. 14.14. Photometric water analyser. Note the treated water samples in the test tubes and the wash bottle beside it.

alkalinity discussed earlier apply to the analyses of irrigation water also. For detailed procedures for water quality analyses, reference is invited to the United Sates Salinity Laboratory Staff (1954) Handbook No. 60, listed at the end of this chapter.

Water Quality Criteria

Of the several factors influencing water quality, the generally accepted criteria for judging the quality are: (*i*) total salt concentration, as measured by electrical conductivity, (*ii*) relative proportion of cations as expressed by sodium adsorption ratio (SAR), and (*iii*) bicarbonate and boron contents.

The suitability of irrigation water (SIW) can be expressed as

$$SIW = f\,(QSPCD) \qquad \ldots(14.3)$$

in which,

Q = quality of irrigation water
S = soil type
P = salt tolerance characteristics of the plant
C = climate
D = drainage characteristics of the soil

Some other factors like the depth of water table, presence of a hard pan of lime or clay, calcium carbonate content in the soil and potassium and nitrate ions in irrigation water also indirectly affect the suitability of irrigation water. This is probably the main reason for the several classifications, varying in limits of salinity and other chemical indices. The soil type, major crops of the area, climate and drainage characteristics profoundly influence the suitability of particular water for irrigation. Highly saline water may be suitable in a well-drained, light textured, fertile soil while a much less saline water may be more harmful for the same crop grown on a heavy textured soil with impeded drainage. It is the actual salt concentration near the root zone, which determines the suitability of irrigation water rather than the chemical properties of irrigation water alone.

The quality of irrigation water is generally judged by its total salt concentration, relative proportion of cations or sodium adsorption ratio and the contents of bicarbonate and boron.

Salinity. Irrespective of the ionic composition, the harmful effects of an irrigation water increase with its total salt concentration as it increases the soil salinity significantly. Waters of low salinity ($EC < 3$ mmhos/cm) are generally composed of higher proportions of calcium, magnesium and bicarbonate ions. Highly saline waters ($EC > 10$ mmhos/cm) consist mostly of sodium and chloride ions. Moderately saline ($EC = 3$ to 9 mmhos/cm) waters have varying ionic compositions. Waters containing high concentrations of sodium, bicarbonate and carbonate ions have high pH.

Sodium adsorption ratio (SAR). Any increase in the SAR of irrigation water increases the SAR of the soil solution. This ultimately increases the exchangeable sodium of the soil. Generally, there is a linear relationship between SAR and exchangeable sodium percentage (ESP) of the soil up to moderate ESP levels, and at high ESP levels, the relationship tends to be curvilinear.

In judging the suitability of irrigation water, both salinity and SAR should be kept in view along with the salinity and sodicity developed during the cropping period. Salinity increases the osmotic stress while adsorption of sodium is increased both by salinity and SAR.

Magnesium : Calcium ratio. At the same level of salinity and SAR, but with varying proportions of calcium and magnesium, adsorption of sodium by soils and clay minerals is more at higher Mg : Ca ratios. This is because the bonding energy of magnesium is generally less than that of calcium, allowing more sodium adsorption. It suggests that soil sodicity would increase more at the same SAR if the water contains a higher proportion of magnesium to calcium. Thus, it is desirable to analyse both calcium and magnesium in irrigation water separately in order to predict soil sodicity hazard more accurately. It is more important if the Mg : Ca ratio in irrigation waters happens to be more than 4.

Bicarbonate. Irrigation water rich in bicarbonate content tend to precipitate soluble calcium and magnesium in the soil as insoluble carbonates:

$$2HCO_3 + Ca^{++} \rightleftharpoons Ca\,CO_3 + H_2O + CO_2$$

This leaves a higher proportion of sodium to divalent cations in the soil solution and increases the SAR. This bicarbonate-induced increase in the SAR of the soil solution ultimately results in higher adsorption of sodium on the soil exchange complex.

Boron. Though boron is an essential nutrient for plant growth, it becomes toxic beyond 2 ppm in irrigation water for most of the field crops. It does not affect the physical and chemical properties of the soil, but at high concentrations it affects the metabolic activities of the plant.

Potassium and nitrate. Potassium and nitrate ions are often present in significant amounts in irrigation waters. Being essential nutrients, they act favourably in reducing the harmful effect of saline water on crop growth by way of providing these nutrients regularly, rather than by reducing soil salinity. Among these, the effect of nitrate ion has been found more spectacular than potassium because irrigated soils are themselves deficient in nitrogen status and are generally well supplied with potassium. Regular supply of nitrate form helps in mitigating the salt-induced nitrogen deficiency and in increasing crop productivity.

Classification of Irrigation Waters

Several classifications of irrigation water have been proposed in India and abroad, on the basis of their chemical characteristics and their effect on crop growth. Amongst these, the classification proposed by the United States Soil Salinity Laboratory (USSSL) staff is widely used because it includes both the factors of salinity and sodium hazard (Richards, 1954). However, such a system is not very suitable under the conditions in India, as waters of greater salinity levels than the highest limit proposed by the USSSL staff are being successfully used in some areas of Haryana and Rajasthan. In view of this and also with a view to include soil texture and salt-tolerance behaviour of the crops, the USSSL staff scheme was modified and the higher range of salt concentration has been extended up to 20,250 micromhos/cm by adding one more salinity group and keeping the same geometric relationship. In 1972 the Conference on Water Quality Ratings at the Central Soil Salinity Research Institute, Karnal proposed a water quality rating giving emphasis on salinity and soil texture along with some specific situations (Table 14.3). However, this scheme has yet to be tested for various types of soils and crops.

Table 14.3. Water quality ratings

Nature of soil	*Crops grown*	*Safe maximum limit of EC of irrigation water (micromhos/cm)*
Deep black soils and alluvial soil having clay content of more than 30 per cent: soils that are fairly to moderately well drained.	Semi-tolerant Tolerant	1500 2000
Heavy textured soils having clay content of 20-30 per cent: soils that are well-drained internally and have a good surface drainage system.	Semi-tolerant Tolerant	2000 4000
Medium textured soils having clay content of 10-20 per cent: soils that are well-drained internally and have a good surface drainage system.	Semi-tolerant Tolerant	4000 6000
Light textured soils having clay content: less than 10 per cent soils that have excellent internal and surface drainage.	Semi-tolerant Tolerant	6000 8000

*These limits are subject to the condition that the groundwater table at no time of the year rises within 1.5 metres from the surface. If the water table is likely to rise up to the root zone, the above limits should be reduced to one-half of its proposed values. Further, if the soil has impeded internal drainage due to any reason, again the suitability limits be reduced to one half.

The limited applicability of the various classifications is quite reasonable on the ground that the crop growth is not only influenced by the chemical characteristics of the irrigation water, soil type and crop species but also by the actual salt concentration near the root zone, physical condition of the soil, particularly that of highly sodic soil, fertility status, depth of water table, soil and water management practices and climatic conditions. Moreover, the favourable effect of potassium and nitrate ions as nutrients in irrigation waters has not been included in any of the schemes. However, these schemes act as a guideline in a broad sense.

Under certain conditions, the use of saline irrigation water is inevitable. It can be better utilized for irrigation purposes after obtaining the following information and adopting proper management practices: (*i*) knowledge of the total salt concentration and ionic composition, (*ii*) salt tolerance behavior of the crop variety to be grown, (*iii*) climate data, (*iv*) consumptive use of water by the crop, (*v*) soil factors such as initial soil salinity, fertility and calcium carbonate content, and (*vi*) drainage characteristics and leaching requirement of the soil.

Quality of Water Sources

The quality of irrigation water is influenced primarily by its source.

River waters. The water quality of Indian rivers, with a few exceptions, and those too in the hot season, is generally good. An analysis of 13 important river waters taken at several places during 1955 to 1966 at monthly intervals has revealed that the water quality of Ganga, Gandak, Kosi, Brahmaputra, Kaveri, Beas, Sutlej and Narmada is good as their EC values fall below 350 micromhos/cm (Table 14.4) while salinity of those of Chambal, Yamuna, Tapti, Godavari and Krishna ranged from 450 to 1400 micromhos/cm. These waters are mostly alkaline. The rivers having low EC values show less monthly and yearly variations as compared to those of moderate salinity. The waters of low EC values are suitable for irrigation purposes throughout the year while those of moderate salinity can be used cautiously. In South India, waters of Hagari and Tungabhadra rivers are of moderate salinity. In some arid and salt affected areas of Gujarat, waters of some seasonal rivers show a salt concentration up to 7000 ppm.

Canal waters. The quality of canal water depends mainly on the river from which it originates. Consequently, the quality of all the canal waters in India is good (Table 14.6). The development of salt affected lands in the canal irrigated regions is mainly due to the rise of water table. Extensive salt affected areas in the canal irrigated regions of Punjab, Rajasthan and UP have been developed because of the rise in the depth of water table and accumulation of salts on the soil surface through capillary conductivity and evaporation.

Table 14.4: Chemical composition of some important Indian rivers

River	*EC micro-mpos/cm*	*pH*	*Cations, meq/litre* Na^+	K^+	Ca^{++}	Mg^{++}	*SAR*
Ganga	274	8.2	0.51	0.22	1.36	0.95	0.46
Gandak	253	8.3	0.19	1.10	1.48	0.80	0.18
Kosi	142	7.8	0.20	0.08	0.85	0.32	0.26
Brahmaputra	177	8.1	0.13	0.09	1.17	0.39	0.17
Yamuna	647	8.3	2.40	0.12	1.85	1.76	1.70
Chambal	448	8.4	1.80	0.60	1.12	1.50	0.53
Narmada	339	8.3	0.92	0.04	1.23	0.18	0.74
Tapti	551	8.2	2.60	0.17	1.19	0.18	2.07
Godavari	725	8.4	3.00	0.11	1.35	1.85	2.86
Krishna	1,392	8.3	2.80	0.06	1.27	1.35	1.60
Kaveri	269	8.3	0.90	0.05	0.91	0.90	0.99
Sutlej	325	—	2.43	—	0.97	—	0.80
Beas	260	—	1.52	—	0.93	—	0.50
Hagari	1,640	8.1	10.90	0.06	3.06	1.92	6.94
Behta	592	—	1.96	—	2.19	1.77	—
Kali Nadi	660	8.6	5.56	—	1.8	0.30	—
Sengar	1,048	9.1	5.38	—	1.25	0.85	—
Tungabhadra	1,730	8.5	14.37	0.74	1.76	1.11	12.98

Compiled by Paliwal (1972).

Quality of other surface water sources. Under the minor irrigation projects, some areas are irrigated by the waters of lakes, tanks, reservoirs and drains. In most of the cases, their quality is fairly good, except in case of the drains or the reservoirs, which are fed by small streams flowing over salt-affected areas. Drainage waters of Godavari and drainage channels in UP are of moderate salinity (Agrawal and Gupta, 1966).

Quality of ground waters. Well water forms an important source of irrigation. Its quality is highly variable due to climatological and hydro-geological conditions. Paliwal (1972) classified the quality of groundwater resources of India into three main groups: (*i*) water quality of arid and semi-arid regions having rainfall below 450 mm per annum. (*ii*) water quality as influenced by hydrological conditions as high water table, consisting of some of the areas of Rajasthan, Punjab, Haryana, Delhi, and Agra and Mathura districts of UP in the canal irrigated areas and river basins and (*iii*) water quality of wells in some areas of the coastal regions of the country as influenced by backward flow of sea water and inundation. Of all the factors, aridity is the most important single factor responsible for a very high degree off salinity of well waters.

In addition to the total salt concentration, individual cation and anion compositions and pH of irrigation water are important characteristics. Statistical analysis shows that all saline waters are alkaline in reaction and their pH values are confined mostly within the range of 7.5 to 8.3. Regarding individual ionic composition, highly saline waters are dominated by sodium and chloride ions. Calcium is confined mostly to the narrow range of traces to 20 per cent of the total salt concentration. Another interesting feature of these waters is the predominance of magnesium over calcium. In those waters, carbonate is either absent or is very little and the concentration of bicarbonate ion is mostly confined to the range of 5 to 30 per cent. Sulphate is uniformly distributed in all ranges. In general, well waters of Rajasthan are of Na^+—Mg^{++} —Ca^{++} and Cl^- —SO_4^-—HCO_3^- type (Paliwal, 1972).

The continuous use of poor quality of irrigation waters may not only cause calcium deficiency but also create poor soil physical conditions, as both sodium and magnesium damage the soil structure by their highly dispersive action. However, hazardous effect of magnesium is not often visible in soils or plants, probably due to the counter-balance of excessive calcium. The magnesium in it may precipitate as magnesium carbonate under the arid condition of Rajasthan where bicarbonate ions are also sufficiently present in irrigation waters. The degree of adverse effect of magnesium and the possibility of its precipitation would depend on the actual proportion of Na : Mg : Ca in the soil solution near the root zone and the climatic conditions.

Extensive data on the quality of well waters of Rajasthan, Gujarat, UP, Punjab, Haryana, Delhi and other parts of the country show that waters of high EC value are predominant with Na and Cl ions, while less saline waters have relatively more Ca, Mg and bicarbonate ions. Potassium and carbonate ions, if present, are mostly confined up to a range of 5 per cent of the total salt concentration. It is rather difficult to draw a general conclusion on the ionic composition of the water in relation to geographical conditions. In general, waters in areas of high rainfall, *i.e.*, above 75 cm per annum and with good drainage are of good quality.

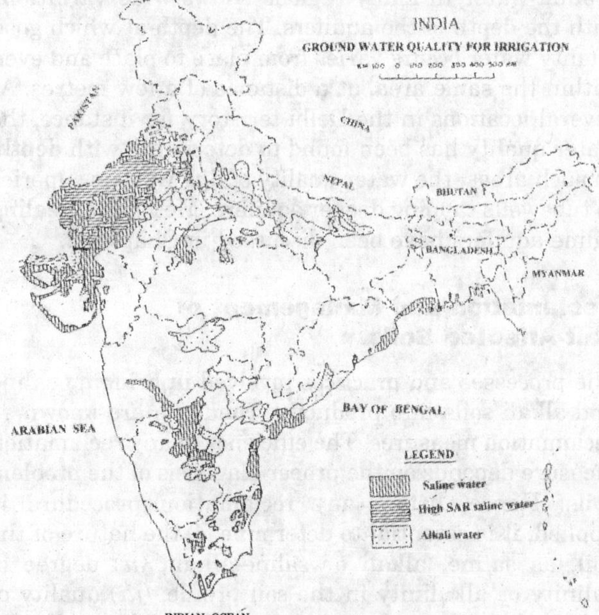

Fig. 14.15. Groundwater quality map of India (1st approximation), 1998.

(***Courtesy:*** Central Soil Salinity Research Institute, Karnal.)

The problem of boron hazard is not serious in the irrigation water sources of India, except in a few cases in the arid areas where its concentration ranges from 2 to 5 ppm. The toxic effect of boron in irrigation water is primarily dependent on its concentration, soil texture, calcium carbonate content, pH and organic matter. It is also influenced by the type and amount of clay in the soil, its adsorption and release characteristics; depth to water table and the drainage condition of the soil. The equilibrium between different forms of boron and their interactions with the above factors and the boron tolerance limit of crops determine the toxic effects of boron on a particular crop species.

The mean chemical composition of well waters of some of the districts of Rajasthan shows that due to low rainfall, the groundwaters of western region are more saline than that of eastern region. However, the adverse effect of poor quality water is not much visible due to the sandy texture of most of the irrigated soils. Similarly, the salt problem is severe in the arid regions of western Gujarat where the quality of well waters is poor.

Seasonal variation in water quality. Salinity of ground water is considerably influenced by the climatic conditions. It increases during the summer and is significantly reduced during the monsoon due to dilution by rain water. The degree of variation in salinity and its ionic composition depends upon the depth of water table, infiltration capacity of the soil and the rainfall characteristics of the area concerned. Besides these factors, some recharge of water takes place in low-lying areas due to flooding by rain water.

Variation in water quality with depth. The quality of ground-water in many regions shows wide variations with the depth of the aquifers. The depth at which good quality water occurs varies from place to place and even within the same area, at a distance of a few metres. At several locations in the Delhi territory, for instance, the water quality has been found to deteriorate with depth. In such areas, the water quality of dug wells is superior to tubewells tapping deeper aquifers. Methods of sealing saline aquifers have been discussed in Chapter 2.

Reclamation and Management of Salt-Affected Soils

The processes and practices involved in bringing saline and alkali soils into productive condition are known as reclamation measures. The efficiency of any reclamation measure depends on the proper diagnosis of the problem soils. Hence, before any reclamation procedure is applied, it is essential to determine (*i*) the nature of the soil, as saline, alkali or saline-alkali, (*ii*) degree of salinity or alkalinity in the soil profile, (*iii*) quality of water available for leaching out the salts and the reaction products, (*iv*) drainage characteristics of the soil, (*v*) topography of the land, and (*vi*) presence of any hard pan of lime or clay in the soil strata.

Reclamation procedures. Reclamation of a soil on a temporary basis can be done by (*i*) removing the salt crust from the surface of the soil, (*ii*) ploughing salt surface crust deep into the soil, and (*iii*) neutralising the effects of certain salts by adding other salts. However, permanent reclamation can be obtained by the following procedures:

(*i*) Lowering of water table (if shallow),
(*ii*) Improving infiltration rate of the soil,
(*iii*) Leaching of salts in saline soils and providing adequate sub-surface drainage,
(*iv*) Replacing excessive exchangeable Na by Ca salts and removing the replaced products, and
(*v*) Suitable management practices.

The reclamation procedures for saline, alkali and saline-alkali soils are discussed separately in the following sections.

Reclamation of Saline Soils by Leaching

The reclamation of a saline soil depends on the efficiency of removal of salts from the upper to lower layers and is thus relatively simple if drainage is not restricted. The simplest procedure is to flood the field after making ridges or levees at the boundaries. The movement of water through the soil profile will carry the dissolved salts into the lower layers below the root zone. These salts are drained away if adequate subsurface drainage is provided (Figs. 14.16, 14.17 and 14.18)

Leaching of saline soils is a process whereby water of low salt concentration is applied to displace the soil solution of relatively high salt concentration. Excess water is allowed to pass through the root zone to move the salts away from the root zone through leaching. Leaching of salts is done to meet either of the following two objectives:

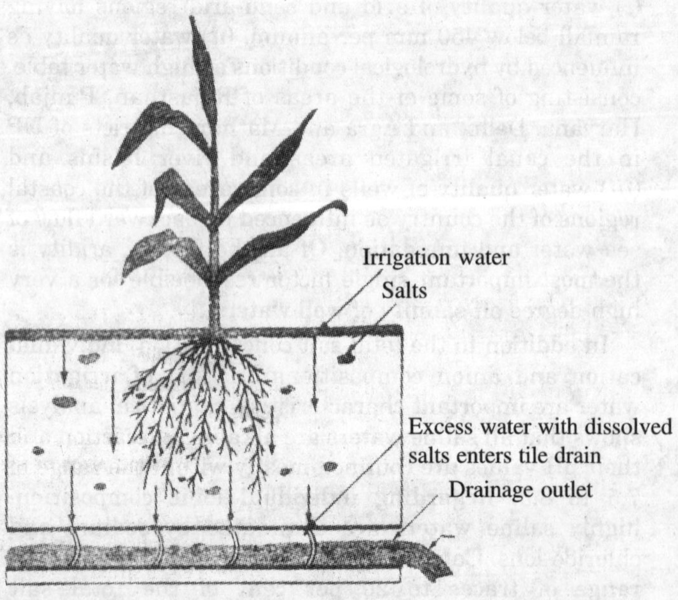

Fig. 14.16. Schematic sketch illustrating the role of tile drains in controlling waterlogging and salt accumulation.

Fig. 14.17. Corrugated PVC drain pipe being laid at the CSSRI Operational Research Project area at Gohana, Rohtak district, Haryana.
(*Courtesy*: CSSRI, Karnal.)

(*i*) To reduce salt concentration in the root zone and reclaim the land for crop production. This is known as one time leaching for reclamation.

(*ii*) Leaching to maintain salt balance for long-term crop production.

Flushing and leaching are the two major operations done to reclaim saline lands. In *flushing* salts are removed by surface washing of the soil as compared to pushing of salts below the root zone. In this method, salt moves along with excess rainfall or irrigation water as the surface water moves down the sloping land. Water along with the salts is removed through a surface drain provided at the tail end. Several studies on flushing have clearly indicated that this technique is useful only under limited situations. Heavy soils are suited to this method of salt management. However, too flat lands and non-availability of outlet drains are the limiting factors in the *flushing* process for land reclamation.

Leaching is the process of dissolving and transporting soluble salts by downward movement of water through the soil. The leaching of soluble salts from the root zone is essential in irrigated soils. When the soils are highly saline, or irrigated with saline water, some more water is to be added to leach out the excessive salts or to avoid accumulation of salts. Without leaching, salts accumulate in direct proportion to the salt content of the irrigation water and the depth of water applied.

Leaching requirement. The depth of irrigation water per unit depth of soil, required to produce any specified increase in soil salinity for any given electrical conductivity of irrigation water can be calculated from the following equation:

$$\frac{D_i}{D_s} = \frac{d_s}{d_w} , \frac{SP}{100} . \frac{\Delta EC_e}{EC_w} \quad ...(14.4)$$

in which,

D_i = depth of irrigation water
D_s = depth of soil
d_s = density of soil (bulk density)
d_w = density of irrigation water
SP = saturation percentage of the soil
ΔEC_e = increase in electrical conductivity of saturation extract of the soil
EC_w = electrical conductivity of irrigation water

Under high water table conditions, the increase in soil salinity by the evaporation of ground water can be determined by the following formula:

$$\Delta EC_e = \frac{D_g}{D_s} . \frac{EC_g}{SP} . \frac{D_w}{D_s} . 100 \quad ...(14.5)$$

in which,

D_w = depth of groundwater evaporated
EC_g = electrical conductivity of groundwater

All other terms are as defined in equation (14.4). It may be noted that equation (14.4) is obtained from the relationship expressed in equation (14.3), after substituting the depth of groundwater evaporated and its EC_g for the depth of irrigation water and its quality.

The fraction of irrigation water that must be leached through the root zone to keep the salinity of the soil below a specific limit is termed as leaching requirement (*LR*). Mathematically, it can be expressed as

$$LR = \frac{D_d}{D_i} = \frac{EC_i}{EC_d} \quad ...(14.6)$$

in which,

LR = leaching requirement, expressed as a ratio or as per cent
EC_i = electrical conductivity of irrigation water, mmhos/cm
EC_d = electrical conductivity of drainage water, mmhos/cm

Information on the consumptive use of water by the crop is necessary if the leaching requirement concept is to be used for determining either the depth of irrigation water that must be applied or the minimum depth of water to be drained, in order to keep the soil salinity from exceeding a specified value. The depth of irrigation water D_i is related to consumptive use D_c and the equivalent depth of drainage water D_d by the equation:

$$D_i = D_c + D_d \quad ...(14.7)$$

Using equation (14.7) to eliminate D_d from equation (14.6)

$$D_i = \frac{D_c}{1-LR} \qquad ...(14.8)$$

Expressing the leaching requirement (LR) in equation (14.8) in terms of the EC ratio of irrigation and drainage waters, as given in equation (14.6), provides the following relationship:

$$D_i = \left(\frac{EC_d}{EC_d - EC_i}\right) D_c \qquad ...(14.9)$$

Equation (14.9) gives the depth of irrigation water, which is required to satisfy both the evapo-transpiration and leaching requirements of the soil. Of the above factors, EC_i is known from the chemical analysis of the irrigation water and EC_d is taken on the basis of permissible salt tolerance limits of the crop. Generally, the salt tolerance limits are expressed in terms of EC of the saturation extract of the soil solution (EC_e). In most cases, EC_d is assumed to be approximately twice as high as the corresponding EC_e.

Example 14.8. *Determine the depth of irrigation water which would change 30 cm depth of loam soil into saline condition, if the EC of irrigation water were 1 millimhos/cm. The bulk density of the soil is 1.2 grams/cu cm and the density of water is 1 gram/cu cm. The saturation percentage of the soil is 40.*

Solution: D_s = 30 cm, d_s = 1.2, d_w = 1.0, SP = 40, EC_e = 4, EC_i = 1

To make the soil saline, EC_e should be raised to 4 millimhos/cm

$$\frac{D_i}{D_s} = \frac{d_s}{d_w} \cdot \frac{SP}{100} \cdot \frac{\Delta EC_e}{EC_i}$$

$$\frac{D_i}{30} = \frac{1.2}{1.0} \cdot \frac{40}{100} \cdot \frac{4}{1}$$

$$\therefore \quad D_i = 57.6 \text{ cm}$$

(***Note:*** Thus, less than 60 cm of reasonably good quality irrigation water contains sufficient salt to change a 30 cm depth of a salt-free loam soil to a saline condition, if there is no leaching or precipitation of salt in the soil.)

Example 14.9. *Determine the change in salinity level of the soil due to the evaporation of groundwater over a period of four months, when the depth of groundwater evaporated is 11 cm, and its electrical conductivity is 10 millimhos/cm. The depth of soil influenced by salt accumulation is 30 cm. The bulk density and saturation per cent of the soil are 1.4 grams/cu cm and 63, respectively. The density of water is assumed to be 1.0 grams/cu cm.*

Solution: Applying equation (14.4)

$$\Delta EC_e = \frac{D_g}{D_s} \cdot \frac{EC_g}{SP} \cdot \frac{d_w}{d_s} . 100$$

$$= \frac{11}{30} \cdot \frac{10}{63} \cdot \frac{1.0}{1.4} . 100$$

$$= 4.1 \text{ millimhos/cm.}$$

Example 14.10. *Estimate the leaching requirement when the EC of the saturation extract of the soil is 11 mmhos/cm at 25 per cent reduction in the yield of cotton. The EC of irrigation water is 1.5 mmhos/cm.*

Solution: $LR = \frac{EC_i}{EC_d}$

$$EC_d = 2EC_e = 2 \times 11 = 22 \text{ mmhos/cm}$$

$$\therefore \quad LR = \frac{1.5}{22} \times 100$$

$$= 6.8 \text{ per cent.}$$

Example 14.11. *In a given region, the monsoon rainfall is 500 mm of which, 50% flows out as runoff, 30% is held in the soil profile and 20% percolates beyond the crop root zone. During the rabi season, 150 mm of groundwater of salinity 4 dS/m is applied for irrigation, as there is no surface water available. At the beginning of a monsoon season, the average salinity of the saturation extract (ECe) of the soil from surface down to crop root zone is 4 dS/m. The runoff has a monsoon average salinity of 1 dS/m and the deep percolated water has a monsoon average salinity of 3 dS/m. Calculate the salt balance and interpret the result.*

Solution: Calculations will be performed for 1 hectare of land. The result will be applicable over the total area that is receiving the saline groundwater for irrigation.

We can write a simple salt balance equation as:

Salt added in a year – Salt removed in a year = Salt storage during the year.

Salt is added by 150 mm of groundwater and salt is removed by runoff and deep percolated water. Runoff water is 250 mm and deep percolated water is 100 mm.

150 mm of groundwater over 1 ha area means 1500 m^3 of water. It has a salinity of 4 dS/m, or a salt concentration of 4 × 460 = 2560 mg/l = 2.56 kg/m^3. Total salt added due to irrigation by groundwater = 1500 × 2.56 = 3840 kg.

250 mm runoff from 1 ha means 2500 m^3 of water. Its salinity is 1dS/m, which corresponds to a salt concentration of 640 mg/l = 0.64 kg/m^3. Total salts removed by runoff water = 2500 × 0.64 = 1600 kg.

100 mm of deep percolated water from 1 ha means 1000 m^3 of water. It has a salinity of 3 dS/m and the corresponding salt concentration is 3 × 640 = 1920 mg/l = 1.92 kg/m^3. Total salt removed by deep percolated water = 1000 × 1.92 = 1920 kg.

Salt storage = 3840 – (1600 + 1920) = 320 kg

The result implies that every year, 320 kg of salt will be net addition to the soil.

Thus, in order to prevent further rise in soil salinity above 11 mmhos/cm in this particular case, it will be necessary to apply approximately 7 per cent more water than the consumptive use requirement.

The leaching requirement, therefore, represents a necessary water use for a permanent irrigated agriculture. When the water needs of a project are planned, the leaching requirements should be considered.

The efficiency of the reclamation per unit volume of water will be maximum with good quality water. The lower the salt content in the irrigation water, the greater is the efficiency in reclaiming saline soils. In practice, where water is inadequate or the soils are slowly permeable, it is desirable to obtain partial reclamation and then a more salt-tolerant crop like barley could be grown. Leaching may be continued during the cropping period also. The effectiveness of leaching varies from soil to soil and depends largely on the quantity of water that passes through the soil. The rate of removal of salt is fast in the beginning but later on the process is slowed down. The efficiency of leaching is also influenced by varying the water flow velocities within a soil. In light soils, higher rate is favourable but in heavy soils intermittent flushing is better as the water gets adequate time to move slowly through the small soil pores.

Leaching methods. Leaching can be done by ponding the required depth of water on the soil surface by means of dikes or ridges, thus establishing a downward water movement through the soil. Leaching can be done at any time of the year, depending on the availability of water. In monsoon Asia, it is advantageous to practice leaching during the monsoon season. The quality of water applied in leaching should be in conformity with the tolerance limits of the crops to be grown. The cost of leaching will be high if salts are leached beyond a threshold value, as no additional benefits will accrue, because yields stabilize at the threshold value. Information on crop tolerance of salt will help in selecting crops. From the initial salinity and the crop tolerance level, fractions of salts to be leached could be derived by the relation:

$$\text{Fraction of salts to be leached} = \frac{\text{Threshold salinity}}{\text{Initial salinity}}$$

For example, to grow wheat in a saline soil with an initial salinity of 20 dS/m, will require leaching of 75 per cent of salts, initially present in the profile. In case any crop is known to be sensitive at the germination stage, the threshold value of this particular stage should be used in deciding the fraction of the salts to be leached. A thumb rule commonly used is that 1 cm of water should be applied for each cm of the soil to be reclaimed. However, the quantity of water required may vary with soil texture and fraction of the salts to be leached. For medium textured soils (sandy loam), 0.5 cm of water for each cm of soil is usually sufficient to remove 80 per cent of the salts present in the soil profile.

Continuous and intermediate leaching. In continuous leaching, water is ponded continuously over the soil surface to allow water to pass through the root zone. In the case of intermittent leaching, water is ponded at intervals. A net saving of 30-35 per cent can be made through intermittent, compared to continuous leaching, in the amount of water needed to leach the salts to the same degree. However, intermittent leaching may not be advantageous for highly degraded alkali soils with restricted infiltration rates. Rectangular check basins and level borders are used when the land is flat. Contour checks can be used when the land slope is considerable. The effectiveness of leaching by furrows is limited. Sprinkling is an effective method for leaching, especially when soils are cracked or are highly permeable.

The efficiency of leaching depends on: (*i*) the amount of water applied, (*ii*) uniformity of water distribution, and (*iii*) the adequacy of drainage. In case of salt-tolerant crops, it is sufficient to leach once or twice in a growing season. But in less tolerant crops, it is advisable to apply irrigation water several times for leaching during the cropping period. Generally, less tolerant crops show higher leaching requirement. In practice, for better crop growth, it is desirable to apply slightly more water than the calculated values. The best practice would be to follow the salinity trends by making regular checks in the cultivated fields irrigated with saline water. Since nutrients are also moved down with water and salts, it is desirable to apply fertilizers after leaching and make up the losses of the nutrients, if any, in leaching. This is especially true in case of nitrogenous fertilizer.

Drainage of Irrigated Lands for Salinity Control in Arid and Semi-Arid Regions

In arid and semi-arid regions, suitable drainage systems are essential for adequate control of soil salinity. This includes both the adequacy of drainage and the quantity of water to be drained. The first factor is more related to the control of water table, preferably at about 1.5 metres below the soil surface, so as to maintain a favourable salt balance near the root zone during the cropping period. While irrigating with saline water, if the water table rises to within 60 cm to 1 metre from the surface, lands often go out of cultivation because of the rapid accumulation of salts in the root zone. Hence, the depth of water table should be kept such that the roots are beyond the zone of capillary action.

Salinity can be controlled by applying requisite amounts of water based on the leaching requirement of the soil. But, water and the salts dissolved in it will be able to move downwards only when the water table is deep. In irrigated areas, the problems of shallow and fluctuating water tables and salinity occur simultaneously in a particular region. Hence, the drainage system of the entire area is to be planned as a single unit, rather than for a particular field. The drainage programme for irrigated land should be initiated and continuously integrated with the development of the irrigation system in order to attain efficient overall water and salinity control.

Shallow water table can be lowered by preventing excessive seepage from canals, efficient water use on the farm and by providing artificial drainage systems in areas where there is inadequate natural drainage. Frequent irrigations with small amounts of water or sprinkler irrigation may help in increasing the efficiency of water use in areas of shallow water table. If the groundwater is of good quality, it could be pumped from shallow water table areas and used in needy areas.

Satisfactory land levelling is the most important step in any reclamation process (Fig. 14.18). The fields should also be provided with 35 to 40 cm high strong bunds all along the field boundary. The whole field is divided into sub-plots. Bunding helps in retaining the entire rainwater and in preventing the outside water from entering the fields. Initial levelling is done with a scraper blade or other levelling equipment. Final levelling can be completed by locating the higher and lower spots in the fields after a shallow irrigation. A gentle slope of about 0.1 per cent towards the drainage channel helps in draining excess water, when required. Proper land levelling helps in proper water distribution and leaching of the soluble salts (Fig. 14.19). Field drains are to be kept clean to facilitate easy disposal of the drainage water (Fig. 14.20).

Fig. 14.18. Layout of the drained area of the Punjab Agricultural University Farm, Ludhiana for raising the *rabi* crop during 1966-67. The construction of outlet drain, followed by an interceptor drain lowered the water table of the area by about 60 to 90 cm.

Fig. 14.19. Leaching for the reclamation of an alkali soil area, after applying gypsum . (*Courtesy:* CSSRI, Karnal.)

Fig. 14.20. Bumper wheat crop in the very first year after reclamation of the low-lying waterlogged area of Punjab Agricultural University, Ludhiana (1967). Note the relief drain being cleaned to facilitate drainage.

Good quality water is an essential input in the reclamation programme. Water may be obtained either from canal or groundwater sources. In canal-irrigated areas, specially in the arid and semi-arid regions, a common feature is that the canal water is of good quality while the groundwater is saline. Mixing groundwater with good quality canal water is a recommended water management practice to utilize the groundwater for irrigation and simultaneously control the water table (Fig. 14.21). For effective land reclamation, provision of adequate drainage is essential. Efforts should be made to conserve the maximum possible amount of available rainfall in the agricultural fields.

Fig. 14.21. Mixing of saline groundwater with good quality canal water in a water course at Narwani in Jind district, Haryana. (*Courtesy:* CSSRI, Karnal.)

Reclamation of Saline-Alkali Soils

When the soil is saline as well as alkali, it is essential to leach out the excessive salts up to an EC value of 6 to 8 mmhos/cm before any attempt is made to remove excessive exchangeable sodium by gypsum. However, some excess salts can be leached by flooding the field after gypsum application for removing the released sodium from the soil exchange phase.

Maintenance of adequate drainage is essential for the successful reclamation of saline, alkali and saline-alkali soils, as leaching out of the salts or the removal of the reaction products is a must for the success of reclamation. In case the sub-soil has lime or clay hard pan, it should be broken by deep ploughing or sub-soiling.

Reclamation of Alkali Soils

The reclamation of alkali soils is more difficult than saline soils because alkali soils have very low permeability and they require the replacement of excessive sodium by calcium for increasing nitrogen availability (Fig. 14.22). It is necessary that the replaced Na ions are leached down to lower layers. The improvement of soil tilth is slow in alkali soils. Excessive exchangeable sodium can be replaced by any low-cost source of calcium and magnesium salts such as gypsum. Soluble calcium and magnesium salts are suitable for the reclamation of alkali soils. Of these salts, calcium and magnesium chlorides are quick in reaction but are too expensive to be used on commercial basis for reclamation. However, gypsum, by virtue of its low cost, is quite suitable and is widely used, though it has a lower solubility than the chlorides. In calcareous alkali soils, some acid forming reagents can also bring about reclamation by dissolving calcium carbonate and creating acidity. These acidifying materials are: (*i*) sulphur, (*ii*) sulphuric acid, (*iii*) iron sulphate, (*iv*) aluminium sulphate, and (*v*) lime sulphur. The choice of the ameliorants depends on the soil pH, the amount of calcium or magnesium carbonate present, cost of the ameliorants, speed of the reaction, desired soil condition, quality of water available and the benefit-cost ratio of reclamation. Sulphuric acid and iron and

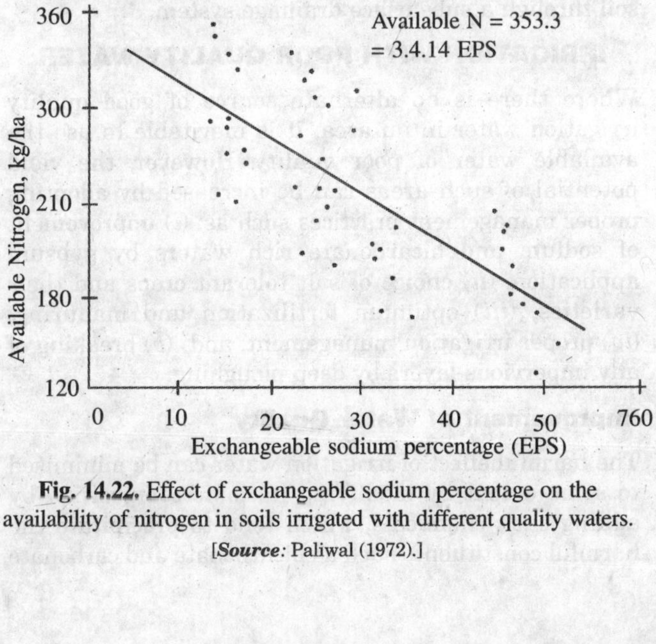

Fig. 14.22. Effect of exchangeable sodium percentage on the availability of nitrogen in soils irrigated with different quality waters. [*Source:* Paliwal (1972).]

aluminium sulphates are quite reactive, whereas sulphur is to be oxidized by microbial activity before it could act as an ameliorating agent.

The following chemical equations illustrate the manner in which some of the ameliorants react with alkali soils. In these equations, the letter X represents the soil exchange complex.

1. *Gypsum*

$$2NaX + CaSO_4 \rightleftharpoons CaX_2 + Na_2SO_4$$

2. *Sulphur*

$$2S + 3O_2 + 2H_2O \rightleftharpoons 2H_2SO_4$$

3. *Sulphuric acid*

$$(a)\ CaCO_3 + H_2SO_4 \rightleftharpoons CaSO_4 + H_2O + CO_2$$

$$(b)\ 2CaCO_3 + H_2SO_4 \rightleftharpoons CaSO_4 + Ca(HCO_3)_2$$

Reaction 3(a) would give only one equivalent of soluble Ca for each equivalent of acid whereas relation 3(b) would furnish two. In practice, about 1½ equivalents would be expected by the application of acid. Released Ca would replace exchangeable Na of alkali soil.

APPLICATION OF SOIL AMENDMENTS

Since alkali soils owe their unproductivity to the presence of excessive amounts of sodium on the exchange complex, application of amendments is necessary for removal of most of the sodium and obtaining higher yields. The harmful effects of alkali soils and poor quality irrigation water can be minimized by modifying its ionic composition by adding such chemicals which tend to precipitate the harmful constituents such as bicarbonate and carbonate in the form of less soluble salts or tend to create a favourable cationic (Ca : Mg : Na) ratio. There appears to be no technique to reduce the total salt concentration of irrigation water for large irrigation projects, except by diluting it and mixing with good quality water, if available nearby (Fig.14.21). If the irrigation water contains excessive amounts of sodium (> 70% soluble sodium), bicarbonate (> 8 meq/1) and magnesium (as Mg: Ca ratio > 4), such a sample is likely to be normally low in calcium content. An unfavourable proportion between calcium and sodium, whether caused by sodium alone or induced by bicarbonate or high Mg: Ca ratio, can be improved by adding a cheap and soluble source of calcium in the irrigation water. This is essential because the lack of calcium below a critical level in the soil-water-plant system is more harmful than the excess of sodium in equivalent amounts.

Gypsum (hydrated calcium sulphate–$CaSO_4.2HO_2$), in spite of low solubility but low cost is a better alternative to use for creating a favourable Ca : Na or Ca : Mg ratio in irrigation water. Improvement in Ca: Na ratio or SAR takes place by increasing the calcium ion concentration, decreasing Mg: Ca ratio and precipitating excessive carbonate ions.

$$2NaHCO_3 + CaSO_4 \rightleftharpoons Ca(HCO_3)_2 + Na_2SO_4$$
$$Ca(HCO_3)_2 \rightleftharpoons CaCO_3 + H_2O + CO_2$$

Success of gypsum application in irrigation water or soil depends upon the choice of the grade of gypsum, its particle size, method of application and quantity applied. In India, large quantities of gypsum are available as deposits in Rajasthan near Bikaner, Rishikesh in Uttaranchal and in parts of Jammu and Kashmir. Gypsum obtained from the mines are of different grades of purity, but, for agricultural proposes, gypsum above 65 per cent purity is quite suitable. However, gypsum of higher grade is preferred for economic reasons. Gypsum is the standard amendment for alkali soils reclamation (Fig. 14.23). It supplies soluble calcium. The quantity of gypsum to be applied in an alkali soil must be determined through proper soil analysis. It is advisable to obtain representative soil samples after the fields have been levelled and bounded. Proper soil tests are necessary for obtaining realistic estimates of the amendment needs of soils. The farmer should establish contact with a soil testing laboratory. Since gypsum application improves both poor quality water and alkali soil, it is desirable to apply gypsum, both in the soil and in water. But, if the soil is good and only the water is bad, gypsum should be applied only in water. The gypsum requirement of water should be calculated, depending upon the relative concentration of sodium, magnesium and calcium ions in irrigation water and the solubility of gypsum. To add 1 meq/1 of calcium, 8.6 quintals of gypsum of 100 per cent purity per hectare-metre of water would be necessary. Hence, the actual amount of gypsum is to be calculated, depending upon the amount of calcium and water to be added for the crop in question.

Fig. 14.23. Use of gypsum in alkali soil reclamation.
(*Courtesy*: CSSRI, Karnal.)

Fig. 14.24. Gypsum application prior to reclamation of alkali land in the Pilot Project area of CSSRI.
(*Courtesy*: CSSRI, Karnal.)

In an alkali soil under reclamation, it is a good practice to grow rice as the first crop, when the climatic conditions permit rice cultivation. Gypsum may be applied in the powder form about 10 to 15 days before transplanting rice. It can be spread evenly in the field with hand when the wind velocity is low (Fig. 14.24). Gypsum should be mixed by light ploughing in the upper 6-8 cm soil. Although many alkali soils contain high levels of exchangeable sodium to depth of 100 cm or more, but usually it would suffice to improve the soil up to a depth of 15 to 20 cm by amendments to grow shallow rooted crops like rice and wheat. For growing deep rooted crops, the quantity of amendments required would be much more.

For getting best results from the field application of gypsum, the soluble salts should first be leached out of the soil as much as practicable. This enhances the efficiency of utilization of the calcium in gypsum for replacing the adsorbed sodium in the soil. Such a step, however, may become infeasible to follow if the soil particles are so dispersed in an alkali soil of fine texture that the soil pores are blocked to permit water movement and leaching. Gypsum in a powdered form is usually broadcast on the field and then mixed with the top soil layer using a suitable implement (*e.g.*, disc plough or offset disc harrow). Following broadcast and mixing, adequate water is added to the field to dissolve the gypsum. The replacement reaction of sodium by calcium (described earlier in this chapter) takes place in the solution phase. It is highly desirable to facilitate leaching out the reaction product in solution from the soil through a subsurface drainage system.

IRRIGATION WITH POOR QUALITY WATER

Where there is no alternate source of good quality irrigation water in an area, it is inevitable to use the available water of poor quality. However, the yield potential of such areas can be increased by adopting proper management practices such as: (*i*) improvement of sodium and bicarbonate rich waters by gypsum application, (*ii*) choice of salt tolerant crops and their varieties, (*iii*) optimum fertilization and manuring, (*iv*) proper irrigation management, and (*v*) breaking of any impervious layers by deep ploughing.

Improvement of Water Quality

The harmful effect of irrigation water can be minimised to some extent by modifying its ionic composition by adding such chemicals, which tend to precipitate the harmful constituents such as bicarbonate and carbonate in the form of less soluble salts or tend to create a favourable cationic (Ca : Mg : Na) ratio.

For all practical purposes, gypsum should be powdered up to 0.5 mm size or passed through a 30 mesh sieve. The gypsum requirement of water should be calculated, depending upon the relative concentration of sodium, magnesium and calcium ions in irrigation water and the solubility of gypsum. Mixing of gypsum directly in irrigation water poses some problems. If it is placed in bags or containers between the entry and outlet of water, it does not dissolve uniformly with increase of time, and most of it is deposited at the place of entry of water in the field and forms bigger granules from powdered gypsum on coming in contact with water. However, some of the above limitations can be overcome by applying its saturated solution by a mechanical device in the flowing water at a constant rate.

Choice of salt-tolerant crops. Some crop and their varieties are more salt-tolerant than others (Table 14.7). Hence, salt-tolerant crops are to be grown in salt affected areas till the soils are improved by vegetation or adopting the reclamation procedures.

Use of fertilizers. Generally, saline and alkali soils or soils irrigated with poor quality waters are low in their fertility status, especially with reference to nitrogen or sometimes to phosphorus. Better crops can be grown by raising their fertility status. Nitrogen response to crops is better when it is applied to soil along with manures. It has been observed that for wheat, barley, *bajra* and maize, the usual does of fertilizers, as applied on normal soils can be applied up to an EC value of 6.5 millimhos/cm and an ESP of about 30. However, excessive fertilization on a highly saline alkali soil is of no value.

Amongst the fertilizers, nitrogen is better than phosphorus, followed by potash. However, up to moderate levels of salinity and alkalinity, phosphorus and potash application may be profitable only in the presence of nitrogen. Application of urea should be avoided due to its higher rate of nitrogen loss in the gaseous form under saline-alkali conditions.

The efficiency of fertilizers on soils irrigated with saline water or salt-affected soils depends largely on the interactive effects of salts, exchangeable sodium and fertilizers. The most important determining factors are the nature and degree of salinity and sodicity, chemical reactivity and solubility of added fertilizers, nutrient requirement of the crops, their availability in the soils and the salt-tolerance behaviour of crops. This indicates that in salt-affected soils, the behaviour of fertilizers is quite complex and efficiency of fertilizers in such soils has to be evaluated, keeping in view the quality of irrigation water, level of soil salinity and sodicity and their interactive effects.

Soil management practices. In order to adopt irrigation with poor quality water on a long-term basis, it is important to have a detailed analysis of representative soil profiles for their physical, chemical and morphological characteristics. Soil analysis should include its structure, texture, pH, lime content, location and amount of gypsum (if present in the profile), total soluble salts, their ionic composition and exchangeable cations. Information on water transmission properties of the soil and depth of water table are to be obtained. Data on the climatic parameters as amount of rainfall, its intensity, distribution, and evaporation are to be obtained, as they control the periodic as well as annual salt build-up in an irrigated area.

Accumulation of excessive amount of salt and development of high sodicity are the main limitations in the safe utilization of poor quality irrigation water. These depend upon the nature and amount of clay minerals in the soil and the quality of irrigation water. Success of irrigation with saline water lies in the degree of accuracy in the predicted values of soil salinity and sodicity expected to be developed during the cropping period. Saline waters can more safely be used in coarse than in fine textured soils.

Table 14.5 presents the effect of gypsum on the yield of some crops. Generally, the gypsum requirement (GR) of an alkali soil is determined by the Schoonover's method. However, studies at the Central Soil Salinity Research Institute, Karnal indicated that on a highly alkali soil, only 25 per cent of the gypsum requirement, as estimated by the Schoonover's method was sufficient to reclaim the soil for crops like rice, wheat and barley (Bhumbla, 1972).

Table 14.5. Effect of gypsum on the yield of some crops

Gypsum (tonnes/ha)	*Paddy*	*Wheat*	*Barley*
	Grain yield (quintals/ha)		
0	2.37	0.43	0.40
7.5	14.31	16.43	22.08
15.0	27.66	28.10	25.85
22.5	30.96	34.05	29.07
30.0	34.17	37.90	33.18

GR = 60 tonnes/ha by Schoonover's method.

[***Source***: Bhumbla (1972).]

The quantity of amendment necessary for the reclamation of a particular field depends upon the quantity of the exchangeable sodium present on the clay surface. This can be calculated on the equivalent basis of sodium to be replaced by calcium. It works out to be 4.1 tonnes of gypsum and 0.78 tonnes of sulphur per hectare for 30 cm depth for each milliequivalent of sodium to be replaced by calcium. Table 14.6 represents the weight of ameliorants equivalent to 1 tonne of sulphur. For a saline-alkali soil, the gypsum requirement is determined after the soluble salts are removed as the salts present interfere with the estimation of calcium. In practice, a quantity of gypsum, about 30 to 50 per cent of the gypsum requirement value as obtained by Schoonover's method, should be thoroughly mixed in an alkali soil and the field flooded with water so that the exchange between calcium and sodium could take place and the released sodium could be leached down by water. By this procedure the physical condition of the soil will improve gradually, as the proportion of exchangeable calcium increases. In some areas gypsum is found in the subsoil < 1 m depth and the same could be mixed by deep ploughing.

Table 14.6. Weight of amendments equivalent to 1 tonne of sulphur

Amendment	*Tonnes equivalent to 1 tonne of sulphur*
Sulphur	1.00
Lime-sulphur (Calcium Polysulphide) solution, 24% sulphur	4.17
Sulphuric acid	3.06
Gypsum ($CaSO_4.2H_2O$)	5.38
Iron sulphate ($FeSO_4.7H_2O$)	8.69
Aluminium sulphate [$Al_2(SO_4)_3$. $18H_2O$]	6.94
Limestone ($CaCO_3$)	3.13

[***Source:*** Richards (1960).]

Highly dispersed alkali soils are devoid of soil organic matter and green maturing with *dhaincha* (Serbiania aculeata) in summer is helpful in reclaiming such soils. Green manuring adds soil organic matter and increases soil fertility. The effectiveness of *dhaincha* is due to the following characteristics: (*i*) it is tolerant to salts and waterlogging conditions, (*ii*) its root system is extensive, (*iii*) it opens compact subsoil and improves soil permeability, (*iv*) its rate of decomposition is fast, and (*v*) its cell sap is acidic and is capable of neutrallising the alkalinity of the soil.

Lime requirement for the reclamation of acid soils. The beneficial effect lime application in soils when soil acidity is a constraint to agricultural production has been discussed earlier. As gypsum requirement for the reclamation of alkali soil is determined on the basis of desired replacement of sodium from the soil complex, the lime requirement is determined based on the requirement of raising soil pH from less than 6 to a desirable higher value. Lime requirement depends on the extent by which the exchangeable H is to be replaced by the Ca of the lime. For detailed information on estimating the lime requirement, reference is invited to Bhattacharya and Michael (2003) listed at the end of this chapter.

A field infested with the problem of salinity should be levelled carefully. This will help in uniform spread of water and facilitate its downward movement, otherwise at high spots (if salts are not leached) it will lead to salt accumulation and will cause heterogeneity in salt concentration.

Medium textured soil with *kankar* layers pose the problem of sodicity. Such soils can be managed better by deep ploughing along with the incorporation of green manure. By virtue of acidic character of cell sap and the penetrating nature of the roots, *green manuring* increases the infiltration rate of the soil. Application of gypsum under conditions of shallow water table may further improve land productivity.

The success of gypsum or flushing of salts lies in the drainability of the soil. Hence, if the improvement of internal drainage by tile drains is not economically feasible, attempts should be made to improve soil physical conditions by using fertilizers and manures in combination. This could be supplemented by improving the surface drainage system.

Irrigation management. While irrigating with poor quality waters, accumulation of salts increases with the fineness of soil texture. Since concentration of salt in the soil solution near the root zone determines the degree of adverse effects on crop growth, rather than the salt concentration of irrigation water alone, it is essential to adopt irrigation practices such that the salinity at the root zone is kept to the minimum. For this purpose the quantity of water and the frequency of irrigation are planned to meet the leaching requirement of the soil and the consumptive use of the crop grown.

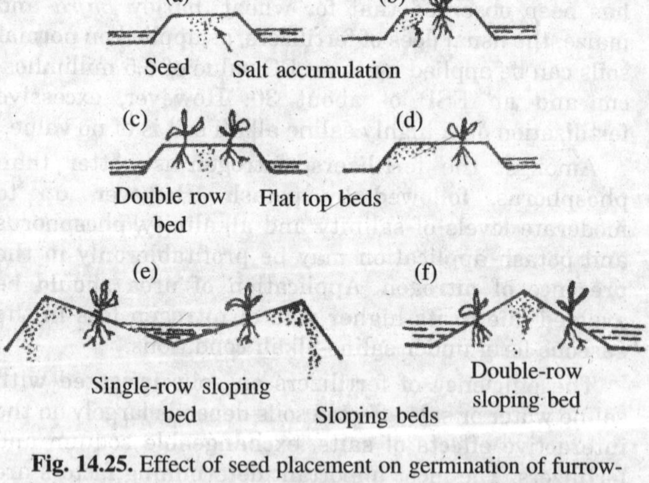

Fig. 14.25. Effect of seed placement on germination of furrow-irrigated crops on saline soils or when using poor quality water.

Salts often accumulate in the top few centimetres of the soil during non-crop periods. Where shallow water tables interfere with salinity control, fallow and idle lands may rapidly accumulate surface salts, particularly in hot arid climates. Under such conditions, both crop germination and yield can be seriously reduced. A heavy pre-sowing irrigation to leach these surface salts will improve germination and early growth and is sometimes an essential practice. It is made far enough in advance of the desired planting date to allow for cultivation to remove weeds and prepare the seedbed.

Seed placement. Obtaining a satisfactory stand of furrow-irrigated crops on saline soils when using poor quality water is important. If salinity is a problem, planting seeds in the centre of a single-row raised bed will place the seed exactly in the area where salts concentrate [Fig. 14.25 (b) & (d)]. A double-row raised planting bed by comparison offers an advantage [Fig.14.25(c)]. The two rows are placed so that each is near a shoulder of the raised bed, thus placing the seed away from the area of greatest salt accumulation. By this method higher soil and water salinities can be tolerated than with the single-row plantings because the water moves the salts through the seed area to the centre of the ridge. Alternate furrow irrigation is often advantageous. If the beds are wetted from both sides, the salts accumulate in the top and the centre of the bed, but if alternate furrows are irrigated, the salt can be moved beyond the single seed row [Fig.14.25(e)], thus reducing the extent of salt accumulation. Off-centre, single-row planting on the shoulder of the bed closest to the watered furrow aids germination under salty soil conditions [Fig. 14.25(f)].

With either single or double row planting, if salts are expected to be a problem, increasing the depth of water in the furrow can also be an aid to improved germination [Figs. 14.25(c) and 14.25(f)]. A better salinity control can be achieved by using sloping beds with seeds planted on the sloping side and the seed row placed just above the water line.

The larger seeded crops, such as maize, have been planted in the water furrow as an aid to salt-control during germination. Grapes, too, have sometimes been grown with problem waters by placing the vine row at the bottom of wide flat furrows or at the bottom of wide, gently sloping V-shaped furrows. Salinity problems have been aggravated when perennial crops such as tree crops and citrus are planted on raised beds and surface irrigated with poor quality water. Salts gradually accumulate in the raised beds to the extent that in a few years crop tolerance level is exceeded.

Plant Response to Saline and Alkali Soils

Excessive salinity usually causes poor stands of crop, stunted growth and reduced yield. It may also cause leaf burns in some crops and deep blue-green colour in other. In some cases, symptoms of excessive salts are similar to those of drought or low fertility status. Soil salinity causes a decrease in the size and vigour of the plant. The germination of seeds is delayed and retarded but the cropping period is slightly reduced with the increase in salinity. However, its effects on the yield and quality of the crop are variable. In general, grain yield is affected more than the height of the plant.

Effect of Salts on Plant Growth

Excessive soil salinity and alkalinity may adversely affect plant growth by: (*i*) increasing the osmotic pressure in the soil solution, (*ii*) accumulating certain ions in toxic concentration in the plant tissue, and (*iii*) by altering the plants mineral nutritional characteristics.

Osmotic pressure and plant growth. It has been observed that there is an inverse relationship between the osmotic pressure of the soil solution and plant growth. The reduced crop growth is due to reduced water availability to plants. Thus, as the osmotic pressure increases with the concentration of salts, it is presumed that salinity inhibits plant growth primarily by reducing the availability of water. There are evidences that the total concentration of salts, rather than their chemical nature, is mainly responsible for the inhibitory effects. On the other hand, recent evidence indicates that the water absorption capacity is relatively unaffected by salinity and plants develop their own internal pressure against salt. Hence, this theory suggests that salt tolerance of the plant may be defined as the degree to which osmotic adjustment can be made without any sacrifice in plant growth.

In addition to the osmotic effect, the force by which water is held by soil particles also affects plant growth. This force, called soil moisture tension, increases as the moisture content of the soil decreases. The osmotic pressure of the soil solution and soil moisture tension are additive in their effect on plant growth. Their combined effect is called total soil moisture stress. It has been observed that plant growth is closely related to the soil moisture stress, regardless whether this stress arises primarily from salinity or soil moisture tension or both.

Through controlled leaching, the osmotic pressure of the soil solution should be maintained at the lowest feasible level. By careful management of the irrigation system, the soil moisture tension in the root zone is maintained in a range that will provide the greatest net return of the crop grown.

Specific ion effect. Certain ions exert specific effects which depress crop growth and yield independent of the salt concentration. These specific ion effects may be toxic or may cause nutritional disorder in the plant, influencing its physiological and metabolic activities. The toxic effect may be due to the presence of an ion in excessive amount in the soil solution, the higher uptake of which damages the plant growth. Nutritional disorders in plants occur when the accumulation of one or more nutrients in excessive amounts inhibits the uptake of other essential nutrients. When the amount of these essential nutrients is below their critical limit which is essential for normal physiological functions of the plant, the growth of the plant is inhibited. Plants vary widely in their nutrient requirement and their ability to absorb specific nutrients. Hence, the effects of salts on nutrition differ markedly from species to species and even between varieties of the same crop.

Salt-tolerance of crops. Salt-tolerance means the ability of a plant to tolerate salt concentrations in the

Fig. 14.26. A CSSRI released rice variety CSR27 growing in the reclaimed saline soil of Sampla Project area, Rohtak district, Haryana. (***Courtesy:*** CSSRI, Karnal.)

Fig. 14.27. Salt-tolerant wheat variety KRL 19 growing under the rice—wheat rotation programme in the Pilot Project area. (***Courtesy:*** CSSRI, Karnal.)

Table 14.7. Relative salt-tolerance of selected crops

Tolerant	*Semi-tolerant*	*Sensitive*
Field Crops		
Barely	Oats	Field beans
Dhaincha (Serbania aculeata)	Rice	Gram (*channa*)
Sugarbeet	Sorghum (*jowar*)	Peas (*mutter*)
Tobacco	Pearl millet (*bajra*)	
Turnips	Maize	
Mustard	Red gram (*arhar*)	
Cotton	Green gram (*mung*)	
Wheat (*Kalyan Sona*)	Sunflower	
Sugarcane	Castor	
Taramira	Sesamum	
	Linseed	
	Fodder Crops	
Saltbush	Turnips (*Senji*)	*Guar*
Dub grass	Sudan Grass	
Rhodes grass	Lucerne	
	Metha	
	Sorghum (*jowar*)	
	Maize	
	Berseem	
	Cowpea (*lobia*)	
Vegetables		
Turnips	Tomato	Celery
Beet-root	Cabbage	Green beans
Asparagus	Cauliflower	
Spinach	Lettuce	
	Potato	
	Carrot	
	Onion	
	Pea	
	Cucumber	
	Pumpkins	
	Bitter gourd (*Karela*)	
Fruits		
Date palm	Pomegranate	
Coconut	Grape	
	Guava	
	Mango	
	Banana	
	Pear	
	Apple	
	Orange	
	Grape fruit	
	Plum	
	Almond	
	Apricot	
	Peach	
	Strawberry	
	Lemon	

root zone. Knowledge of the salt tolerance characteristics of a crop is important in selecting a particular crop or its variety to suit the soil conditions and for determining the leaching requirement.

The effect of soil salinity on the crop grown is intangible when the electrical conductivity of the saturation extract (EC_e) is less than 2 mmhos/cm. Yield of very sensitive crops is reduced when the EC_e value is about 4 mmhos/cm. Many of the common field crops are affected when the EC_e value is in the range of 4 to 8 mmhos/cm. Crops with high salt-tolerance can grow satisfactorily when the EC_e values are between 8 and 16 mmhos/cm. Only a few crops survive at EC_e values beyond 16 mmhos/cm.

Table 14.7 presents the salt-tolerance characteristics of some of the common crops, which are listed in decreasing order of tolerance within each group. In some crops the salt-tolerance limit varies with the stage of growth. For example, rice is quite tolerant during germination but becomes sensitive during seeding stage. Similarly, maize is more tolerant during germination than at later stages of growth. On the other hand, sugarbeet can tolerate salinity only up to 4 mmhos/cm during germination but can sustain itself at much higher levels of salinity after the seedlings are established. Varieties of the same crop may differ significantly in their salt-tolerance behaviour. In the Gangetic Plains of North India, a common practice is to grow rice crop in the monsoon season (*kharif*), followed by wheat in the winter season (*rabi*). In the early few years after reclamation, it is a good practice to grow salt-tolerant varieties of rice and wheat (Figs. 14.26 and 14.27). Soil conditions improve over the years and more high yielding varieties could be adopted. Figure 14.28 shows the progressive increase in the yield of rice and wheat crops over a ten-year period in a recently reclaimed alkali soil area in the Rohtak district of Haryana.

Boron tolerance of crops. Boron, though essential for plant growth, becomes toxic beyond a concentration of a few ppm. It does not seem to affect the germination of seeds up to 20 ppm or even more, but the growth rate and height of seedlings are significantly reduced beyond 1 or 2 ppm. Relatively, legumes are salt-sensitive but can tolerate more boron than cereals. Amongst field crops, wheat and barley can tolerate from 1 to 2 ppm of boron in irrigation water.

Boron is translocated in the leaves and accumulate in the tip and margin. Toxicity symptoms such as yellowing and tip burn is usually visible. If boron toxicity is indicated in the plant, its source should be found. If irrigation water has excessive boron, it should be mixed with good quality water and if soil is the cause, the boron should be reduced by leaching, though its leaching is more difficult than of salts. Table 14.8 lists in decreasing order the boron tolerance of some of the crops.

Table 14.8. Limits of boron in irrigation waters for crops of different degrees of tolerance

Tolerant	*Semi-tolerant*	*Sensitive*
4.0 ppm	*3.0 ppm*	*1.0 ppm*
Palm (Phoenix canariensis)	Sunflower	Plum
Date palm	Potato	Pear
Sugarbeet	Cotton	Apple
Lucerne	Tomato	Grape
Broad bean	Radish	Cherry
Onion	Pea	Peach
Turnip	Barley	Apricot
Cabbage	Wheat	Orange
Lettuce	Maize	Grape fruit
Carrot	Oat	Lemon
	Pumpkin	
	Sweet potato	
2.00 ppm	*1.0 ppm*	*0.3 ppm*

Adapted from Wilcox (1960).

Special Problems in Using Poor Quality Irrigation Water

Some of the special problems encountered in irrigation with poor quality water are the toxicity effects on certain crops due to sprinkler irrigation and the management of sewage effluent for irrigation.

Special toxicity effects due to sprinkler irrigation. Overhead sprinkling of certain crops can cause special toxicity problems, not encountered in surface methods of irrigation. Excess quantities of sodium and chloride can be absorbed through leaves wetted by the sprinkler and can cause leaf burn. This occurs primarily during periods of high temperatures and low humidity. In between rotations of the sprinkler, water evaporates and salts are concentrated. The leaf burn or crop damage seems to be primarily of specific toxicity due to uptake and accumulation of sodium or chloride. Toxicity occurs with sensitive crops at relatively low sodium or chloride concentrations. Toxicity has been reported to occur on citrus at concentrations as low as 3 meq/l of either sodium or chloride in the irrigation water. With surface irrigation, these concentrations caused no toxicity effects or leaf burn. Slight damage has been reported for more tolerant crops such as lucerne under extremely high evaporative conditions at EC_w = 1.3 mmhos/cm containing 6 meq/l of sodium and 7 meq/l of chloride. In contrast, little or no damage occurred with waters having an EC_w value as high as 4 mmhos/cm where sodium was 24 meq/l and chloride was 37 meq/l when evaporation conditions were low (Ayres and Westcot, 1976).

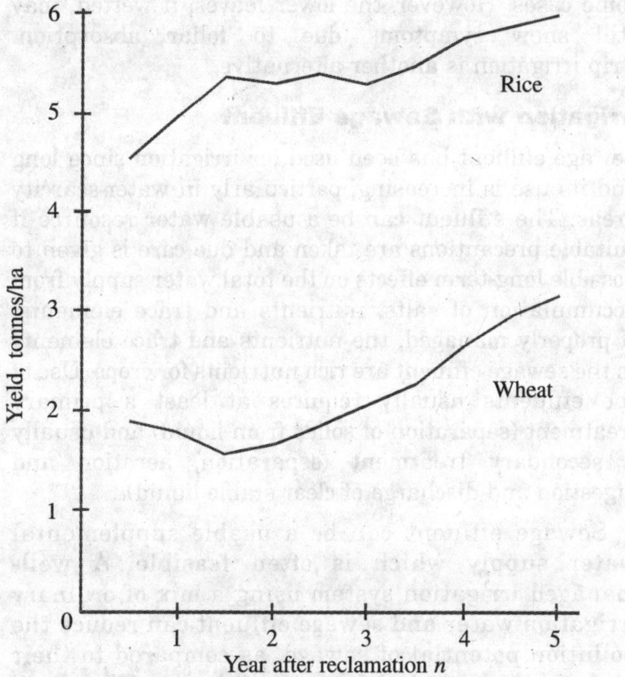

Fig. 14.28. Progressive increase in the productivity of rice and wheat crops in selected demonstration plots of a recently reclaimed alkali soil areas under the CSSRI Operational Research Project in Haryana. [***Source:*** CSSRI Bul. No. 12 (1988).]

Relative crop tolerances to sodium and chloride in sprinkler applied irrigation water are not well established but, in general, crops sensitive to sodium or chloride are most sensitive to foliar absorption from sprinkler applied water. Most annual crops are usually not sensitive. Where foliar absorption has been a problem, the following practices are recommended to reduce the problem:

(1) *Sprinkler irrigation at night.* Night sprinkling has been effective in reducing the sodium and chloride toxicity due to foliar absorption.

(2) *Increasing speed of rotation of sprinkler heads.* Frequent or continuous wetting results in less absorption than intermittent wetting. Use of sprinkler heads that rotate at 1 revolution per minute or more are recommended.

(3) *Sprinkling during periods of low humidity and high evaporative demand.* If weather patterns for an area are known or can be forecast and soil conditions allow for storage of sufficient quantities of water for the crop to use between irrigations, then water applications can be timed to avoid these critical periods as much as possible.

(4) *Crop selection for quality of water.* If overhead sprinklers must be used, it may not be possible to grow certain sensitive crops such as beans or grapes.

(5) *Change of irrigation method.* A change to another irrigation method such as furrow, border or basin may be necessary. Under-tree sprinklers have been used in some cases. However, the lower leaves, if wetted, may still show symptoms due to foliar absorption. Drip irrigation is another alternative.

Irrigation with Sewage Effluent

Sewage effluent has been used for irrigation since long and its use is increasing, particularly in water-scarcity areas. The effluent can be a usable water resource if suitable precautions are taken and due care is given to possible long-term effects on the total water supply from accumulation of salts, nutrients and trace elements. If properly managed, the nutrients and trace elements in the sewage effluent are rich nutrients for crops. Use of the effluents usually requires at least a primary treatment (separation of solids from liquid) and usually a secondary treatment (separation, aeration and digestion and discharge of clear stable liquid).

Sewage effluent can be a usable supplemental water supply which is often feasible. A well-managed irrigation system using a mix of ordinary irrigation water and sewage effluent can reduce the pollution potential of sewage, as compared to their disposal in rivers and other water bodies. The use of sewage effluent, however, requires a higher level of management than in case of ordinary water supplies. This is primarily related to public health and public acceptance. There are certain health precautions required by local public health authorities but these normally are not sufficiently restrictive to prevent the irrigation of selected crops with sewage water.

The main problems that may occur can be grouped into those of physical, chemical and biological characteristics. Each of these are discussed below to the extent needed to point out the problems.

(1) *Physical characteristics.* The physical characteristics of sewage effluent include the total suspended solids, colour, temperature and odours. The total suspended solids include floating and suspended solids and dissolved organic substances. The suspended solids may clog the soil pores, coat the land surface and reduce infiltration and aeration. The degree of the problem due to organic content depends on the extent of treatment given to the sewage prior to its use. Use of treated effluent presents very few problems due to suspended solids. Drying between irrigations and cultivations is the usual field practice used. The organic material can be beneficial to the soil if managed without aeration and odour problems. The desirable management practice will have to be evaluated by location specific trials.

Odours are indicators of lack of aeration and anaerobic decomposition of organic matter. Strong odour may be obnoxious and indicative of operational problems but a normal secondary effluent usually has little odour problem.

(2) *Chemical characteristics.* The chemical characteristics of sewage effluent vary with the source of water and the type of discharge into the system. In effluents, which receive considerable industrial wastes, trace element toxicity may be a problem. Copper, zinc, cadmium and boron content are sometimes high enough to be detrimental. Others of importance include arsenic, chromium, lead, manganese, mercury and nickel. These are to be assessed prior to approval for irrigation use. On the other hand, the trace elements contamination in the effluent may act as a source of certain needed trace elements in deficient soils.

(3) *Biological characteristics.* Biological characteristics are concerned with bacteria, viruses and other disease causing organisms. Raw sewage can be expected to be teeming with micro-organisms, some of which may be pathogenic or disease causing. The degree of disinfection will depend upon the intended use and the health requirements in the area. Usually vegetables, which are eaten raw are not irrigated with sewage water.

SUMMARY

Nature, causes and extent of salt problems in irrigated and unirrigated areas are described. Influence of irrigation practices and soil and crop management on soil salinity and alkali are presented. Chemistry of salt-affected soils is explained. Influence of salts on the physical properties of soils is described. Procedures for the determination of the properties of saline and alkali soils are presented. Quality of irrigation water is described along with the criteria for their classification. Procedures for irrigation with poor quality water are explained. Basic requirement in the reclamation and management of salt-affected soils are presented. Procedures for the reclamation of saline, saline-alkali and alkali soils are explained along with materials and methods. Plant response to saline and alkali soils are explained along with the effect of salts on plant growth. Applicability and limitations of irrigation using sewage effluents are explained.

PROBLEMS

14.1. Express 2100 ppm and 2200 ppm sodium bicarbonate and calcium chloride concentrations in meq/litre

[**Ans.** Sodium bicarbonate 25 meq/l
Calcium chloride 20 meq/l]

14.2. Express the osmotic pressure in atmosphere of a salt solution showing EC values as 3.2, 0.8 and 4.6. mmhos/cm.

[**Ans.** 1.152, 0.288, 1.656 atmospheres]

14.3. Express the salt concentration of solutions in ppm showing their EC value as 1.5 millimhos/cm, 0.2 mhos/cm and 1000 micromhos/cm.

[**Ans.** 960 ppm, 128.000 ppm and 640 ppm]

14.4. Analysis of a 10 grams soil sample showed exchangeable sodium, calcium, magnesium and potassium as 1.46, 1.241, 0.657 and 0.292 meq, respectively. Determine the total cation exchange capacity as well as the degree of saturation of each cation.

[**Ans.** CEC = 36.5; Ex Na% = 40, Ex Ca% = 34,
Ex Mg% = 18 and Ex Na% = 8]

14.5. A soil sample, weighing 5 gram, on shaking with 100 cc of gypsum solution of 31.2 meq/l of calcium concentration showed 26.5 meq/l as Ca + Mg in the filtrate. Determine the gypsum requirement per 100 grams soil and also in tonnes/hectare for 15 and 30 cm depth of soil.

[**Ans.** 9.4 meq/100 gm soil;
8.084 tonnes/ha for 15 cm depth;
16.168 tonnes/ha for 30 cm depth]

14.6. The electrical conductivity of the canal water used for irrigation in a locality is 1 mmhos/cm and its density 1 gram/cu cm. The density of salt-free loam soil of the area is 1.4 grams/cu cm. The saturation per cent is 50. Calculate the volume of water, expressed in depth of application, that will render the soil saline up to a depth of 1 m.

[**Ans.** 2.8 m]

QUESTIONS

1. What is the characteristic of a saline agricultural land. Why is there no constant threshold value of soil salinity to refer to a soil as saline?
2. Which combination of physical and chemical property causes soil disperse in water? What are the problems associated with such soils while practicing agriculture?
3. Why pure water is a bad conductor of electricity? What changes in the water take place by adding a little mineral salt that it starts conducting electricity?
4. Why does a plant show signs of wilting in a waterlogged soil when the soil water is highly saline?
5. Name two crops each of sensitive, semi-tolerant and tolerant class towards root zone soil salinity. What may be a simple method to reduce soil salinity, if soil drainage is not restricted?
6. Distinguish between the salinization processes of the soils of inland and the soils of coastal land.
7. Soil may get salinized due to various factors. Name and briefly explain the controllable and the uncontrollable factors.
8. Discuss the reasons why fine textured soils are more prone to turn alkali than coarse textured soils.
9. Under identical climate conditions, why is an irrigated agricultural land more likely to turn saline that a rain-fed agricultural land?
10. Complete the following chemical equations:

    $2NaHCO_3 + CaSO_4 = Na_2SO_4 + ?$

    $2CaCO_3 + ? = CaSO_4 + Ca(HCO_3)_2$

    $2HCO_3 + Ca = CaCO_3 + ? + ?$
11. Explain how the solubility of sodium, calcium and magnesium salts in water determines the formation of sodic soil in arid and semi-arid regions.
12. Soil salinization of irrigated agricultural lands in semi-arid climate is inevitable. Outline the steps needed to ensure that soil salinization is delayed and kept under control.

13. Write if the following statements are True or False:

(*i*) All Indian rivers have saline water

(*ii*) Canal water salinity depends of the source river and the terrain it passes through

(*iii*) Irrigated land in the semi-arid region is more likely to turn saline than that in the humid region

(*iv*) An alkali soil has a high percentage of exchangeable calcium

(*v*) Alkali soils have pH of 8.5 or higher

(*vi*) If 1 g of NaCL is dissolved in 1 m^3 of water, the salt concentration is 1 ppm

(*vii*) An EC meter works on the principle of Wheatstone Bridge

(*viii*) The larger is the soil particle size, the higher will be its specific surface

(*ix*) In India, the most commonly used chemical for alkali land reclamation is sulphuric acid

(*x*) In furrow irrigation system, salt concentration is higher on the ridge

14. Of the four options given for each of the following questions, select the most appropriate answer:

(*i*) A soil solution of pH 7 is called a:

(a) Saline soil (b) Acid soil
(c) Alkali soil (d) Neutral soil

(*ii*) Electrical conductivity of a soil solution is a measure of:

(a) Soil dryness
(b) Soil iron content
(c) Soil salinity
(d) Clay content of the soil

(*iii*) A dispersed clay soil has:

(a) High infiltration
(b) Low infiltration
(c) Good workability
(d) Good internal drainage

(*iv*) Cations are attached to the surface of the clay particles by:

(a) Absorption (b) Adsorption
(c) Adhesion (d) Cohesion

(*v*) Saline soil can be reclaimed by:

(a) Leaching
(b) Scrapping
(c) Adding gypsum
(d) By growing salt-tolerant crops

(*vi*) In non-saline alkali soil, the exchangeable sodium percentage is:

(a) < 5 (b) 5–10
(c) 10–15 (d) > 15

(*vii*) Clay soils are chemically active because they:

(a) Have high specific surface
(b) Are colloidal
(c) Adsorb cations
(d) Have porosity

(*viii*) An important natural factor causing soil salinity is:

(a) High temperature (b) High evaporation
(c) Low precipitation
(d) Weathering of parent rock

(*ix*) Among the Indian rivers, the water salinity is the lowest and safest in:

(a) Ganga (b) Yamuna
(c) Kosi (d) Tungabhadra

(*x*) In a furrow-irrigated saline land, the seeds should be sown:

(a) On the ridge (b) Facing east
(c) Facing north (d) At the furrow bottom

15. In the following questions, match the entries in the Ist column with the closest entry in the 2nd column:

(*i*)	(A) CEC	(I) Measure of salinity
	(B) SAR	(II) Measure of soil reaction
	(C) pH	(III) Measure of replaceable Na
	(D) EC	(IV) Measure of Na with respect to Ca and Mg
	(E) ESP	(V) Measure of replaceable cations
(*ii*)	(A) Saline soil	(I) Gypsum
	(B) Alkali soil	(II) Poor infiltration
	(C) Acid soil	(III) Leaching
	(D) Dispersed soil	(IV) Salinity prone
	(E) Waterlogged soil	(V) Lime
(*iii*)	(A) Na-clay to Ca-clay	(I) Uptake by plant
	(B) Western Gujarat	(II) Salinity control
	(C) Calcium depletion	(III) Saline groundwater
	(D) Leaching method	(IV) Sodium replacement
	(E) Drainage	(V) Surface ponding
(*iv*)	(A) *Sesbania*	(I) Salt-tolerant tree
	(B) *Accacia*	(II) Soil conditioner
	(C) Mustard	(III) Salt-tolerant crop
	(D) Sorghum	(IV) Sensitive to salinity
	(E) Pea	(V) Semi-tolerant crop
(*v*)	(A) Flame photometer	(I) Soil moisture measurement
	(B) TDR meter	(II) Water transmission property
	(C) Wheatstone Bridge	(III) Detection of elements
	(D) Permeameter	(IV) Extraction by ammonium acetate
	(E) CEC determination	(V) Salinity measurement

FURTHER READING REFERENCES

Abrol, I.P., (1962), *Studies on soil factors affecting the safe utilization of irrigation waters as regards their alkali hazards* Ph. D. Thesis, I.A.R.I., New Delhi.

Abrol, I.P. and Bhumbla D.R., (1971), *Saline and alkali soils in India—their occurrence and management*, World Soil Resources, FAO Report No. 41, p. 42–51.

Aggarwal, R.R. and Gupta, R.N., (1966), *Saline and alkali soils of India,* ICAR Tech. Bull. (Agri. Series) No. 15, New Delhi.

Ayres, R.S. and Westcot, D.W., (1976), *Water quality in agriculture,* Irrigation and Drainage Paper 29, Food and Agriculture Organization of the United Nations, Rome, p. 97.

Bains, S.S., Singh, K.N., Dayanand and Pandey, S.L. (1970), *Crop production techniques on saline soils characterised by high groundwater table*, Paper presented at Seminar on Soil Salinity, U.P. Instt. of Agri. Sciences, Kanpur.

Bernstein, L. (1959), *Salt Tolerance of Vegetable Crops,* USDA Agri. Inf. Bull. No. 206, United States Department of Agriculture, Washington, D.C.

Bernstein, L. (1965), *Salt Tolerance of fruit crops,* Bull. No. 292. United States Department of Agriculture, Agricultural Information Service, Washington, D.C.

Bhattacharya, A.K. and Michael, A.M. (2003), *Land Drainage—Principles, Methods and Applications,* Konark Publishers Pvt. Ltd., Delhi.

Bhumbla, D.R. (1972), *Saline and alkali soils and their management,* Lectures in summer institute held at CSSRI, Karnal.

Biswas, T.D., Das, B. and Verma, H.K.G. (1969), *Quality rating of well waters for irrigation in Delhi area,* Groundwater, Part III, Misc. No. 14. Geological Survey of India, Calcutta, pp. 117–134.

Central Soil Salinity Research Institute (1988), *Field Drainage for the Control of Groundwater Table and Soil Salinity in Irrigated Agricultural Lands of Semi-Arid Regions,* Lecture Notes, Vol. II. CSSRI, Karnal, p. 138.

Central Soil Salinity Research Institute, Karnal (1998), *Management of Salt-Affected Soils and Use of Saline Water in Agriculture,* Biennial Report of I.C.A.R., Coordinated Project, CSSRI, Karnal.

Central Soil Salinity Research Instituite, Karnal (1998), *Annual Report CSSRI, Karnal.*

Finkel, H.J. (1983), *Handbook of Irrigation Technology,* Vol. 1. CRC Press, Boca Raton, Florida, USA, pp.79–92.

Food and Agricultural Organization of The United Nations (1971), *Salinity Seminar Baghdad,* FAO of UN, Rome.

Food and Agriculture Organization of the United Nations (1971) *Drainage of Heavy Soils,* Irrigation and Drainage Paper 6. FAO, of UN, Rome, p. 109.

Food and Agriculture Organization of the United Nations (1973), *Drainage of Salty Soils,* FAO Irrigation and Drainage Paper 15, FAO, of UN Rome, p. 84.

Food and Agriculture Organization of the United Nations (1980), *Drainage Design Factors,* FAO Irrigation and Drainage Paper 38, FAO of UN Rome, p. 52.

Gandhi, A.P. and Paliwal, K.V., (1975), *Effect of different quality of irrigation waters and soil texture on the yield and uptake of nutrients by wheat,* Proc. India Nat. Sci. Acad. 41 (B), pp. 440–451.

Gupta, R.K., (1985), *Soil and Water Sampling for Salinity Appraisal,* Central Soil Salinity Research Institute, Karnal, p. 9.

Gupta, S.K., Dinker, V.S. and Tyagi, N.K. (1995), *Reclamation and Management of Waterlogged and Salt-Affected Areas in Irrigation Commands,* Manual, Central Soil Salinity Research Institute, Karnal.

Israelson, O.W., and Hanson, V.E., (1962), *Irrigation Principles and Practices,* 3rd ed., John Wiley and Sons, New York, pp. 207–230.

Johnson Division (1967), *The Chemistry of Water,* Bull. No. 1237; Johnson Division, St. Paul Mannesota, USA.

Kanwar, J.S., and Kanwar, B.S., (1969), *Quality of irrigation waters.* Trans. 9th Intl. Cong. Soil Sci., 1, pp. 891–903.

Kovda, V.A., (1973), *Irrigation, drainage and salinity—an international source book,* Food and Agriculture Organization of the United Nations, Paris, Hutchinson & Co. Ltd.

Louis Berger International and Water and Power Consultancy Services (1988), *Handbook of Drainage of Irrigated Areas in India,* Tech. Report No. 5., Water and Power Consultancy Service (India), New Delhi.

Maliwal, G.L., (1968), *Effect of irrigation waters of different qualities on soils and crops,* Ph.D. thesis, Udaipur University, Udaipur.

Maliwal, G.L. and Paliwal, K.V., (1972), *Effect of fertilizer on the growth and chemical composition of wheat irrigated with saline water,* Agrochemica, 16, pp. 450–468.

Mehta, K.K. (1970), *Studies on the effect of irrigation water on soil properties and crop growth with special reference to boron,* Ph.D., Thesis, Udaipur University, Udaipur.

Mehta, K.K. (1985), *Alkali Soils—Steps for Reclamation,* Central Soil Salinity Research Institute, Karnal, p. 8.

Ministry of Water Resources, Govt. of India (1996), *Reclamation of Saline and Alkali Soils,* Proc. National Workshop, Dec. 1996.

Minhas, P.S. (1991), *Use of Poor Quality Waters,* 50 Years of Natural Resources Management Research, ed. by Singh, G.B. and Sharma, B.R. Indian Council of Agricultural Research, Krishi Bhawan, New Delhi, pp. 327–346.

Minhas, P.S. and Gupta, R.K., (1992), *Quality of Irrigation Water—Assessment and Management,* Publication and Information Division, Indian Council of Agricultural research, KA Bhawan, Pusa, New Delhi, p. 123.

Minhas, P.S., Sharma, O.P. and Patil, S.G. (1998), *25 Years of Research on Management of Salt-Affected Soils and Use of Saline Water in Agriculture,* Central Soil Salinity Research Institute, Karnal, p. 230.

Murthy, V.V.N. (1985), *Land and Water Management Engineering.* Kalyani Publishers, Ludhiana, pp. 352–372.

Narayana, A., (1966), *Physiological studies on salt-tolerance of wheat,* Ph.D. Thesis, I.A.R.I., New Delhi.

Paliwal K.V. (1972), *Irrigation with saline water,* I.A.R.I., Monograph-2, Water Technology Centre, IARI, New Delhi.

Paliwal, K.V. and Gandhi, A.P., (1976), *Effect of salinity, SAR and Ca : Mg ratio in irrigation water and soil texture on the predictability of ESP of soils.* Soil Sci., 122, pp. 85–90.

Paliwal, K.V., Srivastava K. and Prasad, M., (1975), *Magnesium hazard in well waters of Rajasthan,* J. Indian Soc. Soil Sci. 23, pp. 113–121.

Paliwal, K.V., and Maliwal, G.L., (1971), *Effect of fertilizers and manures on the growth and chemical composition of maize plants irrigated with saline waters,* Indian J. Argon. 16, pp. 316–321.

Paliwal K.V. and Yadav B.R., (1976), *Irrigation water quality and crop management in the Union Territory of Delhi,* I.A.R.I., Tech. Bull. 9, Water Technology Centre, I.A.R.I., New Delhi.

Paliwal, K.V., Maliwal, G.l. and Nanawati, G.C., (1975), *Effect of bicarbonate rich irrigation water on the growth, nutrient, uptake and synthesis of proteins and carbo-hydrates in wheat.* Plant and Soil 43, pp. 533–536.

Prasad, M., (1973), *Effect of different quality irrigation waters on soil properties and crop growth with special reference to magnesium,* Ph.D. Thesis, University of Udaipur, Udaipur.

Praving Titovunno (1971), *Quality of leachate from Vernal Project soil as affected by certain conditions in the soil and applied water,* M.S. Thesis, Utah State University, Logan. USA.

Rajasthan Agricultural Drainage Project (1996), *Analysis of Subsurface Design Criteria,* Chambal Command Area Development Authority, Kota, Rajasthan.

Ramamoorthy, B., (1970), *The scope for living with soil salinity and alkali conditions in India* Proc. Seventh NESA Regional Irrigation Practices Seminar, Lahore, Pakistan, September 16–28, 1968.

Rhoades, J.D., Kondiah. A. and Mashali, A.M. (1992), *The Use of Saline Waters for Crop Production,* FAO Irrigation and Drainage Paper 48, Food and Agricultural Organization of the United Nations, Rome, p. 123.

Rhoads, J.D., Chanduvi, F. and Lesli, S., (1999), *Soil Salinity Assessment,* FAO Irrigation and Drainage Paper 57, FAO of UN, Rome, p. 150.

Rycroft, D.W. Amer, M.H., (1995), *Prospects for the Drainage of Clay Soils,* FAO Irrigation and Drainage Paper 51, Food and Agricultural Organization of the United Nations, Rome, p. 134.

Shady, A.M., (1996), *Subsurface Drainage of Irrigated Lands in India,* International Commission on Irrigation and Drainage, New Delhi.

Singh, S.S., (1963), *Studies on some chemical factors in determining the quality of irrigation waters,* Ph.D. Thesis, I.A.R.I., New Delhi.

Singh, K.N., Bains, S.S. and Agrawal, K.N., (1969), *Fertilizer use under saline-cum-high water table conditions,* Indian J. Argon, 14, pp. 26–30.

Tyagi, N.K., and Tyagi, K.C., (1990), *A Framework for Planning Irrigation System Improvement for Salinity Control,* IWM Report 1, Central Soil Saslinity Research Institute, Karnal, p. 110.

Tyagi, N.K., Kamra, R.K, Prihar, P.S. and Singh, T.S. eds., (1993), *Sustainable Irrigation in Saline Environment,* Central Soil Salinity Research Institute, Karnal, p. 229.

Tyagi, N.K., and Minhas, P.S. eds., (1998), *Agricultural Salinity Management in India,* Central Soil Salinity Research Institute, Karnal, p. 519.

Unted States Salinity Laboratory Staff—Editor Richards, L.A.—(1954), *Diagnosis and improvement of saline and alkali soils.* U.S.D.A., Handbook No. 60.

United States Department of Agriculture, Soil Conservation Service (1964), National Engineering Handbook, Section 151, Chapter 1, *Soil-Plant-Water Relationship,* U.S.D.A., Washington, D.C.

United Nations Development Programme and Food and Agriculture Organization of the United Nations (1971), *Land and Water Management in the Chambal Irrigated Area, Rajasthan, India,* AGL.SF/IND 60, Tech. Rep. 4, p. 97.

United Nations Food and Agriulture Organization (UNFAO, 2003): *Land and Plant Nutrition Service Global Network in Integrated Soil Mangement for Sustainable Use of Salt-Affected Soils. FAO Online htt/www.org/ag/ag/agl/agll/spush/topicl.litm* January 12, 2005.

Wagdi, M.A.H.A., (1968), *Some factors influencing the quality of irrigation waters,* Ph.D. Thesis, I.A.R.I., New Delhi.

Wilcox, L.V., (1960), *Boron injury to crops,* US Dept. of Agri. Inform. Bull., 21.

Chapter

15

Economic Evaluation of Irrigation Projects and Water Pricing Policy

LEARNING OBJECTIVES

Basic principles and procedures for the economic and financial analyses of water resources projects. Tangible and intangible benefits and costs of projects. Procedures for the application of discounting methodologies in the economic analyses of water resources projects. Mutually exclusive projects. Criteria and procedures for pricing of irrigation water from canals, wells and lift irrigation schemes. Water pricing scenario in different states of India. Mobilization of financial resources for water resources projects.

Irrigation projects require large investments. Casual economic analyses and unjustified benefit claims are often made for particular irrigation projects for a variety of political and social reasons. The rapid increase in irrigation development, especially in the developing countries and increasing project costs have focussed the attention of governments, financing institutions and users on the need to establish the economic feasibility of project proposals. Each proposed water resource project requires rigorous technical and economic appraisals.

Investment on irrigation projects is economically justifiable only if it generates an economic surplus, including both direct and indirect benefits and costs. It is, therefore, important that the project is analysed from the economic angle and its profitability assessed properly. The returns from the project should have a clearly higher value than the cost of all the inputs. The economic appraisal of an irrigation project must take full account of all attributable costs and benefits. Appropriate design procedures and management practices will be required if the project is to be economically viable.

Water resources development projects, from the economic angle, may be considered as an investment activity where capital resources are spent to create a producing asset from which the benefits are realised over an extended period of time. They include surface and groundwater irrigation and water supply projects, drainage projects and projects on land reclamation, salinity prevention and flood control. They will have relatively well defined time sequences of investment and production activities. Hence, economic analysis should take into account the costs and benefits of the projects covering their entire service periods.

Economic and financial analyses provide a framework within which all aspects of a proposed project can be evaluated in a coordinated and systematic manner. Careful analysis of the economics of projects will identify unrealistic or questionable assumption and indicate ways in which a project can be modified to improve its income-generating capacity.

ECONOMIC AND FINANCIAL FEASIBILITIES OF PROJECTS

Project appraisals include *economic* analysis and *financial* analysis. The economic feasibility evaluation assesses the economic viability of a proposed project for the society as a whole, at the economic values. It helps in selecting a particular one from amongst the possible alternatives. The financial feasibility evaluation assesses the financial conditions that will be encountered in developing and operating a project at the market price. It is prepared to assist the investor in determining the characteristics of cash flow under different types of financing. Financial analysis evaluates possible financing alternatives with respect to interest rates, repayment schedules and the maturity period of the loan.

Basic Concepts and Definitions in Economic Analysis

Economic Rate of Return. The internal rate of return calculated using economic values.

Economic Value. It is the amount by which the return or output from a project changes the national income.

Financial Rate of Return. The internal rate of return calculated using market values.

National Income. National income is the sum of the financial goods or services produced by an economy in a given period of time. In accounting practice, the income shows the revenues, less the expenses. Income may be calculated including the tax or excluding it. Income shown after deducting the expenses, including the tax is termed as net income.

G.D.P. Gross domestic product, national income.

Funds Flow Analysis. It is the analysis of cash inflow and outflow in a project.

Gross Benefit. The incremental value of output from a project shows the gross benefit.

Net Benefit. In the analysis of projects, the amount remaining after all outflows are deducted from all inflows is the net benefit. In the initial years of analysis of irrigation projects, the net benefit is negative.

Incremental Net Benefit. It is the increase in net benefit with the project, as opposed to the situation without the project. It shows the incremental cash flow.

Indirect Benefit. The secondary benefits resulting from a project is called its indirect benefit.

Opportunity Cost. The opportunity cost is the benefit foregone by investing capital in a particular project, instead of its next best alternative use. It usually forms the basis for selecting the *discount rate* for calculating the benefit–cost ratio and the *net present worth* of an investment.

Discount Rate. Discount rate is the interest rate used to determine the present worth of a future value in discounted cash flow analysis. The discount rate is usually reckoned at the cost of the capital employed.

Efficiency Price. It is the economic value used in economic analysis. Efficiency price reflects the opportunity cost or value of a good or service used or generated by a project. The price used may be the market price adjusted for market distortions or *shadow price*. Efficiency price is used in economic analysis when the objective is to maximise the national income. Hence, the procedure is sometimes termed as efficiency analysis.

Imputed Price. It is a price or economic value obtained by some computation rather than by using an observed market price. It is better to avoid using an imputed price in project analysis, to the extent possible.

Inflation. Inflation is an increase in the general price level of an economy. Inflation occurs when the quantity of money in circulation rises relative to the quantity of goods and services which are available. In project analysis, the usual analytical approach is to work on constant prices rather than current prices and to assume that inflation will affect prices of all costs and benefits equally. It is assumed that the prices will affect both the costs and benefits to the same extent, and hence the general relationship between costs of products and inputs will remain the same.

Shadow Price. Shadow price is the value used in economic analysis for a cost or benefit in a project when it is considered that the market price may not provide a realistic estimate of the economic value. Shadow price usually means the *accounting price* under such a situation. In the strict sense, shadow price is derived using mathematical models, such as linear programming. However, the accounting price simply indicates that the price used is not the market price.

Current Price. It is the price or the value which includes the effects of change in the price of a commodity or service due to inflation. It could be a past value or price, as recorded, or a value or price which is expected to occur in normal conditions.

Current Ratio. It is the ratio obtained by dividing the current assets of a project by the current liabilities. Current ratio is used to judge the liquidity or credit worthiness of a project.

Farm Gate Price. The farm gate price is the price a farmer receives for his product or pays for the inputs at the boundary of the farm. It is the price without any transport cost to the market or any marketing service.

Cut-off Rate. It is the rate below which a project cannot be accepted as economically viable. The opportunity cost of the capital is usually taken as the cut-off rate. It is the minimum acceptable *internal rate of return.*

Transfer Payment. It is a payment made without receiving any goods or services in return. In irrigation projects, the most common transfer payments are the taxes of different types and the subsidies received for growing specialised crops and adopting water-saving irrigation methods like sprinkler and drip irrigation systems.

Economic Life. The period during which a fixed asset is capable of yielding services to its owner.

Grace Period. In credit transactions during the grace period allowed, the holder need not repay principal, and sometimes the interest.

Grants. A payment made to an individual or a cooperative enterprise by a government or other agency without expectation of goods or services in return. The purpose is to encourage a specific activity. A grant is a transfer payment in economic appraisal.

Work Day. The time devoted to an activity by one person during one day. In developing countries, the duration in agricultural operations is usually taken as 8 hours per day.

Equity. It is the ownership right in an enterprise. Equity capital is the residual amount left after deducting the total liabilities from the total assets.

Economic and Financial Analyses of Projects

Economic Analysis

Economic analysis is carried out to determine whether a project is capable of contributing to the development of the economy as a whole. It also reveals whether the returns from the project are likely to be useful enough to justify the investment on it. It is done using economic values. Economic value is the amount by which the return of a project or the costs involved in it changes the national income. The value may be the market price adjusted against market distortions and transfer payments, or an estimate of the value in use, as measured by the *willingness to pay*. It helps to identify those projects which make major contributions to overall economic growth. The project which maximizes the returns to the capital investment is given the highest rank in a comparative economic evaluation of alternative project proposals. It will determine the amount of the income stream generated in excess of all attributable costs. However, economic analysis does not specify as to who receives the benefits. It is neutral to capital ownership as well, whether the project is from public revenues, credit agencies or private sources.

Financial Analysis

Financial analysis is done using market prices. It takes into account the income distribution and capital ownership amongst the various participants of the project, the beneficiaries and the investors. In an irrigation project, the participants are the government, mainly through its irrigation department and the farmers. Other participants include credit agencies, agriculture department, agricultural university, voluntary organizations, input agencies like seed, fertilizer and plant protection organizations and marketing agencies.

Financial analysis may examine whether the project entity, say, the irrigation department responsible for the implementation of the project may have receipts large enough to recover the total cost it incurs. Even otherwise, the project may be suitable for implementation if the economic analysis shows that the total returns to the society exceeds the social costs.

It is generally expected that an irrigation project should generate additional incremental income from the returns from the farming enterprise in the command area, so that the project is economically viable. However, if the governmental policy is not to assess the farmers on economic terms but at a subsidised rate, such a proposal will be inoperative, unless the subsidy cost is quantified and added to the cost of the project. In an irrigation project, there could be additional income from hydro-electric power generation, aquaculture, recreation, flood protection downstream of project headworks, and diverting a part of the water supplies for drinking water and industrial use.

Financial analysis is important when the incentive structure associated with an irrigation project is considered. It will not be in the interest of the farmers, if a project which is attractive from the standpoint of the society, is not attractive to an individual farmer in terms of his ability to earn a living from his participation. Timing of returns which the financial analysis will reveal is important for projects like a major irrigation scheme which has a long gestation period to commence generation of income.

Comparative Advantages of Economic and Financial Analyses

In economic analysis, the prices of certain commodities like those of food grain crops where the procurement prices are fixed by the Government, based on various considerations, may have to be changed to reflect their true social or economic value. The adjusted prices are called '*shadow*' or '*accounting*' prices. For an elaborate economic analysis, the shadow prices are derived by adjusting the market prices for transfer payments. In financial analysis, however, no such change in prices are made and the market prices, including taxes and subsidies, are used. In economic analysis, the society as a whole is considered. It is assumed that all the capital is supplied by someone somewhere in the society. The analytical technique takes care of the return of capital in determining the worth of a project. In economic analysis, taxes and subsidies are treated as transfer payments. In agricultural projects, the most common direct transfer payments are taxes and subsidies. In economic analysis, income taxes, sales taxes, customs duties and other taxes, are transfer payments within the society, not payment for resources used in production. Hence, they are not deducted from the benefit stream when deriving the cash flow as the basis on which to compute the productivity of capital. To the economy of a nation, taxes are a benefit available to the society for the use of its capital invested in the project. In financial analysis, however, taxes are a cost which the individual entity must pay before arriving at the amount available for recovering its capital and compensating it for the use of its capital, *i.e.*, its cash flow. Hence, taxes are a cost, just like any other expenditure. Similarly, subsidies are benefits to the projects in financial analysis.

In the economic analysis of water resources projects, payment of interest and repayment of the principal are treated as transfer payments. This is because the loan terms only divide the claims to goods and services between the borrowers and lenders and do not change the returns to the capital. They also do not increase or reduce the national income. However, the incremental benefit generated by the project includes the taxes that the project has to bear during its productive period. However, a subsidy is a cost to the society, since it is an expenditure incurred by the socicty to operate a project.

In financial analysis, it is necessary to account for the outside capital borrowed in undertaking a project. When borrowed capital is received, it is normally entered into the receipts streams as a *benefit*. When a payment of interest or repayment of principal is made to an outside supplier of the capital, it is deducted from the gross return as a *cost*.

The same *discounted cash flow methodology* (described later) is used in both economic and financial analyses. The term economic analysis usually considers the financial aspects as well, even though some major projects may require analysis from both the angles.

Tangible and Intangible Project Costs and Benefits

Benefit-cost investigations are needed in order to ensure that investments in irrigation projects are desirable and economically sound and that they will be utilized to the utmost advantage to the society. Therefore, it is important that the project is analysed from the economic angle and its profitability assessed properly. The incremental return arising from the project should clearly have a higher value than the cost of all the inputs. Economic evaluation includes the determination of *total* benefits and *total* costs.

In project appraisals, *benefit* means any increase in production or any gain in goods, materials and services which emanates from it. *Cost* means the expenditure on the consumption of goods, materials and services for its production and maintenance. Benefits and costs can be grouped into two general categories, *tangible* and *intangible*.

Tangible Benefits and Costs

In project analysis, the cost or benefit that is capable of being appraised at an actual price or economic value is considered tangible. They are real values which reflect physical quantities or goods or amount of services. A financial price or economic value, in money terms, refers to the monetary prices of goods and services. Their values vary with market trends.

Intangible Benefits and Costs

Intangible benefits and costs are those which, although are having values, cannot be realistically assessed in actual or approximate money terms. Intangible benefits include health cover, education, employment generation, recreation, possibility of intensive cropping, larger crop produce, sustainable and reliable crop production, better income distribution amongst the farmers of the project area, national integration, environmental protection (example, disease vector control by proper drainage), better health standards of people and the like. Intangible benefits are sometimes valued as being equal to the estimated cost of the alternative method of providing the same benefit. This method of alternative cost evaluation could be used only when the alternative would be actually undertaken in the absence of the project. The tangible cost of avoiding an intangible cost may be included in the cost of the project. However, the nature of intangible costs and benefits is such that the real value of the cost to the investor on the benefit provided to the project participants cannot be determined. When substantial intangible costs or benefits are encountered in a project, the analyst should identify them and quantify them to the extent possible. A project in which a major amount of the benefit is intangible should be given due weightage, while finally selecting the project.

Tangible benefits may be either *primary* or *direct benefits* and *secondary* or *indirect benefits*. The primary benefits represent the value of the immediate goods and services resulting from a project. The primary benefit of an irrigation project is the increase in agricultural production from a unit area of its command. The secondary or indirect benefits represent the added value over and above the immediate products and services which the project induces. Indirect benefits of irrigation projects include the increase in the gross-cropped area by producing more than one crop per year on the same land and raising of high-value crops. In many cases, projects intended primarily for irrigation also serve as sources for domestic water supply. These benefits, though indirect, are to be quantified and included in the benefit-cost analysis.

Some costs in irrigation projects are very difficult to be quantified and be given a money value. Siltation, waterlogging, salinization and soil erosion are examples. These costs should not be ignored if they are likely to be substantial. They should be treated in the project analysis as intangible costs. Similarly, an adverse environmental impact which may be caused by an irrigation project is an intangible cost. On the other hand, an increase in the standard of living of the people and attainment of a better health standard resulting from an irrigation project, are intangible benefits. When intangible costs or benefits are encountered in project analysis, the analyst should identify and quantify them to the extent possible. This will help in ranking alternative proposals, when the money available for funding is scarce.

Costs can also be classified as *direct* and *indirect*. *Direct costs* are the values of the material and services directly used for the project. *Indirect costs* are those values which are incurred while undertaking the project, such as approach roads to project sites and other communication facilities, social costs incurred on the project staff (establishment of schools/hospitals), and establishment of agro-service centres.

When economic appraisals are made for irrigation schemes, all expected benefits and costs—primary and secondary, tangible and intangible—should be carefully ascertained and examined. Direct tangible benefits and costs can be quantified with a reasonable degree of accuracy. Tangible secondary benefits can be quantified, though precise estimates are difficult.

Analysis of benefits and costs of a project include the following major steps:

(*i*) Determination of the quantity and value of the various kinds of costs and benefits.

(*ii*) Conversion of costs and benefits taking place at different periods to a common time scale.

(*iii*) Comparison of total costs and total benefits.

The quality of the economic analysis of an irrigation project depends on the quality of data used on benefits and costs. Here, the *'GIGO'* principle–"garbage in and garbage out" works with vengeance. Unrealistic assumptions about yields, acceptance and participation by farmers, future market and price projections can render economic analysis a futile exercise. Hence, it is extremely important that realistic assumptions are made on the benefits and costs of the project

Irrigation Project Costs

Irrigation project costs include all the expenditure made to establish, maintain and operate a project. Costs are estimated on an annual basis. The annual cost of a project includes both *fixed* and *variable costs.*

Fixed Costs

Fixed costs, also referred as investment or initial costs, include the following, as applicable:

1. Costs of obtaining water right and permits.
2. Planning and design costs.
3. Land purchase and rehabilitation of the population in the areas affected by the water resources project.
4. Cost of storage reservoirs, head regulator and canal water distribution system, including associated structures and controls.
5. Command area development–surveying, land development operations and on-farm water conveyance and control.
6. Drainage system—main drains, link drains, and on-farm drainage systems–surface and sub-surface.
7. Cost of wells, pumps, electric motors/engines and pumping plant accessories, and their installation.
8. Pump house.
9. Electric power connection, metering and recording equipment.
10. Automation equipment/remote control, if used.
11. Inspection and approach roads.
12. Equipment for water application, sprinkler/drip irrigation equipment, if used.

The above costs are incurred at the initial stages of the project, while others are paid annually. Annual fixed costs include the interest on the total investment on the project.

Variable Costs

Variable costs are of a recurring nature computed on an annual basis. They are O and M (Operation and Maintenance) costs as well as levies and charges on insurance and miscellaneous operating costs of recurring nature. The variable annual costs may be enumerated as follows:

1. Maintenance of structures and water distribution network.
2. Costs of fuel, namely, diesel or other fuels and electricity.
3. Lubricants, minor repairs, and painting.
4. Layout of field for surface irrigation – renewal of borders, ridges, and field channels.
5. Maintenance of drainage system – desilting, weed control.
6. Operating manpower (Manpower costs include salaries, social benefits, housing, insurance, medical treatment, transportation and similar items).

A simple procedure, commonly used for preliminary cost estimates is to calculate the interest on the average value of the installation at the prevailing interest rate:

$$\text{Annual interest cost} = \frac{(\text{Value of installation} - \text{Salvage value}) \times \text{Interest rate}}{2}$$

Depreciation

Depreciation is the anticipated reduction in the value of an asset due to physical use of the equipment/structure or obsolescence. In the conventional analysis, the annual depreciation is computed as follows:

$$\text{Annual depreciation} = \frac{\text{Original cost} - \text{Salvage value}}{\text{Useful life in years}}$$

Depreciation refers to the process of allocating a portion of the original cost of a fixed asset to each accounting period so that the value is gradually written off during the course of the estimated useful life of the asset. Allowance may be made for the asset's estimated resale value, if any, at the end of the useful life of the enterprise.

In discounted cash flow analysis, depreciation is not treated as a cost. The cost of an asset is shown in the year it is incurred and the benefits are shown in the year they are obtained. Since this is done over the entire life of the project, it is not necessary to show the value of the asset apportioned in any given year as depreciation. That would amount to double counting.

Service period of wells and pumps projects. In the case of ground water and lift irrigation projects, when the expected service period of wells and pumps is specified in hours of operation, same in years is calculated by dividing the total hours of operation by the average annual hours of operation.

Variable costs of irrigation projects. If the electrical connection charges are paid in lump sum, the annualised cost may be estimated, assuming an expected service life of about 25 years. If the charges are to be paid annually, the same are to be added to the operation and maintenance cost, to arrive at the annual variable cost.

Variable costs of surface irrigation projects include the costs on regulation of the conveyance system as well as maintenance and repairs. In case of wells and pumps, they include the cost of power/fuel (electricity/diesel), cost of lubricants, labour charges for operating the pumping units and the expenditure on repairs and maintenance of the equipment and accessories.

The cost of power is often the most important component of variable costs in the case of pumping systems. The usual practice is to calculate the requirement of energy per hour of operation from the known discharge rate of the pumping plant, total operating head and its overall efficiency. The requirement of power is expressed in kilowatt-hours per hour for electricity and litres of diesel per hour of operation of engines.

The energy consumption of an electric motor is computed as follows:

$$\text{Energy consumption} = \frac{\text{Brake horse power}}{\text{Motor efficiency}} \times 0.746$$

Efficiencies of electric motors may be obtained from the performance data supplied by the manufacturers. Motor efficiencies usually vary from 75 to 90 per cent.

The demand of electrical power for hourly operation is multiplied by the annual hours of operation to arrive at the total annual energy consumption. The annual power cost is determined by multiplying the annual energy demand by the prevailing cost per unit of electrical energy.

In case of engines, the cost of fuel is computed as follows:

$$\text{Fuel cost} = \text{BHP} \times \text{Specific fuel consumption} \times \text{Cost of fuel per litre}$$

A realistic estimate of the rate of fuel consumption for a given engine can be made if the manufacturer's fuel-consumption curve for the engine is available. Fuel consumption of diesel engines commonly used in irrigation pumping vary from 0.2 litres to 0.29 litres/bhp-hr. An average value of 0.23 litre/bhp-hr can be assumed in the absence of better data.

The consumption of lubricating oil is usually assumed to be 4.5 litres per 1000 bhp-hr. Many manufacturers provide values for the consumption of lubricants of their products. From the cost of lubricants per hour of operation, the annual cost of lubricants is computed.

The repair and maintenance costs of the component parts of an irrigation system may be assumed as per the norms given in Table 15.1, or an average value may be assumed based on field evaluation studies or local experience.

Table 15.1. Guidelines for estimation of service life and annual maintenance and repair costs of irrigation structures and equipment

Component	*Expected duration (operation hours)*	*Expected economic life (years)*	*Annual maintenance and repair charge as% of initial investment*
Dams	–	50–150	–
Masonry wells	–	50–170	0.2–0.5
Ponds and tanks	–	20–50	–
Tubewell screen and casing (mild steel)	–	20–50	0.5–1.5
Pump house and foundation	–	40–50	0.5–1.5
Bowls of turbine pumps (about 50% of the cost of the pump unit)	16,000–20,000	8–10	5–7
Columns of turbine pump	32,000–40,000	16–20	3–5
Centrifugal pump	32,000–50,000	15–20	3–5
Gear head	30,000–36,000	14–20	5–7
V–belt	6,000	3	5–7
Flat belt (leather)	20,000	10	5–7
Electric motor	50,000–70,000	25–35	1.5–2.5
Diesel engine	28,000	15	5–8
Petrol engine	14,000–18,000	8–12	5–10
Galvanised iron pipes	–	25–50	1–2
Concrete pipes	–	20–60	2–4
Portable aluminium pipes	–	10–15	2–4
Plastic pipes (underground)	–	20–40	1.5–2.5
Asbestos cement pipes	–	20–40	1–1.5
Hydrants	–	20–40	1–2
Water meters	–	20–40	2–4
Sprinkler nozzles	–	5–10	2–3
Fittings of portable pipes	–	15	2–3

Estimating Service Period and Annual Maintenance and Repair Costs of Irrigation System Components. The replacement of the components of an irrigation system is based on expected life of the components. Table 15.1 provides the expected life of various irrigation system components, based on field observations and compiled from various sources. Variability in expected life can occur for many of these components due to different physical conditions like the quality of groundwater, the level of repair, operation and maintenance practised and the length of time, the system is used each year. In case of poor quality groundwater and abnormal working conditions like excessive silt-load in water, the expected service life will be less. Sound engineering judgement should be exercised in estimating the economic life during the analysis. Table 15.1 also provides the commonly used values of annual maintenance and repair costs of different components of irrigation systems.

Selecting Proper Values of Benefits. The market price is usually considered as the best approximation of the true value of a good or service when the benefits are from the agricultural sector. The point of first sale at which it is generally desirable to value agricultural produce is at the *farm gate price.*

Predicting Future Price. Project analysis requires judging future returns from present investments. This is generally a matter of judgement. The analyst must take into consideration all the factors that could be obtained and arrive at a conclusion. The best initial estimate about future prices is that they will retain the present relationships or the average relationship which have been existing over the past few years. However, it must be considered whether these average relationships will change in the future with technological innovations, changes in international trade practices and the consequent shifts in demand or supply of goods.

Discounting Rate, Present Worth, Cost Escalation, Capital Recovery Factor and Amortization

The process of determining the present worth of a future value is called *discounting*. The technique of discounting enables us to determine whether to accept for implementation projects which have variously shaped time streams, namely, when patterns of costs and benefits fall differently during the economic life span of a project. The discount rate is the interest rate used to determine the *present worth* of a future value by discounting. The present worth is determined by multiplying the future amount by the expression:

$$\frac{1}{(1+i)^n} \qquad \ldots(15.1)$$

in which i = the discount rate indicating the opportunity cost of the capital. Discount rate is also called *discount factor* or *present worth factor*. (It is the reciprocal of the *compounding factor* for 1). Compounding factor shows what an initial amount of 1 becomes when growing at a compound interest rate. It is also called the *compound interest factor*. The mathematical expression for compounding factor is $(1 + i)^n$, in which i = the interest rate and n = number of years. Numerically, the discount factor shows how much Re. 1 at a future date is worth today. The discount rates chosen for financial and economic analyses differ. For financial analysis, the discount rate is the rate at which the investor is able to borrow or avail the capital. For economic analysis, the most commonly used discount rate is the opportunity cost of the capital. It is usually expressed in the form of an interest rate. *Cut-off rate* is the rate below which a project is considered unacceptable. It is often taken to be the *opportunity cost of the capital*.

A second discount rate in economic analysis is the *borrowing rate* which a country has to pay to finance a project. The usual practice in discounting is that both costs and benefits are discounted, beginning with the first or second year of the project.

The *present worth* of a future value at the end of the nth period at an interest rate of i is computed using the following formula:

$$PV = F\left(\frac{1}{(1+i)^n}\right) \qquad \ldots(15.2)$$

in which,

PV = Present value of the future income stream

F = Future value of the income stream

Example 15.1. *The opportunity cost of capital is 12 per cent per annum. Determine the discount rate for the first, second, third and fourth years.*

Solution : Discount factor = $\dfrac{1}{(1+i)^n}$

The discount rate for the first year

$$= \frac{1}{(1+0.12)^1} = 0.8929$$

Second year $= \dfrac{1}{(1+0.12)^2} = 0.7972$

Third year $= \dfrac{1}{(1+0.12)^3} = 0.7118$

Fourth year $= \dfrac{1}{(1+0.12)^4} = 0.6355$

Annuity

An annuity is a series of periodic payments made at the end of a compounding period. Examples of annuities are insurance payments and payments made to a person from a retirement fund. The future value of an annuity can be determined using the following formula:

$$S_n = V\left[\frac{(1+i)^n - 1}{i}\right] \quad ...(15.3)$$

in which,

S_n = Future value of the annuity after n periods,

V = Value of an annuity,

n = number of annuity payments, and

i = interest rate per period

Example 15.2. *A project proposes to build a reserve fund by depositing Rs. 1000 million at the end of each year, by investing in suspense accounts available for four years. If the investments can earn at the rate of 6 per cent per year, compounded annually, what would be the sum of annuities deposited at the time of the fourth payment?*

Solution: $S_n = V\left[\frac{(1+i)^n - 1}{i}\right]$

$$= 1000\left[\frac{(1+0.06)^3 - 1}{0.06}\right]$$

$$= 1000 \times 4.37462$$

$$= \text{Rs. } 4374.60 \text{ millions.}$$

(*Note:* The first 3 deposits only earn interest. The fourth deposit will earn interest at the end of the fourth year only.)

Present worth of an annuity factor. Present worth of an annuity factor is used to calculate the present worth of a sequence of level payments to be made at the end of n periods of time at an interest rate of i.

The present worth of an annuity is the amount of money today which is equivalent to a series of equal payments in the future. The present value of an annuity is estimated by the following formula:

$$PV(A) = \left\{\frac{V[(1+i)^n - 1]}{i(1+i)^n}\right\} \quad ...(15.4)$$

in which,

$PV(A)$ = present worth of an annuity,

V = value of annuity,

i = discount rate, and

n = numbers of annuity payments

Example 15.3. *A 3-year annuity for Rs. 100.00 has payments received at the end of each year. Determine the present value of the annuity when discounted at 5 per cent.*

Solution: $PV(A) = \frac{[100\{(1+0.05)^3 - 1\}]}{0.05\,(1+0.05)^3}$

$$= \frac{(100)\,(11576 - 1)}{(0.05)\,(1.1576)}$$

$$= \text{Rs. } 272.32$$

The present worth of a sequence of level payments to be made at the end of n periods at the interest rate of i is computed using the following formula:

$$P = A\,\frac{(1+i)^n - 1}{i(1+i)^n}$$

in which,

P = Present worth of a sequence of level payments,

A = Amount of each payment.

It may be seen that the capital recovery factor (CRF) is the reciprocal of the present worth of an annuity factor.

Compounding Factor

The compounding factor is used to calculate the future worth of a present amount at the end of the nth period at an interest rate of i. The factor for, say, Re. $(1 + i)^n$ is determined by the following relationship:

$$\text{Compounding factor} = (1+i)^n \quad ...(15.5)$$

Example 15.4. *If the interest rate (i) is 10 per cent for Re. 1, compute the compounding factor for the first, second and subsequent years.*

Solution : Compounding factor for the first year

$$= \left(1+\frac{10}{100}\right)^1 = 1.10$$

Second year $= \left(1+\frac{10}{100}\right)^2 = 1.21$

Third year $= \left(1+\frac{10}{100}\right)^3 = 1.33$

Fourth year $= \left(1+\frac{10}{100}\right)^4 = 1.46$

The compounding factor is used to calculate the future worth of a present amount at the end of a particular period, using the following relationship:

$$F_V = P\,(1+i)^n \quad ...(15.6)$$

in which,

F_V = future worth,

P = present amount, and

n = period

Example 15.5. *If an investible amount worth Rs. 25,000, has an opportunity of earning at the rate of 10 per cent per annum, compute the future value at the end of the fifth year.*

Solution: $F_V = P\,(1+i)^n$

$$= 25{,}000\left(1+\frac{10}{100}\right)^5$$

$$= \text{Rs. } 40{,}263.00.$$

Cost escalation. The fixed and variable costs of an irrigation project often increase during the project period. The projected rate of escalation during the period of analysis can be incorporated in the present worth and annual cost values. The escalated present worth at the end of n years of the economic life of a component of a project can be computed using the following relationship:

$$P_w f(e) = \frac{(1+e)^n}{(1+i)^n} \quad \text{...(15.7)}$$

in which,

$P_w f(e)$ = Escalated present worth factor,

e = Annual rate of escalation,

i = Interest rate, and

n = Expected life of the component, years.

The escalated present worth factor can be used to evaluate the effect of escalation on cost terms, such as replacement of the pump, motor and accessories. The present worth value, incorporating the effect of cost escalation, can be obtained using the following relationship:

$$PW(e) = PW \times P_w f(e) \quad \text{...(15.8)}$$

in which,

$PW(e)$ = Present worth cost, incorporating the effect of escalation cost in Rs.

The annual cost value can be adjusted for the cost of escalation by multiplying its value with the annual cost escalation factor (Pearson, 1974).

$$Acf(e) = \frac{(1+e)^n - (1+i)^n}{(1+e)-(1+i)} \times \frac{i}{(1+i)^n - 1} \quad \text{...(15.9)}$$

when $e \neq i$

and $A\,c(e) = A\,cf(e) \times AC$ ...(15.10)

in which,

$AC(e)$ = Annual cost, incorporating the effect of cost escalation, Rs.

$Acf(e)$ = Annual cost escalation factor, fraction

Capital Recovery Factor

Capital recovery factor (*CRF*) is used to calculate the amount of each level payment to be made at the end of each n period, to recover the present amount at the end of the period, at a predetermined interest rate of i. The factor is computed as follows:

$$CRF = \frac{i(1+i)^n}{(1+i)^n - 1} \quad \text{...(15.11)}$$

The amount of level (even) payments to be made is computed by the following formula:

$$A = P\frac{i(1+i)^n}{(1+i)^n - 1} \quad \text{...(15.12)}$$

in which,

A = Amount of each level payment to be made at the end of each of n periods

P = Present amount

The annual cost of capital invested in an irrigation system can be determined from the present worth value of the investments, plus the interest during the period of analysis. The annual cost is usually determined by calculating a uniform series of annual values for depreciation and interest over the analysis period which is equivalent to the single present worth value. The value of this uniform series of annual costs is determined by the application of the *capital recovery factor (CRF).*

Amortization

The term amortization literally means "killing by degrees". Hence, amortization of loans refers to the repayment of loan (both principal and interest) in a series of payments usually in equal instalments. The instalment is determined by the amortization factor computed as follows:

$$\text{Amortization factor} = \frac{i}{1-(1+i)^{-n}} \quad \text{... (15.13)}$$

in which,

i = Interest rate and

n = Compounding frequency

Example 15.6. *A loan of Rs. 50,000 carrying an annual interest rate of 12 per cent is to be amortized by equal instalments over the next 5 years. Determine the value of the annual instalments due.*

Solution: $D = \frac{i}{1-(1+i)^n}$

$= 50{,}000 = \frac{0.12}{1-(112)^5}$

$= 50{,}000 \times 0.2774$

$=$ Rs. 13,870.00.

For accounting purposes, it is advisable to prepare a repayment schedule indicating the distribution of instalments between the outstanding principal and interest due, as shown in Table 15.2.

The method of repaying an equal amount at regular intervals is termed as *level* (even) payment as shown in Table 15.2. This is possible because of progressively diminishing interest payments on outstanding balance of the principal, so that the total payments remain constant. Level payments are suitable when the income flow is regular over the entire project life.

Another repayment plan is the *decreasing payment plan*. In this method, the financial amounts are amortised equally so that the interest due declines

Table 15.2. Amortization plan of repayment schedule on level payment basis

Period years	*Outstanding principal at the beginning of the period (Rs.)*	*Interest due at the end of the period (Rs.)*	*Principal repaid at the end of the period (Rs.)*	*Instalment due at the end of the period (Rs.)*
1	50,000	6000	7870	13,870
2	42,130	5055	8815	13,870
3	33,315	3997	9873	13,870
4	23,442	2813	11,057	13,870
5	12,385	1486	12,385	13,870
Total	—	19,350	50,000	69,350

Table 15.3. Amortization plan of repayment schedule of principal amount on progressively diminishing basis

Period years	*Outstanding principal at the beginning of the period (Rs.)*	*Interest at the end of the period (Rs.)*	*Principal repaid at the end of the period (Rs.)*	*Instalment due at the end of the period (Rs.)*
1	50,000	6,000	10,000	16,000
2	40,000	4,800	10,000	14,800
3	30,000	3,600	10,000	13,600
4	20,000	2,400	10,000	12,400
5	10,000	1,200	10,000	11,200
Total	—	18,000	50,000	68,000

progressively. The total instalment due shows a progressively declining pattern as indicated in Table 15.3. The decreasing payment plan is more suitable for projects generating higher income flow in the earlier years.

APPLICATION OF DISCOUNTING METHODOLOGY IN ECONOMIC ANALYSIS OF IRRIGATION PROJECTS

In project analysis, the amount remaining after all *outflows* (costs) are deducted from all *inflows* (benefits) is the *net benefit*. It is also called the *cash flow*. The *incremental net benefit* is the increase in net benefit with the project, in contrast to the case without the project. It is the incremental cash flow, which may be positive or negative. In the early years of a project, the net incremental benefit is usually negative. The method of analysis for conversion of costs and benefits to a common time basis is called *discounted cash flow technique*. The prime aspect of this technique is to compare the costs and returns that occur at different time periods to a common time scale. The costs are entered in the years they are incurred and the benefits are entered in the years they are realized.

When benefit/cost appraisals are made, all expected benefits and costs, whether primary or secondary, and tangible or intangible, should be carefully ascertained and examined. However, only the values of primary and tangible benefits and costs can be determined and calculated with a reasonable degree of accuracy, since no specific value can usually be assigned to the intangibles. Assigning monetary values to secondary benefits and costs are susceptible to wide inaccuracies and subjectiveness.

The project worthiness of investments is assessed using the following measures:

(*i*) Net present worth (often abbreviated as NPW)

(*ii*) Internal rate of return often (often abbreviated as IRR)

(*iii*) Benefit-cost ratio (often abbreviated as B/C ratio) and

(*iv*) Net benefit-investment (often abbreviated N/K) ratio.

The arithmetics of these discounted measures and the way to interpret the measures and their limitations are the same whether they are used for financial analysis or for economic analysis. The difference is only whether the techniques are applied at financial prices or at economic values.

Net Present Worth

The most direct discounted cash flow measure of project worth is the net present worth (NPW). It is also sometimes referred to as net present value (NPV). It is simply the present worth of the incremental net benefit or incremental *cash flow stream*. The cost and returns on year to year basis are estimated first. Then to estimate the worth of the project, the costs are subtracted from the returns, again on year to year basis, to arrive at the net return stream (cash flow). It is then discounted to the present value. The discounted net cash flow has to be added over the life period of the project to get the net present worth.

Expressed mathematically, net present worth can be estimated as follows:

$$\text{Net present worth (NPW)} = \sum_{t=1}^{n} \frac{B_t - C_t}{(1+i)^t} \quad \ldots(15.14)$$

in which,

B_t = Benefit accrued in each year of project period used in analysis, in Rs.

Ct = Cost incurred in each year of project period used in analysis, in Rs.

t = Time period of project life, reckoned in years, i.e., 1, 2, ... n

n = Number of years of anticipated project life

i = Discount rate selected on the basis of cost of capital.

Net present worth may be interpreted as the present worth of the income stream generated by an investment on a project. In financial analysis, it is the present worth of the income stream accruing to the agency or to the individual for whom the analysis is done. In economic analysis, it is the present worth of the incremental national income generated by the investment. The guiding principle regarding the worthiness of a project is that the net present worth shall be at positive levels at the cost of capital. If the NPW is a positive sum, it implies that the investor will get a return in excess of the value of the capital invested at today's prices. If the NPW is zero, then the return to the investor is just equal to the capital invested to recover his capital at today's prices. Projects with positive NPW, when properly discounted, can be considered as financially/ economically feasible and can be taken up for implementation.

Example 15.7. *Determine the net present worth (NPW) of the investment on a shallow tubewell with the cash flow pattern given below, at 12 per cent cost of capital and a project life of 12 years.*

(*Note:* The values of the discount factor may be computed using Equation 15.1.)

Solution:

Year	*Cash outflow (Rs.)*	*Cash inflow (Rs.)*	*Cash flow (Rs.)*	*Discount factor at 12%*	*Discount cash flow (Rs.)*
1	2	3	4	5	6 = 4 × 5
1	85,000	10,000	– 75,000	0.8929	– 66,968
2	1500	12,000	10,500	0.7972	8371
3	1800	14,500	12,700	0.7118	9040
4	1900	16,800	14,900	0.6355	9469
5	2100	22,100	20,000	0.5674	11,348
6	2200	24,700	22,500	0.5066	11,399
7	2300	18,180	15,880	0.4523	7183
8	2500	17,340	14,840	0.4039	5994
9	2600	16,250	13,650	0.3606	4922
10	2800	14,990	12,190	0.3220	3925
11	3000	13,200	10,200	0.2875	2933
12	3200	11,500	8,300	0.2567	2131

Net present worth = Sum of the values of discounted cash flow for the 12-year period with discount factor at 12% (Col. 6) = Rs. 9747

Since NPW > 0, the investment is attractive.

Simplified Procedure for Regular Cash Flows. In a cash flow pattern of the following type, the usual way of determining the present value of cash flows is to multiply the cash flow with the discount factor for the respective years at the cost of capital as shown in the Table 15.4.

Table 15.4. Simplified procedure to determine the net present worth for regular cash flows

(Amount in Rs.)

Year	*Cash Flow*	*Discount Factor 12%*	*Present Value*
1	*2*	*3*	*4 = (2 ´ 3)*
t_1	6500	0.8929	5804
t_2	6500	0.7972	5182
t_3	6500	0.7118	4627
t_4	6500	0.6355	4131
t_5	6500	0.5674	3688
t_6	6500	0.5066	3293
t_7	6500	0.4523	2940
t_8	6500	0.4039	2625
t_9	6500	0.3606	2344
t_{10}	6500	0.3220	2093
Total	65,000	5.6502	36727

Example 15.8. *Work out the net present worth of the following investment on a shallow tubewell project, if the opportunity cost of capital is 12 per cent per annum.*

Solution:

(Amount in Rs.)

Year	*Capital investment*	*Operation and maintenance*	*Cash outflow*	*Cash inflow*	*Cash flow*	*Discount factor*	*Discounted cash flow*
1	2	3	4 = (2 + 3)	5	6 = (5 − 4)	7	8 = (6 × 7)
1	85000	2300	87300	45000	− 42300	0.8929	−37770
2	—	2500	2500	52000	49500	0.7972	39461
3	—	2750	2750	63750	61000	0.7118	43598
4	—	3010	3010	75250	72240	0.6355	45909
5	—	3370	3370	82000	78630	0.5674	44615
6	—	4200	4200	85225	81025	0.5066	41047
7	—	4600	4600	88730	84130	0.4523	38052
8	—	5100	5100	79500	74400	0.4039	30050
9	—	5450	5450	68515	63065	0.3606	22741
10	—	5680	5680	55200	49520	0.3220	15945
11	—	6750	6750	53400	46650	0.2875	13412
12	—	7150	7150	47000	39850	0.2567	10229
Total						NPW	= 3,45,059

Since the NPW > 0, the investment is project-worthy at the opportunity cost of the capital.

However, this is a cumbersome procedure for computation. When the cash flow is regular and uniform, the same result can be arrived at by multiplying the annual cash flow for the designated period per year, in the present case, Rs. 6500, with the sum of the discount factor for the relevant period, namely 5.6502.

Thus, 6500 × 5.6502 = 36727.

Internal Rate of Return

A method of using the net benefit stream (or cash flow) for measuring the worth of a project is to determine the discount rate that makes the net present worth of incremental net benefit stream (or incremental cash flow) equal to zero. This discount rate is called the *internal rate of return* (IRR). It is the maximum interest that a project can pay for the resources used if the project is to recover its investment and operating costs.

Expressed mathematically, the internal rate of return is the discount rate i chosen, such that

$$\sum_{t=1}^{n} \frac{B_t - C_t}{(1+i)^t} = 0 \quad \text{... (15.15)}$$

The IRR is the interest rate or the earning rate which the investment in a project would earn. When using IRR, the selection criterion for projects is to accept those projects with an IRR, which is greater than the opportunity cost of the capital. In computing IRR, a systematic procedure of trial and error is necessary to arrive at the discount rate which will make the present worth of the incremental net benefit stream equal to zero.

It may be interesting to note that the internal rate of return of a series of values such as a cash flow can exist only when at least one value is negative. Evidently, if all the values are positive, no discount rate can make the net present worth of a stream equal to zero. The value of IRR, is only a broad indicator of the economic feasibility of a project, in so far as its contribution to the national economic development pertaining to the investment. Internal rate of return criterion of assessing the economic feasibility of projects is usually used by the World Bank and other international financing agencies.

Example 15.9. *The cash flow pattern of a shallow tubewell irrigation system is given below. Estimate the internal rate of return of the investment and compare it with the opportunity cost of capital at 14 per cent per year.*

Solution:

(All figures in Rs.)

Year	*Capital investment*	*Operation and maintenance*	*Cash outflow*	*Cash inflow*	*Cash flow*	*Discounted cash flow analysis**		
						at 14%	*at 20%*	*at 25%*
1	2	3	4 = (2 + 3)	5	6 = (5 – 4)	7	8	9
1	1,35,000	2,700	1,37,700	16,000	–1,21,700	–1,06,755	–1,01,413	–97,360
2	—	2,915	2,915	22,250	19,335	14,878	13,426	12,374
3	—	3,150	3,150	25,050	21,900	14,783	12,674	11,213
4	—	3,400	3,400	28,910	25,510	15,105	12,304	10,449
5	—	3,675	3,675	34,720	31,045	16,125	13,954	10,173
6	—	3,967	3,967	38,800	34,833	15,870	12,994	9,130
7	—	4,280	4,280	43,550	39,270	15,692	10,960	9,190
8	—	4,625	4,625	48,450	43,825	15,365	10,194	7,354
9	—	4,950	4,950	45,630	40,680	12,509	7,884	5,459
10	—	5,330	5,330	42,185	36,855	9,940	5,952	3,958
11	—	5,865	5,865	33,575	27,710	6,556	3,730	2,380
12	—	6,300	6,300	28,750	22,450	4,660	2,519	1,542
					NPW	= 34728	5178	–14,138

**IRR* = Lower discount rate + Difference between the two discount rates

$$\left\{\frac{\text{NPW at the lower discount rate}}{\text{Absolute difference between the NPWs at the two discount rates}}\right\} = 20 + 5\left\{\frac{5178}{5178+14138}\right\} = 20 + 5\left\{\frac{5178}{19316}\right\}$$

$$= 20 + 5\,(0.2681) = 20 + 1.34 = 21.34 \text{ per cent, i.e., 21 per cent (rounded off)}$$

As the *IRR* > the opportunity cost of capital, the project is worthy of investment.

Benefit-Cost Ratio

The discounted measure of project worth could be expressed by the benefit-cost ratio. The benefit-cost ratio (BCR) is the ratio obtained when the present worth of the benefit stream is divided by the present worth of the cost stream:

Benefit-cost ratio (BCR)

$$= \left\{\frac{\text{Present worth of benefits}}{\text{Present worth of costs}}\right\}$$

Expressed mathematically,

$$\text{Benefit-cost ratio} = \frac{\sum_{t=1}^{n} \frac{B_t}{(1+i)^t}}{\sum_{t=1}^{n} \frac{C_t}{(1+i)^t}} \qquad \text{... (15.16)}$$

Traditionally, benefit–cost analysis has been the most common method of assessing the economic feasibility of irrigation projects. However, in the past, undiscounted values of costs and benefits were applied on an annual basis in the year when the analysis is done to arrive at the *B–C* ratio. The enumeration of project costs and benefits often did not undergo the rigorous economic criteria followed now. Such analysis frequently resulted in erroneous conclusions.

Example 15.10. *The initial investment on a shallow tubewell project, including pumping set and underground pipeline water distribution system, using PVC pipeline, is Rs. 85,000.00. If the annual operating costs and benefit stream follows the pattern given below, work out the benefit-cost ratio at 12 per cent cost of capital, if the project life is 12 years.*

Solution:

(Amount in Rs.)

Year	*Cash outflow*	*Cash inflow*	*Discount factor at 12% (From Tables)*	*Discounted cash outflow*	*Discounted cash inflow*
1	*2*	*3*	*4*	*5 (2 × 4)*	*6 (3 × 4)*
1	85,000	10,000	0.8929	75,897	8,929
2	1,500	12,000	0.7972	1,196	9,566
3	1,800	14,500	0.7118	1,281	10,321
4	1,900	16,800	0.6355	1,207	10,676
5	2,100	22,100	0.5674	1,192	12,540
6	2,200	24,700	0.5066	1,115	12,513
7	2,300	18,180	0.4523	1,040	8,223
8	2,500	17,340	0.4039	1,010	7,004
9	2,600	16,250	0.3606	938	5,860
10	2,800	14,990	0.3220	902	4,827
11	3,000	13,200	0.2875	863	3,795
12	3,200	11,500	0.2567	821	2,952
Total	1,10,900	1,91,560	—	87,462	97,206

$$\text{Undiscounted benefit-cost ratio} = \frac{191560}{110900}, \textit{ i.e.,} = 1.727$$

$$\text{Discounted benefit-cost ratio} = \frac{97203}{87467}, \textit{ i.e.,} = 1.111$$

(***Note***: The estimate given by the undiscounted measure is a grossly overestimated figure.)

Example 15.11. *The initial investment on a shallow tubewell project, including pumping set and underground pipeline water distribution system, using PVC pipeline, is Rs. 85,000. The operation and maintenance cost and the annual benefit are furnished below. Estimate the benefit-cost ratio of the investment. The opportunity cost of capital is 12 per cent per annum.*

Solution:

(Amount in Rs.)

Year	*Capital invest-ment*	*Operation and main-tenance*	*Cash outflow*	*Cash inflow*	*Discount factor 12%*	*Discounted cash outflow*	*Discounted cash inflow*
1	*2*	*3*	*4=(2+3)*	*5*	*6*	*7=(4 × 6)*	*8=(5×6)*
1	85000	2300	87300	45000	0.8929	77950	40185
2	–	2500	2500	52000	0.7972	1993	41454
3	–	2750	2750	63750	0.7118	1958	45377
4	–	3010	3010	75250	0.6355	1913	47821
5	–	3370	3370	82000	0.5674	1912	46527
6	–	4200	4200	8225	0.5066	2128	43175
7	–	4600	4600	88370	0.4523	2081	40133
8	–	5100	5100	79500	0.4039	2060	32110
9	–	5450	5450	68515	0.3636	1965	24707
10	–	5680	5680	55200	0.3220	1829	17774
11	–	6750	6750	53400	0.2875	1941	15353
12	–	7150	7150	47000	0.2567	1835	12065
Total			1,37,860	7,95,000		99,656	4,06,681

Benefit-cost ratio (BCR)

$$= \frac{\text{Discounted benefit stream}}{\text{Discounted cost stream}} = \frac{406681}{99565} = 4.08$$

Note. The undiscounted BCR is $\frac{795570}{137860} = 5.77$, which is an overestimate.

Since BCR > 1, the project is worthwhile at the prevailing opportunity cost of the capital.

Net Benefit-Investment Ratio

It was seen earlier that none of the three discounted measures of project worth, which we have examined, can be relied upon to precisely rank projects. The usual selection criterion for each is to accept all projects that meet the requiremenr of NPW of zero or greater, a B/C ratio of 1 or greater, both at the opportunity cost of the capital, or an IRR equal to or greater than the opportunity cost of capital. In many cases, it is convenient to have a reliable measure to rank projects to decide on the order in which they should be undertaken.

A suitable criterion for ranking independent projects is the net benefit-investment ratio (also called the N/K ratio). This is the present worth of positive cash flows divided by the present worth of negative cash flows.

Expressed mathematically, Net Benefit-Investment (N/K) ratio

$$= \frac{\sum_{t=1}^{n} \frac{N_t}{(1+i)^t}}{\sum_{t=1}^{n} \frac{K_t}{(1+i)^t}} \quad \text{... (15.17)}$$

in which,

N_t = Incremental net benefit in each year after the stream has turned positive.

K_t = Incremental net benefit in initial years when the stream is negative.

Other terms are the same as defined earlier.

When using the N/K ratio, the selection criterion is to accept all projects with a net benefit-investment ratio of 1 or greater, when discounted at a suitable discount rate, which shall ordinarily be the opportunity cost of the capital. Selecting independent projects in the order of their N/K ratio, maximizes the return per unit of investible fund.

There are some limitations in the use of N/K ratio to rank projects. It can be used to rank mutually exclusive projects only when the N/K ratios of all projects in the investment programme are known. In practice, this would often be difficult.

In general, the net benefit-investment ratio is convenient to be used in real life project investment decisions. It is useful to rank projects when sufficient funds are not available to implement all the proposed ones.

Example 15.12. *The cash flow of a multipurpose river valley project follows the pattern given below. Estimate the N/K ratio of the project, if the opportunity cost of capital is 14 per cent.*

Solution:

(Amount in million Rs.)

Year	*Capital investment*	*Operation and maintenance*	*Cash outflow*	*Cash inflow*	*Cash flow*	*Discount factor flow (14%)*	*Discounted cash flow*
1	*2*	*3*	*4=(2+3)*	*5*	*6=(5–4)*	*7*	*8=(6×7)*
1	1634.21	—	1634.21	—	–1634.21	0.8772	–1434
2	2223.85	—	2223.85	—	–2223.85	0.7695	–1711
3	2021.74	—	2021.74	—	–2021.74	0.6750	–1365
4	—	51.03	51.03	1990.95	1939.92	0.5921	+1149
5	—	50.17	50.17	2193.80	2143.63	0.5194	+1113
6–25	—	50.17	50.17	3212.68	3162.51	3.4399	+10879

Using the values in column 8,

$$\text{N/K ratio} = \frac{1149+1113+10879}{1434+1711+1365} = \frac{13141}{4510} = 2.91$$

Since the N/K ratio > 1, the project is worthy of implementation.

Choice of the Discounted Measure to Use in Project Analysis

All the four discounting procedures, namely, NPW, IRR, B/C ratio, and N/K ratio are valuable in assessing projects proposed for funding. Each of them has certain specific applicability in evaluating projects.

BCR. In benefit-cost ratio approach, all independent projects with B/C ratio of 1 or greater, when discounted at opportunity cost of capital, could be accepted. It is not adequate for ranking independent projects directly.

N/K Ratio. Projects with N/K ratio of 1 or greater can be accepted for funding, when discounted at the opportunity cost of the capital. It is a satisfactory measure for ranking independent projects. The project with the highest value of N/K ratio is ranked first. The measure cannot be used directly to assess mutually exclusive projects (discussed later).

NPW. Generally, all independent projects are ranked by the BCR or N/K ratios. NPW is not popular as a measure of ranking independent projects. NPW is useful for ranking mutually exclusive projects. In case of mutually exclusive alternative project proposals, the one which has the largest NPW is accepted.

IRR. Independent projects with IRR equal to or greater than the opportunity cost of the capital are accepted for funding. The measure does not provide satisfactory criteria for ranking alternative project proposals. In case of mutually exclusive projects, the measure cannot be used directly. It is necessary to discount the differences between incremental net benefit flows of mutually exclusive alternate projects. The opportunity cost of the capital is used as cut-off. (The procedure is dealt in detail later).

In project analysis, the practice is to develop the project and analyse it with one of the discounted measures. The project is considered bankable if it has an NPW zero or greater, IRR equal to or greater than the cut-off rate, or B/C ratio and N/K ratio of 1 or greater.

Netting Out, Field Data, Length of Project Period, Replacement Costs and Salvage Value

Netting Out

In project analysis, the process of subtracting *costs* from *benefits*, to obtain the *net benefit* or *cash flow,* is referred to as *netting out*. In benefit-cost analysis, care should be taken to ensure that the netting out is done in exactly the same pattern for all alternative projects. Otherwise, the values of the ratios of B/C cannot be compared directly with one another. However, in computing discounted cash flow measures of net project worth (NPW) and internal rate of return (IRR), it does not make any difference as to where the netting out occurs. Thus, gross costs can be subtracted from gross benefits or the project costs subtracted from the sum of the net benefits realized, or any other combination. This is often considered a great convenience when using NPW and IRR values, as compared to B/C ratio, since data may reach the analyst with costs subtracted from benefits, or netting out in different forms. Hence, when the benefit-cost ratio is used as a criterion for evaluating projects, it is desirable that all analysts working in a country follow a common netting out convention to derive net benefit streams.

The selection criterion for the benefit-cost ratio measure of project worth is to accept all independent projects with a benefit-cost ratio of 1 or greater, when the cost and benefit streams are discounted at the opportunity cost of capital. An advantage of the benefit-cost ratio is that it can be used directly to observe how much the costs could rise without making the project economically unattractive. The benefit-cost ratio is used almost exclusively as a measure of social benefit, which is important in water resources projects.

Collection of field data for project evaluation. The various types of data required for an irrigation project evaluation include irrigation potential of the project, cropping intensity, cropping pattern, crop yields, value of gross produce and cost of cultivation. The irrigation potential is estimated from the total quantity of water available from the project and the irrigation requirements of crops for an assumed cropping system. A model cropping pattern for a farm is assumed on the basis of the data on climate, soil type, habits of farmers and a favourable economic return. The water requirement of crops during the different crop seasons are computed on a unit area basis. The capacity of the irrigation source divided by the irrigation requirement of crops, gives the area which can be irrigated.

Length of the project period for analysis. A question often asked in project analysis is for how long a period should the analyst carry out the economic analysis? The general practice is to select a period of time that will be close to the economic life of the project. If the project requires a sizable initial capital investment, for example, a lift irrigation project, a convenient starting point for establishing the period of analysis is the economic life of the major investment item which is the pumping unit and the rising main. In smaller irrigation projects, where the benefits accrue mainly from increase of agricultural production of the command area, a 25-year period is usually considered adequate for economic analysis. The influence on the present worth of a project by prolonging the period of analysis beyond 25 years, would only be nominal. Hence, even though major irrigation projects have economic project life of 50 years or more, the period of analysis is confined normally to about 25 years.

Replacement Cost

Irrigation projects often require investments which have different periods of effective service. In case of a pump irrigation scheme, for example, the intake structure and pump foundation may be expected to last 25 to 50 years or more, while the pump itself may last for 10 to 20 years. The pump bowl assembly of a turbine pump lasts only 8 to 10 years. Hence, in project analysis, allowance must be made for the replacement cost of pumps during the economic life of the project. Providing for replacement cost is simple. The analyst includes it in the project analysis among the cost items for the particular year in which the replacement is done. This may sometimes make the cash flow for that particular year negative, but that is of no significance in the discounted measure of project worth.

Example 15.13. *The initial cost of an open well and diesel engine-driven pumping set is Rs. 95,000. The expected project life of the well is 40 years and that of the pump set is 15 years. Prepare a cash flow statement with the pumping set being replaced at the end of the 15th year, at a cost of Rs. 40,000, if the annual net benefit is Rs. 25,000. Estimate the NPW of the investment at the end of the project life, at 14 per cent cost of capital.*

Solution:

(Amount in Rs.)

Year	*Capital invest-ment*	*Operation and main-tenance*	*Cash outflow*	*Cash inflow*	*Cash flow*	*Discount factor 14%*	*Discoun-ted cash flow*
1	*2*	*3*	*4 = (2 + 3)*	*5*	*6=(5–4)*	*7*	*8=(6´7)*
1	95000	1900	96900	25000	–71900	0.8772	–63071
2–15	—	1900	1900	25000	23100	5.2651	121624
16	40000	1900	41900	25000	–16900	0.1229	–2077
17–30	—	1900	1900	25000	23100	0.7376	17039
31	40000	1900	41900	25000	–16900	0.0172	–291
32–40	—	1900	1900	25000	23100	0.0852	1968
							NPW = 75189

Salvage Value

Salvage value, also called *residual value* or *terminal value*, is the value of an asset remaining unused at the end of a project. In project analysis, the residual value is generally added to the *benefit stream* in the last year of the project. It is often considered as the resale value of the asset.

Example 15.14. *The initial cost of an open well and diesel engine-driven pumping set is Rs. 95,000. The expected project life of the well is 40 years and that of the pump set is 15 years. Prepare a cash flow statement with the pumping set being replaced at the end of the 15th year at a cost of Rs. 40,000. The annual net benefit is Rs. 25,000, and the salvage value of the 10 year old diesel engine is Rs. 12,050. Estimate the NPW of the investment at the end of the project life, at 14 per cent cost of capital.*

Solution:

(Amount in Rs.)

Year	*Capital invest-ment*	*Operation and main-tenance*	*Cash outflow*	*Cash inflow*	*Cash flow*	*Discount factor 14%*	*Discoun-ted cash flow*
1	*2*	*3*	*4 = (2 + 3)*	*5*	*6=(5–4)*	*7*	*8=(6× 7)*
1	95000	1900	96900	25000	–71900	0.8772	–63071
2–15	—	1900	1900	25000	23100	5.2651	121624
16	40000	1900	41900	25000	–16900	0.1229	–2077
17–30	—	1900	1900	25000	23100	0.7376	17039
31	40000	1900	41900	25000	–16900	0.0172	–291
32–39	—	1900	1900	25000	23000	0.0799	1846
40	—	1900	1900	37050*	35150	0.0053	186
							NPW = 75256

* Includes the annual net benefit of Rs. 25,000 plus the salvage value of Rs. 12,050.

Mutually Exclusive Projects

In project analyses, there are situations to deal with mutually exclusive projects. They are projects which by their nature are such that if one is chosen, the other cannot be undertaken. For example, a lift irrigation project cannot be undertaken as separate from a surface irrigation project if the same water source is to be used for the surface irrigation project and *vice versa*.

The framework of economic analysis discussed so far was meant for dealing with independent projects, *i.e.* projects for which accepting one alternative does not preclude the chance of accepting another project option. However, the methodology for mutually exclusive projects differs to some extent. In case of mutually exclusive projects, the alternative having higher NPW is selected. A modified IRR method can also be used to select mutually exclusive projects. The incremental cash flow is prepared by subtracting the year-by-year cash flow of the less expensive alternative from the cash flow of the more expensive one. The incremental cash flow thus arrived at, is then discounted at the cost of capital to estimate the internal rate of return. If the incremental IRR is greater than the cost of capital, it is economically justified to undertake the bigger investment. Alternatively, one may opt for a less expensive proposal.

Sometimes, what may initially appear as mutually exclusive projects, can be taken up at successive stages of the development programme/plan, when a project can be expanded in phases to become a large one. Here, each phase is to be treated as independent projects because implementing the first phase need not pre-empt a larger or a smaller second phase.

Example 15.15. *A vegetable producers' cooperative society has 2500 enrolled members, who cultivate high value vegetable crops in 9750 ha. of land. The Society procures the produce and arranges the marketing at the nearest market, 10 km away. The Society identifies that irrigation has been a major bottleneck in stabilising production and proposes a project for installing a central pumping station for pumping water from the nearby river and distributing it to each member for a water cess. The Society has the choice of using either a gravity flow on-farm water distribution system or a sprinkler irrigation system. If the cash flow of the alternatives follow the pattern given below, evaluate the project options and recommend the economically justified options when the opportunity cost of capital is 12 per cent.*

Solution:

(Amount in thousand Rs.)

Year	*Cash flow*	
	Gravity flow system	*Sprinkler system*
1	–1125	–1710
2	900	1240
3	1035	1470
4	1200	1675
5	1350	1885
6–15	1650	2200
Total	19860	26560

Under situations of mutually exclusive projects, the NPW of alternative projects are computed at the opportunity cost of capital. The alternative offering the highest NPW is chosen.

(Amount in thousand Rs.)

Year	*Cash flow*		*Discount factor 12%*	*Discounted cash flow*	
	Gravity flow system	*Sprinkler system*		*Gravity flow system*	*Sprinkler system*
1	*2*	*3*	*4*	5 = (2×4)	6 = (3 × 4)
1	–1125	–1710	0.8929	–1005	–1527
2	900	1240	0.7972	718	989
3	1035	1470	0.7118	737	1046
4	1200	1675	0.6355	763	1065
5	1350	1885	0.5674	766	1070
6–15	1650	2200	3.2061	5290	7053
Total	19,860	26,560	6.8109	7269	9696

The sprinkler irrigation system is a better alternative because its NPW (Rs. 96,96,000) is greater than the NPW of the gravity flow system (Rs. 72,69,000).

It is necessary to examine whether the increased investment on the sprinkler system is justified at the opportunity cost of the capital. Hence, the incremental net benefit (cash flow) is to be estimated by subtracting the cheaper cash flow from the expensive one and determining the IRR of the incremental cash flow. The expensive investment is justified only if IRR ≥ the opportunity cost of the capital.

(Amount in thousand Rs.)

Year	*Incremental cash flow*	*Discounted cash flow analysis*	
		@ 12%	*@ 50%*
1	– 585	– 522	– 390
2	340	271	151
3	435	310	129
4	475	302	94
5	535	304	70
6–15	550	1763	142
Total	6700	2428	196

Since the IRR is more than 50%, while the opportunity cost of capital is only 12%, the installation of sprinkler system is recommended on economic grounds.

Sensitivity Analysis

Sensitivity analysis, also called treatment for uncertainty, is an analytical technique to test systematically what happens to the earning capacity of a project, if events differ from the estimates made about them at the time of planning. It is a means of dealing with uncertainty about future events and values. A sensitivity analysis is done by varying one element or a combination of elements in project costs and/or benefits and determining the effect of that change on the measures of project worth. In irrigation project analysis, most projects should be tested at least for the effects on delay in implementation and resulting variations in construction costs. Effects of possible reduction in project life due to excessive silting, waterlogging and soil salinity are also to be considered.

In general, projects tend to be more sensitive to cost overruns that occur in the early stages of its implementation than to changes in the prices of agricultural commodities which occur later. However, major differences in the predicted farm yield and those realised after commissioning the irrigation project will have profound influence on the net project worth. To conduct the sensitivity analysis of a project, it is adequate to simply calculate the measures of project worth over and over again, using new estimates of costs and benefits. Sensitivity analysis has important implications in project management. For example, a project may be highly sensitive to delay. If decision makers are aware of how sensitive a project is and how much will be the financial loss due to lost opportunities to generate benefits, it would be possible to take timely decisions to forestall delays.

Example 15.16. *The initial investment on a deep tubewell project, including pumping set and underground pipeline water distribution system, using PVC pipeline, is Rs. 2,35,000. The cash flow pattern of the investment is as given below. The opportunity cost of the capital is 16 per cent per annum. The water source is expected to last for 30 years. Examine the sensitivity of the measures of project worthiness to:*

(a) 15 per cent cost escalation in operation and maintenance charges,

(b) Water sources drying up in 20 years.

Solution: (*i*) Benefit-cost ratio (BCR) :

(Amount in Rs.)

Year	*Capital invest-ment*	*Operation and mainte-nance charges*	*Cash outflow*	*Cash inflow*	*Discount factor 16%*	*Discoun-ted cash outflow*	*Discoun-ted cash inflow*
1	*2*	*3*	*4 = (2 + 3)*	*5*	*6*	*7 = (4 × 6)*	*8 = (5 × 6)*
1	2,85,000	8500	2,93,500	32,000	0.8621	2,53,017	27,586
2	—	9685	9685	52,000	0.7432	7198	38,644
3	—	10,615	10,615	62,780	0.6407	6801	40,220
4	—	13,080	13,080	72,440	0.5523	7224	40,008
5–14	—	14,730	14,730	90,400	2.6693	39,319	2,41,305
15	—	15,780	15,780	95,000	0.1079	11,956	10,253
16–20	—	16,900	16,900	98,520	0.3534	5973	34,817
21–30	—	18,325	18,325	1,01,250	0.2484	4552	25,151
Total						3,36,040	4,57,984

$$\text{Benefit-cost ratio (BCR)} = \frac{457984}{336040} = 1.36$$

(ii) Internal rate of return (IRR) and net project worth (NPW).

(Amount in Rs.)

Year	Cash flow	Discounted cash flow analysis		
		@ 16%	@ 20%	@ 25%
1	– 2,61,500	– 2,25,429	– 2,17,908	– 2,09,200
2	42,315	31,449	29,384	27,082
3	52,165	33,422	30,188	26,709
4	59,360	32,785	28,629	24,314
5–14	75,670	2,01,986	1,53,005	1,10,667
15	– 15,780	– 1703	– 1024	– 556
16–20	81,620	28,845	15,851	7713
21–30	82,925	20,599	9,072	3400
	NPW	1,21,944	47,197	– 9871

$$\text{IRR} = 20 + 5\,\frac{47197}{57068} = 20 + 5\,(0.83)$$

$$= 20 + 4.15 = 24.15\%$$

The BCR > 1, NPW > 0 and the IRR > 16%, i.e., the opportunity cost of capital. Hence, the project is worthy of implementation.

Alternate Scenario I:

There is 15 per cent cost escalation in the operation and maintenance charges.

(i) Benefit-cost ratio:

(Amount in Rs.)

Year	Capital invest-ment	Operation and main-tenance charges	Cash outflow	Cash inflow	Discount factor 16%	Discoun-ted cash outflow	Discoun-ted cash inflow
1	2	3	4 = (2 + 3)	5	6	7 = (4 ×6)	8 = (5 ×6)
1	2,85,000	9775	2,94,775	32,000	0.8621	2,54,125	27,586
2	—	11,138	11,138	52,000	0.7432	8,278	38,644
3	—	12,207	12,207	62,780	0.6407	7,821	40,220
4	—	15,042	15,042	72,440	0.5523	8,308	40,008
5–14	—	16,940	16,940	90,000	2.6693	45,218	2,41,305
15	95,000	18,147	1,13,137	95,000	0.1079	12,209	10,253
16–20	—	19,435	19,435	98,520	0.3534	6,868	34,817
21–30	—	21,074	21,074	1,01,250	0.2484	5,235	25,151
						3,48,062	4,57,984

$$\text{BCR} = \frac{457984}{348062} = 1.32$$

(ii) Internal rate of return (IRR) and net project worth (NPW).

(Amount in Rs.)

Year	*Cash flow*	*Discounted cash flow analysis*		
		@ 16%	*@ 20%*	*@ 25%*
1	– 2,62,775	– 2,26,538	– 2,18,970	– 2,10,220
2	40,862	30,369	28,375	26,152
3	50,573	32,402	29,267	25,893
4	57,398	31,701	27,683	23,510
5–14	73,460	1,96,087	1,48,536	1,07,435
15	– 18,147	– 1958	– 1178	– 639
16–20	79,085	27,949	15,358	7474
21–30	80,176	19,916	8771	3287
	NPW	1,09,928	37,842	– 17,108

$$\text{IRR} = 20 + 5\left[\frac{37921}{54950}\right] = 20 + 5\,(0.69) = 20 + 3.45$$

$$= 23.45\%$$

When the operation and maintenance charges escalate by 15 percent, the project continues to be viable though the BCR declines from 1.36 to 1.31, NPW declines from Rs. 1,21,944 to Rs. 1.09,928 and the IRR declines from 24.14 per cent to 23.45 per cent.

Alternate Scenario II:

Water source drying up in 20 years:

(i) Benefits-cost ratio (BCR).

(Amount in Rs.)

Year	*Capital invest-ment*	*Operation and main-tenance charges*	*Cash outflow*	*Cash inflow*	*Discount factor (16%)*	*Discoun-ted cash outflow*	*Discoun-ted cash inflow*
1	*2*	*3*	*4 = (2 + 3)*	*5*	*6*	*7=(4 ×6)*	*8=(5 × 6)*
1	2,85,000	9775	2,94,775	32,000	0.8621	2,54,125	27,586
1	2,85,000	8500	2,93,500	32,000	0.8621	2,53,017	27,586
2	—	9685	9685	52,000	0.7432	7198	38,644
3	—	10,615	10,615	62,780	0.6407	6801	40,220
4	—	13,080	13,080	72,440	0.5523	7224	40,008
5–14	—	14,730	14,730	90,400	2.6693	39,319	2,41,305
15	95,000	15,780	1,10,780	95,000	0.1079	11,956	10,253
16–20	—	16,900	16,900	98,520	0.3534	5973	34,817
Total	3,80,000					3,31,488	4,32,833

$$\text{BCR} = \frac{432833}{331488} = 1.31$$

(ii) Net project worth (NPW) and internal rate of return (IRR).

(Amount in Rs.)

Year	*Cash flow*	*Discounted cash flow analysis*		
		@ 16%	*@ 20%*	*@ 25%*
1	– 2,61,500	– 2,25,439	– 2,17,908	– 2,09,200
2	42,315	31,449	29,384	27,082
3	52,165	33,422	30,188	26,709
4	59,360	32,785	28,629	24,314
5–14	75,670	2,01,986	1,53,005	1,10,667
15	– 15,780	– 1703	– 1024	– 556
16–20	81,620	28,845	15,851	7713
	NPW	1,01,345	38,125	– 13271

$$\text{IRR} = 20 + 5\left[\frac{38125}{51396}\right] = 20 + 5\,(0.74)$$

$$20 + 3.71 = 23.71\%$$

Under a situation when the water source dry up in 20 years (instead of the original project life of 30 years), the project continues to be viable, though the BCR declines from 1.36 to 1.31, NPW declines from Rs. 1,21,944 to Rs. 1,01,345 and the IRR declines from 24.14 per cent to 23.71 per cent under the alternate scenario.

WATER PRICING AND COST RECOVERY

Large scale irrigation projects, involving major storage and diversion works are planned, constructed, owned and operated by government agencies, usually the irrigation department. They are financed as part of the national development programme, out of the government's own revenues, market borrowings and loans from financing agencies. Small-scale storage works like irrigation tanks, lift irrigation schemes and deep tubewells projects are undertaken by government agencies, farmers' cooperatives or by individual farmers. Irrigation wells, including open wells and shallow tubewells are almost exclusively owned and operated by farmers. In developing countries, the farmers may avail subsidies and loans from government and financing agencies to meet a part of the initial investment on projects owned by the farmers or farmers' cooperatives.

Projects owned and operated by farmers and farmers' cooperatives are of self-sustaining nature. They are to be financially viable, taking into account both investment costs, and operation and maintenance costs. In case of individual farmer-owned irrigation works, the need for fixing a price for water arises only when the owner sells a part of his water to his neighbours at a price. Such a price can be fixed on the basis of cost recovery, taking total costs into consideration. A similar principle applies to fixing water charges by a cooperative society to its members.

Major water resources projects have an unusual cost structure. They have high fixed costs for storage, diversion and distribution facilities. For any given scheme of over a considerable operational range, the unit cost of irrigation water fall by increasing the size of the scheme. The cost per unit quantity of water will also fall with an increasing level of utilization of the water resource which has been developed. The decreasing cost-benefit relationship has been balanced by the temporal pattern of demand, which is lowest at the beginning of the life of a project.

In addition to high fixed investment costs, there is a large part of the recurring cost which is not influenced by the amount of water used. For example, the cost of staff for canal operation, maintenance and management are independent of the level of use of the project.

Influence of the inclusion of additional purposes in multipurpose projects on economy. Multipurpose irrigation projects may present spillover costs or benefits. In general, in multipurpose projects, it is often possible to provide a group of related goods and services more economically than from an alternative single purpose project. In water resources projects, there could be a danger that a major benefit like flood control may sometimes overweigh the economic rate of return and hide another useful purpose like irrigation or water supply.

Projects may be tested for the desirability of additional purposes by determining the net present worth of the alternative group of purposes at the opportunity cost of capital. If the project has a larger net present worth with additional purposes than without it, then it is advantageous to include the additional purposes. On the other hand, if the value of the net present worth is reduced by including the additional purposes, it would not be justified to include such an additional purpose in the project.

Need for Appropriate Pricing of Water

Major and medium irrigation projects are developed to benefit a large number of farmers of an irrigation command area. They are usually funded by the government. Government funding is from the taxes and other revenues raised from the public. Irrigation projects result in substantial financial benefits to the farmers of the command area. Hence, it is necessary that appropriate cost recovery policies are developed by the government to recover the money spent on a project for investment in other projects which will benefit other members of the society. The extent that the cost of an irrigation project is not recovered from the beneficiaries, represents the subsidy paid by others in the society who are not benefited by the project.

Two basic issues arise in formulating the cost recovery policy of an irrigation project. The first is the cost of the project which is to be repaid. The second is the projection of the incremental benefit received by individuals to be recovered through direct and indirect means as increased tax revenue. Cost recovery estimates, based on sound economic principles, are valuable in providing realistic projections, resulting in better efficiency and equity of cost recovery policies.

A major objective in fixing an appropriate price for irrigation water is to minimize waste and to allocate water optimally to maximise the net benefit from the project. The criteria for assessing the charges on an irrigation project should ensure equitable income from the project and an adequate recovery of the cost incurred. These are possible through charges which the beneficiary farmers can afford to pay and which still leave for them an adequate assured net benefit.

A certain amount of recovery of costs and benefits results from general taxation, like export tax and income tax. However, general taxation is not applicable to situation specific costs, benefits and risks. General taxation also affects those who are not benefited by a project. Hence, measures to recover costs and benefits, in addition to water pricing and general taxation, should be reflected and targeted to the beneficiaries of the project. These measures may be called *benefit taxes*. *Betterment levy*, assessed on the basis of increase in the value of the land benefited by the project is one such measure of the amount which could be varied according to the different crops grown. The procedure for charging water on volumetric basis, *i.e.*, on the volume of water used, is beneficial to avoid waste and project the cost on the price of water. However, it may not be practicable in the case of canal irrigation projects having large number of beneficiaries with small land holdings. This is usually the case in developing countries. In such situations, it may be possible to allot specific quantities of water to group farming societies or farmers' cooperatives.

The second objective of a sound cost recovery policy is to allot charges equitably and in line with the national policy for income distribution. In many developing countries, the small farmers are charged less than the large farmers.

Irrigation charges and benefit taxes should aim at obtaining an acceptable proportion of the benefit generated by the project, while still meeting the criteria of efficiency, income distribution and equity. The level of the charges and benefit taxes for a particular project, when charged separately, should take into consideration the charges and taxes in other areas, the disincentive effect of a benefit tax and the administrative cost of collecting the charges. Volumetric rate of water in a form which is acceptable to the beneficiary farmers and project specific betterment levies, are often desirable.

Cost recovery rates directly affect the cash flow of the farmer, the project agency and the generation of national income. However, if the cost recovery measures impose too heavy a burden on the farmers, it would be a disincentive for them to participate in the project.

The total net benefit of a project is the theoretical upper limit of the amount of revenue to be collected from water charges and benefit taxes. However, the actual amount will be much less in order to provide for variations in yield and risk factors like pests and diseases, variations in price of farm produce, natural calamities like floods and inefficient project management.

The lower limit of taxes to be collected also need careful consideration. A general policy adopted by many governments is to establish water charges and benefit taxes that will at least recover the operation and maintenance (O&M) costs of the project. This will at least avoid continued drain of national revenue on a particular project benefitting only a section of the population.

In projects which are funded by international agencies like the World Bank, it has been observed that the cost recovery, as a percentage of the incremental net cash income, rarely exceed 30 to 35 per cent (Price Gittinger, 1982).

Historical Perspective of Water Pricing and Sanctioning of Irrigation Projects

The practice of collecting water charges in India in the form of land revenue dates back to the erstwhile pre-British kingdoms in Andhra Pradesh, Karnataka, Tamil Nadu and other states (Vani, 1991, Molony and Raju, 1994). The current policy of water charges and taxes in India owes its origin to the water pricing policy of the British colonial administration. The British treated irrigation projects mainly as commercial enterprises. The irrigation rates were based on a rate of return commensurating with the interest rate which prevailed in the London Market. The water rates were periodically revised in relation to cost-price changes.

The rate of financial return, which was 4% till 1919, was increased to 5% during 1919-21 and again revised to 6% after 1921 (Sangal, 1991).

The finance policy of the Government of India during the British period, pertaining to irrigation projects could be traced back to 1866 A.D., when the Secretary of State expressed his willingness to sanction the raising of loans necessary for the construction of such works which could reasonably be expected to be remunerative. In 1879 A.D., the following criteria were adopted by the British Parliament to test the financial results of the irrigation projects in India (Irrigation Commission, 1972):

(*i*) By considering the capital cost of any work as simply the sum actually spent on its construction.

(*ii*) By crediting the revenue account yearly with:
- (*a*) the simple interest of the capital cost of the works at the commencement of the year, and
- (*b*) the working expenses of the year.

(*iii*) By crediting the revenue account yearly with:
- (*a*) direct receipts, and
- (*b*) indirect receipts.

The difference between (*ii*) and (*iii*) for any one year would show the profit or loss for that year. Schemes were to be sanctioned only if they satisfied the test of financial productivity arrived at, in terms of rate of returns calculated, as suggested in (*i*), (*ii*) and (*iii*) above. The test of financial productivity was that the project should be able to show a certain percentage return on the *sum-at-charge* in the tenth year after its beginning. The sum-at-charge was the capital cost plus the arrears of interest upto that year. This principle was followed since then and the rate of return required before a project could be considered as financially productive was fixed differently from time to time.

With Independence, the development of irrigation was stepped up in India. The lowering of the rate of return in the productivity test in 1949 from 6 per cent to 3.75 per cent, enabled a larger number of projects to be accepted for construction. But the indirect benefits derived from irrigation projects were still not recognized adequately. During 1955-61, investigations by a Committee of the Planning Commission on the overall benefits of selected major irrigation projects revealed that large benefits accrued from irrigation in terms of double cropping, diversification and better quality of crops, higher yields and greater employment opportunities for hired labour. However, they were not adequately accounted in determining the benefit of irrigation projects. Indirect benefits include the establishment of processing industries, the expansion of consumer industries, retail trade and transport and communications. It became clear that the total benefits from irrigation were far larger than the direct financial returns accruing to government from irrigation rates. The Committee recommended that the criterion of benefit-cost ratio should be used for assessing the feasibility of new projects instead of the criterion of the direct financial return to government. For the sake of simplicity, it also proposed that the indirect benefits need not be taken into account in the computation. In working out the ratio, the benefits should be taken to comprise the difference in the value of total annual agricultural production less the cost of cultivation, before and after the introduction of irrigation. The total cost should include the annual interest on capital, depreciation and expenditure on maintenance and operation of the project.

The Irrigation Commission (1972) pointed out that projects with a benefit-cost ratio of less than 1.5 were not generally considered for acceptance, although theoretically, a ratio of unity should meet the criterion. The benefit-cost ratio was, however, not determined by using discounted measures of project worth. Hence, it could not stand the financial scrutiny adopted in the present day analysis, using discounted values of benefits and costs. The value of B-C ratio of 1.5 was fixed mainly due to the progressive increase in project costs, while the rise in prices of agricultural commodities had been much less. A benefit-cost ratio lower than 1.5 could, in these circumstances, lead to the acceptance of border line cases which may later prove to be economically unsound. The Commission, however, recommended that this rule should not be rigidly applied in the case of irrigation projects in the drought-prone areas. In such areas, there are often pressing social and human compulsions to accept a lower limit of benefit-cost ratio. In case of multipurpose projects, the benefit-cost ratio has to be worked out for the project as a whole to ascertain its economic feasibility. The benefit-cost ratio also needs to be worked out for each benefit on the basis of apportioned costs, in order to ascertain whether any particular aspect of the project is unjustified on economic grounds.

Basis of Levying Irrigation Charges

Generally, the water charges imposed by governments comprises of one or more of the following elements:

(1) *Water rate*, depending on the kind of crop and the area irrigated, number of irrigations of each crop or the total volume of water used by the farmer or a designated agency like an irrigation cooperative, in a season or a year.

(2) *Increment in land revenue*, based on increased benefit derived annually from providing irrigation facilities.

(3) *Betterment levy*, representing the government's share in the increase in the land value accruing with the provision of irrigation facilities.

(4) *Irrigation cess*, being the annual charge per hectare of the irrigable area, whether water is actually taken for irrigation or not.

Once the basis of charge is decided, some measurable unit for charging has to be selected. This can be the amount of water used, the cropped area, the area and type of crop, or simply the income of the farmer. Volumetric measurement is technically difficult in large irrigation projects. Measurement of discharge over calibrated concrete weirs, for example, depends upon regular and accurate readings. It is not generally practicable for large-scale projects unless the supply is to cooperatives or to farmers having large size farms.

Volumetric supply, however, has been used extensively and successfully in the United States of America and Australia. In Israel and Taiwan, volumetric supplies are made to cooperatives which allocate water among small holdings. In India, large sugarcane farms in Maharashtra are charged on the basis of volumetric supply.

In developing countries, the rate at which water is charged for the different crops varies from country to country and from state to state in a federal set-up. The rates are related neither to the cost of supplying water nor to the quantity of water used. They are charged on the area basis for different crops. These rates are usually quite low in spite of the fact that the price of agricultural produce has gone up substantially. Where the area of land is used as the basis for charging water, there is the advantage of simplicity and relative ease of measurement. But there is no incentive to the farmer to use water judiciously. In areas where water is scarce and there is surplus land, it may discourage development of the optimum area under irrigation.

A device used widely in India, Bangladesh and Pakistan to recover a part of the cost of project is the *"betterment levy"*. This is a tax on a part of the increase in the capital value of land, following provision of assured irrigation supply. It may be a single tax or one spread over a number of years, sometimes as many as 7 years. However, a betterment levy and high annual irrigation charges would result in double-taxing of benefits.

The Irrigation Commission (1972) of the Government of India has recommended the following principles, as guidelines for fixing irrigation rates:

(1) Water rates should be levied on a *crop basis*, except in the case of irrigation from tubewells.
(2) The rate should be related to the gross income from the crop and not to the cost of the project. It should range between 5% and 12% of gross income, the upper limit being applicable to cash crops.
(3) The rates should be within the paying capacity of irrigators and should aim at ensuring full utilization of available supplies.
(4) Between regions with a similar class of supply, there should be minimum disparity, if any, in the rates charged.
(5) For fixing rates, irrigation should be divided into A, B and C categories, on the basis of the quantity and timeliness of supply. Lower rates may be fixed where, on account of good rainfall, the demand for irrigation water is less or where the supply is inadequate and uncertain.
(6) The general level of irrigation rates in a state should be such that, taken as a whole, the irrigation schemes do not impose any burden on the general revenues.

With all charges levied, the canal water charges in most of the developing countries are only nominal. In India, in almost every case, the charges are less than 5% of gross output. In many cases, they are less than 1%. With the apparent reluctance of governments to tax agricultural income effectively, it may be predicted that water rates, as a percentage of the value of gross output, could fall rather than rise over time.

Current Scenario of Water Pricing in Different States of India

Since water is a state subject in India, the water rates are fixed by the state government. As a result, the water rates vary between states and even between projects within a state. Table 15.5 presents the overall and crop-specific canal water rates in selected states of India and the year in which the rates were last revised, as of 1995.

All states in India, except the seven north-eastern states, levy charges on irrigation water, either directly or indirectly. While the states have separate water charges, in case of certain pre-British projects in Andhra Pradesh, Karnataka and Tamil Nadu, there is no separate water charge, but only higher land revenue for *wet lands*. For example, in Tamil Nadu, the land revenue assessment for *wet lands* ranges from Rs. 7.41–54.36 per hectare, as against Rs. 1.24 – 19.77 per hectare for irrigated dry lands (Government of India, 1992). The water rates in West Bengal vary with the different rice crop seasons. However, in Kerala, the water is charged only in terms of gross area irrigated, irrespective of crops and seasons. In case of other states, water rates vary not only with crops and seasons, but also by the category of projects, type of irrigation (flow or lift) and the category of uses, namely, private parties, cooperatives and government lift irrigation schemes (Sangal, 1991; Govt. of India, 1992).

The rates of water charged for different projects within a state depend on the quality of irrigation service, namely, dependability and continuity. Andhra Pradesh, Haryana and Tamil Nadu have classified their canals into 2, 3 and 5 categories, respectively for fixing water rates. Bihar, Orissa and Madhya Pradesh have distinct rates for perennial and non-perennial canals. Bihar charges different rates for long lease (Rs. 74.11 per ha), seasonal lease (Rs. 89.45 per ha) and in case of non-perennial irrigation systems, the rates are only half that of the perennial systems. Maharashtra follows a nearly identical pattern whose rates for single watering are in the range of Rs. 35.90 per ha, depending upon

Table 15.5. Canal water rates for irrigation of paddy, wheat and sugarcane in selected major states of India

(Rs./ha)

State	Range (Rs./ha)		Crop-specific rates			Year in which they were revised
	Minimum	Maximum	Paddy	Wheat	Sugarcane	
Andhra Pradesh	99	222	222	—	370	1986
Bihar	30	158	89		158	1983
Gujarat	40	830	110	110	830	1981
Haryana	17	99	74	61	99	1975
Karnataka	37	556	99	—	556	1985
Maharashtra	65	1000	—	—	750	1989
Madhya Pradesh	15	297	59	76		1990
Orissa	6	185	40		—	1981
Punjab	14	81	49	29	—	1974
Rajasthan	20	143	—	74	—	1982
Tamil Nadu	6	65	49		62	1962
Uttar Pradesh	7	327	143	143	237	1983
West Bengal	74	593	125	—	—	1977

(***Sources:*** Govt. of India, Planning Commission, 1992; Central Water Commission, 1995.)

crops and seasons. In Maharashtra and Madhya Pradesh, there is a distinction between *demand rates* and *agreement rates.* In both the states, the demand rate is almost twice that of the agreement rate (Govt. of India, 1992).

Even though area-based water rates are charged in almost all states, volume-based water rates are prevalent in sugarcane and rice-based cooperatives in Maharashtra and Gujarat (Govt. of India, 1992).

Other charges in canal irrigation. In most states, besides water charges, levies/cess is charged on canal water. Betterment levies are charged on irrigated land in Tamil Nadu and West Bengal. In Tamil Nadu, it is 45 per cent of the land revenue, while in West Bengal a flat rate of Rs. 400/ha is charged. Irrigation cess on water rates is charged in Gujarat, Maharashtra, Tamil Nadu and Kerala. While it is 20 per cent of the water rates in Maharashtra and Gujarat, the rates in Tamil Nadu range from 37-75 per cent. Madhya Pradesh has a crop cess of Rs. 20/ ha for flow system and Rs. 10/ ha for lift system (Maria Saleth, 1996).

Canal water charges for non-irrigation purposes. There is considerable income from canal water when used for non-irrigation purposes (domestic and industrial), lease of fishing rights and sale of timber grown on canal banks and peripheral land strips. The income from them had a three-fold increase during 1976-88 at the all-India level. Two states accounted for over 70% of rise in the total non-irrigation income, namely, Maharashtra (11 times) and Madhya Pradesh (6 times) during the period. There is great scope for further increase in income from non-irrigational uses, especially in the area of providing water for domestic water supply schemes, industries and emerging scientific fish production in perennial canals and storage reservoirs.

Water Rates of Tubewell Supplies

Tubewell supplies differ from canal supplies in several respects. Tubewells usually provide a steady supply of water for irrigation. It is easy to provide irrigation at optimum periods, with reference to the stage of growth of the crops, thus maximizing production. In contrast, canal supplies fluctuate and canals often remain closed for long periods. In the case of canals, the annual working expenses remain the same, in spite of the variations, in the volume of water supplied. On the other hand, in tubewells the cost of power consumed in pumping is related to the volume of water pumped. It is also possible to conveniently measure the discharge rates of tubewells with a simple V-notch or other water measuring devices. The facility for water measurement and the high cost of tubewells and pumps make it desirable that tubewell water is charged on the basis of the volume of water supplied to the individual farmers.

In many countries, the water rates of tubewells are charged on a volume basis. The discharge rates of tubewells are checked periodically by means of weirs or other measuring devices installed with every deep tubewell. To determine the quantity of water supplied, the time for which the tubewell operates for each irrigator is noted. In addition, the electrical units consumed during the period are recorded to provide a check.

The groundwater prices are usually on an hourly basis, though payment by season and number of irrigations is practiced in some states. The hourly prices are in the range of Rs. 3-6 in Andhra Pradesh, Tamil Nadu, and Uttar Pradesh. They are in the range of Rs. 25-45 in Gujarat, where the groundwater markets are better organised and competitive (Shah, 1993). The water rates charged by individual farmers owning irrigation wells (tubewells and open wells) are on the basis of negotiation between buyers and sellers. Such negotiations could reflect on the demand and supply factors and the overall financial benefit to the respective parties.

In the states of Haryana and Punjab, the assessment of the supply of water by a tubewell is made on the basis of the units of electricity consumed in operating the tubewell for each irrigator. The advantages claimed in the assessment by this method are that it is more dependable than the time-discharge basis of assessment. It also takes into account the total head involved in pumping.

When water supply from a tubewell is measured at the well, there may be disparity in the amounts charged from the farmers at their field, in view of the seepage losses in conveyance channels. Consequently, farms located near the wells pay less per unit of water consumed. The unit cost increases with the increase in the distance of the irrigated area from the well. Provision of lined irrigation channels or water distribution through underground pipelines will alleviate this disparity.

Water Rates for Lift Irrigation from Canals

In many canal command areas, there are some areas which cannot be served by gravity flow. Farmers in such areas carry out irrigation by lifting water from canals or the overflow drains of canals by installing pumping sets or other water lifting devices. It is a common practice to charge lower rates for irrigating such areas, in consideration of the extra cost incurred in lifting water.

In case of lift irrigation projects in India, the water rates vary with the ownership of the lift project (private/cooperatives/state). While crop area based charges are applicable to individual farmers, volumetric rates are usually applicable to cooperatives and state-owned lift irrigation schemes. In Gujarat, Haryana, Punjab and Uttar Pradesh, the water rates in privately owned canal lifting is 30–50% of the crop-area water rates in gravity-fed canal irrigation. However, in Maharashtra, Goa and Himachal Pradesh, the rates for lift irrigation water are substantially higher than the corresponding rates in flow irrigation (Sangal, 1991).

Cost Recovery and Financial Performance of Water Resource Projects

Major water resources projects are usually of multipurpose nature, namely, irrigation, hydro-power generation, industrial and drinking water supply, flood control, aquaculture, recreation and environmental management. Hence, only a part of the cost of the project can be attributed to the agricultural sector for water charges. In the agricultural sector also, even though total benefits may be uneconomical in terms of domestic currency, a project may result in an important saving in foreign currency, or an earning of foreign currency through export of the produce. This is a benefit to the general public and not limited to the farmer. Many developing countries have attained near stability in agricultural production in which irrigation has played a major role. Extension of irrigation to arid areas like the Indira Gandhi Canal Project area in western Rajasthan in northern India has created an important barrier to the advance of desertification, even though there are problems of waterlogging and salt accumulation which are yet to be effectively controlled. Successful irrigated agricultural production usually results in a whole chain of related economic activities which benefit a broad spectrum of the population. These include agro-based industries and professional services of various nature. In general, for the stimulation of any economically productive activity such as agriculture, it is justified to allow a major part of the costs to be covered from the general taxes.

An approach usually advocated by the government is that the basic investments in water resource development and distribution, land reclamation and the settlement of displaced persons of project areas shall be borne by the public and the current operating costs of the system should be borne by the farmers as price of water. There are three basic approaches to fix water charges. They can be related to the cost of providing irrigation to the benefits to be derived from it, or to some value judgement on the beneficiaries' ability to pay the charge. If a cost approach is used, water rates can be set to recover capital and recurrent costs, or some proportion of them. In a low-income economy, the main problem with this approach is that it may ignore both the levels of benefit actually achieved, which may often be low and the poverty of the beneficiaries.

Financial Performance of Major and Medium Irrigation Projects in India

It may be seen from the previous sections that the water charges levied by the different states in India are very low. They are also not being revised periodically to reflect the normal changes in the benefits and costs of the project, including capital investment as well as operation and maintenance costs. The water charges in most states do not cover the operation and maintenance (O&M) costs. Case studies in Bihar and Rajasthan revealed that in some years, the expenditure involved in collecting the water charges exceeded the amount collected (Maria Saleth, 1996). Such situations seriously undermine the financial viability of most of the irrigation projects. The magnitude of the financial gap can be seen from Table 15.6.

Table 15.6. Financial performance of major and medium irrigation projects in selected major states of India, 1989–90

(All figures in million Rs.)

States	*Capital outlay*	*Revenue expenditure*	*Revenue receipts*	*Depreciation*[1]	*Financial loss*[2]	
					A	*B*
Andhra Pradesh	28984	3155	352	276	2803	3079
Bihar	30201	584	061	288	523	810
Gujarat	22327	3559	170	210	3389	3599
Himachal Pradesh	213	0004	000	002	004	006
Haryana	8635	1161	136	084	1025	1110
Karnataka	20892	1536	161	198	1375	1573
Kerala	7601	193	016	072	177	249
Madhya Pradesh	25865	579	123	247	456	703
Maharashtra	40052	4389	276	375	4114	4489
Orissa	15952	161	046	153	116	268
Punjab	8874	781	174	088	608	695
Rajasthan	15517	1299	160	149	1139	1288
Tamil Nadu	6519	809	016	064	793	857
Uttar Pradesh	32180	3507	366	310	3141	3451
West Bengal	5553	457	016	053	441	495
Total	269364	22175	2073	2570	20102	22672

(***Note***: In table 15.6, (1) depreciation is allowed at the rate of one per cent on mean capital outlay. For 1987-88, the depreciation has been calculated at the rate of one per cent on capital outlay at the end of the year, (2) *'A'* denotes the excess of expenditure over receipts and *'B'* denotes the excess of expenditure including depreciation over receipts.)

(***Source:*** Government of India, Planning Commission, 1992.)

According to the Vaidyanathan Committee Report (Government of India, Planning Commission, 1992), the total annual cost of major and medium irrigation projects was about Rs. 25,000 million, as of 1989-90. This included the operation and maintenance costs, one per cent interest on the cumulative capital investment and one per cent depreciation allowance. Against this amount, the total receipts from irrigation charges and other receipts had been Rs. 3000 million, which is 12% of the annual cost incurred in the project. This revenue gap amounts to a total annual financial loss of Rs. 22,000 million on account of major and medium irrigation projects. Amongst the states, the financial loss was higher for Andhra Pradesh, Gujarat, Maharashtra and Uttar Pradesh. Maharashtra and Gujarat accounted for over half of the total financial loss.

Traditionally, India had many grassroots level institutions, built around widely accepted social conventions and moral principles. These institutions have been successful in enforcing an orderly use of water at the local level. With the gradual separation of ownership and use of water resources, effected through colonial and post-independence legislations, the indigeneous institutions have become almost non-functional. There has been some effort in the recent years to service these institutions through *Panchayat Raj* (local self-government) and through people's participation. However, these efforts are in the initial stages, even though a few successful cases can be seen in some states.

Economic Aspects of Different Irrigation Systems

Major and medium projects are almost always surface water projects with storage or diversion at headworks and an extensive water distribution system. Minor irrigation projects could be either surface water projects—tanks, diversions or river-canal lift irrigation projects or ground water irrigation projects—tubewells and open wells. In large projects, storage works are intended to store the rainfall-runoff during the rainy seasons and to augment the flow from diversion works, especially during the dry periods. Storage works are usually used for hydro-power generation and flood control, in addition to irrigation. Integrated irrigation development comprises of harnessing surface water sources through storage and diversion works at appropriate locations in a river basin and using the water supplies to augment the water available in the command area from rainfall, soil moisture and ground water.

In an integrated water resources development programme, all components of the water resources are to be developed optimally. Land development and soil conservation measures in the command area not only reduce soil erosion but facilitate the conservation of rainfall-runoff as soil moisture and increased groundwater recharge to meet a good part of the crop water requirement. Surface irrigation development results in substantially higher groundwater recharge obtained from the seepage from the water conveyance system and deep percolation from areas under crop. Hence, surface water development for sustained productivity shall include the simultaneous development of groundwater irrigation works, with a view to operate them conjunctively as a well-integrated water resources system.

Large land areas lying close to but at higher elevations of canals could be brought under irrigation by having lift irrigation projects to operate conjunctively with the gravity flow systems. This will result in major economic advantage to the integrated system. The feasibility for such conjunctively operated flow and lift systems is highly location–specific.

Economics of a water resources project are profoundly influenced by its design, operation and maintenance. Integrated water resources planning, comprising of conjunctive use of surface and groundwater, provide major economic advantages. Such systems facilitate prevention of environmental degradation. They ensure a favourable hydrological balance, especially groundwater balance, in the command area. Sustainable agricultural production in the command area demands effective control of water. Only the required quantity of canal water supplies is to be released at optimum periods to the different segments of the command area.

Generally, the financial advantage of major water resources projects are low when the benefits arising from increased agricultural production in the irrigation command area alone are considered. Hence, a realistic appraisal of a major project will require attribution of benefits from the different uses such as hydro-power generation, reduction in flood damages in the downstream areas, providing water supply to industrial and domestic consumers, recreation and others. Major projects require much greater care in operation and maintenance in order to obtain the desired level of benefits from various sources. An ill-managed system could prove to be more of a curse than a boon. The irrigation component of the project, to be economically feasible, will require careful attention to agricultural production system, especially soil and water management practices, and supply of the different inputs in agricultural production. The requirements of ancillary inputs are to be predicted at the time of project planning and timely arrangements made to forestall deficiencies in any sector.

Small-scale projects, including groundwater and surface water pumping projects for irrigation and water supply, can provide high benefit-cost ratios from single operation alone. However, including water supplies for domestic and industrial supplies to a predominantly irrigation project is often economically more advantageous. The cost of initial planning of the project by professionally competent agencies will substantially enhance the returns from the project. In the developing country situations, it will be desirable to own and operate the comparatively large size projects in the minor irrigation category, like tank irrigation, lift irrigation and deep tubewells by farmers' cooperatives which should have access to effective technical support in planning and management. The relatively smaller ones like shallow tubewells and open wells could be owned by individual farmers. It is advantageous to design and operate a water well to provide the utilization of its full potential, rather than limiting its design yield to meet only the requirements of the owner. The owner could sell his additional water to other users, including neighbouring farmers who do not possess their own irrigation sources. Such supplies are to be appropriately priced.

Availability of Funds for Investments, Operation and Maintenance

Small scale irrigation projects are financed by the individual farmer from his own resources or by availing credit facilities from banks or other agencies. The usual requirement of banks is that a part of the investment is to be met by the farmer, while the remaining is provided as credit, to be repaid in installments over a period of time, at a specified interest rate. The National Bank for Agriculture and Rural Development (NABARD) is the institution at the national level for financing projects in agriculture and rural development, including minor irrigation and land development, through refinancing arrangement with commercial banks. The NABARD has developed criteria and guidelines for financing of various types of projects. The financing agency insists on a proper techno-economic feasibility assessment of the project before undertaking to provide the credit. The Government of India provides financial support to small-scale irrigation works under the Integrated Rural Development Programme, Drought-Prone Area Programme and special programmes to benefit socially and economically backward sections of the population. Irrigation projects in the cooperative sector, which are intended to benefit a group of farmers forming the cooperative, generate financial resources through shares from its members and availing credit facilities from financing agencies. Financing agencies usually encourage cooperative effort in minor irrigation system development. Elements of subsidy and tax concessions are also available in many countries to encourage cooperative venture of farmers in irrigation development.

Financing Major Water Resources Projects

Major water resources projects require high financial investments, most of which will be required during the construction phase of the project. However, they contribute greatly to national development. They provide benefits which are beyond those that are simply financial and economic. The major financial burden of such projects is borne by the government. The usual practice is for the governmental agency to bear the cost of project feasibility studies, project planning, land acquisition, rehabilitation of the population displaced by the project and construction and management of the headworks, water conveyance system and associated structures as well as the cost of the equipment, their installation and operation and maintenance. The cost of development of the irrigation command area, on-farm water distribution system and on-farm drainage and soil conservation works are usually borne by the farmer and/or their cooperatives with the assistance of the government. The government assistance is usually limited to the cost of engineering and planning involved in on-farm works and subsidies and/or loans through government controlled financing agencies. Non-governmental organizations also provide technical and financial assistance to small-scale water resources projects. Ancillary works like approach roads and agricultural advisory services are provided by the government. Input services like seeds, seedlings, fertilizers, plant-protection chemicals and equipment, crop product processing facilities, storage, transportation and marketing facilities are organized by farmers' cooperatives, public undertakings and private agencies, with the encouragement and support from the government.

Financial requirements of major water resource projects are borne by the government from its own receipts, through market borrowing and credits and aids from national and international financing agencies. Finances from international funding agencies are available in the form of aids and/or loans.

Bilateral assistance. Some governments have a special interest in assisting developing countries to develop water resources projects. It will be advantageous to avail bilateral assistance when such possibilities exist.

Multilateral assistance. Many international agencies provide assistance in water resources development project planning and implementation. They include the European Economic Community (EEC), the United Nations Development Programme (UNDP), the World Bank and a number of other agencies. The EEC provides financial aid and technical assistance to the developing countries which are signatories to the Lome Convention as well as other countries with which it has special agreements.

Within the United Nations system, the main source of technical assistance, including project preparation, is the United Nations Development Programme (UNDP). The UNDP finances pre-investment activities, including techno-economic feasibility study of projects. The recipient country, wishing to avail UNDP assistance, develops a country programme which has to be agreed to by the UNDP. The country programme identifies the sectors and activities for which the assistance will be used and estimates the financial requirements. Most of the activities undertaken with the UNDP assistance require contribution from the recipient country, either in kind or in local currency. Once the assistance for preparing a particular project is included in the country programme, a tripartite agreement is made between the recipient country, the UNDP and the executing agency. The tripartite agreement specifies the details of the works to be undertaken. Detailed information about UNDP assistance for preparing particular projects could be obtained from the UNDP Resident Representative, stationed in most of the developing countries.

The Food and Agriculture Orgnization (FAO) of the United Nations has established an investment centre with a multi-disciplinary team of experts who specialize in the formulation of investment projects. The centre helps to identify and prepare agricultural development projects in collaboration with the World Bank, the Inter-American Development Bank, the Asian Development Bank, the African Development Bank, the International Fund for Agricultural Development and several agencies in the Arab countries. For assistance under the FAO-bank cooperative programmes, the FAO could be contacted through its representative or the appropriate bank could be approached. World Bank provides assistance in the form of long-term loans for irrigation project preparation and implementation.

SUMMARY

Need for appropriate procedures for the economic and financial analyses of water resources projects are highlighted. Basic concepts in the economic and financial analyses of projects are presented, taking into account tangible and intangible benefits and costs. Application and relative merits of different discounting procedures in the discounted cash flow analyses of water resources projects are described. Importance of appropriate pricing of irrigation water supplies from canals, wells and lift irrigation schemes are described along with historical perspective and current scenario. Mobilization of funds for financing irrigation projects of varying magnitude are explained.

QUESTIONS

1. Distinguish between:

(i) Economic analysis and Financial analysis

(ii) Economic rate of return and financial rate of return

(iii) Gross benefit and net benefit

(iv) Shadow price and current price

(v) Efficiency price and imputed price

(vi) Tangible and intangible costs and benefits

(vii) Fixed costs and variable costs

(viii) Discount factor and compounding factor

(ix) Capital recovery factor and amortization

(x) Benefit-cost ratio and net benefit-investment ratio

(xi) Betterment levy and irrigation cess

(xii) Bilateral assistance and multilateral assistance

2. What are the objectives of techno-economic evaluation of irrigation projects?

3. What is opportunity cost of the capital economic analysis?

4. What is transfer payment in economic analysis?

5. How does economic analysis contribute to the development of economy as a whole?

6. What are the comparative advantages of economic and financial analyses?

7. How are tangible and intangible benefits and costs enumerated in project analysis?

8. What is depreciation? How is it computed in project analysis?

9. What are the variable costs in the operation of irrigation projects?

10. What is discounting in the analysis of a project?

11. What is the present worth of an annuity factor?

12. How is the escalated present worth of the component of a project computed at the end of a particular period of time?

13. Define amortization. How are the instruments of repayment computed under the amortization scheme?

14. How is the amount of each level (even) payment determined at the end of each fixed time intervals with a view to recover the present amount at the end of a predetermined time period at the interest rate, as applicable?

15. Define the term internal rate of return. How is it different from the net present worth concept in discounted cash flow analysis?

16. Define netting out in project analysis.

17. What is the special requirement in adopting the netting out procedure in analysing alternative projects, adopting the benefit-cost ratio approach?

18. What are mutually exclusive projects? How does the methodology used in analysing mutually exclusive projects differ from those of independent projects?

19. How do the inclusion of additional purposes influence the economy of multipurpose water resources projects?

20. What are the usually adopted components of water charges by governments?

21. What are the distinguishing factors influencing the water charges of tubewell supplies, as compared to those of canal water supplies?

22. What are the common means of financing major water resources projects?

23. Identify the *true* and *false* statements from the following:

(i) Financial analysis is done using market prices.

(ii) In financial analysis, when borrowed capital is received, it is usually entered into the receipt stream as a cost.

(iii) In discounted cash flow analysis, depreciation is treated as a cost.

(iv) The discount rates chosen in financial and economic analyses differ.

(v) The compounding factor is used to calculate the future worth of a present amount at the end of a particular time period at a predetermined rate of interest.

(vi) The benefit-cost ratio is the rate obtained when the present worth of the cost stream is divided by the present worth of the benefit stream.

(vii) The discounting procedure benefit-cost ratio (BCR) is valuable in ranking independent projects directly.

(viii) The net present worth is the present worth of the net benefit.

(ix) Sensitivity analysis is the means of dealing with uncertainty about future events and values in a project.

(x) Betterment levy is the annual charge per hectare of the irrigated area, whether water is actually used for irrigation or not.

24. Identify the most appropriate answer of the following questions from the four options:

(i) The current price is:

(a) The value or price as recorded

(b) The value or price that is expected to occur under normal conditions

(c) Shadow price

(d) Imputed price

(ii) The discounting procedure which is satisfactory for ranking independent projects directly is:

(a) Benefit-cost ratio

(b) Net benefit investment ratio

(c) Internal rate of return

(d) Net present worth

(*iii*) The discounting procedure suitable for ranking mutually exclusive projects is:
(a) Benefit-cost ratio
(b) Net benefit investment ratio
(c) Internal rate of return
(d) Net present worth

(*iv*) In case of irrigation projects, the usually adopted value of the length of the project period for economic analysis is:
(a) 15 years (b) 25 years
(c) 40 years (d) 50 years

(*v*) The most significant component in the water charges levied by the government is:
(a) Betterment levy
(b) Incremental land revenue
(c) Water rate
(d) Irrigation cess

25. In the following, match each of the entries of the 1st column with the most closely linked entry in the 2nd column:

(*i*)	(A) Economic analysis	(I) Accounting price
	(B) Grants	(II) Liquidity of a project
	(C) Shadow price	(III) Opportunity cost of the capital
	(D) Current ratio	(IV) Encouraging a specific activity
	(E) Cut-off rate	(V) Willingness to pay
(*ii*)	(A) Annuity	(I) The most direct cash flow measure
	(B) Compounding factor	(II) The maximum interest that a project can pay to recover its investment and operating costs
	(C) Net present worth	(III) Suitable method for ranking independent projects
	(D) Internal rate of return	(IV) Future worth of a present amount, at the end of a specified period
	(E) Net benefit investment ratio	(V) Series of periodic payments
(*iii*)	(A) Netting out	(I) Water pricing policy
	(B) Sensitivity analysis	(II) Government's share in the increase in land revenue occurring with the provision of irrigation
	(C) Irrigation Commission (1972)	(III) Financing water resources projects
	(D) Bilateral assistance	(IV) Subtracting costs from benefits to obtain cash flow
	(E) Betterment levy	(V) Treatment for uncertainty

FURTHER READING REFERENCES

Babu, R., Dhyani, B.L., Agarwal, M.C. and Samra, J.S. (1997), *Economic Evaluation of Watershed Management Projects*, Pub. Central Soil and Water Conservation Research and Training Institute, Dehra Dun, No. T-33/D-23, p. 83.

Caruthers, L. and Clark, C. (1981), *The Economics of Irrigation*, Pub. Liverpool University Press, Greves Street, Liverpool L7 7AF, UK, p. 300.

Central Water Commission (1995), *Water and Related Statistics*, Ministry of Water Resources, Government of India, New Delhi, p. 314.

Dhawan, B.D. (1993), *Trends and New Tendencies in Indian Irrigated Agriculture,* Pub. Commonwealth Publishers, New Delhi, p. 294.

Dixon, J.A., Carpenter, R.A., Fallom, L.A., Sherman, P.B., and Menopemok, S. (1986), *Economic Analysis of the Environmental Impacts of Development Projects,* Pub. The Asian Development Bank, Manila, Philippines, p. 100.

Dov Nir (1983), *Economics and Costing of Irrigation,* in Fomlel, H.J., *Handbook of Irrigation Technology*, Vol. 2. Pub. CRC Press Inc. Boca Raten, Florida, USA, pp. 61–75.

Government of India (1992), *Report of the Committee on Pricing Irrigation Water*, Planning Commission, New Delhi.

Indian Agricultural Research Institute, New Delhi (1985), *Economic Criteria for Fixation of Irrigation Charges*, Division of Agrl. Economics, I.A.R.I., New Delhi, Res. Bull. No. 9, p. 178.

James, L.O., and Lee, R.R. (1971), *Economics of Water Resources Planning,* Pub. Tata McGraw Hill Publishing Co. (TMH Edition), New Delhi, p. 615.

Jensen, M.E. (ed.) (1983), *Design and Operation of Farm Irrigation Systems,* ASAE Monograph No.3, American Society of Agricultural Engineers, St. Joseph, Michigan USA, pp. 55–69.

Maloney, C. and Raju, K.V. (1994), *Managing Irrigation Together: Practice and Policy in India*, Pub. Sage Publications, New Delhi.

Maria Saleth, R. (1996), Water *Institutions in India: Economics, Law and Policy,* Pub. Commonwealth Publishers, New Delhi for Institute of Economic Growth, New Delhi, p. 299.

Michael, A.M. and Khepar, S.D., (1992), *Water Well and Pump Engineering,* Pub. Tata McGraw Hill Publishing Co. Ltd., New Delhi, pp. 872–911.

Ministry of Irrigation and Power, Government of India, New Delhi (1972), *Report of Irrigation Commission,* Vol. I, p. 930.

Mruthyunjaya (1986), *Benefit-Cost Analysis of Management Systems,* Mimeographed. Division of Agricultural Economics, I.A.R.I., New Delhi.

National Bank for Agriculture and Rural Development (1995), *Lift Irrigation Schemes in Maharashtra—A Techno-Economic Evaluation*, NBARD, Mumbai, p. 129

Pearson, G.F. (1974), *Life Cycle Costing in an Energy Crisis Era,* Professional Engineer 44(7), 22–29, Washington, D.C., USA.

Price Gittinger, J. (1989), *Economic Analysis of Agricultural Projects,* 2nd ed. Pub. for Economic Development Institute, World Bank, Johns Hopkins Univ. Press, Baltimore, USA, p. 505.

Sangal, S.P., (1991), *Pricing of Irrigation Water in India*, Economic and Political Weekly, Vol. 26, No. 46.

Seagraves, J.A. and Easter, K.W., (1983), *Pricing of Irrigation Water in Developing Countries,* Water Resources Bulletin, Vol.19, No.4.

Shah, T. (1993), *Ground Water Market and Irrigation Development,* Pub. Oxford University Press, New Delhi.

Vani, M.S., (1992), *The Role of Panchayat Institutions in Irrigation Management,* Pub. Indian Law Institute, New Delhi.

Appendices

APPENDIX A

Theis' well function $W(u)$ corresponding to values of u and $1/u$

[***Source***: Walton (1962)]

		$1/u = n$	$n(1)$	$n(2)$	$n(3)$	$n(4)$	$n(5)$	$n(6)$	$n(7)$	$n(8)$	$n(9)$	$n(10)$
n	N	$u = N$	$N(-1)$	$N(-2)$	$N(-3)$	$N(-4)$	$N(-5)$	$N(-6)$	$N(-7)$	$N(-8)$	$N(-9)$	$N(-10)$
1.000	1.0	$W(u)$ =2.194(− 1)	1.823	4.038	6.332	8.633	1.094(1)	1.324(1)	1.554(1)	1.784(1)	2.015(1)	2.245(1)
0.833	1.2	1.584(− 1)	1.660	3.858	6.149	8.451	1.075(1)	1.306(1)	1.536(1)	1.766(1)	1.996(1)	2.227(1)
0.666	1.5	1.000(− 1)	1.465	3.637	5.927	8.228	1.053(1)	1.283(1)	1.514(1)	1.744(1)	1.974(1)	2.204(1)
0.500	2.0	4.890 (− 2)	1.223	3.355	5.639	7.940	1.024(1)	1.255(1)	1.485(1)	1.715(1)	1.945(1)	2.176(1)
0.400	2.5	2.491(− 2)	1.044	3.137	5.417	7.717	1.002(1)	1.232(1)	1.462(1)	1.693(1)	1.923(1)	2.153(1)
0.333	3.0	1.305(− 2)	9.057(− 1)	2.959	5.235	7.535	9.837	1.214(1)	1.444(1)	1.674(1)	1.905(1)	2.135(1)
0.286	3.5	6.970(− 3)	7.942(− 1)	2.810	5.081	7.381	9.683	1.199(1)	1.429(1)	1.659(1)	1.889(1)	2.120(1)
0.250	4.0	3.779(− 3)	7.024(− 1)	2.681	4.948	7.247	9.550	1.185(1)	1.415(1)	1.646(1)	1.876(1)	2.106(1)
0.222	4.5	2.073(− 3)	6.253(− 1)	2.568	4.831	7.130	9.432	1.173(1)	1.404(1)	1.634(1)	1.864(1)	2.094(1)
0.200	5.0	1.148(− 3)	5.598(− 1)	2.468	4.726	7.024	9.326	1.163(1)	1.393(1)	1.623(1)	1.854(1)	2.084(1)
0.166	6.0	3.601(− 4)	4.544(− 1)	2.295	4.545	6.842	9.144	1.145(1)	1.375(1)	1.605 (1)	1.835(1)	2.066(1)
0.142	7.0	1.155(− 4)	3.738(− 1)	2.151	4.392	6.688	8.990	1.129(1)	1.360(1)	1.590(1)	1.820(1)	2.050(1)
0.125	8.0	3.767(− 5)	3.106(− 1)	2.027	4.259	6.555	8.856	1.116(1)	1.346(1)	1.576(1)	1.807(1)	2.037(1)
0.111	9.0	1.245(− 5)	2.602(− 1)	1.919	4.142	6.437	8.739	1.104(1)	1.334(1)	1.565(1)	1.795(1)	2.025(1)

APPENDIX B

Table of the functions, e^{rb}, $e^{-r/B}$, $K_0\ (r/B)$ and $e^{r/B}$, $K_0\ (r/B)$ for preparing De-Glees' type curves for estimating hydraulic properties of semi-confined aquifers

[***Source:*** Hantush (1956)]

r/B	$e^{r/B}$	$e^{-r/B}$	$K_0(r/B)$	$e^{r/B}K_0(r/B)$	r/B	$e^{r/B}$	$e^{-r/B}$	$K_0(r/B)$	$e^{r/B}K_0(r/B)$
0.010	1.010	0.990	4.721	4.769	0.020	1.020	0.980	4.028	4.110
0.011	1.011	0.989	4.626	4.677	0.021	1.021	0.979	3.980	4.064
0.012	1.012	0.988	4.539	4.594	0.022	1.022	0.978	3.933	4.021
0.013	1.013	0.987	4.459	4.517	0.023	1.023	0.977	3.889	3.979
0.014	1.014	0.986	4.385	4.447	0.024	1.024	0.976	3.846	3.940
0.015	1.015	0.985	4.316	4.381	0.025	1.025	0.975	3.806	3.902
0.016	1.016	0.984	4.251	4.320	0.026	1.026	0.974	3.766	3.866
0.017	1.017	0.983	4.191	4.263	0.027	1.027	0.973	3.729	3.831
0.018	1.018	0.982	4.134	4.209	0.028	1.028	0.972	3.692	3.797
0.019	1.019	0.981	4.080	4.158	0.029	1.029	0.971	3.657	3.765

(Contd...)

r/B	$e^{r/B}$	$e^{-r/B}$	$K_0(r/B)$	$e^{r/B}K_0(r/B)$
0.030	1.030	0.970	3.623	3.734
0.031	1.031	0.969	3.591	3.704
0.032	1.032	0.968	3.559	3.675
0.033	1.034	0.967	3.528	0.647
0.034	1.035	0.967	3.499	3.620
0.035	1.036	0.966	3.470	3.593
0.036	1.037	0.965	3.442	3.568
0.037	1.038	0.964	3.414	3.543
0.038	1.039	0.963	3.388	3.519
0.039	1.040	0.962	3.362	3.495
0.040	1.041	0.961	3.336	3.473
0.041	1.042	0.960	3.312	3.450
0.042	1.043	0.959	3.288	3.429
0.043	1.044	0.958	3.264	3.408
0.044	1.045	0.957	3.241	3.387
0.045	1.046	0.956	3.219	3.367
0.046	1.047	0.955	3.197	3.348
0.047	1.048	0.954	3.176	3.329
0.048	1.049	0.953	3.155	3.310
0.049	1.050	0.952	3.134	3.292
0.050	1.051	0.951	3.114	3.274
0.051	1.052	0.950	3.094	3.256
0.052	1.053	0.949	3.075	3.239
0.053	1.054	0.948	3.056	3.223
0.054	1.055	0.947	3.038	3.206
0.055	1.056	0.946	3.019	3.190
0.056	1.058	0.945	3.001	3.174
0.057	1.059	0.945	2.984	3.159
0.058	1.060	0.944	2.967	3.144
0.059	1.061	0.943	2.950	3.129
0.060	1.062	0.942	2.933	3.114
0.061	1.063	0.941	2.916	3.100
0.062	1.064	0.940	2.900	3.086
0.063	1.065	0.939	2.884	3.072
0.064	1.066	0.938	2.869	3.058
0.065	1.067	0.937	2.853	3.045
0.066	1.068	0.936	2.838	3.032
0.067	1.069	0.935	2.823	3.019
0.068	1.070	8.934	2.809	3.006
0.069	1.071	0.933	2.794	2.994
0.070	1.072	0.932	2.780	2.981
0.071	1.074	0.931	2.766	2.969
0.072	1.075	0.930	2.752	2.957
0.073	1.076	0.930	2.738	2.945
0.074	1.077	0.929	2.725	2.934
0.075	1.078	0.928	2.711	2.923
0.076	1.079	0.927	2.698	2.911
0.077	1.080	0.926	2.685	2.900
0.078	1.081	0.925	2.673	2.889
0.079	1.082	0.924	2.660	2.879

r/B	$e^{r/B}$	$e^{-r/B}$	$K_0(r/B)$	$e^{r/B}K_0(r/B)$
0.080	1.083	0.923	2.647	2.868
0.081	1.084	0.922	2.635	2.857
0.082	1.085	0.921	2.623	2.847
0.083	1.086	0.920	2.611	2.837
0.084	1.088	0.919	2.599	2.827
0.085	1.089	0.918	2.587	2.817
0.086	1.090	0.918	2.576	2.807
0.087	1.091	0.917	2.564	2.798
0.088	1.092	0.916	2.553	2.788
0.089	1.093	0.915	2.542	2.779
0.090	1.094	0.914	2.531	2.769
0.091	1.095	0.913	2.520	2.760
0.092	1.096	0.912	2.509	2.751
0.093	1.097	0.911	2.499	2.742
0.094	1.099	0.910	2.488	2.733
0.095	1.100	0.909	2.478	2.725
0.096	1.101	0.908	2.467	2.716
0.097	1.102	0.908	2.457	2.707
0.098	1.103	0.907	2.447	2.699
0.099	1.104	0.906	2.437	2.691
0.10	1.105	0.905	2.427	2.682
0.11	1.116	0.896	2.333	2.605
0.12	1.127	0.887	2.248	2.534
0.13	1.139	0.878	2.169	2.471
0.14	1.150	0.869	2.097	2.412
0.15	1.162	0.861	2.030	2.358
0.16	1.173	0.852	1.967	2.309
0.17	1.185	0.844	1.909	2.262
0.18	1.197	0.835	1.854	2.219
0.19	1.209	0.827	1.802	2.179
0.20	1.221	0.819	1.753	2.141
0.21	1.234	0.811	1.706	2.105
0.22	1.246	0.802	1.662	2.071
0.23	1.259	0.794	1.620	2.039
0.24	1.271	0.787	1.580	2.008
0.25	1.284	0.779	1.541	1.979
0.26	1.297	0.771	1.505	1.952
0.27	1.310	0.763	1.470	1.925
0.28	1.323	0.756	1.436	1.900
0.29	1.336	0.748	1.404	1.876
0.30	1.350	0.741	1.372	1.853
0.31	1.363	0.733	1.342	1.830
0.32	1.377	0.726	1.314	1.809
0.33	1.391	0.719	1.286	1.788
0.34	1.405	0.712	1.259	1.768
0.35	1.419	0.705	1.233	1.749
0.36	1.433	0.698	1.207	1.731
0.37	1.448	0.691	1.183	1.713
0.38	1.462	0.684	1.160	1.696
0.39	1.477	0.677	1.137	1.679

(Contd...)

r/B	$e^{r/B}$	$e^{-r/B}$	$K_0(r/B)$	$e^{r/B}K_0(r/B)$
0.40	1.492	0.670	1.114	1.663
0.41	1.507	0.664	1.093	1.647
0.42	1.522	0.657	1.072	1.632
0.43	1.537	0.650	1.052	1.617
0.44	1.553	0.644	1.032	1.602
0.45	1.568	0.638	1.013	1.589
0.46	1.584	0.631	0.994	1.575
0.47	1.600	0.625	0.976	1.562
0.48	1.616	0.619	0.958	1.549
0.49	1.632	0.613	0.941	1.536
0.50	1.649	0.606	0.924	1.524
0.51	1.665	0.600	0.908	1.512
0.52	1.682	0.594	0.892	1.501
0.53	1.699	0.589	0.877	1.489
0.54	1.716	0.583	0.861	1.478
0.55	1.733	0.577	0.847	1.467
0.56	1.751	0.571	0.832	1.457
0.57	1.768	0.565	0.818	1.446
0.58	1.786	0.560	0.804	1.436
0.59	1.804	0.554	0.791	1.426
0.60	1.822	0.549	0.777	1.417
0.61	1.840	0.543	0.765	1.407
0.62	1.859	0.538	0.752	1.398
0.63	1.878	0.533	0.740	1.389
0.64	1.896	0.527	0.728	1.380
0.65	1.915	0.522	0.716	1.371
0.66	1.935	0.517	0.704	1.363
0.67	1.954	0.512	0.693	1.354
0.68	1.974	0.507	0.682	1.346
0.69	1.994	0.502	0.671	1.338
0.70	2.014	0.497	0.660	1.330
0.71	2.034	0.492	0.650	1.322
0.72	2.054	0.487	0.640	1.315
0.73	2.075	0.482	0.630	1.307
0.74	2.096	0.477	0.620	1.300
0.75	2.117	0.472	0.611	1.293
0.76	2.138	0.468	0.601	1.285
0.77	2.160	0.463	0.592	1.278
0.78	2.181	0.458	0.583	1.272
0.79	2.203	0.454	0.574	1.265
0.80	2.225	0.449	0.565	1.258
0.81	2.248	0.445	0.557	1.252
0.82	2.270	0.440	0.548	1.245
0.83	2.293	0.436	0.540	1.239
0.84	2.316	0.432	0.532	1.233
0.85	2.340	0.427	0.524	1.226
0.86	2.363	0.423	0.516	1.220
0.87	2.387	0.419	0.509	1.214
0.88	2.411	0.415	0.501	1.209
0.89	2.435	0.411	0.494	1.203
0.90	2.460	0.407	0.487	1.197
0.91	2.484	0.402	0.480	1.192
0.92	2.509	0.398	0.473	1.186
0.93	2.534	0.395	0.466	1.181
0.94	2.560	0.391	0.459	1.175
0.95	2.586	0.387	0.452	1.170
0.96	2.612	0.383	0.446	1.165
0.97	2.638	0.379	0.440	1.159
0.98	2.664	0.375	0.433	1.154
0.99	2.691	0.372	0.427	1.149
1.0	2.718	0.368	0.421	1.144
1.1	3.004	0.333	0.366	1.098
1.2	3.320	0.301	0.318	1.057
1.3	3.669	0.272	0.278	1.021
1.4	4.055	0.247	0.244	0.988
1.5	4.482	0.223	0.214	0.958
1.6	4.953	0.202	0.188	0.931
1.7	5.474	0.183	0.165	0.906
1.8	6.050	0.165	0.146	0.883
1.9	6.686	0.150	0.129	0.861
2.0	7.389	0.135	0.114	0.842
2.1	8.166	0.122	0.101	0.823
2.2	9.025	0.111	8.93(– 2)	0.806
2.3	9.974	0.100	7.91(– 2)	0.789
2.4	1.102(1)	9.07(– 2)	7.02– 2)	0.774
2.5	1.218(1)	8.21(– 2)	6.23(– 2)	0.760
2.6	1.346(1)	7.43(– 2)	5.54(– 2)	0.746
2.7	1.488(1)	6.72(– 2)	4.93(– 2)	0.733
2.8	1.644(1)	6.08(– 2)	4.38(– 2)	0.721
2.9	1.817(1)	5.50(– 2)	3.90(– 2)	0.709
3.0	2.009(1)	4.98(– 2)	3.47(– 2)	0.698
3.1	2.220(1)	4.50(– 2)	3.10(– 2)	0.687
3.2	2.453(1)	4.08(– 2)	2.76(– 2)	0.677
3.3	2.711(1)	3.69(– 2)	2.46(– 2)	0.667
3.4	2.996(1)	3.34(– 2)	2.20(– 2)	0.658
3.5	3.312(1)	3.02(– 2)	1.96(– 2)	0.649
3.6	3.660(1)	2.73(– 2)	1.75(– 2)	0.640
3.7	4.045(1)	2.47(– 2)	1.56(– 2)	0.632
3.8	4.470(1)	2.24(– 2)	1.40(– 2)	0.624
3.9	4.940(1)	2.02(– 2)	1.25(– 2)	0.617
4.0	5.460(1)	1.83(– 2)	1.12(– 2)	0.609
4.1	6.034(1)	1.00(– 2)	1.00(– 2)	0.602
4.2	6.669(1)		8.9(– 3)	0.595
4.3	7.370(1)		8.0(– 3)	0.589
4.4	8.145(1)		7.1(– 3)	0.582
4.5	9.002(1)		6.4(– 3)	0.576
4.6	9.948(1)		5.7(– 3)	0.570
4.7	1.099(2)		5.1(– 3)	0.564
4.8	1.215(2)		4.6(– 3)	0.559
4.9	1.343(2)		4.1(– 3)	0.553
5.0	1.484(2)		3.7(– 3)	0.548

APPENDIX C

Table C-1. Table of values of the function *W (u, r/B)* for detemining hydraulic properties of confined aquifers

[***Source***: Hantush (1956)]

u	*1/u* *r/B*	*0*	*0.005*	*0.01*	*0.02*	*0.03*	*0.04*	*0.05*	*0.06*	*0.07*	*0.08*	*0.09*
0		∞	1.08(1)	9.44	8.06	7.25	6.67	6.23	5.87	5.56	5.29	5.06
1(– 6)	1.00(6)	1.32(1)										
2(– 6)	5.00(5)	1.25(1)										
4(– 6)	2.50(5)	1.18(1)	1.07(1)									
6(– 6)	1.66(5)	1.14(1)	1.06(1)									
8(– 6)	1.25(5)	1.12(1)	1.05(1)	9.43	*W (u, r/B) = W (0, r/B)*							
1(– 5)	1.00(5)	1.09(1)	1.04(1)	9.42								
2(– 5)	5.00(4)	1.02(1)	9.95	9.30								
4(– 5)	2.50(4)	9.55	9.40	9.01	8.03							
6(– 5)	1.66(4)	9.14	9.04	8.77	7.98	7.24						
8(– 5)	1.25(4)	8.86	8.78	8.57	7.91	7.23						
1(– 4)	1.00(4)	8.63	8.57	8.40	7.84	7.21						
2(– 4)	5.00(3)	7.94	7.91	7.82	7.50	7.07	6.62	6.22	5.86			
4(– 4)	2.50(3)	7.25	7.23	7.19	7.01	6.76	6.45	6.14	5.83	5.55		
6(– 4)	1.66(3)	6.84	6.83	6.80	6.68	6.50	6.27	6.02	5.77	5.51	5.27	5.05
8(– 4)	1.25(3)	6.55		6.52	6.43	6.29	6.11	5.91	5.69	5.46	5.25	5.04
1(– 3)	1.00(3)	6.33		6.31	6.23	6.12	5.97	5.80	5.61	5.41	5.21	5.01
2(– 3)	5.00(2)	5.64		5.63	5.59	5.53	5.45	5.35	5.24	5.12	4.89	4.85
4(– 3)	2.50(2)	4.95		4.94	4.92	4.89	4.85	4.80	4.74	4.67	4.59	4.51
6(– 3)	1.66(2)	4.54			4.53	4.51	4.48	4.45	4.40	4.36	4.30	4.24
8(– 3)	1.25(2)	4.26			4.25	4.23	4.21	4.19	4.15	4.12	4.08	4.03
1(– 2)	1.00(2)	4.04			4.03	4.02	4.00	3.98	3.95	3.92	3.89	3.85
2(– 2)	5.00(1)	3.35				3.34	3.34	3.33	3.31	3.30	3.28	3.26
4(– 2)	2.50(1)	2.68					2.67	2.67	2.66	2.65	2.65	2.64
6(– 2)	1.66(1)	2.29							2.28	2.28	2.27	2.27
8(– 2)	1.25(1)	2.03							2.02	2.01	2.01	2.01
			W (u, r/B) =W (u, 0)									
1(– 1)	1.00(1)	1.82								1.81	1.81	1.81
2(– 1)	5.00(1)	1.22										1.22
4(– 1)	2.50(1)	7.02(– 1)										7.00(–1)
6(– 1)	1.66(1)	4.54(– 1)										
8(– 1)	1.25(1)	3.11(– 1)										

u	*1/ur/B*	*0*	*0.1*	*0.2*	*0.3*	*0.4*	*0.5*	*0.6*
		∞	4.85	3.50	2.74	2.23	1.55	
1(– 4)	1.00(4)	8.63						
2(– 4)	5.00(3)	7.94						
4(– 4)	2.55(3)	7.25						
6(– 4)	1.66(3)	6.84						
8(– 4)	1.25(3)	6.55	4.84					

(Contd...)

u	*1/ur/B*	*0*	*0.1*	*0.2*	*0.3*	*0.4*	*0.5*	*0.6*
					*W (u, r/*B) = *W (0, r/B)*			
1(– 3)	1.00(3)	6.33	4.83					
2(– 3)	5.00(2)	5.64	4.71					
4(– 3)	2.50(2)	4.95	4.42	3.48				
6(– 3)	1.66(2)	4.54	4.18	3.43				
8(– 3)	1.25(2)	4.26	3.98	3.36	2.73			
1(– 2)	1.00(2)	4.04	3.81	3.29	2.71	2.22		
2(– 2)	5.00(1)	3.35	3.24	2.95	2.57	2.18		
4(– 2)	2.50(1)	2.68	2.63	2.48	2.27	2.02	1.52	
6(– 2)	1.66(1)	2.29	2.26	2.17	2.02	1.84	1.46	1.11
8(– 2)	1.25(1)	2.03	2.00	1.93	1.83	1.69	1.39	1.08
1(– 1)	1.00(1)	1.82	1.80	1.75	1.67	1.56	1.31	1.05
2(– 1)	5.00	1.22	1.21	1.19	1.16	1.11	9.96(– 1)	8.58(– 1)
4(– 1)	2.50	7.02(– 1)	7.00(– 1)	6.93(– 1)	6.81(– 1)	6.65(– 1)	6.21(– 1)	5.65(– 1)
6(– 1)	1.66	4.54(– 1)	4.53(– 1)	4.50(– 1)	4.44(– 1)	4.36(– 1)	4.15(– 1)	3.87(– 1)
8(– 1)	1.25	3.11(– 1)	3.10(– 1)	3.08(– 1)	3.05(– 1)	3.01(– 1)	2.89(– 1)	2.73(– 1)
1	1.00	2.19(– 1)		2.18(– 1)	2.16(– 1)	2.14(– 1)	2.07(– 1)	1.97(– 1)
2	5.00(– 1)	4.88(– 2)		4.87(– 2)	4.85(– 2)	4.82(– 2)	4.73(– 2)	4.60(– 2)
4	2.50(– 1)	3.77(– 3)		3.77(– 3)	3.77(– 3)	3.76(– 3)	3.74(– 3)	3.73(– 3)
0		∞	8.42(– 1)	2.28(– 1)	6.95(– 2)	2.23(– 2)	7.4(– 3)	2.5(– 3)
1(– 2)	1.00(2)	4.04						
2(– 2)	5.00(1)	3.35						
4(– 2)	2.50(1)	2.68						
6(– 2)	1.66(1)	2.29	8.39(–1)	*W (u, r/B)* = *W* (0, *r/B*)				
8(– 2)	1.25(1)	2.03	8.32(– 1)					
1(– 1)	1.00(1)	1.82	8.19(– 1)					
2(– 1)	5.00	1.22	7.15(– 1)	2.27(– 1)				
4(– 1)	2.50	7.02(– 1)	5.02(– 1)	2.10(– 1)	6.91(– 2)			
6(– 1)	1.66	4.54(– 1)	3.54(– 1)	1.77(– 1)	6.64(– 2)	2.22(– 2)		
8(– 1)	1.25	3.11(– 1)	2.54(– 1)	1.44(– 1)	6.07(– 2)	2.18(– 2)		
1	1.00	2.19(– 1)	1.85(– 1)	1.14(– 1)	5.34(– 2)	2.07(– 2)	7.3(– 3)	
2	5.00(– 1)	4.89(– 2)	4.44(– 2)	3.35(– 2)	2.10(– 2)	1.12(– 2)	5.1(– 3)	2.1(– 3)
4	2.50(– 1)	3.78(– 3)	3.6(– 3)	3.1(– 3)	2.4(– 3)	1.60(– 3)	1.0(– 3)	6.0(– 4)

(***Note:*** The number in parenthesis is the power of ten by which the other number is multiplied. For example, 1.08 (7) = 1.08 × 10^7 1.00 (5) = 1 × 10^5. The negative signs in the parenthesis represent negative power of ten. For example, 1 (– 6) = 1 × 10^{-6}, 8 (– 1) = 8 × 10^{-1} etc.)

Table C-2. Table of values of the functions $W(u_A, r/B)$ and $W(u_Y, r/B)$ for determining the hydraulic properties of semi-confined aquifers with delayed yield

[*Source:* Boulton (1963)]

r/B = 0.01		r/B = 0.1		r/B = 0.2		r/B = 0.316	
$1/u_A$	$W(u_A, r/B)$	$1/u_A$	$W(u_A, r/B)$	$1/u_A$	$W(u_A, r/B)$	$1/u_A$	$W(u_A, r/B)$
1(1)	1.82	1(1)	1.80	5(0)	1.19	1(0)	2.16(−1)
1(2)	4.04	5(1)	3.24	1(1)	1.75	2(0)	5.44(−1)
1(3)	6.31	1(2)	3.81	5(1)	2.95	5(0)	1.15
5(3)	7.82	2(2)	4.30	1(2)	3.29	1(1)	1.65
1(4)	8.40	5(2)	4.71	5(2)	3.50	5(1)	2.50
1(5)	9.42	1(3)	4.83	1(3)	3.51	2.65	2.62
1(6)	9.44	1(4)	4.85				2.65

r/B = 0.01		r/B = 0.1		r/B = 0.2		r/B = 0.316	
$1/u_Y$	$W(u_Y, r/B)$	$1/u_Y$	$W(u_Y, r/B)$	$1/u_Y$	$W(u_Y, r/B)$	$1/u_Y$	$W(u_Y, r/B)$
4(2)	9.45	4(0)	4.86	4(−1)	3.51	4(0)	2.66
4(3)	9.54	4(1)	4.95	4(0)	3.54	4(0)	2.74
4(4)	1.02(1)	4(2)	5.64	2(1)	3.69	4(1)	3.38
4(5)	1.23(1)	4(3)	7.72	4(1)	3.85	4(2)	5.42
4(6)	1.46(1)	4(4)	1.00(1)	1.5(2)	4.55	4(3)	7.72
				4(2)	5.42		

r/B = 0.4		r/B = 0.6		r/B = 0.8		r/B = 1.0	
$1/u_A$	$W(u_A, r/B)$	$1/u_A$	$W(u_A, r/B)$	$1/u_A$	$W(u_A, r/B)$	$1/u_A$	$W(u_A, r/B)$
1(0)	2.13(−1)	1(0)	2.06(−1)	5(−1)	4.60(−2)	5(−1)	4.44(−2)
2(0)	5.34(−1)	2(0)	5.04(−1)	1	1.97(−1)	1(0)	1.85(−1)
5(0)	1.11	5(0)	9.96(−1)	2	4.66(−1)	2(0)	4.21(−1)
1(1)	1.56	1(1)	1.31	5	8.57(−1)	5(0)	7.15(−1)
5(1)	2.18	2(1)	1.49	1(1)	1.05	1(1)	8.19(−1)
1(2)	2.22	5(1)	1.55	2(1)	1.12	2(1)	8.41(−1)

r/B = 0.4		r/B = 0.6		r/B = 0.8		r/B = 1.0	
$1/u_A$	$W(u_A, r/B)$	$1/u_A$	$W(u_A, r/B)$	$1/u_A$	$W(u_A, r/B)$	$1/u_A$	$W(u_A, r/B)$
1(−1)	2.23	4.44(−1)	1.59	2.5(−2)	1.13	4(−2)	8.44(−1)
1(0)	2.26	2.22(0)	1.71	2.5(−1)	1.16	4(−1)	9.01(−1)
5(0)	2.40	4.44(0)	1.84	1.25(0)	1.26	4(0)	1.36
1(1)	2.55	1.67(1)	2.45	2.5(0)	1.39	4(1)	3.14
3.75(1)	3.20	4.44(1)	3.26	9.37(0)	1.94		
1(2)	4.05			2.5(1)	2.70		

(Contd...)

r/B = 1.5		r/B = 2.0		r/B = 2.5		r/B = 3.0	
$1/u_A$	$W(u_A, r/B)$	$1/u_A$	$W(u_A, r/B)$	$1/u_A$	$W(u_A, r/B)$	$1/u_A$	$W(u_A, r/B)$
5(– 1)	3.94(– 2)	3.33(– 1)	1.00(– 2)	5(– 1)	2.71(– 2)	5(– 1)	2.10(– 2)
1(0)	1.51(– 1)	5(– 1)	3.35(– 2)	1(0)	8.03(– 2)	1(0)	5.34(– 2)
1.25(0)	1.99(– 1)	1(0)	1.14(– 1)	1.25(0)	9.61(– 2)	1.25(0)	6.07(– 2)
2(0)	3.01(– 1)	1.25(0)	1.44(– 1)	2(0)	1.17(– 1)	2(0)	6.81(– 2)
5(0)	4.13(– 1)	2(0)	1.94(– 1)	5(0)	1.25(– 1)	5(0)	6.95(– 2)
1(1)	4.27(– 1)	5(0)	2.27(– 1)	1(1)	1.25(– 1)	1(1)	6.95(– 2)
2(1)	4.19(– 1)	1(1)	2.28(– 1)				

r/B = 1.5		r/B = 2.0		r/B = 2.5		r/B = 3.0	
$1/u_A$	$W(u_A, r/B)$	$1/u_A$	$W(u_A, r/B)$	$1/u_A$	$W(u_A, r/B)$	$1/u_A$	$W(u_A, r/B)$
7.11(– 2)	4.44(– 1)	4(– 2)	2.39(– 1)	2.56(– 2)	1.32(– 1)	1.78(– 2)	7.43(– 2)
3.55(– 1)	5.09(– 1)	2(– 1)	2.83(– 1)	1.28(– 1)	1.62(– 1)	8.89(– 2)	9.39(– 2)
7.11(– 1)	5.87(– 1)	4(– 1)	3.37(– 1)	2.56(– 1)	1.99(– 1)	1.78(– 1)	1.19(– 1)
2.67(0)	9.63(– 1)	1.5(0)	6.14(– 1)	9.6(– 1)	3.99(– 1)	6.67(– 1)	2.62(– 1)
7.11(0)	1.57	4(0)	1.11	2.56(0)	7.98(– 1)	1.78(0)	5.77(– 1)

APPENDIX D

Table D-1. Monthly percentage of day, time, hours in the year for latitudes 0° to 65°, north of the equator

[Adapted from USDA, ARS Tech. Bull. No. 1275, (1962)]

Latitude (degrees) North	*Jan.*	*Feb.*	*Mar.*	*Apr.*	*May*	*June*	*July*	*Aug.*	*Sep.*	*Oct.*	*Nov.*	*Dec.*
(1)	*(2)*	*(3)*	*(4)*	*(5)*	*(6)*	*(7)*	*(8)*	*(9)*	*(10)*	*(11)*	*(12)*	*(13)*
65	3.45	5.14	7.90	9.92	12.65	14.12	13.66	11.25	8.55	6.60	4.12	2.64
64	3.75	5.30	7.93	9.87	12.42	13.60	13.31	11.15	8.58	6.70	4.35	3.04
63	4.01	5.40	7.95	9.83	12.22	13.22	13.02	11.04	8.60	6.79	4.55	3.37
62	4.25	5.52	7.99	9.75	12.03	12.91	12.79	10.92	8.50	6.86	4.72	3.67
61	4.46	5.61	8.01	9.71	11.88	12.63	12.55	10.84	8.55	6.94	3.89	3.93
60	4.67	5.70	8.05	9.66	11.72	12.39	12.33	10.72	8.57	7.00	5.04	4.15
59	4.81	5.78	8.05	9.60	11.61	12.23	12.21	10.60	8.56	7.07	5.09	4.31
58	4.99	5.85	8.06	9.55	11.44	12.00	12.00	10.55	8.56	7.13	5.13	4.55
57	5.14	5.93	8.07	9.51	11.32	11.77	11.87	10.47	8.54	7.19	5.27	4.69
56	5.29	6.00	8.10	9.45	11.20	11.67	11.69	10.40	8.52	7.25	5.54	4.89
55	5.39	6.06	8.12	9.41	11.11	11.53	11.59	10.32	8.51	7.30	5.62	5.01
54	5.53	6.12	8.15	9.36	11.00	11.40	11.43	10.27	8.50	7.33	5.74	5.17
53	5.64	6.19	8.16	9.32	10.88	11.31	11.34	10.19	8.52	7.38	5.83	5.31
52	5.75	6.23	8.17	9.28	10.81	11.13	11.22	10.15	8.49	7.40	5.94	5.43
51	5.87	6.25	8.21	9.26	10.76	11.07	11.13	10.05	8.48	7.41	5.97	5.46
50	5.98	6.32	8.25	9.25	10.69	10.93	10.90	10.00	8.44	7.43	6.07	5.65
48	6.13	6.42	8.22	9.15	10.50	10.72	10.83	9.92	8.45	7.56	6.24	5.86
46	6.30	6.50	8.24	9.09	10.37	10.54	10.66	9.82	8.44	7.61	6.38	6.05
44	6.45	6.59	8.25	9.04	10.22	10.38	10.50	9.73	8.43	7.67	6.51	6.23
42	6.60	6.66	8.28	8.97	10.10	10.21	10.37	9.64	8.42	7.73	6.63	6.39
40	6.73	6.73	8.30	8.92	9.99	10.08	10.34	9.56	8.41	7.78	6.73	6.53
38	6.87	6.79	8.34	8.90	9.92	9.95	10.10	9.47	8.38	7.80	6.82	6.66
36	6.99	6.86	8.35	8.85	9.31	9.83	9.99	9.40	8.36	7.85	6.92	6.79

(Contd...)

Latitude (degrees) North	*Jan.*	*Feb.*	*Mar.*	*Apr.*	*May*	*June*	*July*	*Aug.*	*Sep.*	*Oct.*	*Nov.*	*Dec.*
(1)	*(2)*	*(3)*	*(4)*	*(5)*	*(6)*	*(7)*	*(8)*	*(9)*	*(10)*	*(11)*	*(12)*	*(13)*
34	7.10	6.91	8.36	8.80	9.72	9.70	9.88	9.33	8.36	7.90	7.02	6.92
32	7.20	6.97	8.37	8.72	9.63	9.60	9.77	9.28	8.34	7.93	7.11	7.05
30	7.30	7.03	8.38	8.72	9.53	9.49	9.67	9.22	8.34	7.99	7.19	7.14
28	7.40	7.02	8.39	8.68	9.46	9.38	9.58	9.56	8.32	8.02	7.27	7.27
26	7.49	7.12	8.40	8.64	9.37	9.30	9.49	9.10	8.32	8.06	7.36	7.35
24	7.58	7.17	8.40	8.60	9.30	9.19	9.41	9.05	8.31	8.10	7.43	7.46
22	7.76	7.22	8.41	8.57	9.22	9.12	9.31	9.00	8.30	8.13	7.50	7.56
20	7.73	7.26	8.20	8.52	9.14	9.22	9.25	8.95	8.30	8.19	7.58	7.88
18	7.88	7.26	8.40	8.46	9.06	8.99	9.20	8.81	8.29	8.24	7.67	7.89
16	7.94	7.30	8.42	8.45	8.98	8.98	9.07	8.80	8.28	8.24	7.72	7.90
14	7.08	7.39	8.43	8.44	8.90	8.73	8.99	8.79	8.28	8.28	7.85	8.04
12	8.08	7.40	8.44	8.43	8.84	8.64	8.90	8.78	8.27	8.28	7.85	8.05
10	8.11	7.40	8.44	8.43	8.81	8.57	8.84	8.74	8.26	8.29	7.89	8.08
8	8.13	7.41	8.45	8.39	8.75	8.51	8.77	8.70	8.25	8.31	7.89	8.11
6	8.19	7.49	8.45	8.39	8.73	8.48	8.75	8.69	8.25	8.41	7.95	8.19
4	8.20	7.58	8.46	8.33	8.65	8.40	8.67	8.63	8.21	8.43	7.95	8.20
2	8.43	7.62	8.47	8.22	8.51	8.25	8.52	8.50	8.20	8.45	8.16	8.42
0	8.49	7.67	8.49	8.22	8.49	8.22	8.49	8.49	8.19	8.49	8.22	8.49

For latitudes, 0° to 50° south of the equator

(1) Latitude (degrees) South

(1)	*(2)*	*(3)*	*(4)*	*(5)*	*(6)*	*(7)*	*(8)*	*(9)*	*(10)*	*(11)*	*(12)*	*(13)*
0	8.49	7.67	8.49	8.22	8.49	8.22	8.49	8.49	8.19	8.49	8.22	8.49
2	8.55	7.71	8.49	8.19	8.44	8.17	8.43	8.44	8.19	8.52	8.27	8.55
4	8.64	7.76	8.50	8.17	8.39	8.08	8.20	8.41	8.19	8.56	8.33	8.65
6	8.71	7.81	8.50	8.12	8.30	8.00	8.19	8.37	8.18	8.59	8.38	8.74
8	8.79	8.84	8.51	8.11	8.24	7.91	8.13	8.32	8.18	8.62	8.47	8.84
10	8.85	7.86	8.52	8.09	8.18	7.84	8.11	8.28	8.18	8.65	8.52	8.90
12	8.91	7.91	8.53	8.06	8.15	7.79	8.08	8.26	7.17	8.67	8.58	8.95
14	8.97	7.97	8.54	8.03	8.07	7.70	7.08	8.19	8.16	8.69	8.65	9.01
16	9.09	8.02	8.56	7.98	7.96	7.57	7.94	8.14	8.14	8.76	8.72	9.17
18	9.18	8.06	8.57	7.93	7.99	7.50	7.88	8.90	8.14	8.80	8.80	9.24
20	9.25	8.09	8.58	7.92	7.83	7.41	7.73	8.05	8.13	8.83	8.85	9.32
22	9.36	8.12	8.58	7.89	7.74	7.30	7.76	8.03	8.13	8.86	8.90	9.38
24	9.44	8.17	8.59	7.87	7.60	7.24	7.58	7.99	8.12	8.89	8.96	9.47
26	9.52	8.28	8.00	7.81	7.56	7.07	7.49	7.87	8.11	8.94	9.10	9.61
28	9.61	8.31	8.61	7.79	7.49	6.99	7.40	7.85	8.10	8.97	9.19	9.74
30	9.69	8.33	8.63	7.75	7.43	6.94	7.30	7.80	8.09	9.00	9.24	9.80
32	9.76	8.36	8.64	7.70	7.39	6.85	7.20	7.73	8.08	9.04	9.31	9.87
34	9.88	8.41	8.65	7.68	7.30	6.73	7.10	7.69	8.06	9.07	9.38	9.99
36	10.06	8.53	8.67	7.61	7.10	6.59	6.99	7.59	8.06	9.15	9.51	10.21
38	10.14	8.61	8.68	7.59	7.03	6.46	6.87	7.51	8.05	9.19	9.60	10.34
40	10.24	8.65	8.70	7.54	6.96	6.33	6.73	7.46	8.04	9.23	9.69	10.42
42	10.39	8.72	8.71	7.49	6.85	6.20	6.60	7.39	8.01	9.27	9.79	10.57
44	10.52	8.81	8.72	7.44	6.73	6.04	6.45	7.30	8.00	9.34	9.91	10.72
46	10.68	8.88	8.73	7.39	6.61	5.87	6.30	7.21	7.98	9.41	10.03	10.90
48	10.85	9.98	8.76	7.32	6.45	5.69	6.13	7.12	7.96	9.47	10.17	11.09
50	11.03	9.06	8.77	7.25	6.31	5.48	5.98	7.03	7.95	9.53	10.32	11.30

Table D-2. Correction factors for length and number of days in a month, for use in the Thornthwaite formula. (Mean possible duration of sunlight expressed in units of 30 days of 12 hours each)

[*Source:* Thornthwaite, (1948)]

North (Lat.)	*Jan.*	*Feb.*	*Mar.*	*April*	*May*	*June*	*July*	*Aug.*	*Sept.*	*Oct.*	*Nov.*	*Dec.*
0	1.04	0.94	1.04	1.01	1.04	1.01	1.04	1.04	1.01	1.04	1.01	1.04
5	1.02	0.93	1.03	1.02	1.06	1.03	1.06	1.05	1.01	1.03	0.99	1.02
10	1.00	0.91	1.03	1.03	1.08	1.06	1.08	1.07	1.02	1.02	0.98	0.99
15	0.97	0.91	1.03	1.04	1.11	1.08	1.12	1.08	1.02	1.01	0.95	0.97
20	0.95	0.90	1.03	1.05	1.13	1.11	1.14	1.11	1.02	1.00	0.93	0.94
25	0.93	0.89	1.03	1.06	1.15	1.14	1.17	1.12	1.02	0.99	0.91	0.91
26	0.92	0.88	1.03	1.06	1.15	1.15	1.17	1.12	1.02	0.99	0.91	0.91
27	0.92	0.88	1.03	1.07	1.16	1.15	1.18	1.13	1.02	0.99	0.90	0.90
28	0.91	0.88	1.03	1.07	1.16	1.16	1.18	1.13	1.02	0.98	0.90	0.90
29	0.91	0.87	1.03	1.07	1.17	1.16	1.19	1.13	1.03	0.98	0.90	0.89
30	0.90	0.87	1.03	1.08	1.18	1.17	1.20	1.14	1.03	0.98	0.89	0.88
31	0.90	0.87	1.03	1.08	1.18	1.18	1.20	1.14	1.03	0.98	0.89	0.88
32	0.89	0.86	1.03	1.08	1.19	1.19	1.21	1.15	1.03	0.98	0.88	0.87
33	0.88	0.86	1.03	1.09	1.19	1.20	1.22	1.15	1.03	0.97	0.88	0.86
34	0.88	0.85	1.03	1.09	1.20	1.20	1.22	1.16	1.03	0.97	0.87	0.86
35	0.87	0.85	1.03	1.09	1.21	1.21	1.23	1.16	1.03	0.97	0.86	0.85
36	0.87	0.85	1.03	1.10	1.21	1.22	1.24	1.16	1.03	0.97	0.86	0.84
37	0.86	0.84	1.03	1.10	1.22	1.23	1.25	1.17	1.03	0.97	0.85	0.83
38	0.85	0.84	1.03	1.10	1.23	1.24	1.25	1.17	1.04	0.96	0.84	0.83
39	0.85	0.84	1.03	1.11	1.23	1.24	1.26	1.18	1.04	0.96	0.84	0.82
40	0.84	0.83	1.03	1.11	1.24	1.25	1.27	1.18	1.04	0.96	0.83	0.81
41	0.83	0.83	1.03	1.11	1.25	1.26	1.27	1.19	1.04	0.96	0.82	0.80
42	0.82	0.83	1.03	1.12	1.26	1.27	1.28	1.19	1.04	0.95	0.82	0.79
43	0.81	0.82	1.02	1.12	1.26	1.28	1.29	1.20	1.04	0.95	0.81	0.77
44	0.81	0.82	1.02	1.13	1.27	1.29	1.30	1.20	1.04	0.95	0.80	0.76
45	0.80	0.81	1.02	1.13	1.28	1.29	1.31	1.21	1.04	0.94	0.79	0.75
46	0.79	0.81	1.02	1.13	1.29	1.31	1.32	1.22	1.04	0.94	0.79	0.74
47	0.77	0.80	1.02	1.14	1.30	1.32	1.33	1.22	1.04	0.93	0.78	0.73
48	0.76	0.80	1.02	1.14	1.31	1.33	1.34	1.23	1.05	0.93	0.77	0.72
49	0.75	0.79	1.02	1.14	1.32	1.34	1.35	1.24	1.05	0.93	0.76	0.71
50	0.74	0.78	1.02	1.15	1.33	1.36	1.37	1.25	1.06	0.92	0.76	0.70
South (Lat.)												
5	1.06	0.95	1.04	1.00	1.02	0.99	1.02	1.03	1.00	1.05	1.03	1.06
10	1.08	0.97	1.05	0.99	1.01	0.96	1.00	1.01	1.00	1.06	1.05	1.10
15	1.12	0.98	1.05	0.98	0.98	0.94	0.97	1.00	1.00	1.07	1.07	1.12
20	1.14	1.00	1.05	0.97	0.96	0.91	0.95	0.99	1.00	1.08	1.09	1.15
25	1.17	1.01	1.05	0.96	0.94	0.88	0.93	0.98	1.00	1.10	1.11	1.18
30	1.20	1.03	1.06	0.95	0.92	0.85	0.90	0.96	1.00	1.12	1.14	1.21
35	1.23	1.01	1.06	0.94	0.89	0.82	0.87	0.94	1.00	1.13	1.17	1.25
40	1.27	1.06	1.07	0.93	0.86	0.78	0.84	0.92	1.00	1.15	1.20	1.29
42	1.28	1.07	1.07	0.92	0.85	0.76	0.82	0.92	1.00	1.16	1.22	1.31
44	1.30	1.08	1.07	0.92	0.83	0.74	0.81	0.91	0.99	1.17	1.23	1.33
46	1.32	1.10	1.07	0.91	0.82	0.72	0.79	0.90	0.99	1.17	1.25	1.35
48	1.34	1.11	1.08	0.90	0.80	0.70	0.76	0.89	0.99	1.18	1.27	1.37
50	1.37	1.12	1.08	0.89	0.77	0.67	0.74	0.88	0.99	1.19	1.29	1.41

Table D-3. Monthly values of 'I' in Thornthwaite formula corresponding to the mean monthly temperature in °C

[*Source:* Thornthwaite, (1948)]

T° C	0.0	0.1	0.2	0.3	0.4	0.5	0.6	0.7	0.8	0.9
0			0.01	0.01	0.02	0.03	0.04	0.05	0.06	0.07
1	0.09	0.10	0.12	0.13	0.15	0.16	0.18	0.20	0.21	0.23
2	0.25	0.27	0.29	0.31	0.33	0.35	0.37	0.39	0.42	0.44
3	0.46	0.48	0.51	0.53	0.56	0.58	0.61	0.63	0.66	0.69
4	0.71	0.74	0.77	0.80	0.82	0.85	0.88	0.91	0.94	0.97
5	1.00	1.03	1.06	1.09	1.12	1.16	1.19	1.22	1.25	1.29
6	1.32	1.35	1.39	1.42	1.45	1.49	1.52	1.56	1.59	1.63
7	1.66	1.70	1.74	1.77	1.81	1.85	1.89	1.92	1.96	2.00
8	2.04	2.08	2.12	2.15	2.19	2.23	2.27	2.31	2.35	2.39
9	2.44	2.48	2.52	2.56	2.60	2.64	2.69	2.73	2.77	2.81
10	2.86	2.90	2.94	2.99	3.03	3.08	3.12	3.16	3.21	3.26
11	3.30	3.34	3.39	3.44	3.48	3.53	3.58	3.62	3.67	3.72
12	3.76	3.81	3.86	3.91	3.96	4.00	4.05	4.10	4.15	4.20
13	4.25	4.30	4.35	4.40	4.45	4.50	4.55	4.60	4.65	4.70
14	4.75	4.81	4.86	4.91	4.96	5.01	5.07	5.12	5.17	5.22
15	5.28	5.33	5.38	5.44	5.49	5.55	5.60	5.65	5.71	5.76
16	5.82	5.87	5.93	5.98	6.04	6.10	6.15	6.21	6.26	6.32
17	6.38	6.44	6.49	6.55	6.61	6.66	6.72	6.78	6.84	6.90
18	6.95	7.01	7.07	7.13	7.19	7.25	7.31	7.37	7.43	7.49
19	7.55	7.61	7.67	7.73	7.79	7.85	7.91	7.97	8.03	8.10
20	8.16	8.22	8.28	8.34	8.41	8.47	8.53	8.59	8.66	8.72
21	8.78	8.85	8.91	8.97	9.04	9.10	9.17	9.23	9.29	9.36
22	9.42	9.49	9.55	9.62	9.68	9.75	9.82	9.88	9.95	10.01
23	10.08	10.15	10.25	10.28	10.35	10.41	10.48	10.55	10.62	10.68
24	10.75	10.82	10.89	10.95	11.02	11.09	11.16	11.23	11.30	11.37
25	11.44	11.50	11.57	11.64	11.71	11.78	11.85	11.92	11.99	12.06
26	12.13	12.21	12.28	12.35	12.42	12.49	12.56	12.63	12.70	12.78
27	12.85	12.92	12.99	13.07	13.14	13.21	13.28	13.36	13.43	13.50
28	13.58	13.65	13.72	13.80	13.87	13.94	14.02	14.09	14.17	14.24
29	14.32	14.39	14.47	14.54	14.62	14.69	14.77	14.87	14.92	14.99
30	15.07	15.15	15.22	15.30	15.38	15.45	15.53	15.61	15.68	15.76
31	15.84	15.92	15.99	16.07	16.15	16.23	16.30	16.38	16.46	16.54
32	16.62	16.70	16.78	16.85	16.93	17.01	17.09	17.17	17.25	17.33
33	17.41	17.49	17.57	17.65	17.73	17.81	17.89	17.97	18.05	18.13
34	18.22	18.30	18.38	18.46	18.54	18.62	18.70	18.79	18.87	18.95
35	19.03	19.11	19.20	19.28	19.36	19.45	19.53	19.61	19.69	19.78
36	19.86	19.95	20.03	20.11	20.20	20.28	20.36	20.45	20.53	20.62
37	20.70	20.79	20.87	20.96	21.04	21.13	21.21	21.30	21.38	21.47
38	21.56	21.64	21.73	21.81	21.90	21.99	22.07	22.16	22.25	22.33
39	22.42	22.51	22.59	22.68	22.77	22.86	22.95	23.03	23.12	23.21
40	23.30									

Table D-4. Mean monthly extra-terrestrial radiation, R_a, expressed in equivalent evaporation: mm/day

(Angot's values)

Northern Hemisphere													Southern Hemisphere											
Jan.	*Feb.*	*March*	*April*	*May*	*June*	*July*	*Aug.*	*Sep.*	*Oct.*	*Nov.*	*Dec.*	*Lat.*	*Jan.*	*Feb.*	*March*	*April*	*May*	*June*	*July*	*Aug.*	*Sep.*	*Oct*	*Nov.*	*Dec.*
3.8	6.1	9.1	12.7	15.8	17.1	16.4	14.1	10.9	7.4	4.5	3.2	50°	17.5	14.7	10.9	7.0	4.2	3.1	3.5	5.5	8.9	12.9	16.5	18.2
4.3	6.6	9.1	13.0	15.9	17.2	16.5	14.3	11.2	7.8	5.0	3.7	48	17.6	14.9	11.2	7.5	4.7	3.5	4.0	6.0	9.3	13.2	16.6	18.2
4.9	7.1	10.1	13.3	16.0	17.2	16.6	14.5	11.5	8.3	5.5	4.3	46	17.7	15.1	11.5	7.9	5.2	4.0	4.4	6.5	9.7	13.4	16.7	18.3
5.3	7.6	10.1	13.7	16.1	17.2	16.6	14.7	11.9	8.7	6.0	4.7	44	17.8	15.3	11.9	8.4	5.7	4.4	4.9	6.9	10.2	13.7	16.7	18.3
5.9	8.1	11.0	14.0	16.2	17.3	16.7	15.0	12.2	9.1	6.5	5.2	42	17.8	15.5	12.2	8.8	6.1	4.9	5.4	7.4	10.6	14.0	16.8	18.3
6.4	8.6	11.4	14.3	16.4	17.3	16.7	15.2	12.5	9.6	7.0	5.7	40	17.9	15.7	12.5	9.2	6.6	5.3	5.9	7.9	11.0	14.2	16.9	18.3
6.9	9.0	11.8	14.5	16.4	17.2	16.7	15.3	12.8	10.0	7.5	6.1	38	17.9	15.8	12.8	9.6	7.1	5.8	6.3	8.3	11.4	14.4	17.0	18.3
7.4	9.4	12.1	14.7	16.4	17.2	16.7	15.4	13.1	10.6	8.0	6.6	36	17.9	16.0	13.2	10.1	7.5	6.3	6.8	8.8	11.7	14.6	17.0	18.2
7.9	9.8	12.4	14.8	16.5	17.1	16.8	15.5	13.4	10.8	8.5	7.2	34	17.8	16.1	13.5	10.5	8.0	6.8*	7.2	9.2	12.0	14.9	17.1	18.2
8.3	10.2	12.8	15.0	16.5	17.0	16.8	15.6	13.6	11.2	9.0	7.8	32	17.8	16.2	13.8	10.9	8.5	7.3	7.7	9.6	12.4	15.1	17.2	18.1
8.8	10.7	13.1	15.2	16.5	17.0	16.8*	15.7	13.9	11.6	9.5	8.3	30	17.8	16.4	14.0	11.3	8.9	7.8	8.1	10.1	12.7	15.3	17.3	18.1
9.3	11.1	13.4	15.3	16.5	16.8	16.7	15.7	14.1	12.0	9.9	8.8	28	17.7	16.4	14.3	11.6	9.3	8.2	8.6	10.4	13.0	15.4	17.2	17.9
9.8	11.5	13.7	15.3	16.4	16.7	16.6	15.7	14.3	12.3	10.3	9.3	26	17.6	16.4	14.4	12.0	9.7	8.7	9.1	10.9	13.2	15.5	17.2	17.8
10.2	11.9	13.9	15.4	16.4	16.6	16.5	15.8	14.5	12.6	10.7	9.7	24	17.5	16.5	14.6	12.3	10.2	9.1	9.5	11.2	13.4	15.6	17.1	17.7
10.7	12.3	14.2	15.5	16.3	16.4	16.4	15.8	14.6	13.0	11.1	10.2	22	17.4	16.5	14.8	12.6	10.6	9.6	10.0	11.6	13.7	15.7	17.0	17.5
11.2	12.7	14.4	15.6	16.3	16.4	16.3	15.9	14.8	13.3	11.6	10.7	20	17.3	16.5	15.0	13.0	11.0	10.0	10.4	12.0	13.9	15.8	17.0	17.4
11.6	13.0	14.1	16.6	16.1	16.1	16.1	15.8	14.9	13.6	12.0	11.1	18	17.1	16.5	15.1	13.2	11.4	10.4	10.8	12.3	14.1	15.8	16.8	17.1
12.0	13.3	14.7	15.6	16.0	15.9	15.9	15.7	15.0	13.9	12.4	11.6	16	16.9	16.4	15.2	13.5	11.7	10.8	11.2	12.6	14.3	15.8	16.7	16.8
12.4	13.6	14.9	15.7	15.8	15.7	15.7	15.7	15.1	14.1	12.8	12.0	14	16.7	16.4	15.3	13.7	12.1	11.2	11.6	12.9	14.5	15.8	16.5	16.6
12.8	13.9	15.1	15.7	15.7	15.5	15.5	15.6	15.2	14.4	13.3	12.5	12	16.6	16.3	15.4	14.0	12.5	11.6	12.0	13.2	14.7	15.8	16.4	16.5
13.2	14.2	15.3	15.7	15.5	15.3	15.3	15.5	15.3	14.7	13.6	12.9	10	16.4	16.3	15.5	14.2	12.8	12.0	12.4	13.5	14.8	15.9	16.2	16.2
13.6	14.5	15.3	15.6	15.3	15.0	15.1	15.4	15.3	14.8	13.9	13.3	8	16.1	16.1	15.5	14.4	13.1	12.4	12.7	13.7	14.9	15.8	16.0	16.0
13.9	14.8	15.4	15.4	15.1	14.7	14.9	15.2	15.3	15.0	14.2	13.7	6	15.8	16.0	15.6	14.7	13.4	12.8	13.1	14.0	15.0	15.7	15.8	15.7
14.3	15.0	15.5	15.5	14.9	14.4	14.6	15.1	15.3	15.1	14.5	14.1	4	15.5	15.8	15.6	14.9	13.8	13.2	13.4	14.3	15.1	15.6	15.5	15.4
14.7	15.3	15.6	15.3	14.6	14.2	14.3	14.9	15.3	15.3	14.8	14.4	2	15.3	15.7	15.7	15.1	14.1	13.5	13.7	14.5	15.2	15.5	15.3	15.1
15.0	15.5	15.7	15.3	14.4	13.9	14.1	14.8	15.3	15.4	15.1	14.8	0	15.0	15.5	15.7	15.3	14.4	13.9	14.1	14.8	15.3	15.4	15.1	14.8

(***Note***: By multiplying the values given in this table, which is in mm/day, by 2.45 gives the equivalent values of R_a in megajoules/square metre/day. (Reference, FAO Irrigation and Drainage Paper 56, *Crop Evapo-transpiration*, FAO (1998), Table 2.6, p. 219.)

Table D-5. Mean daily maximum duration of bright sunshine hours *N* for difference in months and latitudes

[***Source:*** Doorenbos and Pruitt, (1975)]

Northern Latitudes	*Jan.*	*Feb.*	*March*	*April*	*May*	*June*	*July*	*Aug.*	*Sept.*	*Oct.*	*Nov.*	*Dec.*
Southern Latitudes	*July*	*Aug.*	*Sept.*	*Oct.*	*Nov.*	*Dec.*	*Jan.*	*Feb.*	*March*	*April*	*May*	*June*
50°	8.5	10.1	11.8	13.8	15.4	16.3	15.9	14.5	12.7	10.8	9.1	8.1
48°	8.8	10.2	11.8	13.6	15.2	16.0	15.6	14.3	12.6	10.9	9.3	8.3
46°	9.1	10.4	11.9	13.5	14.9	15.7	15.4	14.2	12.6	10.9	9.5	8.7
44°	9.3	10.5	11.9	13.4	14.7	15.4	15.2	14.0	12.6	11.0	9.7	8.9
42°	9.4	10.6	11.9	13.4	14.6	15.2	14.9	13.9	12.9	11.1	9.8	9.1
40°	9.6	10.7	11.9	13.3	14.4	15.0	14.7	13.7	12.5	11.2	10.0	9.3
35°	10.1	11.0	11.9	13.1	14.0	14.5	14.3*	13.5	12.4	11.3	10.3	9.8
30°	10.4	11.1	12.0	12.9	13.6	14.0	13.9	13.2	12.4	11.5	10.6	10.2
25°	10.7	11.3	12.0	12.7	13.3	13.7	13.5	13.0	12.3	11.6	10.9	10.6
20°	11.0	11.5	12.0	12.6	13.1	13.3	13.2	12.8	12.3	11.7	11.2	10.9
15°	11.3	11.6	12.0	12.5	12.8	13.0	12.9	12.6	12.2	11.8	11.4	11.2
10°	11.6	11.8	12.0	12.3	12.6	12.7	12.6	12.4	12.1	11.8	11.6	11.5
5°	11.8	11.9	12.0	12.2	12.3	12.4	12.3	12.3	12.1	12.0	11.9	11.8
0°	12.1	12.1	12.1	12.1	12.1	12.1	12.1	12.1	12.1	12.1	12.1	12.1

Table D-6. Saturation vapour pressure e_a in *m bar* as function of mean air temperature *t* in °C

Temperature 0°C	e_a *m bar*	*Temperature 0°C*	e_a *m bar*
0	6.1	20	23.4
1	6.6	21	24.9
2	7.1	22	26.4
3	7.6	23	28.1
4	8.1	24	29.8
5	8.7	25	31.7
6	9.4	26	33.6
7	10.0	27	35.7
8	10.7	28	37.8
9	11.5	29	40.1
10	12.3	30	42.4
11	13.1	31	44.9
12	14.0	32	47.6
13	15.0	33	50.3
14	16.1	34	53.2
15	17.0	35	56.2
16	18.2	36	59.4
17	19.4	37	62.8
18	20.6	38	66.3
19	22.0	39	69.9

Table D-7. Values of wind function, $f(u) = 0.27\ (1 + W_2/100)$ for wind run at a 2 m height in km/day

Wind speed (km/day)	*0*	*10*	*20*	*30*	*40*	*50*	*60*	*70*	*80*	*90*
		0.30	0.32	0.35	0.38	0.41	0.43	0.46	0.49	0.51
100	0.54	0.54	0.59	0.62	0.65	0.67	0.70	0.73	0.76	0.78
200	0.81	0.84	0.86	0.89*	0.92	0.94	0.97	1.00	1.03	1.05
300	1.08	1.11	1.13	1.16	1.19	1.21	1.24	1.27	1.30	1.32
400	1.35	1.38	1.40	1.43	1.46	1.39	1.51	1.54	1.57	1.59
500	1.62	1.65	1.67	1.70	1.73	1.76	1.78	1.81	1.84	1.80
600	1.89	1.92	1.94	1.97	2.00	2.02	2.05	2.08	2.11	2.15
700	2.16	2.19	2.21	2.24	2.27	2.29	2.32	2.35	2.38	2.40
800	2.43	2.46	2.48	2.51	2.54	2.56	2.59	2.62	2.65	2.65
900	2.70									

*Where wind data are not collected at 2 metre height, the appropriate corrections for wind measurements taken at different heights are given below:

Measurement height, m	0.5	1.0	1.5	2.0	3.0	4.0	5.0	6.0	10.0	
Correction factor		1.35	1.15	1.06	1.00	0.93*	0.88	0.85	0.83	0.77

Table D-8. Values of weighting factor (1 – W) for the effect of wind and humidity of ET_0 at different temperatures and altitudes

Temperature °C		2	4	6	8	10	12	14	16	18	20	22	24	26	28	30	32	34	36	38	40
(1–W) at altitude	m																				
	0	0.57	0.54	0.51	0.48	0.45	0.42	0.39	0.36	0.34	0.32	0.29	0.27	0.25	0.23*	0.22	0.20	0.18	0.17	0.16	0.15
	500	0.56	0.52	0.49	0.46	0.43	0.45	0.38	0.35	0.33	0.30	0.28	0.26	0.24	0.22	0.21	0.19	0.18	0.16	0.15	0.14
	1000	0.54	0.51	0.48	0.45	0.42	0.39	0.36	0.34	0.31	0.29	0.27	0.25	0.23	0.21	0.20	0.18	0.17	0.15	0.14	0.13
	2000	0.51	0.48	0.45	0.42	0.39	0.36	0.34	0.31	0.29	0.27	0.25	0.23	0.21	0.19	0.18	0.16	0.15	0.14	0.13	0.12
	3000	0.48	0.45	0.42	0.39	0.36	0.34	0.31	0.29	0.27	0.25	0.23	0.21	0.19	0.18	0.16	0.15	0.14	0.13	0.12	0.11
	4000	0.46	0.42	0.39	0.36	0.34	0.31	0.29	0.27	0.25	0.23	0.21	0.19	0.18	0.16	0.15	0.14	0.13	0.12	0.11	0.10

$W = \Delta/(\Delta + r)$ in which Δ is the rate of change of the saturation vapour pressure with temperature and r is the psychrometric constant.

Table D-9. Value of weighting factor W for the effect of radiation of ET_0 at different temperatures and altitudes

Temperature °C		2	4	6	8	10	12	14	16	18	20	22	24	26	28	30	32	34	36	38	40
W at altitude	m																				
	0	0.43	0.46	0.49	0.52	0.55	0.58	0.61	0.64	0.66	0.68	0.71	0.73	0.75	0.77*	0.78	0.80	0.82	0.83	0.84	0.85
	500	0.44	0.48	0.51	0.54	0.57	0.60	0.62	0.65	0.67	0.70	0.72	0.74	0.76	0.78	0.79	0.81	0.82	0.84	0.85	0.86
	1000	0.46	0.49	0.52	0.55	0.58	0.61	0.64	0.66	0.69	0.71	0.73	0.75	0.77	0.79	0.80	0.82	0.83	0.85	0.86	0.87
	2000	0.49	0.52	0.55	0.58	0.61	0.64	0.66	0.69	0.71	0.73	0.75	0.77	0.79	0.81	0.82	0.84	0.85	0.86	0.87	0.88
	3000	0.52	0.55	0.58	0.61	0.64	0.66	0.69	0.71	0.73	0.75	0.77	0.79	0.81	0.82	0.84	0.85	0.86	0.87	0.88	0.89
	4000	0.54	0.58	0.61	0.64	0.66	0.69	0.71	0.73	0.75	0.77	0.79	0.81	0.82	0.64	0.85	0.86	0.87	0.89	0.90	0.90

Table-10. Conversion factor for extra-terrestrial radiation to net solar radiation R_{ns} for a given reflection a of 25% and different ratios of actual to maximum sunshine hours (1 – a) (0.25 + 0.50 n/N)

n/N	0.00	0.05	0.10	0.15	0.20	0.25	0.30	0.35	0.40	0.45	0.50	0.55	0.60	0.65	0.70	0.75	0.80	0.85	0.90	0.95	1.00
(1–a) (0.25+ 0.50 n/N)	0.19	0.21	0.22	0.24	0.26	0.28	0.30	0.32	0.34	0.36	0.37	0.39	0.41	0.43	0.45	0.47	0.49*	0.51	0.52	0.54	0.56

Table D-11. Correction for temperature $f(t)$ on long-wave radiation R_{nl}

$t°C$	0	2	4	6	8	10	12	14	16	18	20	22	24	26	28	30	32	34	36
$f(t) = 0Tk^4$	11.0	11.4	11.7	12.0	12.4	12.7	13.1	13.5	13.8	14.2	14.6	15.0	15.4	15.9	16.3*	16.7	17.2	17.7	18.1

Table D-12. Correction for vapour pressure $f(e_d)$ on long-wave radiation R_{nl}

e_d *m bar*	6	8	10	12	14	16	18	20	22	24	26	28	30	32	34	36	38	40
$f(e_d) = 0.34 - 0.444/\sqrt{e_d}$	0.23	0.22	0.20	0.19	0.18	0.16	0.15	0.14*	0.13	0.12	0.12	0.11	0.10	0.09	0.08	0.08	0.07	0.06

Table D-13. Correction for the ratio of actual and maximum bright sunshine hours $f(n/N)$ on long-wave radiation R_{nl}.

n/N	0.00	0.05	0.10	0.15	0.20	0.25	0.30	0.35	0.40	0.45	0.50	0.55	0.60	0.65	0.70	0.75	0.80	0.85	0.90	0.95	1.00
$f(n/N) = 0.1 + 0.9\, n/N$	0.10	0.15	0.19	0.24	0.28	0.33	0.37	0.42	0.46	0.51	0.55	0.60	0.64	0.69	0.73	0.78	0.82*	0.87	0.91	0.96	1.00

APPENDIX E

1. Values of gamma function

x	$\Gamma(x)$	x	$\Gamma(x)$
1.000	1.00000 00000	1.225	0.91191 56071
1.005	0.99713 85354	1.230	0.91075 48564
1.010	0.99432 58512	1.235	0.90962 34274
1.015	0.99156 12888	1.240	0.90852 10583
1.020	0.98884 42033	1.245	0.90744 74922
1.025	0.98617 39633	1.250	0.90640 24771
1.030	0.98354 99506	1.255	0.90538 57663
1.035	0.98097 15606	1.260	0.90439 71178
1.040	0.97843 82009	1.265	0.90343 62946
1.045	0.97594 92919	1.270	0.90250 30645
1.050	0.97350 42656	1.275	0.90159 71994
1.055	0.97110 25663	1.280	0.90071 84765
1.060	0.96874 36495	1.285	0.89986 66769
1.065	0.96642 69823	1.290	0.89904 15863
1.070	0.96415 20425	1.295	0.89824 29947
1.075	0.96191 83189	1.300	0.89747 06963
1.080	0.95972 53107	1.305	0.89672 44895
1.085	0.95757 25273	1.310	0.89600 41767
1.090	0.95545 94882	1.315	0.89530 95644
1.095	0.95338 57227	1.320	0.89464 04630
1.100	0.95135 07699	1.325	0.89399 66866
1.105	0.94935 41778	1.330	0.89337 80535
1.110	0.94739 55040	1.335	0.89278 43850
1.115	0.94547 43149	1.340	0.89221 55072
1.120	0.94359 01856	1.345	0.89167 12485
1.125	0.94174 26997	1.350	0.89115 14420
1.130	0.93993 14497	1.355	0.89065 59235
1.135	0.93815 60356	1.360	0.89018 45324
1.140	0.93641 60657	1.365	0.88973 71116
1.145	0.93471 11562	1.370	0.88931 35074
1.150	0.93304 09311	1.375	0.88891 35692
1.155	0.93140 50217	1.380	0.88853 71494
1.160	0.92980 30666	1.385	0.88818 41041
1.165	0.92823 47120	1.390	0.88785 42918
1.170	0.92669 96106	1.395	0.88754 75748
1.175	0.92519 74225	1.400	0.88726 38175
1.180	0.92372 78143	1.405	0.88700 28884
1.185	0.92229 04591	1.410	0.88676 46576
1.190	0.92088 50371	1.415	0.88654 89993
1.195	0.91951 12341	1.420	0.88635 57896
1.200	0.91816 87424	1.425	0.88618 49081
1.205	0.91685 72606	1.430	0.88603 62361
1.210	0.91557 64930	1.435	0.88590 96587
1.215	0.91432 61500	1.440	0.88580 50635
1.220	0.91310 59475	1.445	0.88572 23397

(Contd...)

x	Γ(x)		x	Γ(x)	
1.450	0.88566	13803	1.725	0.91361	34904
1.455	0.88562	20800	1.730	0.91466	53712
1.460	0.88560	43364	1.735	0.91573	62171
1.465	0.88560	80495	1.740	0.91682	60252
1.470	0.88563	31217	1.745	0.91793	47950
1.475	0.88567	94579	1.750	0.91906	25268
1.480	0.88574	69646	1.755	0.92020	92224
1.485	0.88583	55520	1.760	0.92137	48846
1.490	0.88594	51316	1.765	0.92255	95178
1.495	0.88607	56174	1.770	0.92376	31277
1.500	0.88622	69255	1.775	0.92498	57211
1.505	0.88639	89344	1.780	0.92622	73062
1.510	0.88659	16850	1.785	0.92748	78926
1.515	0.88680	49997	1.790	0.92876	74904
1.520	0.88703	87833	1.795	0.93006	61123
1.525	0.88729	30231	1.800	0.93138	37710
1.530	0.88756	76278	1.805	0.93272	04811
1.535	0.88786	25287	1.810	0.93407	62585
1.540	0.88817	76586	1.815	0.93545	11198
1.545	0.88851	29527	1.820	0.93684	50832
1.550	0.88886	83478	1.825	0.93825	81682
1.555	0.88924	37830	1.830	0.93969	03951
1.560	0.88963	91990	1.835	0.94114	17859
1.565	0.89005	45387	1.840	0.94261	23634
1.570	0.89048	97463	1.845	0.94410	21519
1.575	0.89094	47686	1.850	0.94561	11764
1.580	0.89141	95537	1.855	0.94713	94637
1.585	0.89191	40515	1.860	0.94868	70417
1.590	0.89242	82141	1.865	0.95025	39389
1.595	0.89296	19949	1.870	0.95184	01855
1.600	0.89351	53493	1.875	0.95344	58127
1.605	0.19408	82352	1.880	0.95507	08530
1.610	0.89468	06085	1.885	0.95671	53398
1.615	0.89529	24327	1.890	0.95837	93077
1.620	0.89592	36685	1.895	0.96006	27927
1.625	0.89657	42800	1.900	0.96176	58319
1.630	0.89724	42326	1.905	0.96348	84632
1.635	0.89793	34930	1.910	0.96523	07261
1.640	0.89864	20302	1.915	0.96699	26608
1.645	0.89936	98138	1.920	0.96877	43090
1.650	0.90011	68163	1.925	0.97057	57134
1.655	0.90088	30104	1.930	0.97239	69178
1.660	0.90166	83712	1.935	0.97423	79672
1.665	0.90247	28748	1.940	0.97609	89075
1.670	0.90329	64995	1.945	0.97797	97861
1.675	0.90413	92243	1.950	0.97988	06513
1.680	0.90500	10302	1.955	0.98180	15524
1.685	0.90588	18996	1.960	0.98374	25404
1.690	0.90678	18160	1.965	0.98590	36664
1.695	0.90770	07650	1.970	0.98768	49838
1.700	0.90863	87329	1.975	0.98968	65462
1.705	0.90959	57079	1.980	0.99170	84087
1.710	0.91057	16796	1.985	0.99375	06274
1.715	0.91156	66390	1.990	0.99581	32598
1.720	0.91258	05779	1.995	0.99789	63643
			2.000	1.00000	00000

APPENDIX F

Friction head loss in irrigation pipes

Table F-1. Friction loss in metres per 100 metres in lateral line of portable aluminium pipe with couplings

(Based on Scobey's formula and 9 metres pipe length)

[Adapted from Hurd (1969)]

Flow (litres/sec)	*Diameter of pipe*				
	5.0 cm Ks 0.34	*7.5 cm Ks 0.33*	*10.0 cm Ks 0.32*	*12.5 cm Ks 0.32*	*15.0 cm Ks 0.32*
1.26	0.32				
1.89	2.53				
2.52	4.49	0.565	0.130		
3.15	6.85	0.858	0.198		
3.79	9.67	1.21	0.280		
4.42	12.9	1.63	0.376	0.122	
5.05	16.7	2.10	0.484	0.157	
5.68	20.8	2.63	0.605	0.196	
6.31	25.4	3.20	0.738	0.240	0.099
7.57		4.54	1.04	0.339	0.140
8.83		6.09	1.40	0.454	0.188
10.10		7.85	1.80	0.590	0.242
11.36		9.82	2.26	0.733	0.302
12.62		12.0	2.76	0.896	0.370
13.88		14.4	3.30	1.07	0.443
15.14		16.9	3.90	1.26	0.522
16.41		19.7	4.54	1.47	0.608
17.67		22.8	5.22	1.70	0.700
18.93		25.9	5.96	1.93	0.798
20.19		29.3	6.74	2.18	0.904
21.45		32.8	7.56	2.45	1.02
22.72		36.6	8.40	2.74	1.13
23.98		40.6	9.36	3.03	1.26
25.24		44.7	10.3	3.34	1.38
26.50			11.3	3.66	1.51
27.76			12.3	4.00	1.66
29.03			13.4	4.35	1.80
30.29			14.6	4.72	1.95
31.55			15.8	5.10	2.12
34.70			18.9	6.12	2.52
37.86			22.2	7.22	2.98
41.01			25.9	8.40	3.46
44.17			29.8	9.68	3.99
47.32			33.8	11.0	4.54
50.48				12.5	5.15
53.63				14.0	5.78
56.79				15.6	6.44
59.94				17.3	7.14
63.10				19.0	7.86

(***Note***: For 6 metres pipe length, *increase* values in the Table by 7.0 per cent and for 12 metres length *decrease* values by 3.0 per cent.)

Table F-2. Scobey's pipe friction coefficient K_s

[Adapted from Hurd (1969)]

Type of pipe material	K_s
Asbestos-cement, plastic	0.32
Cement-lined cast iron, dipped and wrapped steel	0.36
Unprotected steel (new), aluminium pipe with couplings	0.40
Slightly used steel	0.44
Fifteen year old steel or iron	0.48
Rough interior	0.54
Very rough interior	0.60

Table F-3. Friction loss in metres per 100 metres in main lines of portable aluminium pipes with couplings

(Based on Scobey's formulae, K_s 0.40 and 9 metres pipe lengths)

[Adapted from Hurd (1969)]

Flow litres / sec.	*Diameter of pipe*						
	7.5 cm	*10.0 cm*	*12.5 cm*	*15.0 cm*	*17.5 cm*	*20.0 cm*	*25.0 cm*
2.52	0.658	0.157					
3.15	1.006	0.239					
3.79	1.423	0.339					
4.42	1.906	0.449	0.150				
5.05	2.457	0.584	0.193				
5.68	3.073	0.731	0.242				
6.31	3.754	0.893	0.295	0.120			
7.57	5.307	1.263	0.417	0.170			
8.83	7.113	1.693	0.560	0.227			
10.10	9.169	2.182	0.721	0.293			
11.36	11.47	2.729	0.967	0.366			
12.62	14.01	3.333	1.102	0.448	0.209		
13.88	16.79	3.996	1.321	0.537	0.251		
15.14	19.81	4.713	1.558	0.633	0.296		
16.41	23.06	5.488	1.814	0.737	0.344		
17.67	26.55	6.316	2.089	0.849	0.397		
18.93	30.27	7.203	2.381	0.967	0.452	0.235	
20.19	34.22	8.142	2.092	1.094	0.511	0.265	
21.45	38.39	9.133	3.020	1.227	0.573	0.298	
22.72	42.80	10.18	3.366	1.368	0.639	0.332	
23.98	47.43	11.29	3.731	1.516	0.708	0.368	
25.24	52.28	12.44	4.113	1.671	0.781	0.399	0.136
26.50	13.65		4.513	1.833	0.857	0.445	0.149
27.76	14.57		4.930	1.988	0.936	0.486	0.163
29.03	16.23		5.364	2.179	1.019	0.529	0.177
30.29	17.59		5.815	2.363	1.104	0.573	0.192
31.55	19.01		6.284	2.554	1.193	0.620	0.208
34.70	22.79		7.532	3.060	1.430	0.742	0.249
37.86	26.88		8.886	3.611	1.687	0.876	0.294
41.01	31.30		10.35	4.204	1.965	1.020	0.342
44.17	36.04		11.91	4.839	2.262	1.174	0.394
47.32		41.08	13.58	5.517	2.520	1.339	0.449
50.48			15.35	6.237	2.915	1.513	0.507
53.63			17.22	6.999	3.271	1.698	0.569
56.79			19.20	7.801	3.666	1.893	0.635
59.94			21.28	8.645	4.041	2.097	0.703
63.10			23.45	9.530	4.454	2.312	0.775
69.49			28.11	11.42	5.338	2.771	0.929
75.72			31.75	13.58	6.298	3.269	1.096
82.03				15.69	7.333	3.886	1.277
88.34				18.06	8.441	4.382	1.470
94.65				20.59	9.624	4.996	1.675
101.0				23.28	10.88	5.648	1.894
107.0				26.12	12.21	6.337	2.125
114.0					13.63	7.064	2.369
120.0					15.08	7.829	2.625
126.0					16.62	8.630	2.894

(***Note***: Where 6.1 m sections of pipes are used, *increase* values shown in table by 7.0 per cent. Where 12.2 m sections of pipe are used, *decrease* values shown in the table by .0 per cent.)

Table F-4. Friction head loss in semi-rigid plastic irrigation pipelines manufactured of PVC or asbestos compounds
(Standard dimension ratio No., SDR 21)

[Adapted from Hurd (1969)]

Flow litres/sec.	*Friction loss in metres/100 metres*						
	Diameter of pipe						
	2.5 cm	*3.18 cm*	*3.8 cm*	*5.08 cm*	*6.25 cm*	*7.61 cm*	*8.9 cm*
0.126	0.15	0.04	0.02				
0.252	0.54	0.17	0.09	0.03	0.01		
0.378	1.15	0.37	0.19	0.06	0.02		
0.503	1.97	0.63	0.32	0.11	0.04	0.01	
0.631	2.98	0.95	0.49	0.16	0.06	0.02	0.01
0.946	6.32	2.03	1.04	0.35	0.14	0.05	0.02
1.26	10.79	3.46	1.78	0.60	0.23	0.09	0.04
1.58	16.30	5.22	2.70	0.91	0.36	0.12	0.07
1.89	22.86	7.32	3.78	1.27	0.50	0.19	0.10
2.21		9.75	5.03	1.70	0.67	0.25	0.13
2.52		12.46	6.46	2.18	0.86	0.32	0.17
2.84		15.51	8.02	2.71	1.07	0.40	0.21
3.15		18.87	9.75	3.30	1.30	0.49	0.25
3.47		22.48	11.64	3.94	1.54	0.59	0.30
3.79			13.64	4.62	1.81	0.69	0.36
4.10			15.85	5.36	2.10	0.80	0.41
4.42			18.19	6.14	2.42	0.92	0.47
4.73			20.65	6.99	2.75	1.06	0.55
5.05			23.28	7.86	3.10	1.19	0.62
5.36				8.81	3.47	1.33	0.69
5.68				9.79	3.85	1.48	0.77
5.99				10.82	4.25	1.64	0.85
6.31				11.89	4.69	1.80	0.93
6.94				14.21	5.59	2.14	1.11
7.59				16.69	6.56	2.52	1.31
8.20				19.35	7.63	2.92	1.53
8.83				22.21	8.73	3.36	1.75
9.46					9.94	3.82	1.99
10.10					11.20	4.29	2.24
10.73					12.51	4.80	2.50
11.36					13.90	5.35	2.79
11.99					15.39	5.92	3.08
12.62					16.91	6.50	3.38
13.88					20.19	7.77	4.04
15.14					23.73	9.12	4.76
16.41						10.57	5.51
17.67						12.11	6.32
18.93						13.78	7.18
20.19						15.52	8.10
21.45						17.37	9.07
22.72						19.27	10.08
23.98						21.33	11.13
25.24						23.45	12.22
26.50							13.40
27.76							14.59
29.03							15.86
30.29							17.15
31.55							18.50

(***Note:*** Table based on Hazen William's equation, C_i=150.)

1. To find friction head loss in PVC or asbestos pipe having a standard dimension ratio other than 21, the values in the table should be multiplied by the appropriate conversion factor shown below:

SDR No.	*Conversion factor*
13.5	1.35
17	1.13
21	1.00
26	0.91
32.5	0.84

Table F-4. Friction head loss in semi-rigid plastic irrigation pipelines manufactured of PVC or asbestos compounds

(Standard dimension ratio, SDR No. 21)

[Adapted from Hurd (1969)]

Flow litres/sec.	Friction loss in metres/100 metres					
	Diameter of pipe					
	10.0 cm	12.5 cm	15.0 cm	20.0 cm	25.0 cm	30.0 cm
0.946	0.01					
1.26	0.02					
1.58	0.04	0.01				
1.89	0.05	0.02				
2.21	0.07	0.02	0.01			
2.52	0.09	0.03	0.01			
2.84	0.12	0.04	0.01			
3.15	0.14	0.05	0.02			
3.47	0.17	0.06	0.02			
3.79	0.20	0.07	0.03			
4.10	0.23	0.08	0.03	0.01		
4.42	0.27	0.09	0.04	0.01		
4.73	0.31	0.11	0.04	0.01		
5.05	0.35	0.12	0.05	0.01		
5.36	0.39	0.14	0.05	0.01		
5.08	0.43	0.15	0.06	0.01		
5.99	0.48	0.17	0.07	0.02		
6.31	0.52	0.19	0.07	0.02		
6.84	0.63	0.22	0.09	0.02		
7.57	0.74	0.26	0.10	0.03	0.01	
8.20	0.85	0.30	0.12	0.03	0.01	
8.83	0.98	0.35	0.14	0.04	0.01	
9.46	1.11	0.40	0.16	0.05	0.01	
10.10	1.26	0.44	0.19	0.05	0.01	
10.73	1.41	0.49	0.21	0.06	0.02	
11.36	1.57	0.55	0.24	0.07	0.02	0.01
11.99	1.73	0.61	0.26	0.07	0.02	0.01
12.62	1.90	0.67	0.29	0.08	0.02	0.01
13.88	2.28	0.81	0.34	0.09	0.03	0.01
15.14	2.67	0.95	0.40	0.10	0.03	0.01
16.41	3.10	1.10	0.46	0.12	0.04	0.02
17.67	3.56	1.26	0.54	0.14	0.05	0.02
18.93	4.04	1.43	0.61	0.17	0.05	0.02
20.19	4.56	1.62	0.69	0.19	0.06	0.03
21.45	5.10	1.82	0.77	0.21	0.07	0.03
22.72	5.67	2.02	0.86	0.24	0.08	0.03
23.98	6.26	2.22	0.95	0.26	0.09	0.04
25.24	6.90	2.45	1.04	0.28	0.10	0.04
26.50	7.55	2.69	1.14	0.31	0.10	0.05
27.76	8.23	2.92	1.25	0.34	0.11	0.05
29.03	8.94	3.18	1.35	0.37	0.12	0.06
30.29	9.67	3.44	1.46	0.41	0.14	0.06
31.55	10.42	3.70	1.58	0.43	0.15	0.06
34.70	12.44	4.42	1.89	0.52	0.18	0.07
37.86	14.61	5.21	2.22	0.61	0.21	0.09

(Contd...)

SDR No.	Factor
13.5	1.35
17	1.13
21	1.00
26	.91
32.5	.84

Flow litres/sec.	Friction loss in metres/100 metres					
	Diameter of pipe					
	10.0 cm	12.5 cm	15.0 cm	20.0 cm	25.0 cm	30.0 cm
37.86	14.61	5.21	2.22	0.61	0.21	0.09
41.01	16.94	6.04	2.58	0.71	0.24	0.10
44.17	19.45	6.92	2.96	0.81	0.28	0.12
47.32	22.08	7.87	3.36	0.93	0.32	0.14
50.48		8.88	3.78	1.04	0.36	0.16
53.63		9.93	4.24	1.17	0.40	0.17
56.79		11.05	4.71	1.30	0.44	0.19
59.84		12.18	5.21	1.44	0.49	0.21
63.10		13.40	5.73	1.58	0.54	0.23
66.25		14.67	6.27	1.73	0.59	0.26
69.41		16.00	6.83	1.88	0.65	0.28
72.56		17.39	7.41	2.05	0.70	0.30
75.72		18.80	8.02	2.21	0.76	0.33
78.87		20.27	8.66	2.39	0.82	0.35
82.03		21.78	9.32	2.57	0.88	0.37
85.18			9.99	2.76	0.95	0.40
88.34			10.66	2.95	1.01	0.43
91.49			11.40	3.16	1.08	0.47
94.65			12.13	3.35	1.15	0.50
101			13.68	3.78	1.30	0.56
107			15.29	4.23	1.45	0.62
114			16.99	4.70	1.62	0.70
120			18.81	5.20	1.79	0.77
126			20.66	5.72	1.97	0.84
133			22.61	6.26	2.15	0.93
139			24.67	6.83	2.34	1.01
145				7.42	2.55	1.10
151				8.02	2.76	1.19
158				8.67	2.97	1.29
164				9.31	3.20	1.39
170				9.98	3.43	1.49
177				10.67	3.67	1.59
183				11.39	3.92	1.69
189				12.10	4.17	1.81
197				12.89	4.43	1.92
202				13.66	4.71	2.04
208				14.46	4.97	2.15
215				15.29	5.27	2.28
221				16.11	5.56	2.41
227				16.99	5.85	2.53
233				17.89	6.17	2.67
240				18.76	6.47	2.80
246				19.69	6.79	2.94
252				20.67	7.11	3.08

SDR No.	*Factor*
13.5	1.35
17	1.13
21	1.00
26	.91
32.5	.84

Table F-5. Friction loss in metres per 100 metres in asbestos cement pressure pipe

[Adapted from Hurd (1969)]

Flow litres/sec.	Nominal pipe diameter				
	10 cm	*15 cm*	*20 cm*	*25 cm*	*30 cm*
6.31	0.677				
7.57	0.954				
8.83	1.28				
10.10	1.65				
11.36	2.06				
12.62	2.53	0.372			
13.88	3.03	0.447			
15.14	3.56	0.525			
16.41	4.16	0.611			
17.67	4.77	0.705			
18.93	5.44	0.803			
20.19	6.16	0.910			
21.45	6.91	1.02			
22.72	7.70	1.14			
23.48	8.54	1.26			
25.24	9.40	1.39	0.324		
26.50	10.3	1.52	0.355		
27.76	11.3	1.66	0.389		
29.03	12.3	1.81	0.423		
30.29	13.3	1.96	0.458		
31.55	14.4	2.12	0.495		
34.70	17.2	2.55	0.594		
37.86	20.3	2.99	0.701	0.214	
41.01	23.7	3.49	0.818	0.249	
44.17	27.3	4.02	0.935	0.287	
47.32	31.1	4.57	1.07	0.328	
50.48		5.18	1.21	0.370	0.152
53.63		5.81	1.36	0.415	0.170
56.79		6.46	1.51	0.464	0.190
59.94		7.17	1.68	0.511	0.210
63.10		7.91	1.85	0.564	0.232
69.49		9.45	2.21	0.675	0.278
75.72		11.2	2.62	0.800	0.328
82.03		13.0	3.04	0.932	0.384
88.34		15.0	3.50	1.07	0.438
94.65		17.1	3.99	1.22	0.502
101.0		19.3	4.52	1.38	0.566
107.3			5.06	1.55	0.637
114.0			5.67	1.73	0.710
120.0			6.26	1.91	0.787
126.0			6.90	2.11	0.864
139.0			8.27	2.53	1.04
151.0			9.75	2.98	1.23
164.0			11.4	3.47	1.43
177.0			13.1	4.00	1.64
189.4			14.9	4.56	1.87

*Scobey's coefficient $K_s = 0.32$.

APPENDIX G

CONVERSION FACTORS

Length

1 metre	= 3.2808 feet
1 metre	= 39.37 inches
1 centimetre	= 0.3937 inch
1 kilometre	= 0.6214 mile
1 foot	= 0.3048 metre
1 inch	= 2.54 centimetres
1 mile	= 5,280 feet
1 mile	= 1.609 kilometres

Area

1 square metre	= 10.764 square feet
1 square centimetre	= 0.155 square inch
1 square kilometre	= 100 hectares
1 square kilometre	= 0.3861 sqare mile
1 hectare	= 10,000 square metres
1 hectare	= 107.640 square feet
1 hectare	= 2.471 acres
1 square foot	= 0.0929 square metre
1 square inch	= 6.452 square centimetres
1 acre	= 43.560 square feet
1 acre	= 0.4047 hectare
1 square mile	= 640 acres
1 square mile	= 258.99 hectares
1 square mile	= 2.59 square kilo-metres

Volume

1 cubic metre	= 35.314 cubic feet
1 cubic metre	= 1.308 cubic yards
1 cubic metre	= 1,000 litres
1 litre	= 0.0353 cubic foot
1 litre	= 0.2642 U.S. gallon
1 litre	= 0.2201 Imperial gallon
1 cubic centimetre	= 0.061 cubic inch
1 cubic foot	= 0.0283 cubic metre
1 cubic foot	= 23.32 litres
1 cubic foot	= 7.40 U.S. gallons
1 cubic foot	= 6.23 Imperial gallons
1 cubic inch	= 16.39 cubic centi-metres
1 cubic yard	= 0.7645 cubic metre
1 U.S. gallon	= 3.7854 litres
1 U.S. gallon	= 0.833 Imperial gallon
1 Imperial gallon	= 1.201 U.S. gallon
1 Imperial gallon	= 4.5436 litres
1 acre-foot	= 43,560 cubic foot
1 acre-foot	= 1,233.5 cubic metres
1 acre inch	= 3,630 cubic feet
1 acre inch	= 102.8 cubic metres

Rate of Flow

1 cubic metre per second	= 35.314 cubic feet per second
1 cubic metre per hour	= 0.278 litres per second
1 cubic metre per hour	= 4.403 U.S. gallons per minute
1 cubic metre per hour	= 3.668 Imperial gallons per minute
1 litre per second	= 0.0353 cubic foot per second
1 litre per second	= 15.852 U.S. gallons per minute
1 litre per second	= 13.206 Imperial gallons per minute
1 litre per second	= 3.6 cubic metres per hour
1 cubic foot per second (second-foot or cusec)	= 0.0283 cubic metres per second
1 cubic foot per second	= 28.32 litres per second
1 cubic foot per second	= 448.8 U.S. gallons per minute
1 cubic foot per second	= 373.8 Imperial gallons per minute
1 cubic foot per second	= 1 acre inch per hour (approx.)
1 cubic foot per second	= 2 acre foot per day (approx.)
1 U.S. gallon per minute	= 0.06309 litres per second
1 Imperial gallon per minute	= 0.07573 litre per second

Useful Approximations

5 cm	= 2 inches
10 cm	= 4 inches
30 cm	= 1 ft
10 metres	= 33 ft
10 km	= 6 miles
16 km	= 10 miles
1 hectare	= 2.5 acres
1 square inch	= 6.5 square cm
1 square metre	= 11 square feet
16.4 cubic centimetres	= 1 cubic inch
35 cubic feet	= 1 cu metre

APPENDIX H

English (FPL) to Metric Conversion Constants

Measure	*FPS* Unit	*FPS to Metric*	*METRIC* Unit	*Metric to FPS*
Length	inch (″)	× 25.4	mm = $\frac{1}{1000}$ m	× 0.03937
	foot (′) =12″	× 0.3048	m	× 3.28
	yard = 3′	× 0.9144	m	× 1.09361
	mile =1,760 yd	× 1.6093	km =1,000 m	× 0.621
	nautical mile	× 1.852	km = 1,000 m	× 2.540
Area	sq in	× 6.451	cm^2	× 0.1550
	sq ft	× 0.0929	m^2	× 10.764
	acres	× 0.4047	hectare	× 2.471
Volume	cu in	× 16.3871	cm^3	× 0.0610
	Imp. gallon (=1.2 American gallon)	× 4.546	litre = $(dm)^3$	× 0.220
	Imp. gallon (=10 lb)	× 0.004546	m^3	× 2.20
	American gallon	× 3.785	litre	× 0.2642
	Petroleum barrel (= 42 American gallons)	× 1.59	hl	× 0.63
	cu ft	× 28.316	litre	× 0.0353
	cu ft	× 0.0283	m^3	× 35.38
Weight	grain = $\frac{1}{7000}$	× 0.0648	gr	× 15.42
	ounce = $\frac{1}{16}$ lb	× 28.35	gr	× 0.035
	pound = 1 lb	× 0.4536	kg	× 2.205
	cwt = 112 lb	× 50.802	kg	× 0.0197
	short ton or American ton = 2,000 lb	× 0.907	tonne	× 1.102
	long ton = 2,240 lb	× 1.016	tonne	× 0.98421
Pressure and head	lb/sq in = $\frac{\text{ft of water}}{2.31}$	× 0.0703	$kg/cm^2 = \frac{\text{m of water}}{10}$	× 14.22
	lb/sq in	× 0.068	atm	× 14.7
	long ton/sq in	× 157.5	kg/cm^2	× 0.00635
	in mercury = $\frac{\text{ft of water}}{1.13}$	× 345	mm of water	× 0.0029
	in mercury	× 25.4	mm of mercury	× 0.03937
	1 atm (= 34 ft of water =14.7 lb/sq in)	× 1.033	1atm. (=10 m of water or 1 kg/cm^2)	× 0.967
Density	lb/cu ft	× 16.02	kg/m^3	× 0.0625
	lb/cu ft	× 0.01602	kg/dm^3 or kg/litre	× 62.424
	lb/cu in	× 27.6799	kg/cm^3	× 0.0362
Velocity	ft/sec	× 0.3048	m/sec	× 3.28
	ft/min	× $\frac{5.08}{1000}$	m/sec	× 196.85
Rate of discharge	cu ft/sec (cusec)	× 0.0283	m^3/sec	× 35.38
	cu ft/sec (cusec)	× 28.316	lit/sec (=dm^3/sec)	× 0.0353
	cu ft/min	× 0.472	lit/sec (=dm^3/sec)	× 2.22
	gal/min (gpm)	× 0.0757	lit/sec (=dm^3/sec)	× 13.22
	gal/min (gpm)	× $\frac{0.757}{10.000}$	m^3/sec	× 13.22

APPENDIX I

Answers to numerical and objective type questions

Chapter 1:

Q.No.79: 11 hours 27 minutes

Q.No. 82: (*i*) T; (*ii*) F; (*iii*) T; (*iv*) T; (*v*) T; (*vi*) T; (*vii*) T; (*viii*) T; (*ix*) F; (*x*) F; (*xi*) T; (*xii*) F.

Q.No.83: (*i*) Dublin, Ireland; (*ii*) Rio de Janeiro, Brazil ... worldwide reforms;
(*iii*) Terrestrial ... solar ... climatic cycle; (*iv*) Hydrometry;
(*v*) (*i*) orographical rain (*ii*) frontal rainfall (*iii*) convective rainfall
(*vi*) Europe ... Asia ... North America; (*vii*) Africa, Asia ... South America (*viii*) 268 million;
(*ix*) 501; (*x*) 200; (*xi*) 84,000 ... 60%; (*xii*) 67 ... 81; (*xiii*) auxiliary; (*xiv*) 700;
(*xv*) 46 ... Karikala Chola; (*xvi*) Colorado ... 1930s; (*xvii*) Rajang River,
(*xviii*) Al-Tabka ... Euphratus; (*xix*) Tahama; (*xx*) 17; (*xxi*) The Gulf of Guinea ... Eastern;
(*xxii*) 2.5; (*xxiii*) Inland basins ... Sahara; (*xxiv*) Egypt ... Libya ... Algeria; (*xxv*) Algeria;
(*xxvi*) Niger ... Senegal ... Volta; (*xxvii*) Ethiopia; (*xxviii*) Sugarcane; (*xxix*) Juba ... Shebelly;
(*xxx*) Grand Anicut ... Kaveri; (*xxxi*) 1355 ... Ferozshah Tughlaq; (*xxxii*) Sir Arthur Cotton;
(*xxxiii*) Anantaraja Sagar ... Cuddapah; (*xxxiv*) Sir Arthur Cotton ... Major P.T. Cautley ... Bharat Ratna M. Visveshwariah; (*xxxv*) Maharashtra ... Gujarat; (*xxxvi*) 1953;
(*xxxvii*) Great ... Middle ... Shivalik; (*xxxviii*) Andhra Pradesh; (*xxxix*) Uttar Pradesh;
(*xl*) Unconsolidated formations ... Semi-consolidated formations ... Consolidated formations;
(*xli*) rice ... cotton; (*xlii*) space ... time ... augmentation; (*xliii*) 75 ... 20

Q.No. 84: (*i*) (A) (IV); (B) (V); (C) (I); (D) (III); (E) (II) (*ii*) (A) (V); (B) (IV); (C) (II); (D) (III); (E) (I)
(*iii*) (A) (IV); (B) (V); (C) (I); (D) (II); (E) (III) (*iv*) (A) (V); (B) (IV); (C) (II); (D) (III); (E) (I)
(*v*) (A) (IV); (B) (V); (C) (I); (D) (II); (E) (III) (*vi*) (A) (IV); (B) (V); (C) (I); (D) (II); (E) (III)
(*vii*) (A) (V); (B) (IV); (C) (I); (D) (II); (E) (III)

Chapter 2:

Q.No.2: 12%;

Q.No.3: 11%, 9%

Q.No.39: (*i*) T; (*ii*) F; (*iii*) T; (*iv*) F; (*v*) T; (*vi*) T; (*vii*) T; (*viii*) F; (*ix*) T; (*x*) T; (*xi*) F; (*xii*) F

Q.No.40: (*i*) (c); (*ii*) (c); (*iii*) (c); (*iv*) (b); (*v*) (a); (*vi*) (b); (*vii*) (b); (*viii*) (d); (*ix*) (c); (*x*) (d); (*xi*) (b);
(*xii*) (b); (*xiii*) (a); (*xiv*) (b); (*xv*) (d); (*xvi*) (d); (*xvii*) (b)

Q.No.41: (*i*) (A) (V); (B) (IV); (C) (II); (D) (III); (E) (I) (*ii*) (A) (IV); (B) (V); (C) (II); (D) (III); (E) (I)
(*iii*) (A) (IV); (B) (V); (C) (I); (D) (II); (E) (III) (*iv*) (A) (V); (B) (III); (C) (I); (D) (II); (E) (IV)
(*v*) (A) (V); (B) (IV); (C) (I); (D) (II); (E) (III)

Chapter 3:

Q.No.2: 20 kg

Q.No.39: (*i*) T; (*ii*) T; (*iii*) F; (*iv*) F; (*v*) T; (*vi*) F; (*vii*) F; (*viii*) T; (*ix*) F; (*x*) T; (*xi*) T; (*xii*) T;
(*xiii*) F; (*xiv*) T; (*xv*) F; (*xvi*) T; (*xvii*) F.

Q.No.40: (*i*) (a); (*ii*) (b); (*iii*) (b); (*iv*) (c); (*v*) (c); (*vi*) (b).

Q.No.41: (*i*) (A) (V); (B) (IV); (C) (I); (D) (II); (E) (III) (*ii*) (A) (V); (B) (IV); (C) (I); (D) (II); (E) (III)
(*iii*) (A) (IV); (B) (V); (C) (I); (D) (II); (E) (III) (*iv*) (A) (V); (B) (IV); (C) (I); (D) (II); (E) (III)
(*v*) (A) (IV); (B) (V); (C) (1); (D) (II); (E) (III)

Chapter 4:

Q.No.36: (*i*) T; (*ii*) F; (*iii*) T; (*iv*) T; (*v*) F; (*vi*) T; (*vii*) T; (*viii*) F; (*ix*) T; (*x*) T; (*xi*) F;
(*xii*) T; (*xiii*) F; (*xiv*) T.

Q.No.37: (*i*) (c); (*ii*) (b); (*iii*) (b); (*iv*) (b); (*v*) (b); (*vi*) (b); (*vii*) (b); (*viii*) (b);
(*ix*) (c); (*x*) (a); (*xi*) (c); (*xii*) (c).

Q.No.38: (*i*) (b), (a), (c), (d); (*ii*) (a), (c), (b), (d); (*iii*) (b), (a), (c), (d); (*iv*) (c), (a), (b) (d); (*v*) (b), (a), (d), (c)

Q.No.39: (*i*) (A) (V); (B) (IV); (C) (I); (D) (II); (E) (III) (*ii*) (A) (IV); (B) (V); (C) (I); (D) (II); (E) (III)
(*iii*) (A) (V); (B) (IV); (C) (I); (D) (II); (E) (III) (*iv*) (A) (V); (B) (IV); (C) (I); (D) (II); (E) (III)

Chapter 5:

Q.No.21: (*i*) T; (*ii*) F; (*iii*) T; (*iv*) F; (*v*) T; (*vi*) F; (*vii*) T; (*viii*) T; (*ix*) T; (*x*) T;
(*xi*) T; (*xii*) T; (*xiii*) F; (*xiv*) F; (*xv*) T; (*xvi*) T; (*xvii*) T

Q.No.22: (*i*) (b); (*ii*) (b); (*iii*) (c); (*iv*) (c); (*v*) (c); (*vi*) (b)

Q.No.23: (*i*) (A) (IV); (B) (V); (C) (I); (D) (III); (E) (II) (*ii*) (A) (IV); (B) (V); (C) (I); (D) (II); (E) (III)
(*iii*) (A) (V); (B) (IV); (C) (I); (D) (II); (E) (III)

Chapter 6:

Q.No.24: (*i*) F; (*ii*) T; (*iii*) F; (*iv*) T; (*v*) T; (*vi*) F; (*vii*) T; (*viii*) T; (*ix*) T; (*x*) F; (*xi*) T;
(*xii*) F; (*xiii*) F; (*xiv*) T; (*xv*) T

Q.No.25: (*i*) (b); (*ii*) (d); (*iii*) (b); (*iv*) (a); (*v*) (a)

Q.No.26: (*i*) (b), (a), (d), (c); (*ii*) (c), (b), (a), (d); (*iii*) (d), (b), (a), (d); (*iv*) (b), (a), (d), (c); (*v*) (b), (a), (d), (c)

Q.No.27: (*i*) (A) (V); (B) (IV); (C) (I); (D) (II); (E) (III) (*ii*) (A) (V); (B) (III); (C) (I); (D) (II); (E) (IV)
(*iii*) (A) (IV); (B) (V); (C) (I); (D) (II); (E) (III)

Chapter 7:

Q.No.33: (*i*) F; (*ii*) T; (*iii*) T; (*iv*) F; (*v*) F; (*vi*) F; (*vii*) T; (*viii*) T; (*ix*) F; (*x*) T; (*xi*) T;
(*xii*) T; (*xiii*) F; (*xiv*) T; (*xv*) T; (*xvi*) F.

Q.No.34: (*i*) (b); (*ii*) (b); (*iii*) (d); (*iv*) (b); (*v*) (c)

Q.No.35: (*i*) (b), (a), (c), (d); (*ii*) (a), (c), (b), (d), (e), (f); (*iii*) (b), (a), (d), (c); (*iv*) (a), (c), (b); (*v*) (b), (e), (a), (c), (d)

Q.No.36: (*i*) (A) (V); (B) (IV); (C) (I); (D) (II); (E) (III) (*ii*) (A) (V); (B) (IV); (C) (I); (D) (II); (E) (III)
(*iii*) (A) (V); (B) (IV); (C) (I); (D) (III); (E) (II)

Chapter 8:

Q.No.41: (*i*) T; (*ii*) F; (*iii*) T; (*iv*) T; (*v*) T; (*vi*) F; (*vii*) T; (*viii*) T; (*ix*) T; (*x*) F; (*xi*) T;
(*xii*) T; (*xiii*) F; (*xiv*) T; (*xv*) T; (*xvi*) T.

Q.No.42: (*i*) (c); (*ii*) (c); (*iii*) (c); (*iv*) (e); (*v*) (c); (*vi*) (c); (*vii*) (b).

Q.No.43: (*i*) (c), (b), (a), (d), (e); (*ii*) (b), (a), (c), (d), (e), (f); (*iii*) (b), (a), (d), (c);
(*iv*) (b), (a), (d), (c), (e); (*v*) (c), (a), (b), (d)

Q.No.44: (*i*) (A) (V); (B) (IV); (C) (I); (D) (II); (E) (III) (*ii*) (A) (V); (B) (I); (C) (IV); (D) (II); (E) (III)
(*iii*) (A) (IV); (B) (V); (C) (I); (D) (II); (E) (III)

Chapter 9:

Q.No.57: (*i*) T; (*ii*) F; (*iii*) T; (*iv*) F; (*v*) T; (*vi*) T; (*vii*) T; (*viii*) F; (*ix*) F; (*x*) T; (*xi*) T; (*xii*) F;
(*xiii*) T; (*xiv*) T; (*xv*) T; (*xvi*) T; (*xvii*) F; (*xviii*) F; (*xix*) T; (*xx*) T; (*xxi*) T; (*xxii*) F; (*xxiii*) T;
(*xxiv*) T; (*xxv*) T; (*xxvi*) F; (*xxvii*) T; (*xxviii*) T; (*xxix*) T; (*xxx*) T

Q.No.58: (*i*) (a); (*ii*) (c); (*iii*) (c); (*iv*) (c); (*v*) (c); (*vi*) (d)

Q.No.59: (*i*) (A) (V); (B) (IV); (C) (I); (D) (II); (E) (III) (*ii*) (A) (IV); (B) (V); (C) (II); (D) (I); (E) (III)
(*iii*) (A) (IV); (B) (V); (C) (I); (D) (II); (E) (III) (*iv*) (A) (V); (B) (IV); (C) (I); (D) (II); (E) (III)
(*v*) (A) (V); (B) (IV); (C) (I); (D) (II); (E) (III)

Chapter 10:

Q.No.25: (*i*) T; (*ii*) F; (*iii*) F; (*iv*) T; (*v*) F; (*vi*) T; (*vii*) T; (*viii*) F; (*x*) T; (*xi*) F; (*xii*) F; (*xiii*) F;
(*xiv*) F; (*xv*) F; (*xvi*) T; (*xvii*) T; (*xviii*) F; (*xix*) T; (*xx*) T

Q.No.26: (*i*) (b); (*ii*) (a); (*iii*) (d); (*iv*) (c); (*v*) (b); (*vi*) (b); (*vii*) (c); (*viii*) (d)

Q.No.27: (*i*) (A) (V); (B) (IV); (C) (II); (D) (III); (E) (I) (*ii*) (A) (IV); (B) (V); (C) (I); (D) (II); (E) (III);
(*iii*) (A) (V); (B) (IV); (C) (II); (D) (III); (E) (I) (*iv*) (A) (IV); (B) (V); (C) (III); (D) (II); (E) (I)
(*v*) (A) (V); (B) (IV); (C) (III); (D) (II); (E) (I)

Chapter 11:

Q.No.22: (*i*) F; (*ii*) F; (*iii*) T; (*iv*) T; (*v*) T; (*vi*) F; (*vii*) F; (*viii*) F; (*ix*) T; (*x*) F; (*xi*) T; (*xii*) T; (*xiii*) F; (*xiv*) T; (*xv*) T

Q.No.23: (*i*) (c); (*ii*) (d); (*iii*) (c); (*iv*) (b); (*v*) (d); (*vi*) (c)

Q.No.24: (*i*) (A) (IV); (B) (V); (C) (I); (D) (II); (E) (III) (*ii*) (A) (IV); (B) (V); (C) (I); (D) (II); (E) (III) (*iii*) (A) (IV); (B) (V); (C) (II); (D) (III); (E) (I)

Chapter 12:

Q.No.26: (*i*) T; (*ii*) T; (*iii*) F; (*iv*) T; (*v*) T; (*vi*) T; (*vii*) F; (*viii*) T

Q.No.27: (*i*) (c); (*ii*) (c); (*iii*) (b); (*iv*) (c); (*v*) (c); (*vi*) (c); (*vii*) (b); (*viii*) (c)

Q.No. 28: (*i*) (A) (V); (B) (IV); (C) (I); (D) (II); (E) (III) (*ii*) (A) (V); (B) (IV); (C) (I); (D) (II); (E) (III) (*iii*) (A) (V); (B) (IV); (C) (I); (D) (II); (E) (III)

Chapter 13:

Q.No.51: (*i*) T; (*ii*) T; (*iii*) F; (*iv*) F; (*v*) T; (*vi*) F; (*vii*) T; (*viii*) T; (*ix*) F; (*x*) T; (*xi*) F; (*xii*) T; (*xiii*) T; (*xiv*) T; (*xv*) F; (*xvi*) T; (*xvii*) F; (*xviii*) F; (*xix*) T; (*xx*) T; (*xxi*) T; (*xxii*) F

Q.No.52: (*i*) (d); (*ii*) (c); (*iii*) (a); (*iv*) (b); (*v*) (b); (*vi*) (c); (*vii*) (b); (*viii*) (c); (*ix*) (b); (*x*) (b); (*xi*) (c); (*xii*) (b); (*xiii*) (b); (*xiv*) (c)

Q.No.53: (*i*) (A) (V); (B) (IV); (C) (I); (D) (II); (E) (III) (*ii*) (A) (V); (B) (IV); (C) (I); (D) (II); (E) (III) (*iii*) (A) (V); (B) (IV); (C) (III); (D) (II); (E) (I) (*iv*) (A) (IV); (B) (V); (C) (II); (D) (III); (E) (I) (*v*) (A) (V); (B) (I); (C) (IV); (D) (III); (E) (II) (*vi*) (A) (IV); (B) (V); (C) (II); (D) (III); (E) (I)

Chapter 14:

Q.No.13: (*i*) F; (*ii*) T; (*iii*) T; (*iv*) F; (*v*) T; (*vi*) T; (*vii*) T; (*viii*) F; (*ix*) F; (*x*) T

Q.No.14: (*i*) (d); (*ii*) (c); (*iii*) (b); (*iv*) (b); (*v*) (a); (*vi*) (d); (*vii*) (a); (*viii*) (b); (*ix*) (c); (*x*) (d)

Q.No.15: (*i*) (A) (V); (B) (IV); (C) (II); (D) (I); (E) (III); (*ii*) (A) (III); (B) (I); (C) (V); (D) (II); (E) (IV) (*iii*) (A) (IV); (B) (III); (C) (I); (D) (V); (E) (II) (*iv*) (A) (II); (B) (I); (C) (III); (D) (V); (E) (IV) (*v*) (A) (III); (B) (I); (C) (V); (D) (II); (E) (IV)

Chapter 15:

Q.No.23: (*i*) T; (*ii*) F; (*iii*) T; (*iv*) T; (*v*) T; (*vi*) F; (*vii*) F; (*viii*) T; (*ix*) T; (*x*) T

Q.No.24: (*i*) (b); (*ii*) (b); (*iii*) (d); (*iv*) (b); (*v*) (c)

Q.No.25: (*i*) (A) (V); (B) (IV); (C) (I); (D) (II); (E) (III) (*ii*) (A) (V); (B) (IV); (C) (I); (D) (II); (E) (III) (*iii*) (A) (IV); (B) (V); (C) (III); (D) (I); (E) (II)

Index

2003 World Water Forum, Kyoto (Japan), 2

Acid soil, reclamation of, 686
Adhesion, 423, 428, 445, 450-52
Afforestation, 7, 31
Afghanistan,
  irrigation efficiency of, 32
  irrigation systems of, 32
Africa, 3-4, 47-57
  agro-climatic regions of, 40
  climatic zones of, 40
  Consultation Group of FAO on Irrigation in, 40
  diversity of water management in, 39
  groundwater resources of, 44
  high-risk zones of, 56
  irrigated crops of, 42
  irrigation development in, 39, 44, 55
  roles of NGOs, 57
  irrigation potential in, utilization of, 55
  Lake Chad system of, 43
  large versus small-scale irrigation systems, 56
  major issues in management, 55
  rainfall in, 55
  river basins of, 43
  semi-arid zone of, 56
  soils of, 42
  surface water sources of, 43
  water quality of, 44
  water-related diseases in, 57
African Development Bank, 724
Aggregation, 425-26, 437
Agricultural Finance Corporation, 344
Agricultural production 5, 661
Agricultural strainers, 152
Agricultural water withdrawal, 16
Agriculture, 2, 5, 11, 17, 18, 19, 50, 53, 421, 661, 665, 670, 672, 681
  role of women in, 57
  salt problems in, 7, 661
Agro-hydrology, 3
Air humidity, 487
Air rotary drilling, 171
Air temperature, 486, 487
Air vent, 364
Air-lift pump, 235, 236
Air-operated down-the-hole drill, 171
Akbar, Emperor, 59
Albedo, 485, 511
Alfalfa, 521, 522, 523, 528, 541
Alfisols, 434, 436, 437, 438, 440, 442
Algeria, 42, 48, 579
Alkali soil, 434, 661-62, 665, 668-70, 672, 678, 681-87, 689
  determination of properties of, 670
  plant response to, 687
  reclamation of, 683
Alkalinity, 665
Alkalinization, 7
Alluvial plain, 253-54, 263
  characteristics of rivers in, 254
  formation of, 254
Alluvial river, 254, 259-61, 263
  intake wells in, 263
Alluvial soil, 430, 432-36
Alluvial streams, meandering streams in, 257
Amortization, 701, 703
Andhra Pradesh, 431, 435-36, 438, 620, 662, 666, 668, 719-20, 722
  salt-affected soils in, 661
  water pricing in, 719
Angle of repose, 315
Angola, 53
Anicut, 19, 57-59; *see also* Diversion weir
Animals, water requirement of, 6
Annual crops, 16
Annuity factor, present worth of, 702
Annular velocity, 169
Apparent resistivity, 129
Apple, 545
Aqueduct, 13
Aquifer, 10-11, 23-24, 44, 47, 50, 52, 83, 124-58, 160
  alluvial, 131
  artesian, 47, 48, 125-26
  homogeneous, 147
  compressibility of, 131, 141
  confined, 126, 131, 133, 141-42, 148
  functions of, 129
  homogeneous, 142
  hydraulic characteristics of, 143
  isotropic, 142
  leaky, 126
  limestone, 53
  multiple deep tube well, 148
  phreatic, 133
  semi-confined, 126, 129-32, 139, 141-42, 148
  shallow, 50
  thickness of, 130, 134
  transmissibility of, 127
  type of, 141
  unconfined, 125, 130, 133, 137, 141, 148
Arch dam, 13
Area of influence, 133, 137
Arecanut, 545
Argentina, 2
Aridisols, 436
Artesian well, 126
Artificial drainage, 34
Artificial gravel packing, 153
Artificial lakes, 33-34
Artificial recharge, 125
  methods of, 127
Asbestos cement pipe, 352, 354
Ashim filters, 151
Asia, 3, 4, 28
Asian Development Bank, 724
Asian Institute of Technology, Bangkok, Thailand, 25
Assam, 431-32, 435-36, 620
Atmospheric pressure, 3, 125-26, 193, 502
Australia, 3, 28, 620
  irrigated area of, 5
Austria, 579
Auxiliary reservoir, 13
Available water (AW), 455, 523
Ayacut, 14

Babylon, 17
Badiah, 38
Bail plug, 176
Balancing reservoir, 13
Balancing tank, 13
Bali, 28
Baluchistan, 22, 76
Bamboo drip irrigation system, 618
Bamboo strainers, 151
Banana, 42-43, 513-14, 522, 528-30, 543-44, 620
Bangladesh, 2, 5, 10, 20-24, 61, 81-82
  Coastal Embankment Project, 23
  flood protection plan of, 23
  Ganga-Brahmaputra-Meghana River system, 23
  geomorphology of, 22
  National Minor Irrigation Development Project, 23
  network of drainage canals in, 23
  RDRS, 204
  treaty with India
    Farraka barrage, 61
Barley, 90, 513-14, 521-22, 536
Barrage, 13
Base exchange capacity, 422
Bean, 512-14, 521-23, 528-29, 536, 541-42, 545-46
Bench mark (BM), 387
Benefit-cost analysis, 707
Benefit-cost ratio (BCR), 710
Betterment levy, 718
Bhutan, 21, 22
Bicarbonate, 675
Bihar, 432-34, 438, 666-67, 719, 721-22
  public tubewells of World Bank supported projects in, 375
  water pricing in, 719
Biodiversity, 110
Biomass, 202

Black gram, 437
Black pepper, 545
Black sand desert, 32
Black soil, 430-31, 435-36, 438-42
  groups of, 431
Blaney-Criddle formula, 492
Blue top levelling, 402
Booster pump, 281, 585
Boring method, choice of, 174
Borneo Island, 27, 28
Boron deficiency, 441, 443, 673, 675, 678, 689-90
Botswana, 53, 54, 56
Brazil, 20
British India, irrigated area in, 20
Brocolli, 542
Brown hill soil, 433
Brussels sprouts, 542
Bubbler irrigation, 600
Bulk density, 427-28, 439, 444, 446, 459
Bulldozers, 410
Bundhs; See Dam
Bureau of Indian Standards, 353, 354, 599
Burkina Faso, 56
Burma; *See* Myanmar
Burundi, 42

Cabbage, 512, 514, 522, 528-29, 541-42
Cable size, 275
Cable tool percussion method, 174
Calcareous alluvial soil, 433
Calcareous sierozemic soil, 433
Calcium, 442
California, 20
Calyx core drilling, 174
Cambodia, 25, 26, 27
Canada, 5, 579
Canal, 7, 13, 14, 96
Canal pumping system, 11, 91, 99, 259, 312, 318, 720
Canalet irrigation distribution system, components of, 328
Canopy resistance, 511
Capillarity, 131, 428, 450-52
  viscosity, 450
Capillary fringe, 124
Capillary pores, 428
Capital recovery factor, 701
Capital recovery factor (CRF), 701, 703
CAR Coordinated Project on Wells and Pumps, 151
Carbon, 184
Carrots, 512, 542
Cash flow, 705
  analysis, 699
  procedure for regular, 705
Cashewnuts, 43
Cassava, 43
Castor, 437
Catchment area, 14
Cation exchange capacity (CEC), 439, 662-63, 668-72
Cauliflower, 512, 514, 522, 541-42
Cavitation, 223
Cavity wells, 133
  development of, 181
Celebes, 28
Central Soil Conservation Research Institute, Dehradun, 343
Centralised grid power system, 198
Centre for Water Resources, Anna University, Chennai, 104
Centrifugal filters, 639
Centrifugal force, 193
Centrifugal pump, 199, 201, 209-17, 219-24, 226-27, 232-35, 237-38, 240, 242
  pump house for, 271
  structure of, 272
  width of, 271
Centrifugal sand separators, 639
Cereals, 33, 90
Chad, 56
Chandragupta, the legendary Gupta ruler, 19
Channel crossings, structures at, 341
Channel flow, patterns of, 255
Characteristic curve, 211, 237
Chezy's formula, 315
Chhattisgarh, 431, 435-36
China, People's Republic of, 5, 17, 20, 22, 30, 620
  irrigation systems' management in, 103
  Yellow River basin of, 17
Chlorine, 443
Chola, Karikala, 19
Chute spillways, 334
Circle of influence, 133
Citrus, 514-15, 522, 529-30, 543-44, 620
Clay, 316, 423-29, 431-34, 436-37, 439, 441-43, 453-55, 457-58, 462, 662, 666, 668-70, 675-76, 678, 683, 685
Clay formation, 125
Clay micelles, 424
Clay mineral, types of, 424
Clay pipe, 352
Clayey soil, 521
Clear-sky solar radiation, 503
Climatic cycle, 3
Climatic data, 490
Clogging, 149, 615, 621, 623-24, 626-27, 629, 631, 635, 638, 641-42, 644-47
Coalescence (fusion), 3
Coastal alluvium, 432
Coastal sand, 433
Cobalt, 443
Cocoa, 27, 42-43, 545
Coconut, 25, 545, 620
Coefficient of storage, 131
Coffee, 42-43, 514, 522, 545
Cohesion, 423, 428, 445, 451
Cohesiveness, 426
Coir strainers, 151
Collar joints, 357
College of Agriculture, Jobner, Rajasthan Agriculture University, 343
Columbia, irrigation systems' management in, 103
Command area, 14, 259, 344, 721
Command Area Development Authorities (CADA), 256
Command Area Development Programme, 344
Communal irrigation system, 15
Comoros, 40
Compensating reservoir, 13
Compounding factor, 702
Compressibility, 141
Concrete pipe, 352-59, 364-65, 369-71, 373
Concrete pipeline, 352, 582
Condensation, 3
Condiments, 545
Conduction of heat, 137
Conductivity, 662-63, 670-76, 679-80, 688
  hydraulic, 126
Cone of depression, 133, 141
Conference on Water Quality Ratings at the Central Soil Salinity Research Institute, Karnal, Haryana, 676
Conjunctive irrigation planning, 14
Consumption, 4
Consumptive use (CU), 479, 483, 516
  peak period, 484
  seasonal, 484
Continuous cultivation, 441
Contour benches, 404-9
Control stand, water surface elevations in, 370
Conventional energy source, 198
Cooperative irrigation systems, 352
Copper, 443
Core drilling, 174
Corrosion, 148, 150, 183, 184, 640
Cost escalation, 701, 703
Cotton, 30, 33, 43, 47, 49, 59, 437, 513-14, 522, 528-30, 538-39, 546
CRIDA, 439
Crop
  boron tolerance of, 689
  effect of soil salinity, 688
  evapo-transpiration of, 516
  field irrigation requirement of, 516
  growth span of, 421
  irrigation requirement of, 516
  moisture requirement of, 5
  moisture use of, 470
  nutrient requirement of, 441
  potential yield of, 478
  productivity of, 6, 479, 527, 661, 679
    basic resources for, 478
    decline in, 80
  restricted root development of, 615
  root system in, 423
  rooting characteristics of, 470
  salt-tolerance of, 687
  selection of, 545
  water requirement of, 478, 529, 532
  yield response factor of, 528
Crop coefficient
  values of, 490, 514
    dual approach, 515
    single approach, 515
Crop evapo-transpiration, 485, 510
  calculation of
    crop coefficient approach, 489
  concept of, 511
  under non-standard conditions, 485
Crop growth, 511
Crop height, 511
Crop management under rain-fed agriculture, 545
Crop selection, 661, 690
Crop water requirement, 500, 501, 515
Crop yield, 444, 527
Crop-soil surface, 511
Cropping pattern, 530
Cropping systems, 98
Cryology, 3
Cucurbits, 541, 543

Cultivation, 351, 352
Culturable command area (CCA), 15
Culvert, 343
  formula, 265
Cumulative infiltration, 445
Current price, 696
Current ratio, 696
Curve fitting, 447
Cut flowers, 42
Cut-fill ratio, 391
Cut-off rate, 696
Cutting drilling, 173
Cyclonic depressions, 8
Cylindrical surface, 133
Cyprus, 620

Dams, 32, 37, 47-48, 50, 53-54, 58, 62-63, 81
Darcy, Henri, 130
  equation, 354
  experiments, 130
  law, 130, 457
  theory, 133
Date palm, 42, 545
Daylight hours, 503
De-Glee's method, 139
Debris screens, 585
Deep percolation, 518
Deficit irrigation, 520
Degree of dispersion, 670
Degree of saturation, 428
Delhi, 667-78
Delivery chamber, 282
Delivery pipe, 278
Delta, development of, 254
Deltaic alluvium, 433
Democratic People's Republic of Korea, 30, 31
Depreciation, 699
Depth of penetration, 142
Desalination, 38
Desert, 22
Desilting basins, 585
Deterioration, 148, 150
Dew point temperature, 487
Dezincification, 184
Diamond core drilling, 174
Diffusion, 483
Diffusion pressure deficit (DPD), 470
Disc filters, 639
Discount rate, 696, 701
Dispersed soil, 662, 670
Distribution chamber
  design of, 375
  specifications of, 283
Distribution of leakage, 132
Diversion dams, 20
Diversion scheme, 13
Diversion structure, 127
Diversion weir, 13, 19, 31
Djibouti, 40
Domestic water supply, 193, 207, 208
Down-the-hole hammer (DTH) drilling, 171-72, 174
  adaptability of, 174
Drainable porosity, 131
Drainage, 36-37, 45, 382-85, 387, 391-95, 404-05, 409, 415, 422, 424, 426-28, 431-34, 438-41, 445, 465, 468, 599, 661-62, 665-67, 669, 672-73, 675-84, 686, 724
  basin, 14
  on-farm, 7
  outflow of water, 33-34
  planning of, 385
  problem, 2
  surface, 383
  system, 384
Drawdown, 133
Drill stem test, 178
Drilling
  methods of
  auger method, 174
  percussion method, 174
  operation of, 169
Drilling fluid, 169, 171
Drilling mud circulation, 169
Drinking water, 7, 8, 35, 83, 108, 111
Drip irrigation, 53, 611-16, 618-26, 629-33, 635-55
Drip irrigation systems, 612, 615, 616, 621, 642, 643
Dripper 626-29, 644
Drought-Prone Area Programme, 723
Droughts, 8
Dry and wet bulb thermometer, 487
Dry bulk density, 427
Dublin Conference (1992), 2
Dug wells, 145
Dummy pipe, 361
Dupuit, 133, 134, 135
Dupuit-Thiem theory, 134
Dykes, 144

Earth channel, 317, 318
Earthquake, 113
Ecological balance, 16
Ecological diversity, 110
Ecological impact, 16
Ecological succession, 16
Ecology, 7, 16
Economic life, 696
Economic prosperity, 1
Economic rate of return, 695
Economic value, 696
Economical development, 4
Ecosystem, 16
Effective rainfall, 516-18, 520, 523, 531, 534
Effective root zone, 471
Effective size, 153
Efficiency price, 696
Effluents, 16, 102
Egypt, 17-20, 40, 42, 45, 103, 620
  1959 Nile Waters Agreement, 45
Electric logging, 128
Electric resistance blocks, 464
Electrical conductivity (EC), 662-63, 670-75, 679-80, 688
Electrical resistivity, 128-29
Electricity generation, conventional sources for, 6
Electrode spacing, 129
Elevated water tank, 375
Embankment dam, 13
Emission uniformity (EU)., 646
Emitter, 623, 625-29, 646, 649
End plug, function of, 365
Energy consumption, 700
Entisols, 434, 436
Entrance velocity, 149
Environment, 7, 16, 23, 34, 35, 198, 631
Equity, 696
Equivalent weight, 663, 671
Eritrea, 51, 52
Ethiopia, 51, 56
Europe, 3, 4
Evaporation, 3, 4, 5, 6, 29, 35, 36, 38, 43, 48, 79, 107, 479-84, 486-90, 494-502, 504, 511
Evapo-transpiration (ET), 28, 51, 80, 478-79, 448-86, 488-91, 493-97, 499-501, 504, 506-11, 516, 521, 667
Exchange complex, 662-63, 668-69, 675, 683
Exchangeable cation, 663, 668-70, 672, 685
Exchangeable sodium percentage (ESP), 662-63, 669, 671, 675
Expert Consultation of FAO Methodologies for Crop, 501
Extraterrestrial radiation, 503

Food and Agriculture Organization (FAO), 2, 5, 16, 22-23, 28, 31, 32, 38, 40-43, 56, 108, 481, 484, 488, 490, 494-95, 497-501, 504, 507-11, 724
  Consultation Group on Irrigation in Africa, 40
  International Action Programme on Water and Sustainable Agricultural Development, 2
  Irrigation and Drainage Papers 490, 501
  FAO Plant Production and Protection Series, 490
  FAOCLIM, 490
Farm gate price, 696
Farmer joint management, 15
Fatigue, 184
Faucet end pipe, 356-57
Fertigation, 601
Fertile Crescent, 17
Fertility, 478
Fertilizer, 579, 598, 601-06, 612-14, 616, 622, 630, 635, 640-42, 644, 685
Fertilizer dissolver, 603
  fertilizer injection systems, location of, 606
Fertilizer tank, 603
Field
  capacity, 454
  channel, 384
    lining of, 322
    materials for, 318
  command, 14
  elevation of, 388
  gains of water in, 516
  layout, 382, 384-85
    equipment for, 409
  water balance of, 516
Field water balance, measurements of, 516
Fiji, irrigated area of, 5
Fills, 409
Filration systems, 635
Filter cake, 169
Filter points, 133
Filters, selection of, 640

Financial rate of return, 696
Finger millet, 437, 535
Fishing tool, 170
Flexible rubber pipe, 368
Flow control valves, 632
Flow irrigation, 81
Flow measurement, principle of, 293
Flow-cum-lift irrigation system, 250
Fluid, properties of, 169
Flume, 291, 296, 299-304, 314, 341-42
Fodder, 33
Fodder crops, 540
Food production, 4, 19
  sustainability of, 2
Food security, 2, 23, 83
Force of gravity, 131
Forests, 110
Formation materials, 154
Formation resistivity, 128
Formation samples, analysis of, 152
Fossil fuel, 198
France, 620
Fresh water, 1-2, 4-5, 667
Frontal rainfall, 3
Fruit crops, 543
Full irrigation, 520
Funds flow analysis, 696

Gamma-ray logging, 54, 128
Ghors, 38
Glaciers, 4
Global Water Partnership (GWP), 2
Goa, 435-36
Gobi Desert, 30
Grade stake, 402
Gram, 512, 514, 536-37, 546
Grand Trigonometrical Survey (G.T.S.), 387
Grape, 514, 522, 528, 529, 530, 543, 544
Grapefruit, 53
Grapes, 620
Graphitization, 184
Gravel, size of, 183
Gravel pack, 152-55, 177
Gravel pack gradation, design theory of, 155
Gravitational potential, 452
Gravity drainage, 131
Gravity flow, 252
Gravity weir, 13
Green gram, 437, 512, 514, 536
Grey brown soil, 433
Grid point, 385-92, 394-99, 402, 407
  reduced levels (RL) of, 387
Gross benefit, 696
Gross command area (GCA), 14
Gross domestic product (G.D.P.), 696
Gross irrigated area, 14
Gross irrigation requirement, 518
Ground surface, topography of, 126
Ground water projects, service period of, 699
Groundnut, 43, 437, 513-14, 522, 528-30, 537, 546
Groundwater
  artificial recharge, 127
  availability of, 11, 83
  capillary rise of, 23
  components of, 83
  conjunctive use of, 7
  contribution of seepage to, 87
  depletion of, 16
  development in
    China, 30
    Korea, Republic of, 31
  exploitation of, 11, 80, 84, 102
  exploration of, 127
    geophysical methods, 128
    test drilling, 127
  importance of, 90
  integrated use of, 64
  jurisdiction of, 101
  legislative measures for utilization of, 88
  minor irrigation schemes in India, 65
  fundamentals of movement of, 133
  rainfall contribution to, 87
  rate of flow of, 129
  rational use of, 90
  renewable resources of, 50
  rate of replenishment of, 126
  resources development in
    Philippines, 29
  resources development in Myanmar, 25
  resources development in Sri Lanka, 24
  safe yield of, 16
  sources of, 144
  storage in Bangladesh, 23
  use of groundwater in, 46
  use of resources in Cambodia, 27
  utilisation of, 20, 90, 127
Groundwater flow, 133
Groundwater movement, 130
Groundwater recharge, 127
Groundwater reservoir, 125-27
Groundwater resources, 127
Groundwater storage, 127
Groundwater utilization, 127
Gujarat, 321, 431-32, 435, 438-40, 620, 722
Gupta dynasty, 58
Gypsum requirement (GR), Schoonover's method, 672
Hammurabi, King, 17
  Code of Hammurabi, 17
Han Dynasty, 18
Hand-made pipe, 352
Harappa, 18
Hargreaves' Equation, 494
Hargreaves' Radiation Formula, 494
HariharaI, King, 59
Harvest index, 528
Haryana, 321, 344, 433, 435, 440, 620, 662, 666-67, 676-77, 689, 719, 721-22
  water pricing in, 719
Heating mirror, 361
Heating plate, 361
High density polyethylene (HDPE) pipes, 620, 629
High specific speed pump, 277
Himachal, 722
Himachal Pradesh, 429
Hindu Kush, 22
Histosols, 436
Hohokam, 20
Hokkaido, 31
Honshu, 31
Human civilization, development of, 1
Humidity, 79, 487, 505
Huns, 31
Hydrant, 367
Hydraulic conductivity, 126, 129, 130, 132, 134, 141, 142, 456
Hydraulic formulae, use of, 650
Hydraulic gradient, 129, 130, 133, 136, 137, 456
Hydraulic head, 125, 456
  types of, 126
Hydraulic pressure, 125, 157
Hydraulic radius, 315
Hydraulic ram, 200
Hydraulic resistance, 129, 132, 139, 141
Hydraulic slope, 315
Hydro-electric power generation, 5, 12
Hydrocyclones, 639
Hydrologic cycle, 3, 4
  concept of, 3
Hydrologic equation, 3
Hydrologic process, 254
Hydrologic regime, 3
Hydrology, 3, 19, 88
  river, 250
  in alluvial plains, 253
  in plateau regions, 255
Hydrometry, 3
Hydrosphere, 3
Hydrostatic pressure, 126, 171, 354

ICAR, 344
  Coordinated Project on Wells, 144
Ice caps, 4
ICRISAT, 439
Impact drive sprinklers, 601
Impeller, 215, 230, 232, 233, 234, 235
  types of, 215
Impermeable layer
  depth of, 127
Impulse pump, 200
Imputed price, 696
Inamgaon, 18
Inceptisols, 434, 436
Incremental net benefit, 696
Incrustation, 148, 149, 183
  forms of, 184
  rate of, 149
India,
  Agra Canal in Uttar Pradesh, 60
  agriculture in
    dependence on South-West monsoon, 18
  agro-climatic zones and regions, 72
  agro-ecological regions of, 74
  Alfisols and associated soils, 436
  Alfisols in semi-arid tropics of, 436
  All India Scheme on Soil Surveys, 429
  All India Soil and Land Use Survey (AISLUS), 441
  All India Soil Survey, 431
  alluvial soils in, 432
  application of drip irrigation system in, 620
  assessment of resources in, 11, 83
  assessment of surface water resources of, 8
  availability of the water resources of, 83
  Bhakra Nangal projects, 62
  Bhavani Project in Tamil Nadu, 98
  black soils in, 431
  canal systems in, 91
  Central Electricity Authority, 6

Central Groundwater Board,
Central Groundwater Board (1989), 11, 64, 84, 87
Central Pollution Control Board, 6, 61, 111
Central Public Health and Engineering Organization, 6
Central Water Commission, 10, 61 63, 64, 107, 109, 112
chain tanks of Karnataka and Andhra Pradesh, 58
Chambal Canal project, 99
Chola empire, 58
classification of river systems of, 81
climate of, 8, 71
Command Area Development Programme (CADP), 64, 104
Community water supply in, 6
Constitution of India, 61, 101, 108, 109
 Annual Plans, 62, 64
 Control and Regulation Bill, 1970, 90
 Famine Commission, 59
 Five Year Plans, 61, 62, 64, 102, 103
 Forest Conservation Act, 1980, 109
 Ganga Action Plan, 111
 Geological Survey of India, 87
 Indian Council of Agricultural Research (ICAR), 61, 620
 Indira Gandhi Nahar Pariyojana, 104
 Indus Water Treaty, 1960, 61
 Inter-State Water Disputes Act, 1956, 61
 Integrated Rural Development Programme (IRDP), 723
 Irrigation Commission, 8, 10, 59-60, 90, 107, 718-19
 jurisdiction of water resources, 61
 National Bank for Agriculture and Rural Development (NABARD), 65, 250, 257, 344, 723
 National Bureau of Soil Survey and Land Use Planning, 72, 429, 431, 436
 National Commission for Integrated Water Resources, 63, 64, 68
 National Commission in Agriculture (1976), 10, 664
 National Committee on Plastics in Agriculture (NCPA), 621
 National Committee on Plasticulture Applications, 621
 National Committee on Use of Plastics in Agriculture, 620
 National Flood Commission, 107
 National Perspective Plan, 112
 National Programme of Flood Management, 106
 National Water Development Agency (NWDA), 112
 National Water Policy (NWP), 6, 102, 107
 Pipe Committees, 104
 Planning Commission, 61
 State Cooperative Acts, 103
 Tamil Nadu Society's Registration Act, 1975, 104
 Water (Prevention and Control) of Pollution (Cess), 109
consumption of macro-nutrients (NPK) in, 441
consumptive use of water by thermal power plants, 6
cropping pattern in, 90
cultivable area of, 69, 80
Damodar Valley Corporation, 60, 106
deficits in food production in, 62
dependence on irrigated agriculture, 19
development of irrigation in, 718
development of private irrigation works, 60
distribution of soil in, 434
distribution of water resources of, 8, 10
diversity of climate in, 69
diversity of land forms in, 69
Eastern Yamuna Canal in Uttar Pradesh, 60
environmental impact on assessment of irrigation projects, 108
evaporation in, 80
evapo-transpiration of different states of, 80
exploitation of groundwater in, 11
famine in, 59, 60
Farraka barrage, treaty with Bangladesh, 61
finance policy of, 718
financial performance of irigation projects in, 721
floods and droughts in, 104
floods in, 105
foodgrain consumption in, 90
hydro-power potential of electricity generation in, 6
Gandhi Sagar projects, 62
Ganga Canal, 60
Ganga Kaveri Link, 112
Ganga-Brahmaputra-Meghna system, 81
geographical area of, 81
geomorphology of, 22
Ghataprabha project, 62
groundwater minor irrigation schemes, 67
groundwater potential in the states and union territories, 89
groundwater regions of, 84
groundwater utilization in, 65
growth of population in, 1
Gupta era, 58
Hirakud dam project, 62
Hirakud reservoir, 106
hydro-power development in, 6
industrial development in, 6
industrial water requirement in, 6
irrigated agriculture in, 22, 661
irrigated area of, 5, 61, 80
productivity of, 80
irrigation and water management in, 18
irrigation development in, 57, 59-62, 64
 since ages, 59-61
 present-day projection of, 64
irrigation potential under surface water minor irrigation, 65
irrigation pumping in, 236
Irrigation Utilisation and Command Area Development, 103
Jawahar Sagar projects, 62
Jayakwadi project, 62
Kadna project, 62
Kangrabati projects, 62
Kaveri project, 104
Ken Canal, 60
Khadakvasla Dam in Maharashtra, 60
Kharung Canal Project, 60
Kolhapur Type Weir (KT weirs), 69
Kosi projects, 62
Krishnarajasagar Project, 60
Krishnarajasagar Scheme, 60
Krishnas project, 62
Kurnool-Cudappa Canal Project, 112
land degradation in extent of, 440
land utilisation, 81
Laterite soils in, 432
lift irrigation projects in, 721
Lower Bhavani project, 104
Lower Chenab Canal in Punjab, 60
Lower Ganga Canal in Uttar Pradesh, 60
Lower Sohang and Para Canals, 60
Mahanadi Canal, 60
Malaprabha project, 62
Mettur Dam project, 60
Ministry of Agriculture, 61, 90
Ministry of Environment and Forests, 109
Ministry of Irrigation, 112
Ministry of Water Resources, 61, 103
Mohini Coperative Distributory Society, 104
monsoon in, 73
 north-east, 75
mountain ranges of, 70
Mughal Period, 59
Mutha Canals in Maharashtra, 60
Nagarjuna Sagar project, 62
National income, 696
natural drainage in, 667
Nira Left Bank Canal, 60
Nizamsagar Project, 60
Orissa Canal system, 59
Pallava Kingdom, 58
Parambikulam Aliyar project, 104
Pawara Canal, 60
Periyar Dam in Kerala, 60
Periyar Vaigai Canal system in Tamil Nadu, 60, 104, 112
physiographic regions of, 71, 73
pipes used in, 353
plateau region of peninsular, 256
plateau regions of, 255
population density of, 70
potential evapo-transpiration in, 80
potential groundwater reservoirs of, 87
power generation capacity of, 6
production of major crops in, 90
rainfall in, 8, 72, 77, 104
 in hot weather season, 76
 regions of, 73
rainfed crop area of, 70
Ranapratap Sagar projects, 62
red soils, 431
rehabilitation and modernization of irrigation scheme, 64
Right Bank Canal, 60
river lift irrigation schemes in, 250

river valley projects, 10
role of irrigated agriculture in, 6
run-off water in, 63
run-off, 108
run-off water in, 83
salt affected areas of, 667-68
Sangam literature, 57
Sarda Canal in Uttar Pradesh, 60
SAT Alfisols of, 437
Sathanpur irrigation project, 104
scientific assessment of groundwater resources of, 11, 83
Sharada Canal Command, Uttar Pradesh, 104
Sirhind Canal in Punjab, 60
soil of, 80, 430, 436, 661
Sone Canal in Bihar, 59
Srisailam Right Bank Canals, 104
statewise structures in, 67
statewise irrigation development in, 68
statewise total groundwater structures in, 67
status of ground water reservoirs of, 85
sunshine hours in a year in, 79
surface water flow in, 83
surface water minor irrigation schemes in, 64
surface water sources of, 81
tank irrigation in, 58, 59, 68
Telugu Ganga Project, 112
temperature regime in different seasons in, 78
topography of, 8
treaty with Nepal, 61
Triple Canal Project, 60
Tungabhadra Project, 62
Ukai Kakrapar Irrigation Project, 104
Ukai project, 62
Upper Ganga Canal system, 59
user's participation in major and medium irrigation, 64
Vertisols and associated soils in, 438
Vijayanagara Kingdom, 59
waste water treatment plants in, 8
water budget in, 8
water management in, 19, 661
Water Users' Associations (WUAs), 103, 104
Water Users' Farmers' Associations, 104
common types of well screens used in, 150
western disturbances, 76
Western Yamuna Canal in Haryana, 59, 100
wind speed and other weather parameters in, 79
yield of crops in, 661
Indian Agricultural Research Institute, Water Technology Centre, 344
Indian Meteorological Department, Pune, 8, 490
Indirect benefit, 696
Indonesia, 5, 18-19, 20, 27, 28, 57, 103
Indus Basin Irrigation System, 22
Indus River system, 20
Indus system of rivers, 22
Indus Valley civilization, 18
Mohenjodaro civilization of, 18
ruins of, 18
Industrial effluents, 6-8, 35, 667
recycling and treatment of, 7
Industrial waste, 7
Industrialisation, 80
Infiltration, 445
accumulated, 445
measurement in irrigation furrows, 450
methods of estimating, 446
opportunity time, 127
rate, 445, 446
factors affecting, 446
velocity, 445
Inflation, 696
Inland water transport, 7
Insect pest, 527
Instream storage, 127
Integrated lift irrigation projects, 250
Integrated river basin planning, 15
Inter-American Development Bank, 724
Interflow, 456
Internal rate of return (IRR), 706
International Commission on Large Dams, 12
International Conference on Land Resources Management, 441
International Fund for Agricultural Development, 724
International Rice Research Institute, Los Banos,, 25
International Soil Science Society (ISSS), 423
Interstices, 125
Intrinsic permeability, 128, 130
Inundation canal, 14
Inverted syphon, 13, 343
Iran, Islamic Republic of, 5, 18, 20, 35, 579
Iraq, 17, 19, 20, 36, 37, 249
Ireland, 2
Iron, 443
IRR, 709, 710
Irrigation,
ancient works in Sri Lanka, 24
depth of, 590
development from 1st to 20th Century A.D, 19
developments in ancient times, 17
economic efficiency in, 526
effect of, 665
efficiency level of water use in, 7
energy requirement in, 102
expansion of, 4
faulty methods, 19
frequency of, 665
Hammurabi's period, 17
harvested crops under, 16
influence of, 17
intensity of, 15
mainstay of food security, 23
planning of, 11
progressive growth of, 19
sewage, 111
suitability of water for, 673
terminology, 12
traditional systems of, 102
utilization of ground water in, 256
water requirement for, 6
Irrigation and Farm Drainage Equipment System, 599, 645
Irrigation Association in USA, 599
Irrigation cess, 718
Irrigation command area, 331, 332, 343, 344, 346, 528, 530, 546
Irrigation Commission, 8, 10, 719
Irrigation conveyance system
water loss in, 312
Irrigation demand, 249
Irrigation depth, 521
Irrigation efficiency, 523, 587
Irrigation frequency, 519
Irrigation interval, 590
Irrigation management, 686
soil properties influencing, 422
Irrigation period, 519
Irrigation pipe
cost comparison of various types of, 372
standard tests for, 354
Irrigation potential, 15
utilisation of, 16
Irrigation project, 14, 695
analysis of benefits and costs of, 699
analysis of discounted measure to use in, 710
availability of funds for investments, 723
benefits of, 698
design and operation of, 530
economic analysis evaluation, 697
concepts and definitions in, 695
economic and financial analyses, 695
economic and financial feasibilities of, 695
economic feasibility evaluation, 695
evaluation of, 710
financial analysis evaluation, 697
advantages of, 697
financial feasibility evaluation, 695
fixed costs, 699
investment on, 695
maintenance of, 723
mutually exclusive, 712
operation of, 723
tangible, 698
variable costs, 699, 700
Irrigation Research Institute, Roorkee, 146
Irrigation scheduling, 15, 520
based on pan evaporation, 521
based on prediction of ET, 521
soil depth influencing, 521
Irrigation system
economic aspects of, 722
mismanagement of, 19, 96
participatory management of, 102
revolution in, 19
Irrigation tank, 311
Irrigation technology, 11
Irrigation water, 21, 290, 291, 293, 298, 305, 351, 362, 367
chemical analysis of, 674
classifications of, 676
control of, 311
depth of, 679
diversion of, 311
measurement of, 290, 291, 302, 306, 307

quality of, 673
requirement of, 648
source of, 5, 684
supply system, 384
Irrigation well, 143, 144, 156, 157, 236, 258-80
Irrigation-based civilizations, 17
Israel, 36, 37, 100, 620
Italy, 20, 103, 579, 620

J-turbo line, 629
Jain Irrigation Systems Ltd., Jalgaon, India, 627, 629
Japan, 2, 18, 20, 21, 29, 31, 32, 103
Java and Madura, 28
Jet mechanism, 234
Jet pump, 234, 235
Jharkhand, 431, 436
Jordan, Royal Kingdom of, 37, 620
Central Water Authority, 38, 103
Jordan Valley, 38
Jute, 539, 662

Kalabagh Barrage, 22
Kalimantan, 28, 29
Kamaran, 39
Kampuchea; See Cambodia
Kanatas, 18, 32, 35
Kara Kum basin, 35
Kara Kum canal, 33
Kara Kum Desert, 33
Kareze *see* Kanats
Karnataka, 431, 432, 435, 436, 438, 439, 620, 722
water pricing in, 719
Kautilya
Arthashastra, 19
Kazakhstan, 32, 34
Kenya, 42, 50, 51, 56, 579
Kerala, 431, 432, 434, 436, 620, 662, 666, 667, 719, 720, 722
salt-affected soils in, 661
Kereze systems, 32
Kharif, 15
Khosla, Dr. A. N., 10
Khosla formula, 10
King, Chola, 57
Klamath Project in California, USA, 321
Korea, 31, 57
Korea, Republic of, 31
Kotri Barrage, 22
Krishnaraja, King, 59
KT weir see Weirs
Kyrgyztan, 32-34
Kyushu, 31
Kyzylkum desert, 34

Lake Chad, 47
Lake Issyk-Kul, 33
Lake Moeris, 18
Lake Tonle, 26
Lake Victoria, 45
Land, productivity of, 11
Land clearing, 384
Land degradation, 17, 80, 439-41
Land development programme, 384
Land grading, 382, 383, 384, 385, 391, 392, 402, 409, 415
staking of, 385
Land levelling, 383, 384, 387, 395-98, 402, 579
Land plane, 411
Land revenue, 718
Land use planning, 434
Land Use Survey of ICAR, 429
Lao PDR, 25-26
underground water resources of, 26
Laplace's equation, 457
Large-scale informal surface water systems, 32
Lateral punching of, 644
Lateral line, 377
Laterals, 630
flushing of, 644
Laterite soil, 430, 432, 435, 436, 440, 441
Leaching, 663, 665, 666, 668, 669, 670, 673, 676, 678, 679, 680, 681, 682, 683, 684, 686, 687, 688, 689
Leakage factor, 129, 132, 141
Lebanon, 579
Lesser Sunda Islands, 28
Levying irrigation charges, basis of, 718
Libya Arab Jamahiriya, 49, 50
Lift irrigation, 193, 721
Lift irrigation project, 249
design of, 249, 260
intake structure, 260
economic feasibility of, 251
economic viability of, 251
environmental considerations of, 252
norms for fxing economic water rates in, 283
service period of, 699
techno-economic feasibility report of, 251
water storage tanks in use of, 283
Lining, 318-23, 630
Lining irrigation channel, costs of, 329
Lithology, 124
Lithosphere, 124
Loam soil, 424, 432, 455, 461, 464
Local Self Government (Panchayat), 7
Location, 143
Lombok, 28
Longwave radiation, 504
Lothal, 18
Low Density Polyethylene (LDPE), 619-21, 630
Lower Sind Inundation Canals, 22
Lucerne, 521, 528, 529, 540, 541, 542
Luzon, 29
Lysimeter experiments, 488

Madagascar, 40, 46, 47
Madhya Pradesh, 250, 321, 431, 432, 435, 436, 438, 620, 719, 720, 722
Magasthenes, 19
Magnesium, 442
hazard, 669
Mahadukhole Canal, 24
Maharashtra, 250, 256, 261, 283, 284, 285, 431, 432, 435, 436, 438, 439, 620, 661, 662, 666, 668, 719, 720, 722
Maize, 30, 31, 513, 514, 521, 522, 523, 524, 528, 529, 530, 531, 532, 535, 540, 542
Malasia, 25
Malay Peninsula, 27
Malaysia, 27, 57
Maldives, 21, 22
Mali, Republic of, 49, 56
Manganese, 443
Mangla Dam, 22
Mango, 620
Manifold, 630
Masonry, 13
Mass density, 128
Mass wetness, 427
Mauritania, 42, 56
Mauritius, 54, 55
Maximum available deficiency (MAD), 523
Mean sea level (MSL), 387
Mekong, 26-27
Mekong River Committee, 26
Mesopotamian Civilization, 17-19
Meter gate, 296, 304, 305, 306
Metric potential, 453
Metric suction, calibration for, 464
Mexico, 5, 17, 18, 20, 103
Michael, A.M., 373
Micro-irrigation, 612, 616, 619, 620, 621, 623, 652
Micro-sprinkler system, 601
Middle East, 1, 2, 5, 35-36, 39
Millet, 30, 42, 90
Mining, 84, 441
Minor irrigation, 23
Missing Wind speed Data, 505
Mixed-flow pump, 232-34
Modified Penman formula, 497
Mohenjodaro, 18
Moisture, 8, 29, 127
circulation of, 3
content calibration for, 464
cycle, 3
distribution pattern, 587
equivalent, 454
movement of,
under unsaturated conditions, 457
Moisture stress, 469
Molar solution, 663
Molecular attraction, 131
Mollisols, 436
Moluccas and Nussa Tenggara, 28
Molybdenum, 443
Mongolia, 30, 31
Monsoon, 1, 2, 5, 7, 8, 10, 15, 18, 20-32
Monsoon Asia, characteristics of, 21
Monsoon rain, 666
Morocco, 40, 42, 43, 47, 48, 103
Mountain meadow soil, 433
Mud pump, 169
Muddy water, 183
Mulching, 444
Mung bean, 437
Municipal effluents, 667
Mustard, 527, 530, 537, 546
Myanmar, 5, 10, 25, 26

N/K ratio., 710
Nam Bo, 27
Natural depressions, 33, 34
Nepal, 22, 24, 103
Net benefit, 696
Net benefit-investment ratio, 709

Net irrigation requirement, 518
Net positive suction head (NPSH), 210
Net present value (NPV), 705
Net present worth (NPW), 705, 709, 710
Netting out, 710
Neutron moisture meter, 465
New Zealand, irrigated area in, 5
Niger, 56
Nigeria, 47
Night irrigation, practice of, 252
Nile Valley, 45
Nitrogen, 441
Non-capillary pores, 428
Non-conventional energy source, 199
Non-government organisations (NGOs), 2, 26, 57, 104
  activities in water resources, 2
Non-pressure pipe, 354
Non-saline alkali soil, 662
Non-steady-state analysis, Theis' equation, 137
North Africa, 2, 40
North America, 3, 4
North Pacific Ocean, 27
North-West Frontier Province (NWFP), 22
Norway, 579
Nozzle operating pressure, 599

Oil palm, 27
Oilseeds, 42, 90, 523, 536, 546
Ojha, T.P., 373
Okinawa, 31
On-farm development, 15
Onion, 512, 514, 522, 523, 529, 541, 542
Open channel, 311, 314, 315, 317, 342, 343, 344
Opportunity cost, 696
Organic fertilizer, 202
Orifices, 182, 291, 296, 304, 305
Orissa, 431, 432, 435, 436, 438, 620, 719, 722
Orographical rain, 3
Osmotic potential, 453, 469, 664, 670, 671, 687
  and plant growth, 687
Osmotic pressure (OP), 664
Osrabandi, 97
Outlet command area, 345
Outlet valve, 377
Oxisols, 436
Pacific Ocean, 31
Paddy, 61, 250
Pakistan, 5, 17-20, 22, 23, 60, 61, 81, 103, 600
Pan coefficient, 481
Panduwasa, King, 24
Papau, New Guinea, irrigated area of, 5
Parshall flumes, 182
Participatory irrigation management, 25
Parts per million (ppm), 663
Pea, 512, 513, 514, 519, 521, 522, 523, 526, 528, 529, 531, 535, 536, 537, 540, 541, 542, 546
Pearl milet (bajra), 90
Pearl millet, 437
Peaty, 434
Pedology, 422
Pedosphere, 422
Peninsular region, 25
Penman equation, 496
Penman Formula, 495
Penman-Montieth equation, 499
Percolation, 456
  tank, 68, 127
Percussive crushing, 173
Perennial irrigation, 17
Performance curve, 211
Performance improvement, 15
Permanent crops, 16
Permanent wilting percentage, 454
Permeability, 126, 144, 178, 426, 446, 456, 457, 466, 469, 662, 666, 667, 668, 670, 674, 683, 686
  coefficient of, 130
Permeable weir, 13
Permissible velocity, 280
Peru, 17, 18
  Chimu empire, 18
pH, 662, 664, 668, 670, 671, 674, 675, 677, 678, 683, 685, 686
Philippines, 18, 27, 29, 57, 103
Phosphorus, 442
Photo-voltaic power generation, 201
Phreatic level, 125
Pick-up-weir, 13
Piezometer, 136
Piezometric heads, 139
Piezometric level, 126
Piezometric surface, 126, 130, 133
  equation of, 135
Pigeon pea, 437, 536
Pineapple, 53, 513, 514, 530, 543, 544, 545
Ping, Li, 18
Pipe
  choice of, 279, 352
  economic size selection of, 243
  polyethylene (HDPE) pipe, 353, 360-61, 631
  intake, 263, 267
    hydrostatic head on, 265
    jointing in, 360
  laying and jointing of, 357
  plastic, 352
  poly vinyl chloride (PVC), 184, 352, 353, 360, 368, 375, 582, 595, 629, 631
  reinforced cement concrete (RCC), 263, 264, 267, 279, 353, 377, 595
    velocity of flow in, 368
    welded joints in, 360
  system
    friction head in, 238
  vent pipe
    area of, 364
  vertical load on, 265
  with tongue and groove joints, 357
Pipeline
  accessories and fittings, 631
  depth below ground, 376-77
  discharge capacity of, 354
  features of, 376
  laying of, 376
    testing of, 378
  prevention of damage of, 355
  surges in, 372
  underground, 384
Piston, 204
Pit, 127
Pitcher irrigation, modified form of,, 617
Pitting, 184
Plant response, 469
Plant species adaptation, 527
Plant structure, 466
Plant temperature, 521
Plant transpiration, 4
Plant water content, 521
Plant-water relationships, 465
  atmospheric condition influencing, 468
Planting date, 527
Planting pattern, 527
Plants
  drought tolerance of, 469
  nutrients in, 470
Plunger, 204
Point source drippers, 624
Poiseuille's law, 456
Polluted water, 7
Pollution, 5, 6, 7, 16, 17, 80, 111
Pollution load, 17
Polythene strainers, 151
Polyvinyle chloride (PVC) laterals, 615
Pomegranate, 620
Population, 1, 2, 4, 5, 6, 16, 17, 21, 22, 31, 32, 35, 38, 39, 44, 45, 46
Pore space, 125, 137
Pores, 125
  distribution of, 131
  size and shape of, 125, 130-31
Porosity, 128, 178, 426, 427, 428, 444, 446, 456, 458
Positive displacement, 193
Potable water, source of, 250
Potassium, 442, 675
Potato, 43, 512, 514, 522, 523, 528, 529, 541, 542, 543
Potential evapo-transpiration (PET), 80, 479, 491
  alternative Hargreaves' equation for, 494
  concept of, 479
  method of computing, 492
Precast concrete channel, 324
Precipitation, 3, 4, 8, 10, 11, 14, 22, 27, 30, 31, 32, 33, 34, 37, 40, 100, 103, 126, 144, 478
Present worth, 701
Pressure differential (PD), 604
Pressure potential, 453
Pressure regulator, 632
Pressurised irrigation systems, 601, 603-04
Priming, 214
Private irrigation, 19
Productivity, 661
Pulses, 43, 90, 514, 536
Pump, 318, 595
  chamber, 275
    hydraulic efficiency of, 275
    piping system in, 278
  efficiency of, 211
  irrigation, 29
  performance of
    effect of speed and impeller diameter on, 212
  propeller, 232, 233, 234, 238
  selection of, 236

discharge capacity of, 237
procedure for, 237
stand, 361, 375
water surface elevations in, 370
Pump house, 250, 253, 258, 260, 262, 267, 268, 269, 270, 271, 272, 273, 274, 275, 276, 279
circular, 271
design of, 273
rectangular, 271
square, 271
vertical driven
pump house for, 273
specific requirements for, 274
vertical turbine, 230, 233, 236, 238
features of, 233
water, 193, 199, 209, 231, 233
specific speed of, 208
Pumping, 127
Pumping plant
design of, 240
economics of, 241
efficiency of, 240
power requirement for, 240
Pumping set, 581, 583, 590, 591, 598, 599
Pumping test
duration of, 143
Pumping water level, 133
Pumping well, 267
Punjab, 321, 344, 429, 433, 435, 440, 620, 662, 666, 667, 676, 677, 722
birth place of modern irrigation, 20
Land of Five rivers, 20
salt-affected soils in, 661

Quality of life, 1
Quicksand, 157

Rabi, 15
Radiation, 3
Radish, 512, 514, 522, 523, 541, 542
Radius of influence, 133, 141
Rain water
management of, 80
Rainfall,
amount of, 5
convective, 3
distribution of, 421
effective, 517
frontal, 3
intensity of, 10
measurement of, 426
orographical, 3
timing of, 5
variability of, 8
variations in, 22
Rajasthan, 321, 431, 433, 435, 438, 440, 620, 662, 666, 667, 669, 676, 677, 678, 684, 721, 722
salt-affected soils in, 661
Rao, Dr. K. L.
National Water Grid, 112
Rapeseed, 514, 530, 537
Reaction drive sprinklers, 601
Readily available water (RAW), 523
Recession cropping, 40
Reciprocal leakage factor, 132
Reciprocating pump, 204
force required to, 206
types of, 205
Red soil, 38, 39, 425, 431, 432, 436, 439, 440, 442, 451, 454, 455, 456, 458, 461, 471
Reference Evapo-transpiration
estimation of FAO Penman-Montieth based combination formula for, 501
Reference line, 388
Regulating pondage weir, 13
Regulating weir, 13
Renewable water, 2
Replacement cost, 710
Republic of Yemen, 39
Reservoirs, 311
pumping from, 263
Retention dam, 13
Return velocity, 169
Reverse circulation rotary drilling, 170
Rice, 30, 31, 40, 47, 49, 54, 90, 514, 529, 532, 533, 535, 545, 548, 662
double cropping in cultivation of, 27
Rice culture, components of water loss in, 533
Rice-based farming systems, 25, 28
Ridge, 383, 385, 387, 409, 413
equipment for making of, 413
Ridge canal, 14
Riparian Rights, 251
Riser valve, 366
Rising main, 279-80
River, 254-58
River basin, 14
River lift irrigation system, growth of, 250
River lift pumping stations
site selection for, 257
River pumping system, 259-60
Rock
drillability of, 172
porosity of, 125
Rock fracture, 124
Rodents, damage from, 615
Roman Empire, 40
Romania, 20
Roosevelt Dam, 20
Root injector, 617
Root proliferation, 438
Rotary crushing, 173
Rotary drilling, 175-76
Rotary rig, capacity range of, 170
Rotating head sprinklers, testing water distribution patterns of, 588
Rough grading, 383
Rubber, 25, 27
Rubber ring, 357, 360
Run-off water,
accumulation of, 666
Rural water supply scheme, 250
Russia, 20, 21
Rwanda, 42

Sabmontane soil, 433
Safflower, 513, 514, 522, 523, 528, 530, 537, 538, 546
Sailaba cultivation, 23
Saline aquifer, 178
Saline soil, 434, 662, 678, 681, 683, 687
peaty, 434
plant response to, 687
properties of
determination of, 670
reclamation of, 678
Saline water, 178
Saline-alkali soil, 662, 668, 670, 678, 68-83, 685
Salinity,
control, 681
hazard, 669
Salinization, 7, 33, 34, 37, 56
Salt
absorption of, 470
accumulation, 19, 615
affliction, 441
balance, 666, 672, 673, 679, 680, 681
concentration, 662
effect on plant growth, 687
problem
causes of, 664
nature and extent of, 666
Salt River Valley in Arizona (USA), 320
Salt water
intrusion of, 23
Salt-affected soil, chemistry of, 668
reclamation and management of, 678
reclamation procedures, 678
Salt-tolerant crop, choice of, 685
Salvage value, 711
Sand, 183
Sand cushion, 321
Sandy soil, 521
Sanitation, 6, 108
Saturation capacity, 454
Saturation extract, 662, 663, 669, 670, 671, 672, 679, 680, 688
Saturation percentage, 663, 672, 679, 680
Saudi Arabia, Kingdom of, 37
desalination of water, 38
irrigation development in, 38
Ministry of Agriculture and Water, 38
physiographic units of, 38
sprinkler irrigation system in, 38
Wadi Jizan Dam, 38
Scraper
animal-drawn buck, 410
tractor-drawn, 410
Screen diameter, 149
Screen filter, 637
Screen opening, 149
Sea water, intrusion of, 30, 127
Sedimentary rocks, 144
Seed placement, 687
Seepage, 133, 312-14
Seismic refraction surveying, 129
Seismic waves, 129
Seismograph, 129
Semi-arid tropics (SAT), 434, 436-39
Alfisols of, 437, 438
fields of, 437
rainfed farming in, 437
red soils of, 436
Semi-pervious layer, 139, 142
Senegal
cultivated area of, 54
irrigation potential of, 54

irrigation systems' management in, 103
topographic feature of, 54
Sensitivity analysis, 713
Sesame seed, 43
Settling basins, 635
Sewage, 7, 111
Shadow price, 696
Shah, EmperorMohammad, 59
Shahjahan, Emperor, 59
Shallow water, , 682
Shejpali, 94, 98
Shifting cultivation, 441
Shikoku, 31
Shock point, 129
Shock waves, 129
Shortwave radiation, 485, 504
Siberia, 30
Silicon, 443
Silt, 10, 18, 106, 183
removal, 271, 275
Single-nozzle sprinkler, 584
Siwalik valleys, 24
Skeletal soil, 434
Slope weir, 13
Slot opening, 149
Small-scale irrigation schemes, 32, 56, 193
Snow-melt, 5
Snowfall, 8
Sodium, 443
Sodium adsorption ratio (SAR), 661, 662, 663, 667, 668, 669, 675, 677, 679, 681, 683, 684, 685, 690
Sodium hazard, 669, 676
Soil, 421, 423, 425-28, 465
chemical degradation of, 440
classification of, 434
depth and permeability of, 10
effect of salts on, 669
formation of, 434
infiltration characteristics of, 383, 446
mineral and organic compounds of, 444
moisture content of, 450
pF of, 452
physical degradation of, 440
reclamation of, 678
salt build-up in, 665
water in
hydraulic equilibrium of, 456
movement of, 445
Soil amendment, 683
Soil conservation, 10
Soil Conservation Society of India, 441
Soil erosion, 56, 80, 113
Soil evaporation, 511
Soil exchange complex, 668
Soil extract, 663, 664
Soil fertility, 440, 441, 442, 444, 465, 466
Soil genesis, 422
Soil heat flux, 486
Soil horizons, 428
Soil, loamy, 521
Soil management, 661, 665, 685
Soil map, 422
Soil microbiology, 422
Soil moisture, 3-10, 40, 423, 459, 478, 483, 484
characteristics, 453
constants, 454
content, 465
potential
laboratory measurements of, 462
retention, 450
stress, 453
tension, 451, 452, 453
Soil morphology, 422
Soil profile, 422, 423, 428, 429, 434, 436, 437, 439, 442, 446, 452, 467, 471
Soil resources, 80
Soil resources survey, 429
Soil salinity, 46, 56, 465, 661, 663, 665, 666, 667, 668, 669, 674, 675, 676, 679, 681, 685, 687, 688
effect of, 688
crop management on, 665
irrigation on, 665
soil management on, 665
Soil, sandy, 521
Soil series, 422
Soil solid
density of, 427
Soil solution, 423
Soil structure, 422, 423, 425, 426, 440, 443, 445, 465
classification based on shape of aggregates, 425
management of, 426
Soil survey, 422
Soil taxonomy, 434
soil groups based on, 436
Soil temperature, 504
Soil texture, 422, 424, 425, 428, 439, 442, 443, 454, 464, 465, 546
Soil thermometer, 487
Soil tilth, 422
Soil type, 422
Soil water
kinds of, 445
metric potential of, 453
movement of
in unsaturated soils, 458
potential energy of, 453
specific potential energy of, 452
Soil wetness, 427
Soil-plant-atmosphere system, 468
Soil-plant-water relationships, 421
Soil-water potential, 452
Soil-water regime, 521
Soil-water-plant-atmospheric relationship, 521
Solar cell, efficiency of, 201
Solar energy, 201, 485
Solar radiation, 485, 503, 505
measurement of, 486
Solenium, 444
Solid slope masonry weir, 13
Soluble salt, 662, 665, 669, 670, 672, 673, 679, 682, 683, 684, 685
Solute potential, measurement of, 463
Somalia, 42, 54, 56
Sorghum, 42, 90, 437, 514, 521, 522, 523, 528, 529, 530, 535, 540, 546
South Africa, 42, 620
South Africa, Republic of, 46
South America, 3, 4
South China Sea, 27
South Korea, irrigation systems' management in, 103
Southern California, 620
Soviet Union, 32, 34, 35
Spain, 20, 620
Spanish colonization, 29
Spate irrigation, 14, 22, 46
Specific capacity, 133
Specific gravity, 193, 427, 428, 459, 460
Specific ion effect, 664, 687
Specific potential energy, 452
Specific retention, 131
Specific surface, 664
Specific yield, 129, 131, 141
Spices, 25, 545
Spigot end pipes, 356
Spodosols, 436
Spray irrigation, 600
Sprinkler, selection of, 591
Sprinkler irrigation,
adaptability of, 578
limitations of, 579
Sprinkler irrigation system, 578, 580, 581, 582, 583, 584, 585, 589, 591, 592, 593, 596, 599, 600, 601
Sprinkler lateral
design of, 594
Sprinkler spacing, 590
Sprinkler system, 593, 595, 597, 598
Sri Lanka, 20-25, 100, 103
Staking, 409
Standard USBR Class A Pan Evaporometer, 480
State of Utah, 20
States of Sabah and Sarawak, 27
Static water level, 132
Steady-state flow, 133, 139
Stop logs, operation of, 267
Storage capacity, 127
Storage coefficient, 129, 141
Stratification, 142
Stream channel, 127
Sub-soil injector, 617
Subaks, 28
Submerged flow analysis, 301
Submersible pump, 230, 231, 232, 236, 238
Suction irrigation, 617
Suction pipe, 278
Sudan, 40, 42, 43, 45, 46, 56, 103
1959 Nile Waters Agreement, 45
Sugarbeet, 512, 514, 522, 530, 539, 540
Sugarcane, 42, 43, 47, 53, 54, 61, 513, 514, 522, 523, 528, 529, 530, 539
Sulawesi, 28
Sulphur, 184, 442
Sumatra, Kalimantan, 28, 29
Sumbawa, 28
Sunflower, 513, 514, 522, 523, 528, 529, 530, 537
Sunken screen pan evaporimeter, 481
Sunshine duration, 485
Surface crust, 669
Surface drip irrigation systems, 621
Surface irrigation, 33, 49, 53, 621
cost analysis of distribution, 329
variable costs of, 700
Surface tension, 428, 445, 450, 451, 452, 456
nature and dimensions of, 450
Surface water,
assessment of, 8

integrated use of, 64
jurisdiction of, 101
large-scale informal systems, 32
minor irrigation schemes in India, 64
small-scale informal systems, 32
Surface water and groundwater, conjunctive use of, 100
Surface water flows, 10
Surface water irrigation scheme, 23, 249
Surface water resources, procedure for estimating, 10
Sustainability, 80
Swaziland, 53
Sweet potato, 43
Syrian Arab Republic, 37
System storage, 311

Taiwan (Republic of China), 27, 29, 30, 57
Tajikistan, 32-34
Tamil Nadu, 431-38, 620, 662, 666, 668, 719, 720, 722
development of lift irrigation projects in, 250
salt-affected soils in, 661
water pricing in, 719
Tank irrigation, 24
Tanzania, United Republic of, 49, 56, 103
Taunsa Barrage, 22
Tea, 514, 522, 533, 545, 546
Technology Mission on Drinking Water, 6
Tensile strength, 146
Tensiometers, 460
Terai soil, 433
Terhela Dam, 22
Test site, hydro-geological conditions of, 141
Thailand, 5, 18, 20, 26, 57
Theis' equation, 137
Thermal and nuclear power generation, 5
Thiem, 135
Thiem, Adolph, 134-35
Thiem-Dupuit equations, assumptions, 137
Thornthwaite Formula, 493-94
Thunderstorm, 3
Tibet (China), 81
Tikabharable Canal, 24
Tillage, 444
Tobacco, 42, 522, 528, 530, 540
Togo, 40
Tomato, 512, 522, 523, 528, 529, 541, 543
Topography, 24, 29, 31, 32, 52, 54
Toxicity, 17, 674, 689, 690
Tractor-operated blade terracer, 355
Traditional canal irrigation system, feature of, 250
Transfer payment, 696
Transmissibility, 129, 130, 134, 136, 139, 141
Transpiration,
rate of, 483
cuticular, 483
ratio, 483
stomatal, 483
Trash screen, 267
Treatment for uncertainty, 713
Tree planting, 7
Trickle irrigation, 611, 612, 646
Tsao-Kung canal, 29
Tube wells, 133, 145-48, 183
Tubewell supplies, water rates of, 720
Tughlaq, Ferozshah, 59
Tunisia, 48, 49, 579
Turbulence, 142
Turkey, 5, 20, 36, 579
irrigation systems' management in, 103
Turkmenistan, 32-34
Turnout, 338
box turnout, 338
siphon tube, 340
spile, 340

Uganda, 40, 52
UK, 579
Ukraine
irrigation development in, 21
Ultimate wilting point, 454
Ultisols, 436
UNICEF, 108
Uniformity coefficient, 153, 589, 647
United Nations Conference, Agenda 21
United Nations Conference on Environment, 2, 108
United Nations Development Programme (UNDP), 108, 724
United Nations Environment Programme, 35
United Nations Environment Programme (UNEP), 108
United Nations Food and Agricultural Organization (UNFAO), 501, 664, 666
United Nations Water Conference, 2
United Sates Salinity Laboratory Staff Handbook No, 675
United States Department of Agriculture (USDA), 423
United States of America, 5, 19, 620
irrigated agriculture in, 18
United States Soil Salinity Laboratory (USSSL), 676
Unlined field channels, 316
Unsteady-state flow, 137
Upper Bari Doab Canal, 312
Urban sewage, 7
Urbanisation, 5
USA, 20, 100, 103, 320, 321, 579, 620
Bureau of Reclamation of USA, 20
Uttar Pradesh, 429, 431, 433, 435, 438, 620, 662, 666,, 676, 677, 720, 722
public tubewells of
World Bank supported projects in, 375
salt-affected soils in, 661
UV degradation, 631
Uzbekistan, 5, 20, 32, 34
V-ditcher, 415
Vaidyanathan Committee Report, 722
Valves, 633
Vapour pressure, 79, 487, 502
deficit, 502
measurement of, 487
Vegetable crops, 541
Vegetation, 3, 5, 8, 10, 16, 44
Venturi injection system, 606
Vertic inceptisols, management of, 439
Verticality, 177
Vertisols, 434, 436, 438, 439, 440
Vertisols and associated soils, 438-39
Vietnam, 5, 20, 25, 26, 27, 103
Viscosity, 450
Visual plant symptoms, 521
Visveshwariah, Sir M., Bharata Rathna, 60
Visveshwariah, SirM., 60
Void ratio, 427
Void space, 125
Voids, 125
gravity drain-age of, 131
Volume wetness, 428
Volute centrifugal pump
fertilizer injection through, 604

Wadi, 14
Wadi Jizan Dam, 38
Warabandi, 94, 97, 98, 99
Washing, 6
Waste dumps, 7
Waste water, 5, 33
Wastewater, treated, 50
Water 1-7, 11, 23, 38
amount in plants, 466
appropriate pricing of, 717
availability of, 2, 3, 4, 256
assessment of, 251
chemical potential of, 469
inter-basin transfer of, 111
historical perspective, 112
national perspective, 112
movement of, 470
rational analysis of, 130
movement, retention and use of, 421
properties of, 444
quality of
factors influencing, 675
sic requirement of, 479
vertical distribution of, 124
Water absorption, 354, 468
Water application efficiency, 525
Water application rate, 590
Water bearing formation, 132
hydraulic properties of, 135
Water conservation, 4, 127
Water consumption, 517
Water control and diversion, 338
structures for, 334
check gate, 334
plate steel gate, 335
portable check dams, 337
precast check gate, 334
watertight seal for check gate, 335
Water conveyance efficiency, 525
Water conveyance system
loss of water in, 318
pre-fabricated canalets, 326
structure for, 331
Water distribution
underground pipeline system, 351
advantages of, 351, 373
design of, 369
efficient use of modern farm machinery, 351
gravity inlet, 362
in community irrigation project, 373
inlet structure, 361
long life and low maintenance cost, 351
loop network, 375

outlet structure, 365
pipes for, 352
rotational delivery system, 378
specialised structures for, 361
survey procedures of the command area, 373
Water distribution efficiency, 525
Water distribution system, 250, 311, 312, 328, 329, 344, 345, 591, 724
design of, 345
in unconsolidated holdings, 345
irrigation command area, 344
planning of, 344
survey of, 345
Water hammer, 224
Water harvesting, 10
Water intake, 456
Water level measurement, 143
Water lift, 193-99
belt-driven pump, 216
biogas engine powered plant, 202
centrifugal pump
classification of, 214
principles of operation of, 213
close-coupled pump, 216
components of manually-operated twin treadle, 204
direct connected pump, 216
feature of variable displaccment pump, 204
installation of mechanically powered deep well reciprocating, 205
medium lift (submersible) pump, 217
multi-stage pump, 215
Persian wheel, 196, 204
portable pumping set, 220
installation of, 259
positive displacement pump, 203
rotary pump, 208
animal powered shallow wellreciprocating pump, 204
vertical turbine pump, 224
advantages of, 226
characteristics of, 226
construction and characteristics of, 224
maintenance of, 228
method of lubrication of, 225
trouble-shooting of, 228, 229
Water lifting
application of non-conventional energy sources in, 198
from canals, 252
permission for, 256
principles involved in, 193
selection of device, 193
Water loss, 23, 533
Water management, 2, 98
women's role in, 2, 7
Water meter, 292-95
Water pollution, 7
Water potential, 470, 521
Water pressure, 352
Water proof channel lining, 322
Water quality, 7, 10, 34, 44, 113, 678, 684
Water rate, 718
Water requirements (WR), 515
Water resource projects
cost recovery, 721
financial performance of, 721
Water resources, 1-3, 6-8, 10, 15, 25
availability in India, 83
drying of, 715
quality of, 676-77
Water resources development projects, 695
financing of, 724
Water right, 14
Water scarcity, 611
Water science and technology, scope of, 2
Water shortage, 5, 478
Water spreading, 127
Water storage efficiency, 525
Water supply, 2, 5, 6, 198, 249, 250, 251, 253, 255, 256, 277, 279
adequate, 531
limited, 531
location, 591
intake points for rver
characteristics influencing, 253
national average of, 6
nature of, 591
urban, 7
Water supply project
economic feasibility of, 251
techno-economic feasibility report of, 251
Water swivel, 169
Water table, 125, 126, 127, 130, 132, 133, 134, 137, 141, 144, 148, 156, 163
perched, 126
shape of, 126
Water use
aspects of, 7
efficiency of, 7, 525, 527-28
low efficiency of, 80
Water vapour, 3
movement, 458
Water-bearing formations
characteristics of, 141
types of, 124
Water-holding capacity, 125
Water-nutrient interactions, 438
Water-plant relationship, areas of, 466
Water-saving technology., 6
Waterlogging, 7, 19, 23, 32, 33, 34, 37, 56, 80, 441, 661, 662, 664, 665, 667, 668, 674, 686
causes of, 664
problem of, 665
problems of, 21
Watershed, 14, 15
Watershed management, 15
Wave radiation, 485
Weed control, 527
Weir, 250, 297-299, 302, 303
Weir-controlled canals, 22
Well, 127, 256, 257, 261, 263, 268, 269, 270, 274, 279, 311, 318
effect of partial penetration of, 147
geologic conditions of, 128
ground-water conditions of, 128
hydraulics of, 132
hydro-geological properties of, 141
measurement of discharge and drawdown, 182
pumping test, 141
step-by-step procedure, 141
rawhiding, 179
testing for drawdown and yield, 181
yield potential of, 143
unconfined, 133
Well development, 178
methods for
compressed air, 180
dispersing, 181
jetting, 181
pumping, 179
surging, 179, 180
Well drilling, 203
Well equations, 133
Well failure, 175
photographic evaluation of. 185
Well, irrigation, 130, 132
Well logging, 127, 175
Well screen, 142, 148-150, 176, 183, 184
Well yield, 133
West Bengal, 321, 431, 432, 433, 436, 620, 722
public tubewells of World Bank supported projects in, 375
West New Guinea, 28
Wetted perimeter, 315
Wheat, 30, 47, 90, 513, 514, 521, 522, 523, 524, 528, 529, 530, 533, 534, 535, 536, 546
Wilting range, 454
Wind erosion, 430, 440, 441
Wind pressure, 502
Wind profile, 486
Wind speed, 481, 485, 486, 490, 494, 496, 497, 498, 499, 501, 502, 504, 505, 506, 507, 508, 510, 511
measurement of, 486
Windmill, 199
Work day, 696
World average, 3
World Bank, 2, 8, 108, 375, 724
Policy Paper in 1993 on goals for the water, 2
World Health Organization (WHO), 6, 108
World War II, 29
World water infrastructure, 2
World water resource, 4

Yellow soil, 432, 435
Yemen, Republic of, 39
Yield capacity, 141
Yugoslavia, 579

Zaire, 40
Zambia 53
Zimbabwe, 52-54, 56, 579
Zinc, 443
Zone of aeration, 124
Zone of saturation, 124-26